BIOCHEMISTRY

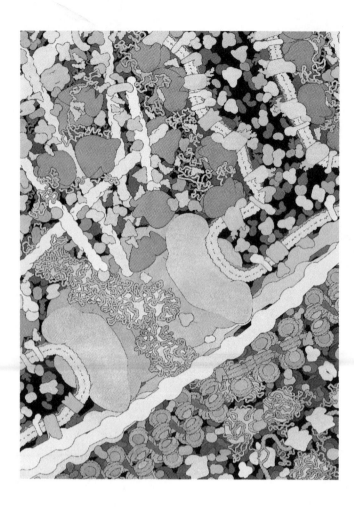

In this cross section through a portion of a typical eukaryotic cell—the image printed on the hard-cover edition of this text—many biochemical functions may be seen. The artist, David Goodsell, reveals all macromolecules by magnifying cell components 500 000 times. Proteins are shown in blue, nucleic acids in orange and red, lipids in yellow, and polysaccharides in green. Small molecules are not shown—imagine the spaces between each macromolecule filled with thousands of cofactors, amino acids, nucleotides, ions, and water.

Biological information flow begins in the nucleus (bottom) with the transcription of genes to produce RNA. The RNA is processed and mature messenger RNAs, complexed with protein, are exported to the cytoplasm through nuclear pores (center, left). These messages associate with ribosomes (purple) and are translated in the cytoplasm (upper left) or on the surface of the endoplasmic reticulum (upper right). The cytoplasm, braced by structural fibers of actin (upper left), is filled with enzymes performing myriad metabolic tasks.

BIOCHEMISTRY

Laurence A. Moran
University of Toronto

Second Edition

K. Gray Scrimgeour
University of Toronto

H. Robert Horton
North Carolina State University

Raymond S. Ochs
Case Western Reserve University

J. David Rawn
Towson State University

NEIL PATTERSON PUBLISHERS
PRENTICE HALL
Englewood Cliffs NJ 07632

Credits

Publisher:	Neil Patterson
Editorial Director:	Sherri Foster
Principal Editor:	Charlotte Pratt
Contributing Editors:	John Challice Terri O'Quin Morgan Ryan
Editorial Assistants:	Donna Curasi Deborah Kirschner
Production Manager:	Donna Young
Production Assistants:	Kevin Cournoyer Kate Dunlap
Principal Artists:	Lisa Shoemaker George Sauer

Contributing Artists:	David Goodsell
	IMS Creative Communications
	    University of Toronto
	Sarah McQueen
	Molecular Graphics and Modelling
	    Duke University
	Andrea Weaver

Designer:	Neil Patterson

*For a full listing of illustration credits, please turn to
the last few pages of this text.*

Biochemistry

© 1994 by Neil Patterson Publishers/Prentice-Hall, Inc.
ISBN 0–13–814443–5

Printed in the United States of America: February, 1994

Library of Congress Cataloging-in-Publication Data

Biochemistry / [edited by] Laurence A. Moran, K. Gray Scrimgeour.—
2nd ed.

p.   cm.

Rev. ed. of: Biochemistry / J. David Rawn. c1989.

Includes bibliographical references and index.

ISBN 0–13–814443–5

1. Biochemistry.   I. Moran, Laurence A.

II. Scrimgeour, K. G.   III. Rawn, J. David. Biochemistry.

QP514.2.R39   1994   574.19′2—dc20   94-134

CIP

NEIL PATTERSON PUBLISHERS
PRENTICE HALL
Englewood Cliffs, NJ 07632

Prentice-Hall International (UK) Limited, *London*
Prentice-Hall of Australia Pty, Limited, *Sydney*
Prentice-Hall Canada Inc., *Toronto*
Prentice-Hall Hispanoamerica, S. A., *Mexico*
Prentice-Hall of India Private Limited, *New Delhi*
Prentice-Hall of Japan, Inc., *Tokyo*
Simon & Schuster Asia Pte. Ltd., *Singapore*
Editora Prentice-Hall do Brasil, Ltda., *Rio de Janeiro*
Prentice-Hall, Inc., *Englewood Cliffs, New Jersey*

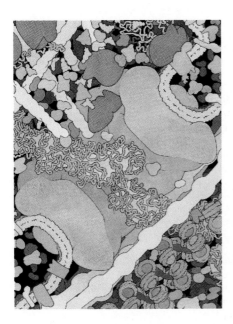

The Authors

Laurence A. Moran

After earning his Ph.D. from Princeton University in 1974, Professor Moran spent four years at the Université de Genève in Switzerland. He has been a member of the faculty at the University of Toronto since 1978, specializing in molecular biology. His research findings on heat-shock genes have been published in many scholarly journals. Professor Moran and K. Gray Scrimgeour drafted most of the chapters for this edition and read and edited all chapters.

K. Gray Scrimgeour

Professor Scrimgeour received his doctorate from the University of Washington in 1961. He has been a faculty member at the University of Toronto since 1967 and is Visiting Scientist at the University of Victoria. He is the author of *The Chemistry and Control of Enzyme Reactions* (1977, Academic Press), and his work on enzymatic systems has been published in more than 50 professional journal articles. From 1984 through 1993, he was editor of the journal *Biochemistry and Cell Biology*. Professor Scrimgeour and Laurence A. Moran drafted most of the chapters for this edition and read and edited all chapters.

H. Robert Horton

Dr. Horton, who received his Ph.D. from the University of Missouri in 1962, is William Neal Reynolds Professor and Undergraduate Coordinator in the Department of Biochemistry at North Carolina State University. He was appointed Alumni Distinguished Professor in 1979. Professor Horton's research includes protein conformation and enzyme mechanisms. He did not write chapters for *Biochemistry*, but his textbook *Principles of Biochemistry* (1993, Neil Patterson Publishers/Prentice Hall) was a substantial source of information, ideas, and narrative. He also prepared the dictionary of biochemical terms for the *Biochemistry Resource Book* that accompanies this new edition.

Raymond S. Ochs

Professor Ochs, who earned his Ph.D. from Indiana University, is an expert on metabolic regulation and has edited a monograph on this topic—Ochs, R. S., Hanson, R. W., and Hall, J., eds. *Metabolic Regulation* (1985, Elsevier)—and authored and co-authored numerous research papers and reports. He did not actively contribute to *Biochemistry*, but his work for Horton *Principles of Biochemistry* (1993, Neil Patterson Publishers/Prentice Hall) was a source of material for several chapters of this edition.

J. David Rawn

Professor Rawn, who earned his Ph.D. from Ohio State University in 1971, has taught and done research in the Department of Chemistry at Towson State University for the past 20 years. He did not write chapters for this edition of *Biochemistry*, but his first edition (1989, Neil Patterson Publishers) served as a source of information and ideas concerning content and organization.

The following experts wrote drafts of the following chapters:
R. Roy Baker (University of Toronto): sections of Lipids (11) and Lipid Metabolism (20);
Roger W. Brownsey (University of British Columbia): sections of Introduction to Metabolism (14);
Willy Kalt (Agriculture Canada): Photosynthesis (19); Robert K. Murray (University of Toronto): Carbohydrates and Glycoconjugates (9); Frances J. Sharom (University of Guelph): Biological Membranes (12); Malcolm Watford (The State University of New Jersey, Rutgers): Integration of Fuel Metabolism in Mammals (23).

Inside a Eukaryotic Cell

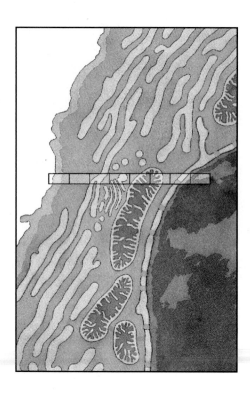

Cells are packed with a collection of diverse macromolecules, all performing specific metabolic tasks required for the maintenance of life. The eight illustrations, beginning below and running along the bottom of the Preface, simulate what we would see if we could magnify a narrow strip of a eukaryotic cell 1 000 000 times. At this magnification, atoms are about the size of a grain of salt, cells are the size of buildings, and you would be roughly one-seventh the size of the earth in height, allowing you to walk from San Francisco to New York in five or six steps.

A long narrow section is shown from the cell surface through the endomembrane system, through a mitochondrion, and into the nucleus.

At the molecular level, the composition of a living cell is complex but curiously homogeneous. Thousands of individual tasks are performed: chemical reactions for energy production, synthesis of new molecules, building and repair of scaffolds and walls, and storage, utilization, and reproduction of genetic information. Each task involves a specific macromolecule or set of macromolecules tailored in size, shape, and chemical composition to perform the job accurately and efficiently. Despite their diverse metabolic roles, however, all of these molecular machines are built according to four simple molecular plans—they are constructed of linear chains of protein, linear chains of nucleic acid, aggregates of lipid, or branching chains of polysaccharide.

The illustrations mirror the homogeneity and diversity of biochemistry. The colors highlight the homogeneity: proteins are drawn in shades of blue, nucleic acids in shades of red, lipids in yellow, and polysaccharides in green. Ribosomes are a special case—they are complexes of protein and nucleic acid and are drawn in purple. The diversity of biochemistry is apparent in the sizes, shapes, and distribution of the myriad molecules.

For clarity, only macromolecules are shown. In reality, the interstices are filled with small molecules, along with ions and water.

David S. Goodsell

The Cell Surface

The surface of the cell has the dual responsibility of protecting the cell from a possibly hostile environment while permitting communication and exchange between the environment and the interior. The primary barrier between the cell and the environment is a lipid bilayer (left), buttressed from the inside by a scaffold of structural fibers, such as spectrin and actin. Embedded in the bilayer are integral membrane proteins that perform the tasks of communication and transport across the impermeable membrane. Attached to the outer surfaces of these proteins are viscous chains of polysaccharide.

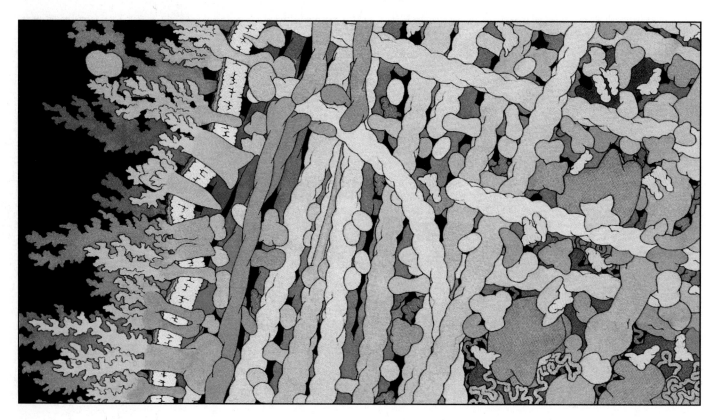

*Science should be as simple as
possible but not simpler.*

Albert Einstein

Preface

Einstein's dictum applies as much to teaching as to doing science, particularly when the subject is as richly complex and swiftly paced as biochemistry, where discoveries at the edge of research—sometimes overnight—reshape our understanding of basic principles and change the way much of the science is done.

One time-honored way to simplify a science as ornate as biochemistry is to present it visually. A clearly rendered illustration often helps students visualize what otherwise would remain abstract and vaguely, or perhaps even wrongly, imagined. We offer three fresh contributions to the art of biochemical illustration. Running along the bottom of this preface is one example—Dr. David Goodsell's informative

The Cytoplasm
Eukaryotic cytoplasm is crisscrossed with the structural fibers of the cytoskeleton. The largest are microtubules. A knobbed intermediate filament is seen below right, and numerous actin filaments run horizontally. The spaces between filaments are teeming with enzymes and ribosomes that orchestrate the degradative and synthetic reactions of the cell.

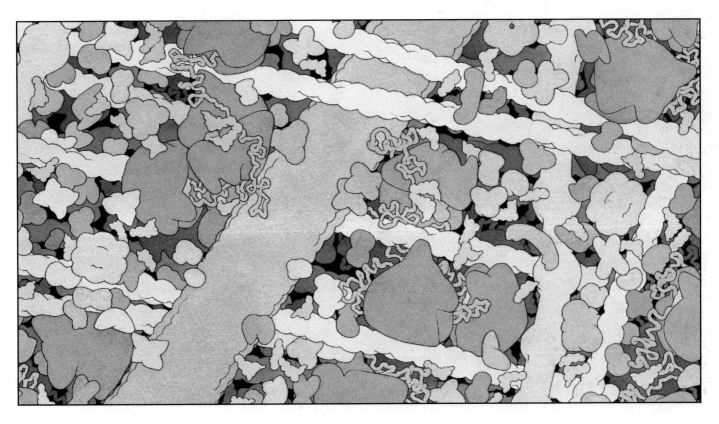

set of eight panels showing the interior of a eukaryotic cell at 1 000 000 times magnification. This series of paintings gives a sharp sense of the sizes and shapes of cellular components and how they collaborate to form a densely packed, gloriously integrated system of molecular machines. You will find throughout the text similarly well-crafted depictions of basic processes that visually clarify the concepts presented in words.

A second kind of clarifying illustration is our set of stereo images showing molecular structures. These images convey an understanding of molecular architecture and its relation to function in a way that cannot be communicated by two-dimensional renderings. Almost all of these computer-generated illustrations have been newly produced at the Richardson Laboratory (Duke University) specifically for this edition. We include, with each new copy of the book, a free stereo viewer that allows the reader to see these pictures in three dimensions.

A third new graphic device, also created by Jane and David Richardson at Duke University, is an interactive computer program, *Exploring Molecular Structure*. This two-disk supplement (available for Macintosh or Windows) gives students an opportunity to manipulate three-dimensional depictions of biomolecules on screen. We have keyed our text to inform the reader of related material on the disks. (Please turn to page *xviii* for a full description of this and other supplements that accompany Moran·Scrimgeour *Biochemistry*.)

Another classic way to reinforce the learning process is to give students practice in applying knowledge while it is still fresh in their minds. We provide with

The Endoplasmic Reticulum

The endomembrane system in eukaryotic cells, composed of the endoplasmic reticulum and the Golgi apparatus, is involved in the processing and transport of lysosomal and cell-surface proteins and lipids. The first step of protein processing is shown below— ribosomes bind to specific receptor proteins on the membrane of the endoplasmic reticulum, and new protein is injected into the lumen as it is synthesized.

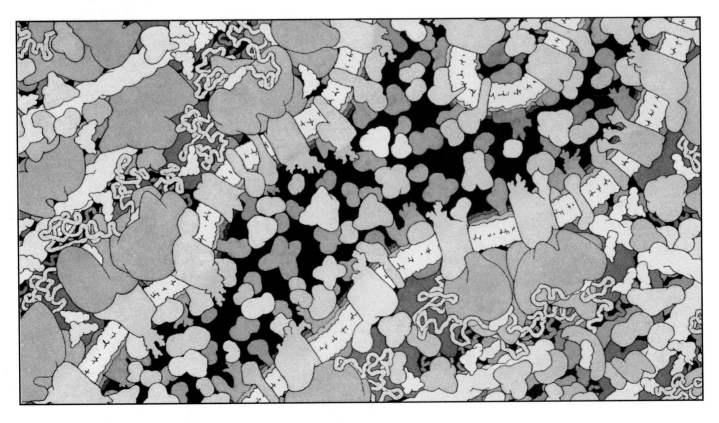

each new copy of our text a free *Biochemistry Resource Book,* which includes a set of problems for each chapter and detailed, step-by-step solutions to every problem. This *Resource Book* also contains a dictionary of biochemical terms, chapter summaries, standard abbreviations, tables of essential data, and "study pages" showing key molecular structures and metabolic pathways, a truly useful set of materials in handy paperback format.

Textbooks can also assist students in their journey through the details (where angels and devils dwell) by stopping now and then for a look at the big picture. Eight of the 33 chapters pause to help students understand, in broad terms, what they have just learned or are about to learn.

Probably the most considerable way to simplify a subject, while avoiding the simplistic, is to present its story in a clear, coherent, well-integrated manner, building at each stage the background needed for the next stage. This brings us to the organization of our book—arranged in four parts, each building on the ones that came before. (Please see pages *xvi* and *xvii* for a list of chapters, which reveals the pattern we follow in presenting *Biochemistry*.) At first glance, this volume may seem big enough to be all-inclusive. It is not. It holds an unwavering fix on basic principles, each one supported by carefully selected examples. We built the book explicitly for the beginning student taking a first course in this burgeoning subject. Parts One and Two establish at the outset a solid foundation of chemical information that will help students understand, rather than merely memorize, the dynamics of metabolic and genetic processes. These sections assume a rudimentary knowledge of the organic

The Golgi Apparatus
Proteins are processed and sorted in the layered compartments of the Golgi apparatus. Here, small digestive proteins destined for lysosomes bind to specific receptors in the membrane, and integral membrane proteins are glycosylated. The completed proteins are transported to their ultimate destination in a coated vesicle (below right). Three-armed triskelions provide the microscopic leverage required to pull a bubble of membrane out of the Golgi stack.

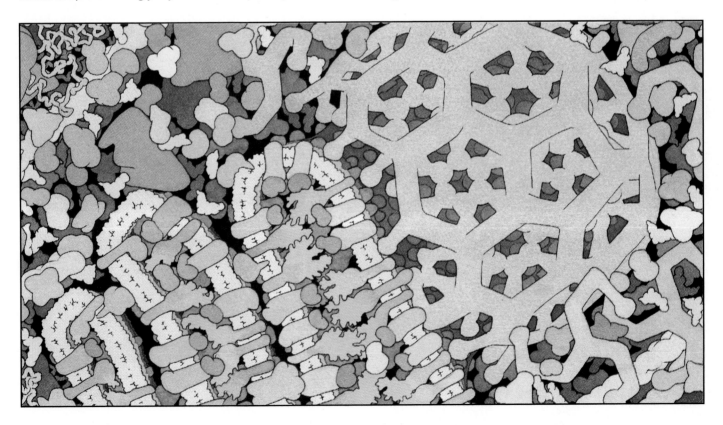

chemistry of carboxylic acids, amines, and aldehydes—knowledge generally gained in prerequisite courses in general and organic chemistry. Even so, key functional groups and chemical properties of each type of biomolecule are carefully explained as their structures and functions are presented.

Part One forms an introduction to all that follows. Chapter 1 gives a brief survey of thermodynamics and kinetics and describes the four classes of biomolecules—proteins, lipids, carbohydrates, and nucleotides—which are covered extensively in Part Two. We hope students will not skip or skim this first chapter; it is essential preparation for the rest of the narrative, a summary of basic biomolecules and a first look at the energy transformations critical to life on earth. In this chapter, and throughout the book as a whole, we sustain a consistent pattern of presentation, from basic chemistry to biochemical function. We first show the chemistry of monomeric units and then explore the properties and functions of the biopolymers formed from those monomers—amino acids to proteins, monosaccharides to polysaccharides and glycoconjugates, lipids to membranes, and nucleotides to nucleic acids. Part Two includes, among its 10 chapters, 3 chapters on enzyme properties, enzyme mechanisms, and coenzymes (Chapters 6, 7, and 8). We encourage a firm grasp of these critical subjects, prerequisite to later appreciation of the role of enzymes in metabolic pathways. Part Two features new material on a host of topics, including glycoconjugates, a subject at the frontier of current biochemical research. We end this part with a survey of digestion (Chapter 13), which forms a bridge between biomolecules and the biochemical dynamics of metabolism.

A Mitochondrion

The conversion of food into usable chemical energy ultimately involves the mitochondrion. The convoluted inner membrane is studded with the proteins of electron transport and oxidative phosphorylation. This protein ensemble creates a concentration gradient of protons across the lipid bilayer and uses the energy of the gradient to drive the formation of ATP, the energy molecule of the cell. The mitochondrion also holds a complement of DNA, mRNA, tRNA, and ribosomes and carries out synthesis of many of its own proteins. (Continued on next page.)

Part Three features 10 chapters that focus on metabolic pathways. In Chapter 14, we introduce the intricate molecular symphonies of metabolism by considering how pathways are energized, interrelated, and regulated. Next comes glycolysis (Chapter 15), which we describe in detail. At this stage, we establish a format that holds in subsequent metabolic chapters—first describing the basic pathway in chemical and enzymatic terms, then demonstrating the bioenergetic requirements and sources of energy, and following with an account of regulatory mechanisms. Part Three includes a comprehensive chapter on photosynthesis (Chapter 19), a process whose early steps we relate to oxidative phosphorylation and whose later steps we relate to the pentose phosphate pathway. We wind up Part Three with a chapter on the integration of fuel metabolism in mammals (Chapter 23). Here we summarize mammalian metabolic states under diverse nutritional circumstances, revealing the complexity of metabolic integration and regulation and the efficiency with which these organisms gain nourishment from a wide array of fuels under widely varying conditions—a suitable coda to our metabolic concert.

Part Four concludes the book with 10 chapters on the flow of biological information, a more comprehensive, up-to-date treatment of molecular biology than is usual in an introductory biochemistry text. Our main emphasis throughout this series of chapters is on the basic processes that govern pathways of information flow. We take pains to incorporate within our story of genes and gene expression the principles of biochemistry taught in preceding chapters. Among the many new topics covered in this part of the book are an extensive treatment of eukaryotic DNA replication

and the mechanisms for replicating the ends of linear chromosomes, new material on eukaryotic repair enzymes, and the coupling of transcription and repair. A separate chapter (Chapter 31) examines a variety of simple strategies for regulating gene expression, emphasizing how these processes are integrated to allow the kinds of cellular decision-making we witness during prokaryotic and eukaryotic development. From development, we move to a unique chapter, Genes and Genomes (Chapter 32), where we explore how genomes are organized. The final chapter (33) is a substantial presentation of techniques used in recombinant DNA technology.

Science textbook publishing, like much of science, is increasingly a collaborative effort. Laurence A. Moran and K. Gray Scrimgeour, seasoned teachers and researchers at the University of Toronto, wrote most of the chapters, and each read and edited all of them, thus assuring a unified whole. The authors also painstakingly reviewed the literature to assure that material presented reflects the most current thinking on each topic. Much of the groundwork for many of the chapters was earlier laid by those experts (including Moran and Scrimgeour) who contributed to the 1989 edition of Rawn *Biochemistry* and the 1993 edition of Horton *Principles of Biochemistry*, a short text (700 pages) for serious courses that must get the job done sooner, or with less detail, than is usual in two-semester courses. Strictly speaking, this new text is the second edition of Rawn *Biochemistry*. The taxonomically inclined would consider this version a new species, not just a variant, since it is so different from its parent. But there remains strong evidence of ancestral forms. It is most like Rawn *Biochemistry* in its use of stereo views and its presentation of

The Nuclear Membrane
The nucleus is surrounded by a two-layered membrane that is continuous with the endoplasmic reticulum. The wall is pierced by large nuclear pore complexes that control the passage of molecules into and out of the nucleus. A layer of intermediate filaments braces the nuclear membrane from the inside.

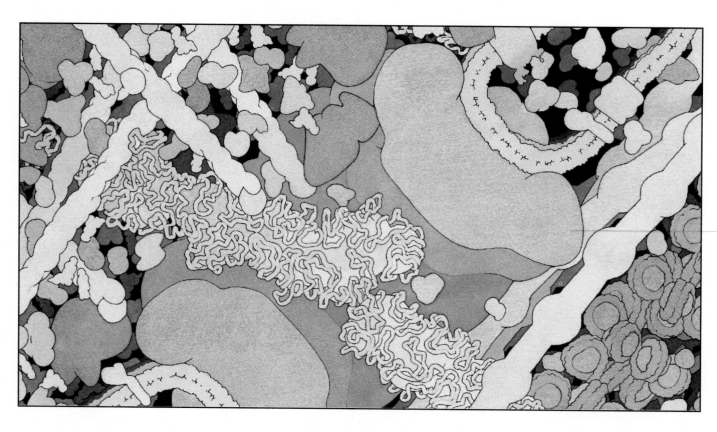

chemical mechanisms where these can deepen a student's understanding. Please turn to "The Authors" (page *v*) for a brief biographical sketch of the two main authors, Professors Moran and Scrimgeour, as well as an account of the authors of background material derived from Rawn or Horton and the authors who wrote single chapters specifically for this edition. Please turn to the "Reviewers and Advisors" (pages *xiv–xv*) for a list of the many scholars who gave us precious advice during development of this text and its precursors. For those intrigued by nuance, we draw attention to page *iv*, where credit is given to editors and illustrators, those crucial catalysts of the publishing process.

The story of biochemistry may seem daunting to the novice—it is a long one, textured with complexity and an elaborate cast of characters—but the student may find solace in Mark Twain's comment about another complicated cultural enterprise: "classical music isn't as bad as it sounds." Neither is biochemistry. Once into it, you will be moved by its harmonies, its sweeping power to explain so much of life itself.

Neil Patterson

Chapel Hill

February 1994

The Nucleus
The nucleus is filled with DNA, some stored in the form of chromatin (upper left, lower right), some in active use (center). DNA is transcribed to RNA by RNA polymerases. The RNA is then processed in large complexes of RNA and protein and eventually transported to the cytoplasm for translation into protein. A host of small proteins protect and repair the DNA and regulate its use.

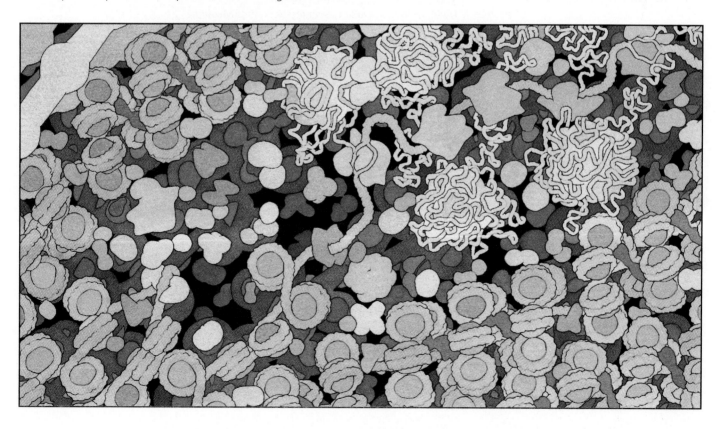

Reviewers and Advisors*

Deborah A. Adams, Agouron Institute
George C. Allen, Stanford University Medical Center
V. S. Ananthanarayanan, McMaster University
G. Harvey Anderson, University of Toronto
Laurens Anderson, University of Wisconsin
Rashid A. Anwar, University of Toronto
Dipali V. Apte, University of Illinois at Urbana-Champaign
James R. Bamburg, Colorado State University
Robert W. Baughman, Harvard Medical School
Bruce P. Bean, Harvard Medical School
G. Vann Bennett, Duke University School of Medicine
Ludwig Brand, Johns Hopkins University
John T. Brosnan, Memorial University of Newfoundland
Greg G. Brown, McGill University
John M. Buchanan, Massachusetts Institute of Technology
Kent O. Burkey, U.S. Department of Agriculture
P. Jonathan G. Butler, Medical Research Council, Cambridge, England
Judith L. Campbell, California Institute of Technology
Philip Carl, University of North Carolina at Chapel Hill
Nicholas C. Carpita, Purdue University
Frank C. Church, University of North Carolina School of Medicine
Steven G. Clarke, University of California, Los Angeles
William J. Coleman, Dupont Central Research & Development
James E. Darnell, Jr., The Rockefeller University
Christian de Duve, Rockefeller University and International Institute of Cellular and
    Molecular Pathology, Belgium
Daniel V. DerVartanian, University of Georgia
Richard E. Dickerson, University of California, Los Angeles
Walter J. Dobrogosz, North Carolina State University
David H. Dressler, Harvard Medical School
Paul T. Englund, Johns Hopkins University School of Medicine
Shelagh M. Ferguson-Miller, Michigan State University
Robert J. Fogelsong, Duke University
Irwin Fridovich, Duke University School of Medicine
Robert B. Gennis, University of Illinois
Naba K. Gupta, University of Nebraska, Lincoln
James H. Hageman, New Mexico State University
Nancy V. Hamlett, Swarthmore College
Franklin M. Harold, National Jewish Center for Immunology and Respiratory
    Medicine
Edward D. Harris, Texas A & M University
Robert A. Harris, Indiana University School of Medicine
Ari H. Helenius, Yale University School of Medicine
Kenneth Hellman, Smith College
Jens M. Hemmingsen, Duke University School of Medicine
Henri G. Hers, de l'Universite Catholique de Louvain, Belgium
Robert L. Hill, Duke University School of Medicine
Peter C. Hinkle, Cornell University
George E. Hoch, University of Rochester
Jan B. Hoek, Thomas Jefferson University
Johns W. Hopkins, III, Washington Univeristy
Ching-hsien Huang, University of Virginia School of Medicine
Evan E. Jones, North Carolina State University
Kenneth M. Jones, University of Leicester, England
Mary Ellen Jones, University of North Carolina at Chapel Hill
Manfred L. Karnovsky, Harvard Medical School

*The affiliations shown are those that obtained when these reviewers and advisors made their
  contributions.

Robert A. B. Keates, University of Guelph
Judith A. Kelly, The University of Connecticut
Debra A. Kendall, The University of Connecticut
Sue C. Kinnamon, Colorado State University
Aaron Klug, Medical Research Council, Cambridge, England
Ronald Kluger, University of Toronto
Ryzard Kole, University of North Carolina at Chapel Hill
Arthur Kornberg, Stanford University Medical School
Nicholas A. Kredich, Duke University School of Medicine
Monty Krieger, Massachusetts Institute of Technology
LeRoy R. Kuehl, University of Utah
James A. Lake, University of California, Los Angeles
M. Daniel Lane, Johns Hopkins University School of Medicine
Robert N. Lindquist, San Francisco State University
Rose G. Mage, National Institutes of Health
Lynn A. Margulis, University of Massachusetts
William F. Marzluff, University of North Carolina at Chapel Hill
Steve W. Matson, University of North Carolina at Chapel Hill
James B. Meade, University of North Carolina School of Medicine
William C. Merrick, Case Western Reserve University School of Medicine
Richard Ogden, Agouron Institute
Philip Oliver, University of Cambridge, England
Juris Ozols, University of Connecticut School of Medicine
Marvin R. Paule, Colorado State University
Henry P. Paulus, Boston Biomedical Research Institute
Norma R. Pedersen, North Carolina State University
Tom D. Petes, University of North Carolina at Chapel Hill
Allen T. Phillips, Pennsylvania State University
Simon J. Pilkis, State University of New York at Stony Brook
Huntington Potter, Harvard Medical School
Jane H. Potter, University of Maryland
Gary L. Powell, Clemson University
Margaret Rand, University of Toronto
Charles C. Richardson, Harvard Medical School
Phillips W. Robbins, Massachusetts Institute of Technology
Nadia A. Rosenthal, Children's Hospital/Harvard Medical School
Milton H. Saier, Jr., University of California, San Diego
Marvin L. Salin, Mississippi State University
Vern G. Schirch, Medical College of Virginia
B. Trevor Sewell, University of Cape Town, South Africa
Gordon C. Shore, McGill University
James N. Siedow, Duke University
Lewis M. Siegel, Duke University School of Medicine
Gerald R. Smith, Fred Hutchinson Cancer Research Center
Jaro Sodek, University of Toronto
Ronald L. Somerville, Purdue University
Trevor Spencer, Victoria, B.C.
Franklin W. Stahl, University of Oregon
Deborah A. Steege, Duke University School of Medicine
Clarence H. Suelter, Michigan State University
Mary C. Sugden, Queen Mary and Westfield College, England
Judith A. Swan, Duke University
George Taborsky, University of California at Santa Barbara
Keith E. Taylor, University of Windsor
Dean R. Tolan, Boston University
George Tomlinson, University of Winnipeg
Thomas W. Traut, University of North Carolina at Chapel Hill
Peter H. von Hippel, University of Oregon
Robert W. Wheat, Duke University
George B. Witman, Worcester Foundation for Experimental Biology
Owen N. Witte, University of California, Los Angeles
Charles Yanofsky, Stanford University
Douglas C. Youvan, Massachusetts Institute of Technology
James K. Zimmerman, Clemson University

List of Chapters

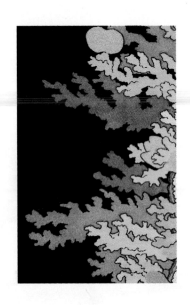

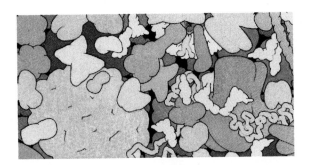

Part Three
Metabolism and Bioenergetics

Part Four
Biological Information Flow

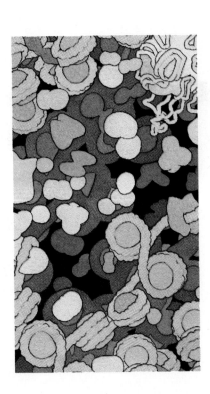

Biochemistry Resource Book

This *Biochemistry Resource Book* comes *free* with each new copy of Moran·Scrimgeour *Biochemistry.*

Supplements

Biochemistry Resource Book

- Problems for each chapter
- Comprehensive, step-by-step solutions to all problems
- Dictionary of biochemical terms
- Summaries of all chapters
- Study pages showing key molecular structures and metabolic pathways
- List of biochemical abbreviations
- Tables of essential data
- Stereo viewer

Overhead Transparencies

- 250 full-color overhead transparencies showing key illustrations from this text
- *Free* to qualified adopting professors

Exploring Molecular Structure

An interactive computer supplement that allows students to manipulate three-dimensional depictions of biomolecules on-screen.

Based on *Kinemages,* dynamic models of molecular structure developed by Jane and David Richardson at Duke University.

- 10 sets of exercises (two disks) that span key topics in biochemistry
- Available for Macintosh or Windows
- On-screen questions presented side-by-side with images
- Moran·Scrimgeour text is keyed to inform students of related material on disks.
- *Free* to qualified adopting professors

Gallery of Macromolecular Structures

Beginning at the bottom of this page and spread across the other pages of the Table of Contents is a collection of macromolecules, each shown at a magnification of 5 000 000. These models were built from coordinates on file in the Brookhaven Protein Data Bank. The structures of the molecules were determined by X-ray diffraction studies.

David S. Goodsell

Contents

**Part One
Introduction**

CRP-cAMP

DNA

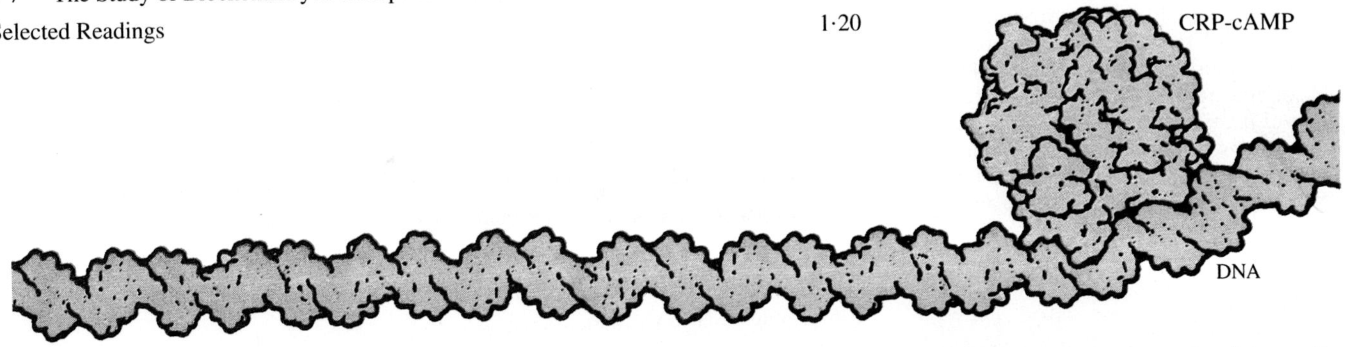

A double strand of DNA carries the genetic information of the cell. Proteins can bind to the DNA helix to control the expression of each gene. cAMP regulatory protein (CRP), bound to cAMP, is shown here in contact with DNA.

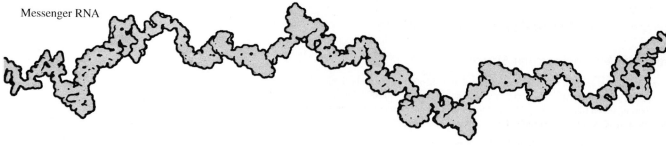

Messenger RNA

Transfer RNA
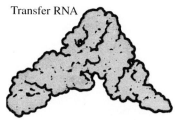

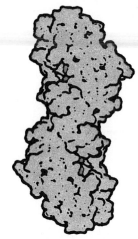

Tyrosyl-tRNA synthetase

RNA is a less stable molecule than DNA. A single strand of messenger RNA carries the information to build proteins from the archive of DNA to the engine of protein synthesis, the ribosome. Short strands of RNA fold to form transfer RNAs, which match each triplet of bases in the messenger RNA to the proper amino acid in a growing protein. Aminoacyl-tRNA synthetases, such as tyrosyl-tRNA synthetase, match the proper amino acid to the proper transfer RNA molecule.

Chapter 2
Cells

Chapter 3
Water

Part Two
Structures and Functions of Biomolecules

Chapter 4
Amino Acids and the Primary Structures of Proteins

4·1

Photosynthetic reaction center

Lipid bilayer

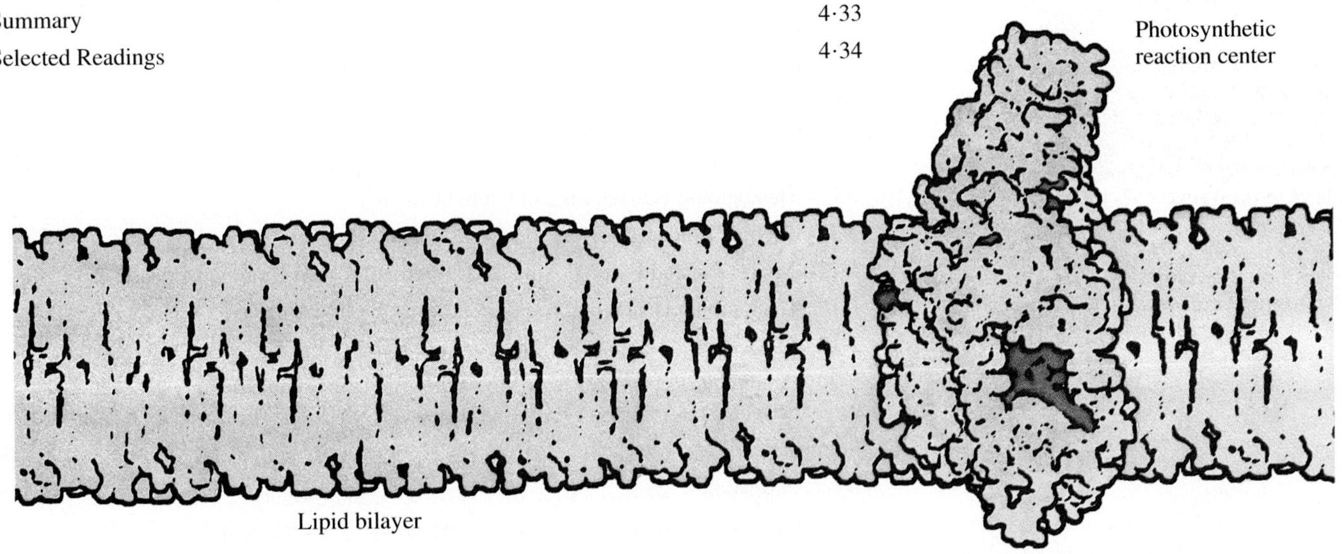

Lipid bilayers form the waterproof skin of cells and delineate compartments within eukaryotic cells. Embedded proteins allow molecules, information, and energy to cross the bilayer.

Papain

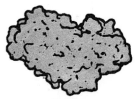

Pepsin

HIV-I protease

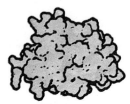

Rhodanese

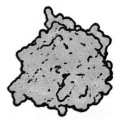

Carboxypeptidase A

The simplest enzymes catalyze a cutting reaction, using water to sever a protein chain.

Chapter 5
Proteins: Three-Dimensional Structure and Function

Chapter 6
Properties of Enzymes

β-Trypsin–pancreatic trypsin inhibitor

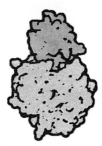

Subtilisin carlsberg–eglin C

Some cells protect themselves from invasive proteases by building specific inhibitor proteins, which block the active sites of the proteases.

Chapter 7
Mechanisms of Enzymes

Some simple enzymes use water to cleave molecules other than proteins, including RNA, polysaccharides, lipids, and penicillins.

Lysozyme

Staphylococcus nuclease

Ribonuclease A

Chapter 8
Coenzymes

Phospholipase A_2

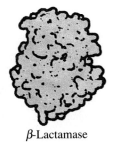

β-Lactamase

Dihydrofolate
reductase

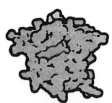

Elongation factor Tu

Carbonic anhydrase

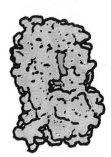

p-Hydroxybenzoate
hydroxylase

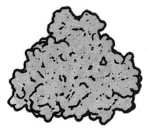

Cytochrome P$_{450cam}$

Some enzymes are composed of a single
polypeptide chain, which folds to form an
active-site pocket on one side.

Chapter 9
Carbohydrates and Glycoconjugates

Chapter 10
Nucleotides   10·1

Chapter 11
Lipids   11·1

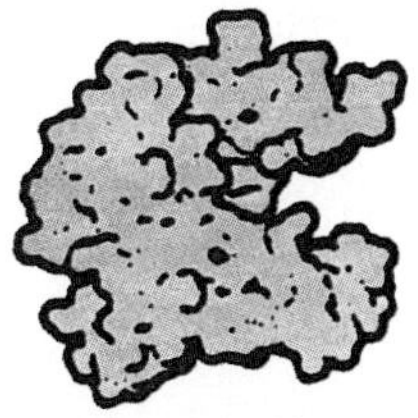

Adenylate kinase

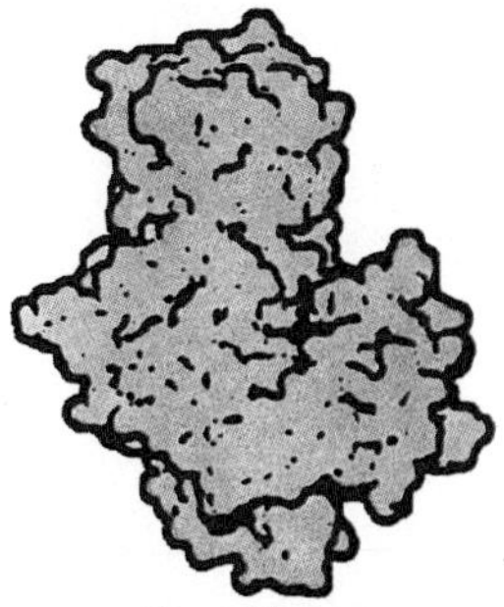

Hexokinase

Phosphoglycerate
kinase

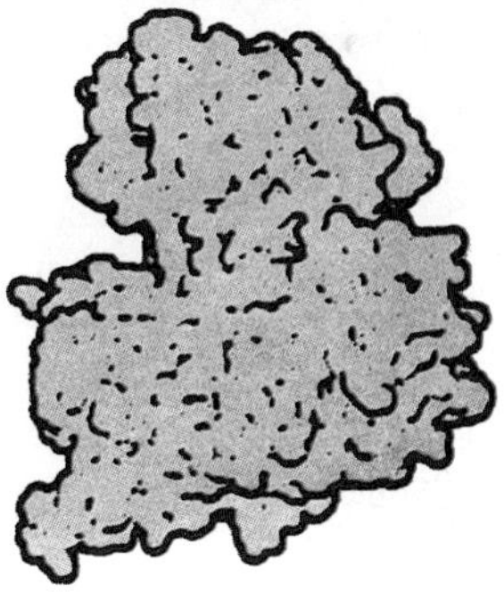

Aconitase

Some enzymes are formed of two domains
connected by a flexible hinge, which may
close to trap a substrate.

Chapter 12
Biological Membranes

Chapter 13
Digestion

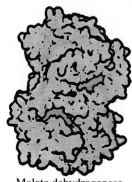

Malate dehydrogenase

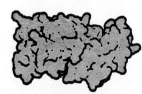

Superoxide dismutase

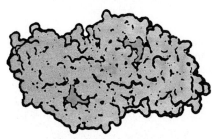

Alcohol dehydrogenase

Some enzymes are composed of two subunits.

Part Three
Metabolism and Bioenergetics

Chapter 14
Introduction to Metabolism   **14·1**

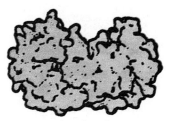

Triose phosphate isomerase

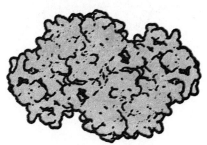

Aspartate transaminase

Chapter 15
Glycolysis   **15·1**

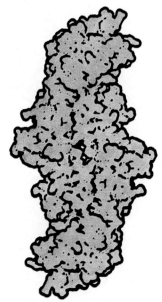

Tryptophan synthase

Chapter 16
The Citric Acid Cycle

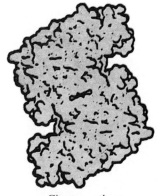

Citrate synthase

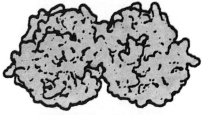

Inorganic pyrophosphatase

Chapter 17
Additional Pathways in Carbohydrate Metabolism

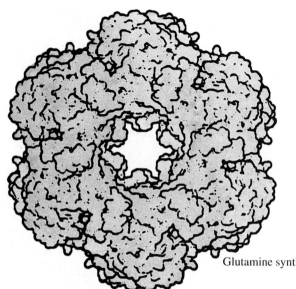

Glutamine synthetase

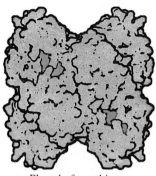

Phosphofructokinase

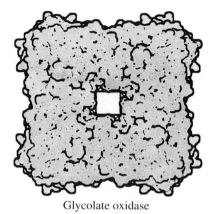

Glycolate oxidase

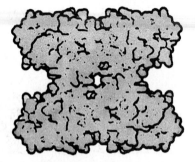

Phosphoglycerate mutase

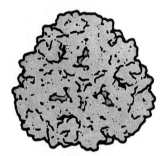

Chloramphenicol
acetyltransferase

Allosteric enzymes typically are composed of several subunits. Subtle shifting between subunits allows a wide range of control of the enzyme's activity.

Chapter 18
Electron Transport and Oxidative Phosphorylation 18·1

Catalase

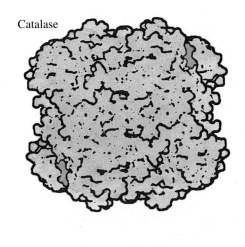

Aspartate transcarbamoylase

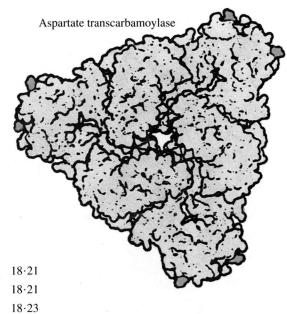

Chapter 19
Photosynthesis   19·1

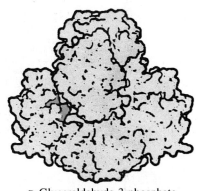

D-Glyceraldehyde-3-phosphate dehydrogenase

Chapter 20
Lipid Metabolism

Erythrocruorin

Myoglobin

Leghemoglobin

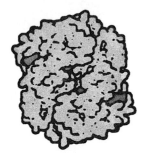

Hemoglobin

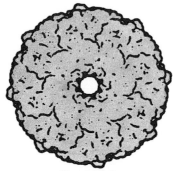

Hemerythrin

A heme surrounded by a protein chain forms a reversible carrier of oxygen. Allosteric oxygen carriers allow fine control of oxygen uptake and release.

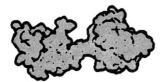
Troponin C

Calcium-binding
parvalbumin

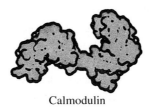
Calmodulin

Calcium is used as a signal molecule in many physiological processes. Proteins are used to sense these signals.

Chapter 21
Amino Acid Metabolism   21·1

Uteroglobin

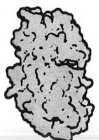

Arabinose-binding
protein

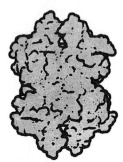

Prealbumin

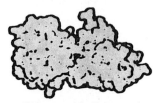

Galactose-binding protein

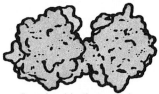

Leucine-binding protein

Protein carriers often transport small mole-cules to regions of need.

Chapter 22
Nucleotide Metabolism

22·1

Chapter 23
Integration of Fuel Metabolism in Mammals

Cytochrome c_5

Cytochrome c_{551}

Cytochrome b_5

Cytochrome b_{562}

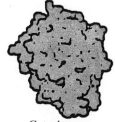

Cytochrome c
peroxidase

Cytochrome c'

Cytochrome c_3

Cytochrome c

Proteins with heme are used to carry electrons.

High-potential
iron protein

Pseudoazurin

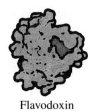
Azurin

Flavodoxin

Plastocyanin

Ferredoxin

Rubredoxin

Metal atoms or small molecules held by proteins are also used to carry electrons.

Part Four
Biological Information Flow

Chapter 24
Nucleic Acids
24·1

Chapter 25
DNA Replication

25·1

Glucagon

Insulin

Interleukin 1β

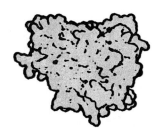
Tumor necrosis factor

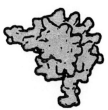
Interleukin 8

Small proteins are used as messengers to warn of blood sugar variation. They are also used to send messages among the various cells of the immune system.

Chapter 26
DNA Repair and Recombination

26·1

Chapter 27
Transcription

27·1

Chapter 28
RNA Processing

Antibodies, which bind to any foreign substance, have two specific binding sites, one at the tip of each arm. Lectins, from plants, also have two binding sites, although they are specific for binding to only a particular sugar.

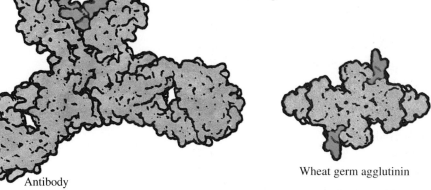

Antibody

Wheat germ agglutinin

α-Cobratoxin

Mellitin

Hirudin

Actinoxanthin

Erabutoxin β

Scorpion
neurotoxin

Many organisms make small, stable proteins that actively attack other cells.

Chapter 31
Gene Expression and Development

Chapter 32
Genes and Genomes   32·1

Other proteins are built for specific functions.

Ubiquitin

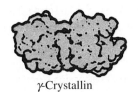

γ-Crystallin

Crambin

Water

Glucose

Alanine

Tryptophan

Heme

ATP

NAD⊕

FAD

Small molecules.

Chapter 33
Recombinant DNA Technology

33·1

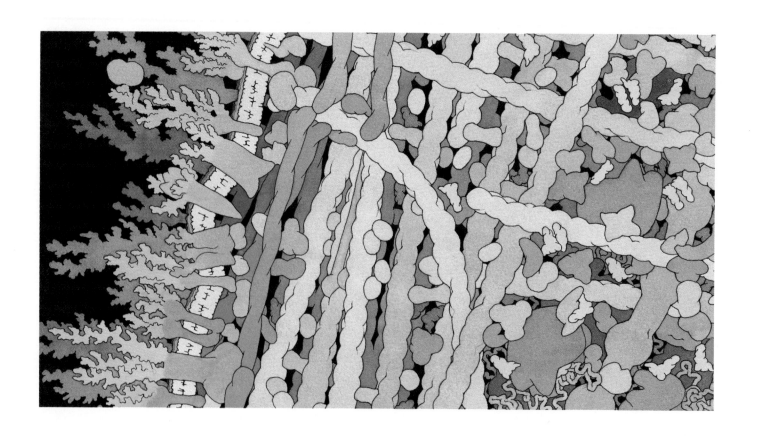

Part One

Introduction

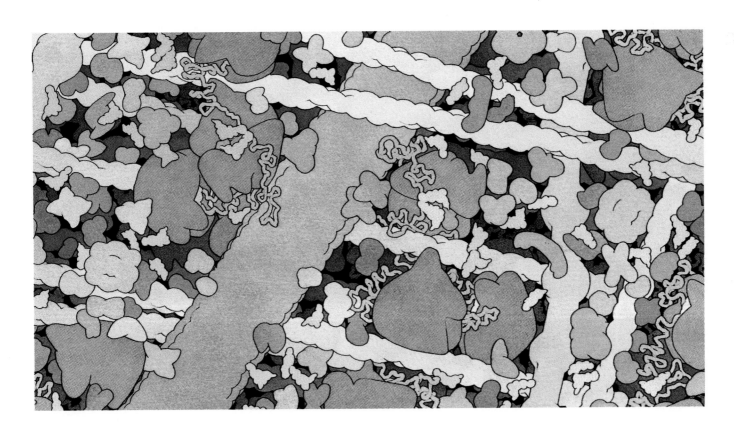

1

Introduction to Biochemistry

Biochemistry is the science of the molecules and chemical reactions of life. One might at first assume that biochemistry is merely a combination of two major sciences, chemistry and biology. However, the defining feature of biochemistry is that it uses the principles and language of one science—chemistry—to explain the other science—biology—at the molecular level. This book introduces the names and structures of the elementary biochemicals and then, building upon this framework, explains the fascinating network of chemical reactions that support life.

1·1 Biochemistry Is a 20th Century Science

Biochemistry has emerged as a dynamic science only within the past hundred years. The first biochemistry departments were founded in Europe and in North America around the turn of the century, and building on a few key discoveries, biochemists have since elucidated many of the basic chemical processes of life. The 20th century has witnessed an explosion of knowledge in the field of biochemistry.

The foundation for modern biochemistry was laid by experiments performed during the 19th century. In 1828, Friedrich Wöhler (Figure 1·1) synthesized the organic compound urea by heating the inorganic compound ammonium cyanate.

$$NH_4(OCN) \xrightarrow{\text{Heat}} H_2N-\overset{\overset{\displaystyle O}{\|}}{C}-NH_2 \qquad \textbf{(1·1)}$$

Wöhler's experiment showed for the first time that organic compounds could be prepared from inorganic substances without the aid of a vital force (i.e., living organism). We now take for granted that both organic ("living") and inorganic ("nonliving") chemicals obey the same chemical and physical laws.

In the last third of the 19th century, physiological chemists analyzed body fluids and tissues and characterized the small, stable compounds they found. However,

Figure 1·1
Friedrich Wöhler (1800–1882). By synthesizing urea, Wöhler showed that compounds occurring in living organisms could be made in the laboratory from inorganic substances.

Figure 1·2
Emil Fischer (1852–1919). Fischer made many contributions to our understanding of the structures and functions of biological molecules.

it took two major breakthroughs—the elucidation of the roles of enzymes and of nucleic acids—to establish biochemistry as the dynamic science it is today.

An understanding of the role of enzymes as the catalysts of biological reactions emerged around the turn of the century. This breakthrough was based in part on the research of Emil Fischer (Figure 1·2), who studied the catalytic effect of yeast enzymes on a simple reaction, the hydrolysis (breakdown by water) of disaccharides (molecules containing two sugars). In 1894, Fischer proposed that during catalysis an enzyme and its reactant, or substrate, combine to form an intermediate compound. He also proposed that only molecules with suitable structures can serve as substrates for the enzyme. Fischer described an enzyme as a rigid template, or lock, and the substrate as a matching key. The lock-and-key theory of enzyme action remains a central tenet of modern biochemistry.

The research of Eduard Büchner further advanced the field of biochemistry. In 1897, Büchner showed that extracts of yeast cells could catalyze the fermentation of the sugar glucose to alcohol and carbon dioxide. It was previously believed that only living cells could catalyze such complex biological reactions. The science of biochemistry developed as researchers realized that virtually all of the reactions of life are catalyzed by enzymes.

The second major breakthrough, elucidation of the biological role of nucleic acids as information molecules, came 50 years after Fischer's theory and Büchner's experiments. In 1944, Oswald Avery, Colin MacLeod, and Maclyn McCarty extracted deoxyribonucleic acid (DNA) from a toxic strain of the bacterium *Streptococcus pneumoniae* and mixed the DNA with a nonpathogenic strain. As a result, the nonpathogenic strain was permanently transformed to a toxic strain. This experiment provided the first conclusive evidence that DNA is the genetic material. In 1953, the three-dimensional structure of DNA was deduced by James D. Watson (Figure 1·3) and Francis H. C. Crick (Figure 1·4). The structure of DNA immediately suggested to Watson and Crick a method whereby DNA could reproduce, or replicate, and thus transmit biological information to succeeding generations.

The study of genetics at the level of nucleic acid molecules has come to be known variously as molecular biology, molecular genetics, or molecular cell biology. However, molecular biology under any name cannot be truly separated from the rest of biochemistry. In order to understand how nucleic acids store and transmit genetic information, one must understand what nucleic acids are and how they are made. One must also understand their role in encoding the enzymes that catalyze the synthesis and degradation of biomolecules, including the nucleic acids themselves. Nucleic acids are intricately enmeshed in the overall metabolism of the cell, and therefore an understanding of their roles is critical to an understanding of biochemistry as a whole.

Figure 1·3
James D. Watson (1928–).

Figure 1·4
Francis H. C. Crick (1916–). Watson and Crick revolutionized the biological sciences with their work on the three-dimensional structure of DNA.

1·2 Many Chemical Elements Are Required for Life

Most complex biomolecules are composed of only a few chemical elements. In fact, over 97% of the weight of most organisms is due to six elements—oxygen, carbon, hydrogen, nitrogen, phosphorus, and sulfur. The relative amounts of these six elements vary among organisms, as shown in Table 1·1.

Water (H_2O) is the most common compound in living organisms, accounting for at least 70% of the weight of most cells. The prevalence of water is responsible in part for the large amount of oxygen in cells. Because water is both the major solvent of organisms and a reactant in many biochemical reactions, an entire chapter (Chapter 3) is devoted to the properties of water.

Carbon is the most prevalent element in the solid material of cells. Carbon atoms can bond to other carbon atoms or to atoms of other elements, principally hydrogen, nitrogen, oxygen, and sulfur. Furthermore, carbon atoms can form up to four covalent bonds, producing an almost infinite number of compounds.

During the 20th century, the definition of organic chemistry changed from the chemistry of compounds of living organisms to the chemistry of carbon-containing compounds (with the exception of a few compounds such as carbon dioxide and

Table 1·1 Elemental composition of three organisms

Element	Human	Alfalfa	Bacterium
Oxygen	62.81%	77.90%	73.68%
Carbon	19.37	11.34	12.14
Hydrogen	9.31	8.72	9.94
Nitrogen	5.14	0.83	3.04
Phosphorus	0.63	0.71	0.60
Sulfur	0.64	0.10	0.32
Total	97.90%	99.60%	99.72%

Composition is given as percentage by weight.
[Adapted from Curtis, H., and Barnes, S. (1989). *Biology*, 5th ed. (New York: Worth Publishers), p. 35.]

Figure 1·5
General formulas of (a) organic compounds, (b) important functional groups, and (c) linkages common in biochemistry. *R* represents an alkyl group ($CH_3(CH_2)_n$—).

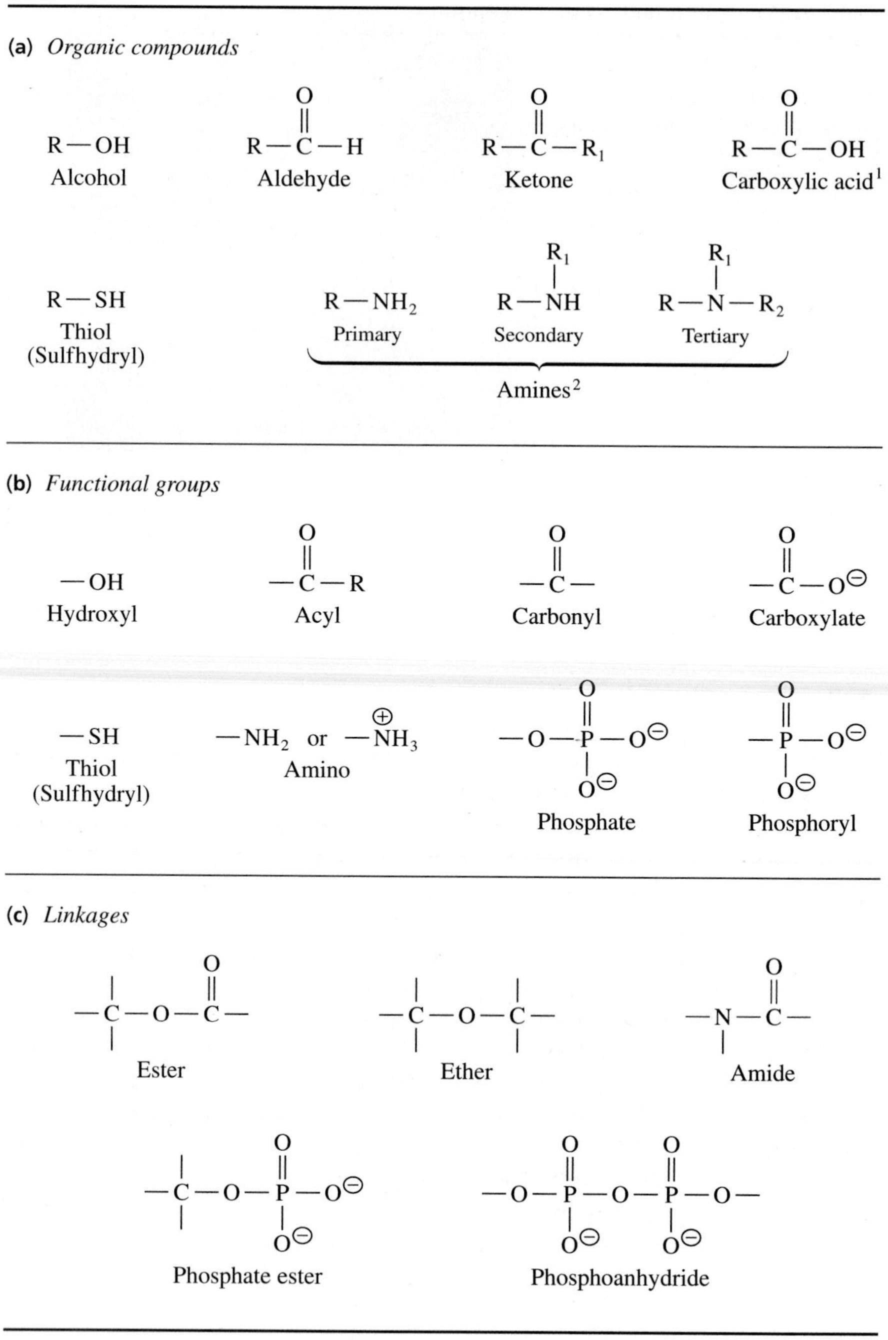

(a) *Organic compounds*

R—OH	R—C(=O)—H	R—C(=O)—R_1	R—C(=O)—OH
Alcohol	Aldehyde	Ketone	Carboxylic acid[1]

R—SH	R—NH_2	R—NH(R_1)	R_2—N(R_1)—R
Thiol (Sulfhydryl)	Primary	Secondary	Tertiary

Amines[2]

(b) *Functional groups*

—OH	—C(=O)—R	—C(=O)—	—C(=O)—$O^{\ominus}$
Hydroxyl	Acyl	Carbonyl	Carboxylate

—SH	—NH_2 or —$\overset{\oplus}{NH_3}$	—O—P(=O)($O^{\ominus}$)—$O^{\ominus}$	—P(=O)($O^{\ominus}$)—$O^{\ominus}$
Thiol (Sulfhydryl)	Amino	Phosphate	Phosphoryl

(c) *Linkages*

—C—O—C(=O)—	—C—O—C—	—N—C(=O)—
Ester	Ether	Amide

—C—O—P(=O)($O^{\ominus}$)—$O^{\ominus}$	—O—P(=O)($O^{\ominus}$)—O—P(=O)($O^{\ominus}$)—O—
Phosphate ester	Phosphoanhydride

[1] Under most biological conditions, carboxylic acids exist as carboxylate anions: R—C(=O)—$O^{\ominus}$.

[2] Amines can also be protonated: R—$\overset{\oplus}{NH_3}$, R—$\overset{\oplus}{NH_2}$(R_1), and R_2—$\overset{\oplus}{NH}$(R_1)—R.

potassium cyanide). By either the classical or the modern definition, the carbon-containing species in biochemistry are known as organic compounds. The types of organic compounds commonly encountered in biochemistry are shown in Figure 1·5a.

Biochemical reactions involve specific chemical bonds and parts of molecules called functional groups. The most common of these sites of reactivity are shown in

Table 1·2 Elements found in living organisms

Major components		Essential ions		Trace elements required by all organisms		Trace elements required by some organisms	
Oxygen	O	Calcium	Ca^{2+}	Cobalt	Co	Aluminum	Al
Carbon	C	Chloride	Cl^{-}	Copper	Cu	Arsenic	As
Hydrogen	H	Magnesium	Mg^{2+}	Iron	Fe	Boron	B
Nitrogen	N	Potassium	K^{+}	Manganese	Mn	Chromium	Cr
Phosphorus	P	Sodium	Na^{+}	Zinc	Zn	Fluorine	F
Sulfur	S					Gallium	Ga
						Iodine	I
						Molybdenum	Mo
						Nickel	Ni
						Selenium	Se
						Silicon	Si
						Tin	Sn
						Vanadium	V

Figure 1·5b. Figure 1·5c shows some of the types of linkages present in derivatives of the compounds shown in Figure 1·5a. We will encounter these compounds, functional groups, and linkages throughout this book.

Biochemistry is not restricted to compounds containing only the six elements listed in Table 1·1. An additional 23 elements have been found in living organisms (Table 1·2). Five of these additional elements are ions that are essential for all organisms. Calcium ions, for example, form crystalline phosphates in bone, function as intracellular signals, and stabilize some proteins. Potassium and sodium ions are common counterions in biological fluids. Many multicellular organisms possess enzymes whose activities maintain high concentrations of potassium and low concentrations of sodium inside cells. The remaining elements, called trace elements, are present in very small quantities. Five trace elements are required by virtually all organisms, and thirteen others are found in some organisms. Despite their low quantities, the trace elements are important components of living organisms. For example, hemoglobin requires iron to bind molecular oxygen. Ions of many trace elements play their roles at the active sites of particular enzymes.

1·3 Many Important Biomolecules Are Polymers

A large part of the chemistry of life is the chemistry of **biopolymers,** macromolecules in which multiple small molecules are linked to each other to form long chains. Although small molecules are familiar to students of chemistry, students of biochemistry must become familiar with macromolecules whose molecular weights range from the tens of thousands to the millions. Aside from water, the most abundant compounds in cells are biopolymers—in order of their abundance, proteins, nucleic acids, polysaccharides (the polymers of carbohydrates), and lipids (Table 1·3). Lipids do not exactly fit the definition of polymer, but since lipid molecules are most often found in association with each other, it is sometimes useful to treat lipids as polymers. Low-molecular-weight **metabolites** (intermediates in the formation or degradation of the biopolymers and their component units) and inorganic ions together make up little more than 1% of the weight of a cell.

Table 1·3 Components of a rapidly growing cell of the bacterium *Escherichia coli*

Type of compound	Percentage of total cell weight
Water	70
Proteins	15
Nucleic acids	7
Polysaccharides	3
Lipids	2
Inorganic ions	1
Metabolites	0.2

The organic molecules from which polymers of living cells are formed are called monomers. The formation of a polymer from its monomers requires many successive condensation reactions; that is, removal of one molecule of water for each monomeric unit added to the growing chain. After a monomer is incorporated into a polymer, it is called a **residue** of the polymer. In some cases, such as some carbohydrates, a single residue is repeated many times; in other cases, such as proteins and nucleic acids, different residues are connected in a particular order that is critical for the function of the molecule.

Biopolymers have properties distinct from those of their constituent monomers. This observation has led to the general principle of the hierarchical organization of life. With each new level of organization comes a new level of complexity that cannot readily be predicted from the properties of the previous level. The levels of organization, in order of increasing biological complexity, are atoms, molecules, biopolymers, organelles, cells, tissues and organs in more complex species, and whole organisms. Proteins and nucleic acids, two types of large polymers that are subjects of much of this book, illustrate the hierarchical principle. Proteins are linear polymers that are folded into unique three-dimensional shapes, or conformations. The activities of the proteins that are enzymes (the cellular catalysts) are dependent on their conformations. Consider the ability of an enzyme to select only certain reactants and to catalyze their conversion to specific products with extremely high efficiency. These life-supporting properties cannot be explained by the chemistry of the individual monomeric components of the protein, only by the organized three-dimensional shape of the enzyme. Similarly, DNA, a nucleic acid, is a polymer that stores genetic information. Individually, the four nucleotide components of DNA have no information capacity. However, when the nucleotides are joined into a linear polymer of DNA, their sequence encodes and directs the synthesis of all of the proteins of a cell and those of future generations.

The following sections briefly introduce the principal types of biopolymers. For each polymer, the discussion includes the structure and chemistry of the monomers, the structure of the polymer, and the relationship between polymer function and its composition and conformation. These discussions serve as both an introduction and a reference to consult if necessary until detailed material on each polymer is presented later in the book.

A. Proteins Are Polymers of Amino Acids

There are 20 common amino acids that are incorporated into proteins in all cells. Amino acids are synthesized from simpler precursor molecules or are absorbed as nutrients. All amino acids contain at least two functional groups—an amino group and a carboxylate group (Figure 1·6). The side chains (R groups) differ among the 20 common amino acids.

The simplest amino acid is glycine, whose side chain is a hydrogen atom (Figure 1·7a). Three amino acid molecules can condense to form a linear polymer called a tripeptide, as shown in Figure 1·7b. This specific tripeptide, containing three residues of glycine, is called glycylglycylglycine. In every polypeptide, the carbonyl group of one amino acid residue is covalently attached to the amino group of its neighbor by an amide bond called a peptide bond. Proteins are much larger than this tripeptide, typically containing 50 or more amino acid residues. Variations in the amino acid composition and in the sequence of amino acid residues are responsible for the great variety of proteins found in nature.

The three-dimensional structure of a protein is determined largely by the sequence of its amino acid residues, which is encoded by a gene. The function of a protein depends on its three-dimensional shape. The structures of many proteins have been determined, and several principles governing the relationship between structure and function have become clear. For example, most proteins that function as enzymes contain a cleft or groove that binds the substrates of a reaction. Figure

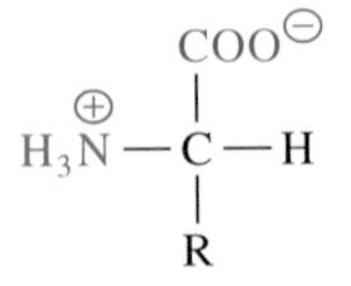

$$\overset{\oplus}{H_3N} - \overset{\overset{\displaystyle COO^{\ominus}}{|}}{\underset{\underset{\displaystyle R}{|}}{C}} - H$$

Figure 1·6
General structure of an amino acid. All amino acids contain at least two functional groups—a protonated amino group (blue) and a carboxylate group (red)—in addition to a side chain (R).

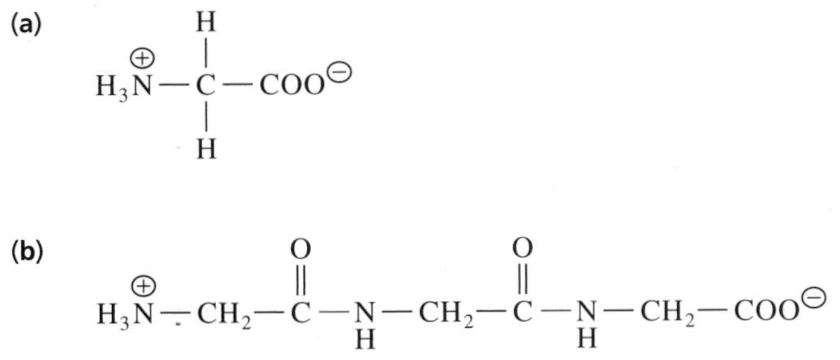

(a)

$$H_3\overset{\oplus}{N} - \overset{\overset{\displaystyle H}{|}}{\underset{\underset{\displaystyle H}{|}}{C}} - COO^{\ominus}$$

(b)

$$H_3\overset{\oplus}{N} - CH_2 - \overset{\overset{\displaystyle O}{||}}{C} - \underset{\underset{\displaystyle H}{|}}{N} - CH_2 - \overset{\overset{\displaystyle O}{||}}{C} - \underset{\underset{\displaystyle H}{|}}{N} - CH_2 - COO^{\ominus}$$

Figure 1·7
Structures of **(a)** glycine and **(b)** a tripeptide containing three residues of glycine. The peptide bonds are shown in red.

(a)   **(b)**

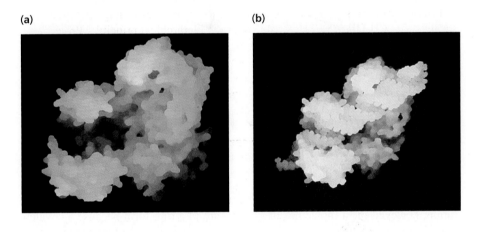

Figure 1·8
Computer-generated images of HIV I reverse transcriptase. The three-dimensional shape of the enzyme allows it to bind specifically to viral RNA. **(a)** Structure of the enzyme. Note the large groove where the substrate binds. **(b)** A model of the enzyme with bound substrate—a DNA/RNA hybrid molecule. (Courtesy of Thomas A. Steitz.)

1·8 shows the structure of the enzyme reverse transcriptase that is encoded by the human immunodeficiency virus I (HIV I) genome. During infection, reverse transcriptase catalyzes synthesis of DNA by promoting the copying of the viral ribonucleic acid (RNA) molecule. The resulting DNA-RNA hybrid lies in a large groove in the middle of the enzyme. Knowledge of the structure of this enzyme has assisted in development of specific drugs that can inhibit its function and possibly prevent the spread of acquired immune deficiency syndrome (AIDS).

Some proteins can associate to form larger structures. This next level of organization produces complexes that can carry out a series of reactions. We will consider protein complexes in our descriptions of **metabolism,** the reactions whereby monomers and macromolecules are synthesized and degraded.

B. Polysaccharides Are Carbohydrate Polymers

Carbohydrates (also called saccharides) are composed primarily of carbon, oxygen, and hydrogen. This group of biomolecules includes simple sugars (monosaccharides), polymers (polysaccharides), and other sugar derivatives. All monosaccharides (and all residues of polysaccharides) contain several hydroxyl groups and are therefore polyalcohols. The most common simple sugars contain either five or six carbon atoms. Glucose, the major six-carbon sugar, is a cellular nutrient and is the monomeric unit of the storage polysaccharides glycogen and starch. Cellulose, the structural polysaccharide of plant cell walls, is also a polymer of glucose.

The structure of glucose can be represented in several ways. Glucose can be shown as a linear molecule containing five hydroxyl groups and one aldehyde

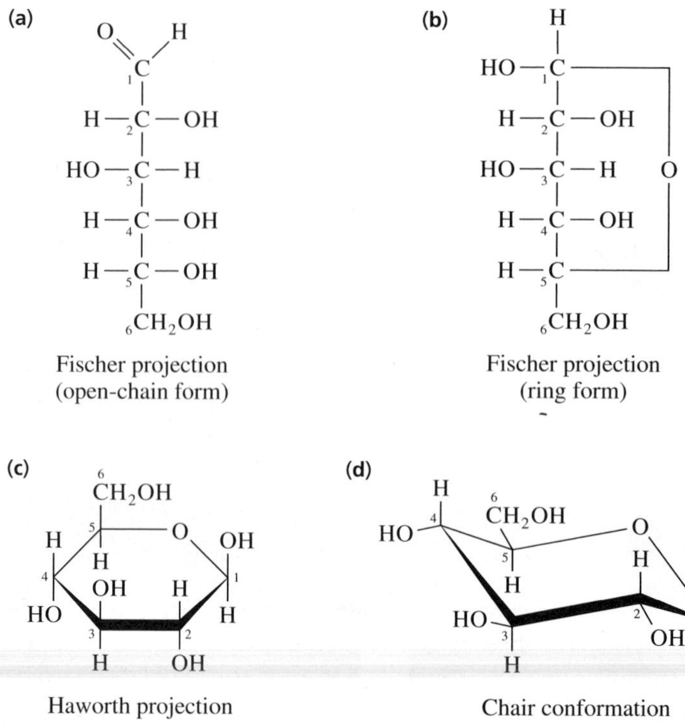

Figure 1·9
Representations of the structure of glucose.
(a) In the Fischer projection, glucose is drawn as a linear molecule. **(b)** The glucose molecule can form a ring structure in solution, shown here as a Fischer projection. **(c)** Ring structures are often drawn as Haworth projections. In a Haworth projection, the ring is almost perpendicular to the page as indicated by the thick lines, which represent bonds closest to the viewer. **(d)** In reality, the ring can exist in several different conformations, including the chair form shown. Each of these representations emphasizes different aspects of the structure of glucose. Sugars will usually be drawn as Haworth projections in this text.

(a)

Fischer projection
(open-chain form)

(b)

Fischer projection
(ring form)

(c)

Haworth projection

(d)

Chair conformation

group (Figure 1·9a). This linear representation is called a Fischer projection. In solution, however, most molecules of glucose are in a ring form because a covalent bond forms between the carbon of the aldehyde group (C-1) and the oxygen of the C-5 hydroxyl group. The ring form can be drawn as a Fischer projection (Figure 1·9b), but in this structure the C—O bond lengths are distorted. The ring form is most commonly shown as a Haworth projection (Figure 1·9c). The Haworth projection is rotated clockwise 90° from the Fischer projection. It portrays the carbohydrate ring as a plane with one edge projecting out of the page; the heavy lines represent the parts of the molecule extending toward the viewer. Hydroxyl groups pointing upward lie above the plane of the ring; those pointing downward lie below

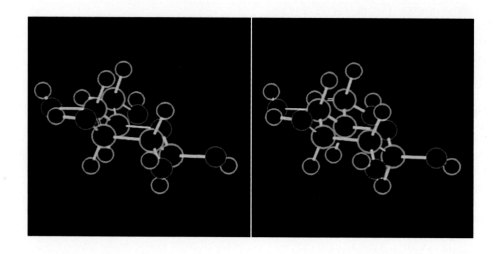

Figure 1·10
Stereo view of glucose in the chair conformation. Color key: carbon, green; hydrogen, blue; oxygen, red.

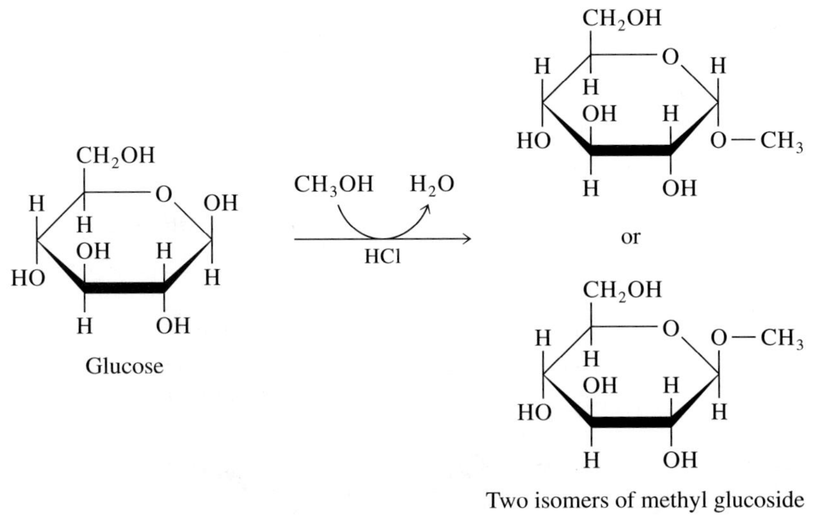

Figure 1·11
Reaction of methanol with glucose. A mixture of two isomers of methyl glucoside is produced.

the plane. Even the Haworth projection, commonly drawn in this text, does not accurately represent the true three-dimensional structure of glucose. The molecule can adopt several conformations, including the chair conformation shown in Figures 1·9d and 1·10.

The hydroxyl group at C-1 of the glucose ring can readily be replaced by alcohols, amines, or thiols. For example, the reaction of methanol with glucose yields methyl glucoside (Figure 1·11). These types of compounds are known generally as glycosides (*glucoside* is specific for glucose). Note that two isomers of methyl glucoside are formed in this nonenzymatic reaction.

Polysaccharides are produced when monosaccharides condense by formation of glycosidic bonds. For example, in cellobiose, the repeating unit of cellulose, C-1 of one glucose residue is connected to the C-4 hydroxyl group of an adjacent glucose residue (Figure 1·12). In cellulose, one glucose residue of each cellobiose unit is rotated 180° relative to the other glucose residue. This arrangement produces a rigid polymer that is further stabilized by hydrogen bonds, which are electrostatic interactions between hydroxyl groups and oxygen atoms of adjacent glucose residues in the same chain. Cellulose polymers interact with one another, forming

Figure 1·12
Structure of cellobiose. The glycosidic bond is shown in red.

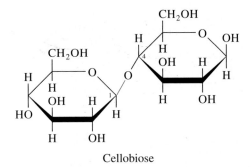

Cellobiose

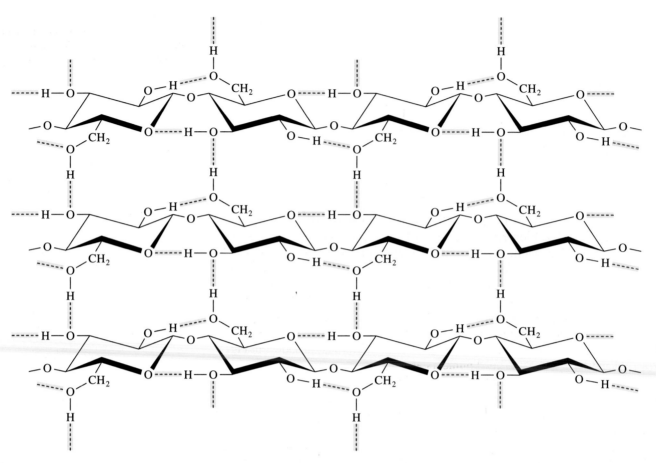

Figure 1·13
Structure of cellulose. Cellulose is a linear polymer of many glucose residues that are connected by glycosidic bonds. Cellulose polymers are stabilized by intrachain hydrogen bonds, and fibers are formed through interchain hydrogen bonding. The hydrogen bonds are shaded yellow.

fibers that are held together by interchain hydrogen bonds (Figure 1·13). Starch and glycogen differ from cellulose in the orientation of their glycosidic bonds at C-1.

The most common five-carbon sugar, ribose (Figure 1·14a), is the carbohydrate component of RNA. 2-Deoxyribose (Figure 1·14b) is found in DNA. As we will see, ribose and deoxyribose are part of nucleotides, which are the monomeric units of nucleic acids.

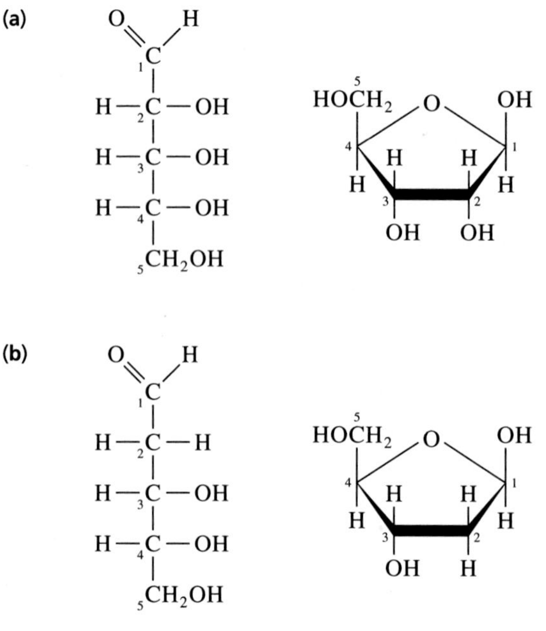

(a)

Figure 1·14
Fischer and Haworth projections of **(a)** ribose and **(b)** 2-deoxyribose.

(b)

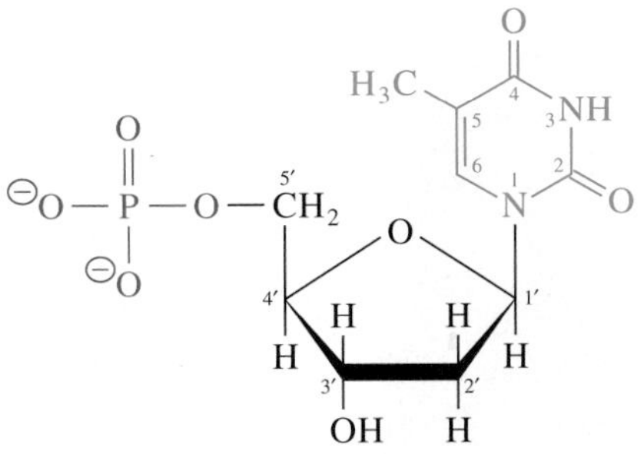

Figure 1·15
Structure of thymidylate (dTMP). This nucleotide contains thymine (blue), deoxyribose (black), and a phosphate group (red).

C. Nucleic Acids Are Polymers of Nucleotides

Nucleotides contain a five-carbon carbohydrate (usually either ribose or deoxyribose), a heterocyclic, nitrogenous base, and at least one phosphate group. The nitrogenous bases of nucleotides are purines and pyrimidines. The major bases are adenine (A), guanine (G), cytosine (C), thymine (T), and uracil (U). In a nucleotide, an amino group of a base is joined to C-1′ of the five-carbon sugar, and the phosphate group is attached to C-5′ of the sugar. Thymidylate, a pyrimidine deoxyribonucleotide from DNA, is shown in Figure 1·15. Thymidylate consists of a thymine ring linked via an *N*-glycosidic bond to 2-deoxyribose, which is phosphorylated at the C-5′ hydroxyl group. Thymidylate is abbreviated dTMP, for deoxythymidine monophosphate. Note that primed numbers are used to designate the five carbon atoms of deoxyribose in dTMP to distinguish them from the carbon and nitrogen atoms of the base.

In nucleic acids, nucleotides are linked by a covalent bond between the 5′-phosphate group of one nucleotide and C-3′ of another. This linkage is called a phosphodiester linkage. The dinucleotide shown in Figure 1·16 illustrates this type of linkage. Nucleic acid polymers contain many nucleotide residues and are characterized by a backbone consisting of alternating sugars and phosphates. The

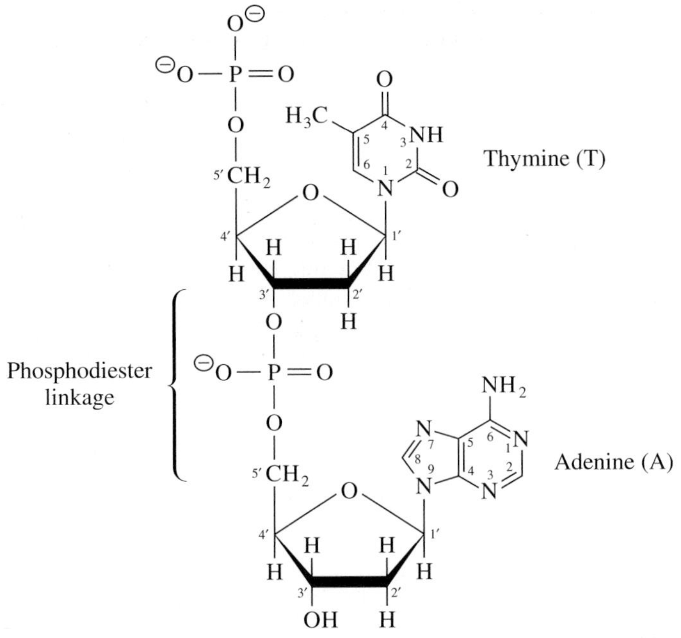

Figure 1·16
Structure of a dinucleotide. This dinucleotide is composed of thymidylate (top) and deoxyadenylate, the deoxyribonucleotide containing the purine adenine (bottom). The phosphodiester linkage involves esterification of phosphate with the hydroxyl group at C-3′ of thymidylate and the hydroxyl group at C-5′ of the deoxyribose attached to adenine.

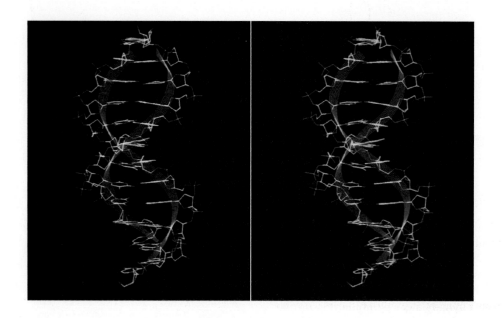

Figure 1·17
Stereo view of a DNA double helix. The sugar-phosphate backbone is highlighted with a purple ribbon. This view shows how the bases (seen edge on) point inward, allowing them to interact in the center of the helix. The specific sequence of nucleotides in DNA carries genetic information. Color key: carbon, green; hydrogen, light blue; nitrogen, dark blue; oxygen, red; phosphorus, orange. (Based on coordinates provided by H. Drew and R. E. Dickerson.)

purine and pyrimidine bases lie nearly perpendicular to the sugar-phosphate backbone. In DNA, the bases of two strands of nucleic acid interact as pairs, and the entire macromolecule forms a helical structure (Figure 1·17). The specific sequence of base pairs carries genetic information. Because A always pairs with T and G always pairs with C, the two strands of DNA are complementary. The double-stranded nature of the DNA polymer is critical for transmitting genetic information, since identical molecules of DNA must be supplied to the progeny when cells divide. When DNA is replicated, the two strands separate, and each acts as a template for the synthesis of a new complementary strand. In this manner, two helices identical to the original DNA duplex are synthesized.

Nucleic acids can be enormous. Some human chromosomes, for example, contain DNA molecules of over one hundred million nucleotide residues. Bacteria, which have fewer genes than mammals, usually contain smaller, circular DNA molecules.

RNA is a single-stranded polynucleotide that is synthesized from a DNA template. There are many different kinds of RNA molecules, including messenger RNA, which is involved directly in the transfer of information from DNA to protein. In messenger RNA, each set of three nucleotides, called a codon, corresponds to a particular amino acid. Proteins are synthesized by "reading" the codons of messenger RNA. Transfer RNAs are smaller molecules required for protein synthesis, and ribosomal RNAs are major components of ribosomes, the RNA-protein complexes on which proteins are synthesized.

D. Lipids Are a Diverse Group of Molecules

Strictly speaking, lipids do not form polymers, but their association with each other creates structures that exhibit properties not shared by individual lipid molecules. Lipids are a diverse group of compounds that are best defined as water-insoluble organic molecules. Some lipids are energy-storage molecules, some are structural elements of membranes, and others are involved in communication within and between cells. The simplest lipids are the fatty acids, which are long-chain hydrocarbons with a carboxylate group at one end. Fatty acids are most commonly found

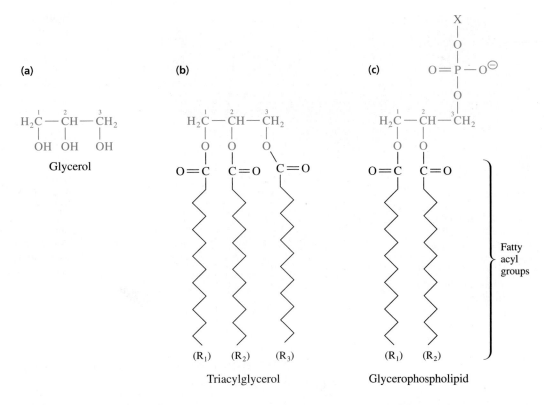

(a)

$$\overset{1}{H_2C} - \overset{2}{CH} - \overset{3}{CH_2}$$
$$\quad | \qquad | \qquad |$$
$$\quad OH \quad OH \quad OH$$

Glycerol

(b)

Triacylglycerol

(c)

Glycerophospholipid

Fatty acyl groups

Figure 1·18
Structures of triacylglycerol and glycero-phospholipid. **(a)** Structure of glycerol. **(b)** Three long-chain fatty acyl groups (designated R_1, R_2, and R_3) are bound to glycerol through ester linkages to form triacylglycerol. **(c)** In glycerophospholipids, two fatty acyl groups are bound to glycerol 3-phosphate, and often a polar group (designated X) is attached to the phosphate.

as part of larger molecules called triacylglycerols (or fats) and glycerophospholipids (Figure 1·18). Triacylglycerols, which contain three fatty acid moieties esterified to a molecule of glycerol, are the most abundant type of lipid in mammals. Glycerophospholipids consist of glycerol 3-phosphate esterified to two fatty acyl groups and usually have a polar group (such as ethanolamine) attached to the phosphate. These lipids are major components of biological membranes.

Membrane lipids usually have a polar, hydrophilic (water-loving) head and a nonpolar, hydrophobic (water-fearing) tail (Figure 1·19). In an aqueous medium, such as that found in cells, the hydrophobic tails of membrane lipids associate, forming two sheets called a lipid bilayer. Lipid bilayers form the structural basis of all biological membranes. Biological membranes separate a cell or a compartment

Figure 1·19
General structure of a membrane lipid. The molecule consists of a polar head (blue) and a nonpolar tail (yellow).

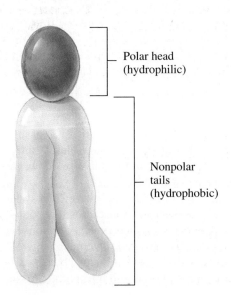

Polar head
(hydrophilic)

Nonpolar tails
(hydrophobic)

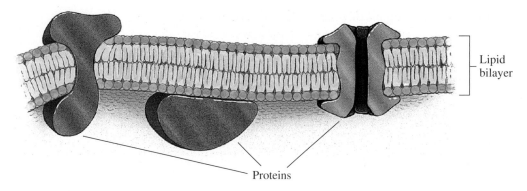

Lipid bilayer

Proteins

Figure 1·20
General structure of a biological membrane. Biological membranes consist of a lipid bilayer with associated proteins. The hydrophobic tails of individual lipid molecules associate to form the core of the membrane; the hydrophilic heads are in contact with the aqueous medium on either side of the membrane. Some proteins span the bilayer; others are attached to the surface in various ways.

within a cell from its surroundings. Most biological membranes have proteins embedded within or attached to the bilayer (Figure 1·20). Some membrane proteins serve as channels for the entry of nutrients and exit of waste, whereas others catalyze reactions that occur specifically at the membrane surface. Membranes are among the largest and most complex cellular structures and are the sites of many important biochemical reactions.

1·4 Thermodynamics, Bioenergetics, and Kinetics Are Important in Biochemistry

There is much more to biochemistry than just understanding the structures and functions of biomolecules. Biochemistry is the study of life at the chemical level—including all the reactions in which biomolecules are synthesized and degraded. Organisms require energy (ultimately derived from sunlight) to sustain their activities, to grow, and to reproduce.

The study of energy in molecules and of changes in energy during biochemical reactions is called **bioenergetics.** Bioenergetics is part of the field of **thermodynamics,** a branch of physical science that deals with energy changes. The principles of thermodynamics, developed in the 19th century, were empirically derived from the study of heat engines. However, the basic thermodynamic principles that apply to energy flow in nonliving systems also apply to the chemistry of life. An appreciation of these principles is one of the keys to understanding biochemistry.

For biochemists, the most useful thermodynamic quantity is the **free-energy change** (ΔG) of a reaction. From ΔG, one can tell whether a reaction is spontaneous and how much energy is given off or required by the reaction. One can also calculate the equilibrium constant (K_{eq}) for a reaction. Free-energy change has two components, **enthalpy** (H) and **entropy** (S). These thermodynamic terms, introduced here, will be encountered throughout the text and are explained more fully in Chapter 14.

A chemical reaction, a cell, or an entire organism can be thought of as a thermodynamic system. Everything that is not part of the defined system constitutes the surroundings of the system. The first law of thermodynamics states that the energy of the universe—that is, the system plus its surroundings—is constant. There can be no net energy change in the universe. However, net energy changes are possible within a given system.

The change in the internal energy of a system, ΔE, is an accounting of the heat or work transactions that occur across the boundary between the system and its surroundings. For isolated biochemical reactions at constant pressure, work effectively equals 0; thus, the change in internal energy of a system is equal to the change in heat, a quantity defined as enthalpy change (ΔH). A biochemical process may generate heat ($\Delta H < 0$) or absorb it from the surroundings ($\Delta H > 0$).

Enthalpy changes alone do not determine the spontaneity of a reaction. Spontaneity also depends on the entropy change (ΔS) of a system. Entropy is a measure of disorder or randomness; usually the entropy of a system tends to increase. For example, when water is carefully layered over a solution of colored solute, the solute eventually diffuses into the water. Through diffusion, the solute achieves a more random distribution; its entropy has increased. Biochemical reactions are more likely to occur spontaneously if they are associated with an increase in entropy. Biochemical reactions in which entropy decreases do occur, but because the total amount of entropy in the universe is increasing (according to the second law of thermodynamics), any localized decrease in entropy must be offset by an increase in entropy elsewhere.

The potential for a reaction to proceed is expressed in terms of the free-energy change of the reaction, which includes terms for changes in enthalpy and entropy (T represents absolute temperature).

$$\Delta G = \Delta H - T\Delta S \qquad\qquad \textbf{(1·2)}$$

ΔG for a reaction is the difference between the free energy of the products and the free energy of the reactants. The sign and magnitude of ΔG are an indication of the extent to which a reaction can proceed. When $\Delta G < 0$, the reaction is spontaneous; that is, it can proceed in the absence of energy provided from outside the system. Such processes are said to be **exergonic.** When $\Delta G > 0$, the reaction is not spontaneous and requires energy from outside the system to proceed. Such processes are termed **endergonic.** When $\Delta G = 0$, the reaction is at equilibrium. For a reaction at equilibrium, the concentrations of products and reactants are no longer changing. The ratio of reactants to products at equilibrium defines the equilibrium constant, K_{eq}. No further net reaction can occur in a system at equilibrium without the input of energy or matter.

Thermodynamic considerations can reveal which direction of a reaction is favored but cannot tell how quickly a reaction will occur. The **kinetic** behavior of a reaction, or its specific rate, depends on the energy content of individual molecules. Potential reactants must have sufficient energy to overcome an energy barrier, called the energy of activation, in order for the reaction to proceed. Molecules in cells usually are stable until they encounter and are acted upon by the appropriate enzyme, a protein catalyst that provides a reaction pathway with a lower energy of activation. Enzymes catalyze virtually all of the reactions that occur in cells. The rate of an enzyme-catalyzed reaction can be up to 10^{17}-fold greater than the rate of the corresponding uncatalyzed reaction. Each enzyme catalyzes a specific reaction by having a binding site that accommodates only substrates of the appropriate size, shape, and charge. The kinetic behavior of many enzymatic reactions is regulated. For example, the activity of the first enzyme of a metabolic pathway may decrease when there is a high concentration of the end product of the pathway. Such regulation conserves energy and building-block material.

The chemical reactions of living organisms are not at equilibrium but rather are in a dynamic steady state. The substrate (A) and the product (B) of a reaction are in equilibrium when the rate of conversion of A to B is equal to the rate of conversion of B to A.

$$A \rightleftharpoons B \qquad\qquad \textbf{(1·3)}$$

A steady state occurs when the rate of formation of B from A is equal to the rate of utilization of B to form C.

$$A \longrightarrow B \longrightarrow C \qquad\qquad \textbf{(1·4)}$$

In this simple steady state, the concentration of B does not change over time, but A is depleted and C accumulates. In living organisms, A is replenished and C is consumed or eliminated.

Organisms maintain a relatively constant chemical composition, but individual molecules in the organism are constantly being degraded and replaced. To synthesize biopolymers, cells require sources of energy as well as the chemical precursors of biopolymers.

1·5 ATP Is the Principal Carrier of Biochemical Energy

Endless numbers of biochemical reactions depend on the chemical potential energy of the nucleotide adenosine triphosphate, ATP (Figure 1·21). The purine adenine is linked to ribose to produce the nucleoside adenosine. In ATP, three phosphoryl groups, designated α, β, and γ, are esterified to the 5′-hydroxyl group of adenosine.

The linkage between adenosine and the α-phosphoryl group is a phosphoester linkage, which like most ester linkages requires relatively little energy to form and releases little energy upon hydrolysis. For example, the ester ethyl acetate partially hydrolyzes in acid solution to form acetic acid and ethanol.

$$H_3C-\overset{\overset{\displaystyle O}{\|}}{C}-O-CH_2CH_3 \ + \ H_2O \ \rightleftharpoons \ H_3C-\overset{\overset{\displaystyle O}{\|}}{C}-OH \ + \ HO-CH_2CH_3$$

(1·5)

In contrast, an acid anhydride such as acetic anhydride is more reactive. It hydrolyzes completely when added to water and releases considerable energy in the form of heat.

$$H_3C-\overset{\overset{\displaystyle O}{\|}}{C}-O-\overset{\overset{\displaystyle O}{\|}}{C}-CH_3 \ + \ H_2O \ \longrightarrow \ 2 \ H_3C-\overset{\overset{\displaystyle O}{\|}}{C}-OH$$

(1·6)

The β- and γ-phosphoryl groups of ATP, which are present as phosphoanhydrides, resemble other acid anhydrides and release considerable free energy when they are hydrolyzed. ATP can be hydrolyzed to adenosine diphosphate (ADP) and inorganic phosphate (abbreviated P_i), and ADP can be hydrolyzed to adenosine monophosphate (AMP) and P_i. The structures of ADP and AMP are also shown in Figure 1·21.

Because the anhydride linkages of ATP contain considerable chemical potential energy, ATP is termed an energy-rich compound. ΔG for hydrolysis of ATP to ADP and P_i is a large negative value, which is used as a standard for comparing energy-yielding reactions. Phosphate compounds having a ΔG for hydrolysis at least as negative as that of ATP are also classified as energy-rich. The ΔG for hydrolysis is a measure of the ability of the phosphate compound to donate its phosphoryl group, that is, its **group-transfer potential.** Simple phosphate monoesters are poor donors of phosphoryl groups; in contrast, phosphoanhydrides have high group-transfer potential. By donating a phosphoryl group, an energy-rich compound can provide free energy to drive biosynthetic reactions.

Many condensation reactions in biosynthetic pathways involve ATP. The release of water during a condensation reaction is energetically unfavorable since the reaction takes place in a medium already rich in water. In fact, hydrolysis of the condensation products is energetically favored. The energy of ATP can be coupled to biosynthetic reactions by transfer of one of the phosphoryl groups of the nucleotide, resulting in formation of a high-energy intermediate that then participates in the synthetic reaction. For example, in the formation of an amide, the carboxylate group reacts with ATP to form a reactive acyl-phosphate intermediate.

$$R-\overset{\overset{\displaystyle O}{\|}}{C}-O^{\ominus} \ + \ ATP \ \longrightarrow \ R-\overset{\overset{\displaystyle O}{\|}}{C}-OPO_3^{\,\textcircled{\scriptsize 2-}} \ + \ ADP$$

(1·7)

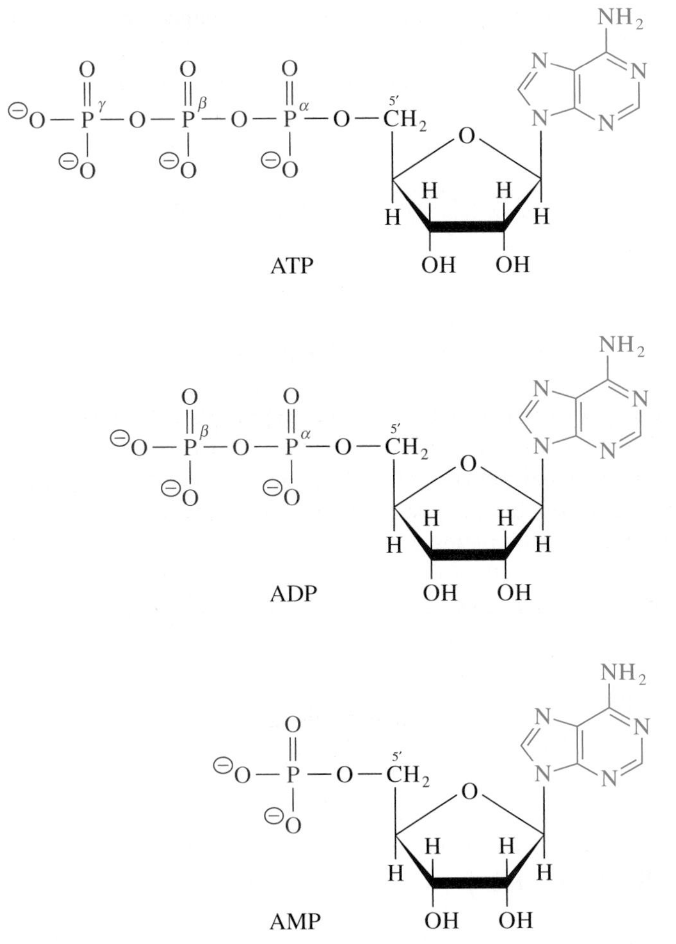

Figure 1·21
Structures of adenosine triphosphate (ATP), adenosine diphosphate (ADP), and adenosine monophosphate (AMP). The base adenine is shown in blue. Phosphoryl groups (red) are bound to the 5′ oxygen of the ribose (black).

The energy associated with the terminal phosphoryl group of ATP is transferred to the intermediate. The acyl phosphate then reacts with the amine to form the amide, releasing phosphate.

$$R-\overset{\overset{\displaystyle O}{\|}}{C}-OPO_3^{\,2-} + H_2N-R_1 \longrightarrow R-\overset{\overset{\displaystyle O}{\|}}{C}-\underset{H}{N}-R_1 + P_i \qquad (1\cdot8)$$

The ATP consumed by biosynthetic reactions is continuously resynthesized from ADP and P_i. This resynthesis is driven by oxidation reactions. Oxidation of energy sources such as fats and sugars proceeds through multistep reaction pathways. In a few instances, oxidation produces a phosphorylated intermediate with a high group-transfer potential. The phosphoryl group of this intermediate can be transferred to ADP to form ATP. However, oxidation pathways often involve a complex series of electron-transferring biomolecules. At certain locations in these electron-transfer pathways, the ΔG of an oxidation step is more negative than the ΔG for the hydrolysis of ATP to ADP + P_i. Such highly exergonic reactions can be coupled to formation of ATP from ADP and P_i (the reverse of the ATP hydrolysis reaction); the coupled processes are called **oxidative phosphorylation.**

Oxidation of carbon-containing fuel molecules, such as carbohydrates, fats, and to a lesser extent amino acids, requires the reduction of other compounds that are biological oxidizing agents. Measurement of the equilibrium constants of these oxidation-reduction reactions has allowed classification of compounds by their **reduction potential,** that is, their potential ability to reduce other compounds. Just as with group-transfer potentials, reduction potentials are related to ΔG values. A

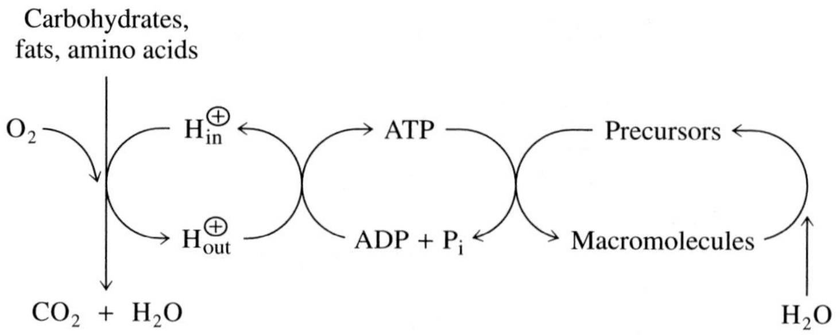

Figure 1·22
The role of ATP. When fuel molecules are oxidized to carbon dioxide and water, protons are translocated across membranes. The re-entry of protons drives the phosphorylation of ADP. The ATP produced by oxidative phosphorylation is coupled to the synthesis of macromolecules and to other energy-requiring reactions.

compound with a large negative reduction potential is said to be a strong reducing agent; it has the potential to produce more ATP than a compound with a less negative reduction potential. In contrast, a compound like oxygen—the ultimate oxidizing agent in many biological processes—has a large positive reduction potential and is termed a strong oxidizing agent. During their oxidation, many compounds—either directly or indirectly—donate protons that contribute to a transmembrane proton gradient. It is the flow of protons back across the membrane that drives the phosphorylation of ADP to ATP (Figure 1·22). The ATP formed by oxidative phosphorylation is available for use in synthetic reactions and other energy-requiring biochemical processes.

1·6 Special Terminology and Abbreviations Are Used in Biochemistry

A full understanding of biochemistry—or any field—depends on a familiarity with the terminology used. Throughout this book, key terms are highlighted in bold when they are first introduced. Most biochemical quantities are specified using SI (Système International) units. Some of the common SI units are listed in Table 1·4. Many biochemists still use more traditional units. For example, the angstrom (Å) is commonly used by protein chemists to report interatomic distances; 1 Å is equal to 0.1 nm, the preferred SI unit. Calories are sometimes used instead of joules (J); one calorie is equal to 4.184 J. Very large or small numerical values for some of the SI units can be indicated by an appropriate prefix. With units of mass, the prefixes are applied to the gram (g) rather than the basic unit, the kilogram. The commonly used prefixes are listed in Table 1·5.

In this book, we will often refer to the molecular weight of a compound. Molecular weight (abbreviated M_r, for relative molecular mass) is the mass of a molecule relative to 1/12th the mass of an atom of the carbon isotope ^{12}C. (The atomic weight of this atom has been defined as exactly 12 atomic mass units.) Because M_r is a relative quantity, it is dimensionless. In some other sources, you may encounter the related term molar mass. Molar mass is the mass (measured in grams) of one mole of that compound. Less often, you may see the use of the dalton as a unit of mass (1 dalton = 1 atomic mass unit). No matter which reference system is used, the values are numerically equivalent.

Biochemistry makes use of many abbreviations. You have already been introduced to the abbreviations ATP and DNA, which are derived from the letters of the full names adenosine triphosphate and deoxyribonucleic acid. Abbreviations like these are convenient and save the laborious repetition of names that are often very long. Learning to associate the abbreviations with their corresponding chemical structures is a necessary step in mastering biochemistry. Abbreviations are given as each class of compounds is introduced.

Table 1·4 Units commonly used in biochemistry

Physical quantity	SI unit	Symbol
Length	Meter	m
Mass	Kilogram	kg
Energy	Joule	J
Electric potential	Volt	V
Time	Second	s
Temperature	Kelvin*	K

*273 K = 0°C.

Table 1·5 Prefixes commonly used with SI units

Prefix	Symbol	Multiplication factor
Mega	M	10^6
Kilo	k	10^3
Milli	m	10^{-3}
Micro	μ	10^{-6}
Nano	n	10^{-9}
Pico	p	10^{-12}
Femto	f	10^{-15}

1·7 The Study of Biochemistry Is a Stepwise Process

This text has been carefully designed to present biochemistry in a logical fashion. The sequence of topics has been chosen so that the material in later chapters builds on the material in the preceding chapters. For example, the structure of a biochemical monomer is presented first, followed by the structure of its respective polymer. Function and biosynthesis are discussed later. This stepwise approach to biochemistry involves several plateaus that require a certain effort to reach. The first challenge is learning the structures of each group of biomolecules. Chemical structures are the vocabulary of biochemistry, and a large vocabulary is built only through practice. At times, the material may seem daunting, but some memorization of molecular structures will pay dividends throughout the course. For example, the structures of 20 amino acids presented near the beginning of Chapter 4 will be encountered in many subsequent chapters.

The study of metabolism represents a second plateau in biochemistry. In contrast to the static structures of molecules, the chemical transformations in cells are dynamic. Almost all of the reactions in metabolism, whether synthetic (anabolic) or degradative (catabolic), are catalyzed by enzymes. For this reason, enzymes are discussed early in the book.

The final plateau in biochemistry is reached with an understanding of biosynthesis of macromolecules. The presentation of biosynthetic reactions must be preceded by the presentation of metabolic pathways that supply the precursor molecules. The biosynthesis of proteins and nucleic acids in particular are discussed last because proper appreciation requires a knowledge of all of the chemical components of cells.

The chapters of this book are organized in four parts. Part One contains 3 introductory chapters that provide an overview of biochemistry and describe the structures of cells and the properties of water.

Part Two contains 10 chapters that describe the structure of biomolecules and the relationship between molecular architecture, chemical reactivity, and biological function. Amino acids and proteins are encountered first, followed by 3 chapters on the properties and mechanisms of enzymes. Chapters on carbohydrates, nucleotides, lipids, and biological membranes follow, providing structural and chemical information that is needed for later chapters. The final chapter in Part Two discusses the physiology and chemistry of human digestion, the process by which biopolymers are hydrolyzed to smaller compounds that are required for the metabolic processes described in Part Three.

Part Three contains 10 chapters devoted to transformations of biomolecules. An introductory chapter gives an overview of metabolism, which serves as a transition from Part Two to Part Three. Chapters on carbohydrate metabolism, the citric acid cycle, oxidative phosphorylation, and photosynthesis are followed by chapters on the metabolic transformations of lipids, amino acids, and nucleotides. A final chapter, *Integration of Mammalian Metabolism,* explains how the various pathways already presented interact in response to changing circumstances in a living organism.

Part Four consists of 10 chapters covering molecular biology in detail. The overall organization follows the flow of biological information within the cell. DNA replication is covered before transcription, which is followed by protein synthesis. Focus on regulation of gene expression in the final chapters provides an appropriate counterpoint to the focus on enzymes in the early part of the book. By the time regulation of gene expression is discussed, the importance of this topic to all the previously encountered metabolic pathways and biosynthetic reactions can be fully appreciated.

Selected Readings

Ingraham, L. L., and Pardee, A. B. (1967). Free energy and entropy in metabolism. In *Metabolic Pathways,* Vol. 1, 3rd ed., D. M. Greenberg, ed. (New York: Academic Press), pp. 1–46. A clear presentation of thermodynamics for biochemistry.

Hanson, R. W. (1989). The role of ATP in metabolism. *Biochem. Ed.* 17:86–92. An excellent introduction to ATP as an energy-transducing agent.

Kohler, R. E., Jr. (1973). The enzyme theory and the origin of biochemistry. *Isis* 64:181–196. Presents the argument that the science of biochemistry originated from an awareness of the importance of enzymes.

Kornberg, A. (1987). The two cultures: chemistry and biology. *Biochemistry* 26:6888–6891. A plea for the use of the language of chemistry as a rational basis for the explanation of life.

Watson, J. D., and Crick, F. H. C. (1953). Molecular structure of nucleic acids. A structure for deoxyribose nucleic acid. *Nature* 171:737–738.

2

Cells

We can make three broad observations that direct further inquiry into the nature and chemical mechanisms of life. First, life obeys the laws of physics and chemistry. For centuries, it was taken for granted that living material possessed some ineffable quality that conferred "life." This unique quality is now regarded by scientists as arising from the ability of living systems to maintain their chemical constituents in highly complex states of spatial and temporal organization. After centuries of probing, we conclude that life is a purely chemical phenomenon. Living material has been dissected at ever finer resolutions, and today the examination and manipulation of biological systems at the molecular level is routine. In no cases have phenomena been shown to violate the known laws of physics and chemistry.

We also observe that life gives rise to life. The origin of life remains a mystery, but the propagation of life is one of the most obvious features of the biological realm. Living systems reproduce themselves.

Finally, we observe that the unit of life is the **cell**—a dollop of water containing dissolved and suspended material, enclosed by a surrounding structure, the **plasma membrane.** Membranes are formed of lipids that aggregate to form sheets. Membranes are flexible, self-sealing, and present impermeable barriers to large molecules and charged species. Embedded in the plasma membrane are a large number of proteins that carry out a variety of functions, including selective transport of materials in and out of the cell. All of the material enclosed by the surrounding membrane of cells, with the exception of the nucleus, is called the **cytoplasm,** which may include large macromolecules and subcellular membrane-bounded structures. The aqueous portion of the cytoplasm minus the subcellular structures is called the **cytosol.** Dissolved in the cytosol and in membranes are enzymes, the protein catalysts whose activities direct the busy chemical commerce of the cell.

The notion that the cell is the unit of life is linked to the observation that life is a chemical phenomenon. Biochemical reaction systems depend on the presence of specific molecules in regulated amounts, usually at high concentrations compared to the external medium. In living organisms, given molecules arise predictably from specific precursors and are converted into specific products. Sequences of reactions in cells, called metabolic pathways, contain branches, loops, and intersections. A map of all of the connections between the thousands of reactions in a cell is analogous to a road map. Like the routes on a road map, few biochemical pathways begin and end without some connection to other pathways in the cell. Yet for all its complexity, the chemical traffic in the cell is controlled and orderly. The traffic patterns of the cell could not be maintained without the isolation of the biochemical reaction system from its environment. Essential products cannot be permitted to diffuse away, and unwanted compounds cannot be permitted to diffuse into the reaction system where they might interfere with essential reactions. However, traffic between the interior and the exterior of the cell must occur. The matter and energy that flow through the cell are sculpted to the needs of the organism by enzyme-directed chemical transformations. The cell is nature's mechanism for achieving an isolated chemical system capable of controlled interaction with its environment. The cell is a unit of biochemical discreteness.

All cells on earth appear to have evolved from a common ancestor that existed nearly four billion years ago. The evidence for a common ancestor includes the appearance in all living organisms of common biochemical building blocks, the same general patterns of metabolism, a common genetic code (with rare, slight variations), and other similarities that point irrefutably to common ancestry. The basic plan of the ancestor cell has been elaborated upon with spectacular inventiveness by billions of years of evolution. This chapter is a broad introduction to the types of cells found in nature.

2·1 Cells Are Either Prokaryotic or Eukaryotic

There is a clear-cut division of cells into two types. **Eukaryotic cells** possess complex internal architecture. Membranous structures divide the cell into compartments with distinct functions. The functional compartments of the cell are called **organelles.** (Some scientists extend this term to include other large subcellular structures, such as ribosomes and transiently formed macromolecular structures involved in cell division. The broader use of the term recalls usage before the emergence of electron microscopy, when *organelle* was a general term for just-visible particles of indeterminate function detected by light microscopists. In this book, we limit the term *organelle* to subcellular structures enclosed by membranes.) In addition to the internal membrane system, a mesh of protein fibers called the **cytoskeleton** contributes to the structure and organization of the eukaryotic cell.

(a)

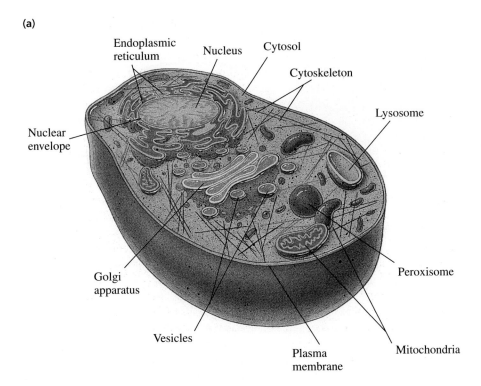

Figure 2·1
Eukaryotic cells. **(a)** Composite animal cell.
(b) Composite plant cell. Unique features of
plant cells include chloroplasts, the sites of
photosynthesis in green plants and algae;
rigid cell walls composed of cellulose; and
vacuoles, large, fluid-filled spaces containing
solutes and cellular wastes.

(b)

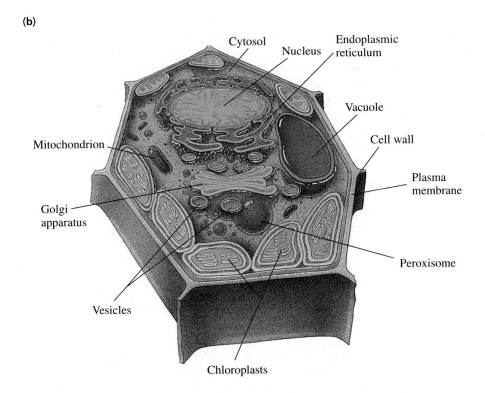

The most prominent subcellular feature of eukaryotic cells is the nucleus, a membrane-bounded structure that contains the genetic material. Plants, animals, fungi, and many microorganisms collectively termed protists are eukaryotes. Figure 2·1 shows composite animal and plant cells exhibiting the principal structures commonly found in each. In upcoming sections, we will examine all of the major intracellular entities and describe their roles.

Figure 2·2
Ultrastructure of an *Escherichia coli* cell.

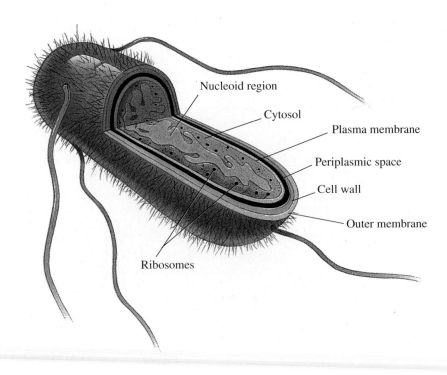

Nucleoid region

Cytosol

Plasma membrane

Periplasmic space

Cell wall

Outer membrane

Ribosomes

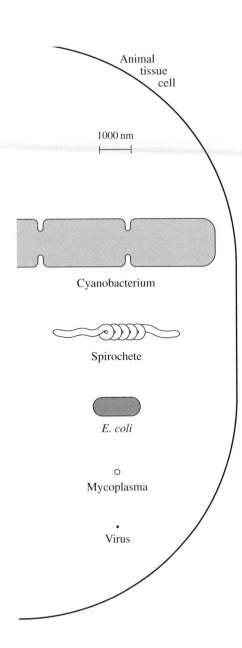

Animal tissue cell

1000 nm

Cyanobacterium

Spirochete

E. coli

Mycoplasma

Virus

Figure 2·3
Relative sizes of a eukaryotic tissue cell, several types of prokaryotic cells, and a virus. Note that extreme deviations from average size are common.

Prokaryotic cells lack a nucleus, organelles, and cytoskeleton and are usually single-celled organisms. Bacteria are prokaryotes. Figure 2·2 illustrates the best-studied of all living organisms, the bacterium *Escherichia coli*. This organism has served as a model biological system, and many of the biochemical reactions described later in this book were first discovered in *E. coli*. The design of the prokaryotic cell has proven immensely successful through evolutionary history. The first cells were prokaryotes, and at present the world is swarming with prokaryotic life. Prokaryotes have been found in almost every conceivable environment on the earth, from hot sulfur springs to the insides of larger cells. They have been detected in the atmosphere, in the deepest parts of the ocean, and several kilometers underground. We often think of biological history in terms of the age of dinosaurs, for instance, or the age of mammals. Life began with the age of bacteria, and to biologists not deceived by the camouflage of extreme smallness, the reign of the prokaryotes has never ended. At present, prokaryotes account for, by some estimates, half of all the biomass on the earth.

In general, eukaryotic cells are much larger than prokaryotic cells (Figure 2·3). A typical eukaryotic tissue cell has a diameter of about 25 μm (25 000 nm). Prokaryotes are about one-tenth that size. Yet evolution has produced tremendous diversity, and trends in the microbiological realm are best described in terms of distributions, with the understanding that extreme deviations from "typical" values for parameters such as size are common. Some eukaryotic unicellular organisms are large enough to be visible to the naked eye, for example, and some nerve cells in the spinal columns of vertebrates can be several feet long, although of microscopic dimensions in cross section.

The smallness of prokaryotes has been assumed to reflect a physical barrier imposed by the simplicity of their cell structures and the constraints of geometry. As the size of a cell increases, volume increases as the cube of the diameter (assuming a roughly spherical cell), whereas surface area increases only as the square of the diameter. Above a certain size, the ratio of surface area to volume becomes limiting for further growth of the cell—imagine an island city whose population grows by cubes as its port facilities, and thus its access to supplies, grow only by squares. Furthermore, although diffusion is an effective delivery mechanism across short distances, the time required for a molecule to diffuse any given distance increases with

Table 2·1 Comparison of prokaryotes and eukaryotes

	Prokaryotes	Eukaryotes
Organisms	Eubacteria, archaebacteria	Animals, plants, fungi, protists
Organization	Unicellular, some colonial	Unicellular, multicellular
Cell size (diameter)	~1–10 μm	~10–100 μm
Membranous organelles	No	Yes
Cytoskeleton	No	Yes
Peptidoglycan cell walls	Yes	No
Endo- and exocytosis	No	Yes
Chromosomes		
Number	1	>1
Location	Nucleoid	Nucleus
Topology	Circular	Linear
Chromosome segregation	Mechanism uncertain	Mitotic spindle

[Adapted from Neidhardt, F. C., Ingraham, J. L., and Schaecter, M. (1990). *Physiology of the Bacterial Cell: A Molecular Approach* (Sunderland, Massachusetts: Sinauer Associates), p. 4, and Stanier, R. Y., Ingraham, J. L., Wheelis, M. L., and Painter, P. R. (1986). *The Microbial World*, 5th ed. (Englewood Cliffs, New Jersey: Prentice Hall), p. 74.]

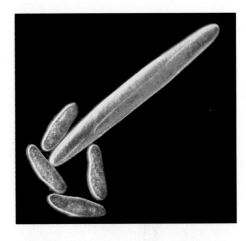

Figure 2·4
The largest prokaryote. With some cells over 600 μm (over 1/50th of an inch long), the huge bacterium *Epulopiscium fishelsoni* is causing previously held ideas about limits on the size of bacteria to be reconsidered. *E. fishelsoni* dwarfs several paramecia, which are relatively large unicellular eukaryotes. The discovery of this titanic prokaryote indicates how much remains to be discovered about microbial diversity. (Courtesy of Esther R. Angert and Norman R. Pace.)

the square of the distance. When the diameter of a cell increases by a factor of 10, the time required for a molecule to diffuse across the cell increases by a factor of 100. These limitations have been thought to limit the size of prokaryotes; the boundary is overcome in eukaryotes by complex internal structures and mechanisms that allow rapid transport and communication from the external medium deep into the cell interior.

A surprising discovery is causing ideas about the limitations on the sizes of prokaryotic cells to be reconsidered. In the intestines of the brown surgeonfish lives a large unicellular organism, *Epulopiscium* (guest at a fish's banquet) *fishelsoni* (Figure 2·4). When discovered in the 1980s, this organism was assumed to be eukaryotic solely because of its size. In 1993, it was announced that molecular biological studies proved *E. fishelsoni* to be a prokaryote. *E. fishelsoni* is monstrously large by prokaryotic standards—commonly 250 μm × 40 μm, with some cells longer than 600 μm. Visible to the naked eye, *E. fishelsoni* is far larger than any previously known prokaryote. The mechanisms by which this bacterium sustains itself despite the apparent constraints of geometry discussed above are the subject of active investigation.

Table 2·1 summarizes the principal differences between prokaryotes and eukaryotes. It should be understood that, in nearly every aspect of cell architecture and biochemical mechanism, there exist some species that have explored peculiar evolutionary paths in ways that defy generalization.

2·2 Prokaryotes Are Structurally Simple, Biochemically Diverse

Prokaryotes possess a variety of fairly regular shapes, the most common being spherical (cocci), rod-shaped (bacilli), and spiral (spirilla). Shape confers a degree of distinction among prokaryotic species, although the developmental options are far less broad than among eukaryotes. Internally, bacteria may appear quite simple, with a relatively homogeneous cytosol and a nucleoid region. The nucleoid region, which contains the genetic material, appears as a light area in electron micrographs.

NO CELL WALL

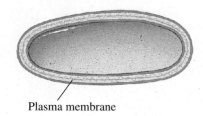

Plasma membrane

GRAM-POSITIVE BACTERIUM

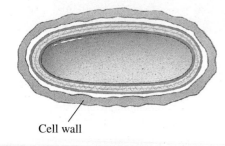

Cell wall

GRAM-NEGATIVE BACTERIUM

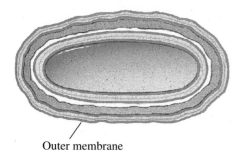

Outer membrane

Figure 2·5
General structures of bacterial cell surfaces. Some bacteria, such as mycoplasmas, are surrounded by a simple membrane. Gram-positive bacteria possess a cell wall formed of peptidoglycan. Gram-negative bacteria possess an additional lipopolysaccharide membrane.

Some prokaryotic species contain regions of infolded plasma membrane, increasing the amount of surface available for membrane-associated functions. While adhering to relatively simple structural patterns, prokaryotes have adapted to countless environments, and natural selection has cobbled together an endlessly inventive variety of prokaryotic solutions to biochemical challenges.

The smallest living organisms are the prokaryotic mycoplasmas, which are parasites of other organisms. Some mycoplasmas are as small as 0.1 μm in diameter and contain only enough genetic material to encode 750 proteins. For a sense of scale, consider that a mycoplasma is one-billionth as massive as a protozoan, a ratio equal to the ratio in mass between a protozoan and a cat! It has been suggested that mycoplasmas contain the minimum equipment necessary to permit the organism to sustain itself and reproduce. Given that mycoplasmas may represent the simplest possible living systems, they would seem to be ideal subjects for investigations into the mechanisms of life. However, mycoplasmas are extremely difficult to culture in the laboratory, probably because they are exclusively adapted to very stable environments—such as the interior of an animal cell—that are difficult to reproduce in the laboratory. Consequently, mycoplasma biochemistry is known only poorly. We might note here that many microorganisms have proven resistant to laboratory domestication—less than 1% of terrestrial microbes can be coaxed to thrive outside of their natural environments, and as few as 0.01% of marine microbial species can be cultured. Many surprises on the order of the gargantuan *E. fishelsoni* remain to be discovered in the prokaryotic realm.

Mycoplasmas are surrounded by a simple plasma membrane with the expected complement of membrane-associated proteins. In nearly all other bacteria, the plasma membrane is surrounded by a **cell wall,** commonly composed of polymers of substituted sugar molecules that are covalently linked by short peptides to form a stiff, porous coat called the peptidoglycan layer. The cell wall confers shape and rigidity. Some bacteria possess an additional outer membranous layer consisting of lipids, proteins, and lipopolysaccharides (lipids covalently linked to carbohydrates) surrounding the cell wall. Bacteria without this outer membranous layer are dyed when stained by the Gram method and are called Gram-positive. Bacteria that possess the outer membranous layer do not stain and are called Gram-negative (Figure 2·5). Gram status has historically been an important diagnostic tool in the identification and characterization of bacteria.

Cell walls provide the cell with mechanical strength and protection from osmotic stress. The concentrations of low-molecular-weight solutes in most bacterial cells greatly exceed the concentrations of solutes in the surrounding medium, and this imbalance can result in high internal osmotic pressure. Without the cell wall, a bacterial cell subjected to such osmotic imbalances would absorb water, swell, and burst. Penicillin and related antibiotics interfere with the action of enzymes involved in the formation of cell walls, causing pathogenic microorganisms to suffer osmotic rupture. Because the host animals do not form cell walls, they are not harmed by the target-specific drugs.

Free-living prokaryotes often possess whiplike appendages that allow for motility. Other structures extending from the cell body allow the bacterium to attach itself to surfaces and to interact with other bacteria. Interactions can include colony formation and the exchange of genetic material.

Subcellular sensory complexes distributed on the plasma membrane of a bacterium allow the organism to detect substances that are potentially either nutritious or toxic. Sensory devices are frequently arrayed in specific regions of the outer membrane, permitting complex responses to chemical information received at the cell surface. This arrangement has led to the question in one published report, "Does *E. coli* have a nose?" The conclusion based on experiment is that *E. coli* does indeed have localized chemosensory apparatus. The differential distribution of sensory devices on the surface of microorganisms conveys a hint of the complex construction of the plasma membrane.

Despite the morphological simplicity of prokaryotes, it is among these microorganisms that most of the metabolic diversity of the natural world is found. In metabolism, molecules are transformed by small steps into other molecules necessary to sustain life, with the reactions catalyzed by a multitude of specific enzymes. Biosynthetic metabolism (or anabolism) utilizes chemical precursors and energy to build up complex molecules, and degradative metabolism (catabolism) breaks down molecules to recover energy and biochemical building blocks. Bacteria exist that can thrive on almost any organic molecule as a source of carbon and energy, and some bacteria exploit a variety of inorganic molecules as energy sources. Furthermore, some bacteria are capable of photosynthesis, the conversion of light energy to chemical energy. This biochemical mechanism originated in the bacterial realm. Bacteria and other organisms that derive energy from light are called **phototrophs.** Those that require chemical sources of energy are called **chemotrophs. Autotrophs** are those that can survive using only CO_2 as a carbon source. **Heterotrophs** require at least one organic nutrient, such as glucose, as a carbon source. Some heterotrophs have extremely elaborate nutritional requirements, reflecting extensive adaptation to lush environments (such as the gastrointestinal tract of an animal) where loss of certain biosynthetic capabilities has carried no penalty.

The role of oxygen in metabolism represents another division in the bacterial realm. **Obligate aerobes** cannot survive except in the presence of oxygen. **Obligate anaerobes** are poisoned by oxygen and live only in environments where oxygen is not present. **Facultative anaerobes** can thrive in either environment. Facultative anaerobes may be merely tolerant of the presence of oxygen while continuing to carry out nonoxidative metabolism, or they may respond to the presence of oxygen by synthesizing the metabolic machinery to support oxidative metabolism using oxygen as the final electron acceptor, which allows far more efficient use of resources.

Bacteria grow rapidly when conditions are suitable. When *E. coli* are grown under optimal conditions, replication of the genome (the organism's complete complement of genetic material) begins even before the previous cycle of genome replication is completed. In fact, we find that rapidly growing *E. coli* cells contain genetic material equal in amount to slightly more than three complete genomes, indicating overlapping cycles of cell reproduction. The bacterial strategy of rapid growth is also reflected in the high rates of metabolic turnover in bacteria. Turnover refers to the continuous synthesis and degradation of molecules in a cell. Turnover in bacterial cells may be from 10^4 to 10^6 times faster than turnover in animal cells. If this seems surprisingly high, consider that a bacterium under ideal growth conditions can double its mass in approximately 20 minutes, a feat clearly beyond the capacity of a large animal. Rapid reproduction and metabolic opportunism have been important aspects of the evolutionary success of bacteria.

The bacterial genome consists of a single, circular molecule of double-stranded DNA localized in a central nucleoid region that, unlike the eukaryotic nucleus, has no enclosing membrane. The nucleoid region appears lighter than the surrounding cytoplasm in electron micrographs because the electron density is low due to the absence of ribosomes, the macromolecular protein-synthesizing machinery of the cell. The bacterial chromosome is thousands of times the length of the bacterium. The DNA molecule is highly folded and densely packed in complexes with small proteins. Information directing the synthesis of proteins is constantly peeled off the DNA molecule in the form of messenger RNA molecules that duplicate the information encoded in DNA. In addition to chromosomal DNA in the nucleoid, most bacteria contain smaller circular DNA molecules, called **plasmids,** located in the cytosol.

During cell reproduction, the bacterial genome is replicated and the prokaryotic cell divides by **binary fission.** Each strand of the double-stranded DNA molecule directs the synthesis of a complementary strand, the two newly created duplex molecules are separated, and the cell divides, with each daughter cell bearing one

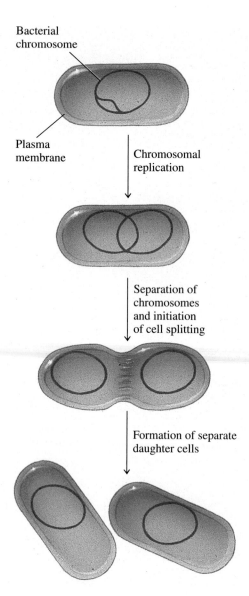

Bacterial chromosome

Plasma membrane

Chromosomal replication

Separation of chromosomes and initiation of cell splitting

Formation of separate daughter cells

Figure 2·6
Binary fission of a prokaryotic cell.

copy of the genetic material (Figure 2·6). Until recently, synthesis of membrane lipids in the middle of the bacterial cell was thought to drive cell growth and fission, but that model has been shown to be inadequate. There is some evidence for the involvement of a system of filaments on the inside surface of the plasma membrane, like the cytoskeleton of eukaryotes, but as yet no filaments have been detected, and the mechanism that drives binary fission remains unexplained.

2·3 Biochemical Studies Are Unveiling the Evolutionary History of Prokaryotes

Prokaryotes have evolved within a fairly confined range of ultrastructural options, their deceptively similar shapes masking their biochemical diversity. For most of the history of microbiology, bacteria have been assigned to a single kingdom, Monera, with inclusion determined by the coarse criteria of unicellularity, smallness, and lack of a nucleus. Recent developments in molecular biology have led to far more refined techniques for the examination and classification of organisms, with important repercussions for the classification of bacteria and other organisms.

Genes evolve. Specifically, the sequences of nucleotides in DNA molecules change over time. The ability to determine the exact nucleotide sequence of DNA by chemical methods provides a new window on the evolutionary relationships of organisms. Parent and offspring of a single species have essentially identical DNA sequences encoding specific proteins. When we compare different species, we find differences in the DNA sequences of the corresponding genes, with a greater number of differences in more distantly related species. The accumulation of these differences, in both DNA and the proteins encoded by DNA, can be used as a gauge to determine how long ago two species branched from a common ancestor. The detection of similarities in the sequences of DNA and proteins across the entire spectrum of living species constitutes the most decisive evidence for a common ancestor for all organisms.

Molecular data have clarified the relationships among bacteria previously grouped as Monera. On the molecular level, the outline of two quite distinct groups emerged in the moneran kingdom, which has led to the separate classification of the archaebacteria, prokaryotes that live in apparently primeval environments. The archaebacteria include the methanogens (methane producers), extreme thermophiles (heat lovers), and extreme halophiles (salt lovers). The remaining bacteria were then classified as eubacteria. Further investigation, led by the microbiologist Carl Woese and coworkers, has shown that the differences between archaebacteria and eubacteria are so striking that the new name *Archaea* has been proposed to avoid the perception that archaebacteria and eubacteria are closely related. By some molecular criteria, archaebacteria, or archaea, seem to be more closely related to eukaryotes than to other prokaryotes; the prokaryotic ultrastructure of archaebacteria appears to be misleading rather than diagnostic of the family relationships.

Organisms have traditionally been divided into five kingdoms (bacteria, fungi, animals, plants, and protists), based on similarities and differences detectable at the scale of overall organism morphology. In a newly proposed system based on molecular data, the kingdom is superseded by a higher level, the domain, of which there are three—the Bacteria, the Archaea, and the Eucarya, which includes all eukaryotes (Figure 2·7). In each domain, organisms are assigned to kingdoms and lower levels, with all members in each level having arisen from the same branch point in evolutionary history. Such an arrangement is termed monophyletic. Note that this is not the case for the traditional five-kingdom classification. The core notion of the

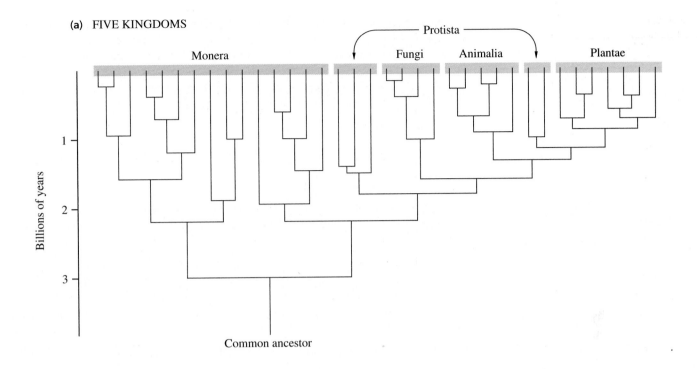

(a) FIVE KINGDOMS

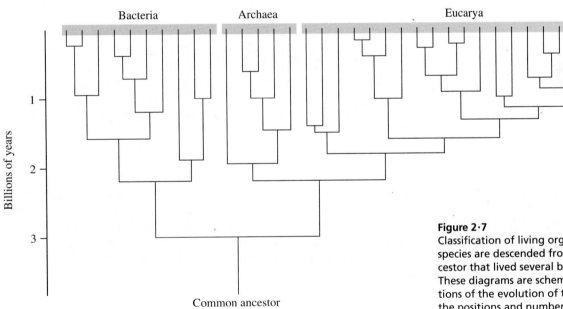

(b) THREE DOMAINS

Figure 2·7
Classification of living organisms. All modern species are descended from a common ancestor that lived several billion years ago. These diagrams are schematic representations of the evolution of the major groups; the positions and numbers of branch points are somewhat arbitrary. **(a)** The five-kingdom classification system emphasizes morphological distinctions among eukaryotes. The Protista are all eukaryotes that do not fit neatly into the fungi, animal, or plant kingdoms, and all prokaryotes are grouped into a single kingdom, the Monera. **(b)** The three-domain classification system corresponds more closely to the actual pattern of evolution. The Archaea represent a single group of prokaryotes that some present evidence suggests are more closely related to eukaryotes than to other prokaryotes.

three-domain system is that a "natural classification system" can be devised using molecular data as the ordering principle for a scheme that accurately expresses evolutionary relationships. The five-kingdom system is based upon homologies and differences at a considerably higher level of organization. The ability to determine evolutionary relationships from the interpretation of molecular data is one of the many exciting, recent advances in biochemistry, and the discussion of a revised system of biological classification reveals the broad implications of biochemical research.

2·4 Eukaryotic Cells Possess an Internal Membrane System and Cytoskeleton

Eukaryotes include most multicellular organisms and many unicellular species. Eukaryotic cells are typically much larger than bacterial cells, commonly 1000-fold greater in volume. The large size of eukaryotic cells requires adaptations that overcome the restrictions imposed by a low ratio of surface area to volume and an internal volume too great for diffusion to be effective as the sole means of internal transport of materials. The special requirements imposed by large size are met in eukaryotic cells by complex internal architecture. The principal structural components unique to the eukaryotic cell are its organelles and cytoskeleton.

In many eukaryotic cells, the amount of internal membrane can account for over 90% of the total membrane of the cell (Table 2·2). The most obvious membrane-defined structure is often the cell nucleus, which is visible through the light microscope and was for decades the criterion for characterization of an organism as a eukaryote. With the development of electron microscopy, beginning in the 1940s, the rest of a labyrinthine membrane system was discovered. Gradually, the characteristic features of distinguishable structures were determined, and the structures were labelled organelles, defined as units of structural and functional discreteness. Many of the distinct membrane-bounded structures are now known to be phases of a single dynamic internal membrane system. Materials flow through the system in paths defined by the membranes. Sheets of membrane burgeon from sites of membrane lipid synthesis, and portions pinch off as vesicles that diffuse through the cell to fuse with membranous structures at other sites. The various membrane regions have unique characteristics and can be separated and identified in the laboratory by testing for enzymatic activities that serve as markers identifying the site of origin.

Internal membranes fill a role analogous to that of the plasma membrane—they enclose regions of biochemical discreteness. In both instances, passage of molecules and signals across the membrane is mediated by transmembrane proteins, and the environments on opposite sides of the membrane are quite distinct.

The ability to internalize parts of the plasma membrane may have been the initial stage in the development of the modern internal membrane system. At some

Table 2·2 Membranes of the rat liver cell

Membrane type	Percent of total membrane in cell
Plasma membrane	5
Rough endoplasmic reticulum	30
Smooth endoplasmic reticulum	15
Nuclear membrane	1
Golgi apparatus	6
Lysosomes, peroxisomes, other compartments	6
Mitochondria	
Outer membrane	7
Inner membrane	30

Figure 2·8
Endocytosis and exocytosis.
The two processes, shown
schematically here, are carried out by separate mechanisms.

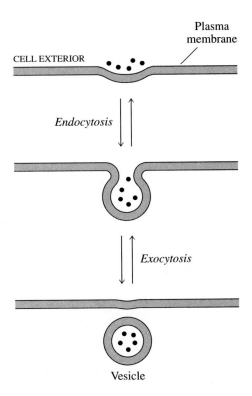

early point in the eukaryotic line, protoeukaryotic cells developed the mechanisms
of endocytosis and exocytosis. In **endocytosis,** a region of the plasma membrane invaginates and is pinched off, forming an internal membranous vesicle engulfing
material from the cell exterior (Figure 2·8). **Exocytosis** is a separate process that
achieves the reverse result, with release of material to the outside of the cell. The
initiation of vesicular traffic between the deep interior of the cell and the cell surface may have provided considerable selective advantages, and selective pressure
favoring the elaboration of this capability could have led to a permanent internal
membrane system.

The organelles in the eukaryotic cytosol are not all of common evolutionary
origin. When we examine the mitochondrion and chloroplast, we will see that these
organelles arose independently of the other internal membrane structures. Though
integrated with other cellular functions, they have retained a considerable degree of
separateness in structure and role.

Our understanding of the other principal characteristic of the eukaryotic cell,
the cytoskeleton, is developing rapidly. The cytoskeleton is a delicate mesh of fibrous proteins that extends throughout the cell. It contributes to cell shape, the
management of vesicular traffic, and the directed movement of organelles. Parts of
the cytoskeleton are also critically involved in eukaryotic cell division. There is ongoing speculation that many of the molecular constituents of the cell, including enzymes and important metabolites, bind to the cytoskeleton; this binding, if widespread, would represent a powerful ordering influence in the cell. At present, the
evidence is inconclusive. We will return to the subject of the degree of organization
in eukaryotic cells after examining the nature and roles of the principal subcellular
structures.

Figure 2·9
Nucleus and endoplasmic reticulum of a eukaryotic cell. (a) Diagram. (b) Electron micrograph of a portion of a rat liver cell. The major ultrastructural features shown are the nucleus (N), the endoplasmic reticulum (ER), and mitochondria (M). (Courtesy of Keith R. Porter.)

(a)

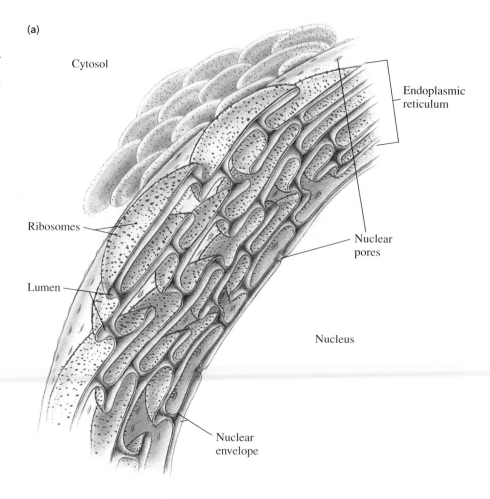

Cytosol

Endoplasmic reticulum

Ribosomes

Nuclear pores

Lumen

Nucleus

Nuclear envelope

(b)

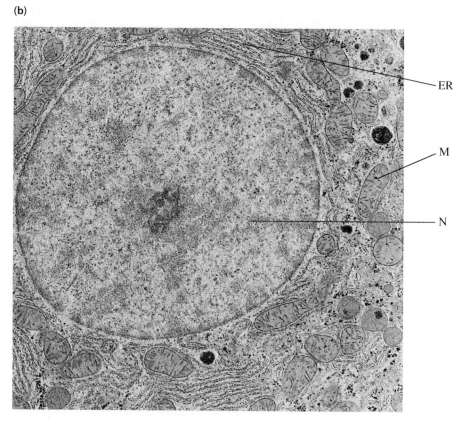

ER

M

N

A. The Nucleus Contains the Genetic Material

The **nucleus** is structurally defined by the **nuclear envelope,** a double membrane that is topologically a single sheet with two layers that join seamlessly at protein-lined nuclear pore complexes (Figure 2·9). The outer membrane of the nuclear envelope is continuous with the endoplasmic reticulum, which consists of highly folded sheets of membrane that extend away from the nucleus over parts of its surface. The endoplasmic reticulum is described in the next section. The inner membrane of the nuclear envelope is lined with filamentous proteins that constitute the nuclear lamina. The lamina has a structural role and also appears to interact with the genetic material, although the nature of the interaction remains ambiguous.

The large nuclear pore complexes (M_r 10^8) control the transport of proteins into the nucleus and RNA molecules out of the nucleus by an ATP-dependent mechanism that is not yet fully understood. Approximately 120 nm in diameter (larger than many viruses), the pore complexes surround an aqueous channel. The channel permits free diffusion of molecules as large as 9 nm, and the complexes can carry out facilitated transport of molecules as large as 25 nm (the size of a ribosome). In a metabolically active cell, such as a maturing oocyte (egg cell), as many as 50 million pore complexes can offer passage across the nuclear envelope.

The nucleus is the control center of the cell, containing 95% of its DNA. Most eukaryotes contain far more DNA than prokaryotes, and much of the nuclear interior is occupied by the eukaryotic genome. If the nucleus were 1 m across, the dimensions of the DNA of a typical eukaryote would be 0.2 mm by 20 km. A human cell may contain 1000 times the DNA of a bacterial cell, and some organisms with unusually large genomes contain tens of times the DNA of human cells. DNA replication and the transcription of DNA into RNA occur in the nucleus. A dense mass in the nucleus, called the nucleolus, is visible in electron micrographs. It is the major site of RNA synthesis and is the site where ribosome subunits are assembled.

DNA in the nucleus is tightly packed with positively charged proteins called histones and coiled into a dense mass called **chromatin** (so named because it stains darkly when treated with certain dyes; *chroma* is Greek for color). Histones account for about half the mass of chromatin. Whereas the main genetic material of prokaryotes is usually a single, circular molecule of DNA, the eukaryotic genome is organized as multiple linear chromosomes. The number of chromosomes can range from few to many. Humans, for example, have 23 pairs of chromosomes.

Division of eukaryotic cells occurs by a coordinated cycle of nucleus-centered events including DNA replication, segregation of chromosomes by mitosis, and cytokinesis, or cell splitting. Mitosis is an elaborate procedure that probably evolved in response to the increased amount of DNA in eukaryotic cells and the necessity for flawlessly duplicating and separating multiple chromosomes. In preparation for cell division, new DNA and histones are synthesized, and then the chromosomal material condenses and separates into two identical sets of chromosomes aligned at the equator of the cell (Figure 2·10). In most eukaryotes, the nuclear envelope disassembles completely during this process, breaking up into thousands of small vesicles. Thus, the contents of the nucleus merge with the cytosol, which is never the case for the contents of other organelles. The alignment and separation of the chromosomes is performed by the mitotic spindle, which consists of paired structures of protein filaments that draw the chromosomes to opposite sides of the cell. Mitosis ends when the chromosomes are separated, nuclear envelopes form, and the chromosomes decondense. At this point, the cell is pinched in two.

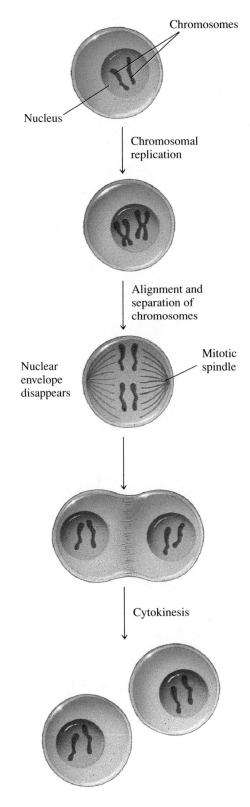

Figure 2·10
Eukaryotic cell division. The chromosomes replicate, and the two sets of chromosomes align at the equator of the cell and are pulled apart by the mitotic spindle. Each daughter cell receives one complete set of chromosomes.

Ancestor
cell

Plasma
membrane

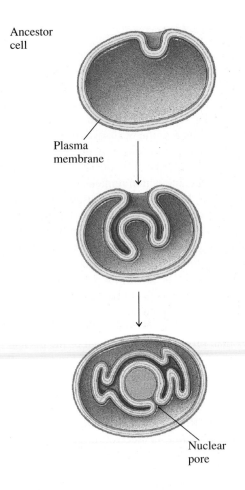

Nuclear
pore

Figure 2·11
Proposed origin of the internal membrane system, including the nucleus. An endocytotic event introduces sheets of membrane continuous with the plasma membrane into the interior of the cell. Stabilization and elaboration of the internal structure produces the nucleus and surrounding endoplasmic reticulum.

The structure of the nuclear membrane, a single sheet folded to form a double layer, suggests an explanation for its evolutionary origin as well as for the origin of the endoplasmic reticulum. Infolding of the plasma membrane in an ancestor cell would have introduced membrane sheets with the topology of interior versus cytosolic faces correctly matching the topology we find in eukaryotic cells (Figure 2·11). Note that the spaces completely enclosed by the internal membrane system are topologically equivalent to the exterior of the cell, in much the same way that the digestive tract of humans is topologically equivalent to the outside of the body.

B. The Endoplasmic Reticulum and Golgi Apparatus Form a Single, Multistage System of Compartments

Continuous with parts of the outer membrane of the nucleus is a network of sheets and tubules called the **endoplasmic reticulum,** already shown in Figure 2·9. The aqueous regions enclosed within the endoplasmic reticulum are known as the lumen. Regions of the endoplasmic reticulum near the nucleus are coated with ribosomes and are called the rough endoplasmic reticulum, reflecting the stubbled appearance of ribosome-bearing membranes in electron micrographs (Figure 2·12). Ribosomes begin translation of RNA into protein in the cytosol; for certain proteins, the first part of the newly synthesized protein to emerge from the ribosome binds to receptors on the endoplasmic reticulum, thereby attaching the ribosome to the membrane. As synthesis continues, the protein is translocated through the membrane into the lumen. If the protein is destined to serve as one of the many membrane-spanning proteins of the cell, part of it remains embedded in the membrane after it is released from the ribosome. A similar process occurs in bacteria, except that the ribosomes attach to the plasma membrane.

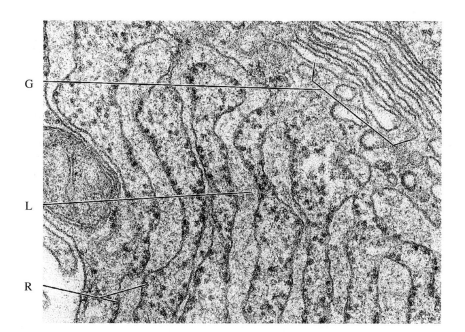

G

L

R

Figure 2·12
Electron micrograph of rough endoplasmic reticulum in a rat liver cell. The dark, electron-dense regions are ribosomes (R). Also labelled are the lumen (L) and a Golgi apparatus (G). Ribosomes bound to the surface of the rough endoplasmic reticulum carry out synthesis of proteins that either pass into the lumen or remain embedded in the membrane. The rough endoplasmic reticulum is continuous with the outer membrane of the nucleus. (Courtesy of Keith R. Porter.)

Many proteins destined for export from the cell are synthesized on the rough endoplasmic reticulum, where they are inserted through the endoplasmic reticulum membrane and into the lumen. As proteins to be exported proceed through the phases of the endoplasmic reticulum system, they are modified and packaged in membranous vesicles that pass through the cell and fuse with the plasma membrane, releasing the contents of the vesicle into the extracellular space. Cells that are specialized for the production of secretory proteins have a great abundance of rough endoplasmic reticulum.

Regions of the endoplasmic reticulum having no attached ribosomes are known as the smooth endoplasmic reticulum (Figure 2·13, next page). The biosynthesis of nearly all of the cell's phospholipids (the main constituents of membranes) is concentrated here, as are other enzyme systems involved in the metabolism of lipids.

The **Golgi apparatus** (Figure 2·14, Page 2·17) is a complex of flattened, fluid-filled membranous sacs often found in proximity to the endoplasmic reticulum and the nucleus. Vesicles that bud off from the endoplasmic reticulum fuse with the Golgi, and the contents of the vesicles may be chemically modified as they pass through the system. The Golgi is a specifically oriented organelle; vesicles bearing materials from the endoplasmic reticulum merge with the Golgi on the *cis* side, and modified products are released in vesicles that bud off the *trans* side. Considerable traffic flows between the endoplasmic reticulum and the *cis* side of the Golgi apparatus. The term *intermediate compartments* has recently been coined for the vesicles in the intervening region to emphasize the continuity of the system between the endoplasmic reticulum and the Golgi apparatus.

Proteins are enzymatically modified as they proceed through the Golgi apparatus. Carbohydrate and lipid groups may be added to them, and then the modified products are sorted and prepared for transport to specific destinations throughout the cell. At the *trans* face of the Golgi, vesicles containing sorted products bud off and travel to a variety of targets, with the destination dictated by docking proteins embedded in the vesicle membrane. The docking proteins must bind to specific receptors before the vesicles can fuse with target membranes. Secretory proteins, for

Figure 2·13
Smooth endoplasmic reticulum. (**a**) Diagram. (**b**) Electron micrograph of smooth endoplasmic reticulum (SER) in a rat liver cell. (Courtesy of George E. Palade.)

(a)

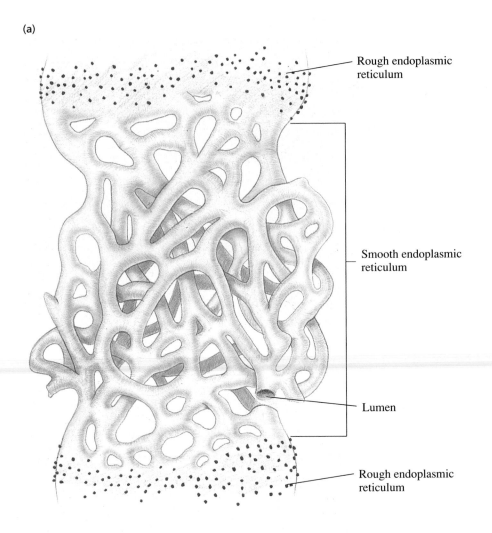

Rough endoplasmic reticulum

Smooth endoplasmic reticulum

Lumen

Rough endoplasmic reticulum

(b)

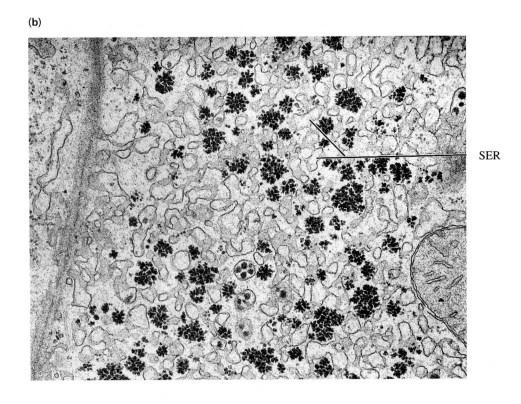

SER

(a)

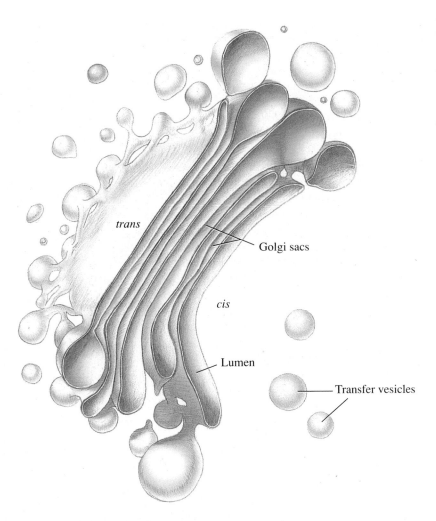

trans

Golgi sacs

cis

Lumen

Transfer vesicles

Figure 2·14
Golgi apparatus. Often associated with the endoplasmic reticulum, the Golgi apparatus is responsible for protein modification, sorting, and targeting to other cellular destinations. **(a)** Diagram. **(b)** Electron micrograph of a rat plasma cell. Note the rough endoplasmic reticulum (RER) adjacent to the *cis* region (CR) of the Golgi apparatus (GA). Transfer vesicles (TV) from the rough endoplasmic reticulum fuse with the Golgi on the *cis* side. (Courtesy of Keith R. Porter.)

(b)

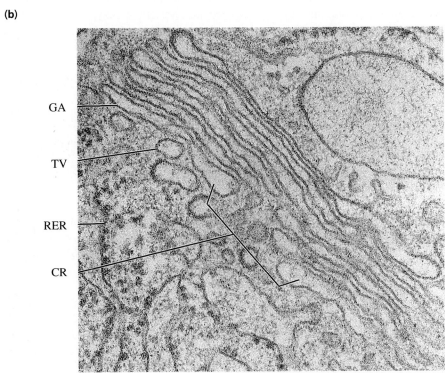

GA

TV

RER

CR

example, are packaged in secretory vesicles. The vesicles may fuse immediately with the plasma membrane (exocytosis), expelling their contents to the extracellular space, or they may remain just beneath the surface awaiting an extracellular signal that triggers fusion with the plasma membrane. Stacked-up vesicles may fuse with each other to form larger vesicles that release the combined contents of precursor vesicles in swift response to one signalling event.

Vesicles budding off from the Golgi may also be destined for sites in the interior of the cell. Targeting follows the same pattern as exocytotic delivery, with specific docking proteins on the vesicle and the intracellular target mediating the binding and release of the vesicle contents. Vesicular transport and fusion with target membranes has the useful feature that proteins and other materials can be transported from the interior of one membranous structure to the interior of another without ever crossing a membrane.

Material must be returned to the Golgi apparatus and endoplasmic reticulum in order for this vesicle delivery system to run continuously. Vesicular traffic returning to the Golgi and endoplasmic reticulum is called retrograde flow; transport *from* these organelles to remote sites is called anterograde flow. The unified nature of the Golgi apparatus and the endoplasmic reticulum is revealed by the striking effect of a compound called Brefeldin A, whose effects demonstrate how delicately this tubulovesicular system is balanced between states. When Brefeldin A is added to cells, swift retrograde flow ensues, vesicles rapidly converge on the Golgi apparatus, and the Golgi itself flows in retrograde fashion into the endoplasmic reticulum until no Golgi apparatus is distinguishable. The molecular mechanisms that maintain the internal membrane structure in its steady state are currently a subject of intense investigation.

C. Lysosomes and Peroxisomes Sequester Potentially Destructive Metabolic Reactions

All eukaryotic cells contain specialized digestive organelles called **lysosomes,** single-membrane vesicles with acidic interiors. As many as several hundred lysosomes may be present in a cell. The low pH is maintained by proton pumps embedded in the surrounding membrane. Enclosed are a variety of enzymes that catalyze the breakdown of cellular biopolymers such as proteins and nucleic acids, whose constituents can be recycled. Lysosomes can also digest large particles, such as retired mitochondria and bacteria ingested by the cell.

The low pH of lysosomes contributes to the digestion of proteins and other polymers but does not affect the specially adapted lysosomal proteins. Lysosomal enzymes are much less active at cytosolic pH. Their pH sensitivity prevents accidentally released lysosomal enzymes from catalyzing disastrous degradation of biopolymers in the cytosol. Thus, compartmentation in lysosomes provides an environment that is specialized for activity and segregated for safety.

Peroxisomes are significant sites of oxygen utilization in all animal cells and many plant cells. Like lysosomes, they are surrounded by a single membrane. In peroxisomes, molecular oxygen is used in a variety of reactions, including many that remove hydrogen atoms from organic molecules, with the formation of the toxic compound hydrogen peroxide, H_2O_2. Some of the hydrogen peroxide is used for oxidation of other compounds. Excess H_2O_2 is destroyed by the action of the peroxisomal enzyme catalase, which catalyzes the conversion of H_2O_2 to water and O_2.

$$2\,H_2O_2 \longrightarrow 2\,H_2O + O_2 \tag{2·1}$$

The sequestering of peroxide-generating reactions in specialized organelles protects the rest of the cell from the toxic by-products of oxidation reactions—another example of the usefulness of intracellular compartmentation.

D. Mitochondria and Chloroplasts Are Sites of Energy Transduction

Mitochondria and chloroplasts have central roles in the bioenergetics of cells in which they are found. **Mitochondria** are the main sites of oxidative energy metabolism and are found in nearly all eukaryotic cells. **Chloroplasts** are the sites of photosynthesis in green plants and algae.

The mitochondrion is surrounded by a double membrane (Figure 2·15). The outer mitochondrial membrane has large pores that permit the passage of molecules having a molecular weight less than ~10 000. The inner membrane is highly folded,

(a)

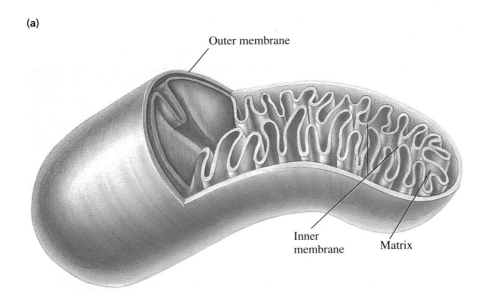

Outer membrane

Inner membrane

Matrix

Figure 2·15
Mitochondrion. Mitochondria are the main sites of energy transduction in aerobic eukaryotic cells. Carbohydrates, fatty acids, and amino acids are metabolized in this organelle. **(a)** Diagram. **(b)** Electron micrograph of a longitudinal section of mitochondria from a bat pancreas cell. The outer membrane (OM) and inner membrane (IM) are indicated. The inner membrane is highly folded, yielding a greatly increased surface area. The space enclosed by the inner membrane is called the matrix (M). (Courtesy of Keith R. Porter.)

(b)

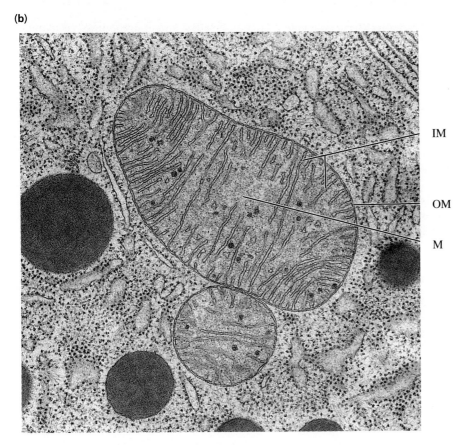

IM

OM

M

resulting in a surface area three to five times that of the outer membrane, and is impermeable to ions and most metabolites. The aqueous phase enclosed by the inner membrane is called the mitochondrial **matrix.** The inner membrane and the matrix contain many of the enzymes involved in aerobic energy metabolism. As carbohydrates, fatty acids, and amino acids are oxidized to carbon dioxide and water in the mitochondria, much of the released energy is conserved in the buildup of a proton concentration gradient across the inner mitochondrial membrane. This stored energy is used to drive the phosphorylation of ADP to produce the energy-rich molecule ATP in a process that will be described in detail in Chapter 18. ATP is then used by the cell for such energy-requiring processes as biosynthesis, transport of certain molecules and ions against concentration and charge gradients, and the generation of mechanical force for such purposes as locomotion and muscle contraction. The number of mitochondria found in cells varies widely. Some eukaryotic cells contain only a few mitochondria, others have thousands. The name *mitochondrion* means filamentous body, and these organelles range in structure from roughly spherical to highly elongated or branched.

Figure 2·16
Chloroplast. This organelle is the site of photosynthesis in plants. Light energy is captured by pigments associated with thylakoid membranes and is utilized to reduce carbon dioxide to carbohydrates. (a) Diagram. (b) Electron micrograph of a cross section of a chloroplast from a spinach leaf. The major features shown are grana (G), the thylakoid membrane (T), and the stroma (S). (Courtesy of A. D. Greenwood.)

(a)

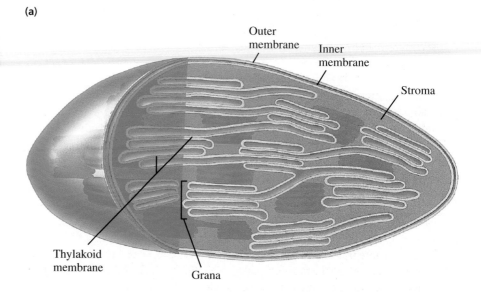

(b)

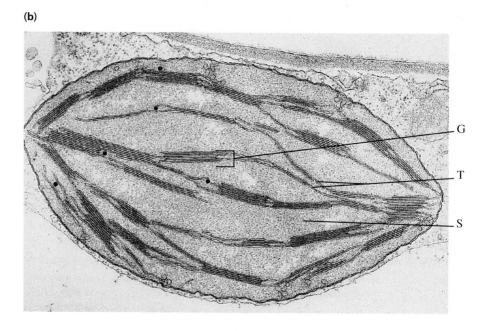

Chloroplasts are surrounded by a double membrane enclosing a third, highly folded internal membrane that forms a system of flattened sacs referred to collectively as the thylakoid membrane (Figure 2·16). The thylakoid membrane, which is suspended in the aqueous stroma, contains chlorophyll and other pigments involved in the absorption of light energy. In many species, the thylakoid membrane is organized in stacks called grana. Also suspended in the stroma are ribosomes and several circular DNA molecules. In the chloroplast, captured light energy is used to drive the formation of carbohydrates from carbon dioxide and water.

Mitochondria and chloroplasts are both almost certainly derived from bacteria that entered into endosymbiotic relationships with primitive eukaryotic cells over one billion years ago. Endosymbionts are symbiotic partners that dwell within their hosts. Evidence for the endosymbiotic origin of mitochondria and chloroplasts includes similarities in membrane composition between these organelles and prokaryotes, the presence of separate, small genomes in both organelles, and ribosomes specific to the organelles that resemble the ribosomes of bacteria. The endosymbiotic event that led to mitochondria occurred first; chloroplasts were later established in cells already colonized by protomitochondria. The few eukaryotes that do not contain mitochondria, such as diplomonads, represent modern species arising from branches deep in the eukaryotic evolutionary tree that have persisted to the present without benefit of the advantages conferred by mitochondria.

The notion of an endosymbiotic origin for mitochondria and chloroplasts has a long history; decisive evidence has accumulated only recently. In fact, a version of the process has been demonstrated in the laboratory of Kwang Jeon. Jeon observed that a batch of amoebas in his lab was stricken with a bacterial invasion lethal to nearly all of the amoebas. He noted that a few amoebas, though critically debilitated and reproducing extremely slowly, were nevertheless viable. Jeon cultivated the amoebas, selecting for the most robust of each generation. After several years, he had a thriving strain in which the number of bacteria in each amoeba had stabilized at about 40 000 and which reproduced at nearly the same rate as the original uninfected amoebas. Most strikingly, he showed that symbiosis had progressed to the point that neither host nor endosymbiont was viable except in the symbiotic state. The metabolism of both organisms had become irreversibly integrated.

The evolution of mitochondria and chloroplasts has progressed far beyond the early stages of integration demonstrated by Kwang Jeon. Most of the proteins of mitochondria and chloroplasts are imported from the cytosol, and only a faint echo of the independent beginnings of these organelles remains.

E. Eukaryotic Cells Possess Extensive Filamentous Architecture

The cytoskeleton suffuses the interior of eukaryotic cells (Figure 2·17). The name is something of a misnomer—the structure is not rigidly skeletal but is instead a dynamic system, assembling and disassembling to meet changing requirements for support, internal organization, and even movement of the cell. The cytoskeleton consists of three types of protein filaments: actin filaments, microtubules, and intermediate filaments. The three types are distinguished by their diameter and their protein composition. All three types of filaments are built of subunits, individual protein molecules that combine to form threadlike fibers. Assemblies of protein subunits are called oligomers (if there are only a few subunits) or multimers (if there are many subunits). The cell is thick with cytoskeletal components, and they have been implicated in a large number of cellular processes.

The most abundant cytoskeletal components, **actin filaments** (sometimes called microfilaments) are assemblies of the protein actin. Actin is a globular protein with two distinct lobes. Molecules of actin assemble into a twisted, double-stranded rope with a diameter of about 7 nm, making actin filaments the most

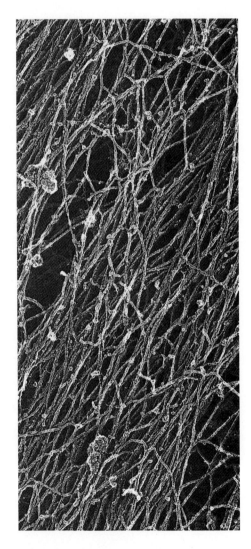

Figure 2·17
Electron micrograph of the cytoskeleton of a fibroblast. This protein scaffold consists of a network of actin filaments, intermediate filaments, and microtubules. (Courtesy of J. E. Heuser and M. Kirschner.)

7 nm diameter

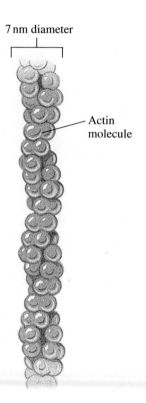

Figure 2·18
Arrangement of actin sub-
units in actin filaments.

Actin
molecule

slender of the cytoskeletal filaments (Figure 2·18). Actin filaments can associate to form bundles, or they can cross-link to form elaborate networks. The type of association is mediated by small actin-binding proteins. Several dozen such proteins are known, and they contribute to the formation of a large number of structures having a variety of roles, some of which are essential for cell viability.

Actin has been found in all eukaryotic cells and is frequently the most abundant protein in plant and animal cells. An important and well-studied function of actin in many organisms is its role in contractile systems. Actin filaments can associate with the protein myosin, and actin and myosin together interact to produce directional movements at the expense of energy from ATP. Myosin is an oligomeric protein with a head region consisting of two globular domains that bind actin, and a tail region. Myosin molecules associate to form thick myosin filaments (Figure 2·19a). The globular domains of myosin molecules dissociate from molecules of actin upon binding of ATP (Figure 2·19b). The ATP is then hydrolyzed by the enzymatic activity of the myosin molecule, producing a myosin-ADP-P_i complex in an energized conformation. Myosin then binds to a molecule of actin further along the actin filament; ADP and P_i are released, and myosin returns to its former shape in the power-stroke step of the cycle. Repetition of these steps results in rapid progression of the myosin molecule down the actin filament. Muscle contraction is carried out by muscle cells adapted to the single purpose of massive concentration of force generated by the interaction of large numbers of actin and myosin molecules. Less spectacular, though no less important, are the other roles of this contractile system within a variety of cells, such as generating force for the routine movement of organelles. Another example occurs during cytokinesis following mitosis. A belt of actin and myosin filaments attaches to the inside of the plasma membrane at the equator of the dividing cell. Tightening of this contractile ring divides the cell into two. After cell division, the filaments of the contractile ring disassemble.

Microtubules, found in nearly all eukaryotic cells, are strong, rigid fibers frequently packed into bundles. They are thicker than actin filaments, with a diameter

(a)

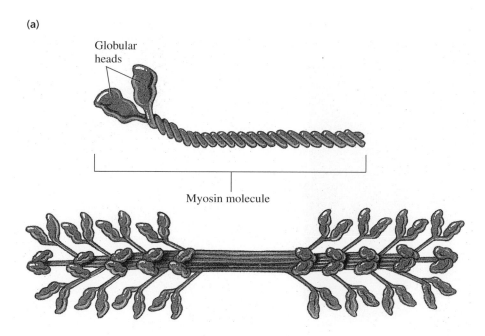

Globular heads

Myosin molecule

Figure 2·19
Actin and myosin in force generation. **(a)** A myosin molecule and a myosin multimer. **(b)** Myosin changes from a low-energy to a high-energy conformation at the expense of ATP. A power stroke results when myosin assumes the high-energy conformation, associates with the actin filament, and changes to the low-energy conformation, causing the actin and myosin filaments to slide past each other. Myosin then binds further down the actin filament.

(b)

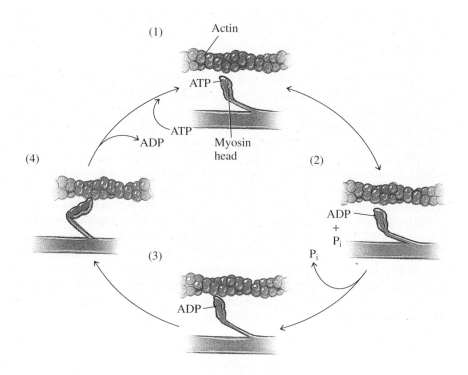

(1) Actin

ATP

ATP

ADP

Myosin head

(4)

(2)

ADP + P$_i$

P$_i$

(3)

ADP

25 nm diameter

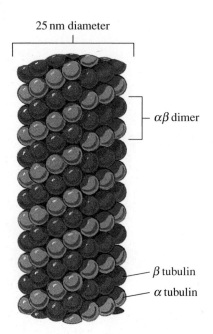

αβ dimer

β tubulin

α tubulin

Segment of a microtubule

Figure 2·20
Structure of microtubules. The α and β tubulin subunits form $\alpha\beta$ heterodimers that assemble around a hollow core, most commonly in 13 vertical rows.

of about 22 nm. Microtubules are composed of two similar protein subunits, α tubulin and β tubulin (M_r 55 000 each). Molecules of each type of subunit combine to form $\alpha\beta$ heterodimers that then assemble around a hollow core in a helical arrangement, usually with 13 vertical rows of tubulin dimers in each microtubule (Figure 2·20).

Microtubules appear to extend radially throughout the cell originating from a microtubular organizing region. Within this region, in many eukaryotes, there is a complex of microtubules and other material called the centrosome. During mitosis, the centrosome divides into two centrosomes that migrate away from each other. Microtubules stretch across the gap between the two centrosomes to form the mitotic spindle, which is responsible for separating newly replicated chromosomes before cell division. The alkaloids vinblastine and vincristine from the periwinkle plant block the assembly of the mitotic spindle, selectively killing cells that divide rapidly. This property has led to the use of these alkaloids as anticancer agents.

Microtubules, like actin filaments, can form structures capable of directed movement. Cilia, which are motile, hairlike structures extending from many eukaryotic cell types, are composed of a bundle of microtubules—nine pairs, or doublets, with a pair of linked singlet microtubules in the center, covered by plasma membrane (Figure 2·21). The outer fibers are linked to the central fiber by radial spokes. At regular intervals along each doublet, arms of the protein dynein protrude from the microtubules. These arms form cross-links between individual microtubules, holding them in position. Driven by the hydrolysis of ATP, dynein changes shape. When the dynein arms move down a microtubule pair on one side and up on the other side, the microtubular structure bends, causing the flexible waving motion characteristic of cilia. Eukaryotic flagella (the long tails that propel sperm cells are an example) are essentially large cilia that also use ATP to drive their waving motion.

Microtubules are also known to form tracks for the directed movement of organelles and vesicles in cells. The rate of movement of lysosomes along microtubular tracks has been measured as high as $2.5\,\mu\text{m s}^{-1}$. At that rate, lysosomes could cross the average tissue cell in just a few seconds. Similar rates are seen for the vesicular traffic between endoplasmic reticulum, Golgi apparatus, and other sites.

(a)

Doublet microtubule

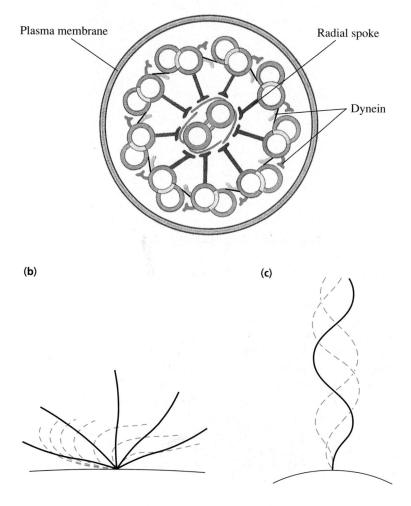

Plasma membrane

Radial spoke

Dynein

(b)   **(c)**

Figure 2·21
Structure and motion of cilia and flagella. **(a)** The cross-sectional diagram shows micro-tubule doublets linked by dynein. Dynein molecules bind and dissociate along the length of microtubules, moving down one side of the structure and up the other side, causing the structure to undulate. The pattern of binding and dissociation gives characteristic beating motions to cilia **(b)** and flagella **(c)**.

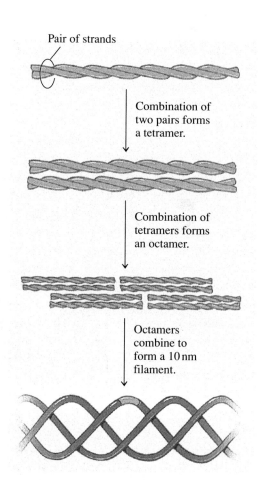

Pair of strands

Combination of two pairs forms a tetramer.

Combination of tetramers forms an octamer.

Octamers combine to form a 10 nm filament.

Figure 2·22
Likely structure of intermediate filaments. Paired strands combine to form tetramers, which combine to form octamers and polymerize end to end, twisted in some way to form ropelike filaments. There are a variety of intermediate filaments, and other structures have been proposed.

Many types of **intermediate filaments** are found in the cytoplasm of most eukaryotic cells (Figure 2·22). Intermediate filaments are essentially aggregations of fibrous protein subunits. The filaments have diameters of approximately 10 nm. Intermediate filaments vary according to the type of subunit. Intermediate filaments line the inside of the nuclear envelope and extend outward from the nucleus to the periphery of the cell. It is possible to dismantle the cytoplasmic network of intermediate filaments by neutralizing a protein involved in its formation, yet cells so treated continue to grow and divide in culture. Nevertheless, since intermediate filaments are found in most eukaryotic cells, they would appear to have an important role, possibly contributing to resistance against the assorted stresses and insults encountered by cells not in culture. Additional roles are suspected since intermediate filaments are represented in great variety.

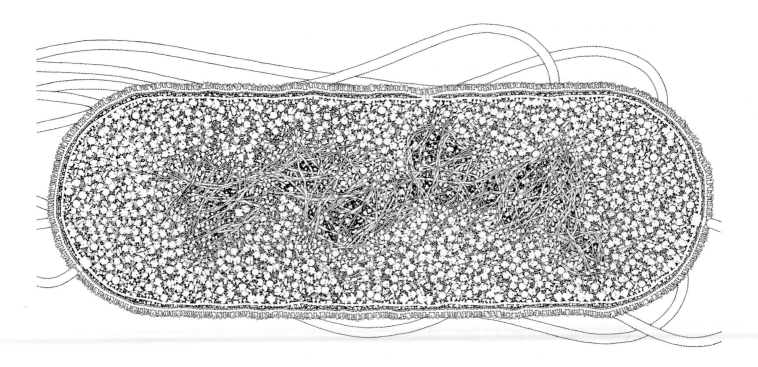

Figure 2·23
E. coli cell, magnified 100 000 times.

2·5 A Picture of the Living Cell

We have discussed the large structures found within cells and described their roles. Keep in mind that even small organelles are immense compared to the molecules and polymers that will be our focus for the rest of this book. Cells contain thousands of different metabolites, many millions of molecules altogether. In every cell, there are hundreds of different enzymes, each acting specifically on only one or possibly a few metabolites. There may be 100 000 copies of some enzymes per cell, yet only a few copies of other enzymes. In the cellular realm, each enzyme is bombarded with potential substrates. The cytosol is dense with biochemical actors.

Molecular biologist and artist David S. Goodsell has produced captivating images that show how an *E. coli* cell would appear on the human scale. The cell in Figure 2·23 is magnified 100 000 times. The square window in Figure 2·24 is a 1 000 000× closeup of the same cell, representing a 100 nm × 100 nm section. Approximately six hundred 100 nm cubes would occupy the volume of the *E. coli* cell. Atoms at this scale are approximately the size of the dot over the letter *i;* small metabolites are smaller than a grain of rice; proteins are the size of a small lima bean. The *E. coli* cell at this scale would be about the size of a bathtub. A human liver cell would be about the size of a large room. Table 2·3 (Page 2·28) lists an inventory of the molecules of the *E. coli* cell.

The still picture of molecules in the cell graphically conveys the density of material in the chemical environment of the cytosol, but it fails to suggest the nature of *action* on the atomic scale. Molecules in the cell are in violent, ceaseless motion. The collisions between molecules are fully elastic—the energy of the collision is conserved in the energy of the rebound. As molecules bounce off each other, they travel a wildly crooked path in space, described as the random walk of diffusion

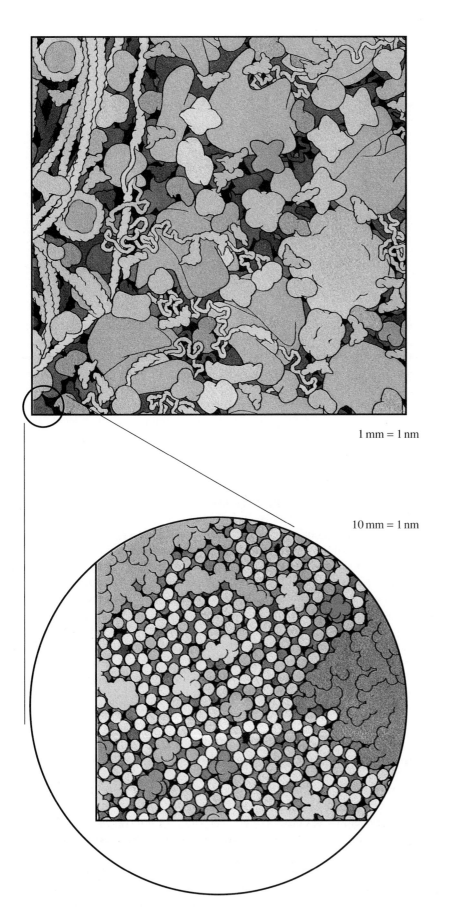

1 mm = 1 nm

10 mm = 1 nm

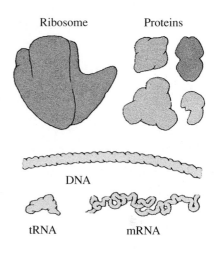

Ribosome   Proteins

DNA

tRNA   mRNA

Figure 2·24
Portion of the cytosol of an *E. coli* cell. The top illustration, magnified one million times, represents a window 100 nm × 100 nm. Proteins are in shades of blue and green. Nucleic acids are in shades of pink. The large ribosomes that pack the cytosol of *E. coli* are shown in purple. Water and small metabolites are not shown. The round inset represents 10 million times magnification and shows small molecules and water.

Table 2·3 Composition of an *E. coli* cell

Component	Molecules per cell	Kinds of molecules
Protein	2 360 000	1050
RNA		
rRNA	56 100	3
tRNA	205 000	60
mRNA	1380	400
Lipid	22 000 000	Four major types
Lipopolysaccharide	1 200 000	1
Metabolites, cofactors, ions	>400 000 000	800+

[Adapted from Neidhardt, F. C., Ingraham, J. L., and Schaecter, M. (1990). *Physiology of the Bacterial Cell: A Molecular Approach* (Sunderland, Massachusetts: Sinauer Associates), p. 4.]

(Figure 2·25). For a small molecule such as water, the mean distance travelled between collisions in a liquid at room temperature is less than the dimensions of the molecule, and the path of travel includes many reversals of direction. Diffusion is thus an extremely indirect mechanism for the transport of a substance between two points. Nevertheless, on the scale of a prokaryotic cell, diffusion is effective. A water molecule can diffuse the length of an *E. coli* cell in one-tenth of a second, visiting many points along the way. In a solution containing a small molecule at a concentration of 1 mM, an enzyme of average size will collide with one of the small molecules one million times per second. Under these conditions, only one in a thousand collisions would have to result in a reaction in order for the enzyme to achieve a catalytic rate of $10^3 \, s^{-1}$. We will see in the upcoming chapter on enzyme mechanisms that some enzymes catalyze reactions with an efficiency far higher than one reaction per thousand collisions. In fact, some enzymes, termed diffusion controlled, catalyze reactions with almost every molecule of substrate with which they collide—an example of the astounding potency of enzyme-directed chemistry.

Lipids in membranes also diffuse vigorously, though only within the two-dimensional plane of the lipid bilayer. Lipid molecules exchange places with neighboring molecules in membranes about six million times per second. Thus, lipid membranes are described as two-dimensional fluids.

Large molecules diffuse more slowly than small ones. In eukaryotic cells, the diffusion of large molecules such as enzymes is retarded even further by the lacy network of the cytoskeleton; large molecules in eukaryotic cells diffuse across a given distance as much as 100 times slower in the cytosol than in pure water.

The full extent of cytosolic organization is not yet known. A number of enzymes are known to form complexes that carry out sequences of reactions. We will encounter several such complexes in our study of metabolism. Cytosolic ribosomes and other factors associated with protein synthesis have been found attached to the cytoskeleton, suggesting that the cytoskeleton plays a role in the organization of the cytosol on a scale as small as individual large molecules. It has also been suggested that many cytosolic enzymes usually considered to be in free solution may in some cases be noncovalently linked to other enzymes that catalyze subsequent reactions of a pathway, and these clusters of enzymes may themselves be noncovalently associated with the cytoskeleton. This arrangement has the advantage that metabolites pass directly from one enzyme to the next without diffusing away into the cytosol. If enzymes are bound to the cytoskeleton, the bonds are noncovalent and so weak that proof of their existence will be extremely difficult to obtain. However, many researchers are sympathetic to the idea that the cytosol is not merely a random mixture of soluble molecules but is highly organized—more machinelike, in

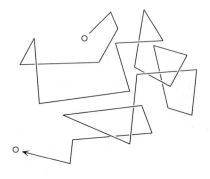

Figure 2·25
Random walk of diffusion. The time required for a molecule to travel a given length by diffusion varies as the square of the distance. In small cells such as bacteria, diffusion is an effective mechanism for delivery of small molecules.

contrast to the long-held impression that simple solution chemistry governs cytosolic activity. At present, the data are ambiguous, but the idea of a highly organized cytosol has generated considerable speculation and represents a research front in biochemistry that may yield important information about how the eukaryotic cell works at the molecular level.

2·6 Viruses Are Quasibiotic Genetic Parasites

Viruses are ubiquitous features of the biological landscape. Are viruses alive? In its simplest form, a virus consists of a nucleic acid molecule surrounded by a protein coat. Viruses lack cellular structure and cannot carry out independent metabolic reactions, commonly regarded as hallmarks of life. They propagate by hijacking the metabolic and genetic machinery of living cells and diverting it to the formation of new viruses. Since viruses contribute nothing to the process but genetic information, one could conclude that viruses are not alive. Yet from another point of view, viruses can be considered extremely efficient forms of life whose life cycle is divided into two stages: the virion stage, when the virus is simply a capsule of genetic material waiting for a host, and the infected-cell stage, when the biochemical machinery of an invaded cell is commandeered for the propagation of the virus. Some bacterial viruses even digest the entire genome of the host bacterium while preserving their own genetic material, which offers a fairly convincing image of a "living" virus whose genetic material is the sole captain of the cell. The question of viral "life" may be mere semantics; certainly viruses have a weighty presence in the biosphere.

Viruses contain a relatively small number of genes, typically between 3 and 100. The size of the viral genome varies with the type of virus. Unlike the genomes of prokaryotic and eukaryotic organisms, which are composed exclusively of DNA, the genomes of some viruses are composed of RNA. The viral genetic material is wrapped in a protein coat, or capsid, which is composed of multiple copies of one or more types of protein. In addition, the capsid of many viruses contains protein spikes that project from its surface (Figure 2·26). These spikes are important in the recognition of host cells and in the transfer of the viral genome into host cells. The capsids of some more complex viruses are surrounded by a membrane envelope composed of membrane lipids and virus-specific membrane-embedded proteins

Figure 2·26
Stereo view of adenovirus. Adenoviruses cause acute respiratory disease, pharyngitis, and conjunctivitis. Note the protein spikes that project from the capsid.

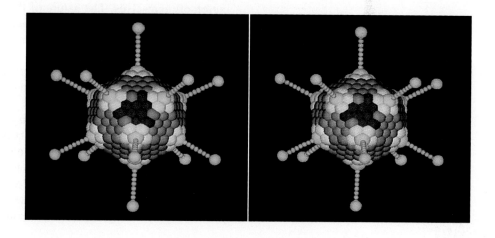

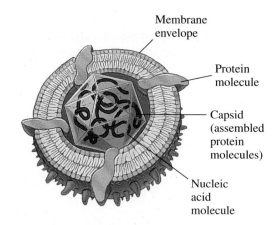

Figure 2·27
General structure of a virus with a membrane envelope.

Membrane envelope

Protein molecule

Capsid (assembled protein molecules)

Nucleic acid molecule

(Figure 2·27). This organization is more characteristic of viruses that infect eukaryotic cells; viruses that infect prokaryotic cells rarely have membrane envelopes.

The structure, assembly, and genetics of viruses are subjects of intense study, in part because they are models for understanding living organisms. Studies of the genetics of bacteriophages, or simply phages, which are viruses that attack bacteria, have provided valuable insights into gene expression and gene regulation in bacteria. Similarly, viruses that infect eukaryotes have been useful as experimental models for eukaryotic genetic mechanisms.

As causative agents of disease in plants and animals, viruses are of immense agricultural and medical importance. Smallpox, chicken pox, and certain tumors are caused by specific DNA-containing viruses, and poliomyelitis, certain leukemias and myelomas, a variety of plant diseases, and acquired immune deficiency syndrome (AIDS) are caused by specific RNA-containing viruses. The name *virus* (Latin for poison) signifies the action of these genetic parasites only too well.

2·7 Life Is a Chemical Chain Reaction That Has Been Proceeding for Nearly Four Billion Years

We noted earlier that all life arises from preexisting life; all cells are produced by the division of older cells. Consider the implications. Life can be considered a chemical chain reaction, begun in the prebiotic vat of the early earth and continuing without interruption ever since. In the most attractive of competing scenarios for the origin of life, there first arose small molecules capable of directing simple replication of themselves. The nature of this chemistry and the identity of the self-replicating molecule is actively disputed. Yet it seems logical that at some point a system arose in which a self-replicating information molecule directed the assembly of additional molecules that contributed to the propagation of the system. The entire system may have been established before or after the appearance of a proto–plasma membrane that enclosed it, keeping the parts together and isolating the interior chemistry.

To the early system, layers of complexity were added, capitalizing on the imperfectness of the system's ability to faithfully replicate itself. Biochemical units most capable of persisting and reproducing gradually monopolized the available resources and prospered.

Branches of the cell line explored structural and chemical innovations, and in many instances tendrils of the chain reaction sputtered and became extinct. One line passed epically through all of the bottlenecks, and in this line appeared the ancestor cell of all life on earth. It was undoubtedly capable of glycolysis (the enzymatic breakdown of glucose) and many of the other fundamental biochemical processes that are common to all cells. It could synthesize amino acids and lipids and almost certainly used ATP as the unit of energy. It used the same standard genetic code that we find in its modern descendants.

This line surged forward for a billion years, proliferating and branching. The cells of the branch that became eukaryotes developed endocytosis, and hence the ability to bring substantial amounts of material into the cell interior. Cells were then able to become predators on other cells. Very likely there was strong selection for larger predator cells and more spectacular assemblages of predatory and digestive functions. The development of the Golgi functions must have been critical—as parts of the cell underwent specialization, sorting of cell constituents would have been indispensable, and dynamic traffic between the cell surface and interior, with the introduction of foreign material, required careful management.

The cells of the branch that became modern prokaryotes explored the advantages of rapid growth. Various species invaded a wide variety of environments and acquired specialized enzymes and pathways. Among the specializations was the ability to harvest the energy of sunlight. In a productive evolutionary leap, separate lines crossed, and prokaryotes were taken into the lush space of the protoeukaryotic interior, where some failure of the digestive function allowed cohabitation. The results, which led to modern-day mitochondria and chloroplasts, opened a vast new developmental space for exploration.

Multicellularity in eukaryotic organisms probably arose a number of times. Incomplete cell division may have led to cell aggregates. Over time, regions of the aggregate took on specialized functions. From such regions arose the organized multicellular structures of tissues and organs, and the need for circulatory systems and intercellular chemical communication via molecules such as hormones. Regulatory mechanisms that allowed single cells to respond to changes in their environment were supplemented by signalling systems that directed the differential expression of genes in different cells of the multicellular organism. Today we can see the existence of multicellularity at a simple stage in colonial prokaryotes, which often form simple aggregates in which some cells take on specialized functions. The development of multicellularity in eukaryotes began a developmental flood of new forms. Emerging large organisms spread to share the surface and waters of the earth with myriad smaller organisms.

2·8 Biochemists Study Model Organisms

Biochemists have studied in great detail the molecular mechanisms of a few model organisms: *E. coli* among prokaryotes; the yeast *Saccharomyces cerevisiae* as the model unicellular eukaryote. The nematode worm *Caenorhabditis elegans* has been used as a model for eukaryotic developmental studies because it has relatively few cells (959 in the mature organism) and its developmental pathway is simple and precise. The fruit fly *Drosophila melanogaster* has been used in tens of thousands of genetics and developmental studies. Biochemical investigations with medical application often use the rat or mouse as proxies for our own system.

How do we use knowledge gained from biological systems as evolutionarily remote from us as *E. coli?* As we probe deeper into the chemistry of life, we find that at the molecular level diversity gives way to unity, and themes emerge that pertain to all life. Knowledge gained from studies on an accommodating organism like *E. coli* can be applied to more recalcitrant model systems, such as the rat. With educated intuition developed from studies of simpler systems, researchers can devise experiments that coax more complex systems to reveal their own intricate mechanisms.

A proper appreciation of the balance between unity and diversity is essential when assessing results of biochemical studies. It is highly unlikely that glycolysis in *E. coli* will be regulated in exactly the same way as glycolysis in rat liver cells. Yet with experience, patterns of regulation are detectable. In upcoming chapters, we report many results for specific organisms with the understanding that the observation is representative of a wide range of organisms.

In this book, we describe the organisms and mechanisms that have been examined most avidly by researchers. The research program of this century has not matched the true pattern of biochemical diversity. For instance, if we were to consider only the total number of individual organisms, we could properly say that, to a first approximation, all life on earth is bacterial. The number of bacteria on earth, if known, would be truly staggering. The number of all other individual creatures combined would be negligible. We might also note that, of all *species* on earth, two-thirds are insects, whereas only 0.4% are vertebrates, yet vertebrate biochemistry has absorbed massive research attention, for obvious reasons. Biochemists are pragmatic. There are things biochemists want to know. It is not possible to know all of the biochemistry of all cells and organisms. But by looking at many organisms, the patterns of an elegant, unified science emerge.

Summary

The unit of life is the cell. Cells are surrounded by a plasma membrane consisting of lipids and protein molecules. Chemical traffic between the cell interior and exterior is strictly controlled by selective transport across the plasma membrane. Cells are of two types, eukaryotic or prokaryotic. Eukaryotic cells contain internal membranous structures termed organelles that divide the cell interior into compartments, and they possess a cytoskeleton consisting of fibers formed from protein subunits. Prokaryotic cells are generally much smaller than eukaryotic cells, and they do not possess internal compartments.

Bacteria are prokaryotes. Most bacteria possess a cell wall surrounding the plasma membrane. If the cell wall is surrounded by an outer membrane, the bacteria are Gram-negative; prokaryotic cells that do not possess an outer membrane surrounding their cell wall are Gram-positive. Much of the biochemical diversity of nature is found in the prokaryotic realm, reflecting adaptation to an enormous range of different environments.

The organelles of eukaryotic cells have specialized functions. The nucleus, surrounded by a double membrane, contains the genetic material. The endoplasmic reticulum is an extensive membrane system continuous with the outer membrane of the nucleus. The rough endoplasmic reticulum is studded with membrane-bound ribosomes. Synthesis of proteins to be exported or proteins that will remain embedded in membranes is one of the principal metabolic activities of the rough endoplasmic reticulum. Ribosomes are absent in regions called smooth endoplasmic reticulum, where much of the lipid synthesis of the cell occurs.

Material newly synthesized in the endoplasmic reticulum is packaged in vesicles that merge with the Golgi apparatus, where modification and sorting of materials occurs. Vesicles then bud off on the far side of the Golgi and carry sorted material to specific cellular destinations.

Lysosomes and peroxisomes contain enzymes that catalyze potentially destructive reactions. Sequestration of these enzymes in lysosomes and peroxisomes protects the cell.

Mitochondria and chloroplasts are organelles involved in energy metabolism. Mitochondria are the centers of oxidative respiration and are the main sites of ATP formation. Chloroplasts are large organelles specialized for the conversion of light energy into chemical energy by the process of photosynthesis. Both organelles almost certainly arose endosymbiotically.

Eukaryotic cells contain elaborate networks of fibrous proteins collectively termed the cytoskeleton. The principal components of the cytoskeleton are actin filaments, microtubules, and intermediate filaments. Actin filaments and microtubules contribute to cell structure and are capable of directed motion. The role of intermediate filaments appears to be primarily structural.

Viruses are genetic parasites. They do not carry out independent metabolism, but depend on the metabolic capacities of host cells to reproduce themselves.

Selected Readings

General References

Bubel, A. (1989). *Microstructure and Function of Cells: Electron Micrographs of Cell Ultrastructure* (Chichester, England: Ellis Horwood). A spectacular collection of annotated electron micrographs.

Darnell, J., Lodish, H., and Baltimore, D. (1990). *Molecular Cell Biology* (New York: Scientific American Books). A masterful introductory textbook.

de Duve, C. (1991). *Blueprint for a Cell: The Nature and Origin of Life* (Burlington, North Carolina: Neil Patterson Publishers). An integrated tour of the history and mechanisms of life.

Goodsell, D. S. (1993). *The Machinery of Life* (New York: Springer-Verlag). Breathtaking vision of the cell on the molecular level.

Schopf, J. W., ed. (1992). *Major Events in the History of Life* (Boston: Jones and Bartlett Publishers). For the inquiring student, this will be a welcome volume. A broad, accessible record of a 1991 symposium for 1700 students and faculty at UCLA covering prebiotic chemistry to the rise of man.

Prokaryotes

Angert, E. R., Clements, K. D., and Pace, N. R. (1993). The largest bacterium. *Nature* 362:239–241. First report of the giant *E. fishelsoni,* with discussion of the constraints on prokaryotic size.

Ingraham, J. L., Brooks Low, K., Magasanik, B., Schaechter, M., and Umbarger, H. E. (1987). *Escherichia Coli and Salmonella Typhimurium: Cellular and Molecular Biology,* Vols. I and II. (Washington, DC: American Society for Microbiology). Massive focus on two prokaryotic organisms—a pair of volumes for students to explore.

Neidhardt, F. C., Ingraham, J. L., and Schaechter, M. (1990). *Physiology of the Bacterial Cell: A Molecular Approach* (Sunderland, Massachusetts: Sinauer Associates). Swiftly becoming a standard reference.

Genomes

Funnell, B. E. (1993). Participation of the bacterial membrane in DNA replication and chromosome partition. *Trends Cell Biol.* 3:20–24. Examines current models for segregation of bacterial chromosomes during division.

Kellenberger, E. (1990). Intracellular organization of the bacterial genome. In *The Bacterial Chromosome,* K. Drlica and M. Riley, eds. (Washington, DC: American Society of Microbiology), pp. 173–186. With discussion of the dependability of electron-micrograph images, given harsh preparation techniques.

O'Brien, S. J., and Seuanez, H. N. (1988). Mammalian genome organization: an evolutionary view. *Annu. Rev. Genet.* 22:323–351. An interesting comparative survey of genome organization across an entire class of organisms.

The Nucleus

Dingwall, C., and Laskey, R. (1992). The nuclear membrane. *Science* 258:942–947. Describes levels of regulation made possible by the nuclear envelope.

Internal Membrane System and Plasma Membrane

Edidin, M. (1992). Patches, posts, and fences: proteins and plasma membrane domains. *Trends Cell Biol.* 2:376–380.

Evans, W. H., and Graham, J. M. (1989). *Membrane Structure and Function* (New York: IRL Press at Oxford University Press). Informative and succinct (86 pages).

Klausner, R. D., Donaldson, J. G., and Lippincott-Schwartz, J. (1992). Brefeldin A: insights into the control of membrane traffic and organelle structure. *J. Cell Biol.* 116:1071–1080.

Lippincott-Schwartz, J. (1993). Bidirectional membrane traffic between the endoplasmic reticulum and Golgi apparatus. *Trends Cell Biol.* 3:81–88. Current thinking on the dynamic state of the internal membrane system.

Warren, G. (1993). Bridging the gap. *Nature* 362:297–298. An accessible summary of the specificity and mechanism of the membrane-fusion machinery in cells. A preamble to two significant research papers in the same issue (see especially Sollner, et al., "SNAP receptors implicated in vesicle targeting and fusion," pp. 318–324).

Endosymbiosis

Jeon, K. W. (1991). *Amoeba* and X-bacteria: a case history of symbiont acquisition and possible species change. In *Symbiosis as a Source of Evolutionary Innovation,* L. Margulis and R. Fester, eds. (Cambridge, Massachusetts: MIT Press), pp. 118–131.

Cytoskeleton

Amos, L. A., and Amos, W. B. (1991). *Molecules of the Cytoskeleton* (London: MacMillan). Comprehensive survey at an undergraduate level.

Luna, E. J., and Hitt, A. L. (1992). Cytoskeleton–plasma membrane interactions. *Science* 258:955–964. Reviews cellular functions involving proteins at the boundary between the cytoskeleton and the plasma membrane.

MacRae, T. H. (1992). Towards an understanding of microtubule function and cell organization: an overview. *Biochem. Cell Biol.* 70:835–841. Discusses the dynamic instability of microtubules and the influential roles of microtubule-associated proteins.

Pumplin, D. W., and Bloch, R. J. (1993). The membrane skeleton. *Trends Cell Biol.* 3:113–117. Describes the structure lining the inside of the plasma membrane, with attention to the linkage between cytoskeleton and extracellular matrix.

Biological Classification

Woese, C. R., Kandler, O., and Wheelis, M. L. (1990). Towards a natural system of organisms: proposal for the domains Archaea, Bacteria, and Eucarya. *Proc. Natl. Acad. Sci. USA* 87:4576–4579.

Wheelis, M. L., Kandler, O., and Woese, C. R. (1992). On the nature of global classification. *Proc. Natl. Acad. Sci. USA* 89:2930–2934.

A linked pair of papers; the first introduces the domain system of biological classification. The second is a response to considerable debate generated by the first paper and an argument in favor of molecular data as the basis for organism classification.

3

Water

Life on earth is often described as a carbon-based phenomenon, but it would be equally correct to call life a water-based phenomenon. Life most likely originated in water, with the first organisms appearing about four billion years ago. In most living cells, water is the most abundant molecule, accounting for 60% to 90% of the mass of the cell. There are a few exceptions, such as seeds and spores, from which water is expelled. It is notable that these cells are dormant until they are revived by the reintroduction of water.

The importance of water is manifold. The macromolecular components of cells—proteins, carbohydrates, lipids, and nucleic acids—assume their shapes in response to water. Some types of molecules interact extensively with water; other molecules or parts of molecules that do not dissolve easily in water tend to associate with each other. In addition, much of the metabolic machinery of cells operates in an aqueous environment. Water is an essential solvent as well as a substrate for many cellular reactions.

We begin our detailed study of the chemistry of life by examining the properties of water. The physical properties of water allow it to act as a solvent for ionic and other polar substances. In addition, water molecules can form weak bonds with other compounds, including other water molecules. These interactions are important sources of structural stability in biomolecules. We will also see how water affects the stability of substances that are not soluble in water. The chemical properties of water are also fundamental to the function of biomolecules and entire cells and organisms. We will examine the ionization of water and discuss acid-base chemistry—topics that are the foundation for understanding the molecules and metabolic processes that we will encounter in subsequent chapters.

3·1 The Water Molecule Is Polar

The structure of a molecule of water (H_2O) is not linear but V-shaped (Figure 3·1a). The angle between the two covalent O—H bonds is 104.5°. The unusual properties of water arise from its angled shape and the intermolecular bonds that it can form. Two orbitals of the oxygen atom participate in covalent bonds with the two hydrogen atoms, whereas each of the remaining two orbitals has an unshared electron pair (Figure 3·1b). Bonds involving these unshared electrons and neighboring molecules are at two of the corners of a tetrahedron that surrounds the central atom of oxygen.

Oxygen attracts electrons within a covalent bond more strongly than does hydrogen; that is, it is more electronegative than hydrogen. As a result, an uneven distribution of charge occurs within each O—H bond of the water molecule, with oxygen bearing a partial negative charge ($\delta^{\ominus}$) and hydrogen bearing a partial positive charge ($\delta^{\oplus}$). This uneven distribution of charge within a bond is known as a **dipole,** and the bond is said to be **polar.**

The polarity of a molecule depends both on the polarity of its covalent bonds and the geometry of the molecule. The angled arrangement of the polar O—H bonds of water creates a permanent dipole for the molecule as a whole (Figure 3·2a). Similarly, a molecule of ammonia contains a permanent dipole (Figure 3·2b).

Figure 3·1
(a) Space-filling structure of a water molecule. (b) Angle between the covalent bonds of a water molecule.

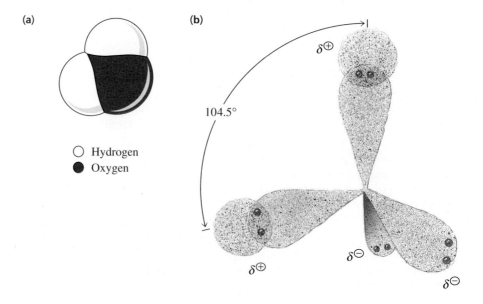

(a)

(b)

104.5°

○ Hydrogen
● Oxygen

$\delta^{\oplus}$

$\delta^{\oplus}$

$\delta^{\ominus}$

$\delta^{\ominus}$

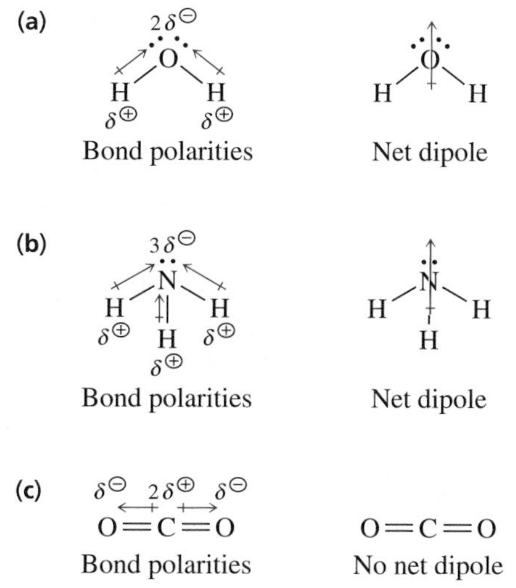

(a)

$2\delta^{\ominus}$

H   O   H
$\delta^{\oplus}$ $\delta^{\oplus}$

Bond polarities

H   O   H

Net dipole

(b)

$3\delta^{\ominus}$

H   N   H
$\delta^{\oplus}$ H $\delta^{\oplus}$
$\delta^{\oplus}$

Bond polarities

H   N   H
   H

Net dipole

(c)

$\delta^{\ominus}$ $2\delta^{\oplus}$ $\delta^{\ominus}$
O $=$ C $=$ O

Bond polarities

O $=$ C $=$ O

No net dipole

Figure 3·2
Polarity of small molecules. **(a)** The geometry of a water molecule is such that the polar covalent bonds create a permanent dipole for the molecule as a whole, with the oxygen bearing a partial negative charge (symbolized by $2\,\delta^{\ominus}$) and each hydrogen bearing a partial positive charge (symbolized by $\delta^{\oplus}$). **(b)** The pyramidal shape of a molecule of ammonia also creates a permanent dipole. **(c)** The polarities of the two covalent bonds in carbon dioxide cancel each other because the bonds are colinear. As a result, the symmetric CO_2 molecule is not polar. (Arrows depicting dipoles point toward the negative charge, with a cross at the positive end.)

Thus, even though water and gaseous ammonia are electrically neutral, both molecules are polar. The high solubility of the polar ammonia molecules in water is facilitated by strong interactions with the polar water molecules. The extreme solubility of ammonia in water is an example of "like dissolves like." Carbon dioxide also contains polar covalent bonds, but because the bonds are colinear, CO_2 is symmetric and the polarities cancel each other (Figure 3·2c). As a result, carbon dioxide has no net dipole and is much less soluble than ammonia in water.

3·2 The Water Molecule Forms Hydrogen Bonds

An important consequence of the polarity of the water molecule is the attraction of water molecules for one another. The attraction between one of the slightly positive hydrogen atoms of one water molecule and the slightly negative oxygen atom of another produces a **hydrogen bond** (Figure 3·3). In a hydrogen bond between two water molecules, the hydrogen atom remains covalently bonded to its oxygen atom, the hydrogen donor. The distance between this hydrogen atom and the other oxygen atom, the hydrogen acceptor, is about twice the length of the covalent bond.

Hydrogen bonds are much weaker than typical covalent bonds. The strength of hydrogen bonds in water and in solutions is difficult to measure directly. However, the energy required to break a hydrogen bond in a vacuum has been estimated to be about 12–25 kJ mol^{-1}, depending on the type of hydrogen bond. In contrast, the energy required to break a covalent O—H bond is about 460 kJ mol^{-1} and to break a covalent C—H bond, about 420 kJ mol^{-1}.

Figure 3·3
Hydrogen bonding between two water molecules. A partially positive ($\delta^{\oplus}$) hydrogen atom of one water molecule attracts the partially negative ($2\,\delta^{\ominus}$) oxygen atom of a second water molecule, forming a hydrogen bond. The distances between atoms of two water molecules in ice are shown.

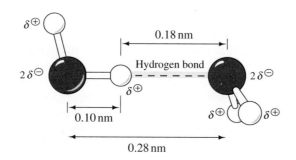

Figure 3·4
Hydrogen bonding by a water molecule. A water molecule can form up to four hydrogen bonds: the oxygen atom of a water molecule is the hydrogen acceptor for two hydrogen atoms, and each O—H group serves as a hydrogen donor. Hydrogen bonds are indicated by dashed lines highlighted in yellow.

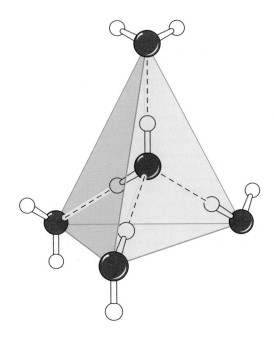

Orientation is important in hydrogen bonding. A hydrogen bond is most stable when the hydrogen atom and the two electronegative atoms associated with it (the two oxygen atoms, in the case of water) are aligned or close to being in line. A single water molecule can form hydrogen bonds with up to four other water molecules (Figure 3·4).

The three-dimensional interactions of liquid water have been difficult to study, but much has been learned by examining the structure of ice crystals (Figure 3·5). In the common form of ice, each molecule of water participates in four hydrogen bonds. Each water molecule is a hydrogen donor in two of these bonds and a hydrogen acceptor in the other two. The hydrogen bonds in ice are arranged tetrahedrally around the oxygen atom of each water molecule. The average energy required to break each hydrogen bond in ice has been estimated to be $23 \, kJ \, mol^{-1}$.

Due to the number of intermolecular interactions in ice, the melting point of ice is much higher than expected for a molecule of its size ($M_r = 18$). In other words, a large amount of energy, in the form of heat, is required to disrupt the hydrogen-bonded lattice of ice. When ice melts, most of the hydrogen bonds are retained by liquid water. It is assumed that each molecule of liquid water can still form up to four hydrogen bonds with its neighbors, but the bonds are distorted relative to those in ice, so that the structure of liquid water is more irregular. The fluidity of liquid water compared to the rigidity of ice is primarily a consequence of the irregular pattern of hydrogen bonding in liquid water, which is constantly fluctuating as hydrogen bonds break and re-form.

An important biological consequence of the different structures of ice and liquid water is that ice is less dense than liquid water. The density of most substances increases upon freezing, as molecular motion slows and tightly packed crystals form. The density of water also increases as it cools—until it reaches a maximum at 4°C (277 K). However, as the temperature drops below 4°C, water expands. This expansion is caused by the formation of the more open hydrogen-bonded ice crystal in which each water molecule is hydrogen bonded rigidly to four others. As a result, ice with its open lattice is less dense than liquid water, whose molecules can move enough to pack more closely. Because ice is less dense than liquid water, ice floats and water freezes from the top down. A layer of ice on a pond thus insulates the creatures below from extreme cold.

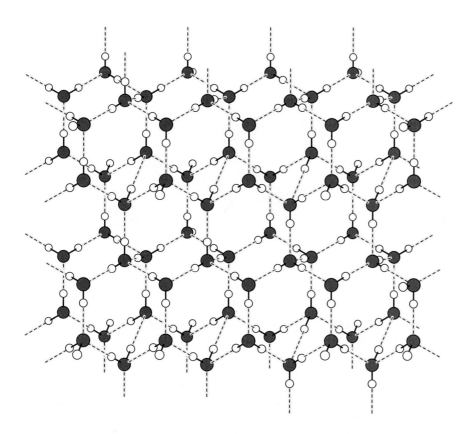

Figure 3·5
Structure of ice. Water molecules in ice form an open, hexagonal lattice in which every water molecule is hydrogen bonded to four others. The geometrical regularity of these hydrogen bonds contributes to the strength of the ice crystal. It is believed that the hydrogen-bonding pattern of liquid water is more irregular than that of ice; the absolute structure of liquid water has not been determined.

Two additional properties of water that are related to its hydrogen-bonding characteristics are its specific heat and its heat of vaporization. Specific heat is the amount of heat needed to raise the temperature of 1 g of a substance by 1°C. A relatively large amount of heat is required to raise the temperature of water because each water molecule participates in multiple hydrogen bonds that must be broken in order for the kinetic energy of the water molecules to increase. Since most of the mass of large multicellular organisms is water and many organisms live in water, temperature fluctuations within cells are thus minimized. This feature is of critical biological importance since the rates of most biochemical reactions are sensitive to temperature.

Like the specific heat, the heat of vaporization of water is much higher than that of many other liquids. As is the case with melting, a large amount of heat is required to evaporate water because hydrogen bonds must be broken to permit water molecules to dissociate from one another and enter the gas phase. Because the evaporation of water absorbs so much heat, perspiration is an effective mechanism for decreasing body temperature. The unusual thermal properties of water make it a stable environment for living organisms, as well as an excellent medium for the chemical processes of life.

3·3 Ionic and Polar Substances Dissolve in Water

Because of its polarity, water readily interacts with compounds that ionize (called electrolytes) and with other polar substances. Substances that readily dissolve in water are said to be **hydrophilic,** or water loving. The interaction of polar molecules with water allows these molecules to dissolve in water. Water molecules align themselves around electrolytes so that their negative ends (the oxygen atoms) are oriented toward the **cations** (the positive ions) and their positive hydrogen atoms

Figure 3·6
Dissolution of sodium chloride (NaCl) in water. **(a)** In the solid state, sodium chloride is a crystalline array held together by electrostatic forces. **(b)** When NaCl is placed in water, the interactions between the positive and negative ions of the crystal are weakened by water. As a result, the crystal dissolves. Each dissolved Na$^\oplus$ and Cl$^\ominus$ is surrounded by a shell of solvent molecules, usually several layers thick, called a solvation sphere. Only one layer of solvent molecules is shown.

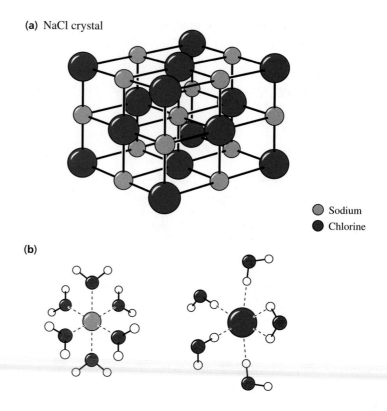

(a) NaCl crystal

○ Sodium
● Chlorine

(b)

Table 3·1 Solubilities of short-chain alcohols in water

Alcohol	Structure	Solubility in water (moles/100 g H$_2$O at 20°C)
Methanol	CH_3OH	∞
Ethanol	CH_3CH_2OH	∞
Propanol	$CH_3(CH_2)_2OH$	∞
Butanol	$CH_3(CH_2)_3OH$	0.11
Pentanol	$CH_3(CH_2)_4OH$	0.030
Hexanol	$CH_3(CH_2)_5OH$	0.0058
Heptanol	$CH_3(CH_2)_6OH$	0.0008

Infinity indicates that there is no limit to the solubility of the alcohol in water.
[Adapted from Brown, T. L., and LeMay, H. E., Jr. (1977). *Chemistry: The Central Science.* (Englewood Cliffs, New Jersey: Prentice-Hall), p. 352.]

are oriented toward the **anions** (the negative ions). Attraction between the polar water molecules and the charged ions of the crystalline electrolyte diminishes the electrostatic interactions between the ions of the solid electrolyte, and the crystal dissolves. For example, the lattice structure of sodium chloride (NaCl) crystals (Figure 3·6a) is disrupted when the crystals are dissolved in water. Because many polar water molecules compete with the relatively few sodium and chloride ions, the interactions between opposite electric charges of dissolved Na$^\oplus$ and Cl$^\ominus$ are much weaker than they are in the intact crystal. As a result, the sodium and chloride ions dissociate from one another. The ions of the crystal continue to dissociate until the solution becomes saturated. At this point, the ions of the dissolved electrolyte are present at high enough concentrations for them to again attach to the solid electrolyte, or crystallize, and an equilibrium is established between dissociation and crystallization. Each dissolved Na$^\oplus$ attracts the negative ends of several water molecules, whereas each dissolved Cl$^\ominus$ attracts the positive ends of several water molecules (Figure 3·6b). Cations are usually smaller than anions and attract more water molecules. The shell of water molecules that surrounds each ion is called a solvation sphere. The solvation sphere usually contains several layers of solvent molecules. A molecule or ion surrounded by solvent molecules is said to be **solvated.** When the solvent is water, such molecules or ions are said to be **hydrated.**

The solubility of organic molecules in water is determined chiefly by their polarity and their ability to form hydrogen bonds with water. Ionic organic compounds such as protonated amines and carboxylates owe their solubility in water to their polar functional groups. Other groups that confer water solubility include amides, alcohols, ketones, and aldehydes. Nonionic organic molecules with polar functional groups, such as some very short-chain alcohols, are very soluble in water. Such molecules disperse among the water molecules, with which their functional groups form intermolecular hydrogen bonds. One-, two-, and three-carbon alcohols are miscible with water, whereas larger hydrocarbons with single hydroxyl groups are much less soluble in water (Table 3·1). In the larger compounds, the properties of the nonpolar hydrocarbon portion of the molecule predominate, thereby limiting solubility.

An increase in the number of polar groups, such as hydroxyl groups, in an organic molecule increases its solubility in water. For example, the carbohydrate glucose, which contains five hydroxyl groups, is very soluble in water: at 17.5°C 83 g of glucose can dissolve in 100 ml of water. Each of the oxygen atoms of glucose can form hydrogen bonds with water (Figure 3·7). The attachment of carbohydrates to some otherwise poorly soluble molecules, including lipids and the bases of nucleosides, increases their solubility.

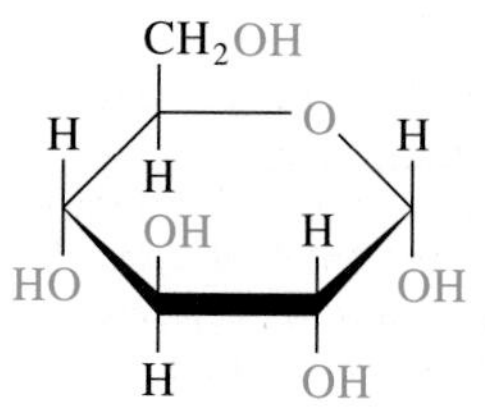

Figure 3·7
Glucose contains five hydroxyl groups and a ring oxygen that can form hydrogen bonds with water.

3·4 Nonpolar Substances Are Insoluble in Water

In contrast to hydrophilic substances, hydrocarbons, such as octane (from gasoline), and other nonpolar substances have very low solubility in water because molecules of water tend to interact with other molecules of water rather than with nonpolar substances. As a result, water molecules exclude nonpolar substances, forcing nonpolar molecules to associate with each other. For example, oil droplets dispersed in water tend to form a single drop, thereby minimizing the area of contact between the two substances. This phenomenon of exclusion of nonpolar substances by water is called the **hydrophobic effect,** and nonpolar molecules are said to be **hydrophobic,** or water fearing. The hydrophobic effect is critical for the folding of proteins and the self-assembly of biological membranes.

Detergents, sometimes called surfactants, are molecules that are both hydrophilic and hydrophobic; they usually have a hydrophobic chain at least 12 carbon atoms in length and an ionic or polar end. Such molecules are said to be **amphipathic.** Soaps, which are alkali metal salts of long-chain fatty acids, are one type of detergent. The soap sodium palmitate, for example, contains a hydrophilic carboxylate group and a hydrophobic tail (Figure 3·8a). One of the synthetic detergents most commonly used in biochemistry is sodium dodecyl sulfate (SDS), which

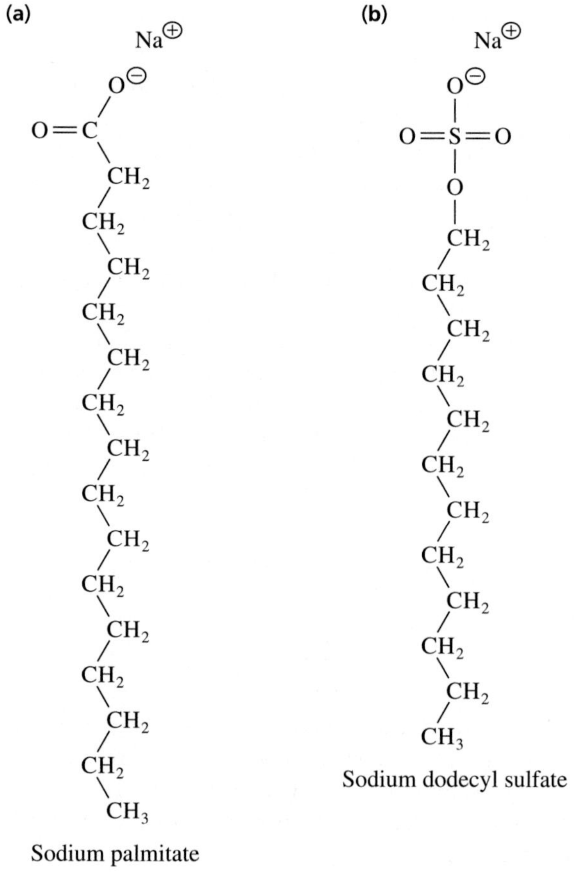

Sodium palmitate

Sodium dodecyl sulfate

Figure 3·8
(a) Sodium palmitate, a soap. (b) Sodium dodecyl sulfate (SDS), a synthetic detergent.

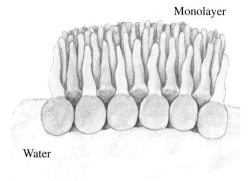

Monolayer

Water

Micelle

Figure 3·9
Cross-sectional views of structures formed by soaps and other detergents in water. Detergents can form monolayers at the air-water interface. They can also form micelles, aggregates of detergent molecules in which the hydrocarbon tails (yellow) associate away from water and the polar head groups (blue) are hydrated.

contains a hydrocarbon chain and a polar sulfate group (Figure 3·8b). The hydrocarbon portion of a detergent is soluble in nonpolar organic substances and its polar group is soluble in water. When a detergent is gently spread on the surface of water, an insoluble monolayer forms in which the hydrophobic, nonpolar tails of the detergent molecules extend into the air, whereas the hydrophilic, ionic heads are hydrated, extending into the water (Figure 3·9). When a sufficiently high concentration of detergent is dispersed in water rather than layered on the surface, groups of detergent molecules aggregate into **micelles.** In one common type of micelle, the nonpolar tails of the detergent molecules associate with one another in the center of the structure, minimizing contact with water molecules. The tails are flexible, not fixed, so that the core of a micelle is liquid hydrocarbon. The ionic heads project into the aqueous solution and are therefore hydrated. Small, compact micelles may contain about 80–100 molecules of detergent. The cleansing action of soaps and other detergents is accomplished by the entrapment of water-insoluble grease and oils within the hydrophobic interiors of micelles. Suspension of nonpolar compounds in water in this fashion is termed solubilization, which is not the same as dissolution. A number of the structures that we will encounter later in this book, including proteins and biological membranes, resemble micelles in having hydrophobic interiors and hydrophilic surfaces.

Some dissolved ions, such as $SCN^\ominus$ (thiocyanate) and $ClO_4^\ominus$ (perchlorate) are called **chaotropes.** These ions are poorly solvated compared to ions such as $NH_4^\oplus$, $SO_4^{2\ominus}$, and $H_2PO_4^\ominus$. Chaotropes enhance the solubility of nonpolar compounds in water by disordering the molecules of water, in effect making them more lipophilic, that is, enhancing their interaction with nonpolar compounds. The exact mechanism of action of chaotropic agents is not known. We will encounter other examples of chaotropic agents, the guanidinium ion and the nonionic compound urea, when we discuss denaturation and the three-dimensional structures of proteins and nucleic acids.

3·5 Noncovalent Interactions Are Important for the Structure and Function of Biomolecules

Two types of noncovalent interactions, hydrogen bonds and hydrophobic interactions, have been introduced so far in this chapter. These weak interactions not only occur in water but also play extremely important roles in the structures and functions of polymers. For example, weak forces are responsible for stabilizing the structures of proteins and nucleic acids. Weak forces are also involved in the recognition of one biopolymer by another and in the binding of reactants to enzymes. There are four major noncovalent bonds or forces involved in the structure and function of biomolecules: electrostatic interactions, hydrogen bonds, van der Waals forces, and hydrophobic interactions.

Electrostatic interactions between two charged particles are potentially the strongest noncovalent forces, but their strengths vary widely. The attractions between ions of opposite charge can extend over greater distances than other attractions. The stabilization of NaCl crystals by ionic interactions is an example of electrostatic force. Weaker electrostatic attractions occur between dipoles and ions. Charge-charge attractions between internal ionic regions of proteins are sometimes called salt linkages or salt bridges. However, these attractions are not as strong as those in salt crystals. The most accurate term for such interactions is **ion pairing.** The strength of an electrostatic interaction depends on the nature of the solvent. Water, as noted earlier, greatly weakens these interactions. Consequently, in an aqueous environment, strong electrostatic forces do not play a major role in the stability of biological polymers. Nevertheless, electrostatic forces do play a role in recognition of one molecule by another. For example, most enzymes have either anionic or cationic sites that attract oppositely charged reactants. Electrostatic forces are also responsible for the mutual repulsion of similarly charged ionic groups. Charge repulsion can influence the structures of individual biomolecules as well as their interactions with other charged molecules.

We have already encountered hydrogen bonds, which are a type of electrostatic interaction, in our discussion of the structure and properties of water. Hydrogen bonds also occur in many macromolecules. They are among the strongest of the common noncovalent bonds in biological systems. The strengths of hydrogen bonds that occur in living cells, such as in the binding of substrates to enzymes and hydrogen bonding between the bases of DNA, have recently been estimated to be about $2–7.5 \, kJ \, mol^{-1}$. These hydrogen bonds are considerably weaker than those of ice, which require $23 \, kJ \, mol^{-1}$ to break, as we saw earlier. Hydrogen bonds are strong enough to confer structural stability but weak enough to be broken readily. In general, a hydrogen bond can form when a hydrogen atom that is covalently bonded to a strongly electronegative atom such as nitrogen, oxygen, or sometimes sulfur (denoted as A in Figure 3·10a) lies approximately 0.2 nm from another strongly electronegative atom (denoted as B) that has an unshared electron pair. The total distance between the two electronegative atoms participating in a hydrogen bond is typically 0.27–0.30 nm. Some common examples of hydrogen bonds are shown in Figure 3·10b. It is believed that the C—H bond is usually insufficiently polar for the hydrogen atom to form a hydrogen bond with an electronegative atom. In rare cases, though, weak C—H------O and C—H------N hydrogen bonds appear in proteins. As we will see, hydrogen bonds are common within (*intra*molecular) and between (*inter*molecular) biological macromolecules such as proteins and nucleic acids. For example, complementary base pairing within DNA molecules depends on precise intramolecular hydrogen bonding (Figure 3·11). Groups or atoms involved in intramolecular hydrogen bonds are also capable of forming hydrogen bonds with the solvent, water. However, under physiological conditions, hydrogen-bonding groups in the interior of a macromolecule do not interact with water. Under certain conditions, the internal groups of a macromolecule may form hydrogen bonds with water, and the structure of the molecule is thereby significantly altered. A biopolymer whose structure has been changed in this way is said to be **denatured.**

In addition to hydrogen bonds, other weak electrostatic forces exist. These forces include the interactions between the dipoles of two uncharged polarized bonds and the interactions between a dipole and a transient dipole induced in a neighboring molecule. These forces are of short range and low magnitude, about $1.3 \, kJ \, mol^{-1}$ and $0.8 \, kJ \, mol^{-1}$, respectively.

Weak intermolecular forces are produced between all neutral atoms by transient electrostatic interactions. These **van der Waals forces,** named after the Dutch physicist Johannes Diderick van der Waals, who discovered them, occur only when atoms are very close together. They originate from the infinitesimal dipole generated in atoms by the random movement of the negatively charged electrons around

(a)

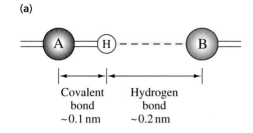

Covalent bond ~0.1 nm   Hydrogen bond ~0.2 nm

(b)

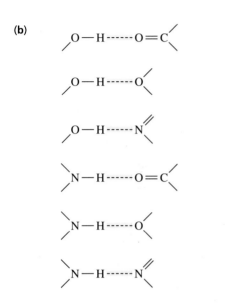

Figure 3·10
(a) Hydrogen bonding between an —A—H group (the hydrogen donor) and an electronegative atom B— (the hydrogen acceptor). A typical hydrogen bond is approximately 0.2 nm long, roughly twice the length of the covalent bond between nitrogen or oxygen and hydrogen. The total distance between the two electronegative atoms participating in a hydrogen bond is approximately 0.30 nm. (b) Examples of biologically important hydrogen bonds.

Figure 3·11
Hydrogen bonding between the complementary bases guanine and cytosine in DNA.

Table 3·2 Van der Waals radii of several atoms

Atom	Radius (nm)
Hydrogen	0.12
Oxygen	0.14
Nitrogen	0.15
Carbon	0.17
Sulfur	0.18
Phosphorus	0.19

the positively charged nucleus. Thus, van der Waals forces are dipolar or electrostatic attractions between the nuclei of atoms or molecules and the electrons of other atoms or molecules. The strength of the interaction between the transiently induced dipoles of nonpolar molecules such as methane is about $0.4 \, kJ \, mol^{-1}$ at an internuclear separation of $0.3 \, nm$. Thus, van der Waals forces are much weaker than hydrogen bonds, although they operate over similar distances. When atoms involved in van der Waals interactions approach one another too closely, there is strong repulsion. Conversely, there is hardly any attraction when the atoms are farther apart than the optimal packing distance, called the van der Waals radius (Table 3·2). When two atoms are separated by the sum of their van der Waals radii, they are said to be in van der Waals contact, and the attractive force between them is maximal (Figure 3·12). The weak van der Waals attractions between nonpolar groups are slightly stronger than the attractions between these groups and water.

Hydrophobic interactions are the association of relatively nonpolar molecules or groups in aqueous solution with other nonpolar molecules rather than with water. Although hydrophobic interactions are sometimes called hydrophobic *bonds,* this description is incorrect. Nonpolar molecules or groups tend to aggregate not because of mutual attraction but because the water molecules trapped around the nonpolar compounds tend to associate with each other. The hydrogen-bonding pattern of water is disrupted by the presence of a nonpolar molecule. Water molecules that surround a less polar molecule in solution are more restricted in their interactions with other water molecules. These restricted water molecules are relatively immobile, or ordered, whereas water molecules in the bulk solvent phase are much more mobile, or disordered. In thermodynamic terms, there is a net gain in the combined entropy of the solvent and the nonpolar solute when the nonpolar groups aggregate and water is freed from its ordered state surrounding the nonpolar groups. Hydrophobic interactions, like hydrogen bonds, are much weaker than covalent bonds. For example, the energy required to transfer a —CH$_2$— group from a hydrophobic to an aqueous environment is about $3 \, kJ \, mol^{-1}$.

Although individual van der Waals forces and hydrophobic interactions are weak, the clustering of nonpolar groups within a protein, nucleic acid, or biological membrane permits formation of a large number of these weak interactions. These cumulative forces play important roles in maintaining the structures of the molecules.

Often, a combination of several weak forces maintains the shape of a biological polymer. For example, the heterocyclic bases of nucleic acids, though polar, are relatively insoluble in water and interact more readily with each other than with water. In DNA, bases are paired by hydrogen bonds (Figure 3·11). Base pairs are also stacked one above another. This stacking arrangement is stabilized by a variety of noncovalent interactions that include van der Waals forces and hydrophobic interactions. These forces are collectively known as stacking interactions.

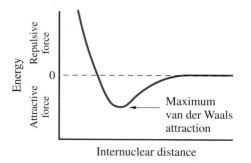

Figure 3·12
Effect of internuclear separation on van der Waals forces. Van der Waals forces are strongly repulsive at short internuclear distances and very weak at long internuclear distances. When two atoms are separated by the sum of their van der Waals radii, the van der Waals attraction is maximal.

3·6 Water Is Nucleophilic

Electron-rich chemicals are called **nucleophiles** (nucleus lovers) because they seek positively charged or electron-deficient species (electrophiles, electron lovers). Nucleophiles are either negatively charged or have unshared pairs of electrons. They attack positive centers in displacement or addition reactions. The most common biological nucleophiles are oxygen, nitrogen, and sulfur atoms. Because water has two unshared pairs of electrons, it too is nucleophilic. Although water is a relatively weak nucleophile, its cellular concentration is so high that one might expect that many biological compounds, such as polymers, would be easily degraded by nucleophilic attack by water. For example, proteins can be hydrolyzed, or degraded by water. In such a reaction, the protein is eventually degraded to its monomeric units, amino acids. This type of reaction involves the transfer of an acyl group from the polymer to the oxygen atom of water. Hydrolysis is essentially an irreversible reaction; that is, the equilibrium for a hydrolytic reaction lies far in the direction of degradation.

Several questions must then be asked. If there is so much water in cells, why are biopolymers not rapidly degraded to their components? Similarly, if the equilibrium lies toward breakdown, how does biosynthesis occur in an aqueous environment? There are several ways whereby cells prevent unwanted hydrolytic reactions or overcome unfavorable equilibria. First, the linkages between the monomeric units of biopolymers, such as the amide bonds in proteins and the ester linkages in DNA, are highly resistant to spontaneous hydrolysis under cellular conditions. Hydrolysis of polymers normally occurs only in the presence of specific enzymes. These enzymes, from the group known as hydrolases, are stored in inactive forms or enclosed in organelles and thus their activity, catalysis of the degradation of the biopolymers, can be limited.

Cells synthesize polymers in an aqueous environment by using the chemical potential energy of ATP to overcome an unfavorable thermodynamic barrier and by excluding water from the regions of the enzymes where the synthetic reactions occur. In chemical terms, biosynthesis of polymers occurs by group-transfer reactions in which an acyl or a carbonyl group is transferred to a nucleophile. Whereas hydrolysis is group transfer to water, synthesis is group transfer to a nucleophile other than water. Because synthetic reactions must proceed against a very unfavorable equilibrium, they follow two-step chemical pathways that differ from the reversal of hydrolysis. In the first step, which is thermodynamically uphill, the molecule to be transferred reacts with ATP to form a reactive intermediate. In the second step, the activated group is readily transferred to the attacking nucleophile in an isoenergetic reaction (i.e., a reaction involving little or no change in free energy). A simple example is the biosynthesis of an amide bond. In this reaction, catalyzed by the enzyme glutamine synthetase, the amino acid glutamate is amidated by reaction with ammonia to form the related amino acid glutamine.

$$\text{Glutamate} + \text{ATP} + \text{NH}_3 \xrightarrow{\text{Glutamine synthetase}} \text{Glutamine} + \text{ADP} + \text{P}_i$$

$$(3\cdot1)$$

In a similar organic chemical reaction, carboxylic acids can be activated to acyl chlorides that are converted to amides (Figure 3·13). In the enzymatic condensation

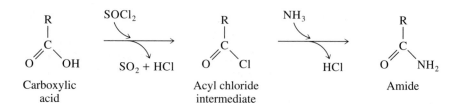

Carboxylic acid

Acyl chloride intermediate

Amide

Figure 3·13
Chemical synthesis of an amide from a carboxylic acid. The acid is activated and the intermediate, an acyl chloride, undergoes nucleophilic attack by ammonia to produce the amide.

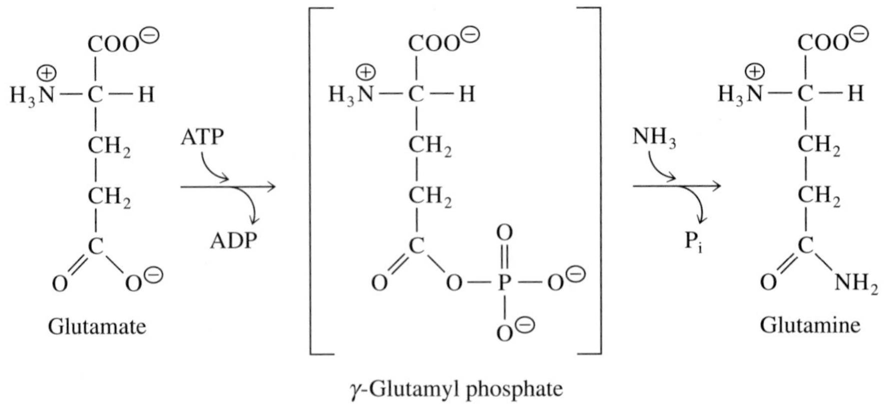

Figure 3·14
Synthesis of glutamine from glutamate, ATP, and ammonia. This two-step reaction is catalyzed by glutamine synthetase. The activated intermediate, which is extremely unstable in aqueous solution, is not hydrolyzed because water is excluded from the active site of the enzyme.

γ-Glutamyl phosphate

of glutamate with ammonia to form glutamine, the chemical potential energy of ATP is used to form a mixed phosphate-carboxylate anhydride called γ-glutamyl phosphate (Figure 3·14). If γ-glutamyl phosphate were in aqueous solution, it would be rapidly hydrolyzed to glutamate and phosphate. However, it remains bound to the enzyme in a cavity called the active site, where it is shielded from water. Because of its extreme reactivity with nucleophiles, γ-glutamyl phosphate readily forms the amide (glutamine) by reacting with ammonia, which is also bound within the active site of the enzyme. Both the activation of a reactant by transfer of a phosphoryl group from ATP and the exclusion of water from active sites of enzymes are characteristic of the synthesis of most biopolymers.

3·7 Water Undergoes Ionization

Among the important properties of water is its slight tendency to ionize. Pure water consists of not only H_2O but also a low concentration of hydronium ions ($H_3O^{\oplus}$) and an equal concentration of hydroxide ions ($OH^{\ominus}$).

$$H_2O \;+\; H_2O \;\rightleftharpoons\; H_3O^{\oplus} \;+\; OH^{\ominus}$$

Although a hydronium ion is often written as $H_3O^{\oplus}$, it actually has several molecules of water associated with it. For convenience, hydronium ions will hereafter be referred to simply as protons and represented as $H^{\oplus}$.

According to the Brønsted-Lowry concept of acids and bases, an **acid** is a substance that can donate protons, and a **base** is a substance that can accept protons. As Equation 3·2 indicates, water can function as either a proton donor (acid) or a proton acceptor (base).

The ionization of water can be analyzed quantitatively. The equilibrium constant (K_{eq}) for the ionization of water is given by

$$K_{eq} = \frac{[H^{\oplus}][OH^{\ominus}]}{[H_2O]} \tag{3·3}$$

One liter of water has a mass of 1000 g, and the molecular weight of water is 18.015. Therefore, water has a concentration of approximately 55.5 M. Thus,

$$K_{eq}(55.5\,M) = [H^{\oplus}][OH^{\ominus}] \tag{3·4}$$

Electrical conductivity measurements have established that the equilibrium constant for the ionization of water is 1.8×10^{-16} M. Because this equilibrium constant is so small, the concentration of water remains virtually unchanged by ionization. Substituting the above equilibrium constant in Equation 3·4 gives

$$1.0 \times 10^{-14} \, \mathrm{M}^2 \;=\; [\mathrm{H}^{\oplus}][\mathrm{OH}^{\ominus}] \qquad \textbf{(3·5)}$$

The value $1.0 \times 10^{-14}\,\mathrm{M}^2$ in Equation 3·5 is called the ion product of water and is designated K_w.

$$K_w \;=\; [\mathrm{H}^{\oplus}][\mathrm{OH}^{\ominus}] \;=\; 1.0 \times 10^{-14} \, \mathrm{M}^2 \qquad \textbf{(3·6)}$$

Since pure water is electrically neutral, its ionization produces an equal number of protons and hydroxide ions ($[\mathrm{H}^{\oplus}] = [\mathrm{OH}^{\ominus}]$). Equation 3·6 can therefore be rewritten as

$$K_w \;=\; [\mathrm{H}^{\oplus}]^2 \;=\; 1.0 \times 10^{-14} \, \mathrm{M}^2 \qquad \textbf{(3·7)}$$

Taking the square root of Equation 3·7 gives

$$[\mathrm{H}^{\oplus}] \;=\; 1.0 \times 10^{-7} \, \mathrm{M} \qquad \textbf{(3·8)}$$

Since $[\mathrm{H}^{\oplus}] = [\mathrm{OH}^{\ominus}]$, the ionization of pure water produces 10^{-7} M $\mathrm{H}^{\oplus}$ and 10^{-7} M $\mathrm{OH}^{\ominus}$. An aqueous solution that contains equal concentrations of $\mathrm{H}^{\oplus}$ and $\mathrm{OH}^{\ominus}$ is said to be neutral. When an acid is dissolved in water, $[\mathrm{H}^{\oplus}]$ increases and the solution is described as acidic. There must be a concomitant decrease in $[\mathrm{OH}^{\ominus}]$, as dictated by Equation 3·6. Dissolving a base in water increases $[\mathrm{OH}^{\ominus}]$ above 1×10^{-7} M, producing a basic, or alkaline, solution, and decreases $[\mathrm{H}^{\oplus}]$ simultaneously.

3·8 The pH Scale Provides a Measure of Acidity and Basicity

Although the concentration of $\mathrm{H}^{\oplus}$ is small relative to the concentration of water, many biochemical processes depend upon the $\mathrm{H}^{\oplus}$ concentration. The transport of oxygen in the blood, the catalysis of reactions by enzymes, and the generation of metabolic energy during respiration or photosynthesis are some of the many biological phenomena that depend upon the concentration of $\mathrm{H}^{\oplus}$. Because the range of $[\mathrm{H}^{\oplus}]$ in aqueous solutions is enormous, it is convenient to use a logarithmic quantity called **pH** to measure the concentration of $\mathrm{H}^{\oplus}$. pH is defined as the negative logarithm of the concentration of $\mathrm{H}^{\oplus}$.

$$\mathrm{pH} \;=\; -\log[\mathrm{H}^{\oplus}] \;=\; \log \frac{1}{[\mathrm{H}^{\oplus}]} \qquad \textbf{(3·9)}$$

A neutral solution has a pH value of 7.0 since $-\log(10^{-7}) = 7.0$. Acidic solutions have pH values less than 7.0, and basic solutions have pH values greater than 7.0. The lower the pH, the more acidic the solution; the higher the pH, the more basic the solution. Since the pH scale is logarithmic, a change in pH of one unit corresponds to a tenfold change in the concentration of $\mathrm{H}^{\oplus}$. Table 3·3 indicates the relationship between pH and the concentrations of $\mathrm{H}^{\oplus}$ and $\mathrm{OH}^{\ominus}$. Figure 3·15, on the next page, shows the pH values of various fluids.

Table 3·3 Relation of $[\mathrm{H}^{\oplus}]$ and $[\mathrm{OH}^{\ominus}]$ to pH

pH	$[\mathrm{H}^{\oplus}]$ (M)	$[\mathrm{OH}^{\ominus}]$ (M)
0	1	10^{-14}
1	10^{-1}	10^{-13}
2	10^{-2}	10^{-12}
3	10^{-3}	10^{-11}
4	10^{-4}	10^{-10}
5	10^{-5}	10^{-9}
6	10^{-6}	10^{-8}
7	10^{-7}	10^{-7}
8	10^{-8}	10^{-6}
9	10^{-9}	10^{-5}
10	10^{-10}	10^{-4}
11	10^{-11}	10^{-3}
12	10^{-12}	10^{-2}
13	10^{-13}	10^{-1}
14	10^{-14}	1

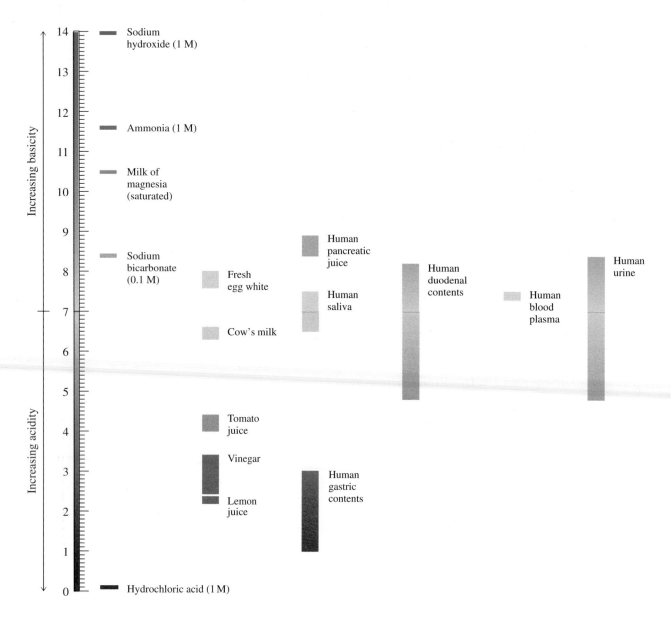

Figure 3·15
pH values for various fluids at 25°C. Lower pH values correspond to acidic fluids; higher pH values correspond to basic fluids. [Adapted from Weast, R. C., Lide, D. R., Astle, M. J., and Beyer, W. H., eds. (1989). *CRC Handbook of Chemistry and Physics,* 70th ed. (Boca Raton, Florida: CRC Press).]

Accurate measurements of pH are routinely made using a pH meter, an instrument that incorporates a selective glass electrode that is sensitive to the concentration of $H^{\oplus}$. Measurement of pH sometimes facilitates the diagnosis of disease. The normal pH of human blood is 7.4, which is frequently referred to as physiological pH. The blood of patients suffering from certain diseases, such as diabetes, can have a lower pH, a condition called acidosis. The condition in which the pH of the blood is higher than 7.4, called alkalosis, can result from persistent, prolonged vomiting (loss of hydrochloric acid from the stomach) or from hyperventilation (excessive loss of carbonic acid as carbon dioxide).

3·9 The Acid Dissociation Constants of Weak Acids Are Determined by Titration

Acids and bases that dissociate completely in water, such as hydrochloric acid and sodium hydroxide, are called strong acids and strong bases. Many acids and bases, such as the amino acids from which proteins are made and the purines and pyrimidines from which DNA and RNA are made, do not dissociate completely in water. These substances are known as weak acids and weak bases.

Let us consider the ionization of acetic acid, the weak acid in vinegar.

$$CH_3COOH \xrightleftharpoons{K_a} H^\oplus + CH_3COO^\ominus \qquad (3 \cdot 10)$$

Acetic acid (weak acid)　　Acetate anion (conjugate base)

As this equation illustrates, an acid can be converted to its **conjugate base** by loss of a proton, and in the reverse reaction, a base can be converted to its **conjugate acid** by addition of a proton.

The equilibrium constant for the dissociation of a proton from an acid is called the acid dissociation constant, designated K_a. To simplify calculations and make easy comparisons, bases are considered in their protonated, or conjugate acid, forms, which are very weak acids. For example, the K_a of the base ammonia (NH_3) is the measure of the acid strength of its conjugate acid, the ammonium ion ($NH_4^\oplus$). For acetic acid

$$K_a = \frac{[H^\oplus][CH_3COO^\ominus]}{[CH_3COOH]} \qquad (3 \cdot 11)$$

The value of K_a for acetic acid at 25°C is 1.76×10^{-5} M. Table 3·4 lists the K_a values for several common acids. Because these values are numerically small and clumsy in calculations, it is useful to place them on a logarithmic scale. Thus, a new parameter, **pKa**, is defined by analogy with pH. A pH value is a measure of acidity, and a pK_a value is a measure of acid strength.

$$pK_a = -\log K_a \qquad (3 \cdot 12)$$

Table 3·4 Dissociation constants and pK_a values of weak acids in aqueous solutions at 25°C

Acid	K_a (M)	pK_a
HCOOH (Formic acid)	1.77×10^{-4}	3.8
CH_3COOH (Acetic acid)	1.76×10^{-5}	4.8
$CH_3CHOHCOOH$ (Lactic acid)	1.37×10^{-4}	3.9
H_3PO_4 (Phosphoric acid)	7.52×10^{-3}	2.2
$H_2PO_4^\ominus$ (Dihydrogen phosphate ion)	6.23×10^{-8}	7.2
$HPO_4^{2\ominus}$ (Monohydrogen phosphate ion)	2.20×10^{-13}	12.7
H_2CO_3 (Carbonic acid)	4.30×10^{-7}	6.4
$HCO_3^\ominus$ (Bicarbonate ion)	5.61×10^{-11}	10.2
$NH_4^\oplus$ (Ammonium ion)	5.62×10^{-10}	9.2
$CH_3NH_3^\oplus$ (Methylammonium ion)	2.70×10^{-11}	10.7

From Equation 3·11, we see that K_a for acetic acid is related to the concentration of $H^{\oplus}$ and to the ratio of the concentrations of the acetate ion and undissociated acetic acid. Taking the logarithm of Equation 3·11 gives

$$\log K_a = \log \frac{[H^{\oplus}][A^{\ominus}]}{[HA]} \tag{3·13}$$

where HA represents the undissociated acid (in this case, acetic acid) and $A^{\ominus}$ represents its conjugate base (in this case, the acetate anion). Since $\log(xy) = \log x + \log y$, Equation 3·13 can be rewritten as

$$\log K_a = \log[H^{\oplus}] + \log \frac{[A^{\ominus}]}{[HA]} \tag{3·14}$$

Rearranging Equation 3·14 gives

$$-\log[H^{\oplus}] = -\log K_a + \log \frac{[A^{\ominus}]}{[HA]} \tag{3·15}$$

The negative logarithms in Equation 3·15 have already been defined as pH and pK_a (Equations 3·9 and 3·12, respectively). Thus,

$$pH = pK_a + \log \frac{[A^{\ominus}]}{[HA]} \tag{3·16}$$

or

$$pH = pK_a + \log \frac{[\text{conjugate base}]}{[\text{weak acid}]} \tag{3·17}$$

Equation 3·17 is called the **Henderson-Hasselbalch equation.** It defines the pH of a solution in terms of the pK_a of the weak acid and the logarithm of the ratio of concentrations of the dissociated species (conjugate base) to the protonated species (weak acid). It is useful to note that the greater the concentration of the conjugate base, the higher the pH, whereas the greater the concentration of the weak acid, the lower the pH. When the concentrations of a weak acid and its conjugate base are exactly the same, the pH of the solution is equal to the pK_a of the acid (since the ratio of concentrations equals 1.0 and the logarithm of 1.0 equals zero).

The pK_a values of weak acids are determined by titration. The titration curve for acetic acid is presented in Figure 3·16. In this example, a solution of acetic acid is titrated by adding small aliquots of a strong base of known concentration. The pH of the solution is measured and plotted versus the number of molar equivalents of strong base added during the titration. Note that since acetic acid has only one ionizable group (its carboxyl group), only one equivalent of a strong base is needed to completely titrate acetic acid to its conjugate base, the acetate anion. When the acid has been titrated with one-half an equivalent of base, the concentration of undissociated acetic acid exactly equals the concentration of the acetate anion. The resulting pH, 4.8, is thus the experimentally determined pK_a for acetic acid.

Similar titration curves can be obtained for each of the five monoprotic acids (acids having only one ionizable group) listed in Table 3·4. All would exhibit the same general shape as Figure 3·16, but the inflection point representing the midpoint of titration (one-half an equivalent titrated) would fall lower on the pH scale for a stronger acid (such as formic acid or lactic acid) and higher for a weaker acid (such as ammonium ion or methylammonium ion).

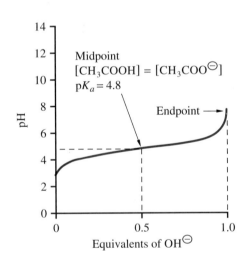

Figure 3·16
Titration of acetic acid (CH_3COOH) with aqueous base ($OH^{\ominus}$). There is an inflection point (a point of minimum slope) at the midpoint of titration, when 0.5 equivalent of base has been added to the solution of acetic acid. This is the point at which $[CH_3COOH] = [CH_3COO^{\ominus}]$ and pH = pK_a. The pK_a of acetic acid is thus 4.8. At the endpoint, all the molecules of acetic acid have been titrated to the conjugate base, acetate.

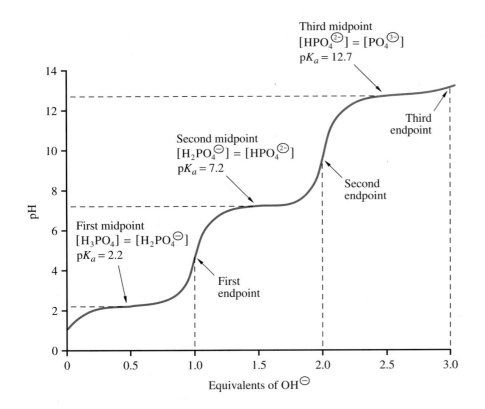

Figure 3·17
Titration curve for phosphoric acid (H_3PO_4). There are three inflection points (at 0.5, 1.5, and 2.5 equivalents of strong base added), corresponding to the three pK_a values for phosphoric acid (2.2, 7.2, and 12.7).

Many biologically important acids and bases have two or more ionizable groups. The number of pK_a values for such substances is equal to the number of ionizable groups, which can be experimentally determined by titration. For example, phosphoric acid requires three equivalents of strong base for complete titration, and three pK_a values are evident from its titration curve (Figure 3·17). The three pK_a values reflect the three equilibrium constants and thus the existence of four possible species (conjugate acids and bases) of inorganic phosphate.

$$(3\cdot18)$$

At physiological pH (7.4), the predominant species of inorganic phosphate are $H_2PO_4^{\ominus}$ and $HPO_4^{2\ominus}$. At pH 7.2, these two species exist in equal concentrations. The concentrations of H_3PO_4 and $PO_4^{3\ominus}$ are so low at pH 7.4 that they can be ignored. This is generally the case for a minor species if the pH is more than two units away from its pK_a.

3·10 Buffered Solutions Resist Changes in pH

If the pH of a solution remains nearly constant when small amounts of strong acid or strong base are added, the solution is said to be **buffered.** The ability of a solution to resist changes in pH is known as its buffer capacity. Inspection of the titration curves of acetic acid (Figure 3·16) and phosphoric acid (Figure 3·17) reveals that the most effective buffering, indicated by the region of minimum slope on the curve, occurs when the concentrations of a weak acid and its conjugate base are equal—in other words, when the pH equals the pK_a. The effective range of buffering by a mixture of a weak acid and its conjugate base is usually considered to be from one pH unit below to one pH unit above the pK_a.

An excellent example of buffer capacity is found in the blood plasma of mammals, which has a remarkably constant pH of 7.4. Consider the results of an experiment that compares the addition of an aliquot of strong acid to a volume of blood plasma with a similar addition of strong acid to either physiological saline (0.154 M NaCl) or water. When 1.0 ml of 10 M HCl (hydrochloric acid) is added to 1000 ml of physiological saline or water that is initially at pH 7.0, the pH is lowered to 2.0 (in other words, $[H^{\oplus}]$ is diluted to 10^{-2} M). However, when 1.0 ml of 10 M HCl is added to 1000 ml of blood plasma at pH 7.4, the pH is again lowered, but only to 7.2—impressive evidence for the effectiveness of physiological buffering.

The pH of the blood is primarily regulated by the carbon dioxide–carbonic acid–bicarbonate buffer system. A plot of the percentages of carbonic acid (H_2CO_3) and each of its conjugate bases as a function of pH is shown in Figure 3·18. Note that the major forms of carbonic acid at pH 7.4 are carbonic acid and the bicarbonate anion ($HCO_3^{\ominus}$).

The buffer capacity of blood depends upon equilibria between gaseous carbon dioxide (which is present in the air spaces of the lungs), aqueous carbon dioxide (which is produced by respiring tissues and dissolved in blood), carbonic acid, and bicarbonate. As shown in Figure 3·18, the equilibrium between bicarbonate and carbonate ($CO_3^{\ominus\ominus}$) does not contribute significantly to the buffer capacity of blood because the pK_a of bicarbonate is 10.2, which is too far from physiological pH to have an effect on buffering of blood.

The first of the three relevant equilibria of the carbon dioxide–carbonic acid–bicarbonate buffer system is the dissociation of carbonic acid to bicarbonate and $H^{\oplus}$ with an effective pK_a of 6.4.

$$H_2CO_3 \; \rightleftarrows \; H^{\oplus} + HCO_3^{\ominus} \tag{3·19}$$

Figure 3·18
Percentages of carbonic acid and its conjugate bases as a function of pH. At pH 7.4 (the pH of blood), the concentrations of carbonic acid (H_2CO_3) and bicarbonate ($HCO_3^{\ominus}$) are substantial, but the concentration of carbonate ($CO_3^{\ominus\ominus}$) is negligible.

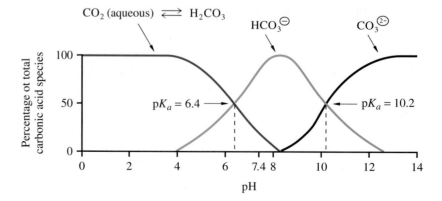

This equilibrium is affected by a second equilibrium in which dissolved carbon dioxide, CO_2(aqueous), is in equilibrium with its hydrated form, carbonic acid.

$$CO_2(\text{aqueous}) + H_2O \rightleftharpoons H_2CO_3 \qquad (3\cdot20)$$

Finally, CO_2(gaseous) is in equilibrium with CO_2(aqueous).

$$CO_2(\text{gaseous}) \rightleftharpoons CO_2(\text{aqueous}) \qquad (3\cdot21)$$

The regulation of the pH of blood afforded by these three equilibria is pictured schematically in Figure 3·19. When the pH of blood falls due to a metabolic process that produces excess $H^\oplus$, the concentration of H_2CO_3 increases momentarily, but H_2CO_3 rapidly loses water to form dissolved CO_2(aqueous), which enters the gaseous phase in the lungs and is expired as CO_2(gaseous). An increase in the partial pressure of CO_2 (pCO_2) in the air expired from the lungs thus compensates for the increased hydrogen ions. Conversely, if the pH of the blood rises, the concentration of $HCO_3^\ominus$ increases transiently, but the pH is rapidly restored as the breathing rate changes and the reservoir of CO_2(gaseous) in the lungs is converted to CO_2(aqueous) and then to H_2CO_3 in the capillaries of the lungs. Again, the equilibrium of the blood buffer system is rapidly restored by changing the partial pressure of CO_2 in the lungs.

Within cells, both proteins and inorganic phosphate contribute to intracellular buffering. Hemoglobin is the strongest buffer in blood other than the carbon dioxide–carbonic acid–bicarbonate buffer. As mentioned earlier, the major species of inorganic phosphate present at physiological pH are $H_2PO_4^\ominus$ and $HPO_4^{\circled{2}}$, reflecting the second pK_a (pK_2) value for phosphoric acid, 7.2.

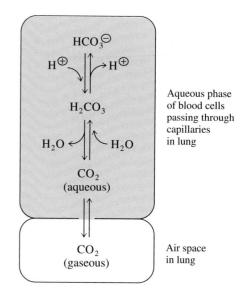

Figure 3·19
Regulation of the pH of blood in mammals. The pH of blood is closely controlled by the ratio of [$HCO_3^\ominus$] and pCO_2 in the air spaces of the lungs. If the pH of blood decreases due to the presence of excess $H^\oplus$, pCO_2 increases in the lungs, restoring the equilibrium. If, on the other hand, the concentration of $HCO_3^\ominus$ rises because the pH of blood increases, CO_2(gaseous) dissolves in the blood, again restoring the equilibrium.

3·11 Natural and Synthetic Compounds Are Biochemically Useful Buffers

Virtually all in vitro biochemical experiments involving purified biomolecules, cell extracts, or intact cells are performed in the presence of a suitable buffer so that a stable pH value can be maintained. Buffers with a variety of pK_a values are available (Table 3·5). Some commonly used buffers are naturally occurring compounds. For example, mixtures of acetic acid and sodium acetate can be used for the pH range from 4 to 6, and mixtures of KH_2PO_4 and K_2HPO_4 can be used in the range from 6 to 8. Some other commonly used buffers are synthetic amines, such as MES, PIPES, HEPES, and Tris. The structures of both the protonated and unprotonated forms of these compounds are shown in Figure 3·20 on the next page. MES, PIPES, and HEPES are tertiary amines and as such have lower pK_a values than aliphatic amines, whose pK_a values are typically 10.6–10.8. The lower pK_a values of MES, PIPES, and HEPES make these compounds suitable for buffering near physiological pH. The primary amine Tris was one of the earliest synthetic buffers for the pH range from 7 to 9. Its pK_a value is lowered by the inductive effect of the three —OH groups. Tris is still widely used, although it can react with aldehydes and ketones and is toxic to some cells. HEPES is also popular because it is an effective buffer at physiological pH and is less reactive and less toxic than Tris.

A buffered solution is usually prepared by one of two methods. The buffer compound can be dissolved and then carefully titrated to the desired pH. Alternatively, stock solutions of both the acidic and unprotonated forms of a buffer can be prepared and mixed in predetermined ratios to achieve exact pH values. For an experiment designed to test the effect of a wide range of pH values, a mixture of several buffers can be prepared to cover the range of interest. For example, to examine the effect of pH values from 5 to 10 on the activity of an enzyme, a mixture of MES ($pK_a = 6.1$), HEPES ($pK_a = 7.5$), and glycine ($pK_a = 9.8$) might be prepared.

Table 3·5 pK_a values of some commonly used buffers

Buffer	pK_a at 25°C
Phosphate (pK_1)	2.2
Acetate	4.8
MES[1]	6.1
Citrate (pK_3)	6.4
PIPES[2]	6.8
Phosphate (pK_2)	7.2
HEPES[3]	7.5
Tris[4]	8.1
Glycylglycine	8.2
Glycine (pK_2)	9.8

[1] 2-(*N*-Morpholino)ethanesulfonic acid
[2] Piperazine-*N,N'-bis*(2-ethanesulfonic acid)
[3] *N*-2-Hydroxyethylpiperazine-*N'*-2-ethanesulfonic acid
[4] *Tris*(hydroxymethyl)aminomethane
[Adapted from Stoll, V. S., and Blanchard, J. S. (1990). Buffers: principles and practice. *Methods Enzymol.* 182:24–38.]

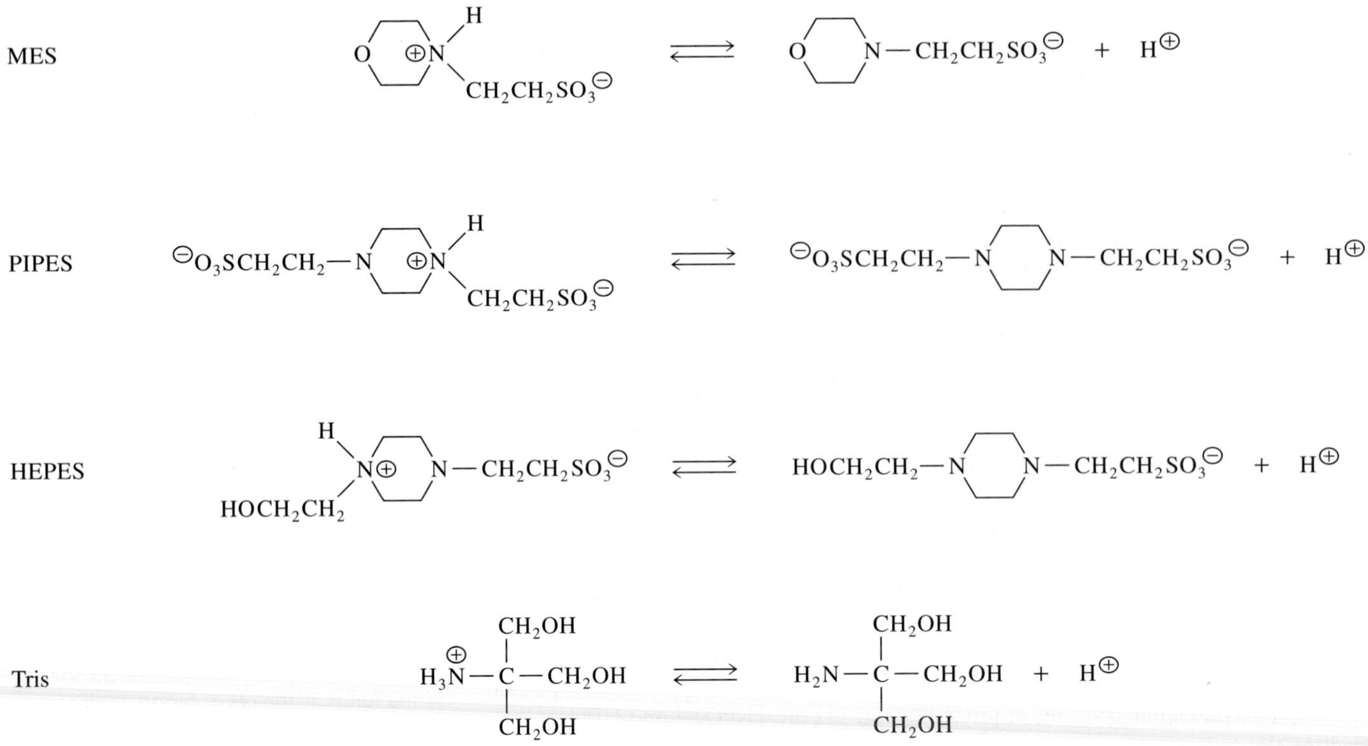

MES

PIPES

HEPES

Tris

Figure 3·20
Structures of some common buffers. The protonated (weak acid) and unprotonated (conjugate base) forms are shown for each compound.

Summary

Water is a nonlinear molecule whose H—O—H bond angle is 104.5°. Because oxygen is more electronegative than hydrogen, an uneven distribution of charge occurs within each O—H bond of the water molecule (that is, each O—H bond is polar). The polarity of the covalent bonds of the water molecule and the geometry of the molecule are such that the molecule has a positive end and a negative end and thus has a permanent dipole.

A water molecule forms four hydrogen bonds in ice and up to four hydrogen bonds in liquid water. In biological systems, intermolecular hydrogen bonds form between water molecules and many types of biomolecules. Some biomolecules also form intramolecular hydrogen bonds.

Ionic substances, such as sodium chloride, and highly polar nonionic compounds, such as short-chain alcohols and glucose, readily dissolve in water and are said to be hydrophilic. Ionic and polar molecules are surrounded by water molecules that form a solvation sphere. Nonpolar substances, such as hydrocarbons, are essentially insoluble in water and are called hydrophobic. The phenomenon of exclusion of nonpolar substances by water is called the hydrophobic effect. Detergents are amphipathic, meaning they have both hydrophobic and hydrophilic groups. These compounds may form monolayers on the surface of aqueous solutions and micelles when dispersed in aqueous media. Chaotropes enhance the solubility of nonpolar compounds in water.

The major noncovalent interactions in cells are electrostatic attractions, hydrogen bonds, van der Waals forces, and hydrophobic interactions. These weak forces stabilize the structures of proteins, nucleic acids, and membranes.

Although water is nucleophilic and present in large amounts, cells use several strategies to prevent degradative hydrolytic reactions and to allow condensation reactions. Unwanted hydrolysis of biopolymers is prevented by storing some hydrolases in inactive forms or sequestering them in organelles. Cells use the chemical potential energy of ATP to overcome the unfavorable equilibria of biosynthetic reactions. Furthermore, water is often excluded from the active sites of enzymes, where biosynthetic reactions are catalyzed.

Pure water, which ionizes slightly, contains 10^{-7} M protons and 10^{-7} M hydroxide ions. The concentration of $H^{\oplus}$ in aqueous solutions is measured on the logarithmic pH scale. Neutral solutions have a pH value of 7.0, acidic solutions have pH values less than 7.0, and basic (or alkaline) solutions have pH values greater than 7.0. Weak acids only partially dissociate when dissolved in water. The pK_a values of weak acids are determined by titration. These values are the pH values at the midpoints of titrations. The Henderson-Hasselbalch equation quantitatively describes the relationship between the pH of a solution, the pK_a of a weak acid, and the ratio of concentrations of the weak acid and its conjugate base.

A solution that resists changes in pH when small amounts of acid or base are added to it is said to be buffered. Maximum buffering is afforded when a weak acid and its conjugate base are present in equal concentrations (in other words, when the pH is equal to the pK_a). In humans, the pH of blood, 7.4 (sometimes called physiological pH), is maintained by the carbon dioxide–carbonic acid–bicarbonate buffer system, which depends upon both the concentration of plasma bicarbonate and the partial pressure of carbon dioxide in the lungs. Proteins and inorganic phosphate contribute to intracellular buffering. Natural and synthetic buffer compounds are used to maintain a constant pH during biochemical experiments.

Selected Readings

Water

Stillinger, F. H. (1980). Water revisited. *Science* 209:451–457. Summary and evaluation of experiments designed to determine the structure of liquid water.

Noncovalent Interactions

Fersht, A. R. (1987). The hydrogen bond in molecular recognition. *Trends Biochem. Sci.* 12:301–304. Discussion of the strength of binding interactions via hydrogen bonds in enzyme-substrate complexes and in DNA.

Frieden, E. (1975). Non-covalent interactions. *J. Chem. Ed.* 52:754–761. A review of the types of noncovalent interactions found in biological systems, with a variety of examples.

Tanford, C. (1980). *The Hydrophobic Effect: Formation of Micelles and Biological Membranes,* 2nd ed. (New York: John Wiley & Sons). A complete description of micelles and hydrophobic interactions.

Ionization of Water

Montgomery, R., and Swenson, C. A. (1976). *Quantitative Problems in Biochemical Sciences,* 2nd ed. (San Francisco: W. H. Freeman and Company).

Segel, I. H. (1976). *Biochemical Calculations: How to Solve Mathematical Problems in General Biochemistry,* 2nd ed. (New York: John Wiley & Sons).

Zumdahl, S. S. (1989). *Chemistry,* 2nd ed. (Lexington, Massachusetts: D. C. Heath and Company). This and other widely available general chemistry texts contain expanded discussions of pH, buffers, and equilibria.

Buffers

Good, N. E., Winget, G. D., Winter, W., Connolly, T. N., Izawa, S., and Singh, R. M. M. (1966). Hydrogen ion buffers for biological research. *Biochemistry* 5:467–477. Description of a variety of synthetic buffers.

Stoll, V. S., and Blanchard, J. S. (1990). Buffers: principles and practice. *Methods Enzymol.* 182:24–38. A practical guide to selecting and preparing buffered solutions.

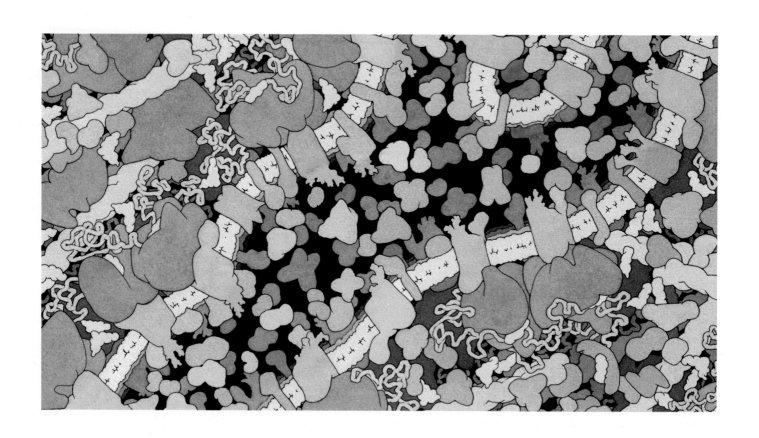

Part Two

Structures and Functions
of Biomolecules

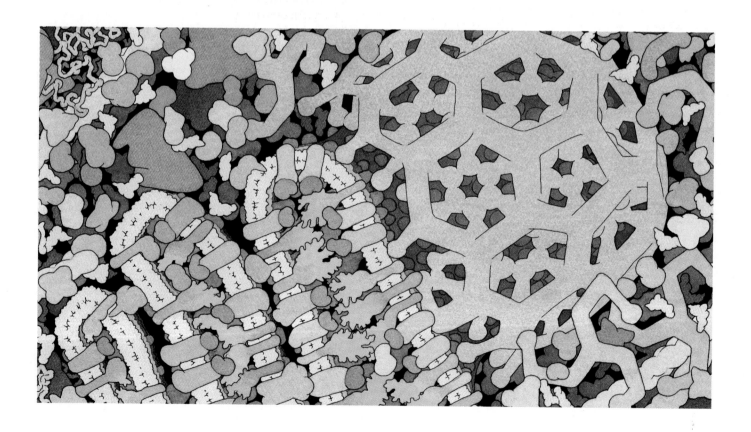

4

Amino Acids and the Primary Structures of Proteins

In the 1830s, the Dutch chemist Gerardus Johannes Mulder was investigating the properties of albumins, substances from milk and eggs that coagulate upon heating. When he measured the elemental composition of albumins, Mulder found that they contain—in order of decreasing amount—carbon, oxygen, nitrogen, and hydrogen. In 1838, the Swedish scientist Jöns Jacob Berzelius suggested to Mulder that these substances be called **proteins,** derived from the Greek *proteios* meaning first or primary, because he suspected that these compounds might be the most important of all biological substances. The name was prophetic. Proteins are now known to be involved in a wide range of biological functions, including, but not limited to, the following:

1. Many proteins function as enzymes, the biological catalysts. Enzymes catalyze nearly all reactions that occur in living organisms.

2. Some proteins bind other molecules for the purposes of storage and transport. For example, myoglobin binds oxygen in skeletal and cardiac muscle cells, and hemoglobin binds and transports O_2 and CO_2 in red blood cells.

3. Structural proteins provide mechanical support and shape to cells and hence to tissues and organisms.

4. Assemblies of proteins can do mechanical work, such as movement of flagella, separation of chromosomes at mitosis, and contraction of muscles.

5. Many proteins play a role in decoding information in the cell. Some are involved in translation, whereas others play a role in regulating gene expression by binding to nucleic acids.

6. Some proteins are hormones, which regulate biochemical activities in target cells or tissues; other proteins serve as receptors for hormones.

7. Some proteins serve other specialized functions. For example, immunoglobulins, one of the classes of proteins within the immune system, defend against bacterial and viral infections in vertebrates.

4·1 Proteins Are Made from 20 Different Amino Acids

Proteins are biopolymers that consist of linear chains of amino acid residues. All organisms use the same 20 amino acids as building blocks for the assembly of protein molecules. These 20 amino acids are therefore often cited as the common or standard amino acids. Despite the limited number of amino acid types, the variations in the order in which they are connected and in the numbers of amino acids per protein allow an almost limitless variety of proteins.

The **primary structure** of a protein is the sequence in which amino acids are covalently connected to form the polypeptide chain. We will consider higher levels of protein structure in Chapter 5, after we have examined the nature of amino acids and the formation of polypeptides.

The 20 common amino acids are termed α-amino acids because they have an amino group and an acidic carboxyl group attached to C-2, which is also known as the α-carbon.

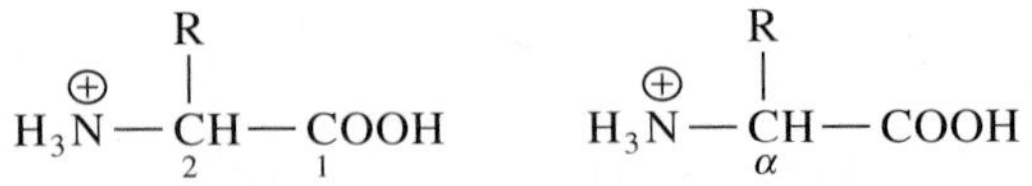

$$\overset{\oplus}{H_3N} - \underset{2}{\overset{R}{\underset{|}{C}H}} - \underset{1}{COOH} \qquad \overset{\oplus}{H_3N} - \underset{\alpha}{\overset{R}{\underset{|}{C}H}} - COOH \qquad (4\cdot1)$$

In addition, a hydrogen atom and a side chain are attached to the α-carbon. The side chain, or R group, is distinctive for each amino acid. Figure 4·1a shows the general structure of an amino acid in perspective; Figure 4·1b shows a ball-and-stick drawing of a representative amino acid, serine, which has —CH_2OH as its side chain. The carbons of a side chain can be lettered sequentially as β, γ, δ, and ε, which refer to carbons 3, 4, 5, and 6, respectively.

At neutral pH, the amino group is protonated (—$NH_3^{\oplus}$) and the carboxyl group is ionized (—$COO^{\ominus}$). The pK_a values of α-carboxyl groups range from 1.8 to 2.5, and the pK_a values of α-amino groups range from 8.7 to 10.7. Thus, at physiological pH (7.4), amino acids are zwitterions, or dipolar ions, even though their net charge may be zero. We will see in Section 4·4 that some side chains can also ionize.

4·2 19 Common Amino Acids Are Stereoisomers

In 19 of the 20 amino acids used for biosynthesis of proteins, the α-carbon atom is **chiral,** or asymmetric, since it has four different groups bonded to it. (The exception is glycine, in which the R group is simply a hydrogen atom, so that two hydrogen atoms are bonded to the α-carbon.) Therefore, amino acids can exist as stereoisomers, compounds with the same molecular formula that differ in the arrangement, or **configuration,** of their atoms in space. A change in configuration requires the breaking of one or more bonds. Two stereoisomers that are nonsuperimposable mirror images can exist for each chiral amino acid. Such stereoisomers are called enantiomers. By convention, the mirror-image pairs of amino acids are designated D (for dextro, from the Latin *dexter,* right) and L (for levo, from the Latin *laevus,* left). The configuration of the amino acid structure in Figure 4·1a is L; its mirror image would be D. To assign the stereochemical designation properly, one must draw the amino acid vertically with its most oxidized group, the α-carboxylate, at the top. It is then compared to the stereochemical reference compound, glyceraldehyde, which will be further described in Chapter 9. In this orientation, the α-amino group of the L isomer is on the left of the α-carbon, and that of the D isomer is on the right. Figure 4·2a (next page) shows the general structure of L- and D-amino acids, and Figure 4·2b shows a particular example, L- and D-serine. These can be compared to the carbohydrates L- and D-glyceraldehyde shown in Figure 4·2c. Figure 4·3 (next page) shows a stereo view of the L and D configurations of serine.

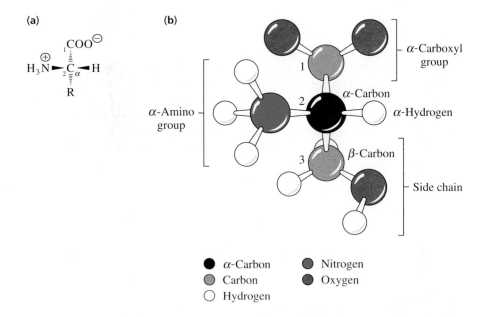

(a)

(b)

α-Amino group

α-Carboxyl group

α-Carbon

α-Hydrogen

β-Carbon

Side chain

● α-Carbon   ● Nitrogen
● Carbon   ● Oxygen
○ Hydrogen

Figure 4·1
Two representations of an α-amino acid at neutral pH. **(a)** General structure. An amino acid has a carboxyl group (designated C-1), an amino group, a hydrogen atom, and a side chain, or R group, all attached to C-2 (the α-carbon). Solid wedges indicate bonds above the plane of the paper; dashed wedges, bonds below the plane of the paper. The blunt ends of wedges are nearer the viewer than are the pointed ends. **(b)** Three-dimensional ball-and-stick drawing of serine (in which —R = —CH$_2$OH). Note the alternative numbering and lettering systems for designating carbons.

Figure 4·2
Mirror-image pairs of compounds containing a single chiral carbon atom. (a) General structures of L- and D-amino acids. (b) L-Serine and D-serine. (c) The historical reference compounds L-glyceraldehyde and D-glyceraldehyde.

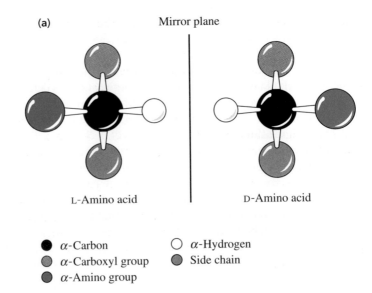

(a)   Mirror plane

L-Amino acid   D-Amino acid

● α-Carbon   ○ α-Hydrogen
◐ α-Carboxyl group   ● Side chain
● α-Amino group

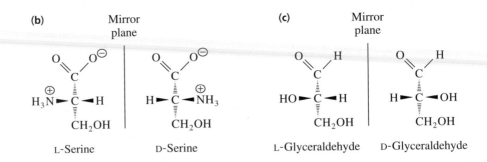

(b)   Mirror plane

L-Serine   D-Serine

(c)   Mirror plane

L-Glyceraldehyde   D-Glyceraldehyde

Although a few D-amino acids occur in nature, the 19 chiral amino acids used in the assembly of proteins are all of the L configuration. Therefore, by convention, amino acids are assumed to be in the L configuration unless specifically designated D. Often, it is convenient to draw the structures of L-amino acids in a form that is stereochemically uncommitted. For example, alanine—the amino acid with a methyl side chain—could be drawn simply as

$$\overset{\oplus}{H_3N} - \overset{\overset{\displaystyle CH_3}{|}}{CH} - COOH \qquad\qquad (4\cdot2)$$

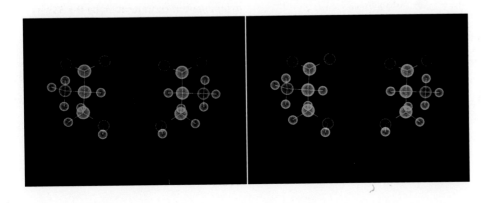

Figure 4·3
Stereo view of L-serine (left) and D-serine (right). Color key: carbon, green; hydrogen, light blue; nitrogen, dark blue; oxygen, red.

This format is particularly useful for short peptides. Uncommitted structures for chemical compounds are commonly used when a correct representation of stereochemistry is not critical to a given discussion.

Two of the 19 chiral amino acids of proteins have two chiral carbon atoms each. The DL system does not completely describe the configurations of these two amino acids, only their chirality at the α-carbon. In these cases, it is necessary to assign each isomer a different name. In addition, assignment of D or L requires comparison to the reference compound glyceraldehyde. The RS system of configurational nomenclature, however, can be applied to each chiral center of these molecules without changing the names of the parent compounds and without comparing them to a reference compound. The RS system, devised by Robert S. Cahn, Christopher K. Ingold, and Vladimir Prelog, is based on assignment of a priority sequence to the four groups bound to a chiral carbon atom. Once assigned, the priority sequence is used to establish the configuration of the molecule. Priorities are numbered one through four and are assigned to groups according to the following rules:

1. For atoms directly attached to the chiral carbon, the one with the lowest atomic mass is assigned the lowest priority, 1.

2. If there are two identical substituent atoms bound to the chiral carbon, the priority is decided by the atomic masses of the next atoms bound. Thus, a —CH$_3$ group has a lower priority than a —CH$_2$Br group because bromine has a greater atomic mass than hydrogen.

3. If an atom is bound by a double or triple bond, the atom is counted once for each formal bond. Thus, —CHO, with a double-bonded oxygen, has a higher priority than —CH$_2$OH. The order of priority for the most common groups, from lowest to highest, is —H, —CH$_3$, —C$_6$H$_5$, —CH$_2$OH, —CHO, —COOH, —COOR, —NH$_2$, —NHR, —OH, —OR, and —SH.

With these rules in mind, imagine the molecule as the steering column of a car, with the group of lowest priority, numbered 1, pointing away from you, and the other three groups arrayed around the rim of the steering wheel. Trace the rim of the wheel moving from the group of highest priority to the group of lowest priority (4, 3, 2). If the movement is clockwise, the molecule is defined as R (Latin, *rectus,* right-handed). If the movement is counterclockwise, the molecule is defined as S (Latin, *sinister,* left-handed). Figure 4·4 demonstrates the assignment of configuration to L-serine by the RS system. L-Cysteine, although it is also an L-amino acid, has the opposite configuration in the RS system, R. Because of this type of inconsistency in the RS system, the DL system is used more often in biochemistry.

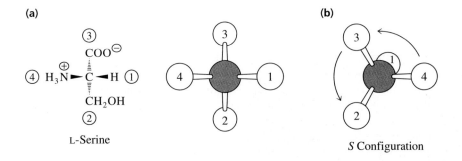

(a)

L-Serine

(b)

S Configuration

Figure 4·4
Assignment of configuration by the RS system. **(a)** Using rules described in the text, each group attached to a chiral carbon is assigned a priority based on atomic mass, 1 being the lowest priority. **(b)** By orienting the molecule with the priority 1 group pointing away (behind the chiral carbon) and tracing the path from the highest priority group to the lowest, the absolute configuration can be established. If the sequence 4, 3, 2 is clockwise, the configuration is R. If the sequence 4, 3, 2 is counterclockwise, the configuration is S. L-Serine has the S configuration.

4·3 The 20 Common Amino Acids Have Different Side Chains

The structures of the 20 amino acids commonly found in proteins are shown in Figure 4·5 as Fischer projections. In Fischer projections, horizontal bonds at a chiral center extend toward the viewer, and vertical bonds extend away. Examination of the structures in Figure 4·5 reveals considerable variation in the side chains of the 20 amino acids. Some side chains are nonpolar and thus hydrophobic, whereas others are polar or ionized at neutral pH and are therefore hydrophilic. Since individual amino acids are zwitterions, they are water soluble. However, it is essential to realize that when amino acids are combined by polymerization into peptides and proteins, the ionic charges on the α-amino and α-carboxyl groups that form the links are lost. As a result, the nonpolar, polar, or ionic characteristics of the side chains greatly influence the overall three-dimensional shape, or conformation, of a protein. For example, water-soluble globular proteins typically contain hundreds of amino acid residues in a compact conformation. Most of the hydrophilic side chains of such a protein are on the surface of the macromolecule and are thus exposed to water, whereas many of the hydrophobic side chains are in the interior of the protein and are shielded from water.

To understand the functions of individual amino acids in proteins, it is important to learn their structures and to be able to relate these structures to the one- and three-letter abbreviations commonly used in referring to them. Learning the names of the amino acids is difficult because biochemists use trivial rather than systematic names for these compounds. In the following sections, the 20 amino acids are organized by the general properties and chemical structures of their side chains. You will notice that some amino acids have nonpolar and chemically inert side chains. Other amino acids have side chains with some polar character (one has a heterocyclic ring system, three have hydroxyl groups, and two have sulfur atoms). Finally, there are amino acids with decidedly polar R groups—five charged (protonated or ionized at physiological pH) and two uncharged. The 20 amino acid side chains fall into the following chemical classes: six aliphatic, three aromatic, two sulfur-containing, two alcohols, three bases, two acids, and two amides. Five of the 20 amino acids are further classified as highly hydrophobic, and seven are classified as highly polar. Understanding the classification of the R groups will simplify memorizing the structures and names.

A. Aliphatic R Groups

Glycine (Gly, G) has the least complex structure of the 20 amino acids, since its side chain is simply a hydrogen atom. As a result, the α-carbon of glycine is not chiral. The two hydrogen atoms of the α-carbon of glycine impart little hydrophobic character to the molecule. Although glycine could be listed in a separate category, we view it as the simplest of the aliphatic group because of its relatively unreactive R group. Glycine has a unique role within the conformations of many proteins because its side chain is small enough to fit into niches that accommodate no other amino acid. The name *glycine* is a good example of the use of trivial names; the proper name, 2-aminoacetic acid, describes the compound exactly but is not commonly used.

Four amino acids, alanine (Ala, A), valine (Val, V), leucine (Leu, L), and the structural isomer of leucine, isoleucine (Ile, I), have saturated aliphatic side chains. The side chain of alanine is simply a methyl group, whereas valine has a three-carbon branched side chain, and leucine and isoleucine each contain a four-carbon branched side chain. Note that both the α- and β-carbon atoms of isoleucine are asymmetric. Because isoleucine has two chiral centers, it has four possible stereoisomers: L-isoleucine, its enantiomer D-isoleucine, L-*allo*isoleucine, and D-*allo*isoleucine. (The prefix *allo-* simply indicates an isomeric form or variety.)

Figure 4·5 (next page)
Fischer projections of the 20 common amino acids at pH 7, their names, and one- and three-letter abbreviations. Amino acids are classified by their hydrophobic or polar characteristics and by the functional groups of their side chains. There are five highly hydrophobic side chains (blue) and seven strongly polar side chains (red).

Aliphatic

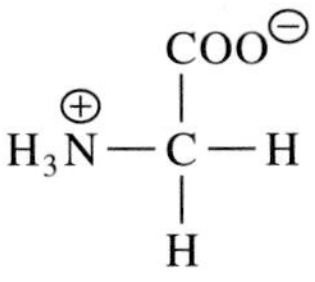

Glycine [G]
(Gly)

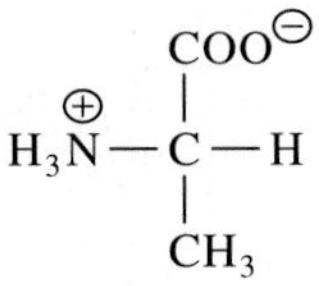

Alanine [A]
(Ala)

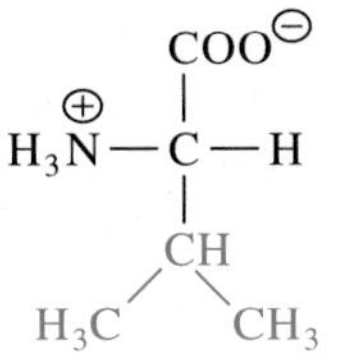

Valine [V]
(Val)

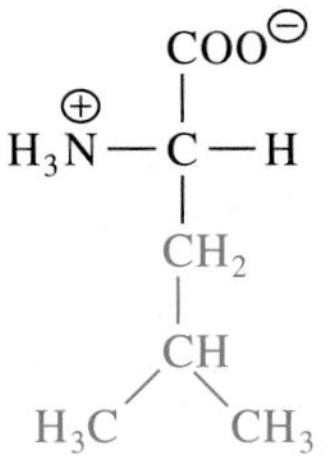

Leucine [L]
(Leu)

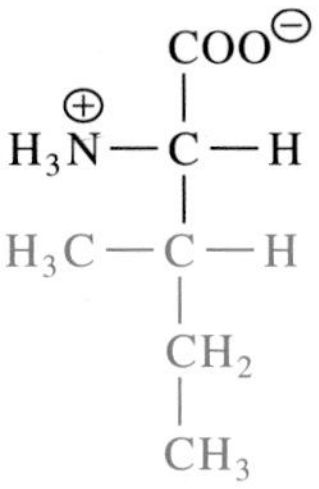

Isoleucine [I]
(Ile)

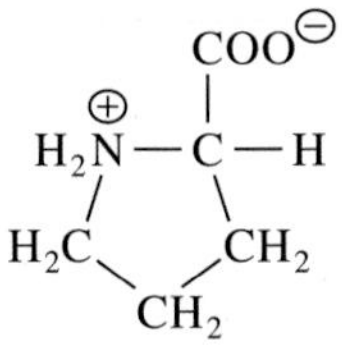

Proline [P]
(Pro)

Aromatic

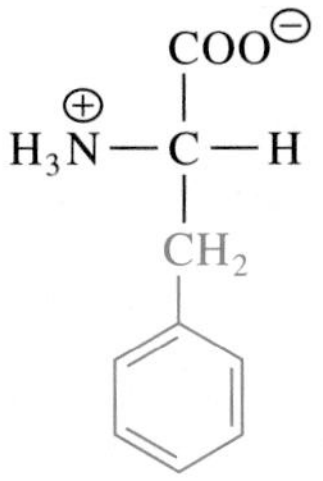

Phenylalanine [F]
(Phe)

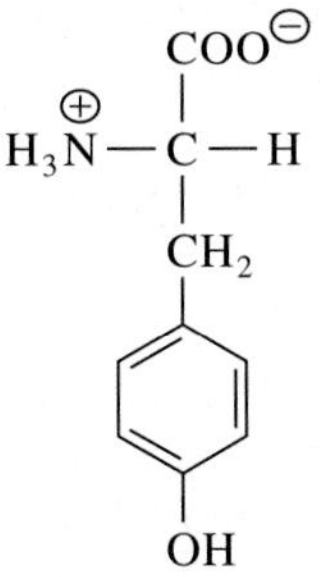

Tyrosine [Y]
(Tyr)

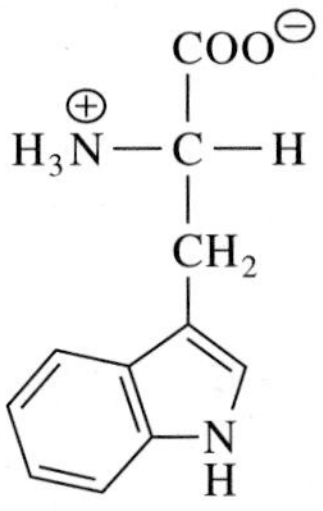

Tryptophan [W]
(Trp)

Sulfur-containing

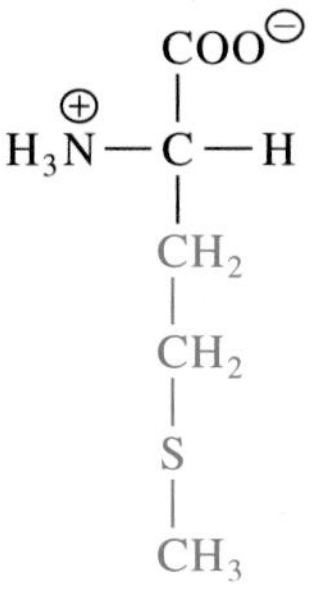

Methionine [M]
(Met)

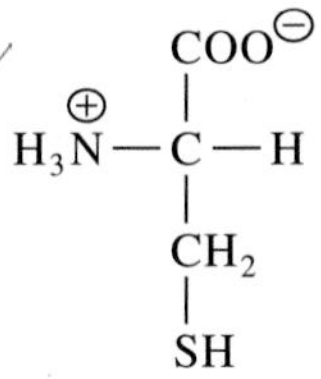

Cysteine [C]
(Cys)

Alcohols

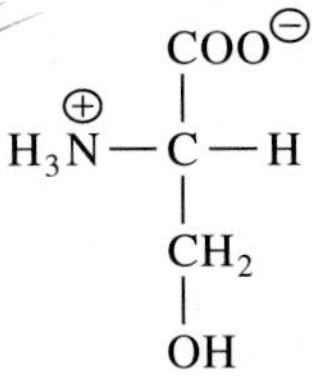

Serine [S]
(Ser)

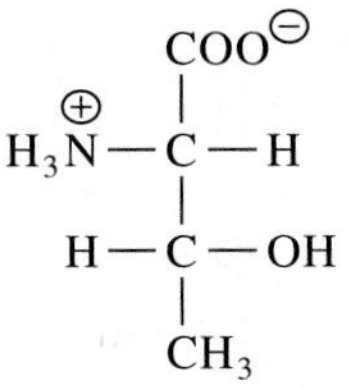

Threonine [T]
(Thr)

Acids

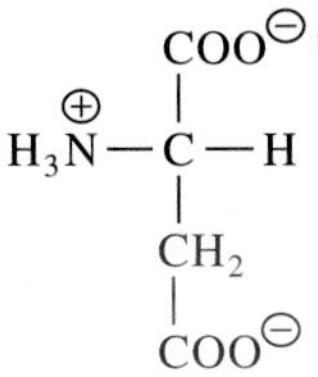

Aspartate [D]
(Asp)

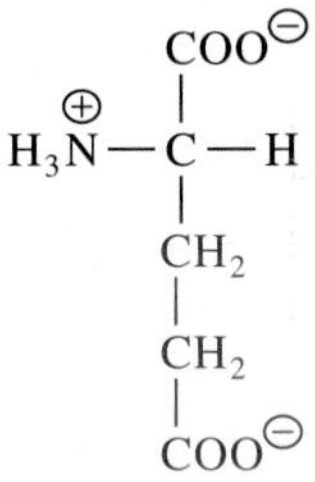

Glutamate [E]
(Glu)

Bases

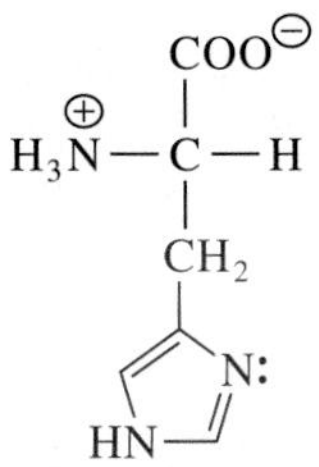

Histidine [H]
(His)

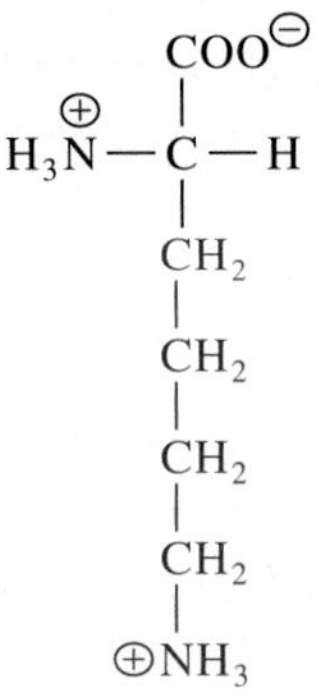

Lysine [K]
(Lys)

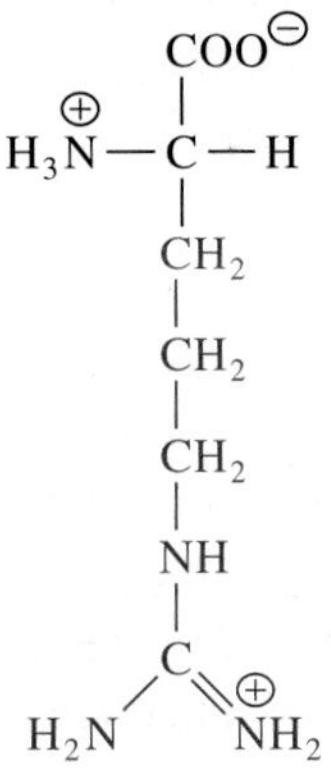

Arginine [R]
(Arg)

Amides

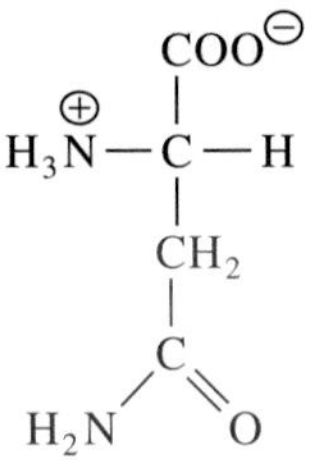

Asparagine [N]
(Asn)

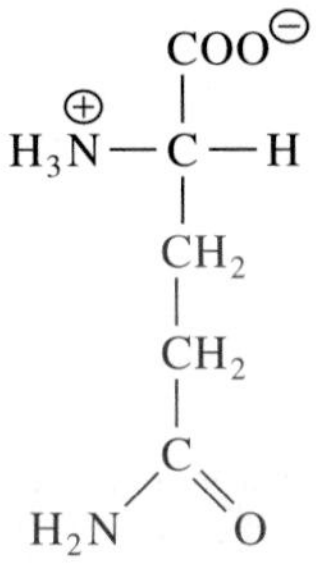

Glutamine [Q]
(Gln)

L-Isoleucine is the only one of the four isomers normally found in proteins. (By the *RS* system, L-isoleucine is (2*S*,3*S*)-isoleucine.) Although the side chains of alanine, valine, leucine, and isoleucine have no reactive functional groups, these amino acids play an important role in establishing and maintaining the three-dimensional structures of proteins because of their tendency to cluster away from water. Valine, leucine, and isoleucine are highly hydrophobic.

Proline (Pro, P) differs markedly from the other 19 amino acids in that its cyclic side chain, a saturated hydrocarbon, is bonded to the nitrogen of its α-amino group as well as to the α-carbon. Thus, strictly speaking, proline is an α-imino acid since it contains a secondary rather than a primary amino group. Nonetheless, the proline residues found in proteins all have the same chirality about their α-carbons as the other L-amino acids. The heterocyclic pyrrolidine ring of proline restricts the geometry of polypeptides, sometimes introducing abrupt changes in the direction of the peptide chain.

B. Aromatic R Groups

Phenylalanine (Phe, F), tyrosine (Tyr, Y), and tryptophan (Trp, W) have side chains with aromatic groups. Phenylalanine, the phenyl analog of alanine, has a benzyl side chain. The benzene ring makes phenylalanine residues highly hydrophobic. Tyrosine is structurally similar to phenylalanine; the *para*-hydrogen of phenylalanine is replaced in tyrosine by a hydroxyl group (—OH), making tyrosine a phenol. Recall that phenols are very weak acids; the pK_a of the side chain of tyrosine is 10.5. The bicyclic structure in the side chain of tryptophan is indole; thus, tryptophan can be regarded as indolealanine. Due to the presence of polar groups on their side chains, tyrosine and tryptophan are not as hydrophobic as phenylalanine.

The three aromatic amino acids absorb ultraviolet (UV) light. At neutral pH, both tryptophan and tyrosine absorb UV light at 280 nm, whereas phenylalanine is almost transparent at 280 nm and absorbs weakly at 260 nm. Since most proteins contain tryptophan and/or tyrosine, absorbance of solutions at 280 nm is routinely used to estimate protein concentration (Section 4·7).

C. Sulfur-Containing R Groups

Methionine (Met, M) and cysteine (Cys, C) are the two sulfur-containing amino acids. Methionine, which is the fifth highly hydrophobic amino acid, contains a nonpolar methyl thioether group in its side chain. Nonetheless, its sulfur atom is nucleophilic. Cysteine resembles alanine in which a hydrogen atom is replaced by a sulfhydryl group (—SH). Since compounds containing sulfhydryl groups are known as thiols or mercaptans, cysteine can also be regarded as thioserine or mercaptoalanine.

Although the side chain of cysteine is somewhat hydrophobic, it is also highly reactive. Because the sulfur atom is polarizable, the sulfhydryl group of cysteine can form weak hydrogen bonds with oxygen and nitrogen. Moreover, the sulfhydryl group of cysteine is a weak acid (pK_a = 8.4); thus, it can lose its proton to become a negatively charged thiolate ion. When some proteins are hydrolyzed, a compound called cystine can be isolated. Cystine consists of two cysteine molecules linked through oxidation by a disulfide bond (Figure 4·6). Disulfide bonds, or bridges, play an important role in stabilizing the three-dimensional structures of some proteins by serving as covalent cross-links between oxidized cysteine residues at different positions in peptide chains.

D. Side Chains with Alcohol Groups

There are four additional amino acids with polar uncharged side chains—two alcohols and two amides. The amides will be discussed in Section 4·3F. Serine (Ser, S) and threonine (Thr, T) have uncharged polar side chains containing β-hydroxyl

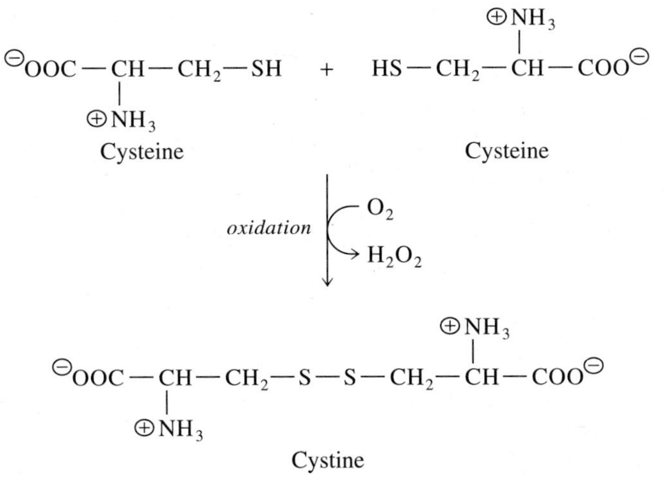

$$^{\ominus}OOC-\underset{\underset{\oplus NH_3}{|}}{CH}-CH_2-SH \quad + \quad HS-CH_2-\underset{\overset{\oplus NH_3}{|}}{CH}-COO^{\ominus}$$

Cysteine   Cysteine

oxidation $\Big\{ \begin{array}{l} -O_2 \\ \rightarrow H_2O_2 \end{array}$

$$^{\ominus}OOC-\underset{\underset{\oplus NH_3}{|}}{CH}-CH_2-S-S-CH_2-\underset{\overset{\oplus NH_3}{|}}{CH}-COO^{\ominus}$$

Cystine

Figure 4·6
Oxidation of the sulfhydryl groups of two cysteine molecules. The oxidation of cysteine proceeds most readily at alkaline pH values at which the thiol group is ionized. When oxidation links the sulfhydryl groups of two cysteine molecules, the resulting compound is a disulfide called cystine.

groups; these alcohol groups give hydrophilic character to the aliphatic side chains. Unlike the more acidic phenolic side chain of tyrosine, the hydroxyl groups of serine and threonine have the weak ionization properties of primary and secondary alcohols. Although the hydroxymethyl group of serine ($-CH_2OH$) does not appreciably ionize in aqueous solutions, this alcohol can react within the active sites of a number of enzymes as though it were ionized. Note that threonine, like isoleucine, has two chiral centers, the α- and β-carbon atoms. L-Threonine, or (2S,3R)-threonine, (shown in Figure 4·5) is the only one of the four stereoisomers that commonly occurs in proteins.

E. Basic R Groups

Five amino acids have side chains that are charged at pH 7; three are basic (positively charged) and two are acidic (negatively charged). Histidine (His, H), lysine (Lys, K), and arginine (Arg, R) have hydrophilic side chains that are nitrogenous bases. Histidine, with an imidazole ring in its side chain, can be regarded as imidazolealanine. The imidazole group is ionizable ($pK_a = 6.0$), and the protonated form of this ring is called an imidazolium ion. Lysine is a diamino acid, having both α- and ε-amino groups. The ε-amino group exists as an alkylammonium ion ($-CH_2-NH_3^{\oplus}$) at neutral pH and confers a positive charge on proteins. Arginine is the most basic of the 20 amino acids; that is, its side-chain guanidinium ion has the highest pK_a value ($pK_a = 12.5$). Thus, arginine side chains also contribute positive charges in proteins.

F. Acidic R Groups and Their Amide Derivatives

Aspartate (Asp, D) and glutamate (Glu, E) are dicarboxylic amino acids. In addition to their α-carboxyl groups, aspartate possesses a β-carboxyl group, and glutamate possesses a γ-carboxyl group. Because the side chains of aspartate and glutamate are ionized at pH 7, they confer negative charges on proteins and are usually found on the surfaces of protein molecules. Aspartate and glutamate are sometimes referred to as aspartic acid and glutamic acid. However, under most physiological conditions, they are found as the conjugate bases and, like other salts, have the *-ate* suffix. Glutamate is probably best known as its monosodium salt, monosodium glutamate (MSG), which is used in food as a flavor enhancer.

   Asparagine (Asn, N) and glutamine (Gln, Q) are the amides of aspartic acid and glutamic acid, respectively. Although the side chains of asparagine and glutamine are uncharged, these amino acids are highly polar and are often found on the surfaces of proteins, where they can interact with water molecules. In addition, the polar amide groups of asparagine and glutamine can form hydrogen bonds with atoms in the side chains of other polar amino acids.

Table 4·1 Comparison of two scales of hydropathy of amino acid residues

Amino acid	Scale #1	Scale #2
Highly hydrophobic:		
Isoleucine	1.4	1.7
Valine	1.1	1.6
Leucine	1.1	1.4
Phenylalanine	1.2	1.1
Methionine	0.64	0.80
Less hydrophobic:		
Alanine	0.62	0.77
Glycine	0.48	0.03
Cysteine	0.29	1.0
Tryptophan	0.81	−0.14
Tyrosine	0.26	−0.27
Proline	0.12	−0.37
Threonine	−0.05	−0.07
Serine	−0.18	−0.10
Highly hydrophilic:		
Histidine	−0.40	−0.91
Glutamate	−0.74	−1.0
Asparagine	−0.78	−1.0
Glutamine	−0.85	−1.0
Aspartate	−0.90	−1.0
Lysine	−1.5	−1.1
Arginine	−2.5	−1.3

[Adapted from Eisenberg, D. (1984). Three-dimensional structure of membrane and surface proteins. *Annu. Rev. Biochem.* 53:595–623.]

G. Hydropathy Is a Measure of the Hydrophobicity of Amino Acid Side Chains

As one can see from Figure 4·5, the properties of the various side chains of amino acids cover the spectrum from highly hydrophobic through weakly polar to highly hydrophilic. There have been numerous attempts to quantitate the hydrophobic character or the polarity of the side chains of the 20 common amino acids. The relative hydrophobic or hydrophilic tendency of each amino acid has been termed its **hydropathy.** Hydropathy values have been estimated experimentally by comparing the water/organic solvent partitioning of amino acids or by measuring the distribution of side-chain analogs between aqueous solutions and the vapor phase. Hydropathy values have also been empirically derived from the relative positions of the 20 types of residues in the three-dimensional structures of a large number of proteins. Large hydropathy values are assigned to amino acid residues that are most often found in the interior of proteins. Two commonly used hydropathy scales, both based upon a combination of experimental and empirical data, are compared in Table 4·1. Amino acids with highly positive hydropathy values are considered hydrophobic; those with the largest negative values are hydrophilic. Both scales list the five highly hydrophobic amino acids at the top, or hydrophobic end, of the list and the five charged groups and the two amides at the hydrophilic end of the list. Alanine, with its methyl group, has some hydrophobic character; other scales ascribe to it much less hydrophobicity than the two scales shown in Table 4·1. It is also difficult to ascertain the hydropathy values of some other amino acid residues in the central group. For example, there is disagreement over the hydrophilicity of the indole group of tryptophan and of the side chain of cysteine (which is often present in disulfide bonds). Although hydropathy is an important determinant of protein-chain folding, it is not possible to accurately predict whether a given residue will be found in the nonaqueous interior of a protein or on the solvent-exposed surface. Indexes such as those in Table 4·1 have been used to predict which segments of membrane-spanning proteins are likely to be embedded in the lipid bilayer (Section 12·4B).

4·4 The Ionic States of Amino Acids Depend on the Ambient pH

The physical properties of amino acids are influenced by the ionic states of the α-carboxyl and α-amino groups and any ionizable groups in the side chains. Seven of the common amino acids have ionizable side chains with measurable pK_a values. Each amino acid has either two or three pK_a values, and these values differ among the amino acids. Consequently, at a given pH, amino acids frequently have different net charges. Even slight differences in net charges on amino acids and proteins can be exploited in order to separate and purify them, as we will see in Section 4·7. The ionic states of amino acid side chains strongly influence the three-dimensional structures and biochemical functions of proteins. In particular, a number of ionizable amino acid residues are involved in catalysis by enzymes; thus, an understanding of the ionic properties of amino acids is necessary in order to understand enzyme mechanisms.

The α-COOH group of an amino acid is a weak acid. Accordingly, we can use the Henderson-Hasselbalch equation (Section 3·9) to calculate the fractions of these groups that are ionized at any given pH.

$$\text{pH} \;=\; pK_a \;+\; \log \frac{[\text{conjugate base}]}{[\text{weak acid}]} \qquad \textbf{(4·3)}$$

As shown in Table 4·2, the pK_a values of the α-carboxyl groups of free amino acids range from 1.8 to 2.5. These values are lower than those of typical carboxylic acids,

Table 4·2 pK_a values of acidic and basic constituents of free amino acids at 25°C*

Amino acid	pK_a values		
	α-Carboxyl group	α-Amino group	Side chain
Glycine	2.4	9.8	
Alanine	2.4	9.9	
Valine	2.3	9.7	
Leucine	2.3	9.7	
Isoleucine	2.3	9.8	
Methionine	2.1	9.3	
Proline	2.0	10.6	
Phenylalanine	2.2	9.3	
Tryptophan	2.5	9.4	
Serine	2.2	9.2	
Threonine	2.1	9.1	
Cysteine	1.9	10.7	8.4
Tyrosine	2.2	9.2	10.5
Asparagine	2.1	8.7	
Glutamine	2.2	9.1	
Aspartic acid	2.0	9.9	3.9
Glutamic acid	2.1	9.5	4.1
Lysine	2.2	9.1	10.5
Arginine	1.8	9.0	12.5
Histidine	1.8	9.3	6.0

*Values have been rounded.
[Values from Dawson, R. M. C., Elliott, D. C., Elliott, W. H., and Jones, K. M. (1986). *Data for Biochemical Research*, 3rd ed. (Oxford: Clarendon Press).]

such as acetic acid ($pK_a = 4.8$), due to the inductive influence of the neighboring —$NH_3^{\oplus}$ group. For a typical amino acid whose α-COOH group has a pK_a of about 2.0, the ratio of conjugate base (carboxylate anion, salt, or the ionized species) to weak acid (carboxylic acid, or the protonated species) at pH 7.0 can be calculated.

$$7.0 = 2.0 + \log \frac{[\text{RCOO}^{\ominus}]}{[\text{RCOOH}]} \qquad \textbf{(4·4)}$$

Therefore, the ratio of carboxylate anion to carboxylic acid is 100 000:1. The predominant species of the α-carboxyl group of amino acids at neutral pH is thus the carboxylate anion.

Next, we consider the ionization of the α-amino groups of free amino acids, whose pK_a values range from 8.7 to 10.7. For an amino acid whose α-amino group has a pK_a of about 10.0, the ratio of conjugate base, or free amine (—NH_2), to conjugate acid, or protonated amine (—$NH_3^{\oplus}$), at pH 7.0 would be 1:1000. Such calculations verify our earlier statement that free amino acids exist predominantly as zwitterions at neutral pH. Thus, it is inappropriate to draw the structure of an amino acid with both —COOH and —NH_2 groups, since there is no pH at which the carboxyl group will be predominantly protonated while the amino group is in its unprotonated, basic form. Note that the imino group of proline ($pK_a = 10.6$) is also protonated at neutral pH, so that proline, despite the bonding of the side chain to the α-amino group, is also zwitterionic at pH 7, as shown in Figure 4·5.

Ionization of the side chains of those amino acids having three readily ionizable groups—aspartate, glutamate, tyrosine, cysteine, lysine, arginine, and histidine—obeys the same principles as the ionizations of the α-carboxyl and α-amino

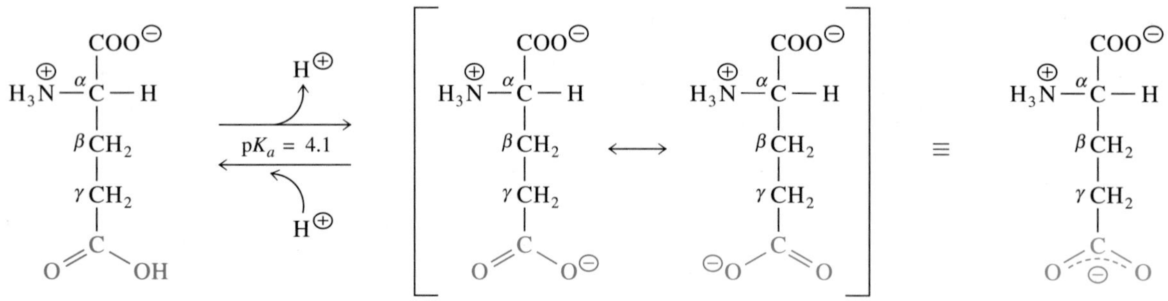

Figure 4·7
Ionization of the protonated γ-carboxyl group of glutamate. The measured pK_a values of 2.1 and 4.1, obtained from titration of fully protonated glutamic acid, reflect the first and second ionizations. Actually, at a pH midway between 2.1 and 4.1, 93% of the α-COOH group is deprotonated to —COO⊖ and 7% of the γ-COOH is deprotonated to —COO⊖. Such overlapping pK_a values are thus not strictly assignable to individual groups. The negative charge of the γ-carboxylate ion is delocalized.

groups. Thus, the Henderson-Hasselbalch equation can be applied to each ionization. Figure 4·7 depicts the ionization of the γ-carboxyl group of glutamate (pK_a = 4.1). Note that the negative charge of the γ-carboxylate anion is delocalized. The γ-carboxyl group, being further removed from the influence of the α-ammonium ion, is a weak acid, similar in strength to acetic acid (pK_a = 4.8), whereas the α-carboxyl group is a stronger acid (pK_a = 2.1).

Figure 4·8 shows the deprotonation of the imidazolium ion of the side chain of histidine and depicts charge delocalization in the imidazolium ion, which predominates below pH 6.0. At pH 7.0, the ratio of imidazole (conjugate base) to imidazolium ion (conjugate acid) is 10:1. Thus, the protonated and neutral forms of the side chain of histidine are both present in significant concentrations near physiological pH. A given histidine side chain in a protein may be either protonated or unprotonated, depending on its immediate environment within the protein. This property makes the side chain of histidine ideal for the transfer of protons within the catalytic sites of enzymes.

Figure 4·9 shows the deprotonation of the guanidinium group of the side chain of arginine in strong alkali. Charge delocalization in the guanidinium ion contributes to its very high pK_a of 12.5.

Titration curves such as those shown for weak acids in Section 3·9 provide data for determining the pK_a values of amino acids. Two examples are shown in Figure 4·10. Alanine, which has two ionizable groups, exhibits two pK_a values, 2.4 and 9.9, each of which is at the center of a buffering region. Histidine, which has three ionizable groups, exhibits pK_a values of 1.8, 6.0, and 9.3, each of which is associated with a buffering zone.

Figure 4·8
Dissociation of a proton from the imidazolium ring of the side chain of histidine. Charge delocalization occurs in the imidazolium ion.

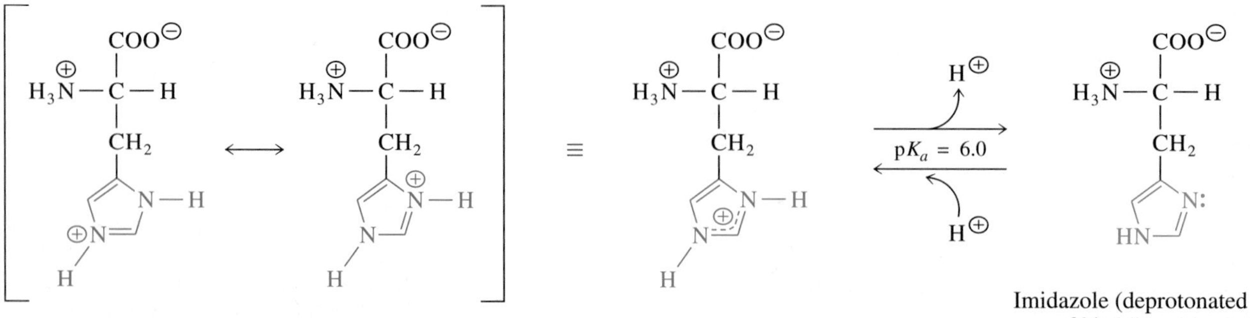

Imidazolium ion (protonated form) of histidine side chain

Imidazole (deprotonated form) of histidine side chain

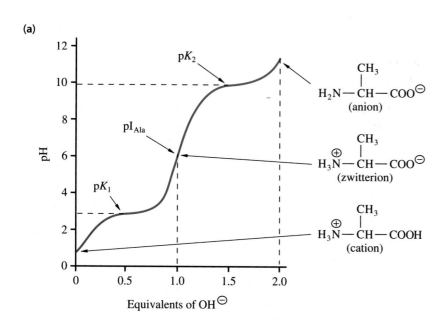

Guanidinium ion (protonated form)
of arginine side chain

Guanidine group
(deprotonated form)
of arginine side chain

Figure 4·9
Deprotonation of the side chain of arginine in strongly basic solution. The equilibrium between the guanidinium ion with its delocalized charge and the unprotonated guanidine group of arginine lies overwhelmingly in the direction of the protonated form at pH 7.

(a)

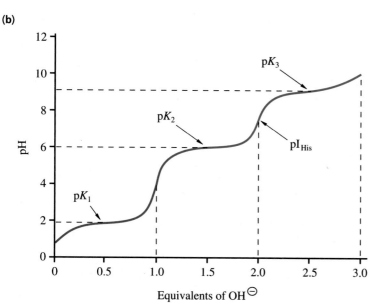

(b)

Figure 4·10
Titration of amino acids. **(a)** Titration curve for alanine. The first pK_a value is 2.4; the second is 9.9. **(b)** Titration curve for histidine. The three pK_a values are 1.8, 6.0, and 9.3. pI_{Ala} and pI_{His} represent the isoelectric points of the two amino acids.

Recall from Section 3·9 that a pK_a value equals the pH at the midpoint in the titration of an ionizable group. In other words, it is the pH at which the concentration of a weak acid exactly equals the concentration of its conjugate base. One can deduce that the net charge on alanine molecules at pH 2.4 averages +0.5 and that the net charge at pH 9.9 averages −0.5. Midway between pH 2.4 and pH 9.9, at pH 6.15, the average net charge on alanine molecules in solution is zero. For this reason, pH 6.15 is referred to as the **isoelectric point** (pI), or isoelectric pH, of alanine. If alanine were placed in an electric field, such as that generated for electrophoresis (Section 4·7), at a pH below its pI, it would carry a net positive charge (in other words, its cationic form would predominate), and it would therefore migrate toward the **cathode** (the negative electrode of the electrophoresis apparatus). At a pH higher than its pI, alanine would carry a net negative charge and would migrate toward the **anode** (the positive electrode). At its isoelectric point (pH = 6.15), alanine would not migrate in either direction. The isoelectric point of an amino acid that, like alanine, contains only two ionizable groups (the α-amino and the α-carboxyl groups) is the arithmetic mean of its two pK_a values. However, for an amino acid such as histidine, which contains three ionizable groups, one must assess the average net charge at each of the three pK_a values in order to identify the isoelectric point correctly. For histidine at pH 1.8, the net charge averages +1.5; at pH 6.0, +0.5; and at pH 9.3, −0.5. Thus, the isoelectric point for histidine is midway between 6.0 and 9.3, or 7.65.

In proteins, the pK_a values of ionizable side chains can vary from those of the free amino acids. Two factors cause this perturbation of ionization constants. First, because α-amino and α-carboxyl groups lose their charges once they are linked by peptide bonds in proteins, they no longer exert strong inductive effects on their neighboring side chains. Second, the position of an ionizable side chain within the three-dimensional structure of a protein can affect its pK_a. As an example, the enzyme ribonuclease A has four histidine residues, but the side chain of each has a slightly different pK_a value as a result of differences in their microenvironments.

4·5 Amino Acids in Proteins Are Linked by Peptide Bonds

Amino acids are linked in a chain by condensation of the α-carboxyl group of one amino acid with the α-amino group of another. The linkage formed between the amino acids is a secondary amide bond called a **peptide bond** (Figure 4·11). Note that a water molecule is lost from the condensing amino acids in the reaction. Unlike the carboxyl and amino groups of free amino acids in solution, the groups involved in peptide bonds carry no ionic charges.

The linked amino acid moieties are called amino acid residues. The names of residues in a polypeptide chain are formed by replacing the ending -*ine* or -*ate* with -*yl*. Thus, a glycine residue in a polypeptide is called glycyl; a glutamate residue is called glutamyl. In the cases of asparagine, glutamine, and cysteine, -*yl* replaces the final -*e* to form asparaginyl, glutaminyl, and cysteinyl, respectively. The -*yl* ending indicates that the residue is an acyl unit (a structure that lacks the hydroxyl of the carboxyl group). Hence, the dipeptide in Figure 4·11 is called alanylserine because the amino acid serine, with the free carboxylate, is substituted by an alanyl (acyl) group.

The free amino group and free carboxyl group at opposite ends of a peptide chain are called the **N-terminus** (amino terminus) and the **C-terminus** (carboxyl terminus), respectively. At neutral pH, each terminus carries an ionic charge. By convention, amino acid residues in a peptide chain are numbered from the N-terminus to the C-terminus and are usually written from left to right.

The **primary structure** of a protein is the linear sequence of amino acid residues linked by peptide bonds. Since both the standard three-letter abbreviations for

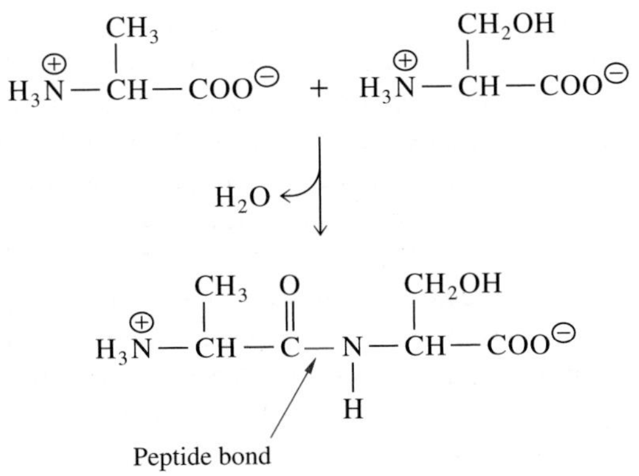

Figure 4·11
Peptide bond between two amino acids. The α-carboxyl group of one amino acid condenses with the α-amino group of another, with loss of a water molecule. The result is a dipeptide in which the amino acids are linked by a peptide bond. Here, alanine is condensed with serine to form alanylserine. (The chemical potential energy of ATP drives the reactions of peptide-bond synthesis, as shown in Chapter 30.)

the amino acids (for example, Gly–Arg–Phe–Ala–Lys) and the one-letter abbreviations (for example, GRFAK) are used to denote the primary structures of peptide chains, it is important to know both abbreviation systems. The terms *dipeptide, tripeptide, oligopeptide,* and *polypeptide* refer to chains of two, three, several (up to about 20), and many (usually more than 20) amino acid residues, respectively. Note that a dipeptide (two residues) contains one peptide bond, a tripeptide contains two peptide bonds, a pentapeptide contains four peptide bonds, and so on. As a general rule, each peptide chain, whatever its length, possesses one free α-amino group and one free α-carboxyl group. (Exceptions include covalently modified terminal residues and circular peptide chains.) Most of the ionic charges associated with a protein molecule are contributed by the side chains of the constituent amino acids, and thus the ionic properties of a protein, as well as its solubility, depend on its amino acid composition. Furthermore, as we will see in Chapter 5, interactions between side chains contribute to the stabilization of the three-dimensional structure of a protein molecule.

Peptides are not merely prototypes of the larger protein molecules but are themselves important biological compounds. The chemistry of peptides is an active area of research. For example, analogs of neuropeptides such as endorphins, the natural pain killers, can be developed for use as drugs. Some very simple peptides are useful as food additives. The sweetening agent aspartame is the methyl ester of aspartylphenylalanine (Figure 4·12). Aspartame, which is about 200 times sweeter than table sugar, is widely used in diet drinks. Many other dipeptides are being tested for their taste, so perhaps a salty-tasting peptide will someday replace sodium chloride for those persons who require a low-sodium diet.

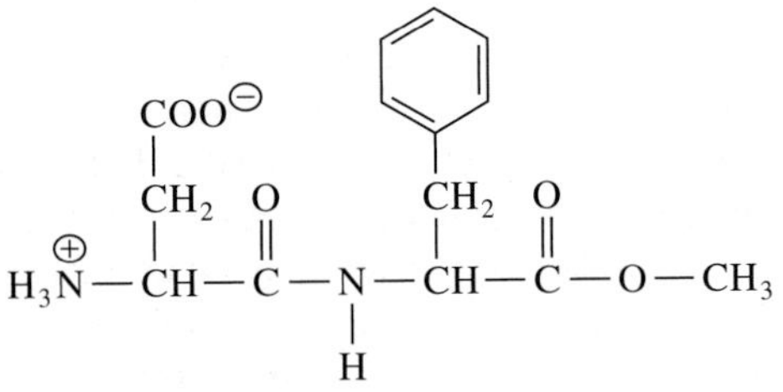

Figure 4·12
Structure of aspartame.

4·6 Polypeptides Can Be Chemically Synthesized

The chemical synthesis of peptides and small proteins is useful for determining the effects of variation of peptide structure on biological function. In addition, synthetic peptides have many uses in the pharmaceutical and food industries. The basic procedure for peptide synthesis was designed by Emil Fischer at the turn of the century. First, the carboxyl group of one amino acid and the amino group of another are substituted (blocked) by protecting groups. Next, the carboxyl group that is to participate in the peptide bond is activated, for example, as an acyl chloride or an acid anhydride. The activated carboxylate can undergo nucleophilic attack by the free amino group (as in Figure 3·13) to form a blocked dipeptide. Finally, the protecting groups are selectively removed by a hydrolysis procedure that leaves the peptide bond intact (Figure 4·13).

A major advance in peptide synthesis was made by R. Bruce Merrifield, who designed and refined the solid-phase procedure. (Merrifield was awarded the Nobel prize in 1984 for his work in solid-phase peptide synthesis.) In this method, an amino acid is covalently attached to a solid support; the new polypeptide chain is grown from this amino acid by sequential addition of other amino acids. The advantages of having the initial reactant and the product bound to an insoluble support are the ability to carry out all reactions in one vessel (making the reactions suitable for automation), the opportunity to add excess second reactant to produce high yields of product, and the easy isolation of the product. The sequence of reactions is shown in Figure 4·14. The amino acid that is to become the C-terminal residue of the synthetic peptide, blocked at its amino group, reacts with a chloromethyl group on a bead of polystyrene. The blocking group is removed in a nonaqueous acidic solution. Dicyclohexylcarbodiimide—an activating agent that in effect removes water from the amine and the carboxylic acid—is used to couple the next amino acid to the amino group of the support-bound amino acid. The blocking group on the new dipeptide is removed, so that the product can react with another amino acid

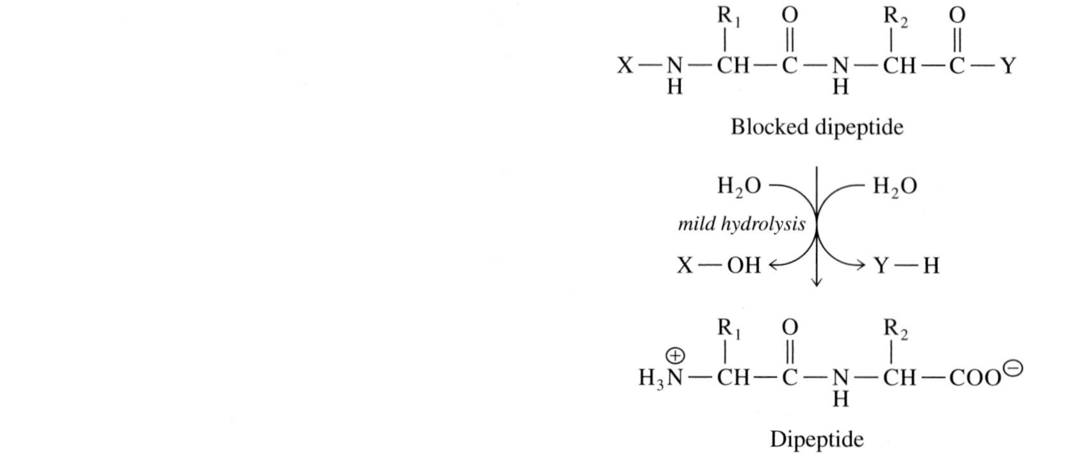

Figure 4·13
Synthesis of a dipeptide from suitably blocked and activated amino acids. X and Y represent blocking groups.

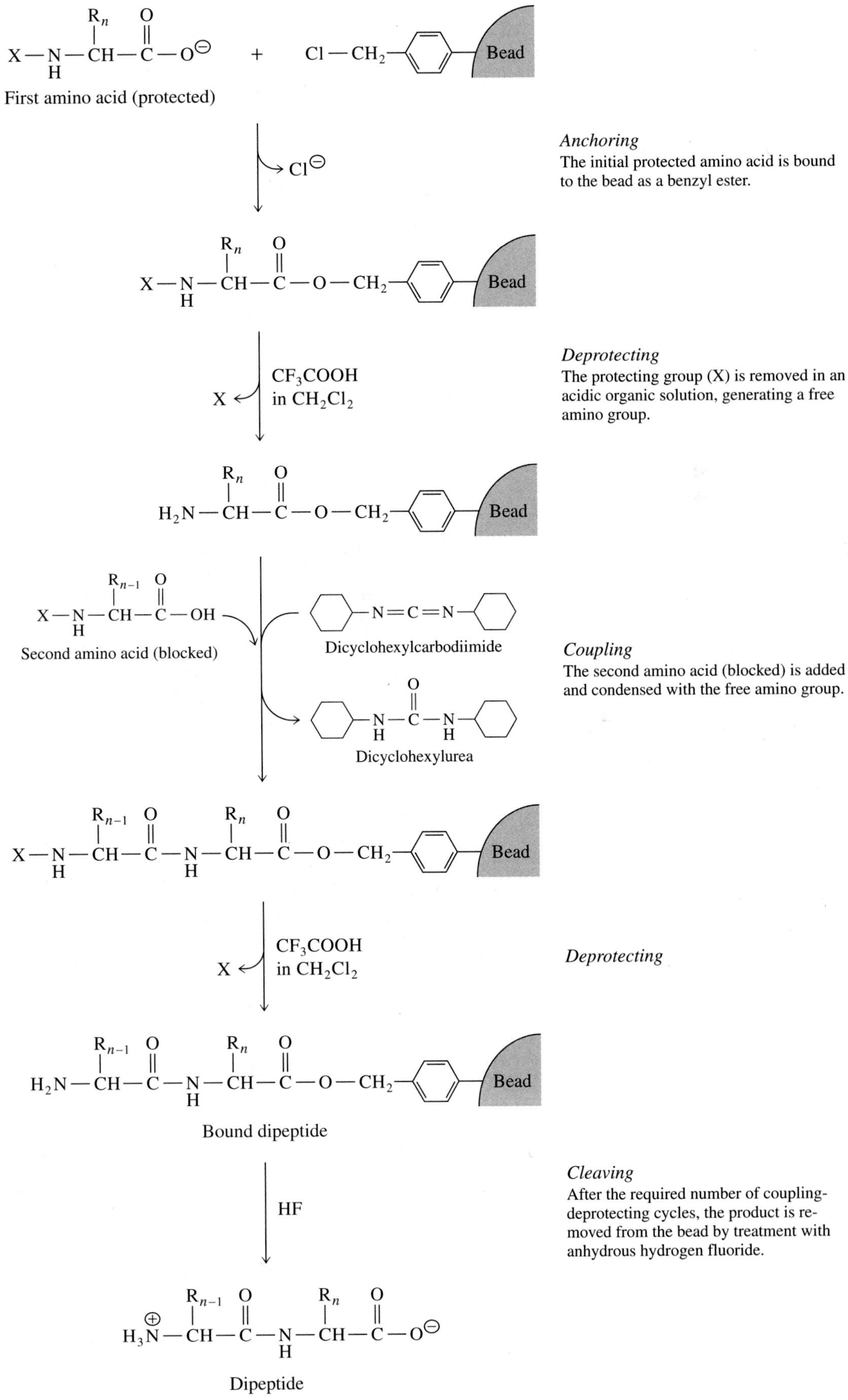

Figure 4·14
Solid-phase peptide synthesis. R represents a side chain, and X represents a blocking group.

First amino acid (protected)

Anchoring
The initial protected amino acid is bound to the bead as a benzyl ester.

CF_3COOH in CH_2Cl_2

Deprotecting
The protecting group (X) is removed in an acidic organic solution, generating a free amino group.

Second amino acid (blocked)

Dicyclohexylcarbodiimide

Dicyclohexylurea

Coupling
The second amino acid (blocked) is added and condensed with the free amino group.

CF_3COOH in CH_2Cl_2

Deprotecting

Bound dipeptide

HF

Cleaving
After the required number of coupling-deprotecting cycles, the product is removed from the bead by treatment with anhydrous hydrogen fluoride.

Dipeptide

or be released from the solid matrix. One of the most notable uses of the solid-phase technique has been chemical synthesis of ribonuclease, a very stable protein that is 124 residues long. However, the solid-phase method is not suitable for synthesis of proteins the size of most enzymes.

4·7 Proteins Can Be Purified by a Variety of Biochemical Techniques

In order to study a particular protein in the laboratory, one must often separate that protein from all other cell components, including other, often very similar, proteins. The steps of purification vary for different proteins but usually involve similar biochemical techniques. These techniques exploit minor differences in the solubilities, net charges, sizes, and binding specificities of proteins. In this section, we will consider some of the common methods of protein purification. For most proteins, these methods must be applied at low temperatures, in the range from 0 to 4°C. Low temperatures minimize protein degradation during purification by impeding the activity of proteases (enzymes that cleave peptide bonds) and by reducing the likelihood that proteins will denature, or unfold (many proteins are extremely heat sensitive).

The first step in protein purification is to obtain a solution of proteins. The source of a protein is often whole cells or tissues in which the target protein may account for less than 0.1% of the total dry weight. Isolation of an intracellular protein requires that cells or chopped tissue be suspended in buffer and homogenized, or disrupted into cell fragments. Disruption of cells can be accomplished mechanically, chemically, or enzymatically. Large cell debris may then be removed, often by filtration through cheesecloth, and subcellular components may be separated by differential centrifugation, which separates principally on the basis of mass and density. To obtain proteins from subcellular organelles, the membrane-bound proteins must be solubilized, usually by treatment of the organelles with a solution containing buffer and detergent. Most proteins are dissolved in the buffer solution in which the cells were suspended for homogenization. For the following discussion, let us assume that the desired protein is one in a mixture of many proteins in this buffered solution.

Usually, the next step in protein purification is a relatively crude separation, or fractionation, procedure that makes use of the different solubilities of proteins in solutions of salts. Ammonium sulfate, a protein-stabilizing salt, is used most often in fractionation. Enough ammonium sulfate is mixed with the solution of proteins to precipitate the less soluble impurities. The target protein and other, more soluble proteins are recovered by centrifugation and remain in the fluid, or supernatant, fraction. Then, more ammonium sulfate is added to the supernatant fluid until the desired protein is precipitated. The mixture is centrifuged, the fluid removed, and the precipitate dissolved in a minimal volume of buffer. Typically, fractionation using ammonium sulfate gives a two- to threefold purification (that is, one-half to two-thirds of the impurities have been removed from the resulting enriched protein fraction). Although the extent of purification is not large, this procedure concentrates the target protein in a solution of smaller volume, which is more suitable for column chromatography. At this point, the solvent is usually exchanged by dialysis for a buffer solution that has no residual ammonium sulfate and has the composition needed for chromatography. In dialysis, a protein solution is placed in a cylinder of cellophane tubing that is knotted at both ends. This sealed sack is suspended in a very large volume of buffer. The cellophane is a semipermeable barrier, so the proteins remain inside the sack because of their high molecular weights, while the buffer and its low-molecular-weight solutes surrounding the protein are gradually exchanged for the medium in which the sack is suspended.

Column chromatography is used next to fractionate the mixture of proteins in solution. A cylindrical column is filled with an insoluble matrix, often consisting of beads of synthetic resin or substituted cellulose fibers. A protein mixture is applied to the column and washed through the matrix by the addition of solvent. As solvent flows through the column, the exiting liquid, or eluate, is collected in many fractions. Figure 4·15a illustrates the general technique of column chromatography. The rate at which proteins travel through the matrix depends on interactions between matrix and protein. For a given column, different proteins will be eluted at different rates. The concentration of protein in each fraction can be determined by measuring the spectrophotometric absorbance of the eluate at 280 nm (Figure 4·15b). (Recall from Section 4·3B that at neutral pH, tyrosine and tryptophan absorb UV light at 280 nm.) The absorbance near 210 nm, the wavelength absorbed by peptide bonds, is sometimes used to measure protein concentrations. To locate the target protein, the fractions containing protein must then be assayed for biological activity or some other characteristic property.

Several types of column chromatography are classified according to the type of matrix. In **ion-exchange chromatography,** the matrix consists of beads or fibers carrying positive charges (anion-exchange resins) or negative charges (cation-exchange resins). Anion-exchange matrices bind negatively charged proteins, retaining them in the matrix for subsequent elution. Conversely, cation-exchange materials bind positively charged proteins. The bound proteins can be serially eluted by gradually increasing the salt concentration in the solvent, since like-charged salt ions and proteins bind to the matrix competitively.

(a)

Steady flow of solvent

Protein mixture

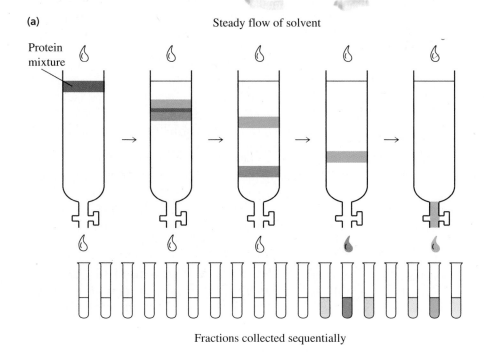

Fractions collected sequentially

(b)

A_{280}

Fraction number

Figure 4·15
Column chromatography. **(a)** Schematic view of fractionation using a chromatographic column. Initially, a mixture of proteins in solution is added to the column, which houses an insoluble, porous matrix. Solvent then flows steadily into the column from a reservoir. Washed by solvent, different proteins (represented by red and blue bands) travel through the column at different rates, depending on their interactions with the matrix. Eluate is collected in a series of fractions, a few of which are shown. **(b)** The protein concentration of each fraction is determined by measuring spectrophotometric absorbance at 280 nm. The peaks correspond to the elution of the protein bands shown in (a). The fractions are then tested for the presence of the target protein.

Gel-filtration chromatography, which separates proteins on the basis of molecular size, utilizes a gel resin consisting of porous beads. Proteins that are smaller than the average pore size penetrate much of the internal volume of the beads and are therefore retarded by the matrix. The smaller the protein, the later it is eluted from the column. Fewer of the pores are accessible to larger protein molecules. Consequently, the largest proteins flow past the beads and are eluted first. Because of this separation of proteins on the basis of molecular size, gel-filtration chromatography is also known as molecular-exclusion chromatography. The choice of pore size of a gel-filtration matrix depends on the molecular weight of the protein to be purified.

Affinity chromatography, the most selective type of column chromatography, relies on binding interactions between a protein and a ligand. A **ligand** is a molecule, group, ion, or atom that binds—usually noncovalently—to another molecule or atom. In affinity chromatography, the ligand is covalently attached to the matrix. It may be a reactant or product to which an enzyme binds in vivo, or it may be an antibody that recognizes the protein of interest. As a mixture of proteins passes through the column, only the target protein specifically binds to the matrix. The column is then washed several times with buffer to rid the column of nonspecifically bound proteins. Finally, the target protein is eluted by washing the column with a solvent that contains a high concentration of the free ligand, or perhaps a concentrated solution of salt. Affinity chromatography alone can sometimes purify a protein 1000- to 10 000-fold.

The resolution of conventional chromatographic methods is greatly increased by the use of **high-pressure liquid chromatography** (HPLC). The matrix in HPLC consists of beads that are smaller, more uniform in size, and more tightly packed than the beads used in conventional chromatographic columns. High pressure is applied to pump a protein mixture and solvent through the column. Columns made of stainless steel can withstand pressures of hundreds of pounds per square inch; however, glass columns, which must operate at lower pressures, are more hospitable to most proteins. Computer-regulated flow rates for the application of samples and solvents allow researchers to select the optimum elution conditions and ensure highly reproducible chromatographic runs. Although speed, resolution, and sensitivity are greatly increased using HPLC, the separations are still based on the same principles that govern conventional ion-exchange, gel-filtration, and affinity chromatography.

Electrophoresis separates proteins based on their migration in an electric field. In **polyacrylamide gel electrophoresis** (PAGE), protein samples are placed on a highly cross-linked gel matrix, and an electric field is applied. The matrix is buffered to a mildly alkaline pH so that most proteins are anionic and migrate toward the anode. Typically, several samples are run at once, together with a reference sample. The gel matrix retards the migration of large molecules as they move in the electric field. Hence, proteins are fractionated on the basis of both charge and mass.

A modification of the standard electrophoresis technique uses the negatively charged detergent sodium dodecyl sulfate (SDS) to overwhelm the native charge on proteins so that they are separated only on the basis of mass. SDS-polyacrylamide gel electrophoresis (SDS-PAGE) is used to assess the purity and to estimate the molecular weight of a protein. In SDS-PAGE, the detergent is added to the polyacrylamide gel as well as to the protein samples. Before loading the protein samples on the gel, 2-mercaptoethanol may be added to the sample, which is heated. 2-Mercaptoethanol breaks any disulfide bonds while SDS and heat denature the proteins for optimal electrophoretic separation. The subunits of proteins that contain multiple peptide chains are also dissociated by this treatment. The dodecyl sulfate anion, which has a long hydrophobic tail (Figure 4·16), binds to hydrophobic side chains of amino acid residues in the polypeptide chain. It binds at a ratio of approximately one molecule of SDS for every two residues of a typical protein. Since

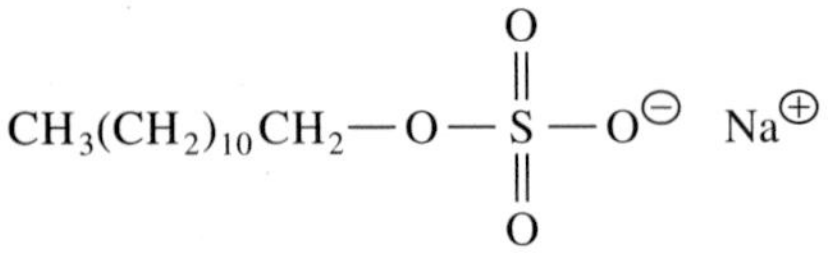

$$CH_3(CH_2)_{10}CH_2-O-\overset{\overset{\displaystyle O}{\|}}{\underset{\underset{\displaystyle O}{\|}}{S}}-O^{\ominus} \quad Na^{\oplus}$$

Figure 4·16
Sodium dodecyl sulfate.

larger proteins bind proportionately more SDS, the charge-to-mass ratios of all treated proteins are approximately the same, and all of the SDS-protein complexes are highly negatively charged.

After the protein samples are loaded onto the gel and an electric field is applied, all SDS-protein complexes move toward the anode, as diagrammed in Figure 4·17a. However, their rate of migration through the gel is inversely proportional to the logarithm of their molecular weight—larger proteins encounter more resistance and therefore migrate more slowly than smaller proteins. This sieving effect differs from gel-filtration chromatography. In gel filtration, larger molecules are excluded from the pores of the gel and thus travel faster. In SDS-PAGE, all molecules must penetrate the pores of the continuous gel; thus, the largest proteins travel most slowly. The protein bands that result from this differential migration, as pictured in Figure 4·17b, can be visualized by staining. Molecular weights of unknown proteins can be estimated by comparing their migrations to the migrations of reference proteins electrophoresed on the same gel.

SDS-PAGE is primarily an analytical tool, although it can be adapted for purifying proteins. Denatured proteins can be recovered from SDS-PAGE by cutting out the bands of a gel. The protein is then electroeluted by applying an electric current to allow the protein to migrate into a buffer solution. After concentration and the removal of salts, such protein preparations can be used for structural analysis or antibody production.

Isoelectric focusing, a modified form of electrophoresis, employs buffers to create a pH gradient within a polyacrylamide gel. When protein samples are electrophoresed in such a gel, each protein migrates to the point in the pH gradient at which it is no longer charged. In other words, each protein bands at its isoelectric point. The separated proteins can then be visualized as in SDS-PAGE.

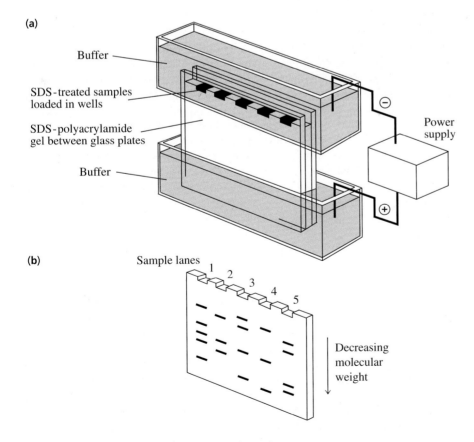

(a)

Buffer

SDS-treated samples loaded in wells

SDS-polyacrylamide gel between glass plates

Buffer

Power supply

(b)

Sample lanes

1 2 3 4 5

Decreasing molecular weight

Figure 4·17
SDS-PAGE. **(a)** An electrophoresis apparatus includes an SDS-polyacrylamide gel between two glass plates and buffer in upper and lower reservoirs, as shown. Samples treated with SDS and 2-mercaptoethanol, heated, and cooled to ambient temperature are loaded into the wells of the gel, and voltage is applied. Since proteins complexed with SDS are negatively charged, they migrate toward the anode. **(b)** The banding pattern of the proteins after electrophoresis can be visualized by staining. Each of the five samples shown contains three or more protein components. Since the smallest proteins migrate fastest, the proteins of lowest molecular weight are at the bottom of the gel.

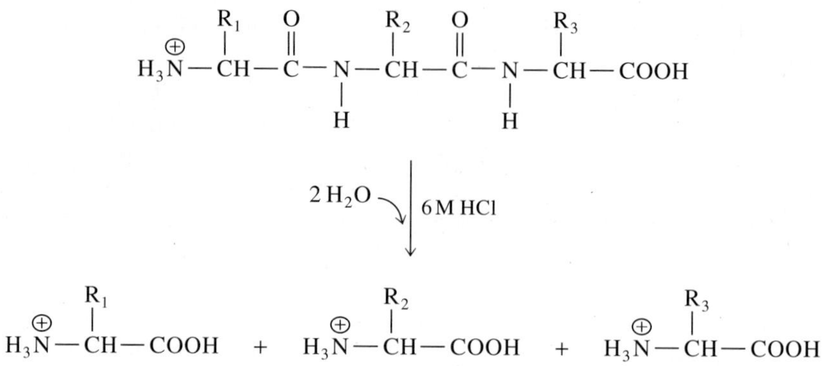

Figure 4·18
Acid-catalyzed hydrolysis of a peptide. Incubation with 6 M HCl at 110°C for 16 to 72 hours releases the constituent amino acids of a peptide.

4·8 The Amino Acid Composition of Proteins Can Be Determined Quantitatively

Once a protein has been isolated in a purified state, its amino acid composition can be determined. First, the peptide bonds of the protein are cleaved by acid hydrolysis, typically using 6 M HCl at 110°C in vacuo for 16 to 72 hours (Figure 4·18). Next, the hydrolyzed mixture, or hydrolysate, is subjected to a chromatographic procedure during which each of the amino acids is separated and quantitated, a process called **amino acid analysis.** One method of amino acid analysis involves treatment of the protein hydrolysate with phenylisothiocyanate (PITC) at pH 9.0 to generate phenylthiocarbamoyl (PTC)–amino acid derivatives (Figure 4·19). The PTC–amino acid mixture is then subjected to HPLC in a column of fine silica beads to which short hydrocarbon chains have been attached. The amino acids are separated according to the hydrophobic properties of their side chains. As each PTC–amino acid derivative is eluted, it is detected, and its concentration is determined by measuring the absorbance of the eluate at 254 nm (the wavelength of peak absorbance of the PTC moiety). A plot showing the absorbance of the eluate from an HPLC column as a function of time is given in Figure 4·20. The peaks, which correspond to amino acids, are identified by standard one-letter abbreviations.

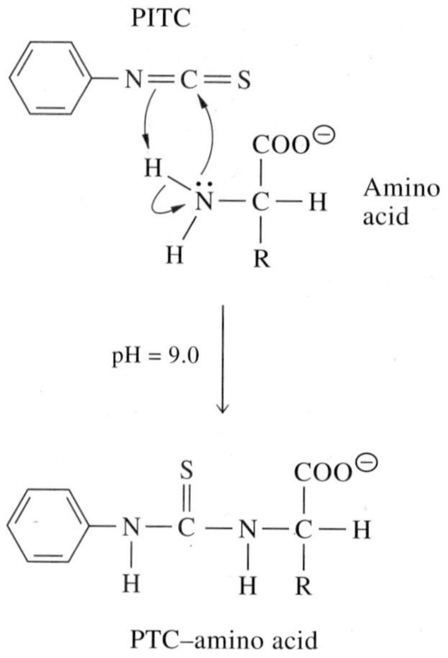

Figure 4·19
Amino acid treated with phenylisothiocyanate (PITC). The α-amino group of an amino acid reacts with phenylisothiocyanate to give a phenylthiocarbamoyl–amino acid (PTC–amino acid).

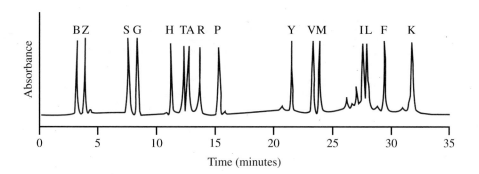

Figure 4·20
Chromatogram obtained from HPLC separation of PTC–amino acids. PTC–amino acids in the column eluate are detected by their absorbance of ultraviolet light at a wavelength of 254 nm. Peaks are labelled with one-letter abbreviations. The letters B and Z indicate totals of asparagine + aspartate and glutamine + glutamate, respectively. [Adapted from Hunkapiller, M. W., Strickler, J. E., and Wilson, K. J. (1984). Contemporary methodology for protein structure determination. *Science* 226:304–311.]

Since different PTC–amino acid derivatives are eluted at different rates, the timing of the peaks identifies the amino acids. The amount of each amino acid present in the aliquot of the hydrolysate subjected to HPLC is proportional to the area under its peak. With this method, amino acid analysis can be performed on samples as small as 1 picomole (10^{-12} mol) of a protein that contains approximately 200 residues.

Despite its usefulness, acid hydrolysis under one set of hydrolytic conditions cannot yield a complete amino acid analysis. Since the side chains of asparagine and glutamine contain amide bonds, the acid used to cleave the peptide bonds of the protein also converts asparagine to aspartic acid plus ammonium ion (with a chloride counterion if HCl is used) and glutamine to glutamic acid plus ammonium ion. When acid hydrolysis is used, the combined totals of glutamate + glutamine are designated with the abbreviations Glx or Z, and the combined totals of aspartate + asparagine are designated Asx or B, as in the chromatogram shown in Figure 4·20.

At elevated temperatures, the indole side chain of tryptophan is especially sensitive to oxidation by air. During acid hydrolysis of proteins, which typically employs temperatures of about 110°C, the side chain of tryptophan is almost totally destroyed, even in evacuated sealed tubes. Thus, the tryptophan content of a protein is often estimated on the basis of its ultraviolet spectrum. Alternatively, it may be analyzed following alkaline hydrolysis or, more often, by including an antioxidant in acid hydrolysis.

Small losses of serine (averaging 10% per 20 hours) and threonine and tyrosine (5% per 20 hours) are also experienced during conventional acid hydrolysis. Conversely, the peptide bonds of valine and isoleucine, which are sterically shielded by hydrocarbon branching at their β-carbons, are slower to hydrolyze than other peptide bonds. For these reasons, several samples of a purified protein are subjected to acid hydrolysis for periods ranging from 16 to 72 hours, and the yields of various amino acid analyses are usually determined by extrapolation of the data obtained.

Cysteine cannot be measured accurately as a component of the acid hydrolysate. Therefore, cysteine residues are oxidized or carboxymethylated (Section 4·10) before hydrolysis of the protein, thereby forming derivatives that can be measured quantitatively after acid hydrolysis.

The amino acid compositions of many proteins have been determined, and dramatic differences in these compositions have been found, illustrating the tremendous potential for diversity based on different combinations of the 20 amino acids. The amino acid compositions of five relatively small proteins are given in Table 4·3 (next page). The data given are based on the complete primary structures that have been obtained for these proteins. Note that proteins need not contain all 20 of the common amino acids.

Table 4·3 Amino acid composition of several proteins

Amino acid	Number of residues per molecule of protein				
	Lysozyme (hen egg white)	Cytochrome c (human)	Ferredoxin (spinach)	Insulin (bovine)	Hemoglobin, α chain (human)
Highly hydrophobic:					
Ile	6	8	4	1	0
Val	6	3	7	5	13
Leu	8	6	8	6	18
Phe	3	3	2	3	7
Met	2	3	0	0	2
Less hydrophobic:					
Ala	12	6	9	3	21
Gly	12	13	6	4	7
Cys	8	2	5	6	1
Trp	6	1	1	0	1
Tyr	3	5	4	4	3
Pro	2	4	4	1	7
Thr	7	7	8	1	9
Ser	10	2	7	3	11
Highly hydrophilic:					
Asn	13	5	2	3	4
Gln	3	2	4	3	1
Acidic					
Asp	8	3	11	0	8
Glu	2	8	9	4	4
Basic					
His	1	3	1	2	10
Lys	6	18	4	1	11
Arg	11	2	1	1	3
Total residues	129	104	97	51	141

A surprising discovery was that a 21st amino acid, selenocysteine (which contains selenium in place of the sulfur of cysteine), is incorporated into a few proteins. In addition, some of the amino acid residues incorporated into certain proteins are biochemically altered after they have been assembled into polypeptide chains. One example, discussed earlier in this chapter, is the formation of cystine residues from two cysteine residues. Another common modification is the attachment of carbohydrate chains to asparagine, serine, or threonine side chains in proteins known as glycoproteins (described in Chapter 9). Reversible phosphorylation of serine, threonine, or tyrosine hydroxyl groups appears to play a role in the regulation of certain intracellular activities, as we will see elsewhere in this book. Thus, although only 20 amino acids are used in the biosynthesis of most proteins, more than 20 species of amino acid residues are found in a number of proteins. Also, many other amino acids occur in cells as free compounds or as components of molecules other than proteins.

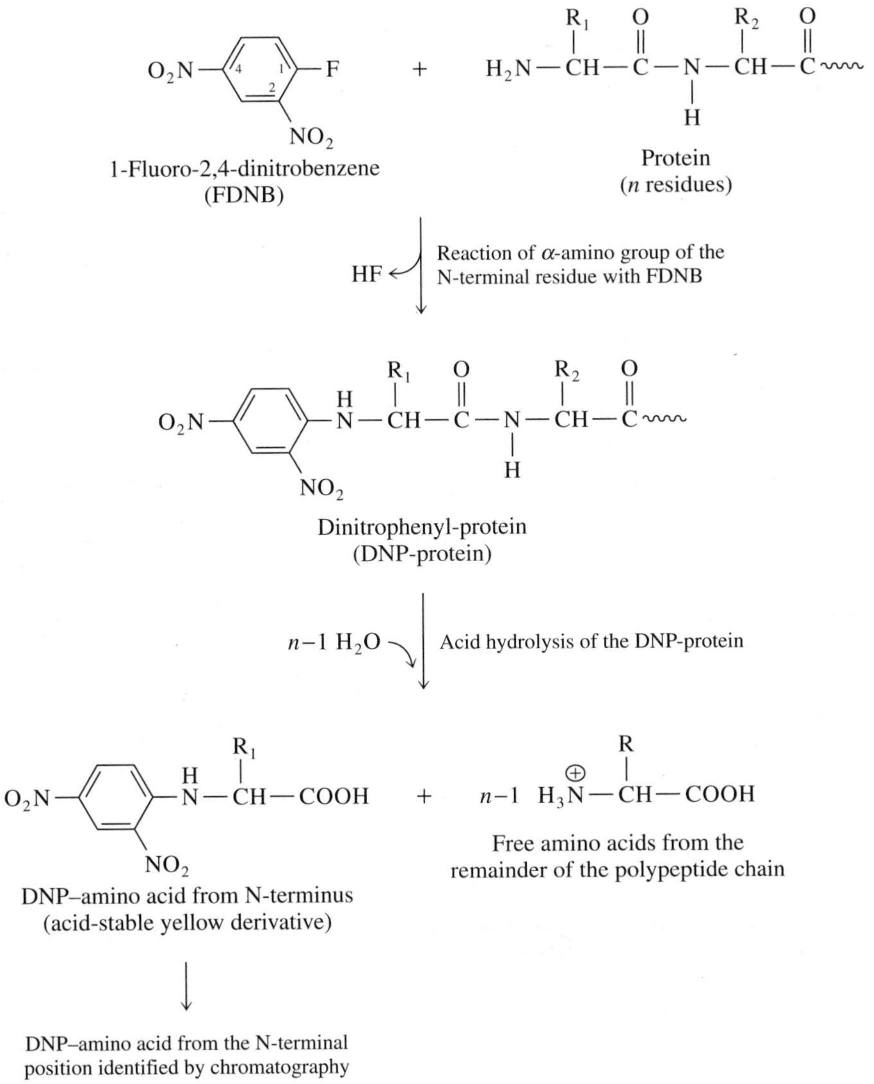

Figure 4·21
Sanger procedure for identifying the N-terminal amino acid residue. The protein is treated with 1-fluoro-2,4-dinitrobenzene (FDNB) under alkaline conditions to produce a dinitrophenyl-protein in which the N-terminal amino acid residue is modified. Hydrolysis in aqueous acid yields free amino acids, including the DNP–amino acid. The labelled amino acid is identified by chromatography.

4·9 Insulin Was the First Protein to Be Sequenced

Amino acid analysis provides information on the composition of a protein but not on the primary structure (sequence of residues). The procedure for sequencing proteins is based on the research of Frederick Sanger and his colleagues, performed between 1943 and 1955. These workers used a technique called end-group analysis to sequence insulin. The Sanger experiments relied on identification of residues with free α-amino groups (i.e., the N-terminal residue(s), along with ε-amino groups of lysine residues). Sanger used 1-fluoro-2,4-dinitrobenzene (FDNB), which reacts with amino groups under mild conditions to form a yellow, acid-stable dinitrophenyl (DNP) derivative of the protein chain (Figure 4·21). During acid hydrolysis, the DNP-protein is cleaved to a mixture of free amino acids and a DNP derivative of the N-terminal amino acid residue. The modified residue is identified by comparison to standard DNP–amino acids by chromatography. Later, other workers made this method more sensitive by replacing FDNB with dansyl chloride, a compound that produces strongly fluorescent derivatives (Figure 4·22).

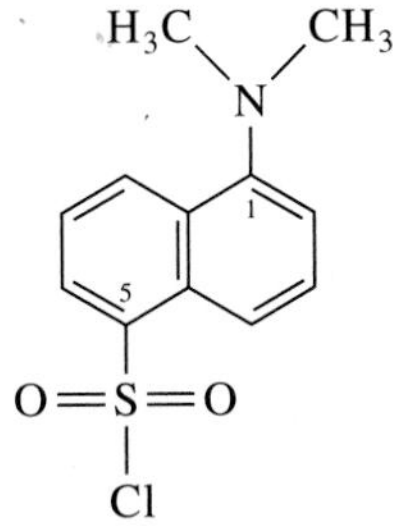

Figure 4·22
Structure of dansyl chloride (1-dimethyl-aminonaphthalene-5-sulfonyl chloride).

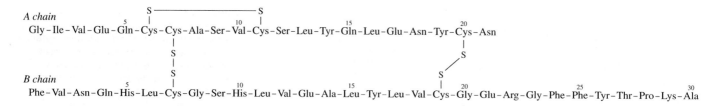

Figure 4·23
Amino acid sequence of bovine insulin.

When the FDNB-based procedure was applied to bovine insulin, one residue each of DNP-glycine and DNP-phenylalanine was formed, suggesting that insulin contains two peptide chains. The amino acid composition indicated the presence of six cysteine residues that presumably form disulfide bridges joining the two chains. By following a procedure for cleaving disulfide bonds, Sanger purified the two chains of insulin: one with glycine and the other with phenylalanine as the N-terminal residue. Sequence work was then performed on each of these chains. Milder conditions for protein hydrolysis were used, so that DNP-peptides (rather than DNP–amino acids) were obtained. For example, partial hydrolysis of the dinitrophenylated phenylalanine chain produced DNP-Phe, DNP-Phe–Val, DNP-Phe–Val–Asp, and DNP-Phe–Val–Asp–Glu. This experiment established the sequence of the first four residues of the phenylalanine chain (residues 3 and 4 were later shown by enzymatic rather than acid cleavage to be asparagine and glutamine, respectively). Data from the analysis of dozens of small overlapping peptides obtained by partial acid hydrolysis or enzymatic cleavage of the chains were required to obtain the complete sequence of insulin (Figure 4·23).

4·10 The Edman Degradation Procedure Is Now Used to Determine the Sequence of Amino Acid Residues

The FDNB method used to sequence insulin, which is a small protein (M_r 5733), is not suitable for larger proteins. In 1950, Pehr Edman described a technique that permits removal and identification of one residue at a time from the N-terminus of a protein. This technique revolutionized protein-sequence analysis. The Edman degradation procedure involves treating a protein or polypeptide with phenylisothiocyanate (PITC) at pH 9.0. PITC reacts under these conditions with the free N-terminus of the chain to form a phenylthiocarbamoyl derivative, or PTC-peptide (Figure 4·24). (Recall that PITC is also used in the measurement of free amino acids, as shown in Figure 4·19.) When the PTC-peptide is treated with an anhydrous acid, such as trifluoroacetic acid, the peptide bond of the N-terminal residue is selectively cleaved, releasing an anilinothiazolinone derivative of the residue. This derivative can be extracted with an organic solvent, such as butyl chloride, leaving the remaining peptide in the aqueous phase. The unstable anilinothiazolinone derivative is treated with aqueous acid, which converts it to a stable phenylthiohydantoin derivative of the amino acid that had been the N-terminal residue (PTH–amino acid). The polypeptide chain in the aqueous phase, now one residue shorter (residue 2 of the original protein is now the N-terminus), can be adjusted back to pH 9.0 and treated again with PITC. The entire procedure (coupling, cleavage, extraction, and conversion) is repeated serially using an automated instrument known as a sequenator. Each cycle yields a PTH–amino acid that can be identified by HPLC. Although it still depends on end-group analysis for sequencing, the Edman degradation procedure has the major advantage of not destroying the polypeptide chain during hydrolytic removal of the labelled N-terminal residue.

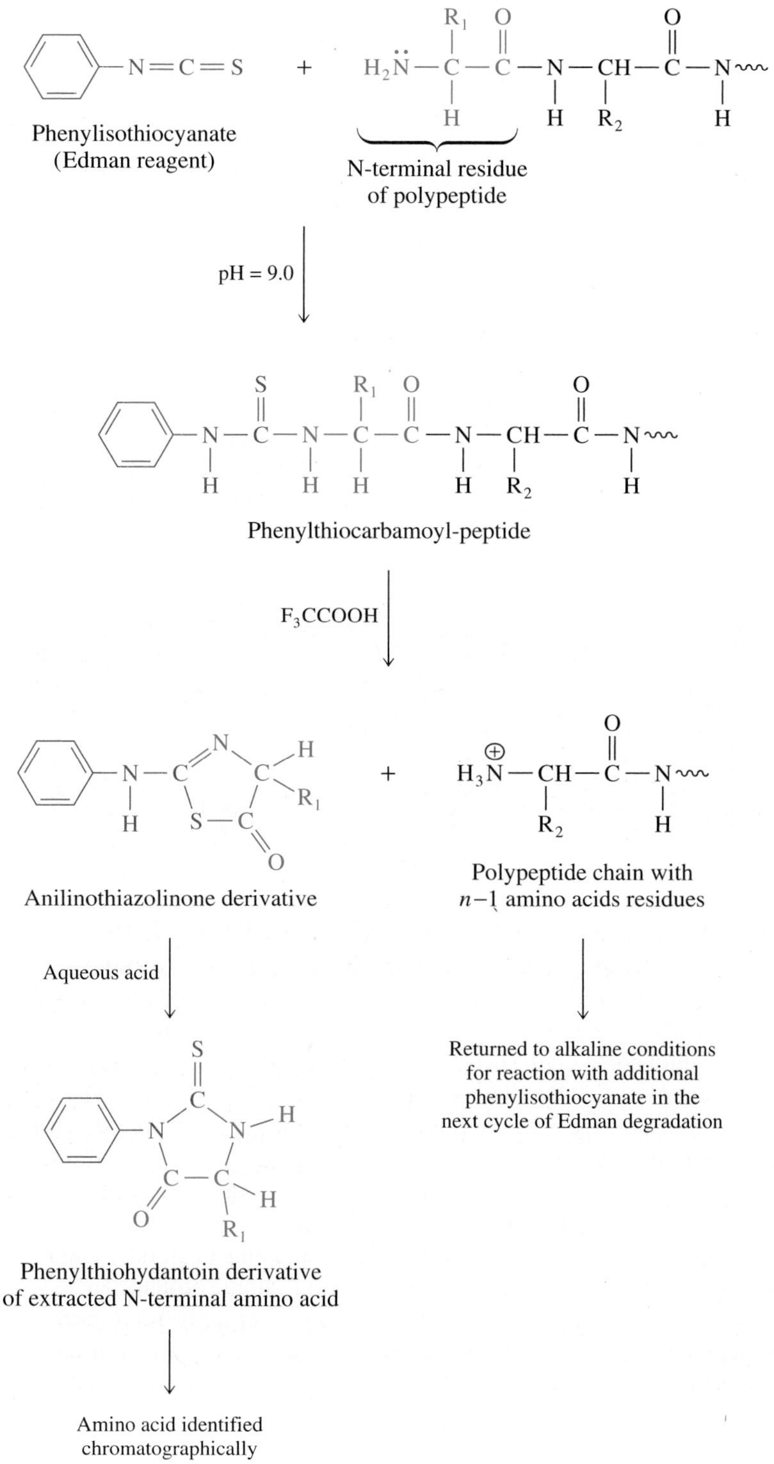

Figure 4·24
Edman degradation procedure for protein sequencing. At pH 9.0, the N-terminal residue of a polypeptide chain reacts with phenylisothiocyanate to give a phenylthiocarbamoyl-peptide. Treating this derivative with trifluoroacetic acid (F_3CCOOH) releases an anilinothiazolinone derivative of the N-terminal amino acid residue without cleaving the other peptide bonds of the polypeptide chain. The anilinothiazolinone is extracted and treated with aqueous acid, which rearranges the derivative so that it forms a stable phenylthiohydantoin derivative. This derivative can then be identified chromatographically. The remainder of the polypeptide chain is returned to alkaline conditions, and the amino acid residue formerly in the second position (now the new N-terminal residue) is subjected to the next cycle of Edman degradation.

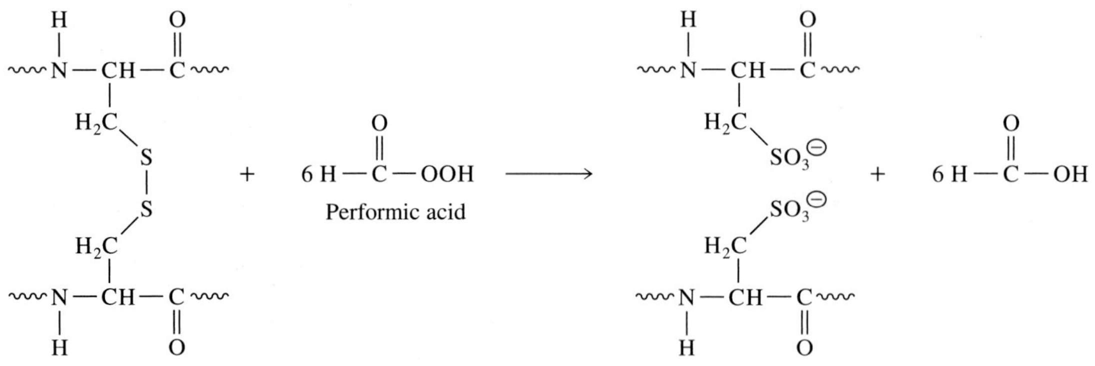

Figure 4·25
Disulfide-bond cleavage by performic acid. Performic acid cleaves a disulfide bond of a protein and prevents re-formation of the bond by oxidizing the cystine residue to two cysteic acid residues.

When a protein contains one or more cystine residues (pairs of cysteine residues cross-linked by disulfide bonds), the disulfide bonds must be cleaved to permit release of the half-cystine residues as PTH–amino acids during the appropriate cycles of Edman degradation. One method of cleavage involves treating the protein with performic acid, which oxidizes cystine to two cysteic acid residues (Figure 4·25). Performic-acid oxidation also oxidizes cysteine residues to cysteic acid and methionine residues to methionine sulfone, each of which is stable and can either be quantitated after acid hydrolysis or converted to a PTH–amino acid by Edman degradation. However, performic acid destroys tryptophan and is therefore not favored for most protein-sequencing tasks. (Of historical interest, tryptophan is not present in either of the first two proteins successfully sequenced, bovine insulin and bovine ribonuclease A. Performic-acid oxidation was thus a suitable choice for cleavage of the disulfide bonds in those proteins.) A preferred method for cleaving disulfide bonds without destroying tryptophan involves treating the protein with an excess of a thiol compound, such as 2-mercaptoethanol, which reduces cystine residues to pairs of cysteine residues (Figure 4·26a). The reactive sulfhydryl groups of the cysteine residues are then blocked by treatment with an alkylating agent, such as iodoacetate, which converts oxidizable cysteine residues to stable *S*-carboxymethylcysteine residues, thereby preventing the re-formation of disulfide bonds in the presence of oxygen (Figure 4·26b).

The yield of the Edman degradation procedure under carefully controlled conditions approaches 100%, and a few picomoles of sample protein can be sequenced for 30 residues or more (higher for larger samples of a given protein) before the yields of additional cycles are obscured by the increasing concentration of unrecovered sample from previous cycles of the procedure. Edman degradation can be performed on proteins electroeluted from SDS-PAGE, provided that the protein sample is extensively desalted. Generally, salts are not soluble in the organic solvents used in Edman degradation and, unless removed, interfere with the chemistry of the degradation. Despite the power of the Edman degradation procedure, it must be supplemented by fragmentation procedures—similar to the partial hydrolyses used by Sanger with insulin—to obtain the primary structure of large proteins.

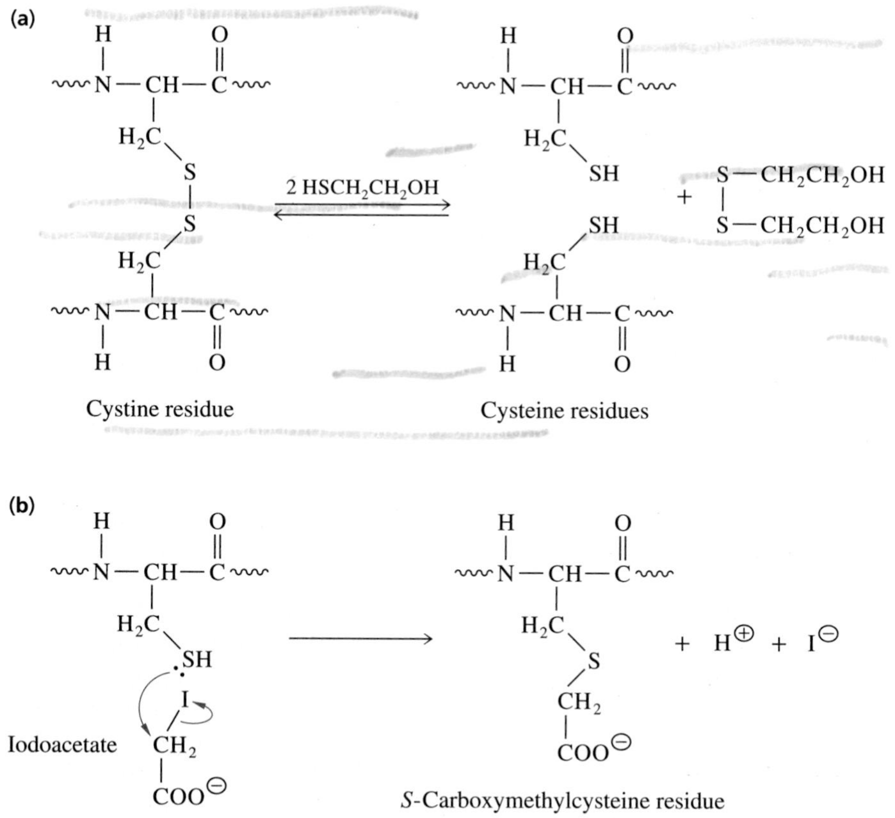

(a)

Cystine residue   Cysteine residues

Figure 4·26
Reduction of cystine and carboxymethylation of cysteine. **(a)** When a protein is treated with excess 2-mercaptoethanol ($HSCH_2CH_2OH$), a disulfide-exchange reaction occurs in which each cystine residue is reduced to two cysteine residues and 2-mercaptoethanol is oxidized to a disulfide. **(b)** Treating the reduced protein with the alkylating agent iodoacetate converts all free cysteine residues to stable S-carboxymethylcysteine residues, thus preventing the re-formation of disulfide bonds in the presence of oxygen.

(b)

S-Carboxymethylcysteine residue

4·11 Large Proteins Are Cleaved to Form Peptides That Can Be Sequenced

Most proteins contain too many residues to be sequenced in their entirety by Edman degradation proceeding from the N-terminus. Therefore, proteases or certain chemical reagents can be used to selectively cleave some of the peptide bonds of a protein. The smaller peptides formed can then be isolated and subjected to sequencing by the Edman degradation procedure.

The chemical reagent cyanogen bromide (BrCN), for example, cleaves proteins by reacting specifically with methionine residues to produce peptides with C-terminal homoserine lactone residues and new N-terminal residues (Figure 4·27). Since most proteins contain relatively few methionine residues, treatment with BrCN usually produces only a few peptide fragments. For example, reaction of

Figure 4·27
Protein cleavage by cyanogen bromide (BrCN). Cyanogen bromide cleaves polypeptide chains on the C-terminal side of methionine residues. The reaction produces a peptidyl homoserine lactone and generates a new amino terminus.

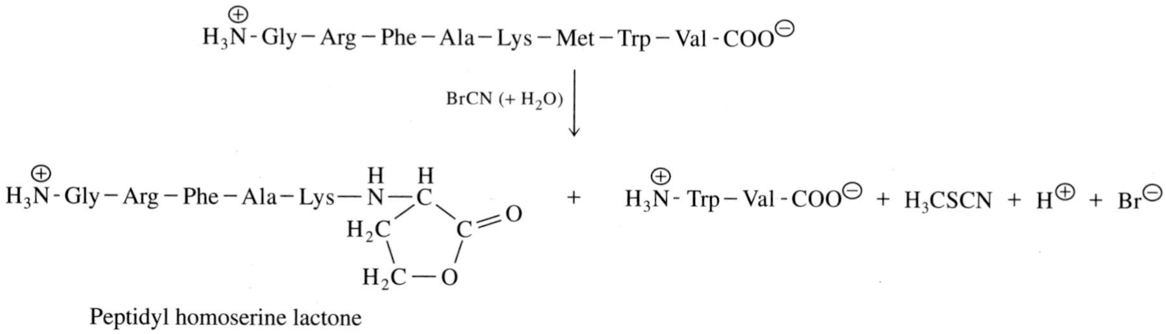

Peptidyl homoserine lactone

BrCN with a polypeptide chain containing three internal methionine residues and a C-terminal alanine residue should generate four peptide fragments, three that possess a C-terminal homoserine lactone residue and one that possesses a C-terminal alanine residue (the original C-terminus of the protein).

Many proteases are available, and trypsin, *Staphylococcus aureus* V8 protease, and chymotrypsin are often used in protein sequencing. Trypsin specifically catalyzes the hydrolysis of peptide bonds on the carbonyl side of lysine residues (lysyl bonds) and arginine residues (arginyl bonds), both of which bear positively charged side chains (Figure 4·28a). *S. aureus* V8 protease catalyzes cleavage of peptide bonds on the carbonyl side of negatively charged residues (glutamyl and aspartyl bonds) and, under appropriate conditions (50 mM ammonium bicarbonate), selectively cleaves only glutamyl bonds. Chymotrypsin, the least specific of the three proteases, preferentially catalyzes the hydrolysis of peptide bonds on the carbonyl side of uncharged residues with aromatic or bulky hydrophobic side chains, for example, phenylalanyl, tyrosyl, and tryptophanyl bonds (Figure 4·28b).

By judicious application of cyanogen bromide, trypsin, *S. aureus* V8 protease, and chymotrypsin to individual samples of a large protein whose disulfide bonds have been reduced and alkylated, one can generate many peptide fragments of various sizes, which can then be isolated and sequenced by Edman degradation. In the final stage of sequence determination, the amino acid sequence of a large polypeptide chain can be deduced by lining up matching sequences of overlapping peptide fragments, as illustrated with a short peptide in Figure 4·28c. When referring to an amino acid residue whose position in the sequence is known, it is customary to follow the residue abbreviation with a number. For example, the third residue of the peptide shown in Figure 4·28 is called Ala-3.

Pure peptides now are often obtained from chemical and enzymatic digests by reverse-phase HPLC, in which elution is effected with an increasing concentration of an organic solvent. The power of this separation method is exemplified by the experiment illustrated in Figure 4·29. Although reverse-phase HPLC is a powerful method for obtaining pure peptides, the successful resolution of a digest is not a routine procedure but a research project with a number of uncertainties.

Figure 4·28
Cleavage of an oligopeptide and sequencing of its amino acids. (a) Trypsin catalyzes cleavage of peptides on the carbonyl side of the basic residues arginine and lysine. (b) Chymotrypsin catalyzes cleavage of peptides on the carbonyl side of aromatic residues, including phenylalanine, tyrosine, and tryptophan, and some other residues with bulky side chains. (c) By using the Edman degradation procedure to determine the sequence of each fragment (highlighted in boxes) and then lining up the matching sequences of overlapping fragments, one can find the order of the fragments and thus deduce the sequence of the entire oligopeptide chain, seen at the top of both (a) and (b).

(a) $H_3\overset{\oplus}{N}$-Gly$-$Arg$-$Ala$-$Ser$-$Phe$-$Gly$-$Asn$-$Lys$-$Trp$-$Glu$-$Val-COO$^{\ominus}$

↓ Trypsin

$H_3\overset{\oplus}{N}$-Gly$-$Arg-COO$^{\ominus}$ + $H_3\overset{\oplus}{N}$-Ala$-$Ser$-$Phe$-$Gly$-$Asn$-$Lys-COO$^{\ominus}$ + $H_3\overset{\oplus}{N}$-Trp$-$Glu$-$Val-COO$^{\ominus}$

(b) $H_3\overset{\oplus}{N}$-Gly$-$Arg$-$Ala$-$Ser$-$Phe$-$Gly$-$Asn$-$Lys$-$Trp$-$Glu$-$Val-COO$^{\ominus}$

↓ Chymotrypsin

$H_3\overset{\oplus}{N}$-Gly$-$Arg$-$Ala$-$Ser$-$Phe-COO$^{\ominus}$ + $H_3\overset{\oplus}{N}$-Gly$-$Asn$-$Lys$-$Trp-COO$^{\ominus}$ + $H_3\overset{\oplus}{N}$-Glu$-$Val-COO$^{\ominus}$

(c)

| Gly – Arg | Ala – Ser – Phe – Gly – Asn – Lys | Trp – Glu – Val |

| Gly – Arg – Ala – Ser – Phe | Gly – Asn – Lys – Trp | Glu – Val |

(a)

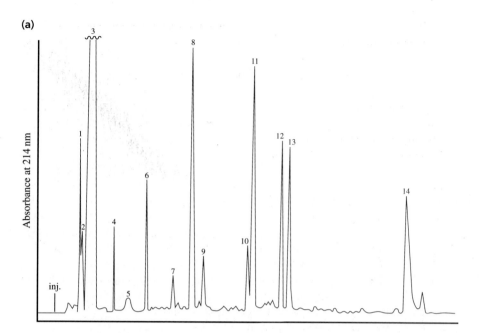

Figure 4·29
Use of reverse-phase HPLC in protein sequencing. **(a)** Separation by reverse-phase HPLC of peptides in a trypsin digest of cytochrome b_5. The eluting solvent was a solution of potassium phosphate (pH 6.0) with a gradually increasing concentration of acetonitrile. The sample was injected into the column at the point marked *inj*. **(b)** The sequence of the peptide in each of the peaks and its location in cytochrome b_5. *Ac* indicates that the first residue is acetylated. The most hydrophobic peptides are eluted last. (Based on data provided by Juris Ozols.)

(b)

Peak	Position	Peptide sequence
1	89−90	Ser-Lys
2	19−23	His-Lys-Asp-Ser-Lys
3	7−9	Asp-Val-Lys
4	1−6	Ac-Ala-Glu-Gln-Ser-Asp-Lys
5	73−76	Glu-Leu-Ser-Lys
6	1−9	Ac-Ala-Glu-Gln-Ser-Asp-Lys-Asp-Val-Lys
7	128−133	Leu-Tyr-Met-SO$_3$-Ala-Glu-Asp
8	52−72	Glu-Gln-Ala-Gly-Gly-Asp-Ala-Thr-Glu-Asn-Phe-Glu-Asp-Val-Gly-His-Ser-Thr-Asp-Ala-Arg
9	128−133	Leu-Tyr-Met-Ala-Glu-Asp
10	33−38	Val-Tyr-Asp-Leu-Thr-Lys
11	39−51	Phe-Leu-Glu-Glu-His-Pro-Gly-Gly-Glu-Glu-Val-Leu-Arg
12	10−18	Tyr-Tyr-Thr-Leu-Glu-Glu-Ile-Gln-Lys
13	77−88	Thr-Tyr-Ile-Ile-Gly-Glu-Leu-His-Pro-Asp-Asp-Arg
14	24−32	Ser-Thr-Trp-Val-Ile-Leu-His-His-Lys

The process of generating and sequencing peptide fragments is especially important for obtaining information about the sequences of proteins whose N-termini are blocked. For example, the N-terminal α-amino groups of many enzymes, especially those of mammalian tissues, are acetylated. These substituted amines do not react at all when subjected to the Edman degradation procedure. Peptide fragments with unblocked N-termini can be produced by selective cleavage, then separated and sequenced, so at least some of the internal sequence of the protein can be obtained. A similar approach, called peptide mapping, can be used to identify an amino acid residue that has been chemically modified. The protein containing the modified residue is chemically or enzymatically fragmented, and the segment containing the modified residue is identified and sequenced.

For proteins that contain disulfide bonds, the complete covalent structure is not fully resolved until the positions of the disulfide bonds have been established. Determining the positions of the disulfide cross-links requires a multistep procedure.

1. Free sulfhydryl groups of the native protein are blocked by treatment with iodoacetate.

2. The modified protein is cleaved using a specific reagent or enzyme, such as cyanogen bromide or trypsin.

3. Each peptide fragment is isolated, and its disulfide bonds are oxidized with performic acid, which splits the cross-linked peptide fragments into two chains containing cysteic acid residues.

4. The sequences of the individual fragments obtained by Edman degradation are compared with the established sequence of the complete polypeptide chain. Positions occupied by cysteic acid residues were originally involved in disulfide bonds, and positions that contain *S*-carboxymethylcysteine originally were free sulfhydryl groups.

In recent years, it has become relatively easy to deduce the amino acid sequence of a protein by determining the sequence of nucleotides in the gene that encodes the protein. (DNA sequencing is described in Chapter 25.) In some cases, especially when a protein is difficult to purify or very scarce, it is more efficient to sequence its gene. For example, the sequence of the insoluble protein elastin (Section 5·8) was determined in this manner. DNA sequences sometimes resolve uncertainties in protein sequences that reflect the lower resolving power of the Edman degradation procedure. However, direct protein sequencing retains its importance because DNA sequences do not reveal the locations of disulfide bonds or the presence of amino acid residues that are modified after synthesis of the protein. Direct knowledge of the amino acid sequences of proteins has also been instrumental in revealing variations in the genetic code and modifications to messenger RNA.

In 1956, Frederick Sanger was awarded a Nobel prize for his work on the sequencing of insulin. Twenty-four years later, Sanger also won a Nobel prize for pioneering the sequencing of nucleic acids. Millions of residues of proteins and nucleic acids have been sequenced to date. These sequences not only reveal details of the structure of individual proteins but allow researchers to identify families of related proteins and to predict the structure and sometimes the function of newly discovered proteins. The rapidly accelerating accumulation of information has also indicated that sequencing every gene in a bacterium or a simple eukaryote is feasible. The task of sequencing the human genome—some three billion base pairs—has likewise begun to bear fruit, with the aid of advanced instruments, computer data bases, and the collaboration of many researchers.

4·12 Comparisons of the Primary Structures of Proteins Can Reveal Evolutionary Relationships

The amino acid sequence of a protein is determined by the gene that encodes it. Therefore, differences among primary structures of proteins reflect evolutionary change.

The sequences of amino acids in proteins from closely related species are likely to be quite similar. The number of differences within the amino acid sequences of these proteins gives an idea of how far various species have diverged in the course of evolution. In general, distantly related species contain proteins that show many differences.

The protein cytochrome *c,* which consists of a single polypeptide chain of 104 to 111 amino acid residues, provides an excellent opportunity for evolutionary comparisons at the molecular level because it is found in all aerobic organisms. Figure 4·30 illustrates the similarities between cytochrome *c* sequences in different species by depicting them as a tree whose branches are proportional in length to the number of differences in the amino acid sequences of the protein. Species that are closely related have relatively few differences in the primary structures of their cytochrome *c* molecules. For example, the penguin and chicken proteins differ at fewer than six positions. At great evolutionary distances, the number of differences

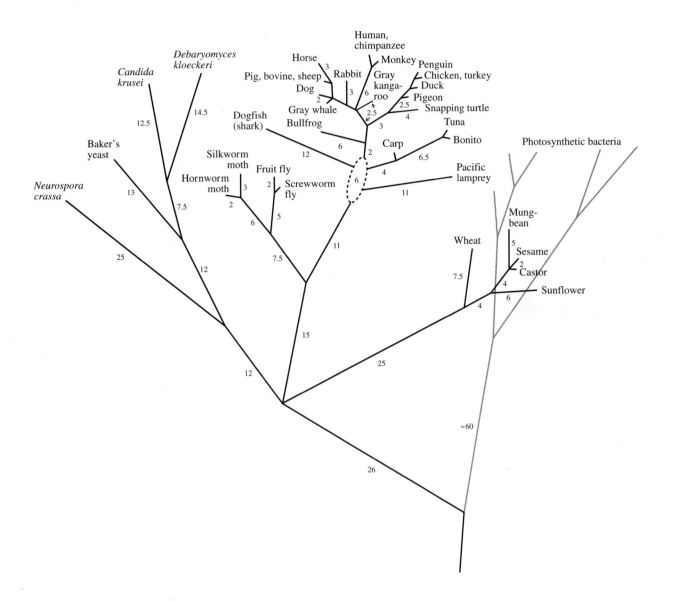

Figure 4·30
Phylogenetic tree for cytochrome c. The numbers next to the branches are changes per 100 amino acid residues. The branching is a close match to classical taxonomic trees, and it shows that eukaryotic cytochromes are related to those of photosynthetic bacteria. In some cases, such as the divergence of fish and lampreys, precise branch points cannot be determined. [Adapted from Dayhoff, M. O., Park, C. M., and McLaughlin, P. J. (1972). Building a phylogenetic tree: cytochrome c. In *Atlas of Protein Sequence and Structure*, Vol. 5, M. O. Dayhoff, ed. (Washington, DC: National Biomedical Research Foundation), pp. 7–16, and Schwartz, R. M., and Dayhoff, M. O. (1978). Origins of prokaryotes, eukaryotes, mitochondria and chloroplasts. *Science* 199:395–403.]

in the same protein may be fairly large; nevertheless, many amino acids in cytochrome *c* are identical in every species. These conserved positions are essential for the proper function of the protein. Proteins that exhibit sequence similarity are said to be homologous if they have evolved from a common ancestor. In general, the more closely related the species, the greater the sequence similarity.

The phylogenetic trees constructed from a comparison of protein sequences closely resemble those constructed by evolutionary biologists using morphological data. Similar trees can also be constructed from nucleotide-sequence information.

Summary

Proteins are made from 20 amino acids, some of which may be modified after synthesis of the protein. Differences in the properties of amino acids reflect differences among their side chains, or R groups. Except for glycine, which has no chiral carbon, all amino acids in proteins are of the L configuration. The side chains of amino acids can be classified according to their chemical structures: aliphatic, aromatic, sulfur-containing, alcohols, bases, acids, and amides. Some amino acids are further classified as highly hydrophobic or highly polar. The properties of the side chains of amino acids are important factors in stabilizing the conformations and determining the functions of proteins.

The ionic state of the acidic and basic groups of amino acids and polypeptides depends on the pH. At pH 7, the α-carboxyl group is anionic ($-COO^{\ominus}$) and the α-amino group is cationic ($-NH_3^{\oplus}$). The charges of ionizable side chains depend on both the pH and their pK_a values. Differences in charges can be used to separate amino acids and proteins.

Amino acid residues in proteins are linked by peptide bonds. The sequence of residues is called the primary structure. Peptides and small proteins can be chemically synthesized. Proteins from biological sources are purified by methods that take advantage of the differences in solubility, net charge, size, and binding properties of individual proteins. The amino acid composition of a protein can be determined quantitatively by hydrolyzing the peptide bonds and analyzing the hydrolysate chromatographically.

The sequence of a polypeptide chain can be determined by the Edman degradation procedure. In each cycle of this procedure, the N-terminal residue of the protein reacts with phenylisothiocyanate to form a phenylthiocarbamoyl-peptide. The modified N-terminal residue is then cleaved and treated with aqueous acid to form a phenylthiohydantoin derivative, which can be identified chromatographically. The polypeptide chain, now one amino acid shorter, is again treated with phenylisothiocyanate. Often, sequences of 30 or more residues can be determined by using the Edman degradation procedure.

Sequences of larger polypeptides can be determined by selective cleavage of peptide bonds using proteases or chemical reagents, followed by Edman degradation of the resulting fragments. The amino acid sequence of the entire polypeptide is then deduced by lining up matching sequences of overlapping peptide fragments.

Comparisons of the primary structures of proteins reveal evolutionary relationships. As species diverge, the primary structures of their common proteins also diverge. By comparing differences in protein structures, we may reach a clearer understanding of evolutionary history.

Selected Readings

General References

Creighton, T. E. (1993). *Proteins: Structures and Molecular Principles,* 2nd ed. (New York: W. H. Freeman and Company), pp. 1–48. This section of Creighton's monograph presents an excellent description of the chemistry of polypeptides.

Dickerson, R. E., and Geis, I. (1969). *The Structure and Action of Proteins* (Menlo Park, California: Benjamin/Cummings Publishing Company). An excellent review on the early characterization of protein molecules.

Haschemeyer, R. H., and Haschemeyer, A. E. V. (1973). *Proteins. A Guide to Study by Physical and Chemical Methods* (New York: John Wiley & Sons). Still an excellent source of general information about proteins.

Meister, A. (1965). *Biochemistry of the Amino Acids,* 2nd ed. (New York: Academic Press). Volume I of this two-volume set is the standard reference for the properties of amino acids.

Synthesis of Polypeptides

Merrifield, B. (1986). Solid phase synthesis. *Science* 232:341–347. Describes the development and use of the solid-phase synthetic method.

Purification of Proteins

Hearn, M. T. W. (1987). General strategies in the separation of proteins by high-performance liquid chromatographic methods. *J. Chromatography* 418:3–26. Discusses general parameters of protein purification as well as the use of HPLC.

Scrimgeour, K. G. (1977). *Chemistry and Control of Enzyme Reactions.* (New York: Academic Press), Chapter 3. Procedures for isolation of enzymes.

Sherman, L. S., and Goodrich, J. A. (1985). The historical development of sodium dodecyl sulphate-polyacrylamide gel electrophoresis. *Chem. Soc. Rev.* 14:225–236.

Stellwagen, E. (1990). Gel filtration. *Methods Enzymol.* 182:317–328.

Quantitation and Amino Acid Composition of Proteins

Hill, R. L. (1965). Hydrolysis of proteins. *Adv. Protein Chem.* 20:37–107.

Ozols, J. (1990). Amino acid analysis. *Methods Enzymol.* 182:587–601.

Stoscheck, C. M. (1990). Quantitation of protein. *Methods Enzymol.* 182:50–68.

Selenocysteine

Leinfelder, W., Zehelein, E., Mandrand-Bethelot, M.-A., and Böck, A. (1988). Gene for a novel tRNA species that accepts L-serine and cotranslationally inserts selenocysteine. *Nature* 331:723–725. Tells how protein synthesis incorporates the 21st amino acid.

Stadtman, T. C. (1991). Biosynthesis and function of selenocysteine-containing enzymes. *J. Biol. Chem.* 266:16 257–16 260. Reviews the role and incorporation of selenium.

Determination of Amino Acid Sequences

Han, K.-K., Belaiche, D., Moreau, O., and Briand, G. (1985). Current developments in stepwise Edman degradation of peptides and proteins. *Int. J. Biochem.* 17:429–445.

Hunkapiller, M. W., Strickler, J. E., and Wilson, K. J. (1984). Contemporary methodology for protein structure determination. *Science* 226:304–311.

Ozols, J. (1990). Covalent structure of liver microsomal flavin-containing monooxygenase form 1. *J. Biol. Chem.* 265:10 289–10 299. Describes current methodology for determination of the amino acid sequence of a large protein.

Sanger, F. (1988). Sequences, sequences, and sequences. *Annu. Rev. Biochem.* 57:1–28. The story of the sequencing of protein, RNA, and DNA.

Walsh, K. A., Ericsson, L. H., Parmelee, D. C., and Titani, K. (1981). Advances in protein sequencing. *Annu. Rev. Biochem.* 50:261–284.

Molecular Evolution

Doolittle, R. F. (1981). Similar amino acid sequences: chance or common ancestry? *Science* 214:149–159. This paper describes the ways in which evolutionary information can be gleaned from comparisons of the primary structures of proteins and the pitfalls that are to be avoided in making such comparisons.

Doolittle, R. F. (1989). Similar amino acid sequences revisited. *Trends Biochem. Sci.* 14:244–245. An update of the 1981 paper, illustrating Darwin's notion of "descent with modification" in terms of similarities of amino acid sequences.

Wilson, A. C. (1985). The molecular basis of evolution. *Sci. Am.* 253(4):164–173. Tells how phylogenetic trees are drawn from comparative protein and DNA structures.

5

Proteins: Three-Dimensional Structure and Function

Proteins are chains of amino acids joined by peptide bonds in a linear sequence specified by DNA. Having examined amino acids and the primary structure of proteins, we move on to the next level of protein organization, three-dimensional structure. A polypeptide chain of a protein is not simply linear but folds into a biologically active shape. The study of proteins has reached such a sophisticated level that three-dimensional structures of hundreds of proteins are now known at atomic resolution. Furthermore, the functions of many of these proteins can be explained on the basis of their conformations.

A **conformation** is a spatial arrangement of atoms that depends on rotation of a bond or bonds. Thus, unlike the configuration of a molecule, the conformation of a protein can change without the breaking of covalent bonds. Considering all possible rotations of the bonds in each amino acid, the number of potential conformations for a protein molecule seems astronomical. However, the actual conformational freedom of any amino acid residue in a protein is surprisingly limited. Furthermore, under physiological conditions, each protein assumes a single stable shape known as its native conformation. It has been proposed that this conformation is the most energetically favored shape possible under normal physiological conditions. We will describe the noncovalent interactions that are responsible for maintaining the properly folded, active states of proteins.

 Related material appears in Exercise 1 of the computer-disk supplement *Exploring Molecular Structure.*

Figure 5·1
Bighorn sheep. The skin, wool, and horns of this animal are composed largely of fibrous proteins that aggregate to form rigid or flexible structures.

On the basis of both their physical characteristics and functions, most proteins can be divided into two major classes, fibrous and globular. These two classes differ markedly in their behavior: one is usually static, and the other dynamic. **Fibrous proteins** are water-insoluble, elongated molecules that are usually physically tough. They provide mechanical support to individual cells and to entire organisms. Typically, fibrous proteins are built upon a single, repetitive structure, assembled into cables or threads. Examples of fibrous proteins are α-keratin, the major component of hair and nails, and collagen, the major protein component of tendons, skin, bones, and teeth (Figure 5·1).

Most **globular proteins** are water-soluble, compact, roughly spherical macromolecules whose polypeptide chains are tightly folded. Globular proteins characteristically have a hydrophobic interior and a hydrophilic surface. They possess indentations or clefts, which specifically recognize and transiently bind other compounds. By selectively binding other molecules, globular proteins serve as dynamic agents of biological action. Globular proteins include enzymes, which are the biochemical catalysts of cells (described in detail in Chapters 6 and 7), and a large number of proteins that serve noncatalytic roles.

Proteins that are part of biological membranes have hydrophobic regions on their surfaces, allowing them to be stable in their nonpolar environment. Some membrane proteins are structural proteins (e.g., part of the cytoskeleton) and many others (such as membrane-bound enzymes and receptors) closely resemble water-soluble globular proteins. Membrane proteins are discussed in Chapter 12. A few proteins have the characteristics of both fibrous and globular proteins. For example, the soluble blood-plasma protein fibrinogen and muscle myosin (Section 2·4E) have fibrous regions and globular ends. However, because globular proteins far outnumber the other types of proteins in variety, most of this chapter is devoted to their description.

In this chapter, we will consider the molecular architecture of proteins, from the simple linkage of amino acids by peptide bonds to the three-dimensional shape of fully formed protein molecules. We will learn that two simple shapes, the α helix and the β sheet—the only conformations of certain fibrous proteins—occur as common elements of structure in the more complex globular proteins. Through several examples, we will examine how the biological functions of proteins relate to and depend on their structures. We will see that the binding sites, which are often clefts, of globular proteins are complementary in shape and chemical properties to the molecules with which they specifically interact. The following two chapters emphasize that this complementarity is a universal property of enzyme proteins. Above all, we will learn that proteins have properties beyond those of free amino acids.

5·1 There Are Four Levels of Protein Structure

Depending upon their complexity, individual protein molecules may be described as having up to four levels of structure (Figure 5·2). As noted in Chapter 4, **primary structure** is the sequence of covalently linked amino acid residues; it describes the linear, or one-dimensional, structure of a protein. The three-dimensional structure of a protein is described by three additional levels: secondary structure, tertiary structure, and quaternary structure. The forces responsible for the formation

(a) Primary structure

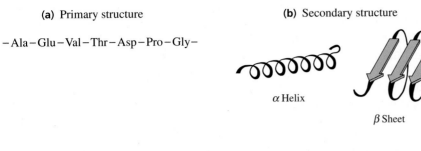

(b) Secondary structure

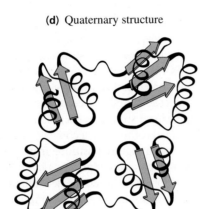

(c) Tertiary structure

(d) Quaternary structure

Figure 5·2
Levels of protein structure. **(a)** The linear sequence of amino acid residues defines the primary structure. **(b)** Secondary structure consists of regions of regularly repeating conformations of the peptide chain, such as α helices and β sheets. **(c)** Tertiary structure refers to the folding of a polypeptide chain into its compact, globular shape. Two domains are shown. **(d)** Quaternary structure refers to the arrangement of two or more separate polypeptide chains into a multi-subunit molecule.

and maintenance, or **stabilization,** of these three levels are primarily noncovalent. **Secondary structure** involves regularities in local conformations, maintained by hydrogen bonds formed between amide nitrogens and carbonyl oxygens of peptide bonds. The common secondary structures are the α helix and the β sheet; we will examine these structures in detail later. **Tertiary structure** is the compacting of a polypeptide into one or more globular units, or domains. Tertiary structures are stabilized by interactions of side chains of non-neighboring amino acid residues. The formation of tertiary structure brings distant portions of the primary structure close together. The active-site cleft of an enzyme, for example, may be composed of residues widely separated in the sequence. Some proteins possess **quaternary structure,** the association of two or more polypeptide chains into a multisubunit protein. The chains may be the same or different. These four structural levels are most clearly evident in globular proteins. In general, fibrous proteins can be characterized almost entirely by their secondary structure. The biological activity of globular proteins is observed only at the tertiary and, in some proteins, the quaternary levels of organization.

The three-dimensional structure of proteins can be studied most easily by examining the general properties of the peptide bond and then studying each level of protein structure in turn—secondary structure, first as seen in several fibrous proteins; tertiary structure and the forces responsible for overall folding; and quaternary structure, which introduces regulatory properties of proteins. In addition to the four fundamental levels of structure, there are recurring patterns of secondary and tertiary structure. Combinations of α helices and β sheets are called supersecondary structure. Some protein chains have several folded units, or domains, with each domain having a specific role. Given supersecondary structures or recognizable domains may appear in a number of proteins with diverse functions, suggesting that these structural units may be evolutionary building blocks for protein molecules.

5·2 X-Ray Crystallography and Nuclear Magnetic Resonance Spectroscopy Are Used to Determine the Three-Dimensional Structures of Proteins

Chemical methods such as Edman degradation are useful for determining the primary structures of proteins. However, X-ray crystallography, a far more powerful tool, is used for determining the secondary, tertiary, and quaternary structures of biological macromolecules. It is possible to construct an electron-density map that shows the arrangement in space of the atoms in a protein using the information obtained from X-ray diffraction patterns of suitable crystals (Figure 5·3). By combining this map with knowledge of the amino acid sequence, it is possible to determine the three-dimensional structure of the protein.

X-ray diffraction studies of fibrous proteins were first attempted in the 1930s; measurements of the simple repeating units of fibrous proteins led to the proposal of the α helix and β sheet as secondary structures. However, determining the three-dimensional structure of a globular protein molecule containing thousands of atoms presented formidable technical difficulties. Chief among these was the difficulty of calculating atomic positions from the positions and intensities of diffracted X-ray beams. Not surprisingly, the development of X-ray crystallography of macromolecules closely followed the development of computers. In 1959, scientists in the laboratory of John C. Kendrew (Figure 5·4a) obtained crystals of sperm-whale myoglobin suitable for X-ray crystallographic analysis. Kendrew ultimately elucidated the structure of this globular protein at a resolution of 0.2 nm. Elucidation of the structure of myoglobin by Kendrew, and of hemoglobin a few years later by Kendrew's colleague Max Perutz (Figure 5·4b), provided the first insights into the nature of the tertiary structure of globular proteins. Their efforts earned these scientists the Nobel prize in 1962. Since then, the structures of many proteins have been revealed by X-ray crystallography. Although analysis of X-ray diffraction patterns has been facilitated by major technical advances, including the use of high-speed computers, determination of protein structures is limited by the difficulty of preparing crystals of a quality suitable for X-ray diffraction.

Crystals of macromolecules have large spaces between neighboring molecules; the spaces are filled with solvent (called mother liquor). The presence of solvent in the crystals limits the resolution of diffraction. However, there are two advantages of the high solvent content. First, the native conformation of the protein is retained in the crystal. Second, the crystal can be suffused with mother liquor containing ligands such as substrate analogs or inhibitors, so that structures of protein complexes can be obtained.

(a)

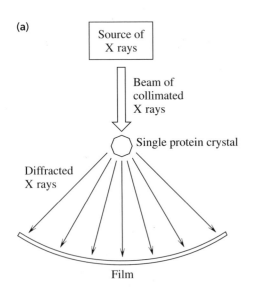

(b)

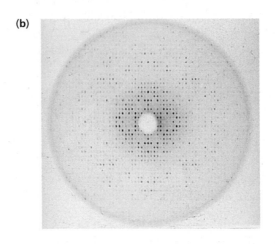

Figure 5·3
X-ray crystallography.
(a) A beam of collimated, or parallel, X rays impinges on a protein crystal, which diffracts the rays onto cylindrical film. The atomic structure is then deduced by mathematical analysis of the diffraction pattern. **(b)** X-ray diffraction pattern of a crystal of adult human deoxyhemoglobin. The location and intensity of these spots are used to determine the protein structure. (Courtesy of Eduardo Padlan.)

Figure 5·4
Pioneers of X-ray crystallography of proteins. **(a)** John C. Kendrew, who determined the structure of myoglobin. **(b)** Max Perutz, who determined the structure of hemoglobin.

(a)

(b)

Unlike X-ray crystallography, nuclear magnetic resonance (NMR) spectroscopy permits the study of proteins in solution and therefore does not require the painstaking preparation of crystals. NMR spectroscopy is a technique that uses the absorption of electromagnetic radiation by molecules in magnetic fields of varying frequencies to determine the spin states of certain atomic nuclei. In the study of protein structure, NMR is usually used to measure the spin states of hydrogen atoms. Specialized NMR techniques record interactions between hydrogen atoms that are close together. Combining these results with a knowledge of the protein sequence allows determination of conformations. The complexity of NMR spectra has thus far precluded determination by NMR spectroscopy of the structures of proteins of molecular weights substantially greater than 15 000.

Both X-ray crystallography and NMR spectroscopy require large amounts of purified protein, and high-resolution structural analysis of a protein may take years. Once the three-dimensional coordinates of the atoms of a macromolecule have been determined, they are customarily deposited in a data bank, where they are available to other researchers. Such data banks have provided assistance in modelling the three-dimensional structures of proteins that are homologous to proteins whose three-dimensional structures are known in detail. Many of the images in this text were created using data files from the Protein Data Bank at Brookhaven National Laboratory.

Protein molecules are both rigid and flexible. This statement may seem contradictory, but it is true. The entire molecule or major parts of it demonstrate relative rigidity. In general, the core of a protein molecule is the most rigid, whereas some regions at the surface, particularly the ends of chains or polypeptide loops, are more mobile. At the atomic level, there is constant motion under physiological conditions. Most of the motion is due to minor bending and stretching fluctuations of no apparent significance; this movement has been called "breathing." Some motions, however—particularly those triggered by ligand binding—cause conformational changes that have functional importance. X-ray and NMR experiments are supplemented by simulations of the atomic motions of macromolecules using calculations made on supercomputers. These calculations, termed molecular dynamics, take into account the known structures and the laws of interaction and motion. They indicate how the relative positions of atoms in a protein can change over even very short time periods. Molecular-dynamics studies have been used to explain features of proteins that cannot be observed when the protein is immobilized in the crystalline state.

Three-dimensional structures of proteins can be depicted in several ways. Each representation has advantages, and the three-dimensional characteristics of each type of representation can be enhanced by the use of stereoscopic views. Ribbon

(a)　　　　　　　　　　　　　(b)　　　　　　　　　　　　　(c)

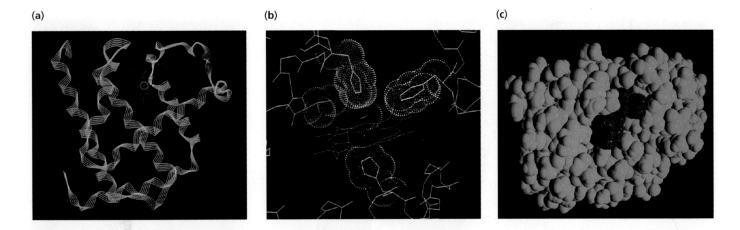

Figure 5·5
Myoglobin structure. **(a)** The polypeptide chain is shown as a ribbon. **(b)** Expanded view of the oxygen-binding site, showing bonds and van der Waals radii of atoms. **(c)** Space-filling model.

models of globular proteins (such as Figure 5·5a) permit tracing of the three-dimensional winding of the polypeptide. Ribbon structures are instructive in explaining the forces stabilizing a protein. An expanded view showing individual atoms depicts active-site regions of proteins (Figure 5·5b). Detailed models of this type can be used to relate a structure to its chemical mechanism. A space-filling model (shown in Figure 5·5c) reflects the dense, closely packed nature of globular proteins. Globular proteins are actually very compact, with side chains nestled together in such a way that the interior of the protein is nearly impenetrable, even by a small molecule such as water. Indeed, we will see in Chapter 7 that the relative rigidity of enzyme proteins is essential for their catalytic action. Large changes in conformation occur only in special circumstances. In these instances, the proteins act as molecules with rigid units joined by flexible hinges. Space-filling models therefore are often suitable representations of enzymes and their specific binding of substrates.

5·3 The Peptide Group Is Polar and Planar

Let us begin our detailed examination of the three-dimensional structure of proteins with the peptide bond. As described in Chapter 4, amino acid residues are linked by peptide bonds to form linear polypeptide chains. The **backbone** of a polypeptide chain consists of repeating $N-C_\alpha-C$ units connected by peptide bonds (Figure 5·6a). Attached to the backbone are the amide hydrogens, carbonyl oxygens, and

Figure 5·6
Generalized structure of a polypeptide chain. Arrows point from the N-terminus to the C-terminus. **(a)** Repeating $N-C_\alpha-C$ units connected by peptide bonds (red) form the backbone of the polypeptide chain. **(b)** The peptide group consists of the N—H and C=O groups involved in formation of the peptide bond, as well as the α-carbons on each side of the peptide bond. Two peptide groups are highlighted in the diagram.

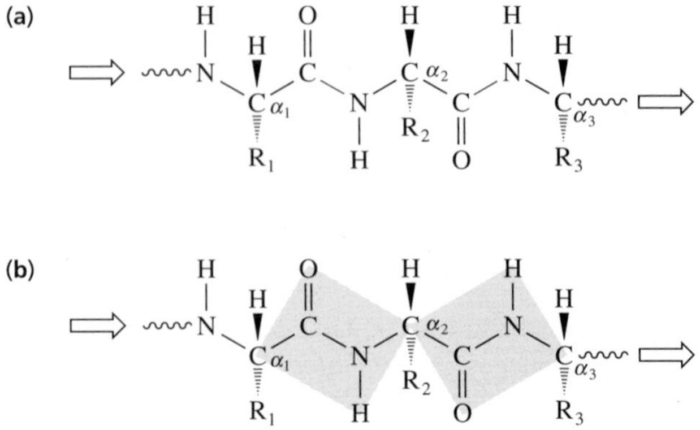

various side chains connected to the α-carbons. It is often helpful to discuss not only the two atoms involved in the peptide bond but also their four substituents: the carbonyl oxygen atom, the amide hydrogen atom, and the two adjacent α-carbon atoms. These six atoms constitute the **peptide group** (Figure 5·6b).

Although it is customary to draw the carbonyl group of a peptide or amide with a double bond between the carbon and oxygen, as shown in Figure 5·7a, the actual nature of the atoms involved in the peptide bond and their substituents lies between this structure and the one shown in Figure 5·7b, in which the electron pair of the carbonyl bond is shifted to the oxygen atom and the unshared electron pair on the amide nitrogen contributes to a double bond. A truer representation is depicted in the resonance hybrid shown in Figure 5·7c. Measurements reveal that the carbonyl carbon–nitrogen bond of a peptide group is shorter than typical carbon-nitrogen single bonds but longer than typical carbon-nitrogen double bonds. The bond appears to have about 40% double-bond character. Since oxygen is more electronegative than nitrogen, the delocalized electrons of the peptide bond are shifted toward oxygen. For this reason, the peptide group is polar. The carbonyl oxygen has a partial negative charge and can serve as a hydrogen acceptor in hydrogen bonds. The nitrogen has a partial positive charge, so that the weakly acidic —NH group can serve as a hydrogen donor in hydrogen bonds.

The partial double-bond character of peptide bonds is sufficient to prevent free rotation around the C—N bond. As a result, the peptide group is planar. However, rotation freely occurs around each N—C_α bond and each C_α—C bond in proteins.

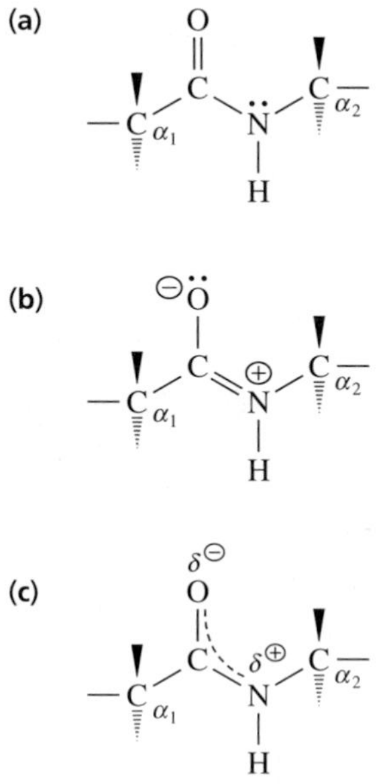

(a)

(b)

(c)

Figure 5·7
Structure of the peptide bond. **(a)** This structure shows the peptide bond as a single C—N bond; the bond between the carbonyl carbon and oxygen is a double bond, and the amide nitrogen has an unshared electron pair. **(b)** In this structure, the peptide bond is a double bond, and the bond between the carbonyl carbon and the oxygen is a single bond, with the carbonyl oxygen having an unshared electron pair. **(c)** This resonance hybrid provides a truer representation of the peptide bond. Electrons are shared by the carbonyl oxygen, the carbonyl carbon, and the amide nitrogen. All six atoms of the peptide group are in a plane because of the partial double-bond character of the peptide bond. The peptide group is polar since the carbonyl oxygen has a partial negative charge ($\delta^\ominus$) and the amide nitrogen has a partial positive charge ($\delta^\oplus$).

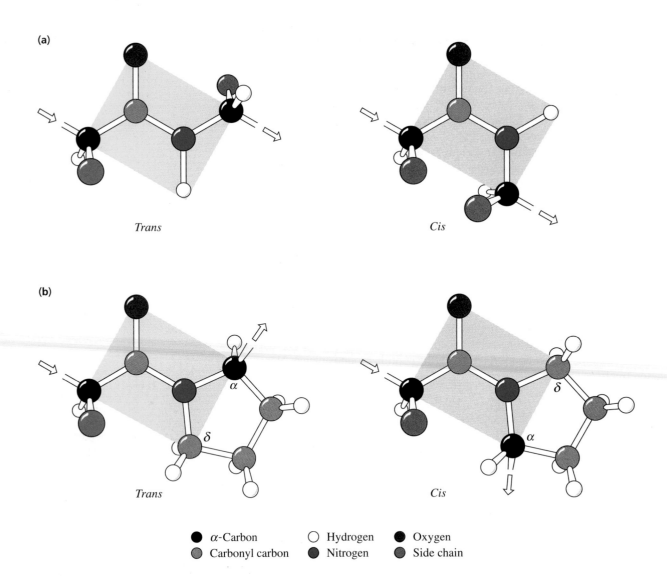

(a)

Trans   Cis

(b)

Trans   Cis

● α-Carbon   ○ Hydrogen   ● Oxygen
● Carbonyl carbon ● Nitrogen   ● Side chain

Figure 5·8
Conformations of a peptide group. (a) *Trans* and *cis* conformations. Nearly all peptide groups in proteins are in the *trans* conformation, which minimizes steric interference between adjacent side chains. (b) *Trans* and *cis* conformations of peptide groups involving proline. In the *trans* conformation, steric interference between the δ position of the pyrrolidine ring and the side chain of the adjacent amino acid residue is only slightly less than the interference, in the *cis* conformation, between the α-carbon of the pyrrolidine ring and the side chain of the adjacent amino acid residue. Most *cis* peptide groups in proteins involve proline residues. The arrows indicate the direction from the N- to the C-terminus.

Despite a large barrier to rotation around the peptide bond, the peptide group can have one of two possible conformations, either the *trans* or *cis* geometric isomer (Figure 5·8a). In the *trans* conformation, the two α-carbons of adjacent amino acid residues are on opposite sides of the peptide bond and at opposite corners of the rectangle formed by the planar peptide group. In the *cis* conformation, the two α-carbons are on the same side of the peptide bond and are closer together. Steric interference between the side chains attached to the two α-carbons makes the *cis* conformation less favorable than the extended *trans* conformation. Consequently, nearly all peptide groups in proteins are *trans*. Rare exceptions occur, however, usually at bonds involving the amide nitrogen of proline, for which the *cis* conformation creates only slightly more steric interference than the *trans* conformation (Figure 5·8b). About 6% of proline residues in proteins analyzed by X-ray crystallography have been found to be in the *cis* conformation.

5·4 Peptide Chains Are Restricted in Conformation

The conformation of a protein depends on rotation around the N—C_α and the C_α—C bonds, which link the rigid peptide groups in a polypeptide chain. This rotation itself is limited by steric interference between main-chain and side-chain atoms of adjacent residues (Figure 5·9). The rotation angle around the N—C_α bond of a peptide group is designated ϕ (phi), and that around the C_α—C bond is designated ψ (psi). In the case of proline, the N—C_α bond is fixed by its inclusion in the pyrrolidine ring of the side chain.

(a)

(b)

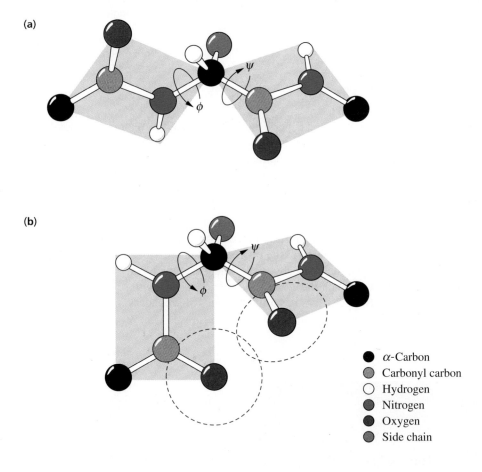

- ● α-Carbon
- ◐ Carbonyl carbon
- ○ Hydrogen
- ● Nitrogen
- ● Oxygen
- ● Side chain

Figure 5·9
Rotation around the N—C_α and the C_α—C bonds, which link peptide groups in a polypeptide chain. **(a)** Peptide groups in an extended conformation. **(b)** Peptide groups in an unstable conformation caused by steric interference between carbonyl oxygens of adjacent residues. The van der Waals radii of the carbonyl oxygen atoms are shown by the dotted lines. The rotation angle around the N—C_α bond is called ϕ and that around the C_α—C bond is called ψ. The planes of the peptide groups are shaded, and the substituents of the outer α-carbons have been omitted for clarity.

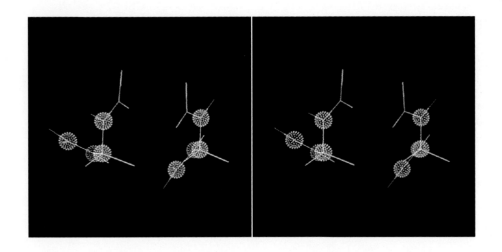

Figure 5·10
Stereo view of measurement of ϕ in peptides. To measure ϕ, the peptide is viewed down the length of the C_α—N bond in the C- to N-terminal direction, and the angle between the C—C_α bond and the next peptide bond toward the N-terminus is measured. The peptide on the left shows ϕ typical of a residue in an α helix; the peptide on the right shows ϕ typical of a residue in a parallel β sheet. Counterclockwise angles, such as these, are negative. Color key: backbone bonds, pink; carbon, green; nitrogen, blue; oxygen, red.

Because rotation around peptide bonds is prevented by their double-bond character, the conformation of each peptide group of a polypeptide can be described by ϕ and ψ. Each of these angles is defined by the relative positions of four atoms of the backbone. Figure 5·10 shows how ϕ is measured, and Figure 5·11 shows how ψ is measured. Clockwise angles are positive, and counterclockwise angles are negative, with each having a 180° sweep.

The biophysicist G. N. Ramachandran and his colleagues constructed space-filling models of peptides and made calculations to determine which values of ϕ and ψ are sterically permitted in the polypeptide chain. Plots of ψ versus ϕ for the residues in a polypeptide chain are called Ramachandran plots. Figure 5·12 is a Ramachandran plot showing the ϕ and ψ values associated with several recognizable conformations. Table 5·1 lists the ϕ and ψ values of secondary structures and other conformations discussed in this chapter.

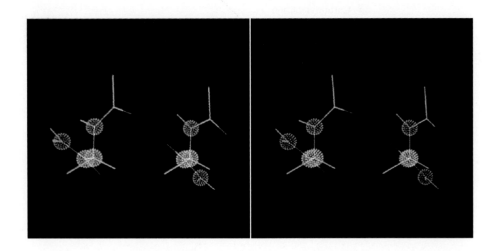

Figure 5·11
Stereo view of measurement of ψ in peptides. To measure ψ, the peptide is viewed down the length of the C_α—C bond in the N- to C-terminal direction, and the angle between the N—C_α bond and the next peptide bond toward the C-terminus is measured. The peptide on the left shows ψ typical of a residue in an α helix; the peptide on the right shows ψ typical of a residue in a parallel β sheet. The counterclockwise angle of the peptide on the left is negative, and the clockwise angle of the peptide on the right is positive.

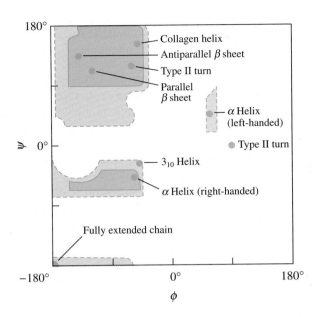

Figure 5·12
Ramachandran plot showing a variety of conformations. Solid lines indicate the range of commonly observed ϕ and ψ values. Dotted lines give the outer limits for an alanine residue. Large blue dots correspond to values of ϕ and ψ that produce recognizable conformations such as the α helix and the β sheet. The positions shown for the type II turn are for the second and third residues. The white portions of the plot correspond to values of ϕ and ψ that rarely or never occur.

Secondary structures result only when a considerable number of consecutive amino acid residues have similar ϕ and ψ values. In native protein conformations, slight distortions in these values often occur. Substantial variation from these values, however, usually disrupts the secondary structure. Vacant areas on the Ramachandran plot represent conformations that are impossible or rare because atoms or groups would be too close together. Most amino acid residues fall within the shaded, or permitted, areas shown on the plot, which are based on alanine as a typical amino acid. Some bulky amino acids have smaller permitted areas. Proline is restricted to a ϕ value of about $-60°$ to $-77°$ because its N—C_α bond is constrained by inclusion in the pyrrolidine ring of the side chain. In contrast, glycine residues are exempt from many steric restrictions because they lack β-carbons; thus, they are very flexible and have ϕ and ψ values that often fall outside the shaded regions of the plot.

Table 5·1 Ideal ϕ and ψ values for some recognizable conformations

Conformation	ϕ	ψ
α Helix (right-handed)	$-57°$	$-47°$
α Helix (left-handed)	$+57°$	$+47°$
3_{10} Helix (right-handed)	$-49°$	$-26°$
Antiparallel β sheet	$-139°$	$+135°$
Parallel β sheet	$-119°$	$+113°$
Collagen helix	$-51°$	$+153°$
Type II turn (second residue)	$-60°$	$+120°$
Type II turn (third residue)	$+90°$	$0°$
Fully extended chain	$-180°$	$-180°$

Figure 5·13
Linus Pauling, who won the Nobel prize in chemistry in 1954.

5·5 The α Helix Is a Common Secondary Structure

In the early 1950s, using data from X-ray crystallographic studies of simple compounds such as amino acids and di- and tripeptides, Linus Pauling (Figure 5·13) and Robert Corey proposed several types of secondary structures. Their proposals took into account possible steric constraints and opportunities for stabilization by formation of hydrogen bonds. It is now known that two of these proposed structures are the major secondary structures of many fibrous and globular proteins. The existence of these two secondary structures explains the earlier classification by William Astbury of fibrous proteins into two categories, the α class and the β class. The elastic proteins in the α class have a structural feature that repeats every 0.5–0.55 nm. The inelastic, or extended, β class proteins have a repeat unit of 0.7 nm. For the α class, Pauling and Corey proposed a structure called the α helix; for the β class, they proposed a structure called the β sheet.

A right-handed **α helix** is depicted in Figures 5·14 and 5·15. (The handedness of a helix is easy to demonstrate. If you wrap your right hand, with your thumb up, around a right-handed helix and move your hand in the direction your thumb is pointing, your fingers will curl the same way the helix is coiled.) Theoretically, an

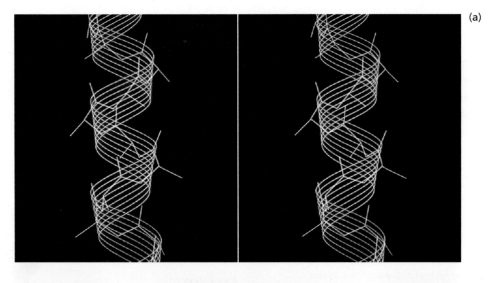

(a)

(b)

Figure 5·14
Stereo views of a right-handed α helix. (a) The ribbon highlights the shape of the helix. All side chains (orange) have been clipped at the β-carbon to make the backbone and carbonyl oxygens (both green) more clearly visible. (b) Space-filling model. Note how tightly packed the atoms are in the α-helical conformation. Color key: carbon, green; β-carbon, orange; nitrogen, blue; oxygen, red. (Based on coordinates provided by B. Shaanan.)

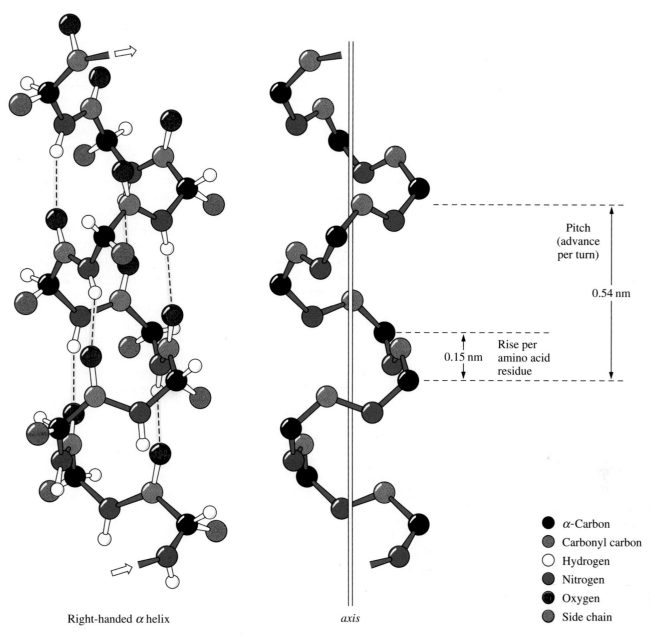

Right-handed α helix   *axis*

- ● α-Carbon
- ● Carbonyl carbon
- ○ Hydrogen
- ● Nitrogen
- ● Oxygen
- ● Side chain

Figure 5·15
Right-handed α helix. Each carbonyl oxygen forms a hydrogen bond with the amide hydrogen of the fourth residue further toward the C-terminus of the polypeptide chain. The N, O, and H atoms of the hydrogen bonds are in line, forming bonds that are approximately parallel to the long axis of the helix. Note that all the —C═O groups point toward the C-terminus. In an ideal α helix, equivalent positions recur every 0.54 nm (this distance is called the pitch of the helix), each amino acid residue advances the helix by 0.15 nm along the long axis of the helix (a dimension called the rise), and there are 3.6 amino acid residues per turn. The arrows at the ends of the helix indicate the direction from the N- to the C-terminus.

α helix could be right-handed or left-handed, but for L amino acid residues, the left-handed conformation is destabilized by steric interference between carbonyl oxygens and side chains. Hence, the α helices found in proteins are nearly always right-handed. Some residues, usually glycines, have been found in left-handed α-helical conformations, but only in stretches not longer than four residues.

In an ideal α helix, the rise, or the distance each residue advances a helix along its axis, is 0.15 nm, and the number of amino acid residues required for one complete turn is 3.6 (i.e., approximately 3⅔; 1 carbonyl group, 3 N—C$_\alpha$—C units, and 1 nitrogen). In proteins, α helices may be slightly distorted, but they still have 3.5 to 3.7 residues per turn. Equivalent positions recur approximately every 0.54 nm, a distance called the pitch of the helix.

Within an α helix, each carbonyl oxygen (residue *n*) of the polypeptide backbone is hydrogen bonded to the α-amino nitrogen of the fourth residue toward the C-terminus (residue *n* + 4). The chain of atoms closed by the hydrogen bond can be regarded as a ring structure of 13 atoms: the carbonyl oxygen, 11 backbone atoms, and the amide hydrogen (an α helix can also be called a 3.6_{13} helix, based on its pitch and hydrogen-bonded loop size). Note that the intrahelical hydrogen bonds are nearly parallel to the long axis of the helix, with the carbonyl groups all pointing toward the C-terminus. Since the orientation of the polar peptide groups is preserved throughout the regularly repeating conformation of the helix, the entire helix is a dipole with a positive N-terminus and a negative C-terminus. Although a single intrahelical hydrogen bond would not provide appreciable structural stability, the cumulative effect of many hydrogen bonds within an α helix stabilizes this conformation, especially in hydrophobic regions within the interior of a protein where water molecules do not compete for hydrogen bonds. In fact, for many amino acid sequences, the α helix is the most stable secondary structure for a polypeptide chain.

As shown in Figures 5·14 and 5·15, all side chains of the amino acids point outward from the cylinder of the helix, thereby minimizing steric interference. Nevertheless, the stability of an α-helical structure is affected by the identity of the side chains, and some amino acid residues are found in α-helical conformations more often than others. Moreover, amino acid residues may have preferred positions within a helix (Table 5·2). For example, alanine, which has a small, uncharged side chain, fits well into the α-helical conformation and is prevalent in the helices of globular proteins. In contrast, glycine, whose side chain consists of a single hydrogen atom, destabilizes α-helical structures by allowing greater freedom of rotation around its α-carbon than is possible for other residues. For this reason, many α helices begin or end with glycine. Proline is the least common residue in an α helix because its rigid, cyclic side chain disrupts or bends the right-handed helical conformation by occupying space that a neighboring residue of the helix would otherwise occupy. In addition, the lack of a hydrogen atom on its amide nitrogen keeps

Table 5·2 Positional preferences of amino acids in helices

Amino acid	N-terminal end	Middle	C-terminal end
Pro	0.8	0.3	0.7
Gly	**1.8**	0.5	**3.9**
Ser	**2.3**	0.6	0.8
Thr	1.6	1.0	0.3
Asn	**3.5**	0.9	1.6
Gln	0.4	1.3	0.9
Asp	**2.1**	1.0	0.7
Glu	0.4	0.8	0.3
Lys	0.7	1.1	1.3
Arg	0.4	1.3	0.9
His	1.1	1.0	1.3
Ala	0.5	**1.8**	0.8
Leu	0.2	1.2	0.7
Val	0.1	1.2	0.2
Ile	0.2	1.2	0.7
Phe	0.2	1.3	0.5
Tyr	0.8	0.8	0.8
Met	0.8	1.5	0.8
Trp	0.3	1.5	0
Cys	0.6	0.7	0.4

Data were obtained from 215 helices in crystal structures of 45 proteins. The values are reported as relative preferences, that is, the ratio of observed occurrences to the number of occurrences expected based upon the percentage composition of the proteins examined. Boldfaced values are statistically higher than expected, and underlined values statistically lower than expected. The middle group tabulates the positions of residues more than five residues from either end of the helix. The end residues are the residues at the N- and C-terminal positions of the helices. [Adapted from Richardson, J. S., and Richardson, D. C. (1989). Principles and patterns of protein conformation. In *Prediction of Protein Structure and the Principles of Protein Conformation*, G. D. Fasman, ed. (New York: Plenum Publishing Corporation), pp. 15–16.]

proline from fully participating in the hydrogen-bond network of an α helix. Thus, proline is found more often at the ends of α helices than in the interior.

Soon after Pauling and Corey proposed the α-helical conformation, Max Perutz verified its existence when he observed in the X-ray diffraction pattern of the fibrous protein α-keratin a minor repeating unit of 0.15 nm, corresponding to the rise of the α helix. (Note that division of the ideal pitch of an α helix, 0.54 nm, by the rise, 0.15 nm, gives the number of residues per turn, 3.6 or approximately $3\frac{2}{3}$. Actually, in α-keratin, the pitch is slightly less than 0.54 nm because the helices are distorted.) α-Keratin, a major component of wool, hair, skin, and fingernails, is a fibrous protein composed almost entirely of α helices. The basic unit of α-keratin is the protofibril, consisting of four right-handed α helices sometimes interrupted by intervening nonhelical sections (Figure 5·16a). The helices are wound around each other to form a left-handed supercoil. Protofibrils are in turn arrayed in a larger structure called a microfibril, which appears to consist of nine protofibrils surrounding a core of two other protofibrils. In hair, microfibrils are packed into macrofibrils, which line up in a parallel fashion (Figure 5·16b).

Protofibrils and microfibrils are cross-linked by disulfide bonds that increase the stability of the overall structure. Keratins with many disulfide bonds, such as those of nails, are hard and rather inflexible. In contrast, keratins with relatively few disulfide bonds, such as those of wool, are flexible and stretch easily. Hair can be

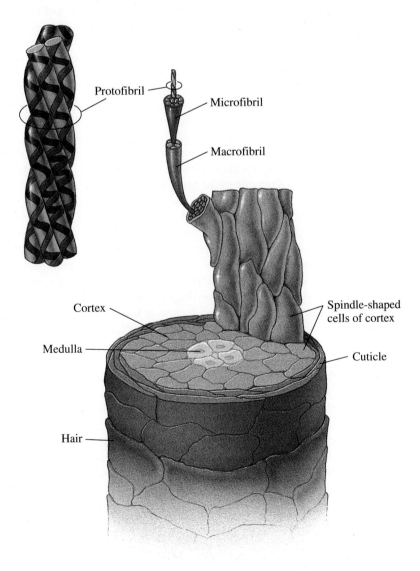

Protofibril

Microfibril

Macrofibril

Cortex

Medulla

Hair

Spindle-shaped cells of cortex

Cuticle

Figure 5·16
Structure of a hair. Four right-handed helices of α-keratin are intertwined to form a protofibril. Keratin protofibrils are assembled into microfibrils, which are in turn assembled into macrofibrils. A single hair contains elongated cells that are packed with macrofibrils.

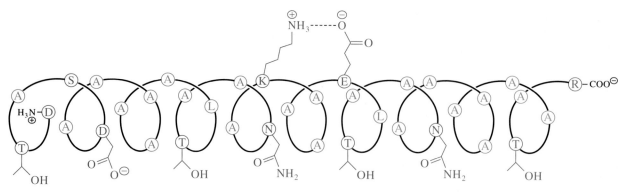

Figure 5·17
Antifreeze protein from the winter flounder. The alanine residues (blue) are on the hydrophobic face of an α helix. The hydrophilic face contains residues (red) that can form hydrogen bonds. A lysine residue and a glutamate residue on the hydrophobic face of the helix form an ion pair. [Adapted from Davies, P. L., and Hew, C. L. (1990). Biochemistry of fish antifreeze proteins. *FASEB J.* 4:2460–2468.]

(a)

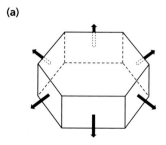

(b)

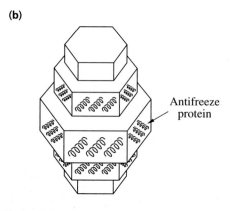

Antifreeze protein

Figure 5·18
Interaction of antifreeze protein with ice. **(a)** In the absence of antifreeze protein, ice crystal growth occurs primarily on the six vertical faces. The direction of crystal growth is indicated by arrows. **(b)** The hydrophilic sides of the antifreeze proteins interact with the vertical faces of the ice crystal, and the hydrophobic sides of the antifreeze proteins inhibit the addition of other water molecules to the crystal. The resulting crystal is smaller and bipyramidal.

curled or uncurled by bending it (e.g., in curlers), treating it with a solution that reduces disulfide bonds, and then treating it with a solution that oxidizes cysteine residues so that they form new disulfide bonds.

Globular proteins vary in their α-helical content. Some, such as the oxygen-binding proteins myoglobin and hemoglobin, contain over 75% of their residues in α helices; others, such as the enzyme chymotrypsin, contain very little α-helical structure. The average content of α helix in the globular proteins that have been examined is 26%.

The regular geometry of the α helix serves an interesting purpose in some proteins in polar fish. These fish live in water whose temperature is low enough to freeze the serum of most fish. The serum of polar fish freezes at a lower temperature than expected, because it contains polypeptides known as antifreeze proteins. Isolated antifreeze proteins have no unusual effect on the melting temperature of ice, but they lower the freezing point of water to a greater extent than predicted by their concentration alone (i.e., their colligative properties). The enhanced depression of the freezing point by antifreeze proteins can be explained by examination of the structure of the antifreeze protein from the winter flounder. This antifreeze protein is a simple α helix. One side of the helix contains many alanine residues and is mostly hydrophobic. The other side of the helix is hydrophilic, with nine side chains that can potentially form hydrogen bonds with the faces of ice crystals (Figure 5·17). In the absence of antifreeze protein, ice forms hexagonal crystals (see Figure 3·5). In the presence of antifreeze protein, the crystals grow only at a lower temperature and are bipyramidal. It has been proposed that the proteins bind to the faces of ice nuclei where crystal growth normally occurs, thereby limiting further crystal growth (Figure 5·18). Thus, any ice crystals that form in the serum of the winter flounder remain small and cannot grow to a size that can damage the fish. The amphipathic helix of the flounder antifreeze protein is not unusual. A number of helix-containing proteins have amphipathic helices whose hydrophobic faces pack together in the interior of the protein.

Some globular proteins contain a few short regions of 3_{10} helix. Like the α helix, the 3_{10} helix is right-handed. Since the carbonyl oxygen forms a hydrogen bond with the amide nitrogen of residue $n + 3$, the 3_{10} helix has a tighter hydrogen-bonded ring structure than the α helix—10 atoms rather than 13—and has fewer residues per turn (3.0) and a longer pitch (0.60 nm). The 3_{10} helix is slightly less stable than the α helix because of steric hindrances and the awkward geometry of its hydrogen bonds. When it occurs, it is usually only a few residues in length and often is the last turn at the C-terminal end of an α helix.

5·6 β Sheets Are Composed of Extended Polypeptide Chains

For the β class of fibrous proteins, Pauling and Corey proposed the **β sheet,** a secondary structure that consists of extended polypeptide chains (called **β strands**). Recall that, in the compact coil of an α helix, each residue corresponds to 0.15 nm of the overall length. In contrast, each residue in a β conformation accounts for about 0.32–0.34 nm. β strands are stabilized by hydrogen bonds between carbonyl oxygens and amide hydrogens. These bonds may link two or more adjacent polypeptide chains or different segments of the same chain. As shown in Figure 5·19, such hydrogen bonds are nearly perpendicular to the extended polypeptide chains, which may be either parallel (running in the same N- to C-terminal direction, as shown in Figure 5·19a) or antiparallel (running in opposite N- to C-terminal directions, as shown in Figure 5·19b). The β sheet is sometimes called a pleated sheet since the side chains point alternately above and below the plane of the sheet due to the bond angles between peptide groups (Figure 5·20).

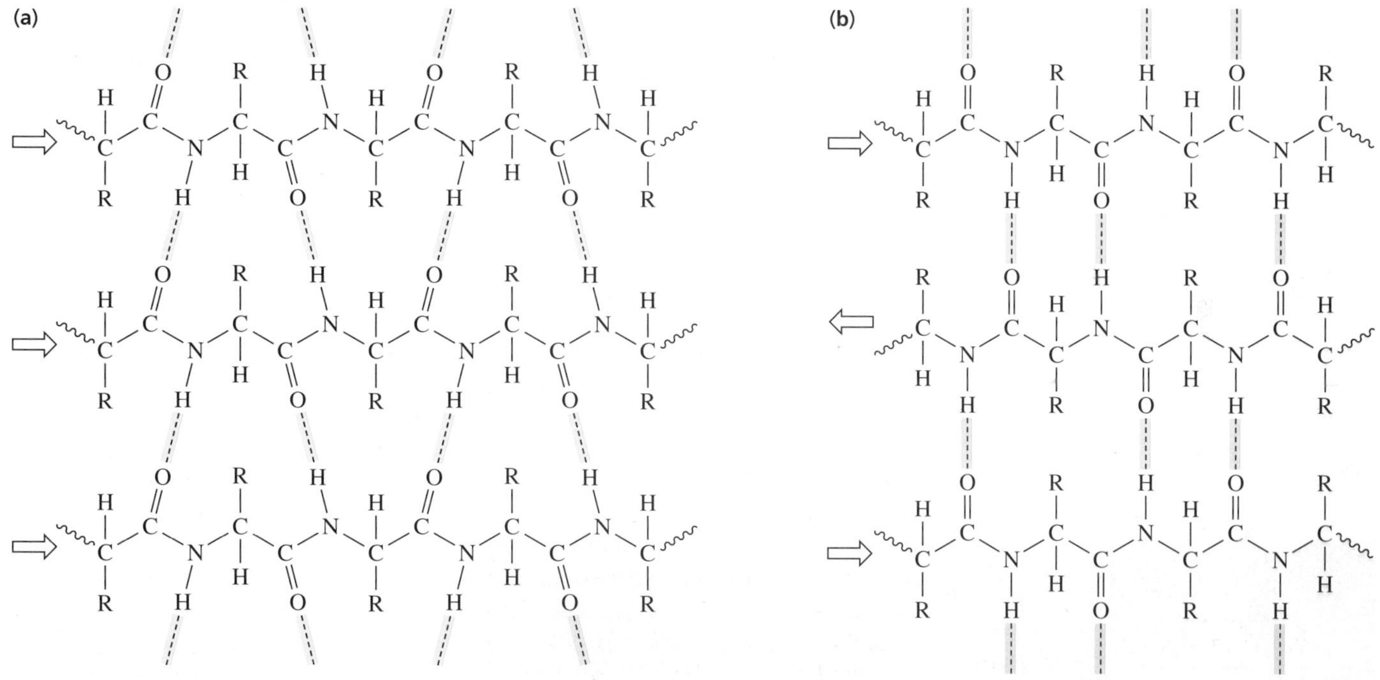

Figure 5·19
β sheets. The arrows point in the N- to C-terminal direction. **(a)** Structure of a parallel β sheet. The hydrogen bonds are evenly spaced but slanted. **(b)** Structure of an antiparallel β sheet. The hydrogen bonds are essentially perpendicular to the strands, and the space between hydrogen-bonded pairs is alternately wide and narrow.

(a) **(b)**

Figure 5·20
Stereo view of the side of a parallel β sheet. In this view, the pleated nature of the parallel β sheet can be seen. Note that the side chains (here clipped at the β-carbon) point alternately above and below the plane of the sheet. Color key: carbon, green; hydrogen, light blue; nitrogen, dark blue; oxygen, red. (Based on coordinates provided by Karl D. Hardman.)

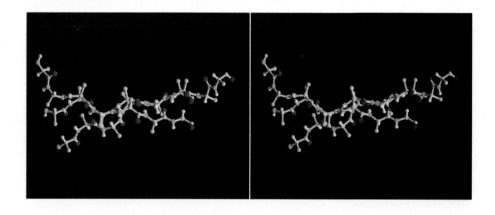

Figure 5·21
Bombyx mori with cocoon.

Measurements from X-ray diffraction patterns of the simple fibrous protein silk fibroin, produced by the silkworm *Bombyx mori* (Figure 5·21), helped Pauling and Corey propose the β sheet. Silk fibroin contains polypeptide chains arrayed in antiparallel β sheets (Figure 5·22). The primary structures of most silk fibroins contain long stretches of the repeating sequence –Gly–Ser–Gly–Ala–Gly–Ala–. Since the side chains of amino acid residues in β sheets extend alternately above and below the plane of the sheet, the side-chain hydrogens of glycine residues lie on one side and the methyl and hydroxymethyl side chains of the alanine and serine residues lie on the other side, allowing a close stacking of sheets. Silk fibroin is flexible because the stacked sheets are held together only by van der Waals forces between the side chains.

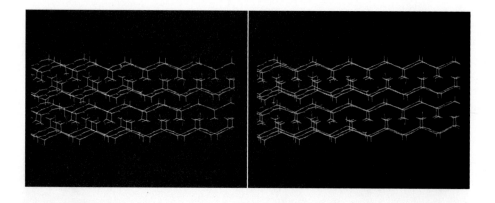

Figure 5·22
Stereo view of silk fibroin, an array of antiparallel β sheets. Side-chain hydrogen atoms of glycine residues lie on one side of each sheet and methyl and hydroxymethyl side chains of the alanine and serine residues, respectively, lie on the other side. Sheets are stacked such that side chains of glycine residues of neighboring sheets face one another, as do the side chains of alanine and serine. Color key: carbon, green; hydrogen, light blue; nitrogen, dark blue; oxygen, red.

Figure 5·23
Stereo view of a domain of Bence-Jones protein. This domain consists largely of two antiparallel β sheets (backbone in blue) linked by intervening nonrepetitive sections (backbone in yellow). (Based on coordinates provided by O. Epp and R. E. Huber.)

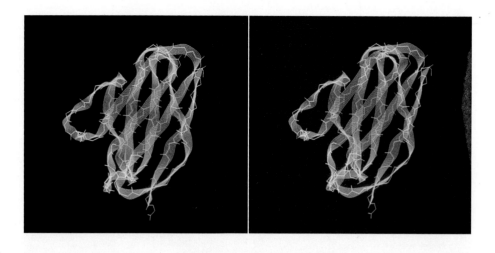

The secondary structures of many globular proteins contain regions of β-sheet conformation. β sheets account for an average of 19% of the residues of the proteins examined. Bence-Jones protein, which is found in high concentrations in patients suffering from multiple myeloma, is a dimer of immunoglobulin light chains (Section 5·20). Each polypeptide consists of two domains that are largely composed of antiparallel β sheets. One domain is shown in Figure 5·23. The right-handed twist that occurs in β-sheet regions of the Bence-Jones protein is typical of the β-sheet regions of globular proteins.

In globular proteins, almost all β strands and β sheets are gently twisted in a right-hand direction. In addition, a disruption called a β bulge occurs fairly often in antiparallel sheets. In a β bulge, one strand contains an extra residue that does not form hydrogen bonds; this causes an abrupt local twist.

5·7 A Different Helical Structure Is Found in Collagen

We now turn to two other fibrous proteins, collagen (described in this section) and elastin (described in the next section). Collagen has a helical structure that differs from the common α helix. The three-dimensional structure of elastin has not been elucidated but it does not appear to have a regularly repeating fibrous structure. The polypeptide chains of both collagen and elastin are held together by covalent bonds. This stabilization makes collagen and elastin suitable for their roles as extracellular components of animal tissues.

Collagen is the major protein component of the connective tissue of most animals and the most abundant vertebrate protein, constituting about 25% to 35% of the total protein in mammals. There are at least 15 different types of collagen proteins with remarkably diverse functions and forms. The hard substance of bone contains collagen and a calcium-phosphate polymer. Collagen in tendons forms stiff, ropelike fibers of tremendous tensile strength. In skin, collagen takes the form of loosely woven fibers, permitting expansion in all directions, and in blood vessels, collagen fibers are arranged in elastic networks. Familiar products derived from collagen include gelatin and glue; collagen is also a major component of leather.

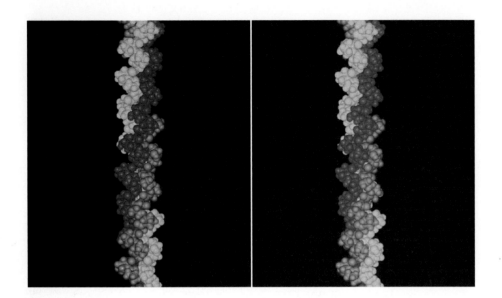

Figure 5·24
Stereo view of the collagen triple helix. Three left-handed helices are coiled around one another in a right-handed supercoil. (Based on coordinates provided by Barbara Brodsky and Cynthia G. Long.)

Native collagen is a molecule consisting of three chains having left-handed helices coiled around each other to form a right-handed supercoil, as shown in Figure 5·24. (Contrast this structure with the structure of keratin, Figure 5·16.) The collagen supercoil is stabilized by interchain hydrogen bonding and by the opposing twist of the helices and the supercoil. A typical collagen molecule is a rod 300 nm long and 1.5 nm in diameter. Within each collagen chain, the left-handed helix has 3.0 amino acid residues per turn and a pitch of 0.94 nm, giving a distance along the axis of 0.31 nm per residue. Thus, the collagen helix is not as highly coiled as an α helix. The left-handedness of the collagen helix and much of the rigidity of collagen arises from steric constraints imposed by many proline residues (Figure 5·25). The ends of collagen molecules contain nonhelical sections, which appear to be necessary for proper alignment and cross-linking of the collagen molecules in the formation of fibrils.

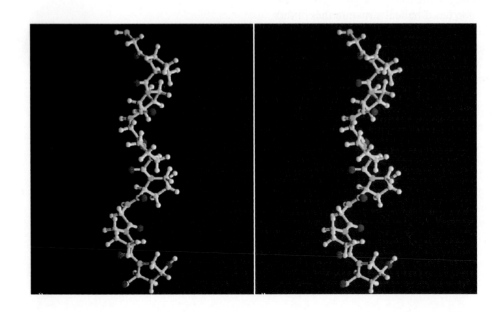

Figure 5·25
Stereo view of one collagen helix. The polypeptide is an extended left-handed helix in which the pyrrolidine rings of proline project away from the long axis of the chain. Color key: carbon, green; hydrogen, light blue; nitrogen, dark blue; oxygen, red.

$\overset{\oplus}{H_3N}$- Glu – Met – Ser – Tyr – Gly – Tyr – Asp – Glu – Lys – Ser – Ala – Gly – Val – Ser – Val –   15
Pro – Gly – Pro – Met – Gly – Pro – Ser – Gly – Pro – Arg – Gly – Leu – Hyp – Gly – Pro –   30
Hyp – Gly – Ala – Hyp – Gly – Pro – Gln – Gly – Phe – Gln – Gly – Pro – Hyp – Gly – Glu –   45
Hyp – Gly – Glu – Hyp – Gly – Ala – Ser – Gly – Pro – Met – Gly – Pro – Arg – Gly – Pro –   60
Hyp – Gly – Pro – Hyp – Gly – Lys – Asn – Gly – Asp – Asp – Gly – Glu – Ala – Gly – Lys –   75
Pro – Gly – Arg – Hyp – Gly – Gln – Arg – Gly – Pro – Hyp – Gly – Pro – Gln – Gly – Ala –   90
Arg – Gly – Leu – Hyp – Gly – Thr – Ala – Gly – Leu – Hyp – Gly – Met – Hyl – Gly – His –   105
Arg – Gly – Phe – Ser – Gly – Leu – Asp – Gly – Ala – Lys – Gly – Asn – Thr – Gly – Pro –   120
Ala – Gly – Pro – Lys – Gly – Glu – Hyp – Gly – Ser – Hyp – Gly – Glx – Asx – Gly – Ala –   135
Hyp – Gly – Gln – Met –

Figure 5·26
Sequence of the first 139 residues of a chain of collagen from rat skin. Beginning with residue 17 from the N-terminus, glycine residues are found at every third position. Proline or hydroxyproline (Hyp) often precedes or follows glycine. *Hyl* indicates a hydroxylysine residue.

Since the collagen helix has 3.0 residues per turn, every third residue of a given strand in a collagen triple helix makes close contact with the other two strands along the central axis of the triple helix. Since only the small side chains of glycine residues can fit at these positions, collagen helices typically contain glycine residues in every third position. There are no intrachain hydrogen bonds in the collagen helix. Characteristic of collagen is the repeating sequence –Gly–X–Y–, where X is often proline and Y is often 4-hydroxyproline (Hyp), a covalently modified derivative of proline (Figure 5·26). Together, proline and hydroxyproline account for about one-fourth of the residues in collagen molecules. For each –Gly–X–Y– triplet, one intermolecular hydrogen bond forms between the amide hydrogen atom of a glycine in one chain and the carbonyl oxygen atom of residue X in an adjacent chain (Figure 5·27).

In addition to hydroxyproline, 5-hydroxylysine (Hyl) residues are found in collagen molecules. Some hydroxylysine residues are covalently bonded to carbohydrate residues, making collagen a glycoprotein. The role of this glycosylation is not known.

Hydroxyproline and hydroxylysine residues are formed in the presence of α-ketoglutarate by reactions in which specific proline and lysine residues are hydroxylated after incorporation into the collagen helix (Figure 5·28, next page). The enzyme-catalyzed hydroxylation of proline and lysine residues requires ascorbic acid (vitamin C). Proline hydroxylation increases the number of groups available for interchain hydrogen bonding, which helps stabilize the triple helix of collagen. In vitamin C deficiency, hydroxylation is impaired, and the triple helix of collagen is not assembled properly, causing blood vessels and skin to become weak. Persons deprived of vitamin C develop scurvy, a disease whose symptoms include skin lesions, fragile blood vessels, loose teeth, and bleeding gums.

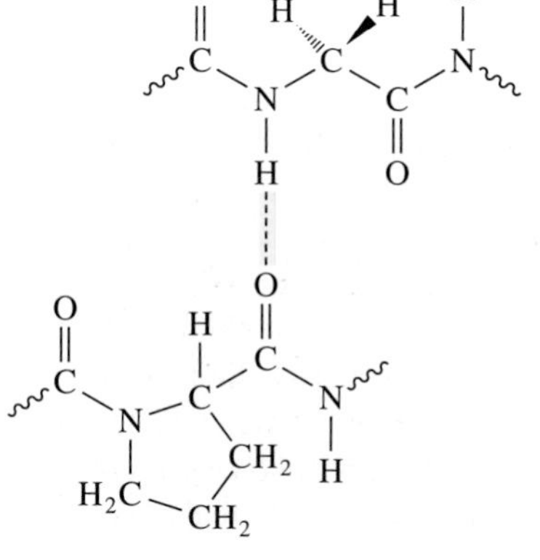

Figure 5·27
An interchain hydrogen bond in collagen. The amide nitrogen of glycine residues lying along the central axis of the triple helix is hydrogen bonded to the carbonyl oxygen of residues, often proline, in an adjacent chain.

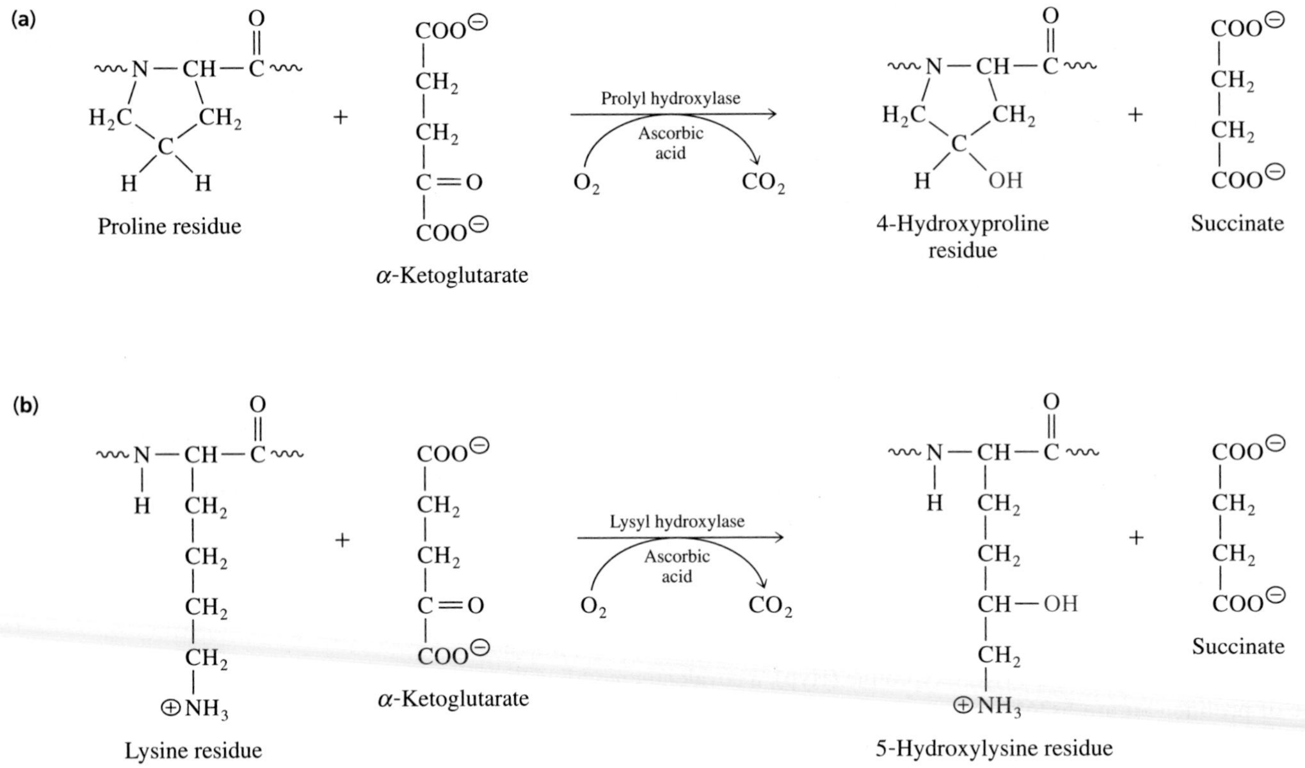

(a)

Proline residue + α-Ketoglutarate $\xrightarrow[\text{Ascorbic acid}]{\text{Prolyl hydroxylase}}$ ($O_2 \longrightarrow CO_2$) 4-Hydroxyproline residue + Succinate

(b)

Lysine residue + α-Ketoglutarate $\xrightarrow[\text{Ascorbic acid}]{\text{Lysyl hydroxylase}}$ ($O_2 \longrightarrow CO_2$) 5-Hydroxylysine residue + Succinate

Figure 5·28
Hydroxylation of proline and lysine. **(a)** Conversion of proline to hydroxyproline. The hydroxylation of specific proline residues, catalyzed by prolyl hydroxylase, requires the presence of α-ketoglutarate. In the reaction, molecular oxygen (O_2) is activated. One oxygen atom from molecular oxygen attaches to proline; the other is incorporated into succinate during the decarboxylation of α-ketoglutarate. Ascorbic acid (vitamin C) acts as a reducing agent in the reaction. **(b)** Conversion of lysine to hydroxylysine. The hydroxylation of specific lysine residues, catalyzed by lysyl hydroxylase, is similar to the reaction described in (a).

Both proline and hydroxyproline residues play a vital role in the structure of collagen. The limited conformational flexibility of these residues not only prevents the formation of α helices but also makes collagen helices and the collagen fiber itself somewhat rigid. Further stabilization of the collagen triple helix seems to be provided by hydrogen bonds involving the hydroxyl group of hydroxyproline, as noted earlier.

Covalent cross-links also contribute to the strength and rigidity of collagen fibers. The ε-$CH_2NH_3^{\oplus}$ groups of the side chains of some lysine residues are converted enzymatically to aldehyde groups (—CHO), producing allysine. Allysine residues and their hydroxy derivatives react with the side chains of lysine and hydroxylysine residues to form **Schiff bases,** complexes formed between carbonyl groups and amines (Figure 5·29). These Schiff bases usually form *between* collagen molecules. Allysine residues also react with other allysine residues by aldol condensation to form cross-links, usually *within* collagen molecules (Figure 5·30). Both types of cross-links are converted to more stable bonds during the maturation of tissues, but the chemistry of these conversions is unknown. Reduction of the double bonds, which would make the cross-links more stable, does not occur.

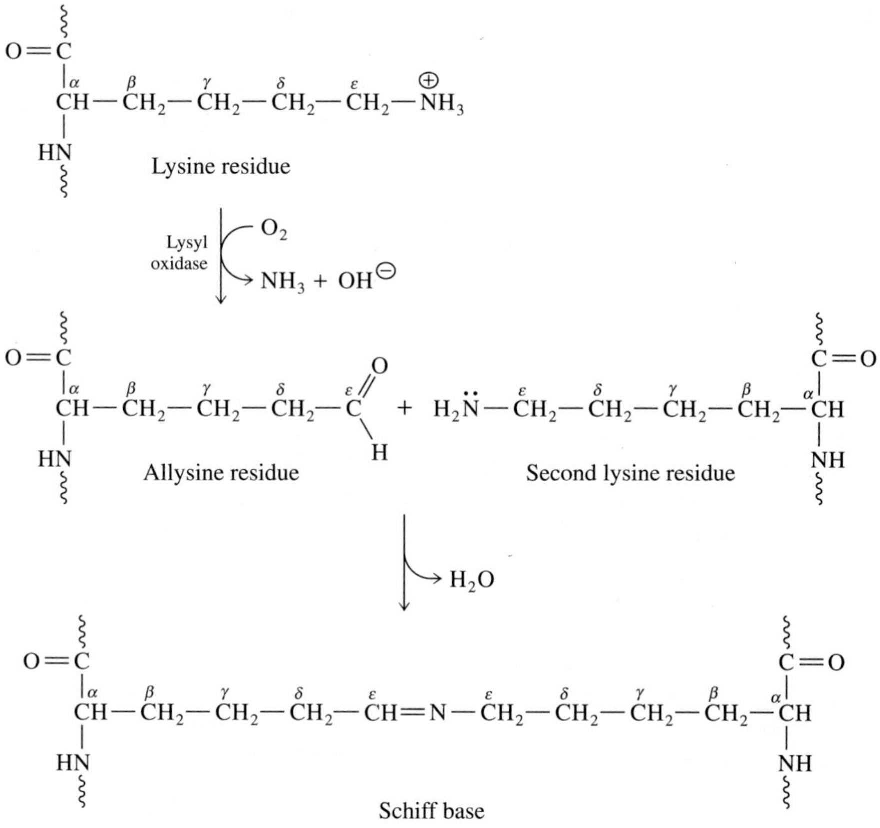

O=C
|α   β   γ   δ   ε
CH—CH₂—CH₂—CH₂—CH₂—NH₃⁺
|
HN   Lysine residue

Lysyl oxidase ⎰ O₂
            ⎱ NH₃ + OH⁻

O=C
|α   β   γ   δ   ε O
CH—CH₂—CH₂—CH₂—C   +   H₂N̈—CH₂—CH₂—CH₂—CH₂—CH
| H |
HN   Allysine residue   Second lysine residue   NH

C=O

→ H₂O

O=C   C=O
|α   β   γ   δ   ε   ε   δ   γ   β   α
CH—CH₂—CH₂—CH₂—CH=N—CH₂—CH₂—CH₂—CH₂—CH
| |
HN   Schiff base   NH

Figure 5·29
Oxidation of a lysine residue and condensation with another lysine residue. Lysyl oxidase catalyzes the oxidation of lysine to produce allysine. Allysine then condenses with a lysine residue to form a Schiff-base cross-link. (Lysyl oxidase also catalyzes the oxidation of hydroxylysine residues to produce hydroxyallysine.)

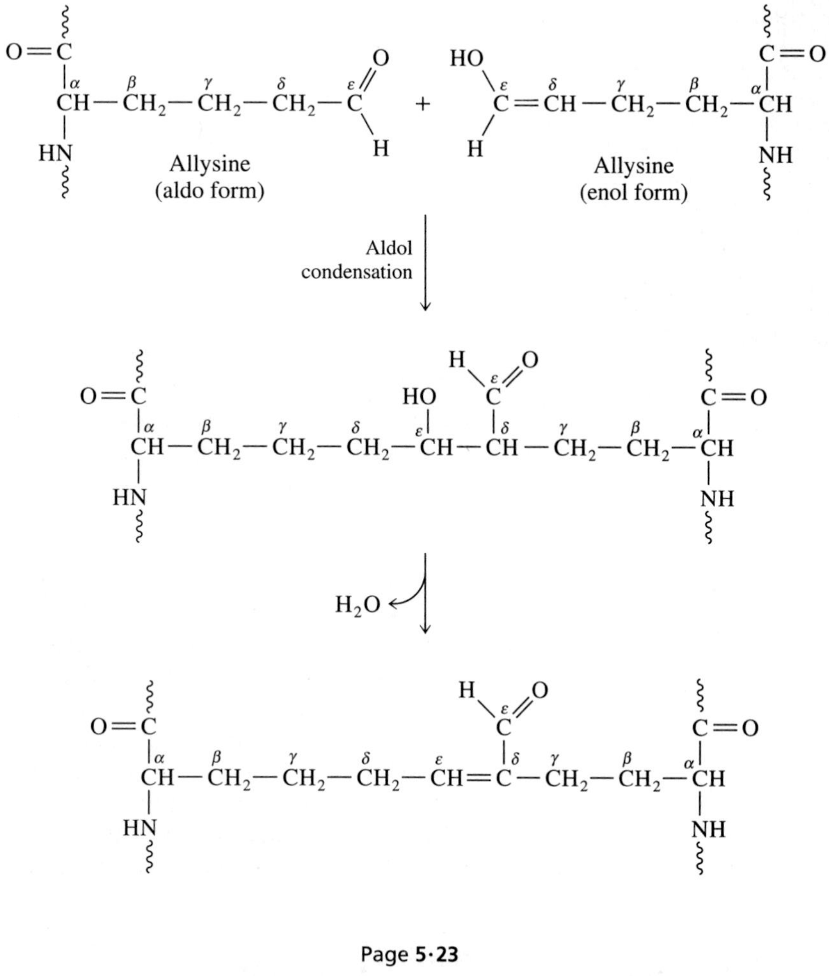

Figure 5·30
Cross-linking in collagen via aldol condensation between two allysine residues. Both the hydrated and dehydrated forms shown probably occur in vivo.

O=C   O   HO   C=O
|α   β   γ   δ   ε   ε   δ   γ   β   α
CH—CH₂—CH₂—CH₂—C   +   C=CH—CH₂—CH₂—CH
| H   H |
HN   Allysine   Allysine   NH
         (aldo form)   (enol form)

Aldol condensation ↓

O=C   H   O   C=O
|α   β   γ   δ   ε HO   C   δ   γ   β   α
CH—CH₂—CH₂—CH₂—CH—CH—CH₂—CH₂—CH
| |
HN   NH

H₂O ↙

O=C   H   O   C=O
|α   β   γ   δ   ε   C   δ   γ   β   α
CH—CH₂—CH₂—CH₂—CH=C—CH₂—CH₂—CH
| |
HN   NH

5·8 Networks of Elastin Can Extend and Relax

Elastin is the major structural protein found in tissues whose functions require rapid extension and complete recovery upon relaxation. Whereas collagens and α-keratins impart structural integrity and rigidity, elastin provides elasticity and flexibility to tissues such as those found in the lungs, skin, and arteries. Elastin is a three-dimensional network of cross-linked molecules that lack a well-defined secondary structure. Elastin is extremely durable, and single fibers can last the lifetime of an individual. The aorta (the large artery leading from the heart) contains roughly twice as much elastin as collagen, giving this vessel the elasticity required to undergo a billion or so stretch-relaxation cycles during a lifetime of heartbeats.

Elastin is a highly water-insoluble protein with a largely hydrophobic amino acid composition, containing slightly less than 50% valine (V) + proline (P) + alanine (A), approximately 33% glycine (G), and smaller amounts of isoleucine (I), leucine (L), and phenylalanine (F). Elastin is usually purified using a variety of solvents that do not hydrolyze its peptide bonds but do dissolve and remove all other proteins and biomolecules. Elastin contains a number of cross-linked residues, including modified lysine residues like those found in collagen, and related cross-linked structures called desmosine and isodesmosine, which are derived from three allysine residues and one lysine residue, as shown in Figure 5·31. Normally, 15 to 17 such cross-links occur within the long (~800-residue) elastin chains. The cross-links of elastin are spaced widely enough to allow extension but closely enough to give the fibers strength. Lysyl oxidase is a copper-containing enzyme that catalyzes the oxidative deamination of lysine residues to allysine residues involved in covalent cross-linking (Figure 5·29). (The suffix -*ase* signifies an enzyme activity.)

Figure 5·31
Desmosine and isodesmosine. These cross-links result from the condensation of three allysine side chains and one lysine side chain.

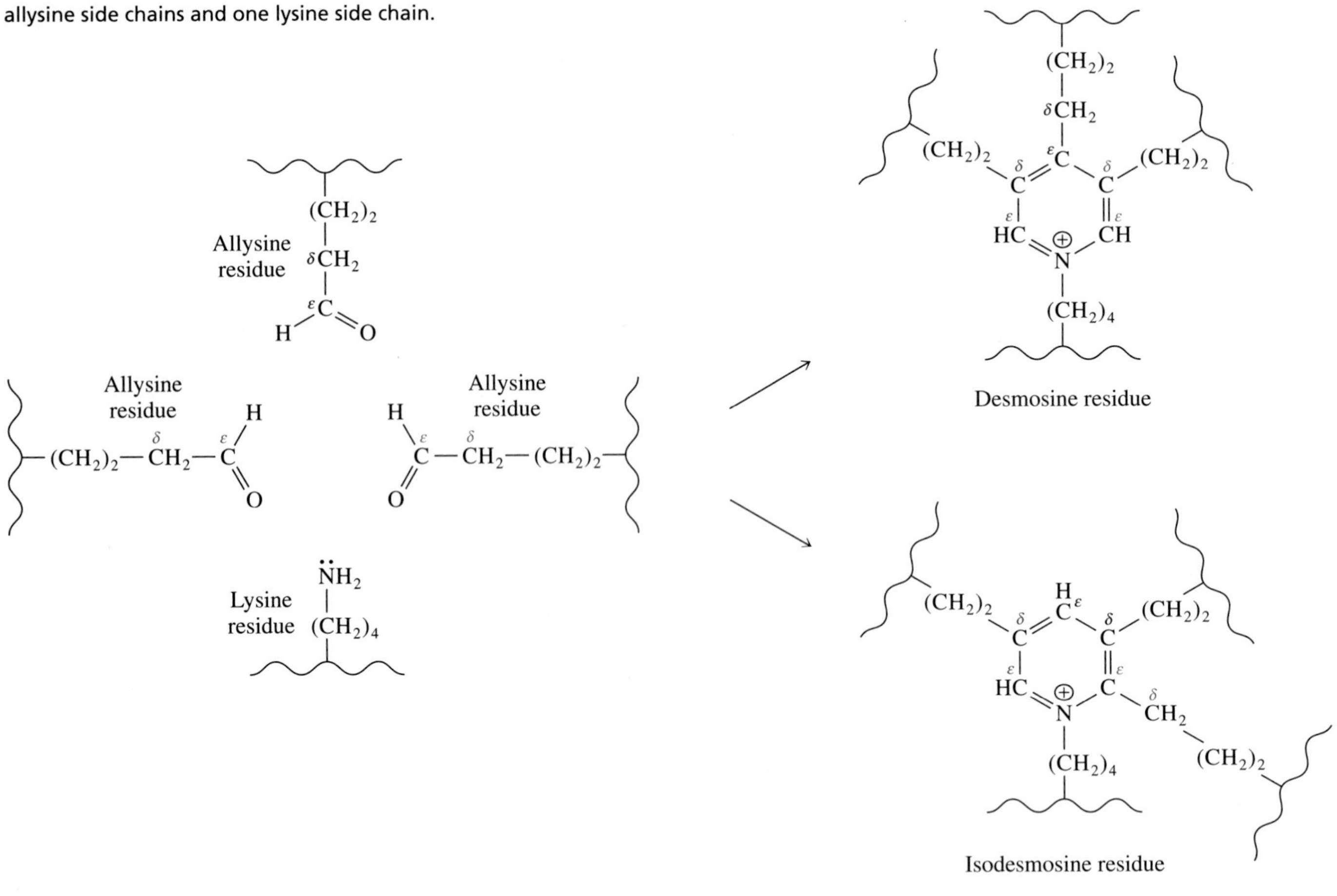

Copper deficiency may diminish the activity of lysyl oxidase, leading to the accumulation of a precursor of elastin, tropoelastin (M_r 72 000), which lacks desmosine and isodesmosine cross-links.

Because of its insolubility, elastin could not be sequenced using the usual methods of protein sequencing, although some soluble fragments of elastin were sequenced. The complete sequence of elastin was deduced by isolating and sequencing the DNA that encodes elastin. Tropoelastin contains a number of repeating oligopeptide sequences. For example, the pentapeptide sequence PGVGV (–Pro–Gly–Val–Gly–Val–) is repeated 11 times in a row. This stretch of residues is flanked by two alanine-rich sections containing lysine (K), AAAAAAAAAKAAKF and AKAAAKAAKF, that are involved in desmosine and isodesmosine cross-links. Other hydrophobic sections of the polypeptide chain that lie between cross-links contain repeating tetrapeptide (PGGV), hexapeptide (VVPGVG), and even nonapeptide (VPGLGVGVG) sequences.

Whereas the cross-linked areas of elastin have a rigid structure, the intervening sequences are responsible for the elastic properties of elastin. It has been suggested that the proline and glycine residues of the intervening sequences cause the polypeptide to turn and adopt a spiral conformation in the relaxed state. However, it is more likely that these hydrophobic regions between the cross-links have an amorphous or nonrepetitive structure.

The loss of elasticity—in skin, arteries, and other tissues—that is associated with aging is a complex and poorly understood process. The loss of lung elasticity characteristic of pulmonary emphysema arises from the degradation of the elastin fibers of the lungs. Tobacco smoke inactivates a protein that normally inhibits elastase, the protease that catalyzes the degradation of elastin. For this reason, smokers are more likely to develop emphysema.

5·9 The Tertiary Structure of Globular Proteins Allows Them to Bind Other Molecules Selectively and Transiently

Tertiary structure refers to the three-dimensional structure of an entire globular protein in its fully folded, biologically active—or native—conformation. Tertiary structure results from the folding of a polypeptide, which may already possess some regions of α helix and/or β sheet, into a closely packed, nearly spherical shape. In the tertiary structure of a protein, regions of secondary structure are connected by segments that we will describe later.

An important feature of tertiary structure is that amino acid residues that are far apart in the primary structure are brought together, permitting interactions among their side chains. Secondary structure is stabilized by hydrogen bonding between amide and carbonyl groups of the backbone; tertiary structure is also stabilized by noncovalent interactions (mostly hydrophobic) between side chains of amino acid residues. Any disulfide bonds that are essential to the structural integrity of a protein are also considered forces that stabilize tertiary structure.

The variety of protein tertiary structures is enormous—but not surprising, given the tremendous variability in protein size and amino acid sequence. However, all globular proteins contain recognizable secondary structure. A sampling of protein conformation is given in Figure 5·32 (next page). As you examine these structures, note the presence of α helices and β sheets. For better visualization of the polypeptide backbone, amino acid side chains are not shown. For additional views of proteins drawn to scale, please see the opening pages of this book.

(a)   **(b)**   **(c)**   **(d)**

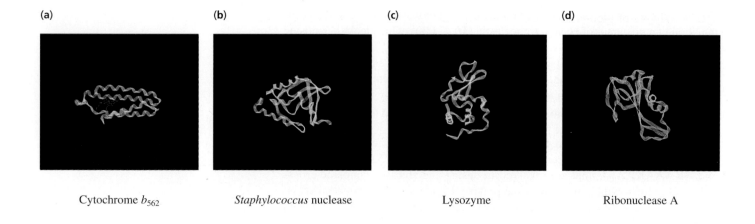

Cytochrome b_{562} *Staphylococcus* nuclease   Lysozyme   Ribonuclease A

(e)   **(f)**

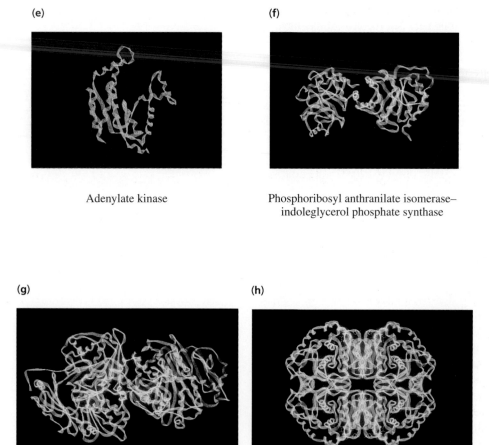

Adenylate kinase

Phosphoribosyl anthranilate isomerase–
indoleglycerol phosphate synthase

Figure 5·32
Ribbon models of proteins. All proteins are drawn in the same scale. **(a)** Cytochrome b_{562}, **(b)** *Staphylococcus* nuclease, **(c)** lysozyme, **(d)** ribonuclease A, **(e)** adenylate kinase, **(f)** phosphoribosyl anthranilate isomerase–indoleglycerol phosphate synthase (a bifunctional protein), **(g)** phosphoglucomutase (a single chain with four domains), and **(h)** lactate dehydrogenase (a protein with four identical subunits). (Structures based on coordinates provided by **(a)** T. Hamada, P. H. Bethge, and F. S. Mathews; **(b)** M. Legg, F. A. Cotton, and E. Hazen; **(c)** C. Kundrot and F. Richards; **(d)** J. Nachman and A. Wlodaver; **(e)** G. E. Schulz, C. W. Müller, and K. Diederichs; **(f)** M. Wilmanns, J. P. Priestle, and J. N. Jansonius; **(g)** W. J. Ray, J. B. Dai, Y. Liu, and M. Konno; **(h)** J. Griffith and M. Rossman.)

(g)   **(h)**

Phosphoglucomutase

Lactate dehydrogenase (tetramer)

The shapes of globular proteins, with their indentations, interdomain interfaces, and other crevices, allow them to fulfill dynamic functions by selectively and transiently binding other molecules. This property is best exemplified by the interaction between specific reactants (substrates) and enzymes at substrate-binding sites, or active sites, but it is also characteristic of other globular proteins. Because many binding sites are positioned toward the interior of a protein, they are relatively free of water. When substrates bind to an active site, they fit so well that water molecules in the cleft are excluded. In some cases, binding at one site can change the conformation of another site and therefore affect binding at the distant site. The diversity of folding and of side-chain structures among proteins allows for diversity in conformation, specificity, and chemical interactions at the binding site.

As part of their binding sites, some globular proteins contain cofactors, relatively small, typically nonprotein molecules or ions that are required for the protein to function, as we will see in Chapter 8. Some cofactors, called prosthetic groups, remain firmly bound to a protein. Others, called cosubstrates, dissociate and reassociate during the course of a physiological reaction. A protein with a prosthetic group in its active site is shown in Figure 5·32a, and a protein with a cosubstrate bound in each of its active sites is shown in Figure 5·32h (both cofactors are in red). An active protein possessing all of its cofactors is often referred to as a **holoprotein;** a protein whose cofactors are absent is called an **apoprotein.**

5·10 Myoglobin Was the First Protein to Have Its Tertiary Structure Determined

Elucidation of the structure of the oxygen-binding protein myoglobin laid the foundation for our present knowledge of the tertiary structures of proteins. Until the structure of myoglobin was deduced, the three-dimensional shapes of globular proteins were merely speculation. The complexity of the structure surprised many protein chemists, who were expecting more of the simple regularity seen in fibrous proteins and the double helix of DNA. But myoglobin does contain regular secondary structure, and we now know that it has a relatively simple conformation.

Like most globular proteins, myoglobin (Mb) and the related protein hemoglobin (Hb), discussed later in this chapter, carry out their biological functions by selectively binding other molecules—in this case, molecular oxygen (O_2). Myoglobin binds oxygen and facilitates its diffusion within muscle. Myoglobin accounts for about 8% of the total protein in the muscles of diving mammals, such as seals and whales, that store large amounts of oxygen. A relatively small protein ($4.4 \times 3.5 \times 2.5$ nm), myoglobin consists of a single polypeptide chain in addition to a heme prosthetic group.

The red color associated with the oxygenated forms of myoglobin and hemoglobin (for instance, the red color of skeletal and cardiac muscle and oxygenated blood) is due to the prosthetic group, heme, where molecular oxygen binds (Figure 5·33). Heme consists of a tetrapyrrole ring system called protoporphyrin IX complexed with ionic iron. The four pyrrole rings of this system are linked by methene (—CH=) bridges, so that the entire porphyrin structure is unsaturated, highly conjugated, and planar. Four methyl groups, two vinyl groups, and two carboxyethyl (or propionate) substituents are attached to the tetrapyrrole ring system. The iron, which is in the reduced—or ferrous (Fe^{2+})—oxidation state, replaces two protons of protoporphyrin IX. The Fe-protoporphyrin IX complex is a resonance hybrid in which the iron is bound equally to the four nitrogens of protoporphyrin IX.

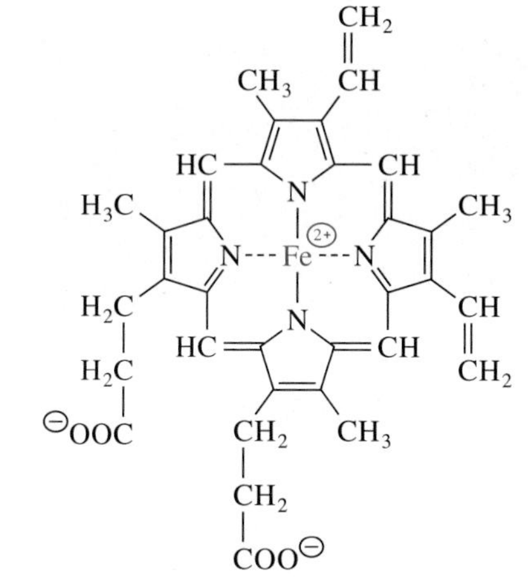

Figure 5·33
Structure of heme (Fe-protoporphyrin IX). The iron atom is in the ferrous oxidation state. Although the figure shows iron covalently bound to only two nitrogen atoms, iron is actually bound equally to all four nitrogen atoms.

Figure 5·34
Diagram of sperm-whale myoglobin. Myoglobin consists of eight α helices (in blue) connected by short nonrepetitive segments (in green). The heme prosthetic group (in red), which is wedged between two helices, binds oxygen. His-64 forms a hydrogen bond with oxygen, and His-93 is complexed to the iron atom within the heme.

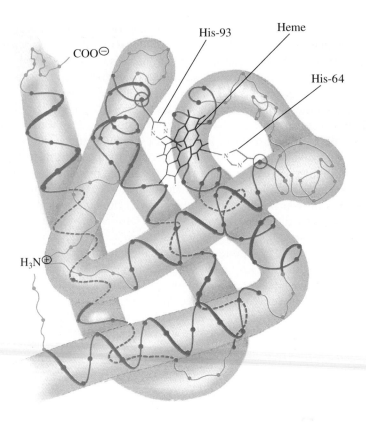

The polypeptide component, or apoprotein, of myoglobin is called globin and consists of 153 amino acid residues. Figure 5·34, which depicts the tertiary structure of sperm-whale myoglobin, shows that the protein consists of a bundle of eight α helices. Over three-quarters of the residues of the protein are in the α helices. Although such a high helix content is atypical of globular proteins, similar folding patterns are found in a large number of related proteins. Figure 5·35a shows the helical structure of myoglobin, and Figure 5·35b reveals the compactness of the molecule.

Several observations regarding the structure of myoglobin are typical of globular proteins. The interior of myoglobin is made up almost exclusively of hydrophobic amino acid residues, particularly those that are highly hydrophobic (valine, leucine, isoleucine, phenylalanine, and methionine). The surface contains both hydrophilic and hydrophobic residues (Figure 5·36). In general, water molecules are excluded from the interior of globular proteins, and most of the ionizable residues of proteins are located on the surface. However, several ionizable residues may be located internally in clefts accessible to molecules that the proteins specifically bind; these exceptional residues are important functional units of proteins. For example, internal ionizable residues play an important role in the chemical mechanisms of many enzymes, as we will see in Chapter 7.

The heme prosthetic group is wedged into a hydrophobic, cagelike cleft formed by the protein. The iron atom of the heme is the site of oxygen binding. Oxygen-free myoglobin is called deoxymyoglobin; the oxygen-bearing molecule is called oxymyoglobin. The reversible binding of oxygen is termed oxygenation. Accessibility of the heme group to molecular oxygen depends on slight movement of nearby amino acid side chains. We will see later that the chemical properties of the hydrophobic crevice of myoglobin (and hemoglobin) are essential for the reversible binding of oxygen.

Figure 5·35
Stereo views of myoglobin. The heme group, shown in red, is almost completely buried. (a) Ribbon model. (b) Space-filling model. (Based on coordinates provided by H. C. Watson and J. C. Kendrew.)

(a)
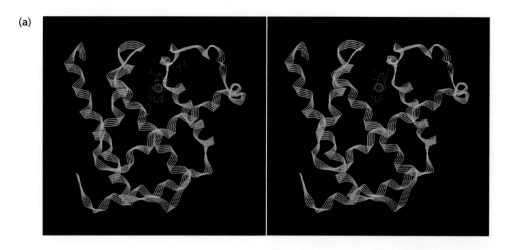

(b)
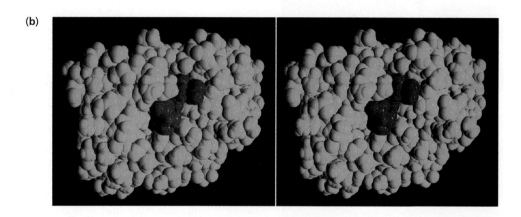

Figure 5·36
Stereo view of a space-filling model of myoglobin. Highly hydrophobic residues (valine, leucine, isoleucine, phenylalanine, and methionine) are blue; highly hydrophilic residues (aspartate, glutamate, histidine, lysine, arginine, asparagine, and glutamine) are pink. The heme group is red. (Based on coordinates provided by T. Takano.)

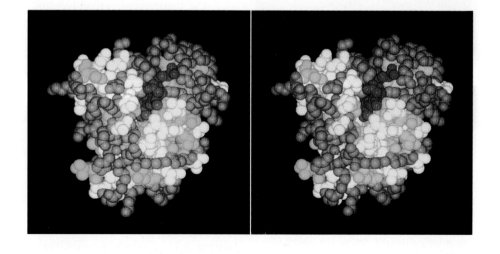

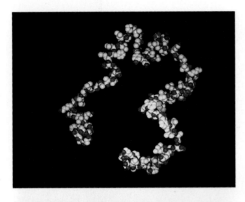

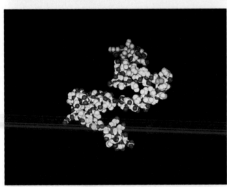

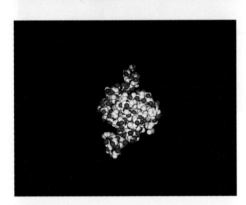

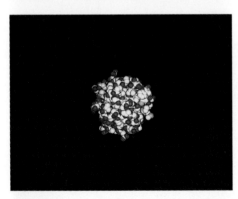

Figure 5·37
Hypothetical folding pathway of thioredoxin. The initially extended polypeptide chain gradually becomes more compact as the tertiary structure forms. The native conformation is roughly spherical, as in most globular proteins. (Courtesy of Frederic M. Richards.)

5·11 Folding of Globular Proteins Depends on a Variety of Interactions

The folding of proteins and the stabilization of the biologically functional conformations of proteins depend on a number of noncovalent factors, including the hydrophobic effect, hydrogen bonding, van der Waals interactions, and ionic interactions. It is thought that, as a protein folds, the first few interactions initiate subsequent interactions by assisting in the alignment of groups. This process is known as cooperativity of folding, the phenomenon whereby the formation of one part of a structure leads to the formation of the remaining parts of the structure. The process of protein folding cannot be described in detail because little is known about the intermediates (there may be several), which are rapidly converted to the native conformation in vivo. It is likely that, as a linear polypeptide—probably already possessing short regions of helix or sheet—folds, it becomes increasingly compact and spherical. A hypothetical folding pathway for the small protein thioredoxin (M_r 12 000) is shown in Figure 5·37. Many weak noncovalent interactions must be involved in the folding process. Although noncovalent interactions are individually weak, the sum of these interactions stabilizes the native conformations of proteins. Furthermore, the weakness of each noncovalent interaction gives proteins the resilience and flexibility to undergo small conformational changes needed for the functioning of the protein.

A. The Hydrophobic Effect Is the Principal Driving Force in Protein Folding

Proteins are more stable in water when their hydrophobic side chains are aggregated in the protein interior rather than solvated by water. The tendency of hydrophobic groups to associate with one another by virtue of their exclusion from water is called the hydrophobic effect (Section 3·4). This effect is presumed to be the major driving force in protein folding. Because water molecules interact more strongly with each other than with nonpolar groups, the nonpolar side chains aggregate, causing the protein to fold. Nonpolar side chains are driven into the interior of the protein, and most polar side chains remain in contact with water on the surface of the protein. The sections of the polar backbone that are forced into the interior of a protein neutralize their polarity by hydrogen bonding to each other and forming secondary structures. Thus, the hydrophobic nature of the interior not only accounts for the association of hydrophobic residues but also the stabilization of helices and sheets.

The hydrophobic effect has traditionally been explained in thermodynamic terms. The water molecules that surround nonpolar groups are relatively well ordered, forming an enclosing structure, or cage. The water molecules are hydrogen bonded to each other but only weakly attracted to the group within the cage. When nonpolar groups are removed from contact with water, the disruption of the cage structure is accompanied by an increase in the entropy of the water molecules as they leave the ordered cages to become part of the bulk solvent. The increase in solvent entropy more than offsets the decreased entropy of the folded protein and provides the most significant driving force for protein folding.

The quantitative aspects of this traditional explanation of the hydrophobic effect have been challenged. For example, there are claims that the change in solvent ordering is not a driving force for the protein-folding process. Reversible unfolding of the protein ubiquitin occurs in a methanol-water solvent system in which the effect of solvent ordering is negligible. In this case, solvent ordering cannot be a component of the hydrophobic effect and is not involved in protein folding. Rather, the optimization of interactions such as hydrogen bonds between amino acid residues and, in particular, van der Waals interactions in the folded state may be the driving force for correct protein folding.

Table 5·3 Examples of hydrogen bonds in proteins

Type of hydrogen bond		Typical distance between donor and acceptor atom (nm)
Hydroxyl-hydroxyl	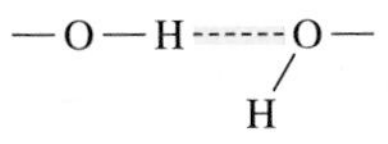 $-O-H\cdots\cdots O-$ $\diagup$ H	0.28
Hydroxyl-carbonyl	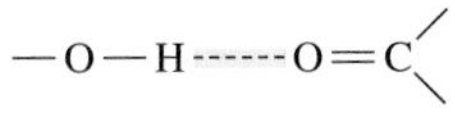 $-O-H\cdots\cdots O=C\Big\langle$	0.28
Amide-carbonyl	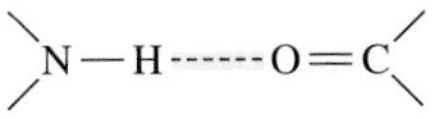 $\diagdown$N$-$H$\cdots\cdots$O$=$C$\Big\langle$	0.29
Amide-hydroxyl	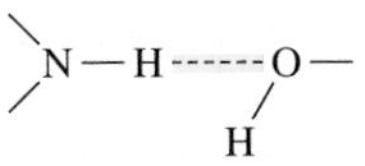 $\diagdown$N$-$H$\cdots\cdots$O$-$ $\diagup$ H	0.30
Amide-imidazole nitrogen	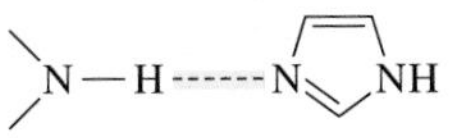 $\diagdown$N$-$H$\cdots\cdots$N⟋NH	0.31

B. Hydrogen Bonds and van der Waals Forces Stabilize Globular Proteins

Hydrogen bonds contribute to the cooperativity of folding and help stabilize the native conformations of globular proteins. As noted previously, the carbonyl and amide groups of the polypeptide backbone, especially those in the interior of a globular protein, often form hydrogen bonds with each other to produce α helices and β sheets. In addition, hydrogen bonds can form between the polypeptide backbone and water, between the polypeptide backbone and polar side chains, between two polar side chains, and between polar side chains and water. Table 5·3 shows some of the many types of hydrogen bonds found in proteins, along with their typical bond lengths. Most hydrogen bonds in proteins are of the N—H------O type. These bonds vary in length from 0.26 to 0.34 nm and may deviate from linearity by up to 40°.

Efficient packing that maximizes van der Waals contacts between nonpolar residues also contributes to the stability of globular proteins. As noted above, the contribution of van der Waals forces to the hydrophobic effect may have been underestimated in the past.

C. Covalent Cross-Links and Ionic Interactions Sometimes Help Stabilize Globular Proteins

In addition to hydrogen bonds, covalent cross-links such as disulfide bonds help stabilize the native conformations of some globular proteins. Although disulfide bonds are not usually found in intracellular proteins, they are sometimes found in proteins that are secreted from cells. When these proteins leave the intracellular environment, the presence of disulfide bonds makes the proteins less susceptible to unfolding and subsequent degradation. Disulfide bonds play no part in determining the native conformation but form spontaneously where two cysteine residues are appropriately located. Formation of a disulfide bond requires oxidation of the thiol groups of the cysteine residues, probably by disulfide exchange reactions involving oxidized glutathione, a small cysteine-containing peptide.

To a small extent, ionic interactions between oppositely charged side chains may help stabilize globular proteins. Ionic side chains typically occur on the surface and are thus solvated and contribute minimally to the overall stabilization of the protein. However, two oppositely charged ions occasionally form an ion pair in the interior of a protein. These ion pairs are typically associated with a special function in the molecular architecture of the protein. For example, in chymotrypsin, an ion pair forms between an α-amino group (Ile-16) and a β-carboxyl group (Asp-194) during activation of the enzyme precursor chymotrypsinogen (Section 6·11A).

D. Protein Folding Is Assisted by Chaperones

Protein-folding experiments using simple proteins, such as ribonuclease A, which we will discuss in the next section, have allowed us to make some general observations regarding the folding of polypeptides into biologically active proteins. First, proper folding of a polypeptide chain does not occur by a random search. Rather, protein folding appears to be a cooperative, sequential process, in which formation of the first few structural elements assists in the alignment of subsequent structural features. Second, the folding pattern and the final conformation of a protein are dependent upon its primary structure. In the test tube, simple proteins may undergo self-assembly, that is, fold into their native conformations without any energy input or assistance. However, in many cases, newly formed protein chains adopt incorrect conformations unless certain other proteins are present. Two types of proteins—enzymes and molecular chaperones—are required for the proper folding of proteins in cells.

There are two enzyme-catalyzed isomerization reactions that could limit the rate of folding of proteins. Peptide bonds involving the amino group of proline residues undergo *cis-trans* isomerization very slowly. The action of peptidyl prolyl *cis-trans* isomerase accelerates the folding of proteins by allowing the formation of the correct proline isomers. The isomerase is present in virtually all organisms and tissues. In eukaryotic cells, the folding of proteins containing disulfide bonds occurs mainly in the endoplasmic reticulum, where proteins are processed for export from the cell. An enzyme, protein-disulfide isomerase, facilitates the formation of the correct disulfide bonds by catalyzing the reshuffling of disulfide bonds via thiol-disulfide interchanges.

Chaperones are proteins that have the net effect of increasing the rate of correct folding by binding newly synthesized polypeptides before they are completely folded. Chaperones prevent the formation of incorrectly folded intermediates that may trap the polypeptide in an aberrant form. Chaperones can also bind to unassembled protein subunits, preventing them from aggregating incorrectly and precipitating before they are assembled into a complete multisubunit protein.

Chaperones appear to inhibit incorrect folding and assembly pathways by interacting with surfaces on the newly synthesized polypeptide chain that are exposed only during folding and assembly. Chaperones form stable complexes with the polypeptide chain, often through interaction with hydrophobic regions. In some cases, the action of chaperones is ATP dependent: apparently, unfolded proteins bind to chaperones, and the release of more compact forms is accompanied by hydrolysis of ATP. Folding that requires molecular chaperones has been aptly described as assisted self-assembly.

Other roles for the chaperones within the cell include assisting in the translocation of polypeptide chains across membranes and the assembly and disassembly of large multiprotein structures. Chaperones are also involved in escorting some proteins to specific locations within the cell (hence their name). There are many different kinds of chaperones. Their synthesis is often increased after cells have been subjected to stress. This response is related to the role of chaperones in refolding proteins that were denatured or damaged when the cell was stressed. Some of the best-studied chaperones are the heat-shock proteins, whose rate of synthesis is greatly increased when cells are subjected to above-normal temperatures. These proteins appear to aid in the recovery of misfolded proteins. Hsp70 (M_r 70 000) is one example of such a chaperone. This protein is a member of a family of closely related proteins in eukaryotic cells. In addition to the heat-shock–inducible members, the family includes an abundant cytoplasmic chaperone (Hsc70) and proteins that are found in the endoplasmic reticulum (BiP), mitochondria, and chloroplasts. Homologous proteins are also found in bacteria; in fact, the Hsp70 proteins are the most highly conserved proteins known, suggesting that chaperones have played an important role since the time of the first cells. Hsp90 and Hsp60 (the latter is also known as chaperonin, or GroE) are additional examples of chaperones that are heat-shock proteins. Hsp90 and Hsp60 are present in all organisms and, like Hsp70, have highly conserved amino acid sequences.

Although there have been many attempts to predict the secondary or tertiary structures of proteins from their primary structures, these attempts have met with limited success, primarily because so many variables determine the conformation of the polypeptide chain, including interactions between residues that are widely separated in the primary structure of the protein and the active role of chaperones in the protein-folding process. Protein folding therefore continues to be a major area of research in protein chemistry.

5·12 Denaturing Agents Cause Proteins to Unfold

Environmental changes or chemical treatments may cause a disruption in the native conformation of a protein, with concomitant loss of biological activity. Such a disruption is called **denaturation.** Because the native conformation is only marginally stable, the energy needed to cause denaturation is often small, perhaps equivalent to the disruption of three or four hydrogen bonds. Although some proteins may completely unfold when denatured to form what is called a random coil (a fluctuating chain considered to be totally disordered), other proteins retain considerable internal structure when similarly treated. It is possible to find conditions under which some denatured proteins—especially certain small proteins—renature, or refold.

Several methods can be used to denature a protein. Raising or lowering the pH can change the ionic state of ionizable side chains in a protein, thereby breaking hydrogen bonds, creating regions of charge repulsion, and disrupting ion pairs; all of these disruptions can contribute to denaturation. Heating a protein solution causes an increase in vibrational and rotational energy that can upset the delicate balance of weak interactions stabilizing the functional, folded conformation. As well as disrupting noncovalent interactions, the harsh conditions of high temperatures or treatment with strong acid or alkali can cause irreversible inactivation through covalent changes, such as deamidation of asparagine or glutamine residues and hydrolysis of peptide bonds involving aspartyl groups.

Two types of chemicals, chaotropic agents (chaos-promoting ions; Section 3·4) and detergents, cause denaturation of proteins under less harsh conditions. Since these chemicals do not cleave covalent bonds, they disrupt only secondary, tertiary, and quaternary structures, not primary structure. Consequently, the effects of these chemicals can sometimes be reversed, and studies using these denaturing agents can provide insight into protein folding. High concentrations of chaotropic agents such as urea and guanidinium salts (Figure 5·38) allow water molecules to penetrate into the interior of proteins and solvate nonpolar side chains, thereby disrupting the hydrophobic interactions that normally stabilize the native conformation. Detergents such as sodium dodecyl sulfate (SDS; shown in Figure 3·8b) denature proteins at lower concentrations than chaotropic agents. The hydrophobic tails of such molecules penetrate the hydrophobic interior of a protein, disrupting the hydrophobic interactions and thereby denaturing the protein.

Complete denaturation of proteins that contain disulfide bonds requires the cleavage of these bonds in addition to disruption of hydrophobic interactions and

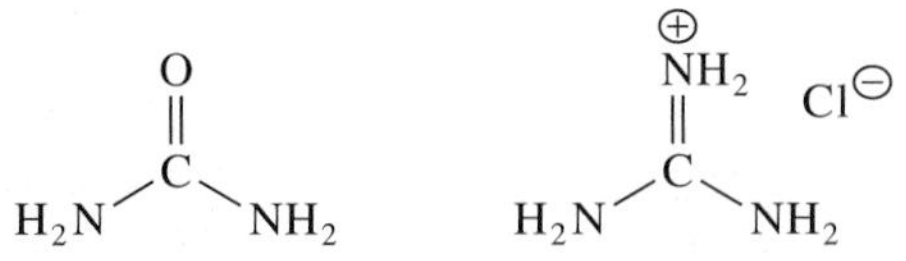

Figure 5·38
Urea and guanidinium chloride.

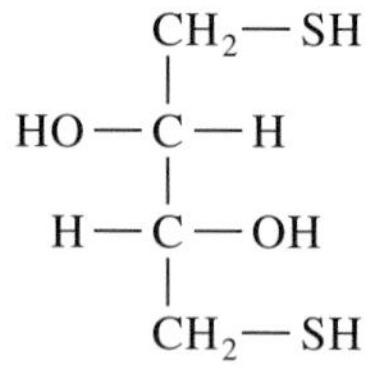

CH₂—SH

HO—C—H

H—C—OH

CH₂—SH

Dithiothreitol
(DTT)

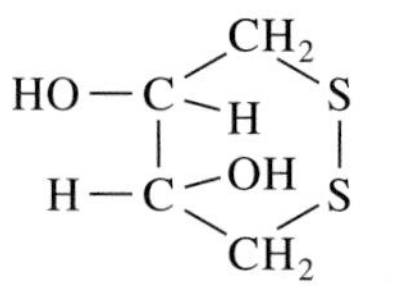

HO—C—H—S

H—C—OH—S

Oxidized dithiothreitol
(a cyclic disulfide)

Figure 5·39
Dithiothreitol.

hydrogen bonds. 2-Mercaptoethanol and other thiol reagents, such as dithiothreitol (DTT), shown in Figure 5·39, can be added to a denaturing medium (8 M urea, 5 M guanidinium chloride, or 1% SDS) in order to reduce any disulfide bonds to sulfhydryl groups. The reduction of the disulfide bonds of the protein is accompanied by oxidation of the thiol reagent.

One early study of protein folding described the regeneration of the native conformation of a protein from its denatured state. This study, conducted by Christian B. Anfinsen and his coworkers, used the protein ribonuclease A, a pancreatic enzyme that catalyzes digestion of ribonucleic acids. Ribonuclease A consists of a single chain of 124 amino acid residues, cross-linked by four disulfide bonds (Figure 5·40). Anfinsen demonstrated the importance of primary structure in directing folding of a simple protein into its native conformation (Figure 5·41).

Denaturation of ribonuclease A with 8 M urea containing 2-mercaptoethanol results in complete loss of tertiary structure and enzymatic activity and yields polypeptide chains containing eight sulfhydryl groups. If the 8 sulfhydryl groups paired randomly during refolding and oxidation, 105 possible disulfide-bonded structures could be produced (7 possible pairings for the first bond, 5 for the second, 3 for the third, and 1 for the fourth; $7 \times 5 \times 3 \times 1 = 105$), with only 1 out of 105 molecules being active. In fact, when reductant is removed and oxidation is allowed to occur in the presence of denaturant (8 M urea), disulfide bonds form between incorrect partners in about 99% of the protein population, generating a solution of protein that has about 1% of its original enzymatic activity. However, when urea and the reductant are removed simultaneously and dilute solutions of the reduced protein are then exposed to air at physiological pH, ribonuclease A spontaneously regains its native conformation, its correct set of disulfide bonds, and its full enzymatic activity. A solution of the inactive forms, containing scrambled disulfide bonds, can be renatured if a small amount of reducing agent (2-mercaptoethanol) is added and the solution gently warmed. These experiments demonstrate that the correct disulfide bonds can form only after the protein folds into its native conformation. Anfinsen concluded that the recovery of the native shape of ribonuclease A is driven entirely by the free energy gained in changing to the stable, physiological conformation. He also concluded that this conformation is determined by the primary structure.

The renaturation of ribonuclease A is a slow process, sometimes taking as long as several hours. Anfinsen discovered that there is an enzyme in the endoplasmic reticulum—now known as protein-disulfide isomerase—that catalyzes the recovery of scrambled ribonuclease in less than two minutes in vitro. The time needed for biosynthesis of a polypeptide the size of ribonuclease is about two minutes. Complete folding, assisted by chaperones and protein-disulfide isomerase in vivo, is likely to require no more than a few additional minutes.

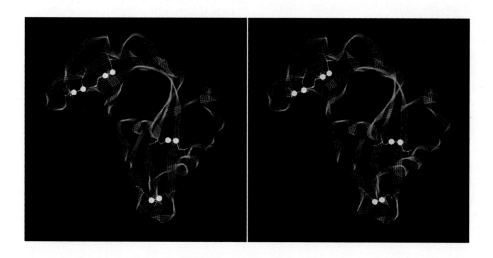

Figure 5·40
Stereo view of ribonuclease A. There are eight cysteine residues (with sulfur atoms highlighted in yellow) and four disulfide bonds in ribonuclease A. If the polypeptide is reductively denatured, the cysteine residues must be paired correctly during renaturation for the enzyme to become active. (Based on coordinates provided by J. Nachman and A. Wlodaver.)

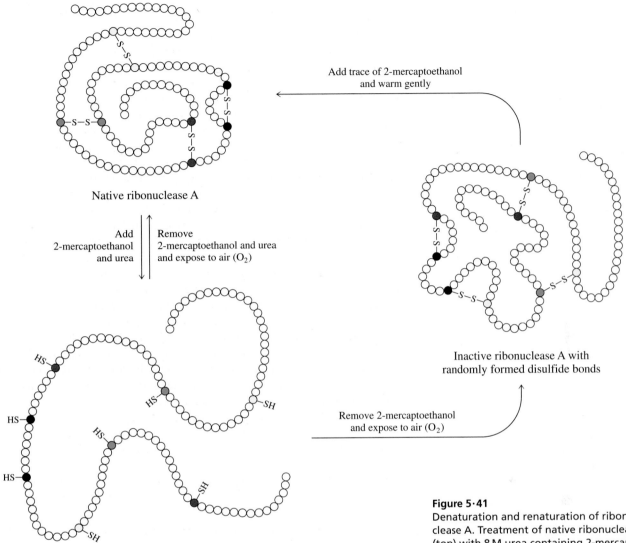

Native ribonuclease A

Add trace of 2-mercaptoethanol
and warm gently

Add
2-mercaptoethanol
and urea

Remove
2-mercaptoethanol and urea
and expose to air (O_2)

Inactive ribonuclease A with
randomly formed disulfide bonds

Remove 2-mercaptoethanol
and expose to air (O_2)

Reversibly denatured ribonuclease A;
disulfide bonds have been reduced

Figure 5·41
Denaturation and renaturation of ribonuclease A. Treatment of native ribonuclease A (top) with 8 M urea containing 2-mercaptoethanol unfolds the protein and disrupts disulfide bonds to produce reduced, reversibly denatured ribonuclease A (bottom). If 2-mercaptoethanol alone is removed, ribonuclease A reoxidizes in the presence of air (O_2), but the disulfide bonds form randomly to produce a scrambled, inactive protein (such as the form shown on the right). If urea and 2-mercaptoethanol are simultaneously removed from reversibly denatured ribonuclease A under appropriate conditions, the protein returns to its native conformation, and in the presence of air, the correct disulfide bonds are formed (top). When a trace of 2-mercaptoethanol is added to a solution of the scrambled protein, the disulfide bonds break and re-form correctly to produce native ribonuclease A. The results of these renaturation experiments demonstrate the importance of primary structure in directing the folding of this simple protein.

5·13 Globular Proteins Possess Conformations Other than Helices and Sheets

The remarkable accumulation of data from X-ray crystallographic studies of proteins has greatly extended the early observations of helices and sheets as secondary structures (such as in α-keratin and silk fibroin) and the existence of compact tertiary structures (as exemplified by myoglobin). Many subtle features of protein structure have been described only after a large number of proteins had been analyzed. Proteins do not exist only as slender rods or chains; small, stabilized loops allow proteins to assume globular shapes. Segments of α helix or β strands are connected by loops, often in recurring patterns. Larger loops, or flaps, have been found on the surfaces of proteins. These loops were once thought to have a random, or disordered, conformation, but it is now known that the loops have specific structures. This section describes some additional features of protein tertiary structure that help explain how the complex shapes of globular proteins are achieved.

A. Nonrepetitive Regions Are Essential in Globular Proteins

As we have noted, globular proteins contain elements of α helix and β strands, that is, regions in which consecutive residues have a single repeating conformation. In addition, globular proteins contain stretches of nonrepetitive three-dimensional structure. These regions have often been called random coils, but they are stable, ordered structures. These nonrepetitive regions connect secondary structures and provide directional changes necessary for a protein to attain its globular shape. These changes occur at **loops,** connecting regions with nonrepetitive structure ranging from 2 to 16 residues in length. Loops often contain many hydrophilic residues and are usually found on the surfaces of globular proteins, where they are exposed to solvent and form hydrogen bonds with water. The term **turn** is applied to loops having only a few (up to five) residues. The two most common types of turns, type I (or common turn) and type II (or glycine turn), are loops that span four amino acid residues and are stabilized by hydrogen bonding between the carbonyl oxygen of the first residue and the amide nitrogen of the fourth residue (Figure 5·42). In the type II turn, the third residue is glycine about 60% of the time; in both types of turns, proline is often the second residue. The tight turns connecting two adjacent antiparallel β strands are often called **hairpin loops.**

Most globular proteins have many residues in nonrepeating conformations. For example, in cytochrome c, a heme-containing globular protein that participates in electron transfer, nearly 50% of the residues of the protein are in regions of nonrepetitive structure (Figure 5·43). Examination of 67 known protein structures shows that, on average, there are almost as many residues in turns (15%) and simple loops (21%) as in helices (26%) and sheets (19%). In addition, approximately 10% of the residues are in more complex loops.

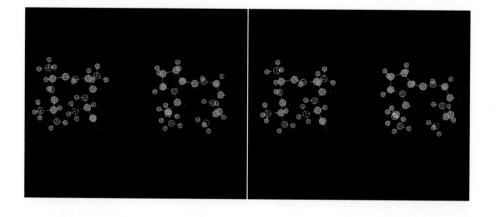

Figure 5·42
Stereo view of two turns. Type I (on the left) is the most common turn. Type II (on the right) often has glycine as the third residue. Both turns are stabilized by hydrogen bonding between the carbonyl oxygen of the first residue and the amide hydrogen of the fourth residue. Color key: carbon, green; β-carbon, orange; hydrogen, light blue; nitrogen, dark blue; oxygen, red. (Type I is based on coordinates provided by Brian W. Matthews and M. A. Holmes; type II is based on coordinates provided by M. N. G. James and A. R. Sielecki.)

Figure 5·43
Stereo view of tuna heart cytochrome c. The heme group (red) is surrounded by a number of nonrepetitive regions and several α-helical regions. (Based on coordinates provided by T. Takano.)

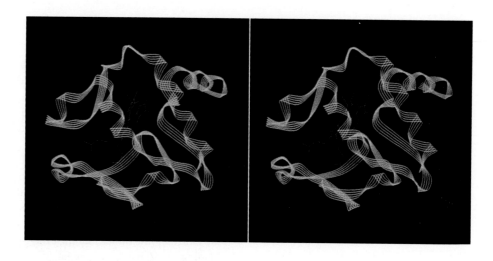

B. Conformations Are Severely Restricted in Globular Proteins

Observations of the three-dimensional structures of globular proteins reveal that the conformational freedom of individual residues in these proteins is quite restricted. Figure 5·44 is a Ramachandran plot showing the ϕ and ψ values for all residues except proline and glycine found in 53 globular proteins. Note that the values for the angles cluster in two major regions: $\phi = -60°$, $\psi = -40°$, the angles for residues in the α-helical conformation, and $\phi = -150°$ to $-60°$, $\psi = 90°$ to $180°$, the angles for residues in the β-sheet conformation. We have just seen that many of the residues plotted are in nonrepetitive structures, not in compact α-helical or extended β structures. Nonetheless, these residues have bond angles characteristic of α helices or β strands.

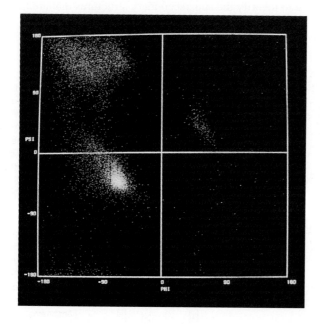

Figure 5·44
Ramachandran plot of the conformational angles observed for all residues except glycine and proline in the structures of 53 globular proteins. As the number of residues per ϕ and ψ value increases, the dots lighten, with orange indicating one residue and white indicating 30 or more. (Courtesy of Jane S. Richardson, David C. Richardson, Kim M. Gernert, and Larry D. Bergman.)

C. Supersecondary Structures, or Motifs, Are Combinations of Secondary Structures

Supersecondary structures, also called **motifs,** are combinations of secondary structure that appear in a number of different proteins. Supersecondary structures may have a particular function or may simply occur as part of a larger functional unit, the domain. Similar supersecondary units may have different functions in different proteins. Some common supersecondary structures are shown in Figure 5·45.

One of the simplest supersecondary structures is the helix-loop-helix (Figure 5·45a). This type of structure occurs in a number of calcium-binding proteins, where glutamate and aspartate residues in the loop provide the calcium-binding site. In certain DNA-binding proteins, this structure is called a helix-turn-helix since the loop is usually only a few residues in length; in these proteins, typically the entire motif, rather than just the loop, provides the binding site. Another DNA-binding motif is the zinc finger, a zinc-containing projection. Zinc fingers are described in more detail in Chapter 27.

The $\beta\alpha\beta$ unit is a supersecondary structure consisting of two parallel β strands linked to an intervening α helix by two loops (Figure 5·45b). The helix connects the carboxy end of one β strand to the amino end of the next and usually runs parallel to the two strands. A hairpin consists of two adjacent antiparallel β strands connected by a hairpin loop (Figure 5·45c).

The Greek key, which takes its name from a design found on classical Greek pottery, is a common supersecondary structure that links four or more antiparallel β strands. Figure 5·45d shows a Greek key in which the antiparallel β strands are linked by two hairpin loops and one long arch connecting the outer strands.

One recently discovered supersecondary structure consists of parallel β strands coiled in a large right-handed helix; the motif is called a parallel β helix. This structure is found in the microbial enzyme pectate lyase. The enzyme is a cylinder composed of seven complete helical turns, with an average of 22 residues per turn (Figure 5·46). The parallel β helix, which is quite distinct from an α helix, illustrates how elements of secondary structure can be combined in interesting ways.

Based on the large number of known protein and peptide structures, attempts have been made to identify the supersecondary structures that are likely to form in

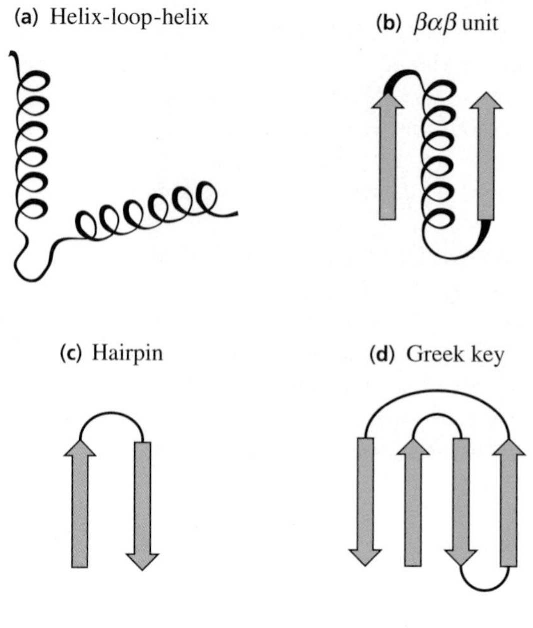

(a) Helix-loop-helix

(b) $\beta\alpha\beta$ unit

(c) Hairpin

(d) Greek key

Figure 5·45
Four supersecondary structures. α Helices and β sheets are commonly connected in globular proteins by loops to form supersecondary structures. Some particularly common supersecondary structures, shown here as two-dimensional representations, are: **(a)** the helix-loop-helix, **(b)** the $\beta\alpha\beta$ unit, **(c)** the hairpin, and **(d)** the Greek key. Arrows indicate the N- to C-terminal direction of the peptide chain.

Figure 5·46
Stereo view of pectate lyase from *Erwinia chrysan-themi.* This protein is an example of a parallel β helix. (Based on coordinates provided by M. D. Yoder, N. T. Keen, and F. Jurnak.)

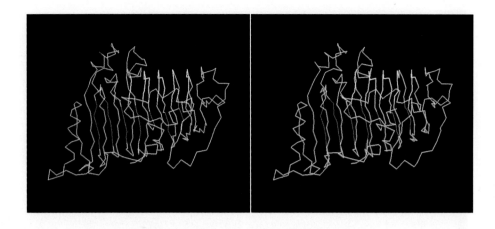

a polypeptide with a given sequence. It may be possible someday to predict the structure—and therefore the function—of a protein known only by the sequence of its gene. However, because so little is understood about the variety of possible side-chain interactions in proteins, such predictive studies are in their infancy.

It has been suggested that motifs may be encoded by distinct segments of DNA that are combined in different ways to assemble a wide variety of proteins. Motifs have roughly 10–40 residues each. Fusion and mutations of ancestral DNA fragments might have been the process by which present proteins evolved from a fairly limited number of fundamental conformational or functional units. Two or more such units, or modules, could have been fused to make the next larger unit, a domain, and several domains could have been assembled to form primitive enzymes that could eventually evolve into more specialized proteins. Although the suggestion seems plausible, it is not yet fully supported by structural data.

D. Domains Are Globular Units Within Tertiary Structure

Discrete, independent folding units within the tertiary structure are known as domains, or lobes. Domains can usually be distinguished easily in the three-dimensional structures of globular proteins. Domains consist of combinations of several units of supersecondary structure. These motifs are usually adjacent in the primary structure. The size of domains varies greatly, from about 25–30 to about 300 amino acid residues, with an average of about 100 residues. Typically, domains have a particular function, such as the binding of a small molecule. Each domain may have binding sites for one to three ligands. Enzymes that require ATP as a reactant may have completely different sequences yet have nucleotide-binding domains of similar three-dimensional structure. In many enzymes that require coenzymes that are dinucleotides, two mononucleotide-binding domains are combined into one dinucleotide-binding domain.

Some domains have easily recognizable structures. We have already seen the globin structure of myoglobin, a domain structure present in a number of proteins. Another recognizable domain structure is the β meander, or up-and-down sheets. It contains antiparallel β strands connected by hairpin loops (Figure 5·47a). This structure sometimes forms a barrel. When β strands form the staves of the barrel, the structure is called a β barrel. The most common recognizable domain structure is an α/β barrel, which is made up of repeating αβ units (Figure 5·47b). Close packing and suitable amino acid composition of the parallel β strands in an α/β barrel often make the core of the barrel a hydrophobic binding or reaction site. Triose phosphate isomerase (Figure 5·48, next page) is a good example of an α/β-barrel domain.

(a) β meander

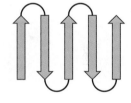

(b) α/β barrel

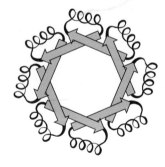

Figure 5·47
Two examples of domain structures. **(a)** The β meander is composed of adjacent anti-parallel β strands connected in a simple, direct pattern. It can be formed into a barrel shape. **(b)** The α/β barrel shown here (top view) forms a structure with eight parallel β strands connected by eight α helices. Arrows indicate the N- to C-terminal direction of the peptide chain.

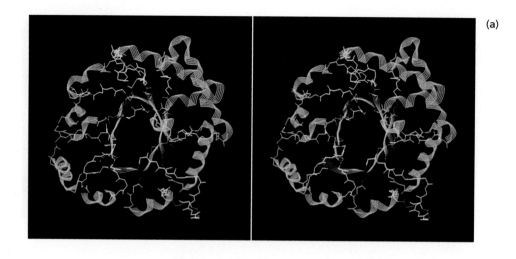

(a)

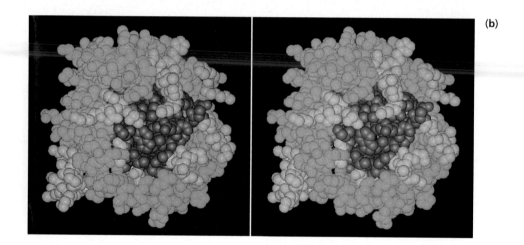

(b)

Figure 5·48
Stereo views of triose phosphate isomerase, an enzyme involved in the metabolism of sugars. **(a)** A cylinder of β strands (in blue) is surrounded by a wreath of α helices (green), forming an α/β barrel. Regions of non-repetitive structure are shown in yellow. **(b)** Space-filling model. Color key: α helix, green; β barrel, blue; nonrepetitive structure, yellow. (Based on coordinates provided by T. Alber, G. A. Petsko, and E. Lolis.)

A small protein may consist entirely of one domain, as triose phosphate isomerase does. Larger proteins usually contain more than one domain. Adenylate kinase is an example of an enzyme with two domains, one containing both helices and a parallel β sheet and the other containing helical rods (Figure 5·49). During catalysis, the domains move toward each other.

The extent of contact between domains varies from protein to protein. At one extreme, domains may exist as separate subunits of a protein that has quaternary structure. In two-lobed proteins, the domains may be connected by a single peptide loop, sometimes called a hinge. The flexibility of this hinge allows one domain to move relative to the other, as in adenylate kinase. Sometimes the hinge can be cleaved so that individual domains can be isolated. At the other extreme, domains may be joined by extensive and close contact. For example, the enzyme papain, which catalyzes selective hydrolytic cleavage of peptide bonds, consists of two interlocking domains (Figure 5·50). The region or regions between domains often form crevices that may serve as binding sites for other molecules, such as in the catalytic sites of enzymes. For instance, the cleft between the two domains of papain functions as a binding site for certain polypeptide substrates.

In some species, certain large multidomain proteins are multifunctional enzymes, with each catalytic activity associated with a separate section (one or several domains) of the single polypeptide chain. One example is shown in Figure

Figure 5·49
Stereo views of porcine adenylate kinase, an enzyme that catalyzes the phosphorylation of AMP by ATP. (**a**) Ribbon model. (**b**) Space-filling model. (Based on coordinates provided by G. E. Schulz, C. W. Müller, and K. Diederichs.)

(a)

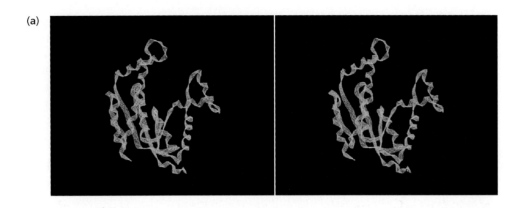

(b)

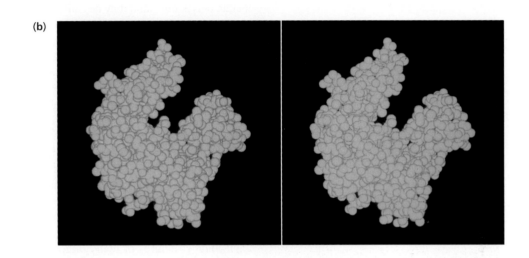

Figure 5·50
Stereo view of papain. Two domains interlock in the native conformation of the protein. One domain contains three α helices (red), and the other contains a single α helix and an antiparallel β sheet (blue). The ribbon model contains a break, but the polypeptide backbone is actually continuous. (Based on coordinates provided by I. Kamphuis and J. Drenth.)

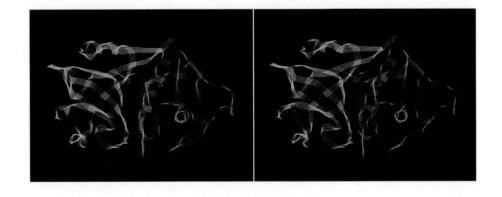

5·32f. (In other species, similar enzyme activities may be present on separate protein molecules or on nonidentical subunits.) Often, the domains of multifunctional proteins catalyze sequential reactions in metabolic pathways. Possible advantages of this type of organization include coordinated regulation of enzyme synthesis and enzymatic activity, direct passage of reactants between active sites (called channelling), and protection of unstable intermediates.

5·14 Proteins with Quaternary Structure Are Assemblies of Globular Subunits

Some—in fact, probably most—globular proteins have an additional level of organization called quaternary structure. Quaternary structure is limited to proteins with multiple subunits and refers to the organization of the subunits. Each subunit is a separate polypeptide and may be called either a monomer or simply a chain. A multisubunit protein is referred to as an **oligomer.** The arrangement of the subunits within an oligomeric protein always has a defined stoichiometry and almost always displays symmetry. The monomers of a multisubunit protein may be identical or different. When the monomers are identical, dimers and tetramers predominate. When the monomers differ, each type of subunit usually has a different function. When two or more reactions are catalyzed by the oligomer, it is called a multienzyme complex. Multienzyme complexes have the same metabolic advantages as multifunctional enzymes.

The subunits of oligomeric proteins are noncovalently associated. Hydrophobic interactions are the principal forces holding the subunits together, although electrostatic forces may contribute to the proper alignment of the subunits. Because intersubunit forces are usually rather weak compared to the forces stabilizing tertiary structure, the subunits of an oligomeric protein can often be isolated in a laboratory. However, the subunits usually remain tightly associated in vivo. The three-dimensional structures of many oligomeric proteins are altered when the proteins bind ligands. When this happens, both the tertiary structures of the subunits and the quaternary structures (i.e., the contacts between subunits) change. Such changes are often key elements in the regulation of the biological activity of oligomeric proteins.

Determination of the subunit composition of an oligomeric protein is an essential step in the physical description of a protein. Although there are a number of procedures that can be used, in typical experiments the molecular weight of the native oligomer is estimated by gel-filtration chromatography, and then the molecular weight of each chain is determined by SDS-polyacrylamide gel electrophoresis (Section 4·7). For a protein having only one type of chain, the ratio of the two values provides the number of chains per oligomer.

The first example we will encounter of a protein having quaternary structure is hemoglobin, which has four polypeptide chains of two slightly different types.

5·15 Hemoglobin Is a Tetramer of Globin Chains

Small aerobic organisms can obtain oxygen required for metabolism by simple diffusion from the environment. Larger animals, which cannot rely on diffusion for an adequate supply of oxygen, have circulatory systems that transport oxygen from the lungs or gills to other tissues. Oxygen is only sparingly soluble in aqueous solution. Therefore, in mammals and birds, oxygen is bound to molecules of hemoglobin for transport in red blood cells, or erythrocytes. Viewed under a microscope, a mature mammalian erythrocyte is a biconcave disk that lacks a nucleus or other internal

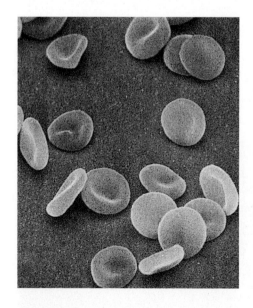

Figure 5·51
Scanning electron micrograph of red blood cells. Mature red blood cells are biconcave disks. Each red blood cell contains approximately 300 million hemoglobin molecules. (Courtesy of Keith R. Porter.)

membrane-enclosed compartments (Figure 5·51). A human erythrocyte is filled with approximately 3×10^8 molecules of hemoglobin.

In 1865, long before any protein had been isolated as a pure substance, Wilhelm Kühne proposed that the oxygen-binding proteins of muscle and blood (now known to be myoglobin and hemoglobin) were related, if not identical. A century later, Max Perutz determined the structure of horse hemoglobin by X-ray crystallography, a task that required over two decades of effort.

Adult hemoglobin is a tetramer composed of two each of two types of globin chains, α and β, both similar to, but slightly shorter than, the single chain of myoglobin. The α and β chains face each other across a central cavity (Figures 5·52 and 5·53). Each of the two α chains consists of 141 amino acid residues; each of the two β chains consists of 146 residues. The tertiary structure of each of the four chains is almost identical to that of myoglobin. The striking similarity in conformation between a myoglobin molecule and a hemoglobin chain can be seen by comparing Figures 5·35a (myoglobin) and 5·54 (hemoglobin). In fact, Perutz described hemoglobin as "just four myoglobin molecules put together." Hemoglobin, however, is not simply a tetramer of myoglobin molecules. Because an α chain interacts with a β chain much more strongly than α interacts with α or β with β, hemoglobin is actually a dimer of $\alpha\beta$ subunits. In addition, as we will see below, the presence of quaternary structure is responsible for oxygen-binding properties not possible with single-chain myoglobin.

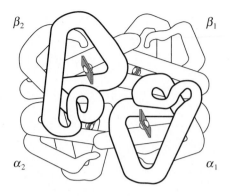

Figure 5·52
Hemoglobin tetramer. The heme groups are shown as red squares. [Adapted from Dickerson, R. E., and Geis, I. (1969). *The Structure and Action of Proteins* (Menlo Park, California: Benjamin/Cummings Publishing Company), p. 56.]

Figure 5·53
Stereo view of the hemoglobin tetramer. Color key: α_1, green; α_2, yellow; β_1, purple; β_2, blue; heme group, red. (Based on coordinates provided by M. Perutz and G. Fermi.)

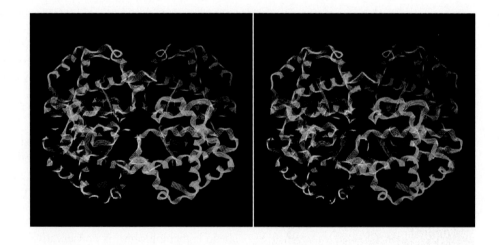

Figure 5·54
Stereo view of the α subunit of horse hemoglobin. The heme group is shown in red. (Based on coordinates provided by B. Shaanan.)

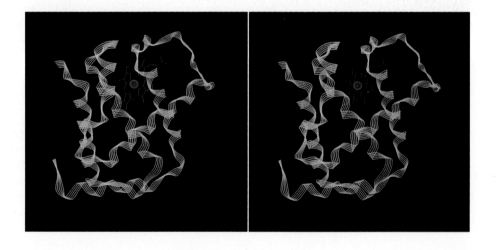

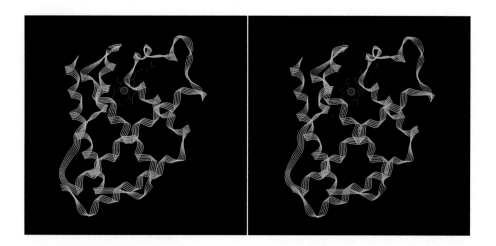

Figure 5·55
Stereo view of leghemoglobin. The heme group is shown in red. (Based on coordinates provided by B. K. Vainshtein and E. H. Harutyunyan.)

The structures of globins in many species are very similar. For example, leghemoglobin, an oxygen-binding monomeric protein found in leguminous plants (Figure 5·55), has a three-dimensional structure much the same as mammalian myoglobin. The determination of the amino acid sequences of globins from dozens of species has revealed the evolutionary relationships of this family of globin genes. For example, the mammalian α and β globins of hemoglobin are more similar to each other than either one is to the globin of myoglobin. This observation suggests that there was a duplication of the globin gene in early ancestors of mammals; one gene gave rise to modern myoglobin and the other to α and β globins. A more recent duplication gave rise to the separate α-globin and β-globin genes. Some invariant residues in the primary structures of globin molecules can be related to their function. For example, the globins of both hemoglobin and myoglobin contain the proximal histidine residue (the fifth ligand to the iron of the heme) and the distal histidine residue (near the oxygen-binding site). Similarly, the phenylalanine residue that contributes to the hydrophobic cage surrounding the heme is invariant in all known vertebrate globins.

An evolutionary relationship based on differences in amino acid sequence inadequately conveys the similarity of tertiary structures because the substitution of one residue by a similar residue may have little effect on the conformation and function of a polypeptide chain. Substitutions that involve similar amino acids are said to be conservative. For example, substitution of a valine (present in most globins) by isoleucine (in the α chains of kangaroo hemoglobin) or by leucine (in soybean leghemoglobin) is conservative. Conservative substitutions are most often found in conserved regions of proteins. In contrast, substitution of a hydrophilic residue (such as glutamate) by a hydrophobic one (such as valine) is nonconservative. Nonconservative substitutions may have a profound effect on the conformation or function of a protein, as is the case with persons suffering from sickle-cell anemia (Section 5·19).

5·16 Oxygen Binds Reversibly to Myoglobin and Hemoglobin

Let us now compare the mechanisms of oxygen binding to hemoglobin and myoglobin. Oxygen binds to the heme prosthetic group, which is the same in the two proteins. However, hemoglobin, which transports oxygen in the blood of vertebrates, is a tetramer; myoglobin, which stores oxygen and facilitates its diffusion within muscle, is a monomer. We will first examine oxygen binding to the heme molecule.

As we saw earlier, nearly all polar residues are located on the surface of the globin molecule. The interior is composed almost entirely of nonpolar residues, with the exception of two histidine residues, His-64 and His-93. We will refer to myoglobin in the following discussion; the same principles apply to hemoglobin. The state of oxygenation of hemoglobin, like myoglobin, is designated by the prefixes *oxy-* and *deoxy-*. The two anionic carboxyl groups of the propionate substituents of heme are exposed on the surface of the globin molecule and are thus hydrated, whereas the nonpolar methyl and vinyl groups are buried in the hydrophobic interior. In oxymyoglobin, the ferrous iron is coordinated by six ligands, making the complex octahedral (Figure 5·56). Four of the ligands of the iron in oxymyoglobin are the four nitrogen atoms of the tetrapyrrole ring system; the fifth ligand is an imidazole nitrogen from His-93 (referred to as the proximal, or near, histidine); and the sixth is molecular oxygen bound between the iron and the imidazole side chain of His-64 (referred to as the distal, or distant, histidine, since it is slightly too far from the iron to be part of its coordination sphere). In deoxymyoglobin, the oxygen-binding position of the iron is not occupied. The nonpolar side chains of Val-68 and Phe-43, shown in Figure 5·57, contribute to the steric constraint and hydrophobicity of the oxygen-binding pocket and help hold the heme group in place. For example, the valine side chain discriminates against the binding of the toxic compound carbon monoxide (CO), apparently by steric hindrance. X-ray analysis has shown that the side chain of His-64 blocks the entrance to the heme-containing pocket in both oxymyoglobin and deoxymyoglobin. Therefore, the side chains that block the entrance to this pocket must move rapidly (breathe) to allow oxygen to bind, a conclusion supported by molecular-dynamics calculations and other indirect experiments.

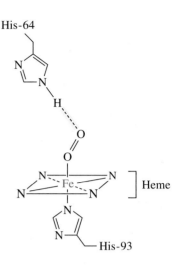

Figure 5·56
Oxygen-binding site of oxymyoglobin. The heme prosthetic group is represented by a parallelogram with a nitrogen atom at each corner.

Figure 5·57
Stereo view of the oxygen-binding site of oxymyoglobin. Fe☺ (red dot), lying almost in the heme plane (also red), is bound to oxygen (pink) and is flanked by His-64 (green) and His-93 (orange). Val-68 (dark blue) and Phe-43 (yellow) contribute to the hydrophobic environment of the oxygen-binding site. The backbones of neighboring residues are shown in light blue. (a) Stick model. (b) Space-filling model. (Based on coordinates provided by T. Takano.)

(a)

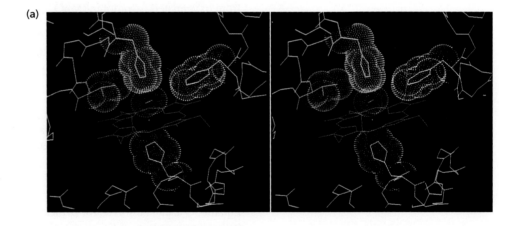

(b)

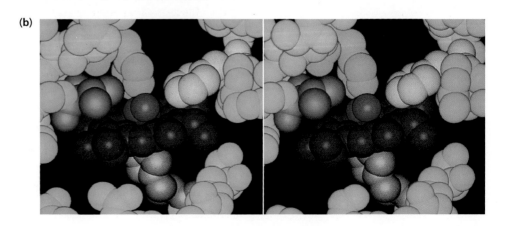

The hydrophobic crevice of the globin polypeptide holds the key to the ability of myoglobin (and hemoglobin) to suitably bind and release oxygen. In aqueous solution, free heme does not reversibly bind oxygen; instead, the Fe^{2+} of the heme is almost instantly oxidized to Fe^{3+}. The structure of myoglobin and hemoglobin prevents formal transfer of an electron and precludes irreversible oxidation, thereby assuring the reversible binding of molecular oxygen for transport. In fact, the ferrous iron atom of heme in hemoglobin is partially oxidized when O_2 is bound. An electron is transferred toward the oxygen atom that is attached to the iron, so that the molecule of dioxygen is partially reduced. If the electron were transferred completely to the oxygen, the complex of oxyhemoglobin would be Fe^{3+}—O_2^{-} (a superoxide anion attached to ferric iron). The globin cage prevents complete electron transfer and enforces return of the electron to the iron atom when O_2 dissociates.

5·17 Myoglobin and Hemoglobin Have Different Oxygen-Binding Curves

The physiological functions of myoglobin and hemoglobin depend on reversible binding of oxygen to the heme prosthetic groups, a process that can be depicted by oxygen-binding curves. Biochemists usually measure the strength of ligand binding to a macromolecule as the equilibrium constant for its dissociation from the macromolecule.

The equilibrium constant (K_{diss}) for the dissociation of molecular oxygen (O_2) from oxygenated myoglobin (MbO_2)

$$MbO_2 \; \underset{\longleftarrow}{\overset{K_{diss}}{\longrightarrow}} \; Mb \; + \; O_2 \qquad (5·1)$$

is calculated as

$$K_{diss} = \frac{[Mb][O_2]}{[MbO_2]} \qquad (5·2)$$

The fractional saturation (Y) of myoglobin is the fraction of the total number of myoglobin molecules that is oxygenated.

$$Y = \frac{[MbO_2]}{[MbO_2] + [Mb]} \qquad (5·3)$$

Equation 5·2 can be solved for $[MbO_2]$ and this value substituted into Equation 5·3. If the concentration of O_2 is measured as the partial pressure of gaseous oxygen, pO_2, an equation that describes Y in terms of pO_2 only can be obtained.

$$Y = \frac{pO_2}{pO_2 + K_{diss}} \qquad (5·4)$$

At half-saturation, when Y = 0.5,

$$0.5 = \frac{pO_2}{pO_2 + K_{diss}} \qquad (5·5)$$

the constant for dissociation of oxygen from the MbO_2 complex becomes another constant, the half-saturation pressure, P_{50}.

$$K_{diss} = (pO_2)_{0.5} = P_{50} \qquad (5·6)$$

A low value for P_{50} indicates high affinity for oxygen; a high P_{50} signifies low affinity.

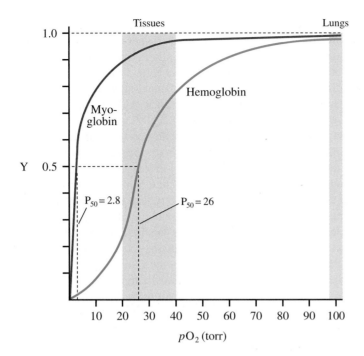

Figure 5·58
Oxygen-binding curves of myoglobin and hemoglobin. The plot shows the fractional saturation, Y, of each protein versus the concentration of oxygen measured as the partial pressure, pO_2 (torr). When Y = 0.5, the protein is half-saturated with oxygen. The oxygen-binding curve of myoglobin is hyperbolic, with half-saturation at an oxygen pressure of 2.8 torr. The oxygen-binding curve of hemoglobin is sigmoidal, with half-saturation at an oxygen pressure of 26 torr. The sigmoidal shape of the oxygen-binding curve indicates that there is positive cooperativity in the binding of oxygen by hemoglobin.

A comparison of the two plots shows that myoglobin has a greater affinity than hemoglobin for oxygen at all oxygen pressures. Nevertheless, in the lungs, where the oxygen pressure is high, hemoglobin is nearly saturated with oxygen. In tissues, where the partial pressure of oxygen is low, oxygen is released from oxygenated hemoglobin and transferred to myoglobin.

When the fractional saturation of a solution containing myoglobin is plotted versus pO_2, a hyperbolic curve described by Equation 5·4 is obtained (Figure 5·58). The hyperbolic shape of the plot indicates that there is a single equilibrium constant for the binding of the ligand to the macromolecule. Whereas only one molecule of oxygen binds to each molecule of myoglobin, up to four molecules of oxygen bind to hemoglobin, one per heme group of the tetrameric protein. The equation that describes oxygen binding to hemoglobin is more complicated than the equation for myoglobin. Binding of more than one molecule of ligand to a macromolecule often results in a sigmoidal (S-shaped) plot of fractional saturation versus concentration of ligand. The sigmoidal curve describing oxygen binding to hemoglobin (Figure 5·58) indicates that the oxygen-binding sites of hemoglobin interact such that binding of one molecule of oxygen to one heme group is not favored but, once it occurs, it facilitates binding of additional molecules of oxygen to the other three hemes. The oxygen affinity of hemoglobin increases as each oxygen molecule is bound. This interactive binding phenomenon is termed **positive cooperativity of binding.** Similarly, loss of oxygen from one heme of fully oxygenated hemoglobin decreases the affinity of the remaining hemes for oxygen.

The myoglobin molecules are half-saturated at a pO_2 of 2.8 torr (1 torr = 133.3 pascal; 1 atmosphere = 760 torr). The P_{50} for myoglobin is much lower than the P_{50} for hemoglobin (26 torr) under similar conditions, reflecting the higher affinity of myoglobin for oxygen. The physiological roles of myoglobin and hemoglobin are directly related to their relative affinities for oxygen at low oxygen pressures. As Figure 5·58 shows, at the high pO_2 found in the lungs (about 100 torr), myoglobin and hemoglobin have a high affinity for oxygen and both are nearly saturated. However, at all pO_2 values below ~50 torr, myoglobin has a significantly greater affinity than hemoglobin for oxygen. Within the capillaries of tissues such as muscles, where pO_2 is low (ranging from 20 to 40 torr), much of the oxygen carried by hemoglobin in erythrocytes is released because of the lower affinity of hemoglobin for oxygen. At this concentration of oxygen, myoglobin in the muscle tissue binds the oxygen released by hemoglobin. The differential affinities of myoglobin and hemoglobin for oxygen thus lead to an efficient system for oxygen delivery from the lungs to muscle.

The same heme prosthetic group is present in myoglobin and hemoglobin, but its affinity for oxygen differs because the environments provided by the protein cages of myoglobin and hemoglobin are slightly different. Protein environments can also change the chemical pathways of a cofactor-requiring reaction. As we will see in our discussion of cytochromes in Chapter 18, under the influence of different apoproteins, the ionic iron of heme undergoes oxidation and reduction instead of oxygenation, cycling between the ferrous (Fe^{2+}) and ferric (Fe^{3+}) states. During reactions catalyzed by peroxidases, the ferric iron of the heme cofactor changes to a more oxidized state. The reaction specificity for these very different types of reactions resides in the protein moiety of the holoprotein.

The cooperative binding of oxygen by hemoglobin can be related to changes in the protein conformation that occur upon oxygenation. Deoxyhemoglobin is stabilized by intra- and intersubunit ion pairs involving the C-terminal residues of each chain. These ion pairs are not present in oxyhemoglobin. Binding of oxygen causes a significant conformational change that disrupts these ion pairs and favors a conformation that has a higher affinity for oxygen. This change is triggered by the behavior of the iron atom of the heme. In deoxyhemoglobin, the iron—binding only five ligands—is somewhat above the plane of the porphyrin ring (about 0.06 nm toward the proximal histidine). When O_2—a sixth ligand—binds, the electronic structure of the iron changes, bringing the iron into the plane of the porphyrin ring (Figure 5·59). The bond between the iron and the proximal histidine residue becomes shorter, pulling the entire helix that contains this histidine. Interactions between the globin chains resist this movement, and it appears that it is only after at least one oxygen molecule binds to each $\alpha\beta$ subunit that the tertiary and quaternary structures change from the deoxy to the oxy conformation. (The regulation and pathway of oxygen binding to hemoglobin chains are discussed further in the next section and in Section 6·13.) The differences in quaternary structure of deoxyhemoglobin and oxyhemoglobin are illustrated in Figure 5·60. One $\alpha\beta$ subunit of hemoglobin rotates about 15° with respect to the other $\alpha\beta$ subunit. Release of the oxygen molecules allows the hemoglobin molecule to re-form the ion pairs and resume the deoxy conformation. The positive cooperativity of oxygen binding by hemoglobin thus promotes its full oxygenation at the high pO_2 of the lungs and the efficient unloading or dissociation of oxygen at the low pO_2 of the capillary beds of other tissues.

Figure 5·59
Conformational changes induced by oxygenation. When the heme iron of hemoglobin is oxygenated, the proximal histidine residue is pulled toward the porphyrin ring. The remainder of the helix containing the histidine also shifts position, disrupting ion pairs that cross-link the subunits of deoxyhemoglobin. Deoxyhemoglobin is shown in light blue; oxyhemoglobin, in dark blue. [Adapted from Baldwin, J., and Chothia, C. (1979). Haemoglobin: the structural changes related to ligand binding and its allosteric mechanism. *J. Mol. Biol.* 129:175–220.]

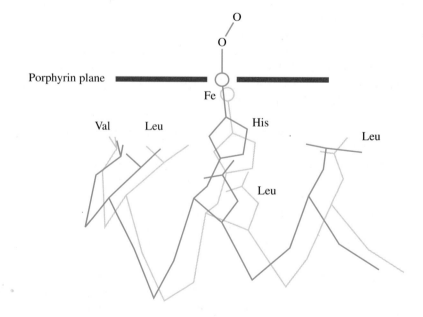

Figure 5·60
Oxygen-induced change in the quaternary structure of hemoglobin. When oxygen binds to both $\alpha\beta$ subunits, the disruption of ion pairs causes one $\alpha\beta$ subunit ($\alpha_1\beta_1$) to rotate 15° relative to the other subunit ($\alpha_2\beta_2$), leading to an increase in the oxygen-binding affinity of the remaining unoxygenated heme groups. Deoxygenated $\alpha_1\beta_1$, light blue; oxygenated $\alpha_1\beta_1$, dark blue.

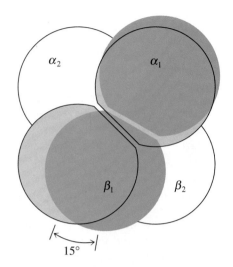

5·18 Hemoglobin Is an Allosteric Protein

In addition to positive cooperativity of binding, **allosteric interactions** (from the Greek, other site) are important in regulation of the binding and release of oxygen by hemoglobin. In this respect, hemoglobin—a transport protein, not an enzyme—resembles certain regulatory enzymes (Section 6·12). Allosteric interactions occur when a specific small molecule, called an **allosteric effector** or modulator, binds to a protein (usually an enzyme) and modulates its activity. The allosteric effector binds reversibly at a site separate from the functional binding site of the protein (i.e., at a binding site or cleft other than the active site of an enzyme). Binding of an allosteric effector transmits information about effector concentration to the functional portion of the protein. A protein whose activity is modulated by allosteric effectors is called an allosteric protein. An effector may be an allosteric activator or an allosteric inhibitor, depending on its effect on the allosteric protein.

Allosteric regulation is accomplished by small but significant changes in the native conformations of allosteric proteins. The binding of an allosteric inhibitor causes the allosteric protein to rapidly change from its active shape (the R state) to its inactive shape (the T state). The binding of an allosteric activator causes the reverse change. The R and T states are in dynamic equilibrium. The change in conformation of an allosteric protein caused by binding or release of an effector, termed the **allosteric transition,** extends from the allosteric site to the functional binding site. A substrate is bound most avidly when the enzyme is in the R state. The activity level of an allosteric protein depends on the relative proportions of molecules in the R and T forms, and these in turn depend on the relative concentrations of the substrates, effectors, or other ligands that bind to each form. In hemoglobin, the deoxy conformation, which resists oxygen binding, is considered the inactive (T) state and the oxy conformation, which facilitates oxygen binding, is considered the active (R) state. The term *T state* was used originally to describe a tense, or taut, state in which the protein was inactive; *R state* referred to the relaxed, active state. Although the terms *tense* and *relaxed* may not strictly apply to all allosteric proteins, *T* and *R* are still used to designate inactive and active forms.

The molecule 2,3-*bis*phospho-D-glycerate (2,3BPG), shown in Figure 5·61, is an allosteric effector of hemoglobin within erythrocytes. The presence of 2,3BPG raises the P_{50} for binding of oxygen to adult hemoglobin in whole blood to about 26 torr—much higher than the P_{50} for oxygen binding to pure hemoglobin in aqueous solution, about 12 torr. In other words, 2,3BPG in erythrocytes substantially lowers the affinity of deoxyhemoglobin for oxygen. The concentrations of 2,3BPG and hemoglobin within erythrocytes are nearly equal (about 4.7 mM).

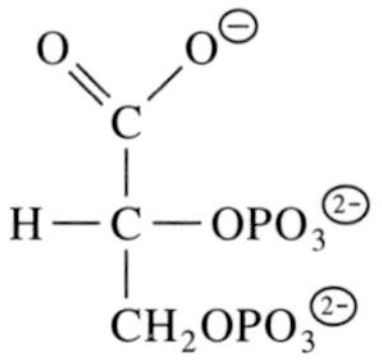

Figure 5·61
2,3-*Bis*phospho-D-glycerate (2,3BPG).

Figure 5·62
Binding of 2,3BPG to deoxyhemoglobin. The central cavity of deoxyhemoglobin is lined with positively charged groups that are complementary to the carboxylate and phosphate groups of 2,3BPG. Both 2,3BPG and the ion pairs shown help stabilize the deoxy conformation. [Adapted from Dickerson, R. E. (1972). X-ray studies of protein mechanisms. *Annu. Rev. Biochem.* 41:815–842.]

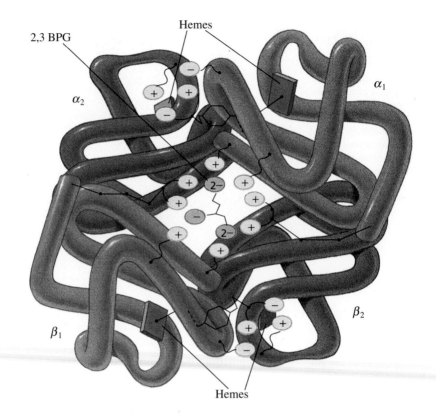

In the central cavity of hemoglobin, between the two β chains, six positively charged side chains and the N-terminal α-amino group of each β chain form a cationic binding site (Figure 5·62). In the deoxy conformation of hemoglobin, these positively charged groups are complementary in position to the five negative charges of 2,3BPG, which electrostatically interacts with these residues. When 2,3BPG is bound, the deoxy conformation is stabilized. In oxyhemoglobin, the β chains are closer together and the allosteric binding site is too small to accommodate 2,3BPG, which dissociates and is expelled from the protein upon oxygenation. Oxygen and 2,3BPG have opposite effects on the R $\rightleftharpoons$ T equilibrium. Binding of O_2 increases the proportion of hemoglobin molecules in the oxy (R) conformation, and binding of 2,3BPG increases the proportion of hemoglobin molecules in the deoxy (T) conformation. Because oxygen and 2,3BPG have different binding sites, 2,3BPG is called an allosteric effector.

2,3BPG has an important physiological role. In the absence of 2,3BPG, hemoglobin is nearly saturated at an oxygen pressure of about 20 torr. Thus, at the low partial pressure of oxygen that prevails in the tissues (20–40 torr), hemoglobin without 2,3BPG would not unload its oxygen to myoglobin. In the presence of equimolar 2,3BPG, however, hemoglobin is only about one-third saturated at 20 torr. The allosteric effect of 2,3BPG allows hemoglobin to transfer oxygen to myoglobin at the low partial pressures of oxygen in the tissues.

In mammals, the regulation of hemoglobin oxygenation by 2,3BPG has a special role in the delivery of oxygen to the fetus from the placenta. A unique hemoglobin, designated hemoglobin F (Hb F), is produced in fetuses. Like adult hemoglobin, Hb F is tetrameric and contains two α chains. However, the other two chains are γ globins, not the β globins found in adult hemoglobin. In γ chains, His-143 of β globin is replaced by serine. Consequently, Hb F has two fewer positive charges in its central cavity, giving fetal hemoglobin a lower affinity for 2,3BPG and a

Figure 5·63
Bohr effect. Lowering the pH decreases the affinity of hemoglobin for oxygen.

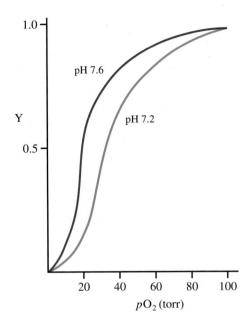

greater affinity for oxygen (lower P_{50}) than maternal hemoglobin. The higher affinity for oxygen of Hb F facilitates transfer of O_2 from the maternal circulation to fetal tissues. Shortly before birth, the synthesis of β globin sharply increases and that of γ globin sharply decreases. Normally, by the end of the first year, γ globin synthesis has completely ceased.

Additional regulation of the binding of oxygen to hemoglobin involves carbon dioxide and protons, both of which are products of aerobic metabolism. Early in this century, Christian Bohr (father of the celebrated physicist Niels Bohr) observed that CO_2 decreases the affinity of hemoglobin for oxygen. Carbon dioxide does this indirectly by lowering the pH inside the blood cells. Enzyme-catalyzed hydration of CO_2 within red blood cells produces carbonic acid (H_2CO_3), a weak acid that dissociates to form bicarbonate and a proton, thereby lowering pH. The lower pH leads to protonation of several groups in hemoglobin that can then form ion pairs that help stabilize the deoxy conformation. The increase in the concentration of carbon dioxide and the concomitant decrease in pH raise the P_{50} of hemoglobin (Figure 5·63). This phenomenon, called the **Bohr effect,** increases the efficiency of the oxygen-delivery system: in the inhaling lung, where the carbon dioxide level is low, oxygen is readily picked up by hemoglobin, which simultaneously releases protons; in metabolizing tissues, where the carbon dioxide level is relatively high and the pH is relatively low, O_2 is readily unloaded from oxyhemoglobin.

Carbon dioxide is transported from the tissues to the lungs in two ways. Most of the CO_2 produced by metabolism is transported as dissolved bicarbonate ions to the lungs, where it reassociates with protons (some of which are released from hemoglobin) and is exhaled as CO_2. Some carbon dioxide, however, is carried via hemoglobin itself, in the form of carbamate adducts (Figure 5·64). At the pH of blood cells (7.2) and at high concentrations of CO_2, the unprotonated amino groups of the four N-terminal α-amino acid residues of deoxyhemoglobin (pK_a values between 7 and 8) can react reversibly with CO_2 to form carbamate adducts. The carbamates of oxyhemoglobin are less stable than those of deoxyhemoglobin. When hemoglobin reaches the lungs, where the partial pressure of CO_2 is low and the partial pressure of O_2 is high, hemoglobin is converted to its oxygenated state, and the CO_2 that was bound is released.

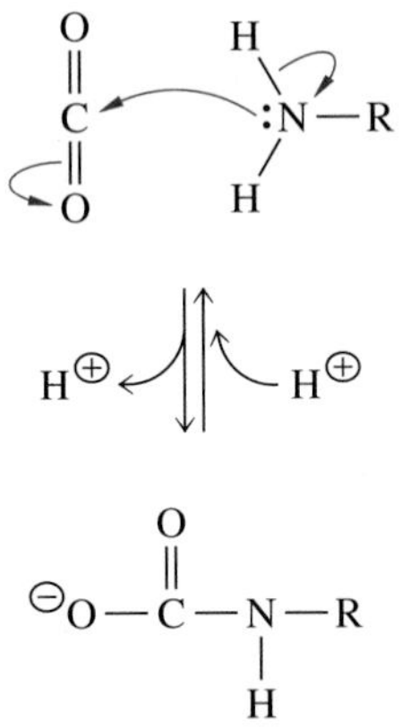

Figure 5·64
Carbamate adduct. Carbon dioxide produced by metabolizing tissues can react reversibly with the N-terminal residues of the globin chains of hemoglobin, converting them to carbamate adducts.

Figure 5·65
Scanning electron micrograph of normal (N) and sickled (S) red blood cells. (Courtesy of Susan Shyne.)

5·19 Sickle-Cell Anemia Is a Molecular Disease

In 1904, a severely anemic student visited a Chicago physician named James Herrick. Examining the student's red blood cells under a microscope, Herrick discovered that among the normal cells were many that had a highly unusual sickle shape (Figure 5·65). He called the abnormal blood condition sickle-cell anemia. Unlike normal erythrocytes, sickled cells cannot pass efficiently through the capillaries. Consequently, circulation is impaired and serious tissue damage can occur. Moreover, sickled cells tend to rupture, resulting in fewer red blood cells (anemia).

It took over four decades to establish that sickle-cell anemia is the result of a molecular alteration in hemoglobin transmitted by a mutated recessive gene on an autosomal (nonsex) chromosome. Individuals who inherit two mutated genes are homozygous and develop sickle-cell anemia, whereas those who inherit one mutated gene and one normal gene are heterozygous and are said to have sickle-cell trait. Individuals with sickle-cell trait do not generally suffer from the symptoms of sickle-cell anemia.

The first clue to the nature of the molecular alteration of sickle-cell hemoglobin (Hb S) was obtained by Linus Pauling and his coworkers, who used electrophoresis to compare Hb S with normal adult hemoglobin, Hb A. The results indicated that Hb S has a greater net positive charge than Hb A. Sequence analysis by Vernon Ingram revealed that the β chains of Hb S molecules contain a nonpolar Val-6 in place of the negatively charged Glu-6 of the β chains of Hb A. This substitution, caused by a single nucleotide substitution in the gene encoding the β chain, accounts for the difference in ionic charge and is responsible for the disease.

In Hb A, the anionic Glu-6 of the β chain lies on the surface of hemoglobin away from either the oxygen-binding or 2,3BPG-binding sites. Why should replacement of this glutamate residue by a valine residue have such profound consequences? The concentration of hemoglobin in red blood cells is so high that it is on the verge of crystallization even under normal conditions. In the deoxy conformation of Hb S, Val-6 on each of the two α chains forms a hydrophobic contact with a pocket in a neighboring Hb S molecule (in the oxy conformation, the pocket is inaccessible). This interaction leads to polymerization of deoxy Hb S at low oxygen pressures typical of capillary beds in metabolically active tissues. The resulting double-stranded polymers aggregate into long helical fibers containing 14 to 16

Figure 5·66
Structure of a deoxyhemoglobin S fiber. Each circle represents a hemoglobin molecule. Strands of hemoglobin tetramers are twisted into helical fibers of 14 to 16 strands. A cutaway view shows an inner core of 4 strands surrounded by 10 outer strands.

strands (Figure 5·66). These insoluble aggregates deform the red blood cells into the sickle shape. Sickle-cell crisis results when the deformed erythrocytes block small blood vessels, causing oxygen deprivation that can lead to permanent tissue damage and even death. Heterozygotes do not develop the disease because the rate of fiber formation in heterozygotes is about 1000 times slower than that in homozygotes.

Sickle-cell trait and sickle-cell anemia are most prevalent in populations of humans who live in or whose origins are the tropics, where malaria is prevalent. Individuals who carry the sickle-cell trait have increased resistance to malaria and, as a consequence, constitute a larger portion of the population. In regions of Africa where the incidence of malaria is particularly high, 20% or more of the population carry the sickle-cell trait. Individuals with the sickle-cell gene are resistant to malaria because the malarial parasite spends part of its life cycle in the red blood cell, and the alteration in red blood cells in persons with the sickle-cell gene prevents the malarial parasite from thriving. Heterozygotes, with only one sickle-cell gene, thus have an advantage over homozygotes with the sickle-cell gene and homozygotes with two Hb A β genes since they do not suffer severely from either sickle-cell anemia or malaria.

5·20 Antibodies Recognize Other Molecules by Their Complementary Shapes

Vertebrates possess a complex immune system that eliminates foreign substances, including infecting organisms. As part of this defense system, vertebrates synthesize proteins called **antibodies,** also known as immunoglobulins, that specifically recognize foreign compounds, called **antigens.** Antigens may be proteins, polysaccharides, or nucleic acids. Few small molecules are antigens, but antibodies to small molecules can be elicited by attaching the small molecules to a biopolymer carrier. Immunization, or exposure to antigen, leads to the immune response, in which a large number of antibodies directed against different regions on the surface of the antigen are produced. Antibodies are glycoproteins synthesized by white blood cells called lymphocytes; each lymphocyte and its descendants synthesize the same antibody. Because animals are exposed to many foreign substances over their lifetime, they develop a huge array of antibody-producing lymphocytes that persist for many years and can later respond to the antigen during reinfection. The "memory" of the immune system is the reason why certain infections do not recur in an individual despite repeated exposure. Vaccines administered to children are effective because immunity established in childhood lasts into adulthood.

When an antigen—either novel or previously encountered—binds to antibodies on the surface of lymphocytes, these cells are stimulated to proliferate and produce soluble antibodies for secretion into the bloodstream. The soluble antibodies bind to the foreign organism or substance, forming antibody-antigen complexes that precipitate and mark the antigen for destruction by a series of interacting proteases or by lymphocytes that engulf the antigen and digest it intracellularly.

The most abundant antibodies in the bloodstream are of the immunoglobulin G class, Y-shaped oligomers composed of four polypeptide chains: two identical light chains and two identical heavy chains (Figure 5·67). Carbohydrate is attached to the heavy chains. The chains are connected by disulfide bonds so that the N-termini

(a)

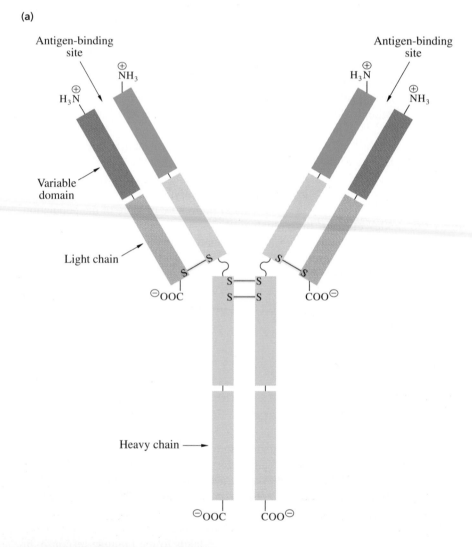

Figure 5·67
Antibody structure. **(a)** The two heavy chains (blue) and two light chains (red) of antibodies of the immunoglobulin G class consist of similar domains. The chains are joined by disulfide bonds (yellow). The variable domains of both the light and heavy chains (where antigen binds) are colored more darkly. **(b)** Stereo view of an antibody. Note how the subunits twist around each other. (Courtesy of Alexander McPherson.)

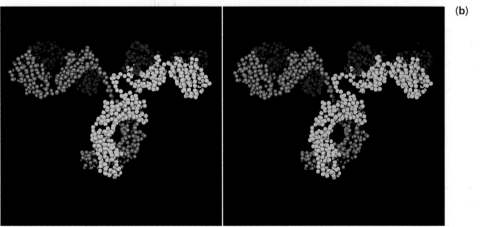

Figure 5·68
Immunoglobulin fold of the variable domain of a light chain. Two antiparallel β sheets lie close to each other. The three hypervariable loops at the N-terminus are colored red.

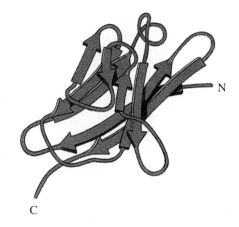

of pairs of light and heavy chains are close together. Each chain is composed of domains of about 110 amino acid residues; light chains contain two domains, and heavy chains contain four domains. All of the domains have a similar supersecondary structure termed the immunoglobulin fold, which consists of a pair of antiparallel β sheets (Figure 5·68). The immunoglobulin fold can also be seen in the stereo view of a domain from the Bence-Jones protein (Figure 5·23), which is a dimer of light chains.

Determination of amino acid sequences has shown that the N-terminal domains of antibodies, termed the variable domains because of their sequence diversity, determine the specificity of antigen binding. X-ray crystallographic experiments have provided additional details about antigen-antibody interactions. The antigen-binding sites of the variable domains are three loops, called hypervariable regions, that show great variation in size and sequence. The loops from both a light chain and a heavy chain combine to form a barrel. The top of the barrel is a surface that is complementary to the shape and polarity of a specific antigen. The match between the antigen and antibody is so close that there is no space for water molecules between the two. The forces that stabilize the interaction of antigen with antibody are noncovalent; there are numerous hydrogen bonds and electrostatic interactions because the antibody is complementary to the polar surface of the antigen. An example of the interaction of antibodies with an antigen is shown in Figure 5·69.

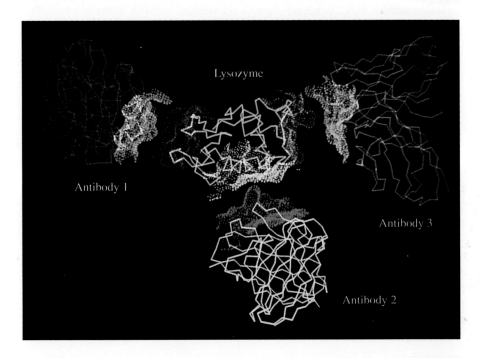

Figure 5·69
Binding of three different antibodies to an antigen (the protein lysozyme). The structures of the three antigen-antibody complexes have been determined by X-ray crystallography. This composite view, in which the antigen and antibody have been separated, shows the surfaces on the antigen and antibody that interact. Only parts of the three antibodies are shown. (Courtesy of David R. Davies.)

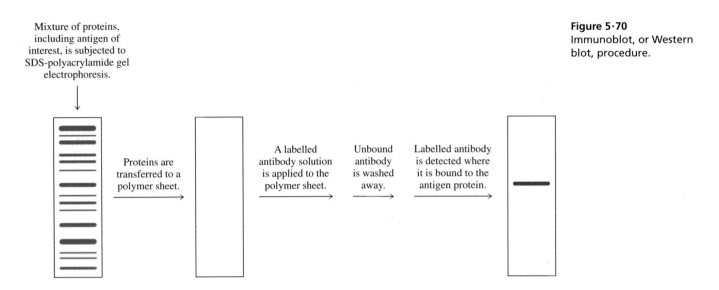

Mixture of proteins, including antigen of interest, is subjected to SDS-polyacrylamide gel electrophoresis.

Proteins are transferred to a polymer sheet.

A labelled antibody solution is applied to the polymer sheet.

Unbound antibody is washed away.

Labelled antibody is detected where it is bound to the antigen protein.

Figure 5·70
Immunoblot, or Western blot, procedure.

Because of their remarkable antigen-binding specificity, antibodies are used in the laboratory for the detection or isolation of small quantities of a biopolymer. In a common type of immunoassay, fluid containing an unknown amount of antigen is mixed with a solution of labelled antibody, and the amount of antibody-antigen complex formed is measured. The sensitivity of these assays, sometimes enhanced by a variety of procedures, makes them suitable for diagnostic tests. For example, immunoassay of a woman's urine for trace amounts of the hormone human chorionic gonadotropin can be used to detect pregnancy within a few days of conception. In a procedure called an immunoblot, or Western blot, a specific protein can be detected in a mixture that has been separated by SDS-polyacrylamide gel electrophoresis. Following electrophoretic separation, proteins are transferred to a sheet of polymer to which they bind avidly. The polymer is then treated with a labelled antibody specific for the desired protein. After unbound antibody is washed away, the antibody remaining bound to the antigen is located (Figure 5·70). Antibodies can also be used as specific ligands for purification of a protein by affinity chromatography. A solution containing a mixture of proteins is applied to a column of matrix-bound antibodies. Next, unbound proteins are washed out with buffer solution. Finally, the desired protein is eluted by increasing the salt concentration or changing the pH to disrupt the antigen-antibody interaction.

Summary

Proteins can be classified as fibrous or globular. Fibrous proteins are generally water insoluble, physically tough, and built of repetitive structures. They usually have static functions. Globular proteins, most of which are soluble in aqueous solutions, are compact, roughly spherical macromolecules whose polypeptide chains are tightly folded. They act dynamically by transiently binding specific molecules.

Adjacent amino acid residues are joined by peptide bonds. Peptide bonds are polar and planar and have some double-bond character. Because of steric restraints, peptide groups are usually in the *trans* conformation. Rotation around the $N-C_\alpha$ and $C_\alpha-C$ bonds gives polypeptide chains conformational flexibility.

Proteins may have up to four levels of structure: primary (sequence of amino acid residues), secondary (local conformation, stabilized by hydrogen bonds), tertiary (the compacted globular structure of an entire polypeptide chain), and quaternary (assembly of two or more polypeptide chains into a multisubunit protein). The highly complex three-dimensional structures of proteins, many of which have been determined by X-ray crystallography, must be preserved to maintain biological activity.

The right-handed α helix is a common secondary structure found in some fibrous and globular proteins. It contains 3.6 amino acid residues per turn and has a pitch of 0.54 nm. The other major type of secondary structure is the β sheet, either parallel or antiparallel, in which the polypeptide chain is extended.

A different helical structure is found in collagen, a fibrous protein of connective tissue. A collagen molecule consists of three left-handed polypeptide helices intertwined to form a right-handed supercoil. Interchain hydrogen bonding and covalent cross-linking through modification of proline and lysine residues stabilizes the protein.

The elastic protein elastin can rapidly extend and relax. Its strong cross-linked structural network is required for the proper functions of organs such as heart and lungs.

Folding of a globular protein into its biologically active state is a sequential, cooperative process involving the hydrophobic effect and, to a lesser extent, hydrogen bonding, van der Waals interactions, and ion pairing. In cells, enzymes and chaperones assist folding. The native conformation of a protein can be disrupted by the addition of denaturing agents.

The compact, folded structures of globular proteins allow them to selectively bind other molecules. For example, the globular structures of the heme-containing proteins myoglobin and hemoglobin allow these proteins to bind and release oxygen in a manner that facilitates delivery of oxygen to respiring tissues.

A monomeric protein, myoglobin contains a single polypeptide chain of 153 residues that is folded into a compact globular structure composed of eight α helices. Its heme prosthetic group, which binds oxygen, is shielded from water in a cleft, or hydrophobic cage, formed by the protein.

Most globular proteins have considerable stretches of residues in nonrepeating conformations. These regions include turns and loops needed to connect α helices and β strands. The secondary structural elements are often connected into recognizable combinations called supersecondary structures, or motifs. Larger globular units called domains, or lobes, are usually associated with a particular function.

Hemoglobin consists of four chains (two α and two β chains in adult hemoglobin), each similar to the globin chain of myoglobin. The deoxy (T) and oxy (R) conformations of hemoglobin differ in their affinity for oxygen. Due to structural interactions associated with its tertiary and quaternary structure, hemoglobin displays positive cooperativity in the binding of oxygen and is subject to allosteric regulation.

Slight differences in the primary structures of hemoglobin molecules can result in significant functional differences: substitution of a serine residue in fetal hemoglobin (Hb F) for a histidine residue found in normal adult hemoglobin (Hb A) increases the affinity of Hb F for oxygen, thereby increasing the efficiency of oxygen delivery from maternal blood to the fetus. A mutation leading to replacement of a glutamate residue at position 6 in the β chains of normal adult hemoglobin with a valine residue results in Hb S, the hemoglobin responsible for sickle-cell anemia.

Antibodies are multidomain proteins that bind foreign substances. The variable domains at the ends of the heavy and light chains of the antibody interact with the antigen. Immunoassays take advantage of the binding specificity of antibodies.

Selected Readings

General References

Creighton, T. E. (1993). *Proteins: Structures and Molecular Properties,* 2nd ed. (New York: W. H. Freeman and Company). Chapters 4–7.

Doolittle, R. F. (1985). Proteins. *Sci. Am.* 253(4):88–99. A brief but clear review of proteins and their binding of ligands.

Goodsell, D., and Olson, A. J. (1993). Soluble proteins: size, shape, and function. *Trends Biochem. Sci.* 18:65–68.

X-Ray Crystallography of Proteins

Kendrew, J. C. (1961). The three-dimensional structure of a protein molecule. *Sci. Am.* 205(6):96–110. Describes the techniques used in determining the structure of myoglobin.

McPherson, A. (1989). Macromolecular crystals. *Sci. Am.* 260(3):62–69. Describes the growth of crystals, the limiting factor in X-ray diffraction studies with macromolecules.

α Helix

Pauling, L., Corey, R. B., and Branson, H. R. (1951). The structure of proteins: two hydrogen-bonded helical configurations of the polypeptide chains. *Proc. Natl. Acad. Sci. USA* 37:205–211. A classic paper in which the structure of the α helix was defined for the first time.

Collagen

Eyre, D. R., Paz, M. A., and Gallop, P. M. (1984). Cross-linking in collagen and elastin. *Annu. Rev. Biochem.* 53:717–748.

Linsenmayer, T. F. (1991). Collagen. In *Cell Biology of Extracellular Matrix,* E. D. Hay, ed. (New York: Plenum Publishing Corporation). A review of the structures and functions of various classes of collagen.

Structure of Globular Proteins

Branden, C., and Tooze, J. (1991). *Introduction to Protein Structure* (New York: Garland Publishing). A comprehensive discussion of three-dimensional structures of proteins, based upon X-ray crystallographic analyses.

Burley, S. K., and Petsko, G. A. (1988). Weakly polar interactions in proteins. *Adv. Protein Chem.* 39:125–189.

Chothia, C., and Finkelstein, A. V. (1990). The classification and origins of protein folding patterns. *Annu. Rev. Biochem.* 59:1007–1039. Describes how α helices and β sheets pack together in supersecondary structures.

Creighton, T. E. (1988). Disulphide bonds and protein stability. *BioEssays* 8:57–63. Discusses the stabilities and roles of dithiol bonds in proteins.

Karplus, M., and Petsko, G. A. (1990). Molecular dynamics simulations in biology. *Nature* 347:631–639. Reviews the use of molecular-dynamics calculations with globular proteins.

Leszczynski, J. F., and Rose, G. D. (1986). Loops in globular proteins: a novel category of secondary structure. *Science* 234:849–855.

Richardson, J. S., and Richardson, D. C. (1989). Principles and patterns of protein conformation. In *Prediction of Protein Structure and the Principles of Protein Conformation,* G. D. Fasman, ed. (New York: Plenum Publishing Corporation), pp. 1–98. An excellent review of secondary and tertiary structures of globular proteins.

Rose, G. D., Geselowitz, A. R., Lesser, G. J., Lee, R. H., and Zehfus, M. H. (1985). Hydrophobicity of amino acid residues in globular proteins. *Science* 229:834–838.

Rose, G. D., Gierasch, L. M., and Smith, J. A. (1985). Turns in peptides and proteins. *Adv. Protein Chem.* 37:1–109.

Traut, T. W. (1988). Do exons code for structural or functional units in proteins? *Proc. Natl. Acad. Sci. USA* 85:2944–2948. Discusses how exons in existing genes do not consistently encode units of proteins, but perhaps their ancestral forms did.

Yoder, M. D., Keen, N. T., and Jurnak, F. (1993). New domain motif: the structure of pectate lyase C, a secreted plant virulence factor. *Science* 260:1503–1507. Describes the parallel β helix.

Denaturation

Ahern, T. J., and Klibanov, A. M. (1985). The mechanism of irreversible enzyme inactivation at 100°C. *Science* 228:1280–1284.

Protein Folding

Anfinsen, C. B. (1973). Principles that govern the folding of protein chains. *Science* 181:223–230. Describes the self-assembly of ribonuclease.

Dill, K. A. (1990). Dominant forces in protein folding. *Biochemistry* 29:7133–7155.

Ellis, R. J., and von der Vies, S. M. (1991). Molecular chaperones. *Annu. Rev. Biochem.* 60:321–347.

Gething, M. J., and Sambrook, J. (1992). Protein folding in the cell. *Nature* 355:33–45.

Hendrick, J. P., and Hartl, F.-U. (1993). Molecular chaperone functions of heat-shock proteins. *Annu. Rev. Biochem.* 62:349–384.

Kim, P. S., and Baldwin, R. L. (1990). Intermediates in the folding reactions of small proteins. *Annu. Rev. Biochem.* 59:631–660.

Matthews, B. W. (1993). Structural and genetic analysis of protein stability. *Annu. Rev. Biochem.* 62:139–160.

Matthews, C. R. (1993). Pathways of protein folding. *Annu. Rev. Biochem.* 62:653–683.

Nilsson, B., and Anderson, S. (1991). Proper and improper folding of proteins in the cellular environment. *Annu. Rev. Microbiol.* 45:607–35.

Richards, F. M. (1991). The protein folding problem. *Sci. Am.* 264(1):54–63.

Woolfson, D. N., Cooper, A., Harding, M. M., Williams, D. H., and Evans, P. A. (1993). Protein folding in the absence of the solvent ordering contribution to the hydrophobic interaction. *J. Mol. Biol.* 229:502–511. Describes how folding of the protein ubiquitin occurs in methanol-water mixtures in which solvent ordering is negligible.

Myoglobin

Phillips, S. E. V. (1980). Structure and refinement of oxymyoglobin at 1.6 Å resolution. *J. Mol. Biol.* 142:531–554.

Hemoglobin

Ackers, G. K., Doyle, M. L., Myers, D., and Daugherty, M. A. (1992). Molecular code for cooperativity in hemoglobin. *Science* 255:54–63.

Baldwin, J., and Chothia, C. (1979). Haemoglobin: the structural changes related to ligand binding and its allosteric mechanism. *J. Mol. Biol.* 129:175–179.

Dickerson, R. E., and Geis, I. (1983). *Hemoglobin: Structure, Function, Evolution, and Pathology* (Menlo Park, California: Benjamin/Cummings Publishing Company).

Perutz, M. F. (1978). Hemoglobin structure and respiratory transport. *Sci. Am.* 239(6):92–125.

Sickle-Cell Hemoglobin

Eaton, W. A., and Hofrichter, J. (1988). Sickle cell hemoglobin polymerization. *Adv. Protein Chem.* 40:63–279.

Ingram, V. M. (1957). Gene mutations in human hemoglobin: the chemical difference between normal and sickle cell hemoglobin. *Nature* 180:326–328.

Pauling, L., Itano, H. A., Singer, S. J., and Wells, I. C. (1949). Sickle cell anemia, a molecular disease. *Science* 110:543–548.

Antibodies

Davies, D. R., Padlan, E. A., and Sheriff, S. (1990). Antibody-antigen complexes. *Annu. Rev. Biochem.* 59:439–473. Describes the three-dimensional structures of antibody-antigen complexes.

Tonegawa, S. (1985). The molecules of the immune system. *Sci. Am.* 253(4):122–131. A good biochemical description of the immune system.

6

Properties of Enzymes

We have seen how the three-dimensional shapes of proteins allow them to serve structural and transport roles; we will now discuss their roles as enzymes. Enzymes are catalysts of extraordinary efficiency and specificity. You may recall from earlier studies that a catalyst accelerates the approach of a reaction toward equilibrium without changing the position of that equilibrium. Enzymes are extremely effective as biological catalysts—most of the reactions they catalyze would not proceed in their absence in a reasonable time without extremes of temperature, pressure, or pH. Enzyme-catalyzed reactions, or **enzymatic reactions,** are 10^3 to 10^{17} times faster than the corresponding uncatalyzed reactions. Enzymes also typically catalyze reactions orders of magnitude faster than nonenzymatic catalysts, such as those used in chemical synthesis.

Enzymes are highly specific for the reactants, or **substrates,** they act upon. The degree of substrate specificity varies. One enzyme might act upon a group of substrates of closely related structures, whereas another might act only upon a single molecular species. Many enzymes exhibit **stereospecificity,** meaning they act upon only a single stereoisomer of the substrate. Perhaps the most important general aspect of enzyme specificity is **reaction specificity,** that is, the lack of formation of wasteful by-products. Reaction specificity is reflected in exceptional product yields, which are essentially 100%. The efficiency of enzymes not only saves energy for the living cell but also precludes buildup of potentially toxic metabolic by-products.

Some enzyme-catalyzed reactions function as control points in metabolism. As we will see, metabolism is regulated in a variety of ways, including alterations in the concentrations of enzymes, substrates, and enzyme inhibitors, and modulation of the activity levels of certain enzymes.

Even the simplest living organisms contain multiple copies of nearly a thousand different enzymes. In multicellular organisms, it is the particular complement of enzymes present that differentiates one cell type from another. Most of the enzymes discussed in this book are among the several hundred enzymes common to virtually all cells. These enzymes catalyze the reactions of the central metabolic pathways necessary for the maintenance of life.

 Related material appears in Exercise 2 of the computer-disk supplement *Exploring Molecular Structure.*

The first enzyme to be isolated was urease, crystallized by James B. Sumner in 1926. Sumner proved that urease is a protein, and since then, almost all enzymes have been shown to be proteins or proteins plus cofactors. However, certain RNA molecules also exhibit catalytic activity. Some of these catalytic RNA molecules are discussed in more detail in Chapters 28 and 30. The enzymes described in this chapter are globular proteins and therefore possess the main functional characteristic of globular proteins—the ability to specifically bind one or several molecules. Substrate molecules are bound in a somewhat hydrophobic cleft known as the **active site,** which is often located between two domains or subunits of a protein or in an indentation, such as a pocket at one end of a barrel. In general, enzymes are much larger than their substrates. Even when the substrate is a macromolecule, only a small portion of it is a reactant—a particular group, bond, or linkage. Enzymes regulated by the cell during metabolism are generally more complex than unregulated enzymes; with few exceptions, they are oligomeric molecules that have separate binding sites for substrates and modulators, the compounds that act as regulatory signals.

This chapter presents a general description of enzymes and their properties. First, the classification and nomenclature of enzymes is presented. Next, kinetic analysis (measurements of reaction rates) is discussed, with emphasis on how kinetic experiments reveal the properties of the enzyme and the nature of the complexes it forms with substrates and inhibitors. The relationship between protein structure and enzymatic function is conveyed by an examination of the serine proteases. Finally, the principles of inhibition and activation of regulatory enzymes are described. The next chapter explains how enzymes work at the chemical level. Chapter 8 is devoted to the biochemistry of coenzymes, the organic molecules that assist some enzymes with their catalytic roles by providing reactive groups not found on amino acid side chains.

6·1 Enzymes Are Named and Classified According to the Reactions They Catalyze

Most enzymes are named by adding the suffix -ase to the name of the substrate they act upon or to a descriptive term for the reactions they catalyze. Thus, urease has urea as a substrate. Alcohol dehydrogenase catalyzes the removal of hydrogen from alcohols (i.e., the oxidation of alcohols). However, a few enzymes, such as trypsin and chymotrypsin, are known by their historic names.

In an effort to standardize nomenclature, a committee of the International Union of Biochemistry (IUB) has published and maintains a classification scheme that categorizes enzymes into six major groups according to the general class of organic chemical reactions they catalyze:

1. **Oxidoreductases** catalyze oxidation-reduction reactions. Most of these enzymes are known as **dehydrogenases,** but some are called oxidases, peroxidases, oxygenases, or reductases.

2. **Transferases** catalyze group-transfer reactions. Many require the presence of coenzymes. A portion of the substrate molecule usually binds covalently to these enzymes or their coenzymes. This group includes the kinases.

3. **Hydrolases** catalyze hydrolysis. They are a special class of transferases, with water serving as the acceptor of the group transferred.

4. **Lyases** catalyze nonhydrolytic and nonoxidative elimination reactions, or lysis, of a substrate, generating a double bond. In the reverse direction, lyases catalyze addition of one substrate to a double bond of a second substrate. A lyase that catalyzes an addition reaction in cells is often termed a **synthase.**

5. **Isomerases** catalyze isomerization reactions. Because these reactions have only one substrate and one product, they are among the simplest enzymatic reactions.

6. **Ligases** catalyze ligation, or joining, of two substrates. These reactions require the input of the chemical potential energy of a nucleoside triphosphate such as ATP. Ligases are usually referred to as **synthetases.**

The IUB classification scheme assigns a unique code number to each enzyme. These numbers identify the major class and subclasses of an enzyme as well as its specific reaction. The formal IUB name and the corresponding number are usually presented near the beginning of scientific reports, after which a shorter common name is used. In this book, enzymes are usually referred to by their common names.

Table 6·1 (next page) lists a representative enzyme of each class with the reaction it catalyzes. Note that most enzymes have more than one substrate, even if the second substrate is only a molecule of water.

6·2 Kinetic Experiments Reveal the General Properties of Enzymes

Enzyme kinetics is the study of the rates of enzyme-catalyzed reactions. Kinetic studies provide indirect information concerning the specificities and catalytic mechanisms of enzymes. Clinically, kinetic experiments are used to detect alterations in the concentration or activity of enzymes, which may be symptomatic of a disease.

One of the first great advances in biochemistry—the explanation that enzyme catalysis involves transient binding of a sterically suitable substrate by an enzyme—resulted from investigation of enzyme kinetics. Early studies also revealed that enzyme activity is stereospecific. The theory put forward to explain the stereospecificity was the lock-and-key theory, proposed by Emil Fischer in 1894. Fischer described an enzyme as a rigid template, or lock, and the substrate as a matching key. Only one or a very few substrates could fit into any given enzyme. The lock-and-key theory led to the concepts that there is an active-site region and that complexes form between substrates and enzymes. For the first half of the 20th century, most research on enzymes involved kinetic experiments. These experiments describe the activity of enzymes by showing how the rates of reactions are affected by variations in concentrations and experimental conditions. For example, it is possible to describe how efficient enzymes are as catalysts and the conditions for their maximal or optimal activity. Early experiments yielded strong but indirect evidence that an enzyme (E) binds a substrate (S) to form an **enzyme-substrate complex** (ES). ES complexes are formed when ligands bind noncovalently in their proper places in the active-site cleft. In the crevice of a globular protein, the substrate reacts transiently with the protein catalyst (and with other molecules of substrate in a multisubstrate reaction) to form the **product** (P) of the reaction.

Before discussing enzyme kinetics in depth, we will review the principles of kinetics for nonenzymatic chemical systems. These principles will then be applied to enzymatic reactions.

A. Chemical Kinetics

Kinetic experiments examine the relationship between the amount of reaction product (P) formed in a unit of time, $\Delta[P]/\Delta t$, and the experimental conditions under which the reaction takes place. The basis of most kinetic measurements is the observation that the rate, or **velocity** (v), of a reaction varies directly with the concentration of each substrate or catalyst. This observation is expressed in an equation

Table 6·1 Examples of enzymes from the six major classes

Major class	Enzyme	Reaction description
1. Oxidoreductase	Lactate dehydrogenase	Oxidation of the secondary alcohol L-lactate to pyruvate, a ketone
2. Transferase	Alanine transaminase (Alanine aminotransferase)	Transfer of an amino group
3. Hydrolase	Trypsin	Hydrolysis of Lys−Y (or Arg−Y) peptide bonds, where Y ≠ Pro
4. Lyase	Pyruvate decarboxylase	Decarboxylation of pyruvate
5. Isomerase	Alanine racemase	Interconversion of D and L isomers of alanine
6. Ligase	Glutamine synthetase	ATP-dependent synthesis of L-glutamine

Example of reaction catalyzed	Coenzyme involved
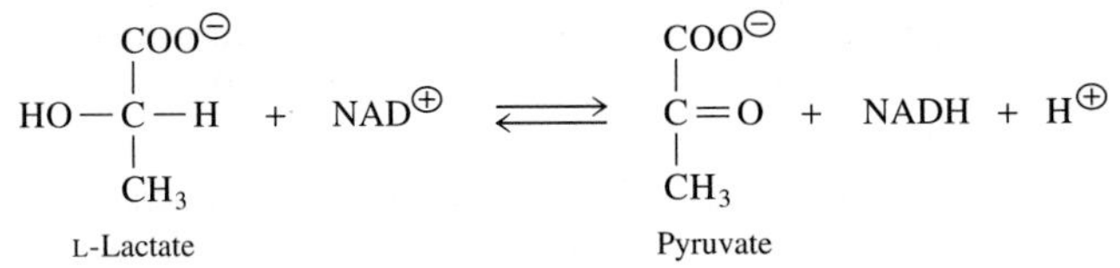 L-Lactate + $NAD^\oplus$ ⇌ Pyruvate + $NADH$ + $H^\oplus$	$NAD^\oplus$ (Nicotinamide adenine dinucleotide)
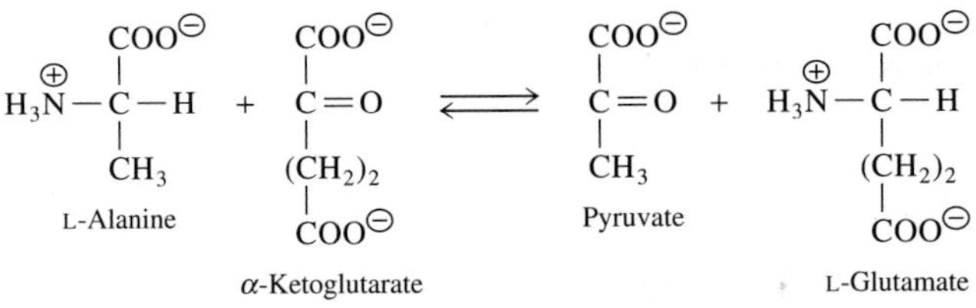 L-Alanine + α-Ketoglutarate ⇌ Pyruvate + L-Glutamate	Pyridoxal phosphate
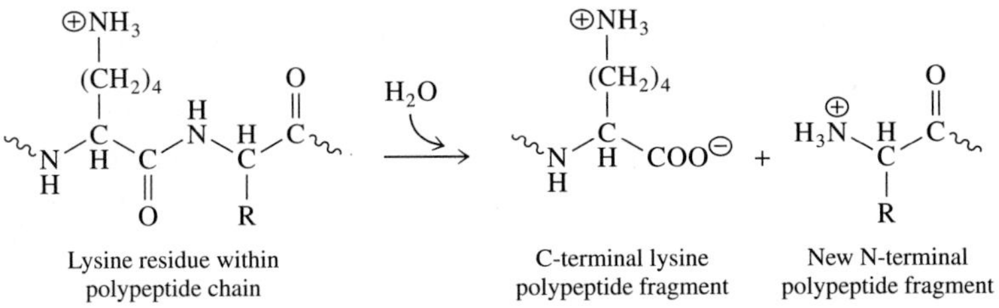 Lysine residue within polypeptide chain + H_2O → C-terminal lysine polypeptide fragment + New N-terminal polypeptide fragment	None
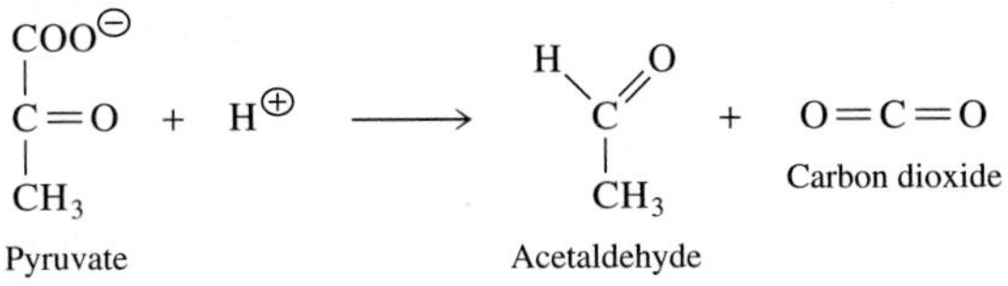 Pyruvate + $H^\oplus$ → Acetaldehyde + $O=C=O$ Carbon dioxide	Thiamine pyrophosphate
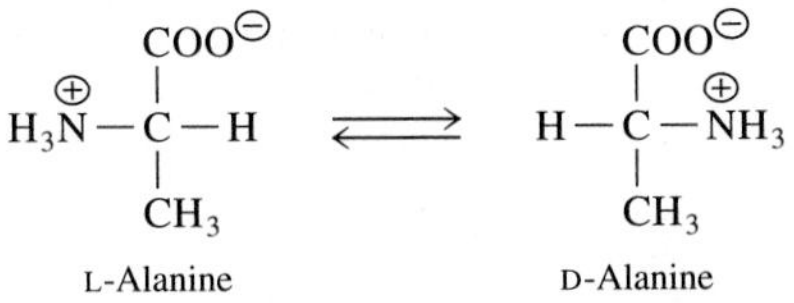 L-Alanine ⇌ D-Alanine	Pyridoxal phosphate
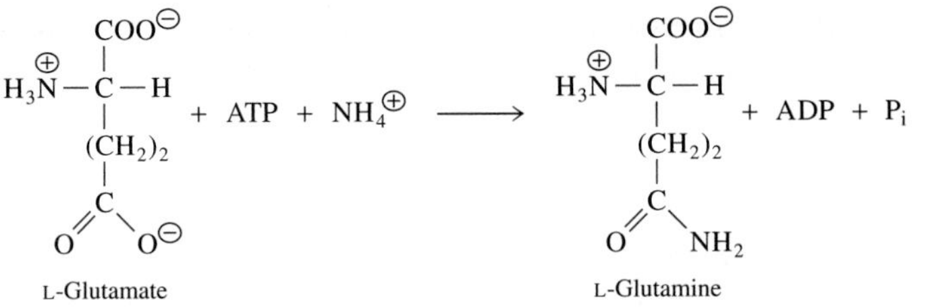 L-Glutamate + ATP + $NH_4^\oplus$ → L-Glutamine + ADP + P_i	ATP

called a rate equation. For example, the rate equation for the nonenzymatic conversion of S to P in an isomerization reaction is written as

$$\frac{\Delta[P]}{\Delta t} = v = k[S] \tag{6·1}$$

The symbol k is the rate constant; each reaction has a different rate constant. Rate constants indicate the speed or efficiency of a reaction. A graph of velocity versus [S] for this reaction would be a straight line that increases linearly with the concentration of substrate. The overall **kinetic order** of a reaction is the sum of the exponents in the rate equation and tells how many molecules are reacting in the slowest step of the reaction. Equation 6·1 is the rate equation for a first-order reaction, in which only one component is reacting. The rate constant for a first-order reaction is expressed in reciprocal time units (s^{-1}).

For a more complicated reaction, such as the reaction $S_1 + S_2 \longrightarrow P_1 + P_2$, the rate is determined by the concentrations of both substrates. If both substrates are present at similar concentrations, the rate equation is

$$v = k[S_1]^1[S_2]^1 \tag{6·2}$$

This reaction is first order with respect to each reactant, and since the sum of the exponents is 2, the reaction is second order overall. The reaction is also termed bimolecular because two molecules react to form the products. The rate constants for second-order reactions have the units $M^{-1}s^{-1}$. Reactions with more than two reactants are often studied, but because the calculations can become quite complex, experimental conditions are arranged so that the reactions proceed in several steps, with each step being either first order or second order.

When the concentration of one reactant is so high that it remains essentially constant during the reaction, the reaction is said to be zero order with respect to that reactant, and the term for the reactant can be eliminated from the rate equation. The reaction then becomes an artificial, or pseudo, first-order reaction.

$$v = k[S_1]^1[S_2]^0 = k'[S_1] \tag{6·3}$$

An example of a pseudo first-order reaction is the nonenzymatic hydrolysis of the glycosidic bond of table sugar, sucrose, in aqueous acid solution. Sucrose is a disaccharide composed of a residue each of the sugars glucose and fructose.

$$\begin{array}{cccccc} \text{Sucrose} & + & \text{Water} & \longrightarrow & \text{Glucose} & + & \text{Fructose} \\ (C_{12}H_{22}O_{11}) & & (H_2O) & & (C_6H_{12}O_6) & & (C_6H_{12}O_6) \end{array} \tag{6·4}$$

The concentration of water, which is both the solvent and a substrate, is so high that it remains effectively constant during the reaction. Under these reaction conditions, the reaction rate depends only on the concentration of sucrose. To prove that the original reaction is not truly first order, the concentration of water can be made a limiting factor by replacing much of the water with a nonreacting solvent.

B. Enzyme Kinetics

Pseudo first-order conditions are used in analyses called **enzyme assays** that determine concentrations of enzymes. Enzyme assays measure the amount of product that is formed in a given time period. In some assay methods, a recording spectrophotometer can be used to record data continuously; in other methods (known as discontinuous, or point, assays), samples are removed and analyzed at intervals. In enzyme assays, all other reactants are present in excess of the enzyme to ensure that reaction conditions are pseudo first order. The assay is performed at a constant temperature that is high enough to produce high activity but not so high that the rate of denaturation of the enzyme is appreciable. The straight line in Figure 6·1 illustrates the effect of enzyme concentration on reaction velocity in a pseudo first-order reaction. The more enzyme present, the faster the reaction. Under the experimental conditions, there are sufficient substrate molecules so that every enzyme molecule

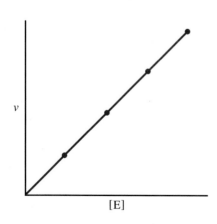

Figure 6·1
Effect of enzyme concentration, [E], on the velocity, v, of an enzyme-catalyzed reaction at a fixed, saturating [S]. Because the reaction rate is affected by the concentration of enzyme but not by the concentration of the other reactant, S, the bimolecular reaction is pseudo first order.

(E) has bound a molecule of substrate to form an ES complex, a condition called saturation of E with S. Enzymes are normally present in assays at very low concentrations compared to the concentrations of substrates. However, enzymes are not necessarily saturated inside cells. The concentration of enzyme in a test sample can be easily determined by comparing its activity to a reference curve similar to the model curve in Figure 6·1.

Let us now consider the kinetic variables for a simple enzymatic reaction, the conversion of a substrate (S) to a product (P), catalyzed by an enzyme (E). Although most enzymatic reactions have two or more substrates, the general principles of enzyme kinetics can be elucidated by assuming the simple case of one substrate and one product. The initial moments of this bimolecular reaction can be written as

$$E + S \underset{k_{-1}}{\overset{k_1}{\rightleftharpoons}} ES \xrightarrow{k_{cat}} E + P \qquad \textbf{(6·5)}$$

The rate constants k_1 and k_{-1} in Equation 6·5 govern the rates of association of S with E and dissociation of S from ES, respectively. The rate constant for the second step is k_{cat}, the **catalytic constant** (or turnover number), which is the number of catalytic events per second per active site or enzyme molecule. Note that the conversion of the ES complex into free enzyme and product is shown by a one-way arrow. During the initial period when measurements are made, little product has been formed, so the rate of the reverse reaction (E + P $\longrightarrow$ EP) is negligible. The velocity measured during this short period is called the **initial velocity, v_0**. The use of v_0 measurements simplifies the interpretation of kinetic data and avoids complications that may arise as the reaction progresses, such as product inhibition and slow denaturation of enzyme. The formation and dissociation of ES complexes are usually very rapid reactions because only noncovalent bonds are formed and broken. In contrast, the conversion of substrate to product is usually rate limiting. It is during this step that the substrate is chemically altered.

Initial velocities are obtained from progress curves, graphs of either the increase in product concentration or the decrease in substrate concentration over time (Figure 6·2). The initial velocity is the tangent or slope $\Delta[P]/\Delta t$ at the origin of a progress curve.

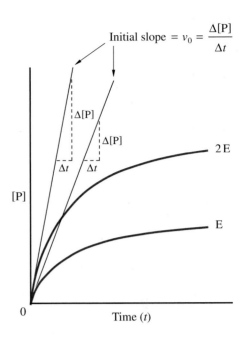

Figure 6·2
Progress curve for an enzyme-catalyzed reaction in which a substrate is converted to product. [P], the concentration of product, increases as the reaction proceeds. The initial velocity of the reaction (v_0) is the slope of the initial linear portion of the curve. Note that the rate of the reaction doubles when twice as much enzyme (2 E, the upper curve) is added to an otherwise identical reaction mixture.

6·3 The Michaelis-Menten Equation Is a Rate Equation for Enzymatic Catalysis

In the early 1900s, workers in several laboratories were examining the effects of variations in substrate concentration in order to derive rate equations for enzymatic processes. Most were using extracts of yeast to catalyze the hydrolysis of sucrose to a molecule each of glucose and fructose (Equation 6·4). Some scientists believed that enzymes acted as catalysts merely by their presence. For example, the radiation theory proposed that enzymes emitted radiations that were absorbed by substrate molecules, leading to the conversion of substrate to product. Other scientists found evidence for complex formation, such as increased heat stability of an enzyme in the presence of its substrate and the stereochemical specificity of enzymes observed by Fischer. Two major kinetic observations supported the theory involving the formation of complexes. First, at high concentrations of substrate, E is saturated by S, and the reaction rate is independent of the concentration of substrate. The value of v_0 for a solution of E that is saturated with S is called the **maximum velocity, V_{max}**. Second, at low concentrations of substrate, the reaction is first order with respect to substrate (making the reaction second order overall: first order with respect to S and first order with respect to E). At intermediate substrate concentrations, the order with respect to S is fractional, decreasing from first order toward zero order as [S]

(a)

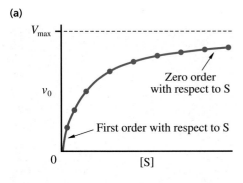

(b)

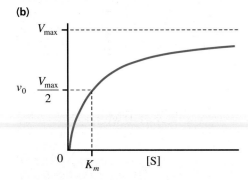

Figure 6·3
Plots of initial velocity, v_0, versus substrate concentration, [S], for an enzyme-catalyzed reaction. **(a)** Each experimental point was obtained from a separate progress curve. The shape of the curve is hyperbolic. At low substrate concentrations, the curve approximates a straight line that rises steeply. In this region of the curve, the reaction is first order with respect to substrate. At high concentrations of substrate, the enzyme is saturated, and the reaction is zero order with respect to substrate. **(b)** The concentration of substrate that corresponds to half-maximum velocity is called the Michaelis constant, K_m. The enzyme is half-saturated when [S] = K_m.

increases. The shape of the v_0 versus [S] curve from low to high [S] is a rectangular hyperbola (Figure 6·3a). The rate equation for Equation 6·5 describes a hyperbola of this type. Hyperbolic curves are indicative of processes involving simple dissociation, as we saw for the dissociation of oxygen from oxymyoglobin (Section 5·17).

The equation for a rectangular hyperbola is

$$y = \frac{ax}{b + x} \tag{6·6}$$

where a is the asymptote of the curve (the value of y at an infinite value of x) and b is the point on the x axis corresponding to a value of $y_{infinity}/2$. We can obtain the rate equation for the simple bimolecular reaction $E + S \rightleftharpoons ES \longrightarrow E + P$ by substituting four terms from enzyme kinetics into the general equation for a rectangular hyperbola. Three of the terms have already been introduced: $y = v_0$, $x = [S]$, and $a = V_{max}$. The fourth term, b in the general equation, is the **Michaelis constant, K_m,** defined as the concentration of substrate when v_0 is equal to one-half V_{max} (Figure 6·3b). The complete rate equation is written

$$v_0 = \frac{V_{max}[S]}{K_m + [S]} \tag{6·7}$$

This is called the **Michaelis-Menten equation,** named after Leonor Michaelis and Maud Menten, who provided strong evidence for this enzymatic relationship (Figure 6·4). Before deriving the Michaelis-Menten equation by a kinetic approach, let us consider the meaning of the constants V_{max} and K_m.

A. The Maximum Velocity, V_{max}, Is Achieved When E Is Saturated with S

As shown in Figure 6·3a, the reaction velocity approaches a constant value as the substrate concentration, [S], becomes very large. In this region of the curve, the enzyme is nearing saturation with S and the reaction is zero order with respect to S (Equation 6·3). Adding more S has virtually no effect. Only the addition of more enzyme can increase the velocity when [S] is very large. The constant velocity that is asymptotically approached is the maximum velocity, V_{max}. V_{max} is defined for any given amount of enzyme as the initial velocity (v_0) at saturating concentrations of substrate. The rate equation for this pseudo first-order region of the curve can be written in these equivalent forms:

$$v_0 \text{ (at saturation)} = V_{max} = k[E][S]^0 = k[E]_{total} = k_{cat}[ES] \tag{6·8}$$

The rate constant k is equal to the catalytic constant k_{cat} (the rate constant for conversion of ES to free enzyme and product) because, at saturation, essentially all molecules of E are present as ES. Another simple relationship shown in Equation 6·8 is that $V_{max} = k_{cat}[E]_{total}$. By rearranging this equation, we obtain the definition for k_{cat}.

$$k_{cat} = \frac{V_{max}}{[E]_{total}} \tag{6·9}$$

The catalytic constant is the maximum velocity, V_{max}, divided by the original concentration of enzyme present, $[E]_{total}$, or the number of moles of substrate converted to product under saturating conditions per second per mole of enzyme (or per mole of active site for an oligomeric enzyme). The unit for k_{cat} is s^{-1}. The reciprocal of k_{cat} is the time required for one catalytic event. Therefore, the catalytic constant is a measure of how quickly an enzyme can catalyze a reaction.

B. The Michaelis Constant, K_m, Is the Concentration of Substrate That Half-Saturates E

The Michaelis constant has a number of meanings. As we saw above, K_m is the initial concentration of substrate at half-maximum velocity, or at half-saturation of E with S. This can be verified by substituting K_m for [S] in Equation 6·7; v_0 is found to be $V_{max}/2$. For some enzymes, K_m can be further defined in terms of the constants for the formation and breakdown of the ES complex, as we will see in Sections 6·3C and 6·3D.

C. The Michaelis-Menten Equation Can Be Derived by Assuming Steady-State Reaction Conditions

The Michaelis-Menten equation has been derived in several ways. One common derivation, provided by George E. Briggs and J. B. S. Haldane, is termed the steady-state derivation. This derivation postulates that there is a period of time during which the substrate binds to the enzyme at the same rate at which it is removed, as may happen in metabolic pathways in the cell. [S] is assumed to be constant but not necessarily saturating. For example, soon after a small amount of enzyme is mixed with substrate, there is a period called the steady state when [ES] is constant because the rates of decomposition of ES to either E + S or E + P are equal to the rate of formation of the ES complex from E + S. The concentration of enzyme-substrate complex reaches a level that remains constant for as long as the concentration of substrate remains much greater than the concentration of enzyme. The rate of formation of ES from E + S depends on the concentration of free enzyme (enzyme molecules not in the form of ES), which is ($[E]_{total}$ − [ES]). Expressing these statements about the steady state algebraically provides this equivalence:

$$(k_{-1} + k_{cat})[ES] = k_1([E]_{total} - [ES])[S] \tag{6·10}$$

Equation 6·10 is rearranged to collect the rate constants, and the ratio of constants obtained defines the Michaelis constant, K_m.

$$\frac{k_{-1} + k_{cat}}{k_1} = K_m = \frac{([E]_{total} - [ES])[S]}{[ES]} \tag{6·11}$$

Next, this equation is solved for [ES] in several steps.

$$([E]_{total} - [ES])[S] = [ES]K_m \tag{6·12}$$

Expanding,

$$[ES]K_m = ([E]_{total}[S]) - ([ES][S]) \tag{6·13}$$

Collecting [ES] terms,

$$[ES](K_m + [S]) = [E]_{total}[S] \tag{6·14}$$

and

$$[ES] = \frac{[E]_{total}[S]}{K_m + [S]} \tag{6·15}$$

Since the velocity of an enzyme-catalyzed reaction depends on the rate of conversion of ES to E + P (Equation 6·8),

$$v = k_{cat}[ES] \tag{6·16}$$

The velocity, and hence the rate equation, can be obtained by substituting the value of [ES] from Equation 6·15 into Equation 6·16, with $v = v_0$.

$$v_0 = \frac{k_{cat}[E]_{total}[S]}{K_m + [S]} \tag{6·17}$$

(a)

(b)

Figure 6·4
(a) Leonor Michaelis and (b) Maud Menten, who performed experiments confirming the Michaelis-Menten equation.

Also, because $k_{cat}[E]_{total} = V_{max}$ (Equation 6·8), Equation 6·17 can be rewritten in the most familiar form of the Michaelis-Menten equation:

$$v_0 = \frac{V_{max}[S]}{K_m + [S]}$$

(6·18)

D. K_m Can Be the Dissociation Constant for the ES Complex

From the steady-state derivation, we see that K_m is the ratio of the constants for breakdown of ES divided by the constant for its formation (Equation 6·11). If the rate constant for the chemical step, k_{cat}, is much smaller than either k_1 or k_{-1}, as is often the case, k_{cat} can be neglected, and K_m becomes k_{-1}/k_1, the equilibrium constant for the dissociation of the ES complex to E + S. Similarly, in their restrictive derivation, Michaelis and Menten assumed that ES is in rapid equilibrium with E + S and that K_m is the dissociation constant of ES. K_m is thus a measure of the affinity of E for S. The lower the value of K_m, the more tightly the substrate is bound. K_m values are sometimes used to differentiate enzymes having the same function. For example, there are several different forms of lactate dehydrogenase in mammals, each with distinct K_m values for its substrates.

The K_m of an enzyme for a substrate is often near the concentration of that substrate in a cell. Because the K_m value falls near the middle of the steeply rising portion of the rate curve for the enzyme, the enzyme is most able to respond proportionally to changes in [S]. Physiologically, this can help prevent the accumulation of substrate under changing metabolic conditions.

E. The Ratio k_{cat}/K_m Describes Rates and Substrate Specificity

In the region of the hyperbolic curve where the concentration of S is very low and the curve still approximates a straight line, reactions are second order: first order with respect to S and first order with respect to E. The rate equation for this region is

$$v_0 = k[E][S]$$

(6·19)

When Michaelis and Menten first wrote the full rate equation, they used the form that included $k_{cat}[E]$ (Equation 6·17) rather than V_{max}. If we consider only the region of the Michaelis-Menten curve at very low [S], Equation 6·17 can be simplified by neglecting the term for [S] in the denominator, a very small value.

$$v_0 = \frac{k_{cat}}{K_m}[E][S]$$

(6·20)

Therefore, the rate constant for Equation 6·19 is k_{cat}/K_m. This rate constant is a measure of two features of enzymatic catalysis. First, it is a measure of the specificity, or preference of an enzyme for different substrates, which can be assessed by comparing the k_{cat}/K_m values of the enzyme-substrate pairs. Suppose that two different substrates, A and B, are competing for binding and catalysis by the same enzyme. When the concentrations of A and B are equal, the ratio of the rates of their conversion to product by the enzyme is equal to the ratio of their k_{cat}/K_m values. For this reason, the ratio k_{cat}/K_m is often called the specificity constant. Second, k_{cat}/K_m is the rate constant for the reaction of dilute concentrations of substrate—an indication of the overall rate of the enzymatic reaction $E + S \longrightarrow E + P$ at very low substrate concentrations. The full significance and application of this meaning for k_{cat}/K_m will be discussed in the next section.

6·4 Rate Constants Indicate the Catalytic Efficiency of Enzymes

The kinetic constants of enzymatic reactions, K_m and k_{cat}, can be used to gauge the catalytic efficiency of enzymes. K_m is a measure of the stability of the ES complex. It is equal to the ratio of $[E]_{free}[S]$ divided by $[ES]$ under steady-state reaction conditions (Equation 6·11). k_{cat} is the first-order rate constant for the conversion of ES to $E + P$. It is a measure of the catalytic activity of an enzyme, telling how many reactions per second a molecule of enzyme can catalyze. Values for k_{cat} of about $10^3 \, s^{-1}$ are typical. The ratio k_{cat}/K_m is an apparent second-order rate constant for the formation of $E + P$ from $E + S$ when the overall reaction is limited by the encounter of S with E. This constant approaches 10^8 to $10^9 \, M^{-1} s^{-1}$, the fastest rate at which two uncharged solutes can approach each other by diffusion at physiological temperature. Enzymes that can catalyze reactions at this extremely rapid rate are discussed in Section 7·8. We will use both k_{cat}/K_m and k_{cat} in Chapter 7 to discuss the rates of enzymatic reactions. The meanings of the terms are summarized in Figure 6·5. Comparisons of enzymatic reactions with nonenzymatic reactions can be made with these constants. The absolute values of these constants tell how powerful each enzyme is as a catalyst.

Enzymes are extremely efficient catalysts. Their efficiency is measured by the **rate acceleration** that they provide. This value is the ratio of the rate constant for a reaction in the presence of the enzyme (k_{cat}) divided by the rate constant for the same reaction in the absence of enzyme (k_n). Surprisingly few values of rate acceleration are known because most cellular reactions occur extremely slowly in the absence of enzymes—so slowly that their nonenzymatic rates are difficult to measure.

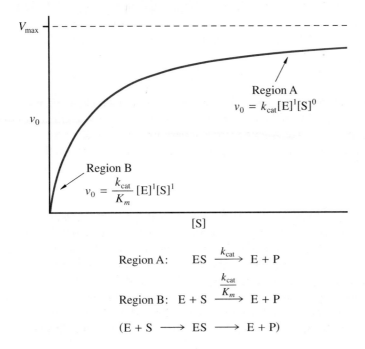

Figure 6·5
Meanings of the two rate constants, k_{cat} and k_{cat}/K_m. The catalytic constant, k_{cat}, is the first-order rate constant for the conversion of the ES complex to $E + P$. It is measured most easily when the enzyme is saturated with substrate (region A on the Michaelis-Menten curve shown). The ratio k_{cat}/K_m is the second-order rate constant for conversion of $E + S$ to $E + P$ at very low concentrations of substrate (region B). The reactions measured by these rate constants are summarized below the graph.

Table 6·2 Rate accelerations of some enzymes

	Nonenzymatic rate constant (k_n in s^{-1})	Enzymatic rate constant (k_{cat} in s^{-1})	Rate acceleration (k_{cat}/k_n)
Chymotrypsin	4×10^{-9}	4×10^{-2}	10^7
Lysozyme	3×10^{-9}	5×10^{-1}	2×10^8
Triose phosphate isomerase	6×10^{-7}	2×10^3	3×10^9
Fumarase	2×10^{-8}	2×10^3	10^{11}
β-Amylase	3×10^{-9}	1×10^3	3×10^{11}
Urease	3×10^{-10}	3×10^4	10^{14}
Adenosine deaminase	10^{-12}	10^2	10^{14}
Alkaline phosphatase	10^{-15}	10^2	10^{17}

Table 6·3 Examples of turnover numbers

Enzyme	Turnover number (k_{cat} in s^{-1})
Papain	10
Ribonuclease	10^2
Carboxypeptidase	10^2
Trypsin	10^2 (to 10^3)
Acetylcholinesterase	10^3
Kinases	10^3
Dehydrogenases	10^3
Transaminases	10^3
Carbonic anhydrase	10^6
Superoxide dismutase	10^6
Catalase	10^7

The values of k_{cat} are the first-order rate constants for the reaction ES → E + P. They are given only as orders of magnitude. [Most values obtained from Eigen, M., and Hammes, G. G. (1963). Elementary steps in enzyme reactions (as studied by relaxation spectrometry). *Adv. Enzymol.* 25:1–38.]

Several examples of rate acceleration are provided in Table 6·2. Typical values are in the range of 10^8 to 10^{12}, but some are quite a bit higher (up to 10^{17}). The difficulty in obtaining rate constants for nonenzymatic reactions is exemplified by the half-time for deamination of adenosine at 20°C and pH 7, about 20 000 years! Because the nonenzymatic rate of this reaction is so slow, adenosine deaminase has an extraordinarily high rate acceleration despite a moderate k_{cat} of 10^2 s^{-1}.

The catalytic constant k_{cat}, the number of catalytic events that an enzyme molecule can catalyze per second, is also called the turnover number. The catalytic constant is a measure of the absolute, not relative, activity of an enzyme. Table 6·3 lists representative values of k_{cat}. Most enzymes are potent catalysts having k_{cat} values of 10^2 to 10^3 s^{-1}. Some enzymes, however, are much faster. For example, the last three examples in Table 6·3 have k_{cat} values of 10^6 s^{-1} or greater. Extremely rapid catalysis is essential for the physiological function of certain enzymes. Carbonic anhydrase, for example, must act very rapidly in order to maintain equilibrium between aqueous CO_2 and bicarbonate (Section 3·10). As we will see in Section 7·8B, both superoxide dismutase and catalase are responsible for rapid removal of the toxic oxygen metabolites superoxide anion and hydrogen peroxide.

6·5 K_m and V_{max} Are Easily Measured

K_m and V_{max} for an enzyme-catalyzed reaction can be measured in several ways. Both values are obtained by analysis of initial velocities at a series of substrate concentrations and a fixed concentration of enzyme. In order to obtain reliable values for the kinetic constants, the [S] points must be spread out, both below and above K_m, to produce a hyperbola. It is difficult to determine either K_m or V_{max} directly from a graph of initial velocity versus concentration because the curve approaches V_{max} asymptotically. However, using a suitable computer program, accurate values are determined by fitting the experimental results to the equation for the hyperbola.

Before computers were widely available, values for V_{max} and K_m were obtained by using graphic transformations of the Michaelis-Menten equation that give straight lines. From these lines, one can extrapolate or obtain a slope to estimate the value of the asymptote (V_{max}) of the v_0 versus [S] plot. Linear transformations are still encountered. The most often used transformation of the Michaelis-Menten equation is the Lineweaver-Burk, or double-reciprocal, plot (Figure 6·6). The values of $1/v_0$ are plotted against $1/[S]$. The absolute value of $1/K_m$ is obtained from the intercept of the line at the x axis, and the value of $1/V_{max}$ is obtained from the y intercept. Although values obtained from double-reciprocal plots are less precise than computer-generated values, double-reciprocal plots are easily understood and provide recognizable patterns for the study of enzyme inhibition, an extremely important aspect of enzymology that we will examine in the next section.

Values of k_{cat} can be obtained from measurements of V_{max} only when the absolute concentration of the enzyme is known. Values of K_m can be determined even when enzymes have not been purified, provided that only one enzyme in the impure preparation can catalyze the observed reaction.

6·6 Reversible Inhibitors Are Bound to Enzymes by Noncovalent Forces

An **inhibitor** (I) is a compound that binds to an enzyme and interferes with its activity by preventing either the formation of the ES complex or its breakdown to E + P. Inhibitors are used experimentally to investigate enzyme mechanisms and to decipher metabolic pathways. Natural inhibitors serve as regulators of metabolism, and many medicinal drugs are enzyme inhibitors.

Inhibitors can be either irreversible or reversible. Irreversible inhibitors are bound to enzymes by covalent bonds. They will be described in Section 6·9. Reversible inhibitors are bound to enzymes by the same noncovalent forces that bind substrates and products. Because they bind noncovalently, reversible inhibitors are easily removed from solutions of enzymes by dialysis or gel filtration (Section 4·7). The constant for the dissociation of I from the EI complex, called the **inhibition constant**, K_i, is described by the equation

$$K_i = \frac{[E][I]}{[EI]} \tag{6·21}$$

There are several types of reversible inhibition, which can be distinguished experimentally by their effects on the kinetic behavior of enzymes. These effects are summarized in Table 6·4. The rate laws for the reactions in the presence of reversible inhibitors are slightly expanded forms of the Michaelis-Menten equation with K_i and [I] incorporated. Schematic views of reversible enzyme inhibition are shown in Figure 6·7 (next page).

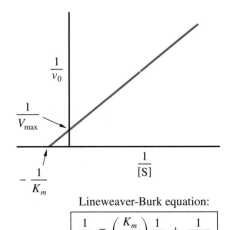

Lineweaver-Burk equation:

$$\frac{1}{v_0} = \left(\frac{K_m}{V_{max}}\right)\frac{1}{[S]} + \frac{1}{V_{max}}$$

Figure 6·6
Lineweaver-Burk (double-reciprocal) plot. This plot is derived from a linear transformation of the Michaelis-Menten equation. Values of $1/v_0$ are plotted as a function of $1/[S]$ values.

Table 6·4 Effects of reversible inhibitors on kinetic constants

Type of inhibitor	Effect	Rate law
Competitive (I binds to E only)	Raises K_m; V_{max} remains unchanged	$v_0 = \dfrac{V_{max}[S]}{K_m\left(1 + \dfrac{[I]}{K_i}\right) + [S]}$
Uncompetitive (I binds to ES only)	Lowers V_{max} and K_m; ratio of V_{max}/K_m remains unchanged	$v_0 = \dfrac{V_{max}[S]}{K_m + [S]\left(1 + \dfrac{[I]}{K_i}\right)}$
Noncompetitive (I binds to E and ES)		$v_0 = \dfrac{V_{max}[S]}{K_m\left(1 + \dfrac{[I]}{K_i}\right) + [S]\left(1 + \dfrac{[I]}{K_i'}\right)}$
Pure noncompetitive (I binds to E and ES equally)	Lowers V_{max}; K_m remains unchanged	$E + I \underset{}{\overset{K_i}{\rightleftarrows}} EI$
Mixed noncompetitive (I binds to E and ES unequally)	Lowers V_{max}; raises or lowers K_m	$ES + I \underset{}{\overset{K_i'}{\rightleftarrows}} ESI$

Figure 6·7
Schematic views of reversible enzyme inhibition. (a) Classical competitive inhibition. S or I binds at the active site, or catalytic center (C), in a mutually exclusive manner. (b) Nonclassical competitive inhibition. S binds to the R (active) conformation, preventing binding of I. I binds to the T (inactive) conformation, preventing binding of S. (c) Uncompetitive inhibition. I binds only to the ES complex. The conformation of the enzyme changes to an inactive shape when I binds. (d) Noncompetitive inhibition. Binding of I changes the conformation of the catalytic center, making the enzyme inactive. Although the inactive conformation can still bind S, no product is formed. [Adapted from Segel, I. H. (1975). *Enzyme Kinetics: Behavior and Analysis of Rapid Equilibrium and Steady-State Enzyme Systems* (New York: Wiley-Interscience), pp. 102–137.]

(a) Classical competitive inhibition

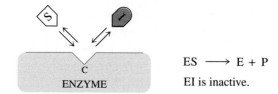

ES ⟶ E + P

EI is inactive.

(b) Nonclassical competitive inhibition

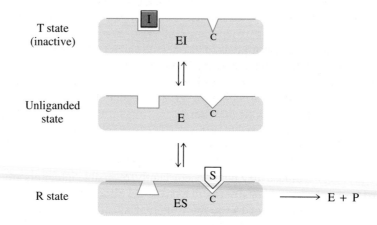

(c) Uncompetitive inhibition

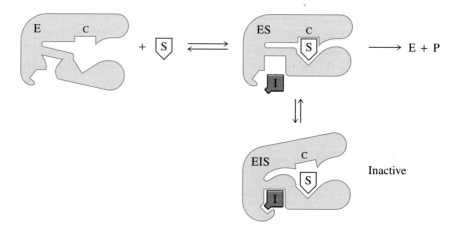

(d) Noncompetitive inhibition

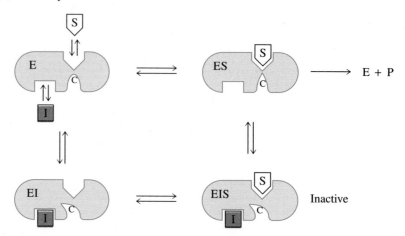

A. Competitive Inhibitors Bind Only to Free E

Competitive inhibitors are the most commonly encountered inhibitors in biochemistry. In competitive inhibition, I can bind only to unliganded molecules of enzyme, E. Competitive inhibition can be described by a diagram (Figure 6·7a and b) and by the kinetic scheme given in Figure 6·8a, which is an expansion of Equation 6·5 that includes formation of the EI complex. In this scheme, only ES can lead to the formation of product.

When a competitive inhibitor is bound to a molecule of enzyme, it prevents the binding of substrate to that enzyme. Conversely, binding of substrate prevents binding of the inhibitor; that is, S and I compete for binding to the enzyme. Most commonly, S and I bind at the same site, the active site. This type of inhibition (Figure 6·7a) is termed classical competitive inhibition. However, with some allosteric enzymes, inhibitors bind at a different site on the enzyme (Figure 6·7b), although the inhibition exhibits competitive characteristics. This type of inhibition is called nonclassical competitive inhibition. When both I and S are present in a solution, the proportion of the enzyme that is able to form ES complexes depends on the relative concentrations of S and I and the relative affinities of the enzyme for them.

The formation of EI can be reversed by increasing the concentration of S. At sufficiently high concentrations of S, saturation of E with S can still be achieved. Therefore, the maximum velocity has the same value as in the absence of inhibitor. The more competitive inhibitor present, the more substrate is needed for half-saturation. We have seen that the concentration of substrate at half-saturation is K_m. Thus, in the presence of increasing concentrations of a competitive inhibitor, K_m increases. The new value is usually referred to as the observed or apparent K_m, K_m^{app}. On a double-reciprocal plot, the effect of a competitive inhibitor is a decrease in the absolute value of the intercept at the x axis ($1/K_m$), whereas the y intercept ($1/V_{max}$) remains the same at any concentration of I (Figure 6·8b).

Substrate analogs are compounds that can bind to enzymes because of their similarity to substrates but that react very slowly or not at all. These compounds are usually competitive inhibitors. Two examples of nonmetabolizable substrate

(a)

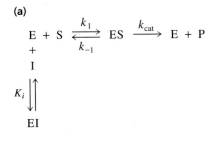

(b)

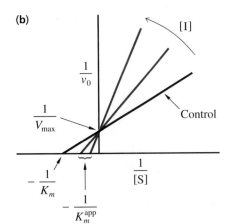

Figure 6·8
Competitive inhibition. **(a)** Kinetic scheme illustrating binding of I to E in competitive inhibition. **(b)** Double-reciprocal plot for competitive inhibition. V_{max} remains unchanged, and K_m increases. The black line labelled "Control" is the result in the absence of inhibitor. The red lines are results in the presence of inhibitor, with the arrow showing the direction of increasing [I].

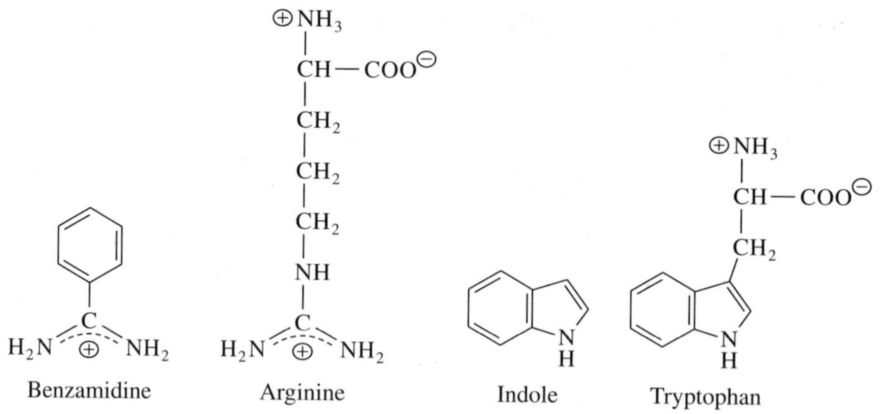

Figure 6·9
Structures of benzamidine and indole. Benzamidine is an analog of an arginine residue in a peptide, and indole is an analog of a tryptophan residue.

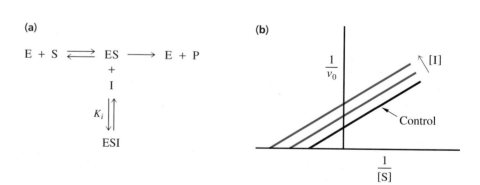

Benzamidine   Arginine   Indole   Tryptophan

analogs are benzamidine, a competitive inhibitor of trypsin, and indole, a competitive inhibitor of chymotrypsin (Figure 6·9). Trypsin catalyzes hydrolysis of peptide bonds on the carboxyl side of arginine and lysine, and benzamidine is an analog of arginine that resembles the alkylguanidyl side chain of the amino acid. Chymotrypsin is specific for bulky aromatic and hydrophobic amino acids, including tryptophan. Recall that tryptophan can be regarded as indolealanine. Benzamidine and indole act as competitive inhibitors by competing with their analogous substrates for the binding pocket of an active site. By systematically varying the structures of substrate analogs and measuring their potencies as inhibitors, one can sometimes draw conclusions about the structure of the active site of an enzyme.

B. Uncompetitive Inhibitors Bind Only to ES

Uncompetitive inhibitors bind only to ES, not to free enzyme (Figures 6·7c and 6·10). In uncompetitive inhibition, V_{max} is decreased ($1/V_{max}$ is increased) by the conversion of some molecules of E to the inactive form ESI. Since it is the ES complex that binds I, the decrease in V_{max} is not reversed by the addition of more substrate. Uncompetitive inhibitors also decrease the K_m (seen as an increase in the absolute value of $1/K_m$ on a double-reciprocal plot) because the equilibria for the formation of both ES and ESI are shifted toward the complexes by the binding of I. Experimentally, the lines on a reciprocal plot representing varying concentrations of an uncompetitive inhibitor all have the same slope, indicating proportionally decreased values for K_m and V_{max} (Figure 6·10b). This type of inhibition usually occurs only with multisubstrate reactions.

Figure 6·10
Uncompetitive inhibition. (a) Kinetic scheme illustrating binding of I to ES in uncompetitive inhibition. (b) Double-reciprocal plot for uncompetitive inhibition. V_{max} and K_m both decrease (i.e., the absolute values of $1/V_{max}$ and $1/K_m$, obtained from the y and x intercepts, respectively, both increase). The ratio of V_{max}/K_m remains unchanged.

(a)

$$E + S \rightleftharpoons ES \longrightarrow E + P$$
$$+$$
$$I$$
$$K_i \updownarrow$$
$$ESI$$

(b)

C. Noncompetitive Inhibitors Bind to Both E and ES

Noncompetitive inhibitors bind to both E and ES, so that EI and ESI complexes—both inactive—can be formed (Figures 6·7d and 6·11). When these inhibitors, which do not structurally resemble substrates, bind to E and ES with equal affinity ($K_i = K_i'$; K_i' is the inhibition constant for ES $\rightleftharpoons$ ESI, as shown in Figure 6·11a), the result is termed pure noncompetitive inhibition. This type of inhibition is characterized by a decrease in V_{max} (increase in $1/V_{max}$) with no change in K_m. On a double-reciprocal plot, the lines for pure noncompetitive inhibition intersect at a point on the 1/[S] axis (Figure 6·11b). The effect of pure noncompetitive inhibition is to reversibly titrate E and ES with I, in essence removing active enzyme molecules from solution. This inhibition cannot be overcome by the addition of S. Pure noncompetitive inhibition is rare, but examples are known among allosteric enzymes. In these cases, the noncompetitive inhibitor probably alters the conformation of the enzyme to a shape that can still bind S but cannot catalyze any reaction.

When an inhibitor binds to E and ES with unequal affinity, the result is termed mixed noncompetitive inhibition. In this case, V_{max} decreases and K_m may either increase or decrease, depending on whether K_i in Figure 6·11a is lower or higher, respectively, than K_i'. On a double-reciprocal plot, the lines for mixed noncompetitive inhibition intersect to the left of the $1/v_0$ axis, anywhere except on the 1/[S] axis (Figure 6·11c). Mixed noncompetitive inhibitors will be encountered in Section 6·8.

(a)

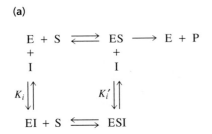

(b)

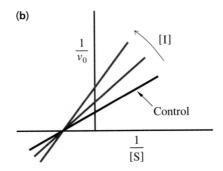

Pure noncompetitive inhibition
($K_i = K_i'$, no change in K_m)

Figure 6·11
Noncompetitive inhibition. **(a)** Kinetic scheme illustrating binding of I to E or ES in noncompetitive inhibition. **(b)** Double-reciprocal plot for pure noncompetitive inhibition, in which I binds to E and ES with equal affinity. V_{max} decreases, but K_m remains the same. **(c)** Double-reciprocal plots for mixed noncompetitive inhibition, in which I binds to E and ES with different affinity: (1) when I has greater affinity for E than ES, K_m increases; (2) when I has greater affinity for ES than E, K_m decreases.

(c) (1)

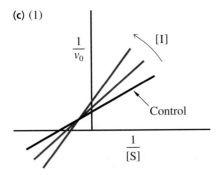

Mixed noncompetitive inhibition
($K_i < K_i'$, K_m increases)

(2)

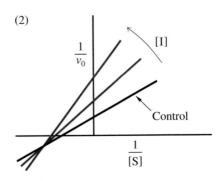

Mixed noncompetitive inhibition
($K_i > K_i'$, K_m decreases)

6·7 Reversible Enzyme Inhibitors Have Both Experimental and Clinical Uses

Reversible enzyme inhibition is not merely a biochemical curiosity but provides a powerful tool for probing enzyme activity and for altering it in the treatment of disease. Information about the shape and chemical reactivity of the active site of an enzyme can be obtained from experiments involving a series of competitive inhibitors with systematically altered structures. One of the most successful inhibition studies has been performed with the enzyme acetylcholinesterase. This enzyme catalyzes the hydrolysis of the neurotransmitter acetylcholine.

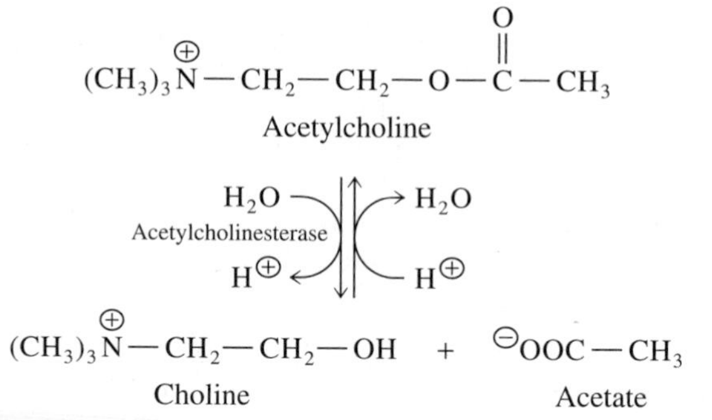

$$(CH_3)_3 \overset{\oplus}{N} - CH_2 - CH_2 - O - \overset{\overset{\textstyle O}{\|}}{C} - CH_3$$

Acetylcholine

$$\text{Acetylcholinesterase}$$

$$(CH_3)_3 \overset{\oplus}{N} - CH_2 - CH_2 - OH \quad + \quad \overset{\ominus}{O}OC - CH_3$$

Choline　　　　　　　　　Acetate

(6·22)

Nerve impulses are relayed at certain junctions between nerves or between a nerve and a muscle by release of the messenger acetylcholine from an activated nerve cell (the presynaptic cell). Acetylcholine diffuses and binds to receptors on a second nerve cell or muscle cell (the postsynaptic cell). When acetylcholine binds to its receptor, it triggers a nerve impulse, a transient wave of membrane depolarization, in the postsynaptic cell. To restore the cell to its resting state, the acetylcholine is rapidly hydrolyzed by the action of acetylcholinesterase.

The active site of acetylcholinesterase has two subsites, an anionic binding site and a catalytic center called the esteratic site (Figure 6·12a). The strength of the electrostatic interaction between the substrate acetylcholine and the anionic site has been measured using competitive inhibitors. For example, dimethylaminoethanol (Figure 6·12b) binds 30 times more tightly than isoamyl alcohol and is therefore a 30-fold more potent inhibitor. The role of the methyl groups of acetylcholine in binding has been studied using ammonium ion and methyl-, dimethyl-, and trimethylammonium ions as inhibitors. Each additional methyl group increases the binding sevenfold, suggesting that hydrophobic binding of the three methyl groups

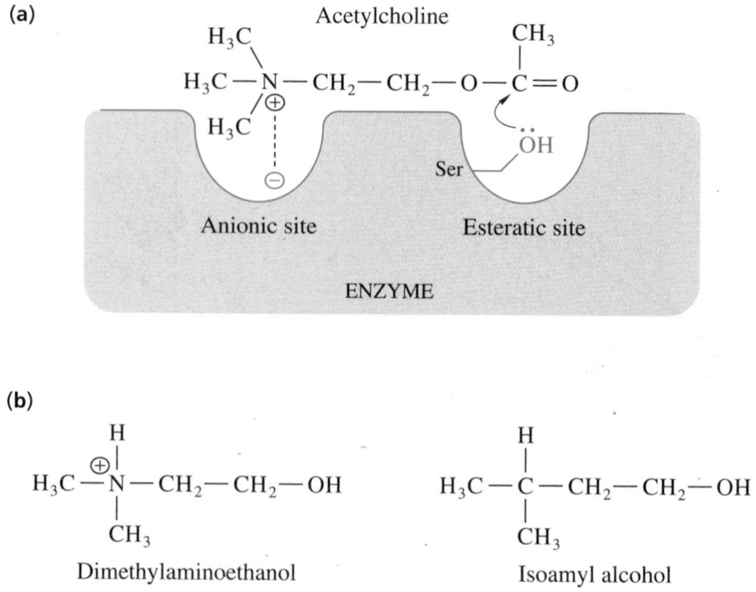

Figure 6·12
Active site of acetylcholinesterase. (a) Schematic map of the active site based on inhibition studies. (b) Dimethylaminoethanol and isoamyl alcohol, two competitive inhibitors of acetylcholinesterase.

(a)

Acetylcholine

Anionic site　　　Esteratic site

ENZYME

(b)

Dimethylaminoethanol　　　　Isoamyl alcohol

also occurs in the anionic site. The esteratic site has also been characterized by inhibition studies, using the irreversible inhibitor diisopropyl fluorophosphate, described in Section 6·9.

The pharmaceutical industry has taken advantage of enzyme-inhibition studies to design clinically useful drugs. In many cases, a naturally occurring enzyme inhibitor is used as the starting point for drug design. Over the years, choices of compounds for synthesis as potential drugs have become less empirical and more rational. Theoretically, with the greatly expanded bank of knowledge about enzyme structure, inhibitors can now be rationally designed to fit the active site of a target enzyme. The effects of a synthetic compound are tested first on isolated enzymes and then in biological systems. Even if a compound has suitable inhibitory activity, other obstacles may be encountered. The drug must reach and permeate the target cells, must not be rapidly metabolized to an inactive compound, and must not be toxic to the host organism. Another difficulty is that organisms often develop resistance to a drug, so that the positive effect of the drug may last only a relatively short time.

In cancer and microbiological or viral infections, the aim of drug treatment is to prevent reproduction of the invading cells or virus without harming host cells. This approach is called selective toxicity. For example, viral infection can be slowed by inhibiting an enzyme essential to the synthesis of viral DNA. The nucleoside 3′-azido-2′,3′-deoxythymidine, or AZT (Figure 6·13a), one of the first drugs used for treatment of acquired immune deficiency syndrome (AIDS), is metabolized to its corresponding 5′-triphosphate; the triphosphate inhibits the enzyme that is responsible for synthesizing DNA from the viral RNA template (Section 25·10C). Methotrexate and trimethoprim (Figure 6·13b) both are competitive inhibitors of dihydrofolate reductase, an enzyme that is essential for synthesis of thymidine monophosphate (TMP). Methotrexate inhibits vertebrate dihydrofolate reductase and can be used to selectively kill rapidly dividing cells; it is one of the most successful cancer chemotherapeutic agents (Section 22·10). Trimethoprim, a potent selective inhibitor of dihydrofolate reductase of prokaryotes but a very poor inhibitor of the human reductase, is used in the treatment of some bacterial infections and malaria.

(a)

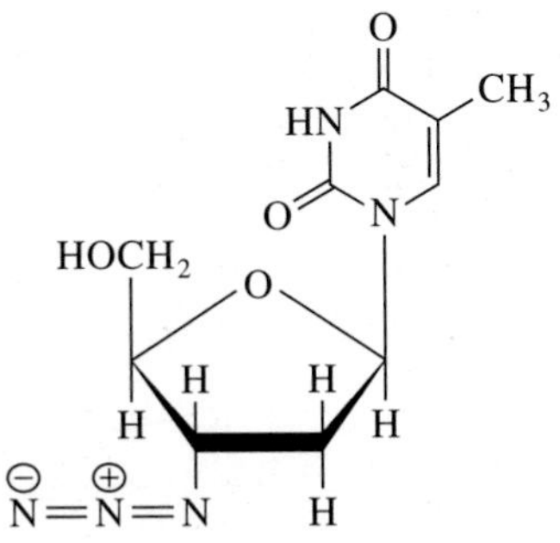

3′-Azido-2′,3′-deoxythmidine
(AZT)

(b)

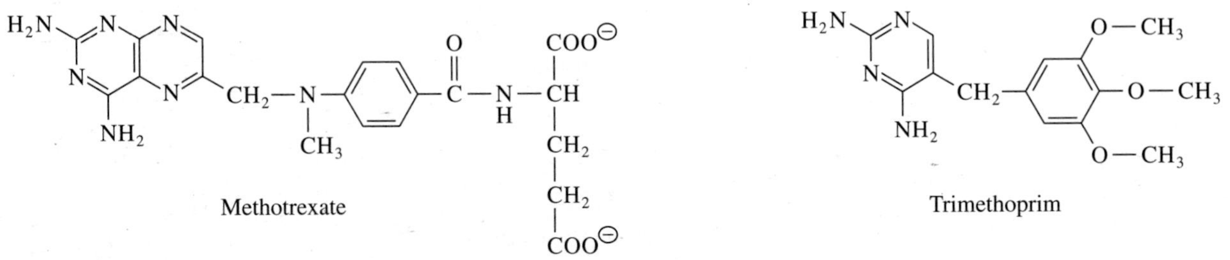

Methotrexate

Trimethoprim

Figure 6·13
Reversible enzyme inhibitors used as drugs.
(a) AZT is metabolized to a 5′-triphosphate that inhibits DNA synthesis from viral RNA.
(b) Methotrexate and trimethoprim are competitive inhibitors of dihydrofolate reductase, an enzyme involved in nucleotide synthesis.

6·8 Kinetic Measurements Can Determine the Order of Binding of Substrates and Release of Products in Multisubstrate Reactions

Kinetic measurements of multisubstrate reactions are a little more complicated than simple one-substrate enzyme kinetics. However, for many purposes, such as designing an enzyme assay, it is sufficient simply to determine the K_m of each substrate in the presence of saturating amounts of each of the other substrates. Furthermore, the simple enzyme kinetics discussed in this chapter can be extended to distinguish among several mechanistic possibilities for multisubstrate reactions, such as group-transfer reactions. This is done by measuring the effect of variations in the concentration of one substrate on the kinetic results obtained for the other. Additional information is obtained by determining the type of reversible inhibition caused by each product of the reaction.

There are several different kinetic schemes by which a multisubstrate reaction could occur. W. W. Cleland introduced a notation that abbreviates a kinetic scheme to a horizontal line or group of lines (Figure 6·14). The sequence of steps proceeds from left to right. The addition of substrate molecules (A, B, C...) to the enzyme and the release of products (P, Q, R...) from the enzyme are indicated by vertical arrows. The various forms of the enzyme, free E or ES complexes, are written under the line. The ES complexes that undergo chemical transformation when the active site is filled are written in parentheses. **Sequential reactions** (Figure 6·14a) are those that require all of the substrates to be present before any product is released. Sequential reactions can be either **ordered,** in which there is an obligatory order for addition of substrates and release of products, or **random,** in which there is no obligatory order of binding or release. In **ping-pong reactions** (Figure 6·14b), a product is released before all of the substrates are bound. For a bisubstrate ping-pong reaction, the first substrate is bound, the enzyme is altered by substitution, and

Figure 6·14
Notation devised by W. W. Cleland for bisubstrate reactions. **(a)** In sequential reactions, all substrates are bound before a product is released. The binding of substrates may be either ordered or random. **(b)** In ping-pong reactions, one substrate is bound and a product is released, leaving a substituted enzyme that binds a second substrate and releases a second product, restoring the enzyme to its original form.

(a) Sequential reactions

Ordered

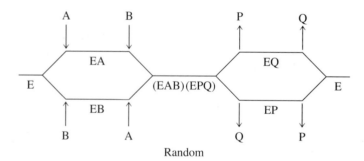

Random

(b) Ping-pong reaction

(a) Sequential

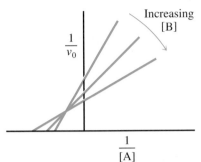

(b) Ping-pong

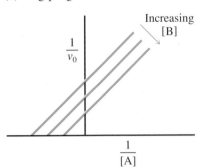

Figure 6·15
Double-reciprocal plots for sequential and ping-pong reactions. In the experiment, the concentration of one substrate (A in the figure) is varied at several fixed concentrations of the second substrate (B), and the initial velocity of each reaction is measured. **(a)** Intersecting lines indicate a sequential kinetic mechanism. **(b)** Parallel lines indicate a ping-pong kinetic mechanism.

the first product is released, after which the second substrate is bound, the altered enzyme is restored to its original form, and the second product is released. The binding and release of ligands in a ping-pong mechanism are usually indicated by slanted lines. The two forms of the enzyme are indicated by E (unsubstituted) and F (substituted).

Sequential and ping-pong reactions can be distinguished by initial-velocity experiments. The concentration of one substrate is varied at several fixed levels of the other substrate, and double-reciprocal plots are drawn, with each line representing a different level of the second substrate. The lines intersect for a sequential mechanism and are parallel for a ping-pong mechanism (Figure 6·15).

To complete the kinetic mechanism (that is, to distinguish ordered from random reactions and to find the order of ligand binding and release), the effect of products as reversible inhibitors with respect to each substrate is measured. This straightforward method was first suggested by Robert Alberty in 1953. Because products are the substrates of the reverse reaction, they act as inhibitors of the forward reaction by binding to the enzyme and preventing the reaction in the forward direction. Each product is tested as a reversible inhibitor at several concentrations, with variable concentrations of one substrate and fixed concentrations of the other. The results can be analyzed as shown in Table 6·5. For example, if a product affects only K_m^{app} of the variable substrate (competitive inhibition), the product and the substrate are competing for the same form of the enzyme. Product Q in the ordered reaction in Figure 6·14 can only bind to free E and therefore affects only the K_m^{app} of the leading substrate, A. Product P in the same reaction can only bind to EQ and thus is a mixed noncompetitive inhibitor of both A and B. Fitting the observations of inhibition studies to Table 6·5 reveals the order of ligand binding and release by identifying the ligands as A or B, P or Q, and distinguishes the type of kinetic mechanism.

Kinetic-mechanism studies like these can be performed with very little enzyme. They supplement information found by other experiments and can assist in the elucidation of the chemical mechanisms of enzymes.

Table 6·5 Determination of kinetic mechanism from types of product inhibition for the general reaction $A + B \rightleftharpoons P + Q$

Product tested as reversible inhibitor:	P	Q	P	Q	
Variable substrate:	A	A	B	B	**Kinetic mechanism**
Observations	Mixed noncompetitive	Competitive	Mixed noncompetitive	Mixed noncompetitive	ORDERED
	Competitive	Competitive	Competitive	Competitive	RANDOM
	Mixed noncompetitive	Competitive	Competitive	Mixed noncompetitive	PING-PONG

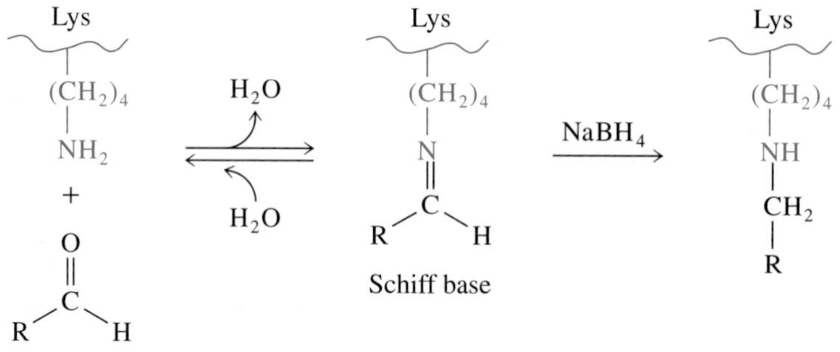

Figure 6·16

Reaction of the ε-amino group of a lysine residue with an aldehyde. Reduction of the Schiff base with sodium borohydride (NaBH$_4$) forms a stable substituted enzyme.

Diisopropyl fluorophosphate (DFP)

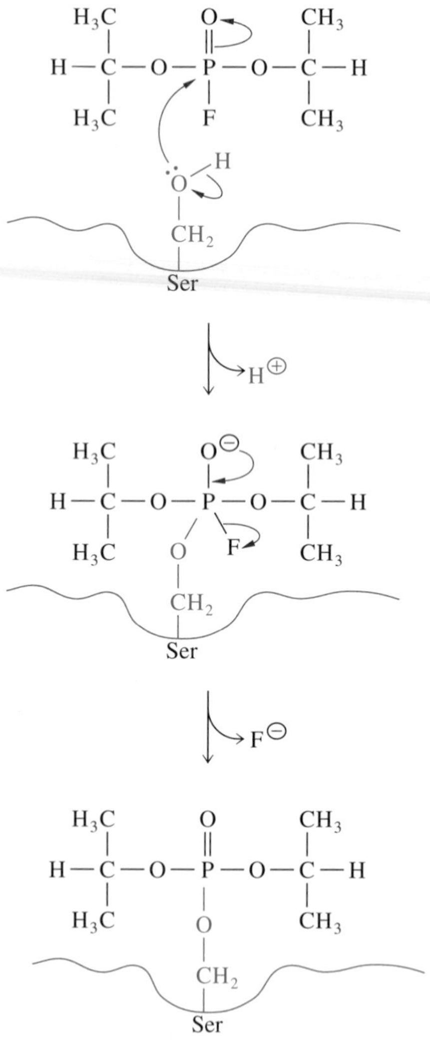

Diisopropylphosphoryl-hydrolase

Figure 6·17

Reaction of diisopropyl fluorophosphate (DFP) with a single, highly nucleophilic serine residue at the active site of a hydrolase, producing inactive diisopropylphosphoryl-hydrolase. DFP inactivates serine proteases and serine esterases. DFP reacts with Ser-195 of chymotrypsin.

6·9 Irreversible Inhibitors Covalently Modify Enzymes

So far, we have discussed the kinetics and applications of reversible enzyme inhibitors. Irreversible inhibition is also experimentally useful. An irreversible inhibitor forms a stable covalent bond with an enzyme molecule. In effect, it inactivates an enzyme by titrating the active site. Formation of a covalent E–I complex occurs in two steps.

$$E^{\ominus} + I-X \;\rightleftharpoons\; E^{\ominus}\cdots I-X \;\rightleftharpoons\; E-I + X^{\ominus} \qquad (6·23)$$

First, a reversible complex is formed between the enzyme ($E^{\ominus}$) and the inhibitor (I–X). A reactive nucleophilic group of the enzyme attacks the inhibitor and is substituted, releasing the leaving group $X^{\ominus}$. An irreversible inhibitor usually substitutes the side chain of an active-site residue; alternatively, it can alter the conformation of an enzyme or block entry of substrate to the active site. Irreversible inhibitors, when analyzed by double-reciprocal plots, mimic noncovalent inhibitors, but since there is no dissociation of E–I, the apparent K_i values obtained are meaningless. Noncompetitive inhibition can be differentiated from irreversible inhibition by testing for reactivation of the reversibly inhibited enzyme by dialysis or dilution. An enzyme may be protected from irreversible inhibition by a substrate or substrate analog. Further evidence for irreversible inhibition can be obtained by observing the time-dependence of inhibition; the extent of irreversible inhibition increases with time. (In contrast, reversible enzyme inhibition rapidly reaches a steady level.)

An important use of irreversible inhibitors is the identification of amino acid residues at the active site by specific substitution of their reactive side chains. A covalent inhibitor that reacts with only one type of amino acid is added to a solution of enzyme, which is tested for loss of activity after incubation. Ionizable side chains are modified by acylation or alkylation reactions. For example, free amino groups such as the ε-amino group of lysine react with an aldehyde to form a Schiff base that can be stabilized by reduction with sodium borohydride, NaBH$_4$ (Figure 6·16).

The nerve gas diisopropyl fluorophosphate (DFP) is one of a group of organic phosphorus compounds that inactivate hydrolases possessing a reactive serine as part of the active site (Figure 6·17). These enzymes are termed serine proteases or serine esterases, depending on reaction specificity. DFP reacts with serine residues to produce diisopropylphosphoryl-serine. Some of the organophosphorus inhibitors are used in agriculture as insecticides; others, such as DFP, are useful reagents for enzyme research. The original nerve gases are extremely toxic poisons developed for military uses. For all of these compounds, the major biological action is irreversible inhibition of acetylcholinesterase, which can cause paralysis.

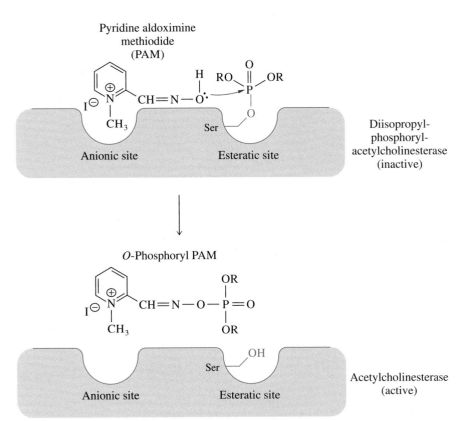

Pyridine aldoximine
methiodide
(PAM)

Diisopropyl-
phosphoryl-
acetylcholinesterase
(inactive)

Anionic site

Esteratic site

O-Phosphoryl PAM

Acetylcholinesterase
(active)

Anionic site

Esteratic site

Figure 6·18
Reactivation of diisopropylphosphoryl-
acetylcholinesterase. Pyridine aldoximine
methiodide (PAM) binds to the anionic site
and displaces the phosphoryl group at the
esteratic site. *R* represents an isopropyl
group.

Knowledge of the structure of the active site of acetylcholinesterase assisted in design of an antidote to treat nerve-gas poisoning. Strong nucleophiles such as hydroxylamine (NH_2OH) can displace phosphate and reactivate the substituted enzyme. Low concentrations of pyridine aldoximine methiodide are as effective as high concentrations of NH_2OH because the positively charged methylpyridine group binds to the anionic site of acetylcholinesterase, placing its nucleophilic oxygen close to the phosphorus atom of the organic ester (Figure 6·18).

DFP was used to identify a serine residue as an active-site component of chymotrypsin. Only 1 serine residue out of the 27 in the enzyme is sufficiently nucleophilic to attack the fluorophosphate of DFP and form a stable phosphoester. When DFP labelled with radioactive (^{32}P) phosphate was used in this reaction, one mole of the phosphate compound was incorporated per mole of chymotrypsin, and the enzyme was inactivated. After the resulting substituted chymotrypsin was partially hydrolyzed in acid, a radioactive peptide with the sequence Asp–^{32}P-Ser–Gly was isolated. When the enzyme was completely sequenced, the substituted serine was identified as Ser-195. Inactivation by DFP and similar compounds is used as a test for active-site serine hydroxyl groups in hydrolytic enzymes.

More useful than general substituting reagents are irreversible inhibitors with structures that allow them to bind specifically to an active site. These inhibitors are referred to as active-site directed reagents, or **affinity labels.**

An affinity label was synthesized to study the active-site residues of chymotrypsin. Chymotrypsin catalyzes the hydrolysis of peptide bonds whose carbonyl group is contributed by a bulky, hydrophobic residue, although it can also catalyze

(a)

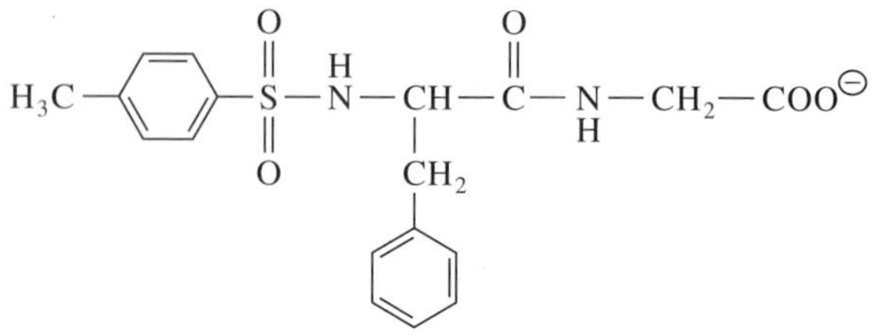

Tosylphenylalanylglycine

(b)

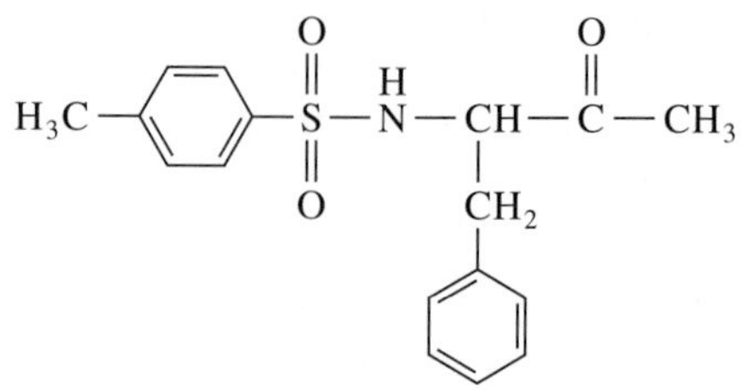

N-Tosylamidophenylethyl methyl ketone
(TPMK)

(c)

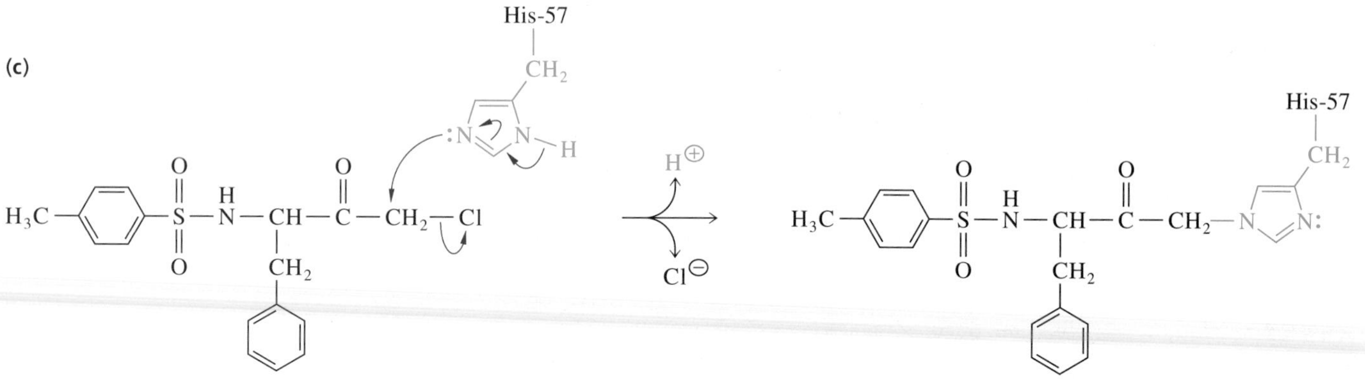

N-Tosylamidophenylethyl chloromethyl ketone
(TPCK)

Alkylated derivative of His-57 at active site of chymotrypsin
(inactive)

Figure 6·19
Structural comparison of a substrate, a reversible competitive inhibitor, and an irreversible inactivator of chymotrypsin.
(a) Structure of tosylphenylalanylglycine, an artificial substrate of chymotrypsin. The peptide bond cleaved by the action of chymotrypsin is shown in red. (b) Structure of *N*-tosylamidophenylethyl methyl ketone, TPMK. This substrate analog acts as a competitive inhibitor of chymotrypsin. (c) Structure and reactivity of *N*-tosylamidophenylethyl chloromethyl ketone, TPCK, an irreversible inhibitor of chymotrypsin. The phenylalanyl side chain of TPCK has affinity for the substrate-binding site of chymotrypsin, while its chloromethyl ketone group is susceptible to attack by nucleophiles such as the imidazole side chain of His-57 at the reactive site. Chymotrypsin activity is lost as His-57 becomes alkylated by reaction with TPCK.

the hydrolysis of esters or simple amides. Tosylphenylalanylglycine (Figure 6·19a) is a substrate for which chymotrypsin has shown specificity. Replacement of the nitrogen-containing leaving group of this substrate with a hydrocarbon group results in a ketone, *N*-1-tosylamido-2-phenylethyl methyl ketone, or TPMK (Figure 6·19b). This compound is a competitive inhibitor of chymotrypsin because it binds to the hydrophobic specificity site of the enzyme but is not subject to nucleophilic attack by the catalytic residues of the enzyme. The ketone can be further modified by substituting a very electronegative group for a hydrogen of its methyl group, making it susceptible to attack by a nucleophilic residue of chymotrypsin but not subject to subsequent hydrolysis. Hence, the chloromethyl ketone (*N*-1-tosylamido-2-phenylethyl chloromethyl ketone, TPCK), synthesized by Guenther Schoellman and Elliott Shaw in the 1960s and pictured in Figure 6·19c, reacts irreversibly with chymotrypsin, forming a covalent bond with a nucleophilic residue in the active site. When Schoellman and Shaw analyzed the covalently modified enzyme, they found that the chymotrypsin was inactive and that His-57 was modified, suggesting its involvement in catalysis. In some hydrolases, a nucleophilic cysteine residue at the active site, rather than a histidine residue, is alkylated by TPCK.

6·10 Site-Directed Mutagenesis Is Used to Produce Modified Enzymes

The latest technique for individually testing the functions of the amino acid side chains of an enzyme is **site-directed mutagenesis.** Mutagenesis is more specific than irreversible inhibition and also allows the testing of amino acids for which there are no specific chemical inactivators. In this procedure, one amino acid is specifically replaced by another through the biosynthesis of a modified enzyme. To begin, the gene that encodes an enzyme is isolated and sequenced. The enzyme can be synthesized in bacterial cells by inserting the gene into a suitable vector, such as a virus, that can be used to transform bacteria. After verifying that the unmodified, or wild-type, enzyme is synthesized by the bacteria, DNA is prepared with the base sequence altered to encode a modified protein. The bacteria are infected with the new DNA, just as they were infected in the control experiment. The mutant protein is synthesized, isolated, and tested for enzymatic activity. If possible, the enzyme is purified and its structure checked by X-ray crystallography to make sure that the mutation has not altered its overall conformation.

A successful example of site-directed mutagenesis is the alteration of the bacterial peptidase subtilisin to make it more resistant to chemical oxidation. Subtilisin has been added to detergent powders to help remove protein stains such as chocolate and blood. Resistance to oxidation, by bleach for instance, increases the suitability of subtilisin as a detergent additive. Subtilisin has a methionine residue at position 222 in the active-site cleft that readily oxidizes, leading to inactivation of the enzyme. In a series of mutagenic experiments, Met-222 was systematically replaced by each of the other common amino acids. All 19 possible mutant subtilisins were isolated and tested for peptidase activity. Most had greatly diminished activity. The Cys-222 mutant had high activity but was also subject to oxidation. The Ala-222 and Ser-222 mutants, with side chains that are nonoxidizable, were not inactivated by oxidation and had relatively high activity. They were the only active and oxygen-stable mutant subtilisin variants. Several studies of active sites by site-directed mutagenesis will be discussed in later sections.

6·11 Serine Proteases Exemplify Many Features of Enzyme Activity

Having examined the general properties of enzymes as revealed by kinetic methods, we now turn to the specific example of serine proteases to explore the relationship between protein structure and catalytic function. We will see how the activity of serine proteases is regulated by zymogen activation and inhibitor proteins, and we will examine the structural basis for substrate specificity. The chemical mechanism of catalysis will be covered in Chapter 7.

A. Chymotrypsin Is Synthesized as an Inactive Precursor

All members of the family of serine proteases, which includes the coagulation factors and the digestive enzymes chymotrypsin, trypsin, and elastase, have a functional serine residue in their active sites and a common catalytic mechanism. In mammals, serine proteases act outside the cells in which they are synthesized. This poses a problem that was raised earlier (Section 3·6): how are the proteins of the cells in which proteases are assembled protected from hydrolytic destruction catalyzed by these enzymes? Cells have solved this problem by a simple type of irreversible regulation, the synthesis of the proteases (and certain other potentially destructive enzymes) as **zymogens,** or proenzymes, inactive precursors that must be covalently modified to become active. The pancreatic zymogens trypsinogen, chymotrypsinogen, proelastase, and procarboxypeptidase are activated extracellularly under appropriate physiological conditions by selective proteolysis—enzymatic cleavage of one or a few specific peptide bonds. This activation process effectively controls the activity of these proteases. We will discuss the metabolic aspects of activation of digestive enzymes in detail in Chapter 13.

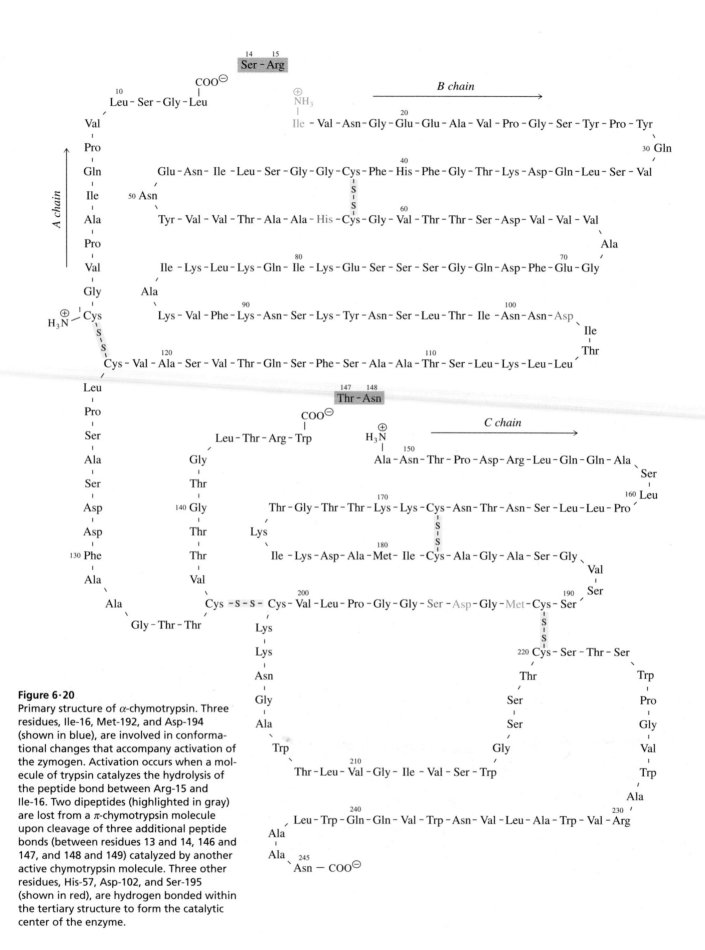

Figure 6·20
Primary structure of α-chymotrypsin. Three residues, Ile-16, Met-192, and Asp-194 (shown in blue), are involved in conformational changes that accompany activation of the zymogen. Activation occurs when a molecule of trypsin catalyzes the hydrolysis of the peptide bond between Arg-15 and Ile-16. Two dipeptides (highlighted in gray) are lost from a π-chymotrypsin molecule upon cleavage of three additional peptide bonds (between residues 13 and 14, 146 and 147, and 148 and 149) catalyzed by another active chymotrypsin molecule. Three other residues, His-57, Asp-102, and Ser-195 (shown in red), are hydrogen bonded within the tertiary structure to form the catalytic center of the enzyme.

As an example of the covalent changes that take place upon activation of a zymogen, let us examine the conversion by sequential cleavages of chymotrypsinogen to active α-chymotrypsin, whose primary structure is shown in Figure 6·20. For convenience, the numbering of the chymotrypsinogen peptide sequence is retained here for chymotrypsin. Trypsin-catalyzed cleavage of a single peptide bond in chymotrypsinogen between Arg-15 and Ile-16 results in a fully active but rather unstable enzyme known as π-chymotrypsin, in which the 15-residue N-terminal peptide remains bound by a disulfide bond between Cys-1 and Cys-122. Three additional peptide bonds of π-chymotrypsin, between residues 13 and 14, 146 and 147, and 148 and 149, are susceptible to cleavage by active chymotrypsin molecules, a process termed autoactivation. Examination of the three-dimensional structure of chymotrypsinogen showed that these three bonds are available for hydrolysis because they are on the surface of π-chymotrypsin. Their cleavage results in the excision of dipeptides Ser-14–Arg-15 and Thr-147–Asn-148. Hence, the most stable form of the enzyme, known as α-chymotrypsin or simply chymotrypsin, is a three-chained protein containing 241 amino acid residues. The tertiary structure of chymotrypsin is stabilized by its five disulfide bonds joining residues 1 and 122, 42 and 58, 136 and 201, 168 and 182, and 191 and 220.

X-ray crystallography has revealed one major difference between the conformation of chymotrypsinogen and chymotrypsin: the lack of a hydrophobic substrate-binding pocket in the zymogen. The differences in structure are shown in Figure 6·21. Upon zymogen activation, the newly generated α-amino group of Ile-16 turns inward and interacts with the β-carboxyl group of Asp-194 to form an ion pair. This local conformational change pulls the side chain of Met-192 away from its hydrophobic contacts in the zymogen conformation, thereby generating a relatively hydrophobic substrate-binding pocket. Near this substrate-binding site are three ionizable side chains (Asp-102, His-57, and Ser-195) that participate in catalysis.

A number of zymogens participate in **enzyme cascades.** A small number of initial zymogen molecules are activated, and each sequentially activates many more zymogens, ultimately leading to a large concentration, and thus high activity, of an active primary enzyme. Enzyme cascades result in rapid signal amplification, as

Figure 6·21
Stereo overlay of the backbones of chymotrypsinogen (purple) and α-chymotrypsin (blue). Ile-16, Asp-194, and the catalytic-site residues (Asp-102, His-57, and Ser-195) in both the zymogen and the active enzyme are shown in red. A portion of the substrate is shown in green. (Chymotrypsin structure based on coordinates provided by A. Tulinsky and R. Blevins; chymotrypsinogen structure based on coordinates provided by D. Wang, W. Bode, and R. Huber.)

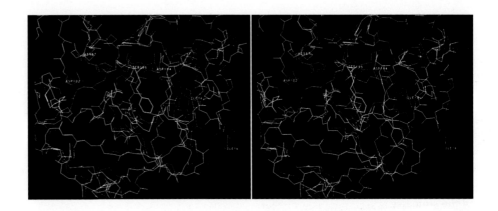

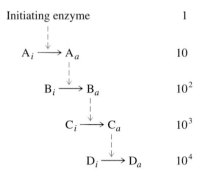

	Number of active enzyme molecules
Initiating enzyme	1
$A_i \longrightarrow A_a$	10
$B_i \longrightarrow B_a$	10^2
$C_i \longrightarrow C_a$	10^3
$D_i \longrightarrow D_a$	10^4

Figure 6·22
Principle of enzyme cascades. A cascade of enzyme activity results when several enzymes are sequentially activated. A very small concentration of the initiating enzyme converts inactive enzyme A_i to active A_a, which in turn converts inactive B_i to active B_a, and so forth. The number of activated enzymes increases exponentially (arbitrarily shown here as 10-fold), so that the effect of the original activating enzyme is enormously amplified.

illustrated in Figure 6·22. The pancreatic zymogens and some coagulation factors are activated in enzyme cascades. We will see in Section 12·9 that enzyme cascades are an important feature of hormone-responsive intracellular signalling pathways. In addition to amplifying signals or biochemical processes, multistep cascades afford the opportunity for regulation at many points so that the final result of the cascade is finely tuned to the needs of the cell or organism.

B. Blood Coagulation Involves a Cascade of Zymogen Activation

In mammals, an elaborate enzyme cascade is responsible for blood clotting, or coagulation. Coagulation involves more than a dozen proteins, some of which circulate in the bloodstream as zymogens of serine proteases. These proteins are thus available at any time but inactive until needed. Blood coagulation is normally initiated by damage to a blood vessel, but it may also occur in response to bacterial infection, cancer, or other disease. Coagulation leads to a blood clot, a mass of insoluble fibrous protein molecules and specialized blood cells called platelets. Blood clots may form in vessels with low blood flow, such as in the lower limbs of bedridden individuals. Blood clots that form in the coronary arteries, the vessels that serve the heart muscle, may cause heart attacks.

The coagulation zymogens are rapidly activated in a cascade fashion. Each serine protease is highly specific for its substrate, catalyzing the hydrolysis of one or two peptide bonds. Sequential activation of zymogens ultimately generates thrombin, the protease that catalyzes the conversion of the soluble circulating protein fibrinogen to insoluble fibrin (Figure 6·23). The inactive forms of the enzymes of the coagulation proteins, or factors, are indicated by Roman numerals that roughly parallel their order of discovery; the subscript a signifies an active factor. Thrombin and factors VII_a, IX_a, X_a, XI_a, and XII_a are serine proteases that are homologous to other serine proteases such as trypsin. Prothrombin and factors VII, IX, and X contain calcium-binding sites; Ca^{2+} is essential for localizing these coagulation factors to the membrane surfaces of platelets. Tissue factor and factors V_a and VIII_a are nonenzymatic cofactor proteins that promote the activity of particular proteases. The common form of hemophilia (an inherited disorder involving defective blood clotting), hemophilia A, results from a deficiency of factor VIII; the less common hemophilia B arises from a deficiency of factor IX. The genes for both factors VIII

Figure 6·23
Mammalian blood coagulation. A cascade involving the activities of serine-protease zymogens ultimately generates thrombin, the enzyme responsible for the proteolytic conversion of fibrinogen to fibrin, which forms a fibrous clot. Factors VII_a, IX_a, X_a, XI_a, and XII_a are also serine proteases; the subscript a indicates the active enzyme. Tissue factor and factors V_a and VIII_a are nonenzymatic proteins that promote the activity of some of the proteases. The conversion of precursors to active forms is indicated by black arrows; the blue arrows indicate the activities of the proteases.

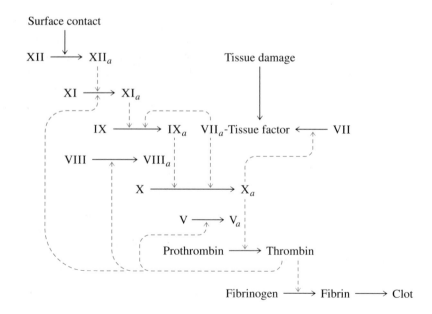

Figure 6·24
Fibrinolytic system. Plasminogen activators proteolytically activate the zymogen plasminogen. Plasmin, the active protease, catalyzes the degradation of fibrin, thereby helping to dissolve blood clots. The blue arrows indicate the activity of the proteases.

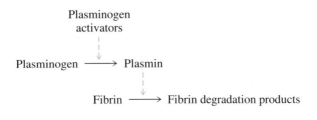

and IX are on the X chromosome, making hemophilia an X-linked disease. Individuals with hemophilia A can now be treated by regular injection of a factor VIII preparation made by cloning and expressing the DNA that codes for the protein.

The spontaneous association of fibrin molecules produces a meshwork of fibers that traps circulating blood cells, gradually blocking the vessel and preventing further blood flow. The classical cascade hypothesis for coagulation has undergone modification as the molecular details of the proteases, their substrates, and their cofactor proteins have been illuminated, but its general features are still valid.

There are two notable features of the coagulation reactions outlined in Figure 6·23. First, the cascade is not merely a linear arrangement of zymogen activation reactions, but it is branched and contains a number of positive-feedback loops. In fact, tissue factor released from injured cells may be the normal physiological initiator of coagulation. Note also that thrombin, the ultimate protease of coagulation, activates factor XI, a protease further upstream in the cascade. Thrombin also activates factors V and VIII, which are accessory proteins in the reactions catalyzed by factors X_a and IX_a, respectively. Thus, thrombin promotes its own activation by accelerating previous steps of the cascade. This feature of the coagulation process allows minute amounts of thrombin, if not checked, to create an explosive response greater than the amplification allowed by a strictly linear cascade.

The second important feature of the coagulation cascade is the presence of multiple proteins and catalytic steps, each of which can be regulated. For virtually every step of coagulation, a control mechanism is known. The activities of the proteases are limited by specific proteins that inhibit their proteolytic activity, and the cofactor proteins are regulated by their susceptibility to degradation. The careful control of the coagulation reactions ensures that blood clots form only where they are needed and only for as long as they are needed.

The counterpart of coagulation is the fibrinolytic system in blood, in which the zymogen plasminogen is activated to the serine protease plasmin by the proteolytic activity of plasminogen activators (Figure 6·24). Plasmin catalyzes the degradation of fibrin to soluble fragments and helps limit the physical extent as well as the duration of blood clots. Plasminogen activators are often administered to patients following heart attacks to help restore blood flow to clotted vessels.

C. X-Ray Crystallographic Analysis Revealed the Basis for the Substrate Specificity of Serine Proteases

Many serine proteases, including those involved in digestion and coagulation, share similarities in primary, secondary, and tertiary structure, as revealed by sequence analysis and X-ray crystallography. Similarities in the backbone conformations and active-site residues of chymotrypsin, trypsin, and elastase may be seen by comparing Figures 6·25a, b, and c (next page). Each enzyme has a two-lobed structure with the active site located in a cleft between the two lobes. In addition to the active-site serine and histidine residues found by chemical substitution reactions, an aspartate residue was found by X-ray crystallographic experiments. The side chains of these three amino acid residues are hydrogen bonded. The alignment of the serine, histidine, and aspartate residues is highlighted in Figures 6·25a, b, and c.

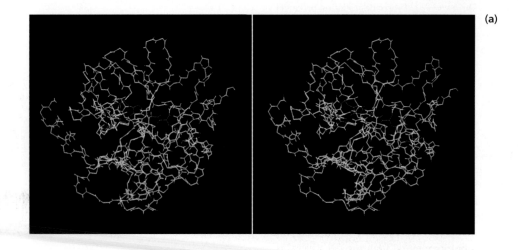

(a)

Figure 6·25
Stereo views of backbone models of chymotrypsin, trypsin, and elastase with residues at the catalytic center shown in red.
(a) Chymotrypsin. (Based on coordinates provided by A. Tulinsky and R. Blevins.) (b) Trypsin. (Based on coordinates provided by J. Walter, R. Huber, and W. Bode.) (c) Elastase. (Based on coordinates provided by T. Prange and I. L. de la Sierra.)

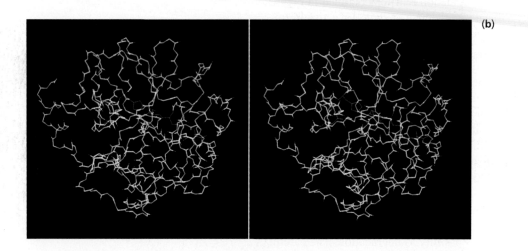

(b)

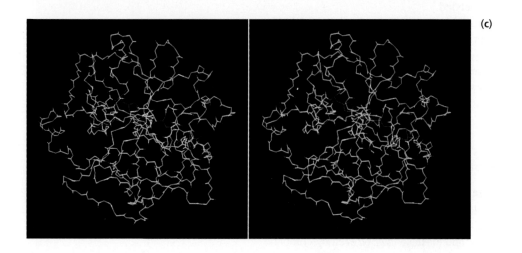

(c)

The substrate specificities of chymotrypsin, trypsin, and elastase can be accounted for by relatively small structural differences in the enzymes. Recall that trypsin catalyzes hydrolysis of peptide bonds whose carbonyl groups are contributed by arginine or lysine. Like chymotrypsin, trypsin contains a binding pocket that correctly positions its substrates for nucleophilic attack by the active-site residue Ser-195. Each protease has a similar extended region into which polypeptides fit, but the so-called specificity pocket near the active-site serine is markedly different for each enzyme. The key difference in structure between trypsin and chymotrypsin is that an uncharged amino acid residue at the base of the hydrophobic binding pocket of chymotrypsin (Figure 6·26a) is replaced in trypsin by an aspartate residue (Figure 6·26b). This negatively charged aspartate residue is responsible for the substrate specificity of trypsin. In the ES complex, it forms an ion pair with the positively charged side chains of arginine and lysine residues of substrates.

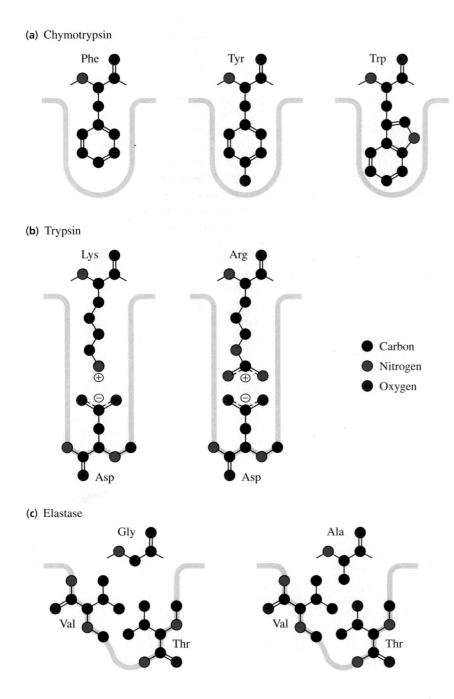

(a) Chymotrypsin

Phe   Tyr   Trp

(b) Trypsin

Lys   Arg

⊕   ⊕
⊖   ⊖

Asp   Asp

● Carbon
● Nitrogen
● Oxygen

(c) Elastase

Gly   Ala

Val   Val
    Thr   Thr

Figure 6·26
Binding sites of chymotrypsin, trypsin, and elastase. The differences in the substrate specificities of these three serine proteases result from differences in their binding sites. **(a)** Chymotrypsin has a hydrophobic pocket that binds the side chains of aromatic amino acid residues. **(b)** A negatively charged aspartate residue sits at the bottom of the binding pocket of trypsin. As a result, trypsin binds the positively charged side chains of lysine and arginine residues. Because the side chain of lysine is one atom shorter than that of arginine, its ion pair is mediated through a molecule of water (not shown). **(c)** The binding pocket of elastase is more shallow due to the presence of valine and threonine residues. Thus, elastase binds only amino acids with small side chains, especially glycine and alanine. Residues with larger side chains cannot be positioned properly for catalysis.

Most of the coagulation proteases resemble trypsin in substrate specificity and probably in tertiary structure as well.

Elastase catalyzes the degradation of elastin, a fibrous protein that is rich in glycine and alanine (Section 5·8), by specifically cleaving peptide bonds in which the carbonyl group belongs to a small residue with an uncharged side chain. Elastase is similar in tertiary structure to chymotrypsin except that the binding pocket of elastase is much shallower. Two glycine residues found at the entrance of the binding site of chymotrypsin and trypsin are replaced in elastase by much larger valine and threonine residues (Figure 6·26c). These residues keep potential substrates with large side chains away from the catalytic center.

A substrate-binding pocket near the catalytic center of an enzyme is not the only determinant of substrate specificity. Many of the peptide bonds hydrolyzed by the action of the coagulation enzymes involve lysine or arginine residues, yet coagulation enzymes do not cleave all bonds involving lysine or arginine residues. For example, the active site of factor IX_a, which catalyzes hydrolysis of an Arg—Ile bond, is so exquisitely tuned that the enzyme has absolute specificity for its natural substrate, factor X; factor IX_a is virtually unreactive toward synthetic oligopeptides containing the same Arg—Ile bond.

6·12 Regulatory Enzymes Are Usually Oligomers

Coordination of the metabolism of an organism requires a number of control mechanisms. One important metabolic control involves the rapid and reversible modulation of the activity of **regulatory enzymes** located at critical points in metabolic pathways. The activities of these enzymes rise when the concentrations of their substrates increase and fall when there is accumulation of the products of their metabolic pathways. Sometimes regulatory enzymes are activated or inhibited by metabolites. When you study metabolism, you will find that these modulations are easy to remember because they are simple and often predictable. By responding to metabolic signals, regulatory enzymes adjust the flux of reactants through entire metabolic pathways. Generally, allosteric phenomena (Section 5·18) are responsible for the reversible control of the activity of these enzymes. Regulatory enzymes are often located at the first step that is unique to a metabolic pathway, called the first committed step of the pathway. This location allows the entire pathway to be regulated by the activity of the first enzyme. Inhibition of the first enzyme of a pathway conserves both material and energy by preventing the accumulation of intermediates and the ultimate end product.

As an example of such regulation, consider the following linear multienzyme pathway, which produces P from A. The end-product P is an allosteric inhibitor of the regulatory enzyme E_1, which catalyzes the first committed step.

$$A \xrightarrow{E_1} B \xrightarrow{E_2} C \xrightarrow{E_3} D \xrightarrow{E_4} \longrightarrow \longrightarrow \longrightarrow P \qquad (6·24)$$

This type of regulation is called **feedback inhibition.** When pathways are branched, regulation may be slightly more complicated. For example, there may be several enzymes catalyzing the first committed step, each one regulated by a different modulator, or complete inhibition of a regulatory enzyme may require binding of several modulators.

We have seen how the conformation of hemoglobin and its affinity for oxygen change when 2,3-*bis*phosphoglycerate is bound (Section 5·18). Regulatory enzymes also undergo allosteric transitions between active R states and inactive

T states. A regulatory enzyme has a second ligand-binding site away from its catalytic center. This second site is called the **regulatory site.** The conformational change associated with the binding of an allosteric activator or inhibitor to its regulatory site is transmitted to the active site of the enzyme, which changes shape sufficiently to alter its activity. The regulatory and catalytic sites are physically distinct regions of the protein, usually on separate domains or separate subunits. Regulatory enzymes are often larger than simple nonregulatory enzymes. For this reason, less is known about the three-dimensional structures of regulatory proteins.

Examination of regulatory enzymes has shown that they have a number of general features.

1. The activities of regulatory enzymes are sensitive to metabolic inhibitors and activators. These modulators, or effectors, seldom resemble the substrates or products of the controlled enzyme. In fact, consideration of the structural differences between substrates and metabolic inhibitors originally led to the conclusion that regulatory compounds are bound to regulatory sites separate from the catalytic sites.

2. Typical allosteric modulators bind noncovalently to the enzymes they regulate. Many modulators alter the K_m of the enzyme for a substrate, others the V_{max} of the enzyme. Modulators themselves are not altered chemically by the enzyme. (There is a special group of regulatory enzymes whose activity is controlled by covalent modification.)

3. Often, a regulatory enzyme can be partially denatured in vitro with loss or impairment of its regulatory properties but no loss of catalytic activity. This process, called desensitization, reinforces the conclusion that catalytic and regulatory events occur at different sites on the enzyme.

4. A regulatory enzyme usually has at least one substrate for which the v_0 versus [S] curve is sigmoidal rather than hyperbolic. A sigmoidal curve is caused by cooperative binding of substrate (Section 5·17), which in turn indicates the presence of multiple substrate-binding sites in such an enzyme.

   Figure 6·27 illustrates the regulatory role that cooperative binding can play. Addition of an activator can shift the sigmoidal curve toward a hyperbolic shape, lowering the apparent K_m, the concentration of substrate required for half-saturation, and raising the activity at a given [S]. Addition of an inhibitor can raise the apparent K_m of the enzyme and lower its activity at any particular concentration of substrate.

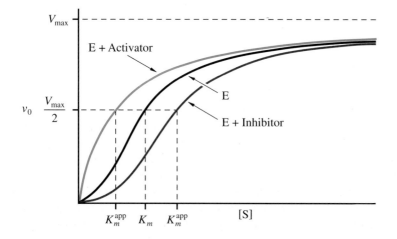

Figure 6·27
Role of cooperativity of binding in regulation. The activity of an allosteric enzyme that has a sigmoidal binding curve can be altered markedly when either an activator or an inhibitor is bound to the enzyme. Addition of an activator can lower the apparent K_m, raising the activity at a given [S]. Conversely, addition of inhibitor can raise the apparent K_m, producing less activity at any given [S].

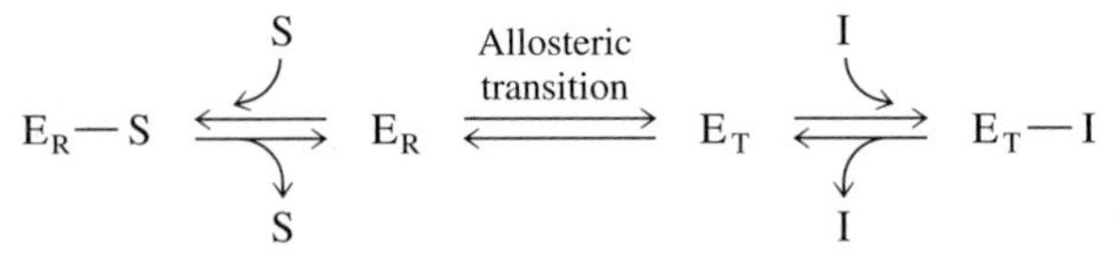

.or on the
ꞏreferentially
ꞏnd I binds pref-
ꞏtion (E_T). Note
ꞏ does not show
ꞏeracting binding

$$E_R\text{—}S \;\rightleftharpoons\; E_R \;\overset{\text{Allosteric}}{\underset{\text{transition}}{\rightleftharpoons}}\; E_T \;\rightleftharpoons\; E_T\text{—}I$$

5. With few exceptions, regulatory enzymes possess quaternary structure. Not all enzymes composed of subunits are regulatory enzymes, however. The individual polypeptide chains of a regulatory enzyme may be identical or different. For those with identical subunits, each polypeptide chain contains both the catalytic and regulatory sites, and the oligomer is a simple symmetric complex—most often a dimer or a tetramer. The oligomers of regulatory enzymes that contain nonidentical subunits have a variety of more complex but still symmetric aggregations.

The allosteric R $\rightleftharpoons$ T transition between the active and the inactive conformations of a regulatory enzyme is rapid. The ratio of the R to T conformations is controlled by the concentrations of the various ligands and the relative affinities of each conformation for those ligands. In the simplest cases, substrate and activator molecules bind only to the R state, and inhibitor molecules bind only to the T state. As outlined in Figure 6·28, addition of S leads to an increase in the concentration of enzyme in the R conformation. Conversely, addition of I increases the proportion of the T species. Activator molecules bind preferentially to the R conformation, leading to an increase in the R:T ratio.

Some allosteric inhibitors are competitive inhibitors even though they are not substrate analogs and do not bind at the active site (i.e., they are nonclassical competitive inhibitors). For example, in the presence of the allosteric inhibitor in Figure 6·27, the enzyme has a higher apparent K_m for its substrate but an unaltered V_{max}. Therefore, this allosteric effector is a competitive inhibitor.

Some regulatory enzymes exhibit noncompetitive inhibition patterns. Binding of a modulator at the allosteric site does not prevent substrate from binding, but it appears to distort the conformation of the active site sufficiently to decrease the activity of the enzyme.

6·13 Two Models Have Been Proposed to Describe Allosteric Regulation

Two models that account for the cooperativity of binding of ligands to oligomeric proteins have gained general recognition. Both the **concerted theory** and the **sequential theory** describe the cooperative transitions in simple quantitative terms. The latter is a more general theory; the former is adequate to explain many allosteric enzymes.

The concerted theory, or symmetry-driven theory, was devised to explain the cooperative binding of identical ligands, such as substrates. It supposes that there is one binding site per subunit for each ligand, that the conformation of each subunit is constrained by its association with other subunits, and that when the protein changes conformation, it retains its molecular symmetry (Figure 6·29a). Thus, there are two conformations in equilibrium, R (with a high affinity for substrate) and T (a low-affinity state). The binding of substrate shifts the equilibrium. When

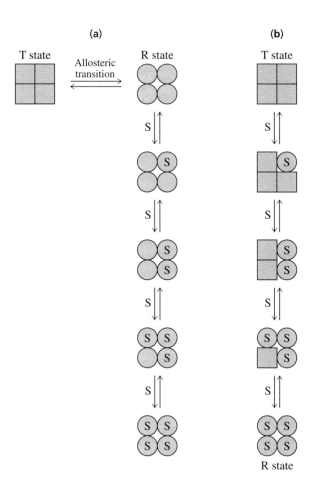

(a)   (b)

Figure 6·29
Two models for cooperativity of binding of substrate (S) to a tetrameric protein. **(a)** In the concerted model, subunits are all either R or T, and S binds only to the R state. **(b)** In the sequential model, binding of S to a subunit converts only that subunit to the R conformation. Neighboring subunits might remain in the T state or might assume conformations between T and R.

the conformation of the protein changes, the affinity of its ligand-binding sites also changes. The concerted theory was extended to include the binding of allosteric modulators, and it can be simplified quantitatively by assuming that the substrate (S) binds only to the R state and the allosteric inhibitor (I) binds only to the T state. The concerted theory is based on the observed structural symmetry of regulatory enzymes. It suggests that all subunits of a given protein molecule have the same conformation, either all R or all T; that is, they retain their symmetry. Experimental data obtained with a number of enzymes can be explained by this simple theory.

The sequential theory, or ligand-induced theory, is a later and more general proposal. It is based on the idea that binding of a ligand may induce a change in the tertiary structure of the subunit to which it binds. This subunit-ligand complex may change the conformations of neighboring subunits to varying extents. This theory too assumes that only one shape has high affinity for the ligand, but it differs from the concerted theory in allowing the existence of both high- and low-affinity subunits in an oligomeric molecule that has fractional saturation (Figure 6·29b). The sequential theory can account for negative cooperativity, a decrease in affinity as subsequent molecules of ligand are bound to an oligomer. Negative cooperativity occurs with a relatively small number of enzymes. The sequential theory treats the concerted theory, with its symmetry requirement, as a limiting and simple case.

One of the most extensive tests of the concerted and sequential theories involved the conformational changes of the oxygen-binding protein hemoglobin (Section 5·17). The composite results from many types of experiments indicate that when hemoglobin is partially oxygenated, the tertiary structure of a monomeric chain changes when it binds oxygen. The change in quaternary structure—the relative movement of the $\alpha\beta$ dimers (Figure 5·60)—occurs only after oxygen has

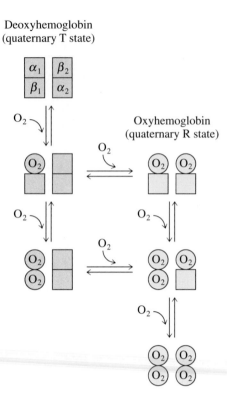

Deoxyhemoglobin
(quaternary T state)

Oxyhemoglobin
(quaternary R state)

Figure 6·30
Conformational changes during oxygen binding to hemoglobin. The tertiary structure of a single chain changes as oxygen is bound. The quaternary structure of hemoglobin changes from the T state to the R state only when both dimers ($\alpha\beta$ subunits) are oxygenated. Only four of the eight possible partially oxygenated species are shown (e.g., oxygen could bind initially to either an α or a β chain, and so on). [Adapted from Ackers, G. K., Doyle, M. L., Myers, D., and Daugherty, M. A. (1992). Molecular code for cooperativity in hemoglobin. *Science* 255:54–63.]

bound to at least one chain in each dimeric subunit. These changes in tertiary and quaternary structure are depicted in Figure 6·30. The process of oxygen binding to hemoglobin has aspects of *both* the sequential and concerted theories: independent alteration of tertiary structure within subunits and a concerted transition in quaternary structure.

6·14 Aspartate Transcarbamoylase Was the First Allosteric Enzyme to Be Thoroughly Characterized

Aspartate transcarbamoylase (ATCase) from the bacterium *Escherichia coli* is the most thoroughly studied regulatory enzyme. It provides an excellent example of allosteric feedback inhibition in the regulation of a biosynthetic pathway. In *E. coli,* ATCase catalyzes the first committed step of the biosynthesis of pyrimidine nucleotides (Figure 6·31). (Nucleotide biosynthesis is discussed in detail in Chapter 22.) This reaction forms carbamoyl aspartate from carbamoyl phosphate and aspartate. The nucleophilic amino group of aspartate attacks the carbonyl group of carbamoyl phosphate, releasing the phosphate and forming a compound that can cyclize to form the six-membered pyrimidine ring. Arthur Pardee and his coworkers showed that the most potent metabolic inhibitor of ATCase is the end product of the pathway, cytidine triphosphate (CTP). The same workers also found that ATP is an activator of ATCase. ATP is an end product of the pathway by which purine nucleotides are synthesized. Activation of ATCase by ATP presumably enhances the production of pyrimidine nucleotides to balance the supply of both types of nucleotides needed for the synthesis of nucleic acids. Both CTP and ATP affect the binding of the substrate aspartate. The velocity versus [aspartate] plot for the ATCase

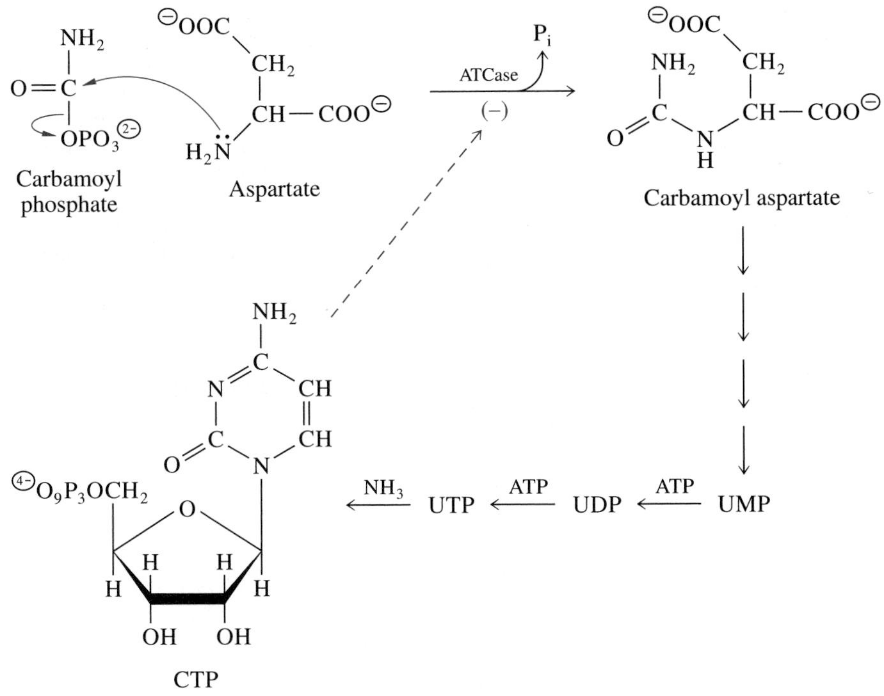

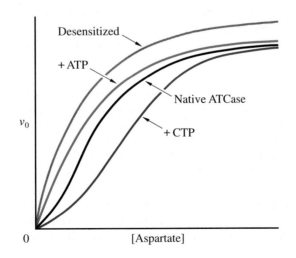

Figure 6·31
Abbreviated pathway for biosynthesis of pyrimidine nucleotides. Aspartate transcarbamoylase (ATCase) catalyzes the formation of carbamoyl aspartate from carbamoyl phosphate and aspartate. Carbamoyl aspartate is converted to the nucleotides UMP, UDP, and UTP, and the ultimate product, CTP. Aspartate and carbamoyl aspartate have been drawn in a configuration that corresponds to the position of the atoms in the pyrimidine ring that is formed. CTP is an allosteric inhibitor of ATCase.

reaction is sigmoidal (Figure 6·32). CTP raises the apparent K_m for aspartate without changing the V_{max} of the enzyme; thus, CTP is a competitive inhibitor, one that binds away from the active site. CTP makes the curve more sigmoidal, indicating greater cooperativity in the binding of aspartate to ATCase. The presence of ATP shifts the curve toward a hyperbolic shape, decreasing the cooperativity of binding to the enzyme.

Several methods of mild denaturation or chemical modification desensitize ATCase. The selective destruction of the regulatory activity is accompanied by conversion of the sigmoidal aspartate-saturation curve for ATCase to a normal hyperbola (the upper curve in Figure 6·32). In addition, there is a slight increase in the activity of ATCase, as though some physical restraint were removed.

Figure 6·32
Allosteric regulation of ATCase from *E. coli*. ATP is an activator and CTP is an inhibitor of the native enzyme. The desensitized ATCase is unaffected by addition of ATP or CTP.

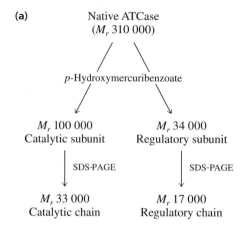

(a) Native ATCase
(M_r 310 000)

p-Hydroxymercuribenzoate

M_r 100 000
Catalytic subunit

M_r 34 000
Regulatory subunit

SDS-PAGE

SDS-PAGE

M_r 33 000
Catalytic chain

M_r 17 000
Regulatory chain

Figure 6·33
Quaternary structure of ATCase. **(a)** The stoichiometry of the quaternary structure of ATCase was arrived at by determining the molecular weights of the protein fragments formed after desensitization and after complete denaturation. **(b)** ATCase is desensitized by *p*-hydroxymercuribenzoate, which reacts with the thiol groups of cysteine residues.

(b)

p-Hydroxymercuribenzoate

Studies of purified ATCase have revealed both the quaternary structure and the path of desensitization of the enzyme (Figure 6·33a). The native form of ATCase has a molecular weight of 310 000. ATCase is desensitized by treatment with *p*-hydroxymercuribenzoate, a compound that reacts specifically with thiol groups (Figure 6·33b). In ATCase, six zinc ions, each complexed to four cysteine residues, stabilize a conformation necessary for proper subunit interaction. Treatment with *p*-hydroxymercuribenzoate releases the Zn^{2+} and results in the formation of subunits of two sizes, having molecular weights of about 100 000 and 34 000. The larger subunit possesses all of the catalytic activity of ATCase and therefore is called the catalytic subunit. It is not inhibited by and does not bind CTP. Only the smaller subunit, the regulatory subunit, can bind CTP or ATP. When analyzed by SDS-polyacrylamide gel electrophoresis (Section 4·7), the components of the catalytic subunit migrate as one chain with a molecular weight of 33 000. The regulatory subunit, when similarly tested, is denatured to a single polypeptide chain of molecular weight 17 000. By considering the molecular weights of the chains and the amounts of protein in each subunit fraction, it was deduced that the native form of ATCase contains 12 polypeptide chains: six catalytic (C) chains and six regulatory (R) chains, arranged as two trimers of C chains and three dimers of R chains. The quaternary structure of ATCase is summarized in Table 6·6, and the structure determined by X-ray crystallography is illustrated in Figure 6·34. Each chain of a catalytic trimer is connected to a chain of the second catalytic trimer through a regulatory dimer.

Table 6·6 Quaternary structure of aspartate transcarbamoylase from *E. coli*

	Molecular weight	Number of chains	Composition
Catalytic chain	33 000	1	C
Catalytic subunit	100 000	3	C_3
Regulatory chain	17 000	1	R
Regulatory subunit	34 000	2	R_2
Native enzyme	310 000	12	$(C_3)_2(R_2)_3$

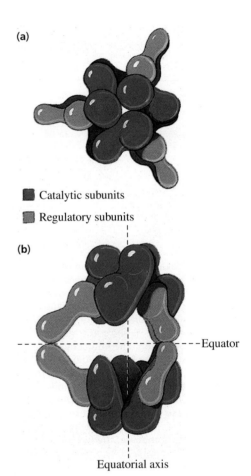

(a)

■ Catalytic subunits

■ Regulatory subunits

(b)

Equator

Equatorial axis

Figure 6·34
Structure of ATCase from *E. coli*. ATCase contains both catalytic and regulatory subunits. **(a)** Top view. The two catalytic subunits are stacked, nearly eclipsed trimers that are held together by three regulatory dimers. **(b)** Side view. Only two regulatory dimers are visible. Each regulatory dimer is bound to two catalytic chains. Note that each chain, both catalytic and regulatory, is composed of two domains. [Adapted from Krause, K. L., Volz, K. W., and Lipscomb, W. N. (1985). Structure at 2.9-Å resolution of aspartate carbamoyltransferase complexed with the bisubstrate analogue *N*-(phosphonacetyl)-L-aspartate. *Proc. Natl. Acad. Sci. USA* 82:1643–1647.]

The most important finding from studies of ATCase is that the active site is physically distinct from the regulatory site but coupled to it. In this case, the regulatory and catalytic sites are even located on different polypeptide chains. Recent X-ray crystallographic studies of ATCase and several enzyme-ligand complexes by William Lipscomb and his colleagues have shown that the active site is composed of amino acids from two adjacent C chains of the catalytic subunit; that is, the active sites are at the interfaces between C chains. They have also shown that the T $\longrightarrow$ R transition involves movement of one trimer away from the other, with the CRRC interchain interactions being maintained during this conformational change. They suggest that binding of the first molecule of aspartate, in the presence of the second substrate, carbamoyl phosphate, causes all six of the catalytic chains to move into a conformation having high affinity for substrate and increased catalytic activity. Thus, ATCase is converted from the T to R state in a concerted manner. The two rigid domains of each catalytic chain move closer together so that the active site in the R form is partially closed, and thus more suitable for reaction, compared to its shape in the T form.

While the major conclusion from the studies of ATCase—that regulatory sites are distinct from catalytic sites—is applicable to almost all regulatory enzymes, you will learn that there is great diversity in the tertiary and quaternary structures of these complex enzymes. For example, the reaction catalyzed by glutamine synthetase (Section 21·4) is an important control point in amino acid metabolism; this enzyme is composed of 12 identical subunits. Ribonucleotide reductase catalyzes a major regulated reaction in DNA biosynthesis (Section 22·9); this reductase contains two chains each of two different types. A fairly common structure for regulatory enzymes is a tetramer, with each identical subunit having a catalytic domain fused to a regulatory domain.

6·15 Some Regulatory Enzymes Undergo Phosphorylation

In addition to simple allosteric interactions, there are other ways that enzyme activity can be regulated. Zymogen activation can control the activity of enzymes, as we saw earlier. The amount of an enzyme can also be controlled by regulating the rate of its synthesis or degradation. This type of control is much slower than the allosteric R $\rightleftharpoons$ T transition—on the scale of hours versus seconds. For some enzymes, an intermediate rate of control, on the scale of minutes, is provided by covalent modification of the enzyme. Phosphorylation is the most common type of regulatory covalent modification. The variety of regulatory strategies allows cells to respond to changing metabolic conditions, whether short-term or long-term. In the following sections, we will describe some of the other mechanisms that have evolved for tailoring enzymatic activity to the needs of the cell.

(a)

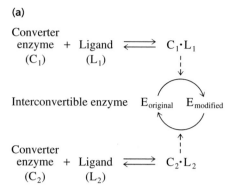

(b)

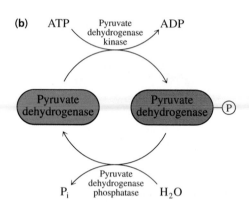

Figure 6·35
Pathway of covalent modification of interconvertible enzymes. **(a)** In the general scheme for covalent modification, two allosterically controlled converter enzymes, C_1 and C_2, catalyze modifications of the interconvertible enzyme, E, the substrate of the covalent substitution. The interconvertible enzyme exists in two forms, original and modified. C_1 and C_2 are activated when they bind their respective modulator ligands, L_1 and L_2. **(b)** Control of the activity of pyruvate dehydrogenase is an example of regulation by covalent modification. Phosphorylation of pyruvate dehydrogenase, the interconvertible enzyme, is catalyzed by pyruvate dehydrogenase kinase. Pyruvate dehydrogenase is inactivated by phosphorylation. It is reactivated by hydrolysis of its phosphoserine residue, catalyzed by an allosteric hydrolase called pyruvate dehydrogenase phosphatase.

Regulation of enzyme activity by covalent modification, including phosphorylation, is reversible, but it usually requires one enzyme for activation and another for inactivation. Enzymes controlled by covalent modification, termed **interconvertible enzymes,** are believed to generally undergo R $\rightleftharpoons$ T transitions. They may be frozen in one conformation or the other by a covalent substitution. The substitution reaction is catalyzed by an accessory enzyme called a **converter enzyme,** and the release of the substituent is catalyzed by another converter enzyme (Figure 6·35a). Converter enzymes are usually controlled by allosteric effectors, although in some cases, converter enzymes within a regulatory sequence are themselves subject to covalent modification, with the entire sequence under allosteric control.

The most common type of covalent modification is phosphorylation of a specific serine residue of an interconvertible enzyme. It has been estimated that up to one-third of the proteins in a typical mammalian cell may contain covalently bound phosphate. An enzyme called a protein kinase catalyzes the transfer of the terminal phosphoryl group from ATP to the appropriate serine residue of the regulated enzyme. The activity of the protein kinase is itself regulated. The phosphoserine of the interconvertible enzyme is hydrolyzed by the activity of a protein phosphatase, releasing phosphate and returning the interconvertible enzyme to its dephosphorylated state. Individual enzymes differ as to whether it is their phospho- or dephospho- forms that are active. Interconvertible enzymes in catabolic pathways are generally activated by phosphorylation and inactivated by dephosphorylation; most interconvertible enzymes in anabolic pathways are inactivated by phosphorylation and reactivated by dephosphorylation.

The reactions involved in the regulation of pyruvate dehydrogenase by covalent modification are shown in Figure 6·35b. Pyruvate dehydrogenase catalyzes a key step in the breakdown of glucose, the decarboxylation of pyruvate leading to the formation of acetyl coenzyme A and CO_2. This reaction connects the pathway of glycolysis to the citric acid cycle. Phosphorylation of pyruvate dehydrogenase, catalyzed by the allosteric enzyme pyruvate dehydrogenase kinase, inactivates the dehydrogenase. The kinase can be activated by any of several metabolites. Phosphorylated pyruvate dehydrogenase is reactivated under different metabolic conditions by hydrolysis of its phosphoserine residue, catalyzed by pyruvate dehydrogenase phosphatase. This phosphatase is activated by Ca^{2+}.

The three-dimensional structures of the phospho- and dephospho- forms of glycogen phosphorylase and isocitrate dehydrogenase, determined by X-ray crystallography, demonstrate two different ways in which covalent modification can alter catalytic activity. A specific kinase catalyzes the phosphorylation of Ser-14 of inactive glycogen phosphorylase b to form active glycogen phosphorylase a. Phosphorylation causes a previously disordered segment of the protein to become helical, a shape stabilized by ion pairs formed between the phosphate group and two arginine residues and by the formation of several new hydrogen bonds. The conformational changes that occur after the covalent modification of glycogen phosphorylase alter the sites where the substrates and modulators are bound. Phosphorylation of this enzyme produces, as expected, a T $\longrightarrow$ R (inactive $\longrightarrow$ active) transition.

An alternate mechanism for regulation by phosphorylation has been found. When isocitrate dehydrogenase from *E. coli* is inactivated by phosphorylation, the binding of its substrate, isocitrate, is prevented. The one covalently modified residue of the enzyme is a serine that, in the unsubstituted enzyme, forms a hydrogen bond with a carboxylate group of isocitrate. When the enzyme is covalently modified, the hydrogen bond cannot be formed, and there is electrostatic repulsion between the carboxylate group and phosphate. X-ray crystallography has shown that the shape of the enzyme does not change appreciably upon covalent modification. Thus, kinase-catalyzed covalent modification is known at both regulatory and catalytic sites of interconvertible enzymes.

6·16 Enzymes May Be Tissue or Organelle Specific

Some reactions are catalyzed by several different enzymes. Multiple forms of enzymes may differ in only one or a few residues (genetic variants) or may be entirely different proteins, for example, in the cytosol and mitochondria. Different proteins from a single biological species that catalyze the same reaction are called **isozymes,** or isoenzymes. Isozymes, which may have different K_m and V_{max} values, are chemically distinct and therefore can be distinguished by physical methods such as electrophoresis. In multisubunit isozymes, the subunits may be similar enough in structure to assemble into hybrid oligomers. For example, lactate dehydrogenase is a tetramer. Extracts of some tissues have up to five separable fractions of lactate dehydrogenase. Two major types of subunits have been isolated, those that predominate in heart (the H type) and muscle (the M type). The five isozyme forms arise from hybridization of the two types of subunits: H_4, H_3M, H_2M_2, HM_3, and M_4.

Measurement of the amount or activity of isozymes in the blood can be used to diagnose medical conditions. For instance, human heart contains mostly the H_4 and H_3M isozymes of lactate dehydrogenase. After a heart attack, heart tissue undergoes partial breakdown, releasing enzymes into the bloodstream. The increase in the ratio of H_4 to other isozymes of lactate dehydrogenase in blood is diagnostic of myocardial infarction (heart attack). When the liver is injured, the amount of M_4 increases. This increase can indicate liver congestion after myocardial infarction or other liver damage such as from cirrhosis or hepatitis.

6·17 Multienzyme Complexes and Multifunctional Enzymes Have Increased Efficiency

In a number of cases, enzymes that catalyze sequential reactions in the same metabolic pathway have been found to be physically associated. This phenomenon offers several potential advantages. Enzyme complexes allow **metabolite channelling.** Channelling of reactants between active sites occurs when the product of one reaction is transferred directly to the next active site without entering the bulk solvent. Channelling can speed up a reaction of a pathway by decreasing the transit time for an intermediate to reach the next enzyme and by producing local high concentrations of the intermediate. Channelling can also protect chemically labile intermediates from degradation by solvent.

There are several ways in which enzymes are associated. In some cases, several active sites are found on a single, multifunctional polypeptide chain. The fusion of related enzyme activities on one protein allows organisms to regulate a series of reactions by coordinating the expression of a single gene. In some pathways, several individual enzyme proteins are noncovalently but fairly tightly associated. Examples of these multienzyme complexes will be encountered later in this text, for example, when we discuss the biosynthesis of macromolecules. Enzymes of some other pathways are associated by weaker interactions, mainly hydrophobic. Because such complexes dissociate easily, it has been difficult to demonstrate their association, and only a few complexes have been incontrovertibly established. Attachment to membranes is another way that enzymes may be associated. Some of the enzymes in coagulation, for example, associate with substrates and specific cofactor proteins on the surfaces of cells. The search for enzyme complexes and the evaluation of their catalytic and regulatory roles is an extremely active area of research in enzyme chemistry.

Summary

Enzymes, the catalysts of living organisms, are remarkable for their catalytic efficiency and their substrate and reaction specificity. With the exception of certain catalytic RNA molecules, enzymes are proteins, or proteins plus auxiliary compounds or ions called cofactors. Enzymes can be grouped into six major classes according to the nature of the reactions they catalyze: oxidoreductases (dehydrogenases), transferases, hydrolases, lyases, isomerases, and ligases (synthetases).

Chemical-kinetic experiments involve the systematic variation of reaction conditions, especially the concentration of substrate, and measurement of the alteration in the rate of formation of the product. Enzyme-kinetic measurements are usually made by measuring initial velocities, which are obtained from progress curves that show the amount of product formed over time. The first step in an enzyme-catalyzed reaction is the formation of a noncovalent enzyme-substrate complex. As a result, enzymatic reactions are characteristically first order with respect to enzyme concentration and typically show hyperbolic dependence on substrate concentration. Maximum velocity (V_{max}) is reached when the substrate concentrations are saturating. The Michaelis-Menten equation describes such kinetic behavior. The Michaelis constant (K_m) is equal to the concentration of substrate that gives half-maximum reaction velocity—that is, half saturation of E with S. Values for the kinetic constants K_m and V_{max} can be determined by computer analysis of kinetic data, or they can be estimated graphically. The double-reciprocal, or Lineweaver-Burk, plot of $1/v_0$ versus $1/[S]$ represents a linear transformation of the hyperbolic Michaelis-Menten equation.

The catalytic constant (k_{cat}), or turnover number for an enzyme, is the maximum number of molecules of substrate that can be transformed into product per molecule of enzyme (or per active site) per second and thus is equal to V_{max} divided by enzyme concentration. The unit for k_{cat} is s^{-1}. The ratio k_{cat}/K_m is an apparent second-order rate constant that governs the reaction of an enzyme with a substrate in dilute, nonsaturating solutions. Its value can provide a measure of the catalytic efficiency and substrate specificity of an enzyme.

The rates of enzyme-catalyzed reactions are also affected by the presence of inhibitors. Knowledge of enzyme-inhibition mechanisms aids in understanding the control of metabolism and the design of clinically useful drugs. Enzyme inhibitors may be classified as reversible or irreversible. Reversible inhibitors are bound noncovalently by the enzyme. They can be competitive (those that increase the apparent value of K_m, with no change in V_{max}), uncompetitive (those that decrease K_m and V_{max} proportionally), or noncompetitive (those that decrease V_{max}, with either no change in K_m—pure noncompetitive—or an increase or decrease in K_m—mixed noncompetitive). Inhibition studies can be used to determine the kinetic scheme of multisubstrate reactions (whether they are ordered, random, or ping-pong) and to determine the order of binding of substrates to an enzyme and the order of release of products.

Irreversible inhibitors form covalent bonds with the enzyme. By treating an enzyme with an irreversible inhibitor and then sequencing a segment of the protein, it is often possible to determine the identity of reactive amino acid residues, which can contribute to deciphering catalytic mechanisms. Site-directed mutagenesis represents another approach to investigating enzymes, including their catalytic mechanisms. Site-directed mutagenesis can be used to change the identity of a single amino acid residue within a protein, generating a new protein whose properties may reveal the role of the original amino acid residue.

Many proteases are synthesized as inactive zymogens that are activated extracellularly under appropriate conditions by selective proteolysis, the specific enzymatic cleavage of one or a few peptide bonds. Blood clotting depends on a cascade of zymogen activation involving over a dozen coagulation factors. Such cascades provide enormous signal amplification. The structures of proteins, as determined by

X-ray crystallography, can reveal information about the active sites, including the binding of specific substrates.

Regulation of metabolism is often provided by allosteric modulators that act on certain key enzymes, almost all of which are oligomers. Allosteric modulators bind to a site other than the active site. They alter either the apparent K_m or V_{max} values of regulatory enzymes through control of the R:T ratio. Aspartate transcarbamoylase catalyzes the first committed step in the biosynthesis of pyrimidine nucleotides by *E. coli*. Its control properties have been elucidated at the molecular level.

Enzyme activity is also regulated by other mechanisms. For example, covalent modification of certain regulatory enzymes provides a control mechanism that is slower than allosteric interactions but more rapid than control achieved by changes in the concentration of enzyme, which is regulated at the level of gene expression. Some organisms or tissues contain isozymes, different proteins catalyzing the same chemical reaction. Isozymes can often be easily distinguished by kinetic or physical properties. The evolution of multienzyme complexes and multifunctional enzymes offers the advantages of metabolite channelling and coordinated regulation of multireaction processes.

Selected Readings

Enzyme Catalysis

Boyer, P. D., ed. (1970–1990). *The Enzymes,* Vols. 1, 2, 3, and 19, 3rd ed. (New York: Academic Press). These volumes, from a comprehensive series, cover general topics of catalysis.

Fersht, A. (1985). *Enzyme Structure and Mechanism,* 2nd ed. (New York: W. H. Freeman and Company). A description of the function of enzymes, with many original proposals.

Lipscomb, W. N. (1983). Structure and catalysis of enzymes. *Annu. Rev. Biochem.* 52:17–34.

Scrimgeour, K. G. (1977). *Chemistry and Control of Enzyme Reactions* (London: Academic Press). An outline of the principles of modern enzymology.

Enzyme Kinetics

Cornish-Bowden, A., and Wharton, C. W. (1988). *Enzyme Kinetics* (Oxford: IRL Press). A brief outline of enzyme kinetics.

Piszkiewicz, D. (1977). *Kinetics of Chemical and Enzyme-Catalyzed Reactions* (New York: Oxford University Press). An introductory text on kinetics.

Segal, H. L. (1959). The development of enzyme kinetics. In *The Enzymes,* Vol. 1, 2nd ed., P. D. Boyer, H. Lardy, and K. Myrbäck, eds. (New York: Academic Press), pp. 1–48. A history of the development of enzyme kinetics.

Segel, I. H. (1975). *Enzyme Kinetics: Behavior and Analysis of Rapid Equilibrium and Steady State Enzyme Systems* (New York: Wiley Interscience). A comprehensive description of kinetic behavior and analysis.

Wong, J. T.-F. (1975). *Kinetics of Enzyme Mechanisms* (London: Academic Press). An advanced mathematical examination of kinetic mechanisms.

Site-Directed Mutagenesis

Johnson, K. A., and Benkovic, S. J. (1990). Analysis of protein function by mutagenesis. In *The Enzymes,* Vol. 19, 3rd ed., P. D. Boyer, ed. (New York: Academic Press), pp. 159–211.

Serine Proteases

Light, A., and Janska, H. (1989). Enterokinase (enteropeptidase): comparative aspects. *Trends Biochem. Sci.* 14:110–112.

Navia, M. A., McKeever, B. M., Springer, J. P., Lin, T.-Y., Williams, H. R., Fluder, E. M., Dorn, C. P., and Hoogsteen, K. (1989). Structure of human neutrophil elastase in complex with a peptide chloromethyl ketone inhibitor at 1.84-Å resolution. *Proc. Natl. Acad. Sci. USA* 86:7–11.

Neurath, H. (1984). Evolution of proteolytic enzymes. *Science* 224:350–357.

Steitz, T. A., and Shulman, R. G. (1982). Crystallographic and NMR studies of the serine proteases. *Annu. Rev. Biophys. Bioeng.* 11:419–444.

Zymogen Activation

Huber, R., and Bode, W. (1978). Structural basis of the activation and action of trypsin. *Acc. Chem. Res.* 11:114–122.

Sharma, S. K., and Hopkins, T. R. (1981). Recent developments in the activation process of bovine chymotrypsinogen A. *Bioorg. Chem.* 10:357–374.

Blood Coagulation

Davie, E. W., Fujikawa, K., and Kisiel, W. (1991). The coagulation cascade: initiation, maintenance, and regulation. *Biochemistry* 30:10363–10370. Provides an overview of the coagulation process.

Mann, K. G., Nesheim, M. E., Church, W. R., Haley, P., and Krishnaswamy, S. (1990). Surface-dependent reactions of the vitamin K-dependent enzyme complexes. *Blood* 76:1–13. Explores the molecular details of coagulation.

Regulatory Enzymes

Ackers, G. K., Doyle, M. L., Myers, D., and Daugherty, M. A. (1992). Molecular code for cooperativity in hemoglobin. *Science* 255:54–63.

Barford, D. (1991). Molecular mechanisms for the control of enzymic activity by protein phosphorylation. *Biochim. Biophys. Acta* 1133:55–62.

Edelman, A. M., Blumenthal, D. K., and Krebs, E. G. (1987). Protein serine/threonine kinases. *Annu. Rev. Biochem.* 56:567–613. Describes properties of the kinases that catalyze phosphorylation of interconvertible enzymes.

Hubbard, M. J., and Cohen, P. (1993). On target with a new mechanism for the regulation of protein phosphorylation. *Trends Biochem. Sci.* 18:172–177. An explanation of the substrate specificity for covalent modification.

Hurley, J. H., Dean, A. M., Sohl, J. L., Koshland, D. E., Jr., and Stroud, R. M. (1990). Regulation of an enzyme by phosphorylation at the active site. *Science* 249:1012–1016. Describes control of the activity of isocitrate dehydrogenase by covalent modification of a binding residue.

Kantrowitz, E. R., and Lipscomb, W. N. (1988). *Escherichia coli* aspartate transcarbamylase: the relation between structure and function. *Science* 241:669–674.

7

Mechanisms of Enzymes

The previous three chapters have described the composition, structures, and actions of enzyme proteins. In this chapter, we examine *how* enzymes catalyze reactions. This involves studying their **mechanisms,** the details of the molecular events that occur during reactions. Individual enzyme mechanisms have been deduced by a variety of methods, including kinetic experiments and—more recently—protein structural studies, especially X-ray crystallography. The general principles of catalysis by enzymes have been elucidated by integrating mechanistic information derived from individual enzymes with studies of suitable nonenzymatic model reactions. A complete explanation of enzymatic catalysis requires quantitative evaluations of mechanisms by interpretation of three kinetic terms that were defined in Chapter 6: k_{cat}, K_m, and the ratio k_{cat}/K_m.

The extraordinary catalytic ability of enzymes is a result of simple physical and chemical properties, the chief of which is proper binding of reactants in the active sites of particular globular-protein molecules. The sciences of chemistry, physics, and biochemistry have combined to take much of the mystery out of enzymes, and now molecular genetics is being applied to test the theories proposed by enzyme chemists. Observations for which there were no explanations just a half century ago are now easily understood.

We begin this chapter with a review of simple chemical mechanisms, followed by a brief discussion of catalysis and transition states. The four major modes of enzymatic catalysis—acid-base and covalent catalysis (classified as chemical effects) and proximity and transition-state stabilization (binding effects)—are then discussed.

 Related material appears in Exercises 2 and 3 of the computer-disk supplement *Exploring Molecular Structure.*

7·1 Enzymatic Mechanisms Are Described with the Terminology of Chemistry

The mechanism of either a nonenzymatic reaction or an enzyme-catalyzed reaction is a description of the atomic or molecular events that occur during the reaction, in as much useful detail as possible. A combination of analytical approaches is required. The reactants, products, and any intermediates must be identified. The pathways of all atoms may be traced by using isotopically labelled reactants. Changes in the chemical bonds of substrate and solvent during the reaction are measured, often by kinetic techniques. Finally, a three-dimensional view of the reaction is obtained through a study of the stereochemical changes that occur during the reaction. In any enzyme mechanism, the mechanistic information about the reactants must be coordinated with what is known about the three-dimensional structure of the enzyme. The mechanisms of many enzymes are well established; in fact, there is now sufficient experimental evidence to support a set of general mechanisms. It is these general modes of catalysis, illustrated by specific examples, that will be the major topic of this chapter.

The same symbolism used in organic chemistry to represent chemical-bond transformations is employed in representing enzymatic mechanisms. With the short review of terms given in this section and the specific points mentioned in discussions of individual mechanisms, you will be able to understand all of the enzyme-catalyzed reactions that are presented throughout this book.

Since covalent chemical bonds consist of pairs of electrons shared between two atoms, bonds can be cleaved in two ways. In most reactions, both electrons stay with one atom, so that an ionic intermediate and a **leaving group** are formed. In ionic reactions, two species are involved: one is electron rich, or **nucleophilic,** and the other is electron poor, or **electrophilic** (Section 3·6). A nucleophile, which has a negative charge or unshared electron pair, attacks the electrophilic center of the other reactant. The convention in mechanistic chemistry is to show a curved arrow pointing from a nucleophile to the electron-deficient position of an electrophile. Hence, transfer of an acyl group can be written as this general mechanism:

$$R-\overset{\displaystyle O}{\underset{\displaystyle X}{\overset{\|}{C}}} \quad \longrightarrow \quad R-\overset{\displaystyle O^{\ominus}}{\underset{\displaystyle Y}{\overset{|}{C}}}-X \quad \longrightarrow \quad \overset{\displaystyle O}{\underset{\displaystyle Y}{\overset{\|}{C}}} \; + \; X^{\ominus}$$

$$\overset{\cdot\cdot}{Y}{}^{\ominus}$$

$$(7\cdot1)$$

The nucleophile $Y^{\ominus}$ attacks the carbonyl carbon (adds to the $C=O$ bond) to form a tetrahedral addition intermediate from which the group $X^{\ominus}$ is eliminated. $X^{\ominus}$ is the leaving group. In hydrolysis of an amide bond, for example, the nucleophile is $OH^{\ominus}$ from water, and $X^{\ominus}$ is an amine. Several examples of this type of mechanism are described in this chapter.

Ionization of carbons involved in covalent bonds occurs in two ways. If a carbon atom retains both electrons, a **carbanion** is produced.

$$R_3C-X \; \rightleftharpoons \; R_3C{:}^{\ominus} \; + \; X^{\oplus} \qquad (7\cdot2)$$

If the bond broken in Equation 7·2 is a $C-H$ bond, a proton is released, and the carbanion is the conjugate base of the organic compound. Formation of a carbanion is the most common pathway for cleavage of carbon bonds in biology.

If the carbon atom loses both electrons, a cationic **carbonium ion,** or **carbocation,** is formed.

$$R_3C-Y \; \rightleftharpoons \; R_3C^{\oplus} \; + \; {:}Y^{\ominus} \qquad (7\cdot3)$$

If Y in Equation 7·3 is a hydrogen atom, $Y^{\ominus}$ is a hydride ion, $H^{\ominus}$ (a proton accompanied by two electrons). Many oxidation reactions proceed by this mechanism. Substitution reactions can also occur in this manner, with $Y^{\ominus}$ (the leaving group) being replaced by another anion. In this substitution mechanism, the ionization step is slow compared to addition of the other anion. Therefore, the overall rate of substitution is first order because it depends on the concentration of only *one* compound, the original reactant.

Another type of substitution, the direct displacement mechanism, involves *two* molecules reacting in the initial slower step. The rate of this type of reaction depends on the concentrations of both reactants. The attacking group or molecule adds to the face of the carbon atom opposite the leaving group to form a transition state that has five groups attached to the central carbon atom. This **transition state,** shown below in square brackets, is an unstable high-energy state. It has a structure between that of the reactant and the product.

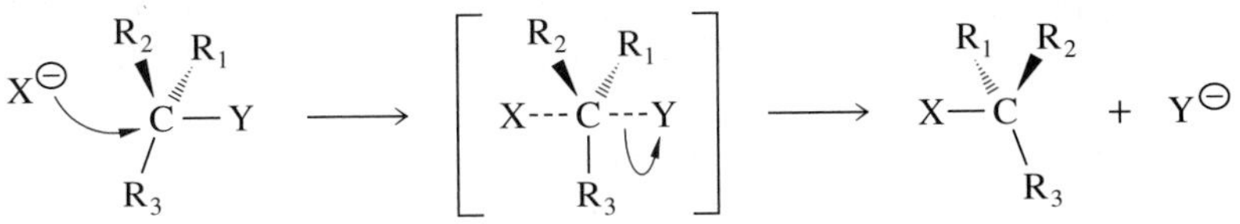

Transition state

(7·4)

In the less common type of bond cleavage, one electron remains with each product to form two free radicals that are usually very unstable.

$$R_1O-OR_2 \longrightarrow R_1O\cdot + \cdot OR_2 \qquad (7\cdot5)$$

Oxidation-reduction reactions are central to the supply of biological energy. Oxidations can take several forms: addition of oxygen, removal of hydrogen, or removal of electrons. In the oxidation of formaldehyde to formic acid, an atom of oxygen is added.

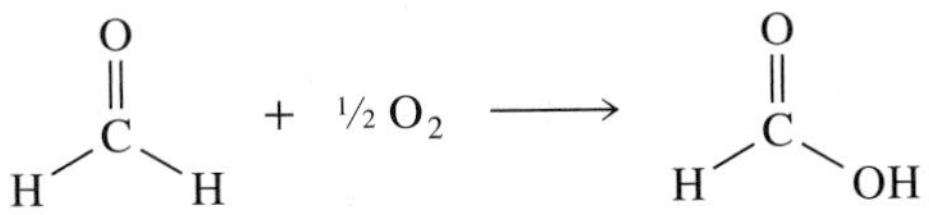

(7·6)

In the oxidation of lactate to pyruvate (Table 6·1), two hydrogen atoms are removed from the alcohol group of lactate. Mechanistically, they are removed as a proton and a hydride ion. Dehydrogenations of this type are the most common biological oxidation-reduction reactions. In the third type of oxidation, the valence of metal ions is increased by removal of electrons. For example, the heme-containing protein cytochrome *c* undergoes cyclic oxidation and reduction between its reduced ferrous (Fe^{2+}) form and its oxidized ferric (Fe^{3+}) form.

7·2 Catalysts Stabilize Transition States

The rate of a chemical reaction depends on the rate of effective collisions of the reacting molecules; that is, it depends on how effectively the reactants collide to form a transition state. For the collisions to lead to product, the colliding substances must be in the correct orientation and must possess sufficient energy to approach the physical configuration of the atoms and bonds of the product. The transition state is an unstable, energized arrangement of atoms in which chemical bonds are in the process of being formed or broken. The energy required to reach the transition state from the ground state of the reactants is called the **energy of activation** of the reaction and is often referred to as the activation barrier.

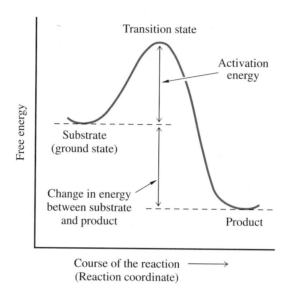

Figure 7·1
Reaction diagram for a single-step reaction. This curve shows the lowest energy path between the substrate and the product. The upper arrow shows the energy of activation for the forward reaction. Molecules of substrate that have more free energy than the activation energy pass over the activation barrier and become molecules of product. Because the substrate is at a higher free-energy level than the product, this is a spontaneous, or exergonic, reaction.

The progress of a reaction is usually represented by a graph called a reaction diagram. An example showing conversion of a substrate to a product in a single step is given in Figure 7·1. The y axis shows the free energies of the reacting species. The x axis, called the reaction coordinate, measures the progress of the reaction. The transition state occurs at the peak of the activation barrier, the energy level that must be exceeded for the reaction to proceed. The energy of activation can be calculated from rate measurements. The lower the value of this energy change, the more stable the transition state, the higher the concentration of the transition state, and the faster the reaction.

Transition states are not yet detectable experimentally, but their structures, which are somewhat similar to those of unstable intermediates, can be predicted. Like unstable intermediates, the structures of transition states may be drawn inside square brackets. Often the structures drawn for transition states indicate bonds that are destabilized (in the process of breaking) or partial (in the process of forming).

Intermediates differ from transition states. They are compounds that are sufficiently stable to be detected or isolated. When there is an intermediate in a reaction, the energy diagram has a trough (Figure 7·2). There are therefore two transition states in the reaction, one preceding formation of the intermediate and one preceding its conversion to product. The slowest step, the **rate-determining** or the rate-limiting step, in the formation of the product (P) from the substrate (S) is the step

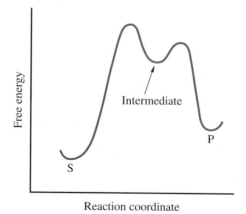

Figure 7·2
Reaction diagram for a reaction with an intermediate. The metastable intermediate occurs in the trough between the two transition states. The rate-determining step in the forward direction is formation of the first transition state, the step that has the higher activation energy. In this reaction, the equilibrium favors the formation of substrate from product because the substrate is at a lower energy level than the product.

(a) Uncatalyzed reaction

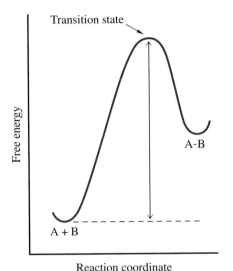

(b) Effect of reactants bound by enzyme

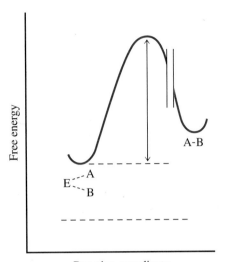

(c) Effect of reactants and transition state bound by enzyme

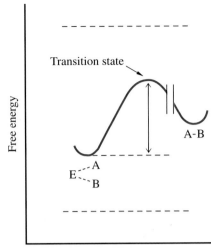

with the highest activation energy. In Figure 7·2, the rate-determining step is the formation of the intermediate. Because relatively little energy is required for the intermediate either to continue to product or revert to the original reactant, the intermediate is not particularly stable.

Catalysts allow reactions to proceed along pathways that have lower activation energies than uncatalyzed reactions. Catalysts directly participate in reactions by stabilizing the transition states along the reaction pathways. An enzyme lowers the overall activation energy by providing a multistep pathway in which each step has a lower energy of activation than the corresponding stage in the nonenzymatic reaction.

The first step in an enzymatic reaction is the formation of a noncovalent enzyme-substrate complex, ES. When reactants bind to enzymes, the reactants lose a great deal of entropy. However, in a reaction between A and B, formation of the EAB complex collects and positions the reactants, making the probability of reaction much higher for the enzyme-catalyzed than for the uncatalyzed reaction (Figure 7·3a and b). Bringing the reactants together accounts for a large part of the catalytic power of enzymes. The active sites of enzymes bind not only substrates and products but also transition states. In fact, transition states bind to active sites much more tightly than substrates bind. The extra binding interactions stabilize the transition state, further lowering the energy of activation (Figure 7·3c). As discussed later in this chapter, the binding by enzymes first of substrates and then of transition states provides the greatest effect in enzyme catalysis.

Figure 7·3
Enzymatic catalysis of the reaction A + B ⟶ A-B. **(a)** Reaction diagram for an uncatalyzed reaction. **(b)** Effect of reactant proximity. By suitably binding A and B in its active site to form the EAB complex, the enzyme brings the two reactants closer to the transition state, making formation of the transition state more frequent. A comparison of the activation arrows in (a) and (b) shows that the collecting of reactants lowers the energy of activation for the reaction of A with B. **(c)** Effect of transition-state stabilization. In addition to the effect in (b), an enzyme binds the transition state more tightly than it binds substrates, further lowering the energy of activation. Thus, the enzymatic reaction has a much lower activation energy than the uncatalyzed reaction. The enzymatic reaction curves include a break to indicate that the enzyme provides a multistep pathway.

7·3 Polar Amino Acids Form the Catalytic Centers of Enzymes

After formation of the ES complex, active-site amino acid residues act as chemical catalysts. Their direct involvement in enzymatic catalysis is relatively easy to observe and understand. We will now examine the chemical processes that contribute to the function of enzymes. Structural studies have verified that, although the active-site cavity of an enzyme—as part of the interior of the protein—is lined with hydrophobic amino acids, a few polar, ionizable amino acids (and a few molecules of water) are usually also found in the cleft. The polar amino acids that undergo chemical changes during enzymatic catalysis are called the **catalytic center** of the enzyme.

Table 7·1 Catalytic functions of ionizable amino acids

Amino acid	Reactive group	Net charge at pH 7	Principal functions
Aspartate	$-COO^{\ominus}$	−1	Cation binding; proton transfer
Glutamate	$-COO^{\ominus}$	−1	Cation binding; proton transfer
Histidine	Imidazole	Near 0	Proton transfer
Cysteine	$-S^{\ominus}$	Near 0	Covalent binding of acyl groups
Tyrosine	$-OH$	0	Hydrogen bonding to ligands
Lysine	$-NH_3^{\oplus}$	+1	Anion binding
Arginine	Guanidinium	+1	Anion binding
Serine	$-CH_2OH$	0	Covalent binding of acyl groups

Table 7·2 Typical pK_a values of ionizable groups of amino acids in proteins

Group	pK_a
Terminal α-carboxyl	3–4
Side-chain carboxyl	4–5
Imidazole	6–7
Terminal α-amino	7.5–9
Thiol	8–9.5
Phenol	9.5–10
ε-Amino	~10
Guanidine	~12
Hydroxymethyl	~16

Table 7·1 lists the ionizable and reactive residues found in the active sites of enzymes and some of their roles. Histidine, which has a pK_a value of about 6 to 7 in proteins, is often an acceptor or donor of protons. Aspartate, glutamate, and occasionally lysine also can participate in the transfer of protons. Certain amino acids such as serine and cysteine have reactivity suitable for covalently transferring groups from one substrate to a second substrate. At neutral pH, aspartate and glutamate usually have negative charges, and lysine and arginine have positive charges. These anions and cations can serve as sites for electrostatic binding of oppositely charged groups on substrates.

As discussed in Section 4·4, the pK_a values of the ionizable groups of amino acid residues in proteins may differ—usually slightly—from the values of the same groups in free amino acids. Table 7·2 lists the typical pK_a values of ionizable groups of amino acids in proteins. Compare these ranges to the exact values given for free amino acids in Table 4·2. Because of differences in their microenvironments, individual groups can differ from similar groups elsewhere in a protein. Occasionally, the side chain of a catalytic amino acid exhibits a pK_a value quite different from the one shown in Table 7·2. Bearing in mind that pK_a values may be perturbed, one method of indirectly testing for involvement of individual amino acids in a reaction is to examine the effect of pH on the rate of the reaction. If the change in rate correlates with the pK_a value of a certain ionic amino acid (Section 7·7), a residue of that amino acid may be involved in the catalytic mechanism.

Years before NMR spectroscopy was used for structural determinations, it was used for titration of ionizable groups of enzymes. Since the observed frequency of energy absorption for an atomic nucleus depends on its chemical environment, identification of spin states can be used in combination with NMR spectra of amino acids and peptides to determine pK_a values of individual ionic residues. These pK_a values can sometimes indicate whether a residue is located in the hydrophobic interior or on the solvated exterior of the protein. In this way, pK_a values provide evidence relating to the roles of ionizable residues at the active sites of enzymes.

7·4 Almost All Enzyme Mechanisms Include Acid-Base Catalysis

In active sites, ionic side chains participate in two kinds of chemical catalysis: acid-base catalysis and covalent catalysis. These types of catalysis are the two major chemical modes of catalysis. Later in this chapter, we will discuss in detail the binding modes of catalysis—proximity effects and transition-state stabilization, which are conceptually more difficult. In **acid-base catalysis,** the acceleration of a reaction is achieved by catalytic transfer of a proton. Many nonenzymatic chemical reactions proceed only when they are performed in strong acid because a proton must be added to the reactant, or they proceed only in strong alkali because a proton must be removed. Indeed, the most common forms of nonenzymatic catalysis involve transfer of protons—proton addition by an acid, proton removal by a base. Not surprisingly, acid-base catalysis is also common in enzymatic reactions. Enzymes use the side chains of amino acids that can donate and accept protons under the nearly neutral pH conditions of cells. In this manner, the active site of an enzyme can provide the biological equivalent of a solution of strong acid or strong base.

It is convenient to use B: to represent a base, or proton acceptor, and $BH^{\oplus}$ to represent its conjugate acid, a proton donor. A proton acceptor can assist reactions in two ways. It can cleave a C—H bond by removing the proton to produce a carbanion.

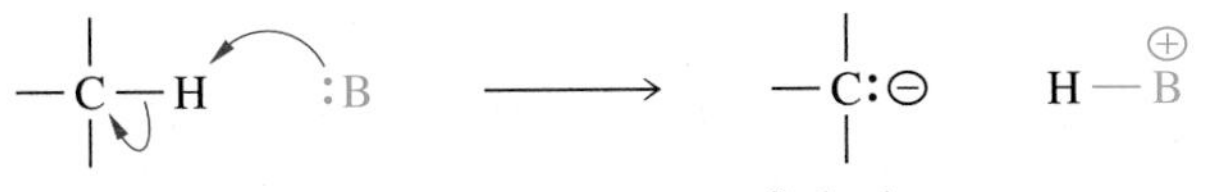

$$(7 \cdot 7)$$

Carbanion

It can also participate in the cleavage of other bonds involving carbon, such as a C—N bond, by generating the equivalent of $OH^{\ominus}$ in neutral solution through removal of a proton from a molecule of water.

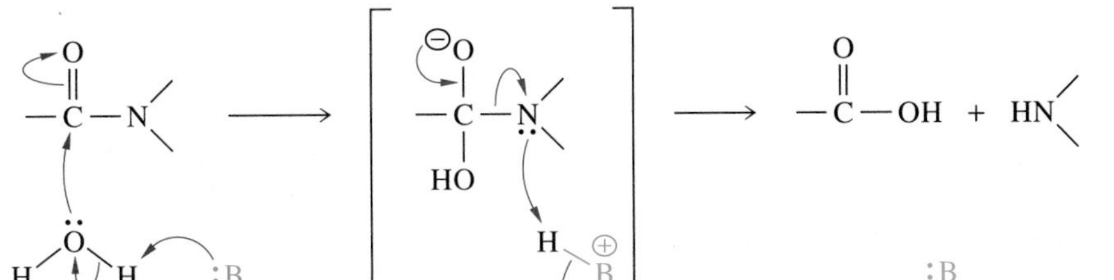

$$(7 \cdot 8)$$

Conversely, $BH^{\oplus}$ can donate a proton when the reaction proceeds in the opposite direction. Because the imidazole/imidazolium of the side chain of histidine has a pK_a of about 6 to 7 in most proteins, it is an ideal group for proton transfer at neutral pH values. The carboxyl groups of aspartate and glutamate residues, which have pK_a values of approximately 4 (but can be perturbed to lower or higher pK_a values), are also acid-base catalysts in some enzymes.

7·5 Many Enzymatic Group-Transfer Reactions Proceed by Covalent Catalysis

Covalent catalysis, the second mode of chemical catalysis, refers to catalyzed group-transfer reactions in which one substrate or part of it forms a covalent bond with the catalyst. In enzymatic covalent catalysis, a portion of one substrate is transferred first to the enzyme and then to a second substrate. For example, the group X can be transferred from molecule A-X to molecule B in the following two steps via the covalent ES complex X-E:

$$A\text{-}X + E \;\rightleftharpoons\; A + X\text{-}E \qquad\qquad (7 \cdot 9)$$

and

$$X\text{-}E + B \;\rightleftharpoons\; B\text{-}X + E \qquad\qquad (7 \cdot 10)$$

The reaction catalyzed by bacterial sucrose phosphorylase is an example of group transfer by covalent catalysis.

$$\text{Sucrose} + \text{P}_i \rightleftharpoons \text{Glucose 1-phosphate} + \text{Fructose} \qquad \textbf{(7·11)}$$

This type of reaction is called phosphorolysis because the glucosyl group from sucrose is transferred to phosphate, rather than to water as in the hydrolysis of sucrose (Equation 6·4). The first chemical step in the sucrose phosphorylase reaction is formation of a covalent glucosyl-enzyme intermediate. In this case, sucrose is equivalent to A-X and glucose is equivalent to X in Equation 7·9.

$$\text{Sucrose} + \text{Enzyme} \rightleftharpoons \text{Glucosyl-Enzyme} + \text{Fructose} \qquad \textbf{(7·12)}$$

The covalent ES intermediate can either donate the glucose unit to another molecule of fructose, in the reverse of Equation 7·12, or to phosphate (which is equivalent to B in Equation 7·10).

$$\text{Glucosyl-Enzyme} + \text{P}_i \rightleftharpoons \text{Glucose 1-phosphate} + \text{Enzyme} \qquad \textbf{(7·13)}$$

The reaction catalyzed by acetoacetate decarboxylase from *Clostridium acetobutilicum,* an anaerobic microorganism, is another excellent example of covalent

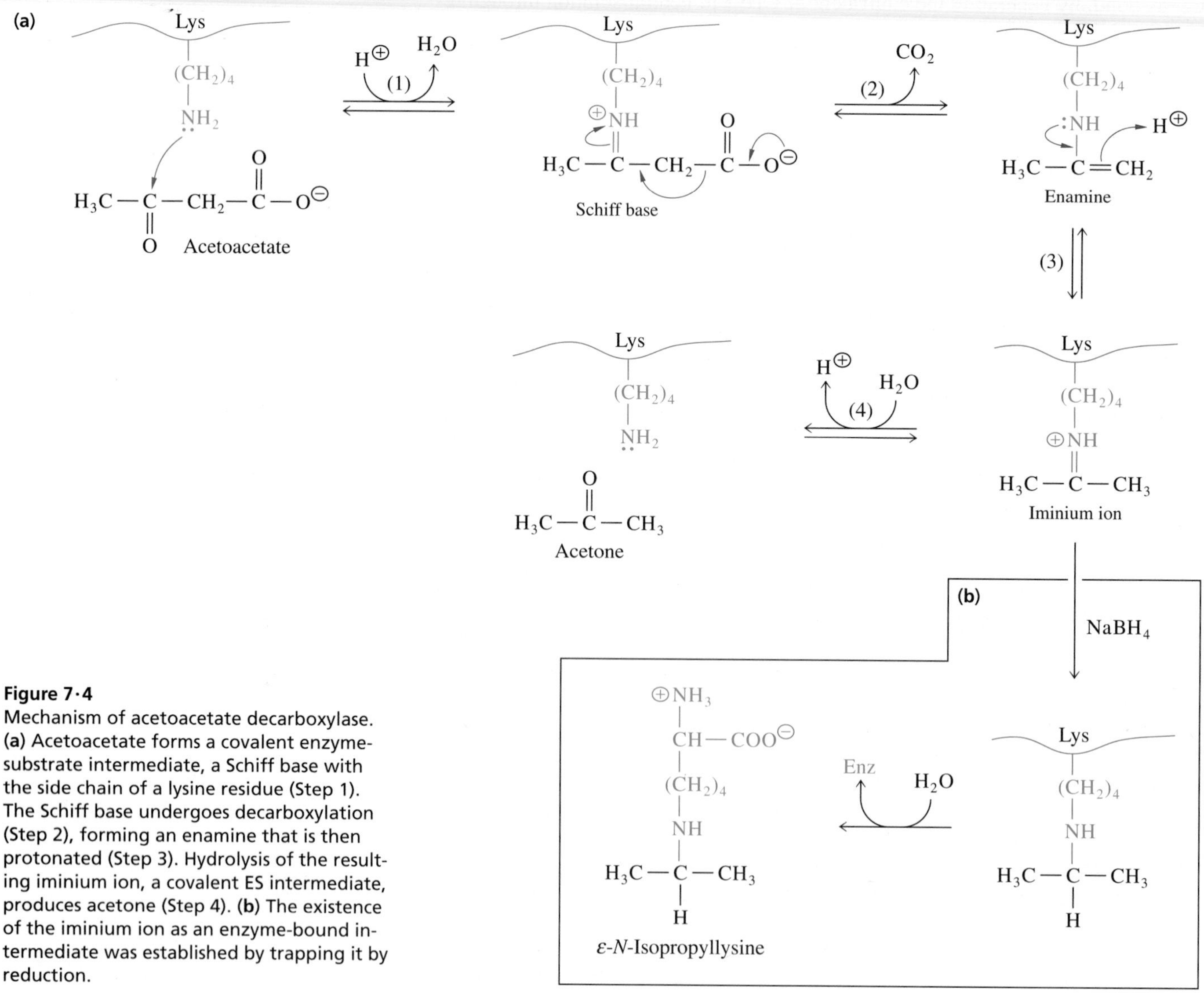

Figure 7·4
Mechanism of acetoacetate decarboxylase. **(a)** Acetoacetate forms a covalent enzyme-substrate intermediate, a Schiff base with the side chain of a lysine residue (Step 1). The Schiff base undergoes decarboxylation (Step 2), forming an enamine that is then protonated (Step 3). Hydrolysis of the resulting iminium ion, a covalent ES intermediate, produces acetone (Step 4). **(b)** The existence of the iminium ion as an enzyme-bound intermediate was established by trapping it by reduction.

catalysis. The enzyme catalyzes the breakdown of acetoacetate. In this reaction, acetone and CO_2 are formed after the substrate reacts with the ε-amino group of an active-site lysine residue (Figure 7·4a). This amino acid has unusual reactivity because the environment of the active site lowers its pK_a from about 10 to about 6. The ε-amino group becomes such a strong nucleophile at neutral pH that it can rapidly react with acetoacetate. The Schiff base that is formed between the β-carbonyl group of acetoacetate and the lysine is decarboxylated to produce an enamine of acetone. This reaction is enhanced by attraction of electrons from the carboxyl group toward the nitrogen of the Schiff base. Decarboxylation is the rate-limiting step in the reaction. In the next step, the enamine is protonated, and finally the iminium-ion complex with the enzyme is hydrolyzed to form acetone and free enzyme. The existence of the covalent ES complex was proven by reduction of a reaction mixture by sodium borohydride, $NaBH_4$. Borohydride reduces the $C=N$ bond of the iminium-ion ES intermediate. The reduced intermediate was hydrolyzed by acid, and the stable compound ε-N-isopropyllysine (Figure 7·4b) was isolated, showing that the active-site lysine is the covalent catalytic group.

It appears that about 20% of enzymes use the covalent mode of catalysis. Proof of covalent catalysis in an enzyme mechanism relies on the isolation or demonstration of the reactive intermediate. The observation of ping-pong kinetics (Section 6·8) is another indication of covalent catalysis.

7·6 Nonenzymatic Reactions Help Elucidate Enzymatic Reactions

Kinetic studies of nonenzymatic model-reaction systems have often been performed to simulate aspects of enzymatic mechanisms. It is possible to examine the kinetics of nonenzymatic reactions in a systematic fashion. For example, the reaction rate can be measured under changing conditions of temperature and solvent composition. Kinetic measurements can also be made in the presence of catalysts, especially buffers, or reactants whose structures are systematically altered. The reaction rates obtained under changing experimental conditions often provide clues to the chemical processes that occur during the reaction. Some model reactions provide information about general catalytic processes that apply to enzymes as a whole. Other reactions, devised as specific models of enzyme-catalyzed reactions, involve nonenzymatic catalysts that contain the same reactive groups as the enzyme and reactants that have chemical properties similar to the substrates of the enzyme.

Here we describe one specific model designed to test a mechanism proposed for peptidases and esterases. In this experiment, imidazole (the reactive group of the side chain of histidine) was tested as the catalyst, and the reactive ester p-nitrophenyl acetate was used as the substrate. Because p-nitrophenyl esters form a colored product when they are hydrolyzed, they are useful substrates for chemical kinetic experiments. The model reaction proceeds in two stages, the first involving covalent catalysis and the second, acid-base catalysis (Figure 7·5, next page). First, the acetyl group is transferred from the substrate to the imidazole catalyst, forming an acetylimidazole intermediate. The rate of this step, which coincides with the rate of formation of the colored phenolate, was measured at several pH values. The results showed that only the conjugate base form of imidazole, which is present at high pH values, not the imidazolium ion, is a covalent catalyst for this reaction.

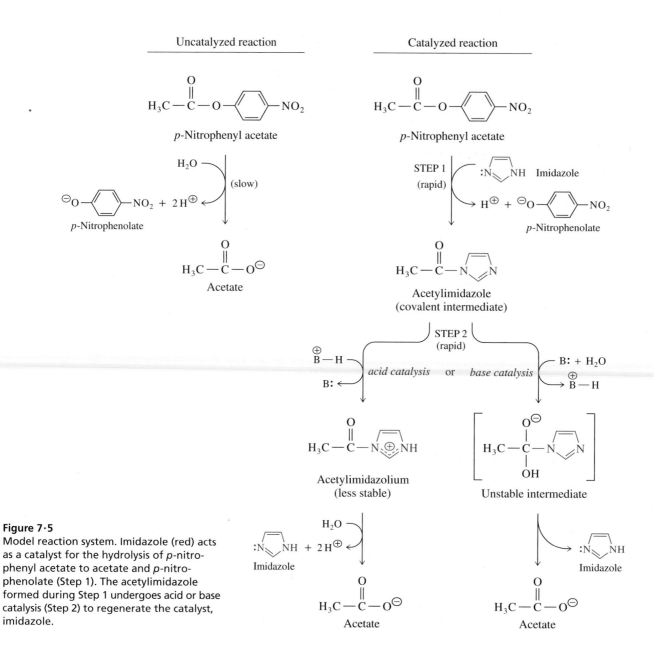

Figure 7·5
Model reaction system. Imidazole (red) acts as a catalyst for the hydrolysis of *p*-nitrophenyl acetate to acetate and *p*-nitrophenolate (Step 1). The acetylimidazole formed during Step 1 undergoes acid or base catalysis (Step 2) to regenerate the catalyst, imidazole.

The second step of the reaction, the hydrolysis of acetylimidazole, is faster than the first step. The second step can be accelerated by imidazole itself and by a buffer such as acetic acid, an acid catalyst, or Tris, a base catalyst. The reaction rate is proportional to the concentration of the acetic acid or the free base form of Tris. The catalytic effect of imidazole in Step 2 must involve acid-base catalysis and not covalent catalysis. Covalent catalysis by imidazole would only produce another molecule of acetylimidazole, not a molecule each of acetate and free imidazole.

Imidazole is a catalyst that allows the two-step hydrolysis of *p*-nitrophenyl acetate to occur more quickly than in the absence of a catalyst. Imidazole is a stronger nucleophile than water and therefore reacts with *p*-nitrophenyl acetate more rapidly than water does. The product of this reaction, acetylimidazole, then reacts with water in the presence of an acid-base catalyst faster than does the original substrate, *p*-nitrophenyl acetate. This model reaction, although it proceeds much more slowly than enzymatic hydrolysis reactions, demonstrates how rate enhancement occurs in both covalent and acid-base catalytic modes.

7·7 The Rates of Enzymatic Reactions Are Affected by pH

The effect of pH on the reaction rate of an enzyme can suggest which ionic amino acid residues are in its active site. The sensitivity to pH usually reflects an alteration in the ionization state of one or more residues involved in catalysis and occasionally in substrate binding. Of course, extremes of pH can cause denaturation of enzymes. However, under pH conditions that do not denature the enzyme, when the initial velocity of the reaction (or in some cases the maximum velocity) is plotted against pH, a bell-shaped curve is often obtained—a curve with information that can help delineate a mechanism. The pH at the point of maximum activity is called the **pH optimum** of the enzyme. Routine assays are usually performed at this pH, which is maintained using appropriate buffers. The pH-rate profile can often be explained by assuming that the ascending sigmoidal curve is caused by the deprotonation of an active-site amino acid residue (B) and that the sigmoidal curve descending from the pH optimum is caused by the deprotonation of a second active-site amino acid residue (A). Thus, a simple bell-shaped curve is the result of two overlapping titrations. The side chain of A must be protonated for activity, and the side chain of B must be unprotonated.

$$\overset{\oplus}{H}A\cdots BH^{\oplus} \;\rightleftharpoons\; \overset{\oplus}{H}A\cdots B \;\rightleftharpoons\; A\cdots B \qquad (7\cdot14)$$
$$\text{Inactive} \qquad\qquad \text{Active} \qquad\qquad \text{Inactive}$$

The inflection points of the two curves approximate the pK_a values of the two ionizable residues. At the pH optimum, midway between the two pK_a values, the greatest number of enzyme molecules are in the active $\overset{\oplus}{H}A\cdots B$ form. Ribonuclease (Section 24·9B), which has two active-site histidine residues, is an example of an enzyme that exhibits a bell-shaped pH-rate profile; the two histidines of ribonuclease have pK_a values of 5.8 (B) and 6.2 ($\overset{\oplus}{H}A$), and the pH optimum is 6.0. The pH-rate profile is not bell shaped if only one or if more than two ionizable amino acids participate in the catalytic mechanism.

Papain is a protease found in papaya fruit and used in meat tenderizers. Its mechanism exhibits both acid-base catalysis and covalent catalysis. It has a typical pH-rate profile with inflection points at pH 4.2 and pH 8.2 (Figure 7·6). This profile suggests that the activity of papain depends on two active-site amino acid residues that have pK_a values of about 4 and 8.

Many proteases, including papain, catalyze the hydrolysis of peptides by the general mechanism shown in Figure 7·7 (next page), which is an expansion of Equation 7·1. After the noncovalent ES complex is formed by binding of the substrate in the active site, an active-site nucleophilic group (—X:, the covalent catalyst) attacks the carbonyl carbon of the peptide bond to form an unstable tetrahedral

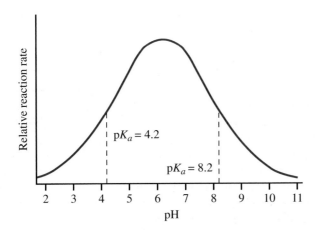

Figure 7·6
pH-rate profile for papain. The pH-rate profile for papain is bell shaped. The left and right segments of the pH-rate curve correspond to titration curves for the side chains of two active-site amino acids. The inflection point at pH 4.2 corresponds to the considerably perturbed pK_a of Cys-25, and the inflection point at pH 8.2 corresponds to the pK_a of His-159. The enzyme is active only when these ionic groups are present as the thiolate-imidazolium (Im) ion pair [—$S^{\ominus}\cdots\overset{\oplus}{H}$Im—].

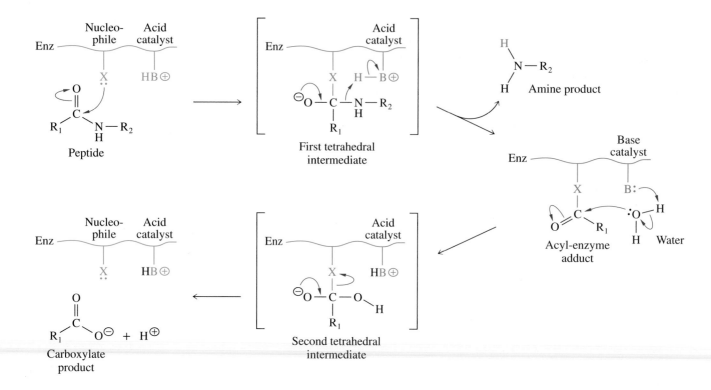

Figure 7·7
General mechanism for enzymatic hydrolysis of a peptide. In this mechanism, the enzyme reacts with the acyl group of the substrate to form a covalent enzyme-substrate intermediate, the acyl-nucleophile adduct. The nucleophile (—X:) is the side chain of an ionizable amino acid residue in the active site. A second active-site residue (—B) acts as a proton donor and acceptor.

intermediate. Cleavage of the peptide bond occurs when a proton is donated by another active-site amino acid ($\oplus$HB), the acid-base catalyst, to the nitrogen atom of the tetrahedral intermediate to produce the amine leaving group. The amine diffuses out of the active site and is replaced by a molecule of water. The other portion of the substrate, the acyl group, has substituted the nucleophilic group of the enzyme, producing the covalent acyl-enzyme adduct. As depicted in the lower half of Figure 7·7, the carboxylic acid from the peptide substrate is released from the substituted enzyme by nucleophilic attack on the carbonyl group by $OH^\ominus$ from water. The tetrahedral intermediate formed as a result of this addition of $OH^\ominus$ decomposes to form the carboxylate product, a proton, and regenerated free enzyme.

Kinetic analysis shows that papain follows a ping-pong kinetic mechanism.

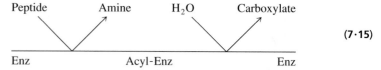

(7·15)

This kinetic behavior is in accord with the general chemical mechanism shown in Figure 7·7. The acyl group of the peptide substrate becomes bound to the enzyme and is then transferred to a molecule of water.

In papain, the two active-site ionizable residues are a nucleophilic cysteine and a proton-donating imidazolium group of histidine. Although the side chain of cysteine normally has a pK_a value of 8 to 9.5, it forms a thiolate-imidazolium ion pair with histidine in the cleft of papain, causing perturbation of its pK_a value to about 4 and raising the pK_a of the histidine residue to about 8. The three ionic forms of the catalytic center of papain are shown in Figure 7·8. Only the upper tautomer of the intermediate form is active. It is only in this form that papain has *both* the nucleophilic cysteine to accept the acyl group from the substrate and the protonated histidine required for acid-base catalysis of the hydrolysis reaction.

Figure 7·8
Three ionic forms of active-site residues of papain. Only the upper tautomer of the intermediate form is active.

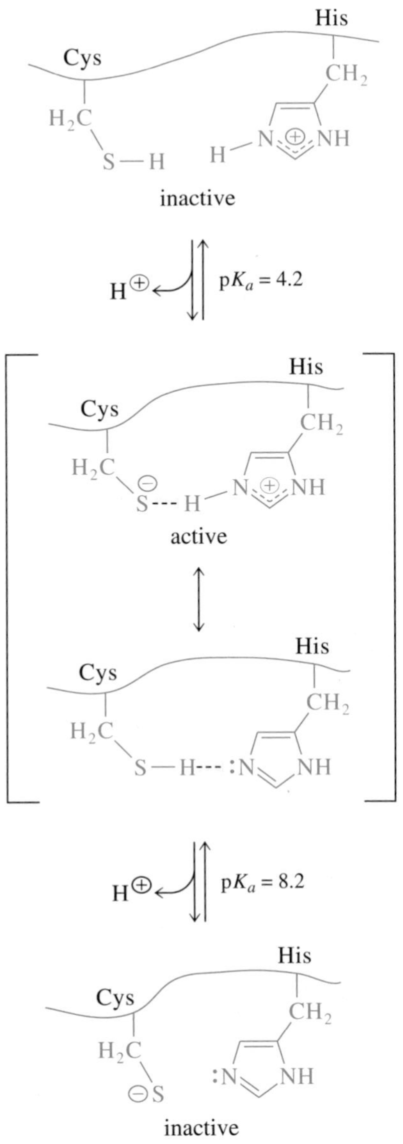

7·8 The Upper Limit for Catalytic Rates Is the Rate of Diffusion

A few enzymes catalyze reactions at rates approaching the upper physical limit of reactions in solution, the rate of diffusion of reactants toward each other. A reaction that occurs with every collision between reactant molecules is termed a **diffusion-controlled reaction.** The frequency of encounter between a typical solute and the active site of an enzyme has been calculated to be about 10^8 to $10^9\,M^{-1}\,s^{-1}$ under physiological conditions. (For reference, the upper limit for a molecular vibration—the fastest unimolecular reaction—is about 10^{12} to $10^{13}\,s^{-1}$.) The frequency of encounter can be higher if there is electrostatic attraction between the reactants. The apparent second-order rate constants for six very fast enzymes are listed in order of increasing rate in Table 7·3 (next page). As you can see, only one of these constants—for superoxide dismutase—greatly exceeds a rate of $10^8\,M^{-1}\,s^{-1}$. The explanation for faster catalysis by superoxide dismutase is discussed in Section 7·8B.

Table 7·3 Enzymes with second-order rate constants near the upper limit

Enzyme	Substrate	k_{cat}/K_m (M^{-1} s^{-1})
Catalase	H_2O_2	4×10^7
Carbonic anhydrase	CO_2	8.3×10^7
Acetylcholinesterase	Acetylcholine	1.6×10^8
Fumarase	Fumarate	1.6×10^8
Triose phosphate isomerase	Glyceraldehyde 3-phosphate	4×10^8
Superoxide dismutase	O_2	2×10^9

The apparent second-order rate constants for the enzyme-catalyzed reaction E + S → E + P were obtained from the ratio k_{cat}/K_m. For these enzymes, the formation of the ES complex is the rate-determining step.

The binding of a substrate to an enzyme is a very fast second-order reaction. If this step is also the rate-determining step, and if the rest of the reaction is simple and fast, the overall rate of the reaction may approach the upper limit for catalysis. Only a limited number of types of chemical reactions can proceed this quickly. These include association reactions, some proton transfers, electron transfers, and conformational changes in proteins. All of the reactions described in Table 7·3 are so simple that their rate-determining steps are roughly as fast as the binding of their substrate molecules by the enzymes.

A. Triose Phosphate Isomerase Catalyzes a Reaction That Is Close to the Diffusion-Controlled Limit

The reaction catalyzed by triose phosphate isomerase is listed as the second fastest in Table 7·3. This enzyme, found in the pathway of glycolysis (Chapter 15), catalyzes the rapid interconversion of dihydroxyacetone phosphate (DHAP) and glyceraldehyde 3-phosphate (G3P).

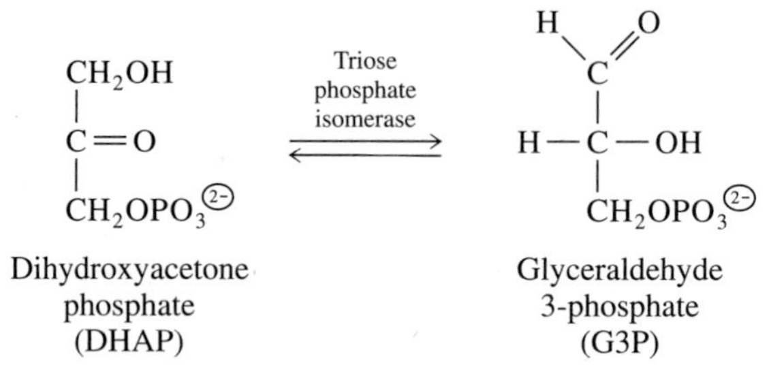

$$
\begin{array}{ccc}
\text{CH}_2\text{OH} & & \text{H} \diagdown \quad \diagup \text{O} \\
| & \text{Triose} & \text{C} \\
\text{C}=\text{O} & \xrightarrow[\phantom{\text{isomerase}}]{\text{phosphate isomerase}} & | \\
| & & \text{H}-\text{C}-\text{OH} \\
\text{CH}_2\text{OPO}_3^{2\ominus} & & \text{CH}_2\text{OPO}_3^{2\ominus}
\end{array} \qquad (7\cdot16)
$$

Dihydroxyacetone phosphate (DHAP)   Glyceraldehyde 3-phosphate (G3P)

The reaction proceeds by a 1,2-proton shift (Figure 7·9). An analogous nonenzymatic interconversion in alkaline solution is a two-step process. The enzymatic interconversion involves two extra steps, the formation of the noncovalent E-DHAP complex and the dissociation of the E-G3P complex. Triose phosphate isomerase has two ionizable active-site residues, glutamate and histidine. Upon binding of substrate, the carbonyl oxygen of dihydroxyacetone phosphate forms a hydrogen bond with the neutral imidazole group of the histidine residue. The carboxylate group of the glutamate residue removes a proton from C-1 of dihydroxyacetone phosphate to form an enediolate intermediate. The oxygen at C-2 is protonated by the histidine residue, which then abstracts a proton from the oxygen at C-1 to form another unstable enediolate intermediate. In this proton transfer, the conjugate acid form of histidine appears to be the neutral species, and the conjugate base appears to be the imidazolate.

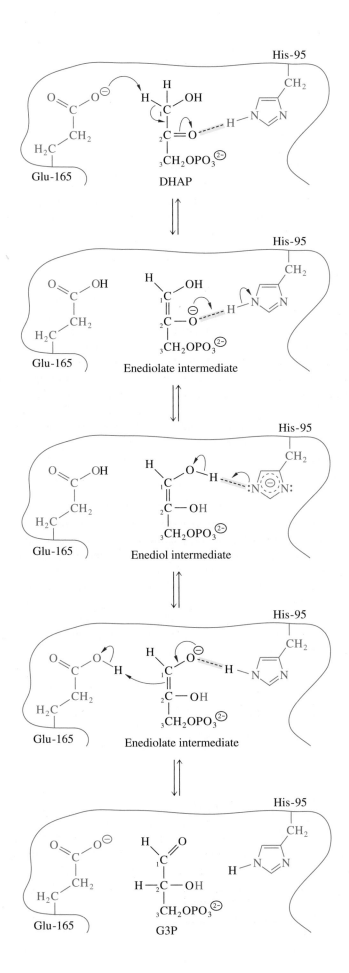

Figure 7·9
Mechanism of the reaction catalyzed by triose phosphate isomerase. The active-site glutamate residue of the enzyme removes a proton from C-1 of the enzyme-bound dihydroxyacetone phosphate (DHAP) and donates a proton to C-2 to form enzyme-bound glyceraldehyde 3-phosphate (G3P). The neutral histidine residue in the active site transfers a proton between the oxygen atoms on C-1 and C-2.

$$\text{(7·17)}$$

This ionization of a histidine residue is unusual and has not been implicated in the catalytic mechanism of any other enzyme.

Elaborate kinetic measurements performed in the laboratory of Jeremy Knowles have determined the rate constants of all four kinetically measurable enzymatic steps in both directions.

$$\text{E + DHAP} \rightleftarrows \text{E-DHAP} \rightleftarrows \text{E-Intermediate} \rightleftarrows \text{E-G3P} \rightleftarrows \text{E + G3P}$$

$$\text{(7·18)}$$

The reaction diagram constructed from these rate constants is shown in Figure 7·10, along with the profile of a mutant triose phosphate isomerase (dashed lines) to be discussed below. Note that all of the barriers for the nonmutant enzyme are approximately the same height; that is, the steps are balanced. This is an ideal strategy for enzymatic catalysis. No single step is rate limiting, so that enzymatic catalysis is not slowed down by accumulation of an intermediate. The value of the second-order rate constant k_{cat}/K_m for the conversion of glyceraldehyde 3-phosphate to dihydroxyacetone phosphate is $4 \times 10^8 \, \text{M}^{-1}\text{s}^{-1}$, close to the rate of a diffusion-controlled reaction. The physical step of S binding to E is rapid but not much faster than the subsequent chemical steps in the reaction sequence. However, improvement of any of the chemical steps would not greatly increase the efficiency of the enzyme. It appears that through evolution this isomerase has virtually reached its maximum possible efficiency as a catalyst.

Knowles and his colleagues have converted the active-site glutamate residue of triose phosphate isomerase to an aspartate residue by site-directed mutagenesis (Section 6·10). In the modified isomerase, the same reactive group ($-\text{COO}^{\ominus}$) is present as in the wild-type enzyme, but it is slightly farther from the substrate. The wild-type and mutant enzymes have fairly similar K_m values. However, the values for k_{cat}, measured for both the forward and reverse reactions catalyzed by the mutant enzyme, are greatly decreased. This experiment verified that the glutamate residue in the active site is essential for full activity. This experiment also changed the enzyme from diffusion controlled to a more typical enzyme that exhibits rapid binding of reactants and slower chemical catalysis steps.

Figure 7·10
Reaction diagram for the reaction catalyzed by triose phosphate isomerase. The solid line represents the profile for the wild-type (naturally occurring) enzyme. The dotted line shows the profile for a mutant enzyme in which the active-site glutamate residue has been replaced by an aspartate residue. In the case of the mutant enzyme, the activation energies for the proton-transfer reactions (Steps 2 and 3) are significantly higher. The wild-type enzyme catalyzes the reaction about one thousand times faster than the mutant enzyme. [Adapted from Raines, R. T., Sutton, E. L., Strauss, D. R., Gilbert, W., and Knowles, J. R. (1986). Reaction energetics of a mutant triosephosphate isomerase in which the active-site glutamate has been changed to aspartate. *Biochemistry* 25:7142–7154.]

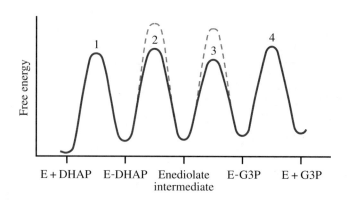

B. Can Diffusion Be Accelerated?

Can an enzymatic rate be faster than the rate of simple diffusion? Yes, when the mechanism involves the **Circe effect.** This effect, proposed by William Jencks in 1975, refers to the strong attractive forces that lure a substrate into an active site. Binding of substrates in the active site by electrostatic forces is well known, but there is little evidence that these interactions direct and attract substrates to the active site. For the Circe effect to increase the rate of a reaction, diffusion of substrate to enzyme must be the rate-limiting step. Also, the substrate molecules must be charged and the product molecules uncharged. Currently, there is only one known enzymatic example of accelerated diffusion: attraction of substrate to the active site of the metal-containing enzyme superoxide dismutase by an electric field at the mouth of the active site.

Superoxide dismutase, a homodimer, catalyzes the very rapid removal of the toxic superoxide-radical anion, $\cdot O_2^{\ominus}$, a by-product of oxidative metabolism. The enzyme catalyzes the conversion of superoxide to molecular oxygen and hydrogen peroxide, which is rapidly removed by the subsequent action of enzymes such as catalase.

$$4 \cdot O_2^{\ominus} \xrightarrow[\substack{\text{Superoxide} \\ \text{dismutase}}]{\substack{4 H^{\oplus} \quad 2 O_2}} 2 H_2O_2 \xrightarrow{\text{Catalase}} 2 H_2O + O_2 \tag{7·19}$$

The reaction catalyzed by superoxide dismutase proceeds in two steps, in which an atom of copper bound to the enzyme is oxidized and then reduced.

$$\text{E-Cu}^{\oplus 2+} + \cdot O_2^{\ominus} \longrightarrow \text{E-Cu}^{\oplus} + O_2 \tag{7·20}$$

and

$$\text{E-Cu}^{\oplus} + \cdot O_2^{\ominus} + 2 H^{\oplus} \longrightarrow \text{E-Cu}^{\oplus 2+} + H_2O_2 \tag{7·21}$$

The overall reaction includes binding of the anionic substrate molecules, transfer of electrons and protons, and release of the uncharged products—all very rapid reactions. The value of k_{cat}/K_m for superoxide dismutase at 25°C is near $2 \times 10^9 \, M^{-1} \, s^{-1}$, *faster* than expected for association of the substrate with the enzyme based on typical diffusion rates. The active-site copper atom is at the bottom of a deep channel in the protein. Four hydrophilic amino acid residues are involved in the electrostatic guidance of $\cdot O_2^{\ominus}$ to the positive side chain of Arg-143 in the substrate-binding pocket. Glu-132, Glu-133, Lys-136, and Thr-137 form a hydrogen-bonded network at the rim of the active site (Figure 7·11). The three charged side chains of this network orient the electrostatic field: the glutamate residues repelling and the lysine

Figure 7·11
Superoxide dismutase. (a) Residues involved in substrate binding. The hydrogen bonds that stabilize the guidance group of superoxide dismutase are shown in yellow. $\cdot O_2^{\ominus}$ binds between Arg-143 and the copper atom. His-63, positioned between the copper (orange) and zinc (blue) atoms, is the proton donor in the reaction. (b) The active-site channel. The calculated electrostatic forces on $\cdot O_2^{\ominus}$ are shown as colored vectors: red, most negative; yellow, negative; green, neutral; light blue, positive; blue, most positive. (Courtesy of E. D. Getzoff, G. H. Liao, and C. L. Fisher.)

(a)

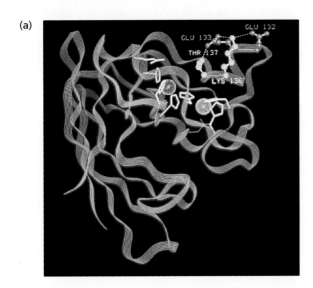

(b)

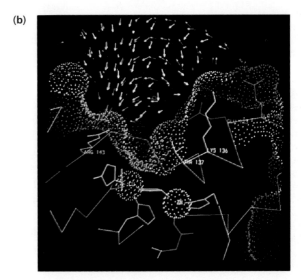

residue attracting $\cdot O_2^{\ominus}$. The electric field around the superoxide dismutase active site enhances the rate of formation of the ES complex about 30-fold. The Circe effect allows superoxide dismutase, and perhaps other enzymes, to catalyze reactions much faster than simple collision would allow.

7·9 The Binding of Reactants Plays the Most Important Role in Enzymatic Catalysis

Hypothetical explanations for enzymatic catalysis have been in circulation for many years, but sufficient scientific evidence to support these ideas has been obtained only recently. Even now, the assignment of quantitative acceleration values to each catalytic mode is difficult. We have already seen two chemical modes by which enzymes catalyze reactions, acid-base catalysis and covalent catalysis. Based upon the effects of acid-base catalysts on nonenzymatic reactions, it is estimated that acid-base catalytic residues probably accelerate enzymatic reactions by about 10 to 100 times. From similar considerations, covalent catalysis provides about the same rate acceleration as does proton transfer.

As important as these chemical modes are, they are far from accounting for the observed rate accelerations achieved by enzymes. The ability of globular proteins to specifically bind and orient ligands can completely account for the remainder. The proper binding of reactants in the active sites of enzymes provides not only substrate and reaction specificity but also most of the catalytic power of enzymes. Two interrelated catalytic modes operate, both based on the interaction of reactant molecules with a matching enzyme. For multisubstrate reactions, the collecting of substrate molecules in the active site raises their effective concentrations over their concentrations in free solution. In the same way, binding of a substrate near a catalytic active-site residue increases the effective concentrations of these two reactants. High effective concentrations favor the more frequent formation of transition states. This phenomenon has had many names, the most widely accepted being the **proximity effect.** The proximity effect requires *weak* binding of reactants to enzymes, since extremely tight binding would defeat catalysis. The second major catalytic mode arising from the ligand-enzyme interaction is the increased binding of transition states to enzymes compared to the binding of substrates or products. This catalytic mode has been referred to as strain or distortion but is more precisely called **transition-state stabilization.** There is an equilibrium (*not* the reaction equilibrium) between ES and the enzymatic transition state, $ES^{\ddagger}$. Interaction between the enzyme and its ligands in the transition state shifts this equilibrium toward $ES^{\ddagger}$ and lowers the energy of activation.

The effects of proximity and transition-state stabilization were outlined in Figure 7·3. Experiments, some of which we will describe, have suggested that proximity contributes to enzymatic catalysis an enhancing effect of about 10^4 to 10^5, and transition-state stabilization at least that much. When these effects are multiplied together with chemical catalytic effects, we can see clearly how enzymes achieve their extraordinary rate accelerations.

The binding forces responsible for formation of ES complexes and for stabilization of $ES^{\ddagger}$ are familiar from Chapters 3 and 5: electrostatic forces, hydrogen bonds, hydrophobic interactions, and van der Waals forces. Electrostatic interactions are stronger in nonpolar environments than in water. Because active sites are largely nonpolar, electrostatic forces in the clefts of enzymes can be quite strong.

Table 7·4 Potential interactions between substrates and amino acid residues of proteins. The binding characteristics shown arise from the side chains. In addition, all amino acid residues can participate in hydrogen bonding through the peptide backbone.

| | Type of interaction | | |
Amino acid	Electrostatic	Hydrogen bonding	Hydrophobic
Ionizable:			
Aspartate	+	+	
Glutamate	+	+	
Lysine	+	+	
Arginine	+	+	
Histidine	+	+	
Cysteine		+	
Polar, nonionizable:			
Asparagine		+	
Glutamine		+	
Serine		+	
Threonine		+	
Polypeptide backbone		+	
Some hydrophobic character:			
Tyrosine		+	+
Tryptophan		+	+
Methionine		+	+
Glycine (little binding, other than through the backbone)			
Hydrophobic:			
Alanine			+
Valine			+
Leucine			+
Isoleucine			+
Phenylalanine			+
Proline			+

The plus sign indicates that the side chain of this amino acid can participate in the type of interaction in that column. [Adapted from Baker, B. R. (1967). *Design of Active-Site-Directed Irreversible Enzyme Inhibitors* (New York: John Wiley & Sons), p. 25.]

Hydrogen bonds, next in bond strength, are often formed between substrates and enzymes. Although hydrogen bonds are fairly stable, they are weak enough that bound reactants can dissociate readily. There are hydrophobic interactions between nonpolar groups of substrates and hydrophobic regions in the clefts of enzymes. Large numbers of weak van der Waals interactions also help to bind substrates. Keep in mind that both the chemical properties of the amino acid components and the shape of the active site of a globular protein also determine the substrate specificity of the enzyme. Table 7·4 provides a list of the interactions possible between individual amino acids and ligands such as substrates.

During the process of the binding of a ligand to an enzyme, the molecules of water that participate in solvation of both the ligand and the complementary part of the active site are lost. Presumably, the released water molecules gain entropy and thus provide some of the energy needed for the ligand-enzyme association. The cleft of an ES complex can be quite nonpolar. However, there is little direct evidence that desolvation plays a major catalytic role.

7·10 The Proximity Effect Is an Antientropy Phenomenon

Two molecules coming together from dilute solution to form a transition state is an improbable event, accompanied by the loss of a great deal of entropy. William Jencks and his colleagues made a conceptual advance by considering the reaction of two molecules at the active site as equivalent to a nonenzymatic bimolecular reaction that was converted to an intramolecular or unimolecular reaction. Correct positioning of two reactants produces a large loss of entropy, sufficient to account for a large rate acceleration. The acceleration is expressed in terms of the enhanced relative concentration, called the **effective molarity,** of the reacting groups in the unimolecular reaction. Experiments have been performed to compare the rate of reaction of a compound containing the reactants correctly preassembled into a single molecule (k_1) with the rate of the corresponding bimolecular reaction (k_2). The effective molarity can be obtained from this ratio:

$$\text{Effective molarity} \quad = \quad \frac{k_1 \ (\text{in s}^{-1})}{k_2 \ (\text{in M}^{-1}\text{s}^{-1})} \qquad (7\cdot22)$$

All of the units in this equation except M cancel, so the ratio is expressed in molar units. Because the effective molarities of suitable chemical model reactions often greatly exceed concentrations that can be attained in solution, the large rate increases characteristic of enzymatic catalysis must reflect factors in addition to high local concentration, such as very favorable geometry of the reacting groups.

The results of an early experiment demonstrate how proximity effects are measured and interpreted. The amine-catalyzed hydrolysis of phenyl esters proceeds through a covalent mechanism, with formation of an N-acyl intermediate and release of a colored phenolate, as we saw in Section 7·6. The rate of this reaction was measured by spectrophotometrically determining the rate of formation of the

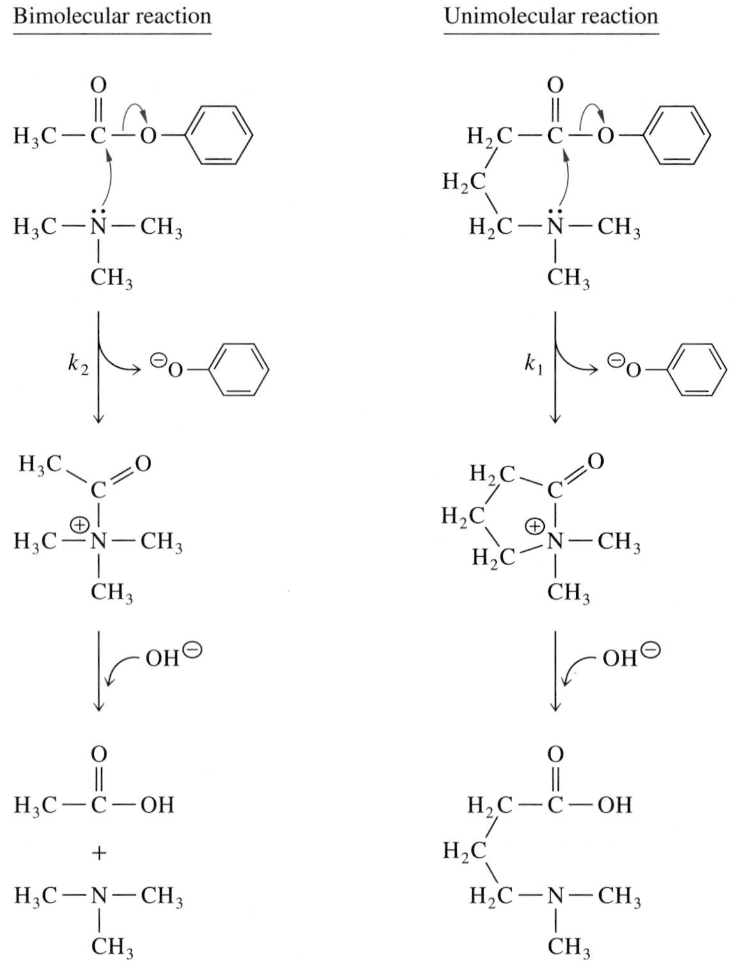

Figure 7·12
Amine-catalyzed hydrolysis of a phenyl ester. The unimolecular reaction is 1260 times faster than the bimolecular reaction.

colored phenolate. *N,N*-Dimethyl aminobutyrate phenyl ester combines the molecules trimethylamine and phenyl acetate in one molecule. It reacts by the same mechanism as trimethylamine and phenyl acetate (Figure 7·12). The unimolecular reaction rate (k_1) exceeds the rate of the bimolecular reaction (k_2) by 1260 M. These results indicate that the effective concentration of trimethylamine in the neighborhood of the acyl group in the ring-closure reaction is 1260 M, a physically impossible concentration. Therefore, the rate enhancement measured for the unimolecular substrate may be the result of more favorable geometry for its reacting groups. Analysis of the kinetic measurements of this experiment showed that, although the two reactions have similar mechanisms, the reactants in the bimolecular reaction require a much greater loss of entropy to reach the transition state.

The proximity effect in a chemical model system is also illustrated by experiments performed by Thomas Bruice and Upendra K. Pandit. In these experiments, the reactivities of a series of intramolecular reactants were compared to the reactivity of bimolecular reactants (Figure 7·13). The unimolecular substrates had progressively greater restriction of rotation around a bridging arm and progressively higher effective molarities. The bimolecular reaction was the two-step hydrolysis of

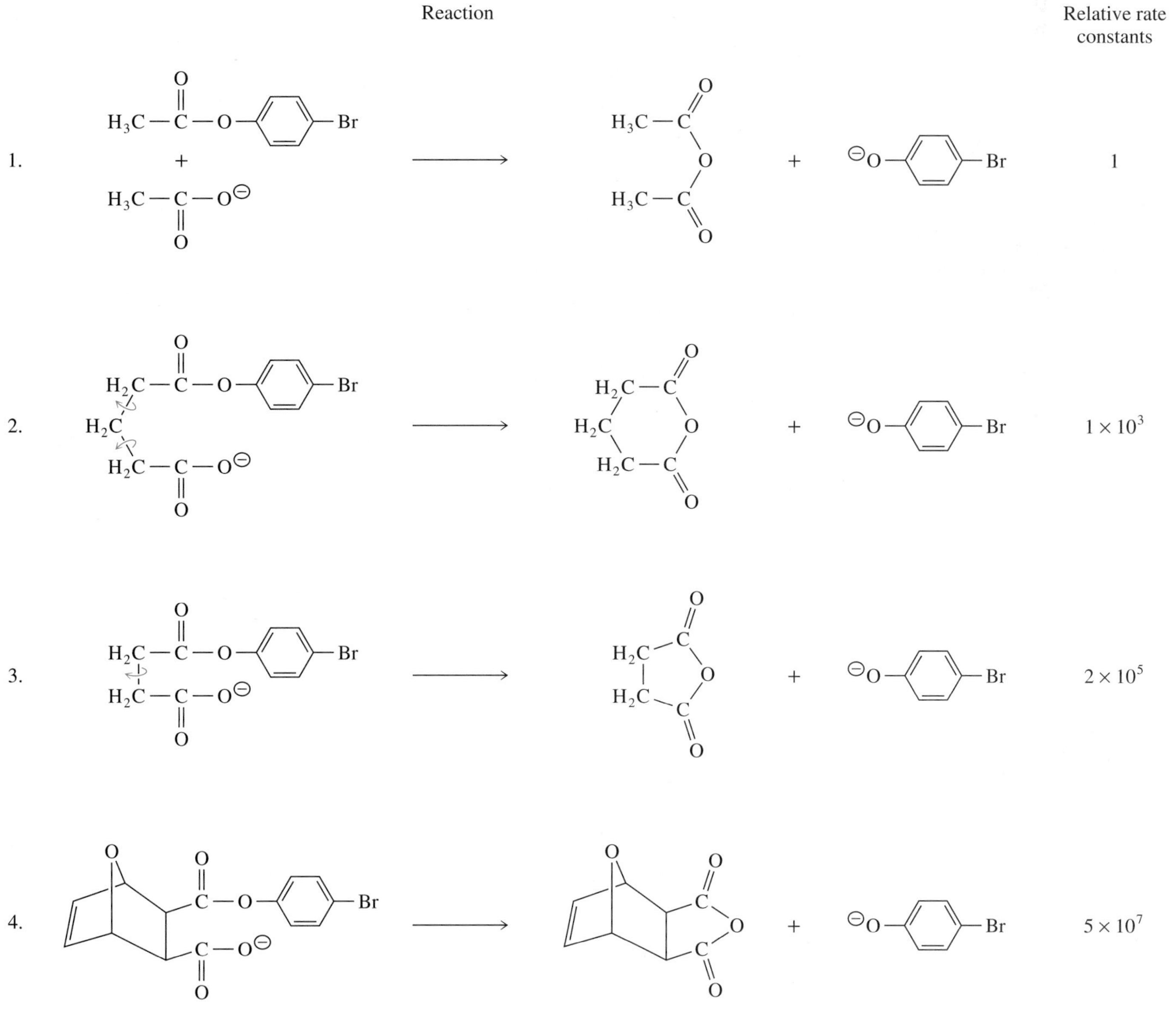

Figure 7·13
Reactions of a series of carboxylates with substituted phenyl esters. The proximity effect is illustrated by the increase in rate observed when the reactants are held more rigidly in proximity. Reaction 4 is 50 million times faster than Reaction 1, the bimolecular reaction.

p-bromophenyl acetate, catalyzed by acetate and proceeding via the formation of acetic anhydride. (The second step, hydrolysis of acetic anhydride, is not shown in Figure 7·13.) The rate of the first step was determined by measuring the release of a colored phenolate. Each of the unimolecular reactions involves formation of either a five- or six-membered ring in the rate-limiting step, covalent catalysis by formation of an acid anhydride. With each restriction placed upon the substrate molecule, the relative rate constant—measured by k_1/k_2—increased markedly. Note that the glutarate ester, compound 2, has two bonds that allow rotational freedom, whereas the succinate ester, compound 3, has only one. The most restricted compound, the rigid bicyclic compound 4, has no rotational freedom. Its rigid structure positions the carboxylate close to the ester. With an extremely high probability of reaction, this compound showed an effective molarity of the carboxylate group of $5 \times 10^7\,M$. Theoretical considerations suggest that the greatest rate acceleration that can be expected from the proximity effect is about 10^8. All of this rate acceleration can be attributed to loss of entropy by the two reactants.

7·11 Extremely Tight Binding of Substrate to an Enzyme Would Interfere with Catalysis

The insight gained from the previous section about unimolecular nonenzymatic reactions can be applied to enzymes. Reactions of ES complexes are analogous to unimolecular reactions. The pre-positioning of substrates in an enzyme active site produces a large rate acceleration. However, the maximum 10^8-fold acceleration that can be generated by the proximity effect cannot be achieved by enzymes. Typically, the change in entropy upon binding of substrate provides an acceleration of roughly 10^4. In ES complexes, the reactants are brought toward but not extremely close to the transition state. This conclusion is based upon both mechanistic reasoning and measurements of the tightness of binding of substrates to enzymes, usually employing K_m values. Because measuring actual dissociation constants is more difficult than measuring K_m values, K_m is assumed to represent the dissociation constant. The following discussion explains why the binding of substrates to enzymes cannot be extremely tight, that is, why K_m values cannot be extremely low.

The reaction diagram in Figure 7·14, for a simple unimolecular reaction, compares the energy path of a nonenzymatic reaction with the multistep path followed in enzymatic catalysis. The transition state ES‡ in the enzymatic reaction is stabilized by the amount shown by the upper arrow (1). If the ES complex were

Figure 7·14
Energy of substrate binding. In this hypothetical reaction, the enzyme accelerates the rate of the reaction by stabilizing the transition state. The amount of stabilization is indicated by the upper arrow (1). The activation barrier for the formation of the transition state ES‡ from ES (2) is low, so this enzyme is a good catalyst. If the enzyme were to bind the substrate tightly, lower dashed line and arrow (3), the activation barrier from the stabilized ES complex would be close to the activation barrier of the nonenzymatic reaction. Tight binding of this type is termed a thermodynamic pit.

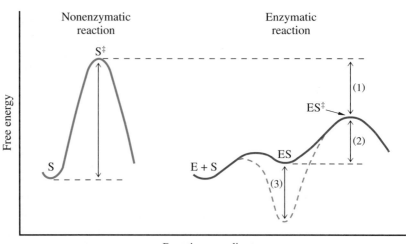

Reaction coordinates

stabilized by a similar amount, shown by the lower arrow (3), the enzyme would not be a catalyst because it would take just as much energy to reach $ES^{\ddagger}$ from ES as is required for the nonenzymatic reaction. In other words, if substrate were bound extremely tightly, there would be little or no catalysis. The situation of excessive ES stability is termed a thermodynamic pit. Because the energy difference between ES and $ES^{\ddagger}$ is significantly less than the energy difference between S and $S^{\ddagger}$ (the transition state in the nonenzymatic reaction), k_{cat} is greater than k_n (the rate constant for the nonenzymatic reaction). Therefore, enzymes must position reactants fairly rigidly before the transition state is reached but not so rigidly that they have lost all of their entropy by forming an ES complex that is too stable. This phenomenon of weak binding of substrates is called destabilization of the ES complex.

Consideration of the K_m values of a wide variety of enzymes for their substrates shows that enzymes avoid the thermodynamic pit. Most K_m values are fairly high, on the order of 10^{-4} M. They would be lower, indicating stronger binding, if all parts of the substrates were bound to the enzymes. For example, the K_m values observed for the binding of ATP to most ATP-requiring enzymes are about 10^{-4} M or greater, but there is a protein (not an enzyme) that binds ATP with a dissociation constant of 10^{-13} M. The billionfold difference in binding indicates that in an ES complex, in which the ground state of a substrate is bound to the enzyme, not all parts of the substrate are bound. This property is an important feature of the fourth major force that drives enzymatic catalysis, increased binding of reactants in the transition state, $ES^{\ddagger}$.

Enzymes that are nonspecific exhibit relatively high K_m values for their substrates. Enzymes specific for small substrate molecules such as urea, carbon dioxide, and superoxide anion exhibit high K_m values for these compounds because the substrates are so small that they can form few noncovalent bonds with enzymes. In contrast, larger substrate molecules and enzyme-bound intermediates are bound relatively tightly to enzymes. Typically, enzymes have lower K_m values for coenzymes—molecules larger than many substrates. Prosthetic groups (coenzymes that remain bound to the enzyme as a permanent component of the active site) bind to enzymes with very low K_m values. For example, the K_m of a typical transaminase for pyridoxal phosphate (a prosthetic group) is 4×10^{-6} M, reflecting binding that is at least 25-fold tighter than the binding of ATP to most of its enzymes.

An enzyme with an extremely low K_m for its substrate must also have a low k_{cat}. This is because k_{cat}/K_m, the second-order rate constant for the association of E and S, cannot exceed the diffusion-controlled value of about 10^8. Very low values of K_m would raise this ratio close to its limit. If the K_m value were 10^{-8} M, k_{cat} could not be higher than about 1 s^{-1}, making the enzyme a relatively slow catalyst.

When the concentration of a substrate inside a cell is below the K_m value of its corresponding enzyme, the equilibrium of the binding reaction $E + S \rightleftharpoons ES$ favors $E + S$. In other words, the formation of the ES complex is slightly uphill energetically (Figures 7·3 and 7·14), and the ES complex is closer to the energy of the transition state than is the ground state. Kinetically, this weak binding of substrates accelerates reactions. K_m values are optimized by evolution for effective catalysis— low enough that proximity is achieved, but high enough that the ES complex is not too stable.

7·12 Transition States Bind Tightly to Enzymes

Since the 1920s, enzyme chemists have suggested that enzymes bind reactants in the transition state with great affinity. Evidence in favor of this theory came in the early 1970s from experiments by Richard Wolfenden and Gustav Lienhard. They showed that chemical analogs of transition states are potent inhibitors of enzymes.

Other experimental data are now available to support transition-state stabilization as a major catalytic mode.

Strain and distortion are terms that have been used for years to describe enzymatic catalysis. It was suggested that enzymes, like solid catalysts, catalyze reactions by physically or electronically distorting substrates. Transition-state stabilization, with its increased interactions with the substrate in the transition state, is just a gentler or more subtle way of explaining strain.

Recall Emil Fischer's lock-and-key theory of enzyme specificity (Section 6·2). Fischer proposed that enzymes were rigid templates that accepted only certain keys—substrates. This theory lost some favor during the 1950s and 1960s as people sought more complex answers to the mechanisms of enzymes. However, the lock-and-key theory is now a generally accepted explanation for enzymatic specificity and also for enzymatic catalysis, with one stipulation—that it is the transition state (or sometimes an unstable intermediate) that is the key, not the substrate molecule. When a substrate binds to an enzyme, the enzyme causes a distortion of the substrate that forces it toward the transition state. There can be maximal interaction with the substrate molecule only in $ES^{\ddagger}$. A portion of this binding in $ES^{\ddagger}$ can be between the enzyme and *nonreacting* portions of the substrate. The effects of binding nonreacting portions of substrates can be seen by comparing chymotrypsin-catalyzed hydrolysis of the amide bond of phenylalanine in N-acetyl-Phe–Gly-NH$_2$ with hydrolysis of N-acetyl-Phe–Ala-NH$_2$. These synthetic substrates have similar K_m values, but the latter substrate, with a methyl side chain replacing a hydrogen atom, has a k_{cat} that is 20 times larger.

For catalysis to occur, the transition state must be stabilized. The enzyme must be complementary in shape and chemical character to the transition state. Figure 7·14 shows that the energy of activation is lowered by the tight binding of the transition state to an enzyme. At the same time, the binding of S to E is weak. The task, then, has been to find examples of differences in the interactions in ES and $ES^{\ddagger}$.

There are several ways in which the comparative stabilization of $ES^{\ddagger}$ could occur, principally:

1. An enzyme could have an active site with a shape that is a closer match to the transition state than to the substrate. An undistorted substrate molecule would not be fully bound.

2. An enzyme could have sites that bind the partial charges present only in the transition state. In a related manner, a nonpolar active site could stabilize a transition state in which there is a decrease in localized charge compared to the substrate.

Examples of these phenomena will be discussed in the next three sections.

7·13 Transition-State Analogs Are Potent Inhibitors

Transition-state analogs are compounds that resemble the transition state. The transition state itself, with a half-life of about 10^{-13} s, is too short-lived to be isolated in the laboratory. However, chemicals can be synthesized that resemble these activated species. If the modified lock-and-key theory is correct, a transition-state analog should bind extremely tightly to the appropriate enzyme and thus be a potent inhibitor. One of the first examples of a successfully designed transition-state analog was 2-phosphoglycolate (Figure 7·15), a compound proposed to be an analog of a transition state in the reaction catalyzed by triose phosphate isomerase (Section 7·8A). This analog binds to the isomerase at least 100 times more tightly than either of the substrates of the enzyme. Some of the additional binding is through a partially negative oxygen atom of the carboxylate group of 2-phosphoglycolate, a feature shared with the transition state but not the substrates.

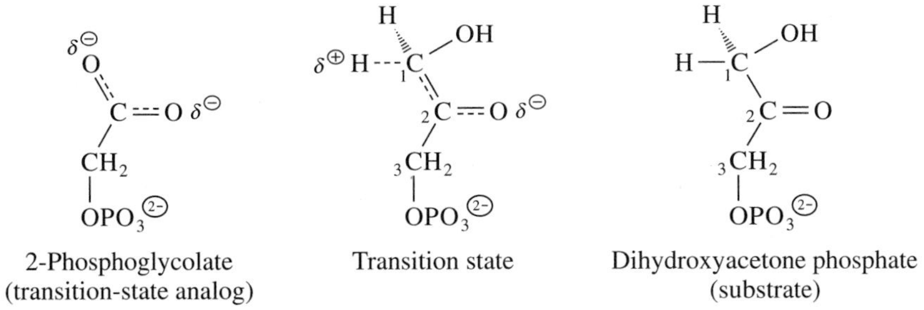

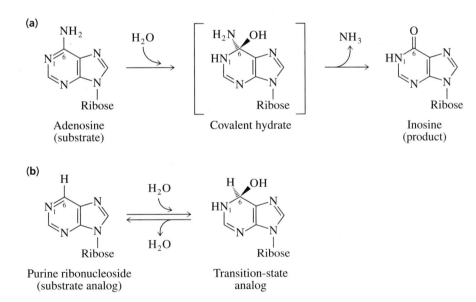

Figure 7·15
2-Phosphoglycolate, a transition-state analog for the enzyme triose phosphate isomerase. 2-Phosphoglycolate is presumed to be an analog of C-2 and C-3 of the transition state (center) between dihydroxyacetone phosphate (right) and the initial enediolate intermediate in the reaction.

Experiments with adenosine deaminase have identified a transition-state analog that binds to the enzyme with amazing affinity because it resembles the transition state very closely. Adenosine deaminase catalyzes the hydrolytic conversion of the purine nucleoside adenosine to inosine. The first step of this reaction, before the loss of the amino group as ammonia, is the addition of a molecule of water (Figure 7·16a). The complex with water, called a covalent hydrate, does not form in aqueous solution but does form as soon as adenosine is bound to the enzyme. The hydrate is a tetrahedral intermediate that quickly decomposes to products. Adenosine deaminase has fairly broad substrate specificity and consequently catalyzes the hydrolytic removal of various groups from the 6-position of purine nucleosides. However, the inhibitor purine ribonucleoside (Figure 7·16b) has no 6-substituent and only undergoes the first enzymatic step of hydrolysis, the addition of the water molecule. The covalent hydrate formed is a transition-state analog, a competitive inhibitor having a K_i of 3×10^{-13} M. The binding of this analog exceeds the binding of either the substrate or the product by a factor of more than 10^8. A very similar reduced inhibitor, 1,6-dihydropurine ribonucleoside (Figure 7·16c), cannot undergo

Figure 7·16
Inhibition of adenosine deaminase by a transition-state analog. **(a)** In the deamination of adenosine, a proton is added to N-1 and a hydroxide ion to C-6 to form an unstable covalent hydrate, which decomposes to produce inosine and ammonia. **(b)** When the inhibitor purine ribonucleoside is added to a solution of adenosine deaminase, it also rapidly forms a covalent hydrate, 6-hydroxy-1,6-dihydropurine ribonucleoside. This covalent hydrate is a transition-state analog that binds over a million times more avidly than another competitive inhibitor, 1,6-dihydropurine ribonucleoside **(c)**, which differs from the transition-state analog only by lacking the 6-hydroxyl group.

a hydration reaction and thus lacks the hydroxyl group at C-6; it has a K_i of only 5×10^{-6} M. Therefore, one can conclude that the enzyme must specifically and avidly bind the transition-state analog—and also the transition state—through interaction with the hydroxyl group at C-6.

7·14 Catalytic Antibodies Have Many Properties of Enzymes

Recently, transition-state analogs bound to proteins have been used as antigens to induce the formation of antibodies that have catalytic activity. Normally, antibodies do not exhibit catalytic activity. It was proposed that antibodies raised to transition-state analogs would bind the transition states well and would therefore be catalysts. It is possible to test this proposal by using pure, or monoclonal, antibodies. In these experiments, mice are immunized with the antigen, and antibody-secreting cells from the spleen are harvested and cloned. Each cell makes just one kind of antibody, which is tested for catalytic activity. Only a few of the antibodies have catalytic activity. These antibodies have the predicted substrate specificity and show a relatively modest rate enhancement, up to about 10^5.

Figure 7·17 illustrates one of the more successful examples of catalysis by an antibody. Antibodies were raised to the protein conjugate of a synthetic phosphonamidate (Figure 7·17a). The structure of the phosphonamidate is thought to resemble the tetrahedral transition state in the hydrolysis of the analogous amide. An antibody with catalytic activity was isolated and found to catalyze the hydrolysis of the synthetic amide (Figure 7·17b), with the greatest rates at pH values of 9 or above. At 37°C, the catalytic rate was about 1/25th the rate of the chymotrypsin-catalyzed hydrolysis of an amide at 25°C and pH 7, which shows that the antibody is a potent catalyst. Like chymotrypsin, the catalytic antibody exhibits fairly narrow substrate specificity. As well as stabilizing the transition state, this antibody uses covalent catalysis, possibly through acylation of a histidine residue.

One possible practical application of catalytic antibodies, raised to the appropriate antigens, is the preparation of catalysts for industrially useful reactions that do not occur in nature. The most important conclusion that can be made from the

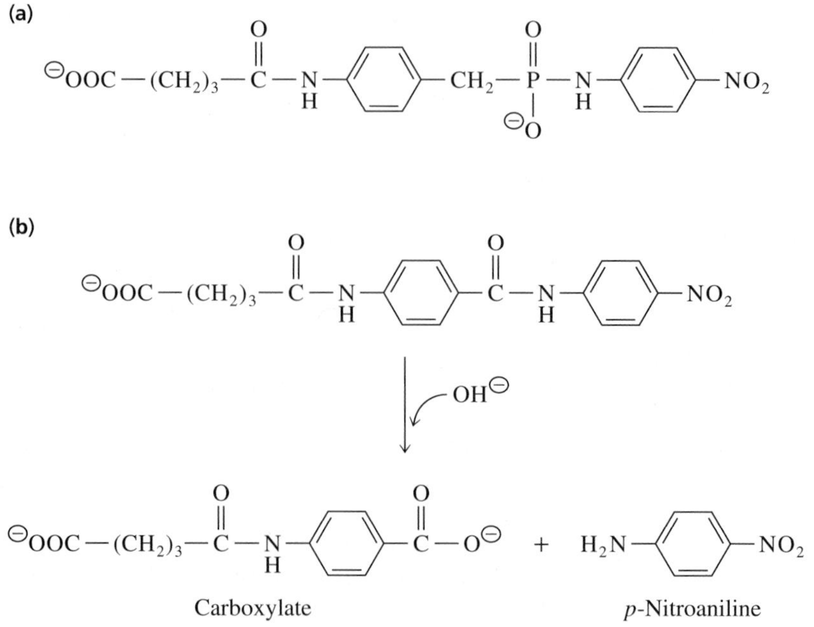

Figure 7·17
Amide hydrolysis catalyzed by an antibody.
(a) The antigen used to induce the antibody was a phosphonamidate coupled to a carrier protein. (b) The substrate of the reaction catalyzed by the antibody is a *p*-nitroanilide.

experiments with catalytic antibodies is that the results support the theory that transition-state stabilization is a major factor in catalysis.

There is great variation in quantitative estimates of the rate enhancement provided by transition-state stabilization. However, enzyme chemists agree that this binding phenomenon is important. Some enzymologists have estimated that transition-state stabilization explains nearly all of the rate acceleration of enzymes. Indeed, by simplifying the equations of transition-state theory, it can be shown that the rate enhancement (the ratio of k_{cat} to k_n, where k_n is the nonenzymatic rate constant) is roughly as large as the ratio of the dissociation constants for the ES and $ES^{\ddagger}$ complexes (K_S and K_T, respectively).

$$k_{cat}/k_n \approx K_S/K_T \tag{7·23}$$

The rate acceleration, therefore, is roughly determined by the increased strength of binding of the transition state ($S^{\ddagger}$) compared to the substrate (S). The assumptions made in deriving this simple expression are not unreasonable. However, as important as transition-state stabilization is, other factors contribute to the catalytic effectiveness of enzymes.

Transition-state analogs often bind slowly to enzymes. Most classical competitive inhibitors bind rapidly to enzymes. The slower attachment of the extra-potent analogs may occur because they must form more bonds with the enzyme. Perhaps a change in the shape of the enzyme is required for their full binding. It is possible that, for full binding, the enzymes sometimes surround transition-state analogs. This leads us to a consideration of the role of enzyme flexibility in catalysis.

7·15 Some Enzymes Undergo Large Conformational Changes

It is important for an enzyme to have an ordered structure, complementary to the transition-state configuration of its substrates. In this respect, enzymes resemble solid catalysts. However, globular proteins are not rigid; molecular-dynamics studies indicate that the atoms of proteins make small and rapid motions. Furthermore, comparison of the three-dimensional structure of an enzyme with the structure of the enzyme-ligand complex shows that small conformational adjustments occur upon binding of the ligand. Some enzymes, in fact, undergo major changes in shape when substrate molecules bind. These enzymes change from an inactive conformation to an active conformation. Activation of an enzyme by a substrate-initiated conformation change is called **induced fit.**

In the late 1950s, before information about the three-dimensional structures of proteins was available, Daniel Koshland suggested that enzyme flexibility is involved in catalytic activity and substrate specificity. His proposal of the induced-fit phenomenon was based upon kinetic experiments with hexokinase.

Hexokinase (Chapter 15) catalyzes the phosphorylation of glucose by ATP.

$$\text{Glucose + ATP} \rightleftharpoons \text{Glucose 6-phosphate + ADP} \tag{7·24}$$

Water (HOH), which resembles the alcoholic group at C-6 of glucose (ROH), is small enough to fit into the active site and therefore should be a good substrate for hexokinase. However, the hydrolysis of ATP is slower than the phosphorylation of glucose by a factor of 40 000. Addition of the unreactive five-carbon sugar lyxose (it has no hydroxymethyl group analogous to C-6 of glucose and thus cannot undergo phosphorylation) increases the rate of hydrolysis of ATP 20-fold. This experiment suggested that the presence of a specific, or natural, substrate in the active site of hexokinase aligns the catalytic groups to make the enzyme active. X-ray crystallographic experiments show that hexokinase exists in two conformations, an open

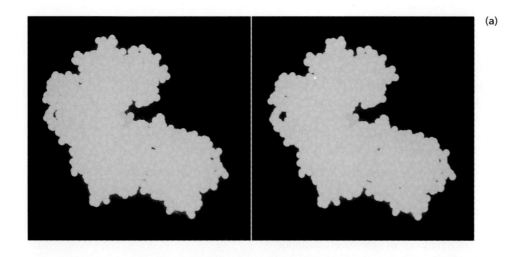

(a)

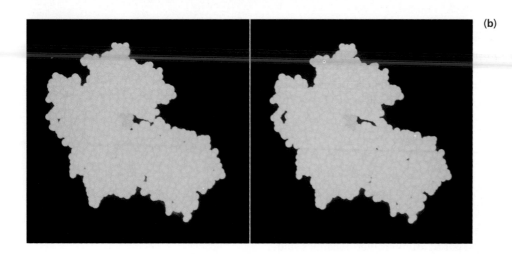

(b)

Figure 7·18
Stereo views of yeast hexokinase. Yeast hexokinase contains two structural domains connected by a hinge region. Upon binding of glucose (green), these domains close, shielding the active site from water. **(a)** Open conformation. **(b)** Closed conformation. (Part (a) based on coordinates provided by T. Steitz, C. Anderson, and R. Stenkamp; part (b) based on coordinates provided by W. Bennett, Jr. and T. Steitz.)

form when glucose is absent and a closed form when glucose is bound (Figure 7·18). The binding of a ligand such as glucose changes the angle between the two domains of hexokinase by $12°$ and closes the cleft in the enzyme-glucose complex. This type of relative motion of two rigid domains is called hinge bending. Closure of the hexokinase active site prevents wasteful hydrolysis of ATP; in this case, induced fit plays a role in conserving cellular energy. A number of other kinases follow induced-fit mechanisms. X-ray crystallographic analysis has shown that the open-to-closed conformational change in adenylate kinase occurs in two steps associated with the binding of each of its two substrates (Figure 10·17).

The more stable shape of the enzyme hexokinase is the inactive form, E_i. Binding of a good substrate such as glucose, but not a poor substrate such as water, forces the enzyme to assume the active enzyme-substrate complex, E_aS, as shown in the following equilibria:

$$E_i + S \; \rightleftharpoons \; E_iS \; \rightleftharpoons \; E_aS \; \rightleftharpoons \; ES^{\ddagger} \qquad (7·25)$$

Because E_a has a less stable conformation than E_i, energy is required to maintain the E_a state. The energy used in the binding of S to form E_aS is energy that cannot be used for catalysis. Consequently, an enzyme that uses an induced-fit mechanism would be less effective than a hypothetical enzyme that is always in its active conformation. The difference in efficiency is the ratio of $[E_a]$ to $[E_i]$, estimated to be 1:40 000 for hexokinase. The catalytic cost of induced fit slows kinases so that their k_{cat} values are approximately $10^3 \, s^{-1}$ (Table 6·3).

Other enzymes undergo hinge bending for reasons other than induced-fit activation. Obviously, allosteric regulatory enzymes undergo conformational changes during the R $\rightleftharpoons$ T transition (Section 6·12). These enzymes have evolved to optimize their regulatory function, not to maximize catalytic rate. Their activity is often controlled by modulation of K_m values. A low K_m value for a critical substrate corresponds to a low rate of catalysis, making the regulatory enzyme the rate-limiting enzyme for a metabolic pathway and limiting the entry of precursors into the pathway. Knowledge of the three-dimensional structures of additional enzymes may show that hinge bending is a common phenomenon. For example, a simple hydrolase, a mutant bacteriophage T4 lysozyme, crystallizes as four independent molecules in one crystal lattice. Each molecule has a slightly different conformation, depending on the angle between its two domains (Figure 7·19). There appears to be little difference in energy between the four conformations of the lysozyme, and hinge bending probably occurs during catalysis.

Certain enzymes have loops or flaps (not full domains) that close after their substrates are bound. These enzymes have increased catalytic ability because of their conformational changes. The flap of the protein strengthens binding to the transition state. Triose phosphate isomerase and tyrosyl-tRNA synthetase (next section) are examples of this special case of transition-state stabilization. In triose phosphate isomerase, a protein loop of 10 residues closes off the active site after the substrate has bound. The active site no longer has access to bulk solvent. The closing of this flap explains why release of product by this diffusion-controlled enzyme is one of the slowest steps of catalysis. Mutant enzymes in which the loop is disabled exhibit lower activity, confirming that the loop assists in catalysis. Some dehydrogenases undergo hinge bending as well as loop closure to enclose their substrates. Both processes may strengthen binding to the transition state.

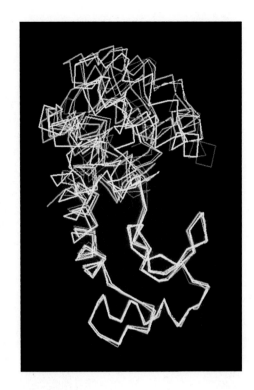

Figure 7·19
Four forms of mutant T4 lysozyme. These forms are conformational variants and do not differ in energy. (Based on coordinates provided by H. R. Faber and B. W. Matthews.)

7·16 Experiments with a tRNA Synthetase Have Shown the Role of Hydrogen Bonding in Catalysis

In this and the following sections, we will use several examples of enzyme mechanisms to illustrate the roles of chemical and binding modes of catalysis. Aminoacyl-tRNA synthetases catalyze the covalent attachment of amino acids to specific transfer RNA (tRNA) molecules, preceding incorporation of the amino acid into a protein (Chapter 29). Tyrosyl-tRNA synthetase catalyzes the attachment of the amino acid tyrosine. The enzymatic "charging" of this tRNA proceeds in two steps, the formation of the energy-rich enzyme-bound intermediate tyrosyl adenylate (Tyr-AMP) and then the isoenergetic transfer of tyrosine to its tRNA.

$$E + Tyr + ATP \rightleftharpoons E\text{-}Tyr\text{-}AMP + PP_i \qquad (7\cdot26)$$

and

$$E\text{-}Tyr\text{-}AMP + tRNA \rightleftharpoons E + Tyr\text{-}tRNA + AMP \qquad (7\cdot27)$$

The mechanism of this enzyme has been studied in the laboratory of Alan Fersht by systematic site-directed mutagenesis accompanied by kinetic measurements, assisted by knowledge of the three-dimensional structure of the synthetase (with and without tyrosine) and the enzyme-bound Tyr-AMP complex. Each side chain suspected of forming a hydrogen bond with the substrate, the intermediate, or the transition state was replaced, one at a time or in various combinations, by a smaller side chain, usually one that does not form hydrogen bonds. The binding contribution of each residue was estimated from the kinetic effect of each alteration, measured by determinations of k_{cat} and K_m values.

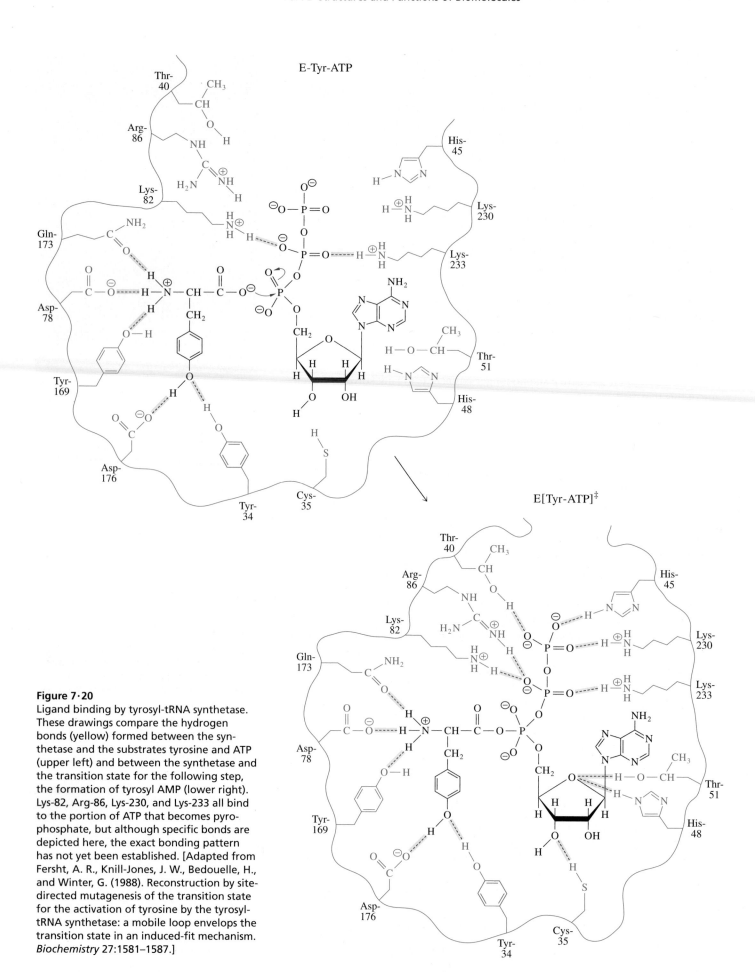

Figure 7·20
Ligand binding by tyrosyl-tRNA synthetase. These drawings compare the hydrogen bonds (yellow) formed between the synthetase and the substrates tyrosine and ATP (upper left) and between the synthetase and the transition state for the following step, the formation of tyrosyl AMP (lower right). Lys-82, Arg-86, Lys-230, and Lys-233 all bind to the portion of ATP that becomes pyrophosphate, but although specific bonds are depicted here, the exact bonding pattern has not yet been established. [Adapted from Fersht, A. R., Knill-Jones, J. W., Bedouelle, H., and Winter, G. (1988). Reconstruction by site-directed mutagenesis of the transition state for the activation of tyrosine by the tyrosyl-tRNA synthetase: a mobile loop envelops the transition state in an induced-fit mechanism. *Biochemistry* 27:1581–1587.]

Because no ionizable amino acid residues that could serve as chemical catalysts have been found to be properly located in the active site of the tyrosyl-tRNA synthetase, it has been proposed that this enzyme uses transition-state stabilization as its principal mechanism of catalysis. Mutagenic alteration of a side chain that binds to a substrate only in the transition state should lower k_{cat} but leave K_m unchanged. This result was obtained when Thr-40 was altered to Ala-40 and when His-45 was changed to Gly-45. It was concluded that these side chains must bind the γ-phosphate of ATP in the transition state but not in the E-Tyr-ATP enzyme-substrate complex. Similarly, Cys-35, His-48, and Thr-51 interact with ribose in the transition state but not in the ES complex. Figure 7·20 illustrates the differences in binding between the E-Tyr-ATP complex and the E[Tyr-ATP]‡ transition state. There are at least seven additional hydrogen bonds in the transition state.

The conformation of tyrosyl-tRNA synthetase changes during catalysis, in an induced-fit mechanism. There are two mobile, exposed loops in the protein structure of this enzyme. Lys-82 and Arg-86 are in one loop, and Lys-230 and Lys-233 are in the other. When ATP binds to the enzyme, it attracts Lys-82 and Lys-233, pulling the loops inward. Arg-86 and Lys-230 from the loops bind to the transition state and also to the E-Tyr-AMP-PP$_i$ intermediate. The mobile loops open to allow substrate to enter or product to leave, but in ES‡, they completely enclose the reactants to provide transition-state stabilization.

7·17 The Mechanism of Lysozyme Involves Tight Binding of an Ionic Intermediate

In the 1960s, David C. Phillips and his colleagues reported the first X-ray crystallographic determination of the structure of an enzyme, lysozyme. Lysozyme catalyzes the hydrolysis of the cell walls of bacteria. Many secretions such as tears, saliva, and nasal mucus contain lysozyme activity. The best studied lysozyme is from hen egg white. The substrate of lysozyme is a polysaccharide composed of alternating residues of *N*-acetylglucosamine (GlcNAc) and *N*-acetylmuramic acid (MurNAc) connected by glycosidic bonds (Figure 7·21). Lysozyme specifically catalyzes hydrolysis of the glycosidic bond containing C-1 of MurNAc residues.

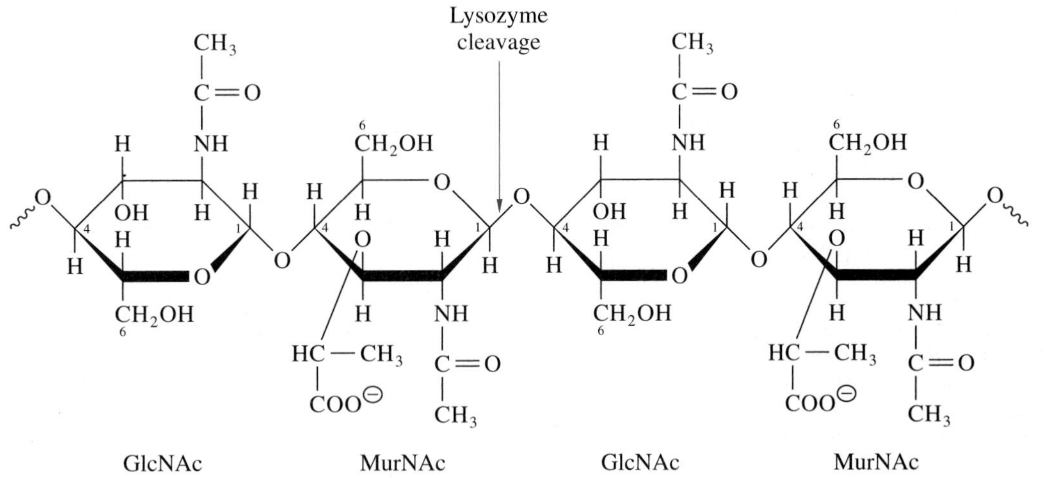

Figure 7·21
Structure of a portion of a bacterial cell wall. Lysozyme catalyzes hydrolytic cleavage of the glycosidic bond between C-1 of MurNAc and the oxygen atom involved in the glycosidic bond.

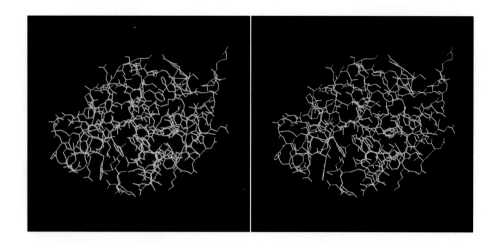

Figure 7·22
Stereo view of hen egg white lysozyme with a trisaccharide substrate (purple). (Based on coordinates provided by J. C. Cheetham, P. J. Artymiuk, and D. C. Phillips.)

Models of lysozyme and its complexes with GlcNAc or a trisaccharide were obtained by detailed X-ray crystallographic analysis (Figure 7·22). The substrate-binding cleft of lysozyme has six sites (designated A through F) for binding six saccharide residues. The sugar residues of the trisaccharide bind to sites A, B, and C. Two additional GlcNAc molecules fit easily into sites E and F of the structural model, but a GlcNAc molecule cannot be fitted into the model at site D unless it is distorted into a half-chair conformation (Figure 7·23). Two ionic amino acid residues were found close to C-1 of the distorted sugar molecule in the D binding site: Glu-35 and Asp-52. Glu-35 is in a nonpolar region of the cleft and has a perturbed pK_a near 6.5. Asp-52, in a more polar environment, has a pK_a near 3.5. The pH optimum of lysozyme is near 5, between these two pK_a values.

The proposed mechanism of lysozyme, shown in Figure 7·24, is based on a combination of X-ray data and chemical evidence. When a molecule of polysaccharide binds to lysozyme, the MurNAc residues bind to sites B, D, and F (there is no cavity for the lactyl side chain in site A, C, or E). The extensive binding of the oligosaccharide appears to distort the MurNAc residue in the D site into the half-chair conformation. Glu-35, which is protonated at pH 5, acts as an acid catalyst, donating its proton to the oxygen involved in the glycosidic bond between the D and E residues. Asp-52, which is negatively charged at pH 5, forms a strong ion pair with the carbonium ion that is formed as an intermediate. This near-covalent bond is an example of transition-state stabilization. The portion of the substrate bound in sites E and F diffuses out of the cleft and is replaced by a molecule of water. A proton from the water molecule is transferred to the conjugate base of Glu-35, and the resultant hydroxide ion adds to the carbonium ion of the D residue, yielding the second product. Evidence for the distortion of the sugar molecule bound to the D site was obtained using a compound that is a transition-state analog. This compound, the δ-lactone analog of $(GlcNAc)_4$ (Figure 7·25), has a slightly distorted half-chair conformation and is a much more potent inhibitor of lysozyme than $(GlcNAc)_4$. In summary, the catalytic modes known for lysozyme include proximity (binding of the substrate close to the active-site residues), acid-base catalysis, and transition-state stabilization, as well as distortion of the substrate toward the transition state.

(a) Chair conformation

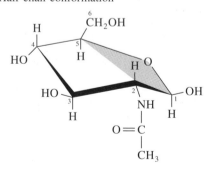

(b) Half-chair conformation

Figure 7·23
Conformations of *N*-acetylglucosamine.
(a) Chair conformation. **(b)** Half-chair conformation proposed for the sugar bound in site D.

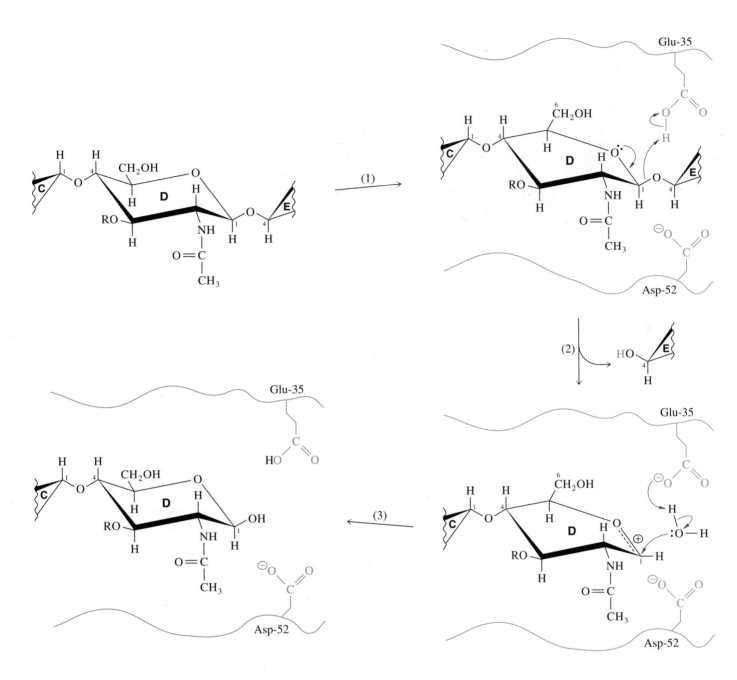

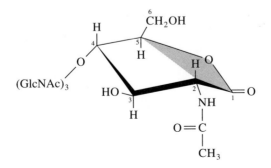

Figure 7·25
Transition-state analog for the reaction catalyzed by lysozyme.

Figure 7·24
Mechanism of lysozyme. Glu-35 donates a proton to the oxygen of the scissile glycosidic bond (the bond to be cleaved), producing an alcohol leaving group (HO–E) and a carbonium ion intermediate. The carbonium ion is stabilized by formation of a strong ion pair with the carboxylate group of Asp-52. The leaving group is replaced in the active site by a molecule of water. Glu-35 then acts as a base, removing a proton from the water. The hydroxide ion that is formed adds to the carbonium ion, completing the hydrolysis. *R* represents the lactyl group of MurNAc.

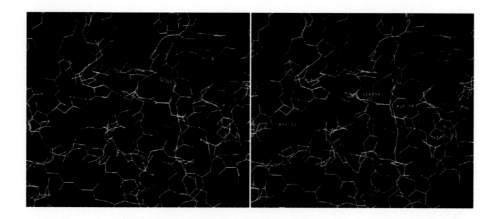

Figure 7·26
Stereo view of the catalytic site of chymotrypsin. Active-site residues Asp-102, His-57, and Ser-195 are arrayed in a hydrogen-bonded network. The conformation of these three residues is stabilized by a hydrogen bond between the carbonyl oxygen of the carboxylate side chain of Asp-102 and the peptide-bond nitrogen of His-57 and by formation of an ion pair between the β-carboxyl group of Asp-194 and the α-amino group of Ile-16. (Based on coordinates provided by A. Tulinsky and R. Blevins.)

7·18 The Mechanisms of Serine Proteases Illustrate Both Chemical and Binding Modes of Catalysis

Chymotrypsin and the other serine proteases are probably the most thoroughly studied enzymes. Because many of the mechanistic properties of serine proteases apply to enzymes in general, they are suitable examples of how the molecular structures of proteins are responsible for their enzymatic functions.

Covalent-modification experiments have shown that both serine and histidine occur in the active site of chymotrypsin (Section 6·9), and X-ray crystallographic studies have revealed that an aspartate residue is also present in the active site (Figure 6·21). The only other charged amino acid side chains in the interior of the protein are the α-amino group of Ile-16 and the β-carboxylate group of Asp-194, which form an ion pair involved in the activation of chymotrypsinogen (Section 6·11A). The discovery that Asp-102, buried in a rather hydrophobic environment, is hydrogen bonded to His-57, which in turn is hydrogen bonded to Ser-195 (Figure 7·26), was particularly exciting. In this grouping, called the **catalytic triad,** His-57—stabilized by Asp-102—abstracts a proton from Ser-195 (Figure 7·27). A common feature of all serine proteases is, in fact, the presence of a Ser-His-Asp catalytic triad.

Figure 7·27
Catalytic triad of chymotrypsin. The imidazole ring of His-57 removes the proton from the hydroxymethyl side chain of Ser-195 (to which it is hydrogen bonded), thereby making Ser-195 a powerful nucleophile. This interaction is facilitated by interaction of the imidazolium ion with its other hydrogen-bonded partner, the buried β-carboxyl group of Asp-102.

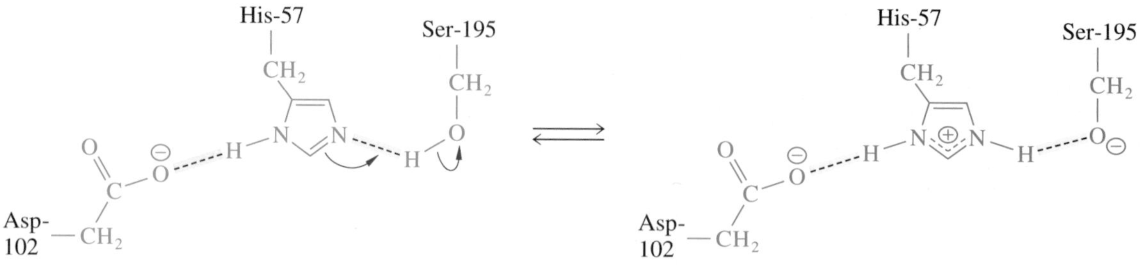

The discovery that Ser-195 is a catalytic residue of chymotrypsin was surprising because the side chain of serine is usually not a strong enough nucleophile to attack an amide bond. The pK_a of the hydroxymethyl group of a serine residue is usually ~16, similar in reactivity to the hydroxyl group of ethanol. You may recall from organic chemistry that, although ethanol can form ethoxides such as sodium ethoxide, this ionization requires the presence of an extremely strong base, such as metallic sodium.

Augmented by later experiments that showed that Asp-102 does not become protonated during catalysis, a fairly detailed mechanism for chymotrypsin and related serine proteases has been proposed. The six steps of the proposed mechanism are illustrated in Figure 7·28 (next page):

1. The enzyme-substrate complex is formed, orienting the substrate for reaction. Interactions holding the substrate in place include binding of the R_1 group in the specificity pocket. The binding interactions position the carbonyl carbon of the scissile peptide bond (the bond susceptible to cleavage) next to the oxygen of Ser-195.

2. A proton is removed from the hydroxyl group of Ser-195 by the basic His-57, and the nucleophilic oxygen attacks the carbonyl carbon of the peptide bond to produce a tetrahedral intermediate (E-TI$_1$), which is believed to be similar to the transition state for this step of the reaction. When the tetrahedral intermediate is formed, the C—O bond changes from a double bond to a longer single bond. This allows the negatively charged oxygen (the oxyanion) of the tetrahedral intermediate to move to a previously vacant position, called the oxyanion hole, where it can form hydrogen bonds with the peptide-chain —NH groups of Gly-193 and Ser-195. These hydrogen bonds stabilize the transition state, the oxyanion form of the substrate, by binding it more tightly to the enzyme than the substrate was bound.

3. Next, His-57, as an imidazolium ion, acts as an acid catalyst, donating a proton to the nitrogen of the scissile peptide bond, thus cleaving it. The first product (P$_1$), the amine, is released. The carboxyl group from the peptide forms a covalent bond with the enzyme, producing an acyl-enzyme intermediate. Steps 1 to 3 are called the acylation steps.

4. As the peptide P$_1$ with the new amino group leaves the active site, water enters. Hydrolysis (deacylation) of the acyl-enzyme intermediate starts when water donates a proton to His-57 (a basic imidazole in this step) and provides an OH$^\ominus$ group to attack the carbonyl group. A second tetrahedral intermediate (E-TI$_2$) is formed and stabilized by the oxyanion hole.

5. His-57, once again an imidazolium ion, donates a proton, leading to collapse of the second tetrahedral intermediate and formation of the second product (P$_2$)—a polypeptide with a new carboxylate group.

6. The carboxylate product is released from the active site, and free chymotrypsin is regenerated.

It was once suggested that Asp-102 becomes protonated when Ser-195 is ionized. However, physical measurements have shown that the proton stays on His-57, and Asp-102 remains negatively charged throughout the catalytic process. Nonetheless, Asp-102 is essential for optimal catalysis by serine proteases. It probably stabilizes and orients His-57 so that it can properly accept the proton from Ser-195.

All serine proteases—not only the pancreatic enzymes but also the blood-coagulation enzymes, the blood-clot–dissolving enzymes, the spermatozoan enzyme acrosin, and the insect enzyme cocoonase—share the same basic mechanism.

Figure 7·28
Mechanism of chymotrypsin-catalyzed cleavage of a peptide bond. In Step 1, the initial, noncovalent enzyme-substrate complex forms. The specificity pocket is shaded. In Step 2, the oxygen of the side chain of Ser-195 attacks the carbonyl carbon of the peptide bond of the substrate to form a tetrahedral intermediate. In Step 3, the negatively charged oxygen is stabilized in the oxyanion hole, and the peptide bond of the substrate breaks, forming an amine (P_1) and an acyl-enzyme intermediate. In Step 4, a water molecule that has replaced the amine at the active site attacks the acyl-enzyme intermediate to give a second tetrahedral intermediate. In Step 5, the tetrahedral intermediate collapses, forming the carboxylate product (P_2) and regenerating active chymotrypsin. Finally, in Step 6, P_2 leaves the active site.

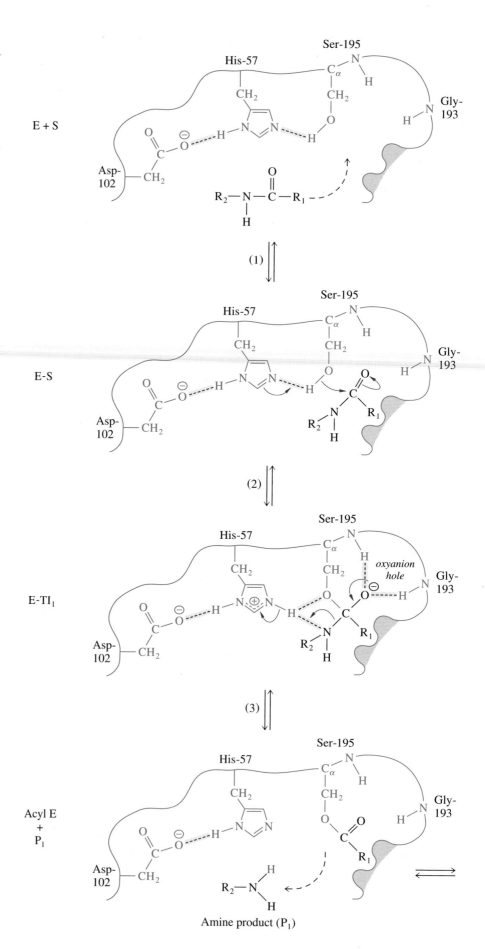

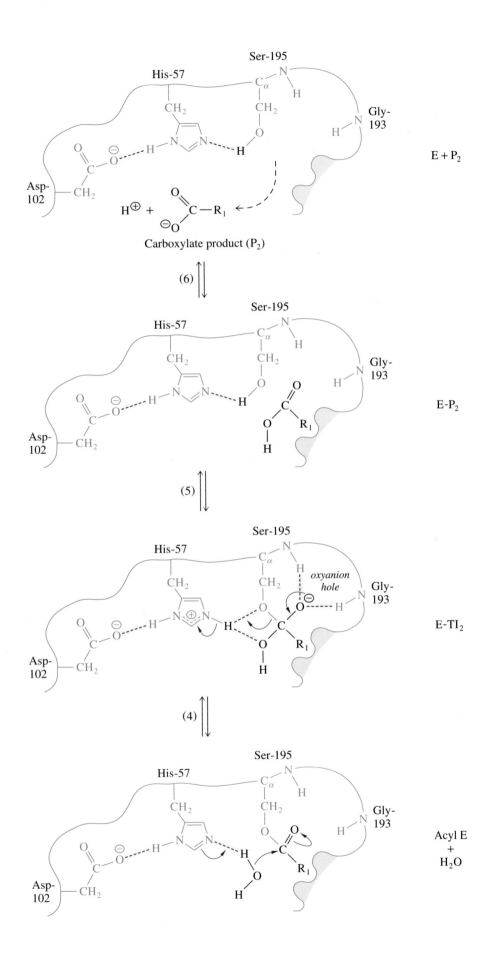

Carboxylate product (P$_2$)

E + P$_2$

(6)

E-P$_2$

(5)

oxyanion hole

E-TI$_2$

(4)

Acyl E
+
H$_2$O

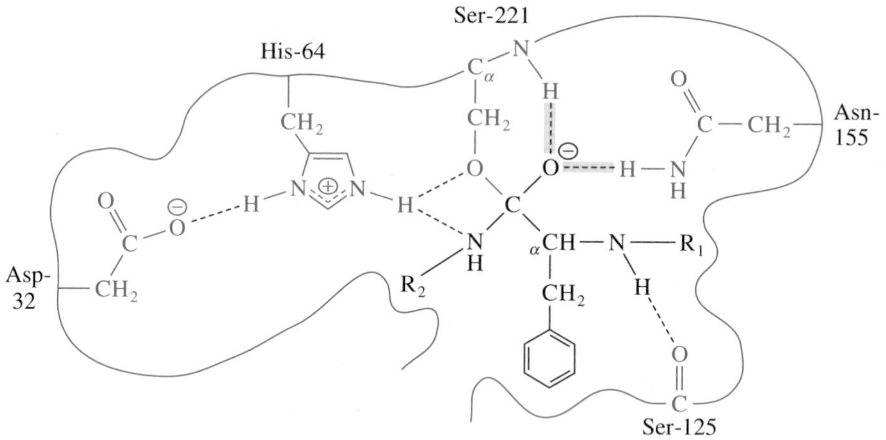

Figure 7·29
Binding of the tetrahedral intermediate by subtilisin. The oxyanion of the tetrahedral intermediate is bound by two hydrogen bonds to subtilisin, one to the backbone amide of Ser-221 and the other to the side chain of Asn-155. These bonds do not exist in the enzyme-substrate complex. This structure is an example of transition-state stabilization.

Part of the evidence for the similarity in mechanism among the serine proteases is the observation that they are all inactivated by treatment with diisopropyl fluorophosphate (DFP) (Section 6·9). Many of the serine proteases are homologous in primary, secondary, and tertiary structure, as revealed by sequence analysis and X-ray crystallography. Presumably, they have evolved from a common ancestral protein.

Subtilisin is one of a family of bacterial proteases that can be inactivated by exposure to DFP but whose primary structures show no homology to the vertebrate family of serine proteases. X-ray crystallography has revealed that even though the tertiary structure of subtilisin bears no resemblance to that of chymotrypsin, the reactive serine in position 221 of subtilisin is hydrogen bonded to a histidine (residue 64) which, in turn, is hydrogen bonded to a buried aspartate (residue 32) (Figure 7·29). Thus, the mechanism outlined for chymotrypsin also applies to the subtilisin family of enzymes. The similarities between subtilisin and chymotrypsin in catalytic mechanism and in the structure of the catalytic triad are an example of convergent evolution, a phenomenon in which proteins take different evolutionary paths to a common mechanism and function.

Site-directed mutagenesis has been used to assess the catalytic importance of the amino acids in the catalytic triad. The active-site aspartate-102 of trypsin was changed to asparagine and the kinetic properties of the mutated and unmutated enzymes were compared. At neutral pH, the wild-type trypsin had values of k_{cat} and k_{cat}/K_m that were about 10^4 greater than those of the mutant enzyme. The rate of inactivation of the mutant enzyme by the inhibitor DFP decreased to the same extent, showing that conversion of the aspartate to asparagine decreased the nucleophilic properties of the active-site serine.

Subtilisin exhibits a rate acceleration of approximately 10^{10}. The catalytic triad of subtilisin was completely dissected by mutagenesis experiments carried out by Paul Carter and James Wells. Each of the three amino acid components was changed, in turn, to an alanine residue. Alanine was chosen as the replacement residue to avoid any unfavorable steric, charge, or hydrogen-bond effects. Replacement of either the covalent catalyst, serine, or the acid-base catalyst, histidine, had the largest effect, a 10^6-fold loss in activity in either case. Replacement of aspartate decreased activity 10^4-fold. The triple mutant, in which all three active-site amino acids were converted to alanine residues, had activity similar to either the serine or histidine single mutants. The combined action of the three amino acids of the triad accelerates the hydrolysis of the substrate by a factor of just over 10^6. The least active mutant forms of subtilisin had a residual catalytic rate approximately 3000 times the nonenzymatic rate. This appreciable residual catalysis is ascribed to binding effects such as transition-state stabilization.

The oxyanion hole, the site that binds the oxyanion of the tetrahedral intermediate, cannot be modified by mutagenesis of either trypsin or chymotrypsin because the hydrogen-bond donors in both these enzymes are amide groups of the protein backbones and so cannot be changed by mutation. In subtilisin, the amide side chain of an asparagine residue participates in the stabilization of the tetrahedral intermediate (Figure 7·29). The catalytic importance of the extra hydrogen bonds has been verified by mutagenesis experiments in which Asn-155 was replaced by amino acids that cannot form hydrogen bonds. The removal of one hydrogen-bonding side chain in subtilisin lowered the activity of the enzyme by a factor of about 10^3, with no change in the K_m for its substrates. This experiment verifies that transition-state stabilization by the oxyanion hole provides considerable enhancement to the rate of hydrolysis catalyzed by subtilisin.

All of the catalytic modes described in this chapter are used in the mechanisms of serine proteases. In the reaction scheme shown in Figure 7·28, Steps 1 and 4 in the forward direction use the proximity effect, the gathering of reactants. For example, when a water molecule replaces the amine P_1, it is held by histidine in Step 4, providing a proximity effect. Acid-base catalysis by histidine lowers the energy barriers for Steps 2 and 4. Covalent catalysis using the $-CH_2OH$ of serine occurs in Steps 2 through 5. The extremely unstable tetrahedral intermediates at Steps 2 and 4 are stabilized by the oxyanion hole. The chemical catalysis and binding effects—both proximity and transition-state stabilization—all make major contributions to the enzymatic activity of serine proteases.

Having gained insight into the general mechanisms of action of enzymes, we can now examine reactions that involve coenzymes. These reactions require reactive groups that cannot be supplied by the side chains of amino acids. Chapter 8 shows how coenzymes provide additional reactive groups to the active sites of enzymes.

Summary

Reaction mechanisms describe in atomic detail how a reaction proceeds. Kinetic experiments—measurements of reaction rates under varying conditions—offer an indirect approach to examining reactions that can contribute to the elucidation of mechanisms. Studies of the structures of enzymes supply additional mechanistic information.

The rate of a reaction depends on the rate of effective collisions between reactants. For each step in a reaction, there is a transition state, or energized configuration, that the reactants must pass through. The amount of energy needed to form the transition state, called the activation energy, affects the rate of the reaction. Catalysis provides a faster reaction pathway by lowering the energy of activation.

Ionizable and reactive amino acid residues in the active site of an enzyme form its catalytic center. Two major chemical modes of enzymatic catalysis are acid-base catalysis and covalent catalysis. In acid-base catalysis, proton transfer contributes to the acceleration of the reaction, with protons either donated by a weak acid or accepted by a base in the active site. The effect of pH on the rate of an enzymatic reaction can suggest the identities of active-site components. In covalent catalysis, the substrate or a portion of it is attached to the enzyme covalently to form a reactive intermediate. Nonenzymatic model-reaction systems are useful for simulating the mechanistic features of enzyme-catalyzed reactions.

The rates of catalysis for a few enzymes are so high that they approach the upper limit set by the rate at which reactants approach each other by diffusion. For enzymatic catalysis to be so rapid, each step of the reaction must be rapid, and the activation energies for the various steps must be balanced. The Circe effect allows catalysis by superoxide dismutase to exceed the diffusion-controlled limit.

Although acid-base catalysis and covalent catalysis are important, the greatest part of the rate acceleration achieved by an enzyme generally arises from the binding of reacting ligands to the enzyme. The initial formation of a noncovalent enzyme-substrate complex (ES) collects and orients reactants, which alone produces an acceleration of the reaction, a phenomenon termed the proximity effect. The binding of reactants must be relatively weak yet strong enough that the entropy of the reactants is considerably decreased. The energy of activation is further lowered by the binding of transition states with greater affinity than the binding of substrates. Support for transition-state stabilization as a major catalytic mode comes from the potent inhibitory activity of transition-state analogs, synthetic compounds that structurally resemble transition states. Furthermore, antibodies with catalytic activity can be induced by using transition-state analogs as antigens.

Enzymes that rely on induced-fit mechanisms use some of the substrate-binding energy for activating the enzyme. Conformational changes in some enzymes help to stabilize the transition state. Enzymes, therefore, catalyze reactions by assisting first in the formation, and then in the stabilization, of transition states.

The role of binding in catalysis has been demonstrated with tyrosyl-tRNA synthetase using structural information, kinetic experiments, and site-directed mutagenesis. The results of these experiments suggest that the transition state is stabilized by at least seven hydrogen bonds that form only with the reactant in the transition state, not with the reactant in its ground state.

Lysozyme presumably binds an unstable carbonium ion, stabilizing it. In addition, it uses proximity, acid-base catalysis, and substrate distortion as catalytic modes.

The serine proteases, exemplified by chymotrypsin, use both chemical and binding modes of catalysis. All serine proteases possess a hydrogen-bonded Ser-His-Asp catalytic triad in their active sites. The serine residue serves as a covalent catalyst, and the histidine residue serves as an acid-base catalyst. The aspartate residue, which is essential for maximum activity, aligns the histidine residue and stabilizes its protonated form. Anionic tetrahedral intermediates form additional hydrogen bonds with the enzyme; these bonds contribute to catalysis by stabilizing the transition state.

Selected Readings

General References

Fersht, A. (1985). *Enzyme Structure and Mechanism,* 2nd ed. (New York: W. H. Freeman and Company).

Jencks, W. P. (1969). *Catalysis in Chemistry and Enzymology* (New York: McGraw-Hill). This monograph explains the basis of the chemistry of catalysis in great detail.

Page, M. I. (1987). Theories of enzyme catalysis. In *Enzyme Mechanisms,* M. I. Page and A. Williams, eds. (London: Royal Society of Chemistry), pp. 1–13. A short review of catalysis.

Scrimgeour, K. G. (1977). *Chemistry and Control of Enzyme Reactions* (London: Academic Press). Chapters 4, 6, 7, and 10.

Walsh, C. (1979). *Enzymatic Reaction Mechanisms* (San Francisco: W. H. Freeman and Company).

Diffusion-Controlled Enzymes

Getzoff, E. D., Cabelli, D. E., Fisher, C. L., Parge, H. E., Viezzoli, M. S., Banci, L., and Hallewell, R. A. (1992). Faster superoxide dismutase mutants designed by enhancing electrostatic guidance. *Nature* 358:347–351. Describes the only good enzymatic example of the Circe effect.

Knowles, J. R., and Albery, W. J. (1977). Perfection in enzyme catalysis: the energetics of triosephosphate isomerase. *Acc. Chem. Res.* 10:105–111.

Lodi, P. J., and Knowles, J. R. (1991). Neutral imidazole is the electrophile in the reaction catalyzed by triosephosphate isomerase: structural origins and catalytic implications. *Biochemistry* 30:6948–6956.

Binding and Catalysis

Hansen, D. E., and Raines, R. T. (1990). Binding energy and enzymatic catalysis. *J. Chem. Ed.* 67:483–489. A concise review of catalysis by proximity and transition-state stabilization.

Herschlag, D. (1988). The role of induced fit and conformational changes of enzymes in specificity and catalysis. *Bioorg. Chem.* 16:62–96. A theoretical examination of various conformational changes on catalytic activity.

Jencks, W. P. (1975). Binding energy, specificity, and enzymatic catalysis: the Circe effect. *Adv. Enzymol.* 43:219–410. A thorough view of binding and catalysis.

Jencks, W. P. (1987). Economics of enzyme catalysis. *Cold Spring Harbor Symp. Quant. Biol.* 52:65–73. A summary of recent experiments on enzymatic catalysis.

Hackney, D. D. (1990). Binding energy and catalysis. In *The Enzymes,* Vol. 19, 3rd ed., P. D. Boyer, ed. (New York: Academic Press), pp. 1–36. Section II of this article clearly explains the interrelation of binding and catalysis.

Transition-State Stabilization

Benkovic, S. J. (1992). Catalytic antibodies. *Annu. Rev. Biochem.* 61:29–54.

Fersht, A. R., Knill-Jones, J. W., Bedouelle, H., and Winter, G. (1988). Reconstruction by site-directed mutagenesis of the transition state for the activation of tyrosine by the tyrosyl-tRNA synthetase: a mobile loop envelops the transition state in an induced-fit mechanism. *Biochemistry* 27:1581–1587.

Kraut, J. (1988). How do enzymes work? *Science* 242:533–540. A review with strong emphasis on transition-state stabilization.

Lolis, E., and Petsko, G. A. (1990). Transition-state analogues in protein crystallography: probes of the structural source of enzyme catalysis. *Annu. Rev. Biochem.* 59:597–630. Discusses the catalytic role of the binding power of enzymes for transition states.

Wolfenden, R. (1972). Analog approaches to the structure of the transition state in enzyme reactions. *Acc. Chem. Res.* 5:10–18.

Serine Proteases

Carter, P., and Wells, J. A. (1988). Dissecting the catalytic triad of a serine protease. *Nature* 332:564–568. Use of site-directed mutagenesis to study the individual active-site residues of subtilisin.

Kossiakoff, A. A., and Spencer, S. A. (1981). Direct determination of the protonation states of aspartic acid 102 and histidine 57 in the tetrahedral intermediate of the serine proteases: neutron structure of trypsin. *Biochemistry* 20:6462–6467.

8

Coenzymes

We have seen how enzymes catalyze myriad biochemical reactions by means of unique arrangements of particular amino acids at their active sites. However, despite the wealth of chemical and conformational possibilities, proteins alone do not account for the entire catalytic repertoire of biochemistry. In this chapter, we will see how the catalytic activities of many enzymes depend on the presence of components called **cofactors.** Cofactors are chemicals required by inactive apoenzymes (proteins only) to convert them to active holoenzymes. There are two types of cofactors: **essential ions** and organic compounds known as **coenzymes** (Figure 8·1). Both inorganic and organic cofactors become essential portions of the active sites to which they specifically bind. Some essential ions, called activator ions, are reversibly bound and often participate in the binding of substrates, whereas tightly bound metal ions frequently participate directly in catalytic reactions.

Coenzymes, which are larger than most metabolites, act as group-transfer reagents. They are specific for the chemical groups, called mobile metabolic groups, that they accept and donate. For some coenzymes, the mobile metabolic group is hydrogen or an electron; other coenzymes carry larger, covalently attached chemical groups. Mobile metabolic groups are attached at the **reactive center** of the coenzyme (colored red in the structures presented in this chapter). Focusing on the chemical reactivity of the reactive center simplifies the study of these rather complex molecules. We begin this chapter with a discussion of metal-ion cofactors.

Related material appears in Exercise 3 of the computer-disk supplement *Exploring Molecular Structure.*

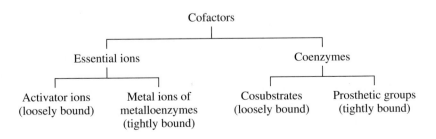

Figure 8·1
Types of cofactors. Cofactors can be divided into two types—essential ions and coenzymes—which can be further distinguished by the strength of interaction with their apoenzymes. An essential ion may be an activator ion, which binds loosely to an enzyme, or a metal ion that is bound tightly within a metalloenzyme. Similarly, a coenzyme may be a loosely bound cosubstrate or a tightly bound prosthetic group.

Next, we describe metabolite coenzymes, most of which are nucleotides or nucleotide derivatives. The bulk of the chapter is devoted to vitamin-derived coenzymes—compounds that are synthesized from dietary precursors. Many vitamins and coenzymes defy easy classification; in a few cases, their exact mechanisms of action are not known. Nevertheless, the study of coenzymes provides a glimpse into the richness of biochemical strategies that organisms have evolved to deal with various metabolic demands. We will encounter many of the structures and reactions presented here in later chapters, when we discuss particular metabolic pathways.

8·1 Many Enzymes Require Inorganic Cations for Activity

Over a quarter of all known enzymes require metallic cations to achieve full catalytic activity. **Metal-activated enzymes** either have an absolute requirement for metal ions or are stimulated by the addition of metal ions. Some of these enzymes require monovalent cations such as $K^{\oplus}$, and others require divalent cations such as $Mg^{2\oplus}$. **Metalloenzymes** contain firmly bound metal ions at their active sites. Although the differences between these two groups of enzymes appear only quantitative, the metal ions that bind more avidly often also play more intimate roles in catalysis. Many of the minerals required by all organisms are essential because they are cofactors.

A. Metal-Activated Enzymes Bind Ions Weakly

Potassium, magnesium, and calcium ions are the most common activator ions. $K^{\oplus}$ is the most abundant cation in the cells of all organisms; an enzyme in the plasma membrane catalyzes the transport of $K^{\oplus}$ into cells. The functions of potassium ions, which bind very weakly to enzymes, are not fully understood, despite the fact that over 60 enzymes are known to depend on the presence of $K^{\oplus}$ for full activity. Potassium ions may stabilize an active conformation of the enzyme, or they may assist in binding a substrate to the enzyme. In some cases, $K^{\oplus}$ may perform both of these functions.

In contrast to $K^{\oplus}$, which binds weakly to proteins, the divalent cations $Mg^{2\oplus}$ and $Ca^{2\oplus}$ form complexes of moderate strength. Magnesium is required by kinases, enzymes that utilize a magnesium-ATP complex as a phosphoryl-group–donating substrate. Most of the magnesium-activated enzymes form E–S–M (or substrate-bridge) complexes, in which the metal ion (M) binds to a substrate (S) but not the enzyme (E). For example, magnesium shields the negatively charged phosphate groups of ATP, making them more susceptible to nucleophilic attack (Section 10·4).

Calcium is a major structural component of bones, teeth, and shells. However, it also functions in stabilizing enzymes and regulating cellular activities. Calcium is bound to some extracellular enzymes, apparently holding the proteins in conformations that are more resistant to thermal denaturation or proteolytic degradation. Calcium ions regulate the activity of many intracellular enzymes, sometimes in association with a small (M_r 16 700) calcium-binding protein called calmodulin (Figure 8·2). When calmodulin binds $Ca^{2\oplus}$, it can activate a variety of enzymes. Calmodulin, which is found in all eukaryotic cells, and other intracellular calcium-binding proteins possess a common helix-loop-helix supersecondary structure that is involved in calcium binding.

Figure 8·2
Stereo view of calmodulin. Four bound calcium ions are shown in white. (Based on coordinates provided by Y. Babu, C. Bugg, and W. Cook.)

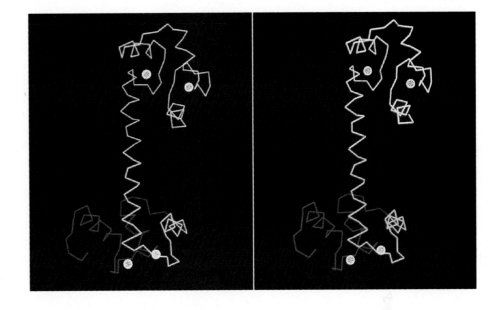

B. Metalloenzymes Bind Their Ions Tightly

Metalloenzymes are easier to characterize than metal-activated enzymes because their metal ions are more firmly attached. The most common ions of metalloenzymes are transition metals such as iron, zinc, copper, and cobalt. Even nickel has been found in the active sites of a few enzymes. Most metalloenzymes form E–M–S (or metal-bridge) complexes in which the metal ion is bound to both the enzyme and a substrate. Rather than playing relatively passive roles in catalysis like most activator ions, the ions bound to metalloenzymes are active participants in catalysis. Because some metal ions have positive charges greater than 1 at neutral pH, they can react much like a proton or an electrophilic catalyst and are sometimes described as super acids. In other reactions, metal ions undergo reversible oxidation and reduction, transferring electrons from a reduced substrate to an oxidized substrate.

Iron—the most common transition metal in mammals—has already been encountered in heme, the site of oxygen binding in hemoglobin and myoglobin. Heme groups also occur in the cytochromes, which are electron-transferring proteins found in mitochondria and chloroplasts. Nonheme iron is often found in the form of iron-sulfur clusters, or centers (Figure 8·3). Several types of iron-sulfur clusters have been described; the most common are the [2 Fe–2 S] and [4 Fe–4 S] clusters, in which the iron is complexed with an equal number of sulfide ions from H_2S and with the —$S^{\ominus}$ groups of cysteine residues. Iron-sulfur clusters are prosthetic groups that mediate oxidation-reduction reactions in complex oxidoreductases and other enzymes.

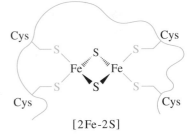

[2Fe-2S]

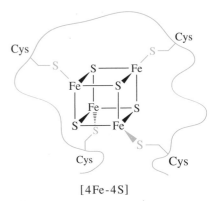

[4Fe-4S]

Figure 8·3
Iron-sulfur clusters, or centers. In each type of iron-sulfur cluster, the iron is complexed to sulfur ($S^{\ominus}$ and the thiolate of a side chain of a cysteine residue).

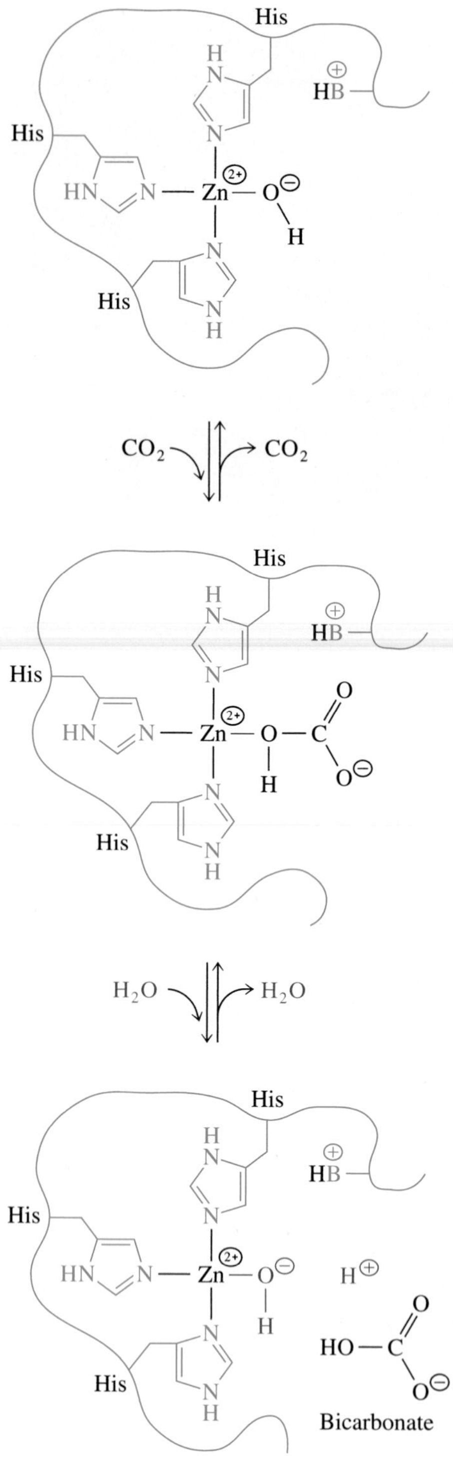

There are over 300 known zinc-containing enzymes, including most RNA polymerases and the digestive enzymes carboxypeptidase A and leucine aminopeptidase. Many proteins and enzymes that bind to DNA or RNA contain a structural motif known as a zinc finger. In this structure, the zinc ion stabilizes the protein "finger" by coordinating with the side chains of histidine and cysteine residues.

The atom of zinc bound to carbonic anhydrase acts as a super-acid catalyst. Zn^{2+} is bound to the side chains of three histidine residues and to a molecule of water. This water molecule is made more acidic by binding to Zn^{2+} and therefore ionizes more readily. A basic carboxylate residue of the enzyme removes a proton from the bound water molecule, producing a nucleophilic hydroxyl group (Figure 8·4). This $OH^{\ominus}$ attacks carbon dioxide, converting it to bicarbonate, which is released from the enzyme. The simplicity of each step allows this enzyme to have an extremely high catalytic rate.

Carbonic anhydrase provides a good example of how inorganic cofactors function as catalysts at the active sites of enzymes. Additional examples of the roles of zinc, iron, and other metals—such as copper in cytochrome c oxidase and cobalt in vitamin B_{12}—will be encountered later in this book.

8·2 Coenzymes Can Be Classified by Their Mechanism or by Their Source

Coenzymes can be separated into two types based upon how they interact with the apoenzyme (see Figure 8·1). Coenzymes of one type—often called **cosubstrates**—are actually substrates in enzyme-catalyzed reactions. A cosubstrate is altered in the course of the reaction and dissociates from the active site. The original structure of the cosubstrate is regenerated in a subsequent enzyme-catalyzed reaction so that the cosubstrate is recycled repeatedly within the cell, unlike the product derived from a simple substrate, which typically undergoes further transformation. Cosubstrates are extremely important in cellular metabolism, serving to shuttle metabolic groups among different enzyme-catalyzed reactions.

The distinction between a cosubstrate and a regular substrate can be illustrated by comparing the fates of the coenzyme ATP and the substrate glucose following the reaction catalyzed by hexokinase. This reaction is the first step in the pathway of glycolysis (Chapter 15), which converts glucose to pyruvate.

$$ATP + Glucose \longrightarrow ADP + Glucose\ 6\text{-phosphate} \qquad (8·1)$$

Although both reactants are converted to products in this reaction, ADP is usually reconverted to ATP by a subsequent enzymatic reaction, whereas most glucose 6-phosphate is either degraded through a series of enzymatic reactions to the three-carbon acids pyruvate and lactate or used by certain cells for the biosynthesis of sucrose, starch, or glycogen.

Figure 8·4
Action of carbonic anhydrase. The zinc ion bound in the active site assists the bound water molecule to become a nucleophile. The hydroxide ion attacks the carbon atom of carbon dioxide to produce bicarbonate.

The second type of coenzyme is referred to as a **prosthetic group.** Whereas a cosubstrate associates reversibly with the coenzyme-binding site of an enzyme, a prosthetic group remains bound to one protein. Like the ionic amino acid residues of the active site, a prosthetic group must be regenerated to its original form during each full catalytic event; otherwise the holoenzyme cannot remain catalytically active. Both types of coenzymes supply active sites with reactive groups that are not present on the side chains of amino acid residues.

Coenzymes can also be classified by their source. Some coenzymes are synthesized from common metabolites and are thus referred to as **metabolite coenzymes.** Other coenzymes, termed **vitamin-derived coenzymes,** are derivatives of **vitamins,** which are compounds that cannot be synthesized by animals and must be obtained as nutrients. Most metabolite coenzymes are nucleotides or their derivatives. Some vitamin-derived coenzymes contain nucleotide moieties.

8·3 ATP Is the Most Common and Versatile Metabolite Coenzyme

Adenosine triphosphate (ATP) is by far the most abundant metabolite coenzyme, although other nucleoside triphosphates sometimes serve as coenzymes. The structure of ATP is shown in Figure 8·5. All of the major components of the ATP molecule can be donated in group-transfer reactions, making ATP a versatile reactant. ATP can donate either a phosphoryl, pyrophosphoryl, adenylyl (AMP), or adenosyl group. Examples of each of these four types of reactions are shown in Figure 8·6 (next page).

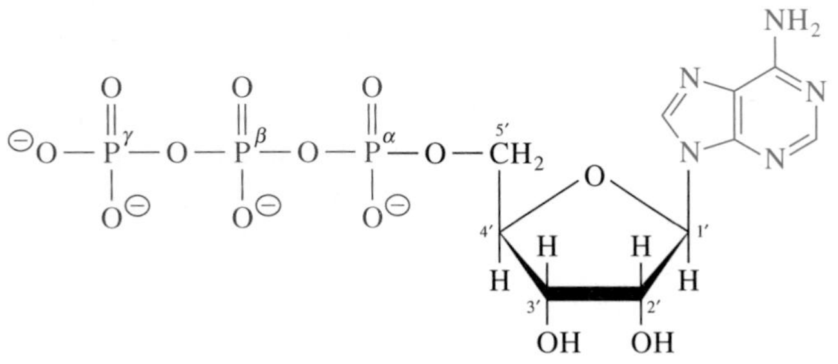

Figure 8·5
Adenosine triphosphate (ATP). The nitrogenous base adenine (blue) is linked to ribose (black) to which three phosphoryl groups (red) are attached. Removal of one phosphoryl group generates adenosine diphosphate (ADP), and removal of two phosphoryl groups as PP$_i$ generates adenosine monophosphate (AMP).

Figure 8·6
Examples of the four types of group-transfer reactions involving ATP. In each case, the two products of ATP—the group transferred and the by-product—are shaded red.

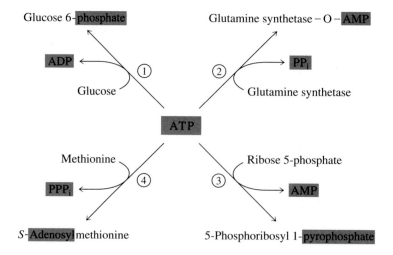

Figure 8·7
Major sites of cleavage in ATP. The bond marked A is cleaved during phosphoryl-group transfer; the bond marked B is cleaved during nucleotidyl-group transfer. *Ado* represents the adenine and ribose groups (the nucleoside adenosine).

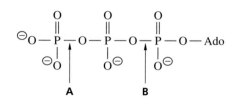

The most common of these four reactions is phosphoryl-group transfer, in which the γ-phosphorus of ATP is transferred to a nucleophile. For example, the 6-hydroxyl group of glucose is converted by ATP to a phosphate ester in the presence of hexokinase (Reaction 1 in Figure 8·6). The other product of this reaction is adenosine diphosphate (ADP). Mildred Cohn showed in 1959 that, in kinase-catalyzed reactions, ATP always transfers its terminal phosphoryl group by breaking the P—O bond marked A in Figure 8·7.

The second most common of the four group-transfer reactions involving ATP is transfer of the AMP moiety (nucleotidyl-group transfer). For example, Reaction 2 of Figure 8·6 shows the adenylylation of the enzyme glutamine synthetase, with concomitant formation of inorganic pyrophosphate (PP_i). The synthesis of DNA and RNA from nucleoside triphosphates also depends on nucleotidyl-group–transfer reactions in which the nucleophile that is being substituted attacks the α-phosphorus atom of ATP. In nucleotidyl-group–transfer reactions, the P—O bond marked B in Figure 8·7 is cleaved.

There are few examples of pyrophosphoryl-group transfer. In the example shown in Figure 8·6 (Reaction 3), a pyrophosphoryl group is donated to ribose 5-phosphate to produce AMP and 5-phosphoribosyl 1-pyrophosphate (PRPP), an intermediate in the biosynthesis of nucleotides, histidine, and tryptophan.

Finally, there are only two reactions known in which ATP donates its adenosyl group. Both of these reactions occur in the biosynthesis of coenzymes. Vitamin B_{12} (also known as B_{12a}, or cobalamin) is a cobalt-containing (Co^{3+}) compound. During its conversion to the coenzyme adenosylcobalamin, vitamin B_{12a} is reduced to vitamin B_{12s} (its Co^{+} form) by two one-electron transfers and then adenosylated by ATP. Inorganic triphosphate (PPP_i) is released in the second step.

$$B_{12a}(Co^{3+}) \xrightarrow{2\,e^{-}} B_{12s}(Co^{+}) \xrightarrow{ATP \quad PPP_i} \text{Adenosyl } B_{12} \qquad (8·2)$$

The other reaction in which ATP donates an adenosyl group is the conversion of methionine to the coenzyme *S*-adenosylmethionine (Reaction 4 in Figure 8·6).

8·4 There Are Several Other Nucleotide or Nucleotide-Derived Metabolite Coenzymes

Several metabolite coenzymes are formed from ATP. These adenosine-containing coenzymes include *S*-adenosylmethionine, phosphoadenosine phosphosulfate, and adenosine diphosphate glucose. As one might expect from their structures, these coenzymes transfer a methyl, a sulfate, and a glucosyl group, respectively. The nucleotide metabolite coenzymes also include compounds other than adenosine nucleotides. For example, uridine nucleotides are involved in the transfer of sugar groups, and cytidine nucleotides are involved in the transfer of alkyl phosphates.

A. *S*-Adenosylmethionine Donates Methyl Groups

The metabolite coenzyme *S*-adenosylmethionine (Figure 8·8) is synthesized by the reaction of methionine with ATP.

$$\text{Methionine} \;+\; \text{ATP} \;\longrightarrow\; \textit{S}\text{-Adenosylmethionine} \;+\; \text{PPP}_i \qquad \textbf{(8·3)}$$

In this reaction, ATP donates an adenosyl group to the sulfur of methionine to form a sulfonium compound. The inorganic triphosphate formed is rapidly hydrolyzed to pyrophosphate and phosphate by the action of the same enzyme that catalyzes the formation of *S*-adenosylmethionine. The thiomethyl group of methionine is unreactive, but the positively charged sulfonium of *S*-adenosylmethionine is a potent donor of methyl groups. Nucleophilic acceptors readily react with *S*-adenosylmethionine, which is the donor of virtually all of the methyl groups used in biosynthetic reactions (except that used in its own biosynthesis).

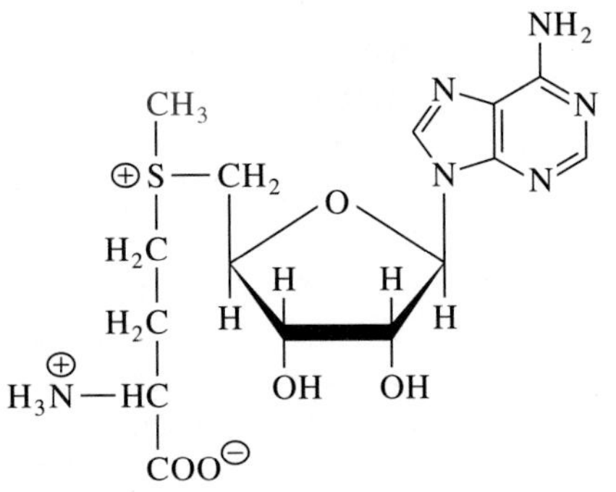

Figure 8·8
Structure of *S*-adenosylmethionine. The activated methyl group of this coenzyme is shown in red.

B. Phosphoadenosine Phosphosulfate Donates Sulfate

Some naturally occurring carbohydrates have sulfate attached to hydroxyl or amino groups. In addition, some phenolic compounds that must be detoxified (i.e., made less biologically active and more water soluble for excretion from the organism) are sulfated. Sulfation reactions require an activated donor compound because free sulfate, like phosphate, is unreactive under physiological conditions. The activated compound that effects the biological transfer of a sulfuryl group ($-SO_3^{\ominus}$) is another ATP-derived metabolite coenzyme, 3'-phosphoadenosine 5'-phosphosulfate (PAPS). The sulfate group of PAPS is attached to a phosphate group as a mixed anhydride. As shown in Figure 8·9, PAPS reacts to produce the appropriate sulfated compound and adenosine 3',5'-*bis*phosphate (phosphoadenosine phosphate, or PAP). The prefix *bis-* denotes two groups attached to separate sites (the prefix *di-* indicates groups joined to each other).

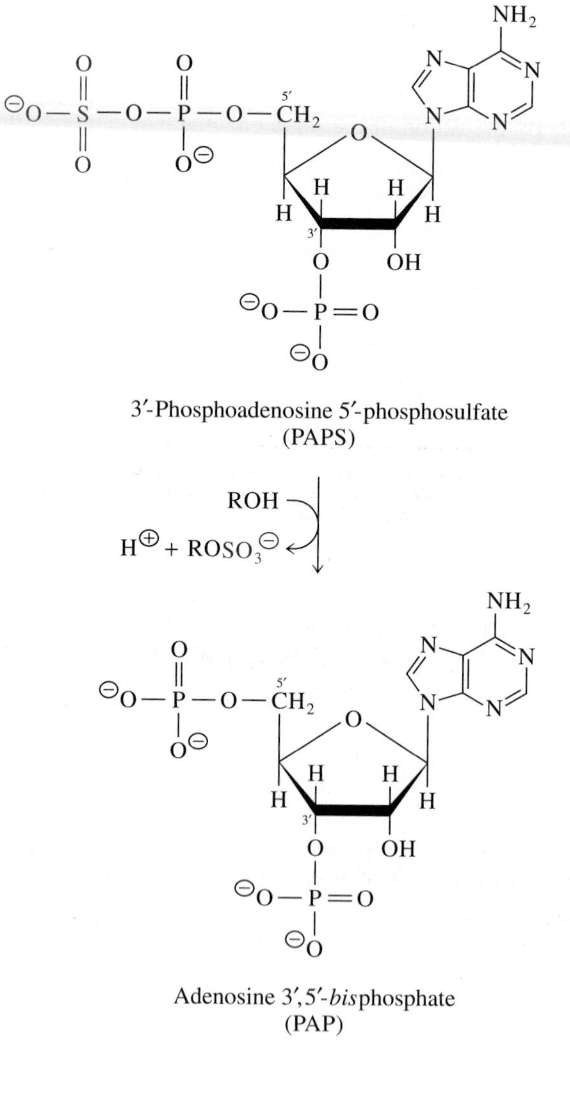

Figure 8·9
Donation of a sulfuryl group to a hydroxyl group by PAPS. *ROH* represents an alcohol.

3'-Phosphoadenosine 5'-phosphosulfate
(PAPS)

Adenosine 3',5'-*bis*phosphate
(PAP)

PAPS is synthesized in two steps. First, an adenylyl group is enzymatically transferred from ATP to sulfate.

$$SO_4^{\ominus 2} + ATP \; \underset{\longleftarrow}{\longrightarrow} \; {}^{\ominus}O-\overset{\overset{O}{\|}}{\underset{\underset{O}{\|}}{S}}-O-\overset{\overset{O}{\|}}{\underset{\underset{O^{\ominus}}{|}}{P}}-O-Ado \; + \; PP_i \qquad \textbf{(8·4)}$$

The reaction is not energetically favored, but hydrolysis of the inorganic pyrophosphate helps to drive the reaction toward completion. Next, the 3′-hydroxyl group of the product, adenosine 5′-phosphosulfate (APS), is phosphorylated by ATP in a reaction catalyzed by a specific kinase.

$$APS \; + \; ATP \; \longrightarrow \; PAPS \; + \; ADP \qquad \textbf{(8·5)}$$

As well as being involved in sulfation reactions, APS and PAPS are intermediates in the formation of the H_2S required for iron-sulfur proteins and the amino acid cysteine. Plants and many microorganisms (but not mammals) reduce sulfate by one of two pathways. The sulfate group of either APS (in yeast and plants) or PAPS (in *Escherichia coli*) undergoes an eight-electron reduction. In the first step, a sulfite group ($-SO_3^{\ominus}$) from APS or free bisulfite ($HSO_3^{\ominus}$) from PAPS is formed; the sulfite is then reduced to sulfide ($S^{\ominus 2}$).

C. ADP-Glucose Donates Glucose

In polysaccharide synthesis in plants and some bacteria, glucose monomers are donated by an adenosine-containing metabolite coenzyme called adenosine diphosphate glucose (ADP-glucose), shown in Figure 8·10. ADP-glucose is formed from ATP and phosphorylated glucose in the following reaction, catalyzed by ADP-glucose pyrophosphorylase.

$$Glucose \; 1\text{-phosphate} \; + \; ATP \; \rightleftharpoons \; ADP\text{-glucose} \; + \; PP_i \qquad \textbf{(8·6)}$$

ADP-glucose is the glucose donor for the biosynthesis of the polysaccharide glycogen in bacteria. We will encounter ADP-glucose again when we discuss the biosynthesis of starch in plants. ADP-glucose is just one of a fairly large number of nucleotide-sugar coenzymes, most of which contain pyrimidines rather than adenine.

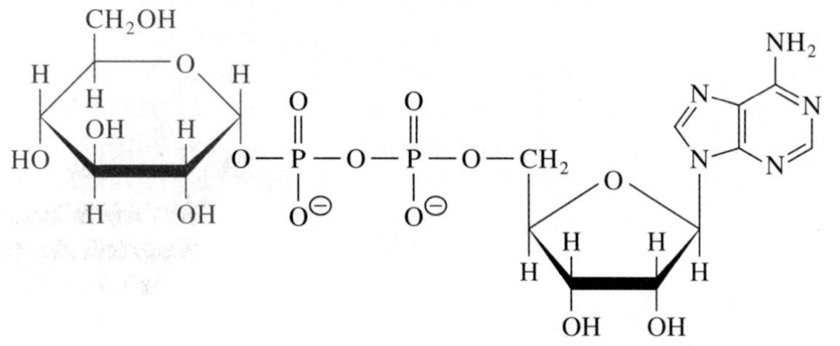

Figure 8·10
Structure of ADP-glucose.

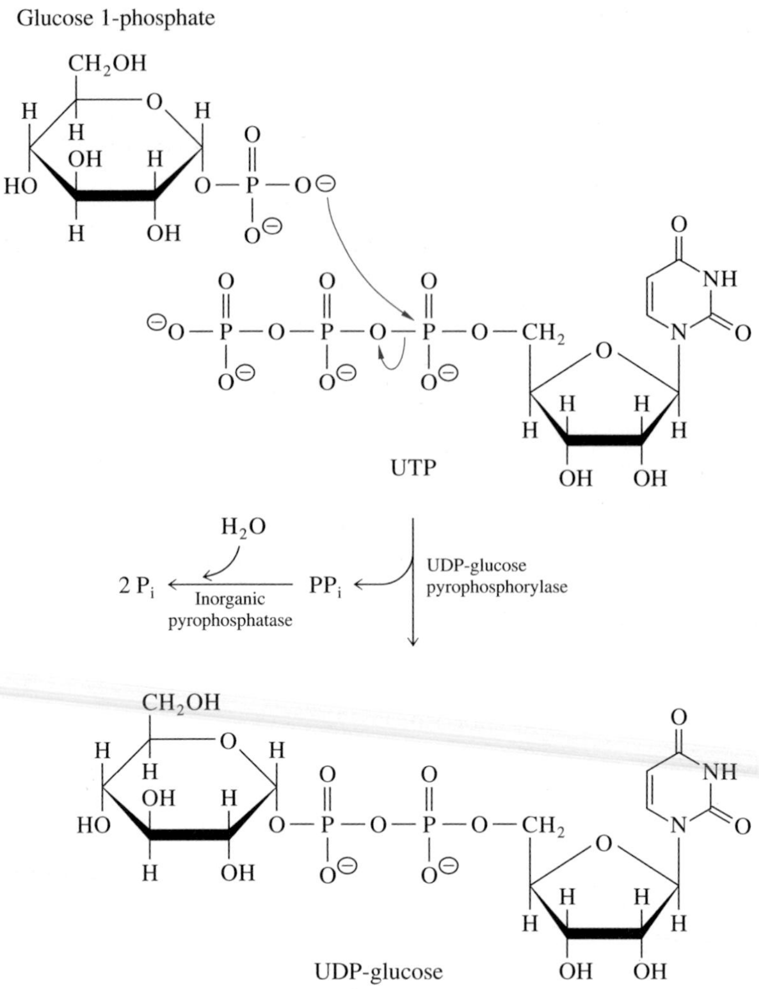

Figure 8·11
Formation of UDP-glucose catalyzed by UDP-glucose pyrophosphorylase.

Glucose 1-phosphate

UTP

UDP-glucose

D. Uridine Nucleotide Coenzymes Are Used in Polysaccharide Synthesis

Polysaccharides, or sugar polymers, are a diverse group of molecules with multiple functions as energy-storage and structural components of cells. A group of glycosyl esters of nucleoside diphosphates are donors of the sugar moiety in polysaccharide biosynthetic reactions. A major finding relating to carbohydrate metabolism and the biosynthesis of many polysaccharides was made in 1949 by Luis F. Leloir and his colleagues when they discovered uridine diphosphate glucose (UDP-glucose), the most commonly encountered of these nucleotide-sugar coenzymes. UDP-glucose is formed from uridine triphosphate (UTP) and glucose 1-phosphate in a reaction catalyzed by UDP-glucose pyrophosphorylase (Figure 8·11). In this reaction, the oxygen of the phosphoryl group of glucose 1-phosphate attacks the α-phosphorus of UTP. The released PP_i is rapidly hydrolyzed to $2\,P_i$ by the action of inorganic pyrophosphatase. This hydrolysis helps drive the pyrophosphorylase-catalyzed reaction toward completion.

UDP-glucose is the source of glucose for the biosynthesis of the storage carbohydrate of animals (glycogen) and the structural polysaccharide of plants (cellulose). UDP-glucose also undergoes conversion to several related sugar nucleotides,

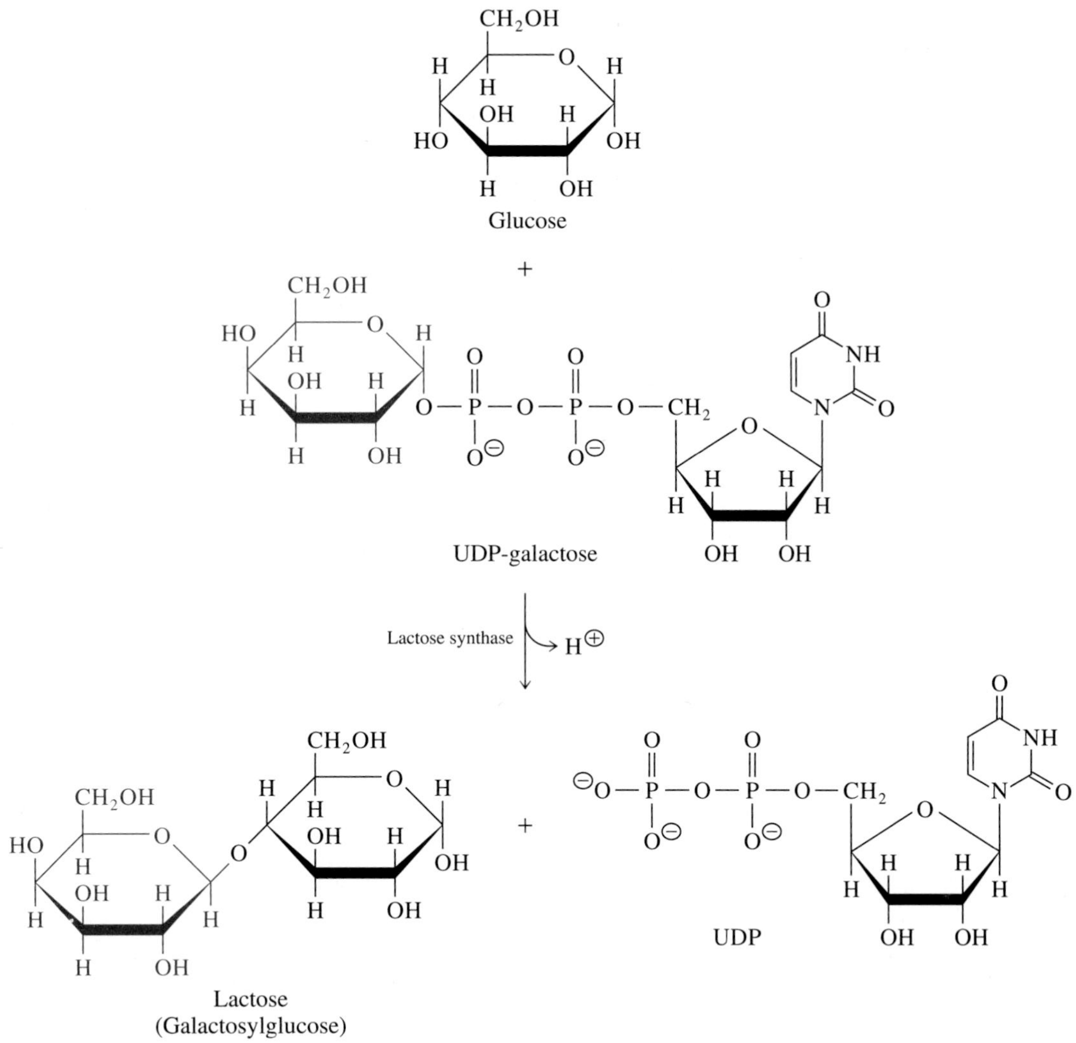

Figure 8·12
Synthesis of lactose.

UDP-galactose and UDP-glucuronic acid. UDP-glucuronic acid is formed by oxidation of C-6 (the —CH_2OH group) in the glucose moiety of UDP-glucose to a carboxylate. UDP-glucuronic acid donates glucuronic acid in the biosynthesis of the complex carbohydrates hyaluronic acid, chondroitin sulfate, and certain polysaccharides in the cell wall of pneumococci.

The UDP-sugar coenzymes donate their sugar groups in reactions in which the glycosyl-phosphate bonds are cleaved. For example, the disaccharide lactose—the most abundant carbohydrate in milk—is formed by transfer of the galactosyl group from UDP-galactose to the 4-hydroxyl group of glucose (Figure 8·12).

E. Cytidine Nucleotide Coenzymes Participate in Lipid Metabolism

Three alcohol derivatives of cytidine diphosphate (CDP) are nucleotide coenzymes that participate in lipid biosynthesis. These derivatives are CDP-choline and CDP-ethanolamine, shown in Figure 8·13 (next page), and CDP-diacylglycerol. All three coenzymes are formed from cytidine triphosphate (CTP) by the general reaction

$$\text{CTP} + \text{Alcohol phosphate} \longrightarrow \text{CDP-alcohol} + \text{PP}_i \qquad (8\cdot7)$$

The three coenzymes each donate an alkyl-phosphoryl group, resulting in cleavage of the pyrophosphate linkage of the cytidine coenzyme. For example, in *E. coli*, phosphatidylserine (a membrane lipid) is synthesized from CDP-diacylglycerol and serine, releasing a molecule of CMP (Figure 8·14).

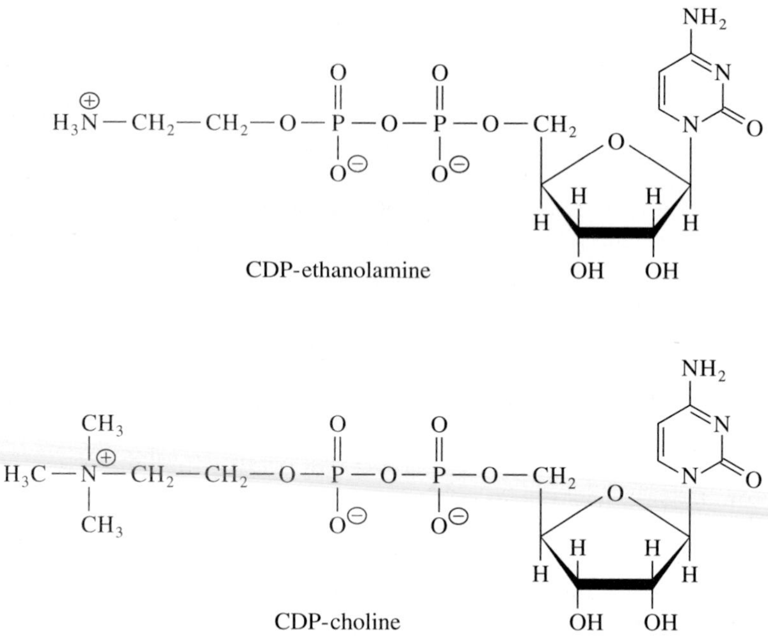

Figure 8·13
CDP-ethanolamine and CDP-choline.

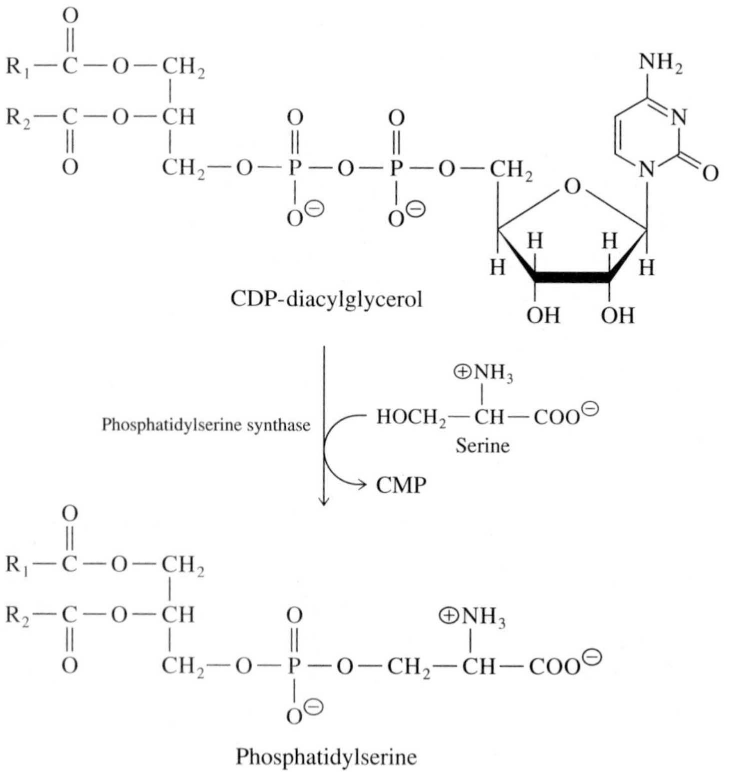

CDP-ethanolamine

CDP-choline

Figure 8·14
Synthesis of phosphatidylserine from CDP-diacylglycerol and serine. *R* represents the hydrocarbon tail of the acyl groups attached to glycerol.

CDP-diacylglycerol

Phosphatidylserine synthase

HOCH$_2$—CH—COO$^\ominus$ Serine with $\oplus$NH$_3$

CMP

Phosphatidylserine

8·5 In Animals, Many Coenzymes Are Derived from B Vitamins

Vitamins are defined as organic substances that must be obtained by an animal as nutrients, usually in small amounts (mg or μg quantities per day). It is believed that mammals and other animals that have vitamin requirements conserve chemical energy by relying on other organisms to supply these micronutrients, thereby avoiding the cost of biosynthesizing the numerous enzymes that would be required for the production of the vitamins. Vitamin-derived coenzymes are formed from precursors that must be obtained as nutrients. In animal cells, many coenzymes are synthesized from dietary precursors known as **B vitamins.** The B vitamins are all water soluble. Although the ultimate sources of B vitamins are usually plants and microorganisms, carnivorous animals can obtain B vitamins from meat.

When a vitamin is lacking from the human diet, the result is a nutritional-deficiency disease, such as scurvy, beriberi, or pellagra, which can be prevented or cured by eating the appropriate vitamin. The definition of vitamins was established early in this century, yet the link between scurvy and nutrition was recognized four centuries ago. British navy physicians discovered that citrus juice was a remedy for scurvy in sailors, whose diet lacked fresh fruits and vegetables. It was not until 1930, however, that ascorbic acid (vitamin C), shown in Figure 8·15, was isolated and proven to be the essential dietary component supplied by citrus juices. The carbohydrate ascorbic acid is a lactone, an internal ester in which the C-1 carboxylate group is condensed with the C-4 hydroxyl group, forming a ring structure. We now know that one of the functions of ascorbic acid is to act as a reducing agent during the hydroxylation of collagen (Section 5·7). Although most animals can synthesize ascorbic acid in the course of carbohydrate metabolism, guinea pigs and primates (including humans) lack this ability and must therefore rely on dietary sources. Unlike most vitamins, ascorbic acid is used in its vitamin form; it does not undergo chemical modification in order to fulfill its biological role.

The term *vitamin* (originally spelled "vitamine") was coined by Casimir Funk in 1912 to describe a "vital amine" from rice husks that cured beriberi, a nutritional-deficiency disease that results in neural degeneration (polyneuritis). Beriberi was first described in fowl, then in humans whose diets consisted largely of polished rice. When more of these organic micronutrients were discovered, the name *vitamin* was applied to the entire group of compounds. Two broad classes were recognized: water-soluble vitamins and fat-soluble, or lipid, vitamins. The antiberiberi vitamin (thiamine) became known as vitamin B_1, and the other B vitamins were referred to as vitamin B_2, vitamin B_6, and so on. Even though many of these substances proved not to be amines, the term *vitamin* has been retained.

Since water-soluble vitamins are readily excreted in the urine, they are required daily in small amounts. Conversely, **lipid vitamins,** which include vitamins A, D, E, and K, are stored by animals, and excessive intakes can result in toxic conditions known as hypervitaminoses. Not all lipid vitamins serve as coenzymes.

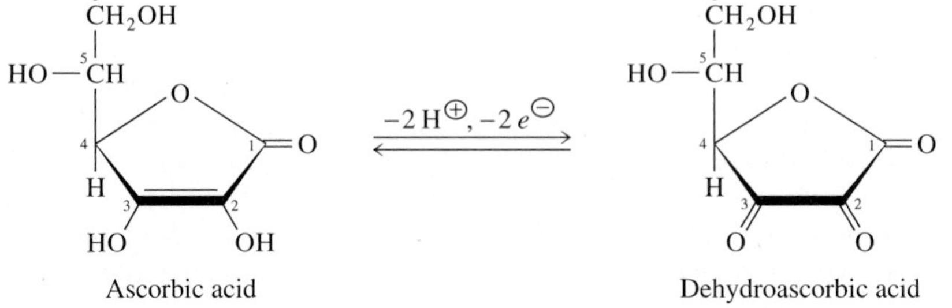

Figure 8·15
Structure of ascorbic acid (vitamin C) and its dehydro, or oxidized, form.

Table 8·1 Major coenzymes

Coenzyme	Vitamin source	Major metabolic roles	Mechanistic role
Adenosine triphosphate (ATP)	No	Transfer of phosphoryl or nucleotidyl groups	Cosubstrate
S-Adenosylmethionine	No	Transfer of methyl groups	Cosubstrate
Phosphoadenosine phosphosulfate (PAPS)	No	Transfer of sulfuryl groups	Cosubstrate
Nucleotide sugars	No	Transfer of carbohydrate groups	Cosubstrate
Cytidine diphosphate (CDP) alcohols	No	Transfer of alcohols in lipid synthesis	Cosubstrate
Nicotinamide adenine dinucleotide (NAD⊕) and nicotinamide adenine dinucleotide phosphate (NADP⊕)	Niacin	Oxidation-reduction reactions involving two-electron transfers	Cosubstrate
Flavin mononucleotide (FMN) and flavin adenine dinucleotide (FAD)	Riboflavin (B$_2$)	Oxidation-reduction reactions involving one- and two-electron transfers	Prosthetic group
Coenzyme A (CoA)	Pantothenic acid (B$_3$)	Transfer of acyl groups	Cosubstrate
Thiamine pyrophosphate (TPP)	Thiamine (B$_1$)	Transfer of aldehyde groups	Prosthetic group
Pyridoxal phosphate (PLP)	Pyridoxine (B$_6$)	Transfer of groups to and from amino acids	Prosthetic group
Biocytin (biotin bound to ε-amino group in a biotinylated enzyme)	Biotin (H)	ATP-dependent carboxylation of substrates or carboxyl-group transfer between substrates	Prosthetic group
Tetrahydrofolate	Folic acid (B$_c$)	Transfer of one-carbon substituents, especially formyl and hydroxymethyl groups; provides the methyl group for thymine in DNA	Cosubstrate
Adenosylcobalamin and methylcobalamin	Cobalamin (B$_{12}$)	Intramolecular rearrangements and transfer of methyl groups	Prosthetic group
Lipoamide residue (lipoyl group bound to ε-amino group in a protein)	No	Oxidation of a hydroxyalkyl group from TPP and subsequent transfer as an acyl group	Prosthetic group
cis-Retinal	Vitamin A	Vision	Prosthetic group
Vitamin K	Vitamin K	Carboxylation of some glutamate residues	Prosthetic group
Ubiquinone	No	Lipid-soluble electron carrier	Cosubstrate
Plastoquinone	No	Lipid-soluble electron carrier	Cosubstrate

Table 8·1 lists the major coenzymes and for vitamin-derived coenzymes, the vitamins that are their precursors. The table also briefly describes the metabolic role of each coenzyme. We have already discussed the metabolite coenzymes; the following sections describe each of the other important coenzymes.

8·6 NAD⊕ and NADP⊕ Are Nucleotides Derived from Niacin

The nicotinamide coenzymes were the first coenzymes to be recognized. By the mid-1930s, nicotinamide adenine dinucleotide (NAD⊕) and the closely related nicotinamide adenine dinucleotide phosphate (NADP⊕) had been isolated and structurally characterized. Both coenzymes contain nicotinamide, the amide of nicotinic acid (Figure 8·16). At about the time that NAD⊕ and NADP⊕ were isolated, nicotinic acid (often referred to as niacin) was identified as the factor missing in the vitamin-deficiency disease pellagra. It was established that nicotinic acid or nicotinamide is essential in the diets of mammals and serves as a precursor of the coenzymes NAD⊕ and NADP⊕. (In many species, metabolism of tryptophan can also lead to NAD⊕. Thus, dietary tryptophan can spare some of the requirement for niacin or nicotinamide.)

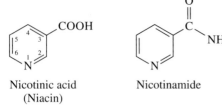

Figure 8·16
Structures of nicotinic acid (niacin) and nicotinamide.

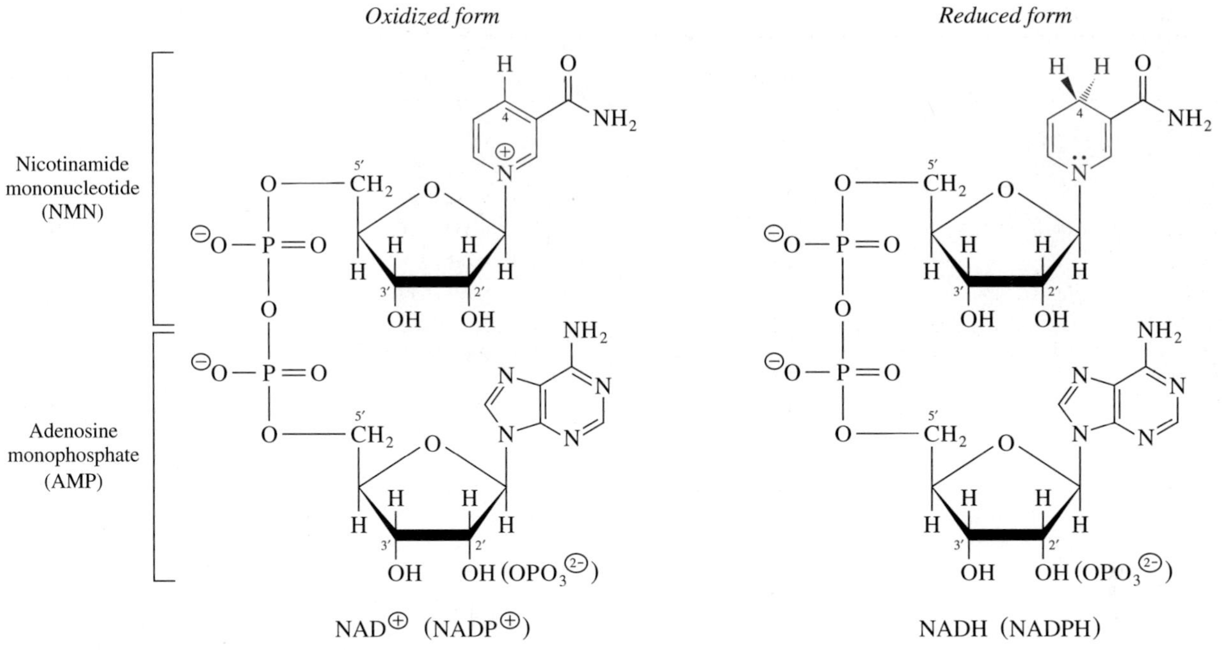

Oxidized form — *Reduced form*

Nicotinamide mononucleotide (NMN)

Adenosine monophosphate (AMP)

NAD$^{\oplus}$ (NADP$^{\oplus}$)   NADH (NADPH)

Figure 8·17
Oxidized and reduced forms of NAD (and NADP). NAD$^{\oplus}$ is the major biological oxidizing agent in catabolic processes. C-4 of the nicotinamide group is reduced when NAD$^{\oplus}$ is converted to NADH. Oxidation of NADH in mitochondria provides energy that is recovered as ATP. In NADP$^{\oplus}$, the 2'-hydroxyl group of the sugar ring of adenosine is phosphorylated. NADP$^{\oplus}$ is reduced to NADPH, largely by reactions of the pentose phosphate pathway (Chapter 17) or by photosynthesis (Chapter 19), and is the reducing agent for many biosynthetic processes.

Because nicotinic acid is the 3-carboxyl derivative of pyridine, nucleotide coenzymes containing nicotinamide are often referred to as pyridine nucleotide coenzymes. The structures of NAD$^{\oplus}$ and NADP$^{\oplus}$ and their reduced forms, NADH and NADPH, are shown in Figure 8·17. Note that both coenzymes contain a phosphoanhydride linkage that joins two 5'-nucleotides: adenosine monophosphate and the ribonucleotide of nicotinamide, called nicotinamide mononucleotide (NMN). In the case of NADP$^{\oplus}$ and NADPH, a phosphoryl group is present on the 2'-oxygen atom of the adenylate moiety. Pyridine nucleotide–dependent dehydrogenases catalyze the oxidation of their substrates by transferring two electrons and a proton in the form of a hydride ion (H$^{\ominus}$) to C-4 of the nicotinamide group of NAD$^{\oplus}$ or NADP$^{\oplus}$, generating the reduced forms, NADH or NADPH. Thus, reduction and oxidation reactions involving the pyridine nucleotides always occur two electrons at a time. NADH and NADPH are said to possess reducing power. NADPH usually supplies energy and hydrogen for reduction reactions, whereas most NADH is formed in catabolic reactions and is oxidized in mitochondria, leading to the production of chemical potential energy in the form of ATP.

NADH and NADPH exhibit an absorbance peak at 340 nm contributed by the dihydropyridine ring, whereas NAD$^{\oplus}$ and NADP$^{\oplus}$ do not absorb light at this wavelength. The appearance and disappearance of the absorbance band at 340 nm is useful for measuring the rates of enzymatic pyridine nucleotide–linked oxidations and reductions.

Lactate dehydrogenase is an NAD$^{\oplus}$-dependent enzyme that catalyzes the reversible oxidation of lactate, the end product of glucose degradation under anaerobic (oxygen-free) conditions. Note that a proton is released from a substrate when either NAD$^{\oplus}$ or NADP$^{\oplus}$ is reduced.

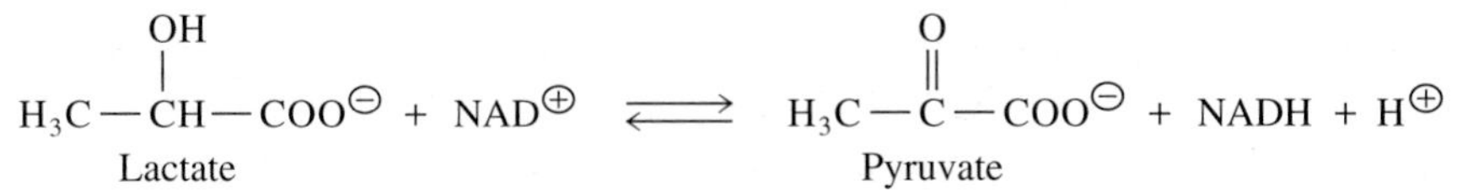

$$\underset{\text{Lactate}}{H_3C-\underset{\underset{\displaystyle OH}{|}}{CH}-COO^{\ominus}} + NAD^{\oplus} \rightleftharpoons \underset{\text{Pyruvate}}{H_3C-\underset{\overset{\displaystyle O}{\|}}{C}-COO^{\ominus}} + NADH + H^{\oplus}$$

(8·8)

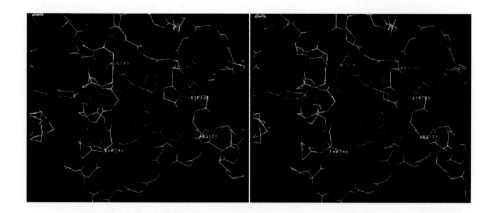

Figure 8·18
Stereo view of the active site of lactate dehydrogenase. Active site residues are shown in red. The co-substrate NAD⊕ is orange, and the substrate lactate is magenta. (Based on coordinates provided by U. M. Grant and M. G. Rossman.)

As shown in Figure 8·18, a histidine residue, His-195, is located within the active site of the enzyme. Figure 8·19 depicts the mechanism by which this histidine residue, serving as a base catalyst, participates in the catalytic action of lactate dehydrogenase. It abstracts a proton from the C-2 hydroxyl group of lactate, facilitating transfer of the hydride ion from C-2 of the substrate to C-4 of the bound NAD⊕. In this mechanism, we see that both the enzyme and the coenzyme are involved in catalyzing the oxidation of lactate to pyruvate. Like most dehydrogenases, lactate dehydrogenase must bind the pyridine nucleotide cosubstrate to form the holoenzyme before it binds its simple substrate (Figure 8·20).

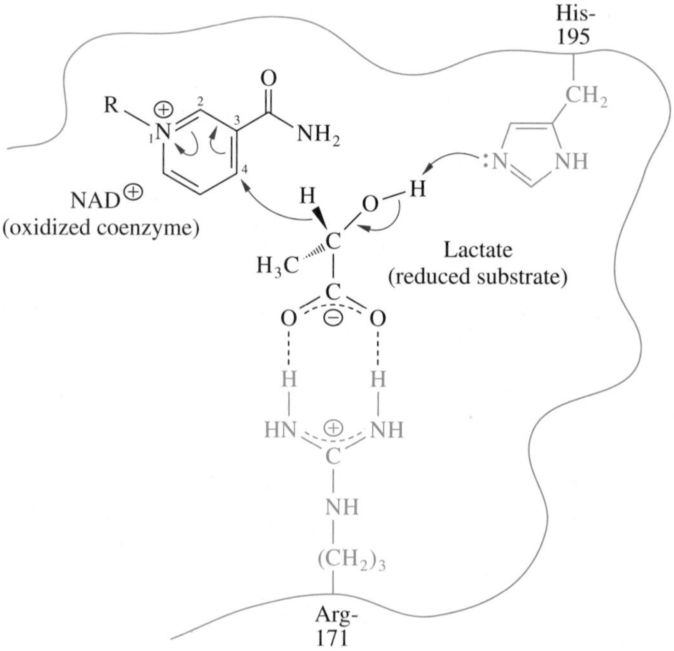

Figure 8·19
Mechanism of lactate dehydrogenase leading to the formation of pyruvate from lactate. The reversible reaction catalyzed by lactate dehydrogenase involves transfer of a hydride ion (H⊖) from the reduced substrate, lactate, to NAD⊕ (as shown here) or from the reduced coenzyme, NADH, to the oxidized substrate, pyruvate.

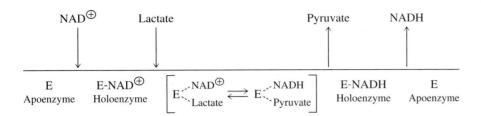

Figure 8·20
Ordered kinetic mechanism for lactate dehydrogenase. Note that, in the forward direction, the coenzyme NAD⊕ is bound first and its reduced form, NADH, is released last.

With all NAD⊕- or NADP⊕-dependent dehydrogenases, reduction of the coenzyme is strictly stereospecific. Any given dehydrogenase transfers a hydride ion exclusively to one side (face) of the pyridine ring and not to the other. In the reverse reaction catalyzed by the same enzyme (in the case of lactate dehydrogenase, for example, reduction of pyruvate to lactate), the same hydrogen is transferred (as a hydride ion) from the reduced coenzyme to the substrate.

Although NAD⊕-dependent dehydrogenases differ in some structural aspects, many have several common features. A number of the dehydrogenases that have been studied have two domains per subunit. One domain binds the pyridine nucleotide cofactor, and the other—the catalytic domain—binds the substrate. The active-site cleft lies between the two domains. In lactate dehydrogenase, the NAD⊕-binding domain includes a six-stranded β sheet and four α helices of the enzyme. The adenine moiety of NAD⊕ is buried in a hydrophobic pocket, whereas the pyrophosphate bridge is bound electrostatically to an arginine and a lysine residue. The nicotinamide ring is in a cleft in the interior of the protein. The carbonyl group of the 3-carboxamide of nicotinamide forms a specific hydrogen bond with the apoenzyme, allowing hydride addition to only one face of the pyridine ring. In contrast to the similar structures of the nucleotide-binding domains, the structures of the catalytic domains of NAD⊕-dependent dehydrogenases vary widely.

8·7 FAD and FMN Are Nucleotides Containing Riboflavin

During their studies of the nicotinamide coenzymes, Otto Warburg and W. Christian observed that NADPH could be reoxidized by molecular oxygen, provided that a protein from yeast was present. When purified from brewer's yeast, this protein was yellow and consequently became known as the "old yellow enzyme." Several research groups succeeded in separating the yellow prosthetic group from the colorless apoenzyme and showed that it was the phosphate ester of riboflavin. Subsequent recombination of the two fractions restored catalytic activity. This was the first demonstration that an apoenzyme and its prosthetic group could be reversibly separated.

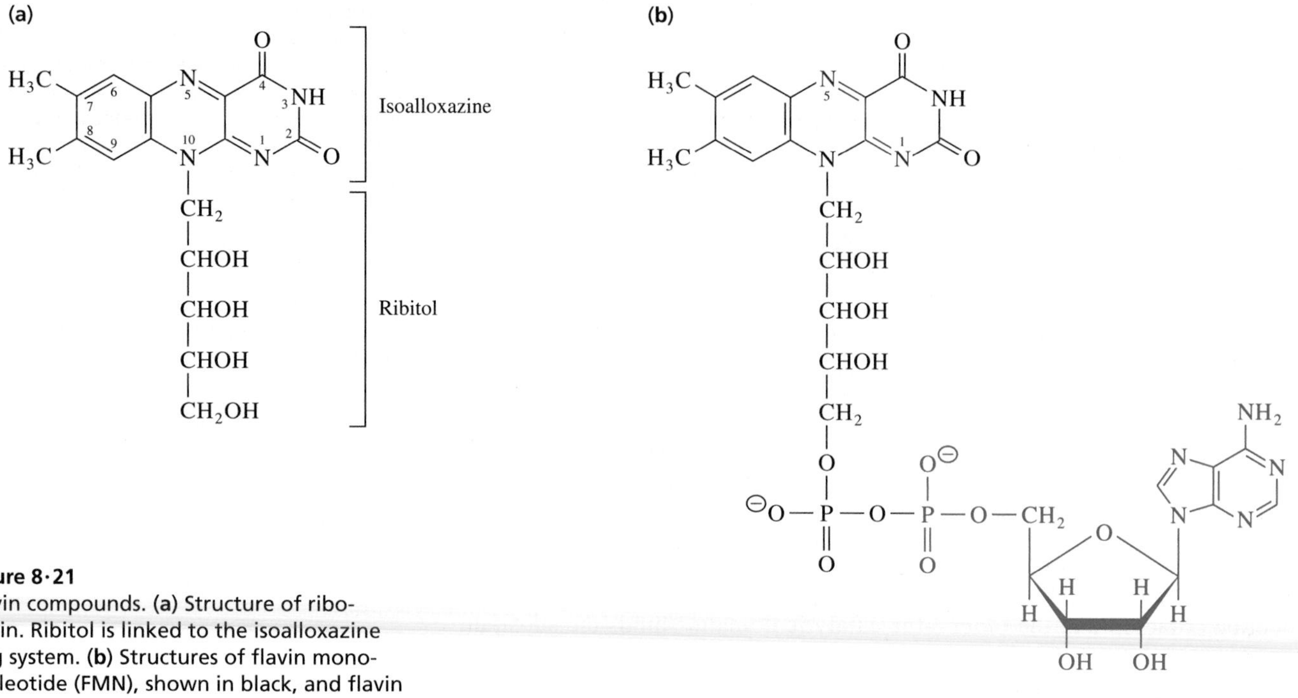

(a)

(b)

Figure 8·21
Flavin compounds. (a) Structure of ribo-flavin. Ribitol is linked to the isoalloxazine ring system. (b) Structures of flavin mono-nucleotide (FMN), shown in black, and flavin adenine dinucleotide (FAD), shown in black and blue. The reactive center is shown in red.

Riboflavin, or vitamin B_2, consists of the five-carbon alcohol ribitol (a reduced form of ribose) linked to a nitrogen atom of isoalloxazine, the characteristic feature of flavins (Figure 8·21a). The yellow color of the prosthetic group is imparted by the conjugated double-bond system of isoalloxazine. Riboflavin is abundant in milk, whole grains, and liver.

The two nucleotide-coenzyme forms of riboflavin are flavin mononucleotide (FMN, riboflavin 5′-phosphate) and flavin adenine dinucleotide (FAD), whose structures are shown in Figure 8·21b. Like $NAD^{\oplus}$ and $NADP^{\oplus}$, FAD contains adenine and a pyrophosphate linkage; however, unlike the pyridine nucleotides, FAD bears no positive charge.

Many enzymes, referred to as flavoenzymes or flavoproteins, require FAD or FMN as a prosthetic group that participates in the catalysis of oxidation-reduction reactions. Characteristically, the nucleotide is noncovalently but very tightly bound, although in a few instances, FAD is actually covalently bound to the protein. Unlike NADH, which is chemically stable in solution at pH 7, $FMNH_2$ and $FADH_2$ rapidly oxidize in solutions containing oxygen. By binding these prosthetic groups tightly, the apoenzymes protect the reduced forms from wasteful reoxidation by O_2. A number of oxidoreductases require, in addition to their flavin prosthetic group, one or more metal ions (for instance, iron or molybdenum); such enzymes are known as metalloflavoproteins.

Whereas the oxidized forms of riboflavin, FAD, and FMN all absorb visible light at 445–450 nm and appear yellow, $FADH_2$ and $FMNH_2$ are colorless because the conjugated double-bond system of the isoalloxazine ring system is lost upon re-duction. In Chapter 18, we will find that FMN is reduced to $FMNH_2$ in mitochon-dria by transfer of a hydride ion from NADH. However, unlike NADH and NADPH, which can only participate in two-electron transfers, $FMNH_2$ and $FADH_2$ can donate electrons either one or two at a time. A partially oxidized compound, FADH· or FMNH·, is formed when the one-electron pathway is followed (Figure

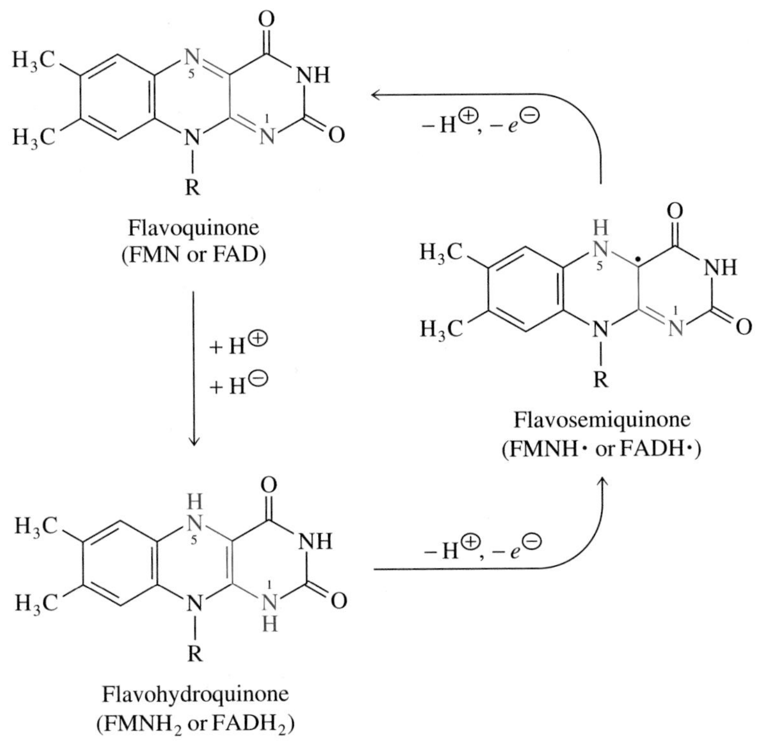

Flavoquinone
(FMN or FAD)

$-H^{\oplus}, -e^{\ominus}$

$+H^{\oplus}$
$+H^{\ominus}$

Flavosemiquinone
(FMNH· or FADH·)

Flavohydroquinone
(FMNH$_2$ or FADH$_2$)

$-H^{\oplus}, -e^{\ominus}$

Figure 8·22
Reduction and reoxidation of FMN or FAD. The conjugated double bonds between N-1 and N-5 are reduced by addition of a hydride ion and a proton to form FMNH$_2$ or FADH$_2$, the hydroquinone form of the coenzymes. Oxidation occurs with an intermediate step. A single electron is removed by a one-electron oxidizing agent such as Fe$^{\oplus}$, with loss of a proton, to form a fairly stable free-radical intermediate. This flavosemiquinone is then oxidized by removal of a proton and one electron to form the fully oxidized FMN or FAD.

8·22). These intermediates are relatively stable free radicals called flavosemiquinones. In mitochondria, the protein that is reduced by FMNH$_2$ has Fe$^{\circled{3+}}$ as its electron acceptor. Because Fe$^{\circled{3+}}$ can only accept one electron, forming Fe$^{\circled{2+}}$, the reduced flavin must be oxidized in two one-electron steps via the semiquinone intermediate. The coupling through FMN of two-electron transfers with one-electron transfers is important in cellular respiration, because further steps in the multistep mitochondrial process by which NADH is oxidized involve one-electron mechanisms. Thus, the oxidation-reduction versatility of the flavins is metabolically significant.

8·8 Coenzyme A Is Derived from Pantothenic Acid

A large number of important metabolic processes, including the oxidation of fuel molecules and the biosynthesis of some carbohydrates and lipids, depend on coenzyme A. This coenzyme was first identified in the mid-1940s as a cofactor required for biological acetylations. It is now recognized as the coenzyme most prominently involved in acyl-group–transfer reactions in which simple carboxylic acids and fatty acids are the mobile metabolic groups. Indeed, the name *coenzyme A* (often abbreviated CoA or CoASH) was derived from its role as the acetylation (now, more generally, acylation) coenzyme. Its structure, determined in the late 1940s by Fritz Lipmann, has three major components: a 2-mercaptoethylamine unit that bears a free —SH group (the site for acylation and deacylation of the coenzyme), a pantothenic acid unit (which is an amide of β-alanine and pantoic acid), and an ADP moiety whose 3′-hydroxyl group is esterified with a third phosphate group (Figure 8·23, next page). Pantothenate, a B vitamin that occurs widely in nature, was extensively purified in the 1930s by Roger J. Williams, who isolated it as an essential growth factor (vitamin B$_3$) for yeast.

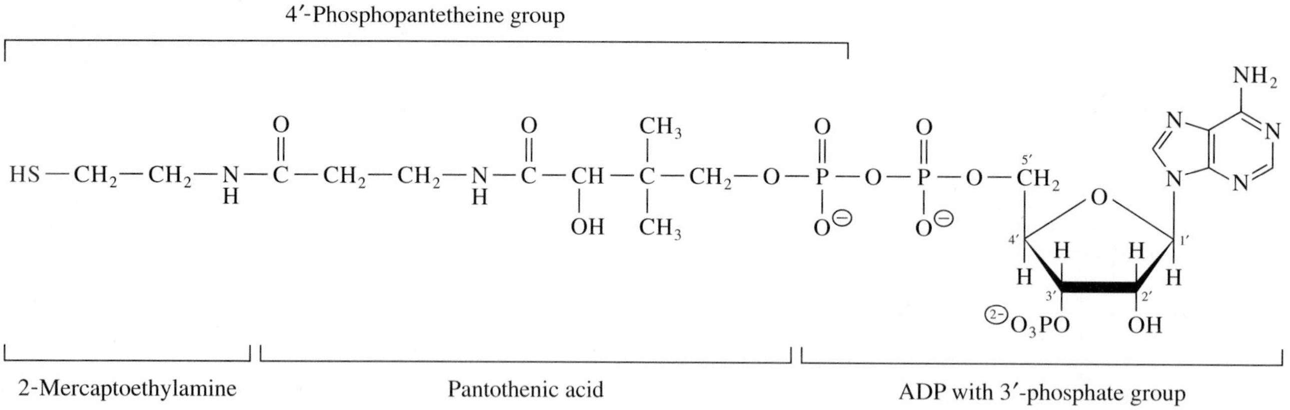

Figure 8·23
Structure of coenzyme A. The molecule consists of 2-mercaptoethylamine bound to the vitamin pantothenic acid, which is bound in turn by a phosphoester linkage to an ADP group that has an additional 3'-phosphate group. The reactive center of coenzyme A is the thiol group (red).

Acetyl CoA has been called "active acetate." The energy of the thioester linkage formed between acetate and the —SH group of CoA is similar to the energy of each of the phosphoanhydride linkages of ATP. Thioesters, unlike ordinary oxygen esters of carboxylic acids, resemble oxygen-acid anhydrides and thus are energy-rich metabolites. Despite the energy associated with acetyl CoA, it is quite resistant to nonenzymatic hydrolysis at neutral pH values. We will see acetyl CoA frequently when we discuss metabolism. The breakdown of carbohydrates, fatty acids, and amino acids produces acetyl CoA, which enters the citric acid cycle, the hub of the metabolic activity of the cell.

When the 3'-phospho-AMP moiety is removed from the CoA structure shown in Figure 8·23 by splitting the molecule between the two phosphoryl groups, the remaining portion, a phosphate ester containing the 2-mercaptoethylamine and pantothenate residues, is known as phosphopantetheine. Phosphopantetheine serves as the prosthetic group of a small protein of 77 amino acid residues known as the acyl carrier protein (ACP), first isolated from *E. coli* by Roy Vagelos and colleagues. The prosthetic group is esterified to ACP via the side-chain oxygen atom of Ser-36 (Figure 8·24). The —SH of the prosthetic group of ACP serves as the site of acylation for intermediates in the biosynthesis of fatty acids (Chapter 20). Thus, in both CoA and ACP, it is the phosphopantetheine group that provides the structure needed for biological transfers of acyl groups activated as thioesters.

Figure 8·24
Phosphopantetheine prosthetic group esterified to acyl carrier protein (ACP) via Ser-36.

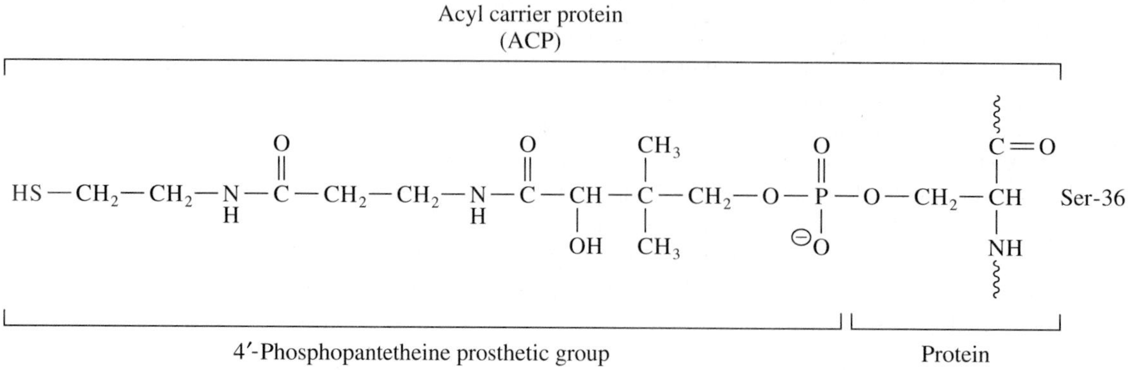

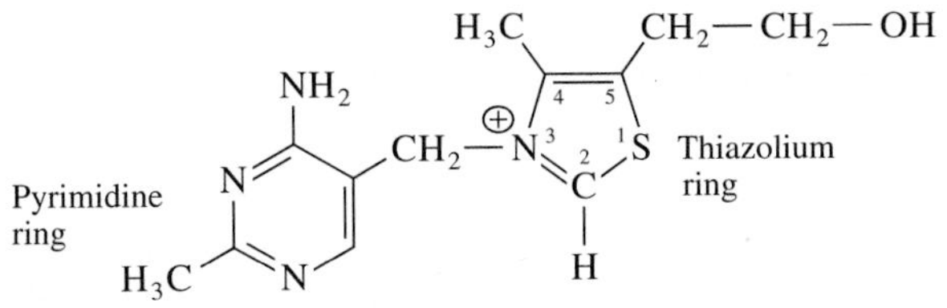

Figure 8·25
Structure of thiamine (vitamin B$_1$).

8·9 Thiamine Pyrophosphate Is a Derivative of Vitamin B$_1$

The structure of thiamine, the antiberiberi vitamin, was elucidated in 1935 by Robert R. Williams, brother of Roger J. Williams, who purified pantothenate. Thiamine includes a pyrimidine ring and a positively charged thiazolium ring, as shown in Figure 8·25. Thiamine (vitamin B$_1$) is abundant in the husks of rice, in other cereals, and in liver. Beriberi is most often found in parts of the world where polished rice is a staple of the diet, since removal of the rice husk also removes most of the essential thiamine.

The coenzyme form of the vitamin is thiamine pyrophosphate (TPP). Within animal cells, the coenzyme is synthesized from dietary thiamine by enzymatic transfer of pyrophosphate from ATP (Figure 8·26). The thiazolium ring contains the reactive center of the coenzyme.

About half a dozen known decarboxylases (carboxy-lyases) require TPP as a coenzyme. In fact, TPP was once referred to as "cocarboxylase." The first successful isolation of thiamine pyrophosphate was from yeast, where it was shown to be the prosthetic group of pyruvate decarboxylase. The activity of pyruvate decarboxylase allows yeast to convert pyruvate to acetaldehyde, which is subsequently reduced to ethanol. The mechanistic role of TPP can be illustrated by examining the

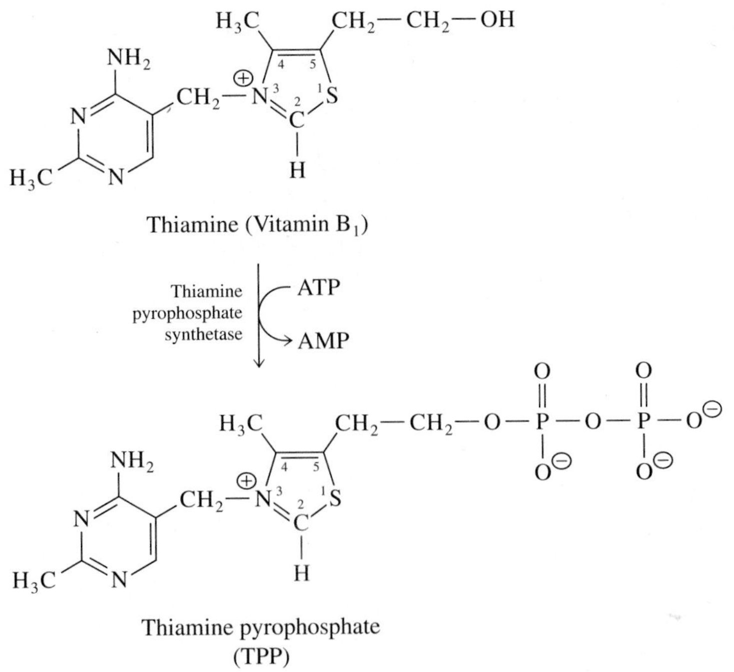

Thiamine (Vitamin B$_1$)

Thiamine pyrophosphate
(TPP)

Figure 8·26
Formation of thiamine pyrophosphate. Thiamine pyrophosphate synthetase catalyzes pyrophosphoryl-group transfer from ATP to thiamine (vitamin B$_1$), converting it to the coenzyme thiamine pyrophosphate (TPP). The thiazolium ring contains the reactive center (red) of the coenzyme.

reaction catalyzed by pyruvate decarboxylase (Figure 8·27). C-2 of TPP has unusual reactivity; it is acidic but has an extremely high pK_a, estimated at 18 in aqueous solution. This pK_a value may be lower in the active site of the enzyme. The positive charge of the thiazolium ring attracts electrons, weakening the bond between C-2 and hydrogen, thus allowing its ionization. The proton that dissociates from C-2 of the thiazolium ring is presumably removed by a basic residue of the enzyme. Ionization generates a resonance-stabilized dipolar carbanion known as an ylid. (Ylids are molecules that have opposite charges on adjacent atoms.) The negatively charged C-2 attacks the electron-deficient carbonyl carbon of the substrate pyruvate. The first product (CO_2) is released, shortening to two carbons the carbon skeleton of the pyruvate attached to the thiazole ring and resulting in the formation of a resonance-stabilized carbanion. In the following step, protonation of this carbanion produces the intermediate hydroxyethylthiamine pyrophosphate (HETPP),

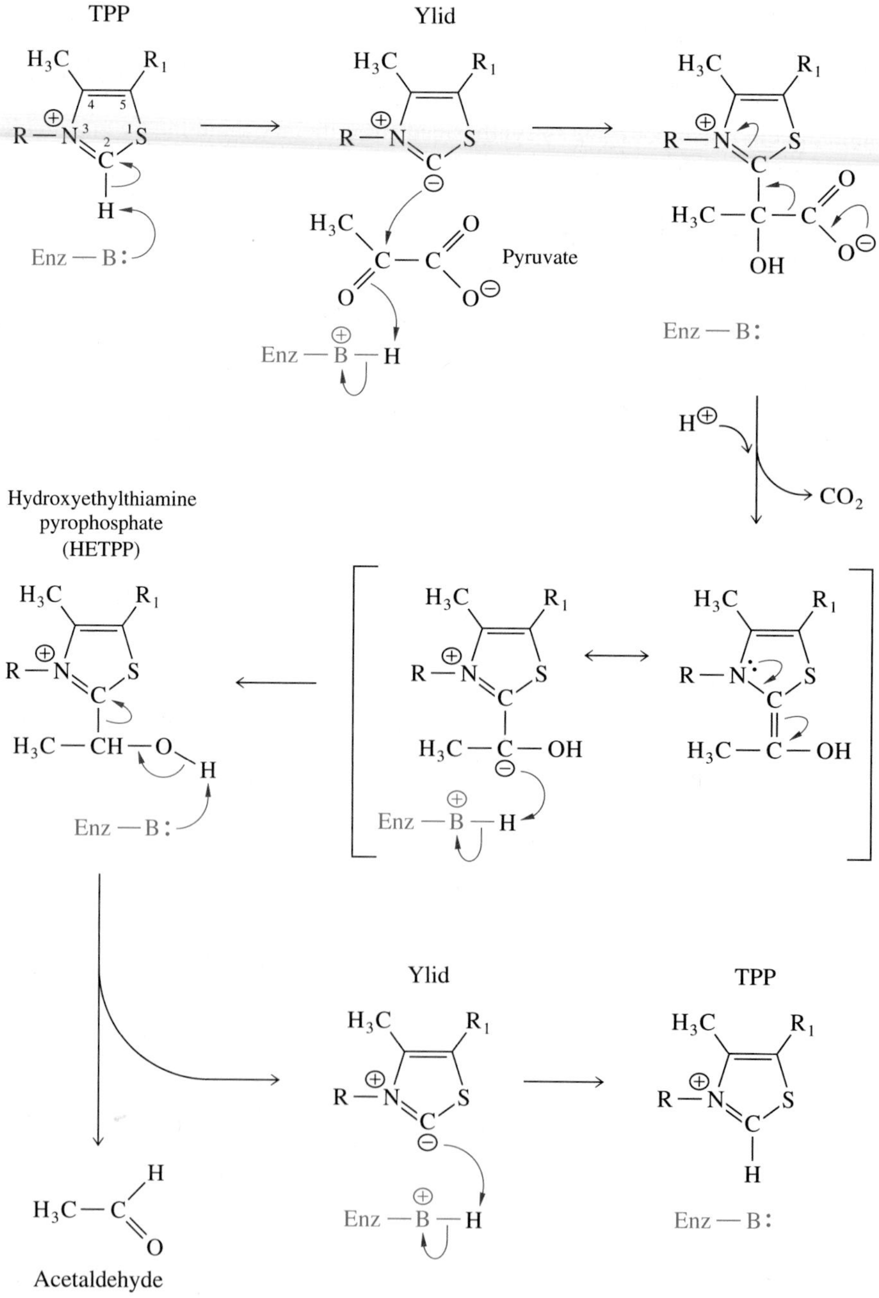

Figure 8·27

Mechanism of pyruvate decarboxylase showing the participation of thiamine pyrophosphate (TPP). Deprotonation of the thiazolium ring of TPP generates a dipolar carbanion known as an ylid. The negatively charged C-2 of the ylid attacks the carbonyl carbon of pyruvate, releasing CO_2 and resulting in the formation of a resonance-stabilized carbanion. Protonation of the carbanion produces HETPP, which is cleaved, releasing acetaldehyde. Protonation of the regenerated ylid form of TPP completes the catalytic cycle.

often called "active acetaldehyde." HETPP is cleaved, releasing acetaldehyde (the second product) and regenerating the ylid form of the enzyme-TPP complex. TPP is then re-formed as the ylid is protonated by the enzyme. TPP will be encountered as a coenzyme required for the oxidation of α-keto acids (Chapter 16). The first steps in the oxidative decarboxylation of pyruvate and of α-ketoglutarate proceed by the mechanism shown in Figure 8·27. In addition to its role as a coenzyme for α-keto acid decarboxylations, TPP serves as a prosthetic group for enzymes known as transketolases, which catalyze transfer of two-carbon keto groups between sugar molecules.

8·10 Pyridoxal Phosphate Is Derived from Vitamin B$_6$

The B$_6$ family of water-soluble vitamins consists of three closely related molecules. The first to be characterized was pyridoxine. However, most naturally occurring vitamin B$_6$ occurs as either pyridoxal or pyridoxamine—usually in a phosphorylated form. These three compounds differ only in the state of oxidation or amination of the carbon bound to position 4 of the pyridine ring (Figure 8·28). Vitamin B$_6$ is widely available from plant and animal sources. Induced deficiencies in rats result in dermatitis and a number of other disorders related to protein metabolism. Once inside a cell, dietary pyridoxine is converted to pyridoxine 5′-phosphate by enzymatic transfer of the γ-phosphoryl group of ATP and then oxidized to form the coenzyme pyridoxal 5′-phosphate (PLP), shown in Figure 8·29.

Figure 8·28
Structures of the vitamins of the B$_6$ family: pyridoxine, pyridoxal, and pyridoxamine.

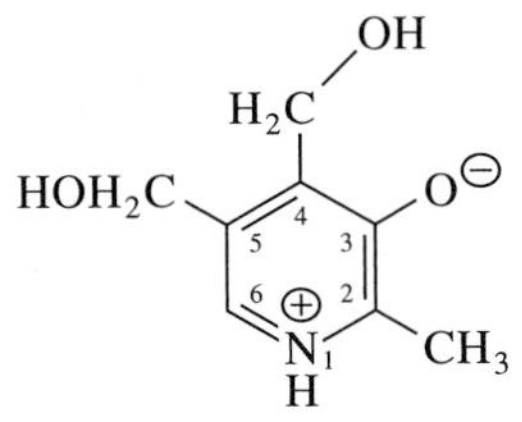

Pyridoxine

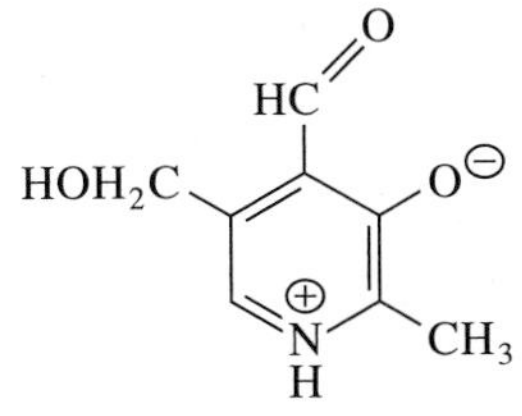

Pyridoxal

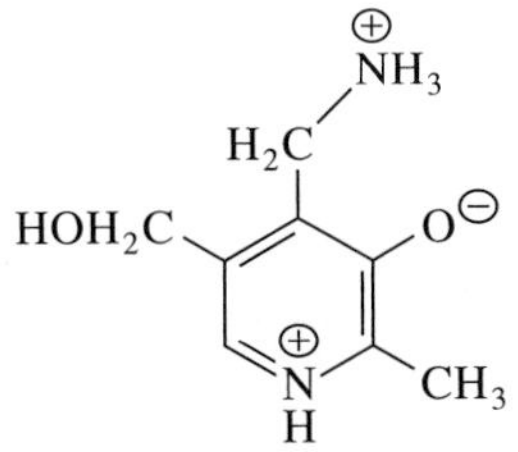

Pyridoxamine

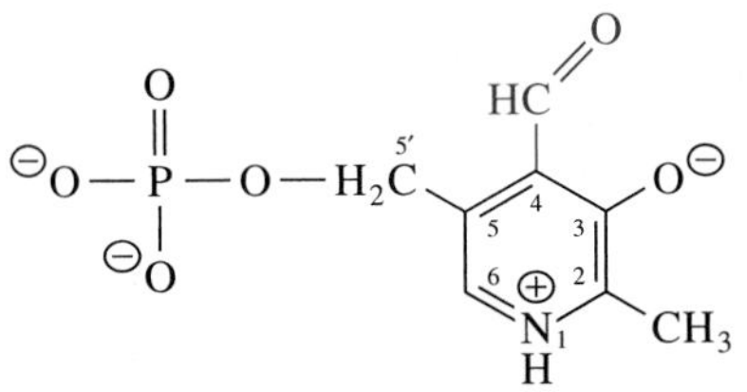

Figure 8·29
Structure of the coenzyme pyridoxal 5′-phosphate (PLP). The reactive center of PLP is the aldehyde group (red).

Pyridoxal phosphate is the prosthetic group for a large number of enzymes that catalyze a variety of reactions involving amino acids, including isomerization, decarboxylation, and side-chain elimination or replacement reactions.

In PLP-requiring enzymes, the prosthetic group is bound as a Schiff base to the ε-amino group of a lysine residue at the active site (Figure 8·30). A Schiff base consists of an imine linkage formed by the reversible condensation of a primary amine with an aldehyde or a ketone. These Schiff bases are sometimes referred to as aldimines or ketimines, respectively. The enzyme-coenzyme Schiff base shown on the right in Figure 8·30 is sometimes referred to as an internal aldimine. During reactions, the coenzyme forms an external aldimine with the substrate. PLP is bound to the enzyme at all times by many weak noncovalent interactions; the covalent but reversible linkage of the internal aldimine gives added strength to the binding of the scarce, vitamin-derived coenzyme to the apoenzyme when the enzyme is not functioning.

The initial step in all PLP-dependent enzymatic reactions involving amino acids is the formation of an external aldimine linking PLP to the α-amino group of the amino acid. When an amino acid substrate binds to a PLP-enzyme that is in the internal-aldimine form, a transimination reaction takes place (Figure 8·31). This transfer reaction proceeds via a geminal-diamine intermediate rather than via formation of the free-aldehyde form of PLP. Note that the Schiff bases contain a system of conjugated double bonds leading to a positive charge on N-1. During subsequent steps in the mechanism of a PLP-enzyme–catalyzed reaction, the prosthetic group serves as an electron sink. Once an α-amino acid forms a Schiff base with PLP, electron withdrawal toward N-1 weakens the three bonds to the α-carbon. In other words, the Schiff base with PLP provides a means for stabilizing a carbanion that can be formed by enzyme-directed loss of one of the three groups attached to the α-carbon of the amino acid. Of the three bonds, the one that breaks depends on the nature and location of amino acid residues contained within the active site of the

Figure 8·30
Binding of PLP to a PLP-dependent enzyme. Pyridoxal phosphate is bound to the apoenzyme by numerous noncovalent interactions and by formation of a Schiff base involving the aldehyde group of the coenzyme and the ε-amino group of a lysine residue in the active site.

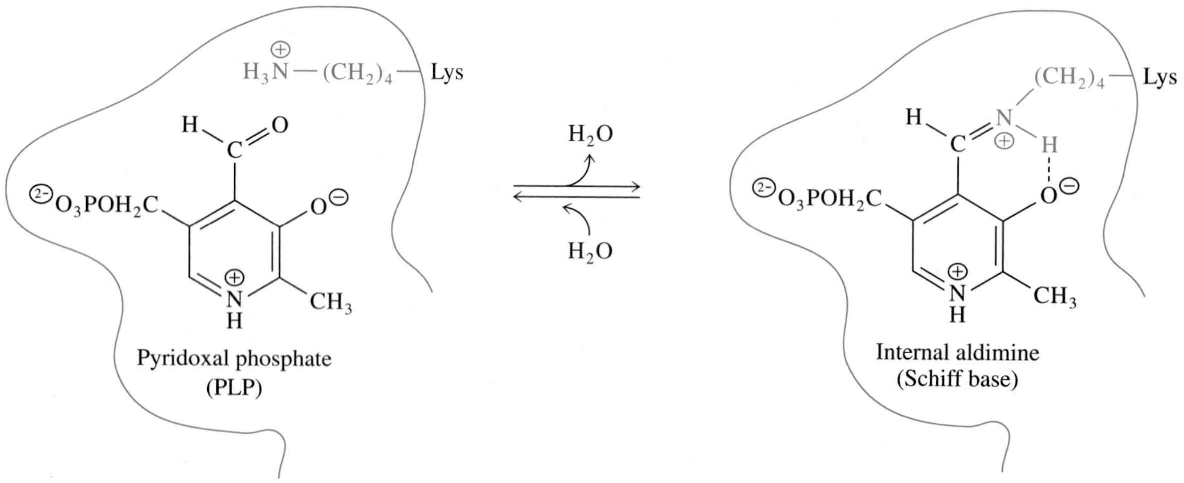

Pyridoxal phosphate
(PLP)

Internal aldimine
(Schiff base)

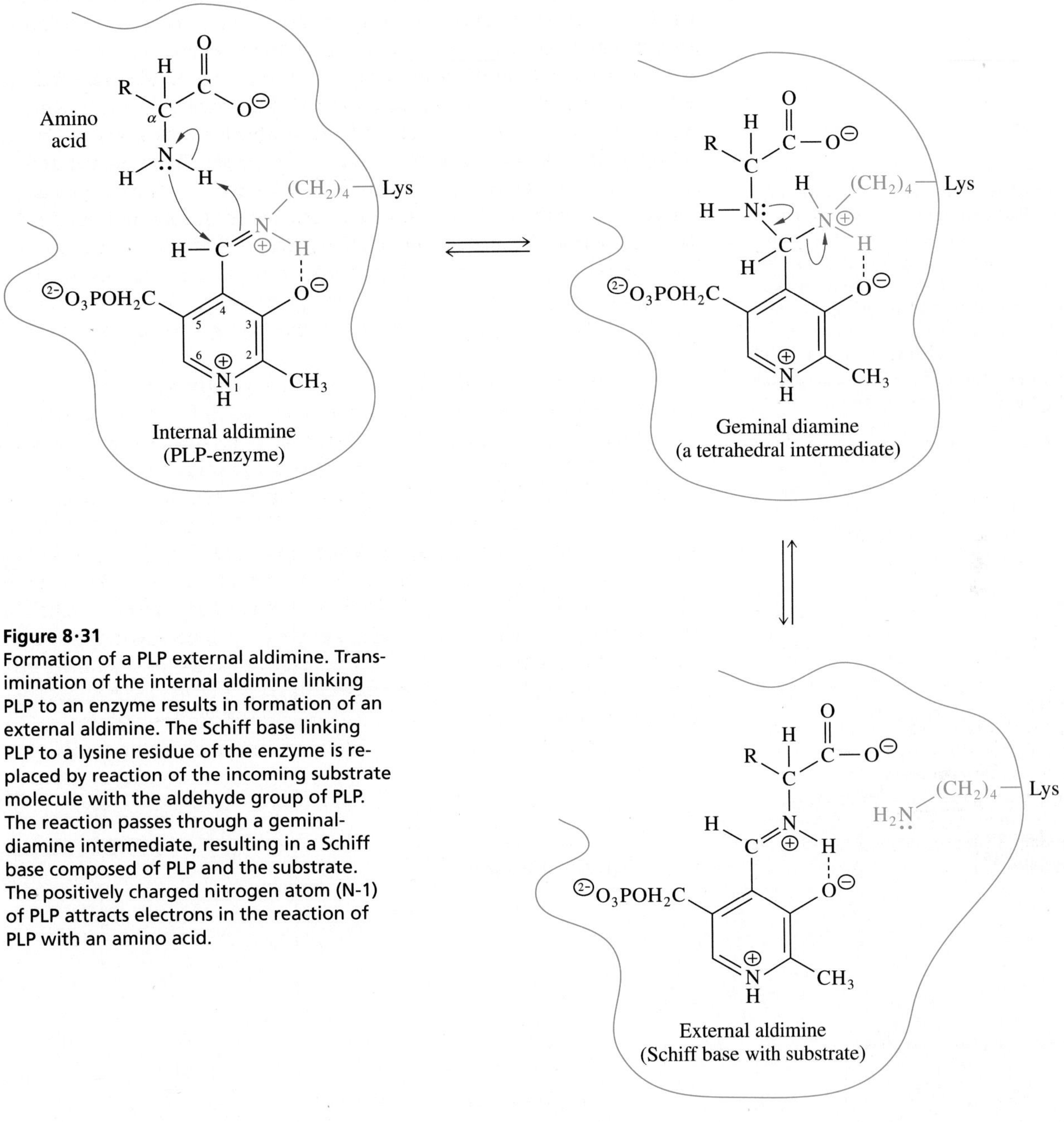

Figure 8·31
Formation of a PLP external aldimine. Trans-imination of the internal aldimine linking PLP to an enzyme results in formation of an external aldimine. The Schiff base linking PLP to a lysine residue of the enzyme is replaced by reaction of the incoming substrate molecule with the aldehyde group of PLP. The reaction passes through a geminal-diamine intermediate, resulting in a Schiff base composed of PLP and the substrate. The positively charged nitrogen atom (N-1) of PLP attracts electrons in the reaction of PLP with an amino acid.

enzyme. A simple illustration of the role of PLP is the racemization of an amino acid (its stereoisomerization at the α-carbon). The cell walls of many bacteria contain D-alanine. These bacteria synthesize the less common D isomer from L-alanine in a reaction catalyzed by alanine racemase. Figure 8·32 illustrates the reversible racemization of D-alanine. The α-hydrogen of the amino acid in the Schiff base is removed, generating a resonance-stabilized carbanion intermediate. The carbanion is then protonated on a different face of the α-carbon. The new Schiff base is hydrolyzed (or cleaved by transimination with the active-site lysine) to generate the opposite stereoisomer. PLP will be encountered frequently in Chapter 21, during the discussion of the metabolism of amino acids.

Figure 8·32
Mechanism of racemization of alanine catalyzed by a PLP-dependent racemase. The bonds of the α-carbon atom of the alanine molecule are weakened by formation of the Schiff base. A proton can then be removed from the α-carbon by a basic group of the enzyme. Two resonance forms of the carbanion intermediate are shown. The proton is returned to a different face of the α-carbon, producing the opposite stereoisomer of alanine.

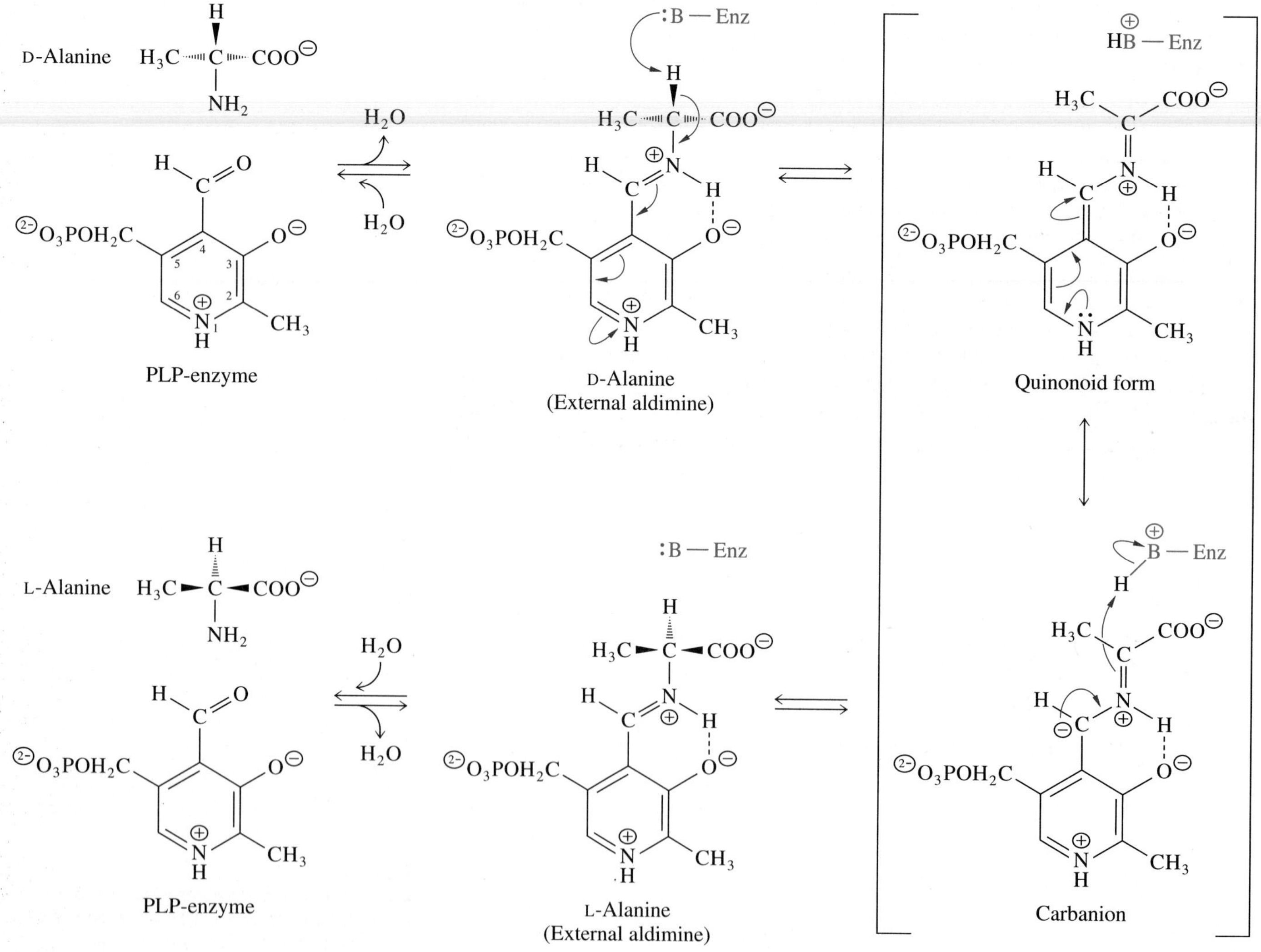

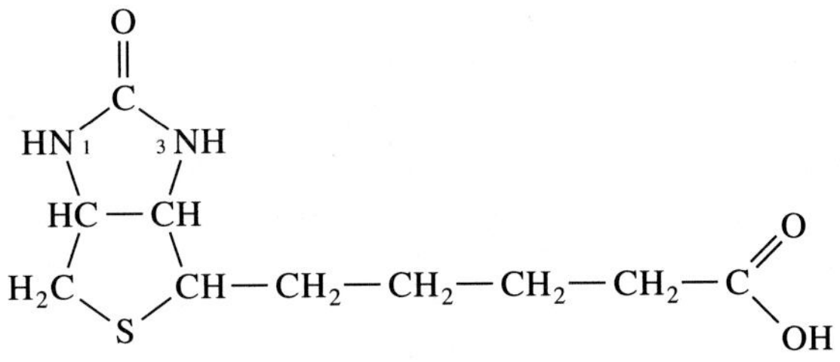

Figure 8·33
Structure of biotin.

8·11 Biotin Serves as a Prosthetic Group for Some Carboxylases

Biotin (once referred to as vitamin H) was initially identified as a necessary factor for the growth of yeast. In 1936, it was successfully isolated from dried egg yolks (1 mg was obtained from 500 lbs) and, in 1940, from liver concentrates. The structure of biotin (Figure 8·33) was proposed in 1942 and confirmed by synthesis a year later. Because biotin is synthesized by intestinal bacteria and is required only in very small (μg) amounts each day, biotin deficiency is rare in humans or animals fed normal diets. A biotin deficiency can be induced, however, by ingestion of raw egg whites. The protein avidin (M_r 70 000), a tetramer of identical subunits, is contained in egg whites. Each subunit of avidin binds biotin tightly (the dissociation constant is about 10^{-15} M), making it unavailable for absorption from the intestinal tract. Biotin deficiency (egg-white toxicity) has been found in people who eat large amounts of raw eggs. When eggs are cooked, avidin is denatured, abolishing its affinity for biotin and eliminating its toxicity.

Biotin serves as a cofactor for enzymes that catalyze carboxyl-group–transfer reactions and ATP-dependent carboxylation reactions. Biotin is covalently linked to the active site of its host enzyme by an amide bond to the ε-amino group of a lysine residue (Figure 8·34). The biotinyl-lysine residue is sometimes referred to as biocytin.

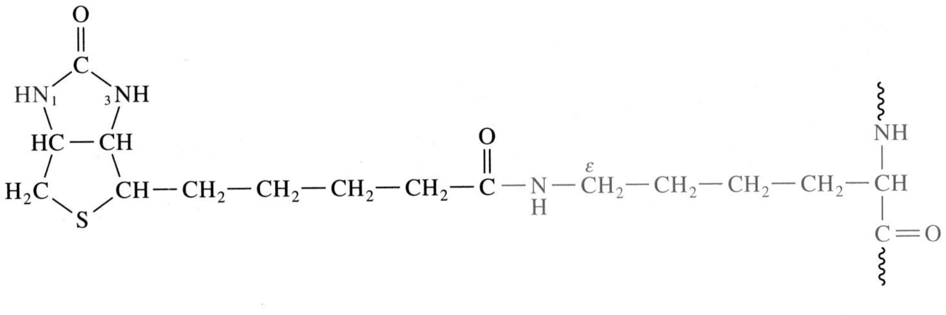

Figure 8·34
Biocytin. Biotin is covalently bound to enzymes by an amide linkage between the carboxylate group of biotin and the ε-amino group of a lysine residue (blue) at the active site of the enzyme. The biotinyl-lysyl adduct is sometimes called biocytin. The reactive center of biocytin is N-1, shown in red.

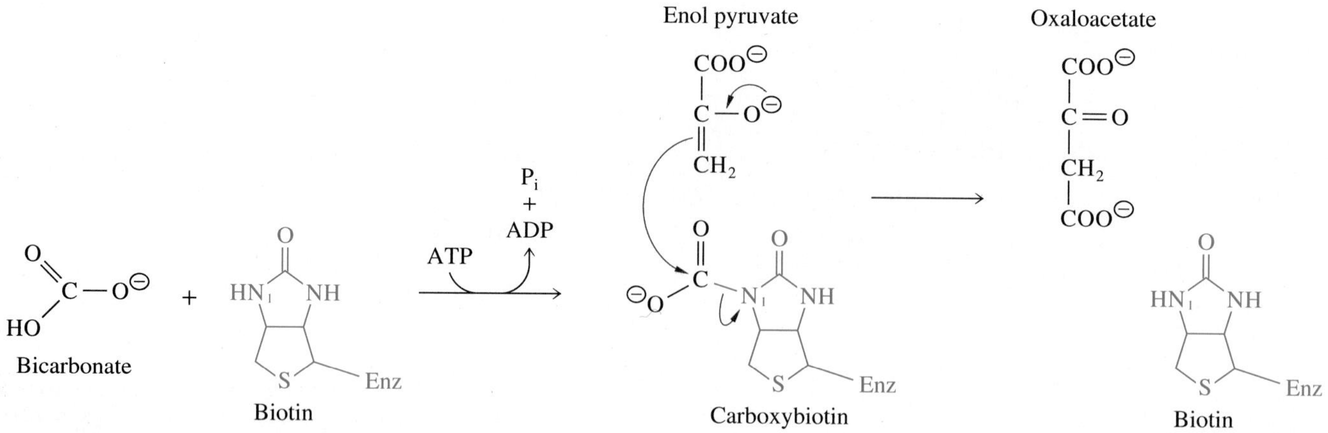

Figure 8·35
Involvement of biotin in the reaction catalyzed by pyruvate carboxylase. In the first step, carboxybiotin is formed from biotin and bicarbonate. In the second step, carboxyl-group transfer from carboxybiotin to the enol form of pyruvate yields oxaloacetate and regenerates biotin.

Biotin acts as a carrier of carbon dioxide, as illustrated by its role in the pyruvate carboxylase reaction. Pyruvate carboxylase catalyzes the carboxylation of the three-carbon acid pyruvate by bicarbonate to form the four-carbon acid oxaloacetate, the substrate for several metabolic reactions. The reaction occurs in two steps: formation of carboxybiotin and transfer of CO_2 from carboxybiotin to pyruvate. As shown in Figure 8·35, the enzyme-bound biotin serves as the intermediate carrier of the mobile carboxyl metabolic group. The mechanism by which biotin is carboxylated is not yet known. It has been suggested that carbonylphosphate is formed from bicarbonate and ATP and that this reactive compound (or CO_2 formed from it) carboxylates N-1 of biotin.

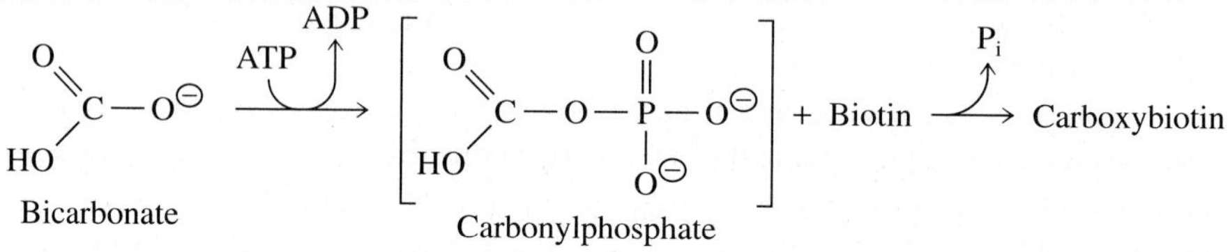

(8·9)

The N-1 carboxybiotinyl-enzyme provides a stable activated form of CO_2 that can be transferred to a second substrate, pyruvate. The enolate form of pyruvate attacks the carboxyl group of carboxybiotin to form oxaloacetate.

8·12 Tetrahydrofolate and Tetrahydrobiopterin Are Both Pterins

The vitamin folate—also known as pteroylglutamate—was isolated in the early 1940s from green leaves, liver, and yeast. Analysis of the isolated vitamin showed that it has three main components: pterin (2-amino-4-oxo-substituted pteridine), a p-aminobenzoic acid moiety, and a glutamate residue (Figure 8·36). However, it was discovered that the coenzyme form of folate was partially degraded during its isolation. The coenzyme differs from the isolated vitamin in two respects: it is reduced to form tetrahydrofolate, and it is modified by the addition of glutamate residues bound to one another through γ-glutamyl amide linkages (Figure 8·37). The anionic polyglutamyl moiety, usually five to six residues long, participates in the binding of tetrahydrofolate to enzymes and assists with its retention inside cells, since lipid membranes are impermeable to charged molecules.

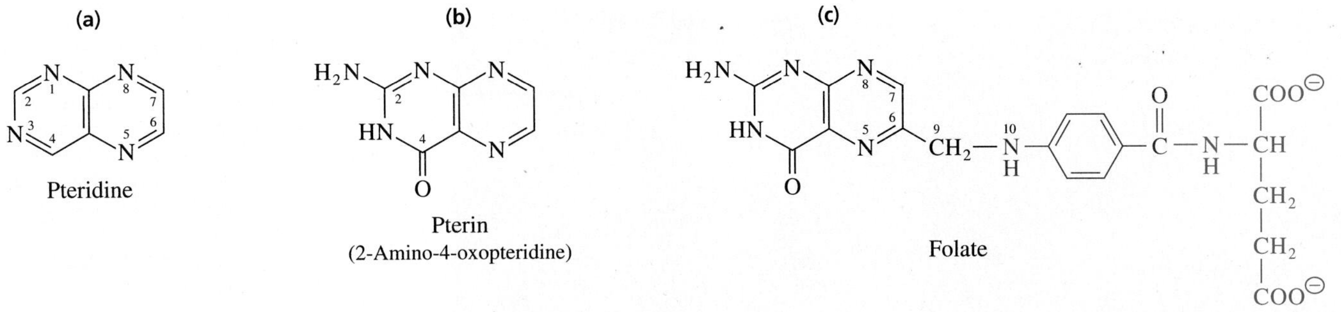

(a)

Pteridine

(b)

Pterin
(2-Amino-4-oxopteridine)

(c)

Folate

Figure 8·36
Structures of **(a)** the pteridine ring system, **(b)** pterin, and **(c)** folate, a conjugated pterin containing *p*-aminobenzoate (red) and glutamate (blue).

Tetrahydrofolate is formed from folate by the addition of hydrogen to positions 5, 6, 7, and 8 of the pterin ring system. Folate is reduced in two steps, both involving NADPH, in a reaction catalyzed by dihydrofolate reductase.

Folate   7,8-Dihydrofolate   5,6,7,8-Tetrahydrofolate

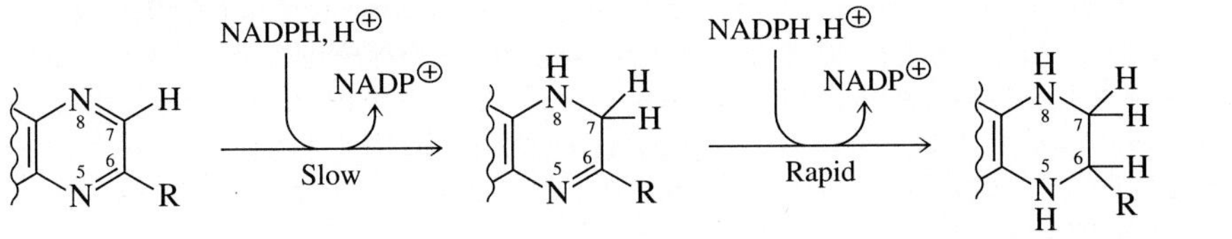

$$(8·10)$$

The primary metabolic function of dihydrofolate reductase is the reduction of dihydrofolate produced during the formation of the methyl group of deoxythymidine monophosphate (dTMP, Section 22·10). This reaction, which utilizes a derivative of tetrahydrofolate, is an essential step in the biosynthesis of DNA and hence is required for cell division. Because of its role in cell division, dihydrofolate reductase has been extensively studied as a target for chemotherapy in the treatment of cancer.

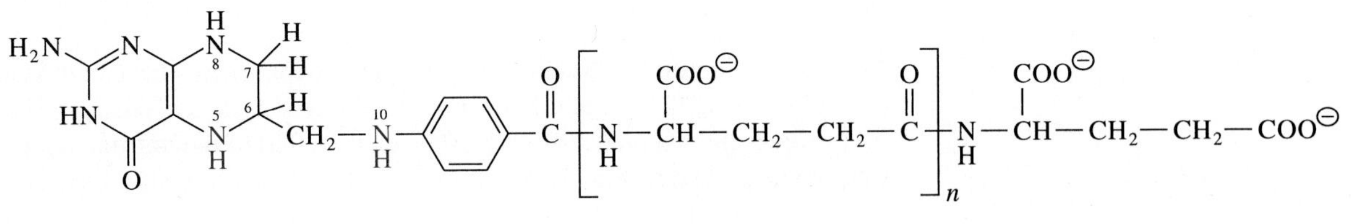

Figure 8·37
Structure of the poly-γ-glutamyl form of tetrahydrofolate. Tetrahydrofolate usually contains five or six glutamate residues. The reactive centers of the coenzyme, N-5 and N-10, are shown in red.

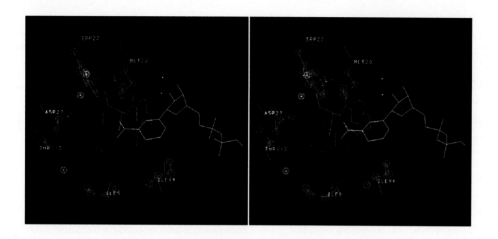

Figure 8·38
Stereo view of the active site of dihydrofolate reductase. The oxidized substrate folate is purple, and the oxidized coenzyme NADP$^\oplus$ is green. The side chains of the active-site residues (red) are labelled. Note how C-4 of the nicotinamide ring of NADP$^\oplus$ is positioned close to the ring of folate that must be reduced. Rather than having a histidine residue as a proton-transferring group, this reductase has an aspartate residue (Asp-27) that transfers a proton to the pterin substrate via intervening water molecules in the active site. One water molecule thought to be involved in this proton transfer is shown as a blue sphere above Asp-27. (Based on coordinates provided by C. Bystroff, S. J. Oatley, and J. Kraut.)

The mechanism of reduction of 7,8-dihydrofolate by NADPH in the dihydrofolate reductase reaction exhibits a remarkable role for water in the active site. The reduction occurs by addition of a proton to N-5, giving C-6 a partial positive charge, thereby assisting the nucleophilic addition of a hydride ion to C-6.

$$\xrightarrow{\quad +\text{H}^\oplus \quad} \qquad \xrightarrow{\quad +\text{H}^\ominus \quad}$$

7,8-Dihydrofolate

5,6,7,8-Tetrahydrofolate

(8·11)

X-ray crystallographic, kinetic, and mutagenic experiments have demonstrated that Asp-27 in the active site of dihydrofolate reductase is an acid catalyst. Examination of the three-dimensional structure of the enzyme led to the unusual conclusion that since the aspartate residue is not close enough to N-5 to transfer the proton directly, proton transfer must involve the participation of water molecules present in the active site. The architecture of the active site, with folate and NADP$^\oplus$ bound, is shown in Figure 8·38.

The conversion of folate to its active coenzyme requires not only reduction but also formation of the polyglutamate tail. The enzyme folylpolyglutamate synthetase catalyzes the condensation of the monoglutamate form of tetrahydrofolate with glutamate; it also catalyzes the addition of each succeeding glutamyl group until the appropriate chain length is reached. The general equation for this process is

$$\text{Tetrahydrofolate}-(\text{Glu})_n \;+\; \text{Glutamate} \xrightarrow[\qquad\qquad]{\text{ATP} \quad \text{ADP}+\text{P}_i} \text{Tetrahydrofolate}-(\text{Glu})_{n+1}$$

(8·12)

Mammals have very low levels of activity of this enzyme; prokaryotes have 50 to 100 times more activity than eukaryotes. The regulation of the length of the poly-glutamate chain is not yet fully understood. Specific hydrolases in the intestine cat-alyze cleavage of the γ-glutamate amide bonds of polyglutamates; these enzymes are involved in digestion of dietary folates. Hydrolases present in lysosomes may be involved in regulation of the chain length of intracellular folylpolyglutamates. We will refer to tetrahydrofolates in later discussions, but keep in mind that these com-pounds have polyglutamate tails.

5,6,7,8-Tetrahydrofolate is required by enzymes that catalyze biochemical transfers of one-carbon units at the oxidation levels of methanol (CH_3OH), form-aldehyde (HCHO), and formic acid (HCOOH). Thus, the fundamental groups bound to tetrahydrofolate are methyl, methylene, or formyl groups. The structures of several one-carbon derivatives of tetrahydrofolate and the enzymatic intercon-versions that occur among the various substituted forms are shown in Figure 8·39.

Figure 8·39
Structures of one-carbon derivatives of tetrahydro-folate. The derivatives can be interconverted enzy-matically by the routes shown. (R denotes the benzoyl oligoglutamate portion of tetrahydro-folate.)

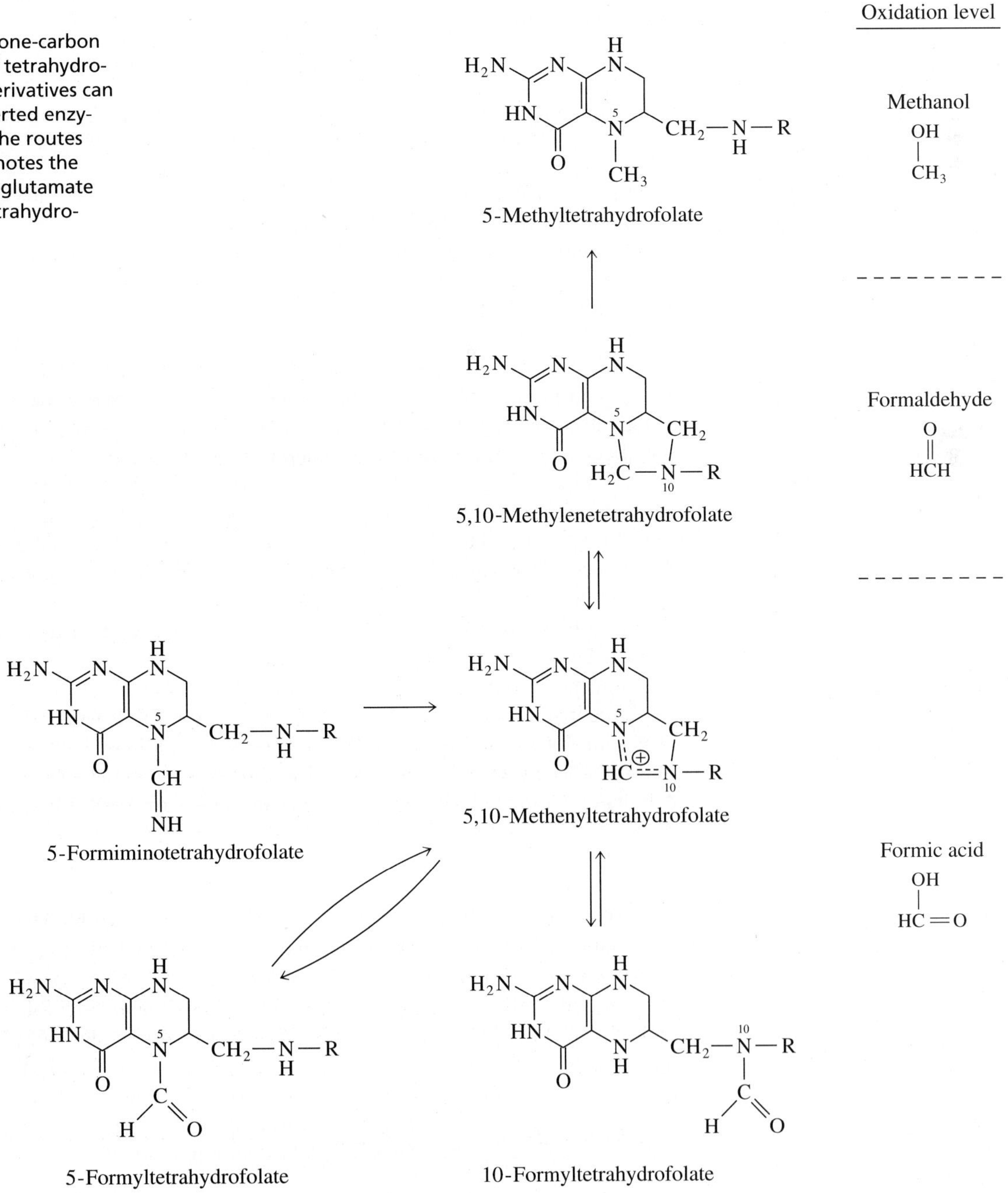

Oxidation level

5-Methyltetrahydrofolate — Methanol (OH / CH_3)

5,10-Methylenetetrahydrofolate — Formaldehyde (O / HCH)

5-Formiminotetrahydrofolate

5,10-Methenyltetrahydrofolate — Formic acid (OH / HC=O)

5-Formyltetrahydrofolate

10-Formyltetrahydrofolate

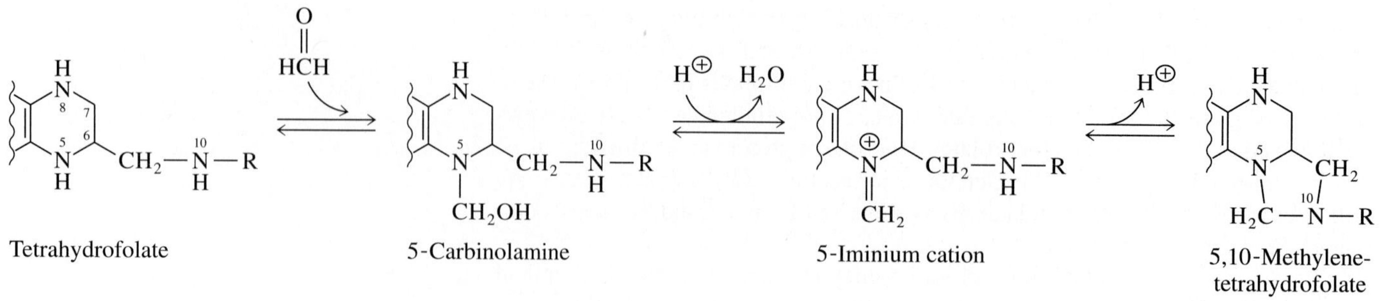

Figure 8·40
Reaction of formaldehyde with tetrahydro-folate.

The one-carbon metabolic groups are covalently bound to the secondary amine N-5 or N-10 of tetrahydrofolate, or to both in a ring form. 10-Formyltetrahydrofolate has been referred to as "active formate" and 5,10-methylenetetrahydrofolate as "active formaldehyde." The addition of formate to tetrahydrofolate is catalyzed by formyltetrahydrofolate synthetase.

$$\text{Tetrahydrofolate} + \text{HCOO}^{\ominus} \underset{}{\overset{\text{ATP} \quad \text{ADP} + \text{P}_i}{\rightleftharpoons}} \text{10-Formyltetrahydrofolate}$$

(8·13)

In several strains of purine-fermenting *Clostridia* (a bacterium), this reaction operates in the reverse direction and serves as a primary source of ATP.

Formaldehyde can react nonenzymatically with tetrahydrofolate through a pathway that involves a carbinolamine that undergoes acid-catalyzed dehydration to a rapidly cyclized 5-iminium cation (Figure 8·40). However, the major metabolic source of the one-carbon group of methylenetetrahydrofolate is not free formaldehyde but the β-carbon of serine, transferred to tetrahydrofolate in the reaction catalyzed by serine hydroxymethyltransferase. We will see in later chapters that tetrahydrofolate-dependent delivery of one-carbon units is important in a number of reactions in nucleic acid metabolism and in protein synthesis.

Vertebrates depend on dietary folate as the precursor of tetrahydrofolate because they have lost the ability to join a pterin to *p*-aminobenzoate. Bacteria retain this ability. Sulfanilamide and other sulfa drugs are antibacterial agents because they inhibit the joining of *p*-aminobenzoate to pterin in certain pathogenic bacteria. The structure of sulfanilamide is shown in Figure 8·41. Since the target of the drug is a reaction catalyzed by bacteria but not their hosts, sulfa drugs are selectively toxic, attacking only the pathogens.

Significant deficiency of folate is not common in North American diets because the major food groups (cereals, meats, and vegetables) contain adequate concentrations of folate. The vitamin folic acid is not routinely included in multivitamin pills because it can mask the symptoms of pernicious anemia, vitamin B_{12} deficiency. Untreated pernicious anemia can lead to irreversible brain damage. The requirement for folate increases during pregnancy, when folate is supplied to the fetus. Recent research indicates that supplementary folate ingested during the first weeks of pregnancy can reduce the occurrence of spina bifida and other birth defects associated with defective development of the neural tube. However, the administration of folate for the prevention of neural-tube defects presents a dilemma. It may not be prudent to recommend that *all* women of child-bearing age consume an increased amount of folate each day—perhaps by making the folate available in multivitamin pills or in fortified foods—because borderline B_{12} deficiency is common. Ideally, additional folate could be administered only to pregnant women, but in practice this is difficult because the extra folate is needed during the first weeks of pregnancy, before many women realize they are pregnant. Researchers and health agencies are working hard to find a solution to the dilemma.

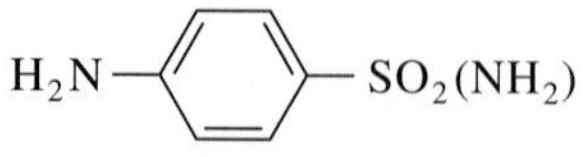

$$\text{H}_2\text{N} - \underset{}{\bigcirc} - \text{SO}_2(\text{NH}_2)$$

Figure 8·41
Structure of sulfanilamide.

5,6,7,8-Tetrahydrobiopterin is another pterin-derived coenzyme. It has a three-carbon side chain at C-6 of tetrahydropterin in place of the large side chain found in tetrahydrofolate and acts by a different mechanism. It is synthesized by animals as well as other organisms. The final step in the biosynthesis of 5,6,7,8-tetrahydrobiopterin, its formation from 7,8-dihydrobiopterin and NADPH, is catalyzed by dihydrofolate reductase. Tetrahydrobiopterin functions as the cofactor for several hydroxylases. The reaction catalyzed by phenylalanine hydroxylase, the conversion of phenylalanine to tyrosine, provides an example of the role of tetrahydrobiopterin (Figure 8·42). Both the coenzyme and phenylalanine are oxidized: one atom of oxygen combines with the aromatic ring of phenylalanine to form tyrosine, and the other combines with the carbon atom at position 4 of tetrahydrobiopterin to form a carbinolamine intermediate. Dehydration of the 4a-carbinolamine, catalyzed by 4a-carbinolamine dehydratase, produces a quinonoid dihydro form of biopterin that has lost hydrogen from N-3 and N-5. The quinonoid form is restored to active tetrahydrobiopterin in an NADH-dependent reaction catalyzed by dihydropteridine reductase. Tetrahydrobiopterin supplies electrons to reduce O_2, which is a substrate for this reaction. One oxygen atom is incorporated into phenylalanine as a hydroxyl group, producing tyrosine, and the other oxygen atom is reduced to H_2O.

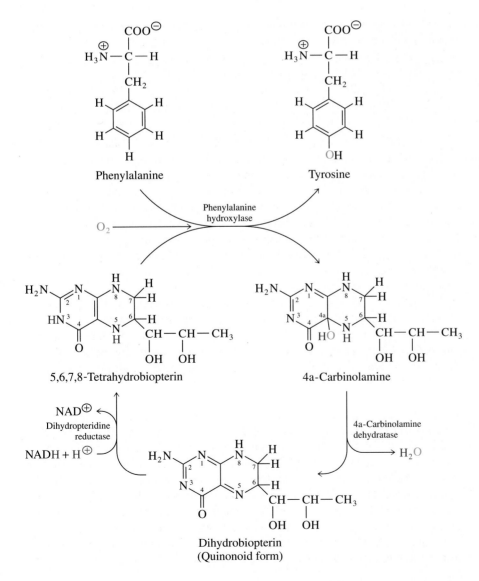

Figure 8·42
Oxidation and regeneration of tetrahydrobiopterin. The active coenzyme 5,6,7,8-tetrahydrobiopterin is oxidized in the phenylalanine hydroxylase reaction, resulting in the formation of first the 4a-carbinolamine of tetrahydrobiopterin and then the quinonoid dihydrobiopterin. The dihydro form is restored to the active tetrahydro form in an NADH-dependent reaction catalyzed by dihydropteridine reductase.

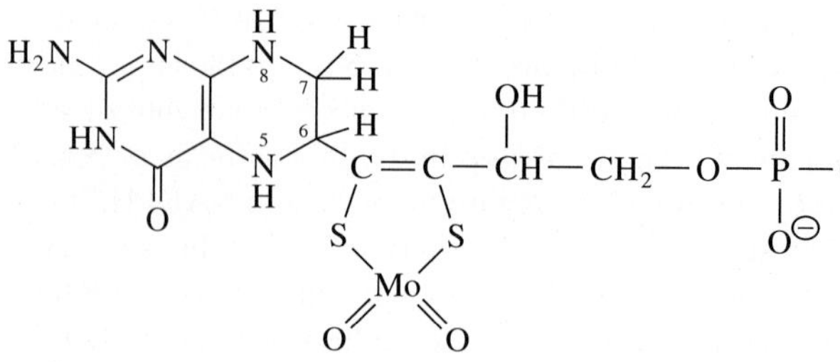

Figure 8·43
Proposed structure of the molybdenum cofactor of xanthine oxidase. The cofactor is shown as a complex of molybdenum with the two thiol groups of tetrahydromolybdopterin, but the pterin ring may be in a dihydro form.

The pterin moiety also occurs as part of a molybdenum-containing cofactor, molybdopterin (Figure 8·43), which is bound tightly to enzymes such as xanthine oxidase and aldehyde oxidase. The molybdenum ion of this cofactor undergoes cyclic oxidation and reduction.

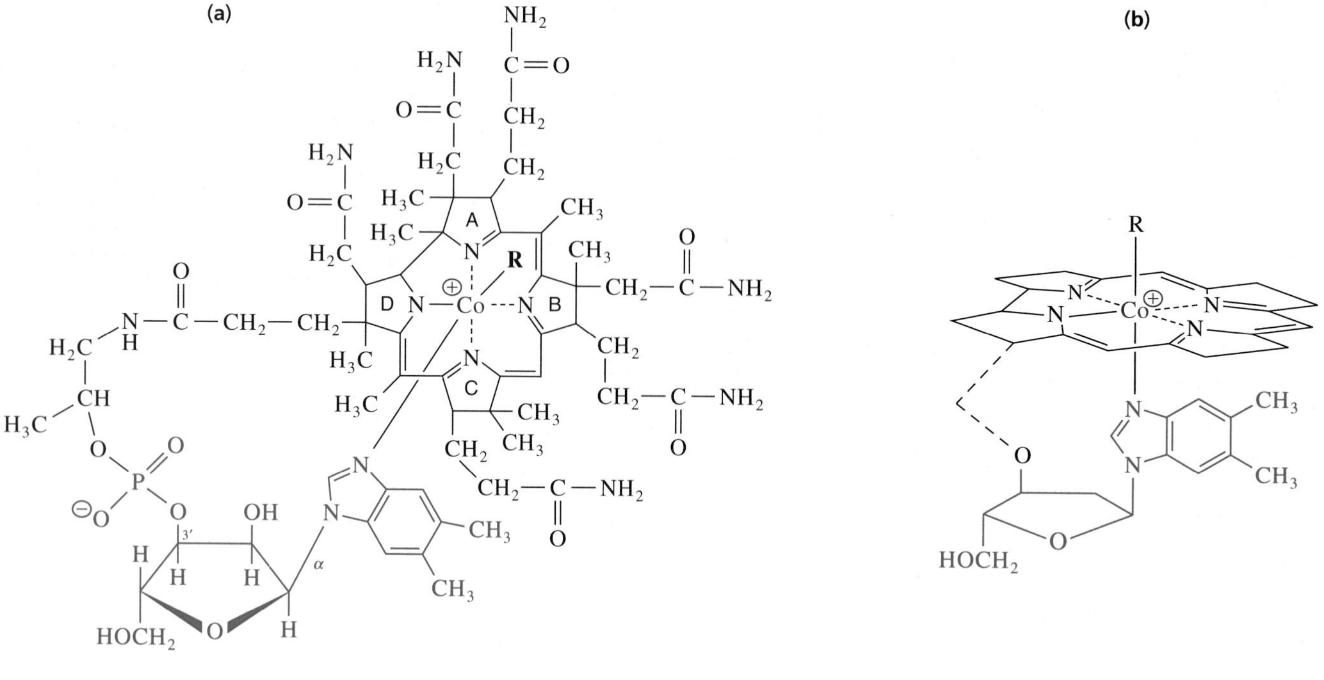

Figure 8·44
Structure of the vitamin B_{12} compounds. (a) Vitamin B_{12} has two characteristic components: the corrin ring system (black) and the 5,6-dimethylbenzimidazole ribonucleotide (blue). The corrin ring system contains four pyrrole rings and resembles the porphyrin ring system of heme. The major difference is that pyrrole rings A and D are joined directly rather than by a methene bridge. Furthermore, the metal coordinated by the four pyrrole nitrogens of corrin is cobalt (shown in red) rather than iron. The second component of vitamin B_{12} is the 5,6-dimethyl-benzimidazole ribonucleotide group. A nitrogen atom of the benzimidazole ring is coordinated with the cobalt of the corrin ring. The benzimidazole ribonucleotide is also bound via an ester linkage to a side chain of the corrin ring system. (b) This abbreviated structure shows the flat corrin ring of vitamin B_{12} with an R group above the ring and the benzimidazole below the ring. In the purified vitamin, the R group is either —CN or —OH. In the primary coenzymatic form of the vitamin, the R group is 5′-deoxyadenosine; in another coenzyme form, it is a methyl group.

8·13 Vitamin B_{12} and Its Coenzyme Forms Contain Cobalt

In 1926, George Minot and William Murphy discovered that pernicious anemia—previously incurable and usually fatal—was improved in patients fed a diet that included large amounts of liver. The factor responsible for relieving the symptoms was vitamin B_{12}, the largest B vitamin and the last to be isolated.

Victims of pernicious anemia lack normal secretion by the stomach mucosa of a glycoprotein called intrinsic factor. This protein specifically binds vitamin B_{12}, and it is the B_{12}–intrinsic factor complex that is absorbed by cells of the small intestine. Currently, impaired absorption is treated by injections of vitamin B_{12} at regular intervals.

(a)

(b)

In 1948, vitamin B_{12}, also called cobalamin, was obtained from liver in the form of a red crystalline cyanide derivative. The cyanide ion is not part of vitamin B_{12} but comes from charcoal used in the purification of the vitamin. The complex structure of cyanocobalamin, the cyanide form of vitamin B_{12}, was determined by Dorothy C. Hodgkin in 1956 by means of X-ray crystallographic and chemical studies. The detailed structure of the vitamin B_{12} compounds is shown in Figure 8·44a. Note the resemblance of the corrin ring system to the porphyrin ring system (such as occurs in heme, Figure 5·33). Differences include the lack of a methene ($—CH{=}$) bridge between rings A and D of the corrin ring system and the presence of trivalent cobalt rather than the divalent iron found in heme. The abbreviated structure shown in Figure 8·44b emphasizes the positions of two axial ligands bound to the cobalt, a benzimidazole ribonucleotide below the cobalt atom and an R group above it. Table 8·2 illustrates the various R groups that occur as upper axial ligands of the cobalt atom of cobalamin.

Vitamin B_{12} is not synthesized in plant or animal cells but is required by all animals and by some bacteria and algae. It functions in the two unstable coenzyme forms adenosylcobalamin and methylcobalamin, so called because a 5′-deoxyadenosyl or methyl group, respectively, is bound to the cobalt atom. The vitamin is synthesized by only a few microorganisms. Ruminants have microorganisms in their forestomach that synthesize vitamin B_{12}, which is then absorbed in the lower gut. Carnivores acquire B_{12} from their diet. After absorption, most of the vitamin is converted via enzymatic reduction and reaction with ATP to the adenosyl-coenzyme derivative. During isolation, the 5′-deoxyadenosyl moiety of adenosyl-cobalamin, linked to the rest of the molecule by the unusual C-5′—Co bond, is readily lost, replaced by $—OH^{\ominus}$ or $—CN^{\ominus}$.

The role of adenosylcobalamin is related to the lability of its unique C—Co bond. The coenzyme participates in several enzyme-catalyzed intramolecular rearrangements in which a hydrogen atom and a second group, bound to adjacent carbon atoms within a substrate, exchange positions (Figure 8·45a). An example is the methylmalonyl-CoA mutase reaction (Figure 8·45b), which is important in the metabolism of odd-chain fatty acids and leads to the formation of succinyl CoA, an intermediate of the citric acid cycle (Chapter 16). In partnership with tetrahydrofolate, methylcobalamin is involved in the transfer of methyl groups, for instance,

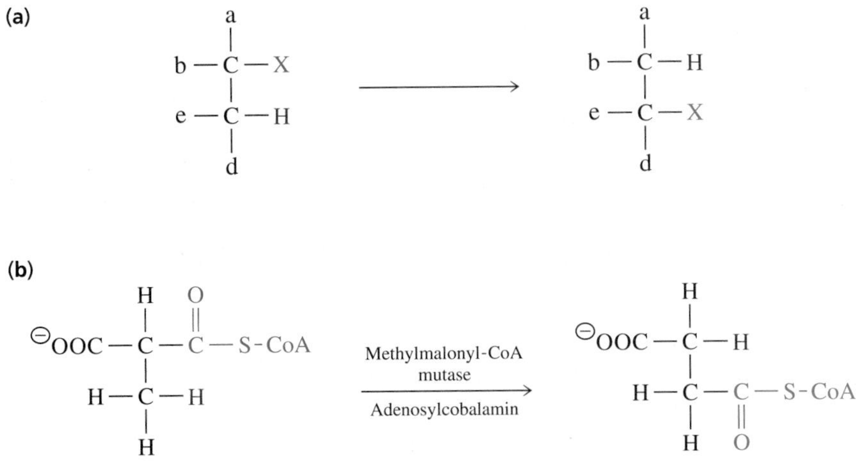

Table 8·2 Forms of vitamin B_{12} and its coenzymes. The various forms of vitamin B_{12} and its coenzymes are distinguished by the R group that is bound to the cobalt ion.

—R	Name
—CN	Cyanocobalamin (Vitamin B_{12})
—OH	Hydroxocobalamin (Vitamin B_{12a})
—CH_3	Methylcobalamin (Methyl B_{12})
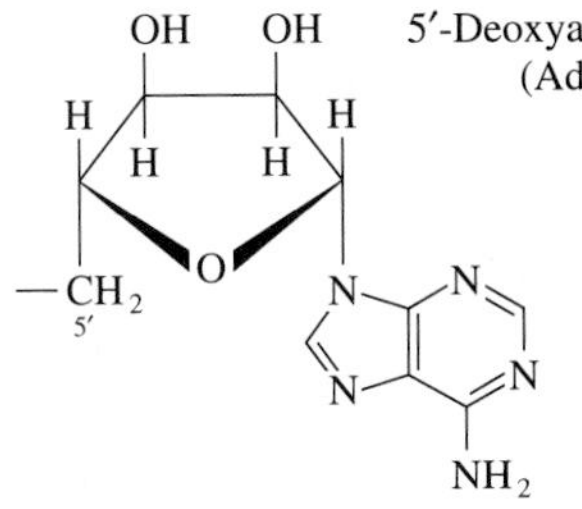	5′-Deoxyadenosylcobalamin (Adenosyl B_{12})

(a)

(b)

Methylmalonyl CoA → (Methylmalonyl-CoA mutase / Adenosylcobalamin) → Succinyl CoA

Figure 8·45
Intramolecular rearrangement catalyzed by adenosylcobalamin-dependent enzymes. **(a)** Adenosylcobalamin participates in intramolecular rearrangements in which a hydrogen atom and a substituent on an adjacent carbon atom exchange places. **(b)** Rearrangement of methylmalonyl CoA to succinyl CoA is catalyzed by the vitamin B_{12}–dependent methylmalonyl-CoA mutase.

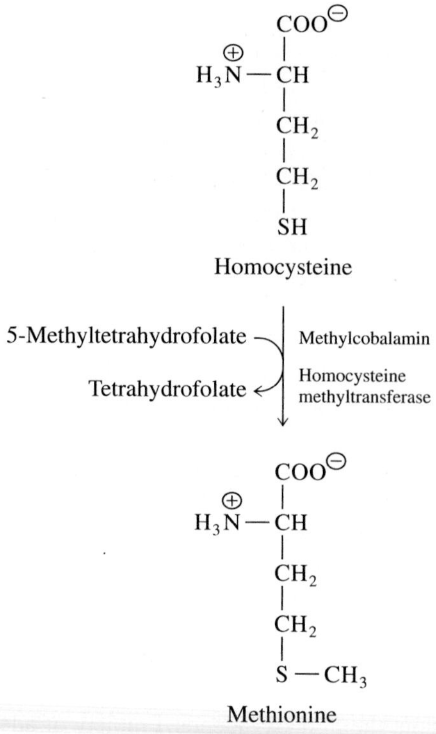

$$\begin{array}{c} COO^{\ominus} \\ | \\ \overset{\oplus}{H_3N} - CH \\ | \\ CH_2 \\ | \\ CH_2 \\ | \\ SH \end{array}$$

Homocysteine

5-Methyltetrahydrofolate ⟍ Methylcobalamin

Tetrahydrofolate ⟍ Homocysteine
methyltransferase

$$\begin{array}{c} COO^{\ominus} \\ | \\ \overset{\oplus}{H_3N} - CH \\ | \\ CH_2 \\ | \\ CH_2 \\ | \\ S - CH_3 \end{array}$$

Methionine

Figure 8·46
Biosynthesis of methionine
from homocysteine.

in the regeneration of methionine from homocysteine in mammals (Figure 8·46). In this reaction, the methyl group of 5-methyltetrahydrofolate is passed to a reactive, reduced form of vitamin B_{12} to form methylcobalamin, which can transfer the methyl group to the thiol side chain of homocysteine.

The mechanism proposed for adenosylcobalamin-mediated intramolecular rearrangements (Figure 8·47) involves some rather unusual chemistry. This mechanism is supported by evidence for the existence of free-radical intermediates and transfer of hydrogen from the substrate to an intermediate formed from the coenzyme, deoxyadenosine. The first step in the rearrangement reactions, the dissociation of the enzyme-bound coenzyme into two free radicals—the $Co^{2\oplus}$ form of cobalamin (vitamin B_{12r}) and the radical of 5′-deoxyadenosine, occurs when the substrate binds to the enzyme. Next, a hydrogen atom is abstracted from the substrate by the 5′-deoxyadenosyl radical. The substrate radical then undergoes rearrangement to the product radical, which accepts a hydrogen atom from deoxyadenosine to re-form the 5′-deoxyadenosyl radical. Finally, the product dissociates from the enzyme, and the 5′-deoxyadenosyl free radical recombines with enzyme-bound vitamin B_{12r}. The use of this free-radical mechanism permits the chemically difficult alkyl-hydrogen exchange reactions to occur.

In humans, vitamin B_{12} deficiency affects the production of both red and white blood cells. The production of blood cells can be partially or temporarily restored by ingestion of large doses of folate. However, in addition to the hematologic problems of pernicious anemia, prolonged vitamin B_{12} deficiency (even when folate supplements are taken) leads to neurological disorders, possibly related to incorporation of methylmalonyl CoA in place of malonyl CoA into long-chain fatty acids of nerve-cell membranes.

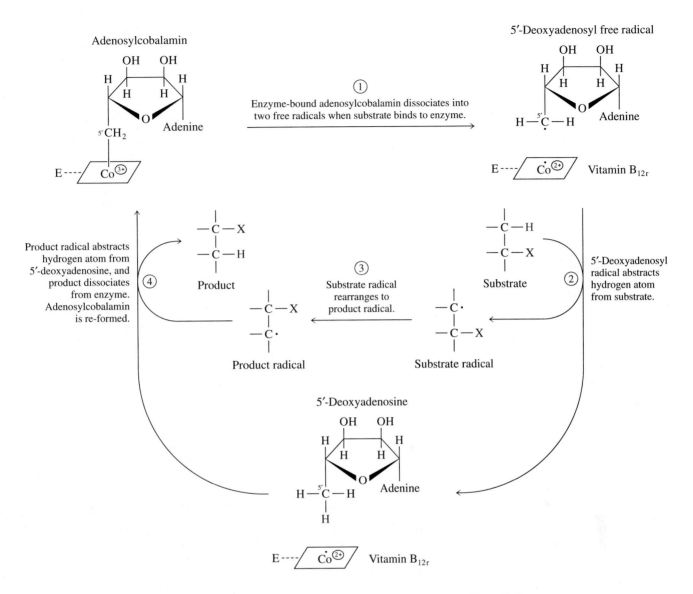

Figure 8·47
Proposed mechanism for adenosylcobalamin-mediated intramolecular rearrangements.

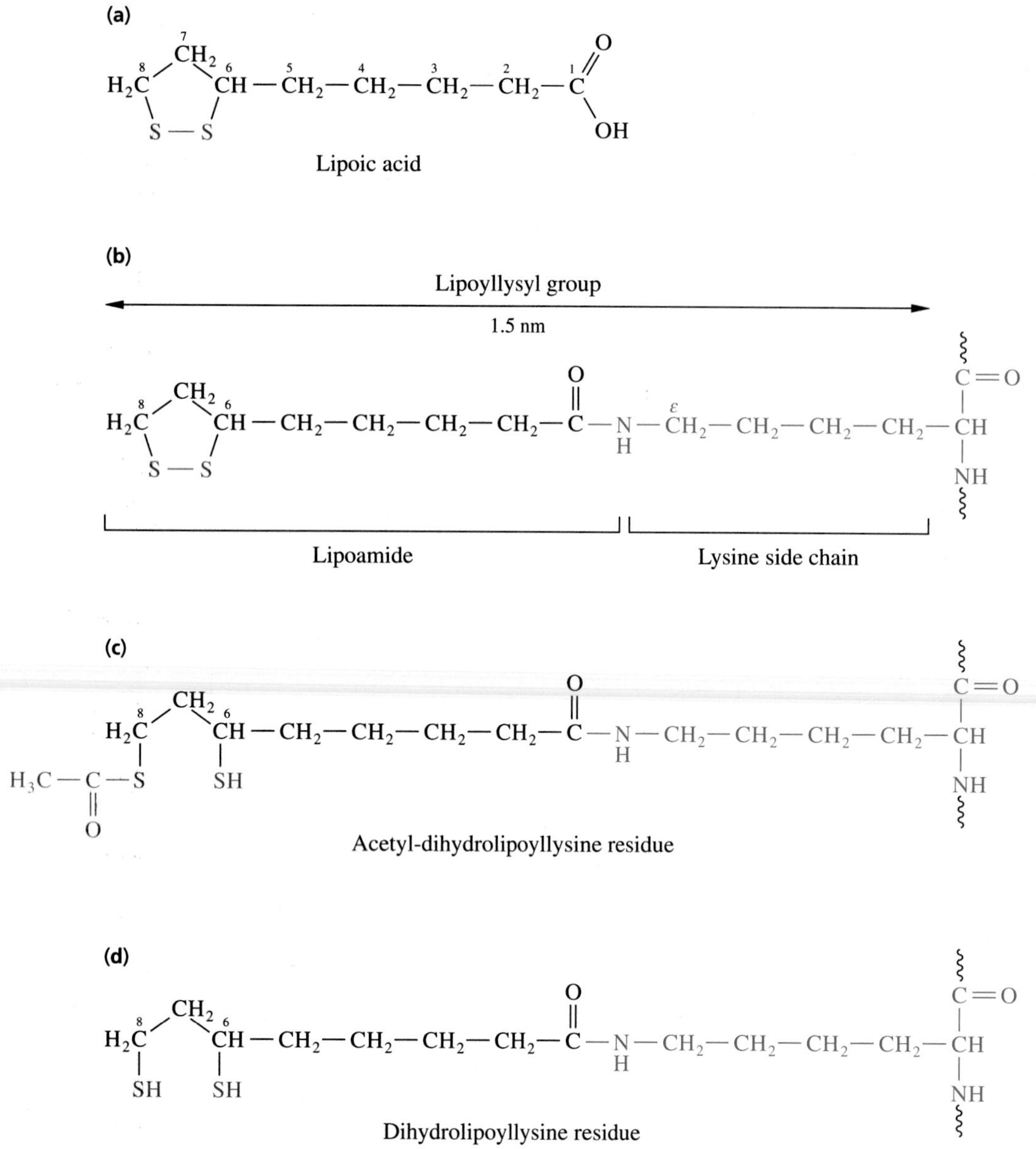

(a)

Lipoic acid

(b)

Lipoyllysyl group

1.5 nm

Lipoamide

Lysine side chain

(c)

Acetyl-dihydrolipoyllysine residue

(d)

Dihydrolipoyllysine residue

Figure 8·48
Lipoic acid. **(a)** Structure of lipoic acid. **(b)** Lipoyllysyl group of dihydrolipoamide acyltransferases. Lipoic acid is bound in amide linkage to the ε-amino group of a lysine residue (blue) of dihydrolipoamide acyltransferases, forming a lipoyllysyl group. The dithiolane ring of the flexible lipoamide is extended 1.5 nm from the polypeptide backbone. **(c)** Structure of an acetyl-dihydrolipoyllysine residue, a thioester. (The acetyl group can be rapidly transferred to the sulfur atom attached to C-6 by non-enzymatic isomerization.) **(d)** Structure of a dihydrolipoyllysine residue.

8·14 Lipoic Acid Is a Cofactor for Some Acyl-Transfer Reactions

Lipoic acid was first isolated in 1951 by Lester Reed and his coworkers, who obtained 30 mg of the coenzyme from 10 *tons* of beef liver. Lipoic acid is often listed with the B vitamins. However, although it is required for growth by a few microorganisms, animals appear to be able to synthesize it from available precursors. Lipoic acid is an eight-carbon carboxylic acid (octanoic acid) in which two hydrogen atoms, on C-6 and C-8, have been replaced by a disulfide linkage, as shown in Figure 8·48a. In Figure 8·48b, the carboxylate group of lipoic acid is shown covalently bound to the ε-amino group of a lysine residue via an amide linkage. This

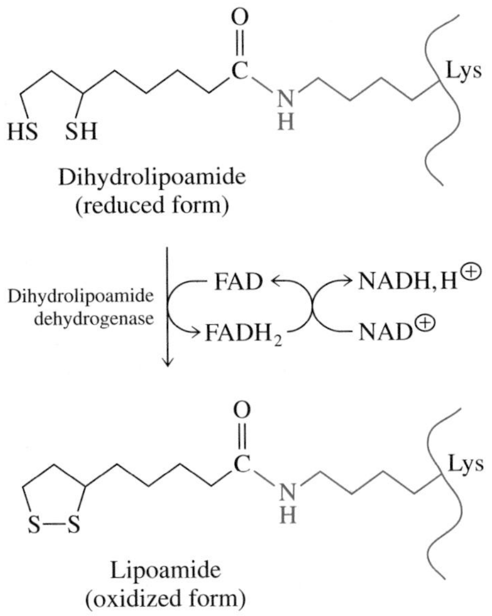

Figure 8·49
Dihydrolipoamide dehydrogenase reaction. The oxidation of the dihydrolipoamide of dihydrolipoamide acyltransferases is accompanied by reduction of an enzyme-bound FAD molecule to FADH$_2$. FADH$_2$ is reoxidized by reaction with NAD$^\oplus$.

Dihydrolipoamide
(reduced form)

Dihydrolipoamide dehydrogenase

FAD ← NADH,H$^\oplus$

FADH$_2$ ← NAD$^\oplus$

Lipoamide
(oxidized form)

structure is found in dihydrolipoamide acyltransferases, protein components of the pyruvate dehydrogenase complex and the α-ketoglutarate dehydrogenase complex, which are multienzyme complexes associated with the citric acid cycle. Both complexes function as α-keto acid dehydrogenases.

Lipoic acid transiently carries acyl groups, such as an acetyl group formed from pyruvate or a succinyl group from α-ketoglutarate. The lipoic acid moiety is believed to function as a swinging arm that carries substrates between active sites in the multienzyme complexes. For example, the disulfide ring of the lipoamide prosthetic group of the pyruvate dehydrogenase complex reacts with hydroxyethyl TPP (HETPP, Section 8·9), binding the acetyl group of HETPP to the sulfur atom attached to C-8 of lipoamide and forming a thioester (Figure 8·48c). The acyl group is then transferred to the sulfur atom of a coenzyme A molecule, generating the reduced (dihydrolipoamide) form of the prosthetic group (Figure 8·48d). Another component of α-keto acid dehydrogenase complexes is the flavin-containing enzyme dihydrolipoamide dehydrogenase, which catalyzes the FAD- and NAD$^\oplus$-dependent reoxidation of the reduced lipoamide prosthetic group (Figure 8·49). The interplay of five coenzymes in the reactions catalyzed by the α-keto acid dehydrogenase complexes, further described in Chapter 16, exemplifies the role of coenzymes as suppliers of reactive groups that augment the catalytic versatility of proteins.

8·15 Some Vitamins Are Lipid Soluble

The structures of the four fat-soluble, or lipid, vitamins (A, D, E, and K) contain rings as well as long, aliphatic side chains. Each of these vitamins is highly hydrophobic, although each possesses at least one polar group. The functions of these vitamins are varied. Furthermore, because they are hydrophobic, the lipid vitamins have proven difficult to study, and less is known about their roles than about the B vitamins. Some of the lipid vitamins are not precursors of coenzymes and are used without chemical alteration.

Figure 8·50

Formation and function of retinal. β-Carotene is cleaved to form two molecules of vitamin A, which are then converted to *cis*-retinal. In rhodopsin, *cis*-retinal is bound to opsin. In the eye, excitation of *cis*-retinal by visible light generates a nerve impulse and converts *cis*-retinal to *trans*-retinal, which dissociates from opsin. *cis*-Retinal is re-formed by the action of retinal isomerase.

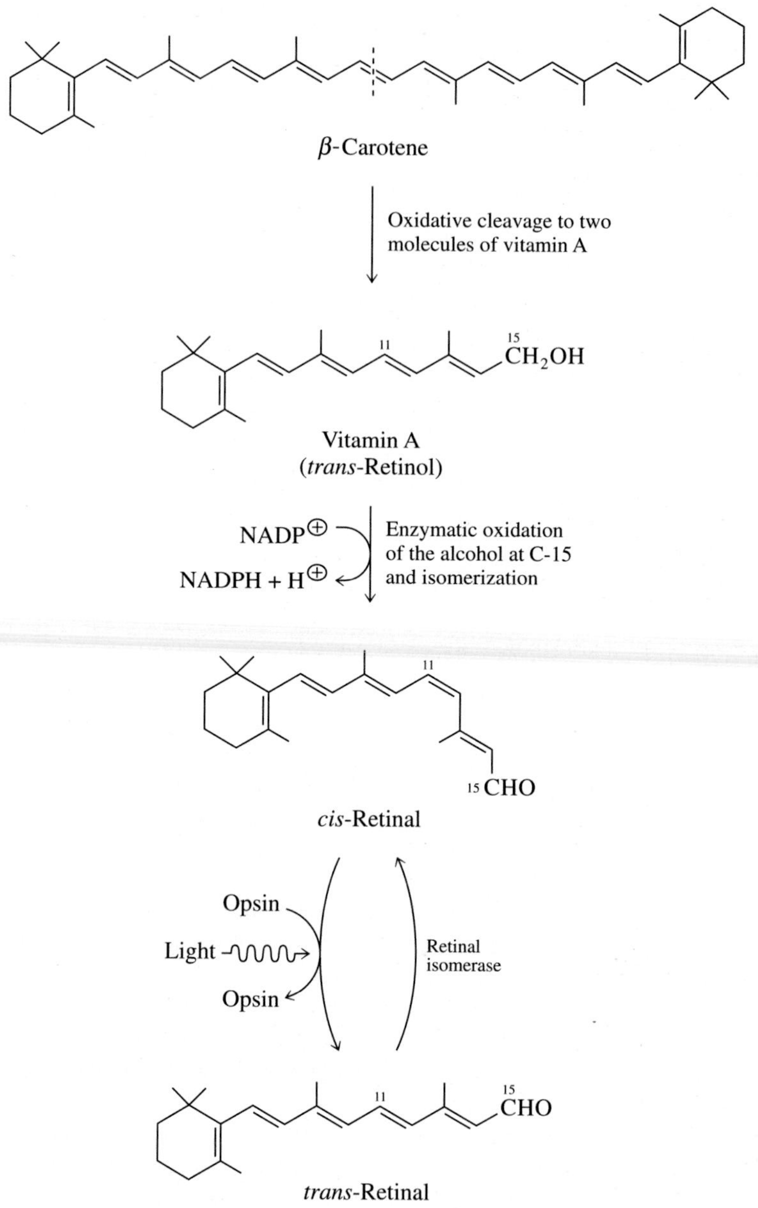

β-Carotene

Oxidative cleavage to two molecules of vitamin A

$^{15}CH_2OH$

Vitamin A
(*trans*-Retinol)

$NADP^{\oplus}$

$NADPH + H^{\oplus}$

Enzymatic oxidation of the alcohol at C-15 and isomerization

cis-Retinal

$_{15}CHO$

Opsin

Light

Opsin

Retinal isomerase

trans-Retinal

$_{15}CHO$

Figure 8·51

Carrots. These vegetables are rich in β-carotene, which is converted to vitamin A.

A. Vitamin A

Vitamin A, or retinol, is a 15-carbon lipid molecule obtained in the diet either by enzymatic oxidative cleavage of the 30-carbon plant lipid β-carotene (Figure 8·50) or directly from liver, egg yolk, or milk products. Carrots and other yellow vegetables are rich in β-carotene (Figure 8·51). Actually, there are three forms of vitamin A, based on the oxidation state of the terminal functional group: the stable alcohol retinol, the aldehyde retinal, and retinoic acid. All three compounds have important biological functions. Retinol and retinoic acid bind to receptor proteins inside cells; the ligand-receptor complexes then bind to chromosomes and can regulate gene expression during cell differentiation.

The aldehyde retinal—the light-sensitive compound required for vision—is formed by oxidation of the alcohol group of retinol or reduction of retinoic acid. Retinal is the prosthetic group of rhodopsin, a membrane-bound protein composed

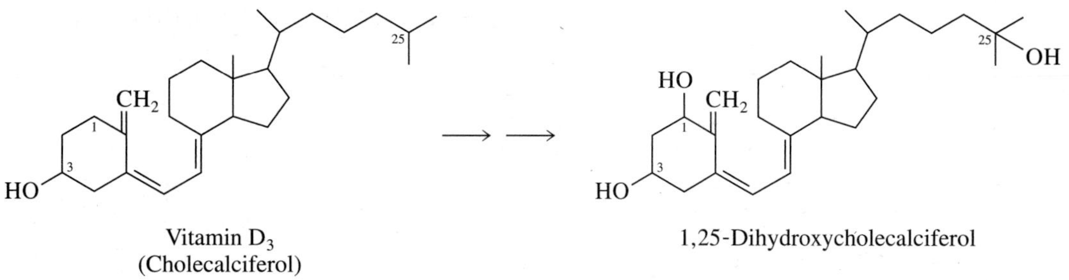

Vitamin D$_3$
(Cholecalciferol)

1,25-Dihydroxycholecalciferol

Figure 8·52
Vitamin D$_3$ (cholecalciferol) and
1,25-dihydroxycholecalciferol.

of the polypeptide opsin and the lipid retinal. The *cis* form of retinal (in which the double bond between C-11 and C-12 has isomerized) binds to opsin. When *cis*-retinal absorbs light, it isomerizes to *trans*-retinal, initiating a conformational change in rhodopsin that triggers a nerve impulse to the brain. To complete the cycle of reactions, the enzyme retinal isomerase catalyzes the re-formation of *cis*-retinal.

B. Vitamin D

Vitamin D is a group of related lipids that are necessary for the proper formation of bone. In vitamin D–deficiency diseases such as rickets (in children) and osteomalacia (in adults), bones are weak because calcium phosphate does not properly crystallize on the collagen matrix of the bones. When a mammal is exposed to sufficient sunlight, vitamin D$_3$ (also called cholecalciferol) is formed nonenzymatically in the skin from the steroid 7-dehydrocholesterol. Vitamin D$_2$, a compound related to vitamin D$_3$, can be obtained in the diet, for example, from fortified milk. The active form of vitamin D, 1,25-dihydroxycholecalciferol, is formed from vitamin D$_3$ by two hydroxylation reactions (Figure 8·52). The active compound is one of several substances that regulate Ca^{2+} utilization in humans.

C. Vitamin E

Vitamin E, or α-tocopherol (Figure 8·53), is one example from a group of closely related tocopherols, compounds having a bicyclic oxygen-containing ring system with a hydrophobic side chain. Vitamin E is believed to function as a scavenger of oxygen and free radicals; this antioxidant action may prevent damage to fatty acids in biological membranes. A deficiency of vitamin E is rare but may lead to fragile red blood cells in humans and sterility in rats.

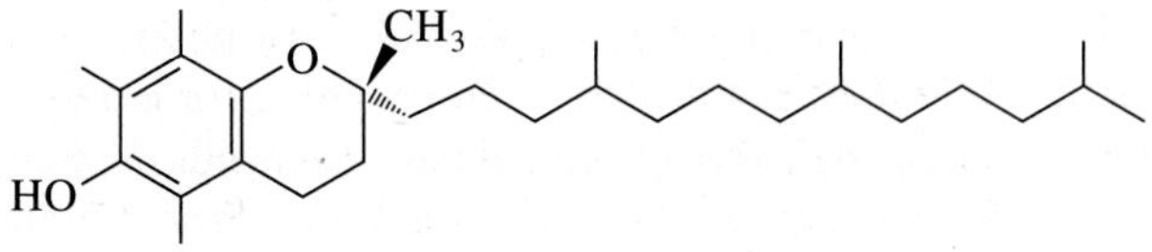

Figure 8·53
Vitamin E (α-tocopherol).

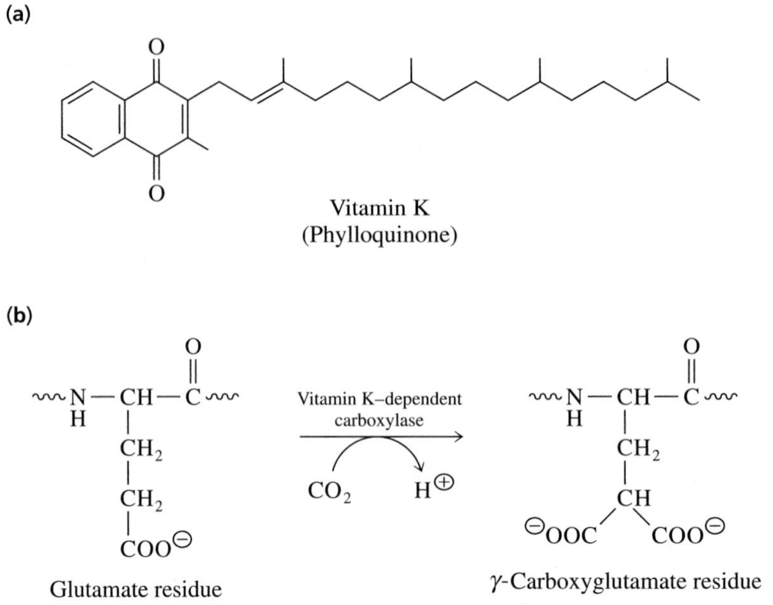

Figure 8·54
(a) Structure of vitamin K. (b) Vitamin K–dependent carboxylase reaction.

(a)

Vitamin K
(Phylloquinone)

(b)

Glutamate residue

Vitamin K–dependent
carboxylase

CO_2 $H^\oplus$

γ-Carboxyglutamate residue

D. Vitamin K

Vitamin K (Figure 8·54a) is a lipid vitamin from plants that is required for the synthesis of some of the proteins involved in blood coagulation. Vitamin K is a cofactor for a carboxylase that catalyzes the conversion of specific glutamate residues to γ-carboxyglutamate residues (Figure 8·54b). The mechanistic role of vitamin K in this reaction is not known. The γ-carboxyglutamate moieties act as chelators of $Ca^{2\oplus}$. Calcium binding to the γ-carboxyglutamate residues of the coagulation proteins allows these proteins to adhere to platelet surfaces where many reactions of the coagulation cascade take place. Vitamin K analogs are sometimes administered to individuals who suffer from thrombotic (blood-clotting) disorders. The analogs cannot participate in the carboxylation of glutamate residues; thus, the coagulation proteins produced are less active in these individuals.

8·16 Ubiquinone Is a Lipid-Soluble Coenzyme

Ubiquinone—also referred to as coenzyme Q and therefore abbreviated Q—is a lipid-soluble coenzyme. It is synthesized by respiring organisms, including mammals, and some photosynthetic bacteria. Ubiquinone is a benzoquinone with four substituents, one of which is a long hydrophobic chain (Figure 8·55a). This tail, which possesses 6 to 10 isoprenoid units, allows ubiquinone to dissolve in lipid membranes. In the inner mitochondrial membrane, ubiquinone transports electrons between membrane-embedded enzyme complexes. An analog of ubiquinone, plastoquinone (Figure 8·55b), serves a similar function in photosynthetic electron transport in chloroplasts (Chapter 19).

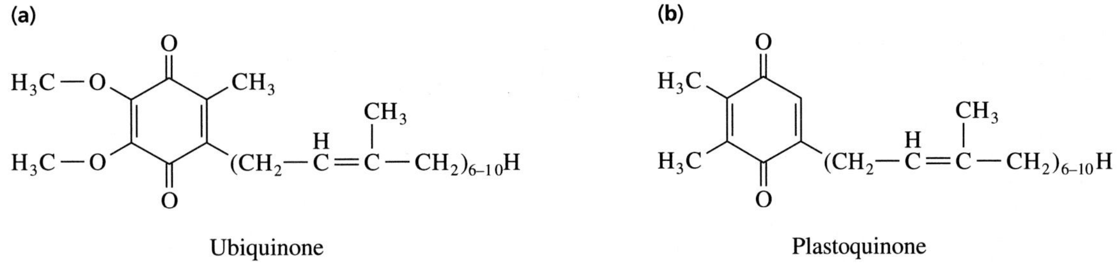

(a)

Ubiquinone

(b)

Plastoquinone

Figure 8·55
Structures of **(a)** ubiquinone and **(b)** plasto-quinone. The hydrophobic tail of each molecule is composed of five-carbon isoprenoid units.

Ubiquinone is an oxidation-reduction compound that is a stronger oxidizing agent than either $NAD^{\oplus}$ or the flavin coenzymes. Consequently, it may be reduced by NADH or $FADH_2$. Like FMN and FAD, ubiquinone can accept or donate either one or two electrons because it has three oxidation states: oxidized Q, a partially reduced semiquinone free radical, and fully reduced QH_2, called ubiquinol (Figure 8·56). In electron transport in mitochondria, the free radical appears to be the anion ($Q^{\cdot\ominus}$).

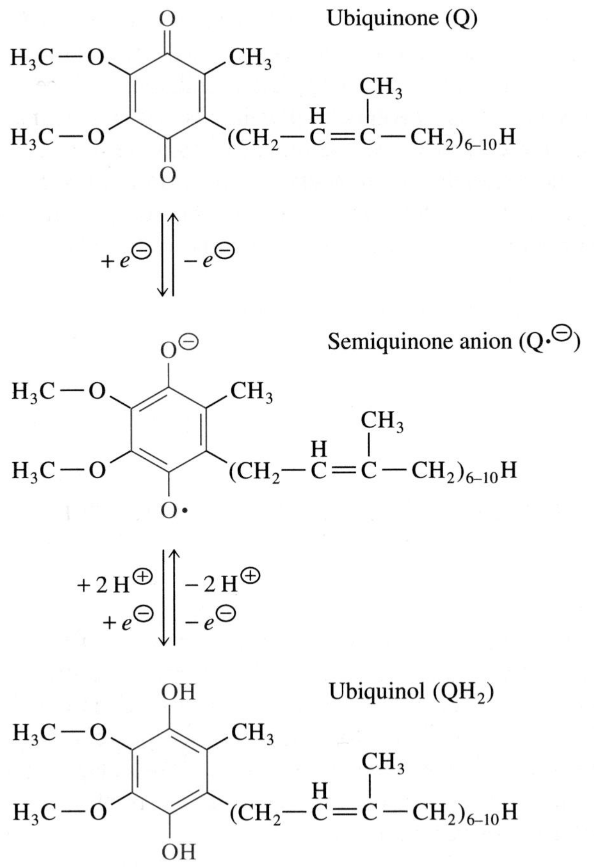

Ubiquinone (Q)

Semiquinone anion ($Q^{\cdot\ominus}$)

Ubiquinol (QH_2)

Figure 8·56
Three levels of oxidation of ubiquinone. Ubiquinone is reduced in two successive one-electron steps via a semiquinone free-radical intermediate.

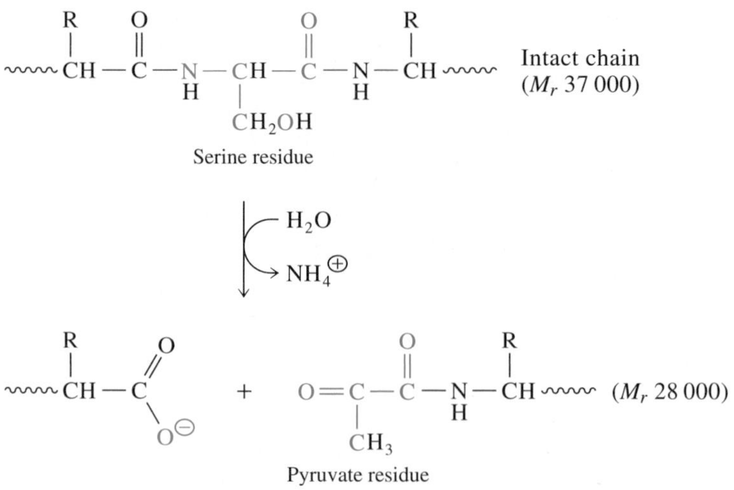

Figure 8·57
Modification of bacterial histidine decarboxylase. The original polypeptide is cleaved and a serine residue of the precursor polypeptide is modified to form a pyruvate residue. This alteration cuts the polypeptide chain into two fragments, with pyruvate at the N-terminus of the larger fragment. An oxygen from the serine residue is incorporated in the carboxylate group of the smaller fragment.

8·17 Some Amino Acid Residues Are Modified to Form Prosthetic Groups

Recently, new catalytic roles have been discovered for amino acid residues in certain enzymes. Posttranslational modification occurs at the active sites of these enzymes, and the modified amino acids then act as prosthetic groups.

The first known and best-characterized example of this phenomenon is the conversion of a serine residue to a pyruvate residue at the active site of bacterial histidine decarboxylase. Cleavage of a polypeptide chain having a molecular weight of 37 000 results in two chains (M_r 9000 and M_r 28 000), the larger of which contains a pyruvate residue at the N-terminus (Figure 8·57). The carbonyl group of the newly formed pyruvate residue reacts with the amino group of the substrate histidine to form a Schiff base analogous to the one formed by pyridoxal phosphate in other enzymes. In fact, it was expected that this enzyme would contain PLP, but analysis revealed instead the amino acid derivative pyruvamide. Pyruvate residues have since been discovered in the active sites of a number of other enzymes.

Three other amino acid–derived prosthetic groups have been discovered (Figure 8·58). They are the trihydroxyphenylalanine residue and its quinone form in plasma amine oxidase—presumably formed by modification of a tyrosine residue; a prosthetic group formed from a residue each of cysteine and tyrosine, present in

Figure 8·58
Three different prosthetic groups derived by modification of amino acid residues. **(a)** The quinone form of a trihydroxyphenylalanine residue. **(b)** A prosthetic group formed by condensation of a cysteine residue and a tyrosine residue. **(c)** A quinone prosthetic group formed from two residues of tryptophan.

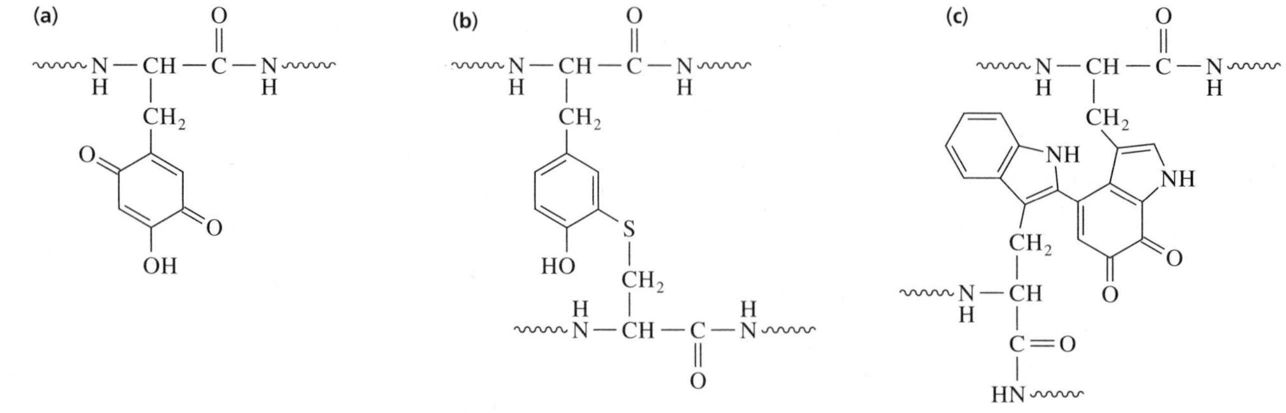

galactose oxidase of the fungus *Dactylium dendroides;* and a quinone cofactor formed from two tryptophan residues in a methylamine dehydrogenase from soil bacteria. The identification of these active-site components adds a new dimension to the diversity of protein structure and function.

8·18 Group-Transfer Proteins May Be Considered Coenzymes

Several proteins serve the same role as coenzymes. They do not themselves catalyze reactions but are required for the action of certain enzymes. These proteins are called **group-transfer proteins,** or protein coenzymes. They often have relatively low molecular weights and are more heat stable than typical enzymes.

Cytochrome *c* (Figure 5·43) and ferredoxin are examples of protein coenzymes involved in oxidation-reduction reactions. There are also several proteins that contain two reactive-center thiol side chains that cycle between their dithiol and disulfide forms, passing reducing power from one substrate to another. Thioredoxins have cysteines three residues apart (–Cys–X–X–Cys–). Their thiol side chains undergo reversible oxidation to form the disulfide bond of a cystine unit. The disulfide reactive center of thioredoxin is on the surface of the protein, rather than in a cleft, so it is accessible to the active sites of appropriate enzymes (Figure 8·59).

Some group-transfer proteins actually contain firmly bound coenzymes or portions of coenzymes. A carboxyl carrier protein containing covalently bound biotin is a component of acetyl-CoA carboxylase in *E. coli,* the enzyme that catalyzes the first committed step of fatty acid synthesis. The acyl carrier protein, introduced in Section 8·8, is a group-transfer protein that contains a phosphopantetheine moiety as its reactive center. A protein cofactor necessary for the degradation of glycine contains a molecule of covalently bound lipoic acid.

Figure 8·59
Stereo view of oxidized thioredoxin. Note that the cystine group is on the surface of the protein. The sulfur atoms are shown in yellow. (Based on coordinates provided by S. K. Katti, D. M. LeMaster, and H. Eklund.)

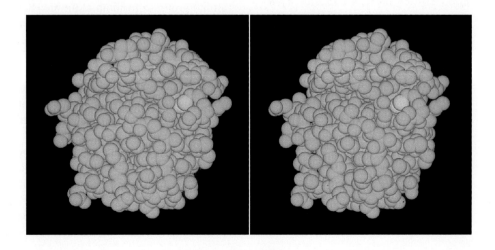

Summary

Many enzyme-catalyzed reactions depend on the presence of cofactors. Cofactors include essential ions and group-transfer reagents called coenzymes. Inorganic ions, such as $K^{\oplus}$, Mg^{2+}, Ca^{2+}, Fe^{2+}, and Zn^{2+}, participate in the binding of substrates, in stabilization of enzymes, and as active catalytic components. Some coenzymes are synthesized from common metabolites, and a number of others are derived from B vitamins. Vitamins are organic compounds that must be supplied in small amounts in the diets of humans and other animals to avoid nutritional-deficiency diseases. Coenzymes that carry mobile metabolic groups may function as cosubstrates of enzymes, or they may be firmly attached to enzymes as prosthetic groups.

ATP is the most common metabolite coenzyme. It can donate a phosphoryl, pyrophosphoryl, adenylyl, or adenosyl group. Other metabolite coenzymes include S-adenosylmethionine, phosphoadenosine phosphosulfate, and the sugar nucleotides. Cytidine nucleotides participate in the biosynthesis of lipids.

The pyridine nucleotide coenzymes, $NAD^{\oplus}$ and $NADP^{\oplus}$, are derived from nicotinic acid (niacin). Pyridine nucleotide–dependent dehydrogenases catalyze transfer of a hydride ion ($H^{\ominus}$) from a specific substrate to position 4 of the pyridine ring of $NAD^{\oplus}$ or $NADP^{\oplus}$, reducing the coenzyme to NADH or NADPH, respectively, with release of a proton. The pyridine nucleotides accept or donate two electrons at a time.

Riboflavin (vitamin B_2) consists of an isoalloxazine ring system and a ribitol residue. Characteristically, the coenzyme forms of riboflavin—FAD and FMN—are tightly bound as prosthetic groups in flavoproteins. FAD and FMN are reduced by hydride (two-electron) transfers to form $FADH_2$ and $FMNH_2$. The reduced flavin coenzymes donate electrons one at a time, passing through the stable free-radical forms FADH· and FMNH·.

Coenzyme A is derived from pantothenic acid. It is the principal coenzyme involved in acyl-group–transfer reactions. The energy of the thioester bond of acetyl CoA is about the same as the energy of the phosphoanhydrides of ATP. The prosthetic group of acyl carrier protein (ACP), 4′-phosphopantetheine, is also derived from pantothenic acid.

The first vitamin isolated, thiamine (vitamin B_1), was found to prevent or cure beriberi in humans subsisting largely on polished rice. Its coenzyme form is thiamine pyrophosphate (TPP), whose thiazole ring binds the aldehyde generated upon decarboxylation of an α-keto acid substrate. A number of α-keto acid dehydrogenases require TPP as a coenzyme or prosthetic group.

Pyridoxal 5′-phosphate (PLP) serves as a prosthetic group for many enzymes involved in amino acid metabolism. It is derived from vitamin B_6, pyridoxine. The aldehyde group at C-4 of the pyridine ring of PLP reversibly reacts with a free amino group to form an aldimine (Schiff base). In many cases, the coenzyme is linked to a lysine residue at the active site of a PLP-dependent enzyme. PLP participates in the catalytic mechanism of such enzymes by forming a Schiff base with an amino acid substrate, through which it stabilizes the carbanion that is generated upon cleavage of a bond to the α-carbon atom of the amino acid.

Biotin serves as a prosthetic group for several ATP-dependent carboxylases and carboxyltransferases. It is covalently linked by an amide bond to the ε-amino group of a lysine residue at the enzyme active site; the biotinyl lysine residue is sometimes referred to as biocytin.

Tetrahydrofolate, the reduced form of folic acid, is involved in the transfer of one-carbon units at the oxidation levels of methanol, formaldehyde, and formic acid. Such transfers are especially important in protein synthesis and in the biosynthesis of purine nucleotides and dTMP. The tetrahydropterin moiety of tetrahydrofolate is also found in tetrahydrobiopterin, a coenzyme involved in O_2-dependent hydroxylation reactions, and as a component of a molybdenum-containing cofactor.

Vitamin B_{12} and its coenzyme forms contain a corrin ring system. The adenosyl-coenzyme derivative of vitamin B_{12}, adenosylcobalamin, is involved in some intramolecular rearrangement reactions. The methyl form, methylcobalamin, is an intermediate in the biosynthesis of methionine from 5-methyltetrahydrofolate and homocysteine.

Lipoic acid, a prosthetic group for α-keto acid dehydrogenase multienzyme complexes, can be synthesized by most organisms. It accepts acyl groups, forming a thioester that passes the acyl group to a second acceptor such as coenzyme A. In this process, dihydrolipoic acid is formed.

The four fat-soluble, or lipid, vitamins are A, D, E, and K. Vitamin A functions as a light-sensitive compound in vision. A derivative of vitamin D regulates Ca^{2+} utilization. Vitamin E helps prevent oxidative damage to membrane lipids. Vitamin K is essential for the conversion of glutamate to γ-carboxyglutamate residues in certain blood-coagulation proteins.

Ubiquinone (coenzyme Q) is a lipid-soluble electron carrier present in the inner membrane of mitochondria. It transfers electrons between enzyme complexes that catalyze the oxidation of NADH and $FADH_2$. Ubiquinone can accept or donate either one or two electrons. Plastoquinone in chloroplasts is similar in structure and function to ubiquinone.

Certain enzymes undergo modifications in which amino acid residues are converted into active-site prosthetic groups. The best-known example is the pyruvamide group of bacterial histidine decarboxylase, which arises from modification of a serine residue. This pyruvate residue plays a mechanistic role similar to that of the aldehyde group of PLP.

Some proteins are not catalysts themselves but function as group-transfer agents, very much like coenzymes. These proteins usually have relatively low molecular weights and are heat stable. Examples are cytochrome *c,* ferredoxin, and acyl carrier protein (ACP).

Selected Readings

General References

Berg, J. M. (1987). Metal ions in proteins: structural and functional roles. *Cold Spring Harbor Symp. Quant. Biol.* 52:579–585. Classifies metalloproteins into four types, based upon the static or dynamic oxidation-reduction and coordination properties of the metal ions.

Scrimgeour, K. G. (1977). *Chemistry and Control of Enzyme Reactions* (London: Academic Press). Chapters 8, 9, and 11.

Wagner, A. F., and Folkers, K. (1964). *Vitamins and Coenzymes* (New York: Wiley-Interscience Publishers). An excellent treatment of the discovery and characterization of the vitamins.

Walsh, C. (1979). *Enzymatic Reaction Mechanisms* (San Francisco: W. H. Freeman and Company).

Specific Cofactors

Blakley, R. L., and Benkovic, S. J., eds. (1985). *Folates and Pterins,* Vol. 1 and Vol. 2. (New York: John Wiley & Sons). The most recent comprehensive coverage of pterin coenzymes.

Coleman, J. E. (1992). Zinc proteins: enzymes, storage proteins, transcription factors, and replication proteins. *Annu. Rev. Biochem.* 61:897–946. Describes the many enzymes and transcription factors that contain zinc.

DiMarco, A. A., Bobik, T. A., and Wolfe, R. S. (1990). Unusual coenzymes of methanogenesis. *Annu. Rev. Biochem.* 59:355–394. Describes the six novel coenzymes required for synthesis of methane by particular bacteria.

Dolphin, D., Poulson, R., and Avramović, O., eds. (1987). *Pyridine Nucleotide Coenzymes.* (New York: John Wiley & Sons). Part A contains chapters on history, nomenclature, chemical and physical properties, and spectroscopic techniques (15 chapters). Part B emphasizes biochemical, nutritional, and medical aspects (17 chapters).

Ghisla, S., and Massey, V. (1989). Mechanisms of flavoprotein-catalyzed reactions. *Eur. J. Biochem.* 181:1–17. Presentation of the best-studied mechanisms of these enzymes.

Golding, B. T., and Rao, D. N. R. (1987). Adenosylcobalamin-dependent enzymic reactions. In *Enzyme Mechanisms,* M. I. Page and A. Williams, eds. (London: Royal Society of Chemistry), pp. 404–428.

Hayashi, H., Wada, H., Yoshimura, T., Esaki, N., and Soda, K. (1990). Recent topics in pyridoxal 5′-phosphate enzyme studies. *Annu. Rev. Biochem.* 59:87–110.

Knowles, J. R. (1989). The mechanism of biotin-dependent enzymes. *Annu. Rev. Biochem.* 58:195–221.

McIntire, W. S., Wemmer, D. E., Chistoserdov, A., and Lidstrom, M. E. (1991). A new cofactor in a prokaryotic enzyme: tryptophan tryptophylquinone as the redox prosthetic group in methylamine dehydrogenase. *Science* 252:817–824. This article describes a new prosthetic group derived from two residues of tryptophan and also reviews other amino acid–derived prosthetic groups.

Popják, G. (1970). Stereospecificity of enzymic reactions. In *The Enzymes,* Vol. 2, 3rd ed., P. D. Boyer, ed. (New York: Academic Press), pp. 115–215. Description of the stereochemistry of reduction reactions involving NADH and NADPH.

Rajagopalan, K. V., and Johnson, J. L. (1992). The pterin molybdenum cofactors. *J. Biol. Chem.* 267:10 199–10 202.

Rhodes, D., and Klug, A. (1993). Zinc fingers. *Sci. Am.* 268(2):56–65. Review of proteins that possess zinc fingers to specifically bind DNA.

Schweitzer, B. I., Dicker, A. P., and Bertino, J. R. (1990). Dihydrofolate reductase as a therapeutic target. *FASEB J.* 4:2441–2452. Describes folate antagonists as therapeutic compounds.

Stover, P., and Schirch, V. (1993). The metabolic role of leucovorin. *Trends Biochem. Sci.* 18:102–106. Report of the route for the biosynthesis of 5-formyltetrahydrofolate.

Strynadka, N. C. J., and James, M. N. G. (1989). Crystal structures of the helix-loop-helix calcium-binding proteins. *Annu. Rev. Biochem.* 58:951–998. Explains the two types of calcium-binding proteins: extracellular enzymes that are stabilized by Ca^{2+} and intracellular proteins whose activities are modulated by reversible binding of Ca^{2+}.

Suttie, J. W. (1985). Vitamin-K dependent carboxylase. *Annu. Rev. Biochem.* 54:459–477. Reviews the posttranslational conversion of glutamate residues to γ-carboxylglutamate residues.

van Poelje, P. D., and Snell, E. E. (1990). Pyruvoyl-dependent enzymes. *Annu. Rev. Biochem.* 59:29–59. Describes pyruvate as a prosthetic group for some amino acid decarboxylases and reductases.

Walsh, C. T., and Orme-Johnson, W. H. (1987). Nickel enzymes. *Biochemistry* 26:4901–4906. Describes the four known nickel-containing enzymes.

9

Carbohydrates and Glycoconjugates

Carbohydrates are widely distributed in nature and represent—on the basis of mass—the most abundant class of bioorganic molecules on earth. Most of this carbohydrate material accumulates as a result of photosynthesis, the process by which certain organisms convert solar energy into chemical energy and incorporate atmospheric carbon dioxide into carbohydrates (Chapter 19). Carbohydrates include monomeric sugars and their polymers. Carbohydrate derivatives include both simple metabolites and molecules of protein and lipid that contain covalently attached sugars.

Carbohydrates play several crucial roles in living organisms. As partially reduced molecules, carbohydrates can be oxidized to yield energy to drive metabolic processes; thus, they can act as energy-storage molecules. Polymeric carbohydrates have a variety of functions. For example, they are found in plant cell walls and the protective coatings of many organisms (Figure 9·1, next page). Carbohydrates attached to cell membranes play a role in cellular recognition and in cell-to-cell communication. Derivatives of sugars are found in a number of biological molecules, including antibiotics, coenzymes (Chapter 8), and the nucleic acids DNA and RNA (Chapter 24). The study of the structure and function, in health and disease, of the various types of carbohydrates is called glycobiology, an area of research that is gaining considerable momentum.

Figure 9·1
Redwood. Plant cell walls contain a high percentage of cellulose, a polymeric carbohydrate that accounts for more than 50% of the organic matter in the biosphere.

Carbohydrates, also known as saccharides, can be classified according to the number of monomeric units they contain. **Monosaccharides** are the monomeric units of carbohydrate structure. All monosaccharides share the empirical formula $(CH_2O)_n$, where n is three or greater (n is usually five or six but can be up to nine). The term *carbohydrate* is derived from $(C \cdot H_2O)_n$, or "hydrate of carbon." **Oligosaccharides** are polymers of 2 to about 20 monosaccharide residues. The most common oligosaccharides are the disaccharides, which consist of two linked monosaccharide residues. **Polysaccharides** are polymers that contain many (usually more than 20) monosaccharide residues. Oligosaccharides and polysaccharides do not have the empirical formula $(CH_2O)_n$ because water is eliminated during polymer formation. The term *glycan* is a more general term for carbohydrate polymers of any length. It can refer to a polymer of identical sugars (homoglycan) or of different sugars (heteroglycan). The term *simple carbohydrate,* or *simple sugar,* is used to refer to monosaccharides and disaccharides. The term *complex carbohydrate* is used to refer to polysaccharides and glycoconjugates.

Glycoconjugates are carbohydrate derivatives in which one or more carbohydrate chains are linked covalently to a peptide chain, protein, or lipid. Glycoconjugates include peptidoglycans, proteoglycans, glycoproteins, and glycolipids. The first three of these compounds are discussed later in this chapter, and glycolipids are discussed in Chapter 11.

This chapter is divided into two parts. Part 1 considers the monosaccharides, disaccharides, and the major homoglycans—starch, glycogen, cellulose, and chitin. Part 2 focuses on three types of glycoconjugates—peptidoglycans, glycoproteins, and proteoglycans—all of which contain heteroglycan chains.

Part 1: Monosaccharides, Disaccharides, and Homoglycans

In this first part of the chapter, we describe the properties of monosaccharides, disaccharides, and homoglycans, including aspects of their structure, nomenclature, and function. We begin with a discussion of the monosaccharides, which are important in their own right and also as the building blocks of disaccharides and glycans. We then discuss the disaccharides, including maltose, cellobiose, lactose, and sucrose, all of which are of major biological or commercial importance. We conclude Part 1 with a discussion of important homoglycans such as starch, glycogen, cellulose, and chitin, which are among the most abundant macromolecules on earth.

9·1 Most Monosaccharides Are Chiral Compounds

Pure monosaccharides are water-soluble, white, crystalline solids that have a sweet taste. Examples include glucose and fructose. Disaccharides such as sucrose and lactose have similar properties. As a rule, the suffix *-ose* is used in naming carbohydrates, although there are a number of exceptions.

Monosaccharides are polyhydroxy aldehydes or ketones. They are classified by the type of carbonyl group and by the number of carbon atoms they contain. All monosaccharides contain at least three carbon atoms. One of these is the carbonyl carbon, and each of the remaining carbon atoms bears a hydroxyl group. The two general classes of monosaccharides are the polyhydroxy aldehydes, or **aldoses,** and the polyhydroxy ketones, or **ketoses.** In aldoses, the most oxidized carbon atom is designated C-1. It is drawn at the top of a Fischer projection of the aldose. In ketoses, the most oxidized carbon atom is usually C-2.

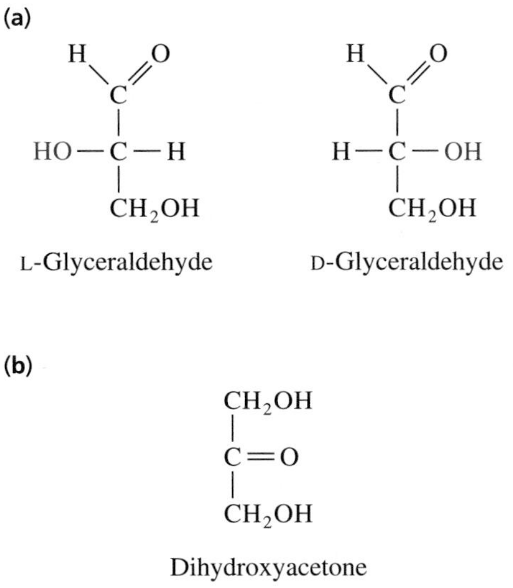

Figure 9·2
Fischer projections of
(a) glyceraldehyde and
(b) dihydroxyacetone. The
designations L (for left)
and D (for right) for glyceraldehyde refer to the configuration of the hydroxyl
group of the chiral carbon
(C-2). Dihydroxyacetone is
achiral.

(a)

L-Glyceraldehyde   D-Glyceraldehyde

(b)

$$CH_2OH$$
$$|$$
$$C=O$$
$$|$$
$$CH_2OH$$

Dihydroxyacetone

The smallest monosaccharides are **trioses,** or three-carbon sugars. One- or two-carbon compounds having the general formula $(CH_2O)_n$ do not have properties typical of carbohydrates. The aldehydic triose, or aldotriose, is glyceraldehyde (Figure 9·2a). Glyceraldehyde is chiral—its central carbon, C-2, is asymmetric (re-call Figure 4·2). The ketonic triose, or ketotriose, is dihydroxyacetone (Figure 9·2b). Dihydroxyacetone is achiral—it has no asymmetric carbon atom. All other monosaccharides can be viewed as longer-chain versions of these two sugars, and as we will soon see, all other monosaccharides are chiral.

The stereoisomers D- and L-glyceraldehyde are shown as stick models in Fig-ure 9·3. Chiral molecules are optically active; that is, they rotate the plane of polar-ized light. The convention for designating D and L isomers was originally based on the optical properties of glyceraldehyde. The form of glyceraldehyde that caused rotation to the right (dextrorotatory) was designated D; the form of glyceraldehyde that caused rotation to the left (levorotatory) was designated L. Structural knowl-edge was limited when this convention was established in the late 19th century, and configurations for the enantiomers of glyceraldehyde were assigned arbitrarily, with a 50% probability of error. It was not until the mid-20th century that X-ray crystallographic experiments proved that the original structural assignments were indeed correct.

Figure 9·3
Stereo view of L-glycer-
aldehyde (left) and
D-glyceraldehyde (right).

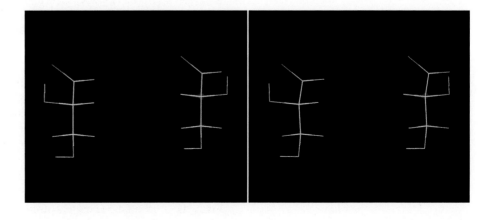

Figure 9·4
Structures of the three- to six-carbon
D-aldoses, shown as Fischer projections.
Those shown in blue are the aldoses of
greatest importance in our study of bio-
chemistry.

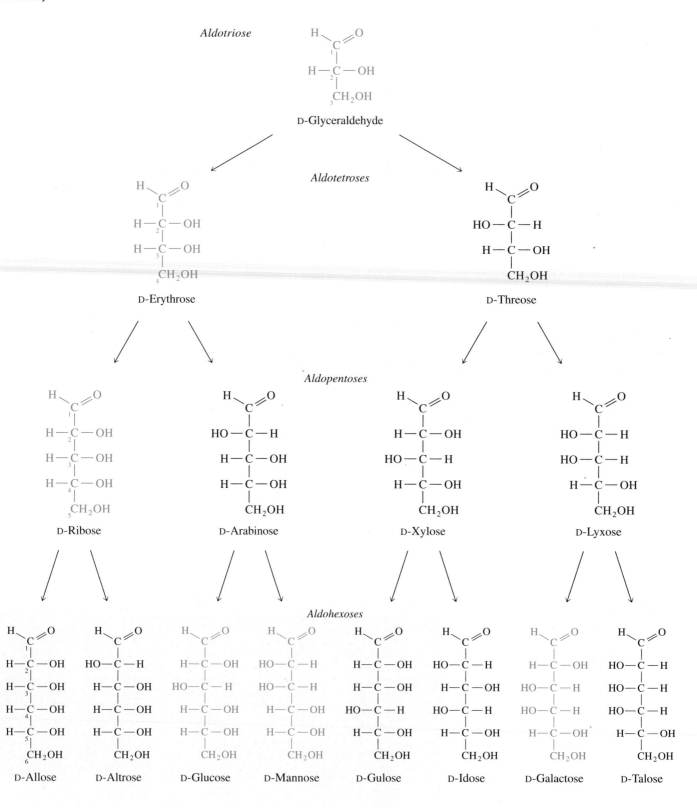

Longer aldoses and ketoses can be regarded as extensions of glyceraldehyde and dihydroxyacetone, respectively, with chiral H—C—OH groups inserted between the carbonyl carbon and the primary alcohol group. Figure 9·4 shows the structures of the tetroses (four-carbon aldoses), pentoses (five-carbon aldoses), and hexoses (six-carbon aldoses) related to D-glyceraldehyde. Note that the numbering of the carbon atoms proceeds from the aldehydic carbon, which is assigned the number 1. By convention, sugars are said to have the D configuration when the configuration of the chiral carbon with the highest number—the chiral carbon most distant from the carbonyl carbon—is the same as that of C-2 of D-glyceraldehyde (i.e., the —OH group attached to this carbon atom is on the right side in a Fischer projection). Except for glyceraldehyde (which was used as the standard), there is no predictable association between the absolute configuration of a sugar and whether it is dextrorotatory or levorotatory. The arrangement of asymmetric carbon atoms is unique for each monosaccharide, giving the sugar its distinctive properties.

Not included in Figure 9·4 are the L enantiomers of the 15 aldoses shown. Note that pairs of enantiomers are mirror images at every chiral carbon; in other words, the configuration at every chiral carbon is opposite. For example, the hydroxyl groups bound to carbon atoms 2, 3, 4, and 5 of D-glucose point right, left, right, and right, respectively, in the Fischer projection; those of L-glucose point left, right, left, and left (Figure 9·5). The D enantiomers of sugars predominate in nature.

Whereas there are two stereoisomers for the one aldotriose (D- and L-glyceraldehyde), which possesses a single chiral carbon atom, there are four stereoisomers for aldotetroses (D- and L-erythrose and D- and L-threose) because erythrose and threose each possess two chiral carbon atoms. In general terms, there are 2^n possible stereoisomers for a compound with n chiral carbons. Aldohexoses, which possess four chiral carbons, have a total of 2^4, or 16, stereoisomers (the eight D aldohexoses in Figure 9·4 and their L enantiomers).

Sugar molecules that differ in configuration at only one of several chiral centers are called **epimers.** For example, D-mannose and D-galactose are epimers of D-glucose (at C-2 and C-4, respectively), although they are not epimers of each other (see Figure 9·4). The three-dimensional structures of the skeletal models of D-glucose and D-mannose can be compared in Figure 9·6.

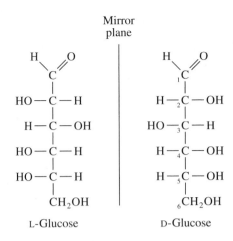

Figure 9·5
Fischer projections of L- and D-glucose.

Figure 9·6
Stereo view of D-glucose (left) and D-mannose (right). D-Glucose and D-mannose are epimers, that is, sugar molecules that differ in configuration at only one carbon (in this case, C-2).

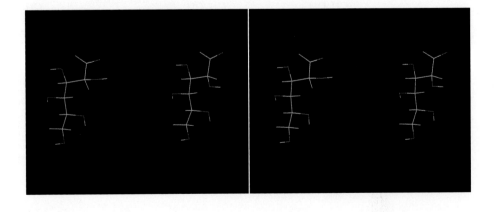

Figure 9·7
Structures of the three- to six-carbon
D-ketoses. Those shown in blue are the
ketoses of greatest importance in our
study of biochemistry.

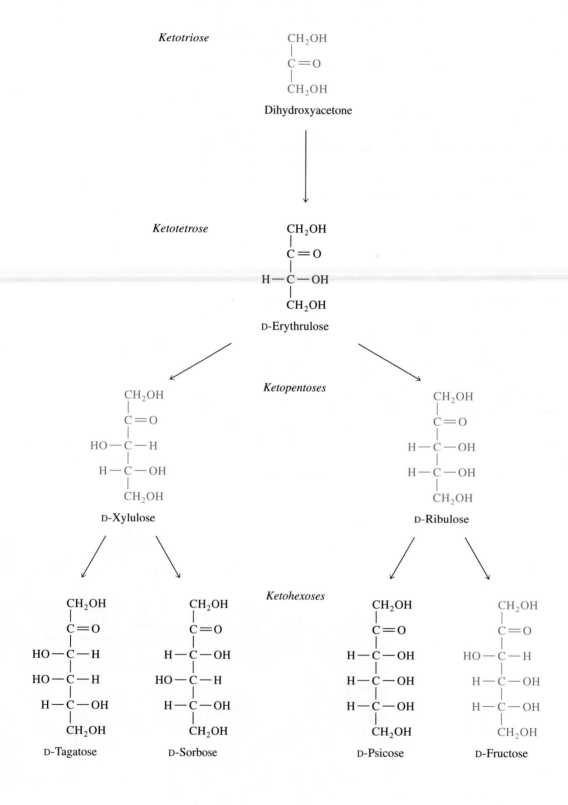

Longer-chain ketoses are related to dihydroxyacetone in the same way that longer-chain aldoses are related to glyceraldehyde (Figure 9·7). Note that a ketose has one fewer chiral carbon atom than the aldose of the same empirical formula. For example, there are only two stereoisomers for the one ketotetrose (D- and L-erythrulose) and four stereoisomers for ketopentoses (D- and L-xylulose and D- and L-ribulose). The ketotetrose and ketopentoses are named by inserting -*ul*- in the name of the corresponding aldose. For example, the ketose xylulose is analogous to the aldose xylose. However, this naming scheme does not apply to the ketohexoses (tagatose, sorbose, psicose, and fructose), which bear trivial names unrelated to the names of the corresponding aldohexoses.

Glucose occurs naturally as a monosaccharide. Other aldoses are usually found only as components of oligosaccharides, polysaccharides, or other sugar derivatives. Early experiments on the hydrolysis of sucrose (table sugar, obtained from plants such as beets and sugarcane) led to the discovery that its monosaccharide components, D-glucose and D-fructose, were dextrorotatory and levorotatory, respectively. As a result, glucose and fructose were initially called *dextrose* and *levulose*. Those names are still encountered occasionally.

9·2 Aldoses and Ketoses Can Form Cyclic Hemiacetals

Decades ago, it was discovered that aldopentoses and aldohexoses behave as though they have one more chiral carbon atom than is evident from the structures shown in Figure 9·4. It was discovered, for example, that D-glucose exists in two forms that contain *five* (not four) asymmetric carbons. The source of this additional asymmetry is an intramolecular cyclization reaction that results in a new chiral center. The two new stereoisomers are called **anomers.**

When an alcohol reacts with an aldehyde (to form a hemiacetal) or with a ketone (to form a hemiketal), a chiral sp^3-hybridized carbon atom is formed from the achiral sp^2-hybridized carbon atom of the carbonyl group. Figure 9·8 shows the formation of chiral hemiacetal and hemiketal structures. The nomenclature committees of the International Union of Pure and Applied Chemistry (IUPAC) and the International Union of Biochemistry (IUB) have recommended abandoning the term *hemiketal;* in carbohydrate chemistry, both products are called *hemiacetals.*

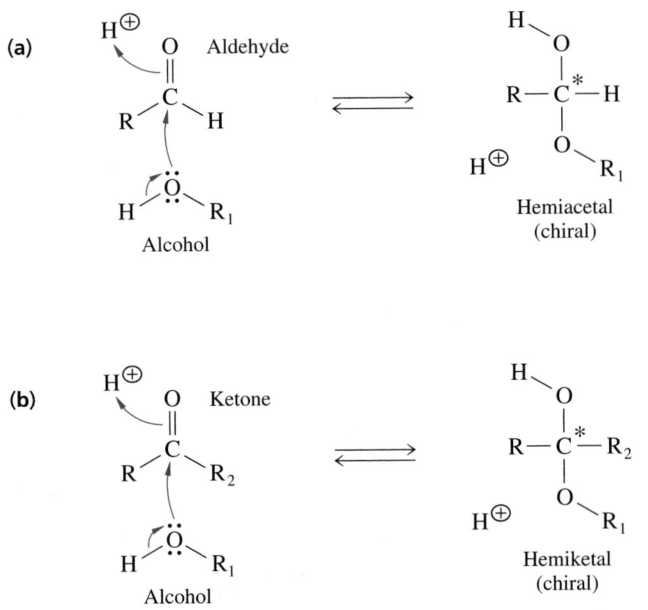

Figure 9·8
(a) Reaction of an alcohol with an aldehyde to form a hemiacetal. (b) Reaction of an alcohol with a ketone to form a hemiketal. In carbohydrate chemistry, both products are called hemiacetals, based on the recommendations of the IUPAC and IUB nomenclature committees. The asterisks indicate the position of the newly formed chiral center.

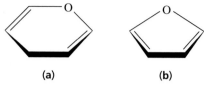

Figure 9·9
(a) Pyran. (b) Furan.

In monosaccharides, the carbonyl carbon of an aldose containing five or more carbon atoms and of a ketose containing six or more carbon atoms can react with an intramolecular hydroxyl group to form a cyclic hemiacetal in solution. Cyclic hemiacetals exist as either five- or six-membered ring structures in which one of the substituents of the ring is the oxygen atom from the hydroxyl group that reacted to form the hemiacetal. (The hemiacetal is thus heterocyclic.) Because it resembles the six-membered ring pyran (Figure 9·9a), the six-membered heterocyclic hemiacetal of a monosaccharide is called a **pyranose.** Similarly, because the five-membered heterocyclic hemiacetal of a monosaccharide resembles furan (Figure 9·9b), it is called a **furanose.** Note, however, that, unlike pyran and furan, the rings of carbohydrates do not contain double bonds.

The most oxidized carbon of a cyclized monosaccharide (which, like the carbonyl carbon of an aldehyde or a ketone, shares a total of four electrons with oxygen) is referred to as the **anomeric carbon.** It can be recognized in a cyclized monosaccharide as the only carbon atom attached to two oxygen atoms. In ring structures, the anomeric carbon is chiral. Thus, the cyclized aldose or ketose can adopt either one of two anomeric configurations (designated α or β), as illustrated for D-glucose in Figure 9·10.

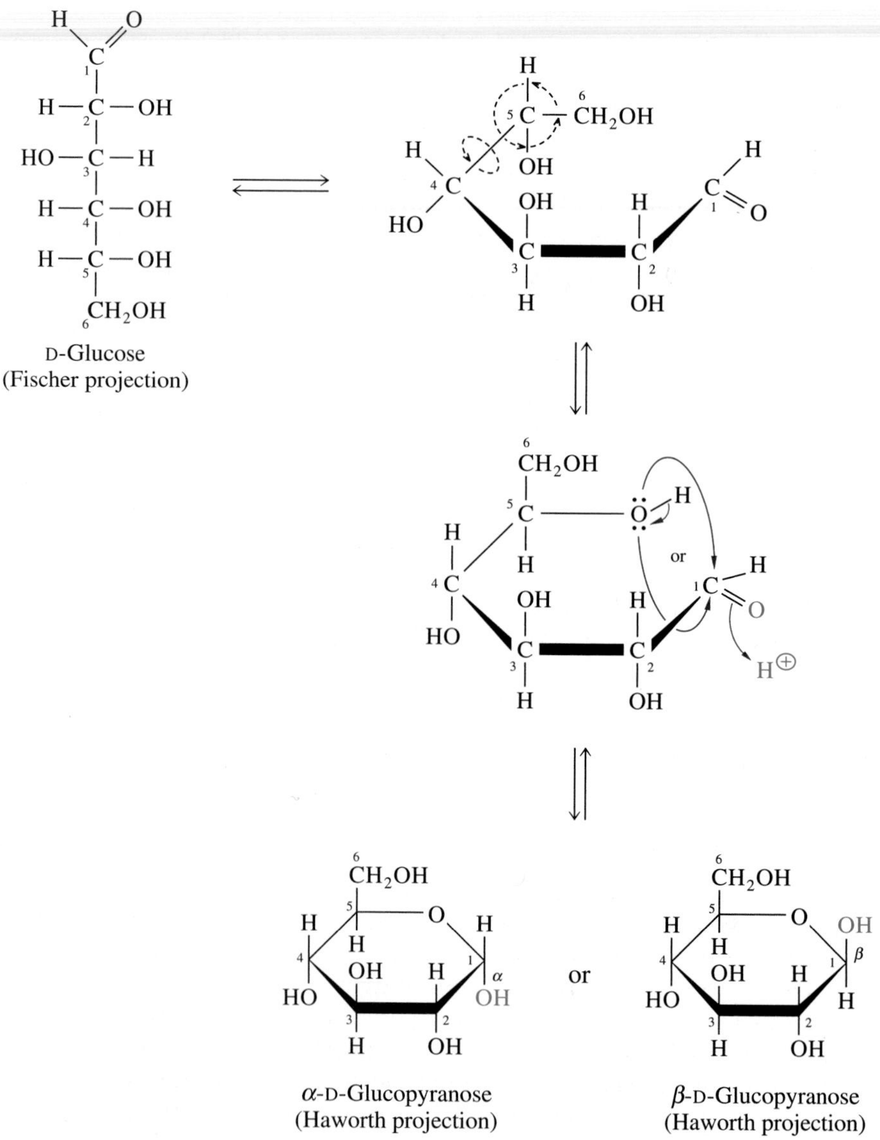

Figure 9·10
Cyclization of D-glucose to form glucopyranose. The Fischer projection (top left) is rearranged into a three-dimensional representation (top right). Rotation of the bond between C-4 and C-5 brings the C-5 hydroxyl group in proximity to the C-1 aldehyde group. Reaction of the hydroxyl group at C-5 with one side of C-1 gives α-D-glucopyranose; reaction of the hydroxyl group with the other side gives β-D-glucopyranose. The α- and β-D-glucopyranose products are shown as Haworth projections. In Haworth projections, the lower edges of the ring (heavy lines) project in front of the plane of the paper, and the upper edges project behind the plane of the paper. In Haworth projections of D sugars, in which the anomeric carbon is depicted on the right and numbered carbons increase in the clockwise direction, the —CH$_2$OH group is always oriented up. Hydroxyl groups to the *right* of the carbon skeleton in the Fischer projection are *down* in the Haworth projection; hydroxyl groups to the *left* (Fischer) are *up* (Haworth). In the α-D anomer of glucose, the hydroxyl group at C-1 is down; in the β-D anomer, the hydroxyl group at C-1 is up.

D-Glucose
(Fischer projection)

α-D-Glucopyranose
(Haworth projection)

β-D-Glucopyranose
(Haworth projection)

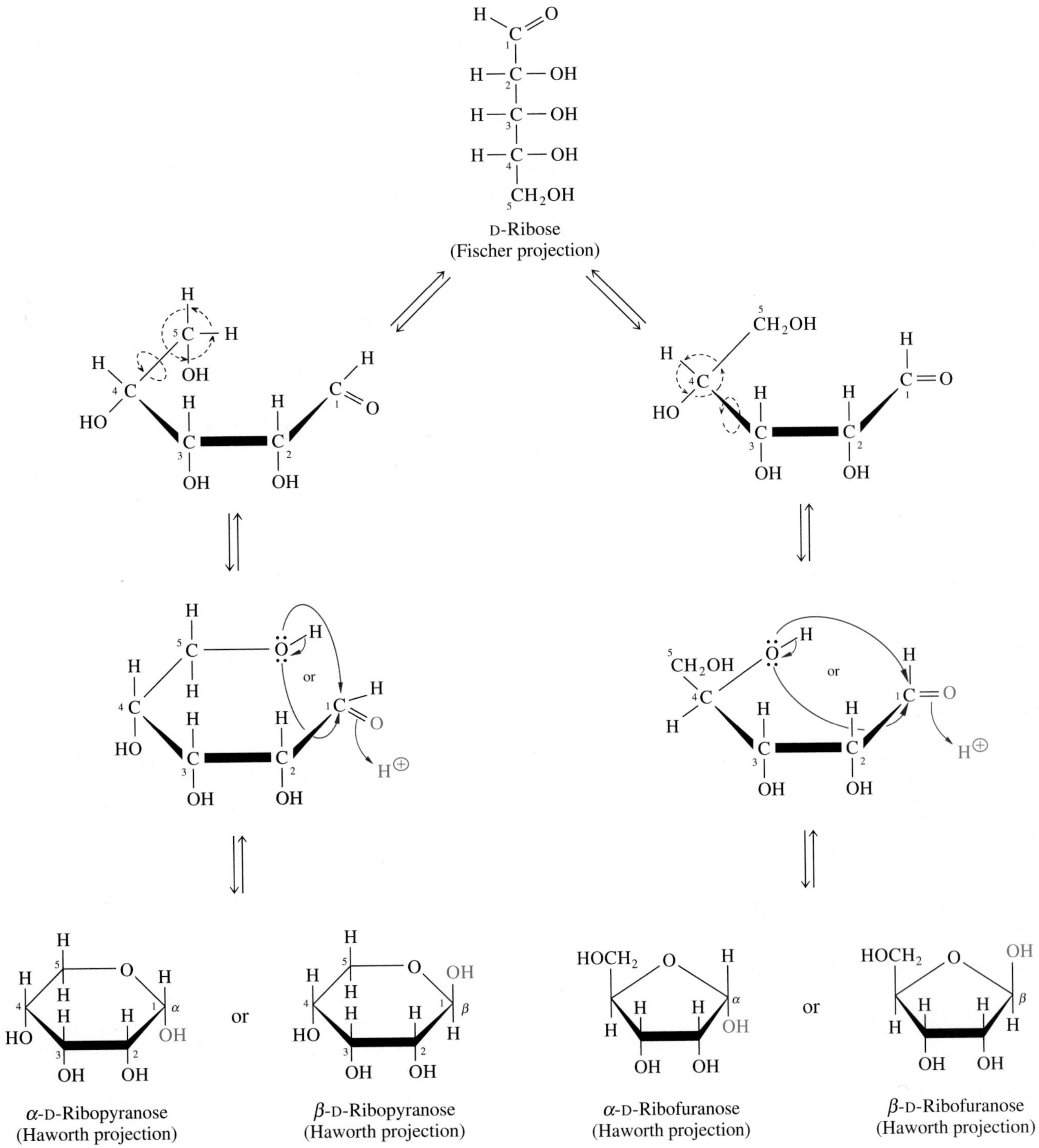

Figure 9·11
Cyclization of D-ribose to form α- and β-D-ribopyranose and α- and β-D-ribofuranose.

In solution, aldoses and ketoses capable of forming ring structures (that is, aldoses containing five or more carbon atoms and ketoses containing six or more carbon atoms) equilibrate among their various cyclic and open-chain forms. At 31°C, for example, D-glucose exists in an equilibrium mixture of approximately 64% β-D-glucopyranose and 36% α-D-glucopyranose, with only a tiny fraction in either the furanose or open-chain forms. Similarly, D-ribose exists as a mixture of 58.5% β-D-ribopyranose, 21.5% α-D-ribopyranose, 13.5% β-D-ribofuranose, and 6.5% α-D-ribofuranose, with a tiny fraction in the open-chain form (Figure 9·11).

The relative abundance of the various forms of monosaccharides at equilibrium reflects the relative stabilities of each form. Although unsubstituted D-ribose is most stable as the β-pyranose, its structure in nucleotides (Chapter 10) is the β-furanoside.

The ring drawings shown in Figures 9·10 and 9·11 are known as Haworth projections, or Haworth perspectives. For most purposes, Haworth projections are adequate representations of stereochemistry. They have the advantage that they can be related easily to Fischer projections. Imagine a cyclic monosaccharide drawn such that the anomeric carbon is on the right and the other carbons are numbered in ascending order in the clockwise direction. Hydroxyl groups pointing *down* in the Haworth projection point to the *right* of the carbon skeleton in the Fischer projection, whereas hydroxyl groups pointing *up* in the Haworth projection point to the *left* in the Fischer projection. The configuration of the anomeric carbon atom in a Haworth projection of a D sugar is designated α if it has the D-like configuration (hydroxyl group pointing down) and β if it has the L-like configuration (hydroxyl group pointing up). Note, however, that enantiomers are mirror images of each other, and therefore the configuration of functional groups at every carbon is opposite in a pair of enantiomers. Accordingly, the configuration of the anomeric carbon atom of an L sugar by convention is designated α if it has the L-like configuration (hydroxyl group pointing up) and β if it has the D-like configuration (hydroxyl group pointing down).

Monosaccharides are often drawn in either the α- or β-D-pyranose form. However, you should remember that the anomeric forms of five- and six-carbon sugars exist in rapid equilibrium. Throughout this chapter and the rest of the book, we will draw sugars in the correct anomeric form if it is known. We will refer to sugars in a nonspecific way (e.g., glucose) when we are referring to an equilibrium mixture of the various anomeric forms as well as the open-chain forms. When we are referring to a specific form of a sugar, however, we will refer to it precisely. For example, if we mean only β-D-glucopyranose, we will use that name. Also, since the D enantiomers of carbohydrates predominate in nature, we will always assume a carbohydrate to be of the D configuration unless specified otherwise. Carbohydrates that exist as L-enantiomers (e.g., ascorbic acid and fucose) will be so designated in this chapter when they are introduced. At other times, their configuration may not be mentioned.

9·3 Monosaccharides Can Adopt Different Conformations

Because of their general clarity and the ease with which they are drawn, Haworth projections are commonly used in biochemistry and in this textbook. However, ring structures containing sp^3-hybridized (tetrahedral) carbons are not actually planar. More realistic views of ring structures are presented in Figures 9·12 and 9·13.

Furanose rings can adopt either an envelope conformation, in which one of the five ring atoms is out-of-plane with the remaining four approximately coplanar, or a twist conformation, in which two of the five ring atoms are out-of-plane, one on either side of the plane formed by the other three atoms (Figures 9·12 and 9·14). For

Figure 9·12
Conformations of β-D-ribofuranose. **(a)** Haworth projection. **(b)** One of 10 possible envelope conformers, in which one ring atom (in this case, C-2) lies above the plane defined by the four remaining ring atoms. **(c)** One of 10 possible twist conformers, in which one ring atom (in this case, C-3) lies above and one ring atom (in this case, C-2) lies below the plane defined by the remaining three ring atoms.

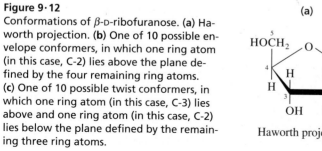

(a)

Haworth projection

(b)

Envelope
conformation

(c)

Twist
conformation

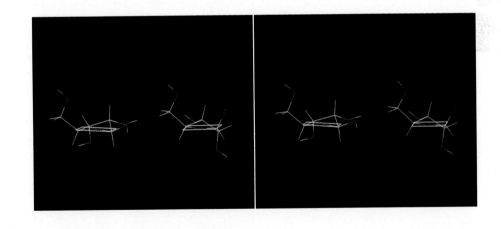

(a)

Haworth projection

(b)

Chair conformation

(c)

Boat conformation

Figure 9·13
Conformations of β-D-glucopyranose.
(a) Haworth projection. **(b)** One of two possible chair conformers. **(c)** One of six possible boat conformers. This boat conformation is destabilized by steric repulsion between the hydroxyl group on C-1 and the hydrogen attached to C-4. There is also repulsion between the hydroxyl group on C-2 and the hydrogen atoms on C-3 and C-5.

each furanose, there are 10 possible envelope conformers and 10 possible twist conformers. All of these forms rapidly interconvert.

Pyranose rings tend to assume one of two distinct conformations, the chair conformation or the boat conformation (Figures 9·13 and 9·15). For each pyranose, there are six distinct boat conformers and two distinct chair conformers. Since steric repulsion among the ring substituents is minimized in the chair conformation, chair conformations are generally more stable than boat conformations. Note that there are two different positions occupied by the substituents of a pyranose ring in the chair conformation. Axial substituents are those that extend perpendicular to the plane of the ring, whereas equatorial substituents are those that extend along the plane of the ring. In the case of cyclohexane derivatives, six ring substituents are always axial (three above and three below the plane of the ring) and six are always equatorial. In the case of pyranoses, five substituents are axial and five are equatorial. However, whether a group is axial or equatorial depends on which carbon atom (C-1 or C-4) extends above the plane of the ring when the ring is in the chair

Figure 9·14
Stereo view of envelope (left) and twist (right) conformations of β-D-ribofuranose. In the envelope conformer shown, C-2 lies above the plane defined by C-1, C-3, C-4, and the ring oxygen. In the twist conformer shown, C-3 lies above and C-2 lies below the plane defined by C-1, C-4, and the ring oxygen. The plane is shown in yellow.

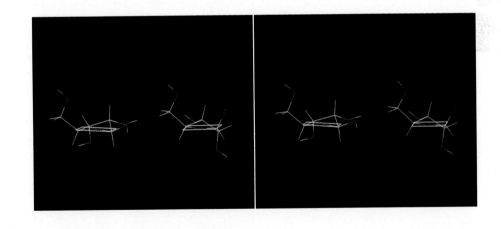

Figure 9·15
Stereo view of chair (left) and boat (right) conformations of β-D-glucopyranose.

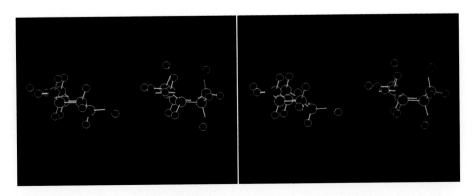

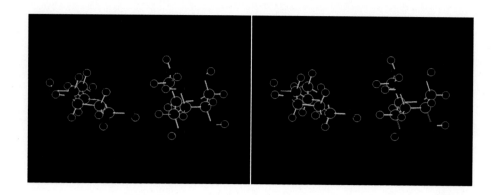

Figure 9·16
Stereo view of the two chair conformers of β-D-glucopyranose. The conformation on the left is more stable.

Figure 9·17
The two chair conformers of β-D-glucopyranose. The top conformer is more stable.

Table 9·1 Abbreviations for some monosaccharides and their derivatives

Monosaccharide or derivative	Abbreviation
Pentoses	
Arabinose	Ara
Ribose	Rib
Xylose	Xyl
Hexoses	
Fructose	Fru
Galactose	Gal
Glucose	Glc
Mannose	Man
Deoxy sugars	
Abequose	Abe
Fucose	Fuc
Rhamnose	Rha
Amino sugars	
Glucosamine	GlcN
Galactosamine	GalN
N-Acetylglucosamine	GlcNAc
N-Acetylgalactosamine	GalNAc
N-Acetylneuraminic acid	NeuNAc
N-Acetylmuramic acid	MurNAc
N-Acetylglucosamine 6-sulfate	GlcNAc-6-SO$_4$
Sugar acids	
Glucuronic acid	GlcA
Iduronic acid	IdoA

conformation. Figures 9·16 and 9·17 show the two different chair conformers of β-D-glucopyranose. The conformation that is most stable is the one in which the bulkiest ring substituents are equatorial. Pyranose rings are occasionally forced to adopt slightly different conformations, such as the half-chair adopted by a polysaccharide residue in the active site of lysozyme (Section 7·17).

9·4 There Are a Variety of Biologically Important Derivatives of Monosaccharides

In addition to the monosaccharides, there are many monosaccharide derivatives in biological systems. These include polymerized monosaccharides, such as oligosaccharides and polysaccharides, as well as several classes of nonpolymerized derivatives. We will consider oligosaccharides and polysaccharides in Sections 9·5 and 9·8. Here, we will briefly introduce a few of the other monosaccharide derivatives important to living systems, including sugar phosphates, the deoxy and amino sugars, the sugar alcohols, the sugar acids, and ascorbic acid (vitamin C). The nucleotide-sugar cosubstrates, which are also monosaccharide derivatives, have already been described in Section 8·4D.

Like other polymer-forming biomolecules, monosaccharides and their various derivatives must be suitably abbreviated so that their polymers can be described easily. The accepted abbreviations contain three letters, with suffixes added for some derivatives. The abbreviations for some pentoses and hexoses and their major derivatives are listed in Table 9·1. We will use these abbreviations extensively in the second part of this chapter.

Figure 9·18
Structures of several metabolically important sugar phosphates.

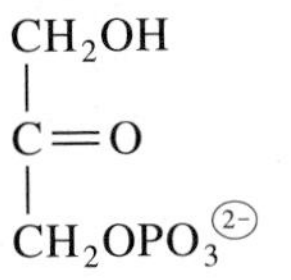

Dihydroxyacetone phosphate

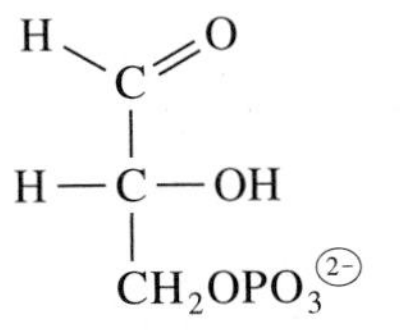

D-Glyceraldehyde 3-phosphate

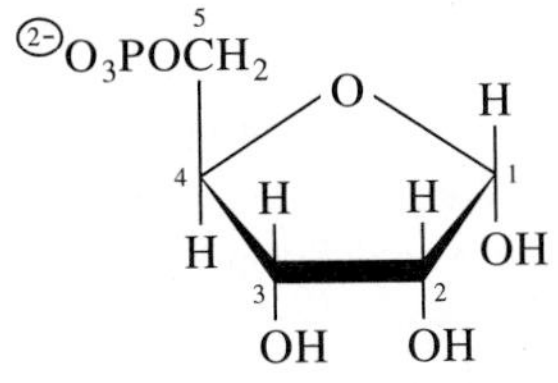

α-D-Ribose 5-phosphate

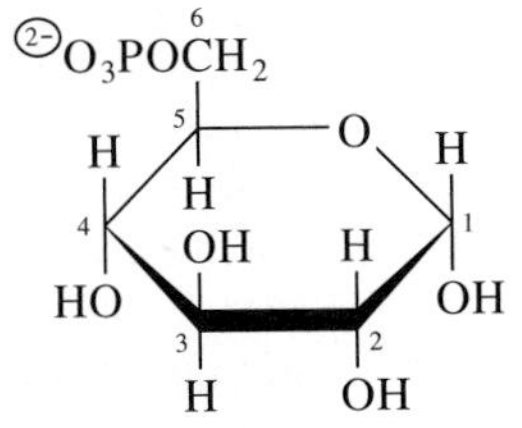

α-D-Glucose 6-phosphate

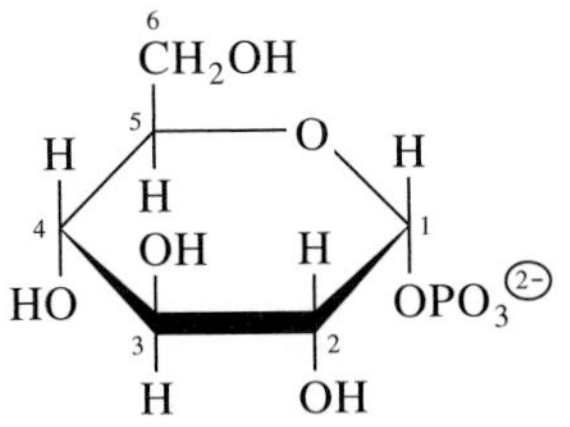

α-D-Glucose 1-phosphate

A. Sugar Phosphates

When monosaccharides are used as fuel, they are metabolized as phosphate esters. Figure 9·18 shows the structures of several sugar phosphates we will encounter in our study of carbohydrate metabolism. The triose phosphates and ribose 5-phosphate are simple phosphate esters, as is glucose 6-phosphate. In glucose 1-phosphate, however, the phosphoryl group is attached to the oxygen of the anomeric carbon, so that the phosphate formed is a hemiacetal phosphate rather than an alcohol phosphate. Because of this chemical difference, the ΔG for hydrolysis of glucose 1-phosphate is more negative than that for the hydrolysis of glucose 6-phosphate.

B. Deoxy Sugars

The structures of four deoxy sugars are shown in Figure 9·19. In three of these, a hydrogen atom replaces one of the hydroxyl groups in the parent monosaccharide. 2-Deoxy-D-ribose is an important building block for DNA. The 6-deoxyhexoses L-fucose (6-deoxy-L-galactose) and L-rhamnose (6-deoxy-L-mannose) are widely distributed in plants, animals, and microorganisms and are frequently found in oligo- and polysaccharides. These sugars are examples of naturally occurring L enantiomers. Despite their unusual L configuration, fucose and rhamnose are derived metabolically from D sugars—D-mannose and D-glucose, respectively.

Figure 9·19
Structures of the deoxy sugars 2-deoxy-D-ribose, L-fucose, L-rhamnose, and D-abequose.

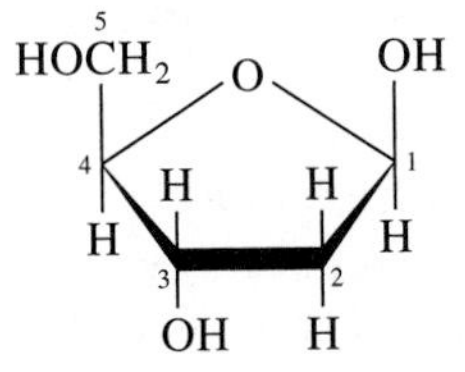

β-2-Deoxy-D-ribose

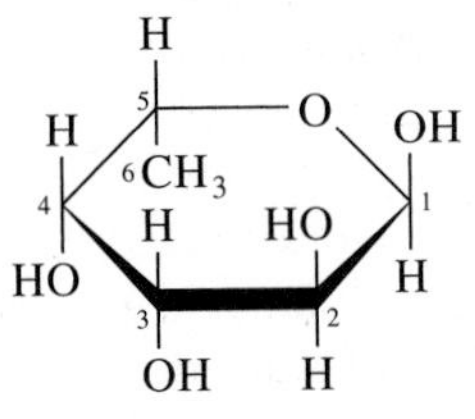

α-L-Fucose
(6-Deoxy-L-galactose)

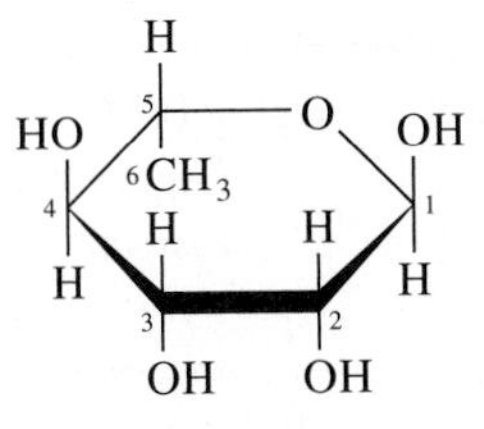

α-L-Rhamnose
(6-Deoxy-L-mannose)

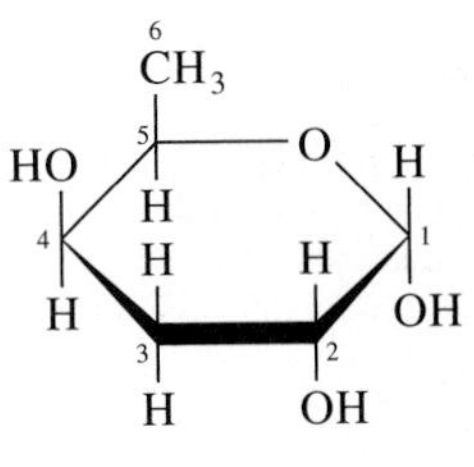

α-D-Abequose
(3,6-Dideoxy-D-*xylo*-hexose)

Abequose is a comparatively rare deoxy sugar that occurs in the outer membranes of certain Gram-negative bacteria. In abequose, two hydrogen atoms replace two of the hydroxyl groups in the parent monosaccharide, thus making it a dideoxy sugar.

C. Amino Sugars

There are a number of sugars in which an amino group replaces one of the hydroxyl groups in the parent monosaccharide. Sometimes the amino group is acetylated. Three examples of amino sugars are shown in Figure 9·20. Amino sugars formed from glucose and galactose commonly occur in glycoconjugates. N-Acetylglucosamine is also the monomer of the homoglycan chitin. N-Acetylneuraminic acid is an acid formed from N-acetylmannosamine and pyruvate. When this compound cyclizes to form a pyranose, the carbonyl group at C-2 (from the pyruvate moiety) reacts with the hydroxyl group of C-6. N-Acetylneuraminic acid is an important constituent of many glycoproteins and of a family of lipids called gangliosides (Section 11·6). Neuraminic acid and its derivatives, including N-acetylneuraminic acid, are collectively known as sialic acids.

D. Sugar Alcohols

In sugar alcohols, the carbonyl oxygen of the parent monosaccharide has been reduced, resulting in polyhydroxy alcohols. Figure 9·21 shows five examples of sugar alcohols. Both glycerol and myo-inositol are important components of lipids (Chapter 11). Ribitol is a component of flavin mononucleotide (FMN) and flavin adenine dinucleotide (FAD), discussed in Chapter 8, and of the teichoic acids, complex polymers found in the cell walls of certain Gram-positive bacteria (Section 9·9C). Xylitol is derived from xylose and is a common constituent of sugarless chewing gum. D-Sorbitol is an intermediate in a metabolic pathway from glucose to fructose that occurs in certain tissues. Note that, in general, sugar alcohols are named by replacing the -ose suffix of the parent monosaccharides with -itol.

Figure 9·20
Structures of the amino sugars glucosamine, N-acetylgalactosamine, and N-acetylneuraminic acid. The amino and acetylamino groups are shown in red.

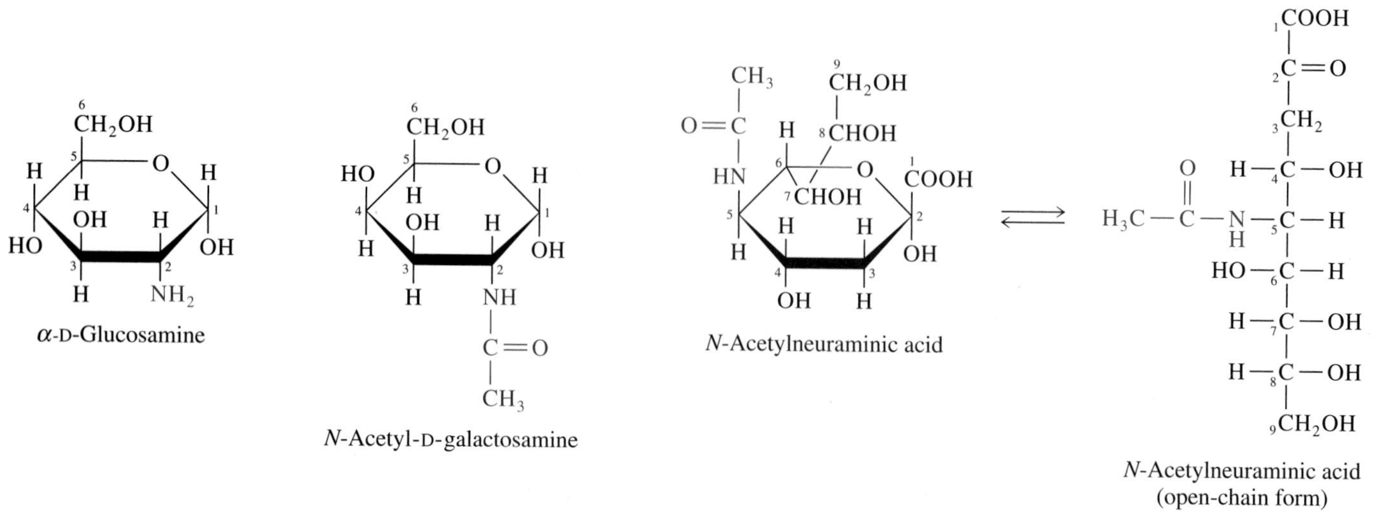

α-D-Glucosamine

N-Acetyl-D-galactosamine

N-Acetylneuraminic acid

N-Acetylneuraminic acid
(open-chain form)

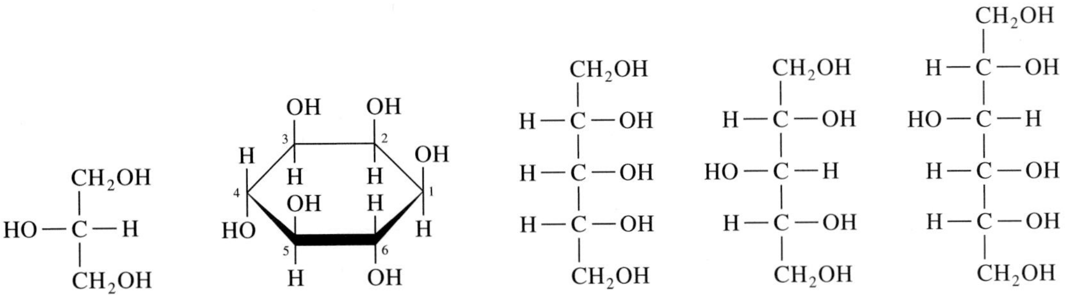

Glycerol   *myo*-Inositol   D-Ribitol   D-Xylitol   D-Sorbitol

Figure 9·21
Structures of several sugar alcohols. Glycerol (a reduced form of glyceraldehyde) and *myo*-inositol (metabolically derived from glucose) are important constituents of many lipids. Ribitol (a reduced form of ribose) is a constituent of the vitamin riboflavin and its coenzymes and of the teichoic acids. Xylitol is a reduced form of xylose. Sorbitol is a metabolite of glucose.

E. Sugar Acids

Sugar acids are carboxylic acids derived from aldoses, either by oxidation of C-1 (the aldehydic carbon) to yield an aldonic acid or by oxidation of the highest-numbered carbon (the carbon bearing the primary alcohol) to yield an alduronic acid. The structures of the aldonic and alduronic derivatives of glucose—gluconate and glucuronate—are shown in Figure 9·22. Aldonic acids exist in the open-chain form in alkaline solution and form lactones (intramolecular esters) upon acidification. Alduronic acids can exist as pyranoses and therefore possess an anomeric carbon. Sugar acids are important components of many polysaccharides.

Figure 9·22
Structures of sugar acids derived from D-glucose. (a) Gluconate and its δ-lactone. (b) The open-chain and pyranose forms of glucuronate.

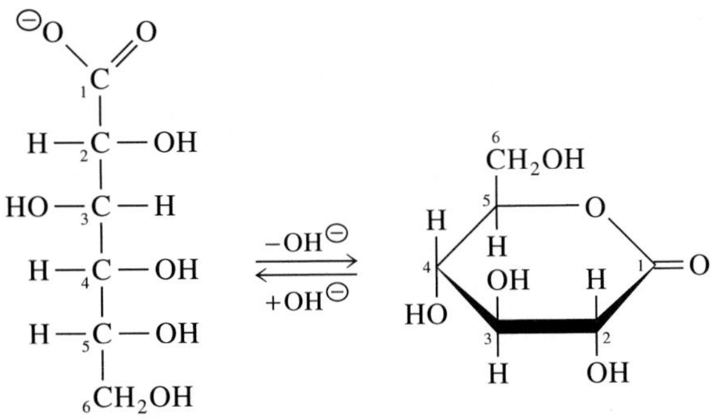

D-Gluconate (open-chain form)   D-Glucono-δ-lactone

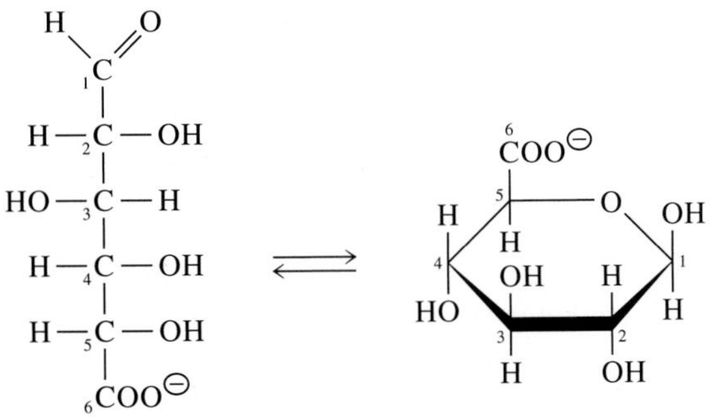

D-Glucuronate (open-chain form)   D-Glucuronate (β pyranose anomer)

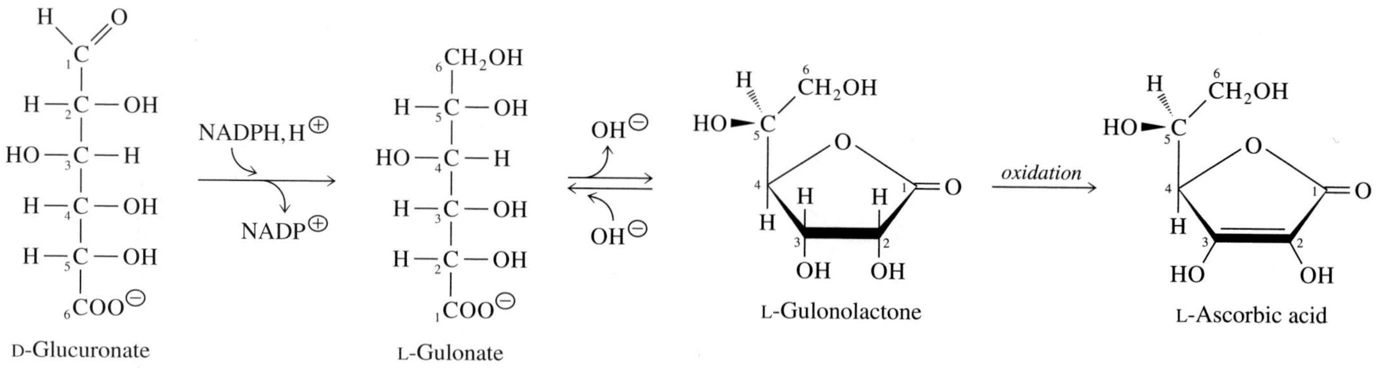

Figure 9·23
Pathway for the synthesis of L-ascorbic acid in mammals. Primates cannot carry out the final step. When the aldehyde group at C-1 of D-glucuronate is reduced, the carbon atom is renumbered to C-6. The change in numbering places the product in the L series.

F. Ascorbic Acid

L-Ascorbic acid, or vitamin C, is an enediol of a lactone derived from glucuronate. Glucuronate is formed by hydrolysis of uridine diphosphate glucuronate (UDP-glucuronate), discussed in Section 8·4D. Glucuronate is reduced to L-gulonate, which forms a lactone before it is oxidized to ascorbic acid (Figure 9·23). Note that the conversion of D-glucuronate to L-gulonate does not involve a change in configuration but merely a reduction that causes a renumbering of the carbon skeleton. Primates cannot carry out the final step, the oxidation of gulonolactone, and must therefore obtain ascorbic acid from the diet. The C-3 hydroxyl group of ascorbic acid has a relatively low pK_a of 4.2 due to resonance stabilization of the conjugate base. Ascorbic acid is an essential cofactor in the catalytic hydroxylation reactions involving proline and lysine residues during collagen synthesis (Sections 5·7 and 8·5).

9·5 Disaccharides Consist of Two Monosaccharide Residues Linked by a Glycosidic Bond

The primary structural linkage in all polymers of monosaccharides is known as the **glycosidic bond.** A glycosidic bond is an acetal linkage involving condensation of the anomeric carbon of a sugar molecule with an alcohol, an amine, or a thiol. Compounds containing glycosidic bonds are called **glycosides.** In a disaccharide, the anomeric carbon of one sugar can interact with one of several hydroxyl groups in the other sugar. Thus, in considering the structure of an oligosaccharide, it is important to note not only which types of monosaccharide residues are involved, but also which of their atoms are involved in glycosidic bonds. Precise nomenclature is needed to specify the exact bonding. In the systematic description of a disaccharide, for example, the linking atoms, the configuration of the glycosidic bond, and the name of each monosaccharide residue (including its designation as a pyranosyl or furanosyl structure) must be specified. Figure 9·24 illustrates structures and nomenclature for four common disaccharides.

Maltose (Figure 9·24a) is a disaccharide that is released during the hydrolysis of starch, which is a polymer of glucose residues. Maltose is composed of two D-glucose residues joined by an α-glycosidic bond. The glycosidic bond links C-1 of one residue (left in Figure 9·24a) to the O atom attached to C-4 of the second residue (right). Maltose is therefore α-D-glucopyranosyl-$(1 \rightarrow 4)$-D-glucose. It is important to note that the anomeric carbon linked by the glycosidic bond is not free to isomerize but is fixed in the α configuration, whereas the glucose residue on the right (the reducing end, as explained in Section 9·6) freely equilibrates among the α, β, and open-chain structures, the latter present only in very small amounts. The structure shown in Figure 9·24a is the β pyranose anomer of maltose (the anomer whose reducing end is in the β configuration, the predominant anomeric form).

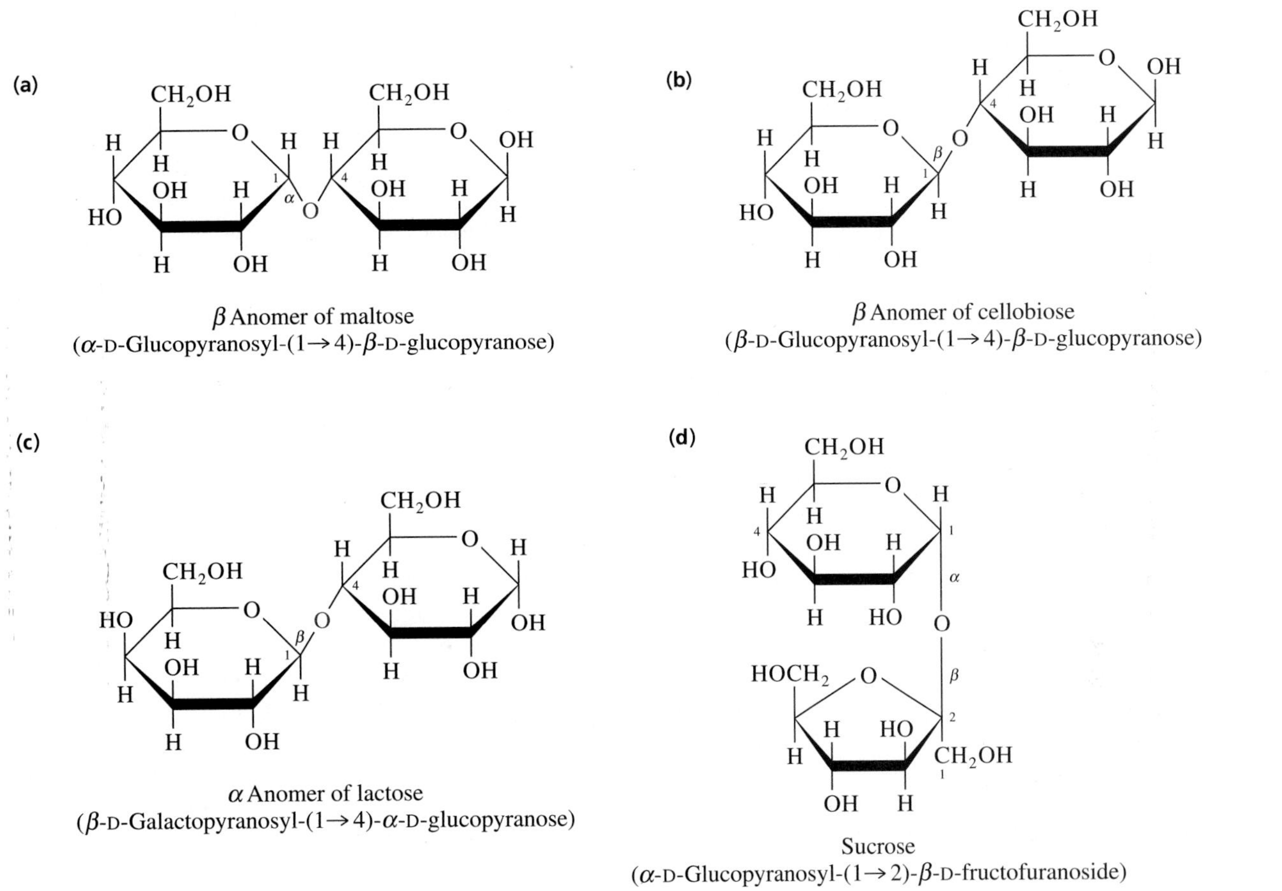

(a)

β Anomer of maltose
(α-D-Glucopyranosyl-(1→4)-β-D-glucopyranose)

(b)

β Anomer of cellobiose
(β-D-Glucopyranosyl-(1→4)-β-D-glucopyranose)

(c)

α Anomer of lactose
(β-D-Galactopyranosyl-(1→4)-α-D-glucopyranose)

(d)

Sucrose
(α-D-Glucopyranosyl-(1→2)-β-D-fructofuranoside)

Figure 9·24
Structures of (a) maltose, (b) cellobiose, (c) lactose, and (d) sucrose. The oxygen atom of each glycosidic bond is shown in red.

Cellobiose (β-D-glucopyranosyl-(1→4)-D-glucose) is another dimer of glucose (Figure 9·24b). Cellobiose is the repeating disaccharide seen in the structure of cellulose and is released during the degradation of cellulose. The only difference between cellobiose and maltose is that the glycosidic linkage in cellobiose is β (it is α in maltose). The glucose residue on the right in Figure 9·24b, like the residue on the right in 9·24a, equilibrates among the α, β, and open-chain structures.

Lactose (β-D-galactopyranosyl-(1→4)-D-glucose), a major carbohydrate found in milk, is a disaccharide that is synthesized only in lactating mammary glands (Figure 9·24c). Note from its structure that lactose is an epimer of cellobiose. The naturally occurring α anomer of lactose is sweeter and more soluble than the β anomer. The β anomer can be found in ice cream, where it crystallizes during storage and gives a gritty texture to the ice cream.

Sucrose (α-D-glucopyranosyl-(1→2)-β-D-fructofuranoside), or table sugar, is the most abundant disaccharide found in nature and is synthesized only in plants (Figure 9·24d). Sucrose is distinguished from the other three disaccharides in Figure 9·24 in that its glycosidic bond links the anomeric carbon atoms of two monosaccharide residues. The configurations of both the glucopyranose and fructofuranose residues in sucrose are fixed, and neither residue is free to equilibrate between α and β anomers.

An important derivative of sucrose is the trisaccharide raffinose (Figure 9·25). This compound is an α-galactosyl derivative of sucrose and is synthesized by transfer of a galactose residue from uridine diphosphate galactose (UDP-galactose) to the 6-hydroxyl group of the glucose residue of sucrose. Plants have a greater variety of oligosaccharides than animal tissues do, and next to sucrose, raffinose is the most common plant oligosaccharide.

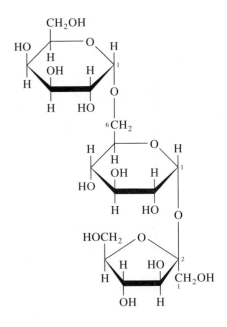

Figure 9·25
Structure of raffinose.

9·6 Monosaccharides and Most Disaccharides Are Reducing Sugars

Because monosaccharides and most disaccharides contain a reactive carbonyl group, they are readily oxidized to diverse products, a property often used in their analysis. Such carbohydrates, which include glucose, maltose, cellobiose, and lactose, are sometimes classified as reducing sugars. Carbohydrates such as sucrose, which are not readily oxidized because both anomeric carbon atoms are fixed in a glycosidic linkage, are classified as nonreducing sugars.

The reducing ability of a sugar polymer is of more than analytical interest. The polymeric chains of oligosaccharides and polysaccharides show directionality, based upon their reducing and nonreducing ends. In a linear polymer, there is usually one reducing-end residue (the residue containing the free anomeric carbon) and one nonreducing-end residue. A branched polysaccharide has a number of nonreducing ends but only one reducing end.

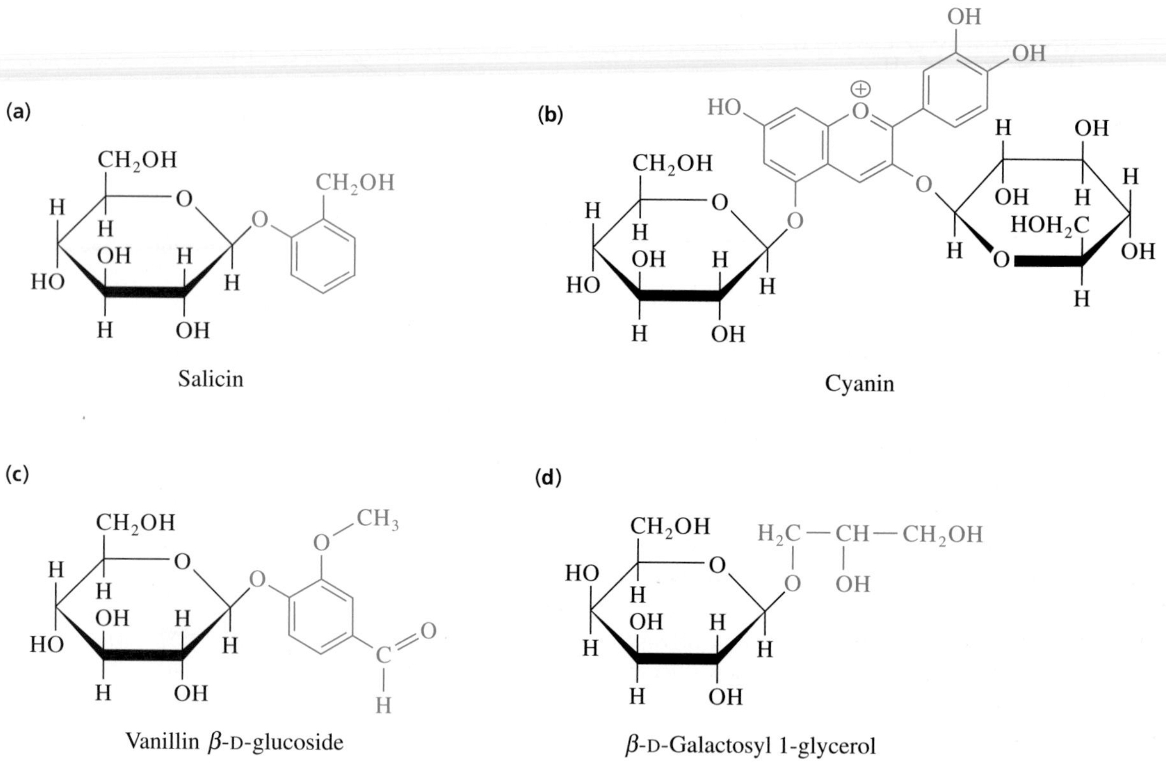

(a)

Salicin

(b)

Cyanin

(c)

Vanillin β-D-glucoside

(d)

β-D-Galactosyl 1-glycerol

Figure 9·26
Structures of four naturally occurring glycosides. The aglycones are shown in blue.
(a) Salicin, a component of willow and poplar bark that has analgesic properties.
(b) Cyanin, the red pigment of various flowers, including roses. (c) Vanillin glucoside, the flavored compound in vanilla extract. (d) β-D-Galactosyl 1-glycerol, derivatives of which are common in eukaryotic cell membranes.

9·7 Carbohydrates Form Glycosides with a Variety of Compounds

The anomeric carbons of sugars form glycosidic linkages with a variety of alcohols, amines, and thiols. These compounds (often organic nonsugar molecules) linked to sugars are referred to as **aglycones.** Other than oligosaccharides and polysaccharides, the most commonly encountered glycosides are the nucleotides, in which a secondary amino group of a purine or pyrimidine is the aglycone of a β-D-ribofuranose or β-D-deoxyribofuranose moiety (Chapter 10). Four other examples of naturally occurring glycosides are shown in Figure 9·26. Hydroxybenzyl alcohol is the aglycone that combines with D-glucose to form salicin, a compound derived from willow and poplar bark that has been used to treat headaches. Cyanin, the pigment that gives roses and certain other flowers their red color, is a glycoside containing cyanidin and two residues of D-glucose. Vanillin glucoside is the flavored compound in natural vanilla extract.

The β-galactosides constitute an abundant class of glycosides. In these compounds, a variety of nonsugar molecules are joined in a β linkage to galactose. For example, derivatives of β-D-galactosyl 1-glycerol (Figure 9·26d) are common in eukaryotic cell membranes and can be hydrolyzed readily by enzymes called β-galactosidases.

9·8 Polysaccharides Are Large Polymers of Monosaccharide Residues

Because the cell walls of plants are made of polysaccharides, polysaccharides are the most abundant organic biomolecules on earth on the basis of mass. They are frequently divided into two broad classes: homopolysaccharides, or homoglycans, which are polymers containing only one type of monosaccharide residue, and heteropolysaccharides, or heteroglycans, which are polymers containing more than one type of monosaccharide residue. Unlike proteins, whose primary structures are encoded by the genome and thus have specified lengths, polysaccharides are created without a template by the addition of particular monosaccharide and oligosaccharide residues. As a result, the lengths and compositions of particular polysaccharide molecules may vary within a population of similar molecules. Such a population is said to be polydisperse.

Most polysaccharides can also be classified according to their biological roles as storage polysaccharides, such as starch and glycogen, or structural polysaccharides, such as cellulose and chitin. There are many different polysaccharides in nature. We will consider the structures of several representative examples in this section.

A. Starch and Glycogen Are Storage Homopolysaccharides of Glucose

D-Glucose, the chief source of metabolic energy for many organisms, is stored intracellularly in polymeric form, thereby avoiding the excessive osmotic pressures that would result from large accumulations of the free monosaccharide. The most common storage homopolysaccharide of glucose in plants and fungi is called starch and in animals, glycogen. Both types of polysaccharides occur in bacteria.

In plant cells, starch is present as a mixture of amylose and amylopectin and is stored in granules whose diameters range from 3 to $100\,\mu$m. Amylose is an unbranched polymer of about 100 to 1000 D-glucose residues connected by α-$(1\rightarrow4)$ glycosidic linkages, specifically termed α-$(1\rightarrow4)$ glucosidic bonds because the

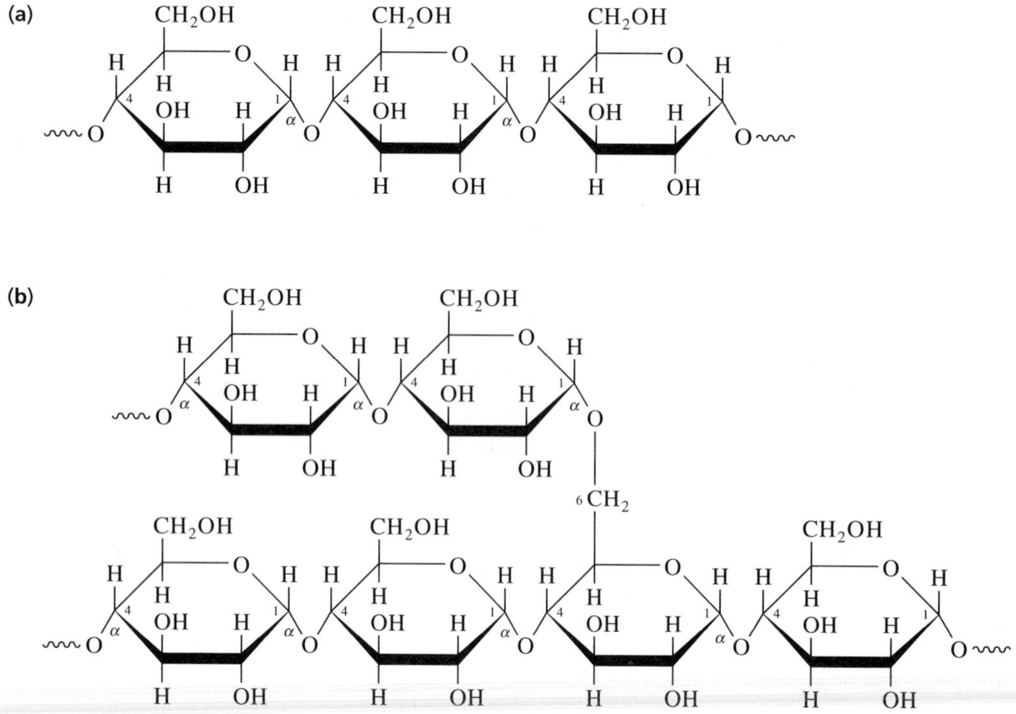

(a)

(b)

Figure 9·27
Structures of amylose and amylopectin. **(a)** Amylose, one form of starch, is a linear polymer of glucose residues linked by α-(1→4)-D-glucosidic bonds. **(b)** Amylopectin, a second form of starch, is a branched polymer. The linear glucose residues of the main chain and the side chains of amylopectin are linked by α-(1→4)-D-glucosidic bonds, and the side chains are linked to the main chain by α-(1→6)-D-glucosidic bonds.

anomeric carbon of a glucose residue is involved in the bond (Figures 9·27a). These same linkages connect glucose monomers in the disaccharide maltose. Although it is not truly soluble in water, amylose forms hydrated aggregates in water and can form a helical structure under some conditions (Figure 9·28).

Amylopectin is essentially a branched version of amylose (Figure 9·27b). In addition to α-(1→4) linkages, amylopectin contains branch points at which α-(1→6) linkages occur. Branching occurs, on average, once every 25 residues, and the branches, or side chains, contain about 15 to 25 glucose residues. Some side chains themselves contain branches. When isolated from living cells, amylopectin molecules range in size from 300 to 6000 glucose residues.

An adult human consumes about 300 g of carbohydrate per day, much of which is supplied by starch. Dietary starch is degraded in the gastrointestinal tract by the actions of α-amylase and a debranching enzyme, each of which catalyzes hydrolysis of certain α-D-glycosidic bonds (Chapter 13). α-Amylase is present in both animals and plants. Another hydrolase, known as β-amylase, exists in the seeds and

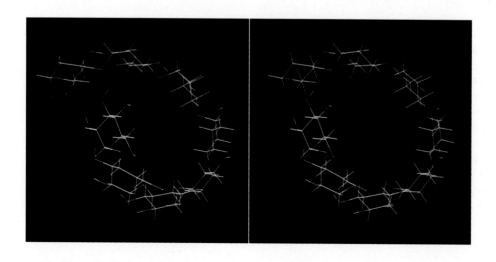

Figure 9·28
Stereo view of amylose. Amylose can assume a left-handed helical conformation, which is hydrated on the inside as well as on the outer surface.

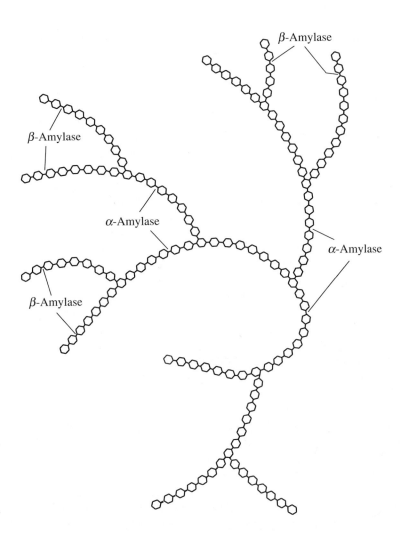

Figure 9·29
Points of hydrolysis of amylopectin by animal and plant amylases. This diagram shows points at which amylopectin can be hydrolyzed by the initial events catalyzed by α-amylase and β-amylase. The position of attack by α-amylase is internal and random; β-amylase always acts on the nonreducing ends. Each hexagon represents a glucose residue; the red hexagon represents the single reducing end of the branched polymer. (An actual amylopectin molecule contains many more glucose residues than are represented here.)

tubers of some higher plants. In the case of α- and β-amylase, the α and β designations refer to types of amylases, not to the configurations of the glycosidic bonds of the substrate; both types of amylases act only on α-(1→4)-D-glycosidic bonds.

Figure 9·29 depicts the sites of action of α-amylase and β-amylase on amylopectin. α-Amylase is an endoglycosidase that catalyzes random hydrolysis of the α-(1→4)-D-glucosidic bonds of amylose and amylopectin. β-Amylase is an exoglycosidase that catalyzes sequential hydrolysis of maltose from the free, nonreducing ends of amylopectin in plants, shortening them one maltose unit at a time. The α-(1→6) linkages at branch points are not substrates for either α- or β-amylase. After amylase-catalyzed hydrolysis of amylopectin, highly branched cores resistant to further hydrolysis, called **limit dextrins,** remain. Limit dextrins can be further degraded only after debranching enzymes have catalyzed hydrolysis of the α-(1→6) linkages at branch points.

Glycogen, a storage polysaccharide found in animals and bacteria, is also a branched polymer of glucose residues. Glycogen contains the same α-(1→6) branching as amylopectin, but the branches occur with greater frequency, and the side chains of glycogen contain fewer glucose residues. In general, glycogen molecules tend to be larger than starch molecules, containing up to several hundred thousand glucose residues. In mammals, depending on the nutritional state, glycogen can account for up to 10% of the mass of the liver and 1% of the mass of muscle.

We will consider enzymes that catalyze the intracellular synthesis and breakdown of glycogen in Chapter 17. At this point, it is sufficient to note that the branched structures of amylopectin and glycogen molecules possess only one reducing end but many nonreducing ends. It is at those nonreducing ends that most enzymatic lengthening and degradation occurs.

B. Cellulose and Chitin Are Structural Homopolysaccharides

Plant cell walls contain a high percentage of the structural homopolysaccharide cellulose, which accounts for over 50% of the organic matter in the biosphere. Unlike storage polysaccharides, cellulose and other structural polysaccharides are extracellular molecules extruded by the cells in which they are synthesized. Like amylose, cellulose is a linear homopolysaccharide of glucose residues, but in cellulose the glucose residues are joined by β-(1→4) linkages rather than α-(1→4) linkages. The glucose monomers of cellulose are therefore joined by the same type of linkage that joins the two glucose residues of cellobiose (Figure 9·24b). The β linkages of cellulose result in a rather rigid, extended conformation in which each successive glucose residue is rotated 180° relative to its neighbors (Figure 9·30). The equatorial hydroxyl groups are extensively hydrogen bonded, forming bundles of polymeric chains, or fibrils (Figure 9·31). The fibrils are insoluble in water and confer strength and rigidity. Cotton fibers are almost entirely cellulose, and wood is about half cellulose.

Cellulose molecules vary greatly in size, ranging from about 300 to over 15 000 glucose residues. Because of the strength it imparts, cellulose is used in a variety of commercial applications and is a component of a number of synthetic materials, including cellophane and the fabric rayon.

Enzymes that catalyze the hydrolysis of α-D-glucosidic bonds (α-glucosidases, such as α- and β-amylase) do not catalyze the hydrolysis of β-D-glucosidic bonds. Similarly, β-glucosidases (such as cellulase) do not catalyze hydrolysis of α-D-glucosidic bonds. Humans and other mammals that can metabolize starch, glycogen, lactose, and sucrose as energy sources cannot themselves metabolize cellulose because they lack enzymes capable of catalyzing the hydrolysis of β-glucosidic linkages. Ruminants such as cows and sheep have microorganisms in their rumens (multichambered stomachs) that produce β-glucosidases. Thus, ruminants can obtain glucose from eating grass and other plants that are rich in cellulose. The digestion of carbohydrates in humans, as well as the fate of bulk materials such as cellulose, is discussed in detail in Chapter 13.

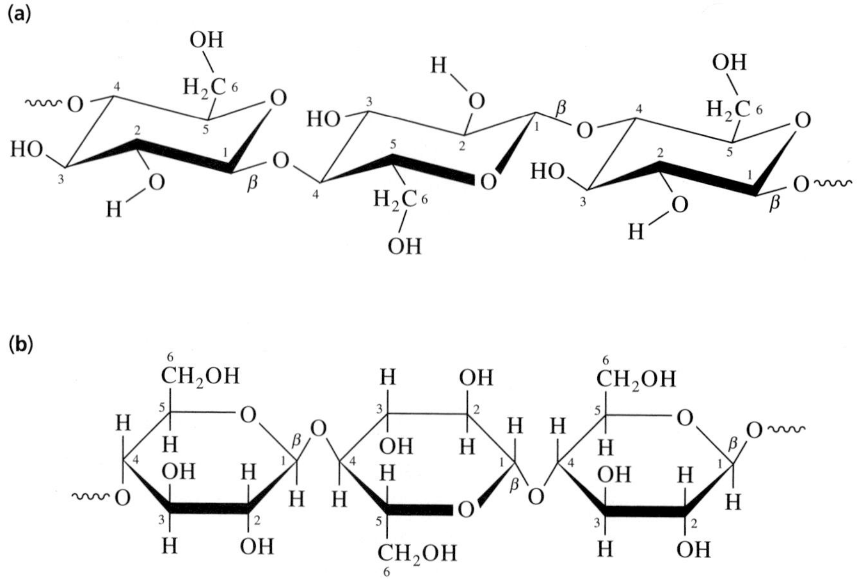

Figure 9·30
(a) Chair conformation of β-(1→4)-linked D-glucose residues in cellulose. (b) Modified Haworth projection of β-(1→4)-linked D-glucose residues in cellulose, emphasizing the alternating orientation of successive glucose residues in cellulose chains.

Figure 9·31
Stereo view of cellulose fibrils. Hydrogen bonding among individual linear chains gives cellulose its strength and rigidity.

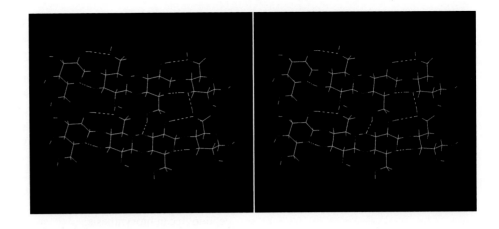

Chitin is a structural homopolysaccharide found in the exoskeletons of insects and crustaceans and also in the cell walls of most fungi and many algae (Figure 9·32). It is probably the second most abundant organic compound on earth. Chitin, a linear polymer similar to cellulose, consists of β-(1→4)-linked *N*-acetylglucosamine (GlcNAc) residues rather than glucose residues (Figure 9·33). Each GlcNAc residue is rotated 180° relative to its neighbors. The GlcNAc residues in adjacent strands form hydrogen bonds with each other, resulting in linear microfibrils that confer great structural strength. Chitin is often closely associated with nonpolysaccharide compounds, such as proteins and inorganic material. Partial deacetylation of chitin produces chitosan, a positively charged, nontoxic polymer. There is growing interest in developing commercial uses for chitosan, which is available as a waste product from shellfish processing. Chitosan may be useful as an absorbent for treatment of waste water or industrial liquids and as a coating for food preservation or cosmetics.

Figure 9·32
Beetle. The exoskeletons of insects contain the structural homopolysaccharide chitin.

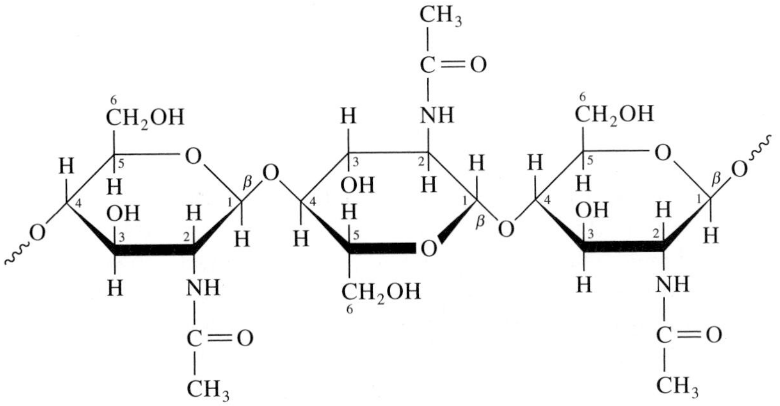

Figure 9·33
Chitin. Chitin is a cellulose-like homopolysaccharide consisting of repeating units of β-(1→4)-linked *N*-acetylglucosamine (GlcNAc) residues. Each GlcNAc residue is rotated 180° relative to its neighbors.

Part 2: Peptidoglycans, Glycoproteins, and Proteoglycans

Having discussed the structure, nomenclature, and function of monosaccharides, disaccharides, and homoglycans, we will now turn our attention to three types of glycoconjugates: peptidoglycans, glycoproteins, and proteoglycans.

Peptidoglycans are composed of a heteroglycan chain made up of alternating units of *N*-acetylglucosamine (GlcNAc) and *N*-acetylmuramic acid (MurNAc, a nine-carbon sugar) linked to peptides of varied composition. These giant macromolecules form the backbone of the cell walls of many bacteria. Interest in peptidoglycans was greatly stimulated when it was discovered that a number of antibiotics, including penicillin, kill microorganisms by inhibiting specific steps in peptidoglycan biosynthesis.

Glycoproteins, proteins to which carbohydrate chains are attached, include an enormous variety of compounds, such as plasma proteins, membrane proteins, certain enzymes, and a number of hormones. The carbohydrate chains of glycoproteins vary in length from 1 to over 30 sugar residues. As is the case for other glycoconjugates, the oligosaccharide chains of glycoproteins help determine both the structures and biological roles of these proteins.

Proteoglycans are proteins to which glycosaminoglycan chains are attached. Proteoglycans differ from glycoproteins in the amount and type of carbohydrate attached to the protein. Proteoglycans are important components of the extracellular matrix, in which they perform a variety of functions.

9·9 Peptidoglycan Is the Major Component of Bacterial Cell Walls

The cell wall of a bacterium helps determine its shape and protects its delicate plasma membrane from fluctuations in osmotic pressure. It is a relatively rigid structure that surrounds the entire bacterium and is generally much thicker in Gram-positive than in Gram-negative bacteria. The observation, made in the late 1950s, that the antibacterial action of penicillin results from inhibition of one of the steps in the synthesis of cell walls gave a major impetus to efforts to elucidate the chemical structure and biosynthetic pathway of bacterial cell walls. Work from many laboratories, notably those of James T. Park and Jack L. Strominger, quickly showed that a glycan linked to a peptide—that is, a **peptidoglycan**—is the major component of the cell walls of both Gram-positive and Gram-negative bacteria. Once the structure was established, the pathway for synthesis of peptidoglycan was then elucidated. Other carbohydrate-containing molecules are present in the cell walls and outer membranes of certain bacteria. For example, teichoic acids are components of the cell walls of certain Gram-positive bacteria, and lipopolysaccharides are components of the outer membranes of Gram-negative bacteria.

A. Peptidoglycan Is Composed of a Glycan Moiety to Which Small Peptides Are Linked

The glycan moiety of peptidoglycan is a polymer composed of alternating *N*-acetylglucosamine (GlcNAc) and *N*-acetylmuramic acid (MurNAc) residues joined in β-$(1 \rightarrow 4)$ linkage (Figure 9·34). MurNAc, a nine-carbon sugar specific to bacteria, consists of D-lactate, a three-carbon acid, joined by an ether linkage to C-3 of GlcNAc. The glycan moiety resembles chitin, except that alternating molecules of GlcNAc are modified by addition of the lactyl moiety.

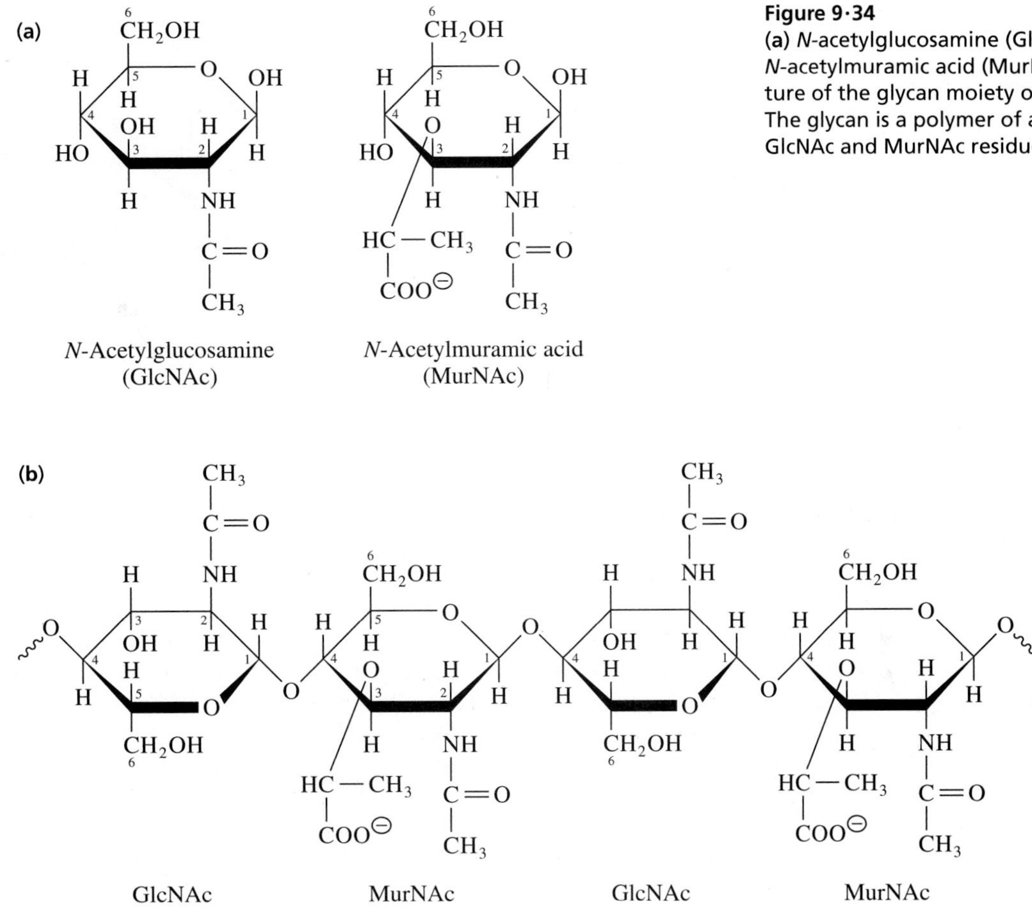

Figure 9·34
(a) *N*-acetylglucosamine (GlcNAc) and *N*-acetylmuramic acid (MurNAc). (b) Structure of the glycan moiety of peptidoglycan. The glycan is a polymer of alternating GlcNAc and MurNAc residues.

In *Staphylococcus aureus,* a Gram-positive pathogen, the peptide component of peptidoglycan is a tetrapeptide with the sequence L-Ala–D-Isoglu–L-Lys–D-Ala (*Isoglu* represents isoglutamate). The amino group of the L-alanine residue is linked to the lactyl carboxylate group of a MurNAc residue of the glycan polymer through an amide bond (Figure 9·35, next page). Noteworthy is the presence of alternating L and D amino acids, one of which is designated D-isoglutamate rather than D-glutamate because it is linked via its γ-carboxyl group to the L-lysine residue. Variations in the amino acid composition of the tetrapeptide occur among bacterial species. The tetrapeptide is cross-linked to another tetrapeptide on a neighboring peptidoglycan molecule by a linker peptide consisting of five glycine residues (pentaglycine). Pentaglycine joins the L-lysine residue of one tetrapeptide to the carboxyl group of the D-alanine residue of the other tetrapeptide. The extensive cross-linking of peptidoglycan essentially converts the peptidoglycan into one giant macromolecule and confers appreciable rigidity on the cell wall.

The enzyme lysozyme catalyzes the hydrolysis of the β-$(1 \rightarrow 4)$ linkages between the MurNAc and GlcNAc residues of peptidoglycan (Section 7·17); this action degrades the bacterial cell wall. Lysozyme is present in a variety of secretions, where it functions as an antibacterial agent, as first described by Alexander Fleming in 1922. Treatment of bacteria with lysozyme can produce protoplasts, forms of bacteria without cell walls, which are viable as long as the osmotic pressure to which they are exposed is carefully adjusted. Even small changes in osmotic pressure may cause the protoplasts to burst.

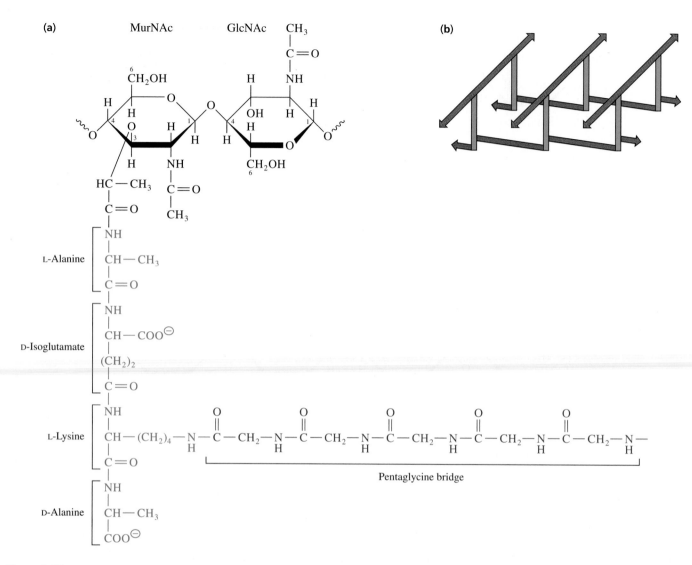

Figure 9·35
Diagram of the structure of the peptidoglycan of *Staphylococcus aureus*. **(a)** Tetrapeptide linked to a MurNAc residue of a
glycan chain. The γ-carboxyl group of
D-isoglutamate is involved in the peptide
linkage to L-lysine. The ε-amino group of the
L-lysine residue of one tetrapeptide is cross-
linked to the α-carboxyl group of the
D-alanine residue of another tetrapeptide on
a neighboring peptidoglycan molecule via a
pentaglycine segment. **(b)** Overall structure
of the peptidoglycan macromolecule and its
cross-linking. The polysaccharide is gray, the
tetrapeptide is blue, and the pentaglycine
cross-link is red.

B. The Synthesis of Peptidoglycan Is Complex, Involving Many Sequential Reactions

Peptidoglycans are synthesized by the pathway shown in Figure 9·36. The peptide
chain is assembled by addition of the amino acids L-alanine, D-glutamate, and
L-lysine in a stepwise fashion to uridine diphosphate MurNAc (UDP-MurNAc), a
nucleotide sugar formed from UDP-GlcNAc and phosphoenolpyruvate. The dipeptide D-Ala–D-Ala is added next. The phosphate-MurNAc-peptide portion of the
nucleotide sugar is then transferred to the lipid bactoprenol phosphate. In Step 6, a
GlcNAc residue from UDP-GlcNAc is added to the MurNAc moiety. Five glycine
residues are then sequentially added to the ε-amino group of the lysine residue. This
is an unusual series of reactions because each glycine residue is donated by a
glycyl–transfer RNA molecule. The only other process involving transfer RNA
molecules is protein synthesis (Chapter 30). The bactoprenol-GlcNAc-MurNAc-
peptide crosses the plasma membrane of the bacterium, an event facilitated by the
bactoprenol moiety. The GlcNAc-MurNAc-peptide is then attached to a preexisting
strand of peptidoglycan in the cell wall, releasing bactoprenol pyrophosphate,
which is hydrolyzed to bactoprenol phosphate and P_i. The new MurNAc residue is
linked to a terminal reducing GlcNAc residue in the cell wall. In the final step, catalyzed by a transpeptidase, the penultimate D-alanine residue is joined in peptide
linkage to a terminal glycine residue of a neighboring peptidoglycan strand. This
reaction is driven by the release of the terminal D-alanine residue.

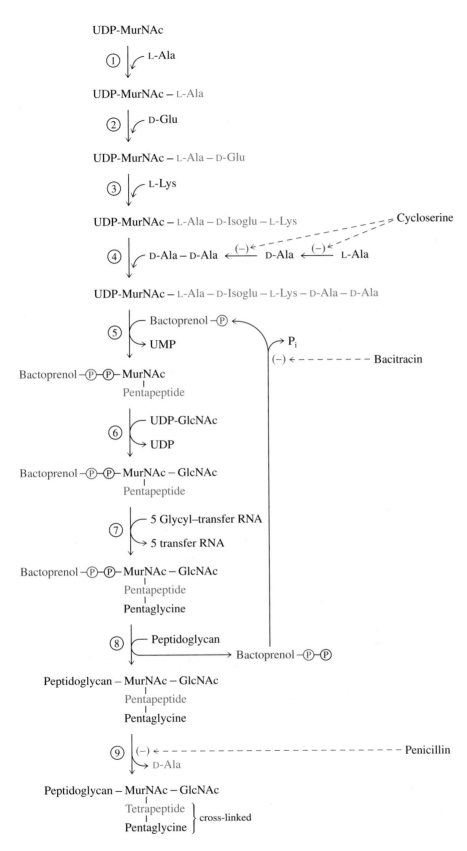

Figure 9·36
Biosynthesis of the peptidoglycan of *S. aureus*. The sites of action of the antibiotics cycloserine, bacitracin, and penicillin are shown. *Isoglu* represents isoglutamate.

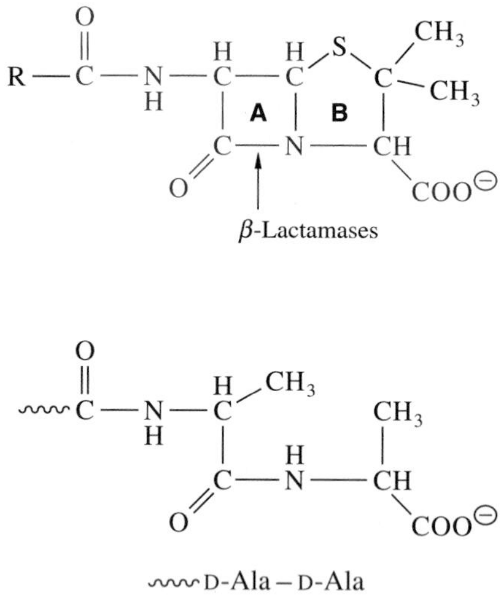

Figure 9·37
Structures of penicillin and D-Ala–D-Ala. The portion of penicillin that resembles the dipeptide is colored red. Ring A is a β-lactam ring, and ring B is a thiazolidine ring. The site of cleavage by β-lactamases (penicillinases) is indicated. R can be a variety of substituents; in the case of benzylpenicillin, it is a benzyl group.

Alexander Fleming observed in 1928 that the mold *Penicillium notatum* produces a compound that inhibits the growth of certain bacteria; this led to the isolation of penicillin and its use in clinical trials in 1941 by Howard Florey and Ernest Chain. The structure of penicillin (Figure 9·37) resembles that of the terminal D-Ala–D-Ala dipeptide, which is a substrate of the transpeptidase that catalyzes the final reaction of peptidoglycan synthesis. Penicillin binds—probably irreversibly—to the active center of the transpeptidase, inhibiting the activity of the enzyme and thereby blocking further peptidoglycan synthesis. A number of other antibiotics also inhibit specific reactions in the biosynthesis of peptidoglycan. The sites of action of cycloserine and bacitracin are shown in Figure 9·36.

Penicillin is selectively toxic to bacteria because the reaction it affects occurs only in certain bacteria, not in eukaryotic cells. It is more effective against Gram-positive than Gram-negative bacteria because Gram-negative bacteria rely more on other compounds for cell-wall integrity. Adverse reactions in humans to penicillin are usually due to immunologic factors (e.g., allergy) and not to direct toxicity caused by the drug.

Bacterial resistance to penicillin can result from the presence in resistant cells of a β-lactamase (penicillinase). β-Lactamases catalyze opening of the β-lactam ring of penicillin, leading to its inactivation. The genes for β-lactamases are generally carried by plasmids harbored by resistant bacteria.

C. Other Carbohydrate-Containing Molecules Are Components of the Cell Walls and Outer Membranes of Certain Bacteria

Other important molecules containing carbohydrates also occur in the cell walls and outer membranes of certain bacteria. For instance, the cell walls of certain Gram-positive bacteria contain large amounts of **teichoic acids**, which contribute to the stability of the cell wall. These are polymers of as many as 30 glycerol phosphate or ribitol phosphate moieties joined by phosphodiester linkages (Figure 9·38). Lipoteichoic acids are linked covalently to lipids in the plasma membrane, and wall teichoic acids are linked covalently to the MurNAc residues of peptidoglycan.

Complex components known as **lipopolysaccharides** are found in the outer membranes of Gram-negative bacteria. Lipopolysaccharides contain lipid A (a disaccharide of phosphorylated glucosamine residues with attached fatty acids) and a polysaccharide. The polysaccharide is composed of a core of approximately 10 sugars and a repeating series of 3–5 sugars (called the O side chain), which varies among different species. The general structure of the lipopolysaccharide of *Salmonella typhimurium* is shown in Figure 9·39. Lipopolysaccharides are released from bacteria undergoing lysis and are toxic to animals and humans. They are often referred to as endotoxins; the toxicity is associated with the lipid A moiety and manifests itself as fever, shock, intravascular coagulation, or even death.

Figure 9·38
Structure of a teichoic acid. This type of teichoic acid is composed of ribitol residues joined by $1 \rightarrow 5$ phosphodiester linkages and is found in the cell wall of *S. aureus* and other Gram-positive bacteria. The number of repeat units is usually 6–10 but can be up to 30 in some cases. D-Alanine, GlcNAc, and other substituents can be attached to specific hydroxyl groups. This type of teichoic acid can be attached covalently to a MurNAc residue of peptidoglycan or can be free.

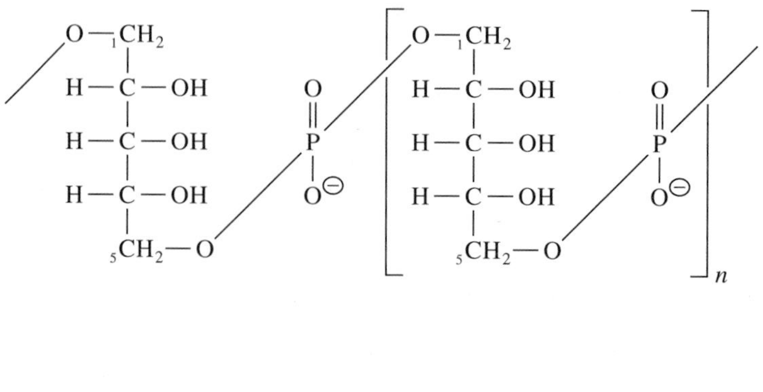

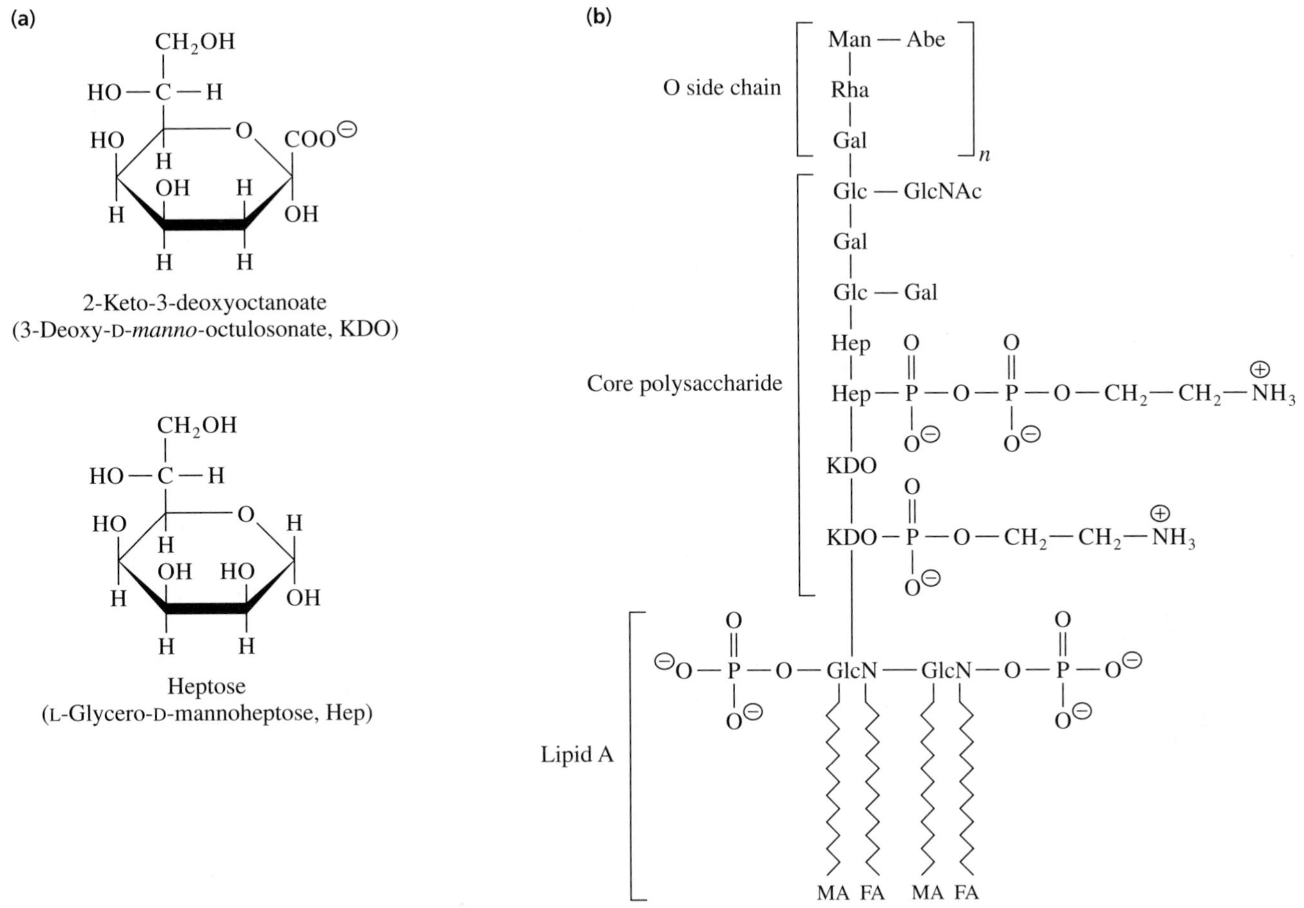

(a)

CH$_2$OH

HO—C—H

2-Keto-3-deoxyoctanoate
(3-Deoxy-D-*manno*-octulosonate, KDO)

CH$_2$OH

HO—C—H

Heptose
(L-Glycero-D-mannoheptose, Hep)

(b)

O side chain

$$\begin{bmatrix} Man—Abe \\ | \\ Rha \\ | \\ Gal \end{bmatrix}_n$$

Core polysaccharide

Glc — GlcNAc
|
Gal
|
Glc — Gal
|
Hep
|
Hep—P—O—P—O—CH$_2$—CH$_2$—NH$_3$
|
KDO
|
KDO—P—O—CH$_2$—CH$_2$—NH$_3$

Lipid A

$^-$O—P—O—GlcN—GlcN—O—P—O$^-$

MA FA   MA FA

Figure 9·39
Lipopolysaccharide from *Salmonella typhimurium*. **(a)** Structure of two sugars found in the lipopolysaccharide. **(b)** Structure of the lipopolysaccharide. The number of repeating units in the O side chain varies from 10 to 40. The sugars found in the O side chain vary among species, whereas the composition of the core polysaccharide is relatively constant. One molecule of β-hydroxymyristic acid (MA), a C$_{14}$ fatty acid, is attached to each GlcN residue. Other fatty acids (FA) are attached to these residues and may be attached to the β-hydroxymyristate groups.

9·10 Glycoproteins Are Ubiquitous and Their Oligosaccharide Chains Exhibit Great Diversity

Glycoproteins, which contain covalently bound carbohydrate, are ubiquitous in nature. The carbohydrate portion contributes from about 1% to more than 80% of the mass of glycoproteins. Glycoproteins are important constituents of plasma membranes; the oligosaccharide chains of plasma-membrane glycoproteins are almost always located on the extracellular surface, forming a thick carbohydrate coat called the glycocalyx (Section 12·5B). Other glycoproteins occur in soluble form, for example, in the blood plasma or in secretions. Glycoproteins are involved in a multitude of biological processes. As we will see in Section 9·13, some of the specific functions of glycoproteins are mediated by their oligosaccharide chains. In this and the following sections, we consider aspects of the structure, analysis, and function of glycoproteins.

The oligosaccharide chains found in glycoproteins show an enormous range of structures. Even among molecules of the same protein, the structure of oligosaccharide chains can vary, a phenomenon called microheterogeneity. Proteins with identical amino acid sequences but different oligosaccharide-chain compositions are called **glycoforms.**

Table 9·2 Principal sugars in eukaryotic glycoproteins and their corresponding nucleotide sugars

Sugar	Nucleotide sugar
L-Fucose	GDP-fucose
D-Galactose	UDP-galactose
D-Glucose	UDP-glucose
D-Mannose	GDP-mannose
N-Acetyl-D-galactosamine	UDP-N-acetylgalactosamine
N-Acetyl-D-glucosamine	UDP-N-acetylglucosamine
N-Acetylneuraminic acid	CMP-N-acetylneuraminic acid
D-Xylose	UDP-xylose

Several factors contribute to the structural diversity of the oligosaccharide chains of glycoproteins.

1. A number of different sugars can occur in an oligosaccharide chain. Eight sugars predominate in eukaryotic glycoproteins. These are the hexoses L-fucose, D-galactose, D-glucose, and D-mannose; the hexosamines N-acetyl-D-galactosamine and N-acetyl-D-glucosamine; the nine-carbon-containing sialic acids (usually N-acetylneuraminic acid); and the pentose D-xylose. Many different combinations of these sugars are possible. In many cases, the sugars are donated to oligosaccharide chains by the corresponding nucleotide sugars, listed in Table 9·2.

2. The glycosidic linkages joining the various sugars may involve either the α or β anomeric configuration.

3. Various carbon atoms can be involved in glycosidic linkages between sugars. In the case of hexoses and hexosamines, the glycosidic linkages always involve C-1 of one sugar but may involve C-2, -3, -4, or -6 of another hexose or C-3, -4, or -6 of an amino sugar (C-2 is usually N-acetylated in this class of sugar). In the case of a sialic acid, it is C-2, not C-1, that is involved in the linkage to other sugars.

4. Oligosaccharide chains of glycoproteins may be branched. For example, two, three, or four sugars can attach to a hexose such as mannose, making possible the formation of chains with two, three, or four branches. Such structures are referred to as bi-, tri-, or tetra-antennary structures, respectively.

The tremendous diversity of oligosaccharide structures afforded by these factors is made evident by the calculation that four different sugars can theoretically be linked to form approximately 36 000 unique tetrasaccharides. This theoretical number of structures is not found in nature, however, because cells do not possess the enormous battery of enzymes of appropriate specificities to synthesize all the linkages necessary to make so many products.

9·11 There Are Three Major Classes of Glycoproteins

Based on the nature of the linkage between their polypeptides and oligosaccharides, glycoproteins are divided into three major classes: O-linked glycoproteins, N-linked glycoproteins, and phosphatidylinositol-glycan–linked glycoproteins. In most **O-linked glycoproteins,** a GalNAc residue is attached via an O-glycosidic

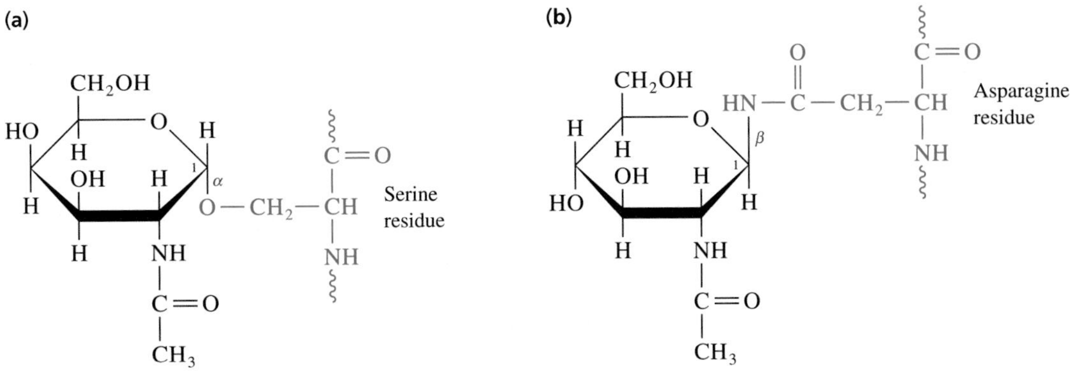

(a)

(b)

Figure 9·40
Comparison of an O-glycosidic linkage and an N-glycosidic linkage. **(a)** N-Acetylgalactosamine–serine linkage, the major O-glycosidic linkage found in glycoproteins. **(b)** N-Acetylglucosamine–asparagine linkage, which characterizes N-linked glycoproteins. The O-glycosidic linkage is α, whereas the N-glycosidic linkage is β.

linkage to a serine or threonine residue. This arrangement is abbreviated GalNAc-Ser/Thr. In **N-linked glycoproteins,** a GlcNAc residue is linked to an asparagine residue via an N-glycosidic linkage, abbreviated GlcNAc-Asn. The structures of an O-glycosidic and N-glycosidic linkage are compared in Figure 9·40. Note that additional sugar residues may be attached to the GalNAc or GlcNAc residue. In **phosphatidylinositol-glycan–linked glycoproteins,** the protein is attached to ethanolamine, which is linked to a branched oligosaccharide to which lipid is also attached.

Members of the O- and N-linked classes of glycoproteins are found in many locations, such as in cell membranes and in blood plasma, whereas members of the phosphatidylinositol-glycan–linked class appear to be confined to the extracellular surface of plasma membranes. An individual glycoprotein may contain both O-linked and N-linked oligosaccharides; for instance, glycophorin A, a major constituent of the membrane of human red blood cells, contains 1 N-linked and 15 O-linked oligosaccharide chains (Section 12·4A). The length of the oligosaccharide chains among O- and N-linked glycoproteins varies; in some cases, the number of sugars can be 30 or more. We will now examine in detail the structures of the three classes of glycoproteins.

A. O-Linked Glycoproteins

Four important subclasses of O-glycosidic linkages in proteins are discussed here. They are listed below and are shown in Figure 9·41 (next page).

1. The most common linkage is the GalNAc-Ser/Thr linkage mentioned above. Other sugars—for example, galactose and sialic acid—are frequently linked to the GalNAc residue. An example of an O-linked oligosaccharide is shown in Figure 9·41a.

2. Some of the 5-hydroxylysine (Hyl) residues of collagen (Section 5·7) are joined to D-galactose via an O-glycosidic linkage (Figure 9·41b). This structure is unique to collagen.

3. The link proteins of certain proteoglycans (see Section 9·14) are joined to their glycosaminoglycans via a Gal-Gal-Xyl-Ser structure (Figure 9·41c). The carbohydrate component is known as a link trisaccharide.

4. A variety of nuclear and cytosolic proteins contain GlcNAc-Ser/Thr moieties in which GlcNAc is the sole sugar attached to the protein (Figure 9·41d).

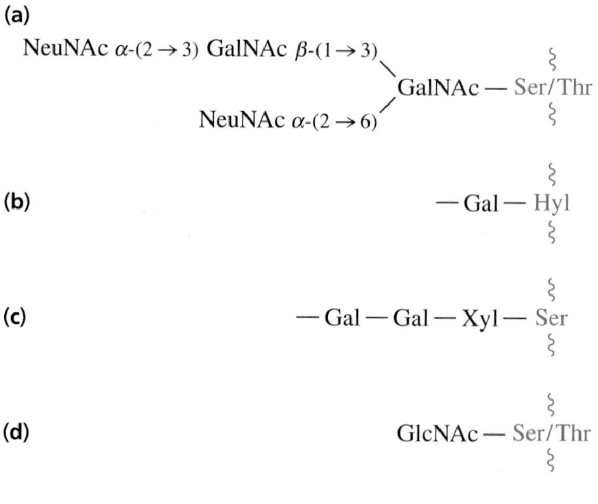

(a)

NeuNAc α-(2 → 3) GalNAc β-(1 → 3)

GalNAc — Ser/Thr

NeuNAc α-(2 → 6)

(b)

— Gal — Hyl

(c)

— Gal — Gal — Xyl — Ser

(d)

GlcNAc — Ser/Thr

Figure 9·41
Four subclasses of *O*-glycosidic linkages. **(a)** Example of a typical linkage in which GalNAc with attached residues is linked to a serine or threonine residue. **(b)** Linkage found in collagen, where a galactose residue, usually attached to a glucose residue, is linked to hydroxylysine (Hyl). **(c)** Link trisaccharide found in certain proteoglycans. **(d)** *O*-GlcNAc linkage found in a variety of nuclear and cytosolic proteins.

Mucins, major members of the *O*-linked class of glycoproteins, are generally of high molecular weight and often contain as much as 80% carbohydrate, in many oligosaccharide chains. They usually contain high amounts of NeuNAc and of sulfate attached to specific sugars. The negative charges of these constituents are in part responsible for the extended shapes of mucins, which confer a high viscosity to fluids containing these glycoproteins. Mucins are found in mucus, the viscous fluid that lines the epithelium of the gastrointestinal, genitourinary, and respiratory tracts and other sites, where they perform protective and lubricatory functions.

The biosynthesis of *O*-linked glycoproteins is complex, involving a battery of specific enzymes in distinct compartments of the cell. In the stepwise synthesis of an oligosaccharide chain, nucleotide sugars play an important role as donors of glycosyl groups in reactions catalyzed by glycosyltransferases. The structure of uridine diphosphate GlcNAc (UDP-GlcNAc) is shown in Figure 9·42. C-1 of the sugar is bound to the terminal phosphate group of the nucleotide in a high-energy linkage. Nucleotide sugars thus have a high group-transfer potential, so that the sugar moiety can be donated to a suitable acceptor. Nucleotide sugars are employed as glycosyl-group donors in a number of biosynthetic reactions, including the synthesis of the oligosaccharide chains of glycolipids, proteoglycans, and homopolysaccharides such as starch and glycogen.

Glycosyltransferases are usually quite specific for their nucleotide sugars, their acceptor substrates, and the linkages they direct (α or β). However, under certain conditions, they may be less specific and accept other nucleotide sugars as substrates. This behavior may account in part for the microheterogeneity that is observed in oligosaccharide chains.

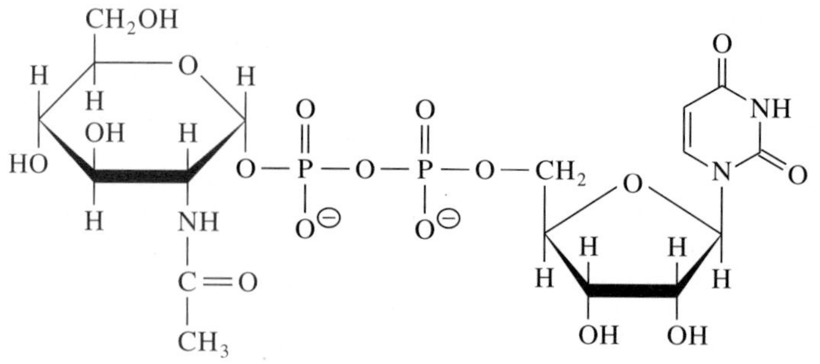

Figure 9·42
Structure of uridine diphosphate *N*-acetylglucosamine (UDP-GlcNAc). GlcNAc is shown in red.

B. *N*-Linked Glycoproteins

N-Linked glycoproteins, like *O*-linked glycoproteins, exhibit a great variety of oligosaccharide chains. Most of them can be divided into three subclasses: high mannose (also known as oligomannose), complex, and hybrid (Figure 9·43). Each of these three classes contains a common core pentasaccharide (GlcNAc$_2$Man$_3$), attached by its innermost GlcNAc residue via an *N*-glycosidic linkage to an asparagine residue, reflecting the fact that all three subclasses share an initial common pathway of biosynthesis. In the biosynthesis of *N*-linked glycoproteins, a branched saccharide chain is transferred to the protein from a lipid-sugar conjugate called dolichol pyrophosphate–oligosaccharide; this compound is not involved in the biosynthesis of *O*-linked glycoproteins. High-mannose chains represent an early stage in the biosynthesis of *N*-linked oligosaccharides. Complex oligosaccharide chains result from the removal of sugar residues from high-mannose chains by the action of specific glycosidases and the addition of other sugar residues, such as fucose, galactose, GlcNAc, and sialic acid (a phenomenon termed oligosaccharide processing). As in the synthesis of *O*-linked glycoproteins, these additional sugar residues are donated by nucleotide-sugar cosubstrates in reactions catalyzed by glycosyltransferases. In certain cases, a glycoprotein may contain a hybrid oligosaccharide chain, a branched oligosaccharide in which one branch is of the high-mannose type and the other is of the complex type. Also, certain oligosaccharide chains may contain three or four (or even more) complex branches; these are called tri- or tetra-antennary structures. These branches are initiated by addition of a GlcNAc residue to various positions on the two exterior mannose residues of the core pentasaccharide.

   Additional details of the biosynthesis of the oligosaccharide chains of *O*- and *N*-linked glycoproteins are presented in Chapter 30 where protein synthesis is described.

Figure 9·43
Structures of *N*-linked oligosaccharides. **(a)** High-mannose chain. **(b)** Complex chain. **(c)** Hybrid chain. The pentasaccharide core common to all *N*-linked structures is shown in red. *SA* represents sialic acid, usually NeuNAc.

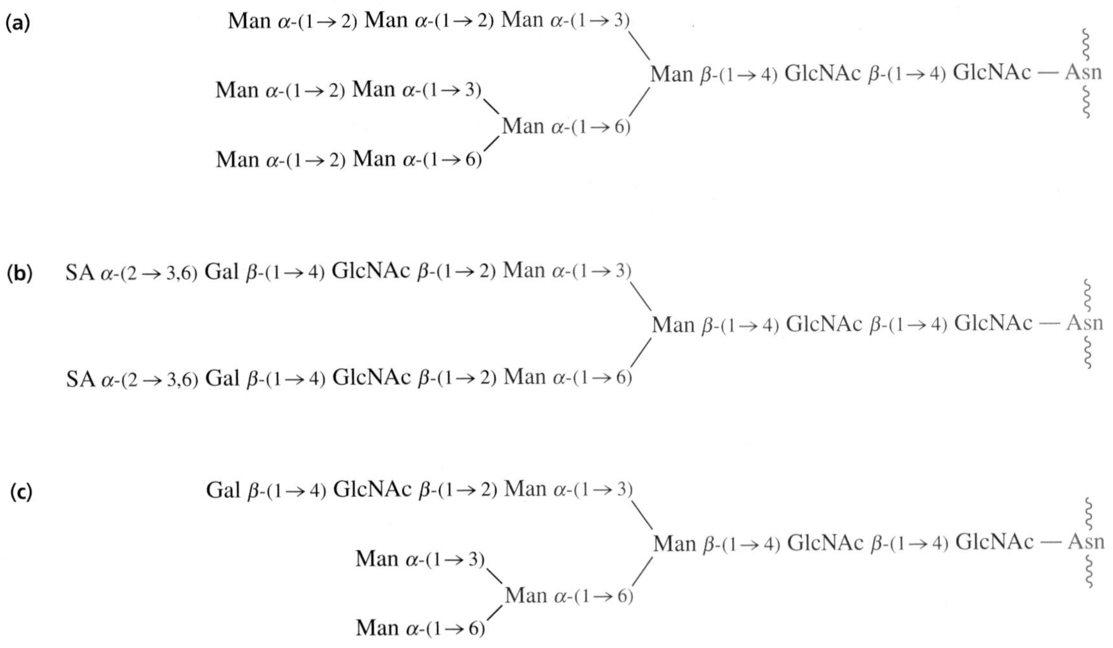

(a)

Man α-(1→2) Man α-(1→2) Man α-(1→3)

Man α-(1→2) Man α-(1→3)

Man β-(1→4) GlcNAc β-(1→4) GlcNAc — Asn

Man α-(1→6)

Man α-(1→2) Man α-(1→6)

(b)

SA α-(2→3,6) Gal β-(1→4) GlcNAc β-(1→2) Man α-(1→3)

Man β-(1→4) GlcNAc β-(1→4) GlcNAc — Asn

SA α-(2→3,6) Gal β-(1→4) GlcNAc β-(1→2) Man α-(1→6)

(c)

Gal β-(1→4) GlcNAc β-(1→2) Man α-(1→3)

Man β-(1→4) GlcNAc β-(1→4) GlcNAc — Asn

Man α-(1→3)

Man α-(1→6)

Man α-(1→6)

C. Phosphatidylinositol-Glycan–Linked Glycoproteins

The third major class of glycoproteins includes proteins linked to surface membranes via phosphatidylinositol-glycan structures, also referred to as glycosylphosphatidylinositol (GPI) membrane anchors, or "sticky feet." These glycoproteins, recognized quite recently, have been identified only in eukaryotic organisms and include certain cell-adhesion molecules and some enzymes. The GPI membrane anchor of the variant surface glycoprotein of *Trypanosoma brucei,* the parasite that causes sleeping sickness, is shown in Figure 9·44. In this glycoprotein, the carboxyl group of the C-terminal aspartate residue of the protein is joined to the amino group of phosphoethanolamine, the phosphate group of which is in turn linked to a mannose residue. This is one of three mannose residues, the innermost of which is linked to a glucosamine residue. The latter is attached to phosphatidylinositol, a glycerophospholipid (Section 11·5). Also attached to the innermost mannose residue is a branched chain containing four galactose residues. The structures of phosphatidylinositol-glycan–linked glycoproteins differ among species. For instance, the amino acid residue to which the GPI group is attached is variable, some GPI groups contain additional molecules of phosphoethanolamine, and there is considerable heterogeneity of the galactosyl oligosaccharides and of the fatty acid composition of the phosphatidylinositol moiety. GPI membrane anchors are embedded in the outer surface of the membrane via the fatty acyl chains of the phosphatidylinositol moiety (Section 12·3).

To test whether particular plasma membranes contain proteins anchored by GPI linkages, a GPI-specific phospholipase C (PIPLC) can be used to hydrolyze the linkage shown in Figure 9·44. If the membrane under study is incubated with PIPLC in a suitable buffer, the bond may be hydrolyzed. Proteins anchored by a

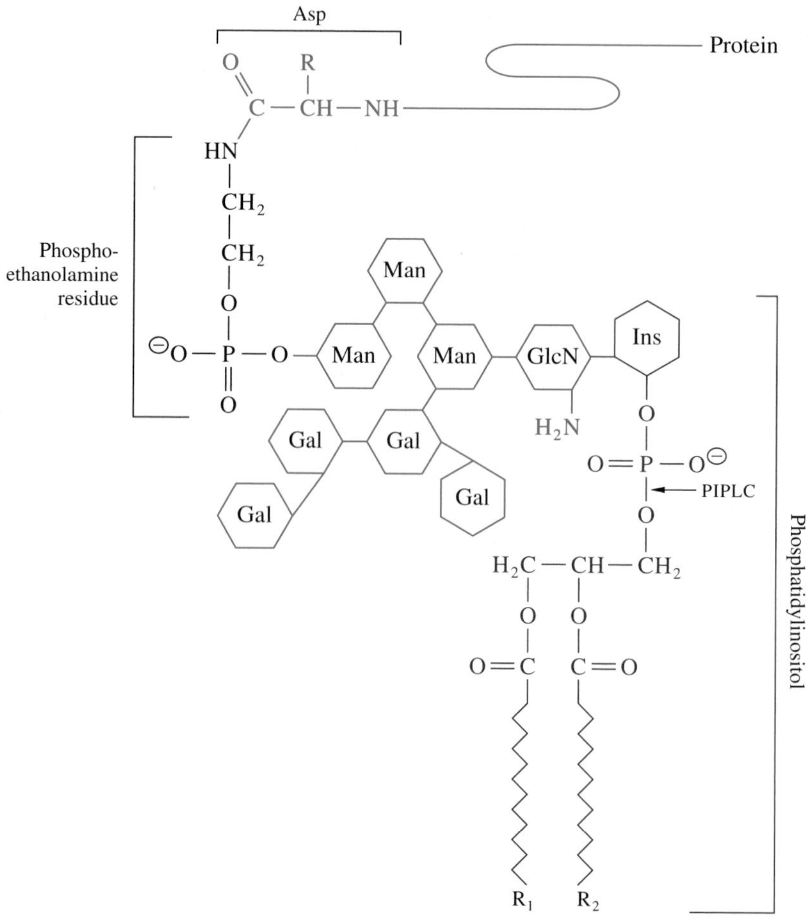

Figure 9·44
Structure of the glycosylphosphatidylinositol membrane anchor of *Trypanosoma brucei.* The C-terminal aspartate residue of the protein is linked to phosphoethanolamine. The innermost mannose residue is linked to a glucosamine residue, which in turn is linked to the inositol group (Ins) of phosphatidylinositol. The fatty acyl chains (R_1 and R_2) of phosphatidylinositol are embedded in the plasma membrane. The site of cleavage by PIPLC is indicated by the arrow.

GPI linkage are thereby released into the aqueous phase, where they can be detected. However, this test is not infallible; in some instances, structural variations in the GPI anchor prevent the action of PIPLC.

The functions of GPI structures are not known with certainty except that they anchor certain proteins to the cell surface. There is evidence that in some epithelial cells they may serve to target certain proteins to the apical region of the plasma membrane (i.e., the top of the cell). Defects in the biosynthesis of GPI anchors may cause certain diseases. Paroxysmal nocturnal hemoglobinuria is a rare condition in which red blood cells hemolyze, particularly at night, releasing hemoglobin into the circulation and hence into the urine. Recent studies have indicated that this condition may be due to a defect in the synthesis of the mannose residues of the GPI anchors of red blood cells, which presumably renders the membrane more susceptible to rupture.

9·12 Many Methods Are Available to Detect, Purify, and Characterize Glycoproteins

A wide variety of methods are available to detect and purify glycoproteins. The first problem is often establishing whether a particular protein contains carbohydrate. One approach is to separate it from other proteins by SDS-PAGE (Section 4·7) and to stain it with periodic acid–Schiff (PAS) reagent. Periodic acid oxidizes the *cis* diols of sugars to aldehydes, which in turn reduce fuchsin (the dye present in the PAS reagent) to form a red product. A highly sensitive staining protocol has been developed that permits detection of as little as one nanogram of glycoprotein.

If cells that synthesize the protein of interest are available, one can show that the protein is probably a glycoprotein by demonstrating that it contains radioactivity after incubation of the cells with a suitable radioactive sugar, such as galactose, mannose, or *N*-acetylglucosamine. However, this approach is not infallible because radioactive sugars can be metabolically converted to other compounds (e.g., amino acids) that are incorporated into proteins.

Further evidence that a protein contains carbohydrate is provided by finding that its electrophoretic migration is affected when it is treated with a purified glycosidase that catalyzes the hydrolysis of its oligosaccharides (Figure 9·45). An enzyme that is widely used for this purpose is peptide *N*-glycosidase F; its action cleaves the linkage between the polypeptide chain and the innermost GlcNAc residue present in an *N*-linked glycoprotein, thus reducing the molecular weight of the protein and increasing its rate of migration in SDS-PAGE. This procedure also helps to distinguish an *N*-linked from an *O*-linked glycoprotein. Similarly, the use of *O*-glycanase, which cleaves GalNAc-Ser/Thr linkages, helps determine the presence of a glycoprotein containing *O*-glycosidic linkages. Endoglycosidase H catalyzes cleavage of the core GlcNAc-GlcNAc bond in high-mannose chains, but not in complex chains, and can therefore be used to reveal information about the structure of the oligosaccharide chains.

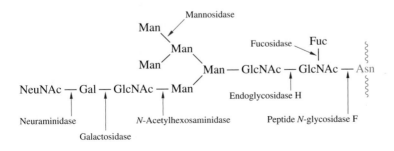

Figure 9·45
Sites of action of various glycosidases on an *N*-linked hybrid chain. For the sake of simplicity, the configurations of the linkages are not indicated, but the various enzymes shown are generally specific for either α or β linkages. Endoglycosidase H catalyzes the cleavage of the bond indicated in high-mannose chains but not in complex chains.

Table 9·3 Some lectins used in the study of glycoproteins

Lectin	Sugar(s) to which lectin binds
Concanavalin A	α-D-Man, α-D-Glc
Ricin	β-D-Gal
Wheat germ agglutinin	β-D-GlcNAc, NeuNAc

Chemical tests can also indicate the presence of carbohydrate in a protein. For instance, oligosaccharide chains of both N- and O-linked glycoproteins are degraded by exposure to anhydrous trifluoromethanesulfonic acid for a suitable period of time. Also, selective exposure of glycoproteins to mild alkali causes β elimination at O-glycosidic linkages.

Certain plant proteins, known as lectins, bind specific sugars present in glycoproteins (Table 9·3). A mixture of proteins can be separated by electrophoresis, and the resultant electrophoretogram can be incubated in a buffer containing a radioactively labelled lectin. Excess lectin is then washed away, and the electrophoretogram is subjected to autoradiography (exposure of film). Glycoproteins that bind the radioactive lectin are revealed as black bands on the developed autoradiogram. Lectins labelled with fluorescent dyes can also be used.

Glycoproteins can be purified by the same procedures used for other proteins (Section 4·7). Membrane-bound glycoproteins are solubilized with mild detergents, which must be present during purification. Affinity chromatography using a lectin covalently attached to the matrix is particularly useful. For example, affinity chromatography using matrix-bound concanavalin A allows the separation of glycoproteins containing high-mannose N-linked chains from those containing complex N-linked chains or O-linked chains.

Once a glycoprotein has been purified, its oligosaccharide moieties can be characterized using a number of methods. A compositional analysis is first performed. This involves quantitative release of the constituent sugars, followed by the separation of these sugars and the determination of their stoichiometry. For instance, the released sugars can be reduced to their corresponding alcohols using sodium borohydride (compare Section 9·4D), acetylated, and then separated and quantitated by gas chromatography or gas chromatography–mass spectrometry (GC-MS). Instruments suitable for compositional analysis of the major sugars present in eukaryotic glycoproteins are now available commercially.

After establishing which sugars are present in the glycoprotein, entire oligosaccharides can be cleaved from their polypeptide chains by the application of hydrazine ($HN_2 = NH_2$), which cleaves N- and O-glycosidic linkages. Alternatively, specific glycosidases can be used to catalyze the hydrolysis of particular oligosaccharide chains from the polypeptide chains to which they are attached (Figure 9·45).

The oligosaccharide chains liberated by some of the above procedures can be purified and then analyzed by methods such as fast atom bombardment–mass spectrometry (FAB-MS) and ^{1}H and ^{13}C nuclear magnetic resonance (NMR) spectroscopy (Section 5·2). In FAB-MS, oligosaccharide chains are fragmented and ionized in the presence of a liquid matrix such as glycerol by bombardment with a beam of neutral atoms, usually argon or xenon, or with a beam of cesium ions. The major fragment ions produced from oligosaccharides have been described and cataloged. Analysis by FAB-MS can yield information on the molecular weight of an oligosaccharide, its composition, the sequence of its sugars, and sometimes on branching and specific linkages. It cannot distinguish epimers (e.g., galactose from mannose) or identify the anomeric nature of glycosidic linkages. Under appropriate circumstances, NMR spectroscopy can be used to identify the specific sugars present in an oligosaccharide, their sequence, their linkages, and the anomeric nature of these linkages (α or β). Information on anomeric configuration can also be obtained by the use of glycosidases that are specific for α or β linkages.

A technique called methylation analysis has proven invaluable in determining the linkages between sugars in oligosaccharide chains. This technique involves treatment of the chain with a methylating agent such as methyl iodide. This methylates all primary or secondary alcohol groups on sugars not involved in glycosidic linkages (Figure 9·46). The various methylated sugars can be released by hydrolysis—after reduction and acetylation—and quantitated by GC-MS. Identification of

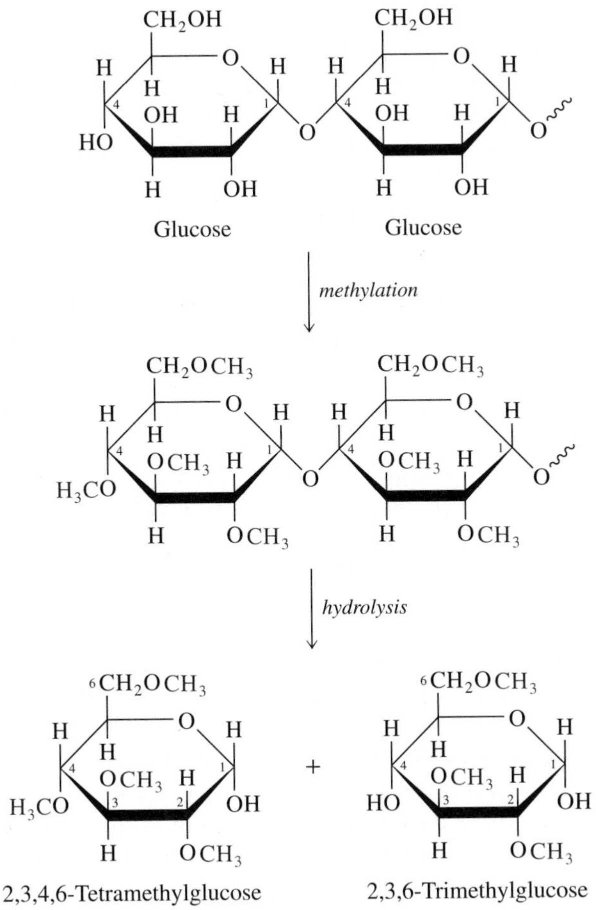

Figure 9·46
Principle of methylation analysis. Two glucose units, joined in α-$(1 \rightarrow 4)$ glycosidic linkage and forming part of a longer oligosaccharide chain, are methylated. The methyl groups are shown in red. When subjected to hydrolysis, one molecule of 2,3,4,6-tetramethylglucose and one molecule of 2,3,6-trimethylglucose are liberated. Note that the hydroxyl groups that were involved in glycosidic linkage are not methylated. The partially methylated products can then be identified and quantified by GC-MS.

the various methylated products permits deduction of the carbon atoms that were involved in glycosidic linkage. However, methylation analysis does not reveal the anomeric nature of glycosidic linkages.

9·13 The Oligosaccharide Chains of Glycoproteins Contribute to Their Physical and Biological Properties

Many types of oligosaccharide chains can be attached covalently to proteins. Proteins of all types (e.g., enzymes, hormones, structural proteins, and transport proteins) may be glycosylated. The presence of one or more oligosaccharide chains on a protein can alter both its physical and biological properties. Various physical properties that may be altered include overall size, solubility, heat stability, conformation, tendency to aggregate, and resistance to proteases. Biological properties that may be altered include rate of secretion, half-life in the circulation, activity, and immunogenicity. Increasing attention is being paid to the effects of glycosylation on the physical and biological properties of glycoproteins, due in part to the number of glycoproteins that are being made available for therapeutic use via recombinant DNA technology. The extent and type of glycosylation of such proteins vary depending upon the host cells in which the gene under study is being expressed, and differences in glycosylation may affect the therapeutic usefulness of the proteins.

In a few cases, specific roles of oligosaccharide chains of glycoproteins have been identified. For example, a number of mammalian hormones are dimeric glycoproteins whose oligosaccharide chains appear to facilitate interaction of the protein subunits and assembly of active hormone, thus preventing proteolysis and degradation. Studies of egg fertilization in mice have revealed that α-galactose residues on a specific glycoprotein in the coat surrounding the oocyte mediate the ability of oocytes to bind sperm. These results suggest that it may be possible to inhibit fertilization by synthesizing compounds that interfere with sperm binding to the α-galactose residues.

The following three subsections shed additional light on the possible functions of oligosaccharide chains in glycoproteins and also show why there is currently great interest in the roles of glycoproteins in normal biological phenomena and in certain disease states.

A. Removal of Terminal Sialic Acid Residues from the Oligosaccharides of Many Plasma Proteins Markedly Shortens Their Half-Lives

Starting in the mid-1960s, a pioneering series of investigations into the possible functions of oligosaccharides in glycoproteins was performed by Gilbert Ashwell and Anatol Morell. They studied the removal, or clearance, of certain proteins from the circulation. It was known that most plasma proteins, with the notable exception of albumin, are glycoproteins, usually containing complex oligosaccharide chains terminating in a sialic acid residue attached to a β-galactose residue. To determine the normal half-life ($t_{1/2}$) of a number of plasma glycoproteins in rabbits, Ashwell and Morell radiolabelled the purified molecules in vitro. The individual radiolabelled proteins were then injected into the bloodstream of the rabbits, blood samples were taken at various times thereafter, and the $t_{1/2}$ of each glycoprotein was calculated. In this way, a normal value for $t_{1/2}$ was obtained for each protein. They next incubated the individual labelled proteins with neuraminidase, which catalyzes removal of terminal sialic acid residues, and studied the $t_{1/2}$ of the resulting asialoglycoproteins. In most cases, the $t_{1/2}$ of the asialoglycoprotein was found to be markedly shortened compared to the normal protein. For example, the half-life of ceruloplasmin was shortened from 56 hours to about 5 minutes (Figure 9·47). Taking the experiments one step further, Ashwell and Morell also removed the terminal galactose residues of the asialoglycoproteins by treatment with β-galactosidase.

Figure 9·47
Half-lives of the plasma protein ceruloplasmin. The untreated protein has a half-life of approximately 56 h. After treatment with neuraminidase to remove terminal sialic acid residues, the protein has a half-life of approximately 5 min. Subsequent treatment with β-galactosidase to remove terminal galactose residues restores the half-life to near-normal values. This experiment suggested that terminal sialic acid residues are involved in the plasma clearance of ceruloplasmin. [Adapted from Ashwell, G., and Morell, A. G. (1974). The role of surface carbohydrates in the hepatic recognition and transport of circulating glycoproteins. In *Advances in Enzymology*, Vol. 41, A. Meister, ed. (New York: Wiley Interscience), pp. 99–128.]

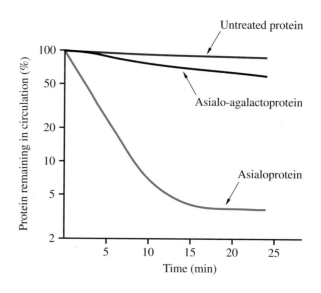

Surprisingly, the half-lives of the glycoproteins lacking both sialic acid and galactose residues were approximately the same as the values for the normal, untreated proteins. From these results, Ashwell and Morell concluded that exposure of subterminal galactose residues acts as a signal for the clearance of many glycoproteins from the plasma. In further studies, Ashwell and Morell showed that hepatocytes possess a receptor that recognizes glycoproteins with terminal galactose residues and removes them from the plasma. This receptor, called the asialoglycoprotein receptor, consists of at least three subunits, two of which are identical. Each subunit can bind the terminal galactose residues of glycoproteins, but the binding is much tighter when the three subunits interact in the native receptor and also when the oligosaccharide chain of the asialoglycoprotein is branched (i.e., the magnitude of binding is triantennary > biantennary > monoantennary). It has been calculated that each hepatocyte can bind approximately 500 000 molecules of certain asialoglycoproteins. After binding to the receptor, the asialoglycoprotein is brought into the cell by endocytosis and subsequently degraded.

The studies of sugar-mediated glycoprotein clearance raised the possibility, since confirmed, that similar interactions of one or more specific sugars with receptors or other target molecules also operate in many other biological phenomena. The activity of the asialoglycoprotein receptor also underscores the importance of proper glycosylation of glycoproteins, including those produced by recombinant DNA technology. For example, erythropoietin, a glycoprotein produced by cells of the kidney that is crucial in regulating the production of red blood cells, is useful in the treatment of certain types of anemia, particularly that associated with chronic kidney disease. Only small amounts of erythropoietin were available for therapeutic use before recombinant DNA technology made feasible its large-scale production. If erythropoietin is not glycosylated in its normal manner, it will be removed from the circulation rapidly and be relatively ineffective. Therefore, it is produced in cells that carry out the appropriate glycosylation reactions.

B. The Plasma Membranes of Cancer Cells Exhibit Abnormal Oligosaccharide Structures

Many changes in the structures of oligosaccharides in glycoproteins and glycosphingolipids, particularly those situated in the plasma membrane, have been detected in cancer cells. These alterations are often due to the reappearance of structures that were expressed in fetal tissues. During development, the expression of some genes for glycosyltransferases fluctuates, with a number of the genes being silenced at specific stages. However, for unknown reasons, certain genes for glycosyltransferases may be reactivated in cancer cells, resulting in the reappearance of oligosaccharide structures that were present in the fetus. The abnormal expression of these structures may be implicated in the unregulated growth and metastasis that are characteristic of cancer cells. Some cancer cells with a high ability to metastasize (spread to distant parts of the body) possess highly branched oligosaccharides (e.g., tetra-antennary structures) on their surfaces. This appears to be due to increased expression in these cells of a GlcNAc transferase that plays a key role in the synthesis of tetra-antennary complex structures. It is possible that inhibition of this particular enzyme, for example, by specifically designed drugs, could diminish the ability of certain cancer cells to metastasize.

Another potential anticancer drug is swainsonine (Figure 9·48). This compound inhibits a mannosidase involved in the trimming of high-mannose oligosaccharide chains, which results in cancer cells with glycoproteins enriched in high-mannose chains and deficient in complex chains. Treatment of mice with swainsonine inhibits the spontaneous metastasis of certain subcutaneous tumors and also inhibits the growth of certain tumor cells.

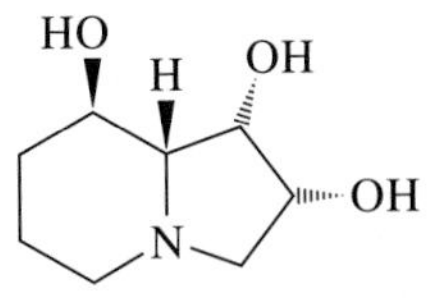

Figure 9·48
Structure of swainsonine, an inhibitor of α-mannosidase.

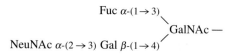

Figure 9·49
Structure of 3'-sialyl-Lewis^x, the oligosaccharide on leukocytes that interacts with E-selectin on endothelial cells.

C. Endothelial Cell-Surface Proteins Recognize Certain Oligosaccharides

Acute inflammation plays a role in protecting animals against bacterial infection. One of its important characteristics is the accumulation of white blood cells (leukocytes) in the inflamed area. In order for leukocytes to accumulate, they must first adhere to a specific protein present on the plasma membranes of the endothelial cells of the local blood vessels. The protein to which leukocytes bind is called E-selectin (*E* because it was originally recognized on the surface of activated endothelial cells that line the blood vessels).

Selectins constitute a family of $Ca^{2 \oplus}$-dependent membrane glycoproteins. They generally have molecular weights of 90 000–140 000 and have a number of distinct domains, including an N-terminal lectinlike domain. Selectins bind specific oligosaccharide groups. The oligosaccharide group to which E-selectin binds is known as 3'-sialyl-Lewis^x (Figure 9·49). (*Lewis* refers to a group of specific blood-group antigens.) 3'-Sialyl-Lewis^x is present in certain glycoproteins and glycolipids found in the plasma membrane of leukocytes.

When tissue is injured, it releases cytokines, which are protein hormones that locally stimulate inflammatory responses, including the expression of E-selectin on endothelial cells. Adhesion of leukocytes to nearby endothelial cells is thereby increased. In addition, cytokines signal the endothelial cells to change shape, resulting in the opening of gaps between the cells. Adherent leukocytes can leave the circulation and enter the tissues through these gaps.

Two other major selectins are P-selectin (first described in activated platelets, also involved in the adhesion of leukocytes to endothelial cells) and L-selectin, also known as homing receptor. L-Selectin, present on certain circulating lymphocytes, directs these cells to peripheral lymph nodes. These lectins also recognize sialylated ligands.

New information available on selectins and on the carbohydrate groups with which they associate suggests that the interactions of a variety of cells, particularly those involved in inflammation, can be altered to produce beneficial therapeutic effects. For instance, 3'-sialyl-Lewis^x or a more potent synthetic analog could be administered in vivo in order to occupy the binding sites of E-selectin molecules on the surfaces of endothelial cells, thus preventing them from interacting with leukocytes. This would inhibit the inflammatory reaction, which in some cases can prove harmful to the host. 3'-Sialyl-Lewis^x is expressed on the surfaces of many tumor cells, raising the possibility that it may also be involved in the altered adhesiveness and metastatic properties shown by such cells.

9·14 Proteoglycans Are Major Components of the Extracellular Matrix

Among its many functions, the extracellular matrix (or connective tissue) acts as a scaffolding for cells and participates in intercellular communication and in the regulation of cellular migration. Its major components are collagen, elastin, and proteoglycans, the third major type of glycoconjugate. Collagen and elastin were discussed in Chapter 5. **Proteoglycans** are complexes of glycosaminoglycans and specific proteins. The glycosaminoglycans may account for up to 95% of the mass of proteoglycans. Many of the same techniques used to detect, purify, and characterize glycoproteins can be applied to proteoglycans.

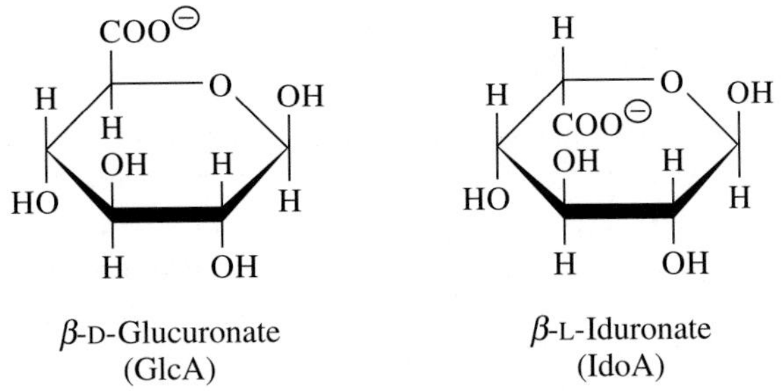

Figure 9·50
Structures of D-glucuronate (GlcA) and L-iduronate (IdoA).

β-D-Glucuronate
(GlcA)

β-L-Iduronate
(IdoA)

A. Glycosaminoglycans Are Unbranched Polysaccharides Composed of Repeating Disaccharides

Each glycosaminoglycan is an unbranched polysaccharide made up of repeating disaccharide units. The number of disaccharide units in these polymers can be very large, and some proteoglycans have extremely high molecular weights. As the name *glycosaminoglycan* indicates, one component of the disaccharide is an amino sugar, either D-galactosamine or D-glucosamine. The amino groups of these two amino sugars can be acetylated, forming GalNAc and GlcNAc, respectively. The other component of the disaccharide is usually a uronic acid: D-glucuronic acid (GlcA) or L-iduronic acid (IdoA), a 5′ epimer of the former. Haworth projections of D-glucuronate and L-iduronate are shown in Figure 9·50. Many glycosaminoglycans also contain sulfate groups, located on various ring hydroxyl groups of the sugars (O-sulfates) or on the amino group of one of the amino sugars (N-sulfates).

At least six different glycosaminoglycans have been isolated and characterized. These include chondroitin sulfate, dermatan sulfate, heparan sulfate, heparin, hyaluronic acid, and keratan sulfate. The structures of the disaccharides that occur in two of these glycosaminoglycans—hyaluronic acid and keratan sulfate—are shown in Figure 9·51. Table 9·4 (next page) lists some of the main features of the six glycosaminoglycans that have been well characterized. Each has its own characteristic sugar composition, linkages, attached proteins, tissue distribution, and function. Note that, unlike the other glycosaminoglycans, hyaluronic acid is not bound covalently to protein and does not contain sulfate. Note also that keratan sulfate does not contain a uronic acid but instead contains galactose.

Figure 9·51
Structures of the repeating disaccharide units in hyaluronic acid and keratan sulfate. (a) Hyaluronic acid consists of alternating D-glucuronic acid (GlcA) and N-acetylglucosamine (GlcNAc) residues. Each GlcA residue is linked to a GlcNAc residue through a β-(1→3) linkage; each GlcNAc residue is in turn linked to the next GlcA residue through a β-(1→4) linkage. (b) Keratan sulfate is composed chiefly of alternating N-acetylglucosamine 6-sulfate (GlcNAc-6-SO$_4$) and galactose residues. Each GlcNAc-6-SO$_4$ residue is linked to a galactose residue through a β-(1→3) linkage; each galactose residue is in turn linked to the next GlcNAc-6-SO$_4$ residue through a β-(1→4) linkage.

(a)

GlcA   GlcNAc

(b)

GlcNAc-6-SO$_4$ Gal

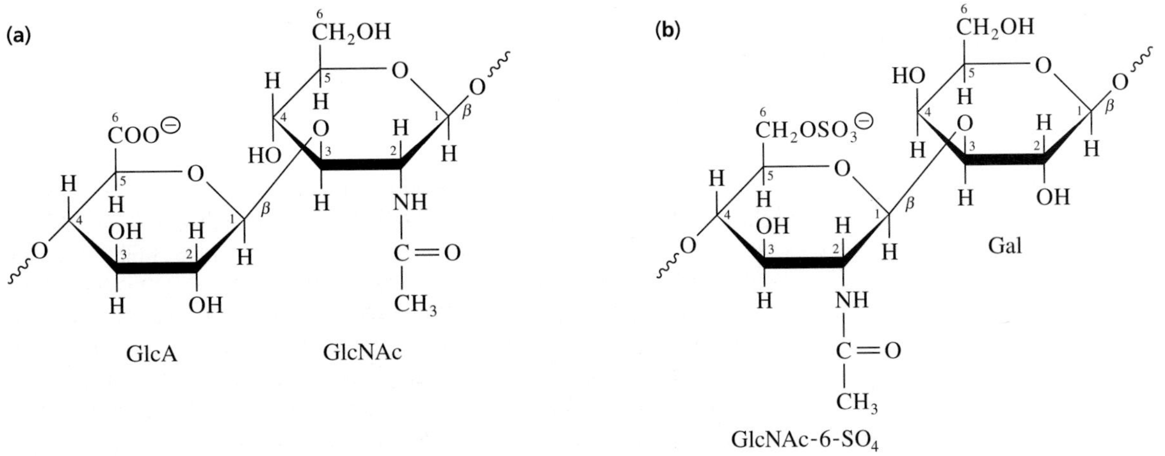

Table 9·4 Summary of the main features of six glycosaminoglycans

Glycosaminoglycans	Uronic acid	Amino sugar	Hexose	Sulfated residue	Protein linkage	Tissue distribution
Chondroitin sulfate	GlcA	GalNAc	—	GalNAc	Xyl-Ser	Cartilage, bone, cornea
Dermatan sulfate	IdoA, GlcA	GalNAc	—	GalNAc, IdoA	Xyl-Ser	Wide distribution (e.g., skin, blood vessels)
Heparan sulfate	GlcA	GlcN	—	GlcN	Xyl-Ser	Skin fibroblasts, aortic wall
Heparin	IdoA, GlcA	GlcN, GlcNAc	—	GlcN, IdoA	Gal-Gal-Xyl-Ser	Mast cells
Hyaluronic acid	GlcA	GlcNAc	—	—	—	Synovial fluid, vitreous humor, loose connective tissues
Keratan sulfate	—	GlcNAc	Gal	GlcNAc, Gal	GlcNAc-Asn	Cornea

B. Cartilage Contains a Proteoglycan Composed of Core and Link Proteins and Several Glycosaminoglycans

Glycosaminoglycans interact with specific core and link proteins to form proteoglycans. Consideration of the structure of the major proteoglycan of cartilage illustrates some important features of proteoglycans. The cartilage proteoglycan aggregate has a very high molecular weight ($M_r \sim 2 \times 10^8$) and contains hyaluronic acid, keratan sulfate, chondroitin sulfate, link proteins, core proteins, and a number of oligosaccharide chains. This complex macromolecular aggregate assumes a characteristic shape resembling a bottle brush (Figure 9·52). A central strand of hyaluronic acid runs through the aggregate, and many core proteins with glycosaminoglycan chains attached branch from its sides. The hyaluronic acid interacts noncovalently—mostly by electrostatic interactions—with the core proteins. These interactions are in turn stabilized by additional interactions—again mainly electrostatic—with a number of link proteins. Each core protein has approximately 30 molecules of keratan sulfate and approximately 100 molecules of chondroitin sulfate attached covalently, the molecular weight of such a proteoglycan monomer being $\sim 2 \times 10^6$. The core and link proteins initially proved difficult to isolate and study, partly because of their insolubility. However, the application of recombinant DNA technology has greatly facilitated the study of these proteins, and the sequences of many of these proteins are becoming available. Already it is known that these proteins differ among various tissues and proteoglycan complexes.

Figure 9·53 shows the structure of a core protein of bovine nasal cartilage in more detail. The protein with attached glycosaminoglycan chains has a molecular weight of ~210 000. It can be subdivided into at least three domains. A specific region of the N-terminal domain (domain A), called the hyaluronate-binding region, interacts noncovalently with five repeating disaccharide units of hyaluronate. The hyaluronate–core protein interaction is stabilized by additional noncovalent interactions between both of these molecules and the link protein. The link protein shows considerable homology with the hyaluronate-binding region of the core protein. Domain B bears about 30 keratan sulfate molecules, attached via O-glycosidic Ser-GalNAc linkages (see Figure 9·40a). The middle and C-terminal regions of the core protein (domain C) bear about 100 molecules of chondroitin sulfate, the density of these molecules diminishing toward the C-terminal end. These chains are attached via O-glycosidic linkages involving the Gal-Gal-Xyl link trisaccharide joined to serine (see Figure 9·41c). Each molecule of chondroitin sulfate is attached via a terminal glucuronate residue to the terminal galactose residue of the link

(a)

Core proteins with glycosaminoglycan chains attached

Central strand of hyaluronic acid

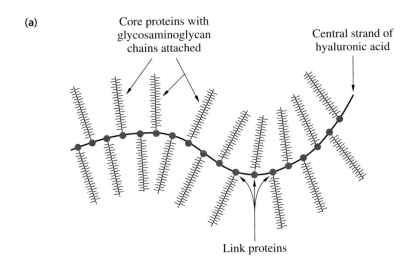

Link proteins

(b)

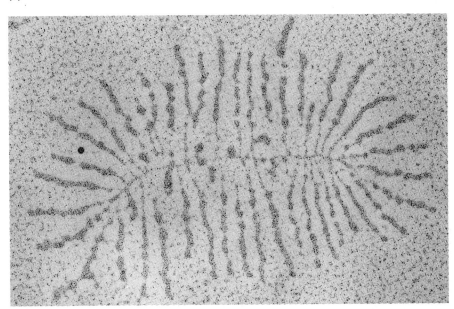

Figure 9·52
Structure of the proteoglycan of cartilage.
(a) The proteoglycan consists of a central strand of hyaluronic acid to which core proteins are attached noncovalently. The core proteins have covalently attached glycosaminoglycan chains (e.g., chondroitin sulfate and keratan sulfate). The interactions of the core proteins with hyaluronic acid are stabilized by link proteins, which interact noncovalently with both types of molecules.
(b) Under the electron microscope, the whole structure has the appearance of a bottle brush. (Courtesy of Joseph A. Buckwalter.)

Domain A   Domain B   Domain C

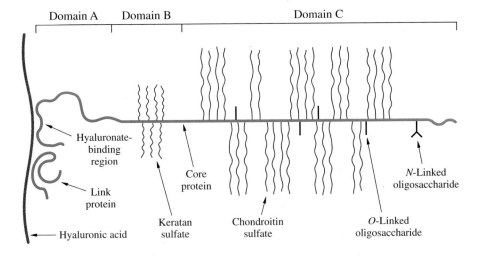

Hyaluronate-binding region

Link protein

Hyaluronic acid

Core protein

Keratan sulfate

Chondroitin sulfate

N-Linked oligosaccharide

O-Linked oligosaccharide

Figure 9·53
Organization of the major proteoglycan of bovine nasal cartilage. A strand of hyaluronic acid is shown on the left. The core protein (M_r ~210 000) has three major domains. Domain A, the N-terminal end of the core protein, interacts with approximately five repeating disaccharides in hyaluronate. The link protein interacts with both hyaluronate and domain A, stabilizing their interactions. Approximately 30 keratan sulfate chains are attached, via GalNAc-Ser linkages, to domain B of the core protein. Domain C contains ~100 chondroitin sulfate chains attached via Gal-Gal-Xyl-Ser linkages and ~40 *O*-linked oligosaccharide chains. An *N*-linked glycan chain is also found near the C-terminus of the core protein.

trisaccharide. Approximately 40 *O*-linked oligosaccharides are also attached to domain C. A fourth type of oligosaccharide chain, an *N*-linked glycan (see Figure 9·40b), is also found attached near the C-terminus of the core protein.

Studies of many other proteoglycans have been performed. Most are smaller in size than the proteoglycan of cartilage, and each has a characteristic complement of glycosaminoglycans, core proteins, and link proteins.

C. The Biosynthesis and Degradation of Glycosaminoglycans Are Catalyzed by Specific Enzymes

Glycosaminoglycans can be attached to core proteins via Gal-Gal-Xyl-Ser linkages, GalNAc-Ser/Thr linkages, and GlcNAc-Asn linkages. The sugar residues added to the growing glycosaminoglycan chain are donated by their corresponding nucleotide sugars (e.g., UDP-GlcNAc, UDP-glucuronate, etc.) in reactions catalyzed by a battery of specific glycosyltransferases. The overall process of chain elongation occurs with high fidelity. The factors determining chain termination are not clearly understood but may include the level of sulfation and the length of the chain.

Sulfation can occur at a variety of locations on individual sugars. For example, hydroxyl groups on carbons 2, 3, 4, and 6, as well as nitrogen moieties of the amino groups of GalN and GlcN, may be sulfated by the action of specific sulfotransferases. In all cases, the sulfate donor is 3′-phosphoadenosine 5′-phosphosulfate (PAPS), shown in Figure 8·9. PAPS is the sulfate donor in almost all biological reactions involving transfer of sulfate to an acceptor.

During the biosynthesis of certain proteoglycans, some D-GlcA residues are epimerized to L-IdoA residues.

$$\text{\small{$\sim\!\!\sim$}} \text{D-GlcA} \text{\small{$\sim\!\!\sim$}} \xrightarrow{\text{5′ Epimerase}} \text{\small{$\sim\!\!\sim$}} \text{L-IdoA} \text{\small{$\sim\!\!\sim$}} \qquad (9·1)$$

L-Iduronic acid is the 5′ epimer of D-glucuronic acid (see Figure 9·50), and the reaction is catalyzed by a 5′ epimerase. Note that the epimerization occurs *after* glucuronate has been incorporated into a glycosaminoglycan, not before. The number of glucuronate residues that are epimerized varies, so that a particular glycosaminoglycan may contain both glucuronate and iduronate residues.

Lysosomes contain enzymes that catalyze the hydrolytic cleavage of specific constituents from glycosaminoglycans. These hydrolases include various sulfatases, hexosaminidases, β-D-galactosidase, β-D-glucuronidase, α-L-iduronidase, etc., the total complement being able to break all the types of linkages present in glycosaminoglycans. These degradative enzymes participate in the normal metabolic turnover of glycosaminoglycans.

Mutations in the genes encoding lysosomal enzymes in animals and humans may result in serious diseases. One example is Hurler's syndrome, caused by mutations in the gene encoding α-L-iduronidase that result in defective activity of the enzyme. Accumulation of the substrates of the enzyme, mainly dermatan sulfate and heparan sulfate, results in early clouding of the cornea, a characteristic coarse facial appearance, abnormal development, and usually early death. Hurler's syndrome can be diagnosed by measuring the activity of α-L-iduronidase in white blood cells or cultured fibroblasts. Increased amounts of dermatan sulfate and heparan sulfate can also be detected in urine.

Diseases affecting the degradation of glycosaminoglycans are examples of lysosomal storage diseases. Mutation in a gene encoding a specific lysosomal hydrolase leads to an accumulation (storage) of its substrate in specific cells or tissues of the body, causing the various signs and symptoms of each disease. Other examples of lysosomal storage diseases are Gaucher's disease, Niemann-Pick disease, and Tay-Sachs disease (see Section 20·17); these diseases are due to mutations in the genes encoding enzymes involved in the catabolism of sphingolipids. Many attempts are being made to develop new treatments for these serious disorders; most attention is focused on enzyme replacement and gene therapy.

D. Proteoglycans Have Many Diverse Functions

The functions of the many proteoglycans found in nature have been relatively difficult to determine, partly because isolating intact proteoglycans is difficult and because proteoglycan structures are complex. It is important to appreciate that proteoglycans are present in the extracellular matrix and also on some cell surfaces. Recent studies show that they interact with many other components of the extracellular matrix, such as collagen, elastin, and fibronectin, probably modulating the functions of these proteins. Proteoglycans usually occupy a very large volume and attract considerable amounts of water, and many of them are polyanions because of their negatively charged sulfate and carboxylate groups. Thus, they help maintain the shape of tissues, bind many inorganic cations such as $Na^{\oplus}$ and $K^{\oplus}$, and act as sieves in the extracellular matrix.

Chondroitin sulfate and hyaluronic acid modulate the compressibility of cartilage in weight-bearing joints. Dermatan sulfate and keratan sulfate contribute to corneal transparency. Heparan sulfate is a component of the plasma membrane of certain cells, where it may act as a receptor and participate in certain cell-cell interactions. Heparin acts as an anticoagulant, and it has been widely used in clinical situations requiring antithrombotic therapy. It is present in mast cells (large connective-tissue cells), from which it can be isolated relatively easily. Heparin inhibits blood clotting primarily by binding to the plasma protein antithrombin III. This interaction induces a conformational change in antithrombin III, increasing the rate at which it binds and inactivates thrombin, the ultimate enzyme of the coagulation cascade (Section 6·11B). Hyaluronic acid appears to play an important role in permitting migration of certain cells through the extracellular matrix. Current research on glycosaminoglycans and proteoglycans is focusing on their possible roles in intercellular communication and transmembrane signalling and on their interactions with specialized proteins and growth factors found in the extracellular matrix.

Summary

Carbohydrates consist of hydroxyaldehydes (aldoses) and hydroxyketones (ketoses) and their derivatives. They include monosaccharides and disaccharides (simple sugars), oligosaccharides, and polysaccharides. Except for the simplest ketose, dihydroxyacetone, carbohydrates are chiral and therefore exhibit optical activity. For a given monosaccharide, there are 2^n possible stereoisomers, where n is the number of chiral carbon atoms. Among these stereoisomers, any two that are nonsuperimposable mirror images of each other are referred to as enantiomers. Two that differ in configuration at only one of several chiral centers are known as epimers. A monosaccharide is designated D or L, depending on the configuration of the chiral carbon farthest from the aldehydic (C-1) or ketonic (usually C-2) carbon atom.

Aldopentoses, aldohexoses, and ketohexoses exist principally as cyclic hemiacetals known as furanoses and pyranoses. In sugar hemiacetals, the anomeric carbon (the carbonyl carbon in the open-chain form) has four substituents, giving these structures an additional asymmetric center. The chirality of the anomeric carbon is designated either α or β. Two optical isomers that differ in configuration only at their anomeric carbon atoms are referred to as anomers. Although furanoses adopt envelope or twist conformations and pyranoses exist primarily in chair conformations, they are often depicted in Haworth projections, which reveal the chirality of each asymmetric carbon atom including the anomeric carbon.

There are several classes of biologically important nonpolymerized derivatives of monosaccharides. These include sugar phosphates, deoxy sugars, amino sugars, sugar alcohols, and sugar acids. Ascorbic acid is also an important monosaccharide derivative.

Monosaccharide residues can be linked via glycosidic bonds to form oligosaccharides and polysaccharides. Four important disaccharides are maltose, cellobiose, lactose, and sucrose. Lactose, an epimer of cellobiose, is the major carbohydrate in milk. Sucrose is synthesized in many plants and is the most abundant disaccharide found in nature. The anomeric carbons of both monosaccharide residues in sucrose are involved in the glycosidic linkage of the disaccharide.

Glycosides are compounds formed when the anomeric carbons of sugars form glycosidic linkages with hydroxyl groups of other sugar molecules or with organic nonsugar molecules. Nucleotides are commonly encountered glycosides.

Glucose is the repeating monomeric unit of the storage polysaccharides amylose, amylopectin, and glycogen and of the structural polysaccharide cellulose, which is the most abundant organic substance in the biosphere. Chitin, the second most abundant organic compound on earth, is another example of a storage homopolysaccharide. Its monomeric unit is β-(1→4)-linked N-acetylglucosamine.

Important glycoconjugates include peptidoglycans, glycoproteins, and proteoglycans. A peptidoglycan is a major component of bacterial cell walls, which maintain the shape and functional integrity of bacteria. The glycan moiety of this peptidoglycan is a polymer of a repeating disaccharide, composed of N-acetylglucosamine and N-acetylmuramic acid joined by β-(1→4) linkages. The action of lysozyme cleaves the linkage between N-acetylmuramic acid and N-acetylglucosamine, depolymerizing the glycan moiety. The carboxyl group of N-acetylmuramic acid is linked to a tetrapeptide containing a mixture of L and D amino acids. The C-terminal D-alanine of this tetrapeptide is in turn cross-linked to the penultimate residue of a tetrapeptide on a neighboring peptidoglycan molecule via a linker peptide consisting of five glycine residues. This cross-linking of peptidoglycans results in a giant macromolecule that provides great overall rigidity to the cell wall.

The pathway of biosynthesis of peptidoglycans features stepwise addition of amino acids to MurNAc and involvement of glycyl–transfer RNA. The final step involves linkage of a D-alanine residue to a pentaglycine cross-link of a neighboring peptidoglycan molecule, catalyzed by a transpeptidase. This step is inhibited by penicillin. Certain other antibiotics also inhibit various steps in peptidoglycan synthesis.

Other important carbohydrate-containing molecules are components of cell walls and outer membranes of certain bacteria. The teichoic acids (polymers of glycerol phosphate or ribitol phosphate) are found in the cell walls of certain Gram-positive bacteria, and lipopolysaccharides are found in the outer membranes of Gram-negative bacteria.

Many proteins with a wide variety of functions are glycosylated. The glycan chains exhibit great diversity in structure due to the variations in the monosaccharides added, the atoms involved in the glycosidic bond, and the potential for branched structures. Most glycoproteins can be classified according to the nature of the linkage joining the protein to its carbohydrate moiety. The three major classes exhibit either an *O*-glycosidic linkage, an *N*-glycosidic linkage, or a linkage to a phosphatidylinositol-glycan structure. The major linkage in *O*-linked glycoproteins is between *N*-acetylgalactosamine and the hydroxyl group of serine or threonine. In *N*-linked glycoproteins, the linkage involves *N*-acetylglucosamine and the nitrogen of the amide group of asparagine. A number of proteins that are present on the extracellular surface of plasma membranes are anchored to its outer leaflet by phosphatidylinositol-glycan structures.

The biosynthesis of *O*-linked glycoproteins uses nucleotide sugars as glycosyl-group donors. Glycosyltransferases sequentially catalyze the addition of individual sugars. In the biosynthesis of *N*-linked glycoproteins, the initial saccharide chain donated by dolichol pyrophosphate–oligosaccharide is processed by specific glycosidases and glycosyltransferases to produce high-mannose chains or complex chains. In certain cases, the result is a hybrid oligosaccharide chain, a branched chain in which one branch is of the high-mannose type and the other is of the complex type.

A variety of methods are available to detect and purify glycoproteins. The principal methods used to characterize their oligosaccharide chains are binding to lectins, compositional analysis, the use of specific glycosidases, mass spectrometry, NMR spectroscopy, and methylation analysis.

The oligosaccharide chains of glycoproteins play roles in many biological processes, including the clearance of certain proteins from the plasma, the spread of cancer cells, and certain aspects of inflammation.

Proteoglycans are composed of specific proteins (core and link) and glycosaminoglycans. Glycosaminoglycans are usually composed of repeating disaccharides of amino sugars (e.g., GlcNAc and GalNAc) and uronic acids (glucuronic and iduronic acids). At least six glycosaminoglycans have been isolated and characterized: chondroitin sulfate, dermatan sulfate, heparan sulfate, heparin, hyaluronic acid, and keratan sulfate.

The principal proteoglycan of cartilage is an enormous molecular aggregate, composed of hyaluronic acid, chondroitin sulfate, keratan sulfate, core proteins, and link proteins, all of which are organized in a specific manner. Like other polysaccharides, the glycan chains are synthesized by the addition of sugar residues donated by nucleotide sugars. Sulfation and epimerization of specific residues may also occur. Certain diseases are due to defects in the degradation of glycosaminoglycans. Proteoglycans have a variety of functions including interactions with other components of the extracellular matrix.

Selected Readings

General References

Candy, D. S. (1980). *Biological Functions of Carbohydrates* (New York: Halsted Press).

Collins, P. M., ed. (1987). *Carbohydrates* (London and New York: Chapman and Hall). A source book that provides names, structural and empirical formulas, physical and chemical properties, and references to the literature for several thousand carbohydrates.

El Khadem, H. S. (1988). *Carbohydrate Chemistry: Monosaccharides and Their Derivatives* (Orlando, Florida: Academic Press).

Oxford GlycoSystems. (1992). *Tools for Glycobiology.* (New York: Oxford GlycoSystems). Contains much valuable information on various topics, procedures, and reagents relating to glycobiology.

Sharon, N. (1980). Carbohydrates. *Sci. Am.* 243(5):90–116.

Chitin

Pennisi, E. (1993). Chitin craze. *Sci. News* 144:72–74.

Glycoprotein Structure

Dwek, R. A., Edge, C. J., Harvey, D. J., Wormald, M. R., and Parekh, R. B. (1993). Analysis of glycoprotein-associated oligosaccharides. *Annu. Rev. Biochem.* 62:65–100.

Kornfeld, R., and Kornfeld, S. (1985). Assembly of asparagine-linked oligosaccharides. *Annu. Rev. Biochem.* 54:631–664.

Lechner, J., and Wieland, F. (1989). Structure and biosynthesis of prokaryotic glycoproteins. *Annu. Rev. Biochem.* 58:173–194.

Schachter, H. (1986). Biosynthetic controls that determine the branching and microheterogeneity of protein-bound oligosaccharides. *Biochem. Cell Biol.* 64:163–181.

Schachter, H. (1991). Enzymes associated with glycosylation. *Curr. Opin. Struct. Biol.* 1:755–765.

Schachter, H., and Brockhausen, I. (1992). The biosynthesis of serine (threonine)-*N*-acetyl-galactosamine-linked carbohydrate moieties. In *Glycoconjugates: Composition, Structure, and Function*, H. J. Allen and E. C. Kisailus, eds. (New York: Marcel Dekker, Inc.), pp. 263–332.

Strous, G. J., and Dekker, J. (1992). Mucin-type glycoproteins. *Crit. Rev. Biochem. Mol. Biol.* 27:57–92.

Glycosylphosphatidylinositols

Englund, P. T. (1993). The structure and biosynthesis of glycosyl phosphatidylinositol protein anchors. *Annu. Rev. Biochem.* 62:121–138.

Ferguson, M. A. J. (1991). Lipid anchors on membrane proteins. *Curr. Opin. Struct. Biol.* 1:522–529.

Thomas, J. R., Dwek, R. A., and Rademacher, T. W. (1990). Structure, biosynthesis, and functions of glycosylphosphatidylinositols. *Biochemistry* 29:5413–5422.

Glycoprotein Function

Borman, S. (1992). Race is on to develop sugar-based anti-inflammatory, antitumor drugs. *Chem. Eng. News* 70(49):25–28.

Hartree, A. S., and Renwick, A. G. C. (1992). Molecular structures of glycoprotein hormones and functions of their carbohydrate components. *Biochem. J.* 287:665–679.

Hughes, R. C. (1992). Lectins as cell adhesion molecules. *Curr. Opin. Struct. Biol.* 2:687–692.

Lasky, L. A. (1992). Selectins: interpreters of cell-specific carbohydrate information during inflammation. *Science* 258:964–969.

Parekh, R. B. (1991). Effects of glycosylation on protein function. *Curr. Opin. Struct. Biol.* 1:750–754.

Rademacher, T. W., Parekh, R. B., and Dwek, R. A. (1988). Glycobiology. *Annu. Rev. Biochem.* 57:785–838. A review of oligosaccharides linked to proteins and lipids, including changes in glycosylation in several diseases.

Rasmussen, J. R. (1992). Effect of glycosylation on protein function *Curr. Opin. Struct. Biol.* 2:682–686.

Proteoglycans

Heinegård, D., and Oldberg, Å. (1989). Structure and biology of cartilage and bone matrix noncollagenous macromolecules. *FASEB J.* 3:2042–2051.

Kjellén, L., and Lindahl, U. (1991). Proteoglycans: structures and interactions. *Annu. Rev. Biochem.* 60:443–475.

10

Nucleotides

Nucleotides are perhaps most familiar as the building blocks of DNA and RNA, just as amino acids are the building blocks of proteins. However, nucleotides either by themselves or in combination with other molecules are players in almost all activities of the cell, including catalysis, transfer of energy, and mediation of hormone signals. We have already seen that some nucleotides function as coenzymes in biosynthetic reactions or are components of other coenzymes (Chapter 8). In this chapter, we will examine in detail the structure and nomenclature of nucleotides, as well as some of their regulatory functions. The role of nucleotides in the transfer of biological energy will be covered in Chapter 14 (Introduction to Metabolism), and the three-dimensional structures and the functions of nucleotide polymers, the nucleic acids, will be covered in the final section of this book.

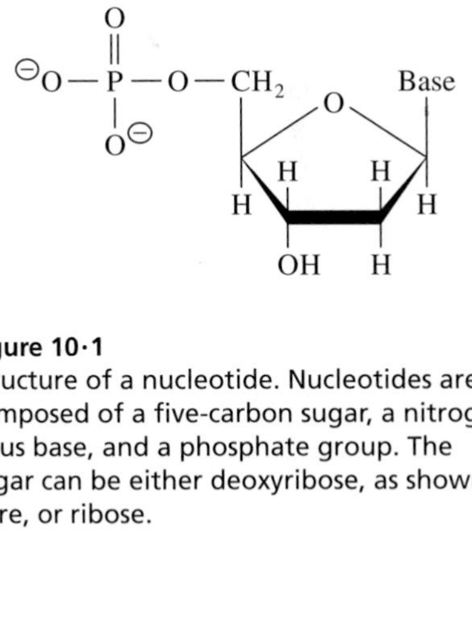

Hydrolysis of nucleotides reveals that they are composed of three kinds of molecules: a weakly basic nitrogenous compound, a five-carbon sugar, and phosphate (Figure 10·1). The nitrogenous bases found in nucleotides are substituted pyrimidines and purines. The pentose is usually either ribose (D-ribofuranose) or 2-deoxyribose (2-deoxy-D-ribofuranose) as shown in Figure 10·2. The pyrimidine or purine *N*-glycosides of these sugars are called **nucleosides.** Nucleotides are the phosphate esters of nucleosides; the common nucleotides contain from one to three phosphate groups. Nucleotides containing ribose are called **ribonucleotides,** and nucleotides containing deoxyribose are called **deoxyribonucleotides.**

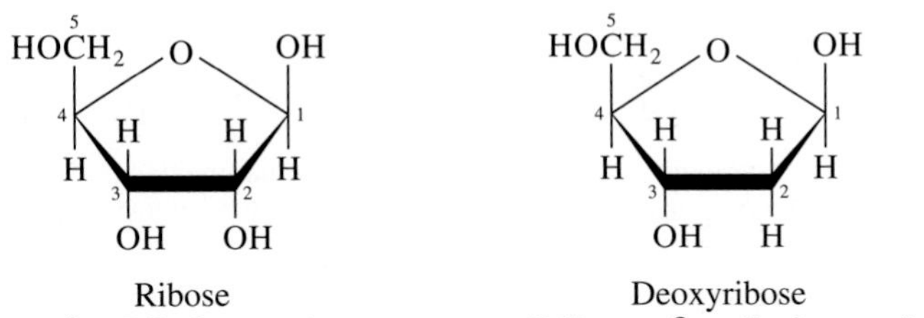

Figure 10·1
Structure of a nucleotide. Nucleotides are composed of a five-carbon sugar, a nitrogenous base, and a phosphate group. The sugar can be either deoxyribose, as shown here, or ribose.

Ribose
(β-D-Ribofuranose)

Deoxyribose
(2-Deoxy-β-D-ribofuranose)

Figure 10·2
Structures of the two sugars found in nucleosides.

Pyrimidine

Purine

Figure 10·3
Structures of pyrimidine and purine. Pyrimidines and purines are usually drawn with C-5 of the pyrimidine and C-8 of the purine to the right.

10·1 Nucleotides Contain Two Classes of Bases

All of the bases found in nucleotides are derivatives of either pyrimidine or purine. The structures of these heterocyclic compounds and the systems for numbering the carbon and nitrogen atoms of each are shown in Figure 10·3. Pyrimidine is a heterocyclic compound that contains four carbon and two nitrogen atoms. Purine is a bicyclic structure consisting of pyrimidine fused to an imidazole ring. Note that the ring structures of both types of bases are unsaturated, with conjugated double bonds. This feature makes the rings planar and also accounts for their ability to absorb ultraviolet light.

Unsubstituted purine and pyrimidine are not commonly found in biological systems, but a number of substituted derivatives are. These derivatives are classified as pyrimidines or purines, depending on the molecule to which they are related. The major pyrimidines found in nucleotides are uracil (2,4-dioxopyrimidine, U), thymine (2,4-dioxo-5-methylpyrimidine, T), and cytosine (2-oxo-4-aminopyrimidine, C). The major purines are adenine (6-aminopurine, A) and guanine (2-amino-6-oxopurine, G). The structures of these five bases are shown in Figure 10·4.

Adenine, guanine, and cytosine are found in both ribonucleotides and deoxyribonucleotides. In contrast, uracil is found mainly in ribonucleotides and thymine in deoxyribonucleotides. Note that thymine is a substituted form of uracil and can be called 5-methyluracil.

In addition to the common bases described above, many other pyrimidines and purines occur naturally. Some examples are shown in Figure 10·5. For example, transfer ribonucleic acid (tRNA) molecules contain a number of bases that have been covalently modified, such as 5-methylcytosine and N^6-methyladenine. These modified bases will be described more fully in Chapter 28. Other important bases include orotic acid and hypoxanthine, intermediates in the metabolism of nucleotides. Caffeine and theophylline, two methylated purines, are stimulants present in the beverages prepared from coffee beans and tea leaves.

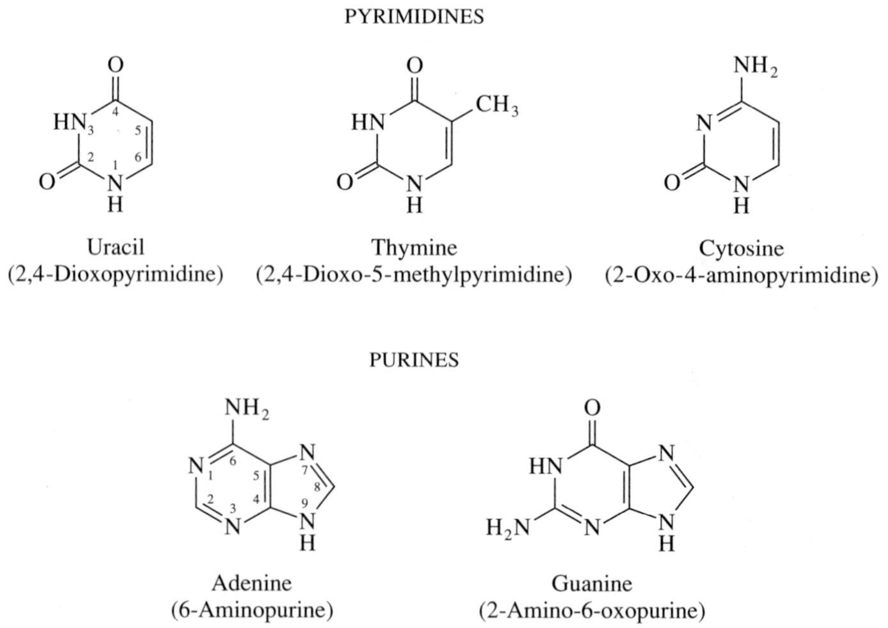

PYRIMIDINES

Uracil
(2,4-Dioxopyrimidine)

Thymine
(2,4-Dioxo-5-methylpyrimidine)

Cytosine
(2-Oxo-4-aminopyrimidine)

PURINES

Adenine
(6-Aminopurine)

Guanine
(2-Amino-6-oxopurine)

Figure 10·4
Structures of the major pyrimidines (uracil, thymine, and cytosine) and of the major purines (adenine and guanine).

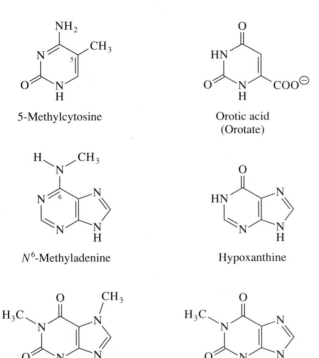

Figure 10·5
Structures of some other naturally occurring purines and pyrimidines.

5-Methylcytosine

Orotic acid
(Orotate)

N^6-Methyladenine

Hypoxanthine

Caffeine

Theophylline

A number of synthetic purines and pyrimidines have medical applications. These compounds can replace the natural pyrimidine or purine substrates in the active sites of certain enzymes. Synthetic molecules that masquerade as natural metabolites are sometimes called antimetabolites. In some cases, the antimetabolites are simply alternate substrates for the enzymes; in other cases, they act directly as enzyme inhibitors. Figure 10·6 shows the structures of 5-fluorouracil and 6-mercaptopurine, two synthetic bases that are used to treat some forms of cancer. Both of these compounds are converted intracellularly to products that are effective enzyme inhibitors. Fluorouracil, a substrate for several enzymes, is converted to its corresponding nucleotide, which mimics thymidylate and is a potent inhibitor of thymidylate synthase, an enzyme essential for DNA synthesis. Mercaptopurine is metabolically converted to several nucleotides that have inhibitory action. When these compounds are given to patients, they are administered as bases or nucleosides rather than as nucleotides, because nucleotides, which are negatively charged, do not readily cross cell membranes.

Purines and pyrimidines, despite being weak bases, are relatively insoluble in water at physiological pH. Within cells, however, most pyrimidine and purine bases occur as constituents of nucleotides and polynucleotides, compounds that are highly soluble.

Figure 10·6
5-Fluorouracil and
6-mercaptopurine.

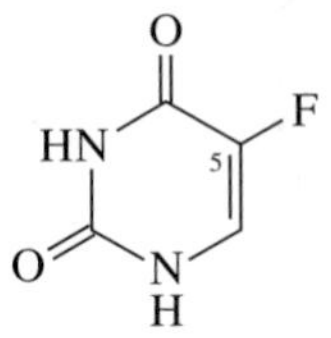

5-Fluorouracil

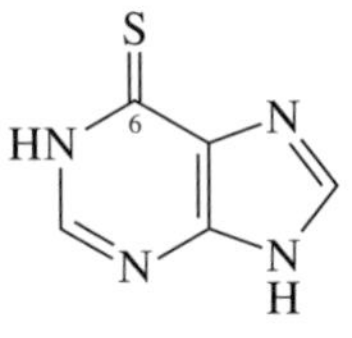

6-Mercaptopurine

Each heterocyclic base of the common nucleosides can exist in at least two tautomeric forms. Adenine and cytosine (which are cyclic amidines) can exist in either amino or imino forms, and guanine, thymine, and uracil (which are cyclic amides) can exist in either lactam (keto) or lactim (enol) forms (Figure 10·7). The tautomeric forms of each base exist in equilibrium, but under the conditions found inside most cells, the amino and lactam tautomers are more stable and therefore predominate. Note that the rings remain unsaturated and are thus planar in each tautomer. As we will see in Chapter 24, the hydrogen-bonding patterns of the amino and lactam tautomers of the bases have important consequences for the three-dimensional structure of nucleic acids.

Figure 10·7
Tautomers of adenine, cytosine, guanine, thymine, and uracil. At physiological pH, the equilibria of these tautomerization reactions lie far in the direction of the amino and lactam forms.

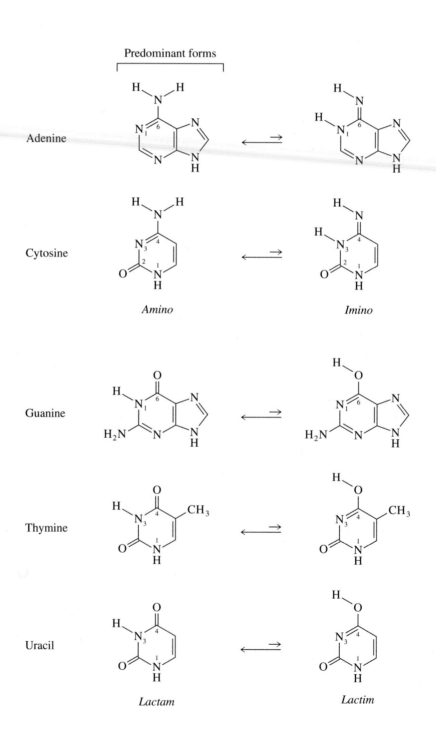

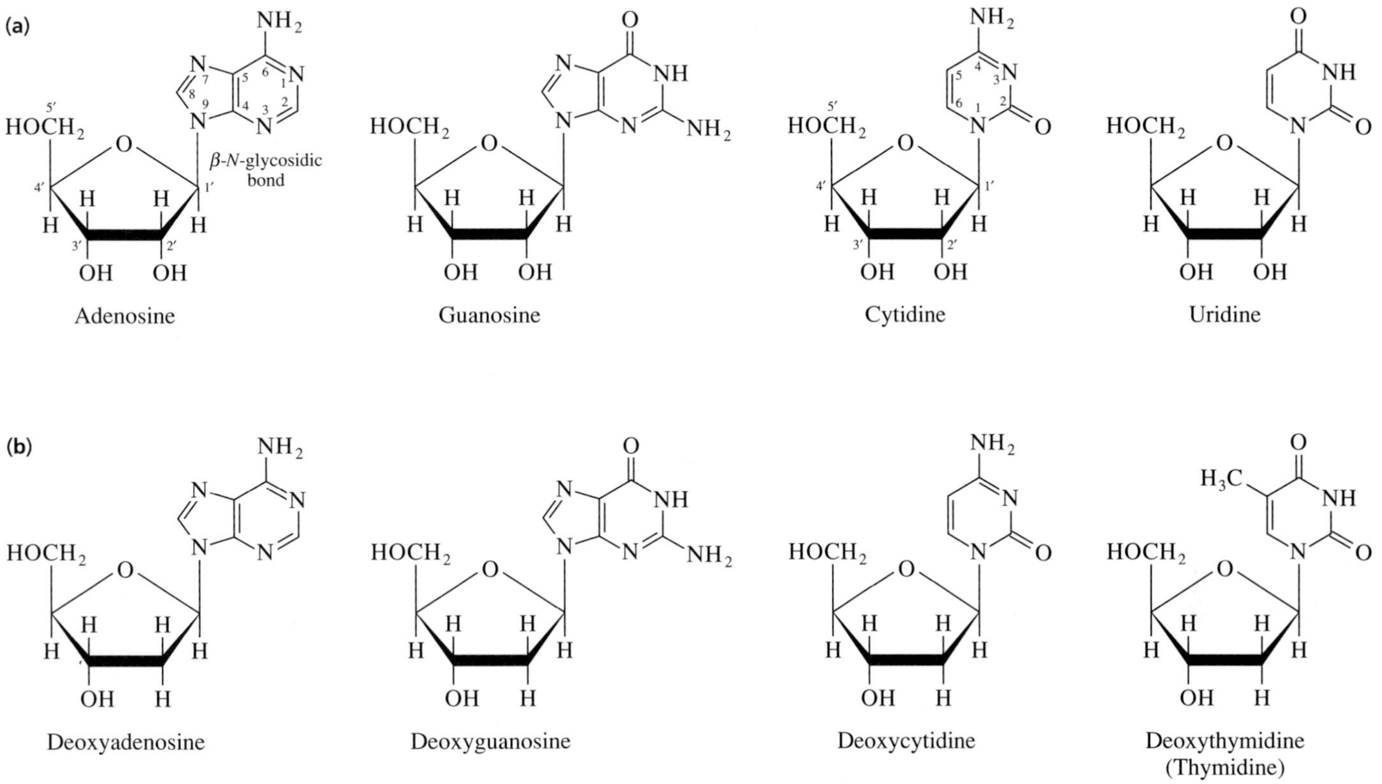

(a)

Adenosine   Guanosine   Cytidine   Uridine

(b)

Deoxyadenosine   Deoxyguanosine   Deoxycytidine   Deoxythymidine
(Thymidine)

Figure 10·8
Structures of some important nucleosides.
(a) Ribonucleosides. The β-N-glycosidic
bond of adenosine is shown in red.
(b) Deoxyribonucleosides.

10·2 Nucleosides Are *N*-Glycosides of Ribose or Deoxyribose

Purine and pyrimidine bases are joined to ribose or deoxyribose to form nucleosides. The structures of the major ribonucleosides and deoxyribonucleosides are
given in Figure 10·8. In each nucleoside, the sugar is connected to the base by a
β-*N*-glycosidic bond between the anomeric carbon of the sugar and N-1 of the
pyrimidine (Figure 10·9) or N-9 of the purine. Nucleosides are therefore *N*-ribosyl
or *N*-deoxyribosyl derivatives of pyrimidines or purines. The numbers of the carbon
atoms of the sugars are accompanied by primes to distinguish them from the numbered atoms of the bases.

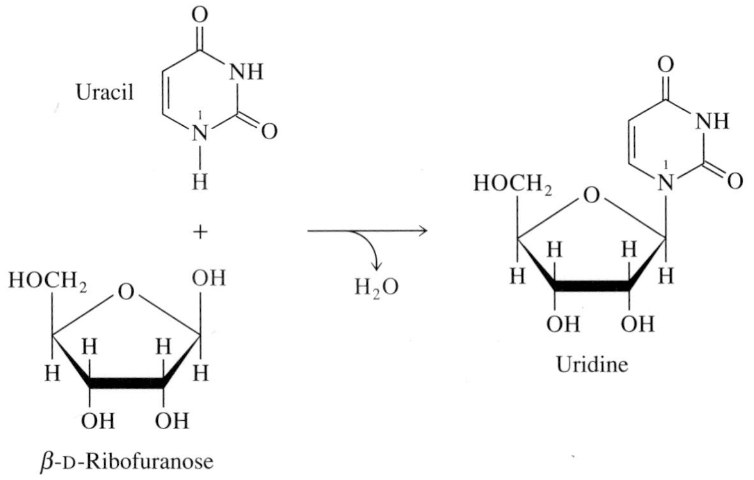

β-D-Ribofuranose

Figure 10·9
Chemical principle of the formation of a
nucleoside from a base and ribose. In uridine, the elements of a molecule of water
(shown in red) are removed from the aglycone (from N-1 of uracil) and from the
anomeric carbon of β-D-ribofuranose. This is
not a physiological reaction, but it illustrates
how the base is joined to the sugar.

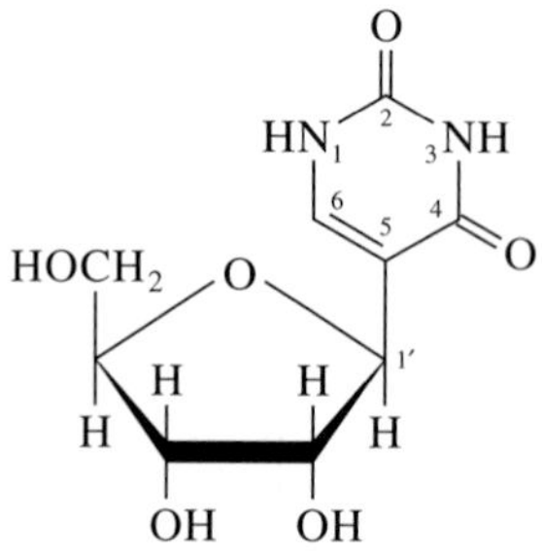

Figure 10·10
Structure of pseudouridine, a nucleoside in which the β-glycosidic bond is formed between two carbon atoms.

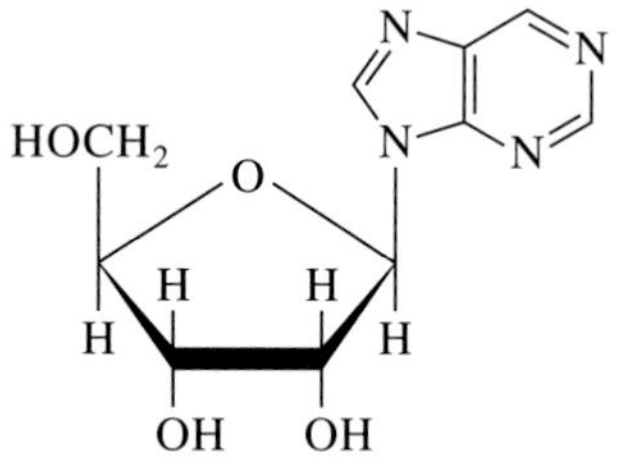

Pseudouridine
(5-(1'-Ribosyl)-uracil)

Nebularine
(9-β-D-Ribofuranosylpurine)

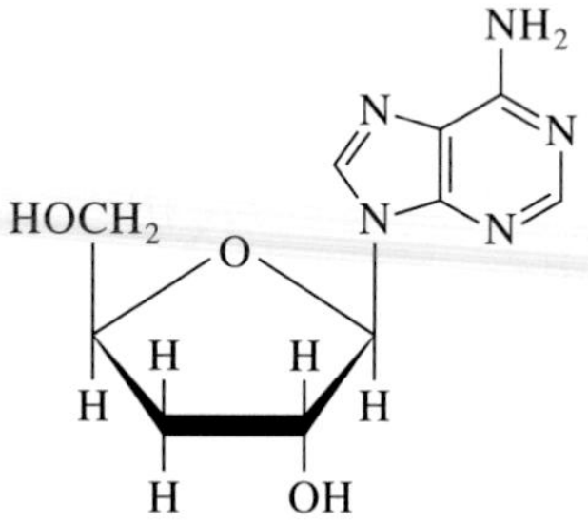

Cordycepin
(9-β-(3'-Deoxy-D-ribofuranosyl)adenine)

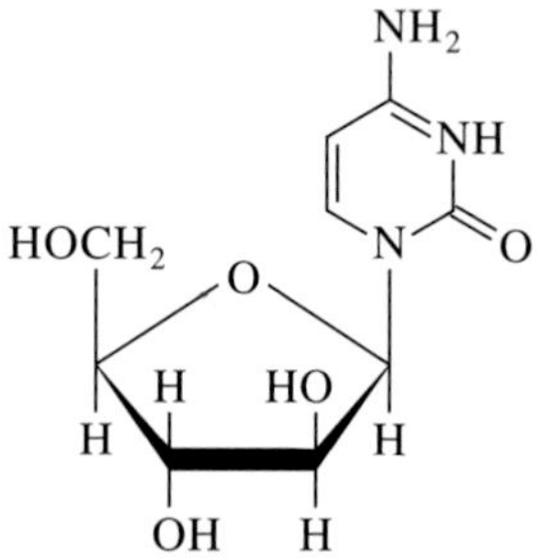

Cytosine arabinoside
(1-β-D-Arabinofuranosylcytosine)

Figure 10·11
Structures of some other nucleosides.

One exception to the typical structure of nucleosides is pseudouridine (Figure 10·10), a component of tRNA. In pseudouridine, a molecule of uracil is attached to a molecule of ribose but not via an *N*-glycosidic bond. Instead, C-5 of the base is linked to C-1' of the sugar.

There are many other naturally occurring nucleosides; some examples are given in Figure 10·11. Several of these compounds are produced by bacteria, molds, or fungi and are antibiotics that prevent growth of other organisms. Nebularine and cordycepin are examples of nucleoside antibiotics. There is also a group of nucleosides that can be isolated from sponges. The best known of these is cytosine arabinoside, which is used to treat several forms of cancer.

The names of nucleosides are derived from the names of the bases they contain. The ribonucleoside containing adenine is called adenosine. (The systematic name, 9-β-D-ribofuranosyladenine, is seldom used.) Its deoxy counterpart is called deoxyadenosine. Similarly, the ribonucleosides of guanine, cytosine, and uracil are guanosine, cytidine, and uridine. Because thymine rarely occurs in ribonucleosides, deoxythymidine is often simply called thymidine. Single-letter abbreviations for bases are also commonly used to designate ribonucleosides: A, G, C, and U (for adenosine, guanosine, cytidine, and uridine, respectively). Three-letter abbreviations for these nucleosides are Ado, Guo, Cyd, and Urd. The deoxyribonucleosides are abbreviated dA or dAdo, dG or dGuo, dC or dCyd, and dT or dThd.

Rotation about the glycosidic bonds of nucleosides and nucleotides is sometimes hindered. In purine nucleosides, the *syn* and *anti* conformations are in rapid equilibrium (Figure 10·12). The *anti* conformation predominates in the common pyrimidine nucleosides. The *anti* conformations of purines and pyrimidines, shown in Figure 10·8, predominate in nucleic acids, the polymers of nucleotides. The furanose ring of the five-carbon sugar moiety of nucleosides can adopt several different conformations, as discussed in Chapter 24.

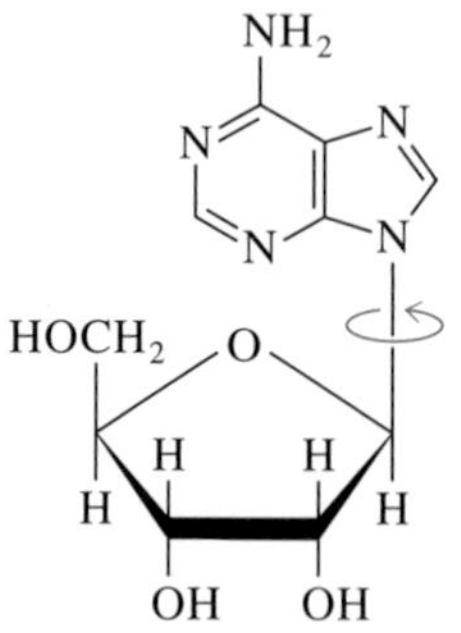

Figure 10·12
Syn and *anti* conformations of adenosine. Some nucleosides can assume either the *syn* or *anti* conformation, but the *anti* form is usually more stable in pyrimidine nucleosides.

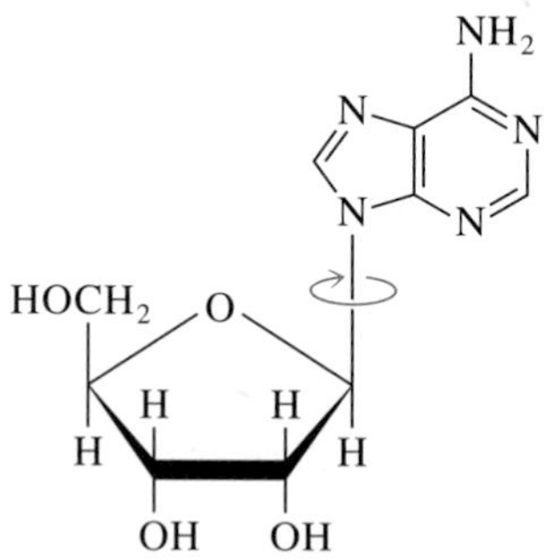

syn Adenosine

anti Adenosine

10·3 Nucleotides Are Phosphate Esters of Nucleosides

Ribonucleosides contain three hydroxyl groups to which phosphate can be esterified (2′, 3′, and 5′), whereas deoxyribonucleosides contain two such hydroxyl groups (3′ and 5′). In naturally occurring nucleotides, the phosphoryl groups are most commonly attached to the oxygen atom of the 5′-hydroxyl group; thus, a nucleotide is always assumed to be a 5′-phosphate ester unless otherwise designated.

The systematic names for nucleotides indicate the number of phosphate groups present. For example, the 5′-monophosphate ester of adenosine is called adenosine monophosphate (AMP). It is also simply called adenylate. Similarly, the 5′-monophosphate ester of deoxycytidine can be referred to as deoxycytidine monophosphate (dCMP) or deoxycytidylate. The 5′-monophosphate ester of the deoxyribonucleoside of thymine is usually called thymidylate but is sometimes called deoxythymidylate to avoid ambiguity. An overview of the nomenclature of bases, nucleosides, and the 5′-nucleotides is presented in Table 10·1. Note that nucleotides with the phosphate esterified at the 5′ position are abbreviated as AMP, dCMP, and so on. Nucleotides with the phosphate esterified to a position other than 5′ are given similar abbreviations but with position numbers designated (for example, 3′-AMP). Nucleoside 3′-phosphates are products of some hydrolytic reactions. 3′,5′-Cyclic monophosphates of adenosine and guanosine will be described in Section 10·8A. Ribonucleoside 2′,3′-cyclic phosphates are intermediates in the enzymatic hydrolysis of RNA.

Nucleoside monophosphates, which can be considered as derivatives of phosphoric acid, are anionic at physiological pH. They are dibasic acids with pK_a values of approximately 1 and 6. The nitrogen atoms of the heterocyclic rings can also ionize.

Nucleoside monophosphates can be further phosphorylated to form nucleoside diphosphates and nucleoside triphosphates. These additional phosphoryl groups are present as phosphoanhydrides. The structures of adenosine monophosphate

Table 10·1 Nomenclature of bases, nucleosides, and nucleotides

Base	Ribonucleoside	Ribonucleotide (5′-monophosphate)
Adenine (A)	Adenosine	Adenosine 5′-monophosphate (AMP); adenylate*
Guanine (G)	Guanosine	Guanosine 5′-monophosphate (GMP); guanylate*
Cytosine (C)	Cytidine	Cytidine 5′-monophosphate (CMP); cytidylate*
Uracil (U)	Uridine	Uridine 5′-monophosphate (UMP); uridylate*

Base	Deoxyribonucleoside	Deoxyribonucleotide (5′-monophosphate)
Adenine (A)	Deoxyadenosine	Deoxyadenosine 5′-monophosphate (dAMP); deoxyadenylate*
Guanine (G)	Deoxyguanosine	Deoxyguanosine 5′-monophosphate (dGMP); deoxyguanylate*
Cytosine (C)	Deoxycytidine	Deoxycytidine 5′-monophosphate (dCMP); deoxycytidylate*
Thymine (T)	Deoxythymidine or thymidine	Deoxythymidine 5′-monophosphate (dTMP); deoxythymidylate* or thymidylate*

*Anionic forms of phosphate esters predominant at pH 7.4

Figure 10·13
Structures of three common adenine ribonu-cleotides, adenosine 5′-monophosphate (AMP), adenosine 5′-diphosphate (ADP), and adenosine 5′-triphosphate (ATP). Each of the three nucleotides consists of an adenine (blue), a ribose (black), and one or more phosphoryl groups (red).

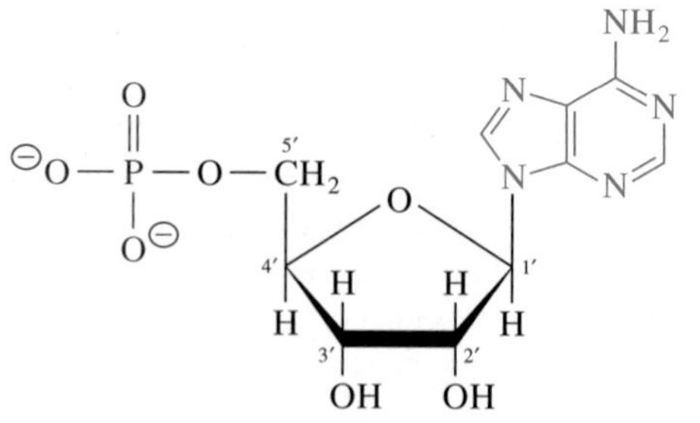

Adenosine 5′-monophosphate
(AMP)

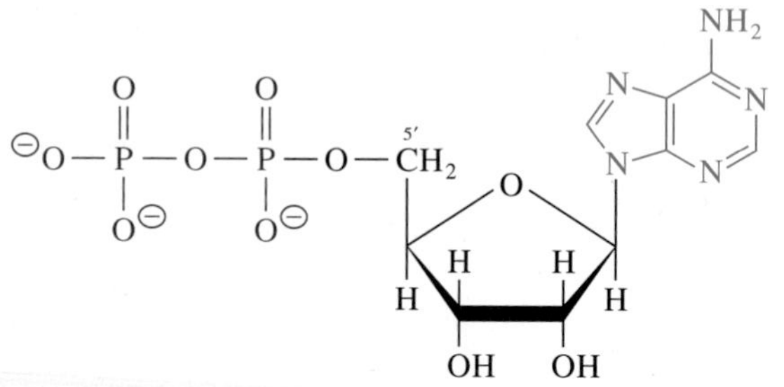

Adenosine 5′-diphosphate
(ADP)

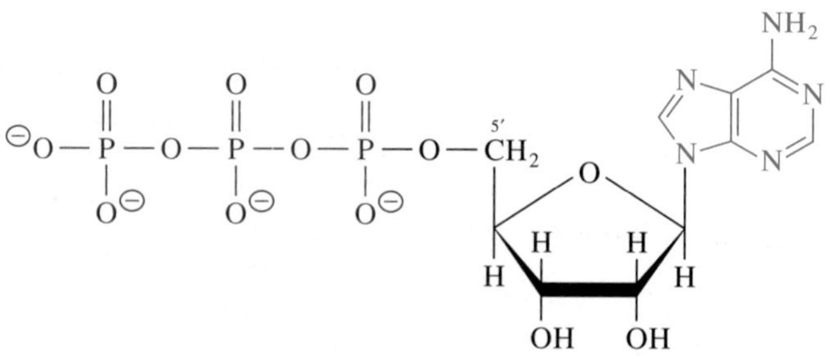

Adenosine 5′-triphosphate
(ATP)

(AMP), adenosine diphosphate (ADP), and adenosine triphosphate (ATP) are compared in Figure 10·13, and two three-dimensional views of ATP are shown in Figure 10·14.

Nucleoside polyphosphates and polymers of nucleotides can also be abbreviated using a scheme in which phosphate groups are abbreviated as p and the ribonucleosides are represented by single letters. The position of the p relative to the nucleoside abbreviation tells its position; a 5′ substitution precedes the nucleoside, and a 3′ substitution follows it. Thus, 5′-adenylate (AMP) can be abbreviated as pA, deoxyadenylate 3′-phosphate as dAp, and ATP as pppA.

10·4 Some Nucleotides Form Complexes with Magnesium Ions

Nucleoside diphosphates and triphosphates, both in aqueous solution and at the active sites of enzymes, are usually present not as free nucleotides but as complexes with magnesium (or sometimes manganese) ions. These cations coordinate with oxygen atoms of the phosphate groups, forming six-membered rings, as shown in Figure 10·15 for ADP and ATP. A magnesium ion can form several different complexes with ATP; the complexes involving the α and β and the β and γ phosphate

Figure 10·14
Stereo views of adenosine
5'-triphosphate (ATP).
(a) Ball-and-stick model.
Some of the hydrogen
atoms have been omitted
for clarity. **(b)** Space-filling
model. Color key: carbon,
green; hydrogen, light
blue; nitrogen, dark blue;
oxygen, red; phosphorus,
orange.

(a)

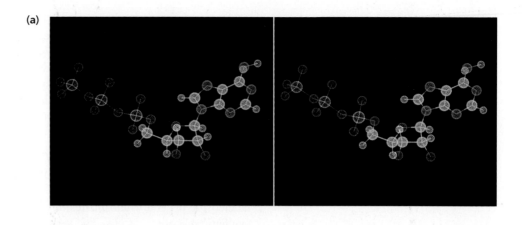

(b)

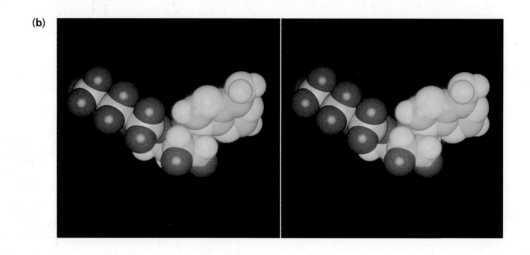

(a)

MgADP

(b)

α,β Complex of MgATP

β,γ Complex of MgATP

Figure 10·15
Complexes between **(a)** ADP and Mg$^{2\oplus}$ and
(b) ATP and Mg$^{2\oplus}$. (*Ado* in the figure represents the adenosyl moieties of ADP and ATP.)

groups are shown in Figure 10·15b. (The phosphate groups are assigned Greek letters according to their position relative to the adenosyl group.) In solution, formation of the β,γ complex is favored. We will see later that nucleic acids are also usually complexed to counterions such as Mg^{2+} or cationic proteins. For convenience, we often refer to the nucleoside triphosphates as ATP, GTP, CTP, and UTP, but keep in mind that, in cells, these molecules actually exist as complexes with Mg^{2+}.

10·5 Specific Kinases Catalyze the Formation of Nucleoside Di- and Triphosphates

Nucleoside di- and triphosphates, not monophosphates, are used in many of the synthetic reactions of cells. The monophosphates must undergo a two-step alteration to become triphosphates. Transfer of a phosphoryl group, usually the terminal phosphate of ATP, is catalyzed by enzymes known as **kinases.** The acceptors of the phosphoryl group include hydroxyl or carboxylate groups, nitrogen, or other phosphate groups. Kinase-catalyzed reactions are very common in biochemistry. Let us examine their role in the formation of nucleoside di- and triphosphates.

A. Nucleoside Monophosphate Kinases

A group of enzymes called nucleoside monophosphate kinases catalyzes the conversion of monophosphates to diphosphates. For example, GMP is converted to GDP by the action of guanylate kinase, which is specific for GMP or dGMP as the phosphoryl-group acceptor and uses ATP or dATP as the phosphoryl-group donor.

$$\text{GMP} + \text{ATP} \rightleftharpoons \text{GDP} + \text{ADP} \qquad \text{(10·1)}$$

Although the equilibrium constant for this reaction is near 1.0, the formation of GDP is favored in cells because the concentration of ATP is high.

Three other kinases act upon dTMP; dCMP, CMP, and UMP; and AMP and dAMP, respectively. Adenylate kinase, which acts on AMP and dAMP, is usually purified from muscle tissue and has been particularly well studied. Adenylate kinase catalyzes the phosphorylation of AMP by ATP.

$$\text{AMP} + \text{ATP} \rightleftharpoons 2\,\text{ADP} \qquad \text{(10·2)}$$

Energy for muscle contraction is generated by hydrolysis of ATP to ADP. When ATP is depleted and ADP accumulates, two molecules of ADP can dismutate by the reverse of Reaction 10·2 to form ATP. When AMP concentrations are relatively high, ADP can be formed from AMP (and ATP); the ADP can then undergo phosphorylation (described fully in Chapter 18) to re-form ATP. In this manner, the activity of adenylate kinase maintains the required levels of the different adenosine nucleotides under varying metabolic conditions.

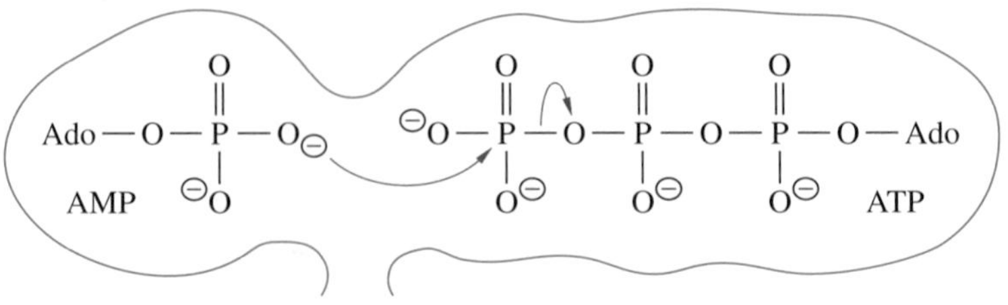

Figure 10·16
Mechanism postulated for adenylate kinase. Note that there are separate binding sites for AMP and ATP and that there is direct nucleophilic attack, not formation of a covalent phosphoryl-enzyme intermediate.

Studies of the three-dimensional structure of adenylate kinase have shown that there are separate binding sites for AMP and ATP. Both substrates must be bound and properly aligned for the reaction to occur. The mechanism for this reaction probably involves nucleophilic attack on the γ-phosphorus atom of ATP by an oxygen atom of AMP (Figure 10·16). The changes in the conformation of the enzyme have been revealed by the X-ray crystallographic structures of three homologous adenylate kinase molecules, each studied as a different form: free enzyme, an enzyme-AMP complex, and a complex of the enzyme with an analog of both substrates (AppppppA, which is AMP and ATP connected by an extra phosphate). When AMP binds, one domain of the enzyme moves about 40°. When both substrate-binding sites are filled, the enzyme moves even more, so that it encloses the substrates (Figure 10·17). This mechanism is an outstanding example of how induced fit in kinases protects the ATP substrate from wasteful hydrolysis.

Figure 10·17
Movement of domains when substrates bind to adenylate kinase. **(a)** The substrate-free enzyme. **(b)** When AMP (red) binds to the enzyme, the domain on the right rotates about 40°. **(c)** In this view, the E-AMP complex has been rotated 90° around a vertical axis. **(d)** Binding of ATP (pink) to the E-AMP complex causes a very large movement in both domains of the enzyme. The upper domain closes down over the substrates, protecting the substrates and products from hydrolysis. (Based on coordinates provided by Georg E. Schulz, Christopher W. Müller, and Kay Diederichs.)

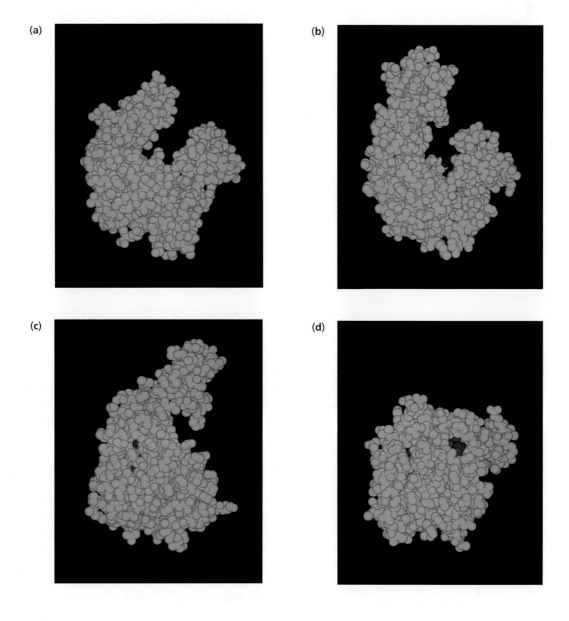

B. Nucleoside Diphosphate Kinase

Nucleoside diphosphates are converted to triphosphates by the action of the enzyme nucleoside diphosphate kinase. This enzyme, present in both the cytosol and mitochondria of eukaryotes, is much less specific than the nucleoside monophosphate kinases. All nucleoside diphosphates, regardless of the purine or pyrimidine base, are substrates for nucleoside diphosphate kinase. Nucleoside monophosphates are not substrates. ATP, due to its relative abundance, is usually the phosphoryl-group donor in vivo. Nucleoside diphosphate kinase is much more active in cells than the nucleoside monophosphate kinases. The mechanism of nucleoside diphosphate kinase is quite different from that of the monophosphate kinases. The reaction follows a ping-pong kinetic scheme, as shown for the phosphorylation of GDP:

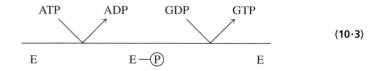

(10·3)

The terminal phosphoryl group of ATP is transferred to an active-site histidine residue of the enzyme to form an energy-rich phosphohistidine intermediate. The phosphoryl group can then be transferred isoenergetically to the diphosphate substrate, in this case GDP.

10·6 Nucleotides Are the Monomeric Units of Nucleic Acids

As described briefly in Chapter 1, nucleotides can be joined together to form nucleic acids, the linear polynucleotides that store and transmit genetic information. Alternating sugar and phosphate moieties form the covalent backbones of these polymers. The bases that are attached to the pentose units provide the variation in sequences of the nucleic acids. Eight nucleoside triphosphates (ATP, GTP, CTP, UTP, dATP, dGTP, dCTP, and dTTP) serve as substrates for RNA and DNA polymerases, the enzymes that catalyze the synthesis of cellular RNA and DNA molecules. The polymerases can add a nucleoside monophosphate to the growing polynucleotide chain by catalyzing the formation of a phosphodiester linkage between the 3′-hydroxyl group of one nucleotide and the 5′-phosphate group of another. The elongation of a DNA molecule is depicted in Figure 10·18. In this reaction, the nucleophilic 3′-hydroxyl group of the nucleic acid attacks the α-phosphorus atom of a nucleoside triphosphate, forming a 3′–5′-phosphodiester and releasing inorganic pyrophosphate (PP_i). The same type of reaction covalently links the monomeric units of RNA. The structures of polynucleotide chains are discussed in Chapter 24.

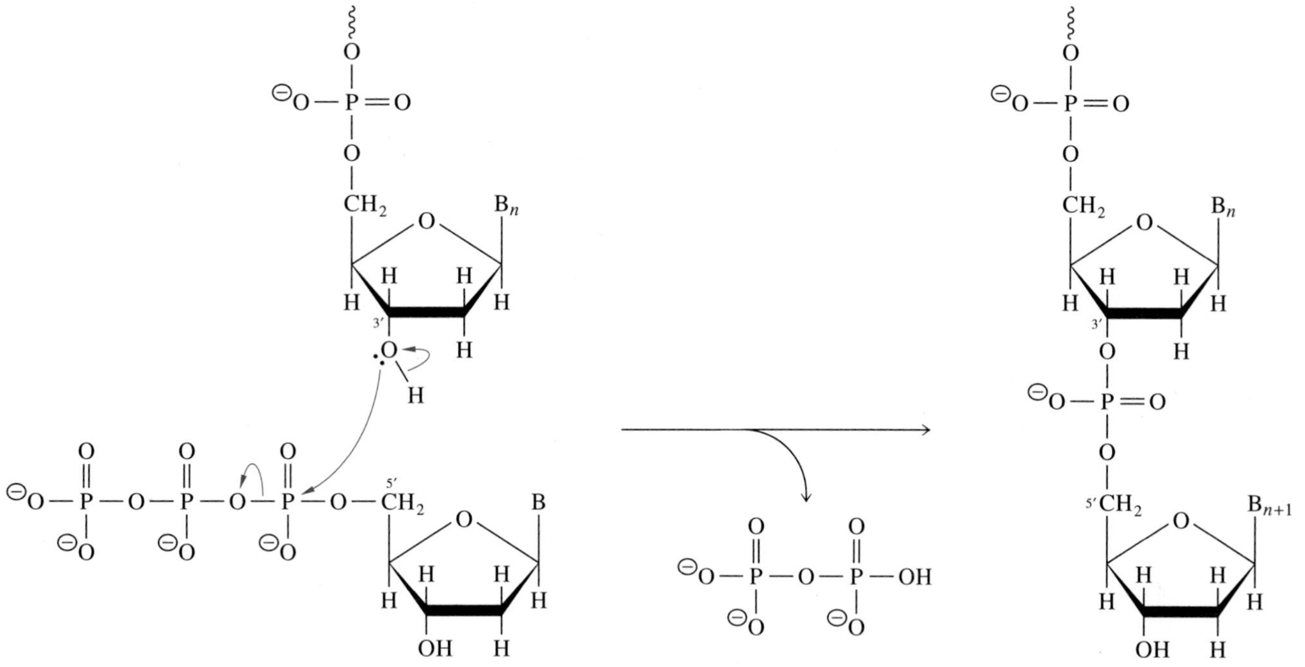

Figure 10·18
Formation of the phosphodiester linkage in DNA. The 3'-hydroxyl group of the polynu-cleotide condenses with the 5'-phosphate group of a nucleoside triphosphate, releasing pyrophosphate (PP$_i$). *B* represents a base.

10·7 Oligodeoxynucleotides Can Be Chemically Synthesized

Oligodeoxynucleotides of about 50 or 60 residues can be chemically synthesized. As with oligopeptides, automated solid-phase methods and anhydrous solvents are used. A deoxynucleoside whose reactive groups are substituted is covalently anchored to a solid support through its 3'-hydroxyl group. The support matrix is usually a column of silica beads. The 3'-hydroxyl group of the nucleotide to be added is activated, and the other reactive groups are substituted so that they will not react. The activated 3'-hydroxyl group of the free nucleotide is condensed with the 5'-hydroxyl group of the matrix-bound nucleoside. The substituted nucleotide is added in excess to ensure a high yield of product; the excess nucleotide is easily removed by washing the column. Next, the protecting group is removed from the 5'-hydroxyl group of the product by mild hydrolysis. The cycle of synthetic reactions—condensation and deblocking—is repeated until an oligonucleotide of the desired sequence has been synthesized. After the product is released from the solid support by mild hydrolysis, all the remaining protecting groups are removed and the synthetic DNA is purified.

One method of oligodeoxynucleotide synthesis is outlined in Figure 10·19 (next page). In this procedure, the nucleotide to be added is converted to an activated derivative, a 3'-phosphoramidite. This compound condenses with the 5'-hydroxyl group, a free primary alcohol, of the bound nucleoside. The resulting phosphite triester is oxidized to a phosphate, thus completing the synthesis of the new 3'–5'-phosphodiester linkage. As we will see, synthetic oligodeoxynucleotides are useful for investigating the structure of nucleic acids and are essential for many laboratory procedures, such as site-directed mutagenesis and the polymerase chain reaction.

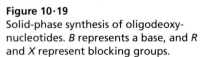

Figure 10·19
Solid-phase synthesis of oligodeoxy-nucleotides. *B* represents a base, and *R* and *X* represent blocking groups.

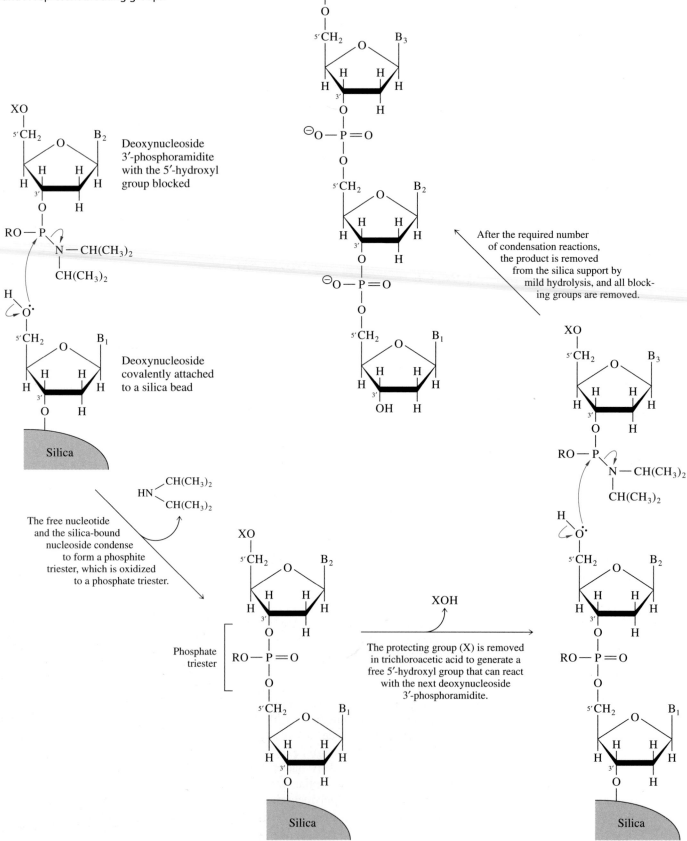

Deoxynucleoside 3'-phosphoramidite with the 5'-hydroxyl group blocked

Deoxynucleoside covalently attached to a silica bead

The free nucleotide and the silica-bound nucleoside condense to form a phosphite triester, which is oxidized to a phosphate triester.

Phosphate triester

The protecting group (X) is removed in trichloroacetic acid to generate a free 5'-hydroxyl group that can react with the next deoxynucleoside 3'-phosphoramidite.

After the required number of condensation reactions, the product is removed from the silica support by mild hydrolysis, and all blocking groups are removed.

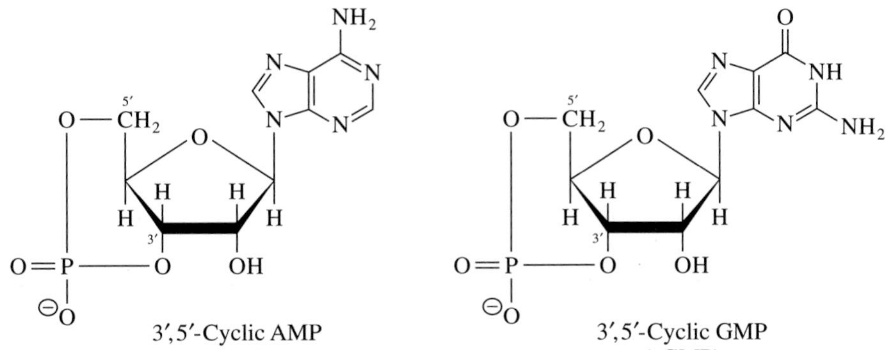

Figure 10·20
Structures of cyclic AMP (cAMP) and cyclic GMP (cGMP).

3′,5′-Cyclic AMP
(cAMP)

3′,5′-Cyclic GMP
(cGMP)

10·8 Some Nucleotides and Nucleosides Are Regulatory Molecules

Nucleotides not only are components of nucleic acids and coenzymes, but they also perform a variety of other functions in prokaryotic and eukaryotic cells. The potential energy of the phosphoanhydride groups of ATP drives many enzymatic reactions that would not otherwise occur spontaneously. The thermodynamics of phosphoryl-group transfer will be discussed fully in Chapter 14. GTP is also an important source of energy, especially during protein synthesis (Chapter 30). Below we discuss some regulatory functions of nucleosides and nucleotides.

A. Cyclic Nucleotides Are Second Messengers

In eukaryotes, the cyclic nucleotides 3′,5′-cyclic adenosine monophosphate (cAMP) and 3′,5′-cyclic guanosine monophosphate (cGMP) are involved in transmitting information from extracellular hormones to intracellular enzymes. The structures of cAMP and cGMP are shown in Figure 10·20; a stereo view of cAMP is shown in Figure 10·21. Because the five-membered cyclic phosphate rings of cAMP and cGMP are strained, these compounds are high-energy compounds. The energy required for the biosynthesis of the 3′,5′-cyclic phosphate bonds is supplied by the precursors of cAMP and cGMP—ATP and GTP, respectively. cAMP is produced from ATP by the action of the enzyme adenylate cyclase.

$$\text{ATP} \longrightarrow \text{cAMP} + \text{PP}_i \qquad (10\cdot4)$$

The formation of cGMP and inorganic pyrophosphate from GTP is catalyzed by guanylate cyclase.

Figure 10·21
Stereo view of cyclic AMP (cAMP). Some of the hydrogen atoms have been omitted for clarity.

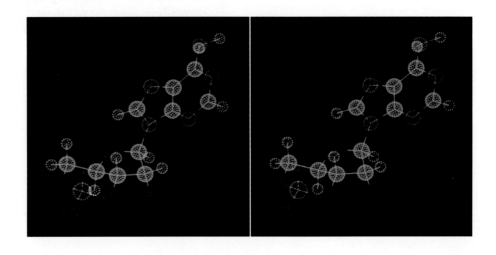

Hormones, most of which are polypeptides, modified amino acids, or steroids, are molecules that allow cells in one part of an organism to communicate with cells in another part of the same organism. In mammals, hormones are synthesized in endocrine glands and travel through the bloodstream to reach responsive tissues. Cells with the appropriate hormone receptors can respond to the hormone, which is called the first (or extracellular) messenger, by producing **second messengers.** cAMP and cGMP represent one type of second messenger. These compounds act intracellularly to alter the activity of certain enzymes. Increased levels of cAMP and cGMP are separately induced by a variety of different hormones. In fact, these two second messengers seem to mediate opposing intracellular effects. cGMP is present at much lower concentrations than cAMP and consequently has not been as well studied. The metabolic roles of cyclic nucleotides in eukaryotes will be discussed in the metabolism section of this book. Cyclic nucleotides are also involved in intracellular signalling in prokaryotes. For example, in *Escherichia coli,* cAMP activates a regulatory DNA-binding protein, which in turn stimulates the synthesis of several enzymes.

B. Adenosine Has Hormonelike Properties

Adenosine administered to mammals causes many physiological responses, including increased blood flow, decreased blood pressure, and muscle relaxation. Many tissues produce adenosine from ADP and AMP when ATP is consumed. This endogenous adenosine, when released from cells, appears to elicit the same responses as exogenous adenosine. Adenosine acts as a messenger that regulates energy metabolism by lowering the utilization of ATP. In this respect, it resembles a hormone. However, adenosine differs from hormones in that no specialized cells or tissues synthesize and store it.

Adenosine receptors are present on the cell surface. When adenosine binds, the receptors are activated, and the activity of adenylate cyclase is altered. Caffeine and theophylline increase blood pressure by binding to adenosine receptors but not activating them; they inhibit the binding of adenosine. Adenosine receptors appear to be the major site of caffeine and theophylline action. At higher concentrations, these purines have additional effects. The activity of adenosine is transient; adenosine is either deaminated or phosphorylated to produce compounds with less messenger activity or none at all.

C. GTP-Binding Proteins Are Signal Transducers

The nucleotide GTP plays an extremely important role in the regulation of many cellular processes. GTP binds intracellularly to proteins known as G proteins, or guanine nucleotide–binding proteins, which serve as transducing agents that transmit signals from hormone receptors to enzymes within the cell. G proteins will be discussed further in Chapter 12. Adenosine receptors are coupled to adenylate cyclase by G proteins. In addition, several G proteins are essential for protein synthesis, although they are not coupled to hormone receptors.

D. Specific Nucleotides Act as Intracellular Signals

Guanosine and adenosine triphosphates can be converted to compounds that act as intracellular signals called **alarmones** (or signal nucleotides). Alarmones are synthesized and accumulate during times of metabolic stress. Elevated levels of alarmones influence gene expression and regulate certain enzyme activities in a manner that tends to overcome the stress. Alarmones have been described as the intracellular hormones of prokaryotes, carrying information from one metabolic pathway to another.

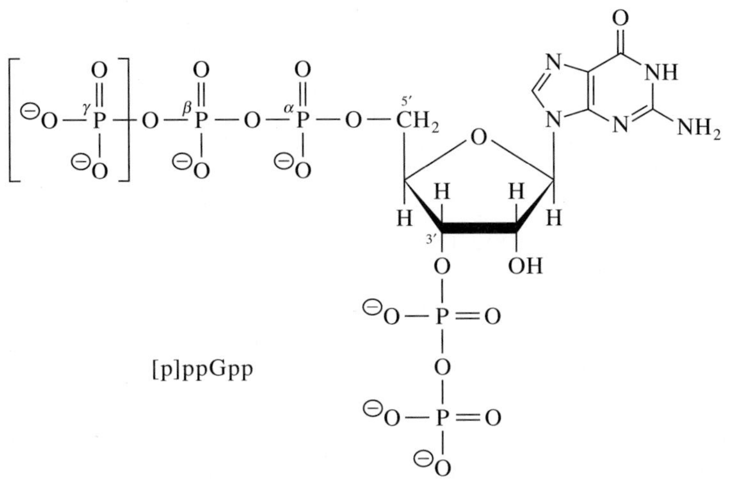

[p]ppGpp

Figure 10·22
The magic spot nucleotides, ppGpp and pppGpp. The γ-phosphoryl group shown in brackets is present in pppGpp.

The first alarmones to be investigated in detail were two guanosine phosphates that accumulate in bacteria during amino acid deficiency. These compounds were first called "magic spot 1" and "magic spot 2" but were later identified as guanosine 5'-diphosphate 3'-diphosphate (ppGpp) and guanosine 5'-triphosphate 3'-diphosphate (pppGpp), shown in Figure 10·22. pppGpp, the minor magic spot, is formed by transfer of pyrophosphate from ATP to the 3'-hydroxyl group of GTP.

$$\text{GTP} + \text{ATP} \longrightarrow \text{pppGpp} + \text{AMP} \qquad \textbf{(10·5)}$$

The production of the major magic spot, ppGpp, appears to involve hydrolysis of pppGpp.

$$\text{pppGpp} + \text{H}_2\text{O} \longrightarrow \text{ppGpp} + \text{P}_i \qquad \textbf{(10·6)}$$

ppGpp is constantly formed and degraded, and a basal level of ppGpp is maintained in bacteria at all times. In response to amino acid deficiency, ppGpp synthesis is increased and its degradation is decreased. By a complex process that is not fully understood, the increased ppGpp concentration causes a decrease in synthesis of RNA, an increase in protein turnover, and other major metabolic adjustments, including synthesis of amino acids. These adjustments allow the cells to overcome the deficiency that triggered the response.

Both prokaryotes and eukaryotes synthesize adenylylated nucleotide alarmones. Diadenosine tetraphosphate (AppppA), shown in Figure 10·23 on the next page, is the major adenylylated signal compound. These compounds are synthesized by side reactions of aminoacyl-tRNA synthetases. For example, AppppA is formed when ATP (pppA) replaces tRNA in Step 2 of the reaction catalyzed by aminoacyl-tRNA synthetases.

$$\text{Amino acid} + \text{ATP} \xrightarrow[\text{Step 1}]{\text{PP}_i} \text{Aminoacyl-AMP} \xrightarrow[\text{Step 2}]{\text{ATP}} \text{Amino acid} + \text{AppppA}$$
$$\textbf{(10·7)}$$

In the bacterium *Salmonella typhimurium*, AppppA and other adenylylated nucleotides accumulate when oxidizing agents are added to the growth medium. The alarmones are believed to stimulate the synthesis of enzymes and other proteins that protect the cells from oxidative damage. Some mammalian cells grown in tissue culture have greatly increased concentrations of AppppA just prior to cell division. In these cells, the alarmone appears to trigger cell proliferation, possibly by stimulating synthesis of DNA. AppppA is degraded to either AMP + ATP or 2 ADP (depending on the tissue or organism) by the action of hydrolytic enzymes. The full physiological importance of these naturally occurring compounds is not yet known.

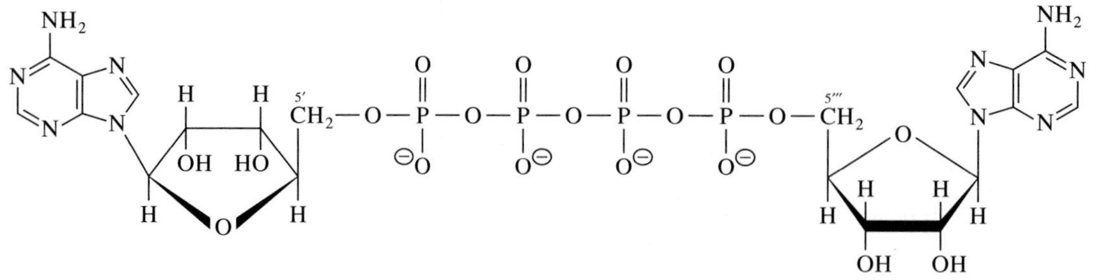

Figure 10·23
Diadenosine 5′,5‴-P¹,P⁴-tetraphosphate (AppppA).

Summary

A nucleotide consists of a heterocyclic base linked to a phosphorylated pentose. The nitrogenous bases of nucleotides are derivatives of pyrimidine and purine. The most common pyrimidines are cytosine (C), thymine (T), and uracil (U); the most common purines are adenine (A) and guanine (G). The amino and lactam tautomers of the bases predominate.

In a nucleoside, the base is linked to a carbohydrate via a *β-N*-glycosidic bond. In ribonucleosides, the sugar is ribose; in deoxyribonucleosides, it is deoxyribose. The glycosidic bond is more often found in the *anti* conformation than in the *syn* conformation in nucleic acids.

Nucleosides can be phosphorylated, usually at the 5′-hydroxyl group, to form mono-, di-, and triphosphates, called nucleotides. Magnesium ions form complexes with the phosphate groups of nucleoside di- and triphosphates.

Transfer of phosphoryl groups among nucleotides is catalyzed by two types of kinases. Phosphoryl groups originating in ATP are ultimately transferred to mono- and diphosphates to convert them to the triphosphates needed for synthetic reactions. Nucleotides are linked by 3′–5′-phosphodiester linkages in nucleic acids.

Nucleosides and nucleotides have additional roles as regulatory molecules. Cyclic AMP and cyclic GMP are second messengers, molecules that play a role in the transmission of information from extracellular hormones to intracellular enzymes. Adenosine, formed from the catabolism of adenine nucleotides, has some properties of a hormone. GTP binds to intracellular G proteins that are involved in many essential cell processes. Several alarmones, or intracellular signal nucleotides, have been isolated, including ppGpp and AppppA.

Selected Readings

General Reference

Westheimer, F. H. (1987). Why nature chose phosphates. *Science* 235:1173–1178. Interesting speculation on the suitability of phosphate esters as metabolites, coenzymes, and information-storage molecules.

Chemical Synthesis of Oligodeoxyribonucleotides

Itakura, K., Rossi, J. J., and Wallace, R. B. (1984). Synthesis and use of synthetic oligo-nucleotides. *Annu. Rev. Biochem.* 53:323–356. Procedures and uses for making synthetic oligomers and polymers.

Khorana, H. G. (1979). Total synthesis of a gene. *Science* 203:614–625. A summary of synthesis of the DNA that encodes a tRNA.

Adenosine Receptors

Daly, J. W. (1982). Adenosine receptors: targets for future drugs. *J. Med. Chem.* 25:197–207. A general review of adenosine receptors and compounds that bind to them.

Olsson, R. A., and Pearson, J. D. (1990). Cardiovascular purinoceptors. *Physiol. Rev.* 70:761–845. A comprehensive review of the role of adenosine in heart tissue.

Stiles, G. L. (1992). Adenosine receptors. *J. Biol. Chem.* 267:6451–6454. A brief review on the molecular properties of individual adenosine receptors.

Cyclic Nucleotides

Berridge, M. J. (1985). The molecular basis of communication within the cell. *Sci. Am.* 253(4):142–152.

Chinkers, M., and Garbers, D. L. (1991). Signal transduction by guanylyl cyclases. *Annu. Rev. Biochem.* 60:553–575. Roles of cGMP as a second messenger in animal cells.

Nucleotide Kinases

Frey, P. A. (1992). Nucleotidyl transferases and phosphotransferases: stereochemistry and covalent intermediates. In *The Enzymes,* Vol. XX, 3rd ed., D. S. Sigma, ed. (San Diego: Academic Press), pp. 141–186.

Schulz, G. E., Müller, C. W., and Diederichs, K. (1990). Induced-fit movements in adenylate kinases. *J. Mol. Biol.* 213:627–630.

Alarmones

Bochner, B. R., Lee, P. C., Wilson, S. W., Cutler, C. W., and Ames, B. N. (1984). ApppppA and related adenylylated nucleotides are synthesized as a consequence of oxidation stress. *Cell* 37:225–232.

Gallant, J. A. (1979). Stringent control in *E. coli. Annu. Rev. Genet.* 13:393–415. A paper that relates the effects of ppGpp to those of hormones in eukaryotes.

Stephens, J. C., Artz, S. W., and Ames, B. N. (1975). Guanosine 5′-diphosphate 3′-diphosphate (ppGpp): a positive effector for histidine operon transcription and a general signal for amino acid deficiency. *Proc. Natl. Acad. Sci. USA* 72:4389–4393.

Zamecnik, P. (1983). Diadenosine 5′, 5‴-P^1, P^4-tetraphosphate (Ap_4A): its role in cellular metabolism. *Anal. Biochem.* 134:1–10. A review of the distribution and functions of Ap_4A.

11

Lipids

Having studied proteins and carbohydrates, we are now ready to consider a third major class of biomolecules, the **lipids.** Like proteins and carbohydrates, lipids (*lipo-,* fat) are found in all known living organisms and play an essential role in the maintenance of life. Unlike proteins and carbohydrates, however, lipids are highly polymorphic and difficult to define structurally. Rather, they are often defined operationally, as water-insoluble (or only sparingly soluble) organic compounds found in biological systems. Lipids are either hydrophobic (nonpolar) or amphipathic (containing both nonpolar and polar groups). In this chapter, we will discuss the structures and functions of the different classes of lipids. The following chapter describes biological membranes, whose properties are directly dependent on the properties of their constituent lipids.

11·1 Lipids Exhibit Great Diversity in Structure and Function

Although lipid structures are often complex, there are some common architectural themes. Among the simplest lipids are the fatty acids, monocarboxylic acids of the general formula R—COOH, where R represents a hydrocarbon tail. Fatty acids are components of many more complex types of lipids, including triacylglycerols (fats and oils), glycerophospholipids (also called phosphoglycerides), and sphingolipids, as well as waxes. Eicosanoids are hormone-like compounds derived from 20-carbon unsaturated fatty acids. Steroids and lipid vitamins are structurally distinct lipids derived from a five-carbon molecule called isoprene; steroids are based

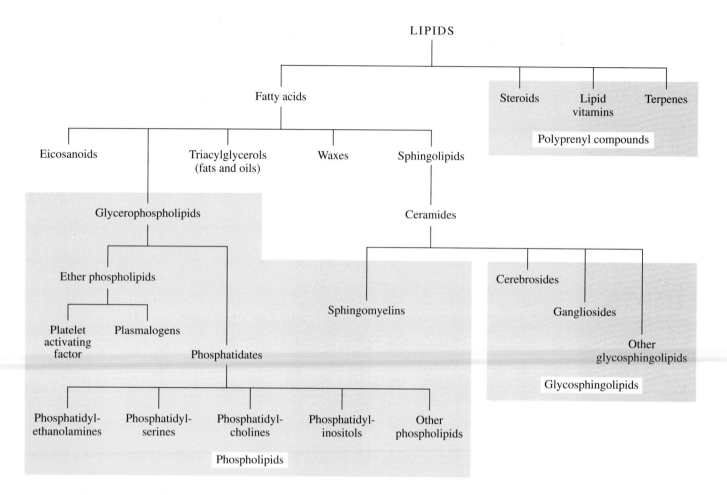

Figure 11·1

Organization of the major types of lipids based on structural relationships. Fatty acids are the simplest lipids in terms of structure. A number of other types of lipids either contain or are derived from fatty acids. These include the triacylglycerols, the glycerophospholipids, and the sphingolipids, as well as the eicosanoids and the waxes. The glycerophospholipids and the sphingomyelins all contain phosphate and are thus classified as phospholipids. Cerebrosides and gangliosides contain monosaccharides or their derivatives and are thus classified as glycosphingolipids. Steroids, lipid vitamins, and terpenes are derived from the five-carbon molecule isoprene and are classified as polyprenyl compounds, or isoprenoids.

on a fused, four-membered cyclic ring structure, and lipid vitamins are composed primarily of long hydrocarbon chains or fused rings. Figure 11·1 shows the major types of lipids and their relationships to one another. Note that lipids containing phosphate moieties are grouped into the general category of phospholipids, lipids containing sphingosine and carbohydrate moieties are called glycosphingolipids, and lipids derived from isoprene are called polyprenyl compounds, or isoprenoids. The name *terpenes* has been applied to all polyprenyl compounds but usually is restricted to those that occur in plants.

In addition to diverse structures, lipids have diverse biological functions. A variety of amphipathic lipids, including glycerophospholipids and sphingolipids, serve as important structural components of all biological membranes. In some organisms, fats and oils (triacylglycerols) function as intracellular storage molecules for metabolic energy. Fats also provide animals with thermal insulation and padding. Waxes provide living organisms with protective surface barriers as components of cell walls, exoskeletons, and skins. A variety of specialized functions are associated with specific lipids; for example, in animals, the steroid hormones regulate and integrate a host of metabolic activities, and in mammals, eicosanoids are used to regulate blood pressure, body temperature, and smooth-muscle contraction. Gangliosides and other glycosphingolipids are located at the cell surface and may participate in cellular recognition.

In some cases, lipids perform their biological functions as individual molecules; in other cases, they interact with other biomolecules to function as part of a complex or aggregate. Lipid complexes include lipoproteins (particles composed of lipid and protein), and lipid aggregates include biological membranes (thin bilayers containing lipid, protein, and sometimes carbohydrate).

Table 11·1 Some common fatty acids (anionic forms) incorporated in membrane lipids

Number of carbons	Number of double bonds	Common name	IUPAC name	Melting point, °C	Molecular formula
12	0	Laurate	Dodecanoate	44	$CH_3(CH_2)_{10}COO^\ominus$
14	0	Myristate	Tetradecanoate	52	$CH_3(CH_2)_{12}COO^\ominus$
16	0	Palmitate	Hexadecanoate	63	$CH_3(CH_2)_{14}COO^\ominus$
18	0	Stearate	Octadecanoate	70	$CH_3(CH_2)_{16}COO^\ominus$
20	0	Arachidate	Eicosanoate	75	$CH_3(CH_2)_{18}COO^\ominus$
22	0	Behenate	Docosanoate	81	$CH_3(CH_2)_{20}COO^\ominus$
24	0	Lignocerate	Tetracosanoate	84	$CH_3(CH_2)_{22}COO^\ominus$
16	1	Palmitoleate	cis-Δ^9-Hexadecenoate	-0.5	$CH_3(CH_2)_5CH{=}CH(CH_2)_7COO^\ominus$
18	1	Oleate	cis-Δ^9-Octadecenoate	13	$CH_3(CH_2)_7CH{=}CH(CH_2)_7COO^\ominus$
18	2	Linoleate	cis, cis-$\Delta^{9,12}$-Octadecadienoate	-9	$CH_3(CH_2)_4(CH{=}CHCH_2)_2(CH_2)_6COO^\ominus$
18	3	Linolenate	all cis-$\Delta^{9,12,15}$-Octadecatrienoate	-17	$CH_3CH_2(CH{=}CHCH_2)_3(CH_2)_6COO^\ominus$
20	4	Arachidonate	all cis-$\Delta^{5,8,11,14}$-Eicosatetraenoate	-49	$CH_3(CH_2)_4(CH{=}CHCH_2)_4(CH_2)_2COO^\ominus$

[Values from Dawson, R. M. C., Elliott, D. C., Elliott, W. H., and Jones, K. M. (1986). *Data for Biochemical Research,* 3rd ed. (Oxford: Clarendon Press).]

11·2 Fatty Acids Are Components of Many Lipids

More than 100 different fatty acids have been identified in the lipids of microorganisms, plants, and animals. These fatty acids differ from one another by the lengths of their hydrocarbon tails, their degree of unsaturation (the number of carbon-carbon double bonds in the hydrocarbon tail), and the positions of those double bonds in their fatty acid chains.

Some of the fatty acids commonly found in mammals are shown in Table 11·1. Most fatty acids have a pK_a of about 4.5–5.0 and are therefore ionized at physiological pH. Fatty acids can be referred to by either IUPAC (International Union of Pure and Applied Chemistry) or common names. The common names of the frequently encountered fatty acids are used most often.

The number of carbon atoms in the most abundant fatty acids usually ranges from 12 to 20 and is almost always even, since fatty acids are synthesized by the sequential addition of two-carbon units. (Fatty acid biosynthesis is discussed in Chapter 20.) In IUPAC nomenclature, the carboxyl carbon is labelled C-1 and the remaining carbon atoms are numbered sequentially. In common nomenclature, Greek letters are used to identify the carbon atoms. The carbon adjacent to the carboxyl carbon (C-2 in IUPAC nomenclature) is designated α, and the other carbons are lettered β, γ, δ, ε, and so on. The Greek letter ω is used to specify the carbon atom farthest from the carboxyl group, whatever the length of the hydrocarbon tail (Figure 11·2).

The physical properties of saturated and unsaturated fatty acids differ considerably. Typically, saturated fatty acids are waxy solids at room temperature (22°C), whereas unsaturated fatty acids are oils at this temperature. Unsaturated fatty acids with only one double bond are called **monounsaturated,** and those with two or more are called **polyunsaturated.** The configuration of the double bonds in unsaturated fatty acids is generally *cis*. In IUPAC nomenclature, the positions of double bonds are indicated by the symbol Δ^N, where the superscript N indicates the lower-numbered carbon atom of each double-bonded pair (see Table 11·1). Note that the carbon atoms involved in the double bonds of most polyunsaturated fatty acids are separated by a methylene group, and thus the double bonds are not conjugated.

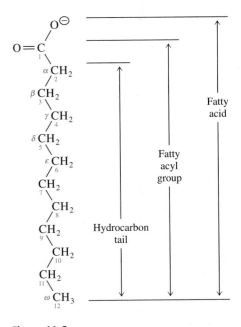

Figure 11·2
Basic structure and nomenclature of fatty acids. Fatty acids consist of a long hydrocarbon tail bound to a carboxyl group. Since the pK_a of the carboxyl group is approximately 4.5–5.0, fatty acids are anionic at physiological pH. In IUPAC nomenclature, the carboxyl carbon is numbered C-1 and other carbon atoms are numbered sequentially from that point. In common nomenclature, the carbon atom adjacent to the carboxyl carbon is designated α, and other carbons are lettered β, γ, δ, ε, and so on. The carbon atom farthest from the carboxyl carbon is referred to as the ω carbon, whatever the length of the tail. The fatty acid shown, laurate (or dodecanoate), has 12 carbon atoms and is saturated since it contains no carbon-carbon double bonds.

A shorthand notation often used to identify fatty acids is two numbers separated by a colon; the first number refers to the number of carbon atoms in the fatty acid, and the second refers to the number of carbon-carbon double bonds. Using this shorthand, palmitate can be written as 16:0, oleate as 18:1, and arachidonate as 20:4, with the double bond positions indicated as a superscript following the Δ symbol (i.e., 20:4 $\Delta^{5,8,11,14}$). Unsaturated fatty acids can also be described by the location of the last double bond in the chain. This double bond is usually found 3, 6, or 9 carbon atoms from the end of the chain. Such fatty acids are called ω-3 (e.g., 18:3 $\Delta^{9,12,15}$), ω-6 (e.g., 18:2 $\Delta^{9,12}$) or ω-9 (e.g., 18:1 Δ^{9}). Note that both linoleate and arachidonate belong to the ω-6 family; in fact, linoleate is the precursor of the longer-chain, more highly unsaturated arachidonate.

The length of the hydrocarbon chain of a fatty acid and its degree of unsaturation influence the melting point of the acid. Compare the melting points listed in Table 11·1 for the saturated fatty acids laurate (12:0), myristate (14:0), and palmitate (16:0). As the length of the hydrocarbon tail increases, the melting point of the saturated fatty acid also increases. Van der Waals interactions among neighboring hydrocarbon tails increase as the tails get longer, and more energy is required to disrupt the increased van der Waals interactions during melting.

Compare also the structures of stearate (18:0), oleate (18:1), and linolenate (18:3) in Figures 11·3 and 11·4. Although represented here in an extended conformation, the saturated hydrocarbon tail of stearate is flexible since every carbon-carbon bond is free to rotate. Because rotation around double bonds is hindered, the presence of *cis* double bonds in oleate and linolenate produces a pronounced bend in the hydrocarbon chain. This bend prevents the formation of the closely packed, well-ordered crystals formed by saturated fatty acids and hence decreases van der Waals interactions among the hydrocarbon chains. Consequently, *cis* unsaturated fatty acids have lower melting points than saturated fatty acids. As the degree of unsaturation increases, lipids become more fluid. Note that stearate (melting point 70°C) is a solid at body temperature, whereas oleate (melting point 13°C) and linolenate (melting point −17°C) are both liquids.

Polyunsaturated fatty acids are readily oxidized by exposure to air. Oxygen reacts with double bonds to form peroxides and free radicals (compounds that have an unpaired electron and are extremely reactive). These compounds then can damage other lipids, proteins, and nucleic acids. Thus, oxidized oils are quite toxic. Spanish toxic oil syndrome was first described when a number of people in Spain died after accidentally eating an oil that had been commercially oxidized.

Although fatty acids are important components of many lipids, free fatty acids occur only in trace amounts in living cells. It is fortunate that free fatty acids are scarce because, as anions, they are detergents and at high concentrations could disrupt membrane structure. Some free fatty acids are found associated with the protein albumin in blood. Most fatty acids are esterified to form more complex lipid molecules. In esters and other derivatives of carboxylic acids, the $RC=O$ moiety contributed by the acid is called the acyl group. In common lipid nomenclature, complex lipids that contain specific fatty acyl groups are named according to the length of the hydrocarbon chain and the number of double bonds of the acyl groups they contain. For example, esters based on the 12-carbon saturated fatty acid laurate are called lauroyl esters, and those based on the 18-carbon unsaturated fatty acid with double bonds at carbons 9 and 12 are called linoleoyl esters.

The relative abundance of particular fatty acids varies with type of organism, type of organ (in multicellular organisms), and food source. The most abundant fatty acids in animals are usually oleate (18:1), palmitate (16:0), and stearate (18:0), although polyunsaturated fatty acids are also prevalent. Mammals require in their diets certain polyunsaturated fatty acids that they cannot synthesize, such as linoleate (18:2, which is abundant in plant oils) and linolenate (18:3, which is abundant in fish oils). Such fatty acids are called **essential fatty acids.** The early work

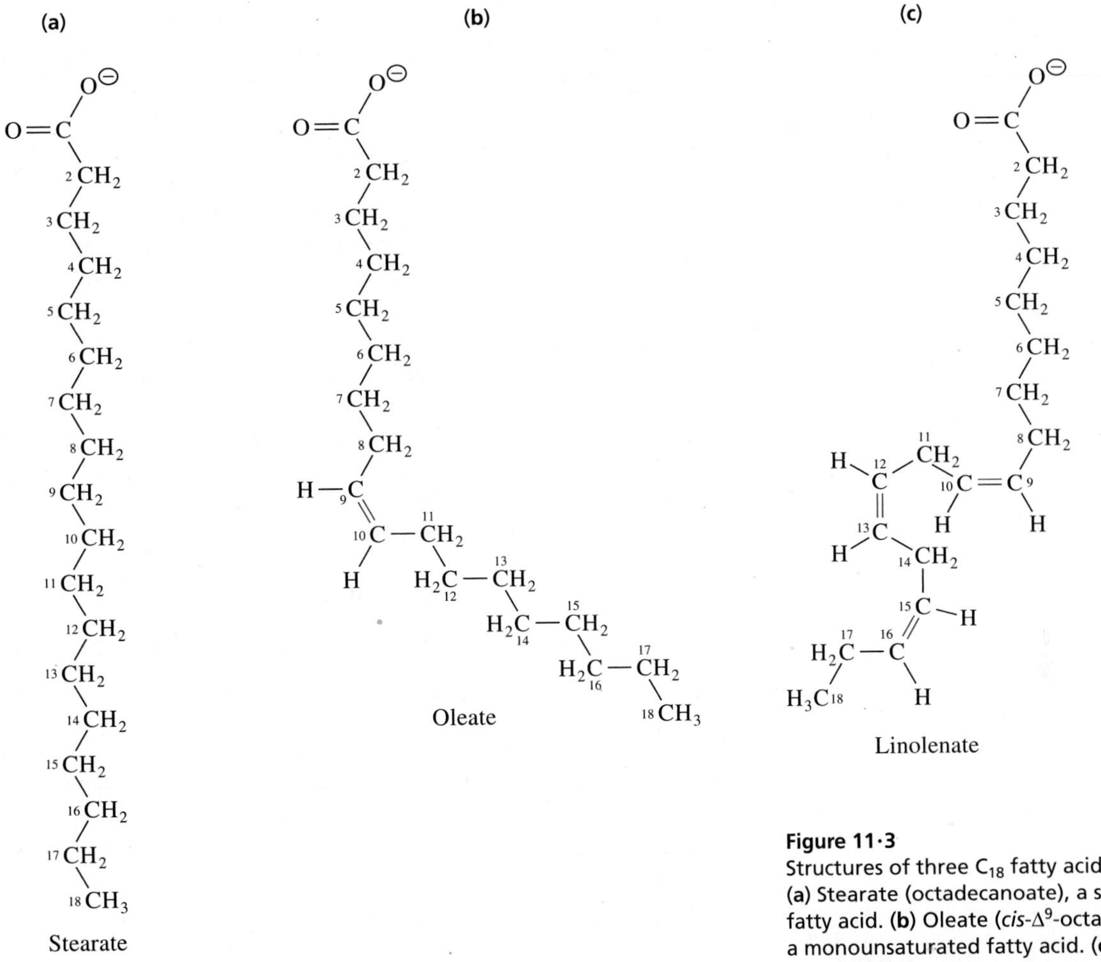

(a)

$O=C$
$O^{\ominus}$
$_2 CH_2$
$_3 CH_2$
$_4 CH_2$
$_5 CH_2$
$_6 CH_2$
$_7 CH_2$
$_8 CH_2$
$_9 CH_2$
$_{10} CH_2$
$_{11} CH_2$
$_{12} CH_2$
$_{13} CH_2$
$_{14} CH_2$
$_{15} CH_2$
$_{16} CH_2$
$_{17} CH_2$
$_{18} CH_3$

Stearate

(b)

$O=C$
$O^{\ominus}$
$_2 CH_2$
$_3 CH_2$
$_4 CH_2$
$_5 CH_2$
$_6 CH_2$
$_7 CH_2$
$_8 CH_2$
$H-C_9$
$_{10}C-CH_2 {}^{11}$
H
$H_2C-CH_2 {}^{13}$
${}_{12}$
$H_2C-CH_2 {}^{15}$
${}_{14}$
$H_2C-CH_2 {}^{17}$
${}_{16}$
$_{18} CH_3$

Oleate

(c)

$O=C$
$O^{\ominus}$
$_2 CH_2$
$_3 CH_2$
$_4 CH_2$
$_5 CH_2$
$_6 CH_2$
$_7 CH_2$
$_8 CH_2$
$H {}_{12}C \quad {}^{11}CH_2$
$\quad {}_{10}C={}^{9}C$
$H \qquad H$
$_{13}C \quad CH_2 {}_{14}$
H
$_{15}C-H$
$H_2C-C {}^{16}$
${}_{17}$
$H_3C_{18} \qquad H$

Linolenate

Figure 11·3
Structures of three C_{18} fatty acids.
(a) Stearate (octadecanoate), a saturated fatty acid. **(b)** Oleate (cis-Δ^9-octadecenoate), a monounsaturated fatty acid. **(c)** Linolenate (all cis-$\Delta^{9,12,15}$-octadecatrienoate), a polyunsaturated fatty acid. The cis double bonds produce kinks in the tails of the unsaturated fatty acids.

Figure 11·4
Stereo views of stearate (left), oleate (middle), and linolenate (right). Color key: carbon, green; hydrogen, blue; oxygen, red.

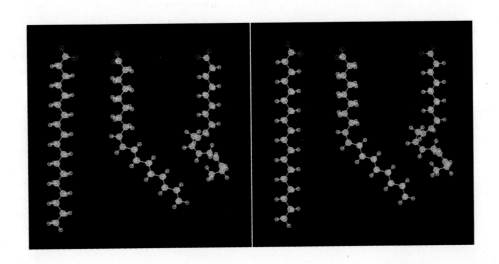

of G. O. Burr and M. M. Burr showed that rats fed fat-free diets grew slowly, developed scaly skin and tail necrosis, and died prematurely. In the absence of the essential fatty acids linoleate and linolenate, mammals can make only saturated fatty acids, oleic acid, and polyunsaturated derivatives of oleic acid (which belong to the ω-9 family), but not the ω-6 and ω-3 fatty acids that are required for the normal function of biological membranes.

A number of fatty acids in addition to those listed in Table 11·1 are found in nature. For example, fatty acids containing cyclopropane or cyclopropene rings are structural components of the cell walls of certain bacteria, and branched-chain fatty acids occur on the feathers of ducks. Phytanic acid, a C_{20} fatty acid with four methyl-group branches, is found in plants and is part of the human diet. In certain individuals who cannot degrade phytanic acid, this compound accumulates within cells, damaging peripheral nerves and the retina. The disorder is known as Refsum's disease.

11·3 Eicosanoids Are Biologically Active Derivatives of Polyunsaturated Fatty Acids

Aspirin (acetylsalicylic acid) is a well known drug that alleviates pain, fever, swelling, and inflammation. It does so by inhibiting the synthesis of prostaglandins, which belong to a group of compounds known as **eicosanoids.** Eicosanoids are oxygenated derivatives of C_{20} polyunsaturated fatty acids such as arachidonic acid; **prostaglandins** are eicosanoids that have a cyclopentane ring. Some examples are shown in Figure 11·5. Eicosanoids are involved in a wide variety of physiological processes and can also mediate a number of potentially pathological responses. For example, prostaglandin E_2 can cause the constriction of blood vessels, and thromboxane A_2 is involved in the formation of blood clots, or thrombi (aggregates of

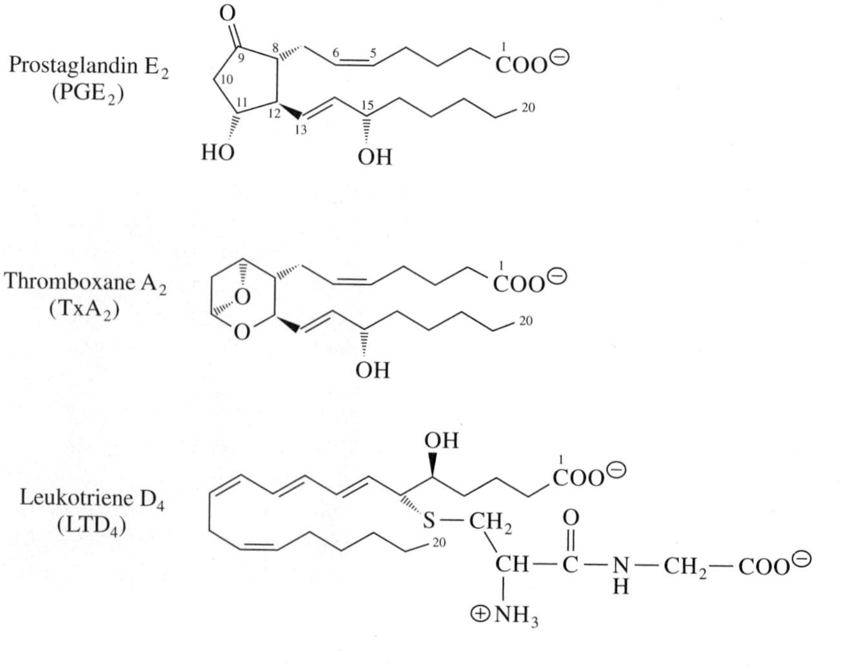

Figure 11·5
Prostaglandin E_2, thromboxane A_2, and leukotriene D_4. These are examples of eicosanoids, which are derived metabolically from arachidonate, a C_{20} polyunsaturated fatty acid with four *cis* double bonds.

cells and protein), which in some cases can block the flow of blood to the heart or brain. Leukotriene D_4, a mediator of smooth-muscle contraction, also provokes the bronchial constriction seen in asthmatics. Interestingly, the Greenland Eskimo have a very low incidence of heart disease, possibly because their diet is enriched in marine oils such as fish and seal oils that have relatively high proportions of ω-3 fatty acids. Some of these fatty acids are converted to eicosanoids that are physiologically less potent than the arachidonate-derived eicosanoids. The presence of the less-active ω-3 compounds has been correlated with a decreased incidence of heart attacks.

11·4 Triacylglycerols Are Neutral, Nonpolar Lipids

Fatty acids can serve as important fuel molecules during metabolism. Oxidation of fatty acids yields more energy ($\sim$37 kJ g^{-1}) than the oxidation of either proteins or carbohydrates ($\sim$16 kJ g^{-1} each). Fatty acids are generally stored as neutral lipids called **triacylglycerols.** As their name implies, triacylglycerols (historically referred to as triglycerides) are composed of three fatty acyl residues esterified to glycerol, a three-carbon alcohol (Figure 11·6). Triacylglycerols are neutral (nonionic), nonpolar (and hence hydrophobic) lipids. Being hydrophobic (unlike carbohydrates), triacylglycerols can be stored in an anhydrous environment and are not solvated by water, which takes up space and adds mass, reducing the efficiency of energy storage. The effect of hydration on polysaccharides is easily illustrated. For example, macaroni in its dry state is quite hard and compact, but when fully hydrated, it expands to several times its original size.

Fats and oils are mixtures of triacylglycerols. Whether a triacylglycerol mixture is a solid (fat) or a liquid (oil) depends on its fatty acid composition and on the temperature. Triacylglycerols that contain only saturated, long-chain fatty acyl groups tend to be solids at body temperature, whereas those that contain unsaturated or short-chain fatty acyl groups tend to be liquids. A sample of a naturally occurring triacylglycerol may contain as many as 20–30 different molecular species that differ in their fatty acid constituents. One molecular species, tripalmitin, is found in animal fat and contains three residues of palmitic acid. Triolein, which contains three oleic acid residues, is the principal molecular species of triacylglycerol found in olive oil.

Many people try to limit their intake of animal fat because it contains the steroid cholesterol. Cholesterol accumulates in deposits of lipid, or plaques, in the walls of blood vessels. These atherosclerotic plaques have been implicated in cardiovascular disease, which can precipitate heart attacks or strokes. Because plant oils usually contain no cholesterol, many people prefer them over lard for cooking. Plant oils such as corn oil and sunflower oil can be converted into "spreadable" semi-solid substances known as margarines. Margarines can be produced by the partial or complete hydrogenation of double bonds in plant oils. The hydrogenation process itself not only saturates the carbon-carbon double bonds of fatty acid esters, it can also change the configuration of the remaining double bonds from *cis* to *trans*. The physical properties of these *trans* fatty acids are similar to those of the saturated fatty acids. Because saturated fatty acids are considered to be unhealthy, many margarines are now produced from plant oils without hydrogenation by adding other edible components such as skim milk powder. The North American diet is still believed to be excessively high in fat, and nutritionists recommend that fat from any source be eaten in moderation.

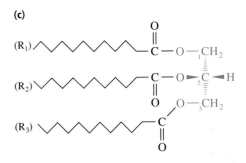

Figure 11·6
Structure of triacylglycerol. Glycerol **(a)** forms the backbone to which three fatty acyl residues are esterified **(b)**. Although glycerol is nonchiral, C-2 of a triacylglycerol is chiral if the acyl groups bound to C-1 and C-3 (shown schematically as R_1 and R_3) are not identical. The general structure of a triacylglycerol is shown in **(c)** with perspective bonds around the chiral carbon; the molecule is oriented for ease of comparison with the structure of L-glyceraldehyde shown in Chapter 4. This orientation allows stereospecific numbering of glycerol derivatives, designated *sn*, with C-1 at the top and C-3 at the bottom.

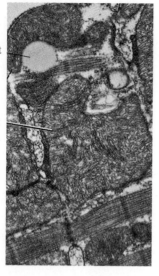

Fat droplet

Mitochondrion

Figure 11·7
Fat droplet in cardiac muscle cell. Fat droplets are often found close to mitochondria in cells that rely on fat to supply energy for cellular activities. Hydrolysis of triacylglycerols in the fat droplets of cardiac muscle cells releases fatty acids. The oxidation of these fatty acids in the mitochondria produces ATP, which is required for muscle contraction. (Courtesy of Mary C. Reedy.)

Triacylglycerols, because of their hydrophobicity, coalesce as fat droplets in cells. Fat droplets are sometimes seen near mitochondria in cells that rely on fatty acids to supply energy for cellular activities (Figure 11·7). In mammals, most fat is stored in adipose tissue, which is composed of specialized cells known as adipocytes. Each adipocyte contains a large fat droplet that accounts for nearly the entire volume of the cell. Although widely distributed throughout the bodies of mammals, most adipose tissue occurs just under the skin and in the abdominal cavity. This subcutaneous fat serves both as a storage depot for energy and as thermal insulation and therefore tends to be more extensive in warm-blooded aquatic mammals.

Fatty acids are released from triacylglycerols through the catalytic activity of lipases, a family of hydrolases. In humans, lipids ingested in food are broken down by lipases in the small intestine. Because lipids are not water soluble, lipid digestion requires the presence of strong detergents called bile salts. Bile salts, which are amphipathic derivatives of cholesterol, aid digestion by emulsifying lipids in the intestine.

11·5 Glycerophospholipids Are Major Components of Biological Membranes

Although triacylglycerols are the most abundant type of lipid in mammals on the basis of weight, they are not structural components of biological membranes because they are not amphipathic and therefore do not form lipid bilayers. The most abundant lipids in membranes are the **glycerophospholipids** (also called phosphoglycerides), which have a glycerol backbone (Figure 11·8). The simplest type of

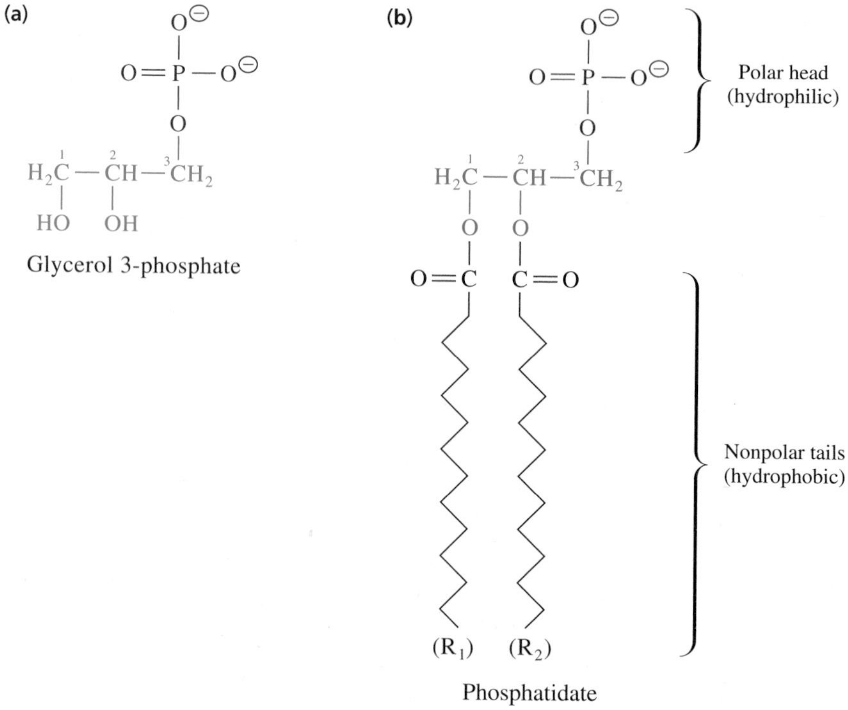

Figure 11·8
Glycerophospholipids. **(a)** Glycerol 3-phosphate. **(b)** Phosphatidate. A phosphatidate consists of glycerol 3-phosphate with two fatty acyl groups (R_1 and R_2) esterified to its C-1 and C-2 hydroxyl groups, respectively. Phosphatidates are the simplest form of glycerophospholipid. Like other glycerophospholipids, phosphatidates have a polar head group and nonpolar tails. Note that C-2 of the glycerol backbone is chiral because the substituents at C-1 and C-3 are different. For simplicity, we will show glycerophospholipids in the stereochemically uncommitted form.

(a)

Glycerol 3-phosphate

(b)

Polar head (hydrophilic)

Nonpolar tails (hydrophobic)

(R_1) (R_2)

Phosphatidate

glycerophospholipid, known as phosphatidate, consists of two fatty acyl groups esterified to C-1 and C-2 of glycerol 3-phosphate. Phosphatidates usually occur only in small concentrations as metabolic intermediates in the biosynthesis or breakdown of more complex glycerophospholipids. The structures of glycerophospholipids can be drawn as derivatives of L-glycerol 3-phosphate in which the substituent esterified to C-2 is shown on the left in a Fischer projection; to save space, we will usually show these compounds as stereochemically uncommitted structures.

In more complex glycerophospholipids, the phosphate is esterified to both glycerol and another compound bearing an —OH group. Table 11·2 identifies some of the families of glycerophospholipids. The structures of three of the most common

Table 11·2 Some common substituents attached to the phosphate group of glycerophospholipids

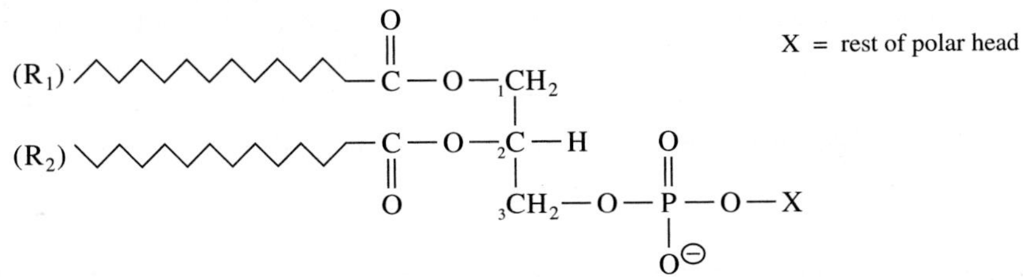

Precursor of X (HO—X)	Formula of X	Name of resulting glycerophospholipid family
Water	—H	Phosphatidate
Choline	—CH$_2$CH$_2$N(CH$_3$)$_3$ ⊕	Phosphatidylcholine
Ethanolamine	—CH$_2$CH$_2$NH$_3$ ⊕	Phosphatidylethanolamine
Serine	—CH$_2$—CH with ⊕NH$_3$ and COO⊖	Phosphatidylserine
Glycerol	—CH$_2$CH—CH$_2$OH with OH	Phosphatidylglycerol
Phosphatidyl-glycerol	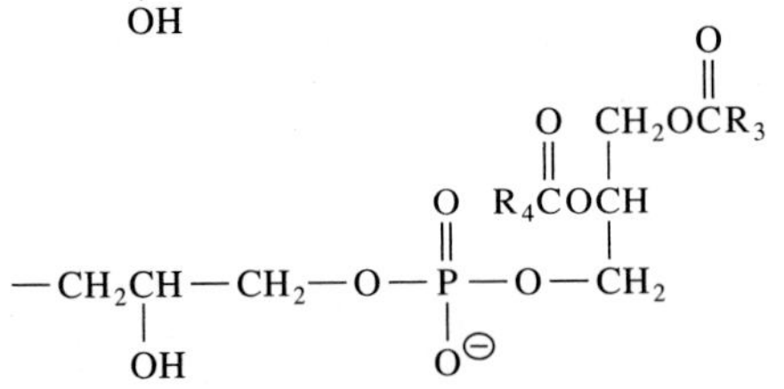 —CH$_2$CH—CH$_2$—O—P—O—CH$_2$ structure	Diphosphatidylglycerol (Cardiolipin)
myo-Inositol	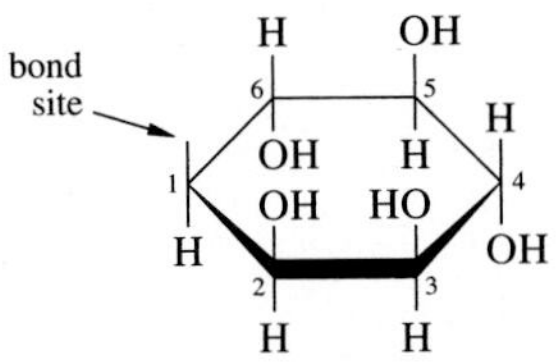 inositol ring structure, bond site	Phosphatidylinositol

families of membrane glycerophospholipids—phosphatidylethanolamine, phosphatidylserine, and phosphatidylcholine—are shown in Figure 11·9.

The glycerophospholipids listed in Table 11·2 are not single compounds but families of molecules that have the same polar head group but different fatty acyl chains. For example, human red blood cell membranes are known to contain at least 21 different species of phosphatidylcholine that differ from each other in the fatty acyl chains esterified at C-1 and C-2 of the glycerol backbone. In general, the type of fatty acid at each position is not random—saturated fatty acids are usually esterified to C-1 and unsaturated fatty acids to C-2 in glycerophospholipids. One notable exception to this rule is dipalmitoylphosphatidylcholine (DPPC), in which palmitic acid is esterified to both C-1 and C-2 (Figure 11·10).

The specific positions of fatty acids in glycerophospholipids can be determined by using enzymes called phospholipase A_1 and phospholipase A_2, which specifically catalyze the hydrolysis of the fatty ester bonds at C-1 and C-2, respectively (Figure 11·11). High concentrations of **lysophosphoglycerides** resulting from phospholipase A action can disrupt cellular membranes. Snake, bee, and wasp venoms are good sources of phospholipase A_2. Injection of snake venom into the blood can result in life-threatening hemolysis, or lysis of the membranes of red blood cells. There are two other classes of phospholipase, namely phospholipase C, which catalyzes hydrolysis of the bond between glycerol and phosphate to liberate diacylglycerol, and phospholipase D, which catalyzes the production of phosphatidates from glycerophospholipids.

Figure 11·9
General structures of **(a)** phosphatidylethanolamine, **(b)** phosphatidylserine, and **(c)** phosphatidylcholine. Functional groups derived from the esterified alcohol are shown in blue. Since each of these lipids can contain many combinations of fatty acyl groups, the general name refers to a family of compounds, not to a single molecule. Note that all three of these molecules have a polar head group and nonpolar tails.

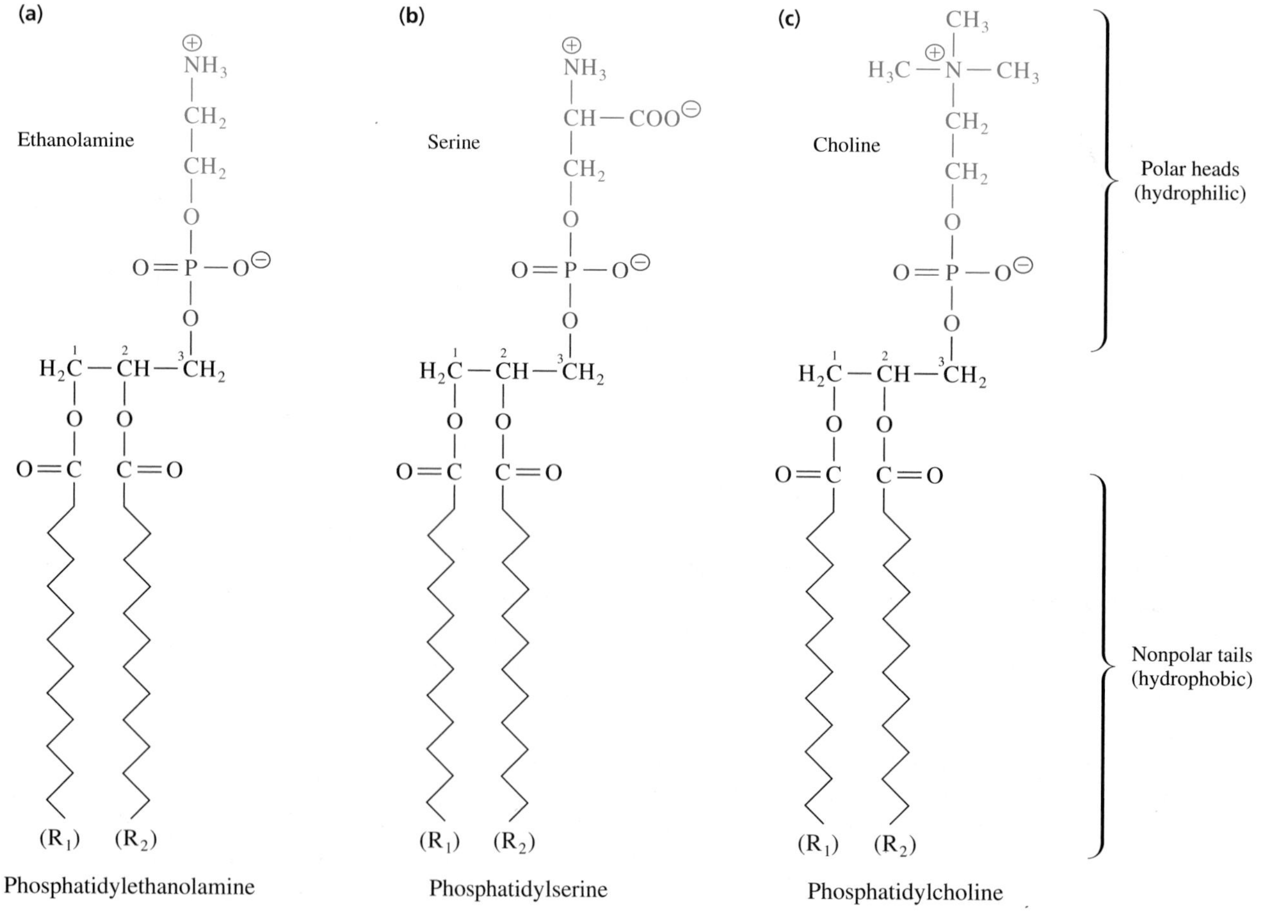

(a) Phosphatidylethanolamine

(b) Phosphatidylserine

(c) Phosphatidylcholine

Figure 11·10
Stereo view of dipalmitoylphosphatidylcholine. Color key: carbon, light green; hydrogen, blue; oxygen, red; phosphorus, dark green.

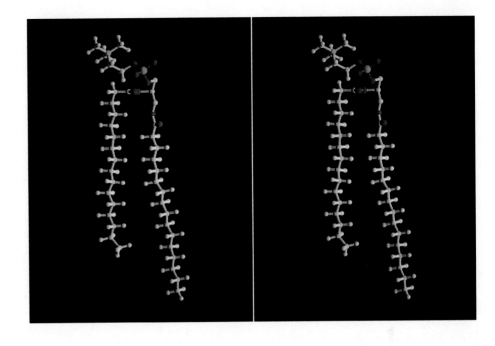

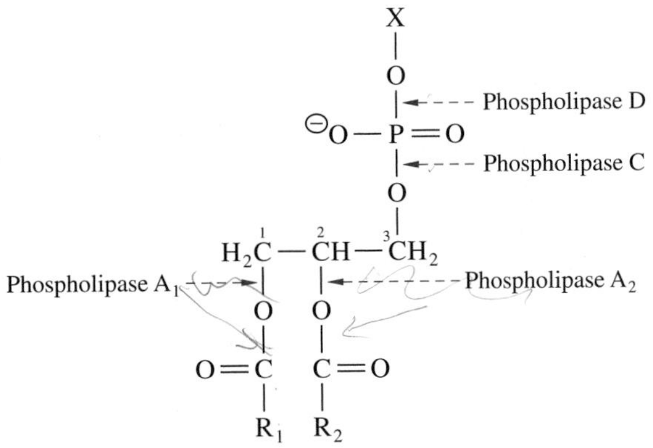

Figure 11·11
Action of four phospholipases. Phospholipases A$_1$, A$_2$, C, and D are used to dissect glycerophospholipid structure. Phospholipases catalyze the selective removal of fatty acids from C-1 or C-2 or convert glycerophospholipids into diacylglycerols or phosphatidic acids.

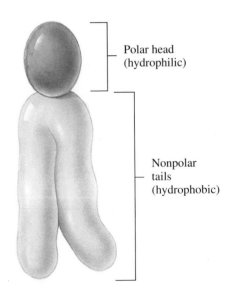

Polar head
(hydrophilic)

Nonpolar
tails
(hydrophobic)

Note that glycerophospholipids are amphipathic molecules, having a polar head (with an anionic phosphate and often one or two other charged groups) and long, nonpolar tails (Figure 11·12). DPPC, shown in Figure 11·10, is a glycerophospholipid that constitutes over 50% of a secretion known as lung surfactant. Lung surfactant reduces the surface tension at the air-water interface of the alveoli (the bubble-like walls of the lungs), which enables oxygen to pass efficiently from the airways into the cells. In recent years, DPPC has been used to treat respiratory distress syndrome, which can be fatal if not treated. This condition, characterized by shallow breathing, affects premature infants whose immature lungs cannot produce sufficient surfactant. DPPC is given as an aerosol to these babies to help them breathe.

Figure 11·12
General depiction of an amphipathic glycerophospholipid molecule. The lipid illustrated here has two nonpolar tails. This icon is used extensively throughout the book.

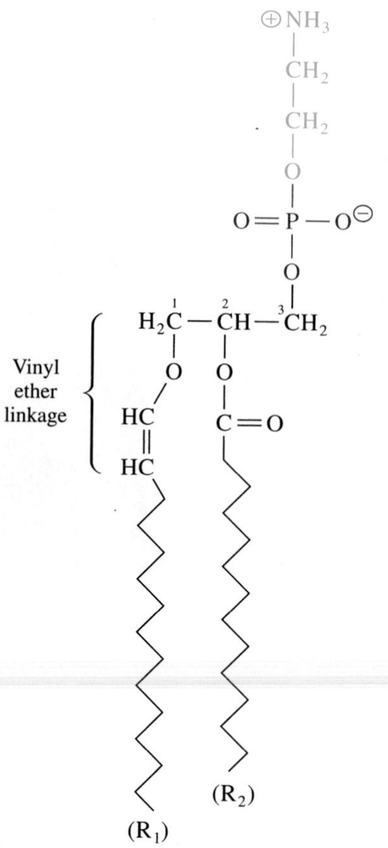

Figure 11·13
Structure of an ethanolamine plasmalogen. A hydrocarbon tail is linked to C-1 of glycerol as a vinyl ether.

Although most glycerophospholipids consist of fatty acids linked to glycerol by esterification, members of the **plasmalogen** family have a hydrocarbon chain linked to C-1 of the glycerol backbone by a vinyl ether linkage (Figure 11·13). The most common compounds esterified to the phosphate group of plasmalogens are ethanolamine and choline. Plasmalogens account for about 23% of the glycerophospholipids in the human central nervous system and are also found in the membranes of cells of peripheral nerve and muscle tissue.

Platelet activating factor (PAF) is a choline glycerophospholipid with an alkyl ether group at C-1 and an acetyl group at C-2 of the glycerol backbone (Figure 11·14). PAF is a biologically active phospholipid. At very low concentrations (0.1 nM), it can cause platelets to aggregate and form thrombi. PAF also functions in smooth-muscle contraction. There is evidence that abnormally high levels of PAF are involved in the morbidity associated with stroke and in toxic shock produced by bacterial infections.

Phosphatidylinositol (PI) is the precursor of lipids that are even more polar, which are formed by the addition of phosphate groups to the inositol ring. These derivatives are phosphatidylinositol 4-phosphate (PIP) and phosphatidylinositol 4,5-*bis*phosphate (PIP_2). The structures of PI, PIP, and PIP_2 are shown in Figure 11·15. PIP_2 is of considerable interest because it is involved in signalling events across the plasma membrane. For example, binding of the neurotransmitter acetylcholine to its receptor at a nerve ending activates a phospholipase C, which catalyzes degradation of PIP_2 to diacylglycerol and inositol 1,4,5-*tris*phosphate. Diacylglycerol and inositol 1,4,5-*tris*phosphate act as second messengers within cells by activating a protein kinase and triggering the release of calcium ions, thereby eliciting a variety of cellular metabolic responses (see Chapter 12). Note that the prefix *tris* is used to designate three phosphate groups on three separate carbon atoms in the inositol ring, whereas the prefix *tri* indicates three phosphates joined in tandem by phosphoanhydride linkages.

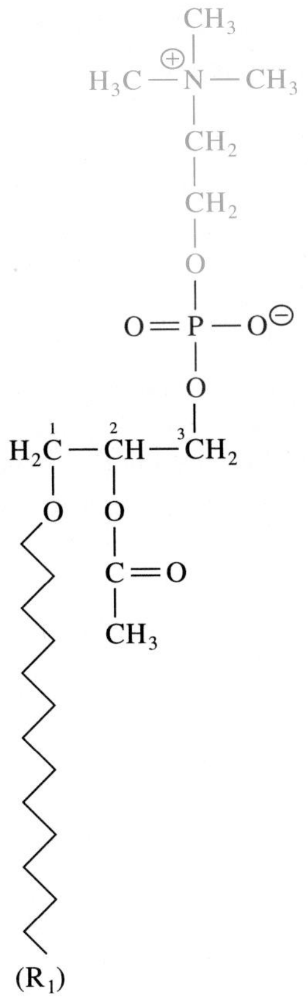

Figure 11·14
Structure of platelet activating factor. An alkyl ether is linked to C-1 of glycerol, and an acetyl group is linked to C-2.

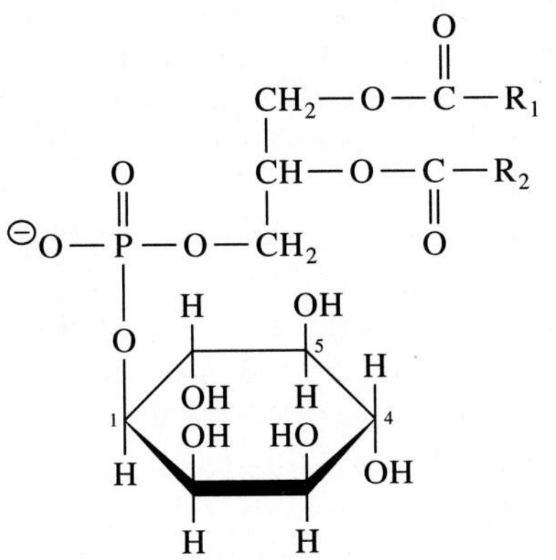

Phosphatidylinositol
(PI)

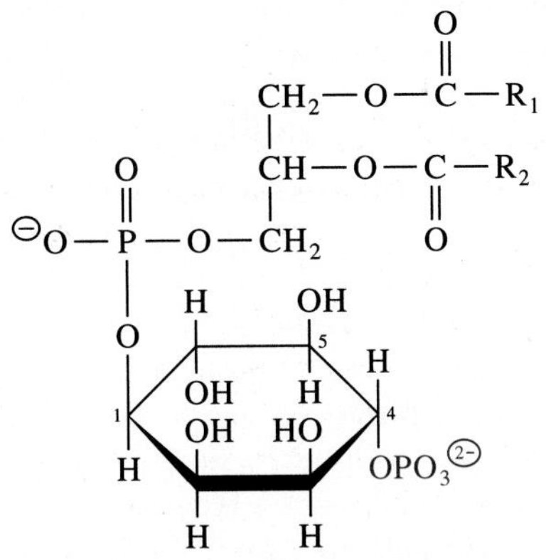

Phosphatidylinositol 4-phosphate
(PIP)

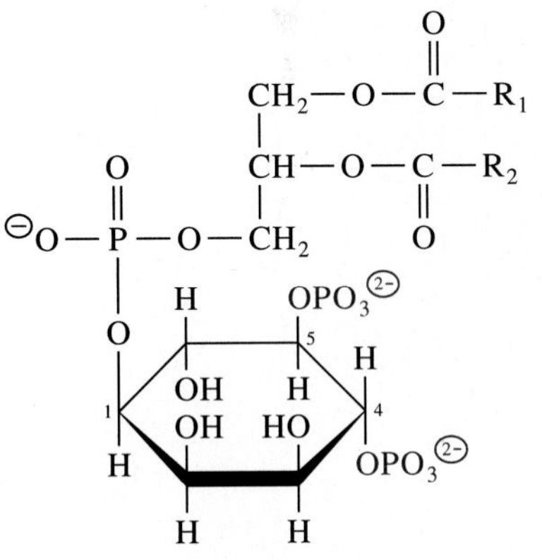

Phosphatidylinositol 4,5-*bis*phosphate
(PIP$_2$)

Figure 11·15
Structures of phosphatidylinositol (PI), phosphatidylinositol 4-phosphate (PIP), and phosphatidylinositol 4,5-*bis*phosphate (PIP$_2$). The additional phosphate groups attached to the inositol ring make PIP and PIP$_2$ highly polar.

11·6 Sphingolipids Constitute a Second Class of Lipids in Biological Membranes

Although glycerophospholipids account for the bulk of the lipids in most membrane systems, other amphipathic lipids, called **sphingolipids,** are also present in plant and animal membranes. In mammals, they are particularly abundant in tissues of the central nervous system.

The structural backbone of sphingolipids is sphingosine (*trans*-4-sphingenine), an unbranched C_{18} alcohol with a *trans* double bond between C-4 and C-5, an amino group at C-2, and hydroxyl groups at C-1 and C-3 (Figure 11·16a). **Ceramide** consists of a fatty acid linked to the C-2 amino group of sphingosine by an amide bond (Figure 11·16b). Ceramides are the metabolic precursors of all sphingolipids. The three major families of sphingolipids are the sphingomyelins, the

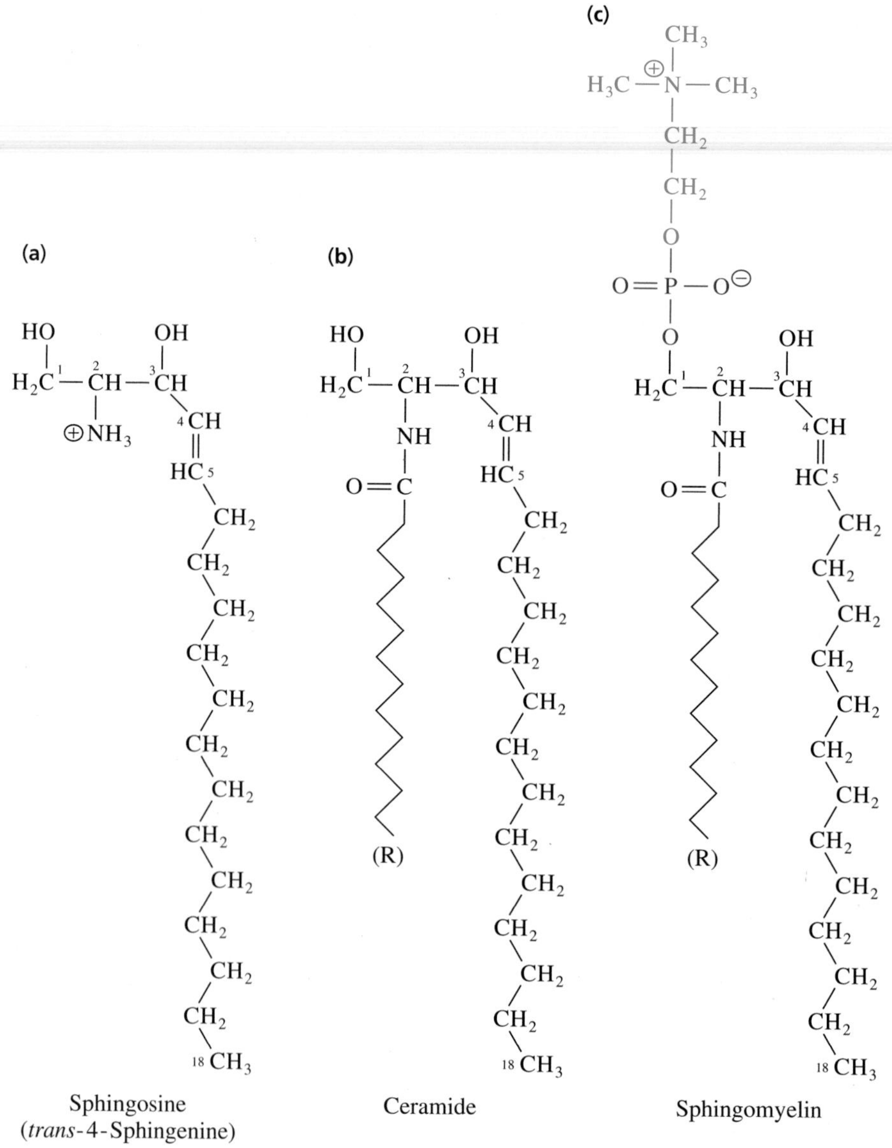

Figure 11·16
Structures of sphingosine, ceramide, and sphingomyelin. **(a)** Sphingosine is a long-chain alcohol with an amino group at C-2. Sphingosine serves as the backbone for sphingolipids, a second class of lipid molecules found in biological membranes. **(b)** Ceramides have a long-chain fatty acyl group attached to the amino group of sphingosine. **(c)** Sphingomyelins have a phosphate group (red) attached to the C-1 hydroxyl group of a ceramide and a choline group (blue) attached to the phosphate.

Sphingosine
(*trans*-4-Sphingenine)

Ceramide

Sphingomyelin

cerebrosides, and the gangliosides. Of these, only the sphingomyelins contain phosphate and are therefore classified as phospholipids; cerebrosides and gangliosides contain carbohydrate residues and are therefore classified as glycosphingolipids.

Sphingomyelin consists of phosphocholine attached to the C-1 hydroxyl group of a ceramide (Figure 11·16c). Note the resemblance of the amphipathic lipid molecule sphingomyelin to a phosphatidylcholine molecule (Figure 11·9c); both are zwitterions containing choline, phosphate, and two long hydrophobic tails. Sphingomyelins are present in the plasma membranes of most mammalian cells and are a major component of the myelin sheaths that surround certain nerve cells.

Cerebrosides are glycosphingolipids that contain one monosaccharide residue attached via a β-glycosidic linkage to C-1 of a ceramide. Galactocerebrosides, also known as galactosylceramides, are cerebrosides whose polar head groups are single β-D-galactosyl residues (Figure 11·17). Galactocerebrosides are abundant in nerve tissue and account for about 15% of the lipids of myelin sheaths. Many other mammalian tissues contain glucocerebrosides, in which a β-D-glucosyl residue, rather than a β-D-galactosyl residue, is bound to C-1 of a ceramide. In some related glycosphingolipids, a linear chain of up to three additional monosaccharide residues is attached to the galactosyl residue of a galactocerebroside or to the glucosyl residue of a glucocerebroside. Because they contain more than one monosaccharide residue, such glycosphingolipids are not called cerebrosides but have a variety of names.

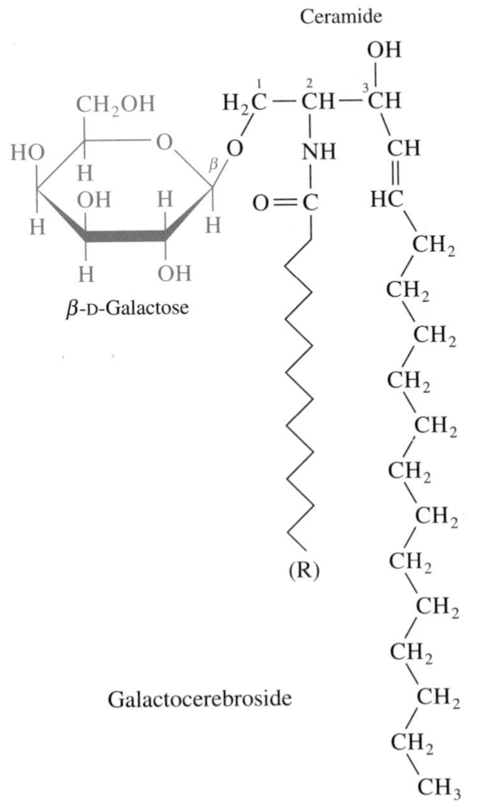

Figure 11·17
General structure of galactocerebroside. β-D-Galactose (blue) is attached to the C-1 hydroxyl group of a ceramide (black).

Gangliosides are more complex glycosphingolipids in which oligosaccharide chains containing *N*-acetylneuraminic acid (NeuNAc) are attached to a ceramide. NeuNAc is the acetyl derivative of a complex nine-carbon amino sugar, a member of the sialic acid family (Figure 11·18a and Figure 11·19). The sugar residues provide the large polar head groups of gangliosides. Gangliosides are anionic due to the presence of the carboxyl group of NeuNAc.

Figure 11·18
Ganglioside structure. (**a**) *N*-Acetylneuraminic acid (NeuNAc) is a component of gangliosides. (**b**) G_{M2} and (**c**) G_{M1} are representative gangliosides.

(a)

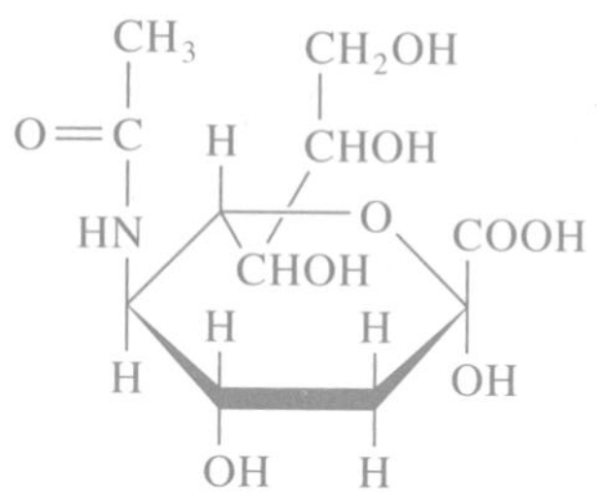

N-Acetylneuraminic acid
(NeuNAc)

(b)

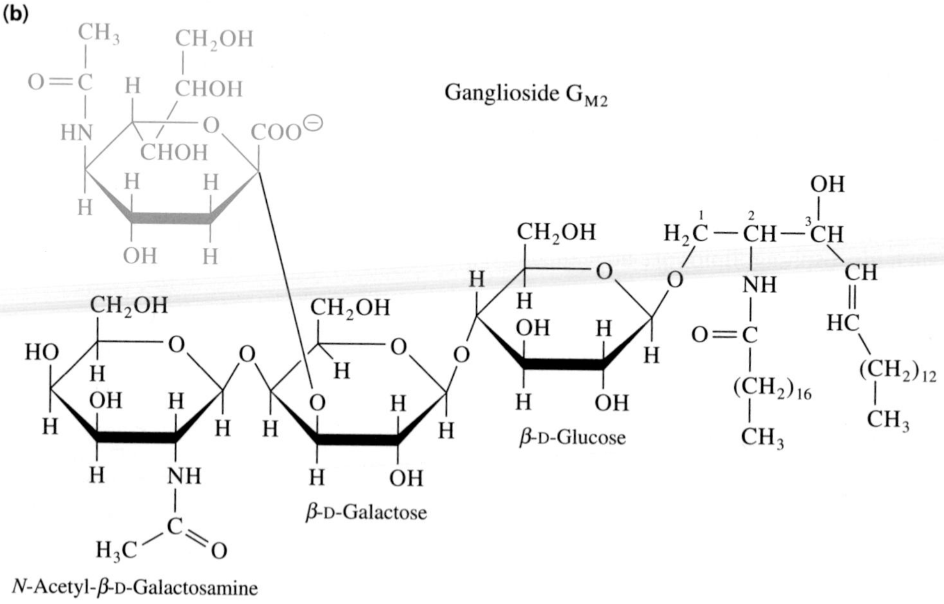

Ganglioside G_{M2}

β-D-Glucose

β-D-Galactose

N-Acetyl-β-D-Galactosamine

(c)

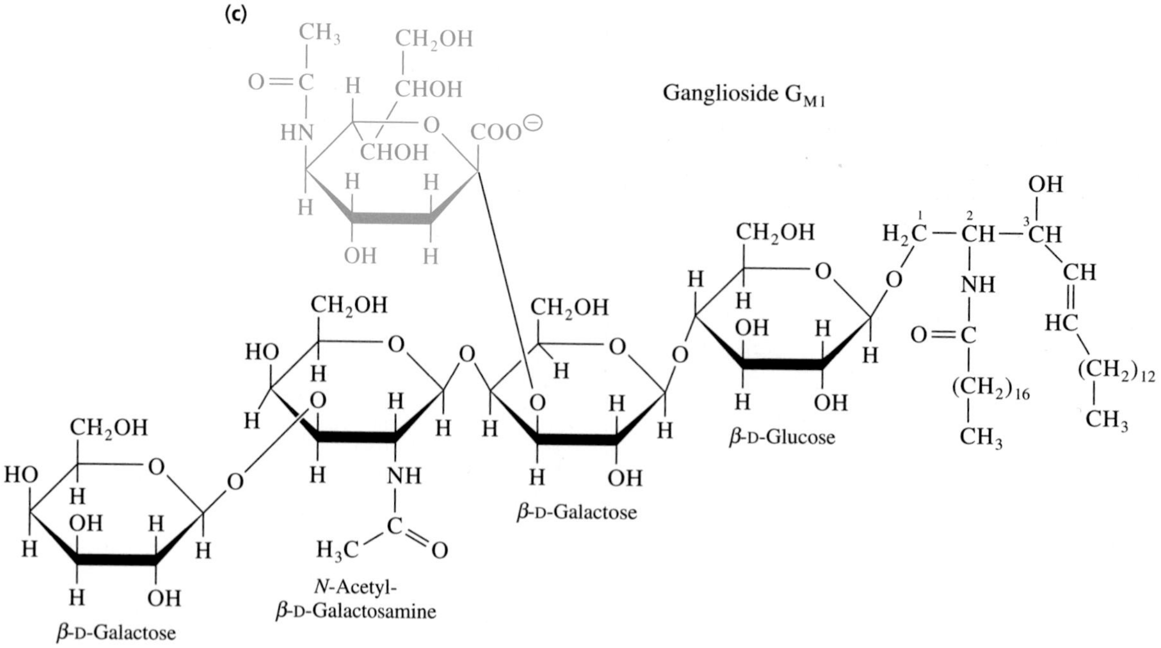

Ganglioside G_{M1}

β-D-Glucose

β-D-Galactose

N-Acetyl-
β-D-Galactosamine

β-D-Galactose

Figure 11·19
Stereo view of *N*-acetyl-neuraminic acid.

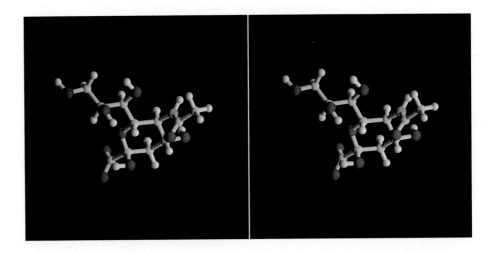

The considerable diversity in the structures of gangliosides results from variations in the composition and sequence of sugar residues; well over 60 varieties of gangliosides have been characterized. Gangliosides are present on cell surfaces, with the two hydrocarbon chains of the ceramide moiety embedded in the hydrophobic part of the plasma membrane and the complex oligosaccharide structures forming an outside coat. For example, the ABO blood group antigens are based on the oligosaccharide chains of glycosphingolipids and glycoproteins. Gangliosides and other glycosphingolipids provide cells with distinguishing surface markers that may serve in cellular recognition and cell-to-cell communication.

The ganglioside G_{M1} serves as a receptor for the cholera toxin produced by the bacterium *Vibrio cholerae*. When this microorganism enters the small intestine, the bacterial toxin can enter the cells lining the intestine and trigger a massive loss of fluid, producing diarrhea and ultimately death by dehydration. The cholera toxin is actually a protein that binds specifically to G_{M1} at the cell surface and is taken into the intestinal epithelial cells. Once inside the cell, the toxin stimulates the release of large quantities of fluid by permanently switching on a cellular signalling pathway.

Although highly varied in their molecular structures, gangliosides can be classified according to the identity of core oligosaccharide structures that are bound to the ceramide. In each of the four recognized core structures, the ceramide is linked through C-1 to a β-glucosyl residue, which is in turn bound to a β-galactosyl residue. The core structures differ in size and in the number and position of their NeuNAc residues. One of the four core structures is exemplified by ganglioside G_{M2}, which is depicted in Figure 11·18b. The *M* in G_{M1} and G_{M2} stands for monosialo, indicating that these lipids have only one NeuNAc residue in their core structure (G_{M2} was the second monosialo ganglioside characterized, thus the subscript 2.)

Ganglioside metabolism is an active area of medical research. Recent investigations have revealed that the composition of membrane glycosphingolipids can change dramatically during the development of malignant tumors. It is also known that genetically inherited defects in ganglioside metabolism are responsible for a number of debilitating and often lethal diseases, such as Tay-Sachs disease and generalized gangliosidosis.

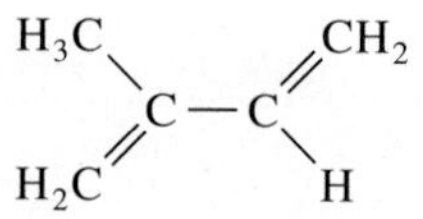

Figure 11·20
Isoprene. Isoprene forms the basic structural unit of all polyprenyl compounds, which include the steroids and lipid vitamins.

11·7 Steroids Constitute a Third Class of Membrane Lipids

The steroids constitute a third class of lipids commonly found in the membranes of eukaryotes. Steroids and the lipid vitamins are more broadly classified as polyprenyl compounds, compounds whose structures are related to the five-carbon molecule isoprene (Figure 11·20). Steroids have a characteristic cyclic nucleus consisting of four fused rings: three six-carbon rings designated A, B, and C and a five-carbon D ring, as shown in Figure 11·21. Cholesterol is one example of a steroid. It is an important component of the plasma membranes of mammals but is only rarely found in plants and never in prokaryotes. Cholesterol can be termed more specifically a **sterol** (a steroid alcohol) because of the —OH group at C-3. Substituents of the ring system can point either down (termed α) or up (termed β) from the plane of the rings. Note the methyl groups at C-10 and C-13 and the eight-carbon side chain attached to C-17.

Other steroids include the sterols of plants and fungi (which contain at least 27 carbon atoms and, like cholesterol, have a hydroxyl group at C-3); the mammalian steroid hormones (such as the C_{18} estrogens, the C_{19} androgens, and the C_{21} progestins and adrenal corticosteroids); and the C_{24} bile salts. The structures of these

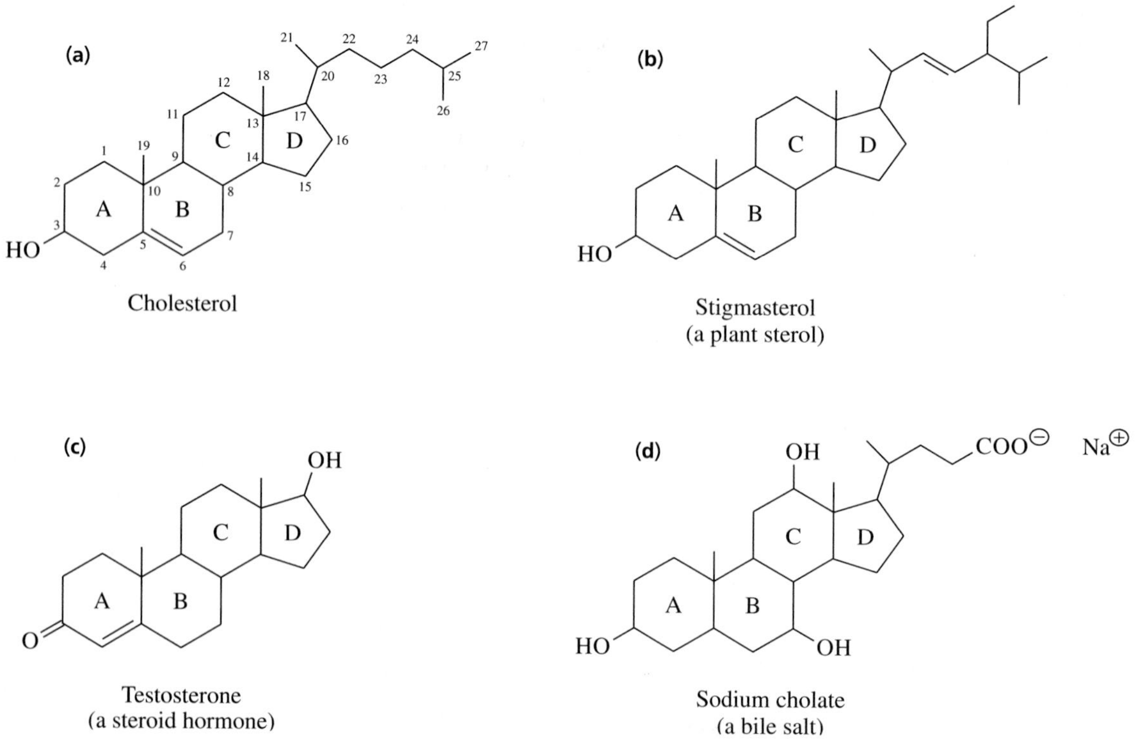

Figure 11·21
Structures of several steroids. The fused ring system of steroids consists of four rings (lettered A, B, C, and D). (a) Cholesterol. (b) Stigmasterol, a common sterol component of plant membranes. (c) Testosterone, a steroid hormone involved in male development in animals. (d) Sodium cholate, a bile salt. Cholate (the conjugate base of cholic acid) aids in the digestion of lipids.

Cholesterol

Stigmasterol
(a plant sterol)

Testosterone
(a steroid hormone)

Sodium cholate
(a bile salt)

Figure 11·22
Stereo view of cholesterol. The polar hydroxyl group is on the left. Note that cholesterol is nearly planar due to its fused ring system.

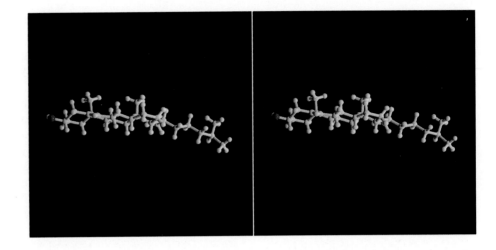

steroids differ in the length of the side chain attached to C-17 of the fused ring system and in the number and placement of methyl groups, double bonds, hydroxyl groups, and in some cases, keto groups.

Despite its implication in cardiovascular disease, cholesterol plays an essential role in mammalian biochemistry. Cholesterol, which is synthesized by mammalian cells, is not only a component of certain membranes, but it is also a precursor of the steroid hormones and bile salts. Steroid hormones include testosterone (the male sex hormone) and estradiol (one of the female hormones) as well as aldosterone (from the adrenal gland, which regulates salt excretion). It is evident from the structure shown in Figure 11·21a and Figure 11·22 that cholesterol is far more hydrophobic than the glycerophospholipids and sphingolipids we have previously encountered as membrane lipids. In fact, with the hydroxyl group at C-3 as its only polar constituent, free cholesterol has a maximal concentration in water of 10^{-8} M. The ring structure of cholesterol is rigid; thus, it is less flexible than most other lipids. As a result, the presence of cholesterol serves to modulate the fluidity of mammalian cell membranes, as we will see in the following chapter.

The esterification of fatty acids to the C-3 hydroxyl group of cholesterol forms cholesteryl esters (Figure 11·23). Because the 3-acyl group of the ester is a nonpolar substituent, a cholesteryl ester is even more hydrophobic than cholesterol itself. Cholesteryl esters are found in lipoproteins in the blood and are formed when cholesterol is to be stored within cells. Lipoproteins will be discussed in Chapter 13.

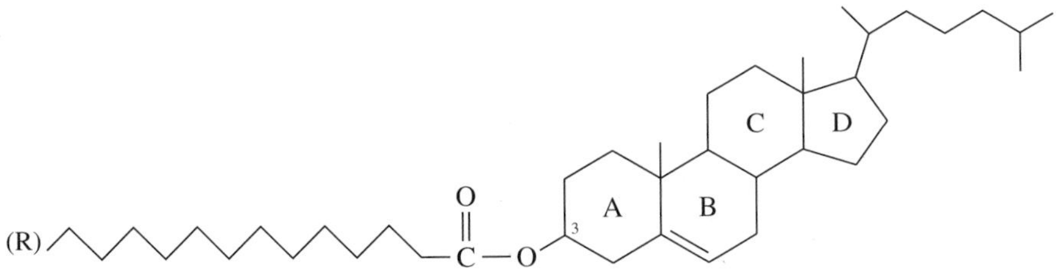

Figure 11·23
A cholesteryl ester. The presence of a fatty acyl group at C-3 makes cholesteryl esters even more hydrophobic than cholesterol.

A cholesteryl ester

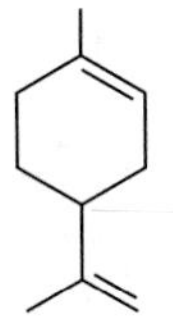

Limonene

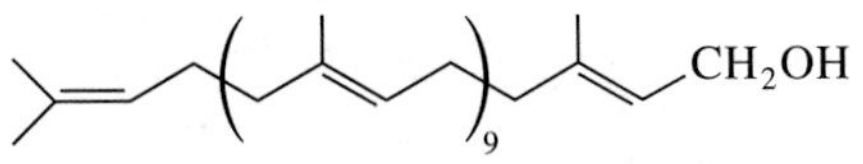

Bactoprenol
(Undecaprenyl alcohol)

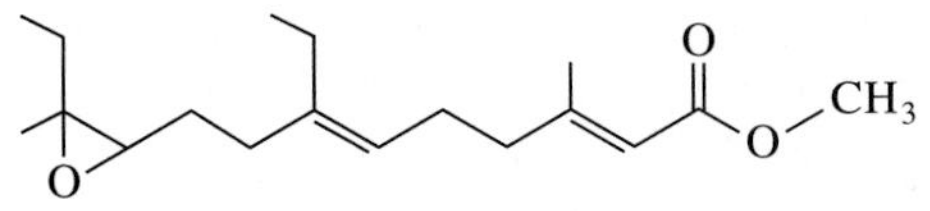

Juvenile hormone I

Figure 11·26
Some biologically important polyprenyl compounds. Limonene is the lipid component of lemons that gives them their characteristic smell. Bactoprenol is a bacterial lipid that plays a role during the synthesis of cell walls. Juvenile hormone I regulates larval development in insects.

The fat-soluble, or lipid, vitamins are vitamins A, D, E, and K. Like steroids, the lipid vitamins are polyprenyl compounds, composed primarily of long hydrocarbon chains or fused rings (Figure 11·25). Even though the lipid vitamins contain polar groups, they are relatively hydrophobic and are more soluble in a lipid environment than in an aqueous one. All serve important roles in human metabolism, as we saw in Chapter 8.

A variety of other polyprenyl compounds also occurs in biological systems (Figure 11·26). **Terpenes** are polyprenyl compounds found in plants. Limonene, for example, is a cyclic terpene, the lipid component of lemons chiefly responsible for their distinctive smell. The polyprenyl compound bactoprenol is a bacterial lipid that plays a role during synthesis of cell walls, and juvenile hormone I is a polyprenyl compound that regulates larval development in insects.

11·9 Special Nonaqueous Techniques Must Be Used to Study Lipids

The study of lipids in the laboratory is rather unusual because it involves the use of organic solvents. Unlike the water-soluble or hydrophilic components of tissues or cells such as carbohydrates and most proteins, lipids generally have very low solubility in water. In order to analyze the lipid components of organelles, cells, or tissues, the sample must be homogenized or dispersed in an organic solvent solution such as methanol in chloroform. The methanol precipitates proteins while the chloroform effectively dissolves the lipids of membranes. The resulting lipid solution can be manipulated further to separate the major lipid classes and then to resolve the various components within each class. Plastic vessels must be avoided since many plastics can also dissolve in chloroform. Several new "lipids" have been discovered after placing chloroform extracts in plastic tubes. A schematic view of lipid purification is shown in Figure 11·27 (next page) and is described in the following paragraphs.

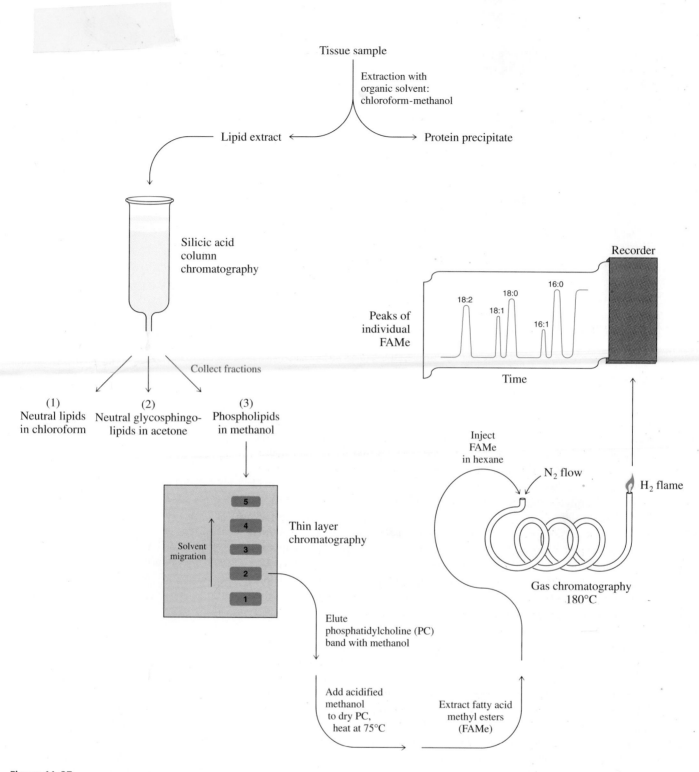

Figure 11·27
Flow chart of lipid extraction and purification. Silicic acid column chromatography, thin layer chromatography, and gas chromatography are used to separate and analyze lipid classes and component fatty acids.

Silicic acid, a chromatographic medium that resembles fine white beach sand, is useful for lipid purification. For example, a lipid extract prepared from red blood cells can be layered on the top of a silicic acid column equilibrated with chloroform. By running chloroform through the column, neutral lipids such as cholesterol, triacylglycerols, and diacylglycerols can be eluted from the column, leaving the more polar lipids bound to the silicic acid. Changing the solvent to acetone allows the elution of neutral glycosphingolipids, and the application of methanol elutes the phospholipids.

Finer resolution of the compounds within each of these groups can be accomplished by thin-layer chromatography. A very thin layer (0.2–0.5 mm) of silicic acid is placed upon a glass plate, and the lipid mixture (for example, the phospholipids) is spotted on the thin layer near the bottom of the plate. The plate is then placed upright in a covered, shallow tank containing a small volume of a solution of chloroform, methanol, water, and acetic acid. The solvent slowly migrates up the plate, resolving the less polar lipid components (which move more quickly with the solvent) from the slower, more polar compounds. In this way, bands of phospholipids such as sphingomyelin, phosphatidylcholine, phosphatidylserine, phosphatidylinositol, phosphatidylethanolamine, and others are separated. The bands can be located by exposing the plate to iodine vapor or by spraying the plate with a fluorescent dye that permits the visualization of the lipid bands under ultraviolet light. Each band can be scraped off the plate and the phospholipid eluted from the silicic acid with solvent. A convenient way to quantitate phospholipids is by acid hydrolysis and measurement of the released inorganic phosphate by a simple color assay.

Luckily for lipid chemists, many fats and their derivatives are volatile at high temperature. When you pass a fast-food restaurant, you can smell fats that have been volatilized. Component fatty acids of phospholipids that have been separated by thin layer chromatography can be analyzed by a technique known as **gas chromatography.** The phospholipids are treated with acidified methanol to produce fatty acid methyl esters; the mixture of fatty acid methyl esters is injected onto the top of a long heated column, which contains a solid adsorptive medium. The methyl esters are carried by the flow of nitrogen gas through the column. The longer and more highly unsaturated fatty acids are separated from the faster-moving shorter and more saturated chains. Since lipids burn at high temperatures, the methyl esters can be detected by a hydrogen flame positioned at the end of the column that burns each methyl ester as it emerges. This combustion is detected electronically and the signal inscribed by a recorder, permitting identification and quantitation of each peak.

The fatty acid profile in Figure 11·27 indicates the diversity of molecular species within a phospholipid class. The individual species of phosphatidylcholine could also be resolved by treating the phospholipid with phospholipase C to produce the various diacylglycerol components. These diacylglycerols can be separated on the basis of chain length and degree of unsaturation of their fatty acids by high-pressure liquid chromatography (HPLC). In HPLC, the mixture of diacylglycerols is resolved by the flow of solvents under pressure through a column. It is common to see 20–30 different species in one phospholipid sample.

Summary

Lipids are water-insoluble organic compounds that can be extracted from biological samples with nonpolar organic solvents. Lipids are quite diverse, both structurally and functionally.

Fatty acids are relatively long-chain monocarboxylic acids. The majority of naturally occurring fatty acids contain an even number of carbon atoms, ranging from 12 to 20. Fatty acids that contain no carbon-carbon double bonds are classified as saturated fatty acids; those that contain one carbon-carbon double bond are classified as monounsaturated, and those that contain more than one carbon-carbon double bond are classified as polyunsaturated. Most of the double bonds found in unsaturated fatty acids have the *cis* configuration. As esters, saturated and unsaturated fatty acids are constituents of a wide variety of lipids.

Fatty acids are generally stored as complex lipids called triacylglycerols (fats and oils). Triacylglycerols are neutral and nonpolar. Waxes, also neutral, nonpolar lipids, are esters of long-chain aliphatic alcohols and fatty acids. Eicosanoids are physiologically important derivatives of C_{20} fatty acids such as arachidonate.

Glycerophospholipids are among the major amphipathic lipid components of biological membranes. They include phosphatidylcholine, phosphatidylethanolamine, phosphatidylserine, and phosphatidylinositol. Their polar heads include an anionic phosphodiester group that links C-3 of the glycerol backbone to another water-soluble component, whereas their nonpolar tails are made up of fatty acyl residues esterified to C-1 and C-2 of the glycerol moiety. Plasmalogens are glycerophospholipids in which the C-1 oxygen of the glycerol 3-phosphate moiety is bound to a hydrocarbon chain as a vinyl ether. Platelet activating factor is a biologically active lipid that has an alkyl ether group at C-1 and an acetyl group at C-2. Phosphatidylinositol 4,5-*bis*phosphate, a polar derivative of phosphatidylinositol, is involved in transmembrane signalling.

Other major classes of lipids include the sphingolipids, steroids, and fat-soluble, or lipid, vitamins. The long-chain amino alcohol sphingosine provides the backbone for sphingolipids. Three major classes of sphingolipids are sphingomyelins, cerebrosides, and gangliosides. The lipid vitamins are examples of polyprenyl compounds, or isoprenoids, lipids synthesized from a five-carbon compound related to isoprene. Cholesterol, a steroid, is an important component of animal membranes and serves as the precursor to a variety of hormones. Other steroids are found in plants and other eukaryotes.

Because they are poorly soluble in water, lipids are extracted and purified in organic solvents.

Selected Readings

General References

Gurr, M. I., and Harwood, J. L. (1991). *Lipid Biochemistry: An Introduction,* 4th ed. (London: Chapman and Hall). A general reference for lipid structure and metabolism.

Mead, J. F., Alfin-Slater, R. B., Howton, D. R., and Popják, G. (1986). *Lipids: Chemistry, Biochemistry and Nutrition* (New York: Plenum Publishing Corporation).

Vance, D. E., and Vance, J. E., eds. (1991). *Biochemistry of Lipids, Lipoproteins, and Membranes* (New York: Elsevier Science Publishing Company). Contains up-to-date reviews.

Specific Lipids

Caminiti, S. P., and Young, S. L. (1991). The pulmonary surfactant system. *Hosp. Practice* 26(1):87–100.

Fisher, S. K., Heacock, A. M., and Agranoff, B. W. (1992). Inositol lipids and signal transduction in the nervous system: an update. *J. Neurochem.* 58:18–38.

Hakomori, S. (1986). Glycosphingolipids. *Sci. Am.* 254(5):44–53. Discussion of the clinical importance of gangliosides in blood- and tissue-donor compatibility and their use in diagnosis and treatment of cancer.

Holmgren, J. (1981). Actions of choleratoxin and the prevention and management of cholera. *Nature* 292:413–417.

Keough, K. M. W. (1985). Lipid fluidity and respiratory distress syndrome. In *Membrane Fluidity in Biology,* Vol. 3. R. C. Aloia and J. M. Boggs, eds. (New York: Academic Press), pp. 39–84.

Lai, C-Y. (1980). The chemistry and biology of cholera toxin. *Crit. Rev. Biochem.* 9:171–206.

Shimizu, T., and Wolfe, L. S. (1990). Arachidonic acid cascade and signal transduction. *J. Neurochem.* 55:1–15. Includes the biochemistry of prostaglandins.

12

Biological Membranes

A lipid membrane barrier is essential for the existence of the cell as an independent entity. Indeed, the presence of an external membrane surrounding the cell was first suggested by William Bowman in 1840, only a year after the cell doctrine had been proposed. Over the course of the 19th century, it became clear that this barrier was relatively impermeable and had the electrical properties of an oil film. Elucidation of the molecular details of membrane structure had to await the application in the 1950s of electron microscopy, which made it possible to visualize cellular membranes. The "triple-track" appearance of many different types of biological membranes stained with heavy metals implied that all membranes might have a common structure (Figure 12·1).

Membranes define the external boundary of the cell as well as separate compartments within the cell. As we saw in Chapter 2, all living cells possess at least one membrane. Most prokaryotes lack internal membranes, and in many cases, their cytoplasmic membrane is surrounded by a rigid cell wall of peptidoglycan. An additional outer membrane is present in Gram-negative bacteria. Eukaryotes possess an exterior plasma membrane as well as internal organelles that are surrounded and supported by membranes having unique components and functions. Enveloped viruses, such as the influenza virus, consist of nucleic acid surrounded by a simple membrane.

Biological membranes are not passive barriers; they have a huge variety of complex functions. In addition to compartmentalizing the cell, membranes contain selective pumps that strictly control the entry and exit of ions and small molecules and also generate the proton concentration gradients essential for production of ATP by oxidative phosphorylation. Receptors found in the membrane recognize extracellular signals and communicate them to the cell interior. In this chapter, we will explore the structures of biological membranes, including the lipid and protein components, and then examine important aspects of membrane function.

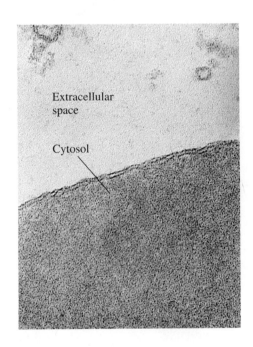

Extracellular space

Cytosol

Figure 12·1
Electron micrograph of a portion of the human erythrocyte membrane stained with osmium. Note that the structure has a trilaminar (three-layer) appearance, which is also characteristic of other biological membranes.

Table 12·1 Lipid, protein, and carbohydrate content of some membranes

Membrane	Percentage by mass			Protein-to-lipid ratio
	Protein	Lipid	Carbohydrate	
Myelin	18	79	3	0.2
Human erythrocyte (plasma membrane)	49	43	8	1.1
Bovine retinal rod	51	49	0	1.0
Mitochondria (outer membrane)	52	48	0	1.1
Amoeba (plasma membrane)	54	42	4	1.3
Sarcoplasmic reticulum (muscle cells)	67	33	0	2.0
Chloroplast lamellae	70	30	0	2.3
Gram-positive bacteria	75	25	0	3.0
Mitochondria (inner membrane)	76	24	0	3.2

[Adapted from Guidotti, G. (1972). Membrane proteins. *Annu. Rev. Biochem.* 41:731–752.]

12·1 Biological Membranes Are Composed of Lipids and Proteins

Related material appears in Exercise 4 of the computer-disk supplement *Exploring Molecular Structure.*

A typical membrane consists of about 40% lipid and 60% protein (by mass) and up to 10% carbohydrate (as a component of glycolipids and glycoproteins). However, the compositions of biological membranes vary considerably between species and between cells of multicellular organisms (Table 12·1). For example, the myelin membrane, which insulates nerve fibers, has a high lipid content and contains little protein. For some other membranes, the reverse is true; the mitochondrial inner membrane and some bacterial cytoplasmic membranes are exceptionally rich in protein, reflecting their high level of protein-dependent metabolic activity. In addition to having a characteristic lipid-to-protein ratio, each biological membrane has a distinctive lipid composition, which again can be highly variable.

A. Lipid Bilayers Form the Structural Basis for Membranes

A typical membrane contains a complex mixture of phospholipids, glycosphingolipids, and (in some eukaryotes) cholesterol. The structures and properties of these molecules were discussed in Chapter 11. The lipid components of membranes have one feature in common: they are all amphipathic, with both polar and nonpolar moieties. In Section 3·4, we saw that soaps, being amphipathic, can form monolayers or micelles. Like soaps, phospholipids and glycosphingolipids can form monolayers under certain conditions. In vivo, however, these lipids tend to assemble into a **lipid bilayer.** Because they have two hydrocarbon tails, phospholipids and glycosphingolipids do not pack well into micelles but fit nicely into lipid bilayers (Figure 12·2).

Not all amphipathic lipids can form bilayers. Cholesterol, for example, is amphipathic but cannot form a bilayer by itself, since the polar —OH group is too small relative to the hydrophobic, fused-hydrocarbon ring system. In biological membranes, cholesterol and other lipids that do not form bilayers (about 30% of the total) are stabilized in a bilayer arrangement by the presence of the other 70% of the lipids.

A lipid bilayer is typically about 5–6 nm thick. Lipid molecules within the bilayer are oriented with their hydrophobic tails pointing toward the interior of the bilayer and their hydrophilic heads in contact with the aqueous solution on each surface. The positive and negative charges of the head groups of phospholipids provide both **leaflets** (or layers) of the bilayer with an ionic surface. The interior of the bilayer is highly nonpolar and can be considered hydrocarbon-like. An increase in the entropy of the solvent water provides the major driving force for the formation of lipid bilayers, as it does for protein folding (Section 5·11A). Lipid bilayers tend to close up to form spherical structures; this property minimizes unfavorable contact between the hydrophobic edge of the bilayer and the aqueous solution.

Synthetic vesicles consisting of phospholipid bilayers that enclose an aqueous compartment can be formed in high yield in the laboratory. Such structures, called **liposomes,** are generally quite stable and impermeable to many substances. Liposomes whose aqueous inner compartment contains drug molecules can be used to deliver drugs to particular tissues in the body, provided that specific targeting proteins are present in the liposome membrane. Synthetic bilayers are an important experimental tool in the investigation of cellular membranes.

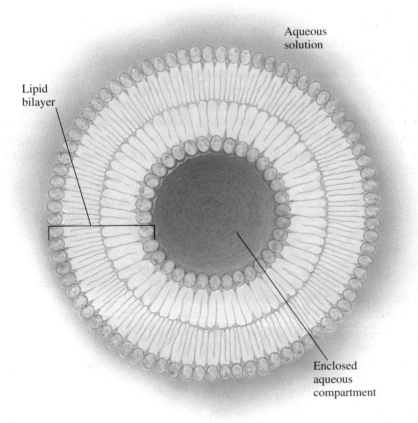

Lipid bilayer

Aqueous solution

Enclosed aqueous compartment

Figure 12·2
Schematic cross-section of a lipid bilayer vesicle, or liposome. The bilayer is made up of two leaflets. In each leaflet, the polar head groups of the amphipathic lipids extend into the aqueous medium, and the nonpolar hydrocarbon tails point inward and are in van der Waals contact with each other.

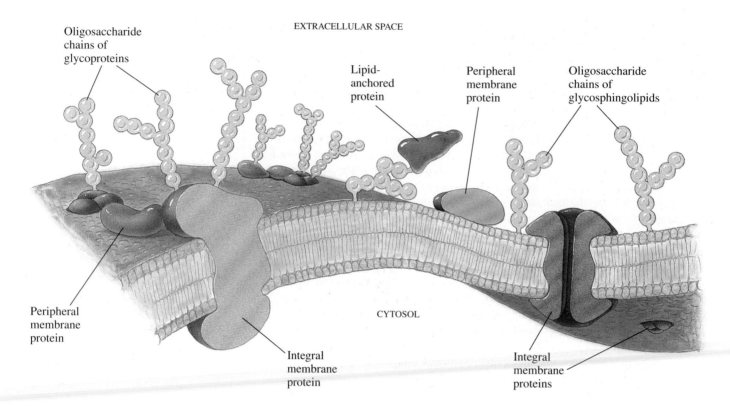

EXTRACELLULAR SPACE

CYTOSOL

Oligosaccharide chains of glycoproteins

Lipid-anchored protein

Peripheral membrane protein

Oligosaccharide chains of glycosphingolipids

Peripheral membrane protein

Integral membrane protein

Integral membrane proteins

Figure 12·3
Structure of a typical eukaryotic plasma membrane. A lipid bilayer forms the basic matrix of biological membranes, and proteins (some of which are glycoproteins) are associated with it in various ways. Integral membrane proteins are embedded in the lipid bilayer and usually span both leaflets. Hydrophobic amino acid residues in a transmembrane segment of these proteins form van der Waals contacts with the hydrophobic tails of lipid molecules. Peripheral membrane proteins are weakly associated with the surface of the membrane through ionic interactions and hydrogen bonds with either the polar heads of the lipid molecules or with integral membrane proteins. Lipid-anchored membrane proteins are linked to the membrane via covalent attachment to several types of lipid molecules.

B. The Fluid Mosaic Model Describes Biological Membranes

Lipid bilayers form the basis of all biological membranes, including plasma membranes and intracellular membranes, and are responsible for many of their physical properties. Lipid bilayers can exist in the absence of protein, but proteins are essential components of biological membranes. Membrane proteins are directly involved in the transport of molecules across the lipid bilayer, the transduction of signals to the cell interior, and interactions between the plasma membrane and the cytoskeleton. Whereas a lipid bilayer that contains no proteins is about 5–6 nm thick, biological membranes are typically 6–10 nm thick, due to the presence of proteins embedded in or associated with the bilayer.

In 1972, S. Jonathan Singer and Garth L. Nicolson proposed the **fluid mosaic model** for the structure of biological membranes. Membrane proteins are visualized as globular icebergs floating in a highly fluid lipid bilayer sea (Figure 12·3). Integral membrane proteins penetrate and span the bilayer, in contact with the hydrophobic interior. Peripheral membrane proteins are more loosely associated with the membrane surface. According to the fluid mosaic model, the membrane is a dynamic structure in which both proteins and lipids undergo rapid lateral diffusion, that is, diffusion within one leaflet of the bilayer. Although some aspects of the original fluid mosaic model have been adjusted and new features have been added, the model is generally valid today.

C. Membranes Can Be Isolated from Cells

Much of our knowledge of the structure and function of biological membranes has been obtained by study of membranes isolated from human red blood cells. Because mammalian erythrocytes contain no nucleus or internal organelles, these cells are a convenient source of plasma membrane. Lysis of the red blood cells in a hypotonic solution (a solution with a salt concentration lower than that of the cytosol) followed by washing to remove cytosolic proteins, such as hemoglobin, produces large quantities of plasma membrane (erythrocyte "ghosts") free of contamination with other membranes. Several erythrocyte membrane proteins have been purified, and their structures and functions are known in some detail. As we will see throughout this chapter, the erythrocyte membrane is useful as a prototype for other membranes about which we know much less.

The isolation of plasma membranes and intracellular membranes from eukaryotic cells other than erythrocytes involves differential centrifugation. Cells are first broken open mechanically, often by homogenization. During this process, organelles such as mitochondria, chloroplasts, and nuclei are released, while the plasma membrane and the membranes of the endoplasmic reticulum and Golgi apparatus pinch off to form small vesicles less than 0.5 μm in diameter. The intact organelles are removed by low-speed centrifugation, and membrane vesicles are collected from the supernatant fluid by high-speed centrifugation. Further separation of the different types of membranes, each of which has a characteristic density, can be accomplished by centrifugation on a sucrose density gradient. Cytoplasmic membranes from Gram-negative bacteria are recovered by disrupting the cells and collecting membrane vesicles by centrifugation.

The electron microscope is a powerful tool for studying cells and macromolecules. One technique that has been used specifically for membranes is **freeze-fracture electron microscopy.** In this technique, a droplet of membrane sample is rapidly frozen to the temperature of liquid nitrogen and then fractured with a knife. The membrane splits along the interface between the leaflets of the lipid bilayer, where the intermolecular interactions are weakest (Figure 12·4). Ice is evaporated in a vacuum, and the exposed internal surface of the membrane is then coated with a thin film of platinum to make a metal replica of the surface for examination in the electron microscope. The leaflets of membranes rich in integral membrane proteins, such as the plasma membrane of a red blood cell (shown in Figure 12·5, next page), resemble a pitted lunar landscape. In contrast, the leaflets of liposomes—which contain no proteins—are smooth.

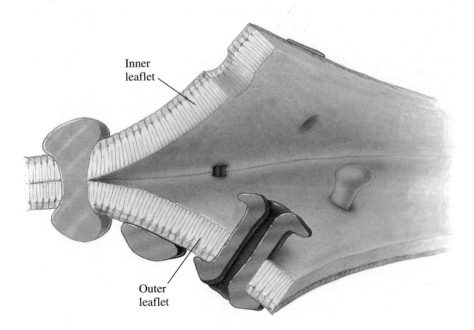

Inner
leaflet

Outer
leaflet

Figure 12·4
Freeze fracturing of a biological membrane. The lipid bilayer splits along the interface of the two leaflets, where intermolecular attractions are weakest. The exposed internal surface of the membrane is then coated with a thin film of platinum to make a metal replica that is examined in the electron microscope. Integral membrane proteins are detected as protrusions or cavities in the replica.

(a) Lateral diffusion

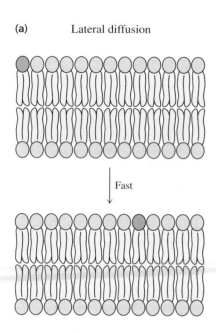

Fast

(b) Transverse diffusion

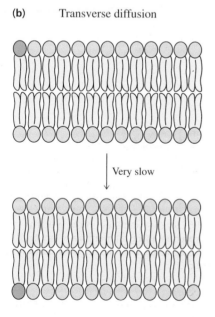

Very slow

Figure 12·6
Diffusion of lipids within a bilayer. **(a)** Lateral diffusion of lipids (diffusion in the plane of one leaflet of the bilayer) is relatively rapid. **(b)** Transverse diffusion, or flip-flop, of lipids (diffusion from one leaflet of the membrane to the other) is relatively slow, occurring at about one-billionth the rate of lateral diffusion.

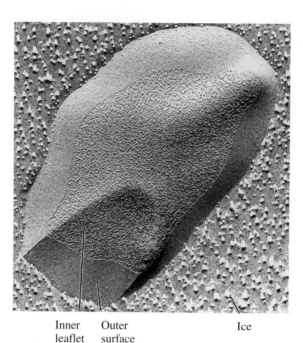

Figure 12·5
Electron micrograph of a freeze-fractured erythrocyte membrane. The bumps on the surface show where integral membrane proteins are. (Courtesy of Vincent T. Marchesi.)

Inner leaflet   Outer surface   Ice

12·2 Lipid Bilayers Are Dynamic Structures

The lipids in a bilayer are in constant motion, thereby conferring on lipid bilayers many of the properties of fluids. Lipids in a bilayer membrane undergo several different types of molecular motion. **Lateral diffusion,** the motion of lipids within the plane of one leaflet of the bilayer, is very rapid. In a bacterial cell about 2 μm long, a phospholipid molecule can diffuse from one end to the other in about one second at 37°C. A lipid bilayer can therefore be regarded as a two-dimensional solution.

In contrast to lateral diffusion, **transverse diffusion** (or flip-flop), the passage of lipids from one leaflet of the bilayer to the other, occurs very slowly because of the large activation energy for this process. The polar head of a phospholipid molecule is highly solvated and must shed its solvation sphere and penetrate the hydrocarbon interior of the bilayer in order to move from one leaflet to the other. The energy barrier associated with such movement is so high that transverse diffusion of phospholipids in a bilayer occurs 10^9 times more slowly than exchange of any two lipids within the same leaflet (Figure 12·6).

The very slow rate of transverse diffusion of membrane lipids allows the inner and outer leaflets of biological membranes to have a different lipid composition. Lipid asymmetry has been documented for plasma membranes as well as internal membranes. As shown in Figure 12·7, sphingomyelin and phosphatidylcholine each account for almost half of the phospholipid molecules in the outer leaflet of the human erythrocyte plasma membrane, but together they represent only a very small fraction of the phospholipid in the inner leaflet, where phosphatidylethanolamine and phosphatidylserine predominate. Lipid asymmetry is also found in the cytoplasmic membranes of bacteria. In mammals, asymmetry is generated and maintained by the activity of a membrane-bound protein that moves phosphatidylserine and phosphatidylethanolamine, but not phosphatidylcholine or sphingomyelin, from the outer to the inner leaflet, using the energy of ATP. This protein, called flippase or translocase, is present in many cell types. Serine and ethanolamine head groups of erythrocyte membrane lipids appear to interact with cytoskeletal proteins present on the inner face of the membrane, augmenting the effect of the flippase.

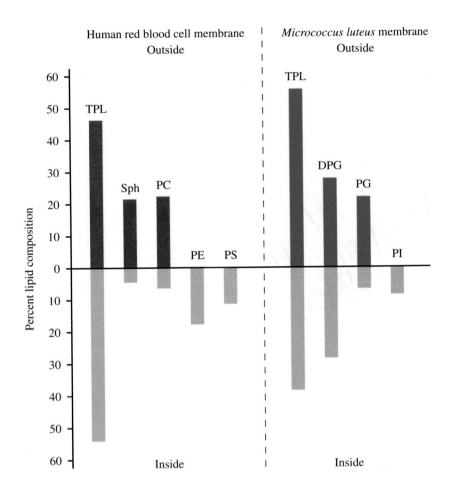

Human red blood cell membrane
Outside

Micrococcus luteus membrane
Outside

Figure 12·7
Asymmetrical distribution of phospholipids in the inner and outer leaflets of the human erythrocyte membrane and the membrane of the bacterium *Micrococcus luteus.* Abbreviations: TPL, total phospholipid; Sph, sphingomyelin; PC, phosphatidylcholine; PE, phosphatidylethanolamine; PS, phosphatidylserine; DPG, diphosphatidylglycerol; PG, phosphatidylglycerol; PI, phosphatidylinositol. [Adapted from Bergelson, L. D., and Barsukov, L. I. (1977). Topological asymmetry of phospholipids in membranes. *Science* 197:224–230.]

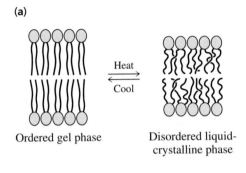

(a)

Ordered gel phase

Disordered liquid-crystalline phase

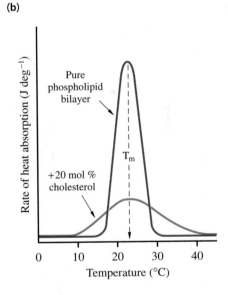

(b)

The fluid properties of bilayers also depend on intrachain rotations of lipid fatty acyl chains and their ability to bend and flex. Saturated acyl chains can exist in two types of conformation. At lower temperatures, there is little rotational motion of the carbon-carbon bonds in the chains, which are stiffly extended. At higher temperatures, constant molecular motion produces short-lived kinks in the chains. A lipid bilayer made from a phospholipid species with a single type of acyl chain exists in a well-ordered gel phase at lower temperatures. In this state, the acyl chains adopt the extended conformation and pack together like stiff rods, with maximal van der Waals contact, to form a crystalline array. When the lipid bilayer is heated, a phase transition (analogous to the melting of a crystalline solid) occurs. The lipids in the resulting liquid-crystalline phase have highly disordered and loosely packed acyl chains. Because the hydrocarbon tails of fatty acyl groups become less extended, the thickness of the bilayer decreases by about 15% during the phase transition (Figure 12·8a). Synthetic bilayer membranes composed of a single type of lipid undergo a phase transition at a distinct temperature called the **phase-transition temperature,** or melting point (T_m), of that particular lipid. Naturally occurring biological membranes, which contain a heterogeneous population of lipids, exhibit a gradual change from the gel to the liquid-crystalline phase, often at temperatures between 30°C and 40°C.

Phase transitions in lipid bilayers can be monitored using a technique called **differential-scanning calorimetry.** To make measurements of this type, a small sample of aqueous lipid dispersion and a sample of water are warmed simultaneously at the same rate, and the calorimeter measures the difference in energy absorption between the two samples. While the temperature is below the phase transition, the rate of heat absorption by the two samples is similar. However, when the

Figure 12·8
Phase transition of a lipid bilayer. **(a)** In the ordered gel state, the hydrocarbon chains adopt an extended conformation. Above the phase-transition temperature (T_m), the fatty acyl chains undergo rapid rotational motion about C—C bonds, which results in kink formation and disorders the liquid-crystalline phase. **(b)** Differential-scanning calorimetry trace observed during the melting of dimyristoylphosphatidylcholine, which has a phase-transition temperature of 23°C. In the presence of cholesterol, the phospholipid melts over a broad temperature range.

phase-transition temperature of the bilayer is reached, heat energy is absorbed at a much higher rate by the bilayer as it melts to the liquid-crystalline phase. This change is detected by the calorimeter, and a "spike" is produced in the trace of differential heat absorption with temperature (Figure 12·8b).

The molecular structure of a phospholipid has dramatic effects on its fluidity and phase-transition temperature. As we saw in Section 11·2, the presence of a *cis* double bond in an unsaturated fatty acid residue introduces a permanent kink into the acyl chain. This kink disrupts packing, increases bilayer fluidity, and greatly lowers the phase-transition temperature. For example, the phase-transition temperature of an 18:1 dioleoylphosphatidylcholine bilayer is –20°C, compared to 54°C for the 18:0 distearoyl species. Increasing the chain length of a saturated fatty acid residue, in contrast, stabilizes the gel phase by increasing the van der Waals interactions between chains in the extended conformation. This results in a higher phase-transition temperature and a reduction in the fluidity of the bilayer. For example, 16:0 dipalmitoylphosphatidylcholine has a phase-transition temperature of 41°C, whereas that of the 14:0 dimyristoyl species is 23°C.

In many organisms, membrane fluidity is relatively constant under different conditions. The maintenance of constant membrane fluidity is important, since changes in fluidity affect the catalytic functions of membrane proteins. The fluidity of membranes can be regulated through changes in the ratio of unsaturated to saturated fatty acyl groups in membrane lipids. When bacteria are grown at low temperatures, the proportion of unsaturated fatty acyl groups increases, thereby maintaining membrane fluidity.

In most warm-blooded organisms, the ratio of unsaturated to saturated fatty acyl groups within membranes varies less because a constant body temperature is usually maintained. An interesting exception is the reindeer leg (Figure 12·9), where the proportion of unsaturated fatty acyl chains in membrane lipids increases closer to the hoof. The lower melting point and greater fluidity of unsaturated fatty acyl groups in these membrane lipids permit the membranes to remain fluid and functional even at the low temperatures to which the lower leg is exposed.

Cholesterol makes up a large proportion (20–25% by mass) of the lipids of a typical mammalian plasma membrane and plays a major role in regulating membrane fluidity. The intercalation of cholesterol molecules between the hydrocarbon tails of the lipids in a bilayer greatly broadens the phase-transition temperature range (Figure 12·8b). The rigid ring structure of cholesterol restricts the mobility of fatty acyl chains of lipids in the liquid-crystalline state and thus decreases fluidity. When added to gel-phase lipid, cholesterol disrupts the ordered packing of the extended fatty acyl chains and thereby increases fluidity. The presence of cholesterol in animal cell membranes thus helps to maintain fairly constant fluidity despite fluctuations in temperature or degree of fatty acid saturation.

12·3 Membrane Proteins Can Be Integral, Peripheral, or Lipid Anchored

There are three broad classes of membrane proteins: **integral membrane proteins, peripheral membrane proteins,** and **lipid-anchored proteins** (Figure 12·3). Specialized membrane proteins are characteristic of particular cellular and intracellular membranes. We will consider the functions of some of these membrane proteins and protein assemblies later in this chapter. We will also see membrane proteins in other chapters, including those covering oxidative phosphorylation, photosynthesis, and protein synthesis.

Integral membrane proteins, also referred to as intrinsic proteins, contain hydrophobic regions that are embedded in the hydrophobic core of the lipid bilayer. Integral membrane proteins usually span the bilayer completely, although at least

Figure 12·9
Reindeer. The proportion of lipids with low melting points is higher near the hooves in these animals, allowing the cell membranes to remain fluid at low temperatures.

one protein, cytochrome b_5, appears to be anchored in the membrane by a hydrophobic protein tail that may not traverse the entire bilayer. Because integral membrane proteins are embedded in the membrane bilayer, their isolation requires disruption of the hydrophobic interactions between the protein and the membrane lipids. This disruption is most often accomplished by using a detergent, which replaces most of the membrane lipids during the solubilization process, forming soluble mixed micelles (micelles containing protein, lipid, and detergent). The commonly used detergents shown in Figure 12·10 are relatively mild and are less likely than strong detergents to denature the protein. Many integral membrane proteins bind one or more layers of lipid around them very tightly. This **annular** (or boundary) **lipid** appears to be critical for maintaining membrane protein integrity and often cannot be removed without denaturing the protein. Individual proteins from a detergent-solubilized membrane extract are purified using the same techniques used for soluble proteins, including gel-filtration, ion-exchange, and affinity chromatography. The integral membrane protein is usually kept in detergent at all times,

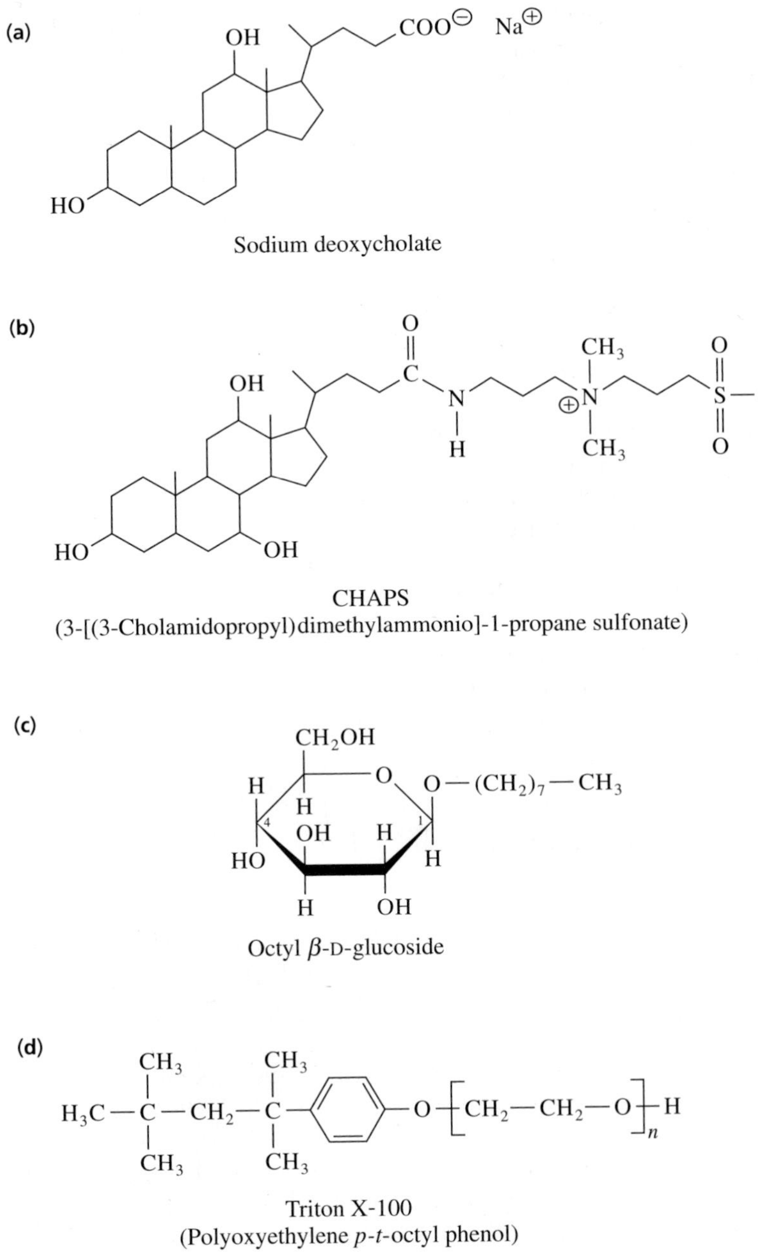

(a)

Sodium deoxycholate

Figure 12·10
Detergents commonly used to solubilize membrane proteins. (**a**) Sodium deoxycholate, (**b**) CHAPS (like deoxycholate, a bile salt derivative), (**c**) Octyl β-D-glucoside, and (**d**) Triton X-100 ($n = 9$–10). Although not all of these detergents are ionic, they all contain polar as well as nonpolar regions.

(b)

CHAPS
(3-[(3-Cholamidopropyl)dimethylammonio]-1-propane sulfonate)

(c)

Octyl β-D-glucoside

(d)

Triton X-100
(Polyoxyethylene *p-t*-octyl phenol)

since removal of the detergent often leads to irreversible protein aggregation and precipitation.

Unlike integral membrane proteins, peripheral membrane proteins are more weakly associated with either face of the membrane through ionic and hydrogen-bonding interactions with the polar head groups of membrane lipids or with integral membrane proteins. Because they are neither covalently attached to the lipid bilayer nor embedded within the lipid matrix, peripheral membrane proteins are more readily dissociated from membranes by procedures that do not require cleavage of covalent bonds or disruption of the membrane itself. For example, these proteins can be removed by changing the ionic strength or pH.

Lipid-anchored membrane proteins are tethered to membranes through covalent linkage to a lipid molecule (Figure 12·11). Members of the simplest class of lipid-anchored membrane proteins are modified by covalent attachment of a fatty acid, often myristate or palmitate, to an amino acid residue via an amide or ester linkage. The fatty acid is inserted into the cytoplasmic leaflet of the bilayer, linking the protein to the membrane. Proteins of this type are found in viruses and eukaryotic cells.

Many eukaryotic lipid-anchored proteins are linked to a molecule of glycosyl-phosphatidylinositol. In this case, the anchor is embedded in the outer leaflet of the plasma membrane by the 1,2-diacylglycerol portion of the glycosylphosphatidyl-inositol. A glycan of varied composition (discussed in more detail in Chapter 9) is attached to the inositol via a glucosamine residue, a mannose residue links the glycan to a phosphoethanolamine residue, and the C-terminal α-carboxyl group of the

Figure 12·11
Structure of fatty acyl-, glycosylphosphatidyl-inositol-, and prenyl-anchored membrane proteins. (a) A myristoylated protein anchored in the inner leaflet. (b) The anchor of the variant surface glycoprotein of the parasitic protozoan *Trypanosoma brucei*. The protein is covalently bound to a phospho-ethanolamine residue that is in turn bound to a glycan. The glycan can be of varied composition (blue hexoses) but usually features a mannose residue to which the phospho-ethanolamine residue is attached and a glucosamine residue, which is attached to the inositol group (Ins, in red) of phosphatidyl-inositol. The diacylglycerol portion of the phosphatidylinositol anchors the protein in the membrane. (c) The farnesyl anchor of a prenylated membrane protein. The iso-prenoid chain is covalently linked to the protein via the —SH group of a cysteine residue close to the C-terminus of the protein. The three types of anchors can be found in the same membrane, but they do not form a complex as shown here. Abbreviations: glucosamine, GlcN; mannose, Man. [Adapted from Ferguson, M. A. J., and Williams, A. F. (1988). Cell-surface anchoring of proteins via glycosyl-phosphatidylinositol structures. *Annu. Rev. Biochem.* 57:285–320.]

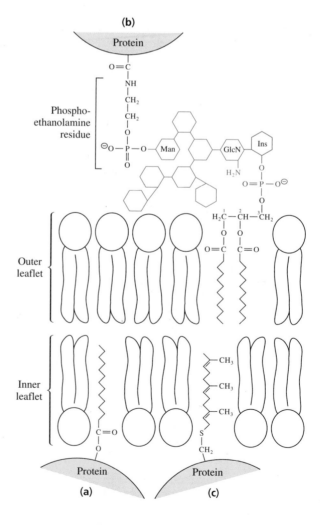

protein is linked to the ethanolamine via an amide bond. Glycosylphosphatidyl-inositol-anchored proteins can be released from the membrane surface by the action of phospholipases C and D that specifically recognize the anchor structure. Over 100 different proteins are now known to be associated with membranes via a glycosylphosphatidylinositol anchor. These proteins, which have a variety of different functions, are present only in the outer leaflet of the membrane. The surface coat proteins of many parasitic protozoa, several adhesion proteins (which are involved in cell-cell contacts), and some hydrolytic enzymes are anchored to the membrane in this way.

Members of a third class of lipid-anchored membrane proteins are covalently linked to an isoprenoid chain (either farnesyl or geranylgeranyl) via the —SH group of a cysteine residue at the C-terminus of the protein. These **prenylated proteins** are found anchored to the cytoplasmic face of both the plasma membrane and intracellular membranes. All three types of lipid anchors are covalently linked to amino acid residues posttranslationally, that is, after the protein is synthesized. Like integral membrane proteins, lipid-anchored proteins are stably associated with membranes, although the proteins themselves do not interact with the membrane, and once released, they behave like soluble proteins.

12·4 Transmembrane Proteins Contain Hydrophobic Regions That Span the Bilayer

Integral membrane proteins usually possess regions of polypeptide that span the membrane; that is, they are transmembrane proteins. Indeed, this is an essential feature for proteins involved in transport or the transmission of signals across the membrane. **Monotopic proteins** are anchored in the bilayer by only a single membrane-spanning segment, which may make up only about 5–10% of the total protein mass. Receptor proteins are often of this type. **Polytopic proteins** possess several membrane-spanning segments, which are separated by loops at the membrane surface. In this case, a larger fraction of the protein mass, typically 50–70%, is embedded in the bilayer. Transport proteins are always polytopic. All transmembrane proteins have a three-part structure: two hydrophilic regions (including the N- and C-termini, which may lie on the same or opposite sides of the membrane) and a hydrophobic region that consists of the membrane-spanning segment(s). The polypeptide chain of monotopic proteins appears to fold independently in these three parts of the protein, a feature that has important implications for the way proteins function in the membrane.

A. Nearly All Integral Membrane Proteins Contain α Helices

The secondary structure of the membrane-spanning segments of integral membrane proteins, with only one known exception, appears to be an α helix. A stretch of approximately 20 amino acid residues in an α-helical conformation is sufficient to completely span the bilayer. Monotopic proteins have a single putative α-helical segment, which is often composed entirely of hydrophobic or uncharged residues. In polytopic proteins, several putative α-helical segments are connected by turns, loops, or larger hydrophilic structures to form a bundle. Although the helix amino acids are usually predominantly hydrophobic, they may include some charged residues, which are often important in maintaining protein structure and function. The arrangement of the polypeptide chain within the membrane, that is, the location of the spanning segments and connecting loops, is referred to as the **topology** of a membrane protein.

Figure 12·12
Membrane topology of human erythrocyte glycophorin A. The N-terminal extracellular domain is highly hydrophilic and carries 1 N-linked and 15 O-linked oligosaccharide chains. A sequence of 23 hydrophobic amino acids makes up the single membrane-spanning segment. The C-terminal intra-cellular domain is rich in charged residues and is the point of attachment of the glyco-protein to the cytoskeleton.

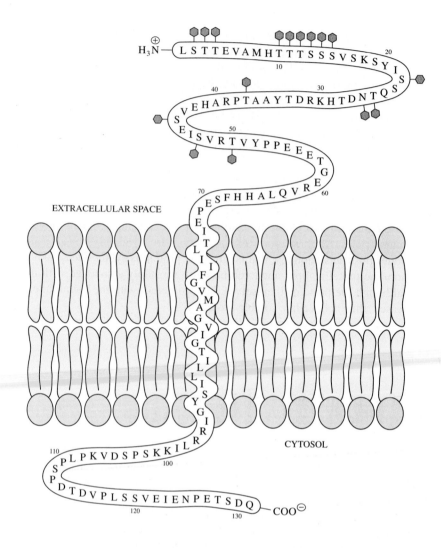

One of the best-studied monotopic integral membrane proteins is glycophorin A, the major acidic glycoprotein of the human erythrocyte membrane (Figure 12·12). This 131-residue glycoprotein (M_r 31 000), which contains 60% carbohydrate by mass, was one of the first integral membrane proteins to be examined biochemically. Glycophorin A has no known enzymatic or transport function but is believed to protect the surface of the erythrocyte and prevent adhesion to other cells and tissues. One N-linked and 15 O-linked oligosaccharides, which carry many negatively charged N-acetylneuraminic acid residues, are bound to the large hydrophilic extracellular domain of glycophorin A. The transmembrane domain of glycophorin A is very likely an α helix containing 23 hydrophobic amino acid residues. The C-terminal intracellular domain is rich in positively charged lysine and arginine and negatively charged aspartate and glutamate residues. This region interacts with the red blood cell cytoskeleton via a linker protein.

Bacteriorhodopsin is one of the best-characterized polytopic membrane proteins. This protein is a light-driven proton pump found in the cytoplasmic membrane of the halophilic (salt-loving) bacterium *Halobacterium halobium*. Bacteriorhodopsin uses the energy of light to generate a transmembrane proton concentration gradient that drives the synthesis of ATP. The three-dimensional structure of bacteriorhodopsin has been determined at high resolution by electron microscopy. Seven α-helical segments, each about 25 residues long, span the membrane (Figure 12·13). These helices are arranged as a bundle that is tilted slightly relative to the bilayer. However, not all of the amino acids in the transmembrane segments of bacteriorhodopsin are hydrophobic. For example, the side chain of

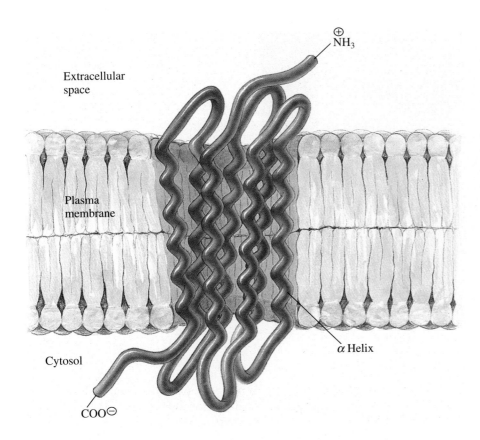

$\overset{\oplus}{N}H_3$

Extracellular
space

Plasma
membrane

Cytosol

α Helix

$\overset{\ominus}{C}OO$

Figure 12·13
Membrane topology of bacteriorhodopsin from *Halobacterium halobium.* The polypeptide chain folds to form seven membrane-spanning α helices, separated by loops at the membrane surface. The transmembrane segments are grouped together to form a helix bundle that is tilted slightly relative to the bilayer. The few charged amino acid side chains face the interior of the bundle, and hydrophobic residues contact the membrane lipids.

Lys-216 carries a covalently bound *cis*-retinal moiety (Section 8·15A), which is the light-absorbing prosthetic group of the protein. Several other charged residues are thought to be oriented toward the interior of the helix bundle. These residues may be involved in specific electrostatic interactions between different helices, or they may provide a hydrophilic passage through which protons can move across the membrane. The exterior faces of the α helices consist of hydrophobic amino acids that contact the acyl chains of membrane lipids. This arrangement accommodates charged residues within the bilayer in a thermodynamically favorable manner.

The sole known exceptions to the α-helical motif are the porin proteins, which form pores in the outer membranes of mitochondria and Gram-negative bacteria. The membrane-spanning region of porins consists of a 16-stranded β barrel (Section 5·13).

B. The Topology of Membrane Proteins Can Be Determined

The detailed three-dimensional atomic structures of many soluble proteins have been investigated by X-ray crystallography (Section 5·2). However, because water-insoluble integral membrane proteins do not readily form crystals, only a few membrane proteins have been successfully crystallized, including bacterial porins and the photosynthetic reaction centers of some bacteria. The *Rhodopseudomonas viridis* reaction center consists of three integral membrane proteins (one monotopic and two polytopic) and one peripheral membrane protein (Figure 12·14, next page). The architecture of the complex is consistent with the principles described for glycophorin A and bacteriorhodopsin: the integral membrane protein components are anchored in the membrane via hydrophobic α-helical segments, and the regions of the proteins outside the membrane are relatively hydrophilic.

Electron microscopy is a useful tool for examining the structures of some membrane proteins. For example, bacteriorhodopsin aggregates in the cytoplasmic membrane of the bacterium, forming large patches termed purple membrane, which

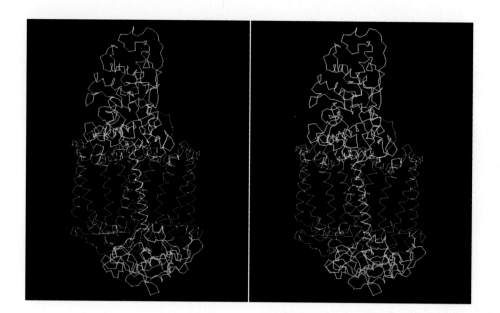

Figure 12·14
Stereo view of the photosynthetic reaction center from *Rhodopseudomonas viridis*. (Based on coordinates provided by J. Deisenhofer.)

are essentially two-dimensional crystals of tightly packed protein molecules in a regular array. When purple membrane patches are examined in the electron microscope at several different angles, computerized image enhancement and reconstruction can generate a high-resolution image of the protein in which the positions of the seven α helices can be established. This method took 25 years to perfect and is now being applied to other proteins that form two-dimensional arrays in the membrane.

The structure and topology of membrane proteins can also be determined by biochemical methods termed vectorial labelling. This approach relies on the fact that biological membranes are impermeable to many hydrophilic or charged reagents that can be used to chemically modify proteins. For example, tyrosine residues in the extracellular domains of integral membrane proteins can be labelled with radioactive iodine by the action of lactoperoxidase, an enzyme that is membrane impermeable. Tyrosine residues in the intracellular protein domains are not labelled unless the cell is first treated with detergent. The locations of the modified tyrosine residues can be found by peptide mapping techniques. The extracellular domains of integral membrane proteins can also be identified by their susceptibility to proteolytic cleavage or by binding to antibodies.

As we have seen, the most common structural motif in integral membrane proteins is predicted to be a hydrophobic α helix, which is favored for thermodynamic reasons. It is possible in principle to identify potential membrane-spanning regions of proteins from their amino acid sequences. The amino acid sequences of hydrophobic proteins are difficult to determine directly, but because DNA sequences of the genes for a large number of membrane proteins are known, the amino acid sequences can be deduced.

Insertion of an unfolded polypeptide chain into a hydrophobic membrane bilayer is energetically unfavorable because the polar carbonyl and amide moieties of the peptide bonds can no longer form hydrogen bonds with water molecules. Formation of an α helix fully satisfies the hydrogen-bonding requirements of the peptide groups. Furthermore, an α helix whose side chains are hydrophobic is likely to be more stable in the membrane interior than in water. In other words, the free-energy change (ΔG) for transfer of the helix from the membrane to water is positive (unfavorable). This thermodynamic principle is the basis for several related methods for predicting the existence of membrane-spanning α helices from an amino acid sequence. As we saw in Section 4·3G, the free-energy change for transfer of an amino acid residue from a hydrocarbon-like solvent to water has been estimated for each of the 20 amino acids commonly found in proteins. This free-energy change has a large positive value for hydrophobic residues such as isoleucine and phenylalanine and a large negative value for charged and polar residues such as lysine and glutamine (Table 12·2). For a stretch of amino acids, the sum of the ΔG values for the individual amino acids is an estimate of the overall polarity, or **hydropathy index,** of that sequence. A large positive hydropathy index indicates that a putative helical segment is more stable in the membrane, whereas a large negative hydropathy index means that the helix is more stable in water. To search for potential membrane-spanning segments, a protein sequence is examined by moving a "window" 7 to 20 amino acids long, one residue at a time. The hydropathy index for the segment is calculated at each position. The result is a **hydropathy plot,** which maps the summed hydropathy index at each window position along the sequence. Segments that are hydrophobic appear as positive peaks on the plot, and if these segments are over 20 residues long, they are considered to be potentially membrane spanning. This method has been validated by the few membrane proteins whose complete topology is known. For example, the presence of transmembrane helices in glycophorin A and bacteriorhodopsin is predicted reasonably well by hydropathy plots (Figure 12·15, next page).

Table 12·2 Polarity scale for amino acid residues

Amino acid	Free-energy change for transfer* (kJ mol^{-1})
Isoleucine	3.1
Phenylalanine	2.5
Valine	2.3
Leucine	2.2
Tryptophan	1.5
Methionine	1.1
Alanine	1.0
Glycine	0.67
Cysteine	0.17
Tyrosine	0.08
Proline	−0.29
Threonine	−0.75
Serine	−1.1
Histidine	−1.7
Glutamate	−2.6
Asparagine	−2.7
Glutamine	−2.9
Aspartate	−3.0
Lysine	−4.6
Arginine	−7.5

*The free-energy change is for transfer of an amino acid residue in a stretch of α helix from the interior of a lipid bilayer to water.
[Adapted from Eisenberg, D., Weiss, R. M., Terwilliger, T. C., Wilcox, W. (1982). Hydrophobic moments in protein structure. *Faraday Symp. Chem. Soc.* 17:109–120.]

Figure 12·15
Hydropathy plots for glycophorin A and bacteriorhodopsin. **(a)** Glycophorin A. One potential membrane-spanning helix is indicated. **(b)** Bacteriorhodopsin. The seven positive peaks correspond to the seven membrane-spanning helices.

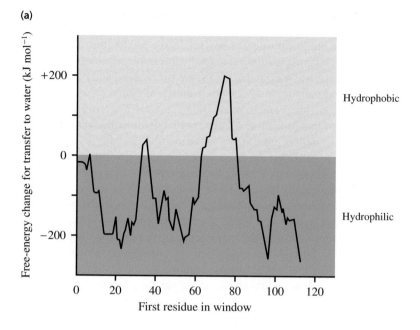

(a)

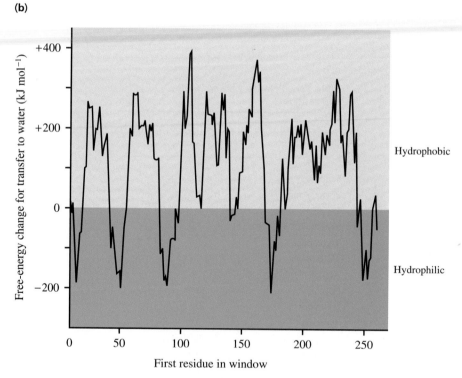

(b)

C. Some Membrane Proteins Can Diffuse Rapidly in the Bilayer

L. D. Frye and Michael A. Edidin devised an elegant experiment to test the hypothesis that integral membrane proteins diffuse laterally within the lipid bilayer. These scientists fused mouse cells with human cells to form heterokaryons (hybrid cells). By using red fluorescent–labelled antibodies that specifically bind to certain proteins in human plasma membranes and green fluorescent–labelled antibodies that specifically bind to certain proteins in mouse plasma membranes, they were able to observe by immunofluorescence microscopy the changes in the distribution of integral membrane proteins over time. The surface antigens were intermixed within 40 minutes after cell fusion (Figure 12·16). This experiment demonstrated that at least some integral membrane proteins diffuse freely within biological membranes.

The lateral-diffusion rate of proteins in the plasma membrane of an intact cell can be measured experimentally by a technique called fluorescence recovery after photobleaching. The membrane protein of interest is first made fluorescent, usually by binding of a specific fluorescent-labelled antibody. The cell is then viewed in a fluorescence microscope, and a small circular patch of the membrane is illuminated by a short, intense pulse of laser light. This exposure destroys the fluorescent groups within the illuminated patch, a process known as photobleaching. The fluorescence within the illuminated region increases with time as unbleached protein molecules diffuse into the bleached patch, a process referred to as recovery. The rate of lateral diffusion of the labelled protein can be calculated from the rate of recovery.

A few integral membrane proteins move laterally very rapidly. For example, the pigment protein rhodopsin in the rod cells of the eye diffuses almost as rapidly as a lipid. However, the majority of membrane proteins diffuse about 100–500 times more slowly than a membrane lipid. The diffusion of other proteins appears to be severely restricted. Many cell membranes are stabilized by the cytoskeleton, a scaffolding of intracellular proteins. Some membrane proteins are anchored to this cytoskeleton and are therefore relatively immobile. Contractile filaments of the cytoskeleton may also actively pull membrane proteins that are attached to them. Other membrane proteins, such as bacteriorhodopsin, cluster together to form large protein aggregates that are not able to move freely in the bilayer. Integral membrane proteins do not readily undergo transverse diffusion. The existence of hydrophilic protein domains on one or both sides of the membrane as well as the presence of carbohydrate chains on many membrane proteins create very large activation-energy barriers to transverse movement.

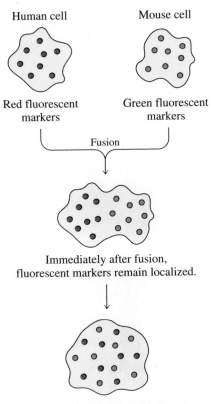

Figure 12·16
Diffusion of integral membrane proteins. Human cells whose integral membrane proteins have been labelled with a red fluorescent marker are fused with mouse cells whose integral membrane proteins have been labelled with a green fluorescent marker. The initially localized markers become dispersed over the entire surface of the fused cell within 40 minutes.

12·5 The Cytoskeleton and Glycocalyx Are Additional Features of Biological Membranes

In keeping with its multiple roles, the biological membrane is not merely an inert barrier between the contents of the cell or organelle and the external environment. The ability of the membrane to define the boundaries of the cell may require additional features in eukaryotes: an internal cytoskeleton and an external glycocalyx.

A. The Cytoskeleton Provides Mechanical Strength

Because lipid bilayers alone are thin structures, some reinforcement of the plasma membrane is needed to increase its mechanical strength. Reinforcement is provided by the **cytoskeleton,** a scaffolding of proteins that extends throughout the cell and is attached to the inner face of the membrane. The cytoskeleton of the human erythrocyte resembles a hairnet spread out underneath the plasma membrane (Figure 12·17). The erythrocyte cytoskeleton provides the cell with sufficient mechanical strength to withstand the powerful shearing forces in the circulation and maintains the characteristic biconcave shape of the cell. The erythrocyte cytoskeletal network contains several peripheral membrane proteins: spectrin, actin, ankyrin, and band 4.1 protein (named for its position in gel electrophoresis). The spectrin α and β chains are large extended proteins (M_r 240 000 and 220 000, respectively), which are twisted around each other to form a long, flexible rod about 100 nm long and 5 nm wide. Two of these $\alpha\beta$ dimers are joined in head-to-head fashion to form spectrin $\alpha_2\beta_2$ tetramers, which are connected in a two-dimensional network by additional cytoskeletal proteins. Short actin filaments are double stranded and contain 15–20 actin monomers. Actin filaments and band 4.1 protein act together as the glue that links the spectrin tetramers. The entire cytoskeleton is connected to the plasma membrane via linker proteins, which mediate interactions between the spectrin–actin–band 4.1 network and integral membrane proteins. Ankyrin is a large peripheral membrane protein with two domains: one binds to the spectrin network and the other to the cytoplasmic domain of the anion exchange protein (also

Figure 12·17
Erythrocyte cytoskeleton. The protein cytoskeletal network lining the inner face of the erythrocyte membrane is formed by spectrin, actin, and band 4.1 protein. The meshwork is attached to the plasma membrane by the linker proteins ankyrin, which binds to the cytoplasmic domain of the anion exchange protein (band 3), and band 4.1, which binds to the cytoplasmic domain of glycophorin A.

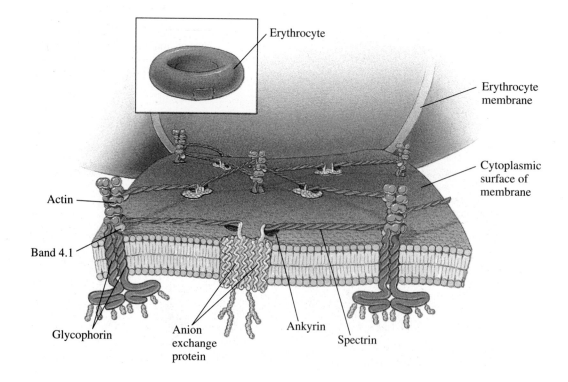

called band 3) of the red blood cell. Band 4.1 also acts as a peripheral linker protein; it connects the spectrin-actin junction to the cytoplasmic domain of glycophorin A.

Hereditary spherocytosis and hereditary elliptocytosis are diseases that result from mutations in the genes encoding erythrocyte cytoskeletal proteins. Depending on the exact molecular defect, the amounts of cytoskeletal proteins may be reduced, or the proteins themselves may be defective. The red blood cells have an abnormal shape (spherical or ellipsoid), are very fragile, and have a shorter life span than normal erythrocytes. Individuals suffering from these diseases have too few circulating red blood cells and are therefore anemic.

Proteins structurally and functionally similar to spectrin, band 4.1, and ankyrin have been identified in many eukaryotic cells. However, the cytoskeleton in nonerythroid cells is substantially more complex and usually contains three different protein networks based on actin filaments (also called microfilaments), intermediate filaments, and microtubules (Section 2·4E). Actin filaments (about 7 nm in diameter) consist of two chains of polymerized actin subunits twisted around each other in a helical fashion. These actin filaments are much longer than the oligomers found in the red blood cell. Intermediate filaments are double- or triple-stranded structures about 10 nm in diameter and are composed of several different protein subunits, depending on the cell type. The intermediate filament proteins (vimentin, keratin, desmin, and others) are encoded by a large multigene family and have similar structural properties. Microtubules are hollow tubes about 25 nm in diameter, consisting of 13 parallel protofilaments, which are in turn made up of heterodimers of α and β tubulin. Each of these cytoskeletal components forms a complex network that extends throughout the interior of the cell. Specialized linker proteins connect the actin filament and microtubule networks to integral membrane proteins at the cytoplasmic face of the plasma membrane or to organelles. All three networks are highly dynamic structures that are constantly assembled and disassembled, depending on the needs of the cell.

The cytoskeletal network can be visualized in the intact cell by fluorescence microscopy. The cells are first fixed and the plasma membrane is made permeable with methanol. The cells are then incubated with fluorescent antibodies or other compounds that bind to particular cytoskeletal proteins. Fluorescence microscopy provides a dramatic (although static) image of the entire cytoskeletal network (Figure 12·18). A more challenging technique for visualizing the cytoskeleton in a living cell involves microinjection of fluorescent-labelled actin or tubulin monomers, which become incorporated into the cellular network over time. This approach allows detailed examination of the dynamic changes that occur in the cytoskeleton in response to a variety of stimuli.

Figure 12·18
Fluorescence microscopy of the actin filament and microtubule networks of mouse embryo fibroblasts. **(a)** Microtubules are visualized by immunofluorescence, using fluorescent antibodies to tubulin. **(b)** Actin filaments are visualized using a fluorescent derivative of phallicidin, which binds to actin. (Courtesy of Vitauls Kalnins.)

(a)

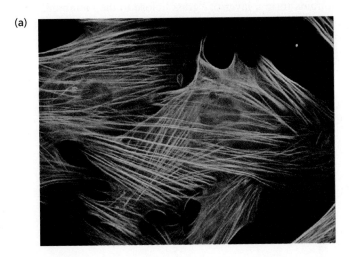

(b)

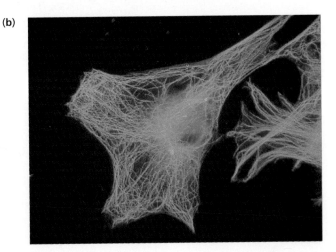

B. The Plasma Membrane Is Covered by a Thick Glycocalyx That Confers Surface Antigenicity

As indicated in Chapter 9 and illustrated in Figure 12·3, the oligosaccharide chains of eukaryotic plasma membrane glycoproteins and glycolipids appear to be located exclusively on the extracellular surface. Glycosphingolipids do not appear to undergo transverse diffusion at any significant rate, since the presence of a hydrophilic carbohydrate chain imposes a very large activation-energy barrier to passage through the membrane interior. The cell-surface carbohydrate forms a **glycocalyx,** or fuzzy coat, which extends up to 150 nm from the plasma membrane of some cells and is visible in electron micrographs. This sugar coat displays a distinctive "fingerprint" of the cell to the exterior.

In the erythrocyte membrane, a variety of oligosaccharide structures are present on glycosphingolipids and on the extracellular domains of glycophorin A and the anion exchange protein. Glycophorin A carries 15 O-linked tetrasaccharides, each of which contains two N-acetylneuraminic acid (NeuNAc) residues, as shown in Figure 12·19a. These sugar chains confer a high negative charge on the surface of the red blood cell. An additional complex N-linked oligosaccharide chain is attached to an asparagine residue. The five N-terminal amino acid residues, together with their O-linked oligosaccharide chains, make up the MN blood group antigens (Figure 12·19b). The two common alleles (alternate forms of the same gene), M and N, are found with approximately equal frequency in human populations and are of no consequence in blood transfusions.

The erythrocyte anion exchange protein carries a single large N-linked oligosaccharide, which contains a characteristic repeating unit of N-acetyllactosamine (Gal β-(1→4) GlcNAc). These polylactosaminyl chains extend a considerable distance from the cell surface.

The oligosaccharides of erythrocyte glycosphingolipids (Section 11·6) carry the ABO blood group antigens. Individuals with different blood types have different terminal oligosaccharide structures (Figure 12·20). The differences arise as a result of genetic variation in the glycosyltransferase enzymes that are responsible for biosynthesis of the sugar chains. Type A individuals possess a transferase that catalyzes the addition of an N-acetylgalactosamine residue to the core structure, whereas the enzyme of B-type individuals adds a galactose residue. In type O cells, the enzyme is inactive, and no modifications are made to the core structure. The enzymes of the A and B blood types differ in only four amino acid residues. The presence of oligosaccharide variants in the glycocalyx does not appear to affect the function of the cells.

Figure 12·21 depicts the glycocalyx of the human erythrocyte. Because it carries many short sugar chains, glycophorin A is represented as a short, prickly "bush" nestled among the taller "trees" of the anion exchange protein oligosaccharides. The glycosphingolipids form a layer of "grass" close to the membrane surface. Surface oligosaccharides are exploited by many infectious agents, parasites,

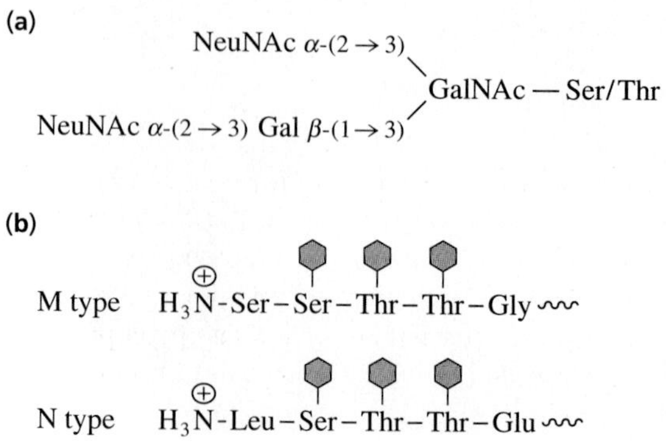

Figure 12·19
MN blood group antigens. (a) Structure of the O-linked oligosaccharides of glycophorin A. Abbreviations: Gal, galactose; GalNAc, N-acetylgalactosamine; NeuNAc, N-acetyl-neuraminic acid. (b) The five N-terminal residues, together with their oligosaccharide chains (represented by hexagons), make up the MN blood group antigens.

(a)

NeuNAc α-(2 → 3)
$\qquad$ GalNAc — Ser/Thr
NeuNAc α-(2 → 3) Gal β-(1 → 3)

(b)

M type H_3N-Ser–Ser–Thr–Thr–Gly ⌇

N type H_3N-Leu–Ser–Thr–Thr–Glu ⌇

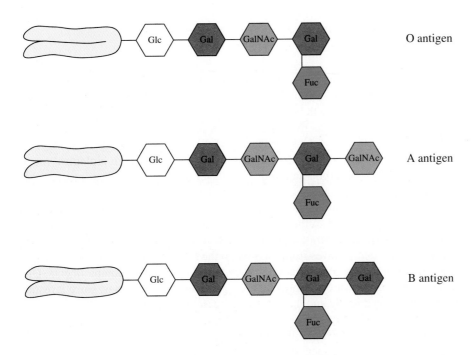

Figure 12·20
Structures of the O-, A-, and B-type glycosphingolipid oligosaccharides. Abbreviations: Gal, galactose; GalNAc, *N*-acetylgalactosamine; Glc, glucose; Fuc, fucose.

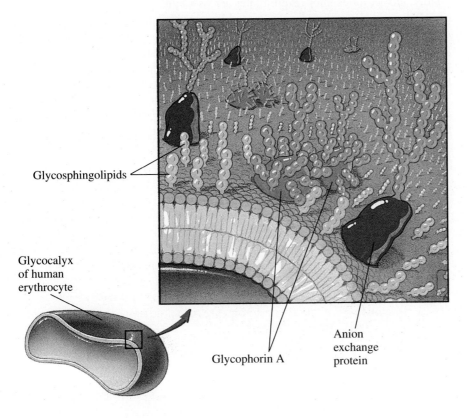

Figure 12·21
Surface view of the glycocalyx of the human erythrocyte. [Adapted from Viitala, J., and Järnefelt, J. (1985). The red cell surface revisited. *Trends Biochem. Sci.* 14:392–395.]

and protein toxins that must enter the cells of the host. For example, *O*-linked oligosaccharides of the type found in glycophorin A act as receptors for certain enveloped viruses, such as the influenza virus, allowing them to enter and infect the cell. The malarial parasite (a protozoan) invades the erythrocyte by binding to the sugar chains of both the anion exchange protein and glycophorin A. Certain pathogenic strains of bacteria attach to glycosphingolipid oligosaccharides as the first step in infection.

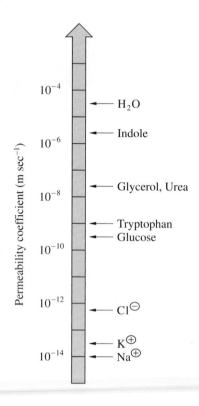

Figure 12·22
Permeability coefficients for diffusion of various species across a lipid bilayer.

12·6 Lipid Bilayers and Membranes Are Selectively Permeable Barriers

While it is essential that a biological membrane physically separate a living cell from its environment, it is equally important that water, oxygen, and all the other nutrients required for growth be able to enter the cell and that products generated by the cell for export (such as hormones, certain digestive enzymes, and toxins), as well as waste products (such as carbon dioxide and urea), be able to leave. Hydrophobic or small, uncharged molecules can freely diffuse through biological membranes, but the hydrocarbon nature of the bilayer presents an almost impenetrable barrier to most charged species.

A small nonpolar molecule can diffuse across a membrane from the side with the higher concentration to the side with the lower concentration. Equilibrium is reached when the concentrations on each side of the membrane are the same. The rate at which a solute moves from one side to the other depends on the difference in concentrations, or the concentration gradient, between the two compartments. Diffusion of a molecule down a concentration gradient is spontaneous, since it results in an increase in entropy and therefore a decrease in free energy.

The ability of many ions and small molecules to diffuse across lipid bilayers, as measured by their **permeability coefficients,** varies from 10^{-4} to 10^{-14} m sec^{-1}. The magnitude of the permeability coefficient roughly parallels the solubility of the species in nonpolar solvents (Figure 12·22). Hydrophobic compounds diffuse rapidly; indole, for example, equilibrates across a lipid bilayer in less than one second. Polar and charged molecules cross bilayers much more slowly because they must shed the water molecules in their solvation sphere while they move through gaps between lipid molecules. This process is thermodynamically unfavorable for polar molecules, especially ions. For example, glucose (a polar molecule) equilibrates across a membrane in minutes to hours, while sodium ions require days to weeks. Water itself is an exception to the general pattern; individual water molecules diffuse across bilayers and membranes extremely rapidly, reaching equilibrium in milliseconds. Many biologically important molecules enter and leave the

(a)

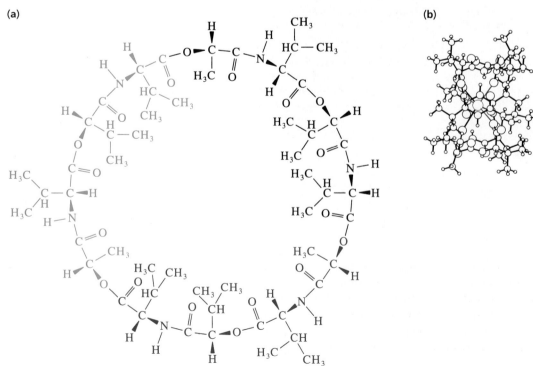

(b)

Figure 12·23
Valinomycin. **(a)** Covalent structure. **(b)** Ball and stick model with K$^{\oplus}$ bound in the central cavity.

Figure 12·24
Gramicidin D. **(a)** Covalent
structure. **(b)** A dimer of
gramicidin D forms a channel in the membrane.

(a)

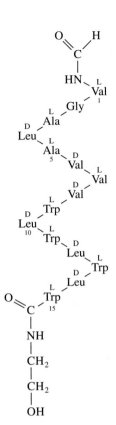

(b)

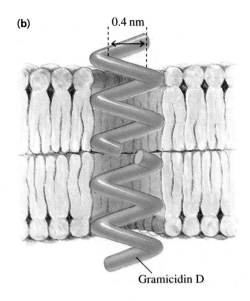

Gramicidin D

cell by simple diffusion, including nonpolar gases such as O_2 and CO_2 and hydrophobic molecules such as steroid hormones, lipid vitamins, and some drugs.

The rate of diffusion of ions across bilayers and membranes is greatly enhanced by **ionophores** (ion bearers). Many of these compounds are toxins that are
synthesized by fungi or bacteria. There are two types of ionophores, mobile carriers and channel formers; both move ions down a concentration gradient. *Streptomyces* species synthesize the mobile carrier valinomycin, which has a cyclic structure made up of L-lactate, D- and L-valine, and D-α-hydroxyisovalerate (Figure
12·23). This doughnut-shaped ionophore binds a potassium ion tightly in its central
cavity by coordination to carbonyl groups, thereby shielding the charge of the ion.
Valinomycin is remarkably ion selective; $K^\oplus$ binds 1000 times more tightly than
$Na^\oplus$. Valinomycin diffuses across the membrane, shuttling the ion to the other
side, where it is released. The nonpolar side chains of the ionophore are in contact
with the acyl chains of the membrane lipids, making the entire valinomycin-$K^\oplus$
complex lipid soluble.

Gramicidin D, a channel-forming ionophore made by the bacterium *Bacillus
brevis*, is a 15-residue hydrophobic peptide of alternating D- and L-amino acids.
Gramicidin D inserts into membranes to form a helical structure, wider than an α
helix, with a central pore (Figure 12·24). A single gramicidin D helix is able to span
only one leaflet of the bilayer; in order to cross the membrane completely, two helices form a head-to-head dimer in which the two N-termini interact. The central
pore of the membrane-spanning dimer, which is about 0.4 nm in diameter, allows
the rapid passage of monovalent cations, such as $H^\oplus$, $Na^\oplus$, and $K^\oplus$. Gramicidin
D channels constantly assemble and dissociate, remaining in the ion-conducting
state for about one second. Because channel formers such as gramicidin D do not
need to diffuse through the bilayer, as mobile carriers do, the rate of transmembrane
ion diffusion via gramicidin D is 10^4 times faster than diffusion via valinomycin.
Ionophores such as valinomycin and gramicidin D are powerful antibiotics; they
kill other cells by dissipating the cation concentration gradients essential for generating ATP and driving secondary active transport.

Table 12·3 Characteristics of different types of membrane transport

	Protein carrier	Saturable with substrate	Movement relative to concentration gradient	Energy input required
Simple diffusion	No	No	Down	No
Channels and pores	Yes	No	Down	No
Passive transport	Yes	Yes	Down	No
Active transport				
Primary	Yes	Yes	Up	Yes (direct source)
Secondary	Yes	Yes	Up	Yes (ion gradient)

(a)

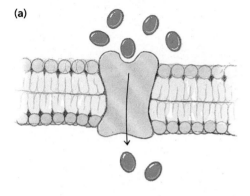

(b)

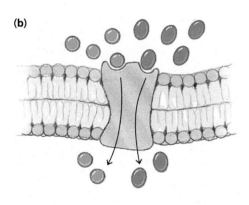

(c)

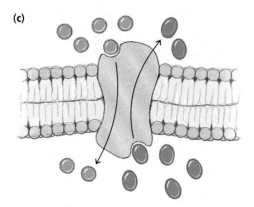

Figure 12·25
Types of membrane transport. **(a)** Uniport.
(b) Symport. **(c)** Antiport.

12·7 Membrane Transport Systems Are Essential to the Cell

To move materials across membranes, living cells use a variety of mechanisms, including transport proteins (for small molecules and ions) and endocytosis and exocytosis (for macromolecules). In this section, we will discuss the general features of membrane transport systems and the energetics of transport, as well as specific transport proteins. Some of these proteins regulate the ionic composition of both the cytosol and the internal compartments of cellular organelles. Some are responsible for the import of fuel molecules (such as sugars) and building-block materials (such as amino acids and nucleosides). Transport proteins also generate the ion concentration gradients essential for ATP production and other important cellular functions.

A. Overview of Membrane Transport

The traffic of small molecules and ions across membranes is mediated by three types of integral membrane proteins: channels and pores, passive transporters, and active transporters. **Channels** and **pores** act much like ionophores, providing an alternate route for the passage of ions and small molecules moving down a concentration gradient. No energy is required for this process, which can be thought of as diffusion through a protein rather than through the lipid bilayer, and the rate of movement of solute across the membrane is in general not saturable at high concentrations. In fact, some channels allow extremely rapid flow of solutes (for example, 10^7 sodium ions per second for the acetylcholine-receptor ion channel), and the rate often approaches the diffusion-controlled limit.

   Passive and **active transporters,** in contrast to channels and pores, specifically bind and transport solutes across the membrane. These proteins function much like enzymes, except that instead of catalyzing a chemical change in the substrate, they move it from one side of the membrane to the other. Solutes can be moved against a concentration gradient by active transport, a process that requires the input of energy. The rate of transport by an active or passive carrier protein, like the rate of an enzyme-catalyzed reaction, is saturable and approaches a maximum value at high substrate concentration. This kinetic property can be used to distinguish pores and channels from active and passive transporters. The characteristics of membrane transport are summarized in Table 12·3.

   Transport proteins are usually specific for a certain molecule or group of structurally similar molecules. The transporter is usually stereochemically specific, transporting only the biologically important stereoisomer. The simplest sort of membrane transporters carry out **uniport;** that is, they carry only a single type of solute across the membrane (Figure 12·25a). Many transporters carry out **cotransport** of two solutes, either in the same direction, **symport** (Figure 12·25b), or in opposite directions, **antiport** (Figure 12·25c). Some transporters are **electroneutral;** that is, there is no net transfer of charge across the membrane as a result of their activity. Transport by **electrogenic** transporters results in the net transfer of charge across the membrane and may thereby create changes in membrane potential (the difference in electrical potential on opposite sides of the membrane).

All transport proteins operate as gates. The transporter can adopt an outward-facing or an inward-facing conformation, depending on whether the substrate-binding site is accessible from the external or cytoplasmic side of the membrane, respectively. When the protein in its outward-facing conformation binds a specific molecule or ion, it undergoes a conformational change to the inward-facing state. The transported molecule is then released on the inner face of the membrane, and the transporter reverts to the outward-facing state (Figure 12·26). The conformational change in the transporter is often triggered by binding of the transported species, as in the induced fit of certain enzymes to their substrates. In active transport, the conformational change may be driven by ATP hydrolysis or other sources of energy. The protein conformational change involved in switching between the two states is likely to be relatively subtle and probably entails the shift of only a few atoms.

B. Pores and Channels Allow Solutes to Diffuse Across the Membrane

Pores and channels are transmembrane proteins with a central hydrophilic passage. (In general, the term *pore* is used for bacteria, and *channel* for animals.) Solutes of the appropriate size, charge, and geometry can pass rapidly through the passage in either direction by diffusion down a concentration gradient (Figure 12·27). The outer membranes of Gram-negative bacteria are rich in porins, a family of pore proteins that allow ions and many small molecules to gain access to specific transporters in the cytoplasmic membrane. Similar pores are found in the outer membranes of mitochondria. X-ray crystallography of the OmpF and PhoE porins from *Escherichia coli* has revealed that individual porin polypeptides form 16-stranded β-barrel structures, three of which assemble to form the trimeric pore (Figure 12·28, next page). The central aqueous channel within each porin subunit has a cross-sectional area of 0.7 nm × 1.1 nm.

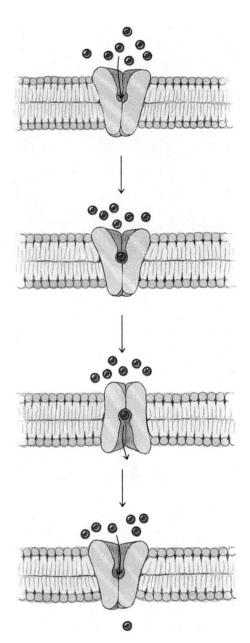

Figure 12·27
Membrane transport through a pore or channel. These proteins contain a hydrophilic passage that allows molecules and ions of the appropriate size, charge, and geometry to traverse the membrane in either direction.

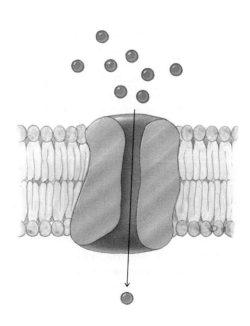

Figure 12·26
Transport protein function. Transport carrier proteins operate as gates. The protein binds its specific substrate and then undergoes a conformational change that allows the molecule or ion to be released on the other side of the membrane. Cotransporters have specific binding sites for each transported species.

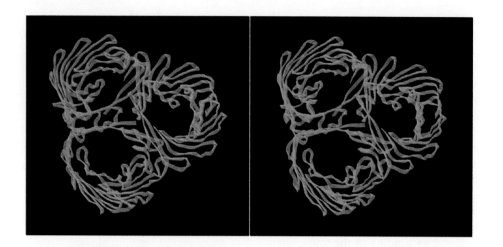

Figure 12·28
Stereo view of a porin from *Rhodobacter capsulatus.* (Based on coordinates provided by M. Weiss and G. Schulz.)

Porins are only weakly solute selective and act as a sieve, usually remaining permanently open. In contrast, animal cell membranes contain many channel proteins that are highly specific for certain ions. Some of these channels are open continuously, whereas others are "gated" and open or close in response to certain stimuli. **Ligand-gated** ion channels open in response to binding of a specific signal molecule. Other channels are **voltage-gated** and open or close only when the electrical properties of the membrane change. **Stretch-gated** channels respond to changes in tension or turgor pressure within the membrane.

One very important ligand-gated ion channel is the acetylcholine receptor, which is found in the plasma membrane of neurons. The acetylcholine receptor is involved in the transmission of nerve impulses from one nerve cell to the next at the synapse. When the electrical impulse arrives at the presynaptic neuron, the neurotransmitter acetylcholine is released into the synaptic cleft, the narrow space between the two nerve cells at their junction. Acetylcholine diffuses across the synaptic cleft and binds to high-affinity sites on the acetylcholine-receptor ion channel in the plasma membrane of the postsynaptic neuron. Ligand binding causes the channel to open, which allows sodium ions to diffuse inward down their concentration gradient (Figure 12·29). There is also a much smaller outflow of potassium ions from the cell through the channel, which is specific for only these two ions. The net inward flow of cations reduces the membrane potential from $-75\,mV$ to $0\,mV$. This depolarization of the plasma membrane of the postsynaptic neuron propagates the electrical impulse. The acetylcholine-receptor ion channel becomes desensitized (unresponsive) to its ligand within a fraction of a second and closes automatically. This response ensures that the nerve impulse leads to only transient depolarization of the neuron.

The natural product curare from the bark of *Chondodendron tomentosum* (used to make poison arrows by aboriginal people of the Amazon) and the snake-venom component bungarotoxin block the action of the acetylcholine-receptor ion channel, leading to muscle paralysis and death.

C. Passive Transport Occurs Spontaneously

No energy is required to drive passive transport, which is also called facilitated diffusion. In this case, the transport protein serves to accelerate greatly the attainment of equilibrium, which would occur only very slowly by diffusion alone. As with enzymes, examination of the kinetics of transport can provide useful information about the properties and mechanisms of the carrier protein. The use of isotopically labelled materials affords a convenient way to follow the transport of small molecules or ions in both intact cells and membrane vesicles. After addition of labelled substrate, cells or vesicles are collected at various times by centrifugation or filtration, and the radioactivity present in the cytosol of the cells or the lumen of the vesicles is measured. When making kinetic measurements, it is important that the substrate not be chemically altered after it reaches the cytosol. For this reason, nonmetabolizable substrate analogs are often used in intact cells. For a simple passive uniport system, the initial rate of entry of substrate into a cell, like the initial rate of an enzyme-catalyzed reaction, depends on the external substrate concentration. The equation describing this dependence is analogous to the Michaelis-Menten equation for enzyme catalysis.

$$v_0 = \frac{V_{max}[S]_{out}}{K_{tr} + [S]_{out}} \tag{12·1}$$

where v_0 is the initial rate of transport of the substrate at an external concentration $[S]_{out}$, V_{max} is the maximum rate of transport of the substrate, and K_{tr} is a constant analogous to the Michaelis constant, K_m. The lower the value of K_{tr}, the higher the affinity of the transporter for the substrate. Equation 12·1 can be represented graphically, as shown in Figure 12·30. As substrate accumulates in the interior of the cell, the rate of outward transport increases until it balances the rate of inward transport, and $[S]_{in}$ equals $[S]_{out}$. At this point, dynamic equilibrium is reached: there is no net change in the concentration of substrate on either side of the membrane, but substrate continues to move across the membrane in both directions.

Erythrocytes rely largely on glucose as a source of metabolic energy. D-Glucose moves from the blood (where its concentration is about 5 mM) down its concentration gradient into the erythrocyte by passive transport via the glucose transporter. This membrane glycoprotein (M_r 55 000) belongs to a large family of transport proteins, all of which possess 12 membrane-spanning segments. Glucose binds to the outward-facing conformation of the transporter and moves across the bilayer when the transporter conformation changes. Glucose dissociates from the protein on the cytoplasmic side, and the transporter reverts to its original conformation. Glucose transporters similar to the erythrocyte protein are present in all mammalian cells, but their activity may be highly regulated.

Another example of a passive transporter is the anion exchange protein (also called band 3) of the human erythrocyte membrane. The anion exchange protein (M_r 95 000) is an antiport protein that carries out the one-for-one electroneutral exchange of bicarbonate ions and chloride ions. In both cases, the ions are transported down their concentration gradients. The anion exchange protein plays an important physiological role in the transport of carbon dioxide from the tissues to the lungs. Waste CO_2, which must be converted to a more soluble form for transport in the blood, enters the red blood cell by diffusion and is transformed to the more soluble

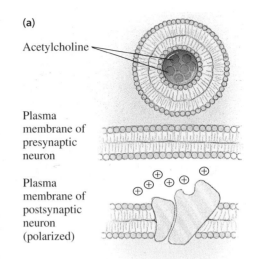

(a)

Acetylcholine

Plasma membrane of presynaptic neuron

Plasma membrane of postsynaptic neuron (polarized)

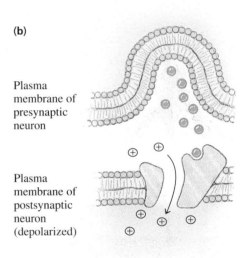

(b)

Plasma membrane of presynaptic neuron

Plasma membrane of postsynaptic neuron (depolarized)

Figure 12·29
Acetylcholine-receptor ion channel. **(a)** In the resting state, the channel is closed and the interior of the postsynaptic neuron is more negatively charged than the exterior, generating a negative membrane potential. **(b)** Acetylcholine released from the presynaptic neuron binds to its receptor, which then allows the entry of sodium ions. The membrane is depolarized (the electrical potential is reduced) as the ions enter the cell.

Figure 12·30
Kinetics of transport via a simple passive transporter. The initial rate of transport (v_0) increases with substrate concentration until a maximum rate of transport, V_{max}, is reached. The K_{tr} for transport, which is analogous to the Michaelis constant, K_m, is the concentration of transported substrate at which the initial rate of transport is half maximal.

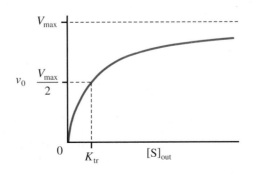

Figure 12·31
Function of the human erythrocyte anion exchange protein. The protein carries out the one-for-one exchange of $Cl^{\ominus}$ for $HCO_3^{\ominus}$, thus allowing CO_2 to be efficiently transported in the blood from the tissues to the lungs as $HCO_3^{\ominus}$ ions.

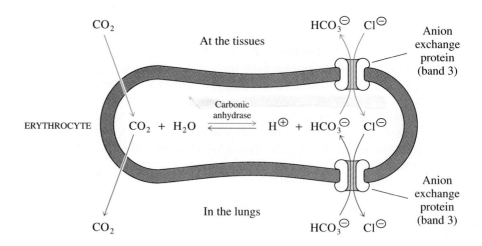

bicarbonate by the cytosolic enzyme carbonic anhydrase (Figure 12·31). The bicarbonate then exits the erythrocyte via the anion exchange protein in exchange for $Cl^{\ominus}$ from the blood. When bicarbonate dissolved in the blood reaches the lungs, the reverse process occurs. $HCO_3^{\ominus}$ reenters the red blood cell via the anion exchange protein and intracellular $Cl^{\ominus}$ exits in exchange. Carbonic anhydrase then catalyzes the formation of CO_2, which diffuses out into the blood and is ultimately exhaled. The anion exchange protein allows extremely rapid exchange of chloride and bicarbonate—on the order of seconds.

The anion exchange protein is an abundant erythrocyte protein (accounting for about 25% of the protein mass in the membrane) and exists as a dimer or tetramer. Anion exchange proteins related to that of the erythrocyte can be found in the plasma membranes of many other mammalian cell types. The erythrocyte anion exchange protein can also transport nonphysiological substrates such as $SO_4^{\ominus\ominus}$ and $NO_3^{\ominus}$. After the carrier has delivered an anion to the cytosol, it must bind another anion in order to switch back to the outward-facing conformation. Thus, the anion exchange protein mediates exchange rather than one-way transport. The membrane topology of the transporter has not been unequivocally established, but hydropathy analysis and biochemical studies suggest that it spans the bilayer 12 times. Several lysine and arginine residues appear to participate in attracting anions into a binding "funnel," and lysine residues are situated in the passage through which the anion moves across the membrane. This feature is reminiscent of substrate binding by superoxide dismutase (Section 7·8B).

D. Active Transport Requires the Input of Energy

Whereas passive transport involves the movement of a solute down a concentration gradient, active transport involves the movement of a solute up a concentration gradient. Active transport therefore requires the input of energy, which may be supplied in various forms.

A simple equation describes the thermodynamics of membrane transport. The free-energy change, ΔG, for transport of an uncharged solute from a concentration on one side of the membrane, c_1, to a concentration on the other side of the membrane, c_2, is

$$\Delta G = RT \ln \frac{c_2}{c_1} \tag{12·2}$$

where R is the universal gas constant ($8.315 \, J \, mol^{-1} \, K^{-1}$) and T is the absolute temperature. When c_1 is larger than c_2 (i.e., transport down a concentration gradient), the value of c_2/c_1 is less than 1. Thus, the values of $\ln c_2/c_1$ and ΔG are negative, and

Table 12·4 Energy sources for active transport

Type of transport	Energy source	Transporter	Transported species
Primary	Light	Bacteriorhodopsin	Protons
	ATP	ATPase	Ions
	ATP	P-Glycoprotein	Nonpolar compounds
	Substrate oxidation (electron transfer)	Electron transport proteins	Protons
Secondary	Proton gradient	Lactose permease	Lactose
	Sodium gradient	Active glucose transporter	Glucose

transport occurs spontaneously. On the other hand, when c_1 is smaller than c_2 (i.e., transport up a concentration gradient), the value of c_2/c_1 is greater than 1. In this case, the values of $\ln c_2/c_1$ and ΔG are positive, and the transport process requires the input of energy.

An **electrical potential** ($\Delta\psi$), or charge separation, often exists across membranes in living cells; a typical potential across a plasma membrane is -60 mV (negative inside). When a transported solute is charged, an additional term must be added to Equation 12·2 to account for the effects of this membrane polarization.

$$\Delta G = RT \ln \frac{c_2}{c_1} + z\mathcal{F}\Delta\psi \qquad (12\cdot3)$$

where z is the unit charge on the transported solute, $\mathcal{F}$ is the Faraday constant (96.48 kJ V^{-1} mol^{-1}), and $\Delta\psi$ is the transmembrane electrical potential in volts. Intuitively, it is obvious that transport of a positively charged solute from the positively polarized side of the membrane to the negatively polarized side results in a release of free energy compared to transport of an uncharged solute. Conversely, transport of a negatively charged solute in the same situation has an energy cost.

Active transporters use a variety of energy sources. The most common is ATP. A large class of ATP-driven ion transporters, the **ion-transporting ATPases,** are found in all organisms. These active transporters, which include the Na$^\oplus$-K$^\oplus$ ATPase and the Ca$^{2\oplus}$ ATPase, play an essential role in creating and maintaining ion concentration gradients across both the plasma membrane and the membranes of internal organelles. Light is the energy source for some active transporters, such as bacteriorhodopsin. **Primary active transport** is powered by a direct source of energy, such as ATP, light, or electron transport (Chapter 18). **Secondary active transport** is driven by ion concentration gradients. The "uphill" transport of the first solute is coupled to the "downhill" transport of a second solute. The free energy stored in the gradient of the second solute drives the transport of the first solute. In most cases, primary active transport is used to create the gradient in the second solute. The sources of energy for active transport are summarized in Table 12·4.

In bacteria, electron flow through the proteins of the electron-transport chain (located in the cytoplasmic membrane) generates a transmembrane proton concentration gradient. Such a proton concentration gradient is often used to power secondary active transport. For example, the uptake of lactose into *E. coli* cells is coupled to the inward flow of H$^\oplus$ (Figure 12·32). Lactose transport is mediated by lactose permease, a transmembrane protein (M_r 46 500) that carries out 1:1 symport of H$^\oplus$ and lactose. The influx of H$^\oplus$ down its concentration gradient provides a source of energy to pull lactose into the cell up its concentration gradient.

Many mammalian proteins that carry out secondary active transport are powered by a sodium ion gradient. In animal cells, the cytosolic sodium ion concentration is

Extracellular space

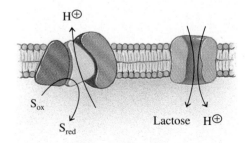

Cytosol

Figure 12·32
Secondary active transport. In *E. coli*, a proton concentration gradient is established by the proteins of the electron-transport chain as oxidized substrates (S$_{ox}$) are reduced (S$_{red}$). The energy released by protons moving down their concentration gradient drives the transport of lactose into the cell via lactose permease.

Figure 12·33
Na$^\oplus$-K$^\oplus$ ATPase generates a sodium ion gradient that drives secondary active transport of glucose.

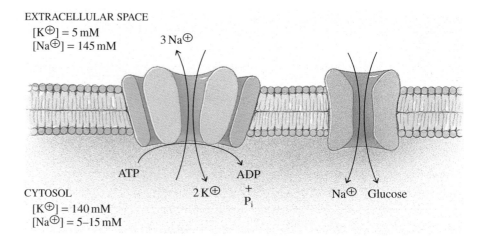

EXTRACELLULAR SPACE
$[K^\oplus] = 5\,mM$
$[Na^\oplus] = 145\,mM$

$3\,Na^\oplus$

ATP

ADP
+
P_i

$2\,K^\oplus$

$Na^\oplus$ Glucose

CYTOSOL
$[K^\oplus] = 140\,mM$
$[Na^\oplus] = 5–15\,mM$

low, whereas the extracellular concentration is high; the reverse is true for potassium ions. The active transporter for glucose in the intestinal cells of animals uses the free energy available in the sodium concentration gradient, importing one sodium ion with each glucose molecule. The energy released by the movement of Na$^\oplus$ down its concentration gradient powers the uphill transport of glucose.

E. The Na$^\oplus$-K$^\oplus$ ATPase Is an ATP-Driven Active Transporter

In large multicellular animals, most cells maintain an intracellular potassium ion concentration of about 140 mM in the presence of an extracellular K$^\oplus$ concentration of about 5 mM. The cytosolic concentration of sodium ions is maintained at about 5–15 mM in the presence of an extracellular concentration of about 145 mM. The pump that maintains these ion concentration gradients is the Na$^\oplus$-K$^\oplus$ ATPase, an ATP-driven antiport system that pumps two K$^\oplus$ into the cell and ejects three Na$^\oplus$ for every molecule of ATP that is hydrolyzed. Each Na$^\oplus$-K$^\oplus$ ATPase catalyzes the hydrolysis of about 100 molecules of ATP per minute under optimal cellular conditions, a significant portion (up to one-third) of the total energy consumption of a typical animal cell. The Na$^\oplus$ gradient generated by the Na$^\oplus$-K$^\oplus$ ATPase is the major source of energy for secondary active transport—including glucose transport—in animal cells (Figure 12·33).

The Na$^\oplus$-K$^\oplus$ ATPase is an integral membrane protein composed of two types of subunits with a stoichiometry of $\alpha_2\beta_2$ (Figure 12·34). The α subunits (subunit

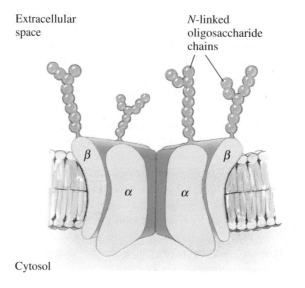

Extracellular
space

N-linked
oligosaccharide
chains

β

β

α

α

Cytosol

Figure 12·34
Structure and orientation of the Na$^\oplus$-K$^\oplus$ ATPase. The protein consists of two α subunits and two β subunits. The catalytic α subunits catalyze hydrolysis of ATP and transport Na$^\oplus$ and K$^\oplus$ across the cell membrane. The smaller β subunits are glycoproteins whose function is unknown.

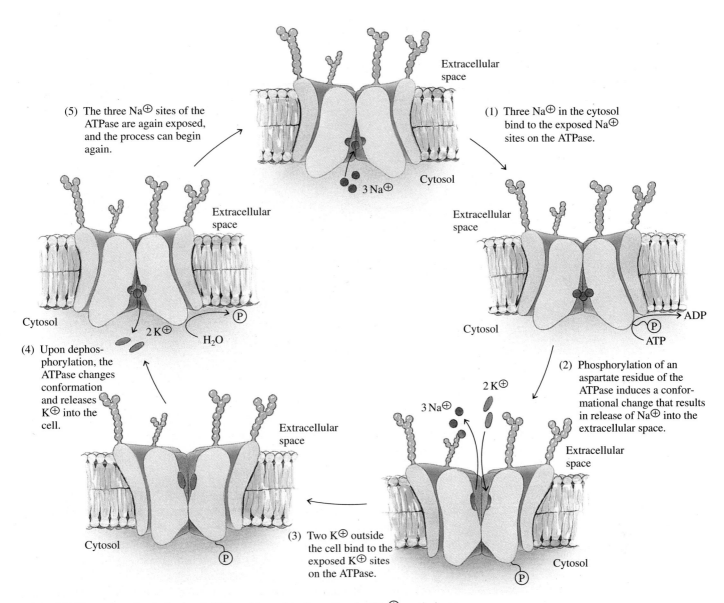

(5) The three Na$^\oplus$ sites of the ATPase are again exposed, and the process can begin again.

Extracellular space

Cytosol

3 Na$^\oplus$

(1) Three Na$^\oplus$ in the cytosol bind to the exposed Na$^\oplus$ sites on the ATPase.

Extracellular space

Cytosol

Extracellular space

Cytosol

2 K$^\oplus$ H$_2$O

(4) Upon dephosphorylation, the ATPase changes conformation and releases K$^\oplus$ into the cell.

Extracellular space

Cytosol

ADP
ATP

(2) Phosphorylation of an aspartate residue of the ATPase induces a conformational change that results in release of Na$^\oplus$ into the extracellular space.

2 K$^\oplus$

3 Na$^\oplus$

Extracellular space

Cytosol

(3) Two K$^\oplus$ outside the cell bind to the exposed K$^\oplus$ sites on the ATPase.

M_r 110 000) catalyze hydrolysis of ATP and have binding sites for Na$^\oplus$ on their cytosolic face and sites for K$^\oplus$ on their extracellular face. The β subunits (subunit M_r 55 000) are glycoproteins whose function within the complex is still unknown.

The mechanism of ion transport by the Na$^\oplus$-K$^\oplus$ ATPase depends on the existence of two different protein conformations (Figure 12·35). The inward-facing state has a high affinity for Na$^\oplus$ and a low affinity for K$^\oplus$. The outward-facing state is phosphorylated and has a high affinity for K$^\oplus$ and a low affinity for Na$^\oplus$. Three sodium ions bind to the Na$^\oplus$-K$^\oplus$ ATPase on the cytosolic side of the membrane. A phosphoryl group is then transferred from ATP to the side chain of an aspartate residue on the cytosolic side of the ATPase, generating an energy-rich aspartyl-γ-phosphate intermediate. Phosphorylation induces a conformational change, during which the three sodium ions are moved across the membrane and released outside the cell. Two potassium ions then bind to the outward-facing Na$^\oplus$-K$^\oplus$ ATPase. Hydrolysis of the aspartyl-phosphate bond results in a conformational switch, and the two potassium ions are brought across the membrane and released inside the cell. Both sodium and potassium ions are moved up their concentration gradients. Because there is a net movement of charge across the membrane, the Na$^\oplus$-K$^\oplus$ ATPase is electrogenic; ion movement generates an electrical potential across the membrane of about -60 mV. In neurons, this potential is essential for transmission of nervous impulses.

Figure 12·35
Mechanism of transport by the Na$^\oplus$-K$^\oplus$ ATPase. The Na$^\oplus$-K$^\oplus$ ATPase pumps out three sodium ions for every two potassium ions that it imports into the cell. Both ions move up their concentration gradients. Energy to drive the process is supplied by the hydrolysis of ATP.

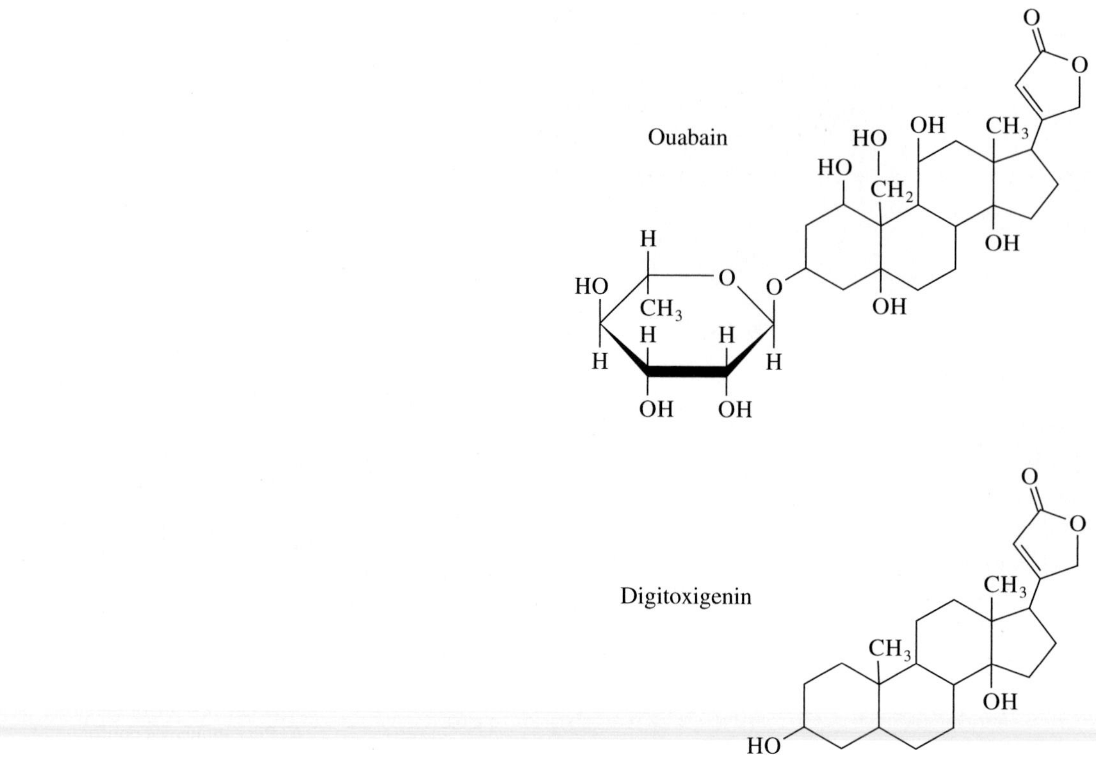

Figure 12·36
Structures of the Na⊕-K⊕ ATPase inhibitors ouabain and digitoxigenin. Both are present in the heart drug digitalis, which is extracted from the purple foxglove.

The highly toxic compounds ouabain and digitoxigenin (extracted from the purple foxglove) are powerful inhibitors of the Na⊕-K⊕ ATPase (Figure 12·36). These molecules, known as cardiotonic steroids, bind to the extracellular domain of the ATPase and lock it into the outward-facing phosphorylated conformation. Digitalis, an extract containing both compounds, is used in tiny doses as a heart stimulant. It increases the intracellular sodium level in heart muscle, which in turn activates the Na⊕-Ca²⊕ antiport system. This transporter exports Na⊕ and imports Ca²⊕. The higher calcium levels increase the strength of the heart muscle contractions.

The Na⊕-K⊕ ATPase is a member of a very large family of ion-transporting ATPases, which fall into three classes. The P-type ATPases become phosphorylated on an aspartate residue during the transport cycle. P-type ATPases share considerable amino acid sequence homology, especially in the region of the modified aspartate. They are all believed to operate by a similar mechanism. All P-type ATPases are susceptible to inhibition by vanadate, which is an analog of phosphate. The plasma membrane Na⊕-K⊕ ATPase and the K⊕ ATPase of the *E. coli* cytoplasmic membrane are representative members of this class of ATPase. One P-type ATPase, the Ca²⊕ ATPase of the endoplasmic reticulum, plays an essential role in signal transduction (Section 12·9). This Ca²⊕ ATPase pumps calcium ions from the cytosol into the lumen of the endoplasmic reticulum for storage, using ATP as an energy source. The net result is that the cytosolic concentration of calcium is very low (10^{-7} M), while the concentration in the lumen approaches the millimolar range.

The vacuolar, or V-type, ATPases are found in the membranes of vacuoles in fungi and plants as well as in endosomes and lysosomes in all eukaryotic cells. These transporters maintain the low pH in the interior compartments of the organelles. Acidification is accomplished by pumping protons from the cytosol into the organelle up the concentration gradient, using ATP as an energy source.

Members of the third class, F-type ATPases, are essential components of the oxidative phosphorylation systems of mitochondria, chloroplasts, and bacteria. Rather than hydrolyzing ATP, they are responsible for its generation and are thus actually ATP synthases. The F-type ATPases all have a characteristic knob-and-stalk structure attached to an integral membrane protein complex. The flow of $H^{\oplus}$ down its concentration gradient through a transmembrane proton channel is coupled to synthesis of ATP from ADP and P_i by the proteins of the knob (Chapter 18).

F. Transporters and Channels Are Involved in Human Disease

A unique membrane transporter named P-glycoprotein appears to play a major role in the resistance of tumor cells to multiple chemotherapeutic drugs. Such multidrug resistance is the leading cause of failure in the clinical treatment of human cancers. P-Glycoprotein is an integral membrane glycoprotein (M_r 170 000) that is over-expressed in the plasma membrane of drug-resistant cells. P-Glycoprotein belongs to a large group of related membrane transport proteins known as the ABC (ATP-binding cassette) superfamily, whose members are present in all organisms. Using ATP as an energy source, P-glycoprotein pumps a large variety of structurally un-related, nonpolar compounds, such as drugs, out of the cell up a concentration gradient. The cytosolic drug concentration is thus maintained at a level low enough to circumvent cell death. The normal physiological function of P-glycoprotein is currently unclear; it has been proposed that the transporter acts as a "vacuum cleaner" for toxic hydrophobic compounds ingested by the organism.

Cystic fibrosis is the most common inherited disorder in Caucasian populations. The disease is characterized by abnormal chloride ion secretion by epithelial cells in the lungs and gastrointestinal tract. This defect is responsible for the production of very thick mucus, which results in infection and tissue damage. A protein named CFTR (cystic fibrosis transmembrane [conductance] regulator) is defective in the disease. Like P-glycoprotein, CFTR is also a member of the ABC superfamily of proteins. CFTR appears to function as a chloride channel that is regulated by 3′,5′-cyclic adenosine monophosphate (cAMP) and protein kinase A and is opened by ATP. About 1 in 20 individuals carries a mutation within the gene that encodes CFTR, but only homozygous individuals show symptoms of disease. Current research is aimed at using gene therapy to correct the defect in CFTR in the lungs of cystic fibrosis patients.

G. Group Translocation Is a Special Type of Active Transport

Group translocation is a type of primary active transport in which the translocated species is chemically modified during the transport process. An example can be found in the phosphoenolpyruvate (PEP)-dependent sugar phosphotransferase system of *E. coli,* which transports at least 10 different sugar species across the cytoplasmic membrane. The sugar phosphotransferase system uses PEP rather than ATP as the energy-rich phosphoryl-group donor. The translocation process involves three or four proteins, depending on the sugar being transported. These proteins are enzymes I, II, III, and HPr. Enzyme I and HPr are both soluble proteins found in the cytosol. Enzyme III is a peripheral membrane protein that forms a complex with enzyme II, an integral membrane protein.

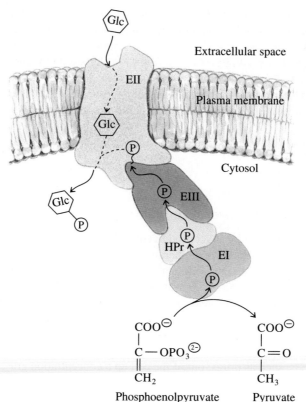

Figure 12·37
Group translocation. During group translocation, a molecule is covalently modified as it is translocated across the membrane. In the case of the phosphoenolpyruvate (PEP)-dependent sugar phosphotransferase system of *E. coli*, glucose is converted to glucose 6-phosphate. The phosphate group (Ⓟ) is ultimately derived from PEP.

The transport of glucose into an *E. coli* cell by the phosphotransferase system involves at least five steps (Figure 12·37). First, PEP donates its phosphoryl group to enzyme I. The phosphorylated enzyme in turn transfers this phosphoryl group to HPr. HPr then donates the phosphoryl group to enzyme III. When enzyme II binds a glucose molecule from the extracellular medium, a complex of enzyme II-glucose-phosphoenzyme III is formed. Enzyme III transfers its phosphoryl group to glucose, which is released into the cytosol as glucose 6-phosphate.

12·8 Endocytosis and Exocytosis Involve the Formation of Lipid Vesicles

The transport we have discussed so far involves the passage of molecules or ions across an intact membrane. Cells also need to import and export large molecules whose sizes preclude direct movement across the membrane via pores or transport proteins. Prokaryotes possess specialized multicomponent export systems in their cytoplasmic and outer membranes, which allow them to secrete certain proteins (often toxins or enzymes) into the extracellular medium. In eukaryotic cells, proteins are moved into and out of the cell by **endocytosis** and **exocytosis,** respectively. In both cases, transport involves formation of a specialized type of lipid vesicle.

Endocytosis is the process by which proteins (and certain other large substances) are engulfed by the plasma membrane and brought into the cell within a lipid vesicle. Receptor-mediated endocytosis begins with the binding of macromolecules to receptor proteins in the plasma membrane of the cell. The membrane then invaginates, forming a vesicle that contains the molecules of interest. As shown in Figure 12·38, the inside of such a membrane vesicle is topologically equivalent to the outside of a cell. Thus, proteins inside the vesicle have not actually crossed the plasma membrane. The major protein of the vesicles is known as clathrin, reflecting its latticelike, or clathrate, structure. Clathrin-coated pits on the cell surface facilitate the endocytosis of receptor-bound macromolecules. Once inside the cell, clathrin-coated vesicles may fuse with endosomes (smooth vesicles that lack clathrin) and then with lysosomes. This fusion process is the reverse of membrane vesicle invagination. Once inside a lysosome, the endocytosed protein and receptor components may be degraded to their constituent amino acids. Alternatively, the protein or the receptor or both may be recycled from the endosome back to the plasma membrane.

Exocytosis is a process similar to endocytosis, except that the direction of transport is reversed. During exocytosis, proteins destined for secretion from the cell are enclosed in vesicles by the Golgi apparatus. The vesicles transporting the proteins then fuse with the plasma membrane, releasing the vesicle contents into the extracellular space. The zymogens of digestive enzymes are exported from pancreatic cells in this manner (Section 13·2C).

Figure 12·38
Endocytosis of coated vesicles. Endocytosis begins with the binding of macromolecules to the plasma membrane of the cell. The membrane then invaginates, forming a clathrin-coated vesicle that contains the molecules of interest. The inside of the vesicle is topologically equivalent to the outside of the cell. (**a**) Electron micrographs. (Courtesy of Margaret M. Perry.) (**b**) Schematic views.

(a)

(b)

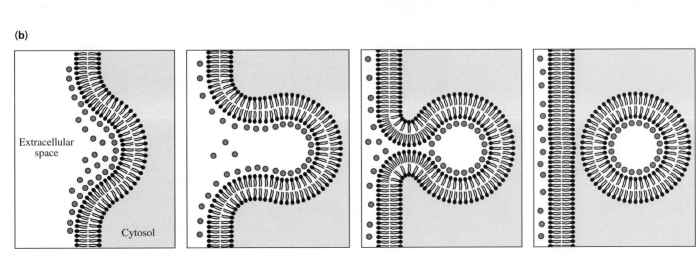

Extracellular space

Cytosol

12·9 Signals Are Transduced Across Membranes by Specific Receptor and Effector Systems

The cells of all organisms must be able to react to stimuli from the external environment. The responses of cells to such molecules vary according to the stimulus and may include movement toward or away from the stimulus, altered metabolism, secretion, differentiation, and cell growth and division. Many stimuli, such as some hormones and pheromones, do not cross the plasma membrane but interact with specific receptor proteins in the plasma membrane of the target cell. This interaction represents the first step in the process of **signal transduction,** whereby the initial external stimulus is converted into an intracellular signal, eventually producing an appropriate cellular response. A cell responds to a particular stimulus only if the appropriate specific membrane receptors are displayed on its surface, thus ensuring selectivity of responses.

A conceptual scheme for signal transduction is presented in Figure 12·39. After the ligand (also called the first messenger), for example, a hormone or growth factor, binds to its specific receptor, the signal is passed through a membrane protein **transducer** to a membrane-bound **effector enzyme.** The action of the effector enzyme generates a **second messenger,** which is usually a small molecule or ion. The second messenger is responsible for carrying the signal to its ultimate destination, which may be the nucleus, an intracellular compartment, or the cytosol. A vast diversity of ligands, membrane receptors, and transducers exist, but very few effector enzymes and second messengers are known. Cells thus possess exquisitely specific sensing systems to detect extracellular signals at their surface but make use of common pathways to transduce these signals to the interior of the cell. In the following pages we will discuss the three main signalling pathways in mammals; these involve adenylate cyclase, inositol phospholipids, and receptor tyrosine kinases.

Steroid hormones, which do not bind to plasma membrane receptors, are an exception to the general scheme of signal transduction shown in Figure 12·39. Because steroids are hydrophobic, they are able to diffuse across the plasma membrane into the cell, where they bind to specific receptor proteins in the cytoplasm.

Figure 12·39
Signal transduction across the plasma membrane of a cell.

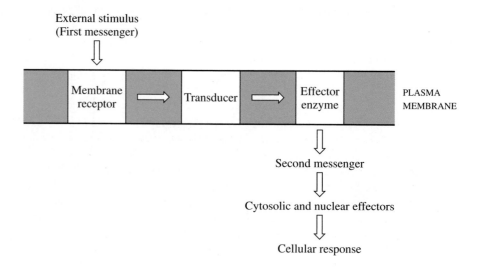

The steroid-receptor complexes are then transferred to the nucleus. The thyroid hormones (thyroxine and triiodothyronine, Section 21·14D) and vitamin D (Section 8·15B) also act directly at the nuclear level. The hormone-receptor complexes bind to specific DNA sequences called **hormone response elements** and thereby enhance or suppress the expression of adjacent genes.

In the case of cell surface–mediated signal transduction, our knowledge of the molecular mechanisms that give rise to the ultimate response of the cell is sketchy. However, one common thread runs through all signalling events: the production of second messengers almost invariably results, either directly or indirectly, in the activation of protein kinases. Enzymes in this large and diverse family catalyze the phosphorylation of various protein substrates in the membrane, the cytosol, or the nucleus. Although the identity and molecular properties of only a few of these phosphorylated proteins are known, many of them appear to play critical roles in metabolic regulation and the control of cell growth and division. In some cases, the proteins are activated by phosphorylation; in other cases, they are inactivated.

Protein kinases fall into two main classes, both of which use ATP as the phosphoryl-group donor. The first group, the serine-threonine protein kinases, catalyzes the phosphorylation of the hydroxyl group of specific serine and threonine residues in proteins. Tyrosine protein kinases are specific for the phenolic hydroxyl group of tyrosine residues. The specificity of a particular protein kinase can be readily determined by analysis of the phospho–amino acids present in a target protein after phosphorylation. Phosphorylation of amino acid residues in the kinase proteins themselves often dramatically alters kinase catalytic activity.

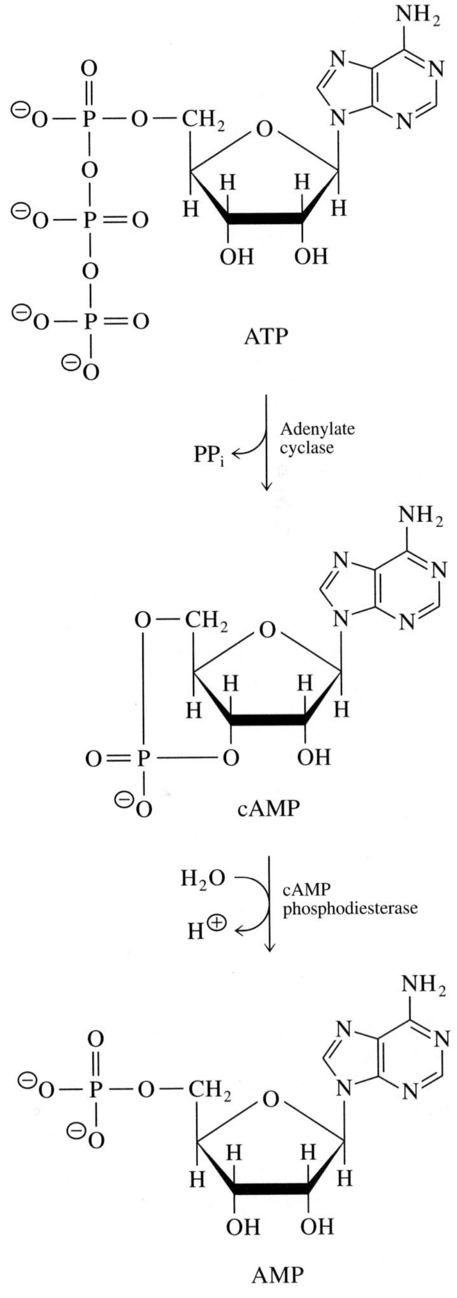

A. Many Hormones Activate the Adenylate Cyclase Signalling Pathway

Many of the hormones that regulate intracellular metabolism exert their effects on target cells by activating the adenylate cyclase signalling pathway. For example, epinephrine (a catecholamine also known as adrenaline) plays a central regulatory role in many pathways, including those of glycogen metabolism. Epinephrine is released from the adrenal gland and acts on all cells whose plasma membrane contains the β-adrenergic receptor. When epinephrine binds, the conformation of the receptor changes, promoting interaction between the receptor and a specific protein transducer, the G protein G_s, which is located on the cytosolic face of the plasma membrane. The receptor-ligand complex activates the G protein, which in turn binds to and activates the effector enzyme adenylate cyclase.

Adenylate cyclase is an integral transmembrane enzyme whose active site faces the cytosol. It catalyzes the formation of the second messenger cAMP from ATP (Figure 12·40). cAMP then diffuses from the membrane surface into the cytosol and activates a cAMP-dependent enzyme known as protein kinase A. This kinase is made up of a dimeric regulatory subunit and two catalytic subunits and is inactive in its fully assembled state. When the cytosolic concentration of cAMP increases as a result of signal transduction through adenylate cyclase, four molecules of cAMP bind to the regulatory subunit of protein kinase A, releasing the two

Figure 12·40
Production and destruction of cAMP. ATP is converted to cAMP by the transmembrane enzyme adenylate cyclase. The second messenger is subsequently converted to 5′-AMP by the action of a cytosolic phosphodiesterase.

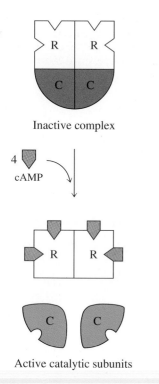

Figure 12·41
Activation of protein kinase A by cAMP. The assembled complex is inactive. When four molecules of cAMP bind to the regulatory subunit (R) dimer, the catalytic subunits (C) are released.

catalytic subunits, which are enzymatically active (Figure 12·41). Protein kinase A, a member of the serine-threonine class of protein kinases, can profoundly alter cellular metabolism by catalyzing the phosphorylation of a number of target enzymes.

Phosphorylation of amino acid side chains on the target enzymes is reversed by the action of protein phosphatases, which catalyze the hydrolytic removal of the phosphoryl groups. Hormone-sensitive lipase is a well-known example of an enzyme whose activity is regulated by a phosphorylation-dephosphorylation cycle (Figure 12·42). In this case, lipase is activated in response to hormone signals; the activated enzyme converts triacylglycerols to fatty acids that can be oxidized to generate energy. Many other enzymes that play key metabolic roles are also regulated by protein kinase A–dependent phosphorylation. We will encounter some of these in our discussions of metabolic pathways.

An important feature of the adenylate cyclase pathway (and many other signalling pathways) is **amplification.** A single hormone-receptor complex is able to interact with and activate many G proteins as a result of rapid lateral diffusion of both receptors and G proteins within the membrane. A single activated adenylate cyclase enzyme may in turn produce many molecules of the second messenger cAMP. Further downstream in the signalling pathway, a single protein kinase A molecule can phosphorylate many target proteins. This series of amplification events is designated a **cascade.**

The ability to turn off a signal-transduction pathway is an essential element of all signalling processes. For example, the cAMP concentration in the cytosol increases only transiently. A soluble phosphodiesterase catalyzes the hydrolysis of cAMP to AMP (Figure 12·40), thus limiting the effects of the second messenger. Caffeine and theophylline (Figure 10·5) at high concentrations inhibit cAMP phosphodiesterase, thereby decreasing the rate of conversion of cAMP to AMP. The effects of cAMP, and thus the activating effects of the stimulatory hormones, are prolonged and intensified by the inhibitors.

B. G Proteins Act as Signal Transducers

A family of intracellular **guanine nucleotide–binding proteins,** or G proteins, serves as transducers between hormone receptors in the plasma membrane and effector enzymes. G proteins bind guanine nucleotides (either GTP or GDP) and have GTPase activity; that is, they slowly catalyze hydrolysis of bound GTP to GDP. G proteins are peripheral membrane proteins on the inner surface of the plasma membrane. There are two classes of G proteins: heterotrimeric and monomeric. Only the heterotrimeric class is involved in signalling via hormone receptors. G proteins act

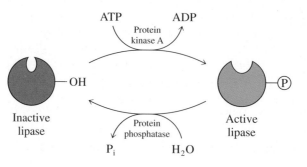

Figure 12·42
Regulation of the catalytic activity of hormone-sensitive lipase. Phosphorylation of hydroxyl groups catalyzed by protein kinase A activates the lipase, and dephosphorylation catalyzed by protein phosphatases inactivates it.

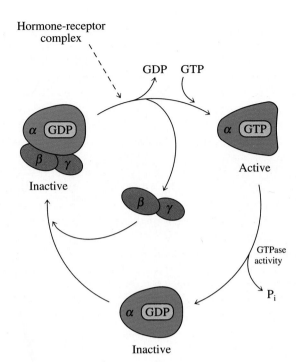

Figure 12·43
G-protein cycle. G proteins undergo a̶
tion after binding a receptor-ligand co̶
plex and are inactivated by their own
GTPase activity.

as molecular switches and exist in two interconvertible states: the GDP-bound form (which is inactive) and the GTP-bound form (which is active). The cyclic activation and deactivation of G proteins is illustrated in Figure 12·43. Heterotrimeric G proteins consist of an α, β, and γ subunit. $G_{\alpha\beta\gamma}$ normally contains GDP, which is bound to the α subunit; this complex is inactive. When a hormone-receptor complex interacts with $G_{\alpha\beta\gamma}$, the G protein undergoes a conformational change, becoming active. Bound GDP is exchanged for GTP, promoting the dissociation of G_α-GTP from $G_{\beta\gamma}$. G_α-GTP then interacts with adenylate cyclase in the membrane, greatly increasing cAMP production. The GTPase activity of the G protein acts as a built-in, timed off switch, since GTP is slowly hydrolyzed to GDP. The hydrolysis deactivates the G protein and allows G_α-GDP to reassemble with $G_{\beta\gamma}$, halting the effect of the G protein on adenylate cyclase.

Many different hormones use the cAMP signalling pathway. Hormones that bind to stimulatory receptors activate adenylate cyclase and raise intracellular cAMP levels, whereas hormones that bind to inhibitory receptors inhibit adenylate cyclase activity. The two types of receptors are distinguished by their associated G proteins. Stimulatory G proteins (G_s) activate adenylate cyclase, and inhibitory G proteins (G_i) inhibit the enzyme. Somatostatin is a polypeptide hormone from the hypothalamus that binds to receptors coupled to G_i. The action of somatostatin decreases cellular cAMP levels and reduces protein phosphorylation by protein kinase A. When epinephrine binds to the β-adrenergic receptor, which interacts with G_s, adenylate cyclase activity is stimulated. However, epinephrine can also bind to another type of receptor, the α_2-adrenergic receptor, which interacts with G_i, resulting in inhibition of adenylate cyclase. (The two classes of adrenergic receptors are found in different cell types.) Thus, the ultimate response of a cell to a hormone depends on the type of receptors present and the type of G protein to which they are coupled.

The amino acid sequences are known for several receptors that transduce signals via the adenylate cyclase pathway. The receptors are all homologous to the

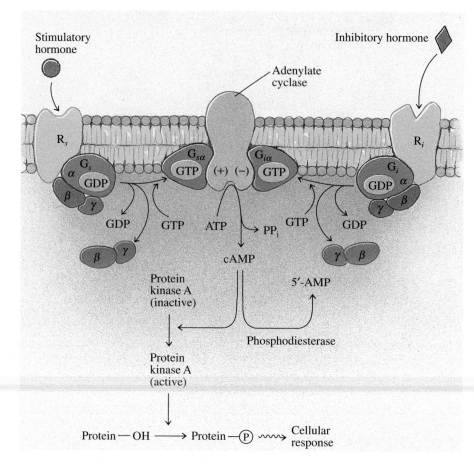

...ase signalling
...to a stimula-
... leads to
...otein G$_s$.
...ibitory re-
...oupled to adenylate
...ibitory G protein G$_i$. G$_s$ ac-
...e integral membrane enzyme
...enylate cyclase, whereas G$_i$ inhibits it.
cAMP activates protein kinase A, resulting in
the phosphorylation of cellular proteins.

β-adrenergic receptor and share a common structural motif of seven membrane-spanning segments. This similarity is perhaps not surprising, since all these receptors must interact with the same set of G proteins. The main features of the adenylate cyclase signalling pathway, including G proteins, are summarized in Figure 12·44.

Heterotrimeric G proteins play a central role as transducers in other signalling pathways. The inositol-phospholipid signalling pathway (discussed in the next section) also involves a G protein (G$_P$), whereas G$_t$ (transducin) takes part in visual excitation in the rod cells of the eye. Another G protein, G$_K$, controls the activity of potassium channels in heart muscle. These G proteins are structurally and mechanistically similar. They possess distinct α subunits, but their β and γ subunits are similar and often are interchangeable.

GTP hydrolysis is an essential step in the cyclic activation and deactivation of G proteins. Synthetic nonhydrolyzable GTP analogs, such as GTPγ-S and guanosine [$\beta\gamma$-imido]triphosphate (GMP-PNP), shown in Figure 12·45, interfere with the normal cycle by competing with GTP for the nucleotide-binding site of the G protein. Since the analogs cannot be hydrolyzed, they lock G$_\alpha$ into the activated state, leading to permanent stimulation of the signalling pathway.

G proteins are the biological targets of cholera and pertussis (whooping cough) toxins, which are secreted by the disease-producing bacteria *Vibrio cholerae* and *Bordetella pertussis*, respectively. Cholera toxin is made up of α and β subunits and exists as an $\alpha\beta_5$ complex. Following binding of the β subunits to the oligosaccharide chains of ganglioside G$_{M1}$ molecules on the surface of cells (Section 11·6), the α subunit crosses the plasma membrane and enters the cytosol. The α subunit is an ADP-ribosylase that catalyzes the transfer of an ADP-ribosyl moiety from the cofactor NAD$^\oplus$ to an arginine residue in the α subunit of the G protein G$_s$ (Figure 12·46). This covalent modification inactivates the GTPase activity of G$_{\alpha s}$. Thus,

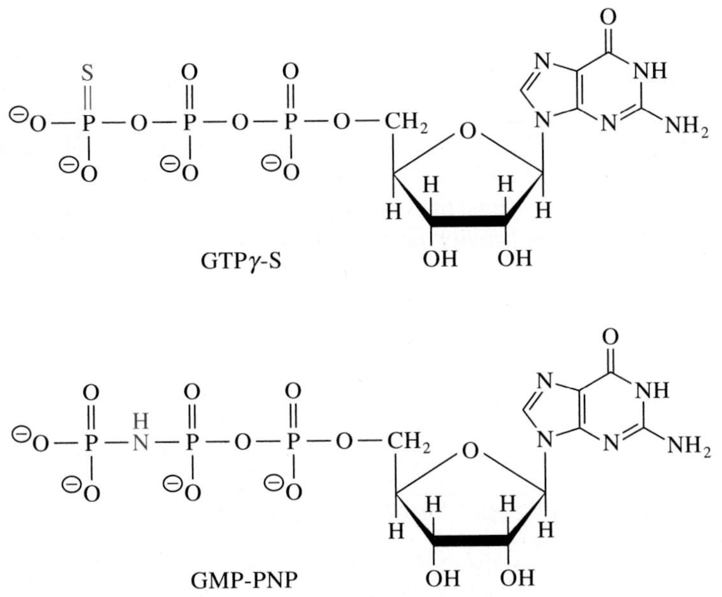

GTPγ-S

GMP-PNP

Figure 12·45
Nonhydrolyzable GTP analogs, GTPγ-S and guanosine [βγ-imido]triphosphate (GMP-PNP).

Figure 12·46
Effect of cholera toxin on the G protein G_s. The α subunit of cholera toxin enters the cytosol and catalyzes ADP-ribosylation of an arginine residue on the α subunit of G_s. The modification abolishes the GTPase activity of the G protein, and as a result, adenylate cyclase is continuously activated.

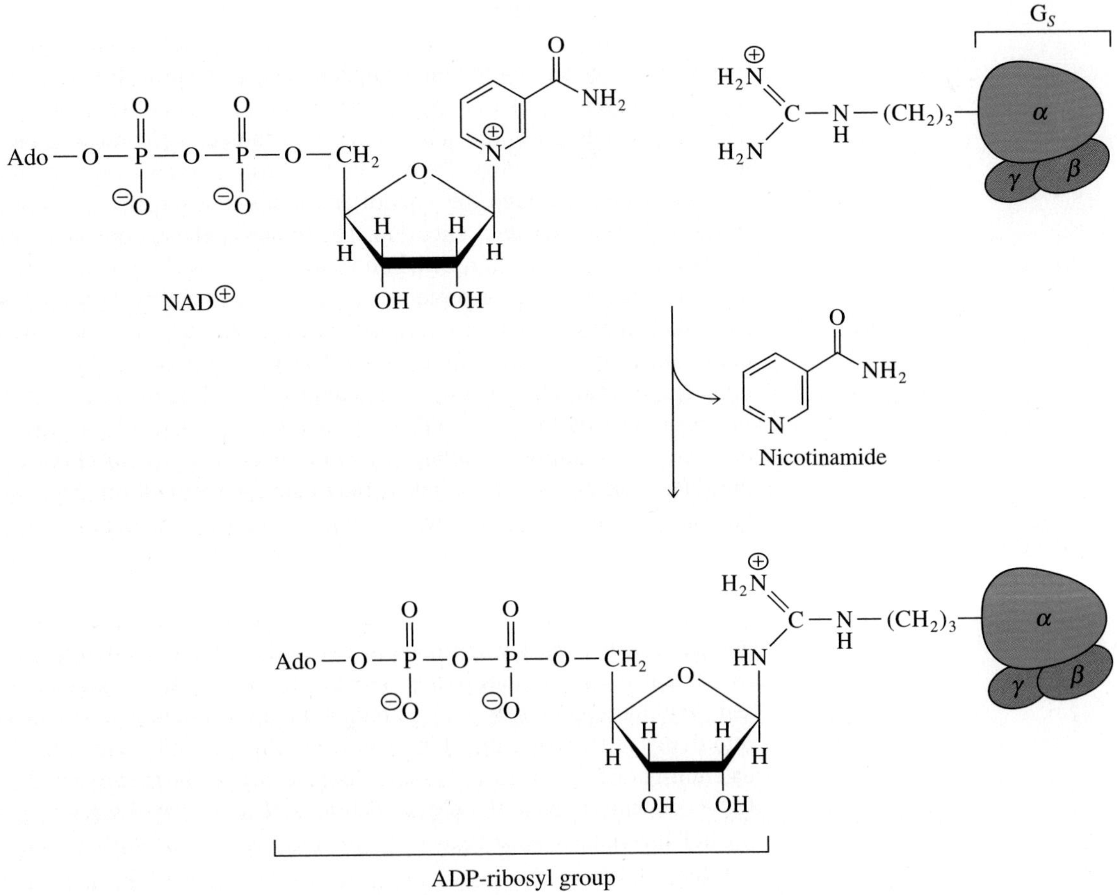

$NAD^\oplus$

Nicotinamide

ADP-ribosyl group

when cells are treated with cholera toxin, adenylate cyclase is continuously activated, and cAMP levels remain high. In organisms infected with *V. cholerae*, cAMP stimulates certain transporters in the plasma membrane of the intestinal cells, leading to massive secretion of ions and water into the gut lumen. The resulting rapid dehydration can be fatal. Pertussis toxin catalyzes the ADP-ribosylation of $G_{\alpha i}$. In this case, the modified $G_{\alpha i}$ is unable to associate with an inhibitory ligand–receptor complex. Thus, the G protein cannot be activated by ligand binding, and adenylate cyclase activity cannot be reduced via inhibitory receptors.

C. The Inositol-Phospholipid Signalling Pathway Produces Two Second Messengers

Another major signal-transduction pathway used by some hormones, growth factors, and other ligands involves the production of two different second messengers, both of which arise from a plasma membrane phospholipid, phosphatidylinositol 4,5-*bis*phosphate (PIP_2). A minor component of plasma membranes, PIP_2 is located in the inner leaflet (an example of lipid asymmetry). It is synthesized from phosphatidylinositol by two successive phosphorylation steps catalyzed by kinases that require ATP.

Following binding of a ligand to a specific receptor, the signal is transduced through a G protein, G_P, which undergoes conversion to the GTP-bound form. G_P subsequently activates the effector enzyme phosphoinositide-specific phospholipase C, which is bound to the cytoplasmic face of the plasma membrane. Phospholipase C catalyzes the hydrolysis of PIP_2 to two products: inositol 1,4,5-*tris*phosphate (IP_3), a water-soluble molecule, and diacylglycerol (Figure 12·47). Both IP_3 and diacylglycerol, like cAMP, are second messengers that transmit the original signal to the interior of the cell. IP_3 action results in an increase in the calcium-ion concentration of the cytosol, and diacylglycerol activates a protein kinase.

IP_3 diffuses through the cytosol and binds to a specific receptor in the membrane of the endoplasmic reticulum. The lumen of the endoplasmic reticulum contains a high concentration of calcium ions as a result of the action of the Ca^{2+} ATPase. The IP_3 receptor protein is a ligand-gated calcium channel, which opens for a short time and releases sequestered Ca^{2+} into the cytosol. The cytosolic calcium concentration rises rapidly, from less than 500 nM to about 1 μM. This calcium flux is transient, since Ca^{2+} is pumped back into the lumen of the endoplasmic reticulum by the Ca^{2+} ATPase as soon as the channel closes. The increased calcium concentration stimulates calcium-dependent protein kinases, which phosphorylate various protein targets. Many calcium-calmodulin–dependent enzymes and protein kinases are also activated under these conditions. Recall that calmodulin is a small (M_r 17 000) acidic calcium-binding protein, which is found both free in the cytosol and as a regulatory subunit of several enzymes (Figure 8·2). Binding of Ca^{2+} to calmodulin induces a conformational change in the protein, which then associates with a variety of enzymes and modulates their activity.

The level of calcium in the cytosol can be measured using special calcium-sensitive fluorescent dyes, such as Indo-1 and Fura-2. These dyes can be introduced into living cells, where they bind Ca^{2+} with high affinity. The fluorescence of the calcium-bound dye differs from the fluorescence of the free dye and is sensitive to Ca^{2+} concentrations in the nM to μM range. Changes in fluorescence inside a living cell can thus be used to measure the increase in [Ca^{2+}] that occurs during signal transduction.

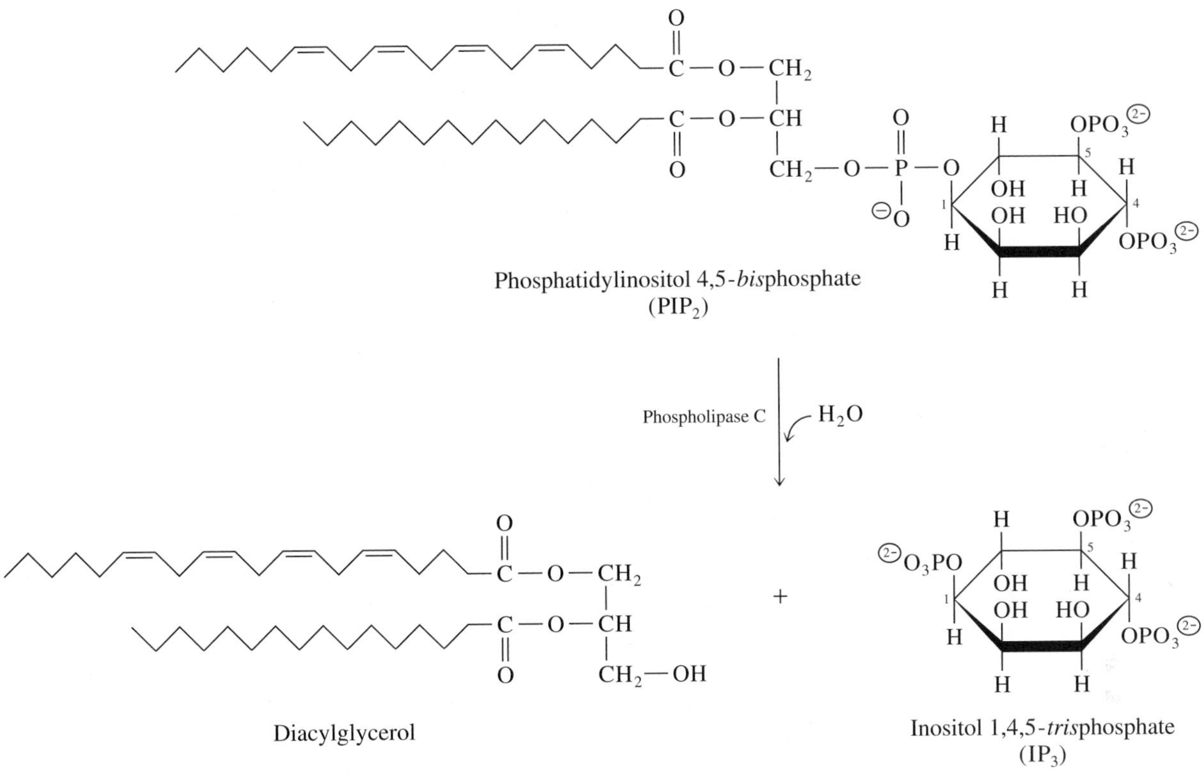

Phosphatidylinositol 4,5-*bis*phosphate
(PIP$_2$)

Phospholipase C — H$_2$O

Diacylglycerol

+

Inositol 1,4,5-*tris*phosphate
(IP$_3$)

Figure 12·47
Phospholipase C activity. Phosphatidylinositol 4,5-*bis*phosphate (PIP$_2$) is hydrolyzed to inositol 1,4,5-*tris*phosphate (IP$_3$) and diacylglycerol by the action of a phosphoinositide-specific phospholipase C.

Calcium-specific ionophores, such as ionomycin and A23187, are also useful tools for examining the importance of Ca^{2+} in signalling events. Addition of these compounds to cells in a calcium-rich medium results in a rapid influx of Ca^{2+} into the cell in the absence of an external stimulus. Such ionophores are used to calibrate the calcium levels measured using fluorescent dye binding. They are also able to mimic the actions of the IP$_3$ arm of the inositol-phospholipid signalling pathway, bypassing membrane receptors.

The other product of PIP$_2$ hydrolysis, diacylglycerol, remains in the plasma membrane and activates a calcium- and phospholipid-dependent protein kinase known as protein kinase C. This kinase is active only when both phosphatidylserine and Ca^{2+} are bound to it. Protein kinase C is normally inactive in the presence of low intracellular concentrations of calcium. However, when diacylglycerol binds to the kinase, the affinity for calcium is greatly increased, so that the kinase can bind Ca^{2+} at low concentrations and become active. Protein kinase C exists in equilibrium between a soluble cytosolic form and a peripheral membrane form. Activation by diacylglycerol results in the translocation of the kinase to the inner face of the plasma membrane, where it remains associated for a period of time. Protein kinase C, a member of the serine-threonine kinase family, phosphorylates many target proteins, both cytosolic and membrane-bound, altering their catalytic activity. Several protein kinase C isozymes exist, each with different catalytic properties and tissue distribution.

Signalling via the inositol-phospholipid pathway is turned off in a number of ways. First, when GTP is hydrolyzed, G$_P$ returns to its inactive form and no longer stimulates phospholipase C. Diacylglycerol is a short-lived second messenger; it is rapidly converted to phosphatidate by the action of enzymes responsible for phospholipid biosynthesis and is eventually recycled back to phosphatidylinositol. IP$_3$ is also a transient second messenger; a series of cytosolic phosphatase enzymes rapidly convert it to inositol, which is also recycled back into phosphatidylinositol.

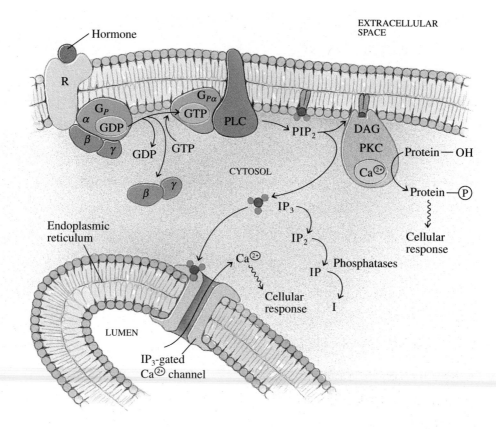

Figure 12·48

Inositol-phospholipid signalling pathway. Binding of a hormone to its transmembrane receptor (R) activates the G protein G_P. This in turn stimulates a specific membrane-bound phospholipase C (PLC), which catalyzes the hydrolysis of the phospholipid PIP_2 in the inner leaflet of the plasma membrane. The second messengers IP_3 and diacylglycerol (DAG) are responsible for carrying the signal to the interior of the cell. IP_3 diffuses to the endoplasmic reticulum, where it binds to and opens an IP_3-gated Ca^{2+} channel in the membrane, releasing stored Ca^{2+}. Diacylglycerol remains in the plasma membrane, where it activates the calcium- and phospholipid-dependent enzyme protein kinase C (PKC).

Thus, the net effect of signalling through this pathway is an increase in the **metabolic turnover** of inositol phospholipids. The main features of the inositol-phospholipid signalling pathway are summarized in Figure 12·48.

Phorbol myristate acetate, a phorbol ester found in the leaves of the plant *Croton tiglium*, mimics diacylglycerol and activates protein kinase C (Figure 12·49). However, unlike diacylglycerol, this compound is not readily metabolized, and it leads to persistent activation and membrane association of protein kinase C. Phorbol myristate acetate is a powerful tumor promoter; that is, it greatly increases the development of tumors following exposure to a carcinogen. Prolonged activation of signalling by protein kinase C may thus be involved in the uncontrolled cell growth of cancer.

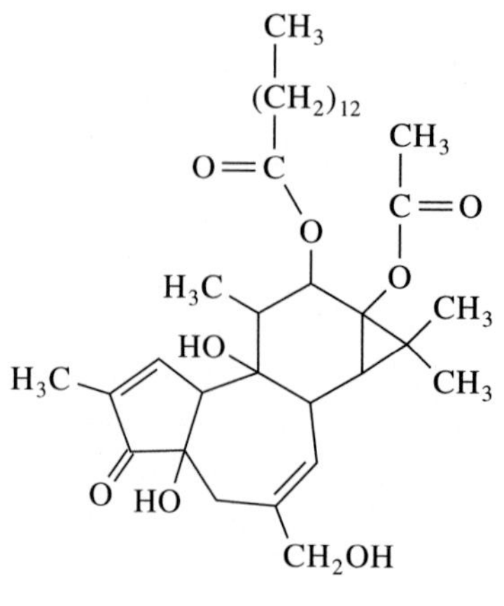

Figure 12·49
Phorbol myristate acetate, a tumor promoter.

It should be noted that activation of both arms of the inositol-phospholipid signalling pathway is necessary for a cellular response; in other words, the actions of IP$_3$ and diacylglycerol are synergistic. Neither an increase in Ca^{2+} concentration alone nor the activation of protein kinase C alone leads to cell stimulation. However, treatment of cells with a combination of a calcium ionophore (which increases the cytosolic Ca^{2+} concentration) and phorbol myristate acetate (which activates protein kinase C) mimics the effect of an external growth factor or hormone.

D. Many Growth Factors Activate Receptor Tyrosine Kinases

Growth factors are proteins that regulate cell proliferation by stimulating resting cells to undergo cell division. Growth factors are specific for cell type, acting only on a cell that displays the appropriate cell-surface receptors. Growth factors that act only on a few cell types include epidermal growth factor, platelet-derived growth factor, nerve growth factor, fibroblast growth factor, and the lymphocyte growth factors known as the interleukins. Other growth factors, such as insulin, have effects on many different cell types because their receptors are ubiquitous.

A common signalling pathway operates for many growth factors. The primary event, binding of the growth factor to its cell-surface receptor, is followed by activation of tyrosine-kinase catalytic activity of the intracellular domain of the receptor. Isolation and sequencing of the genes for many receptor tyrosine kinases suggests that these proteins have a similar molecular topology. Common features include a large glycosylated extracellular domain to which the growth factor binds, a single hydrophobic membrane-spanning domain, and an intracellular tyrosine-kinase catalytic domain. For many growth-factor receptors, activation of the catalytic activity of the kinase domain is thought to occur as a result of receptor dimerization. In the case of the insulin receptor, which is a dimer to begin with (Figure 12·50), binding of insulin to the α subunit induces a conformational change that brings together the tyrosine-kinase domains of the β subunits, which are thereby activated. Many of the activated kinase domains of growth-factor receptors use ATP

Figure 12·50
Molecular topology of the insulin receptor. The receptor consists of two extracellular α chains, each of which contains an insulin-binding site, and two transmembrane β chains with cytosolic tyrosine-kinase domains. Following insulin binding to the α chains, the β chains catalyze autophosphorylation of tyrosine residues in the adjacent intracellular domains. Other cellular proteins are also phosphorylated by the tyrosine kinase, which sets off a cascade of events in the cell.

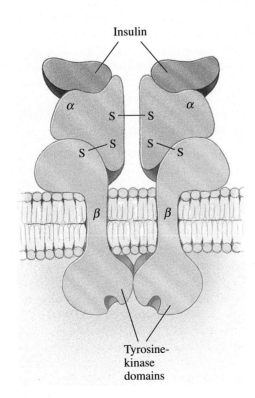

Insulin

Tyrosine-
kinase
domains

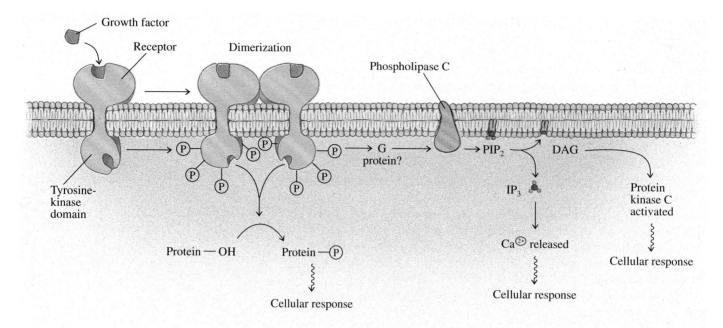

Figure 12·51
General scheme for signal transduction by membrane-receptor tyrosine kinases. A growth factor binds to its receptor, activating the intracellular tyrosine-kinase domain. The receptor is autophosphorylated, and tyrosine residues on other cellular proteins are phosphorylated. Other signalling pathways are activated by processes that are not well understood. Hydrolysis of PIP_2 catalyzed by phospholipase C and release of IP_3 and diacylglycerol (DAG) often occur.

to phosphorylate the hydroxyl group of tyrosine residues in the receptor itself, a process known as **autophosphorylation.** It is likely that each receptor in the dimer catalyzes the phosphorylation of its partner. Other proteins are also phosphorylated by the activated receptor tyrosine kinase, leading to the stimulation of secondary signalling pathways. Hydrolysis of PIP_2 often occurs, with subsequent activation of protein kinase C and an increase in the cytosolic calcium concentration. Cellular stimulation by growth factors is a complex process that involves the triggering of multiple signal-transduction pathways (Figure 12·51).

Phosphoryl groups are removed from both the growth-factor receptors and their protein targets by another class of enzymes, the protein tyrosine phosphatases. Although only a few enzymes in this class have been studied so far, they appear to play a very important role in regulation of signalling via protein tyrosine kinases.

E. Some Oncogene Products Are Defective Proteins of Signal-Transduction Pathways

Tumor cells show uncontrolled proliferation, suggesting that regulation of their signalling pathways may be defective. The discovery of **oncogenes** has provided some support for this hypothesis. Oncogenes are genes that transform normal cells into cancer cells. These genes are often mutated or inappropriately expressed versions of normal cellular genes that encode proteins involved in the signalling pathways leading to cell proliferation. Oncogenes were first discovered in tumor viruses, RNA-containing viruses that cause cancer. The cancer-causing gene in many of these viruses was apparently acquired from a host organism during the course of evolution and integrated into the viral genome. The original host gene, a **proto-oncogene,** may have encoded a growth factor, a growth-factor receptor, or some downstream intracellular component of the signalling pathway, such as a nuclear transcription factor that is involved in gene expression (Table 12·5). Infection of a host cell by the tumor virus leads to expression of the defective protein and uncontrolled stimulation of cell proliferation. For example, the avian erythroblastosis

Table 12·5 Viral oncogenes, their protein products, and the corresponding cellular proto-oncogenes

Oncogene	Source	Proto-oncogene product
Growth factor		
v-sis	Simian sarcoma	Platelet-derived growth factor
Growth-factor receptor		
*v-erb*B	Avian erythroblastosis	Epidermal growth-factor receptor
v-fms	Feline sarcoma	Colony stimulating-factor 1 receptor
v-ros	UR II avian sarcoma	Insulin receptor
Transducer		
ras	Human tumors	Monomeric G protein
Cytosolic signalling component		
v-src	Rous avian sarcoma	Cytosolic protein tyrosine kinase
v-abl	Abelson murine leukemia	Cytosolic protein tyrosine kinase
Nuclear transcription factor		
v-jun	Avian sarcoma virus 17	Transcription factor AP-1 subunit
v-fos	FBJ osteosarcoma	Transcription factor AP-1 subunit
v-myc	Avian myelocytomatosis	Involved in transcription

virus, which causes cancer in chickens, contains the oncogene *v-erb*B (*v* indicates a viral gene). The protein product of this oncogene is homologous to the epidermal growth-factor receptor, whose gene is the proto-oncogene *c-erb*B (*c* indicates a cellular gene). As shown in Figure 12·52, the *v-erb*B product is a truncated version of the receptor for epidermal growth factor. Amino acids 551 to 1154 of the epidermal growth-factor receptor are 90% identical to those of the *v-erb*B protein, which lacks the external ligand-binding domain. The oncogenic protein is permanently activated in the absence of growth factor and continuously transduces a signal for cell division. More than half the known viral oncogenic proteins are altered receptor tyrosine kinases.

Only a small number of human cancers arise from oncogenic viruses. Many tumors develop following spontaneous or chemically induced mutations in cellular proto-oncogenes. Mutated versions of genes encoding proteins involved in the signalling pathways regulating cell growth have been identified in a significant proportion of human tumors. For example, altered *ras* proteins have been found in lung, colon, and pancreatic tumors. The study of oncogenes and signalling pathways will lead to a better understanding of the causes and possible means for prevention and cure of human cancer.

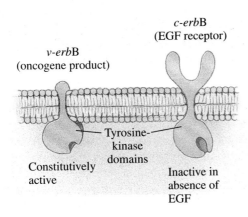

*c-erb*B
(EGF receptor)

*v-erb*B
(oncogene product)

Tyrosine-
kinase
domains

Constitutively
active

Inactive in
absence of
EGF

Figure 12·52
Protein products of the viral oncogene *v-erb*B and the proto-oncogene *c-erb*B. Note that the extracellular ligand-binding domain is absent in the viral protein, whose amino acid sequence is over 90% identical to the corresponding region of the cellular product, the epidermal growth-factor (EGF) receptor. The oncogenic protein is believed to transduce a signal in the absence of activation by growth factor.

Summary

Biological membranes define the external boundary of the cell and separate compartments within the cell. A typical membrane consists of lipids and proteins, with small amounts of carbohydrate on glycosphingolipids and glycoproteins. Membranes are a fluid mosaic of proteins in a lipid bilayer matrix. Amphipathic lipids, such as glycerophospholipids and sphingolipids, assemble into bilayers spontaneously. Lateral diffusion of lipids in the bilayer is rapid, whereas transverse diffusion from one leaflet to the other takes place very slowly. Specific lipids are asymmetrically distributed between the inner and outer leaflets of biological membranes. At low temperatures, a lipid bilayer exists in an ordered gel state in which the acyl chains are extended. The bilayer undergoes a phase transition on warming and adopts the fluid liquid-crystalline phase in which the acyl chains constantly bend and flex. *Cis* double bonds create kinks in the acyl chain, thus decreasing the phase-transition temperature and increasing fluidity. Cholesterol modulates membrane fluidity in animal cell membranes by disrupting the packing of gel-phase lipids and restricting the motion of acyl chains of liquid-crystalline–phase lipids.

Most integral membrane proteins span the hydrophobic interior of the bilayer, whereas peripheral membrane proteins are more loosely associated with the membrane surface. Lipid-anchored proteins are tethered to the bilayer by covalent linkage to either a fatty acyl chain, a molecule of glycosylphosphatidylinositol, or an isoprenoid group. Nearly all integral membrane proteins contain α-helical segments that span the lipid bilayer. Receptor proteins often possess only a single α-helical region, whereas transport proteins always have multiple membrane-spanning segments connected by loops at the membrane surface. A stretch of about 20 amino acid residues is sufficient to completely span the bilayer. Although these amino acids are usually predominantly hydrophobic, they may include some charged residues. The topology of a membrane protein can be determined by experimental methods such as X-ray crystallography, electron microscopy, and vectorial labelling and may be predicted using hydropathy plots.

Although many proteins are free to diffuse laterally in the membrane, others are anchored to the cytoskeletal network, which is attached at the inner face of the plasma membrane and provides mechanical strength. In the human erythrocyte, a meshwork of spectrin and actin is attached to the cytoplasmic domains of integral membrane proteins by peripheral linker proteins. In other cell types, the cytoskeleton consists of actin microfilaments, microtubules, and intermediate filaments. All three networks are highly dynamic structures that are constantly assembled and disassembled, depending on the needs of the cell.

The oligosaccharide chains of glycolipids and glycoproteins appear to be located exclusively on the external surface of the cell. Glycosphingolipid sugar chains and the *N*- and *O*-linked oligosaccharides of membrane glycoproteins form a sugar coat, called the glycocalyx, which displays a "fingerprint" of the cell to the exterior. Oligosaccharides make up blood group antigens and act as receptors for viruses, parasites, and protein toxins.

A lipid bilayer is a selectively permeable barrier that is impenetrable to most charged species but allows water and hydrophobic molecules to diffuse freely across it. The rate of diffusion of ions across a membrane may be greatly enhanced by certain ionophores. Specific transport, channel, or pore proteins mediate the movement of ions and polar molecules across membranes. Channel proteins allow rapid diffusion of large numbers of specific ions or small molecules through a central pore, down a concentration gradient. Transport proteins bind a substrate and move it across the membrane by alternating between an outward-facing and an inward-facing conformation. Passive transporters move molecules down a concentration gradient and do not require energy. Active transporters, which move substrates up a concentration gradient, require energy input. In primary active transport, energy is supplied directly from ATP hydrolysis, light, or electron transport. Secondary active transport is driven by an ion gradient; "uphill" transport of the substrate is coupled to "downhill" transport of the ion. Large protein molecules can be moved into and out of the cell by the processes of endocytosis and exocytosis, respectively, which involve creation and fusion of lipid vesicles.

Hormones and growth factors bind to specific receptors in the plasma membrane, which convert the external stimulus to an intracellular signal. After ligand binding to the receptor, a G-protein transducer is often activated by exchange of bound GDP for GTP. The activated G protein passes the signal to a membrane effector enzyme, which produces one or more second messengers. These small molecules or ions then carry the signal to the cell interior. These second messengers often activate protein kinases. Phosphorylation of amino acid side chains by these kinases alters the function of target proteins and produces a cellular response. Only a few common routes are used for signal transduction. One signalling pathway results in the G protein–mediated activation of membrane-bound adenylate cyclase, which produces the second messenger cAMP. This in turn activates protein kinase A, a serine-threonine protein kinase. In another major pathway, a G protein activates phospholipase C, which catalyzes the hydrolysis of the lipid phosphatidylinositol 4,5-*bis*phosphate. One product (inositol 1,4,5-*tris*phosphate) increases the cytoplasmic Ca^{2+} concentration, whereas the other product (diacylglycerol) activates protein kinase C. Many growth factors activate a protein tyrosine-kinase domain on the cytoplasmic side of their membrane receptors. This activation results in phosphorylation of tyrosine residues of the receptor as well as other target proteins. Oncogenes, genes that transform normal cells to cancer cells, often encode altered versions of proteins involved in the signalling pathways leading to cell proliferation.

Selected Readings

General References

Bretscher, M. S. (1985). The molecules of the cell membrane. *Sci. Am.* 253(4):100–108. A clear introduction to the structure and function of biological membranes.

Frye, L. D., and Edidin, M. (1970). The rapid intermixing of cell surface antigens after formation of mouse-human heterokaryons. *J. Cell Sci.* 7:319–335.

Gennis, R. B. (1989). *Biomembranes* (New York: Springer-Verlag).

Jain, M. K. (1988). *Introduction to Biological Membranes* (New York: John Wiley & Sons).

Singer, S. J., and Nicolson, G. L. (1972). The fluid mosaic model of the structure of cell membranes. *Science* 175:720–731. This paper discusses the properties of membranes that led the authors to propose their model.

Singer, S. J. (1992). The structure and function of membranes: a personal memoir. *J. Membr. Biol.* 129:3–12. This article revisits the main highlights of the fluid mosaic model on its 20th anniversary and discusses how recently discovered features of membranes fit into the model.

Membrane Lipids

Devaux, P. F. (1992). Protein involvement in transmembrane lipid asymmetry. *Annu. Rev. Biophys. Biomol. Struct.* 21:417–439. A detailed review of the role of phospholipid flippases and cytoskeletal proteins in membrane lipid asymmetry.

Storch, J., and Kleinfeld, A. M. (1985). The lipid structure of biological membranes. *Trends Biochem. Sci.* 10:418–421. A review of the structures of biological membranes, including areas of interest and debate.

Membrane Proteins

Englund, P. T. (1993). The structure and biosynthesis of glycosylphosphatidylinositol protein anchors. *Annu. Rev. Biochem.* 62:121–38. This article reviews recent progress in biosynthesis of glycosylphosphatidylinositol anchors.

Fasman, G. D., and Gilbert, W. A. (1990). The prediction of transmembrane protein sequences and their conformations: an evaluation. *Trends Biochem. Sci.* 15:89–92. A concise assessment of the current methods used for prediction of membrane protein topography from the amino acid sequence.

Jennings, M. L. (1989). Topography of membrane proteins. *Annu. Rev. Biochem.* 58:999–1027. A detailed review of membrane protein topography, its experimental determination, and prediction.

Marshall, C. J. (1993). Protein prenylation: a mediator of protein-protein interactions. *Science* 259:1865–1866. A short review of the structure, biosynthesis, and possible functions of protein prenylation.

Glycocalyx

Gahmberg, C. G., and Hermonen, J. (1988). The human red cell sialoglycoprotein glycophorin A: biosynthesis, glycosylation, and interaction with external ligands. *Ind. J. Biochem. Biophys.* 25:133–136.

Viitala, J., and Järnefelt, J. (1985). The red cell surface revisited. *Trends Biochem. Sci.* 10:392–395. An interesting article describing the glycocalyx of the human erythrocyte.

Cytoskeleton

Bennett, V. (1985). The membrane skeleton of human erythrocytes and its implications for more complex cells. *Annu. Rev. Biochem.* 54:273–304. A detailed description of the structure and interactions of the proteins of the red blood cell cytoskeleton.

Marchesi, V. T. (1985). Stabilizing infrastructure of cell membranes. *Annu. Rev. Cell Biol.* 1:531–561. A review of components of the cytoskeleton.

Membrane Transport

Collins, F. S. (1992). Cystic fibrosis: molecular biology and therapeutic implications. *Science* 256:774–779. One of the discoverers of the gene reviews current knowledge of the CFTR protein and discusses some possible therapies for the disease.

Gottesman, M. M., and Pastan, I. (1993). Biochemistry of multidrug resistance mediated by the multidrug transporter. *Annu. Rev. Biochem.* 62:385–427. A comprehensive review of the molecular biology, biochemical properties, and clinical significance of P-glycoprotein.

Hille, B. (1992). *Ionic Channels of Excitable Membranes,* 2nd ed. (Sunderland, Massachusetts: Sinauer Associates).

Jennings, M. L. (1989). Structure and function of the red blood cell anion transport protein. *Annu. Rev. Biophys. Biophys. Chem.* 18:397–430. An in-depth review of the band 3 transport protein.

Kaback, H. R., Bibi, E., and Roepe, P. D. (1990). β-Galactoside transport in *E. coli*: a functional dissection of *lac* permease. *Trends Biochem. Sci.* 15:309–314. A comprehensive review of the lactose permease transport protein.

Pedersen, P. L., and Carafoli, E. (1987). Ion motive-ATPases. I. Ubiquity, properties, and significance to cell function. *Trends Biochem. Sci.* 12:146–150. II. Energy coupling and work output. *Trends Biochem. Sci.* 12:186–189. Comprehensive reviews of the ion-translocating ATPases, including the Ca^{2+} and Na^{+}-K^{+} ATPases.

Signal Transduction

Berridge, M. J. (1985). The molecular basis of communication within the cell. *Sci. Am.* 253(10):142–152. An introduction to the adenylate cyclase and inositol phospholipid signalling pathways.

Fantl, W. J., Johnson, D. E., and Williams, L. T. (1993). Signalling by receptor tyrosine kinases. *Annu. Rev. Biochem.* 62:453–481. This review article provides comprehensive information on the biochemical mechanisms of signalling through many tyrosine kinase growth-factor receptors.

Farago, A., and Nishizuka, Y. (1990). Protein kinase C in transmembrane signalling. *FEBS Lett.* 268:350–354. This short review discusses several aspects of the structure and role of protein kinase C in signal transduction.

Hepler, J. R., and Gilman, A. G. (1992). G-Proteins. *Trends Biochem. Sci.* 17:383–387. An advanced review of the G-protein family.

Linder, M. E., and Gilman, A. G. (1992). G-Proteins. *Sci. Am.* 267(7):56–65. An introduction to G proteins.

Michell, R. H. (1992). Inositol lipids in cellular signalling mechanisms. *Trends Biochem. Sci.* 17:274–276. A short review of recent advances in the field of signalling via inositol phospholipids.

Walton, K. M., and Dixon, J. E. (1993). Protein tyrosine phosphatases. *Annu. Rev. Biochem.* 62:101–120. A detailed review of intracellular and receptor tyrosine phosphatases.

13

Digestion

Living organisms universally obtain nourishment from their environment. Phototrophs, of course, thrive on little more than water, sunlight, and air. The needs of other organisms are more complex, and a variety of means have evolved for obtaining energy and molecular building blocks from the environment (Figure 13·1). The pitcher plant, for example, traps insects and digests them. Ruminants, such as cows, rely on the metabolic capacity of endosymbiotic microorganisms to subsist on grass, which is otherwise indigestible. The young of many species are sustained by food processed by their parents. In some cases, offspring are fed partially digested material from the parent's diet; in other cases, mammary glands produce a specific mixture of nutrients, delivered in the form of milk. We saw in Chapter 1 that a variety of inorganic ions are required by all living things. We have also seen that certain organic compounds—vitamins—are required by some species (Chapter 8).

Figure 13·1
Mechanisms for obtaining nutrients. (a) The pitcher plant traps insects and digests them. (b) Cows digest grass with the aid of endosymbiotic microorganisms. (c) Many birds, including penguins, regurgitate partially digested food to feed their offspring.

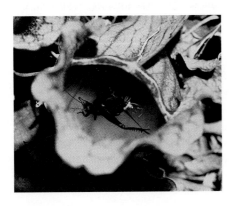

This chapter is not concerned with micronutrients such as essential ions and vitamins but with the dietary compounds that are major nutrients.

Regardless of its origin, the food we eat is converted into energy and precursors of macromolecules. Much of our food is in the form of biopolymers. In the digestive system, or **gastrointestinal tract,** biopolymers such as starch and proteins are hydrolyzed enzymatically to their monomeric units. Only small molecules, such as monosaccharides and amino acids, can be absorbed by the cells of the intestinal wall. Once absorbed, the smaller compounds are transported through the bloodstream and lymphatic system to the cells that use them. Dietary fats (triacylglycerols) are also broken down to smaller units—fatty acids and monoacylglycerols—that are absorbed and then reassembled into triacylglycerols for transport. **Digestion** refers to the process of hydrolyzing dietary macromolecules to smaller molecules that can be absorbed by the body.

Many of the first biochemical processes to be described were reactions of digestion. For example, digestive enzymes, such as chymotrypsin (Sections 6·11 and 7·18), were among the first enzymes characterized. Kinetic experiments with digestive enzymes delineated their substrate specificities and the optimal conditions for their activity, which match the substrates and the physiological conditions that the enzymes meet in the digestive process. The relatively simple biochemistry of digestion serves as a good conclusion to our discussion of the structures of biomolecules and an introduction to the more complex features of cellular metabolism, which is the subject of Part 3 of this book. Our examination of digestion will show how essential end products are obtained through coordination of sequential enzymatic reactions.

The enzymatic reactions and the organs described in this chapter are those involved in the conversion of dietary macromolecules to smaller molecules and their absorption by the human body. Similar enzymatic processes occur in many other species. Fungi and bacteria, for example, also secrete enzymes that catalyze the hydrolysis of biopolymers to monomeric compounds that can be absorbed.

13·1 Digestion Involves Catalyzed Hydrolysis of Biopolymers

Digestive hydrolysis of biopolymers is necessary because polymers themselves usually cannot cross cell membranes. Dietary polymers must therefore be broken down into small molecules that can enter intestinal cells. An organism can then synthesize its own biopolymers from a pool of monomers or their precursors that are acquired from food.

In Chapter 1, we noted that considerable energy in the form of ATP is required for synthesis of biopolymers. However, hydrolysis of biopolymers does not require the input of energy, nor does it yield any usable energy. Cellular energy is only obtained by further metabolism of the products of digestion. Although the hydrolysis of biological macromolecules is spontaneous (thermodynamically favored), the molecules are stable in the absence of the proper hydrolases.

13·2 Digestion Requires Enzymes from the Mouth, Stomach, Pancreas, and Intestine

The major macromolecular components of food are carbohydrates, proteins, and lipids. Although enzymes that catalyze the hydrolysis of DNA and RNA are present in the digestive tract, nucleic acids are not major nutrients. Starch, usually the principal dietary carbohydrate, is hydrolyzed to glucose units; proteins are hydrolyzed

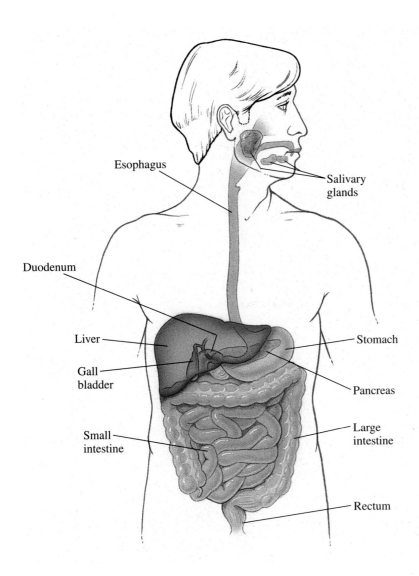

Figure 13·2
Gastrointestinal tract.

Esophagus

Salivary glands

Duodenum

Liver

Gall bladder

Small intestine

Stomach

Pancreas

Large intestine

Rectum

to their constituent amino acids; and fats are hydrolyzed to fatty acids and mono-acylglycerols. The products of digestion may be further metabolized to release energy or to be incorporated into new macromolecules.

In this chapter, we will follow the steps of digestion from the entry of food into the mouth through hydrolysis and absorption, and we will see the specialized role of each digestive organ in the overall process (Figure 13·2 and Table 13·1). The end products of digestion are the ultimate sources of the substrates for many of the pathways described in later chapters. After we have examined all of the major pathways of cellular metabolism, we will return to the subject of interorgan cooperation in Chapter 23, when mammalian metabolism is summarized.

A. Overview of the Gastrointestinal Tract

In the mouth, food is mechanically disrupted by chewing and is moistened by saliva. Saliva is an aqueous solution of ions (mainly $Na^{\oplus}$, $K^{\oplus}$, $Cl^{\ominus}$, and $HCO_3^{\ominus}$), mucins, and other proteins. (Mucins are large, carbohydrate-rich glycoproteins that increase the viscosity of the fluids in which they are dissolved; Section 9·11A.) A salivary amylase and a lipase from glands in the tongue begin the hydrolysis of starch and fats, respectively. Once food is swallowed, it is mixed with hydrochloric acid (HCl) in the stomach. The HCl (usually 0.01–0.1 M, or pH 1–2) denatures

Table 13·1 Functions of the digestive organs

Organ	Major digestive roles
Salivary glands	Provide fluid and amylase
Tongue	Provides lipase
Stomach	Provides HCl and pepsin
Pancreas	Provides enzymes, zymogens, and $NaHCO_3$
Liver	Synthesizes bile salts
Gall bladder	Stores bile
Small intestine	Completes digestion; absorbs end products
Large intestine	Absorbs water, ions, and short-chain acids

Figure 13·3
Villi of the small intestine. (**a**) Cross section of the small intestine. Villi of the intestinal mucosa project into the lumen of the small intestine. (**b**) Longitudinal section of part of a villus, showing columnar epithelial cells and microvilli. Each villus consists of a layer of epithelial cells and a central core of connective tissue. Microvilli, consisting of infolded membranes that provide a large surface area, project from columnar epithelial cells into the lumen of the small intestine, collectively forming an absorptive brush border. Embedded in the connective core are lymphatic vessels and blood capillaries.

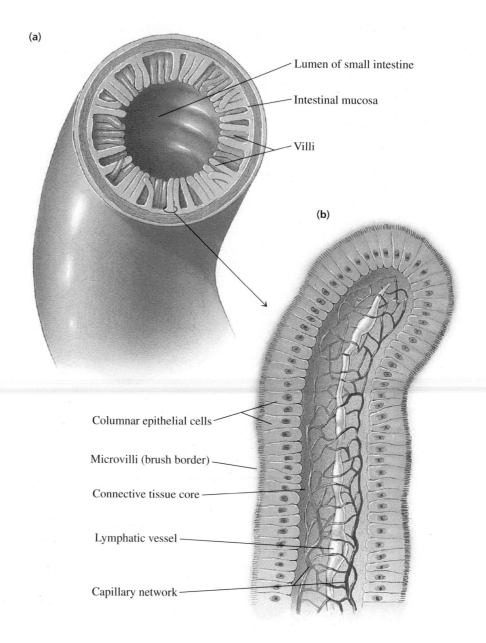

(a)

Lumen of small intestine

Intestinal mucosa

Villi

(b)

Columnar epithelial cells

Microvilli (brush border)

Connective tissue core

Lymphatic vessel

Capillary network

dietary protein and also helps kill ingested bacteria. Pepsin, a protease from the stomach, has high activity in this acidic medium. Pepsin catalyzes the hydrolysis of proteins to large peptide fragments and a few amino acids. The resulting mixture of food and secretions, called **chyme,** is liquefied by gastric juice and the kneading action of the stomach. At this point, hydrolysis of all three food components has begun.

Chyme then passes to the first segment of the small intestine, the duodenum, where it encounters secretions from the pancreas and the liver. The pancreas supplies two major secretions to the duodenum: large quantities of hydrolases and zymogens that are activated to hydrolases, and sufficient bicarbonate to neutralize the chyme. Bile, a suspension originating in the liver and stored in the gall bladder, also enters the duodenum. Bile contains bile salts, natural detergents that aid in the absorption of lipids. In the small intestine, a glycosidase, a lipase, a phospholipase, and several proteases catalyze the major hydrolytic reactions of digestion but do not complete the job. Epithelial cells of the small intestine possess hydrolases on their outer surface that virtually complete the digestion of polymers to monomers. Most digestive hydrolysis and absorption takes place in the upper small intestine.

The small intestine is long (4–6 meters) and has a huge surface area since its inner surface, also called the mucosal surface, is covered with many protrusions termed villi. The surface area is further increased by projections of the plasma membranes of epithelial cells, called microvilli, which form a brushlike coat called the brush border on the surface of villi (Figure 13·3). The large surface area of the small intestine, on the order of $300 \, m^2$, allows rapid absorption of the end products of digestion by the epithelial cells. Monosaccharides, amino acids, and small peptides are transported to the blood, and fatty acids and monoacylglycerols are transported to the lymphatic vessels. Some water is also absorbed in the small intestine. After undigested material leaves the small intestine and enters the large intestine, bacteria effect further degradation of the remaining food polymers, producing gases such as CH_4, CO_2, and H_2, as well as short-chain carboxylic acids that can be absorbed. Water, bile salts, and some ions are also absorbed by the large intestine, and then the remaining solid material passes from the gastrointestinal tract in the feces.

Before examining the digestion of each group of nutrients, we will describe several biochemical aspects of gastrointestinal function, namely production of acid by the stomach, secretion of pancreatic juice and its role in digestion, and the bacterial flora of the large intestine.

B. Gastric Juice

When stimulated, the parietal cells in the stomach wall (also called oxyntic cells) secrete gastric juice, an aqueous solution containing mucus, the zymogen pepsinogen, ions, and $0.15 \, M$ HCl, into a channel that leads to the lumen of the stomach. Several transport proteins take part in the release of the strong acid HCl, including a P-type $H^{\oplus}$-$K^{\oplus}$ ATPase and an anion exchange protein that exchanges extracellular $Cl^{\ominus}$ for intracellular $HCO_3^{\ominus}$. The actions of these transporters resemble the actions of the $Na^{\oplus}$-$K^{\oplus}$ ATPase and the anion exchange protein of erythrocytes, respectively (Section 12·7). The sequence of reactions involved in HCl secretion is shown in Figure 13·4. Parietal cells are rich in mitochondria, which generate the energy and CO_2 required for HCl secretion.

Three stimuli—acetylcholine, gastrin, and histamine—can activate the $H^{\oplus}$-$K^{\oplus}$ ATPase and thus initiate the secretion of HCl from the parietal cells. The neurotransmitter acetylcholine is released after stimulation of the vagus nerve by the sight, smell, or thought of food. Acetylcholine triggers the release of gastric juice that contains all of its solutes. Both the 17-residue polypeptide hormone gastrin and

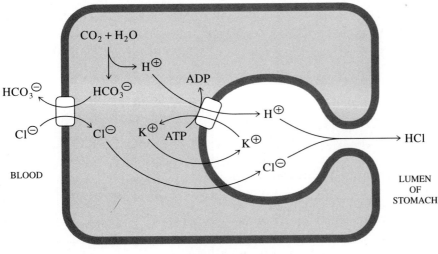

PARIETAL CELL

Figure 13·4
Production of HCl by a parietal cell. $H^{\oplus}$ and $HCO_3^{\ominus}$ are derived from CO_2 and H_2O inside the cell. $HCO_3^{\ominus}$ exits to the blood plasma in exchange for $Cl^{\ominus}$. $K^{\oplus}$ and $Cl^{\ominus}$, both at high concentrations in the parietal cell, enter the lumen of the stomach. $K^{\oplus}$ is transported back into the cell in exchange for $H^{\oplus}$ in an ATP-dependent transport reaction. The net reaction of the ion-transport system is the release of HCl into the lumen of the stomach.

(a)

$$CH_2-CH_2-\overset{\oplus}{N}H_3$$

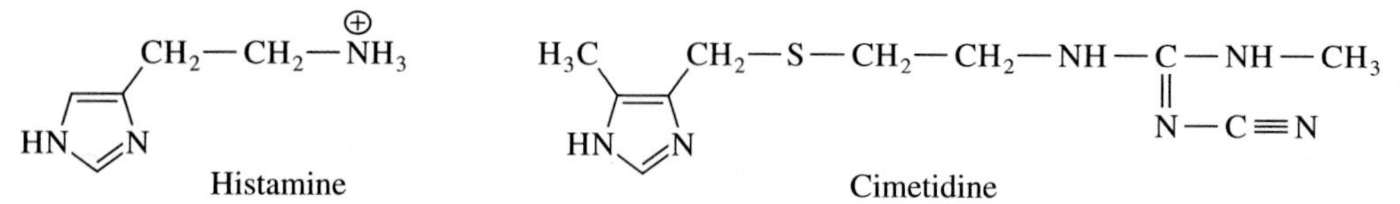

HN—N

Histamine

(b)

$$H_3C \qquad CH_2-S-CH_2-CH_2-NH-\underset{\underset{N-C\equiv N}{\|}}{C}-NH-CH_3$$

HN—N

Cimetidine

Figure 13·5
Histamine and cimetidine. **(a)** Histamine stimulates secretion of acid. **(b)** Cimetidine competes with histamine for binding to the histamine receptor and is used as a drug to diminish acid secretion.

the amine histamine stimulate release mostly of HCl and water from parietal cells. Histamine (Figure 13·5a) is the most potent stimulant of acid secretion.

The stomach tissue itself is normally resistant to the hydrochloric acid secreted in the gastric juice. A layer of mucus (a viscous mixture of water and mucins, containing secreted bicarbonate to neutralize the HCl) that coats the cells that line the stomach is a barrier to attack by $H^\oplus$. Mucus also covers most of the other surfaces of the gastrointestinal tract. However, the strong acid in the stomach may cause or irritate peptic ulcers, which are holes in the stomach tissue. Recent research suggests that ulcers may form as a result of chronic infection by the bacterium *Helicobacter pylori*. *Helicobacter* thrives in the nutrient-rich mucus, while the host's chronically activated immune system leads to inflammation and damage of nearby tissue. A common treatment for peptic ulcers is administration of the drug cimetidine (Figure 13·5b), a histamine analog that competitively inhibits the binding of histamine to its receptors on the parietal cells. Cimetidine binds to but does not activate these receptors and therefore interferes with secretion of HCl. A recent and apparently more effective treatment for some ulcers is administration of antibiotics to eliminate the bacterial infection and chronic inflammation that may ultimately be responsible for the ulcers. The stomach wall can also be injured by exposure to acetylsalicylic acid (aspirin) and ethanol and by bile salts that come up the gastrointestinal tract from the intestine. The ulcerogenic actions of aspirin include disruption of the mucus barrier, damage to the epithelial cells, and inhibition of bicarbonate secretion. To avoid irritating the stomach, many individuals now use slow-acting, coated aspirins that do not dissolve until they reach the intestine.

Figure 13·6
Pancreas. **(a)** Location of the pancreas. Special cells of the duodenum, the first part of the small intestine, secrete hormones that stimulate secretion of zymogens and bicarbonate from the pancreas. **(b)** Arrangement of the pancreatic acini, groups of cells surrounding tiny tubes that lead to the pancreatic duct system. The acinar cells synthesize zymogens and enzymes and release them to the pancreatic duct.

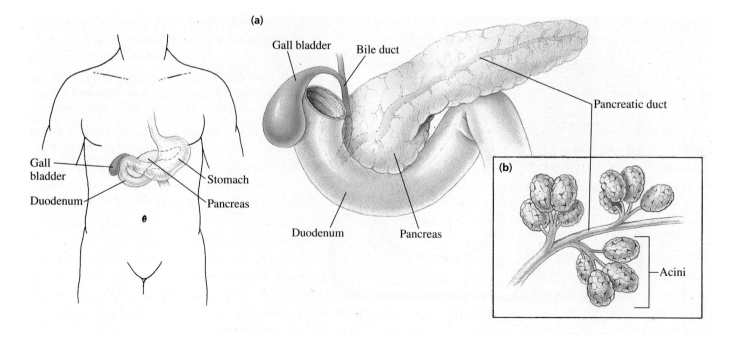

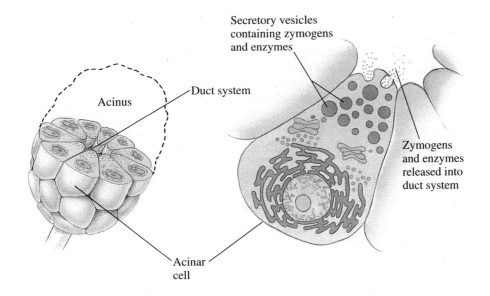

C. Pancreatic Juice

In humans, the pancreas lies near the back wall of the abdominal cavity. Many pancreatic cells are organized in bundles, called acini, that are shaped like bunches of grapes. The acinar cells are clustered to form tiny tubes that lead to larger channels and eventually to the pancreatic duct. The pancreatic duct extends the length of the pancreas and empties into the duodenum (Figure 13·6).

In the acinar cells, zymogens and enzymes are synthesized and then sequestered in secretory vesicles called zymogen granules. When the cells are stimulated, the vesicles fuse with the plasma membrane and release a solution containing sodium chloride, zymogens, and enzymes into the pancreatic duct system (Figure 13·7). Table 13·2 lists the enzymes that originate in the pancreas. Note that those enzymes whose action could damage the pancreatic cells (proteases and phospholipase) are stored and released as zymogens.

The pancreas is also responsible for secretion of sodium bicarbonate. The solution that carries the electrolytes, zymogens, and enzymes through the pancreatic duct to the duodenum is called the pancreatic juice. $NaHCO_3$, which can reach concentrations of $0.12\,M$, is needed to neutralize the HCl from the stomach. The duodenum also secretes $NaHCO_3$. The pH of the fluid in the duodenum is usually between 6 and 8.

Table 13·2 Pancreatic enzymes

Zymogen	Active enzyme	Substrate
	α-Amylase	Starch and glycogen
	Ribonuclease	RNA
	Deoxyribonuclease	DNA
Chymotrypsinogen	Chymotrypsin	Peptides
Trypsinogen	Trypsin	Peptides
Proelastase	Elastase	Peptides
Procarboxypeptidases	Carboxypeptidases	Peptides
	Lipase	Triacylglycerols
Prophospholipase A_2	Phospholipase A_2	Glycerophospholipids

Two polypeptide hormones produced by endocrine cells of the duodenum enter the bloodstream and trigger the pancreatic secretions. Cholecystokinin stimulates the release of $NaHCO_3$, zymogens, and enzymes, as well as the emptying of bile from the gall bladder. Cholecystokinin is released from the small intestine in response to the presence of peptides and fatty acids. Secretin, the other hormone, is released when the pH of the extracellular fluid drops below 5. Secretin stimulates the secretion of pancreatic juice that is rich in $NaHCO_3$. The presence of peptides and fatty acids, as well as the low pH, thus coordinates the arrival of partially digested food from the stomach with the release of hydrolases and $NaHCO_3$ into the small intestine.

The lumen of the duodenum is the site of most of the digestive hydrolytic reactions. The pH of the contents of the small intestine is near neutral, suitable for the activity of the enzymes produced by the pancreas. In the fluid of the lumen, pancreatic enzymes—including activated zymogens—catalyze the hydrolysis of starch, proteins, and triacylglycerols.

The final hydrolytic reactions of digestion occur at the intestinal brush border. The brush-border cells have a few oligosaccharidases that catalyze digestion of dietary disaccharides and any oligosaccharides produced by the action of amylase. The brush-border cells also contain a great variety of peptidases, most bound to the luminal surface of the cells and some located in the cytoplasm. The concerted action of these enzymes results in the formation of monosaccharides, free amino acids, and di- and tripeptides that can be absorbed by the intestinal cells.

D. Bacteria of the Large Intestine

The large intestine absorbs water and salts and salvages a few nutrients from undigested food. Unlike other digestive organs, the large intestine does not secrete any enzymes. Instead, enzymes produced by bacteria living in the large intestine are responsible for metabolizing some of the remaining food.

There are more than 400 bacterial species living in the large intestine of an adult human. Most of the bacteria, called enteric bacteria, are facultative anaerobes. The most familiar of these organisms is *Escherichia coli*. *E. coli* cells can divide every 20 minutes under optimal laboratory conditions, but the bacterial generation time in the intestine is much longer, probably about one to four divisions per day. Nonetheless, the concentration of bacteria in the large intestine is high. From one-third to one-half the dry weight of the feces is derived from viable bacteria. The tremendous growth potential of enteric bacteria is limited in the upper portion of the small intestine by gastric juice, whose low pH is inhospitable to bacterial growth. In the lower portion of the small intestine, the bacterial population is limited by the constant forward movement of material by peristalsis—the rhythmic, propulsive action of the intestinal smooth muscle.

The bacterial flora of the large intestine degrade undigested disaccharides to monosaccharides and thence to carboxylic acids such as acetate, propionate, and butyrate, which are then absorbed. Some ammonia, produced by microbial degradation of protein, is also absorbed by the host. Vitamin K (Section 8·15D), which is required for the proper function of the blood-coagulation enzymes, is synthesized by intestinal microorganisms. Because the establishment of intestinal flora in newborn infants requires several days, and because vitamin K does not cross the placental barrier in large amounts, newborns have little of the vitamin in their bodies at birth and are susceptible to hemorrhagic disorders. For this reason, vitamin K is often administered soon after birth.

We will now examine the specific chemical reactions that occur during digestion and absorption of individual nutrients.

13·3 Digestion of Carbohydrates Occurs in Stages

Carbohydrates supply about half the calories in the typical North American diet. Starch, the major polysaccharide of plants, supplies about half of the carbohydrates. Sucrose and lactose are the other major dietary carbohydrates. Glycogen is mostly degraded during food preparation, so it does not concern us here. Enzymatic digestion of carbohydrates occurs in stages at three locations: in the mouth, in the lumen of the small intestine, and at the brush border. The first steps of carbohydrate digestion in the mouth and intestinal lumen involve only starch. The final stage is the digestion of oligosaccharides at the brush border.

Digestion of starch commences in the mouth with the action of salivary α-amylase. The size of starch polymers is greatly diminished by hydrolysis in the mouth. Some further degradation occurs by nonenzymatic hydrolysis in the stomach. In the lumen of the small intestine, starch becomes a mixture of maltose, maltotriose, and limit dextrins as a result of the action of pancreatic α-amylase. The salivary and pancreatic amylases catalyze the same reaction and have the same substrate specificity, and both have a pH optimum near neutrality. The reaction catalyzed by the amylases, hydrolysis of the glycosidic bond of α-$(1 \rightarrow 4)$-linked glucose polymers, is shown in Figure 13·8. α-Amylases are endoglycosidases; that is, the bonds cleaved are internal bonds of the polymer. Amylases do not catalyze the cleavage of the α-$(1 \rightarrow 6)$ bonds that occur at the branch points of amylopectin (Section 9·8A). The products of amylase-catalyzed hydrolysis are oligosaccharides. For example, the final products of hydrolysis of the linear starch amylose are the disaccharide maltose and the trisaccharide maltotriose. The action of amylase on the branched substrate amylopectin also produces maltose and maltotriose. However, because the α-$(1 \rightarrow 4)$ glycosidic bonds near the α-$(1 \rightarrow 6)$ branch points are somewhat resistant to hydrolysis, a mixture containing limit dextrins—varying in size from about 5 to 10 glucose residues—is eventually formed. Figure 13·9 (next page) illustrates the action of amylase on the two types of starch.

The final digestion of oligosaccharides takes place at the brush border of the small intestine. The surface of the brush border contains a series of membrane-bound oligosaccharidases (Table 13·3, next page). These hydrolases, present in the glycocalyx of the cell, are all large glycoproteins (M_r 90 000–220 000). Together, their actions convert sucrose, lactose, and all the starch-derived oligosaccharides to monosaccharides. The enzyme α-limit dextrinase catalyzes the cleavage of larger limit dextrins to smaller limit dextrins containing three or four saccharide units. The

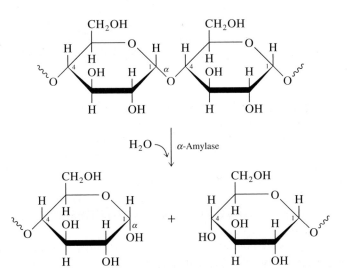

Figure 13·8
Reaction catalyzed by α-amylase. The substrate is an α-$(1 \rightarrow 4)$-linked glucose polymer. Hydrolysis generates a new reducing end (the residue on the left), an α anomer.

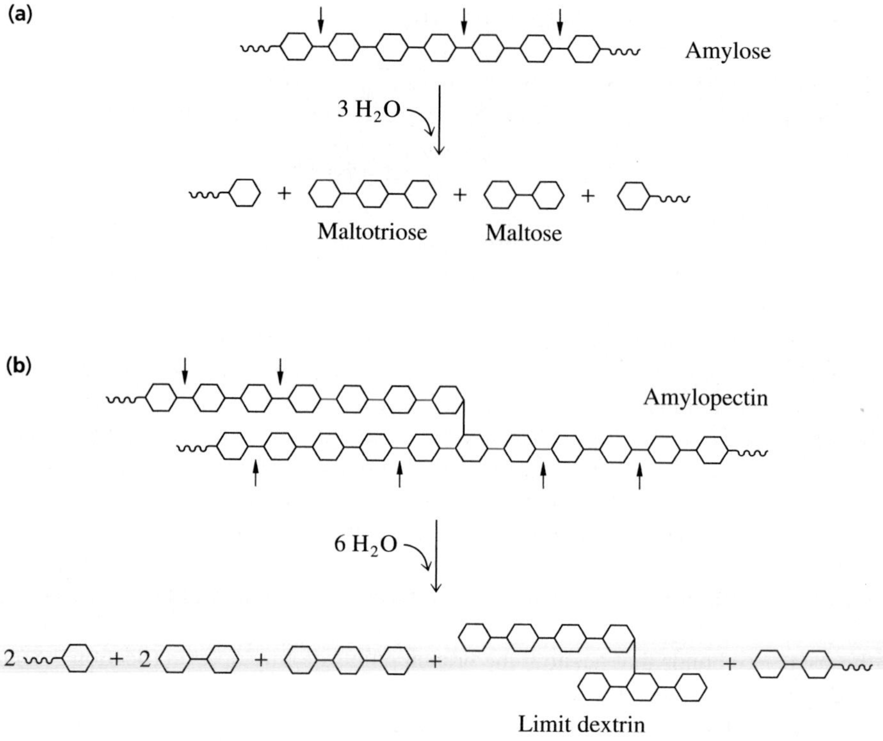

Figure 13·9
Action of α-amylase on amylose and amylo-pectin. Glucose residues are indicated by hexagons. In this figure, α-(1→4) glycosidic bonds are horizontal and α-(1→6) glycosidic bonds are vertical. Only small portions of the substrates are shown. **(a)** In amylose, α-(1→4) bonds are hydrolyzed to produce maltotriose and maltose. **(b)** In amylopectin, all α-(1→4) bonds are easily hydrolyzed except the seven shown in red, which are more resistant to attack. The α-(1→6) bond is not hydrolyzed.

(a)

Amylose

$3\,H_2O$

Maltotriose   Maltose

(b)

Amylopectin

$6\,H_2O$

Limit dextrin

Table 13·3 Oligosaccharidases of the brush border

Enzyme	Substrate	Products
α-Limit dextrinase	Larger limit dextrins	Glucose, residual oligosaccharides
Sucrase-isomaltase	Maltose, maltotriose, tri- or tetrasaccharide limit dextrins	Glucose
Sucrase	Sucrose	Glucose, fructose
Lactase	Lactose	Glucose, galactose
Trehalase	Trehalose	Glucose

hydrolysis of the smaller limit dextrins, maltotriose, and maltose to glucose is catalyzed by sucrase-isomaltase, a complex enzyme with two active sites. This enzyme can complete the digestion of starch to glucose. Other less well characterized glucosidases may also contribute to the final hydrolysis of oligosaccharides.

Sucrose undergoes sucrase-catalyzed hydrolysis to glucose and fructose, and lactose is hydrolyzed to glucose and galactose by the β-galactosidase activity of lactase. The brush border also contains the enzyme trehalase, which catalyzes the hydrolysis of the nonreducing disaccharide trehalose, two glucose molecules linked by an α-(1→α-1) bond. Trehalose occurs in mushrooms and insects. The glucose monomers, but not trehalose, are absorbed by cells of the villi and enter the bloodstream.

13·4 Certain Carbohydrates Are Not Digested

Some carbohydrates are not digested in the small intestine. These indigestible carbohydrates fall into two categories: dietary fiber and small oligosaccharides. They are often termed unavailable carbohydrates because they do not become available to us as monosaccharides. Any metabolism of indigestible carbohydrates, such as hydrolysis of polymers to monosaccharides or smaller compounds, is performed by enteric bacteria. Because cells of the large intestine have no transport system for the absorption of monosaccharides, these compounds do not contribute significantly to human nutrition.

A. Dietary Fiber

In the past two decades, an interest in dietary fiber has arisen from suggestions that adding fiber to the diet may protect against conditions including chronic constipation, colon cancer, and obesity. Dietary fiber, which is not hydrolyzed during digestion, consists primarily of polysaccharides from plant cell walls and includes cellulose, hemicelluloses (polymers containing β-$(1 \rightarrow 4)$-linked galactose and mannose residues), and pectins (α-$(1 \rightarrow 4)$ chains of polygalacturonic acid, containing some rhamnose). Lignin, a nonsaccharide phenolic polymer, is also a component of dietary fiber. The typical fiber content of edible plants is about 25% cellulose and almost 70% noncellulose polysaccharide. These biopolymers have traditionally been considered digestively inert and were described as roughage. Actually, some cellulose and appreciable amounts of hemicelluloses and pectins are hydrolyzed by bacteria in the lumen of the large intestine and are metabolized to gases and short-chain acids that are absorbed by the intestinal cells. Certain forms of starch that escape digestion in the small intestine are similarly degraded by bacteria in the large intestine.

The effects of dietary fiber on the intestine are quite complex and not yet fully understood. Dietary fiber causes changes in the morphology of the villi of the small intestine, increases the rate of turnover of some cells, increases the rate of mucin synthesis, and alters the activities of intestinal surface disaccharidases. Insoluble fibers such as cellulose, which increase the bulk of feces and decrease the time needed for the stool to travel through the large intestine, probably affect digestion or absorption of nutrients and may also decrease the time of exposure to any dietary carcinogens that may cause intestinal cancer. Other fibers, such as pectin and lignin, have the additional property of binding or sequestering lipid compounds such as bile acids and cholesterol, thereby decreasing their absorption. Dietary fiber also appears to decrease the absorption of water-soluble nutrients such as monosaccharides and trace elements. The purported benefits of dietary fiber may therefore depend on additional dietary factors and the overall nutritional state of the individual.

Fresh fruits and vegetables, whole wheat, and bran are good sources of dietary fiber. In addition, some people consume an indigestible pentose polymer prepared from *Psyllium* seeds to add bulk to the stool.

B. Indigestible Oligosaccharides

Some oligosaccharides that are not substrates for the hydrolases of the small intestinal mucosa are fermented by bacteria in the large intestine. These fermentation reactions can result in flatulence (gas in the intestine) and sometimes diarrhea (from the osmotic effect of the short-chain acids, which draw water from mucosal cells into the intestinal lumen). Indigestible oligosaccharides that may elicit digestive disturbances include lactose and certain plant oligosaccharides. The disturbances can now be treated by enzyme supplements.

Most adults are deficient in lactase, the enzyme that catalyzes the conversion of lactose to glucose and galactose, resulting in lactose intolerance. The prevalence of lactase deficiency is 70–95% in most parts of the world. Northern European populations and their descendants are among the notable exceptions. In lactase-deficient

Figure 13·10
Structures of sucrose, raffinose, stachyose, and verbascose. Except for sucrose, these plant oligosaccharides are indigestible by mammals. (Abbreviations: Fru, fructose; Gal, galactose; Glc, glucose.)

Sucrose	Glc α-(1→β2) Fru
Raffinose	Gal α-(1→6) Glc α-(1→β2) Fru
Stachyose	Gal α-(1→6) Gal α-(1→6) Glc α-(1→β2) Fru
Verbascose	Gal α-(1→6) Gal α-(1→6) Gal α-(1→6) Glc α-(1→β2) Fru

individuals, lactose is metabolized by bacteria in the large intestine, with the production of gases such as CO_2 and H_2 and short-chain acids. The acids can cause diarrhea by increasing the ionic strength of the intestinal fluid. Although some individuals retain the ability to digest lactose throughout life, the majority of humans undergo a reduction in the level of lactase at about five to seven years of age.

One treatment for lactose intolerance is a commercially prepared enzyme supplement that contains β-galactosidase from a microorganism. This enzyme is specific for the same glycosidic bond hydrolyzed by lactase. β-Galactosidase supplements can be taken when milk or foods that contain milk products are ingested. Alternatively, milk pretreated with β-galactosidase can be purchased. Some lactose-intolerant individuals can tolerate yogurt, in which the lactose has been partially hydrolyzed by the action of the endogenous β-galactosidase of the microorganism *Lactobacillus bulgaricus* in the yogurt culture.

Considerable amounts of the trisaccharide raffinose and its relatives, the tetrasaccharide stachyose and the pentasaccharide verbascose, are present in certain plants (Figure 13·10). These three oligosaccharides are α-(1→6) derivatives of sucrose and are especially abundant in peas and beans. Because they have no endogenous α-galactosidase, mammals are incapable of digesting these oligosaccharides. Intestinal bacteria that contain this enzyme activity may ferment the monosaccharides. Ingestion of the raffinose family of oligosaccharides therefore leads to formation of the gases H_2 and CO_2 in the large intestine. Stachyose and verbascose produce the most gas. A commercial preparation of α-galactosidase from the mold *Aspergillus niger* can be eaten along with the first few bites of offending foods so that the substrates of α-galactosidase are degraded in the early stages of digestion. Some people are thus able to eat beans, cabbage, and onions without intestinal distress.

One indigestible oligosaccharide, sucralose, is marketed as a low-calorie sweetener. Sucralose is synthesized from sucrose by substitution of three hydroxyl groups by chlorine atoms (Figure 13·11). Sucralose is approximately 600 times sweeter than sucrose but has a similar taste. It is heat stable and therefore can be used in cooked foods. Sucralose is calorie free because it cannot be hydrolyzed as it passes through the body.

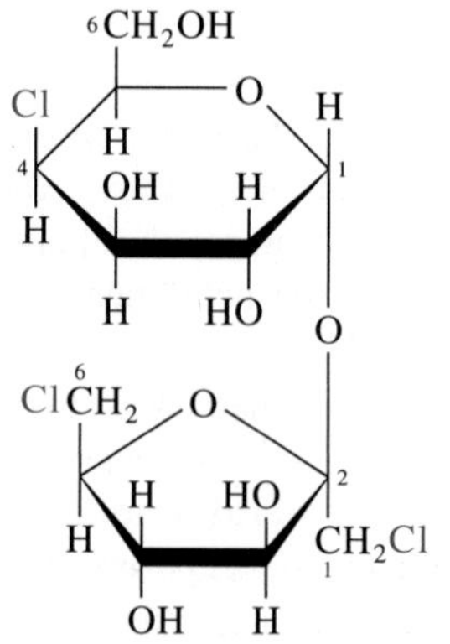

Figure 13·11
Structure of sucralose (1,6-dichloro-1,6-dideoxy-β-D-fructofuranosyl-4-chloro-4-deoxy-α-D-galactopyranoside).

13·5 Proteins Are Digested in Stages

After protein-containing foods are disrupted by chewing and moistened with saliva, they mix with the hydrochloric acid in the stomach and are denatured. Protein digestion begins in the stomach under acidic conditions and continues in the small intestine.

A. Hydrolysis of Proteins Begins in the Stomach

Digestion of dietary proteins is initiated by hydrolysis catalyzed by pepsin, a protease stored by gastric mucosa cells in the form of the zymogen pepsinogen. Pepsinogen is released in response to stimulation by neurotransmitters and a number of other agents. When exposed to the acidic pH of the gastric juice, pepsinogen undergoes slow autoactivation to the active enzyme. Activation involves proteolytic cleavage of a peptide from the N-terminus of the zymogen. This proteolysis is catalyzed by another molecule of pepsinogen, which has some catalytic activity under acidic conditions. The rate of pepsinogen activation increases with time as fully active pepsin molecules catalyze formation of more pepsin molecules from pepsinogen.

Pepsin has a pH optimum near 2; the two midpoints of the bell-shaped pH-rate profile coincide with pK_a values of 1 and 4.5. These values correspond to two catalytically important residues: Asp-32, which is unusually acidic, and Asp-215. Pepsin is one example of a group of proteases—termed **aspartic proteases,** or acid proteases—that have similar catalytic centers and similar mechanisms. All of the aspartic proteases have a pH optimum of about 2 to 4. The involvement of aspartyl groups at the active site of pepsin originally suggested that covalent catalysis was likely, with formation of an acid anhydride composed of an aspartate carboxyl group and the carbonyl group of the scissile peptide bond. However, no covalent enzyme-substrate intermediate has been detected, and a number of experiments indicate that the mechanism includes acid-base catalysis. A proposed mechanism is shown in Figure 13·12. A water molecule is held in the active site by hydrogen bonds to the side chains of the aspartic acid and aspartate residues. The ionized aspartate acts as a base catalyst, removing a proton from the water molecule. This leads to formation of an unstable intermediate, which decomposes to the amine and carboxylic acid products.

Pepsin exhibits fairly broad substrate specificity. It has a preference for peptide bonds involving the amino group of leucine, tryptophan, phenylalanine, or tyrosine residues, but it slowly catalyzes the hydrolysis of some other peptide bonds. Collagen, which is present in large amounts in meat, is a particularly good substrate for pepsin. The product of pepsin action in the stomach is a mixture of peptides that passes into the small intestine. The peptides have a regulatory role in digestion. As noted earlier, their presence in the small intestine stimulates release of the hormone cholecystokinin, which triggers pancreatic secretion.

B. Hydrolysis of Proteins Continues in the Small Intestine

The pancreatic zymogens, after delivery from the acinar cells of the pancreas to the duodenum, become hydrolases that catalyze the further digestion of dietary proteins. The peptide bonds of dietary proteins are accessible to the intestinal proteases since the proteins were denatured by the stomach acid. All of the proteases formed from pancreatic zymogens are active only at neutral pH. Therefore, it is important that the chyme from the stomach be neutralized by $NaHCO_3$. When stimulated by acid in the lumen, the epithelial cells of the duodenum secrete bicarbonate. The increase in pH in the duodenum relative to the stomach also inactivates pepsin.

In addition to the serine proteases trypsin, chymotrypsin, and elastase (whose specificities were discussed in Section 6·11C), two carboxypeptidases from the pancreatic juice are active in the duodenum. These exopeptidases are zinc-containing

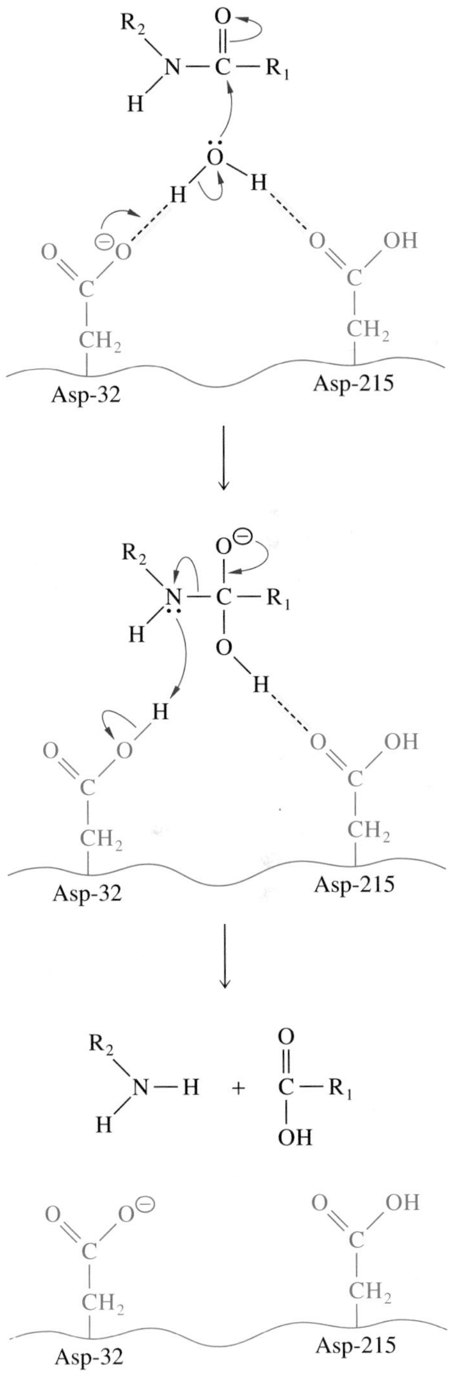

Figure 13·12
Proposed mechanism for peptide-bond cleavage catalyzed by pepsin. The carboxylate group of Asp-32 removes a proton from the bound water molecule. The resulting hydroxide ion attacks the carbonyl carbon atom of the peptide bond to form an unstable tetrahedral intermediate. After the proton is transferred to the nitrogen atom of the peptide bond, the tetrahedral intermediate decomposes to form the amine and carboxylic acid products.

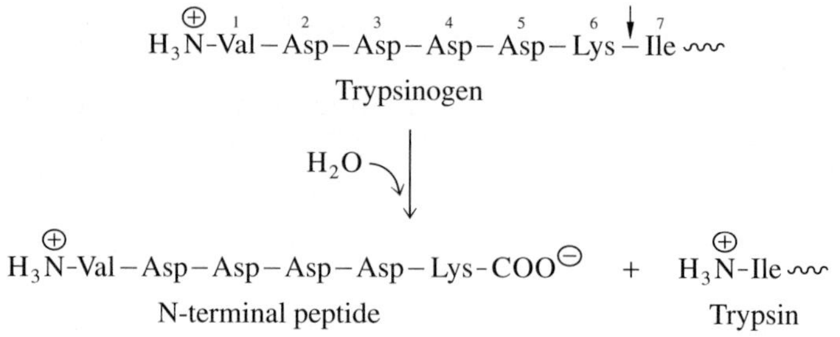

Figure 13·13
Activation of trypsinogen. Hydrolysis of the Lys-6—Ile-7 bond is catalyzed by enteropeptidase or trypsin.

$$\overset{\oplus}{H_3}N-\overset{1}{Val}-\overset{2}{Asp}-\overset{3}{Asp}-\overset{4}{Asp}-\overset{5}{Asp}-\overset{6}{Lys}\overset{\downarrow}{-}\overset{7}{Ile}\sim$$

Trypsinogen

$$H_2O \searrow \downarrow$$

$$\overset{\oplus}{H_3}N-Val-Asp-Asp-Asp-Asp-Lys-COO^{\ominus} \quad + \quad \overset{\oplus}{H_3}N\text{-}Ile\sim$$

N-terminal peptide Trypsin

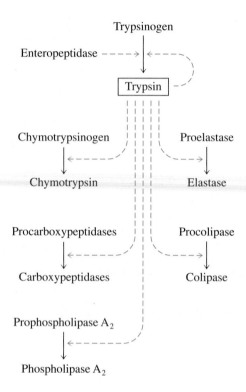

Trypsinogen

Enteropeptidase

Trypsin

Chymotrypsinogen Proelastase

Chymotrypsin Elastase

Procarboxypeptidases Procolipase

Carboxypeptidases Colipase

Prophospholipase A$_2$

Phospholipase A$_2$

Figure 13·14
Activation of pancreatic zymogens in the duodenum. Enteropeptidase initiates this process by catalyzing the activation of trypsinogen to trypsin. Trypsin is the common activator of several pancreatic zymogens, including trypsinogen. Trypsin also activates procolipase, releasing colipase, an activator of lipase (Section 13·6).

metalloenzymes that catalyze the hydrolysis of the C-terminal peptide bond of protein substrates. The two carboxypeptidases are specific for different peptide bonds. The combined activities of the pancreatic proteases produce a mixture of amino acids and peptides. About 60–70% of the protein-derived compounds in the small intestine are peptides containing from two to six residues; the remaining compounds are free amino acids.

In the duodenum, the brush-border protease enteropeptidase (formerly called enterokinase) is activated in response to the arrival of bile salts. Enteropeptidase, like trypsin, catalyzes cleavage of peptide bonds on the carboxyl side of lysine and arginine residues. It can therefore catalyze the specific cleavage of the Lys-6—Ile-7 bond of trypsinogen, which leads to active trypsin (Figure 13·13). Once activated by the removal of its N-terminal hexapeptide, trypsin proteolytically activates the other pancreatic zymogens, including additional trypsinogen molecules (Figure 13·14). The formation of active trypsin by the action of enteropeptidase can be viewed as the master activation step. (The molecular events in the activation of chymotrypsinogen, which leads to the formation of its substrate-binding site, are described in Section 6·11A.)

The amount of enzyme protein secreted by the pancreas is large: 20–30 grams per day in a healthy adult. The enzymes, after they serve their catalytic purposes, are themselves digested, and most of their amino acids are absorbed by the body for reuse. Another endogenous source of digestible protein is dead intestinal mucosa cells. These cells, which are replaced very rapidly, account for about 30 grams of protein per day. Since only about 70–100 grams of dietary (exogenous) protein are digested per day, recycled enzymes and dead cells represent a significant portion of absorbed amino acids.

C. Oligopeptides Are Hydrolyzed by Brush-Border and Intestinal Cytoplasmic Peptidases

The oligopeptides and amino acids from the lumen of the small intestine diffuse to the intestinal mucosa cells, where peptides larger than tripeptides are further hydrolyzed by the action of a series of brush-border peptidases. These membrane-bound aminopeptidases and dipeptidases appear to complement the pancreatic proteases in specificity. For example, the brush-border peptidases include enzymes that catalyze the hydrolysis of proline residues from the N- or C-termini of peptides. The cytosol of the mucosal cells also contains several dipeptidases and a tripeptidase that catalyze hydrolysis of peptides absorbed by the cells. Tripeptides are hydrolyzed by the action of both brush-border and cytoplasmic enzymes. The actions of the battery of gastric, pancreatic, and mucosal proteases thus reduce ingested proteins to a mixture of mostly amino acids and some dipeptides and tripeptides.

D. Inhibitor Proteins Block the Activity of Serine Proteases

A general feature of serine proteases is their regulation by inhibitor proteins. For example, premature activation of the pancreatic zymogens is prevented by a molecule called trypsin inhibitor. This protein, synthesized by pancreatic cells, protects

Figure 13·15
Stereo views of the trypsin–trypsin inhibitor complex. **(a)** Space-filling model showing trypsin in gold and the inhibitor in blue. **(b)** Backbone model of trypsin (gold) and complete structure of trypsin inhibitor (blue), with dot surfaces showing van der Waals contacts between residues at the enzyme-inhibitor interface (color key: carbon, green; nitrogen, blue; oxygen, red). Lys-15 of the inhibitor binds to trypsin at the same site that a normal substrate would bind. The very close match between the two proteins makes the binding of the pancreatic trypsin inhibitor to trypsin one of the strongest known noncovalent interactions between protein molecules. (Based on coordinates provided by R. Huber and J. Deisenhofer.)

(a)

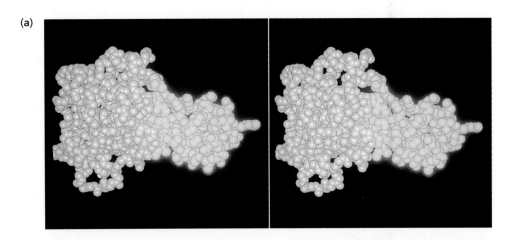

(b)

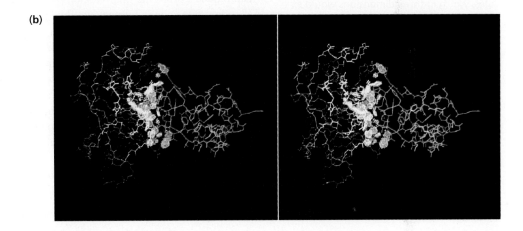

the cells from the destruction that might occur if even a single molecule of trypsinogen were prematurely activated. The potency of trypsin stems not only from its own proteolytic activity but also from the proteolytic activity of the zymogens it activates. Trypsin inhibitor, a small protein (M_r 6000), binds noncovalently but extremely tightly to the active site of trypsin, forming an enzyme-inhibitor complex with a dissociation constant of 10^{-13} M. This is one of the strongest known noncovalent interactions between protein molecules. The inhibitor protein, an analog of the peptide substrates of trypsin, is bound by one of its lysine residues to the active site of trypsin. The molecular structure of the inhibitor matches the shape of the active site of the enzyme so well that no water can enter the cleft to cause hydrolysis of the inhibitor (Figure 13·15). Other low-molecular-weight protease inhibitors, as well as larger inhibitors, are found in virtually all mammalian tissues. These inhibitors are believed to limit the activity of proteases that are active in the tissues under certain circumstances, such as during tissue growth, wound repair, and infection.

13·6 Lipases Catalyze the Digestion of Fats

Most of the lipids in a typical North American diet—about 90%—are triacylglycerols; phospholipids and cholesterol account for about 10%. Digestion of fats starts slowly in the mouth, where fat-containing foods are broken up by chewing and where an acid-stable lipase secreted by the tongue catalyzes limited hydrolysis. This enzyme has a pH optimum of 4 and retains some activity in the stomach.

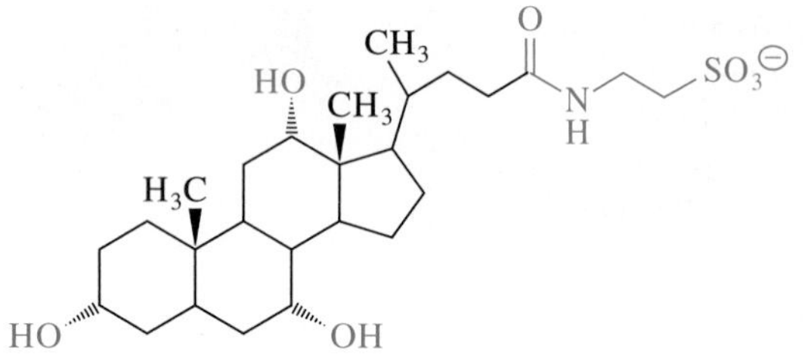

Taurocholate

Glycocholate

Figure 13·16
Bile salts. Taurocholate and glycocholate, conjugates of cholic acid with taurine and glycine, respectively, are the most abundant bile salts in humans. Bile salts are amphipathic. The hydrophilic parts of the molecules, which in three dimensions would extend below the plane of the page, are shown in blue. The other parts of the molecules are hydrophobic.

Digestion of dietary lipids occurs predominantly in the intestine when chyme, which includes suspended fat particles, arrives from the stomach. In the small intestine, the fat particles are coated with bile salts.

Bile salts are synthesized from cholesterol in the liver and are conveyed to the gall bladder. From the gall bladder, they are secreted into the duodenum. The most abundant bile salts in humans are the amides taurocholate and glycocholate, which are the conjugates of cholic acid with taurine (3-sulfinyl-2-aminopropionate) and glycine, respectively (Figure 13·16). Bile salts are amphipathic lipids that play an important role in the absorption of dietary lipids. Because triacylglycerols have limited solubility in bile-salt micelles, bile salts coat rather than dissolve fat particles. However, the products of triacylglycerol hydrolysis (free fatty acids and monoacylglycerols) are more soluble in bile-salt micelles and form **mixed micelles**—micelles formed from more than one type of amphipathic molecule. These micelles transport the fatty acids and monoacylglycerols to the intestinal wall, where they are absorbed.

Triacylglycerols in the chyme emulsion are subjected to degradation by the action of pancreatic lipase. This protein is synthesized and stored as an active enzyme

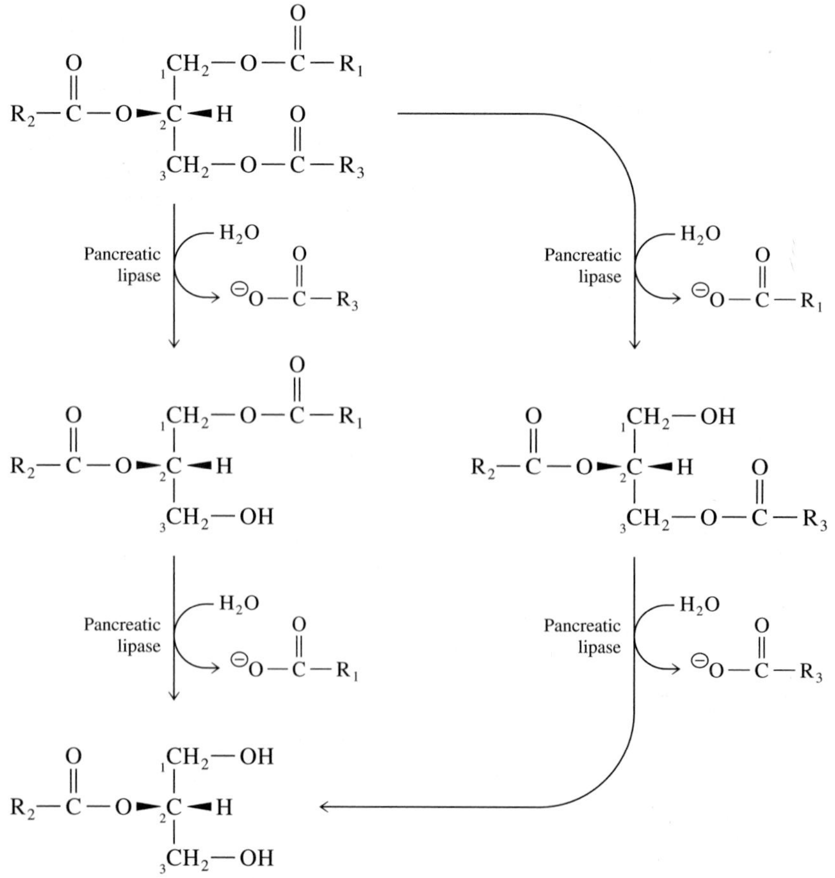

Figure 13·17
Action of pancreatic lipase. The enzyme catalyzes removal of C-1 and C-3 acyl chains, producing free fatty acids and 2-monoacylglycerol. 1,2-Diacylglycerol and 2,3-diacylglycerol are intermediates in the pathway.

by the acinar cells of the pancreas and enters the duodenum in the pancreatic juice. Pancreatic lipase is a serine esterase that catalyzes the hydrolysis of the ester bonds at the C-1 and C-3 positions of triacylglycerols, producing free fatty acids, the intermediates 1,2-diacylglycerol and 2,3-diacylglycerol, and finally, 2-monoacylglycerol (Figure 13·17). Pancreatic lipase has maximal activity near pH 7. The enzyme has two domains: a 133-residue C-terminal domain and a 333-residue N-terminal domain that contains the active site. The bile salts coating the fat particles prevent the direct binding of lipase. However, the soluble enzyme is able to bind and act upon its lipid substrate in the presence of the protein cofactor colipase. Colipase is synthesized in the pancreas as its precursor procolipase, which is activated when a short N-terminal peptide is removed by the action of trypsin in the intestine.

X-ray crystallographic analyses of the lipase-procolipase complex, both alone and in the presence of phosphatidylcholine and bile salt, have been performed. The studies show that one face of procolipase binds to the C-terminal domain of lipase by noncovalent bonds. The opposite face of procolipase has three mobile hydrophobic loops that can interact with a lipid surface. This hydrophobic interaction brings the lipase into contact with its lipid substrate and also induces a conformational change in the lipase, thereby opening an α-helical region that acts as a lid that covers the active site (Figure 13·18). When the lid is opened, a hydrophobic surface

(a)

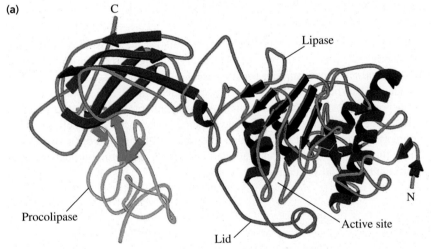

(b)

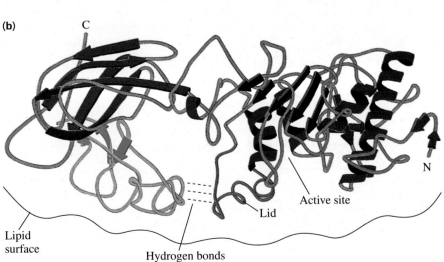

Figure 13·18
Lipase-procolipase complex. **(a)** Procolipase (blue) binds to the C-terminal domain of pancreatic lipase. The enzyme is inactive because a lid (red) covers the active site. **(b)** When three loops of the procolipase protein interact with a lipid surface, lipase undergoes a conformational change in which the lid is opened by formation of hydrogen bonds with procolipase. This movement exposes the active site and makes it fully functional. These models are based on X-ray analysis of a lipase-procolipase complex; it is believed that the lipase-colipase complex is similar.

is created at the entrance of the active site. In addition, lid opening exposes a serine-containing catalytic triad and an oxyanion hole similar to those of the serine proteases (Section 7·18). In this manner, the nonpolar lipid substrate is able to enter the active-site cleft and be hydrolyzed by the action of the water-soluble enzyme. Figure 13·18 is based on a complex containing procolipase, but the complex containing colipase is probably similar.

13·7 The Products of Digestion Are Absorbed by the Small Intestine

The final phase of nutrient assimilation is the absorption of the digestion products—monosaccharides, amino acids, small peptides, and lipids. The water-soluble nutrients are absorbed by active transport. Transport of lipids within cells and between tissues via the bloodstream must overcome the physical difficulties associated with the extremely low solubility of lipids in aqueous solutions.

A. Hexoses, Amino Acids, and Peptides Are Absorbed by Specific Transport Proteins

Several membrane transport proteins are responsible for the transport of monosaccharides across the plasma membrane of the microvilli. One transporter is specific for glucose and galactose, another for fructose. Glucose, the most abundant monosaccharide, is absorbed by cotransport with $Na^{\oplus}$ (Section 12·7D). Sodium ions are then transported back into the lumen in an ATP-dependent manner by the action of $Na^{\oplus}$-$K^{\oplus}$ ATPase. Monosaccharides leave the opposite face of the cell by a passive, uniport sugar-transport system (also referred to as facilitated diffusion) that is driven by the sugar concentration gradient. The sugars enter the bloodstream and flow through the portal vein directly to the liver.

Figure 13·19
Pathways for the absorption of hexoses, amino acids, and peptides. The small molecules are transported from the intestinal lumen into mucosal cells via transporters (shown as squares). Absorbed hexoses exit to the blood by passive transport. Absorbed di- and tripeptides are largely converted to amino acids for export to the blood, also by passive transport. [Adapted from Alpers, D. H. (1987). Digestion and absorption of carbohydrates and proteins. In *Physiology of the Gastrointestinal Tract,* 2nd ed., L. R. Johnson, ed. (New York: Raven Press), pp. 1469–1483.]

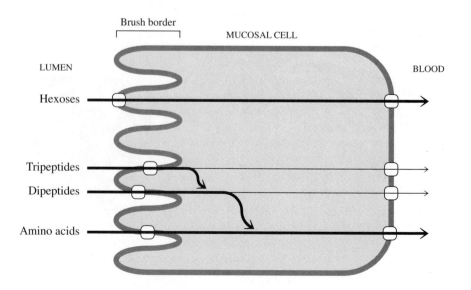

About a dozen amino acid transport systems have been recognized. Most involve cotransport of amino acids with sodium ions. These multiple transporters have overlapping specificities. For example, at least four transporters are specific for the neutral amino acids. In addition, dipeptides and tripeptides are absorbed by transport systems distinct from those for amino acids. After their absorption, most peptides are hydrolyzed by the action of the cytoplasmic di- and tripeptidases. The amino acids and remaining di- and tripeptides then enter the portal bloodstream by passive transport. Figure 13·19 summarizes the pathways for absorption of the water-soluble nutrients.

Several genetic disorders are associated with defects in amino acid transport. In individuals with these disorders, the concentrations of certain amino acids are elevated in the urine. Many small compounds, including nutrients and metabolic wastes, are filtered out of the blood as it passes through the kidneys. Some nutrients, such as amino acids, are recovered by the action of specific transport mechanisms before the urine leaves the kidney. A defective amino acid transporter does not function properly during either absorption by the intestinal mucosa or reabsorption by the tubules of the kidney. Hartnup's disease results from a defect in a transporter of neutral amino acids. The disease is characterized by general aminoaciduria, and the concentration of tryptophan and other neutral amino acids in the urine is especially high. The symptoms of Hartnup's disease resemble those of the vitamin-deficiency disease pellagra (Section 8·6), probably because there is not sufficient tryptophan available for biosynthesis of the coenzymes $NAD^{\oplus}$ and $NADP^{\oplus}$. Individuals with the disorder respond well to administration of nicotinic acid.

In cystinuria, the transporter of cysteine and the basic amino acids is defective. The defect leads to elevated excretion of the amino acids arginine, lysine, and ornithine, and cystine formed from the easily oxidized cysteine. Cystine, which is less soluble than cysteine, may crystallize in the form of kidney stones. Because there are multiple transporters for amino acids and because peptides can still be transported, malnutrition is not necessarily a symptom of amino acid transporter defects.

B. Absorption of Lipid Nutrients Requires Bile Salts

In the lumen of the small intestine, free fatty acids and monoacylglycerols produced by the hydrolysis of triacylglycerols mix with bile salts to form mixed micelles. The products of fat hydrolysis are delivered in these micelles to the intestinal wall, where the micelles dissociate and the fatty acid and monoacylglycerol molecules enter the cell by permeating the plasma membrane, although a transporter may also be involved. Cholesterol is absorbed by diffusion. Inside the cell, a fatty acid carrier protein (M_r 12 000) binds fatty acids that are absorbed from the lumen of the intestine. The lipid-protein complex fuses with the endoplasmic reticulum, where the fatty acids are converted to fatty acyl CoA molecules.

$$\text{Fatty acid} + \text{CoASH} + \text{ATP} \longrightarrow \text{Acyl CoA} + \text{AMP} + \text{PP}_i$$

(13·1)

A monoacylglycerol may combine with two acyl CoA molecules to form a triacylglycerol (Figure 13·20a, next page). Alternatively, the fatty acyl CoA derivatives can be incorporated into triacylglycerols by a series of reactions that uses glycerol 3-phosphate as the acyl acceptor (Figure 13·20b, next page).

The water-insoluble triacylglycerols combine with absorbed cholesterol (some of which is esterified) and specific apoproteins to form aggregates known as chylomicrons. The chylomicrons accumulate in the Golgi apparatus, are packaged into secretory vesicles, and leave the cell by exocytosis. Chylomicrons transported out

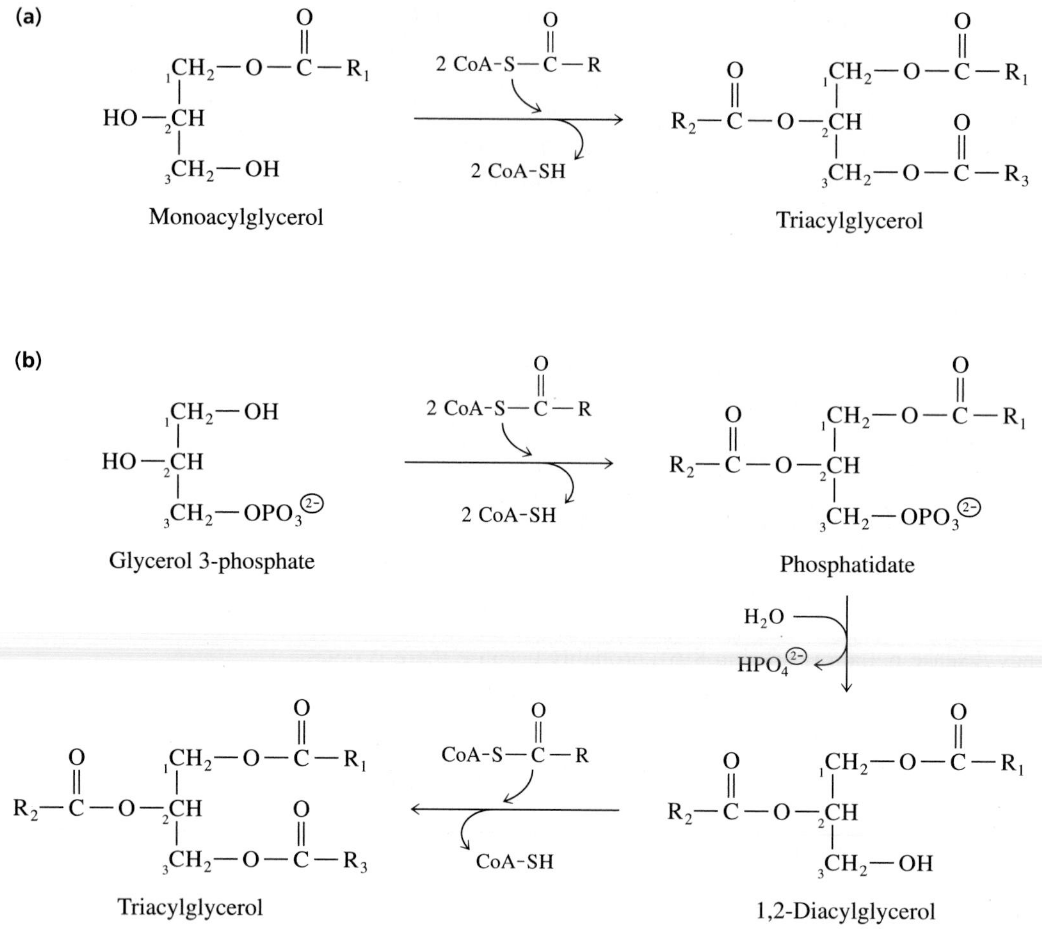

(a)

(b)

Figure 13·20
Re-esterification of fatty acyl CoA molecules and monoacylglycerols. (a) Formation of triacylglycerol from monoacylglycerol and two acyl CoA molecules. (b) Formation of triacylglycerol from three fatty acyl CoA molecules and glycerol 3-phosphate.

of intestinal cells pass through the lymphatic system before entering the bloodstream for delivery to the tissues. The steps of fatty acid and monoacylglycerol absorption are summarized in Figure 13·21.

The fate of dietary phospholipids is similar to the fate of triacylglycerols. Pancreatic phospholipases secreted into the intestine catalyze removal of fatty acids from phospholipids present in food. The major phospholipase present in the pancreatic juice is phospholipase A_2, which catalyzes hydrolysis of the ester bond at C-2 of a glycerophospholipid to form a lysophosphoglyceride and a fatty acid (Figure 13·22). Phospholipase A_2 is secreted by the pancreas in the form of a zymogen, prophospholipase A_2, which is proteolytically activated by trypsin. Phospholipase A_2 requires the presence of bile salts in order to catalyze hydrolysis of glycerophospholipids. Lysophosphoglycerides are absorbed through the brush-border membrane and are re-esterified to glycerophospholipids in the intestinal epithelial cells.

After the products of lipid hydrolysis are released from the mixed micelles, bile-salt micelles reenter the intestinal lumen. Most of the bile salts (over 90%) are absorbed from the material in the lower portion of the small intestine. The bile salts are conserved by recirculation through the intestine, portal blood, and liver in what is called the enterohepatic circulation. This routing involves several membrane-bound transporters in both the intestinal and liver cells. Bile salts circulate through the liver and intestine several times during the digestion of a single meal. The small

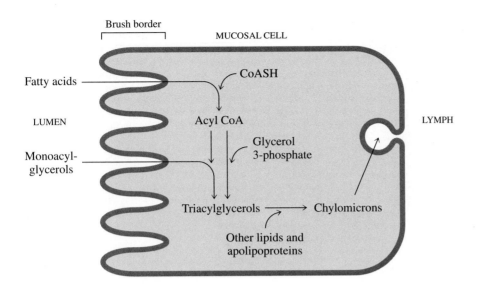

Figure 13·21
Pathway of absorption of fatty acids and monoacylglycerols. Fatty acids and monoacylglycerols are absorbed and used to synthesize triacylglycerols. Triacylglycerols, other lipids, and proteins are assembled into chylomicrons, which exit the cell by exocytosis.

amounts of bile salts that are not extracted from the intestine are excreted in the feces. This elimination of bile salts represents a major route for disposal of their precursor, cholesterol. In addition to bile salts, bile contains free cholesterol, some of which is also excreted in the feces. Some dietary cholesterol is excreted when epithelial cells in the intestine, after absorbing cholesterol, slough off from the mucosal wall.

C. Some Lipids Are Transported in Lipoproteins

Because they are essentially insoluble in water, cholesterol and its esters, as well as the nonpolar triacylglycerols, cannot be transported in blood as free molecules. Instead, these lipids are complexed with phospholipids and amphipathic apolipoproteins to form particles known as **lipoproteins.** A variety of lipoproteins are found in human blood plasma. Lipoproteins are macromolecular assemblies with hydrophobic cores and hydrophilic surfaces. The core contains triacylglycerols and cholesteryl esters; the surface consists of amphipathic molecules: cholesterol, phospholipids, and the apolipoproteins (Figure 13·23, next page).

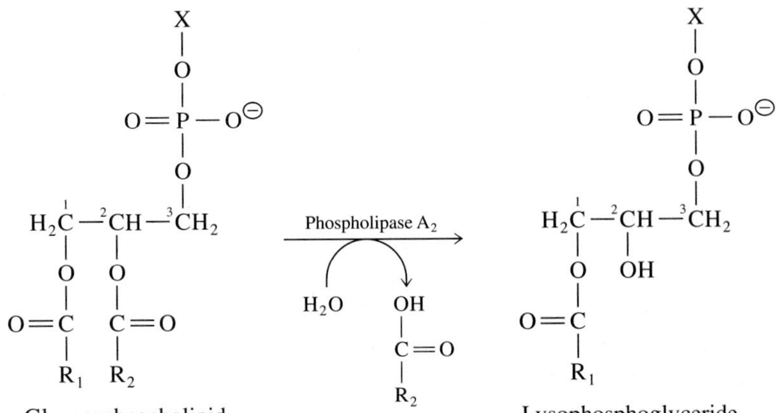

Figure 13·22
Action of phospholipase A_2. The reaction produces a lysophosphoglyceride and a free fatty acid.

Glycerophospholipid → Phospholipase A_2 → Lysophosphoglyceride

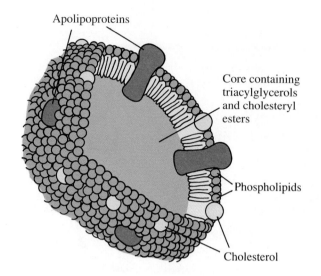

Apolipoproteins

Core containing
triacylglycerols
and cholesteryl
esters

Phospholipids

Cholesterol

Figure 13·23
Structure of a lipoprotein.
A lipoprotein contains a
core of neutral lipids,
which include triacylglycerols and cholesteryl es-
ters. The core is coated
with a monolayer of phos-
pholipids in which apolipoproteins and cholesterol
are embedded.

Lipoproteins are classified according to their relative densities. Lipids are less dense than proteins; therefore, the greater the lipid content of a lipoprotein, the lower its density (Table 13·4). The major classes of human lipoproteins are the chylomicrons, which carry triacylglycerols and cholesterol from the small intestine to the tissues; the very low density lipoproteins (VLDL), which carry endogenous triacylglycerols, cholesterol, and cholesteryl esters from the liver (their primary site of synthesis) to the tissues; the low density lipoproteins (LDL), which are formed during the breakdown of VLDL and are enriched in cholesterol and cholesteryl esters; and the high density lipoproteins (HDL), which are enriched in protein and transport cholesterol and cholesteryl esters from tissues back to the liver.

About a dozen apolipoproteins have been identified as constituents of lipoprotein complexes. The concentrations of particular apolipoproteins may vary by as much as 1000-fold between individuals. This variation has implications for human health, since high concentrations of certain lipoproteins are associated with increased risk of cardiovascular disease. There are also multiple genetic forms of some apolipoproteins. Much attention has been focused on these variants, since one of them, an apolipoprotein E variant, is associated with increased incidence of Alzheimer's disease, a neurodegenerative disorder.

Table 13·4 Characteristics of lipoproteins in human plasma

	Chylomicrons	VLDL	LDL	HDL
Molecular weight $\times 10^{-6}$	>400	5–6	2.3	0.18–0.36
Density (g cm^{-3})	<1.006	0.95–1.006	1.006–1.063	1.063–1.210
Chemical composition (%)				
Triacylglycerol	85	50	10	4
Free cholesterol	1	7	8	2
Cholesteryl ester	3	12	37	15
Phospholipid	9	18	20	24
Protein	2	10	23	55

[Adapted from Kritchevsky, D. (1986). Atherosclerosis and nutrition. *Nutr. Int.* 2:290–297.]

Summary

The major human nutrients are carbohydrates (mostly in the form of starch), proteins, and fats. Each of these biomolecules must be hydrolyzed so that its components can be assimilated. The hydrolysis, or digestion, of food polymers occurs at several locations in the gastrointestinal tract, mediated by enzymes that function under conditions that vary among the different digestive organs. The greatest amount of hydrolysis takes place in the small intestine.

Starch is partially hydrolyzed in the mouth by the action of salivary α-amylase. The action of a second α-amylase, which is synthesized in the pancreas and secreted into the small intestine, converts the oligosaccharides from starch into a mixture of maltose, maltotriose, and limit dextrins. Enzymatic hydrolysis of these intermediates and of the dietary disaccharides sucrose and lactose to monosaccharides is catalyzed by a series of glycosidases that are bound to the plasma membrane of intestinal cells. Dietary fiber and certain oligosaccharides pass through the small intestine undigested. Enzymes produced by bacteria growing in the large intestine are involved in the metabolism of some of these undigested carbohydrates.

The initial step in the digestive hydrolysis of proteins occurs in the stomach. There, the zymogen pepsinogen is activated to pepsin, an aspartic protease with an acidic pH optimum. The pancreas secretes a group of zymogens that, upon activation by limited proteolysis in the small intestine, become trypsin, chymotrypsin, elastase, and two carboxypeptidases. The product of the combined actions of these enzymes is a mixture of amino acids and peptides containing from two to six residues. Peptidases of the intestinal mucosa catalyze further hydrolysis of peptides.

Fats (triacylglycerols) are hydrolyzed to fatty acids and 2-monoacylglycerols by the action of a lipase in the mouth and pancreatic lipase. When colipase—which forms a complex with pancreatic lipase—binds to lipid, the active site of lipase becomes accessible to its substrates.

Monosaccharides, amino acids, and di- and tripeptides are absorbed by intestinal mucosa cells by cotransport with $Na^{\oplus}$. Most of the di- and tripeptides are hydrolyzed in the mucosal cells. The sugar and amino acid monomers then enter the portal blood. In the intestine, the products of fat hydrolysis are taken up by bile-salt micelles and delivered by these carriers to the mucosa, where the hydrolyzed products are absorbed. The products of fat hydrolysis are reassembled into triacylglycerols. Lipids are packaged in lipoprotein particles for transport from mucosal cells through the lymph.

Selected Readings

General References

Caspary, W. F. (1992). Physiology and pathophysiology of intestinal absorption. *Am. J. Clin. Nutr.* 55:299S–308S.

Johnson, L. R., ed. (1987). *Physiology of the Gastrointestinal Tract,* 2nd ed. (New York: Raven Press). An excellent two-volume review of the physiology and biochemistry of digestion and absorption of nutrients.

Specific Topics

Büller, H. A., and Grand, R. J. (1990). Lactose intolerance. *Annu. Rev. Med.* 41:141–148.

Cristofaro, E., Mottu, F., and Wuhrmann, J. J. (1974). Involvement of the raffinose family of oligosaccharides in flatulence. In *Sugars in Nutrition,* H. L. Sipple and K. W. McNutt, eds. (New York: Academic Press), pp. 313–336.

Erickson, R. H., and Kim, Y. S. (1990). Digestion and absorption of dietary protein. *Annu. Rev. Med.* 41:133–139.

Kassell, B., and Kay, J. (1973). Zymogens of proteolytic enzymes. *Science* 180:1022–1027. Describes the inherent proteolytic activity of zymogens such as pepsinogen.

Keil, B. (1971). Trypsin. In *The Enzymes,* Vol. 3, 3rd ed., P. D. Boyer, ed. (New York: Academic Press), pp. 249–275.

Thoma, J. A., Spradlin, J. E., and Dygert, S. (1971). Plant and animal amylases. In *The Enzymes,* Vol. 5, 3rd ed., P. D. Boyer, ed. (New York: Academic Press), pp. 115–189.

van Tilbeurgh, H., Egloff, M.-P., Martinez, C., Rugani, N., Verger, R., and Cambillau, C. (1993). Interfacial activation of the lipase-procolipase complex by mixed micelles revealed by X-ray crystallography. *Nature* 362:814–820.

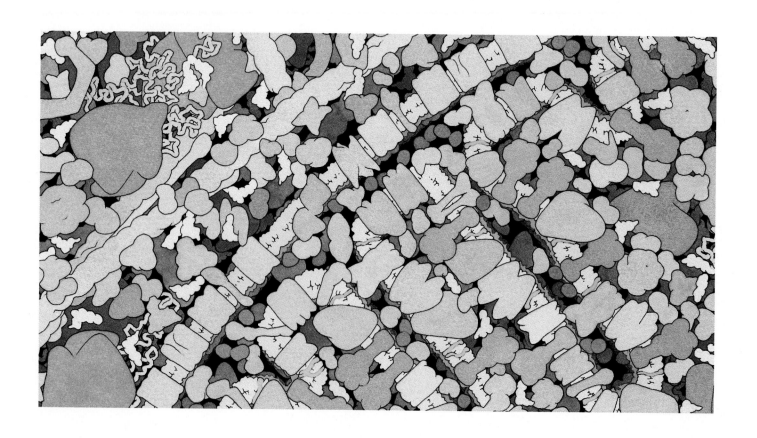

Part Three

Metabolism and Bioenergetics

14

Introduction to Metabolism

The major emphasis of the first two parts of this book has been description of the structures and functions of the major components of living cells—from small molecules to polymers to larger aggregates such as membranes.

But living cells are not static assemblies of molecules. Although some cell components are stable, others turn over rapidly. By turnover, we mean they are continually synthesized and degraded even though their concentrations may remain constant. This third part of the book focuses on the biochemical activities that account for the assimilation, transformation, degradation, and synthesis of many of the cellular components already described. In short, we will move from a static view of the structure of the cell to consideration of the dynamics of cell function. Understanding the biochemical basis for the many functions that set living cells apart from nonliving structures might be considered a daunting task. We can meet this challenge by taking a stepwise approach that builds upon the foundations established in the first two sections. In this chapter, we will discuss some general themes of metabolism and the thermodynamic principles that underlie cellular activities.

Figure 14·1
Anabolism and catabolism. Biopolymers, including those obtained from food, are catabolized to release monomeric building blocks and energy. Chemical or solar energy is used in the synthesis of macromolecules and in the performance of cellular work.

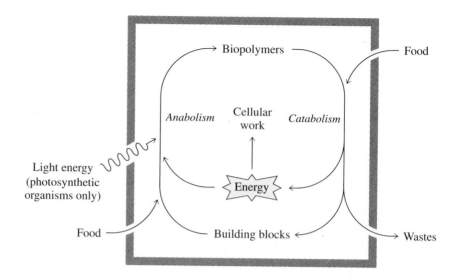

14·1 Metabolism Is the Sum of Cellular Reactions

In the broadest terms, **metabolism** may be described as all of the chemical reactions that occur within living cells. (The term *intermediary metabolism* is sometimes used to refer to the reactions involving the small molecules of cells.) It is convenient to consider separately those reactions that degrade complex molecules (catabolic reactions) and those that synthesize molecules (anabolic reactions). Catabolic reactions degrade molecules to liberate smaller molecular building blocks as well as energy (Figure 14·1). Living cells harness the released energy to drive anabolic reactions, which synthesize the molecules needed for cell maintenance and growth. Cells also use captured energy for the performance of other kinds of cellular work, such as transport across membranes and cell movement.

Whether we observe microbes or large multicellular organisms, we find a bewildering variety of biological adaptations. It is fascinating to think that there may be more than 10 million species alive on the earth—and that several hundred million species may have come and gone through the course of evolution. Within

(a)

(b)

Figure 14·2
Hummingbird and tortoise. Many metabolic activities occur in all living organisms, including the small and swift hummingbird (**a**) and the large, slow-moving tortoise (**b**).

multicellular organisms, there is striking specialization of cell types or tissues. Despite this extraordinary diversity, the biochemistry of living cells reveals a surprising degree of similarity not only in the chemical composition and structure of cellular components but also in the metabolic routes by which the components are modified (Figure 14·2).

In general terms, the common themes of cellular function include the following:

1. Cells maintain specific internal concentrations of ions, metabolic intermediates, and enzymes. Cell membranes universally provide the physical barrier that segregates cell components from the environment.

2. Cells extract energy from external sources in order to drive energy-consuming reactions. Energy may be derived from the conversion of solar energy to chemical energy or from the ingestion of energy-yielding compounds.

3. Cells grow and reproduce according to a blueprint encoded in the genetic material.

4. Cells respond to environmental influences. Cell activities must be geared to the availability of energy. When the supply of energy from the environment is limited, energy demands may be met by using internal stores or by slowing metabolic rates as in hibernation, stasis, sporulation, or seed formation.

Since the vast majority of metabolic reactions are facilitated by enzymatic catalysis, a complete description of metabolism must include not only the reactants, intermediates, and products of cellular reactions but also the characteristics of the enzymes that catalyze the reactions. Most cells are able to perform hundreds to thousands of reactions. In the face of such complexity, how do we structure our approach so that we can deal with the information in manageable pieces and in an orderly manner? Our strategy is to subdivide metabolism into segments or branches. We begin by considering separately the metabolism of the four major groups of biomolecules—amino acids, carbohydrates, lipids, and nucleotides. Furthermore, within each of the four branches of metabolism, we recognize a number of distinct sequences of metabolic reactions, which we call **pathways.** A pathway can be regarded as the biological equivalent of a synthetic or degradative scheme in organic chemistry. However, defining the limits of a metabolic pathway is not straightforward. Unlike chemical synthesis in the laboratory, which has an obvious conclusion, metabolism usually leads to products that undergo further change. It is important to consider the physiological context of the sequence of reactions in order to establish plausible start and end points for the pathway. For example, you might examine the pathway of glycolysis (Chapter 15) and ask, where does glycolysis—the degradation of sugar—begin and end? Does it begin with polysaccharides (such as glycogen and starch), extracellular glucose, or intracellular glucose? Does the pathway end with pyruvate, lactate, or ethanol? Be aware that start and end points may be assigned somewhat arbitrarily, often according to tradition or for ease of study. Indeed, in Chapter 23 we will see examples in which pathways are linked to form extended metabolic routes that are more difficult to characterize.

We will also see that individual metabolic pathways can take different forms. A linear metabolic pathway, such as the biosynthesis of proline, consists of a series of enzyme-catalyzed reactions in which the product of one reaction (i.e., a metabolite) is the substrate for the next reaction (Figure 14·3, next page). A cyclic metabolic pathway, such as the citric acid cycle (Chapter 16), also consists of sequential

Glutamate

↓

γ-Glutamyl phosphate

↓

Glutamate γ-semialdehyde

↓

Δ^1-Pyrroline 5-carboxylate

↓

Proline

Figure 14·3
Linear metabolic pathway. The biosynthesis of proline is an example of a linear metabolic pathway. The product of each step is the substrate for the next step in the pathway.

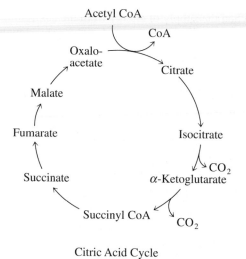

Citric Acid Cycle

Figure 14·4
Cyclic metabolic pathway. The sequence of reactions in a cyclic pathway forms a closed loop. In the citric acid cycle, an acetyl group is metabolized to carbon dioxide via reactions that regenerate the intermediates of the cycle.

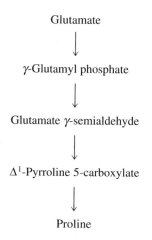

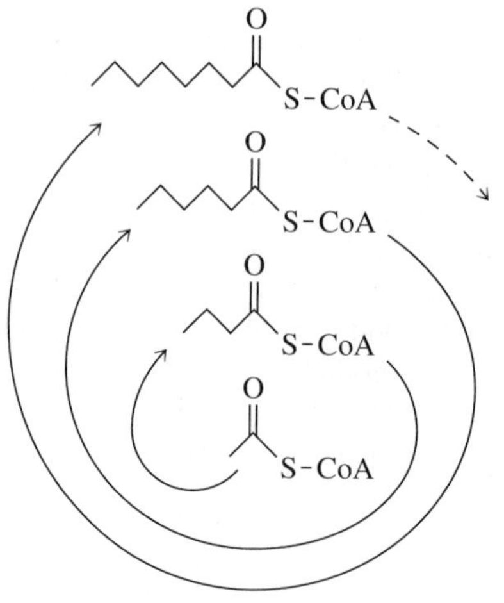

Figure 14·5
Spiral metabolic pathway. In fatty acid biosynthesis, the same set of enzymes catalyzes the progressive lengthening of the acyl chain.

enzyme-catalyzed steps, but the sequence forms a closed loop, with the intermediates being regenerated with every turn of the cycle (Figure 14·4). In a spiral metabolic pathway, such as the biosynthesis of fatty acids (Section 20·9), the same set of enzymes is used repeatedly for chain-lengthening or degradation of a given molecule (Figure 14·5).

14·2 Metabolism Proceeds by Discrete, Controlled Steps

Why are so many distinct reactions carried out in living cells? In principle, it should be possible to carry out the synthesis and especially the degradation of complex organic molecules with far fewer reactions. Indeed, it has been argued that in the early terrestrial environment—rich in nitrogen, ammonia, methane, and sulfurous gases—ambient temperatures and frequent electrical discharges provided the conditions for synthesis of hydrocarbons, amino acids, and even nucleotides without enzymatic catalysts.

The primeval environment contrasts sharply with the intracellular environment, which is for the most part a haven of comparative stability. Reactions in cells must proceed at moderate temperatures and pressures, at rather low reactant concentrations, and at near-neutral pH. All these factors contribute to the requirement for a multitude of efficient enzymatic catalysts.

Furthermore, reactions in vivo must proceed in a manner such that energy input or extraction can be achieved in a controlled way. Energy flow is mediated by energy donors and acceptors that carry discrete quanta of energy. As we will see, the energy transferred in a single reaction seldom exceeds $60 \, kJ \, mol^{-1}$. The energy released during a catabolic process (such as glucose oxidation to carbon dioxide and water, which releases $\sim 2800 \, kJ \, mol^{-1}$) is transferred to individual acceptors one step at a time rather than released in one grand explosion (Figure 14·6). The transfer of energy at each step is never 100%, but it appears that the energy loss is offset by the greater manageability of small quanta of energy. Like catabolic pathways, pathways for the synthesis of a variety of biomolecules require the transfer of discrete quanta of energy at multiple points. Energy carriers that accept and donate energy, such as adenine nucleotides and nicotinamide coenzymes, are found in all life forms.

Figure 14·6
Single-step versus multistep pathways. The uncontrolled combustion of glucose would release a large amount of energy all at once. In contrast, a multistep, enzyme-catalyzed pathway allows conservation of manageable amounts of energy.

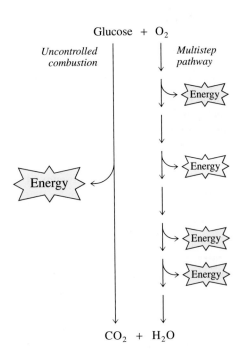

Some compounds can be substrates or products of more than one enzyme (i.e., have two or more metabolic functions). Intermediates shared by different pathways permit the exchange of energy and chemical entities between pathways. The interdependence of metabolic pathways presents an additional challenge in understanding metabolism—it is important to understand not only the critical individual pathways but also their interrelationships.

Finally, multistep pathways provide opportunities to establish control points. The regulation of metabolism is the very essence of cell function. As we discuss various metabolic pathways, we will see the exquisite balance of energy supply and demand in living cells, as well as the ability of cells to mount timely and extensive responses to stimulation or challenge from the environment or from internal signals.

14·3 Overview of the Metabolic Pathways in Cells

It is worth taking a bird's-eye view of metabolism before embarking on the detailed coverage given in subsequent chapters. In this section, we will examine some general features of the organization and function of metabolic pathways. Although a considerable portion of cellular metabolic activity is devoted to biosynthetic reactions, we will first concentrate on the energy-producing pathways. An overview of the major catabolic pathways is given in Figure 14·7.

Complex macromolecules can provide substrates for energy-yielding metabolism by first being broken down into smaller units. These compounds and other small energy-yielding metabolites that are easily transported between tissues in multicellular organisms are called **metabolic fuels.** Degradation of macromolecules may begin extracellularly, as we saw in Chapter 13. Alternatively, the initial source of substrates for energy metabolism may be endogenous compounds, including glycogen in animals and some bacteria, starch in plants, and triacylglycerols, found in large amounts in adipose tissue and seeds. In addition, amino acids liberated by protein breakdown are a significant source of metabolic fuel, especially when exogenous fuels are not available. The monomeric units (monosaccharides, fatty acids, and amino acids) released from the biopolymers may be metabolized within the cell of origin or may be delivered to other cells. Note that catabolism of nucleic acids is not shown in Figure 14·7. These molecules are the repository of biological information, and although they are synthesized and degraded at considerable rates in certain circumstances, they do not contribute significantly to the production of cellular energy as the other three types of molecules do. Of course, the catabolic reactions shown in Figure 14·7 exist not only to generate energy. The reactions also yield small molecules that serve as the building blocks for the synthesis of macromolecular cellular components.

The catabolism of the major energy-yielding monomers produces three types of compounds that mediate the release of energy: acetyl CoA, nucleoside triphosphates, and reduced coenzymes. Glucose, fatty acids, and some amino acids are oxidized to generate two-carbon (acetyl) units linked to coenzyme A. Acetyl CoA enters the common pathway of oxidative metabolism, the citric acid cycle, and upon

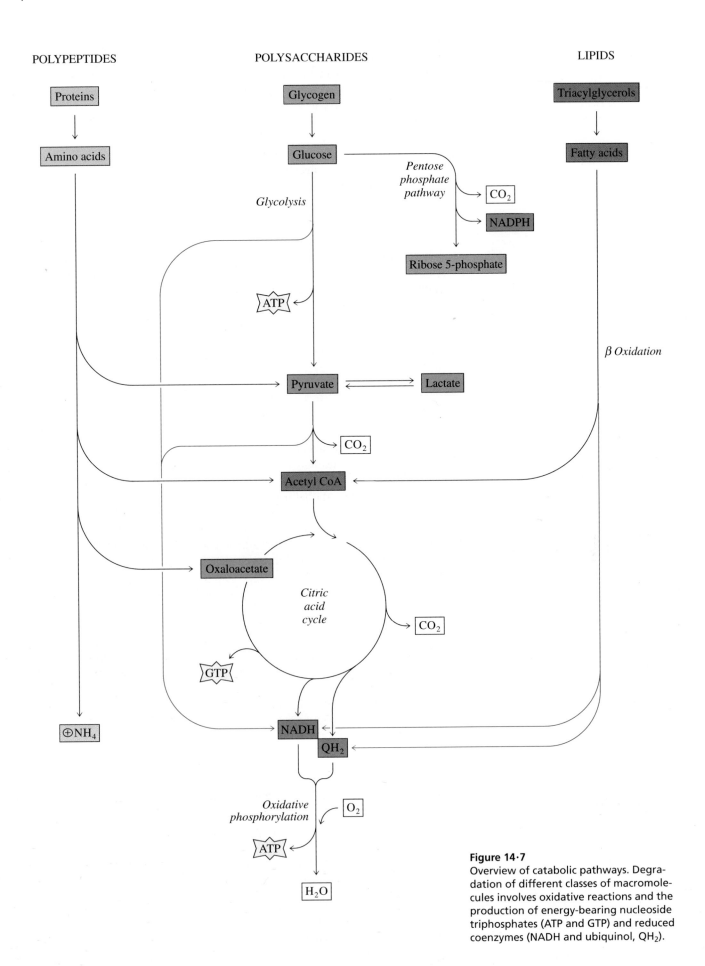

POLYPEPTIDES POLYSACCHARIDES LIPIDS

Figure 14·7
Overview of catabolic pathways. Degradation of different classes of macromolecules involves oxidative reactions and the production of energy-bearing nucleoside triphosphates (ATP and GTP) and reduced coenzymes (NADH and ubiquinol, QH_2).

oxidation to carbon dioxide, releases energy. The energy released in catabolic reactions is conserved in nucleoside triphosphates and reduced coenzymes. The general principle for energy conservation in phosphate-containing compounds, such as ATP or GTP, can be shown as

$$\text{Energy} \longleftarrow \underset{X\text{—}\textcircled{P}}{\overset{X + P_i}{\bigcirc}} \longrightarrow \text{Energy}$$

(14·1)

where X is the phosphoryl-group acceptor. The energy of oxidation reactions is also conserved in the reducing power of reduced coenzymes, such as NADH and ubiquinol, QH_2.

$$2\,H^{\oplus}, 2\,e^{\ominus} \longleftarrow \underset{YH_2}{\overset{Y}{\bigcirc}} \longrightarrow 2\,H^{\oplus}, 2\,e^{\ominus}$$

(14·2)

Y represents the oxidized molecule, and YH_2 its reduced counterpart. Energy-carrying intermediates may be used directly to drive energy-dependent reactions. Alternatively, the energy of the reduced coenzyme can be transferred to the energy carrier ATP by the terminal pathway of oxidative phosphorylation. The properties of ATP as a high-energy compound are described in Section 14·9.

In the next few chapters, we will examine energy-transducing processes in detail. Our discussion of metabolic pathways will begin in Chapter 15 with **glycolysis,** a ubiquitous pathway for glucose catabolism. In glycolysis, the hexose is split

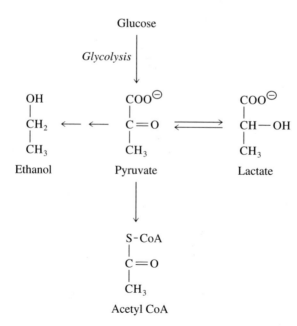

Figure 14·8
Products of the glycolytic pathway. Pyruvate may be converted to either lactate or ethanol. Pyruvate may also be converted to acetyl CoA for further oxidation.

into two three-carbon metabolites. This pathway can generate ATP without consuming oxygen. Three common products of this pathway are lactic acid, ethanol, and pyruvate, as shown in Figure 14·8. Pyruvate can be converted to acetyl CoA for further oxidation.

The **citric acid cycle** (Chapter 16) operates exclusively under aerobic conditions and metabolizes acetyl CoA derived from pyruvate and other compounds. The citric acid cycle facilitates the complete oxidation of the acetate carbons of acetyl CoA to carbon dioxide, and the energy released is conserved in the formation of NADH, QH_2, and ATP or GTP.

In Chapter 17, we will discuss other pathways of glucose metabolism, including the synthesis and degradation of glycogen, a glucose-storage polymer. The synthesis of glycogen is not simply the reverse of its degradation. These opposing pathways involve distinct enzymes that respond rapidly to metabolic signals in such a way that the rates of glycogen breakdown or synthesis can be carefully matched to the energy demands of the individual cell or to those of the organism as a whole. We will also see that glucose can be synthesized from three-carbon precursors such as pyruvate, lactate, and glycerol—a process called **gluconeogenesis.** Oxidation of glucose by the **pentose phosphate pathway** produces not ATP but ribose (necessary for the synthesis of nucleotides and nucleic acids) and NADPH (required in many reductive biosynthetic reactions). For this reason, the pentose phosphate pathway can be considered both anabolic and catabolic.

The final phase of energy-yielding catabolism is tied to the reduction of molecular oxygen by reduced coenzymes, principally NADH and QH_2 (Chapter 18). The energy of the reduced compounds is used to generate an electrochemical gradient of protons across a cell membrane. The potential energy of this gradient is harnessed to drive the phosphorylation of ADP to ATP.

$$ADP + P_i \longrightarrow ATP + H_2O \qquad \textbf{(14·3)}$$

This process, called **oxidative phosphorylation,** involves the plasma membrane of prokaryotes and the inner mitochondrial membrane of eukaryotes. Electrons are passed between electron carriers in order of decreasing reduction potential (discussed in Section 14·13). As electrons flow toward the terminal oxidizing agent, usually O_2, protons are pumped across the membrane from inside to outside, thereby creating a transmembrane proton concentration gradient. We will see that the reactions of oxidative phosphorylation and the reactions involved in capturing light energy during photosynthesis (discussed in Chapter 19) are quite similar.

Three additional chapters examine the metabolism—both anabolic and catabolic—of lipids, amino acids, and nucleotides. Some lipids can be catabolized to acetyl CoA, which enters the citric acid cycle and leads to the generation of ATP. Lipid metabolism (Chapter 20) also includes the synthesis of energy-storage molecules (the triacylglycerols), membrane lipids, and other compounds such as steroids and eicosanoids.

Amino acid metabolism (Chapter 21) encompasses a rich array of enzymes and intermediates, due to the diversity of the 20 common amino acids. Although amino acids were introduced as the building blocks of proteins, some of them also play important roles as metabolic fuels and biosynthetic precursors.

Nucleotide biosynthesis and degradation are considered (Chapter 22). Nucleotides are synthesized from small precursor molecules by de novo pathways or from preformed purines and pyrimidines by salvage pathways. Unlike the other three classes of biomolecules, however, nucleotides are catabolized primarily for excretion rather than for energy production. Nucleotides, like amino acids, contain nitrogen, an element not usually found in carbohydrates and lipids; specific pathways exist for the assimilation and excretion of nitrogen.

Finally, the reactions for biosynthesis of nucleic acids and proteins are discussed in Chapters 24–30.

14·4 Metabolic Pathways Are Regulated

In spite of the plethora of enzymes and metabolites in each cell, metabolism is not random. Rather, it is highly regulated. If each of the possible metabolic reactions were to occur at a fixed rate all of the time, organisms would be incapable of reacting to changes in their environment. For example, the intake of energy may be sporadic (e.g., meals), yet organisms expend energy continuously. Metabolism must therefore be regulated so that an organism can respond efficiently to the availability of energy or food. When there is no intake of food but a continued need for energy expenditure, metabolic fuels are mobilized from storage depots at a rate sufficient to supply cells with oxidizable substrates. Humans, for example, can survive periods of starvation as long as five to six weeks when provided with water. Keep in mind that any organism responds not only to external demands but also to genetically programmed instructions. For example, during embryogenesis or reproductive cycles, the metabolism of individual cells may change dramatically.

A. Metabolic Pathways Are Flexible

The responses of organisms to environmental factors may involve fine-tuning or dramatic reorganization of metabolic processes. A brief glance at some extremes of activity in the biological world illustrates the flexibility of metabolic systems. Consider the speed of animals such as the cheetah (which can run faster than 100 kilometers per hour, albeit briefly) or the American pronghorn antelope (which can sustain speeds of more than 60 kilometers per hour for many minutes). At the other extreme, consider the remarkable suppression of metabolic activity in seeds and spores, which may remain viable for decades.

The responses of organisms to changing demands may involve alterations in many pathways or only a few and may occur on a time scale ranging from less than a second to hours or longer. Perhaps the most rapid biological responses involve changes in the passage of small ions (sodium, potassium, and calcium, for example) through cell membranes. Transmission of nerve impulses and muscle contraction are events that depend on ion movement. Rapid change in ion movement is achieved by regulating the opening and closing of membrane transport proteins on a time scale of milliseconds. In general, the most rapid effects involve a preexisting transmembrane gradient in the concentration of an ion; opening a channel allows flow down the concentration gradient. A considerable portion of the energy expended by cells is devoted to the maintenance of transmembrane concentration gradients.

B. Regulation Alters the Flux Through Pathways

The ultimate response of an organism to shifting physiological demands must involve alterations in the rates of individual metabolic pathways that govern the synthesis and degradation of molecules as well as the generation and consumption of energy. We have already mentioned that multiple steps in pathways can facilitate regulation. Metabolic regulation is accomplished in a variety of ways. Many pathways are irreversible under physiological conditions. This means that a metabolite enters a pathway and can proceed through each step in sequence, without backing up and wasting cellular materials or energy. It is a general rule that biomolecules are synthesized and degraded by distinct routes, although some steps may be common to both the anabolic and catabolic reaction sequences.

In many pathways, substrates enter at multiple points. It is necessary, therefore, to consider a given pathway in the context of the pathways that supply all the substrates. Furthermore, the accumulation of products can affect both specific reactions and entire pathways. For example, lactic acid formed by anaerobic metabolism in rapidly contracting muscle must be removed for continued muscle activity. If the removal of lactic acid is restricted—as in poorly vascularized muscle—the capacity of cells for anaerobic energy production is severely curtailed by the ensuing acidification. This is humbling for an undertrained athlete who attempts to sprint 300 meters or more and, after a promising start, dramatically slows down or even collapses. The same phenomenon can have devastating consequences for a patient suffering from ischemic heart disease, who may experience debilitating pain after only modest exertion.

The flow of material through a metabolic pathway, or **flux,** depends not only on the supply of substrates and removal of products but also on the capabilities of the catalysts that facilitate the individual reactions. The concept of a "rate-limiting" or "pacemaker" enzyme has been a popular and attractive hypothesis. It is tempting to visualize regulation of a pathway by mechanisms that involve the efficient manipulation of a single point of highest sensitivity, sometimes likened to the narrow part of an hourglass. In practice, however, this may be an oversimplification. Flux through a pathway often depends on controls at several steps. These steps are generally irreversible reactions in the pathway, that is, those far from thermodynamic equilibrium. An enzyme is not rate limiting for a given pathway but rather has a particular control strength, or contribution, to the overall flux of the pathway. Because intermediates or cosubstrates from disparate sources may feed into or out of the pathway, multiple control points are the norm; an isolated, linear pathway is a rarity.

C. Regulation Is Accomplished in Different Ways

Two patterns of metabolic regulation frequently encountered are feedback inhibition and feed-forward activation. **Feedback inhibition** occurs when a product (usually the ultimate product) of a pathway controls the rate of its own synthesis through inhibition of an early step, usually the first committed step (the first reaction that is unique to the pathway).

$$A \longrightarrow B \longrightarrow C \longrightarrow D \longrightarrow E \longrightarrow P \qquad (14\cdot4)$$

The value of such a regulatory pattern in a biosynthetic pathway is obvious. When a sufficient amount of the product (P) is available, flux in the pathway is inhibited. Flux is restored when P is depleted. It is logical to inhibit the pathway in the early steps; otherwise, metabolic intermediates would accumulate unnecessarily.

Feed-forward activation occurs when a metabolite produced early in a pathway activates an enzyme that catalyzes a reaction further down the pathway.

$$A \longrightarrow B \longrightarrow C \longrightarrow D \longrightarrow E \longrightarrow P \qquad (14\cdot5)$$

In this example, the activity of one enzyme (which converts A to B) is coordinated with the activity of another enzyme (which converts D to E).

Figure 14·9
Regulatory role of a protein kinase. The effect of the initial signal is amplified by the signalling cascade. Simultaneous phosphorylation of different cellular proteins by the activated kinase results in coordinated regulation of different metabolic pathways. Particular proteins may become more active or less active when phosphorylated. Protein phosphatases catalyze the removal of phosphoryl groups.

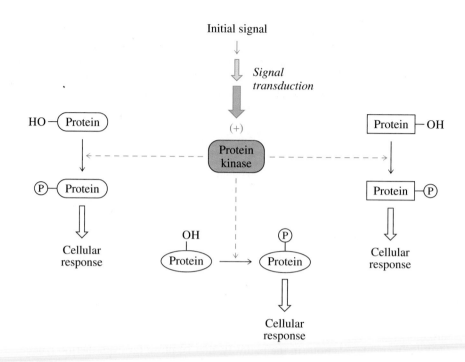

In Chapter 6, we discussed the ways in which individual enzymes are regulated. Allosteric activators and inhibitors, which are usually small molecules, can rapidly alter the activity of enzymes by inducing conformational changes that affect catalytic activity. We will see numerous examples of allosteric modulation in the coming chapters.

The activity of enzymes can also be rapidly and reversibly altered by covalent modification. The most common modifications are the addition and removal of phosphoryl groups. Phosphorylation usually occurs at serine, tyrosine, or threonine residues of enzymes or other proteins such as metabolite transporters or cell-signalling components. Phosphorylation, catalyzed by enzymes known as **protein kinases,** may stabilize a particular conformation in the target protein, much as in allosteric modification, or it may block access of substrates or other ligands to their binding sites on the protein. The effects of phosphorylation are reversed by the action of **protein phosphatases,** which catalyze the removal of phosphoryl groups. The activation of kinases with multiple specificities allows the coordinated regulation of more than one metabolic pathway by one signal—some pathways may be activated while some are inhibited. The cascade nature of intracellular signalling pathways, described in Section 12·9, also means that the initial signal is amplified (Figure 14·9).

Although we have mentioned numerous examples from mammalian metabolism, highly efficient metabolic regulation occurs in all organisms. For example, yeast cells are capable of completely oxidizing glucose to carbon dioxide when oxygen is available but readily switch to fermentation of glucose, with the production of ethanol, when oxygen is depleted.

The amounts of specific enzymes can be altered by increasing protein synthesis or degradation. This is usually a slow process relative to allosteric or covalent activation or inhibition. However, the turnover of certain enzymes may be rapid. In Part 4 of this book, we will examine more closely a variety of mechanisms involved in regulating gene expression and protein synthesis. Keep in mind that several modes of regulation may be operating simultaneously.

The importance of understanding metabolic control mechanisms is relevant not only to researchers but also to those who exploit the metabolic machinery of organisms—mostly microbial—for commercial purposes. Metabolic engineering is the manipulation of cellular activities by altering enzymatic, transport, and regulatory functions of the cell through the use of recombinant DNA technology (which is described in more detail in Chapter 33). Existing metabolic pathways can be extended to obtain new products, or the overall yield of a synthetic pathway can be maximized, a process termed metabolite overproduction. Examples of products of engineered metabolism are some antibiotics, vitamins, industrial chemicals such as acetone and butanol, and products used in animal feed.

14·5 In Multicellular Organisms, Metabolic Tasks Are Divided Among Different Tissues

In eukaryotes, one factor that contributes to metabolic complexity and efficiency is the sequestering of metabolic processes in specialized subcellular compartments. For example, the enzymes that catalyze fatty acid synthesis in vertebrates are located in the cytosol, whereas the enzymes that catalyze fatty acid oxidation (breakdown) are located inside the mitochondria. Within a prokaryotic cell, different metabolic processes may be localized in distinct regions. In multicellular organisms, compartmentation can also take the form of specialization of tissues.

The division of labor among tissues also allows the site-specific regulation of metabolic processes. Cells from different tissues may be distinguished by their complement of enzymes, including isozymes (Section 6·16). Individual cells maintain different concentrations of metabolites, depending in part on the presence of specific transporters that facilitate the entry and exit of metabolites. Finally, depending on the cell-surface receptors and signal-transduction mechanisms that are present, individual cells in multicellular organisms respond differently to hormonal or neuronal signals. We know that the glycolytic pathway is very active in muscle and brain tissue, where glucose is metabolized for energy. However, in liver and adipose tissue, excess glucose can be converted to fat. The control of glucose metabolism is obviously quite different in these four tissues.

Some major metabolic processes involve reactions in a number of tissues and require the transport of substances between tissues. Consider as an example an athlete running a race (Figure 14·10). In a short race, almost all of the energy in the

Figure 14·10
Athletes running a race. Initially, glucose derived from muscle glycogen is used for fuel. In longer races, glycogen in liver and triacylglycerols in adipose tissue are broken down for fuel.

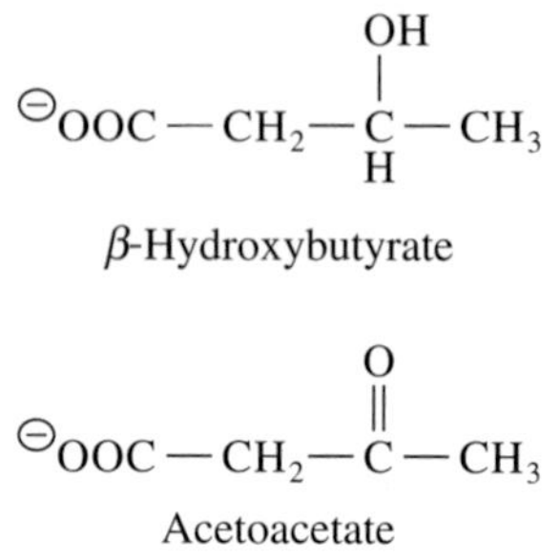

$$^\ominus OOC - CH_2 - \overset{\overset{\displaystyle OH}{|}}{\underset{\underset{\displaystyle H}{|}}{C}} - CH_3$$

β-Hydroxybutyrate

$$^\ominus OOC - CH_2 - \overset{\overset{\displaystyle O}{||}}{C} - CH_3$$

Acetoacetate

Figure 14·11
Structures of two ketone bodies.

muscles may be supplied by the metabolism of intracellular fuels—principally glucose derived from glycogen stored inside muscle cells. However, in longer races, or during particularly strenuous exercise, interorgan movement of fuels becomes important. In a marathon, glucose derived from the breakdown of liver glycogen and fatty acids produced by lipolysis of triacylglycerols stored in adipose tissue are released into the blood and carried to the muscles.

The changes in metabolism that occur between the fed and fasted states in mammals also illustrate the importance of interorgan metabolism. The pattern of metabolism is dictated by two factors: the metabolic fuels available within the body and the fuel specificity of particular tissues. The brain, for example, must be supplied with glucose. During starvation, glucose stores are depleted within 12–24 hours. Survival depends on the ability of most tissues to oxidize fatty acids released from adipose tissue and on the ability of the liver to convert certain amino acids to glucose for use by tissues such as the brain. The demand for amino acids, which are mostly derived from muscle protein catabolism, decreases after about one week, when ketone bodies derived from fatty acids become important fuels (Figure 14·11).

Upon refeeding after a fast, the mobilization of internal energy stores ceases, and ingested compounds are metabolized for immediate energy needs and for storage. The pattern of storage is not simply a reflection of what is ingested. Large quantities of energy can be stored as triacylglycerols, but there is only a small store of carbohydrate (glycogen). There is no purely storage form of free amino acids; rather, amino acids are incorporated into functional proteins. The dramatic metabolic changes during feeding and fasting in mammals are orchestrated primarily by the hormones insulin and glucagon. Understanding hormone-dependent regulatory networks is a major challenge in metabolic regulation. We will return to the subject of interorgan metabolism in Chapter 23.

14·6 Thermodynamic Principles Underlie the Study of Metabolism

For a firm understanding of the nature of equilibrium and flux in metabolism, we turn to the basic principles of thermodynamics. Each of the reactions of metabolism involves the transfer of both matter and energy. The transfer of matter is intuitive, represented by the movement of atoms among reacting species—movements that are captured in the familiar symbols of chemistry. Energy transfer in chemical reactions is less intuitive, and information about energy does not appear overtly in chemical equations. A quantitative understanding of energy transfer is needed to determine whether a reaction will proceed in a particular direction in a test tube and whether it will do so in a living cell.

An important point to keep in mind as you study metabolism is that many metabolic pathways are not at equilibrium but are maintained in a steady state. A figurative description of the steady state is the "leaking bucket" model: a bucket has a small hole in the bottom, but a faucet adds water to it at a rate exactly equal to the rate of loss through the hole. Over time, the water level in the bucket remains constant, although the water itself is flowing. In metabolic pathways consisting of several reactions, all of the intermediates are in a steady state. The initial reactant in the pathway is steadily replenished and the end product steadily removed. The concentrations of intermediates remain unchanged even as flux occurs.

The laws of thermodynamics apply to biological systems, as we saw in Chapter 1. The free-energy change (ΔG) is a measure of the energy available within a system to do work. The ability of a system to do work decreases as equilibrium is approached; at equilibrium, no energy is available to do work, and there is no net conversion of reactants to products. The free-energy change defines the equilibrium condition in terms of the enthalpy (heat) and entropy (randomness) of the system at constant pressure.

$$\Delta G = \Delta H - T\Delta S \tag{14·6}$$

where ΔH is the change in enthalpy, ΔS is the change in entropy, and T is the absolute temperature.

If ΔG for a process is *negative,* the process is spontaneous; that is, it can proceed in the absence of energy provided from outside the system. Such processes are said to be **exergonic.** If ΔG is *positive,* the process cannot proceed spontaneously. For such a process to proceed, enough energy must be supplied from outside the system to make the free-energy change negative. Processes with a positive free-energy change are said to be **endergonic,** which is a way of saying the reverse process is thermodynamically favored. When the process is at equilibrium, no change in free energy occurs, and $\Delta G = 0$.

Because ΔG is composed of both enthalpy and entropy contributions, only the sum of these at a given temperature—as indicated in Equation 14·6—must be negative for a reaction to be spontaneous. Thus, even if ΔS for the system during a particular process is negative, a sufficiently negative ΔH can overcome the decrease in entropy, resulting in a ΔG that is less than zero. Similarly, even if ΔH is positive, a sufficiently positive ΔS can overcome the increase in enthalpy, resulting in a negative ΔG. Spontaneous processes that proceed due to a large positive ΔS are said to be "entropy-driven." Examples of entropy-driven processes include protein folding (Section 5·11) and the formation of lipid bilayers (Section 12·1A), both of which are associated with the hydrophobic effect (Section 3·5). The processes of protein folding and lipid-bilayer formation both result in states of decreased entropy for the protein molecule and bilayer components, respectively. However, the decrease in entropy is offset by a large increase in the entropy of surrounding water molecules.

The free-energy change of a reaction depends on the conditions under which the reaction occurs. Therefore, it is useful to have a set of reference reaction conditions established by convention. These reference conditions are referred to as the **standard state.** Chemists define standard temperature as 298 K (25°C), standard pressure as 1 atmosphere, and standard solute concentration as 1.0 M. The notation ΔG° indicates the change in free energy under standard conditions as defined by chemists. The chemical standard state has been modified slightly for biological chemistry. Since most biochemical reactions occur at a pH near 7, the standard concentration of hydrogen ions in the biological standard state is 10^{-7} M (pH = 7.0) rather than 1.0 M (pH = 0.0). Free-energy change under the biological standard state is indicated by $\Delta G^{\circ\prime}$.

14·7 The Equilibrium Constant of a Reaction Is Related to the Standard Free-Energy Change

An important relationship exists between the actual free-energy change, or free-energy change under nonstandard conditions (ΔG), the standard free-energy change ($\Delta G^{\circ\prime}$), and the equilibrium constant (K_{eq}). To illustrate the relationship, we will first consider the free energy of each component of a reaction. The free energy of a solution of substance A is related to its standard free energy by

$$G_A = G_A^{\circ\prime} + RT \ln[A] \tag{14·7}$$

where R is the universal gas constant (8.315 J K^{-1} mol^{-1}). Free energy is expressed in units of kJ mol^{-1}. The term RT $\ln[A]$ is sometimes given as 2.303 RT $\log[A]$.

For the reaction

$$A + B \rightleftharpoons C + D \qquad (14\cdot8)$$

the actual free-energy change is the sum of the free energies of the products minus the free energies of the reactants.

$$\Delta G_{reaction} = (G_C + G_D) - (G_A + G_B) \qquad (14\cdot9)$$

By substituting the relationship of actual free energy to standard free energy (Equation 14·7), we can obtain

$$\Delta G_{reaction} = (G_C^{\circ\prime} + G_D^{\circ\prime} - G_A^{\circ\prime} - G_B^{\circ\prime}) + RT \ln \frac{[C][D]}{[A][B]} \qquad (14\cdot10)$$

or

$$\Delta G_{reaction} = \Delta G_{reaction}^{\circ\prime} + RT \ln \frac{[C][D]}{[A][B]} \qquad (14\cdot11)$$

If the reaction has reached equilibrium, the ratio of concentrations in the last term of Equation 14·11 is, by definition, the equilibrium constant, K_{eq}. When the concentrations of the reactants and products are at equilibrium, the rates of the forward and reverse reactions are the same, and $\Delta G_{reaction} = 0$. Thus,

$$\Delta G_{reaction}^{\circ\prime} = -RT \ln K_{eq} \qquad (14\cdot12)$$

If we know the value of $\Delta G^{\circ\prime}$ for a reaction, Equation 14·12 allows us to calculate K_{eq} and vice versa. Note that the relationship between $\Delta G^{\circ\prime}$ and K_{eq} is logarithmic. Small changes in $\Delta G^{\circ\prime}$ therefore represent large changes in K_{eq}.

14·8 Actual Free-Energy Change, Not Standard Free-Energy Change, Is the Criterion for Spontaneity in Cellular Reactions

For any enzymatic reaction within a living organism, the *actual* free-energy change (the free-energy change under cellular conditions) must be less than zero for the reaction to proceed. However, many metabolic reactions have *standard* free-energy changes that are positive. The difference between ΔG and $\Delta G^{\circ\prime}$ depends upon cellular conditions. The most important condition affecting free-energy change in cells is the concentration of substrates and products of a reaction. Consider Reaction 14·8. At equilibrium, the ratio of substrates and products is by definition the equilibrium constant, K_{eq}, and the free-energy change of the reaction is zero.

$$K_{eq} = \frac{[C][D]}{[A][B]} \qquad \Delta G = 0 \qquad (14\cdot13)$$

However, when this reaction is not at equilibrium, a different ratio of products to substrates is observed, and the free-energy change is derived using Equation 14·11.

$$Q = \frac{[C]'[D]'}{[A]'[B]'} \qquad \Delta G = \Delta G^{\circ\prime} + RT \ln Q \qquad (14\cdot14)$$

where Q is the **mass action ratio.** It is this ratio relative to the ratio of products to substrates at equilibrium that determines the free-energy change for a reaction. In other words, the free-energy change is a measure of how far from equilibrium the reacting system is poised. Consequently, it is ΔG, and not $\Delta G^{\circ\prime}$, that is the criterion for assessing the spontaneity of a reaction and therefore its direction within a particular system.

In general, we can divide metabolic reactions into two types. Let us take Q to represent the steady-state ratio of reactant and product concentrations in a living cell. Reactions for which Q is close to K_{eq} are called **near-equilibrium reactions.** The free-energy changes associated with near-equilibrium reactions are quite small, and these reactions are readily reversible. Reactions for which Q is far from K_{eq} are called **metabolically irreversible reactions.** These reactions are greatly displaced from equilibrium, with Q usually two or more orders of magnitude from K_{eq}. Whereas ΔG must be at least slightly negative for all reactions in living cells, ΔG is a large negative number for metabolically irreversible reactions.

When flux through a pathway changes, the intracellular concentrations of metabolites vary but usually over a narrow range of no more than two- or threefold. Due to the large amount of enzymatic activity present, enzymes that catalyze near-equilibrium reactions are able to quickly restore levels of substrate and product to near-equilibrium status. These enzymes accommodate flux in either direction.

In contrast, enzymes that catalyze metabolically irreversible reactions are usually present in limiting amounts in cells, insufficient to achieve near-equilibrium status for their reactions. The control points of pathways are generally metabolically irreversible reactions. The enzymes that catalyze these reactions are usually regulated in some way. Metabolically irreversible reactions may act as bottlenecks in metabolic traffic, controlling the flux through reactions further along the pathway.

Near-equilibrium reactions are not usually suitable control points. Flux through a near-equilibrium step cannot be significantly increased, since it is already virtually at equilibrium. Near-equilibrium reactions are therefore controlled only by changes in substrate and product concentrations. In contrast, flux through metabolically irreversible reactions is relatively unaffected by changes in metabolite concentration; flux through these reactions is primarily controlled by effectors that modulate the catalytic rate.

14·9 ATP Is the Principal Carrier of Biological Energy

The energy produced by one biological reaction or process is often used to drive a second reaction that would not otherwise occur spontaneously. The two reactions are linked by a shared energized intermediate (B-X) in the overall reaction sequence.

$$A\text{-}X + B \longrightarrow A + B\text{-}X$$
$$B\text{-}X + C \longrightarrow B + C\text{-}X$$
(14·15)

The sum of the free-energy changes for the coupled reactions must be negative for the reactions to proceed. Energy flow in metabolism depends on many coupled reactions in which ATP is the shared energy-carrying intermediate.

$$A\text{—}\textcircled{P} + ADP \longrightarrow A + ATP$$
$$ATP + C \longrightarrow ADP + C\text{—}\textcircled{P}$$
(14·16)

ATP, a nucleoside triphosphate, contains one phosphate ester formed by linkage of the α-phosphate to the 5' oxygen of ribose and two phosphoanhydrides formed by the α,β and β,γ linkages between phosphate groups. ATP is a donor of mobile metabolic groups, usually donating a phosphoryl group, leaving adenosine diphosphate (ADP), or donating a nucleotidyl (AMP) group, leaving inorganic pyrophosphate (PP_i), as described in Section 8·3. Transfer of a phosphoryl group or a

Table 14·1 Standard free energies of hydrolysis for ATP and AMP

Major ionic form of reactants and products	$\Delta G^{\circ\prime}_{\text{hydrolysis}}$ (kJ mol^{-1})
$ATP^{4-} + H_2O \longrightarrow ADP^{3-} + HPO_4^{2-} + H^{+}$	-30
$ATP^{4-} + H_2O \longrightarrow AMP^{2-} + HP_2O_7^{3-} + H^{+}$	-32
$AMP^{2-} + H_2O \longrightarrow \text{Adenosine} + HPO_4^{2-}$	-14

nucleotidyl group, both of which participate in phosphoanhydride linkages in ATP, to a substrate activates that substrate (i.e., raises its free energy). The activated compound, which may be either a metabolite or the side chain of an amino acid residue of a synthetase, has the increased reactivity needed for completion of the coupled reaction. Although the various groups of ATP are usually transferred to acceptors other than water, hydrolysis reactions provide useful estimates of the free-energy changes involved. Table 14·1 lists the standard free energies of hydrolysis ($\Delta G^{\circ\prime}_{\text{hydrolysis}}$) for ATP and AMP, and Figure 14·12 depicts the hydrolytic cleavage of each of the phosphoanhydrides of ATP. Note from Table 14·1 that whereas cleavage of the ester releases only 14 kJ mol^{-1}, cleavage of either of the phosphoanhydrides releases at least 30 kJ mol^{-1} under standard conditions.

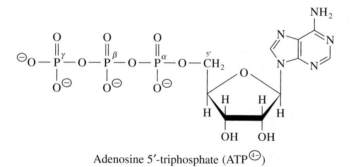

Figure 14·12
Hydrolysis of ATP to
(1) ADP and inorganic phosphate (P$_i$) and
(2) AMP and inorganic pyrophosphate (PP$_i$).

Adenosine 5′-triphosphate (ATP^{4-})

Adenosine 5′-diphosphate (ADP^{3-})

Adenosine 5′-monophosphate (AMP^{2-})

Inorganic phosphate (P$_i$)

Inorganic pyrophosphate (PP$_i$)

Several factors contribute to the large amount of energy released during hydrolysis of the phosphoanhydride linkages of ATP under both standard and cellular conditions. First, electrostatic repulsion among the negatively charged oxygen atoms of the phosphoanhydride groups of ATP is eliminated by hydrolysis. Recall that virtually all of the ATP and ADP molecules in a cell actually exist in complexes with Mg^{2+} (Figure 10·15). Magnesium, by partially neutralizing the negative charges on the oxygen atoms of ATP, diminishes electrostatic repulsion and decreases $\Delta G_{hydrolysis}$. The amount of electrostatic repulsion is also affected by pH. The reactants and products of ATP hydrolysis have ionizable phosphate groups with pK_a values between 6 and 7. Changes in pH alter the ionic states of all reactants and products in the reaction and strongly affect the free energy of hydrolysis. At high pH, when the phosphate groups are ionized, electrostatic repulsion is increased, and $\Delta G_{hydrolysis}$ increases.

Solvation effects also contribute significantly to the negative free energy of ATP hydrolysis. The products of hydrolysis—ADP and inorganic phosphate or AMP and pyrophosphate—are better solvated than ATP itself. When ions are solvated, they are electrically shielded from each other. The decrease in the repulsion between phosphate groups helps drive the reaction toward hydrolysis.

Finally, there is more resonance stabilization of the products of ATP hydrolysis than of ATP itself. The phosphoryl groups of ATP compete for the unpaired electrons on the bridging oxygen atoms of the phosphoanhydrides, whereas the products of hydrolysis assume a more stable electronic structure. Resonance stabilization of the products provides a large driving force for ATP hydrolysis.

The factors listed above combine to provide the large free-energy change when ATP is hydrolyzed. Because of the free-energy change associated with cleavage of their phosphoanhydrides, ATP and the other nucleoside triphosphates—uridine triphosphate (UTP), guanosine triphosphate (GTP), and cytidine triphosphate (CTP)—are often referred to as **energy-rich metabolites.** We will occasionally express the consumption of the phosphoanhydride linkages of nucleoside triphosphates in terms of ATP equivalents.

Since all the phosphoanhydrides of nucleoside phosphates have nearly equal standard free energies of hydrolysis (or formation), phosphoryl-group transfers between nucleoside phosphates have equilibrium constants close to 1.0. As we saw in Section 10·5, ATP is usually the phosphoryl-group donor in reactions whereby nucleoside monophosphates and diphosphates are phosphorylated. Of course, the intracellular concentrations of individual nucleoside mono-, di-, and triphosphates differ, depending on regulatory mechanisms designed to meet different metabolic needs. For example, the intracellular levels of ATP are far greater than dTTP levels. ATP is involved in many biochemical reactions and metabolic functions, whereas dTTP has only one known function, to serve as a substrate for DNA synthesis.

In vivo, the concentrations of ATP and its hydrolysis products are such that the free-energy change for ATP hydrolysis is actually greater than the standard value of $-30 \, kJ \, mol^{-1}$. Although the concentration of ATP varies among cell types, the intracellular ATP concentration fluctuates little within a particular cell. Intracellular ATP concentrations are maintained in part by the action of adenylate kinase (Section 10·5A), which catalyzes the following near-equilibrium reaction:

$$AMP \; + \; ATP \; \rightleftharpoons \; 2 \, ADP \qquad \textbf{(14·17)}$$

This reaction allows the formation of ADP from AMP when concentrations of AMP are high; some of the ADP is then converted to ATP by oxidative phosphorylation.

ATP concentrations are greater than ADP or AMP concentrations, and relatively minor changes in ATP concentration may result in large changes in the concentrations of the di- and monophosphate forms. Table 14·2 shows the theoretical

Table 14·2 Theoretical changes in concentrations of adenine nucleotides

ATP (mM)	ADP (mM)	AMP (mM)
4.8	0.2	0.004
4.5	0.5	0.02
3.9	1.0	0.11
3.2	1.5	0.31

[Adapted from Newsholme, E. A., and Leech, A. R. (1986). *Biochemistry for the Medical Sciences* (New York: John Wiley & Sons), p. 315.]

increases in [ADP] and [AMP] under conditions where ATP is consumed, assuming that the total adenine nucleotide concentration remains 5.0 mM. In fact, when cells are well supplied with oxidizable fuels and oxygen, they maintain a balance of adenine nucleotides in which ATP is present at a steady concentration in the 2–10 mM range, [ADP] is less than 1 mM, and [AMP] is even lower. As we will see, ADP and AMP are often effective allosteric regulators of some energy-yielding metabolic processes. ATP, whose concentration is relatively constant, is generally not an important regulator under physiological conditions.

14·10 Consumption of ATP Can Be Coupled to Biosynthesis of Other Molecules

Enzymes known as kinases (also called phosphotransferases) catalyze transfer of the γ-phosphoryl group from ATP (or, less frequently, from another nucleoside triphosphate) to another substrate. Kinase reactions are typically unidirectional (that is, kinases catalyze metabolically irreversible reactions). A few kinase reactions, however, such as those catalyzed by adenylate kinase (Section 14·9) and creatine kinase (Section 14·11), are near-equilibrium reactions.

Although the reactions they catalyze are sometimes described as phosphate-group–transfer reactions, kinases actually transfer a phosphoryl group ($—PO_3^{2-}$) to their acceptors. This transfer of the electrophilic phosphoryl group from one nucleophile to another is presumed to occur by a mechanism called **in-line nucleophilic displacement** (Figure 14·13). Most kinases catalyze direct transfer to the nucleophilic acceptor substrate; in a few reactions, such as the one catalyzed by nucleoside diphosphate kinase, the phosphoryl group is first transferred to a nucleophilic group on the enzyme (Section 10·5B).

The ability of a phosphate compound to transfer its phosphoryl group(s) is termed its phosphoryl-group–transfer potential, or simply **group-transfer potential**. Some compounds, such as phosphoanhydrides, are excellent group donors; others, such as phosphoesters, are poor donors (i.e., have a group-transfer potential less than that of ATP). The energetics of various kinase reactions can be compared by assessing the standard free-energy changes of the reactions given in Table 14·3. (Under standard conditions, group-transfer potentials have the same values as the standard free energies of hydrolysis but are opposite in sign. Thus, the group-transfer potential is a measure of the free energy required for the formation of the phosphorylated compound.) Most often, one kinase catalyzes transfer of a phosphoryl group from an excellent donor to ADP to form ATP, which then acts as a donor in a second kinase-catalyzed reaction.

Figure 14·13
Mechanism of phosphoryl-group transfer.
(a) In-line nucleophilic displacement.
(b) Pentacovalent intermediate of the in-line mechanism, in which the approach of the nucleophile and the departure of the leaving group occur at apical positions.

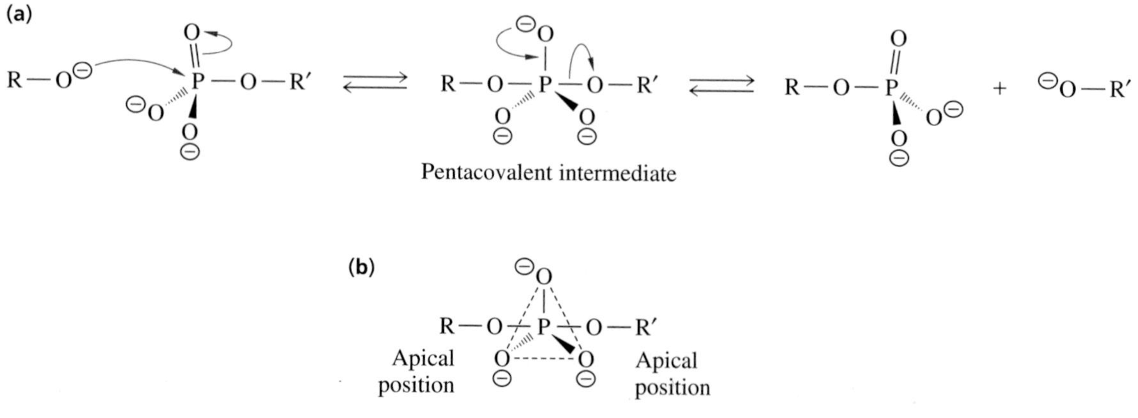

The synthesis of glutamine from glutamate and ammonia illustrates how a cell can use the free energy of ATP to drive a biosynthetic reaction (Figure 14·14). This reaction is catalyzed by glutamine synthetase and provides an important means by which organisms can incorporate inorganic nitrogen into biomolecules as carbon-bound nitrogen. In this reaction—the enzymatic synthesis of an amide bond—the carboxyl group of the substrate is activated via synthesis of an anhydride intermediate. Glutamine synthetase catalyzes the nucleophilic displacement of the γ-phosphoryl group of ATP by the γ-carboxylate of glutamate, generating enzyme-bound γ-glutamyl phosphate, a high-energy intermediate, and releasing ADP. γ-Glutamyl phosphate is unstable in aqueous solution but is protected from water in the active site of glutamine synthetase. In the second step of the mechanism, ammonia acts as an attacking nucleophile, displacing the phosphate (a good leaving group) from the carbonyl carbon to generate the product, glutamine. Overall, one molecule of ATP is converted to ADP + P_i for every molecule of glutamine formed from glutamate and ammonia.

Table 14·3 Standard free energies of hydrolysis for common metabolites

Metabolite	$\Delta G^{\circ\prime}_{hydrolysis}$ (kJ mol^{-1})
Phosphoenolpyruvate	−62
1,3-*Bis*phosphoglycerate	−49
Acetyl phosphate	−43
Phosphocreatine	−43
Pyrophosphate	−33
Phosphoarginine	−32
ATP to AMP + PP$_i$	−32
ATP to ADP + P$_i$	−30
Glucose 1-phosphate	−21
Glucose 6-phosphate	−14
Glycerol 3-phosphate	− 9

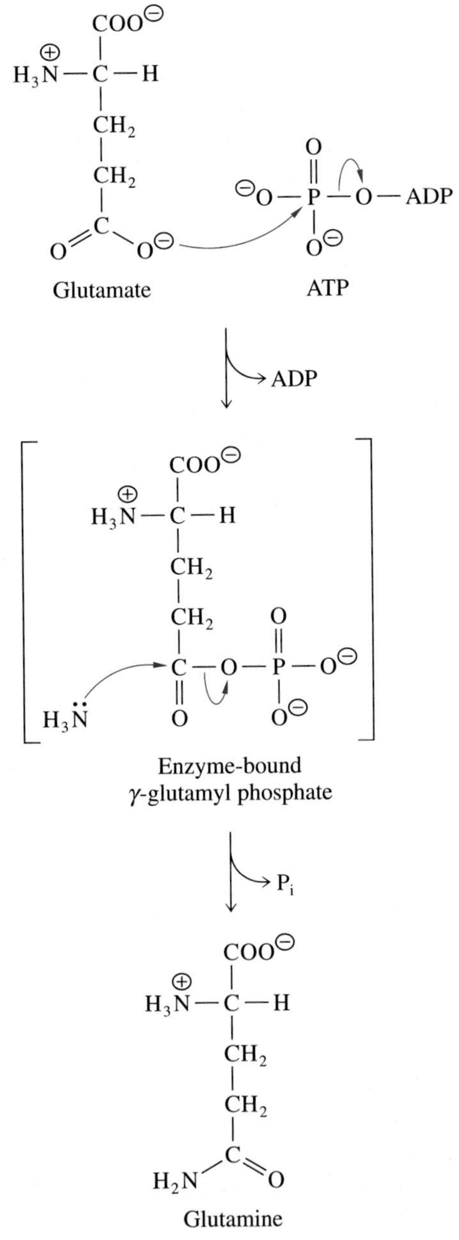

Figure 14·14
Conversion of glutamate to glutamine, catalyzed by ATP-dependent glutamine synthetase. Binding of the anhydride intermediate within the active site of the enzyme prevents reaction with water (hydrolysis).

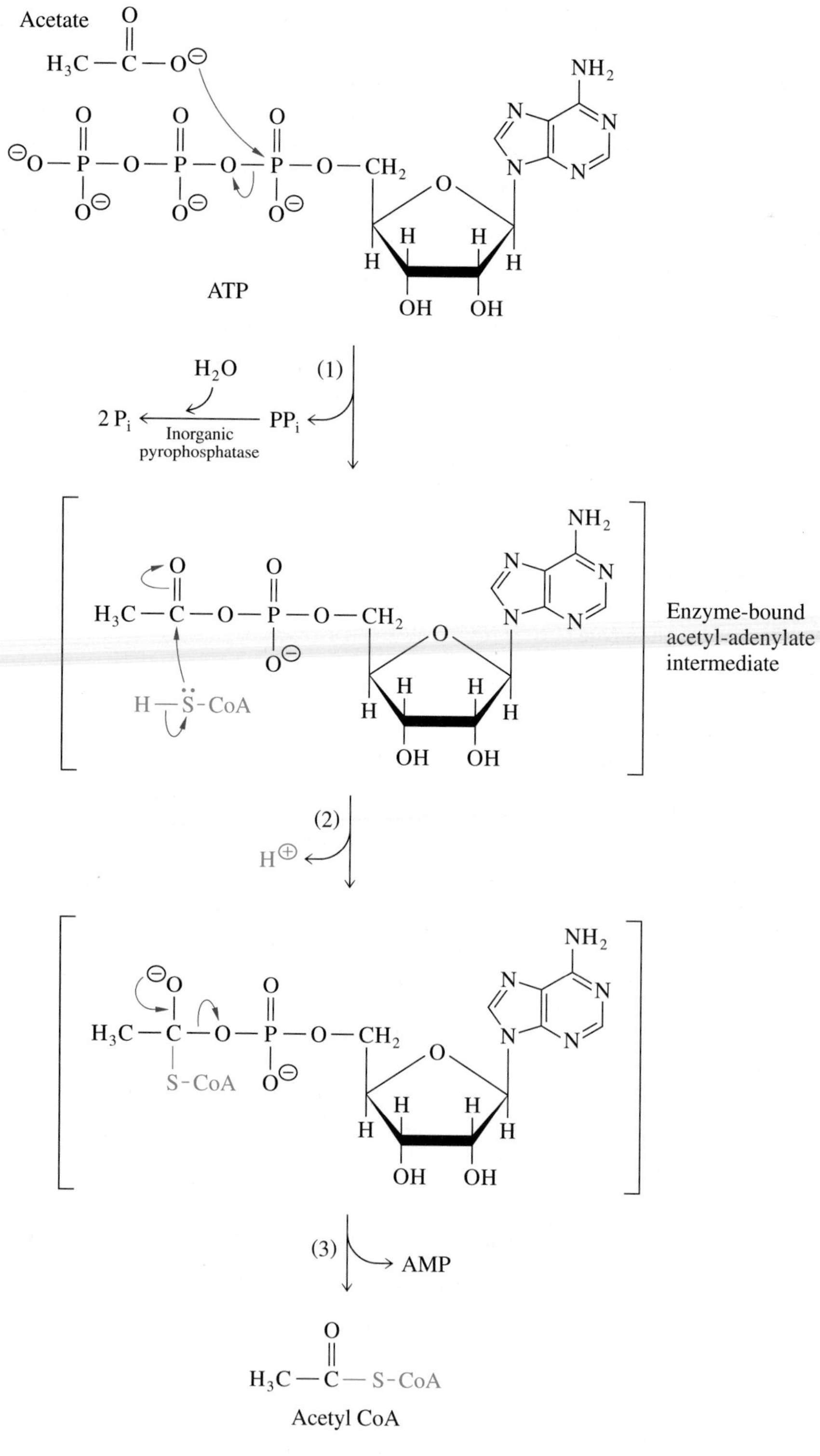

Figure 14·15
Synthesis of acetyl CoA from acetate, catalyzed by acetyl-CoA synthetase. Hydrolysis of the acetyl-adenylate intermediate is prevented by tight binding in the active site of the enzyme.

The other common group-transfer reaction involving ATP is transfer of the nucleotidyl group. An example is the synthesis of acetyl CoA, catalyzed by acetyl-CoA synthetase. In this reaction, the AMP moiety of ATP is transferred to the nucleophilic carboxylate group of acetate to form acetyl-adenylate as an intermediate (Figure 14·15). Note that pyrophosphate (PP_i) is released in this step. Like the glutamyl-phosphate intermediate in the previous figure, the high-energy intermediate is shielded from nonenzymatic hydrolysis by tight binding in the active site of the enzyme. The reaction is completed by transfer of the acetyl group to the nucleophilic sulfur atom of coenzyme A, leading to the formation of acetyl CoA and AMP.

The synthesis of acetyl CoA also illustrates how removal of a product can cause a metabolic reaction to approach completion, just as formation of a precipitate or a gas can drive an inorganic reaction toward completion. In Figure 14·15, for example, a molecule of the product PP_i is hydrolyzed to two molecules of P_i by the action of inorganic pyrophosphatase. Almost all cells have high levels of activity of this enzyme, hence the concentration of PP_i in cells is generally very low (less than 10^{-6} M). The additional hydrolytic reaction adds the energy cost of one additional phosphoanhydride linkage to the overall synthetic process. We will see that hydrolysis of pyrophosphate accompanies many synthetic reactions in metabolism.

14·11 The Energy of Other Metabolites Can Be Coupled to the Synthesis of ATP

Several energy-rich metabolites release more free energy upon hydrolysis than ATP does under the same conditions. A variety of specific kinases catalyze the transfer of phosphoryl groups from such molecules to ADP, generating ATP.

Certain bacteria generate the mixed anhydride acetyl phosphate during anaerobic fermentation by reaction of acetyl CoA with phosphate.

$$\text{Acetyl CoA} + P_i \rightleftharpoons \text{Acetyl phosphate} + \text{CoASH} \qquad (14\cdot18)$$

In these cells, the phosphoryl group of acetyl phosphate can be used to generate ATP. Hydrolysis of acetyl phosphate yields acetate and inorganic phosphate (Figure 14·16). Under standard conditions, the group-transfer potential of acetyl phosphate exceeds that of ATP. Reaction 14·18 can proceed under cellular conditions, with the transfer of the activated phosphoryl group from acetyl phosphate to ADP providing the bacteria with ATP needed for anaerobic metabolism. In other bacteria, it appears that the same reactions occur in reverse. Acetate is activated to acetyl phosphate by reaction with ATP, and the resulting acetyl phosphate is converted to acetyl CoA, which is required for fatty acid synthesis and other anabolic reactions.

The **phosphagens,** including phosphocreatine and phosphoarginine, are high-energy phosphate-storage molecules found in animal muscle cells (Figure 14·17). Phosphagens are phosphoamides (rather than phosphoanhydrides) and have a higher group-transfer potential than ATP. In the muscles of vertebrates, large amounts of phosphocreatine are formed during times of ample ATP supply. In resting muscle, the concentration of phosphocreatine is about five-fold higher than that of ATP. When ATP is needed, creatine kinase catalyzes rapid replenishment of ATP

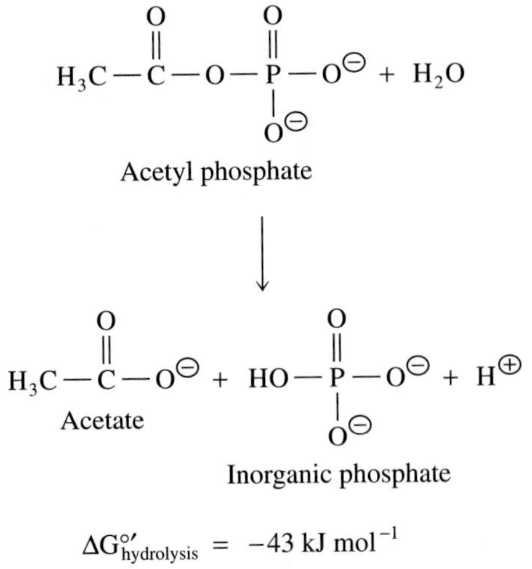

Figure 14·16
Hydrolysis of acetyl phosphate to acetate and inorganic phosphate.

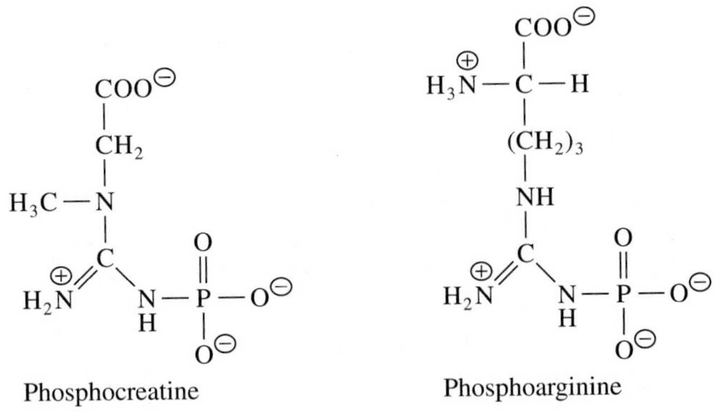

Figure 14·17
Structures of phosphocreatine and phosphoarginine.

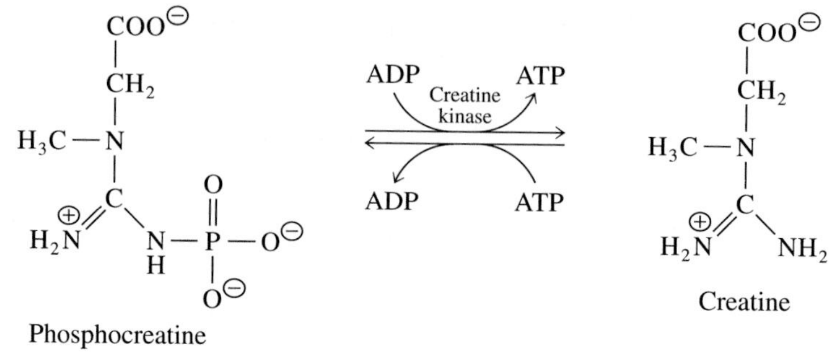

Part 3 Metabolism and Bioenergetics

Figure 14·18
Phosphoryl-group transfer from phospho-creatine to ADP with formation of ATP and creatine. Under cellular conditions, the substrate and product concentrations and the activity of creatine kinase are such that this is a near-equilibrium reaction.

Phosphocreatine

Creatine

through transfer of the activated phosphoryl group from phosphocreatine to ADP (Figure 14·18). In many invertebrates—notably mollusks and arthropods—phosphoarginine is the source of the activated phosphoryl group.

Phosphoenolpyruvate, an intermediate of the glycolytic pathway, has the most energy-rich phosphate linkage known. The standard free energy of phosphoenolpyruvate hydrolysis is $-62\,\text{kJ mol}^{-1}$. Phosphoenolpyruvate is neither a phosphoanhydride nor a phosphoamide but an enol ester. The hydrolysis of phosphoenolpyruvate can be divided into three steps (Figure 14·19). First, cleavage of the P—O bond of the phosphate ester produces inorganic phosphate and a resonance-stabilized enolate anion. The unstable enolate anion is then protonated to yield enolpyruvate. Finally, tautomerization of enolpyruvate yields pyruvate. The free energy associated with phosphoenolpyruvate can be understood by considering the molecule as an enol whose structure is locked by attachment of the phosphoryl group. When the phosphoryl group is removed, the molecule can assume the much more stable keto form.

Transfer of a phosphoryl group from phosphoenolpyruvate to ADP is catalyzed by the enzyme pyruvate kinase. $\Delta G^{\circ\prime}$ for the reaction is about $-32\,\text{kJ mol}^{-1}$ ($-62 + 30\,\text{kJ mol}^{-1}$). Thus, the equilibrium for this reaction under standard conditions lies far in the direction of transfer of the phosphoryl group from phosphoenolpyruvate to ADP, with formation of pyruvate and ATP. In cells, this reaction is an important source of ATP.

Figure 14·19
Hydrolysis of phosphoenolpyruvate. The reaction can be divided into three steps. (1) The hydrolysis step produces a resonance-stabilized enolate anion and inorganic phosphate. (2) The enolate anion is protonated. (3) Tautomerization converts the enol to a ketone.

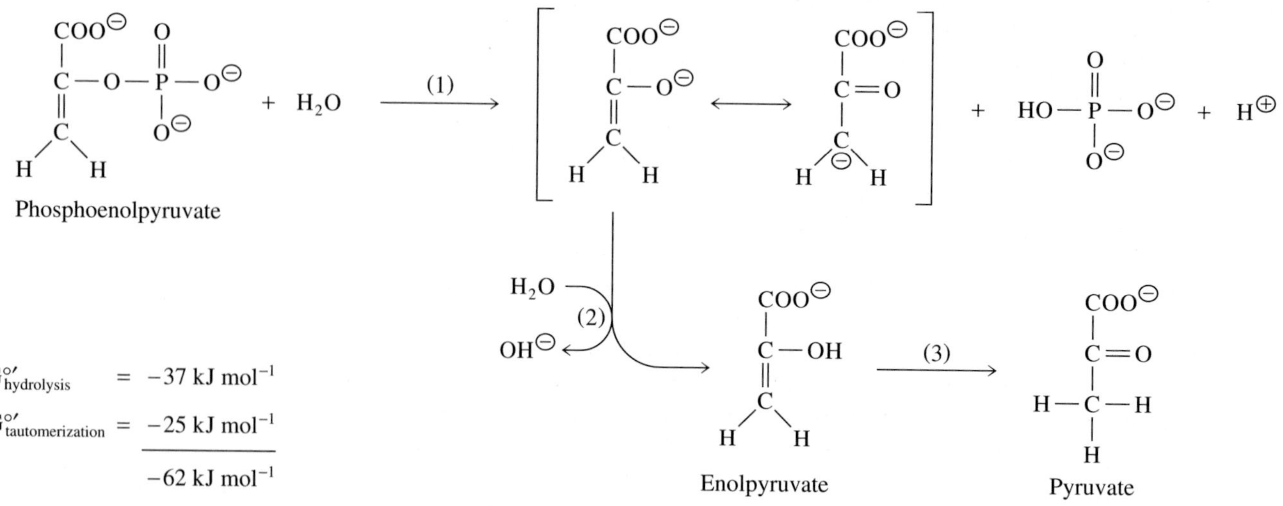

Phosphoenolpyruvate

Resonance-stabilized enolate anion

Enolpyruvate

Pyruvate

(1) $\Delta G^{\circ\prime}_{\text{hydrolysis}} = -37\,\text{kJ mol}^{-1}$

(3) $\Delta G^{\circ\prime}_{\text{tautomerization}} = -25\,\text{kJ mol}^{-1}$

$-62\,\text{kJ mol}^{-1}$

14·12 Acyl-Group Transfer Is Also Important in Metabolism

In addition to phosphoryl-group transfer, a number of metabolic reactions involve the transfer of acyl groups from an acyl coenzyme A (or acyl carrier protein) to an acceptor molecule. Recall that acyl groups are attached to the coenzyme via a thioester linkage (Section 8·8).

$$R-\overset{\overset{\displaystyle O}{\|}}{C}-S\text{-Coenzyme} \qquad (14\cdot19)$$

Sulfur is in the same group of the periodic table as oxygen; however, thioesters are less stable than oxygen esters because resonance structures that contribute to the stability of oxygen esters are not available to thioesters. The standard free energy of hydrolysis of acetyl CoA molecules is -31 kJ mol^{-1}, about the same as ATP hydrolysis (Figure 14·20).

The high energy of the acyl coenzyme is used in the transfer of acyl groups, for instance, when two-carbon acetate units are donated to the four-carbon compound oxaloacetate in the first step of the citric acid cycle and when acetyl CoA condenses with a growing acyl chain during the biosynthesis of fatty acids.

$$H_3C-\overset{\overset{\displaystyle O}{\|}}{C}-S\text{-CoA}$$
Acetyl coenzyme A

$$\Big\downarrow \begin{array}{l} -H_2O \\ \searrow HS\text{-CoA} \end{array}$$

$$H_3C-\overset{\overset{\displaystyle O}{\|}}{C}-O^{\ominus} + H^{\oplus}$$
Acetate

$$\Delta G^{\circ\prime}_{hydrolysis} = -31 \text{ kJ mol}^{-1}$$

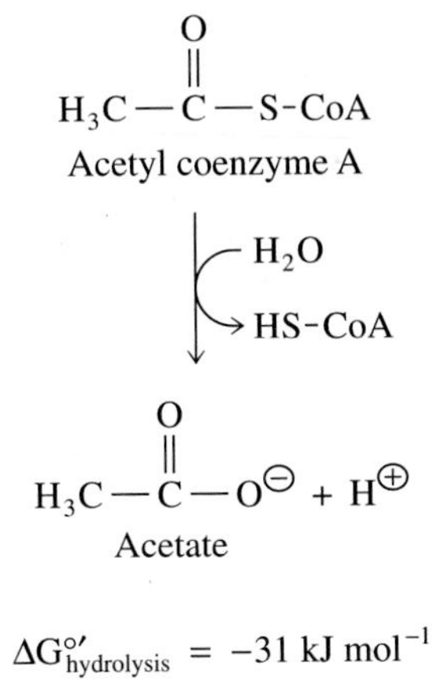

Figure 14·20
Hydrolysis of acetyl coenzyme A to coenzyme A and acetate.

14·13 The Free Energy of Biological Oxidation Reactions Can Be Captured in the Form of Reduced Coenzymes

Carbohydrates, lipids, and amino acids are oxidized in catabolic reactions. The oxidation of one molecule is necessarily coupled with the reduction of another molecule. A molecule that accepts electrons and is reduced is an **oxidizing agent.** A molecule that loses electrons and is oxidized is a **reducing agent.** The net oxidation-reduction reaction is

$$A_{red} + B_{ox} \rightleftharpoons A_{ox} + B_{red} \qquad (14\cdot20)$$

The electrons released in biological oxidation reactions are most often transferred enzymatically to one of two oxidizing agents, either nicotinamide adenine dinucleotide (NAD$^{\oplus}$) or nicotinamide adenine dinucleotide phosphate (NADP$^{\oplus}$) in reactions catalyzed by dehydrogenases. The structures and functional mechanisms of these two coenzymes were discussed in Section 8·6. When NAD$^{\oplus}$ and NADP$^{\oplus}$ are reduced, their nicotinamide groups accept a hydride ion (Figure 8·17). Lipid-soluble quinones are also important electron carriers (Section 8·16).

The reduced coenzymes NADH and NADPH are similar structurally, yet they have sharply different functions. During oxidative phosphorylation, NADH—produced during catabolic oxidation reactions—is converted to NAD$^{\oplus}$ with concomitant production of ATP. NADPH—generated by oxidation reactions in specialized pathways such as the pentose phosphate pathway (Section 17·11)—provides hydride ions for reductive reactions such as those required for the synthesis of fatty acids, amino acids, and nucleotides. The role of both of these coenzymes can be described as supplying reducing power, which is measured quantitatively as **reduction potential.**

Figure 14·21
Schematic diagram of an electrochemical cell. Electrons flow through the external circuit from the zinc anode to the copper cathode. The salt bridge permits the flow of counterions (sulfate ions in this example) without extensive mixing of the two solutions. The electromotive force is measured by the voltmeter connected across the two electrodes.

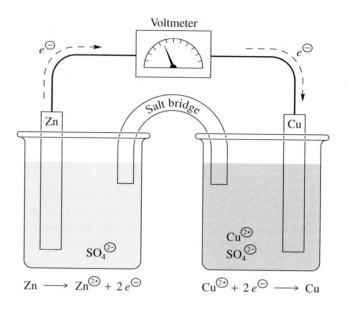

$$Zn \longrightarrow Zn^{2+} + 2\,e^{-} \qquad\qquad Cu^{2+} + 2\,e^{-} \longrightarrow Cu$$

A. Standard Reduction Potential Is Related to Standard Free Energy

Reduction potential can be measured quantitatively in electrochemical cells. A simple oxidation-reduction reaction involving the transfer of a pair of electrons from a zinc atom (Zn) to a copper ion (Cu^{2+}) illustrates the principle.

$$Zn + Cu^{2+} \rightleftharpoons Zn^{2+} + Cu \qquad\qquad \textbf{(14·21)}$$

This reaction can be carried out in two separate solutions that divide the overall reaction into two half-reactions (Figure 14·21). At the anode, two electrons are given up by each zinc atom that reacts (the reducing agent or reductant). The electrons flow through a wire to the cathode, where they reduce Cu^{2+} (the oxidizing agent or oxidant) to metallic copper. A salt bridge—consisting of a tube with a porous partition, filled with electrolyte—preserves electroneutrality by providing an aqueous path for the flow of nonreactive counterions between the two solutions. This separation of ion flow from flow of electrons (i.e., electrical energy) permits measurement of the latter, using a voltmeter.

The direction of the current through the circuit in Figure 14·21 indicates that Zn^{2+} is more easily oxidized than Cu^{2+}. The reading on the voltmeter represents a potential difference—the difference between the reduction potential of the reaction on the left and that on the right. The measured potential is the **electromotive force.**

As was the case with measurements of free energy, it is useful to have a reference standard for measurements of reduction potential. In the case of reduction potential, however, the reference is not simply a set of reaction conditions but a reference half-reaction to which all other half-reactions can be compared. The chosen reference half-reaction is the oxidation of hydrogen gas (H_2), whose reduction potential under standard conditions ($E°$) is arbitrarily set equal to 0.0 volts (V). The standard reduction potential of a given half-reaction is measured with an oxidation-reduction couple in which the reference half-cell contains a solution of 1 M $H^{\oplus}$ and 1 atm $H_2(g)$, and the sample half-cell contains 1 M each of the oxidized and reduced species of the substance whose reduction potential is to be determined. Under standard conditions for biological measurements, the hydrogen ion concentration in the sample half-cell is 10^{-7} M (pH 7.0). The voltmeter across the oxidation-reduction couple measures the difference in the reduction potential between the reference and

Table 14·4 Standard reduction potentials for some important biological half-reactions

Reduction half-reaction	$E^{\circ\prime}$ (V)
Acetyl CoA $+ CO_2 + H^{\oplus} + 2\,e^{\ominus} \longrightarrow$ Pyruvate $+$ CoA	-0.48
Ferredoxin (spinach), $Fe^{3+} + e^{\ominus} \longrightarrow Fe^{2+}$	-0.43
$2\,H^{\oplus} + 2\,e^{\ominus} \longrightarrow H_2$	-0.42
α-Ketoglutarate $+ CO_2 + 2\,H^{\oplus} + 2\,e^{\ominus} \longrightarrow$ Isocitrate	-0.38
Lipoyl dehydrogenase (FAD) $+ 2\,H^{\oplus} + 2\,e^{\ominus} \longrightarrow$ Lipoyl dehydrogenase (FADH$_2$)	-0.34
$NADP^{\oplus} + 2\,H^{\oplus} + 2\,e^{\ominus} \longrightarrow$ NADPH $+ H^{\oplus}$	-0.32
$NAD^{\oplus} + 2\,H^{\oplus} + 2\,e^{\ominus} \longrightarrow$ NADH $+ H^{\oplus}$	-0.32
Lipoic acid $+ 2\,H^{\oplus} + 2\,e^{\ominus} \longrightarrow$ Dihydrolipoic acid	-0.29
Glutathione (oxidized) $+ 2\,H^{\oplus} + 2\,e^{\ominus} \longrightarrow$ 2 Glutathione (reduced)	-0.23
FAD $+ 2\,H^{\oplus} + 2\,e^{\ominus} \longrightarrow$ FADH$_2$	-0.22
FMN $+ 2\,H^{\oplus} + 2\,e^{\ominus} \longrightarrow$ FMNH$_2$	-0.22
Acetaldehyde $+ 2\,H^{\oplus} + 2\,e^{\ominus} \longrightarrow$ Ethanol	-0.20
Pyruvate $+ 2\,H^{\oplus} + 2\,e^{\ominus} \longrightarrow$ Lactate	-0.18
Oxaloacetate $+ 2\,H^{\oplus} + 2\,e^{\ominus} \longrightarrow$ Malate	-0.17
Cytochrome b_5 (microsomal), $Fe^{3+} + e^{\ominus} \longrightarrow Fe^{2+}$	0.02
Fumarate $+ 2\,H^{\oplus} + 2\,e^{\ominus} \longrightarrow$ Succinate	0.03
Ubiquinone (Q) $+ 2\,H^{\oplus} + 2\,e^{\ominus} \longrightarrow$ QH$_2$	0.04
Cytochrome b (mitochondrial), $Fe^{3+} + e^{\ominus} \longrightarrow Fe^{2+}$	0.08
Cytochrome c_1, $Fe^{3+} + e^{\ominus} \longrightarrow Fe^{2+}$	0.22
Cytochrome c, $Fe^{3+} + e^{\ominus} \longrightarrow Fe^{2+}$	0.23
Cytochrome a, $Fe^{3+} + e^{\ominus} \longrightarrow Fe^{2+}$	0.29
Cytochrome f, $Fe^{3+} + e^{\ominus} \longrightarrow Fe^{2+}$	0.36
$NO_3^{\ominus} + e^{\ominus} \longrightarrow NO_2^{\ominus}$	0.42
Photosystem P700	0.43
$Fe^{3+} + e^{\ominus} \longrightarrow Fe^{2+}$	0.77
$\frac{1}{2} O_2 + 2\,H^{\oplus} + 2\,e^{\ominus} \longrightarrow H_2O$	0.82

[Most values are from Loach, P. A. (1968). Oxidation-reduction potentials, absorbance bands and molar absorbance of compounds used in biochemical studies. In *Handbook of Biochemistry: Selected Data for Molecular Biology*, H. A. Sober, ed. (Cleveland, Ohio: CRC Press).]

sample half-reactions. Since the standard reduction potential of the reference half-reaction is 0.0 V, the measured potential is that of the sample half-reaction. The standard reduction potentials ($E^{\circ\prime}$) of some important biological half-reactions are given in Table 14·4. More negative potentials are assigned to reaction systems that have an increasing tendency to donate electrons. Thus, electrons flow spontaneously from more negative to more positive reduction potentials. (This table is analogous to the well-known electromotive-force series, which lists metals in the order of their tendency to reduce other ions. For example, Zn is listed above Cu in the electromotive-force series.)

The standard reduction potential for the transfer of electrons from one molecular species to another is related to the standard free-energy change by the equation

$$\Delta G^{\circ\prime} = -n\mathcal{F}\Delta E^{\circ\prime} \qquad \text{(14·22)}$$

where n is the number of electrons transferred; $\mathcal{F}$ is Faraday's constant (96.48 kJ V^{-1} mol^{-1}); and $\Delta E^{\circ\prime}$ is the difference in volts between the standard reduction potentials of the oxidized and reduced species.

Recall from Equation 14·12 that $\Delta G^{\circ\prime} = -RT \ln K_{eq}$. Combining this equation with Equation 14·22, we get

$$\Delta E^{\circ\prime} = -\frac{RT}{n\mathcal{F}} \ln K_{eq} \qquad (14 \cdot 23)$$

Under biological conditions, the reactants in a system will not be present in standard concentrations of 1 M. Just as the actual free-energy change for a reaction is related to the standard free-energy change by Equation 14·11, observed reduction potential (ΔE) is related to the change in the standard reduction potential ($\Delta E^{\circ\prime}$) by the Nernst equation. For Reaction 14·20 on Page 14·25,

$$\Delta E = \Delta E^{\circ\prime} - \frac{RT}{n\mathcal{F}} \ln \frac{[A_{ox}][B_{red}]}{[A_{red}][B_{ox}]} \qquad (14 \cdot 24)$$

At 298 K, Equation 14·24 reduces to

$$\Delta E = \Delta E^{\circ\prime} - \frac{0.026}{n} \ln Q \qquad (14 \cdot 25)$$

where Q represents the actual concentrations of reduced and oxidized species. If we wish to calculate the electromotive force of a reaction under nonstandard conditions, we use the Nernst equation and substitute the actual concentrations of reactants and products. Keep in mind that a positive ΔE value indicates that a reaction is spontaneous.

B. Electron Transfer from NADH Is an Important Source of Free Energy

In living cells, most of the reduced coenzyme NADH formed in metabolic reactions is oxidized by the respiratory electron-transport chain, with concomitant production of ATP from ADP + P_i. NADH, and to a lesser extent QH_2, acts as a conduit delivering the energy of biological oxidation reactions to the site of oxidative phosphorylation. The ultimate acceptor of the electrons from NADH is oxygen. It is possible to calculate the free-energy change associated with this overall oxidation-reduction reaction under standard conditions by adding the standard reduction potentials of the two half-reactions and using Equation 14·22. The two half-reactions, from Table 14·4, are

$$NAD^{\oplus} + 2 H^{\oplus} + 2 e^{\ominus} \longrightarrow NADH + H^{\oplus} \qquad E^{\circ\prime} = -0.32 \text{ V} \qquad (14 \cdot 26)$$

and

$$\tfrac{1}{2}O_2 + 2 H^{\oplus} + 2 e^{\ominus} \longrightarrow H_2O \qquad E^{\circ\prime} = 0.82 \text{ V} \qquad (14 \cdot 27)$$

Since the NADH half-reaction has the more negative standard reduction potential, NADH will be oxidized whereas oxygen will be reduced. The net reaction is

$$NADH + \tfrac{1}{2}O_2 + H^{\oplus} \longrightarrow NAD^{\oplus} + H_2O \qquad \Delta E^{\circ\prime} = 1.14 \text{ V} \qquad (14 \cdot 28)$$

Using Equation 14·22,

$$\Delta G^{\circ\prime} = -(2)(96.48 \text{ kJ V}^{-1}\text{mol}^{-1})(1.14 \text{ V}) = -220 \text{ kJ mol}^{-1} \qquad (14 \cdot 29)$$

The standard free-energy change for the formation of ATP from ADP + P_i is 30 kJ mol^{-1} (greater under the conditions of the living cell, as noted earlier). The energy released during the oxidation of NADH under cellular conditions is sufficient to drive the formation of several molecules of ATP. A more precise examination of the ratio of NADH oxidized to ATP formed will be given in Chapter 18, where other features of the process of oxidative phosphorylation that affect this ratio will be taken into account.

14·14 Metabolic Pathways Are Studied by a Variety of Methods

The complexity of many metabolic pathways makes them difficult to study. Reaction conditions used with isolated reactants in the test tube (in vitro) are often very different from the reaction conditions in the intact cell (in vivo). The study of the chemical events of metabolism is one of the oldest branches of biochemistry, and a variety of approaches have been developed to characterize the enzymes, intermediates, flux, and regulation of metabolic pathways.

A classical approach to unravelling metabolic pathways is to add a substrate to preparations of tissues, cells, or subcellular fractions and then follow the emergence of intermediates and end products. The fate of a substrate is easier to trace when the substrate has been specifically labelled. Since the advent of nuclear chemistry, isotopic tracers have been used to map the transformations of metabolites. For example, compounds containing atoms of radioactive isotopes such as ^3H or ^{14}C can be added to cells or other preparations, and the radioactive compounds produced by anabolic or catabolic reactions can be purified and identified. Verification of the steps of a particular pathway can be accomplished by reproducing in vitro the separate reactions using isolated substrates and enzymes. Individual enzymes for almost all known metabolic steps have been isolated. By determining the substrate specificity and kinetic properties of a purified enzyme, it is possible to draw some conclusions regarding the regulatory role of that enzyme. However, a complete assessment of the regulation of a pathway requires analysis of metabolite concentrations in the intact cell or organism under various conditions.

A valuable source of information is provided by studying mutations in single genes that are associated with the abnormality of single proteins. While some mutations are lethal and not transmitted to subsequent generations, others can be tolerated by the descendants. Investigation of these mutant organisms has helped identify enzymes and intermediates of metabolic pathways. Typically, a defective enzyme results in deficiency of its product and accumulation of its substrate or a product derived from the substrate by a branch pathway. This approach has been extremely successful in elucidating metabolic pathways in simple organisms such as bacteria and yeast. In humans, enzyme defects are manifested in metabolic diseases. Over 300 single-gene diseases are known. Some are extremely rare and others fairly common; some are tragically severe. In cases where the metabolic disorder produces mild symptoms, it appears that the network of metabolic reactions contains enough overlap and redundancy to allow near-normal development of the organism.

In a similar fashion, investigation of the actions of metabolic inhibitors has aided in the identification of individual steps in metabolic pathways. Inhibition of one step of a pathway affects the entire pathway. Because the substrate of the inhibited enzyme accumulates, it can be isolated and characterized more easily. You will see in Chapter 18 how inhibitors were used to decipher the sequence of electron carriers in the respiratory electron-transport chain. The effects of a medicinal drug on cellular metabolism indicate the mechanism of action of the drug and often suggest ways in which the drug can be improved. There are a number of compounds—particularly toxins—whose biological effects have been known for centuries but only recently characterized at the biochemical level.

In instances where natural mutations are not available, mutants have been generated by treatment with radiation or chemical agents that cause mutations (Chapter 26). By producing a series of mutants, isolating them, and examining their nutritional requirements and accumulated metabolites, entire pathways have been delineated. More recently, site-directed mutagenesis (Section 6·10) has proven valuable in defining the roles of enzymes. Bacterial and yeast systems have been most widely used for introduction of mutations because these organisms can be grown in large numbers in a short period of time. It is possible to produce animal models—particularly insects and nematodes—in which certain genes are not expressed. It is also possible to delete certain genes in vertebrates. "Gene knockout" mice, for instance, provide an experimental system for investigating the complexities of mammalian metabolism.

Summary

The chemical reactions that occur in cells are collectively called metabolism. These reactions can be classified as catabolic (degradative) or anabolic (synthetic) reactions. Metabolic activities allow cells to maintain intracellular conditions different from those of the environment, to extract energy from external sources, to grow and reproduce, and to respond to internal or external influences.

Sequences of reactions are called pathways. Although the start and end points are sometimes arbitrary, metabolic pathways may be linear, cyclic, or spiral; the pathways may also branch. Enzymes are required for cells to carry out reactions under conditions of moderate temperature, pressure, and pH. Degradative and synthetic pathways proceed in stepwise fashion, with the participation of energy carriers such as nucleoside triphosphates and nicotinamide coenzymes.

The major catabolic pathways in cells convert macromolecules to smaller energy-yielding metabolites, or fuels. The smaller compounds also serve as building blocks for the synthesis of new macromolecules. Glucose, fatty acids, and some amino acids are oxidized to form acetyl CoA, which enters the citric acid cycle, the common pathway of oxidative metabolism. The energy released in catabolic reactions is conserved in the form of nucleoside triphosphates and reduced coenzymes. The energy of the reduced compounds is used to synthesize ATP from ADP and P_i by the process of oxidative phosphorylation.

Metabolic pathways are regulated to allow the organism to use fuel sources efficiently and to respond to changing demands. Responses may involve many pathways or only a few. The flux, or flow of material through a pathway, usually depends on regulation of multiple steps, each with a particular control strength for the overall pathway. Feedback inhibition and feed-forward activation are commonly encountered. Regulation of particular enzymes may be accomplished by allosteric modulation, reversible covalent modification, and changes in the rate of enzyme synthesis or degradation.

In multicellular organisms, tissues are specialized for different metabolic tasks. Interorgan metabolism may be coordinated by hormones.

The laws of thermodynamics apply to metabolic reactions, which are in a steady state, not at equilibrium. The direction of a chemical or enzyme-catalyzed reaction depends on the change in free energy. Reactions occur spontaneously only when the free-energy change is negative. The standard free-energy change of a reaction is related to the equilibrium constant of the reaction by the formula $\Delta G^{\circ\prime} = -RT \ln K_{eq}$.

In cells, the change in free energy (ΔG) that occurs during a given reaction depends primarily on the concentrations of reactants and products and usually differs from the standard free-energy change ($\Delta G^{\circ\prime}$). Each reaction of a metabolic pathway proceeds with a negative free-energy change. The concentrations of reactants and products in many cellular reactions approach the equilibrium state; such reactions are called near-equilibrium reactions. Reactions for which the steady-state concentrations of reactants are far from equilibrium are called metabolically irreversible reactions.

ATP plays a central role in energy metabolism. The energy released by one biological process is often conserved in the form of ATP, to be used by other, energy-requiring processes. The energy of ATP is made available when a terminal phosphoryl group or a nucleotidyl group is transferred. The action of adenylate kinase helps maintain a constant concentration of ATP in cells. The concentrations of ADP and AMP are much lower.

Phosphoryl-group transfer from another energy-rich substrate to ADP forms ATP. In addition to nucleoside triphosphates, there are several other metabolites with activated phosphoryl groups, including acetyl phosphate, the phosphagens, and phosphoenolpyruvate.

Acyl-group transfer is an important metabolic reaction. The transfer of acyl groups from coenzyme A proceeds with a large negative free-energy change.

The free energy of biological oxidation reactions can be captured in the form of reduced coenzymes. This form of energy is measured as reduction potential, the quantitative measure of the ability of a molecule to donate electrons. Standard reduction potential is related to standard free-energy change by the formula $\Delta G^{\circ\prime} = -n \mathcal{F} \Delta E^{\circ\prime}$. Under nonstandard conditions, reduction potential is given by the Nernst equation.

Selected Readings

Bridger, W. A., and Henderson, J. F. (1983). *Cell ATP* (New York: John Wiley & Sons). A lucid treatment of the chemistry and metabolic function of ATP, including in-depth discussion of the reasons for constant cellular ATP concentrations and pathologies associated with elevated or depressed ATP concentrations.

Hanson, R. W. (1989). The role of ATP in metabolism. *Biochem. Ed.* 17:86–92.

Harold, F. M. (1986). *The Vital Force: A Study of Bioenergetics* (New York: W. H. Freeman and Company). Chapters 1 and 2.

Ingraham, L. L., and Pardee, A. B. (1967). Free energy and entropy in metabolism. In *Metabolic Pathways*, 3rd ed., Vol. 1, D. M. Greenberg, ed. (New York: Academic Press). The basic thermodynamics of living organisms.

Klotz, I. M. (1967). *Energy Changes in Biochemical Reactions* (New York: Academic Press).

Newsholme, E. A., and Leech, A. R. (1986). *Biochemistry for the Medical Sciences* (New York: John Wiley & Sons). Chapters 1 and 2. An introduction to thermodynamics and methods used in the study of metabolism.

Stephanopoulos, G., and Vallino, J. J. (1991). Network rigidity and metabolic engineering in metabolite overproduction. *Science* 252:1675–1681. Discusses why control functions must be altered as well as enzymes amplified for metabolite overproduction.

van Holde, K. E. (1985). *Physical Biochemistry*, 2nd ed. (Englewood Cliffs, New Jersey: Prentice Hall).

15

Glycolysis

The first two pathways we will examine in our study of metabolism carry out reactions of carbohydrate metabolism: glycolysis is examined in this chapter, and the citric acid cycle is presented in the chapter that follows. Both pathways have major roles in energy metabolism, and both are also involved in the formation and degradation of other types of molecules, such as amino acids and lipids. The amount of energy obtained from carbohydrate catabolites is maximized when they proceed through the glycolytic pathway and are then further acted upon by the citric acid cycle.

The relationship between the two pathways is shown in Figure 15·1 (next page). In **glycolysis,** glucose is converted to the three-carbon acid pyruvate. Pyruvate has a number of possible fates, of which three will principally concern us. Pyruvate may undergo oxidative decarboxylation to form acetyl CoA, the metabolite that enters the citric acid cycle. In some microorganisms, pyruvate is converted to ethanol, the end product of an energy-producing process called alcoholic fermentation. And in some circumstances, pyruvate can be reversibly reduced to lactate. These pathways, seemingly different, are simply variations on the glycolytic theme.

The citric acid cycle is the route by which the acetate moiety of acetyl CoA is oxidized to carbon dioxide and water. The complete oxidation of acetate is exergonic, with energy conserved in the reduced forms of the oxidizing agents $NAD^{\oplus}$ and ubiquinone (Q). Recovery of some of this energy as ATP is described in Chapter 18, which examines the terminal oxidation reactions of aerobic metabolism (oxidation of NADH and QH_2 by O_2) and how these oxidation reactions are coupled to the phosphorylation of ADP.

In the summer of 1897, the doctrine of vitalism was discredited by experiments that contributed to the birth of biochemistry. A German chemist, Martin Hahn, was attempting to extract proteins from yeast by grinding yeast cells in a mortar with fine sand and diatomaceous earth. The mixture was wrapped in cheesecloth and placed in a press. The resulting yeast extract was difficult to preserve. Remembering that fruit preserves are made by adding sugar, Hans Büchner, a colleague, suggested mixing sucrose with the yeast extracts. The idea was tested by Hans's brother Eduard (Figure 15·2, next page), who discovered that bubbles of carbon dioxide evolved from the mixture. Eduard Büchner concluded that fermentation was occurring. More than 20 years earlier, Louis Pasteur had shown that yeast ferment

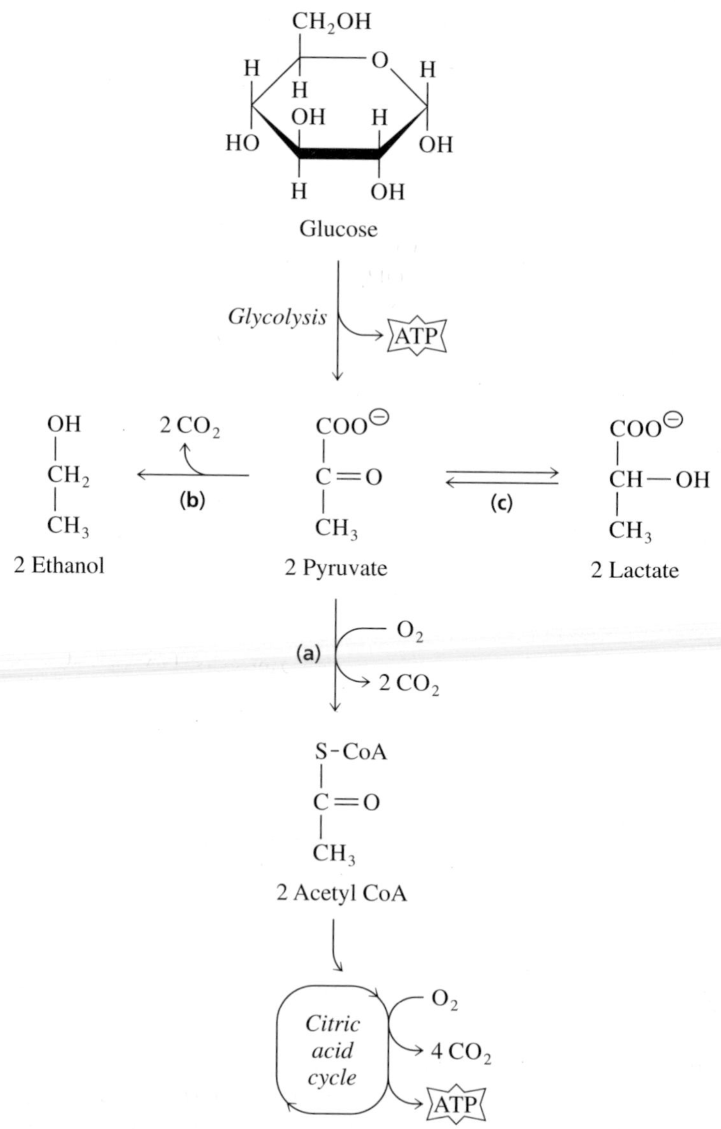

Figure 15·1

Catabolism of glucose. One molecule of glucose is converted to two molecules of pyruvate by the glycolytic pathway. Three major fates of pyruvate are shown. **(a)** Under aerobic conditions, pyruvate is oxidized to the acetyl group of acetyl CoA, which can be oxidized by the citric acid cycle. A limited amount of ATP is generated in glycolysis; much more ATP is formed from products of the citric acid cycle. **(b)** Glucose is fermented to ethanol by certain microorganisms under anaerobic conditions. **(c)** Glucose undergoes anaerobic glycolysis to lactate in certain cells.

Glucose

Glycolysis → ATP

2 Ethanol ← (b) ← 2 CO_2 — 2 Pyruvate ⇌ (c) ⇌ 2 Lactate

(a) O_2 → 2 CO_2

2 Acetyl CoA

Citric acid cycle — O_2 → 4 CO_2 → ATP

Figure 15·2
Eduard Büchner (1860–1917).

sugar to alcohol in the absence of oxygen but not when grown in air. He characterized fermentation as "life without air." Pasteur extended his results to conclude that fermentation requires *living* organisms. In a manuscript published in 1898, Eduard Büchner concluded that "such a complicated apparatus as the yeast cell presents is not required. It is considered that the bearer of the fermenting action of the press juice is much more probably a dissolved substance, doubtless a protein; this will be called *zymase*." Today, we recognize that the zymase of yeast extracts is not a single enzyme but actually a mixture of enzymes that together catalyze the reactions of glycolysis. For his discovery of cell-free fermentation, Eduard Büchner was awarded the Nobel prize in 1907.

The reactions of the glycolytic pathway were gradually elucidated by analyzing the reactions catalyzed by yeast juice and muscle extracts. In 1905, Arthur Harden and William John Young found that the rate of fermentation of glucose by yeast juice decreased over time but could be restored by adding inorganic phosphate. Harden and Young assumed that phosphate derivatives of glucose were being formed, so they sought such compounds and succeeded in isolating fructose 1,6-*bis*phosphate, which they referred to as hexose diphosphate. The role of fructose 1,6-*bis*phosphate as an intermediate in the fermentation of glucose was established by the demonstration that it too was fermented by cell-free yeast extracts.

Harden and Young also found that yeast juice lost its fermentative activity when it was dialyzed. Their demonstration that boiled yeast juice (which contains heat-stable cofactors) restored activity led to the later discovery of the coenzyme $NAD^{\oplus}$.

Many researchers helped establish the details of the glycolytic pathway. At one time it was called the Embden-Meyerhof pathway, and sometimes even the Embden-Meyerhof-Warburg-Christian-Parnas pathway, to honor some of the pioneer investigators of the first pathway to be completely elucidated. By the 1940s, the complete pathway—including its enzymes, intermediates, and coenzymes—was known. The further characterization of its individual enzymes and investigations into the regulation of glycolysis and its integration with other pathways have taken many more years.

In the beginning of this book, we discussed plateaus in the study of biochemistry (Section 1·7). We have reached the second plateau, the study of metabolic interconversions. You will find in upcoming chapters that many biomolecules and many enzyme names are introduced. If a systematic approach is taken, these pathways will not seem overwhelming. It is essential to learn the structures of the metabolites and the names of the enzymes, since in many cases they will be encountered again in connection with other pathways. Keep in mind that the chemical structures of the metabolites prompt the enzyme names, and the names of the enzymes reflect the substrate specificity and the type of reaction catalyzed. With a confident grasp of terminology, you will be prepared to enjoy the chemical elegance of metabolism. However, it is important not to lose sight of the important concepts and general strategies of metabolism while memorizing the details. The names of particular enzymes might fade from memory after the final exam, but we hope you will retain an understanding of the patterns and purposes behind the interconversion of metabolites in cells.

15·1 Glycolysis Is a Ubiquitous Cellular Pathway

The pathway of glycolysis is a sequence of 10 enzyme-catalyzed reactions by which glucose is converted to pyruvate (Figure 15·3, Pages 15·4 and 15·5). The enzymes of this pathway are located in the cytosol. The conversion of one molecule of glucose to two molecules of pyruvate is accompanied by the net conversion of two molecules of ADP to ATP. The glycolytic pathway is found in virtually all cells; for some it is the only ATP-producing pathway.

The net reaction of glycolysis is shown in Equation 15·1. In addition to the production of ATP, two molecules of $NAD^{\oplus}$ are reduced to NADH.

$$\text{Glucose} + 2\,\text{ADP} + 2\,\text{NAD}^{\oplus} + 2\,\text{P}_i \longrightarrow$$
$$2\,\text{Pyruvate} + 2\,\text{ATP} + 2\,\text{NADH} + 2\,\text{H}^{\oplus} + 2\,\text{H}_2\text{O} \tag{15·1}$$

In the first and third reactions of glycolysis, a phosphoryl group is transferred to the hexose substrate. Phosphorylation helps retain glucose inside the cell, since phosphorylated glucose is not a substrate for the transporter that conveys glucose in and out of the cell. The phosphoryl groups are also central to the mechanism of energy conservation. Two intermediates of glycolysis have sufficient group-transfer potential to allow transfer of a phosphoryl group to ADP, producing ATP.

The enzymes of the glycolytic pathway that act on phosphorylated (and therefore anionic) substrates possess anion-binding residues in their active sites. In some enzymes, arginine or lysine residues fill this role. In addition, the positively charged N-terminus of a helix can be positioned in such a way that it binds the phosphate group of a substrate. (Recall that an α helix is dipolar, positive at the N-terminus and negative at the C-terminus.)

Figure 15·3
Conversion of glucose to pyruvate by glycolysis. Following interconversion of glyceraldehyde 3-phosphate and dihydroxyacetone phosphate, the remaining reactions of glycolysis are traversed by two triose molecules for each hexose molecule metabolized. ATP is consumed in the hexose stage and generated in the triose stage.

(2 ATP Consuming Part)

Transfer of a phosphoryl group from ATP to glucose

① Hexokinase, glucokinase
(irreversible)

not comitted step because this can be converted to other products.

Glucose

Glucose 6-phosphate

Isomerization

Phosphoglucoisomerase
② Glucose 6-phosphate isomerase
(Reversible)

Fructose 6-phosphate

Transfer of a second phosphoryl group from ATP to fructose 6-phosphate

③ Phosphofructokinase-1
or 6-phosphofructo 1 kinase.
Regalatory enzyme.
V. import

comitted step.

Fructose 1,6-bisphosphate

C-3—C-4 bond cleavage, yielding two triose phosphates

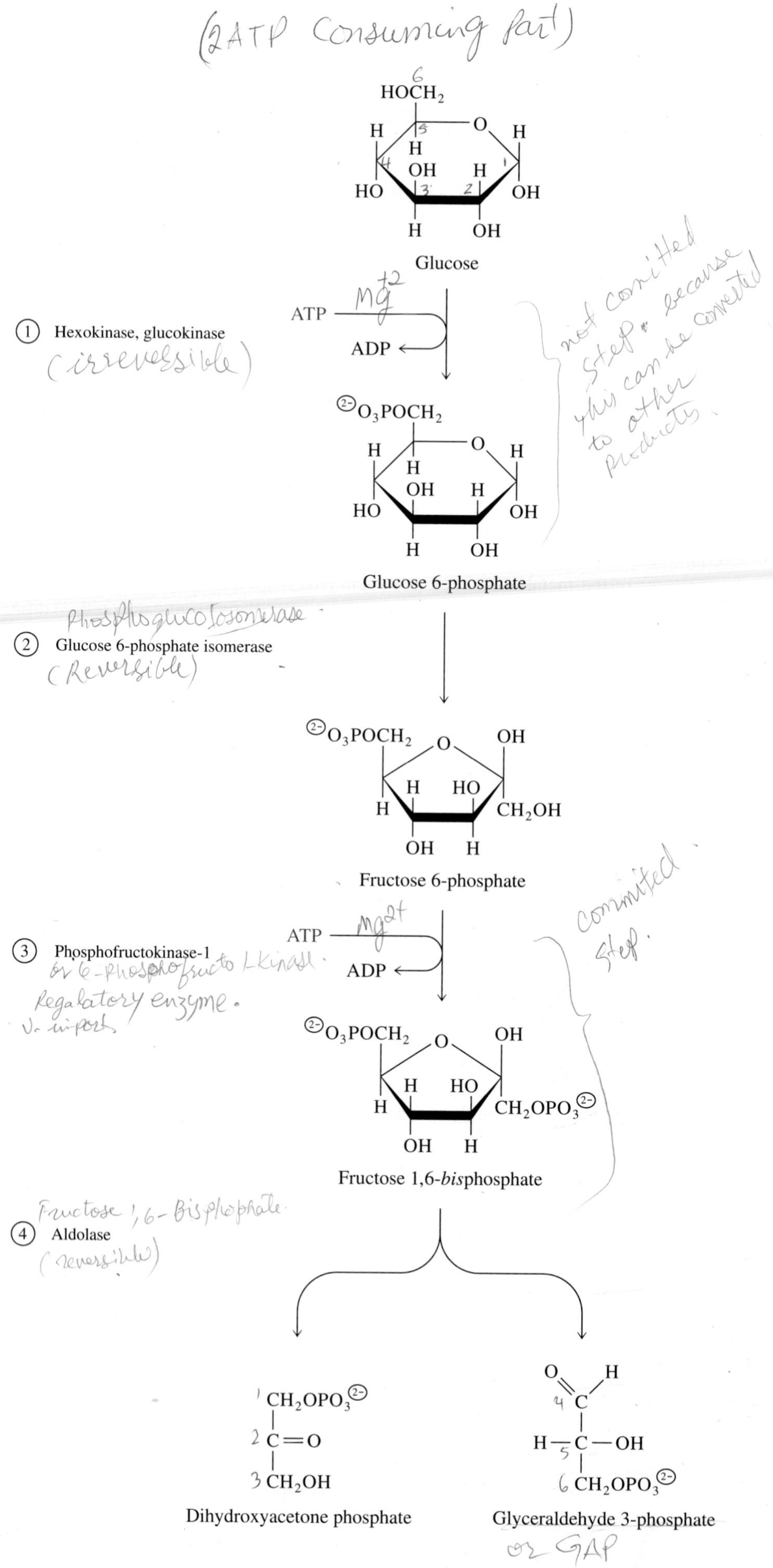

Fructose 1,6-Bisphosphate.
④ Aldolase
(reversible)

Dihydroxyacetone phosphate   Glyceraldehyde 3-phosphate
or GAP

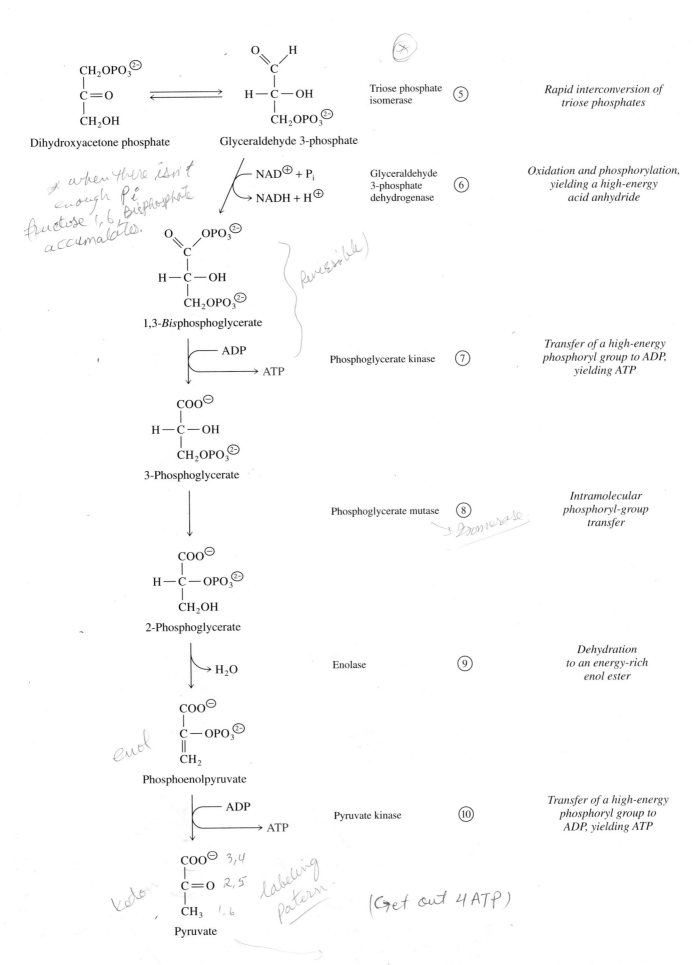

Dihydroxyacetone phosphate

Glyceraldehyde 3-phosphate

Triose phosphate isomerase ⑤

Rapid interconversion of triose phosphates

when there isn't enough Pi fructose 1,6 Bisphosphate accumulates.

NAD⊕ + Pᵢ → NADH + H⊕

Glyceraldehyde 3-phosphate dehydrogenase ⑥

Oxidation and phosphorylation, yielding a high-energy acid anhydride

1,3-*Bis*phosphoglycerate

(Reversible)

ADP → ATP

Phosphoglycerate kinase ⑦

Transfer of a high-energy phosphoryl group to ADP, yielding ATP

3-Phosphoglycerate

Phosphoglycerate mutase ⑧

→ Isomerase

Intramolecular phosphoryl-group transfer

2-Phosphoglycerate

H₂O

Enolase ⑨

Dehydration to an energy-rich enol ester

enol

Phosphoenolpyruvate

ADP → ATP

Pyruvate kinase ⑩

Transfer of a high-energy phosphoryl group to ADP, yielding ATP

keto

COO⊖ 3,4
C=O 2,5
CH₃ 1,6

labeling pattern

(Get out 4 ATP)

Pyruvate

Table 15·1 The enzymatic reactions of glycolysis

Reaction	Enzyme
1. Glucose + ATP $\longrightarrow$ Glucose 6-phosphate + ADP + $H^{\oplus}$	Hexokinase, glucokinase
2. Glucose 6-phosphate $\rightleftarrows$ Fructose 6-phosphate	Glucose 6-phosphate isomerase
3. Fructose 6-phosphate + ATP $\longrightarrow$ Fructose 1,6-*bis*phosphate + ADP + $H^{\oplus}$	Phosphofructokinase-1
4. Fructose 1,6-*bis*phosphate $\rightleftarrows$ Dihydroxyacetone phosphate + Glyceraldehyde 3-phosphate	Aldolase
5. Dihydroxyacetone phosphate $\rightleftarrows$ Glyceraldehyde 3-phosphate	Triose phosphate isomerase
6. Glyceraldehyde 3-phosphate + $NAD^{\oplus}$ + P_i $\rightleftarrows$ 1,3-*Bis*phosphoglycerate + NADH + $H^{\oplus}$	Glyceraldehyde 3-phosphate dehydrogenase
7. 1,3-*Bis*phosphoglycerate + ADP $\rightleftarrows$ 3-Phosphoglycerate + ATP	Phosphoglycerate kinase
8. 3-Phosphoglycerate $\rightleftarrows$ 2-Phosphoglycerate	Phosphoglycerate mutase
9. 2-Phosphoglycerate $\rightleftarrows$ Phosphoenolpyruvate + H_2O	Enolase
10. Phosphoenolpyruvate + ADP + $H^{\oplus}$ $\longrightarrow$ Pyruvate + ATP	Pyruvate kinase

Pyruvate produced by glycolysis has a number of fates that vary among cells or species, and among the same cells under different metabolic conditions. Under anaerobic conditions in muscle, pyruvate is converted to lactate, accompanied by the regeneration of $NAD^{\oplus}$. Lactate is also produced from pyruvate by some microorganisms under anaerobic conditions, although in the microbiological realm, pyruvate is also converted into a number of other products. For instance, in yeast, pyruvate is converted to ethanol. In the presence of oxygen, pyruvate can be completely oxidized to CO_2 and H_2O. Pyruvate is also a precursor for the biosynthesis of the amino acid alanine in most organisms and of valine and leucine in plants and bacteria.

The 10 reactions of glycolysis can be divided into two phases: the hexose stage and the triose stage. The left page of Figure 15·3 shows the hexose stage. At Step 4, a C—C bond of fructose 1,6-*bis*phosphate is cleaved, and thereafter the intermediates of the pathway are triose phosphates. Note that the two triose phosphates formed from fructose 1,6-*bis*phosphate undergo interconversion, with one triose phosphate—glyceraldehyde 3-phosphate—continuing through the pathway. Therefore, all subsequent steps of the triose stage of glycolysis are traversed by two molecules for each molecule of glucose metabolized. The reactions of glycolysis are summarized in Table 15·1.

In the hexose stage of glycolysis, two molecules of ATP are converted to ADP. In the triose stage, four molecules of ATP are formed from ADP for each molecule of glucose metabolized. Note that the pathway contains two ATP-forming steps, each of which is traversed by both molecules arising from the split hexose molecule. Thus, the net production of ATP by glycolysis is two molecules of ATP per molecule of glucose.

$$
\begin{array}{lll}
\text{ATP consumed per glucose:} & 2 & \text{(hexose stage)} \\
\text{ATP produced per glucose:} & \underline{4} & \text{(triose stage)} \\
\text{Net ATP production per glucose:} & 2 &
\end{array}
\qquad \textbf{(15·2)}
$$

15·2 Glycolysis Has 10 Enzyme-Catalyzed Steps

We will now examine the chemistry and enzymes of each of the reactions of glycolysis in turn. As you read, pay attention to the chemical logic and economy of the pathway. Consider how each chemical reaction prepares a substrate for the next step in the process. Note, for example, that a cleavage reaction converts a hexose to two trioses, not a two-carbon compound and a tetrose. The two trioses are in rapid equilibrium, and this allows both products of the cleavage reaction to be further metabolized by the action of one set of enzymes, not two. Finally, be aware of how ATP energy is both used and produced in glycolysis. We have already seen a number of examples of the transfer of the chemical potential energy of ATP (e.g., in Sections 8·3 and 8·4), but the reactions in this chapter are our first detailed examples of how the energy released by oxidation reactions is conserved.

1. Hexokinase Catalyzes the Phosphorylation of Glucose to Form Glucose 6-Phosphate, Consuming One Molecule of ATP

In the first reaction of glycolysis, glucose is phosphorylated at position 6, producing glucose 6-phosphate. This phosphoryl-group–transfer reaction is catalyzed by hexokinase and consumes a molecule of ATP (Figure 15·4). Kinases, which catalyze phosphoryl-group–transfer reactions, as described in Section 10·5, catalyze four of the reactions in the glycolytic pathway—Steps 1, 3, 7, and 10. All four kinases have mechanisms that include direct nucleophilic attack of a hydroxyl group on the terminal phosphoryl group of ATP. (For comparison, see the mechanism of adenylate kinase in Figure 10·16.)

Hexokinases from yeast and mammalian tissues are the most thoroughly studied. The kinases from both sources have a fairly broad substrate specificity, catalyzing phosphorylation of glucose, mannose, and fructose when it is present at high concentrations. As described in Section 7·15, yeast hexokinase undergoes an induced-fit conformational change when glucose binds. This conformational change helps prevent wasteful hydrolysis of ATP.

Multiple forms of hexokinase occur in both yeast and mammalian tissues. For example, four isozymes—numbered I, II, III, and IV based on their order of appearance in several separation techniques—have been isolated from mammalian liver. All four isozymes are found in other mammalian tissues, in varying proportions. The isozymes differ in kinetic properties, particularly in their K_m values for glucose. Hexokinases I, II, and III have moderately low K_m values (about 10^{-6} to 10^{-4} M), whereas hexokinase IV, also called glucokinase, has a K_m value for glucose of about 10^{-2} M. In the laboratory, glucokinase can catalyze the phosphorylation of fructose as well as glucose, but only when fructose is present at extremely high concentrations; at physiological concentrations, the activity of glucokinase with fructose is negligible.

The reaction catalyzed by hexokinase is a site of regulation of glycolysis. Hexokinase is allosterically inhibited by its immediate product, glucose 6-phosphate. The regulation of glycolysis will be discussed in detail in Section 15·7.

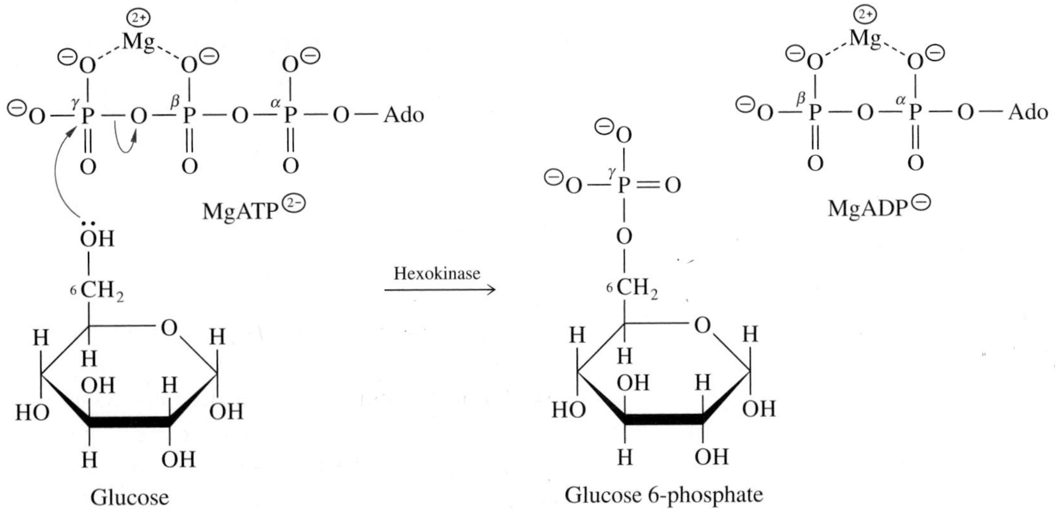

Figure 15·4
Phosphoryl-group–transfer reaction catalyzed by hexokinase. This reaction occurs by attack of the C-6 hydroxyl group of glucose on the γ-phosphorus of MgATP^{2-}. MgADP$^{\ominus}$ is displaced and glucose 6-phosphate is generated. Mg$^{2\oplus}$, shown explicitly here, is also required in the other kinase reactions in this chapter, although it is not shown. (*Ado* represents the adenosyl moiety of ATP and ADP.)

Glucose

Hexokinase

Glucose 6-phosphate

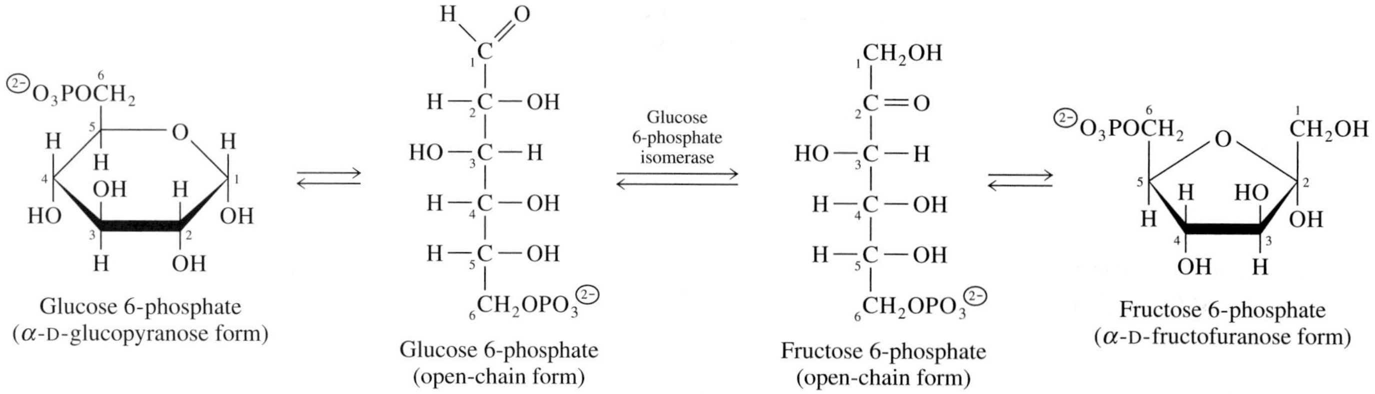

Glucose 6-phosphate
(α-D-glucopyranose form)

Glucose 6-phosphate
(open-chain form)

Glucose
6-phosphate
isomerase

Fructose 6-phosphate
(open-chain form)

Fructose 6-phosphate
(α-D-fructofuranose form)

Figure 15·5
Conversion of glucose 6-phosphate to fructose 6-phosphate. The reaction is an aldose-ketose isomerization catalyzed by glucose 6-phosphate isomerase.

2. Glucose 6-Phosphate Isomerase Catalyzes the Conversion of Glucose 6-Phosphate to Fructose 6-Phosphate

In the second step of glycolysis, glucose 6-phosphate isomerase catalyzes the conversion of glucose 6-phosphate to fructose 6-phosphate (Figure 15·5). This reaction is an example of an aldose-ketose isomerization; its mechanism is similar to that of triose phosphate isomerase (Section 7·8A). The α anomer of glucose 6-phosphate (α-D-glucopyranose 6-phosphate) preferentially binds to glucose 6-phosphate isomerase. The open-chain form of glucose 6-phosphate is then generated within the active site of the enzyme, and an aldose-to-ketose conversion occurs. The open-chain form of fructose 6-phosphate cyclizes to form α-D-fructofuranose 6-phosphate.

Glucose 6-phosphate isomerase exhibits absolute stereospecificity. When the reaction catalyzed by this enzyme runs in the reverse direction, fructose 6-phosphate (in which C-2 is not chiral) is converted only to glucose 6-phosphate. No mannose 6-phosphate, the C-2 epimer of glucose 6-phosphate, is formed; mannose 6-phosphate isomerase specifically catalyzes isomerization of the epimer. Glucose 6-phosphate isomerase catalyzes a near-equilibrium reaction in cells and therefore is not a control point for glycolysis. (The relationship between distance from equilibrium and regulation is discussed in Chapter 14.)

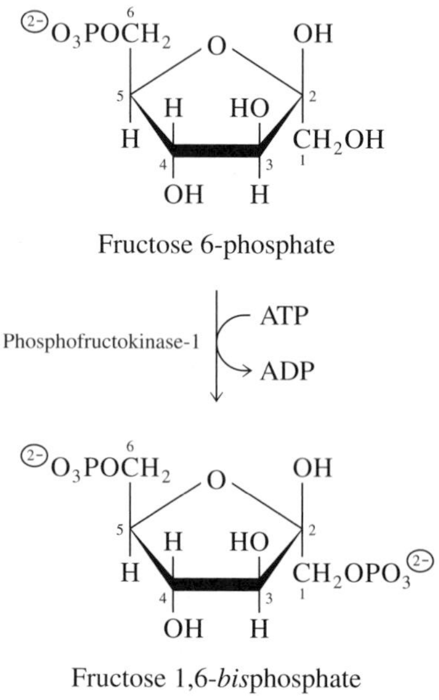

Fructose 6-phosphate

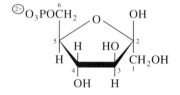

Fructose 1,6-*bis*phosphate

Figure 15·6
Conversion of fructose 6-phosphate to fructose 1,6-*bis*phosphate. The reaction is catalyzed by phosphofructokinase-1. In most cells, this reaction is a critical regulatory step of glycolysis.

3. The Reaction Catalyzed by Phosphofructokinase-1 Consumes a Second Molecule of ATP

Phosphofructokinase-1 (PFK-1) catalyzes the transfer of a phosphoryl group from ATP to the C-1 hydroxyl group of fructose 6-phosphate, producing fructose 1,6-*bis*phosphate (Figure 15·6). Note that although the reaction catalyzed by glucose 6-phosphate isomerase produces α-D-fructose 6-phosphate, it is the β-D anomer that is the substrate for phosphofructokinase-1. The α and β anomers of fructose 6-phosphate equilibrate extremely rapidly in aqueous solution by a nonenzymatic reaction.

PFK-1 is a large, oligomeric enzyme with a molecular weight ranging in different species from about 130 000 to 600 000. PFK-1 catalyzes a metabolically irreversible reaction that is a critical regulatory point for glycolysis in most cells. The logic of regulation at both the first and third steps of glycolysis relates to the source of substrate for the pathway. Glucose enters glycolysis via the regulated step at hexokinase, but hexoses other than glucose can enter glycolysis by conversion to fructose 6-phosphate. For substrates entering the pathway at this step, the reaction catalyzed by PFK-1 is the first committed step of the pathway.

The numerical designation phosphofructokinase-1 is applied to this enzyme because there is a second, more recently discovered phosphofructokinase that catalyzes the synthesis of fructose 2,6-*bis*phosphate. This enzyme, which we will encounter later in this chapter, is known as PFK-2.

A variation in glycolytic metabolism has been observed in the amoeba. This organism possesses a phosphofructokinase that catalyzes transfer of a phosphoryl group to fructose 6-phosphate from inorganic pyrophosphate rather than from ATP.

$$\text{Fructose 6-phosphate} \xrightarrow[\quad]{\text{PP}_i \quad \text{P}_i} \text{Fructose 1,6-}bis\text{phosphate} \qquad \textbf{(15·3)}$$

The amoeba utilizes inorganic pyrophosphate in place of nucleoside triphosphates in several other reactions of carbohydrate metabolism.

4. Aldolase Catalyzes the Cleavage of Fructose 1,6-*Bis*phosphate, Forming Two Triose Phosphates

The first three steps of glycolysis prepare the hexose for cleavage into two triose phosphates, glyceraldehyde 3-phosphate and dihydroxyacetone phosphate. The enzyme that catalyzes this reaction is fructose 1,6-*bis*phosphate aldolase, commonly shortened to aldolase (Figure 15·7). Aldolase catalyzes a near-equilibrium reaction in cells and therefore is not a control point for glycolysis.

The reaction catalyzed by aldolase is an aldol cleavage, frequently encountered in biological systems as a mechanism for cleaving C—C bonds. The reverse, C—C bond formation, is commonly carried out by aldol condensation (base-catalyzed addition of a carbon atom to a carbonyl carbon). The substrate of the aldolase cleavage reaction is an aldol (aldehyde-alcohol), in this case a β-hydroxy-ketone, with C-2 of the fructose skeleton being the carbonyl group and C-4, the

Figure 15·7
Aldolase-catalyzed cleavage of fructose 1,6-*bis*phosphate to dihydroxyacetone phosphate and glyceraldehyde 3-phosphate. Note that dihydroxyacetone phosphate is derived from C-1 to C-3 of fructose 1,6-*bis*phosphate, and glyceraldehyde 3-phosphate is derived from C-4 to C-6 of the hexose.

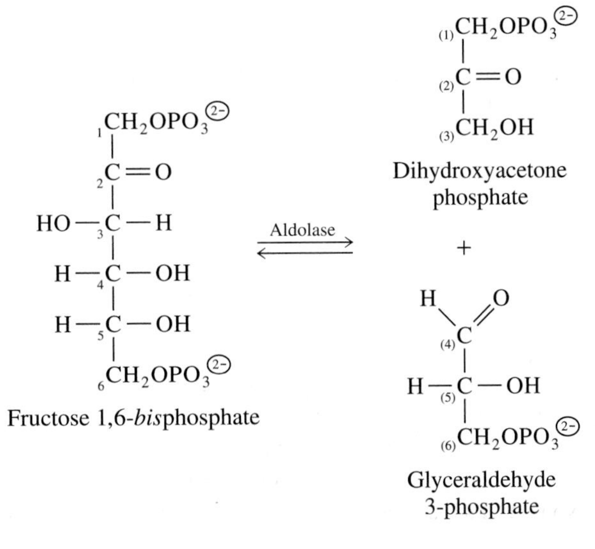

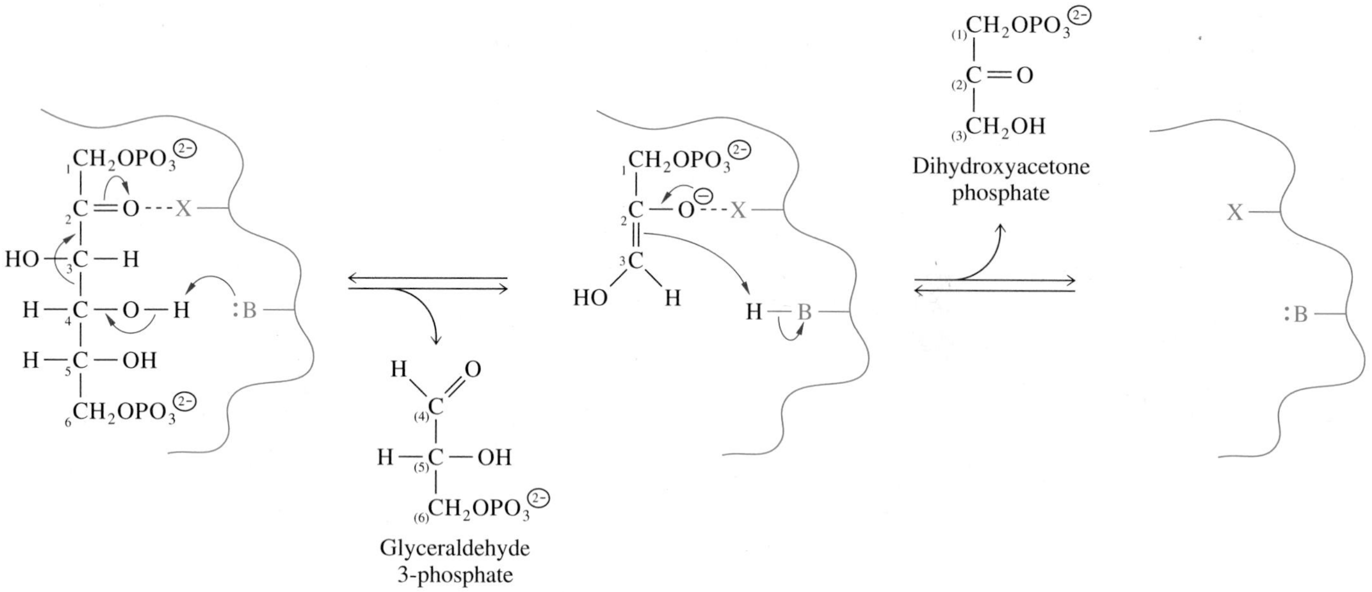

CH$_2$OPO$_3^{2-}$
C=O---X
HO—C—H
H—C—O—H   :B
H—C—OH
CH$_2$OPO$_3^{2-}$

H   O
C
H—C—OH
CH$_2$OPO$_3^{2-}$

Glyceraldehyde
3-phosphate

CH$_2$OPO$_3^{2-}$
C---O$^{\ominus}$---X
C
HO   H
H—B

CH$_2$OPO$_3^{2-}$
C=O
CH$_2$OH

Dihydroxyacetone
phosphate

X
:B

Figure 15·8
Mechanism of the aldol cleavage catalyzed by aldolases. The substrate, fructose 1,6-*bis*-phosphate, is an aldol (not a β-hydroxy-aldehyde in this case but a ketone). Aldolases possess an electron-withdrawing group (X—) that polarizes the C-2 carbonyl group. One class of aldolase has an amino group and the other has Zn$^{2\oplus}$ for this purpose. There is also a basic residue (designated :B—) that removes a proton from the C-4 hydroxyl group.

alcohol. Aldolase possesses both a basic catalytic residue and an electron-withdrawing group (Figure 15·8). In aldolases from plant and animal tissues, a lysine residue at the active site reacts with the carbonyl group of the substrate to form a Schiff base that withdraws electrons. Aldolases from fungi, algae, and some microorganisms use a zinc ion as the electron sink.

The sequence of reactions leading from glucose 6-phosphate to cleavage of fructose 1,6-*bis*phosphate into two triose phosphates makes chemical sense. The aldol grouping required for aldol cleavage is present in both glucose 6-phosphate and fructose 1,6-*bis*phosphate. However, given the position of its β-hydroxy carbonyl group, aldol cleavage of glucose 6-phosphate would yield a two-carbon and a four-carbon fragment, with only one bearing a phosphoryl group. Isomerization of glucose 6-phosphate in Step 2 of glycolysis positions the carbonyl group properly for cleavage into two trioses; phosphorylation of fructose 6-phosphate prepares for the production of two interchangeable triose phosphates that can follow a common route through glycolysis.

5. Triose Phosphate Isomerase Catalyzes the Interconversion of Glyceraldehyde 3-Phosphate and Dihydroxyacetone Phosphate

Of the two molecules produced by the splitting of fructose 1,6-*bis*phosphate, only glyceraldehyde 3-phosphate is a substrate for the next reaction in the glycolytic pathway. The other product, dihydroxyacetone phosphate, is converted to glyceraldehyde 3-phosphate (Figure 15·9). This near-equilibrium reaction is catalyzed by triose phosphate isomerase. As glyceraldehyde 3-phosphate is consumed in Step 6, its steady-state concentration is maintained by flux from dihydroxyacetone phosphate. In this way, two molecules of glyceraldehyde 3-phosphate are supplied to the glycolytic pathway for each molecule of fructose 1,6-*bis*phosphate that is split.

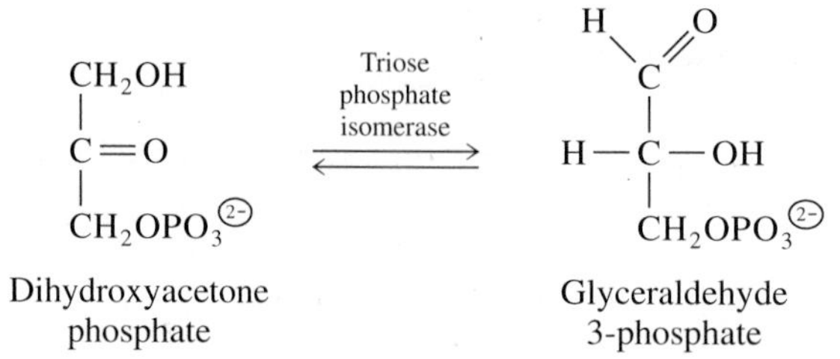

CH$_2$OH
C=O
CH$_2$OPO$_3^{2-}$

Dihydroxyacetone
phosphate

Triose
phosphate
isomerase

H   O
C
H—C—OH
CH$_2$OPO$_3^{2-}$

Glyceraldehyde
3-phosphate

Figure 15·9
Interconversion of dihydroxyacetone phosphate and glyceraldehyde 3-phosphate, catalyzed by triose phosphate isomerase.

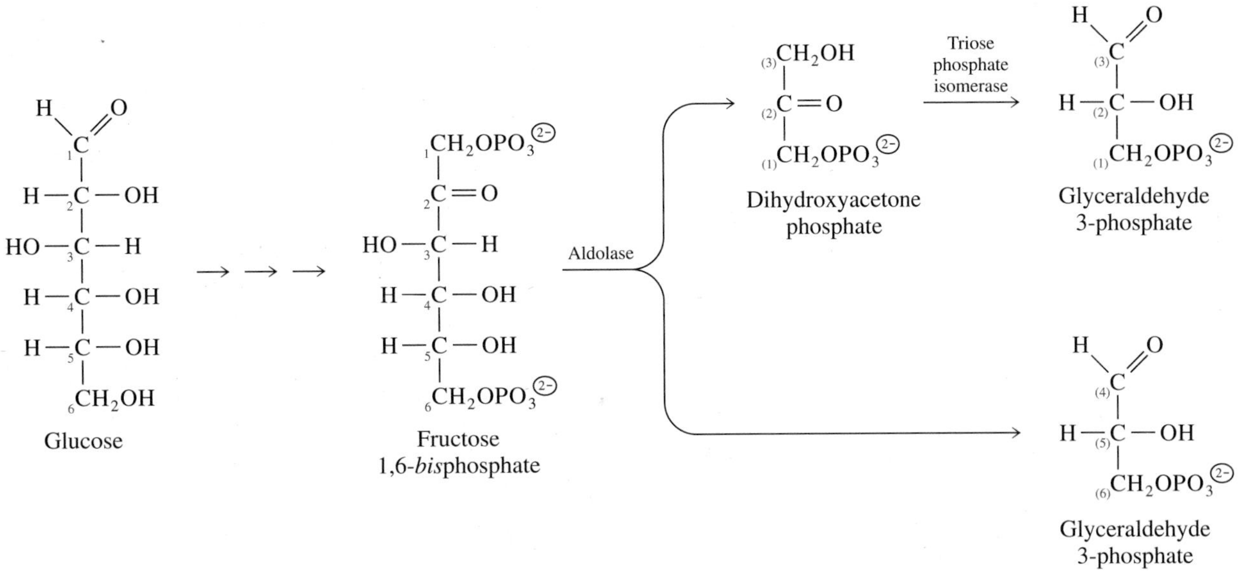

Figure 15·10
Fate of carbons from the hexose stage to the triose stage of glycolysis. All numbers refer to the carbon atoms in the original glucose molecule.

Triose phosphate isomerase, like glucose 6-phosphate isomerase, catalyzes an aldose-to-ketose conversion. The mechanism of the reaction catalyzed by triose phosphate isomerase is described in Section 7·8A and shown in Figure 7·9. The catalytic mechanisms of aldose-ketose isomerases have been studied extensively, and the formation of enzyme-bound enediolate intermediates appears to be a common feature.

We noted earlier that the reaction rates of a few enzymes appear to be diffusion controlled (Section 7·8). Such enzymes catalyze a reaction each time a substrate encounters the active site. Triose phosphate isomerase was given as an example; it catalyzes the interconversion of dihydroxyacetone phosphate and glyceraldehyde 3-phosphate with efficiency near the maximum possible.

The fate of the individual carbon atoms of a molecule of glucose metabolized to two molecules of glyceraldehyde 3-phosphate, depicted in Figure 15·10, has been determined by radioisotopic tracer studies in a variety of organisms. Note that carbons 1, 2, and 3 of one molecule of glyceraldehyde 3-phosphate are derived from carbons 4, 5, and 6 of glucose, whereas carbons 1, 2, and 3 of the second molecule of glyceraldehyde 3-phosphate (converted from dihydroxyacetone phosphate) originate as carbons 3, 2, and 1 of glucose. When these molecules of glyceraldehyde 3-phosphate mix to form a single pool of metabolites, a carbon atom from C-1 of glucose can no longer be distinguished from a carbon atom from C-6 of glucose.

6. Glyceraldehyde 3-Phosphate Dehydrogenase Catalyzes the Only Oxidation Reaction of Glycolysis

The recovery of energy from triose phosphates begins with the reaction catalyzed by glyceraldehyde 3-phosphate dehydrogenase, in which glyceraldehyde 3-phosphate is oxidized and phosphorylated to produce 1,3-*bis*phosphoglycerate (Figure 15·11). In the process, a molecule of $NAD^{\oplus}$ is reduced to NADH. The

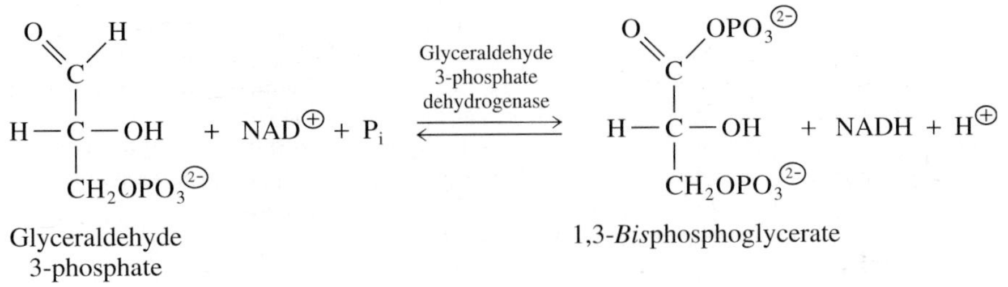

Figure 15·11
Conversion of glyceraldehyde 3-phosphate to 1,3-*bis*phosphoglycerate, catalyzed by glyceraldehyde 3-phosphate dehydrogenase.

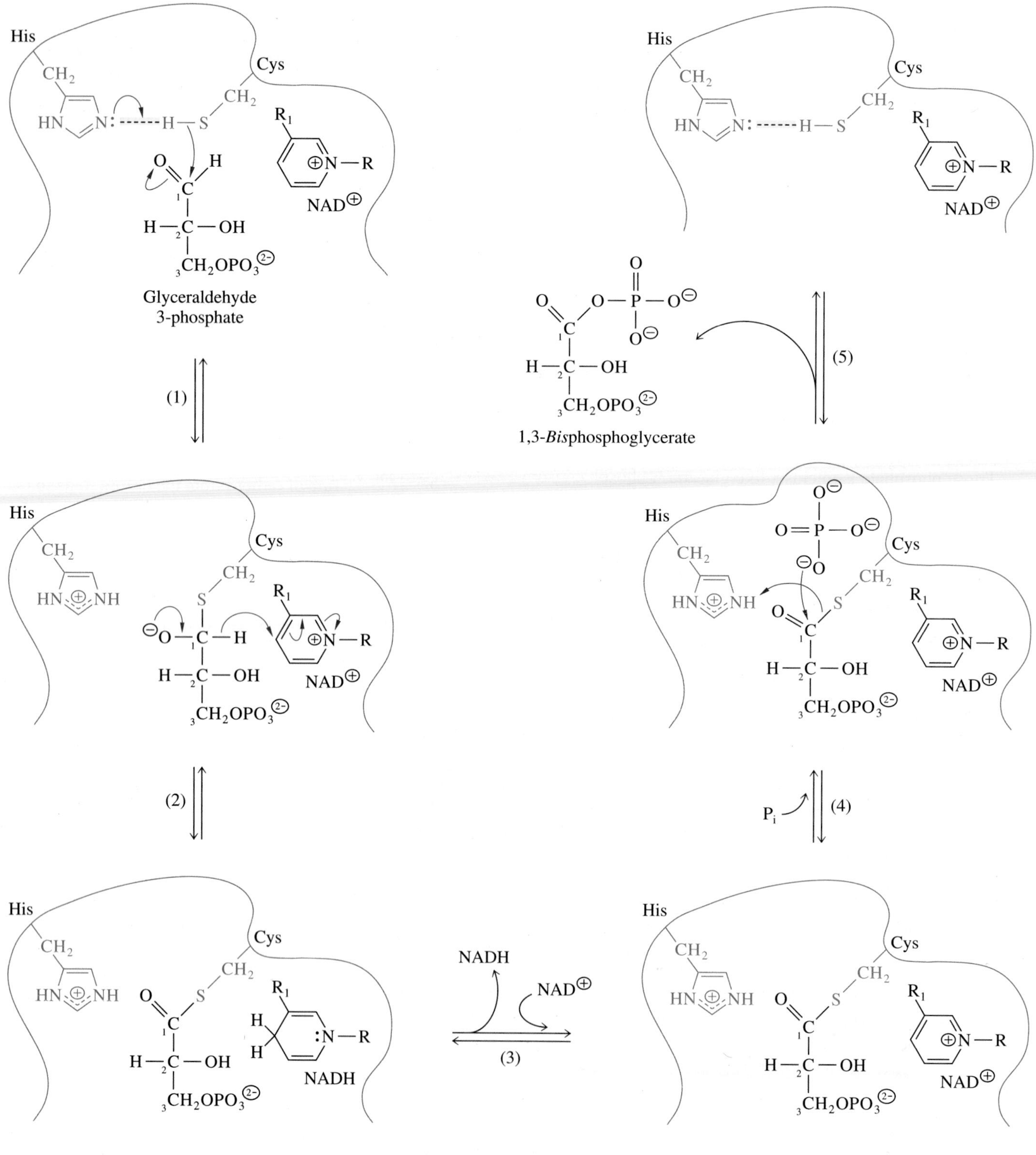

Figure 15·12

Reaction mechanism of glyceraldehyde 3-phosphate dehydrogenase. (1) An ionized cysteine sulfhydryl group attacks C-1 of glyceraldehyde 3-phosphate, forming a thiohemiacetal. (2) A hydride ion from the thiohemiacetal reduces NAD⊕ and forms a high-energy thioacyl-enzyme intermediate. (3) NADH dissociates and is replaced by NAD⊕. (4) Phosphate binds and displaces the thioacyl-enzyme intermediate, resulting in the formation of 1,3-*bis*phosphoglycerate, which dissociates from the enzyme (5).

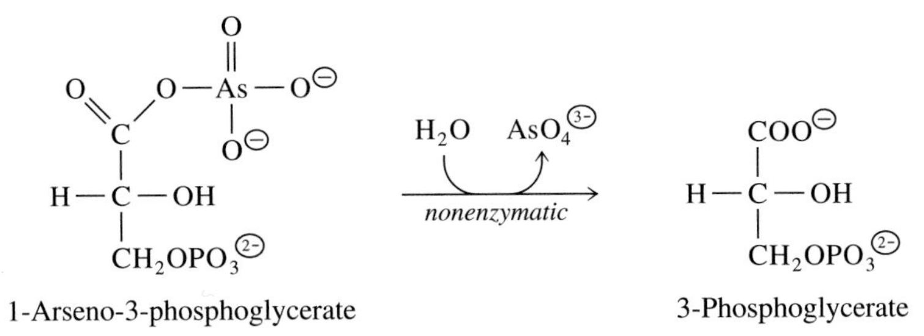

Figure 15·13
Spontaneous hydrolysis of 1-arseno-3-phosphoglycerate. Inorganic arsenate can replace inorganic phosphate as a substrate for glyceraldehyde 3-phosphate dehydrogenase, forming the unstable 1-arseno analog of 1,3-*bis*phosphoglycerate.

overall reaction is near equilibrium in cells. Nevertheless, the reaction generates a high-energy bond in 1,3-*bis*phosphoglycerate; the energy content of the aldehyde group of glyceraldehyde 3-phosphate decreases when it is oxidized to its acid, and some of that energy is conserved in the acid-anhydride linkage of 1,3-*bis*phosphoglycerate. In the next step of glycolysis, the energy of this acid-anhydride compound contributes to the formation of ATP from ADP.

The catalytic mechanism of glyceraldehyde 3-phosphate dehydrogenase can be depicted in five steps (Figure 15·12). In Step 1, the —$S^{\ominus}$ group of a cysteine residue attacks the carbonyl group of glyceraldehyde 3-phosphate, resulting in formation of a covalently bound thiohemiacetal. In Step 2, the thiohemiacetal is oxidized by $NAD^{\oplus}$ to form NADH and a thioacyl group covalently attached to the enzyme. The thioacyl group possesses high group-transfer potential. In Step 3, NADH dissociates from the enzyme and is replaced by $NAD^{\oplus}$. In Step 4, inorganic phosphate binds in the active site and attacks the carbonyl group of the thioacyl-enzyme intermediate to form a phosphoanhydride group. In the final step, the product, 1,3-*bis*phosphoglycerate, dissociates from the active site of the enzyme, completing the catalytic cycle. In the next reaction of glycolysis, this product isoenergetically transfers a phosphoryl group to ADP to form ATP.

The NADH formed in the glyceraldehyde 3-phosphate dehydrogenase reaction is reoxidized, either by the respiratory electron-transport chain (Chapter 18) or in other reactions where NADH serves as a reductant, such as the reduction of acetaldehyde to ethanol (Section 15·3) or of pyruvate to lactate (Section 15·4).

Arsenic, like phosphorus, is in Group V of the periodic table, and arsenate ($AsO_4^{\ominus}$) is an analog of inorganic phosphate. Arsenate competes with phosphate for its binding site in glyceraldehyde 3-phosphate dehydrogenase. Like phosphate, arsenate cleaves the energy-rich thioacyl-enzyme intermediate, a process called arsenolysis. The arsenolysis reaction parallels Step 4 in Figure 15·12. However, arsenolysis produces an unstable analog of 1,3-*bis*phosphoglycerate, 1-arseno-3-phosphoglycerate, which is rapidly hydrolyzed upon contact with water (Figure 15·13). This nonenzymatic hydrolysis produces 3-phosphoglycerate and regenerates inorganic arsenate, which can again react with a thioacyl-enzyme intermediate. In the presence of arsenate, glycolysis can proceed from 3-phosphoglycerate, but the ATP-producing reaction involving 1,3-*bis*phosphoglycerate is bypassed. As a result, there is no net formation of ATP from glycolysis, with potentially lethal consequences.

7. ATP Is Generated by the Action of Phosphoglycerate Kinase

Phosphoglycerate kinase catalyzes phosphoryl-group transfer from the energy-rich mixed anhydride 1,3-*bis*phosphoglycerate to ADP, generating ATP and 3-phosphoglycerate (Figure 15·14). Steps 6 and 7 together couple the oxidation of an aldehyde to a carboxylic acid with the phosphorylation of ADP to ATP. The formation of ATP by transfer of a phosphoryl group from a high-energy compound such as 1,3-*bis*phosphoglycerate to ADP is termed **substrate-level phosphorylation.** This reaction is the first ATP-generating step of glycolysis, yet it is a near-equilibrium

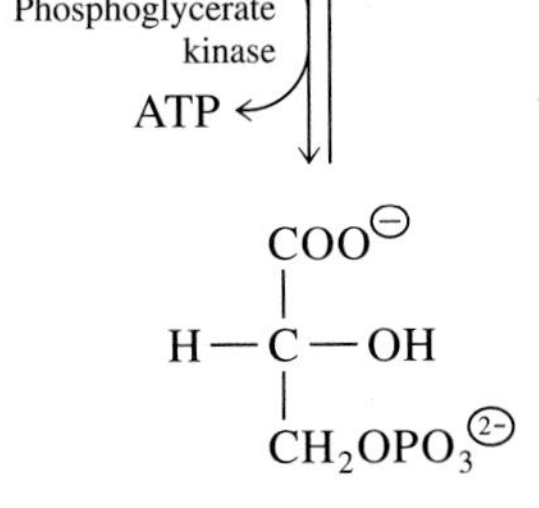

1,3-*Bis*phosphoglycerate

Figure 15·14
Phosphoryl-group transfer from the phosphoanhydride group of 1,3-*bis*phosphoglycerate to ADP, catalyzed by phosphoglycerate kinase. This reaction is the first ATP-producing step of glycolysis.

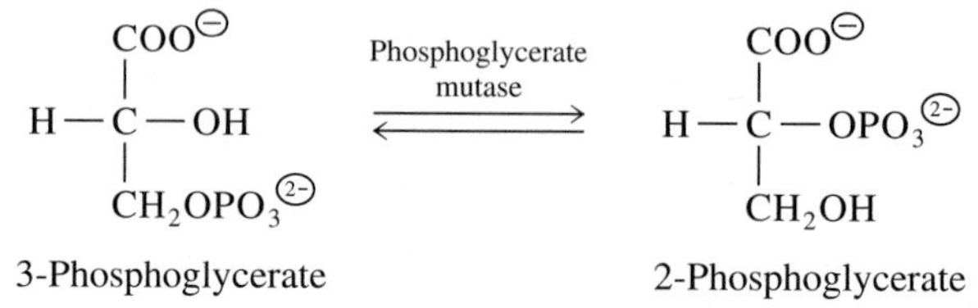

Figure 15·15
Conversion of 3-phosphoglycerate to 2-phosphoglycerate, catalyzed by phosphoglycerate mutase.

3-Phosphoglycerate ⇌ (Phosphoglycerate mutase) 2-Phosphoglycerate

reaction in cells and therefore is not a regulatory step. The three other glycolytic reactions that feature the production or consumption of ATP are each regulated steps of the pathway.

8. Phosphoglycerate Mutase Catalyzes the Conversion of 3-Phosphoglycerate to 2-Phosphoglycerate

Phosphoglycerate mutase catalyzes a near-equilibrium reaction in which 3-phosphoglycerate and 2-phosphoglycerate are interconverted (Figure 15·15). Mutases are isomerases that catalyze transfer of a group from one part of a substrate molecule to another. This type of reaction was initially a conundrum to biochemists. How could a phosphoryl group be removed from one hydroxyl group and transferred to another hydroxyl group of the same substrate molecule without the input of energy from ATP? Evolution has generated two distinct answers to this mechanistic riddle. One is associated with animal muscle and yeast phosphoglycerate mutases, the other with plant phosphoglycerate mutase. Both involve two phosphoryl-group–transfer steps.

The mechanism of the muscle and yeast enzymes is characterized by formation of a 2,3-*bis*phosphoglycerate (2,3BPG) intermediate, as shown in Figure 15·16. The active enzyme is phosphorylated at a histidine residue prior to binding 3-phosphoglycerate. After substrate binding occurs, the enzyme transfers its phosphoryl group to the substrate to form 2,3BPG. In the next step, the phosphate from position 3 of 2,3BPG is transferred back to the enzyme, leaving 2-phosphoglycerate. The original phosphorylation of phosphoglycerate mutase requires 2,3BPG itself as a phosphorylating cofactor; small amounts of 2,3BPG must be available to prime the enzyme if the 2,3BPG intermediate escapes from the active site.

Plant phosphoglycerate mutase does not form a 2,3BPG intermediate. Instead, 3-phosphoglycerate binds to plant phosphoglycerate mutase and transfers its phosphoryl group to the enzyme. The phosphoryl group is transferred back to the substrate at position 2 to form 2-phosphoglycerate. Note that despite their differences, both types of phosphoglycerate mutase have an intermediate that is poised to produce either 3-phosphoglycerate or 2-phosphoglycerate.

9. Enolase Catalyzes the Conversion of 2-Phosphoglycerate to Phosphoenolpyruvate

2-Phosphoglycerate is dehydrated to phosphoenolpyruvate in a near-equilibrium reaction catalyzed by enolase, formally termed 2-phosphoglycerate dehydratase. Enolase requires Mg^{2+} for activity. Two magnesium ions participate in this reaction: a "conformational" ion binds to the hydroxyl group of the substrate, and a

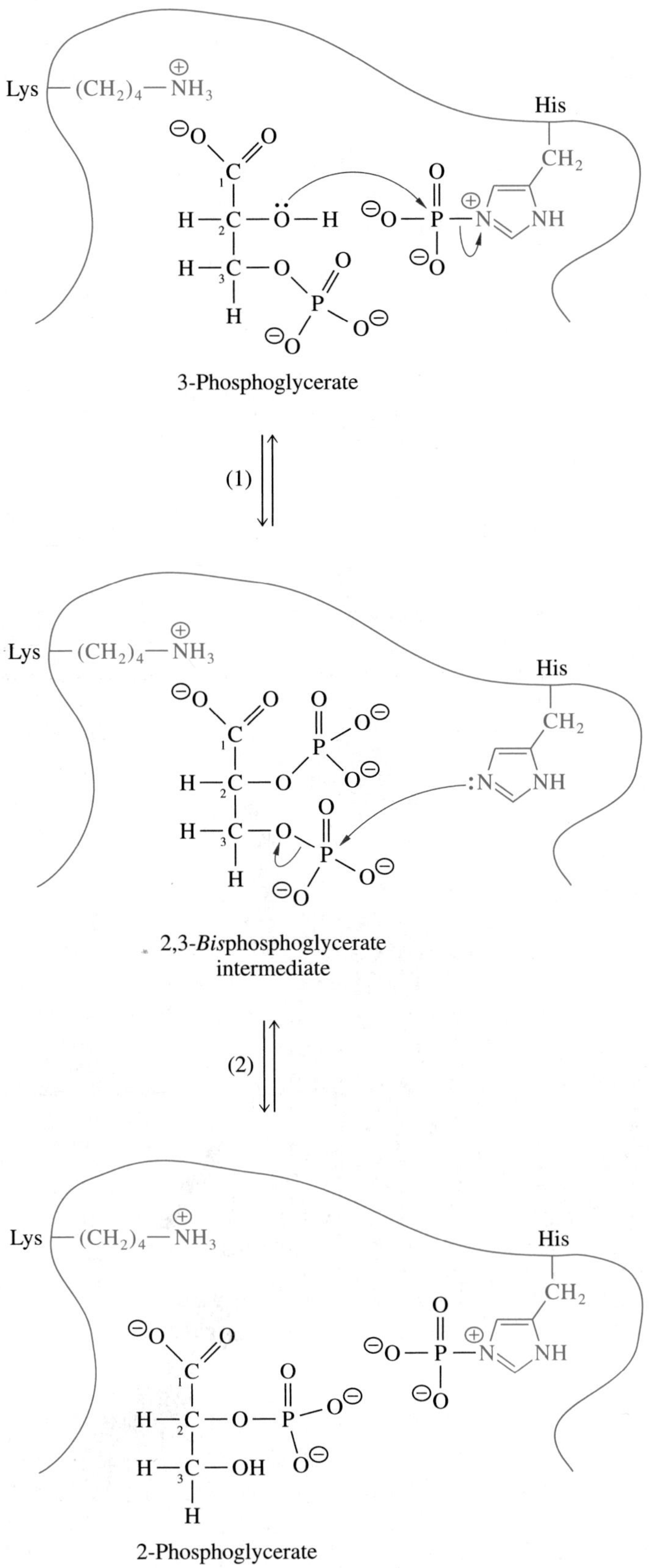

Figure 15·16
Mechanism of conversion of 3-phospho-glycerate to 2-phosphoglycerate in animals and yeast. A lysine residue at the active site of phosphoglycerate mutase binds the car-boxylate anion of the substrate. A histidine residue in the enzyme, which must be ini-tially phosphorylated in order for 3-phospho-glycerate to bind, donates its phosphoryl group to form the *bis*phosphoglycerate inter-mediate. Rephosphorylation of the enzyme yields 2-phosphoglycerate.

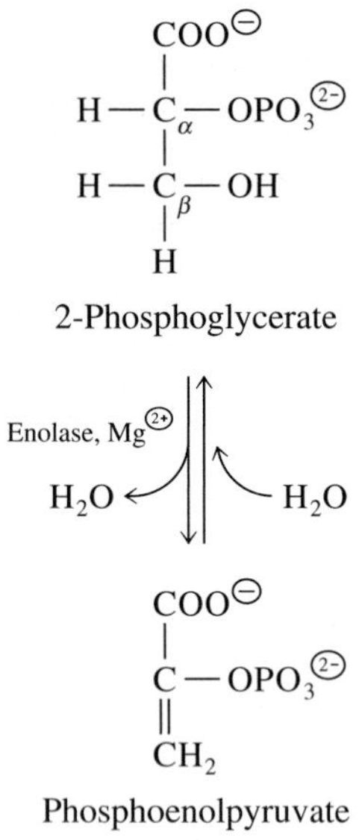

COO$^{\ominus}$

H—C$_\alpha$—OPO$_3^{2\ominus}$

H—C$_\beta$—OH

H

2-Phosphoglycerate

Enolase, Mg^{2+}

H$_2$O ←   ← H$_2$O

COO$^{\ominus}$

C—OPO$_3^{2\ominus}$

‖

CH$_2$

Phosphoenolpyruvate

Figure 15·17
Conversion of 2-phosphoglycerate to phosphoenolpyruvate. This reaction, catalyzed by enolase, converts the phosphomonoester 2-phosphoglycerate to the energy-rich enol-phosphate ester phosphoenolpyruvate.

"catalytic" ion participates in the dehydration reaction. Enolase catalyzes the conversion of 2-phosphoglycerate to phosphoenolpyruvate by the reversible α,β elimination of water (Figure 15·17). Phosphoenolpyruvate has an extremely high phosphoryl-group–transfer potential because the phosphoryl group holds pyruvate in its unstable enol tautomeric form.

In the presence of inorganic phosphate, fluoride ions (F$^{\ominus}$) inhibit enolase by forming magnesium fluorophosphate with Mg^{2+} from the active site. In early laboratory investigations, the inhibition of glyceraldehyde 3-phosphate metabolism in yeast extracts treated with F$^{\ominus}$ resulted in the accumulation of 2-phosphoglycerate and 3-phosphoglycerate, suggesting that these compounds are intermediates in glycolysis.

10. Pyruvate Kinase Catalyzes Phosphoryl-Group Transfer from Phosphoenolpyruvate to ADP, Forming Pyruvate and ATP

The second substrate-level phosphorylation of glycolysis is catalyzed by pyruvate kinase (Figure 15·18). This step is the third metabolically irreversible reaction of glycolysis and is another site of regulation. When the phosphoryl group of phosphoenolpyruvate is transferred to the β-phosphate group of ADP, enolpyruvate is formed. This enzyme-bound intermediate isomerizes to the more stable keto form of pyruvate. Pyruvate kinase is subject to regulation by both allosteric effectors and covalent modification.

The reaction catalyzed by pyruvate kinase is the last step of the glycolytic pathway. Next, we will examine subsequent steps of anaerobic glycolysis that result in the conversion of pyruvate to ethanol or lactate. The complete oxidation of pyruvate under aerobic conditions will be described in Chapter 16, when we examine the citric acid cycle.

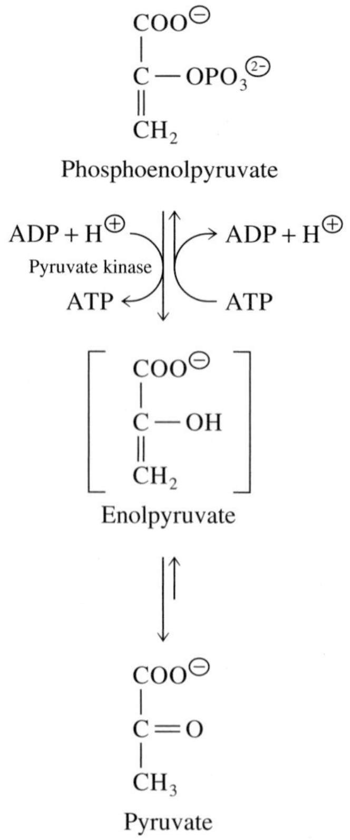

COO$^{\ominus}$

C—OPO$_3^{2\ominus}$

‖

CH$_2$

Phosphoenolpyruvate

ADP + H$^{\oplus}$ ←   → ADP + H$^{\oplus}$

Pyruvate kinase

ATP ←   ← ATP

COO$^{\ominus}$

C—OH

‖

CH$_2$

Enolpyruvate

COO$^{\ominus}$

C=O

CH$_3$

Pyruvate

Figure 15·18
Formation of pyruvate from phosphoenolpyruvate, catalyzed by pyruvate kinase. Phosphoryl-group transfer from phosphoenolpyruvate to ADP generates ATP in this metabolically irreversible reaction. The unstable enol tautomer of pyruvate is an enzyme-bound intermediate.

15·3 Pyruvate Can Be Anaerobically Metabolized to Ethanol in Yeast

The conversion of glucose to pyruvate is accompanied not only by the synthesis of ATP but also by the reduction of $NAD^{\oplus}$ to NADH at the glyceraldehyde 3-phosphate dehydrogenase reaction (Step 6 of glycolysis). In order for glycolysis to operate continuously, the cell must have a means of regenerating $NAD^{\oplus}$. Without a reaction in which NADH is oxidized, all of the coenzyme would rapidly accumulate in the reduced form, and glycolysis would cease. Under *aerobic* conditions, oxidation of NADH is accomplished primarily by the process of oxidative phosphorylation (Chapter 18), which requires molecular oxygen. Under *anaerobic* conditions, the synthesis of ethanol or lactate consumes NADH and regenerates the $NAD^{\oplus}$ that is essential for continued glycolysis.

In the anaerobic state, yeast cells convert pyruvate to ethanol and CO_2, in the process oxidizing NADH to $NAD^{\oplus}$. Two reactions are involved. First, pyruvate is decarboxylated to acetaldehyde in a reaction catalyzed by pyruvate decarboxylase (Section 8·9; the mechanism is shown in Figure 8·27). Next, alcohol dehydrogenase catalyzes the reduction of acetaldehyde to ethanol by NADH. These reactions and the cycle of $NAD^{\oplus}$/NADH reduction and oxidation in alcoholic fermentation are shown in Figure 15·19.

The sum of the glycolytic reactions and the conversion of pyruvate to ethanol is

$$\text{Glucose} + 2\,P_i + 2\,\text{ADP} + 2\,H^{\oplus} \longrightarrow 2\,\text{Ethanol} + 2\,CO_2 + 2\,\text{ATP} + 2\,H_2O$$

$$(15 \cdot 4)$$

These reactions have familiar commercial roles in the manufacture of beer and bread (Figure 15·20). In the brewery, the carbon dioxide produced during the conversion of pyruvate to ethanol can be captured and used to carbonate the final alcoholic brew; this gas produces the foamy head. In the bakery, carbon dioxide is the agent that causes bread dough to rise.

Ethanol is a neutral end product of fermentation and is excreted by yeast. Even at reasonably high concentrations, ethanol is innocuous to the yeast. In contrast, the formation of lactic acid by anaerobic glycolysis, which we will examine in the next section, can be accompanied by a drastic decrease in pH.

Figure 15·20
For centuries, bakeries and breweries have exploited the conversion of pyruvate to ethanol by yeast.

Figure 15·19
Anaerobic conversion of pyruvate to ethanol in yeast. Pyruvate is first decarboxylated by the action of pyruvate decarboxylase. Subsequently, NADH produced by the glyceraldehyde 3-phosphate dehydrogenase reaction can be reoxidized to $NAD^{\oplus}$ by the action of alcohol dehydrogenase, which catalyzes reduction of acetaldehyde to ethanol. Regeneration of $NAD^{\oplus}$ allows fermentation to continue under anaerobic conditions.

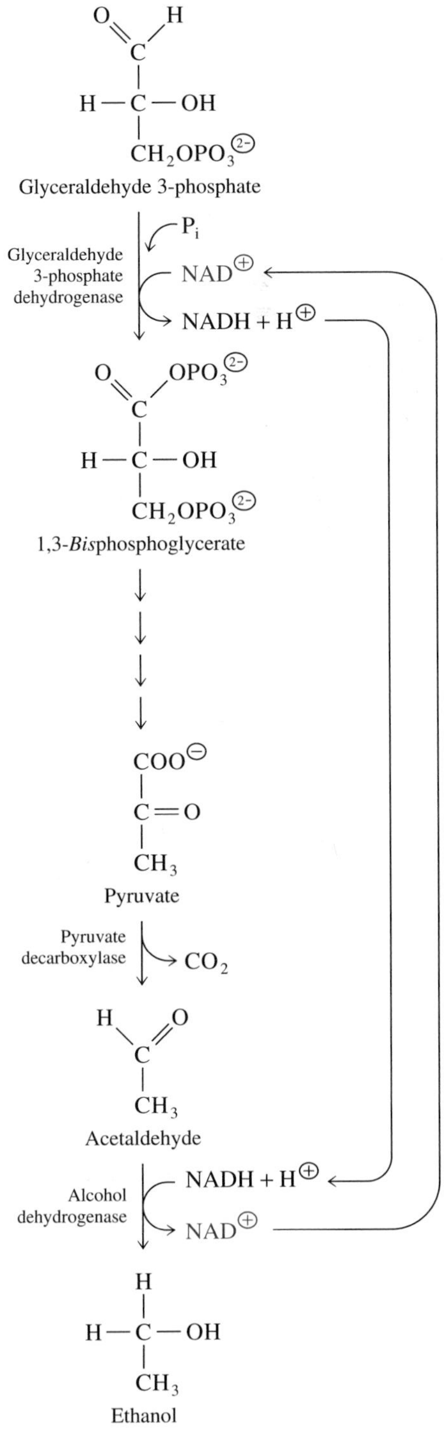

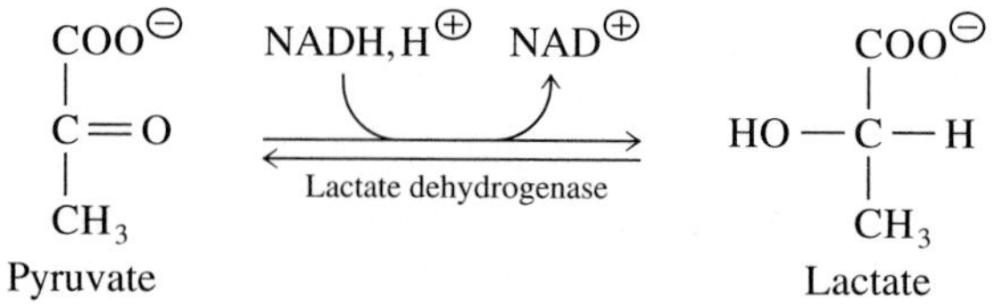

$$
\begin{array}{l}
\text{COO}^{\ominus} \\
| \\
\text{C}=\text{O} \\
| \\
\text{CH}_3
\end{array}
\quad
\xrightleftharpoons[\text{Lactate dehydrogenase}]{\text{NADH,H}^{\oplus} \quad \text{NAD}^{\oplus}}
\quad
\begin{array}{l}
\text{COO}^{\ominus} \\
| \\
\text{HO}-\text{C}-\text{H} \\
| \\
\text{CH}_3
\end{array}
$$

Pyruvate   Lactate

Figure 15·21
Anaerobic conversion of pyruvate to lactate in muscle. The NADH produced by the glyceraldehyde 3-phosphate dehydrogenase reaction of glycolysis can be reoxidized to NAD$^{\oplus}$ under anaerobic conditions by this reaction, catalyzed by lactate dehydrogenase.

15·4 Pyruvate Can Be Converted to Lactate in Most Cells

Unlike yeast, most organisms lack pyruvate decarboxylase and therefore cannot produce ethanol from pyruvate. Instead, pyruvate is reduced to lactate in a reversible reaction catalyzed by lactate dehydrogenase (Figure 15·21). Once formed, lactate has no other metabolic fate than reconversion to pyruvate. Hence, lactate is a metabolic dead end. Since lactate formation catalyzed by lactate dehydrogenase regenerates NAD$^{\oplus}$ from NADH, the pathway of glycolysis is complete, with NAD$^{\oplus}$ becoming available for the glyceraldehyde 3-phosphate dehydrogenase reaction, in the same manner as shown for alcoholic fermentation (Figure 15·19).

In mammals, production of lactate is usually accompanied by its reconversion to pyruvate. Lactate formed in muscles during exercise is transported out of muscle cells and carried via the bloodstream to the liver, where it is converted to pyruvate by the action of hepatic lactate dehydrogenase. Further metabolism of pyruvate requires oxygen. When the supply of oxygen to tissues is inadequate, all tissues produce lactate by anaerobic glycolysis. Under these conditions, lactate accumulates, causing an elevation of lactic acid in blood, a condition termed lactic acidosis. During lactic acidosis, the pH of the blood may drop to dangerously acidic levels.

The overall reaction for glucose degradation to lactate is

$$\text{Glucose} + 2\,\text{P}_i + 2\,\text{ADP} \longrightarrow 2\,\text{Lactate} + 2\,\text{ATP} + 2\,\text{H}_2\text{O} \qquad \textbf{(15·5)}$$

The ATP produced by glycolysis is utilized by cells as a source of energy. Like NAD$^{\oplus}$, ADP must be regenerated if glycolysis is to continue. Thus, if we add to Equation 15·5 the reaction for consumption of ATP,

$$2\,\text{ATP} + 2\,\text{H}_2\text{O} \longrightarrow 2\,\text{ADP} + 2\,\text{P}_i + 2\,\text{H}^{\oplus} \qquad \textbf{(15·6)}$$

we see that lactic acid, not lactate, is formed from glucose.

$$\text{Glucose} \longrightarrow 2\,\text{Lactate} + 2\,\text{H}^{\oplus} \qquad \textbf{(15·7)}$$

Lactic acid is the substance that causes muscles to ache during and after exercise. When certain bacteria ferment the sugars in milk to lactic acid, the acid denatures the proteins in milk, causing the curdling necessary for cheese and yogurt production.

Regardless of the final product—lactate or ethanol—glycolysis generates two molecules of ATP per molecule of glucose consumed. Oxygen is not required in either case. This feature is not only essential for anaerobic organisms but also for some specialized cells in multicellular organisms. In most cells, the majority of ATP is produced by oxidative phosphorylation, which is a strictly oxygen-dependent process. However, some tissues, termed obligatory glycolytic tissues, rely on glycolysis for all of their energy. In the cornea of the eye, for instance, oxygen availability is limited by poor blood circulation. Anaerobic glycolysis provides the necessary ATP for such tissues in the absence of sufficient oxygen for oxidative phosphorylation.

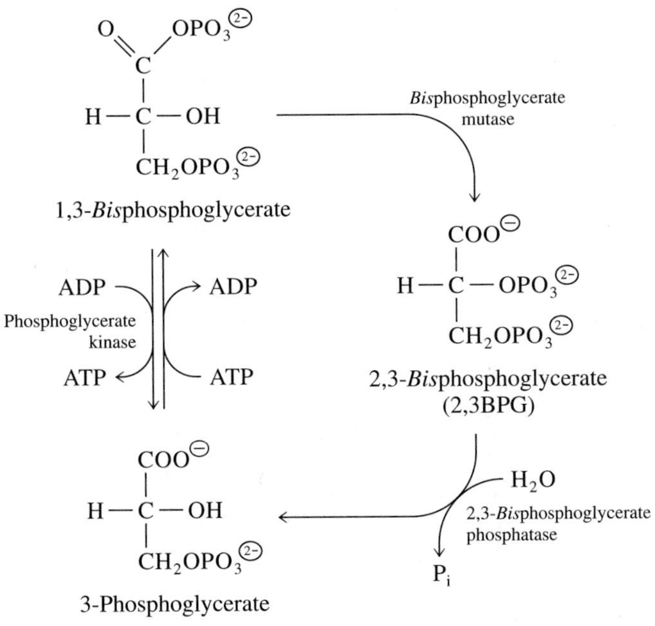

Figure 15·22
Formation of 2,3-*bis*phosphoglycerate (2,3BPG). In red blood cells, *bis*phosphoglycerate mutase catalyzes conversion of 1,3-*bis*phosphoglycerate to 2,3BPG, an allosteric inhibitor of the binding of oxygen to hemoglobin. 2,3-*Bis*phosphoglycerate phosphatase catalyzes the conversion of 2,3BPG to 3-phosphoglycerate, returning the intermediate to the glycolytic pathway. This shunt bypasses the generation of ATP by phosphoglycerate kinase.

15·5 1,3-*Bis*phosphoglycerate Can Be Converted to 2,3-*Bis*phosphoglycerate in Red Blood Cells

An important function of glycolysis in red blood cells is the production of 2,3-*bis*-phosphoglycerate (2,3BPG), an allosteric inhibitor of the oxygenation of hemoglobin (Section 5·18). We encountered this metabolite earlier as a reaction intermediate and cofactor in Step 8 of glycolysis.

Erythrocytes contain the enzyme *bis*phosphoglycerate mutase, which catalyzes the transfer of a phosphoryl group from C-1 to C-2 of 1,3-*bis*phosphoglycerate, forming 2,3-*bis*phosphoglycerate. As shown in Figure 15·22, 2,3-*bis*phospho-glycerate phosphatase catalyzes the hydrolysis of 2,3-*bis*phosphoglycerate to 3-phosphoglycerate, which can reenter the glycolytic pathway and be catabolized to pyruvate.

The shunting of 1,3-*bis*phosphoglycerate through these two enzymes bypasses phosphoglycerate kinase, which catalyzes Step 7 of glycolysis, one of the two ATP-generating steps. However, only a small portion of glycolytic flux in blood cells—about 10%—is diverted through the mutase and phosphatase. Accumulation of free 2,3-*bis*phosphoglycerate (i.e., 2,3BPG not bound to hemoglobin) inhibits *bis*-phosphoglycerate mutase. In exchange for diminished ATP generation, this bypass provides a regulated supply of 2,3-*bis*phosphoglycerate, which is necessary for the efficient release of O_2 from oxyhemoglobin.

15·6 The Overall Free-Energy Change of Glycolysis Is Highly Negative

As explained in the discussion of thermodynamics in Chapter 14, the free-energy change of a reaction must be negative for the reaction to occur. Obviously, all of the reactions of glycolysis in cells must have negative free-energy changes for the pathway to proceed toward product. However, the free-energy changes of the individual reactions in the glycolytic pathway are not equal: they are, by definition, large for the metabolically irreversible reactions and near zero for the near-equilibrium reactions.

Figure 15·23
Cumulative standard and actual free-energy changes for the reactions of glycolysis. The vertical axis measures free-energy changes in kJ mol⁻¹. The reactions of glycolysis are plotted in sequence horizontally. The upper plot (red) tracks the standard free-energy changes, and the bottom plot (blue) shows actual free-energy changes under cellular conditions. The data for actual free-energy changes are taken from erythrocytes. The interconversion reaction catalyzed by triose phosphate isomerase (reaction 5) is not shown. [Adapted from Hamori, E. (1975). Illustration of free energy changes in chemical reactions. *J. Chem. Ed.* 52:370–373.]

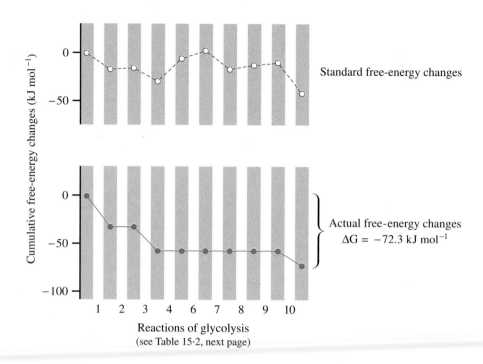

In Figure 15·23, the cumulative standard free-energy changes for the glycolytic reactions and the actual free-energy changes for the steps of glycolysis in erythrocytes are diagrammed. Values for the free-energy changes of the individual reactions are given in Table 15·2. The vertical axis of Figure 15·23 measures free-energy changes. The glycolytic reactions are arranged sequentially along the horizontal axis. The figure rewards careful study by imparting a visual sense of the difference between free-energy changes under standard conditions ($\Delta G^{\circ\prime}$) and free-energy changes under cellular conditions (ΔG).

It can be readily seen from the blue plot, which tracks the actual free-energy changes for the reactions of glycolysis, that each reaction has a free-energy change that is negative or zero, which must be so for the reaction to proceed toward product. It follows that the overall pathway, which is the sum of the individual reactions, must also be negative. This sum is depicted by the bracket on the right side of the graph.

Note that the actual free-energy changes are large only for the reactions catalyzed by hexokinase, phosphofructokinase-1, and pyruvate kinase—the metabolically irreversible steps of the pathway and also the steps with regulatory roles. The ΔG values for the other steps are very close to zero. In other words, they are near-equilibrium reactions in cells.

In contrast, the standard free-energy changes for the same reactions exhibit no consistent pattern. Although there are large negative standard free-energy changes at the same steps that are highly exergonic under cellular conditions, it is apparent that this is coincidental, since there are equally large standard free-energy changes for reactions known to be near-equilibrium in the cell. Furthermore, some of the standard free-energy changes for the reactions of glycolysis are *positive*, indicating that under standard conditions, flux through these reactions would occur in the direction of substrate rather than product. Finally, it is apparent that there is a large difference between the free energy of the intermediates at their standard-state concentrations and at the much lower concentrations that exist in cells.

Table 15·2 Comparison of standard free-energy changes ($\Delta G^{\circ\prime}$) and actual free-energy changes (ΔG) in erythrocytes for the enzyme-catalyzed reactions of glycolysis

Enzyme	$\Delta G^{\circ\prime}$ $(kJ\ mol^{-1})$	ΔG $(kJ\ mol^{-1})$
1. Hexokinase	− 17	−33
2. Glucose 6-phosphate isomerase	+ 2	near equilibrium
3. Phosphofructokinase-1	− 14	−22
4. Aldolase	+ 24	near equilibrium
5. Triose phosphate isomerase	+ 8	near equilibrium
6. Glyceraldehyde 3-phosphate dehydrogenase	+ 6	near equilibrium
7. Phosphoglycerate kinase	− 19	near equilibrium
8. Phosphoglycerate mutase	+ 5	near equilibrium
9. Enolase	+ 2	near equilibrium
10. Pyruvate kinase	− 32	−17

15·7 The Degradation of Glucose Is a Regulated Process

The catabolism of glucose is regulated by a variety of cellular mechanisms. Here we will emphasize the regulation of the enzymes of glycolysis and only mention in passing the regulation of glucose transport into cells and the regulatory effects of hormones, which are discussed in more detail in Chapter 23. Glycolysis is regulated in all cells, but the mechanisms of regulation may differ in detail among cell types.

The regulation of the glycolytic pathway has been examined more thoroughly than that of any other pathway. Data about enzymatic regulation come primarily from two types of biochemical research: enzymology and metabolic biochemistry. In the enzymological approaches, metabolites are tested for their effects on isolated enzymes, and information is sought on the structure and regulatory mechanisms of enzymes. In metabolic biochemistry, the concentrations of pathway intermediates in vivo are analyzed, and pathway dynamics under cellular conditions are stressed. We sometimes find that in vitro studies are deceptive indicators of pathway dynamics in vivo. For instance, a compound may be an effector of an enzyme in vitro, but only at concentrations not found in the cell. Accurate interpretation of biochemical data greatly benefits from a combination of enzymological and metabolic expertise.

As discussed in Section 6·12, regulatory enzymes are almost always oligomeric proteins. However, not all oligomeric enzymes have regulatory properties. For example, of the 10 enzymes that catalyze reactions in the glycolytic pathway, only one is monomeric (phosphoglycerate kinase). All of the other enzymes are either dimers or tetramers (hexokinase exists in both monomeric and dimeric forms; the dimer is more active). Five of the enzymes show cooperativity of substrate binding, including the only three enzymes that catalyze regulated reactions in glycolysis—hexokinase, phosphofructokinase-1, and pyruvate kinase. Each of these three enzymes has regulatory sites located in regions between subunits. Each catalyzes a nonequilibrium reaction in glycolytically active tissues such as red blood cells (see Table 15·2).

The seven enzymes catalyzing near-equilibrium reactions of glycolysis have catalytic properties suitable for maintaining a mixture of their substrates and products at or near equilibrium. In Chapter 7, we saw that for an enzyme to have maximum catalytic efficiency, the binding of substrates to the enzyme should be moderately weak. For an enzyme to respond maximally to changes in substrate concentration and quickly reestablish equilibrium, the K_m value of the enzyme for substrate should be near the cellular concentration (Section 6·3D). Allosterically regulated enzymes, on the contrary, may have low K_m values for substrate, and the K_m may be modulated allosterically (Sections 6·12 and 7·15).

Table 15·3 Ratio of $K_m/[S]$ for the enzymes catalyzing near-equilibrium reactions of glycolysis in mammals

Enzyme	Substrate	Tissue	K_m (μM)	$K_m/[S]$
1. Hexokinase, glucokinase				
2. Glucose 6-phosphate isomerase	Glucose 6-phosphate	Brain	210	1.6
		Muscle	700	1.6
3. Phosphofructokinase-1				
4. Aldolase	Fructose 1,6-*bis*phosphate	Brain	12	0.06
		Muscle	100	3.1
5. Triose phosphate isomerase	Dihydroxyacetone phosphate	Muscle	870	17
6. Glyceraldehyde 3-phosphate dehydrogenase	Glyceraldehyde 3-phosphate	Brain	44	15
		Muscle	70	23
7. Phosphoglycerate kinase	1,3-*Bis*phosphoglycerate	Brain	9	>9
8. Phosphoglycerate mutase	3-Phosphoglycerate	Brain	240	6
		Muscle	5000	83
9. Enolase	2-Phosphoglycerate	Brain	33	7
		Muscle	70	10
10. Pyruvate kinase				

[Adapted from Fersht, A. (1985). *Enzyme Structure and Mechanism*, 2nd ed. (New York: W. H. Freeman), p. 328.]

Table 15·3 lists the ratio of K_m to cellular concentration for the substrates of the near-equilibrium reactions of glycolysis in mammals. The data in the table, derived from several species, show that for all but one enzyme, the concentration of substrate is below the K_m value. Metabolic studies also indicate that these enzymes are all present in sufficiently high concentrations to maintain the concentrations of substrates and products near equilibrium in vivo. Under suitable physiological conditions, all seven of these reactions may proceed in the reverse direction, as steps in a pathway of glucose synthesis from pyruvate (Chapter 17).

Regulatory enzymes can be exceptions to the high-K_m principle. For example, hexokinase in erythrocytes has a K_m of about 0.1 mM for glucose, but the concentration of glucose is much higher. In this situation, alteration of the glucose concentration does not alter the flux through the hexokinase-catalyzed reaction. However, an allosteric inhibitor or activator that alters the K_m value for glucose would have a profound effect.

We will now examine each of the sites of regulation of glycolysis. Our primary focus is the regulation of glycolysis in mammalian cells. Variations on the regulatory themes we discuss can be found in other species.

A. Regulation of Hexose Transporters

The first potential site for the regulation of glycolysis is the transport of glucose into the cell. Bacteria possess the phosphoenolpyruvate-dependent sugar phosphotransferase system, discussed in Section 12·7G. The activity of this sugar transporter is regulated by the concentration of sugar molecules inside the cell. All mammalian cells possess membrane-spanning glucose transporters. Intestinal and kidney cells have a $Na^{\oplus}$-dependent cotransport system for absorption of dietary glucose and urinary glucose, respectively. Five members of the GLUT family of glucose transporters, discussed in detail in Chapter 23, have also been found in mammalian cells. Each has different properties suitable for the metabolic activities of the tissues in which they are found. These transporters function by passive transport. In most mammalian cells, the intracellular glucose concentration is far lower than the blood glucose concentration, and the glucose transporters facilitate the movement of glucose down its concentration gradient. Glucose transport into cells represents a crossover point between intracellular and interorgan metabolism. Our focus for now is on intracellular control of the pathway of glycolysis, and therefore we will not examine regulation of glucose transporters in detail. We will return to the regulation of glucose transport in Chapter 23, when we discuss interorgan metabolism in mammals and the regulation of sugar transporters by hormones.

B. Regulation of Hexokinase

The reaction catalyzed by hexokinase is metabolically irreversible. The key regulatory factor is the concentration of the product, glucose 6-phosphate, which allosterically inhibits hexokinase isozymes I, II, and III, but not isozyme IV, glucokinase, which predominates in liver and pancreas. The concentration of glucose 6-phosphate increases when glycolysis is inhibited at sites further along the pathway. The product inhibition of hexokinase by glucose 6-phosphate therefore coordinates the regulation of hexokinase with the regulation of subsequent enzymes of glycolysis.

   Glucokinase has characteristics suited to the physiological role of the liver and pancreas in managing the supply of glucose for the entire body. In most cells, glucose concentrations are maintained far below the concentrations in blood. However, glucose freely enters the liver and pancreas, and the concentration of glucose in the cells of these organs matches the concentration in blood. The blood glucose concentration is typically 5 mM, though after a meal it may rise as high as 10 mM. Since the K_m of glucokinase for glucose is 10 mM, glucokinase is never saturated with glucose. Therefore, liver and pancreas tissue can respond to increases in blood glucose concentration with proportionate increases in the phosphorylation of glucose. The role of glucokinase in the metabolism of mammals is detailed in Chapter 23.

C. Regulation of Phosphofructokinase-1

The second site of allosteric regulation of glycolysis is the reaction catalyzed by phosphofructokinase-1. The quaternary structure of PFK-1 varies among species. The bacterial and mammalian enzymes are both tetramers; the mammalian enzyme has a much higher molecular weight. In yeast the enzyme is an octamer with four subunits each of two types. Figure 15·24 (next page) shows the structure of the *E. coli* phosphofructokinase-1 complexed with its products, fructose 1,6-*bis*-phosphate and ADP, and its allosteric activator, ADP. Four elongated subunits associate symmetrically to form the native oligomer. Note the physical separation of the active site and the regulatory site of each subunit. Note also the alignment of the products bound in the active site.

   ATP is both a substrate of PFK-1 and, in most species, an allosteric inhibitor of the enzyme. ATP causes a decrease in the affinity of PFK-1 for its glycolytic substrate, fructose 6-phosphate (i.e., raises its K_m value). In mammalian cells, AMP is an allosteric activator of PFK-1 that acts by relieving the inhibition caused by ATP (Figure 15·25, next page). ADP activates mammalian PFK-1 but inhibits the plant kinase; in microorganisms, the regulatory effects of purine nucleotides vary among species.

   The concentration of ATP varies little in most mammalian cells, despite large changes in the rate of its formation and utilization. However, as discussed in Chapter 14, significant changes in the concentrations of ADP and AMP do occur because they are present in cells in much smaller concentrations than ATP, and small changes in the level of ATP cause proportionally larger changes in the levels of ADP and AMP. The steady-state concentrations of these compounds have a large role in determining the rate of the reaction catalyzed by PFK-1.

   Citrate, an intermediate of the citric acid cycle, is another physiologically important inhibitor of PFK-1. The citric acid cycle connects the partial oxidation of glucose to pyruvate with the complete oxidation of pyruvate to CO_2 and H_2O. An elevated concentration of citrate indicates that ample substrate is entering the citric acid cycle. The regulatory effect of citrate on PFK-1 is an example of feedback inhibition that regulates the supply of pyruvate to the citric acid cycle.

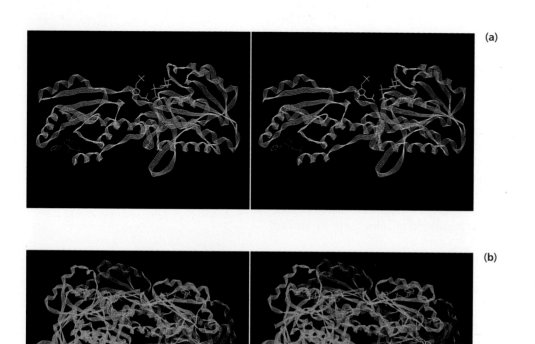

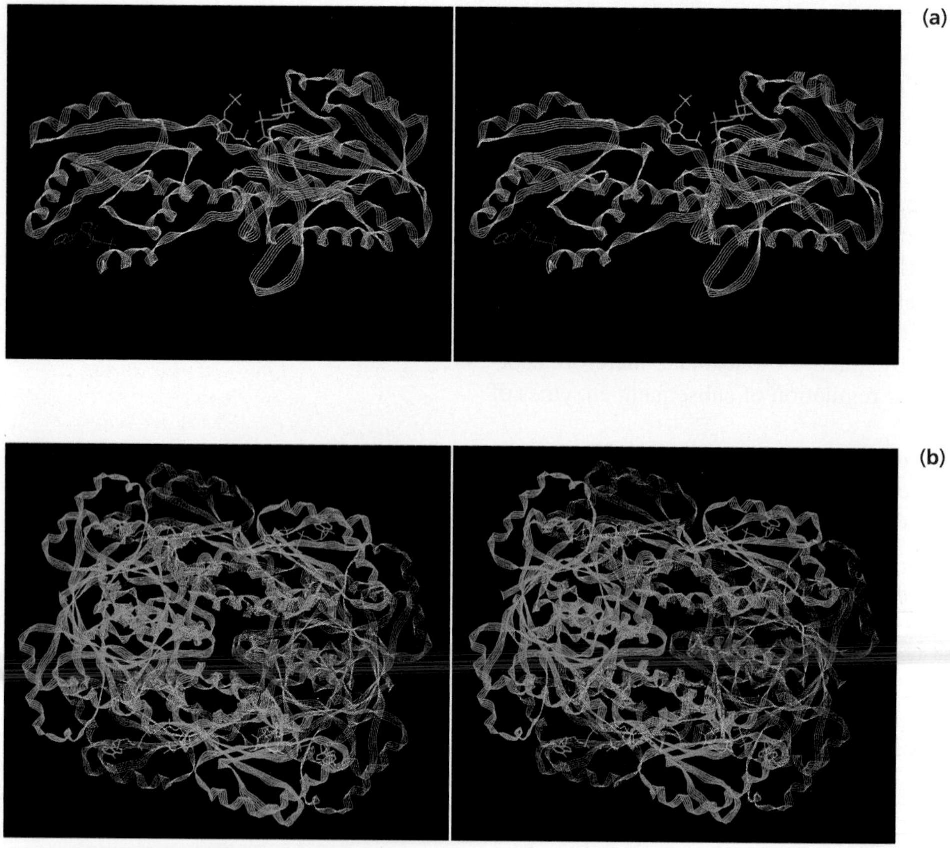

(a)

(b)

Figure 15·24
Stereo views of the structure of phosphofructo-kinase-1 from *E. coli.* The holoenzyme is a tetramer of identical subunits. **(a)** Single subunit. The products, fructose 1,6-*bis*phosphate (yellow) and ADP (green), are bound in the active site. The allosteric activator ADP (orange) is bound in the regulatory site.
(b) Tetramer. Two subunits are blue, two are purple. The white dots are Mg^{2+} ions.

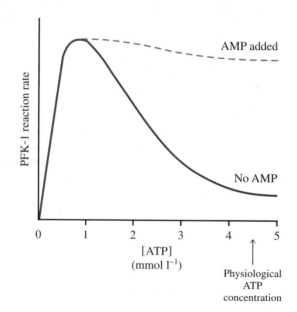

Figure 15·25
Regulation of PFK-1 by ATP and AMP. In the absence of AMP, PFK-1 is almost completely inhibited by physiological concentrations of ATP. In the range of concentrations of AMP found in the cell, the inhibition of PFK-1 by ATP is relieved. In this way, flux through PFK-1 is regulated by changes in the concentration of AMP. [Adapted from Martin, B. R. (1987). *Metabolic Regulation: A Molecular Approach* (Oxford, England: Blackwell Scientific Publications), p. 222.]

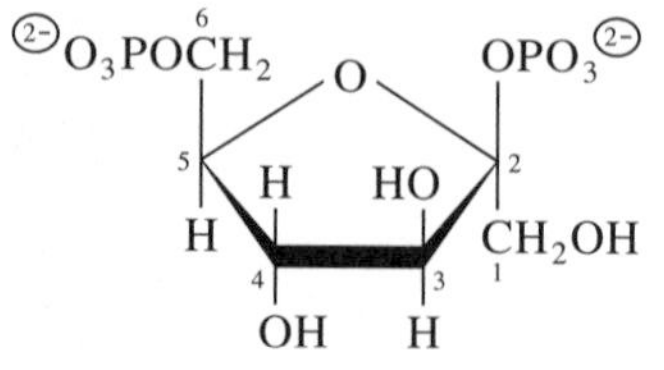

Figure 15·26
Structure of β-D-fructose 2,6-*bis*phosphate, the predominant form in cells.

Fructose 2,6-*bis*phosphate (Figure 15·26), discovered in 1980, is a potent activator of PFK-1, effective in the micromolar range. This compound is present in mammals, fungi, and plants but not prokaryotes. Fructose 2,6-*bis*phosphate is formed from fructose 6-phosphate through the action of the enzyme phosphofructokinase-2 (PFK-2). PFK-2 is stimulated by inorganic phosphate and inhibited by citrate. Surprisingly, in mammalian liver a different active site of the same protein catalyzes the dephosphorylation of fructose 2,6-*bis*phosphate, re-forming fructose 6-phosphate. This activity of the enzyme is called fructose 2,6-*bis*phosphatase.

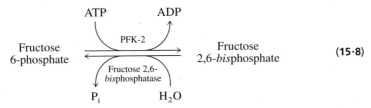

(15·8)

The dual activities of PFK-2 control the steady-state concentration of fructose 2,6-*bis*phosphate.

In liver, the activity of PFK-2 is linked to the action of glucagon, a hormone produced by the pancreas in response to low blood sugar. An elevation of the concentration of glucagon in the blood triggers the adenylate cyclase signalling pathway (Section 12·9A) in liver cells, culminating in the phosphorylation of a serine residue in PFK-2 (Figure 15·27). This phosphorylation is catalyzed by protein kinase A, the cAMP-dependent protein kinase. Phosphorylation inactivates the kinase activity of the bifunctional enzyme and activates its phosphatase activity.

Figure 15·27
Effect of glucagon on glycolysis. (1) The glucagon transducer system includes the glucagon receptor, coupling proteins, and the enzyme adenylate cyclase. When glucagon binds to its receptor, adenylate cyclase is activated and the second messenger cyclic AMP (cAMP) is formed. (2) cAMP binds to the regulatory subunit (R) of protein kinase A, which releases activated catalytic subunits (C). (3) The C subunits catalyze the phosphorylation of the bifunctional enzyme PFK-2, inhibiting its kinase activity and stimulating its fructose 2,6-*bis*phosphatase activity. When the concentration of glucagon is high, formation of the potent PFK-1 activator fructose 2,6-*bis*phosphate is decreased, and degradation of the activator is increased. As a result, the rate of glycolysis is slowed.

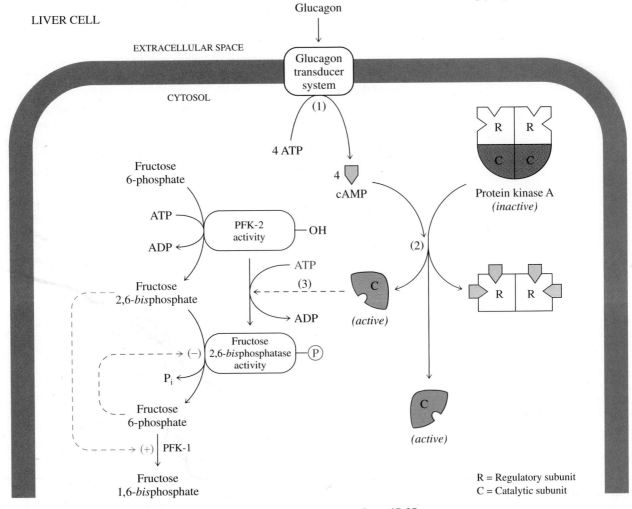

Under these conditions, the level of fructose 2,6-*bis*phosphate falls, PFK-1 becomes less active, and glycolysis is depressed. Under conditions in which glucose is rapidly metabolized, the concentration of fructose 6-phosphate increases, and more fructose 2,6-*bis*phosphate is formed, since fructose 6-phosphate is both a substrate of PFK-2 and a potent inhibitor of fructose 2,6-*bis*phosphatase. A phosphoprotein phosphatase catalyzes the dephosphorylation of PFK-2. Thus, in liver cells, control of glycolysis by glucagon and glucose is accomplished through control of the bifunctional enzyme whose activity establishes the steady-state concentration of fructose 2,6-*bis*phosphate.

D. Regulation of Pyruvate Kinase

Four different isozymes of pyruvate kinase are present in mammalian tissues. For some of these, such as the isozymes found in liver, kidney, and red blood cells, a sigmoidal curve is obtained when initial velocity is plotted against phosphoenolpyruvate concentration (Figure 15·28a). These isozymes are allosterically activated by fructose 1,6-*bis*phosphate and inhibited by ATP. In the absence of fructose 1,6-*bis*phosphate, physiological concentrations of ATP almost completely inhibit the isolated enzyme. The presence of fructose 1,6-*bis*phosphate—considered to be the most important modulator in vivo—shifts the curve to the left. With sufficient fructose 1,6-*bis*phosphate, the curve becomes hyperbolic. The figure shows that at a constant concentration of substrate, enzyme activity is greater in the presence of the allosteric activator, up to a saturating concentration of phosphoenolpyruvate. Recall that fructose 1,6-*bis*phosphate is the product of the reaction catalyzed by PFK-1. Its concentration can be expected to increase when the activity of PFK-1 is increased. Since fructose 1,6-*bis*phosphate is an activator of pyruvate kinase, activation of PFK-1 (which catalyzes Step 3 of the glycolytic pathway) causes subsequent activation of pyruvate kinase (the last enzyme in the pathway). This type of regulation is called feed-forward activation.

The isozyme of pyruvate kinase found in mammalian liver and intestinal cells is subject to an additional type of regulation, covalent modification by phosphorylation of the enzyme. Protein kinase A, which also catalyzes the phosphorylation of PFK-2, catalyzes the phosphorylation of pyruvate kinase. Pyruvate kinase is less active in the phosphorylated state. The change in kinetic behavior when pyruvate kinase is phosphorylated is depicted in Figure 15·28b, which shows the response of isolated liver cells incubated with glucagon, a stimulator of protein kinase A. Dephosphorylation of pyruvate kinase is catalyzed by a protein phosphatase.

The general effects of regulatory metabolites on glycolysis are summarized in Figure 15·29. Activation of the energy-producing glycolytic pathway is desirable

Figure 15·28
Regulation of pyruvate kinase. (a) A plot of phosphoenolpyruvate concentration versus enzyme activity for isozymes in some cells yields a sigmoidal curve, which indicates positive cooperativity of binding of phosphoenolpyruvate. The presence of fructose 1,6-*bis*phosphate shifts the curve to the left, showing that fructose 1,6-*bis*phosphate is an activator of the enzyme. (b) Incubation of liver cells with glucagon, which results in phosphorylation of pyruvate kinase, shifts the curve to the right, indicating a relatively less active pyruvate kinase.

(a)

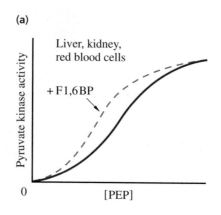

(b)

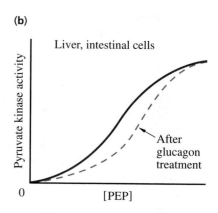

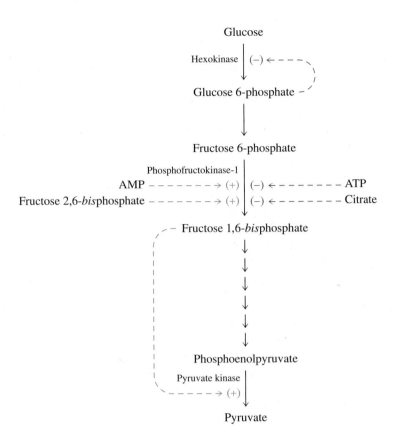

Figure 15·29
Summary of the metabolic regulation glycolytic pathway. Not shown are the effects of ADP on PFK-1, which vary among species.

under conditions when ATP is required, such as during exercise by muscle. Use of ATP leads to accumulation of AMP, which relieves the inhibition of PFK-1 by ATP. Fructose 2,6-*bis*phosphate also relieves this inhibition. Formation of fructose 1,6-*bis*phosphate then increases, which in certain tissues activates pyruvate kinase. The activity of the pathway decreases when its products are no longer required. Hexokinase is inhibited by excess glucose 6-phosphate, and PFK-1 by accumulation of ATP and citrate (an intermediate in the energy-producing citric acid cycle). ATP and citrate both signal an adequate energy supply.

E. The Pasteur Effect

Louis Pasteur observed that when yeast cells grow anaerobically, they produce much more ethanol and consume much more glucose than when they grow aerobically. In a similar manner, muscle accumulates lactic acid under anaerobic conditions but not when it metabolizes glucose aerobically. In both yeast and muscle, the rate of conversion of glucose to pyruvate is much higher under anaerobic conditions. The slowing of glycolysis in the presence of oxygen is called the **Pasteur effect.** As we will see in the next chapter, aerobic metabolism of glucose produces much more ATP than the two molecules per molecule of glucose produced by glycolysis. Therefore, for any given ATP requirement, less glucose must be consumed under aerobic conditions. Cells sense the state of ATP supply and demand and modulate glycolysis by means of a variety of regulatory systems. For example, availability of oxygen leads to the inhibition of PFK-1 (and thus glycolysis), probably through an increase in the ATP:AMP ratio. In the next chapter, we examine the energy-yielding metabolism of pyruvate under aerobic conditions, which can be seen as a second stage in the process of sugar degradation that begins with glycolysis.

Summary

Glycolysis is a ubiquitous pathway for the catabolism of monosaccharides. The 10 enzymes that catalyze the individual steps of glycolysis are located in the cytosol. For each molecule of hexose that is converted to pyruvate, there is a net production of two molecules of ATP from ADP + P_i, and two molecules of $NAD^\oplus$ are reduced to NADH. Glycolysis can be divided into two stages: a hexose stage, in which ATP is consumed, and a triose stage, in which a net gain of ATP is realized.

Anaerobic reoxidation of glycolytically produced NADH to $NAD^\oplus$ can occur through reductive metabolism of pyruvate. During alcoholic fermentation in yeast, pyruvate is cleaved to acetaldehyde in a reaction catalyzed by pyruvate decarboxylase, and acetaldehyde is reduced to ethanol, with concomitant oxidation of NADH to $NAD^\oplus$, in a reaction catalyzed by alcohol dehydrogenase. In anaerobic glycolysis to lactate, NADH is oxidized to $NAD^\oplus$ during the reduction of pyruvate to lactate, which is catalyzed by lactate dehydrogenase.

Three of the glycolytic reactions are metabolically irreversible in cells. These are the steps catalyzed by hexokinase, phosphofructokinase-1, and pyruvate kinase. Regulation of these enzymes, and thus of glycolysis, involves both allosteric interactions and covalent modifications. Regulation by covalent modification is mediated by enzymes that are not part of the glycolytic pathway but that act on enzymes of the pathway.

Selected Readings

Cullis, P. M. (1987). Acyl group transfer–phosphoryl group transfer. In *Enzyme Mechanisms*, M. I. Page and A. Williams, eds. (London: Royal Society of Chemistry), pp. 178–220. Summarizes the mechanisms of kinases, mutases, and phosphatases.

Engström, L., Ekman, P., Humble, E., and Zetterqvist, Ö. (1987). Pyruvate kinase. In *The Enzymes,* Vol. 18, P. D. Boyer and E. Krebs, eds. (New York: Academic Press), pp. 47–75. Describes the regulation of pyruvate kinase from mammalian liver.

Fothergill-Gilmore, L. A. (1986). Domains of glycolytic enzymes. In *Multidomain Proteins: Structure and Evolution.* D. G. Hardie and J. R. Coggins, eds. (Amsterdam: Elsevier Science Publishers), pp. 85–174. Discusses the X-ray crystallographic studies of the enzymes of the glycolytic pathway.

Hamori, E. (1975). Illustration of free energy changes in chemical reactions. *J. Chem. Ed.* 52:370–373.

Hers, H.-G., and Van Schaftingen, E. (1982). Fructose 2,6-*bis*phosphate 2 years after its discovery. *Biochem. J.* 206:1–12.

Hoffmann-Ostenhof, O., ed. (1987). *Intermediary Metabolism* (New York: Van Nostrand Reinhold Company). This collection of classic papers in metabolism contains the 1898 paper by Eduard Büchner mentioned in the introduction.

Lienhard, G. E., Slot, J. W., James, D. E., and Mueckler, M. M. (1992). How cells absorb glucose. *Sci. Am.* 266(1):86–91. Describes how glucose enters mammalian cells.

Pilkis, S. J., El-Maghrabi, M. R., and Claus, T. H. (1988). Hormonal regulation of hepatic gluconeogenesis and glycolysis. *Annu. Rev. Biochem.* 57:755–783.

Pilkis, S. J., and Granner, D. K. (1992). Molecular physiology of the regulation of hepatic gluconeogenesis and glycolysis. *Annu. Rev. Physiol.* 54:885–909.

Saier, M. H. (1987). *Enzymes in Metabolic Pathways* (New York: Harper & Row). Includes a readily accessible account of the enzymes of carbohydrate metabolism.

Seeholzer, S. H., Jaworowski, A., and Rose, I. A. (1991). Enolpyruvate: chemical determination as a pyruvate kinase intermediate. *Biochemistry* 30:727–732. First direct evidence for enolpyruvate in the reaction catalyzed by pyruvate kinase.

16

The Citric Acid Cycle

In the laboratory, organic compounds composed of carbon, hydrogen, and oxygen can be completely oxidized to CO_2 and H_2O by combustion. During this process, nearly all of the energy in the organic compounds is converted to heat. Combustion is an uncontrolled, often violent process. In biological systems, energy is released from molecules not by combustion but by the controlled process of respiration, with considerable conservation of the energy released. Compounds are oxidized in discrete steps, with enzymatically catalyzed transfer of electrons to molecular oxygen. The energy released is stored in the form of ATP. One major respiratory pathway is the oxidation of glucose to pyruvate followed by the complete oxidation of pyruvate coupled to the formation of ATP by oxidative phosphorylation. This chapter describes the oxidation of the carbon chain of pyruvate, forming ATP and reduced coenzymes. Chapter 18 describes the reoxidation of the reduced coenzymes and the conservation of the energy released as large amounts of ATP.

In the cells of aerobic organisms, pyruvate formed by glycolysis can be oxidized to CO_2 and H_2O through a series of enzymatic steps, during which much of the energy released is captured in the form of energy-rich compounds. The first enzymatic step in the conversion of pyruvate to CO_2 and H_2O is an oxidative decarboxylation reaction involving coenzyme A (CoASH). The products of this reaction are

 Related material appears in Exercise 5 of the computer-disk supplement *Exploring Molecular Structure.*

Figure 16·1
Hans Krebs (1900–1981). Krebs discovered the citric acid cycle, urea cycle, and glyoxylate cycle.

CO_2 and acetyl CoA, a molecule consisting of a two-carbon moiety attached to the group carrier coenzyme A (Figure 8·23). Subsequent oxidation of the acetyl group of acetyl CoA is carried out by the **citric acid cycle.** The energy released in the oxidation reactions of the citric acid cycle is conserved as reducing power when the coenzymes $NAD^{\oplus}$ and ubiquinone, Q (Section 8·16), are reduced to form NADH and ubiquinol (QH_2).

The citric acid cycle is also known as the tricarboxylic acid cycle, because there are tricarboxylate intermediates, and as the Krebs cycle, after the biochemist Hans Krebs who discovered it (Figure 16·1). In 1937, Krebs and W. A. Johnson proposed the citric acid cycle to explain a number of puzzling observations. It had been found by Albert Szent-Györgyi that addition of succinate, fumarate, or oxaloacetate to a suspension of minced muscle stimulated the consumption of O_2. The substrate of the oxidation was carbohydrate, either glucose or glycogen. Especially intriguing was the observation that addition of small amounts of these substances caused larger amounts of oxygen to be consumed than is required for their own oxidation, indicating that the substances had catalytic effects. Krebs and Johnson observed that the six-carbon compound citrate and the five-carbon compound α-ketoglutarate also had a catalytic effect on the carbohydrate-supported respiration of muscle. Krebs and Johnson proposed that citrate was formed from a four-carbon cycle intermediate and a derivative of glucose, an unknown two-carbon compound that was later shown to be acetyl CoA. The cyclic nature of the pathway explained how its intermediates could act catalytically, that is, lead to the consumption of more O_2 than is required for their own combustion. The citric acid cycle was only the second metabolic cycle to be discovered. The first was the urea cycle (Section 21·10), outlined by Krebs in 1933. A third metabolic cycle, the glyoxylate cycle, was proposed by Krebs and Hans Kornberg in 1957. The glyoxylate cycle is a variation of the citric acid cycle and is described later in this chapter.

The citric acid cycle is the hub of aerobic metabolism. The aerobic catabolism of carbohydrates, fats, and amino acids merges at the reactions of the citric acid cycle, and intermediates of the citric acid cycle are the starting points for many biosynthetic pathways. The citric acid cycle is thus both catabolic and anabolic, or **amphibolic.**

The enzymes of the citric acid cycle are found in the cytosol of prokaryotes and in eukaryotic mitochondria (which are evolutionary descendants of bacteria). Before pyruvate produced in the cytosol by glycolysis can enter the citric acid cycle, it must be converted to acetyl CoA. In eukaryotes, enzymatic conversion to acetyl CoA occurs after transport of pyruvate into the mitochondrion.

16·1 Pyruvate Enters the Mitochondrion via a Transport Protein Embedded in the Inner Mitochondrial Membrane

The mitochondrion is delimited by a double membrane, but only the inner membrane presents a barrier to the passage of small molecules such as pyruvate. Small molecules pass through the outer membrane via an aqueous channel formed by a transmembrane protein called porin (Section 12·7B) that allows free diffusion of molecules having a molecular weight less than 10 000. Embedded in the inner mitochondrial membrane is a protein, pyruvate translocase, that specifically transports pyruvate from the intermembrane space to the interior space of the mitochondrion,

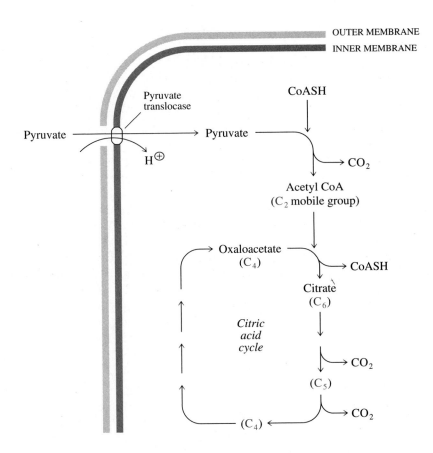

Figure 16·2
Entry of pyruvate into the citric acid cycle. Pyruvate is transported into the mitochondrion via pyruvate translocase. It is then converted to acetyl CoA and CO_2. For every acetyl moiety that enters the citric acid cycle, two molecules of CO_2 are released.

known as the mitochondrial matrix. In Figure 16·2, pyruvate is shown crossing the membrane in symport with $H^{\oplus}$. Once inside the mitochondrion, pyruvate is converted to CO_2 and acetyl CoA, which is further oxidized by the reactions of the citric acid cycle.

16·2 The Pyruvate Dehydrogenase Complex Converts Pyruvate to Acetyl CoA

In both prokaryotes and eukaryotes, the conversion of pyruvate to acetyl CoA and CO_2 is catalyzed by a complex of enzymes and cofactors known as the pyruvate dehydrogenase complex. The overall reaction is

$$
\begin{array}{ccc}
COO^{\ominus} & & S\text{-}CoA \\
| & & | \\
C=O & \xrightarrow{\text{Pyruvate dehydrogenase complex}} & C=O \\
| & & | \\
CH_3 & & CH_3 \\
\text{Pyruvate} & & \text{Acetyl CoA}
\end{array}
\qquad (16\cdot1)
$$

HS-CoA → CO₂, NAD$^{\oplus}$ → NADH

The pyruvate dehydrogenase complex is a multienzyme complex, that is, a noncovalently linked assembly of enzyme molecules that catalyze successive reactions. The product formed by one reaction of the enzyme complex does not diffuse into the medium but is immediately acted upon by the next component of the system. This channelling of metabolites is accomplished by covalent binding of the metabolite to a flexible prosthetic group of one enzyme of the complex. (Channelling and multienzyme complexes were discussed in Section 6·17.)

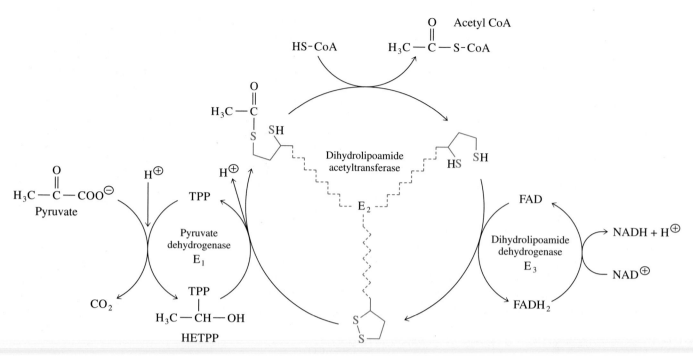

Figure 16·3
Reactions of the pyruvate dehydrogenase complex. A product is released at each step: CO_2 at pyruvate dehydrogenase (E_1), acetyl CoA at dihydrolipoamide acetyltransferase (E_2), and NADH at dihydrolipoamide dehydrogenase (E_3). The swinging arm of E_2 is a lipoamide prosthetic group formed by amide linkage of lipoic acid to a lysine residue of the enzyme.

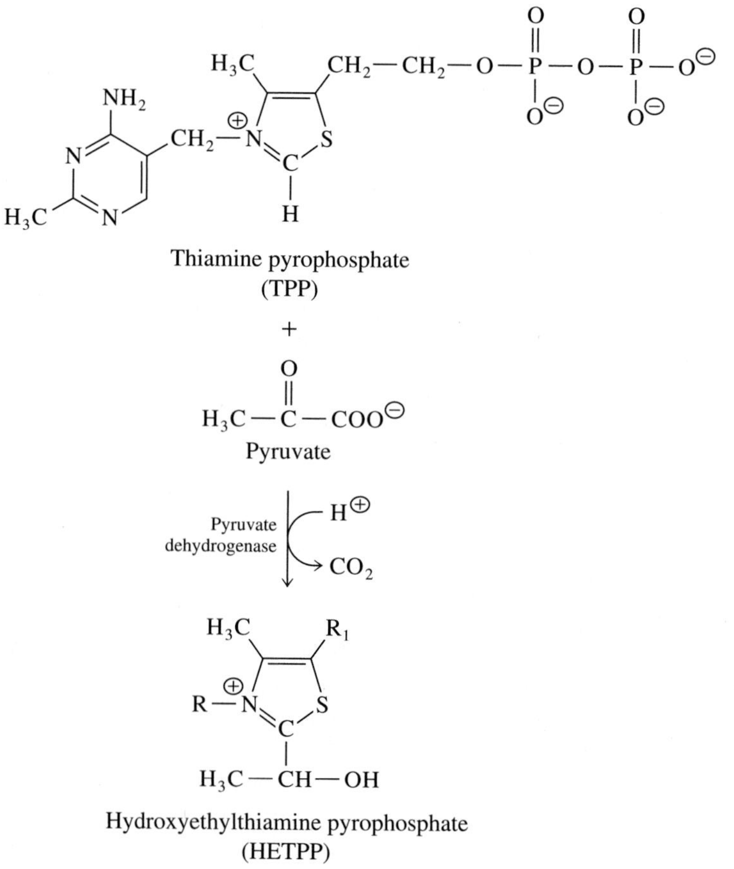

Thiamine pyrophosphate
(TPP)

+

Pyruvate

Pyruvate dehydrogenase

Hydroxyethylthiamine pyrophosphate
(HETPP)

Figure 16·4
Formation of hydroxyethylthiamine pyrophosphate (HETPP) in the pyruvate dehydrogenase reaction.

The individual components and the reaction sequence of the pyruvate dehydrogenase complex are shown in Figure 16·3. The complex consists of multiple copies of three enzymes: pyruvate dehydrogenase (E_1), dihydrolipoamide acetyltransferase (E_2), and dihydrolipoamide dehydrogenase (E_3), as well as several other proteins.

E_1 catalyzes the decarboxylation of pyruvate and the transfer of the remaining two-carbon fragment to the lipoamide moiety of E_2. Pyruvate first reacts with the prosthetic group of E_1, thiamine pyrophosphate (TPP, Section 8·9); after CO_2 is released, the resulting intermediate is hydroxyethylthiamine pyrophosphate (HETPP), as shown in Figure 16·4. The mechanism of this reaction is similar to that of pyruvate decarboxylase, diagrammed in Figure 8·27 as an example of the role of TPP. In that reaction, HETPP is cleaved, releasing acetaldehyde. In the reaction catalyzed by E_1, the two-carbon hydroxyethyl fragment is instead transferred to the lipoamide prosthetic group of E_2. This prosthetic group consists of lipoic acid covalently bound by an amide linkage to a lysine residue of E_2 (Figure 16·5).

The lipoamide prosthetic group acts as a swinging arm that moves between the reactive sites of E_1 and E_3. The transfer of the two-carbon hydroxyethyl fragment from E_1 to the lipoamide prosthetic group of E_2 is thought to involve the oxidation of HETPP by the disulfide form of the lipoamide group to form acetyl-TPP, followed by transfer of the acetyl group to the dihydro form of lipoamide (Figure 16·6). Next, CoASH reacts with the acetyl group, forming acetyl CoA and leaving the lipoamide in the reduced, sulfhydryl form (top of Figure 16·3).

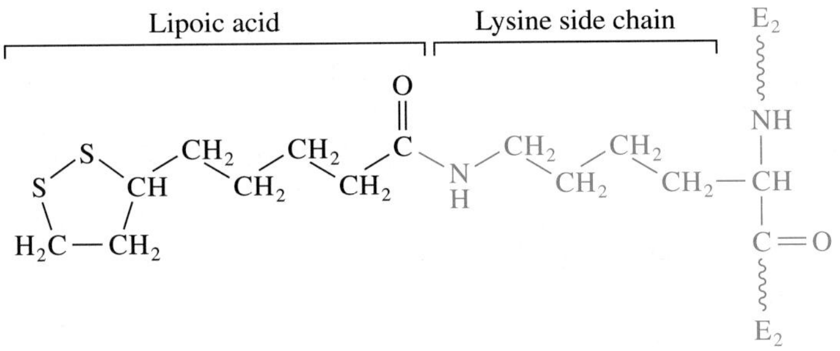

Figure 16·5
Lipoamide prosthetic group of E_2 of the pyruvate dehydrogenase complex.

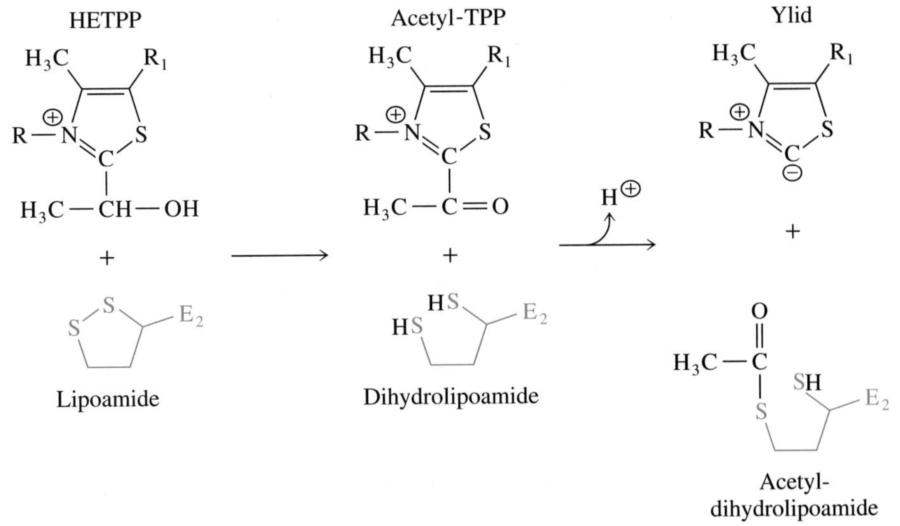

Figure 16·6
Proposed mechanism for conversion of the hydroxyethyl group of HETPP to the acetyl group of acetyl-dihydrolipoamide. In this reaction, oxidation of HETPP to acetyl-TPP is coupled to reduction of the disulfide of the lipoamide group. The acetyl group is then transferred to a sulfhydryl group of dihydrolipoamide.

Table 16·1 Components of the pyruvate dehydrogenase complex in mammals and *E. coli*

Enzyme	Coenzyme	Oligomeric form		Number of oligomers per complex	
		Mammals	*E. coli*	Mammals	*E. coli*
Pyruvate dehydrogenase (E_1)	TPP	$\alpha_2\beta_2$	α_2	20–30	12
Dihydrolipoamide acetyltransferase (E_2)	Lipoic acid, CoASH	α_{60}	α_{24}	1	1
Dihydrolipoamide dehydrogenase (E_3)	FAD, $NAD^\oplus$	α_2	α_2	6	6
X	Lipoic acid	Monomer	—	6	—
Kinase	None	$\alpha\beta$	—	2–3	—
Phosphatase	None	$\alpha\beta$	—	>3	—

[Adapted from Patel, M. S., and Roche, T. E. (1990). Molecular biology and biochemistry of pyruvate dehydrogenase complexes. *FASEB J.* 4:3224–3233.]

E_3 catalyzes the reoxidation of the reduced lipoamide of E_2, allowing E_2 to participate in another round of catalysis. The prosthetic group of E_3, flavin adenine dinucleotide (FAD, Figure 8·21), oxidizes the reduced lipoamide, resulting in formation of the reduced coenzyme, E_3-$FADH_2$. $NAD^\oplus$ is then reduced by E_3-$FADH_2$, so that E_3-FAD and NADH are formed and the catalytic cycle is completed.

An important distinction between FAD and $NAD^\oplus$ is that FAD remains tightly bound to the enzyme with which it is associated, whereas $NAD^\oplus$ is released to solution. In other words, FAD is a prosthetic group, whereas $NAD^\oplus$ is a cosubstrate (Section 8·2). In the pyruvate dehydrogenase complex, the E_3-$FADH_2$ complex must be reoxidized by reaction with $NAD^\oplus$ to regenerate the original holoenzyme and complete the catalytic cycle. Reduced NADH dissociates into the mitochondrial matrix (or cytosol of bacteria), where it serves as a mobile carrier of reducing power.

The pyruvate dehydrogenase complex is a large assembly with multiple copies of E_1 and E_3 surrounding a core consisting of many E_2 monomers. There are two types of cores, those with 60 E_2 chains and those with 24 E_2 chains. Table 16·1 lists the components and stoichiometry of the complexes in mammals and *Escherichia coli*. The molecular weight of the complex in *E. coli* is about five million; in mammals, it is between seven and nine million, depending on the tissue of origin. The pyruvate dehydrogenase complex from the Gram-negative bacterium *Azotobacter vinelandii* has the same subunit stoichiometry as the enzyme from *E. coli*. The structure of the core in the *A. vinelandii* complex has been determined by X-ray crystallography, and the structure of the entire complex has been deduced from related experiments. The 24 E_2 monomers of the core are arranged in a cube to which the 12 E_1 dimers and the 6 E_3 dimers are bound noncovalently (Figure 16·7). A protein species present only in mammalian pyruvate dehydrogenase complexes, protein X, contains a bound lipoyl group. Befitting its name, the function of this protein is unknown. The eukaryotic complex also has associated with it several copies of a protein kinase and a protein phosphatase. These enzymes have regulatory roles that will be examined in Section 16·5.

We will encounter two other α-keto acid dehydrogenase complexes that have properties very much like those of the pyruvate dehydrogenase complex. One is specific for α-ketoglutarate (Section 16·3, part 4), and the other for branched-chain α-keto acids (Section 21·12E). In each complex, E_1 is a specific dehydrogenase and E_2 a specific dihydrolipoamide acyltransferase; E_3 is the same enzyme in all three complexes.

Figure 16·7
Structure of the pyruvate dehydrogenase complex. **(a)** Core of the pyruvate dehydrogenase complex from *A. vinelandii*. The core is composed of 24 E_2 chains. **(b)** Model of the pyruvate dehydrogenase complex, showing 12 E_1 dimers surrounding the core. (Courtesy of Andrea Mattevi and Wim G. J. Hol.)

(a)

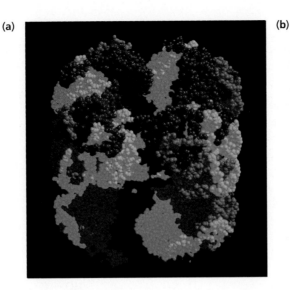

(b)

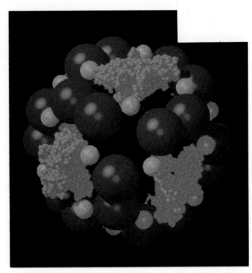

16·3 The Citric Acid Cycle Generates Energy-Rich Molecules

Acetyl CoA formed from pyruvate or as the product of other pathways (such as the catabolism of fatty acids or amino acids) can be oxidized by the citric acid cycle. The eight enzyme-catalyzed reactions of the citric acid cycle are listed in Table 16·2 and shown in Figure 16·8 (next page). Before examining the reactions individually, we will consider two general features of the pathway: the flow of carbon and the production of energy-rich molecules.

In the first reaction of the citric acid cycle, the two-carbon acetyl group of acetyl CoA condenses with the four-carbon molecule oxaloacetate to form the six-carbon intermediate citrate. Two molecules of CO_2 are released when a six-carbon acid and a five-carbon acid undergo oxidative decarboxylation. The decarboxylations are followed by several steps that lead to regeneration of oxaloacetate, the original condensation partner for acetyl CoA. Because oxaloacetate is regenerated, the citric acid cycle can be seen as a multistep catalyst, oxidizing acetyl CoA to CO_2 but returning to its original state with each complete round of reactions.

Table 16·2 The enzymatic reactions of the citric acid cycle

Reaction	Enzyme
1. Acetyl CoA + Oxaloacetate + H_2O $\longrightarrow$ Citrate + CoASH + $H^\oplus$	Citrate synthase
2. Citrate $\rightleftarrows$ Isocitrate	Aconitase (Aconitate hydratase)
3. Isocitrate + $NAD^\oplus$ $\longrightarrow$ α-Ketoglutarate + NADH + CO_2	Isocitrate dehydrogenase
4. α-Ketoglutarate + CoASH + $NAD^\oplus$ $\longrightarrow$ Succinyl CoA + NADH + CO_2	α-Ketoglutarate dehydrogenase complex
5. Succinyl CoA + GDP (or ADP) + P_i $\rightleftarrows$ Succinate + GTP (or ATP) + CoASH	Succinyl-CoA synthetase
6. Succinate + Q $\rightleftarrows$ Fumarate + QH_2	Succinate dehydrogenase complex
7. Fumarate + H_2O $\rightleftarrows$ L-Malate	Fumarase (Fumarate hydratase)
8. L-Malate + $NAD^\oplus$ $\rightleftarrows$ Oxaloacetate + NADH + $H^\oplus$	Malate dehydrogenase

Net equation:
Acetyl CoA + 3 $NAD^\oplus$ + Q + GDP (or ADP) + P_i + 2 H_2O $\longrightarrow$ CoASH + 3 NADH + QH_2 + GTP (or ATP) + 2 CO_2 + 2 $H^\oplus$

Figure 16·8
Citric acid cycle. In each turn of the cyclic pathway, the two-carbon acetyl group of an acetyl CoA molecule enters the cycle by condensing with oxaloacetate. Two molecules of CO_2 are subsequently released, the mobile coenzymes $NAD^⊕$ and Q are reduced, a phosphoryl group is transferred to GDP (or ADP), and the original acceptor molecule, oxaloacetate, is re-formed.

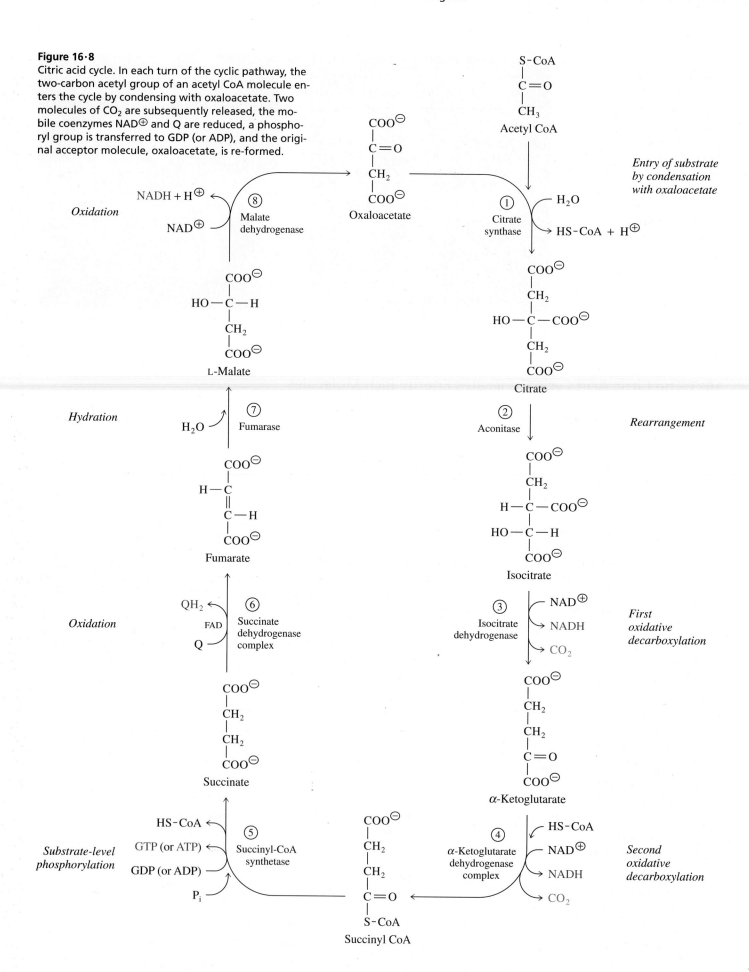

Note that the two carbon atoms that enter the cycle by condensation with oxaloacetate are not the same carbon atoms subsequently lost as CO_2 (Figure 16·9). However, the *balance* of carbon is such that for each two-carbon group from acetyl CoA that enters the cycle, two molecules of CO_2 are released during a complete turn of the cycle. The two carbons of acetyl CoA that entered the cycle become half of the symmetrical four-carbon molecule succinate in the fifth reaction. The two halves of this symmetrical molecule are chemically equivalent so that, in terms of carbon tracing, carbons arising from acetyl CoA are evenly distributed in molecules arising from succinate.

Most of the energy released in the reactions of the citric acid cycle is conserved in the form of the reduced coenzymes NADH and QH_2 (Figure 16·10). NADH is formed by reduction of $NAD^{\oplus}$ during two oxidative decarboxylation steps; QH_2 is formed when succinate is oxidized to fumarate. Oxidation of the reduced coenzymes by the electron-transport chain leads to the production of ATP by the process of oxidative phosphorylation. Most of the ATP produced by aerobically metabolizing cells is generated by oxidative phosphorylation. In addition, one reaction of the citric acid cycle produces a nucleoside triphosphate via substrate-level phosphorylation. The triphosphate produced may be either GTP or ATP, depending on the cell type.

We will now examine each of the eight enzymatic steps of the citric acid cycle in detail.

1. Acetyl CoA Enters the Citric Acid Cycle by Condensing with Oxaloacetate to Form Citrate

In the first reaction of the citric acid cycle, acetyl CoA reacts with oxaloacetate to form citrate and CoASH.

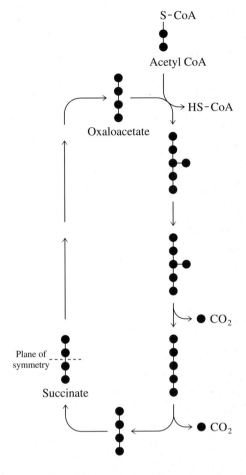

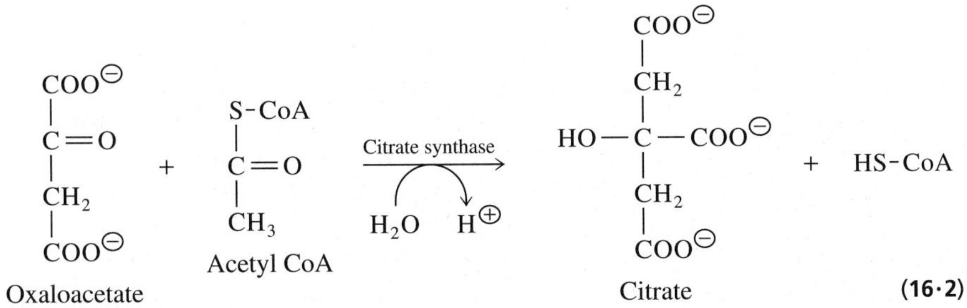

Figure 16·9
Fates of the carbon atoms from oxaloacetate and acetyl CoA during one turn of the citric acid cycle. The plane of symmetry of succinate means that the two halves of the molecule are chemically equivalent; thus, carbon from acetyl CoA (red) is uniformly distributed in the four-carbon intermediates leading to oxaloacetate. Carbon from acetyl CoA that enters in one turn of the cycle is thus lost as CO_2 only in the second and subsequent turns.

$$\text{Oxaloacetate} + \text{Acetyl CoA} \xrightarrow[\text{H}_2\text{O} \quad \text{H}^{\oplus}]{\text{Citrate synthase}} \text{Citrate} + \text{HS-CoA} \qquad (16\cdot2)$$

This reaction is catalyzed by citrate synthase. In most tissues, citrate synthase has a molecular weight of about 100 000 and is composed of two identical subunits. The reaction it catalyzes appears to be metabolically irreversible. The uncertainty about

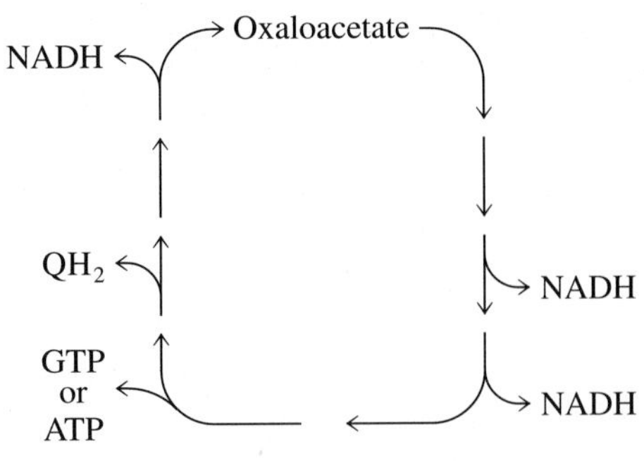

Figure 16·10
Production of energy-rich compounds by the citric acid cycle. The reduced coenzymes NADH and QH_2 are oxidized by the respiratory electron-transport chain, and some of the energy released from their oxidation is used for the formation of ATP from ADP and P_i via oxidative phosphorylation. One nucleoside triphosphate—either GTP or ATP, depending on the cell type—is produced by substrate-level phosphorylation.

its steady-state condition arises from the technical difficulty of measuring precisely the concentrations of the reactants and products in the mitochondrial compartment, especially the concentration of oxaloacetate, which is present in very small quantities.

The reaction catalyzed by citrate synthase is the only reaction of the citric acid cycle in which a C—C bond is formed. In the first step, a proton is abstracted from C-2 of acetyl CoA by a histidine residue of the enzyme. Thioesters enolize, a property that facilitates the proton abstraction. The enol form of the substrate is stabilized by the positive charge on the imidazolium group of the histidine residue.

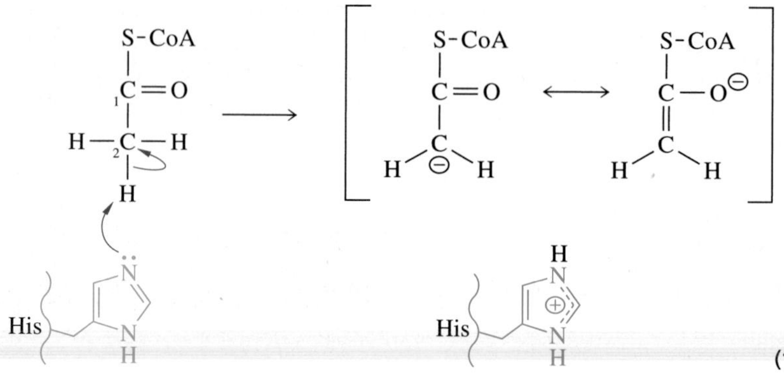

(16·3)

As shown in Figure 16·11, the carbanion (a nucleophile) attacks the carbonyl carbon of oxaloacetate to form an enzyme-bound intermediate, citryl CoA. This energy-rich thioester is hydrolyzed to release the products, citrate and CoASH.

Citrate synthase (Figure 16·12) undergoes several conformational changes upon binding of oxaloacetate and formation of the intermediate citryl CoA. First, the binding of oxaloacetate induces hinge bending, so that the two domains move toward each other by 18° and form a binding site for acetyl CoA. Acetyl CoA is then bound, its proton abstracted, and citryl CoA formed. Citryl CoA induces the enzyme to close completely; only the closed conformation catalyzes hydrolysis of citryl CoA by a bound water molecule. After hydrolysis, the enzyme opens, allowing the products to leave. The closing of the enzyme in the presence of reactive ligands is an example of induced fit, a property that citrate synthase shares with hexokinase (Figure 7·18). The conformational changes undergone by citrate synthase and hexokinase prevent side reactions by shielding the reactants from water.

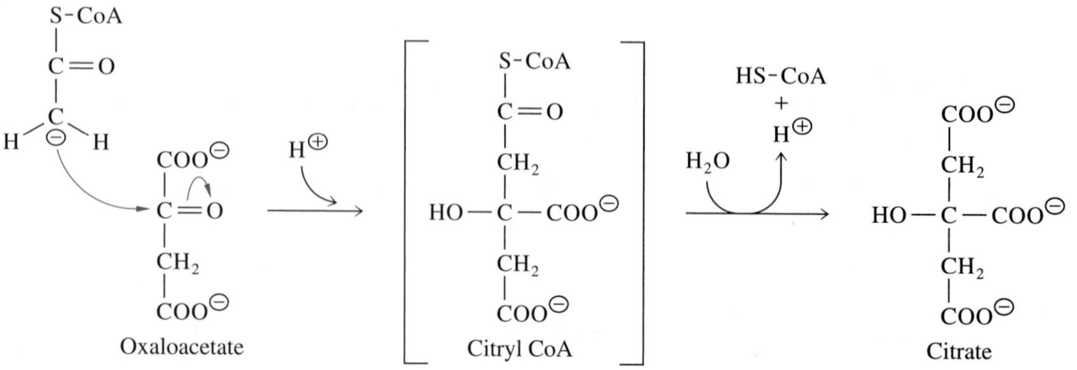

Figure 16·11
Reaction catalyzed by citrate synthase.

Figure 16·12
Stereo views of citrate synthase. **(a)** Open conformation. **(b)** Fully closed conformation. The product citrate (green) is positioned in the active site. (Based on coordinates provided by S. Remington, G. Weigand, and R. Huber.)

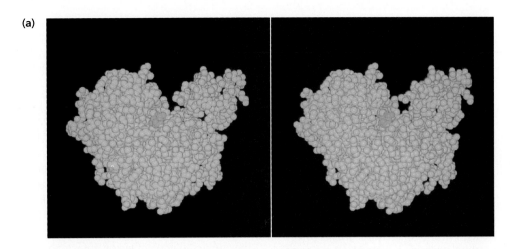

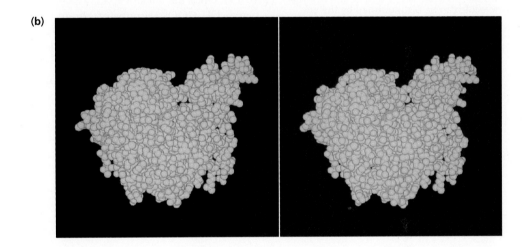

2. Aconitase Catalyzes the Conversion of Prochiral Citrate to Chiral Isocitrate

Aconitase (formally, aconitate hydratase) catalyzes the near-equilibrium conversion of citrate to isocitrate. Citrate is a tertiary alcohol and thus cannot be oxidized to a keto acid. The action of aconitase converts citrate to an oxidizable secondary alcohol. The name of the enzyme is derived from the enzyme-bound intermediate of the reaction, *cis*-aconitate. The reaction proceeds by elimination of water to form a carbon-carbon double bond, followed by stereospecific addition of water to form isocitrate (Figure 16·13).

Citrate   *cis*-Aconitate   2*R*,3*S*-Isocitrate

Figure 16·13
Reaction catalyzed by aconitase. The enzyme is named for its intermediate, *cis*-aconitate.

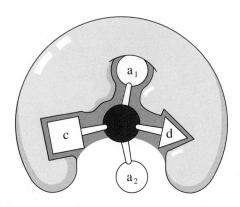

Figure 16·14
Nonequivalence of identical groups when a prochiral molecule is bound by three-point attachment. In a chiral active site, the chemically identical groups a_1 and a_2 of a prochiral molecule can be distinguished by the enzyme. The enzyme cannot bind a_2 in the a_1 binding site when c and d are correctly bound.

Aconitase contains a covalently bound iron-sulfur prosthetic group that assists in correct positioning of citrate by binding both the C-3 carboxylate group and the hydroxyl group. In other iron-sulfur proteins, the iron is involved in electron transfer and undergoes oxidation-reduction changes, as discussed in Chapter 18.

The proper positioning of citrate in the active site of aconitase is essential for the stereospecific reaction that follows. Citrate is a prochiral molecule; that is, it has a carbon atom with three types of substituents (C_{aacd}). A prochiral molecule can become chiral (C_{abcd}) by substituting a fourth type for one of the identical substituents. When the citric acid cycle was proposed by Krebs, the inclusion of the citrate-to-isocitrate reaction was a major barrier to its acceptance because labelling studies indicated that only a single isomer of isocitrate was produced in cells. At the time, conversion of a prochiral molecule to a single chiral isomer was unknown. In a conceptual breakthrough, Alexander Ogston suggested in 1948 how a chiral active site of an enzyme could distinguish between chemically equivalent groups on the citrate molecule. Ogston envisioned citrate binding in a manner he called three-point attachment, with nonidentical groups being the points of attachment (Figure 16·14). Once the substrate is correctly bound to the asymmetric binding site, the two $-CH_2COO^{\ominus}$ groups of citrate have specific orientations and thus are no longer equivalent. Conversion of citrate to isocitrate creates two chiral centers in the isocitrate molecule, 2R,3S-isocitrate. (The RS nomenclature system is discussed in Section 4·2.)

Fluoroacetate, produced by a few plants, is highly poisonous. It is converted to fluoroacetyl CoA and then incorporated into fluorocitrate by the action of citrate synthase.

$$\begin{array}{c} COO^{\ominus} \\ | \\ CH_2F \\ \text{Fluoroacetate} \end{array} \xrightarrow{\text{HS-CoA}} \begin{array}{c} S\text{-CoA} \\ | \\ C=O \\ | \\ CH_2F \\ \text{Fluoroacetyl CoA} \end{array} \xrightarrow[\text{Oxaloacetate}]{\text{Citrate synthase}\quad\text{HS-CoA}} \begin{array}{c} COO^{\ominus} \\ | \\ CH-F \\ | \\ HO-C-COO^{\ominus} \\ | \\ CH_2 \\ | \\ COO^{\ominus} \\ \text{Fluorocitrate} \end{array}$$

$$(16\cdot4)$$

The citrate analog fluorocitrate is a potent inhibitor of animal aconitase, and aerobic metabolism via the citric acid cycle is blocked in its presence. The lethal properties of fluoroacetate have led to its use as a rat poison.

3. Isocitrate Dehydrogenase Catalyzes the Oxidation of Isocitrate to α-Ketoglutarate and CO_2

Isocitrate dehydrogenase catalyzes the oxidative decarboxylation of isocitrate to form α-ketoglutarate. This reaction is the first of four oxidation-reduction reactions within the citric acid cycle. Two steps are involved in this $NAD^{\oplus}$-dependent reaction (Figure 16·15). First, the alcohol group of isocitrate is oxidized via transfer of the hydrogen on C-2 as a hydride ion to $NAD^{\oplus}$, reducing the coenzyme to NADH and forming oxalosuccinate, an unstable β-keto acid. Oxalosuccinate undergoes nonenzymatic β decarboxylation to α-ketoglutarate before it is released from the enzyme. Overall, the hydroxy acid isocitrate is converted to the keto acid

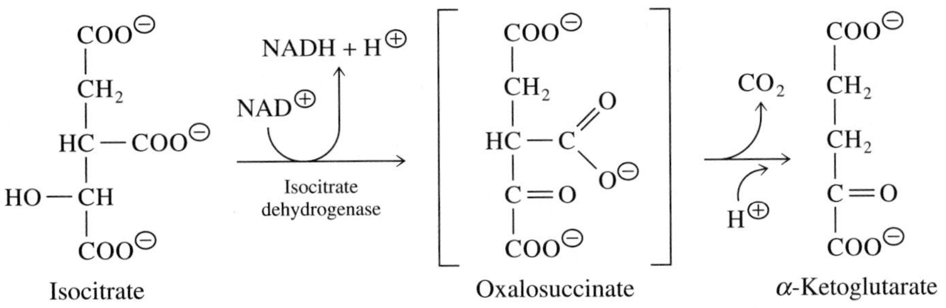

Figure 16·15
Reaction catalyzed by isocitrate dehydrogenase. $NAD^{\oplus}$ is reduced to NADH and carbon dioxide is released in this oxidative decarboxylation reaction. Oxalosuccinate, indicated in brackets, is an unstable enzyme-bound intermediate.

α-ketoglutarate with release of CO_2, and a molecule of $NAD^{\oplus}$ is reduced to NADH. The reaction catalyzed by isocitrate dehydrogenase is metabolically irreversible and can be regulated by phosphorylation in *E. coli,* as discussed in Section 16·5.

4. The α-Ketoglutarate Dehydrogenase Complex Catalyzes the Formation of Succinyl CoA

Like pyruvate, α-ketoglutarate is an α-keto acid. Oxidative decarboxylation of α-ketoglutarate is analogous to the reaction catalyzed by the pyruvate dehydrogenase complex, and the product is again a high-energy thioester, succinyl CoA (Figure 16·16). The multienzyme α-ketoglutarate dehydrogenase complex closely resembles the pyruvate dehydrogenase complex. The same coenzymes are involved, and the mechanisms are identical. In the α-ketoglutarate dehydrogenase complex, the three component enzymes are α-ketoglutarate dehydrogenase (E_1, containing TPP), dihydrolipoamide succinyltransferase (E_2, containing a flexible lipoamide prosthetic group), and dihydrolipoamide dehydrogenase (E_3, the same flavoprotein as E_3 of the pyruvate dehydrogenase complex). The α-ketoglutarate dehydrogenase complex catalyzes a key regulatory step of the citric acid cycle, as discussed in Section 16·5. This is the second of the two CO_2-producing reactions of the citric acid cycle.

The conversion of α-ketoglutarate to succinyl CoA brings us to the stage of the citric acid cycle at which there has been a net oxidation of carbon atoms to CO_2 equal to the number of carbon atoms that entered the cycle at the citrate synthase reaction. In the four remaining reactions of the cycle, the four-carbon succinyl group of succinyl CoA is converted back to oxaloacetate. With the regeneration of oxaloacetate, additional acetyl CoA can enter the citric acid cycle to be oxidized.

Figure 16·16
Reaction catalyzed by the α-ketoglutarate dehydrogenase complex. The reaction is analogous to that catalyzed by the pyruvate dehydrogenase complex. The two complexes are very similar and have identical mechanisms.

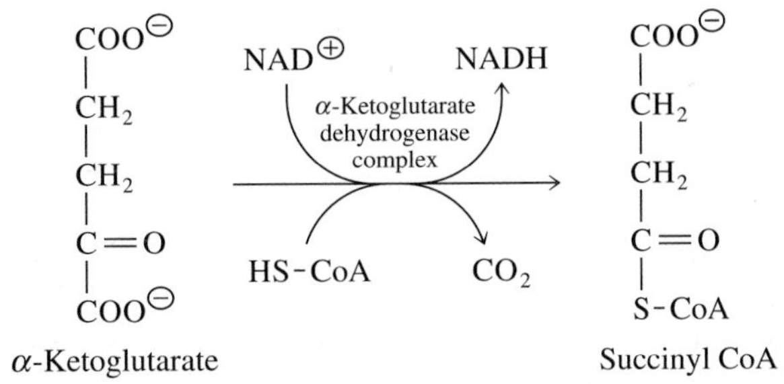

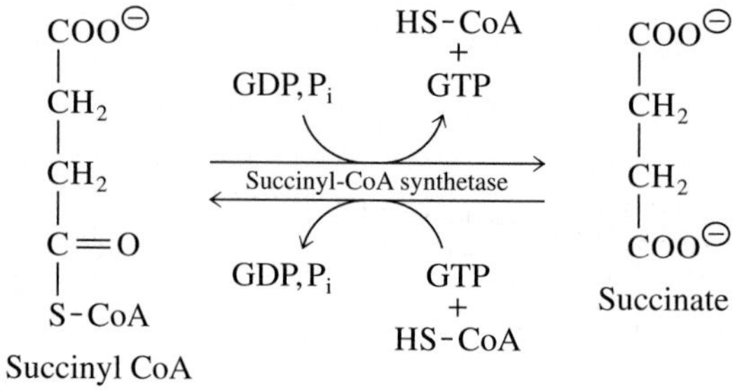

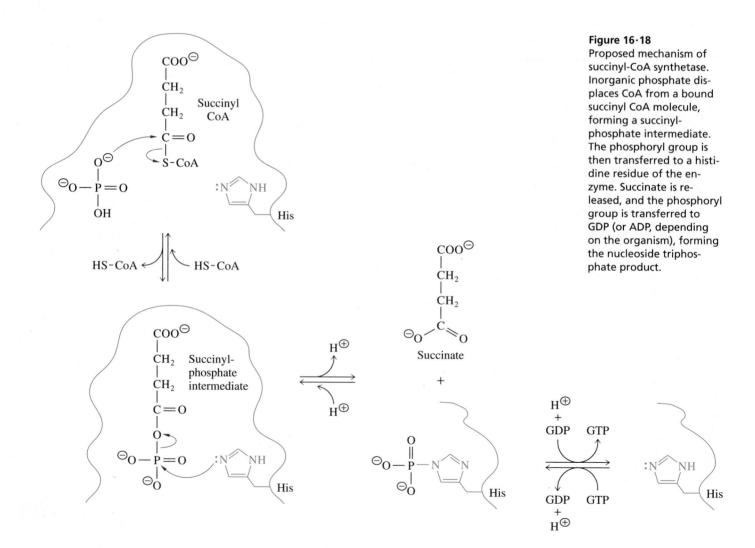

Figure 16·17
Reaction catalyzed by succinyl-CoA synthetase. In plants and several bacteria, ATP is used in place of GTP.

5. Succinyl-CoA Synthetase Catalyzes Substrate-Level Phosphorylation

The free energy stored in the thioester group of succinyl CoA is conserved through the synthesis of a nucleoside triphosphate (usually GTP in mammals and ATP in plants and several bacteria) from the corresponding nucleoside diphosphate and P_i. The reaction is catalyzed by succinyl-CoA synthetase, sometimes referred to as succinate thiokinase (Figure 16·17). This reaction, like the reactions catalyzed by phosphoglycerate kinase and pyruvate kinase in glycolysis, is an example of substrate-level phosphorylation.

Figure 16·18
Proposed mechanism of succinyl-CoA synthetase. Inorganic phosphate displaces CoA from a bound succinyl CoA molecule, forming a succinyl-phosphate intermediate. The phosphoryl group is then transferred to a histidine residue of the enzyme. Succinate is released, and the phosphoryl group is transferred to GDP (or ADP, depending on the organism), forming the nucleoside triphosphate product.

During catalysis, succinyl-CoA synthetase generates a relatively stable phospho-enzyme intermediate in which a phosphoryl group is covalently linked to a histidine residue of the enzyme. In the proposed mechanism, the enzyme binds succinyl CoA, inorganic phosphate displaces CoA to form succinyl phosphate, and the phosphoryl group is subsequently transferred to a histidine residue of the enzyme. The phosphohistidine intermediate then transfers its phosphoryl group to GDP, forming GTP (Figure 16·18).

GTP can be regarded as an ATP equivalent because the γ-phosphoryl group of GTP can be transferred to ADP in a reaction catalyzed by nucleoside diphosphate kinase, producing ATP and GDP.

$$\text{GTP} + \text{ADP} \xrightleftharpoons[\text{}]{\substack{\text{Nucleoside} \\ \text{diphosphate kinase}}} \text{ATP} + \text{GDP} \qquad \textbf{(16·5)}$$

6. The Succinate Dehydrogenase Complex Catalyzes the Conversion of Succinate to Fumarate

The eukaryotic succinate dehydrogenase complex is embedded in the inner mitochondrial membrane, whereas the other components of the citric acid cycle are dissolved in the matrix of the mitochondrion. In aerobic prokaryotes, the succinate dehydrogenase complex is embedded in the plasma membrane, whereas the other components of the citric acid cycle are in the cytosol.

The succinate dehydrogenase complex (often loosely referred to as succinate dehydrogenase) catalyzes the formation of a double bond in the oxidation of succinate to fumarate (Figure 16·19). The complex consists of several polypeptides and includes a covalently bound FAD prosthetic group (Figure 16·20) and iron-sulfur clusters. These electron-transport cofactors are involved in the removal of electrons from succinate and the donation of electrons to ubiquinone (Q), the lipid-soluble mobile carrier of reducing power. Just as $FADH_2$ in E_3 of the pyruvate dehydrogenase complex is reoxidized by water-soluble $NAD^{\oplus}$ to complete the catalytic cycle of that enzyme, $FADH_2$ in the succinate dehydrogenase complex is reoxidized by Q. $FADH_2$ is often shown as the redox product of this reaction, but since FAD is permanently bound to the enzyme, the catalytic cycle is not completed until bound $FADH_2$ is reoxidized and the mobile product QH_2 is released. We will consider the transfer of reducing equivalents to Q again in Chapter 18, where we will see that the succinate dehydrogenase complex has a role in the respiratory electron-transport chain.

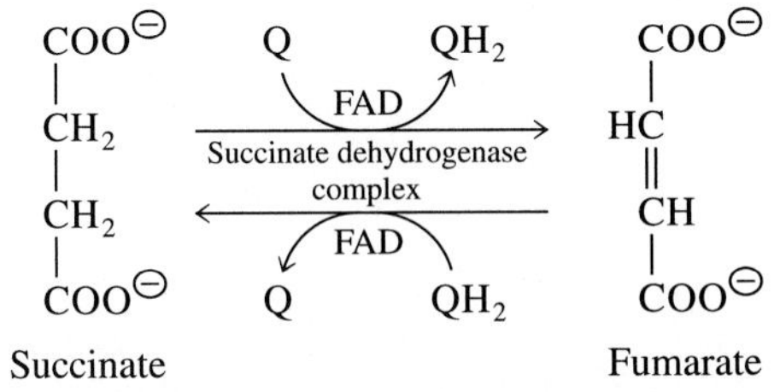

Figure 16·19
Reaction catalyzed by the succinate dehydrogenase complex. The lipid-soluble electron carrier ubiquinone (Q) accepts electrons from the molecule of $FADH_2$ covalently bound to the succinate dehydrogenase complex. The electrons pass from $FADH_2$ to Q via iron-sulfur proteins that are also part of the complex.

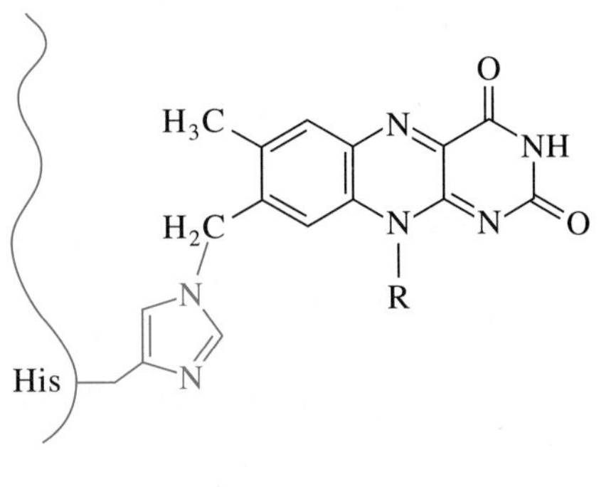

Figure 16·20
FAD covalently linked to the side chain of a histidine residue in the succinate dehydrogenase complex. *R* represents the ribitol-ADP moiety of FAD.

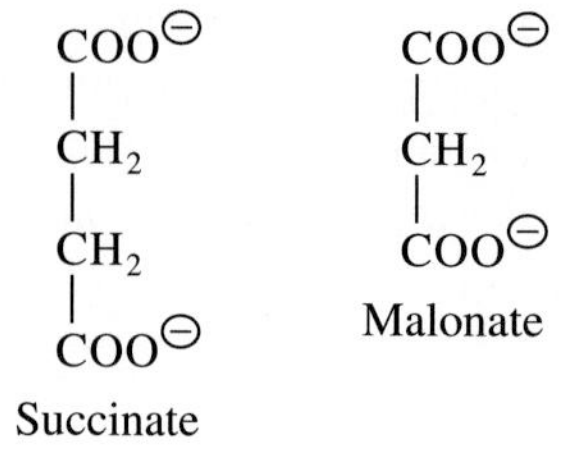

Figure 16·21
Succinate and malonate. Malonate is a structural analog of succinate that binds to the substrate-binding site of the succinate dehydrogenase complex but does not react. Malonate therefore acts as a competitive inhibitor.

The substrate analog malonate is a competitive inhibitor of the succinate dehydrogenase complex (Figure 16·21). Malonate, like succinate, is a dicarboxylate that binds to cationic amino acid residues in the active site of the succinate dehydrogenase complex, but malonate cannot undergo oxidation. In experiments with isolated mitochondria or cell homogenates, the presence of malonate causes succinate, α-ketoglutarate, and citrate to accumulate. Such experiments provided some of the original evidence for the sequence of reactions in the citric acid cycle.

7. Fumarase Catalyzes the Reversible Hydration of Fumarate to Malate

Fumarase (formally, fumarate hydratase) catalyzes the near-equilibrium conversion of fumarate to malate through the stereospecific *trans* addition of water to the double bond of fumarate (Figure 16·22). Fumarate is another example of a prochiral molecule. When fumarate is positioned in the active site of fumarase, the double bond of the substrate can be attacked from only one direction. The product of the reaction is the L stereoisomer of the hydroxy acid malate.

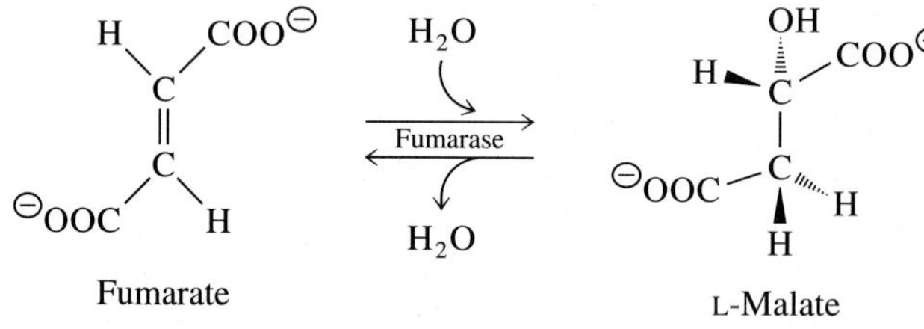

Figure 16·22
Reaction catalyzed by fumarase. Fumarate is prochiral; the stereospecific addition of water forms the chiral molecule L-malate.

8. Malate Dehydrogenase Catalyzes the Oxidation of Malate to Oxaloacetate, Completing the Citric Acid Cycle

The last step in the citric acid cycle is the oxidation of malate to regenerate oxaloacetate, with formation of a molecule of NADH (Figure 16·23). The reaction is catalyzed by $NAD^{\oplus}$-dependent malate dehydrogenase. The near-equilibrium interconversion of the α-hydroxy acid L-malate and the keto acid oxaloacetate is analogous to the reversible reaction catalyzed by lactate dehydrogenase (Figure 15·22).

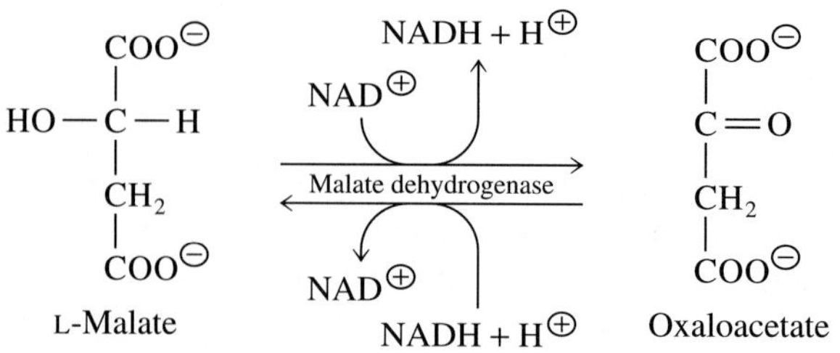

Figure 16·23
Reaction catalyzed by malate dehydrogenase.

16·4 Reduced Coenzymes Produced by the Citric Acid Cycle Fuel the Production of ATP by Oxidative Phosphorylation

In the net reaction of the citric acid cycle, three molecules of NADH, one molecule of QH_2, and one molecule of GTP or ATP are produced for each molecule of acetyl CoA entering the pathway.

$$\text{Acetyl CoA} + 3\,NAD^{\oplus} + Q + GDP\,(\text{or ADP}) + P_i + 2\,H_2O \longrightarrow$$
$$\text{CoASH} + 3\,NADH + QH_2 + GTP\,(\text{or ATP}) + 2\,CO_2 + 2\,H^{\oplus}$$

$$(16·6)$$

As mentioned earlier, NADH and QH_2 can be oxidized by the respiratory electron-transport chain of the inner mitochondrial membrane, with the concomitant production of ATP via oxidative phosphorylation. As we will see when we examine oxidative phosphorylation in Chapter 18, approximately 2.5 molecules of ATP are generated for each molecule of NADH oxidized to $NAD^{\oplus}$ by the electron-transport chain, and up to 1.5 molecules of ATP are produced for each molecule of QH_2 oxidized to Q. Thus, the complete oxidation of 1 molecule of acetyl CoA by the citric acid cycle and oxidative phosphorylation is associated with the production of approximately 10 molecules of ATP.

Reaction	Energy-yielding product	ATP equivalents
Isocitrate dehydrogenase	NADH	2.5
α-Ketoglutarate dehydrogenase complex	NADH	2.5
Succinyl-CoA synthetase	GTP or ATP	1.0
Succinate dehydrogenase complex	QH_2	1.5
Malate dehydrogenase	NADH	2.5
Total		10.0

$$(16·7)$$

The citric acid cycle is the final stage in the catabolism of many major nutrients, oxidizing acetyl CoA produced by the degradation of carbohydrates, lipids, and amino acids. Having covered glycolysis in the previous chapter, we can now give a complete accounting of the ATP produced from the degradation of one molecule of glucose. Recall that for each glucose molecule degraded by glycolysis, there is a net gain of 2 molecules of ATP, with production of 2 molecules of pyruvate. Conversion of both pyruvate molecules to acetyl CoA by the pyruvate dehydrogenase complex yields 2 NADH molecules, or about 5 additional ATP equivalents. When this is combined with the ATP equivalents from the citric acid cycle via oxidation of 2 molecules of acetyl CoA, the total yield is about 27 molecules of ATP.

	ATP equivalents
Glycolysis	2
Pyruvate dehydrogenase complex (Pyruvate × 2)	5
Citric acid cycle (Acetyl CoA × 2)	20
Total	27

$$(16·8)$$

When glycolysis operates *anaerobically,* $NAD^{\oplus}$ converted to NADH in the reaction catalyzed by glyceraldehyde 3-phosphate dehydrogenase is reoxidized when pyruvate is reduced to form lactate. The recycling of $NAD^{\oplus}$ is necessary for glycolysis to continue in the absence of oxygen. Under *aerobic* conditions, NADH is

Figure 16·24
ATP production from the catabolism of one molecule of glucose by glycolysis, the citric acid cycle, and oxidative phosphorylation. The complete oxidation of glucose leads to the formation of approximately 32 molecules of ATP.

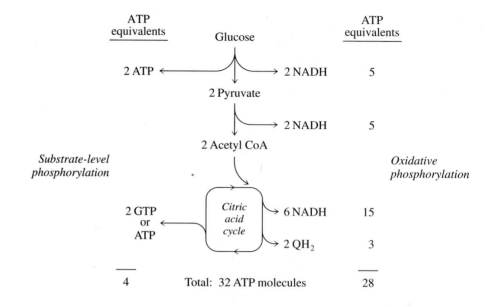

not reoxidized within the glycolytic pathway and is available to fuel ATP formation. However, the NADH produced by glycolysis is located in the cytosol, whereas the electron-transport chain in eukaryotes is located in the mitochondrion. The reducing equivalents from cytosolic NADH may be transported into the mitochondrion by shuttle mechanisms, two of which are described in detail in Chapter 18. Both shuttles regenerate $NAD^{\oplus}$ in the cytosol so aerobic glycolysis can continue. By far the most common shuttle used in cells is the malate-aspartate shuttle. When this system operates, 1 NADH can yield between 2 and 2.5 ATP. Adding an additional 5 ATP equivalents from oxidation of 2 NADH to the total in Equation 16·8 gives a total of approximately 32 molecules of ATP produced by the complete oxidation of glucose (Figure 16·24). When NADH enters the mitochondrion by an alternate mechanism, the glycerol phosphate shuttle, each NADH yields about 1.5 ATP, with an overall yield of about 30 ATP per molecule of glucose.

16·5 The Citric Acid Cycle Is Closely Regulated

Because the citric acid cycle occupies a central position in cellular metabolism, it is not surprising to find that the pathway is stringently controlled. Regulation is mediated by allosteric effectors and by covalent modification of citric acid cycle enzymes. Flux through the pathway is further controlled by regulation of the supply of acetyl CoA.

As noted earlier, acetyl CoA arises from a number of sources, including pathways for the degradation of carbohydrates, lipids, and amino acids. Regulation of the pyruvate dehydrogenase complex controls the supply of acetyl CoA produced from pyruvate, and hence from the degradation of carbohydrates. In general, substrates of the pyruvate dehydrogenase complex are activators of the complex and products are inhibitors. In mammalian cells and in *E. coli,* the E_2 and E_3 components of the pyruvate dehydrogenase complex (dihydrolipoamide acetyltransferase and dihydrolipoamide dehydrogenase, respectively) are regulated by simple mass-action effects when their products accumulate. The activity of the acetyltransferase, E_2, is inhibited when the concentration of acetyl CoA is high, whereas the dehydrogenase, E_3, is inhibited by a high $NADH/NAD^{\oplus}$ ratio (Figure 16·25). In general, the inhibitors are likely to be present in high concentrations when energy resources are plentiful. The converse applies to the activators of the pyruvate dehydrogenase complex.

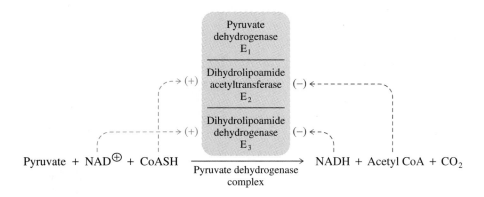

Figure 16·25
Regulation of the E_2 and E_3 ⟨...⟩
the pyruvate dehydrogenase ⟨...⟩
mammalian cells and *E. coli*. Accumulation
of the products acetyl CoA and NADH decreases flux through the reversible reactions
catalyzed by E_2 and E_3.

Mammalian, but not prokaryotic, pyruvate dehydrogenase complexes are subject to further regulation by covalent modification. In Section 16·2, it was noted that
a protein kinase and a protein phosphatase are associated with the mammalian multienzyme complex, though not with the complex in *E. coli*. The kinase and phosphatase both act on the interconvertible enzyme pyruvate dehydrogenase, E_1, which
under most conditions catalyzes the rate-determining step of the complex. Pyruvate
dehydrogenase kinase catalyzes the phosphorylation of E_1, thereby inactivating the
enzyme. Pyruvate dehydrogenase phosphatase catalyzes the dephosphorylation and
activation of E_1 (Figure 16·26).

There are only about two molecules of the pyruvate dehydrogenase kinase in
the complex, both firmly bound to the E_2 core. Somehow, possibly by migrating
along the surface of the core, these few kinase molecules rapidly catalyze the inactivation of the 20–30 E_1 tetramers.

Pyruvate dehydrogenase kinase and phosphatase are themselves regulated. The
kinase is allosterically activated by NADH and acetyl CoA, which are products of
pyruvate oxidation. Accumulation of NADH and acetyl CoA signals energy availability and leads to an increase in phosphorylation of the pyruvate dehydrogenase
subunit and inhibition of the further oxidation of pyruvate. Conversely, pyruvate
and ADP inhibit the kinase, leading to activation of the pyruvate dehydrogenase
subunit. Another regulator of the phosphorylation status of E_1 in vitro is Ca^{2+},
which activates pyruvate dehydrogenase phosphatase. Elevated levels of Ca^{2+} lead
to dephosphorylation of the E_1 subunit and an increase in flux through the pyruvate
dehydrogenase complex. Certain hormones regulate the level of Ca^{2+} in mitochondria, but it has not yet been proven that Ca^{2+} concentrations fluctuate in a range
suitable to affect the phosphatase in vivo.

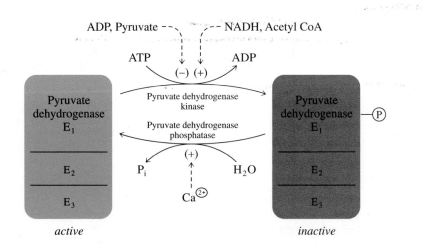

Figure 16·26
Regulation of the mammalian pyruvate dehydrogenase complex by reversible phosphorylation of the E_1 subunit, pyruvate dehydrogenase. The regulatory kinase and
phosphatase enzymes are both part of the
mammalian complex. The kinase is activated
by NADH and acetyl CoA, products of the reaction catalyzed by the pyruvate dehydrogenase complex, and inhibited by ADP and the
reaction substrate pyruvate. Dephosphorylation is stimulated by elevated levels of Ca^{2+}.
E_1, under most conditions, catalyzes the
rate-determining step of the pyruvate dehydrogenase complex, and control of E_1 controls the rate of the entire complex.

Figure 16·27
Regulation of isocitrate dehydrogenase in *E. coli* by covalent modification. A bifunctional enzyme catalyzes phosphorylation and dephosphorylation of isocitrate dehydrogenase. The two activities of the bifunctional enzyme are reciprocally regulated allosterically by intermediates of glycolysis and the citric acid cycle.

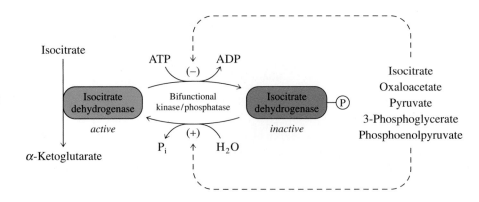

Within the citric acid cycle, three reactions are metabolically irreversible and are therefore potential sites of control. These are the reactions catalyzed by citrate synthase, isocitrate dehydrogenase, and the α-ketoglutarate dehydrogenase complex. Citrate synthase catalyzes the first reaction of the citric acid cycle, which would seem a suitable control point. However, mechanisms for the regulation of this enzyme are not yet well established. ATP inhibits the enzyme in vitro, but significant changes in intramitochondrial ATP concentration are unlikely in vivo, and therefore ATP may not be a physiological regulator.

Mammalian isocitrate dehydrogenase is allosterically activated by Ca^{2+} and ADP and inhibited by NADH. The enzyme in mammals is not subject to covalent modification. In *E. coli,* however, phosphorylation of a serine residue of isocitrate dehydrogenase catalyzed by a protein kinase virtually abolishes the activity of the enzyme. The same protein molecule that contains the kinase activity also has a phosphatase activity located on a separate domain that catalyzes hydrolysis of the phosphoserine residue, reactivating isocitrate dehydrogenase. The kinase and phosphatase activities are reciprocally regulated: isocitrate, oxaloacetate, pyruvate, and the glycolytic intermediates 3-phosphoglycerate and phosphoenolpyruvate allosterically activate the phosphatase and inhibit the kinase (Figure 16·27). Thus, when

Figure 16·28
Regulation of the pyruvate dehydrogenase complex and the citric acid cycle.

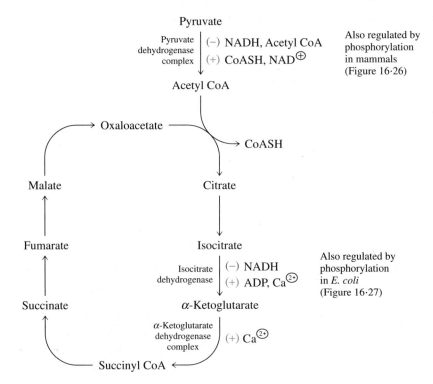

the concentrations of glycolytic and citric acid cycle intermediates in *E. coli* are high, isocitrate dehydrogenase is active. When phosphorylation abolishes the activity of isocitrate dehydrogenase, isocitrate is diverted to the glyoxylate cycle, a specialized pathway present in some microorganisms and plants. The glyoxylate cycle is discussed in Section 16·8.

The α-ketoglutarate dehydrogenase complex catalyzes a reaction that is analogous to that of the pyruvate dehydrogenase complex. The complexes are very similar, yet the α-ketoglutarate dehydrogenase complex has quite different regulatory features. No kinase or phosphatase is associated with the α-ketoglutarate dehydrogenase complex. Instead, calcium ions bind to E_1 of the complex and decrease the K_m of the enzyme for α-ketoglutarate, leading to an increase in the rate of formation of succinyl CoA. NADH and succinyl CoA are inhibitors of the α-ketoglutarate complex in vitro, but it has not been established that they have a significant regulatory role in living cells.

The regulation of the pyruvate dehydrogenase complex and the citric acid cycle is summarized in Figure 16·28.

16·6 The Citric Acid Cycle Functions as a Multistep Catalyst

By definition, a catalyst is an agent that increases the rate of a reaction without itself undergoing net transformation. Catalysis by metabolic cycles bears comparison to catalysis by enzymes. An enzyme goes through a cyclic series of conversions, finishing by returning to the form in which it began. All enzymatic reactions, in fact all catalytic reactions, can be represented as cycles. The citric acid cycle fits the description of a catalyst. Taken as a whole, the citric acid cycle can be viewed as a mechanism for oxidation of the acetyl group of acetyl CoA to CO_2 by $NAD^{\oplus}$ and ubiquinone. When the citric acid cycle operates in isolation, its intermediates are reformed with each full turn of the cycle. Therefore, the citric acid cycle is not a pathway for the net degradation of any of the intermediates of the pathway. None of the intermediates appears in the net reaction of the citric acid cycle as a substrate or product.

$$\text{Acetyl CoA} + 3\,NAD^{\oplus} + Q + \text{GDP (or ADP)} + P_i + 2\,H_2O \longrightarrow$$
$$\text{CoASH} + 3\,NADH + QH_2 + \text{GTP (or ATP)} + 2\,CO_2 + 2\,H^{\oplus}$$

$$(16·9)$$

Only small amounts of a catalyst are needed to participate in the conversion of large quantities of substrate to product. Similarly, only small amounts of citric acid cycle intermediates need to be present for the oxidation of a large number of acetyl CoA molecules. A small change in the amount of a cycle intermediate greatly changes the metabolic capacity of the cycle.

16·7 Metabolites Enter and Exit the Citric Acid Cycle at Several Points

The citric acid cycle is not exclusively a catabolic pathway for the oxidation of acetyl CoA. It also plays a major role in metabolism by serving as the intersection of a number of other pathways, hence its description as the hub of aerobic metabolism. Some intermediates of the cycle are important metabolic precursors, and some

Figure 16·29
Routes leading to and from the citric acid cycle. Intermediates of the citric acid cycle are precursors for carbohydrates, lipids, and amino acids, as well as nucleotides and porphyrins. Reactions feeding into the cycle replenish the pool of cycle intermediates.

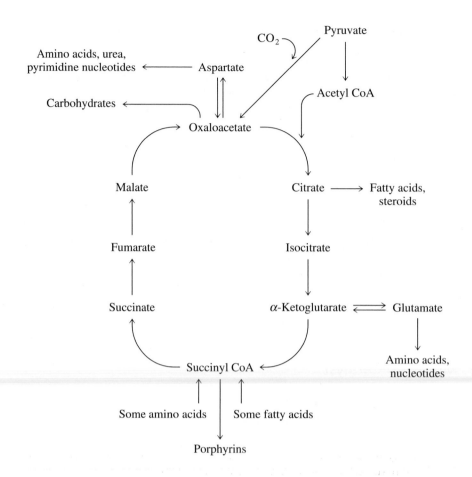

catabolic pathways produce cycle intermediates. As shown in Figure 16·29, citrate, α-ketoglutarate, succinyl CoA, and oxaloacetate all lead to biosynthetic pathways. In adipose tissue, citrate is part of a pathway for formation of fatty acids and steroid molecules. Citrate from the mitochondria is transported to the cytosol, where it undergoes cleavage to form acetyl CoA, the precursor of the lipids. One major metabolic fate of α-ketoglutarate is reversible conversion to glutamate, which can then be incorporated into proteins or serve as a precursor in the biosynthesis of other amino acids or nucleotides. Succinyl CoA can condense with glycine to initiate the biosynthesis of porphyrins. Oxaloacetate is a precursor of carbohydrates formed by the pathway of gluconeogenesis—a pathway that supplies the essential fuel glucose when it is no longer available from physiological stores or exogenous sources. Gluconeogenesis is a major topic in the next chapter. Oxaloacetate is also interconvertible with aspartate, which can be used in urea synthesis, protein synthesis, and the synthesis of pyrimidine nucleotides.

As noted in the previous section, the rate at which the citric acid cycle metabolizes acetyl CoA is extremely sensitive to changes in the concentrations of its intermediates. Metabolites that are removed by entry into biosynthetic pathways must be replenished by **anaplerotic** (Greek, filling up) **reactions.** Because the pathway is cyclic, replenishment of any of the cycle intermediates results in a greater concentration of all intermediates.

The reversible reactions in Figure 16·29 (oxaloacetate $\rightleftharpoons$ aspartate and α-ketoglutarate $\rightleftharpoons$ glutamate) supply intermediates to the citric acid cycle, and pathways for the degradation of some amino acids and fatty acids can contribute succinyl CoA.

An important regulated replenishment reaction, catalyzed by pyruvate carboxylase, produces oxaloacetate from pyruvate.

$$\text{Pyruvate} + CO_2 + ATP + H_2O \longrightarrow \text{Oxaloacetate} + ADP + P_i$$

(16·10)

This is the major anaplerotic reaction in mammals. Pyruvate carboxylase is allosterically activated by acetyl CoA. Accumulation of acetyl CoA signals slow utilization by the citric acid cycle and a need for more intermediates. The activation of pyruvate carboxylase supplies oxaloacetate for the cycle. The metabolic role of pyruvate carboxylase will be described further when we examine gluconeogenesis in the next chapter.

Other organisms have a variety of regulated anaplerotic reactions that keep the intake and output of citric acid cycle intermediates in a delicate balance. For example, many plants and some bacteria supply oxaloacetate to the citric acid cycle via the reaction catalyzed by phosphoenolpyruvate carboxylase.

$$\text{Phosphoenolpyruvate} + HCO_3^{\ominus} \rightleftharpoons \text{Oxaloacetate} + P_i \qquad \text{(16·11)}$$

The interplay of all of these reactions—entry of acetyl CoA from glycolysis and other sources, entry of intermediates from catabolic pathways and from anaplerotic reactions, and exit of intermediates to anabolic pathways—is carefully regulated, permitting the amphibolic citric acid cycle to perform its role as the traffic circle of metabolism.

16·8 The Glyoxylate Cycle Permits Acetyl CoA to Be Incorporated into Glucose

The pathway for the formation of glucose from noncarbohydrate precursors requires that pyruvate or oxaloacetate serve as a precursor (Chapter 17). In animals, acetyl CoA cannot lead to net formation of either pyruvate or oxaloacetate, and therefore acetyl CoA is not a carbon source for net production of glucose. (The carbon atoms of acetyl CoA are incorporated into oxaloacetate by the reactions of the citric acid cycle, but note that for each two carbons so incorporated, two other carbons are released as CO_2 by reactions of the citric acid cycle.) In plants, bacteria, and yeast, but not in animals, there exists a biosynthetic route—the **glyoxylate cycle**—that leads from two-carbon compounds to glucose. The glyoxylate cycle, named for a two-carbon intermediate, is a modification of the citric acid cycle. It provides an anabolic alternative to the citric acid cycle for metabolism of acetyl CoA, leading to the formation of glucose from acetyl CoA via four-carbon compounds. Cells that contain the glyoxylate cycle enzymes can synthesize all their required carbohydrates from any substrate that is a precursor of acetyl CoA. For example, yeast can grow on ethanol because yeast cells can oxidize ethanol to acetyl CoA, which can be metabolized via the glyoxylate cycle to form oxaloacetate. Similarly, certain microorganisms employ the glyoxylate cycle to sustain growth on acetate, which can be incorporated into acetyl CoA in a reaction catalyzed by acetate thiokinase.

$$\underset{\text{Acetate}}{H_3C-COO^{\ominus}} + HS\text{-}CoA \quad \xrightarrow[\text{Acetate thiokinase}]{\overset{AMP, PP_i}{\underset{ATP}{\wedge}}} \quad \underset{\text{Acetyl CoA}}{H_3C-\overset{\overset{\displaystyle O}{\|}}{C}-S\text{-}CoA}$$

(16·12)

The glyoxylate cycle is especially active in oily seed plants. In these plants, stored seed oil is converted to carbohydrates that sustain the plant during germination. We will examine the glyoxylate cycle in these plants as an example of the metabolic role of this cycle.

Figure 16·30
Glyoxylate cycle. Isocitrate lyase and malate synthase are unique to the glyoxylate cycle. When this bypass operates, the carbon of acetyl CoA is converted to glucose rather than CO_2 via the conversion of malate to oxaloacetate and entry of oxaloacetate into the pathway of gluconeogenesis.

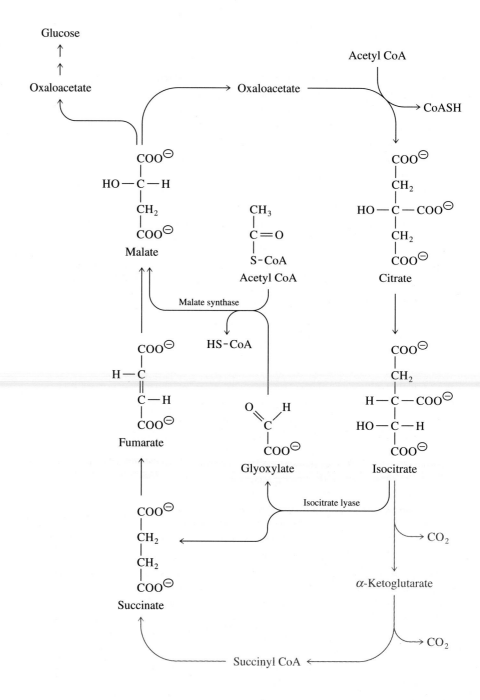

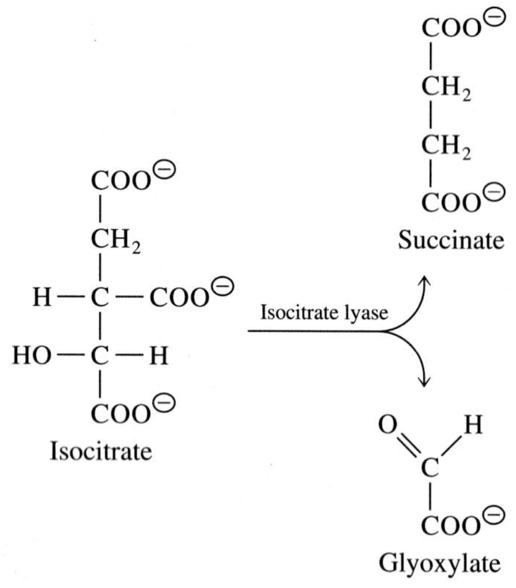

Figure 16·31
Reaction catalyzed by isocitrate lyase, the first bypass enzyme of the glyoxylate cycle.

The glyoxylate cycle can be regarded as a shunt within the citric acid cycle (Figure 16·30). Some reactions are shared with the citric acid cycle; two allow isocitrate to bypass the remainder of the citric acid cycle. The six-carbon intermediate isocitrate, rather than undergoing decarboxylation, is converted to the four-carbon molecule succinate and the two-carbon molecule glyoxylate in a reaction catalyzed by isocitrate lyase, the first of the two bypass enzymes of the glyoxylate cycle (Figure 16·31). Glyoxylate then condenses with acetyl CoA to form the four-carbon molecule malate in a reaction catalyzed by malate synthase (Figure 16·32). No carbon atoms of acetyl CoA are released as CO_2 during operation of the glyoxylate cycle, and the net formation of four-carbon molecules from acetyl CoA supplies

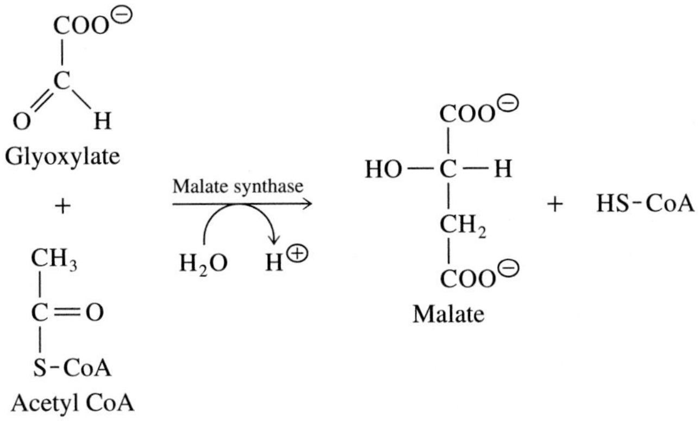

$$COO^{\ominus}$$
$$|$$
$$C$$
$$O \quad \diagup \searrow \quad H$$
Glyoxylate

+

$$CH_3$$
$$|$$
$$C = O$$
$$|$$
$$S\text{-}CoA$$
Acetyl CoA

$\xrightarrow[\quad H_2O \quad H^{\oplus} \quad]{\text{Malate synthase}}$

$$COO^{\ominus}$$
$$|$$
$$HO - C - H$$
$$|$$
$$CH_2$$
$$|$$
$$COO^{\ominus}$$
Malate

+ HS-CoA

Figure 16·32
Reaction catalyzed by malate synthase, the second bypass enzyme of the glyoxylate cycle.

precursors for the formation of glucose by gluconeogenesis. Succinate is oxidized to malate and oxaloacetate to maintain the catalytic amounts of cycle intermediates. The net reaction of the glyoxylate cycle is

$$2\,\text{Acetyl CoA} \;+\; 2\,\text{NAD}^{\oplus} \;+\; Q \;\longrightarrow$$
$$\text{Oxaloacetate} \;+\; 2\,\text{CoASH} \;+\; 2\,\text{NADH} \;+\; \text{QH}_2 \;+\; 2\,\text{H}^{\oplus} \qquad \textbf{(16·13)}$$

In eukaryotes, the glyoxylate cycle requires transfer of metabolites between the mitochondrion, the cytosol, and a separate organelle, the single-membraned glyoxysome. The glyoxylate cycle that exists in germinating seedlings of castor beans is illustrated in Figure 16·33. Glyoxysomes contain enzymes that catalyze the catabolism of the fatty acids of the stored oils. The seed oil of the castor bean is converted to acetyl CoA by the action of these enzymes. Glyoxysomal citrate synthase and aconitase, isozymes of the citric acid cycle enzymes in mitochondria, catalyze

Figure 16·33
Glyoxylate cycle in germinating castor bean seeds. The conversion of acetyl CoA to glucose requires transfer of metabolites between three metabolic compartments: the glyoxysome, the cytosol, and the mitochondrion.

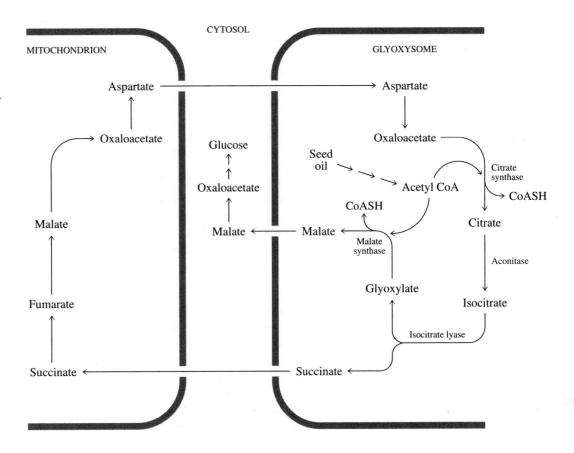

the incorporation of acetyl CoA into citrate and the conversion of citrate to iso-citrate. Isocitrate is cleaved in the reaction catalyzed by isocitrate lyase, forming succinate and glyoxylate. Succinate leaves the glyoxysome and enters the citric acid cycle in the mitochondrion. Glyoxylate in the glyoxysome condenses with acetyl CoA to form malate in the reaction catalyzed by malate synthase. Malate leaves the glyoxysome and enters the cytosol, where a cytosolic malate dehydrogenase catalyzes the formation of oxaloacetate, a precursor for gluconeogenesis.

Malate, the four-carbon precursor of glucose, thus arises from two molecules of acetyl CoA in two condensation reactions catalyzed by citrate synthase and malate synthase. Keep in mind that the pool of acetyl CoA in the glyoxysome, which fuels biosynthesis of glucose, is entirely separate from the mitochondrial pool of acetyl CoA, which fuels the catabolic oxidation reactions of the citric acid cycle.

Summary

The citric acid cycle is the final stage in the aerobic catabolism of carbohydrates, amino acids, and fatty acids. In eukaryotic cells, pyruvate generated by glycolysis in the cytosol is transported into the mitochondrial matrix where a multienzyme complex, the pyruvate dehydrogenase complex, catalyzes its oxidation to acetyl CoA and CO_2. The pyruvate dehydrogenase complex consists of enzyme components E_1 (pyruvate dehydrogenase), E_2 (dihydrolipoamide acetyltransferase), and E_3 (dihydrolipoamide dehydrogenase) and requires as cofactors thiamine pyrophosphate, lipoamide, CoASH, FAD, and $NAD^{\oplus}$.

The citric acid cycle consists of eight enzyme-catalyzed reactions. Citrate synthase catalyzes the condensation of the acetyl group of acetyl CoA with oxaloacetate, generating the tricarboxylic acid citrate. Citrate is a tertiary alcohol that must be converted to a secondary alcohol before further oxidation can occur. Aconitase catalyzes the conversion of prochiral citrate to the chiral molecule $2R,3S$-isocitrate. Successive oxidative decarboxylation reactions catalyzed by isocitrate dehydrogenase and the α-ketoglutarate dehydrogenase complex produce two molecules of NADH and two molecules of CO_2, leading to the formation of succinyl CoA. Succinyl-CoA synthetase catalyzes substrate-level phosphorylation of GDP to GTP (or ADP to ATP, depending on the organism) as the thioester of succinyl CoA is cleaved to succinate and CoASH. The eukaryotic flavoprotein succinate dehydrogenase, embedded in the inner mitochondrial membrane, is a component of a complex that catalyzes oxidation of succinate to fumarate, with electrons transferred from $FADH_2$ to ubiquinone (Q), forming ubiquinol (QH_2), a lipid-soluble mobile carrier of reducing equivalents. The double bond of fumarate is hydrated by the action of fumarase to produce malate, which can be oxidized to oxaloacetate by $NAD^{\oplus}$-dependent malate dehydrogenase. The regeneration of oxaloacetate completes one turn of the citric acid cycle.

For each molecule of acetyl CoA oxidized via the citric acid cycle, three molecules of $NAD^{\oplus}$ are reduced to NADH (in the reactions catalyzed by isocitrate dehydrogenase, the α-ketoglutarate dehydrogenase complex, and malate dehydrogenase), one molecule of Q is reduced to QH_2 (succinate dehydrogenase), and one molecule of GTP is generated from GDP + P_i or one molecule of ATP is generated from ADP + P_i (succinyl-CoA synthetase). Oxidation via the respiratory electron-transport chain of the reduced coenzymes NADH and QH_2 produced during one turn of the citric acid cycle leads to the formation of about 10 ATP molecules per molecule of acetyl CoA that enters the cycle. Complete oxidation of one molecule of glucose by glycolysis, the pyruvate dehydrogenase complex, the citric acid cycle, and the electron-transport chain leads to the formation of approximately 32 molecules of ATP by a combination of substrate-level and oxidative phosphorylation.

The citric acid cycle has several control points. The pyruvate dehydrogenase complex is regulated by the levels of its end products, acetyl CoA and NADH. Isocitrate dehydrogenase and the α-ketoglutarate dehydrogenase complex are allosterically regulated. In mammals, the pyruvate dehydrogenase complex is further controlled by covalent modification. In *E. coli,* isocitrate dehydrogenase is subject to covalent modification.

In addition to its role in oxidative catabolism, the citric acid cycle provides precursors for biosynthetic pathways. Citrate, α-ketoglutarate, succinyl CoA, and oxaloacetate are the principal branch-point metabolites. The pathway is replenished by formation of oxaloacetate from pyruvate, by reversible reactions forming oxaloacetate and α-ketoglutarate, and by pathways for the degradation of some amino acids and fatty acids, which can contribute succinyl CoA.

The glyoxylate cycle is a pathway closely related to the citric acid cycle that allows plants and some microorganisms to use acetyl CoA to generate four-carbon intermediates for gluconeogenesis and other biosynthetic pathways. Two enzymes unique to the glyoxylate cycle, isocitrate lyase and malate synthase, provide a bypass around the CO_2-producing reactions of the citric acid cycle. These enzymes are not present in animal cells; thus, animals cannot synthesize anabolic metabolites from acetyl CoA or two-carbon molecules such as acetate. Isocitrate lyase catalyzes the cleavage of isocitrate to succinate and glyoxylate. Succinate enters the citric acid cycle, and glyoxylate condenses with CoASH to form malate, catalyzed by malate synthase. Malate can serve as a precursor for the formation of glucose.

Selected Readings

Pyruvate Dehydrogenase Complex

Mattevi, A., Obmolova, G., Schulze, E., Kalk, K. H., Westphal, A. H., de Kok, A., and Hol, W. G. J. (1992). Atomic structure of the cubic core of the pyruvate dehydrogenase multienzyme complex. *Science* 255:1544–1550.

Patel, M. S., and Roche, T. E. (1990). Molecular biology and biochemistry of pyruvate dehydrogenase complexes. *FASEB J.* 4:3224–3233.

Reed, L. J., and Yeaman, S. J. (1987). Pyruvate dehydrogenase. In *The Enzymes,* Vol. 18, 3rd ed., P. D. Boyer and E. G. Krebs, eds. (Orlando: Academic Press), pp. 77–95. Describes the regulation of α-keto acid dehydrogenases by covalent modification.

Reed, L. J., and Hackert, M. L. (1990). Structure-function relationships in dihydrolipoamide acyltransferases. *J. Biol. Chem.* 265:8971–8974.

Citric Acid Cycle

Beinert, H., and Kennedy, M. C. (1989). Engineering of protein bound iron-sulfur clusters. *Eur. J. Biochem.* 186:5–15. Explains the mechanism of binding of citrate to aconitase.

Kay, J., and Weitzman, P. D. J., eds. (1987). *Krebs' Citric Acid Cycle—Half a Century and Still Turning.* (London: The Biochemical Society). Intriguing and accessible papers, some provocative, covering a range of approaches to the pathway. Published in celebration of the 50th anniversary of Krebs's discovery.

Krebs, H. A. (1970). The history of the tricarboxylic acid cycle. *Perspect. Biol. Med.* 14:154–170.

LaPorte, D. C., and Koshland, D. E., Jr. (1983). Phosphorylation of isocitrate dehydrogenase as a demonstration of enhanced sensitivity in covalent regulation. *Nature* 305:286–290.

McCormack, J. G., and Denton, R. M. (1988). The regulation of mitochondrial function in mammalian cells by Ca^{2+} ions. *Biochem. Soc. Trans.* 109:523–527.

Ottaway, J. H., McClellan, J. A., and Saunderson, C. L. (1981). Succinic thiokinase and metabolic control. *Int. J. Biochem.* 13:401–410.

Singer, T. P., and Johnson, M. K. (1985). The prosthetic groups of succinate dehydrogenase: 30 years from discovery to identification. *FEBS Lett.* 190:189–198.

Wiegand, G., and Remington, S. J. (1986). Citrate synthase: structure, control, and mechanism. *Annu. Rev. Biophys. Biophys. Chem.* 15:97–117.

Williamson, J. R., and Cooper, R. H. (1980). Regulation of the citric acid cycle in mammalian systems. *FEBS Lett.* 117(Suppl.):K73–K85.

Glyoxylate Cycle

Beevers, H. (1980). The role of the glyoxylate cycle. In *The Biochemistry of Plants: A Comprehensive Treatise,* Vol. 4, P. K. Stumpf and E. E. Conn, eds. (New York: Academic Press), pp. 117–130.

17

Additional Pathways in Carbohydrate Metabolism

By now, you should appreciate that the metabolism of glucose is central to energy metabolism in the cell. It is time to broaden our perspective on carbohydrate metabolism by examining a number of other important pathways linked to glucose or its derivatives. These metabolic processes include

1. the entry of sugars other than glucose into the glycolytic pathway;

2. the entry of glucose residues from intracellular polysaccharides into the glycolytic pathway;

3. the synthesis of storage polysaccharides from glucose;

4. the synthesis of glucose from noncarbohydrate precursors;

5. the oxidation of glucose by the pentose phosphate pathway, which produces NADPH and ribose.

We will also examine the integration and regulation of pathways of carbohydrate metabolism.

In Chapter 13, we saw how humans digest starch and oligosaccharides to form glucose. Glucose is used as a source of energy by many types of organisms, not just mammals. For example, microorganisms growing in the digestive tract of ruminants digest cellulose and ferment the glucose monomers via pyruvate to acetate, lactate, succinate, or the three-carbon acid propionate. The glucose supplies energy to the microorganisms, and the end products of the fermentation pathways supply the host with energy and, to some extent, precursors for glucose synthesis. Chapters 15 and 16 presented the enzymatic reactions by which absorbed sugar is degraded via pyruvate either anaerobically to ethanol or lactate or aerobically to carbon dioxide and water. Other monosaccharides are metabolized by similar routes.

Glucose availability is controlled by regulating the uptake or synthesis of glucose and related molecules and by regulating the synthesis and degradation of storage polysaccharides composed of glucose residues. Glucose is stored in plants as starch and in animals and some microorganisms as glycogen. Starch and glycogen can be degraded to release glucose monomers that fuel energy production via glycolysis, the citric acid cycle, and oxidative phosphorylation.

When glucose or its storage polymers are not available, glucose can be synthesized from noncarbohydrate precursors by the process of gluconeogenesis. This pathway is closely related to glycolysis but runs in the direction of glucose synthesis rather than degradation. Four reactions specific to the gluconeogenic pathway bypass the three metabolically irreversible reactions of glycolysis and permit the formation of glucose when conditions are appropriate.

In addition to its role as a source of energy in living cells, glucose can yield reducing equivalents in the form of NADPH for biosynthetic and other reactions. Glucose is also a precursor of the ribose and deoxyribose moieties of nucleotides and deoxynucleotides. These functions are accomplished by the pentose phosphate pathway.

In mammals, glycogen metabolism, gluconeogenesis, and the pentose phosphate pathway are closely and coordinately regulated in accordance with the moment-to-moment requirements of the organism. In this chapter, we will review these pathways and examine some of the mechanisms for regulation of glucose metabolism in cells.

17·1 Dietary Saccharides Can Be Catabolized via Glycolysis

As we saw in Chapter 13, foods of all kinds are digested to some extent by a variety of chemical and enzymatic means in the digestive tract. Although glucose is the most abundant carbohydrate, some other monosaccharides can be absorbed and used as sources of energy. In the following sections, we will see how fructose, galactose, and mannose can be metabolized by the glycolytic pathway.

A. Fructose Is Converted to Glyceraldehyde 3-Phosphate

The fructose-containing disaccharide sucrose and the monosaccharide fructose, sweetening agents in many foods and beverages, account for a significant fraction of the carbohydrate in human diets. After absorption, almost all of the fructose is metabolized by the liver. However, fructose is not a good substrate of any of the isozymes of hexokinase. The K_m value for fructose is much higher than that for glucose. In liver, fructose is phosphorylated by the action of a specific fructokinase that catalyzes the ATP-dependent formation of fructose 1-phosphate. Next, fructose 1-phosphate aldolase catalyzes cleavage of fructose 1-phosphate to dihydroxyacetone phosphate and glyceraldehyde. The glyceraldehyde is then phosphorylated to glyceraldehyde 3-phosphate in a reaction catalyzed by triose kinase, consuming a second molecule of ATP. These metabolic steps, together with the conversion of dihydroxyacetone phosphate to a second molecule of glyceraldehyde 3-phosphate by triose phosphate isomerase, are shown in Figure 17·1. Both molecules of glyceraldehyde 3-phosphate can then be metabolized to pyruvate by the remaining steps of glycolysis (triose stage).

The metabolism of one molecule of fructose to two molecules of pyruvate produces two molecules of ATP and two molecules of NADH. This is the same yield as the conversion of glucose to pyruvate. However, fructose catabolism bypasses phosphofructokinase-1 and its associated regulation. Diets rich in fructose or sucrose may lead to a fatty liver due to overproduction of pyruvate, which is a precursor for the synthesis of fats and cholesterol.

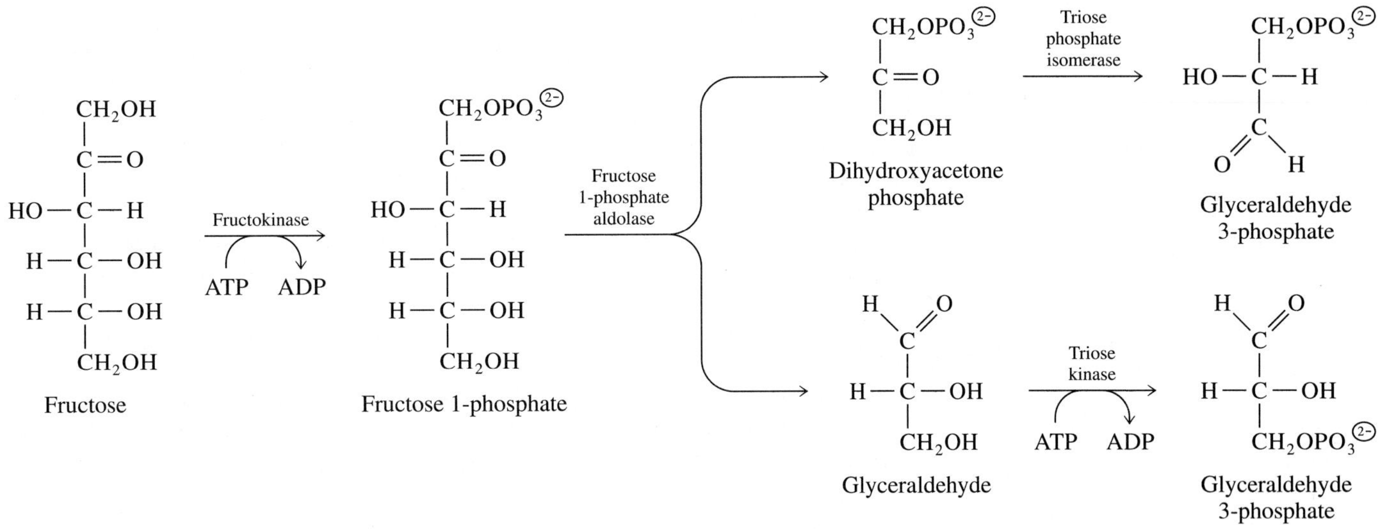

Figure 17·1
Conversion of a fructose molecule to two molecules of glyceraldehyde 3-phosphate by the actions of fructokinase, fructose 1-phosphate aldolase, triose phosphate isomerase, and triose kinase.

B. Lactose-Derived Galactose Is Converted to Glucose 1-Phosphate

The disaccharide lactose, present in milk, provides a major source of energy for nursing mammals, including human infants. Nearly all infants and young children are able to metabolize lactose by the action of intestinal lactase, which catalyzes hydrolysis of lactose to one molecule each of glucose and galactose, both of which are absorbed from the intestine and transported in the circulatory system.

As shown in Figure 17·2, galactose—the C-4 epimer of glucose—can be converted to glucose 1-phosphate by a pathway that involves the recycling of the metabolite coenzyme uridine diphosphate glucose (UDP-glucose). In the liver, galactose is phosphorylated by the action of galactokinase, consuming a molecule of ATP. The product of this reaction is galactose 1-phosphate, which exchanges with the glucose 1-phosphate moiety of UDP-glucose by cleavage of the pyrophosphate bond of UDP-glucose catalyzed by galactose 1-phosphate uridylyltransferase. The products of this reaction are glucose 1-phosphate and UDP-galactose. Glucose 1-phosphate can enter the glycolytic pathway after conversion

Figure 17·2
Conversion of galactose to glucose 6-phosphate. The metabolic intermediate UDP-glucose is recycled in the process. The overall stoichiometry for the pathway is galactose + ATP ⟶ glucose 6-phosphate + ADP.

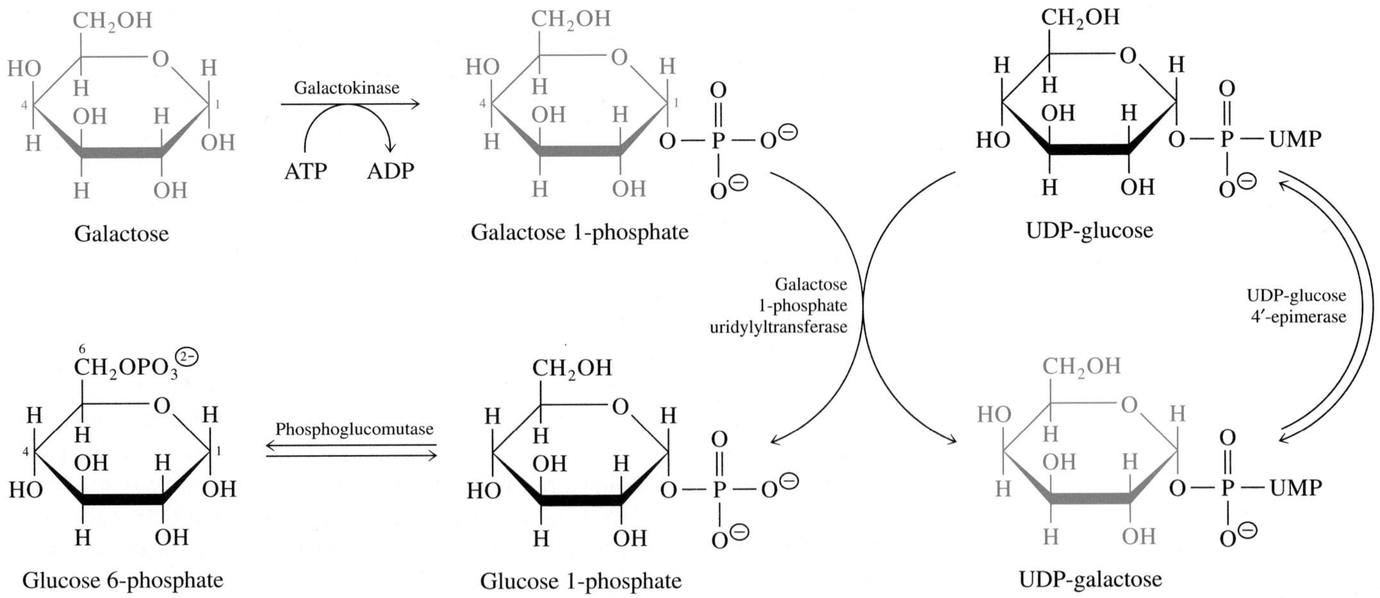

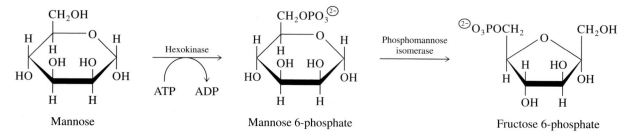

Figure 17·3
Conversion of mannose to fructose 6-phosphate by the actions of hexokinase and phosphomannose isomerase.

to glucose 6-phosphate in a reaction catalyzed by phosphoglucomutase. UDP-galactose, the other product of the reaction catalyzed by galactose 1-phosphate uridylyltransferase, is recycled to UDP-glucose by the action of UDP-glucose 4′-epimerase.

Conversion of one molecule of galactose to two molecules of pyruvate produces two molecules of ATP and two molecules of NADH, the same yield as the conversions of glucose and fructose. Although there is a requirement for UDP-glucose, which is formed from glucose and the ATP equivalent UTP, only small (catalytic) amounts of it are needed since it is recycled.

Infants fed a normal milk diet rely on the pathway of galactose metabolism. In the most common and severe form of the genetic disorder galactosemia (the inability to properly metabolize galactose), infants are usually deficient in galactose 1-phosphate uridylyltransferase. In such cases, galactose 1-phosphate accumulates in the cells. This can compromise liver function, which is recognized by the appearance of jaundice, the yellowing of the skin. The liver damage is potentially fatal. Other effects include damage to the central nervous system. Screening for galactose 1-phosphate uridylyltransferase in the red blood cells of the umbilical cord allows detection of galactosemia at birth, and the severe effects of this genetic deficiency can be avoided by excluding lactose from the diet.

C. Mannose Is Converted to Fructose 6-Phosphate

The aldohexose mannose is obtained in the diet from glycoproteins and certain polysaccharides. Mannose is converted to mannose 6-phosphate by the action of hexokinase. In order to enter the glycolytic pathway, mannose 6-phosphate undergoes isomerization to fructose 6-phosphate in a reaction catalyzed by phosphomannose isomerase. These two reactions are depicted in Figure 17·3.

17·2 Polysaccharide Phosphorylases Catalyze Mobilization of Glucose Residues

As we have already seen, glucose is stored in the intracellular polysaccharides starch and glycogen. Most of the glycogen in vertebrates is found in the cells of muscle and liver. Glycogen in muscle cells appears in electron micrographs as cytosolic granules with a diameter of 10 to 40 nm, similar in size to ribosomes. Glycogen particles in liver are about three times larger. The enzymes required for **glycogenolysis,** the intracellular degradation of glycogen, are similar in muscle and liver, but the pathway has different roles in these two sites. In muscle tissue, glycogen breakdown leads to the formation of glucose 6-phosphate that is metabolized via glycolysis and the citric acid cycle. In liver, most glucose 6-phosphate is converted to glucose that is delivered to the bloodstream to be taken up by other cells, such as brain cells, red blood cells, and adipocytes (fat cells).

The glucose residues of starch and glycogen are released from the storage polymers, or mobilized, through the action of enzymes classified as polysaccharide phosphorylases, specifically, starch phosphorylase (in plants) and glycogen phosphorylase (in many other types of organisms). These enzymes catalyze the removal of glucose residues from the nonreducing ends of starch or glycogen, provided the monomers are attached by α-$(1 \rightarrow 4)$ linkages. As the name implies, the enzymes catalyze **phosphorolysis**—cleavage of a bond by group transfer to an oxygen atom of phosphate. There is an important distinction between hydrolysis (group transfer to water) and phosphorolysis. In phosphorolysis, the residue released from the substrate is a phosphate ester. Thus, the first product of polysaccharide breakdown is α-D-glucose 1-phosphate, not free glucose.

$$\text{Polysaccharide} \atop (n \text{ residues}) \; + \; \text{P}_i \; \xrightarrow{\substack{\text{Polysaccharide} \\ \text{phosphorylase}}} \; \text{Polysaccharide} \atop (n-1 \text{ residues}) \; + \; \text{Glucose 1-phosphate}$$

$$(17 \cdot 1)$$

Glycogen phosphorylase, which catalyzes this major regulatory step of glycogenolysis in vertebrates, is a dimer of identical two-lobed subunits (Figure 17·4). Each subunit has a molecular weight of 97 000. The catalytic sites are located in clefts between the two domains of each subunit, at the ends opposite the subunit interface. Each subunit has a region that can attach to a glycogen particle. The binding sites for the allosteric inhibitors ATP and glucose 6-phosphate and the activator

Figure 17·4
Structure of glycogen phosphorylase. This enzyme is a symmetric dimer. There is a site that binds glycogen or oligosaccharides, a catalytic site that binds glucose 1-phosphate or glucose, and an allosteric site that binds glucose 6-phosphate and the nucleotides AMP and ATP. [Adapted from Sprang, S., Goldsmith, E., and Fletterick, R. (1987). Structure of the nucleotide activation switch in glycogen phosphorylase a. *Science* 237:1012–1019.]

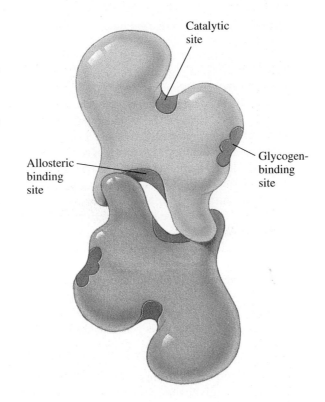

Catalytic site

Allosteric binding site

Glycogen-binding site

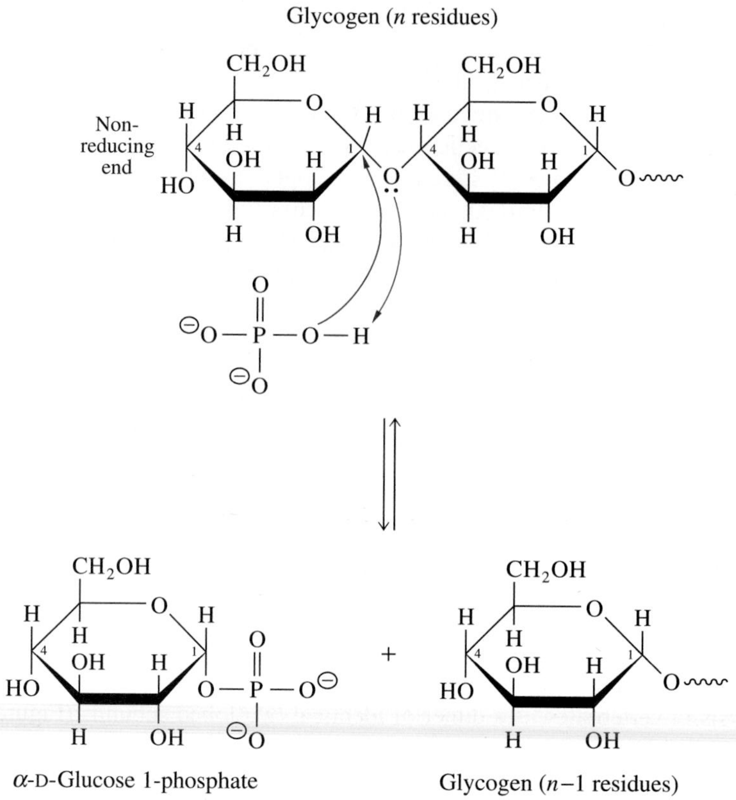

Figure 17·5
Cleavage of the terminal glucose residue from the nonreducing end of a glycogen chain, catalyzed by glycogen phosphorylase. A molecule of inorganic phosphate in the active site donates a proton to the oxygen atom of the terminal α-glucosidic bond of glycogen. An oxygen atom of the deprotonated inorganic phosphate then attacks C-1 to produce glucose 1-phosphate. The other product is a glycogen molecule shortened by one glucose residue.

Glycogen (n residues)

α-D-Glucose 1-phosphate   +   Glycogen ($n-1$ residues)

AMP are close to the interface between the subunits. Glycogen phosphorylase is an interconvertible enzyme that exists in cells in two forms, an active phosphorylated form called phosphorylase a and a less active dephosphorylated form called phosphorylase b.

The mechanism of the phosphorolysis reaction catalyzed by glycogen phosphorylase is shown in Figure 17·5. Inorganic phosphate in the active site donates a proton to the oxygen atom involved in the glycosidic bond of a glucose residue at a nonreducing end of the glycogen molecule. The deprotonated oxygen then attacks C-1, the anomeric carbon. The terminal glucosyl group forms glucose 1-phosphate, leaving a glycogen molecule that is one residue smaller.

Glycogen phosphorylase catalyzes the progressive degradation of glycogen chains from the nonreducing ends. The enzyme stops four glucose residues from a branch point (an α-(1→6) glucosidic bond), leaving a limit dextrin (Section 13·3). The limit dextrin can be further degraded by the action of the glycogen debranching enzyme, which has two separate activities (Figure 17·6). A glucanotransferase activity catalyzes the relocation of three glucose residues from a branch to a free 4′ end of the glycogen molecule. The original linkage and the new linkage are both α-(1→4). The second activity of glycogen debranching enzyme, amylo-1,6-glucosidase, catalyzes hydrolytic (not phosphorolytic) removal of the remaining α-(1→6)-linked glucose residue. One glucose molecule is thereby released for each branch in the original glycogen polymer. A considerably larger number of glucose 1-phosphate molecules is generated by the action of glycogen phosphorylase. Recall that two ATP molecules are obtained when glucose is the substrate for glycolysis. In contrast, three molecules of ATP are obtained for most (about 90%) of the glucose residues of glycogen metabolized by the glycolytic pathway. The energy yield from glycogen is higher because glycogen phosphorylase catalyzes phosphorolysis rather than hydrolysis: no ATP is consumed in the phosphorylation of glucose.

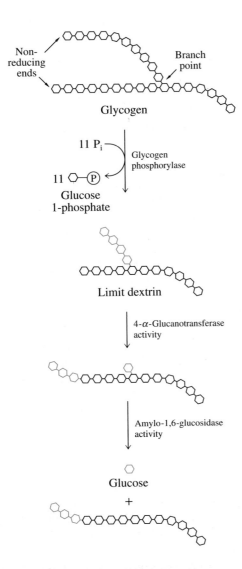

Glycogen

11 P_i

Glycogen phosphorylase

11 (P)

Glucose 1-phosphate

Limit dextrin

4-α-Glucanotransferase activity

Amylo-1,6-glucosidase activity

Glucose

+

Figure 17·6
Degradation of glycogen. Glycogen phosphorylase catalyzes the sequential phosphorolytic cleavage of α-(1→4) glucosidic bonds from the nonreducing ends of glycogen chains, stopping four residues from an α-(1→6) branch point. One molecule of glucose 1-phosphate is formed for each glucose residue mobilized by glycogen phosphorylase. Further degradation is accomplished by the two enzymatic activities of the glycogen debranching enzyme. The 4-α-glucanotransferase activity catalyzes transfer of a trimer from a branch of the limit dextrin to a free 4′ end of the glycogen molecule. The amylo-1,6-glucosidase activity catalyzes hydrolytic release of the remaining α-(1→6)-linked glucose residue. The linear chain that is formed becomes a substrate for glycogen phosphorylase.

17·3 Glucose 1-Phosphate Is Converted to Glucose 6-Phosphate and Glucose

The product of glycogenolysis, the hemiacetal ester glucose 1-phosphate, is rapidly interconverted with glucose 6-phosphate in a near-equilibrium reaction catalyzed by phosphoglucomutase (Figure 17·7). The mechanism of this reaction is similar to that of phosphoglycerate mutase from animals and yeast (Section 15·2, Part 8). Glucose 1-phosphate binds to a phosphoenzyme, and glucose 1,6-*bis*phosphate is formed as an enzyme-bound intermediate. Transfer of phosphate back to the enzyme leaves glucose 6-phosphate. In phosphoglucomutase, the active-site residue that transfers the phosphoryl group is a serine rather than a histidine residue. Glucose 6-phosphate is an intermediate in a number of cellular pathways, including glycolysis, glycogen synthesis, and the pentose phosphate pathway.

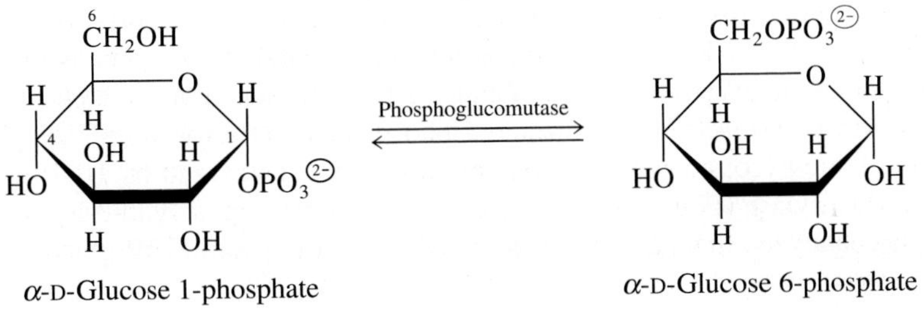

α-D-Glucose 1-phosphate   Phosphoglucomutase   α-D-Glucose 6-phosphate

Figure 17·7
Interconversion of glucose 1-phosphate and glucose 6-phosphate, a near-equilibrium reaction catalyzed by phosphoglucomutase. Glucose 6-phosphate produced by glycogenolysis enters a number of cellular pathways.

In liver, the principal end product of glycogenolysis is glucose, which is formed from glucose 6-phosphate by the action of glucose 6-phosphatase. As we will see, this enzyme also catalyzes the final reaction of gluconeogenesis (discussed in Section 17·7).

$$\text{Glucose 6-phosphate} + H_2O \xrightarrow{\text{Glucose 6-phosphatase}} \text{Glucose} + P_i$$

$$(17\cdot2)$$

The glucose thus formed can be transported via the bloodstream to other tissues. Liver, kidney, pancreas, and small intestine are the only tissues in vertebrates that contain glucose 6-phosphatase activity. A deficiency of this enzyme causes von Gierke's disease, also called type I glycogen storage disease. In afflicted individuals, the concentration of glucose in the blood is low (hypoglycemia), and glycogen accumulates in the liver and kidneys because glucose is not mobilized normally. Accumulation of glucose 6-phosphate inhibits glycogen phosphorylase and activates glycogen synthase. The hypoglycemia can be alleviated by eating many small meals containing carbohydrate that is slowly digested. The ingestion of excess carbohydrate, which is stored as liver glycogen, may result in a greatly enlarged liver.

Von Gierke's disease is the most common of the glycogen-storage diseases. Deficiencies of other enzymes involved in the metabolism of glycogen cause metabolic disorders of differing severity. For example, in Cori's disease (type III glycogen storage disease), there is a deficiency of the debranching enzyme. Only the glucose residues from the outer branches of glycogen are mobilized. The manifestations of this disease are similar to those of von Gierke's disease but milder.

17·4 Glycogen Synthesis and Glycogen Degradation Require Separate Pathways

After a meal, vertebrates absorb dietary glucose and convert some of it to the storage polysaccharide glycogen. Based on the results of in vitro studies, it was widely believed until the late 1950s that glycogen phosphorylase catalyzed both synthesis and breakdown of glycogen. However, when the in vivo concentrations of reactants and products were reliably determined, it became clear that the reverse reaction does not occur in vivo and that glycogen synthesis must occur by another route. The discovery of the nucleotide-sugar coenzymes—including UDP-glucose as a donor of glucosyl groups—and of an enzyme that uses this nucleotide sugar as a substrate for glycogen synthesis changed the prevailing view. We have already seen the role of nucleotide sugars in the biosynthesis of glycoconjugates (Chapter 9, Part 2); in Chapter 19, we will see that UDP-glucose and ADP-glucose are intermediates in the synthesis of sucrose and starch, respectively. It is now known that synthesis and degradation of glycogen require separate enzymatic steps. In fact, it is a general rule of metabolism that different routes are required for opposing degradative and synthetic pathways.

Glucose enters cells from the bloodstream and is phosphorylated to glucose 6-phosphate by the action of hexokinase. Three separate enzyme-catalyzed reactions are required for the incorporation of a molecule of glucose 6-phosphate into glycogen (Figure 17·8). First, phosphoglucomutase catalyzes the near-equilibrium conversion of glucose 6-phosphate to glucose 1-phosphate. Glucose 1-phosphate is then activated by reaction with uridine triphosphate (UTP), forming UDP-glucose and inorganic pyrophosphate (PP_i). This reaction is catalyzed by UDP-glucose pyrophosphorylase (Figure 8·11). In the last step of glycogen synthesis, glycogen synthase catalyzes the addition of the glucose residue from UDP-glucose to the

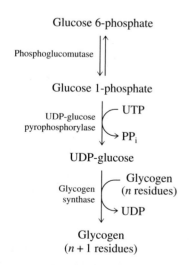

Figure 17·8
Synthesis of glycogen. The three-step pathway is catalyzed by phosphoglucomutase, UDP-glucose pyrophosphorylase, and glycogen synthase.

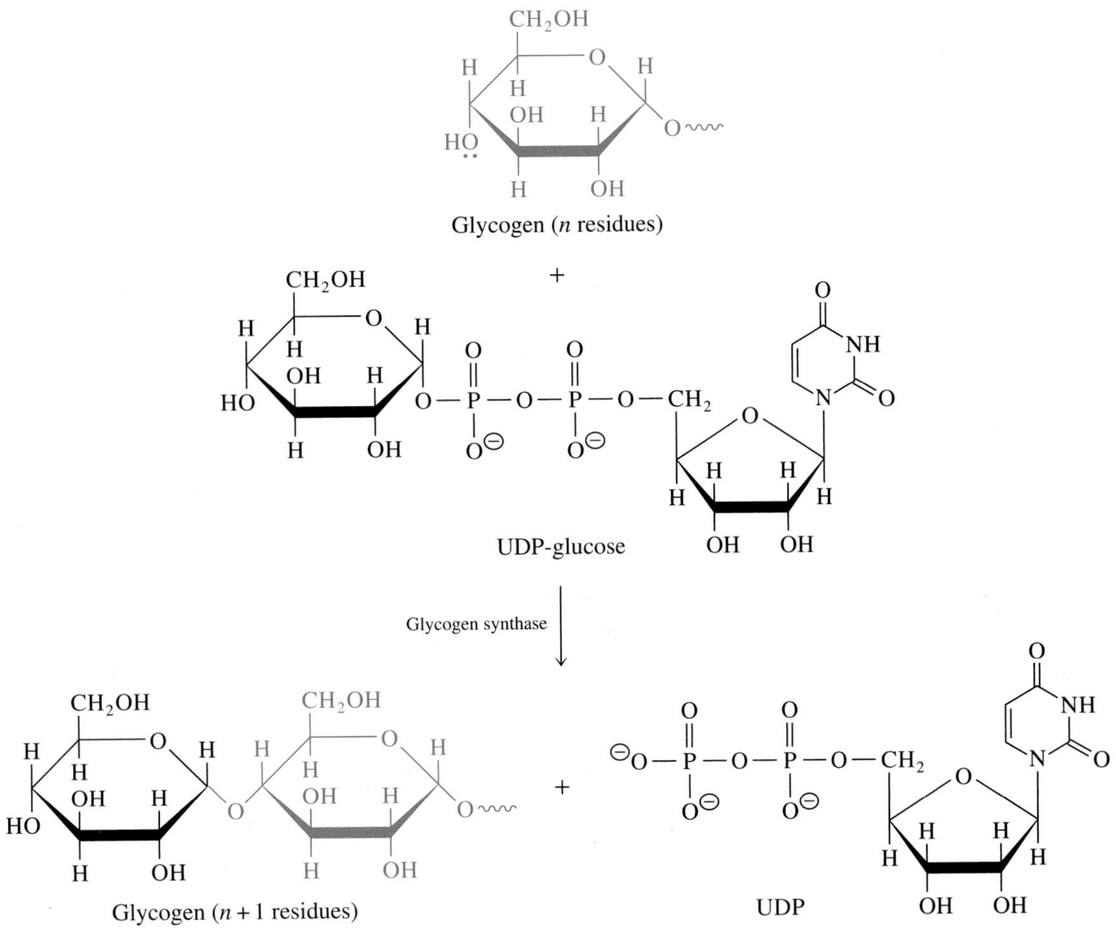

Figure 17·9
Addition of a glucose residue to the non-reducing end of a glycogen primer molecule, catalyzed by glycogen synthase.

nonreducing end of glycogen (Figure 17·9). The glycogen synthase reaction is the major regulatory step of the pathway of glycogen synthesis. Hormones that control the rate of glycogen synthesis do so by regulating the activity of glycogen synthase.

Glycogen synthase requires a preexisting glycogen primer of at least four glucose residues. The primer consists of up to eight α-(1→4)-linked glucose residues attached by the 1'-hydroxyl group of the reducing end in glycosidic linkage to a specific tyrosine residue of the protein glycogenin (M_r 37 000). The primer is formed in two steps. Attachment of the first glucose residue appears to require UDP-glucose as the glucosyl-group donor and a specific glucosyltransferase. Extension of the primer by up to seven additional UDP-glucose molecules is catalyzed by glycogenin itself. Thus, glycogenin is both a protein scaffold for the construction of glycogen as well as an enzyme. Each completed molecule of glycogen (which can contain several hundred thousand glucose residues) contains a single molecule of glycogenin. Further lengthening of the glycogen primer is catalyzed by glycogen synthase. UDP-glucose-requiring glycogen synthases are also present in amoeba, molds, yeast, and insects. Recall that some bacteria synthesize glycogen, but by a reaction that uses ADP-glucose (Section 8·4C).

Another enzyme, amylo-(1,4→1,6)-transglycosylase, catalyzes the formation of the branches in glycogen. This enzyme, also known as the branching enzyme, removes an oligosaccharide of at least six residues from the nonreducing end of an elongated chain and attaches it by an α-(1→6) linkage to a position at least four glucose residues from the nearest α-(1→6) branch point. Branching of glycogen provides many sites for glucose addition or phosphorolysis to occur and contributes to the speed with which glucose can be mobilized.

17·5 Glycogen Metabolism Is Regulated

The role of glycogen in mammals is to store glucose in times of plenty (after feeding) and to supply glucose in times of need (during fasting or "fight-or-flight" situations). The major tissues involved in the storage and use of glycogen are the liver and muscle. Muscle is of quantitative importance in large part because so much of the total mass of mammals consists of muscle. In muscle, glycogen is used as a store of readily available fuel for muscle contraction. In contrast, the glycogen stores in liver are largely converted to glucose that exits liver cells and enters the bloodstream to be transported to other tissues. Other cell types, such as adipocytes, can also take up glucose and synthesize glycogen, but such cells account for a relatively minor portion of overall glycogen metabolism.

The control of glycogen metabolism in mammals was the first metabolic regulatory system to be well understood at the intracellular level. It is now apparent that the regulatory systems of other pathways follow patterns similar to those of glycogen metabolism. In fact, the enzymes that control glycogen metabolism are also the controlling enzymes for several other pathways, including glycolysis, fatty acid synthesis, and gluconeogenesis.

As mentioned in Section 17·2, glycogen phosphorylase is allosterically inhibited by ATP and glucose 6-phosphate and activated by AMP. However, glycogen phosphorylase activity is physiologically regulated primarily by the reversible phosphorylation of a single serine residue, which converts relatively inactive phosphorylase b to active phosphorylase a.

In this section, we will see how both the mobilization and synthesis of glycogen are regulated by hormones, cell-surface receptors, and their associated signal-transduction systems. The principles of hormone regulation were presented in Chapter 12. The regulation of glycogen metabolism demonstrates these principles in action.

A. Hormones Are Messengers at the Level of the Whole Organism

The principal hormones involved in the control of glycogen metabolism in mammals are insulin, glucagon, and epinephrine. Insulin (Section 4·9) is a small protein synthesized by the β cells of the pancreas as a single-chain precursor called proinsulin. In the secretory granules, selective proteolysis of proinsulin releases an internal peptide and the two-chained hormone insulin. Insulin is secreted in response to elevations in blood glucose concentration. Thus, high levels of insulin are associated with the fed state of an organism. Insulin elicits increased uptake and intracellular use or storage of glucose in target cells such as muscle and adipose tissue, resulting in a decrease in blood glucose concentration.

Glucagon, a hormone containing 29 amino acid residues, is secreted by the α cells of the pancreas. Glucagon is released in response to low blood glucose concentration. Glucagon increases the blood glucose concentration by activating glycogen degradation. The effect of glucagon is opposite that of insulin, and an elevated glucagon concentration is associated with the fasted state.

Epinephrine (also known as adrenaline) and its precursor norepinephrine (or noradrenaline, which also has hormone activity) are catecholamines derived from the amino acid tyrosine (Figure 17·10). Epinephrine, and to a lesser extent norepinephrine, is released from the adrenal glands in response to neural signals that trigger the fight-or-flight response. As one of its diverse physiological effects, epinephrine stimulates increased breakdown of glycogen to glucose 1-phosphate, which results in elevated intracellular levels of glucose 6-phosphate. This increase in glucose 6-phosphate leads to enhanced glycolysis in muscle and an increase in the amount of glucose released to the bloodstream from the liver.

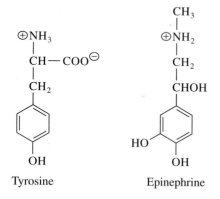

Figure 17·10
Tyrosine and epinephrine. Epinephrine is a catecholamine derived from tyrosine. Norepinephrine lacks the methyl group attached to the amino group of epinephrine.

B. Hormone Action Depends on Cell-Surface Receptors and Second Messengers

When vertebrate hormones are released from their sites of synthesis (endocrine glands), they enter the bloodstream, where they could potentially communicate with all cells of the organism. Selectivity in hormone response is achieved by the presence in given cells of specific hormone receptors. Many cells, for example, possess insulin receptors; these cells are said to be insulin responsive. Insulin causes these cells to increase their uptake of glucose, which results in a decrease in the concentration of blood glucose. By virtue of their *lack* of insulin receptors, red blood cells and brain cells are said to be insulin insensitive.

Liver cells are the only cells rich in glucagon receptors, making glucagon extremely selective in its target. In contrast, a large number of tissues are responsive to epinephrine and norepinephrine. Epinephrine and norepinephrine bind to adrenergic receptors, of which there are a number of subtypes. Two of these—the α_1 and β receptors—are considered in detail here. They elicit distinct and well understood intracellular responses. In cells that contain both receptor subtypes, one is commonly predominant, so that both responses are not evident in the same cell. The distribution of different adrenergic receptors varies between species.

Glucagon and epinephrine are among the hormones that have plasma membrane receptors coupled to a G protein. Glucagon binds to a glucagon receptor that activates the stimulatory G protein, G_s. Binding of epinephrine or norepinephrine to the β-adrenergic receptor also stimulates G_s. G_s in turn stimulates the activity of adenylate cyclase, a membrane-bound enzyme that catalyzes the formation of the second messenger cyclic AMP (Section 10·8A). Cyclic AMP (cAMP) activates protein kinase A (also called cAMP-dependent protein kinase). The steps that trigger the increase in the intracellular concentration of cAMP are summarized in Figure 17·11. (The cascade system involving G proteins, cAMP, and protein kinase A is discussed in detail in Section 12·9B.)

The binding of epinephrine or norepinephrine to the α_1-adrenergic receptor activates the G protein G_P, which invokes an entirely different regulatory cascade. In this case, the membrane-bound enzyme phospholipase C is activated, resulting in the formation of inositol 1,4,5-*tris*phosphate and diacylglycerol. (This process is described in Section 12·9C.) Formation of inositol 1,4,5-*tris*phosphate causes release of calcium from the lumen of the endoplasmic reticulum into the cytosol, where it has diverse physiological effects. The other product, diacylglycerol, is a lipid molecule that activates protein kinase C, a membrane-bound regulatory enzyme. One of the known targets of protein kinase C is the insulin receptor. When

Figure 17·11
Effect of glucagon and epinephrine. Inactive adenylate cyclase (red) is activated by the action of the G protein G_s. The cAMP produced then activates protein kinase A.

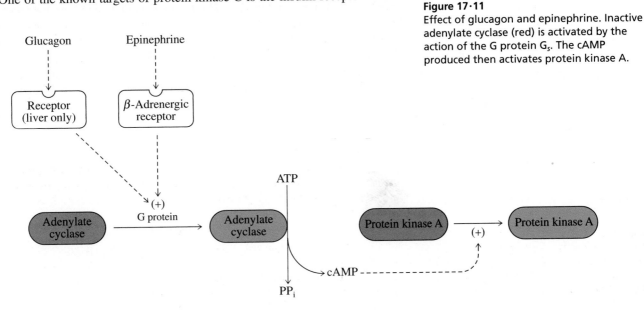

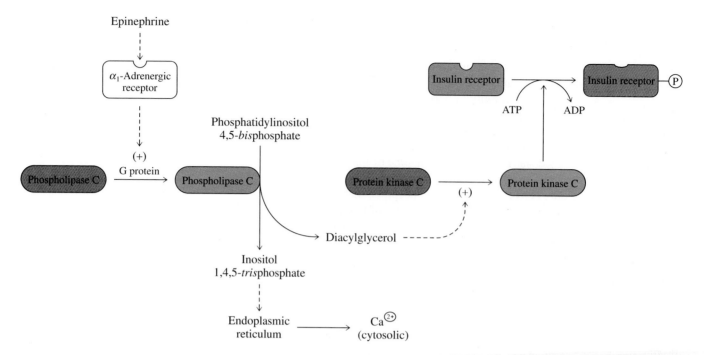

Figure 17·12

Effects of the binding of epinephrine to the α_1-adrenergic receptor. Epinephrine stimulates the activation of phospholipase C via the α_1-adrenergic receptor and the G protein G_P. Phospholipase C catalyzes the formation of the second messengers inositol 1,4,5-*tris*phosphate and diacylglycerol. Diacylglycerol activates protein kinase C, which, among other activities, catalyzes the phosphorylation and inhibition of the insulin receptor. Inositol 1,4,5-*tris*phosphate stimulates release of calcium ions from the lumen of the endoplasmic reticulum to the cytosol.

phosphorylated, the insulin receptor binds insulin poorly. Thus, the action of protein kinase C, triggered by epinephrine, attenuates the effect of insulin. A summary of the effects of epinephrine binding to the α_1-adrenergic receptor is shown in Figure 17·12.

Both inositol 1,4,5-*tris*phosphate and diacylglycerol, like cAMP, are second messengers in that they transmit information from the first messenger—the hormone. The signal molecule calcium could be considered a third messenger, though this terminology is not used. All of these molecules are signal molecules that control the activities of regulatory proteins. Note that the initial signal—binding of a hormone to a receptor on the cell surface—is amplified by the cascade nature of the signalling pathways.

C. Intracellular Regulation of Glycogen Metabolism Involves Interconvertible Enzymes

The enzymes and proteins involved in the intracellular regulation of glycogen metabolism are listed in Table 17·1. These proteins function in both liver and muscle. The core of this control system is the modulation of the two major regulatory steps of glycogen metabolism, namely, the reactions catalyzed by glycogen phosphorylase (degradation) and glycogen synthase (synthesis). These enzymes are regulated reciprocally: when one is active, the other is inactive. The principal mechanism of intracellular regulation in glycogen metabolism is phosphorylation and dephosphorylation of specific residues of interconvertible enzymes. Glycogen phosphorylase and glycogen synthase are both interconvertible enzymes subject to the activity of protein kinases and protein phosphatases. The phosphorylated form of glycogen synthase is the inactive state, whereas the phosphorylated form of glycogen phosphorylase is its more active state (Figure 17·13). As with glycogen phosphorylase, the active form of the synthase is termed glycogen synthase *a* and the inactive form glycogen synthase *b*.

Figure 17·13

Active and inactive states of glycogen synthase and glycogen phosphorylase. Both are interconvertible enzymes regulated by phosphorylation and dephosphorylation of specific serine residues.

	a form (active)	*b* form (inactive)
Glycogen synthase	—OH	—Ⓟ
Glycogen phosphorylase	—Ⓟ	—OH

Table 17·1 Regulatory proteins of intracellular glycogen metabolism

Name	Description
Glycogen phosphorylase	Catalyzes major regulatory reaction of glycogen breakdown
Glycogen synthase	Catalyzes major regulatory reaction of glycogen synthesis
Protein kinase A	Catalyzes phosphorylation of phosphorylase kinase and glycogen synthase
Phosphorylase kinase	Catalyzes phosphorylation of glycogen phosphorylase
Calmodulin	Calcium-binding protein found free in cytosol and as a subunit of phosphorylase kinase
Protein phosphatase-1	Catalyzes dephosphorylation of the phosphoproteins of glycogen metabolism

When the blood glucose concentration is low, epinephrine and glucagon trigger their respective enzyme cascades in muscle and liver, thereby increasing the concentration of intracellular cAMP, which activates protein kinase A. Protein kinase A catalyzes both the phosphorylation and activation of phosphorylase kinase and the phosphorylation and inactivation of glycogen synthase (Figure 17·14). Phosphorylase kinase in turn catalyzes the phosphorylation and activation of glycogen phosphorylase b to glycogen phosphorylase a. Thus, by increasing the production of cAMP in the cells to which they bind, glucagon and epinephrine cause increases in glycogenolysis and simultaneous suppression of glycogen synthesis.

Phosphorylase kinase is a large enzyme (M_r 1.3×10^6 in skeletal muscle) consisting of multiple copies of four distinct subunits. Two of the larger subunits contain serine residues that are phosphorylated when protein kinase A is active. Like most converter enzymes, phosphorylase kinase has noncovalent allosteric effectors. The most important modulator is ionic calcium. The smallest of the subunits of

Figure 17·14
Activation of glycogen phosphorylase and inactivation of glycogen synthase. Protein kinase A catalyzes the phosphorylation and inactivation of glycogen synthase, suppressing glycogen synthesis. Protein kinase A also activates phosphorylase kinase, which catalyzes the phosphorylation and activation of glycogen phosphorylase, leading to glycogen degradation.

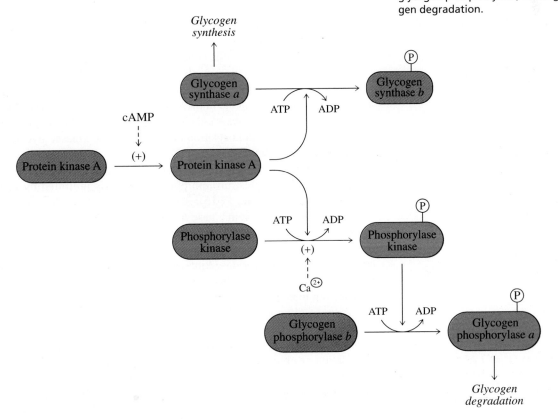

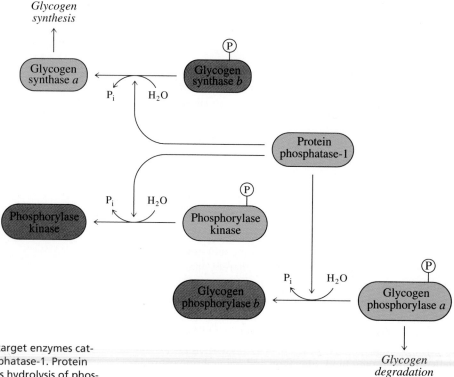

Figure 17·15
Dephosphorylation of target enzymes catalyzed by protein phosphatase-1. Protein phosphatase-1 catalyzes hydrolysis of phosphate ester bonds of the phosphorylated enzymes glycogen synthase, phosphorylase kinase, and glycogen phosphorylase. The result is an increase in glycogen synthesis and a decrease in glycogen degradation.

phosphorylase kinase is the calcium-binding regulatory protein calmodulin (Figure 8·2), which changes conformation upon binding of Ca^{2+}. Calcium binding alone leads to partial activation of phosphorylase kinase. The enzyme is most active when it is both phosphorylated and in the presence of high concentrations of calcium. The fourth subunit of phosphorylase kinase contains the active site that catalyzes the phosphorylation and activation of glycogen phosphorylase, leading to the mobilization of glucose reserves.

The effects of cAMP are abolished by cAMP phosphodiesterase, which catalyzes the rapid hydrolysis of cAMP to AMP, which has no second messenger activity (Figure 12·40). Once the concentration of cAMP falls, the dephosphorylated forms of phosphorylase kinase, glycogen phosphorylase, and glycogen synthase are restored by the action of protein phosphatases. Four major protein phosphatases have been found in cells. One of these, protein phosphatase-1, catalyzes most of the protein dephosphorylations of glycogen metabolism. This single enzyme largely accounts for the dephosphorylation of phosphorylase kinase, glycogen phosphorylase, and glycogen synthase (Figure 17·15).

The activity of protein phosphatase-1 is subject to regulation by cAMP. In muscle and other tissues, protein phosphatase-1 is inhibited by a small protein named inhibitor-1 (Figure 17·16). A threonine residue of inhibitor-1 is phosphorylated by the action of protein kinase A in the presence of cAMP. When it is phosphorylated, inhibitor-1 becomes a potent allosteric inhibitor of protein phosphatase-1. Interconvertible enzymes that would be dephosphorylated by the action of protein phosphatase-1 (phosphorylase kinase, glycogen phosphorylase, and glycogen synthase) remain in their phosphorylated states much longer when inhibitor-1 is active.

In liver, there is no inhibitor-1, but protein phosphatase-1 is inhibited by the active form of glycogen phosphorylase, phosphorylase a, to which it binds very tightly. Since there is much more phosphorylase than phosphatase, inhibition can be almost complete, and no glycogen synthesis can occur (i.e., there is no dephosphorylation of glycogen synthase b to its active form). The presence of glucose in liver promotes glycogen synthesis. Glucose binds to glycogen phosphorylase a, inducing a conformational change that allows the bound protein phosphatase-1 to

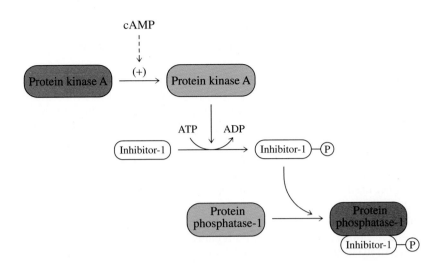

Figure 17·16
Inhibition of protein phosphatase-1 by inhibitor-1. Phosphorylated inhibitor-1 is a potent inhibitor of protein phosphatase-1. The effects of cAMP, ultimately leading to phosphorylation and activation of glycogen phosphorylase and phosphorylation and deactivation of glycogen synthase, are reinforced by inhibitor-1, which is activated by the action of protein kinase A.

become active. Protein phosphatase-1 then catalyzes the phosphorylase *a* to *b* conversion of the enzyme molecule to which it is bound and is released (Figure 17·17). Release of protein phosphatase-1 from its inactive complex with phosphorylase *a* leads to activation of glycogen synthase. This regulatory relationship ensures that glycogen synthase cannot be activated by protein phosphatase-1 until glycogen phosphorylase is nearly all converted to its inactive phosphorylase *b* form. Just as the level of glucose in the blood mediates the initial release of insulin or glucagon, the level of intracellular glucose regulates glycogen metabolism in liver cells.

In summary, the major regulation of glycogen metabolism is accomplished by hormones whose signals are amplified by enzyme-cascade systems. Small concentrations of hormone trigger the formation of larger concentrations of second messengers, principally cAMP. cAMP activates protein kinase A, which results in the phosphorylation and activation of glycogen phosphorylase and the phosphorylation and inactivation of glycogen synthase. Glycogen mobilization is then favored. When glucose accumulates to a sufficiently high level, protein phosphatase-1 reverses the mobilization phase by catalyzing the dephosphorylation and inactivation of glycogen phosphorylase and the dephosphorylation and activation of glycogen synthase.

Figure 17·17
Regulation of glycogen metabolism by glucose in the liver. The presence of glucose in the liver has two effects. First, binding of glucose to glycogen phosphorylase *a* releases the inhibition of the protein phosphatase. Glycogen phosphorylase *a* becomes a substrate for protein phosphatase-1, resulting in inactivation of glycogen phosphorylase. Second, the protein phosphatase catalyzes the activation of glycogen synthase *b*. The net result is decreased glycogen breakdown and increased glycogen synthesis when the concentration of glucose is high.

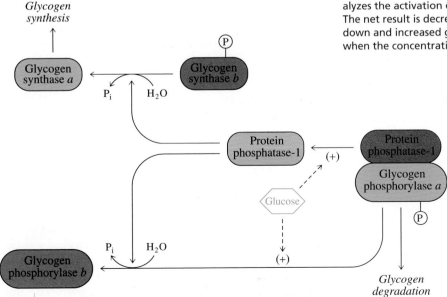

17·6 Glucose Can Be Synthesized from Noncarbohydrate Precursors by Gluconeogenesis

Because glucose may not always be available from exogenous sources or intracellular stores, most organisms have a pathway for glucose biosynthesis. Microorganisms can convert many nutrients to phosphate esters of glucose and to glycogen. Certain mammalian tissues, primarily liver and kidney, synthesize glucose de novo from noncarbohydrate precursors such as lactate and alanine. This process is called **gluconeogenesis.** In our discussion of gluconeogenesis, we will focus on the liver, the primary site of glucose synthesis in mammals. The role of the liver in supplying other tissues with glucose is discussed more fully in Chapter 23.

The pathway for gluconeogenesis from pyruvate to glucose is compared to the glycolytic pathway in Figure 17·18. Note that many of the intermediates are identical; some reactions are common to both pathways and involve the same enzymes. All seven of the near-equilibrium reactions of glycolysis are traversed in the reverse direction during gluconeogenesis. Enzymatic reactions unique to gluconeogenesis are required to bypass the three highly exergonic reactions of glycolysis—the reactions catalyzed by pyruvate kinase, phosphofructokinase-1, and hexokinase.

The synthesis of one molecule of glucose from two molecules of pyruvate requires four ATP and two GTP molecules, as well as two molecules of NADH. The net equation for gluconeogenesis is

$$2\,\text{Pyruvate} \;+\; 2\,\text{NADH} \;+\; 4\,\text{ATP} \;+\; 2\,\text{GTP} \;+\; 6\,H_2O \;+\; 2\,H^{\oplus} \longrightarrow$$

$$\text{Glucose} \;+\; 2\,\text{NAD}^{\oplus} \;+\; 4\,\text{ADP} \;+\; 2\,\text{GDP} \;+\; 6\,P_i \qquad \textbf{(17·3)}$$

Four ATP equivalents are needed to overcome the thermodynamic barrier to formation of two molecules of the high-energy compound phosphoenolpyruvate from two molecules of pyruvate. Recall that the conversion of phosphoenolpyruvate to pyruvate is a metabolically irreversible reaction catalyzed by pyruvate kinase. Two ATP molecules are also required to carry out the reverse of the reaction catalyzed by phosphoglycerate kinase. No energy is recovered in gluconeogenesis between fructose 1,6-*bis*phosphate and glucose because the kinase-catalyzed reactions of glycolysis are metabolically irreversible and are bypassed by hydrolytic reactions in gluconeogenesis. Recall that glycolysis consumes two ATP molecules and generates four, for a net yield of two ATP equivalents. Synthesis of one molecule of glucose by gluconeogenesis consumes a total of six ATP equivalents—four more than are produced by glycolysis.

We will begin our examination of the individual steps in the conversion of pyruvate to glucose with the two enzymes required to bypass the glycolytic reaction catalyzed by pyruvate kinase. First, pyruvate carboxylase catalyzes the conversion of pyruvate to oxaloacetate, coupled to hydrolysis of a molecule of ATP.

$$\textbf{(17·4)}$$

Pyruvate carboxylase has a molecular weight of 520 000 and is composed of four identical subunits. Linked covalently to a lysine residue of each subunit is a biotin prosthetic group. The biotin is required for the addition of bicarbonate to pyruvate. The reaction mechanism for pyruvate carboxylase was described in Section 8·11. This carboxylase catalyzes a metabolically irreversible reaction and can be allosterically activated by acetyl CoA. This is the only regulatory mechanism known for the enzyme. Recall that the pyruvate carboxylase reaction also plays an anaplerotic role by supplying oxaloacetate to the citric acid cycle (Section 16·7).

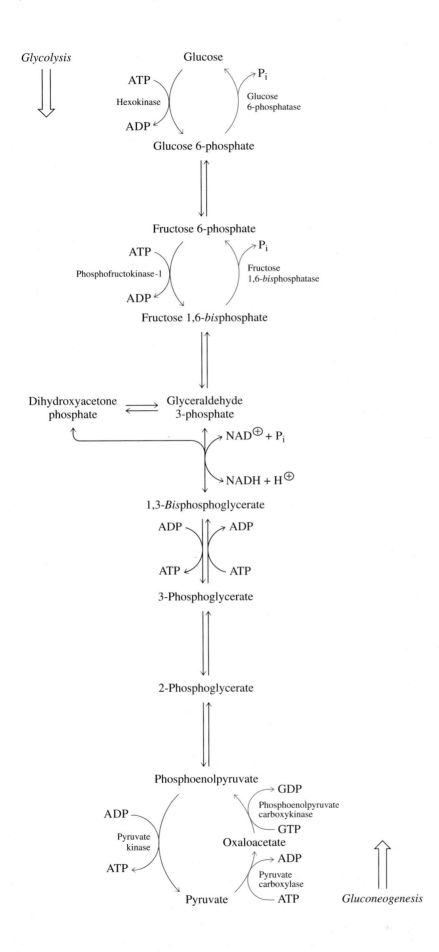

Figure 17·18
Comparison of gluconeoge
sis. The three metabolically
tions of glycolysis highlighte
passed in gluconeogenesis b   ̄ ̄
reactions highlighted in blue. in both pathways, the triose phase is traversed by two
molecules for each glucose molecule.

Following the pyruvate carboxylase reaction, phosphoenolpyruvate (PEP) carboxykinase catalyzes the conversion of oxaloacetate to phosphoenolpyruvate.

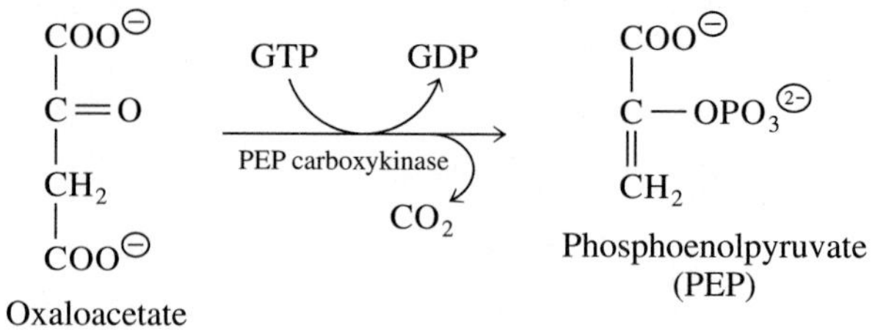

(17·5)

This decarboxylation reaction uses GTP as the donor of a high-energy phosphoryl group (Figure 17·19). PEP carboxykinase is a monomer with a molecular weight of about 70 000. The enzyme displays no allosteric kinetic properties and has no known physiological effectors. Although the reaction catalyzed by isolated PEP carboxykinase is readily reversible in vitro, the reaction is metabolically irreversible. Regulation in vivo is imposed by the amount of PEP carboxykinase synthesized by cells. The level of enzyme sets an upper limit for the rate of gluconeogenesis. During fasting in mammals, chronic production of the hormone glucagon by the pancreas leads to increased synthesis of PEP carboxykinase in liver, a process called hormonal induction. The increased synthesis of PEP carboxykinase is the result of a prolonged elevation of the intracellular concentration of cAMP, which triggers increased transcription of the PEP carboxykinase gene. After several hours, the amount of PEP carboxykinase rises, and there is an increase in the rate of gluconeogenesis. Insulin, abundant in the fed state, acts in opposition to glucagon at the level of the gene, leading to a reduction in the synthesis of PEP carboxykinase.

The reactions of gluconeogenesis between phosphoenolpyruvate and fructose 1,6-*bis*phosphate are simply the reverse of the near-equilibrium reactions of glycolysis. However, the glycolytic reaction catalyzed by phosphofructokinase-1 is metabolically irreversible. This reaction is bypassed by the third enzyme specific to gluconeogenesis, fructose 1,6-*bis*phosphatase, which catalyzes the conversion of fructose 1,6-*bis*phosphate to fructose 6-phosphate (Figure 17·20). The hydrolysis of the phosphate ester in this reaction has a large negative ΔG and is a metabolically

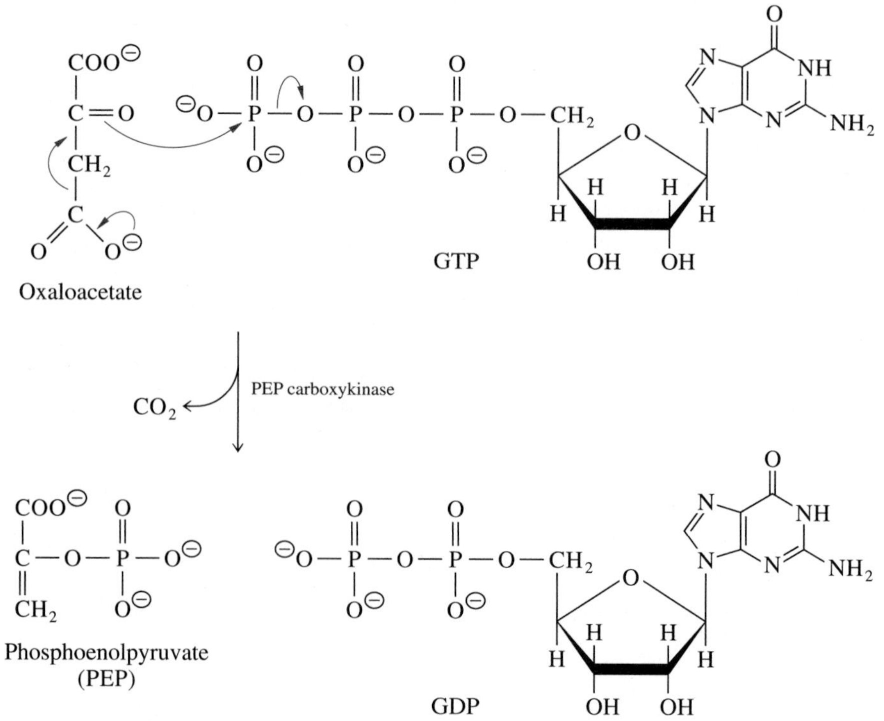

Figure 17·19
Mechanism of the reaction catalyzed by PEP carboxykinase. Decarboxylation of oxaloacetate is followed by nucleophilic attack on the phosphorus atom of the γ-phosphoryl group of GTP, yielding phosphoenolpyruvate, GDP, and CO_2.

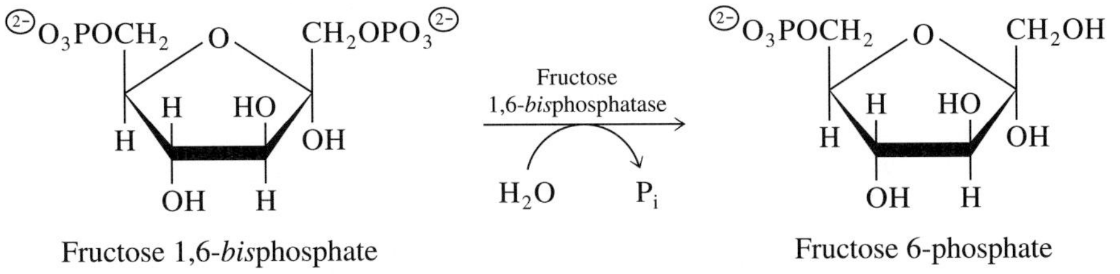

Fructose 1,6-*bis*phosphate

Fructose 6-phosphate

Figure 17·20
Formation of fructose 6-phosphate from fructose 1,6-*bis*phosphate, catalyzed by fructose 1,6-*bis*phosphatase.

irreversible reaction. Fructose 1,6-*bis*phosphatase is a tetrameric enzyme with a molecular weight of 150 000. The mammalian enzyme displays sigmoidal kinetics and is allosterically inhibited by AMP and by the regulatory molecule fructose 2,6-*bis*phosphate. Recall that fructose 2,6-*bis*phosphate is a potent activator of phosphofructokinase-1, the enzyme that catalyzes the formation of fructose 1,6-*bis*phosphate in glycolysis (Section 15·7C). Thus, the two enzymes that catalyze interconversion of fructose 6-phosphate and fructose 1,6-*bis*phosphate are reciprocally controlled by the concentration of fructose 2,6-*bis*phosphate.

Following the near-equilibrium conversion of fructose 6-phosphate to glucose 6-phosphate, the final step of gluconeogenesis is the reaction catalyzed by glucose 6-phosphatase. This enzyme catalyzes the hydrolysis of glucose 6-phosphate, producing glucose and inorganic phosphate (Figure 17·21). The hydrolytic reaction is metabolically irreversible. Glucose 6-phosphatase is bound to the membrane of the endoplasmic reticulum, apparently with its active site in the lumen. It has been proposed that a transporter is required to convey glucose 6-phosphate from the cytosol to the site of hydrolysis. Transporters would also be necessary to return glucose and P_i to the cytosol; the P_i transporter has been isolated. Whereas all of the other enzymes required for gluconeogenesis are found in small amounts in many types of mammalian tissue, glucose 6-phosphatase is found only in cells from the liver, kidney, pancreas, and small intestine.

Many mammalian tissues contain an incomplete set of gluconeogenic enzymes. For example, some muscle cells contain high fructose 1,6-*bis*phosphatase activity, but muscle is not gluconeogenic. In tissues with at least partial gluconeogenic pathways, these enzymes catalyze so-called "futile cycles," reactions whose net formal balance is the hydrolysis of ATP. For example, the net reaction of phosphofructokinase-1 and fructose 1,6-*bis*phosphatase acting simultaneously is the hydrolysis of ATP to ADP plus P_i.

$$\text{Fructose 6-phosphate} + \text{ATP} \longrightarrow \text{Fructose 1,6-}bis\text{phosphate} + \text{ADP}$$

$$\text{H}_2\text{O} + \text{Fructose 1,6-}bis\text{phosphate} \longrightarrow \text{Fructose 6-phosphate} + \text{P}_i$$

$$\text{Net} \qquad \text{H}_2\text{O} + \text{ATP} \longrightarrow \text{ADP} + \text{P}_i \qquad\qquad \textbf{(17·6)}$$

When these reaction sequences were first considered, they were called futile cycles because their operation seemed only to expend ATP with no apparent gain. It was later realized that the presence of opposing, metabolically irreversible reactions that catalyze a cycle between two pathway intermediates provides a sensitive regulatory site. Control of one or both of these reactions can determine the direction and level of flux through the metabolic sequence. These reactions are now termed **substrate cycles,** and it is believed that they serve an important role in cells. In gluconeogenic

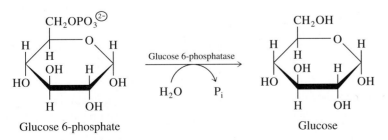

Glucose 6-phosphate

Glucose

Figure 17·21
Formation of glucose from glucose 6-phosphate, catalyzed by glucose 6-phosphatase.

Figure 17·22
Example of a substrate cycle. The direction of flux is toward gluconeogenesis. However, phosphorylation of fructose 6-phosphate (catalyzed by phosphofructokinase-1) and dephosphorylation of fructose 1,6-*bis*phosphate (catalyzed by fructose 1,6-*bis*phosphatase) occur simultaneously. The overall rate of gluconeogenesis can be controlled with great sensitivity by altering the rate of either or both of the enzymes of the substrate cycle.

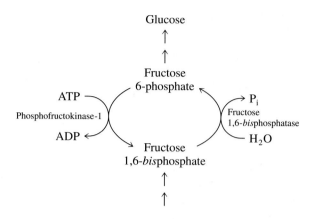

tissues, where actual reversal of flux through the glycolytic/gluconeogenic pathways is possible, both the direction and the amount of flow through the pathway are controlled by substrate cycles (Figure 17·22).

17·7 There Are Several Important Precursors for Gluconeogenesis

Any metabolite that can be converted into pyruvate or a citric acid cycle intermediate can serve as a precursor of glucose. Entry of the appropriate compound into the citric acid cycle leads to increased formation of oxaloacetate, and the conversion of oxaloacetate or pyruvate to glucose is straightforward. The major gluconeogenic precursors are lactate and amino acids, of which alanine is by far the most important. Glycerol from the hydrolysis of triacylglycerols is another important substrate for gluconeogenesis, entering the pathway after conversion to a triose phosphate. Gluconeogenic precursors arise in a variety of tissues and must be transported to the liver to serve as substrates in the formation of glucose.

Glycolysis generates large amounts of lactate in active muscle, and red blood cells generate a steady output of lactate even when muscle is inactive. Lactate from these and other sources enters the bloodstream and travels to the liver, where it is converted to pyruvate by the action of lactate dehydrogenase. Pyruvate can then serve as a substrate for gluconeogenesis. Glucose produced by the liver enters the bloodstream for delivery to peripheral tissues. The sequence of glucose oxidation to lactate in peripheral tissues, followed by delivery of lactate to liver, formation of glucose from lactate, and delivery of glucose back to peripheral tissues, is known as the **Cori cycle.** This metabolic cycle operates with no net loss or gain of carbon. Conversion of lactate to glucose requires energy, most of which is derived from the oxidation of fatty acids in liver. Thus, the Cori cycle is a vehicle for the delivery of chemical potential energy from liver to the peripheral tissues.

The carbon skeletons of most amino acids are catabolized to pyruvate or intermediates of the citric acid cycle. Pyruvate formed from glycolysis or amino acid catabolism in peripheral tissues can accept an amino group from an α-amino acid, forming alanine in a process called transamination (Section 21·5).

$$
\begin{array}{ccc}
\text{COO}^{\ominus} & & \text{COO}^{\ominus} \\
| & \overset{\text{Amino}\quad\alpha\text{-Keto}}{\underset{transamination}{\longrightarrow}} & | \\
\text{C}=\text{O} & & \text{H}_3\overset{\oplus}{\text{N}}-\text{CH} \\
| & & | \\
\text{CH}_3 & & \text{CH}_3 \\
\text{Pyruvate} & & \text{Alanine}
\end{array} \qquad (17\cdot7)
$$

The alanine travels to the liver, where it undergoes transamination with α-ketoglutarate to re-form pyruvate for gluconeogenesis. Subsequent release of glucose from

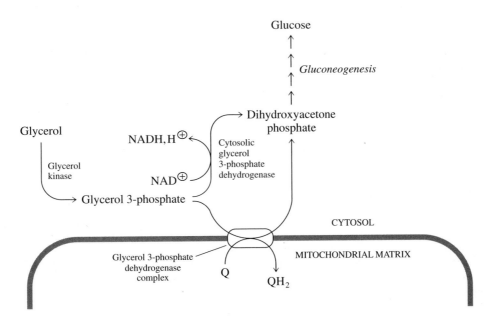

Figure 17·23
Gluconeogenesis from glycerol. Glycerol 3-phosphate can be oxidized in reactions catalyzed by either of two dehydrogenases; both reactions result in the formation of reduced coenzymes.

the liver completes a cycle similar to the Cori cycle that is known as the **glucose-alanine cycle.** We will reconsider the Cori cycle and the glucose-alanine cycle in Chapter 23. We will see then that amino acids become a major source of carbon for gluconeogenesis during fasting, when glycogen supplies are depleted.

The carbon skeleton of aspartate, which is the amino-group donor in the urea cycle (a pathway that eliminates excess nitrogen from the cell, as discussed in Section 21·10), also is a precursor of glucose. Aspartate is converted to fumarate in the urea cycle; fumarate is hydrated to malate, which is oxidized to oxaloacetate.

The catabolism of triacylglycerols produces glycerol and acetyl CoA. As we saw earlier, acetyl CoA cannot contribute to the net formation of glucose in animals, but can in organisms that carry out the reactions of the glyoxylate cycle (Section 16·8). Glycerol, though, can be converted to glucose by a route that begins with phosphorylation to glycerol 3-phosphate, catalyzed by the enzyme glycerol kinase (Figure 17·23). Glycerol 3-phosphate enters the gluconeogenic pathway after conversion to dihydroxyacetone phosphate. This oxidation reaction can be catalyzed by a flavin-containing glycerol 3-phosphate dehydrogenase embedded in the inner mitochondrial membrane. The cytosolic face of this enzyme binds glycerol 3-phosphate, and electrons are passed to ubiquinone (Q) and subsequently to the rest of the mitochondrial respiratory electron-transport chain. Oxidation of glycerol 3-phosphate can also be catalyzed by cytosolic glycerol 3-phosphate dehydrogenase, in which case NADH is a coproduct. In liver, the site of most gluconeogenesis in mammals, both reactions occur, as indicated in Figure 17·23.

In ruminants, which include not just cattle but sheep, giraffes, deer, and camels, the lactate and propionate produced by the microorganisms in their rumen (stomach chambers) is absorbed and in great part metabolized to glucose. Lactate from the rumen is oxidized to pyruvate. Propionate is converted to propionyl CoA and then to succinyl CoA, an intermediate of the citric acid cycle that can be metabolized to oxaloacetate (Section 20·3).

17·8 The Pathway of Glucose Synthesis Depends on Enzyme Localization and Substrate Identity

All the enzymes that catalyze the reactions of gluconeogenesis are cytosolic, with the exception of glucose 6-phosphatase in the endoplasmic reticulum, pyruvate carboxylase in the mitochondria, and PEP carboxykinase, which in different species or

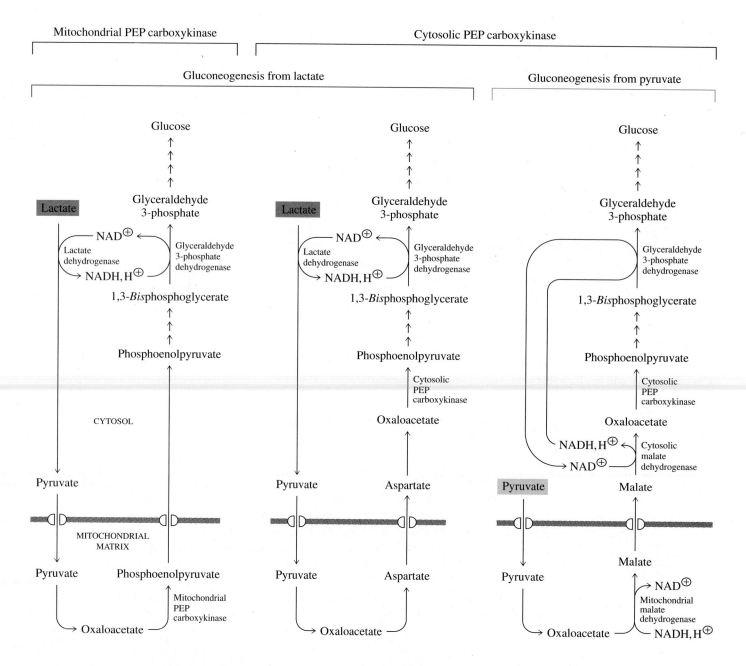

Figure 17·24
Effect of enzyme localization and substrate identity on the pathway for glucose synthesis. When lactate is the substrate for gluconeogenesis and PEP carboxykinase is present in mitochondria, mitochondrial oxaloacetate is converted to phosphoenolpyruvate for transport out of the mitochondrion (left). If PEP carboxykinase is not present in the mitochondrion, oxaloacetate is converted to aspartate for transport (center). In either case, reducing equivalents generated by lactate dehydrogenase are retained in the cytosol. When pyruvate is the substrate for gluconeogenesis, mitochondrial oxaloacetate is converted to malate for transport (right). The reoxidation of malate in the cytosol generates NADH for the glyceraldehyde 3-phosphate reaction.

tissues is found in the cytosol (rat liver), the mitochondria (bird liver), or both. Humans and many other mammals contain almost equal amounts of cytosolic and mitochondrial PEP carboxykinase activity. The localization of pyruvate carboxylase in mitochondria means that gluconeogenesis from pyruvate or lactate requires transport of pyruvate from the cytosol to the mitochondrial matrix. Recall from Chapter 16 that the outer mitochondrial membrane presents no barrier to molecules with molecular weights less than 10000. A specific transporter conveys pyruvate across the inner mitochondrial membrane. In the mitochondria, pyruvate is acted on by pyruvate carboxylase to form oxaloacetate. However, no transporter exists to transport mitochondrial oxaloacetate to the cytosol. Oxaloacetate must therefore be shuttled to the cytosol by conversion to a metabolite for which a transporter exists. Reconversion to oxaloacetate then occurs in the cytosol.

Three options exist for shuttling oxaloacetate into the cytosol, with the route depending on both the localization of enzymes and the identity of the initial substrate for gluconeogenesis—either lactate or pyruvate. When lactate is the initial substrate, the conversion of lactate to pyruvate in the cytosol by the action of lactate dehydrogenase generates NADH that balances the consumption of NADH in the glyceraldehyde 3-phosphate dehydrogenase reaction of gluconeogenesis. Cytosolic reducing equivalents remain in the cytosol. The two pathways on the left of Figure 17·24 show gluconeogenesis from lactate. On the far left is the pathway that is active when PEP carboxykinase is present in the mitochondria. Mitochondrial phosphoenolpyruvate can be transported to the cytosol by a transporter called the tricarboxylate carrier, with straightforward conversion of phosphoenolpyruvate to glucose in the cytosol. When PEP carboxykinase is not present in the mitochondria, oxaloacetate can be converted to aspartate, which can be transported to the cytosol and reoxidized to oxaloacetate (middle pathway of Figure 17·24). The action of cytosolic PEP carboxykinase then converts oxaloacetate to phosphoenolpyruvate.

When pyruvate rather than lactate is the initial substrate for gluconeogenesis, NADH required for the glyceraldehyde 3-phosphate dehydrogenase reaction in the cytosol is not supplied by the action of lactate dehydrogenase. In this case, oxaloacetate in the mitochondria is reduced by mitochondrial NADH to malate for transport to the cytosol. Reoxidation of malate in the cytosol regenerates NADH (pathway on the right of Figure 17·24). The existence of different pathways for the exit of oxaloacetate from the mitochondria allows the cell to match the oxidation state of the gluconeogenic substrate with the requirement for cytosolic NADH consumed in the glyceraldehyde 3-phosphate dehydrogenase reaction of gluconeogenesis.

When the substrate for gluconeogenesis is oxaloacetate derived from aspartate via the urea cycle, a shuttle is also needed to convey aspartate from the mitochondria to the cytosol. In this case, aspartate exits the mitochondria and is converted in the cytosol to fumarate, then malate. The oxidation of malate produces both oxaloacetate for gluconeogenesis and the NADH needed to reduce 1,3-*bis*phosphoglycerate.

17·9 Gluconeogenesis Is Regulated by Hormones and by Substrate Supply

Gluconeogenesis, like glycogen synthesis and degradation, is carefully regulated in vivo. We have seen that glucagon exerts a long-term regulatory effect by causing an increase in the quantity of cytosolic PEP carboxykinase, which enhances gluconeogenesis. Elevated levels of glucagon also lead to inactivation of pyruvate kinase by a cAMP-dependent phosphorylation, as noted in Section 15·7D during our discussion of the regulation of glycolysis.

Figure 17·25
Substrate cycles between fructose 6-phosphate and fructose 1,6-*bis*phosphate and between phosphoenolpyruvate and pyruvate. A change in activity of any of the enzymes involved in the substrate cycles can affect not only the rate of flux but also the direction of flux toward either glycolysis or gluconeogenesis.

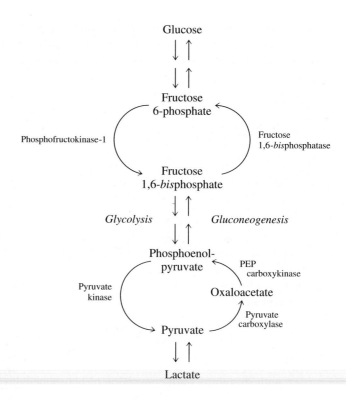

Short-term regulation of gluconeogenesis (regulation that occurs within minutes and does not involve synthesis of new protein) is exerted at two sites in the pathway—the substrate cycles between pyruvate and phosphoenolpyruvate and between fructose 1,6-*bis*phosphate and fructose 6-phosphate (Figure 17·25). We can now appreciate an important feature of substrate cycles: modification of the activity of any enzyme in the cycle can alter the flux through two opposing pathways. For example, inhibition of pyruvate kinase stimulates gluconeogenesis. This can also be viewed as causing more phosphoenolpyruvate to enter the pathway leading to glucose rather than being converted to pyruvate by pyruvate kinase. Glucagon also regulates the fructose 6-phosphate–fructose 1,6-*bis*phosphate cycle. cAMP-activated protein kinase A catalyzes the phosphorylation of phosphofructokinase-2, as described in Section 15·7C. The resulting decrease in the concentration of fructose 2,6-*bis*phosphate removes a potent activator of phosphofructokinase-1 and relieves the inhibition of fructose 1,6-*bis*phosphatase, thereby activating gluconeogenesis.

A different regulator of gluconeogenesis is the concentration of substrate. The principal substrates for the gluconeogenic pathway in mammals are amino acids (mostly alanine) and lactate. The amino acids arise from breakdown of muscle protein and are converted by pathways described in Chapter 21 to intermediates of the citric acid cycle. Under certain metabolic conditions, oxaloacetate produced by the citric acid cycle is diverted to the gluconeogenic pathway. The concentration of amino acids in the blood does not saturate the gluconeogenic pathway, and an increase in the concentration of free amino acids results in a greater conversion to glucose. Similarly, the concentration of lactate in the blood is below the saturation level for gluconeogenesis. Lactate destined for gluconeogenesis is produced largely by muscle, reflecting the large mass and high rate of glycolytic activity of muscle tissue. In this way, the metabolic activities of other tissues can affect the rate of gluconeogenesis in the liver.

The pathways of gluconeogenesis and glycogen metabolism are not mutually exclusive. In liver, substrates may traverse the gluconeogenic pathway and go on to form glycogen rather than glucose if glycogen synthase is active. As glycogen stores become depleted, gluconeogenesis increases in order to maintain a constant blood glucose concentration. This corresponds both to changes in the activity of gluconeogenic enzymes and to the progressive inactivation of glycogen synthase. The pathways of glucose metabolism are not abruptly switched on or off but rather are continuously adjusted according to the minute-to-minute metabolic needs of the organism.

17·10 Glucose Is Sometimes Converted to Sorbitol or Lactose

Glucose, whether obtained from the diet, from the degradation of glycogen, or from de novo synthesis, is usually either oxidized or reincorporated into glycogen. However, some glucose is converted to other compounds. In this section, we will discuss two specialized pathways involving glucose: the sorbitol pathway and the process of lactose synthesis. In the next section, we will discuss the pentose phosphate pathway, which generates ribose and NADPH.

A. The Sorbitol Pathway

A number of mammalian tissues, including the testes, pancreas, brain, and the lens of the eye, contain two enzymes that provide a pathway for the conversion of glucose to fructose (Figure 17·26). This pathway is called the sorbitol, or polyol, pathway. A similar pathway occurs in fruits such as apples, pears, apricots, and peaches; in these fruits, glucose 6-phosphate is reduced to sorbitol 6-phosphate, which is hydrolyzed to sorbitol and then converted to fructose. Aldose reductase catalyzes the reduction of glucose by NADPH to produce sorbitol. The enzyme has a single domain, consisting of an eight-stranded barrel to which the nicotinamide coenzyme is bound (Figure 17·27, next page). Surprisingly, the structure resembles that of triose phosphate isomerase (Figure 5·48) and shows no similarity to the structures of other NAD- or NADP-binding dehydrogenases, in which the coenzyme binds to a site between two domains of the protein.

The second reaction in the sorbitol pathway is oxidation of sorbitol to fructose. This $NAD^{\oplus}$-dependent reaction is catalyzed by polyol dehydrogenase (also called sorbitol dehydrogenase). The sorbitol pathway supplies essential fructose to some cells. For example, fructose is the main fuel of sperm cells. In some mammalian cells, sorbitol may be involved in osmoregulation.

Because aldose reductase has a high K_m value for glucose (approximately 0.1 M), flux through the sorbitol pathway is normally low, and glucose is usually metabolized by glycolysis. However, when the concentration of glucose is higher than usual—for example, in individuals with diabetes—increased amounts of sorbitol are produced in tissues such as the lens. Because there is less polyol dehydrogenase activity than aldose reductase activity, sorbitol accumulates rather than being converted to fructose. Membranes are relatively impermeable to sorbitol. The

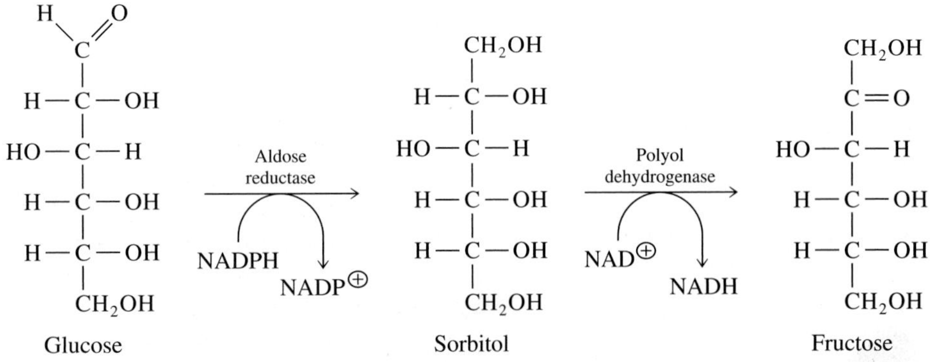

Figure 17·26
Sorbitol, or polyol, pathway. Flux through this pathway increases in some tissues when the concentration of glucose is high.

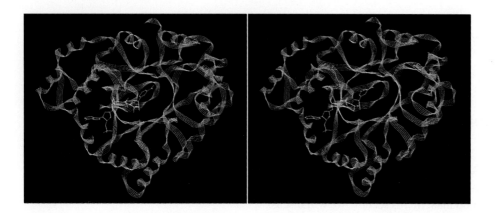

Figure 17·27
Stereo view of aldose reductase. The coenzyme NADPH is orange. (Based on coordinates provided by D. K. Wilson and F. A. Quiocho.)

resulting change in osmolarity of the cells causes aggregation and precipitation of lens proteins, leading to cataracts, opaque regions in the lens. In galactosemia, when levels of galactose are elevated, the sugar is reduced to its sugar alcohol, galactitol (Figure 17·28), which also accumulates and leads to cataracts.

B. Lactose Synthesis

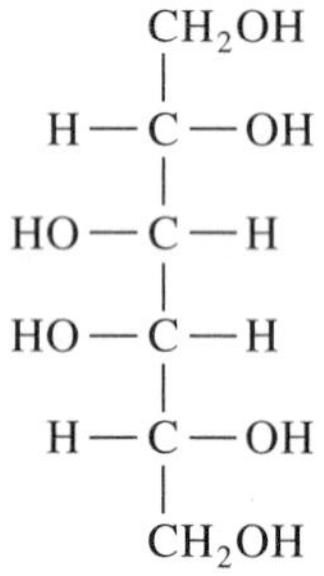

The disaccharide lactose (β-D-galactopyranosyl-$(1 \rightarrow 4)$-β-D-glucopyranose) is present in high concentrations (2–9% by weight) in the milk of most mammals. Both of the monosaccharide residues of lactose are derived from blood glucose. Recall that UDP-glucose, synthesized from UTP and glucose 1-phosphate (Figure 8·11), is converted to UDP-galactose by the action of UDP-glucose 4′-epimerase (Figure 17·2). Lactose synthase catalyzes the transfer of the galactosyl group from UDP-galactose to the 4-hydroxyl group of glucose (Figure 8·12). The enzyme is located in the endoplasmic reticulum of mammary tissue cells. Lactose synthase is composed of two different types of subunits. One is a galactosyltransferase, and the other is the protein α-lactalbumin. For many years, the role of α-lactalbumin was not known. In most tissues, the galactosyltransferase catalyzes the transfer of galactose from UDP-galactose to free N-acetylglucosamine (GlcNAc) or to GlcNAc residues of glycoproteins (Figure 17·29). However, the substrate specificity of the galactosyltransferase is modified during lactation when, under control of the hormone prolactin, mammary cells synthesize α-lactalbumin. α-Lactalbumin binds to the galactosyltransferase to form active lactose synthase and lowers the K_m value for glucose from over 1 M to the millimolar range, making glucose the preferred substrate for the transferase. Because α-lactalbumin is synthesized only in mammary glands, this tissue is the only site of lactose synthesis in mammals.

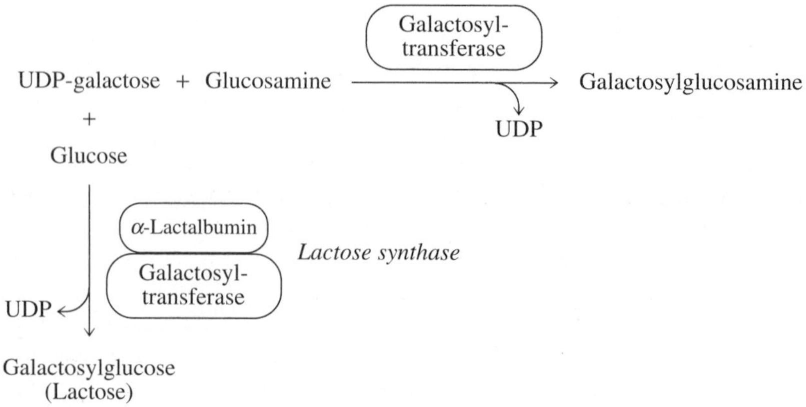

Figure 17·28
Structure of D-galactitol.

$$
\begin{array}{c}
CH_2OH \\
| \\
H - C - OH \\
| \\
HO - C - H \\
| \\
HO - C - H \\
| \\
H - C - OH \\
| \\
CH_2OH
\end{array}
$$

Figure 17·29
Reactions catalyzed by galactosyltransferase. In most tissues, galactosyltransferase catalyzes transfer of a galactose residue to an N-acetylglucosamine acceptor molecule. In mammary tissue during lactation, a complex of α-lactalbumin and galactosyltransferase is formed. In the reaction catalyzed by this oligomeric enzyme, glucose is the major acceptor of the galactosyl group of UDP-galactose.

UDP-galactose + Glucosamine $\xrightarrow{\text{Galactosyltransferase}}$ Galactosylglucosamine + UDP

+

Glucose

α-Lactalbumin / Galactosyltransferase — *Lactose synthase*

UDP

Galactosylglucose (Lactose)

17·11 The Pentose Phosphate Pathway Produces NADPH and Ribose 5-Phosphate

The **pentose phosphate pathway,** sometimes called the hexose monophosphate shunt, is the final pathway in our examination of carbohydrate metabolism. We have already seen that glucose can enter the glycolytic pathway or be incorporated into storage polysaccharides. Once it has been converted to glucose 6-phosphate, glucose can also enter the pentose phosphate pathway. This pathway has two primary functions: production of NADPH and formation of ribose 5-phosphate. NADPH is the pyridine nucleotide coenzyme used for reductive biosynthesis. Ribose 5-phosphate is required for biosynthesis of ribonucleotides and their derivatives, which are incorporated into RNA, DNA, and certain coenzymes. The pentose phosphate pathway is active in tissues that synthesize fatty acids or steroids (e.g., the mammary gland, the liver, the adrenal gland, and adipose tissue) since large amounts of NADPH are consumed in these biosynthetic reactions. In other cells, such as muscle and brain, the pentose phosphate pathway accounts for little of the overall consumption of glucose. The enzymes that catalyze the reactions of this pathway are all found in the cytosol, the site of many of the biosynthetic reactions that require NADPH.

The pentose phosphate pathway can be divided into an oxidative stage and a nonoxidative stage, as illustrated in Figure 17·30 (next page). In the oxidative stage, NADPH is produced as glucose 6-phosphate is converted to the five-carbon compound ribulose 5-phosphate.

$$\text{Glucose 6-phosphate } + \text{ 2 NADP}^{\oplus} + \text{ H}_2\text{O} \longrightarrow$$
$$\text{Ribulose 5-phosphate } + \text{ 2 NADPH } + \text{ CO}_2 + \text{ 3 H}^{\oplus} \qquad \textbf{(17·8)}$$

If the cells carrying out this pathway require considerable amounts of both NADPH and nucleotides, all of the ribulose 5-phosphate can be isomerized to ribose 5-phosphate and the pathway is completed at this stage. Usually, though, more NADPH than ribose 5-phosphate is needed, and most of the pentose phosphates are converted into glycolytic intermediates.

The nonoxidative stage of the pentose phosphate pathway is a means for disposing of the pentose phosphate formed in the oxidative stage by providing a route to glycolysis. In the nonoxidative stage, ribulose 5-phosphate is converted to fructose 6-phosphate and glyceraldehyde 3-phosphate, intermediates of the glycolytic pathway. If all of the pentose phosphate were converted to intermediates of glycolysis, the sum of the nonoxidative reactions would be conversion of three pentose molecules to two hexose molecules plus one triose molecule.

$$\text{3 Ribulose 5-phosphate } \longrightarrow \text{ 2 Fructose 6-phosphate } + \text{ Glyceraldehyde 3-phosphate}$$
$$\textbf{(17·9)}$$

Both fructose 6-phosphate and glyceraldehyde 3-phosphate can be metabolized further through either the glycolytic or gluconeogenic pathways. Let us now take a closer look at the individual reactions of the pentose phosphate pathway.

The oxidative stage of the pentose phosphate pathway is shown in Figure 17·31 (Page 17·29). The first reaction, catalyzed by glucose 6-phosphate dehydrogenase, is the oxidation of glucose 6-phosphate to 6-phosphogluconolactone (equivalent to the oxidation of an aldehyde to a carboxylic acid). This step is the major regulatory site for the entire pentose phosphate pathway. Glucose 6-phosphate dehydrogenase is allosterically inhibited by NADPH. By virtue of this simple regulatory feature, production of NADPH by the pentose phosphate pathway is self-limiting. The next enzyme of the oxidative phase is gluconolactonase, which catalyzes hydrolysis of 6-phosphogluconolactone, an internal ester, to the sugar acid 6-phosphogluconate. Finally, 6-phosphogluconate dehydrogenase catalyzes the oxidative decarboxylation of 6-phosphogluconate, producing a second molecule of NADPH, ribulose

Figure 17·30
Pentose phosphate pathway. The pathway can be divided into an oxidative and a non-oxidative stage. The oxidative stage produces the five-carbon sugar phosphate ribulose 5-phosphate, with concomitant production of NADPH. The nonoxidative stage produces the glycolytic intermediates glyceraldehyde 3-phosphate and fructose 6-phosphate.

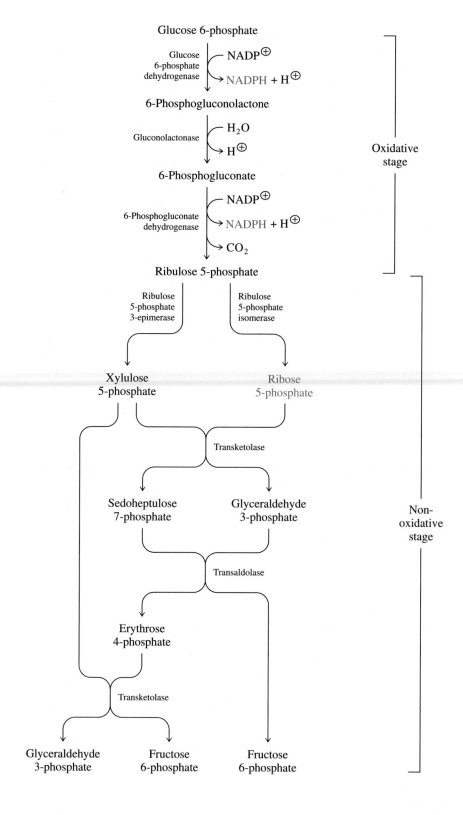

5-phosphate, and CO_2. This oxidation is similar to the citric acid cycle reaction catalyzed by isocitrate dehydrogenase, in which an unstable β-keto acid is formed as an intermediate (Figure 16·15).

The nonoxidative stage of the pentose phosphate pathway in cells that have a high overall flux through the pathway consists entirely of near-equilibrium reactions. This stage of the pathway serves two functions: it provides five-carbon sugars for biosynthesis and introduces sugar phosphates into the pathways of glycolysis or gluconeogenesis. There are two fates for ribulose 5-phosphate: an epimerase can catalyze the formation of xylulose 5-phosphate, or an isomerase can catalyze the conversion of ribulose 5-phosphate into ribose 5-phosphate (Figure 17·32). Ribose 5-phosphate is the precursor of the ribose portion of nucleotides. However, only a very small amount of ribose 5-phosphate is siphoned from the pentose phosphate pathway to fill this need. The remaining steps of the pathway convert the five-carbon sugars into glycolytic intermediates. Rapidly dividing cells, which require both ribose 5-phosphate and NADPH, generally have high pentose phosphate pathway activity. Ribose 5-phosphate is a precursor of deoxyribonucleotides, and NADPH is consumed in the reduction of ribonucleotides to deoxyribonucleotides (Section 22·9).

Figure 17·31
Oxidative stage of the pentose phosphate pathway. Two molecules of NADPH are generated for each molecule of glucose 6-phosphate that enters the pathway.

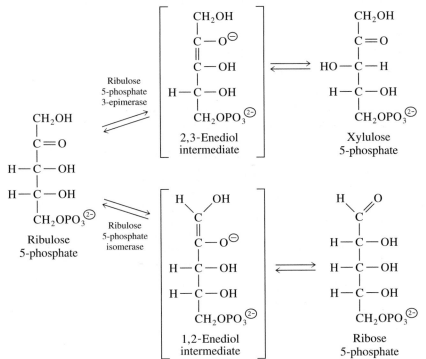

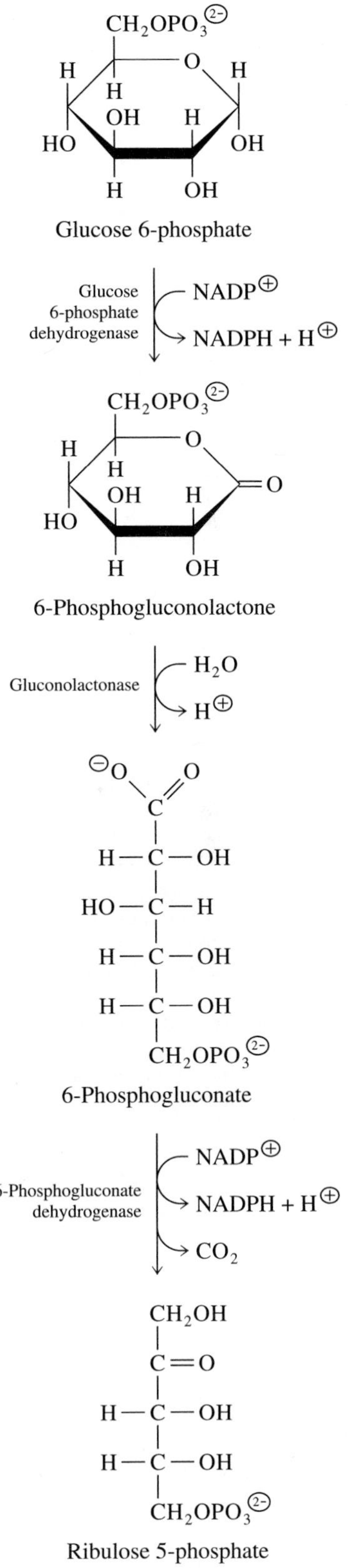

Figure 17·32
Conversion of ribulose 5-phosphate to xylulose 5-phosphate or ribose 5-phosphate. In either case, removal of a proton leads to formation of an enediol intermediate. Reprotonation forms either the ketose xylulose 5-phosphate or the aldose ribose 5-phosphate.

17·12 Transketolase and Transaldolase Catalyze Interconversions Among Sugar-Phosphate Metabolites

Transketolase and transaldolase, enzymes with broad substrate specificities, catalyze the exchange of two- and three-carbon fragments between sugar phosphates. For both enzymes, one substrate is an aldose and the other a ketose. The combined actions of these enzymes convert the five-carbon sugar phosphates to three- and six-carbon sugar phosphates.

Transketolase is a thiamine pyrophosphate–dependent enzyme that catalyzes transfer of a two-carbon keto group from a ketose phosphate to an aldose phosphate. An example of the action of transketolase is shown in Figure 17·33. In terms of carbon transfer, the ketose phosphate is shortened by two carbons, and the aldose phosphate is elongated by two carbons. The mechanism of transketolase, which illustrates the essential role of thiamine pyrophosphate (TPP), is shown in Figure 17·34. First, the carbanion of TPP attacks the carbonyl carbon of the substrate molecule. The resulting TPP adduct undergoes proton extraction by a basic group of the enzyme, and subsequent electron rearrangement leads to fragmentation, releasing the first product of the overall reaction, glyceraldehyde 3-phosphate. The remaining two-carbon fragment bound to TPP is resonance-stabilized; one resonance form is a carbanion. This carbanion attacks the carbonyl carbon of ribose 5-phosphate, forming another TPP adduct. A second fragmentation, similar to the one in the first part of the reaction, yields the original TPP carbanion and a second product, sedoheptulose 7-phosphate.

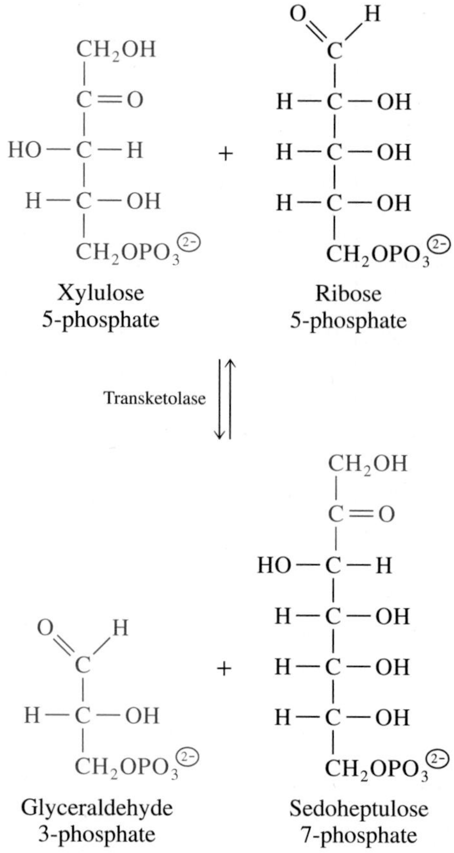

Figure 17·33
Reaction catalyzed by transketolase. Transketolase catalyzes the reversible transfer of a two-carbon fragment (C-1 and C-2) from xylulose 5-phosphate, shown in red, to ribose 5-phosphate, generating glyceraldehyde 3-phosphate and sedoheptulose 7-phosphate. Note that the number of carbon atoms balances and that the ketose-phosphate substrate (in either direction) is shortened by two carbon atoms while the aldose-phosphate substrate is lengthened by two carbon atoms. In this example, $5C + 5C \longrightarrow 3C + 7C$.

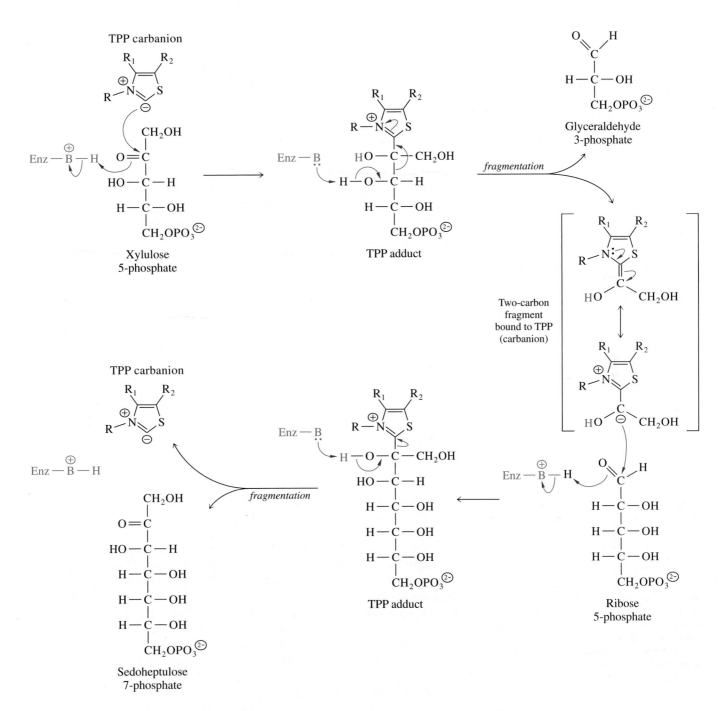

Figure 17·34
Mechanism of the reaction catalyzed by transketolase. Only the thiazolium ring of thiamine pyrophosphate (TPP) is shown.

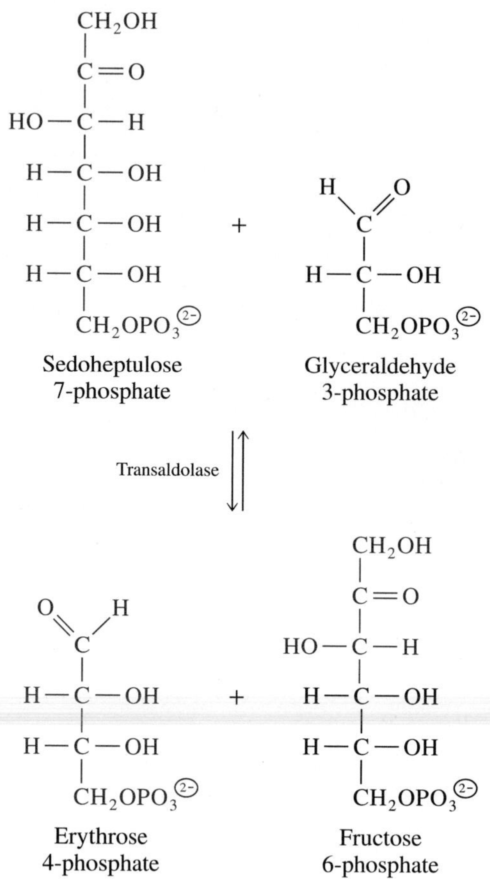

Figure 17·35
Reaction catalyzed by transaldolase. Transaldolase catalyzes the reversible transfer of a three-carbon unit (dihydroxyacetone) from sedoheptulose 7-phosphate, shown in red, to C-1 of glyceraldehyde 3-phosphate, generating a new ketose phosphate, fructose 6-phosphate, and releasing a new aldose phosphate, erythrose 4-phosphate. Note that the carbon atoms balance:
$7 C + 3 C \longrightarrow 6 C + 4 C.$

Transaldolase catalyzes the transfer of a three-carbon fragment from a ketose phosphate to an aldose phosphate. The transaldolase reaction of the pentose phosphate pathway converts sedoheptulose 7-phosphate and glyceraldehyde 3-phosphate to erythrose 4-phosphate and fructose 6-phosphate (Figure 17·35). The mechanism of the transaldolase reaction is shown in Figure 17·36. In the first step, a lysine residue of the enzyme condenses with the carbonyl group of sedoheptulose 7-phosphate to form a Schiff base. Deprotonation at C-4 by a base of the enzyme leads to cleavage of the bond between C-3 and C-4, releasing the first product, erythrose 4-phosphate. A fragment containing the first three carbons of sedoheptulose phosphate remains bound to the enzyme. The carbanion of this fragment, a resonance form, attacks the carbonyl of glyceraldehyde 3-phosphate, and a new six-carbon sugar is formed, still bound to the enzyme as a Schiff base. Finally, hydrolysis of the Schiff base releases the second product, fructose 6-phosphate.

A reexamination of the overall pentose phosphate pathway (Figure 17·30) shows that a six-carbon sugar (glucose 6-phosphate) is converted to a five-carbon sugar (ribulose 5-phosphate) and CO_2. Next, two isomerization reactions generate the substrates for the sequential actions of transketolase and transaldolase. These reactions, along with another transketolase reaction, generate one three-carbon molecule (glyceraldehyde 3-phosphate) and two six-carbon molecules (fructose 6-phosphate). Thus, the carbon-containing products from the passage of three molecules of glucose through the pentose phosphate pathway are glyceraldehyde 3-phosphate, fructose 6-phosphate, and CO_2. The balanced equation for this process is

$$3 \text{ Glucose 6-phosphate} + 6 \text{ NADP}^{\oplus} + 6 \text{ H}_2\text{O} \longrightarrow$$

$$2 \text{ Fructose 6-phosphate} + \text{Glyceraldehyde 3-phosphate} + 6 \text{ NADPH} + 3 \text{ CO}_2 + 9 \text{ H}^{\oplus}$$

(17·10)

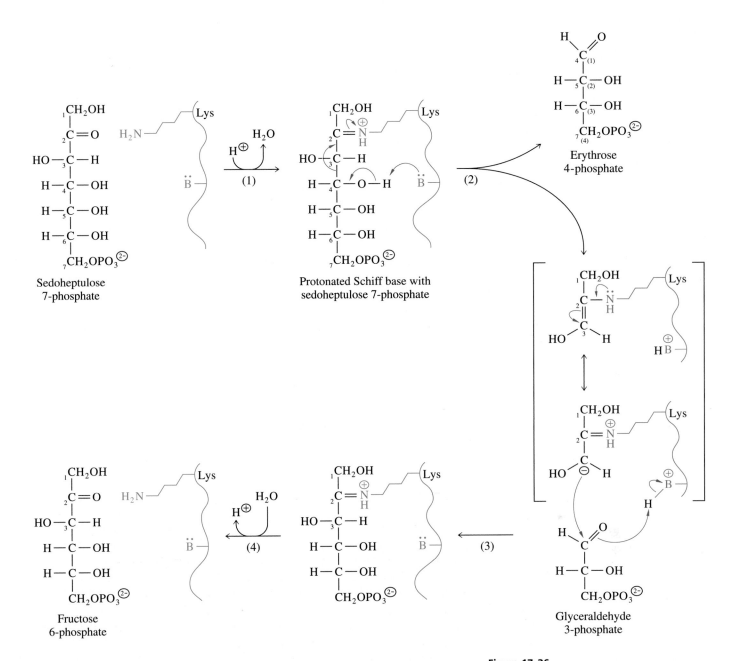

Figure 17·36
Mechanism of the reaction catalyzed by transaldolase, in which a three-carbon fragment (dihydroxyacetone) is transferred from sedoheptulose 7-phosphate to glyceraldehyde 3-phosphate.

In most cells, the primary cellular requirement is for NADPH rather than ribose 5-phosphate, and the formation of glyceraldehyde 3-phosphate and fructose 6-phosphate by the pentose phosphate pathway is followed by resynthesis of glucose 6-phosphate from these intermediates. In this case, the equivalent of one molecule of glucose is completely oxidized to CO_2 by six passages through the pathway. When six molecules of glucose 6-phosphate are oxidized, the six ribulose 5-phosphates produced can be rearranged by the reactions of the pentose phosphate pathway and part of the gluconeogenic pathway to form five glucose 6-phosphate molecules. (Recall that two glyceraldehyde 3-phosphate molecules are equivalent to one fructose 1,6-*bis*phosphate molecule.) Disregarding H_2O and $H^{\oplus}$, the overall stoichiometry for this process is

$$6 \text{ Glucose 6-phosphate} + 12 \text{ NADP}^{\oplus} \longrightarrow$$

$$5 \text{ Glucose 6-phosphate} + 12 \text{ NADPH} + 6 \text{ CO}_2 + \text{P}_i \qquad \textbf{(17·11)}$$

Indeed, an alternative name for the pathway is the *pentose phosphate cycle*. This formulation emphasizes that most of the glucose 6-phosphate that enters the pathway is recycled; one-sixth is converted to CO_2 and P_i, and there is net formation of NADPH.

We will encounter some of the reactions of the pentose phosphate pathway when we discuss the reductive pentose phosphate cycle of photosynthesis (Section 19·12). Ribulose 5-phosphate 3-epimerase, ribulose 5-phosphate isomerase, and transketolase all catalyze reactions in that pathway, but in the opposite direction from the pentose phosphate pathway.

Aside from photosynthesis (Chapter 19), this chapter completes the discussion of carbohydrate metabolism. We have seen how glucose, a major fuel, can be stored in polysaccharide form and mobilized as needed. Glucose can also be synthesized from noncarbohydrate precursors by the reactions of gluconeogenesis. We have seen that glucose can be oxidized by the pentose phosphate pathway to produce NADPH or transformed by the glycolytic pathway into pyruvate. Pyruvate and many noncarbohydrate metabolites feed into the citric acid cycle, which generates the reduced coenzymes NADH and QH_2. We will now see how the energy in the reducing power of these compounds can be recovered. The next chapter describes the major cellular role of molecular oxygen as the oxidant in the conservation of energy as ATP. Indeed, the process of oxidative phosphorylation is the source of most ATP in aerobic organisms.

Summary

Dietary saccharides can be converted to metabolites that can enter the glycolytic pathway. Fructose is converted to glyceraldehyde 3-phosphate, galactose is converted to glucose 1-phosphate, and mannose is converted to fructose 6-phosphate.

Glycogen is the glucose-storage polysaccharide of animals. Glycogen phosphorylase catalyzes degradation of intracellular glycogen to form glucose 1-phosphate, which can be converted to glucose 6-phosphate in a near-equilibrium reaction catalyzed by phosphoglucomutase. In mammalian liver, glucose 6-phosphate is hydrolyzed to glucose and P_i.

In liver and muscle, glycogen degradation and glycogen synthesis are reciprocally regulated pathways that are controlled by the hormones epinephrine, glucagon, and insulin. Both epinephrine and glucagon stimulate the production of cAMP, which activates protein kinase A. This leads to phosphorylation and inactivation of glycogen synthase. In addition, protein kinase A catalyzes phosphorylation of glycogen phosphorylase kinase, which in turn catalyzes phosphorylation of glycogen phosphorylase, converting glycogen phosphorylase to its active form. Epinephrine can also increase cytosolic calcium, which activates phosphorylase kinase, leading to activation of glycogen phosphorylase. The hormonally induced phosphorylations in both liver and muscle are reversed by a protein phosphatase that is itself regulated.

Gluconeogenesis is the pathway for glucose synthesis from noncarbohydrate precursors, such as lactate and amino acids. Many of the gluconeogenic reactions are simply the reverse of the near-equilibrium reactions of glycolysis. Enzymes specific to gluconeogenesis catalyze the metabolically irreversible conversion of pyruvate to phosphoenolpyruvate (pyruvate carboxylase and phosphoenolpyruvate carboxykinase), fructose 1,6-*bis*phosphate to fructose 6-phosphate (fructose 1,6-*bis*phosphatase), and glucose 6-phosphate to glucose (glucose 6-phosphatase). The principal regulatory site of gluconeogenesis is the conversion of pyruvate to phosphoenolpyruvate; a secondary site is the reaction catalyzed by fructose 1,6-*bis*phosphatase. Because pyruvate carboxylase is located in mitochondria, pyruvate must be transported into the mitochondria, and oxaloacetate must be transported to the cytosol via shuttle systems.

Sometimes, glucose can be converted to sorbitol or fructose via the sorbitol pathway. In the mammary gland, glucose reacts with UDP-galactose to form lactose. Lactose synthase consists of galactosyltransferase and α-lactalbumin.

The pentose phosphate pathway provides an alternate pathway for glucose 6-phosphate metabolism, initiated by the action of glucose 6-phosphate dehydrogenase. Glucose 6-phosphate dehydrogenase is also the major regulatory site of the pathway; the enzyme is allosterically inhibited by NADPH. The oxidative stage of the pentose phosphate pathway generates two molecules of NADPH per molecule of glucose 6-phosphate converted to ribulose 5-phosphate and CO_2. The nonoxidative stage of pentose phosphate metabolism includes isomerization of ribulose 5-phosphate to ribose 5-phosphate, a metabolite required for the biosynthesis of nucleotides and nucleic acids. Further metabolism of pentose phosphate molecules via the nonoxidative stage of the pentose phosphate pathway provides a mechanism for their conversion to triose-phosphate and hexose-phosphate intermediates of glycolysis and gluconeogenesis.

Selected Readings

General References

Schaub, J., van Hoof, F., and Vis, H. L. (1991). *Inborn Errors of Metabolism* (New York: Raven Press). Examines many genetic disorders, including the disorders of lactose and galactose metabolism discussed in this chapter.

Hers, H. G., and Hue, L. (1983). Gluconeogenesis and related aspects of glycolysis. *Annu. Rev. Biochem.* 52:617–653.

Pilkis, S. J., El-Maghrabi, M. R., and Claus, T. H. (1988). Hormonal regulation of hepatic gluconeogenesis and glycolysis. *Annu. Rev. Biochem.* 57:755–783.

Pilkis, S. J., and Granner, D. K. (1992). Molecular physiology of the regulation of hepatic gluconeogenesis and glycolysis. *Annu. Rev. Physiol.* 57:885–909.

Specific Topics

Jeffrey, J., and Jörnvall, H. (1988). Sorbitol dehydrogenase. *Adv. Enzymol. Mol. Biol.* 61:47–106.

Sukalski, K. A., and Nordlie, R. C. (1989). Glucose 6-phosphatase: two concepts of membrane-function relationship. *Adv. Enzymol. Mol. Biol.* 62:93–117.

Wood, T. (1985). *The Pentose Phosphate Pathway* (Orlando, Florida: Academic Press).

Wood, T. (1986). Physiological functions of the pentose phosphate pathway. *Cell Biochem. Func.* 4:241–247.

Glycogen Metabolism

Cohen, P. (1989). The structure and regulation of protein phosphatases. *Annu. Rev. Biochem.* 58:453–508.

Hubbard, M. J., and Cohen, P. (1993). On target with a new mechanism for the regulation of protein phosphorylation. *Trends Biochem. Sci.* 18:172–177.

Johnson, L. N., and Barford, D. (1990). Glycogen phosphorylase: the structural basis of the allosteric response and comparison with other allosteric proteins. *J. Biol. Chem.* 265:2409–2412.

Larner, J. (1990). Insulin and the stimulation of glycogen synthesis. The road from glycogen synthase to cyclic AMP-dependent protein kinase to insulin mediators. *Adv. Enzymol. Mol. Biol.* 63:173–231. A review of the control of glycogen metabolism by insulin.

Madsen, N. B. (1986). Glycogen phosphorylase. In *The Enzymes,* Vol. 17, 3rd ed., P. D. Boyer and E. G. Krebs, eds. (New York: Academic Press), pp. 365–394.

Smythe, C., and Cohen, P. (1991). The discovery of glycogenin and the priming mechanism for glycogen biosynthesis. *Eur. J. Biochem.* 200:625–631.

Sprang, S., Goldsmith, E., and Fletterick, R. (1987). Structure of the nucleotide activation switch in glycogen phosphorylase a. *Science* 237:1012–1019.

Taylor, S. S. (1989). cAMP-dependent protein kinase. *J. Biol. Chem.* 264:8443–8446.

18

Electron Transport and Oxidative Phosphorylation

Only a small fraction of the energy potentially available from glucose is extracted by glycolysis. We have seen that the citric acid cycle generates the reduced coenzymes NADH and QH_2. The oxidation of fatty acids and amino acids, examined in ensuing chapters, leads to production of these same energy-rich molecules. We will now examine the process by which the reduced coenzymes are oxidized and ATP is formed. The overall process is called **oxidative phosphorylation.** An analogous process, photosynthesis in plant chloroplasts, also involves electron transfer and conservation of energy (Chapter 19).

Oxidative phosphorylation requires several enzyme complexes embedded in a membrane. Some of the complexes generate a proton concentration gradient, and another complex uses the proton gradient for synthesis of ATP from ADP and P_i. In eukaryotes, oxidative phosphorylation occurs in mitochondria, with the enzymes embedded in the inner mitochondrial membrane. In bacteria, the enzymes are embedded in the plasma membrane. In chloroplasts, the thylakoid membrane contains the components involved in photosynthesis. We will examine in detail oxidative phosphorylation as it occurs in mammalian mitochondria, the most thoroughly studied system.

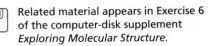

Related material appears in Exercise 6 of the computer-disk supplement *Exploring Molecular Structure.*

Figure 18·1
Overview of oxidative phosphorylation. A proton concentration gradient is produced from reactions catalyzed by the respiratory electron-transport chain. As electrons from reduced substrates flow through the complexes of the electron-transport chain, protons are translocated across the inner mitochondrial membrane from the matrix to the intermembrane space. The free energy stored in the proton concentration gradient is tapped as protons reenter the mitochondrial matrix; their reentry is coupled to the conversion of ADP and P_i to ATP.

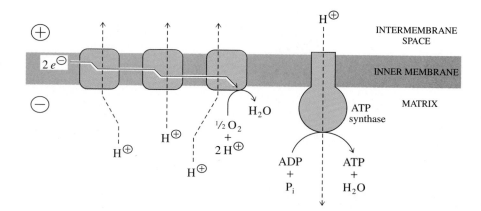

Oxidative phosphorylation in mitochondria consists of two tightly coupled phenomena, summarized in Figure 18·1.

1. Mitochondrial NADH and QH_2 are oxidized by the **respiratory electron-transport chain,** a series of membrane-embedded enzyme complexes that serve as electron carriers. Electrons are passed from the reduced coenzymes to the terminal electron acceptor of aerobic metabolism, molecular oxygen (O_2). As electrons move through the complexes, the energy from the oxidation of NADH and QH_2 is used to transport protons across the inner membrane from the **matrix** (the region enclosed by the inner membrane) to the intermembrane space, generating a proton concentration gradient, with the matrix more alkaline and negatively charged than the intermembrane space.

2. The proton concentration gradient serves as a reservoir of free energy. This energy is tapped when protons are channelled back across the inner membrane through another integral membrane enzyme complex, ATP synthase. This complex catalyzes the phosphorylation of ADP as protons flow down the concentration gradient:

$$ADP + P_i \longrightarrow ATP + H_2O \qquad \textbf{(18·1)}$$

After examining the structure of the mitochondrion, we will consider how the proton concentration gradient is generated and examine the nature of the energy stored in the gradient. We will then examine in detail the ensemble of protein complexes that carry out oxidative phosphorylation. We will conclude with an examination of transport processes closely linked to oxidative phosphorylation and a brief discussion of toxic oxygen species.

18·1 In Eukaryotes, Oxidative Phosphorylation Takes Place in Mitochondria

The final stages in the aerobic oxidation of biomolecules in eukaryotes occur in the mitochondrion. This organelle is the site of the citric acid cycle and fatty acid oxidation, both of which generate reduced coenzymes that are oxidized by the respiratory electron-transport chain. The structure of a typical mitochondrion is shown in Figure 18·2.

The number of mitochondria in cells varies dramatically. Some algae contain only one mitochondrion, whereas the protozoan *Chaos chaos* contains half a million. A mammalian liver cell contains up to 5000 mitochondria. The number of mitochondria is related to the overall energy requirements of the cell. White muscle tissue contains relatively few mitochondria, relying upon anaerobic glycolysis for its energy needs. The rapidly contracting but swiftly exhausted jaw muscles of the alligator are an extreme example of white muscle. Alligators can snap their jaws with astonishing speed and force but cannot continue this motion beyond a very few

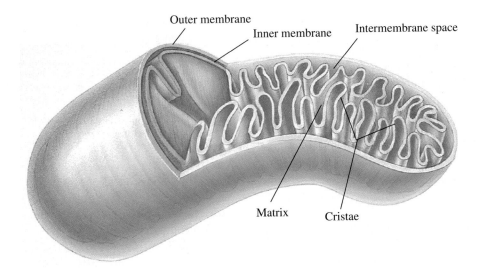

Outer membrane

Inner membrane

Intermembrane space

Matrix

Cristae

Figure 18·2
Structure of the mitochondrion. The outer mitochondrial membrane is freely permeable to small compounds. The inner membrane, which is virtually impermeable to polar and ionic substances, is highly folded into cristae. The protein complexes that catalyze the reactions involved in oxidative phosphorylation are located in the inner membrane.

repetitions (Figure 18·3a). Red muscle tissue has many mitochondria, exemplified by the flight muscles of migratory birds, which must sustain substantial and steady outputs of power (Figure 18·3b).

Mitochondria vary greatly in size and shape among different tissues and even within a cell. A typical mammalian mitochondrion has a diameter of 0.2 to 0.8 μm and a length of 0.5 to 1.5 μm, similar in size to an *Escherichia coli* cell.

Mitochondria are bounded by two membranes with markedly different properties. The outer mitochondrial membrane is relatively protein poor. Embedded in the outer membrane is the transmembrane protein porin, which forms channels that permit free diffusion of ions and water-soluble metabolites having molecular weights less than 10 000 across the membrane. The inner mitochondrial membrane is a barrier to protons. It is permeable to uncharged molecules such as water, molecular oxygen, and carbon dioxide but is virtually impermeable to larger polar and ionic substances. To cross the inner membrane, such substances must be transported by specific membrane-spanning transporters. Entry of anionic metabolites into the negatively charged interior of a mitochondrion is energetically unfavorable;

(a)

(b)

Figure 18·3
(a) The white muscle of the alligator jaw contains few mitochondria. Although fast and strong, the jaw muscles are rapidly exhausted. **(b)** The flight muscles of migratory birds, rich in mitochondria, are adapted for sustained output of work.

these metabolites are usually exchanged for other anions from the interior or are accompanied by protons flowing down the concentration gradient generated by oxidation (Section 12·7D).

The inner mitochondrial membrane is very rich in protein, with a protein-to-lipid ratio of about 4:1 by mass. The inner membrane is often highly folded, resulting in a greatly increased surface area. The folds are called cristae. The components that carry out the oxidative reactions of oxidative phosphorylation are embedded in the inner membrane. The ATP synthase complex is also embedded in the inner membrane, but some of its subunits extend into the matrix.

The area between the inner and outer mitochondrial membranes is called the intermembrane space. Since the outer membrane is freely permeable to small molecules, the intermembrane space has about the same composition of ions and metabolites as the cytosol. Thus, from the standpoint of mitochondrial functions discussed here, the intermembrane space is equivalent to the cytosol.

The contents of the matrix include the pyruvate dehydrogenase complex, the enzymes of the citric acid cycle (except for the succinate dehydrogenase complex, which is embedded in the inner membrane), and most of the enzymes that catalyze fatty acid oxidation. The protein concentration in the matrix is so high, approaching $500\,mg\,ml^{-1}$, that the matrix is a gel-like substance. The matrix also contains metabolites and inorganic ions and a pool of $NAD^{\oplus}$ and $NADP^{\oplus}$ that remains separate from the pyridine nucleotide coenzymes of the cytosol.

18·2 The Chemiosmotic Theory Explains How Electron Transport Is Coupled to Phosphorylation of ADP

The concept that a proton concentration gradient serves as the energy reservoir for driving the formation of ATP is termed the **chemiosmotic theory,** originally formulated by Peter Mitchell in the early 1960s (Figure 18·4). Since the mid-1920s, the mechanism by which cells carry out oxidative phosphorylation has been the subject of intensive research and much controversy. Many early attempts were made to identify an energy-rich phosphorylated metabolite that could participate in phosphoryl-group transfer to ADP, forming ATP. No such intermediate was detected. Instead, Mitchell proposed that energy conservation is linked to membrane transport in mitochondria. Mitchell's revolutionary idea that a proton concentration gradient was the much-sought-after energy source mobilized enormous experimental activity, with the result that the formation and dissipation of ion gradients is now acknowledged as a central motif in bioenergetics. Mitchell was awarded the Nobel prize in 1978 for his research in this area.

Before considering the individual reactions of the electron-transport chain, including electron transport and proton translocation, we will examine Mitchell's chemiosmotic theory and the nature of the energy stored in a proton concentration gradient. By the time Mitchell proposed the chemiosmotic theory, much information had accumulated on the oxidation of substrates and the cyclic oxidation and reduction of mitochondrial electron carriers. It was assumed that these oxidation reactions drove the formation of ATP, but the pathway linking oxidation to phosphorylation of ADP was not known.

In 1956, Britton Chance and Ronald Williams showed that intact isolated mitochondria suspended in phosphate buffer oxidize substrates and consume oxygen only when ADP is added to the suspension. In other words, the oxidation of a substrate is obligatorily *coupled* to the phosphorylation of ADP. Experiments showed not only that respiration proceeded rapidly until all of the ADP was phosphorylated (Figure 18·5a) but also that the amount of O_2 consumed depended on the amount of ADP added. The phenomenon of coupling was puzzling, and no theoretical model adequately explained how oxidation reactions were linked to ATP formation.

Figure 18·4
Peter Mitchell (1920–1992), whose research on transport phenomena led to the chemiosmotic theory.

(a)

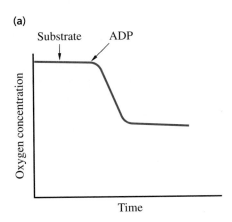

(b)

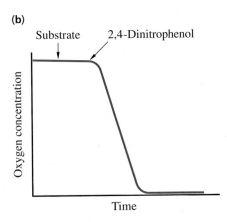

Figure 18·5
Respiration by mitochondria. **(a)** In the presence of excess P_i and substrate, intact mitochondria consume oxygen rapidly only when ADP is added. When all of the ADP is phosphorylated, rapid respiration ceases. **(b)** Addition of the uncoupler 2,4-dinitrophenol allows oxidation of the substrate to proceed in the absence of phosphorylation of ADP. The arrows indicate times when additions were made to the mitochondria.

The phenomenon of **uncoupling** provided an essential piece of the puzzle. It was known that synthetic compounds called uncouplers stimulate oxidation of substrates in the absence of ADP (Figure 18·5b). Respiration in the presence of an uncoupler proceeds until virtually all of the available oxygen is reduced, and rapid oxidation of substrates proceeds with little or no phosphorylation of ADP. Simply put, these compounds uncouple oxidation from phosphorylation. A large number of uncouplers have been discovered. They have little in common chemically except that all are lipid-soluble weak acids. Both their protonated and conjugate-base forms can cross the inner mitochondrial membrane; the anionic conjugate base retains lipid solubility because the negative charge is delocalized. The resonance structures of a typical uncoupler, 2,4-dinitrophenol, are shown in Figure 18·6.

The chemical properties of uncouplers provided a clue to their action and, when combined with a mass of other experimental data, suggested a mechanism for oxidative phosphorylation. In his chemiosmotic theory, Mitchell proposed the radical idea that a proton concentration gradient across the mitochondrial membrane was created by the action of the mitochondrial enzyme complexes and that this gradient provided the energy for phosphorylation of ADP. Mitchell offered several postulates whose confirmation would lend experimental support to his theory.

1. An intact inner mitochondrial membrane is an absolute requirement for coupling. The membrane must be impermeable to charged solutes; otherwise, the proton concentration gradient would collapse. Specific transporters allow ionic metabolites to cross the membrane.

2. Electron transport through the electron-transport chain generates a proton concentration gradient, with the cytosolic side of the inner mitochondrial membrane having the higher concentration of $H^{\oplus}$.

3. A membrane-bound enzyme, ATP synthase, catalyzes ADP phosphorylation in a reaction driven by transfer of protons across the inner mitochondrial membrane.

Figure 18·6
Conjugate acid and conjugate base forms of 2,4-dinitrophenol. The dinitrophenolate anion is resonance stabilized, and its negative ionic charge is broadly distributed over the ring structure of the molecule. Three of the structures contributing to the resonance hybrid are shown. Because the negative charge is delocalized, both the acid and base forms of dinitrophenol are sufficiently hydrophobic to dissolve in the membrane.

The postulates listed on the previous page explained the effect of the lipid-soluble uncoupling agents, suggesting that they bind protons in the cytosol, carry them through the inner membrane, and release them in the matrix, thereby dissipating the proton concentration gradient. The proton carriers uncouple oxidation from phosphorylation since protons enter the matrix without passing through the ATP synthase.

In 1965, Mitchell and Jennifer Moyle provided support for the second postulate by demonstrating that the medium is acidified when mitochondria under anaerobic conditions are given small amounts of oxygen. This type of experiment continues today in an effort to quantitate the proton concentration gradient generated by the electron-transport chain (i.e., how many protons are translocated for each pair of electrons transferred).

ATP synthase activity was first recognized in 1948 as ATPase activity in damaged mitochondria. It was assumed by most workers that the ATPase functions physiologically in reverse. Efraim Racker and his coworkers isolated and characterized this membrane-bound oligomeric enzyme in the 1960s. The reversibility of the ATPase reaction was demonstrated by observing the expulsion of protons upon addition of ATP to mitochondria or to liposomes in which the ATPase was incorporated.

18·3 The Energy Stored in a Proton Concentration Gradient Has Electrical and Chemical Components

The protons that are translocated into the intermembrane space by the respiratory complexes flow back into the matrix via the ATP synthase, forming a circuit that is similar to an electrical circuit. The energy of the proton concentration gradient, the **protonmotive force,** is analogous to the electromotive force of electrochemistry (Section 14·13A). This analogy is depicted in Figure 18·7. In a reaction such as the reduction of molecular oxygen by a reducing agent XH_2 in an electrochemical cell,

$$XH_2 + \tfrac{1}{2}O_2 \rightleftharpoons X + H_2O \tag{18·2}$$

electrons from the reducing agent XH_2 pass along a wire that connects the two electrodes where the oxidation and reduction half-reactions occur. Electrons flow from the anode, where XH_2 is oxidized,

$$XH_2 \rightleftharpoons X + 2H^{\oplus} + 2e^{\ominus} \tag{18·3}$$

to the cathode, where O_2 is reduced.

$$\tfrac{1}{2}O_2 + 2H^{\oplus} + 2e^{\ominus} \rightleftharpoons H_2O \tag{18·4}$$

Protons pass freely from one reaction cell to the other through the solvent in a salt bridge. Electrons move through an external wire due to a potential difference between the cells; the potential, measured in volts, is the electromotive force. The direction of electron flow and the extent of reduction of the oxidizing agent is determined by the difference in free energy between XH_2 and O_2, which in turn depends on their respective reduction potentials.

In mitochondria, protons—not electrons—flow through the external connection, an aqueous circuit connecting the electron-transport chain and ATP synthase that is analogous to the wire of the electrochemical reaction. The electrons still pass from the reducing agent XH_2 to the oxidizing agent O_2, in this case through the membrane-bound electron-transport chain. The free energy of these oxidation-reduction reactions is recovered in the phosphorylation of ADP.

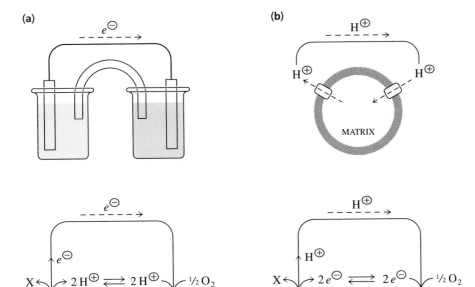

(a)

(b)

MATRIX

$X \leftarrow 2\,H^{\oplus} \rightleftharpoons 2\,H^{\oplus} \quad \tfrac{1}{2}\,O_2$

$XH_2 \qquad\qquad H_2O$

$X \leftarrow 2\,e^{\ominus} \rightleftharpoons 2\,e^{\ominus} \quad \tfrac{1}{2}\,O_2$

$XH_2 \qquad\qquad H_2O$

Figure 18·7
Analogy of electromotive and protonmotive force. **(a)** In an electrochemical cell, electrons pass from the reducing agent (XH_2) to the oxidizing agent (O_2) through a wire that connects the two electrodes. The measured electrical potential between cells is the electromotive force. **(b)** When the configuration is reversed (i.e., the external pathway for electrons is substituted by an aqueous pathway for protons), the potential is the protonmotive force. In the mitochondria, protons are translocated across the membrane when electrons are transported within the membrane by the electron-transport chain. The proton concentration gradient serves as a reservoir of electrical and chemical potential energy. The flow of $H^{\oplus}$ down this gradient is coupled to the formation of ATP.

The movement of protons generates both chemical and electrical potential energy. The chemical contribution to the free-energy change is described by the difference in proton concentration on either side of the membrane (compare Section 12·7D, Equation 12·2).

$$\Delta G_{chem} = nRT \ln \frac{[H^{\oplus}]_{in}}{[H^{\oplus}]_{out}} \qquad (18\text{·}5)$$

where n is the number of protons, R is the universal gas constant ($8.315\,\mathrm{J\,K^{-1}\,mol^{-1}}$), and T is the absolute temperature. Equation 18·5 can be rewritten as

$$\Delta G_{chem} = 2.303\,nRT(pH_{in} - pH_{out}) \qquad (18\text{·}6)$$

The electrical contribution to the free-energy change is described by the change in membrane potential, $\Delta\psi$.

$$\Delta G_{elec} = z\mathcal{F}\Delta\psi \qquad (18\text{·}7)$$

where z is the charge of the transported substance, and $\mathcal{F}$ is Faraday's constant ($96.48\,\mathrm{kJ\,V^{-1}\,mol^{-1}}$). Since the charge ($z$) is 1.0 for each proton translocated, $n = z$. The total free-energy change for proton transport is therefore

$$\Delta G = n\mathcal{F}\Delta\psi + 2.303\,nRT\Delta pH \qquad (18\text{·}8)$$

Dividing Equation 18·8 by $n\mathcal{F}$ yields an expression for the potential that develops between the two sides of the mitochondrial membrane. The term $\Delta G/n\mathcal{F}$ is the protonmotive force, Δp.

$$\Delta p = \frac{\Delta G}{n\mathcal{F}} = \Delta\psi + \frac{2.303\,RT\Delta pH}{\mathcal{F}} \qquad (18\text{·}9)$$

At 25°C, $2.303\,RT/\mathcal{F} = 0.059\,\mathrm{V}$ and

$$\Delta p = \Delta\psi + (0.059\,\mathrm{V})\Delta pH \qquad (18\text{·}10)$$

The proton concentration gradient is often referred to as a pH gradient. This is an oversimplification that omits the charge gradient, the other source of free energy contained in the proton concentration gradient. For example, in liver mitochondria,

$\Delta\psi = 0.17$ V and ΔpH $= 0.5$. Thus, the amount of free energy available from the charge gradient is actually greater than that from the chemical (pH) difference (85% versus 15%).

$$\Delta p = 0.17 \text{ V} + (0.059 \text{ V})(0.5) = 0.20 \text{ V} \qquad \text{(18-11)}$$

Respiring liver mitochondria maintain Δp at this value, keeping the proton concentrations sufficiently far from equilibrium to provide enough free energy to drive the phosphorylation of ADP.

Having examined the theory that explains energy storage in proton concentration gradients, we will now consider the reactions of the electron-transport chain.

18·4 Electron Transport and Oxidative Phosphorylation Depend on Complexes of Proteins

Five assemblies of proteins associated with oxidative phosphorylation are found in the inner membrane of mitochondria. They have been isolated in their active forms by careful solubilization using detergents. Each enzyme assembly catalyzes a separate portion of the energy-transduction process. The numbers I through V are assigned to these complexes. Four of them, I through IV, contain multiple cofactors and are involved in electron transfer. Complex V is the ATP synthase.

Some characteristics of the mammalian complexes I through IV are listed in Table 18·1. The complexes are not physically associated with one another but appear to diffuse independently in the inner mitochondrial membrane. The four enzyme complexes contain a wide variety of oxidation-reduction cofactors. Electron flow occurs via the reduction and oxidation of these cofactors, with flow proceeding from a reducing agent to an oxidizing agent.

Electrons flow through the components of the electron-transport chain roughly in the direction of increasing reduction potential. The reduction potentials of the components fall between that of the strong reducing agent NADH and that of the terminal oxidizing agent O_2. Ubiquinone (Q) and cytochrome c serve as links between complexes of the electron-transport chain. Q transfers electrons from Complexes I and II to Complex III. Cytochrome c links Complexes III and IV. Complex IV uses the electrons to catalyze the reduction of oxygen to water. The order of the

Table 18·1 Characteristics of protein complexes of the mitochondrial respiratory electron-transport chain in bovine heart

Complex	Subunits	Molecular weight	Oxidation-reduction components
I. NADH-ubiquinone oxidoreductase	≥ 41	$\sim 800\,000$	1 FMN 22–24 Fe-S in 5–7 clusters
II. Succinate-ubiquinone oxidoreductase	4	125 000	1 FAD 7–8 Fe-S in 3 clusters Cytochrome b_{560}
III. Ubiquinol-cytochrome c oxidoreductase	9–10	$\sim 250\,000$	1 Fe-S cluster Cytochrome b Cytochrome c_1
IV. Cytochrome c oxidase	13	400 000 (dimer)	Cytochrome a Cytochrome a_3 2 Copper ions

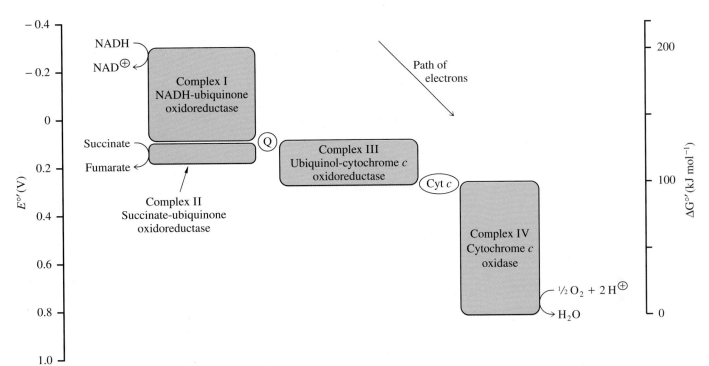

$$NADH \longrightarrow FMN \longrightarrow Fe\text{-}S \longrightarrow Q \longrightarrow \begin{bmatrix} Fe\text{-}S \\ Cyt\ b \end{bmatrix} \longrightarrow Cyt\ c_1 \longrightarrow Cyt\ c \longrightarrow Cyt\ a \longrightarrow Cyt\ a_3 \longrightarrow O_2$$
$$Succinate \longrightarrow FAD \longrightarrow Fe\text{-}S \longrightarrow$$

Figure 18·8
Mitochondrial electron transport. Each of the four complexes of the electron-transport chain, composed of several protein subunits and cofactors, undergoes cyclic reduction and oxidation. The complexes are linked by the mobile carrier ubiquinone (Q) and by cytochrome c. The height of each complex indicates the $\Delta E^{\circ\prime}$ between its reducing agent (substrate) and its oxidizing agent (product). Transfer of electrons from NADH and succinate to O_2 is accompanied by translocation of protons from the matrix to the intermembrane space, resulting in a proton concentration gradient across the membrane.

electron-transport reactions is depicted in Figure 18·8 against a scale of standard reduction potential on the left and a relative scale of standard free-energy change on the right. The standard reduction potential, in units of volts, is directly related to the standard free-energy change, in units of kJ mol⁻¹, by the formula

$$\Delta G^{\circ\prime} = -n\mathscr{F}\Delta E^{\circ\prime} \tag{18·12}$$

As you can see from Figure 18·8, a substantial amount of energy is released during the electron-transfer process. It is this energy, released in discrete steps during electron transfer, that is stored in the form of a proton concentration gradient.

The values shown are strictly true only under standard conditions. The relationship between actual reduction potentials (E) and standard ones ($E^{\circ\prime}$) is similar to the relationship between actual and standard free energy (Section 14·8).

$$E = E^{\circ\prime} + RT \ln\left[\frac{red}{ox}\right] \tag{18·13}$$

(The abbreviations *red* and *ox* represent the two oxidation states of the electron carrier.) The standard reduction potential is also known as the midpoint potential because, under standard conditions, the concentrations of reduced and oxidized carrier molecules are equal; thus, the ratio [red]:[ox] is 1, and the second term in Equation 18·13 is 0. In this case,

$$E = E^{\circ\prime} \tag{18·14}$$

In order for electron carriers to be efficiently reduced and reoxidized in a cyclic fashion, appreciable quantities of both the reduced and oxidized forms of the carriers must be present. This is the situation found in mitochondria. We may therefore assume that, for any given oxidation-reduction reaction in the electron-transport complexes, E and $E^{\circ\prime}$ are fairly similar. Our discussion hereafter refers only to $E^{\circ\prime}$ values.

Table 18·2 Standard reduction potentials of mitochondrial oxidation-reduction components

Substrate or complex	$E°'$ (V)
NADH	−0.32
Complex I	
FMN	−0.30
Fe-S clusters	−0.25 to −0.05
Succinate	+0.03
Complex II	
FAD	0.0
Fe-S clusters	−0.26 to 0.00
QH_2/Q	+0.04
$(Q·^{\ominus}/Q)$	−0.16)
$(QH_2/Q·^{\ominus}$	+0.28)
Complex III	
Fe-S cluster	+0.28
Cytochrome b_{560}	−0.10
Cytochrome b_{566}	+0.05
Cytochrome c_1	+0.22
Cytochrome c	+0.23
Complex IV	
Cytochrome a	+0.21
Cu_A	+0.24
Cytochrome a_3	+0.39
Cu_B	+0.34
O_2	+0.82

Table 18·3 Standard free energy released in the oxidation reaction catalyzed by each complex

Complex	$E°'_{reductant}$ (V)	$E°'_{oxidant}$ (V)	$\Delta E°'$ (V)	$\Delta G°'$ (kJ mol^{-1})
I (NADH/Q)	−0.32	+0.04	+0.36	−70
II (Succinate/Q)	+0.03	+0.04	+0.01	− 2
III (QH$_2$/Cytochrome c)	+0.04	+0.23	+0.19	−37
IV (Cytochrome c/O$_2$)	+0.23	+0.82	+0.59	−110

$\Delta E°'$ was calculated as the difference between $E°'_{reductant}$ and $E°'_{oxidant}$.

The standard free energy obtained by the oxidation of one mole of NADH or the electrons derived from NADH was calculated using Equation 18·12, where $n = 2$ electrons.

The standard reduction potentials of the substrates and cofactors of the electron-transport chain are listed in Table 18·2. Note that the values progress from negative to positive, so that—in general—each substrate or intermediate is oxidized by a cofactor or substrate that has a more positive $E°'$. In fact, one consideration in determining the sequence of the electron carriers was their reduction potentials.

The standard free energy available from the reactions catalyzed by each of the complexes is shown in Table 18·3. Note that only the reactions catalyzed by Complexes I, III, and IV provide substantial energy for the conversion of ADP + P$_i$ to ATP. It is these three complexes that translocate protons across the inner mitochondrial membrane as electrons pass through the complex. Complex II, which is also the succinate dehydrogenase complex that we examined as a component of the citric acid cycle, does not contribute to the formation of the proton concentration gradient. Complex II transfers electrons from succinate to Q and thus represents a tributary of the respiratory chain.

The sequence of electron carriers in the respiratory chain can be demonstrated by the use of inhibitors that bind to specific components of the electron-transport chain and block the transfer of electrons. The resulting changes in oxidation states of the electron carriers can be detected by changes in light absorption. Any compound that causes a block in respiration (as evidenced by a diminished O$_2$ consumption in isolated mitochondria) should cause electron carriers closer to O$_2$ than the blocked site to be oxidized and those closer to NADH to be reduced. For example, the plant toxin rotenone binds to a site on Complex I and blocks the transfer of electrons to Q. In the presence of rotenone, the electron carriers of Complex I are reduced and those of complexes III and IV are oxidized. This suggests the order

$$\text{NADH} \longrightarrow \text{I} \longrightarrow \text{III,IV} \longrightarrow \text{O}_2 \tag{18·15}$$

The state in which some carriers are reduced and others oxidized is called a *crossover*, and experiments of this type are called crossover analyses. Other specific inhibitors are the antibiotic antimycin A, which binds to Complex III, and cyanide, which binds to Complex IV.

18·5 Cofactors Have Special Roles in Electron Transport

As depicted in Figure 18·8 on the previous page, the electrons that flow through Complexes I–IV are actually transferred between coupled cofactors. Electrons enter the respiratory electron-transport chain two at a time from the reduced substrates NADH and succinate. In Complexes I and II, the flavin coenzymes FMN and FAD (Section 8·7) are reduced, respectively. The reduced coenzymes FMNH$_2$ and FADH$_2$ donate one electron at a time, and all of the subsequent steps in the electron-transport chain proceed by one-electron transfers. Iron-sulfur (Fe-S) clusters of both the [2 Fe–2 S] and [4 Fe–4 S] type (Section 8·1B) are present in Complexes I, II, and III. Each iron-sulfur cluster can accept or donate one electron as an iron atom undergoes oxidation and reduction between the ferric (Fe^{3+}) and ferrous (Fe^{2+}) states. The structures of the two main types of mitochondrial iron-sulfur clusters are shown in Figure 8·3. The iron in the mitochondrial clusters is usually complexed with inorganic sulfide (S^{2-}) and thiol groups of cysteine residues. Copper atoms and cytochromes also act as one-electron oxidation-reduction agents.

Cytochromes are heme-containing proteins that undergo reversible oxidation and reduction of their iron atoms, alternating between the ferric and ferrous oxidation states. The structures of two cytochromes were presented in Chapter 5 (Figures 5·32a and 5·43). All except cytochrome *c* are integral membrane proteins. Cytochromes were first discovered by C. A. MacMunn in 1886 and were rediscovered by David Keilin in 1925. It was Keilin who recognized that these hemoproteins are involved in respiration. He classified them on the basis of their visible absorption spectra into the categories *a*, *b*, and *c*. The heme prosthetic groups of the three types of cytochromes have slightly different structures (Figure 18·9). The heme of *b*-type

Figure 18·9
Heme groups of cytochromes *a*, *b*, and *c*, shown in their reduced (ferrous) states. All share a highly conjugated porphyrin ring system, but the substituents of the ring differ.

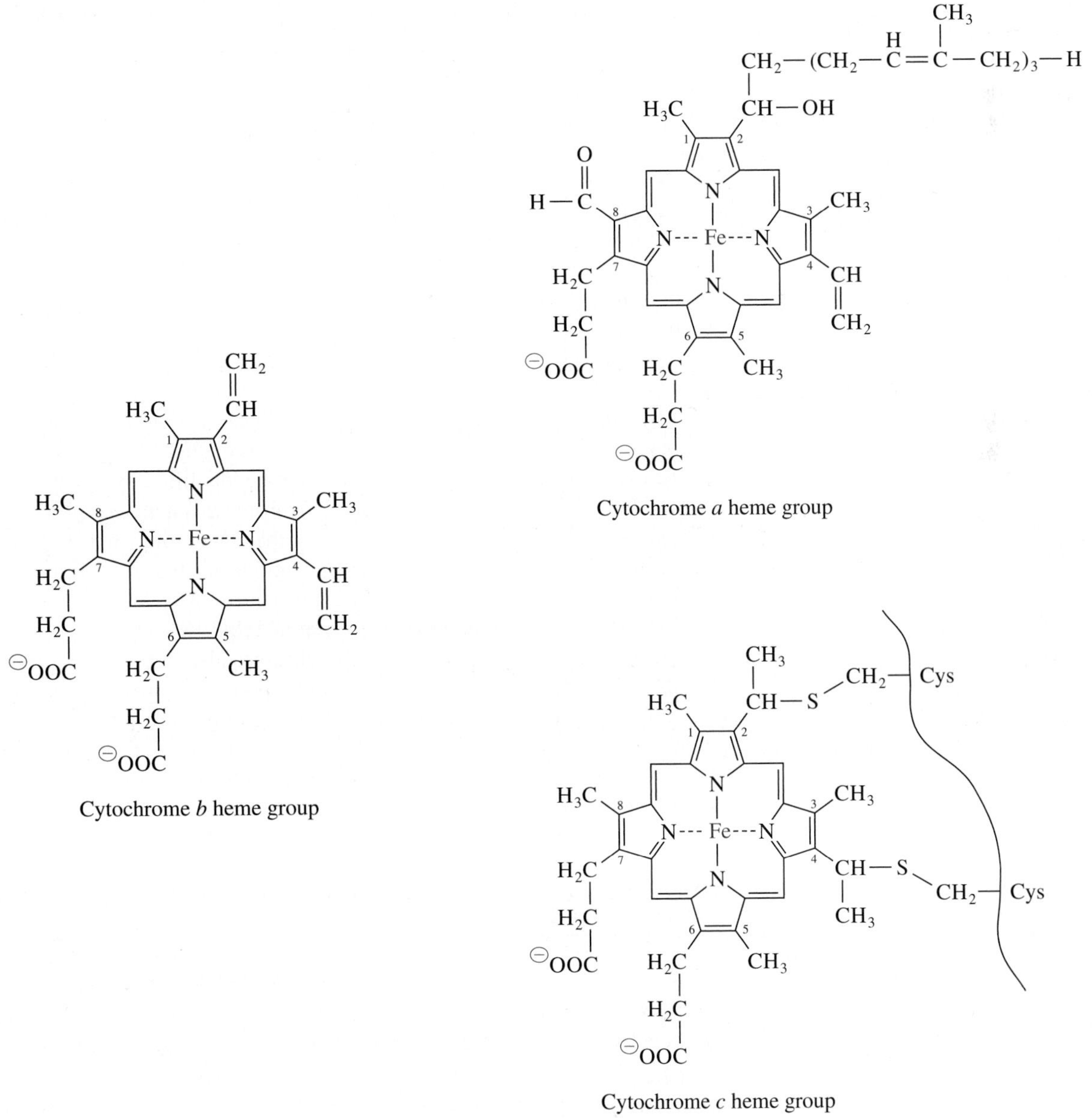

Cytochrome *a* heme group

Cytochrome *b* heme group

Cytochrome *c* heme group

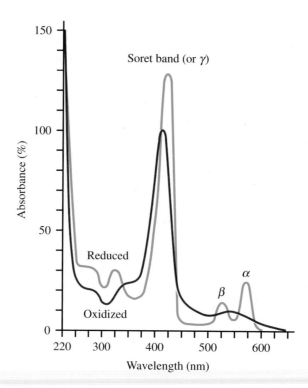

Figure 18·10
Absorption spectra of oxidized and reduced horse cytochrome c. The reduced cytochrome shows three absorbance peaks in the visible portion of the spectrum; these peaks are designated α, β, and γ, starting with the peak of longest wavelength. Upon oxidation, the γ band, also called the Soret band, decreases in intensity and shifts to a slightly shorter wavelength, whereas the α and β peaks disappear, leaving a single broad band of absorbance. Units are absorbances in percent, normalized to the Soret band of the oxidized form.

cytochromes is the same as that of hemoglobin and myoglobin. In the c-type cytochromes, the heme is covalently attached to the apoprotein by two thioether linkages, formed by addition of thiol groups of two cysteine residues to vinyl groups of the heme. In cytochrome a proteins, a 17-carbon hydrophobic chain replaces a vinyl group attached to the porphyrin ring, and a formyl group replaces a methyl group.

Cytochromes can be differentiated by their absorption spectra. Absorption spectra of reduced and oxidized cytochrome c are shown in Figure 18·10. Although the most strongly absorbing band is the Soret band near 400 nm, the band labelled α is used to characterize cytochromes. Within each category, individual cytochromes vary slightly in their spectra and can be identified by the absorption maximum of the α band. Thus, a particular cytochrome b is denoted cytochrome b_{560} to indicate the peak in the α absorption band of the cytochrome in its reduced form. In some cases, nondescriptive subscript designations, such as cytochromes c_1 and a_3, are used to distinguish cytochromes of a given class. Wavelengths of maximum absorption for reduced cytochromes are given in Table 18·4.

Most importantly, the individual cytochromes also differ in their reduction potentials. The difference arises because each apoprotein provides a different environment for its heme prosthetic group. Heme groups with different reduction potentials function as electron carriers at several points in the electron-transport chain. Similarly, the reduction potentials of iron-sulfur clusters vary widely depending on the chemical and physical environment provided by the apoproteins.

Table 18·4 Absorption maxima (in nm) of major spectral bands in the visible absorption spectra of the cytochromes

| Heme protein | Absorption band | | |
	α	β	γ
Cytochrome c	550–558	521–527	415–423
Cytochrome b	555–567	526–546	408–449
Cytochrome a	592–604	Absent	439–443

The electron-transport complexes of the inner mitochondrial membrane are functionally linked by the mobile electron carriers Q and cytochrome c. Ubiquinone (Section 8·16) is a lipid-soluble molecule that can accept and donate two electrons, one at a time. Q diffuses within the lipid bilayer, accepting electrons from Complexes I and II and passing them to Complex III. The other mobile electron carrier is cytochrome c, which is a peripheral membrane protein associated with the outer face of the membrane. Cytochrome c ferries electrons from Complex III to Complex IV. The components and oxidation-reduction reactions of each of the four electron-transporting mitochondrial complexes are examined in detail in the following sections.

18·6 Complex I Transfers Electrons from NADH to Ubiquinone

Complex I, NADH-ubiquinone oxidoreductase (also called NADH dehydrogenase), catalyzes the transfer of two electrons from NADH to Q. Complex I is an extremely complicated enzyme. For example, complex I from bovine mitochondria is composed of at least 41 subunits. Figure 18·11 illustrates the sequence of reactions within Complex I. NADH donates electrons to Complex I two at a time as a hydride ion ($H^\ominus$, an ion that is composed of two electrons and a proton). Q accepts electrons one at a time, passing through a semiquinone anion intermediate ($Q\cdot^\ominus$) before reaching its fully reduced state, ubiquinol (QH_2).

$$Q \xrightarrow{+\,e^\ominus} Q\cdot^\ominus \xrightarrow{+\,e^\ominus,\,+\,2\,H^\oplus} QH_2 \qquad \textbf{(18·16)}$$

In the first step of electron transfer through Complex I, NADH transfers a hydride ion to FMN, forming $FMNH_2$. $FMNH_2$ is oxidized in two steps via a semiquinone intermediate, donating the two electrons to the next oxidizing agent—an iron-sulfur cluster—one at a time.

$$FMN \xrightarrow{+\,H^\oplus,\,+\,H^\ominus} FMNH_2 \xrightarrow{-\,H^\oplus,\,-\,e^\ominus} FMNH\cdot \xrightarrow{-\,H^\oplus,\,-\,e^\ominus} FMN$$

$$\textbf{(18·17)}$$

Thus, FMN is a transducer that converts the flow of electrons from the two-electron transfers of the NAD-linked dehydrogenases to the one-electron steps that characterize the rest of the electron-transport chain. $FMNH_2$ transfers electrons to sequentially linked Fe-S clusters, obligatory one-electron acceptors, which pass the electrons to Q.

During the movement of electrons through Complex I, protons are translocated from the matrix to the intermembrane space. The most recent studies of the stoichiometry of proton translocation have estimated that four protons are moved across the membrane for each electron pair passed from NADH to Q. It is clear that protons originating in the hydride ion of NADH and in the matrix are transferred to FMN to form $FMNH_2$ and that these two protons or their equivalents are consumed in the reduction of Q to QH_2. However, it is not known how protons are transported across the membrane.

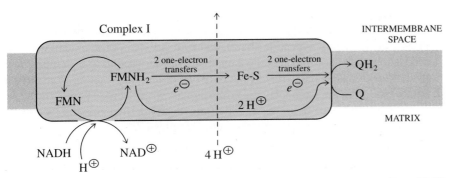

Figure 18·11
Electron transfer and proton flow in Complex I. Electrons are passed from NADH to Q via FMN and a series of Fe-S centers. In addition, two protons are required for the reduction of Q to QH_2. The stoichiometry of protons translocated is about four per pair of electrons transferred. The chemical mechanism for the transfer of protons across the membrane is not known.

Figure 18·12
Electron flow through Complex II. The role of cytochrome b_{560}, which copurifies with the complex, is not clear. Complex II does not contribute to the proton concentration gradient but serves as a tributary that introduces electrons into the electron-transport sequence at the level of QH_2.

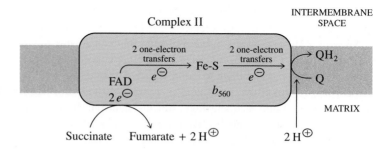

18·7 Complex II Transfers Electrons from Succinate to Ubiquinone

Complex II, succinate-ubiquinone oxidoreductase (also called the succinate dehydrogenase complex), accepts electrons from succinate and, like Complex I, catalyzes the reduction of Q to QH_2. Mammalian complex II is composed of four subunits. The two largest subunits constitute succinate dehydrogenase, which contains the FAD prosthetic group and three iron-sulfur clusters. The other two subunits appear to be required in vivo for binding of succinate dehydrogenase to the membrane and for passage of electrons to the physiological acceptor, Q. We previously considered the succinate dehydrogenase–catalyzed oxidation of succinate to fumarate as one of the reactions of the citric acid cycle (Section 16·3, Part 6).

The sequence of reactions for transfer of two electrons from succinate to ubiquinone is the reduction of FAD by a hydride ion and the transfer of two single electrons from the reduced flavin to the series of three iron-sulfur clusters (Figure 18·12). A cytochrome b_{560} copurifies with the complex, bound to one of the smaller subunits, but its role in electron transfer is not known. Because little free energy is released in the reaction sequence catalyzed by Complex II, the complex cannot contribute to the proton concentration gradient across the inner mitochondrial membrane. Instead, it introduces electrons from the oxidation of succinate into the electron-transport sequence at the level of QH_2. Q can accept electrons from Complex I or II and donate them to Complex III and thence to the rest of the electron-transport chain. Reactions in several other pathways also donate electrons to Q. We saw one of them, the reaction catalyzed by glycerol 3-phosphate dehydrogenase, in the previous chapter (Section 17·7).

18·8 Complex III Transfers Electrons from QH_2 to Cytochrome c

Complex III, ubiquinol-cytochrome c oxidoreductase, contains 9 or 10 distinct subunits, including a [2 Fe-2 S] protein, cytochrome b (which includes two hemes, b_{560} and b_{566}, on a single polypeptide chain), and cytochrome c_1. The oxidation of one molecule of QH_2 is accompanied by the transfer of four protons to the outer side of the inner membrane (Figure 18·13), but only two of these protons are from the matrix; the other two are from QH_2. Electrons are accepted from the complex by the one-electron carrier cytochrome c, which moves along the cytosolic face of the inner membrane and transfers an electron to Complex IV.

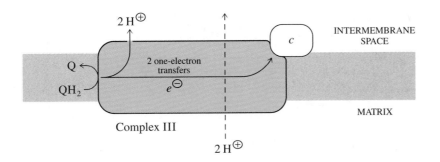

Figure 18·13
Electron transfer and proton flow in Complex III. Overall, electrons are passed from QH_2 to cytochrome c. Four protons are translocated across the membrane: two from the matrix and two from QH_2.

The sequence of electron transfers in Complex III is complicated. The oxidation-reduction sequence and its associated proton movements proved elusive until a cyclic pathway termed the **Q cycle** was suggested in 1975 by Peter Mitchell, who also proposed the chemiosmotic theory. Later investigators, including Bernard Trumpower, have contributed to our present understanding of the Q cycle.

When the electron carriers of Complex III were first identified, the linear sequence

$$Q \longrightarrow b \longrightarrow c_1 \longrightarrow c \qquad \textbf{(18·18)}$$

was proposed, based upon the finding that antimycin A caused a crossover between cytochrome b and the c cytochromes. This sequence was placed in doubt, however, by the finding that mitochondria under near-anaerobic conditions exhibited a transiently *reduced* cytochrome b when pulsed with oxygen, an effect magnified in the presence of antimycin. Based on the linear sequence, one would expect that when oxygen was suddenly available, there would be sequential increases in the oxidation of cytochrome c, cytochrome c_1, cytochrome b, and ubiquinol. The finding that cytochrome b was instead reduced could not be rationalized by the linear sequence.

A recent formulation for the Q cycle, which not only explains this paradox but also describes how electrons are moved between the reduced molecule QH_2 (from either Complex I or II) and cytochrome c, is diagrammed in Figure 18·14, next page. Note that three forms of Q are involved: QH_2, Q, and the semiquinone anion, $Q \cdot \ominus$. The first two forms are soluble in the lipid bilayer and move between the cytosolic and matrix faces of the membrane; $Q \cdot \ominus$, however, exists at two separate binding sites near the aqueous phase at each side of the membrane.

The Q cycle occurs in two steps. First, QH_2 diffuses to the cytosolic face of the membrane, where it is oxidized by transferring an electron to the Fe-S protein and thence to cytochrome c_1 (Figure 18·14a). When QH_2 is oxidized to ubisemiquinone $(Q \cdot \ominus)$, two protons are released to the cytosol. The semiquinone anion $Q \cdot \ominus$ reduces the b_{566} heme group, which transfers the electron to the b_{560} group. Reduced b_{560} then reduces Q on the matrix side to $Q \cdot \ominus$. (Recall that Q is able to diffuse through the membrane.)

In the second step of the Q cycle (Figure 18·14b), a second molecule of QH_2 repeats the process just described, with a second electron donated ultimately to cytochrome c_1, an additional $2 H^\oplus$ released to the cytosol, and another electron arriving at b_{560}, which can be used, along with $2 H^\oplus$ from the matrix, to convert the previously formed matrix-side $Q \cdot \ominus$ to QH_2. The sum of the two cyclic steps (Figure 18·14c) is oxidation of two molecules of QH_2, formation of one molecule of QH_2, reduction of two molecules of cytochrome c, transfer of four protons across the inner membrane of the mitochondrion, and consumption of two protons from the matrix. Because the net charge difference created between the matrix and the cytosol is only 2, the protonmotive force generated by Complex III is less than that from Complex I or Complex IV.

Figure 18·14
Proposed Q cycle. **(a)** QH_2 at the cytosolic face of the membrane donates an electron to the Fe-S protein of Complex III and $2\,H^{\oplus}$ to the cytosol, forming $Q\cdot^{\ominus}$. The Fe-S protein then reduces cytochrome c_1, and the electron continues down the respiratory chain. $Q\cdot^{\ominus}$ donates an electron to the b_{566} heme group, with formation of Q, which diffuses to the matrix face of the membrane. Reduced b_{566} donates an electron via b_{560} to Q at the matrix face of the membrane to form $Q\cdot^{\ominus}$. **(b)** Completion of the cycle requires a second molecule of QH_2 to repeat the first part of the cycle, producing another reduced cytochrome c_1, another $2\,H^{\oplus}$, another Q, and reduced b_{560}, which this time reduces the $Q\cdot^{\ominus}$ formed in the first step of the cycle to QH_2, consuming $2\,H^{\oplus}$ from the matrix. The result is the oxidation of two QH_2 molecules and the formation of one QH_2 molecule, as shown in the overall pathway **(c)**. Electrons from Complex III are donated to the mobile carrier cytochrome c. The site of inhibition by antimycin is indicated in red. (Adapted from Trumpower, B. L. (1990). The protonmotive Q cycle: energy transduction by coupling of proton translocation to electron transfer by the cytochrome bc_1 complex. *J. Biol. Chem.* 265:11 409–11 412.)

(a) First step of Q cycle

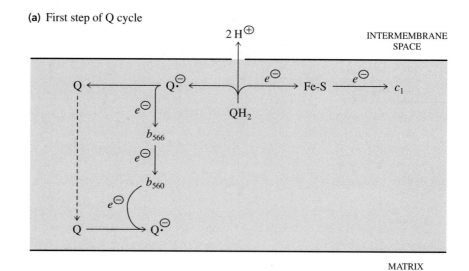

(b) Second step of Q cycle

(c) Complete Q cycle

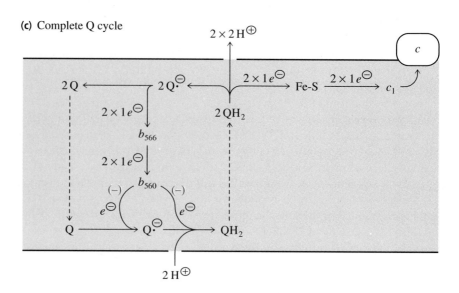

The cyclic flows in Complex III are possible because the uncharged co-enzymes Q and QH_2 can diffuse freely between the two faces of the mitochondrial membrane and because cytochrome b is a transmembrane protein containing two heme groups that together seem to span the membrane to form a path for electrons. We can now understand the paradoxical observations. Providing a pulse of oxygen to oxygen-deprived mitochondria oxidizes the c cytochromes and the Fe-S protein of Complex III, but as QH_2 becomes oxidized, it is converted to $Q\cdot^{\ominus}$ and donates electrons to cytochrome b, at least transiently. Thus, the pulse of oxygen must lead to a *reduction* of b_{566} and b_{560}. It is now known that the exact site of inhibition by antimycin A is between b_{560} and $Q\cdot^{\ominus}$ or QH_2 (Figure 18·14c). Thus, in the presence of antimycin, electrons can go no further than the hemes of the b cytochromes.

18·9 Complex IV Transfers Electrons from Cytochrome c to O_2

Complex IV, cytochrome c oxidase, is the last component of the respiratory electron-transport chain. This complex catalyzes the four-electron reduction of molecular oxygen (O_2) to water ($2\,H_2O$) and translocates protons into the intermembrane space.

The membrane-bound form of the mammalian enzyme is a dimer that contains 13 types of polypeptides with a total molecular weight of about $400\,000$. Among the polypeptide chains of Complex IV are cytochromes a and a_3, whose heme prosthetic groups have identical structures but different standard reduction potentials resulting from different environments within the oligomeric complex. The other oxidation-reduction cofactors of Complex IV are two copper ions, termed Cu_A and Cu_B, that alternate between the Cu^{2+} and $Cu^{\oplus}$ states as they participate in electron transfer.

Figure 18·15 shows that cytochrome c oxidase contributes to the proton concentration gradient in two ways. The first is a mechanism that translocates two $H^{\oplus}$ for each pair of electrons transferred (i.e., for each atom of O_2 reduced). The nature of the proton translocation is not known. The second mechanism by which cytochrome c oxidase contributes to the proton concentration gradient is by consuming matrix $H^{\oplus}$ when oxygen is reduced to water. While not involving the actual translocation of $H^{\oplus}$ across the membrane, this mechanism nevertheless contributes to the formation of Δp. The effect is the same as a net transfer of four $H^{\oplus}$ for each pair of electrons.

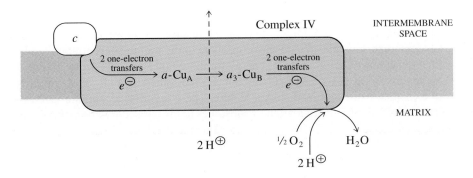

Figure 18·15
Flow of electrons and protons in Complex IV. The iron atoms of the heme groups in the a cytochromes and the copper atoms are both oxidized and reduced as electrons flow. Complex IV contributes to the proton concentration gradient in two ways: proton translocation from the matrix to the intermembrane space occurs in close association with electron transfer between the a cytochromes, and the formation of water subtracts protons from the matrix.

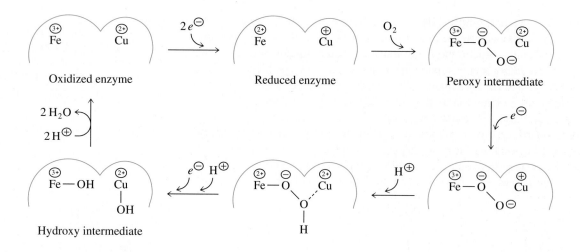

Figure 18·16
Proposed mechanism for reduction of dioxygen by cytochrome c oxidase. O_2 binds between the iron atom of cytochrome a_3 and the Cu_B atom. Two electrons are required to reduce the oxidized (Fe^{3+}-Cu^{2+}) enzyme so that oxygen can bind and form the peroxy intermediate. Subsequent addition of two electrons and two protons leads to the hydroxy intermediate. Protonation of the hydroxy species releases H_2O. [Adapted from Babcock, G. T., and Wikström, M. (1992). Oxygen activation and the conservation of energy in cell respiration. *Nature* 356:301–309.]

The reduction of a molecule of dioxygen requires four electrons (i.e., the equivalent of two molecules of NADH) and four protons.

$$O_2 + 4\,e^{\ominus} + 4\,H^{\oplus} \longrightarrow 2\,H_2O \qquad (18\cdot19)$$

Reduction of O_2 occurs at a catalytic center that includes the heme iron atom of cytochrome a_3 and the neighboring copper atom, Cu_B. It is this so-called binuclear center that binds toxic ligands such as cyanide and carbon monoxide. The other heme and copper cofactors (cytochrome a and Cu_A) transfer electrons one at a time from cytochrome c molecules on the cytosolic side of the complex to the oxygen-reducing site. Partially reduced oxygen species that could cause oxygen toxicity occur as reaction intermediates, but because the electron- and proton-transfer reactions are extremely rapid, the intermediates are generally not released from the enzyme.

A proposed mechanism for the reduction of dioxygen is presented in Figure 18·16. Only the reduced form of the catalytic site (the $Fe_{a3}{}^{2+}$-$Cu_B{}^{\oplus}$ binuclear center) can bind oxygen. The bound oxygen molecule is reduced to a peroxy intermediate by two electrons, one from the iron and one from the copper. A third electron reduces the cupric copper atom. The oxygen nearer the copper atom is then protonated, and the electrons are redistributed between the iron and copper atoms. Upon addition of another proton and a fourth electron, the O—O bond is cleaved, producing a hydroxy intermediate. This intermediate accepts two protons, releasing two molecules of water and oxidizing the enzyme.

18·10 Complex V Couples the Reentry of Protons into the Matrix with the Formation of ATP

Distinct from the other complexes that can be resolved from the inner mitochondrial membrane, Complex V does not contribute to the proton concentration gradient but rather consumes it, using its energy for the synthesis of ATP from ADP and P_i. Complex V is an F-type ATPase (Section 12·7E), called the F_0F_1 ATP synthase, with a characteristic knob-and-stalk structure. The F signifies a coupling factor—ATP synthase couples the oxidation of substrates to the phosphorylation of ADP in the mitochondrion. The F_1 component contains the catalytic subunits; when isolated in solubilized form from membrane preparations, it catalyzes the hydrolysis

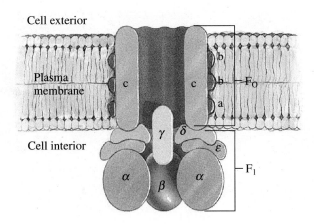

Cell exterior

Plasma
membrane

Cell interior

Figure 18·17
Schematic cross-section of the knob-and-stalk structure of F_OF_1 ATP synthase in *E. coli*. The F_1 component in *E. coli* is found on the inside of the plasma membrane; in mitochondria, F_1 is found on the inside of the inner mitochondrial membrane. The F_O component spans the membrane, forming a proton channel. F_1 is believed to consist of three α and three β subunits associated to form a knoblike hexameric structure, and one γ, one δ, and one ε subunit, which form the stalk that connects F_1 to the F_O channel, or basepiece. The stoichiometry of F_O in *E. coli* is thought to be $a_1b_2c_{10-12}$. Mitochondrial F_O has a more complex structure and subunit composition and is not as well characterized.

of ATP. For this reason, it has traditionally been referred to as F_1 ATPase. The F_O component is a proton channel that spans the membrane. The passage of protons through the channel into the matrix is coupled to the formation of ATP. The stoichiometry of proton entry per ATP synthesized is estimated to be $3\,H^\oplus$ per ATP. The F_O component is named for its sensitivity to oligomycin, an antibiotic that binds in the channel, preventing the entry of protons and thereby inhibiting ATP synthesis.

Figure 18·17 shows the arrangement of the F_1 and F_O components with respect to the bacterial plasma membrane and depicts the apparent subunit composition of the best-studied Complex V, the *E. coli* ATP synthase. The molecular interactions among the subunits within each component have not been fully elucidated. However, the subunit composition of the bacterial F_1 component appears to be $\alpha_3\beta_3\gamma\delta\varepsilon$, and that of the F_O component is thought to be $a_1b_2c_{10-12}$. The c subunits of F_O interact to form the channel for passage of $H^\oplus$ through the membrane.

The F_1 component of the mitochondrial ATP synthase is similar to its bacterial counterpart. Preliminary structures derived from X-ray crystallographic studies of both rat liver and bovine heart F_1 ATPases show that a symmetric $\alpha_3\beta_3$ oligomer is connected to the transmembrane F_O channel by a multisubunit stalk. The orientation of F_1 components in mitochondria—the knobs on the inside of the membrane—parallels the orientation in bacteria. Recall that mitochondria are evolutionary descendants of bacteria (Section 2·4D).

The mechanism of synthesis of ATP from ADP and P_i catalyzed by Complex V has been the target of intensive research for over two decades. In 1979, Paul Boyer proposed the **binding-change mechanism** based on observations that suggested energy input was required not for formation of the phosphoanhydride linkage but for release of ATP from the catalytic site. Evidence continues to accumulate in favor of this mechanism. In the active site, the ΔG for the reaction $ADP + P_i \rightleftharpoons ATP + H_2O$ is near 1.0; thus, ATP formation at the active site is readily accomplished. However, the ATP produced is in a thermodynamic pit (Section 7·11); that is, ATP has such high binding energy that release is prevented. A conformational change driven by proton influx weakens binding of ATP sufficiently to allow the product to be released from the enzyme. The dissociation constant for ATP changes from about 10^{-12} M to 10^{-5} M when the conformation of one of the three $\alpha\beta$ subunits in the ATP synthase changes. Information is not yet available, though, on the nature of the proton-driven conformational change.

The current binding-change proposal suggests that the $\alpha_3\beta_3$ oligomer of ATP synthase contains three catalytic sites. At any given time, each site is in a different

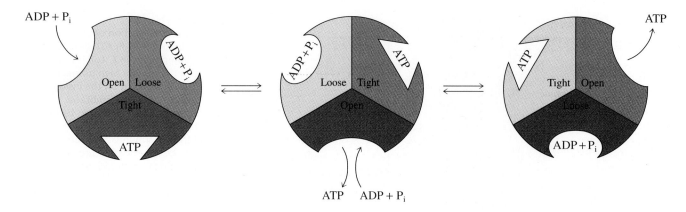

Figure 18·18
Proposed steps in ATP formation and release. The three different binding-site conformations for ATP synthase are shown as different shapes. ADP and P_i bind to an open site. A proton-driven conformational change causes them to bind to the enzyme more tightly, in the loose conformation. Meanwhile, the site that had bound ATP tightly has become an open site, and a loose site containing other molecules of ADP and P_i has become a tight site. ATP is released from the open site, and ATP is synthesized in the tight site. [Adapted from Boyer, P. D. (1989). A perspective of the binding change mechanism for ATP synthase. *FASEB J.* 3:2164–2178.]

conformation: open, loose, or tight. All three of the catalytic sites pass sequentially through these conformations. Formation and release of ATP is believed to involve the following steps, summarized in Figure 18·18:

1. One molecule each of ADP and P_i binds to an open site.

2. Inward passage of protons across the inner mitochondrial membrane causes each of the three catalytic sites to change conformation. The open conformation (containing the newly bound ADP and P_i) becomes a loose site. The loose site, already filled with ADP and P_i, becomes a tight site. The ATP-bearing tight site becomes an open site.

3. ATP is released from the open site, and ADP and P_i condense to form ATP in the tight site.

18·11 ATP, ADP, and P_i Are Actively Transported Across the Mitochondrial Inner Membrane

ATP is synthesized in the matrix of mitochondria, but since most of it is consumed in the cytosol, it must be exported. A transporter is necessary to allow ADP to enter and ATP to leave mitochondria because the inner mitochondrial membrane is impermeable to ATP and ADP. The ADP/ATP carrier, called the adenine nucleotide translocase, carries out the exchange of mitochondrial ATP and cytosolic ADP (Figure 18·19). Although the adenine nucleotides are commonly complexed with Mg^{2+}, this is not the case when they are transported across the membrane. Exchange of ADP^{3-} and ATP^{4-} causes the loss of a net charge of -1 in the matrix and is therefore electrogenic (Section 12·7A). This type of exchange draws on the electrical part of Δp, that is, $\Delta\psi$. Thus, some of the free energy conserved in the proton concentration gradient is expended to drive this transport process.

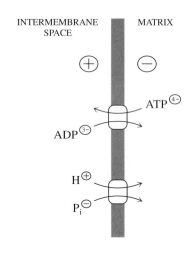

Figure 18·19
Transport of ATP, ADP, and P_i across the inner mitochondrial membrane. The exchange of ATP for ADP, carried out by the adenine nucleotide translocase, is electrogenic and unidirectional; the symport of P_i and $H^{\oplus}$ is electroneutral.

The formation of ATP from ADP and P_i in the mitochondrial matrix also requires a phosphate transporter that is responsible for the uptake of P_i from the cytosol. Phosphate ($H_2PO_4^{\ominus}$) is transported into mitochondria in electroneutral symport with $H^{\oplus}$ (Figure 18·19). This carrier does not involve the electrical component of the proton motive force, $\Delta\psi$, but does draw upon the chemical difference, ΔpH. Thus, the transporters necessary for ATP formation both consume Δp. The combined energy cost of transport of ATP out of the matrix and ADP and P_i into the matrix is approximately equivalent to the influx of one proton. Therefore, the synthesis of one molecule of ATP by oxidative phosphorylation requires the influx of four protons from the intermembrane space (one for transport and three that pass through the F_O component of ATP synthase).

18·12 The Energy-Conserving Complexes Do Not Contribute Equally to ATP Yield

Before the chemiosmotic theory was proposed, when many researchers were searching for an energy-rich intermediate capable of forming ATP by direct phosphoryl-group transfer, it was assumed that Complexes I, III, and IV each contributed to ATP formation with one-to-one stoichiometry. We now know that energy transduction occurs by generation and consumption of a proton concentration gradient. The yield of ATP need not be equivalent for each proton-translocating electron-transport complex, nor must the yield of ATP per molecule of substrate oxidized be an integral number. Of course, half molecules of ATP are not synthesized. Many electron-transport chains in a mitochondrion simultaneously contribute to the proton concentration gradient, which is a common energy reservoir that is drawn upon by many ATP synthase complexes. Recall that the formation of one molecule of ATP from ADP and P_i catalyzed by ATP synthase requires inward passage of about three protons; one additional proton is needed for transport of P_i, ADP, and ATP. The ratio of protons translocated to the intermembrane space per pair of electrons transferred by the coupled electron-transport complexes is 4:1 for Complex I, 2:1 for Complex III, and 4:1 for Complex IV. These measurements can be used to calculate the ratio of molecules of ADP phosphorylated to atoms of oxygen reduced, called the **P:O ratio.** When mitochondrial NADH is the substrate for the respiratory electron-transport chain, 10 protons are exported to the cytosol, and the P:O ratio is $10 \div 4$, or 2.5. The P:O ratio for succinate is $6 \div 4$, or 1.5. These calculated values are close to the P:O ratios that have been observed in experiments similar to that shown in Figure 18·5a, which measure the amount of O_2 reduced when a given amount of ADP is phosphorylated. These P:O values should be considered maximum values—the efficiency of energy conservation is not 100% for several reasons, including the slow leakage of protons back into the mitochondria and the energy used by the adenine nucleotide translocase and other proton-gradient–dependent transport processes (Section 12·7).

   The P:O ratios given above are the most recent estimates and differ from the values traditionally used, which were 3 for NADH and 2 for succinate. The traditional values were derived from early experimental work. However, even after the acceptance of the chemiosmotic theory, the values of 3 and 2 continued to be used in some cases, possibly because of the appeal of whole numbers compared to fractional values. The traditional P:O ratios still appear occasionally in publications.

18·13 Shuttle Mechanisms Permit Aerobic Oxidation of Cytosolic NADH

The reducing equivalents produced by glycolysis in the cytosol must enter the mitochondria in order to participate in electron transport and oxidative phosphorylation. In fact, glycolysis cannot continue unless cytosolic NADH is converted back to $NAD^{\oplus}$, since the glyceraldehyde 3-phosphate dehydrogenase reaction requires a continuous supply of $NAD^{\oplus}$. Because neither NADH nor $NAD^{\oplus}$ can diffuse across the inner mitochondrial membrane, reducing equivalents enter the mitochondrion by shuttle mechanisms. These shuttles are cyclic pathways by which a reduced coenzyme in the cytosol passes its reducing power to a metabolite that is transported into the mitochondrion; the reducing power is passed back to another molecule of the coenzyme in the matrix; and the reoxidized metabolite is returned to the cytosol, where the cycle can be reinitiated.

   The glycerol phosphate shuttle is prominent in insect flight muscles, which sustain very high rates of oxidative phosphorylation. It is also present to a lesser extent in most mammalian cells. Two enzymes are involved: an $NAD^{\oplus}$-dependent cytosolic glycerol 3-phosphate dehydrogenase and a membrane-embedded glycerol 3-phosphate dehydrogenase complex that contains an FAD prosthetic group and has a substrate-binding site on the outer face of the inner mitochondrial membrane. In the cytosol, NADH reduces dihydroxyacetone phosphate in a reaction catalyzed by cytosolic glycerol 3-phosphate dehydrogenase.

$$
\text{NADH} + \text{H}^{\oplus} + \underset{\substack{\text{Dihydroxyacetone} \\ \text{phosphate}}}{\text{O}=\text{C}\begin{smallmatrix}\text{CH}_2\text{OH}\\|\\|\\\text{CH}_2\text{OPO}_3^{\tiny\text{2}\ominus}\end{smallmatrix}} \xrightleftharpoons[\substack{}]{\substack{\text{Glycerol} \\ \text{3-phosphate} \\ \text{dehydrogenase}}} \underset{\text{Glycerol 3-phosphate}}{\text{HO}-\text{C}-\text{H}\begin{smallmatrix}\text{CH}_2\text{OH}\\|\\|\\\text{CH}_2\text{OPO}_3^{\tiny\text{2}\ominus}\end{smallmatrix}} + \text{NAD}^{\oplus}
$$

(18·20)

Glycerol 3-phosphate is then converted back to dihydroxyacetone phosphate by the membrane-embedded glycerol 3-phosphate dehydrogenase complex. In the process, two electrons are transferred to the FAD prosthetic group of the membrane-bound enzyme.

$$\text{Enz-FAD} + \text{Glycerol 3-phosphate} \rightleftharpoons \text{Enz-FADH}_2 + \text{Dihydroxyacetone phosphate}$$

(18·21)

FADH_2 transfers two electrons to the mobile electron carrier Q, which then transports the electrons to ubiquinol-cytochrome c oxidoreductase (Complex III). Overall, the glycerol phosphate shuttle oxidizes cytosolic NADH and generates QH_2 in the inner mitochondrial membrane (Figure 18·20). Oxidation of cytosolic NADH equivalents by this pathway produces less energy (1.5 ATP per molecule of cytosolic NADH) than oxidation of intramitochondrial NADH because the reducing equivalents introduced by the shuttle bypass the mitochondrial NADH-ubiquinone oxidoreductase (Complex I).

The malate-aspartate shuttle is the more active shuttle in liver and some other mammalian tissues. It involves malate dehydrogenase and aspartate transaminase in both the cytosol and the mitochondrial matrix, in addition to transporters in the inner mitochondrial membrane. The operation of the shuttle is diagrammed in Figure 18·21. NADH in the cytosol reduces oxaloacetate to malate in a reaction catalyzed by cytosolic malate dehydrogenase. Malate enters the mitochondrial matrix via the dicarboxylate translocase in electroneutral exchange with α-ketoglutarate. In the matrix, mitochondrial malate dehydrogenase catalyzes the reoxidation of malate to oxaloacetate, with the reduction of mitochondrial $\text{NAD}^{\oplus}$ to NADH. NADH is then oxidized by Complex I of the respiratory electron-transport chain.

Figure 18·20
Glycerol phosphate shuttle. Cytosolic NADH reduces dihydroxyacetone phosphate to glycerol 3-phosphate in a reaction catalyzed by cytosolic glycerol 3-phosphate dehydrogenase. The reverse reaction is catalyzed by the mitochondrial glycerol 3-phosphate dehydrogenase complex, an integral membrane flavoprotein that transfers electrons to ubiquinone.

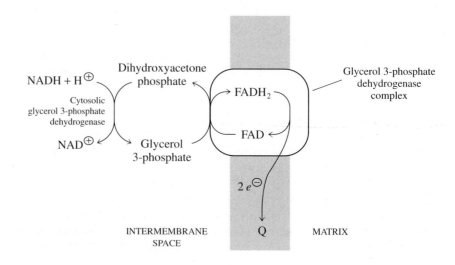

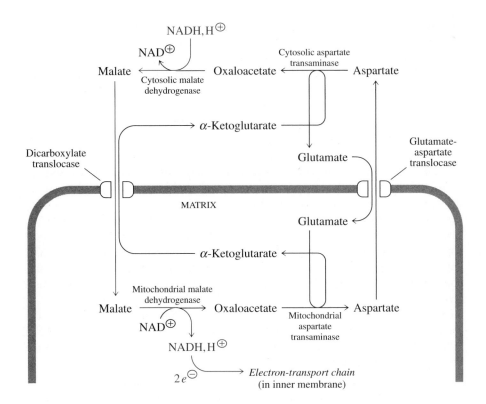

Figure 18·21
Malate-aspartate shuttle. NADH in the cytosol reduces oxaloacetate to malate, which is transported into the mitochondrial matrix. Reoxidation of malate generates NADH that can pass electrons to the electron-transport chain. Completion of the shuttle cycle requires the activities of mitochondrial and cytosolic aspartate transaminase.

Continued operation of the shuttle requires the return of oxaloacetate to the cytosol. This does not proceed directly, since oxaloacetate itself is not transported across the inner mitochondrial membrane. Instead, oxaloacetate reacts with glutamate in a reaction catalyzed by mitochondrial aspartate transaminase, with the production of α-ketoglutarate and aspartate. α-Ketoglutarate exits the mitochondrion via the dicarboxylate translocase in exchange for malate entry. Aspartate exits through the glutamate-aspartate translocase in exchange for entry of glutamate. Once in the cytosol, aspartate and α-ketoglutarate are the substrates for cytosolic aspartate transaminase, which catalyzes the formation of glutamate and oxaloacetate. Glutamate reenters the mitochondrion in antiport with aspartate, and oxaloacetate reacts with another molecule of cytosolic NADH, repeating the cycle. Because reduction of oxaloacetate in the cytosol consumes a proton that is released in the matrix by the oxidation of malate, indirect oxidation of cytosolic NADH via the electron-transport chain contributes fewer protons to the proton concentration gradient than direct oxidation of mitochondrial NADH (9 versus 10). The ATP yield for oxidation of two molecules of cytoplasmic NADH by this shuttle is therefore about 4.5 rather than 5.0.

18·14 Oxidative Phosphorylation Is Controlled by Cellular Energy Demand

Control of oxidative phosphorylation does not occur by simple feedback inhibition or by any allosteric mechanism. The overall control of the rate of oxidative phosphorylation is coupled to substrate availability and to the cellular energy demand. The rate of respiration depends on the concentrations of intramitochondrial NADH, O_2, P_i, and ADP. As discussed in Section 18·2, intact mitochondria suspended in phosphate buffer respire only in the presence of ADP. As ATP is consumed in cytosolic reactions, ADP is made available for translocation through the adenine nucleotide translocase, and electron transport—which is strictly coupled to the availability of ADP—accelerates. Substrate oxidation accelerates only when an increase in the concentration of ADP signals that the ATP pool needs to be replenished.

In certain situations, substrate oxidation is not tied to the generation of ATP. The respiratory burst of phagocytes, in which NADPH is oxidized, is one example (Section 18·15). Another example is the uncoupling that occurs in certain mammals for the purpose of generating heat. This physiological uncoupling occurs in brown adipose tissue, whose brown color is due to its many mitochondria. Brown adipose tissue is found in abundance in newborn mammals and in species that hibernate. In the hairless newborn or upon arousal from hibernation, release of epinephrine triggers the activation of a lipase, leading to release of fatty acids that are then oxidized. The reduced coenzymes that are generated are substrates for the electron-transport chain. However, the free energy of the reduced compounds is not conserved as ATP but is lost as heat because oxidation is uncoupled from phosphorylation. The uncoupling is due to an increase in the activity of thermogenin, a protein that forms a channel for reentry of protons into the mitochondria. Thermogenin is inactive as a channel until the binding of fatty acids induces a conformational change in the proton-transporting structure. When thermogenin is active, fatty acids are oxidized in the mitochondria, and the free energy produced is dissipated as heat, raising the body temperature of the animal.

18·15 Some Oxygen Metabolites Are Toxic

In this chapter, we have seen how the reduction of molecular oxygen is linked to the generation of cellular energy. Certain white blood cells called phagocytes are also capable of reducing molecular oxygen for the purpose of producing microbicidal oxidizing agents. In this section, we will examine this activity and then discuss mechanisms whereby other cells protect themselves from toxic oxygen species.

Phagocytes remove foreign materials such as bacteria from tissues by engulfing and then destroying them using proteolytic enzymes as well as oxygen metabolites. When stimulated, phagocytes undergo a burst of respiratory activity, increasing their oxygen intake up to 50-fold. At the same time, glucose is rapidly metabolized by the pentose phosphate pathway (Section 17·11) to produce NADPH. A plasma membrane–embedded NADPH oxidase catalyzes the one-electron reduction of oxygen by transferring electrons from NADPH to FAD and then apparently to a heterodimeric cytochrome b, which participates in oxygen reduction. The product of the reaction is the free-radical superoxide anion $\cdot O_2^{\ominus}$.

$$\text{NADPH} + 2\,O_2 \longrightarrow \text{NADP}^{\oplus} + 2\,\cdot O_2^{\ominus} + H^{\oplus} \qquad \text{(18·22)}$$

The extracellular $\cdot O_2^{\ominus}$ may spontaneously dismutate to hydrogen peroxide, H_2O_2; these two oxygen species are the starting materials for more toxic compounds, such as the hydroxyl radical, $\cdot OH$.

Small amounts of toxic oxygen species such as $\cdot O_2^{\ominus}$, H_2O_2, and $\cdot OH$ are also generated as by-products of oxidative metabolism in other types of cells. The concentration of superoxide in the liver is about $10^{-11}\,M$, and that of H_2O_2 is about 1000-fold higher. Oxygen radicals and peroxides are capable of damaging lipids, proteins, and nucleic acids. The principal targets of the free radicals are the unsaturated bonds in the lipids of membranes. Peroxidation of these fatty acid residues produces a decrease in membrane fluidity and can lead to lysis of cells. The thiol groups of cysteine side chains of proteins can also be oxidized, thereby cross-linking and inactivating proteins. Oxidative damage to DNA may induce mutations. In mammals, injury to the heart after a heart attack occurs when there is increased formation of free radicals at the site of the tissue damage. It has been suggested that molecular damage by reactive oxygen compounds may also be a factor in a wide variety of diseases, including arthritis, emphysema, and some cancers, and even in the aging process.

A variety of enzymes helps protect cells against oxidative damage. One of these is superoxide dismutase (Section 7·8B), which is found in prokaryotes as well as eukaryotes. Superoxide dismutase catalyzes the dismutation of two superoxide anions to hydrogen peroxide.

$$2 \cdot O_2^{\ominus} + 2 H^{\oplus} \longrightarrow H_2O_2 + O_2 \qquad \text{(18·23)}$$

Hydrogen peroxide can then be converted to H_2O and O_2 by the action of catalase.

$$2 H_2O_2 \longrightarrow 2 H_2O + O_2 \qquad \text{(18·24)}$$

The importance of superoxide dismutase is illustrated by one type of hereditary amyotrophic lateral sclerosis (a neurodegenerative disease, also called Lou Gehrig's disease), which appears to be caused by decreased levels of superoxide dismutase activity.

In red blood cells, where the concentration of oxygen is relatively high, the potential for oxidative damage is great. Superoxide, rather than O_2, is occasionally released from oxyhemoglobin and then converted to H_2O_2. In erythrocytes, glutathione peroxidase is more efficient than catalase in removing H_2O_2. This enzyme is present in other tissues and in some invertebrates but not in bacteria or plants.

Glutathione (GSH), the tripeptide γ-glutamylcysteinylglycine, is the reduced cofactor for oxygen detoxification in erythrocytes.

$$
\underset{\text{Glutathione (GSH)}}{
\overset{\oplus}{H_3}N - \underset{\underset{COO^{\ominus}}{|}}{CH} - CH_2 - CH_2 - \overset{\overset{O}{\|}}{C} - NH - \underset{\underset{\underset{SH}{|}}{\underset{CH_2}{|}}}{CH} - \overset{\overset{O}{\|}}{C} - NH - CH_2 - COO^{\ominus}
} \qquad \text{(18·25)}
$$

In the presence of glutathione peroxidase and H_2O_2, the sulfhydryl groups of two molecules of glutathione react to form a molecule of glutathione disulfide (GSSG).

Glutathione disulfide (GSSG) (18·26)

NADPH is required for the restoration of reduced glutathione by the reaction catalyzed by glutathione reductase.

$$GSSG + NADPH + H^{\oplus} \longrightarrow 2\,GSH + NADP^{\oplus} \qquad \text{(18·27)}$$

To meet the demand for reduced $NADP^{\oplus}$ in the red blood cell, up to 10% of the total consumption of glucose may be mediated by the pentose phosphate pathway. Glutathione plays a role in a variety of cellular activities in addition to oxygen detoxification. For instance, glutathione is involved in leukotriene biosynthesis and the transport of some amino acids into the cell. In addition, reduced glutathione can help restore the conformation of proteins whose free sulfhydryl groups have spontaneously oxidized to form mixed disulfides.

In humans, a mutation of the gene encoding glucose 6-phosphate dehydrogenase (the first enzyme of the pentose phosphate pathway) can lead to hemolytic anemia, the destruction of red blood cells. This is a common condition, affecting about 100 million people. Clinical problems arise only when these individuals ingest compounds (such as the antimalarial drug primaquine) that lead to formation of high levels of peroxides. The demand for NADPH then exceeds the capacity of the mutant cells to metabolize glucose via the pentose phosphate pathway.

It is fascinating that the reducing power of NADPH is used in different cells for opposite purposes: protection against the toxicity of oxygen metabolites and formation of these metabolites as a defensive device.

In addition to enzymes such as superoxide dismutase, catalase, and glutathione peroxidase, essential dietary nutrients protect against oxygen toxicity. These dietary compounds include the antioxidants vitamin E and ascorbic acid. Vitamin E (Figure 8·53) scavenges free radicals by donating a hydrogen atom to a free radical and itself becoming a less reactive vitamin E–O· radical. Ascorbic acid (Figure 9·23) reduces both the radical form of vitamin E and ROO· peroxyl radicals. Roles in the prevention of oxygen toxicity help explain the dietary requirements for a number of micronutrients. For example, selenium is incorporated into the amino acid selenocysteine; there is an essential selenocysteine residue in the active site of glutathione peroxidase. Zinc, copper, and manganese are essential components of isozymes of superoxide dismutase.

Summary

The energy in reduced coenzymes is recovered in the mitochondria by the process of oxidative phosphorylation. This process consists of two tightly coupled phenomena: (1) oxidation of substrates by the respiratory electron-transport chain, accompanied by the translocation of protons across the inner mitochondrial membrane to generate a proton concentration gradient; and (2) formation of ATP driven by the energy of protons flowing into the matrix through a channel in ATP synthase. The chemiosmotic theory explains how ADP phosphorylation is coupled to electron transport. It also explains the effects of uncoupling agents that allow substrate oxidation without ADP phosphorylation. The postulates of the theory include an intact mitochondrial membrane, a proton concentration gradient, and a membrane-bound ATPase that operates in reverse.

As electrons move through the electron-transport complexes in the mitochondrial membrane, proton translocation results in a potential difference across the membrane. The protonmotive force (Δp) depends on the difference in concentration of the protons (ΔpH) and the difference in charge ($\Delta\psi$) on either side of the membrane.

The five assemblies of proteins and cofactors involved in oxidative phosphorylation include the electron-transferring Complexes I–IV and ATP synthase. Electron flow through the complexes generally proceeds according to the relative reduction potentials of the various components. Electrons flow from NADH through Complexes I, III, and IV. Electrons from succinate are introduced via Complex II.

Several cofactors participate in electron transfer, including FMN, FAD, iron-sulfur clusters, copper atoms, cytochromes, and ubiquinone. The cytochromes are differentiated by their absorption spectra and reduction potentials. The mobile carrier ubiquinone links Complexes I and II with Complex III, and cytochrome c links Complex III with Complex IV.

For every two electrons transferred from NADH to Q by Complex I, four protons are translocated to the intermembrane space. Electron transfer by Complex II does not contribute to the proton concentration gradient. The transfer of electrons from QH_2 to cytochrome c by Complex III results in the net translocation of two protons from the matrix. The sequence of electron transfers in Complex III is described by the two-step Q cycle. Complex IV transfers electrons from cytochrome c to O_2, the ultimate oxidizing agent of the electron-transport chain. Electron transfer and reduction of O_2 to H_2O contributes four protons to the gradient.

Protons reenter the mitochondrial matrix by flowing through the multimeric F_O portion of the F_OF_1 ATP synthase. The mechanism by which proton flow drives the synthesis of ATP from ADP + P_i catalyzed by F_1 ATPase is not understood. The binding-change mechanism has been proposed to explain how the flow of three protons is linked to conformational changes in the synthase.

Transport of ADP and P_i into and ATP out of the mitochondrial matrix consumes the equivalent of one proton. The ATP yield per pair of electrons transferred by Complexes I–IV can be calculated from the number of protons translocated. Oxidation of mitochondrial NADH generates 2.5 ATP; oxidation of succinate generates 1.5 ATP. These values are lower than those traditionally used. Cytosolic NADH can contribute to oxidative phosphorylation when the reducing power is transferred to the mitochondria by the action of the glycerol phosphate shuttle or the malate-aspartate shuttle.

In certain cells, oxidation of reduced substrates is uncoupled from phosphorylation to generate heat. Phagocytes use the reducing power of NADPH to generate superoxide for fighting bacterial infections. Cells use reduced molecules such as glutathione, vitamin E, and ascorbic acid to protect against oxidative damage by reactive oxygen species.

Selected Readings

General References

Cramer, W. A., and Knaff, D. B. (1991). *Energy Transduction in Biological Membranes: A Textbook of Bioenergetics* (New York: Springer-Verlag). Text for an advanced course in bioenergetics emphasizing the experimental approach.

Harold, F. M. (1986). *The Vital Force: A Study of Bioenergetics* (New York: W. H. Freeman and Company). Engaging monograph with several examples of different biological solutions to the problem of energy transduction.

Mitchell, P. (1979). Keilin's respiratory chain concept and its chemiosmotic consequences. *Science* 206:1148–1159. Mitchell's Nobel prize lecture, relating the concepts and controversial history of the chemiosmotic theory.

Nicholls, D. G., and Ferguson, S. J. (1992). *Bioenergetics 2* (London: Academic Press). Explains the fundamentals of chemiosmosis.

Racker, E. (1980). From Pasteur to Mitchell: a hundred years of bioenergetics. *Fed. Proc.* 39:210–215.

Rees, D. C., and Farrelly, D. (1990). Biological electron transfer. In *The Enzymes*, Vol. 19, 3rd ed, David S. Sigman and P. D. Boyer, eds. (New York: Academic Press). pp. 38–97. Describes the fundamental thermodynamic and kinetic properties of electron-transfer reactions.

Electron-Transport Complexes

Anraku, Y., and Gennis, R. B. (1987). The aerobic respiratory chain of *Escherichia coli. Trends Biochem. Sci.* 12:262–266. Describes enzyme complexes that catalyze oxidation of QH_2 that are very different from the mammalian complexes.

Babcock, G. T., and Wikström, M. (1992). Oxygen activation and the conservation of energy in cell respiration. *Nature* 356:301–309.

Capaldi, R. A. (1990). Structure and function of cytochrome *c* oxidase. *Annu. Rev. Biochem.* 59:569–596.

Chan, S. I., and Li, P. M. (1990). Cytochrome *c* oxidase: understanding nature's design of a proton pump. *Biochemistry* 29:1–12.

Hatefi, Y. (1985). The mitochondrial electron transport and oxidative phosphorylation system. *Annu. Rev. Biochem.* 54:1015–1069.

Malmström, B. G. (1990). Cytochrome oxidase: some unsolved problems and controversial issues. *Arch. Biochem. Biophys.* 280:233–241.

Trumpower, B. L. (1990). The protonmotive Q cycle: energy transduction by coupling of proton translocation to electron transfer by the cytochrome bc_1 complex. *J. Biol. Chem.* 265:11409–11412.

Walker, J. E. (1992). The NADH:ubiquinone oxidoreductase (complex I) of respiratory chains. *Q. Rev. Biophys.* 25:253–324.

Weiss, H., Friederich, T., Hofhaus, G., and Preis, D. (1991). The respiratory-chain NADH dehydrogenase (complex I) of mitochondria. *Eur. J. Biochem.* 197:563–576.

Wikström, M. (1989). Identification of the electron transfers in cytochrome oxidase that are coupled to proton-pumping. *Nature* 338:776–778.

ATP Synthase and P:O Ratios

Boyer, P. D. (1989). A perspective of the binding change mechanism for ATP synthesis. *FASEB J.* 3:2164–2178.

Ferguson, S. J. (1986). The ups and downs of P/O ratios (and the question of non-integral coupling stoichiometries for oxidative phosphorylation and related processes). *Trends Biochem. Sci.* 11:351–353.

Hinkle, P. C., Kumar, M. A., Resetar, A., and Harris, D. L. (1991). Mechanistic stoichiometry of mitochondrial oxidative phosphorylation. *Biochemistry* 30:3576–3582. Gives most recent estimates of the stoichiometry of oxidative phosphorylation.

Pedersen, P. L., and Amzel, L. M. (1993). ATP synthases: structure, reaction center, mechanism, and regulation of one of nature's most unique machines. *J. Biol. Chem.* 268:9937–9940. Summarizes the structure, mechanism, and regulation of F_OF_1 ATPase and describes future research directions.

Penefsky, H. S., and Cross, R. L. (1991). Structure and mechanism of F_OF_1-type ATP synthases and ATPases. *Adv. Enzymol. Mol. Biol.* 64:173–214.

Oxygen Toxicity

Babior, B. M. (1992). The respiratory burst oxidase. *Adv. Enzymol. Mol. Biol.* 65:49–95.

Cadenas, E. (1989). Biochemistry of oxygen toxicity. *Annu. Rev. Biochem.* 58:79–110.

Cross, A. R., and Jones, O. T. G. (1991). Enzymic mechanisms of superoxide production. *Biochim. Biophys. Acta* 1057:281–298.

Machlin, L. J., and Bendich, A. (1987). Free radical tissue damage: protective role of antioxidant nutrients. *FASEB J.* 1:441–445.

Segal, A. W. (1989). The electron transport chain of the microbicidal oxidase of phagocytic cells and its involvement in the molecular pathology of chronic granulomatous disease. *J. Clin. Invest.* 83:1785–1793.

19

Photosynthesis

As we have seen in earlier chapters, metabolism is fueled by two forms of chemical energy: the phosphoryl-group–transfer potential of ATP and the reducing power of the nicotinamide cofactors NADH and NADPH. ATP and the reduced forms of the nicotinamide cofactors are produced by glycolysis and the citric acid cycle, and additional ATP is produced by oxidative phosphorylation. Together, these pathways result in the complete oxidation of carbohydrates to carbon dioxide and water.

If the formation of ATP from ADP and P_i and the reduction of $NAD^{\oplus}$ and $NADP^{\oplus}$ are dependent on a supply of carbohydrate, what is the ultimate source of carbohydrate? The answer is **photosynthesis.** Through the process of photosynthesis, carbohydrates are synthesized from atmospheric CO_2 and water. Photosynthesis is an energy-requiring process that is powered by light. It completes the global carbon cycle—nonphotosynthetic organisms derive energy by oxidizing carbohydrates, and they give off CO_2; photosynthetic organisms capture and reduce CO_2 to carbohydrates, which are the source of energy and building blocks for all metabolic processes. Photosynthetic carbohydrate formation is the basis of the world's food supply, as well as the source of a vast array of nonfood products.

Organisms capable of photosynthesis are called **phototrophs,** a diverse group that includes certain bacteria, such as green and purple sulfur bacteria, purple non-sulfur bacteria, and cyanobacteria (blue-green algae), algae, and, of course, plants (Figure 19·1, next page). Photosynthesis is intensively studied in terrestrial vascular plants because of its important role in agriculture and our environment. However, much of our fundamental knowledge of photosynthesis has been obtained from research on the simplest kinds of phototrophs (photosynthetic bacteria and unicellular algae). With the exception of anaerobic bacteria, all phototrophs give off oxygen as a product of photosynthesis. It is believed that the atmosphere surrounding the earth was transformed from a reducing to an oxidizing environment more than two billion years ago following the appearance of oxygen-producing phototrophs. This change in the atmosphere represents a pivotal event in the course of biological evolution.

Figure 19·1
Forest understory. Photosynthesis, a process carried out by almost all plants, occurs under a variety of environmental conditions. Plants optimize their photosynthetic productivity as they adapt to diverse conditions.

19·1 Photosynthesis Consists of Two Major Processes

The net reaction of photosynthesis is

$$CO_2 + H_2O \xrightarrow{\text{Light}} (CH_2O) + O_2 \qquad \textbf{(19·1)}$$

where (CH_2O) represents carbohydrate.

Photosynthesis encompasses two major processes that can be described by two partial reactions.

$$H_2O + ADP + P_i + NADP^{\oplus} \xrightarrow{\text{Light}} O_2 + ATP + NADPH + H^{\oplus}$$
$$CO_2 + ATP + NADPH + H^{\oplus} \longrightarrow (CH_2O) + ADP + P_i + NADP^{\oplus}$$

$$\text{Sum:} \quad CO_2 + H_2O \xrightarrow{\text{Light}} (CH_2O) + O_2 \qquad \textbf{(19·2)}$$

In the first process, often called the "light reactions," protons derived from water are used in the chemiosmotic synthesis of ATP from ADP and P_i, while a hydride ion from the aqueous medium is used for the reduction of $NADP^{\oplus}$ to NADPH. The reactions are characterized by the light-dependent production of oxygen gas derived from the splitting of water molecules. Such reactions are possible because photosynthetic organisms can harness light energy and use it to drive metabolic reactions.

The second process of photosynthesis involves the use of NADPH and ATP in carbon assimilation, a process that results in the reduction of gaseous carbon dioxide to carbohydrate. Because these reactions do not directly depend on light but only on a supply of ATP and NADPH, they are often referred to as the "dark reactions." Although the terms *light reactions* and *dark reactions* have been widely used, both processes normally occur simultaneously. We will see later that the dark reactions require light for maximum rates of carbon assimilation.

19·2 Photosynthesis in Algae and Plants Occurs in Chloroplasts

In algae and plants, photosynthesis occurs in specialized organelles called **chloroplasts** (Figure 19·2). The main structural feature of the chloroplast is an internal membranous network called the thylakoid membrane, or thylakoid lamella, which is the site of the light-dependent reactions that lead to the formation of NADPH and ATP. The thylakoid membrane is a highly folded, continuous membrane network suspended in the aqueous matrix of the chloroplast. This aqueous matrix, called the stroma, is the site of the second portion of the photosynthetic process, the reduction of carbon dioxide to carbohydrate. The chloroplast is enclosed by a double membrane that is highly permeable to CO_2 and selectively permeable to other metabolites.

The aqueous space enclosed by the thylakoid membrane is called the lumen. As we will see later, the translocation of protons across the thylakoid membrane into the lumen creates the protonmotive force that drives the synthesis of ATP. The thylakoid membrane is folded into a network of flattened vesicles arranged in stacks called grana (singular, granum) or present as single, unstacked vesicles that traverse the stroma and connect grana. Regions of the thylakoid membrane located within grana and not in contact with the stroma are called granal lamellae, whereas regions in contact with the stroma are called stromal lamellae. We will discuss later how membrane-embedded components involved in photosynthesis are differentially distributed between the granal and stromal lamellae.

(a)

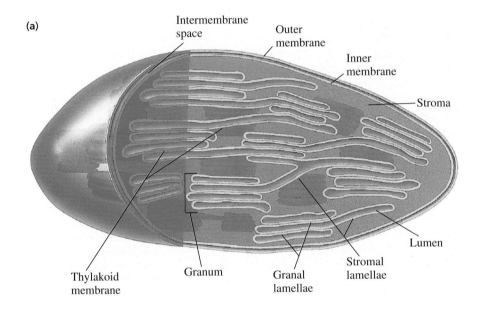

Figure 19·2
Structure of a chloroplast. **(a)** Illustration. The chloroplast possesses a double membrane that surrounds an aqueous region called the stroma. Located within the stroma is the thylakoid membrane, which is folded into flattened, saclike vesicles. These vesicles can occur in stacks called grana or as single vesicles that traverse the stroma and connect grana. Regions of the thylakoid membrane that are located within grana and are not in contact with the stroma are called granal lamellae. Regions exposed to the stroma are called stromal lamellae. The space enclosed by the thylakoid membrane is called the lumen. **(b)** Electron micrograph of a spinach leaf chloroplast. (Courtesy of A. D. Greenwood.)

(b)

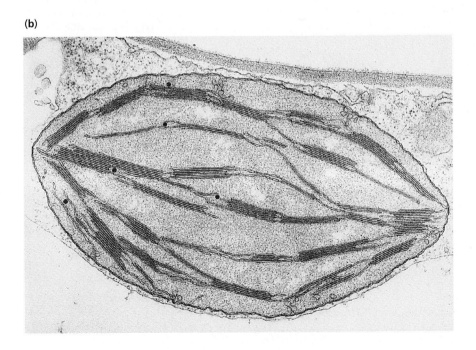

19·3 Chlorophyll and Other Pigments Absorb Light

Among the components embedded in the thylakoid membrane are various pigments that capture light energy for photosynthesis. **Chlorophyll,** a green pigment, is the most abundant and has the most significant role in the harvesting of light. Four different types of chlorophyll commonly occur among phototrophic organisms. These are distinguished by the chemistry of the substituent groups attached to the main part of the molecule, as shown in Figure 19·3. Chlorophyll *a* (Chl *a*) and chlorophyll *b* (Chl *b*) are the two types of chlorophyll found in green plants, with Chl *a* generally more abundant than Chl *b*. The major pigments in photosynthetic bacteria are bacteriochlorophyll *a* (BChl *a*) and bacteriochlorophyll *b* (BChl *b*).

Chlorophyll molecules are specifically oriented in the photosynthetic membranes by both covalent and noncovalent bonds to integral membrane proteins. Features common to all the chlorophylls are the hydrophilic chlorin ring, which contains a light-absorbing network of conjugated double bonds, and the hydrophobic phytol side chain, which helps to make chlorophyll soluble in the membrane. The chlorin ring of the chlorophyll is similar to the porphyrin ring of the heme prosthetic group of hemoglobin, myoglobin, and the cytochromes (Figure 5·33). Unlike heme, which contains Fe^{2+}, chlorophyll contains Mg^{2+} chelated by the pyrrole nitrogen atoms of the ring.

In addition to chlorophyll, several accessory pigments are present in the photosynthetic membranes. These are the carotenoids (yellow to brown), which are present in all phototrophs, and the phycobilins, including phycoerythrin (red) and phycocyanin (blue), which are found in some algae and cyanobacteria. Like the

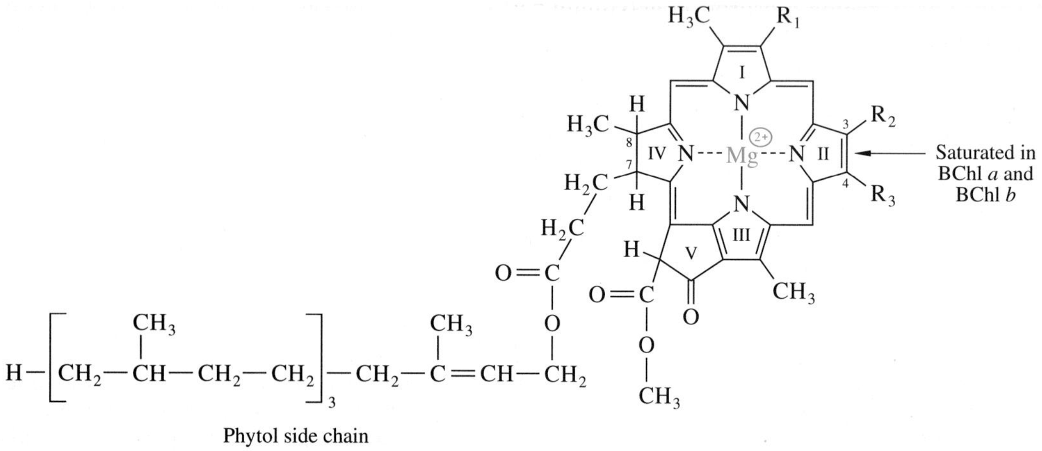

Figure 19·3
Structures of chlorophyll and bacteriochlorophyll pigments. Differences in substituent groups (R_1, R_2, and R_3) are shown in the table. In the bacteriochlorophylls, the bond between C-3 and C-4 in ring II is saturated. The hydrophobic phytol side chain and hydrophilic chlorin ring give chlorophyll amphipathic characteristics.

Chl species	R_1	R_2	R_3
Chl *a*	$-CH=CH_2$	$-CH_3$	$-CH_2-CH_3$
Chl *b*	$-CH=CH_2$	$-\overset{O}{\overset{\|}{C}}-H$	$-CH_2-CH_3$
BChl *a*	$-\overset{O}{\overset{\|}{C}}-CH_3$	$-CH_3$	$-CH_2-CH_3$
BChl *b*	$-\overset{O}{\overset{\|}{C}}-CH_3$	$-CH_3$	$-CH=CH_2$

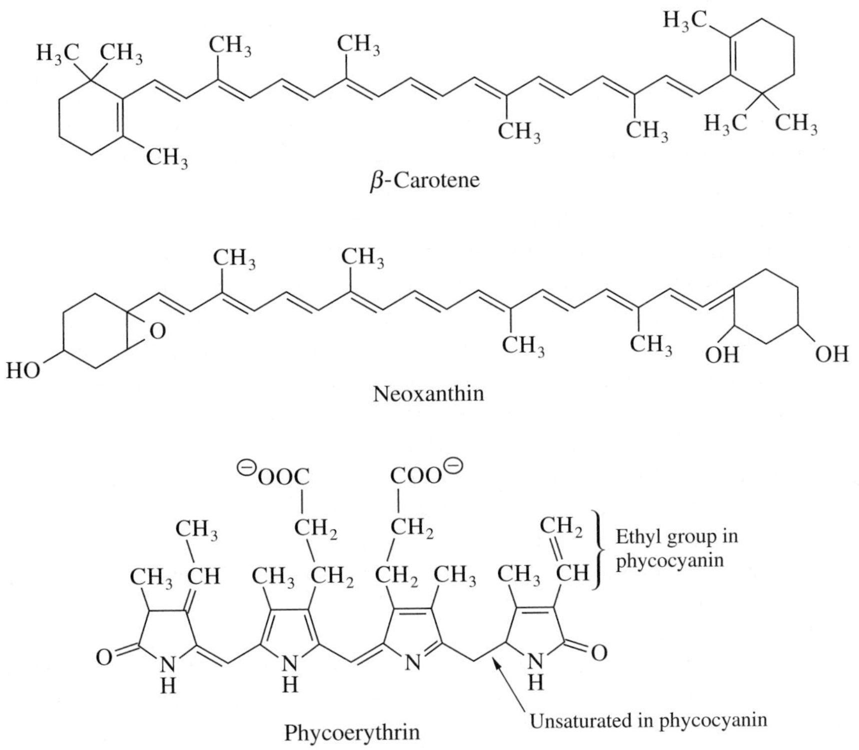

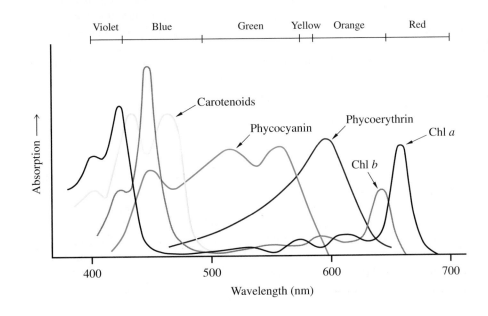

β-Carotene

Neoxanthin

Phycoerythrin

Figure 19·4
Structures of several accessory pigments. β-Carotene and neoxanthin are carotenoids found in green plants. Phycoerythrin and phycocyanin are phycobilins present in certain algae and in cyanobacteria.

chlorophylls, these molecules contain a series of conjugated double bonds that give them the light-absorbing properties characteristic of pigments (Figure 19·4).

Chl a and Chl b both absorb light in the violet-to-blue region (absorption maximum 400–500 nm) and the orange-to-red region (absorption maximum 650–700 nm) of the electromagnetic spectrum. Because the structures of Chl a and Chl b are slightly different, the absorption maxima of these molecules are not identical (Figure 19·5). The absorption maxima of the accessory pigments complement those of the chlorophylls and thus broaden the range of light energy that can be absorbed by phototrophs. Collectively, the pigments absorb a range of radiant energy that spans the spectrum of visible light. Since the absorption spectrum of a pigment determines its color, the amounts and kinds of chlorophylls and accessory pigments determine the characteristic color of a photosynthetic tissue or organism.

Figure 19·5
Absorption spectra of major photosynthetic pigments. Collectively, the pigments absorb radiant energy across the spectrum of visible light. [Adapted from Govindjee and Govindjee, R. (1974). The absorption of light in photosynthesis. *Sci. Am.* 231(6):68–82.]

19·4 Pigments Are Components of Photosystems

The light-absorbing pigments of photosynthesis are components of functional units called **photosystems.** Each photosystem contains a **reaction center,** which forms the core of the photosystem. The reaction center consists of a complex of proteins, specialized electron-transfer molecules, and a pair of Chl *a* molecules (BChl *a* or BChl *b* in bacteria) called the **special pair,** or the primary electron donor. Light-absorbing pigments, called **antenna pigments,** are associated with the reaction center. These pigments may form a separate antenna complex or may be bound directly to the reaction-center proteins. Each photosynthetic unit in green bacteria contains about 2000 pigment molecules, whereas those in plants contain about 200 to 300, and those in purple bacteria contain about 60.

The pigments of a photosystem absorb **photons** (quanta of light energy) and can transfer this absorbed energy to the special pair of chlorophyll molecules in the reaction center. Only the chlorophyll special pair can participate directly in the conversion of photochemical energy to electrochemical energy. The energy absorbed by the special pair can initiate the transfer of an electron through the photosynthetic electron-transport chain; thus, light energy is converted to chemical energy.

19·5 The Thylakoid Membrane Contains Two Types of Photosystems

The thylakoid membrane contains two types of photosystems, designated photosystem I (PSI) and photosystem II (PSII). PSI is located predominantly in the stromal lamellae and is thus exposed to the chloroplast stroma, whereas PSII is located predominantly in the granal lamellae, away from the stroma. The special pair of Chl *a* molecules in the reaction center of PSI has a long-wavelength absorption maximum at 700 nm; this primary donor is sometimes called P700. The special pair of Chl *a* molecules in the reaction center of PSII has an absorption maximum at 680 nm, and so the primary donor is sometimes called P680.

Although the two membrane-spanning photosystems are located in different regions of the thylakoid membrane, they are linked by specific electron carriers and work in series. The spatial separation of PSI and PSII prevents spontaneous transfer of excitation energy between the two photosystems, ensuring that PSI and PSII are linked only by the transfer of electrons. Light energy for photosynthesis is captured by the antenna pigments associated with each reaction center and by a pigment complex called the **chlorophyll *a/b* light-harvesting complex** (LHC). The PSII antenna chlorophylls are part of a pigment-protein complex that is external to the PSII reaction center. In contrast, the PSI antenna chlorophylls are bound to protein subunits of the reaction center. The LHC contains about half of the chlorophyll of the chloroplast. The LHC is distributed in the thylakoid membrane in three fractions: one fraction is bound to PSI, another is bound to PSII, and the third is mobile and can harvest light for either PSI or PSII, as we will see later in the chapter.

In addition to the photosystems and light-harvesting complexes, other components participating in photosynthesis are embedded in or associated with the thylakoid membrane (Figure 19·6). These include the oxygen-evolving complex, the cytochrome *bf* complex, and the chloroplast ATP synthase. The oxygen-evolving complex, composed of several peripheral proteins and four manganese atoms, is associated with PSII on the luminal side of the thylakoid membrane. The manganese atoms make up the active site and are believed to be bound directly to the reaction-center proteins. The peripheral proteins assist in oxygen-evolving reactions and are

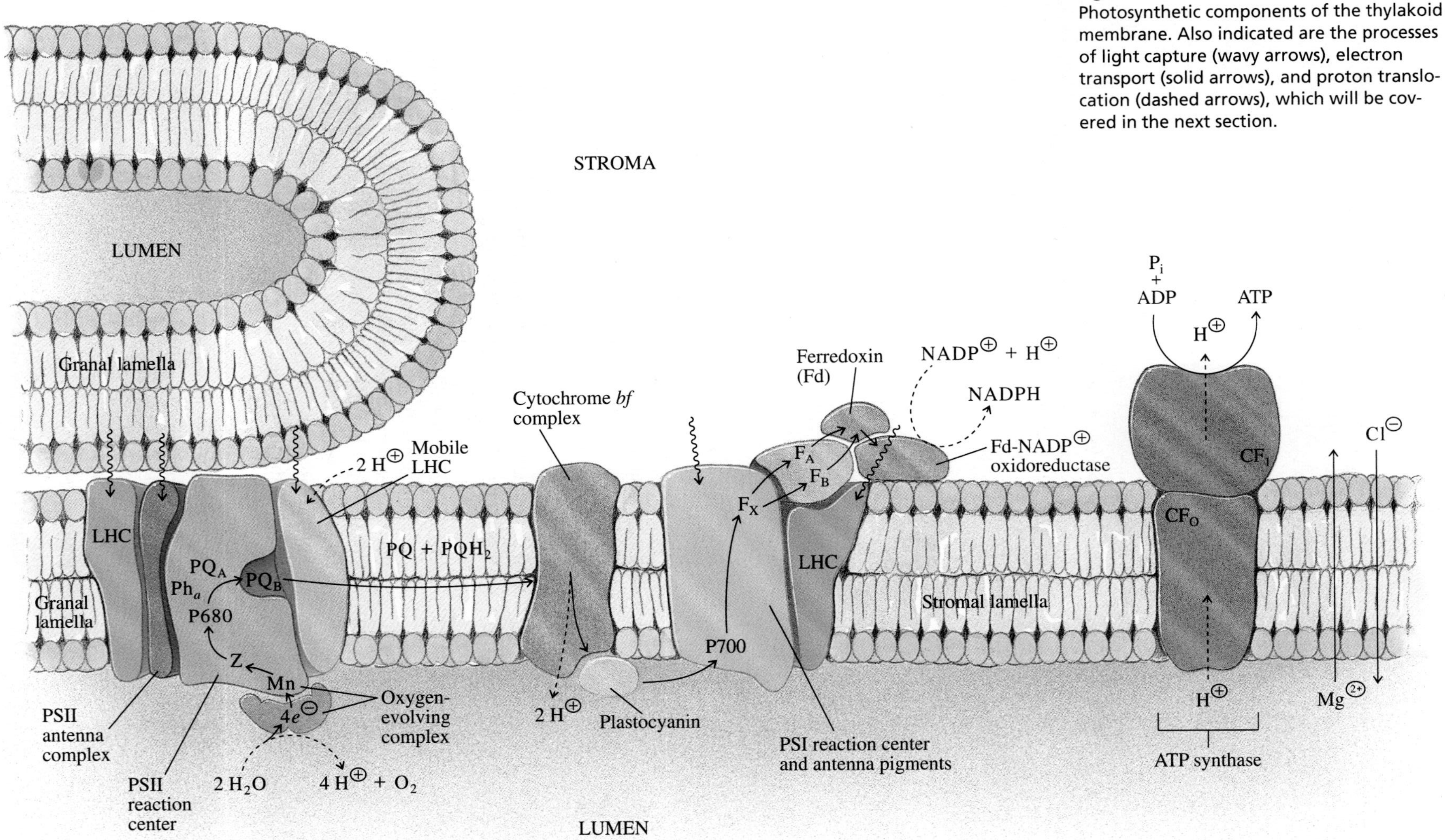

Figure 19·6
Photosynthetic components of the thylakoid membrane. Also indicated are the processes of light capture (wavy arrows), electron transport (solid arrows), and proton translocation (dashed arrows), which will be covered in the next section.

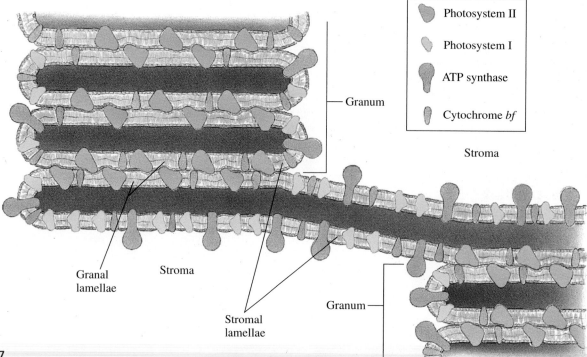

Stroma

Granum

Granal
lamellae

Stroma

Stromal
lamellae

Granum

Figure 19·7
Distribution of photosynthetic components between stromal and granal lamellae. PSI occurs predominantly in the stromal lamellae, PSII occurs predominantly in granal lamellae, and the cytochrome *bf* complex occurs in both the stromal and granal lamellae. ATP synthase occurs exclusively in the stromal lamellae.

noncovalently bound to the reaction center. The cytochrome *bf* complex spans the thylakoid membrane and is located in both the stromal and granal lamellae. The ATP synthase also spans the thylakoid membrane and is located exclusively in the stromal lamellae. Figure 19·7 illustrates the predominant locations within the thylakoid membrane of PSI, PSII, the cytochrome *bf* complex, and ATP synthase.

19·6 Noncyclic Electron Transport Results in the Reduction of NADP$^{\oplus}$ to NADPH

Photosynthetic electron transfer follows one of two routes. Electrons can be transferred linearly through a series of membrane-associated carriers, terminating with the transfer of electrons to NADP$^{\oplus}$, or they can be transferred through many of the same carriers, but in a cyclic manner, so that there is no terminal electron transfer to NADP$^{\oplus}$. During both noncyclic and cyclic electron transport, protons are translocated across the thylakoid membrane from the stroma to the lumen, generating the protonmotive force used to drive the formation of ATP. However, only noncyclic electron flow yields NADPH. In this section, we will discuss noncyclic electron transfer. Cyclic electron transfer will be discussed in Section 19·7.

The initial event preceding photosynthetic electron transfer is the absorption of a photon of light by a pigment molecule. LHC pigments, antenna pigments, or reaction-center pigments can absorb photons. For the sake of illustration, we will follow the transfer of excitation energy from the antenna complex to PSII. When a pigment absorbs a photon, a low-energy, ground-state electron is promoted to a higher-energy molecular orbital, resulting in absorption of excitation energy by the pigment. Molecular orbitals have discrete energy levels, and the difference in energy between the pigment's ground-state and higher-energy orbital must match the photon's energy for the photon to be absorbed. The energy of a photon is inversely

---

Legend:
Photosystem II
Photosystem I
ATP synthase
Cytochrome *bf*

related to wavelength, and therefore pigments absorb only certain wavelengths of light (Figure 19·5).

The energy of an excited pigment molecule can be transferred among pigment molecules and to the reaction center. However, the energy of some excited pigments can decay by means other than excitation-energy transfer. The energy of an absorbed photon can be spent in molecular motion (i.e., kinetic energy, or heat). Alternatively, the excited pigment can reemit a photon of longer wavelength (lower energy), which is detectable as fluorescence. Measurement of fluorescence emission is a research tool widely used to monitor photosynthesis experimentally. Fluorescence can be observed when existing or imposed photosynthetic "bottlenecks" reduce the ability of the reaction center to become excited, forcing more energy to be reemitted from the chlorophyll as fluorescence.

Photon absorption by the LHC or the PSII antenna complex and the subsequent rapid transfer (10^{-15} s) of excitation energy among adjacent pigments can lead to the excitation of the PSII reaction center (Figure 19·8). Pigments transfer excitation energy through interactions among their molecular orbitals, which are facilitated by their orientation in the membrane and their proximity to each other. Specific orientation of the pigment molecules is the result of binding to integral membrane proteins. Transfer of energy to one of the special-pair chlorophyll molecules of PSII (designated P680) raises this pigment to an excited state. The excited reaction center is then designated as P680*. Because the energy level of an excited reaction-center chlorophyll is slightly lower than the energy required to excite an antenna chlorophyll, energy transfer proceeds no further.

Once excited, P680* becomes a relatively strong reductant, and a reaction-center chlorophyll readily donates an electron to pheophytin, the first in a series of electron acceptors. The separation of charge that takes place (due to the oxidation of the reaction center to P680$^\oplus$ and the reduction of pheophytin to Ph$^\ominus$) is the step that converts captured light energy to chemical energy. A similar type of charge separation results from excitation and electron transfer in PSI.

The photooxidized reaction center, P680$^\oplus$, returns to the ground state and becomes a very powerful oxidizing agent. The oxidized reaction-center chlorophyll of P680$^\oplus$ is reduced by an electron that is ultimately derived from the oxygen-evolving complex. The oxygen-evolving complex uses water as a source of electrons. Four electrons from two oxidized water molecules are used for photosynthetic electron transport, while the remaining four protons are released into the lumen and contribute to the transmembrane proton concentration gradient. Molecular oxygen (O_2) from water is released to the atmosphere. Thus, during photosynthesis, all phototrophs (except anaerobic bacteria) give off oxygen.

The manner in which the oxygen-evolving complex catalyzes the oxidation of water—the extraction of its electrons and release of oxygen—is quite interesting. Located on the luminal side of PSII, the oxygen-evolving complex contains a cluster of four manganese ions that are specifically oriented to accumulate and then

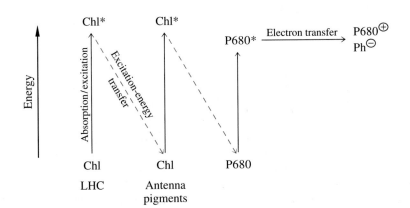

Figure 19·8
Excitation-energy transfer from pigment molecules in the LHC and antenna complex to the P680 reaction center and subsequent electron transfer from P680* to pheophytin (Ph). The same process occurs in P700.

transfer four water-derived electrons, one at a time, to the oxidized P680 (P680$^\oplus$). Once the Mn cluster has given up its four electrons, the following events related to water splitting occur: 1) two molecules of water are oxidized and their four electrons reduce the Mn cluster, 2) four protons are released into the lumen, and 3) one molecule of O_2 is evolved. This stoichiometry can be observed experimentally by exposing chloroplasts in darkness to brief flashes of light that excite PSII. With each flash, P680 is excited to P680*, charge separation takes place, and then P680$^\oplus$ is reduced by an electron from the Mn cluster. After four flashes (when the Mn cluster is fully oxidized), oxygen is given off, and H$^\oplus$ movement across the membrane can be detected. The oxidation states of the Mn cluster are stabilized by various coordination configurations with the proteins and with the oxygen atoms of the bound water molecules. However, re-reduction of P680$^\oplus$ does not occur directly from the Mn cluster. Instead, transfer of an electron from the Mn cluster to P680$^\oplus$ occurs via the oxidation and reduction of Z, a tyrosine residue of a protein subunit of the reaction center. Electron transfer from Z restores P680$^\oplus$ to its neutral state so that it is ready to be excited again.

A difference in reduction potential is the basis of electron transfer between membrane-associated photosynthetic electron carriers. Electron transfer proceeds spontaneously from carriers of lower reduction potential (stronger reductants) to carriers of higher reduction potential (stronger oxidants). This same principle underlies mitochondrial electron transport (Chapter 18). In photosynthesis, however, the absorption of light energy transiently lowers the reduction potential of the primary donor; that is, P680* and P700* have far lower reduction potentials (i.e., they are stronger reductants, or electron donors) than their ground-state equivalents, P680 and P700.

When the reduction potentials of the series of photosynthetic electron-transport components are plotted in series, the result is a zigzag figure called the **Z-scheme**

Figure 19·9
Z-scheme. The Z-scheme, so called because of its shape, is a widely accepted model illustrating the reduction potentials associated with electron flow through photosynthetic electron carriers. Because the reduction potentials of the carriers vary with experimental conditions, the values shown are approximate. Abbreviations: Z, electron donor to P680; Ph a, pheophytin a, electron acceptor of PSII; PQ$_A$, plastoquinone tightly bound to PSII; PQ$_B$, reversibly bound PQ undergoing reduction by PSII; PQ$_{pool}$, plastoquinone pool made up of PQ and PQH$_2$; A$_0$, chlorophyll a, the primary electron acceptor of PSI; A$_1$, phylloquinone, or vitamin K$_1$; F$_X$, F$_B$, and F$_A$, iron-sulfur clusters; and Fd, ferredoxin.

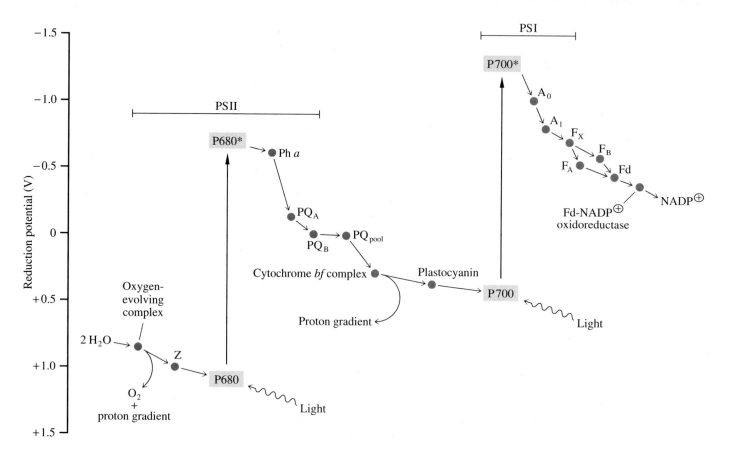

(Figure 19·9). The Z-scheme shows the increase in reducing power resulting from the excitation of the photosystem reaction centers. Also shown are electron carriers between P680 and P700. Note that the order of electron transfer is from PSII to PSI, not from PSI to PSII. (The terminology *PSI* and *PSII* reflects the order of discovery of the photosystems, not the order of electron transfer.) The following discussion of the electron-transport sequence will refer to the Z-scheme in Figure 19·9 and to Figure 19·6, which illustrates the spatial relationships of the electron-transport components within the thylakoid membrane.

Once excited, P680* donates its electron to pheophytin *a* (Ph *a*), which is identical to Chl *a* except that the magnesium ion of Chl *a* is replaced by two protons. Electron transfer then proceeds from reduced pheophytin *a* to PQ_A, a plastoquinone tightly bound to a PSII polypeptide. Like mitochondrial ubiquinone (Figure 8·56), plastoquinone can be reduced by two sequential one-electron transfers (Figure 19·10). During two successive photochemical events, PQ_A transfers two electrons, one at a time, to a second plastoquinone molecule, PQ_B, which is reversibly bound to a PSII polypeptide. PQ_B first receives one electron from PQ_A and is protonated to form a semiquinol (i.e., plastosemiquinol, $\cdot PQ_B H$). Following a second photochemical event, plastosemiquinol is reduced and protonated again to form plastoquinol ($PQ_B H_2$). Fully reduced, protonated $PQ_B H_2$ is then released into the pool of quinone. Its long hydrophobic tail ensures that the plastoquinone pool is soluble in the thylakoid membrane.

The photosynthetic Q cycle, like the mitochondrial Q cycle (Section 18·8), carries out the oxidation of reduced quinone. The oxidation of PQH_2 occurs via the cytochrome *bf* complex. This complex is made up of heme-containing cytochromes and iron-sulfur proteins and is analogous to the mitochondrial Complex III (ubiquinol-cytochrome *c* oxidoreductase). The two-step Q cycle (Figure 19·11, next page) results in the oxidation of two molecules of PQH_2, the reduction of one molecule of PQ to PQH_2, the transfer of two electrons to plastocyanin (one at a time), and the net transfer of four protons into the lumen. In the first step of the Q cycle, PQH_2 is oxidized at a site adjacent to cytochrome *f*, and the two protons from the oxidized PQ are released into the lumen. One of the electrons from PQH_2 oxidation is funnelled through cytochrome *f* and reduces plastocyanin. The second electron, plus one stromal proton, converts a molecule of PQ to $\cdot PQH$ at a site away from the PQH_2 oxidation site. In the second step, a second molecule of PQH_2 is oxidized. Once again, one electron ultimately reduces plastocyanin; the other electron, plus one stromal proton, converts $\cdot PQH$ to PQH_2. The fully reduced PQH_2 is then released into the plastoquinone pool. Since the cytochrome *bf* complex can accept only electrons and not protons from PQH_2, Q cycling by plastoquinone and the cytochrome *bf* complex serves as a proton pump, contributing to the protonmotive force that drives ATP synthesis.

Plastocyanin is a small, copper-containing protein that is reduced by cytochrome *f* of the cytochrome *bf* complex and oxidized by $P700^{\oplus}$ of PSI. Electron transport through plastocyanin occurs via changes in the oxidation state of its copper atom, which is coordinated at the active site by two nitrogen and two sulfur atoms of four amino acid residues. There is a small pool of plastocyanin (fewer than five molecules) per P700. These plastocyanin molecules move laterally in the thylakoid membrane to shuttle electrons between the cytochrome *bf* complex and P700.

Once $P700^{\oplus}$ has been reduced to P700 by plastocyanin, it can become energized to P700* by pigment excitation that migrates from the PSI antenna pigments, the PSI-bound LHC, or, under certain conditions, the mobile pool of LHC. Note that P700*, with a reduction potential of approximately $-1.3\,V$, is the strongest reductant in the chain of electron carriers. The electron from P700* is readily donated to a chlorophyll *a* electron acceptor, known as A_0. The reduced species, $A_0^{\ominus}$, donates its electron in turn to a bound molecule of phylloquinone (vitamin K_1), known as A_1. From A_1, the electron is transferred to a series of iron-sulfur clusters, first to

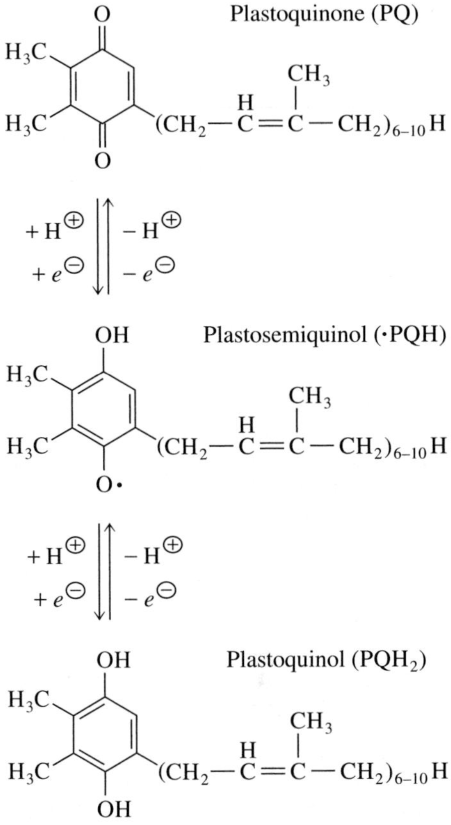

Figure 19·10
Reduction of plastoquinone to plastoquinol by two successive one-electron transfers.

Figure 19·11

Photosynthetic Q cycle. The Q cycle is a two-step process. In the first step, PQH_2 is oxidized to PQ. In the process, two protons are released into the lumen; one electron is funnelled to plastocyanin via cytochrome f; and the other electron, along with a proton from the stroma, converts a molecule of PQ to ·PQH. In the second step, a second molecule of PQH_2 is oxidized. Again, two protons are released into the lumen; one electron is funnelled to plastocyanin; and the other electron reduces ·PQH to PQH_2, which is then released into the plastoquinone pool. Together, the two steps result in the oxidation of two molecules of PQH_2, the reduction of one molecule of PQ to PQH_2, the transfer of two electrons to plastocyanin, and the transfer of four protons into the lumen.

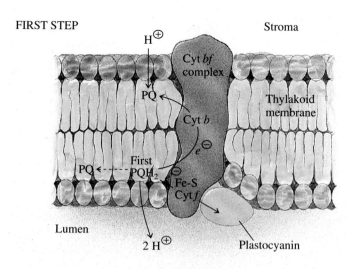

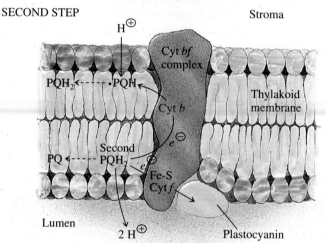

F_X and then to either F_A or F_B. F_X is part of the PSI reaction-center complex, whereas F_A and F_B are located in a peripheral polypeptide on the stromal side of the thylakoid membrane.

The electron delivered to the last iron-sulfur cluster, either F_A or F_B, is transferred to an iron-sulfur protein called ferredoxin (Fd). Ferredoxin is a small water-soluble protein in the chloroplast stroma. The low reduction potential of ferredoxin ($E° = -0.43$ V) allows it to readily reduce $NADP^\oplus$ to NADPH ($E° = -0.32$ V). The reduction of $NADP^\oplus$ by reduced ferredoxin is catalyzed by ferredoxin-$NADP^\oplus$ oxidoreductase, an enzyme loosely bound to the stromal side of the thylakoid membrane. The oxidoreductase contains the prosthetic group FAD, which is reduced to $FADH_2$ by reduced ferredoxin in two one-electron transfers. $FADH_2$ in turn reduces $NADP^\oplus$ by donating two electrons and a proton in the form of a hydride ion. The pH difference across the thylakoid membrane is increased by this step, since a proton from the stroma is consumed in the conversion of $NADP^\oplus$ to NADPH. Formation of NADPH completes the noncyclic electron-transport sequence.

Three separate steps in the photosynthetic electron-transport chain change the distribution of protons between the stroma and lumen. First, protons from water are released into the lumen by the oxygen-evolving complex. Second, protons originating in the stroma are translocated to the lumen during the oxidation of PQH_2 by the cytochrome *bf* complex. Third, the uptake of a proton during reduction of $NADP^{\oplus}$ lowers the proton concentration in the stroma. As protons are moved from the stroma into the lumen, magnesium ions move across the thylakoid membrane from the lumen into the stroma so that charge balance is maintained between the two compartments. Similarly, chloride ions are translocated into the lumen to compensate for the higher concentration of electrons in the stroma resulting from electron transport. As we will see later in the chapter, changes in the level of $Mg^{2\oplus}$ in the stroma during active photosynthesis play a role in the regulation of carbohydrate synthesis.

In the presence of a transmembrane proton concentration gradient, the membrane-spanning ATP synthase catalyzes the formation of ATP from ADP and P_i. Since this process is light-dependent in plants, photosynthetic ATP formation is termed **photophosphorylation.** The ATP synthase of chloroplasts is a multiprotein complex similar to its mitochondrial counterpart in that it consists of two major particles, CF_O and CF_1. (The designations CF_O and CF_1 distinguish the chloroplast ATP synthase components from the mitochondrial F_O and F_1 components.) The CF_O particle spans the thylakoid membrane and forms a channel for protons. CF_1 protrudes into the stroma and catalyzes the formation of ATP from ADP and P_i. The chloroplast ATP synthase is localized exclusively in the stromal lamellae.

19·7 Cyclic Electron Flow Contributes to the Proton Concentration Gradient Across the Thylakoid Membrane

It is estimated that for every two electrons transferred to reduce one molecule of $NADP^{\oplus}$ to NADPH, a protonmotive force develops across the thylakoid membrane sufficient to synthesize one molecule of ATP. This one-to-one ATP:NADPH stoichiometry of the light reactions would create an imbalance with the stoichiometry of the carbohydrate-producing dark reactions, since, as we will see, two molecules of NADPH and three molecules of ATP are consumed in the reduction of one molecule of CO_2 to (CH_2O). Under conditions where the coupling between the light and dark reactions becomes unbalanced, a modified sequence of electron-transport steps operates as a compensatory cycle to form ATP without the simultaneous formation of NADPH. This modified reaction sequence is called **cyclic electron transport.** As in noncyclic electron transport, electrons are transferred via the Q cycle from plastoquinone to the cytochrome *bf* complex and then on to PSI and ferredoxin. In cyclic electron transport, however, the soluble ferredoxin donates its electron back to the PQ pool in a process mediated by the cytochrome *bf* complex via a cytochrome that is involved exclusively in cyclic electron flow. Through Q cycling, cyclic electron transfer supplements the protonmotive force generated during noncyclic electron flow and thus increases the production of ATP without the concomitant production of NADPH.

In general, the relative rates of cyclic and noncyclic electron flow are influenced by the relative amounts of NADPH and $NADP^{\oplus}$. When the stromal ratio of NADPH to $NADP^{\oplus}$ is high, the rate of noncyclic electron transfer is limited, due to the low availability of $NADP^{\oplus}$. Under these circumstances, cyclic electron flow is favored.

Figure 19·12
Regulation of light-energy input to photosystems by phosphorylation of the mobile LHC. Activated by a high ratio of PQH_2 to PQ, a specific LHC kinase catalyzes phosphorylation of the mobile LHC, causing it to migrate to the stromal lamellae. Preferential energy input into PSI increases the rate of PQH_2 oxidation and lowers the ratio of PQH_2 to PQ. Dephosphorylation of the mobile LHC is catalyzed by a phosphatase and causes the LHC to migrate back to PSII in the granal lamellae. Wavy arrows indicate light-energy transfer from pigments to the photosystems.

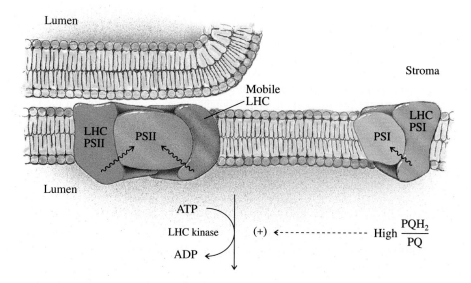

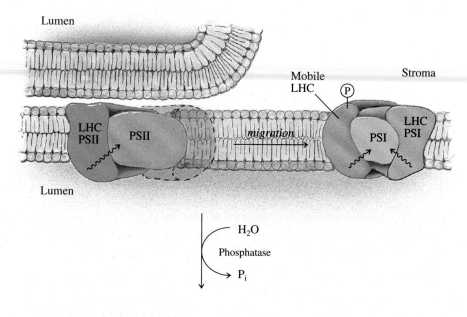

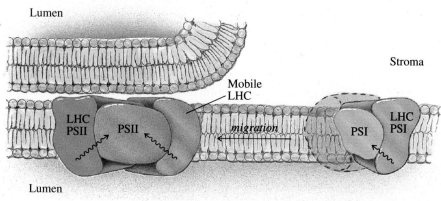

19·8 The Capture of Light Energy by Chloroplasts and the Energy Distribution Between Photosystems Are Regulated

Photosynthesis occurs under a wide range of light conditions. For example, compared to late-day sun, full sunlight has both greater intensity and a higher proportion of shorter-wavelength light. Because of the different absorption maxima of P680 and P700, P680 is preferentially excited in full sun compared to P700. Under varying conditions, the amount of light energy captured by the two photosystems can become imbalanced as a result of daily changes and short-term fluctuations in the spectral quality and intensity of ambient light. Although this imbalance could limit the rate of electron transport, the problem is ameliorated by a short-term regulatory mechanism that affects the location of the mobile LHC, which preferentially surrounds PSII in the granal lamellae. Two tyrosine residues on a surface-exposed segment of the mobile LHC can be phosphorylated in a reaction catalyzed by a specific protein kinase. LHC phosphorylation causes electrostatic repulsion between adjacent mobile LHCs and leads to the migration of the phosphorylated LHC to the stromal lamellae, where PSI is located (Figure 19·12). When this happens, light energy captured by the phosphorylated mobile LHC is funnelled preferentially to PSI.

The kinase that catalyzes LHC phosphorylation is more active when the ratio of PQH_2 to PQ in the PQ pool is high, which occurs when plastoquinone is reduced by PSII more rapidly than it is oxidized by PSI. Phosphorylation of the mobile LHC alleviates this condition by increasing energy input into PSI. A phosphatase catalyzes dephosphorylation of the LHC, causing it to migrate back to PSII in the granal lamellae. Thus, under various light conditions, the equilibrium between phosphorylated and dephosphorylated mobile LHC is determined by the oxidation state of the plastoquinone pool and results in a balanced energy distribution between PSII and PSI.

Long-term changes in photosynthetic capacity can occur in response to long-term changes in light intensities and spectral qualities. For example, the total number and proportions of the various pigment complexes can change and the stoichiometries of PSII, PSI, and electron carriers can be modified by different light conditions. Under conditions of limited light, synthesis of chlorophyll and accessory pigments is increased in order to maximize light-gathering capacity. Conversely, in highly illuminated leaves, photosynthetic capacity is increased by higher proportions of cytochrome *bf* complex, plastoquinone, ferredoxin, and ATP synthase per photosystem and a proportionately lower chlorophyll content of the light-harvesting components. It is not surprising that a process so fundamental to the survival of the plant can adapt over both the short and long term to optimize photosynthetic efficiency under different conditions.

19·9 Photosynthetic Bacteria Have Only One Photosystem

The preceding outline of noncyclic and cyclic electron transport sets the stage for a discussion of bacterial photosynthesis. A major feature that distinguishes photosynthesis in bacteria (except cyanobacteria) from that in algae and plants is that photosynthetic bacteria have only one type of photosystem instead of two. The bacterial photosystem differs among bacterial species. The long-wavelength absorption maximum of the primary electron donor in the reaction center also varies among bacterial species; for example, P870 and P960 are both found among species of the purple bacterium *Rhodospirillum*. Since bacteria do not contain chloroplasts, photosynthesis occurs in the plasma membrane. The plasma membrane of photosynthetic bacteria is a highly invaginated, contiguous membrane creating two cellular compartments—the cytosol and the periplasmic space. The translocation of protons into the periplasmic space creates the protonmotive force for photophosphorylation. The plasma membrane is not homogeneous; respiratory and photosynthetic activities are localized in different portions of the membrane.

A few characteristics distinguish photosynthetic bacterial species. These characteristics include 1) the use of endogenous versus exogenous electrons for electron transport, 2) the specific source of electrons (most photosynthetic bacteria use electron donors other than water, for example, H_2, H_2S, or organic acids, and do not give off oxygen during photosynthesis), and 3) the requirement for other organic compounds in addition to CO_2 as sources of assimilatory carbon.

The focus of this discussion will be on anoxygenic (non–oxygen-producing) bacterial photosynthesis since it is much more widespread among bacteria and has been most thoroughly studied. Anoxygenic photosynthesis occurs in the purple bacteria and the green bacteria. Electron transport may be either noncyclic, in which $NAD^{\oplus}$ is the terminal electron acceptor, or cyclic, in which only a transmembrane proton concentration gradient is formed. In the cyclic process, NADH is formed by harnessing the transmembrane proton concentration gradient.

In both cyclic and noncyclic anoxygenic bacterial photosynthesis, antenna pigments capture light energy and transfer it to the reaction center. In the purple bacterial reaction center, the bacteriochlorophylls of the special pair, once excited, donate an electron to reduce a bacteriopheophytin. The reduced bacteriopheophytin then donates the electron to a quinone molecule (Q_A), which can be either ubiquinone or menaquinone, depending on the bacterial species. This quinone is tightly bound to the reaction center; adjacent to it is a second molecule of quinone (Q_B), which is reversibly bound to the reaction center. In a manner analogous to PQ_A reduction in PSII, the bound semiquinone, $\cdot Q_A$, donates its electron to the reversibly bound quinone, Q_B (see Figure 19·10). A subsequent photochemical event provides a second electron via Q_A to reduce $\cdot Q_B$ to $Q_B^{\ominus}$. Two protons are taken up by $Q^{\ominus}$ from the cytosolic side of the plasma membrane, and then fully reduced, protonated QH_2 is released into the pool of quinone. The removal of protons from the cytosol during the reduction of quinone generates a proton concentration gradient across the membrane.

The manner of quinol oxidation determines whether electrons are cycled back to the reaction center (cyclic flow) or used to reduce $NAD^{\oplus}$ (noncyclic flow). In cyclic electron flow, QH_2 is oxidized by the bacterial cytochrome bc complex via the Q cycle (see Figure 19·11). A second type of cytochrome c then reduces the oxidized special pair in the reaction center. In cyclic flow, NADH is formed by harnessing the proton concentration gradient in the following manner. The bacterial NADH-quinone oxidoreductase, which is equivalent to the mitochondrial Complex I (NADH-ubiquinone oxidoreductase), is responsive to the magnitude of the proton concentration gradient across the membrane. When the gradient is sufficiently large, the complex can catalyze the normal respiratory reaction in the reverse direction. Therefore, instead of undergoing Q cycle oxidation by the cytochrome complex, QH_2 is oxidized by the NADH-quinone oxidoreductase, and NADH is formed. The majority of photosynthetic bacteria carry out this type of cyclic electron transport and indirect nicotinamide reduction.

Noncyclic electron transport, in which $NAD^{\oplus}$ is reduced directly by the electron-transport chain, is not well understood. It appears to involve serial electron transfer through iron-sulfur clusters in the reaction center, Q, soluble ferredoxin, and ferredoxin-nicotinamide oxidoreductase, which catalyzes the reduction of $NAD^{\oplus}$. Figure 19·13 compares the chloroplast electron-transport pathways (noncyclic and cyclic) with the cyclic and noncyclic pathways found in different bacteria.

(a) CHLOROPLAST

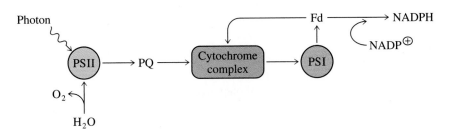

(b) CYCLIC FLOW

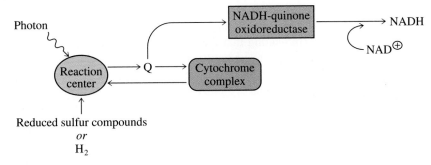

(c) NONCYCLIC FLOW

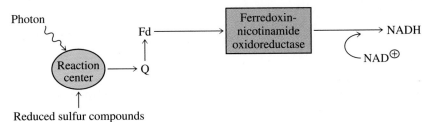

Figure 19·13
Comparison of electron transport in chloroplasts and bacteria. **(a)** Cyclic and noncyclic electron flow in chloroplasts. **(b)** Cyclic electron flow and **(c)** noncyclic electron flow in bacteria. Noncyclic flow occurs in the green sulfur bacteria, whereas cyclic flow is found in the purple sulfur and nonsulfur bacteria.

According to evolutionary theory, the present-day eukaryotic chloroplasts evolved from endosymbiotic photosynthetic bacteria. In the 1980s, X-ray crystallographic analysis of a prokaryotic reaction center from the purple nonsulfur bacterium *Rhodopseudomonas viridis* (Figure 12·14) and a similar analysis of the PSI reaction center from the cyanobacterium *Synechococcus* in 1993 illustrate interesting features with regard to the endosymbiotic theory. Certain electron-transfer components and mechanisms are common to both the *Rhodopseudomonas* reaction center and the PSII reaction center in plants. In both the bacterial and the plant PSII reaction centers, photooxidation of the primary donor reduces a pheophytin, which donates an electron to two quinone molecules in series. Similarly, the reaction center of green sulfur bacteria is similar to the PSI reaction center in the cyanobacterium *Synechococcus* (cyanobacteria, like plants, have two types of photosystems). In both cases, the first stable electron acceptor outside of the photosystem is an iron-sulfur cluster.

The recent characterizations of the three-dimensional atomic-level structure of the *R. viridis* reaction center and cyanobacterial PSI are significant in that they provide detailed information about the specific orientation of reaction-center components and relate this information to the functioning of the reaction center. The *R. viridis* reaction center was in fact the first integral membrane protein to be described in atomic-level detail.

19·10 Reactions of the Reductive Pentose Phosphate Cycle Assimilate CO$_2$ into Carbohydrates

The second major phase of photosynthesis is the reductive conversion of carbon dioxide into carbohydrates, a process that is powered by the ATP and NADPH formed during the light reactions of photosynthesis. Recall that these reactions, although often referred to as the dark reactions because they are not directly dependent on light, normally occur simultaneously with the light reactions. The formation of carbohydrates occurs in the chloroplast stroma and is accomplished by a cycle of enzyme-catalyzed reactions that have three major consequences: 1) the fixation of atmospheric CO$_2$, 2) the reduction of CO$_2$ to carbohydrate, and 3) the regeneration of the molecule that accepts CO$_2$. The metabolic pathway leading to carbon assimilation has several names, including the reductive pentose phosphate cycle; the C$_3$ pathway (indicating that the first intermediate of the pathway is a three-carbon molecule); the photosynthetic carbon reduction cycle; and the Calvin, Calvin-Benson, or Calvin-Bassham cycle. We will refer to the pathway as the **reductive pentose phosphate cycle,** or simply the RPP cycle. This name is appropriate because there are similarities between the RPP cycle and the oxidative pentose phosphate pathway (Section 17·11).

The substrate for the RPP cycle, CO$_2$, diffuses directly into photosynthetic cells. In terrestrial vascular plants, CO$_2$ passes through surface structures called stomata to access photosynthetic cells. Stomata are composed of two adjacent cells on the surface of the leaf surrounding a cavity that is lined with photosynthetic cells. The aperture created by the stomatal cells changes in response to ion fluxes and the resulting osmotic uptake of water. The flux of ions across the stomatal cells is regulated by factors that reflect the suitability of conditions for carbon assimilation, such as temperature and availability of CO$_2$ and water.

19·11 RuBisCO Catalyzes the Initial Step of the RPP Cycle

The fixing of gaseous CO$_2$ into an organic product is accomplished in the first step of the RPP cycle. This step is catalyzed by ribulose 1,5-*bis*phosphate carboxylase-oxygenase, abbreviated RuBisCO. RuBisCO makes up about 50% of the soluble protein in plant leaves; it is thus one of the most abundant enzymes in nature. Although the stromal concentration of RuBisCO active sites has been estimated to be as high as 4 mM, the enzyme is so stringently regulated that the activity of RuBisCO can limit the rate of carbon assimilation.

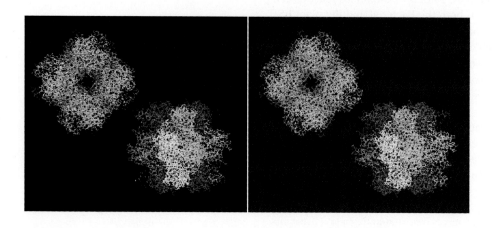

Figure 19·14
Stereo view of the subunit organization of the ribulose 1,5-*bis*phosphate carboxylase-oxygenase (RuBisCO), top and side views. Both views are shown with depth suppressed to emphasize structural organization. Large subunits are shown alternately yellow and blue; small subunits are purple. (Based on coordinates provided by David Eisenberg.)

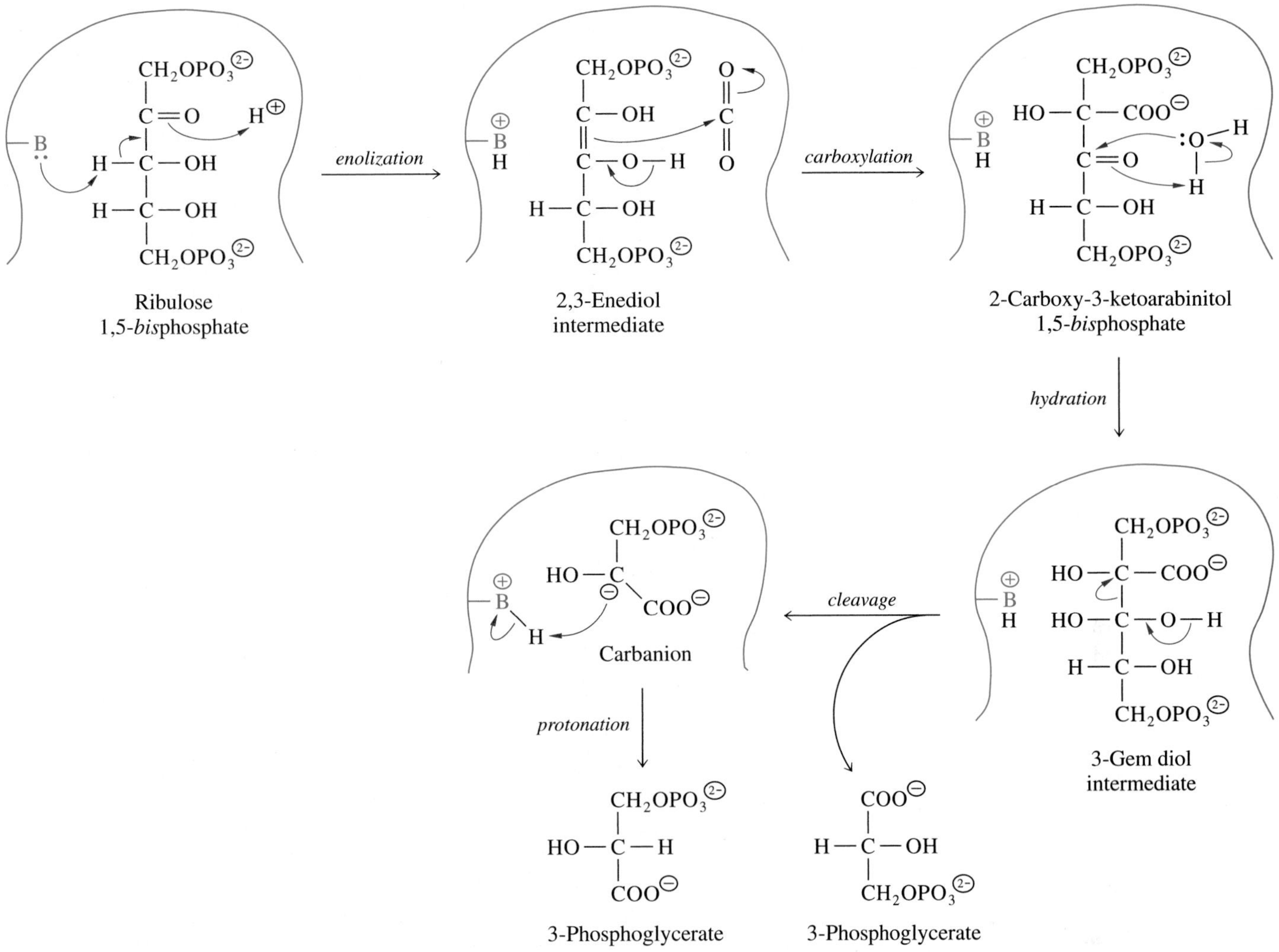

Ribulose
1,5-*bis*phosphate

2,3-Enediol
intermediate

2-Carboxy-3-ketoarabinitol
1,5-*bis*phosphate

enolization

carboxylation

hydration

Carbanion

cleavage

3-Gem diol
intermediate

protonation

3-Phosphoglycerate

3-Phosphoglycerate

Figure 19·15
Mechanism of the carboxylation activity of RuBisCO. In this reaction, ribulose 1,5-*bis*phosphate is carboxylated to form two molecules of 3-phosphoglycerate.

The RuBisCO of plants, algae, and cyanobacteria is composed of eight large subunits (M_r 56 000 each) and eight small subunits (M_r 14 000 each) for a total molecular weight of about 560 000. This composition is denoted L_8S_8. The eight large subunits associate to create the core (L_8) of the molecule (Figure 19·14), and the interfaces between these subunits form eight active sites. Four small subunits are located at each end of the L_8 core. The RuBisCO molecule is simpler in photosynthetic bacteria, having only large subunits. In the purple bacterium *Rhodospirillum rubrum*, RuBisCO is a dimer of large subunits that have about 30% sequence homology with the RuBisCO large subunit in plants.

To catalyze the fixation of CO_2, RuBisCO must be in an activated state. Activation of RuBisCO requires CO_2, Mg^{2+}, and correct stromal pH. In the light, RuBisCO activity increases in response to the higher pH and Mg^{2+} concentration in the stroma developed during proton translocation. The molecule of CO_2 required to activate RuBisCO reacts with the side chain of a lysine residue (Lys-202) that is located away from the active site of RuBisCO. The carbamate adduct formed between CO_2 and the lysine residue is similar to the reaction product of CO_2 with erythrocyte hemoglobin (Figure 5·64). During activation of RuBisCO, Mg^{2+} binds to the CO_2-lysine carbamate adduct and the β-carboxylate group of Asp-203. During catalysis, Mg^{2+} is also coordinated at two sites formed by the bound substrate. In addition to activation by CO_2, Mg^{2+}, and pH, RuBisCO is further activated by light directly.

The reaction mechanism of the carboxylation activity of RuBisCO is shown in Figure 19·15. The substrates for the reaction are a free molecule of CO_2 (not the

CO_2 bound to Lys-202) and the five-carbon phosphorylated sugar ribulose 1,5-*bis*-phosphate. The slow step in the reaction is the abstraction of a proton from ribulose 1,5-*bis*phosphate by a basic residue (—B:) of the enzyme to create the 2,3-enediol intermediate. Reaction of the enediol with CO_2 yields 2-carboxy-3-ketoarabinitol 1,5-*bis*phosphate, which is hydrated to an unstable 3-gem diol intermediate. The C-2—C-3 bond of the intermediate is immediately cleaved, generating a carbanion and one molecule of 3-phosphoglycerate. Stereospecific protonation of the carbanion yields a second molecule of 3-phosphoglycerate. This step completes the carbon-fixation stage of the RPP cycle—two molecules of 3-phosphoglycerate are formed from CO_2 and the five-carbon sugar ribulose 1,5-*bis*phosphate.

19·12 After Fixation, Carbon Is Reduced, and the CO_2 Acceptor Molecule Is Regenerated

Figure 19·16 illustrates all the steps of the RPP cycle. Note that the figure shows the steps for the assimilation of not one but three molecules of carbon dioxide. This is because the smallest carbon intermediate in the RPP cycle is a C_3 molecule. Thus, three CO_2 molecules must be fixed before one C_3 unit can be removed from the cycle without decreasing the size of the pools of RPP-cycle intermediates.

The reductive phase of the RPP cycle occurs in the two reactions following the RuBisCO reaction. First, 3-phosphoglycerate is converted to 1,3-*bis*phosphoglycerate in an ATP-dependent reaction catalyzed by phosphoglycerate kinase. Next, 1,3-*bis*phosphoglycerate is reduced by NADPH in a reaction catalyzed by glyceraldehyde 3-phosphate dehydrogenase (named for the reverse reaction). An equilibrium is maintained between the reaction product, glyceraldehyde 3-phosphate, and its isomer, dihydroxyacetone phosphate, through the activity of triose phosphate isomerase.

The sequence of reactions following the reductive phase of the cycle ends with the regeneration of the CO_2 acceptor molecule, ribulose 1,5-*bis*phosphate. Most of the reactions of the RPP cycle are part of the regeneration phase, which is stoichiometrically complex. Glyceraldehyde 3-phosphate is diverted into three different branches of the pathway and is interconverted between four-, five-, six-, and seven-carbon phosphorylated sugars. This regenerative phase is catalyzed by various isomerases, epimerases, aldolases, phosphatases, and transketolase. The last step is the phosphorylation of ribulose 5-phosphate by ATP to form ribulose 1,5-*bis*phosphate in a reaction catalyzed by phosphoribulokinase. All the substrates, enzymes, and reactions of the RPP cycle are summarized in Table 19·1 (Page 19·22).

After three rounds of the pathway (i.e., after three molecules of CO_2 have been fixed), one triose phosphate can be removed from the cycle. The remaining five triose phosphates are used to regenerate three molecules of ribulose 1,5-*bis*phosphate. This stoichiometry of triose phosphate removal maintains the levels of pathway intermediates. Cycle intermediates are depleted when triose phosphate removal exceeds the rate of CO_2 fixation and are replenished when triose phosphate is diverted from the cycle at a rate less than the rate of CO_2 fixation. The availability of cycle intermediates, which is affected by the rate of triose phosphate removal, strongly influences the capacity of the cycle to assimilate CO_2.

Figure 19·16 (next page)
Reductive pentose phosphate (RPP) cycle. The cycle has three stages: CO_2 fixation, carbon reduction to (CH_2O), and regeneration of the CO_2 acceptor molecule. The concentration of RPP-cycle intermediates is maintained when one molecule of triose phosphate (glyceraldehyde 3-phosphate [G3P] or dihydroxyacetone phosphate [DHAP]) exits the cycle after three molecules of CO_2 are fixed.

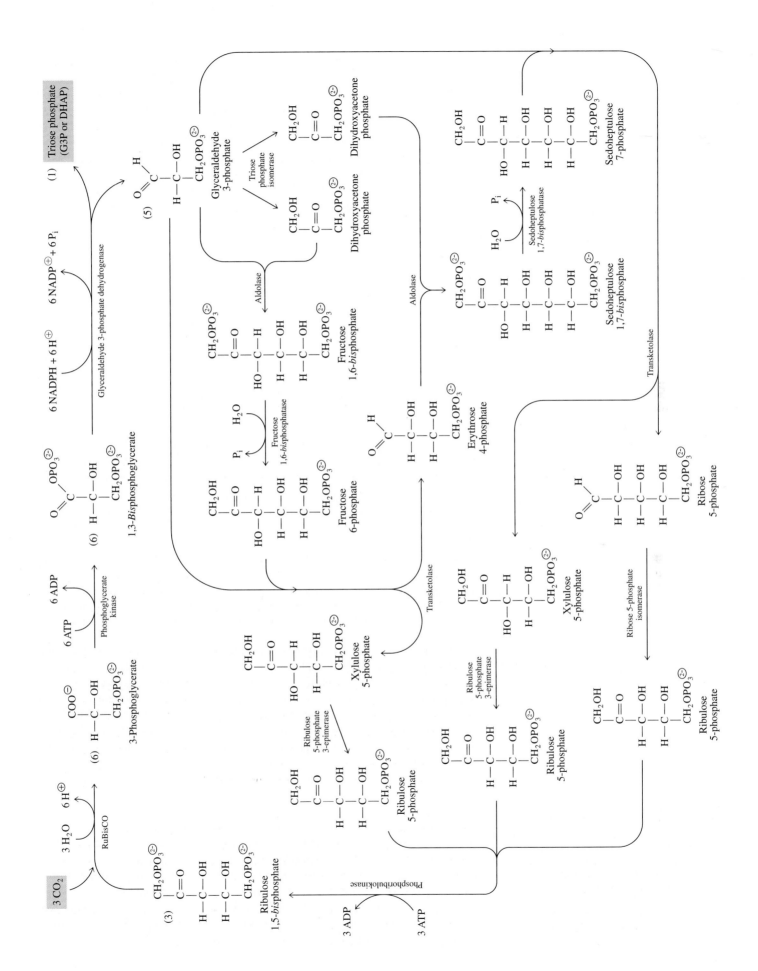

Table 19·1 Reactions of the RPP cycle, involving the fixation of three molecules of carbon dioxide for each molecule of triose phosphate to be used in carbohydrate synthesis

Reaction	Enzyme
3 Ribulose 1,5-*bis*phosphate $+$ 3 CO_2 $+$ 3 H_2O $\longrightarrow$ 6 3-Phosphoglycerate $+$ 6 $H^{\oplus}$	RuBisCO
6 3-Phosphoglycerate $+$ 6 ATP $\longrightarrow$ 6 1,3-*Bis*phosphoglycerate $+$ 6 ADP	Phosphoglycerate kinase
6 1,3-*Bis*phosphoglycerate $+$ 6 NADPH $+$ 6 $H^{\oplus}$ $\longrightarrow$ 6 Glyceraldehyde 3-phosphate $+$ 6 $NADP^{\oplus}$ $+$ 6 P_i	Glyceraldehyde 3-phosphate dehydrogenase
2 Glyceraldehyde 3-phosphate $\rightleftarrows$ 2 Dihydroxyacetone phosphate	Triose phosphate isomerase
Dihydroxyacetone phosphate $+$ Glyceraldehyde 3-phosphate $\rightleftarrows$ Fructose 1,6-*bis*phosphate	Aldolase
Fructose 1,6-*bis*phosphate $+$ H_2O $\longrightarrow$ Fructose 6-phosphate $+$ P_i	Fructose 1,6-*bis*phosphatase
Fructose 6-phosphate $+$ Glyceraldehyde 3-phosphate $\rightleftarrows$ Erythrose 4-phosphate $+$ Xylulose 5-phosphate	Transketolase
Erythrose 4-phosphate $+$ Dihydroxyacetone phosphate $\rightleftarrows$ Sedoheptulose 1,7-*bis*phosphate	Aldolase
Sedoheptulose 1,7-*bis*phosphate $+$ H_2O $\longrightarrow$ Sedoheptulose 7-phosphate $+$ P_i	Sedoheptulose 1,7-*bis*phosphatase
Sedoheptulose 7-phosphate $+$ Glyceraldehyde 3-phosphate $\rightleftarrows$ Xylulose 5-phosphate $+$ Ribose 5-phosphate	Transketolase
2 Xylulose 5-phosphate $\rightleftarrows$ 2 Ribulose 5-phosphate	Ribulose 5-phosphate 3-epimerase
Ribose 5-phosphate $\rightleftarrows$ Ribulose 5-phosphate	Ribose 5-phosphate isomerase
3 Ribulose 5-phosphate $+$ 3 ATP $\longrightarrow$ 3 Ribulose 1,5-*bis*phosphate $+$ 3 ADP	Phosphoribulokinase

Net reaction

$$3\,CO_2 + 9\,ATP + 6\,NADPH + 5\,H_2O \longrightarrow 9\,ADP + 8\,P_i + 6\,NADP^{\oplus} + \text{Triose phosphate (G3P or DHAP)}$$

19·13 Light, pH, and $Mg^{\oplus\oplus}$ Regulate the Activities of Some Enzymes of the RPP Cycle

Regulation of the RPP cycle occurs at the metabolically irreversible steps of the pathway, including the *bis*phosphatase reactions, the steps that consume ATP and NADPH, and the carboxylation reaction. Specifically, these steps include the reactions catalyzed by the fructose 1,6-*bis*phosphatase, sedoheptulose 1,7-*bis*-phosphatase, phosphoglycerate kinase, phosphoribulokinase, glyceraldehyde 3-phosphate dehydrogenase, and RuBisCO. All these enzymes appear to be regulated by one or more of several factors, including light, the concentration of stromal $Mg^{\oplus\oplus}$, and stromal pH.

The activities of all the enzymes mentioned above except phosphoglycerate kinase and RuBisCO are activated by light by a mechanism involving the reduction of surface-exposed disulfides; this reduction increases the activity of the enzyme by causing a change in its tertiary structure. Reducing equivalents are provided by the photosynthetic electron-transport chain for the formation of the —SH groups. Reduced ferredoxin, instead of reducing $NADP^{\oplus}$ as shown in Figure 19·6, donates its electrons to a small stromal protein, thioredoxin (Figure 19·17). Changes in the oxidation state of ferredoxin are between the $Fe^{\oplus\oplus}$ and $Fe^{\oplus\oplus\oplus}$ states of iron, whereas the oxidation and reduction of thioredoxin is due to a pair of sulfhydryl groups. Reduction of a cystine disulfide of thioredoxin to thiols is catalyzed by the enzyme ferredoxin-thioredoxin reductase. The sulfhydryl groups of the soluble thioredoxin

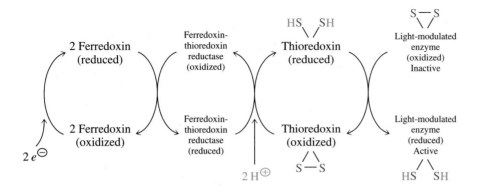

Figure 19·17
General scheme for the light activation of light-modulated enzymes of the RPP cycle. During light activation, electrons are diverted from the photosynthetic electron-transport chain and used to reduce disulfide groups of the inactive enzymes to sulfhydryl groups.

can then undergo spontaneous sulfhydryl-disulfide exchange with a surface disulfide of a light-modulated enzyme, resulting in activation of the enzyme. Thus, enzyme activation is controlled by light through the provision of electrons from light-dependent electron transport.

Recent research indicates that reversible phosphorylation may be involved in the short-term light activation of RuBisCO as well as enzymes of the C_4 pathway (Section 19·15) and sucrose metabolism (Section 19·17). In addition to the short-term light regulation of RuBisCO, a tightly binding inhibitor, present only at night, keeps the enzyme inactive in the dark. The inhibitor, 2-carboxyarabinitol 1-phosphate (Figure 19·18), is very similar in structure to the RuBisCO reaction intermediate 2-carboxy-3-ketoarabinitol 1,5-*bis*phosphate.

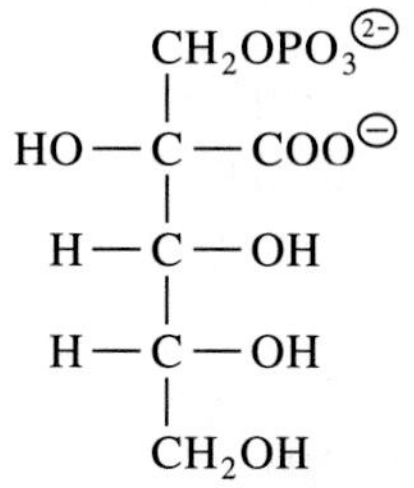

The stromal concentration of Mg^{2+} and the stromal pH, both of which increase as protons are translocated into the lumen, are also regulatory signals for certain enzymes of the RPP cycle. As we saw in Section 19·11, RuBisCO requires Mg^{2+} and CO_2 for activation. Fructose 1,6-*bis*phosphatase and sedoheptulose 1,7-*bis*phosphatase also require an alkaline pH and a high concentration of Mg^{2+} for maximum activity. Phosphoribulokinase is subject to another type of pH regulation through its inhibition by 3-phosphoglycerate (3PG), the RuBisCO reaction product. Phosphoribulokinase is only inhibited by 3-phosphoglycerate in the $3PG^{2-}$ form. During active photosynthesis, when the stroma is alkaline, the predominant ionic form is $3PG^{3-}$. However, when photosynthetic activity decreases and the stromal pH drops, the level of $3PG^{2-}$ increases; consequently, there is increased inhibition of phosphoribulokinase. This inhibition prevents continued ATP consumption by phosphoribulokinase when the rate of photosynthesis declines.

Factors such as pH and $[Mg^{2+}]$ that affect the activities of the regulated RPP-cycle enzymes are significantly different under photosynthetic versus nonphotosynthetic conditions and thereby coordinate the rate of carbohydrate synthesis with the rate of ATP and NADPH formation. Consider the importance of this coordination to a plant leaf that receives intermittent light. Rapid activation of the RPP cycle is important to achieve maximum carbon assimilation during the light. However, in the dark, efficient deactivation is necessary to prevent imbalances between the concentrations of RPP-cycle intermediates that would preclude rapid response to renewed photosynthetic activity.

$$\begin{array}{c} CH_2OPO_3^{2-} \\ | \\ HO-C-COO^{-} \\ | \\ H-C-OH \\ | \\ H-C-OH \\ | \\ CH_2OH \end{array}$$

Figure 19·18
Structure of 2-carboxyarabinitol 1-phosphate, a competitive inhibitor of RuBisCO.

19·14 RuBisCO Also Catalyzes the Oxygenation of Ribulose 1,5-*Bis*phosphate

As its complete name indicates, ribulose 1,5-*bis*phosphate carboxylase-oxygenase catalyzes not only carboxylation but also oxygenation of ribulose 1,5-*bis*phosphate. The two reactions are competitive—CO_2 and O_2 compete for active sites on RuBisCO. The apparent (or measured) affinity of RuBisCO for CO_2 is much

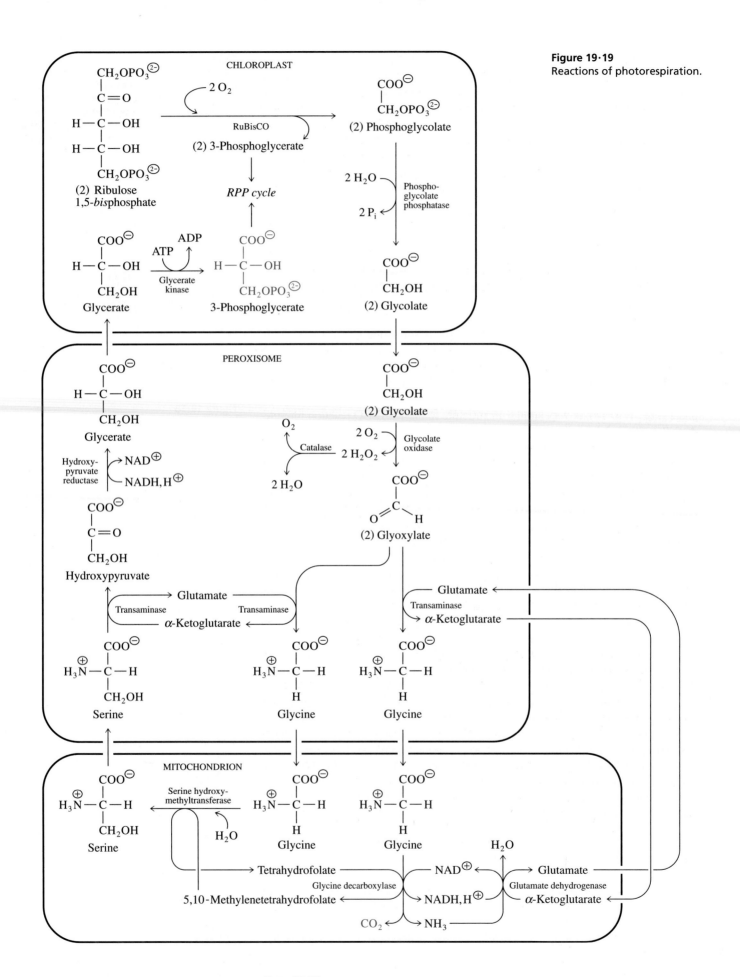

Figure 19·19
Reactions of photorespiration.

greater than for oxygen, and CO_2 is more soluble in the stroma than O_2, but the atmospheric concentration of oxygen (21%) is much higher than that of carbon dioxide (0.03%). As a result, both reactions contribute significantly to the consumption of ribulose 1,5-*bis*phosphate in vivo. Under normal conditions, there is a carboxylation-to-oxygenation ratio of 3:1 to 4:1. In Section 19·12, we discussed the outcome of the carboxylation reaction catalyzed by RuBisCO. We now turn to the metabolic sequence that follows the oxygenation of ribulose 1,5-*bis*phosphate by RuBisCO.

The products of the RuBisCO oxygenation reaction are one molecule of 3-phosphoglycerate and one molecule of phosphoglycolate. Figure 19·19 shows the net result of two oxygenation events. Two molecules of 3-phosphoglycerate enter the RPP cycle directly, and two molecules of 2-phosphoglycolate are metabolized to one molecule of CO_2 and one molecule of 3-phosphoglycerate, which enters the RPP cycle. **Photorespiration,** the light-dependent uptake of O_2 and release of CO_2, is the name given to the processes of oxygen uptake by RuBisCO and the metabolism of phosphoglycolate.

The metabolism of phosphoglycolate occurs in three different cellular compartments—the chloroplast, the peroxisome, and the mitochondrion. Phosphoglycolate is dephosphorylated to glycolate in the chloroplast; glycolate then enters the peroxisome, where it reacts with O_2 to form glyoxylate and hydrogen peroxide. Hydrogen peroxide is readily converted to O_2 and H_2O by the enzyme catalase, and glyoxylate is transaminated to glycine. Glycine exits the peroxisome and enters the mitochondrion, where it can undergo oxidative decarboxylation in a cleavage reaction that yields 5,10-methylenetetrahydrofolate, CO_2, and NH_3 (Section 21·12C). CO_2 is either photosynthetically refixed or released from the leaf, and NH_3 is reassimilated to form glutamate. 5,10-Methylenetetrahydrofolate donates its methylene group to a second molecule of glycine (also from glyoxylate) to form serine. Serine exits the mitochondrion and enters the peroxisome, where it is deaminated to hydroxypyruvate; hydroxypyruvate is then reduced to glycerate. Glycerate enters the chloroplast, where it is phosphorylated by ATP to 3-phosphoglycerate. 3-Phosphoglycerate can be incorporated into the RPP cycle.

Since extensive searches of RuBisCO mutants have not uncovered a mutant enzyme that catalyzes *only* the carboxylation of ribulose 1,5-*bis*phosphate (a "RuBisC"), it is speculated that, although a seemingly wasteful process (one carbon atom is lost for every two O_2 fixed), photorespiration is physiologically important. It may be that photorespiration regenerates ADP and $NADP^{\oplus}$ under conditions of low CO_2 concentration, for example, when stomata are closed and light intensity is high. During rapid photosynthesis, the capacity to form ATP and NADPH can be limited by the supply of stromal ADP and $NADP^{\oplus}$. Under these conditions, electron carriers can become over-reduced, and the proton concentration gradient across the thylakoid membrane can become excessively high. With no means to turn over ATP and NADPH and dissipate the proton concentration gradient, light-sensitive photosynthetic pigments of the LHC and photosystem antenna complexes can be oxidatively damaged. When stomata are closed, reactive oxygen species (e.g., superoxide anions) formed from accumulated oxygen generated by the oxygen-evolving complex can cause severe oxidative damage to pigments and lipids. Photorespiratory oxygen uptake and other reactions that consume oxygen and scavenge reactive oxygen species may reduce the possibility of such damage.

Figure 19·20
Structures of leaves of C_4 and C_3 plants. Note that the large bundle sheath cells of C_4 plants contain chloroplasts and are completely surrounded by mesophyll cells.

C_4 LEAF SECTION

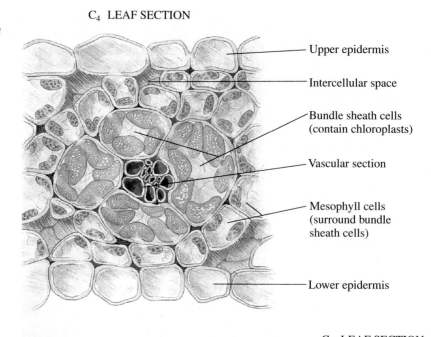

- Upper epidermis
- Intercellular space
- Bundle sheath cells (contain chloroplasts)
- Vascular section
- Mesophyll cells (surround bundle sheath cells)
- Lower epidermis

C_3 LEAF SECTION

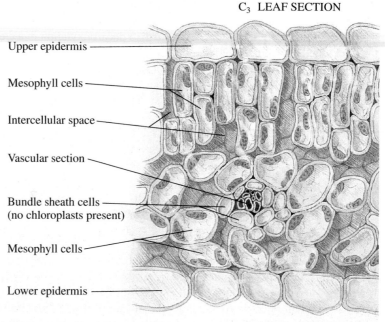

- Upper epidermis
- Mesophyll cells
- Intercellular space
- Vascular section
- Bundle sheath cells (no chloroplasts present)
- Mesophyll cells
- Lower epidermis

19·15 The C_4 Pathway Minimizes the Oxygenase Activity of RuBisCO by Concentrating CO_2

In several plant species, a second pathway for carbon fixation occurs in conjunction with the RPP cycle. In this pathway, the initial product of carbon fixation is a four-carbon acid rather than a three-carbon acid, and two distinct cell types are involved. Since the initial product of CO_2 fixation is a C_4 acid, the metabolic route is called the **C_4 pathway,** just as the RPP cycle is also called the C_3 pathway. C_4 plants include such economically important species as maize (corn), sorghum, and sugarcane, as well as many of the most troublesome weeds. Plants with the C_4 pathway have essentially no photorespiratory activity.

The C_4 pathway, which can be considered a prelude to the C_3 pathway, has two phases, carboxylation and decarboxylation. In the initial phase, CO_2 is fixed to C_4

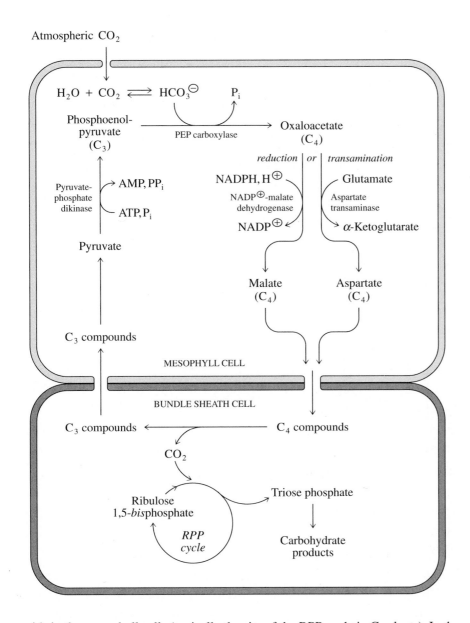

Figure 19·21
General scheme of the C_4 pathway. Four-carbon acids are formed in the mesophyll cell, transported to the bundle sheath cells via plasmodesmata, and decarboxylated in the bundle sheath cells. The CO_2 released is refixed by the RuBisCO reaction and enters the RPP cycle. The remaining three-carbon unit is converted back to the CO_2 acceptor molecule, phosphoenolpyruvate.

acids in the mesophyll cells (typically the site of the RPP cycle in C_3 plants). In the second phase, C_4 acids are decarboxylated in the bundle sheath cells, where CO_2 is released, refixed by RuBisCO, and incorporated into the RPP cycle. Figure 19·20 illustrates the cellular organization of C_4 and C_3 leaves in cross section. Two major structural features related to carbon assimilation distinguish C_4 from C_3 leaf types. First, in the C_4 leaf, mesophyll cells completely surround the bundle sheath cells, which surround the vascular tissue. Second, the C_4 bundle sheath cells are large and contain chloroplasts. In contrast, C_3 bundle sheath cells are nonphotosynthetic.

A schematic representation of the C_4 reaction sequence in the mesophyll and bundle sheath cells is shown in Figure 19·21. CO_2 is hydrated to bicarbonate ($HCO_3^{\ominus}$) in the mesophyll cytosol. Bicarbonate and phosphoenolpyruvate are the substrates for carboxylation catalyzed by phosphoenolpyruvate carboxylase (PEP carboxylase), a cytosolic enzyme that has no oxygenase activity. Note that meso-phyll cells of C_4 plants contain PEP carboxylase but not RuBisCO, whereas bundle sheath cells contain RuBisCO but not PEP carboxylase.

Depending on the C_4 species, the oxaloacetate formed by the action of PEP carboxylase is either reduced to malate or transaminated to aspartate. C_4 acids are transported via plasmodesmata (intercellular connections) to the adjacent bundle sheath cells. Since the cell walls of bundle sheath cells are quite impermeable to

(a)

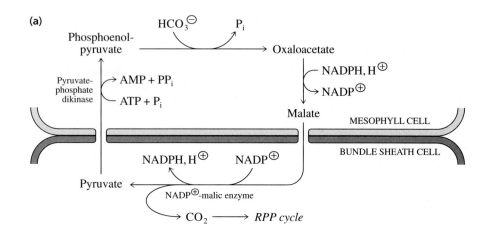

(b)

(c)

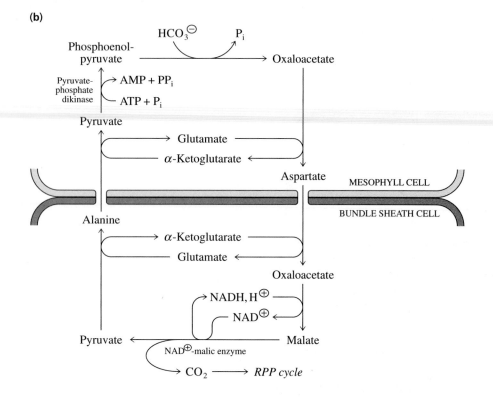

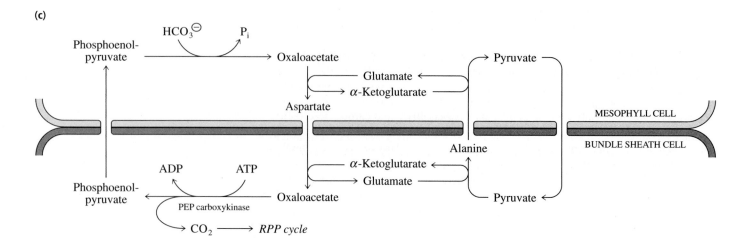

Figure 19·22
The three major C_4 subgroups, based on the predominant decarboxylation enzyme in their bundle sheath cells. **(a)** C_4 pathway in species with $NADP^{\oplus}$-malic enzyme. **(b)** C_4 pathway in species with $NAD^{\oplus}$-malic enzyme. **(c)** C_4 pathway in species with PEP carboxykinase.

gases, C_4 acid decarboxylation greatly increases the CO_2 concentration inside the cells and creates a high ratio of CO_2 to O_2. Since there is no RuBisCO in the mesophyll cells and the ratio of CO_2 to O_2 is extremely high in the bundle sheath cells, the oxygenase activity of RuBisCO is minimized. Thus, there is essentially no photorespiratory activity in C_4 plants.

Three major subgroups of C_4 plants are based on the predominance of one of three decarboxylation enzymes in bundle sheath cells (although not all C_4 species necessarily have one predominant decarboxylating enzyme). The three decarboxylases, which are each associated with somewhat different pathways, are $NADP^\oplus$-malic enzyme, $NAD^\oplus$-malic enzyme, and PEP carboxykinase (Figure 19·22).

$NADP^\oplus$-malic enzyme is located in the cytosol, whereas $NAD^\oplus$-malic enzyme is in the mitochondrion. Both of these decarboxylases catalyze the formation of CO_2 and pyruvate from malate. Pyruvate is subsequently phosphorylated to phosphoenolpyruvate, the bicarbonate acceptor molecule. Recall that in gluconeogenesis, pyruvate is converted to phosphoenolpyruvate via two reactions: first, pyruvate is carboxylated to form oxaloacetate; then oxaloacetate is decarboxylated to form phosphoenolpyruvate (Section 17·6). In the C_4 cycle, carboxylation and decarboxylation of pyruvate does not occur; pyruvate is converted to phosphoenolpyruvate in a single reaction catalyzed by pyruvate-phosphate dikinase. In this reaction, pyruvate, ATP, and P_i react to form phosphoenolpyruvate, AMP, and PP_i.

$$\text{Pyruvate} + \text{ATP} + \text{P}_i \xrightarrow{\substack{\text{Pyruvate-phosphate} \\ \text{dikinase}}} \text{Phosphoenolpyruvate} + \text{AMP} + \text{PP}_i$$

(19·3)

A second molecule of ATP is required to form ADP from AMP.

$$\text{AMP} + \text{ATP} \xrightarrow{\text{Adenylate kinase}} 2\,\text{ADP} \qquad \textbf{(19·4)}$$

The total energy cost for conversion of one molecule of pyruvate to phosphoenolpyruvate is two molecules of ATP.

The third decarboxylase, PEP carboxykinase, is located in the cytosol. This enzyme differs from $NAD^\oplus$-malic enzyme and $NADP^\oplus$-malic enzyme in that, instead of malate, oxaloacetate is decarboxylated and phosphorylated to generate phosphoenolpyruvate. In contrast to the malic enzymes, PEP carboxykinase regenerates phosphoenolpyruvate in a single step that requires only one molecule of ATP.

$$\text{Oxaloacetate} + \text{ATP} \xrightarrow{\substack{\text{PEP} \\ \text{carboxykinase}}} \text{CO}_2 + \text{Phosphoenolpyruvate} + \text{ADP}$$

(19·5)

Phosphoenolpyruvate is transported to the mesophyll cell. Since a C_4 amino acid (aspartate) is imported to the bundle sheath cell and a C_3 acid (phosphoenolpyruvate) is subsequently exported, an additional shuttle of pyruvate and alanine is needed to maintain the pool of glutamate required for oxaloacetate transamination in the mesophyll cell.

Although there is an extra energy cost to form phosphoenolpyruvate for C_4 carbon assimilation, the absence of photorespiration gives C_4 plants a significant advantage over C_3 plants. C_4 plants have significantly higher quantum yields (molecules of CO_2 fixed per photon absorbed) than C_3 plants under their respective optimal photosynthetic conditions. For reasons related to the speed of intercellular metabolite movement, requirements for light activation, and the temperature optima for certain C_4 enzymes, C_4 plants are most photosynthetically efficient under conditions of high light intensity and high temperature.

19·16 Certain Plants Carry Out Carbon Fixation at Night to Conserve Water

As CO_2 enters the plant during photosynthesis, a great deal of water can be lost from the leaf tissues through open stomata. A group of plants found primarily in arid environments and dry microclimates greatly reduce water loss during photosynthesis by carrying out a modified sequence of carbon-assimilation reactions that involves the accumulation of malate at night. Because the reaction sequence associated with the nocturnal accumulation of acid was first discovered in the stonecrop family, *Crassulaceae,* this sequence is called **Crassulacean acid metabolism,** or CAM (Figure 19·23). In addition to the stonecrops, CAM is found in several other plant families as well, including the cacti, bromeliads, and orchids (Figure 19·24). CAM plants of economic importance include pineapple and many ornamental succulent plants.

CAM plants take up CO_2 into mesophyll cells through open stomata at night. Water loss through the stomata is much lower at cooler nighttime temperatures than

Figure 19·23
General scheme of Crassulacean acid metabolism (CAM). During the night, PEP carboxylase and NAD⊕-malate dehydrogenase catalyze the formation of malate. Phosphoenolpyruvate required for malate synthesis is derived from starch. The next day, decarboxylation of malate increases the cellular concentration of CO_2, which can be fixed by the RPP cycle, while NADPH and ATP are formed by the light reactions of photosynthesis. The decarboxylation of malate also yields phosphoenolpyruvate, which is subsequently converted to starch by gluconeogenesis.

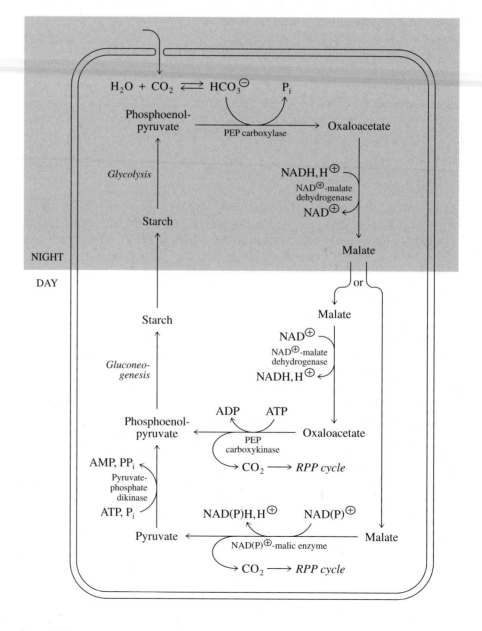

during the day. CO_2 is fixed by the PEP carboxylase reaction, and the oxaloacetate formed is reduced to malate, which is then translocated into the large, central vacuole. Transport of malate into the vacuole is necessary in order to maintain a near-neutral pH in the cytosol, since the cellular concentration of this acid can be as high as 0.2 M by the end of the night. The vacuole of a CAM mesophyll cell is large, generally occupying over 90% of the total cell volume. During the next light period, when ATP and NADPH are formed by the light reactions of photosynthesis, malate is released from the vacuole and decarboxylated. Thus, the large pool of malate accumulated at night supplies CO_2 for carbon assimilation during the day. During decarboxylation of malate, stomata are tightly closed so that neither water nor CO_2 can escape from the leaf, and the level of cellular CO_2 can be much higher than the level of atmospheric CO_2. As in C_4 plants, the higher internal CO_2 concentration greatly reduces photorespiration.

In fact, CAM is analogous to C_4 metabolism in that the RPP cycle is preceded by a reaction sequence involving the formation of C_4 acids, catalyzed by PEP carboxylase. Also, the three alternative decarboxylation pathways (Figure 19·22) are the same in CAM as in C_4 metabolism. CAM and C_4 metabolism differ, however, in that the C_4 pathway involves the spatial separation of the carboxylation and decarboxylation phases of the cycle between mesophyll and bundle sheath cells, whereas CAM, which takes place in the mesophyll cells, involves the temporal separation of these phases into a day and night cycle. As in the C_4 pathway, PEP carboxylase catalyzes the reaction of bicarbonate and phosphoenolpyruvate to yield oxaloacetate, which is reduced to malate. In CAM plants, however, this reaction occurs only at night. The phosphoenolpyruvate required for malate formation is derived from starch, which is glycolytically converted to phosphoenolpyruvate. During the day, the phosphoenolpyruvate formed during malate decarboxylation (either directly by PEP carboxykinase or via malic enzyme and pyruvate-phosphate dikinase) is converted to starch by gluconeogenesis and stored in the chloroplast. Thus, in CAM, not only are there large day/night changes in the pool of malate but also in the pool of starch (Figure 19·25).

An important regulatory feature of the CAM pathway is the inhibition of PEP carboxylase by malate and low pH. During the day, when the cytosolic concentration of malate is high and pH is low, PEP carboxylase is effectively inhibited. This inhibition is essential to prevent futile cycling of CO_2 and malate by PEP carboxylase and to avoid competition between PEP carboxylase and RuBisCO for CO_2.

Figure 19·24
Orchid. Orchids are one of several families of plants that carry out Crassulacean acid metabolism (CAM), a pathway of carbon assimilation that reduces water loss during photosynthesis. Orchids grow in semitropical climates in very loose soil or suspended in trees. Because of their exposed roots, orchids often undergo water stress.

Figure 19·25
Reciprocal changes in starch and malate during the CAM cycle. Starch serves as a source of carbon for the formation of the three-carbon molecule phosphoenolpyruvate, which constitutes three-quarters of the carbon in the malate formed during nocturnal CO_2 fixation.

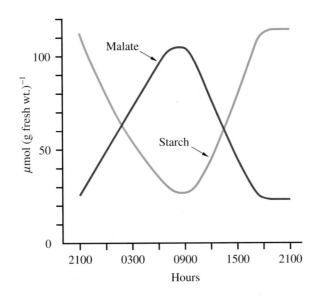

19·17 Sucrose and Starch Are Synthesized from Metabolites of the RPP Cycle

Essentially all plants convert the carbon fixed during the RPP cycle into starch or sucrose. Sucrose, some sugar alcohols, and certain oligosaccharides based on sucrose are translocatable forms of carbohydrate that are moved through the phloem (nutrient-conducting vascular elements) to plant organs that are not photosynthetically self-sustaining. These "sink" organs include roots, seeds, and young, developing leaves. Thus, in the common terminology, sink organs obtain carbohydrate from "source" leaves. Some carbohydrate accumulates in source leaves during the day as starch. In the absence of photosynthesis during the night, starch is broken down, converted to sucrose, and used as the metabolic fuel for the source leaves and other organs of the plant.

Sucrose is synthesized in the cytosol from triose phosphates (glyceraldehyde 3-phosphate and dihydroxyacetone phosphate) that originate either directly from the RPP cycle or from starch. Chloroplast starch is broken down to glucose or glucose monophosphates and glycolytically converted to triose phosphates. The movement of triose phosphates from the chloroplast to the cytosol is mediated by the phosphate translocator, a transport protein that spans the inner membrane of the chloroplast (Figure 19·26). The outer chloroplast membrane is readily permeable to triose phosphates and other small molecules. The translocation of chloroplast triose phosphate occurs in a strict exchange for inorganic phosphate from the cytosol. This stoichiometric requirement of the phosphate translocator regulates the rate of sucrose synthesis in relation to the ambient photosynthetic conditions by controlling the movement of triose phosphates, the substrate for sucrose formation, and inorganic phosphate, a product of sucrose formation.

Chloroplast triose phosphate is available for export to the cytosol during sustained photosynthesis, and inorganic phosphate becomes available during sucrose synthesis. When sucrose synthesis slows, cytosolic P_i is less available to the phosphate translocator, and more triose phosphate is retained in the chloroplast and converted to starch. When photosynthesis slows, triose phosphates are less available for export to the cytosol for sucrose formation.

The reaction sequence for sucrose synthesis is depicted in Figure 19·27. Four molecules of triose phosphate are used to produce one molecule of sucrose. In this sequence, four triose phosphates are converted to one molecule of fructose 6-phosphate and one molecule of glucose 6-phosphate by the action of aldolase,

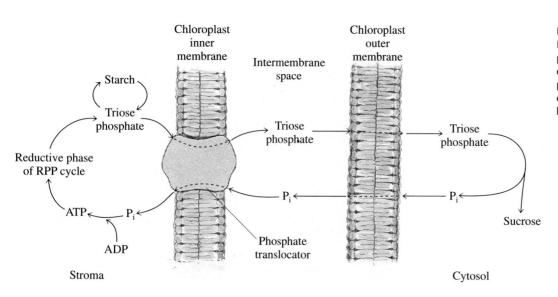

Figure 19·26
Export of stromal triose phosphate in exchange for cytosolic P_i via the phosphate translocator of the chloroplast inner membrane.

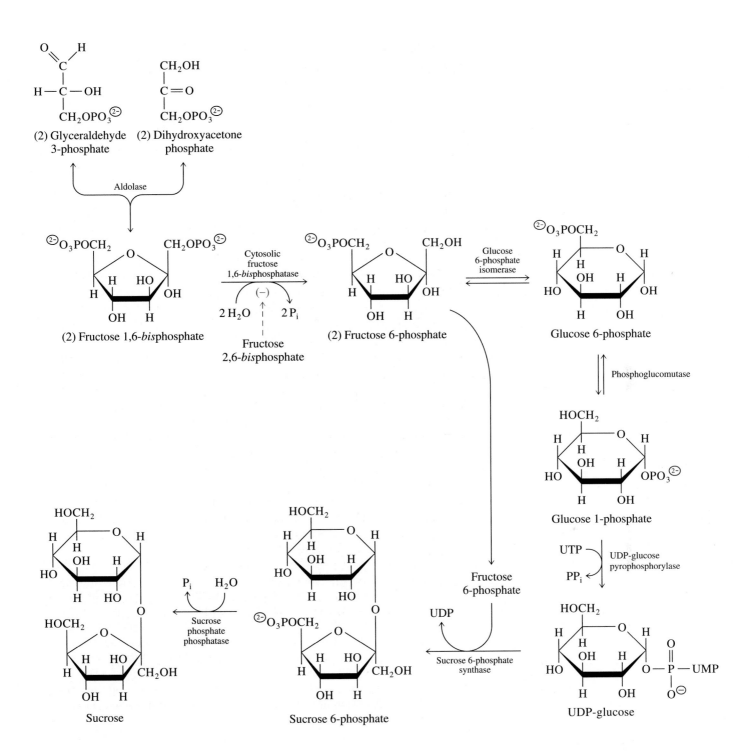

Figure 19·27
Pathway for the biosynthesis of sucrose from triose phosphates (glyceraldehyde 3-phosphate and dihydroxyacetone phosphate) in the cytosol. Four molecules of triose phosphate ($4\,C_3$) from the RPP cycle are consumed for each molecule of sucrose (C_{12}) synthesized. The pathway is regulated via the phosphate translocator and fructose 2,6-*bis*phosphate.

fructose 1,6-*bis*phosphatase, and glucose 6-phosphate isomerase. The glucose 6-phosphate is converted to glucose 1-phosphate, which reacts with UTP to form UDP-glucose (Figure 8·11). Fructose 6-phosphate and UDP-glucose then combine to form sucrose 6-phosphate. Phosphate is cleaved from sucrose 6-phosphate, forming sucrose.

During sucrose synthesis, inorganic phosphate is produced in the reactions catalyzed by fructose 1,6-*bis*phosphatase and sucrose phosphate phosphatase, and inorganic pyrophosphate is produced in the reaction catalyzed by UDP-glucose pyrophosphorylase. The first metabolically irreversible step in the pathway is the hydrolysis of fructose 1,6-*bis*phosphate to yield fructose 6-phosphate and P_i. Not surprisingly, the activity of fructose 1,6-*bis*phosphatase is highly regulated, and interestingly, its regulation is via the allosteric effector molecule fructose 2,6-*bis*phosphate (Figure 15·26), which we encountered in our examinations of glycolysis and gluconeogenesis. Fructose 2,6-*bis*phosphate, although not an intermediate in the sucrose pathway, serves a regulatory role in sucrose metabolism as a potent inhibitor of cytosolic fructose 1,6-*bis*phosphatase. The level of fructose 2,6-*bis*phosphate is determined by the relative rates of its synthesis and degradation, catalyzed in plants by the activities of fructose 2,6-*bis*phosphate kinase and fructose 2,6-*bis*phosphatase, respectively. Both of these enzyme activities are subject to complex regulation by a number of metabolites that reflect the suitability of conditions for sucrose synthesis.

Triose phosphate from the RPP cycle can also be converted to starch in the chloroplast (Figure 19·28). In starch synthesis as in sucrose synthesis, triose phosphate is condensed to fructose 1,6-*bis*phosphate and acted upon by fructose 1,6-*bis*phosphatase to yield fructose 6-phosphate, which is then isomerized to glucose 6-phosphate and converted to glucose 1-phosphate. In starch synthesis, glucose 1-phosphate is then converted to ADP-glucose (versus UDP-glucose in sucrose synthesis). In this reaction, catalyzed by ADP-glucose pyrophosphorylase, one molecule of ATP is consumed, and pyrophosphate is released. Glucose from ADP-glucose is donated to the growing starch polymer in a reaction catalyzed by starch synthase.

The synthesis of ADP-glucose catalyzed by ADP-glucose pyrophosphorylase is the regulated step in the pathway of starch biosynthesis. The activity of this enzyme is strongly influenced by the ratio of 3-phosphoglycerate to inorganic phosphate, which can change under different photosynthetic conditions. During rapid photosynthesis, when the concentration of 3-phosphoglycerate (the product of the RuBisCO reaction) is high and the concentration of inorganic phosphate (used in ATP formation) is low, ADP-glucose pyrophosphorylase becomes activated. Conversely, when the rates of photosynthesis and photophosphorylation decrease and the ratio of 3-phosphoglycerate to inorganic phosphate decreases, ADP-glucose pyrophosphorylase and starch synthesis are inhibited. This regulatory feature avoids excessive diversion of triose phosphates from the RPP cycle to starch synthesis when the rate of photosynthesis declines.

At night, starch serves as a source of carbon for growth and respiration. The starch molecule, a glucose polymer similar to glycogen, is phosphorolytically cleaved by the action of starch phosphorylase to generate glucose 1-phosphate. Glucose 1-phosphate is converted to glucose 6-phosphate and metabolized by glycolysis to triose phosphates, which are exported from the chloroplast for sucrose synthesis and subsequent respiration or transport from the cell. An alternate route for starch breakdown is hydrolysis by amylases to variously sized dextrins and eventually to maltose and then glucose. Glucose formed via this route is phosphorylated by the action of hexokinase and enters the glycolytic and sucrose biosynthetic pathways.

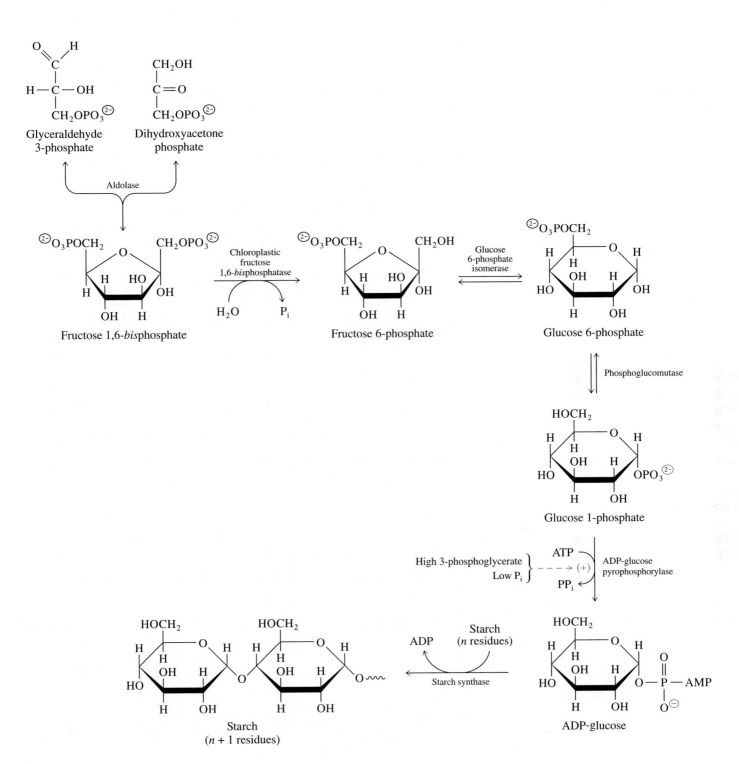

Figure 19·28
Pathway for the biosynthesis of starch in the chloroplast.

Summary

The photosynthetic processes encompass the light-dependent electron-transport reactions that form NADPH and ATP and the subsequent utilization of NADPH and ATP in the conversion of atmospheric CO_2 to carbohydrate (CH_2O).

The chloroplast is the organelle in which photosynthesis occurs in algae and plants. In the chloroplast, the thylakoid membrane contains electron-transport components, including photosystem I (PSI) and photosystem II (PSII), and ATP synthase. The thylakoid membrane is suspended in the aqueous stroma. The stroma contains all the enzymes and metabolites that participate in the conversion of CO_2 to (CH_2O) by the reductive pentose phosphate (RPP) cycle.

Chlorophyll and other pigments capture light energy for photosynthesis. These pigments are components of photosystems, the functional units of the light-dependent reactions of photosynthesis. Each photosystem is composed of light-absorbing antenna pigments and a reaction center. The reaction center contains two chlorophyll molecules called the special pair. Plants have two types of photosystems, designated PSI and PSII, whereas each photosynthetic bacterium has only one type of photosystem, which is different from the plant photosystems.

The absorption of light energy by chlorophyll and accessory pigments culminates in the excitation of the special pair of reaction-center chlorophyll molecules, which lowers their reduction potential. The excitation of the reaction-center primary-donor chlorophylls occurs in each of the two photosystems of plants and drives the noncyclic transfer of electrons from water through the photosystems to $NADP^{\oplus}$. The lowering of the reduction potential of a reaction-center chlorophyll by photoexcitation makes electron transfer to carriers of lower reduction potential possible. Cyclic electron transport, in which electrons are not transferred to $NADP^{\oplus}$ but rather are cycled back to PSI, also occurs. In both noncyclic and cyclic electron transport, protons are translocated across the photosynthetic membrane to generate a proton concentration gradient that drives the chemiosmotic conversion of ADP and P_i to ATP.

An elaborate short-term regulatory mechanism controls the energy distribution between PSI and PSII through phosphorylation and subsequent migration of the mobile LHC. Long-term changes in photosynthetic capacity (e.g., changes in the number and proportions of the various pigment complexes and modifications in the stoichiometries of PSI, PSII, and electron carriers) can occur in response to long-term changes in light intensity and spectral quality.

Each photosynthetic bacterium (except cyanobacteria) has only one type of photosystem, which is located in the plasma membrane. Protons are translocated across the plasma membrane into the periplasmic space. Different species of bacteria carry out either cyclic or noncyclic electron transfer using electrons from various reduced compounds.

The products of the light-dependent reactions, NADPH and ATP, are consumed during the reductive pentose phosphate (RPP) cycle. There are three stages in the RPP cycle: 1) the fixation of CO_2 by RuBisCO, 2) the reduction of CO_2 to (CH_2O), and 3) the regeneration of the CO_2 acceptor molecule, ribulose 1,5-*bis*phosphate. The ATP and NADPH formed during the electron-transport reactions are consumed during the reduction and regeneration phases of the RPP cycle.

Coordination of the RPP cycle with the rate of ATP and NADPH synthesis depends on the regulation of several RPP-cycle enzymes by light, the stromal concentration of Mg^{2+}, and stromal pH. This coordination is essential to maintain balanced pools of RPP-cycle intermediates during changing photosynthetic conditions. The rate of photosynthesis can be limited by one or more of the following factors: light, the supply of CO_2, the rate of photosynthetic electron transport, RuBisCO activity, and the availability of inorganic phosphate for photophosphorylation.

There are three additional metabolic pathways associated with the RPP cycle. The first pathway is directly related to the oxygenase activity of RuBisCO. The reaction products that arise from the oxygenation of ribulose 1,5-*bis*phosphate by RuBisCO are catabolized by the photorespiratory pathway. This reaction sequence, which takes place in three different cellular organelles, recovers half of the carbon that is consumed by the oxygenation of RuBisCO. The other two pathways, the C_4 and CAM pathways, essentially eliminate the oxygenation of ribulose 1,5-*bis*phosphate by concentrating CO_2 at the site of RuBisCO. The C_4 pathway is found in such plants as maize (corn), sorghum, sugarcane, and many weeds, whereas the CAM pathway is found in such plants as cacti, bromeliads, and orchids. Both the C_4 and CAM pathways occur in addition to the RPP cycle. In both pathways, there is a preliminary PEP-carboxylation step, which yields a four-carbon acid. The subsequent decarboxylation of the four-carbon acid releases CO_2, which is refixed by RuBisCO. In both pathways, the preliminary carboxylation reaction has the effect of increasing the ratio of CO_2 to oxygen at the site of RuBisCO. In the C_4 pathway, the processes of CO_2 fixation by PEP carboxylase and CO_2 fixation by RuBisCO occur in two different cell types. In the CAM pathway, these two processes occur in the same cell type but at different times during the day/night cycle.

The end products of carbon assimilation are starch and sucrose. Starch is synthesized in the chloroplast. Sucrose is formed in the cytosol from triose phosphates that are exported from the chloroplast to the cytosol via the phosphate translocator. The major regulatory control for the rate of synthesis of starch and sucrose is the ratio of phosphate esters to inorganic phosphate. This ratio is a general indicator of the rate of photophosphorylation and carbon assimilation.

Selected Readings

General References

Goodwin, T. W., and Mercer, E. I. (1983). *Introduction to Plant Biochemistry,* 2nd ed. (New York: Pergamon Press).

Hatch, M. D., and Boardman, N. K., eds. (1981). *Photosynthesis.* Vol. 8 of *The Biochemistry of Plants: A Comprehensive Treatise.* P. K. Stumpf and E. E. Conn, eds. (New York: Academic Press).

Photosynthetic Electron Transport

Golbeck, J. H. (1992). Structure and function of photosystem I. *Annu. Rev. Plant Physiol. Plant Mol. Biol.* 43:293–324.

Govindjee and Coleman, W. J. (1990). How plants make oxygen. *Sci. Am.* 262(2):50–58.

Staehlin, L. A., and Arntzen, C. J., eds. (1986). *Photosynthesis III: Photosynthetic Membranes and Light Harvesting Systems.* Vol. 19 of *Encyclopedia of Plant Physiology* (New York: Springer-Verlag).

Youvan, D. C., and Marrs, B. L. (1987). Molecular mechanisms of photosynthesis. *Sci. Am.* 256(6):42–48.

Photophosphorylation

Arnon, D. I. (1984). The discovery of photosynthetic phosphorylation. *Trends Biochem. Sci.* 9:258–262.

Bennett, J. (1991). Phosphorylation in green plant chloroplasts. *Annu. Rev. Plant Physiol. Plant Mol. Biol.* 42:281–311.

Photosynthetic Carbon Metabolism

Bassham, J. A., and Calvin, M. (1957). *The Path of Carbon in Photosynthesis* (Englewood Cliffs, New Jersey: Prentice Hall).

Edwards, G. E., and Walker, D. (1983). C_3, C_4: *Mechanisms and Cellular and Environmental Regulation of Photosynthesis* (Berkeley: University of California Press).

Heber, U., and Krause, G. H. (1980). What is the physiological role of photorespiration? *Trends Biochem. Sci.* 5:32–34.

Kelly, G. J., Holtum, J. A. M., and Latzko, E. (1989). Photosynthesis. Carbon metabolism: new regulators of CO_2 fixation, the new importance of pyrophosphate and the old problem of oxygen involvement revisited. *Prog. Bot.* 50:74–101.

Woodrow, I. E., and Berry, J. A. (1988). Enzymatic regulation of photosynthetic CO_2 fixation in C_3 plants. *Annu. Rev. Plant Physiol. Plant Mol. Biol.* 39:533–594.

20

Lipid Metabolism

Consider the flight of the migratory bird: some species can fly 1000 miles nonstop at 25 miles per hour. Sustained work such as this can only be achieved when lipid stores are used for fuel. Glycogen, the other important depot of stored energy, can supply ATP for muscle contraction for at best a fraction of an hour. Other instances of prolonged, intense output of work, such as the migrations of locusts and the record-chasing of marathon runners, are similarly fueled by the metabolism of triacylglycerols. In this chapter, we consider the pathways by which lipids are oxidized for energy and synthesized for storage. We also consider the metabolism of membrane lipids, lipid signal molecules, and cholesterol.

20·1 Adipocytes Are the Main Repositories of Fat in Mammals

An adipocyte is essentially a fat droplet surrounded by a thin shell of cytosol in which the nucleus and other organelles are suspended. Fatty acids are stored in adipocytes as esterified constituents of triacylglycerols. Hormones control a carefully balanced system in which release of fatty acids is inhibited or stimulated depending on metabolic need. Hydrolysis of triacylglycerols, or **lipolysis,** is inhibited by insulin; the rate of lipolysis increases when the level of insulin in blood is low. Lipolysis is further stimulated when the level of epinephrine in blood rises, for instance during fasting or exercise. The pathway for inhibition of fatty acid release by

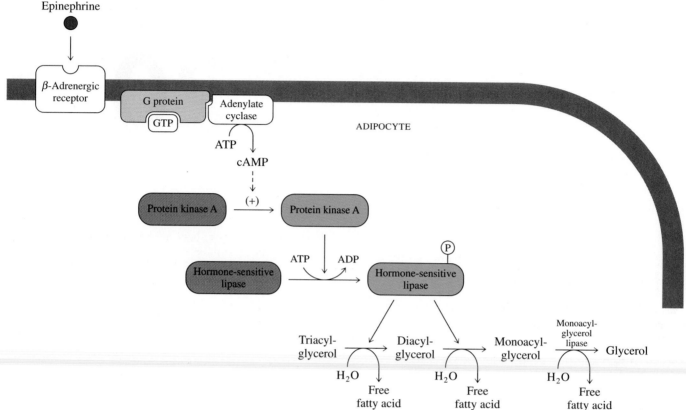

Figure 20·1
Hormonal regulation of triacylglycerol degradation. In adipocytes, epinephrine initiates an enzymatic cascade that results in the activation of protein kinase A. Protein kinase A catalyzes the phosphorylation and activation of hormone-sensitive lipase, which catalyzes hydrolysis of triacylglycerols to monoacylglycerols and free fatty acids. Hydrolysis of monoacylglycerols is catalyzed by monoacylglycerol lipase.

insulin is poorly understood despite huge ongoing research efforts. Epinephrine acts by binding to the β-adrenergic receptors of adipocytes, leading to the activation of adenylate cyclase, which catalyzes the production of cyclic AMP (cAMP). Elevated levels of cAMP in the cell lead to the activation of protein kinase A (cAMP-dependent protein kinase). This activation sequence is described in detail in Section 12·9. In adipocytes, protein kinase A catalyzes the phosphorylation and activation of hormone-sensitive lipase. This enzyme catalyzes the conversion of triacylglycerols to free fatty acids and monoacylglycerols. Although hormone-sensitive lipase can also catalyze conversion of monoacylglycerols to glycerol and free fatty acids, adipocytes contain a more specific and more active monoacylglycerol lipase that probably accounts for most of this catalytic activity. The pathway of triacylglycerol degradation is diagrammed in Figure 20·1. Glycerol and free fatty acids diffuse through the plasma membrane and enter the bloodstream. Glycerol is metabolized by the liver, where most of it is converted to glucose via gluconeogenesis (Section 17·8). Fatty acids are poorly soluble in aqueous solution and travel in blood bound to serum albumin, one of the major serum proteins—serum albumin accounts for about half the mass of serum proteins. Fatty acids are ferried in the bloodstream to a number of tissues, including heart, skeletal muscle, and liver, where they are oxidized in mitochondria to release energy.

20·2 Fatty Acids Are Oxidized by Removal of Two-Carbon Groups

In 1904, Franz Knoop conducted a classic biochemical experiment that led the way to the complete elucidation of the pathway for fatty acid degradation. Knoop fed dogs fatty acid derivatives that contained phenyl groups attached to the terminal

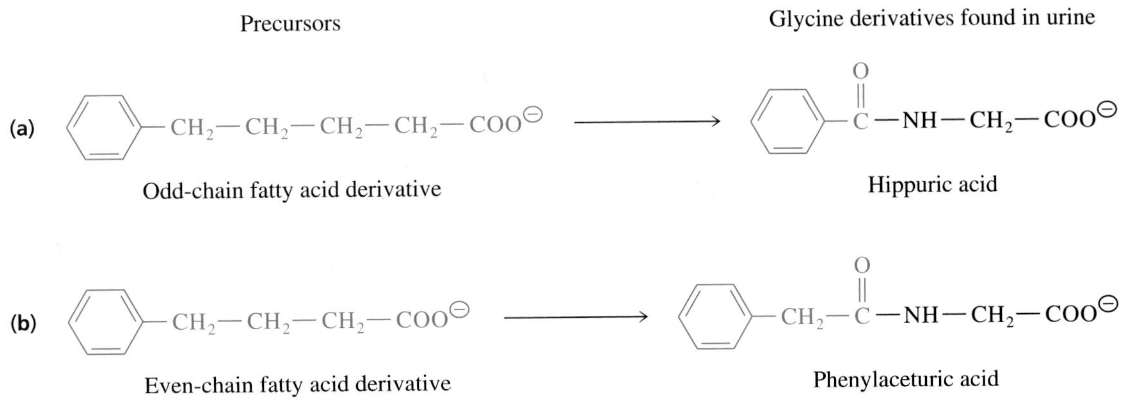

Precursors   Glycine derivatives found in urine

(a) ⟨phenyl⟩—CH_2—CH_2—CH_2—CH_2—$COO^{\ominus}$ ⟶ ⟨phenyl⟩—$\overset{\overset{\displaystyle O}{\|}}{C}$—$NH$—$CH_2$—$COO^{\ominus}$

Odd-chain fatty acid derivative   Hippuric acid

(b) ⟨phenyl⟩—CH_2—CH_2—CH_2—$COO^{\ominus}$ ⟶ ⟨phenyl⟩—CH_2—$\overset{\overset{\displaystyle O}{\|}}{C}$—$NH$—$CH_2$—$COO^{\ominus}$

Even-chain fatty acid derivative   Phenylaceturic acid

Figure 20·2
Knoop experiment. **(a)** Dogs fed a phenylated odd-chain fatty acid derivative always produced a glycine derivative with a single carbon of the acyl chain remaining. **(b)** When an even-chain fatty acid derivative was fed to the dogs, two carbons of the acyl chain were left. Together, the results suggest that two-carbon units are removed during fatty acid oxidation.

carbon, and then he isolated phenyl compounds in the dogs' urine. Knoop's addition of an aromatic group to the fatty acids to allow detection and isolation of the product was the earliest metabolic labelling study. When phenyl derivatives of fatty acids containing an odd number of carbon atoms (called odd-chain fatty acids) were ingested, hippuric acid, the conjugate of benzoate and glycine, was detected (Figure 20·2a). When even-chain fatty acid derivatives were ingested, phenylaceturic acid, the conjugate of phenylacetate and glycine, was recovered (Figure 20·2b). Clearly, the acyl chain was not degraded one carbon at a time, or both types of fatty acids would have produced hippuric acid. If degradation proceeded by groups of more than two carbons, correspondingly larger phenylated derivatives would have been found. Knoop therefore proposed (correctly, it was later shown) that fatty acids are shortened two carbons at a time by oxidation at the β-carbon. The biochemical route consists of several steps and is called the **β-oxidation pathway.**

The two-carbon fragments produced during β oxidation of fatty acids are now known to be transferred to coenzyme A to form acetyl CoA. The overall process of fatty acid degradation in cells can be divided into three stages: activation of fatty acids in the cytosol, transport into mitochondria, and degradation in two-carbon increments. We will consider these stages in detail, focusing first on the oxidation of an even-chain, saturated fatty acid. Specific reactions necessary for the oxidation of odd-chain and unsaturated fatty acids are considered in subsequent sections.

A. Fatty Acids Are Activated by Esterification to Coenzyme A

In the cytosol, fatty acids are activated by conversion to thioesters of CoA, catalyzed by acyl-CoA synthetase.

$$R\text{—}COO^{\ominus} + HS\text{-}CoA \xrightarrow[\text{Acyl-CoA synthetase}]{\overset{\displaystyle ATP \quad AMP}{\overset{\displaystyle \searrow \quad \nearrow}{\underset{}{}}}} R\text{—}\overset{\overset{\displaystyle O}{\|}}{C}\text{—}S\text{-}CoA \qquad (20\cdot1)$$

where PP_i is produced with AMP.

The pyrophosphate product is hydrolyzed by the action of inorganic pyrophosphatase. Thus, two high-energy phosphoanhydride linkages are consumed to form the CoA thioesters of fatty acids. Four different acyl-CoA synthetases are found in cells, with specificities for fatty acids whose chains are short ($<C_6$), medium (C_{6-12}), long ($>C_{12}$), or very long ($>C_{16}$).

The mechanism of the activation reaction involves an acyl-adenylate intermediate formed by the reaction of a fatty acid with ATP (Figure 20·3, next page). Nucleophilic attack by the sulfur atom of coenzyme A on the carbonyl carbon of the acyl group leads to release of AMP and the thioester fatty acyl CoA.

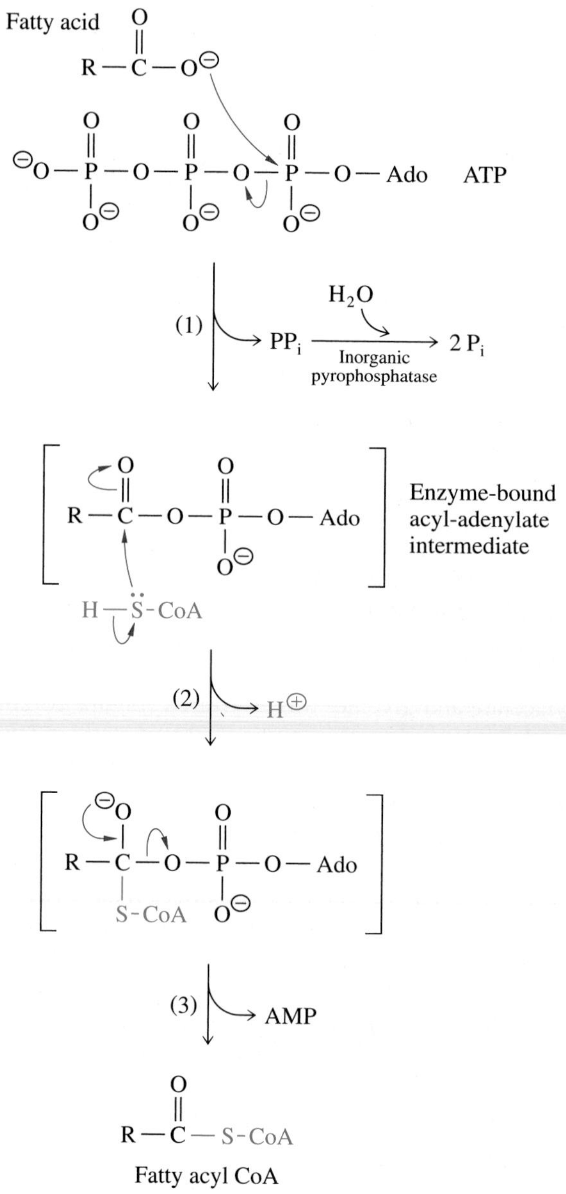

Figure 20·3
Mechanism of acyl-CoA synthetase. (1) An enzyme-bound acyl-adenylate intermediate is formed from a fatty acid and ATP. Pyrophosphate released in the process is hydrolyzed to two molecules of inorganic phosphate, with release of a large amount of free energy. (2) The thiol group of CoASH reacts with the carbonyl carbon of the intermediate. (3) Following cleavage, AMP and fatty acyl CoA are released. (*Ado* represents adenosine.)

B. Fatty Acyl CoA Is Transported into the Mitochondrial Matrix by a Shuttle System

Fatty acyl CoA formed in the cytosol cannot cross the inner mitochondrial membrane and enter the mitochondrial matrix, where the reactions of β oxidation occur. It might therefore seem more efficient to generate fatty acyl CoA from fatty acids in the mitochondrial matrix rather than in the cytosol. However, the production of fatty acyl CoA outside the matrix allows for control of β oxidation by regulation of a transport system that shuttles fatty acid molecules across the inner mitochondrial membrane. The shuttle process is carried out by two acyltransferases and a translocase protein embedded in the inner membrane (Figure 20·4). Transport is initiated when the acyl group of fatty acyl CoA is esterified to L-carnitine in a reaction catalyzed by carnitine acyltransferase I (CAT I), forming acylcarnitine. It appears that CAT I is localized to the inside of the outer mitochondrial membrane. As we will see in Section 20·11, regulation of CAT I is central to the control of fatty acid metabolism.

Figure 20·4
Carnitine shuttle system. A fatty acyl CoA molecule reacts with carnitine to form acylcarnitine and CoASH in a reaction catalyzed by carnitine acyltransferase I. This enzyme is apparently bound to the inside of the outer mitochondrial membrane. Acylcarnitine is transported across the inner mitochondrial membrane by carnitine:acylcarnitine translocase via antiport with carnitine from the matrix. In the matrix, membrane-bound carnitine acyltransferase II catalyzes the regeneration of fatty acyl CoA from acylcarnitine and CoASH. The path of the acyl group through the shuttle system is traced in red.

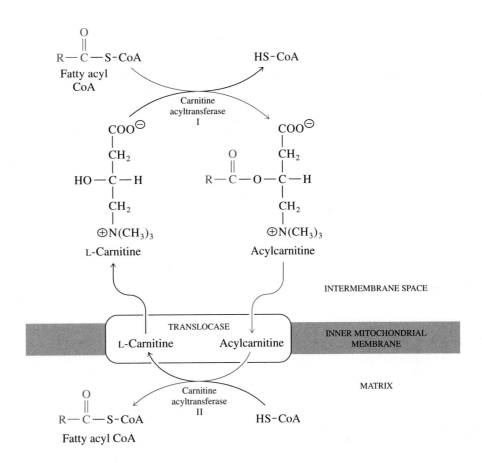

Once formed, acylcarnitine enters the mitochondrial matrix in exchange for free carnitine via the carnitine:acylcarnitine translocase. In the mitochondrial matrix, membrane-bound carnitine acyltransferase II (CAT II), an isozyme of CAT I, catalyzes the reverse of the reaction catalyzed by CAT I. Overall, the shuttle system results in removal of fatty acyl CoA from the cytosol and generation of fatty acyl CoA in the mitochondrial matrix.

C. Fatty Acid Oxidation Produces Acetyl CoA, NADH, and QH_2

Four steps are required to convert fatty acyl CoA to acetyl CoA: oxidation, hydration, further oxidation, and thiolysis (Figure 20·5, next page). In the first step, acyl-CoA dehydrogenase catalyzes the formation of a double bond between the α- and β-carbons of the acyl group, producing *trans*-Δ^2-enoyl CoA. In mammals, four separate acyl-CoA dehydrogenase isozymes catalyze this first step; each has a different chain-length preference: short, medium, long, or very long.

Associated with this first step is a series of electron transfers. Electrons from fatty acyl CoA are transferred to the FAD prosthetic group of acyl-CoA dehydrogenase, then to another FAD prosthetic group bound to a water-soluble protein called electron-transferring flavoprotein (ETF). Electrons are then passed to a membrane-bound iron-sulfur flavoprotein, ETF:ubiquinone reductase, which catalyzes the reduction of ubiquinone (Q). ETF, a heterodimer containing one FAD group, interacts with membrane-bound enzymes, including several flavoproteins not connected with fatty acid metabolism. Thus, ETF, like Q, is a mobile carrier of reducing equivalents.

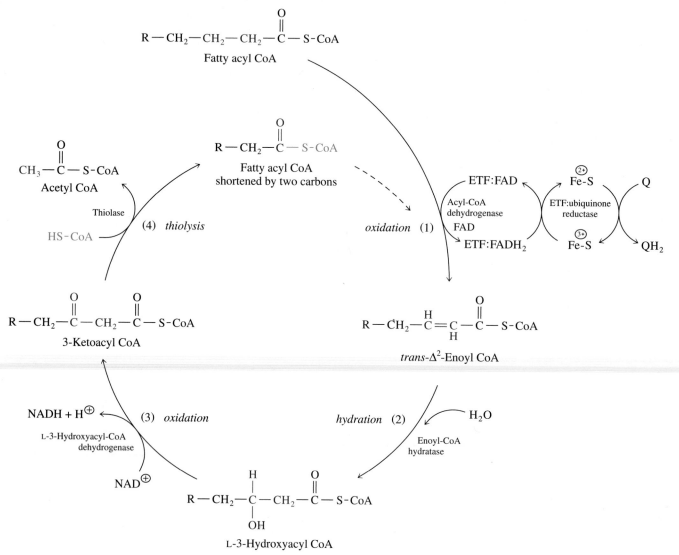

Figure 20·5
β Oxidation of saturated fatty acids. One round of β oxidation consists of four enzyme-catalyzed reactions. Each round generates one molecule each of QH$_2$, NADH, and acetyl CoA and produces a fatty acyl CoA molecule two carbon atoms shorter than the molecule that entered the round.

Enoyl CoA produced in the first step of β oxidation is hydrated to form the L isomer of 3-hydroxyacyl CoA. L-3-Hydroxyacyl CoA is then oxidized in an NAD$^\oplus$-dependent reaction to produce 3-ketoacyl CoA. Finally, the nucleophilic sulfhydryl group of a molecule of CoASH attacks the C-3 carbonyl group of 3-ketoacyl CoA in a cleavage reaction catalyzed by thiolase. The release of acetyl CoA leaves a fatty acyl CoA molecule shortened by two carbons. This acyl CoA molecule is a substrate for the same round of reactions, which continues until the entire molecule has been converted to acetyl CoA. As the fatty acyl chain becomes shorter, the different isozymes of acyl-CoA dehydrogenase (those with preferences for short, medium, long, or very long chains) are the catalysts.

Interestingly, the first three reactions of fatty acid oxidation are chemically parallel to three steps of the citric acid cycle—and are therefore easy to remember! The parallel steps are shown in Figure 20·6. The β-oxidation pathway is not cyclic like the citric acid cycle but is an example of a spiral metabolic pathway (Section 14·1).

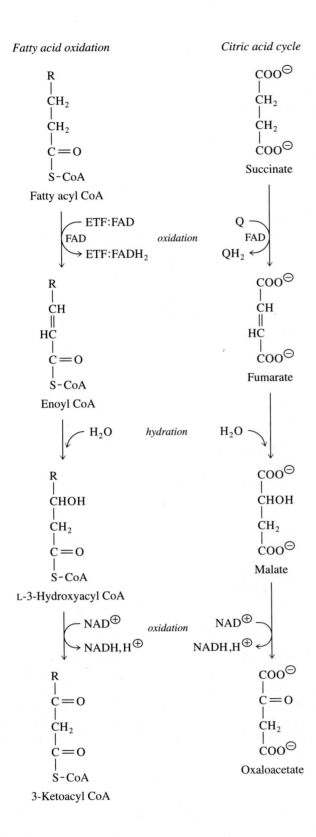

Fatty acid oxidation

Citric acid cycle

Figure 20·6
Chemical similarity between steps of the citric acid cycle and fatty acid oxidation.

D. Fatty Acid Oxidation Generates a Large Amount of ATP

The energy yield of fatty acid oxidation can be estimated from the amount of acetyl CoA, QH_2, and NADH produced. Consider the balanced reaction for the oxidation of one molecule of palmitoyl CoA by the β-oxidation pathway. Oxidation of palmitoyl CoA (C_{16}) to eight molecules of acetyl CoA requires seven cycles of β oxidation. (Remember that to divide a string into eight pieces, you need to cut it only seven times.)

$$\text{Palmitoyl CoA} + 7\,\text{CoASH} + 7\,Q + 7\,\text{NAD}^{\oplus} + 7\,H_2O \longrightarrow$$
$$8\,\text{Acetyl CoA} + 7\,QH_2 + 7\,\text{NADH} + 7\,H^{\oplus} \qquad \textbf{(20·2)}$$

Each acetyl CoA molecule can enter the citric acid cycle, generating more NADH and QH_2 and releasing two molecules of CO_2 per turn of the cycle. NADH and QH_2 produced by both the citric acid cycle and the β-oxidation pathway feed the electron-transport chain, generating ATP by oxidative phosphorylation. Water is also formed as a product of the electron-transport chain reactions and as a product of the formation of ATP from ADP and P_i. The overall equation for the complete oxidation of palmitoyl CoA via β oxidation, the citric acid cycle, electron transport, and oxidative phosphorylation is

$$\text{Palmitoyl CoA} + 23\,O_2 + 108\,P_i + 108\,\text{ADP} \longrightarrow$$
$$\text{CoASH} + 16\,CO_2 + 123\,H_2O + 108\,\text{ATP} \qquad \textbf{(20·3)}$$

(We use the conversion equivalents explained in Section 18·12: oxidation of 1 NADH yields approximately 2.5 ATP, 1 QH_2 yields approximately 1.5 ATP, and 1 acetyl CoA yields approximately 10 ATP). Since 2 ATP equivalents are expended in activating palmitate to palmitoyl CoA, the net yield is 106 ATP equivalents per molecule of palmitate completely oxidized.

By comparison, oxidation of one molecule of glucose to CO_2 and water gives an ATP yield of approximately 32. Since palmitate has 16 carbons and glucose only 6, we normalize the yield of ATP from glucose for direct comparison with fatty acids by multiplying the ATP yield of glucose by 16/6: $16/6 \times 32 = 85$ ATP, which is less than 80% of the ATP yield of palmitate. Fatty acids provide more energy than carbohydrates on a per-carbon basis because carbohydrates are already partially oxidized. Especially important for their role as fuel molecules is the hydrophobic nature of fatty acids, which permits them to be stored in large quantities without large amounts of bound water, as would be found with carbohydrates. Anhydrous storage allows far more energy to be stored on a per-weight basis. It is easy to appreciate why Antarctic explorers like Roald Amundsen came to prefer fat over carbohydrate as a dietary source of calories: they had to carry all their calories with them for the 1500-mile round trip to the South Pole.

When fat is used as a source of energy, the organism also receives as a bonus a remarkable amount of water: 123 water molecules for every molecule of palmitoyl CoA that is oxidized. The hump on a camel is a fat-storage depot that supplies the animal with water as well as energy, allowing the camel to travel long distances without drinking.

20·3 β Oxidation of Odd-Chain Fatty Acids Produces Propionyl CoA

Most fatty acids found in nature have an even number of carbon atoms. Nevertheless, odd-chain fatty acids are formed, for instance by bacteria in the stomachs of ruminants. Odd-chain fatty acids are oxidized by the same sequence of reactions as even-chain fatty acids. However, the product of the final thiolytic cleavage is propionyl CoA (CoA with a C_3 group attached) rather than acetyl CoA (C_2 group). In mammalian liver, three enzymes catalyze the conversion of propionyl CoA to succinyl CoA (Figure 20·7). Propionyl-CoA carboxylase is a biotin-dependent enzyme that catalyzes the incorporation of bicarbonate into propionyl CoA to produce D-methylmalonyl CoA. Methylmalonyl-CoA racemase catalyzes the conversion of D-methylmalonyl CoA to its L isomer. Finally, methylmalonyl-CoA mutase catalyzes the formation of succinyl CoA. Methylmalonyl-CoA mutase is one of the few enzymes that requires the vitamin-B_{12} derivative adenosylcobalamin as a cofactor. Recall from Section 8·13 that adenosylcobalamin-dependent enzymes catalyze intramolecular rearrangements in which a hydrogen atom and a substituent on an adjacent carbon atom exchange places. In the reaction catalyzed by methylmalonyl-CoA mutase, the —C(O)—S-CoA group exchanges positions with a hydrogen atom of the methyl group. This rearrangement is shown in Figure 8·45.

The succinyl CoA formed by the action of methylmalonyl-CoA mutase is metabolized to oxaloacetate. Since oxaloacetate is a substrate for gluconeogenesis, the propionyl group derived from odd-chain fatty acids can be converted on a net basis to glucose. Recall that in organisms without the glyoxylate cycle, carbohydrates can be converted to fatty acids, but fatty acids cannot be converted to carbohydrates. Gluconeogenesis from propionyl CoA is a minor exception to this rule.

Figure 20·7
Conversion of propionyl CoA to succinyl CoA.

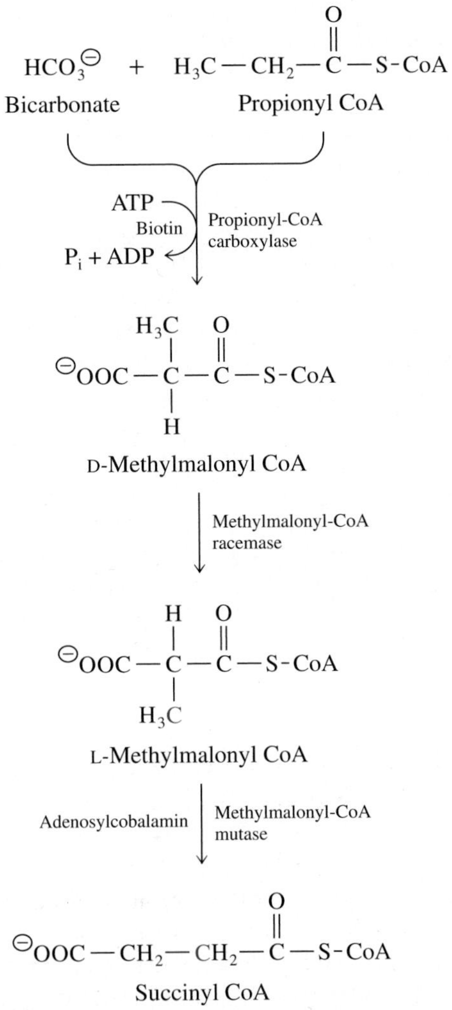

Figure 20·8
Oxidation of linoleoyl CoA. In addition to the enzymes of the β-oxidation pathway, two other enzymes are required, enoyl-CoA isomerase and 2,4-dienoyl-CoA reductase.

Linoleoyl CoA
$(C_{18}, cis, cis\text{-}\Delta^{9,12})$

① Three rounds of β oxidation

$(C_{12}, cis, cis\text{-}\Delta^{3,6})$

② Enoyl-CoA isomerase

$(C_{12}, trans, cis\text{-}\Delta^{2,6})$

③ One round of β oxidation

$(C_{10}, cis\text{-}\Delta^{4})$

④ Acyl-CoA dehydrogenase
(first reaction of β oxidation)

$(C_{10}, trans, cis\text{-}\Delta^{2,4})$

⑤ NADPH, H$^{\oplus}$
2,4-Dienoyl-CoA reductase
NADP$^{\oplus}$

$(C_{10}, trans\text{-}\Delta^{3})$

⑥ Enoyl-CoA isomerase
(same enzyme as Step 2)

$(C_{10}, trans\text{-}\Delta^{2})$

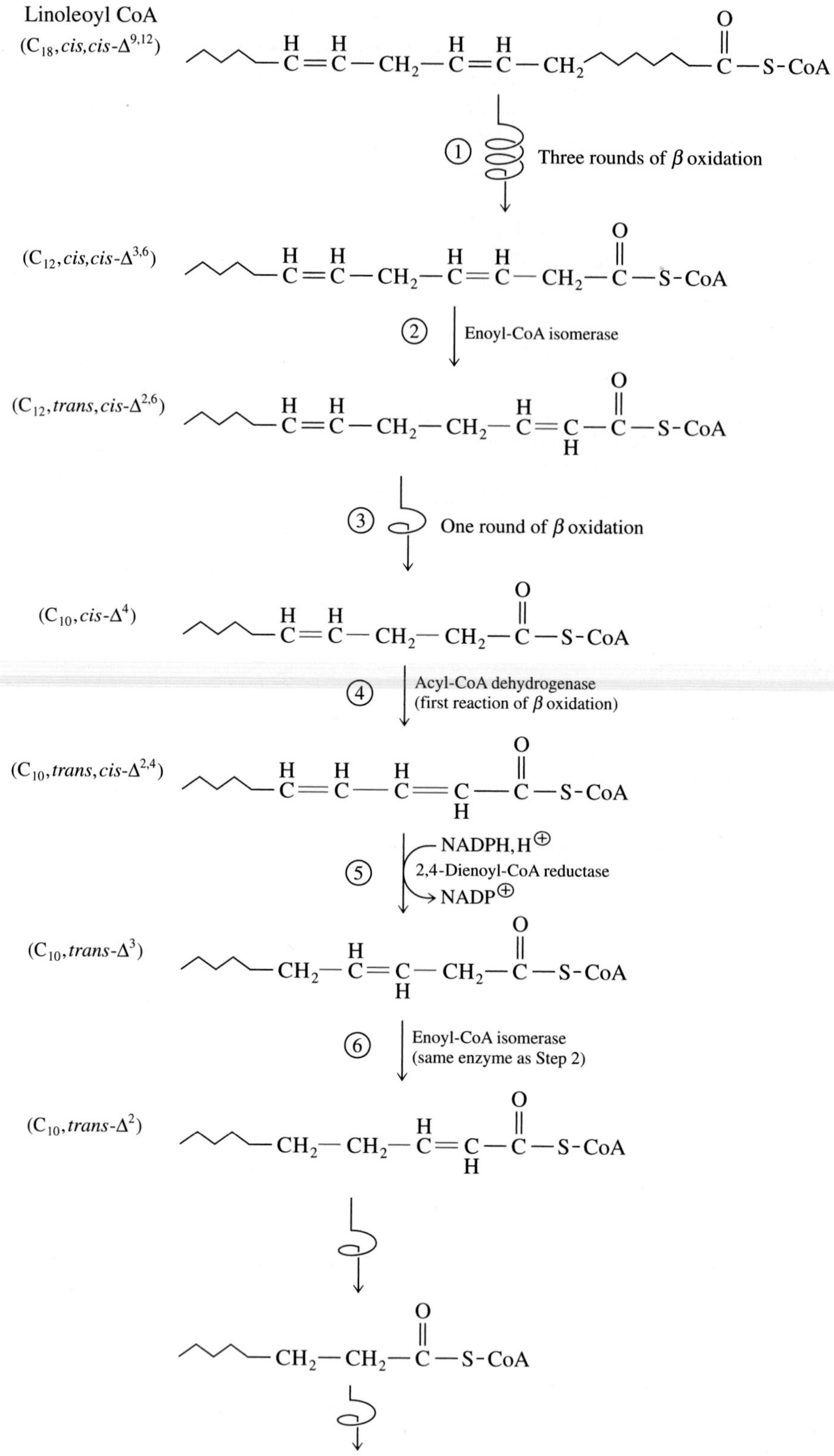

20·4 Oxidation of Unsaturated Fatty Acids Requires Two Enzymes in Addition to Those of the β-Oxidation Pathway

Unsaturated fatty acids are common in nature. Using the oxidation of linoleate (C_{18}, cis,cis-$\Delta^{9,12}$-octadecadienoate) as an example, Figure 20·8 outlines the pathway for catabolism of unsaturated fatty acids. As is the case with most polyunsaturated fatty acids, the double bonds in linoleate are separated by a methylene group. Linoleoyl CoA is a substrate for the enzymes of the β-oxidation pathway until a double bond of the shortened fatty acid chain interferes with catalysis. After three rounds of β oxidation, linoleoyl CoA is converted to a C_{12} molecule, cis,cis-$\Delta^{3,6}$-dienoyl CoA (Step 1). This molecule has a cis-β,γ double bond rather than the usual $trans$-α,β double bond found at this stage of β oxidation and as a result is not a substrate for enoyl-CoA hydratase. In an alternate reaction catalyzed by enoyl-CoA isomerase (Step 2), the double bond is rearranged from Δ^3 to Δ^2 to produce $trans,cis$-$\Delta^{2,6}$-dienoyl CoA. This product can reenter the β-oxidation pathway, and another round of β oxidation can be completed (Step 3). If our example fatty acid had been a monounsaturated species, the double bond would now have been dealt with and oxidation would proceed as usual via the reactions of the β-oxidation pathway. In the pathway from linoleate, an unsaturated C_{10} molecule, cis-Δ^4-acyl CoA, has been generated. This molecule is acted upon by the first enzyme of the β-oxidation pathway, acyl-CoA dehydrogenase, producing the C_{10} molecule $trans,cis$-$\Delta^{2,4}$-dienoyl CoA. Because the diene is resonance stabilized, the double bond is a poor substrate for hydration, unlike an isolated double bond. Hydride addition to the cation and proton addition to the anion of one resonance form, catalyzed by 2,4-dienoyl-CoA reductase, accounts for the position of the double bond in the product, C_{10}, $trans$-Δ^3-enoyl CoA.

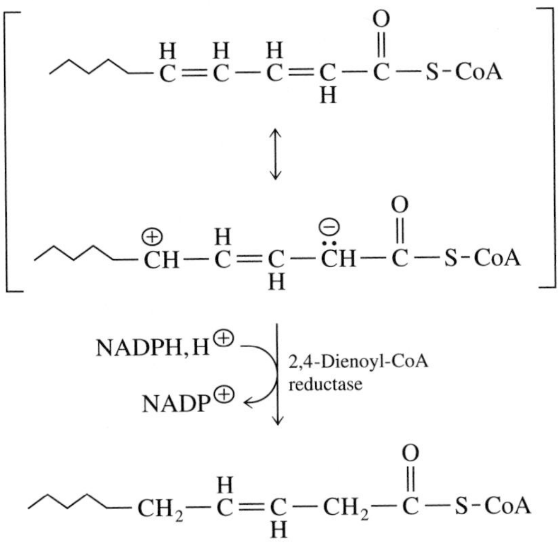

$$C_{10}, trans\text{-}\Delta^3\text{-Enoyl CoA} \qquad\qquad (20\cdot4)$$

This product (like its cis isomer in Step 2) is a substrate for enoyl-CoA isomerase. The product of isomerization, $trans$-Δ^2-enoyl CoA, then continues through the β-oxidation pathway.

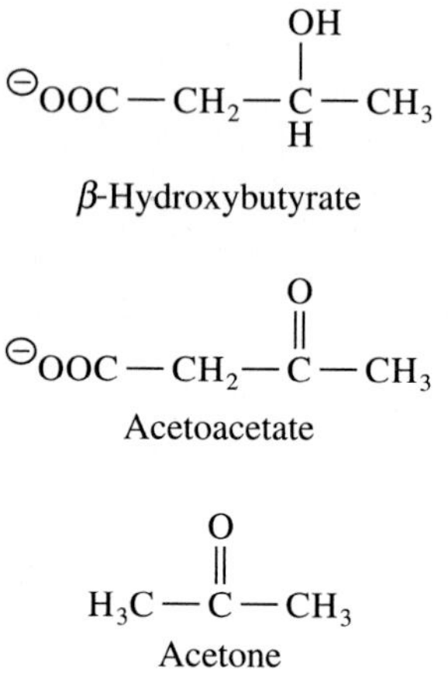

$$^{\ominus}OOC - CH_2 - \underset{\underset{H}{|}}{\overset{\overset{OH}{|}}{C}} - CH_3$$

β-Hydroxybutyrate

$$^{\ominus}OOC - CH_2 - \overset{\overset{O}{\|}}{C} - CH_3$$

Acetoacetate

$$H_3C - \overset{\overset{O}{\|}}{C} - CH_3$$

Acetone

Figure 20·9
The ketone bodies.

20·5 Ketone Bodies Are Fuel Molecules

Most acetyl CoA from fatty acid oxidation is routed to the citric acid cycle. However, when the amount of acetyl CoA increases above the capacity of the citric acid cycle to oxidize it, the excess acetyl CoA is diverted to the formation of ketone bodies—β-hydroxybutyrate, acetoacetate, and acetone. As indicated by their structures (Figure 20·9), not all ketone bodies are ketones. Of the three, only β-hydroxybutyrate and acetoacetate are quantitatively significant; small amounts of acetone are produced by the nonenzymatic decarboxylation of acetoacetate.

The ketone bodies are fuel molecules. Although they represent less potential metabolic energy than the fatty acids from which they are derived, they are able to

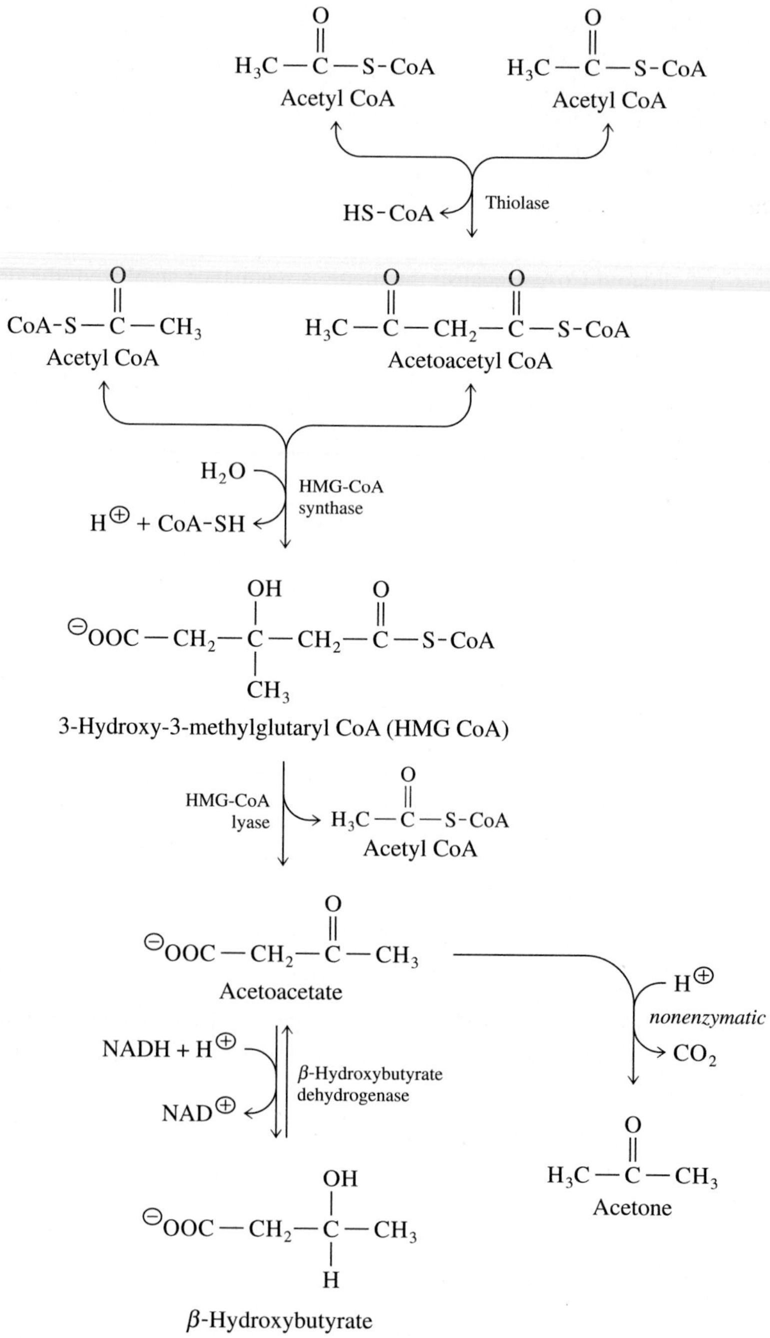

Figure 20·10
Biosynthesis of β-hydroxybutyrate, acetoacetate, and acetone.

serve as "water-soluble lipids" that can be more rapidly oxidized than fatty acids by organs such as the heart and kidneys. During starvation, ketone bodies are produced in large amounts by the liver, increasing their concentration in blood to the point where they become a substitute for glucose as the principal fuel for brain cells. In Chapter 23, we will consider in detail the role of ketone bodies in different metabolic states such as late starvation and also in the aberrant state of diabetes. Here we examine the general pathways by which ketone bodies are produced and consumed.

A. Ketone Bodies Are Synthesized in the Liver

In mammals, ketone bodies are synthesized in the liver and exported for use by other tissues. The pathway for ketone body synthesis, or **ketogenesis,** which occurs in the mitochondrial matrix, is shown in Figure 20·10. First, two molecules of acetyl CoA condense to form acetoacetyl CoA and CoASH in a reaction catalyzed by thiolase. Subsequently, a third molecule of acetyl CoA is added to acetoacetyl CoA to form 3-hydroxy-3-methylglutaryl CoA (HMG CoA) in a reaction catalyzed by HMG-CoA synthase. Next, HMG-CoA lyase catalyzes the cleavage of HMG CoA to acetoacetate and acetyl CoA. NADH-dependent reduction of acetoacetate produces β-hydroxybutyrate in a reaction catalyzed by β-hydroxybutyrate dehydrogenase. Both acetoacetate and β-hydroxybutyrate can be transported across the mitochondrial membrane and the plasma membrane of liver cells, after which they enter the bloodstream to be used as fuel by other cells of the body. In the bloodstream, small amounts of acetoacetate are decarboxylated to acetone.

B. Ketone Bodies Are Oxidized in Mitochondria

In cells that use β-hydroxybutyrate and acetoacetate for fuel, both compounds enter mitochondria, where they are converted to acetyl CoA and oxidized by the citric acid cycle. β-Hydroxybutyrate is converted to acetoacetate in a near-equilibrium reaction catalyzed by β-hydroxybutyrate dehydrogenase. Acetoacetate reacts with succinyl CoA to form acetoacetyl CoA in a reaction catalyzed by succinyl-CoA transferase (Figure 20·11). This reaction siphons some of the succinyl CoA from

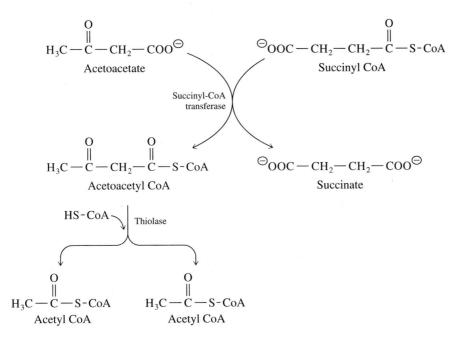

Figure 20·11
Conversion of acetoacetate to acetyl CoA. Succinyl-CoA transferase catalyzes the conversion of acetoacetate and succinyl CoA to acetoacetyl CoA and succinate. Acetoacetyl CoA is then converted to two molecules of acetyl CoA by the action of thiolase.

the citric acid cycle. Energy that would normally be captured as GTP in the substrate-level phosphorylation catalyzed by succinyl-CoA synthetase (Section 16·3, Part 5) is used instead to activate acetoacetate to its CoA ester. Acetoacetyl CoA is then converted to two molecules of acetyl CoA by the action of thiolase.

20·6 Fatty Acid Synthesis Occurs by a Pathway Different from Fatty Acid Oxidation

Fatty acid synthesis in mammals occurs largely in the liver and adipocytes and, to a lesser extent, in specialized cells under specific conditions, such as mammary gland cells during lactation. Fatty acids are synthesized and degraded by completely different pathways. The key differences between the two processes in mammals are summarized in Table 20·1. Fatty acid oxidation takes place principally in mitochondria, whereas synthesis takes place in the cytosol. The active thioesters in fatty acid oxidation are CoA derivatives, whereas in fatty acid synthesis the intermediates are bound as thioesters to **acyl carrier protein** (ACP). Separate enzymes catalyze the reactions of oxidation, whereas a multifunctional protein of two identical polypeptide chains catalyzes most of the biosynthetic reactions in mammals. Synthesis and degradation both proceed in two-carbon steps. However, oxidation results in a two-carbon product, acetyl CoA, whereas synthesis requires a three-carbon substrate, malonyl CoA, which transfers a two-carbon unit to the growing chain. In the process, CO_2 is released. Finally, reducing power for synthesis is supplied by NADPH, whereas oxidation is dependent on $NAD^\oplus$ and ETF (with ETF feeding electrons to Q).

Fatty acid synthesis in eukaryotes depends on three processes. First, mitochondrial acetyl CoA is shuttled into the cytosol. Next, carboxylation of acetyl CoA generates malonyl CoA, the substrate for the elongation reactions that extend the fatty acyl chain. The carboxylation of acetyl CoA is the regulated step of fatty acid synthesis. Finally, assembly of the fatty acid chain is carried out by fatty acid synthase.

Table 20·1 Comparison of fatty acid oxidation and synthesis in mammals

	Oxidation	Synthesis
Localization	Mitochondria	Cytosol
Acyl carrier	CoA	Acyl carrier protein (ACP)
Carbon units	C_2	C_2
Acceptor/donor	CoA (C_2)	Malonyl CoA (C_3, donor reaction evolves CO_2)
Mobile redox cofactors	$NAD^\oplus$, ETF	NADPH
Organization of enzymes	Separate enzymes	Multifunctional enzyme

20·7 The Citrate Transport System Provides Acetyl CoA to the Cytosol

Fatty acid synthesis in the cytosol of eukaryotes requires that acetyl CoA be supplied from mitochondria, where it is produced. In the fed state, fatty acids are synthesized from acetyl CoA produced by carbohydrate metabolism. The net export of acetyl CoA from mitochondria is accomplished by the **citrate transport system.** The transport is indirect, requiring a series of steps (Figure 20·12). In the first step, mitochondrial acetyl CoA condenses with oxaloacetate in a reaction catalyzed by citrate synthase. This is also the first step of the citric acid cycle. Next, citrate is transported out of the mitochondrion by the citric acid–dicarboxylic acid carrier via antiport with a dicarboxylate anion. In the cytosol, citrate is cleaved to form oxaloacetate and acetyl CoA in an ATP-requiring reaction catalyzed by citrate lyase.

Once cytosolic acetyl CoA has been generated, further steps are required to return the remaining carbon of the original citrate to the mitochondrion. Cytosolic malate dehydrogenase, an isozyme of mitochondrial malate dehydrogenase, catalyzes the reduction of oxaloacetate to malate, with conversion of NADH to $NAD^{\oplus}$. Next, malate is decarboxylated to pyruvate in a reaction catalyzed by malic enzyme, with reduction of $NADP^{\oplus}$ to NADPH. Thus, the citrate transport system serves not only to transport acetyl CoA from the mitochondrion to the cytosol but also to generate cytosolic NADPH. About half the NADPH required for fatty acid synthesis is generated by the citrate transport system, with the rest contributed by the pentose phosphate pathway (Section 17·11).

The newly formed pyruvate enters the mitochondrion via pyruvate translocase. Once in the mitochondrion, pyruvate can be carboxylated to form oxaloacetate in an ATP-requiring step, or it can be converted to acetyl CoA by the action of the pyruvate dehydrogenase complex, thus completing the cycle of reactions. Cellular conditions determine the fate of pyruvate.

Figure 20·12
Citrate transport system. The system achieves net transport of acetyl CoA from the mitochondrion to the cytosol (the shuttling of acetyl CoA is shown in blue) and net conversion of cytosolic NADH into NADPH (the path for conversion of reducing equivalents from NADH to NADPH is shown in red). Two molecules of ATP are expended for each round of the cyclic pathway.

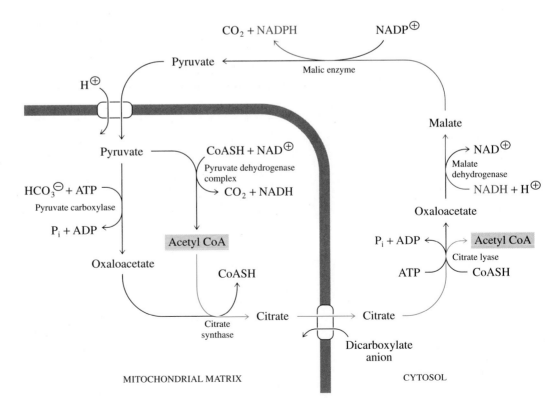

Table 20·2 Components of acetyl-CoA carboxylase

E. coli	Animals and yeast
Biotin carboxyl carrier protein 2 subunits, M_r 22 500 each	Two copies of one polypeptide chain containing all catalytic sites 2 subunits, M_r 230 000–290 000 each
Biotin carboxylase 2 subunits, M_r 51 000 each	Total M_r of complex: 460 000–580 000
Transcarboxylase 2 subunits, M_r 30 000 each 2 subunits, M_r 35 000 each	
Total M_r of complex: 277 000	

20·8 Acetyl CoA Is Carboxylated to Form Malonyl CoA in a Regulated Step

The second stage of fatty acid synthesis consists of the carboxylation of acetyl CoA in the cytosol to form malonyl CoA in a reaction catalyzed by the biotin-dependent enzyme acetyl-CoA carboxylase. This reaction is the key regulatory step of fatty acid synthesis, as we will see in Section 20·11.

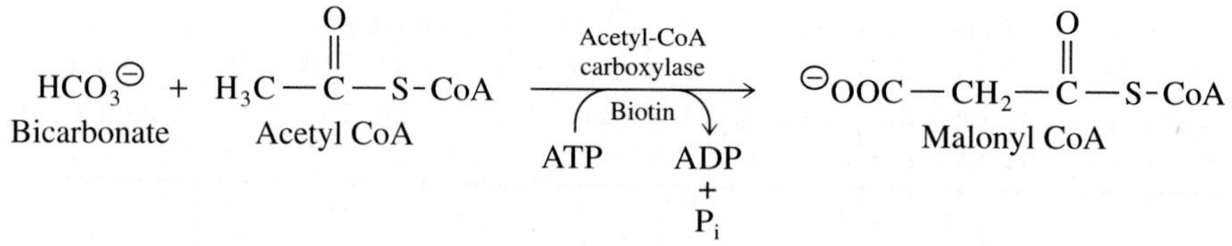

$$HCO_3^{\ominus} + \underset{\text{Acetyl CoA}}{H_3C-\overset{\overset{\textstyle O}{\|}}{C}-S\text{-}CoA} \xrightarrow[\text{Biotin} \quad \underset{\substack{\text{ADP} \\ + \\ P_i}}{\text{ATP}}]{\text{Acetyl-CoA carboxylase}} \underset{\text{Malonyl CoA}}{{}^{\ominus}OOC-CH_2-\overset{\overset{\textstyle O}{\|}}{C}-S\text{-}CoA}$$

$$\textbf{(20·5)}$$

In *Escherichia coli,* three separate protein subunits carry out this conversion: a carrier protein termed biotin carboxyl carrier protein (BCCP) and two enzymes, a biotin carboxylase and a transcarboxylase. In animals and yeast, all of these activities are contained on a single polypeptide chain (Table 20·2).

Carboxylation of acetyl CoA proceeds in two steps. The ATP-dependent activation of $HCO_3^{\ominus}$ forms carboxybiotin.

$$HCO_3^{\ominus} + \text{Enz-Biotin} + \text{ATP} \rightleftharpoons \text{Enz-Biotin}-COO^{\ominus} + \text{ADP} + P_i$$

$$\textbf{(20·6)}$$

This reaction is followed by the transfer of activated CO_2 to acetyl CoA, forming malonyl CoA.

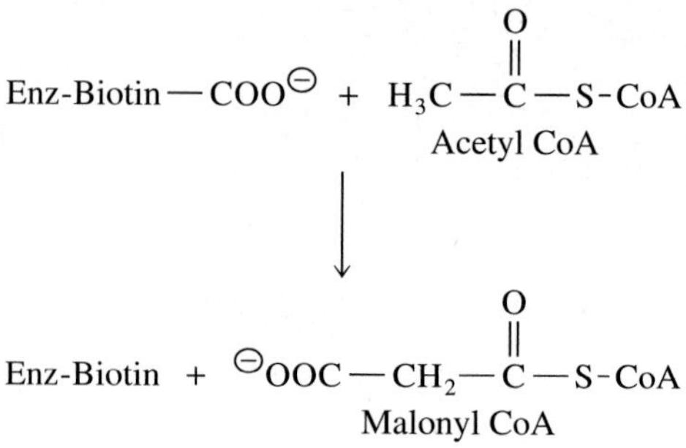

$$\text{Enz-Biotin}-COO^{\ominus} + \underset{\text{Acetyl CoA}}{H_3C-\overset{\overset{\textstyle O}{\|}}{C}-S\text{-}CoA}$$

$$\downarrow$$

$$\textbf{(20·7)}$$

$$\text{Enz-Biotin} + \underset{\text{Malonyl CoA}}{{}^{\ominus}OOC-CH_2-\overset{\overset{\textstyle O}{\|}}{C}-S\text{-}CoA}$$

20·9 Fatty Acid Synthesis Is Catalyzed by a Multienzyme Pathway in *E. coli* and by a Multifunctional Enzyme in Mammals

Fatty acids are synthesized from acetyl CoA and malonyl CoA following transfer of acetyl and malonyl groups to the prosthetic group phosphopantetheine, the same group found in coenzyme A. In *E. coli,* this prosthetic group is attached to ACP (Figure 20·13). Seven enzymes and ACP are required to synthesize fatty acids in *E. coli.* In mammals, a single polypeptide chain contains all of the catalytic activities, and the prosthetic group is incorporated within the multifunctional protein, whose active form is a dimer. We will first examine fatty acid formation in *E. coli* and then compare it to the process in mammals.

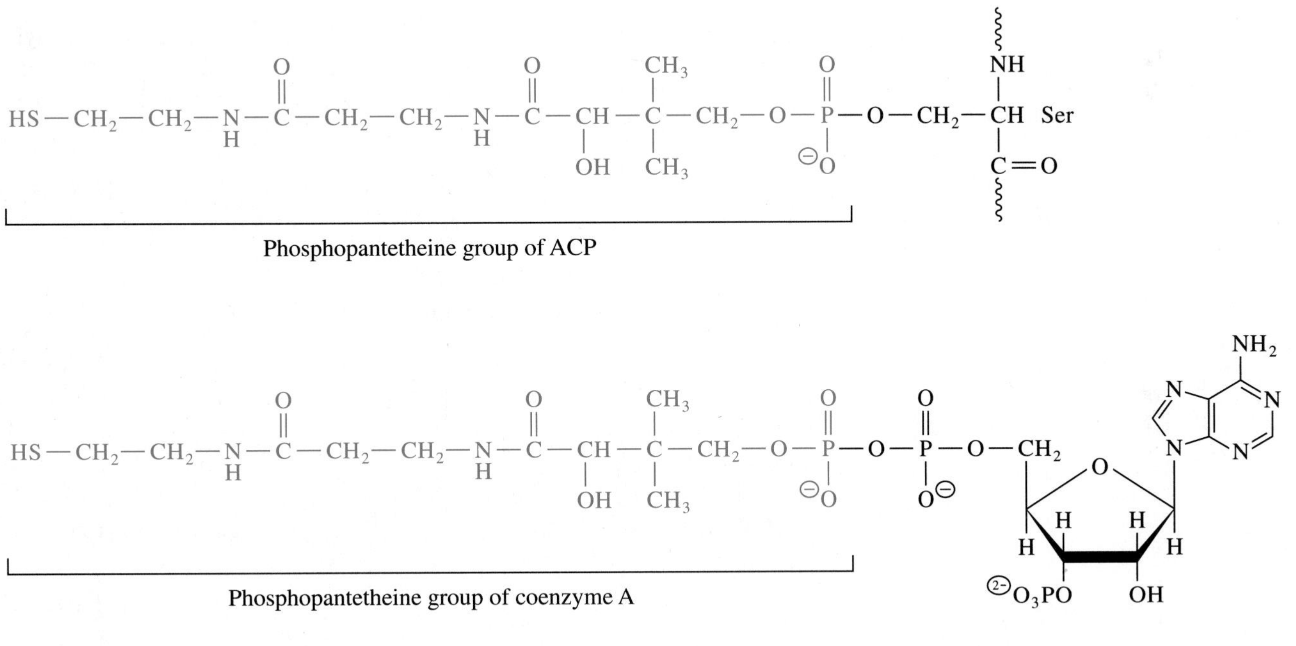

Phosphopantetheine group of ACP

Phosphopantetheine group of coenzyme A

Figure 20·13
Phosphopantetheine in acyl carrier protein (ACP) and in coenzyme A.

Part 3 Metabolism and Bioenergetics

Figure 20·14
Stages in the biosynthesis of fatty acids from acetyl CoA and malonyl CoA in *E. coli*. In the loading stage, acetyl CoA and malonyl CoA are esterified to ACP. In the condensation stage, ketoacyl-ACP synthase (also called the condensing enzyme) accepts an acetyl group from acetyl-ACP, releasing ACP-SH. Ketoacyl-ACP synthase then catalyzes transfer of the acetyl group to malonyl-ACP to form aceto-acetyl-ACP and CO_2. In the first reduction, acetoacetyl-ACP is converted to D-β-hydroxy-butyryl-ACP in a reaction catalyzed by NADPH-dependent ketoacyl-ACP reductase. Dehydration of D-β-hydroxybutyryl-ACP results in the formation of a double bond, pro-ducing *trans*-butenoyl-ACP. Finally, reduction of *trans*-butenoyl-ACP produces butyryl-ACP. Synthesis continues by repeating the last four stages, with butyryl-ACP substituting for acetyl-ACP in the next condensation stage.

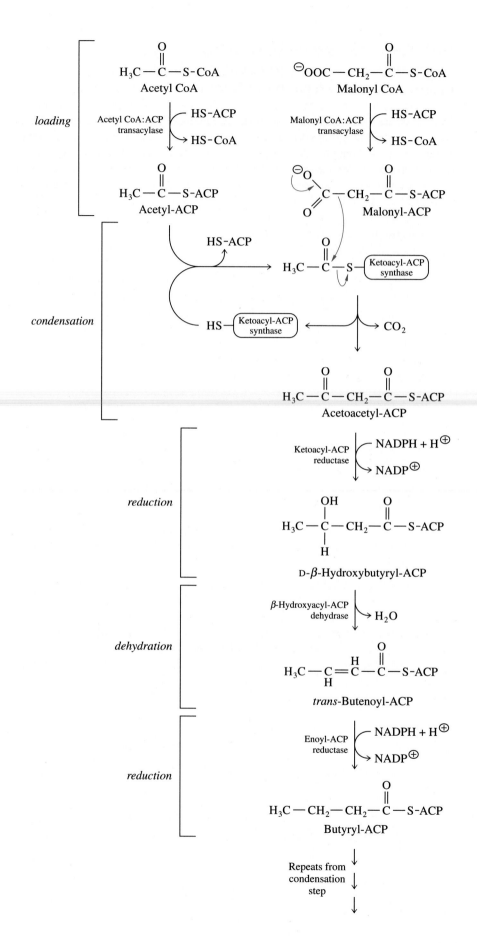

A. Assembly of Fatty Acids Occurs in Five Stages

Fatty acid synthesis can be divided into five separate stages: loading of precursors by formation of two thioester derivatives, condensation of the precursors, reduction, dehydration, and a further reduction. The entire sequence is shown in Figure 20·14. The last four stages are repeated until a long-chain fatty acid is synthesized. Fatty acid synthesis is often called palmitate synthesis because palmitate is the preponderant product of these reactions. Other fatty acids are formed by ancillary reactions that are discussed in Section 20·10.

1. *Loading.* Two enzymes, acetyl CoA:ACP transacylase and malonyl CoA:ACP transacylase, are required for the loading steps, in which acetyl CoA and malonyl CoA are each transesterified to ACP.

2. *Condensation.* Ketoacyl-ACP synthase, also called the condensing enzyme, accepts an acetyl group from acetyl-ACP, and ACP-SH is released. Ketoacyl-ACP synthase then catalyzes transfer of the acetyl group to malonyl-ACP, evolving CO_2 from the latter substrate and forming acetoacetyl-ACP. Recall that the synthesis of malonyl CoA involves ATP-dependent carboxylation. The strategy of first carboxylating and then decarboxylating a compound used for a synthetic reaction results in a favorable free-energy change for the synthetic reaction at the expense of ATP consumed in the carboxylation step. A similar strategy is seen in mammalian gluconeogenesis: pyruvate (C_3) is carboxylated to form oxaloacetate (C_4), which is subsequently decarboxylated to form the C_3 molecule phosphoenolpyruvate (Section 17·7).

3. *Reduction.* The ketone of acetoacetyl-ACP is converted to an alcohol, forming D-β-hydroxybutyryl-ACP, in an NADPH-dependent reaction catalyzed by ketoacyl-ACP reductase.

4. *Dehydration.* A dehydrase catalyzes the removal of water, with formation of a double bond.

5. *Reduction.* The product of dehydration, *trans*-butenoyl-ACP, undergoes reduction to form a four-carbon acyl-ACP, butyryl-ACP, in a reaction catalyzed by NADPH-dependent enoyl-ACP reductase.

Synthesis continues by repeating the process from the condensation stage, with the growing acyl-ACP substituting for acetyl-ACP, and a new molecule of malonyl CoA entering with each round (Figure 20·15, next page). Note that in the condensation stage of the first round, acetyl CoA contributes the two-carbon unit that becomes the penultimate and terminal carbons of the growing molecule; in subsequent rounds, malonyl CoA contributes the two-carbon units. Rounds of synthesis continue until a C_{16} palmitoyl group is formed. Palmitoyl-ACP is a substrate for thiolase, which catalyzes the formation of palmitate and ACP-SH.

$$\text{Palmitoyl-ACP} \xrightarrow[\text{Thiolase}]{\text{H}_2\text{O}} \text{Palmitate} + \text{HS-ACP} \qquad \textbf{(20·8)}$$

The overall stoichiometry of palmitate synthesis from acetyl CoA and malonyl CoA is

$$\begin{aligned} \text{Acetyl CoA} + 7\,\text{Malonyl CoA} + 14\,\text{NADPH} + 14\,\text{H}^{\oplus} \longrightarrow \\ \text{Palmitate} + 7\,CO_2 + 14\,\text{NADP}^{\oplus} + 8\,\text{CoASH} + 6\,H_2O \end{aligned} \qquad \textbf{(20·9)}$$

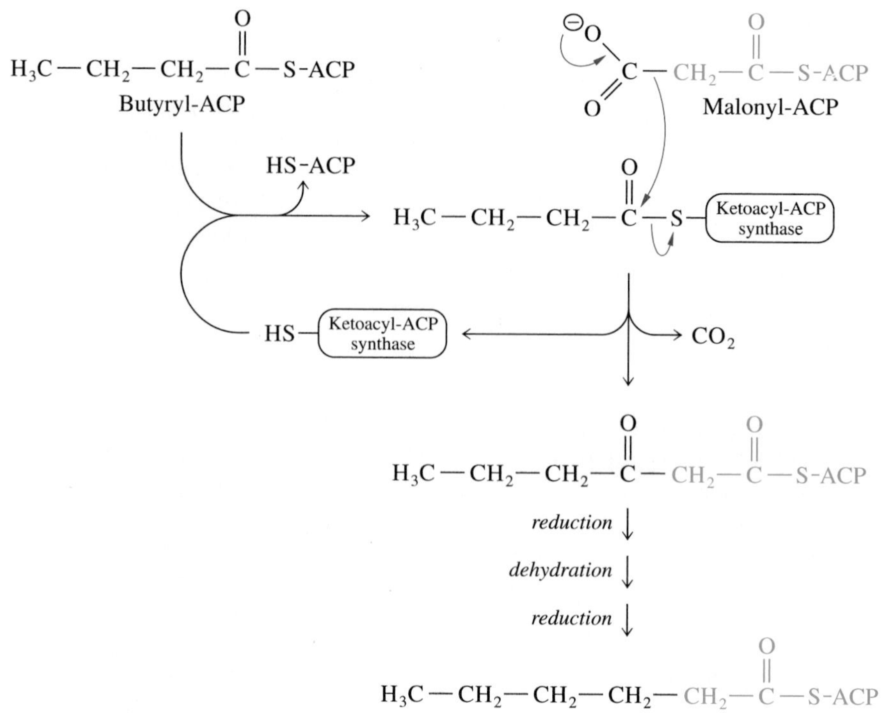

Figure 20·15
Condensation step for the second round of fatty acid biosynthesis. Butyryl-ACP formed by the reactions illustrated in Figure 20·14 reacts with ketoacyl-ACP synthase, which displaces ACP and binds the acyl group. The acyl group is then transferred to malonyl-ACP, with release of CO_2. This reaction is followed by the sequence of reduction, dehydration, and reduction outlined in Figure 20·14, with new molecules of malonyl CoA entering until a C_{16} chain is formed.

B. Mammalian Fatty Acid Synthase Is a Dimer of Identical Multifunctional Polypeptides

The overall sequence of reactions for mammalian fatty acid synthesis is quite similar to the sequence in *E. coli,* yet the structural features of the two enzyme systems are strikingly different. Fatty acid synthase in mammals is a dimer of identical polypeptides. The two subunits of the dimer are arranged in head-to-tail fashion, as indicated schematically in Figure 20·16. Each half of the holoenzyme contains all of the activities for the synthesis of fatty acids, so that two fatty acids are formed simultaneously.

ACP, part of the synthase in mammals, is the site of entry for carbon destined for fatty acids. (Since the acyl-carrying prosthetic group is attached directly to a multifunctional protein in the mammalian enzyme rather than to an independent protein, acyl-carrier domain—ACD—might be a better term in this case than ACP. However, the original term referring to the arrangement in bacteria is widely used and recognized for the acyl-carrier component of all fatty acid synthases.) Acetyl

Figure 20·16
Mammalian fatty acid synthase. Each identical polypeptide chain contains seven enzymatic activities. The dimeric holoenzyme is composed of identical monomers arranged head-to-tail. Three domains can be identified in each monomer. Ketoacyl-ACP synthase, acetyl CoA:ACP transacylase, and malonyl CoA:ACP transacylase constitute a condensation domain. The remaining activities except for thiolase constitute a reduction and dehydration domain. Thiolase, in a separate domain, catalyzes release of fully formed C_{16} fatty acids.

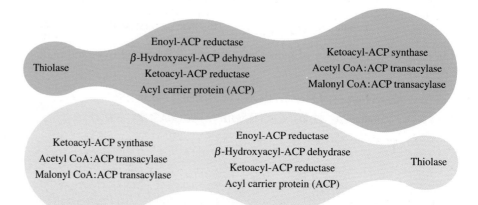

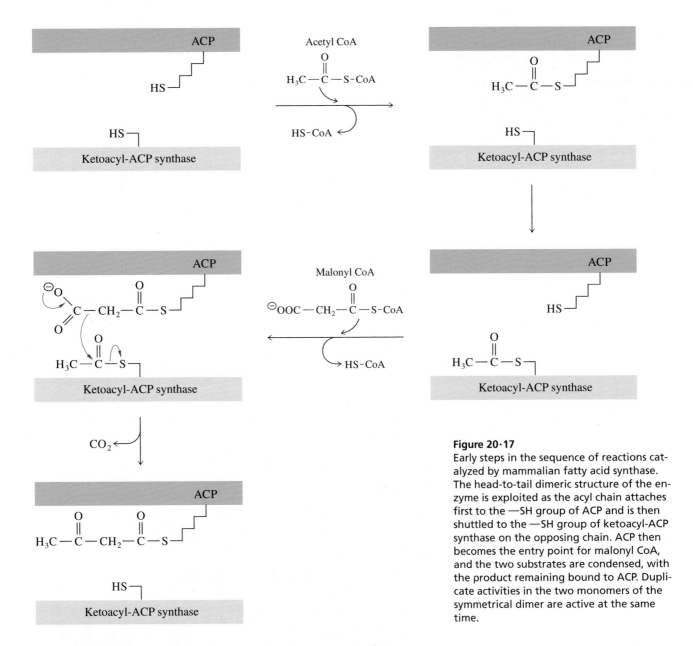

Figure 20·17
Early steps in the sequence of reactions catalyzed by mammalian fatty acid synthase. The head-to-tail dimeric structure of the enzyme is exploited as the acyl chain attaches first to the —SH group of ACP and is then shuttled to the —SH group of ketoacyl-ACP synthase on the opposing chain. ACP then becomes the entry point for malonyl CoA, and the two substrates are condensed, with the product remaining bound to ACP. Duplicate activities in the two monomers of the symmetrical dimer are active at the same time.

CoA reacts with ACP, and the acetyl group is then transferred to a sulfhydryl group of a cysteine residue of ketoacyl-ACP synthase, which resides on the facing subunit (Figure 20·17). ACP then binds malonyl CoA and the condensation step ensues, with the end product joined to ACP. During the remaining reduction, dehydration, and reduction steps, the carbon chain remains attached to ACP. For subsequent rounds of elongation, the reduced four-carbon chain attaches to the —SH group of ketoacyl-ACP synthase, ACP accepts an additional malonyl CoA, and condensation takes place as before.

20·10 Additional Enzymes Are Required for Further Chain Elongation and Desaturation

The most common product of fatty acid synthase in plants and animals is palmitate (16:0). Synthesis of a wide variety of other fatty acids requires a host of enzymes present in the endoplasmic reticulum as well as in mitochondria. For example, animal cells contain a number of desaturases that catalyze the formation of double

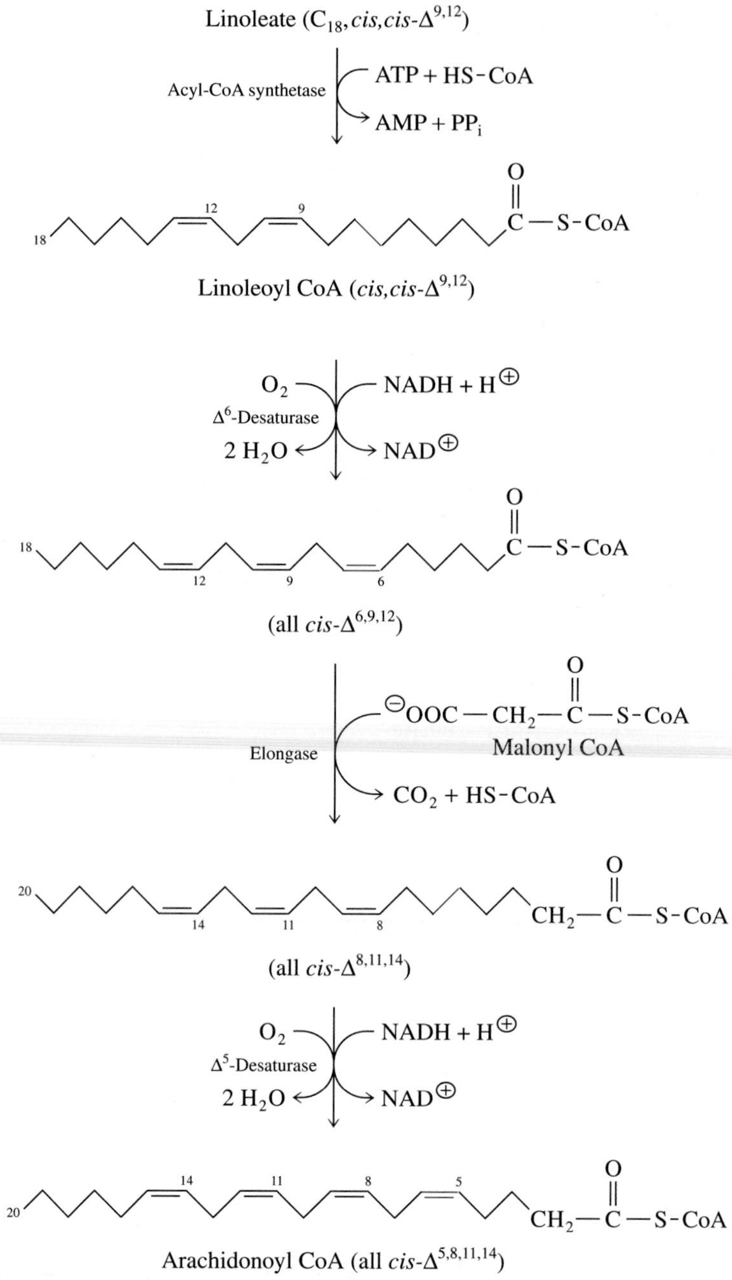

Figure 20·18
Elongation and desaturation reactions in the conversion of linoleate to arachidonoyl CoA. The activation of linoleate as linoleoyl CoA is the same activation step that precedes fatty acid oxidation. Instead of oxidation, further desaturation and elongation leads to arachidonoyl CoA. The arachidonoyl moiety of arachidonoyl CoA is incorporated into triacylglycerols or phospholipids.

Linoleate (C_{18}, cis,cis-$\Delta^{9,12}$)

Acyl-CoA synthetase: ATP + HS–CoA → AMP + PP$_i$

Linoleoyl CoA (cis,cis-$\Delta^{9,12}$)

Δ^6-Desaturase: O_2 + NADH + H$^\oplus$ → 2 H$_2$O + NAD$^\oplus$

(all cis-$\Delta^{6,9,12}$)

Elongase: $^\ominus$OOC–CH$_2$–C(=O)–S–CoA (Malonyl CoA) → CO$_2$ + HS–CoA

(all cis-$\Delta^{8,11,14}$)

Δ^5-Desaturase: O_2 + NADH + H$^\oplus$ → 2 H$_2$O + NAD$^\oplus$

Arachidonoyl CoA (all cis-$\Delta^{5,8,11,14}$)

bonds as far as nine carbons from the carboxyl end of a fatty acid. Desaturation of double bonds further removed than nine carbons can be catalyzed only by plant desaturases. A fatty acid such as linoleate, 18:2 $\Delta^{9,12}$, which animals require but cannot synthesize, is thus an essential fatty acid, one that must be supplied by the diet.

Linoleate is a precursor of arachidonoyl CoA, which is formed in a pathway involving desaturation and elongation reactions (Figure 20·18). The pathway consumes malonyl CoA and thus depends on the activity of acetyl-CoA carboxylase (Section 20·8). However, the pathway is catalyzed by enzymes different from those of fatty acid synthase. The arachidonoyl moiety of arachidonoyl CoA can be incorporated into triacylglycerols or phospholipids. Arachidonate derived from phospholipid cleavage is a precursor of eicosanoids, discussed in Section 20·12.

20·11 Fatty Acid Metabolism in Animals Is Regulated by Hormones

The metabolism of fatty acids in mammals is under hormonal control. Glucagon, epinephrine, and insulin are the principal hormonal regulators of fatty acid metabolism, with glucagon and epinephrine present in high concentrations in the fasted state and insulin present in high concentrations in the fed state. The concentration of circulating glucose must be maintained within fairly narrow limits at all times. (The principles and mechanisms of glucose homeostasis are examined in detail in Chapter 23.) In the fasted state, carbohydrate stores become depleted, and synthesis of carbohydrates must occur to maintain the level of glucose in the blood. To further relieve pressure on the limited supply of glucose, fatty acids are mobilized to serve as fuel, and many tissues undergo regulatory transitions that decrease their use of carbohydrates and increase their use of fatty acids. The opposite occurs in the fed state, when carbohydrates are used as fuel and as precursors for fatty acid synthesis. The important points of control for fatty acid oxidation are release of fatty acids from adipocytes and regulation of carnitine acyltransferase I in the liver. The key regulatory enzyme for fatty acid synthesis is acetyl-CoA carboxylase. Regulation of synthesis and degradation is reciprocal, with increased metabolism by one pathway balanced by decreased activity in the opposing pathway. The hormonal regulation of fatty acid metabolism is summarized in Figure 20·19.

Figure 20·19
Hormonal control of fatty acid metabolism.

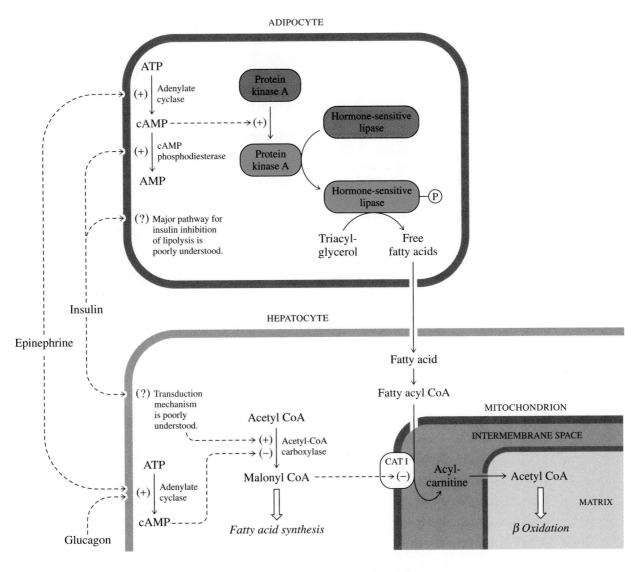

The major regulator of fatty acid mobilization is inhibition by insulin. The rate of lipolysis is extremely sensitive to the levels of insulin. Enormous effort has been devoted to elucidating this regulatory pathway, but the biochemical mechanism for its action remains mostly veiled.

When the concentration of insulin drops, the inhibition of lipolysis is relieved, and triacylglycerols are released. Lipolysis is further stimulated by epinephrine. As we saw in Section 20·1, binding of epinephrine (and norepinephrine, a precursor of epinephrine and also a hormone) to adipocytes leads to elevated production of cAMP, activation of protein kinase A, subsequent activation of hormone-sensitive lipase, and release of free fatty acids.

As the concentration of albumin-bound fatty acids in the blood increases, more fatty acids enter tissues to be oxidized. Fatty acid oxidation is regulated intracellularly by control of the formation of acylcarnitine, and thus control of the transport of fatty acids into mitochondria by the carnitine shuttle system. In liver cells, carnitine acyltransferase I (CAT I) is allosterically inhibited by malonyl CoA, the product of acetyl-CoA carboxylase in the pathway for fatty acid synthesis. The level of malonyl CoA rises during active fatty acid synthesis, and its inhibitory effect on CAT I decreases the rate of fatty acid oxidation.

The inhibition of hepatic fatty acid oxidation by malonyl CoA is relieved when the levels of glucagon and epinephrine rise. These hormones cause an increase in the concentration of cAMP, which leads to the cAMP-dependent phosphorylation and inactivation of acetyl-CoA carboxylase.

In the fed state, a different hormonal scenario emerges, with the level of insulin rising in response to the presence of carbohydrates and dominating the effects of glucagon and epinephrine. When insulin binds to adipocytes, cAMP phosphodiesterase is stimulated and the conversion of cAMP to AMP increases, interrupting the sequence leading to active hormone-sensitive lipase. However, experiments have shown that stimulation of cAMP phosphodiesterase is not the only important route of lipolysis regulation by insulin. Other yet-to-be-discovered mechanisms triggered by insulin are also major regulators of triacylglycerol hydrolysis.

In diabetes, a disease in which either insulin is lacking or its receptors are insensitive to insulin, inhibition of fatty acid release by insulin does not occur, resulting in massive release of fatty acids from adipose tissue. The result is a very high rate of β oxidation in the liver and production of more acetyl CoA than the citric acid cycle can oxidize. Excess acetyl CoA is diverted to the formation of ketone bodies. The levels of ketones in the blood of diabetics can become so high that the sweet smell of acetone can be discerned on the breath. We will return to the subject of diabetes in Chapter 23.

Insulin also stimulates acetyl-CoA carboxylase in liver cells, leading to an increase in the formation of malonyl CoA, which supplies the pathway of fatty acid synthesis and inhibits the activity of CAT I, thereby blocking the delivery of fatty acids to the β-oxidation pathway. The signal transduction pathway leading from the binding of insulin to the activation of acetyl-CoA carboxylase is, once again, poorly understood.

It is known that hormonal regulation of acetyl-CoA carboxylase is exerted through reversible phosphorylation. Glucagon (and less importantly, epinephrine) stimulates phosphorylation and inactivation of the enzyme in liver. Inactivation of acetyl-CoA carboxylase depresses the rate of fatty acid synthesis. Acetyl CoA-carboxylase is further regulated by fatty acyl CoA, which increases the activity of a protein kinase that catalyzes phosphorylation and inhibition of the enzyme. At higher concentrations, fatty acyl CoA is itself an allosteric inhibitor of acetyl-CoA carboxylase. The ability of fatty acid derivatives to regulate acetyl-CoA carboxylase is physiologically appropriate—an increased concentration of fatty acids causes a decrease in the rate of the first committed step of fatty acid synthesis.

The in vitro activation of acetyl-CoA carboxylase by citrate led early investigators to propose a feed-forward mechanism for the activation of fatty acid synthesis, since citrate in the cytosol is a precursor of acetyl CoA, the substrate for acetyl-CoA carboxylase. Another finding regarded as support for this mechanism was the observation that when purified acetyl-CoA carboxylase was exposed to citrate in vitro, polymerization of enzyme molecules into long chains occurred, which appeared to coincide with activation of the enzyme. Electron micrographs of polymerized enzyme were regarded as impressive visual evidence of an important control mechanism (Figure 20·20). The evidence was dramatic and unusual, and the conclusions about the regulation of acetyl-CoA carboxylase became quite well known. Recently, experiments with liver cells have shown that changes in the concentration of citrate in vivo do not correspond to changes in the rate of fatty acid synthesis. In fact, since citrate lyase catalyzes a near-equilibrium reaction, whereas acetyl-CoA carboxylase catalyzes a metabolically irreversible reaction that is a control point for fatty acid synthesis, it may be predicted that stimulation of the activity of acetyl-CoA carboxylase would shift flux through the citrate lyase reaction toward product and decrease the concentration of citrate. Furthermore, recent studies have deflated the importance of the striking polymerization process. Activation of the enzyme has been shown to precede polymerization. Citrate clearly can activate the enzyme, and polymerization clearly occurs in vitro, but these phenomena are apparently not important factors in the cellular control of acetyl-CoA carboxylase.

Having examined the oxidation and synthesis of fatty acids and the regulation of these processes, we now turn to the other pathways of lipid metabolism, including those leading to lipids with roles as signal molecules, membrane constituents, and other specialized functions.

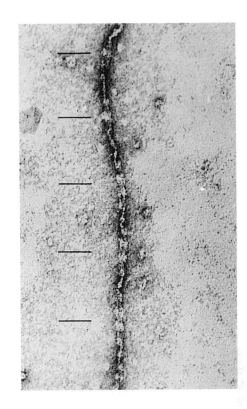

Figure 20·20
Electron micrograph of a filamentous polymer of purified acetyl-CoA carboxylase. The physiological relevance of polymerization is in doubt. (Courtesy of M. Daniel Lane.)

20·12 Eicosanoids, a Large Class of Signal Molecules, Are Derived from Arachidonate

Most arachidonate in cells is found in the inner leaflet of the plasma membrane, esterified at C-2 of the glycerol backbone of phospholipids. Release of arachidonate to the interior of cells is catalyzed by phospholipase A_2 (Figure 11·11).

Arachidonate can serve as the precursor for a large number of products called eicosanoids, a group of long-chain unsaturated fatty acid derivatives that serve as metabolic regulators. There are two general classes of eicosanoid regulatory molecules. One class is derived from cyclization of arachidonate, catalyzed by a bifunctional enzyme with two interacting but separate active sites. The cyclooxygenase activity of the enzyme catalyzes formation of an extremely unstable cyclic compound, a hydroperoxide that is rapidly acted on by the hydroperoxidase activity of the enzyme (Figure 20·21, next page). The product of second step, prostaglandin H_2, is enzymatically converted into an array of regulatory molecules with short lifetimes (on the order of seconds to minutes). These compounds, including prostacyclin, thromboxane, and the prostaglandins, have been termed **local regulators.** Unlike hormones, which are produced by glands and travel in the blood to their sites of action, eicosanoids typically act in the immediate neighborhood of the cell in which they are produced. For example, blood platelets produce thromboxane A_2, the only known naturally occurring thromboxane, which causes platelet aggregation and constriction of smooth muscles of arterial walls and leads to changes in

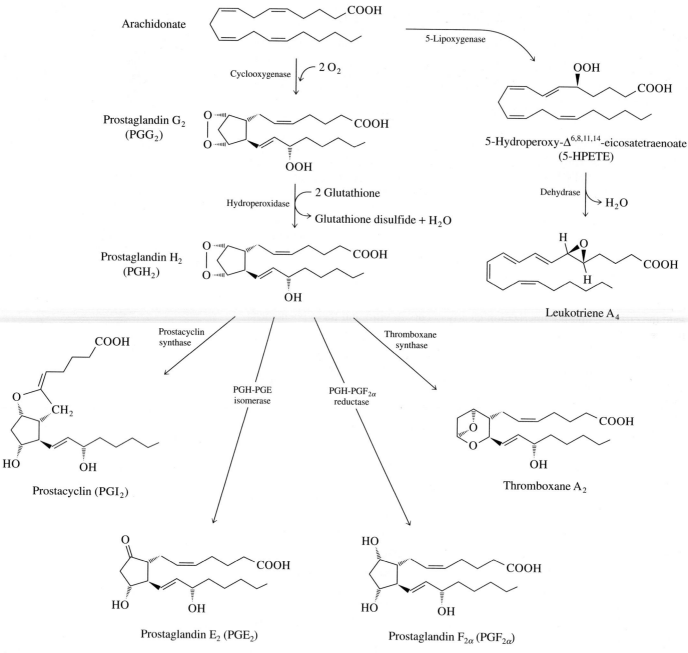

Figure 20·21
Pathways for the formation of eicosanoids. The two major pathways for eicosanoid formation begin with the products of the cyclooxygenase and lipoxygenase reactions. The cyclooxygenase pathway leads to prostacyclin, thromboxane A_2, and a variety of prostaglandins. The lipoxygenase pathway shown produces leukotriene A_4, a precursor of other leukotrienes, some of which trigger allergic responses.

local blood flow. The uterus produces contraction-triggering prostaglandins during labor. Eicosanoids also mediate pain sensitivity, inflammation, and swelling. Aspirin is effective in blocking these effects because its active ingredient, acetylsalicylic acid, irreversibly inhibits cyclooxygenase by transferring an acetyl group to an active-site serine residue of the enzyme. By blocking the activity of cyclooxygenase, aspirin prevents the formation of eicosanoids that are synthesized subsequent to the cyclooxygenase reaction.

The second class of eicosanoids are the products of reactions catalyzed by lipoxygenases. In Figure 20·21, 5-lipoxygenase is shown catalyzing the first step in the pathway leading to leukotriene A_4. Further reactions produce other leukotrienes, including a group of three that constitute what was once called the "slow-reacting substance of anaphylaxis" (allergic response) and are responsible for the occasionally fatal side effects of immunization shots.

20·13 Triacylglycerols and Neutral Phospholipids Are Synthesized from Diacylglycerol

Most fatty acids in cells exist in esterified forms as triacylglycerols or glycerophospholipids. A useful division of phospholipids based on metabolic origins classifies them as **neutral (zwitterionic) phospholipids,** such as phosphatidylcholine and phosphatidylethanolamine, and **acidic (anionic) phospholipids,** such as phosphatidylinositol and phosphatidylserine.

Triacylglycerols and the neutral phospholipids phosphatidylcholine and phosphatidylethanolamine are synthesized by a common pathway. The first portion of this pathway is shown in Figure 20·22. First, dihydroxyacetone phosphate produced by glycolysis is reduced to glycerol 3-phosphate in a reaction catalyzed by

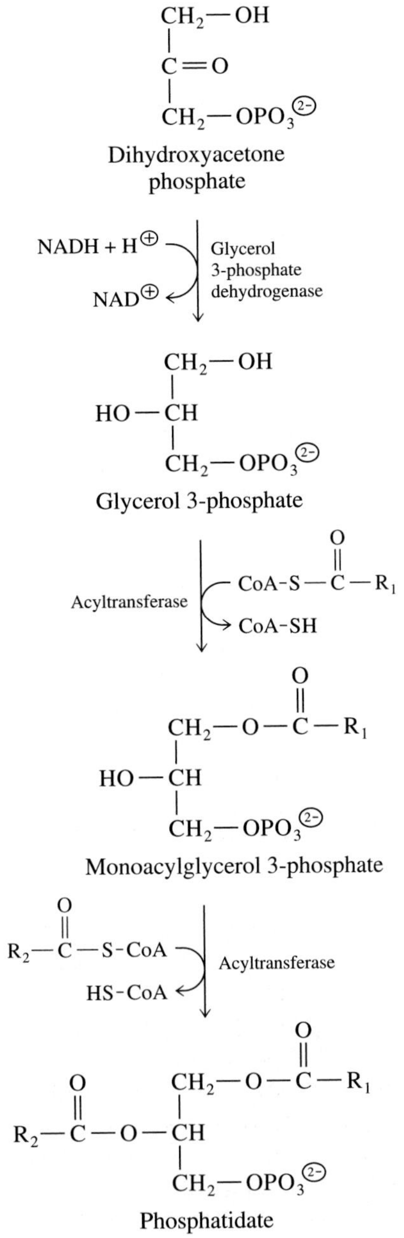

Figure 20·22
Pathway for the formation of phosphatidate. The acyltransferase that catalyzes esterification at C-1 of glycerol 3-phosphate has a preference for saturated acyl chains; the acyltransferase that catalyzes esterification at C-2 has a greater affinity for unsaturated acyl chains.

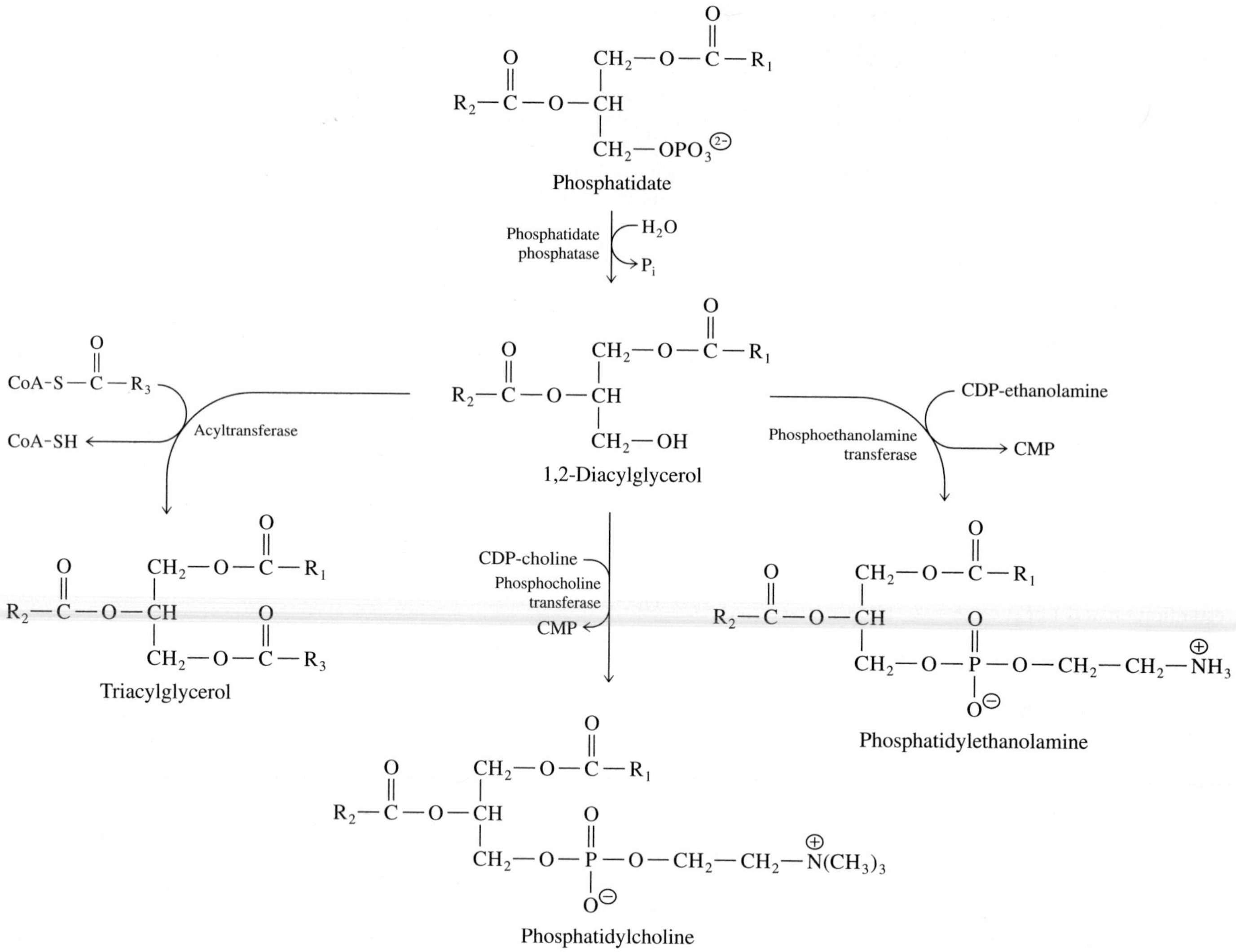

Figure 20·23
Neutral lipid synthesis. Formation of tri-acylglycerols, phosphatidylcholine, and phosphatidylethanolamine proceeds via a diacylglycerol intermediate. A cytidine nucleotide cosubstrate donates the polar head group for the phospholipids.

glycerol 3-phosphate dehydrogenase. Glycerol 3-phosphate then serves as the backbone for acylation reactions catalyzed by two separate acyltransferases, with fatty acyl CoA molecules donating the acyl groups. The first acyltransferase, which has a preference for fatty acyl CoA molecules with saturated acyl chains, catalyzes esterification at C-1 of glycerol 3-phosphate; the second acyltransferase, which has a greater affinity for unsaturated species, catalyzes esterification at C-2 of mono-acylglycerol 3-phosphate. The resulting molecule is a phosphatidate (the term refers to a family of molecules whose specific properties depend upon the attached acyl groups).

The next step in the formation of triacylglycerols and neutral phospholipids is the dephosphorylation of phosphatidate, catalyzed by phosphatidate phosphatase. The product of this reaction, 1,2-diacylglycerol, can be either directly acylated to form a triacylglycerol or it can react with a cytidine triphosphate–derived co-substrate, CDP-choline or CDP-ethanolamine, to form either of two phospholipids, phosphatidylcholine or phosphatidylethanolamine, respectively (Figure 20·23).

Phosphatidylcholine synthesis requires CDP-choline formed through the phosphorylation of choline, catalyzed by choline kinase, followed by condensation with CTP to form CDP-choline, catalyzed by CTP:phosphocholine cytidylyltransferase (Figure 20·24). A parallel series of reactions is employed to form CDP-ethanolamine for phosphatidylethanolamine biosynthesis, with different kinase and transferase enzymes required for the analogous steps.

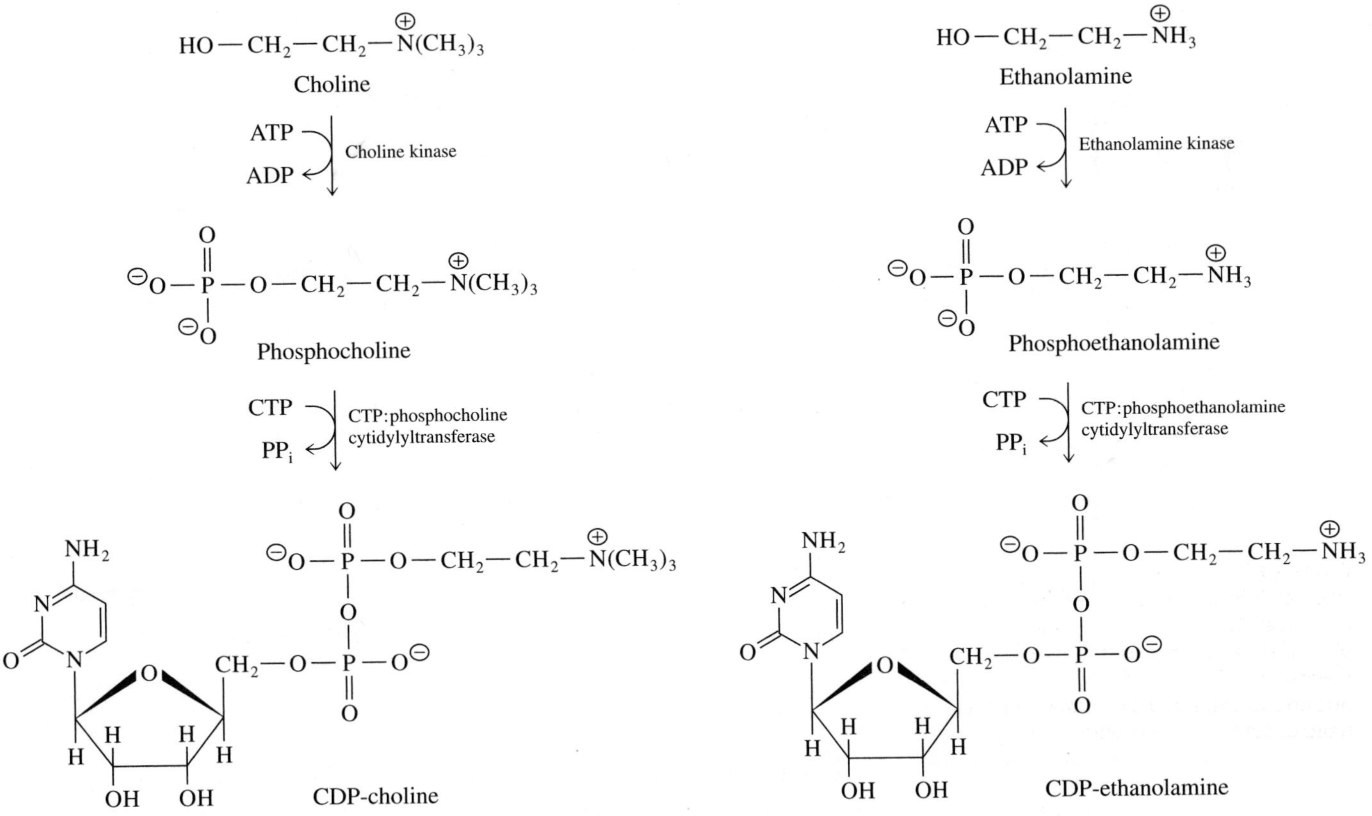

Figure 20·24
Formation of CDP-choline and CDP-ethanolamine.

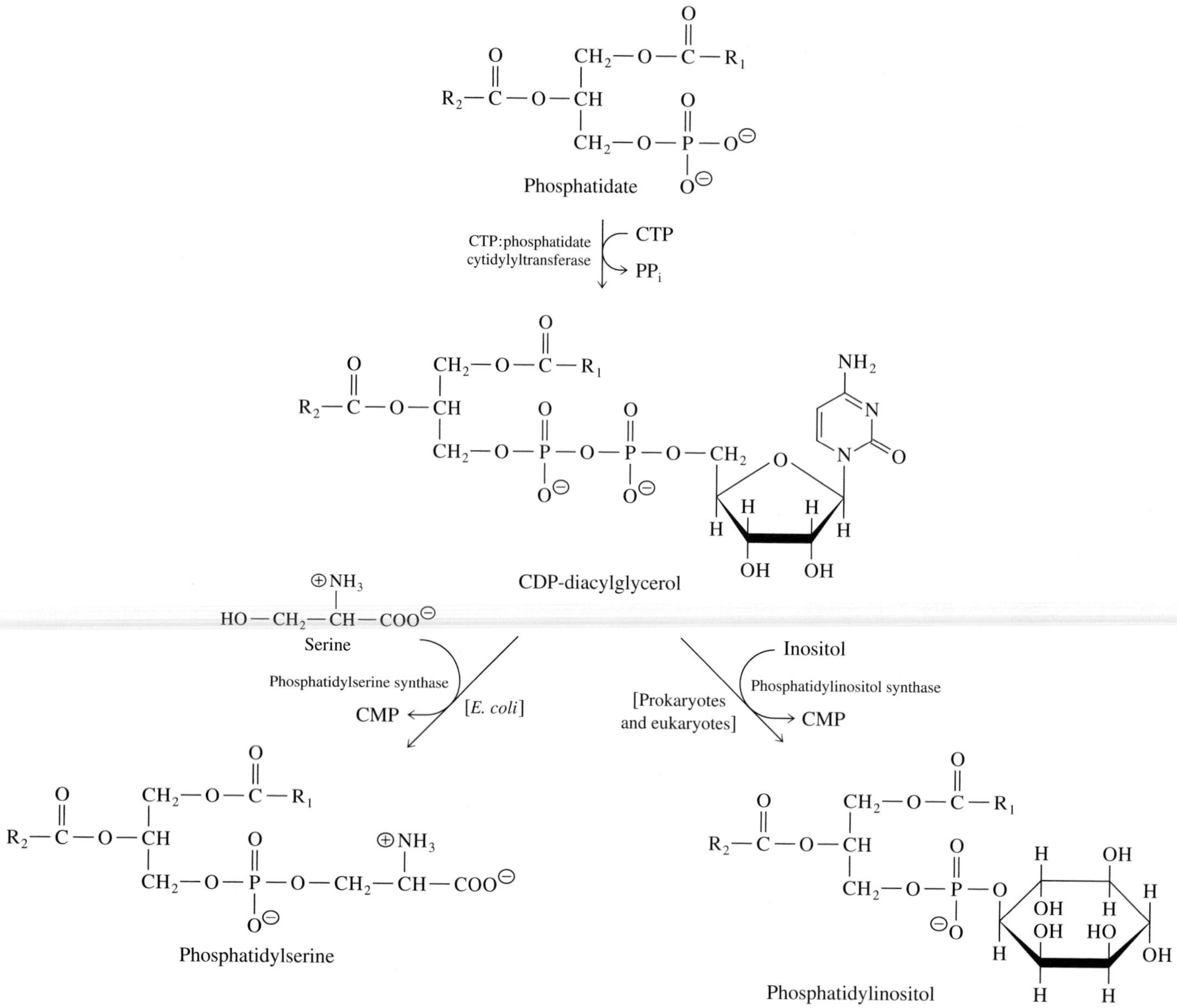

Figure 20·25
Synthesis of the acidic phospholipids phosphatidylserine and phosphatidylinositol. Phosphatidate accepts a cytidylyl group from CTP to form CDP-diacylglycerol. CMP is then displaced by an alcohol group of serine or inositol to form phosphatidylserine or phosphatidylinositol, respectively.

20·14 Biosynthesis of Acidic Phospholipids Proceeds from Phosphatidate

Acidic phospholipids have a net negative charge at physiological pH. The immediate precursor for acidic phospholipids is phosphatidate formed by the pathway shown in Figure 20·22. Phosphatidate condenses with CTP to form CDP-diacylglycerol. Next, displacement of CMP by serine leads to phosphatidylserine in *E. coli*. In both prokaryotes and eukaryotes, displacement of CMP from CDP-diacylglycerol by inositol leads to phosphatidylinositol (Figure 20·25). Through successive phosphorylation reactions, phosphatidylinositol may be converted to phosphatidylinositol 4-phosphate (PIP) and phosphatidylinositol 4,5-*bis*phosphate, or PIP$_2$ (Figure 20·26).

Figure 20·26
Biosynthesis of
phosphatidylinositol
4-phosphate and
phosphatidylinositol
4,5-*bis*phosphate from
phosphatidylinositol.

Phosphatidylinositol

Phosphatidylinositol
kinase

ATP
ADP

Phosphatidylinositol 4-phosphate
(PIP)

Phosphatidylinositol
4-phosphate kinase

ATP
ADP

Phosphatidylinositol 4,5-*bis*phosphate
(PIP$_2$)

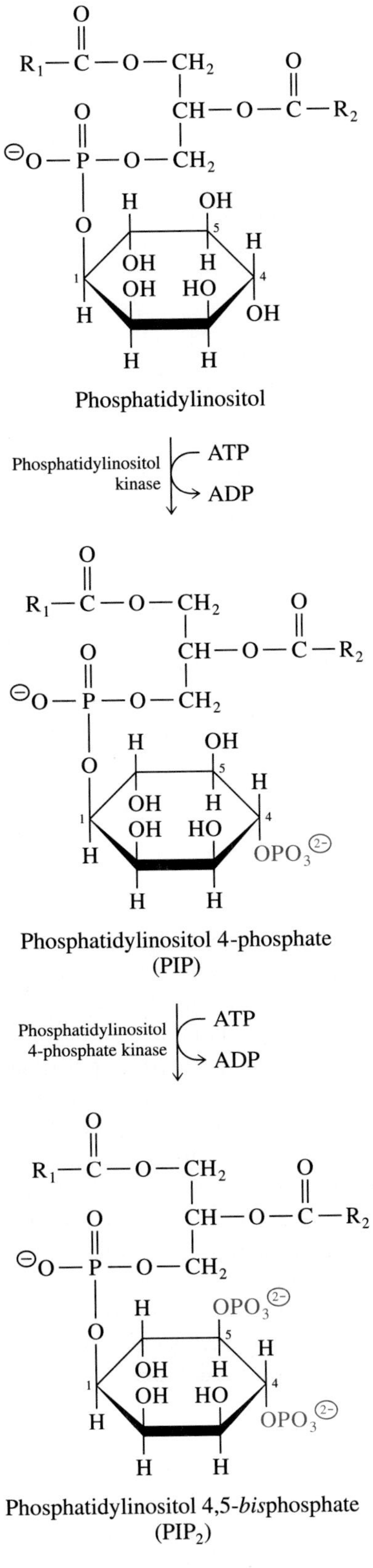

Figure 20·27
Interconversions of phosphatidylethanolamine and phosphatidylserine.

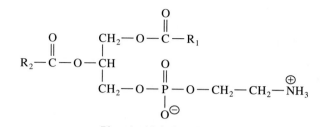

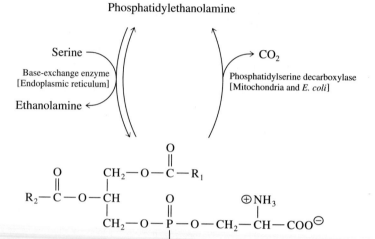

In mammals, phosphatidylserine is formed by the action of the base-exchange enzyme, which catalyzes the reversible displacement of ethanolamine from phosphatidylethanolamine by serine (Figure 20·27). The conversion of phosphatidylserine to phosphatidylethanolamine also occurs via decarboxylation of phosphatidylserine in a reaction catalyzed by phosphatidylserine decarboxylase. The base-exchange reaction takes place in the endoplasmic reticulum of eukaryotes, whereas the decarboxylation reaction takes place in the mitochondria and in *E. coli*.

In *E. coli* and mitochondria, glycerol 3-phosphate itself can serve as a head group of phospholipids. Glycerol 3-phosphate displaces CMP from CDP-diacylglycerol to form phosphatidylglycerol phosphate, which is dephosphorylated to form phosphatidylglycerol (Figure 20·28). Here the pathways in *E. coli* and mitochondria diverge. In mitochondria, phosphatidylglycerol condenses with CDP-diacylglycerol; in *E. coli,* a second molecule of phosphatidylglycerol condenses with the first. The product is the same in both cases, diphosphatidylglycerol, more commonly known as cardiolipin.

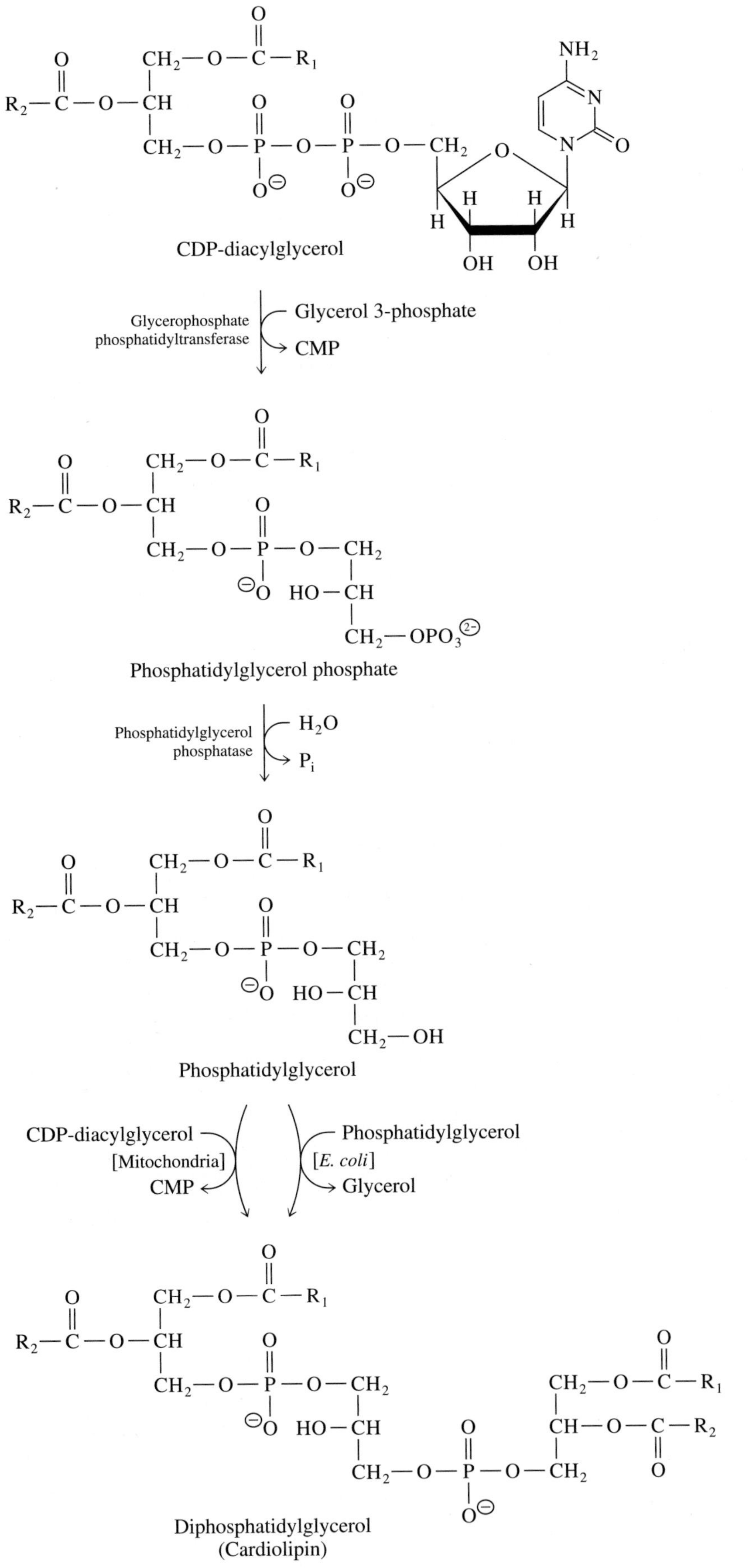

Figure 20·28
Formation of diphosphatidylglycerol. A molecule of glycerol 3-phosphate is incorporated into CDP-diacylglycerol by displacement of CMP to form phosphatidylglycerol phosphate. Next, a phosphatase catalyzes the formation of phosphatidylglycerol. Separate displacement reactions ensue in *E. coli* (where a second molecule of phosphatidylglycerol displaces glycerol) and mitochondria (where phosphatidylglycerol reacts with CDP-diacylglycerol, displacing CMP). Diphosphatidylglycerol, also known as cardiolipin, is found in the plasma membrane of prokaryotes and the inner mitochondrial membrane of eukaryotes.

CDP-diacylglycerol

Glycerophosphate phosphatidyltransferase — Glycerol 3-phosphate → CMP

Phosphatidylglycerol phosphate

Phosphatidylglycerol phosphatase — H_2O → P_i

Phosphatidylglycerol

CDP-diacylglycerol [Mitochondria] CMP — Phosphatidylglycerol [*E. coli*] → Glycerol

Diphosphatidylglycerol (Cardiolipin)

Figure 20·29
Deacylation/reacylation cycle. A deacylation/reacylation mechanism allows for the tailoring of fatty acids attached to newly synthesized phosphatidylinositol. A specific example is shown. Oleate is first removed from C-2 of phosphatidylinositol by the action of phospholipase A_2. The 1-stearoylglycero-3-phosphoinositol product then reacts with arachidonoyl CoA to form the 1-stearoyl-2 arachidonoyl species of glycero-3-phosphoinositol. This reaction, catalyzed by an acyltransferase, yields the characteristic fatty acid pattern seen in mature phosphatidylinositol.

1-Stearoyl-2-oleoylglycero-3-phosphoinositol

Phospholipase A_2 → Oleate

1-Stearoylglycero-3-phosphoinositol

Acyltransferase — Arachidonoyl CoA → CoASH

1-Stearoyl-2-arachidonoylglycero-3-phosphoinositol

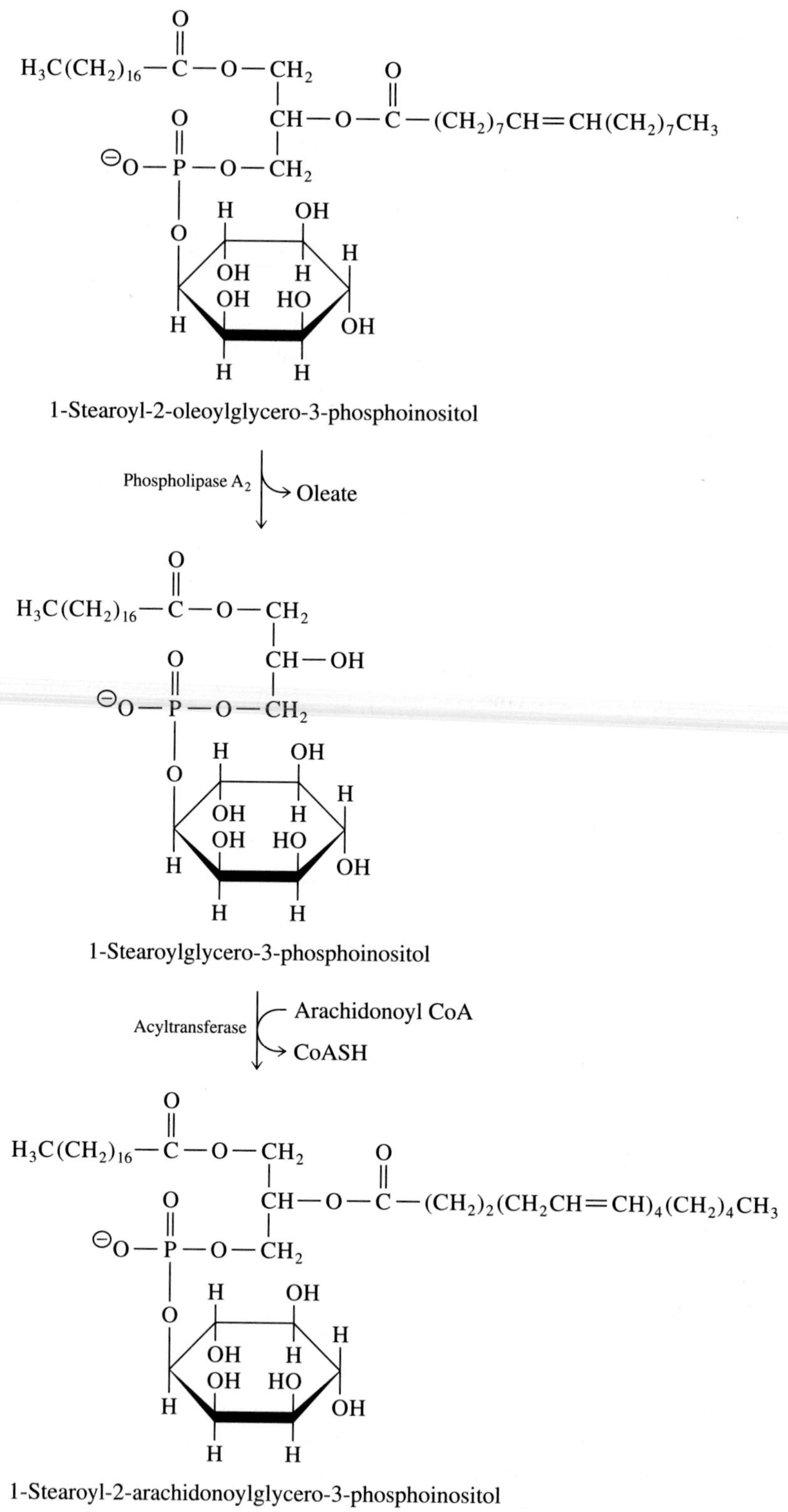

20·15 Fatty Acid Constituents of Phospholipids Can Be Changed by Deacylation/Reacylation Reactions

Neutral and acidic phospholipids are both derived from phosphatidate via 1,2-diacyl-glycerol or CDP-diacylglycerol (Figures 20·23 and 20·25, respectively). However, the combinations of acyl chains in the two classes of phospholipids are quite distinct. This could not be the case if the original fatty acid constituents of phosphatidate were conserved in the phospholipid end products. For example, arachidonate is not a major component of phosphatidate, but in phosphatidylinositol and phosphatidylinositol 4,5-*bis*phosphate, arachidonate can account for more than 40% of the total fatty acids. It does not appear that molecular species of phosphatidate enriched in arachidonate are selected for phosphatidylinositol synthesis. Instead, newly synthesized molecules of phosphatidylinositol are tailored so that fatty acids such as oleate are removed from the C-2 position and arachidonate is attached in their place. This exchange is accomplished by a deacylation/reacylation cycle in which phospholipase A_2 catalyzes the removal of the original C-2 fatty acid, producing 1-stearoylglycero-3-phosphoinositol (Figure 20·29). This lysophospholipid (so-called because when introduced in high concentrations it can cause cells to lyse) is then reacylated by the action of an acyltransferase that uses arachidonoyl CoA as the fatty acyl donor. Other newly made phospholipids can be tailored by the same mechanism; varying the theme slightly, the fatty acid constituents at C-1 can be replaced by the activity of phospholipase A_1 in the deacylation step.

20·16 Ether Lipids Are Synthesized from Dihydroxyacetone Phosphate

Ether lipids have an ether linkage in place of one of the usual ester linkages. Ether lipids are derived from dihydroxyacetone phosphate rather than glycerol 3-phosphate, the precursor for most phospholipids. The overall pathway for the formation of ether lipids is shown in Figure 20·30 (next page). First, an acyl group from fatty acyl CoA is esterified to the alcohol moiety at C-1 of dihydroxyacetone phosphate, producing 1-acyldihydroxyacetone phosphate. Next, a fatty alcohol displaces the fatty acid to produce 1-alkyldihydroxyacetone phosphate. The keto group of this compound is reduced by NADPH to form 1-alkylglycero-3-phosphate. Reduction is followed by esterification at C-2 of the glycerol residue to produce 1-alkyl-2-acylglycerol-3-phosphate. The subsequent reactions—dephosphorylation and transfer of the choline group—are the same as those shown earlier in Figure 20·23.

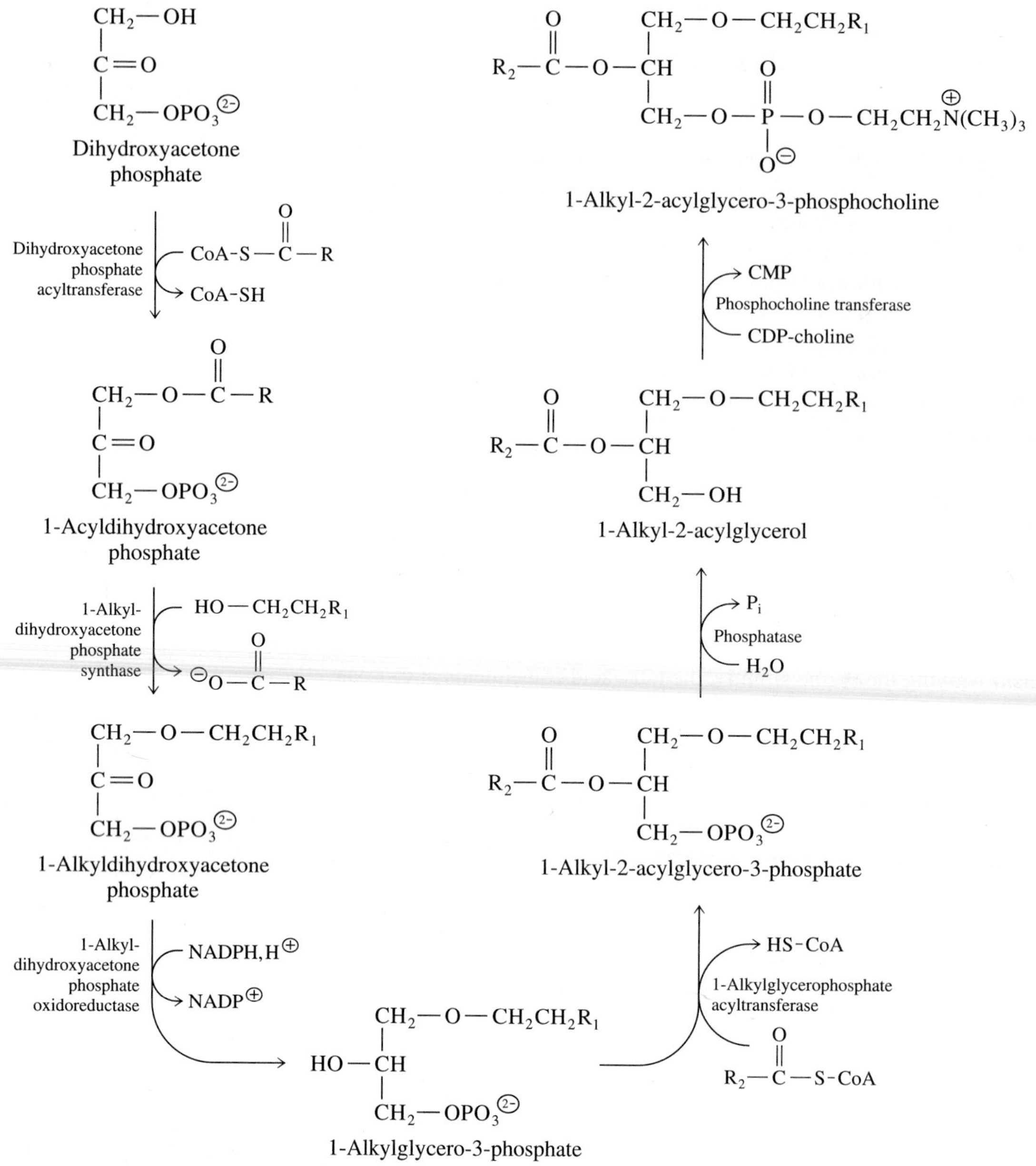

Figure 20·30
Formation of ether lipids. First, an acyl group from fatty acyl CoA is esterified to the alcohol moiety at C-1 of dihydroxyacetone phosphate, producing 1-acyldihydroxyacetone phosphate. A fatty alcohol displaces the fatty acid and the product is reduced by NADPH to produce 1-alkylglycero-3-phosphate. Esterification at C-2 produces 1-alkyl-2-acylglycero-3-phosphate, which undergoes dephosphorylation. Finally, transfer of a choline group from CDP-choline produces 1-alkyl-2-acylglycero-3-phosphocholine.

The displacement of a fatty acid by a fatty alcohol to produce 1-alkyl-dihydroxyacetone phosphate is an unusual reaction. The mechanism is detailed in Figure 20·31. Tautomerization in Step 1 is followed by addition of a proton (Step 2) to produce a carbonium ion. The fatty alcohol adds to the carbonium ion (Step 3), forming an ether and an ester bond at C-1. The fatty acid, a good leaving group, is then displaced in Step 4, and a second carbonium ion is formed. Proton extraction and tautomerization (Steps 5 and 6) form the product, 1-alkyldihydroxyacetone phosphate.

One class of ether lipids called plasmalogens has a vinyl ether linkage at C-1 of the glycerol backbone (see Figure 11·13). Plasmalogens are synthesized using ethanolamine and choline glycerophospholipids with an alkyl ether constituent at C-1. A mixed-function oxidase similar to the fatty acid desaturases catalyzes the conversion of the alkyl ether linkage into a vinyl, or alkenyl, ether linkage (Figure 20·32).

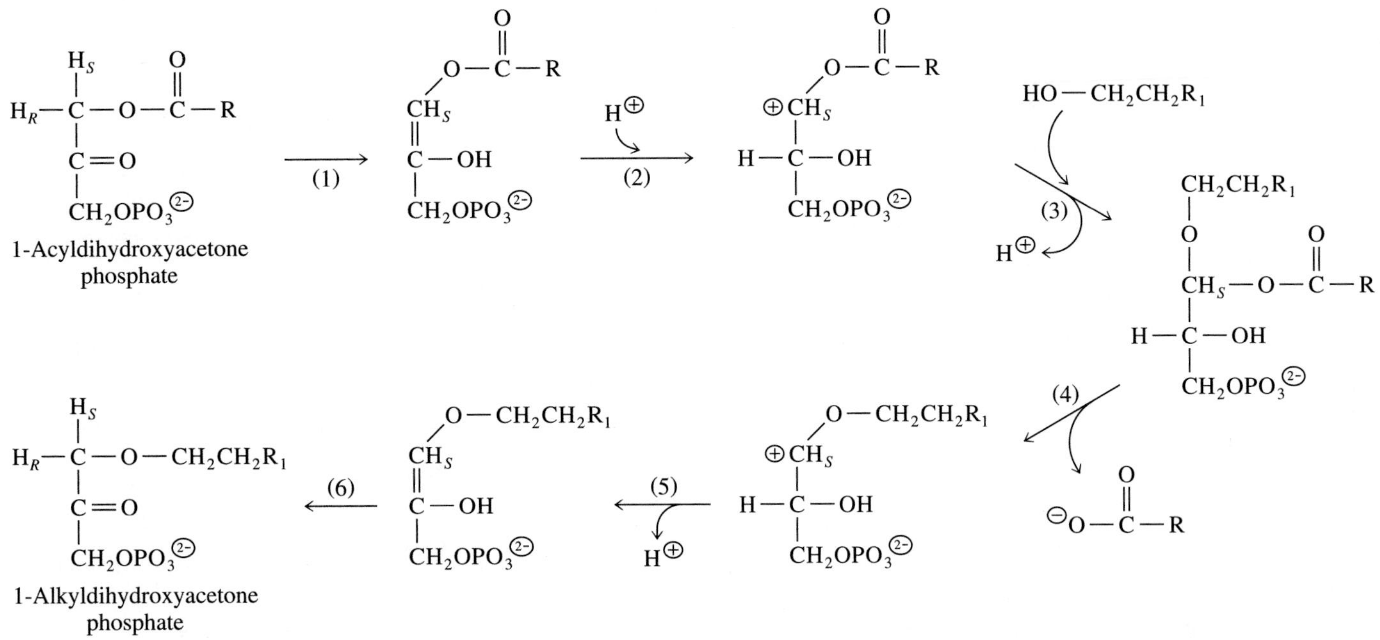

The structures show the conversion of 1-Acyldihydroxyacetone phosphate to 1-Alkyldihydroxyacetone phosphate through steps (1) through (6).

1-Acyldihydroxyacetone phosphate

1-Alkyldihydroxyacetone phosphate

Figure 20·31
Proposed mechanism for ether bond formation. Tautomerization and addition of a proton forms a carbonium ion at C-1 of the dihydroxyacetone phosphate backbone. Addition of an alcohol ($R_1CH_2CH_2$—OH) to this moiety displaces the fatty acid at C-1, and deprotonation and tautomerization forms the 1-ether lipid, 1-alkyldihydroxyacetone phosphate.

The physiological roles of plasmalogens and most other ether lipids are unknown, though the function of one ether lipid called platelet activating factor (PAF) has been determined. An ether lipid with a palmitoyl group at C-1 of the glycerol backbone and an acetyl group at C-2, PAF serves as a signal molecule that stimulates the aggregation of platelets during the process of blood clotting. Platelet activating factor is extremely potent, effective at about 0.1 nM.

The pathway for synthesis of PAF was first elucidated in white blood cells. Following stimulation of these cells with a preparation of yeast particles, 1-alkyl-2-acylglycero-3-phosphocholine molecules are attacked by cellular phospholipase A_2 to yield 1-alkylglycero-3-phosphocholine, which is also known as lyso-PAF (Figure 20·33, next page). An acetyltransferase then catalyzes transfer of an acetyl group from acetyl CoA to form PAF. Transfer of the acetyl group is the rate-limiting step of the pathway.

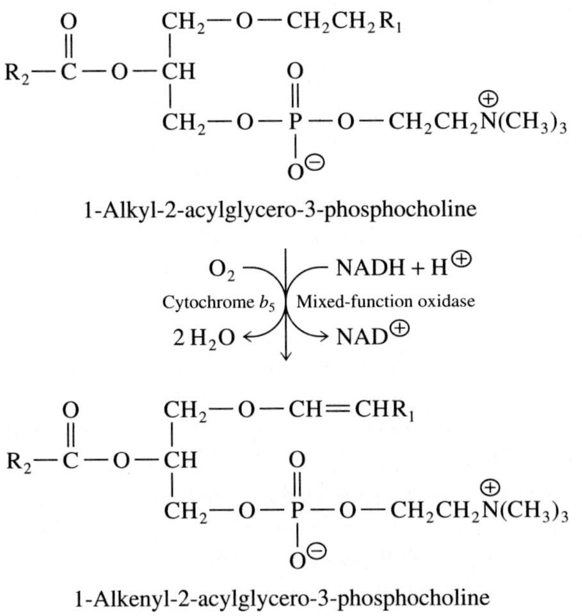

1-Alkyl-2-acylglycero-3-phosphocholine

1-Alkenyl-2-acylglycero-3-phosphocholine

Figure 20·32
Biosynthesis of a choline plasmalogen. A mixed-function oxidase uses oxygen and NADH in a reaction that involves cytochrome b_5 to introduce a double bond into the alkyl chain at C-1 of 1-alkyl-2-acylglycero-3-phosphocholine. This produces the characteristic vinyl (alkenyl) ether linkage found at C-1 of plasmalogens.

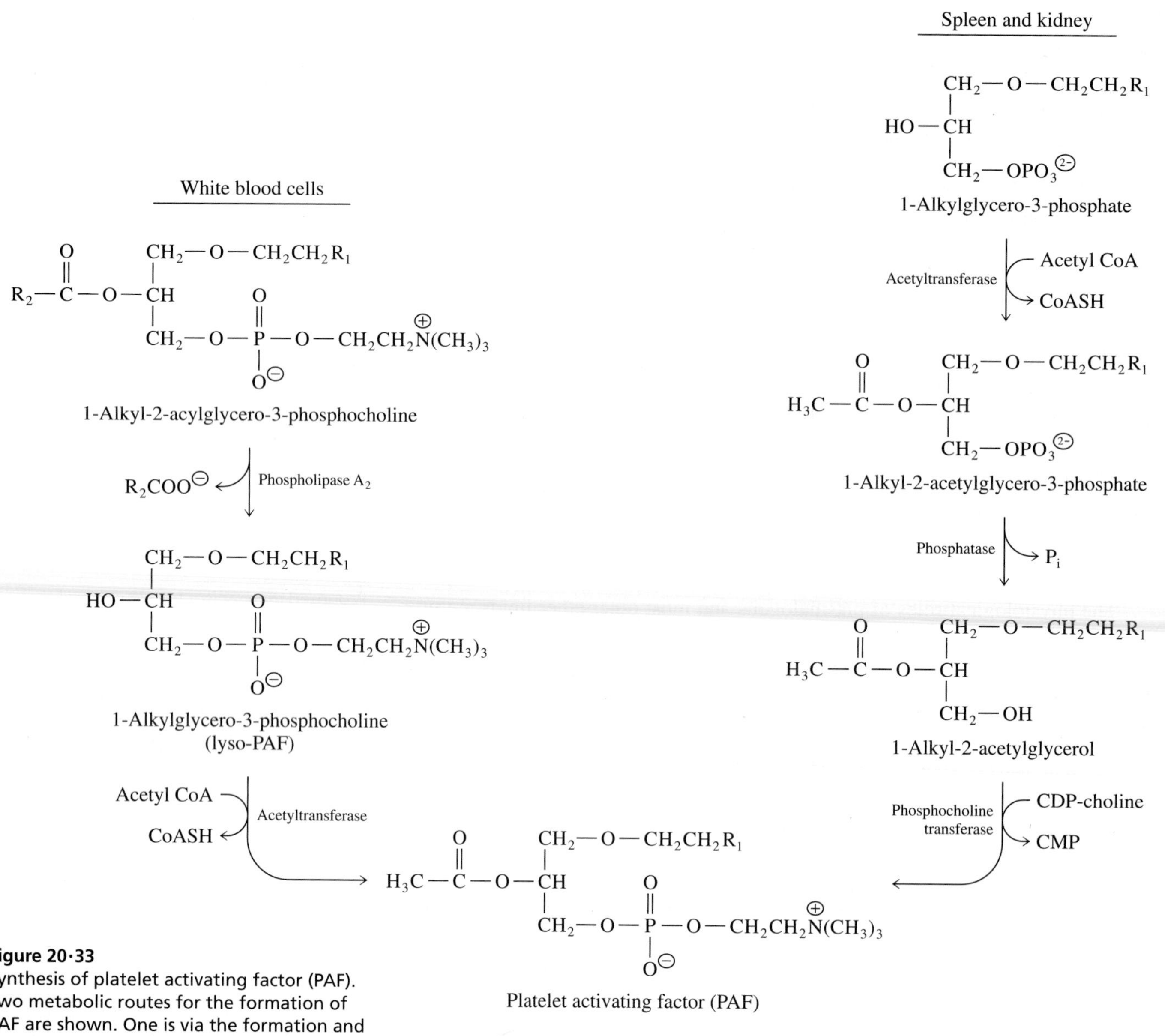

Spleen and kidney

$$CH_2-O-CH_2CH_2R_1$$
$$HO-CH$$
$$CH_2-OPO_3^{2\ominus}$$

1-Alkylglycero-3-phosphate

Acetyltransferase — Acetyl CoA → CoASH

$$H_3C-\overset{O}{\overset{\|}{C}}-O-CH$$
$$CH_2-O-CH_2CH_2R_1$$
$$CH_2-OPO_3^{2\ominus}$$

1-Alkyl-2-acetylglycero-3-phosphate

Phosphatase → P_i

$$H_3C-\overset{O}{\overset{\|}{C}}-O-CH$$
$$CH_2-O-CH_2CH_2R_1$$
$$CH_2-OH$$

1-Alkyl-2-acetylglycerol

White blood cells

$$R_2-\overset{O}{\overset{\|}{C}}-O-CH$$
$$CH_2-O-CH_2CH_2R_1$$
$$CH_2-O-\overset{O}{\overset{\|}{P}}-O-CH_2CH_2\overset{\oplus}{N}(CH_3)_3$$
$$O^\ominus$$

1-Alkyl-2-acylglycero-3-phosphocholine

$R_2COO^\ominus$ ← Phospholipase A_2

$$HO-CH$$
$$CH_2-O-CH_2CH_2R_1$$
$$CH_2-O-\overset{O}{\overset{\|}{P}}-O-CH_2CH_2\overset{\oplus}{N}(CH_3)_3$$
$$O^\ominus$$

1-Alkylglycero-3-phosphocholine
(lyso-PAF)

Acetyl CoA — Acetyltransferase → CoASH

Phosphocholine transferase — CDP-choline → CMP

$$H_3C-\overset{O}{\overset{\|}{C}}-O-CH$$
$$CH_2-O-CH_2CH_2R_1$$
$$CH_2-O-\overset{O}{\overset{\|}{P}}-O-CH_2CH_2\overset{\oplus}{N}(CH_3)_3$$
$$O^\ominus$$

Platelet activating factor (PAF)

Figure 20·33
Synthesis of platelet activating factor (PAF). Two metabolic routes for the formation of PAF are shown. One is via the formation and acetylation of lyso-PAF, which takes place in activated white blood cells. The second, which takes place in spleen and kidney, is a longer process in which 1-alkyl-2-acetyl-glycero-3-phosphate is formed and then converted into 1-alkyl-2-acetylglycerol. This compound is processed in a manner similar to that described for phosphatidylcholine synthesis in Figure 20·23.

Interestingly, PAF is made by a different pathway in tissues such as spleen and kidney. In the alternate synthetic route shown in Figure 20·33, 1-alkylglycero-3-phosphate, derived from 1-alkyldihydroxyacetone phosphate by reduction as shown in Figure 20·30, is acetylated by the action of an acetyltransferase that differs from the enzyme that catalyzes formation of PAF in white blood cells. A phosphatase catalyzes conversion of 1-alkyl-2-acetylglycero-3-phosphate to 1-alkyl-2-acetylglycerol, which in turn serves as a substrate for a phosphocholine transferase activity that produces PAF. This transferase, which uses CDP-choline as a cosubstrate, is distinct from the enzyme that produces phosphatidylcholine from 1,2-diacylglycerol (shown in Figure 20·23).

The two routes of PAF production may serve two distinct purposes: the generation of PAF in response to an extracellular stimulus (as in white blood cells) or the continuous production of PAF for regulation of blood pressure (as in kidney). In either event, PAF can be rapidly degraded by the action of an acetylhydrolase to produce lyso-PAF. The lyso-PAF can be acylated to form 1-alkyl-2-acylglycero-3-phosphocholine, the phospholipid that initiates the pathway of PAF synthesis in white cells by serving as a substrate for phospholipase A_2.

20·17 Sphingolipids Are Derived from Palmitoyl CoA and Serine

The sphingolipids are a class of lipids that have sphingosine as their structural backbone. The backbone is derived from palmitoyl CoA and serine. In the first step of the sphingosine biosynthetic pathway, serine condenses with palmitoyl CoA, producing 3-ketosphinganine (Figure 20·34). Reduction of 3-ketosphinganine, catalyzed by NADPH-dependent 3-ketosphinganine reductase, produces sphinganine. Finally, desaturation produces sphingosine in a reaction catalyzed by the flavin-containing sphinganine dehydrogenase.

Figure 20·34
Biosynthesis of sphingosine.

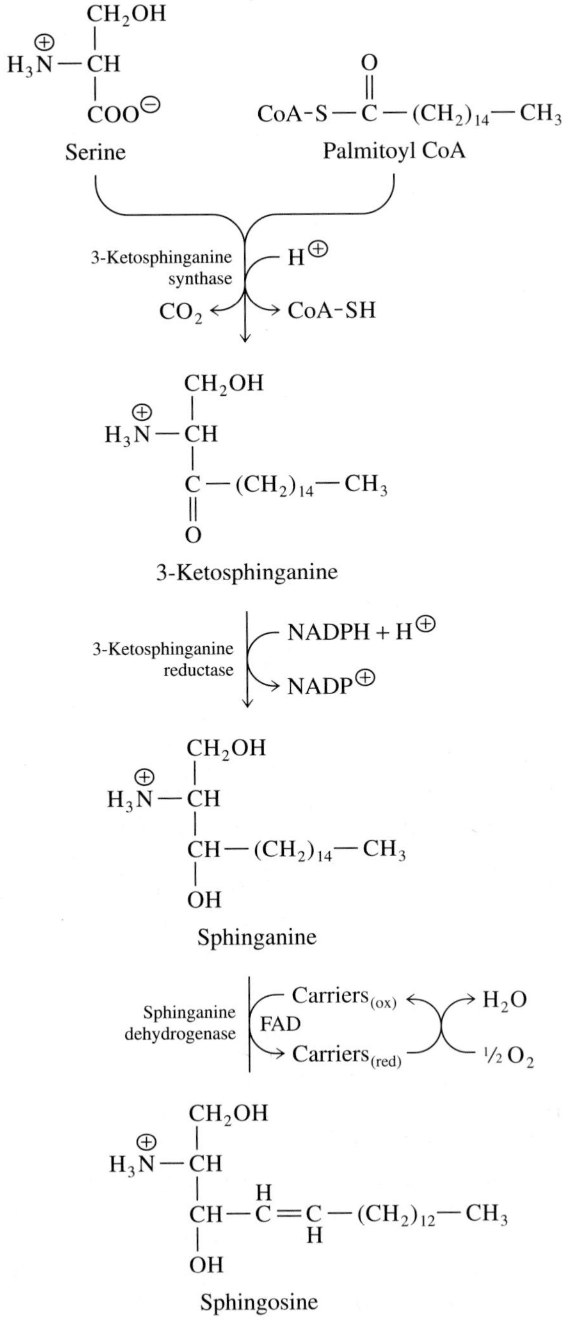

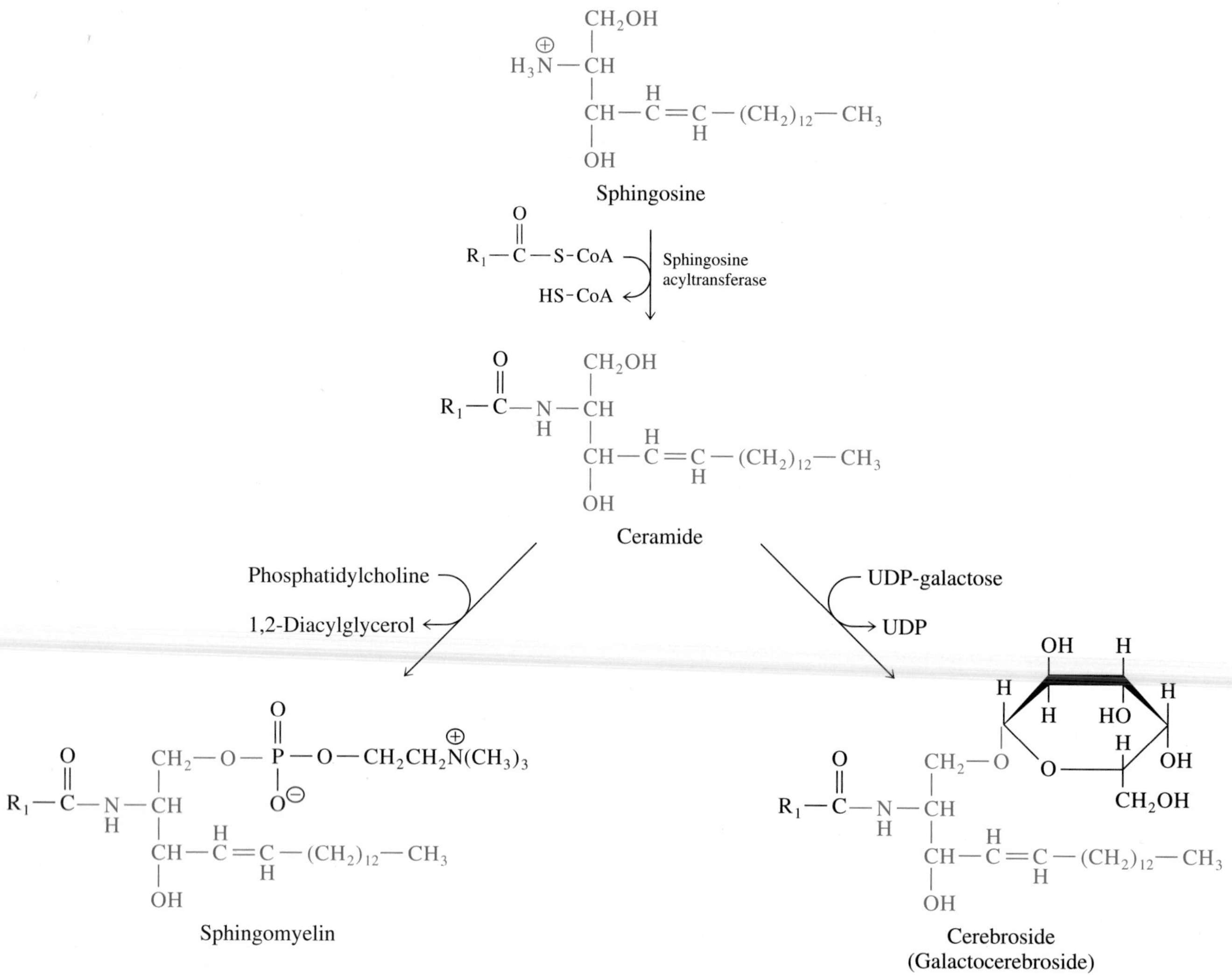

Figure 20·35
Formation of a sphingomyelin and a cerebroside (galactocerebroside), starting from sphingosine.

Like most fatty acid–modifying enzymes in eukaryotes, sphinganine dehydrogenase is embedded in the membrane of the endoplasmic reticulum, with its active site facing the cytosol. Although the details of the electron transfer needed to reoxidize the $FADH_2$ of the enzyme are not clear, it is known that the endoplasmic reticulum contains a number of electron carriers (including heme-containing proteins and Fe-S proteins) that reduce O_2, but at a small fraction of the rate at which mitochondrial O_2 is reduced. Unlike mitochondrial electron carriers, the electron carriers of the endoplasmic reticulum are not involved in ATP production.

Acylation of sphingosine produces a ceramide, which can then be modified by reaction with CDP-choline or phosphatidylcholine to form a sphingomyelin or with a UDP-sugar to form a cerebroside (Figure 20·35). More complex sugar-lipid conjugates can be formed by reaction with additional nucleotide sugars and CMP-N-acetylneuraminic acid. The resulting molecules are gangliosides. The gangliosides are part of the external leaflet of the plasma membrane in mammals, as are most sugar lipids. Complete structures of the ganglioside component N-acetylneuraminic acid (NeuNAc) and the gangliosides G_{M1} and G_{M2} were shown in Figure 11·18. Schematic pathways for the formation and degradation of a variety of sphingolipids (cerebrosides, gangliosides, and globosides) are shown in Figure 20·36.

Defects in sphingolipid degradation can have serious clinical consequences. Sphingolipid catabolism is largely carried out in the lysosomes of cells. Lysosomes contain a variety of glycosidases that catalyze the stepwise hydrolytic removal of

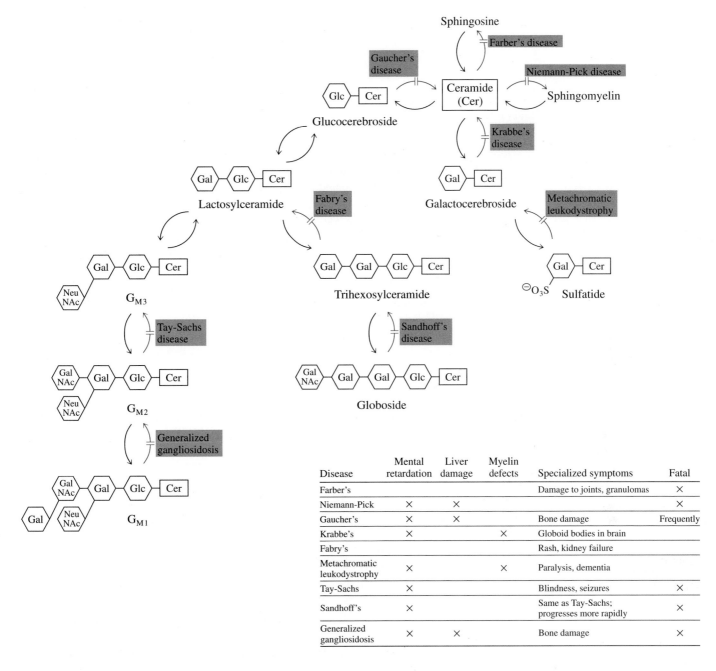

Disease	Mental retardation	Liver damage	Myelin defects	Specialized symptoms	Fatal
Farber's				Damage to joints, granulomas	×
Niemann-Pick	×	×			×
Gaucher's	×	×		Bone damage	Frequently
Krabbe's	×		×	Globoid bodies in brain	
Fabry's				Rash, kidney failure	
Metachromatic leukodystrophy	×		×	Paralysis, dementia	
Tay-Sachs	×			Blindness, seizures	×
Sandhoff's	×			Same as Tay-Sachs; progresses more rapidly	×
Generalized gangliosidosis	×	×		Bone damage	×

Figure 20·36
Pathways for the formation and degradation of a variety of sphingolipids, with hereditary metabolic diseases indicated. The diseases resulting from the enzyme defects are called sphingolipidoses.

sugars from the oligosaccharide chains of sphingolipids. There are certain inborn errors of metabolism in which a genetic defect leads to a deficiency in a particular degradative lysosomal enzyme, resulting in lysosomal storage diseases (Section 9·14C). In Tay-Sachs disease, for instance, there is a deficiency in hexosaminidase A, which catalyzes the removal of N-acetylgalactosamine from the oligosaccharide chain of gangliosides. If removal of this sugar does not occur, the disassembly of gangliosides is blocked, leading to a buildup of a nondegradable by-product, the ganglioside G_{M2}. A number of defects in sphingolipid metabolism, whose clinical manifestations are termed sphingolipidoses, are identified in Figure 20·36. The accumulation of nondegradable lipid by-products can cause lysosomes to swell, leading to cellular and ultimately tissue enlargement. This is particularly deleterious in central nervous tissue, which offers little room for expansion. Swollen lysosomes accumulate in the cell bodies of nerve cells and lead to neuronal death, possibly by leakage of lysosomal enzymes into the cell. As a result, blindness, mental retardation, and death can occur.

Figure 20·37
Isoprene.

20·18 Cholesterol Is Derived from Cytosolic Acetyl CoA

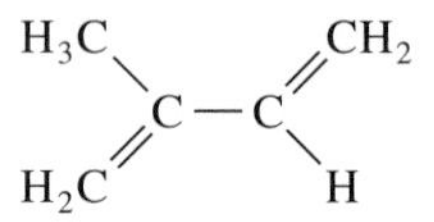

Although most mammalian cells are capable of synthesizing cholesterol, the pathway for cholesterol formation is substantially active only in liver cells. Part of the function of lipoproteins is the delivery of dietary and liver-derived cholesterol to the rest of the body's cells.

The first milestone in the elucidation of cholesterol synthesis was the discovery that all of the carbon in cholesterol arises from acetyl CoA, a fact that emerged from early radioisotope labelling experiments. It was also discovered that squalene, a C_{30} linear hydrocarbon, is an intermediate in the biosynthesis of the 27-carbon cholesterol molecule and that squalene itself is formed from five-carbon units related to isoprene, whose structure is shown in Figure 20·37. Thus, the stages of cholesterol biosynthesis were found to be

$$\text{Acetate (C}_2) \longrightarrow \text{Isoprenoid (C}_5) \longrightarrow \text{Squalene (C}_{30}) \longrightarrow \text{Cholesterol (C}_{27})$$

(20·10)

We will organize our examination of cholesterol synthesis by focusing on the 5-carbon, 30-carbon, and 27-carbon stages.

Figure 20·38
Initial reactions of cholesterol biosynthesis leading to the formation of mevalonate. HMG-CoA reductase catalyzes the major regulatory step for the synthesis of cholesterol.

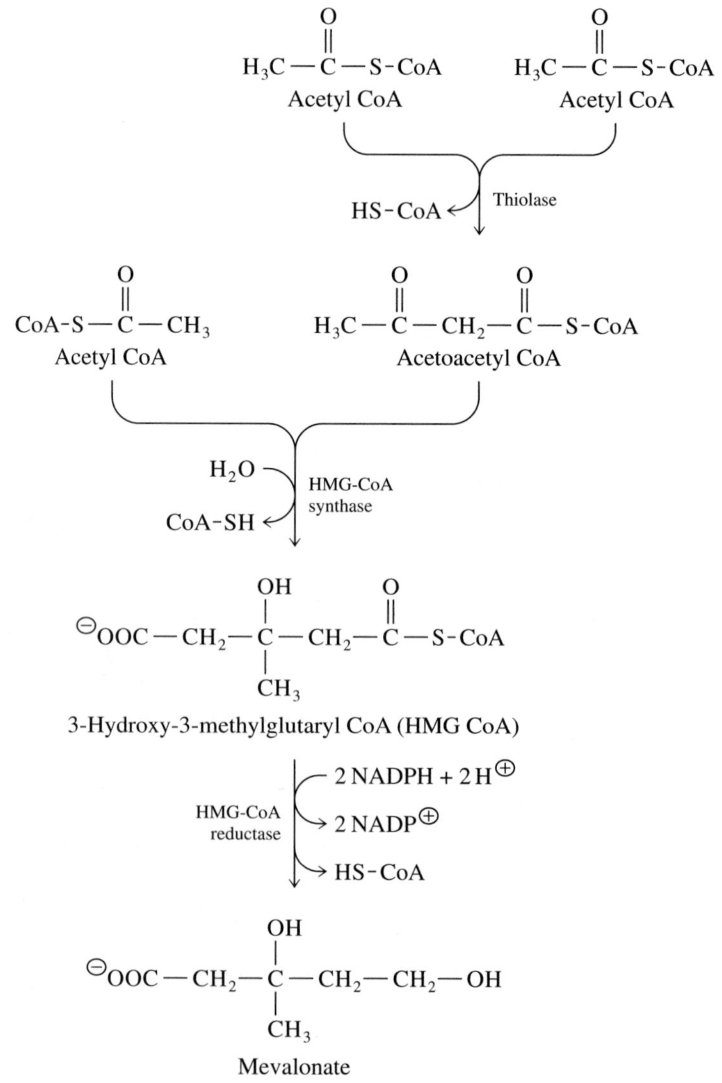

A. Stage 1: Acetyl CoA to Isopentenyl Pyrophosphate

Cholesterol is synthesized from cytosolic acetyl CoA, which is transported from mitochondria via the citrate transport system. Since cytosolic acetyl CoA is also used for fatty acid synthesis, it is the branch-point metabolite of two major pathways of lipid biosynthesis.

The first stage in the formation of cholesterol is the sequential condensation of three molecules of acetyl CoA, with the formation of hydroxymethylglutaryl CoA (HMG CoA) as an intermediate leading to mevalonate (Figure 20·38). The condensation steps are catalyzed by cytosolic isozymes of the mitochondrial enzymes involved in the formation of ketone bodies.

HMG-CoA reductase catalyzes the first committed step in cholesterol biosynthesis, the conversion of HMG CoA to mevalonate. Two molecules of NADPH are consumed in this reaction. HMG-CoA reductase is a large integral membrane protein of the endoplasmic reticulum that has seven membrane-spanning units and a C-terminal domain that extends into the cytosol. The active site of the enzyme is in the cytosolic domain. HMG-CoA reductase is regulated by several mechanisms that will be described in Section 20·19.

Mevalonate is converted to isopentenyl pyrophosphate in three enzyme-catalyzed steps: phosphorylation of mevalonate, a second phosphorylation, and decarboxylation (Figure 20·39).

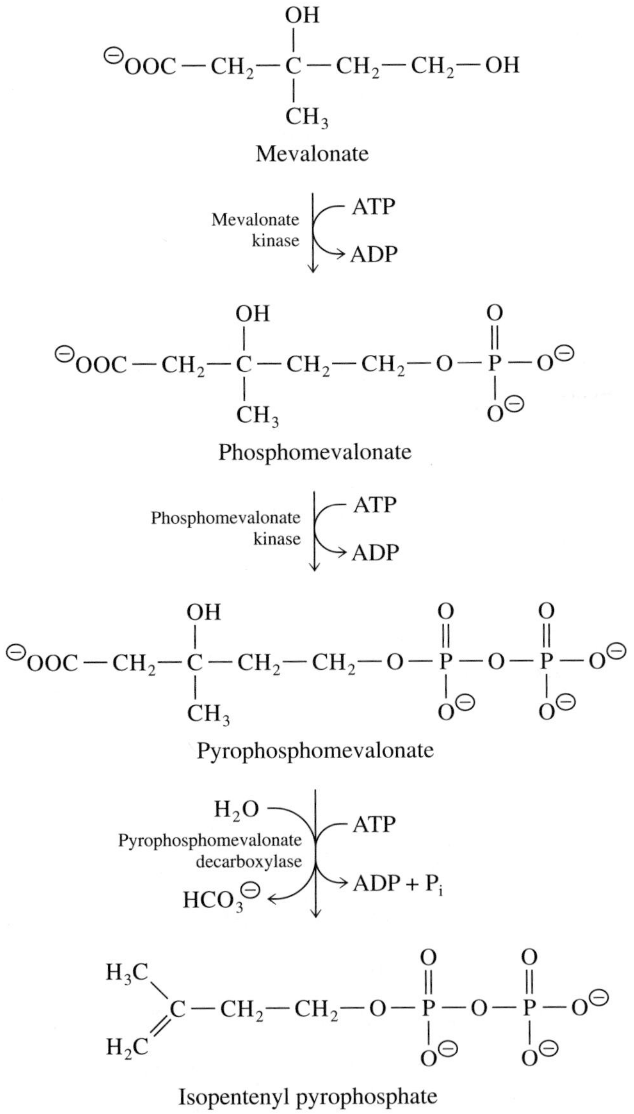

Figure 20·39
Formation of isopentenyl pyrophosphate. Two phosphorylations and a decarboxylation reaction convert mevalonate into the five-carbon molecule isopentenyl pyrophosphate.

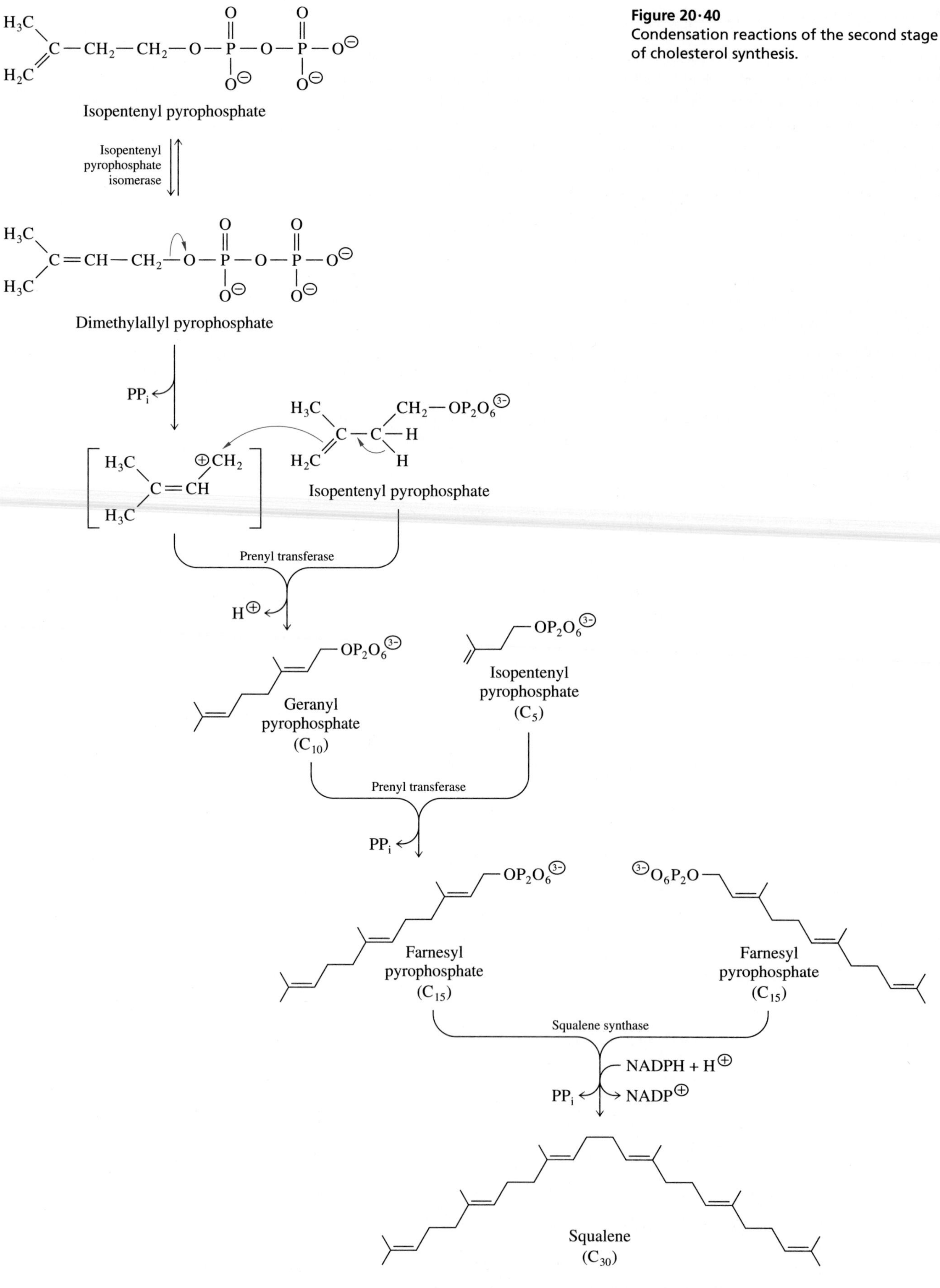

Figure 20·40
Condensation reactions of the second stage
of cholesterol synthesis.

Isopentenyl pyrophosphate

Isopentenyl
pyrophosphate
isomerase

Dimethylallyl pyrophosphate

PP_i

Isopentenyl pyrophosphate

Prenyl transferase

$H^{\oplus}$

Geranyl
pyrophosphate
(C_{10})

Isopentenyl
pyrophosphate
(C_5)

Prenyl transferase

PP_i

Farnesyl
pyrophosphate
(C_{15})

Farnesyl
pyrophosphate
(C_{15})

Squalene synthase

$NADPH + H^{\oplus}$

PP_i

$NADP^{\oplus}$

Squalene
(C_{30})

B. Stage 2: Isopentenyl Pyrophosphate to Squalene

An isomerase catalyzes the conversion of isopentenyl pyrophosphate to its isomer, dimethylallyl pyrophosphate, which then condenses in head-to-tail fashion with isopentenyl pyrophosphate to form the C_{10} molecule geranyl pyrophosphate (Figure 20·40). The condensation of the two C_5 molecules proceeds by formation of an intermediate carbonium ion of dimethylallyl pyrophosphate that is attacked by isopentenyl pyrophosphate. This basic reaction is repeated: the C_{10} molecule geranyl pyrophosphate condenses in head-to-tail fashion with isopentenyl pyrophosphate to form the C_{15} molecule farnesyl pyrophosphate. Two molecules of farnesyl pyrophosphate then condense in head-to-head fashion to form the C_{30} molecule squalene.

C. Stage 3: Squalene to Cholesterol

The steps between squalene and cholesterol are numerous and complex. One intermediate, lanosterol, accumulates in appreciable quantities in cells actively synthesizing cholesterol (Figure 20·41). The steps between squalene and lanosterol involve addition of an oxygen atom followed by cyclization of the chain to form the four-ring steroid nucleus. The cyclization occurs in a single step of concerted rearrangements of electrons from neighboring double bonds. The conversion of lanosterol to cholesterol occurs via a multistep pathway involving methyl-group shifts, oxidations, and decarboxylations. Many different enzyme activities and several possible routes have been implicated in the path between lanosterol and cholesterol. The exact pathway is at present unknown.

Figure 20·41
Final stage of cholesterol synthesis: squalene to cholesterol. The intermediates of this process are numerous, and the exact sequence of steps between lanosterol and cholesterol remains uncertain.

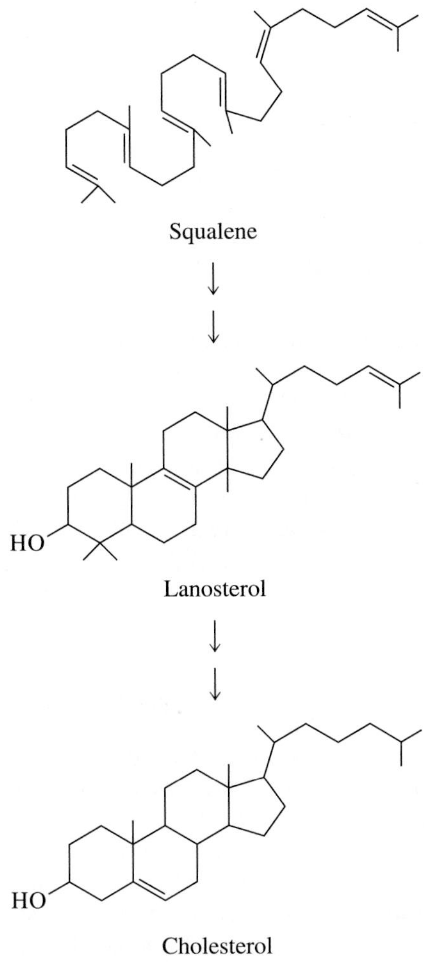

Squalene

Lanosterol

Cholesterol

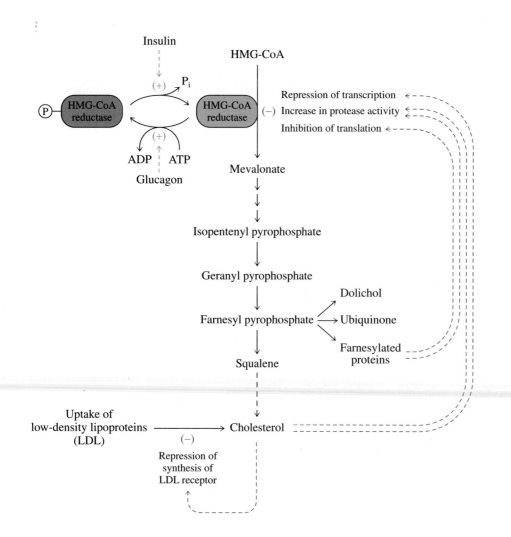

Figure 20·42
Regulation of cholesterol metabolism.

20·19 The Regulation of Cholesterol Biosynthesis Is Remarkably Complex

There are several mechanisms for regulation of the synthesis or uptake of cholesterol in liver cells. Cholesterol can mediate repression of the transcription of genes that encode HMG-CoA reductase, and high levels of cholesterol also inhibit the uptake of cholesterol-rich low-density lipoproteins (LDL) at the plasma membrane by repressing transcription of the gene for the LDL receptor (Figure 20·42). There is further evidence that elevated cholesterol levels in liver cells lead to an increase in the activity of a protease that attacks HMG-CoA reductase.

The hormones insulin and glucagon can control the activity of HMG-CoA reductase by mediating reversible phosphorylation of the enzyme. Binding of glucagon to the cell activates a kinase that catalyzes phosphorylation and inactivation of the reductase. The activity of the phosphorylated reductase can be restored by a phosphatase that is activated when insulin binds at the cell surface.

Yet another regulatory mechanism exists. When mevalonate is introduced into cells, it appears to reduce the activity of the reductase more effectively than sterols. Once it was shown that mevalonate itself was not the effector, a search began for nonsterol metabolites produced from mevalonate that could control the reductase. The path from acetyl CoA to cholesterol has branch points at farnesyl pyrophosphate that lead to a number of nonsterols, including dolichol and ubiquinone. Farnesyl pyrophosphate is also used in the synthesis of farnesylated proteins in which the polyprenyl group provides a membrane anchor for proteins (Section 12·3).

These farnesylated proteins appear to suppress the translation of mRNA species encoding the HMG-CoA reductase, though the regulatory pathway is not known, and they appear to stimulate the degradation of the reductase itself.

Uncontrolled production or ingestion of cholesterol can contribute to cardiovascular disease. Commonly associated with this disease is a buildup of cholesterol in atherosclerotic plaques in the walls of blood vessels. These plaques can block blood flow to the heart, causing heart attack, or to the brain, causing stroke.

20·20 Cholesterol Metabolism Is the Source of a Large Number of Other Cell Constituents

The pathway leading to cholesterol and pathways emanating from cholesterol contribute to a plethora of cellular constituents. Isopentenyl pyrophosphate, the C_5 precursor of squalene, and ultimately of cholesterol, is the precursor for a large number of other products, such as the lipid vitamins A, E, and K, ubiquinone in animal cells, and terpenes, plastoquinone, and the phytol side chain of chlorophyll in plants (Figure 20·43).

Cholesterol is a precursor of the bile salts, which facilitate intestinal absorption of fat; vitamin D, which stimulates Ca^{2+} uptake from the intestines; steroid hormones such as testosterone and β-estradiol that control sex characteristics; and steroids that control salt balance. Finally, it is important to keep in mind that the principal product of sterol synthesis is cholesterol itself, which modulates membrane fluidity and is an essential component of the plasma membrane of animal cells.

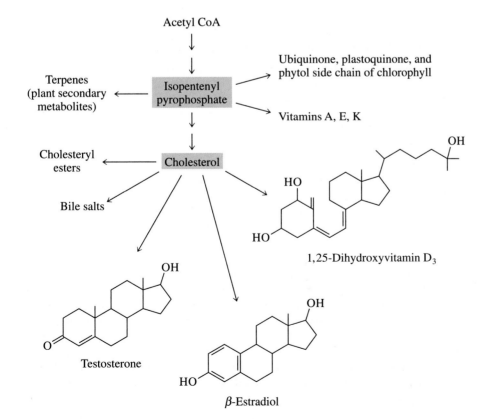

Figure 20·43
Diversity of products related to cholesterol and the pathway of cholesterol synthesis.

20·21 Lipids Are Made at a Variety of Sites Within the Cell

Eukaryotic cells are highly compartmentalized, with membranes serving as the boundaries for these compartments. The compartments can have quite different functions and quite distinct phospholipid and fatty acyl constituents in their surrounding membranes. Most lipid biosynthesis in eukaryotic cells occurs in the endoplasmic reticulum. Here, for example, phosphatidylcholine, phosphatidylethanolamine, phosphatidylinositol, and phosphatidylserine are synthesized using glycerol-3-phosphate, fatty acyl CoA thioesters, CDP-choline, CDP-ethanolamine, CTP, serine, and inositol. The enzymes of lipid synthesis in the endoplasmic reticulum are membrane bound with their active sites oriented toward the cytosol so that they have access to the water-soluble cytosolic compounds. For example, the phosphocholine transferase that catalyzes the formation of phosphatidylcholine is oriented toward the cytosol and can utilize cytosolic CDP-choline. CTP:phosphocholine cytidylyltransferase, which produces CDP-choline for use in the synthesis of phosphatidylcholine, is a cytosolic enzyme that is activated following translocation to the cytosolic face of the endoplasmic reticulum. The major phospholipids are transported to other membranes in the cell in vesicles that travel between the endoplasmic reticulum and Golgi apparatus and between the Golgi and various membrane target sites. There are also soluble transport proteins that carry phospholipids and cholesterol to other membranes.

Although the endoplasmic reticulum is the principal site of lipid metabolism in the cell, there are also lipid-metabolizing enzymes at other locations. For instance, membrane lipids can be tailored to give the lipid profile characteristic of individual cellular organelles. In the plasma membrane, acyltransferase activities catalyze the acylation of lysophospholipids. Mitochondria have the enzyme phosphatidylserine decarboxylase, which catalyzes the conversion of phosphatidylserine to phosphatidylethanolamine (Figure 20·27). Mitochondria also contain the enzymes responsible for the synthesis of diphosphatidylglycerol (cardiolipin), a molecule considered unique to the mitochondrion. Lysosomes possess various acid hydrolases, including phospholipase and glycosidase activities, that degrade phospholipids and sphingolipids. Peroxisomes possess enzymes involved in the early stages of ether lipid synthesis. In Zellweger's disease, a defect exists in the pathway for assembly of peroxisomes. The precise deficiency is not known. Defects in peroxisomal formation can lead to poor plasmalogen synthesis, with potentially fatal consequences. Especially prone to damage are the tissues of the central nervous system, where plasmalogens constitute a substantial portion of the lipids of the myelin sheath.

Summary

Fatty acids are released from adipocytes by epinephrine-stimulated activation of hormone-sensitive lipase, which catalyzes the conversion of triacylglycerols stored in fat cells to free fatty acids and monoacylglycerols. Another more specific and active monoacylglycerol lipase catalyzes the hydrolysis of monoacylglycerol to free fatty acids and glycerol. The free fatty acids and glycerol then enter the bloodstream. Fatty acids bind to albumin and are delivered to tissues where they are oxidized for energy.

Fatty acids are degraded to acetyl CoA by the sequential removal of two-carbon fragments, a process called β oxidation. Before oxidation can take place, fatty acids are activated by esterification to coenzyme A and then transported into mitochondria. In eukaryotic cells, β oxidation occurs in the mitochondrial matrix. Oxidation of fatty acids produces large amounts of ATP.

The β-oxidation pathway for saturated fatty acids consists of four enzyme-catalyzed steps: oxidation, hydration, further oxidation, and thiolysis. Oxidation of unsaturated fatty acids follows the same pathway until a double bond is reached. Then additional enzymes are required, an isomerase and a reductase.

β oxidation of odd-chain fatty acids produces propionyl CoA rather than acetyl CoA in the final reaction of the degradative pathway. Three enzyme-catalyzed reactions convert the propionyl CoA to succinyl CoA, an intermediate of the citric acid cycle.

The ketone bodies β-hydroxybutyrate and acetoacetate are fuel molecules produced in the liver by the condensation of acetyl CoA molecules. They are water soluble and serve as a readily available fuel. A third ketone body, acetone, is produced in trace amounts by nonenzymatic decarboxylation of acetoacetate.

Fatty acid synthesis, which in animal cells takes place in the cytosol, occurs by a different pathway than fatty acid oxidation. Acetyl CoA needed for fatty acid synthesis is shuttled from the mitochondrion, where it is produced, to the cytosol via the citrate transport system. In addition, the citrate transport system, along with the pentose phosphate pathway, provides NADPH needed for the reactions of fatty acid biosynthesis.

The formation of long-chain fatty acids from malonyl CoA and acetyl CoA occurs in five stages: loading, condensation, reduction, dehydration, and further reduction. This sequence is repeated until a long-chain fatty acid is released. In *E. coli,* the reactions of fatty acid synthesis are carried out by separate enzymes; in mammals, they are carried out by a multienzyme complex. The most common product of fatty acid synthesis is palmitate. Longer-chain and unsaturated fatty acids are produced by additional reactions.

Fatty acid oxidation is regulated at the release of fatty acids from adipocytes and the transport of fatty acids into mitochondria. Epinephrine stimulates fatty acid release from adipocytes. The cAMP-linked hormones glucagon and epinephrine stimulate transport of fatty acids into mitochondria for oxidation. The conversion of acetyl CoA to malonyl CoA is the first committed step of fatty acid synthesis and the key regulatory step. The enzyme that catalyzes this reaction, acetyl-CoA carboxylase, is controlled by reversible phosphorylation, responding both to hormone signals and the presence of fatty acyl CoA.

Eicosanoids are metabolic regulators derived from arachidonate by two branched pathways. The cyclooxygenase pathway leads to prostacyclin, prostaglandins, and thromboxane. Among the products of the lipoxygenase pathway are the leukotrienes.

Triacylglycerols and neutral phospholipids are synthesized by a common pathway, both being derived from 1,2-diacylglycerol. Phosphatidate is the precursor of the acidic phospholipids phosphatidylserine and phosphatidylinositol. Members of a separate class of phospholipids called plasmalogens contain an ether bond at C-1 of their glycerol backbone. Sphingolipids have a backbone derived from palmitoyl CoA and serine. Acylation of sphingosine produces a ceramide, which can be modified by addition of phosphatidylcholine to form a sphingomyelin or by addition of a sugar moiety to form a cerebroside. More complex sugar-lipid conjugates constitute the gangliosides.

Cholesterol is an essential component of animal cell membranes and a precursor of a large number of cell constituents. All the carbon atoms of cholesterol arise from acetyl CoA. The major regulatory step in cholesterol biosynthesis is the conversion of 3-hydroxy-3-methylglutaryl CoA to mevalonate.

Selected Readings

Frerman, F. E. (1988). Acyl-CoA dehydrogenases, electron flavoprotein and electron transfer flavoprotein dehydrogenase. *Biochem. Soc. Trans.* 16:416–418.

Hardie, D. G. (1989). Regulation of fatty acid synthesis via phosphorylation of acetyl-CoA carboxylase. *Prog. Lipid Res.* 28:117–146.

Harwood, J. L. (1988). Fatty acid metabolism. *Annu. Rev. Plant Physiol. Plant Mol. Biol.* 39:101–138. Focuses on fatty acid metabolism in plants, with attention given to comparisons with bacteria and animals.

Marinetti, G. V. (1990). *Disorders of Lipid Metabolism* (New York: Plenum Press).

Mead, J. F., Alfin-Slater, R. B., Howton, D. R., and Popják, G. (1986). *Lipids: Chemistry, Biochemistry, and Nutrition* (New York: Plenum Publishing Corporation). Authoritative treatment of the structure and metabolism of lipids.

McGarry, J. D., Brown, N. F., Inthanousay, P. P., Park, D. I., Cook, B. A., and Foster, D. W. (1992). Insights into the topography of mitochondrial carnitine palmitoyltransferase gained from the use of proteases. In *New Developments in Fatty Acid Oxidation,* P. M. Coates and K. Tanaka, eds. (New York: Wiley-Liss, Inc.), pp. 47–61.

Stralfors, P., and Belfrage, P. (1984). Reversible phosphorylation of hormone sensitive lipase/cholesterol ester hydrolyase in the hormone control of adipose tissue lipolysis and of adrenal steroidogenesis. In *Enzyme Regulation by Reversible Phosphorylation: Further Advances,* P. Cohen, ed. (Amsterdam: Elsevier Science Publishing Company).

Vance, D. E., and Vance, J. E., eds. (1991). *Biochemistry of Lipids, Lipoproteins, and Membranes* (Amsterdam: Elsevier Science Publishing Company). Well-organized and highly readable.

Voelker, D. R., and Kennedy, E. P. (1982). Cellular and enzymatic synthesis of sphingomyelin. *Biochemistry* 21:2753–2759. E. P. Kennedy provides supporting evidence for phosphatidylcholine as the immediate donor of the phosphocholine moiety of sphingomyelin, recanting his earlier position that CDP-choline is the donor.

Wakil, S. J. (1989). Fatty acid synthase, a proficient multifunctional enzyme. *Biochemistry* 28:4523–4530.

21

Amino Acid Metabolism

The metabolism of amino acids involves a large number of enzyme-catalyzed interconversions of small molecules that contain the element nitrogen. The relatively inert gas N_2 is the ultimate source of biological nitrogen. Ammonia (NH_3) derived from N_2 is incorporated into metabolites, including the amino acids, via glutamate and glutamine.

The metabolism of amino acids is best considered from two points of view: the origins and fates of the amino acid nitrogen atoms and the origins and fates of their carbon skeletons. Nitrogen metabolism includes the major routes for the enzyme-catalyzed incorporation of nitrogen, beginning with N_2. This compound can only be metabolized, or fixed, by a few species. The product of nitrogen metabolism is ammonia, which can be metabolized by all organisms. We will see how the metabolism of the precursors and degradation products of amino acids converges with many of the major pathways of metabolism discussed in previous chapters.

A few organisms can assimilate N_2 and simple carbon compounds into amino acids. Others can synthesize the carbon chains of the amino acids but require nitrogen in the form of ammonia. Mammals have even less capacity for amino acid biosynthesis. They can synthesize only about half the amino acids they require; the rest—called **essential amino acids**—must be provided by dietary sources. The **nonessential amino acids** are those that mammals can produce in sufficient quantity for growth. Whether amino acids are essential or nonessential is ascertained experimentally for individual organisms and species. The patterns of amino acid metabolism can be somewhat different across species, with variations in the pathways for synthesis of the same molecule. There are also a variety of routes for the degradation and disposal of the nitrogen-containing waste products of amino acid breakdown.

The role of amino acids in metabolism is distinct from that of the other major dietary components (fats and carbohydrates) because amino acids are not primarily stored as a supply of cellular energy. Rather, amino acids are required principally to build proteins, which undergo constant turnover in cells. Each type of protein in the cell is degraded to its constituent amino acids at a different rate. In mammals, there are situations, such as fasting or a dietary excess of protein, during which metabolites from the catabolism of amino acids enter the pathway of gluconeogenesis.

Considering that there are 20 common amino acids, many of which are converted to other metabolites, the number of enzymatic reactions in this chapter may seem overwhelming. However, the pathways of amino acid metabolism possess central features and common themes that reflect an underlying pattern of organization. We begin our examination of amino acid metabolism by examining the incorporation of nitrogen into cellular metabolites.

21·1 Nitrogen Is Cycled Through the Biosphere

Nitrogen in biological systems originates from gaseous N_2, which constitutes about 80% of the atmosphere. The two atoms of molecular nitrogen are held together by an extremely strong triple bond (bond energy = 940 kJ mol^{-1}). This feature makes nitrogen gas chemically unreactive, so it is not surprising that very specific and sophisticated enzyme systems are required to assimilate nitrogen into amino acids and their metabolic derivatives.

A. A Few Organisms Carry Out Nitrogen Fixation

The reduction of N_2 to ammonia is called **nitrogen fixation.** In the chemical industry, nitrogen is fixed for use in plant fertilizer by an energetically expensive process involving the use of high temperature and pressure in concert with special catalysts to drive the reduction of N_2 by H_2 to form ammonia. The availability of biologically useful nitrogen is often a limiting factor for plant growth, and the application of nitrogenous fertilizers is important for obtaining higher crop yields.

Most nitrogen fixation in the biosphere is carried out by a few species of bacteria or algae that have the ability to synthesize the complicated enzyme nitrogenase. This multisubunit protein catalyzes the conversion of N_2 to two molecules of NH_3. Nitrogenase is present in members of the genus *Rhizobium* that live symbiotically in root nodules of many plants, including beans, peas, alfalfa, and clover, all of which are legumes (Figure 21·1). N_2 is also fixed by free-living soil bacteria such as *Azotobacter, Klebsiella,* and *Clostridium* and by cyanobacteria found in aqueous environments. Most plants require a supply of fixed nitrogen from the environment. Sources of this nutrient include excess fixed and oxidized nitrogen excreted by microorganisms, decayed animal and plant tissue, and fertilizer. Vertebrates obtain fixed nitrogen exclusively through ingestion of plant and animal tissue.

The enzyme nitrogenase usually consists of two protein components, one containing iron and the other, iron and molybdenum. Because both metalloproteins are highly susceptible to inactivation by O_2, nitrogenases are protected from oxygen within nitrogen-fixing organisms. For example, strict anaerobes carry out nitrogen fixation only in the absence of O_2. Within the root nodules of leguminous plants, the protein leghemoglobin (Figure 5·55) binds O_2 and thereby keeps its concentration sufficiently low in the immediate environment of the nitrogen-fixing enzymes of rhizobia.

Figure 21·1
Root nodules of a clover plant. Symbiotic bacteria of the genus *Rhizobium* reside in these nodules, which provide a suitable environment for the reduction of atmospheric nitrogen to ammonia.

A strong reducing agent—either reduced ferredoxin or reduced flavodoxin (a flavoprotein electron carrier from microorganisms)—is required for the enzymatic reduction of N_2 to NH_3. An obligatory reduction of $2\,H^{\oplus}$ to H_2 accompanies the reduction of N_2.

$$N_2 + 8\,H^{\oplus} + 8\,e^{\ominus} + 16\,ATP \longrightarrow 2\,NH_3 + H_2 + 16\,ADP + 16\,P_i \quad \text{(21·1)}$$

For each electron transferred in vitro by nitrogenase, two ATP molecules must be converted to ADP and P_i, so that the six-electron reduction of a single molecule of N_2 (plus the two-electron reduction of $2\,H^{\oplus}$) consumes 16 ATP. In order to obtain the reducing power and ATP for this process, symbiotic nitrogen-fixing microorganisms rely on photosynthesis carried out by the plants with which they are associated.

The reduction of N_2 catalyzed by nitrogenase is thought to proceed in three discrete steps, with two electrons transferred at each step. It has been postulated that diimine and hydrazine are reaction intermediates that remain bound to the enzyme.

$$N\equiv N \xrightarrow{2\,e^{\ominus},\,2\,H^{\oplus}} \left[HN = NH \xrightarrow{2\,e^{\ominus},\,2\,H^{\oplus}} H_2N - NH_2 \right] \xrightarrow{2\,e^{\ominus},\,2\,H^{\oplus}} 2\,NH_3$$

Dinitrogen   Diimine   Hydrazine   Ammonia

$$\text{(21·2)}$$

Two other types of nitrogenases have been found. Rather than containing both molybdenum and iron, these enzymes contain either vanadium or iron. They generally have lower catalytic activity than the molybdenum-iron nitrogenases.

B. Plants and Microorganisms Can Convert Nitrate and Nitrite to Ammonia

During lightning storms, high-voltage discharges catalyze the oxidation of N_2. When N_2 reacts with O_2 in the atmosphere, biologically useful nitrate ($NO_3^{\ominus}$) and nitrite ($NO_2^{\ominus}$) are formed and washed into the soil. Other sources of nitrate and nitrite are the oxidation of NH_3 by microorganisms such as *Nitrosomonas* and *Nitrobacter*.

Most plants and microorganisms contain nitrate reductase and nitrite reductase, enzymes that together catalyze the reduction of the nitrogen oxides to ammonia. In higher plants, reduced ferredoxin, formed in the light reactions of photosynthesis, passes its reducing power to either $NAD^{\oplus}$ or $NADP^{\oplus}$. The reduced pyridine nucleotide is a cosubstrate in the reaction, catalyzed by nitrate reductase, that converts nitrate to nitrite.

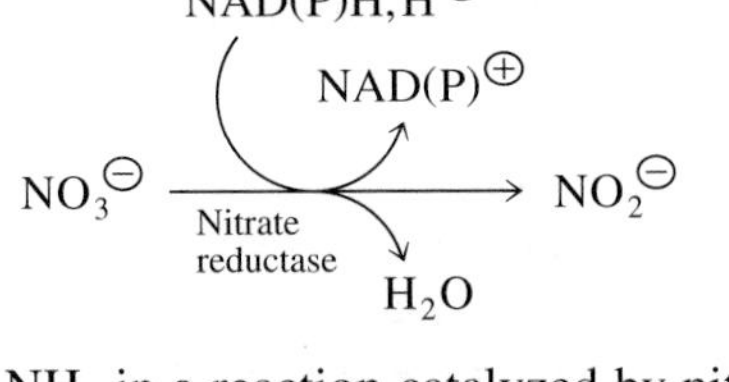

$$NO_3^{\ominus} \xrightarrow[\substack{\text{Nitrate} \\ \text{reductase}}]{} NO_2^{\ominus} \qquad \text{(21·3)}$$

with $NAD(P)H, H^{\oplus} \to NAD(P)^{\oplus}$ and $\to H_2O$

Nitrite is reduced to NH_3 in a reaction catalyzed by nitrite reductase. This reduction uses reduced ferredoxin as a source of electrons. The intermediates of the reaction have not been isolated, but it is presumed that the sequence of steps is

$$NO_2^{\ominus} \longrightarrow \left[NO^{\ominus} \longrightarrow NH_2OH \right] \longrightarrow NH_3 \qquad \text{(21·4)}$$

Figure 21·2
Nitrogen cycle. A few free-living or symbiotic microorganisms can convert N₂ to ammonia. Ammonia is incorporated into biomolecules such as amino acids and proteins, which later are degraded, re-forming ammonia. Many soil bacteria and plants can carry out the reduction of nitrate to ammonia via nitrite. Several genera of bacteria can convert ammonia to nitrite. Others can oxidize nitrite to nitrate. Some other bacterial genera can reduce nitrate to N₂.

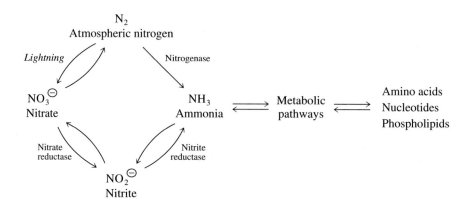

Some bacteria can convert ammonia to nitrite, whereas others can convert nitrite to nitrate. This formation of nitrate is called nitrification. Still other bacteria can reduce nitrate or nitrite to N_2 (denitrification). The overall scheme for interconversion of the major nitrogen-containing chemical species of the biosphere is depicted in Figure 21·2. The biological cycling of N_2 between nitrogen oxides, ammonia, nitrogenous biomolecules, and back to N_2 is termed the nitrogen cycle. By means of this cycle, nitrogen gas is converted to usable forms and assimilated into simple organic compounds, which then transfer nitrogen to a variety of low-molecular-weight metabolites and eventually to macromolecules. The cycle is completed when nitrogenous products arising from excretion or from the death of organisms are broken down to nitrogen oxides and N_2.

21·2 Glutamate Dehydrogenase Catalyzes the Incorporation of Ammonia into Glutamate

Ammonia generated biosynthetically from N_2 or nitrogen oxides is assimilated into a large number of low-molecular-weight metabolites. Ammonia has a pK_a of 9.2 and therefore exists in neutral aqueous solution mainly as the ammonium ion, $NH_4^{\oplus}$. In the catalytic centers of enzymes, however, the unprotonated nucleophile NH_3 is the reactive species. One highly efficient route for the incorporation of ammonia into the central pathways of amino acid metabolism is the reductive amination of α-ketoglutarate to glutamate, catalyzed in both plants and animals by glutamate dehydrogenase.

$$
\begin{array}{c}
COO^{\ominus} \\
| \\
{}_{\alpha}C{=}O \\
| \\
CH_2 \\
| \\
CH_2 \\
| \\
COO^{\ominus}
\end{array}
+ NH_4^{\oplus}
\xrightleftharpoons[\text{Glutamate dehydrogenase}]{NAD(P)H,H^{\oplus} \quad NAD(P)^{\oplus}}
\begin{array}{c}
COO^{\ominus} \\
| \\
H_3N^{\oplus}{-}{}_{\alpha}C{-}H \\
| \\
CH_2 \\
| \\
CH_2 \\
| \\
COO^{\ominus}
\end{array}
+ H_2O
$$

α-Ketoglutarate   L-Glutamate

(21·5)

The glutamate dehydrogenases of some species or tissues are specific for NADH, whereas others are specific for NADPH. Still others can use either cofactor. The reaction involves the condensation of ammonia with the carbonyl group of α-ketoglutarate, forming an enzyme-bound α-iminiumglutarate intermediate. This intermediate is reduced by the transfer of a hydride ion from NADH or NADPH (Figure 21·3).

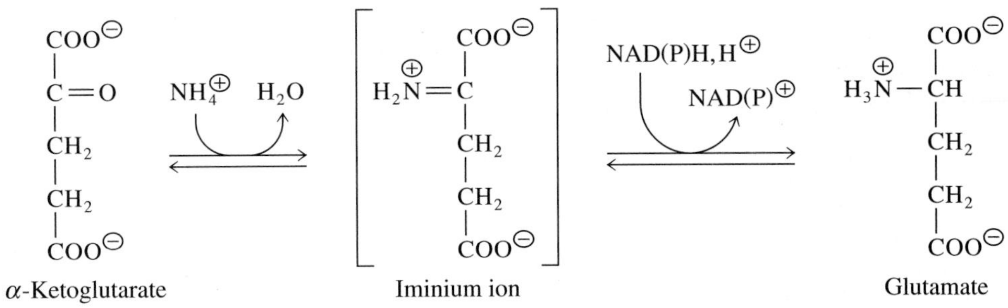

Figure 21·3
Glutamate dehydrogenase reaction.

The glutamate dehydrogenase reaction can have different roles in different organisms. In *Escherichia coli,* its major role is glutamate formation when excess $NH_4^{\oplus}$ is present. In the mold *Neurospora crassa,* separate enzymes with distinct pyridine nucleotide specificities exist; an NADPH-dependent enzyme catalyzes primarily the reductive amination of α-ketoglutarate to glutamate, and an $NAD^{\oplus}$-dependent enzyme functions in oxidative catabolism of glutamate to α-ketoglutarate. In mammals, glutamate dehydrogenase catalyzes a reaction close to equilibrium with net flux usually toward glutamate catabolism. In plants, the main direction of flow in this reaction seems to be toward glutamate production.

Although a number of compounds are allosteric effectors of liver glutamate dehydrogenase in vitro, they probably have little physiological significance because, as we have seen, near-equilibrium reactions are not subject to regulation. Flux through these reactions is controlled simply by changes in the concentrations of substrates and products.

21·3 Glutamine Is an Important Carrier of Ammonia-Derived Nitrogen

Another reaction critical to the assimilation of ammonia in many organisms is the formation of glutamine from glutamate and ammonia, catalyzed by glutamine synthetase.

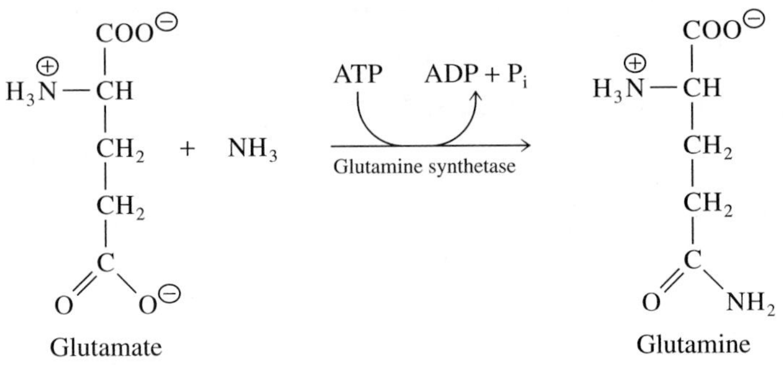

(21·6)

Glutamine is a nitrogen donor in many biosynthetic reactions. For example, the amide nitrogen of glutamine is the direct precursor of several of the nitrogen atoms of the purine and pyrimidine ring systems of nucleotides (Chapter 22).

Glutamine has a special role in mammals, where it functions to transport nitrogen and carbon between tissues and so decreases the circulating levels of toxic $NH_4^{\oplus}$. In mammals, glutamine is mainly synthesized in muscle and transported via the circulatory system to other tissues, such as liver and kidney.

Figure 21·4
Combined action of glutamine synthetase and glutamate synthase in a pathway leading to the transamination of α-keto acids to form amino acids.

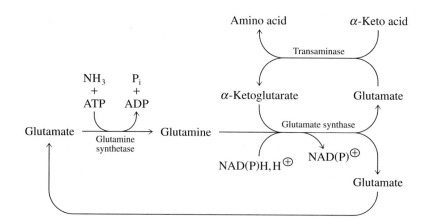

In prokaryotes, the amide nitrogen of glutamine can be transferred to α-ketoglutarate in a reductive amination reaction catalyzed by glutamate synthase. This enzyme requires a reduced pyridine nucleotide as a reducing agent.

$$\text{Glutamine} + \alpha\text{-Ketoglutarate} \xrightarrow[\text{Glutamate synthase}]{\text{NAD(P)H, H}^\oplus \quad \text{NAD(P)}^\oplus} 2\text{ Glutamate} \qquad (21\cdot7)$$

The glutamate synthase reaction has the same outcome as the glutamate dehydrogenase reaction—the assimilation of ammonia into glutamate. Coupled reactions catalyzed by glutamine synthetase and glutamate synthase are an important alternative to the glutamate dehydrogenase reaction in prokaryotes, especially when the concentration of available ammonia is low. Figure 21·4 shows how the combined actions of glutamine synthetase and glutamate synthase lead to the incorporation of ammonia into a variety of amino acids by formation of glutamate, followed by transamination with α-keto acids to form a number of amino acids. (Transamination reactions are examined in detail in Section 21·5.) Note that the glutamate dehydrogenase reaction uses two molecules of reduced pyridine nucleotide per pair of glutamate molecules formed, whereas the coupled reactions of glutamine synthetase and glutamate synthase consume one molecule of reduced pyridine nucleotide and one molecule of ATP.

21·4 Glutamine Synthetase in *E. coli* Is Regulated by a Sophisticated Mechanism

Glutamine synthetase plays a major role in nitrogen metabolism since glutamine is a precursor of many other metabolites. The regulatory properties of *E. coli* glutamine synthetase have been intensively investigated. This enzyme is regulated at three levels: by feedback inhibition (there are at least nine allosteric inhibitors), by covalent modification, and by regulation at the level of protein synthesis.

E. coli glutamine synthetase consists of 12 identical subunits (subunit M_r 51 600). Each subunit has not only a catalytic site but also binding sites for the allosteric inhibitors. The nine recognized inhibitors are AMP, carbamoyl phosphate, CTP, histidine, tryptophan, glucosamine 6-phosphate, alanine, serine, and glycine (Figure 21·5). Six of the inhibitors contain a nitrogen atom obtained directly from the amide nitrogen of glutamine. The nitrogen atoms of alanine, serine, and glycine can be derived from glutamine indirectly. No single inhibitor completely blocks the catalytic activity of glutamine synthetase. Instead, the degree of inhibition increases as more of the inhibitors bind, a process known as **cumulative feedback inhibition.** For

Figure 21·5
Allosteric feedback inhibition of glutamine synthetase in *E. coli*. Several compounds that contain a nitrogen atom ultimately derived from the amide group of glutamine are inhibitors of glutamine synthetase. The presence of a single inhibitor causes only partial inhibition. When all the inhibitors are present, inhibition is nearly complete.

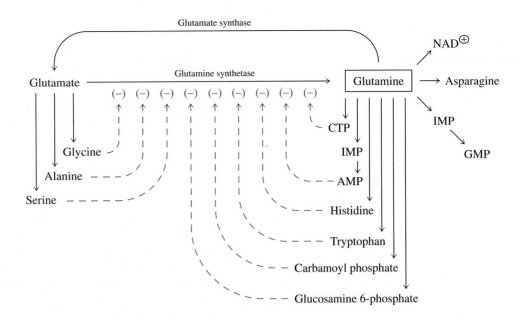

example, when glutamine synthetase was tested with saturating concentrations of tryptophan, CTP, carbamoyl phosphate, or AMP, the residual activities were 84%, 86%, 87%, and 59%, respectively. When all four of these inhibitors were present simultaneously, the residual activity was 37%. The presence of all nine inhibitors blocks essentially all the activity of the enzyme.

Covalent modification of glutamine synthetase in *E. coli* occurs as part of a regulatory cascade that includes the sequential addition of AMP moieties (adenylylation) to specific tyrosine residues, one present in each of the 12 subunits of this enzyme (Figure 21·6). Adenylylation, catalyzed by glutamine synthetase adenylyltransferase, decreases the activity of glutamine synthetase. The removal of the

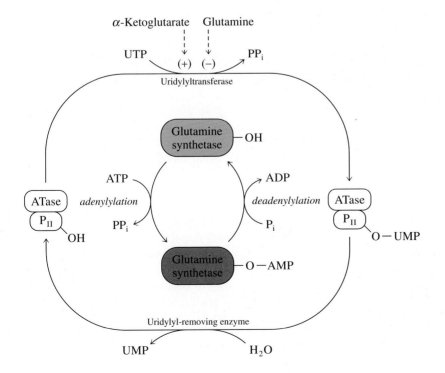

Figure 21·6
Regulation of glutamine synthetase by covalent modification. Glutamine synthetase is inactivated by adenylylation and activated by deadenylylation, both of which are catalyzed by the bifunctional enzyme glutamine synthetase adenylyltransferase (ATase). ATase is complexed to a regulatory protein called P_{II}. The reversible uridylylation of P_{II} controls ATase; UMP incorporation is controlled in turn by the allosteric effectors α-ketoglutarate and glutamine.

AMP group from glutamine synthetase, with concomitant reactivation of the enzyme, is also catalyzed by the adenylyltransferase protein. Which activity predominates is determined by another protein, P_{II}, that can form a complex with the adenylyltransferase. P_{II} itself is subject to covalent modification by the reversible transfer of a UMP group to the phenolic oxygen of a specific tyrosine residue of the protein, a reaction catalyzed by uridylyltransferase. The removal of UMP is catalyzed by uridylyl-removing enzyme. The uridylylation of P_{II} activates glutamine synthetase by stimulating its deadenylylation. Uridylylation is stimulated by α-ketoglutarate and inhibited by glutamine. α-Ketoglutarate appears in the ammonia-assimilation pathway before the glutamine synthetase reaction and thus is a feed-forward activator. Glutamine, the product of the reaction, is a feedback inhibitor.

Allosteric regulation of the activity of an enzyme is extremely rapid, with the conformational change occurring in roughly a millisecond. Changes in enzyme activity through covalent modification are significantly slower, taking closer to a second. Still longer-term adaptation to environmental changes is afforded by regulation at the level of gene expression. When ample quantities of usable nitrogen are available to *E. coli* from its environment, relatively low levels of glutamine synthetase are found within cells. Under these conditions, assimilation of ammonia can occur via the glutamate dehydrogenase reaction. In contrast, when *E. coli* grows in nitrogen-limited media, the alternate pathway of ammonia assimilation using glutamate synthase is followed, and higher levels of glutamine synthetase are produced through increased transcription and translation of the gene encoding it. The levels of glutamine synthetase in *E. coli* can vary over 100-fold.

Mammalian glutamine synthetases differ markedly from the bacterial enzyme in both their physical properties and regulatory behavior. The enzymes of liver and brain have been most thoroughly characterized. These proteins contain eight identical subunits, but they can also exist as tetramers. The mammalian glutamine synthetases are not regulated by covalent modification. Glycine, serine, alanine, and carbamoyl phosphate are inhibitors of mammalian liver glutamine synthetase, and α-ketoglutarate is an activator. However, the range of modulation of the mammalian synthetase activity is not as extensive as that of the *E. coli* enzyme.

21·5 Transaminases Catalyze the Reversible Interconversion of α-Amino Acids and α-Keto Acids

Glutamate is a key intermediate in a large number of the reactions of amino acid metabolism, both synthetic and degradative. The amino group of glutamate can be transferred to many α-keto acids in reactions catalyzed by enzymes known as transaminases, or aminotransferases. The basic reaction scheme for transaminases is shown in Figure 21·7.

In amino acid biosynthesis, the amino group of glutamate is transferred to various α-keto acids, generating the corresponding α-amino acids. Most of the common amino acids can be formed by transamination. However, the α-keto acid corresponding to lysine is unstable, so lysine is not formed by transamination. In addition, because the α-keto acid of threonine is a poor substrate for transamination, little threonine is formed by this type of reaction. In catabolism, one or more transamination reactions generate glutamate or aspartate. The amino groups of glutamate and aspartate then enter pathways leading to nitrogen disposal, such as the urea cycle (Section 21·10) or the synthesis of uric acid (Section 22·12).

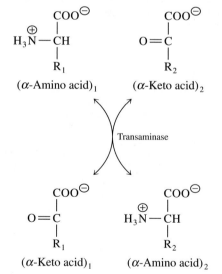

Figure 21·7
Transfer of an amino group from an α-amino acid to an α-keto acid, catalyzed by a transaminase. In biosynthetic reactions, (α-amino acid)$_1$ is often glutamate, with its carbon skeleton producing α-ketoglutarate, (α-keto acid)$_1$. (α-Keto acid)$_2$ represents the precursor of a variety of newly formed amino acids, (α-amino acid)$_2$.

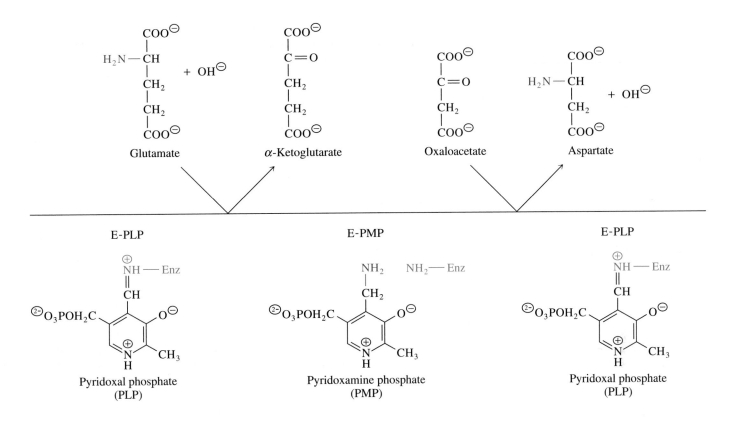

Figure 21·8
Ping-pong kinetic mechanism of aspartate transaminase. Part of the substrate is transferred to the enzyme, and a stable enzyme–pyridoxamine phosphate (E-PMP) intermediate is generated. The second half of the mechanism mirrors the first half.

All known transaminases require the coenzyme pyridoxal phosphate (Section 8·10), and all have ping-pong kinetic mechanisms. The most thoroughly studied is aspartate transaminase, also called aspartate aminotransferase.

$$\text{Glutamate} + \text{Oxaloacetate} \underset{\text{transaminase}}{\overset{\text{Aspartate}}{\rightleftharpoons}} \alpha\text{-Ketoglutarate} + \text{Aspartate}$$

(21·8)

Figure 21·8 shows the kinetic mechanism of this enzyme. The amino group of glutamate is transferred to enzyme-bound pyridoxal phosphate (E-PLP), forming a stable enzyme–pyridoxamine phosphate (E-PMP) intermediate, with release of α-ketoglutarate. At this point, a second substrate, oxaloacetate, binds to the enzyme and reacts with E-PMP, forming aspartate and regenerating the original form of the enzyme. Each step in this reaction sequence is reversible.

The mechanism of the initial half-reaction of transamination is shown in Figure 21·9. The E-PLP holoenzyme exists in the form of an internal aldimine, with PLP covalently linked as a Schiff base to the side chain of an active-site lysine residue of the apoenzyme. PLP is further anchored to the enzyme by many noncovalent bonds. In Step 1, the nucleophilic —NH$_2$ form of the donor amino acid displaces

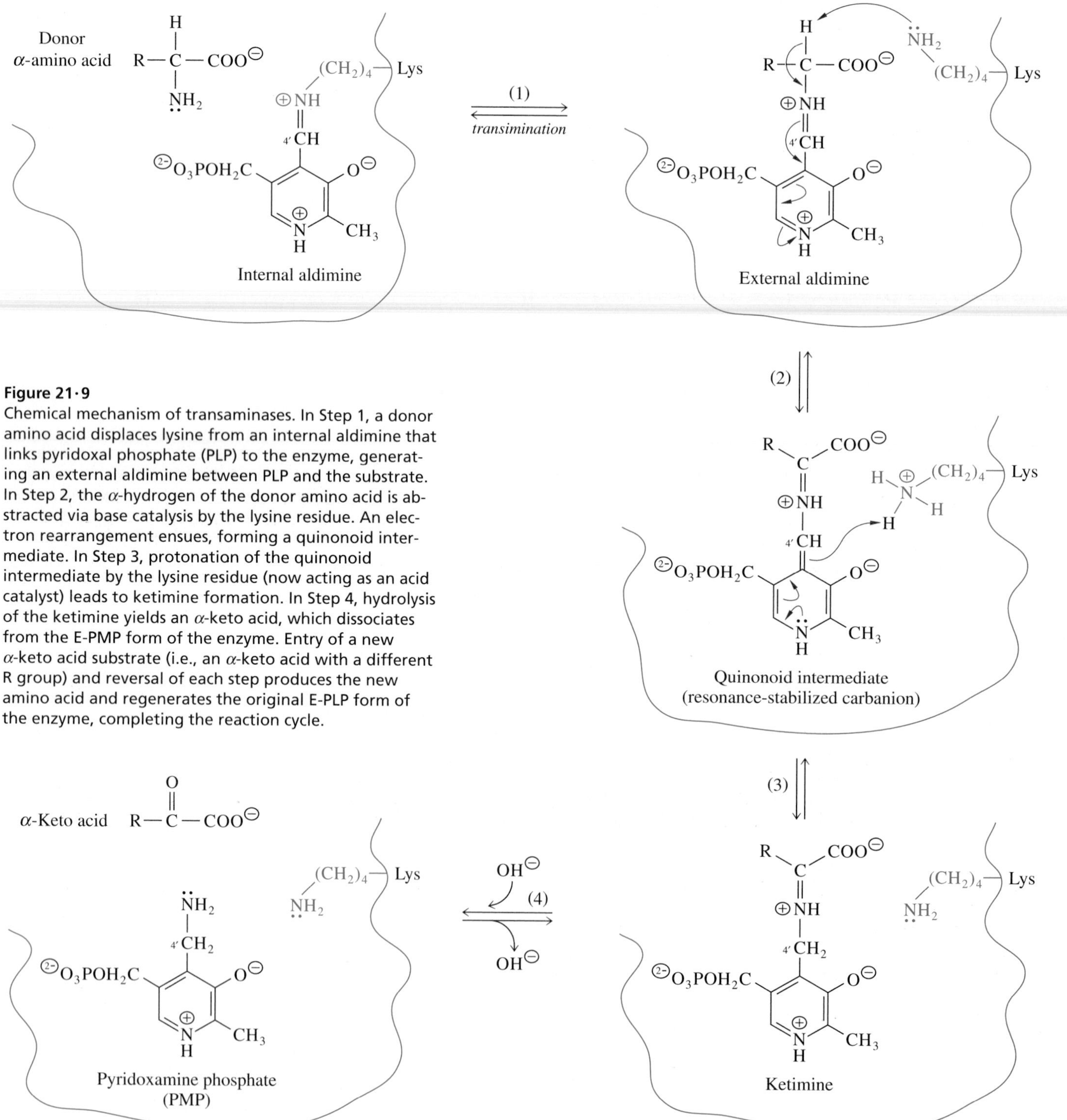

Figure 21·9
Chemical mechanism of transaminases. In Step 1, a donor amino acid displaces lysine from an internal aldimine that links pyridoxal phosphate (PLP) to the enzyme, generating an external aldimine between PLP and the substrate. In Step 2, the α-hydrogen of the donor amino acid is abstracted via base catalysis by the lysine residue. An electron rearrangement ensues, forming a quinonoid intermediate. In Step 3, protonation of the quinonoid intermediate by the lysine residue (now acting as an acid catalyst) leads to ketimine formation. In Step 4, hydrolysis of the ketimine yields an α-keto acid, which dissociates from the E-PMP form of the enzyme. Entry of a new α-keto acid substrate (i.e., an α-keto acid with a different R group) and reversal of each step produces the new amino acid and regenerates the original E-PLP form of the enzyme, completing the reaction cycle.

the lysine residue by transimination (see Figure 8·31). Electron rearrangement, assisted by base catalysis involving the lysine residue, leads in Step 2 to a quinonoid intermediate. Further rearrangement in Step 3 leads to a ketimine, which is hydrolyzed to yield a free α-keto acid and the E-PMP form of the enzyme. In the second phase of the reversible and symmetrical reaction, a different α-keto acid becomes the substrate for the reverse sequence, starting with Step 4. The reaction sequence is completed when Step 1 is reached and E-PLP is formed, with release of an amino acid containing the carbon skeleton of the second α-keto acid.

The transaminases catalyze near-equilibrium reactions, and the direction in which the reactions proceed in vivo depends on the supply of substrates and the removal of products. For example, in cells having an excess of α-amino nitrogen groups, the amino groups can be transferred via one or a series of transamination reactions to α-ketoglutarate to yield glutamate, which can undergo oxidative deamination catalyzed by glutamate dehydrogenase. When amino acids are being actively formed, transamination reactions in the opposite direction occur, with the amino groups donated by glutamate.

21·6 Many Nonessential Amino Acids Are Synthesized Directly from Key Intermediates of Central Metabolism

Having examined the incorporation of nitrogen into amino acids, we will now turn our attention to the origins of their carbon skeletons. Figure 21·10 depicts in schematic form the biosynthetic routes leading to the 20 common amino acids. Keep in mind that most bacteria and plants (but not mammals) synthesize all amino acids. In general, amino acids that are nonessential for mammals are derived by short pathways from intermediates of glycolysis, the pentose phosphate pathway, or the citric acid cycle. The classification of amino acids as either nonessential or essential for mammals roughly parallels the number of steps in their synthetic pathways and the energy required for their synthesis. Amino acids with the largest

Figure 21·10
Biosynthesis of amino acids. Shown are the connections to glycolysis, the pentose phosphate pathway, and the citric acid cycle. Amino acids essential for humans are shaded in red; nonessential amino acids are shaded in blue.

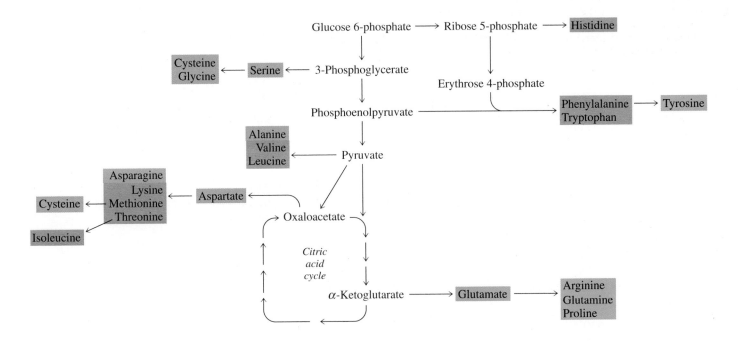

Table 21·1 Essential and nonessential amino acids for mammals, with energetic requirements for their biosynthesis

Amino acid	Moles of ATP required per mole of amino acid produced[a]	
	Nonessential	*Essential*
Glycine	12	
Serine	18	
Cysteine	19[b]	
Alanine	20	
Aspartate	21	
Asparagine	22	
Glutamate	30	
Glutamine	31	
Threonine		31
Proline	39	
Valine		39
Histidine		42
Arginine	44[c]	
Methionine		44
Leucine		47
Lysine		50 or 51
Isoleucine		55
Tyrosine	62[d]	
Phenylalanine		65
Tryptophan		78

[a] Moles of ATP required includes ATP used for synthesis of precursors and conversion of precursors to products.
[b] Formed from serine and homocysteine, a metabolite of the essential amino acid methionine.
[c] Essential for some mammals.
[d] Formed from the essential amino acid phenylalanine.
[Adapted from Atkinson, D. E. (1977). *Cellular Energy Metabolism and Its Regulation* (New York: Academic Press).]

energy requirements for synthesis are generally essential amino acids (Table 21·1). It has been difficult to establish by nutritional experiments whether histidine and arginine are essential for humans. However, since no pathway for de novo synthesis of histidine in humans is known, histidine is classified as an essential amino acid. It had been thought that adult humans could synthesize sufficient arginine but that infants could not. The current evidence suggests that arginine normally is not essential for humans of any age. Other mammals also synthesize arginine but not always in sufficient quantity.

Three amino acids are synthesized from common metabolites by one reaction, transamination. We have seen that glutamate can be formed by either the reductive amination or transamination of α-ketoglutarate. Other amino acids formed by simple transamination are alanine and aspartate. Pyruvate is the amino-group acceptor in the synthesis of alanine.

$$(21\cdot9)$$

Oxaloacetate is the amino-group acceptor in the synthesis of aspartate.

$$(21\cdot10)$$

The two amino acids with amide side chains are synthesized by one additional reaction, amidation. Glutamine, as we have seen, is formed from ammonia and glutamate in an ATP-dependent reaction catalyzed by glutamine synthetase. Asparagine, the other amide-containing amino acid, is synthesized in many mammals in a reaction involving the transfer of the amide nitrogen of glutamine to aspartate, catalyzed by asparagine synthetase.

$$(21\cdot11)$$

In some bacteria and plants, ammonia rather than glutamine is the source of the amide group of asparagine. This ATP-dependent reaction is mechanistically similar to that catalyzed by glutamine synthetase.

We will now examine the biosynthetic pathways for the other six amino acids that are classified as nonessential for humans.

A. Serine, Glycine, and Cysteine Are Derived from 3-Phosphoglycerate

Serine is the major biosynthetic precursor of both glycine and cysteine. Serine formation begins with 3-phosphoglycerate and involves three reactions (Figure 21·11). First, the secondary hydroxyl substituent of 3-phosphoglycerate is oxidized to a keto group by the action of 3-phosphoglycerate dehydrogenase, forming 3-phosphohydroxypyruvate. This compound undergoes transamination with glutamate to form 3-phosphoserine and α-ketoglutarate. Finally, 3-phosphoserine phosphatase catalyzes the formation of serine and P_i. This final step, which is irreversible, is controlled by the concentration of serine, which acts as a feedback inhibitor.

Serine is a major source of glycine via a reversible reaction catalyzed by serine hydroxymethyltransferase (Figure 21·12). The reverse reaction, which occurs in

Figure 21·11
Serine biosynthesis.

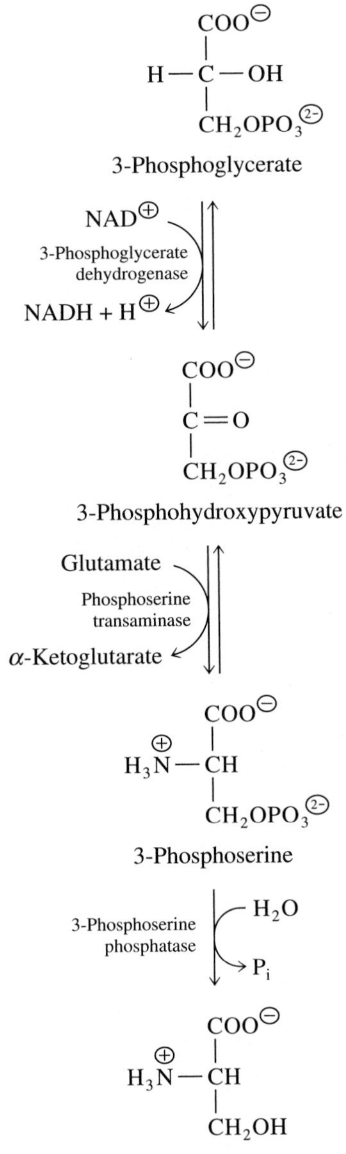

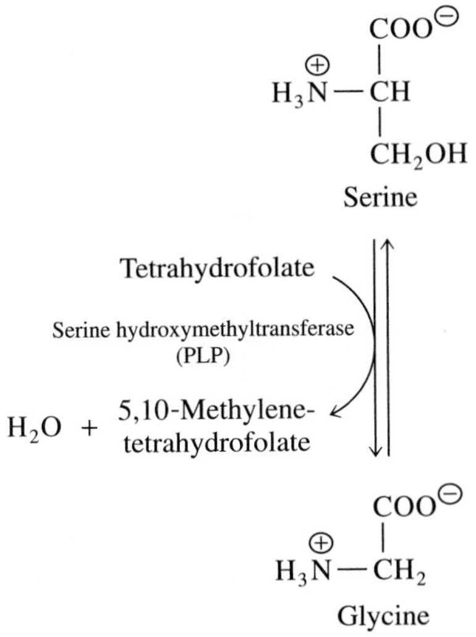

Figure 21·12
Biosynthesis of glycine. Pyridoxal phosphate (PLP) and tetrahydrofolate are both cofactors for this reaction, which is catalyzed by serine hydroxymethyltransferase.

Figure 21·13
Metabolic routes from serine and glycine.

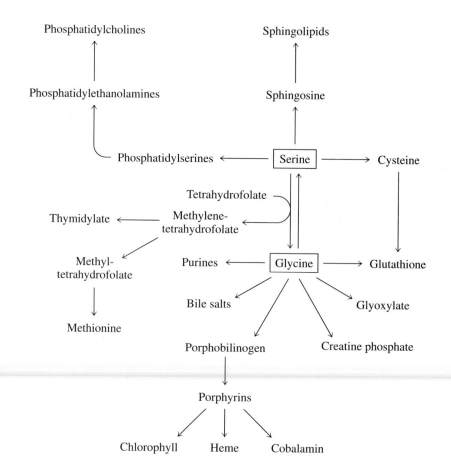

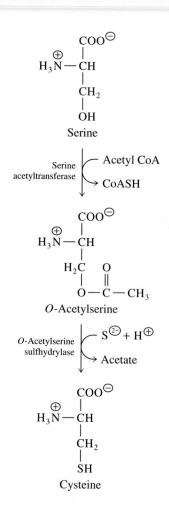

Figure 21·14
Biosynthesis of cysteine from serine in bacteria and plants.

plant mitochondria, represents a different route to serine from that shown in Figure 21·11. The serine hydroxymethyltransferase reaction requires two cofactors: the prosthetic group pyridoxal phosphate and the cosubstrate tetrahydrofolate.

Serine and glycine, as well as 5,10-methylenetetrahydrofolate derived from the hydroxymethyl group of serine, are metabolic precursors of many other compounds (Figure 21·13). These include the amino acids methionine and cysteine, purines, thymidylate, glycerophospholipids, and porphyrins.

The biosynthesis of cysteine from serine in bacteria and plants is carried out in two steps (Figure 21·14). First, an acetyl group from acetyl coenzyme A (acetyl CoA) is transferred to the β-hydroxyl substituent of serine, forming *O*-acetylserine. Next, inorganic sulfide, $S^{2\ominus}$ (formed by a complex reduction of $SO_4^{2\ominus}$, as discussed in Section 8·4B), displaces the acetate group, and cysteine is formed.

Cysteine biosynthesis in mammals also begins with serine (Figure 21·15). In the first step, serine condenses with homocysteine, a metabolite generated during the biosynthesis and catabolism of methionine. Since methionine is an essential amino acid for mammals, the availability of homocysteine for this reaction depends upon an ample supply of methionine in the diet of mammals. The product of the condensation reaction, cystathionine, is cleaved to α-ketobutyrate and cysteine.

B. Proline Is Formed from Glutamate

The pathway for biosynthesis of proline is shown in Figure 21·16. First, γ-glutamate kinase catalyzes the phosphorylation of glutamate to form γ-glutamyl phosphate. This product is converted to glutamate γ-semialdehyde by addition of a hydride ion from NADH and release of P_i, catalyzed by glutamate γ-semialdehyde dehydrogenase. The mechanism of this enzyme is analogous to that of the glycolytic enzyme

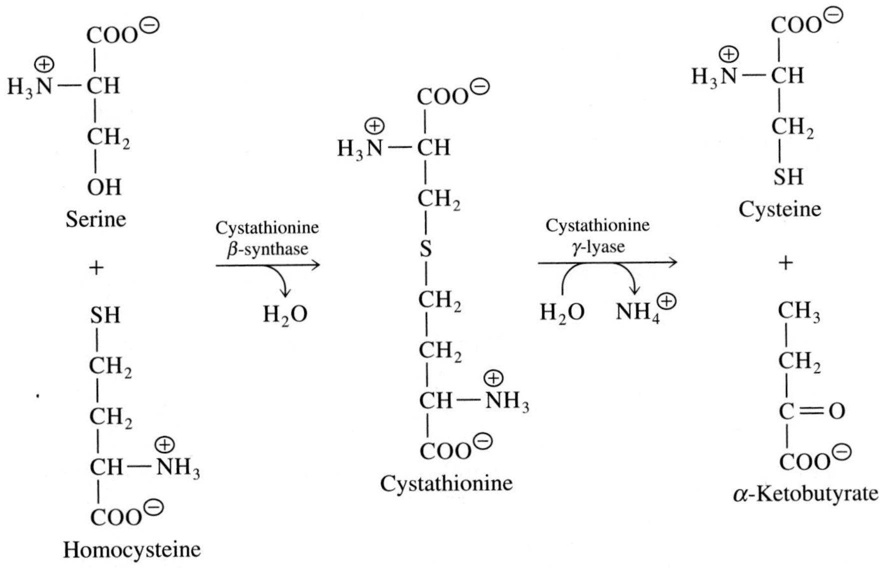

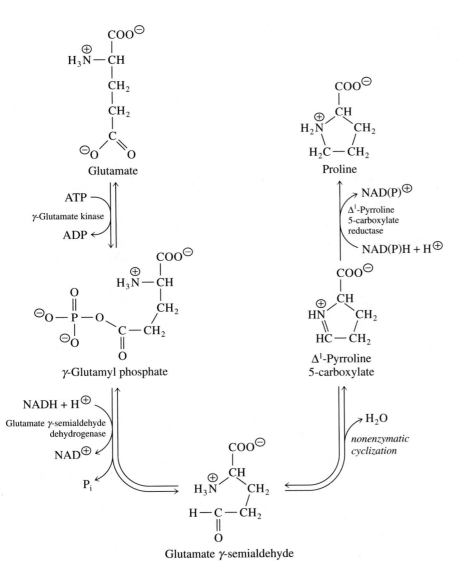

Figure 21·15
Biosynthesis of cysteine in mammals.

Figure 21·16
Biosynthesis of proline.

glyceraldehyde 3-phosphate dehydrogenase. Glutamate γ-semialdehyde undergoes nonenzymatic cyclization by formation of an internal Schiff base to form Δ^1-pyrroline 5-carboxylate. Finally, a reductase catalyzes the formation of proline, with NADH or NADPH serving as a cofactor. The specificity for the pyridine nucleotide varies among organisms.

C. In Mammals, Tyrosine Can Be Formed from Phenylalanine

De novo synthesis of tyrosine follows a multistep, energy-dependent pathway in some organisms, such as bacteria (Section 21·7C); however, in mammals, tyrosine is formed from the essential amino acid phenylalanine in a single step catalyzed by phenylalanine hydroxylase (Figure 21·17). This reaction requires molecular oxygen and the reductant tetrahydrobiopterin (Section 8·12). One oxygen atom from O_2 is incorporated into tyrosine, and the other is converted to water. Regeneration of the reduced form of biopterin, 5,6,7,8-tetrahydrobiopterin, from quinonoid dihydrobiopterin is catalyzed by dihydropteridine reductase in a reaction that requires NADH.

In North America and Europe, one of the most common disorders of amino acid metabolism is phenylketonuria. This condition is caused by a mutation in the gene that encodes phenylalanine hydroxylase. The result is an impairment in the ability of the affected individual to convert phenylalanine to tyrosine. The blood of children with the disease contains very high levels of phenylalanine and low levels of tyrosine. The phenylalanine, rather than being converted to tyrosine, is metabolized to phenylpyruvate by transamination. Elevated levels of phenylpyruvate and its derivatives in young children result in irreversible mental retardation. Phenylketonuria can be detected by testing for elevated levels of phenylalanine in the blood during the first days after birth. If the dietary intake of phenylalanine is strictly limited during the first decade of life, phenylalanine hydroxylase–deficient individuals can show normal mental development. Elevated levels of phenylalanine

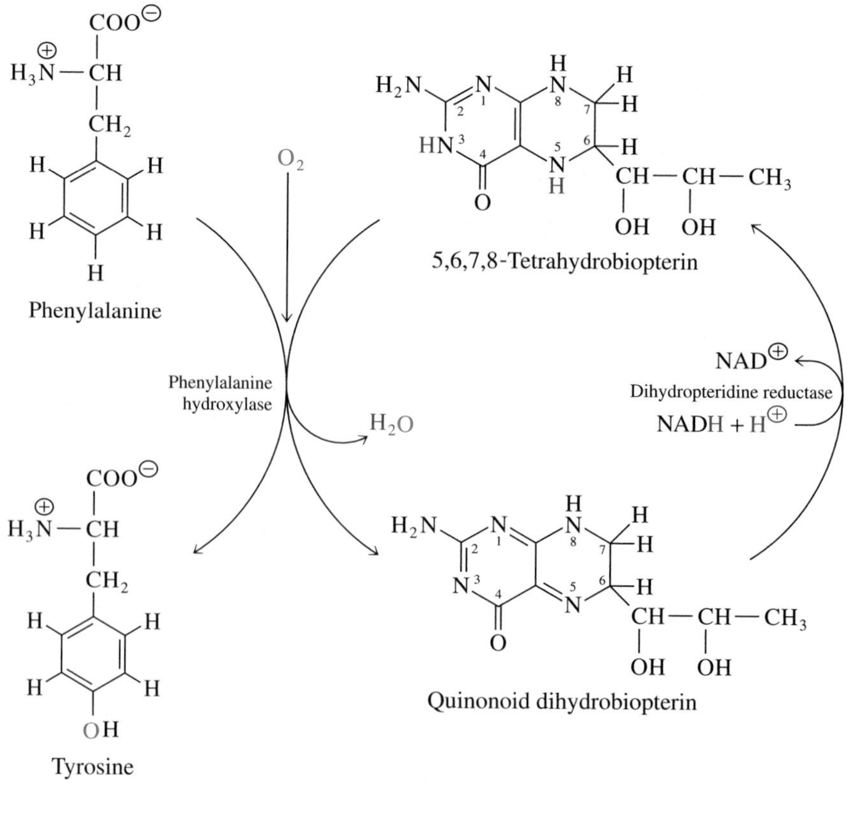

Figure 21·17
Conversion of phenylalanine to tyrosine in mammals. Tyrosine formation catalyzed by phenylalanine hydroxylase involves molecular oxygen and the cofactor tetrahydrobiopterin. The cofactor is regenerated in an NADH-dependent reaction catalyzed by dihydropteridine reductase.

are also observed in individuals with deficiencies in dihydropteridine reductase or defects in the biosynthesis of the coenzyme tetrahydrobiopterin because each of these disorders results in impairment of the hydroxylation of phenylalanine.

D. In Mammals, Synthesis of Arginine Requires Reactions in Two Tissues

Synthesis of arginine in mammals involves a metabolic pathway that occurs in two tissues—the intestine and the kidney (Figure 21·18). In the intestine, ornithine is synthesized by transamination of glutamate γ-semialdehyde (this compound is also an intermediate in the conversion of glutamate to proline, Figure 21·16). Ornithine then reacts with carbamoyl phosphate to form citrulline and P_i. Next, citrulline passes through the bloodstream to the kidney, where the amino group of aspartate condenses with citrulline, producing argininosuccinate. Fumarate is then eliminated from argininosuccinate to form arginine. Most of the arginine synthesized by this pathway is released from the kidney for use by the liver and other tissues in such pathways as protein synthesis and the urea cycle. In the kidney, some arginine is converted to guanidoacetic acid, which is then transported to the liver, where it is the substrate for creatine synthesis.

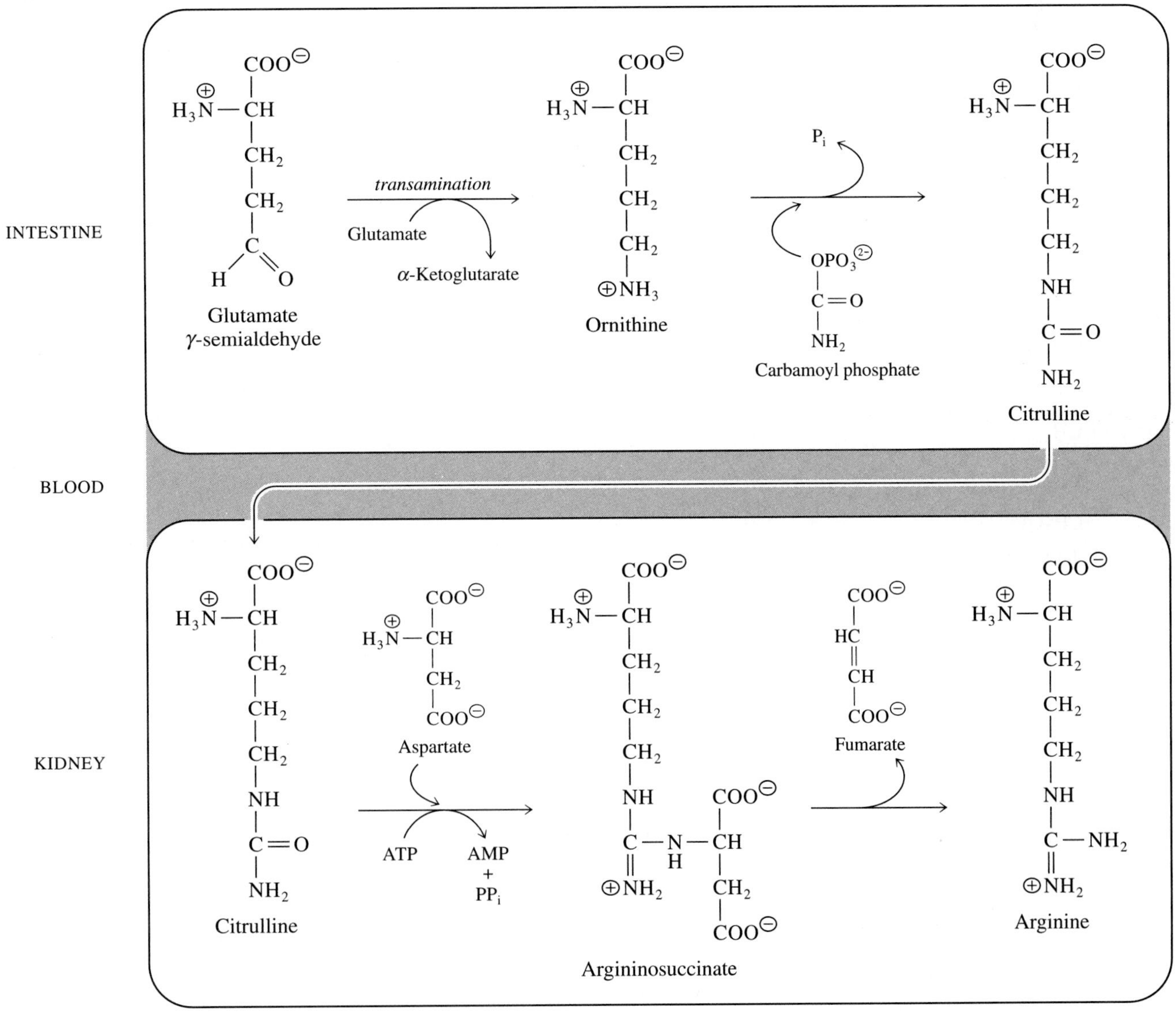

Figure 21·18
De novo synthesis of arginine in mammals. Citrulline is synthesized in the intestine and then passes through the bloodstream to the kidney, where it is converted to arginine.

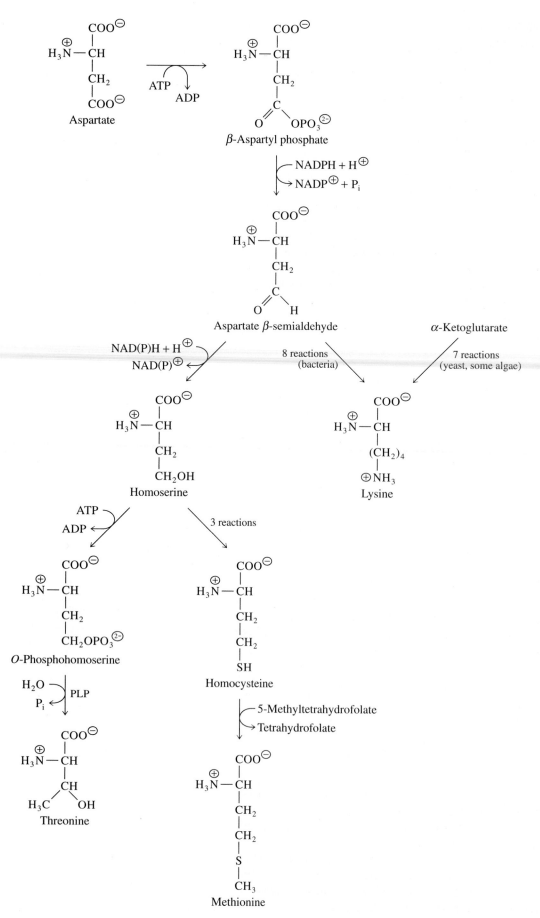

Figure 21·19
Pathways for the biosynthesis of lysine, threonine, and methionine in microorganisms.

21·7 Bacteria and Plants Synthesize the Amino Acids That Are Essential for Mammals

Most of the research on the biosynthesis of amino acids that are essential for mammals has been performed with bacteria. Therefore, much of the following discussion relies on information obtained from experiments with bacterial enzymes. Plants are believed to synthesize these amino acids by similar pathways.

A. Aspartate Is the Precursor of Lysine, Threonine, and Methionine

In most microorganisms, aspartate is the precursor of lysine, threonine, and methionine. Recall that aspartate is formed in a transamination reaction involving oxaloacetate as the amino-group acceptor. The intermediate steps in the conversion of aspartate to lysine, threonine, and methionine are summarized in Figure 21·19. The first two reactions, leading to aspartate β-semialdehyde, are common to the formation of all three amino acids. In the branch leading to lysine, pyruvate is the source of carbon atoms added to the skeleton of aspartate β-semialdehyde, and glutamate is the source of the ε-amino group. An entirely different route to lysine, beginning with α-ketoglutarate, is operative in some eukaryotic microorganisms such as yeasts and some algae.

Homoserine, formed from aspartate β-semialdehyde, is a branch point for the formation of threonine and methionine. Threonine is derived from homoserine after phosphorylation to O-phosphohomoserine. The phosphate group of O-phosphohomoserine is removed and a hydroxyl group is added in a pyridoxal phosphate–dependent reaction to yield threonine. Methionine formation requires the synthesis of homocysteine in three steps from homoserine. Homocysteine is then methylated at its sulfur atom by 5-methyltetrahydrofolate (Section 8·12), forming methionine. An enzyme that catalyzes this reaction is also found in mammals, but its activity is too low to supply enough methionine for the needs of the cell. Therefore, methionine is an essential amino acid for mammals.

B. The Pathways for Synthesis of the Branched-Chain Amino Acids Isoleucine, Valine, and Leucine Share Enzymatic Steps

The biosynthesis of the branched-chain amino acids is depicted in Figure 21·20 (next page). Threonine is converted to isoleucine in five steps. In a very similar series of reactions, pyruvate is converted to valine. In fact, the same four enzymes catalyze the final four steps in isoleucine biosynthesis and the four steps of valine biosynthesis from pyruvate. The pathway for biosynthesis of leucine branches from the pathway leading to valine.

C. A Branched Pathway Leads to the Aromatic Amino Acids

In some classic studies of bacterial mutants, Bernard Davis noted that some single-gene mutants required provision of as many as five compounds for growth—phenylalanine, tyrosine, tryptophan, p-hydroxybenzoate, and p-aminobenzoate (a component of folate). These compounds all contain an aromatic ring system. The inability of Davis's mutants to grow without these compounds was reversed when the compound shikimate was provided, suggesting that shikimate is an intermediate in the biosynthesis of these aromatic compounds. Further experiments by Frank Gibson showed that chorismate is formed in three steps from shikimate and that it is a key intermediate in aromatic amino acid synthesis.

Figure 21·20
Pathways for the biosynthesis of the branched-chain amino acids in microorganisms and plants. The final four enzymes in the pathway from threonine to isoleucine also catalyze the four reactions in the pathway from pyruvate to valine. The carbon chain of α-ketoisovalerate, an intermediate in the formation of valine, is lengthened by one methylene group during the formation of leucine.

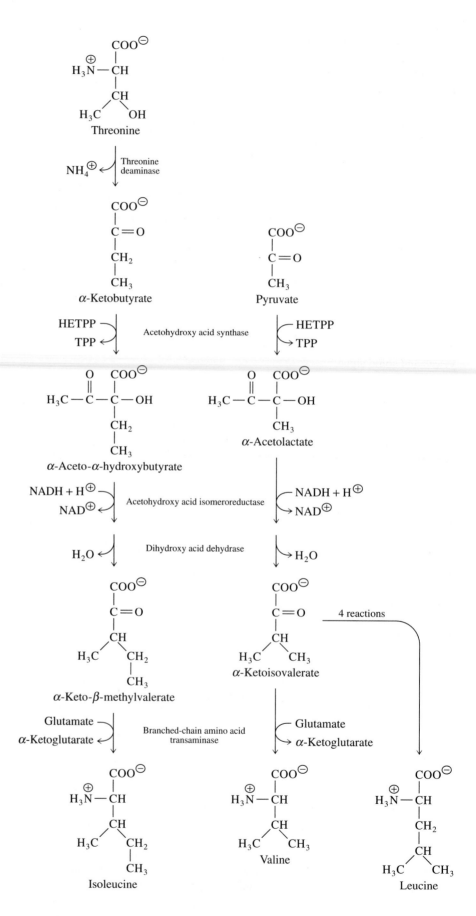

Chorismate is synthesized from phosphoenolpyruvate and erythrose 4-phosphate in seven steps (Figure 21·21). First, phosphoenolpyruvate and erythrose 4-phosphate condense to form a seven-carbon sugar derivative and P_i. In the enteric bacteria, three separately regulated isozymes each catalyze this reaction. The activities of these enzymes are allosterically inhibited by phenylalanine, tyrosine, or tryptophan. Three steps are required for transformation of the seven-carbon sugar phosphate to shikimate. First, cyclization of the open-chain form of the sugar produces 3-dehydroquinate. Dehydration introduces the first double bond of the ring, and an NADPH-dependent reduction forms shikimate. The C-3 hydroxyl group of shikimate undergoes phosphorylation, followed by addition to the C-5 hydroxyl group of a three-carbon moiety from phosphoenolpyruvate. Subsequent elimination of P_i introduces a second double bond to the ring, forming chorismate. This compound is a branch-point metabolite that is converted via different pathways to phenylalanine, tyrosine, and tryptophan.

Figure 21·21
Biosynthesis of chorismate from phospho-enolpyruvate and erythrose 4-phosphate. Chorismate is a branch-point metabolite that leads via different pathways to phenyl-alanine, tyrosine, and tryptophan.

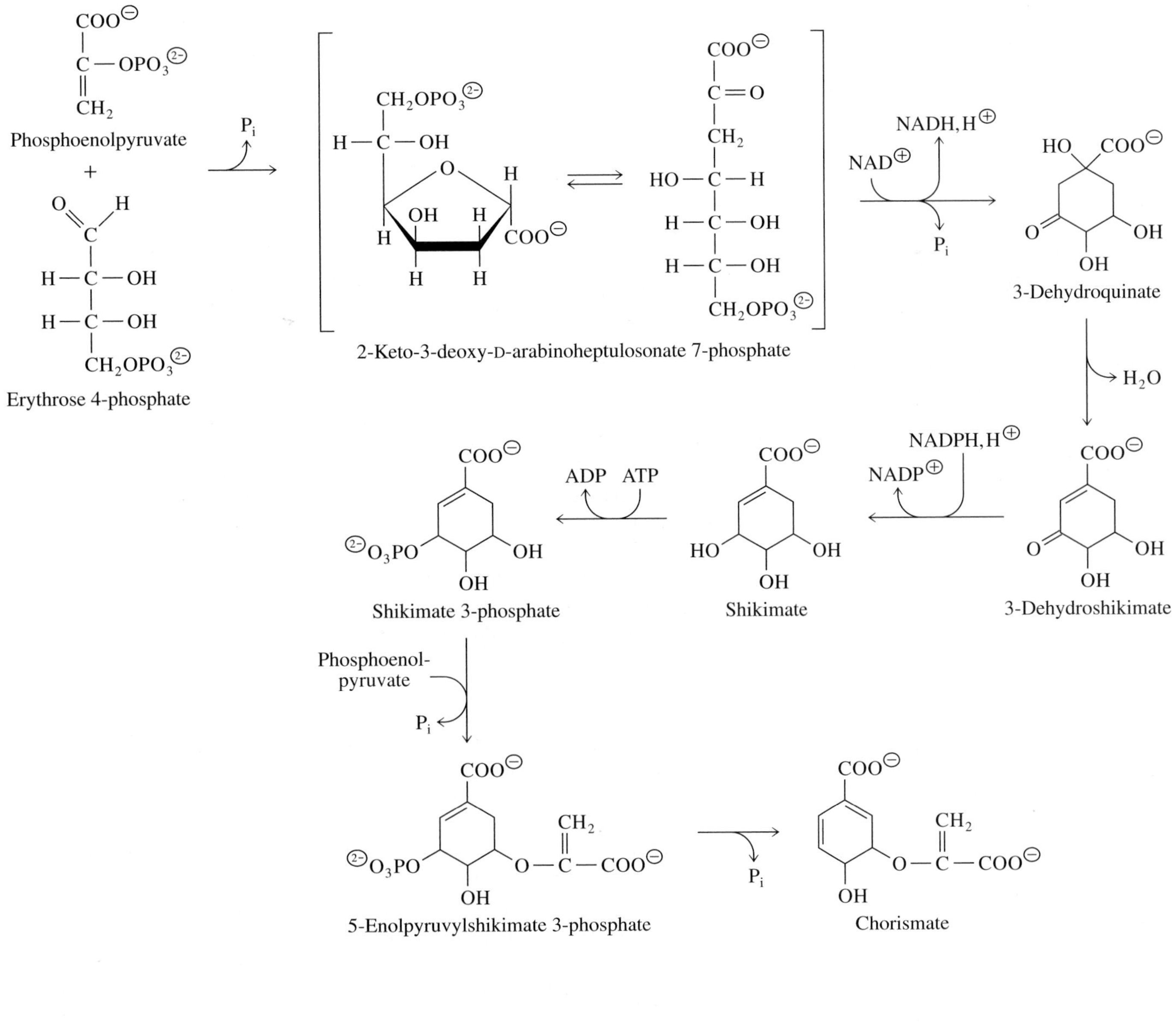

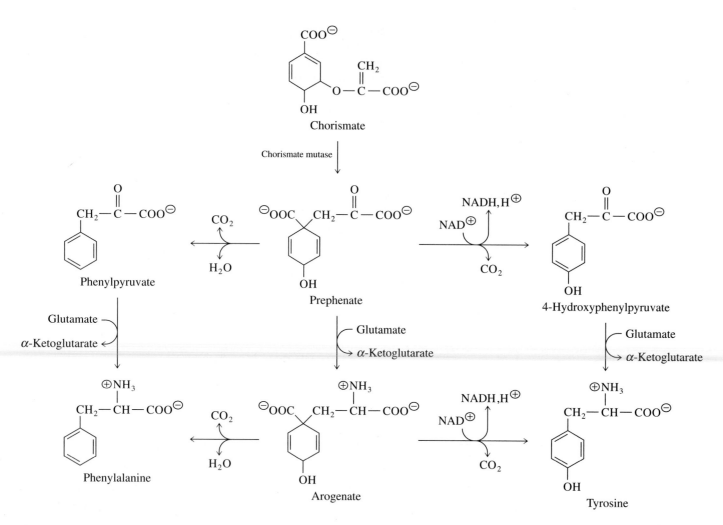

Figure 21·22
Biosynthesis of phenylalanine and tyrosine from chorismate.

There are two pathways for the biosynthesis of phenylalanine from chorismate, with different organisms using one or the other of these routes (Figure 21·22). First, chorismate mutase catalyzes the rearrangement of chorismate to produce prephenate, a highly reactive compound. In *E. coli,* several other bacteria, and some plants, water and CO_2 are eliminated from prephenate to form the fully aromatic product phenylpyruvate, which is then transaminated to phenylalanine. In the alternate pathway, used by some bacteria, prephenate is transaminated to form arogenate, from which water and CO_2 are removed to form phenylalanine.

Two pathways also lead from chorismate to tyrosine, as shown on the right side of Figure 21·22. Following the enzymatic conversion of chorismate to prephenate, some organisms transaminate prephenate to produce arogenate, which is oxidatively decarboxylated to yield tyrosine. In other organisms, oxidative decarboxylation precedes transamination. The product of the oxidation reaction, 4-hydroxyphenylpyruvate, undergoes transamination to tyrosine.

The biosynthesis of tryptophan from chorismate in *E. coli* involves five enzymes, one of which catalyzes two consecutive reactions (Figure 21·23). It appears that tryptophan is synthesized by the same pathway in all organisms for which it is nonessential. In the first step, the amide nitrogen of glutamine is transferred to chorismate. Elimination of the hydroxyl group and the adjacent pyruvate moiety of chorismate produces the aromatic ring. The resulting compound, anthranilate, accepts a ribose 5-phosphate moiety from phosphoribosyl pyrophosphate (PRPP).

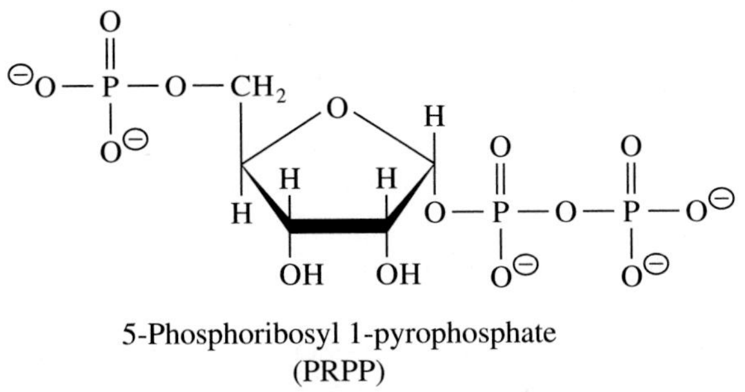

5-Phosphoribosyl 1-pyrophosphate
(PRPP)

(21·12)

Rearrangement of the ribose, decarboxylation, and ring closure generate the indole component of indole glycerol phosphate.

The final stage in tryptophan biosynthesis is catalyzed by tryptophan synthase. In some organisms, the two catalytic domains of tryptophan synthase are fused in a single polypeptide chain. In other organisms, this enzyme contains two types of subunits, α and β, in the form of an $\alpha_2\beta_2$ tetramer. The α subunit or domain catalyzes the cleavage of indole glycerol phosphate to glyceraldehyde 3-phosphate and indole. The β subunit or domain catalyzes the condensation of indole and serine in a reaction that requires pyridoxal phosphate as a cofactor. The indole produced in the reaction catalyzed by the α subunit of $\alpha_2\beta_2$ tetramers does not dissociate to the solvent but is instead transferred directly to the active site of the β

Figure 21·23
Biosynthesis of tryptophan from chorismate in *E. coli*.

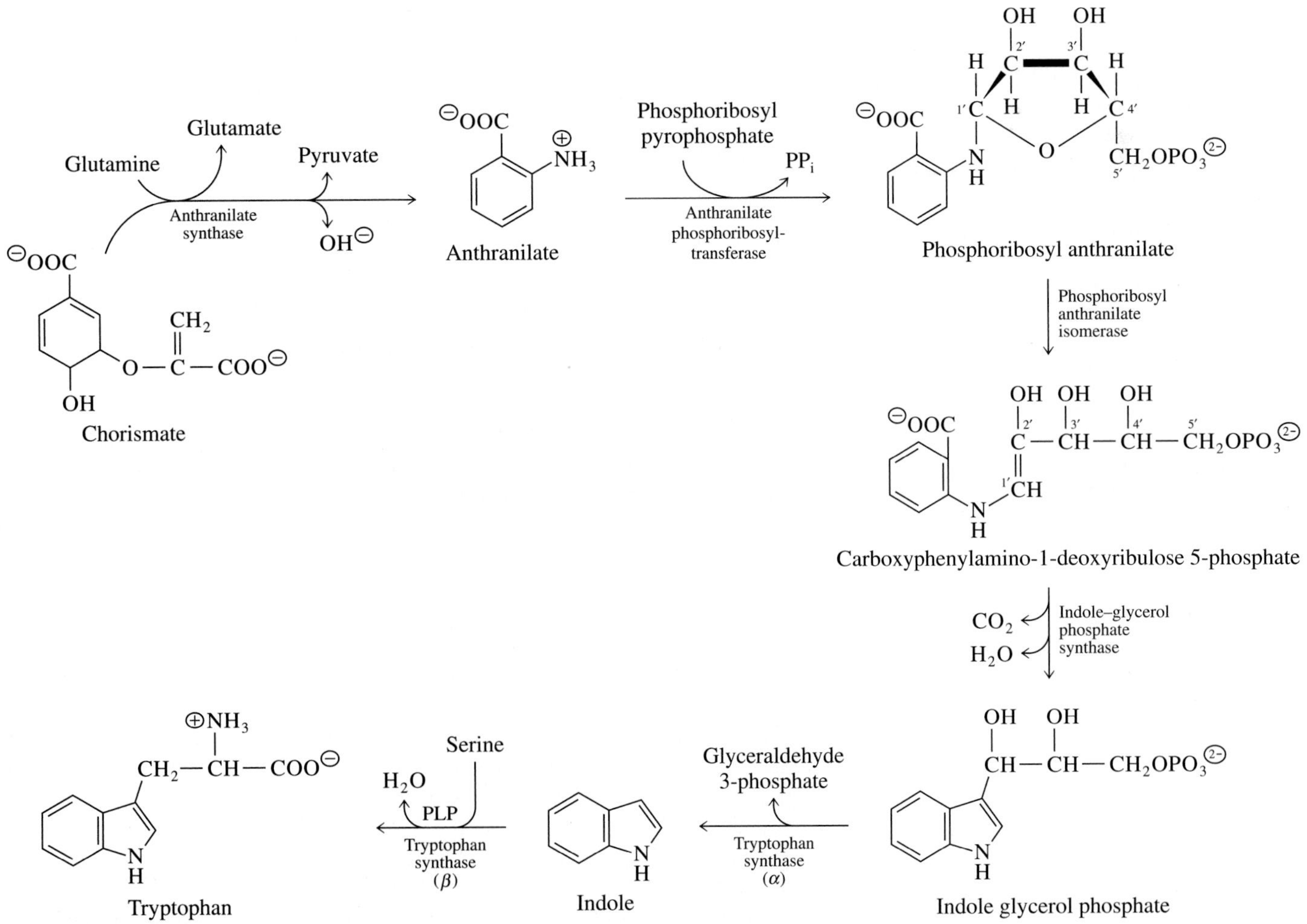

subunit. When the three-dimensional structure of tryptophan synthase from *Salmonella typhimurium* (an organism whose tryptophan synthase has the $\alpha_2\beta_2$ oligomeric structure) was determined by X-ray crystallography, a tunnel joining the α and β active sites was discovered. The presence of a channel through the interior of a globular protein is most unusual. Because the diameter of the tunnel matches the molecular dimensions of indole, passage of indole through the tunnel has been invoked to explain why indole does not enter free solution.

Short-term control of tryptophan biosynthesis is exerted by feedback inhibition of the first step in the pathway (conversion of chorismate to anthranilate) by tryptophan. In bacteria, high concentrations of tryptophan also inhibit the synthesis of the five enzymes of the pathway via transcriptional and translational control mechanisms (Chapter 30).

D. Phosphoribosyl Pyrophosphate, ATP, and Glutamine Are the Precursors of Histidine

Nutritional experiments have suggested that mammals can synthesize small amounts of histidine but not a sufficient quantity for the organism's needs. There are several sources of endogenous histidine that make the interpretation of these experiments difficult. However, no pathway for the biosynthesis of histidine has been found in mammals; therefore, histidine is classified as an essential amino acid.

Figure 21·24

Outline of the biosynthesis of histidine.

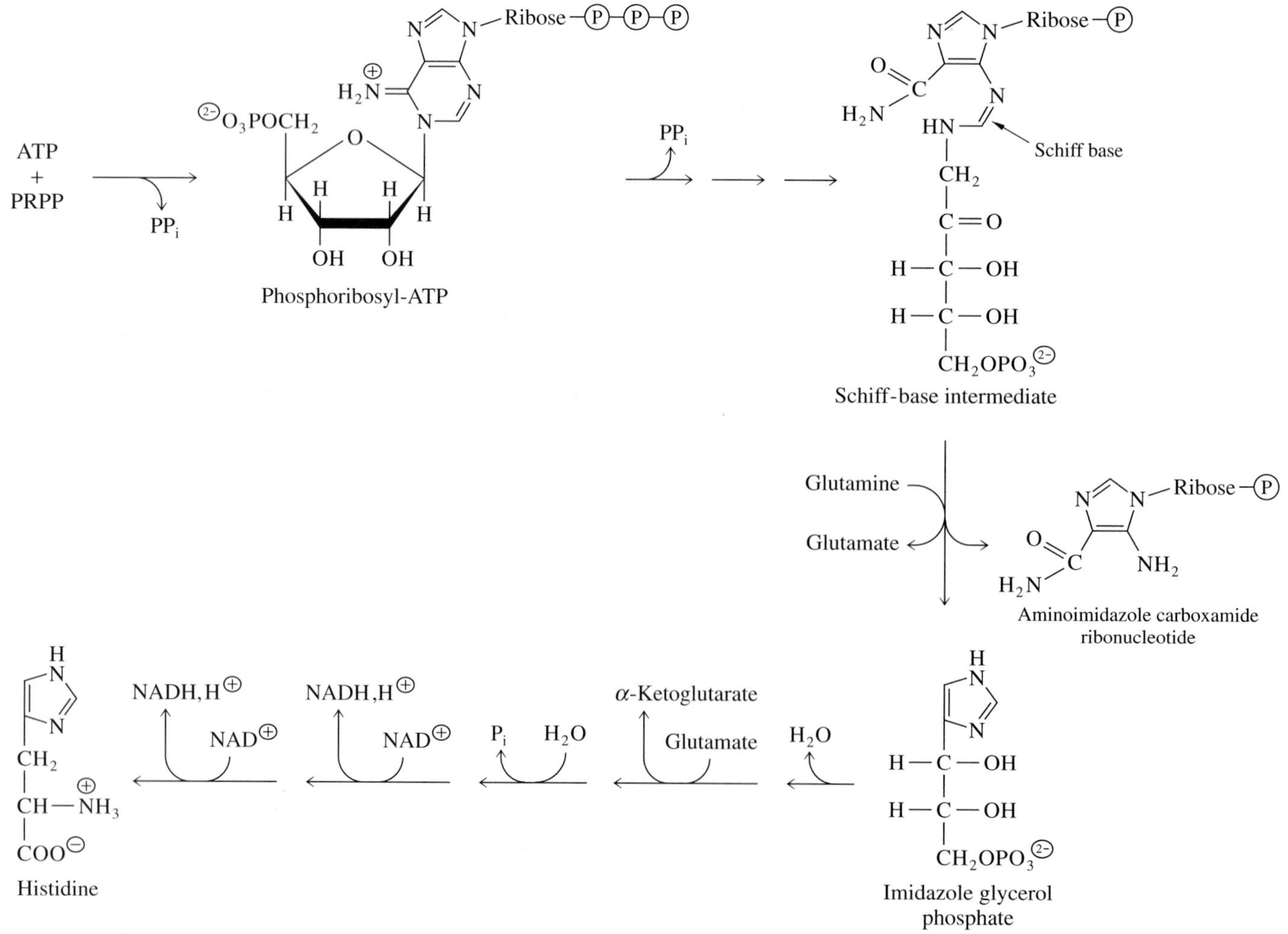

In plants, simpler eukaryotes, and bacteria, histidine is synthesized from PRPP, which is also a precursor of tryptophan and of purine and pyrimidine nucleotides. The pathway for the biosynthesis of histidine begins with a condensation between the pyrimidine ring of ATP and PRPP (Figure 21·24). In subsequent reactions, the six-membered ring of the adenine moiety is cleaved, and a Schiff-base intermediate is formed. Glutamine donates a nitrogen atom to the intermediate, releasing the amine portion of the Schiff base and initiating cyclization. The nitrogen atom from glutamine is incorporated into the imidazole ring of the product, imidazole glycerol phosphate. Imidazole glycerol phosphate then undergoes dehydration, transamination by glutamate, hydrolytic removal of its phosphate, and oxidation from the level of a primary alcohol to that of a carboxylic acid in two sequential $NAD^{\oplus}$-dependent steps, forming histidine.

21·8 Intracellular Proteins Are in a Dynamic Equilibrium

Once amino acids are biosynthesized or ingested, they are incorporated into proteins. One might assume that only growing or reproducing cells would require new protein molecules (and therefore a supply of amino acids), but this is not the case. In organisms of all types, proteins are continuously synthesized and degraded. Let us consider this dynamic process in mammals.

Early experiments demonstrated not only that the growth of young mammals requires greater intake of protein than excretion of protein-derived nitrogen but also that adults must have continued dietary intake of proteins. For healthy adults, the intake of nitrogen balances the excretion of nitrogen. A major advance in these studies was made in the 1940s by Rudolf Schoenheimer and his colleagues. During studies of the metabolism of amino acids in rats, they observed that while some of the label from ingested isotopically labelled amino acids appeared quickly in the urine, much was retained by the animal, presumably as protein. Schoenheimer concluded that existing protein molecules must be replaced by protein synthesized from the ingested amino acids. Schoenheimer's further examination of this dynamic equilibrium, or **turnover,** of proteins showed that proteins in some tissues (e.g., intestinal mucosa, liver, and kidney) are replaced more rapidly than those in other tissues (e.g., skin and bone). Later work has shown that individual proteins turn over at different rates. The half-lives of intracellular proteins can vary from a few minutes to several weeks, but the half-life of a given protein in different organs and species is generally similar. The energetically expensive process of rapid protein turnover protects cells by eliminating abnormal proteins. Rapid protein turnover also ensures that proteins involved in regulatory processes are degraded so that the cell can respond to constantly changing conditions.

Eukaryotic cells degrade proteins to amino acids through lysosomal hydrolysis and other processes, some of which require ATP. In lysosomal hydrolysis, vesicles containing material to be destroyed fuse with lysosomes. Lysosomes hydrolyze the engulfed proteins using a variety of proteases. The collective proteolytic action of lysosomal enzymes is nonselective, causing extensive hydrolysis of all trapped proteins.

In contrast, some intracellular proteins have very short half-lives because they are specifically targeted for degradation. A major pathway for selective hydrolysis of proteins in eukaryotic cells involves covalent linkage of proteins to the protein ubiquitin, followed by hydrolysis of the ubiquitinated protein by the action of a

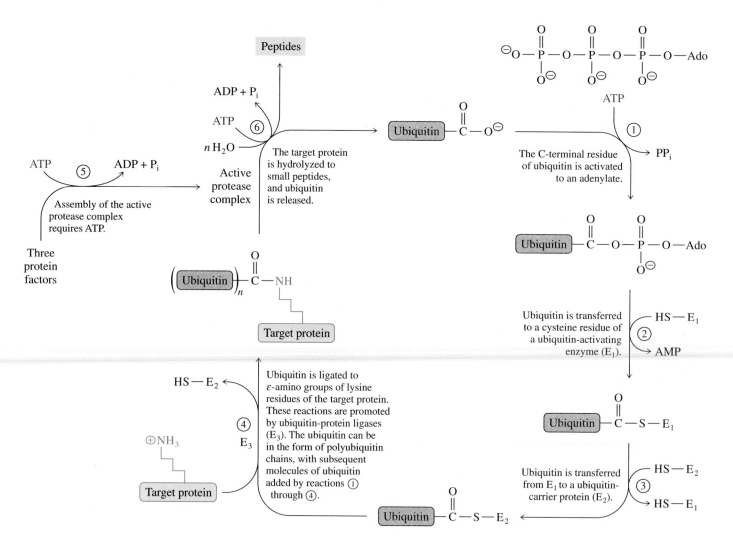

Figure 21·25
Proposed pathway for ubiquitination of proteins and their subsequent hydrolysis.

large multiprotein complex. The pathway proposed for this system is illustrated in Figure 21·25. This process occurs in the cytosol but may also take place in the nucleus.

Ubiquitin, a highly conserved protein having 76 amino acid residues, is present in all eukaryotic cells. The thiol esters formed between ubiquitin and a family of ubiquitin-carrier proteins (designated E_2 in Figure 21·25) donate ubiquitin to the protein to be degraded. The C-terminal carboxyl groups of ubiquitin molecules form isopeptide bonds, that is, bonds with ε-amino groups of lysine residues of the target protein. Additional molecules of ubiquitin are often linked to the ubiquitin already attached to the protein to form polyubiquitin chains. A large complex formed from three protein factors (M_r 250 000, 600 000, and 650 000) catalyzes the hydrolysis of ubiquitinated proteins to inactive peptides. Ubiquitin is then released from these peptides and from other ubiquitin molecules by the action of ubiquitin C-terminal hydrolases. The peptides are subsequently hydrolyzed to amino acids in reactions catalyzed by other cytosolic proteases. ATP is required for activation of ubiquitin, for assembly of the active protease complex, and for hydrolysis of the target protein to peptides; its role in the latter two processes is unknown.

The signals by which a protein is recognized as a target for ubiquitination are not fully understood. One possible signal is the N-terminal amino acid residue of a protein. Proteins with certain N-terminal residues are very stable; those with other N-terminal residues are unstable. The selection of proteins for degradation is based on the binding specificity of one of the ubiquitin-protein ligases (E_3). Other sites of recognition on the target proteins appear to be necessary for degradation to occur. Much remains to be learned both about the reactions and regulation of the hydrolysis of intracellular proteins.

21·9 Catabolism of Amino Acids Often Begins with Deamination, Followed by Degradation of the Remaining Carbon Chains

We will now examine the degradation of amino acids. The amino acids catabolized by organisms are obtained from food and from the degradation of intracellular proteins. The first step in their degradation is often removal of the α-amino group. Next, the carbon chains are altered in specific ways for entry into the central pathways of carbon metabolism. We will first consider the metabolism of the ammonia arising from amino acid degradation, and then we will examine the metabolic fates of the various carbon skeletons.

The removal of the α-amino group occurs in several ways. Most often, the amino acid undergoes transamination with α-ketoglutarate to form an α-keto acid and glutamate. The glutamate is oxidized to α-ketoglutarate and ammonia by the action of glutamate dehydrogenase. The net effect of these two reactions is the release of α-amino groups as ammonia and the formation of NADH and of α-keto acids.

$$\text{Amino acid} + \alpha\text{-Ketoglutarate} \rightleftharpoons \alpha\text{-Keto acid} + \text{Glutamate}$$

$$\text{Glutamate} + \text{NAD}^{\oplus} \rightleftharpoons \alpha\text{-Ketoglutarate} + \text{NADH} + \text{H}^{\oplus} + \text{NH}_4^{\oplus}$$

$$\text{Sum:} \quad \text{Amino acid} + \text{NAD}^{\oplus} \rightleftharpoons \alpha\text{-Keto acid} + \text{NADH} + \text{H}^{\oplus} + \text{NH}_4^{\oplus}$$

$$(21\cdot13)$$

The amide groups of glutamine and asparagine are hydrolyzed by specific enzymes—glutaminase and asparaginase, respectively—to produce ammonia and the corresponding dicarboxylic amino acids glutamate and aspartate.

21·10 The Urea Cycle Converts Ammonia into Urea

Because ammonia is highly toxic to cells, its concentration is usually maintained at low levels. Different organisms have different pathways for the elimination of waste nitrogen. Many aquatic organisms are able to excrete ammonia directly across the cell membranes of gill tissue, to be flushed away by the surrounding water. Most terrestrial vertebrates convert waste nitrogen to the less toxic product urea (Figure 21·26), an uncharged and highly water-soluble compound that can be carried in the blood to the kidney, where it is excreted as the major solute of urine. Birds and many terrestrial reptiles convert surplus ammonia to uric acid, a relatively insoluble compound that precipitates from aqueous solution to form a semisolid slurry. Uric acid is also a product of the degradation of purine nucleotides by birds, reptiles, and primates. The further degradation of uric acid is considered in Section 22·13.

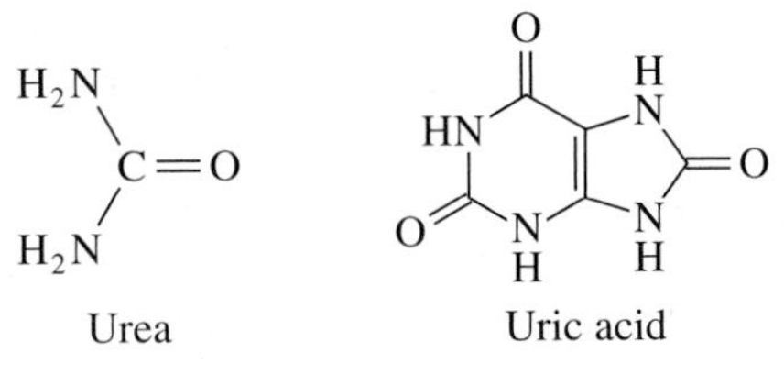

Figure 21·26
Urea and uric acid.

Urea

Uric acid

Figure 21·27
Urea cycle. Two transport proteins connecting the mitochondrial matrix and the cytosol are required for the operation of the urea cycle: the citrulline-ornithine exchanger and the glutamate-aspartate translocase.

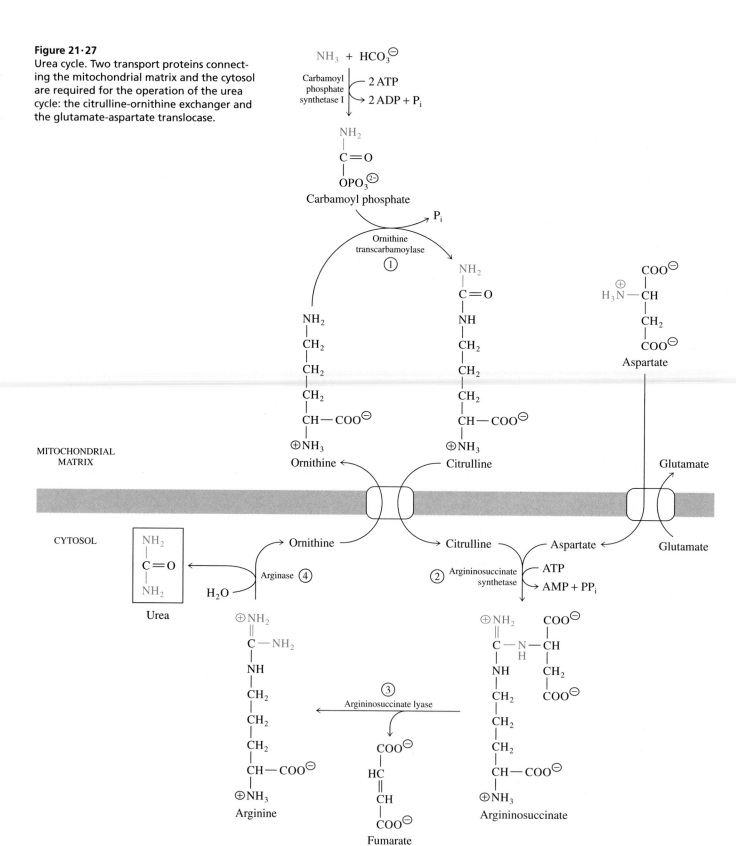

In mammals, the synthesis of urea, or ureogenesis, occurs almost exclusively in the liver. Urea is the product of a set of reactions called the **urea cycle.** This pathway was elucidated by Hans Krebs and Kurt Henseleit in 1932, several years before Krebs discovered the citric acid cycle. Several observations led to the proposal of the urea cycle. High levels of the enzyme arginase occur in the livers of all organisms that synthesize urea. Slices of rat liver can bring about the net conversion of ammonia to urea. Urea synthesis by these preparations is greatly stimulated when the amino acid ornithine is added; the amount of urea synthesized greatly exceeds the amount of added ornithine, suggesting that ornithine acts catalytically.

Ammonia released by oxidative deamination of glutamate, catalyzed by mitochondrial glutamate dehydrogenase, is incorporated into urea in five steps. Carbamoyl phosphate is formed first, and its nitrogen atom is incorporated into urea in the four reactions of the urea cycle (Figure 21·27). Two reactions occur in the mitochondria of liver cells, and the other three occur in the cytosol. The precursors of the two nitrogen atoms of urea are ammonia and aspartate. The carbon atom of urea comes from bicarbonate. The overall reaction for urea synthesis is

$$NH_3 + HCO_3^{\ominus} + Aspartate + 3\,ATP \longrightarrow$$
$$Urea + Fumarate + 2\,ADP + 2\,P_i + AMP + PP_i$$

(21·14)

Four equivalents of ATP are consumed in urea synthesis. Three molecules of ATP are converted to two ADP and one AMP during formation of one molecule of urea, and hydrolysis of the molecule of inorganic pyrophosphate that is formed consumes a fourth ATP equivalent. Urea passes from the liver through the bloodstream to the kidney, where it is excreted in urine.

Carbamoyl phosphate is synthesized from ammonia, bicarbonate, and ATP in a mitochondrial reaction catalyzed by carbamoyl phosphate synthetase I. This enzyme is one of the most abundant in liver mitochondria, accounting for as much as 20% of the protein of the mitochondrial matrix. The mechanism of the reaction catalyzed by carbamoyl phosphate synthetase I involves three steps (Figure 21·28). First, carbonyl phosphate is formed by reaction of bicarbonate with the γ-phosphate of ATP; ammonia then displaces the phosphate to form carbamate, and finally, carbamoyl phosphate is formed by transfer of the γ-phosphate of a second molecule of ATP. A similar enzyme called carbamoyl phosphate synthetase II uses glutamine rather than ammonia as the nitrogen donor. This enzyme, located in the cytosol of liver and most other cells, catalyzes the formation of carbamoyl phosphate destined for the synthesis of pyrimidine nucleotides (Section 22·6).

Carbamoyl phosphate synthetase I is allosterically activated by *N*-acetylglutamate, which is formed by the reaction of glutamate with acetyl CoA. The formation of *N*-acetylglutamate is catalyzed by *N*-acetylglutamate synthase.

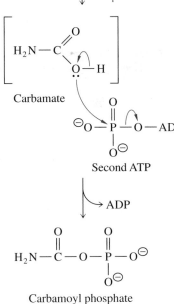

Figure 21·28
Synthesis of carbamoyl phosphate catalyzed by carbamoyl phosphate synthetase I. The reaction involves two phosphoryl-group transfers. In the first, nucleophilic attack by bicarbonate on ATP produces carbonyl phosphate and ADP. Next, ammonia reacts with carbonyl phosphate, forming a tetrahedral intermediate. Elimination of a phosphate group produces carbamate. A second phosphoryl-group transfer from another ATP forms carbamoyl phosphate and ADP. Structures in brackets remain enzyme bound during the reaction.

(21·15)

High rates of amino acid catabolism lead to elevated glutamate concentrations and concomitant increases in the concentration of *N*-acetylglutamate. Activation of carbamoyl phosphate synthetase I by *N*-acetylglutamate increases the supply of substrate for the urea cycle.

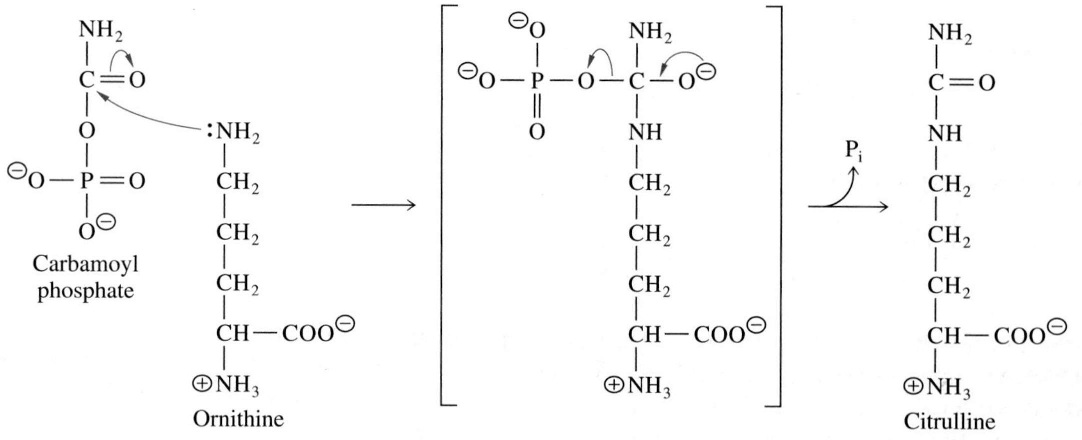

Figure 21·29
Synthesis of citrulline by ornithine transcarbamoylase. The δ-amino group of ornithine attacks the carbonyl carbon of carbamoyl phosphate, releasing the phosphate group and forming citrulline.

The urea cycle itself consists of four reactions. The first reaction occurs in the mitochondrial matrix, and the other three reactions occur in the cytosol.

1. In the first reaction of the urea cycle, carbamoyl phosphate reacts in the mitochondrion with the urea cycle intermediate ornithine to form citrulline in a reaction catalyzed by ornithine transcarbamoylase. This step incorporates the nitrogen atom originating from ammonia into citrulline; citrulline thus contains half the nitrogen destined for urea. The mechanism of ornithine transcarbamoylase is shown in Figure 21·29. The nucleophilic δ-amino group of ornithine reacts with the carbonyl carbon of carbamoyl phosphate, releasing inorganic phosphate and yielding citrulline. Citrulline is then transported out of the mitochondrion in an exchange with cytosolic ornithine, a compound that is produced in a subsequent reaction of the urea cycle.

2. The second nitrogen atom destined for urea is incorporated when citrulline condenses with aspartate to form argininosuccinate in the cytosol. This ATP-dependent reaction is catalyzed by argininosuccinate synthetase. The mechanism of this reaction is shown in Figure 21·30. First, a citrullyl-AMP intermediate is formed when the ureido oxygen of citrulline displaces PP_i from ATP. AMP becomes the leaving group when this oxygen is displaced by the amino group of aspartate.

Figure 21·30
Synthesis of argininosuccinate. This reaction, catalyzed by argininosuccinate synthetase, proceeds by the formation of citrullyl-AMP. The amino group of aspartate displaces AMP. Ado in the figure represents the adenosine moiety.

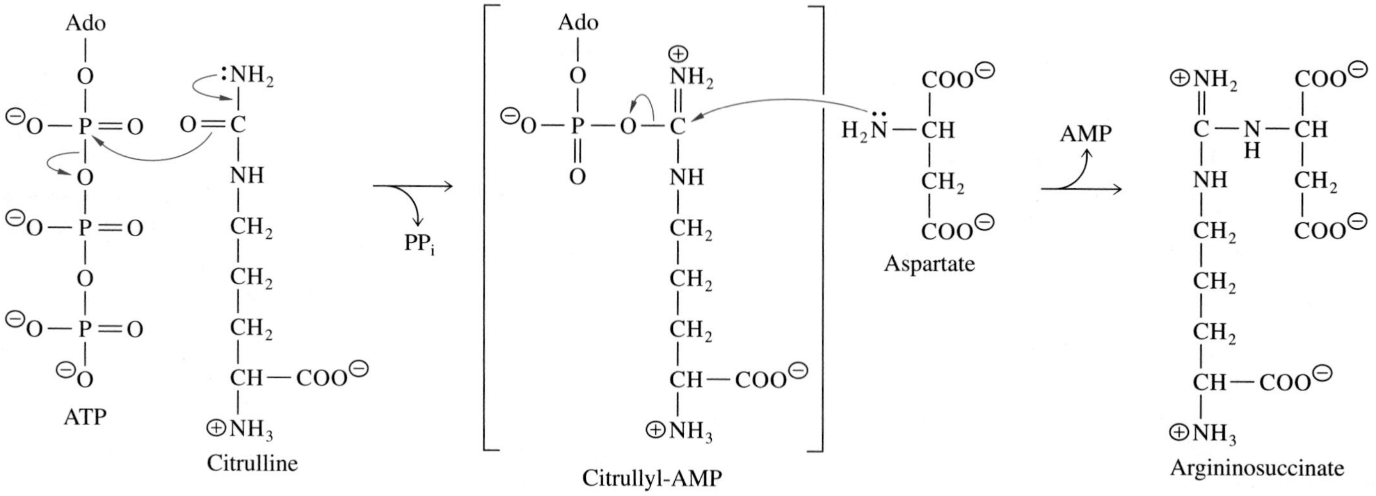

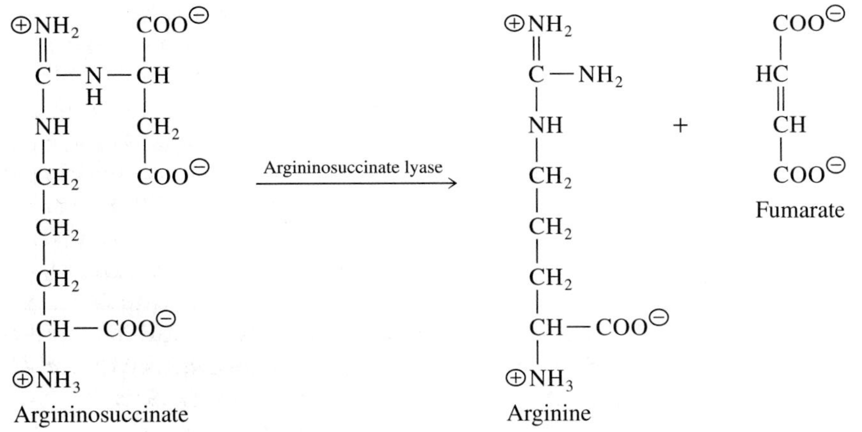

Figure 21·31
Conversion of argininosuccinate to arginine and fumarate, catalyzed by argininosuccinate lyase.

3. Argininosuccinate is cleaved nonhydrolytically to form arginine plus fumarate in an elimination reaction catalyzed by argininosuccinate lyase (Figure 21·31). Together, the second and third steps of the urea cycle exemplify a strategy for donation of the amino group of aspartate that will be encountered twice in the next chapter as part of purine biosynthesis. The key processes are an ATP-dependent condensation, followed by the elimination of fumarate.

4. In the final reaction of the urea cycle, the guanidinium group of arginine is hydrolytically cleaved to form ornithine and urea in a reaction catalyzed by arginase (Figure 21·32). Ornithine generated by the action of arginase is transported into the mitochondrion where it reacts with carbamoyl phosphate to support the continued operation of the urea cycle.

Each round of the urea cycle consumes a molecule of aspartate and releases a molecule of fumarate. The aspartate is derived primarily from the catabolism of other amino acids within the liver cell, and the fumarate formed is converted to glucose. This linkage of the urea cycle to gluconeogenesis is discussed further in Chapter 23.

Studies of mammalian liver cells have shown that they differ in enzyme activity, depending on their location. Those liver cells nearest the inflow of blood (the periportal cells) have the highest capacity for urea synthesis, whereas glutamine synthesis occurs only in cells nearest the outflow (the perivenous cells). The blood sequentially encounters these reciprocally distributed ammonia-removal systems. When the blood first encounters the periportal cells, most ammonia is removed as urea. The glutamine synthetase activity of the perivenous cells then scavenges ammonia that has escaped urea formation and ensures that very little ammonia escapes detoxification in the liver.

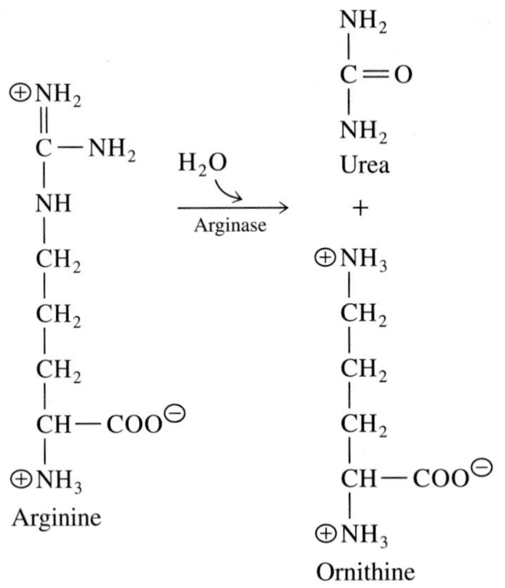

21·11 Ancillary Reactions Balance the Supply of Substrate into the Urea Cycle

The reactions of the urea cycle convert stoichiometric amounts of nitrogen from ammonia and from aspartate into urea. Many amino acids can function as amino-group donors via transamination reactions with α-ketoglutarate, forming glutamate. Glutamate can undergo either transamination with oxaloacetate to form aspartate or deamination to form ammonia. Both glutamate dehydrogenase and aspartate

Figure 21·32
Hydrolysis of arginine to form ornithine and urea, catalyzed by arginase.

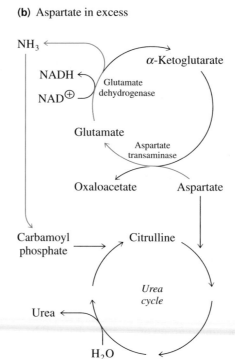

(a) NH₃ in excess

(b) Aspartate in excess

Figure 21·33
Balancing the supply of nitrogen for the urea cycle. Two theoretical situations are described: **(a)** NH₃ in extreme excess and **(b)** aspartate in extreme excess. Note that the directions of the reactions catalyzed by glutamate dehydrogenase and aspartate transaminase are reversed in the two cases. Adjustments to the flux through these reactions meet the requirement for equal amounts of nitrogen from ammonia and aspartate. More significantly, this mechanism allows the liver to synthesize urea when presented with any mixture of ammonia and amino acids, since near-equilibrium transamination reactions leading to either aspartate or glutamate correct imbalances.

transaminase are abundant in liver mitochondria and catalyze near-equilibrium reactions. These reactions can convert any proportion of amino acids and ammonia into the equal amounts of aspartate and ammonia that are consumed during operation of the urea cycle. Consider the theoretical case of a relative surplus of ammonia (Figure 21·33a). In this situation, the near-equilibrium reaction catalyzed by glutamate dehydrogenase would proceed in the direction of glutamate formation; elevated concentrations of glutamate would then result in increased flux to aspartate through aspartate transaminase, also a near-equilibrium step. In contrast, if an excess of aspartate exists, the net flux in the reactions catalyzed by glutamate dehydrogenase and aspartate transaminase would occur in the opposite direction to provide ammonia for urea formation (Figure 21·33b).

21·12 The Pathways for Catabolism of the Carbon Chains of Amino Acids Converge with the Major Pathways of Metabolism

As shown in Figure 21·34, some amino acids are degraded to citric acid cycle intermediates, others to pyruvate, and still others to acetyl CoA. While all of these products can be oxidized to CO_2 and H_2O, they can have other metabolic fates. Amino acids that are degraded to citric acid cycle intermediates can supply the pathway of gluconeogenesis and are called **glucogenic.** Those that form acetyl CoA can contribute to the formation of fatty acids or ketone bodies and are called **ketogenic.** Those that form pyruvate can be metabolized to either oxaloacetate or acetyl CoA and so can be either glucogenic or ketogenic.

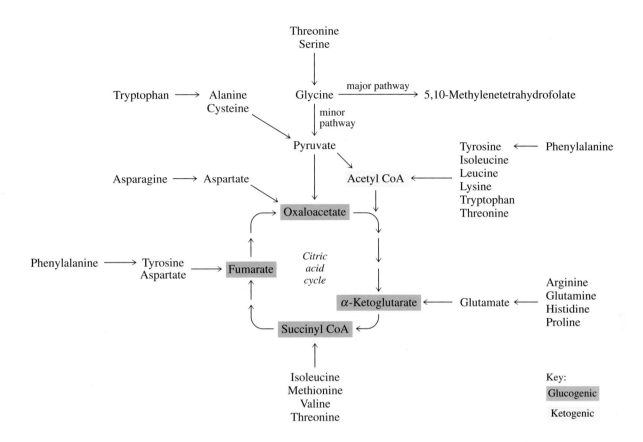

Figure 21·34
Conversion of the carbon skeletons of amino acids to pyruvate, acetyl CoA, or citric acid cycle intermediates for further catabolism.

We will examine the pathways of amino acid degradation in mammals, beginning with the simplest routes. Particular attention is given to common end products and the intersections of pathways at common intermediates.

A. Alanine, Aspartate, Glutamate, Asparagine, and Glutamine Are Degraded by Simple Transformations

Alanine, aspartate, and glutamate are synthesized by reversible transamination reactions, as we have seen. The breakdown of these three amino acids involves their reentry into the pathways from which their carbon skeletons arose. By reversal of the original transamination reactions, alanine gives rise to pyruvate, aspartate to oxaloacetate, and glutamate to α-ketoglutarate. Since aspartate and glutamate are converted to citric acid cycle intermediates, both are glucogenic. Pyruvate formed from alanine can be either glucogenic or ketogenic, as noted above, depending upon whether it is converted to oxaloacetate or acetyl CoA.

The degradation of both glutamine and asparagine begins with their hydrolysis to glutamate and aspartate, respectively. Thus, glutamine and asparagine are both glucogenic. In rapidly dividing tissues such as intestinal mucosa and cancer cells, the carbon skeleton of glutamine is a major fuel. The mitochondria of these cells contain a phosphate-dependent glutaminase. The glutamate produced by the action of this enzyme undergoes transamination, and α-ketoglutarate is then oxidized via the citric acid cycle to malate, which is converted to pyruvate in a reaction catalyzed by a mitochondrial malic enzyme. Glutamine is present at high concentrations in blood; considerable amounts therefore are available for use as a respiratory fuel. We will further examine the roles of amino acids, particularly alanine and glutamine, when we discuss the integration of mammalian metabolism (Chapter 23).

Figure 21·35
Catabolism of proline and arginine.

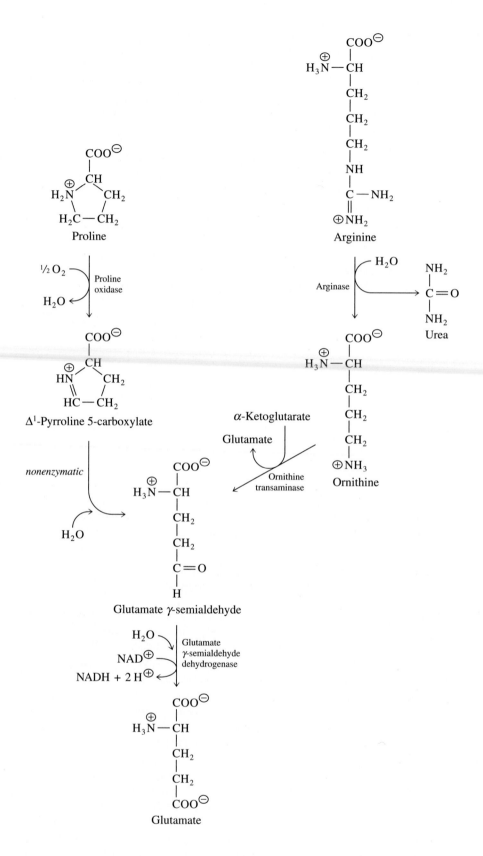

B. The Pathways for the Degradation of Proline, Arginine, and Histidine Lead to Glutamate

Figure 21·35 illustrates the convergence of the degradative pathways of proline and arginine. Proline forms glutamate in three steps. Following oxidation by molecular oxygen, the ring of the Schiff base Δ^1-pyrroline 5-carboxylate is opened by nonenzymatic hydrolysis. The product, glutamate γ-semialdehyde, is oxidized to form glutamate.

Arginine degradation commences with the reaction catalyzed by arginase, an enzyme of the urea cycle. The ornithine produced is transaminated to glutamate γ-semialdehyde, which, as in the pathway for proline degradation, is oxidized to form glutamate. A defect in the activity of ornithine transaminase causes the metabolic disease gyrate atrophy of the choroid and retina of the eye. Gyrate atrophy leads to tunnel vision and later to blindness. The progress of this disorder can be slowed by restricting dietary intake of arginine or administration of pyridoxine.

The major pathway for histidine degradation in mammals is shown in Figure 21·36. In the first step, histidine undergoes nonoxidative deamination to form urocanate and ammonia, catalyzed by histidine ammonia lyase. Next, urocanase catalyzes the addition of water to urocanate. The ensuing hydrolysis reaction opens the ring, forming *N*-formiminoglutamate. The formimino moiety is then transferred to tetrahydrofolate, forming 5-formiminotetrahydrofolate and glutamate. 5-Formiminotetrahydrofolate is enzymatically deaminated to form 5,10-methenyltetrahydrofolate.

Figure 21·36
Major pathway for the catabolism of histidine to glutamate in mammals.

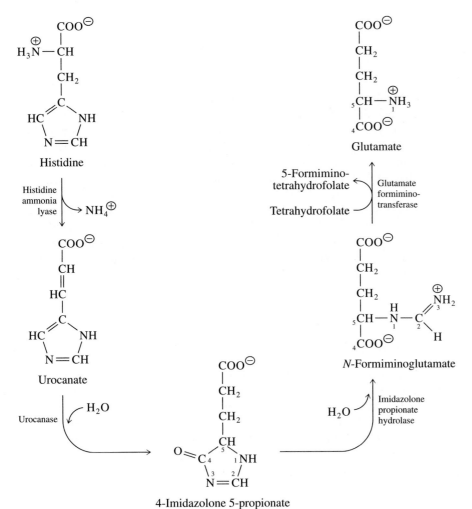

4-Imidazolone 5-propionate

C. Serine Is Converted to Glycine, Which Is Degraded by the Glycine-Cleavage System

Although there are enzymes that catalyze the breakdown of serine and glycine to pyruvate, this is not the major degradative pathway. In some organisms or tissues, a small amount of serine is converted directly to pyruvate by the action of serine dehydratase, a pyridoxal phosphate–dependent enzyme.

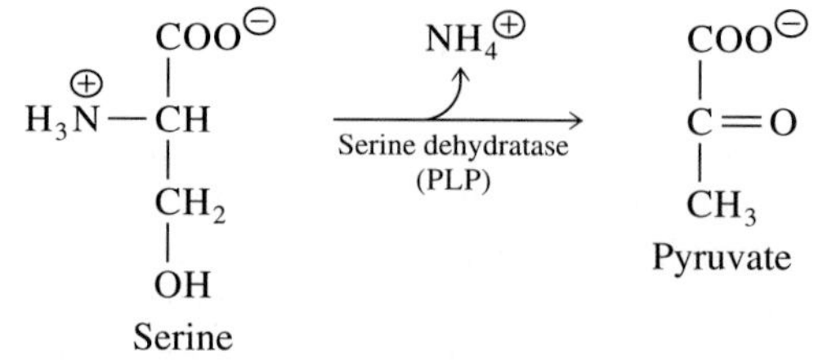

$$(21 \cdot 16)$$

Similarly, a small amount of glycine can be converted to serine in a reaction catalyzed by serine hydroxymethyltransferase (Section 21·6A). Thus, glycine can be catabolized via serine to pyruvate.

The principal pathway of serine catabolism, however, involves the formation of glycine, which is then broken down by the glycine-cleavage system, the major pathway for degradation of glycine in mammals.

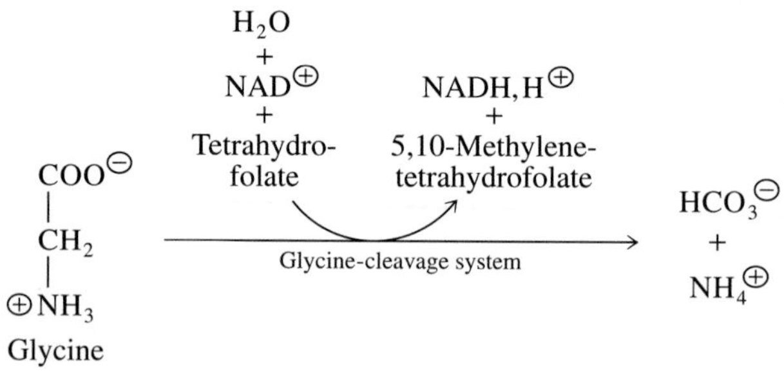

$$(21 \cdot 17)$$

Catalysis by this system, also present in plants and bacteria, requires an enzyme complex containing four nonidentical subunits. Pyridoxal phosphate, lipoamide, and FAD are present as prosthetic groups, and tetrahydrofolate is a cosubstrate. Initially, glycine is decarboxylated (Figure 21·37). Then, $NH_4^{\oplus}$ is released, and the remaining one-carbon group is transferred to tetrahydrofolate to form 5,10-methylenetetrahydrofolate. The reduced lipoamide arm is oxidized by FAD, which in turn reduces the mobile carrier $NAD^{\oplus}$. Although it is reversible in vitro, the glycine-cleavage system catalyzes an irreversible reaction in cells. The irreversibility of the reaction sequence is due in part to the K_m values for the products ammonia and methylenetetrahydrofolate, which are far greater than the concentrations of these compounds in vivo. Defects in components of the enzyme complex that catalyzes glycine cleavage lead to accumulation of large amounts of glycine in body fluids, resulting in a disease called nonketotic hyperglycinemia. Most individuals with this disorder are severely retarded and die in infancy.

D. Two Pathways for Threonine Degradation Lead to Glycine

There are several routes for the degradation of threonine. In the major pathway, threonine is oxidized to 2-amino-3-ketobutyrate in a reaction catalyzed by threonine dehydrogenase (Figure 21·38). 2-Amino-3-ketobutyrate can undergo thiolysis to form acetyl CoA and glycine. Another route for the catabolism of threonine involves cleavage to acetaldehyde and glycine by the action of threonine aldolase,

Figure 21·37
Components and reactions
of the glycine-cleavage
system.

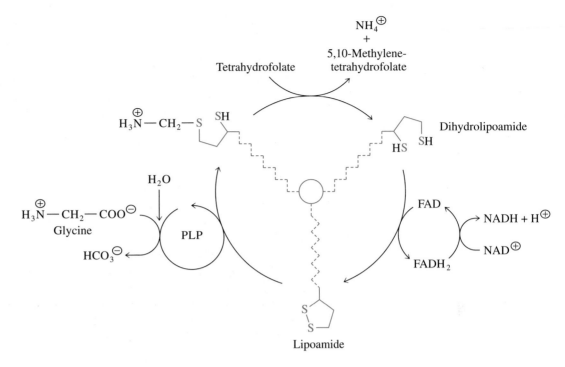

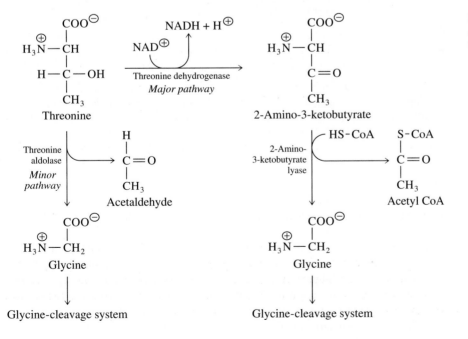

Figure 21·38
Alternate routes for the degradation of
threonine.

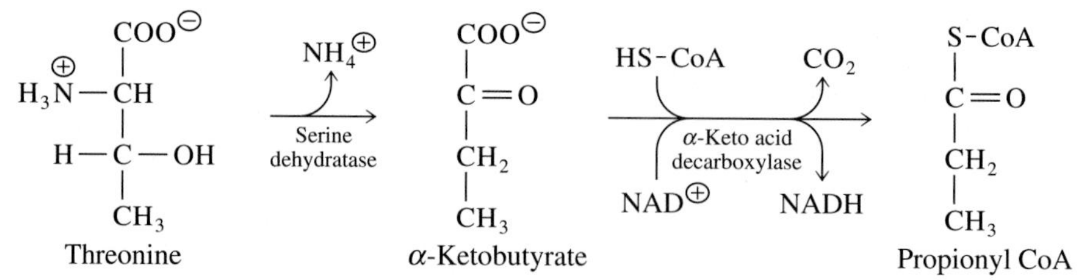

Figure 21·39
Conversion of threonine to propionyl CoA.

which in many tissues and organisms is actually a minor activity of serine hydroxy-methyltransferase. Acetaldehyde can be oxidized to acetate by the action of acetaldehyde dehydrogenase, and acetate can be converted to acetyl CoA by acetyl-CoA synthetase.

A third route for the catabolism of threonine—occurring in human tissues—involves deamination to α-ketobutyrate, catalyzed by serine dehydratase (Figure 21·39), the same enzyme that catalyzes the conversion of serine to pyruvate. α-Ketobutyrate can be converted to propionyl CoA, a precursor of the citric acid cycle intermediate succinyl CoA. The pathway from propionyl CoA to succinyl CoA is shown in detail in Chapter 20.

$$\text{Propionyl CoA} \longrightarrow \text{D-Methylmalonyl CoA} \longrightarrow$$
$$\text{L-Methylmalonyl CoA} \longrightarrow \text{Succinyl CoA} \tag{21·18}$$

Threonine can thus produce either succinyl CoA or glycine and acetyl CoA, depending on the pathway by which it is catabolized.

E. The Branched-Chain Amino Acids—Leucine, Valine, and Isoleucine—Are Degraded by Pathways That Share Common Steps

Leucine, valine, and isoleucine, the branched-chain amino acids, are degraded by related pathways, shown in Figure 21·40. The first three steps are catalyzed by the same three enzymes in all pathways. The first step, transamination, is catalyzed by branched-chain amino acid transaminase. In mammals, transamination of branched-chain amino acids occurs primarily in muscle, whereas the ensuing oxidative steps occur primarily in liver.

The second step in the catabolism of branched-chain amino acids is catalyzed by branched-chain α-keto acid dehydrogenase. In this reaction, the branched-chain α-keto acids undergo oxidative decarboxylation to form branched-chain acyl CoA molecules one carbon atom shorter than the precursor α-keto acids. The branched-chain α-keto acid dehydrogenase is a multienzyme complex that contains lipoamide and thiamine pyrophosphate and requires $NAD^{\oplus}$ and coenzyme A. Its catalytic mechanism is similar to those of the pyruvate dehydrogenase and α-ketoglutarate dehydrogenase complexes. The branched-chain α-keto acid dehydrogenase complex is inhibited by branched-chain acyl CoA molecules and by phosphorylation of one of its subunits. Regulation of this enzyme controls the overall degradation of the branched-chain amino acids.

The branched-chain acyl CoA molecules are oxidized by an FAD-containing acyl-CoA dehydrogenase. The electrons removed in this oxidation step are transferred via the electron-transferring flavoprotein to ubiquinone (Q).

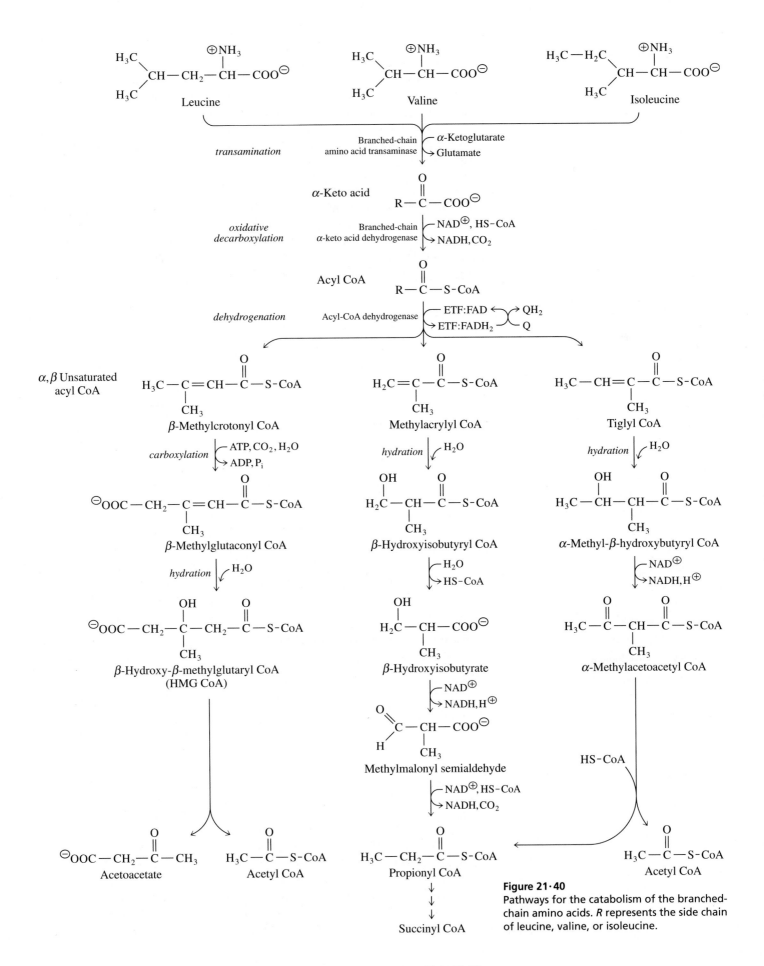

Figure 21·40
Pathways for the catabolism of the branched-chain amino acids. *R* represents the side chain of leucine, valine, or isoleucine.

At this point, the steps in the catabolism of the branched-chain amino acids diverge. In the degradation of leucine, the intermediate formed after the initial three steps is carboxylated and hydrated to form hydroxymethylglutaryl CoA (HMG CoA), an intermediate in ketone-body synthesis in liver mitochondria. All of the carbons of leucine are ultimately converted to acetyl CoA.

In the pathway for degradation of valine in mammals, the product of the first three steps is ultimately converted to propionyl CoA. As in the degradation of threonine, propionyl CoA is converted via methylmalonyl CoA to succinyl CoA, which enters the citric acid cycle.

The isoleucine degradation pathway leads to both propionyl CoA and acetyl CoA. Isoleucine is therefore both glucogenic (succinyl CoA formed from propionyl CoA) and ketogenic (acetyl CoA). In contrast, degradation of valine produces only succinyl CoA, and degradation of leucine produces only acetyl CoA. Thus, although the initial steps in the degradation of the three branched-chain amino acids are similar, their carbon skeletons have different fates.

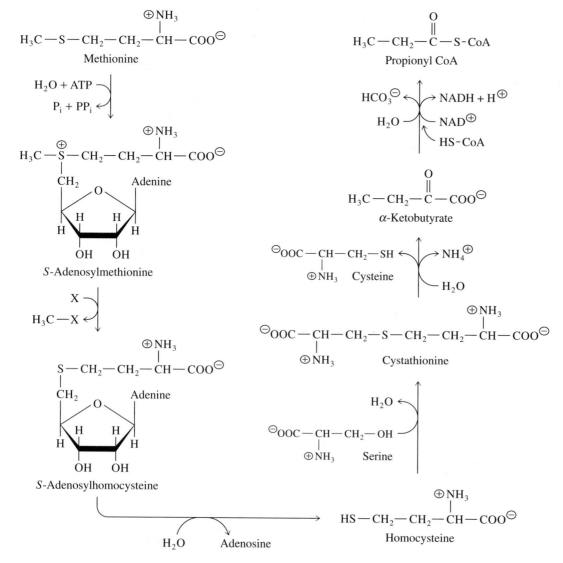

Figure 21·41
Conversion of methionine to cysteine and propionyl CoA. *X* in the second step represents any of a number of methyl-group acceptors.

F. Methionine Degradation Involves Cysteine Synthesis

One major role of methionine is conversion to the activated methyl donor *S*-adenosylmethionine (Section 8·4A). Transfer of the methyl group from *S*-adenosylmethionine to a methyl acceptor leaves *S*-adenosylhomocysteine, which is degraded by hydrolysis to homocysteine and adenosine (Figure 21·41). Homocysteine can either be methylated by 5-methyltetrahydrofolate to form methionine, or it can react with serine to form cystathionine, which can be cleaved to cysteine and α-ketobutyrate. We encountered this series of reactions earlier as part of a pathway for the formation of cysteine (Figure 21·15). By this pathway, mammals can form cysteine using a sulfur atom from the essential amino acid methionine. α-Ketobutyrate is converted to propionyl CoA by the action of an α-keto acid dehydrogenase. Propionyl CoA can be further metabolized to succinyl CoA, as explained earlier.

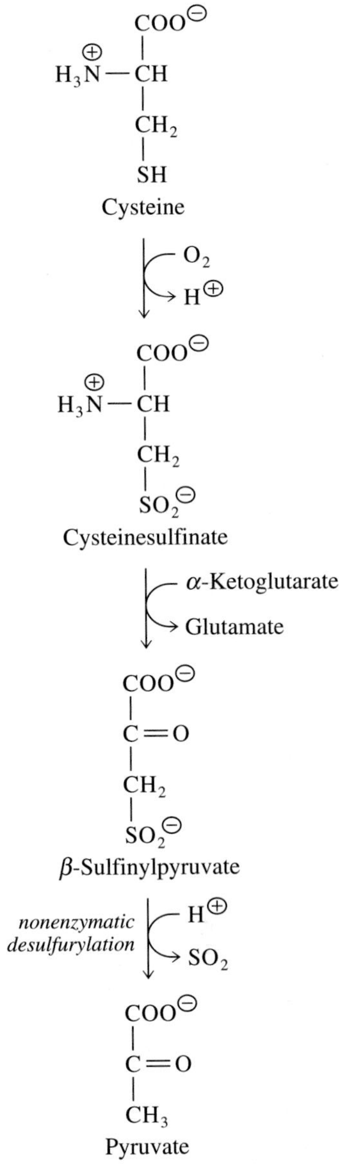

G. Cysteine Is Converted to Pyruvate

In mammals, the major route of cysteine catabolism is a three-step pathway leading to pyruvate (Figure 21·42). Cysteine is first oxidized to cysteinesulfinate, which loses its amino group by transamination to form β-sulfinylpyruvate. Nonenzymatic desulfurylation produces pyruvate. The SO_2 formed in this reaction can be oxidized to sulfate by the action of sulfite oxidase. Cysteinesulfinate is also the major precursor for the synthesis of taurine, a β-amino acid that is involved in bile-salt formation and development of the retina (Figure 21·43). Taurine may be essential in infants.

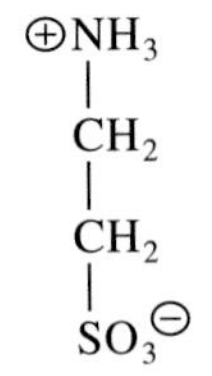

H. Phenylalanine, Tyrosine, and Tryptophan Are Catabolized via Oxidation, Deamination, and Ring-Opening Hydrolysis

The aromatic amino acids share a common pattern of catabolism. There is generally an oxidation at the earliest stage of the pathway, followed by the removal of nitrogen by transamination or hydrolysis, and then ring opening coupled with an oxidation.

The conversion of phenylalanine to tyrosine, catalyzed by phenylalanine hydroxylase (Section 21·6C), is important not only in the biosynthesis of tyrosine but also in the catabolism of phenylalanine. Catabolism of tyrosine begins with the removal of its α-amino group in a transamination reaction with α-ketoglutarate. Subsequent oxidation steps lead to ring opening and eventually to the final products, fumarate and acetoacetate (Figure 21·44, next page). Like the fumarate formed in the urea cycle, this fumarate is cytosolic and is converted to glucose. Acetoacetate is a ketone body. Thus, the degradation of tyrosine is both glucogenic and ketogenic.

The first metabolic disease to be characterized as a genetic defect was alcaptonuria, a rare disease in which one of the intermediates in the catabolism of phenylalanine and tyrosine, homogentisate, accumulates. A deficiency of homogentisate dioxygenase prevents further metabolism of this catabolite. Solutions of homogentisate turn dark upon standing because this compound is converted to a pigment; alcaptonuria was recognized by observing the darkening of urine. Individuals with alcaptonuria develop arthritis, but it is not known how the metabolic defect produces this complication.

Tryptophan, which has an indole ring system, has a more complex pathway of catabolism than phenylalanine or tyrosine in that two ring-opening reactions are required. The major route of tryptophan oxidation in liver and many microorganisms

Figure 21·42
Conversion of cysteine to pyruvate in mammals.

Figure 21·43
Structure of taurine.

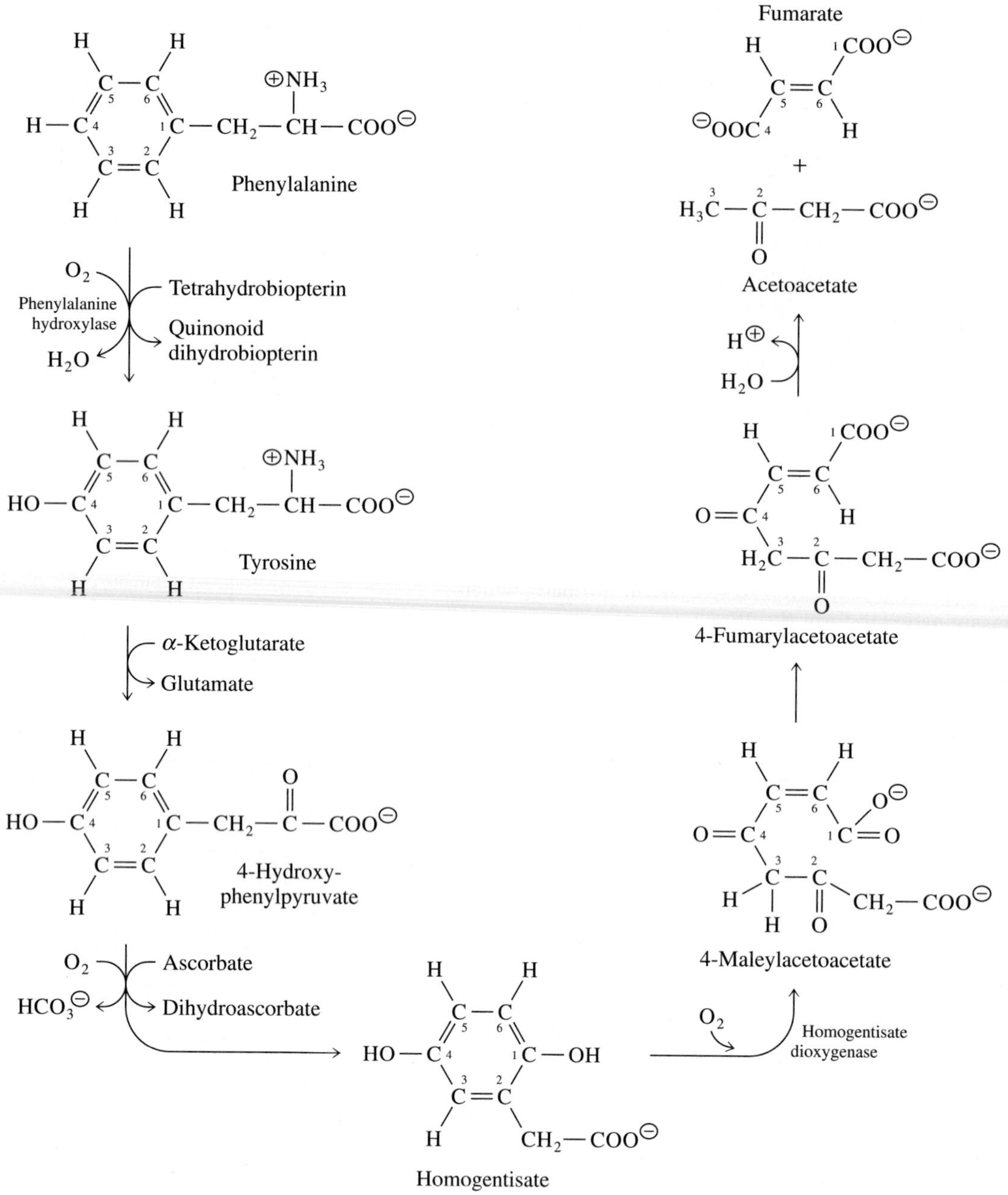

Figure 21·44
Conversion of phenyl-alanine and tyrosine to fumarate and aceto-acetate.

is the kynurenine pathway (Figure 21·45). The first step is catalyzed by tryptophan 2,3-dioxygenase. O_2 is a substrate of the reaction, and both atoms of O_2 are incorporated into the product, N-formylkynurenine. The next three steps are hydrolysis, oxidation, and a second hydrolysis, in which alanine is released. Then another oxidation occurs, in which a second dioxygenase catalyzes the opening of the benzene ring, forming 2-amino-3-carboxymuconate ε-semialdehyde. As shown in Figure 21·45, this intermediate is a branch-point metabolite. The major degradative branch consists of a number of steps leading to α-ketoadipate and ultimately to acetyl CoA. The six steps from α-ketoadipate also occur in lysine catabolism (Section 21·12I). Alanine, formed early in tryptophan catabolism, is transaminated to pyruvate, which in turn is converted to either acetyl CoA or oxaloacetate. Thus, the catabolism of tryptophan is both ketogenic and glucogenic.

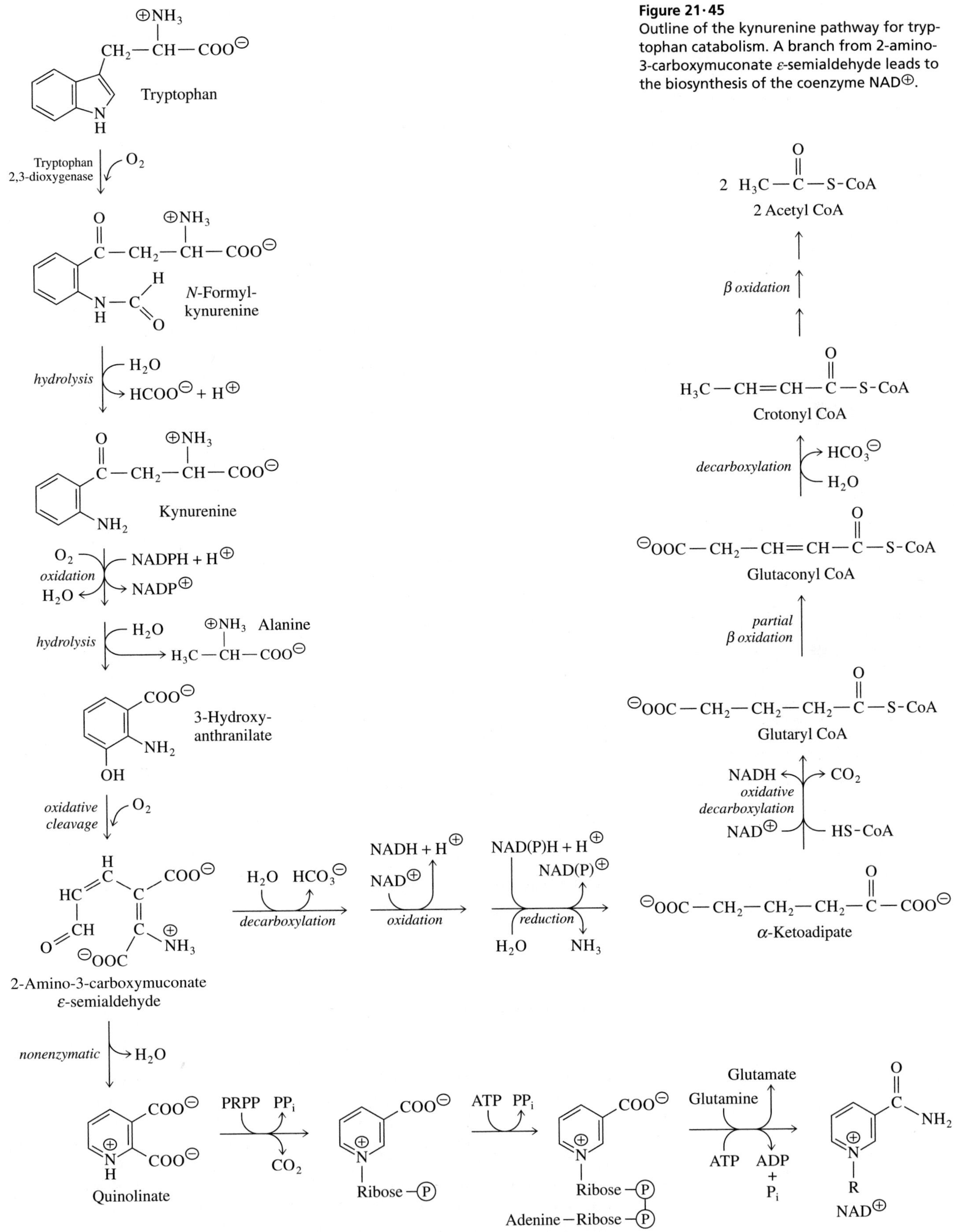

Figure 21·45
Outline of the kynurenine pathway for tryptophan catabolism. A branch from 2-amino-3-carboxymuconate ε-semialdehyde leads to the biosynthesis of the coenzyme $NAD^{\oplus}$.

Proceeding down the other branch from 2-amino-3-carboxymuconate ε-semi-aldehyde, nonenzymatic ring closure forms quinolinate, a precursor of the pyridine nucleotide coenzymes $NAD^{\oplus}$ and $NADP^{\oplus}$. Small but significant amounts of the pyridine nucleotide coenzymes can be formed by this route in humans, sparing some of the dietary requirement for niacin. There are pathological conditions associated with impaired catabolism of tryptophan in which clinical pellagra occurs, strongly suggesting that the metabolic route from tryptophan to $NAD^{\oplus}$ is important under normal conditions.

I. The Pathway for Lysine Catabolism Merges with Tryptophan Catabolism at α-Ketoadipate

The main pathway for the degradation of lysine in both mammals and bacteria involves the intermediate saccharopine (Figure 21·46). Saccharopine is formed by the reductive condensation of lysine with α-ketoglutarate. Cleavage of saccharopine releases glutamate, accomplishing the transfer of the amino group of lysine

Figure 21·46
Saccharopine pathway for lysine degradation in mammalian liver.

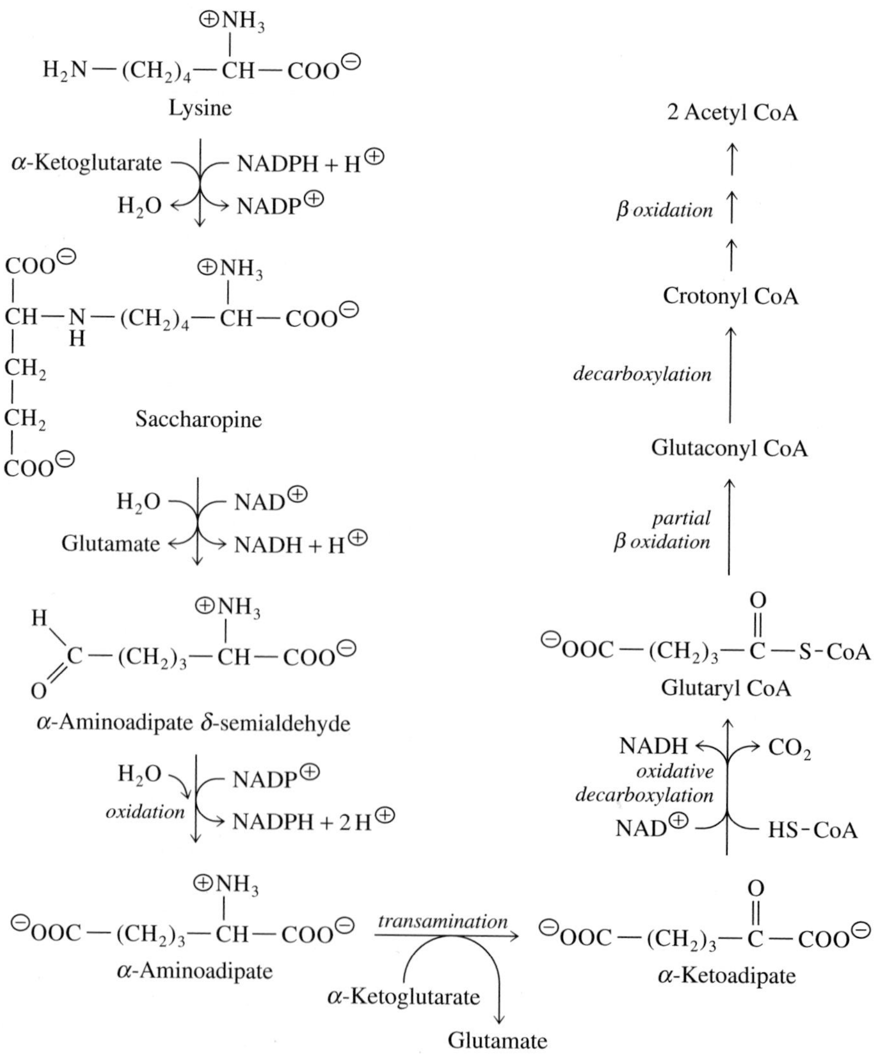

to α-ketoglutarate and generating α-aminoadipate δ-semialdehyde. The latter molecule is oxidized to α-aminoadipate, which loses its amino group by transamination to become α-ketoadipate. α-Ketoadipate is converted to acetyl CoA by the same steps shown in Figure 21·45 for the oxidation of tryptophan.

21·13 Renal Glutamine Metabolism Produces Bicarbonate for Acid-Base Regulation

The body often produces acids as metabolic end products. Acids are eliminated in the urine. One example of this is the massive production of β-hydroxybutyric acid (a ketone body) during uncontrolled diabetes mellitus. Another is the production of sulfuric acid during the catabolism of the sulfur-containing amino acids, cysteine and methionine. These acids, of course, dissociate to give protons and the corresponding anion, β-hydroxybutyrate or sulfate. Blood has an effective buffer system for dealing with the protons; they react with bicarbonate to produce CO_2, which is eliminated by the lungs, and H_2O (Figure 21·47). While this system effectively neutralizes the excess hydrogen ions, it does so at the cost of depleting blood bicarbonate. A bicarbonate repletion mechanism exists in the kidney and requires the amino acid glutamine.

In the kidney, the two nitrogens of glutamine are removed by the sequential action of glutaminase and glutamate dehydrogenase to give α-ketoglutarate$^{\ominus\ominus}$ and $2\,NH_4^{\oplus}$.

$$\text{Glutamine} \longrightarrow \longrightarrow \alpha\text{-Ketoglutarate}^{\ominus\ominus} + 2\,NH_4^{\oplus} \qquad (21·19)$$

Since α-ketoglutarate is a divalent anion, its subsequent metabolism to a neutral product (e.g., glucose) is necessarily accompanied by the production of two molecules of bicarbonate. Two molecules of glutamine are required to produce one molecule of glucose, so the overall transformation can be represented as

$$2\,\text{Glutamine} \longrightarrow \text{Glucose} + 4\,NH_4^{\oplus} + 4\,HCO_3^{\ominus} \qquad (21·20)$$

A key event in the repletion of bicarbonate by the kidney is the segregation of bicarbonate ions from ammonium ions. Complex transport processes in the renal tubules ensure that the $NH_4^{\oplus}$ is excreted in the urine while the $HCO_3^{\ominus}$ is added to the venous blood and returned to the body (Figure 21·48). In this manner, bicarbonate is generated to replace the bicarbonate lost in the initial buffering of the metabolic acid (e.g., β-hydroxybutyric acid or sulfuric acid). The excreted $NH_4^{\oplus}$ is accompanied in the urine by the anion (e.g., β-hydroxybutyrate or sulfate) of the original acid.

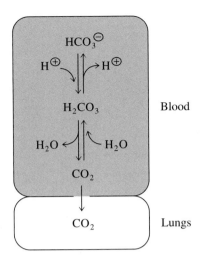

Figure 21·47
$H^{\oplus}$ buffering in blood. The $H^{\oplus}$ buffer system leads to bicarbonate loss.

Figure 21·48
Regeneration of bicarbonate via renal glutamine metabolism.

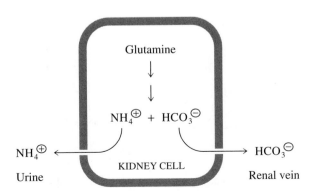

21·14 Many Other Biomolecules Are Derived from Amino Acids

The pathways of amino acid metabolism have a number of prominent synthetic roles, including the incorporation of glycine into purine nucleotides and aspartate into pyrimidine nucleotides (Chapter 22). Glycine is also a precursor of porphyrins, and arginine is a precursor of the messenger molecule nitric oxide. In addition, various amino acids are converted into amines that are hormones or neurotransmitters. A few of the many derivatives of amino acids are discussed in this section.

A. Glycine Is a Precursor of Heme

In mammals, heme prosthetic groups are synthesized primarily in bone marrow and in the liver. The first step in the synthesis of the porphyrins destined for the heme prosthetic groups of hemoglobin, myoglobin, and the cytochromes is condensation of the carbonyl group of succinyl CoA with the α-carbon of glycine to form δ-aminolevulinate (Figure 21·49). Two molecules of δ-aminolevulinate condense to form the substituted pyrrole porphobilinogen. Four molecules of porphobilinogen are converted in several steps to uroporphyrinogen, a porphyrin that is then modified by reactions that include alterations of its side chains. The insertion of Fe^{2+} to produce heme is catalyzed by the enzyme ferrochelatase.

Porphyria is a series of related metabolic diseases in which there is accumulation of porphyrin or its precursors. These diseases are associated with defects in one of several enzymes catalyzing the synthesis of heme. The symptoms of porphyria

Figure 21·49
Biosynthesis of porphobilinogen and uroporphyrinogen. R_1 represents an acetyl group, and R_2 represents a propionyl group.

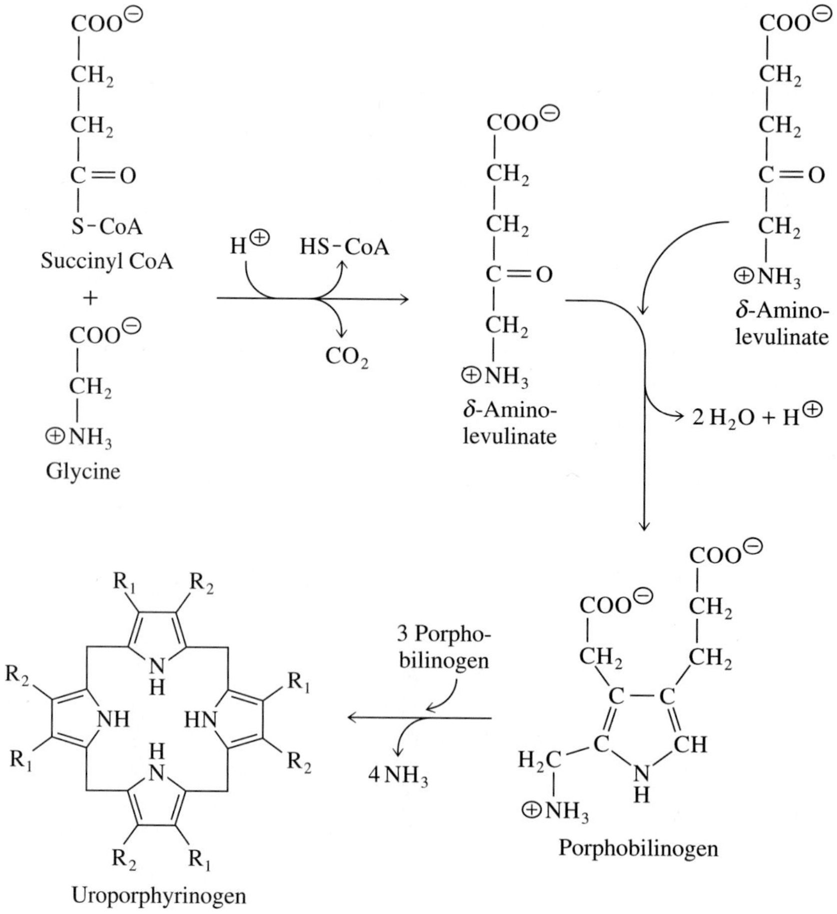

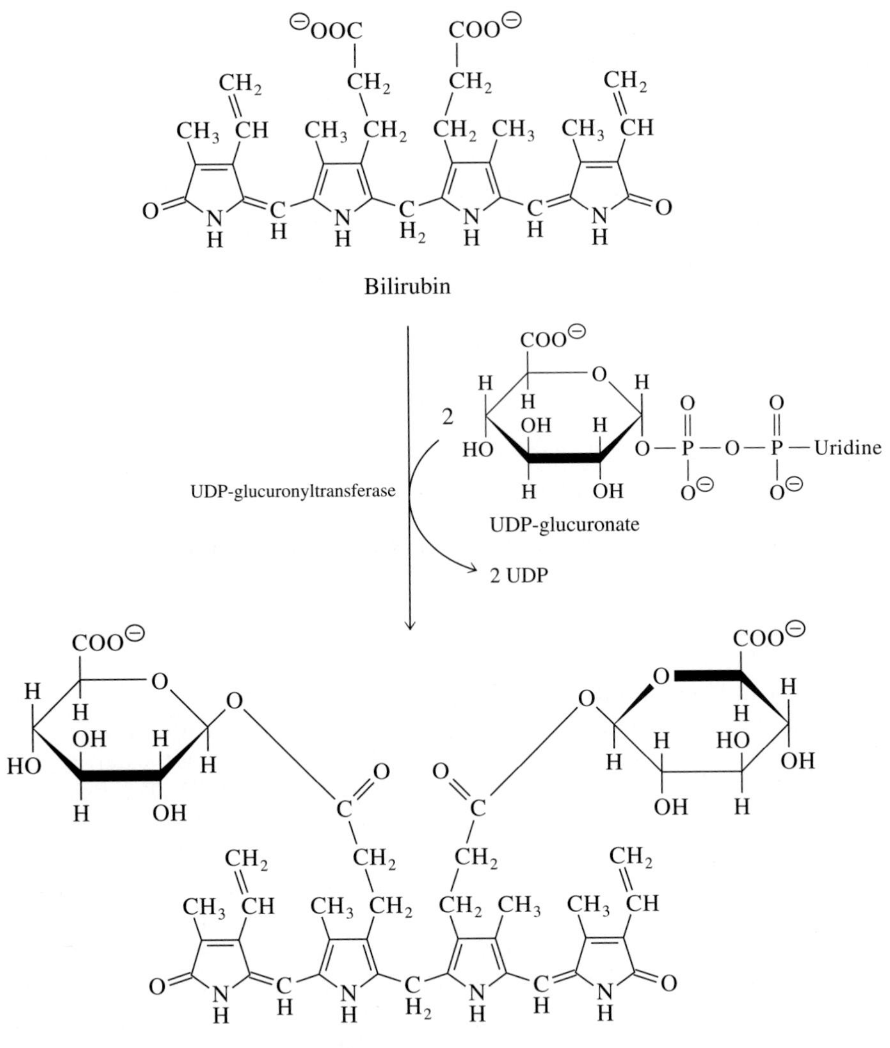

Figure 21·50
Conversion of bilirubin to bilirubin diglucuronide.

include weakness in the arms and legs, constipation and nausea, slightly rapid heart beat, severe agitation, and eventually convulsions. In one form of porphyria, the excretion of an intermediate in the biosynthetic pathway of heme turns the urine dark red. It has been suggested that King George III suffered from porphyria; rather than being mad (as often related in history books), his derangement may have been the result of acute intermittent porphyria.

Red blood cells are continuously being formed in marrow cells and degraded in the spleen. Heme from hemoglobin is catabolized to ionic iron and bilirubin, a lipophilic linear tetrapyrrole. The poorly soluble bilirubin is bound to albumin in the bloodstream for transport to the liver. In the liver, bilirubin is detoxified by addition of two glucuronate molecules donated by UDP-glucuronate (Figure 21·50) so that it becomes water soluble and can be secreted into the bile and thence to the small intestine. Jaundice—caused by liver dysfunction, obstruction of the bile duct, or excessive destruction of erythrocytes—occurs when deposits of bilirubin, which is red-orange in color, turn the skin and the whites of the eyes yellow. The livers of some infants do not have sufficiently high activities of UDP-glucuronyltransferase, the enzyme that catalyzes the formation of the soluble glucuronide of bilirubin. These infants develop jaundice.

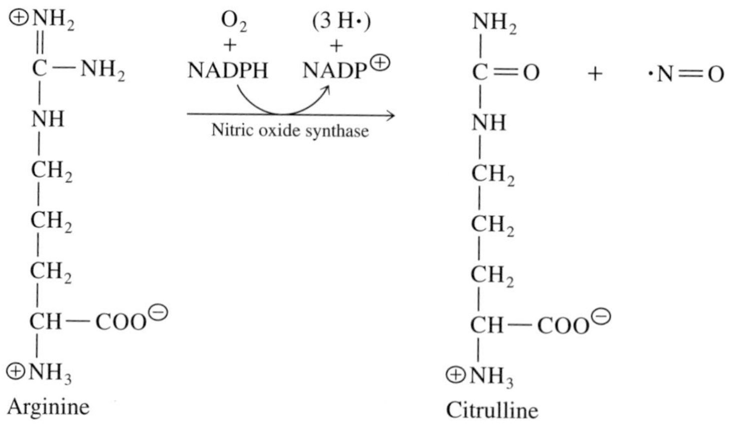

Figure 21·51
Tentative stoichiometry of the reaction catalyzed by nitric oxide synthase.

B. The Messenger Compound Nitric Oxide Is Synthesized from Arginine

Nitric oxide is an unstable gaseous derivative of nitrogen with an odd number of electrons ($\cdot N = O$). Although it is less reactive than other free-radical compounds, it can be toxic. Because nitric oxide in aqueous solution reacts rapidly with oxygen and water to form nitrates and nitrites, it exists in vivo for only a few seconds.

A cytosolic enzyme found in mammals, nitric oxide synthase, catalyzes the formation of nitric oxide and citrulline from arginine (Figure 21·51). This oxidation of arginine requires the cofactors NADPH, tetrahydrobiopterin, FMN, FAD, and a cytochrome. Nitric oxide formed by this reaction has been implicated as a messenger molecule involved in several quite different physiological functions, including resistance to bacterial infection, dilation of blood vessels, and the action of the neurotransmitter glutamate. Nitric oxide synthase is present in two forms, a constitutive calcium-dependent form present in endothelial cells and an inducible calcium-independent form found in macrophages.

When macrophages (a type of white blood cell) are exposed to certain stimulants such as bacterial lipopolysaccharides, they synthesize nitric oxide. The short-lived nitric oxide free radical is one of the weapons used by macrophages to kill bacteria and tumor cells. Nitric oxide may interact with superoxide anions ($\cdot O_2^{\ominus}$) to form more toxic reactants that account for the cell-killing activity.

Blood vessels are tubes of smooth muscle with a thin inner layer of endothelial cells. The endothelial cells contain nitric oxide synthase and cell-surface receptors for binding circulating molecules such as acetylcholine that stimulate dilation of blood vessels. Binding of acetylcholine to the receptors triggers synthesis of nitric oxide, which diffuses to the muscle layer of the blood vessel and causes it to relax. Nitric oxide activates the enzyme guanylate cyclase, which catalyzes the conversion of GTP to cyclic GMP (cGMP) in the muscle cell. cGMP stimulates a protein kinase that causes the muscle to relax by an unknown mechanism. Nitroglycerin, used to dilate coronary arteries in the treatment of angina pectoris, exerts its effect by virtue of its metabolic conversion to nitric oxide.

Nitric oxide also functions in brain tissue. The amino acid glutamate is a neurotransmitter, exciting neurons by binding to specific receptors and causing the opening of channels that allow $Ca^{2\oplus}$ to enter the neuronal cells. The nitric oxide synthase in these cells is rapidly activated by the influx of calcium ions, which bind to molecules of the regulatory protein calmodulin (Figure 8·2) associated with the synthase. Although the exact role of nitric oxide in the brain is not known, it has been shown that it elicits an increase in the concentration of the second messenger cGMP. A sidelight to this research is the finding that nitric oxide formed in the brain of laboratory animals is involved in neural damage during strokes. Release of

excess glutamate during stroke causes formation of abnormally high amounts of nitric oxide that appear to kill some neurons in the same way macrophages kill bacterial cells. Administering an inhibitor of nitric oxide synthase to an animal produces
some protection from stroke damage. The roles of nitric oxide and cGMP in the
brain are being thoroughly examined, and an active search for other simple messenger molecules, such as CO, that may function in ways similar to nitric oxide is
underway.

C. Other Amines Are Formed from Amino Acids by Decarboxylation

Several neurotransmitters and hormones are synthesized from amino acids by short
pathways that involve pyridoxal phosphate–dependent decarboxylation reactions.
In the brain, glutamate is converted to γ-aminobutyrate, an inhibitory neurotransmitter (Figure 21·52a). Histamine (Figure 21·52b), formed from histidine in various animal tissues, controls the constriction of certain blood vessels and also the secretion of hydrochloric acid by the stomach (Chapter 13). There are two forms of
histamine receptors: H1 receptors in smooth muscle and capillaries, which are responsible for blood vessel constriction, and H2 receptors, which are involved in
gastric acid secretion. Conditions that result in the release of excess histamine, such
as asthma and ulcers, can be treated using histamine-receptor–specific inhibitors.

The formation of biological amines from both tryptophan and tyrosine begins
with a tetrahydrobiopterin-dependent hydroxylation reaction, mechanistically similar to the reaction catalyzed by phenylalanine hydroxylase (Section 21·6C). Hydroxylation is followed by decarboxylation. The amine formed from tryptophan is
serotonin (Figure 21·52c), which functions not only as a neurotransmitter but also
as a hormone for some nonneural tissues. Tyrosine is converted to dopamine (Figure 21·52d), which can be further metabolized in the adrenal medulla to the hormone epinephrine (Figure 21·52e). In the brain, this pathway stops at dopamine.
Parkinson's disease, which is associated with a decrease in the production of
dopamine, has been treated by administration of 3,4-dihydroxyphenylalanine, the
intermediate in the formation of dopamine from tyrosine (dopamine itself cannot
enter the brain).

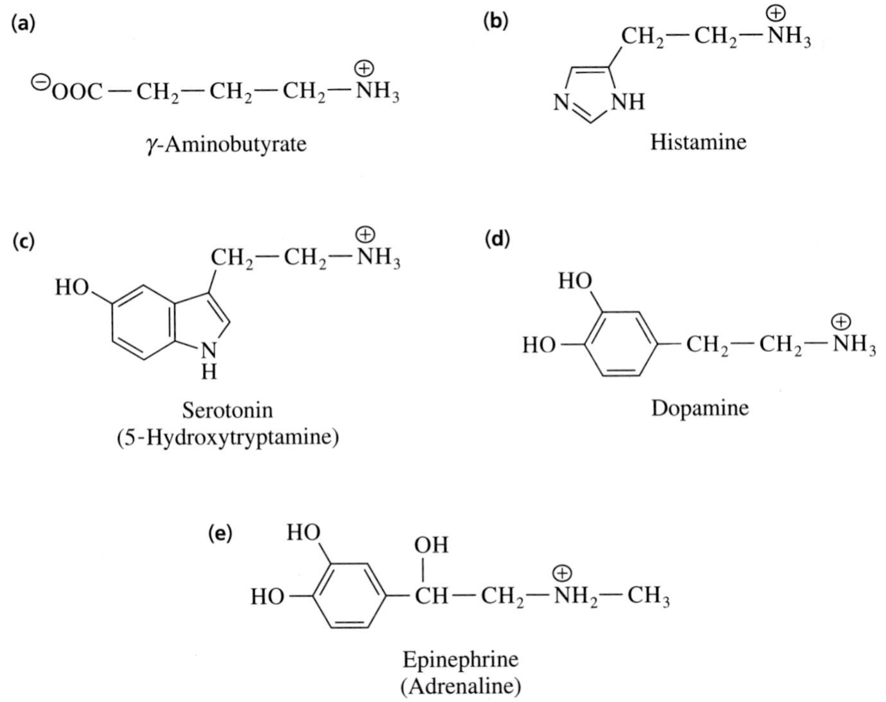

Figure 21·52
Amines formed by decarboxylation of amino acids.

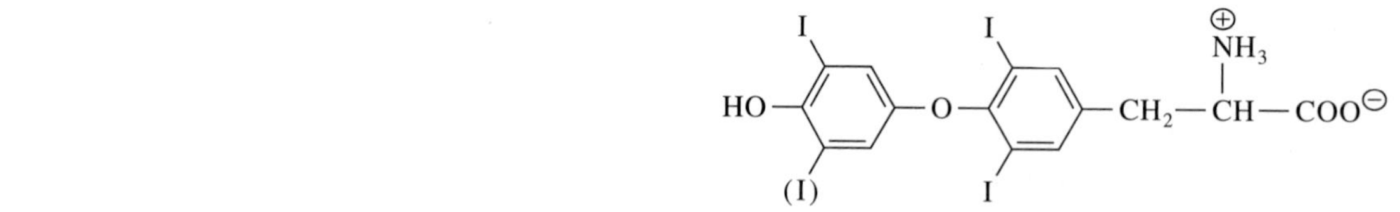

Figure 21·53
Triiodothyronine (T$_3$) and thyroxine (T$_4$). Thyroxine contains one more atom of iodine (in parentheses) than triiodothyronine.

D. The Thyroid Hormones Are Formed from Iodinated Tyrosine Residues of Thyroglobin

Two closely related hormones, triiodothyronine (T$_3$) and thyroxine (T$_4$), are synthesized in the thyroid gland (Figure 21·53). Although tyrosine is a precursor of these hormones, as expected from their structures, it is the tyrosine residues of a specific prohormone protein—not free molecules of tyrosine—that are involved in the synthesis. The precursor protein in the thyroid gland is thyroglobin, whose tyrosine residues are enzymatically iodinated posttranslationally. Pairs of iodinated residues adjacent in the three-dimensional structure of a special region of thyroglobin condense to form T$_3$ and T$_4$ residues, which are released as needed by the action of a lysosomal protease.

The thyroid hormones stimulate metabolism. Individuals with low thyroid hormone levels are lethargic and have cold skin; those with an excess are active and have warm skin. Small amounts of sodium iodide are commonly added to table salt to prevent goiter, a condition of hypothyroidism caused by a lack of iodide in the diet.

E. Several Polyamines Are Derived from Ornithine

Three amino acid derivatives, the naturally occurring polyamines putrescine, spermidine, and spermine (Figure 21·54), are required for the optimal growth of all cells. Because they are strongly cationic, it has been suggested that they interact with polyanions such as nucleic acids. However, the mechanism of their action is not yet known. Putrescine is formed by decarboxylation of ornithine. An aminopropyl group can be transferred to putrescine to form spermidine, which can accept a second aminopropyl group to form spermine. The donor of these two three-carbon fragments is S-adenosylthiopropylamine, the product of the decarboxylation of S-adenosylmethionine (Figure 8·8).

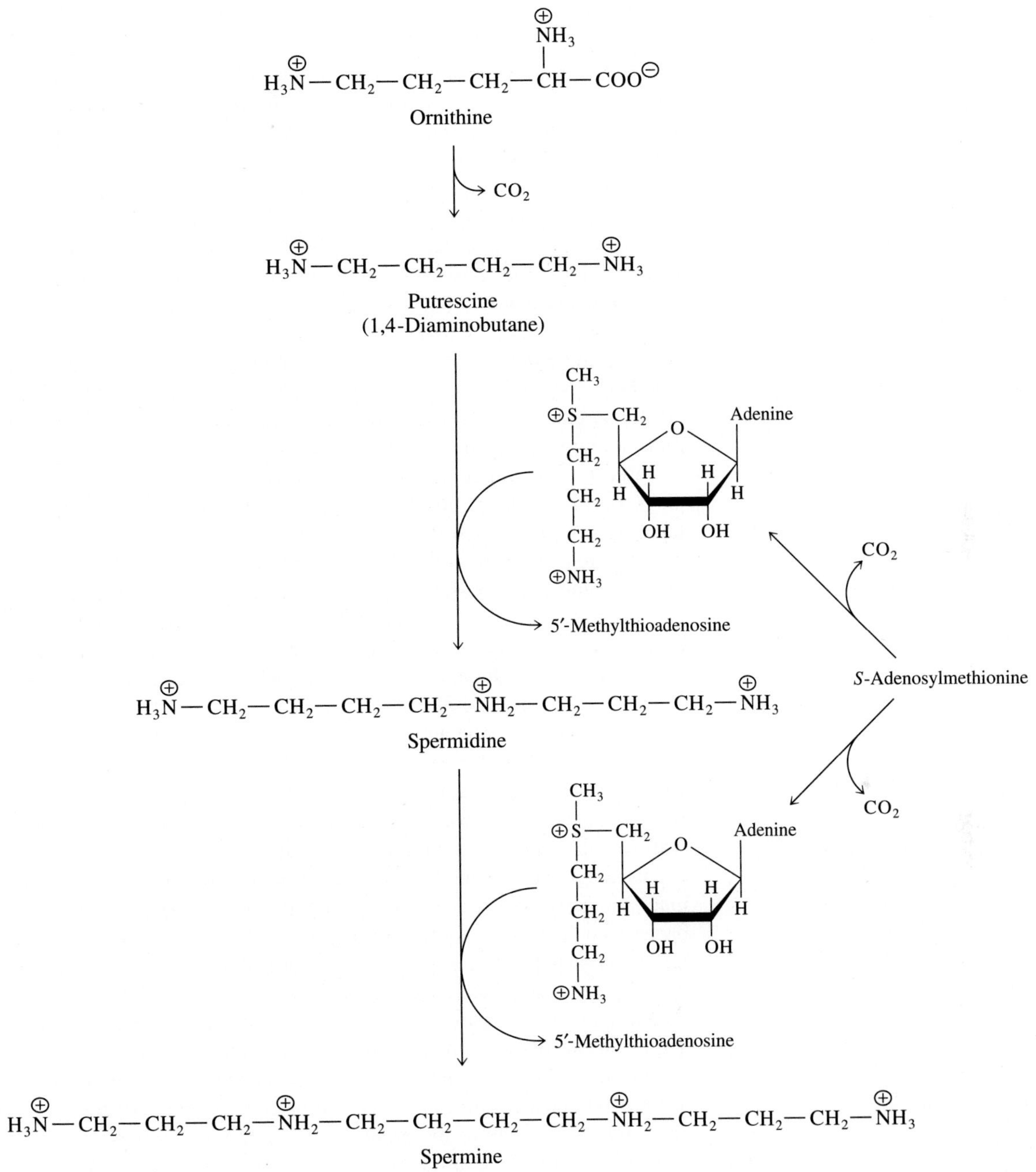

$$\overset{\oplus}{H_3N}-CH_2-CH_2-CH_2-\overset{\overset{\displaystyle\overset{\oplus}{NH_3}}{|}}{CH}-COO^{\ominus}$$

Ornithine

↓ → CO_2

$$\overset{\oplus}{H_3N}-CH_2-CH_2-CH_2-CH_2-\overset{\oplus}{NH_3}$$

Putrescine
(1,4-Diaminobutane)

5′-Methylthioadenosine

S-Adenosylmethionine

→ CO_2

$$\overset{\oplus}{H_3N}-CH_2-CH_2-CH_2-CH_2-\overset{\oplus}{NH_2}-CH_2-CH_2-CH_2-\overset{\oplus}{NH_3}$$

Spermidine

→ CO_2

5′-Methylthioadenosine

$$\overset{\oplus}{H_3N}-CH_2-CH_2-CH_2-\overset{\oplus}{NH_2}-CH_2-CH_2-CH_2-CH_2-\overset{\oplus}{NH_2}-CH_2-CH_2-CH_2-\overset{\oplus}{NH_3}$$

Spermine

Figure 21·54
Formation of putrescine, spermidine, and spermine.

Summary

Amino acid metabolism can be approached from two points of view: the origins and fates of the amino acid nitrogen atoms, and the origins and fates of their carbon skeletons. Nitrogen is introduced into biological systems by reduction of chemically unreactive N_2 from the atmosphere to ammonia. This reaction, catalyzed by nitrogenase, is carried out by a few species of bacteria and algae. An alternate route for the assimilation of nitrogen, carried out by plants and microorganisms, is the reduction of nitrate to nitrite, catalyzed by nitrate reductase. Nitrite is reduced by the action of nitrite reductase to form ammonia, which can be used by all organisms.

Ammonia is assimilated into metabolites by several routes. Glutamate dehydrogenase catalyzes the reductive amination of α-ketoglutarate in a reversible reaction forming glutamate, which can then enter the central pathways of metabolism. Another important carrier of nitrogen is glutamine, which can be formed from glutamate and ammonia by the action of glutamine synthetase. Glutamate and glutamine are nitrogen donors in many reactions.

Because glutamine is a precursor of many metabolites, glutamine synthetase plays a critical role in nitrogen metabolism. *E. coli* glutamine synthetase is regulated by a sophisticated mechanism that includes feedback inhibition, covalent modification, and regulation at the level of protein synthesis.

The amino group of glutamate can be transferred to many α-keto acids in reversible transamination reactions that form α-ketoglutarate and the corresponding α-amino acids. In the reverse direction, transamination reactions convert a number of amino acids to glutamate or aspartate, which contribute amino groups to pathways for nitrogen disposal.

Nonessential amino acids are those that an organism can produce in sufficient quantity for growth. Essential amino acids are those that must be supplied in the diet. The nonessential amino acids are generally those formed by short, energetically inexpensive pathways. Of the nonessential amino acids in mammals, glutamate, alanine, and aspartate are formed by simple transamination, and glutamine and asparagine are formed by transfer of amide groups to the side chains of glutamate and aspartate. Serine and glycine are derived from 3-phosphoglycerate. In bacteria and plants, cysteine is formed from serine; in mammals, it is formed from the breakdown of methionine. Proline is formed from glutamate. Tyrosine is formed in a single reaction from the essential amino acid phenylalanine. In mammals, arginine is synthesized from glutamate γ-semialdehyde in a pathway that involves reactions in two tissues—the intestine and the kidney.

Of the amino acids that are essential in the diets of mammals, lysine, threonine, and methionine are synthesized in bacteria from aspartate. Pathways sharing certain enzymatic steps lead to the branched-chain amino acids isoleucine, valine, and leucine. The aromatic amino acids arise from a pathway in which chorismate is formed in seven steps from phosphoenolpyruvate and erythrose 4-phosphate, followed by conversion of chorismate to phenylalanine, tyrosine, or tryptophan. In plants, simpler eukaryotes, and bacteria, histidine is formed from phosphoribosyl pyrophosphate, ATP, and glutamine.

Protein molecules in all living cells are continuously being synthesized and degraded. The rate of turnover differs for each protein. In eukaryotes, intracellular proteins are degraded by nonselective hydrolysis in lysosomes and by selective ATP-dependent and ATP-independent hydrolysis in the cytosol. A major pathway for ATP-dependent hydrolysis requires that the target proteins be covalently attached to the protein ubiquitin. Proteins are selected for ubiquitination and subsequent degradation by a complex regulatory system that is not fully understood.

Catabolism of amino acids often begins with deamination, followed by modification of the remaining carbon chains for entry into the central pathways of carbon metabolism. Glutamate or aspartate is the amino-group acceptor at the end of one or a series of transamination reactions. In mammals, nitrogen is disposed of via urea formed by the urea cycle in the liver. Urea is then transported to the kidneys and excreted. The carbon atom of urea is derived from bicarbonate. One amino group is derived from ammonia, and the other, from the α-amino group of aspartate. Fumarate is released in the cytosol as a product of the urea cycle and enters the pathway of gluconeogenesis. Aspartate needed for continued operation of the urea cycle arises from the catabolism of other amino acids.

The pathways for degradation of amino acids lead to pyruvate, acetyl CoA, or intermediates of the citric acid cycle. Amino acids that are degraded to citric acid cycle intermediates can supply the pathway of gluconeogenesis and are called glucogenic. Those that form acetyl CoA can contribute to the formation of fatty acids or ketone bodies and are called ketogenic. Amino acids degraded to pyruvate can be metabolized to either oxaloacetate or acetyl CoA and so can be either glucogenic or ketogenic.

Alanine, aspartate, and glutamate are degraded by reversal of the transamination reactions by which they were formed. Degradation of glutamine and asparagine begins with their hydrolysis to glutamate and aspartate, respectively. Proline, arginine, and histidine are degraded to glutamate. Serine is converted to glycine, which is catabolized by the multienzyme glycine-cleavage system. Threonine can be converted by alternate routes to succinyl CoA or glycine and acetyl CoA. The branched-chain amino acids are degraded by pathways that share common steps leading to different fates; leucine is degraded to acetyl CoA, valine to succinyl CoA, and isoleucine to succinyl CoA and acetyl CoA. Methionine is degraded to succinyl CoA, with formation of cysteine as a by-product. Cysteine is degraded to pyruvate. Of the aromatic amino acids, phenylalanine is converted to tyrosine, which is degraded to fumarate and acetoacetate. Tryptophan is degraded to acetyl CoA, with formation of alanine as a by-product; alanine is converted to pyruvate and then to either oxaloacetate or acetyl CoA. Lysine catabolism merges with the pathway of tryptophan catabolism, leading to acetyl CoA.

The metabolism of glutamine in the kidney provides bicarbonate to buffer acids formed in the catabolism of some ketone bodies and amino acids.

Amino acids are converted to a number of important biomolecules including heme prosthetic groups, nitric oxide, and some hormones and neurotransmitters.

Selected Readings

General Reference

Bender, D. A. (1985). *Amino Acid Metabolism,* 2nd ed. (Chichester, England: John Wiley & Sons). A monograph devoted to the pathways of amino acid metabolism, with emphasis on mammalian systems.

Nitrogen Fixation

Burris, R. H. (1991). Nitrogenases. *J. Biol. Chem.* 266:9339–9342. A brief, clear review.

Stacey, G., Burris, R. H., and Evans, H. J., eds. (1992). *Biological Nitrogen Fixation* (New York: Chapman and Hall). A large, thorough, advanced survey, including detailed reviews of nitrogenase structure and function.

Individual Amino Acids

Adams, E., and Frank, L. (1980). Metabolism of proline and the hydroxyprolines. *Annu. Rev. Biochem.* 49:1005–1061.

Cooper, A. J. L. (1983). Biochemistry of sulfur-containing amino acids. *Annu. Rev. Biochem.* 52:187–222.

Dhanakoti, S. N., Brosnan, J. T., Herzberg, G. R., and Brosnan, M. E. (1990). Renal arginine synthesis: studies in vitro and in vivo. *Am. J. Physiol.* 259:E437–E442. Explains how citrulline from the intestine is converted to arginine in the rat kidney cortex.

Harper, A. E., Miller, R. H., and Block, K. P. (1984). Branched-chain amino acid metabolism. *Annu. Rev. Nutr.* 4:409–454.

Häussinger, D. (1990). Nitrogen metabolism in liver: structural and functional organization and physiological relevance. *Biochem. J.* 267:281–290. Describes how different liver cells have different pathways for nitrogen metabolism.

Herrmann, K. M., and Somerville, R. L., eds. (1983). *Amino Acids: Biosynthesis and Genetic Regulation* (Reading, Pennsylvania: Addison-Wesley Publishing Company). A collection of review articles on amino acid biosynthesis, mainly in prokaryotes.

Hyde, C. C., Ahmed, S. A., Padlan, E. A., Miles, E. W., and Davies, D. R. (1988). Three-dimensional structure of the tryptophan synthase $\alpha_2\beta_2$ multienzyme complex from *Salmonella typhimurium. J. Biol. Chem.* 263:17 857–17 871. Includes evidence from X-ray crystallography of a tunnel from one active site to another.

Kaufman, S. (1987). Aromatic amino acid hydroxylases. In *The Enzymes: Control by Phosphorylation, Part B,* Vol. 18, 3rd ed., P. D. Boyer and E. G. Krebs, eds. (New York: Academic Press), pp. 217–282. A description of the hydroxylation of phenylalanine, tyrosine, and tryptophan, with emphasis on regulation.

Kvamme, E., ed. (1988). *Glutamine and Glutamate in Mammals,* Vol. 1. (Boca Raton, Florida: CRC Press).

Stadtman, E. R., and Ginsburg, A. (1974). The glutamine synthetase of *Escherichia coli:* structure and control. In *The Enzymes,* Vol. 10, 3rd ed., P. D. Boyer, ed. (New York: Academic Press), pp. 755–807.

Umbarger, H. E. (1978). Amino acid biosynthesis and its regulation. *Annu. Rev. Biochem.* 47:533–606.

Pyridoxal Phosphate

Hayashi, H., Wada, H., Yoshimura, T., Esaki, N., and Soda, K. (1990). Recent topics in pyridoxal 5′-phosphate enzyme studies. *Annu. Rev. Biochem.* 59:87–110.

Metabolic Diseases

Macalpine, I., and Hunter, R. (1969). Porphyria and King George III. *Sci. Am.* 221(1):38–46. Describes porphyria and suggests that King George III of England was not insane but suffered from porphyria.

Scriver, C. R., Beaudet, A. L., Sly, W. S., and Valle, D., eds. (1989). *The Metabolic Basis of Inherited Disease,* Vols. 1 and 2. (New York: McGraw-Hill). Up-to-date descriptions of over 100 metabolic diseases, including many involving enzymes associated with amino acids, explained through molecular genetics.

Woo, S. L. C. (1989). Molecular basis and population genetics of phenylketonuria. *Biochemistry* 28:1–7.

Intracellular Proteases

Bond, J. S., and Butler, P. E. (1987). Intracellular proteases. *Annu. Rev. Biochem.* 56:333–364.

Hershko, A., and Ciechanover, A. (1992). The ubiquitin system for protein degradation. *Annu. Rev. Biochem.* 61:761–807.

Urea Cycle

Meijer, A. J., Lamers, W. H., and Chamuleau, R. A. F. M. (1990). Nitrogen metabolism and ornithine cycle function. *Physiol. Rev.* 70:701–748.

Nitric Oxide

Snyder, S. H. (1992). Nitric oxide: first in a new class of neurotransmitters? *Science* 257:494–496.

Snyder, S. H., and Bredt, D. S. (1992). Biological roles of nitric oxide. *Sci. Am.* 266(5):68–77.

22

Nucleotide Metabolism

We have seen a variety of roles for nucleotides: as coenzymes, energy transducers, and regulatory compounds. The biosynthesis of DNA and RNA from nucleotide monomers will be examined in Chapters 25 and 27. Virtually all organisms and tissues have the ability to synthesize purine and pyrimidine nucleotides, reflecting the importance of these molecules. De novo synthesis of the ring structures of the heterocyclic bases consumes a considerable amount of ATP. To conserve energy, synthesis of nucleotides is closely regulated to match the needs of cell division and other uses. ATP itself, used in so many reactions, is the most abundant nucleotide. Because nucleotide biosynthesis is closely linked to cell division, its study has been of particular importance to modern medicine; a number of synthetic agents that inhibit nucleotide synthesis are useful as therapeutic agents against cancer.

In this chapter, we will first describe the biosynthesis of purine nucleotides from simpler precursors. We will then discuss simple pathways by which intact purines obtained from the breakdown of nucleic acids within cells or from the external environment (such as from the intestinal tract during the digestion of food) can be incorporated directly into nucleotides, a process called salvage. The ability to catalyze formation of purine nucleotides by the de novo pathway provides a measure of independence from dietary sources, and the salvage pathway conserves energy by recycling the products of nucleic acid breakdown.

We will then describe the pathway for the de novo biosynthesis of pyrimidine nucleotides, a reaction sequence that differs in chemical strategy from that of the purine nucleotides, and we will see that pyrimidines too are salvaged. Next, we will present the conversion of the purine and pyrimidine ribonucleotides to their 2′-deoxy forms, the form in which they are incorporated into DNA.

Finally, we will examine the biological degradation of nucleotides. Breakdown of purines leads to the formation of potentially toxic compounds that are excreted, whereas breakdown of pyrimidines leads to readily metabolized products.

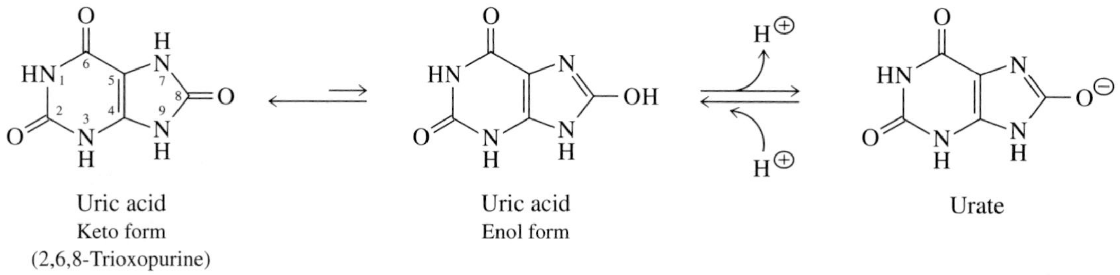

Figure 22·1
Uric acid (2,6,8-trioxopurine). The keto tautomer of uric acid is more stable than the enol tautomer. The enol form can ionize to form urate.

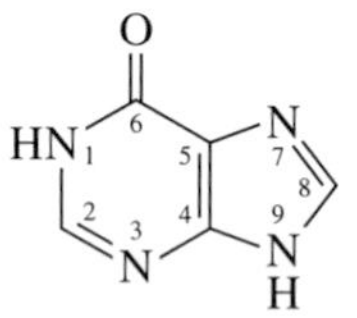

Figure 22·2
Hypoxanthine (6-oxopurine).

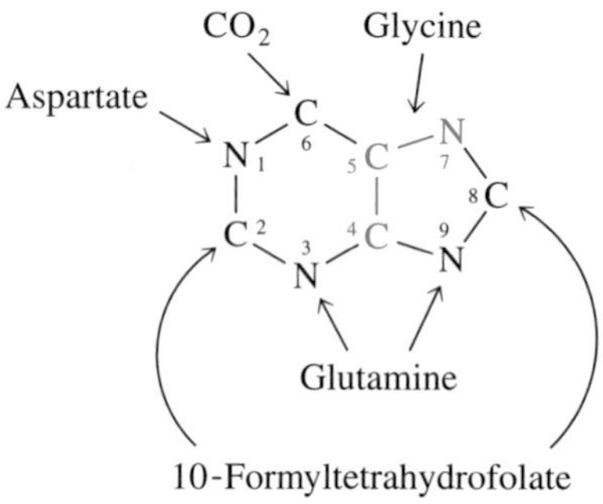

Figure 22·3
Sources of the ring atoms in purines synthesized de novo. Note that the nitrogen atom and both carbon atoms of a glycine molecule are incorporated into positions 4, 5, and 7 of the purine; the amide nitrogen atoms of two glutamine molecules provide N-3 and N-9; C-6 arises from CO_2; aspartate contributes its amino nitrogen to position 1; and C-2 and C-8 originate from formate via 10-formyltetrahydrofolate.

22·1 Early Investigations of Purine Biosynthesis Involved Isotopic Labelling

Uric acid (Figure 22·1) was the first purine discovered. As early as 1776, this compound was found by both Karl W. Scheele and Torbern Bergmann in human urine and bladder stones. A century later, it was shown that most of the nitrogen ingested by adult hens is excreted as uric acid. Chemists of the 19th century established that in birds and many reptiles the major end product of the metabolism of nitrogen is uric acid, analogous to urea as the major end product of nitrogen metabolism in mammals.

In 1936, H. A. Krebs and his colleagues performed experiments that paved the way for the complete elucidation of the pathway leading to purine nucleotides. They showed that in pigeons the incorporation of ammonia (NH_3) into uric acid occurs in two stages. Ammonia is first incorporated into hypoxanthine in the liver. Next, hypoxanthine is oxidized in the kidney to uric acid in reactions catalyzed by xanthine dehydrogenase. The structure of hypoxanthine is shown in Figure 22·2. Until the 1950s, many biochemists believed that the purine bases of nucleic acids arose by a route different from the pathway in birds that leads to the uric acid formed for nitrogen excretion. However, experiments with isotopically labelled precursor molecules demonstrated that nucleic acid purines and uric acid arise from the same precursors and reaction sequence. Homogenates of pigeon liver—a tissue in which purines are actively synthesized—were then used extensively as a convenient source of enzymes for studying the steps in purine biosynthesis. The pathway in avian liver has proven a reliable model for the pathway in other organisms.

Isotopes of carbon and nitrogen became available as tracers for metabolic studies in the 1940s. The stable isotopes ^{13}C and ^{15}N were used at first, and the radioactive ^{14}C a few years later. Simple compounds such as $^{13}CO_2$, $H^{13}COO^{\ominus}$ (formate), and $^{\oplus}H_3N-CH_2-^{13}COO^{\ominus}$ (glycine) were administered to pigeons and rats by John M. Buchanan and his colleagues. They then isolated and chemically degraded excreted uric acid using a procedure that allowed labelled carbon and nitrogen atoms to be traced to their positions in uric acid. They observed that the carbon from carbon dioxide was incorporated into C-6 and carbon from formate into C-2 and C-8 of purines. Ultimately, the sources of the ring atoms were identified as N-1, aspartate; C-2 and C-8, formate via 10-formyltetrahydrofolate (Section 8·12); N-3 and N-9, amide groups from glutamine; C-4, C-5, and N-7, glycine; and C-6, carbon dioxide. These findings are summarized in Figure 22·3.

22·2 Phosphoribosyl Pyrophosphate (PRPP) Is Required for Nucleotide Biosynthesis

G. Robert Greenberg demonstrated that the purine ring structure of hypoxanthine is not synthesized as a free base but as a substituent of ribose phosphate. The end product of the biosynthetic pathway is not hypoxanthine but its 5′-ribonucleotide, inosine 5′-monophosphate (IMP, or inosinate), shown in Figure 22·4. The hypoxanthine detected in avian liver by Krebs and his colleagues was formed by the action of degradative enzymes that catalyze removal of sugar and phosphate groups from ribonucleotides. The de novo pathway of purine synthesis involves a series of sugar-phosphate intermediates with incomplete purine skeletons attached.

The source of ribose 5-phosphate for purine biosynthesis was identified by Buchanan and his colleagues as 5-phosphoribosyl 1-pyrophosphate (PRPP), a compound whose structure was elucidated in the laboratory of Arthur Kornberg in 1954. PRPP is synthesized from ribose 5-phosphate and ATP in a reaction catalyzed by PRPP synthase (Figure 22·5); PRPP then donates its ribose 5-phosphate to serve as the foundation upon which the purine structure is built. PRPP is also a precursor of the pyrimidine nucleotides, although in that pathway it reacts with a preformed pyrimidine to form a nucleotide. In addition, PRPP is utilized in the salvage pathways and in the biosynthesis of histidine and tryptophan.

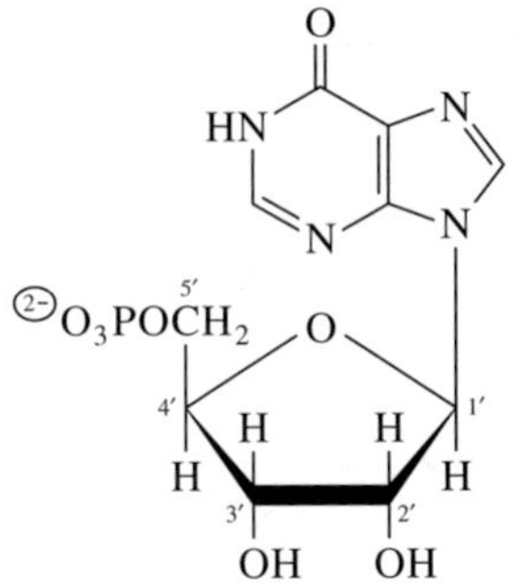

Figure 22·4
Inosine 5′-monophosphate (IMP, or inosinate).

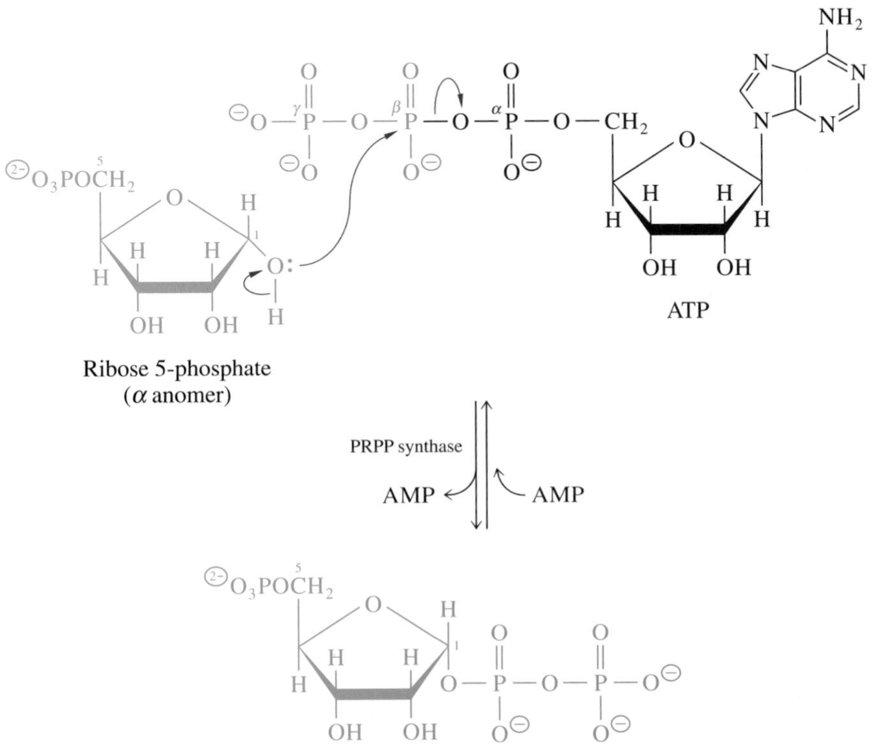

Figure 22·5
Synthesis of 5-phosphoribosyl 1-pyrophosphate (PRPP) from ribose 5-phosphate and ATP. PRPP synthase catalyzes the transfer of a pyrophosphoryl group from ATP to the 1-hydroxyl group of ribose 5-phosphate.

Figure 22·6
The 10-step pathway for the de novo biosynthesis of IMP. *R5′P* stands for ribose 5′-phosphate. The atoms are numbered according to their positions in the completed purine ring structure.

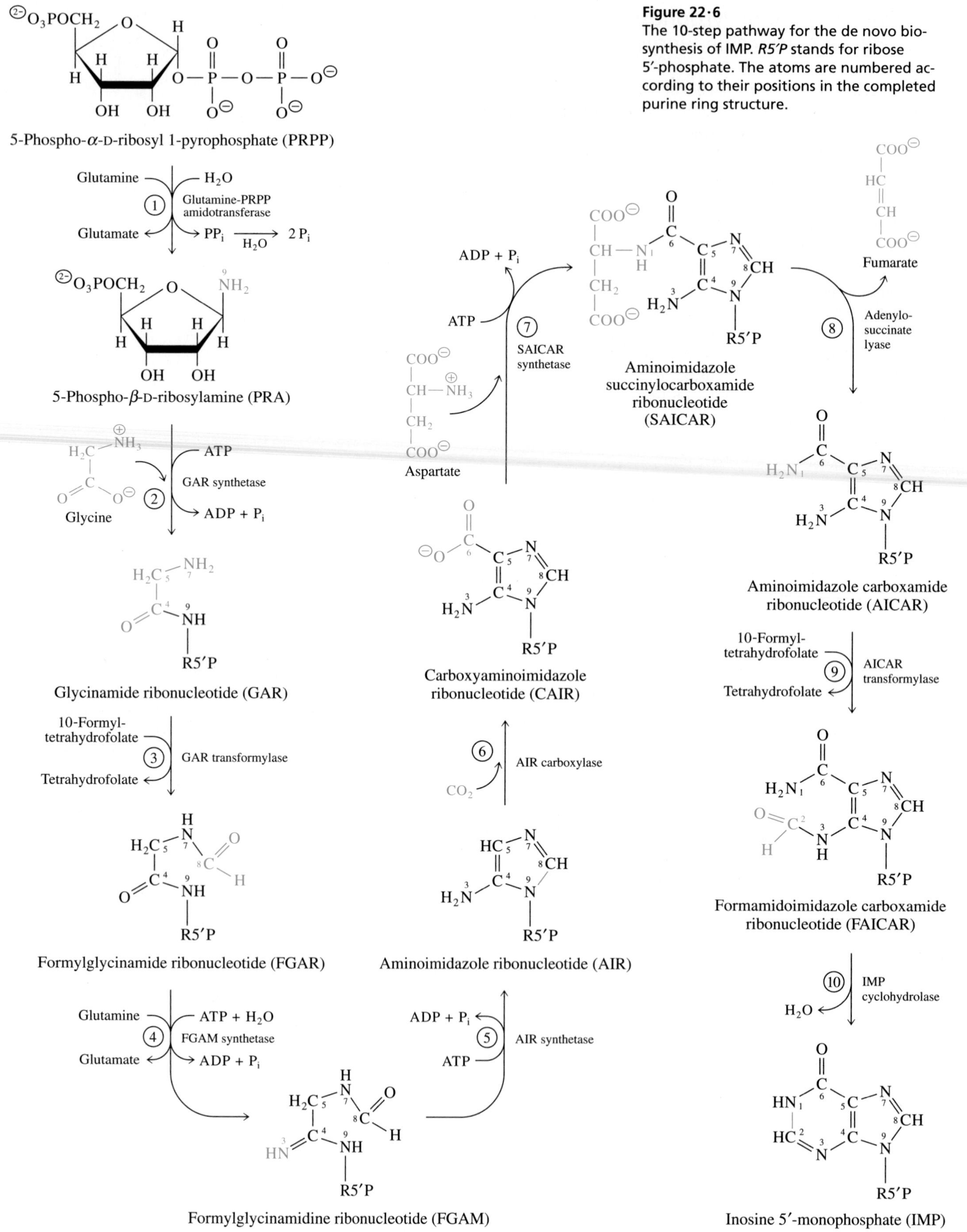
5-Phospho-α-D-ribosyl 1-pyrophosphate (PRPP)
Glutamine
H_2O
Glutamine-PRPP amidotransferase
Glutamate
PP_i
$2 P_i$
5-Phospho-β-D-ribosylamine (PRA)
Glycine
ATP
GAR synthetase
ADP + P_i
Glycinamide ribonucleotide (GAR)
10-Formyl-tetrahydrofolate
GAR transformylase
Tetrahydrofolate
Formylglycinamide ribonucleotide (FGAR)
Glutamine
ATP + H_2O
FGAM synthetase
Glutamate
ADP + P_i
Formylglycinamidine ribonucleotide (FGAM)
ADP + P_i
AIR synthetase
ATP
Aminoimidazole ribonucleotide (AIR)
CO_2
AIR carboxylase
Carboxyaminoimidazole ribonucleotide (CAIR)
Aspartate
ATP
ADP + P_i
SAICAR synthetase
Aminoimidazole succinylocarboxamide ribonucleotide (SAICAR)
Fumarate
Adenylo-succinate lyase
Aminoimidazole carboxamide ribonucleotide (AICAR)
10-Formyl-tetrahydrofolate
AICAR transformylase
Tetrahydrofolate
Formamidoimidazole carboxamide ribonucleotide (FAICAR)
H_2O
IMP cyclohydrolase
Inosine 5′-monophosphate (IMP)

22·3 The Purine Ring System of IMP Is Assembled in 10 Steps

The 10-step pathway for de novo synthesis of IMP and the structures of each intermediate were elucidated in the 1950s by the research groups of Buchanan and Greenberg. These workers used partially purified enzyme preparations from pigeon and chicken liver. The painstaking isolation and structural characterization of the intermediates and enzymes took about 10 years. This work was greatly aided by knowledge of the precursors of all the ring atoms, availability of an enzyme source with high levels of activity—homogenates of avian liver—and identification of the starting substrate (PRPP) and the ultimate product (IMP).

The de novo pathway to IMP is shown in Figure 22·6. The order in which atoms are added to the purine ring structure is N-9 from glutamine; C-4, C-5, and N-7 from glycine; C-8 from 10-formyltetrahydrofolate; N-3 from glutamine; C-6 from CO_2; N-1 from aspartate; and C-2 from 10-formyltetrahydrofolate. Note that the five-membered imidazole ring is completed (Step 5) before the six-membered pyrimidine ring (Step 10). We will discuss only a few of the steps in detail.

The pathway begins with displacement of the pyrophosphoryl group of PRPP by the amide nitrogen of glutamine, catalyzed by glutamine-PRPP amidotransferase. Notice that the anomeric configuration is inverted from α to β in this nucleophilic displacement; the β configuration persists in completed purine nucleotides. The amino group of the product, phosphoribosylamine, is then acylated by glycine to form glycinamide ribonucleotide. The mechanism of this reaction, in which an enzyme-bound glycyl phosphate is formed, resembles that of glutamine synthetase, which has glutamyl phosphate as an intermediate (Section 3·6).

Next, a formyl group is transferred from 10-formyltetrahydrofolate to the amino group destined to become N-7 of IMP. In Step 4, an amide is converted to an amidine, (R)HN—C=NH, in an ATP-dependent reaction that requires glutamine as the nitrogen donor. The enzyme that catalyzes this step, an amidotransferase, is irreversibly inhibited by antibiotics that are analogs of glutamine, such as azaserine and 6-diazo-5-oxo-norleucine (Figure 22·7, next page). These compounds, which are examples of affinity labels, react with a sulfhydryl group of the enzyme.

Step 5 of IMP synthesis is a ring-closure reaction that requires ATP and produces an imidazole derivative. In Step 6, CO_2 is incorporated by attachment to the carbon that becomes C-5 of the purine. This carboxylation is unusual in that neither biotin nor ATP is required. C-5 of the imidazole ring is an activated nucleophile because it is part of an enamine (C=C—NH₂), with —NH₂ being analogous to the —OH group of an enol. The mechanism by which this nucleophile reacts with the electrophilic CO_2 is shown in Figure 22·8 (next page).

In the next two steps (7 and 8), the amino group of aspartate is incorporated into the growing purine ring system. First, the entire aspartate molecule condenses with the newly added carboxylate group in an ATP-dependent reaction to form an amide, specifically, a succinylocarboxamide. Then adenylosuccinate lyase catalyzes a nonhydrolytic cleavage reaction that releases fumarate. This two-step process results in transfer of an amino group containing the nitrogen destined to become N-1 of IMP.

In Step 9, which resembles Step 3, 10-formyltetrahydrofolate donates a formyl group to the nucleophilic amino group of aminoimidazole carboxamide ribonucleotide. The amide nitrogen of the final intermediate condenses with the formyl group in a ring-closure reaction that completes the purine ring system of IMP.

The de novo synthesis of IMP consumes considerable energy. ATP is converted to AMP during synthesis of PRPP. Also, Steps 2, 4, 5, and 7 are driven by the conversion of ATP to ADP. Additional ATP is required for the synthesis of glutamine from glutamate and ammonia (Section 21·3).

Figure 22·7
Glutamine analogs as affinity labels.
(a) Glutamine and the related antibiotics
azaserine and 6-diazo-5-oxo-norleucine.
(b) The diazo-containing antibiotics act as
affinity-labelling reagents, initially binding
to the glutamine-binding site of the amidotransferase that catalyzes Step 4 of purine
biosynthesis (or the binding site of other
glutamine-dependent amidotransferases).
Inactivation results from the alkylation of a
nucleophilic cysteine residue at the active
site.

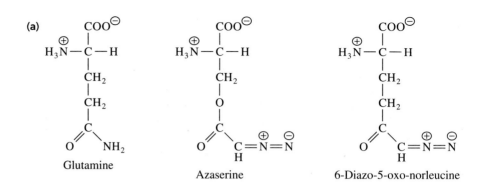

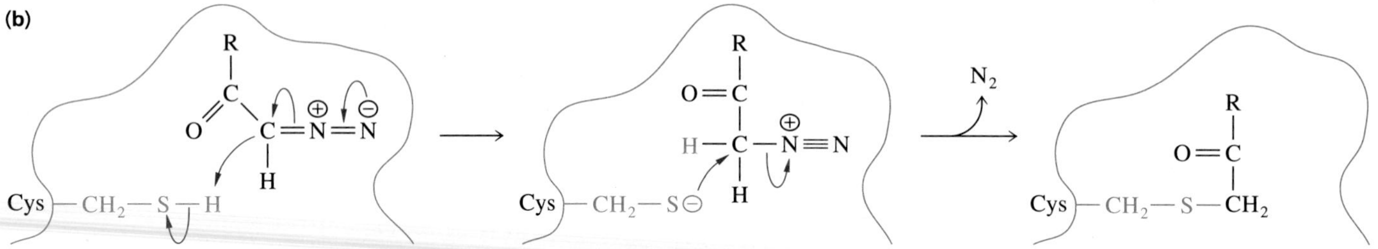

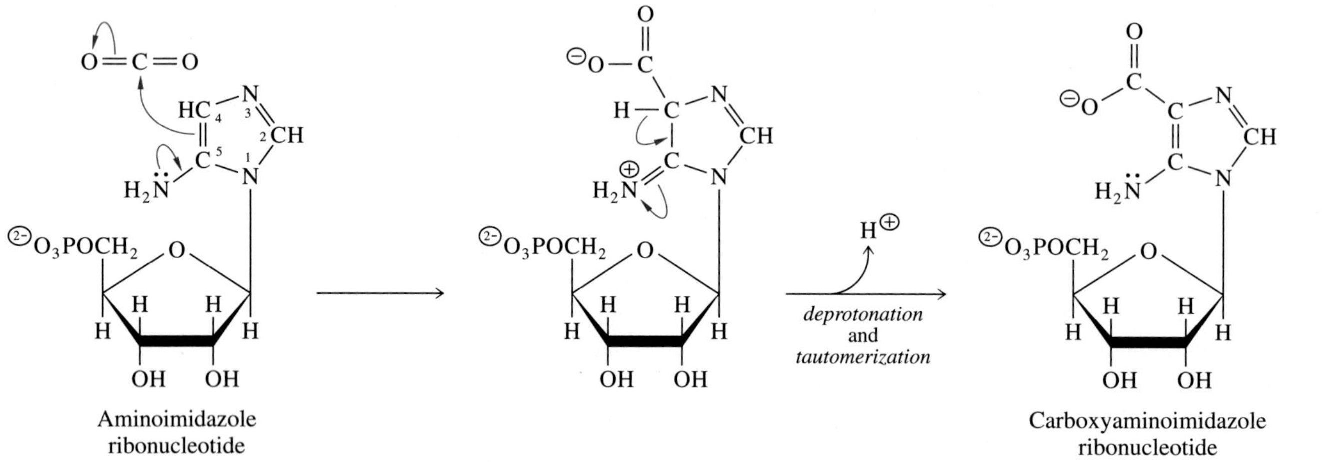

(a)

Glutamine   Azaserine   6-Diazo-5-oxo-norleucine

(b)

Aminoimidazole
ribonucleotide

*deprotonation
and
tautomerization*

Carboxyaminoimidazole
ribonucleotide

Figure 22·8
Mechanism for the addition of carbon dioxide to aminoimidazole ribonucleotide. The
enamine is an activated nucleophile that
attacks the electrophilic carbon of CO_2. Subsequent deprotonation and tautomerization
produces carboxyaminoimidazole ribonucleotide.

22·4 AMP and GMP Are Synthesized from IMP

IMP can be converted to either of the major purine nucleotides, AMP or GMP, as shown in Figure 22·9. Two enzymatic reactions are required for each of these conversions.

The biosynthesis of AMP from IMP involves two steps that closely resemble Steps 7 and 8 in the biosynthesis of IMP. First, the amino group of aspartate condenses with the keto group of IMP in a reaction catalyzed by GTP-dependent adenylosuccinate synthetase. Then the elimination of fumarate from adenylosuccinate is catalyzed by adenylosuccinate lyase, the same enzyme that catalyzes

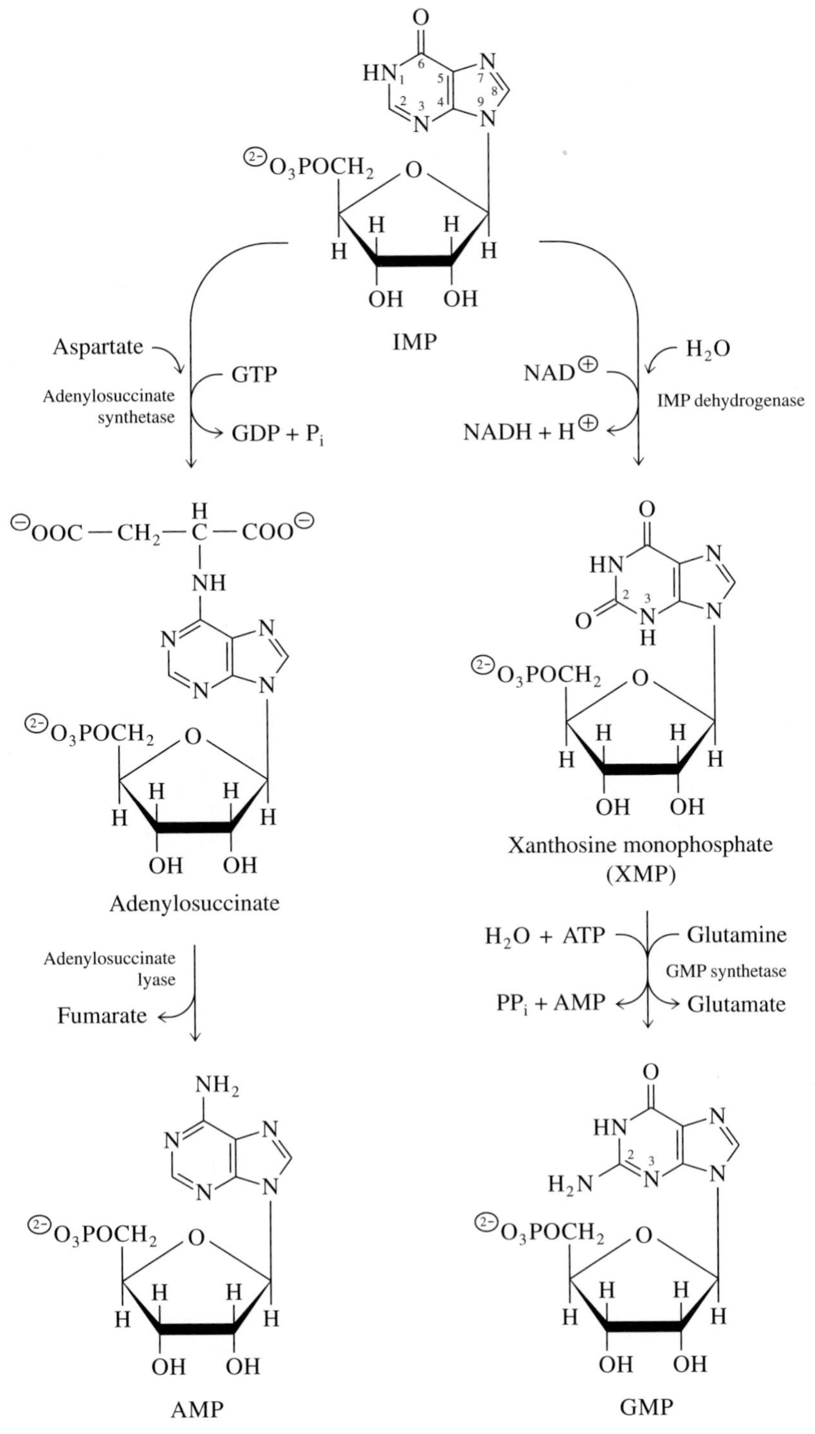

Figure 22·9
Pathways for the conversion of IMP to AMP or GMP.

Step 8 of the de novo synthesis of IMP. Steps similar to these amine-transfer reactions are involved in the biosynthesis of arginine in the urea cycle (Section 21·10).

In the conversion of IMP to GMP, C-2 is oxidized in a reaction catalyzed by IMP dehydrogenase. This reaction proceeds by addition of a molecule of water to the C-2=N-3 double bond and oxidation of the hydrate by $NAD^{\oplus}$. The product of the oxidation is xanthosine monophosphate (XMP). Next, in an ATP-dependent reaction catalyzed by GMP synthetase, the amide nitrogen of glutamine replaces the oxygen at C-2 of XMP. GMP synthetase, like other glutamine-dependent amidotransferases, is inactivated by azaserine (Figure 22·7). The use of GTP as a cosubstrate in the synthesis of AMP from IMP, and of ATP in the synthesis of GMP from IMP, helps to balance the formation of the two products.

Purine nucleotide synthesis is probably regulated in cells by feedback inhibition. Several enzymes that catalyze steps in the biosynthesis of purine nucleotides exhibit allosteric kinetic behavior in vitro. PRPP synthase is inhibited by several purine ribonucleotides, but only at concentrations higher than those usually found in the cell. PRPP is a donor of ribose 5-phosphate in over a dozen reactions, and therefore one would not expect PRPP synthesis to be regulated exclusively by the concentrations of purine nucleotides. The enzyme that catalyzes the first committed step in the pathway of purine nucleotide synthesis, glutamine-PRPP amidotransferase (Step 1 in Figure 22·6), is allosterically inhibited by 5′-ribonucleotide end products at intracellular concentrations. The first committed step appears to be the principal site of regulation of this pathway. None of the enzymes between this step and formation of IMP are known to be regulated.

The paths leading from IMP to AMP and from IMP to GMP are also regulated by feedback inhibition. Adenylosuccinate synthetase is inhibited in vitro by AMP, the product of this two-step branch. IMP dehydrogenase is inhibited by both XMP and GMP. The feedback inhibition patterns of certain enzymes in this branched pathway to AMP and GMP are shown in Figure 22·10. Notice that the end products inhibit two of the initial common steps as well as steps leading from the branch point.

Figure 22·10
Regulation of purine nucleotide biosynthesis as indicated by in vitro studies.

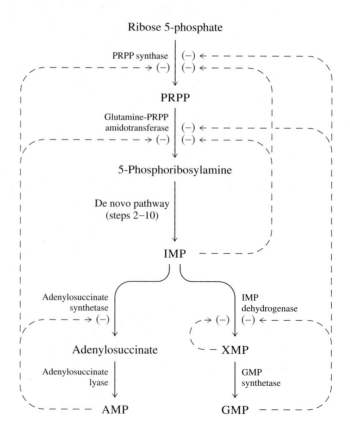

22·5 Purines Arising from the Breakdown of Nucleotides Can Be Salvaged

During cellular metabolism and during digestion in animals, nucleic acids are degraded to mononucleotides, nucleosides, and eventually, heterocyclic bases. These catabolic reactions are catalyzed by enzymes that include ribonucleases, deoxyribonucleases, and a variety of nucleotidases and nucleosidases. Some of the purines formed in this way are further degraded to uric acid or other excretory products, but a considerable fraction is normally salvaged by direct conversion to purine ribonucleotides. Purines formed by intracellular reactions are salvaged to a greater extent than those formed during digestion. The recycling of intact purine molecules conserves cellular energy.

PRPP is the donor of ribose 5-phosphate for the purine salvage reactions. Adenine phosphoribosyltransferase catalyzes the reaction of adenine with PRPP to form AMP and PP_i. Hydrolysis of PP_i catalyzed by inorganic pyrophosphatase renders the reaction metabolically irreversible. Hypoxanthine-guanine phosphoribosyltransferase catalyzes similar reactions—the conversion of hypoxanthine to IMP and guanine to GMP with concomitant formation of PP_i. Hypoxanthine-guanine phosphoribosyltransferase has a lower K_m for PRPP and a higher k_{cat} than glutamine-PRPP amidotransferase; these properties allow the salvage reactions to occur preferentially over de novo purine synthesis at low concentrations of PRPP. The degradation of purine nucleotides to their respective purines and purine salvage through reaction with PRPP are outlined in Figure 22·11.

In 1964, Michael Lesch and William Nyhan described a severe metabolic disease characterized by mental retardation, palsy-like spasticity, and a bizarre tendency toward self-mutilation. Individuals afflicted with this disease, called Lesch-Nyhan syndrome, rarely survive past childhood. Prominent biochemical features of the disease are the excretion of up to six times the normal amount of uric acid and a greatly increased rate of de novo purine biosynthesis. The disease is caused by a hereditary deficiency of the enzyme hypoxanthine-guanine phosphoribosyltransferase and is usually restricted to males because the gene for this enzyme is on the X chromosome. In the absence of hypoxanthine-guanine phosphoribosyltransferase activity, hypoxanthine and guanine are degraded to uric acid instead of being converted to IMP and GMP. Cells devoid of hypoxanthine-guanine phosphoribosyltransferase contain markedly increased concentrations of PRPP, indicating that the salvage reaction normally consumes an appreciable amount of the PRPP produced by the cell. The PRPP normally used for salvage of hypoxanthine and guanine contributes to the de novo synthesis of excessive amounts of IMP, and the surplus IMP is degraded to uric acid. It is not known how this single-enzyme defect causes the various behavioral symptoms. The catastrophic effects of the deficiency indicate that the purine salvage pathway in humans is not just an energy-saving addendum to the central pathways of purine nucleotide metabolism, but that synthesis and salvage are critically integrated.

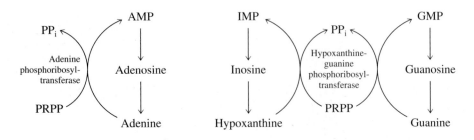

Figure 22·11
Degradation and salvage of purines.

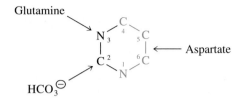

Figure 22·12
Sources of the ring atoms of pyrimidines synthesized de novo. Note that C-4, C-5, C-6, and N-1 are all contributed by aspartate.

22·6 The De Novo Pathway for Pyrimidine Synthesis Leads to UMP

The pathway for de novo synthesis of pyrimidine nucleotides is simpler than the de novo purine pathway and requires less consumption of ATP. There are three metabolic precursors of the pyrimidine ring: bicarbonate, which contributes C-2; the amide group of glutamine (N-3); and aspartate, which contributes the remaining atoms (Figure 22·12).

PRPP is required for biosynthesis of pyrimidine nucleotides, but the sugar-phosphate from PRPP is donated after the ring is formed rather than entering the pathway in the first step as in purine nucleotide biosynthesis. A compound with a completed pyrimidine ring—orotate (6-carboxyuracil)—reacts with PRPP to form a pyrimidine ribonucleotide in the fifth step of the six-step pathway.

The six reactions of the pathway for de novo pyrimidine synthesis are shown in Figure 22·13. The first two steps generate a noncyclic intermediate that contains all of the atoms destined for the pyrimidine ring. This intermediate, carbamoyl aspartate, is enzymatically cyclized. The product, dihydroorotate, is subsequently oxidized to orotate. Orotate is then converted to the ribonucleotide orotidine 5′-monophosphate (OMP, or orotidylate), which undergoes decarboxylation to form uridine 5′-monophosphate (UMP). This pyrimidine nucleotide is the precursor of all other pyrimidine ribo- and deoxyribonucleotides. As we will see, the enzymes required for de novo pyrimidine synthesis are organized differently in prokaryotes and eukaryotes, and the pathways in these cell types are regulated differently.

The first step in the pathway of de novo pyrimidine biosynthesis is the formation of carbamoyl phosphate from bicarbonate, the amide nitrogen of glutamine, and the phosphoryl group of ATP. This reaction, catalyzed by carbamoyl phosphate synthetase, requires two molecules of ATP, one to drive the formation of the C—N bond and the other to donate the phosphoryl group of the product.

Carbamoyl phosphate is also a metabolite in the pathway leading to the biosynthesis of arginine via citrulline (Section 21·6D). In prokaryotes, the same carbamoyl phosphate synthetase is utilized in both pyrimidine and arginine biosynthetic pathways. This enzyme is allosterically inhibited by pyrimidine ribonucleotides such as UMP, the product of the pyrimidine biosynthetic pathway. It is activated by L-ornithine, a precursor of citrulline, and by purine nucleotides, the substrates (along with pyrimidine nucleotides) for the synthesis of nucleic acids.

In eukaryotic cells, there are two distinct carbamoyl phosphate synthetases. A mitochondrial synthetase, called carbamoyl phosphate synthetase I because it was discovered first, catalyzes a reaction that utilizes ammonia as the source of the amide of carbamoyl phosphate.

$$NH_3 \ + \ HCO_3^\ominus \ + \ 2\,ATP$$

$$\Big\downarrow \begin{array}{l} \text{Carbamoyl} \\ \text{phosphate} \\ \text{synthetase I} \end{array}$$

$$\underset{\text{Carbamoyl phosphate}}{H_2N-\overset{\displaystyle O}{\overset{\|}{C}}-OPO_3^{\,2\ominus}} \ + \ 2\,ADP \ + \ P_i$$

(22·1)

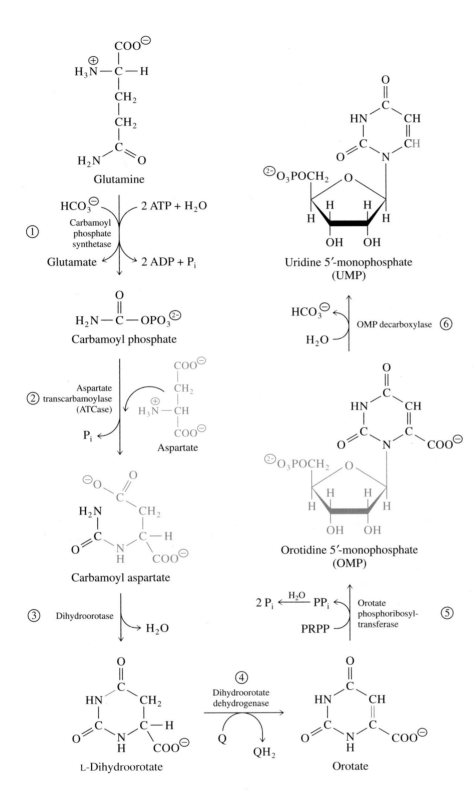

Figure 22·13
The six-step pathway for the de novo synthesis of uridine 5'-monophosphate (UMP, or uridylate) in prokaryotes. In eukaryotes, Steps 1, 2, and 3 are catalyzed by a multifunctional protein called dihydroorotate synthase, and Steps 5 and 6 are catalyzed by a bifunctional enzyme, UMP synthase.

In mammalian liver, carbamoyl phosphate generated in mitochondria is utilized for synthesis of urea. The cytosolic synthetase, carbamoyl phosphate synthetase II, catalyzes the first committed step of pyrimidine synthesis. The amide nitrogen of carbamoyl phosphate in this reaction comes from glutamine.

$$\text{Glutamine} + \text{HCO}_3^{\ominus} + 2\,\text{ATP} + \text{H}_2\text{O}$$

$$\downarrow \quad \text{Carbamoyl phosphate synthetase II}$$

$$\text{Carbamoyl phosphate} + \text{Glutamate} + 2\,\text{ADP} + \text{P}_i$$

(22·2)

Carbamoyl phosphate synthetase II is allosterically regulated. PRPP and IMP activate the enzyme, and several pyrimidine nucleotides inhibit it. The compartmentation of reactions in mitochondria and the cytosol allows separate control of each enzyme and the pathway it serves.

In the second step of UMP biosynthesis, the activated carbamoyl group of carbamoyl phosphate is transferred to aspartate to form carbamoyl aspartate. In this reaction, catalyzed by aspartate transcarbamoylase (ATCase), the nucleophilic nitrogen of aspartate attacks the carbonyl group of carbamoyl phosphate.

ATCase from *Escherichia coli* was the first allosteric enzyme to be fully characterized. In *E. coli,* where carbamoyl phosphate synthetase generates an intermediate that can enter pathways leading either to pyrimidines or to arginine, it is ATCase that catalyzes the first committed step of pyrimidine biosynthesis. This enzyme is inhibited by pyrimidine nucleotides and activated in vitro by ATP, as discussed in Chapter 6. Although ATCase in *E. coli* is only partially inhibited by CTP alone (50–70%), inhibition can be almost total when both CTP and UTP are present. UTP alone does not inhibit the enzyme. The allosteric controls—inhibition by pyrimidine nucleotides and activation by the purine nucleotide ATP—provide a means for carbamoyl phosphate synthetase and ATCase to balance the pyrimidine nucleotide and purine nucleotide pools in *E. coli.* The ratio of the concentrations of the two types of allosteric effectors determines the activity level of ATCase.

Eukaryotic ATCase is not feedback inhibited. Regulation by feedback inhibition is not necessary because the substrate of ATCase in eukaryotes is not a branch-point metabolite—the synthesis of carbamoyl phosphate and citrulline for the urea cycle occurs in the mitochondrion, and the synthesis of carbamoyl phosphate for pyrimidines occurs in the cytosol. These pools of carbamoyl phosphate are separate. Consequently, the pyrimidine pathway can be controlled by regulation of the enzyme preceding ATCase, carbamoyl phosphate synthetase II.

Dihydroorotase catalyzes the third step of UMP biosynthesis, the reversible closure of the pyrimidine ring. The product, dihydroorotate, is then oxidized by the action of dihydroorotate dehydrogenase to form orotate. In eukaryotes, dihydroorotate dehydrogenase is associated with the inner membrane of the mitochondrion. The enzyme is an iron-containing flavoprotein that catalyzes transfer of electrons from dihydroorotate to ubiquinone (Q) and thence to O_2 by the respiratory electron-transport chain.

Once formed, orotate displaces the pyrophosphate group of PRPP, producing OMP in a reaction catalyzed by orotate phosphoribosyltransferase. This enzyme also has a salvage role—it can catalyze the conversion of pyrimidines other than orotate to the corresponding pyrimidine nucleotides. As in the reactions catalyzed by adenine phosphoribosyltransferase and hypoxanthine-guanine phosphoribosyltransferase, the pyrophosphate produced is hydrolyzed, rendering Step 5 irreversible.

Finally, OMP is decarboxylated to UMP. Product inhibition of OMP decarboxylase by UMP in vitro has been reported, but the primary control point in vivo is earlier in the biosynthetic pathway.

22·7 Two Multifunctional Proteins Are Involved in Pyrimidine Biosynthesis in Eukaryotes

Although the six enzymatic steps leading to UMP are the same in prokaryotes and eukaryotes, the structural organization of the enzymes (free versus associated) varies among organisms. For example, in *E. coli,* each of the six reactions is catalyzed by a separate enzyme. In eukaryotes, a multifunctional protein in the cytosol known as dihydroorotate synthase contains separate catalytic sites (carbamoyl phosphate synthetase II, ATCase, and dihydroorotase) for the first three steps of the pathway. Dihydroorotate produced in the cytosol by Steps 1 through 3 passes through the outer mitochondrial membrane prior to being oxidized to orotate by dihydroorotate dehydrogenase. The substrate-binding site of this enzyme is located on the outer surface of the inner mitochondrial membrane. Orotate then moves to the cytosol, where it is converted to UMP. A bifunctional enzyme known as UMP synthase catalyzes both the reaction of orotate with PRPP to form OMP and the rapid decarboxylation of OMP to UMP.

The intermediates formed in Step 1 and Step 2 (carbamoyl phosphate and carbamoyl aspartate) and OMP (from Step 5) are not normally released to the solvent but remain enzyme bound and are channelled from one catalytic center to the next. Several multifunctional proteins, each catalyzing several steps, also occur in the pathway of purine nucleotide biosynthesis in some organisms. This sequestering of labile intermediates within multidomain enzymes prevents nonproductive degradation of intermediates, thereby conserving energy.

22·8 CTP Is Synthesized from UMP

UMP is converted to CTP in three steps. First, uridylate kinase (UMP kinase) catalyzes the transfer of the γ-phosphoryl group of ATP to UMP to generate UDP, and then nucleoside diphosphate kinase catalyzes the transfer of the γ-phosphoryl group of a second ATP molecule to UDP to form UTP. In these two reactions, two molecules of ATP are converted to two molecules of ADP.

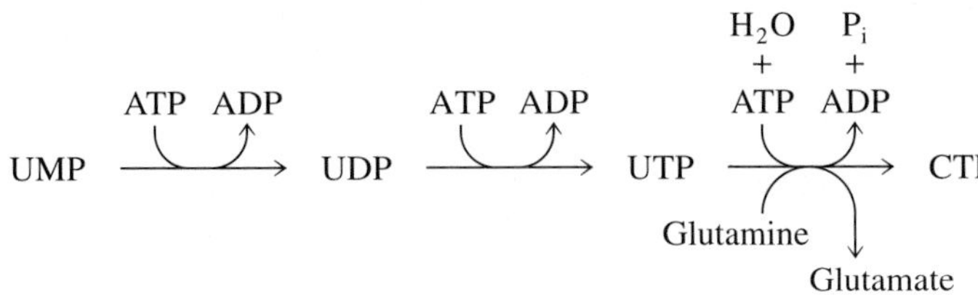

$$\text{(22·3)}$$

CTP synthetase then catalyzes the ATP-dependent transfer of the amide nitrogen from glutamine to C-4 of UTP, forming CTP. This reaction, shown in Figure 22·14, is analogous to Step 4 of purine biosynthesis (Figure 22·6).

CTP synthetase is allosterically inhibited by its product, CTP, and in *E. coli* is allosterically activated by GTP. The activation involves both an increase in V_{max} and a decrease in the K_m of the enzyme for glutamine.

Thymidylate (dTMP) is also a derivative of UMP, but before thymidylate can be formed, UMP must undergo reduction to form dUMP. We will examine the formation of deoxyribonucleotides next and then the formation of dTMP from dUMP.

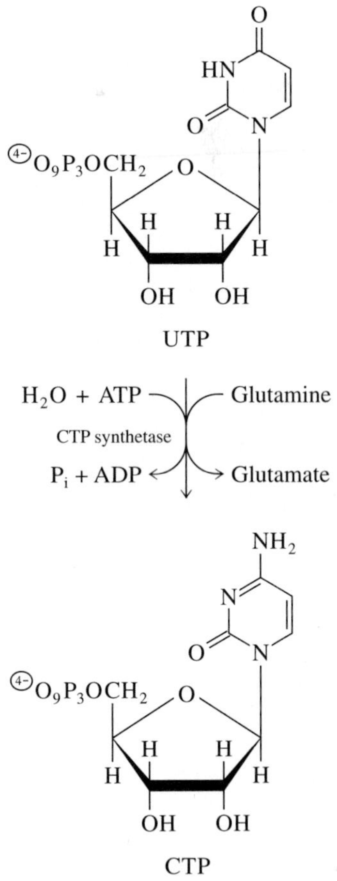

Figure 22·14
Conversion of UTP to CTP, catalyzed by CTP synthetase.

22·9 Deoxyribonucleotides Are Synthesized by Reduction of Ribonucleotides

The 2′-deoxyribonucleotides, whose principal role is to serve as triphosphate substrates for DNA polymerase, are synthesized through enzymatic reduction of ribonucleotides. In most organisms, this reduction occurs at the nucleoside diphosphate level. Peter Reichard and his colleagues, who overcame difficult technical barriers to elucidate this complicated reaction system, showed that all four ribonucleoside diphosphates—ADP, GDP, CDP, and UDP—are substrates of a single, closely regulated ribonucleoside diphosphate reductase. However, in some microorganisms, including species of *Lactobacillus, Clostridium,* and *Rhizobium,* ribonucleoside *tri*phosphates are the substrates for reduction by a cobalamin-dependent reductase. Both types of enzymes are referred to as ribonucleotide reductase, although more precise names are ribonucleoside diphosphate reductase and ribonucleoside triphosphate reductase, respectively.

NADPH provides the reducing power for the synthesis of deoxyribonucleoside diphosphates. A disulfide bond at the active site of ribonucleotide reductase is reduced to two thiol groups, which in turn reduce C-2′ of the ribose moiety of the nucleotide substrate by a free-radical mechanism. As shown in Figure 22·15, electrons are transferred from NADPH to ribonucleotide reductase via the flavoprotein thioredoxin reductase and the dithiol protein coenzyme thioredoxin. In the absence of thioredoxin (for example, in *E. coli* mutants that lack thioredoxin), another small dithiol protein, glutaredoxin, can replace reduced thioredoxin in deoxyribonucleotide formation. Glutaredoxin transfers electrons from reduced glutathione to ribonucleotide reductase. Once formed, dADP, dGDP, and dCDP are phosphorylated to the triphosphate level by the action of nucleoside diphosphate kinases. dUDP, as we will see in the next section, is converted to dTMP via dUMP.

Figure 22·15
Reduction of ribonucleoside diphosphates by electrons derived from NADPH, transferred via thioredoxin. Three proteins are involved: the NADPH-dependent flavoprotein thioredoxin reductase, thioredoxin, and ribonucleotide reductase. In this figure, *B* represents a purine (A, G) or pyrimidine (C, U) base.

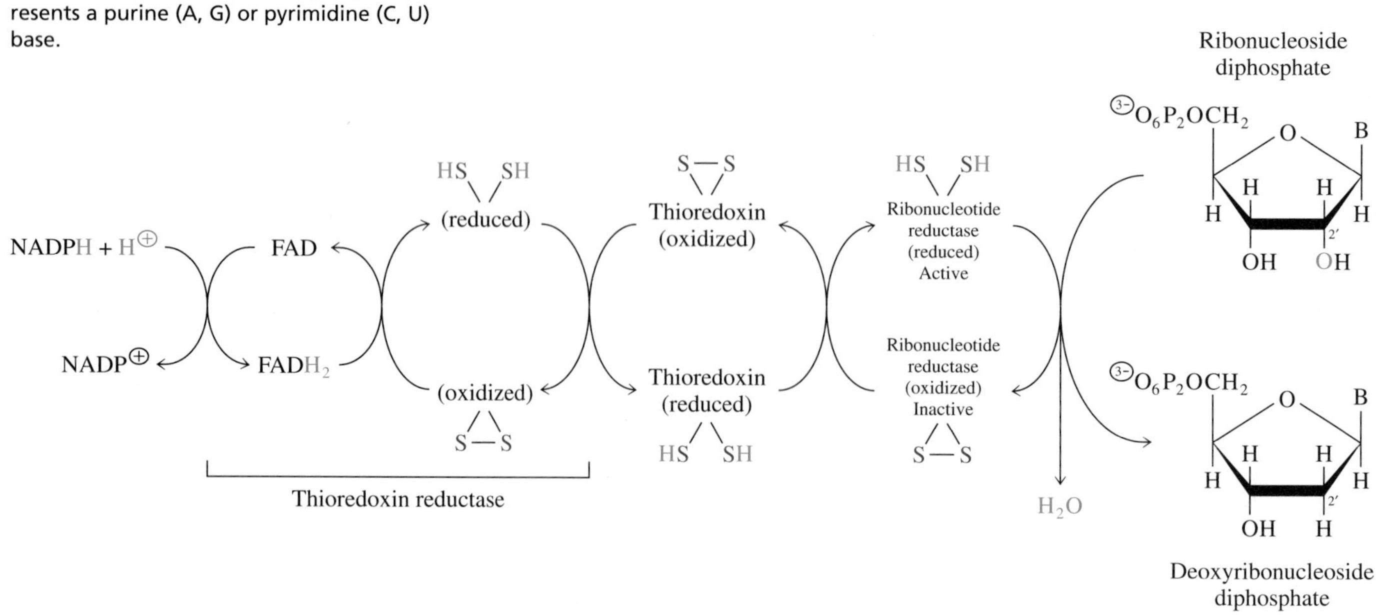

A mechanism postulated for *E. coli* ribonucleotide reductase is shown in Figure 22·16. This mechanism is based on the presence in the active site of a tyrosine residue that forms a phenoxy radical and cysteine thiol groups (not shown) that act as one-electron reducing agents. After a molecule of ribonucleoside diphosphate binds to the enzyme, the protein radical abstracts a hydrogen atom from C-3′, converting the substrate into a free radical. The oxygen atom at C-2′ is then protonated and ejected as a molecule of water. The resulting free radical, with the unpaired electron now at C-2′, is reduced in two one-electron steps and protonated. Finally, the product free radical—differing from the substrate free radical only by reduction at C-2′—abstracts the hydrogen atom from the active site tyrosine. The unusual mechanism presented here, in which a protein radical and then a substrate radical are generated, appears to be a general feature of ribonucleotide reductases.

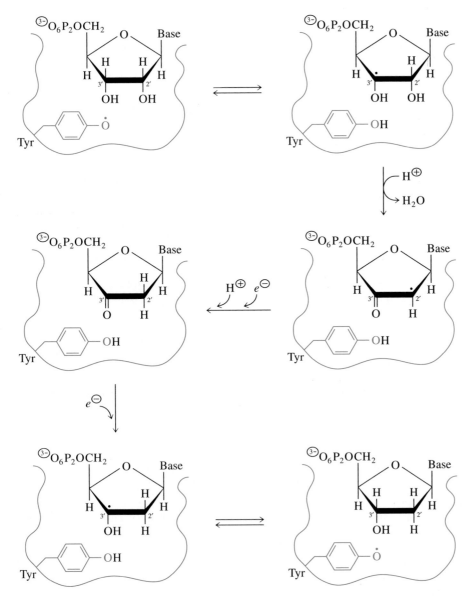

Figure 22·16
Mechanism proposed for the *E. coli* ribonucleotide reductase.

Table 22·1 Allosteric regulation of ribonucleotide reductase

Ligand bound to activity site	Ligand bound to specificity site	Activity of catalytic site
dATP	Enzyme inactive	
ATP	ATP or dATP	Specific for CDP or UDP
ATP	dTTP	Specific for GDP
ATP	dGTP	Specific for ADP

Under some circumstances, ribonucleotide reduction can be the rate-limiting step in DNA synthesis. Ribonucleotide reductases are strictly regulated by allosteric interactions. Both the substrate specificity and the catalytic rates of the enzymes are regulated in cells by the reversible binding of nucleotide metabolites. This regulatory system has features that ensure that a balanced selection of deoxyribonucleotides is available for DNA synthesis. The allosteric modulators—ATP, dATP, dTTP, and dGTP—exert their effects by binding to ribonucleotide reductases at either of two regulatory sites. One allosteric site, termed the *activity site,* controls the activity of the catalytic site. A second allosteric site, called the *specificity site,* determines the substrate specificity of the catalytic site. The binding of ATP to the activity site activates the reductase; the binding of dATP to the activity site inhibits all enzymatic activity. When ATP is bound to the activity site and either ATP or dATP is bound to the specificity site, the reductase becomes pyrimidine specific, catalyzing the reduction of CDP and UDP; binding of dTTP to the specificity site activates the reduction of GDP, and binding of dGTP activates the reduction of ADP. The main features of the allosteric regulation of ribonucleotide reductase are summarized in Table 22·1.

22·10 A Unique Methylation Reaction Produces dTMP from dUMP

Thymidylate (dTMP), which is required for DNA synthesis, is formed from UMP in four steps. UMP is phosphorylated to UDP, which is reduced to dUDP, and dUDP is dephosphorylated to dUMP, which is then methylated.

$$\text{UMP} \longrightarrow \text{UDP} \longrightarrow \text{dUDP} \longrightarrow \text{dUMP} \longrightarrow \text{dTMP} \qquad \textbf{(22·4)}$$

The conversion of dUDP to dUMP can occur by two routes. dUDP can react with ADP in the presence of a nucleoside monophosphate kinase to form dUMP and ATP.

$$\text{dUDP} + \text{ADP} \rightleftarrows \text{dUMP} + \text{ATP} \qquad \textbf{(22·5)}$$

dUDP can also be phosphorylated to dUTP at the expense of ATP through the action of rather nonspecific nucleoside diphosphate kinases. dUTP is then rapidly hydrolyzed to dUMP + PP_i by the action of deoxyuridine triphosphate diphosphohydrolase (dUTPase).

$$\text{dUDP} + \text{ATP} \xrightarrow[\text{ADP}]{} \text{dUTP} \xrightarrow[\text{H}_2\text{O}]{} \text{dUMP} + \text{PP}_i \qquad \textbf{(22·6)}$$

The rapid hydrolysis of dUTP prevents it from being accidentally incorporated into DNA in place of dTTP.

dCMP can also serve as a source of dUMP via hydrolysis catalyzed by dCMP deaminase.

$$\text{dCMP} + \text{H}_2\text{O} \longrightarrow \text{dUMP} + \text{NH}_4^{\oplus} \qquad \textbf{(22·7)}$$

dCMP deaminase in liver is subject to allosteric regulation: dTTP is an allosteric inhibitor of the enzyme, and dCTP is an activator.

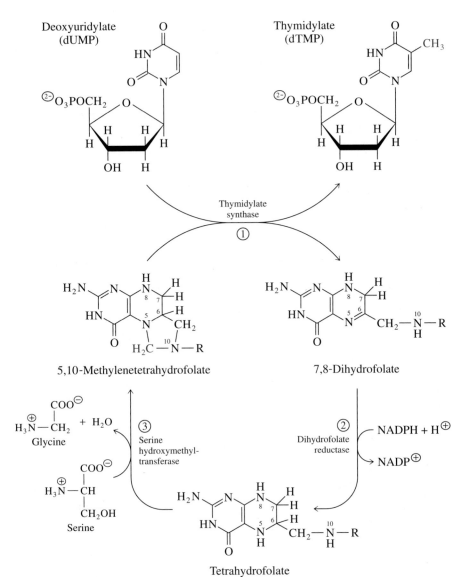

Figure 22·17
Cycle of reactions in the synthesis of thymidylate (dTMP) from dUMP. Thymidylate synthase catalyzes the first reaction of this cycle, producing dTMP. The other product of the reaction, dihydrofolate, must be reduced by NADPH in a reaction catalyzed by dihydrofolate reductase before a methylene group can be added to regenerate 5,10-methylenetetrahydrofolate. Methylene-tetrahydrofolate is regenerated in a reaction catalyzed by serine hydroxymethyltransferase.

The conversion of dUMP to dTMP is catalyzed by thymidylate synthase. In this reaction, 5,10-methylenetetrahydrofolate (Section 8·12) is the donor of the one-carbon group (Figure 22·17). Note that the methyl ($-CH_3$) group in the product, dTMP, is more reduced than the methylene ($-CH_2-$) group in 5,10-methylenetetrahydrofolate, whose oxidation state is equivalent to that of a hydroxy-methyl ($-CH_2OH$) group or formaldehyde. Thus, not only does methylenetetra-hydrofolate serve as a group-transfer coenzyme providing a one-carbon unit, but it also serves as the reducing agent for the reaction, furnishing a hydride ion and being oxidized to 7,8-dihydrofolate in the process. This is the only known reaction in which the transfer of a one-carbon unit from tetrahydrofolate results in oxidation at N-5 and C-6 to produce dihydrofolate. Dihydrofolate must be converted to *tetra*hydrofolate before the coenzyme can accept another one-carbon unit for further transfer reactions. The 5,6 double bond of dihydrofolate is reduced by NADPH in the reaction catalyzed by dihydrofolate reductase (Section 8·12). Serine hydroxymethyltransferase then catalyzes the transfer of the β-CH_2OH group of serine to tetrahydrofolate to regenerate 5,10-methylenetetrahydrofolate.

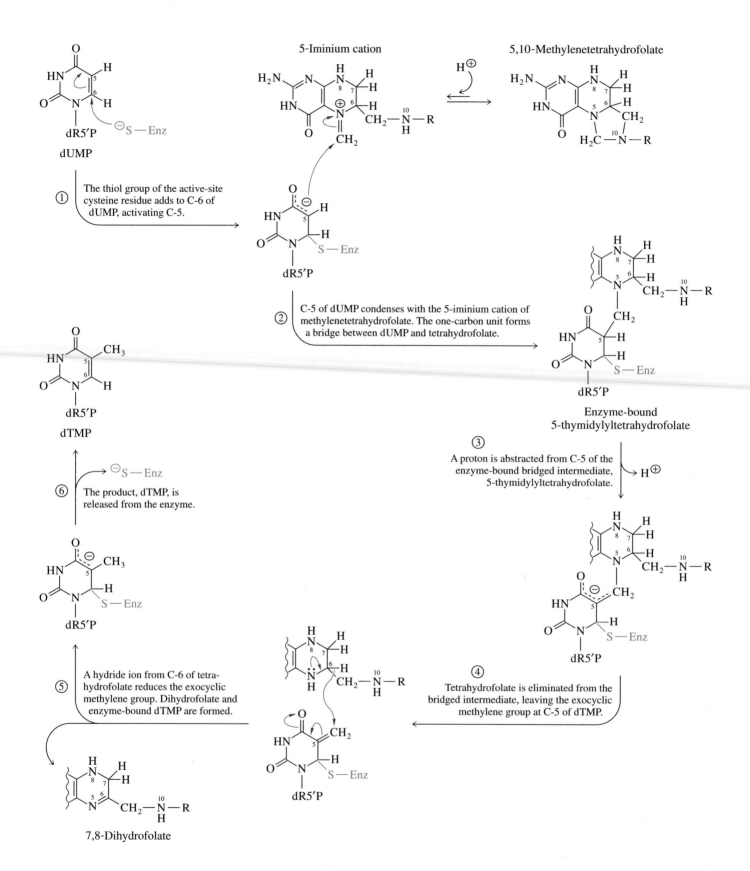

Figure 22·18
Mechanism of thymidylate synthase.

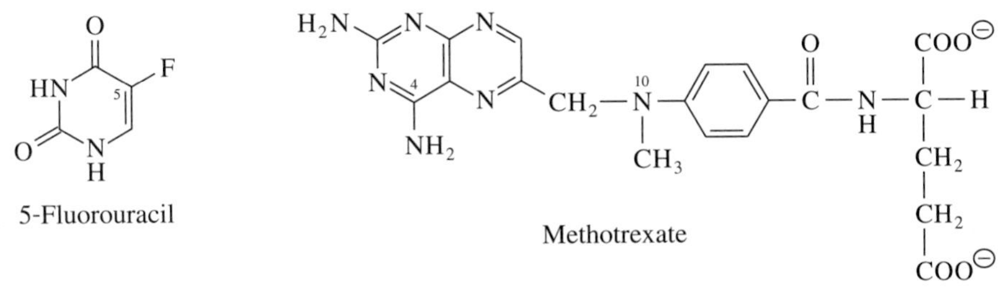

Figure 22·19
5-Fluorouracil and methotrexate, drugs designed to inhibit reproduction of cancer cells.

5-Fluorouracil

Methotrexate

The mechanism proposed for the action of thymidylate synthase has an unusual feature: covalent binding of a substrate with a concomitant increase in reactivity (Figure 22·18). Note that this mechanism is quite different from covalent catalysis in group-transfer reactions, during which a mobile group of the substrate is transiently bound covalently to an enzyme. The first step of the reaction pathway is noncovalent binding of the substrate dUMP, which then is bound covalently by addition of a nucleophilic thiolate of the enzyme ($^\ominus$S—Enz, an ionized cysteine residue). This addition reaction converts dUMP to a carbanion in which C-5 is nucleophilic. Next, the one-carbon group is donated to the carbanion, not by methylenetetrahydrofolate but by the 5-iminium cation with which it is in equilibrium. This electrophile reacts with C-5 of dUMP to form an intermediate called 5-thymidylyltetrahydrofolate in which dUMP and tetrahydrofolate are combined via a methylene bridge. A proton must be eliminated from C-5 of this enzyme-bound intermediate before it decomposes to form tetrahydrofolate and the exocyclic methylene derivative of dUMP. This methylene group is reduced by a hydride ion from tetrahydrofolate, producing enzyme-bound dTMP and dihydrofolate.

Since dTMP serves as an essential precursor of DNA, any agent that lowers dTMP levels drastically affects cell division. Because rapidly dividing cells are particularly dependent on the activities of thymidylate synthase and dihydrofolate reductase, these enzymes have been major targets for anticancer drugs. Inhibition of either or both of these enzymes blocks the synthesis of dTMP and therefore the synthesis of DNA.

5-Fluorouracil and methotrexate, shown in Figure 22·19, have proven effective in combatting some types of cancer. 5-Fluorouracil is converted to its deoxyribonucleotide, 5-fluorodeoxyuridylate, by a pyrimidine salvage pathway. 5-Fluorodeoxyuridylate binds tightly to thymidylate synthase, inhibiting the enzyme and bringing the three-reaction cycle shown in Figure 22·17 to a halt. 5-Fluorodeoxyuridylate accomplishes this inhibition by acting as a substrate for the first few steps of the thymidylate synthase reaction, forming the 5-fluoro analog of enzyme-bound thymidylyltetrahydrofolate (Figure 22·20). At this stage, the normal intermediate releases a proton. The 5-fluoro analog cannot because its C-5—F bond is stable, and the reaction proceeds no further.

Methotrexate, the most commonly used anticancer drug, is an analog of folate with an amino group in place of the oxygen atom at C-4 and a methyl substituent at N-10. Methotrexate is a potent and relatively specific inhibitor of dihydrofolate reductase, which catalyzes Step 2 of the cycle shown in Figure 22·17. The folate analog binds to the reductase extremely tightly by noncovalent interactions only. The resulting decrease in tetrahydrofolate levels greatly diminishes the formation of dTMP, since dTMP synthesis is dependent on adequate concentrations of methylenetetrahydrofolate. Most normal cells undergo cell division more slowly than cancer cells and consequently are less sensitive to methotrexate. However, because methotrexate is toxic to all cells, specific protocols for its use have been developed. Often, 5-formyltetrahydrofolate, which can be converted to methylenetetrahydrofolate (Section 8·12), is given to patients shortly after they have been administered an otherwise lethal dose of methotrexate. This high dose–rescue approach enhances the therapeutic use of the drug.

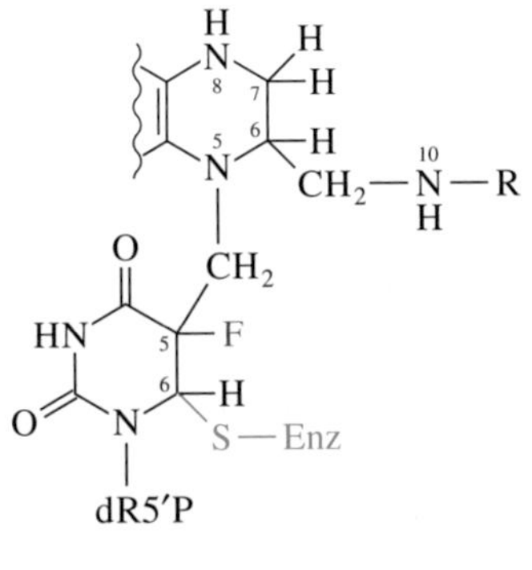

Figure 22·20
Complex formed between 5-fluorodeoxyuridylate and thymidylate synthase.

dR5′P

In most organisms, thymidylate synthase and dihydrofolate reductase exist as distinct polypeptides. However, in protozoa, these two enzyme activities are contained on the same polypeptide chain. Dihydrofolate reductase is located on the N-terminus and thymidylate synthase is located on the C-terminus, with the two domains linked by a short peptide. The native enzyme is a dimer consisting of two of the bifunctional chains. One biological advantage of the bifunctional structure may be channelling of dihydrofolate from the thymidylate synthase active site to the dihydrofolate reductase site without release of dihydrofolate to solvent. The reductase rapidly catalyzes reduction of dihydrofolate as it is formed, preventing both inhibition of dTMP synthase by dihydrofolate and depletion of the pool of tetrahydrofolate.

dTMP can also be synthesized via the salvage of thymidine, which is catalyzed by ATP-dependent thymidine kinase.

$$\text{Thymidine} \xrightarrow[\substack{\text{Thymidine} \\ \text{kinase}}]{\text{ATP} \quad \text{ADP}} \text{dTMP} \tag{22·8}$$

Radioactive thymidine is often used as a highly specific tracer for monitoring intracellular synthesis of DNA because it enters cells readily and its principal metabolic fate is salvage and incorporation into DNA.

Figure 22·21
Interconversions of purine nucleotides and their constituents. [Adapted from Traut, T. W. (1988). Enzymes of nucleotide metabolism: the significance of subunit size and polymer size for biological function and regulatory properties. *Crit. Rev. Biochem.* 23:121–169.]

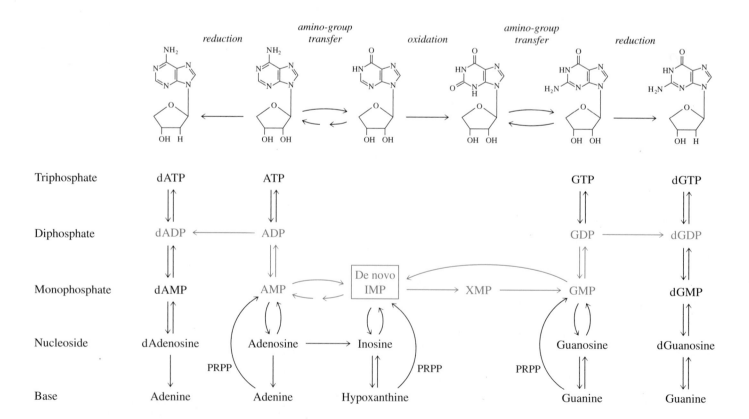

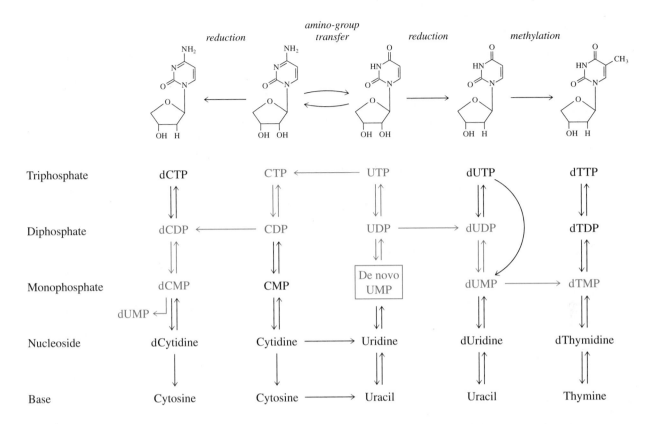

Triphosphate	dCTP	CTP ← UTP		dUTP	dTTP
Diphosphate	dCDP ← CDP		UDP → dUDP		dTDP
Monophosphate	dCMP	CMP	De novo UMP	dUMP → dTMP	
	dUMP ←				
Nucleoside	dCytidine	Cytidine → Uridine		dUridine	dThymidine
Base	Cytosine	Cytosine → Uracil		Uracil	Thymine

reduction   *amino-group transfer*   *reduction*   *methylation*

Figure 22·22
Interconversions of pyrimidine nucleotides and their constituents. [Adapted from Traut, T. W. (1988). Enzymes of nucleotide metabolism: the significance of subunit size and polymer size for biological function and regulatory properties. *Crit. Rev. Biochem.* 23:121–169.]

22·11 Interconversion of Nucleotides: A Summary

Nucleotides and their constituents are interconverted by a number of reactions, many of which we have seen already. The principal routes for purine nucleotide metabolism are shown in Figure 22·21. IMP, the first nucleotide product of the de novo biosynthetic pathway, is readily converted to AMP and GMP, their di- and triphosphates, and the deoxy counterparts of these nucleotides. The kinases responsible for the conversion of monophosphates to di- and triphosphates have been discussed in Section 10·5. As we will see in the next section, nucleotides are not only interconverted but also degraded into their constituents. Salvage reactions can re-form nucleotides from purine bases produced by nucleotide degradation.

The major interconversions of pyrimidine nucleotide metabolism are illustrated in Figure 22·22. UMP formed by the de novo pathway can be converted to cytidine and thymidine phosphates. In addition, free pyrimidines can be salvaged by the action of phosphorylases and kinases that lead to formation of nucleosides and nucleotides, respectively.

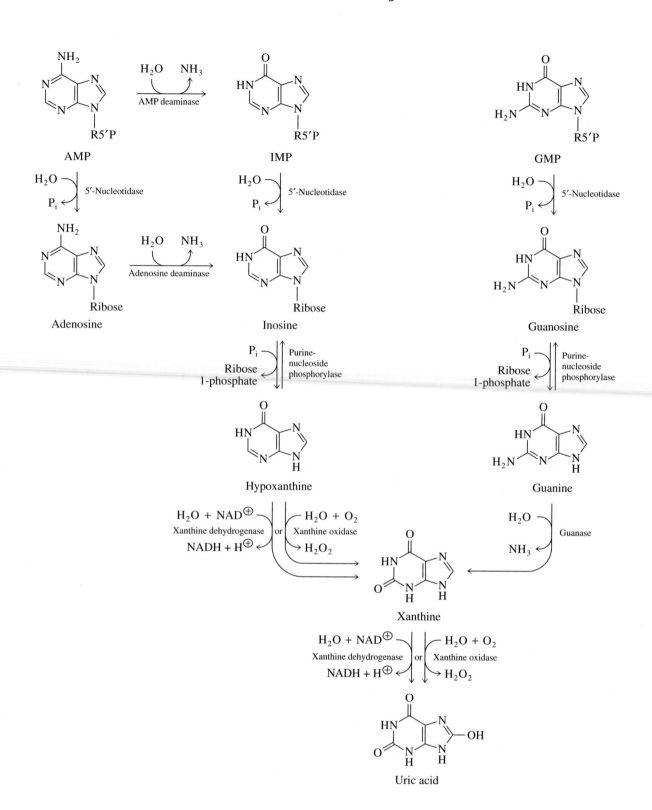

Figure 22·23
Pathways for the breakdown of AMP and
GMP to uric acid.

22·12 Purine Breakdown by Birds, Reptiles, and Primates Leads to Uric Acid

Although most intracellular purine and pyrimidine molecules are salvaged, some are catabolized. These molecules originate from an excess of ingested nucleotides or from routine intracellular turnover of nucleic acids.

In birds, reptiles, and primates (including humans), purine nucleotides are converted to uric acid, which is excreted. In birds and reptiles, amino acid catabolism also leads to uric acid; in mammals, surplus nitrogen from amino acid catabolism is disposed of in the form of urea. Birds and reptiles lack enzymes that catalyze further breakdown of uric acid, but many organisms degrade uric acid to other products, as we will see in the next section.

The pathways leading from AMP and GMP to uric acid are summarized in Figure 22·23. Hydrolytic removal of phosphate from AMP and GMP produces adenosine and guanosine, respectively. Adenosine is deaminated to inosine by the action of adenosine deaminase. Similarly, AMP is deaminated to IMP by the action of AMP deaminase. Hydrolysis of IMP produces inosine, which can be converted to hypoxanthine by phosphorolysis. Phosphorolysis of guanosine produces guanine. Both of these phosphorolysis reactions (as well as phosphorolysis of several deoxynucleosides) are catalyzed by purine-nucleoside phosphorylase and produce ribose 1-phosphate (or deoxyribose 1-phosphate) as well as a base. Adenosine is not a substrate of mammalian purine-nucleoside phosphorylase.

Individuals with deficient activity of the enzyme adenosine deaminase suffer a severe disease of the immune system in which their lymphocytes fail to function properly. It has been suggested that the observed buildup of dATP associated with the disease inhibits ribonucleoside diphosphate reductase, blocking synthesis of DNA and the proliferation of lymphocytes required for the immune response. Lymphocytes arise as differentiated products of bone marrow cells, offering a target for the new technique of **gene replacement therapy.** In this procedure, marrow cells are removed, the gene for wild-type adenosine deaminase is transferred into the cells in culture, and the cells are returned to the patient. The altered marrow cells quickly generate a population of lymphocytes containing the wild-type adenosine deaminase. Under very strict guidelines, the first human trial of this type of gene transfer was begun in 1990, when the gene for adenosine deaminase was introduced into lymphocytes of two individuals having defects in this enzyme. These patients were also receiving injections of purified enzyme. The gene therapy increased the level of enzyme activity well above the activity level associated with injection therapy alone, and the immune function of both patients improved significantly. Development of gene replacement therapy could greatly benefit individuals suffering from inherited metabolic diseases that arise from single-protein defects, such as cystic fibrosis and hemophilia.

Hypoxanthine formed from inosine is oxidized to xanthine, and xanthine is oxidized to uric acid, the final product in Figure 22·23. Either xanthine oxidase or xanthine dehydrogenase can catalyze both reactions. In the reactions catalyzed by xanthine oxidase, electrons are transferred to O_2 to form hydrogen peroxide, H_2O_2. Xanthine oxidase, an extracellular enzyme in mammals, appears to be an altered form of the intracellular enzyme xanthine dehydrogenase, which generates the same products as xanthine oxidase but transfers electrons to $NAD^{\oplus}$ to form NADH. These two enzyme activities occur widely in nature and exhibit broad substrate specificity. Their active sites contain complex electron-transfer systems that include an iron-sulfur center, a molybdopterin, and FAD.

In most cells, guanine is deaminated to xanthine in a reaction catalyzed by guanase. Animals that lack guanase excrete guanine. For example, pigs excrete guanine but metabolize adenine derivatives further to allantoin, the major end product of catabolism of purines in most mammals.

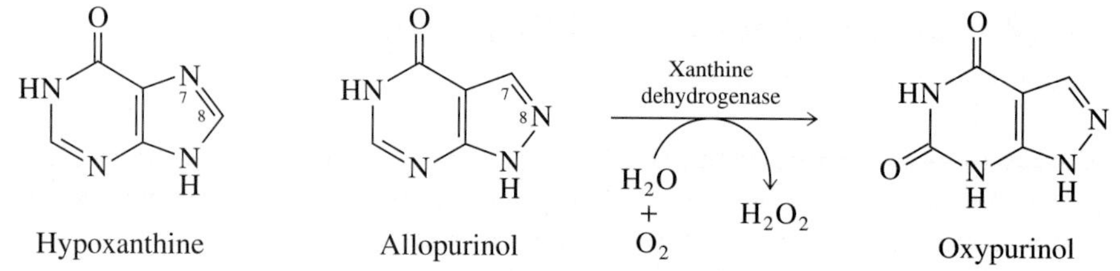

Figure 22·24
Allopurinol and oxypurinol. Xanthine dehydrogenase catalyzes the oxidation of allopurinol. The product of this oxidation, oxypurinol, binds tightly to xanthine dehydrogenase, inhibiting the enzyme.

Hypoxanthine   Allopurinol $\xrightarrow[\substack{H_2O \\ + \\ O_2}]{\text{Xanthine} \atop \text{dehydrogenase}}$ H_2O_2 Oxypurinol

Gout is a disease of humans caused by the overproduction or inadequate excretion of uric acid. Sodium urate in blood is relatively insoluble, and when its concentration is elevated, it can crystallize (sometimes along with uric acid) in cartilage and soft tissues, especially the kidney, and in toes and joints. The deposits in joints can cause severe pain. Gout has several causes, including partial deficiency of hypoxanthine-guanine phosphoribosyltransferase activity, which results in less salvage of purines and more catabolic production of uric acid. Gout can also be caused by defective regulation of purine biosynthesis.

Gout can be treated by administration of allopurinol, a synthetic C-7, N-8 positional isomer of hypoxanthine (Figure 22·24). Allopurinol is converted in cells to oxypurinol, a powerful inhibitor of xanthine dehydrogenase. The administration of allopurinol prevents the formation of abnormally high levels of uric acid, thus preventing deposits and formation of kidney stones. During treatment with allopurinol, neither hypoxanthine nor xanthine accumulates. These compounds are converted to IMP and XMP via the reaction catalyzed by hypoxanthine-guanine phosphoribosyltransferase. Subsequent reactions form AMP and GMP. Hypoxanthine and xanthine are more soluble than sodium urate and uric acid, and if they are not reutilized by salvage reactions, they are excreted by the kidney.

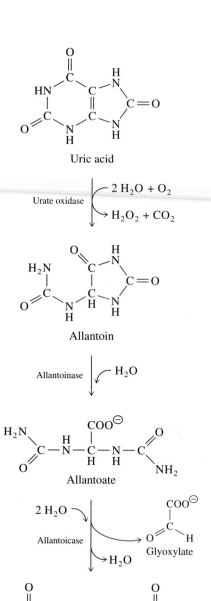

Uric acid

Urate oxidase $2 H_2O + O_2$
                 $H_2O_2 + CO_2$

Allantoin

Allantoinase H_2O

Allantoate

Allantoicase $2 H_2O$ $COO^{\ominus}$
                            Glyoxylate
                 H_2O

Urea   +   Urea

H_2O Urease   Urease H_2O

$CO_2 + 2 NH_3$ $CO_2 + 2 NH_3$

Figure 22·25
Catabolism of uric acid through oxidation and hydrolysis.

22·13 Most Organisms Degrade Uric Acid

In most organisms, uric acid can be further oxidized. Urate oxidase, also called uricase, catalyzes the O_2-dependent conversion of uric acid to allantoin, H_2O_2, and CO_2 (Figure 22·25). In this process, the pyrimidine ring of uric acid is opened. Allantoin is the major end product of purine degradation in most mammals (though not humans, for whom the end product is uric acid). Allantoin is also excreted by turtles, some insects, and gastropods.

The enzyme allantoinase catalyzes the hydrolytic opening of the imidazole ring of allantoin to produce allantoate, the conjugate base of allantoic acid. Some bony fishes (teleosts) possess allantoinase activity but cannot degrade allantoate and therefore excrete it as the end product of purine degradation.

Most fishes, amphibians, and freshwater mollusks metabolize allantoate one step further. These species contain the enzyme allantoicase, which catalyzes the hydrolysis of allantoate to one molecule of glyoxylate and two molecules of urea. Thus, urea is the nitrogenous end product of purine catabolism in these organisms.

Finally, many organisms—including plants, marine crustaceans, and many marine invertebrates—are able to hydrolyze urea in a reaction catalyzed by urease. Carbon dioxide and ammonia are the products of this reaction. Urease is found only in organisms in which hydrolysis of urea does not lead to ammonia toxicity. For example, in plants, ammonia generated from urea is rapidly assimilated by the action of glutamine synthetase. In marine creatures, toxic ammonia is produced in surface organs such as gills, which allows it to be flushed away.

22·14 The Purine Nucleotide Cycle in Muscle Produces Ammonia

Exercising muscle produces ammonia in proportion to the work performed. The pathway for ammonia formation involves one of the enzymes of purine catabolism, AMP deaminase, integrated into a route for energy production.

AMP deaminase catalyzes the deamination of AMP to IMP.

$$AMP + H_2O \longrightarrow IMP + NH_3 \qquad \textbf{(22·9)}$$

In exercising muscle, the IMP is recycled to AMP by the same reactions we saw earlier in the biosynthesis of AMP: GTP-dependent condensation with the amino group of aspartate and elimination of fumarate (Figure 22·9). The cyclic sequence of reactions, termed the **purine nucleotide cycle,** is shown in Figure 22·26. The net reaction for one turn of the cycle is

$$Aspartate + GTP + H_2O \longrightarrow Fumarate + GDP + P_i + NH_3$$
$$\textbf{(22·10)}$$

For continued operation, the purine nucleotide cycle must have a supply of amino nitrogen in the form of aspartate. Amino acids contribute the ammonia as they undergo transamination with α-ketoglutarate to form glutamate, and glutamate transaminates with oxaloacetate to form aspartate, which enters the cycle. When active, the purine nucleotide cycle supports some of the energy requirements of muscular work; the production of fumarate increases the capacity of the citric acid cycle to oxidize acetyl CoA.

A deficiency of AMP deaminase in muscle is characterized by muscle fatigue and cramps after exercise. Individuals with this deficiency do not accumulate ammonia or IMP during exercise. The disorder can cause considerable discomfort but is not life-threatening.

Figure 22·26
The purine nucleotide cycle.

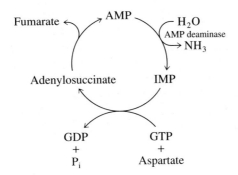

22·15 Pyrimidines Are Catabolized to Acetyl CoA and Succinyl CoA

The catabolism of pyrimidine nucleotides begins with hydrolysis to the corresponding nucleosides and P_i, catalyzed by 5′-nucleotidase. This reaction and subsequent reactions of pyrimidine nucleotide catabolism are shown in Figure 22·27. In the case of CMP, initial hydrolysis to cytidine is followed by deamination to uridine in a reaction catalyzed by cytidine deaminase. The glycosidic bonds of uridine and thymidine are then cleaved by phosphorolysis in reactions catalyzed by uridine phosphorylase and thymidine phosphorylase, respectively. Deoxyuridine can also undergo phosphorolysis catalyzed by uridine phosphorylase. The products of these phosphorolysis reactions are ribose 1-phosphate or deoxyribose 1-phosphate in addition to uracil and thymine.

The catabolism of pyrimidine bases leads to intermediates of central metabolism; thus, no distinctive excretory products are formed. Breakdown of both uracil and thymine involves several steps (Figure 22·28). First, the pyrimidine ring is reduced to a 5,6-dihydropyrimidine in a reaction catalyzed by dihydrouracil dehydrogenase, a reductase different from the one involved in de novo biosynthesis. The reduced ring is then opened by hydrolytic cleavage of the N-3—C-4 bond in a reaction catalyzed by dihydropyrimidinase. The resulting carbamoyl-β-amino acid

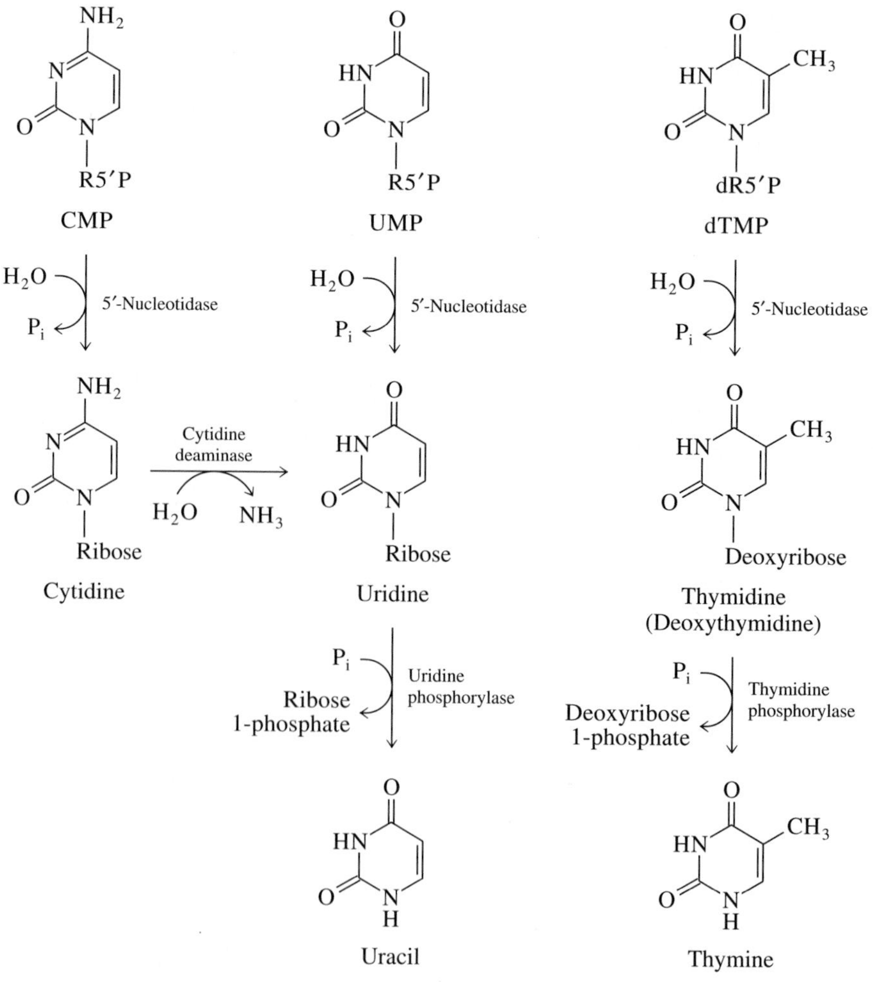

Figure 22·27
Major pathways for the degradation of pyrimidine nucleotides to uracil and thymine.

derivative (ureidopropionate or ureidoisobutyrate) is further hydrolyzed to $NH_4^\oplus$, $HCO_3^\ominus$, and a β-amino acid. β-Alanine (from uracil) and β-aminoisobutyrate (from thymine) can then be converted to acetyl CoA and succinyl CoA, respectively, which can enter the citric acid cycle and be converted to other compounds. In bacteria, β-alanine can also be used in the synthesis of pantothenate, a constituent of coenzyme A.

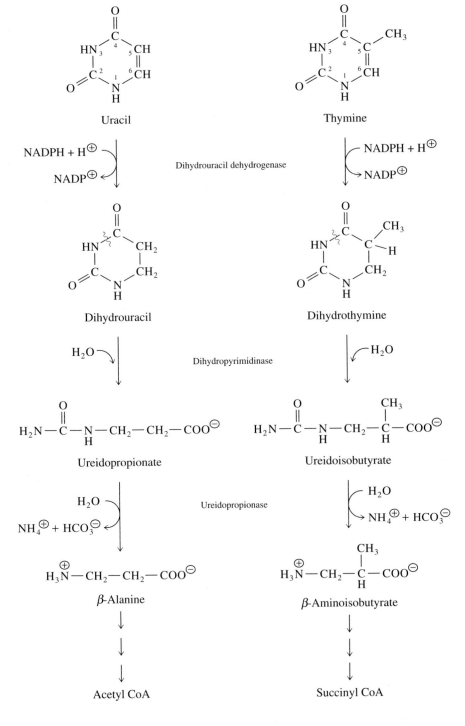

Figure 22·28
Catabolism of uracil to β-alanine and thymine to β-aminoisobutyrate, which can be further degraded to acetyl CoA and succinyl CoA, respectively.

Summary

Most organisms can synthesize purines and pyrimidines de novo. This biosynthesis involves nucleotides as key intermediates.

The de novo synthesis of purine nucleotides is achieved by a 10-step pathway that leads to the formation of IMP (inosinate). Isotopic labelling experiments showed that the carbon and nitrogen atoms of the purine ring are derived from glycine, 10-formyltetrahydrofolate, glutamine, aspartate, and carbon dioxide. The purine molecule is assembled by the stepwise addition of precursor units to a series of intermediates built upon a foundation of ribose 5′-phosphate donated by 5-phosphoribosyl 1-pyrophosphate (PRPP). The first purine-containing product of the de novo pathway, IMP, can be converted to either AMP or GMP. These compounds are precursors of nucleic acids and a variety of nucleotide coenzymes.

PRPP can also react directly with adenine, guanine, or hypoxanthine in salvage reactions to yield AMP, GMP, or IMP, respectively. A deficiency of hypoxanthine-guanine phosphoribosyltransferase, the enzyme that catalyzes the salvage of hypoxanthine and guanine, is the cause of Lesch-Nyhan syndrome, a disease that leads to spasticity and severe mental disability.

In the synthesis of the pyrimidine nucleotide UMP, PRPP enters the pathway *after* completion of the ring structure. The metabolic precursors of the pyrimidine ring are bicarbonate, glutamine, and aspartate. In eukaryotes, the regulation of this pathway occurs at the first step, namely, the formation of carbamoyl phosphate. In prokaryotes, the enzyme that catalyzes the second step—aspartate transcarbamoylase (ATCase)—is inhibited allosterically by the ultimate products CTP and UTP. In eukaryotes, two multifunctional proteins catalyze key steps in the biosynthesis of UMP. CTP is formed by the ATP- and glutamine-dependent amidation of UTP.

Deoxyribonucleotides are synthesized by reduction of ribonucleotides at C-2′ of the ribose moiety. The reaction, catalyzed by ribonucleotide reductase, requires the reducing power of NADPH. Reducing equivalents from NADPH are transferred via a protein coenzyme, either thioredoxin or glutaredoxin. Deoxyribose formation occurs at the nucleoside diphosphate level in most organisms, but in certain microorganisms reduction takes place at the triphosphate level. In both *E. coli* and mammalian cells, ribonucleotide reductase is closely regulated. Allosteric effectors control not only the overall catalytic activity of the enzyme but also its substrate specificity. Through the action of ribonucleotide reductase and nucleoside diphosphate kinases, dATP, dGTP, and dCTP are made available for the synthesis of DNA.

Thymidylate (dTMP) is formed from deoxyuridylate (dUMP) by a methylation reaction in which 5,10-methylenetetrahydrofolate donates both a one-carbon group and a hydride ion. 7,8-Dihydrofolate, the other product of this methylation, is recycled by reduction to the active coenzyme tetrahydrofolate. Inhibition of this reduction by methotrexate can prevent the synthesis of dTMP and thus of DNA.

In birds and reptiles, waste nitrogen from both amino acid and purine nucleotide catabolism is incorporated into IMP and eventually excreted as uric acid. Primates (for whom urea is the main excretion product of amino acid catabolism) also degrade surplus purines to uric acid. Most organisms catabolize uric acid further to allantoin, allantoate, urea, or even ammonia. In humans, overproduction of uric acid leads to gout.

Pyrimidines can be catabolized to ammonia, bicarbonate, and either β-alanine (from cytosine or uracil) or β-aminoisobutyrate (from thymine). Further breakdown produces acetyl CoA and succinyl CoA, respectively.

Selected Readings

General Reference

Traut, T. W. (1988). Enzymes of nucleotide metabolism: the significance of subunit size and polymer size for biological function and regulatory properties. *Crit. Rev. Biochem.* 23:121–169.

Purine Metabolism

Elion, G. B. (1989). The purine path to chemotherapy. *Science* 244:41–47. Description of the use of purine analogs for treatment of leukemia, gout, and viral infections and of azathioprine for immunosuppression in kidney transplantation.

Florkin, M. (1979). Biosynthesis of the purine nucleotides. In *Comprehensive Biochemistry,* Vol. 33A, M. Florkin and E. H. Stotz, eds. (Amsterdam: Elsevier Science Publishing Company), pp. 325–355. An excellent summary of the history of the research that elucidated the purine de novo pathway.

Lowenstein, J. M. (1972). Ammonia production in muscle and other tissues: the purine nucleotide cycle. *Physiol. Rev.* 52:382–414. A review of the pathway for the formation of ammonia in muscle.

Rowe, P. B. (1984). Folates in the biosynthesis and degradation of purines. In *Folates and Pterins,* Vol. 1, R. L. Blakley and S. J. Benkovic, eds. (New York: John Wiley & Sons), pp. 329–344. Description of the role of tetrahydrofolate in purine nucleotide metabolism.

Wyngaarden, J. B. (1976). Regulation of purine biosynthesis and turnover. *Adv. Enzyme Regul.* 14:25–37. Describes how purine nucleotide synthesis de novo is regulated at its first unique step.

Pyrimidine Metabolism

Jones, M. E. (1980). Pyrimidine nucleotide biosynthesis in animals: genes, enzymes, and the regulation of UMP biosynthesis. *Annu. Rev. Biochem.* 49:253–279. Review describing how three structural genes code for the six enzyme centers.

Kantrowitz, E. R., and Lipscomb, W. N. (1988). *Escherichia coli* aspartate transcarbamylase: the relation between structure and function. *Science* 241:669–674. Description of the three-dimensional structure of this enzyme.

Schachman, H. K. (1988). Can a simple model account for the allosteric transition of aspartate transcarbamoylase? *J. Biol. Chem.* 263:18583–18586. A review of the regulation of this enzyme.

Wild, J. R., Loughrey-Chen, S. J., and Corder, T. S. (1989). In the presence of CTP, UTP becomes an allosteric inhibitor of aspartate transcarbamoylase. *Proc. Natl. Acad. Sci. USA* 86:46–50.

Reduction of Ribonucleotides

Stubbe, J. (1990). Ribonucleotide reductases: amazing and confusing. *J. Biol. Chem.* 265:5329–5332. A brief review of the mechanisms of ribonucleotide reductases.

Thelander, L., and Reichard, P. (1979). Reduction of ribonucleotides. *Annu. Rev. Biochem.* 48:133–158. A review of the reaction and its regulation.

Thymidylate Synthase

Douglas, K. T. (1987). The thymidylate synthesis cycle and anticancer drugs. *Medicinal Research Reviews* 7:441–475. A chemical view of inhibitions of thymidylate synthase and dihydrofolate reductase.

Finer-Moore, J. S., Montfort, W. R., and Stroud, R. M. (1990). Pairwise specificity and sequential binding in enzyme catalysis: thymidylate synthase. *Biochemistry* 29:6977–6986. A stereochemical mechanism for the reaction catalyzed by thymidylate synthase based upon the three-dimensional structure of the enzyme.

Ivanetich, K. M., and Santi, D. V. (1990). Bifunctional thymidylate synthase-dihydrofolate reductase in protozoa. *FASEB J.* 4:1591–1597. Considers the usefulness of two enzymes on a single polypeptide chain.

Santi, D. V., and Danenberg, P. V. (1984). Folates in pyrimidine nucleotide biosynthesis. In *Folates and Pterins,* Vol. 1, R. L. Blakley and S. J. Benkovic, eds. (New York: John Wiley & Sons), pp. 345–398. A comprehensive review of the biosynthesis of dTMP.

23

Integration of Fuel Metabolism in Mammals

In previous chapters, we considered many reactions and pathways of biological chemistry. Our presentation was organized by treating pathways as discrete metabolic units. The concept of the pathway has some limitations. Pathways frequently intersect and may even overlap for several reactions. The end of one pathway and the beginning of the next is defined somewhat arbitrarily, since the starting compounds and end products of a pathway are themselves products and precursors. Pathways in cells may be linked by the effects of regulatory molecules, which further blurs the division between one pathway and another. Finally, pathways may extend across organs, and a sequence of reactions in one organ may be operating simultaneously, but in the opposite direction, in another organ. Intermediates flowing between the two organs therefore complete an interorgan cycle. Mapping the flux between different organs under different conditions is an ongoing research effort.

The major catabolic and anabolic pathways are shown in Figure 23·1 (next page). You have already encountered all of these pathways and reactions. We will now reconsider some of these metabolic routes in the context of overall metabolism, including the integration of multiple pathways and their regulation. Using humans as a model mammalian system, we examine the metabolic changes that occur in response to consumption of a meal consisting of carbohydrate, lipid, and protein, followed by prolonged fasting. We will see how the different organs of the body work in concert to maintain a constant supply of fuel to the tissues during and long after the ingestion of a meal. As we proceed, we will consider in passing some of the roles of major organs and their biochemical features during different metabolic states, such as fuel use in muscle during rest and exercise. Ultimately, we will give a sense of the degree to which the pathways of mammalian metabolism are interdependent and integrated, rather than independent and isolated.

23·1 Metabolic Tasks Are Distributed Among Different Tissues

The organs of the mammalian body are distinct not only in gross anatomy but in biochemistry. Different organs, indeed different cells within an organ, possess characteristic complements of enzymes. Not all cells possess the enzymatic machinery

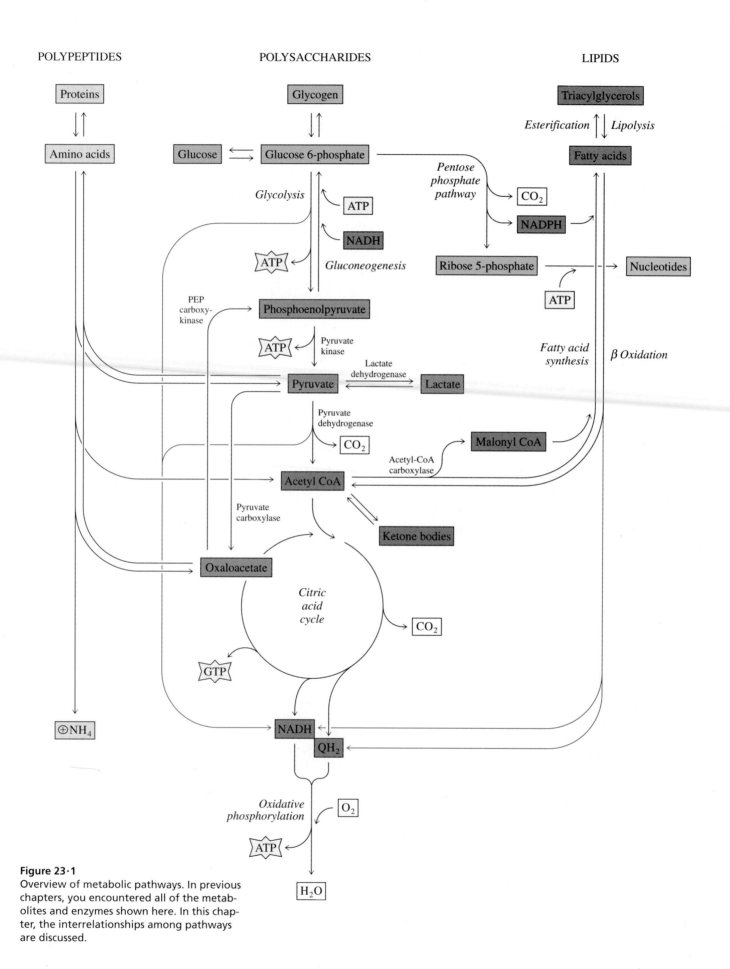

Figure 23·1
Overview of metabolic pathways. In previous chapters, you encountered all of the metabolites and enzymes shown here. In this chapter, the interrelationships among pathways are discussed.

necessary to carry out all metabolic reactions. This differentiation results in tissue-specific metabolic functions. The patterns of energy metabolism in the major tissues are shown in Figure 23·2. Some tissues, such as brain, are simply consumers of fuels. Others, such as liver and adipose tissue, function as fuel stores, taking up excess fuels in times of plenty and releasing fuels at other times.

	Fuel Used	Fuel Released
BRAIN	Glucose Ketone bodies	—
INTESTINE	Glucose Glutamine	Dietary lipids, amino acids, and carbohydrates
LIVER	Glucose Amino acids Fatty acids	Glucose (from glycogen and gluconeogenesis) Triacylglycerols Ketone bodies
ADIPOSE TISSUE	Glucose	Fatty acids Glycerol
MUSCLE	Glucose Fatty acids Lactate Ketone bodies Branched-chain amino acids	Lactate Alanine and glutamine
KIDNEY	NOT SHOWN Glucose Fatty acids Ketone bodies Lactate Glutamine (in acidosis)	Glucose (from gluconeogenesis) *significant during prolonged starvation*

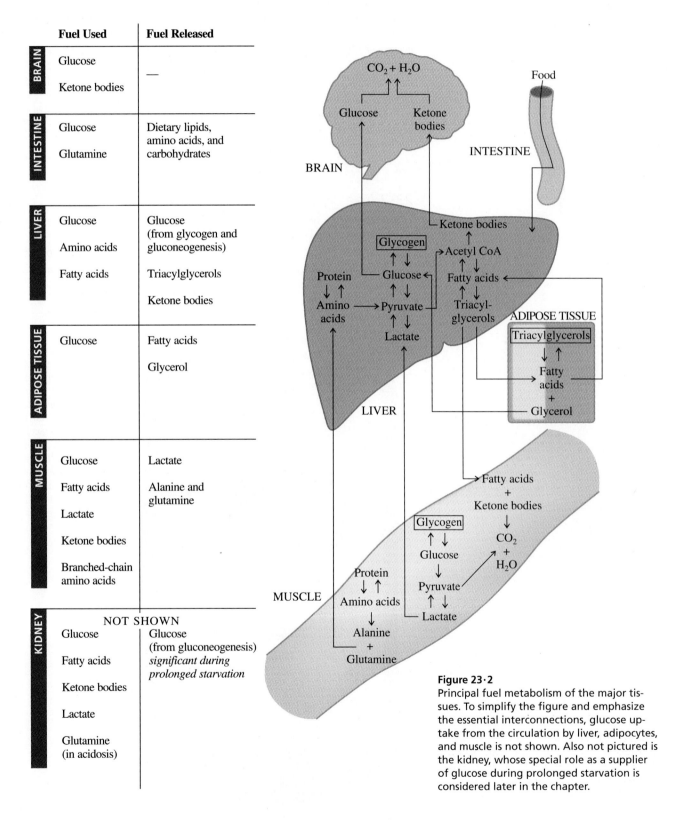

Figure 23·2
Principal fuel metabolism of the major tissues. To simplify the figure and emphasize the essential interconnections, glucose uptake from the circulation by liver, adipocytes, and muscle is not shown. Also not pictured is the kidney, whose special role as a supplier of glucose during prolonged starvation is considered later in the chapter.

Table 23·1 Fuel reserves in a 70 kg man

Fuel	kJ	g
Triacylglycerols	590 000	15 000
Glycogen in liver	1500	90
Glycogen in muscle	6000	350
Free glucose	320	20
Protein	100 000	6000

A fuel is defined as a metabolite that circulates within the body and can be taken up by cells and used to produce ATP. The principal fuels in mammals are glucose, lactate, fatty acids, ketone bodies, triacylglycerols, and amino acids. In addition to these compounds, glycerol and ethanol can be used as fuel, but their contribution to overall energy metabolism is generally minor. Not all cells can use each of the major fuels, and not all fuels are available at all times.

According to the definition of fuels as circulating metabolites, glycogen is not a fuel but a storage molecule. Degradation of glycogen mobilizes the fuel glucose. Triacylglycerols represent an interesting test of the definition since they are clearly energy storage molecules within adipose tissue, muscle, and liver. Degradation releases fatty acids that serve as fuels. However, there is good evidence that muscle derives energy from circulating triacylglycerols—extracellular hydrolysis of triacylglycerols releases fatty acids that are taken up for fuel.

The fuel reserves of a 70 kg man are given in Table 23·1. The carbohydrate and protein reserves are limited, but it appears that there is no limit to the amount of triacylglycerol that can potentially be stored.

23·2 The Liver Plays a Central Role in Mammalian Metabolism

The unique role of the liver is reflected in both its anatomical location and its specialized biochemistry. Blood perfuses most organs in parallel—the arterial system

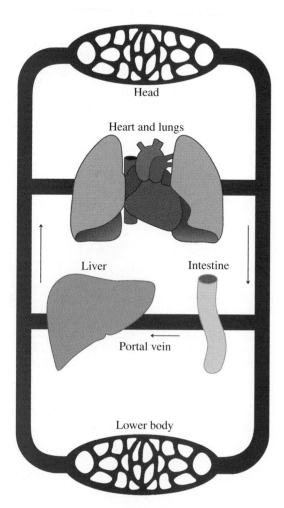

Figure 23·3
Placement of the liver in the circulation. Most tissues are perfused in parallel. The liver is perfused in series with visceral tissue; blood that drains from the intestine flows to the liver via the portal vein. Thus, the liver is ideally placed to regulate the passage of fuels to the peripheral tissues.

supplies oxygenated blood to the organs, and venous blood is returned to the lungs for reoxygenation (Figure 23·3). The liver, however, is perfused in series with the portal-drained visceral tissues—blood from the major visceral tissues (gastrointestinal tract, pancreas, spleen, and adipose tissue) drains into the portal vein and then flows to the liver. This means that after the products of digestion are absorbed by the intestines, they can pass immediately to the liver. The liver is therefore ideally placed to take up surplus nutrients and to regulate the delivery of fuels to the periphery. Peripheral tissues are defined loosely as tissues other than the liver—an indication of the central importance of the liver in interorgan metabolism.

The liver not only regulates the distribution of dietary fuels, it becomes a supplier of fuel from its own reserves when dietary supplies are exhausted. The liver also serves as the site for interconversion of fuels. For instance, energy derived from fatty acid oxidation is used to synthesize glucose in the liver, and glucose is then made available to tissues that cannot directly oxidize fatty acids. The liver is thus a metabolic buffer, ensuring that the levels of circulating fuels are neither too high nor too low and interconverting metabolites so that the fuels in circulation are those that peripheral organs require.

23·3 Blood Glucose Concentration Is Maintained Within Strict Limits

The main product of the digestion of dietary carbohydrates is glucose, the major fuel of the body. Blood glucose levels are highest after ingestion of a meal rich in carbohydrates, but even then the level of glucose is not unregulated. The delivery of glucose to the periphery after a meal is controlled by both the rate of absorption by the intestine and the rate of uptake and release by the liver. The concentration of glucose in the blood seldom exceeds 10 mM, and even after many weeks of starvation the levels of glucose in the blood do not drop below 3 mM. Hypo- and hyperglycemia both have detrimental consequences. For instance, certain parts of the brain have an absolute requirement for glucose. If the concentration of glucose in the blood falls below 2.5 mM, glucose uptake into the brain is compromised, resulting in central nervous system depression and coma. Conversely, when blood glucose levels are very high, glucose is filtered out of the blood by the kidneys at a rate faster than it can be reabsorbed, and glucose is lost in the urine. The loss of glucose is accompanied by osmotic loss of water and electrolytes, which can result in hyperglycemic hyperosmotic coma.

Glucose homeostasis—the maintenance of constant levels of glucose in the circulation—is achieved by balancing glucose synthesis or absorption against utilization. Circulating glucose is turned over very rapidly. Forty grams of glucose are removed from the circulation per hour in the period immediately following a meal, which means that the total supply of glucose in the blood, about 20 g, would last less than 30 minutes. As consumption by tissues removes dietary glucose from the circulation, liver glycogen and gluconeogenesis become the sources of glucose. However, if the body continued to remove glucose from the blood at the high rate of 40 grams per hour, liver glycogen could supply glucose for only another 8–10 hours. Similarly, although gluconeogenesis could in theory yield 3–4 kilograms of glucose from the catabolism of all of the body's protein (1.75 grams of protein is required to make 1 gram of glucose), protein catabolism must be restricted. The body lacks a purely storage form of protein, and even the loss of small amounts of protein to fuel metabolism carries a penalty in lost function. Therefore, as glucose becomes less plentiful, hormones act to restrict the use of glucose to those tissues that

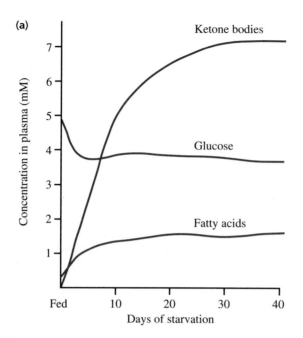

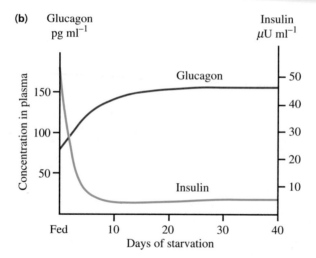

Figure 23·4
Levels of circulating fuels and hormones during progressive starvation. (a) Levels of fuels. (b) Levels of the hormones insulin and glucagon.

absolutely require it. Figure 23·4 illustrates the levels of several fuels and hormones in the bloodstream beginning immediately after a meal and continuing for several weeks of starvation.

Tissues for which glucose is the only usable fuel include kidney medulla, retina, red blood cells, and parts of the brain. These obligatory glycolytic tissues have little or no capacity for oxidative metabolism and therefore depend on the metabolism of glucose to lactate for the generation of ATP. In humans, the brain requires about 120 grams of glucose per day, and the other glycolytic tissues together require about 60 grams of glucose per day. This use of glucose accounts for about half of the daily carbohydrate consumption for a person with a typical intake of 9000 kJ that is 50% carbohydrate.

The consumption of glucose by obligatory glycolytic tissues is essentially constant over time. The supply of glucose from exogenous sources (i.e., the diet) occurs in bursts (meals), with the sudden load spread out somewhat by the rate of absorption, which takes approximately 2–4 hours.

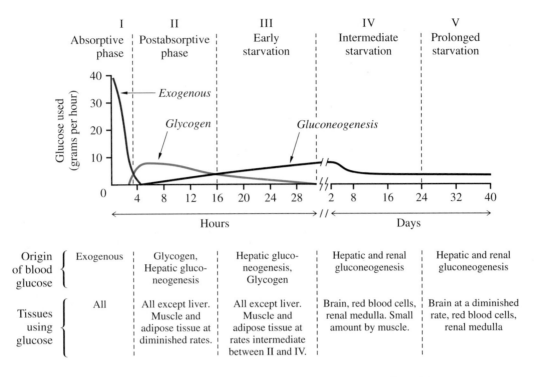

	I	II	III	IV	V
	Absorptive phase	Postabsorptive phase	Early starvation	Intermediate starvation	Prolonged starvation
Origin of blood glucose	Exogenous	Glycogen, Hepatic gluco-neogenesis	Hepatic gluco-neogenesis, Glycogen	Hepatic and renal gluconeogenesis	Hepatic and renal gluconeogenesis
Tissues using glucose	All	All except liver. Muscle and adipose tissue at diminished rates.	All except liver. Muscle and adipose tissue at rates intermediate between II and IV.	Brain, red blood cells, renal medulla. Small amount by muscle.	Brain at a diminished rate, red blood cells, renal medulla

Figure 23·5
The five phases of glucose homeostasis. The graph, based on observations of a number of individuals, illustrates glucose utilization in a 70 kg man consuming a 100 g oral glucose load and then fasting for 40 days. [Adapted from Ruderman, N. B., Aoki, T. T., and Cahill, G. F. (1978). Gluconeogenesis and its disorders in man. In *Gluconeogenesis: Its Regulation in Mammalian Species*, R. W. Hanson and M. A. Mehlman, eds. (New York: John Wiley & Sons), p. 515.]

23·4 There Are Five Phases of Glucose Homeostasis

In the late 1960s, George Cahill, Jr., had the opportunity to examine a number of obese patients as they underwent therapeutic starvation for six weeks. During the starvation period, the subjects received only water and micronutrients (vitamins and minerals). Cahill's observations led to a general description of glucose homeostasis based on five distinct phases (Figure 23·5). During the initial absorptive phase, which lasts up to 4 hours, most cells in the body utilize glucose. As exogenous glucose is used up, the postabsorptive phase begins, and the body mobilizes liver glycogen to maintain circulating glucose levels. After about 24 hours, liver glycogen is depleted, and the only source of circulating glucose is gluconeogenesis in the liver. Gluconeogenesis continues at a high rate for a few days, then decreases during the intermediate starvation phase. In prolonged starvation, the fifth and final phase of glucose homeostasis, glucose utilization is very low, the body reserves are severely decreased, and if refeeding does not occur, death will follow.

We will now examine in detail the strikingly different metabolic states that occur during the period diagrammed in Figure 23·5. As we follow the changes that occur during progressive starvation, a picture will emerge of which interactions are most important in the maintenance of metabolic stability.

23·5 The Level of Glucose in the Blood Rises After a Meal: The Absorptive Phase

The concentration of blood glucose is low after an overnight fast—it does not drop appreciably lower even during prolonged starvation. The consumption of carbohydrate after an overnight fast results in a large increase in the concentration of glucose in the blood entering the liver via the portal vein. Although it is very difficult to measure the change precisely in humans, in rats the concentration of glucose in the portal vein has been shown to rise as high as 15 mM, a considerably higher concentration than is ever reached in the peripheral circulation. Most of the blood delivered to the liver arrives from the portal vein, with about 20% received from the hepatic artery. When the difference in glucose concentration between the portal vein and the hepatic artery is large, the liver takes up glucose. The mechanism for sensing the difference in glucose concentration between the portal vein and the hepatic artery, and for subsequently adjusting liver metabolism, is poorly understood.

The other major products of carbohydrate digestion, fructose and galactose, are almost completely removed from the portal blood by the liver and therefore are not available as fuels for peripheral tissues. Within the liver, fructose and galactose are phosphorylated and ultimately converted to glycolytic intermediates. We considered the metabolism of fructose and galactose in Sections 17·1A and B.

Although the immediate fate of dietary glucose depends on the composition of the diet and the physiological state, in general about one-third is deposited in the liver, one-third goes to muscle, and one-third goes to the brain and obligatory glycolytic tissues. Uptake of glucose by the liver, muscle, and brain is quite efficient, and only a very large load of dietary glucose will overwhelm the capacity of the body to restrict the level of circulating glucose. When this happens, glucose is lost to the urine as the ability of the kidney to reabsorb filtered glucose is exceeded.

A. Insulin Regulates Glucose Uptake

The presence of food in the intestinal lumen causes the secretion into the bloodstream of a number of gastrointestinal hormones, such as gastric inhibitory polypeptide and cholecystokinin, which induce the β cells of the pancreas to begin to secrete insulin. The increase in blood glucose concentration following absorption of dietary glucose causes the β cells of the pancreas to release more insulin. The exact pathway by which metabolism of glucose leads to release of insulin is not clear. It is known that glucose must enter the β cells and be metabolized; glucose does not exert its effects by merely binding to a receptor on the pancreatic cell surface. Other agents circulating in the blood, including certain amino acids, also increase the secretion of insulin.

Insulin has many functions, one of which is to stimulate glucose uptake by peripheral tissues such as skeletal muscle and adipose tissue. In mammals, glucose is taken up into cells by specific transporters (Table 23·2). We have already seen the

Table 23·2 Mammalian hexose transporters

Name	Tissues	Functions
GLUT1	Brain, kidney, colon, placenta	Basal uptake of glucose
GLUT2	Liver, pancreatic β cells, small intestine, kidney	Uptake and release of glucose by hepatocytes and kidney; β cell glucose sensor; release of glucose from intestinal cells
GLUT3	Many tissues	Basal uptake of glucose
GLUT4	Skeletal and cardiac muscle, adipose tissue	Insulin-stimulated glucose uptake
GLUT5 (Fructose transporter)	Small intestine	Fructose transport
SGLT1 (Sodium-linked transporter)	Small intestine and kidney	Uptake of dietary glucose from small intestine; reabsorption of filtered glucose in kidney

[Adapted from Pessin, J. E., and Bell, G. I. (1992). Mammalian facilitative glucose transporter family: structure and molecular regulation. *Annu. Rev. Physiol.* 54:912.]

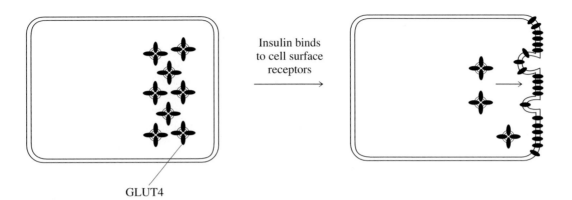

GLUT4

Figure 23·6
Regulation of glucose transport by insulin. A pool of GLUT4 transporters resides in the cell embedded in the membranes of intracellular vesicles. The binding of insulin to the cell surface promotes the translocation of the vesicles to the plasma membrane and fusion of the vesicles with the membrane, thereby increasing the capacity of the cell to transport glucose. [Adapted from Pessin, J. E., and Bell, G. I. (1992). Mammalian facilitative glucose transporter family: structure and molecular regulation. *Annu. Rev. Physiol.* 54:919.]

sodium-linked active transporter of the small intestine, which is responsible for importing glucose from the lumen (Section 12·7D). The same transporter, called SGLT1, is also present in the kidney, where it reabsorbs filtered glucose. The other hexose transporters listed in the table carry out reversible facilitated transport.

The elevated level of blood glucose, together with an elevated level of insulin, results in high rates of glucose uptake into the cells of skeletal muscle and adipose tissue by a transport system called GLUT4. In 1980, Samuel Cushman and Tetsuro Kono independently described the mechanism by which the rate of glucose transport is increased. A preformed pool of glucose transporters exists in the cytoplasm of insulin-targeted cells, with the transporters embedded in the membranes of intracellular vesicles. When insulin binds to receptors on the cell surface, the vesicles fuse with the plasma membrane, thereby delivering the transporters to the cell surface and increasing the capacity of the cell to transport glucose (Figure 23·6).

GLUT4 is found at high levels only in adipose tissue and skeletal muscle, and insulin-regulated uptake of glucose is seen only in these tissues. In these and most other tissues, a basal level of glucose transport in the absence of insulin is accomplished by two other transporters, GLUT1 and GLUT3.

The transporter GLUT2 is found in liver and pancreatic β cells. In liver, GLUT2 is responsible for the traffic of glucose in and out. The remaining transporter, known as GLUT5, appears to be responsible for intestinal fructose transport and is now often referred to as the fructose transporter.

Once inside a cell, glucose is rapidly phosphorylated by the action of a hexokinase. This reaction effectively traps the glucose inside the cell since phosphorylated glucose cannot diffuse across membranes and is not a substrate for the glucose transporters.

As discussed in Section 15·7B, liver and β cells of the pancreas express hexokinase IV, or glucokinase as it is more commonly known. Most hexokinases have K_m values for glucose of about 0.1 mM or less; the K_m of glucokinase for glucose is approximately 2–5 mM. Glucokinase exhibits sigmoidal kinetics towards glucose at physiological concentrations of ATP (Figure 23·7). A protein inhibitor present in liver raises the K_m value of glucokinase for glucose to about 5–10 mM, and activity increases dramatically as the concentration of glucose increases through the steep part of the sigmoidal curve. The kinetic properties indicate that flux through glucokinase is very low when blood glucose levels are in the normal range but can increase when the glucose level in the hepatic portal vein rises after a meal. Glucokinase—in contrast to the other hexokinases—is not subject to inhibition by physiological concentrations of glucose 6-phosphate. Thus, the glucokinase activity in liver is able to respond effectively to high concentrations of glucose and is able to clear glucose from the portal blood. Since glucokinase is the major hexokinase in the β cells of the pancreas, significant glucose utilization in these cells (and insulin secretion triggered by metabolism of glucose in those cells) occurs only when glucose is abundant.

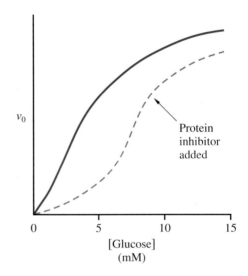

Figure 23·7
Plot of initial velocity, v_0, versus concentration of glucose for glucokinase from rat liver at 4 mM ATP. The addition of a protein inhibitor raises the apparent K_m for glucose.

B. Glycogen Is Formed During the Absorptive Phase

During the absorptive phase, almost all tissues in the body use glucose as a fuel, and the availability of other fuels, such as free fatty acids and ketone bodies, is limited. Excess glucose is stored primarily as glycogen in liver and muscle, but there is a major difference in the function of these two stores and hence in the regulation of their use. Glucose stored as glycogen in the liver is available to the whole body because liver possesses the enzyme glucose 6-phosphatase, which catalyzes the hydrolysis of glucose 6-phosphate to glucose. This glucose can exit the liver via glucose transporters. Skeletal muscle lacks glucose 6-phosphatase, and glucose 6-phosphate obtained from glycogenolysis in muscle is not released as glucose. Instead, it is metabolized in muscle via glycolysis to provide ATP for contraction. Lactate produced by glycolysis in muscle is released into the circulation to be used as a fuel by other tissues or to be used by liver for gluconeogenesis.

There are two pathways leading from dietary glucose to synthesis of glycogen in the liver, known as the direct and indirect pathways (Figure 23·8). In the direct pathway, dietary glucose is taken up by the liver, phosphorylated by the action of glucokinase, and converted to glycogen. A more roundabout route to glycogen also exists, related to the conditions that generally exist just before a meal is consumed. If it has been quite a few hours since the last meal, the level of glucokinase in the liver will be low. The level may increase 10-fold in response to a meal, but this inductive synthesis, stimulated by insulin, takes a considerable amount of time. In addition, hepatic gluconeogenesis, which provides glucose to the body between meals, continues right up to the time that a meal is consumed. It has been proposed that on refeeding, the liver does not immediately begin to take up glucose. Rather, it is suggested that gluconeogenesis in the liver continues, but instead of being released as glucose, the glucose 6-phosphate formed is used for glycogen synthesis. At the same time, peripheral tissues metabolize dietary glucose to lactate, which is released to the circulation, taken up by the liver, and converted to glycogen by gluconeogenesis and glycogen synthesis. This latter route is known as the indirect pathway of glycogen synthesis. The exact contributions of the direct and indirect pathways to glycogen synthesis after a meal have been the subject of much controversy over the past 10 years. The current consensus is that for a short time following refeeding after a fast, the indirect pathway operates and may even predominate, but the direct pathway rapidly becomes the major pathway.

Figure 23·8
Direct and indirect pathways of hepatic glycogen synthesis. In the direct pathway, dietary glucose is taken up by the liver, phosphorylated, and incorporated into glycogen. In the indirect pathway, dietary glucose is taken up by peripheral tissues and metabolized to lactate. Lactate is transported to liver and converted to glucose 6-phosphate by the gluconeogenic pathway, and glucose 6-phosphate is then incorporated into hepatic glycogen.

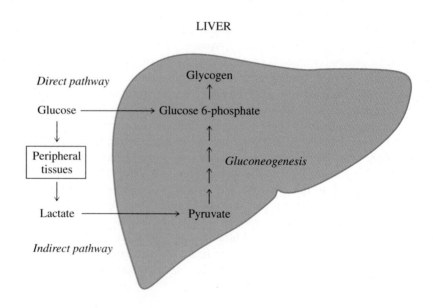

C. Glycogen Synthesis in the Liver Is Regulated by Insulin and Glucagon

Glycogen synthesis in the liver is subject to regulation by insulin and glucagon. In addition to stimulating glycogen synthesis, insulin has numerous other effects on hepatic metabolism, such as stimulation of glycolysis, fatty acid synthesis and esterification, and protein synthesis, as well as inhibition of glycogenolysis, gluconeogenesis, fatty acid oxidation, ketogenesis, and proteolysis. The signalling pathways and second messengers for insulin action are not yet known. Glucagon acts only on liver, where its major effects on hepatic metabolism are, in general, opposite those of insulin. Glucagon acts by stimulating the formation of cyclic AMP, which triggers activation of protein kinase A by a pathway that has been quite well characterized (Section 17·5C).

One enzyme known to be under the control of insulin is glycogen synthase. Glycogen synthase consists of four subunits (subunit M_r 85 000). Each subunit possesses eight potential phosphorylation sites, so the number of possible phosphorylation states is large. The phosphorylation state of glycogen synthase is controlled by insulin, which appears to act by causing dephosphorylation of one particular site, thereby stimulating the enzyme, but again the intracellular signalling pathway is not known. The action of insulin together with a high concentration of glucose 6-phosphate stimulates the formation of glycogen.

When the level of insulin is high, indicating an abundance of fuel, the level of glucagon is low. In the absence of stimulation by glucagon, glycogenolysis in the liver is inhibited. Thus, a futile cycle of glycogen synthesis and breakdown is prevented.

D. Muscle Glycogen Supplies Fuel for Local Requirements

Muscle is usually classified as cardiac, skeletal, or smooth. Cardiac muscle preferentially oxidizes fatty acids and depends heavily on aerobic metabolism. In the absorptive phase, the availability of free fatty acids is very low, whereas glucose is abundant. During this period, even the heart uses glucose as its principal fuel. Skeletal muscle makes up 40% of the body mass but can account for much more than 40% of energy metabolism during exercise. The rate of metabolism in skeletal muscle can increase 100-fold within a fraction of a second. Smooth muscle, exemplified by the muscle lining the walls of intestines and blood vessels, has very low metabolic activity and will not be considered further here.

Skeletal muscle is composed of several types of fibers, with individual muscles possessing a distinctive combination of these types. The biochemical properties of muscle fibers have important bearings on function. Type I fibers, or slow twitch, have a large number of mitochondria and highly active oxidative pathways such as β oxidation of fatty acids and the citric acid cycle. They have an extremely good blood supply and low levels of glycogen. Type II fibers, or fast twitch, are divided into two classes. Type IIA fibers are oxidative but have fewer mitochondria than type I fibers. Type IIB fibers are glycolytic and have relatively few mitochondria. Oxidative metabolism in type IIB fibers is limited by poor blood supply, but these fibers contain large amounts of glycogen and the energy-storage molecule phosphocreatine (Section 14·11). The capacity of the glycolytic pathway in type IIB fibers is very high, and these fibers are capable of considerably greater power output than type I or type IIA fibers, though only for short periods of time.

Despite a marked increase in ATP utilization during exercise, intramuscular ATP levels do not decrease by more than 20%. ATP can be regenerated for a short period by phosphoryl-group transfer from phosphocreatine in a near-equilibrium reaction catalyzed by creatine kinase.

$$
\underset{\text{Phosphocreatine}}{
\begin{array}{c}
\text{COO}^{\ominus} \\
| \\
\text{CH}_2 \\
| \\
\text{H}_3\text{C} - \text{N} \\
| \\
\overset{\oplus}{\underset{\text{H}_2\text{N}}{}}\text{C} \\
\text{N} - \text{P} - \text{O}^{\ominus}
\end{array}
}
\quad
\xrightleftharpoons[\text{ADP} \quad \text{ATP}]{\underset{\text{kinase}}{\text{Creatine}} \quad \text{ADP} \quad \text{ATP}}
\quad
\underset{\text{Creatine}}{
\begin{array}{c}
\text{COO}^{\ominus} \\
| \\
\text{CH}_2 \\
| \\
\text{H}_3\text{C} - \text{N} \\
| \\
\text{C} \\
\overset{\oplus}{\text{H}_2\text{N}} \quad \text{NH}_2
\end{array}
}
$$

(23·1)

Although present in high concentrations (up to 25 mM), phosphocreatine can only replenish ATP for short intervals—about 3–4 seconds. Phosphocreatine also serves to buffer the level of ATP in oxidative tissue during sudden bursts of activity in the short period before delivery of fuel and activation of metabolic pathways catches up with demand.

Once the reservoir of phosphocreatine is exhausted, glycogen is drawn upon for energy. It is present within the cell (meaning it does not have to be transported via the blood and imported into the cell, processes that are too slow to meet the needs of suddenly active muscle), the metabolism of glycogen to lactate does not require additional blood-borne oxygen, and the product of glycogen breakdown is glucose 6-phosphate, which spares a molecule of ATP that would be consumed if glucose were the substrate for glycolysis. Even when muscle glycogen is plentiful, however, the amount of high-intensity exercise that can be fueled by anaerobic metabolism is limited. After a short time, the accumulation of protons causes exhaustion (Section 15·4). Low-intensity exercise can be sustained for much longer periods.

Intramuscular fuels are inadequate for extremely prolonged exertion, such as a 26-mile marathon, and therefore other fuels must be provided. Blood-borne glucose from hepatic glycogen suffices for a while, but this must soon be supplemented. The need is met by the mobilization of fatty acids. Increased perfusion of the muscles increases the delivery of blood-borne fuels such as glucose and fatty acids, along with the molecular oxygen necessary for the oxidative catabolism of these fuels. Type IIB fibers (glycolytic, nonoxidative) make a minor contribution during a marathon. Type I fibers (oxidative, large numbers of mitochondria, and highly active oxidative pathways) are more important for this type of exercise. The intensity of exercise supportable by fatty acid oxidation is considerably less than that supported by anaerobic metabolism (phosphoryl-group transfer from phosphocreatine and metabolism of glycogen to lactate). However, aerobically supported exercise can be maintained for much longer periods of time.

In trained athletes, distance running is supported by a balance of glycolytic and oxidative metabolism. Training increases the ability of muscle to resist acidification

Figure 23·9
Carbohydrate loading—glycogen deposition and utilization in skeletal muscle in response to fasting and exercise. [Adapted from McGilvery, R. W., and Goldstein, G. (1979). *Biochemistry: A Functional Approach*, 2nd ed. (Philadelphia, Pennsylvania: W. B. Saunders Company).]

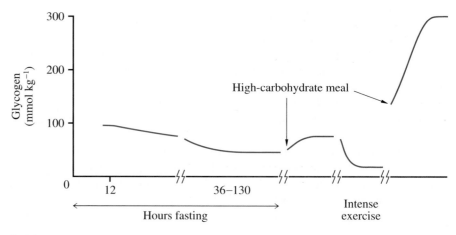

and consequent fatigue, and exercise can be sustained for longer periods. Ultimately, however, depletion of muscle glycogen stores occurs, and fatigue ensues, accompanied by a considerable drop in work output. The use of lipid fuels along with muscle glycogen slows the depletion of glycogen and thus delays fatigue. Similarly, the higher the store of glycogen in muscle at the beginning of exercise, the longer the duration of exercise, and perhaps as important, the longer the duration at a higher intensity. Muscle glycogen levels can be raised above normal levels by the athlete's technique of carbohydrate loading. Typically, glycogen levels are depleted by intense exercise 1–3 days before an athletic event, after which the athlete consumes meals containing large amounts of carbohydrate (Figure 23·9). When glycogen is restored from a state of depletion in muscle, the amount stored overshoots the amount of glycogen originally present. Carbohydrate loading does not cause a marked overshoot in liver glycogen stores. To capitalize on pre-race carbohydrate loading, distance runners attempt to calibrate pace and fuel reserves so that, ideally, exhaustion of muscle glycogen coincides precisely with the end of the race.

E. Glucose Metabolism Is Linked to Fatty Acid Metabolism in Liver

As glycogen reserves in liver are restored following the end of a fast or completion of exercise, the rate of glycogen synthesis decreases, and glucose is diverted into pathways that lead to the synthesis and storage of lipids. However, individuals carry out very little de novo fatty acid synthesis when fed a typical North American diet of ~40% fat. Instead, triacylglycerols are derived from dietary fatty acids, with glucose contributing the glycerol moieties.

In humans consuming a great excess of calories in the form of a high-carbohydrate/low-fat diet, fatty acid synthesis may occur. The conversion of glucose to fatty acids requires flux through glycolysis. However, during active fatty acid synthesis in the liver, the level of cytosolic citrate is high (citrate shuttles acetyl units from mitochondria to the cytosol, as described in Section 20·7), and citrate is an inhibitor of phosphofructokinase-1 (PFK-1). PFK-1 is also inhibited by ATP at physiological concentrations. The activator fructose 2,6-*bis*phosphate effectively relieves the inhibition of PFK-1 by citrate and ATP (Section 15·7C). Fructose 2,6-*bis*phosphate is synthesized in a reaction catalyzed by the bifunctional protein phosphofructokinase-2 (PFK-2). When phosphorylated by the action of protein kinase A, the bifunctional protein shows predominantly fructose 2,6-*bis*phosphatase activity; in its dephosphorylated form, the enzyme catalyzes the conversion of fructose 6-phosphate to fructose 2,6-*bis*phosphate. The regulation of the fructose 2,6-*bis*phosphate control system is shown in Figure 23·10. In the fed state (insulin high, glucagon

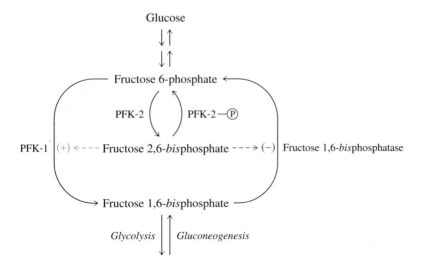

Figure 23·10
Regulation of hepatic carbohydrate metabolism by fructose 2,6-*bis*phosphate.

low), the level of fructose 2,6-*bis*phosphate is high, which stimulates PFK-1 and increases flux through glycolysis. High levels of fructose 2,6-*bis*phosphate also inhibit fructose 1,6-*bis*phosphatase, thus preventing a futile cycle. During a fast, the level of fructose 2,6-*bis*phosphate is low, decreasing the stimulation of PFK-1 and lifting the inhibition of fructose 1,6-*bis*phosphatase, and net flux occurs in the direction of gluconeogenesis.

High rates of insulin-stimulated glycolysis in the liver during the fed state provide large amounts of pyruvate, which is converted to acetyl CoA by the action of the pyruvate dehydrogenase complex. Insulin also stimulates acetyl-CoA carboxylase, resulting in elevated levels of malonyl CoA and an increase in fatty acid synthesis. Flux through the pentose phosphate pathway is also high under these conditions and provides the NADPH necessary for fatty acid synthesis. The fatty acids formed are then esterified, exported from the liver, and deposited in adipose tissue as a store of energy.

23·6 In the Absorptive Phase, Dietary Lipids Are Processed by Different Tissues

Lipids entering the body from the digestive tract are packaged in lipoproteins and transported in the lymph, eventually reaching the bloodstream. In the absorptive period, the amount of lipid in the blood can be high, but the concentration of free fatty

Figure 23·11
Entry of blood-borne dietary lipids into an adipocyte. Lipoprotein lipase (LPL) synthesized in the adipocyte is stored in intracellular vesicles and is secreted into the capillary, where it is anchored by glycosaminoglycans. In the capillary, LPL catalyzes the hydrolysis of triacylglycerols in lipoproteins to release free fatty acids. Fatty acids enter the adipocyte, where they are esterified and stored in fat droplets.

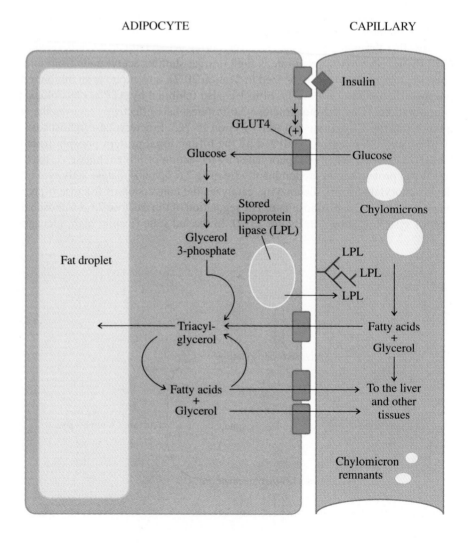

acids is low. Lipoproteins deliver triacylglycerols to peripheral tissues such as muscle and adipose tissue. At these sites, triacylglycerols are not taken up directly but are first hydrolyzed to release free fatty acids. Hydrolysis is catalyzed by lipoprotein lipase, an enzyme found predominantly in adipose tissue and skeletal muscle but expressed in almost every tissue. In adipose tissue, lipoprotein lipase is synthesized intracellularly and secreted into the lumen of a capillary, where it is anchored by glycosaminoglycans on the endothelial cell surface. In the lumen, lipoprotein lipase catalyzes the hydrolysis of triacylglycerols, releasing fatty acids for uptake into the adipocyte (Figure 23·11). Following entry into the cell, the fatty acids are esterified and stored. Glycerol 3-phosphate required for the esterification pathway (Section 20·13) arises from the metabolism of glucose to dihydroxyacetone phosphate via the initial reactions of glycolysis, followed by conversion of dihydroxyacetone phosphate to glycerol 3-phosphate in a reaction catalyzed by glycerol 3-phosphate dehydrogenase.

In the liver, fatty acids can be oxidized or esterified. Oxidation is predominantly a mitochondrial process; activated fatty acids must be transported into the mitochondria via the carnitine shuttle system (Section 20·2B). This system is very sensitive to inhibition by malonyl CoA. During the absorptive phase, hepatic levels of malonyl CoA are high, so fatty acids are prevented from entering the mitochondria and thus are available for esterification to triacylglycerols, a pathway that occurs in the cytosol. In liver, glycerol 3-phosphate needed for the esterification pathway can arise from glycolysis or from glycerol via the reaction catalyzed by glycerol kinase. This enzyme is not present in adipocytes, and glycerol arising from hydrolysis of triacylglycerols in adipocytes cannot serve as a substrate for triacylglycerol synthesis in adipose tissue. Instead, this glycerol is released and transported to the liver for further metabolism.

23·7 Dietary Proteins Are Usually Present in Excess of Requirements

The amount of protein in the average North American diet exceeds the requirement for growth and renewal. However, unlike carbohydrates and lipids, excess proteins and amino acids are not stored. Approximately half of the amino acids supplied by the diet are essential amino acids that are conserved for protein synthesis. Dietary glutamine, glutamate, aspartate, and asparagine, and a considerable amount of arterial glutamine, are metabolized within intestinal cells to provide energy. The nitrogen of these amino acids is released into the portal vein in the form of ammonia, alanine, citrulline, and proline. Once again, the location of the liver in the circulation is important since the large amounts of potentially toxic ammonia produced by the intestinal cells are delivered by the portal vein directly to the liver for detoxification by conversion to urea.

There is a large increase in the amino acid load reaching the liver after a meal. These amino acids can be used for protein synthesis, they can be catabolized, or they can pass through the liver untouched. Nonessential amino acids are generally freely interconverted in the liver due to the very high activities of alanine transaminase, aspartate transaminase, and glutamate dehydrogenase. Essential amino acids are conserved and are only catabolized if they are present in excess. Branched-chain amino acids may be used for hepatic protein synthesis, but unlike most other amino acids, they are not catabolized in the liver, and most pass through to the peripheral tissues. The other essential amino acids can be catabolized by the actions of liver enzymes. Flux through these reactions is controlled mainly by substrate availability. The catabolic enzymes exhibit high K_m values for their amino acid substrates, and there is little catabolism when the amino acids are present in low concentrations. Some of these enzymes, such as phenylalanine hydroxylase, are also regulated by reversible phosphorylation, and many are subject to long-term changes through regulation of gene expression, with the relevant genes being expressed at very high rates when the diet includes large amounts of protein.

In the absorptive phase, the large increase in the amount of amino acids delivered to the liver results in a marked increase in hepatic protein synthesis. The amino acids that are not incorporated into polypeptides are partially broken down. These amino acids are not completely oxidized in liver since this would produce more ATP than the metabolism of the liver could utilize, and the amount of ADP would swiftly become limiting. Instead, the partially oxidized carbon skeletons of amino acids enter the ATP-consuming pathways of urea synthesis and gluconeogenesis. Amino acid oxidation and gluconeogenesis are closely coupled, and the rate of gluconeogenesis may be limiting for the rate of hepatic amino acid catabolism.

The increased rate of hepatic protein synthesis observed in the absorptive phase may represent temporary stockpiling of some amino acids until conditions allow them to be converted to glucose via gluconeogenesis. Protein synthesis also consumes large amounts of ATP and thereby allows more amino acid catabolism. During the absorptive phase after a mixed meal, amino acid catabolism is the major energy-producing pathway in liver. Ammonia released from catabolized amino acids is converted to urea, and the carbon is converted to glucose 6-phosphate, which is then either released as free glucose or used for the synthesis of glycogen. Urea synthesis and gluconeogenesis consume much of the ATP generated by the catabolism of dietary amino acids.

Despite the high rates of amino acid utilization in the liver, the level of amino acids in the blood still rises in the absorptive phase, and most tissues take up amino acids for protein synthesis. As mentioned above, the branched-chain amino acids are not catabolized in the liver. They pass into the bloodstream and are taken up by the peripheral tissues. Both skeletal muscle and adipose tissue have high capacities for the catabolism of branched-chain amino acids. The first step is reversible transamination to form branched-chain α-keto acids. The branched-chain keto acids are then decarboxylated by a multienzyme complex, branched-chain α-keto acid dehydrogenase (Section 21·12E), which is similar to the pyruvate dehydrogenase complex and the α-ketoglutarate dehydrogenase complex. The degree of catabolism that occurs in the peripheral tissues seems to vary among species, with a considerable portion of the branched-chain α-keto acids released into the circulation in rats but apparently less in humans. However, some branched-chain α-keto acids probably undergo complete oxidation in muscle.

Glutamine is the most abundant amino acid in the body, making up over 40% of the free amino acid pool. Free glutamine is found predominantly in the blood plasma and inside skeletal muscle cells. Since glutamine absorbed from the diet is catabolized in intestinal cells, the large amount of glutamine in the body must arise from de novo synthesis. The major sites of glutamine synthesis are skeletal muscle, adipose tissue, the lungs, and under certain conditions, the liver. Glutamine is the major respiratory fuel in some cells, including intestinal cells, cells of the immune system, tumor cells, and fetal tissues. These cells also utilize large amounts of glutamine for the synthesis of nucleotides and proteins. However, these pathways account for only 5% of the glutamine taken up by these cells. The remaining 95% is catabolized for energy production.

23·8 Fuel Selection Gradually Shifts Between the Postabsorptive and Early Starvation Phases

The absorption of nutrients from the intestine takes 2–4 hours. Toward the end of this period, the delivery of dietary glucose slows and the use of glucose from endogenous sources increases. The level of insulin drops, and the level of glucagon rises. The liver stops removing glucose from the circulation, and adipose tissue and resting skeletal muscle no longer take up glucose for storage, although muscle still takes up glucose during contraction. The liver responds to the high glucagon/low insulin profile by breaking down glycogen to maintain the level of circulating glucose. Low levels of insulin also mean that the inhibition of lipolysis in adipose tissue is lifted, and free fatty acids are released into the circulation (Section 20·11). The heart and other tissues begin to use free fatty acids for fuel as their concentration in the blood increases.

A. Lipolysis During Early Starvation Provides Free Fatty Acids

The storage and release of fatty acids in adipose tissue represents a tightly regulated balance between esterification of fatty acids and hydrolysis of triacylglycerols by the action of hormone-sensitive lipase. As the concentration of glucose in the blood falls, insulin levels drop. Inhibition by insulin is the major regulator of lipolysis, and low levels of insulin are accompanied by high rates of lipolysis. Epinephrine and norepinephrine further stimulate lipolysis by increasing the activity of hormone-sensitive lipase via cAMP-dependent phosphorylation. Insulin causes a decrease in the level of cAMP in adipose tissue by stimulating cAMP phosphodiesterase, but this regulatory mechanism does not fully account for the antilipolytic effect of insulin—other signalling pathways yet to be identified are also important.

B. The Glucose–Fatty Acid Cycle Spares Glucose Utilization

When a choice is available, most tissues use fatty acids as fuels before ketone bodies, and both before glucose. This means that glucose is used only when it is abundant and other fuels are scarce. Fatty acids are used in preference to glucose even though the concentration of circulating glucose (>3 mM) is always much higher than the concentration of free fatty acids (<2 mM). The glucose–fatty acid cycle, first proposed by Philip Randle, explains how the preference for glucose or fatty acids is determined by metabolic conditions. As free fatty acids become available, they are taken up by muscle cells and metabolized by β oxidation and the citric acid cycle to yield energy. The oxidation of fatty acids is accompanied by increased levels of citrate, acetyl CoA, and NADH within the cell, which then inhibit glucose oxidation: citrate inhibits PFK-1, and acetyl CoA and NADH inhibit the pyruvate dehydrogenase complex. Thus, fatty acid oxidation effectively provides the energy required by the muscle cell and at the same time inhibits glucose utilization. Operation of the glucose–fatty acid cycle requires that low concentrations of glucose correspond to greater fatty acid availability. As outlined in the previous section, low availability of glucose results in decreased levels of insulin. Release of insulin-mediated inhibition of lipolysis in adipose tissue increases the availability of free fatty acids in the plasma. The glucose–fatty acid cycle is shown in Figure 23·12 (next page).

Figure 23·12
Glucose–fatty acid cycle. When the level of glucose is low, the level of insulin is low, and insulin inhibition of lipolysis is relieved. The oxidation of fatty acids generates inhibitors of glucose metabolism, thus sparing the use of glucose. In this way, the presence of free fatty acids generates a preference for fatty acids as a fuel over glucose.

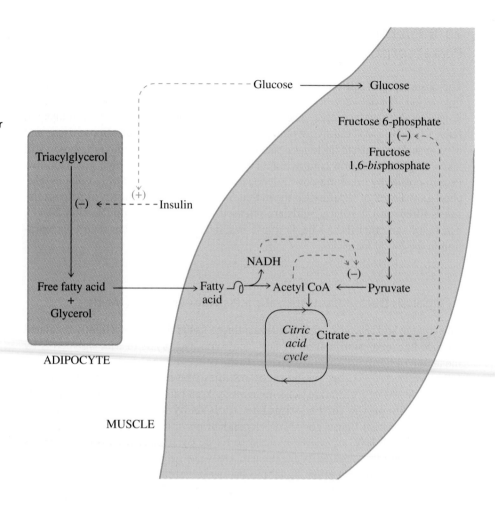

By means of the glucose–fatty acid cycle, fatty acid oxidation contributes to the maintenance of blood glucose levels by sparing the oxidation of glucose in peripheral tissues. This pattern of fuel use is usually observed only when the availability of glucose is low, such as after liver glycogen stores are exhausted, but the cycle predicts that any rise in the level of circulating fatty acids will decrease glucose use. This is indeed true. A drug such as caffeine, which inhibits cAMP phosphodiesterase and thereby raises cAMP levels, causes an increase in lipolysis in adipose tissue. The resulting increase in the level of circulating fatty acids causes a decrease in the oxidation of glucose.

C. Liver Glycogen Provides Circulating Glucose for a Few Hours

The level of glucagon rises gradually for several days as starvation progresses (Figure 23·4b). The binding of glucagon to receptors in the liver stimulates the production of cAMP, which stimulates protein kinase A, as detailed in Section 17·5A. This signalling cascade activates glycogen phosphorylase, leading to increased glycogen degradation. At the same time, glycogen synthase is phosphorylated and inactivated. Other important regulatory enzymes in the liver are also phosphorylated by the action of protein kinase A, including PFK-2 (inhibited), pyruvate kinase (inhibited), and acetyl-CoA carboxylase (inhibited). The result is decreased flux through glycolysis. The glucose 6-phosphate produced by glycogenolysis is hydrolyzed by the action of glucose 6-phosphatase and released into the circulation as free glucose.

23·9 During Intermediate Starvation, Hepatic Gluconeogenesis Is the Only Source of Glucose

Hepatic glycogen stores would last only a few hours if other fuels or sources of glucose were not provided. We saw in Section 23·8B how the glucose–fatty acid cycle operates to provide fatty acids as an alternate fuel while inhibiting the use of glucose, but even with the contribution of fatty acids, the hepatic glycogen reserves are depleted 16–24 hours into starvation. Nevertheless, the supply of glucose must be maintained. After liver glycogen stores are depleted, the only source of circulating glucose is gluconeogenesis.

In mammals, the only organs capable of gluconeogenesis are the liver and kidneys. During the early stages of starvation, the liver is by far the more important. The substrates taken up by the liver for gluconeogenesis are lactate, pyruvate, glycerol, alanine, and other amino acids that can give rise to three-carbon metabolites.

A critical aspect of metabolism at this stage of starvation is regulation of the pyruvate dehydrogenase complex. Mammals lack a bypass for this complex, which catalyzes the conversion of pyruvate (C_3) to acetyl CoA (C_2). Acetyl CoA from this reaction and from the oxidation of fatty acids cannot be converted to glucose in organisms that lack the reactions of the glyoxylate cycle (Section 16·8). Therefore, pyruvate decarboxylated to acetyl CoA represents a loss of gluconeogenic potential, increasing the metabolic pressure on the systems that maintain glucose homeostasis. If the pyruvate is converted to lactate or alanine instead of serving as a substrate for the pyruvate dehydrogenase complex, the pyruvate carbon skeleton can still serve as a precursor of glucose via gluconeogenesis in the liver.

A. The Cori Cycle Recycles Carbon and Transports Energy

The obligatory glycolytic tissues continue to use glucose and produce lactate in all five phases of glucose homeostasis. The lactate is converted to glucose in the liver for release to the peripheral tissues. In other words, a C_6 compound is degraded to two C_3 compounds, which are then used to resynthesize the C_6 compound (Figure 23·13). This interorgan loop, known as the **Cori cycle,** does not represent a supply

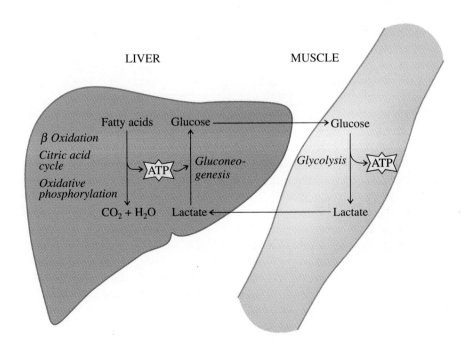

Figure 23·13
Cori cycle.

of new glucose to the body. Instead, the Cori cycle is simply a system for transferring energy from the liver to the peripheral tissues. The synthesis of glucose in the liver requires energy, and the catabolism of glucose to two C_3 compounds by glycolysis yields energy in the peripheral tissues. Much of the energy required for glucose synthesis comes from hepatic fatty acid oxidation. Thus, the cycle makes the energy of fatty acid oxidation available to tissues that lack the β-oxidation pathway.

If the molecules of the Cori cycle were completely oxidized by peripheral tissues, additional carbon would be required to maintain blood glucose levels. Recycling the same carbons permits fatty acid oxidation to contribute to glucose homeostasis even though fatty acids cannot be converted directly to glucose, while at the same time making economical use of available carbon.

B. The Glucose-Alanine Cycle Transports Nitrogen

The glucose-alanine cycle is a variant of the Cori cycle. A major end product of muscle metabolism is alanine, whose carbon skeleton arises from pyruvate produced by glycolysis. Lactate, pyruvate, and alanine are all in equilibrium in muscle by virtue of reactions catalyzed by lactate dehydrogenase and alanine transaminase.

$$
\underset{\text{Lactate}}{\overset{\displaystyle COO^{\ominus}}{\underset{\displaystyle CH_3}{\overset{\displaystyle |}{\underset{\displaystyle |}{CH-OH}}}}}
\xrightleftharpoons[]{\text{Lactate}\atop\text{dehydrogenase}}
\underset{\text{Pyruvate}}{\overset{\displaystyle COO^{\ominus}}{\underset{\displaystyle CH_3}{\overset{\displaystyle |}{\underset{\displaystyle |}{C=O}}}}}
\xrightleftharpoons[]{\text{Alanine}\atop\text{transaminase}}
\underset{\text{Alanine}}{\overset{\displaystyle COO^{\ominus}}{\underset{\displaystyle CH_3}{\overset{\displaystyle |}{\underset{\displaystyle |}{CH-\overset{\oplus}{N}H_3}}}}}
\qquad (23\cdot2)
$$

Therefore, we can redraw the Cori cycle to include alanine synthesis in muscle (in place of lactate), conversion of alanine to glucose (and urea) in liver, and conversion of glucose to alanine in muscle (Figure 23·14).

The glucose-alanine cycle has two advantages over the Cori cycle. First, it allows the transport of nitrogen to the liver (the only site of the urea cycle), and second, it yields more energy in the glucose-utilizing tissue. The reducing equivalents from NADH that would have been consumed in reducing pyruvate to lactate are now available to be transferred into the mitochondria in muscle for oxidation by the respiratory electron-transport chain.

Figure 23·14
Glucose-alanine cycle.

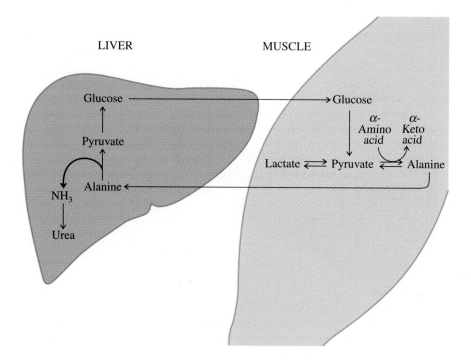

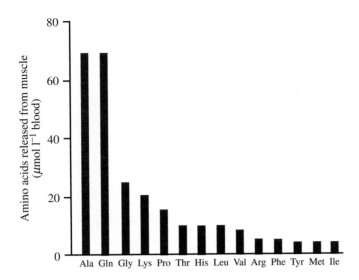

Figure 23·15
Amino acids released from muscle during starvation. [Adapted from Felig, P., and Wahren, J. (1975). Fuel homeostasis in exercise. *N. Engl. J. Med.* 293:1078–1084.]

C. Amino Acids Are the Major Precursors for New Glucose Synthesis

During intermediate starvation, about 100–110 grams of glucose are used per day, of which about 40–50 grams are accounted for by the activity of the Cori cycle and the glucose-alanine cycle, leaving about 50–60 grams that must be obtained by other means. Lipolysis in adipose tissue contributes enough glycerol to account for 15–20 grams of glucose per day, a rate that remains fairly constant throughout starvation. Essentially no esterification of fatty acids occurs in liver during starvation, and the available glycerol is directed entirely to gluconeogenesis. The only other source of gluconeogenic precursors is amino acids.

The rate of gluconeogenesis from amino acids is quite high even in the absorptive phase, when gluconeogenesis serves to dispose of the carbon skeletons of excess dietary amino acids. During early starvation, there is an increase in proteolysis in the liver, which releases considerable amounts of amino acids for hepatic gluconeogenesis. However, this does not last long. As starvation progresses, proteolysis in peripheral tissues increases to provide gluconeogenic precursors to the liver. The production of 100 g of glucose requires the breakdown of about 175 g of amino acids. All proteins in the body have a function, and any loss of protein to provide energy is accompanied by a loss of functional capacity. The rate of gluconeogenesis from amino acids, high during the first few days of starvation, must decline if body protein is to be conserved.

During the intermediate phase of starvation, muscle is in a state of net proteolysis, with release of amino acids for hepatic gluconeogenesis. Muscle protein contains a full complement of amino acids, and there is some release of most amino acids into the circulation (Figure 23·15). As we saw earlier, branched-chain amino acids have a special role in muscle—some are released, but most are metabolized within the muscle cell. Aspartate is also metabolized within muscle. Of the amino acids released, alanine and glutamine make up more than 50%, although they (together with glutamate, the immediate precursor of glutamine) make up only 11–15% of muscle protein. Some of the glutamine released from muscle cells is from a very large, preexisting pool of free glutamine. However, the steady release of glutamine during starvation requires that glutamine be continually synthesized in muscle.

The carbon skeleton required for glutamine synthesis presumably comes from the catabolism of other amino acids within the muscle cell. Muscle takes up some glutamate from the circulation, but the amount is insufficient to account for the amount of glutamine released. The actual pathway in muscle that gives rise to the carbon skeleton of glutamine has not been firmly established. According to the

Figure 23·16
Proposed pathway for de novo synthesis of glutamine in skeletal muscle. The precursors are aspartate and branched-chain amino acids, with acetyl CoA possibly arising from the degradation of branched-chain amino acids.

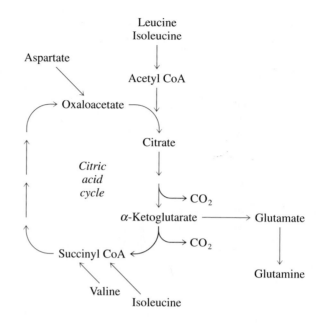

best-accepted proposal, the catabolism of aspartate and two of the branched-chain amino acids, valine and isoleucine, gives rise to oxaloacetate, which then combines with acetyl CoA (possibly derived from the catabolism of isoleucine and leucine) to form citrate. The citrate is then converted to α-ketoglutarate and then glutamate for glutamine synthesis (Figure 23·16). Note that the fixation of carbon from acetyl CoA into glutamine does not represent the net incorporation of a two-carbon unit into a gluconeogenic precursor. When glutamine is metabolized in other tissues (liver, intestine, kidney), the pathways of glutamine metabolism include release of CO_2, and the acetyl CoA carbons are lost, just as they would be if the acetyl CoA carbons had been metabolized by the reactions of the citric acid cycle.

The other amino acids released by muscle proteolysis are taken up by the liver and used for gluconeogenesis. In addition, although the glucose-alanine cycle does not give rise to new glucose, there is some alanine present in muscle protein, and considerable amounts of alanine are synthesized from glutamine in tissues such as the small intestine. Therefore, some of the alanine arriving at the liver represents potential new glucose.

D. Regulation of Hepatic Gluconeogenesis

During starvation, the major sites of regulation of gluconeogenesis in the liver are enzymes involved in the catabolism of amino acids and enzymes that bypass the ir-reversible reactions of glycolysis: phosphoenolpyruvate (PEP) carboxykinase, fruc-tose 1,6-*bis*phosphatase, and glucose 6-phosphatase.

PEP carboxykinase is not known to be subject to short-term regulation, but in most species, the half-life of the cytosolic isozyme is only 6 hours. This short half-life allows relatively rapid changes in the amount of the enzyme through changes in the rate of its synthesis. In the rat, glucagon stimulates and insulin inhibits the tran-scription of the gene encoding this enzyme, with a resultant high activity of PEP carboxykinase during starvation. Conversely, hepatic pyruvate kinase levels de-crease during starvation, and the activity of the enzyme is further lowered by cAMP-dependent phosphorylation and by high levels of alanine, which allosteri-cally inhibit the enzyme. Further up the gluconeogenic pathway, fructose 2,6-*bis*-phosphate levels become low, again the result of mechanisms that depend on the levels of glucagon and cAMP. Low levels of fructose 2,6-*bis*phosphate allow flux through the reaction catalyzed by fructose 1,6-*bis*phosphatase while flux through the opposing reaction, catalyzed by PFK-1, is inhibited, as shown earlier in Figure 23·10. The activity of glucokinase is decreased due to decreased gene expression and a lack of substrate (recall that the K_m of glucokinase for glucose in the presence of its protein inhibitor is about 5–10 mM; the concentration of circulating glucose during intermediate starvation is much lower than this). Little is known about the regulation of glucose 6-phosphatase. This enzyme is present in the endoplasmic reticulum, and regulation of its activity may involve the regulation of other proteins that transport substrates and products to and from the lumen of the endoplasmic reticulum.

E. Ketone Bodies Are Synthesized in the Liver

As starvation progresses, lipolysis in adipocytes continues to release considerable amounts of fatty acids, which are taken up by the liver. The level of glycolytic intermediates is low in liver, as is the level of malonyl CoA, the main inhibitor of fatty acid oxidation. With inhibition of carnitine acyltransferase I (CAT I) by malonyl CoA lifted, fatty acids can enter the mitochondria, where they are partially oxidized. The citric acid cycle in liver can absorb only a fraction of the acetyl CoA produced by β oxidation, and most is converted to the ketone bodies β-hydroxybutyrate and acetoacetate. The ketone bodies are water soluble, and their concentration in the blood increases with the length of starvation, as shown earlier in Figure 23·4a. Ketone bodies that are taken up and metabolized by tissues spare the use of glucose, just as glucose is spared by the operation of the glucose–fatty acid cycle. As ketone bodies initially become available, they provide up to 50% of the energy requirements of skeletal muscle. However, as starvation progresses, skeletal muscle takes up smaller amounts of ketone bodies and returns to the use of fatty acids as the major fuel. The mechanism governing the change in fuel preference from ketone bodies back to fatty acids is not known.

As glucose availability becomes limited, even the brain must use alternate fuels. The blood-brain barrier is impermeable to long-chain fatty acids, which therefore cannot be used by the brain. Ketone bodies cross the blood-brain barrier, enter the brain, and serve as a fuel that lowers the amount of glucose required by the brain. Ketone bodies are also used by the small intestine, thus sparing glutamine. The most important overall effect of the use of fatty acids and ketone bodies as alternatives to glucose is that less demand for glucose means that less proteolysis of muscle protein is required to supply the gluconeogenic pathway.

23·10 In Late Starvation, Renal Gluconeogenesis Gains in Importance

As the late stage of starvation begins, the body has adapted to alternate fuels, and all but the most essential glucose-consuming pathways have been shut off (Figure 23·17, next page). The advantage of using ketone bodies during starvation is that they can be used by the brain, but they have a notable disadvantage: large amounts of ketone bodies are lost in the urine. This wastes valuable energy. In addition, ketone bodies are strong acids, and their dissociation causes a drop in plasma pH, a condition known as metabolic acidosis. As described in Section 21·13, the kidneys respond to metabolic acidosis by metabolizing large amounts of glutamine to drive

Figure 23·17
Pattern of fuel use during the late stage of starvation. The use of glucose as fuel is essentially limited to the brain. Other tissues, including parts of the brain, depend on ketone bodies and fatty acids for fuel.

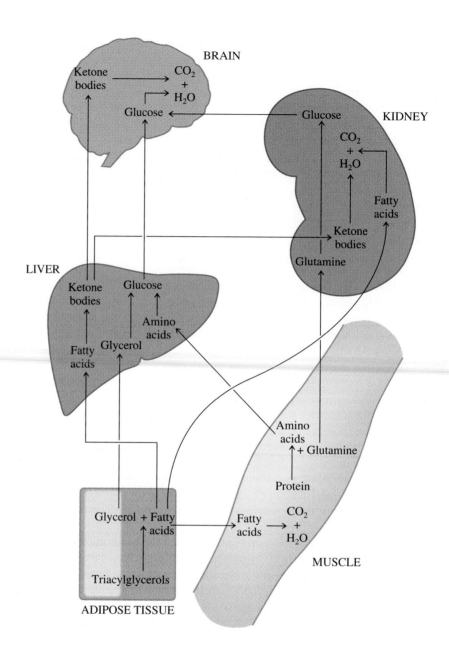

repletion of the bicarbonate buffer system. Renal metabolism of glutamine generates ammonia that is released in urine, and the carbon skeleton of glutamine is used for gluconeogenesis. The accompanying production of bicarbonate helps maintain acid-base homeostasis.

$$2\,\text{Glutamine} \longrightarrow \text{Glucose} + 4\,\text{NH}_4^{\oplus} + 4\,\text{HCO}_3^{\ominus} \qquad \textbf{(23·3)}$$

In the intestine, ketone bodies replace glutamine as the major fuel, sparing glutamine for renal metabolism. During this phase of starvation, gluconeogenesis in the kidneys probably produces more glucose than gluconeogenesis in the liver. However, it should be remembered that the total production of glucose has decreased considerably. Renal gluconeogenesis never operates at a rate close to that seen in liver during early starvation.

23·11 Refeeding After Starvation Restores the Metabolic Profile of the Fed State

In this chapter, we have seen how the body maintains blood glucose levels even during conditions of extreme duress through the regulation of both the synthesis and utilization of glucose. On refeeding, metabolism is quickly restored to the condition of the fed state. The concentration of glucose in the blood rises with refeeding, causing an increase in the release of insulin. The rise in the level of insulin causes the immediate cessation of gluconeogenesis in hepatic tissue and lipolysis in adipose tissue. Insulin also stimulates glucose uptake into muscle and adipose tissue via the GLUT4 glucose transporter. Fuel stores begin to be replenished.

Some long-term adaptations to starvation are reversed less rapidly and may persist for a considerable time after refeeding. For instance, the levels of gluconeogenic enzymes such as PEP carboxykinase are quite high in the liver during starvation, and the enzyme activities may remain high for hours after feeding. Similarly, the level of glucokinase in liver is very low during starvation, and several hours pass before new enzyme synthesis completely restores glucokinase activity. The delay in restoration of normal metabolism has the beneficial effect that the body does not immediately abandon essential adaptations made during starvation at the first refeeding but instead is prepared to discriminate between transient and more sustained changes in food supply.

23·12 Diabetes Results from Lack of Insulin or Insulin Resistance

Diabetes mellitus is a metabolic disease that has been recognized for thousands of years. There are two types of diabetes, each involving a derangement of metabolic control that results in extreme hyperglycemia. The levels of glucose in the blood often rise high enough to exceed the renal reabsorption threshold, and glucose spills over into the urine. The high concentration of glucose in urine draws water osmotically from the body. The disease is named for the symptoms known to the ancients; *diabetes mellitus* means excessive, sweet urine.

The symptoms of diabetes arise from a breakdown in the integration of fuel metabolism due to dysfunctional hormonal control by insulin. In insulin-dependent diabetes mellitus (IDDM), the disease arises from the lack of insulin. Damage to the β cells of the pancreas results in diminished or absent secretion of insulin. Once known as juvenile-onset or Type I diabetes, this form of the disease appears to be autoimmune in nature. It is characterized by early onset (usually before age 15), and patients are usually thin and exhibit hyperglycemia, polyuria (excessive urination), polydipsia (excessive thirst), polyphagia (excessive hunger), and dehydration.

In non–insulin-dependent diabetes (NIDDM), usually associated with obesity, insulin secretion may be normal, and circulating levels of insulin may even be elevated. Thus, the problem is not a shortage of insulin, but insulin resistance resulting from decreased sensitivity, poor responsiveness, or both. Chronic hyperglycemia is the result. A drop in insulin production is often observed in individuals with NIDDM as time passes. There is strong evidence that chronic hyperglycemia damages pancreatic β cells.

NIDDM affects about 5% of the population, and IDDM affects about 1%. In addition, about 1 woman in 120 develops a form of diabetes during pregnancy. There is a trend not to call this diabetes but rather impaired glucose tolerance of pregnancy. About 70% of women who exhibit gestational diabetes return to normal after giving birth.

To understand diabetes, we must consider the functions of insulin. The presence of insulin indicates nutritional sufficiency. It stimulates glucose transport into muscle and adipose tissue cells, increases glycogen synthesis and glycolysis, decreases hepatic glycogen breakdown and gluconeogenesis, inhibits lipolysis in adipocytes, and stimulates amino acid transport and protein synthesis in a number of tissues. In IDDM, when insulin levels are low, glycogen is broken down in liver, and gluconeogenesis occurs regardless of the actual state of glucose supply. At the same time, glucose uptake and use in peripheral tissues is restricted in the absence of stimulation by insulin. Inhibition of lipolysis by insulin is also deficient, while other hormones such as epinephrine trigger release of fatty acids from fat cells. The resulting high rates of hepatic β oxidation produce more acetyl CoA than the citric acid cycle can oxidize. Excess acetyl CoA is diverted to the formation of ketone bodies. In other words, with the exception of hyperglycemia, the body exhibits the metabolic responses characteristic of starvation. Hyperglycemia and ketoacidosis can cause dehydration, coma, and death.

Acute complications such as ketoacidosis are rare in NIDDM, but hyperglycemia is present due to both decreased glucose uptake by peripheral tissues and increased glucose output by liver. Chronic hyperglycemia can lead to a number of complications, such as nonenzymatic glycosylation of proteins, and intracellular accumulation of sorbitol formed via the sorbitol pathway (Section 17·10A), leading to cataracts in the lens and sclerotic lesions in blood vessel walls.

The aim of all regimens for the management of diabetes is control of blood glucose levels. Dietary modifications are often sufficient to control NIDDM, and this type of diabetes may even disappear with moderate weight loss and a program of exercise. If this is not successful, a number of oral drugs are available that act by increasing the secretion of insulin and also by potentiating insulin action at peripheral tissues. Insulin replacement may be needed, however, and is often required as diabetes progresses. IDDM must be treated with injections of insulin. Treatment is complicated by the difficulty of timing the dosage. A number of different forms of insulin are available, including fast- and slow-acting types. Patients learn to monitor their blood glucose level and to adjust their dosage accordingly. In the clinical trial stage are devices that would act as an artificial pancreas, monitoring blood glucose and secreting insulin when appropriate.

Summary

In mammals, metabolic tasks are distributed among different tissues. Some organs, such as the liver, are suppliers of fuels, whereas others, such as the brain, are simply consumers. The principal fuels in mammals are glucose, lactate, fatty acids, ketone bodies, triacylglycerols, and amino acids. Glycerol and ethanol are fuels of minor importance. Not all cells can use each of the major fuels, and not all fuels are available at all times. Glucose is the most important fuel, and control of glucose metabolism is central to overall metabolic control.

The maintenance of constant levels of glucose in the circulation (glucose homeostasis) is achieved by balancing glucose synthesis or absorption against utilization. When large amounts of glucose are present, tissues absorb glucose. As the supply of glucose diminishes, tissues switch to other fuels. The use of glucose is governed by hormones, principally insulin and glucagon. The liver plays a central role in glucose homeostasis.

During starvation, the body responds to the interruption in the supply of exogenous fuels with unified changes in metabolic activity. The coordinated nature of the changes reveals the degree of integration of metabolic pathways. Five phases of glucose homeostasis have been identified in the course of starvation: the absorptive and postabsorptive phases, followed by early, intermediate, and prolonged starvation.

In the absorptive phase, extending from 2 to 4 hours after a meal, the level of glucose in the blood rises. Glucose is rapidly taken up by the liver, skeletal muscle, and brain. If the level of glucose in the blood rises above the ability of the kidney to reabsorb filtered glucose, the excess is lost to the urine.

High blood glucose triggers the release of insulin, which has many physiological effects. In liver, the effects of insulin include stimulation of glycolysis, fatty acid synthesis and esterification, and protein synthesis, as well as inhibition of glycogenolysis, gluconeogenesis, fatty acid oxidation, ketogenesis, and proteolysis. Glucagon, which acts only on liver, has effects that are, in general, opposite those of insulin. Insulin levels are high in the fed state; glucagon levels are high in the fasted state.

Lipids entering the body during the absorptive phase are packaged in chylomicrons. Triacylglycerols in chylomicrons are delivered to peripheral tissues, where triacylglycerols are hydrolyzed outside the cell and fatty acids are taken up, esterified, and stored.

Large amounts of dietary amino acids arrive at the liver during the absorptive phase. These amino acids are either catabolized, used for protein synthesis, or permitted to pass unaltered to peripheral tissues. Oxidation of the amino acids in liver generates a large amount of ATP, much of which is immediately consumed by the pathways of gluconeogenesis and urea synthesis, which remove the carbon and ammonia, respectively, generated by amino acid catabolism.

In the transition to the postabsorptive and early starvation phases, the absorption of dietary glucose slows, the level of insulin drops, and the level of glucagon rises. The liver responds to the hormonal changes by breaking down glycogen and releasing glucose. Low levels of insulin are also accompanied by an increase in the rate of lipolysis in adipose tissue and an increase in the release of free fatty acids into the circulation. The glucose–fatty acid cycle is a mechanism that spares the use of glucose when fatty acids are available: when glucose is low, insulin is low, release and catabolism of fatty acids is high, and products of fatty acid catabolism (NADH, acetyl CoA, and citrate) inhibit glucose degradation and spare the use of glucose.

The Cori cycle is an important aspect of fuel metabolism in which glucose is metabolized to lactate in muscle and other tissues, and lactate is returned to liver for conversion back to glucose. The energy for gluconeogenesis is derived from the catabolism of other fuels, such as fatty acids. The glucose-alanine cycle is a variation of the Cori cycle that includes conversion of pyruvate to alanine in muscle and transport of alanine to liver. This cycle has the advantages of transport of nitrogen to liver and greater energy yield in muscle, since pyruvate is not converted to lactate at the expense of NADH.

Amino acids are major gluconeogenic precursors, although the use of endogenous amino acids requires degradation of proteins and accompanying loss of protein function.

As starvation progresses, fatty acids are converted to ketone bodies, which, unlike fatty acids, can cross the blood-brain barrier and serve as a fuel for the brain, further sparing the use of glucose. Less use of glucose at this stage means less proteolysis of muscle protein for gluconeogenesis.

Gluconeogenesis in kidney becomes important in the late stage of starvation. High levels of ketone bodies result in metabolic acidosis, and gluconeogenesis in kidney is linked to the production of ammonia and bicarbonate. These products of renal gluconeogenesis are used to help maintain acid-base homeostasis.

Refeeding reverses the adaptations that develop during starvation. The concentration of glucose in the blood rises, the level of insulin rises, gluconeogenesis in liver and lipolysis in adipose tissue decreases, glucose is taken up by tissues, and fuel stores are replenished.

Diabetes mellitus is a metabolic disease characterized by extreme hyperglycemia. Insulin-dependent diabetes mellitus (IDDM) arises from a lack of insulin due to damage of the pancreatic β cells. In non–insulin-dependent diabetes mellitus (NIDDM), insulin levels may be normal or even elevated, but tissues are insulin-resistant due to decreased sensitivity, poor responsiveness, or both.

Selected Readings

Bonadonna, R. C., and De Fronzo, R. A. (1992). Glucose metabolism in obesity and Type II diabetes. In *Obesity*, P. Bjorntorp and B. N. Brodoff, eds. (Philadelphia, Pennsylvania: J. B. Lippincott), pp. 474–501.

Cahill, G. F. (1970). Starvation in man. *N. Engl. J. Med.* 282:668–675.

Dinneen, S., Gerich, J., and Rizza, R. (1992). Carbohydrate metabolism in non-insulin dependent diabetes mellitus. *N. Engl. J. Med.* 327:707–713.

Felig, P. (1975). Amino acid metabolism in man. *Annu. Rev. Biochem.* 44:933–955.

Kahn, B. B. (1992). Facilitative glucose transporters: regulatory mechanisms and dysregulation in diabetes. *J. Clin. Invest.* 89:1367–1374.

Larner, J. (1990). Insulin and the stimulation of glycogen synthesis. *Adv. Enzymol. Relat. Areas Mol. Biol.* 63:173–231.

Newsholme, E. A., and Leech, A. R. (1986). *Biochemistry for the Medical Sciences.* (Chichester, England: John Wiley & Sons). See especially Chapter 14, "The integration of metabolism during starvation, refeeding, and injury."

Pessin, J. E., and Bell, G. I. (1992). Mammalian facilitative glucose transporter family: structure and molecular regulation. *Annu. Rev. Physiol.* 54:911–930.

Randle, P. J. (1986). Fuel selection in animals. *Biochem. Soc. Trans.* 14:799–806.

Ruderman, N. B., Aoki, T. T., and Cahill, G. F. (1976). Gluconeogenesis and its disorders in man. In *Gluconeogenesis: Its Regulation in Mammalian Species,* R. W. Hanson and M. A. Mehlman, eds. (New York: Wiley Interscience), pp. 515–558.

Shulman, G. I., and Landau, B. R. (1992). Pathways of glycogen repletion. *Physiol. Rev.* 72:1019–1035.

Taylor, R., and Agius, L. (1988). The biochemistry of diabetes. *Biochem. J.* 250:625–640.

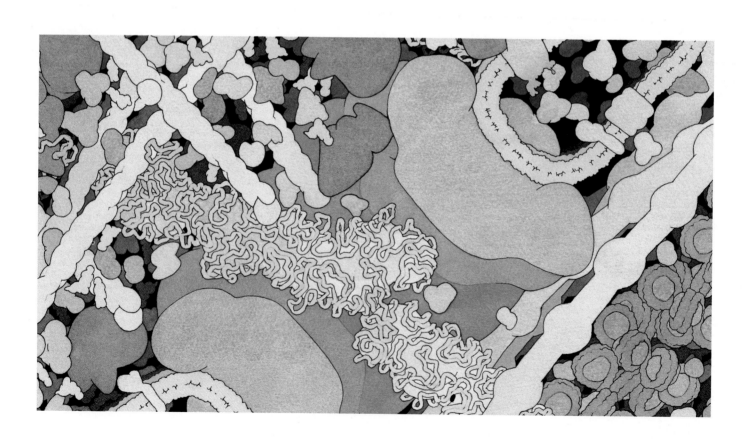

Part Four

Biological Information Flow

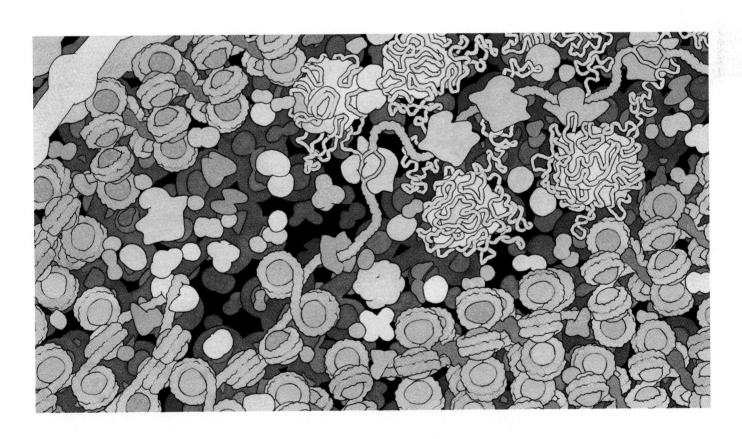

24

Nucleic Acids

Living organisms contain a set of instructions that directs the timely synthesis of ribonucleic acid (RNA) and proteins. The required information resides in the genetic material, or **genome,** which is composed of one or more large molecules of deoxyribonucleic acid, or DNA (Figure 24·1, next page). The genomes of many viruses are also composed of DNA, but some are composed of RNA.

The information that specifies the primary structures of proteins is encoded in the nucleotide sequence of DNA. This information is enzymatically copied during the synthesis of RNA, a process known as transcription. Some of the transcribed RNA molecules are translated by the protein-synthesis machinery, giving rise to polypeptide chains, which subsequently fold and assemble to form protein molecules. Thus, we can generalize that the biological information stored in DNA flows from DNA to RNA to protein.

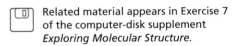
Related material appears in Exercise 7 of the computer-disk supplement *Exploring Molecular Structure.*

(a)

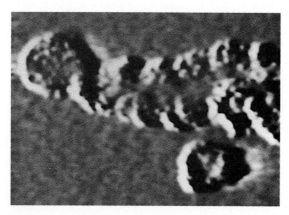

Figure 24·1
(a) Scanning tunneling electron micrograph of DNA. (b) Schematic view, redrawn for clarity. (Courtesy of Lawrence Livermore National Laboratory.)

(b)

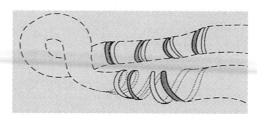

(a)

(b)

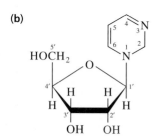

Figure 24·2
General structures of nucleosides. In purine nucleosides (a) and pyrimidine nucleosides (b), the base is joined to the sugar by a β-N-glycosidic bond. The sugar in ribonucleosides is ribose, which contains a hydroxyl group at C-2′, as shown here. In deoxyribonucleosides, there is a hydrogen atom at C-2′ instead of a hydroxyl group. Note that the numbers of the carbon atoms in the sugar residues are designated with primes in order to distinguish them from the numbered atoms of the purine and pyrimidine rings.

The discovery of the substance that proved to be DNA was made in 1869 by Friedrich Miescher, a young Swiss physician working in the laboratory of the German physiological chemist Felix Hoppe-Seyler. Miescher treated white blood cells (which came from the pus on discarded surgical bandages) with hydrochloric acid to obtain nuclei for study. When the nuclei were subsequently treated with acid, a precipitate formed that proved to contain carbon, hydrogen, oxygen, nitrogen, and a high percentage of phosphorus. Because of its occurrence in nuclei, Miescher called the precipitate "nuclein." Later, when it was found to be strongly acidic, its name was changed to **nucleic acid.** Shortly after Miescher's discovery of DNA, Hoppe-Seyler isolated a similar substance from yeast cells; this substance is now known to be RNA.

Nucleic acids represent the third major class of large biopolymers. Like proteins and polysaccharides, nucleic acids are polymers composed of similar monomeric units that are covalently joined to produce macromolecules. In this chapter, we will discuss the molecular structure and topology of nucleic acids and the ways they are packaged in living cells. We will also introduce some of the enzymes involved in the degradation of nucleic acids.

24·1 Nucleoside Monophosphates Are the Monomers in Polynucleotides

Nucleic acids are polymers of nucleotides, or phosphorylated nucleosides. The structures and nomenclature of the major ribonucleosides and deoxyribonucleosides have already been described. Here we briefly review the main features of nucleoside structure. A purine or pyrimidine base is connected to ribose or deoxyribose via a β-N-glycosidic bond (Figure 24·2). The base can rotate around this bond to adopt different conformations. The nucleosides shown in Figure 24·2 are drawn in the *anti* conformation. The *anti* and *syn* conformers are in rapid equilibrium in purine nucleosides, whereas the *anti* conformation predominates in pyrimidine nucleosides. In nucleic acids, nucleoside monophosphate residues are most commonly in the *anti* conformation.

Ribonucleosides contain three hydroxyl groups to which phosphate can be esterified (2′, 3′, and 5′), whereas 2′-deoxyribonucleosides contain two such hydroxyl groups (3′ and 5′). In naturally occurring nucleotides, the phosphoryl groups are most commonly attached to the oxygen atom of the 5′-hydroxyl group; thus, a nucleotide is always assumed to be a 5′-phosphate ester unless otherwise designated. The four nucleoside 5′-monophosphates that are the monomeric units in DNA are shown in Figure 24·3.

The bases of nucleosides can tautomerize, as we saw earlier. The amino tautomers of adenine and cytosine and the lactam (keto) tautomers of guanine and thymine, as shown in Figure 24·3 and subsequent figures, are the forms that predominate under the conditions usually found inside cells.

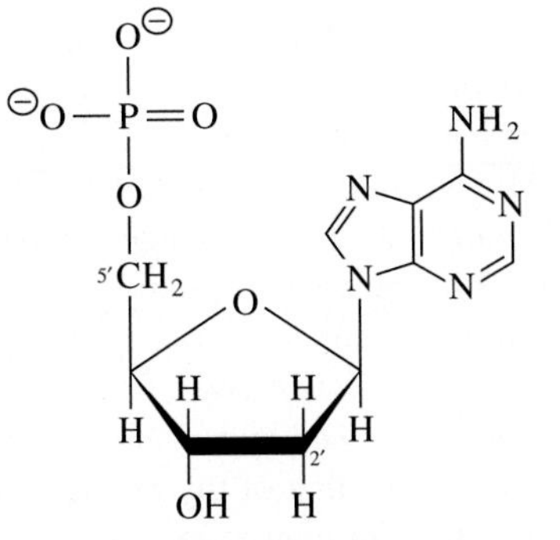

Figure 24·3
Structures of the deoxyribonucleoside 5′-monophosphates.

2′-Deoxyadenosine 5′-monophosphate
(Deoxyadenylate, dAMP)

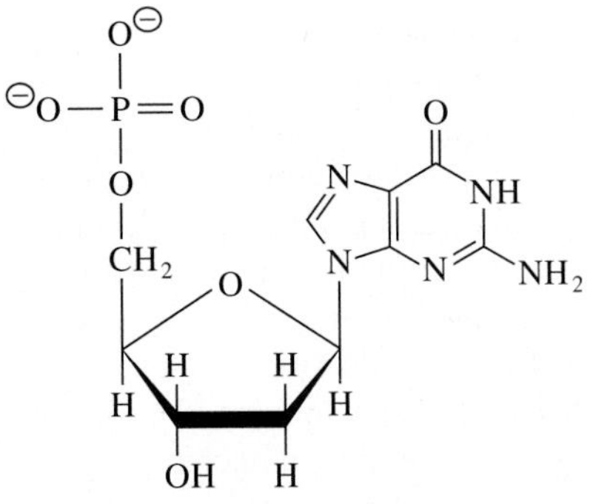

2′-Deoxyguanosine 5′-monophosphate
(Deoxyguanylate, dGMP)

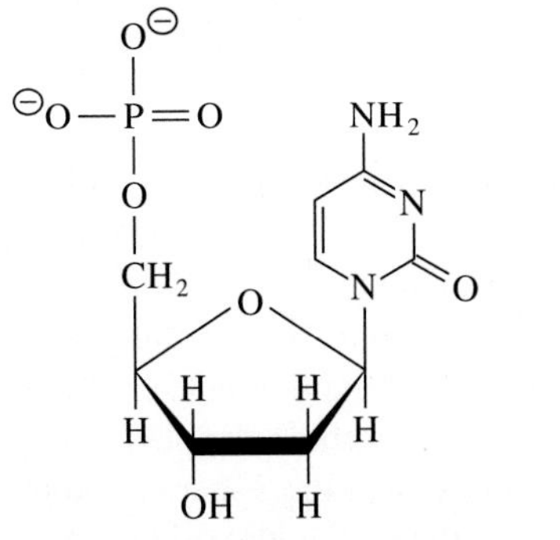

2′-Deoxycytidine 5′-monophosphate
(Deoxycytidylate, dCMP)

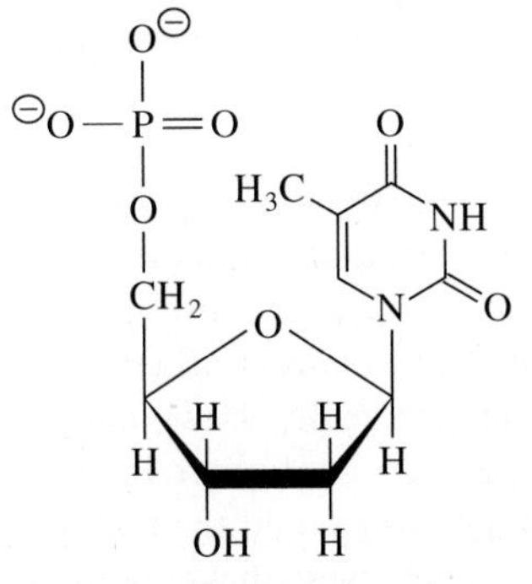

2′-Deoxythymidine 5′-monophosphate
(Deoxythymidylate, dTMP)

Figure 24·4
Hydrogen bond sites of nucleosides in DNA. Each base contains atoms and functional groups that can serve as hydrogen donor or acceptor sites. The common tautomeric forms of the nucleoside bases are shown. Hydrogen donor and acceptor groups differ in the other tautomers. *R* represents deoxyribose.

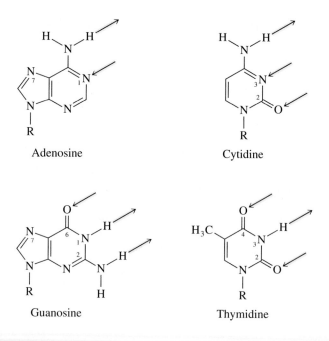

Adenosine

Cytidine

Guanosine

Thymidine

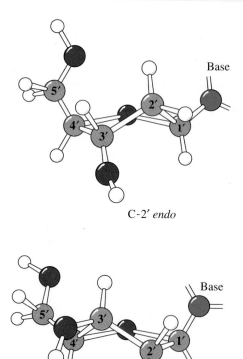

C-2′ endo

C-3′ endo

○ Carbon ● Oxygen
○ Hydrogen ● Nitrogen

Figure 24·5
Common pucker conformations of the furanose ring in deoxynucleosides. The bases are attached at the 1′ position. Either C-2′ or C-3′ can be found above the plane formed by the remaining four atoms of the ring.

The bases of nucleosides can participate in hydrogen bonding. For example, in the common forms of DNA, the amino groups of adenosine and cytidine serve as hydrogen donors, and the ring nitrogen atoms (N-1 in adenosine and N-3 in cytidine) serve as hydrogen acceptors (Figure 24·4). Cytidine also has a hydrogen-acceptor group at C-2. The lactam tautomers of guanosine and thymidine can participate in the formation of three hydrogen bonds. In the case of guanosine, the group at C-6 serves as a hydrogen acceptor; N-1 and the amino group at C-2 serve as hydrogen donors. In the case of thymidine, the groups at C-4 and C-2 serve as hydrogen acceptors, and N-3 serves as a hydrogen donor. The hydrogen-bonding ability of uridine, a nucleoside found in RNA, is similar to that of thymidine. The hydrogen-bonding patterns of nucleosides have important consequences for the three-dimensional structure of nucleic acids. Additional hydrogen bonding occurs in some nucleic acids and in nucleic acid–protein interactions. For example, N-7 of adenosine and guanosine can serve as a hydrogen acceptor, and both amino hydrogen atoms of adenosine, guanosine, and cytidine can be donated to form hydrogen bonds.

The furanose rings in nucleosides and nucleotides can adopt a number of different conformations. The puckered forms are referred to as the envelope and twist (half-chair) conformers. In the twist conformations, the 2′- or 3′-carbon atom is not in the same plane as the rest of the atoms in the five-membered ring (Figure 24·5). Atoms have an *endo* conformation if they lie on the same side of the plane as C-5′, which is usually drawn above the plane. The most common puckered forms are the C-2′ *endo* conformer, in which the 2′-carbon atom is above the plane formed by the other atoms, and the C-3′ *endo* conformer, in which the 3′-carbon atom is above the plane. Note that the relative orientations of the C-3′ hydroxyl groups and C-4′—C-5′ bonds are quite different in the two structures. In nucleosides and nucleotides that are free in solution, the C-2′ and C-3′ *endo* conformations rapidly interconvert.

24·2 DNA Contains Two Linear Polymers of Nucleotides

In 1944, Oswald Avery, Colin MacLeod, and Maclyn McCarty demonstrated that DNA was the molecule that carried genetic information. They showed that purified DNA could transform a harmless strain of *Streptococcus pneumoniae* to an infectious strain that was genetically stable. A few years later, Alfred D. Hershey and Martha Chase proved that the DNA of bacteriophages (viruses that infect bacteria), not the proteins in the viral coat, directed production of new phage particles inside the bacterial cell.

Experiments such as these helped convince scientists that DNA was the genetic material, although it was difficult to envisage how information was stored in this molecule. By 1950, it was clear that DNA was a linear polymer of deoxyribonucleotide residues joined by 3′–5′ phosphodiester linkages, and the first X-ray analysis of DNA fibers by William T. Astbury in 1947 indicated that the bases were probably stacked 0.34 nm apart. Moreover, through analysis of hydrolysates, Erwin Chargaff had deduced certain regularities in the nucleotide compositions of DNA samples obtained from a wide variety of prokaryotes and eukaryotes. Chargaff observed that within the DNA of a given cell, dA and dT were present in equimolar amounts, as were dG and dC (Table 24·1). The ratio of purines to pyrimidines was 1:1 even when the total mole percent of G + C differed considerably from that of A + T. Based on his analysis of the physical properties of DNA, J. M. Gulland suggested in 1947 that the bases were involved in the formation of hydrogen bonds.

The first crystal structure of a nucleoside (cytidine) was solved by S. Furberg in 1951, but it wasn't until 1953—nine years after Avery, MacLeod, and McCarty suggested a genetic role for DNA—that James D. Watson and Francis H. C. Crick proposed a model for the structure of DNA. Their model was based on X-ray diffraction patterns that Rosalind Franklin, Maurice Wilkins, and others had obtained from DNA fibers, as well as on the chemical equivalencies noted by Chargaff, on model building, and on intuition. The model, which we now know to be essentially correct, accounted for the equal amounts of purine and pyrimidine bases by suggesting that DNA was double stranded and that bases on each strand paired specifically with bases on the other strand, A with T and G with C. Watson and Crick's proposed structure is now referred to as the B conformation of DNA, or simply **B-DNA**.

Table 24·1 Base composition of DNA (mole %) and ratios of bases

Source	A	G	C	T	A/T*	G/C*	G + C	Purine/ Pyrimidine*
Escherichia coli	26.0	24.9	25.2	23.9	1.09	0.99	50.1	1.04
Mycobacterium tuberculosis	15.1	34.9	35.4	14.6	1.03	0.99	70.3	1.00
Yeast	31.7	18.3	17.4	32.6	0.97	1.05	35.7	1.00
Ox	29.0	21.2	21.2	28.7	1.01	1.00	42.4	1.01
Pig	29.8	20.7	20.7	29.1	1.02	1.00	41.4	1.01
Human	30.4	19.9	19.9	30.1	1.01	1.00	39.8	1.01

*Deviations from a 1:1 ratio are due to experimental variations.

A. Nucleotides in DNA Are Linked as 3′–5′ Phosphodiesters

We have seen that the primary structure of a protein refers to the sequence of amino acid residues linked by peptide bonds; by analogy, the primary structure of a nucleic acid is the sequence of nucleotide residues connected by 3′–5′ phosphodiester linkages. A tetranucleotide representing a segment of single-stranded DNA is shown in Figure 24·6. Note that all the nucleotide residues within a polynucleotide chain have the same orientation; thus, like polypeptide chains, polynucleotide chains have directionality. One end of a linear polynucleotide chain is said to be 5′ (that is, there is no residue attached to its 5′-carbon atom), and the other is said to be 3′ (no residue attached to its 3′-carbon atom). The forward direction, which is the direction of polymerization in the biosynthesis of both DNA and RNA, is 5′→3′;

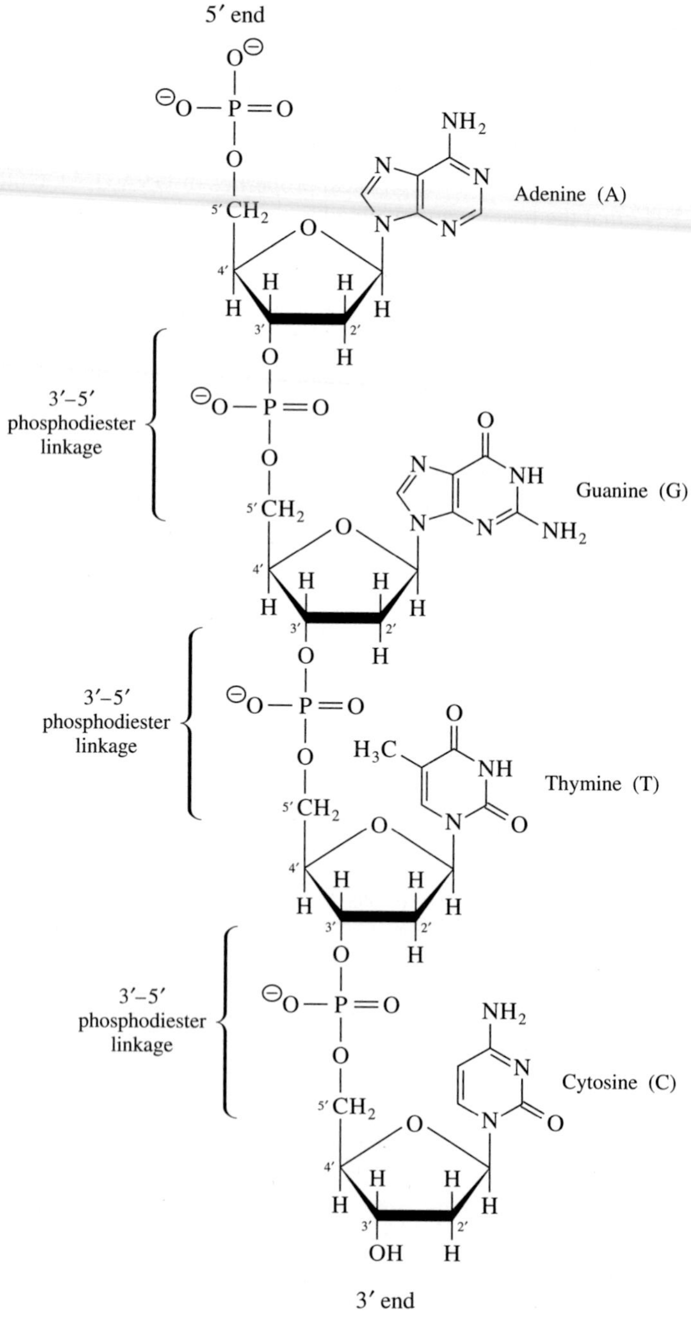

Figure 24·6
Structure of a tetranucleotide. The nucleotide residues are connected by 3′–5′ phosphodiester linkages. This tetranucleotide has a terminal nucleotide with a 5′-phosphoryl group that is not bound to another nucleotide (the 5′ end) and a terminal nucleotide with a free 3′-hydroxyl group (the 3′ end). The forward direction of a polynucleotide chain is defined as 5′→3′, that is, proceeding from the 5′ terminus to the 3′ terminus. This tetranucleotide can be abbreviated as pdApdGpdTpdC, or simply AGTC.

therefore, when not otherwise specified, nucleotide sequences are assumed to read $5' \rightarrow 3'$. Phosphates are often abbreviated p. Thus, 5'-adenylate (AMP) can be abbreviated as pA, deoxyadenosine 3'-monophosphate as dAp, and ATP as pppA. The tetranucleotide in Figure 24·6 can be abbreviated pdApdGpdTpdC, or even AGTC when it is clear that the reference is to DNA and only the sequence of the bases is important.

Each phosphate group that participates in a phosphodiester linkage has a pK_a value of approximately 2 and bears a negative charge at neutral pH. Consequently, nucleic acids are polyanions under physiological conditions. Inside the cell, the phosphate groups are usually neutralized by counterions such as Mg^{2+} or cationic proteins (proteins that contain an abundance of the basic residues arginine and lysine) that form complexes with DNA.

Single-stranded nucleic acid molecules can be quite flexible due to relatively unhindered rotation around many of the bonds in the sugar-phosphate backbone (Figure 24·7). The backbone contains the C-3'—C-4' bond (labelled δ in Figure 24·7), which is part of the furanose ring. Rotation around this bond is somewhat restricted, but some rotation is possible because the sugars can exist in a variety of conformations. The C-2' *endo* conformation of the sugar residue is shown in Figure 24·7.

Steric hindrance and electrostatic repulsion between oxygen atoms bound to the phosphorus atoms limit the number of possible conformations of the polynucleotide chain in solution. In some cases, the torsion angles are restricted because of clashes between the phosphate groups and the bases. This interference limits rotation around the C-5'—O bond (labelled β in Figure 24·7). Single-stranded polynucleotides tend to adopt an extended, or stretched-out, form, which minimizes interaction between phosphate groups. As we will see shortly, rotation around the bonds in the backbone is even more restricted in double-stranded polynucleotides.

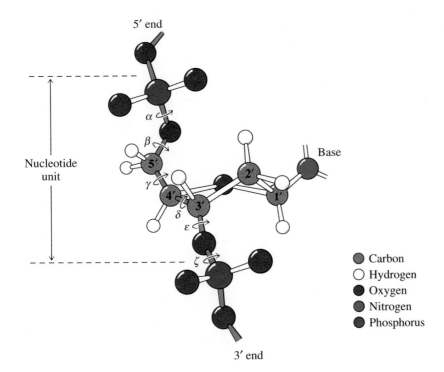

Figure 24·7
Structure of the sugar-phosphate backbone (green) of polynucleotides. A single nucleotide unit is shown with the 5'-phosphate group at the top. Each bond is identified by a letter of the Greek alphabet. The conformation of the polynucleotide is determined by the torsion angles indicated by circular arrows. [Adapted from Saenger, W. (1984). *Principles of Nucleic Acid Structure* (New York: Springer-Verlag), p. 14.]

● Carbon
○ Hydrogen
● Oxygen
● Nitrogen
● Phosphorus

B. A Double Helix Is Formed from Two Antiparallel Polynucleotide Strands

Most genomes are made up of double-stranded DNA, which is composed of two single-stranded polynucleotides. Each base of one strand is hydrogen bonded to a base of the opposite strand, forming a **base pair.** The two strands of this structure run in opposite directions; in other words, they are antiparallel (Figure 24·8).

Note that the lactam and amino tautomers of each base have the appropriate complementary hydrogen donor and acceptor sites to form a base pair. Guanine pairs with cytosine, and adenine with thymine, giving rise to base pairs that maximize hydrogen bonding between potential donor and acceptor sites. G/C base pairs

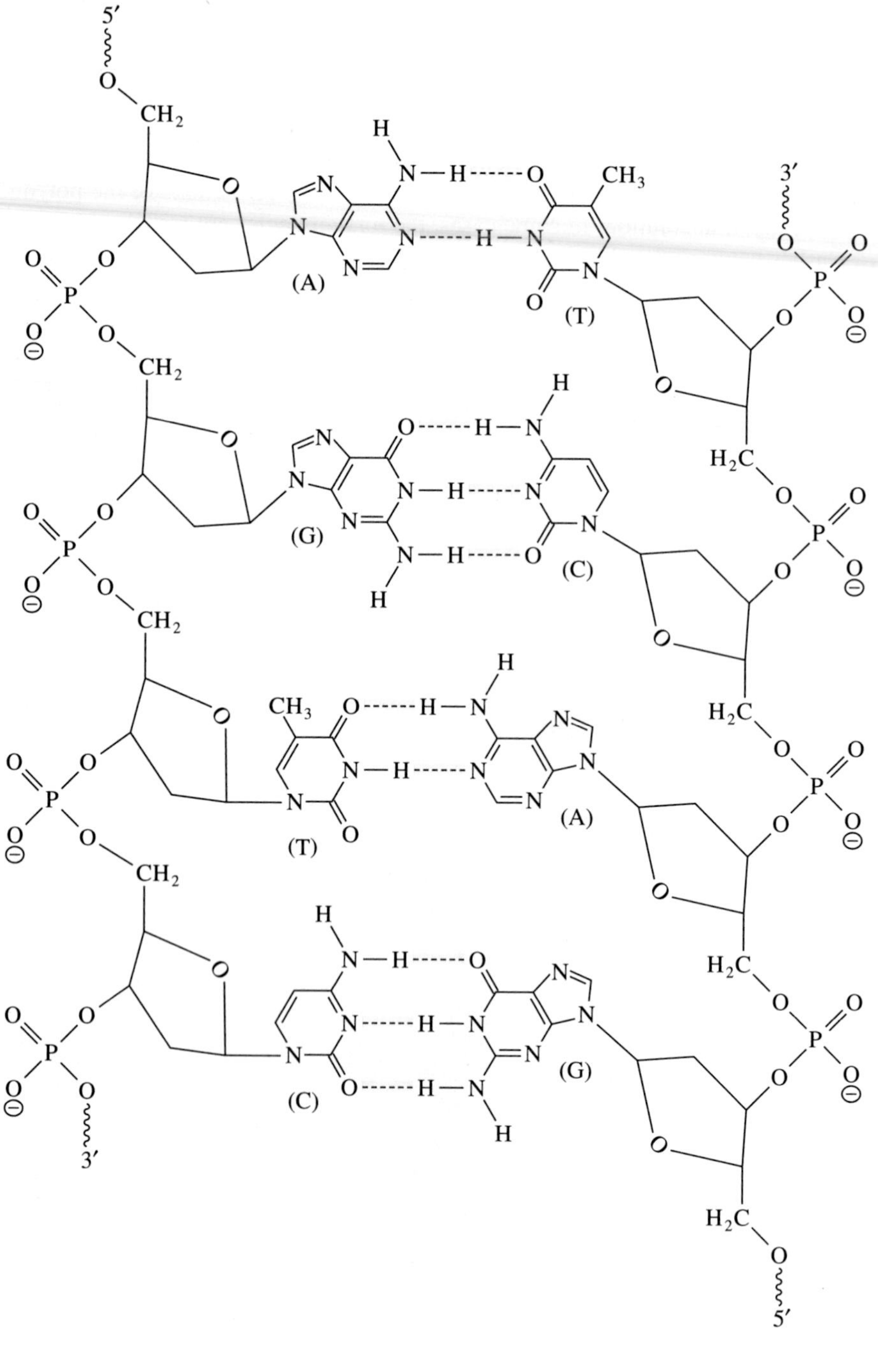

Figure 24·8
Diagram of double-stranded DNA. The two strands run in opposite directions. Adenine in one strand pairs with thymine in the opposite strand, and guanine pairs with cytosine. In actuality, the base pairs are tilted so that the plane of the base pair is perpendicular to the page.

have three hydrogen bonds, and A/T base pairs have two. The base pairs shown in Figure 24·8 are referred to as Watson-Crick base pairs. Other patterns of base pairing in DNA are possible in certain rare structures.

Because A in one strand pairs with T in the other strand and G pairs with C, the strands are complementary and can serve as templates for each other. The lengths of the hydrogen bonds are similar in all cases, and because each base pair is formed from one purine and one pyrimidine, the sizes of the two types of Watson-Crick base pairs are nearly the same (Figure 24·9). In each case, the distance between the 5′-carbon atoms in opposite strands is close to 1.1 nm. This arrangement means that the strands of double-stranded DNA are the same distance apart throughout their length. Thus, DNA has a regular structure that is not greatly affected by base composition.

Because most naturally occurring DNA is double stranded, the length of DNA molecules or segments is often expressed in terms of base pairs (bp). For convenience, longer structures are measured in thousands of base pairs, or **kilobase pairs,** commonly abbreviated kb.

The actual structure of DNA differs from that shown in Figure 24·8 in two important aspects. In a true three-dimensional representation, the base pairs are rotated so that they are perpendicular to the page, and the two strands wrap around each other to form a two-stranded helical structure, or **double helix.**

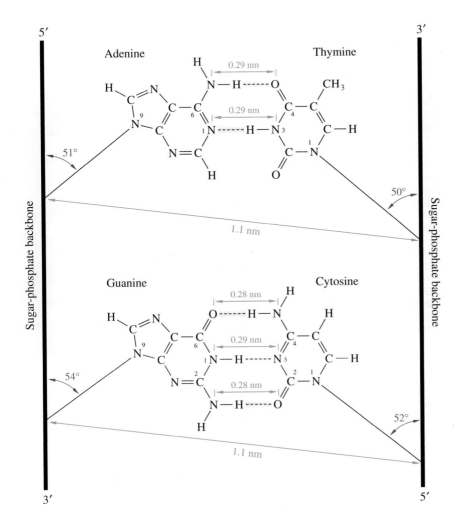

Figure 24·9
Dimensions of base pairs in DNA. An A/T base pair and a G/C base pair are almost the same size. The distance between the two sugar-phosphate backbones and the angles of the β-N-glycosidic bonds relative to the long axis of DNA are similar for all base pairs.

DNA can be visualized as a "ladder" whose ends have been twisted in opposite directions to form a helix. The paired bases represent the rungs of this ladder, and the antiparallel sugar-phosphate backbones represent the supports. Each strand of DNA serves as a perfect template for the other. This templating is responsible for the overall symmetry of double-stranded DNA. However, templating alone does not produce a helix. It is the interaction of adjacent base pairs that brings the base pairs closer together and creates a hydrophobic interior, resulting in twisting of the sugar-phosphate backbone to form a helix. In other words, the helical form of DNA is due to the fact that the base pairs stack on top of each other, leaving no space between the rungs of the ladder. These stacking and hydrophobic interactions make double-stranded DNA much more stable than separated single strands. The templating and stacking components of DNA structure are illustrated in Figure 24·10.

The three-dimensional structure of B-DNA is shown in Figure 24·11. Note that the two strands are completely intertwined and that they run in opposite directions. This figure also shows that the base pairs are not perfectly planar, nor are they exactly perpendicular to the long axis of the molecule.

The B-DNA helix is right-handed. (Imagine that you are descending a spiral staircase with the same helical sense as DNA; if you are making right turns, the helix is right-handed.) According to calculations based on models of DNA, a right-handed helix is somewhat more stable than a left-handed helix, which may explain why the most common form of DNA is right-handed.

Figure 24·10
Templating and stacking of bases in double-stranded DNA.

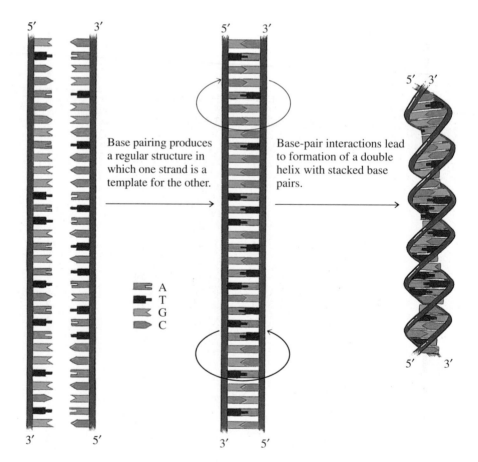

Base pairing produces a regular structure in which one strand is a template for the other.

Base-pair interactions lead to formation of a double helix with stacked base pairs.

A
T
G
C

Figure 24·11
Stereo views of B-DNA.
(a) Stick model. Note that the base pairs are not exactly perpendicular to the sugar-phosphate backbones. **(b)** Space-filling model. Color key: carbon, green; nitrogen, blue; oxygen, red; phosphorus, orange. (Based on coordinates provided by H. Drew and R. E. Dickerson.)

(a)

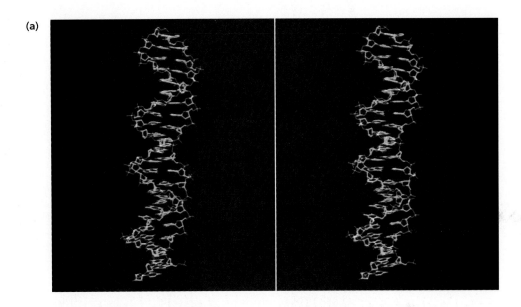

(b)

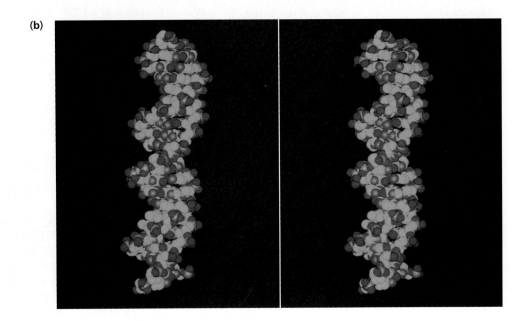

The hydrophilic sugar-phosphate backbones, one from each strand, wind around the outside of the helix, where they are exposed to aqueous solvent. In contrast, the stacked, relatively hydrophobic bases are located in the interior of the helix, where they are largely shielded from aqueous solvent. The space-filling model of DNA illustrates these points (Figure 24·11b). Note that the top and bottom surfaces of adjacent base pairs are quite close. They are, in fact, in van der Waals contact. The exposed phosphate groups are also evident in this representation of the structure of DNA.

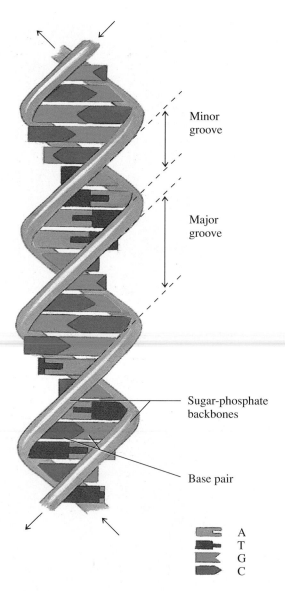

Minor groove

Major groove

Sugar-phosphate backbones

Base pair

A
T
G
C

Figure 24·12
Three-dimensional structure of B-DNA. This stylized model of B-DNA depicts the orientation of base pairs and sugar-phosphate backbones, as well as the relative sizes of the pyrimidine and purine bases. The sugar-phosphate backbones wind around the outside of the helix, and the bases occupy the interior. Stacking of the base pairs creates two grooves of unequal width, the major and the minor groove. For clarity, a slight space has been left between the stacked base pairs, and the interaction between complementary bases has been schematized.

The stacking of base pairs and the resulting twist in the sugar-phosphate backbones lead to the formation of two grooves of unequal width on the surface of the helix. The larger groove is called the **major groove;** the smaller groove, the **minor groove** (Figure 24·12). Within these grooves, the edges of the base pairs are exposed to solvent and are chemically distinguishable; thus, molecules that interact with particular base pairs through this groove can identify the base pairs without disrupting the helix. This is particularly important for proteins that bind to DNA and "read" a specific sequence.

In aqueous solution, B-DNA is a right-handed helix with a diameter of 2.37 nm. The **rise,** or distance between one base pair and the next along the helical axis, is 0.33 nm, and the **pitch** of the helix, or distance to complete one turn, is 3.40 nm. These dimensions are shown in Figure 24·13. Since there are about 10.4 base pairs per turn of the helix, the angle of rotation between adjacent nucleotides within each strand, or the **twist,** is 34.6° (360°/10.4). The original model proposed by Watson and Crick predicted 10 base pairs per turn of the helix; this is the structure that is found in crystalline fibers of the type examined by Franklin and Wilkins. It is now known that B-DNA winds up slightly when it is dehydrated to form crystalline DNA fibers. The actual number of base pairs per turn can vary from less than 10.4 to more than 10.5. The dimensions given above are average values.

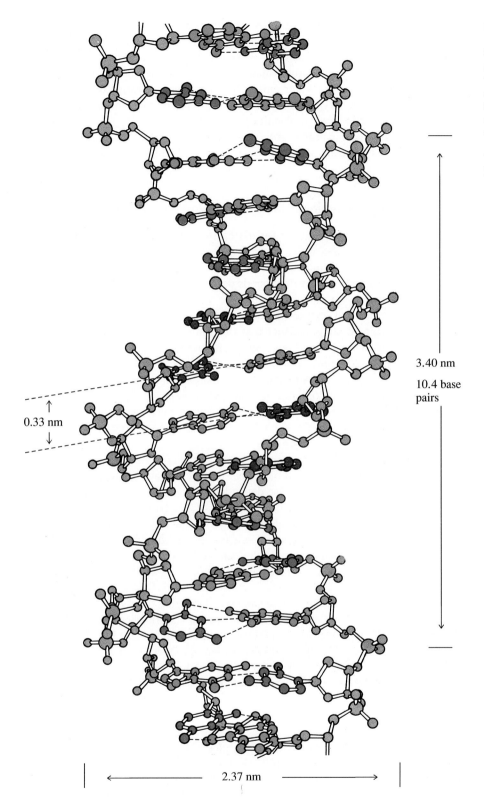

0.33 nm

3.40 nm

10.4 base pairs

2.37 nm

Figure 24·13
Molecular structure of B-DNA. This model shows G/C and A/T base pairing and the anti-parallel orientation of the two polynucleotide chains. The overall dimensions of B-DNA and the distance between the centers of the stacked base pairs are also shown. Color key: adenine, orange; cytosine, blue; guanine, green; thymine, red. [Adapted from a drawing by Irving Geis, in Dickerson, R. E. (1983). The DNA helix and how it is read. *Sci. Am.* 249(6):94–111.]

Table 24·2 Stacking energies for the 10 possible dimers in B-DNA

Stacked dimers		Stacking energies (kJ mol^{-1})
↑C-G↓ / G-C↓		−61.0
↑C-G↓ / A-T↓	↑T-A↓ / G-C↓	−44.0
↑C-G↓ / T-A↓	↑A-T↓ / G-C↓	−41.0
↑G-C↓ / C-G↓		−40.5
↑G-C↓ / G-C↓	↑C-G↓ / C-G↓	−34.6
↑T-A↓ / A-T↓		−27.5
↑G-C↓ / T-A↓	↑A-T↓ / C-G↓	−27.5
↑G-C↓ / T-A↓	↑T-A↓ / C-G↓	−28.4
↑A-T↓ / A-T↓	↑T-A↓ / T-A↓	−22.5
↑A-T↓ / T-A↓		−16.0

Arrows designate the direction of the sugar-phosphate backbone and point from C-3′ of one sugar unit to C-5′ of the next.
[Adapted from Ornstein, R. L., Rein, R., Breen, D. L., and MacElroy, R. D. (1978). An optimized potential function for the calculation of nucleic acid interaction energies: I. Base stacking. *Biopolymers* 17:2341–2360.]

Inside the cell, the grooves of DNA are usually filled with water molecules that form hydrogen bonds with the atoms of the bases and the backbone. In the case of B-DNA, the minor groove is densely packed with water molecules, which can fit easily within this space and form contacts with the exposed edges of the bases and adjacent water molecules. Water molecules in the major groove form hydrogen bonds with phosphate oxygen atoms and the keto and amino groups of the bases. These highly ordered water molecules contribute to the rigidity and stability of B-DNA. There are on average three water molecules bound to each phosphate group. Altogether there are about 20 bound water molecules for each nucleotide in DNA.

C. Many Weak Interactions Stabilize Double-Stranded DNA

Although the forces responsible for maintaining the native conformations of complex cellular structures such as nucleic acids, proteins, and membranes are necessarily strong enough to maintain the structures, they are weak enough to allow conformational flexibility. Covalent bonds define the primary structures of macromolecules, whereas weak forces dominate the folded forms. The major forces that influence the stability of double-stranded DNA are hydrophobic interactions, base stacking, hydrogen bonding, and electrostatic repulsion.

The burying of the hydrophobic purine and pyrimidine rings in the interior of the double helix away from solvent increases the stability of DNA. As in proteins, the hydrophobic interactions are related to an increase in entropy when water molecules are freed from their ordered state around the exposed residues. This gain in entropy is partially negated by highly ordered, bound water molecules that associate with B-DNA. Hydrophobic interactions make an important contribution to the overall stability of double-stranded DNA.

The stacked base pairs form van der Waals contacts in the interior of the molecule. In addition, an induced-dipole effect enhances stacking interactions. The strength of these interactions is base dependent; calculations of stacking energies show that stacking interactions involving dimers of G/C base pairs are stronger than those involving A/T base pairs (Table 24·2). Consequently, DNA rich in G + C is more stable than DNA with a lower G + C content. Although the forces between stacked bases are individually weak, they are additive; thus, in DNA molecules, which may contain millions of base pairs, stacking interactions are an important source of stability.

Hydrogen bonding between base pairs is a stabilizing force in double-stranded DNA, but it does not account for the difference in stability between the double-stranded molecule and separated single strands. This is intuitively obvious since the nonpaired bases of single-stranded DNA could also form hydrogen bonds with water molecules. Base pairing is most important for the formation of a regular structure, or templating. The hydrogen bonds between base pairs are constantly being broken and re-formed at a rapid rate. This process allows the polynucleotide chains to undergo transient small shifts in conformation, a phenomenon known as "breathing."

Electrostatic repulsion of the negatively charged phosphate groups in the backbone is a potential source of instability in the DNA helix. However, repulsion is minimized by electrostatic interactions between the phosphate groups and cations (especially Mg^{2+}) or DNA-binding proteins. In the absence of protein and cations, the overall stability of double-stranded DNA is greatly reduced.

Although duplex DNA is thermodynamically more stable than the separated single strands, the double helix must be disrupted during DNA replication, transcription, and recombination. Thus, it is reasonable to conclude that the forces holding the two strands together are sufficient to confer stability but not strong enough to prevent strand separation during these processes.

24·3 DNA Can Be Denatured and Renatured in Vitro

Complete unwinding of a double helix and separation of the complementary single strands is called **denaturation.** Denaturation occurs only in vitro. The reverse process, whereby two separated strands come together to form double-stranded DNA, is called **renaturation.**

Double-stranded DNA can be disrupted when solutions of DNA are heated above a certain temperature. In studies of thermal denaturation, the temperature of a solution of DNA is slowly increased. As the temperature is raised, more and more of the bases become unstacked, and hydrogen bonds between base pairs are broken. Eventually, the two strands separate completely.

Absorbance of ultraviolet light can be used to measure the extent of denaturation. Measurements are made at 260 nm, close to the absorbance maximum for nucleic acids. At this wavelength, single-stranded DNA absorbs 12–40% more light than double-stranded DNA because stacking interactions between base pairs reduce ultraviolet absorbance (Figure 24·14). The absorbance of an equivalent amount of free nucleotides is even greater than the absorbance of single-stranded DNA, indicating that single-stranded DNA maintains some stacking interactions.

A plot of the change in absorbance versus temperature of a DNA solution is called a melting curve (Figure 24·15). The sharp transition from the double-helical state to the denatured state reflects disruption of the interactions between the stacked, hydrogen-bonded base pairs in double-helical DNA. The midpoint of this transition, where half of the DNA is single stranded, is referred to as the **melting temperature,** or T_m. Organic solvents, such as ethanol, lower the T_m of DNA by reducing the strength of hydrophobic interactions in the interior of the molecule. Chaotropic agents, such as urea, guanidinium chloride, and formamide, also destabilize double-stranded DNA.

As shown in Figure 24·15, the T_m of poly(GC) differs significantly from that of poly(AT). A/T-rich DNA melts more readily than G/C-rich DNA because stacking interactions between base pairs are stronger in DNA that is rich in G + C. The additional hydrogen bond in G/C base pairs compared to A/T base pairs contributes only slightly to overall stability; stronger stacking interactions are largely responsible for the higher melting temperature of G/C-rich DNA.

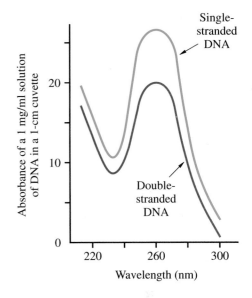

Figure 24·14
Absorbance spectra of double-stranded and single-stranded DNA. At pH 7.0, DNA has an absorbance maximum near 260 nm. Denatured DNA absorbs 12–40% more ultraviolet light than double-stranded DNA.

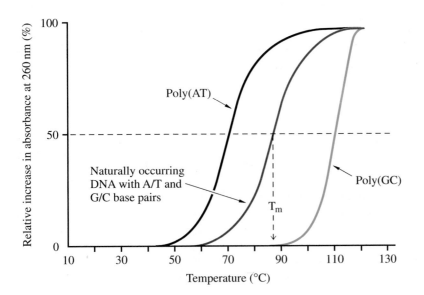

Figure 24·15
Melting curve of DNA. In this experiment, the temperature of a DNA solution is increased while the absorbance at 260 nm is monitored. At T_m, which is determined by the inflection point of the sigmoidal curve, the increase in absorbance of the sample is one-half the increase in absorbance of completely denatured DNA. Poly(AT), whose strands are polymers of AT dimers, melts at a lower temperature than either naturally occurring DNA or poly(GC) since more energy is required to disrupt stacked G/C base pairs.

Denaturation is a two-step process. In the first step, the two strands separate at localized regions rich in A/T base pairs. As the temperature is increased, more of the DNA becomes single stranded until only the most stable, G/C-rich regions remain double stranded. At this point, double-stranded DNA could be rapidly reformed by lowering the temperature. As the temperature continues to rise, the strands separate completely, and the DNA is denatured.

Denatured DNA can be renatured by lowering the temperature below the T_m. Renaturation is a slow process because the complementary strands must first find each other in solution and then come together in the proper orientation to form base pairs between aligned sequences. This nucleation step is rate limiting, which means that the rate of renaturation of DNA is concentration dependent, as are most second-order reactions. Once a short region of double-stranded DNA has formed, the remaining DNA renatures rapidly by a zippering mechanism (Figure 24·16).

The T_m of a given duplex DNA molecule is highly dependent on the ionic strength of the solution. At low salt concentrations, the charges of the phosphate groups are not neutralized, and electrostatic repulsion causes the two strands of DNA to separate at lower temperatures. Conversely, at high salt concentrations, the negative charges of phosphate oxygen atoms are shielded, and double-stranded

Figure 24·16
Denaturation and renaturation of DNA.

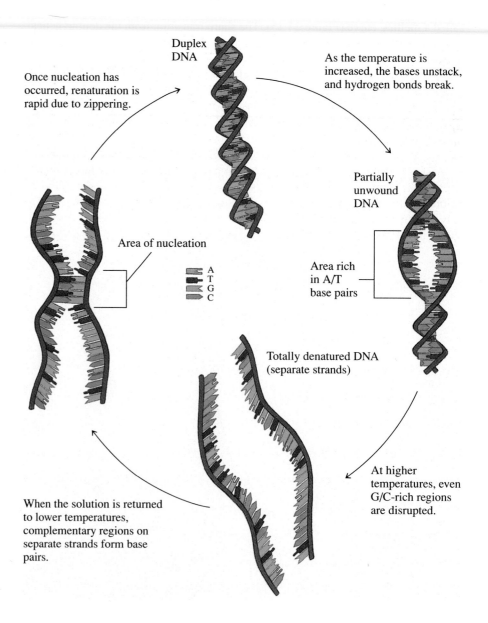

Duplex DNA

As the temperature is increased, the bases unstack, and hydrogen bonds break.

Once nucleation has occurred, renaturation is rapid due to zippering.

Partially unwound DNA

Area of nucleation

A
T
G
C

Area rich in A/T base pairs

Totally denatured DNA (separate strands)

At higher temperatures, even G/C-rich regions are disrupted.

When the solution is returned to lower temperatures, complementary regions on separate strands form base pairs.

DNA denatures at higher temperatures. Renaturation is faster at high salt concentrations because nucleation is more likely to occur when the charges on the single strands are shielded.

24·4 Double-Stranded DNA Can Exist in Several Conformations

In the years that followed publication of the Watson-Crick model for double-helical DNA, X-ray crystallographic studies of various synthetic oligodeoxyribonucleotides of known sequence revealed that different conformations of double-stranded DNA can exist. One of these forms, known as A-DNA, can occur naturally in some DNA molecules of unusual base composition. A-DNA is also commonly observed when purified DNA is dehydrated.

Both A- and B-DNA are right-handed helices with standard Watson-Crick base pairs. The essential differences in A-DNA are that the bases are tilted approximately 20° relative to the helix axis; there are about 11 base pairs per turn; and the helix is wider than in B-DNA (Figure 24·17). Furthermore, A-DNA has a deeper

Figure 24·17
Stereo views of A-DNA.
(a) Stick model. Note that the base pairs (blue) are not perpendicular to the sugar-phosphate backbone. (b) Space-filling model. (Based on coordinates provided by U. Heinemann and H. Lauble.)

(a)

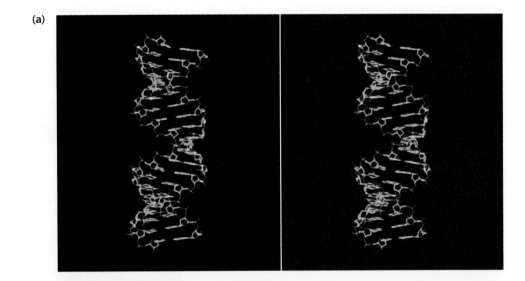

(b)

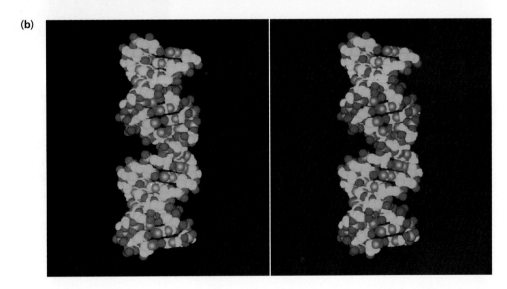

Figure 24·18
Structures of (**a**) A-DNA and (**b**) B-DNA. The two forms of DNA are viewed along the helix axis. The closest base pair (G/C) is highlighted. Dehydration of B-DNA causes a transition to the A form. The grooves of A-DNA do not contain bound water, but the phosphate groups are still hydrated.

(a)

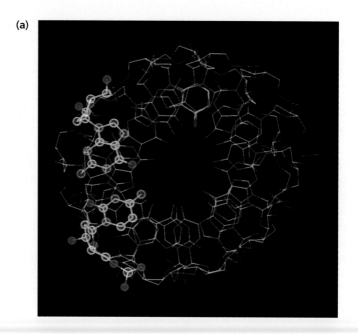

(b)

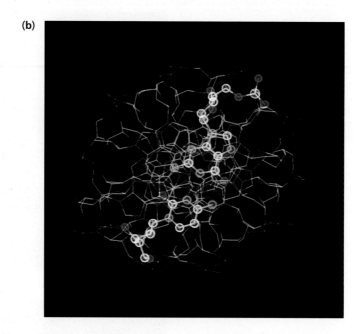

major groove than B-DNA because the base pairs are not centered on the helix axis but instead wind around the center of the molecule. Figure 24·18 is a comparison of A- and B-DNA structures from a perspective looking down the helix axis. In B-DNA, adjacent base pairs are stacked almost directly on top of each other, and the core of the molecule is dense and hydrophobic. In contrast, the base pairs in A-DNA are somewhat offset from each other, and the center of the structure is accessible to solvent.

At the level of individual nucleotides, the most obvious difference between A-DNA and B-DNA is the conformation of the deoxyribose residues. In B-DNA,

(a)

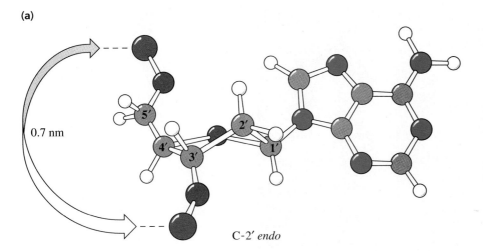

C-2′ *endo*

(b)

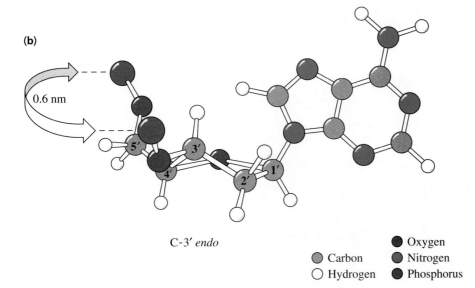

C-3′ *endo*

Carbon • Oxygen
Hydrogen • Nitrogen
• Phosphorus

Figure 24·19
Conformation of deoxyadenylate in B- and A-DNA. **(a)** In B-DNA, the sugar conformation is C-2′ *endo,* and the phosphorus atoms are 0.7 nm apart along the sugar-phosphate backbone. **(b)** The sugar conformation in A-DNA is C-3′ *endo,* and the phosphorus atoms are only 0.6 nm apart. [Adapted from Sundaralingam, M. (1974). Principles governing nucleic acid and polynucleotide conformations. In *Structure and Conformation of Nucleic Acids and Protein-Nucleic Acid Interactions,* M. Sundaralingam and S. T. Rao, eds. (Baltimore: University Park Press), pp. 487–524.]

the C-2′ *endo* form predominates, and the distance between adjacent phosphorus atoms is 0.7 nm, whereas in A-DNA, the sugar adopts a C-3′ *endo* conformation, and the phosphorus atoms are closer together (Figure 24·19). This difference in sugar pucker allows A-DNA to contain about 11 base pairs per turn, compared to about 10.4 in B-DNA. In both forms of DNA, the nucleotides adopt an *anti* conformation, although there is some difference in the torsion angle of the *β-N*-glycosidic bond.

It is now known that DNA inside the cell does not exist in a pure B conformation. Rather, DNA is a dynamic molecule whose conformation changes as the helix bends in solution and is complexed to protein. Most of the DNA appears to be in forms that closely resemble the standard B conformation, with short sequences of A-DNA occurring in certain regions. Some DNA molecules that contain a high proportion of modified nucleotides, such as the DNA of T-even bacteriophages, may adopt additional conformations in vivo.

Some synthetic oligonucleotides crystallize in a form that is radically different from A- or B-DNA. The best-studied examples are polynucleotides containing alternating C and G residues. These synthetic oligonucleotides form a double-stranded, left-handed helix with about 12 base pairs per turn. This form is known as

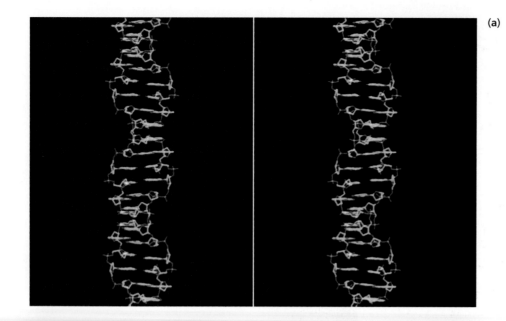

(a)

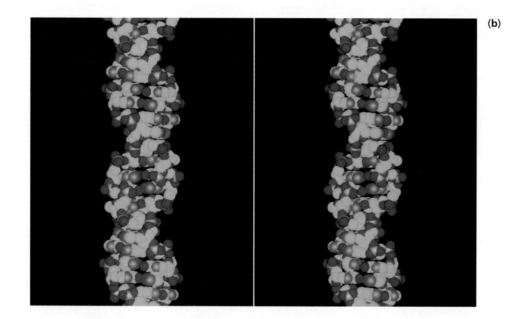

(b)

Figure 24·20
Stereo views of Z-DNA.
(a) Stick model. **(b)** Space-filling model. (Based on coordinates provided by H. Drew and R. E. Dickerson.)

Z-DNA (Figure 24·20). In addition to its left-handed helical structure, one of the main characteristics of Z-DNA is the absence of obvious grooves. The base pairs are only slightly tilted relative to the helix axis, and unlike base pairs in A-DNA, they are centered on the axis. Because there are about 12 base pairs per turn, the distance between each base pair is greater than in the other forms of DNA. Consequently, a stretch of Z-DNA is longer and thinner than either B-DNA or A-DNA (Figure 24·21).

The repeating unit in Z-DNA is two base pairs, unlike the other forms, which have a single–base-pair repeating unit. Furthermore, the individual nucleotides in Z-DNA do not adopt identical conformations. The sugar pucker of the guanylate residue is C-3′ *endo,* and the base is in the *syn* position (Figure 24·22). In contrast,

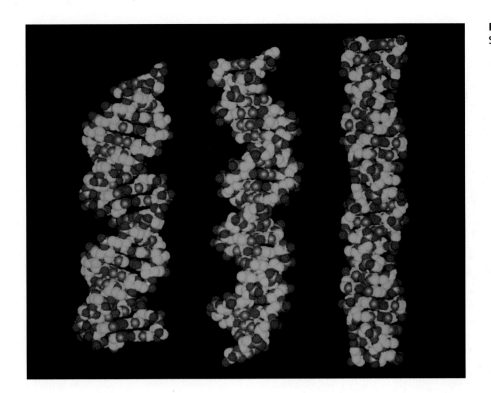

Figure 24·21
Structures of A-, B-, and Z-DNA.

Figure 24·22
Structure of deoxyguanylate in B- and Z-DNA. (a) In B-DNA, the sugar of the nucleotide adopts the C-2' *endo* conformation, and the base is in the *anti* position. (b) In Z-DNA, the sugar pucker is C-3' *endo,* and the base is in the *syn* position. [Adapted from Wang, A. H., Fujii, S., Van Boom, J. H., and Rich, A. (1982). Right-handed and left-handed double-helical DNA: structural studies. *Cold Spring Harbor Symp. Quant. Biol.* 47:33–44.]

● Carbon	● Nitrogen
○ Hydrogen	● Phosphorus
● Oxygen	

(a)

(b)

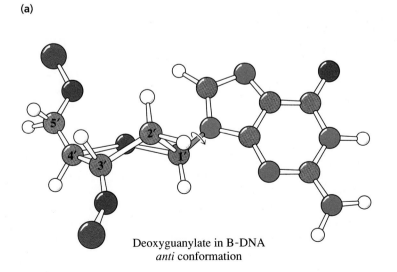

Deoxyguanylate in B-DNA
anti conformation

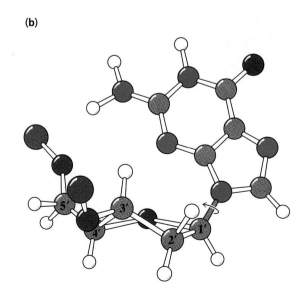

Deoxyguanylate in Z-DNA
syn conformation

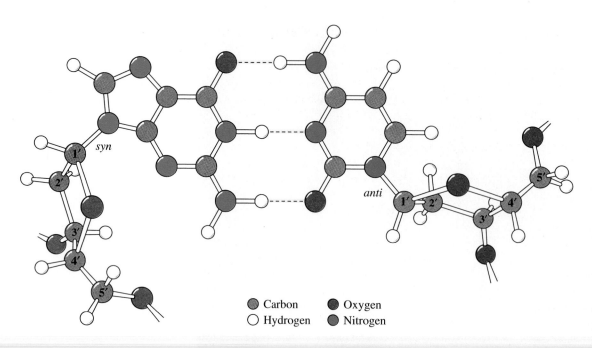

○ Carbon   ● Oxygen
○ Hydrogen   ● Nitrogen

Figure 24·23
Structure of a G/C base pair in Z-DNA. The conformation of the sugar differs for each half of the base pair, as does the orientation around the β-N-glycosidic bond. [Adapted from Saenger, W. (1984). *Principles of Nucleic Acid Structure* (New York: Springer-Verlag), p. 287.]

the cytidylate sugar residue adopts a C-2′ *endo* conformation, and the base is in the *anti* position, as it is in B-DNA (Figure 24·23). Recall that, in solution, the *anti* conformation usually predominates in free nucleosides, but guanine nucleotides such as dGMP can adopt the *syn* conformation, which is favored by electrostatic attraction between the amino group at C-2 and the 5′-phosphate. This interaction is also seen in Z-DNA. The structural parameters of A-, B-, and Z-DNA are compared in Table 24·3.

Although synthetic polynucleotides composed of alternating purine and pyrimidine residues can form Z-DNA, this form of DNA appears to be rare in the genomes of living organisms. The B to Z transition can occur in supercoiled DNA (Section 24·6) and possibly in plants and mammals whose DNA is modified.

Table 24·3 Major structural features of A-, B-, and Z-DNA

Property	A-DNA	B-DNA	Z-DNA
Helix handedness	right	right	left
Repeating unit	1 base pair	1 base pair	2 base pairs
Rotation per base pair	32.7°	34.6°	30°
Base pairs per turn	~11	~10.4	~12
Inclination of base pair to helix axis	19°	1.2°	9°
Rise per base pair along helix axis	0.23 nm	0.33 nm	0.38 nm
Pitch	2.46 nm	3.40 nm	4.56 nm
Diameter	2.55 nm	2.37 nm	1.84 nm
Conformation of glycosidic bond	*anti*	*anti*	*anti* at C *syn* at G
Sugar pucker	C-3′ *endo*	C-2′ *endo*	C-2′ *endo* at C C-3′ *endo* at G

[Adapted from Dickerson, R. E., Drew, H. R., Conner, B. N., Wing, R. M., Fratini, A. V., and Kopka, M. L. (1982). The anatomy of A-, B-, and Z-DNA. *Science* 216:475–485.]

24·5 Some DNA Molecules Contain Modified Nucleotides

Several different modified nucleotides have been identified in DNA from a variety of organisms. By far the most common variants are methylated nucleotides and their derivatives (Figure 24·24). In most cases, the bases are modified after DNA synthesis is completed. Methylases act on particular bases in DNA. The site of methylation is defined by a specific DNA sequence that is recognized by the methylase. The methyl-group donor is *S*-adenosylmethionine.

There are two common methylases in *Escherichia coli*. The Dam methylase acts on adenine residues in the sequence GATC, and the Dcm methylase is responsible for the methylation of cytosine residues on opposite strands in the sequence $CC^A/_TGG$ ($^A/_T$ indicates that both CCAGG and CCTGG are methylation sites). In both cases, the methylated bases serve as protein-recognition sites in processes such as recombination and the initiation of DNA replication. In other cases, modified nucleotides may protect DNA from cleavage by restriction endonucleases, as explained in Section 24·10. The 5-hydroxymethylcytosine base incorporated into the DNA of bacteriophages T2, T4, and T6 protects the phage DNA from degradation by the phage enzyme that catalyzes breakdown of bacterial host DNA containing unmodified cytosine bases. In the case of methylated cytosine, CTP is modified so that 5-hydroxymethylcytosine is incorporated as DNA is synthesized. Many of these hydroxymethylcytosine residues are further modified by glycosylation.

5-Methylcytosine is commonly found in the genomes of mammals and vascular plants, but it is not present in DNA from insects, yeast, and most other eukaryotes. Up to 5% of the cytosine residues in mammalian DNA and as many as one-third of the cytosine residues in flowering plants may be methylated. The methylation site in mammals is the dinucleotide CG, and the cytosine residues on both strands are methylated. In plant genomes, CG as well as CNG (*N* can be any nucleotide) are methylation sites. The formation of Z-DNA is favored in regions that are rich in 5-methylcytidine. The exact role of base methylation in mammals and vascular plants is unknown, but modified nucleotides may play a role in gene regulation.

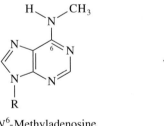

N^6-Methyladenosine
5′-monophosphate

N^2-Methylguanosine
5′-monophosphate

N^4-Methylcytidine
5′-monophosphate

5-Methylcytidine
5′-monophosphate

5-Hydroxymethylcytidine
5′-monophosphate

Figure 24·24
Common nucleotides containing modified bases. *R* represents the phosphoribosyl group.

24·6 DNA Topology

DNA molecules are very large, and they can bend and twist in solution to assume a variety of shapes. DNA topology is the study of the overall morphology of these molecules.

The two ends of some double-stranded DNA molecules, such as bacterial chromosomes and the DNA molecules in mitochondria and chloroplasts, are covalently linked to form closed circles. Many viruses also have circular DNA. Eukaryotic nuclear DNA is linear, as is the DNA in some viruses and bacteriophages. Bacteria and simple eukaryotes may also contain small circular DNA molecules called **plasmids**, which contain a small number of genes. DNA molecules, whether circular or linear, have considerable conformational flexibility.

A. Cruciform Structures in DNA

Some DNA molecules contain **inverted repeats**, sequences of nucleotides that are repeated in the opposite orientation. These inverted repeats are often separated by a short stretch of DNA that is not part of the repeated sequence (Figure 24·25). In such regions, each strand of DNA is capable of forming a base-paired structure with the complementary region on the same strand, giving rise to a conformation known as a **cruciform structure.**

Figure 24·25
Formation of a cruciform structure in double-stranded DNA.

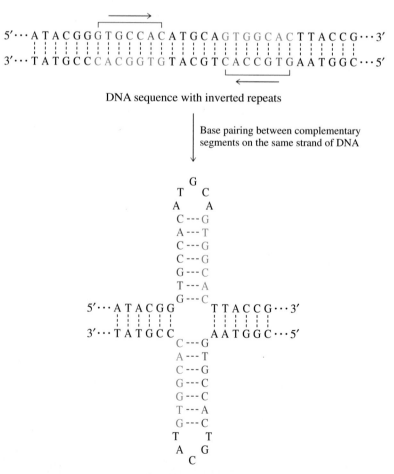

Cruciform structures are less stable than normal B-DNA, but they do occur under certain conditions. Note that there are no unpaired bases in a cruciform structure, with the exception of the bases in the loops where the DNA turns back on itself. Compared to B-DNA, there are fewer turns of the double helix per number of bases in a cruciform because of these unpaired bases.

B. Supercoiled DNA

In the B form of DNA, the helix exhibits an average twist of 34.6° from one base pair to the next, and one turn is completed every 10.4 base pairs. Circular DNA molecules with this structure are said to be relaxed. However, the double helix can be overwound or underwound if the circular DNA is broken and the two ends of the linear molecule are twisted in opposite directions. When the DNA is overwound so that it has a 40° twist, one strand crosses over the other every 9 base pairs; when the helix is underwound so that it has only a 33° twist, one strand crosses over the other every 11 base pairs. When linear overwound or underwound DNA is rejoined to create a circle, the molecule compensates for the change in twist by forming **supercoils** in order to maintain the B conformation (Figure 24·26). Circular DNA molecules that are underwound give rise to negatively supercoiled structures, whereas those that are overwound produce positively supercoiled structures.

The formation of supercoils is an important aspect of the topology of DNA inside the cell. Most DNA molecules in living organisms are negatively supercoiled, and supercoiling has significant consequences for the function of DNA. Positively supercoiled DNA is not found in nature. An example of negatively supercoiled plasmid DNA is shown in Figure 24·27. The number of supercoils in such a molecule can range from 1 to 20 or 30, indicating progressive underwinding of the double helix. Bacterial chromosomes typically contain about 5 supercoils per 1000 base pairs, whereas the DNA of eukaryotes usually has a much higher degree of supercoiling.

Negative supercoiling to maintain the B conformation of DNA introduces torsional strain. The addition of a single negative supercoil increases the standard free energy of formation of DNA by about 38 kJ mol^{-1}. Therefore, elimination of supercoils is thermodynamically favored. One way to eliminate negative supercoils is to create a region of Z-DNA, since this form of DNA is a left-handed helix. Such structures have been detected in supercoiled plasmids when the nucleotide composition favors formation of Z-DNA. Similarly, formation of a cruciform structure relieves negative supercoils because the cruciform contains fewer helical turns.

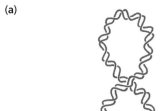

(a)

Negative supercoil

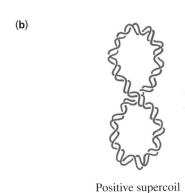

(b)

Positive supercoil

Figure 24·26
Supercoiled DNA. **(a)** A negatively supercoiled molecule is produced when a circular DNA molecule is underwound around its longitudinal axis. The resulting superhelical turn is right-handed. **(b)** A positively supercoiled molecule is produced when DNA is overwound. This superhelical turn is left-handed.

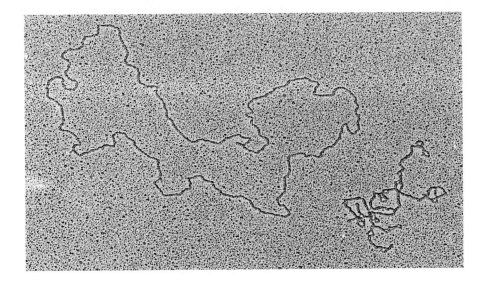

Figure 24·27
Electron micrograph of plasmid DNA. A relaxed, closed, circular molecule is on the left; a supercoiled plasmid is on the right. (Courtesy of Jack Griffith.)

Figure 24·28
Local unwinding in negatively supercoiled DNA.

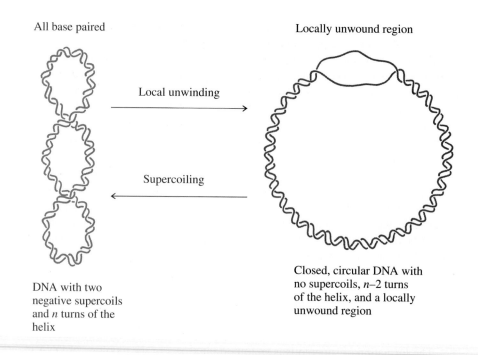

All base paired

Locally unwound region

Local unwinding →

← Supercoiling

DNA with two negative supercoils and n turns of the helix

Closed, circular DNA with no supercoils, $n-2$ turns of the helix, and a locally unwound region

Negative supercoils can also be eliminated through local unwinding of DNA (Figure 24·28). The regions that are unwound tend to be A/T rich, since such regions are inherently less stable than G/C-rich regions in B-DNA. Local unwinding is important during the initiation of DNA replication and transcription, when strands must be separated. These processes are more efficient when DNA is negatively supercoiled.

The amount of supercoiling in a DNA molecule can be defined quantitatively. The number of times that one strand of a circular DNA molecule crosses over another when the molecule is lying flat on a plane is defined as the **linking number** (L). The linking number is a topological property of DNA and is constant for any covalently closed, circular molecule. The linking number is defined as positive in right-handed helices such as B-DNA and negative in left-handed helices such as Z-DNA.

When negative supercoils are added to a circular DNA molecule, the linking number decreases. The easiest way to understand this is to examine Figure 24·28 and note that negatively supercoiled DNA is equivalent to DNA containing no supercoils and a locally unwound region. As an example, consider the circular DNA molecule of the animal virus SV40, which contains 5200 base pairs. Since the double helix of normal B-DNA makes one complete turn every 10.4 base pairs, this DNA molecule has a linking number of 500 (5200/10.4) when it is not supercoiled. However, SV40 DNA isolated from cells contains 25 negative supercoils; its linking number is therefore 475.

C. Topoisomerases Change the Linking Number

The linking number of a DNA molecule can be changed only by cleaving a phosphodiester linkage in one or both strands, rewinding the DNA, and resealing the break. Enzymes that catalyze this process alter the topology of DNA and are therefore called **topoisomerases.** All cells contain similar enzymes that introduce or remove supercoils. Type I topoisomerases catalyze a break in one strand of DNA and change the linking number by 1; type II topoisomerases catalyze a break in both strands of DNA and change the linking number by 2.

   E. coli topoisomerase I is a monomeric protein with a molecular weight of 100 000. It catalyzes the removal of negative supercoils in DNA. Since supercoiled DNA is under torsional strain, the removal of supercoils is a spontaneous process, and no external energy source is needed to drive the topoisomerase reaction. An overview of the reaction is shown in Figure 24·29, and a proposed mechanism for the reaction is presented in Figure 24·30 (next page). The enzyme binds to a region

Figure 24·29
Action of *E. coli* topo-isomerase I.

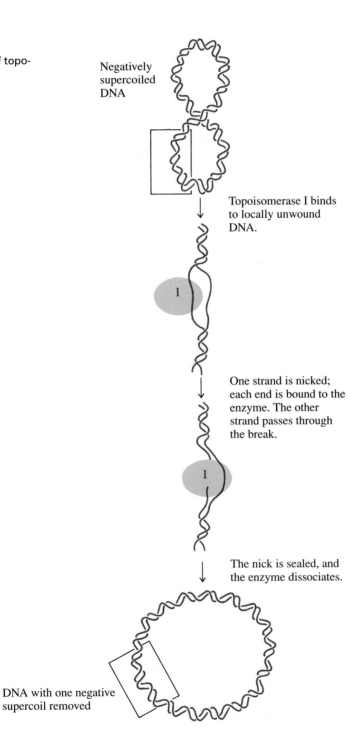

Negatively supercoiled DNA

Topoisomerase I binds to locally unwound DNA.

One strand is nicked; each end is bound to the enzyme. The other strand passes through the break.

The nick is sealed, and the enzyme dissociates.

DNA with one negative supercoil removed

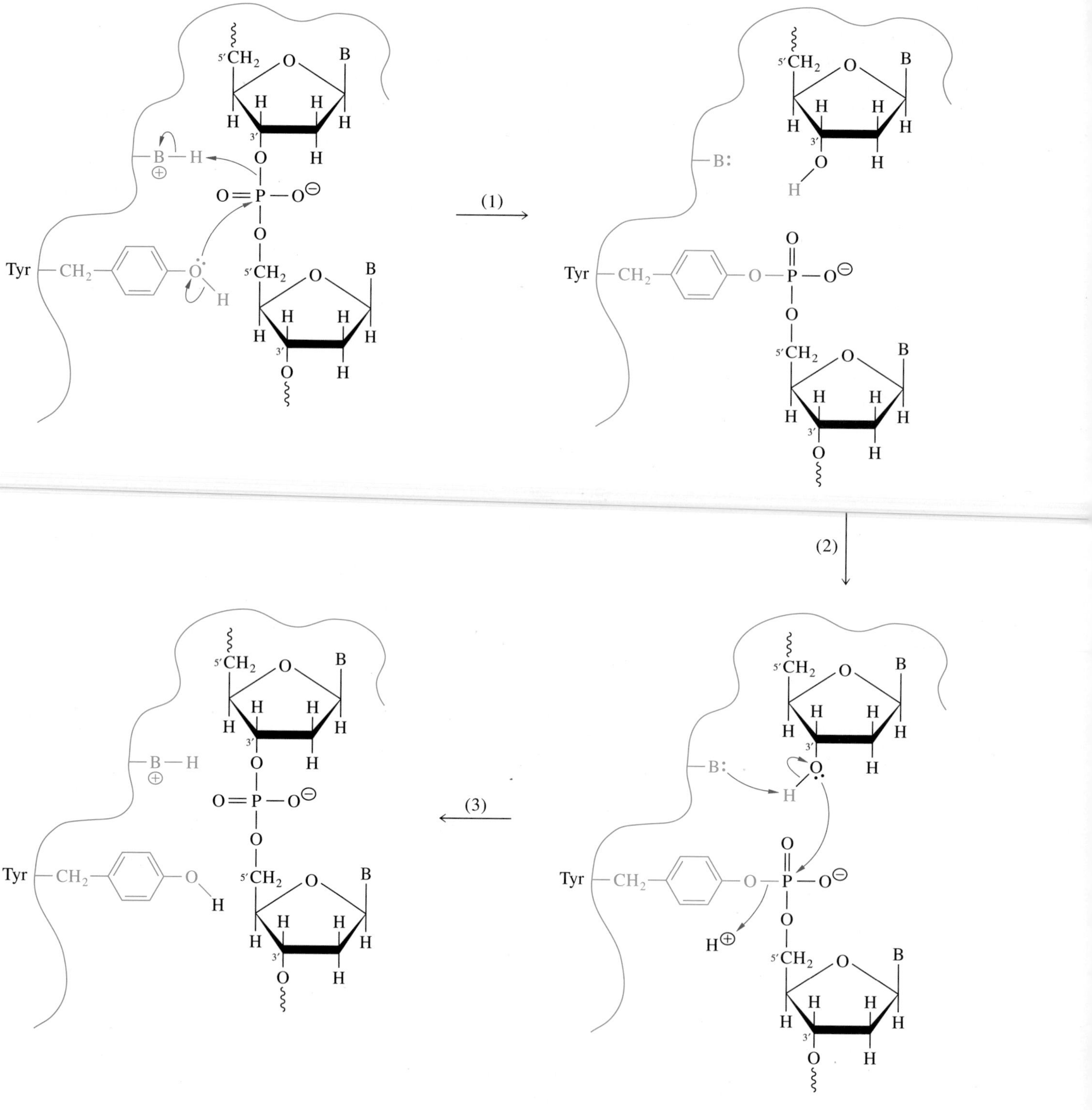

Figure 24·30
Mechanism of *E. coli* topoisomerase I. In Step 1, the hydroxyl group of a tyrosine residue at the active site of the enzyme forms a phosphodiester linkage with a 5′-phosphate group of the DNA chain, breaking the DNA strand. During Step 2, the second strand (not shown) passes through the break in the first strand. In Step 3, the broken phosphodiester linkage of the first strand is re-formed.

of single-stranded DNA that is locally unwound, and a 5′-phosphate group of one of the strands becomes covalently bound to an active-site tyrosine in the enzyme through a transesterification reaction. The 3′ end of the broken strand is also held in place within the active site. A supercoil is removed and the linking number of DNA increases when the second strand is passed through the break. The break is then sealed by a second transesterification reaction that removes the covalent bond between the enzyme and DNA and regenerates the sugar-phosphate backbone. Only one supercoil is removed at a time.

Since positively supercoiled DNA is overwound and does not contain locally unwound regions, and since topoisomerase I binds only to single-stranded regions, only negatively supercoiled DNA is relaxed by this enzyme. Type I topoisomerases are found in eukaryotes as well as prokaryotes, but the mechanism of the eukaryotic enzymes differs in that the protein is covalently bound to the 3′ end of the break in DNA, not to the 5′ end.

E. coli topoisomerase II (also called DNA gyrase) can add negative supercoils to DNA and thereby decrease the linking number. Since supercoiled DNA is under torsional strain, the addition of supercoils requires energy, which is supplied by coupling the reaction to the hydrolysis of ATP. The proposed action of topoisomerase II is shown in Figure 24·31. According to this scheme, DNA wraps around the outside of the protein. The DNA is not supercoiled in this complex. Topoisomerase II then catalyzes the formation of a double-stranded break in one loop of DNA. In the next step, another loop of DNA is passed through the break. After the cleaved strand is resealed, the DNA dissociates. The net effect is to introduce two negative supercoils and decrease the linking number by 2. Note that in the reaction catalyzed by topoisomerase II, one double strand is passed through a double-stranded break, whereas in topoisomerase I–catalyzed reactions, a single strand is passed through a single-stranded break.

Only the bacterial topoisomerase II enzymes can introduce negative supercoils into DNA. However, both the bacterial and eukaryotic type II enzymes can relax positively and negatively supercoiled DNA by a reaction that is the reverse of the one shown in Figure 24·31. In addition to introducing and removing supercoils in DNA, topoisomerase II can untangle knotted DNA inside the cell. For example, when two closed, circular DNA molecules become concatenated, that is, joined like the links in a chain, topoisomerase II can separate the linked DNA molecules.

24·7 Cells Contain Several Kinds of Nucleic Acids

Whereas DNA molecules serve as the repository of genetic information, RNA molecules participate in several processes by which the genetic information is expressed. Within a given cell, RNA molecules are found in multiple copies and in several forms. There are four major classes of RNA in all living cells:

1. **Ribosomal RNA** (rRNA) molecules are an integral part of ribosomes, intracellular complexes of protein and RNA that catalyze protein synthesis. rRNA is the most abundant class of ribonucleic acid, accounting for about 80% of the total cellular RNA.

2. **Transfer RNA** (tRNA) molecules carry activated amino acids to the ribosomes for incorporation into growing peptide chains during protein synthesis. tRNA accounts for about 15% of the total cellular RNA. tRNA molecules are only 73 to 95 nucleotide residues in length.

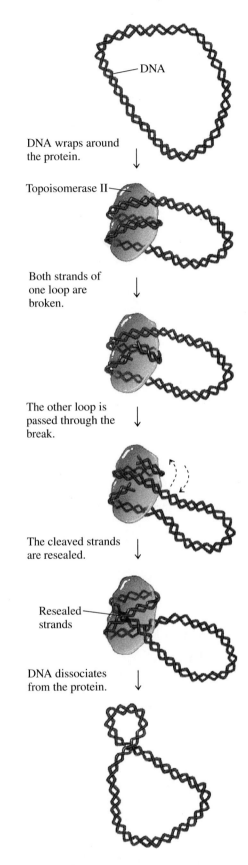

DNA wraps around the protein.

Topoisomerase II

Both strands of one loop are broken.

The other loop is passed through the break.

The cleaved strands are resealed.

Resealed strands

DNA dissociates from the protein.

Figure 24·31
Action of topoisomerase II.

3. **Messenger RNA** (mRNA) molecules encode the sequences of amino acids in proteins. Thus, mRNA molecules serve as the "messengers" that carry information from DNA to the ribosomes. Generally, mRNA accounts for only 3% of the total cellular RNA. These molecules are the least stable of the cellular ribonucleic acids.

4. **Small RNA** molecules are present in all cells. Some of them have catalytic activity or contribute to catalytic activity in association with proteins. Many small RNA molecules are associated with processing events that modify RNA after it is synthesized.

24·8 Some RNA Molecules Have Stable Secondary Structure

RNA polynucleotides differ from DNA polynucleotides by the presence of 2′-hydroxyl groups and by the substitution of the base uracil for thymine found in DNA (Figure 24·32). Note that thymine is actually 5-methyluracil. As in DNA, the nucleotides in RNA are joined by 3′–5′ phosphodiester linkages to produce the polynucleotide. RNA molecules usually exist as single-stranded polyribonucleotides.

Under physiological conditions, most single-stranded polynucleotides fold back on themselves to form duplex regions between complementary base pairs. RNA commonly has extensive secondary structure. One such structure is a **hairpin,** or stem-loop, which arises when short regions of complementary sequence form base pairs (Figure 24·33). Such structures are important in transcription and are common features in both tRNA and rRNA.

Figure 24·32
Structures of the ribonucleotides in RNA.

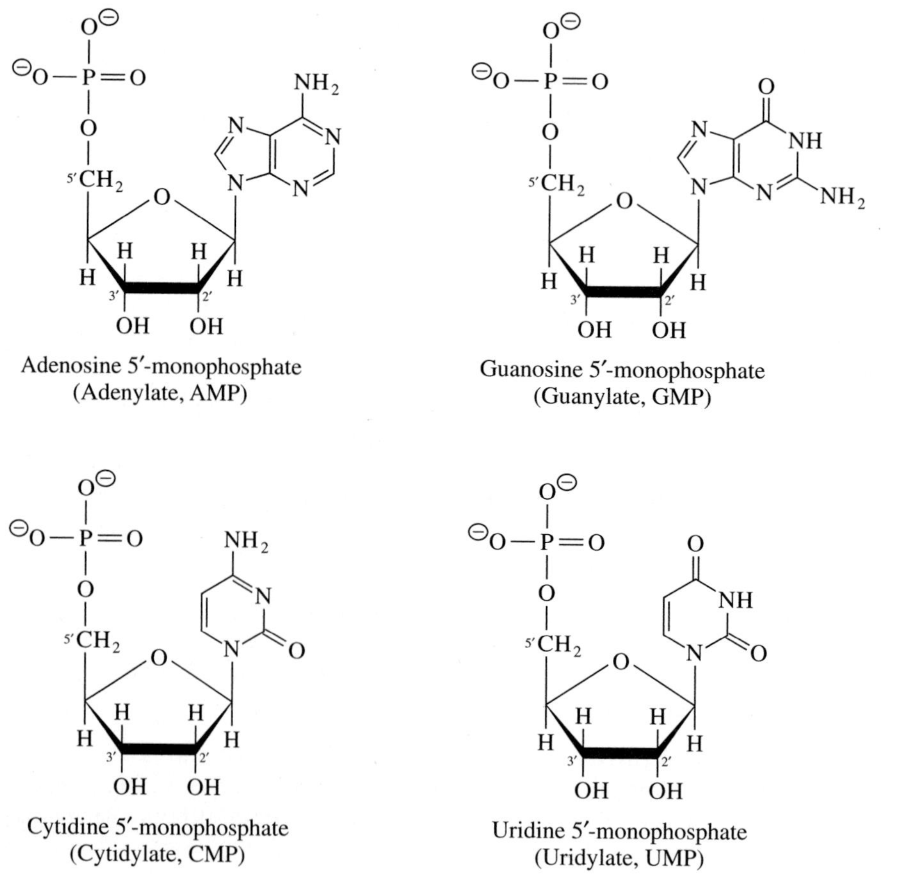

Adenosine 5′-monophosphate
(Adenylate, AMP)

Guanosine 5′-monophosphate
(Guanylate, GMP)

Cytidine 5′-monophosphate
(Cytidylate, CMP)

Uridine 5′-monophosphate
(Uridylate, UMP)

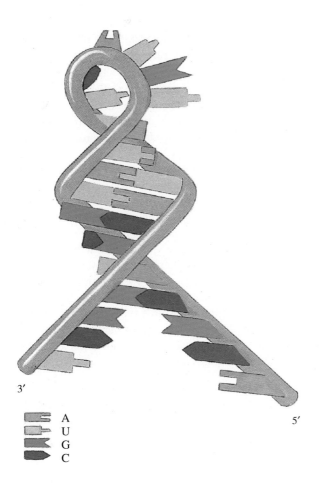

3′

5′

A
U
G
C

Figure 24·33
Hairpin in RNA. Single-stranded polynucleotides such as RNA can form hairpins, or stem-loops, when short regions of complementary sequence form base pairs. The stem of the hairpin consists of base-paired nucleotides, and the loop consists of noncomplementary nucleotides. Note that the strands in the stem are antiparallel.

   The double-stranded regions of RNA adopt a structure that is similar to that of A-DNA. In other words, the sugar is in the C-3′ *endo* form, and the bases are in the *anti* position (see Figure 24·19). Single-stranded RNA molecules can also adopt a similar structure, depending on the base composition. For example, polyA forms a single-stranded helix that is stabilized by extensive base stacking (Figure 24·34). Other single-stranded RNA molecules, such as polyU, do not appear to adopt a regular conformation in solution.

Figure 24·34
Stereo view of polyA, a single-stranded RNA helix.

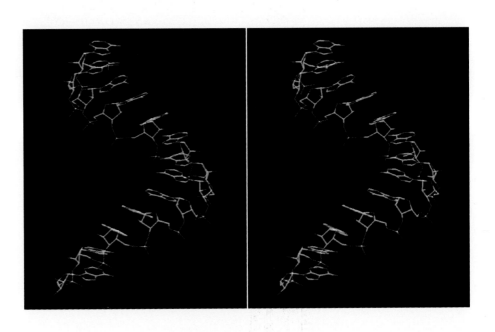

24·9 Nucleases Catalyze Hydrolysis of the Phosphodiester Linkages of Nucleic Acids

Enzymes that catalyze the hydrolysis of phosphodiester linkages in nucleic acids are collectively known as nucleases. There are a variety of different nucleases in all cells. Some of them are required for the synthesis and repair of DNA, as we will see in subsequent chapters, and others are needed for the production and degradation of cellular RNA.

Ribonucleases (RNases) act only on RNA, and **deoxyribonucleases** (DNases) act only on DNA. Some nucleases catalyze hydrolysis of both RNA and DNA. Like proteases, nucleases can be further classified according to their site of cleavage: **exonucleases** catalyze hydrolysis of phosphodiester linkages to release nucleotide residues from the ends of a polynucleotide chain, whereas **endonucleases** catalyze hydrolysis of phosphodiester linkages at various sites within polynucleotide chains. Nucleases exhibit a wide variety of specificities for nucleotide sequences.

In a polynucleotide, either the 3′ or the 5′ ester bond of a 3′–5′ phosphodiester linkage can be cleaved during hydrolysis. One type of hydrolysis yields a 5′-monophosphate and a 3′-hydroxyl group; the other type yields a 3′-monophosphate and a 5′-hydroxyl group (Figure 24·35). Each type of nuclease catalyzes one reaction or the other, but not both. Table 24·4 lists some common nucleases from different organisms, as well as their substrates and products.

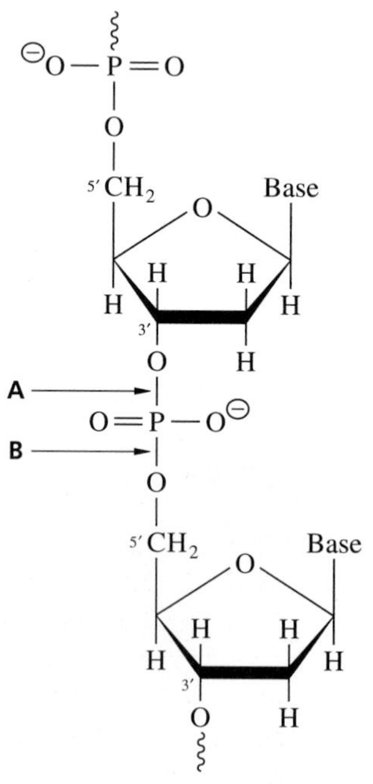

Figure 24·35
Nuclease cleavage sites. Cleavage at bond A generates a 5′-phosphate and a 3′-hydroxyl terminus. Cleavage at bond B generates a 3′-phosphate and a 5′-hydroxyl terminus. Both DNA (shown) and RNA are substrates of nucleases.

Table 24·4 Some representative nucleases

	Substrate	Product
Exonucleases (3′ → 5′)		
Exonuclease I (*E. coli*)	Single-stranded DNA	Nucleoside 5′-phosphates
DNA polymerase I (*E. coli*)	Single-stranded DNA*	Nucleoside 5′-phosphates
Exonuclease III (*E. coli*)	Double-stranded DNA*	Nucleoside 5′-phosphates
Bacillus subtilis exonuclease	Double-stranded DNA, RNA	Nucleoside 3′-phosphates, oligonucleotides
Exonucleases (5′ → 3′)		
Bacillus subtilis exonuclease	Single-stranded DNA	Nucleoside 3′-phosphates (also works from 3′ end)
Exonuclease VII (*E. coli*)	Single-stranded DNA	Oligonucleotides (also works from 3′ end)
Neurospora crassa exonuclease	Single-stranded DNA, RNA	Nucleoside 5′-phosphates
Exonuclease VI or DNA polymerase I (*E. coli*)	Double-stranded DNA	Nucleoside 5′-phosphates
Endonucleases		
Neurospora crassa endonuclease	Single-stranded DNA, RNA	Polynucleotides with 5′-phosphate termini
Endonuclease I (*E. coli*)	Single- or double-stranded DNA	Polynucleotides with 5′-phosphate termini
DNase I (bovine pancreas)	Single- or double-stranded DNA	Polynucleotides with 5′-phosphate termini
DNase II (calf thymus)	Single- or double-stranded DNA	Polynucleotides with 3′-phosphate termini
RNase A (bovine pancreas)	RNA	Polynucleotides with 3′-phosphate termini

*Also catalyzes hydrolysis of nicked and gapped double-stranded DNA.

A. Alkaline Hydrolysis of RNA Involves Formation of a Cyclic Monophosphate

The difference between 2′-deoxyribose in DNA and ribose in RNA may seem small, but it greatly affects the chemical and physical properties of the nucleic acids. The 2′-hydroxyl group of ribose contributes to some of the hydrogen bonding within RNA molecules and is involved in certain chemical and enzyme-catalyzed reactions. The net effect is that single-stranded RNA is less stable than DNA. The greater chemical stability of DNA is probably an important factor in its role as the primary genetic material.

One example illustrating the differences in chemical reactivities that result from the presence of the 2′-hydroxyl groups of RNA is the effect of alkaline solutions on RNA and DNA. RNA treated with 0.1 M NaOH at room temperature is degraded to a mixture of 2′- and 3′-nucleoside monophosphates within a few hours, whereas DNA is quite stable under the same conditions. The mechanism of alkaline hydrolysis of RNA is shown in Figure 24·36 (next page). In both the first and second steps, hydroxide ions act only as catalysts, since removal of a proton from water (to form the 5′-hydroxyl group in the first step or the 2′- or 3′-hydroxyl group in the second) regenerates one hydroxide ion for each hydroxide ion consumed. Note that a 2′,3′-cyclic nucleoside monophosphate intermediate is formed. DNA is not hydrolyzed under these conditions because it lacks the 2′-hydroxyl groups needed to initiate the intramolecular transesterification.

Figure 24·36
Alkaline hydrolysis of RNA. In Step 1, a hydroxide ion in alkaline solution abstracts the proton from the 2′-hydroxyl group of a nucleotide residue in RNA, converting it to an alkoxide. The 2′-alkoxide is a potent nucleophile that attacks the adjacent phosphorus atom, displacing the 5′-oxygen and generating a 2′,3′-cyclic nucleoside monophosphate intermediate. As each phosphodiester of RNA is cleaved, the polyribonucleotide chain rapidly depolymerizes. The 2′,3′-cyclic nucleoside monophosphate is not stable in alkaline solutions, however, and a second hydroxide ion catalyzes its hydrolysis to either a 2′- or a 3′-nucleoside monophosphate (Step 2).

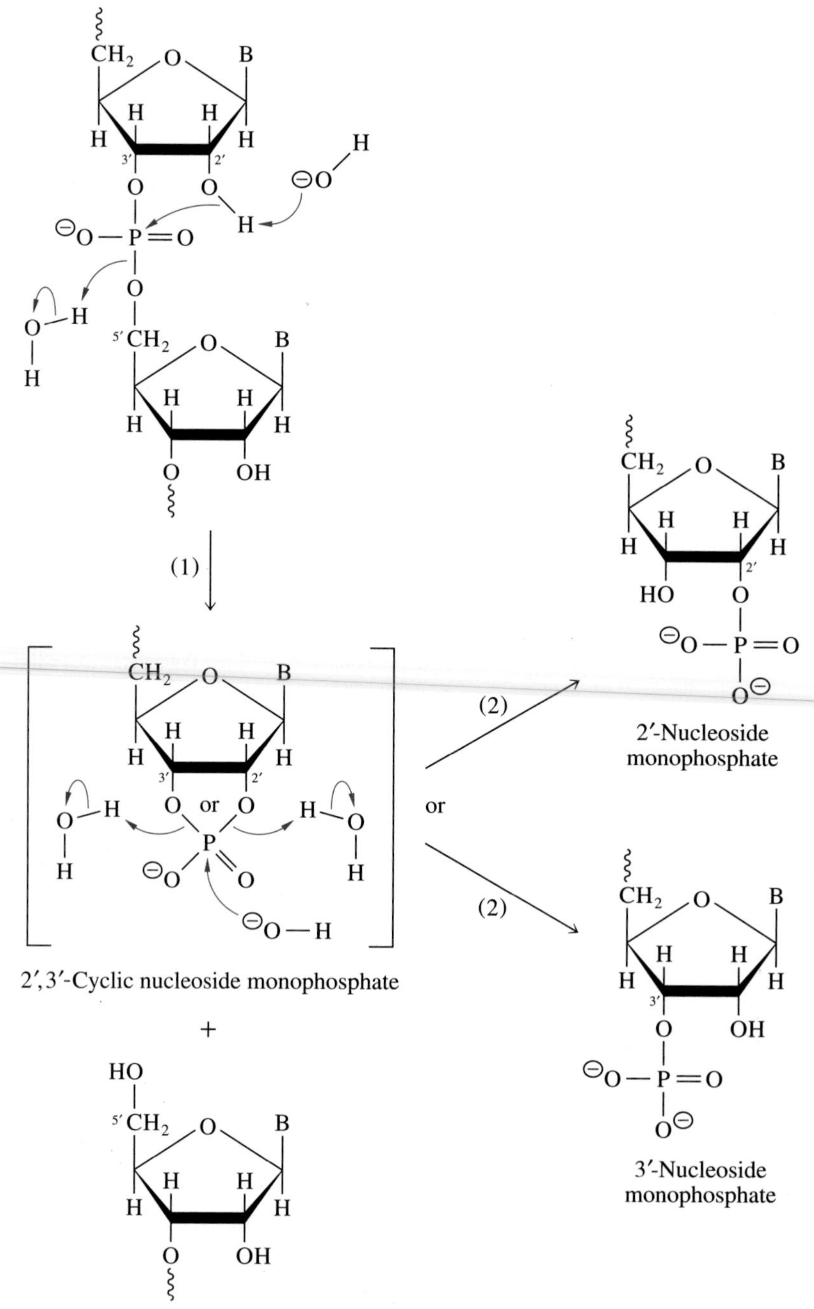

Figure 24·37 (next page)
Mechanism of RNA cleavage by RNase A. In Step 1, His-12 serves as a base catalyst, abstracting the proton from the 2′-hydroxyl group of the pyrimidine nucleotide. The resulting nucleophilic oxygen atom attacks the phosphorus atom. His-119 (as an imidazolium ion) serves as an acid catalyst, ultimately donating a proton to the 5′-oxygen atom of the next nucleotide residue to produce the alcohol leaving group. In the transition state, the phosphorus atom possesses five covalent bonds. In Step 2, a 2′,3′-cyclic nucleoside monophosphate intermediate is formed, and the first product (P_1), which has a 5′-hydroxyl terminus, is released. In Step 3, water enters the active site upon departure of P_1, and His-119 (now in the conjugate-base form) removes a proton from water, generating a hydroxide ion near the phosphodiester of the cyclic intermediate. The hydroxide ion attacks the phosphorus atom to form a second transition state with a pentacovalent phosphorus atom. The imidazolium form of His-12 donates a proton to the 2′-oxygen atom, enabling the transition state to decompose to P_2 (Step 4).

2′,3′-Cyclic nucleoside monophosphate

2′-Nucleoside monophosphate

3′-Nucleoside monophosphate

B. Some Ribonucleases Catalyze Hydrolysis of RNA via a Cyclic Monophosphate Intermediate

Bovine pancreatic ribonuclease A (RNase A) consists of a single polypeptide chain of 124 amino acid residues cross-linked by four disulfide bridges. The enzyme has a pH optimum of about 6. RNase A catalyzes cleavage of phosphodiester linkages in RNA at 5′ ester bonds. In chains drawn in the 5′ → 3′ direction, cleavage occurs to the right of pyrimidine nucleotide residues. Thus, a strand of the sequence pApGpUpApCpGpU would be hydrolyzed by RNase A to form pApGpUp + ApCp + GpU.

RNase A contains three ionic amino acid residues at the active site—Lys-41, His-12, and His-119. Lys-41 contributes to transition-state stabilization; His-12 and His-119 participate as acid-base catalysts in the cleavage of RNA. Many studies have led to the formulation of the mechanism of catalysis shown in Figure 24·37.

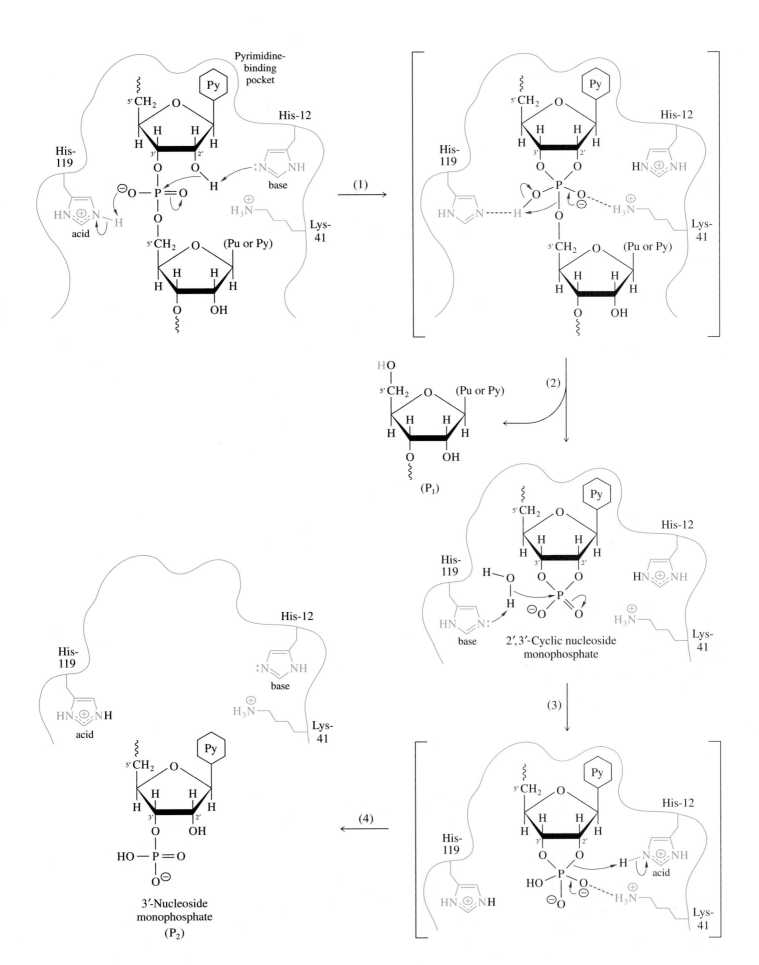

The pyrimidine-binding pocket of the enzyme accounts for the specificity of RNase A (much as the arginine- and lysine-binding pocket of trypsin gives that endopeptidase its specificity). In the first transition state of this reaction, the phosphorus atom possesses five covalent bonds; the 2′-oxygen atom is bonded to the phosphorus atom but the 5′-oxygen has not yet left. Note that the first and second transition states are stabilized by the charged side chain of Lys-41. Hydrolysis of the cyclic phosphate intermediate involves water at the active site, as in the displacement mechanisms of proteases. Upon release of the second product, which has a 3′-monophosphate terminus, the original state of the enzyme is restored—His-12 is a base, and His-119 is an acid. The activity of RNase A illustrates the use of three fundamental catalytic mechanisms: proximity (in the binding and positioning of a suitable phosphodiester between the two histidine residues); acid-base catalysis (by His-119 and His-12); and transition-state stabilization (by Lys-41).

Although alkaline hydrolysis and the reaction catalyzed by RNase A both involve formation of a 2′,3′-cyclic nucleoside monophosphate intermediate, it is important to note three ways in which the reactions differ. First, alkaline hydrolysis can occur at any residue, whereas enzyme-mediated cleavage occurs only at pyrimidine nucleotide residues; second, hydrolysis of the 2′,3′-cyclic monophosphate intermediates is random in the case of alkaline hydrolysis (producing mixtures of 2′- and 3′-nucleotides) but specific in the case of enzyme-mediated cleavage (producing only 3′-nucleotide products); third, enzymatic hydrolysis proceeds at physiological pH, rather than at an extremely basic pH.

C. Cells Contain Many Different Deoxyribonucleases

Exonuclease III is an enzyme that is commonly used in the laboratory to manipulate DNA. The substrate for this enzyme is double-stranded DNA with free ends terminating in 3′-hydroxyl groups. The enzyme acts on the 3′ ends of both strands, catalyzing the progressive removal of single nucleotides as nucleoside 5′-monophosphates. Both strands are degraded until only a small double-stranded region remains (Figure 24·38). Exonuclease III is isolated from *E. coli,* where it plays a role in recombination and repair of DNA.

DNase I is an endonuclease that is secreted by cells of the mammalian pancreas. Its role is to degrade ingested DNA. The enzyme binds to double-stranded DNA and interacts with the bases and the sugar-phosphate backbone in the minor groove (Figure 24·39). Strand cleavage occurs by acid-base catalysis involving a glutamate residue and a histidine residue at the active site. When DNase I binds to DNA, protein contacts cause some distortion, and the DNA is slightly bent. It is quite common for DNA-binding proteins to alter the conformation of DNA to some extent.

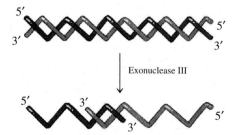

Figure 24·38
Action of exonuclease III.

Figure 24·39
Stereo view of DNase I bound to DNA. The DNA molecule is shown in purple, and the DNase protein is in blue. Amino acid residues in contact with DNA are red. (Based on coordinates provided by D. Suck, A. Lahm, and C. Oefner.)

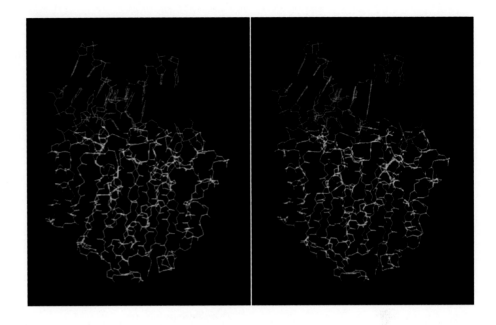

24·10 Restriction Endonucleases Catalyze Hydrolysis of Duplex DNA at Specific Sites

The **restriction endonucleases** are an important class of endonucleases that act on DNA. The term *restriction endonuclease* is derived from the observation that certain bacteria can block phage infections by specifically destroying the incoming viral DNA. Such bacteria are known as restricting hosts, since they restrict the expression of foreign DNA.

Restricting hosts synthesize nucleases that digest foreign DNA at sites corresponding to specific nucleotide sequences. The DNA of the host cell itself is protected from the nucleases because some bases are methylated at the sites recognized by the endonucleases and the nuclease cannot catalyze hydrolysis of the modified substrate. Any DNA that enters the cell and is not methylated at those specific sites is subject to cleavage by restriction endonucleases.

Restriction endonucleases have been divided into types I and II. Type I restriction endonucleases catalyze both the methylation of host DNA and the cleavage of unmethylated DNA. Type II restriction endonucleases are simpler in that they have only nuclease activity; separate enzymes, specific methyltransferases known as restriction methylases, catalyze methylation at the same recognition sequence. Well over 200 type I and type II restriction nucleases have been characterized.

Many species of bacteria have at least one highly specific type II restriction endonuclease as well as a restriction methylase of identical specificity. Normally, all DNA of the host cell is methylated at the potential restriction sites. Following DNA replication, each site is hemimethylated; that is, bases on only one strand are methylated. Hemimethylated sites are high-affinity substrates for the methylase but are not recognized by the restriction endonuclease.

Table 24·5 Specificities of some common restriction endonucleases

Source	Enzyme[1]	Recognition sequence[2]
Bacillus amyloliquefaciens H	*Bam*HI	G↓GATCC
Bacillus globigii	*Bgl*II	A↓GATCT
Escherichia coli RY13	*Eco*RI	G↓A$\overset{*}{A}$TTC
Escherichia coli R245	*Eco*RII	↓C$\overset{*}{C}$TGG
Haemophilus aegyptius	*Hae*III	GG↓CC
Haemophilus influenzae R$_d$	*Hind*II	GTPy↓Pu$\overset{*}{A}$C
Haemophilus influenzae R$_d$	*Hind*III	$\overset{*}{A}$↓AGCTT
Haemophilus parainfluenzae	*Hpa*II	C↓CGG
Nocardia otitidis-caviarum	*Not*I	GC↓GGCCGC
Providencia stuartii 164	*Pst*I	CTGCA↓G
Serratia marcescens S$_b$	*Sma*I	CCC↓GGG

[1] The names of restriction endonucleases are abbreviations of the names of the organisms that produce them. Some abbreviated names are followed by a letter denoting the strain. Roman numerals indicate the order of discovery of the enzyme in that strain.

[2] Recognition sequences are written 5′ to 3′. Only one strand is represented. The arrows indicate cleavage sites. Asterisks represent positions where bases can be methylated. Pu (purine) denotes that either A or G is recognized. Py (pyrimidine) denotes that either C or T is recognized.

The specificities of a few representative restriction endonucleases are listed in Table 24·5. In nearly all cases, the four– to eight–base pair segment that serves as a recognition site has a twofold axis of symmetry; that is, the 5′→3′ sequence of residues is the same in both strands of duplex DNA. Thus, the paired sequences read the same on opposite strands; such sequences are known as palindromes. (Palindromes in English include BIB, DEED, RADAR, and even MADAM I'M ADAM, provided one ignores punctuation and spacing.)

A. *Eco*RI Binds to DNA and Catalyzes Double-Stranded Cleavage

*Eco*RI from *E. coli* was one of the first restriction enzymes discovered. As shown in Table 24·5, *Eco*RI has a six–base pair recognition sequence that is palindromic (each strand being d(GAATTC) in the 5′→3′ direction). The *Eco*RI enzyme is a homodimer and thus, like its substrate, possesses a twofold axis of symmetry. *Eco*RI is a type II restriction endonuclease; the companion methylase converts the second adenine within the recognition sequence to N^6-methyladenine (Figure

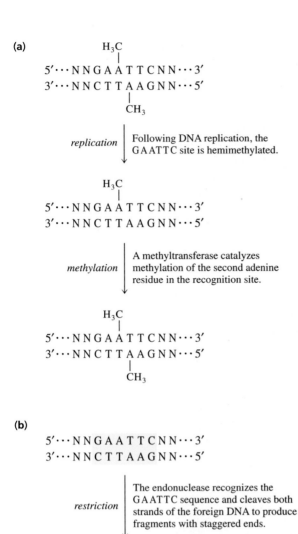

Figure 24·40
Methylation and restriction at the *Eco*RI site.
(a) Methylation of adenine residues.
(b) Cleavage produces staggered, or sticky, ends.

24·40). Although the methylase and the endonuclease recognize the same DNA sequence, the two proteins are not homologous.

As Table 24·5 and Figure 24·40 show, some restriction endonucleases (including *Eco*RI, *Bam*HI, and *Hin*dIII) catalyze staggered cleavage, producing fragments with single-stranded extensions at their ends. These single-stranded regions are called sticky ends because they are complementary and can thus re-form a double-stranded structure. Other enzymes, such as *Hae*III, *Hin*dII, and *Sma*I, catalyze reactions that produce blunt ends, molecules with no single-stranded extensions.

In order to recognize a specific sequence and cleave at a specific site, restriction endonucleases must bind tightly to DNA. In general, the enzymes interact nonspecifically with DNA and then move down the length of the molecule until the specific binding site is encountered. The structure of a complex between *Eco*RI and

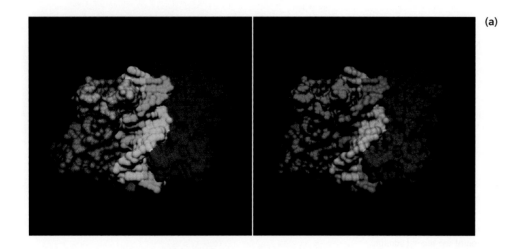

(a)

Figure 24·41
Stereo view of restriction endonuclease *Eco*RI bound to DNA. *Eco*RI is composed of two identical monomers shown here in gold and red. The enzyme is bound to a fragment of DNA with the sequence CGCGAATTCGCG (recognition sequence underlined). The two strands of DNA are colored differently (blue and green) for clarity. (a) Side view. (b) Top view. (Courtesy of John M. Rosenberg.)

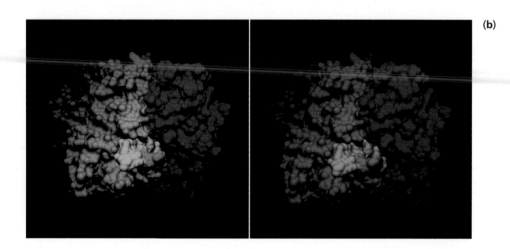

(b)

DNA of the same sequence as the restriction site has been determined by X-ray crystallography. As shown in Figure 24·41, each half of the *Eco*RI homodimer binds one side of the DNA molecule so that the DNA, which retains its helical structure, is almost surrounded. The enzyme recognizes the specific nucleotide sequence by contacting base pairs in the major groove. The minor groove (in the middle of the structure shown in Figure 24·41) is exposed to the aqueous environment. Note that identical regions of each protein monomer bind to the major groove where the recognition sequence GAATTC is located. Since this sequence is palindromic, each monomer interacts with identical base pairs. Contrast this structure with that shown in Figure 24·39. In the case of DNase I, the active enzyme is monomeric, and the enzyme binds nonspecifically to any DNA sequence.

Several basic amino acid residues line the cleft formed by the two *Eco*RI monomers. The side chains of these amino acids interact electrostatically with the sugar-phosphate backbone of DNA. In addition, two arginine residues (Arg-145 and Arg-200) and one glutamate residue (Glu-144) of each *Eco*RI monomer form hydrogen bonds with base pairs in the recognition sequence, thus ensuring specific binding. Other nonspecific interactions with the backbone further stabilize the complex.

The interaction between DNA and the *Eco*RI enzyme is similar to the interactions between DNA and many other DNA-binding proteins that recognize specific sequences. For instance, the structure of the DNA double helix in most protein-DNA complexes is close to the standard B form. Part of the protein penetrates the major groove, where amino acid side chains interact with functional groups on the edges of the base pairs. The recognition of specific base pairs often involves charged amino acid residues in the protein. Finally, many specific DNA-binding proteins are homodimers that recognize specific palindromic sequences.

B. Restriction Enzymes Can Be Used to Manipulate DNA in the Laboratory

Soon after restriction enzymes were first discovered, they were exploited in the laboratory. One of the first uses was in developing restriction maps of DNA, that is, maps that show the specific sites of cleavage. Such maps are useful in identifying fragments of DNA that contain specific genes.

A restriction map of bacteriophage λ DNA is shown in Figure 24·42. The DNA of bacteriophage λ is a linear, double-stranded DNA molecule that consists of approximately 48 400 bp, or 48.4 kb. A restriction digest is made by treating this DNA with various restriction enzymes. By measuring the sizes of the resulting fragments, it is possible to develop a map of the cleavage sites. The DNA fragments in a restriction digest are separated by electrophoresis, as shown in Figure 24·43. In an actual restriction-mapping experiment, the information from many such digests is combined to produce a complete, accurate map.

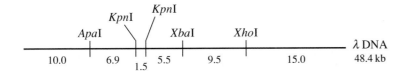

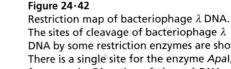

Figure 24·42
Restriction map of bacteriophage λ DNA. The sites of cleavage of bacteriophage λ DNA by some restriction enzymes are shown. There is a single site for the enzyme *Apa*I, for example. Digestion of phage λ DNA with this enzyme produces two fragments: one of 10.0 kb and the other of 38.4 kb, as shown in lane 1 of Figure 24·43.

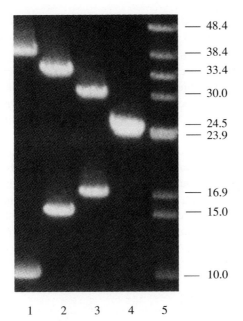

1 2 3 4 5

Figure 24·43
Digestion of bacteriophage λ DNA by various restriction endonucleases. A solution of DNA is treated with the enzyme and then electrophoresed on an agarose gel, which separates fragments according to their size. The smallest fragments move fastest and are found at the bottom of the gel. Lane 1: *Apa*I digestion. Lane 2: *Xho*I digestion. Lane 3: *Kpn*I digestion; the smallest fragment (1.5 kb) is not visible. Lane 4: *Xba*I digestion; the two fragments are approximately the same size and are not well resolved. Lane 5: Intact DNA and a mixture of fragments from the digestions in lanes 1–4. (Copyright 1992 New England Biolabs Catalog. Reprinted with permission.)

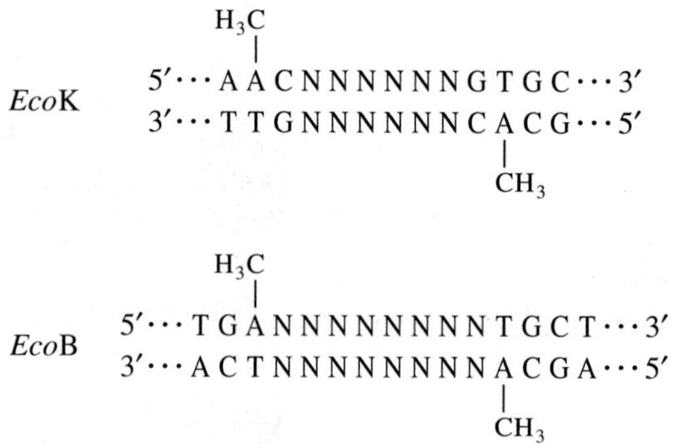

Figure 24·44

EcoB and *EcoK* recognition sites. The recognition sequences are methylated at adenine residues. (*N* can be any nucleotide.)

$$
\begin{array}{c}
\text{H}_3\text{C} \\
|
\end{array}
$$

EcoK
5′···A A C N N N N N N G T G C···3′
3′···T T G N N N N N N C A C G···5′
$$|$$
$$\text{CH}_3$$

$$
\begin{array}{c}
\text{H}_3\text{C} \\
|
\end{array}
$$

EcoB
5′···T G A N N N N N N N N T G C T···3′
3′···A C T N N N N N N N N A C G A···5′
$$|$$
$$\text{CH}_3$$

C. Bacteria Contain Strain-Specific Nucleases

EcoB and *EcoK* are examples of type I restriction enzymes. In the presence of ATP, Mg^{2+}, and *S*-adenosylmethionine, these enzymes catalyze the methylation and cleavage of DNA. It is not known why *S*-adenosylmethionine is required for the cleavage; it does not lose its methyl group in this process, so it may be an allosteric activator. Although the enzymes bind directly to DNA at the sites shown in Figure 24·44, restriction occurs randomly at distant sites. The enzymes cause the formation of loops of DNA—one end of the loop at the recognition site of the enzyme and the other at the active site.

Both enzymes are composed of three subunits encoded by three genes in the *E. coli* genome. The *EcoB* form is found in B strains of *E. coli,* and the *EcoK* form is present in K strains. The presence of different restriction enzymes in these strains inhibits the exchange of DNA.

24·11 DNA Is Packaged into Chromatin in the Nuclei of Eukaryotic Cells

In 1879, 10 years after Miescher's discovery of nuclein, Walter Flemming observed the presence of banded objects in the nuclei of stained eukaryotic cells viewed through a microscope. He called the material **chromatin** after the Greek word *chroma,* which means color. Modern usage of the term chromatin refers specifically to the complex of DNA and protein found in the nuclei of eukaryotic cells. In humans, the nucleus must accommodate 46 such chromatin fibers, or **chromosomes.** Since the largest human chromosome contains 2.4×10^8 base pairs of DNA, it would have a length of about 8.2 cm if the DNA were stretched out in the B conformation. In fact, the length of the largest chromosome during metaphase (when it is in its most condensed state) is about $10\,\mu m$, or about 1/8000 the length it would have if it were a linear rod of B-DNA. One of the primary roles of the proteins in chromatin is to package DNA in a manageable form.

Table 24·6 Basic and acidic residues in mammalian histones

Type	Molecular weight	Number of residues	Number of basic residues	Number of acidic residues
Rabbit thymus H1	21 000	213	65	10
Calf thymus H2A	14 000	129	30	9
Calf thymus H2B	13 800	125	31	10
Calf thymus H3	15 300	135	33	11
Calf thymus H4	11 300	102	27	7

A. Histones and DNA Combine to Form Nucleosomes

The major protein components of chromatin are known as **histones.** Most eukaryotic species contain five different histones, known as H1, H2A, H2B, H3, and H4. All five histones are basic proteins containing numerous lysine and arginine residues whose positive charges promote binding of the protein to the negatively charged sugar-phosphate backbone of DNA (Table 24·6). The primary structures of histones H2A, H2B, H3, and H4 are highly conserved in all species of eukaryotes. For example, bovine histone H4 differs from pea histone H4 in only two residues out of 102. Such similarity in primary structure implies a corresponding conservation of tertiary structure and function.

When chromatin is treated with solutions of low ionic strength (<5 mM), it unfolds, giving rise to a structure that has an appearance in electron micrographs like beads on a string (Figure 24·45). The "beads" are DNA-histone complexes called **nucleosomes,** and the "string" is double-stranded DNA.

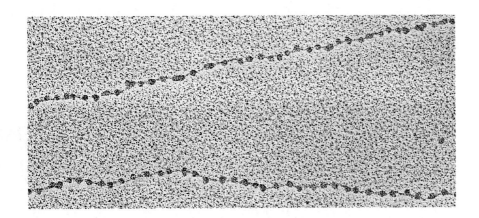

Figure 24·45
Electron micrograph of extended chromatin showing the beads-on-a-string organization. (Courtesy of Victoria E. Foe.)

Figure 24·46
Nucleosome structure. (**a**) A histone octamer. (**b**) Nucleosomes. The nucleosome core particle consists of a histone octamer and about 146 base pairs of DNA; linker DNA consists of about 54 base pairs and links core particles. Histone H1 binds to the nucleosome core particle and linker DNA.

(a)

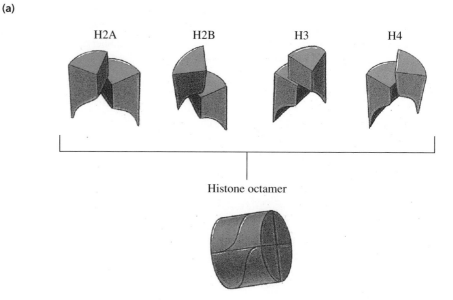

(b)

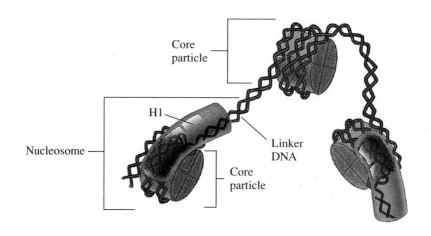

Each nucleosome is composed of two molecules each of histones H2A, H2B, H3, and H4 and about 200 base pairs of DNA (Figure 24·46). The histone molecules form an organized protein complex, and the DNA is wrapped around the outside for about 1.75 turns. Approximately 146 base pairs of DNA are in close contact with the histone octamer, forming a **nucleosome core particle.** The wrapping of DNA around the core nucleosome induces positive supercoils in the remaining DNA. These supercoils are relaxed by eukaryotic topoisomerase II, an enzyme that is bound to chromatin. When nucleosomes are removed, the naked eukaryotic DNA contains about one negative supercoil for each original nucleosome.

The DNA between each core particle (the string in Figure 24·45) is called **linker DNA.** It is about 54 base pairs long. The fifth histone, H1, binds to the linker DNA and to the nucleosome core particle. Histone H1 is involved in higher-order chromatin structures and is depleted in the beads-on-a-string structure.

The amino acid sequences of histone H1 proteins from different organisms are not as conserved as those of the other histones. Many species contain a number of histone H1 variants that may be synthesized in different cell types or at different stages of development. Some organisms, such as vertebrates, contain an additional histone, H5, that is functionally similar to H1. Histone H1 is not present at all in the budding yeast *Saccharomyces cerevisiae*.

B. Nucleosomes Are Packaged to Form Higher Levels of Chromatin Structure

The packaging of DNA into nucleosomes results in about a 10-fold reduction in the length of chromatin compared to B-DNA. Higher levels of DNA packaging lead to further condensation. For example, the beads-on-a-string structure is itself coiled into a **solenoid** form to yield a structure called the 30-nm fiber. One possible solenoid structure is shown in Figure 24·47. The 30-nm fiber forms when every nucleosome contains a molecule of histone H1 and adjacent molecules of H1 bind to each

Figure 24·47
The 30-nm chromatin fiber.

30 nm

H1 histones

other cooperatively, bringing the nucleosomes together into a more compact and stable form of chromatin. The packaging of the beads-on-a-string structure into the 30-nm fiber results in a further fourfold condensation of chromosome length.

The transition between various forms of chromatin is associated with covalent modification of the core histones. One of the most important changes is the acetylation of histones, especially H3 and H4, which occurs when chromatin decondenses to form the extended beads-on-a-string structure. Histone H1 binds less tightly to core nucleosomes containing acetylated histones, causing the 30-nm fiber to unravel.

Figure 24·48
Electron micrograph of a histone-depleted chromosome. (**a**) In this view, the entire protein scaffold is visible. (**b**) In this magnification of a portion of (a), individual loops attached to the protein scaffold can be seen. (Courtesy of Ulrich K. Laemmli.)

(a)

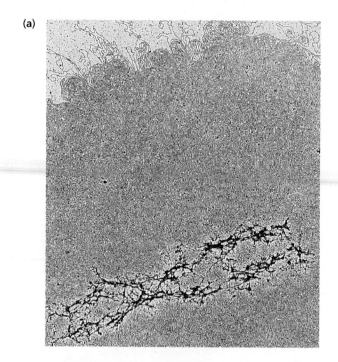

(b)

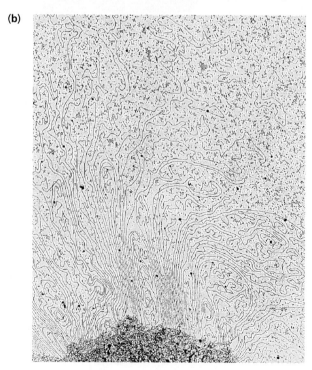

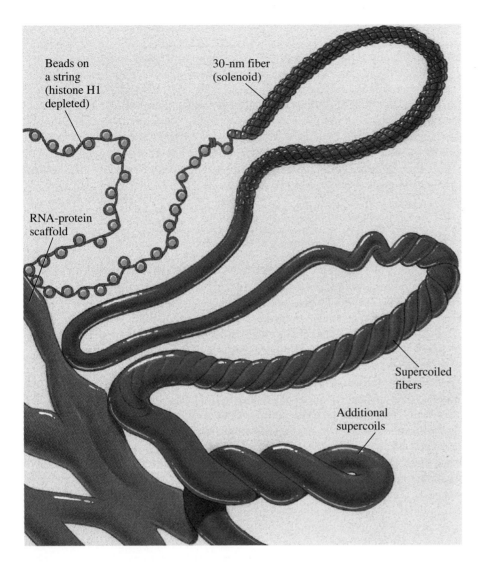

Figure 24·49 labels: Beads on a string (histone H1 depleted); 30-nm fiber (solenoid); RNA-protein scaffold; Supercoiled fibers; Additional supercoils

Figure 24·49
Summary of chromatin structure. [Adapted from Rindt, K.-P., and Nover, L. (1980). Chromatin structure and function. *Biol. Zentralbl.* 99:641–673.]

Active genes (genes that are transcribed) are found in decondensed chromatin, whereas genes in the more compact, higher-order structures are repressed. The switch from compact chromatin to decondensed chromatin is mediated, in part, by proteins known as HMG proteins (high-mobility–group proteins), which preferentially associate with chromatin in the extended beads-on-a-string conformation. HMG proteins also prevent binding of histone H1.

The 30-nm fibers form large loops that are attached to an RNA-protein scaffold. Because the loops are fixed at the base, supercoils can accumulate as they do in circular DNA molecules. There may be as many as 2000 such loops on a large mammalian chromosome. The organization of DNA into loops accounts for an additional 200-fold condensation in the length of DNA. The RNA-protein scaffolding of a chromosome can be seen under an electron microscope when histones have been removed from the DNA (Figure 24·48). Large loops of DNA are clearly visible in such electron micrographs. (The loops are not supercoiled because the DNA was extensively cut by single-strand–specific endonucleases during isolation.) Chromatin is further condensed when 30-nm fibers form supercoiled fibers, which can in turn form additional supercoils (Figure 24·49). The most condensed chromatin structure is found in metaphase chromosomes, which are formed only during cell division (Figure 24·50).

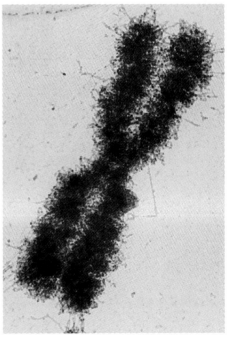

Figure 24·50
Electron micrograph of a metaphase chromosome. (Courtesy of Gunther Bahr.)

C. Prokaryotic DNA Is Also Packaged

Although histones are found only in eukaryotes, the DNA in prokaryotes is also associated with proteins that allow it to be packaged in a condensed form. HU protein is an abundant component of bacterial cells. It binds to DNA and forms a nucleosome-like structure. Other histonelike proteins, such as H-NS, are also abundant and appear to be involved in DNA packaging. Although bacterial chromosomes are circular, the DNA is organized as large loops attached to a protein scaffold as in eukaryotes. These loops of DNA form supercoils independently. In *E. coli,* there are about 40 large loops, each containing approximately 100 kb. This arrangement gives rise to a relatively compact structure known as the nucleoid.

Summary

Nucleic acids are polymers of nucleoside monophosphates, or nucleotides. A purine or pyrimidine ring is attached to ribose or 2′-deoxyribose by a β-N-glycosidic bond. The *anti* conformation of nucleosides predominates in nucleic acids. The furanose rings can pucker, with the C-2′ *endo* and C-3′ *endo* conformations appearing most often. The bases in DNA are often methylated. Hydrogen bonding by the bases of nucleotides is important for the three-dimensional structure and functions of nucleic acids.

DNA is a double-stranded nucleotide polymer whose monomeric units are linked as 3′–5′ phosphodiesters. As originally proposed by Watson and Crick, the two strands are antiparallel. Base pairing between A and G and between C and T allows each strand to serve as a template for the other. Stacking interactions between base pairs are largely responsible for the helical conformation of double-stranded DNA. B-DNA, the most common form, is a right-handed helix stabilized by hydrophobic interactions, stacking interactions, hydrogen bonding, and electrostatic repulsion. The sugar-phosphate backbone winds around the outside of the helix, and the helix surface contains two grooves. DNA can be denatured by heating above its melting temperature, T_m. Separated single strands can renature; nucleation is the rate-limiting step.

Double-stranded DNA can adopt several conformations. A-DNA is right-handed, but compared to B-DNA, the base pairs are tilted and the center of the helix is accessible to solvent. Z-DNA is a left-handed helix with no obvious grooves.

DNA molecules are flexible and assume a variety of shapes. Cruciform structures can form when inverted repeats are present. Naturally occurring DNA is usually underwound, or negatively supercoiled. Topoisomerases catalyze the removal of supercoils and, in some cases, can introduce supercoils.

There are four major classes of RNA: ribosomal RNA, transfer RNA, messenger RNA, and small RNA. RNA is usually single stranded and, like DNA, has considerable secondary structure, including hairpins, or stem-loops. Double-stranded RNA adopts a helical conformation similar to that of A-DNA.

The phosphodiester linkages of nucleic acids are hydrolyzed by nucleases that are classified as DNases or RNases and as exonucleases or endonucleases. Hydrolysis of RNA in alkaline solutions or by the action of RNase A proceeds via a 2′,3′-cyclic phosphate intermediate. Restriction endonucleases cleave DNA at specific sequences. Host DNA is protected from hydrolysis by the presence of bases that have been methylated by the restriction enzyme or by another enzyme. When *Eco*RI binds to DNA, specific residues of the protein contact the nucleotide residues at the recognition site.

In cells, large DNA molecules are condensed and packaged. The histone proteins of eukaryotes bind to DNA to form nucleosomes. Nucleosomes are strung together, and the resulting structure undergoes additional degrees of condensation, forming supercoils that are attached to an RNA-protein scaffold in the nucleus.

Selected Readings

Polynucleotide Structure and Properties

Adams, R. L. P., Knowler, J. T., and Leader, D. P. (1992). *The Biochemistry of the Nucleic Acids,* 11th ed. (New York: Chapman and Hall).

Blackburn, G. M., and Gait, M. J., eds. (1990). *Nucleic Acids in Chemistry and Biology* (Oxford: IRL Press at Oxford University Press).

Crothers, D. M., Haran, T. E., and Nadeau, J. G. (1990). Intrinsically bent DNA. *J. Biol. Chem.* 265:7093–7096.

Dickerson, R. E., Drew, H. R., Conner, B. N., Wing, R. M., Fratini, A. V., and Kopka, M. L. (1982). The anatomy of A-, B-, and Z-DNA. *Science* 216:475–485.

Saenger, W. (1984). *Principles of Nucleic Acid Structure* (New York: Springer-Verlag).

DNA Topology and Topoisomerases

Champoux, J. J. (1990). Mechanistic aspects of type-I topoisomerases. In *DNA Topology and Its Biological Effects,* N. R. Cozzarelli and J. C. Wang, eds. (Cold Spring Harbor, New York: Cold Spring Harbor Laboratory Press), pp. 217–242.

Cozzarelli, N. R., Boles, T. C., and White, J. H. (1990). Primer on the topology and geometry of DNA supercoiling. In *DNA Topology and Its Biological Effects,* N. R. Cozzarelli and J. C. Wang, eds. (Cold Spring Harbor, New York: Cold Spring Harbor Laboratory Press), pp. 139–184.

Hsieh, T.-S. (1990). Mechanistic aspects of type-II DNA topoisomerases. In *DNA Topology and Its Biological Effects,* N. R. Cozzarelli and J. C. Wang, eds. (Cold Spring Harbor, New York: Cold Spring Harbor Laboratory Press), pp. 243–263.

McLean, M. J., and Wells, R. D. (1988). The role of DNA sequences in the formation of Z-DNA versus cruciforms in plasmids. *J. Biol. Chem.* 263:7370–7377.

Pulleyblank, D. E., and Ellison, M. J. (1982). Purification and properties of type I topoisomerase from chicken erythrocytes: mechanism of eukaryotic topoisomerase action. *Biochemistry* 21:1155–1161.

Nucleases

Heitman, J. (1992). How the *Eco* RI endonuclease recognizes and cleaves DNA. *BioEssays* 14:445–454.

Kim, Y., Grable, J. C., Love, R., Greene, P. J., and Rosenberg, J. M. (1990). Refinement of *Eco* RI endonuclease crystal structure: a revised protein chain tracing. *Science* 249:1307–1309.

McClarin, J. A., Frederick, C. A., Wang, B.-C., Greene, P., Boyer, H., Grable, J., and Rosenberg, J. M. (1986). Structure of the DNA-*Eco* RI endonuclease recognition complex at 3 Å resolution. *Science* 234:1526–1541.

Suck, D., Lahm, A., and Oefner, C. (1988). Structure refined to 2 Å of a nicked DNA octanucleotide complex with DNase I. *Nature* 332:464–468.

Chromatin

Beebe, T. P., Jr., Wilson, T. E., Ogletree, D. F., Katz, J. E., Balhorn, R., Salmeron, M. B., and Siekhaus, W. J. (1989). Direct observation of native DNA structures with the scanning tunneling microscope. *Science* 243:370–372.

Burlingame, R. W., Love, W. E., Wang, B.-C., Hamlin, R., Xuong, N.-H., and Moudrianakis, E. N. (1985). Crystallographic structure of the octameric histone core of the nucleosome at a resolution of 3.3 Å. *Science* 228:546–553.

Richmond, T. J., Finch, J. T., Rushton, B., Rhodes, D., and Klug, A. (1984). Structure of the nucleosome core particle at 7 Å resolution. *Nature* 311:532–537.

Schmid, M. B. (1988). Structure and function of the bacterial chromosome. *Trends Biochem. Sci.* 13:131–135.

Schmid, M. B. (1990). More than just "histone-like" proteins. *Cell* 63:451–453.

25

DNA Replication

The transfer of genetic information from one generation to the next has puzzled naturalists since at least the time of Aristotle. We now know that genetic information is encoded in the nucleotide sequence of DNA and that transferring information from a parental cell to two daughter cells requires the faithful duplication of this information. The structure of DNA immediately suggests a method of replication, a point that did not escape the notice of James D. Watson and Francis H. C. Crick when they proposed the structure of DNA in 1953. Since the two strands of double-helical DNA are complementary, the nucleotide sequence of one strand automatically specifies the sequence of the other. During DNA replication, each strand of DNA acts as a template for the synthesis of a complementary strand. In this way, DNA replication produces two daughter DNA duplexes, each of which contains one parental strand and one newly synthesized strand. This mode of replication is termed **semiconservative replication** (Figure 25·1).

Although elucidation of the structure of DNA in 1953 suggested a mode of DNA replication, the mechanism of semiconservative replication continues to be studied today. The overall process of DNA replication can be divided into three stages: initiation, elongation, and termination. Researchers have isolated and characterized enzymes that catalyze these various steps and have, in many cases, identified the genes that encode the enzymes. Analyses of mutations in the genes involved in DNA replication have played a crucial role in deciphering this complex mechanism.

The results of many studies have produced an illuminating picture of how large numbers of polypeptides assemble into complexes that are capable of carrying out many functions simultaneously. The DNA replication complex is like a machine whose parts are made of protein. Some of the component polypeptides are active or partially active in isolation, but others are active only in association with the complete protein machine. The correct assembly of this replication complex at the site where DNA replication begins is an essential part of the initiation of DNA replication. The complex replicates DNA semiconservatively during the subsequent elongation stage, catalyzing the incorporation of nucleotides into the growing strands. Finally, during termination, when replication of the chromosome is complete, the protein machine is disassembled, and the daughter molecules segregate.

Related material appears in Exercise 8 of the computer-disk supplement *Exploring Molecular Structure.*

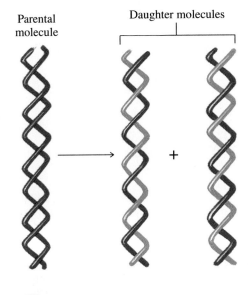

■ Parental strand
■ Newly synthesized strand

Figure 25·1
Semiconservative replication. Each strand of duplex DNA acts as a template for synthesis of a daughter strand. Each daughter molecule of DNA contains one parental strand and one newly synthesized strand.

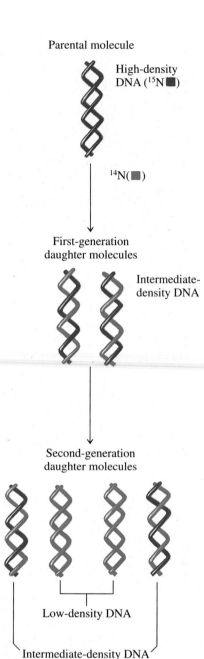

Parental molecule

High-density DNA (^{15}N ■)

^{14}N(■)

First-generation daughter molecules

Intermediate-density DNA

Second-generation daughter molecules

Low-density DNA

Intermediate-density DNA

Figure 25·2
Predicting the results of the Meselson-Stahl experiment. Molecules of parental DNA were synthesized in a medium containing only the heavy isotope of nitrogen, ^{15}N; daughter strands were synthesized in a medium containing only the lighter isotope of nitrogen, ^{14}N. If replication is semiconservative, all fragments of the chromosomal DNA should be of intermediate density after one round of replication in the medium containing ^{14}N. Similarly, after two rounds of replication in the medium containing ^{14}N, half the DNA molecules should be of intermediate density and half should be of low density.

Protein machines that carry out a series of related biochemical reactions are not unusual; such complexes are also involved in transcription, RNA processing, and translation. Furthermore, there is increasing evidence that complexes of weakly associated enzymes and other macromolecules are responsible for other processes of cellular metabolism.

The transmission of genetic information from generation to generation requires that DNA replication be both rapid (because the entire complement of DNA must be replicated before cells can divide) and accurate (because poor fidelity would prevent transmission of the correct genetic information to daughter cells). Several billion years of evolution by natural selection have resulted in modern organisms that balance these somewhat competing requirements.

The overall process of chromosomal DNA replication in prokaryotes and eukaryotes appears to be similar, although the sequence and structure of specific enzymes vary greatly among organisms. We should not be surprised by this: just as Ferraris and Fiats function in a similar way, individual parts from one cannot substitute for parts in the other. The same is true for chromosomal DNA replication: it is similar in all organisms even though the individual enzymes may differ. We will focus on chromosomal DNA replication mechanisms in the bacterium *Escherichia coli* because many of its replication enzymes are known and well characterized. We will also outline the special features of eukaryotic replication. Some replication mechanisms that are unique to eukaryotic viruses and bacteriophages will also be discussed; these mechanisms are variations of the typical cellular processes.

25·1 DNA Replication Is Semiconservative

We indicated above that the very structure of DNA suggests a mechanism for replication in which each strand serves as a template for synthesis of a complementary daughter strand. However, at least two other mechanisms are possible. A **conservative** replication mechanism would preserve the parental molecule and synthesize a completely new, double-stranded daughter molecule. A **dispersive** replication mechanism would require that the parental DNA be broken down and two new DNA molecules made from the released nucleotides.

If DNA were replicated by a semiconservative mechanism, each daughter duplex DNA molecule should contain one strand of the original parental DNA and one strand of newly synthesized DNA. This hypothesis was tested in an elegant experiment designed by Matthew Meselson and Franklin Stahl, using an analytical technique developed by Jerome Vinograd. In their experiment, Meselson and Stahl distinguished between parental DNA and newly synthesized DNA by labelling the different DNA molecules with isotopes of different densities. To do this, they grew *E. coli* cells in a medium in which the sole nitrogen source was a heavy (but nonradioactive) isotope of nitrogen, ^{15}N, in the form of ^{15}NH$_4^{\oplus}$. Under these conditions, all the nucleotides synthesized by the bacteria contained ^{15}N and thus had a high density. These nucleotides were incorporated into parental DNA. Cells that had been grown in ^{15}NH$_4^{\oplus}$ were then shifted to a medium that contained only the naturally abundant, less dense isomer of nitrogen, ^{14}N, supplied as ^{14}NH$_4^{\oplus}$. In this medium, newly synthesized nucleotides contained only the lighter isotope and thus exhibited a low density. Meselson and Stahl reasoned that if DNA replication were semiconservative, DNA molecules isolated from cells after one round of replication should contain a 50/50 mixture of ^{14}N and ^{15}N and thus exhibit an intermediate density. A second round of semiconservative replication in the ^{14}N-containing medium should lead to progeny in which one-half of the isolated DNA contains no ^{15}N and exhibits a low density and the other half contains a 50/50 mixture of ^{14}N and ^{15}N and exhibits an intermediate density. This prediction is summarized in Figure 25·2.

Figure 25·3
Experimental results obtained by Meselson and Stahl. **(a)** Ultraviolet-absorption photographs showing the DNA bands after 0, 1, and 1.9 generations in a medium containing ^{14}N. **(b)** Microdensitometer tracings of the density profiles in (a). Lower density is on the left, higher density on the right. (Courtesy of Matthew Meselson.)

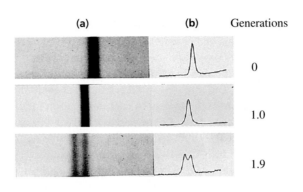

(a)   (b)   Generations

0

1.0

1.9

DNA molecules of different densities can be separated from one another in an analytical ultracentrifuge on the basis of their buoyancy in cesium salt. This technique, called equilibrium density gradient centrifugation, subjects a solution of cesium salt to high-speed ultracentrifugation. The salt solution forms a density gradient with the most dense part of the solution at the outermost part of the centrifuge tube and the least dense part of the solution at the innermost part of the tube. When a molecule is present in such a salt solution during centrifugation, the molecule moves through the gradient until it comes to rest at a position where the density of the molecule is equal to that of the salt solution. Figure 25·3 shows the bands of DNA molecules that result from equilibrium density gradient centrifugation of bacterial lysates sampled at various times after a growing, ^{15}N-containing culture is transferred to a medium containing ^{14}NH$_4^{\oplus}$. Before transfer to the medium containing the light isotope, all the DNA forms bands at the high density. After one generation (one round of replication), all the DNA forms bands at an intermediate density, halfway between the densities of ^{14}N- and ^{15}N-containing DNA molecules. After two generations, half the DNA forms bands at the intermediate density and half at the low density. This banding pattern demonstrated that the mechanism of DNA replication in *E. coli* is indeed semiconservative.

25·2 DNA Replication Is Bidirectional

Semiconservative replication is an elegant means of transferring genetic information, but a variety of mechanistic hurdles must be overcome in order to complete this reaction inside a cell. First, the two DNA strands of the double helix must be unwound and separated, a process that is energetically unfavorable. This in turn creates the second hurdle, which is that the unwinding of the helix leads to a change in the linking number of the helix and the introduction of torsional strain that must be relieved (see Section 24·6). Third, synthesis of a new, complementary DNA strand requires a mechanism for template-directed polymerization. And finally, the mechanism of replication must be regulated so that chromosome duplication occurs only when coupled to cell division.

The *E. coli* chromosome is a large, circular, double-stranded DNA molecule of 4.6×10^3 kilobase pairs (kb) that is tightly coiled into superhelical twists and compressed into a dense nucleoid about $1\,\mu m$ in diameter. When partially replicated DNA is isolated from growing cells and purified, it appears as a large, circular molecule with two junctions where the newly synthesized daughter strands meet the unreplicated parental DNA (Figure 25·4). These junctions are called **replication forks.**

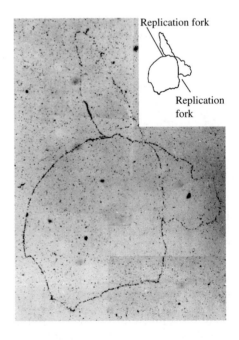

Replication fork

Replication fork

Figure 25·4
Autoradiograph of a replicating *E. coli* chromosome. The DNA has been labelled with [^{3}H]-deoxythymidine and has been detected by overlaying the replicating chromosome with photographic emulsion. The autoradiograph reveals that the *E. coli* chromosome has two replication forks. (Courtesy of John Cairns.)

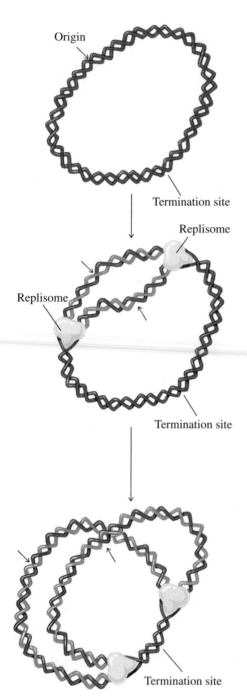

Figure 25·5
Bidirectional DNA replication in *E. coli*. Semi-conservative DNA replication begins at a unique origin and proceeds in both directions. The synthesis of the new strands of DNA (light gray) occurs at the two replication forks, where the replisomes are located. When the replication forks meet at the termination site, the two double-stranded DNA molecules separate.

Replication of the *E. coli* chromosome begins at a unique site called the origin of replication and proceeds bidirectionally during the elongation phase until the two replication complexes meet at a termination site, where replication stops (Figure 25·5). The protein machine that carries out the polymerization reaction is called a **replisome,** and one replisome is located at each of the two replication forks.

The rate of movement of each replication fork in *E. coli* is approximately 1000 base pairs per second. Since each of the two replication forks moves at this rate, the entire 4.6×10^3 kb *E. coli* chromosome can be duplicated in about 38 minutes.

E. coli cells can divide as rapidly as once every half hour or as slowly as once every several days. When a cell divides, its chromosomal DNA is replicated only once, so replication is obviously closely coupled to the cell cycle. DNA replication is regulated largely at the initiation step, so when cells are dividing slowly, DNA replication is initiated less frequently.

25·3 DNA Polymerase III Catalyzes a Polymerization Reaction at the Replication Fork

Over 30 different proteins are needed for DNA replication in *E. coli*. Some of these are specifically required for initiation or termination, but most are needed either directly or indirectly for DNA synthesis at the replication forks. During DNA synthesis, nucleotides are added to the 3′ end of a growing chain in reactions catalyzed by enzymes known as DNA-directed DNA polymerases, or simply **DNA polymerases.**

There are three different DNA polymerases in *E. coli* cells. DNA polymerase I repairs damaged DNA and participates in the synthesis of one of the strands of DNA during replication. DNA polymerase II plays a role in DNA repair. DNA polymerase III is the enzyme responsible for chain elongation at the replication fork.

Structurally, DNA polymerase III is the most complex of the three DNA polymerases. It is a multimeric protein composed of 10 different types of subunits. Table 25·1 lists the subunits of DNA polymerase III, their activities, masses, and the genes that encode them. The minimally active core of the enzyme consists of three subunits: α, ε, and θ. The polymerization activity is localized to the α subunit, and the ε subunit contains a $3′ \rightarrow 5′$ exonuclease activity. The θ subunit is required to

Table 25·1 Major identified subunits of DNA polymerase III holoenzyme

Subunit	M_r		Gene	Activity
α	130 000	core	*pol*C/*dna*E	Polymerase
ε	27 000		*dna*Q/*mut*D	$3′ \rightarrow 5′$ exonuclease
θ	8846		*hol*E	?
β	40 000		*dna*N	Forms sliding clamp
τ	71 000		*dna*X	Enhances dimerization of core; ATPase
γ	47 000	γ complex	*dna*X	Enhance processivity; assist in replisome assembly
δ	38 700		*hol*A	
δ'	36 900		*hol*B	
χ	16 600		*hol*C	
ψ	15 174		*hol*D	

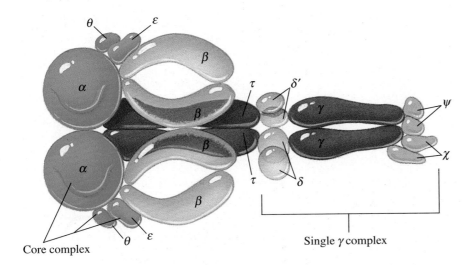

Figure 25·6
Structure of *E. coli* DNA polymerase III holoenzyme. The holoenzyme is a dimer of the core complex (α, ε, θ), a dimer of β and τ subunits, and a single γ complex (two copies each of γ, δ, δ', χ, and ψ). The structure is thus an asymmetric dimer. Other models of the holoenzyme structure have been proposed. (Adapted from O'Donnell, M. (1992). Accessory protein function in the DNA polymerase III holoenzyme from *E. coli*. *BioEssays* 14:105–111.)

form protein-protein contacts with other polypeptides at the replication fork. Two copies of the core complex bind to a dimer of τ subunits to create a larger, dimeric structure. Hydrolysis of ATP is required during assembly of DNA polymerase III; this ATPase activity is localized to the τ dimer. Two β subunits bind to each half of this core-τ complex; as we will see later, these β subunits also bind tightly to DNA. The remaining subunits of DNA polymerase III associate with two copies of the γ subunit to form a γ complex that is composed of two copies each of γ, δ, δ', χ, and ψ. When all the subunits are present, the entire oligomeric protein is called a **holoenzyme.**

One model of the structure of DNA polymerase III is shown in Figure 25·6, and other similar models have been proposed. There are two important features of this proposed structure: first, there are two active sites of polymerization, and second, the holoenzyme is not perfectly symmetrical. As we will see in the next few sections, two daughter DNA strands are synthesized at each replication fork, and the holoenzyme dimer is responsible for synthesizing both of these strands simultaneously.

The DNA replication polymerases in other bacteria are also large enzyme complexes, but the number of subunits and their roles may differ. For example, the *Bacillus subtilis* enzyme does not have a subunit that is homologous to the ε subunit of *E. coli* (the $3' \rightarrow 5'$ exonuclease activity of the *B. subtilis* enzyme is part of the large α subunit). In general, the amino acid sequences of bacterial DNA polymerases are only weakly conserved.

As we mentioned earlier, DNA polymerase III in *E. coli* catalyzes DNA synthesis at a rate of approximately 1000 nucleotides per second—the fastest rate known for any in vivo polymerization reaction. However, *purified* DNA polymerase III catalyzes polymerization at a much slower rate, suggesting that the isolated enzyme lacks some of the components necessary for full activity. We now know that DNA polymerase III inside the cell is associated with a number of other proteins necessary for full activity during replication. The entire complex—DNA polymerase III plus its associated proteins—forms the replisome, and only when a complete replisome is assembled does polymerization in vitro occur at approximately the rate found in vivo.

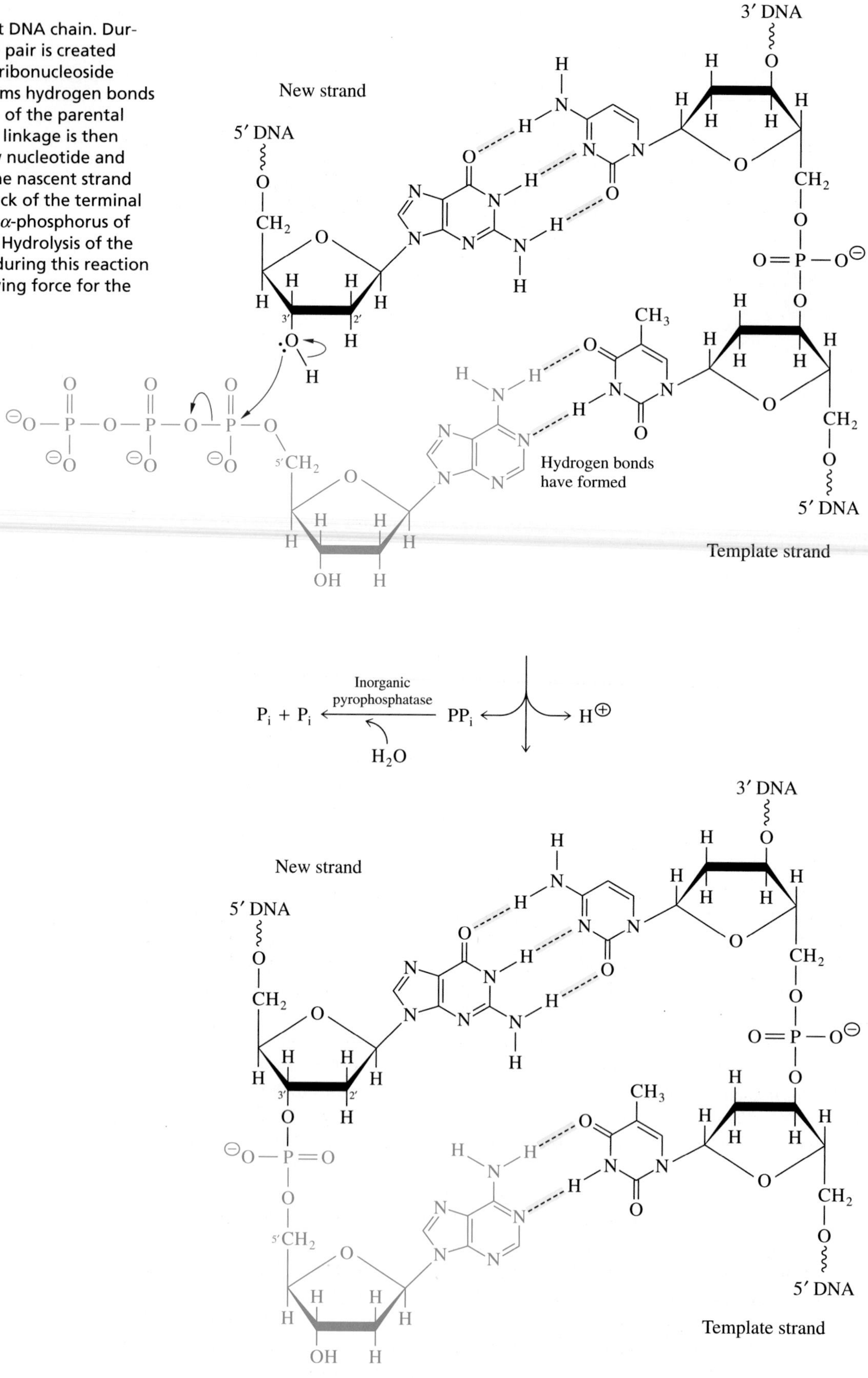

Figure 25·7
Elongation of the nascent DNA chain. During DNA synthesis, a base pair is created when an incoming deoxyribonucleoside 5'-triphosphate (blue) forms hydrogen bonds with a nucleotide residue of the parental strand. A phosphodiester linkage is then formed between the new nucleotide and the terminal residue of the nascent strand through nucleophilic attack of the terminal 3'-hydroxyl group on the α-phosphorus of the incoming nucleotide. Hydrolysis of the pyrophosphate released during this reaction provides much of the driving force for the reaction.

A. Chain Elongation Is a Nucleotidyl-Group–Transfer Reaction

The polymerization of DNA, or chain elongation, is a stepwise reaction in which nucleotides are added one at a time to the 3′ end of a growing chain. This is the reaction catalyzed by DNA polymerase III at the replication fork. The substrate is a deoxyribonucleoside 5′-triphosphate (dNTP), and the correct nucleotide is determined by Watson-Crick base pairing to the template strand.

DNA polymerase III contains a binding site for the DNA template and another for any of the four deoxyribonucleoside triphosphates. Incoming dNTPs enter the active site, where the 3′ end of the growing DNA strand is located. The dNTPs form hydrogen bonds with a residue of the template strand, an interaction that is thought to be stabilized by base stacking. When the correct nucleotide is positioned at the active site, the enzyme-DNA complex undergoes a conformational change, triggering a reaction that leads to formation of a phosphodiester linkage between the incoming dNTP and the nascent chain. The reaction mechanism involves nucleophilic attack on the α-phosphorus of the incoming dNTP by the free 3′-hydroxyl group of the nascent DNA chain (Figure 25·7). This reaction creates a phosphodiester linkage and displaces pyrophosphate. Subsequent hydrolysis of this released pyrophosphate provides much of the driving force for the polymerization reaction.

After each polymerization step, DNA polymerase III advances by one residue, a new nucleotide binds, and another nucleotidyl-group–transfer reaction occurs. The new chain is thus synthesized by the stepwise addition of single nucleotides that are properly aligned by base pairing with the template strand. DNA polymerase III cannot catalyze synthesis of DNA in the absence of a template, nor can it initiate synthesis of DNA in the absence of the 3′ end of a preexisting chain. In other words, DNA polymerase III requires both a template and a primer to synthesize DNA. Because nucleotides are added to the 3′ end of the growing chain, DNA replication is said to proceed in the 5′→3′ direction.

During polymerization, several different dNTP molecules are often bound and released before the correct one is found and positioned in the active site. This binding and testing of dNTPs is very rapid; the release of pyrophosphate and the shifting of the enzyme relative to the template by one base pair are usually the rate-limiting steps during polymerization.

B. DNA Polymerase III Holoenzyme Remains Bound to the Replication Fork During Chain Elongation

Once DNA synthesis has been initiated, the holoenzyme remains bound to the replication fork until replication is complete. The 3′ end of the growing chain is associated with the active site of the enzyme, and about 20 base pairs of double-stranded DNA behind the exposed 3′ end are bound to DNA polymerase III. Enzymes that remain bound to their nascent chains through many polymerization steps are said to be **processive.** (The opposite of processive is **distributive;** a distributive enzyme dissociates from the growing chain after the addition of each monomer.) The high processivity of DNA polymerase III in vivo means that only a small number of DNA polymerase III molecules are needed for DNA replication. Processivity also helps account for the rapid overall rate of chromosome replication.

The processivity of the DNA polymerase III holoenzyme is due in part to the β subunits of the enzyme. These subunits have no enzymatic activity on their own, but when assembled into the holoenzyme, they interact to form a ring that can completely surround a DNA molecule (Figure 25·8, next page). The β subunits thus act as sliding clamps that help make the enzyme extremely processive. The sliding clamp binds only to double-stranded DNA, thus ensuring that the polymerase does not slide forward to the single-stranded region ahead of the growing chain. Other components of the replisome also ensure that DNA polymerase remains associated with the nascent chains during replication.

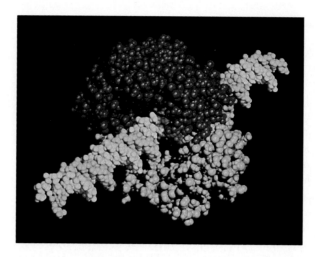

Figure 25·8
Structure of the complex between the β subunits of DNA polymerase III and DNA. Two β subunits associate to form a head-to-tail dimer in the shape of a ring that completely surrounds the DNA. The remaining subunits of DNA polymerase III holoenzyme are bound to this structure. (Courtesy of X.-P. Kong, M. O'Donnell, and J. Kuriyan.)

C. Errors Made by DNA Polymerase III Can Be Corrected by Proofreading

DNA polymerase III holoenzyme is a $3' \rightarrow 5'$ exonuclease as well as a $5' \rightarrow 3'$ polymerase. The exonuclease activity, which lies primarily within the ε subunit, can catalyze hydrolysis of the phosphodiester linkage that binds the $3'$-terminal residue to the rest of the growing polynucleotide chain. The ability of the DNA polymerase III holoenzyme simultaneously to catalyze chain elongation and act as a phosphodiesterase allows the holoenzyme to proofread, or edit, newly synthesized DNA in order to correct mismatched base pairs. When DNA polymerase III recognizes a distortion produced by an incorrectly paired base, the exonuclease activity of the enzyme catalyzes the removal of the mispaired nucleotide. The holoenzyme retreats one residue, and polymerization then continues. This proofreading activity is quite sloppy; some DNA polymerases remove as many as 10% of correct bases as well. Perhaps this expense of energy is part of the price of accurate DNA replication.

DNA polymerase III incorporates the wrong base about once every 10^5 polymerization steps, which corresponds to an error rate of 10^{-5}. The $3' \rightarrow 5'$ proofreading exonuclease activity has an error rate of about 10^{-2}. The combination of these two sequential reactions has an overall error rate for polymerization of 10^{-7}, which is one of the lowest error rates for any enzyme. Some of the errors introduced during DNA replication are subsequently corrected by DNA repair mechanisms (Chapter 26), so the overall error rate is usually less than 10^{-9}. Nevertheless, replication errors are not uncommon when large genomes are duplicated. Mammalian genomes, for example, consist of more than 3×10^9 bp. This translates into about six replication errors per cell division in diploid cells.

Proofreading appears to be essential for accurate DNA synthesis, and with few exceptions, all DNA polymerases possess a $3' \rightarrow 5'$ exonuclease activity. It is possible, however, to isolate strains of *E. coli* (called "mutator" strains) in which replication is less accurate, either because the polymerase activity or the exonuclease activity is impaired. Interestingly, mutant forms of DNA polymerase with enhanced accuracy have also been isolated. These mutant enzymes tend to catalyze polymerization less rapidly than the wild-type forms, suggesting that there is a trade-off between accuracy and speed of replication. The major contribution to the fidelity of the polymerization step appears to be selection of the correct dNTP to be incorporated into the growing chain. It is reasonable to postulate that the genes for modern DNA polymerases are the result of selection for an acceptable balance between accuracy and speed of polymerization.

DNA polymerases are able to catalyze chain elongation only in the $5'\rightarrow3'$ direction. This is due, in part, to the requirement for proofreading. In order to understand this, imagine that DNA synthesis proceeded in the $3'\rightarrow5'$ direction using nucleoside 5'-triphosphates as substrates. The growing end of the DNA chain would contain a 5'-triphosphate group contributed by addition of the previous nucleotide. Each new residue would be added by nucleophilic attack of the 3'-hydroxyl group of the incoming nucleotide on the α-phosphorus of the growing chain. This hypothetical mechanism is an example of **head growth,** which means the high-energy bond broken during the addition of the next monomer is found on the head of the growing chain. Fatty acids and proteins are polymerized in this manner, whereas the mechanism of DNA, RNA, and polysaccharide polymerization involves **tail-growth.**

Now imagine that an incorrect nucleotide is incorporated into the growing chain. If DNA were synthesized in a head-growth mechanism, proofreading would remove the most recently added nucleotide and leave an exposed 5'-hydroxyl or 5'-monophosphate group at the growing end of the chain. In either case, further addition of nucleotides would be impossible because there would be no high-energy bond available. Thus, proofreading is only compatible with a tail-growth mechanism. Perhaps the reason DNA is synthesized exclusively in the $5'\rightarrow3'$ direction is that DNA synthesis in the opposite direction would be less accurate.

25·4 DNA Polymerase III Synthesizes Two Strands Simultaneously

Since the two template strands of DNA are antiparallel, and since DNA is always synthesized in the $5'\rightarrow3'$ direction, synthesis of the two new strands occurs in opposite directions. This in turn means that synthesis of one daughter strand occurs in the *same* direction as replication-fork movement, while synthesis of the other daughter strand occurs in the *opposite* direction (Figure 25·9). The newly synthesized strand formed by $5'\rightarrow3'$ polymerization in the same direction as replication fork movement is called the **leading strand;** the newly synthesized strand formed by $5'\rightarrow3'$ polymerization in the opposite direction is called the **lagging strand.** Each half of the DNA polymerase III holoenzyme dimer is responsible for synthesis of one of these strands. We will discuss the topological arrangement required to achieve bidirectional replication in Section 25·5.

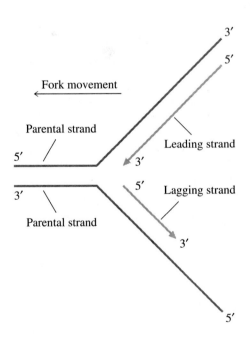

A. DNA Synthesis on the Lagging Strand Is Discontinuous

The leading strand is synthesized continuously from the origin of replication as the replication fork advances. However, because DNA polymerase only replicates DNA in the $5'\rightarrow3'$ direction, the lagging strand is synthesized discontinuously in short pieces that are then joined in a separate process.

Discontinuous DNA synthesis was demonstrated by labelling growing *E. coli* cells with a short pulse of [^{3}H]-deoxythymidine. Synthesis of DNA was then interrupted, and the newly made DNA molecules were isolated, denatured, and separated according to their sizes. Two types of labelled DNA molecules were found: very large DNA molecules that collectively contained about half the radioactivity of the partially replicated DNA, and shorter DNA fragments about 1000 residues in

Figure 25·9
Schematic diagram of the replication fork. The two newly synthesized strands are of opposite polarity. On the leading strand, $5'\rightarrow3'$ synthesis moves in the same direction as the replication fork; on the lagging strand, $5'\rightarrow3'$ synthesis moves in the opposite direction.

length that collectively contained the other half of the radioactivity (Figure 25·10). The large DNA molecules arose from continuous synthesis of the leading strand; the shorter fragments arose from discontinuous synthesis of the lagging strand. The short pieces of DNA were named **Okazaki fragments** in honor of their discoverer, Reiji Okazaki.

B. Each Okazaki Fragment Begins with an RNA Primer

Discontinuous synthesis explains how the lagging strand is synthesized, but it does not explain how synthesis of each Okazaki fragment is initiated. As we mentioned earlier, DNA polymerase III cannot begin polymerization de novo; it can only add nucleotides to existing polynucleotides. In order to synthesize the lagging strand, a series of short RNA primers is made at the replication fork. Each RNA primer is complementary to a portion of the lagging-strand DNA template and is extended from its 3′ end by DNA polymerase III to form an Okazaki fragment (Figure 25·11).

RNA primers are synthesized by an enzyme called primase, the product of the *dna*G gene. Primase is one component of a larger complex called the **primosome,** which also includes helicase, an enzyme involved in unwinding DNA at the replication fork, and other accessory proteins (Table 25·2). The primosome is an integral part of the replisome.

Every second, primase catalyzes synthesis of a short RNA primer approximately 10 nucleotides long. Since the replication fork advances at a rate of approximately 1000 nucleotides per second, one primer is synthesized approximately every thousand nucleotides. DNA polymerase III then catalyzes DNA synthesis in the 5′ → 3′ direction by extending each RNA primer. Continued unwinding of DNA at the replication fork exposes more single-stranded DNA behind the holoenzyme, and additional RNA primers are made as the replisome moves.

Primer synthesis by primase is very error prone, and this probably explains why cells use RNA primers rather than DNA primers during replication. When a cell synthesizes RNA primers, this RNA is incorporated into the nascent strand. The presence of "non-DNA" highlights the primer site, however, and other enzymes in the cell can be directed to remove the RNA and replace it with DNA polymerized by more accurate enzymes. If the cell used DNA primers, this would not be possible, and some of the fidelity of DNA replication would be lost.

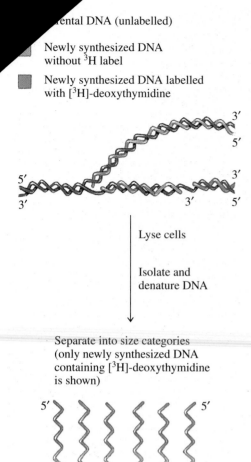

ental DNA (unlabelled)

Newly synthesized DNA without ^{3}H label

Newly synthesized DNA labelled with [^{3}H]-deoxythymidine

Lyse cells

Isolate and denature DNA

Separate into size categories (only newly synthesized DNA containing [^{3}H]-deoxythymidine is shown)

Short fragments from lagging strand

Long fragments from leading strand

Figure 25·10
Discontinuous DNA synthesis demonstrated by analysis of newly synthesized DNA. Nascent DNA molecules were labelled in *E. coli* with a short pulse of [^{3}H]-deoxythymidine. The cells were then lysed, DNA was isolated and denatured, and single strands were separated on the basis of their size. Only the most recently synthesized DNA molecules contained [^{3}H]-deoxythymidine. These labelled molecules fell into two classes: molecules of great length and short fragments of approximately 1000 nucleotides. The long DNA molecules arose from continuous synthesis of the leading strand, and the short fragments from discontinuous synthesis of the lagging strand.

Table 25·2 Components of the primosome

Component	Subunit M_r	Activity
DnaB	50 000	Helicase; assists in initiation
DnaC	29 000	Assists DnaB
Primase (DnaG)	60 000	Synthesizes RNA primers
PriB (n)	12 000	Assist in primosome assembly
PriA (n′)	81 000	
PriC (n″)	21 000	
DnaT (i)	22 000	

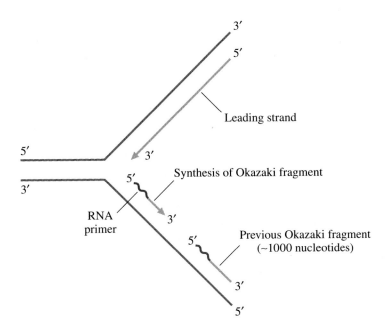

Figure 25·11
Schematic diagram of lagging-strand synthesis. A short piece of RNA (brown) serves as the primer for synthesis of each Okazaki fragment.

C. The RNA Primer Is Removed and Okazaki Fragments Are Extended by DNA Polymerase I

DNA polymerase I of *E. coli* was discovered by Arthur Kornberg in 1957. It is called DNA polymerase I because it was the first enzyme found to be capable of catalyzing DNA synthesis using a template strand. DNA polymerase I consists of a single, large polypeptide chain (M_r 110 000). Like *E. coli* DNA polymerase III, DNA polymerase I has both a polymerization activity and a $3' \rightarrow 5'$ proofreading exonuclease activity, but unlike DNA polymerase III, both activities are found within the same polypeptide chain. DNA polymerase I also has a $5' \rightarrow 3'$ exonuclease activity, an activity not found in DNA polymerase III.

DNA polymerase I can be cleaved with certain proteolytic enzymes to generate a small fragment that contains the $5' \rightarrow 3'$ exonuclease activity and a larger fragment that retains the polymerization and proofreading activities. The larger fragment consists of the C-terminal 605 amino acid residues, and the smaller fragment contains the remaining N-terminal 323 residues. The large fragment, known as the **Klenow fragment,** has been widely used in DNA sequencing and other reactions in the laboratory that require DNA synthesis in the absence of $5' \rightarrow 3'$ degradation. The mechanisms of DNA synthesis and proofreading have been intensively studied using the Klenow fragment as a model for more complicated DNA polymerases. The structure of the Klenow fragment bound to a fragment of DNA with a mismatched terminal base pair shows the nascent strand positioned at the $3' \rightarrow 5'$ exonuclease site of the enzyme (Figure 25·12). At least 10 base pairs of double-stranded DNA are bound by the enzyme. Many of the amino acid residues involved in binding DNA are similar in all DNA polymerases, although the three bacterial DNA polymerases are otherwise quite different in terms of three-dimensional structure and amino acid sequence.

The $3' \rightarrow 5'$ exonuclease and polymerase active sites are separated by 3.5 nm within the enzyme. This suggests that the enzyme must shuffle back and forth as it excises and replaces mismatched nucleotides. In other words, all base pairs move through the proofreading site during processive polymerization, but following excision of any mismatched nucleotide, DNA polymerase I must back up to reposition the 3' end of the primer in the polymerization site (Figure 25·13, next page).

The $3' \rightarrow 5'$ exonuclease active site contains two bound metal ions (which can be either $Mg^{2 \oplus}$, $Zn^{2 \oplus}$, or $Mn^{2 \oplus}$ in the purified enzyme). In one model of the exonuclease reaction mechanism, a hydroxide ion attacks the phosphorus atom,

Figure 25·12
Space-filling model of the Klenow fragment bound to DNA with a terminal mismatch. The enzyme (yellow) is wrapped around the DNA template strand (blue) and the nascent strand (red). The extreme 3' end of the nascent strand (where the mismatch is located) is positioned at the $3' \rightarrow 5'$ exonuclease site. (Courtesy of Thomas A. Steitz and Lorena Beese.)

Figure 25·13
Model for proofreading by the Klenow fragment.

(a) During DNA synthesis, the 3′ end of the growing strand is bound to the polymerization site. The enzyme is clamped to double-stranded DNA by means of a flexible thumb.

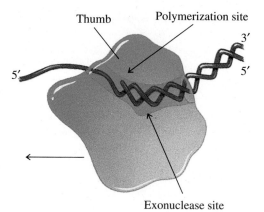

(b) Movement during synthesis brings the most recently added nucleotide into the exonuclease site. If the nucleotide is mismatched, it is excised by the 3′→5′ exonuclease activity of the polymerase. The polymerase must then back up to reposition the 3′ end of the growing strand at the polymerization site.

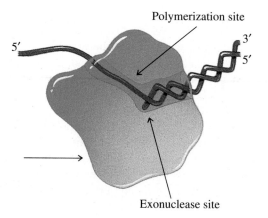

giving rise to a pentacovalent transition state. The hydroxide ion is positioned at the active site by one of the bound metal ions and a nearby tyrosine residue. The second metal ion stabilizes the transition state. The hydrolysis presumably follows an in-line mechanism, leading to inversion of the configuration at the phosphorus atom.

The 5′→3′ exonuclease activity of intact DNA polymerase I is responsible for the removal of the RNA primer at the beginning of each Okazaki fragment. DNA polymerase I also fills in the gap between Okazaki fragments as the RNA primer is degraded. This process of simultaneous synthesis and digestion is called **nick translation** because the net effect is movement of the nick in the direction of synthesis (Figure 25·14). DNA polymerase I recognizes and binds to the nick between the 3′ end of the nascent DNA chain and the 5′ end of the next primer. The 5′→3′ exonuclease activity then catalyzes hydrolytic removal of the first RNA nucleotide while the 5′→3′ polymerase adds a deoxyribonucleotide to the 3′ end of the nascent DNA chain. In this way, the enzyme moves the nick along the lagging strand. After completing 10 or 12 cycles of hydrolysis and polymerization, DNA polymerase I dissociates from the DNA, leaving behind two strands of DNA separated by a nick in the phosphodiester backbone. Note that DNA polymerase I is only slightly processive; it does not bind to accessory proteins that can form a sliding clamp to improve processivity and is not able to catalyze more than about 10 or 12 cycles of hydrolysis and polymerization before it dissociates from the template.

However, this is usually sufficient to remove the short RNA primer and replace it with DNA.

The final product of DNA replication must consist entirely of double-stranded DNA, and the removal of RNA primers by DNA polymerase I is therefore an essential part of DNA replication. In addition to its role in lagging-strand synthesis, DNA polymerase I is also required for many DNA repair reactions. We will discuss DNA repair further in Chapter 26.

(a) Completion of lagging-strand synthesis leaves a nick between the Okazaki fragment and the preceding RNA primer.

Figure 25·14
Joining of Okazaki fragments by the combined action of DNA polymerase I and DNA ligase.

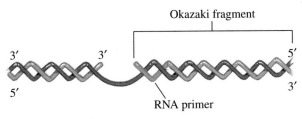

(b) DNA polymerase I extends the Okazaki fragment while its $5'{\rightarrow}3'$ exonuclease activity removes the RNA primer. This process, which results in movement of the nick along the lagging strand, is called nick translation.

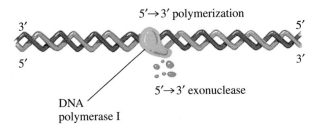

(c) DNA polymerase I dissociates after extending the Okazaki fragment 10–12 nucleotides. DNA ligase binds to the nick between Okazaki fragments.

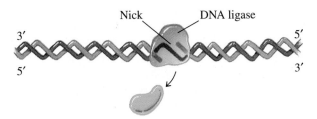

(d) DNA ligase catalyzes formation of a phosphodiester linkage, which seals the nick, creating a continuous lagging strand. The enzyme then dissociates.

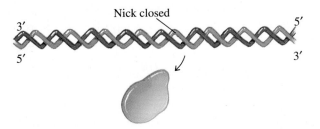

Figure 25·15
Mechanism of ligation by DNA ligase in
E. coli. Using NAD⊕ as a cosubstrate, *E. coli*
DNA ligase catalyzes the formation of a
phosphodiester linkage at nicks in DNA.
The reaction can be divided into three steps.
In Step 1, the ε-amino group of a lysine residue of DNA ligase displaces nicotinamide
mononucleotide (NMN⊕) from NAD⊕. The
result is an AMP–DNA-ligase intermediate.
With DNA ligases that use ATP as the cosubstrate, pyrophosphate is displaced. In
Step 2, the free 5'-phosphate group of the
DNA displaces the enzyme, thereby forming
an ADP-DNA intermediate. In Step 3, the nucleophilic 3'-hydroxyl group on the adjacent
DNA strand displaces AMP, generating a
phosphodiester linkage that seals the nick in
the DNA strand.

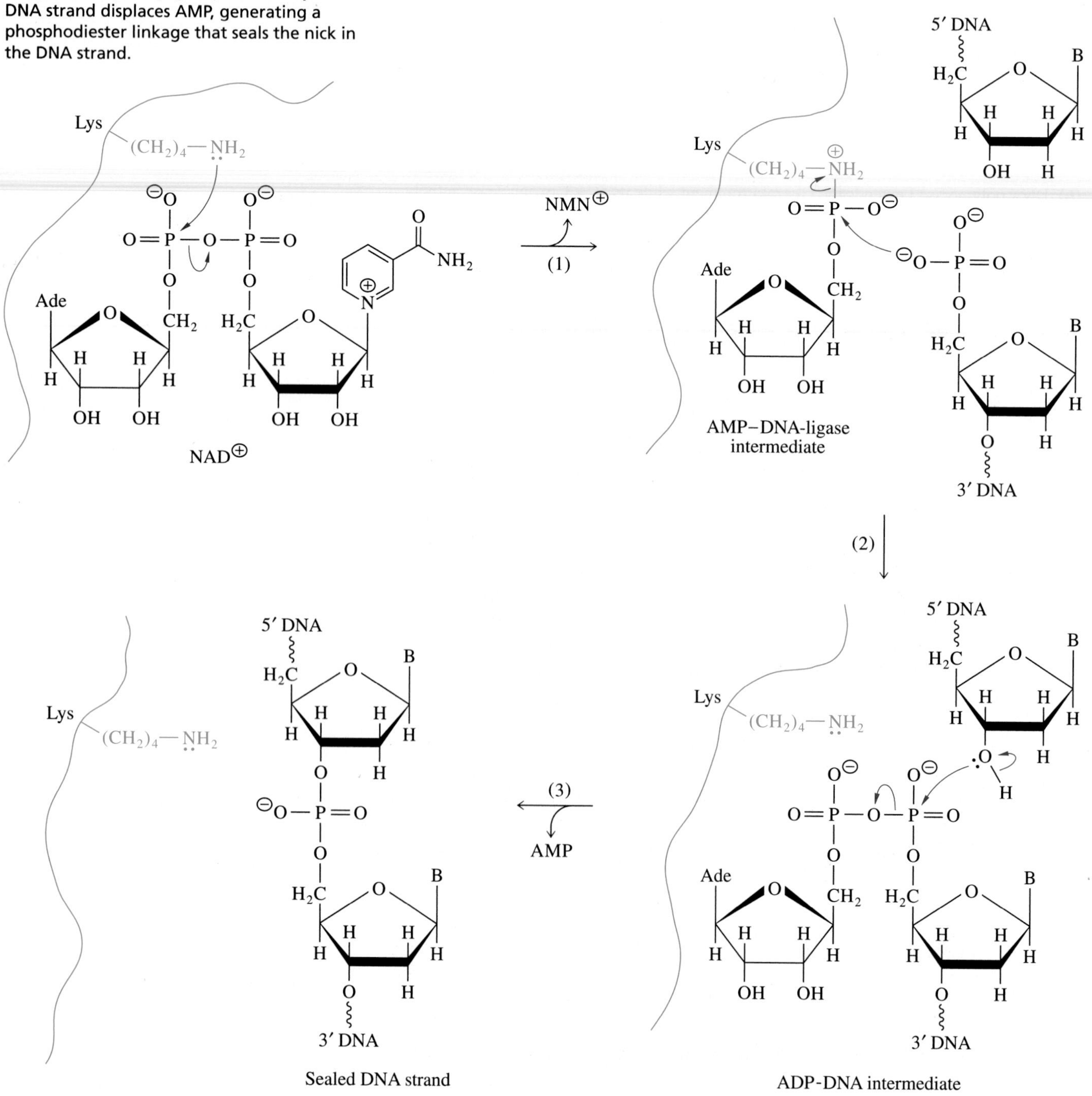

D. Okazaki Fragments Are Joined by DNA Ligase

The last step in the synthesis of the lagging strand of DNA is the formation of a phosphodiester linkage between the 3'-hydroxyl group at the end of one Okazaki fragment and the 5'-phosphate group of an adjacent Okazaki fragment. This step is catalyzed by DNA ligase. Most ligases, including DNA ligases in other species, use ATP as a cosubstrate, but the *E. coli* enzyme requires $NAD^{\oplus}$. This is an unusual role for $NAD^{\oplus}$.

The mechanism of ligation by *E. coli* DNA ligase is summarized in Figure 25·15. In the first step, the phosphorus atom bonded to the 5'-oxygen of the adenosine group of $NAD^{\oplus}$ undergoes nucleophilic attack by the ε-amino group of an active-site lysine residue. Nicotinamide mononucleotide ($NMN^{\oplus}$) is released, and an energy-rich AMP–DNA-ligase intermediate is generated. The free 5'-phosphate group of the DNA then attacks the phosphoryl group of the AMP-enzyme complex, displacing the enzyme and forming an energy-rich ADP-DNA intermediate. The activated 5'-phosphate of the ADP-DNA intermediate then undergoes nucleophilic attack by the free 3'-hydroxyl group on the adjacent DNA chain, resulting in displacement of AMP and sealing of the nick. The overall reaction is

$$DNA_{(nicked)} + NAD^{\oplus} \longrightarrow DNA_{(sealed)} + NMN^{\oplus} + AMP \qquad \textbf{(25·1)}$$

25·5 The DNA Template Is Unwound During Replication

The two strands of the double helix must be unwound and separated during DNA replication. This unwinding is accomplished primarily by a class of proteins called **helicases,** which are found at the replication fork. In *E. coli,* one of the most important helicases is DnaB, a protein tightly associated with primase in the primosome. DnaB unwinds duplex DNA ahead of the replication fork. This reaction is coupled to hydrolysis of nucleoside triphosphates. Since DnaB is part of the replisome, the rate of DNA unwinding is directly coupled to the rate of polymerization as the replisome moves along the chromosome.

Rapid unwinding of DNA generates supercoils ahead of the replication fork. These supercoils could be relieved by rotation of the DNA if the chromosome were linear, but the molecule would have to rotate along its long axis at 6000 revolutions per minute to relieve all the tension! Even if this were possible, the heat generated by friction would kill any cell. In actuality, the action of topoisomerases rapidly relieves supercoils generated by the advancing replication fork. These topoisomerases are not part of the replisome but are nonetheless essential for replication.

As the replication fork advances and DNA is unwound, single-stranded DNA is exposed on the lagging strand. The formation of intrastrand secondary structures, such as hairpin loops, is prevented by **single-strand binding protein** (SSB), also known as helix-destabilizing protein. *E. coli* SSB is a tetrameric protein consisting of identical subunits (subunit M_r 18 000). SSB binds tightly to single-stranded DNA (covering a region of about 32 nucleotides) and prevents the DNA from folding back on itself to form double-stranded regions. The binding of SSB to DNA is cooperative; that is, binding of the first tetramer facilitates binding of the second, and so on. The presence of several adjacent SSB molecules on single-stranded DNA produces an extended, relatively inflexible DNA conformation that is free of secondary structure. Since such secondary structure can inhibit the movement of DNA through the replisome, single-stranded DNA coated with SSB is an ideal template for synthesis of a complementary strand during DNA replication.

An overall model of DNA synthesis by the replisome is shown in Figure 25·16. Helicase is located at the head of the replication fork, followed by the DNA polymerase III holoenzyme and primase. As the helicase unwinds the DNA, primase synthesizes RNA primers on the lagging-strand template approximately once every second. One half of the DNA polymerase III holoenzyme catalyzes synthesis of the leading strand, while the other half is responsible for extending the RNA primers on the lagging-strand template to form Okazaki fragments. The lagging-strand template is thought to be folded back into a large loop. This configuration not only allows both template strands to have the same apparent polarity, but also allows the proteins required for the synthesis of both the leading and lagging strands to be contained in a single replisome at the replication fork.

Although the two halves of the DNA polymerase III holoenzyme are drawn in this model as equivalent, they function slightly differently: one remains firmly bound to the leading-strand template, whereas the other binds the lagging-strand

Figure 25·16
Model showing simultaneous synthesis of leading and lagging strands by the replisome. The replisome contains several elements: the DNA polymerase III holoenzyme; a primosome containing primase, a helicase, and other subunits; and additional components including single-strand binding protein (SSB). One half of the DNA polymerase III holoenzyme catalyzes the synthesis of the leading strand, while the other half catalyzes the synthesis of the lagging strand.

(a) The lagging-strand template is looped back through the replisome so that synthesis of both the leading and the lagging strand proceeds in the same direction.

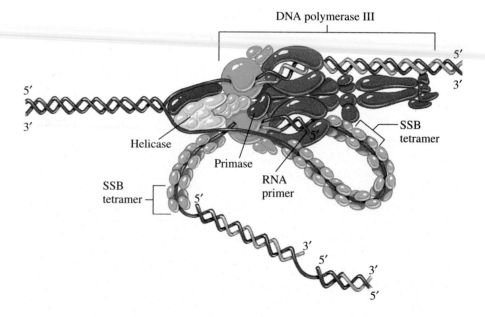

(b) As the helicase unwinds DNA, primase synthesizes the next RNA primer of the lagging strand, and the lagging-strand polymerase completes an Okazaki fragment.

template until it encounters the RNA primer of the previously synthesized Okazaki fragment. At this point, the lagging-strand template is released. It then rebinds to DNA polymerase III at the newly synthesized primer. The γ complex (see Figure 25·6) is thought to play a role in binding and releasing the loop of the lagging strand. The γ complex is also thought to open and close the β subunit ring, allowing half the DNA polymerase III holoenzyme to transfer from one region of DNA to another.

This replisome model explains how syntheses of the leading and lagging strands are coordinated. Note that the composition of the replisome (a protein machine) ensures that all the components necessary for replication are available at the right time, in the right amount, and in the right place. This process of assembling a number of proteins into an interactive unit for carrying out multiple functions appears to be critical for a variety of activities in the flow of information in the cell, including RNA synthesis, RNA processing, and protein synthesis.

(c) After encountering the RNA primer of a previously synthesized Okazaki fragment, the lagging-strand polymerase releases the lagging-strand template while remaining bound to the replisome.

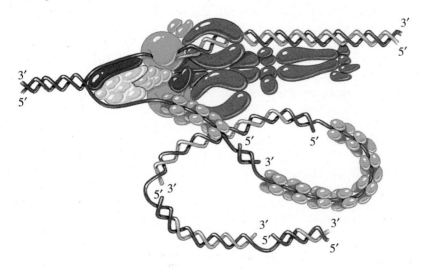

(d) The lagging-strand polymerase binds to a newly synthesized RNA primer and begins synthesis of another Okazaki fragment.

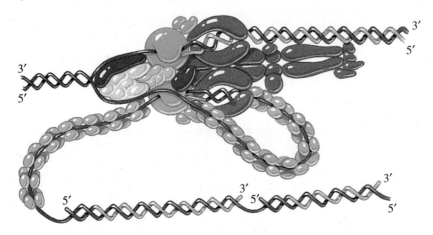

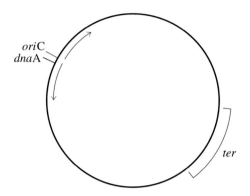

Figure 25·17
Location of the origin (*ori*C) and terminus (*ter*) of DNA replication in *E. coli*. *dna*A is the gene for the protein DnaA, which is involved in the initiation of replication. The distance between *ori*C and *dna*A is about 40 kb. The red arrows indicate the direction of movement of the replication forks.

25·6 Initiation of DNA Replication Occurs at a Unique Site on the Bacterial Chromosome

Genetic analysis has shown that DNA replication begins at a unique site on the *E. coli* chromosome. This site is called the origin and is designated *ori*C. DNA replication begins at *ori*C and proceeds in both directions until the two replication forks meet at the terminus (designated *ter*), a region opposite the origin (Figure 25·17).

A. DnaA Binds to the Origin of Replication

The origin of replication in *E. coli* consists of a minimal sequence of 245 nucleotides. Within this region are four binding sites for DnaA, a protein required for the initiation of DNA replication and the assembly of replisomes at the origin. DnaA is a tetramer of four identical subunits (subunit M_r 50 000). The binding sites for DnaA all have a similar nucleotide sequence (TATT$^C/_A$CA$^C/_A$A), although two of the sites are inverted. In addition to the four DnaA-binding sites, the origin contains three tandem repeats of a 13 bp sequence that is rich in A/T. There are also 11 Dam methylase sites in *ori*C. Figure 25·18 shows the organization and sequence of *ori*C in *E. coli*. The DnaA-binding sites, the methylation sites, and the A/T-rich repeats are common features of many bacterial origins of replication, suggesting that origin sequences have been evolutionarily conserved.

The formation of a replication fork begins when four DnaA molecules bind to the DnaA-binding sites. The binding sites must be fully methylated before DnaA can bind efficiently. Six to sixteen additional DnaA molecules then bind in a highly cooperative process, wrapping the DNA into a structure that resembles a eukaryotic nucleosome (Section 24·11). HU protein (Section 24·11C) is also required to form this structure. At this stage, the DnaA-HU-DNA complex is referred to as a closed complex.

DnaA, in association with HU protein, unwinds *ori*C, creating an open complex that contains a locally unwound region of DNA. Unwinding begins in the region of the A/T-rich repeats, and about 60 base pairs are unwound to form a replication bubble (Figure 25·19). The unwinding of DNA is energetically unfavorable

Figure 25·18
Organization and sequence of *ori*C in *E. coli*. (a) Genetic map showing the organization of DnaA-binding sites at *ori*C. There is a transcription initiation site within this region. (b) Sequence of *ori*C. Nucleotides shown in black occur at the same position in other bacteria, indicating that these sites have been conserved. Nucleotides shown in blue are either partially conserved or not conserved. The four binding sites for DnaA are underlined in black. Methylated adenines (located at the Dam methylase site GATC) are indicated by asterisks. The 13 bp, A/T-rich regions are underlined in red.

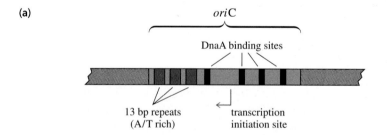

(a)

$$ori\text{C}$$

DnaA binding sites

13 bp repeats
(A/T rich)

transcription
initiation site

(b)

```
 1  *                         *                         *      35
G A T C T A T T T A T T T A G A G A T C T G T T C T A T T G T G A T C
                                        *                        70
T C T T A T T A G G A T C G C A C T G C C C T G T G G A T A A C A A G
                              *                            * 105
G A T C C G G C T T T T A A G A T C A A C A A C C T G G A A A G G A T
                        *                    *               140
C A T T A A C T G T G A A T G A T C G G T G A T C C T G G A C C G T A
                    *                                        175
T A A G C T G G G A T C A G A A T G A G G G G T T A T A C A C A A C T
                                                             210
C A A A A A C T G A A C A A C A G T T G T T C T T T G G A T A A C T A
                *                                            245
C C G G T T G A T C C A A G C T T C C T G A C A G A G T T A T C C A C
```

(a) The DnaA molecules first bind to four sites in *ori*C.

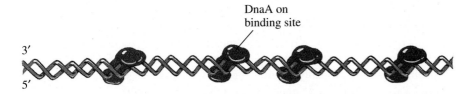

DnaA on
binding site

3′

5′

Figure 25·19
Conceptual model of the initiation of replication in *E. coli*.

(b) The original DnaA molecules are joined by others and by HU protein, which folds the chromosome into a structure similar to a nucleosome. This folding results in unwinding at *ori*C, allowing DnaB-DnaC complexes to enter.

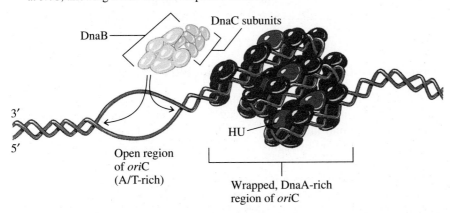

DnaB

DnaC subunits

3′

5′

HU

Open region
of *ori*C
(A/T-rich)

Wrapped, DnaA-rich
region of *ori*C

(c) The helicase activity of DnaB further unwinds the DNA, displacing the DnaA molecules. The unwound single strands are prevented from reannealing by SSB molecules.

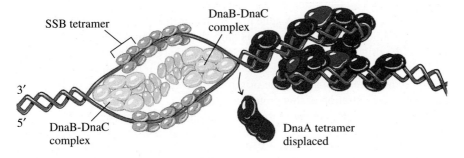

SSB tetramer

DnaB-DnaC
complex

3′

5′

DnaB-DnaC
complex

DnaA tetramer
displaced

but occurs readily if the DNA containing *ori*C is negatively supercoiled. Negative supercoils are introduced by topoisomerase II in a reaction coupled to ATP hydrolysis (Section 24·6). Transcription of adjacent genes also aids formation of the open complex since transcription creates negatively supercoiled regions behind the transcription complex (Chapter 27). Formation of the open complex is probably the rate-limiting step in the initiation of DNA replication.

B. Replisomes Are Assembled at the Open Complex

The next proteins to join the open complex are DnaB and DnaC. Recall that DnaB, a hexamer, is a helicase that forms part of the primosome at the replication fork. The DnaB hexamer associates with six DnaC polypeptides, and it is this dodecamer that binds to the open complex. DnaA interacts with DnaB and guides the DnaB-DnaC complex to the unwound region of DNA. DNA in the *ori*C region is further unwound by the helicase activity of DnaB in association with topoisomerase II and gyrase (Figure 25·20, next page). This unwinding is coupled to hydrolysis of ATP.

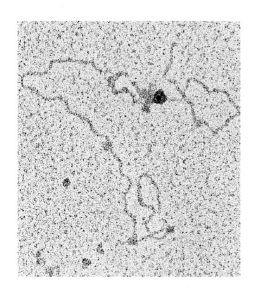

Figure 25·20
Electron micrograph of unwinding at *ori*C. The region around the origin has been unwound to form an open complex. Single-stranded DNA is coated with SSB, and the DnaB-DnaC complex at one of the forks has been made visible by binding gold-labelled antibody to DnaB. This technique is not highly efficient, and the complex at the other fork, while present, was not labelled. The large proteins on the double-stranded DNA are molecules of topoisomerase II, which are also required for unwinding. (Courtesy of Barbara Funnell.)

DnaA, DnaB, DnaC, topoisomerase II, and SSB prepare the origin of replication for the initiation of DNA synthesis by forming a prepriming complex. The remaining components of the replisome are then assembled on this prepriming complex. These components include the various other subunits of the primosome, including primase itself (DnaG). DNA polymerase III holoenzyme is added by stepwise addition of its subunits, beginning with τ, the γ complex, and then the β subunits. Replisome assembly probably requires the aid of chaperones such as DnaK, which is the prokaryotic form of the major heat-shock protein (Hsp70).

Primase is required for initiation of DNA synthesis on the two lagging strands at *ori*C. The initial RNA primers used for leading-strand synthesis may also be made by primase, although in some cases the primers are synthesized by the action of RNA polymerase, which initiates RNA synthesis at several sites in and near *ori*C. One of the most frequently used transcription initiation sites is shown in Figure 25·18.

C. Initiation of DNA Replication Is Regulated

Regulation of the initiation of DNA replication is extremely important since it ensures that DNA synthesis is coupled to cell division. In *E. coli* and most other bacteria, initiation is controlled in two ways: by restricting the availability of DnaA-binding sites at *ori*C and by regulating the concentration of this protein inside the cell.

Immediately after initiation of replication, the DNA near *ori*C binds to the plasma membrane, where it is sequestered from cellular proteins. Because the DNA has just been replicated, the Dam methylase sites within *ori*C are hemimethylated. But because the DNA is sequestered, Dam methylase cannot reach these sites. Since DnaA cannot bind to hemimethylated DNA even if it were available, replication cannot be reinitiated at *ori*C until the DNA is released from the membrane and fully methylated by the action of Dam methylase.

The proximity of the *dna*A gene to *ori*C also influences the rate of initiation of replication. In *E. coli*, the *dna*A gene is only about 40 kb from *ori*C, while in other bacteria, the origin of replication is immediately adjacent to the *dna*A gene. When *ori*C binds to the membrane immediately after initiation of replication, the *dna*A gene is sequestered along with the origin. This prevents production of DnaA and lowers the concentration of this protein in the cell.

DnaA is absolutely required for the initiation of replication on the *E. coli* chromosome, but in other replicating chromosomes, different *ori*-protein pairs occur. For example, in the bacteriophage λ, the origin of replication specifically interacts with a λ replication protein encoded by the λ *O* gene. DnaA protein cannot substitute for the O protein in λ replication, nor can the λ O protein substitute for DnaA in *ori*C-directed replication. But O protein helps direct DnaB and other host proteins to initiate replication at the λ origin. Thus, the combination of a specific DNA sequence and a particular protein directs the replication machinery to begin replication at a unique site. In this way, the bacteriophage subverts the bacterial replication machinery for its own use.

25·7 DNA Replication Terminates Within the *ter* Region

Termination of replication in *E. coli* occurs within the *ter* region, opposite *ori*C on the circular chromosome. The *ter* region acts as a trap, preventing replication forks from passing through. Within this 600 kb region are five DNA sequences (*ter*A to *ter*E) that are binding sites for a protein known as **terminator utilization substance** (Tus). Tus (M_r 36 000) binds *ter* sites very tightly ($K_{diss} = 10^{-13} - 10^{-11}$ M); the half-life of dissociation is 10 hours.

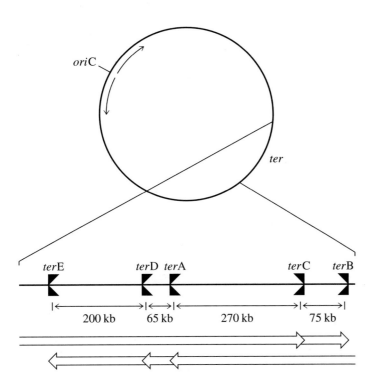

Figure 25·21
Organization of the *ter* region in *E. coli*. The 600 kb *ter* region contains five asymmetric *ter* sites, each of which can be bound by Tus. The *ter* sites stop only a replisome approaching from one direction (the concave sides of the boxes shown here represent the replisome traps). The arrowheads represent the possible termination sites for each of the two converging replisomes. Note that the *ter* sites are oriented to ensure that the replisomes do not arrest at a *ter* site without meeting. (Adapted from Hidaka, M., Kobayashi, T., and Horiuchi, T. (1991). A newly identified DNA replication terminus site, *ter*E, on the *Escherichia coli* chromosome. *J. Bacteriol.* 173:391–393.)

When bound to DNA, Tus blocks the passage of replication forks by preventing the unwinding of DNA by helicase. Each Tus binding site is asymmetric, and Tus arrests only replisomes that approach it from one direction (replisomes that encounter Tus from the other direction are thought to strip Tus from its binding site). The organization of the *ter* region is shown in Figure 25·21. Note that the *ter* sites are oriented so that each replisome must pass over all the sites oriented the opposite way before encountering one that causes termination. This arrangement ensures that the two replisomes entering the *ter* region from opposite directions will always collide, and hence that the chromosome will always be completely replicated. The replisomes disassemble following collision.

DNA replication is complete when one of the replication forks runs into the other. After replication of the entire chromosome, the two circular, duplex DNA molecules are topologically linked as a concatenate (like the links in a chain). The chromosomes are separated by the activity of DNA topoisomerases, including specific topoisomerases that are required for chromosome segregation and partitioning to the two daughter cells.

25·8 DNA Can Be Sequenced Using Dideoxyribonucleotides

In 1976, Frederick Sanger developed a method for sequencing DNA enzymatically using the Klenow fragment of *E. coli* DNA polymerase I. Sanger was awarded his second Nobel prize for this achievement (his first Nobel prize was awarded for developing a method for sequencing proteins). The advantage of using the Klenow fragment for this type of reaction is that the enzyme lacks $5' \rightarrow 3'$ exonuclease activity, which could degrade newly synthesized DNA. However, one of the disadvantages is that the Klenow fragment is not very processive and is easily inhibited by the presence of secondary structure in the single-stranded DNA template. This

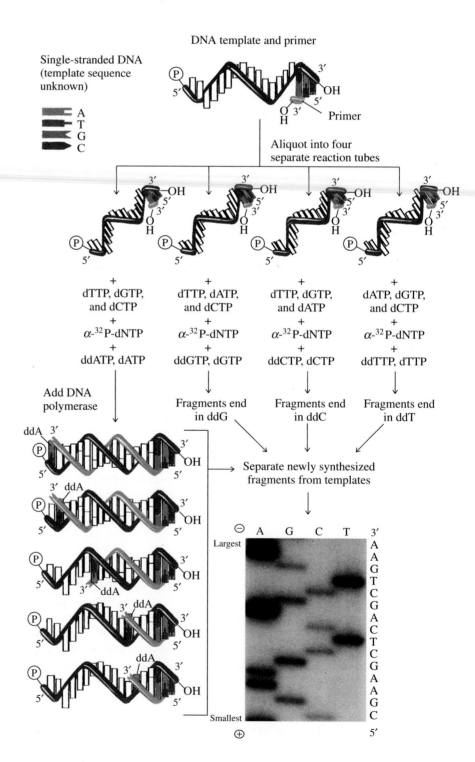

Figure 25·22
Structure of a 2',3'-dideoxy-nucleoside triphosphate. *B* represents any base.

Figure 25·23
Sanger method for sequencing DNA. In the Sanger sequencing method, addition of a small amount of a particular dideoxynucleoside triphosphate (ddNTP) to each reaction mixture causes DNA synthesis to terminate when that ddNTP is incorporated in place of the normal nucleotide. The positions of incorporated ddNTPs, determined by the lengths of the fragments generated, indicate the positions of the corresponding nucleotide in the sequence. The fragments generated during synthesis with each ddNTP are separated according to size using an electrophoretic sequencing gel, and the sequence of the DNA can be read off an autoradiograph of the gel (as shown by the vertical column of letters to the right of the gel).

limitation can be overcome by adding SSB or analogous proteins, or more commonly, by using DNA polymerases from bacteria that grow at high temperatures instead of the Klenow fragment. Such polymerases are active at 60°–70°C, a temperature at which secondary structure in single-stranded DNA is unstable.

The Sanger sequencing method uses 2′,3′-dideoxynucleoside triphosphates (ddNTPs), which differ from the deoxyribonucleotide substrates of DNA synthesis in that the 3′-hydroxyl group is replaced with a hydrogen (Figure 25·22). The dideoxyribonucleotides are recognized by DNA polymerase and are added to the 3′ end of the growing chain. But because these nucleotides lack a 3′-hydroxyl group, subsequent nucleotide additions cannot take place. Their incorporation therefore terminates the growth of the DNA chain. When a low concentration of a particular dideoxyribonucleotide is included in a DNA synthesis reaction, the dideoxyribonucleotide is occasionally incorporated in place of the corresponding dNTP and thus causes immediate termination of replication. The length of the resulting fragment of DNA identifies the position of the corresponding nucleotide that should have been incorporated.

DNA sequencing using ddNTP molecules involves several steps (Figure 25·23). The DNA to be sequenced is prepared as single-stranded molecules and mixed with a short oligonucleotide complementary to one end of the DNA to be sequenced. This oligonucleotide acts as a primer for DNA synthesis catalyzed by DNA polymerase. The oligonucleotide-primed material is then split into four separate reaction tubes. Each tube receives a small amount of an α-^{32}P-labelled dNTP, which allows the newly synthesized DNA to be visualized by autoradiography. Then each tube receives an excess of the four nonradioactive dNTP molecules and a small amount of one of the four ddNTPs. For example, the A reaction tube receives an excess of nonradioactive dTTP, dGTP, dCTP, and dATP mixed with a small amount of ddATP. DNA polymerase is then added to the reaction mixture. As the polymerase replicates the DNA, it occasionally incorporates a ddATP residue instead of a dATP residue, and synthesis of the growing DNA chain is terminated. Random incorporation of ddATP results in the production of newly synthesized DNA fragments of different lengths, each ending with an A. The length of each fragment is therefore a measure of the distance from the primer to one of the adenine residues in the sequence. Adding a different dideoxyribonucleotide to each reaction tube produces a different set of fragments: ddTTP produces fragments that terminate with a T, ddGTP with a G, and ddCTP with a C. The newly synthesized chains from each sequencing reaction are then separated from the template DNA. Finally, the mixtures from each sequencing reaction are subjected to electrophoresis in adjacent lanes on a sequencing gel, where the fragments are resolved by size. The sequence of the DNA molecule can then be read from the gel.

25·9 DNA Replication in Eukaryotes Is Similar to DNA Replication in Prokaryotes

Eukaryotic chromosomes are linear, double-stranded DNA molecules that are usually much larger than the chromosomes of bacteria. For example, the *E. coli* genome is a single chromosome of 4.6×10^3 kb, whereas the genome of the fruit fly *Drosophila melanogaster* is 1.65×10^5 kb, and the average mammalian genome is approximately 3×10^6 kb (haploid DNA content). In addition, eukaryotes usually have more than one chromosome: *D. melanogaster* has three large chromosome pairs and one small pair, and mammals have between 20 and 30 chromosome pairs, depending on the species. But despite this potential for increased genomic complexity, the biochemical mechanism of DNA replication in all organisms is fundamentally similar. Many of the differences in replication between prokaryotes and eukaryotes arise from the greater size of eukaryotic DNA and its being packaged into chromatin.

In eukaryotes, as in *E. coli,* replication is bidirectional. But whereas the *E. coli* chromosome has a unique origin of replication, eukaryotic chromosomes have multiple sites where DNA synthesis is initiated. Although the rate of fork movement in eukaryotes is slower than in bacteria, the presence of a large number of independent origins of replication enables the larger eukaryotic genomes to be copied in approximately the same time as prokaryotic genomes.

In eukaryotes, as in prokaryotes, synthesis of the leading strand is continuous, and synthesis of the lagging strand is discontinuous. The overall process of primer synthesis, Okazaki fragment synthesis, primer hydrolysis, and gap filling in eukaryotes is similar to that in bacteria, but the enzymes that carry out DNA replication in eukaryotes have not been characterized as extensively as those in prokaryotes.

DNA replication in several species of eukaryotes has been studied intensely. These include the yeast *Saccharomyces cerevisiae,* the fruit fly *D. melanogaster,* and cells from several mammals including humans, cows, mice, rats, and hamsters. Although there may be some minor differences in the details of DNA replication among different eukaryotic species, most of what we discuss in this section applies to all known eukaryotes.

A. There Are Several Eukaryotic DNA Polymerases

Eukaryotic cells contain at least five different DNA polymerases. DNA polymerases α, δ, and ε are found in the nucleus and are responsible for catalyzing DNA replication and some repair reactions. These three enzymes are evolutionarily related to each other and share a common ancestor with *E. coli* DNA polymerase II, forming a family known as the B family of DNA polymerases. DNA polymerase β is a small, monomeric protein (M_r 40 000). This enzyme associates tightly with the exonuclease DNase V and is involved in the repair of nuclear DNA. DNA polymerase γ (M_r 140 000) carries out replication of mitochondrial DNA. A sixth DNA polymerase is responsible for replicating DNA in chloroplasts.

DNA polymerase α is the enzyme responsible for lagging-strand synthesis at the replication fork. The largest subunit of this tetrameric protein contains the polymerase active site, and two of the other subunits possess primase activity. The active site of RNA primer synthesis is mostly associated with the smallest subunit. (Note that the primase and polymerase activities of prokaryotes, which are associated in the replisome, separate during purification, whereas in eukaryotes the activities do not separate because the interactions between the relevant subunits are strong enough to withstand the relatively harsh treatments of protein purification.)

Like the DNA polymerase III holoenzyme from *E. coli,* DNA polymerase α cannot begin DNA synthesis de novo and can lengthen DNA chains only when provided with an RNA primer. The primase subunit of DNA polymerase α operates in bursts of activity. An RNA primer containing from 6 to 15 nucleotides is made in each burst. In the presence of dNTP molecules and the other subunits of DNA polymerase α, the RNA primer is extended as DNA.

DNA polymerase α also has an associated $3' \rightarrow 5'$ exonuclease activity that appears to be inhibited or masked in vitro. The significance of this is uncertain since proofreading activity is probably required in vivo. DNA replication in eukaryotic cells is extremely accurate; only one base in 10^9 to 10^{11} is incorrectly incorporated.

DNA polymerase δ catalyzes synthesis of the leading strand at the replication fork (note that the leading- and lagging-strand polymerases in eukaryotes are isolated as separate enzymes). This enzyme is composed of two subunits. The largest subunit contains the active site of polymerization. The enzyme also has a $3' \rightarrow 5'$ exonuclease activity.

The third polymerase involved in chromosomal DNA replication in eukaryotes is DNA polymerase ε. It is a large, monomeric protein. Within this single polypeptide chain is a polymerase activity, a $3' \rightarrow 5'$ proofreading exonuclease activity, and

Table 25·3 Eukaryotic DNA polymerases

DNA polymerase	Organism[a]	Subunit size[b] (M_r in thousands)	Activities	Location	Role
α	Mammals, yeast (pol I), _Drosophila_	160–180 70 55–60 48	Polymerase Primase $3' \rightarrow 5'$ Exonuclease[c]	Nucleus	Lagging-strand synthesis Repair
β	Mammals[d]	39–41	Polymerase	Nucleus	Repair
γ	Mammals, chicken, frog, _Drosophila_	140–144[e]	Polymerase $3' \rightarrow 5'$ Exonuclease	Mitochondrion	Mitochondrial DNA replication
δ	Mammals, yeast (pol III)	125–130 47–50	Polymerase $3' \rightarrow 5'$ Exonuclease	Nucleus	Leading-strand synthesis Repair
ε	Mammals, yeast (pol III)	240–290[e]	Polymerase $3' \rightarrow 5'$ Exonuclease $5' \rightarrow 3'$ Exonuclease	Nucleus	Leading-strand synthesis Repair Gap filling on lagging strand

[a] Alternative names for the polymerase in yeast are given in parentheses.
[b] Ranges refer to subunit sizes reported in different species.
[c] Polymerase α $3' \rightarrow 5'$ exonuclease activity is not detectable in all species.
[d] Polymerase β may be confined to vertebrates.
[e] Smaller associated subunits appear to be proteolytic fragments.

a $5' \rightarrow 3'$ exonuclease activity. Like its functional counterpart in _E. coli_ (DNA polymerase I), this enzyme is mainly required for DNA repair but also fills and seals gaps between Okazaki fragments on the lagging stand. In some cases, DNA polymerase ε may substitute for DNA polymerase δ as the leading-strand polymerase. The characteristics of eukaryotic DNA polymerases are summarized in Table 25·3.

B. Accessory Proteins Are Required at the Eukaryotic Replication Fork

The eukaryotic equivalent of SSB is replication protein A (RPA), sometimes also known as replication factor A. RPA is composed of three subunits (M_r 70 000, 35 000, and 12 000). It binds tightly to single-stranded DNA and increases the efficiency of DNA synthesis on the lagging strand.

DNA polymerase δ by itself acts in a distributive manner in vitro. It requires accessory proteins that act as a clamp to lock the enzyme onto the leading strand, making the enzyme processive. The clamp protein has been identified as proliferating cell nuclear antigen (PCNA), a multimeric protein (subunit M_r 36 000). PCNA is a highly conserved protein found in all eukaryotic cells. It is functionally equivalent to the β subunits of _E. coli_ DNA polymerase III. The prokaryotic and eukaryotic clamps may be structurally similar as well.

Replication factor C (RFC) is a multimeric protein that binds to DNA polymerase δ and helps make it processive. RFC appears to load PCNA onto DNA to form the sliding clamp. The role of RFC is analogous to the roles of the γ complex and the τ subunit of _E. coli_ DNA polymerase III holoenzyme. The amino acid sequences of the RFC subunits are similar to those of the τ, γ, and δ' proteins from bacteria, suggesting that these functionally related proteins may share a common ancestor.

Many different helicases have been isolated from eukaryotic cells, including some that bind specifically to DNA polymerase δ or DNA polymerase α. In contrast to the bacterial helicase, which is part of the primosome, the eukaryotic helicases do not appear to be tightly associated with primase activity. Supercoils induced by unwinding in eukaryotes are relieved by topoisomerases associated with chromatin.

Figure 25·24
Conceptual model of a eukaryotic replication fork.

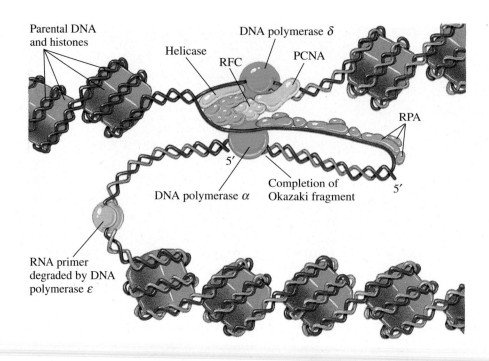

There are many different ligases in eukaryotic cells, and several of them have been implicated in sealing Okazaki fragments on the lagging strand. Bloom's syndrome is a rare genetic disease in humans that is due to low activity of DNA ligase I. The symptoms are slow growth and development, and patients usually die before adulthood. Replication forks move more slowly in the cells of patients with Bloom's syndrome, suggesting that DNA ligase I normally plays a role in DNA replication. The fact that the deficiency is not immediately lethal indicates that other ligases probably compensate for much of the reduced level of DNA ligase I activity.

The presence of many accessory proteins at the replication fork indicates that there is a replisome-like complex that also contains DNA polymerases α and δ. However, there is little direct evidence for the existence of such a complex in eukaryotes. Unlike the case in bacteria, the lagging-strand polymerase in eukaryotes may dissociate from the replication fork after the synthesis of each Okazaki fragment. A diagram of the proteins found at the eukaryotic replication fork is shown in Figure 25·24.

C. Eukaryotic DNA Has Many Origins of Replication

In contrast to the single origin of replication of *E. coli* (*ori*C), DNA replication in eukaryotes originates at many sites. An extreme example is provided by *D. melanogaster:* during early embryogenesis, when DNA is being rapidly replicated, the largest chromosome contains some 6000 replication forks, or one every 10 kb. This is an exceptional case; most of the origins used at this stage of development are not active in the adult, when the rate of chromosome duplication is slower. Replication proceeds bidirectionally from each origin, and the forks from different origins move toward one another, merging to form bubbles of ever-increasing size (Figure 25·25).

In yeast, the origin of replication is referred to as an autonomously replicating sequence, or ARS element. There are two important sequences within each ARS element: a binding site for initiator proteins and a region of DNA that is easily unwound (the DNA unwinding region). The binding site for initiator proteins consists

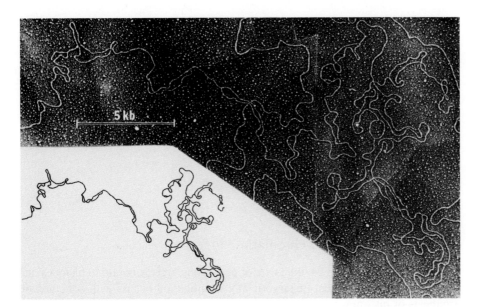

Figure 25·25
Electron micrograph of replicating DNA in the embryo of the fruit fly *Drosophila melanogaster.* Note the large number of replication forks. (Courtesy of David S. Hogness.)

of an 11 bp, A/T-rich sequence known as the ARS consensus sequence (Figure 25·26). In yeast, the protein that initiates DNA replication is a complex of at least six different polypeptides. This protein complex is called the origin replication complex (ORC). ORC binds to the ARS consensus sequence in a manner that is probably similar to the binding of DnaA to *ori*C; that is, the DNA is wrapped around the outside of the protein complex. Binding leads to unwinding of the nearby DNA unwinding element and formation of a replication bubble.

Other protein-binding sites often occur in or near eukaryotic origins. For example, within the yeast origin shown in Figure 25·26 there are weaker binding sites for ORC at *B1* and *B2,* upstream of the ARS consensus sequence. Site *B3* is the binding site for a common yeast protein that normally plays a role in transcription initiation. The binding of such proteins at origins may assist the unwinding of DNA.

As is the case in prokaryotes, the initial unwound region is stabilized by a single-strand binding protein (in this case, RPA) and extended by helicase. Within the region of DNA that is unwound are sites at which DNA polymerase α preferentially synthesizes primers. These are the sites at which DNA synthesis actually begins. Following the addition of DNA polymerase α, RFC, PCNA, and DNA polymerase δ are added to complete formation of the replication fork.

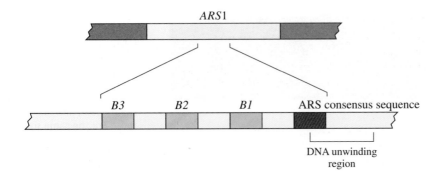

Figure 25·26
Replication origin *ARS*1 in yeast. The *ARS*1 replication origin is located between two genes. The origin replication complex binds to the ARS consensus sequence and more weakly to sites *B1, B2,* and *B3.*

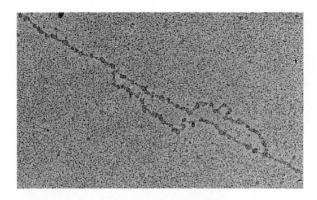

Figure 25·27
Nucleosomes on replicating DNA. Note the nucleosomes on the parental DNA molecule (ahead of the replication fork) and on both daughter molecules. (Courtesy of Oscar L. Miller, Jr.)

D. DNA Replication and Chromatin

As described in Chapter 24, the DNA of eukaryotic cells is bound to histones and packaged in nucleosomes. Eukaryotic DNA replication occurs with concomitant synthesis of histones, a process known as **histone duplication.** DNA replication and histone synthesis involve completely different enzymes acting in different parts of the cell, yet both processes are coordinated. As soon as DNA is replicated, nucleosomes form on each daughter molecule (Figure 25·27).

In some cases, the newly synthesized histone proteins preferentially form nucleosomes on the lagging strand of the replication fork, but in other cases, the original nucleosomes segregate equally to each daughter molecule. The nucleosome core particle may even be split in half during DNA replication, with half going to one strand and half to the other in a manner that is reminiscent of semiconservative DNA replication.

There appears to be an origin of replication associated with many of the large loops of chromatin that are bound to the nuclear scaffold. Some origins contain scaffold attachment sites, suggesting that these origins are located at the base of the loops where DNA is bound to the scaffold. During DNA replication, it is likely that the replication proteins remain associated with the nuclear scaffold and that the DNA is reeled through a large protein complex. This allows all of the several hundred replication forks in each chromosome to be localized in one region of the nucleus. These replication foci within a nucleus can be visualized using fluorescent antibodies against the single-strand binding protein (RPA) (Figure 25·28).

Figure 25·28
Replication foci in the nucleus of the frog *Xenopus laevis.* Antibodies to RPA were tagged with a red fluorescent dye and used to detect regions of high concentrations of RPA (on the left). Newly incorporated nucleotides in the same nucleus were detected with a binding protein attached to a green dye (on the right). Newly replicated DNA forms discrete foci in the nucleus, and many of these are the same sites as the RPA foci (compare the patterns of the left and right images). (Courtesy of Ulrich Laemmli.)

25·10 Other Modes of DNA Replication Have Been Described

The standard mechanism of DNA replication in both eukaryotes and prokaryotes consists of two replication forks moving in opposite directions from an origin. Both the leading- and lagging-strand templates are copied, and the net result is semiconservative replication. However, modified forms of the standard mechanism have been described.

A. Rolling-Circle Replication

In some cases, only one strand of DNA is copied during replication, and the product is a single-stranded DNA molecule. For example, the circular DNA of certain bacteriophages is single stranded. Upon injection into a host bacterium, this single-stranded genome is converted to a double-stranded molecule by the action of primase and DNA polymerase III. The double-stranded circular molecule is, in turn, used as the template for rolling-circle replication to produce new, single-stranded molecules (Figure 25·29). The single-stranded DNA molecules that are produced by such a mechanism are incorporated into new bacteriophage particles. A rolling-circle intermediate in φX174 DNA replication is shown in Figure 25·30.

There are many variations of rolling-circle replication. In some cases, the displaced tails serve as templates for lagging-strand synthesis, and this results in the production of double-stranded DNA molecules that are packaged into phage particles. Bacteriophage λ replicates in this way during the late stages of the infectious cycle. In many single-stranded bacteriophages, including φX174, the endonuclease that makes the nick in the double-stranded intermediate remains attached to the 5′ end of the single-stranded tail. When one round of synthesis is completed, the endonuclease catalyzes cleavage of the single strand and ligation of its ends to form a single-stranded, circular molecule.

The rolling-circle mechanism is also used during conjugation in *E. coli* and related bacteria. Cells of opposite mating types can be joined by sex pili, which form

A primosome assembles on the template (+) strand, and an RNA primer is synthesized.

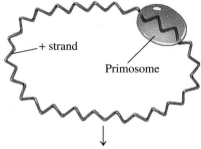

The primer is extended by DNA polymerase III to produce a complementary (–) strand.

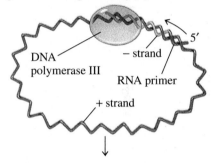

The resulting double-stranded DNA molecule is nicked at a specific site on the + strand by a phage-encoded endonuclease.

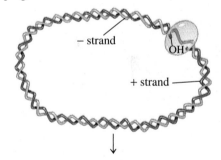

The + strand is extended from the exposed 3′-hydroxyl group by DNA polymerase III, displacing the original + strand.

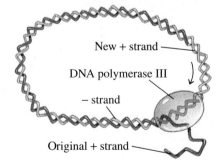

Figure 25·29
Replication of single-stranded bacteriophage DNA.

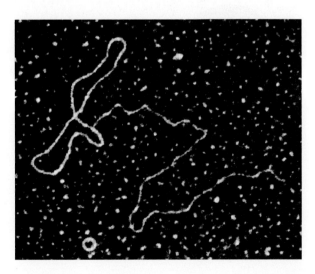

Figure 25·30
Electron micrograph of a rolling-circle intermediate in the replication of bacteriophage φX174 DNA. A long, single-stranded tail extends from a circular, double-stranded molecule. (Courtesy of David Dressler and Kirston Koths.)

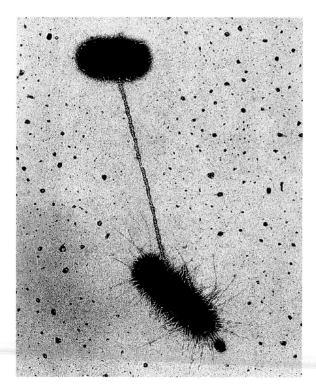

Figure 25·31
Conjugating *E. coli* cells. The cell on the bottom (F⁺) contains an F plasmid that is being replicated by a rolling-circle mechanism. DNA is passed through the long sex pilus that connects this cell to the F⁻ cell. (Courtesy of Charles Brinton, Jr.)

a hollow tube between cells (Figure 25·31). The donor cell (referred to as F^+, or male) contains a 95 kb F plasmid that carries genes for the sex pili and other proteins involved in conjugation. One of these proteins is an endonuclease that nicks the plasmid DNA at a specific site used to initiate rolling-circle replication. Single-stranded DNA is passed through the pilus to a recipient cell (designated F^-, or female), where the complementary strand is synthesized (Figure 25·32). Sometimes the F plasmid becomes integrated into the chromosome of the F^+ cell; if this happens, the entire chromosome is transferred when plasmid-directed replication begins. This process takes over 100 minutes.

Rolling-circle replication is also used in the transfer of a large plasmid from the soil bacterium *Agrobacterium tumefaciens* to plant cells. Infection by the bacterial

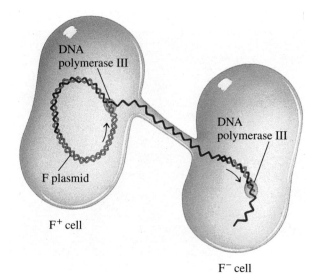

DNA polymerase III

F plasmid

F^+ cell

DNA polymerase III

F^- cell

Figure 25·32
Rolling-circle replication of an F plasmid DNA during conjugation. A nuclease creates a single-stranded cut at a unique site on the plasmid in the F^+ cell. DNA polymerase III uses the exposed 3'-hydroxyl group as a primer, synthesizing a new strand and displacing the existing strand. The existing strand is then transferred through the pili to the F^- cell, where a complementary strand is synthesized.

plasmid causes crown gall tumors (Figure 25·33). This is one of the very few known examples of exchange of genetic information between prokaryotes and eukaryotes.

B. Delayed Synthesis on the Lagging Strand Leads to the Formation of D-Loops

On some plasmids, the initiation of replication involves displacement of one strand of DNA to form a displacement loop, or D-loop. The D-loop is an extended region of unwound DNA in which one of the strands has been copied but the other remains single stranded. The replication of the ColE1 plasmid in *E. coli* is a well-studied example of this mechanism of DNA replication (Figure 25·34). ColE1 replication origins are often present in recombinant DNA plasmids (Chapter 33).

Figure 25·33
Crown gall tumors on a tobacco plant. The tumors form when DNA from the bacterium *Agrobacterium tumefaciens* is inserted into plant cells. Bacterial DNA is transferred by rolling-circle replication. (Courtesy of Eugene Nester.)

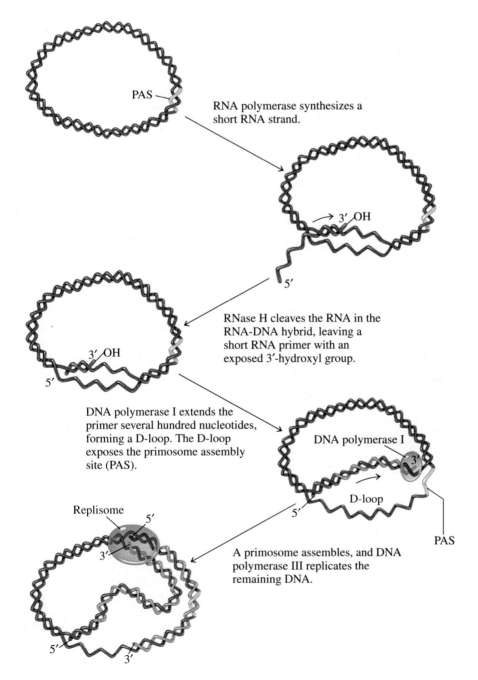

RNA polymerase synthesizes a short RNA strand.

3′ OH

5′

RNase H cleaves the RNA in the RNA-DNA hybrid, leaving a short RNA primer with an exposed 3′-hydroxyl group.

3′ OH

5′

DNA polymerase I extends the primer several hundred nucleotides, forming a D-loop. The D-loop exposes the primosome assembly site (PAS).

DNA polymerase I

D-loop

5′

PAS

Replisome

5′

3′

5′

3′

A primosome assembles, and DNA polymerase III replicates the remaining DNA.

PAS

Figure 25·34
DNA replication by a D-loop mechanism.

Initiation begins with transcription from a site 500 base pairs upstream from the origin. Transcription causes unwinding of the replication origin, and the 5′ end of the RNA transcript forms a stable RNA-DNA hybrid. This hybrid is recognized by RNase H, a ubiquitous enzyme that cleaves RNA in RNA-DNA duplexes. In this case, the enzyme catalyzes an endonucleolytic cleavage at a specific site, giving rise to a short RNA primer with an exposed 3′-hydroxyl group.

This primer is extended by DNA polymerase I for several hundred nucleotides, forming a D-loop. The single-stranded region is coated with SSB. Formation of the D-loop eventually exposes a primosome assembly site (PAS) on the displaced strand. The PAS is adjacent to DnaA-binding sites, and the combined action of DnaA and the primosome components leads to the formation of an active replisome containing DNA polymerase III. The remainder of the plasmid is then replicated unidirectionally by the replisome.

The D-loop mechanism is also used in the replication of mitochondrial DNA in eukaryotic cells. Almost two-thirds of the mitochondrial DNA molecule is displaced before synthesis on the lagging strand begins.

C. Reverse Transcriptase Catalyzes DNA Synthesis Using an RNA Template

The genomes of retroviruses are RNA molecules that are copied into DNA during infection. The DNA can be immediately transcribed to produce RNA for production of more virus particles, or it can combine with host DNA, allowing the virus to remain latent for many generations.

One of the key enzymes required for conversion of retroviral RNA to DNA is a virally encoded reverse transcriptase. This enzyme is a member of the B family of DNA polymerases; it is homologous to eukaryotic DNA polymerases α and δ and to DNA polymerase II from *E. coli*. The structure of the human immunodeficiency virus I (HIV I) reverse transcriptase has been solved; it contains a large groove that accommodates the DNA-RNA hybrid.

The mechanism of retroviral DNA replication is complex. The first step requires formation of a circular RNA molecule that serves as the template for DNA synthesis. The primer for the synthesis reaction is a tRNA molecule that hybridizes to the retroviral RNA. Reverse transcriptases usually have RNase H activity as well as DNA polymerase activity. The RNA component of the DNA-RNA product is degraded by this activity, and the remaining single-stranded DNA is copied to produce a double-stranded DNA molecule.

Reverse transcriptases do not have associated 3′→5′ exonuclease activities or proofreading activities. They have the highest error rates of any DNA polymerases; the HIV I reverse transcriptase, for example, incorporates an incorrect base every 2000 to 4000 nucleotides. This accounts in part for the high mutation rate of HIV I.

Reverse transcriptases are not confined to retroviruses; uninfected cells also possess enzymes capable of copying RNA into DNA. These enzymes are responsible for the formation of processed pseudogenes and for replication of transposable elements (Chapter 32). The key role of reverse transcriptases during the retrovirus life cycle has prompted investigation of specific inhibitors that could prevent or retard infection. One such inhibitor is AZT (3′-azido-2′,3′-dideoxythymidine), which is used as an anti-HIV drug (Figure 25·35). This nucleoside is taken up by cells and phosphorylated; it is then incorporated into DNA, where it acts as a chain terminator much like the dideoxyribonucleotides used in DNA sequencing. Most DNA polymerases have a low affinity for the phosphorylated derivative, but for some unknown reason, HIV reverse transcriptase binds the drug very effectively. Thus, AZT triphosphate is an effective competitor of dTTP in HIV-infected cells.

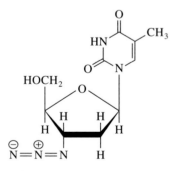

Figure 25·35
Structure of AZT
(3′-azido-2′,3′-dideoxythymidine).

25·11 Special Strategies Exist to Replicate the Ends of Linear Chromosomes

When a replication fork reaches the end of a linear chromosome, the replisome falls apart, and the daughter DNA molecules separate. Synthesis of the leading-strand template is complete because newly synthesized DNA is made continuously in the $5' \to 3'$ direction. On the lagging strand, however, there is a gap whose size depends on the location of the last Okazaki fragment (Figure 25·36). If left untouched during subsequent rounds of synthesis, this gap would become larger and larger, and some of the products of replication would be shorter than the original parental molecule. In addition, the single-stranded $3'$ extensions produced would eventually be degraded by exonucleases. Clearly, there must be a mechanism that prevents such chromosome shortening.

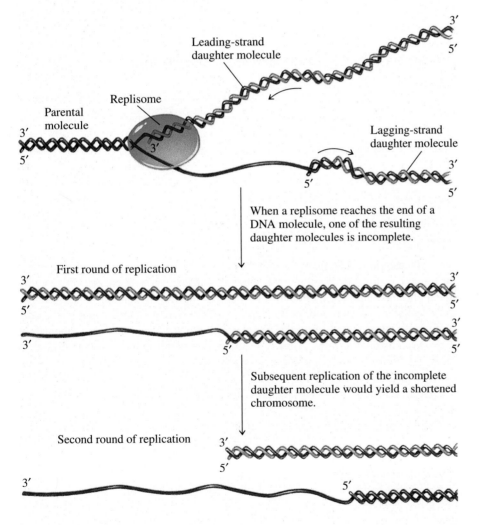

Figure 25·36
DNA replication at the ends of linear chromosomes. In the absence of a mechanism for maintaining the ends of linear chromosomes, one tip of each chromosome would get shorter each time a chromosome was replicated. (Only one end of the chromosome is shown here.)

Figure 25·37
DNA replication of a bacteriophage with terminal repeats.

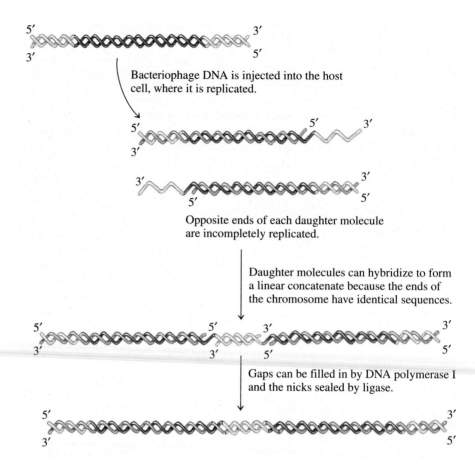

Bacteriophage DNA is injected into the host cell, where it is replicated.

Opposite ends of each daughter molecule are incompletely replicated.

Daughter molecules can hybridize to form a linear concatenate because the ends of the chromosome have identical sequences.

Gaps can be filled in by DNA polymerase I and the nicks sealed by ligase.

A. The Ends of Viral DNA Molecules Are Replicated by a Variety of Mechanisms

Several different mechanisms have evolved for coping with the replication of chromosomal termini. For example, many bacteriophages and eukaryotic viruses with linear chromosomes form circular intermediates during replication. These circular intermediates are then copied by a rolling-circle mechanism, which avoids the problem of unreplicated termini. Other bacteriophages have identical sequences at their chromosomal termini. After replication, the daughter molecules hybridize at their ends to form large, linear concatenates. The gaps at the junctions can be repaired by the combined actions of DNA polymerase I and ligase (Figure 25·37). The large concatenates are then cleaved into phage-sized DNA pieces during packaging.

Bacteriophage ϕ29 and the mammalian virus adenovirus use a different mechanism to ensure that the end of the lagging strand is completely copied. These viruses contain a protein called TP (for terminal protein). Some molecules of TP are covalently bound to the 5′ ends of viral DNA molecules through a phosphodiester linkage to the hydroxyl group of a serine residue (Figure 25·38). Modified versions of TP that contain a covalently bound nucleotide (dCMP in the case of adenovirus and dAMP in ϕ29) also exist in the cell. These modified terminal proteins associate with bound TP to form a complex at the ends of the DNA. The covalently bound nucleotide then pairs with the terminal nucleotide at the 3′ end of each strand. This nucleotide is now able to serve as a primer for replication. Virally encoded DNA polymerases bind specifically to TP and the primer nucleotide and copy the strands from their 3′ to 5′ ends (note that one polymerase copies each strand; there is no bidirectional replication in this case). Since priming occurs at the ends of the parental DNA, both strands are copied completely to produce two double-stranded daughter molecules. When DNA synthesis is finished, the second molecule of TP is

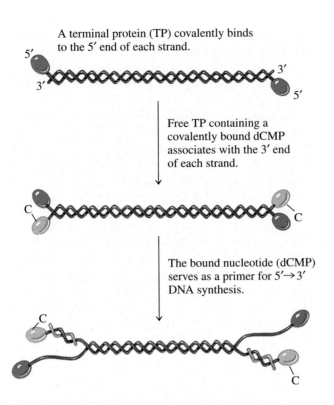

A terminal protein (TP) covalently binds to the 5' end of each strand.

Free TP containing a covalently bound dCMP associates with the 3' end of each strand.

The bound nucleotide (dCMP) serves as a primer for 5'→3' DNA synthesis.

Figure 25·38
Replication of adenovirus DNA.

released by cleavage of the bond between TP and the primer nucleotide. This strategy ensures that the primer is at the very end of the chromosome.

B. Replication of the Ends of Eukaryotic Chromosomes Requires a Sequence-Specific Telomerase

Eukaryotic chromosomes are linear, double-stranded DNA molecules. As is the case with bacteriophage chromosomes, a specific mechanism is required to replicate the ends of eukaryotic chromosomes and prevent the loss of genetic information. Part of the mechanism involves special DNA sequences located at the ends of the chromosomes. The average chromosome terminates in a stretch of up to 10 000 base pairs consisting of short, repetitive sequences. These terminal regions are known as **telomeres** (Figure 25·39). Depending on the species, the repeat can range

Figure 25·39
Telomeric sequences on human chromosomes. Telomeric sequences have been identified using in situ hybridization with a fluorescent probe that specifically detects the human telomeric repeat. (Courtesy of Judy Fantes and Howard Cooke.)

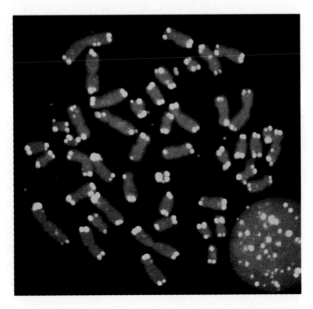

Table 25·4 Repetitive sequences at the ends of eukaryotic chromosomes

Species	Repeat
Tetrahymena (ciliate)	T T G G G G
Euplotes (ciliate)	T T T T G G G G
Paramecium (ciliate)	G G G $^G/_T$ T T
Homo sapiens (vertebrate)	T T A G G G
Saccharomyces cerevisiae (yeast)	[T G]$_{(1-3)}$

from 4–8 bp in length, but in all cases one strand is rich in T and G (Table 25·4). At the very tips of the chromosomes, there is a short stretch of single-stranded DNA ending in a 3′-hydroxyl group. This is the lagging strand that was incompletely replicated when the replication fork reached the end of the chromosome (see Figure 25·36).

Progressive shortening of the chromosome is prevented by telomerase, an enzyme that extends the exposed 3′ end of DNA. Telomerase contains a bound RNA molecule whose sequence is complementary to the telomere repeats. The enzyme is a reverse transcriptase that extends the 3′ end of DNA using the bound RNA as a template (Figure 25·40). Several rounds of addition result in extension of the 3′ end by about one hundred nucleotides, and this length of exposed single-stranded DNA can serve as a template for DNA polymerase α. Recall that DNA polymerase α has an associated primase activity, so it can fill in the opposite strand by making an RNA primer. The combined action of telomerase and DNA polymerase α counteracts the shortening of the chromosome during successive rounds of replication.

Despite the mechanism for protecting against chromosome shortening, there appears to be a gradual loss of DNA at the tips of human chromosomes in replicating somatic cells. Telomere shortening may contribute to cell senescence in mammals; this effect is presumably due to the loss of telomerase activity with age. In the

Figure 25·40
Action of telomerase. The enzyme uses a bound RNA molecule as a template to extend the ends of chromosomes. Once a sufficiently long single-stranded 3′ end has been synthesized, it is recognized by DNA polymerase α. This enzyme then replicates the 5′ end of the molecule using the single-stranded 3′ end as a template.

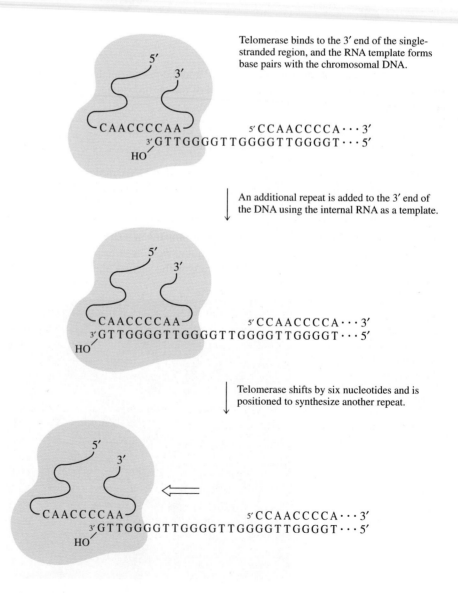

Telomerase binds to the 3′ end of the single-stranded region, and the RNA template forms base pairs with the chromosomal DNA.

An additional repeat is added to the 3′ end of the DNA using the internal RNA as a template.

Telomerase shifts by six nucleotides and is positioned to synthesize another repeat.

yeast *S. cerevisiae,* a gene known as *EST*1 (for ever shorter telomeres) is required for maintaining the ends of chromosomes. Mutations in this gene lead to a senescence phenotype and the progressive shortening of chromosomes. There are, of course, many other factors that contribute to senescence, and it is still unclear whether chromosome shortening is a cause or an effect of aging.

Summary

DNA replication is, in most cases, semiconservative. Each strand of the parental molecule serves as a template for the synthesis of a complementary daughter strand. After one round of replication, each DNA molecule contains one strand from the parent and one newly synthesized strand.

DNA replication in prokaryotes and eukaryotes occurs by similar mechanisms. In all organisms, DNA replication requires a complex protein machine, although the composition of the machine differs among species. In all organisms, replication is bidirectional and begins at particular sites called origins of replication. In *E. coli,* the origin, *ori*C, is recognized by DnaA, which binds to it and allows the other proteins involved in DNA replication to assemble into an active complex. The complex initiates replication by separating the strands of the helix through the concerted action of DNA topoisomerases, SSB, and DnaB, a helicase. The separated strands are then available to the primosome, which includes primase. DNA polymerase III then joins the complex to complete formation of the replication protein machine. Eukaryotic chromosomes contain multiple origins of replication.

Chain elongation is carried out by DNA polymerases. The polymerase responsible for DNA synthesis in *E. coli* is DNA polymerase III; in eukaryotes it is DNA polymerases α and δ. All DNA polymerases catalyze formation of a phosphodiester linkage between an activated deoxyribonucleoside 5′-triphosphate and a nascent DNA chain. In the replication complex, leading-strand DNA polymerases are processive: they do not dissociate from the DNA until the entire chromosome has been replicated. DNA polymerases require a free 3′-hydroxyl group for activity; all catalyze chain growth in the 5′→3′ direction. Most DNA polymerases are also associated with a 3′→5′ exonuclease activity, which recognizes the distortion caused by an improperly paired base and removes the mismatched nucleotide. This exonuclease activity proofreads newly synthesized DNA and greatly increases the fidelity of DNA replication.

Chain growth differs for the two strands of DNA at the replication fork. One new strand is synthesized continuously in a 5′→3′ direction; it is called the leading strand. The other new strand is synthesized discontinuously in the 5′→3′ direction; it is called the lagging strand. Discontinuous synthesis of the lagging strand produces oligonucleotides called Okazaki fragments. Each Okazaki fragment starts with an RNA primer, which is synthesized by primase. The RNA primer is removed by a 5′→3′ exonuclease, and the resulting gap is filled in by the action of a DNA polymerase. Once the gap is filled, the nick in the phosphodiester backbone is closed by DNA ligase.

Syntheses of the leading and lagging strands are closely coupled in vivo and are carried out simultaneously at the replication fork by the replisome. The replisome contains helicases to open the DNA, primase to synthesize primers on the lagging strand, and two DNA polymerases to lengthen the two new DNA chains. DNA polymerase III synthesizes both leading and lagging strands in *E. coli,* whereas two different polymerases are required in eukaryotes. Some of the polypeptides in the replisome exhibit enzymatic activities that can be detected when the individual polypeptides are isolated; others display no activity in isolation but function by stabilizing the complex or enhancing the activity of the other polypeptides.

Part **4** Biological Information Flow

DNA replication terminates when two replication forks moving in opposite directions meet. On circular bacterial chromosomes, this occurs opposite the origin in the *ter* region. Tus binds to sites in the *ter* region and prevents replication forks from passing through.

Not all DNA replication is semiconservative. Some bacteriophage DNA is synthesized by a rolling-circle mechanism, which is also used during conjugation in *E. coli* and related species. Mitochondrial DNA and some plasmid DNA molecules are replicated by the D-loop mechanism. The single-stranded RNA genomes of retroviruses are copied into DNA by reverse transcriptase, a special type of DNA polymerase.

Special mechanisms are required to synthesize the ends of linear DNA molecules. In some cases, this involves the formation of linear concatenates, while in other cases, specific proteins bind to the ends of DNA to ensure that entire chromosomes are replicated. Shortening of the ends of eukaryotic chromosomes is prevented by telomerases.

Selected Readings

General References

Adams, R. L. P., Knowler, J. T., and Leader, D. P. (1992). *The Biochemistry of the Nucleic Acids,* 11th ed. (New York: Chapman and Hall).

Echols, H., and Goodman, M. F. (1991). Fidelity mechanisms in DNA replication. *Annu. Rev. Biochem.* 60:477–511.

Kornberg, A., and Baker, T. (1992). *DNA Replication,* 2nd ed. (New York: W. H. Freeman and Company).

DNA Polymerases

Beese, L. S., Derbyshire, V., and Steitz, T. A. (1993). Structure of DNA polymerase I Klenow fragment bound to duplex DNA. *Science* 260:352–355.

Braithwaite, D. K., and Ito, J. (1993). Compilation, alignment, and phylogenetic relationships of DNA polymerases. *Nucleic Acids Res.* 21:787–802.

Downey, K. M., Tan, C.-K., and So, A. G. (1990). DNA polymerase delta: a second eukaryotic DNA replicase. *BioEssays* 12:231–236.

Freemont, P. S., Friedman, J. M., Beese, L. S., Sanderson, M. R., and Steitz, T. A. (1988). Cocrystal structure of an editing complex of Klenow fragment with DNA. *Proc. Natl. Acad. Sci. USA* 85:8924–8928.

Johnson, K. A. (1993). Conformational coupling in DNA polymerase fidelity. *Annu. Rev. Biochem.* 62:685–713.

Kong, X.-P., Onrust, R., O'Donnell, M., and Kuriyan, J. (1992). Three-dimensional structure of the β subunit of *E. coli* DNA polymerase III holoenzyme: a sliding DNA clamp. *Cell* 69:425–437.

Kunkel, T. A. (1992). DNA replication fidelity. *J. Biol. Chem.* 267:18 251–18 254.

Maki, S., and Kornberg, A. (1988). DNA polymerase III holoenzyme of *Escherichia coli:* III. Distinctive processive polymerases reconstituted from purified subunits. *J. Biol. Chem.* 263:6561–6569.

McHenry, C. S. (1991). DNA polymerase III holoenzyme. *J. Biol. Chem.* 266:19 127–19 130.

O'Donnell, M. (1992). Accessory protein function in the DNA polymerase III holoenzyme from *E. coli. BioEssays* 14:105–111.

Wang, T. S.-F. (1991). Eukaryotic DNA polymerases. *Annu. Rev. Biochem.* 60:513–552.

Prokaryotic DNA Replication

Alberts, B. M. (1984). The DNA enzymology of protein machines. *Cold Spring Harbor Symp. Quant. Biol.* 49:1–12.

Alberts, B. M. (1985). Protein machines mediate the basic genetic processes. *Trends Genet.* 1:26–30.

Baker, T. A., Funnell, B. E., and Kornberg, A. (1987). Helicase action of DnaB protein during replication from the *Escherichia coli* chromosomal origin in vitro. *J. Biol. Chem.* 262:6877–6885.

Baker, T. A., and Wickner, S. H. (1992). Genetics and enzymology of DNA replication in *Escherichia coli. Annu. Rev. Genet.* 26:447–477.

Bramhill, D., and Kornberg, A. (1988). Duplex opening by DnaA protein at novel sequences in initiation of replication at the origin of the *E. coli* chromosome. *Cell* 52:743–755.

Bramhill, D., and Kornberg, A. (1988). A model for initiation at origins of DNA replication. *Cell* 54:915–918.

Hidaka, M., Kobayashi, T., and Horiuchi, T. (1991). A newly identified DNA replication terminus site, *ter*E, on the *Escherichia coli* chromosome. *J. Bacteriol.* 173:391–393.

Hiraga, S. (1992). Chromosome and plasmid partition in *Escherichia coli. Annu. Rev. Biochem.* 61:283–306.

Kuempel, P. L., Pelletier, A. J., and Hill, T. M. (1989). Tus and the terminators: the arrest of replication in prokaryotes. *Cell* 59:581–583.

Marians, K. J. (1992). Prokaryotic DNA replication. *Annu. Rev. Biochem.* 61:673–719.

McMacken, R., Silver, L., and Georgopoulos, C. (1987). DNA replication. In Escherichia coli *and* Salmonella typhimurium: *Cellular and Molecular Biology,* F. C. Neidhardt, ed. (Washington, DC: American Society for Microbiology), pp. 564–612.

von Meyenburg, K., and Hansen, F. G. (1987). Regulation of chromosome replication. In Escherichia coli *and* Salmonella typhimurium: *Cellular and Molecular Biology,* F. C. Neidhardt, ed. (Washington, DC: American Society for Microbiology), pp. 1555–1577.

Eukaryotic DNA Replication

Adachi, Y., and Laemmli, U. K. (1992). Identification of nuclear pre-replication centers poised for DNA synthesis in *Xenopus* egg extracts: immunolocalization study of replication protein A. *J. Cell Biol.* 119:1–15.

Benbow, R. M., Zhao, J., and Larson, D. (1992). On the nature of origins of DNA replication in eukaryotes. *BioEssays* 14:661–670.

Cook, P. R. (1991). The nucleoskeleton and the topology of replication. *Cell* 66:627–635.

dePamphilis, M. L. (1993). Origins of DNA replication that function in eukaryotic cells. *Curr. Opin. Cell Biol.* 5:434–441.

dePamphilis, M. L. (1993). Eukaryotic DNA replication: anatomy of an origin. *Annu. Rev. Biochem.* 62:29–63.

Hamlin, J. L. (1992). Mammalian origins of replication. *BioEssays* 14:651–659.

Heintz, N. H., Dailey, L., Held, P., and Heintz, N. (1992). Eukaryotic replication origins as promoters of bidirectional DNA synthesis. *Trends Genet.* 8:376–381.

Huberman, J. A. (1992). Quest's end and questions' beginning. *Curr. Biol.* 2:351–352.

Kunkel, T. A. (1992). Biological asymmetries and the fidelity of eukaryotic DNA replication. *BioEssays* 14:303–308.

Linn, S. (1991). How many pols does it take to replicate nuclear DNA? *Cell* 66:185–187.

Natale, D. A., Umek, R. M., and Kowalski, D. (1993). Ease of DNA unwinding is a conserved property of yeast replication origins. *Nucleic Acids Res.* 21:555–560.

O'Donnell, M., Onrust, R., Dean, F. B., Chen, M., and Hurwitz, J. (1993). Homology in accessory proteins of replicative polymerases— *E. coli* to humans. *Nucleic Acids Res.* 21:1–3.

Podust, V. N., and Hübscher, U. (1993). Lagging strand DNA synthesis by calf thymus DNA polymerases α, β, δ, and ε in the presence of auxiliary proteins. *Nucleic Acids Res.* 21:841–846.

So, A. G., and Downey, K. M. (1992). Eukaryotic DNA replication. *Crit. Rev. Biochem. Mol. Biol.* 27:129–155.

Telomeres

Blackburn, E. H. (1991). Structure and function of telomeres. *Nature* 350:569–573.

Blackburn, E. H. (1992). Telomerases. *Annu. Rev. Biochem.* 61:113–129.

Boeke, J. D. (1990). Reverse transcriptase, the end of the chromosome, and the end of life. *Cell* 61:193–195.

Lundblad, V., and Szostak, J. W. (1989). A mutant with a defect in telomere elongation leads to senescence in yeast. *Cell* 57:633–643.

26

DNA Repair and Recombination

DNA is the only cellular macromolecule that can be repaired in vivo. This is probably because the cost of damaged DNA is severe, whereas the cost of damage to other macromolecules is slight. For example, if a protein is damaged, little is lost: the protein will eventually be replaced by a new, functional protein (recall the discussion of protein turnover in Section 21·8). If DNA is damaged, on the other hand, the entire organism may be in jeopardy because the instructions for the synthesis of an essential molecule may be permanently altered. In bacteria, unrepaired damage to a gene encoding an essential protein is lethal. Even in multicellular organisms, the accumulation over time of defects in DNA can lead to progressive derangement of cellular functions. DNA repair mechanisms protect not only individual cells but also subsequent generations.

It is important to make the distinction between DNA damage and mutations. As we will see, DNA damage takes a variety of forms, including base modifications, nucleotide deletions and insertions, cross-linking of DNA strands, and breakage of the phosphodiester backbone. While severe DNA damage may be lethal, much of the DNA damage that occurs in vivo is repaired. Only damage that escapes repair becomes a mutation.

Mutations are heritable changes to the sequence of nucleotides in DNA. They are thus permanent alterations of genetic information. Many mutations are due to DNA damage that was not repaired, although mutations can also arise from normal cellular reactions (including repair itself, as we will discuss).

In single-celled organisms, whether prokaryotes or eukaryotes, mutations are passed on directly to the daughter cells following DNA replication and cell division. However, in multicellular organisms, mutations can be passed on to the next generation only if they occur in the germ line (i.e., gamete-producing cells). Germ-line mutations may have no noticeable effect on the organism that contains them but may have profound effects on the progeny, especially if the mutated genes play a role in development and differentiation.

Some forms of DNA repair involve **recombination.** Recombination can be loosely defined as the exchange of pieces of DNA on different molecules or within the same molecule. In addition to its role in repair, recombination has great significance for evolution. Because the information content of chromosomes within a species differs as a result of mutation, recombination is a mechanism that creates new combinations of mutant and wild-type genes on a single chromosome—something like shuffling a deck of cards to get a new hand.

In this chapter, we discuss the common kinds of DNA damage and how they are repaired. We also discuss the biochemistry of recombination. As we proceed, you will notice that many of the enzymes involved in both repair and recombination are the same as those we encountered in the previous chapter.

26·1 Some DNA Damage Is Due to DNA Replication Errors

Occasionally, an incorrect nucleotide is incorporated into DNA during replication. The frequency of misincorporation is very low, but because DNA molecules are so large, this is nonetheless an appreciable source of DNA damage. The relatively high probability of replication errors in organisms with large genomes is part of the **genetic load** that these organisms must tolerate. Genetic load refers to the overall risk of mutation due to all kinds of DNA damage. The high genetic load of mammals, which have large genomes, seems inconsistent with their survival and is one of the reasons why a large fraction of the mammalian genome is assumed to be nonfunctional in terms of providing genetic information. If much of mammalian DNA is "junk," most replication errors will have no effect on the organism.

As we mentioned in the previous chapter, the low overall error rate of DNA replication is due to the combined effects of the fidelity of the polymerization reaction itself and the accuracy of proofreading. The overall error rate appears to involve a trade-off between accuracy and speed, suggesting that there is a cost associated with increased fidelity of replication. We know that there is a cost associated with mutation, especially harmful mutation; these costs must be balanced in living organisms.

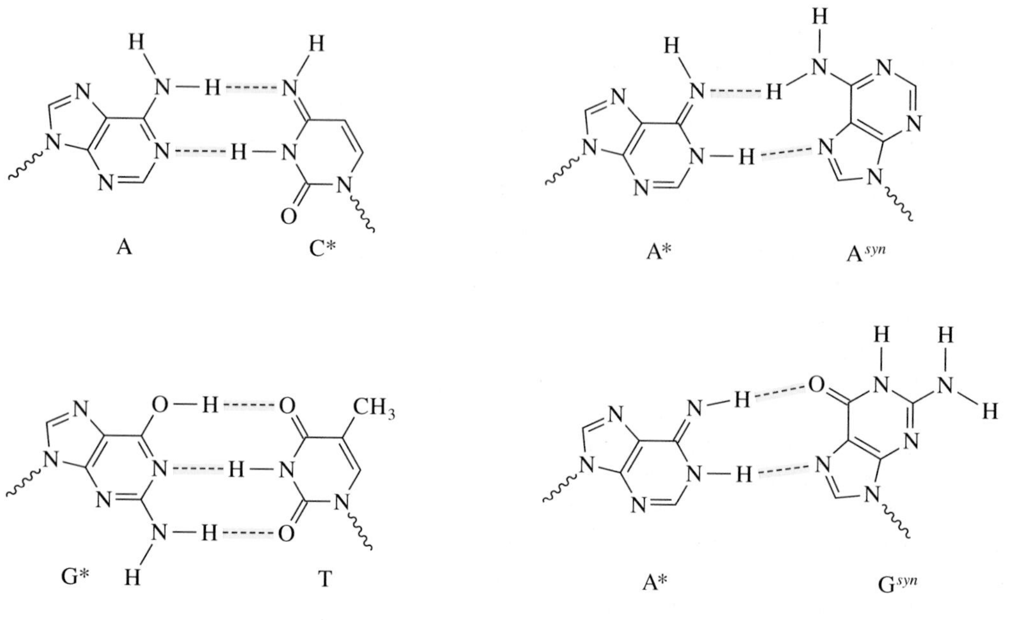

Figure 26·1
Base pairs involving unusual tautomers of bases. The asterisk indicates the rare tautomer; *syn* refers to the conformation of the nucleotide.

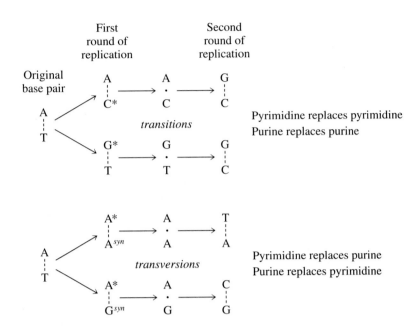

Figure 26·2
Pathways to nucleotide substitutions. Tautomerization of bases can lead to formation of unusual base pairs and the misincorporation of nucleotides during DNA replication. If the damage is not repaired, a second round of replication leads to a transition or transversion in one of the new strands.

Mismatches during DNA replication can be caused by the formation of base pairs between nucleotides and rare tautomers of other nucleotides. Figure 26·1 shows some examples of pairing between unusual tautomers of bases; other mismatches, including pyrimidine/pyrimidine base pairs, can also occur. Such mismatches are frequently repaired before the next round of replication. However, if they are not repaired, replication leads to fixation of the error, resulting in a permanent base substitution (a mutation).

Base substitutions can be transitions or transversions (Figure 26·2). **Transitions** occur when a purine is substituted by the other purine, or when a pyrimidine is substituted by the other pyrimidine. **Transversions** occur when a purine is substituted by a pyrimidine, or vice versa.

26·2 A Variety of Agents Cause DNA Damage

Many different agents, including chemicals and radiation, can damage DNA. The primary targets of such agents are the bases because they contain many reactive groups. However, the sugars and the phosphodiester linkages can also be attacked. Agents that cause DNA damage are often called **mutagens.**

A. Some Bases Can Be Deaminated

A common type of DNA damage is hydrolytic deamination of adenine, cytosine, or guanine (Figure 26·3, next page). (Because thymine does not have an amino group, it cannot be deaminated.) Deaminated bases form incorrect base pairs and, if not removed, result in the incorporation of incorrect nucleotides during the next round of DNA replication. Modified bases that are found in DNA can also be deaminated. Deamination of one of these, methylcytosine, is shown in Figure 26·3.

Whereas deamination of adenine and guanine is rare, deamination of cytosine and methylcytosine is common. Deamination of cytosine, which produces uracil, would give rise to large numbers of mutations if uracil rather than thymine were found in DNA as it is in RNA. In these cases, it would be impossible to distinguish between a correct uracil and a uracil arising from deamination of cytosine. However, uracil is not one of the bases normally found in DNA, and damage arising from the deamination of cytosine can be recognized and repaired. Thus, the presence of thymine in DNA increases the stability of genetic information.

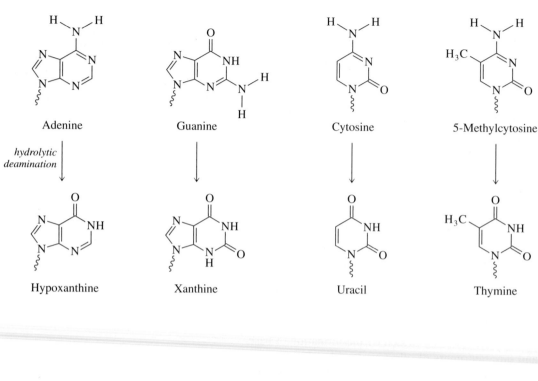

Figure 26·3
Hydrolytic deamination of DNA bases. Hydrolytic deamination produces other nucleotide bases, some of which form incorrect base pairs and lead to a change in the information content of DNA. Deamination of adenine produces hypoxanthine, which pairs with cytosine rather than thymine. Deamination of guanine produces xanthine, which still pairs with cytosine, but less well than guanine does. Deamination of cytosine produces uracil, which pairs with adenine rather than guanine. Deamination of 5-methylcytosine produces thymine, which pairs with adenine.

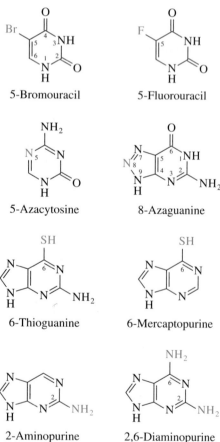

Figure 26·4
Common base analogs. Substituted atoms and functional groups are highlighted in blue.

Even though uracil can be recognized and repaired, deamination of cytosine is still a common cause of spontaneous mutation. Spontaneous mutation due to deamination is especially common in organisms that contain methylcytosine in their DNA since deamination of methylcytosine produces thymine. Thymine arising from deamination cannot be distinguished from that present naturally.

Not all deaminations are spontaneous. Nitrous acid (HNO_2) is a potent mutagen that oxidizes the amino groups of nucleotide bases. In the presence of nitrous acid, cytosine can be converted to uracil, adenine to hypoxanthine, guanine to xanthine, and methylcytosine to thymine. Nitrous acid can be produced from nitrates and nitrites, which are common in nature.

B. Base Analogs Can Be Incorporated into DNA During Replication

There are many base analogs that can be incorporated into DNA by the replication machinery. Some of these analogs occur naturally within the cell as by-products of nucleotide metabolism or because of spontaneous chemical alterations of the nucleotide precursors used in DNA replication. Other base analogs are taken up by the cell from the environment. In the intestines of mammals, for example, base analogs can be formed during digestion and absorbed by the cells lining the intestine or by the many species of bacteria that inhabit the gut. Base analogs imported into a cell are converted to nucleoside triphosphates via the salvage pathways (Section 22·5) and can then be incorporated into DNA. Thus, although we refer here to base analogs, the compounds that are incorporated during DNA replication are actually nucleotide analogs.

Most of the base analogs that are biochemically important are inhibitors of DNA replication or mutagens. Some common base analogs are shown in Figure 26·4. 5-Fluorouracil is a potent drug used to treat cancer. The most important effect of 5-fluorouracil is blockage of thymidylate synthesis by inhibition of thymidylate synthase (Figure 22·20), but the analog is occasionally converted to a nucleoside triphosphate and incorporated into DNA in place of thymidylate. Uracil is also a base analog of thymine; it too can be incorporated into DNA. (Note, however, that whereas deamination of cytosine to uracil results in a U/G base pair, misincorporation of uracil generates a U/A base pair, which probably does not lead to mutation.)

Although it resembles 5-fluorouracil, 5-bromouracil does not inhibit thymidylate synthase. However, it is readily converted to a nucleoside triphosphate that substitutes for thymidylate at high frequency during DNA replication in all species. This analog is often used as a mutagen; 5-bromouracil normally pairs with A in DNA, but the enol form of 5-bromouracil is more stable than that of the other pyrimidines, and consequently it is often incorporated opposite G, leading to a transition (A/T to G/C). In addition to its use as a mutagen, 5-bromouracil is employed in laboratories to label newly synthesized DNA.

There are a number of other base analogs. 8-Azaguanine is an analog of guanine. It is incorporated rarely into DNA but readily into RNA. 2,6-Diaminopurine (2-aminoadenine) is an analog of adenine. When it is incorporated, 2,6-diaminopurine pairs with thymine to form a base pair with three hydrogen bonds. This is not necessarily detrimental; in some bacteriophages (such as S-2L, which infects cyanobacteria), all adenines in the phage DNA are replaced by 2,6-diaminopurine. The mutagenic effect of 2,6-diaminopurine is probably due to its occasional mispairing with cytosine.

2-Aminopurine is mutagenic. Its nucleoside-triphosphate derivative is incorporated during DNA replication, when it pairs with either C or T. If not repaired, this leads to A/T→G/C and G/C→A/T transitions. 6-Mercaptopurine and 6-thioguanine are analogs that resemble hypoxanthine and guanine, respectively (the oxygens of the normal bases are replaced by sulfhydryl groups). Both of these analogs are incorporated into DNA after conversion to nucleoside triphosphates. Both are effective therapeutic agents because they inhibit purine synthesis and also critically damage DNA, which selectively kills rapidly growing cells such as cancer cells.

C. Some Alkylating Agents Can Damage DNA

Alkylating agents are molecules that contain a reactive alkyl group (usually a methyl group). Some alkylating agents can be formed inside the cell as by-products of normal metabolism. For example, nitrous acid can lead to the nitrosation of methylurea and methylnitrosoguanidine, giving rise to N-methyl-N-nitrosourea and N-methyl-N'-nitro-N-nitrosoguanidine (MNNG), respectively. The structures of these and two other alkylating agents are shown in Figure 26·5.

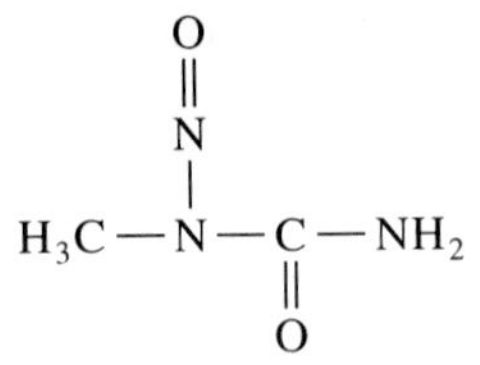

Figure 26·5
Common alkylating agents.

N-Methyl-N-nitrosourea

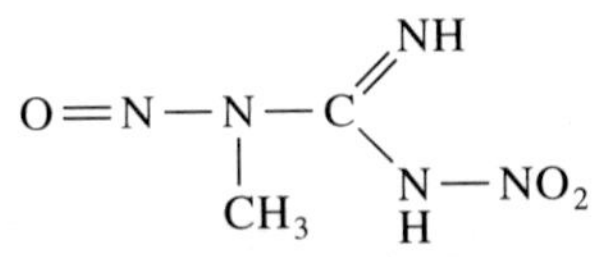

N-Methyl-N'-nitro-N-nitrosoguanidine
(MNNG)

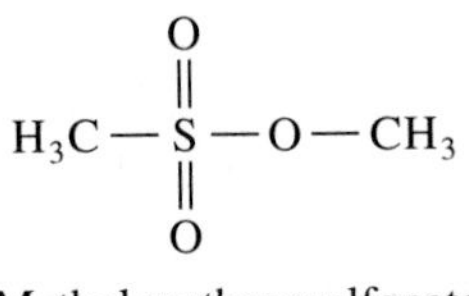

Methyl methanesulfonate
(MMS)

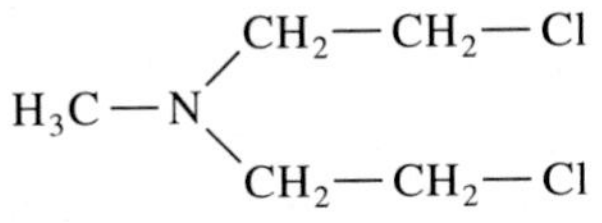

Methyl-bis(β-chloroethyl)amine
(Nitrogen mustard)

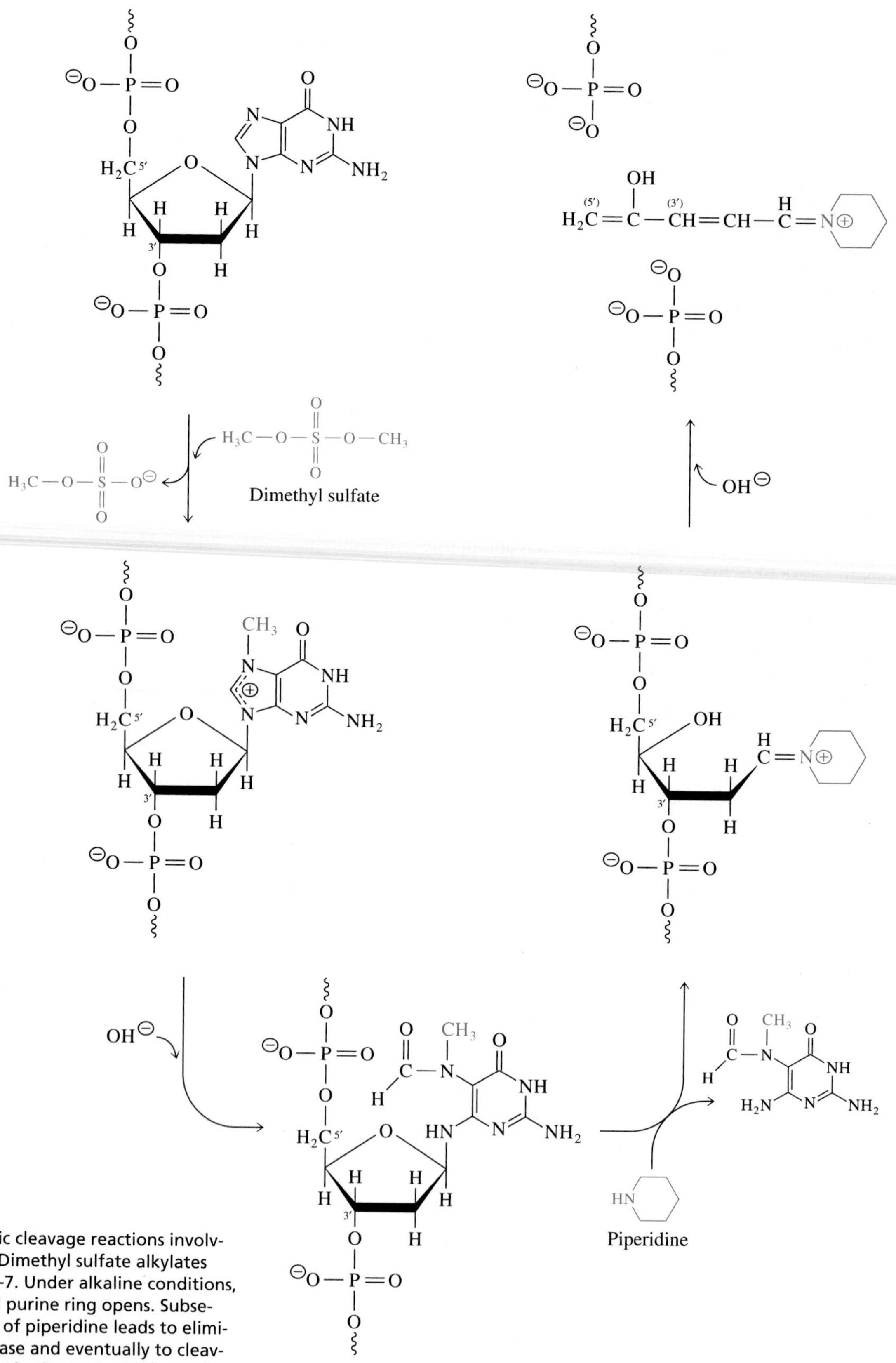

Figure 26·6
Guanine-specific cleavage reactions involving alkylation. Dimethyl sulfate alkylates G residues at N-7. Under alkaline conditions, the methylated purine ring opens. Subsequent addition of piperidine leads to elimination of the base and eventually to cleavage of the phosphodiester backbone. This reaction is used in vitro to cleave DNA specifically at G residues.

Alkylating agents react with nucleophilic atoms such as N-7 of guanine and N-3 of adenine. In some cases, the resulting modified bases in DNA can lead to mispairing, while in other cases the modified bases disrupt the structure of DNA and block DNA replication. Often the alkylated bases contain an unstable quaternary amino group, which can lead to spontaneous cleavage of the β-*N*-glycosidic bond and release of the base from DNA. For example, dimethyl sulfate alkylates N-7 of guanine, leading to excision of the base. This is one of the reactions used in vitro to cleave DNA specifically at G residues. In this procedure, piperidine is used to enhance cleavage of the glycosidic bond, and cleavage of both phosphoester linkages occurs at high pH (Figure 26·6).

D. Intercalating Agents Can Cause Insertions and Deletions

There are a number of complex chemicals that can fit between the stacked base pairs of DNA, thus causing DNA damage. These **intercalating agents** have a planar ring structure that mimics the structure of a base pair. Ethidium, daunomycin, and acridine dyes such as acridine orange and proflavin are examples of synthetic drugs that are effective intercalating agents. Benzopyrene, which is found in soot, is an example of a naturally occurring chemical that can act as an intercalating agent (Figure 26·7).

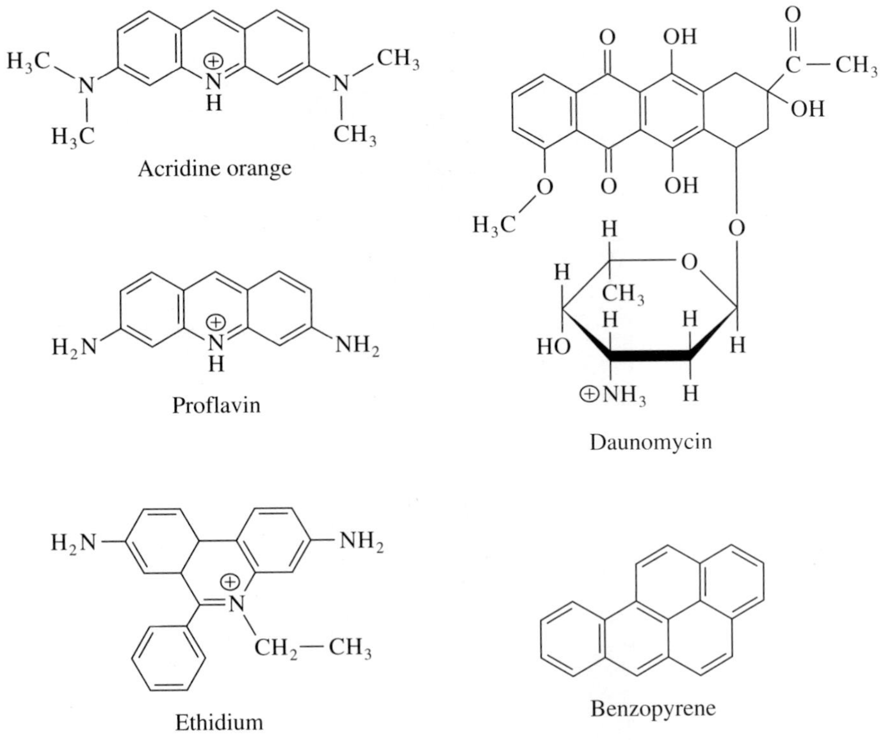

Figure 26·7
Common intercalating agents.

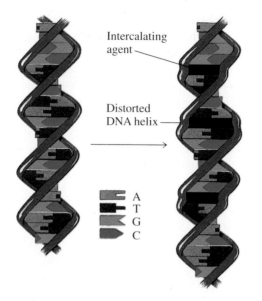

Figure 26·8
Effect of intercalation. Intercalating agents (dark bands) distort the DNA helix, partially unwinding it.

Intercalating agent

Distorted DNA helix

A
T
G
C

Intercalating agents do not disrupt base pairing, but they do interfere with stacking. They distort the double helix by separating adjacent base pairs, unwinding the DNA slightly (Figure 26·8). Linear DNA molecules can bind large amounts of intercalating agents; covalently closed, circular DNA molecules bind much less because the helix cannot unwind easily to accommodate the intercalating molecules. This difference is exploited in the laboratory to separate circular plasmid DNA from linear chromosomal DNA. Insertion of an intercalating agent into DNA reduces the

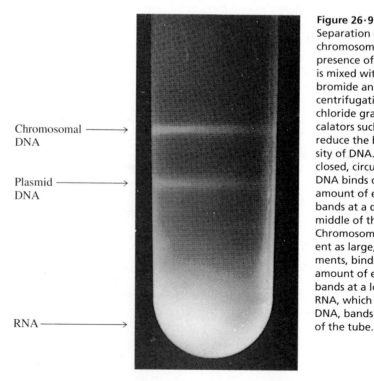

Chromosomal DNA

Plasmid DNA

RNA

Figure 26·9
Separation of plasmid and chromosomal DNA in the presence of ethidium. DNA is mixed with ethidium bromide and subjected to centrifugation in a cesium-chloride gradient. Intercalators such as ethidium reduce the buoyant density of DNA. Covalently closed, circular plasmid DNA binds only a small amount of ethidium and bands at a density near the middle of the gradient. Chromosomal DNA, present as large, linear fragments, binds a large amount of ethidium and bands at a lower density. RNA, which is denser than DNA, bands at the bottom of the tube.

Figure 26·10
Structure of a daunomycin-DNA complex. **(a)** End view. **(b)** Side view. Note that the ring complex of the drug (shaded) penetrates the stacked base pairs and causes unwinding. The keto and hydroxyl groups at the end of the planar ring complex form hydrogen bonds with groups on cytosine and guanine. *W* represents water. (Courtesy of A. H.-J. Wang.)

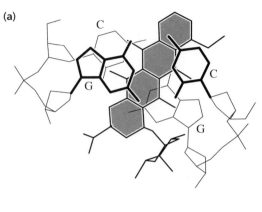

(a)

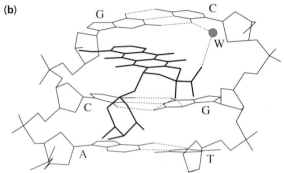

(b)

buoyant density of the DNA. Thus, linear DNA that binds large amounts of an intercalating agent can be separated from covalently closed, circular DNA by equilibrium density gradient centrifugation. Ethidium bromide is often used for such separations because, in addition to being an intercalating agent, it fluoresces under ultraviolet light (Figure 26·9).

The structures of several DNA-intercalator complexes have been examined by X-ray crystallography. In many cases, the chemicals show some preference for particular DNA sequences. The interaction between the intercalating agent and the DNA is often stabilized by stacking interactions as well as hydrogen bonds between the agent and functional groups on the bases or the sugar-phosphate backbone. The structure of a daunomycin-DNA complex is shown in Figure 26·10.

The presence of intercalating agents in DNA usually inhibits both replication and transcription. These effects have been exploited by a number of organisms that synthesize and excrete intercalating agents that act as antibiotics. Actinomycin D, the aflatoxins, and echinomycin are examples of naturally occurring antibiotics that are intercalating agents (Figure 26·11, next page). Actinomycin D is secreted by the bacterium *Streptomyces chrysomallus* and is widely used in the laboratory as an inhibitor of transcription. Aflatoxins are made by species of *Aspergillus,* a mold. In addition to acting as intercalating agents, the aflatoxins also bind covalently to DNA. They are potent carcinogens in mammals. Echinomycin is produced by *Streptomyces* species and is effective as an antibacterial drug and as a chemotherapeutic agent in the treatment of cancer. Echinomycin is an example of a *bis*-intercalator; its has two planar rings that can intercalate at sites on the same DNA molecule that are separated by one turn of the helix.

Occasionally, a replication fork passes over a region of DNA that contains an intercalating agent. This results in small insertions or deletions in the newly synthesized strand as the replication complex attempts to cope with the presence of a large molecule between the stacked bases. In this way, intercalating agents also cause DNA damage.

(a) Actinomycin D

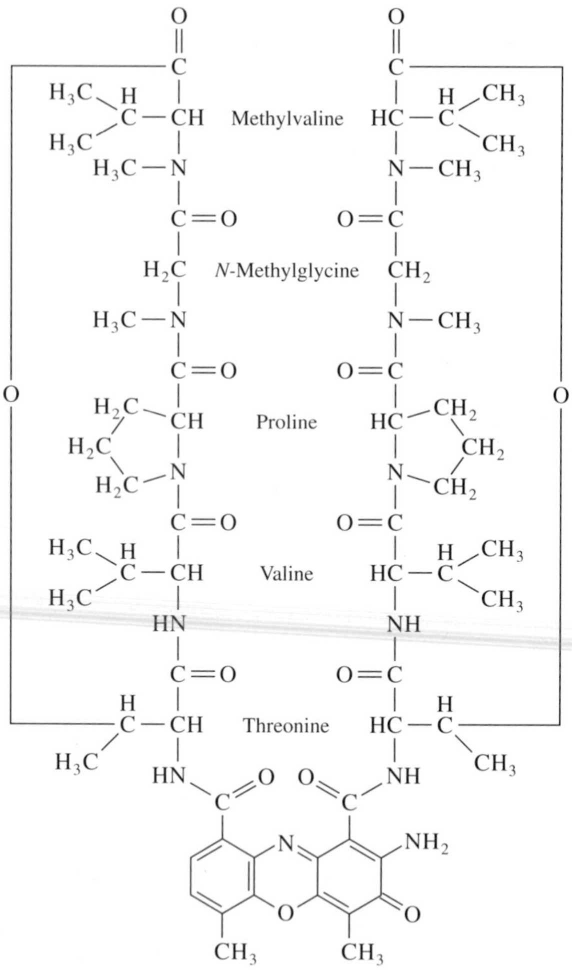

(b) Aflatoxin B1

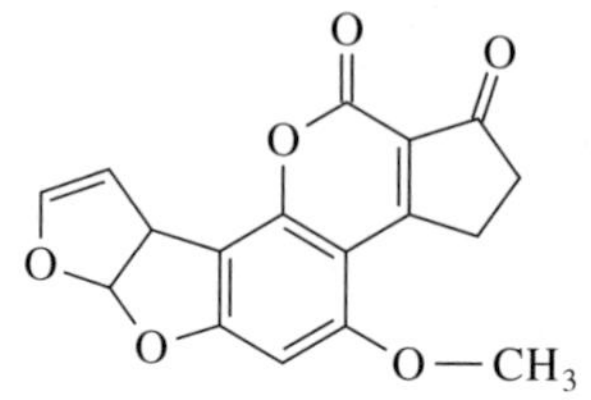

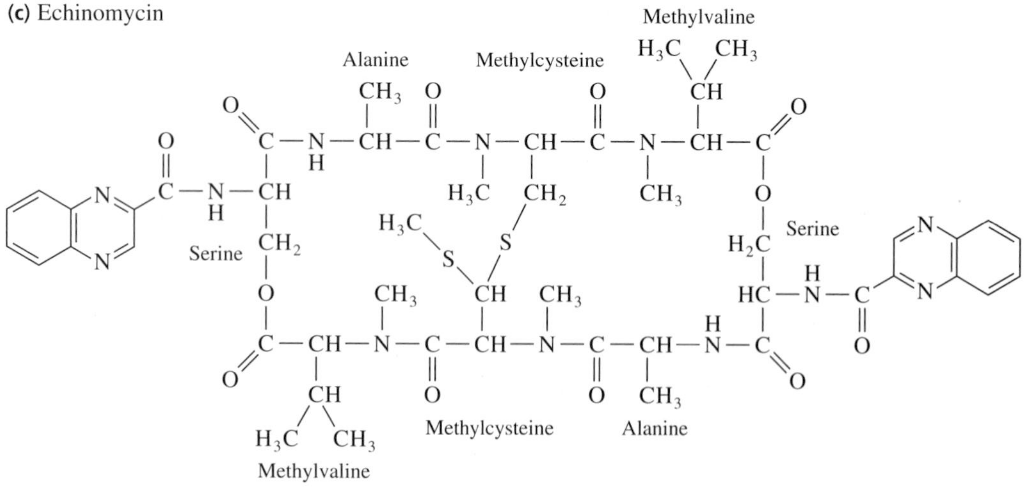

Figure 26·11
Intercalating agents that act as antibiotics. **(a)** Actinomycin D. *N*-Methylglycine is also known as sarcosine. **(b)** Aflatoxin B1. **(c)** Echinomycin. In three dimensions, the two planar rings of echinomycin are positioned to intercalate at two sites on DNA separated by one turn of the helix.

(c) Echinomycin

E. Radiation Damages DNA

X rays, gamma rays, and ultraviolet light can damage DNA. X rays are sometimes used to treat cancer because the damage X rays cause overwhelms the repair capacity of rapidly growing cells and leads to their death. Unfortunately, not only cancer cells grow rapidly; hair follicle cells, bone marrow cells, and the cells lining the intestine also rapidly divide and are also killed by systemic X-ray therapy. In addition, mutations caused by X rays can themselves cause cancer.

The major damage caused by ionizing radiation is the introduction of single-stranded and double-stranded breaks in DNA. Strand breakage is due to a reaction between hydroxyl radicals (formed from water as a result of ionization) and one of the carbons in deoxyribose. The hydroxyl radical abstracts a hydrogen from C-4′, which then reacts with molecular oxygen to form an unstable intermediate. The intermediate spontaneously collapses to a more stable molecule, causing strand breakage. In the absence of DNA replication, single-stranded breaks are tolerated by most cells. During replication, however, unwinding of DNA can result in disruption of the replication fork wherever a break is encountered. Double-stranded breaks lead to fragmentation of chromosomes and thus can be lethal to both growing and resting cells.

Irradiation of DNA can also produce dozens of different modified bases. Many of the products of irradiation involve rearrangement or cleavage of the heterocyclic rings of both pyrimidines and purines. For example, gamma radiation can produce radicals that participate in breakage of the imidazole ring of purines and formation of a formamidopyrimidine ring (Figure 26·12).

The effect of ultraviolet light on DNA is similar to that of X rays except that UV light is less penetrating. The mutagenic effect of ultraviolet light on humans is most often manifested as skin cancer, caused by prolonged exposure of the skin to direct sunlight. Ultraviolet light causes dimerization of stacked pyrimidines. This process is an example of photodimerization and most commonly occurs between adjacent thymines (Figure 26·13), although thymine-cytosine and cytosine-cytosine dimers can also be formed. DNA synthesis is inhibited in the presence of such dimers, probably because they distort the template molecule. Therefore, removal of pyrimidine dimers is important for survival.

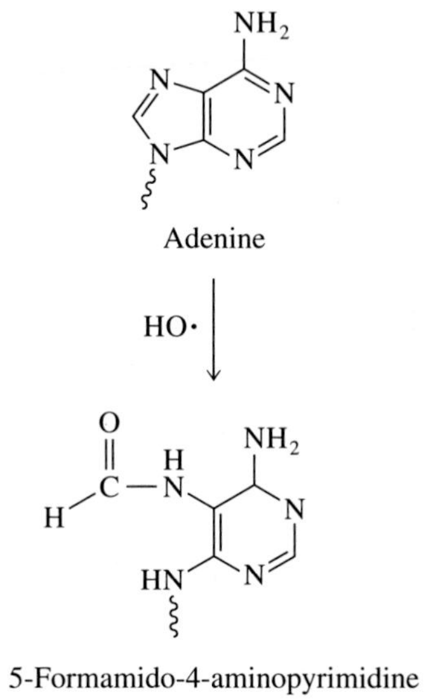

Adenine

HO·

5-Formamido-4-aminopyrimidine

Figure 26·12
Formation of a formamidopyrimidine. Hydroxyl radicals produced by gamma radiation catalyze cleavage of the purine ring.

Figure 26·13
Photodimerization of adjacent thymidylate residues in DNA. Ultraviolet light causes adjacent thymidylate residues in DNA to dimerize, thereby distorting the structure of DNA. In double-stranded DNA, the thymidylate residues are base paired with adenylate residues on the opposite strand. For simplicity, only a single strand is shown.

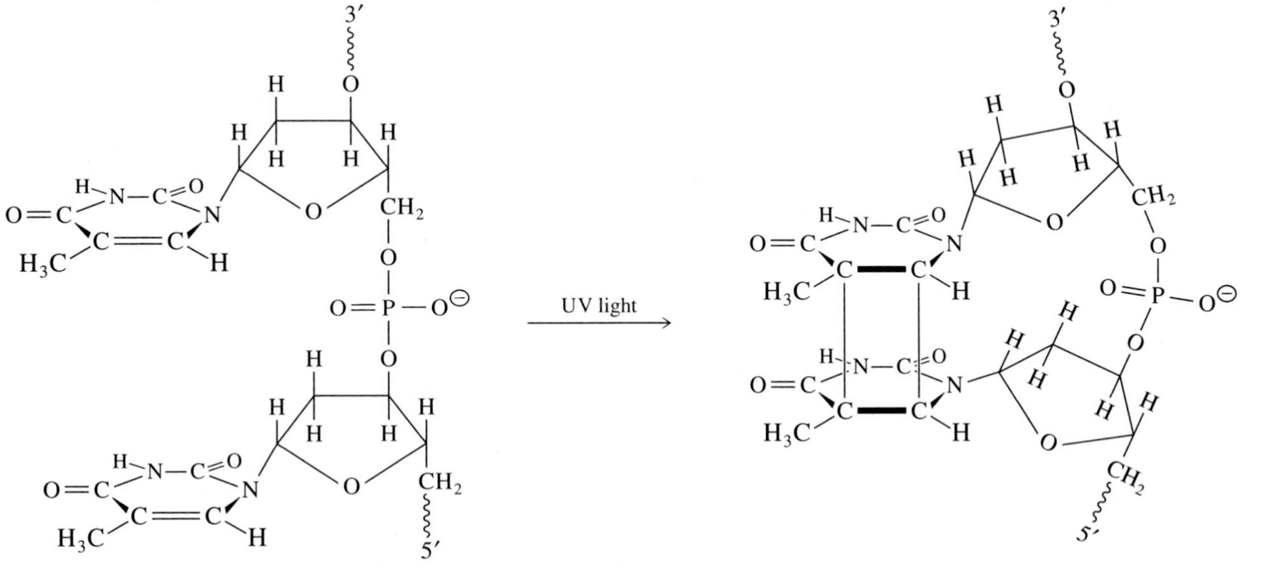

UV light

26·3 Mutagens and Cancer

For decades, geneticists have been using radiation and chemicals to generate mutants. X rays, ultraviolet light, and nitrous acid are favored mutagens. Much of what we know about biochemistry comes from studies that have identified the function of a protein by examining the effect of mutating its gene. The mechanisms of action of many enzymes have been elucidated from a knowledge of the activity, or lack of activity, of mutant proteins.

But mutagens abound in nature, and many mutations occur outside the laboratory. The average rate of mutation in bacteria and simple eukaryotes is about 0.003 mutations per genome per replication. Rates of mutation reflect errors that arise during DNA replication as well as those that are due to other forms of DNA damage. The amount of damage sustained by DNA per generation is not known, but it is known that most damage is repaired before it becomes a mutation.

Many kinds of mutations have been detected in natural populations. Substitutions—either transversions or transitions—are common, and deletions and insertions are also observed. Naturally occurring mutations in the human gene for hypoxanthine-guanine phosphoribosyltransferase (the *hgprt* gene), for example, have been extensively analyzed by many workers, and a large data base of observed mutations has been compiled. The enzyme is required for the purine salvage pathway, but the gene is nonessential in cells in culture, and most mammals show no effects when the gene is deleted. Because this gene is not essential, it is ideal for studies of the spontaneous mutation rate and the effect of mutagens such as cigarette smoke and the drugs used to treat cancer.

More than 1000 mutations in human *hgprt* have been characterized. About half of them are single base-pair substitutions, with transitions and transversions occurring in about equal numbers. About 15% of these mutations are not inherited, and many of these are detected in individuals who have undergone extensive chemotherapy or who have been exposed to harmful chemicals. Based on these and similar data, it appears that environmental mutagens can contribute significantly to mutation in humans as they do in other organisms.

It is important to identify potential mutagens in the environment in order to make informed and intelligent decisions about their possible risk to the ecosystem. To this end, tests have been developed for detecting mutagens. The **Ames test** uses strains of *Salmonella typhimurium,* a relative of *Escherichia coli,* to measure the frequency of mutagenesis caused by various compounds. The *Salmonella* strains used in this test are designated *his⁻* because they are unable to synthesize histidine due to a mutation in one of the genes in the biosynthetic pathway. The strains therefore require histidine in the medium in order to grow. The Ames test detects the ability of a mutagen to cause a reversion of the *his⁻* phenotype, which allows the bacteria to grow in the absence of histidine (i.e., the chemical causes a mutation that restores the ability of the bacterium to synthesize histidine). Several strains carrying different *his⁻* mutations are used in order to assay the effects of various mutagens that may preferentially cause certain types of damage.

An example of the Ames test is shown in Figure 26·14. A small piece of filter paper containing the chemical to be tested is placed in the middle of a culture plate containing millions of *his⁻* bacteria. The culture medium does not contain histidine, so the bacteria are unable to grow unless the chemical causes reversion of the *his⁻* mutation. The presence of a halo of bacterial colonies around the filter indicates that the chemical is a mutagen that promotes reversion. In this example, the chemical is ethyl methanesulfonate, an alkylating agent. Larger colonies distributed randomly over the culture plate are due to spontaneous reversion and not to the action of the mutagen.

Not all mutagens are carcinogens, and not all carcinogens are potent mutagens. Nevertheless, there is a high correlation between mutagenicity and carcinogenicity,

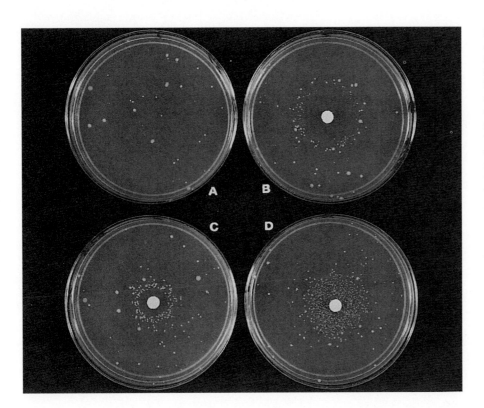

Figure 26·14
Ames test. The colonies on the plates are *his⁻ S. typhimurium,* and the medium does not contain histidine. **(a)** When no mutagen is present, the few colonies that appear are due to spontaneous reversion. These colonies are distributed randomly on the surface of the medium. **(b)** A small piece of filter paper containing a high concentration of mutagen is placed in the center of the plate. The mutagen diffuses outward from the filter paper. Near the center of the plate, the concentration of mutagen is so high that all bacteria are killed. Further away, however, the concentration is low enough to prevent cell death and high enough to induce mutation. Reversion of the *his⁻* phenotype leads to a ring of colonies. **(c)** A lower concentration of mutagen on the filter paper produces a smaller zone of lethality. **(d)** At the lowest concentration of mutagen, a large number of *his⁻* reversions are observed. (Courtesy of Bruce N. Ames.)

and mutagens identified by the Ames test should be considered potential carcinogens until proven otherwise. Powerful mutagens are more likely to induce mutations that kill cells. Cancer, on the contrary, is due to nonlethal mutations that generally alter the level or timing of the expression of certain genes. Some cancers occur fairly frequently because a particular site on a chromosome is a hot spot for mutation. Agents that act selectively at this hot spot may not be potent mutagens in other systems, but they are nonetheless powerful carcinogens. Other cancers are due to chromosomal rearrangements that place a gene in a different environment and lead to inappropriate expression. Agents that induce chromosomal rearrangements can be potent carcinogens but not necessarily potent mutagens.

Some chemicals known to cause cancer in mammals are not effective mutagens. These chemicals are modified in vivo by the action of various enzymes in the liver, and it is the modified product that is mutagenic (and carcinogenic). Liver extracts, which can carry out these modification reactions, are added to the media used in the Ames test in order to explore this possibility.

26·4 Damaged DNA Molecules Can Be Repaired

A number of different repair systems and a plethora of repair enzymes have evolved to ensure efficient repair of DNA in vivo. We will focus on four repair systems that are found in all species: direct repair, excision repair, mismatch repair, and error-prone repair. The distinguishing features of these repair systems in *E. coli* are summarized in Table 26·1 (next page); similar repair systems have been characterized in many other species. A fifth type of repair, recombinational repair, will be discussed in Section 26·6.

Table 26·1 DNA repair systems in *E. coli*

DNA repair system	Types of damage repaired	Enzymes/proteins involved in repair
Direct repair	Pyrimidine dimers Alkylated bases Other modified bases Mismatched bases	Photoreactivating enzyme Methyltransferases
Excision repair	Alkylated bases Deaminated bases Pyrimidine dimers Other lesions that distort the DNA helix	DNA glycosylases AP endonucleases UvrABC endonuclease Helicase II DNA polymerase I DNA ligase Mfd
Mismatch repair	Mismatched bases on newly synthesized DNA strands	Mut proteins Helicases Exonucleases DNA polymerase III DNA ligase
Error-prone repair	Cross-linked bases Sites where intercalators are bound	UmuC UmuD RecA

A. Direct Repair

Some damaged nucleotides and mismatched bases are recognized and repaired by proteins that continually scan DNA in order to detect specific lesions. These proteins effect repair without cleaving the DNA or excising the base. Such reversal of damage is called **direct repair.**

The formation of thymine dimers is an example of DNA damage that can be repaired directly. Recall that thymine dimers are caused by irradiation with ultraviolet light. In all prokaryotes and eukaryotes, there is an enzyme known as photoreactivating enzyme that binds distorted double helices at the site of thymine dimers. In the presence of visible light, the enzyme catalyzes the regeneration of two thymine bases. Photoreactivating enzyme then dissociates from the repaired DNA, and normal A/T base pairs re-form (Figure 26·15).

Methyltransferases can repair DNA damaged by alkylating agents. These proteins recognize and bind specific alkylated bases in DNA and remove the alkyl group. O^6-Methylguanine-DNA methyltransferase, for example, is the product of the *ada* gene in *E. coli*. This protein transfers alkyl groups from the O^6 position of methylguanine to a sulfhydryl group on a cysteine residue in the protein (Figure 26·16). The protein is inactivated in this process and cannot carry out another methyl-group–transfer reaction. However, the methylated form of the protein acts as a regulator of transcription, stimulating expression of the *ada* gene and other genes that encode repair enzymes, thus signalling the cell that there is a need for more repair enzymes. This type of positive feedback at the level of gene expression is common in repair and recombination processes. (We discuss the regulation of transcription in more detail in the next chapter.)

A variety of different methyltransferases are present in all organisms. Some eukaryotic and prokaryotic enzymes are homologous, suggesting that direct repair of alkylated bases is an ancient process. As in *E. coli*, the levels of methyltransferase in other species increase in response to alkylation of DNA. Permanently elevated levels of the enzyme are found in some human cancer cells, which are consequently resistant to therapeutic alkylating agents.

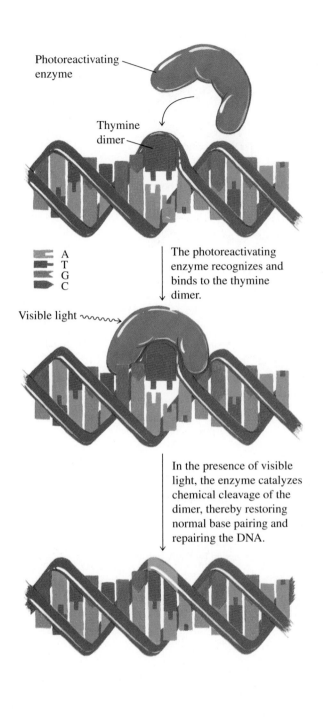

Figure 26·15
Repair of thymine dimers by photoreactivating enzyme in the presence of visible light.

Photoreactivating enzyme

Thymine dimer

A
T
G
C

The photoreactivating enzyme recognizes and binds to the thymine dimer.

Visible light

In the presence of visible light, the enzyme catalyzes chemical cleavage of the dimer, thereby restoring normal base pairing and repairing the DNA.

O^6-Methylguanine

Guanine

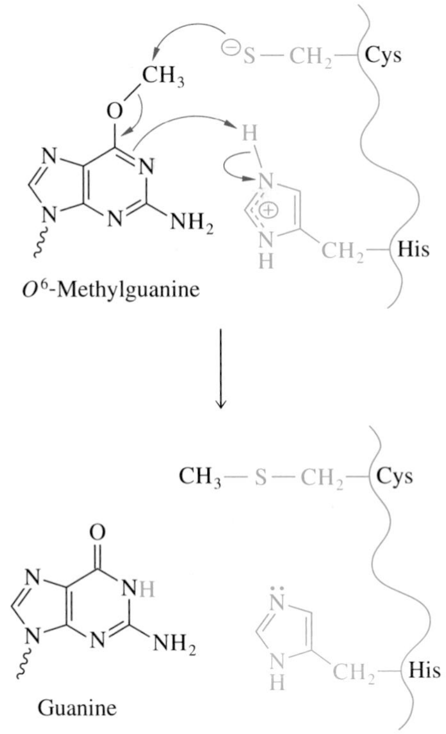

Figure 26·16
Mechanism of O^6-methylguanine-DNA methyltransferase. This protein transfers alkyl groups (in this case a methyl group) from the O^6 position of methylguanine to the sulfhydryl group on a cysteine residue in the protein. [Adapted from Blackburn, G. M., and Gait, M. J., eds. (1990). *Nucleic Acids in Chemistry and Biology* (New York: IRL Press), p. 289.]

B. Excision Repair

Deaminated bases, pyrimidine dimers, and gross lesions that alter the structure of the DNA helix can be repaired by a general excision-repair pathway whose overall features are similar in all species. In the first step of the pathway, the strand of DNA containing the lesion is cleaved on both sides of the damage by excision-repair endonucleases. In some cases, helicase activity is required to remove the DNA fragment between the endonuclease cleavage sites, while in other cases, the fragment is degraded by an exonuclease. The result in all cases is a single-stranded gap. In the next step of the pathway, the gap is filled by the action of DNA polymerase I in prokaryotes and repair DNA polymerases in eukaryotes. The nick is then sealed by

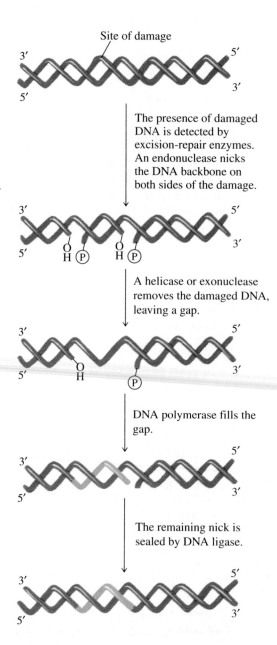

Figure 26·17
General excision-repair pathway.

Site of damage

The presence of damaged DNA is detected by excision-repair enzymes. An endonuclease nicks the DNA backbone on both sides of the damage.

A helicase or exonuclease removes the damaged DNA, leaving a gap.

DNA polymerase fills the gap.

The remaining nick is sealed by DNA ligase.

DNA ligase (Figure 26·17). There are several variations of this general excision-repair mechanism; they differ in the protein that recognizes the DNA damage and in the protein that catalyzes cleavage of DNA.

In *E. coli,* the recognition of damaged DNA and the initial endonucleolytic cleavages can be catalyzed by the UvrABC endonuclease (sometimes referred to as an exinuclease). As its name implies, the genes encoding the subunits of this endonuclease (*uvr*A, *uvr*B, and *uvr*C) were first identified as genes responsible for UV resistance. Two UvrA subunits (M_r 103 900), in association with a single UvrB subunit (M_r 76 100), bind sequentially to DNA. The UvrA$_2$B complex appears to first bind undamaged DNA and then scan for lesions by moving along the double helix in a manner that requires ATP hydrolysis. At the site of DNA damage, the proteins unwind the helix slightly, creating a small bubble. The UvrA$_2$B complex recognizes a variety of DNA damage, including pyrimidine dimers, small insertions and deletions, small gaps, and modified bases with bulky substituents. A common feature of these types of damage is that they interfere with base stacking, and one model of damage recognition suggests that UvrB, a hydrophobic protein, can intercalate at sites where stacking interactions are weak.

Next, the two UvrA subunits dissociate from the DNA and are replaced by UvrC (M_r 66 000). UvrC binds preferentially to single-stranded regions, including those created by the unwinding of the DNA at the site of the lesion. The UvrBC complex then catalyzes cleavage of the DNA strand containing the lesion at sites on either side of the damaged region (Figure 26·18). The endonucleolytic cleavage sites are at the eighth phosphodiester bond 5′ to the lesion and the fourth or fifth bond 3′ to the lesion. The oligonucleotide is removed with the help of helicase II (the product of the *uvr*D gene), and the gap is repaired by the actions of DNA polymerase I and DNA ligase. Helicase II and DNA polymerase I are also responsible for displacing the UvrB and UvrC subunits from DNA following excision.

Some sites of damage are excised more frequently than others by UvrABC. For example, the DNA of genes that are actively transcribed is preferentially repaired

Figure 26·18
Excision repair by UvrABC.

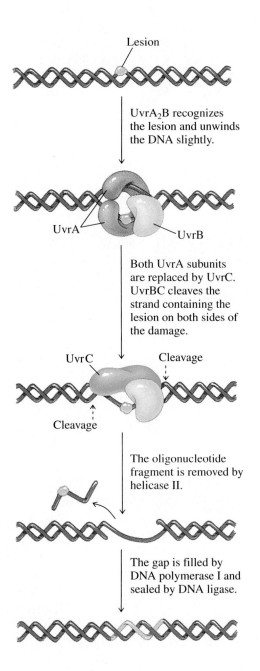

Lesion

UvrA₂B recognizes the lesion and unwinds the DNA slightly.

UvrA
UvrB

Both UvrA subunits are replaced by UvrC. UvrBC cleaves the strand containing the lesion on both sides of the damage.

UvrC
Cleavage
Cleavage

The oligonucleotide fragment is removed by helicase II.

The gap is filled by DNA polymerase I and sealed by DNA ligase.

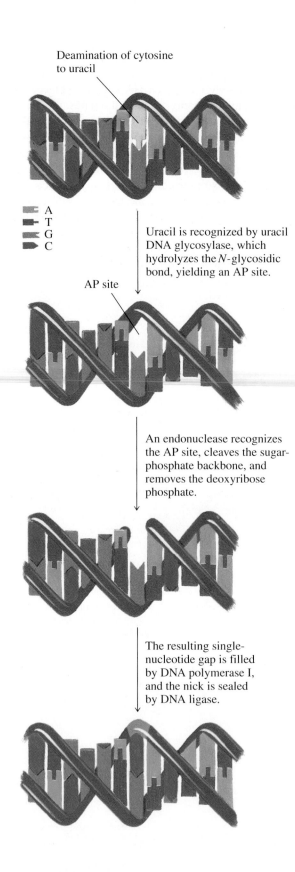

Deamination of cytosine
to uracil

A
T
G
C

AP site

Figure 26·19
Repair of deaminated cyto-
sine by the action of uracil
DNA glycosylase.

Uracil is recognized by uracil
DNA glycosylase, which
hydrolyzes the *N*-glycosidic
bond, yielding an AP site.

An endonuclease recognizes
the AP site, cleaves the sugar-
phosphate backbone, and
removes the deoxyribose
phosphate.

The resulting single-
nucleotide gap is filled
by DNA polymerase I,
and the nick is sealed
by DNA ligase.

over DNA in nontranscribed regions of the genome. The cellular machinery recognizes actively transcribed genes by the presence of RNA polymerase. When
stalled at the site of DNA damage, *E. coli* RNA polymerase is recognized by a
transcription-repair–coupling factor known as Mfd (mutation frequency decline).
Mfd (M_r 126 000) displaces RNA polymerase from the template strand by binding

to the site of damage. UvrA then interacts with bound Mfd at the lesion and initiates excision. The rest of the excision-repair pathway proceeds normally. In this way, the excision-repair enzymes are directed to sites where transcription has been inhibited by DNA damage. This mechanism ensures that actively transcribed genes are efficiently repaired before the lack of a gene product causes cell death. In fact, the main role of excision-repair enzymes may be to fix actively transcribed genes.

Enzymes called DNA glycosylases (sometimes known as *N*-glycosylases) also play a role in excision repair. These enzymes remove specific modified bases by catalyzing hydrolysis of the *N*-glycosidic bonds linking the modified bases to the sugars. For example, one class of DNA glycosylases removes deaminated bases such as uracil and hypoxanthine. By hydrolyzing the *N*-glycosidic bond, uracil DNA glycosylase removes uracil to create an apyrimidinic site (both apyrimidinic and apurinic sites are abbreviated *AP site*). An AP endonuclease then recognizes the AP site and catalyzes cleavage of the sugar-phosphate backbone, leading to excision of deoxyribose phosphate. The single-nucleotide gap in the duplex DNA is repaired by the actions of DNA polymerase I and DNA ligase (Figure 26·19). In some cases, the gap is extended by an exonuclease activity before it is filled by the action of DNA polymerase.

Endonuclease II is the most active AP endonuclease in *E. coli*. It catalyzes cleavage of the backbone on the 5′ side of an AP site, creating a single-stranded break. The abasic residue is then eliminated spontaneously in an alkali-catalyzed reaction, leaving a gap at the site of the original lesion (Figure 26·20). Endonuclease II also possesses a 3′→5′ exonuclease activity that can extend the gap on the 5′ side of the lesion.

Figure 26·20
Cleavage by endonuclease II. The enzyme cleaves on the 5′ side of the AP site. Spontaneous elimination in the presence of alkali then leads to cleavage on the other side of the lesion.

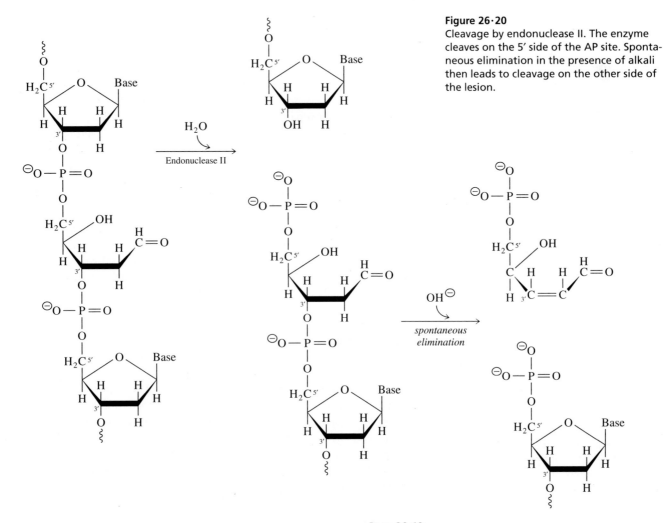

There are many bacterial DNA glycosylases that catalyze the initial reactions in the excision-repair pathway as described for uracil DNA glycosylase. In addition to the enzymes specific for deaminated bases, there are enzymes that remove alkylated bases (for example, 3-methyladenine) and enzymes such as pyrimidine dimer DNA glycosylase that repair thymine dimers. All these enzymes are relatively small proteins ($M_r \sim 30\,000$) that possess no intrinsic nuclease activity. Other DNA glycosylases, such as those that recognize oxidized pyrimidines and oxidized purines, are more complex. In addition to being glycosylases, these enzymes are AP endonucleases; they possess an intrinsic endonucleolytic activity that cleaves on the 3′ side of the AP site, thus eliminating the need for an alkali-catalyzed elimination step. AP endonucleases that are also DNA glycosylases are known as class I AP endonucleases; those that are not associated with glycosylase activity fall into class II. Following cleavage by a class I enzyme, a separate class II AP endonuclease such as endonuclease II is still needed for cleavage on the 5′ side of the lesion and excision of the abasic residue. The other steps of excision repair are identical to those described above. Formamidopyrimidine DNA glycosylase is an example of this second type of DNA glycosylase. This enzyme recognizes damage caused by ionizing radiation (see Figure 26·12).

DNA glycosylases are also found in eukaryotic cells. Eukaryotic uracil DNA glycosylase is homologous to its bacterial equivalent, suggesting that the excision of uracil is a reaction that was carried out in a common ancestral cell. This is presumably a reflection of the high frequency of deamination of cytosine and the need to remove the resulting uracil, which forms a stable base pair with guanine. Recall that the presence of thymine rather than uracil in DNA may be an evolutionary response to the need to detect cytosine deamination.

C. Mismatch Repair

Both direct-repair and excision-repair mechanisms rely on the ability of the cellular machinery to recognize inappropriate bases in DNA. Mismatch repair, on the other hand, replaces bases normally found in DNA and therefore not recognized by direct- or excision-repair systems. There are eight possible base mismatches in double-stranded DNA (A/C, A/G, A/A, C/C, C/T, T/T, T/G, and G/G). These mismatches are formed as a result of misincorporation during DNA replication or by chemical alterations of bases in DNA. Mismatches can also be formed during recombination (Section 26·5).

Mismatches can be repaired in both prokaryotes and eukaryotes by mechanisms that recognize the strand with the correct base and replace the incorrect base on the other strand. Such mechanisms require that the two strands of DNA be distinguishable. Most mismatches arise during DNA replication, when it is possible to distinguish the parental strand from the newly synthesized strand on the basis of methylation (in those species whose DNA is methylated). Immediately after DNA replication in such species, daughter molecules are hemimethylated, a status that persists for many seconds after replication (Section 25·6C). The major mismatch-repair system acts selectively on the undermethylated DNA strand and uses the methylated strand as a template for repair (Figure 26·21). In this way, errors due to misincorporation during DNA replication are corrected.

Mismatch repair has been extensively studied in *E. coli* and related species, where methylation at the Dam methylase sites (GATC) labels the parental strand. These methylation sites are usually separated by about 4000 bp, so the site of the mismatch may be as far as 2000 bp from the nearest hemimethylated site. During mismatch repair, the unmethylated DNA at the Dam methylase site is cleaved by an endonuclease, and a large piece of this strand, extending to the mismatch, is removed. This particular form of mismatch repair is known as long-patch or methyl-directed repair.

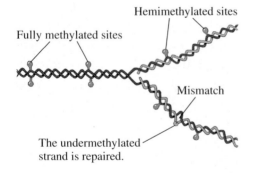

Fully methylated sites

Hemimethylated sites

Mismatch

The undermethylated strand is repaired.

Figure 26·21
Mismatch repair of methylated DNA. Immediately after replication, newly synthesized DNA (light gray) is undermethylated. A mismatch-repair mechanism recognizes the new DNA and selectively repairs the newly synthesized strand using the parental strand as a template. (Methylated bases are identified by orange dots.)

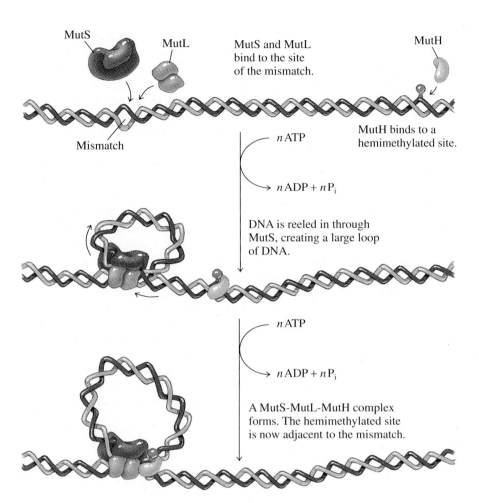

MutS and MutL bind to the site of the mismatch.

MutH binds to a hemimethylated site.

n ATP

n ADP + n P$_i$

DNA is reeled in through MutS, creating a large loop of DNA.

n ATP

n ADP + n P$_i$

A MutS-MutL-MutH complex forms. The hemimethylated site is now adjacent to the mismatch.

Figure 26·22
Assembly of the MutS-MutL-MutH complex during long-patch mismatch repair.

As expected, *E. coli* strains with a defective Dam methylase have high mutation rates because the long-patch–repair system cannot recognize the parental strand. Strains that overproduce the Dam methylase have also been created, and these too have high mutation rates. Overproduction of Dam methylase leads to more rapid methylation of the hemimethylated sites, which doesn't allow enough time for repair.

Long-patch mismatch repair is carried out by proteins that were originally identified as products encoded by mutator genes (*mut*). MutS (M_r 97 000) binds directly to the sites of mismatched base pairs. MutL (two subunits of M_r 70 000 each) associates with MutS, forming a complex that covers about 50 bp on either side of the mismatch. MutH binds to hemimethylated sites. The hemimethylated and mismatch sites are brought together, probably by the action of the MutS-MutL complex, which creates a loop by reeling in the DNA between the mismatch and hemimethylated sites (Figure 26·22). This reaction is associated with hydrolysis of ATP, and the complex is readily observed in vitro (Figure 26·23).

Once the complex has formed, the undermethylated DNA strand is cleaved at the Dam methylase site. Cleavage is catalyzed by an endonuclease activity associated with MutH and requires ATP. The DNA is then unwound from the nick introduced by MutH by the action of helicase II. Helicase II binds to the MutS complex at the site of the nick. This ensures that only DNA at the site of a mismatch is unwound. Once unwinding begins, the complex may dissociate, releasing the DNA loop. The newly synthesized (undermethylated) DNA released by helicase II is degraded by the action of an exonuclease, beginning at the free end. Different exonucleases carry out this degradation, depending on which strand is nicked by MutH.

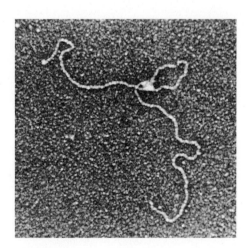

Figure 26·23
Electron micrograph of a DNA loop formed by MutS in vitro. (Courtesy of M. Grilley, P. Modrich, and J. Griffith.)

Exonuclease I degrades DNA in the $3' \rightarrow 5'$ direction, and either exonuclease VII or RecJ degrades DNA in the $5' \rightarrow 3'$ direction. In either case, the result is a large stretch of single-stranded DNA that extends from the Dam methylase site through the formerly mismatched region.

The large, single-stranded gap is coated with SSB, and a new strand of DNA is synthesized by DNA polymerase III. Note that it is the replication polymerase that is responsible for DNA synthesis during long-patch mismatch repair. This is presumably because the other polymerases are not processive enough to synthesize such a long piece of new DNA. Since DNA polymerase III is involved in mismatch repair as well as DNA replication, it is difficult to assess the fidelity of DNA replication alone; mutations in DNA polymerase III that alter its error rate affect both repair and replication. As mentioned in the previous chapter, the error rate of DNA replication is approximately 10^{-7}, but many of the replication errors that occur are subsequently corrected by long-patch mismatch repair. The result is a further 100-fold reduction in the overall rate of misincorporation. Thus, mutations due to mistakes in DNA replication are reduced to a level of 10^{-9}, or one error for every billion nucleotides incorporated.

The repair reaction is completed by the action of DNA ligase, which seals the nick at the site where the single-stranded gap ends. The repair steps are summarized in Figure 26·24. Note that the mismatched base pair is repaired using the parental DNA strand as a template.

Not all mismatches are repaired with equal efficiency by this mechanism. MutS binds preferentially to G/T and A/C mismatched base pairs and only weakly to C/C pairs. Consequently, G/T and A/C pairs are preferentially repaired, and C/C

Figure 26·24
Replacement of the undermethylated DNA strand during long-patch mismatch repair. Helicase II binds originally to the complex formed by MutS, MutL, and MutH.

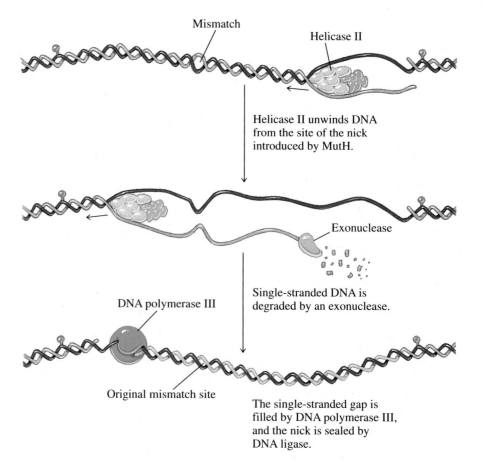

Mismatch

Helicase II

Helicase II unwinds DNA from the site of the nick introduced by MutH.

Exonuclease

Single-stranded DNA is degraded by an exonuclease.

DNA polymerase III

Original mismatch site

The single-stranded gap is filled by DNA polymerase III, and the nick is sealed by DNA ligase.

pairs are often ignored by the long-patch–repair mechanism. Other pairs are corrected at intermediate frequencies. Long-patch repair mechanisms are also capable of correcting small deletions and insertions.

There are several other mismatch repair systems in bacteria. These carry out short-patch mismatch repair over a region of about 10 base pairs. Some of these other systems recognize specific DNA sequences. For example, one short-patch–repair mechanism is responsible for repairing mismatches exclusively at methylation sites.

Mismatch-repair systems are also present in eukaryotic cells. In some cases, the eukaryotic enzymes are homologous to those in bacteria. The mechanism of long-patch mismatch repair in mammals is similar to that in bacteria, except that the methylation sites are at CG sequences in the mammalian genome. However, some eukaryotes, such as yeast and insects, do not contain methylated DNA. Mismatch repair in these organisms also requires an enzyme that preferentially nicks the newly synthesized strand, although the mechanism of strand recognition is not known.

D. Error-Prone Repair

Error-prone repair is a fourth type of repair mechanism that is found in all cells. Error-prone repair is induced when the replisome stalls at a site of DNA damage that cannot be bypassed. Such damage arises from irradiation or the action of mutagens that cross-link DNA bases. The signal for induction of error-prone repair is the presence of a single-stranded gap near the replication fork. The cellular response involves the formation of a modified DNA replication complex that passes over the damage, inserting nucleotides haphazardly opposite unrecognizable bases on the template strands (hence the name *error prone*).

In *E. coli,* stalled replisomes trigger a complex process known as the SOS response. A protein called RecA, which we will encounter in upcoming sections, binds to the single-stranded region. This binding in turn stimulates expression of a number of genes by a mechanism described more fully in Section 27·13. The genes regulated by the SOS response include *uvr*A, *uvr*B, and *uvr*D, which are all involved in excision repair, as well as genes encoding inhibitors of cell division. Expression of the genes *umu*C and *umu*D is also enhanced during the SOS response. These genes encode proteins that bind to RecA and DNA polymerase III. The resulting UmuCD-RecA-DNA polymerase III complex is able to carry out error-prone DNA replication. The roles of UmuC and UmuD in this complex are unclear, but it is thought that these proteins act to increase the processivity of DNA polymerase III and inhibit its proofreading activity.

E. Genetic Defects in Repair Systems Can Cause Disease

DNA repair does not appear to be essential for cell viability in *E. coli* and in those eukaryotes that have been examined. Even organisms with mutations in the genes for repair enzymes still grow (as long as the genes do not encode products that are also needed for DNA replication). Eventually, however, the accumulation of mutations in such organisms leads to death of the descendants.

There are many examples of human diseases that are due to genetic defects in repair pathways. Fanconi's anemia is characterized by malignant tumors and increased susceptibility to mutagens. The symptoms are due to defects in repair pathways, probably low activities of an exonuclease or ligase. Cockayne's syndrome is a repair deficiency that causes increased sensitivity to UV light. Patients with Cockayne's syndrome also grow slowly, leading to cachectic dwarfism.

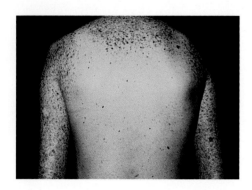

Figure 26·25
Xeroderma pigmentosum. This hereditary disease is associated with extreme sensitivity to UV light and increased frequency of skin cancer, as shown here. Note that most of the cancer occurs on skin that is exposed to sunlight (on the arms and neck). (Courtesy of Dirk Bootsma.)

Xeroderma pigmentosum is a hereditary disease associated with extreme sensitivity to ultraviolet light and increased frequency of skin cancer (Figure 26·25). Excision repair is defective in patients with xeroderma pigmentosum, but the phenotype can be due to mutations in at least eight different genes. One of these genes encodes a photoreactivating enzyme known as ultraviolet light–damaged DNA recognition protein. Another encodes a DNA glycosylase with class I AP-endonuclease activity. Two other genes encode helicases required during repair. One of these helicases is also a transcription factor (TFIIH) required for the initiation of transcription in eukaryotes (Section 27·8A). Proteins similar to both of these human helicases are also found in distantly related eukaryotes, such as yeast, where they also play a role in DNA repair. This is further evidence that DNA repair mechanisms have been conserved from the time of our earliest ancestors.

26·5 Homologous Recombination Involves Exchange of DNA Between Molecules

From a biochemical perspective, recombination is any event that results in the exchange or transfer of pieces of DNA from one chromosome to another or within a chromosome. Most recombinations are examples of **homologous recombination;** they occur between any homologous DNA sequences (for example, exchanges between paired chromosomes during meiosis). Recombinations between unrelated sequences are examples of nonhomologous recombination. Transposons, which are mobile genetic elements, jump from chromosome to chromosome by taking advantage of nonhomologous recombination mechanisms. Recombinations between DNA molecules also occurs when bacteriophages integrate into host chromosomes. If recombination occurs at a specific location, it is called site-specific recombination. Site-specific recombination and transposons are discussed in Chapter 32.

In genetics, the only meaningful recombinations are those that involve exchanges between DNA molecules with different sequences. Such recombinations are biologically significant because they lead to new combinations of genes on a chromosome. (Remember that although the DNA sequences involved in homologous recombination are similar, they need not be identical. That is why we say *homologous recombination* and not *identical recombination*.)

While mutation is one of the ways of creating variation within a population, recombination is another. Variation is required for evolution to occur, and most species have some means of exchanging information between individual organisms. In prokaryotes, which usually contain only a single copy of their genome (i.e., they are haploid), this exchange requires recombination. Some eukaryotes are also haploid, but most are diploid, with one of the two complete sets of chromosomes contributed by each parent. Genetic recombination in diploids mixes the genes on the chromosomes contributed by each parent so that subsequent generations receive very different combinations of genes. None of your children's chromosomes, for example, will be the same as yours, and none of yours are the same as your parents'.

As we will see in later chapters, recombinational events also play a role in the regulation of gene expression. The most important role of recombination, however, is as a form of DNA repair. Certain types of DNA damage cannot be effectively repaired by the mechanisms described earlier. These types of damage require repairs involving recombination between DNA molecules. In this section, we describe the Holliday model of general recombination—a type of recombination that seems to be common in many species.

A. The Holliday Model of General Recombination

Homologous recombination involves introducing either single-stranded breaks or double-stranded breaks into DNA molecules. The mechanism of recombination involving single-stranded breaks is often known as general recombination; recombination involving double-stranded breaks, which occurs in some yeast species, will not be discussed here.

As an example of recombination in prokaryotes, consider general recombination between two molecules of double-stranded bacteriophage DNA or between two molecules of plasmid DNA in bacteria. This exchange of information between two molecules of DNA begins with the alignment of identical or homologous DNA sequences. Next, single-stranded nicks are introduced in the homologous regions of each sequence, and single strands are exchanged in a process called **strand invasion.** The resulting structure contains a region of strand crossover and is known as a Holliday junction after Robin Holliday, who first proposed it in 1964 (Figure 26·26).

Figure 26·26
Holliday model of general recombination. First, nicks are introduced into a homologous region of each sequence. Subsequent strand invasion, DNA cleavage at the crossover junction, and sealing of nicked strands results in exchange of the ends of the chromosomes.

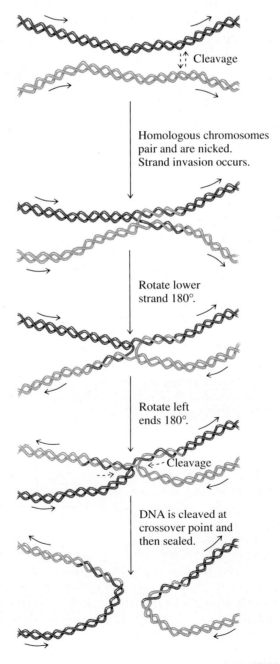

Cleavage

Homologous chromosomes pair and are nicked. Strand invasion occurs.

Rotate lower strand 180°.

Rotate left ends 180°.

Cleavage

DNA is cleaved at crossover point and then sealed.

Figure 26·27
Holliday junction. Note that no base pairs are disrupted and that the two double-stranded helices consist of fully stacked base pairs.

The chromosomes can be separated at this stage by cleaving the two invading strands at the crossover point, but it is important to realize that the Holliday junction can exist in several different conformations. As shown in Figure 26·26, the ends of the homologous DNA molecules can be rotated to generate a structure that, when cleaved at the crossover, produces two chromosomes that have exchanged ends. Recombination in many different organisms probably occurs by a similar process.

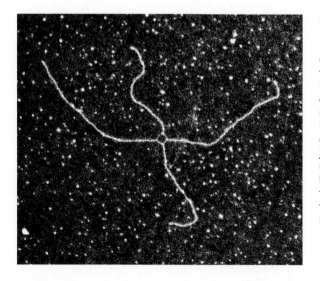

Figure 26·28
Electron micrograph of a Holliday junction formed during recombination between plasmids. The circular plasmid DNA has been cut with a restriction endonuclease in order to form linear molecules. In this electron micrograph, the junction contains partially unwound DNA, which may be an artifact of preparation. (Courtesy of Huntington Potter and David Dressler.)

(a)

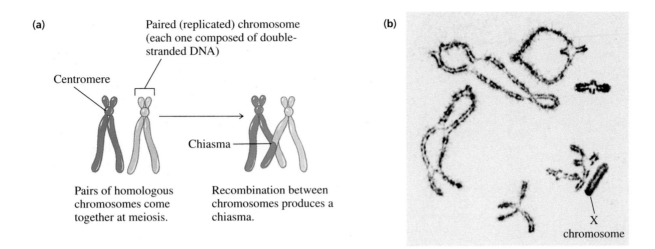

Centromere

Paired (replicated) chromosome (each one composed of double-stranded DNA)

Chiasma

Pairs of homologous chromosomes come together at meiosis.

Recombination between chromosomes produces a chiasma.

(b)

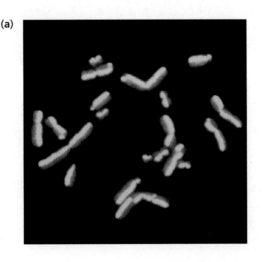

X chromosome

Figure 26·29
Formation of chiasmata during meiosis in eukaryotes. **(a)** Conceptual model. **(b)** Light micrograph of meiotic chromosomes in grasshopper. There are between one and four chiasmata per pair of chromosomes except for the X chromosome, which is unpaired and therefore cannot cross over. (Courtesy of Bernard John.)

A more detailed view of a Holliday junction is shown in Figure 26·27. This view corresponds to the final rotated version seen in the previous figure. Note that no base pairs are disrupted at the crossover point and that the two double helices consist of fully stacked base pairs. The normal breaking and re-forming of base pairs can lead to further strand exchange, a process known as **branch migration.** An example of a Holliday junction formed during plasmid recombination is shown in Figure 26·28.

During meiosis in eukaryotes, crossing over can be observed in the light microscope because chromosomes are condensed and chromosome segregation is prolonged (Figures 26·29 and 26·30). These crossovers, or chiasmata, are not Holliday junctions. Instead, they represent regions where recombination has already been resolved and large pieces of DNA have been exchanged between homologous chromosomes. Note that the scale of the images in Figures 26·26 and 26·29 is vastly different; chiasmata represent regions of thousands of base pairs.

While homologous recombination is similar in all organisms, the details are best understood in *E. coli,* from which most of the enzymes and proteins have been isolated and characterized. In the following sections, we will describe the mechanism of action of the bacterial proteins.

Figure 26·30
Mammalian chromosomes showing regions where DNA fragments have exchanged by homologous recombination. **(a)** One of the original chromosomes is labelled with a fluorescent dye; the other is unlabelled. **(b)** After homologous recombination, both chromosomes display a banded appearance. (Courtesy of Sheldon Wolff and Judy Bodycote-Thomas.)

(a)

(b)

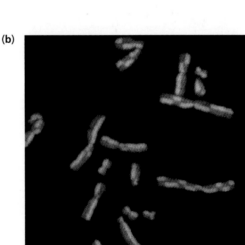

B. RecBCD Is a Travelling Helicase

One of the first steps in recombination is the generation of single-stranded DNA with a free 3′ end. In *E. coli*, this step is carried out by an enzyme with subunits that are encoded by three genes whose products have long been known to play a role in recombination: *rec*B, *rec*C, and *rec*D.

RecBCD (also known as exonuclease V) binds to DNA and acts as a helicase, unwinding the DNA in a process coupled to ATP hydrolysis. When RecBCD encounters a specific sequence known as a *chi* site (GCTGGTGG), it catalyzes cleavage of a single strand. Continued unwinding from this point generates single-stranded DNA with a 3′ terminus. Strand exchange during recombination begins at the *chi* site, which marks the 3′ end of the invading strand. The exchange terminates at various distances from the *chi* site, depending on the extent of branch migration during recombination.

Given the action of RecBCD, *chi* sites are recombinational hot spots. They occur frequently in the *E. coli* genome (about once every 4 kb). The frequency of recombination declines gradually the further one moves away from a *chi* site in the 3′→5′ direction along the single-stranded region, and it drops off precipitously on the 5′ side of the site, where no single-stranded DNA is produced for strand invasion. Strains of *E. coli* that are *rec*B⁻ and *rec*C⁻ exhibit less than 1% of the normal level of recombination, suggesting that most recombinational events in *E. coli* require RecBCD.

C. RecA Promotes Single-Strand Invasion

RecA is a strand-exchange protein. It is essential for homologous recombination and for some forms of repair (such as error-prone repair, as we saw in Section 26·4D). The protein functions as a monomer (M_r 37 800) that binds cooperatively to single-stranded DNA. Each monomer covers about five nucleotides, and each successive monomer binds to the opposite side of the DNA helix. RecA molecules bind to the single-stranded tails produced by the action of RecBCD helicase, creating a well-characterized structure in which the single-stranded DNA forms a right-handed helix that is underwound relative to B-DNA. The single-stranded DNA lies within a deep groove in the RecA protein. SSB is displaced during RecA binding but aids formation of the RecA-coated single strand by removing regions of secondary structure (Figure 26·31).

Figure 26·31
Formation of the complex between RecA and single-stranded DNA. RecA molecules bind cooperatively to the single-stranded tails produced by the action of RecBCD helicase. SSB is displaced during RecA binding.

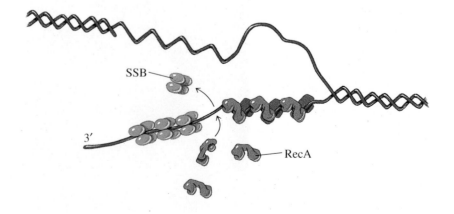

SSB

3′

RecA

The RecA–single-stranded DNA complex is so stable that dissociation of RecA from DNA in vivo may require ATP hydrolysis. In vitro, the RecA–single-stranded DNA complex has a half-life of 30 minutes and can be readily seen in electron micrographs (Figure 26·32). The coated DNA has a diameter of 12 nm, due largely to the presence of bound RecA. It is held in a more extended form than the uncoated single-stranded DNA next to it, which forms a tighter helix due to base-stacking interactions.

One of the key roles of RecA in recombination is to recognize regions of homology. RecA promotes the formation of a triple-stranded intermediate between the coated single strand and a homologous region of double-stranded DNA. RecA then catalyzes strand exchange in which the single strand displaces the homologous strand from the double helix. Strand exchange takes place in two steps: strand invasion, followed by branch migration (Figure 26·33). During the exchange reaction, both the single-stranded and the double-stranded DNA are in an extended conformation. The strands must rotate around each other, a process that is presumably aided by topoisomerases. Coating of single-stranded DNA with RecA is a relatively slow process; formation of a triple-stranded intermediate between homologous DNA molecules is much faster. Strand exchange is also slow, despite the fact that no covalent bonds are broken.

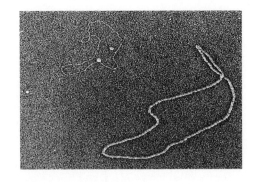

Figure 26·32
Electron micrograph of a RecA–single-stranded DNA complex. Two closed, circular, single-stranded DNA molecules are shown. The one on the right is coated with RecA. An uncoated molecule is shown on the upper left. (Courtesy of Jack Griffith.)

Figure 26·33
Strand exchange catalyzed by RecA.

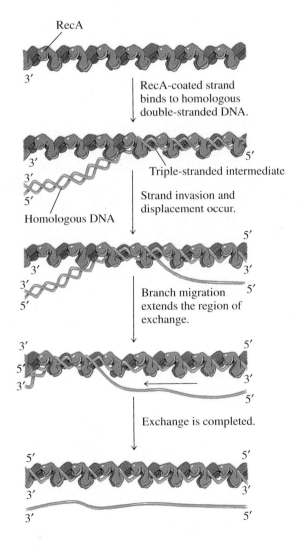

RecA-coated strand binds to homologous double-stranded DNA.

Triple-stranded intermediate

Strand invasion and displacement occur.

Homologous DNA

Branch migration extends the region of exchange.

Exchange is completed.

Figure 26·34
Models of hydrogen-bonded bases in a triple-stranded intermediate. The base from the invading strand is on the top.

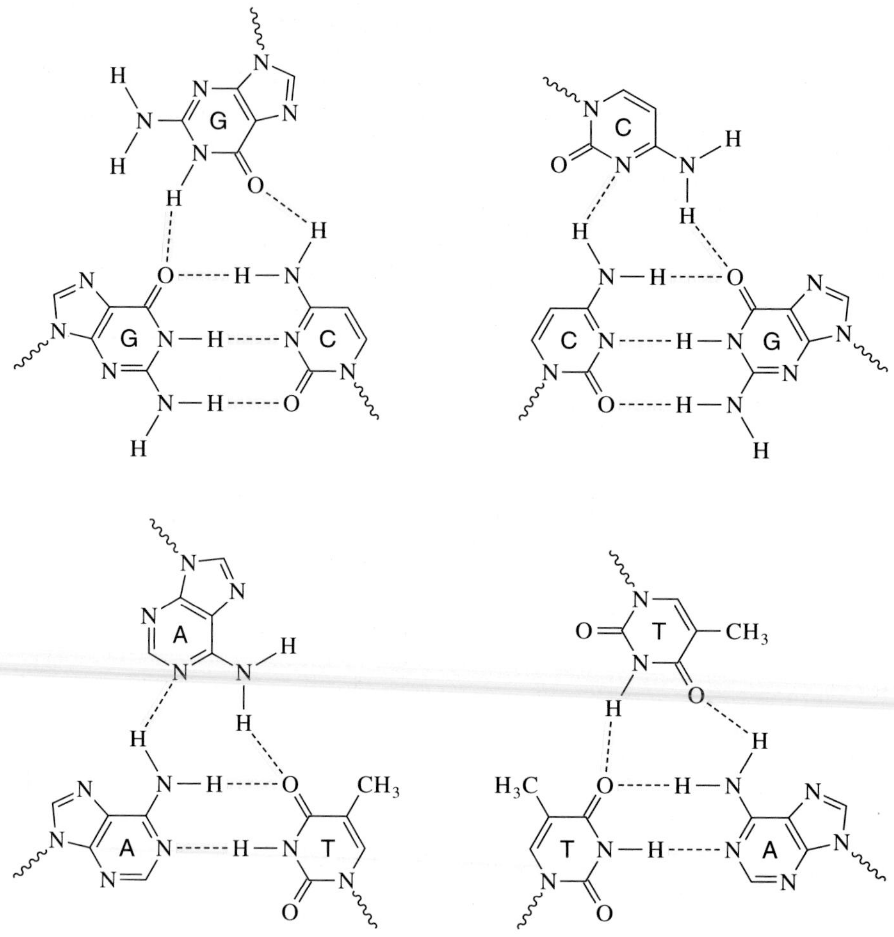

Figure 26·35
Model of RecA-mediated strand exchange. RecA (blue) binds to and compresses the triple-stranded intermediate, rotating the bases from the invading strand (red) and the bases from one of the original strands, and thus bringing them into alignment to form Watson-Crick base pairs. As this happens, the other original strand is displaced.

The triple-stranded intermediate is quite stable. A model that accounts for this stability and also explains how RecA is able to recognize homologous DNA has been proposed. This model suggests that the bases in the single strand form hydrogen bonds with the homologous base pairs in the double helix (Figure 26·34). RecA then catalyzes strand exchange by flipping the base pairing of the two identical strands (Figure 26·35).

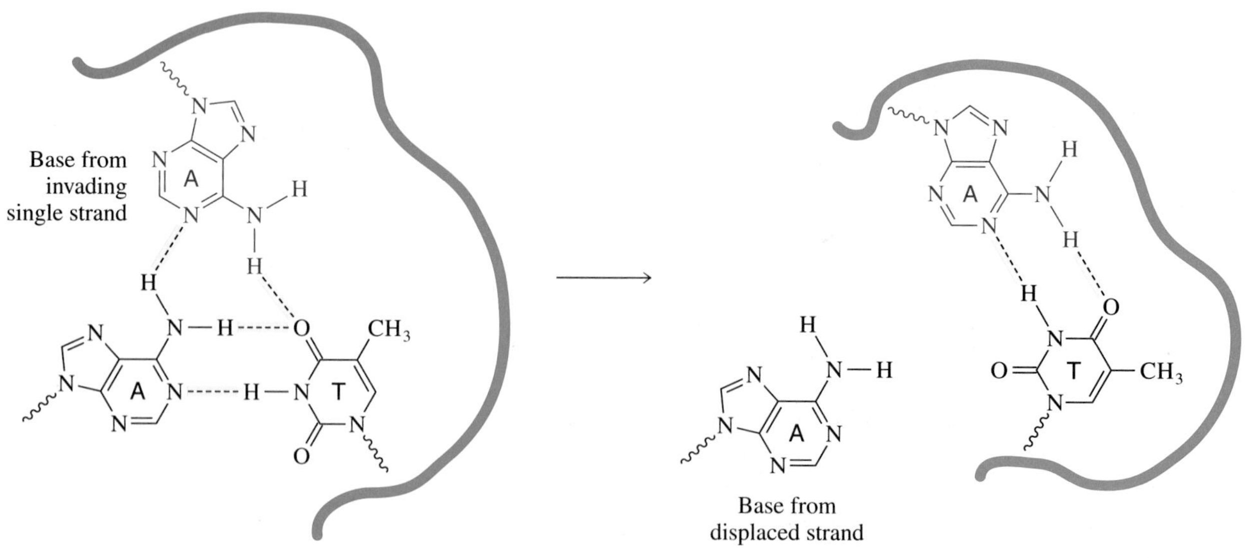

Base from invading single strand

Base from displaced strand

D. The Holliday Junction Is Resolved by Ruv Proteins

RecA can also promote strand invasion between two aligned, double-stranded DNA molecules. Both molecules must contain single-stranded tails bound to RecA that are wound around the corresponding complementary strands in the homologue. This exchange gives rise to a Holliday junction such as that shown in Figure 26·26. Subsequent branch migration can extend the region of strand exchange. Branch migration can continue even after RecA dissociates from the recombination intermediate. A complex of RuvA (M_r 22 000) and RuvB (M_r 37 000) promotes ATP-dependent branch migration at Holliday junctions. The rate of RuvAB-mediated migration is significantly faster than strand invasion.

RuvC (M_r 19 000) catalyzes cleavage of the crossover strands to resolve Holliday junctions (Figure 26·36). Two types of recombinant molecules are produced as a result of this cleavage: those in which only single strands are exchanged, and those in which the ends of the chromosome have been swapped (see Figure 26·26). RuvC can also resolve Holliday junctions that remain associated with RecA.

E. Recombination Occurs During Conjugation in *E. coli*

During conjugation in *E. coli*, a single-stranded copy of the chromosome of one bacterium is passed to another via the sex pilus (Figure 25·32). While this single-stranded DNA can be used directly for recombination via strand invasion mediated by RecA, the DNA is usually converted to double-stranded DNA first. This double-stranded DNA recombines with the chromosome of the recipient (female) bacterium by homologous recombination. Formation of two separate Holliday junctions, or crossovers, leads to replacement of a segment of the female chromosome (Figure 26·37). When single-stranded DNA is recombined with the female chromosome, a heteroduplex consisting of one strand from the male and one strand from the female is formed. Such a heteroduplex is also found at the ends of the recombined regions involving double-stranded DNA.

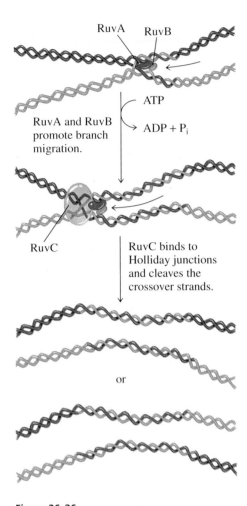

RuvA and RuvB promote branch migration.

RuvC binds to Holliday junctions and cleaves the crossover strands.

or

Figure 26·36
Action of Ruv proteins. RuvA and RuvB promote branch migration in a reaction requiring hydrolysis of ATP. RuvC cleaves Holliday junctions. Two types of recombinant molecules can be generated in this reaction.

Figure 26·37
Recombination during conjugation in *E. coli*. The male chromosome, which is double stranded in this case, forms two separate Holliday junctions with the female chromosome. The result is the replacement of part of the female chromosome with homologous DNA from the male chromosome. The displaced linear DNA is eventually degraded in vivo.

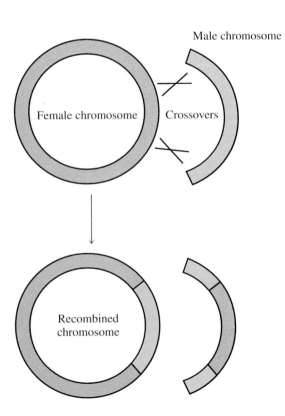

Since recombination of bacterial chromosomes only occurs between homologous DNA sequences, the two strands are most often completely complementary. However, RecA-catalyzed strand exchange is not affected by the presence of short regions of nonhomologous DNA (such as allelic differences) nor by the presence of insertions or deletions. Thus, recombination may generate mismatches or gaps in one strand. Such DNA damage is subsequently corrected by repair enzymes using one strand as a template.

Mismatch repairs that lead to both double-stranded DNA molecules having the same sequence are known as **gene conversion** events (Figure 26·38). One gene is "converted" to the other if the repair enzymes choose DNA strands from the same original molecule as the two "correct" templates. In this case, net reciprocal exchange does not occur; instead, one gene "disappears" and is replaced by the other. Gene conversion is observed in all organisms; it is a consequence of homologous recombination coupled to mismatch repair.

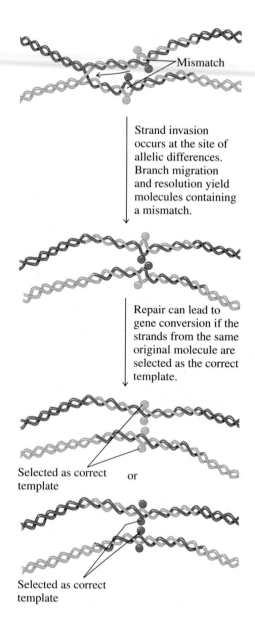

Mismatch

Strand invasion occurs at the site of allelic differences. Branch migration and resolution yield molecules containing a mismatch.

Repair can lead to gene conversion if the strands from the same original molecule are selected as the correct template.

Selected as correct template   or

Selected as correct template

Figure 26·38
Gene conversion. The colored dot above each strand indicates a single-nucleotide allelic difference between two genes; in all other respects, the two strands are identical. Gene conversion occurs if the repair enzymes use DNA strands from the same original molecule as the two correct templates.

26·6 Recombination Can Be a Form of Repair

The enzymes that catalyze recombination can be directly involved in the repair of severe DNA damage. We have already seen an example of such an involvement in the SOS response, which requires RecA. However, some types of damage cannot be overcome even by error-prone repair during DNA replication. In these cases, recombination can generate an intact template strand, and the lesion can then be repaired by mismatch or error-prone repair (Figure 26·39).

Recombination probably evolved as a form of repair. Since natural selection works predominantly at the level of individual organisms, it is difficult to see why recombination per se would have evolved unless it also had a direct role in survival. The ability to repair DNA is of direct benefit to an individual organism, and recombination enzymes likely evolved because they conferred a selective advantage. That recombination also creates new combinations of genes on a chromosome is an added bonus to the population and its chance for evolutionary survival.

Over 100 *E. coli* genes are required for recombination and repair. Most if not all of the genes used in recombination play some role in repair as well. The number of genes required for repair and recombination in eukaryotes is not known, but considering the complexity and number of these pathways, it is likely that many hundreds of genes still await discovery.

Summary

Mutations in DNA can arise as a result of DNA damage that is not repaired. A variety of different agents can cause damage; these include alkylating agents, intercalating agents, and ionizing radiation. DNA can also be damaged as a result of deamination of bases, incorporation of base analogs, or mistakes during DNA replication. Mutations that are caused by DNA damage can lead to cancer in animals; the Ames test is an assay for the mutagenic effects of various chemicals.

There are four important DNA repair pathways in all organisms: direct repair, excision repair, mismatch repair, and error-prone repair. In direct repair, the normal bases in DNA are restored by enzymes that modify the damaged nucleotides. The excision-repair pathway involves enzymes, such as UvrABC endonuclease, that remove a stretch of damaged DNA. The resulting gap is filled by the repair DNA polymerase. Glycosylases scan DNA for specific types of damage and then excise damaged nucleotides. Mismatch repair is a pathway that allows selective repair of errors introduced during DNA replication. The enzymes involved in this pathway recognize newly synthesized strands of DNA, in some cases because they are undermethylated. Base mismatches are corrected using the parental DNA strand as a template. In error-prone repair, the DNA replication complex is modified to enable it to bypass regions of damage that cannot be readily repaired by the other repair mechanisms.

Homologous recombination involves the exchange of DNA between identical or almost identical DNA strands. Exchange is initiated by the formation of single-stranded tails that invade the double helix on the homologous chromosome. Reciprocal single-strand exchange leads to the formation of a Holliday junction that can be resolved in several different ways by cutting the single strands at the crossover point. In *E. coli*, strand exchange is mediated by RecBC nuclease and RecA, and the Holliday junction is resolved by Ruv proteins.

Recombination probably evolved as a form of repair. The general recombination pathway seen in *E. coli* is similar in all organisms. Recombination is important during conjugation in bacteria and during meiosis in eukaryotes.

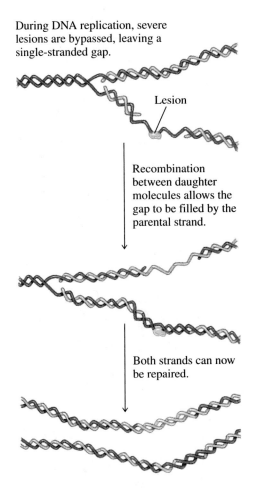

During DNA replication, severe lesions are bypassed, leaving a single-stranded gap.

Lesion

Recombination between daughter molecules allows the gap to be filled by the parental strand.

Both strands can now be repaired.

Figure 26·39
Repair by recombination.

Selected Readings

DNA Repair

Bootsma, D., and Hoeijmakers, J. H. J. (1993). Engagement with transcription. *Nature* 363:114–115.

Buratowski, S. (1993). DNA repair and transcription: the helicase connection. *Science* 260:37–38.

Echols, H., and Goodman, M. F. (1991). Fidelity mechanisms in DNA replication. *Annu. Rev. Biochem.* 60:477–511.

Friedberg, E. C. (1992). Xeroderma pigmentosum, Cockayne's syndrome, helicases, and DNA repair: what's the relationship? *Cell* 71:887–889.

Haseltine, W. A. (1983). Ultraviolet light repair and mutagenesis revisited. *Cell* 33:13–17.

Lindahl, T. (1982). DNA repair enzymes. *Annu. Rev. Biochem.* 51:61–87.

Selby, C. P., and Sancar, A. (1993). Molecular mechanism of transcription-repair coupling. *Science* 260:53–58.

Van Houten, B., and Snowden, A. (1993). Mechanism of action of the *Escherichia coli* UvrABC nuclease: clues to the damage recognition problem. *BioEssays* 15:51–59.

Walker, G. C. (1985). Inducible DNA repair systems. *Annu. Rev. Biochem.* 54:425–457.

Mutagenesis

Drake, J. W. (1991). Spontaneous mutation. *Annu. Rev. Genet.* 25:125–146.

Lindahl, T., Sedgwick, B., Sekiguchi, M., and Nakabeppu, Y. (1988). Regulation and expression of the adaptive response to alkylating agents. *Annu. Rev. Biochem.* 57:133–157.

Recombination

Conley, E. C. (1992). Mechanism and genetic control of recombination in bacteria. *Mutat. Res.* 284:75–96.

Ganesan, S., and Smith, G. R. (1993). Strand-specific binding to duplex DNA ends by the subunits of the *Escherichia coli* RecBCD enzyme. *J. Mol. Biol.* 229:67–78.

Hastings, P. J. (1992). Mechanism and control of recombination in fungi. *Mutat. Res.* 284:97–110.

Rao, B. J., Chiu, S. K., and Radding, C. M. (1993). Homologous recognition and triplex formation promoted by RecA protein between duplex oligonucleotides and single-stranded DNA. *J. Mol. Biol.* 229:328–343.

Stahl, F. W. (1987). Genetic recombination. *Sci. Am.* 256(2):90–101.

West, S. C. (1992). Enzymes and molecular mechanisms of genetic recombination. *Annu. Rev. Biochem.* 61:603–640.

27

Transcription

As we have seen, the structure of DNA proposed by James Watson and Francis Crick in 1953 immediately suggested a means of replicating the DNA and thereby transmitting genetic information from one generation to the next. However, the structure of DNA did not reveal how an organism uses the information stored in its genetic material.

The answer to this question had been emerging gradually over decades as scientists examined mutant organisms and attempted to characterize the changes induced by mutation. In the 1940s, George Beadle and Edward Tatum proposed, based on studies of the bread mold *Neurospora crassa,* that a single unit of heredity, or gene, in some way directed the production of a single enzyme. The full demonstration of the relationship between genes and proteins came when a study of sickle-cell hemoglobin was published by Vernon Ingram in 1956. Ingram showed that hemoglobin from patients with the heritable disease sickle-cell anemia differed from normal hemoglobin by the replacement of a single amino acid. His results indicated that genetic changes can manifest themselves as changes in the primary sequence of a protein. Thus, by extension, the information in the genome specifies the primary structure of each protein in an organism.

The flow of information from gene to protein requires an RNA intermediate. DNA is transcribed to produce RNA, and some RNA molecules, namely messenger RNA, are translated to yield proteins. This pathway is known as the **central dogma** of molecular biology (Figure 27·1). We will define a **gene** as a region of DNA that is transcribed. It is important to keep this definition in mind; it means that there can be genes that do not encode proteins since not all transcripts are messenger RNA.

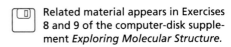

Related material appears in Exercises 8 and 9 of the computer-disk supplement *Exploring Molecular Structure.*

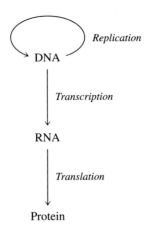

Figure 27·1
Central dogma of molecular biology. DNA directs its own replication, and the flow of cellular information is from DNA to RNA to protein.

Our definition of a gene normally excludes regions of the genome that control transcription but are not themselves transcribed. We will encounter some exceptions to our definition of a gene—surprisingly, there is no definition that is entirely satisfactory.

The genomes of some prokaryotes contain as few as 750 genes, although most have about 3000 genes. Most of the proteins encoded by these genes are involved in metabolism, including the biosynthesis and degradation of macromolecules such as DNA, RNA, and protein. Many of the remaining genes encode proteins required for the structural integrity of the cell and for processes such as repair and recombination. Such genes are often referred to collectively as **housekeeping genes** since their products are essential for the normal activities of all living cells. Simple eukaryotes such as yeast have a genome that consists largely of a set of housekeeping genes similar to those in prokaryotes but includes genes required to maintain the more complex eukaryotic cell.

In addition to housekeeping genes, all cells contain genes that are expressed only in special circumstances, such as during cell division. Multicellular organisms also contain genes that are expressed only in certain types of cells. For example, all cells in a maple tree contain the genes for the enzymes that make chlorophyll, but these genes are only expressed in cells that are exposed to light, such as cells on the surface of a leaf. Similarly, all cells in mammals contain insulin genes, but only certain pancreatic cells produce insulin. The total number of genes in multicellular eukaryotes is still not known with precision, but it is thought that insects contain approximately 7000 genes and mammals between 20 000 and 50 000 genes.

In this and the next five chapters, we will examine the process by which the information stored in DNA directs the synthesis of proteins. In this chapter, we describe the process of transcription and how it is regulated. In the next chapter, we will examine RNA processing, or how RNA molecules essential for the transfer of information are modified following transcription. In Chapter 29, we will encounter the genetic code and learn how the sequence of nucleotides specifies the sequence of amino acids in proteins. We will also discuss transfer RNA molecules, which bring amino acids to the site of protein synthesis. In Chapter 30, we will examine the structure and function of ribosomes and the process of translating the genetic information and synthesizing proteins. In Chapters 31 and 32, we will consider additional examples of the regulation of gene expression and how genes are organized in the genome.

27·1 Several Classes of RNA Molecules Participate in the Transfer of Genetic Information from DNA to Protein

The transfer of information from DNA to protein requires the participation of all four classes of RNA molecules. Messenger RNA (mRNA) was the first class shown to be transcribed from a DNA template. The discovery of mRNA is due largely to the work of François Jacob, Jacques Monod, and their collaborators at the Pasteur Institute in Paris. In the early 1960s, these researchers showed that ribosomes participate in the synthesis of proteins by translating unstable RNA molecules (mRNA). They also made the important discovery that the sequence of an mRNA molecule is complementary to a segment of one of the strands of DNA.

The other three classes of RNA are now also known to be involved in gene expression. Transfer RNA (tRNA) carries amino acids to the translation machinery. Ribosomal RNA (rRNA) makes up much of the ribosome. Several different small RNA molecules that participate in various metabolic events, such as posttranscriptional modifications, are also involved in information flow. Many of these small RNA molecules have catalytic activity.

A large percentage of the total RNA in the cell is rRNA, and only a small percentage is mRNA. But if we compare the rates at which the cell synthesizes RNA rather than the steady-state levels of RNA, we see a different picture (Table 27·1). Even though mRNA accounts for only 3% of the total RNA in *Escherichia coli*, one-third of the cell's capacity to synthesize RNA is devoted to the production of mRNA. In fact, this value can increase to about 60% when the bacterium is growing slowly and does not need to replace ribosomes and tRNA. The discrepancy between steady-state levels of various RNA molecules and the rates at which they are synthesized can be explained by the differing stabilities of the various RNA molecules: rRNA and tRNA molecules are extremely stable, whereas mRNA is rapidly degraded after translation. In bacterial cells, half of all newly synthesized mRNA is degraded by nucleases within three minutes. In eukaryotes, the average half-life of mRNA is about 10 times longer. The relatively high stability of eukaryotic mRNA is a result of processing and modification events that prevent eukaryotic mRNA from being degraded during its transport from the nucleus, where transcription occurs, to the cytosol, where translation occurs.

Table 27·1 The RNA content of an *E. coli* cell

Type	Steady-state level	Synthetic capacity[1]
rRNA	83%	58%
tRNA	14%	10%
mRNA	3%	32%
RNA primers[2]	<1%	<1%
Other RNA molecules[3]	<1%	<1%

[1] Relative amount of each type of RNA being synthesized at any instant.
[2] RNA primers are those used in DNA replication; these are not synthesized by RNA polymerase.
[3] Other RNA molecules include several RNA enzymes, such as the RNA component of RNase P.
[Adapted from Bremer, H., and Dennis, P. P. (1987). Modulation of chemical composition and other parameters of the cell by growth rate. In Escherichia coli *and* Salmonella typhimurium: *Cellular and Molecular Biology*, Vol. 2, F. C. Neidhardt, ed. (Washington, DC: American Society for Microbiology), pp. 1527–1542.]

27·2 RNA Polymerase Catalyzes RNA Synthesis

At about the time that mRNA was identified, researchers in several laboratories independently discovered an enzyme that catalyzed the synthesis of RNA when provided with ATP, UTP, GTP, CTP, and a template DNA molecule. The newly discovered enzyme was RNA polymerase. This enzyme catalyzes DNA-directed RNA synthesis, or **transcription.**

RNA polymerase was initially identified by its ability to catalyze polymerization of ribonucleotides, but further study of the enzyme revealed that it does much more. RNA polymerase is the core of a larger transcription complex (just as DNA polymerase is the core of a larger replication complex). This complex is assembled at the beginning of a gene during the initiation phase of transcription. Initiation is associated with partial unwinding of the template DNA and synthesis of an RNA primer. In the elongation phase, RNA polymerase catalyzes the processive elongation of the RNA chain while DNA is continuously unwound and rewound. Finally, the transcription complex, which includes RNA polymerase, responds to specific transcription termination signals and disassembles.

Although the composition of the transcription complex varies considerably among different organisms, all transcription complexes catalyze essentially the same types of reactions. Therefore, we will introduce the general process of transcription by discussing the reactions catalyzed by the well-characterized transcription complex in *E. coli*. The more complicated eukaryotic RNA polymerases will be presented in Sections 27·7 and 27·8.

A. RNA Polymerase Is an Oligomeric Protein

Free RNA polymerase is isolated from *E. coli* cells as an oligomeric protein composed of six subunits of five different types (Table 27·2). Five of these subunits combine in a stoichiometry of $\alpha_2\beta\beta'\sigma$ to form the holoenzyme; the ω subunit is usually present in less than stoichiometric amounts and may not be part of the functional holoenzyme. The subunits of the holoenzyme function together to carry out the reactions of transcription: the β' subunit contributes to DNA binding, the β subunit contains part of the active site of the enzyme, and the σ subunit plays an important role in the initiation of transcription. The α subunits are the scaffold upon which the other subunits are assembled; in addition, they interact with many proteins that regulate transcription. Several different types of σ subunits, also known as σ factors, exist within any one bacterium. The major form of the RNA polymerase

Table 27·2 Subunits of *E. coli* RNA polymerase holoenzyme

Subunit	M_r
β'[1]	155 600
β	150 600
σ	70 300[2]
α	36 500
ω	10 100

[1] The β and β' subunits are unrelated despite the similarity of their names.
[2] The molecular weight given is for the σ subunit found in the most common form of the holoenzyme.

Figure 27·2
Subunit composition of an *E. coli* RNA polymerase holoenzyme. The holoenzyme is composed of four different subunits with the stoichiometry $\alpha_2\beta\beta'\sigma^{70}$; ω is present in less than stoichiometric amounts. The holoenzyme is depicted here and in subsequent diagrams as the core enzyme plus σ because σ dissociates from the holoenzyme after transcription is initiated. [Adapted from Darst, S. A., Kubalek, E. W., and Kornberg, R. D. (1989). Three-dimensional structure of *Escherichia coli* RNA polymerase holoenzyme determined by electron crystallography. *Nature* 340:730–732.]

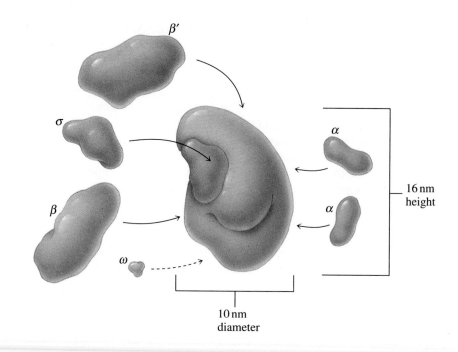

Figure 27·3
Schematic representation of SDS-PAGE patterns of subunits of purified RNA polymerases of different bacterial species. *E. coli*, an enteric bacterium, is distantly related to the anaerobic bacterium *Bacillus subtilis* and the cyanobacterium *Anabaena*. These three types of eubacteria are more distantly related to the archaebacteria *Sulfolobus acidocaldarius* and *Halobacterium halobium*. Identically colored subunits are related by sequence homology. [Adapted from Zillig, W., and Gropp, F. (1987). The distribution of antigenic determinants in the B components of DNA-dependent RNA polymerases of archaebacteria, eubacteria, and eukaryotes suggests the primeval character of the extremely thermophilic archaebacteria. In *RNA Polymerase and the Regulation of Transcription: Proceedings of the Sixteenth Steenbock Symposium,* W. S. Reznikoff, R. R. Burgess, J. E. Dahlberg, C. A. Gross, M. T. Record, and M. P. Wickens, eds. (New York: Elsevier Science Publishing Company), p. 19.]

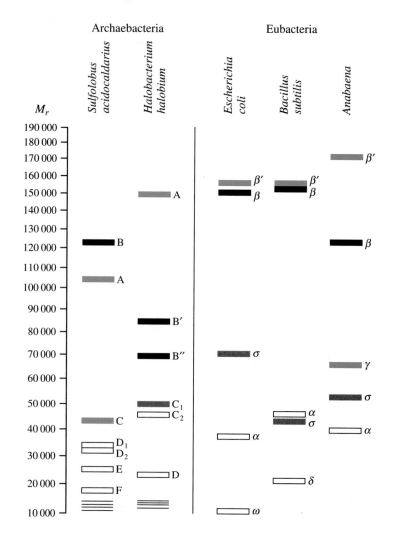

holoenzyme found in *E. coli* contains the subunit σ^{70} (M_r 70 300). This holoenzyme is irregularly shaped, with a groove at one end (Figure 27·2). The groove is large enough to accommodate about 16 base pairs of double-stranded B-DNA, and its structure resembles that of the DNA-binding site of DNA polymerase I.

Under certain conditions, the σ subunit can be dissociated from purified holoenzyme, leaving a core polymerase composed only of the α, β, and β' subunits ($\alpha_2\beta\beta'$). The weaker association of σ with the core polymerase reflects the role of σ in the cell; σ dissociates from the core polymerase after the initiation of transcription.

B. Prokaryotic RNA Polymerases Are Homologous

RNA polymerases from several different bacteria have been purified and compared. Although all function similarly, their subunit compositions vary widely. As shown in Figure 27·3, RNA polymerases from other bacterial species are composed of subunits that differ in size from those of *E. coli* RNA polymerase. Some bacterial RNA polymerases contain a number of tiny subunits (M_r ~10 000) not found in polymerases from the other species. This variation in composition among functionally similar enzymes recalls the variation mentioned in Chapter 25 among replication complexes; it suggests that, like the replisome, the transcription complexes in different organisms are composed of functionally similar but noninterchangeable components. However, in the case of RNA polymerase, these apparent differences mask an underlying sequence homology. The genes for the largest subunits of all bacterial RNA polymerases are descended from common ancestral genes. In some cases, the various domains of these proteins are encoded by separate but adjacent genes. The A and C subunits of RNA polymerase from *Sulfolobus,* for example, are encoded by adjacent genes that in other species have fused to form a single, larger gene.

C. Chain Elongation Is a Nucleotidyl-Group–Transfer Reaction

RNA polymerase catalyzes chain elongation by a mechanism almost identical to that of DNA polymerase (Section 25·3). Part of the growing RNA chain is base-paired to the DNA template strand, and incoming ribonucleoside triphosphates are tested in the active site of the polymerase for correct hydrogen bonding to the next unpaired nucleotide of the template strand. When the incoming nucleotide forms correct hydrogen bonds, RNA polymerase catalyzes nucleophilic attack by the 3′-hydroxyl group of the growing RNA chain on the α-phosphorus of the incoming ribonucleoside triphosphate. The result of this nucleotidyl-group–transfer reaction is the formation of a new phosphodiester linkage and the release of pyrophosphate (Figure 27·4, next page). This reaction is likely to be thermodynamically favorable in vivo since hydrolysis of the terminal phosphoanhydride linkage of a nucleoside triphosphate in vitro yields about 13 kJ mol^{-1} more energy than is required for formation of a new phosphoester linkage in a growing polynucleotide chain.

Like DNA polymerase, RNA polymerase catalyzes addition of nucleotides in the 5′→3′ direction and is highly processive when it is part of the complex bound to DNA. RNA polymerase differs from DNA polymerase in using ribonucleoside triphosphates (UTP, GTP, ATP, and CTP) rather than deoxyribonucleoside triphosphates (dTTP, dGTP, dATP, and dCTP). Another difference is that the growing RNA strand interacts with the template strand over only a short distance. The product of transcription is single-stranded RNA, not an RNA-DNA duplex.

The overall chain-elongation reaction is thermodynamically assisted by the subsequent hydrolysis of pyrophosphate inside the cell. Thus, two phosphoanhydride linkages are ultimately disrupted for every nucleotide added to the growing chain.

Figure 27·4
Reaction catalyzed by RNA polymerase. When an incoming ribonucleoside triphosphate correctly pairs with the next unpaired nucleotide on the DNA template strand, RNA polymerase catalyzes nucleophilic attack by the 3'-hydroxyl group of the lengthening RNA strand on the α-phosphorus of the incoming ribonucleoside triphosphate. As a result, a phosphodiester linkage is formed and pyrophosphate is released. The subsequent hydrolysis of pyrophosphate, catalyzed by inorganic pyrophosphatase, provides additional thermodynamic driving force for the reaction. (For simplicity in this and in subsequent figures, B represents a base and hydrogen bonding between bases is designated by a single dashed line.)

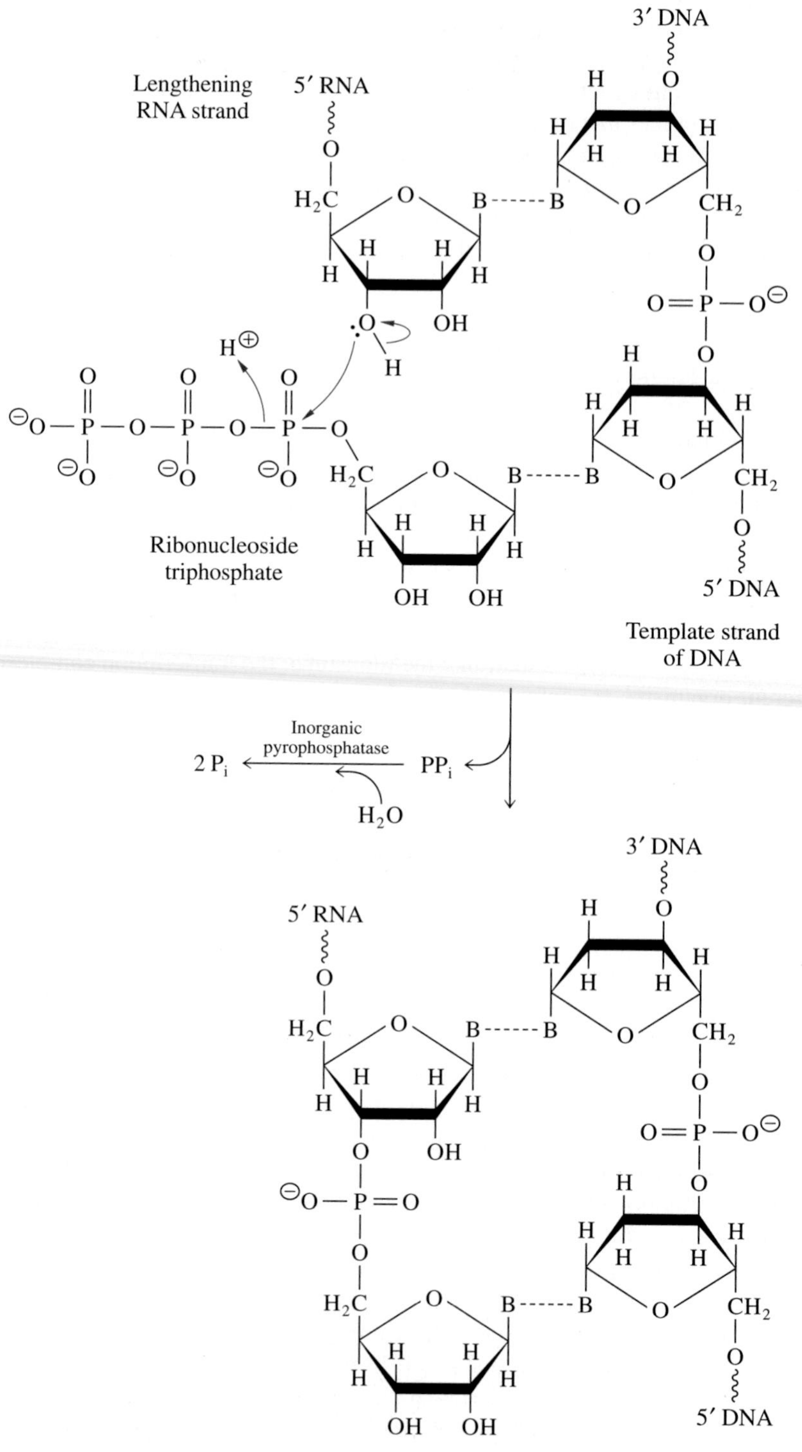

The rate of transcription is rather slow compared to the rate of DNA replication. In *E. coli*, even though the flux of nucleotides into the active site of RNA polymerase is estimated to be greater than 10^4 molecules per second, the rate of transcription ranges from 30 to 85 nucleotides per second, which is less than one-tenth the rate of DNA replication. Part of the reason for the slowness of transcription may be less efficient unwinding of the template, but the polymerization activity of the enzyme also affects the transcription rate. Apparently, most collisions between incoming ribonucleoside triphosphates and the active site are nonproductive, suggesting that RNA polymerase forms a new phosphodiester linkage only when the incoming ribonucleoside triphosphate fits the active site of the enzyme precisely. A precise fit into the active site requires base stacking and hydrogen-bond formation between the incoming ribonucleoside triphosphate and a complementary nucleotide on the template strand.

Despite the requirement for a high-precision fit, RNA polymerase does make mistakes. The error rate of RNA synthesis is 10^{-6} (one mistake for every one million nucleotides incorporated). This rate is higher than the overall error rate of DNA synthesis. Unlike DNA polymerase, RNA polymerase does not possess an exonuclease proofreading activity. Note, however, that while extreme precision in DNA replication is necessary to minimize mutations (which could be passed on to progeny), accuracy in RNA synthesis is not as crucial for survival since many copies of RNA are made from each gene, and the occasional error affects only one molecule of RNA.

27·3 Transcription Is Initiated at Promoter Sequences

The elongation reactions of RNA synthesis are preceded by a distinct initiation step in which a transcription complex is assembled at an initiation site and a ribonucleotide primer is synthesized. As in DNA replication, the assembly of an initiation complex in transcription requires proteins that can recognize and bind to the initiation site.

The sites of transcription initiation are called **promoters.** In bacteria, several genes are often cotranscribed from a single promoter; such a transcriptional unit is called an **operon.** In eukaryotic cells, each gene usually has its own promoter. There are hundreds of promoters in bacterial cells and thousands in eukaryotic cells.

The frequency of transcription initiation is usually related to the need for a gene product—either RNA or protein. For example, in cells that are dividing rapidly, the genes for ribosomal RNA are usually transcribed frequently; every few seconds a new transcription complex begins transcribing at the promoter. This process gives rise to structures such as those in Figure 27·5, which shows multiple transcription complexes on adjacent *E. coli* ribosomal RNA genes. Transcripts of increasing length are arrayed along the genes as many RNA polymerases transcribe the genes at the same time. (A striking image of the same phenomenon in eukaryotes is shown in Figure 27·27.) In contrast, some bacterial genes are transcribed only once every two generations. In these cases, initiation may only occur once every few hours. (Outside the laboratory, the average generation time of most bacteria is many hours.)

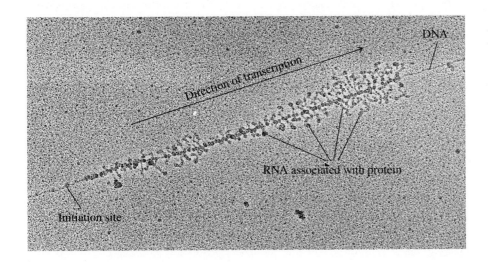

Figure 27·5
Transcription of *E. coli* ribosomal RNA genes. The genes are being transcribed from left to right. The RNA product is cleaved and associated with proteins required for processing, which occurs before transcription is complete. (Courtesy of Oscar L. Miller, Jr.)

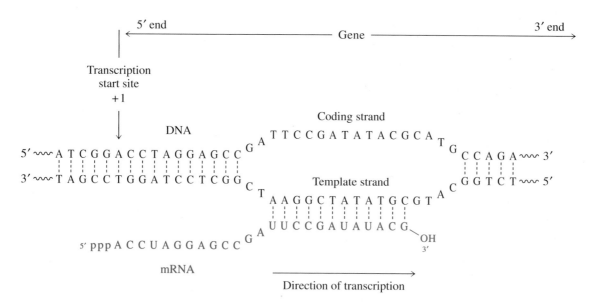

Figure 27·6
Orientation of genes. The sequence of a hypothetical gene and RNA in the process of being transcribed from it are shown. Note that genes are transcribed from the 5′ end to the 3′ end, but the actual template strand of DNA is copied from the 3′ end to the 5′ end. Growth of the ribonucleotide chain proceeds 5′→3′.

A. By Convention, Genes Have a 5′→3′ Orientation

In Section 24·2A, we introduced the convention that nucleic acid sequences are always written in the 5′→3′ direction. When a sequence of double-stranded DNA is displayed, the sequence of the top strand is written 5′→3′, and the sequence of the bottom, antiparallel strand is written 3′→5′.

Since our operational definition of a gene is a DNA sequence that is transcribed, we consider a gene to begin at the point where transcription starts (designated +1) and to end at the point where transcription terminates. RNA polymerization proceeds in the 5′→3′ direction. Consequently, in accordance with the convention established for writing nucleic acid sequences, the start site of a gene is presented on the left of a diagram and the termination site on the right. The top strand is called the **coding strand** and the bottom strand is called the **template strand** (Figure 27·6). Moving along a gene in the 5′→3′ direction is described as moving "downstream"; moving in the 3′→5′ direction is moving "upstream." Note that transcription proceeds in the 5′→3′ direction but that the template strand is copied from the 3′ end to the 5′ end. Also note that the RNA product is identical in sequence to the coding strand except that uridylate has replaced thymidylate.

B. The Transcription Complex Is Assembled at Promoter Sequences

In both prokaryotes and eukaryotes, a competent transcription complex forms when an upstream promoter sequence in DNA is recognized by one or more proteins that bind to the promoter and also to RNA polymerase. These DNA-binding proteins thus assist RNA polymerase in binding tightly to the promoter. In bacteria, the σ subunit of RNA polymerase is required for promoter recognition and formation of the transcription complex. In eukaryotes, a variety of accessory proteins are needed for the assembly of the transcription complex. These eukaryotic proteins are collectively known as **transcription factors.**

The nucleotide sequence of a promoter is one of the most important features affecting the frequency of transcription of a gene. Soon after the development of DNA-sequencing technology, many different promoters were examined. The start site, the point at which transcription actually begins, was identified, and the region upstream of this site was sequenced to find out whether the promoter sequences of different genes were similar. This analysis revealed a common pattern of the type we now refer to as a **consensus sequence,** the sequence of nucleotides found most often in each position.

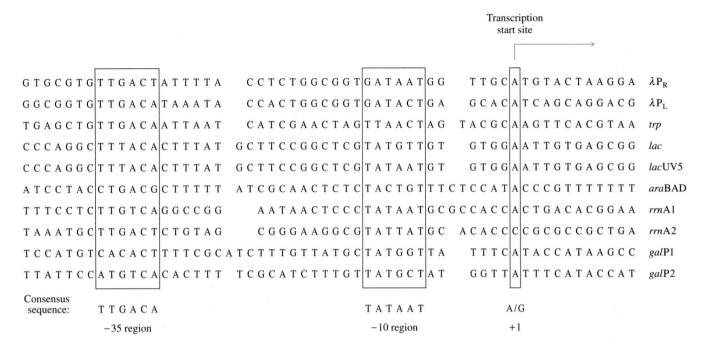

Figure 27·7
Promoter sequences from 10 bacteriophage and bacterial genes. These promoter sequences are all recognized by the σ^{70} subunit in *E. coli*. The nucleotide sequences are aligned so that their +1, −10, and −35 regions are in register. Note the degree of sequence variation at each position. The consensus sequence has been derived from a much larger data base of over 300 well-characterized promoters. In the larger sample, either purine can be found at the start site.

The consensus sequence of the most common type of promoter in *E. coli* is shown in Figure 27·7. This promoter is bipartite: there is a consensus sequence of TATAAT in the −10 region (relative to the transcription start site), and a consensus sequence of TTGACA in the −35 region. The average distance between the two blocks is 17 bp. The promoter sequences in Figure 27·7 are all taken from the top strand of the gene since convention dictates that nucleotide sequences are written in the $5' \rightarrow 3'$ direction. However, since the actual binding site is double-stranded DNA, the complementary sequence is equally important.

The −10 region is known as a **TATA box,** and the −35 region is simply referred to as the **−35 region.** Together, these two regions define the promoter, which is recognized by the *E. coli* RNA polymerase holoenzyme. The DNA consensus sequence given in Figure 27·7 is recognized by σ^{70}, the most common σ subunit in *E. coli* cells. Other σ subunits in *E. coli* recognize and bind to promoters with quite different consensus sequences (Table 27·3). Furthermore, in other species of bacteria, the promoter consensus sequences may also be very different.

Even among those *E. coli* promoters recognized by the σ^{70} subunit, very few have sequences that are exact matches to the consensus sequence shown in Figure 27·7.

Table 27·3 *E. coli* σ subunits

Subunit	Gene	Genes transcribed	Consensus sequence −35	Consensus sequence −10
σ^{70}	*rpo*D	Many	TTGACA	TATAAT
σ^{54}	*rpo*N	Nitrogen metabolism	None	CTGGCACNNNNNTTGCA*
σ^{38}	*rpo*S	Stationary phase	?	TATAAT
σ^{28}	*fla*I	Flagellar synthesis and chemotaxis	TAAA	GCCGATAA
σ^{32}	*rpo*H	Heat shock	CTTGAA	CCCATNTA*
$\sigma^{\text{gp}55}$	gene *55*	Bacteriophage T4	None	TATAAATA

* *N* represents any nucleotide.

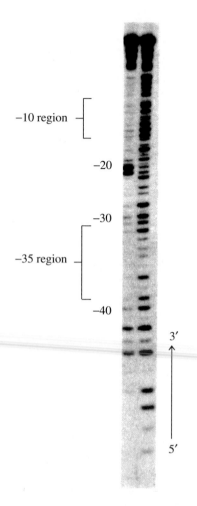

Figure 27·8
Autoradiograph obtained in a footprinting experiment. The lane on the right shows the ladder of fragments obtained in the absence of RNA polymerase; the lane on the left, the ladder of fragments obtained in the presence of RNA polymerase. Note that in the presence of RNA polymerase, the nucleotides in the −10 and −35 regions are protected from nuclease cleavage, indicating that RNA polymerase contacts both of these regions. (Courtesy of D. J. Galas.)

This sequence is called a consensus sequence precisely because it describes the most common nucleotides at given positions and not an exact sequence. In some cases, the match is quite poor, with G and C present at positions normally occupied by A and T. Such promoters are usually associated with genes that are transcribed infrequently. Other promoter sequences, such as the promoters for the ribosomal RNA operons, resemble the consensus sequence quite closely. These genes are generally transcribed very efficiently. Observations such as these suggest that the consensus sequence describes the most efficient initiation site for the RNA polymerase holoenzyme. The promoter sequence of each gene has likely been optimized by natural selection to fit the requirements of the cell. A weak promoter is ideal for a gene whose product is not needed in large quantities, whereas a strong promoter is necessary for the production of large amounts of gene product.

When RNA polymerase and DNA containing a promoter are mixed in a buffered solution of moderate ionic strength, a polymerase-DNA complex forms. The polymerase-binding site on DNA can be determined by an experimental technique known as **footprinting,** or nuclease protection. In footprinting, a DNA fragment containing a promoter is labelled at one end with ^{32}P and digested with just enough endonuclease to produce one or two single-stranded nicks per fragment. Nicks occur randomly in each molecule and, on denaturation, generate a population of single-stranded fragments of various sizes. The fragments can be separated according to size by gel electrophoresis. Autoradiography of this gel reveals only those fragments with ^{32}P-labelled ends. In the absence of RNA polymerase, the result is a ladder of labelled fragments corresponding to an endonuclease cut at each nucleotide along the DNA molecule. When the same experiment is carried out in the presence of RNA polymerase, the polymerase covers some of the DNA, protecting those phosphodiester linkages from the endonuclease and causing the ladder of fragments to contain gaps. The gaps correspond to nucleotides that are covered and protected by RNA polymerase. An example of an autoradiograph from such an experiment is shown in Figure 27·8.

These and other experiments reveal that RNA polymerase binds to an extended region of DNA that includes both the TATA box and the −35 region. The protected sequence covers about 70 base pairs, from position −50 to +20. The length of this binding region is approximately consistent with the size of the RNA polymerase holoenzyme, which is about 16 nm long.

Within the protected sequence, the specific nucleotides that contact RNA polymerase can be determined by another experimental protocol. In this technique, a population of identical DNA molecules is randomly modified at particular nucleotides. For example, the alkylating agent dimethylsulfate methylates both adenylate and guanylate (Figure 27·9). DNA is exposed to dimethylsulfate at a level that results in modification of only a few nucleotides per molecule. The collection of randomly modified DNA molecules is then mixed with RNA polymerase, and enzyme-DNA complexes are isolated. Modification of nucleotides normally in close contact with RNA polymerase inhibits polymerase binding; therefore only DNA molecules containing modified nucleotides that are not normally in contact with RNA polymerase will still bind to the enzyme. The results of such experiments reveal that RNA polymerase binds predominantly to one face of the double-stranded helix and contacts several nucleotides in the major groove (Figure 27·10). Although some of the polymerase-DNA contacts occur in regions that are not part of the consensus sequence, most are localized to the −10 and −35 regions. Because these regions are separated by about 17 nucleotides, they are approximately two turns apart on the same side of the helix.

C. Promoter Recognition in *E. coli* Depends on the σ Subunit

The effect of σ subunits on promoter recognition can best be explained by comparing the binding properties of the core polymerase ($\alpha_2\beta\beta'$) and the holoenzyme

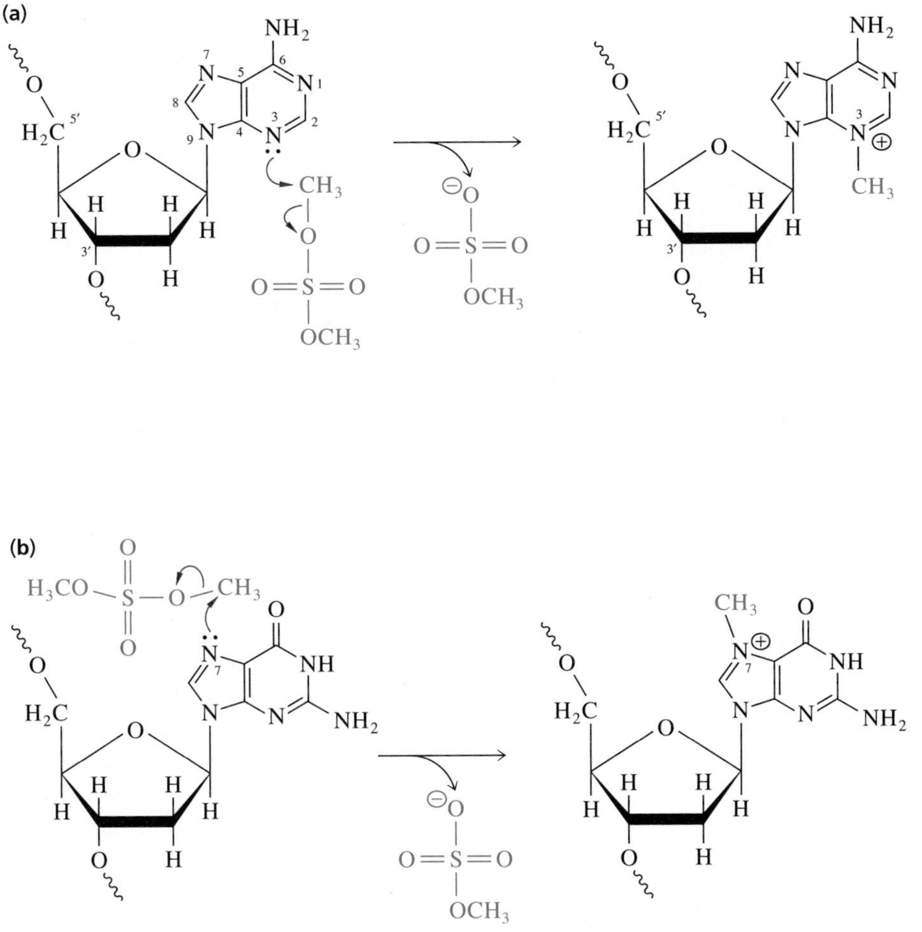

(a)

Figure 27·9
Chemical modification of DNA. DNA can be chemically modified by dimethylsulfate (blue), which methylates **(a)** adenylate and **(b)** guanylate. Methylation occurs at different sites on adenylate and guanylate due to the different chemical reactivity of the nitrogen atoms.

(b)

Figure 27·10
RNA polymerase–promoter contacts in the *lac*UV5 promoter. RNA polymerase binds predominantly to one face of the double-stranded DNA molecule. Here, the DNA helix is marked with a dot pattern to show contact points with RNA polymerase. Most of the contacts are located in the −10 and −35 regions of the DNA molecule. Many of the contacts with the stacked bases occur in the major groove.

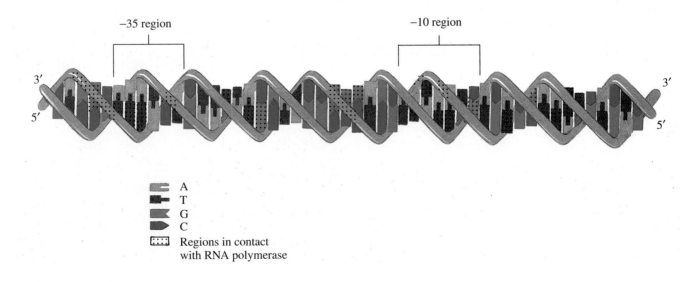

−35 region   −10 region

A
T
G
C
Regions in contact
with RNA polymerase

$(\alpha_2\beta\beta'\sigma^{70})$. The core polymerase, which lacks a σ subunit, binds to DNA non-specifically; it has no greater affinity for promoters than for any other DNA sequence (its association constant, K_{assoc}, is approximately $10^{10}\,M^{-1}$). Once formed, this DNA-protein complex dissociates slowly ($t_{1/2} \sim 60\,min$). In contrast, the holoenzyme, which contains the σ subunit, binds more tightly to promoter sequences than the core polymerase ($K_{assoc} \sim 2 \times 10^{11}\,M^{-1}$) and forms more stable complexes ($t_{1/2} \sim 2$–$3\,h$). Although the holoenzyme binds preferentially to promoter sequences, it also has appreciable affinity for nonpromoter DNA ($K_{assoc} \sim 5 \times 10^6\,M^{-1}$). The complex formed by nonspecific binding of the holoenzyme to DNA dissociates very rapidly ($t_{1/2} \sim 3\,s$). These association constants reveal how the σ subunit affects the binding properties of the core polymerase: 1) σ decreases the affinity of the core polymerase for nonpromoter sequences, and 2) σ increases the affinity of the core polymerase for specific promoter sequences.

The association constants reveal nothing about the mechanism by which the RNA polymerase holoenzyme finds a promoter. We might expect the holoenzyme to search for a promoter by continuously binding and dissociating until it encounters a promoter sequence. Such binding would be a second-order reaction, and its rate would be limited by the rate at which the holoenzyme diffuses in three dimensions. However, the rate constant for promoter binding has been measured experimentally as $k = 10^{10}\,M^{-1}\,s^{-1}$, which is about two orders of magnitude greater than the maximum theoretical value for a diffusion-controlled, second-order reaction.

The remarkable rate of the promoter-binding reaction is achieved by one-dimensional diffusion of RNA polymerase along the length of the DNA molecule. During the short period of time that the enzyme is bound nonspecifically, RNA polymerase can scan 2000 base pairs in its search for a promoter sequence. Several other sequence-specific DNA-binding proteins, such as restriction enzymes, locate their DNA-binding sites in a similar manner.

Once transcription has been initiated and the polymerase proceeds past the promoter, the promoter is said to be cleared, and it can be recognized by another holoenzyme. Subsequent binding of other RNA polymerases may occur rapidly because RNA polymerase molecules in the cell are in excess relative to promoters. An average bacterial cell contains approximately 5000 molecules of RNA polymerase;

Figure 27·11
Estimate of the distribution of the approximately 5000 RNA polymerase molecules typically found in an *E. coli* cell. Very few molecules are free in the cytosol, yet only half of all RNA polymerases are actively transcribing.

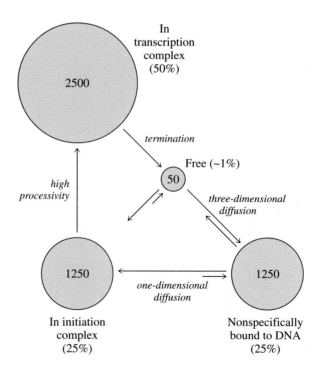

their probable distribution is illustrated in Figure 27·11. Because promoter clearance is slow for some genes, many RNA polymerase molecules are located at promoter sequences. A significant number, however, are found in transcription complexes that are in the process of transcribing genes. Nearly all RNA polymerase molecules that are involved in neither initiation nor elongation are bound nonspecifically to DNA rather than existing free in the cytosol.

D. Formation of the Transcription Bubble and Synthesis of a Primer Are Rate-Limiting Steps in Initiation

Initiation of transcription is slow, even though the holoenzyme may search for and bind to the promoter very quickly. In fact, initiation is often the rate-limiting step in transcription because, like initiation of DNA replication, it requires unwinding of the DNA helix and synthesis of an RNA primer for chain elongation. During DNA replication, these steps are carried out by a helicase and a primase, but during transcription, these steps are carried out by the RNA polymerase holoenzyme itself. RNA polymerase is thus capable of initiating RNA synthesis in the absence of an exogenous primer.

Unwinding of DNA at the initiation site involves an isomerization reaction in which RNA polymerase (R) and the promoter (P) shift from a closed complex (RP_c) to an open complex (RP_o). In the closed complex, the DNA is double stranded; in the open complex, 18 base pairs of DNA are denatured, forming a **transcription bubble.** Formation of the open complex is usually the slowest step of the initiation events.

Once the open complex has formed, the template strand is positioned at the polymerization site of the enzyme. In the next step, a phosphodiester linkage is formed between two ribonucleoside triphosphates that have diffused into the active site and formed hydrogen bonds with the +1 and +2 nucleotides of the template strand. This reaction is slower than the analogous polymerization reaction during chain elongation, in which one of the substrates (the growing RNA chain) is held in position by the formation of a short RNA-DNA helix.

Additional nucleotides are then added to the first dinucleotide to create a short RNA primer that is paired with the template strand. When the primer reaches approximately 10 nucleotides in length, the RNA polymerase holoenzyme undergoes a transition from the initiation to the elongation mode, and the transcription complex moves away from the promoter along the DNA template (promoter clearance). The initiation reactions are summarized in Equation 27·1.

$$R + P \underset{}{\overset{K_{assoc}}{\rightleftharpoons}} RP_c \xrightarrow[isomerization]{} RP_o \xrightarrow[\substack{promoter \\ clearance}]{} \qquad (27\cdot1)$$

As noted earlier, the holoenzyme has a much greater affinity for the promoter than for any other DNA sequence. Because of this tight binding, it resists moving away from the initiation site. On the other hand, during elongation, the core polymerase binds nonspecifically to all DNA sequences to form a highly processive complex. The transition from initiation of transcription to chain elongation is associated with a conformational change in the holoenzyme that causes release of the σ subunit. Without σ, the enzyme no longer binds specifically to the promoter and is able to leave the site of initiation. At this point, several accessory proteins bind to the core polymerase to create the complete protein machine required for elongation.

The binding of one of these accessory proteins, NusA (M_r 55 000), helps convert RNA polymerase to the elongation form. NusA slows down the elongation reaction, especially when the concentration of nucleoside triphosphates is low. NusA also interacts with other accessory proteins and plays a role in termination (Section 27·6). Transcription initiation in *E. coli* is summarized in Figure 27·12.

Figure 27·12
Initiation of transcription.

(a) RNA polymerase holoenzyme has significant nonspecific affinity for DNA.

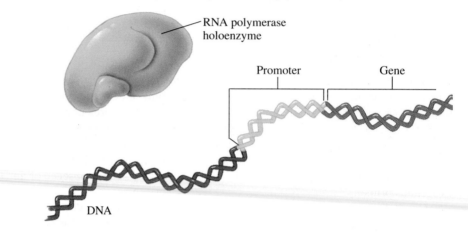

(b) After binding to DNA, the holoenzyme conducts a one-dimensional search for a promoter.

(c) When a promoter is found, the RNA polymerase holoenzyme and the promoter form a closed complex.

(d) Isomerization of the closed complex results in an open complex and the formation of a transcription bubble at the initiation site. A short RNA primer is then synthesized.

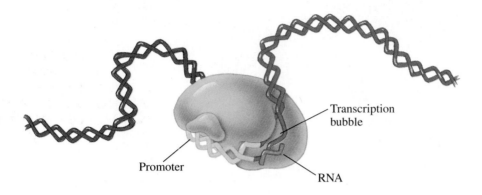

Transcription bubble

Promoter

RNA

(e) The σ subunit dissociates from the core enzyme, and RNA polymerase clears the promoter. Several accessory proteins, including NusA, are part of the elongation complex.

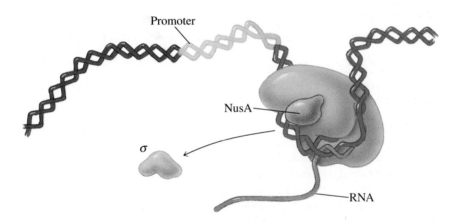

Promoter

NusA

σ

RNA

27·4 RNA Polymerase Catalyzes Several Topological Changes Within the Transcription Bubble

RNA polymerase contacts a stretch of DNA approximately 30 base pairs long during the elongation stage of RNA synthesis. (Note that RNA polymerase covers a shorter stretch of DNA during elongation than during initiation since switching from initiation to elongation is accompanied by a conformational change in the enzyme.) Within the 30 bp region, 18 base pairs of DNA are unwound, exposing the template strand to the enzyme. As many as 12 nucleotides of the template strand form base pairs with newly synthesized RNA to form a short hybrid RNA-DNA helix.

The length of the double-stranded region, up to 12 base pairs, corresponds to one turn or less of a helix in an A conformation (Figure 27·13, next page). (Recall from Section 24·8 that double-stranded RNA preferentially adopts the A conformation.) One turn of the helix allows the maximum number of interactions possible between the template strand and the nascent RNA strand without impeding the elongation reaction. If the RNA-DNA hybrid exceeded one turn of the helix, the newly synthesized RNA strand would wrap more than one full turn around the template DNA strand, pass through the transcription bubble, and become topologically linked to the DNA.

Figure 27·13

Hybrid RNA-DNA helix formed within the transcription bubble during elongation. The RNA-DNA hybrid adopts an A conformation, in which 12 base pairs correspond to one turn of a helix. During transcription, the extent of the RNA-DNA hybrid may at times be less than 12 base pairs.

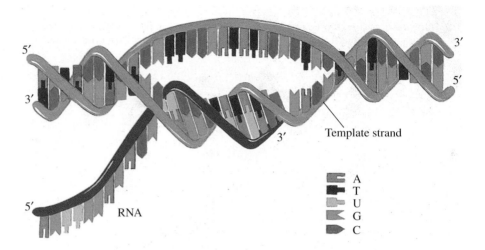

When a new nucleotide joins the 3′ end of the growing RNA chain, the following steps must occur:

1. The length of the RNA-DNA hybrid must be reduced by disrupting one base pair at the upstream end of the transcription bubble.

2. A DNA base pair must be disrupted at the downstream end of the bubble.

3. A DNA base pair must be re-formed at the upstream end of the bubble.

4. RNA polymerase must advance along the DNA by one nucleotide, positioning its active site over the next unpaired nucleotide on the template strand.

These steps are carried out by the combined subunits of RNA polymerase in *E. coli* and do not require additional cofactors or enzymes. The disruption of a base pair in the transcription bubble is coupled to the formation of another base pair elsewhere, thus maintaining an approximate thermodynamic equilibrium.

The elongation reaction has been analyzed by Peter H. von Hippel, who pointed out that if the four steps listed above require energy, it can come only from the cleavage of phosphoanhydride linkages during nucleotidyl-group transfer. (Recall that two phosphoanhydride linkages are broken for each nucleotide added.) However, the enzymatic and topological changes are unlikely to be directly coupled since these activities occur at distinct sites, separated by a considerable distance. Hence, it appears that additional energy is not required either to denature and renature DNA at the ends of the transcription bubble or to form and disrupt a base pair of the RNA-DNA hybrid.

The formation of double-stranded nucleic acid is energetically favorable, and depending on which base pair forms, each of the renaturation events releases between 4 and 12 kJ mol^{-1}. Most of this energy is recovered by RNA polymerase through the coupling of denaturation and renaturation steps. The thermodynamically most favorable case involves forming two G/C base pairs and disrupting an A/T and an A/U base pair within the transcription bubble. This process results in an overall standard free-energy change of −21 kJ mol^{-1}. The least favorable case is disrupting two G/C base pairs and forming an A/T and an A/U base pair; this process is not thermodynamically favored ($\Delta G^{\circ\prime} = +21$ kJ mol^{-1}). The input of energy required to transcribe these sequences is part of the activation energy for the overall reaction catalyzed by RNA polymerase. Since the magnitude of activation energy strongly influences the rate of a reaction, it is not surprising that the nucleotide sequence of a gene affects its rate of transcription.

Transcription presents one other topological problem: as the transcription complex moves along a gene, either RNA polymerase must trace a helical path around

DNA or the DNA must rotate as it passes through the transcription complex. Except for the transcription of the beginning of a gene, it is difficult to envisage only rotation of RNA polymerase since such rotation would be impeded by the frictional drag of the growing RNA chain, especially if the chain were covered with ribosomes. A combination of RNA polymerase rotation and DNA rotation probably occurs in vivo, with the result that some positive supercoiling is introduced ahead of RNA polymerase and some negative supercoiling is introduced behind it. Since the DNA of both bacteria and eukaryotes is held in large loops by a combination of proteins and RNA, the elongation complex encounters increasing resistance unless the supercoils are relieved. As in DNA replication, supercoils formed during transcription are removed by topoisomerases, whose activities are required for efficient transcription.

27·5 The *E. coli* Genome Is Arranged to Prevent Polymerases from Colliding

As we saw in Chapter 25, DNA replication begins at a unique origin on the circular *E. coli* chromosome and proceeds bidirectionally until the termination site is reached at a point opposite the origin. The protein machine for DNA replication is a large structure that is highly processive and moves rapidly (1000 base pairs per second). The transcription complex is also large and highly processive, but it moves 20 times more slowly than a DNA replication fork. What happens when replication forks collide with transcription complexes?

When a replication fork passes through a gene in the same orientation in which it is transcribed ($5' \rightarrow 3'$), it overtakes the transcription complex. In some species, the DNA replication machinery contains a "cowcatcher" protein that strips off proteins ahead of the replication fork. In *E. coli*, the replication fork passes efficiently through an actively transcribed gene at a rate that is close to the rate for DNA that is devoid of transcription complexes (Figure 27·14). On the other hand, if the replication fork enters a gene from the 3' end, DNA polymerase and RNA polymerase collide head-on, which causes the replication complex to pause and disrupts the transcription complex. Although RNA polymerase is eventually dislodged in head-on collisions, dislodging is much slower than in same-orientation collisions.

(a)

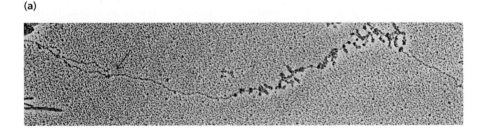

(b)

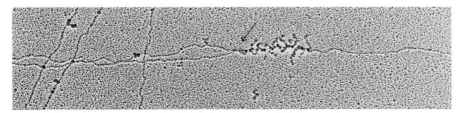

Figure 27·14
Electron micrograph of a replication fork passing through actively transcribed genes in *E. coli*. (a) A replication fork (arrow) is approaching the actively transcribed ribosomal RNA genes from the left, in the same orientation as transcription. (b) The replication fork has entered the transcribed region and stripped off the transcription complexes. (Courtesy of Sarah French.)

Figure 27·15
Transcriptional orientation on the *E. coli* chromosome of genes for the synthesis of DNA, RNA, and protein. Long arrows indicate the direction of replisome movement from the origin (*oriC*); short arrows indicate the direction of transcription of the genes marked. Each arrow outside the circular map refers to a gene encoding a part of the protein synthesis machinery; the sites of seven rRNA (*rrn*) operons in this strain are labelled. Each arrow inside the circular map refers to a gene necessary for DNA or RNA synthesis. [Adapted from Brewer, B. J. (1988). When polymerases collide: replication and the transcriptional organization of the *Escherichia coli* chromosome. *Cell* 53:679–686.]

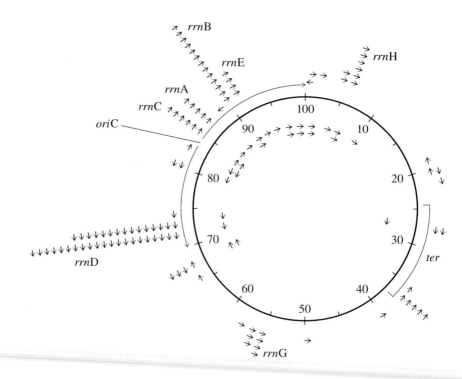

Since some genes may have up to 70 active transcription complexes bound at any one time, such repeated head-on collisions could considerably slow DNA replication to the point that the cell would not survive. However, Bonita Brewer has noted that on many actively transcribed *E. coli* genes, transcription and replication complexes move in the same direction (Figure 27·15). In fact, more than two-thirds of all *E. coli* genes are oriented in the same direction as DNA replication. Thus, the genome has evolved over millions of years to favor an orientation of active genes that avoids head-on collisions with replisomes.

Gene orientation relative to the origin of DNA replication has topological implications. Positively supercoiled DNA is created ahead of both the transcription and DNA replication complexes. This may impede the progress of both complexes as they approach each other head-on. On the other hand, when replication and transcription are codirectional, negative supercoils behind the transcription complex are cancelled by positive supercoils ahead of the replisome. The same considerations apply to transcription of adjacent genes; there may be a selective advantage to transcribing adjacent genes in the same direction.

Eukaryotic genes are not globally aligned in the same manner as those on the bacterial chromosome. Recall that eukaryotic DNA replication is much slower; the rate is comparable to that of transcription. Thus, a slowdown due to head-on collisions may not be detrimental. Furthermore, there are many origins of replication on eukaryotic chromosomes, and genes may be arranged locally with respect to the nearest origin.

27·6 Termination of Transcription Requires Special Terminator Sequences

Termination is a distinct event in transcription because, just as the transcription complex is actively assembled, it must be actively disassembled after elongation is complete. Termination is signalled by information contained at sites in the DNA sequence being transcribed. A priori, we might expect terminators to be sites where a

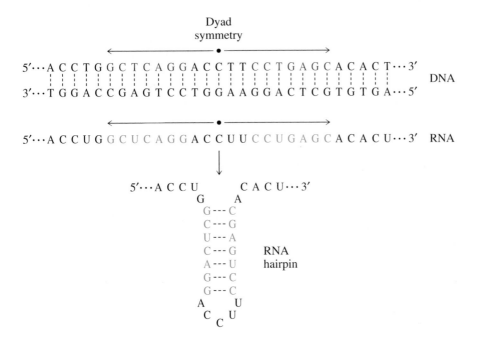

Dyad
symmetry

5′···A C C T G G C T C A G G A C C T T C C T G A G C A C A C T···3′   DNA
3′···T G G A C C G A G T C C T G G A A G G A C T C G T G T G A···5′

5′···A C C U G G C U C A G G A C C U U C C U G A G C A C A C U···3′   RNA

5′···A C C U   C A C U···3′
       G   A
       G --- C
       C --- G
       U --- A
       C --- G   RNA
       A --- U   hairpin
       G --- C
       G --- C
       A   U
        C   U
           C

Figure 27·16
Formation of a hairpin structure in RNA. The transcribed DNA sequence contains a region of dyad symmetry. Complementary sequences in RNA may form a region of secondary structure known as a hairpin. A three-dimensional representation of stacked, double-stranded RNA in such a structure is shown in Figure 24·33.

specific protein binds and disrupts the elongation complex, but this is not the mechanism that has evolved in prokaryotes. Instead, termination often occurs because the elongation complex is less stable when transcribing certain DNA sequences. In those cases where termination requires a specific factor, the factor interacts with the newly synthesized RNA, not with the DNA.

Termination is subject to regulation by factors that can bypass the normal transcription termination signals. We will discuss an example of the coupling between transcription termination and protein synthesis in Chapter 30 and additional examples of the regulation of termination in Chapter 31.

A. RNA Polymerase Pauses While Transcribing Some Sequences

As we saw in the previous section, the rate of RNA synthesis during elongation is somewhat sequence dependent: the topological changes are thermodynamically less favorable when the transcription complex is passing through a sequence rich in G/C base pairs because it is more difficult to open the transcription bubble and to denature the RNA-DNA hybrid. When such sequences are encountered, transcription can be slowed by a factor of 10 to 100. Sites within a gene where transcription slows are called **pause sites.**

Pausing is exaggerated at palindromic sequences. At such sites, the newly synthesized RNA can form a hairpin structure (Figure 27·16). Formation of an RNA hairpin pulls the RNA strand up through the transcription bubble and destabilizes the RNA-DNA hybrid in the elongation complex by prematurely stripping off part of the newly transcribed RNA. This partial disruption of the transcription bubble probably causes the transcription complex to cease elongation until the hybrid re-forms.

The RNA-DNA hybrid usually re-forms spontaneously due to fluctuations that cause the bases in the hairpin structure to separate, thus making them available for base-pairing with the DNA template strand. In some cases, however, the RNA hairpin may interact with NusA protein in the transcription complex. NusA increases pausing at such palindromic sites, perhaps by stabilizing the hairpin structure, as illustrated in Figure 27·17. Depending on the structure of the hairpin, the transcription complex may pause at such sites for anywhere from 10 seconds to 30 minutes.

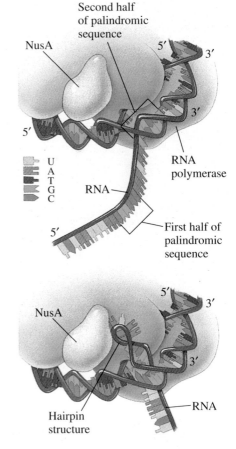

Figure 27·17
RNA polymerase at a pause site. Formation of an RNA hairpin disrupts the normal, 12 bp RNA-DNA hybrid and causes the transcription complex to pause. The RNA-DNA hybrid usually re-forms spontaneously due to slight structural fluctuations. In some cases, however, NusA may interact with the hairpin structure and lengthen the pause.

Pause sites are significant physiologically; by slowing the rate of transcription, they allow other events to occur that may abort transcription. Pausing is also important in the normal termination of transcription at the ends of genes. In *E. coli,* there are two kinds of transcription termination reactions: simple termination, which does not require additional protein factors, and rho-mediated termination, which requires a special termination factor named rho.

B. In Simple Termination, Specific Sequences Reduce the Stability of the RNA-DNA Hybrid

A number of sequences at the ends of genes in *E. coli* play a role in transcription termination. In many of these transcription terminators, a region of dyad symmetry that is rich in G and C is followed by a stretch of polyA on the template strand. As described above, when the palindromic region is transcribed, the newly synthesized RNA forms a hairpin structure and destabilizes the RNA-DNA hybrid. In this case, the RNA-DNA hybrid is maintained only by dA/rU base pairs. Such pairs are the weakest of all hybrid base pairs and are easily disrupted, leading to release of the RNA, renaturation of the transcription bubble, and dissociation of RNA polymerase from the gene. The weakness of a run of dA/rU base pairs is due partly to formation of two, rather than three, hydrogen bonds between these bases. The stacking interactions in such a hybrid helix are also unfavorable because adjacent dA residues are not compatible with the A form of the double helix.

The efficiency of simple termination can be enhanced by other proteins. For example, at some terminators, NusA increases the frequency of termination, probably by interacting with the RNA hairpin as described above. At other terminators, the efficiency of termination is greatly increased when a protein factor called τ is present. The mechanism of action of τ is not yet known.

C. Some Termination Sites Are Rho Dependent

Some genes lack terminators with a self-complementary GC-rich sequence, and others do not have a stretch of polyA. Termination of transcription in many of these cases requires a specific regulatory protein named rho. Terminator sequences that require rho are said to be rho dependent. In the absence of rho, transcription does not stop, and RNA polymerase continues to copy the template strand of DNA.

Rho exists in the cytosol as a hexamer of identical subunits. It is a potent ATPase with an affinity for single-stranded RNA and may act as an RNA-DNA helicase. Rho binds single-stranded RNA exposed behind a paused transcription complex and may actively wrap approximately 80 nucleotides of RNA around itself. In this process, ATP is hydrolyzed to ADP + P_i (Figure 27·18), and the RNA dissociates from the transcription bubble.

Termination appears to be due to both destabilization of the RNA-DNA hybrid and direct contact between the transcription complex and rho as it wraps the RNA strand around itself. Rho can also bind to NusA and other accessory proteins. This binding may cause the transcription complex to change conformation and disassemble.

Rho is usually prevented from binding to newly synthesized mRNA by the presence of ribosomes on the single-stranded RNA molecule. In bacterial cells, transcription of protein-encoding genes is intimately coupled to translation: before transcription is complete, ribosomes bind mRNA and begin synthesizing protein. The presence of the protein synthesis machinery on newly synthesized mRNA masks the single-stranded RNA and prevents rho from binding. When transcription passes beyond the site where protein synthesis terminates, the single-stranded RNA becomes exposed and accessible to rho. The transcription complex stalls at the transcription pause site present at the end of a gene, allowing rho to bind to the RNA transcript and terminate transcription.

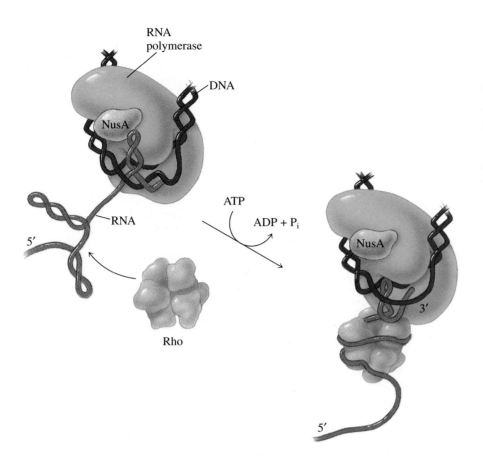

RNA polymerase

DNA

NusA

RNA

5′

Rho

ATP

ADP + P$_i$

NusA

3′

5′

Figure 27·18
Rho-dependent termination of transcription in *E. coli*. RNA polymerase is stalled at a pause site. Rho binds to the exposed, newly synthesized RNA. This binding is accompanied by ATP hydrolysis. Rho probably wraps the nascent RNA chain around itself, thereby destabilizing the RNA-DNA hybrid and terminating transcription. [Adapted from Platt, T. (1986). Transcription termination and the regulation of gene expression. *Annu. Rev. Biochem.* 55:339–372.]

The affinity of rho for RNA that is not covered by ribosomes allows pause sites within a gene to become termination sites when protein synthesis slows due to lack of amino acids or some other component of the translation machinery. Under these conditions, mRNA is no longer protected by ribosomes, and rho binds, terminating transcription. This mechanism prevents the inefficient production of mRNA that cannot be translated.

D. Rho-Dependent Termination Can Be Inhibited by Antiterminators

Pause sites in the genes for ribosomal RNA should lead to transcription termination by rho since the nascent transcripts are not protected by ribosomes. In order to prevent this, the transcription complexes on rRNA genes are modified so that they do not respond to rho. This mechanism is referred to as **antitermination.**

Near the promoters of rRNA genes, there is a DNA sequence known as boxA. Following transcription of this region, the boxA sequence of the RNA transcript is bound by the antiterminators NusB and S10, a ribosomal protein also known as NusE. These factors associate with the elongation complex in a complex that also includes NusG (Figure 27·19). The modified transcription complex is resistant to rho-dependent termination because interaction between rho and RNA polymerase is prevented.

Note that the presence of a binding site at the beginning of the RNA transcript serves as a tether, which restricts the antiterminator factors to a particular transcription complex. NusB, S10, and NusG only act as antiterminators of rRNA genes, and they are only effective against rho-dependent termination. The normal terminators at the ends of these genes are strong, rho-independent terminators.

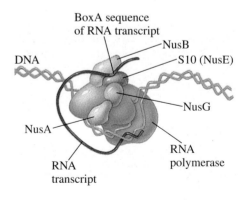

BoxA sequence of RNA transcript

NusB

DNA

S10 (NusE)

NusG

NusA

RNA polymerase

RNA transcript

Figure 27·19
Antiterminators bound to the transcription complex transcribing ribosomal RNA genes. NusB and S10 bind to the boxA sequence on the nascent transcript and to RNA polymerase in association with NusG. The modified complex is resistant to rho-dependent termination.

Table 27·4 Eukaryotic RNA polymerases

Polymerase	Location	Copies per cell	Products	Polymerase activity of cell
RNA polymerase I	Nucleolus	40 000	35–47S pre-rRNA	50–70%
RNA polymerase II	Nucleoplasm	40 000	mRNA precursors U1, U2, U4, and U5 snRNA	20–40%
RNA polymerase III	Nucleoplasm	20 000	5S rRNA tRNA U6 snRNA 7S RNA Other small RNA molecules	10%
Mitochondrial RNA polymerase	Mitochondrion	?	Products of all mitochondrial genes	?
Chloroplast RNA polymerase	Chloroplast	?	Products of all chloroplast genes	?

Figure 27·20

Schematic representation of SDS-PAGE patterns of subunits from purified RNA polymerases from the yeast *S. cerevisiae* and the bacterium *E. coli*. Identically colored subunits are related by sequence homology. Subunits of eukaryotic RNA polymerases connected by lines are identical. [Adapted from Zillig, W., and Gropp, F. (1987). The distribution of antigenic determinants in the B components of DNA-dependent RNA polymerases of archaebacteria, eubacteria, and eukaryotes suggests the primeval character of the extremely thermophilic archaebacteria. In *RNA Polymerase and the Regulation of Transcription: Proceedings of the Sixteenth Steenbock Symposium,* W. S. Reznikoff, R. R. Burgess, J. E. Dahlberg, C. A. Gross, M. T. Record, and M. P. Wickens, eds. (New York: Elsevier Science Publishing Company), p. 19.]

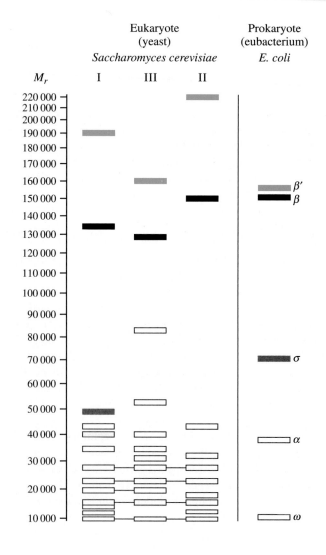

27·7 Eukaryotic and Prokaryotic RNA Polymerases Are Homologous

The same processes carried out by a single RNA polymerase in *E. coli* are carried out in eukaryotes by several similar enzymes. In eukaryotes, three different RNA polymerases transcribe nuclear genes, and other RNA polymerases are found in mitochondria and chloroplasts. The three nuclear enzymes each transcribe different classes of genes: RNA polymerase I transcribes class I genes, those that encode large ribosomal RNA molecules; RNA polymerase II transcribes class II genes, those that encode proteins and a few that encode small RNA molecules; and RNA polymerase III transcribes class III genes, those that encode a number of small RNA molecules including tRNA and 5S rRNA (Table 27·4). (Some of the RNA molecules listed in the table will be discussed in subsequent chapters.)

The mitochondrial RNA polymerase is a monomeric enzyme encoded by the nuclear genome. In yeast, the mitochondrial RNA polymerase is substantially similar in amino acid sequence to the RNA polymerases of T3 and T7 bacteriophages. This similarity suggests that these enzymes share a common ancestor. Unlike mitochondria, chloroplasts encode their own RNA polymerase. The genes encoding the chloroplast RNA polymerase are similar in sequence to those of eubacterial RNA polymerase, consistent with the theory that chloroplasts, like mitochondria, originated from bacterial endosymbionts in ancestral eukaryotic cells.

The three nuclear RNA polymerase enzymes differ in subunit composition, although they have several smaller polypeptides in common (Figure 27·20). The presence of these small subunits makes the eukaryotic nuclear RNA polymerases more complex than the prokaryotic RNA polymerases. The actual number of subunits in each polymerase varies somewhat among organisms, but there are always 2 large subunits and from 7 to 12 smaller ones. The two largest subunits of each polymerase are similar in sequence to the β' and β subunits of *E. coli*. Figure 27·21 shows regions of similarity in the amino acid sequences of the β' subunit of *E. coli* and the largest subunit of RNA polymerase II from the yeast *Saccharomyces cerevisiae* and the fruit fly *Drosophila melanogaster.*

The three nuclear RNA polymerases found in eukaryotic cells transcribe different genes, and each requires different accessory transcription factors in vivo. The partitioning of cellular transcription among three polymerases probably reflects an efficient division of labor. Eukaryotic cells are larger than prokaryotic cells and contain considerably more ribosomes and genes. Conceivably, in an early primitive eukaryote, a novel RNA polymerase arose that could accommodate the heavy

Figure 27·21
Amino acid sequence similarities among the largest subunits of RNA polymerases from different organisms. The genes for the β' subunit of *E. coli* and the largest subunit of RNA polymerase II from the yeast *S. cerevisiae* and the fruit fly *D. melanogaster* have been cloned and sequenced. The deduced amino acid sequences show considerable similarity in spite of their distant evolutionary relationships. The C-terminus of each eukaryotic protein contains multiple repeats of the unusual sequence –Ser–Pro–Tyr–Ser–Pro–Thr. The function of this sequence is not known. [Adapted from Greenleaf, A. L., Jokerst, R. S., Zehring, W. A., Hamilton, B. J., Weeks, J. R., Sluder, A. E., and Price, D. H. (1986). *Drosophila* RNA polymerase II: genetics and in vitro transcription. In *RNA Polymerase and the Regulation of Transcription: Proceedings of the Sixteenth Steenbock Symposium,* W. S. Reznikoff, R. R. Burgess, J. E. Dahlberg, C. A. Gross, M. T. Record, M. P. Wickens, eds. (New York: Elsevier Science Publishing Company), p. 460.]

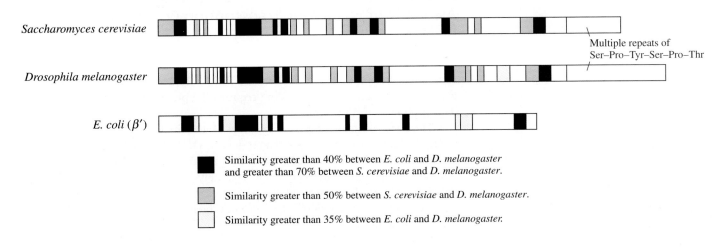

Saccharomyces cerevisiae

Multiple repeats of
Ser–Pro–Tyr–Ser–Pro–Thr

Drosophila melanogaster

E. coli (β')

■ Similarity greater than 40% between *E. coli* and *D. melanogaster* and greater than 70% between *S. cerevisiae* and *D. melanogaster*.

▨ Similarity greater than 50% between *S. cerevisiae* and *D. melanogaster*.

☐ Similarity greater than 35% between *E. coli* and *D. melanogaster*.

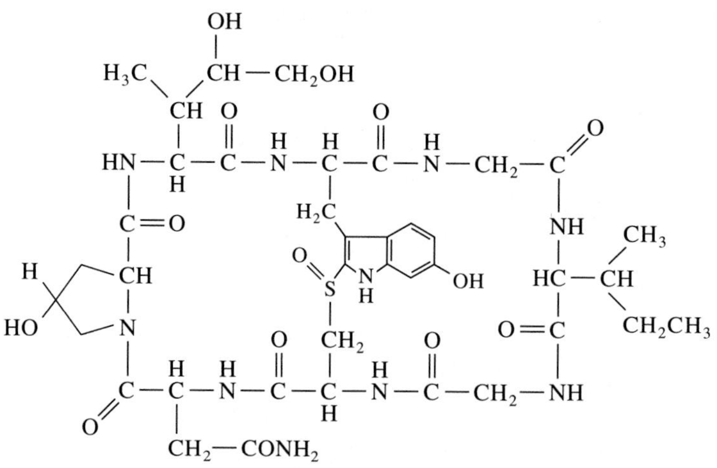

Figure 27·22
α-Amanitin.

Figure 27·23
Amanita phalloides.

demand for ribosomal RNA while all other genes were transcribed by a second enzyme. Later on, the functions of the second RNA polymerase were subdivided further when an RNA polymerase specializing in small RNA molecules arose. The three nuclear polymerases are almost certainly derived from a single ancestral polymerase, given their structural relatedness; in yeast, the largest subunits of each nuclear RNA polymerase appear to be related.

The activities of the eukaryotic RNA polymerases are often distinguished on the basis of their sensitivity to the fungal toxin α-amanitin (Figure 27·22). α-Amanitin binds tightly to mammalian RNA polymerase II and inhibits transcription. The toxin binds less tightly to RNA polymerase III. RNA polymerase I, organellar RNA polymerases, and prokaryotic RNA polymerases are insensitive. RNA polymerases from other eukaryotes, such as yeast, may have different relative sensitivities.

α-Amanitin is produced by the poisonous mushroom *Amanita phalloides* (Figure 27·23), colloquially known as the death cap, or death angel. Ingestion of even one mushroom of this species can induce vomiting and diarrhea within hours. Death follows in a few days.

27·8 Transcription Factors Are Required for the Assembly of Eukaryotic Transcription Complexes

Although the eukaryotic transcription machine catalyzes the same reactions as the prokaryotic machine, assembly of the eukaryotic machine during initiation of transcription is more complex. No eukaryotic RNA polymerase appears to bind specifically to promoters. Instead, binding requires the presence of initiation factors or transcription factors (TF).

The bacterial σ subunit is an example of an initiation factor that binds more tightly to core RNA polymerase than it does to DNA. In contrast, eukaryotic transcription factors are not isolated as part of the multisubunit RNA polymerase but assemble into the preinitiation complex separately by binding to DNA before RNA polymerase binds. General transcription factors are required for transcription of all genes of a class, whereas other transcription factors are required only for certain genes or gene families.

Table 27·5 Some representative RNA polymerase II transcription factors

Factor	Characteristics
TFIIA	Binds to TFIID; can interact with TFIID in the absence of DNA
TFIIB	Interacts with RNA polymerase II
TFIID	RNA polymerase II initiation factor
TBP	TATA binding protein; subunit of TFIID
TAFs	TBP-associated factors; many subunits
TFIIE	Interacts with RNA polymerase II
TFIIH	Required for initiation; helicase activity; couples transcription to DNA repair
TFIIS[1]	Binds to RNA polymerase II; elongation factor
RAP30/74[2]	Binds to RNA polymerase II; two subunits – RAP30 and RAP74
SP1	Binds to GC-rich sequence
CTF[3]	Family of different proteins that recognize the core sequence CCAAT

[1] Also known as sII or RAP38
[2] Also known as Factor 5 or TFIIF
[3] Also known as NF1

A. Transcription Factors for Class II Genes Bind to Promoter Sites Located Upstream of the Transcription Start Site

RNA polymerase II transcribes all genes whose products are destined to be translated into proteins, as well as some genes that encode small RNA molecules involved in posttranscriptional modifications. The protein-coding RNA synthesized by this enzyme was originally called heterogeneous nuclear RNA (hnRNA), but it is now more commonly referred to as **mRNA precursor.** Processing of this precursor into mature mRNA is considered in Chapter 28. About 40 000 molecules of RNA polymerase II are found in large eukaryotic cells; the activity of this enzyme accounts for roughly 20 to 40% of all cellular RNA synthesis.

Analysis of the promoter sequences for RNA polymerase II was attempted in a number of ways, such as by inspecting DNA sequences from class II genes and by footprinting the sites where the polymerase contacts the DNA, but the analyses proved to be complicated. Whereas in *E. coli* the same RNA polymerase (core polymerase plus σ^{70}) transcribes many genes in a single cell under similar conditions, the situation is quite different in eukaryotes. The class II genes sequenced to date come from many species and are transcribed in different tissues under quite different conditions. We now know that the individual class II genes are bound by a variety of transcription factors that recognize specific sequences in each organism. However, in spite of this apparent diversity, all eukaryotes contain a number of similar general transcription factors that interact directly with RNA polymerase II and control initiation at all class II genes (Table 27·5).

Many class II genes contain an AT-rich region that is functionally similar to the prokaryotic TATA box. This eukaryotic AT-rich region, also called a TATA box, is located between 19 and 27 base pairs upstream of the transcription start site and is the site where RNA polymerase II binds to DNA and the initiation complex is assembled.

Transcription initiation begins when the general transcription factor TFIID binds to the promoter. TFIID is a multisubunit factor, and one of the subunits, TATA-binding protein (TBP), is responsible for binding to the region containing the

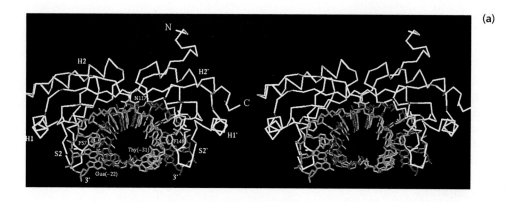

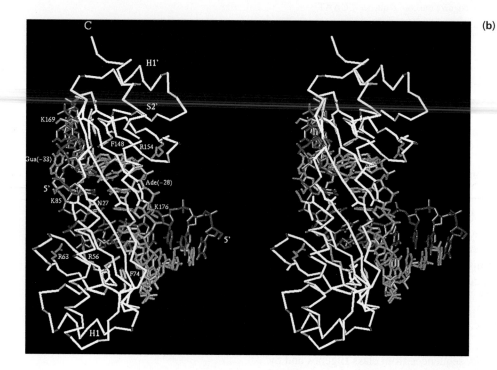

Figure 27·24
Stereo views of *Arabidopsis thaliana* TATA-binding protein (TBP) bound to DNA. (a) Image viewed at 90° to the axis of two-fold symmetry of TBP, shown as a white α-carbon trace. TBP is bound to the TATA element of DNA (red and green). In this view, the cylindrical axis of B-DNA, drawn as an atomic stick model, is nearly perpendicular to the page. The two pairs of phenylalanine side chains that induce sharp kinks in the DNA are colored yellow (Phe-57, Phe-74, Phe-148, and Phe-165). DNA is partially unwound between these two kinks. Two asparagine residues that make hydrogen bonds in the minor groove of the TATA element (Asn-27 and Asn-117) are also colored yellow. Two hydrophobic residues thought to be involved in DNA recognition are colored blue (Ile-152 and Leu-163). (b) Image viewed parallel to the axis of symmetry of TBP. TBP forms a saddle that straddles the DNA. Positively charged amino acid side chains that form ionic bonds with the phosphodiester backbone are shown in blue. (Courtesy of Stephen K. Burley.)

TATA box. The structure of TBP from *Arabidopsis thaliana* is shown in Figure 27·24. It forms a molecular clamp that distorts the structure of DNA at the TATA box by binding to the minor groove. The DNA is sharply kinked and partially unwound. The significance of DNA bending remains to be determined, but it appears to be a common feature of TBPs from different species since the structure of a yeast TBP-DNA complex is similar to that formed by a plant protein.

The next steps in the formation of an initiation complex are the association of TFIIA and TFIIB. Neither of these factors can bind to DNA in the absence of TFIID; they appear to associate with the exposed surface of TBP shown in Figure 27·24. RNA polymerase II then joins the preinitiation complex. RAP30/74 (also known as Factor 5 or TFIIF) binds to RNA polymerase II during initiation (Figure 27·25). This factor plays no direct role in recognizing the promoter, but nevertheless it is analogous to bacterial σ factors in two ways: it decreases the general affinity of RNA polymerase II for nonpromoter DNA, and it assists in the formation of the open complex. TFIIH, TFIIE, and other less well characterized factors are also part of the transcription initiation complex, although they are not shown in the figure. One of the subunits of TFIIH is the helicase that may play a role in coupling transcription to DNA repair (Section 26·4B).

Figure 27·25
Class II transcription initiation complex.

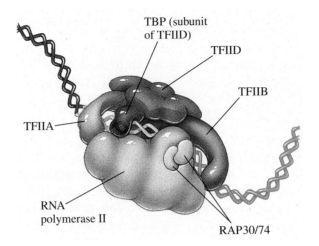

Most of the transcription factors dissociate from DNA and from RNA polymerase II once elongation begins. However, RAP30/74 may remain bound, and a specific elongation factor, TFIIS (also called sII or RAP38) associates with the transcribing polymerase. TFIIS may play a role in pausing and transcription termination that is similar to the role of NusA in bacteria.

Eukaryotic promoter regions commonly contain additional protein-binding sites. A common promoter element in mammals is found between 60 and 80 base pairs upstream of the transcription start site for RNA polymerase II. The core consensus sequence for this region is CCAAT. It is called the CCAAT box (pronounced "cat box"). Although this sequence is not palindromic, it occurs naturally in both orientations. This site is bound by transcription factor CTF (CCAAT box transcription factor, also called NF1). Mammalian class II promoters frequently contain yet another common element whose consensus sequence is GGGGCGGGGC. It is located upstream of the TATA box at variable distances. It is the recognition site for binding of the transcription factor SP1.

SP1 binds directly to one of the components of TFIID, and CTF probably interacts directly with the complex at the TATA box as well. Both SP1 and CTF stimulate transcription initiation at many but not all genes. They most likely contact DNA and the initiation complex simultaneously through formation of a DNA loop (Figure 27·26). The formation of loops of DNA that bring transcription factors into contact is a common feature of initiation in eukaryotes, as it is in prokaryotes. Although drawn as naked DNA in Figure 27·26, the promoter region in reality contains bound nucleosomes that assist in the formation of such long-range contacts.

Figure 27·26
SP1 and CTF interaction with the RNA polymerase II preinitiation complex.

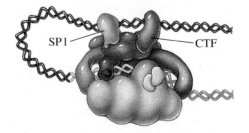

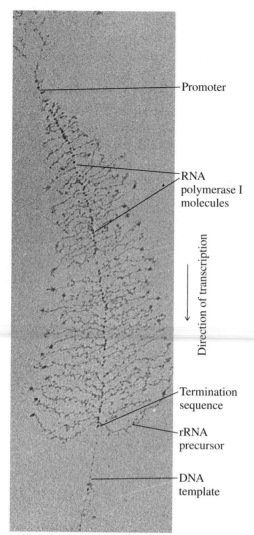

Promoter

RNA
polymerase I
molecules

Direction of transcription

Termination
sequence

rRNA
precursor

DNA
template

Figure 27·27
Electron micrograph of a single active ribosomal RNA gene isolated from a nucleolus of an oocyte of the spotted newt, *Notophthalmus viridescens*. Because transcription initiation is so frequent, there are many molecules of RNA polymerase actively transcribing the gene at the same time. (Courtesy of Oscar L. Miller, Jr. and Barbara R. Beatty.)

B. Transcription Factors for Class I Genes Bind to Sites Upstream of the Promoter

In eukaryotic cells, the genes for three ribosomal RNA molecules—18S, 28S, and 5.8S—are transcribed as a single unit. In most eukaryotes, these units are found in clusters composed of 100 to 5000 repeats. They are transcribed by RNA polymerase I in the nucleolus, a region of the nucleus devoted to the synthesis of rRNA. A eukaryotic cell typically contains 40 000 RNA polymerase I molecules, which may seem like a large number for transcribing as few as 100 genes. But the rRNA precursor is long (6000 to 15 000 nucleotides), and as many as 50 polymerase molecules may transcribe each rRNA gene simultaneously (Figure 27·27). This intense transcriptional activity reflects the need of the cell for rRNA, especially during rapid growth, when millions of copies of rRNA are required per cell division.

The promoter sequences for class I genes are located just upstream of the transcription start site, in a region of the repeating gene unit called the nontranscribed spacer. All class I promoters analyzed consist of several binding sites that can be subdivided into three important regions. The first region spans the transcription initiation site and is required for initiation by RNA polymerase I. The second region, further upstream, is necessary for binding of multisubunit transcription factors. One of these factors in humans is SL1. Binding of SL1 to this region of DNA assists the polymerase in accurately selecting the start site. One of the subunits of SL1 and similar factors in other species is TBP, the same protein that binds to class II promoters. Other general transcription factors, such as TFID, are also required. The third promoter region lies even further upstream, typically 100 to 150 base pairs from the transcription start site, and is also involved in binding transcription factors.

C. Some Transcription Factors for Class III Genes Bind to Internal Control Regions

RNA polymerase III carries out about 10% of cellular transcription in eukaryotes. We know more about transcription by RNA polymerase III than about transcription by any other eukaryotic polymerase, in part because the class III transcription-initiation complex is less complicated than the class I or class II complex. There are three class III transcription factors: TFIIIA, TFIIIB, and TFIIIC, but not all of them are required for transcription of every class III gene. The properties of class III transcription factors are listed in Table 27·6.

The 5S rRNA and tRNA genes are examples of genes transcribed by RNA polymerase III. Surprisingly, the promoter sequences for these genes lie within their transcribed regions. Such promoter elements are called **internal control regions**

Table 27·6 RNA polymerase III transcription factors

Factor	Characteristics
TFIIIA	Binds internal control region of 5S rRNA gene; requires zinc
TFIIIB	Interacts with stable complex via protein-protein and protein-DNA interactions; large multisubunit complex
TFIIIC	Binds internal control region of all class III genes except the 5S rRNA gene; large multisubunit complex

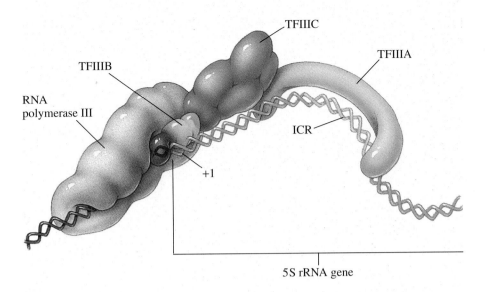

Figure 27·28
Binding of class III transcription factors to the internal control region of a 5S ribosomal RNA gene.

(ICR). The class III transcription factors bind to the internal promoter during initiation. Once formed, the transcription-factor complex remains stably associated with the gene for multiple rounds of initiation and is called a stable transcription complex. Thus, it is not necessary to assemble a new transcription-factor complex each time these genes are transcribed.

Initial transcription of 5S rRNA genes begins with the formation of a stable transcription complex. This complex includes TFIIIA, TFIIIB, and TFIIIC (Figure 27·28). TFIIIA is the first protein to bind DNA. It recognizes and binds tightly to specific nucleotides of the ICR. Once TFIIIA is bound, TFIIIC enters the complex by making both protein-protein contacts with TFIIIA and protein-DNA contacts with the 5S rRNA gene. TFIIIB, a multisubunit factor, then binds to the TFIIIA–TFIIIC–5S rRNA gene complex, again via protein-protein and protein-DNA interactions. RNA polymerase III then joins the stable transcription complex and begins transcription at the start site at the 5′ end of the gene.

Transcription of tRNA genes requires TFIIIB and TFIIIC but not TFIIIA. Other class III genes have slightly more complicated promoters that rely on proteins that bind to upstream sequences as well as those that bind to an internal control region. However, these genes appear to use some of the same transcription factors as the better-understood genes for 5S rRNA. In particular, TFIIIB is required. One of the subunits of TFIIIB is TBP, which is also required for transcription of class I and class II genes.

How do large transcription-factor complexes remain associated with the 5S rRNA genes while RNA polymerase passes through their binding sites? The answer to this question is not entirely clear, but analysis of the interaction of TFIIIA with the 5S rRNA gene has provided some clues. TFIIIA from the frog *Xenopus laevis* (M_r 38 500) was the first eukaryotic transcription factor purified to homogeneity, and much about its structure is now known. The N-terminal domain of this protein binds DNA. The C-terminal domain is the activation domain that interacts with TFIIIC. When TFIIIA binds to the 5S rRNA gene, it recognizes determinants that are located primarily on the coding strand of the 5S rRNA gene. As RNA polymerase III reads through the gene, the template and coding strands separate. While the polymerase interacts with the template strand, the huge transcription-factor complex remains bound to the coding strand. Thus, when the strands reanneal after the polymerase has passed, the transcription-factor complex is still bound to the same site.

Figure 27·29
Two zinc-finger structural motifs in *Xenopus* TFIIIA. Each Zn^{2+} is tetrahedrally coordinated by two conserved histidine and two conserved cysteine residues. Additional conserved residues or conservative substitutions occur in the finger and at the base of the finger. Each zinc finger is connected to its neighbor by a short linker sequence.

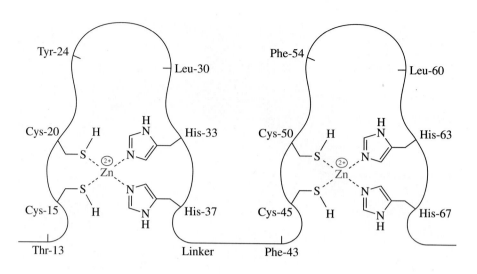

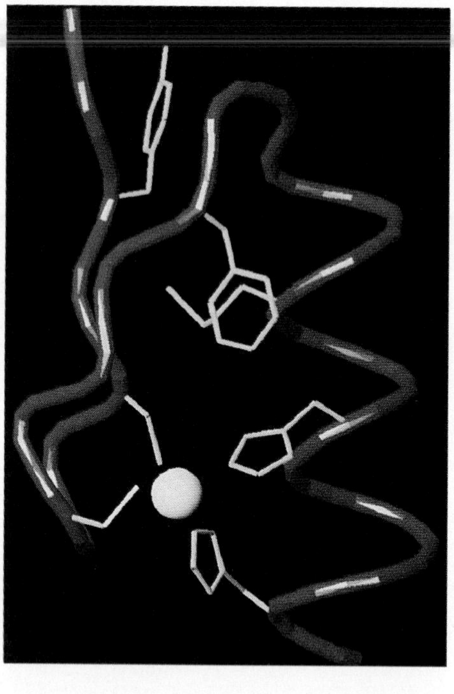

D. TFIIIA Contains Zinc Fingers

Analysis of the amino acid sequence of TFIIIA and several other DNA-binding proteins has identified a common, repetitive amino acid sequence. This sequence of about 30 amino acids folds into a structure that includes tetrahedrally coordinated zinc ions (Zn^{2+}). This structural motif is referred to as a **zinc finger.** In those proteins analyzed, the number of zinc fingers varies from 2 to over 30.

The prototype for zinc-containing DNA-binding proteins is TFIIIA from *Xenopus*. It contains nine tandem repeats of the zinc-finger consensus sequence and nine tetrahedrally coordinated zinc ions. The amino acids that interact with Zn^{2+} are two conserved histidine and two conserved cysteine residues (Figure 27·29).

Zinc-containing motifs have been identified in many eukaryotic DNA-binding proteins, including several transcription factors and many proteins that control developmentally regulated genes in *Drosophila* and yeast. The motifs can be subdivided into several classes. The TFIIIA class is the one usually called a zinc finger, but it is also referred to as C$_2$H$_2$ because the Zn^{2+} is coordinated by two cysteine and two histidine residues. We will discuss another zinc-containing motif, C$_2$C$_2$, in Section 27·14A.

The three-dimensional structure of a single zinc finger consists of an α helix and a β sheet in addition to the region containing the conserved cysteine residues (Figure 27·30). Zinc-finger motifs also contain conserved hydrophobic residues that stabilize the structure via hydrophobic interactions between the α-helical region and the β sheet. Adjacent zinc fingers in proteins such as TFIIIA interact with DNA through the α helices, which lie in the major groove. The mouse regulatory protein Zif268 contains three zinc fingers of the type found in TFIIIA. The structure of a Zif268-DNA complex is shown in Figure 27·31. SP1, which is required for transcription of some class II genes, also contains three zinc fingers of the C$_2$H$_2$ class.

Figure 27·30
Structure of a zinc finger from the fruit fly transcription factor tramtrack. The protein backbone is shown in red. The zinc ion is shown as a white sphere, and zinc ligands are shown in yellow. Hydrophobic residues in the core of the protein are shown in blue. (Courtesy of John Schwabe.)

27·9 Genes Can Be Transcribed at Different Rates

As noted at the beginning of this chapter, there are a large number of genes that are expressed in every cell. Expression of these genes is said to be constitutive. These housekeeping genes encode the proteins required for basic metabolic events such as

Figure 27·31
Three zinc fingers from mouse transcription factor Zif268 in complex with DNA. DNA is shown as a white space-filling model, and the zinc fingers are shown in red. Amino acid side chains in contact with base pairs are shown in purple. Zinc ions and ligands are shown in yellow. (Courtesy of John Schwabe.)

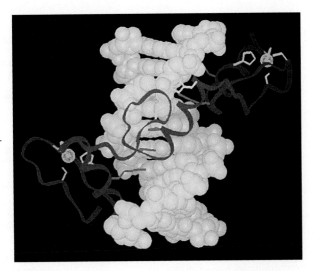

glycolysis and for the biosynthesis of molecules such as amino acids and DNA. In general, housekeeping genes whose products are required continuously at high levels have strong promoters and are transcribed efficiently and continuously. Conversely, housekeeping genes whose products are required at low levels usually have weak promoters and are transcribed infrequently. In addition to housekeeping genes, there are genes that are expressed at high levels in some circumstances and not at all in others. Such genes are said to be regulated.

Regulation of gene expression can take place at any point in the flow of biological information but occurs most often at the level of transcription. Various elaborate mechanisms have evolved that allow cells to program gene expression during differentiation and development, to maintain basic functions, and to respond to environmental stimuli by altering the rate of production of RNA. Since initiation of transcription requires both specific DNA sequences and soluble proteins, either of these elements can be manipulated to modulate transcriptional activity.

The initiation of transcription of regulated genes is controlled by the action of special regulatory proteins that are gene specific. Regulatory proteins are required in addition to the general transcription factors in eukaryotes or the holoenzyme containing σ^{70} in bacteria. These proteins regulate gene expression by binding to DNA, usually near the promoters of the genes they control. Genes that are regulated at the level of transcription initiation by regulatory proteins are of two basic types: negatively regulated genes and positively regulated genes. Transcription of a negatively regulated gene is prevented by a regulatory protein called a **repressor.** A negatively regulated gene can be transcribed only in the absence of active repressor. Transcription of a positively regulated gene can be activated by a regulatory protein called an **activator.** A positively regulated gene is transcribed poorly or not at all in the absence of its activator.

Repressors and activators are often allosteric proteins whose activities are affected by the binding of ligands. Some activators are themselves activated by the binding of ligands. Some repressors controlling the synthesis of enzymes of a catabolic pathway are inactivated by the binding of the first substrate in the pathway. Ligands that bind to and inactivate repressors are called **inducers** because they induce the transcription of the genes controlled by the repressors. Other repressors control the synthesis of enzymes in a biosynthetic pathway and are activated by the binding of the end product of the pathway. Ligands that bind to and activate repressors are called **corepressors.** Four general mechanisms for regulation of allosteric activators and repressors are illustrated in Figure 27·32 (next page).

Figure 27·32
Strategies for the regulation of transcription initiation by regulatory proteins.

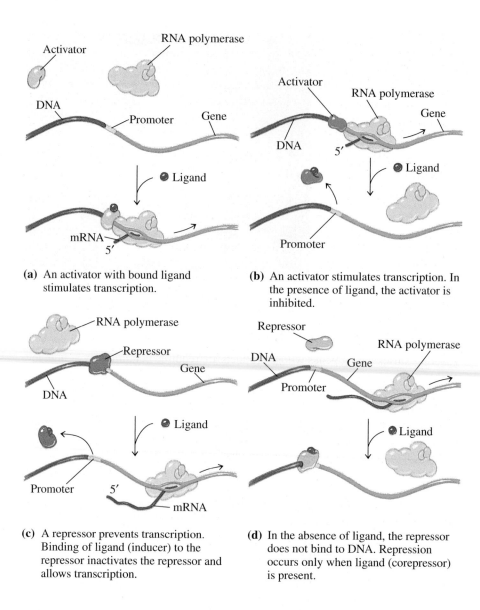

(a) An activator with bound ligand stimulates transcription.

(b) An activator stimulates transcription. In the presence of ligand, the activator is inhibited.

(c) A repressor prevents transcription. Binding of ligand (inducer) to the repressor inactivates the repressor and allows transcription.

(d) In the absence of ligand, the repressor does not bind to DNA. Repression occurs only when ligand (corepressor) is present.

Few regulatory systems are as simple as those described above. For example, the transcription of many genes is regulated by a combination of repressors and activators, and the transcription of many other genes is regulated by multiple activators. These more elaborate mechanisms of transcriptional regulation have evolved to meet the specific requirements of individual organisms. When transcription is regulated by a host of regulatory mechanisms acting together, a more varied response to cellular requirements is possible. By examining how the transcription of a few particular genes is controlled, we can begin to understand how positive and negative regulatory mechanisms can be combined to produce the remarkably sensitive forms of regulation seen in vivo.

27·10 Expression of the *lac* Operon Is a Classic Example of Both Negative and Positive Regulation

Bacteria normally obtain the carbon they need for growth by catabolizing five- or six-carbon sugars via the glycolytic pathway. *E. coli* preferentially uses glucose as a carbon source but can also use other sugars, including β-galactosides such as lactose. The enzymes required for β-galactoside uptake and utilization are not synthesized unless a β-galactoside substrate is available. Furthermore, even in the presence of β-galactosides, these enzymes are synthesized in limited amounts if the preferred carbon source, glucose, is also present. The synthesis of the enzymes required for β-galactoside utilization is regulated at the level of transcription initiation.

A. The Genes Required for the Metabolism of β-Galactosides in *E. coli* Are Controlled by a Repressor

Uptake and utilization of β-galactosides by *E. coli* requires three proteins—lactose permease, β-galactosidase, and thiogalactoside transacetylase (Figure 27·33). β-Galactosides are taken up by *E. coli* in a transport process coupled to the inward flow of protons. The protein responsible for the transport of β-galactosides is lactose permease, also called β-galactoside permease, a membrane-bound protein

Figure 27·33
Proteins required for the utilization of β-galactosides in *E. coli*. Several different β-galactosides are transported with protons into the cell by lactose permease. Most β-galactosides, including lactose, are then hydrolyzed by β-galactosidase to hexoses that enter other catabolic pathways. Nonmetabolizable β-galactosides are acetylated by thiogalactoside transacetylase. These acetylated compounds then diffuse out of the cell.

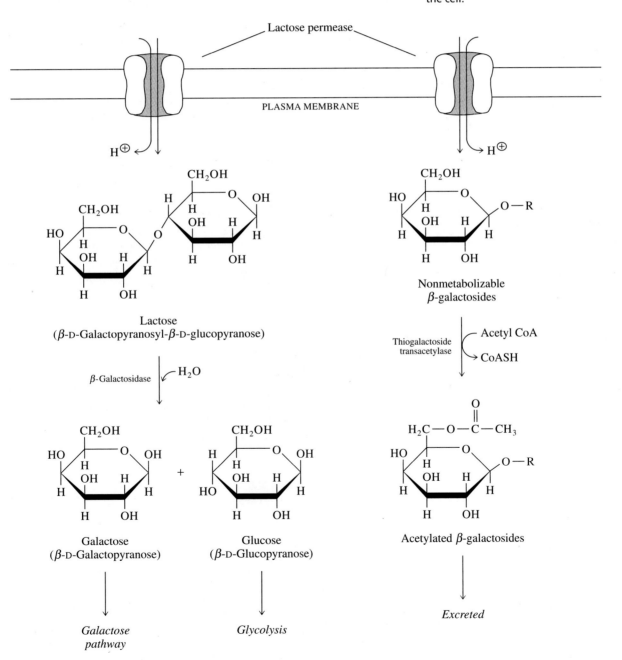

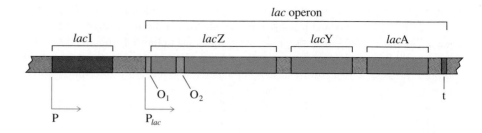

Figure 27·34
Organization of the genes that encode the proteins required for the metabolism of lactose. The coding regions for the three proteins required for the utilization of β-galactosides are cotranscribed from a single promoter (P_{lac}). These three coding regions—lacZ, lacY, and lacA—constitute the lac operon. The gene that encodes lac repressor, lacI, is located upstream of the lac operon. The lacI gene has its own promoter, P. The lac repressor binds to the operators O_1 and O_2 near P_{lac}. The transcription terminator is denoted by t.

(M_r 46 500). Lactose permease, one of the classic examples of a symport transporter (Section 12·7), is the product of the lacY gene. After transport into an E. coli cell, most β-galactosides are hydrolyzed to metabolizable hexoses by the activity of β-galactosidase, a large, tetrameric enzyme consisting of identical subunits (subunit M_r 116 400). β-Galactosidase is encoded by the lacZ gene. Those β-galactosides that cannot be hydrolyzed by the activity of β-galactosidase are acetylated by the activity of thiogalactoside transacetylase. Acetylated β-galactosides can diffuse through the plasma membrane and are thus eliminated from the cell. The transacetylase, a dimer of identical subunits (subunit M_r 22 700), is encoded by the lacA gene.

The three genes encoding these proteins—lacZ, lacY, and lacA—form an operon that is transcribed from a single promoter to produce a large mRNA molecule containing three separate protein-coding regions. In this case, we refer to a protein-coding region as a gene, a definition that differs from our standard use of the term. Operons composed of genes with related functions are efficient because they allow a set of proteins, whose concentrations in the cell would otherwise have to be independently regulated, to be controlled by transcription from a single promoter. Operons composed of protein-coding genes are very common in E. coli and other prokaryotes but are almost never found in eukaryotes.

Expression of the three genes of the lac operon is controlled by a regulatory protein called lac repressor. The lac repressor is a tetramer of identical subunits (subunit M_r 38 600). It is encoded by a fourth gene, lacI, which is located just upstream of the lac operon but is transcribed separately (Figure 27·34).

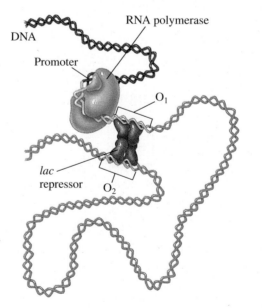

Figure 27·35
Binding of lac repressor to the lac operon. The tetrameric lac repressor interacts simultaneously with two sites near the promoter of the lac operon. As a result, a loop of DNA is formed. RNA polymerase can still bind to the promoter in the presence of the lac repressor–DNA complex.

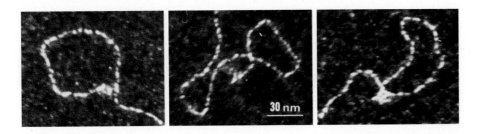

Figure 27·36
Electron micrographs of DNA loops in vitro. These loops were formed by mixing *lac* repressor with a fragment of DNA bearing two synthetic *lac* repressor–binding sites. One binding site is located at one end of the DNA fragment, and the other is located 535 base pairs away. DNA loops of 535 base pairs were formed when *lac* repressor bound simultaneously to the two binding sites. (Courtesy of Michèle Amouyal.)

B. *lac* Repressor Binds to DNA

The *lac* repressor represses transcription by binding simultaneously to two sites near the promoter of the *lac* operon. Repressor-binding sites are called **operators,** and both *lac* operators (O_1 and O_2) are necessary for repression. One operator (O_1) is adjacent to the promoter, and the other is within the coding region of *lac*Z. When bound to the operators, *lac* repressor forms a stable loop of DNA (Figures 27·35 and 27·36). The *lac* repressor was originally believed to block initiation of transcription of the *lac* operon by preventing the binding of RNA polymerase to the *lac* promoter. However, it is now known that *lac* repressor can be bound at O_1 and RNA polymerase can be bound at the *lac* promoter simultaneously. *Lac* repressor appears to block initiation of transcription by preventing formation of the open complex. The gene for the repressor (*lac*I) does not need to be located close to the *lac* operon and could, in principle, be moved to other locations without significantly affecting regulation.

The equilibrium binding constant (K_B) for the binding of *lac* repressor (R) to O_1 (O) in vitro is very high.

$$R + O \underset{k_{-1}}{\overset{k_1}{\rightleftharpoons}} RO$$

$$K_B = \frac{[RO]}{[R][O]} = \frac{k_1}{k_{-1}} = 10^{13}\,M^{-1}$$

(27·2)

The dissociation rate constant (k_{-1}) in vitro is $6 \times 10^{-4}\,s^{-1}$. Thus, the repressor-operator complex has a half-life of approximately 20 minutes, less than the normal generation time of an *E. coli* cell. The association rate constant (k_1) in vitro is $7 \times 10^9\,M^{-1}\,s^{-1}$ when measured using a small fragment of DNA containing an operator sequence. The in vivo association rate constant is estimated to be greater than $10^{10}\,M^{-1}\,s^{-1}$. This rate is faster than the maximum rate expected for a simple diffusion-controlled bimolecular reaction. Consequently, *lac* repressor must locate the operator not by searching in three dimensions but by binding nonspecifically to DNA and searching in one dimension. (Recall from Section 27·3C that RNA polymerase also uses this kind of searching mechanism.) There are only about 10 molecules of *lac* repressor in a bacterial cell, but the repressor searches for and finds O_1 so rapidly that when a repressor dissociates spontaneously from the operator, another finds and occupies the site within a very short time. During this interval, no more than one transcript is likely to be initiated.

K_B for the interaction of *lac* repressor with nonspecific DNA in vitro is $4 \times 10^4\,M^{-1}$. Thus, the repressor binds nearly 10^9 times more strongly to O_1 than it binds to random DNA sequences. The difference in the affinity of *lac* repressor for specific and nonspecific DNA sequences is likely to be close to the maximum possible for DNA-binding proteins. Note, for example, that the difference in affinities is less than 10^5 for RNA polymerase holoenzyme.

Figure 27·37
Regions of dyad symmetry within the *lac* operators. **(a)** Operator O_1. The extended region of imperfect dyad symmetry is highlighted in blue. The region covered by *lac* repressor (as determined by DNase I footprinting) is indicated. **(b)** Operator O_2. The region of dyad symmetry within O_2 is less extensive; *lac* repressor binds less tightly to this operator.

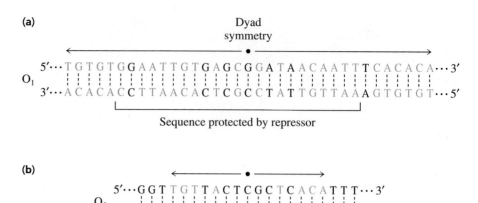

O_1 contains an extended region of imperfect dyad symmetry (Figure 27·37). Most of this region is covered when O_1 is bound by *lac* repressor. The sequence of O_2 is less symmetric; *lac* repressor appears to bind to O_2 with lower affinity than it binds to O_1.

The symmetry of the operator sequences parallels the symmetry of *lac* repressor. Each of the four subunits of *lac* repressor is composed of two domains: an N-terminal domain, which specifically binds to DNA, and a larger C-terminal domain, which interacts with the other subunits. The N-terminal domains of two subunits interact directly with each operator. The binding of a dimeric or tetrameric protein to two DNA sequences possessing dyad symmetry appears to be a general feature of many DNA-binding proteins. The restriction enzyme *Eco*RI is an example of a dimeric protein binding to a symmetrical site (Section 24·10A). The K_B for binding of a dimer to two sites on a molecule is greater than the K_B for binding of two monomers to the same two sites. This phenomenon is known as the **chelate effect.**

The chelate effect applies to the simultaneous binding of *lac* repressor to two operators as well as to the binding of a dimer to a symmetrical binding site. Although in this case the unbound repressor is tetrameric and already possesses two separate DNA-binding sites, in other cases two dimeric repressors bind to separate sites and then interact to form a DNA loop. The presence of two specific binding sites increases the effective concentration of the repressors and promotes interaction between them. The effect works over large distances; some repressor-binding sites are separated by 800 bp, and the resulting DNA loop is larger than the one shown in Figure 27·36.

C. *lac* Repressor Is an Allosteric Protein

In the absence of lactose, *lac* repressor binds to O_1 and O_2 and blocks expression of the *lac* operon by preventing RNA polymerase from initiating transcription. If lactose is present in the medium, however, a small amount of lactose is converted to allolactose in a reaction catalyzed by β-galactosidase (Figure 27·38). Allolactose is an inducer of the *lac* operon: it binds tightly to *lac* repressor ($K_B \sim 10^6\,M^{-1}$) and causes a conformational change that reduces the affinity of the repressor for the operators. In the presence of inducer, *lac* repressor dissociates from the DNA, thereby allowing RNA polymerase to initiate transcription. In this way, expression of the proteins required for the utilization of lactose depends on the presence of allolactose, which in turn depends on the presence of lactose.

When *lac* repressor is bound to the operator sequences, it actually increases the rate of the initial binding of RNA polymerase to the *lac* operon promoter more than 100-fold. This effect is due to contacts between *lac* repressor and RNA polymerase when both proteins are bound. Because it is poised at the promoter, RNA polymerase can transcribe the *lac* operon immediately following induction. Without the

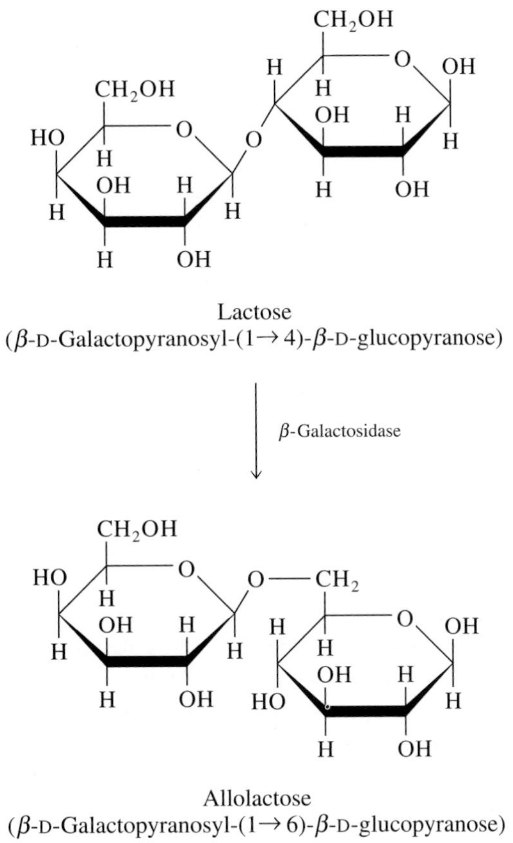

Figure 27·38
Formation of allolactose
from lactose. This reaction
is catalyzed by β-galactosidase.

Lactose
(β-D-Galactopyranosyl-(1→4)-β-D-glucopyranose)

β-Galactosidase

Allolactose
(β-D-Galactopyranosyl-(1→6)-β-D-glucopyranose)

assistance of *lac* repressor, RNA polymerase would more slowly locate the weak *lac* promoter, whose sequence does not closely match the consensus sequence recognized by σ^{70}.

Because RNA polymerase is bound to the *lac* promoter, the spontaneous dissociation of *lac* repressor from the *lac* operators allows a single transcript of the *lac* operon to be produced before a new molecule of *lac* repressor can bind. Thus, even in the absence of induction, the genes of the *lac* operon are expressed at a low level. Such transcription of repressed genes is called **escape synthesis.** Escape synthesis provides the lactose permease and β-galactosidase required to obtain the inducer allolactose from lactose. Bacterial cells growing in the absence of inducer normally contain about 15 molecules of β-galactosidase in the cytosol and a few molecules of lactose permease in the membrane. In the presence of inducer, however, the number of β-galactosidase molecules rises to approximately 15 000, a 1000-fold increase.

Escape synthesis may not be essential for induction of the *lac* operon, however, since there are β-galactosides other than lactose that are substrates for β-galactosidase, and these substrates can be direct inducers. For example, although *E. coli* is best adapted to the intestinal tracts of adult mammals, adult mammals do not normally feed on milk, which is a primary source of lactose. This suggests that *E. coli* rarely have the opportunity to utilize lactose as a carbon source. But adult mammals do commonly eat plants, which contain chloroplasts whose thylakoid membranes are a rich source of galactolipids. Removal of the fatty acids from galactolipids yields β-galactosyl glycerol, an excellent substrate for the enzyme β-galactosidase (Figure 27·39). β-Galactosyl glycerol is one of the most effective inducers of the *lac* operon. In addition, this substrate can readily enter the bacterial cell via a constitutive sugar permease. Thus, β-galactosyl glycerol requires neither β-galactosidase nor lactose permease to function as an inducer. It is likely that β-galactosyl glycerol, and perhaps similar compounds, are much more common substrates for *E. coli* β-galactosidase than is lactose.

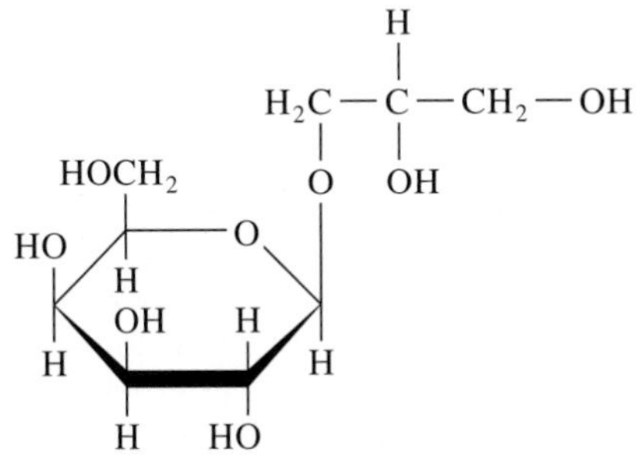

Figure 27·39
Structure of β-galactosyl glycerol.

(a) CRP-cAMP binds to a site near the promoter.

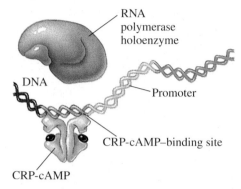

(b) RNA polymerase holoenzyme binds to the promoter and also contacts the bound activator, causing an increase in the rate of transcription initiation.

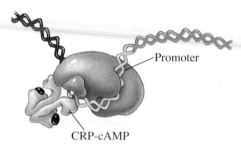

Figure 27·40
Activation of transcription initiation at the *lac* promoter by CRP-cAMP. CRP-cAMP binds to a specific DNA sequence near the *lac* promoter and enables RNA polymerase to initiate transcription more efficiently.

D. Transcription of the *lac* Operon in *E. coli* Is Also Regulated by an Activator

Transcription of the *lac* operon in *E. coli* depends not only upon the presence of lactose but also upon the concentration of glucose in the external medium: the *lac* operon is transcribed maximally when lactose alone is the carbon source, but transcription is reduced 50-fold when glucose is also present in the medium. The decreased rate of transcription from operons such as the *lac* operon when glucose is present is termed **catabolite repression.**

Catabolite repression is a feature of many operons encoding metabolic enzymes. These operons characteristically have weak promoters from which transcription is initiated inefficiently in the presence of glucose. In the absence of glucose, however, the rate of transcription initiation increases dramatically due to binding near the promoter of an activator that converts the relatively weak promoter to a stronger one. There is no repressor involved in the regulation of these operons, despite the use of the term *catabolite repression*. In fact, this is a well-studied example of an activation mechanism.

The activator responsible for the activation of genes in the absence of glucose is cAMP regulatory protein (CRP), a dimeric protein whose activity is controlled by cAMP. In the absence of cAMP, CRP has little affinity for DNA. When cAMP is present, however, it binds to CRP, converting it into a sequence-specific DNA-binding protein. The CRP-cAMP complex binds to specific DNA sequences near the promoters of more than 30 genes, including the *lac* operon. Because there are many more binding sites for CRP-cAMP than there are for *lac* repressor, it is not surprising that there are at least 1000 molecules of CRP per cell compared to only 10 molecules of *lac* repressor. The CRP-cAMP binding sites are often just upstream of the −35 regions of the promoters they activate. While bound to DNA, CRP-cAMP can contact RNA polymerase at the promoter site, leading to increased rates of transcription initiation (Figure 27·40). The effect of CRP-cAMP is to increase the rate of production of enzymes that can make use of substrates other than glucose. In the case of the *lac* operon, the rate of transcription initiation is only affected when β-galactosides are available; otherwise transcription of the operon is repressed.

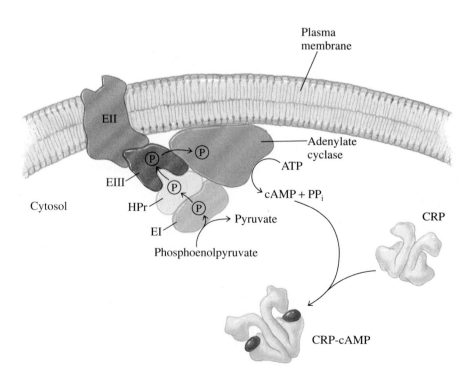

Figure 27·41
cAMP production. In the absence of glucose, enzyme III transfers a phosphoryl group, originating from phosphoenolpyruvate, to membrane-bound adenylate cyclase. Phosphorylated adenylate cyclase catalyzes the conversion of ATP to cAMP. As a result, levels of cAMP in the cell rise and cAMP binds to CRP. CRP-cAMP activates the transcription of a number of genes encoding enzymes that compensate for the lack of glucose as a carbon source.

In *E. coli,* the concentration of cAMP inside the cell is controlled by the presence of glucose outside the cell. When glucose is available, it is imported into the cell and phosphorylated by a complex of transport proteins collectively known as the phosphoenolpyruvate-dependent sugar phosphotransferase system. When glucose is not available, one of the glucose transport enzymes, enzyme III, catalyzes transfer of a phosphoryl group, ultimately derived from phosphoenolpyruvate, to adenylate cyclase, leading to its activation (Figure 27·41). Adenylate cyclase then catalyzes the conversion of ATP to cAMP, thereby increasing the levels of cAMP in the cell. As molecules of cAMP are produced, they bind to CRP, stimulating transcription initiation at promoters that respond to catabolite repression. Very similar mechanisms of response to external stimuli operate in eukaryotes, where molecules such as cAMP act as second messengers.

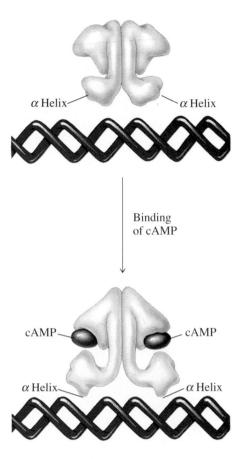

E. CRP Contains a Helix-Turn-Helix DNA-Binding Domain

Each subunit of the CRP dimer contains an α-helical domain, which, in the presence of cAMP, fits into adjacent sections of the major groove of DNA and contacts the nucleotides of the CRP-cAMP–binding site. In the absence of cAMP, the conformation of CRP changes such that the two α helices can no longer bind to the major groove (Figure 27·42). In this way, cAMP regulates the binding activity of CRP. The DNA-binding motif of CRP is common to many transcriptional regulatory proteins. In some cases, such as CRP-cAMP, the DNA is bent slightly when the protein binds (Figure 27·43).

The DNA-binding region of each CRP-cAMP subunit and many similar proteins includes a second α helix in addition to the one that lies in the major groove of DNA. The two helices are separated by a short stretch of polypeptide chain that

Figure 27·42
Allosteric changes in CRP caused by the binding of cAMP. Each monomer of the CRP dimer contains an α helix. In the absence of cAMP, the α helices cannot fit into adjacent sections of the major groove of DNA and hence cannot recognize the CRP-cAMP–binding site. When cAMP binds to CRP, the two α helices assume the proper conformation for binding to DNA.

Figure 27·43
Structure of a complex between CRP-cAMP and DNA. (Courtesy of Steve Schultz and Thomas A. Steitz.)

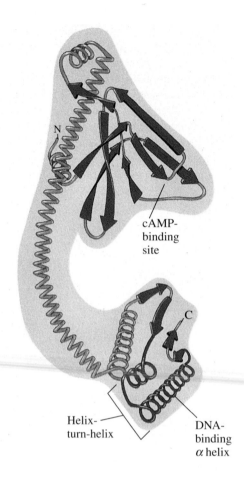

Figure 27·44
Drawing of CRP structure showing the helix-turn-helix motif.

cAMP-binding site

N

C

Helix-turn-helix

DNA-binding α helix

turns sharply, giving rise to the name helix-turn-helix for this type of DNA-binding motif (Figure 27·44). The *lac* repressor, σ^{70}, and the restriction endonuclease *Eco*RI (Section 24·10A) also contain a helix-turn-helix motif and bind DNA in a similar manner. This structure is very different from the zinc-finger motif of TFIIIA.

The actual contacts between amino acid side chains and a specific DNA sequence are similar to those between enzymes and their substrates. For example, CRP-cAMP recognizes adjacent C/G and A/T base pairs through hydrogen bonds between the bases and its glutamate and lysine side chains (Figure 27·45).

CRP-cAMP increases the rate of transcription initiation in different ways at different promoters. At some promoters, CRP-cAMP enhances binding of RNA polymerase, whereas at other promoters, it aids initiation directly by accelerating either formation of the open complex or primer synthesis. In some cases, as we will see later, CRP-cAMP cooperates with other DNA-binding proteins to achieve activation. CRP-cAMP can also act as a repressor. For example, a CRP-cAMP–binding site overlaps the promoter for the *crp* gene itself. When CRP-cAMP binds to this site, it prevents RNA polymerase from interacting with the promoter, thereby blocking transcription of the *crp* gene. Thus, when the concentration of CRP-cAMP inside the cell rises sufficiently, CRP-cAMP represses transcription of its own gene by **autoregulation.** The exact role of CRP-cAMP in activation or repression depends largely on the position of the binding site relative to the promoter. Many transcription factors behave similarly.

Figure 27·45
Hydrogen bonds between amino acid side chains and DNA in the CRP-cAMP–DNA complex.

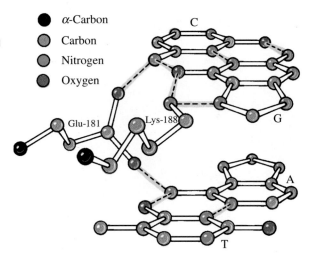

- ● α-Carbon
- ○ Carbon
- ○ Nitrogen
- ● Oxygen

Glu-181 Lys-188

C
G
A
T

27·11 Activators Control the Maltose Regulon

Starch and glycogen are abundant sources of carbon for bacteria that feed on plant and animal remains. In mammals, the activities of the enzymes required for glycogen degradation are regulated by a protein phosphatase and a phosphorylase kinase (Section 17·5C). In bacteria, however, it is transcription of the genes that encode the enzymes that is regulated and not the activities of the enzymes themselves.

The bacterial genes encoding the proteins required for the utilization of maltose are scattered throughout the genome. Some are single genes with their own promoters, and others are organized into several separate operons. All of these genes and operons are controlled by the regulatory protein MalT and, either directly or indirectly, by CRP-cAMP. Such coordinately regulated operons and genes are collectively referred to as a **regulon.**

The regulatory protein MalT is a large, monomeric protein (M_r 103 000) that is normally present in the cell at low concentration (20 to 40 molecules per cell). Transcription of the *mal*T gene is positively regulated by CRP-cAMP—when glucose is absent, the concentration of cAMP rises, and CRP-cAMP activates transcription of the *mal*T gene until the concentration of MalT is several hundred molecules per cell. Since MalT is itself an activator of the other genes of the maltose regulon, the enzymes required to convert starch to glucose are produced only when glucose is unavailable. Note that this mechanism of indirect activation by CRP-cAMP differs from the direct activation by CRP-cAMP at the *lac* operon.

The rate of transcription of the maltose-utilization genes also depends upon the presence of substrate and of ATP. The ability of MalT to activate transcription is increased by maltotriose, a trimer of glucose residues linked by α-$(1 \rightarrow 4)$ glycosidic bonds. Maltotriose is produced by the extracellular enzymatic degradation of amylopectin and amylose (Figure 13·9). Two forms of MalT exist in equilibrium. The inactive form ($MalT_i$) cannot function as an activator, whereas the active form ($MalT_a$) can bind to specific DNA sequences and activate transcription. Since $MalT_i$ and $MalT_a$ exist in equilibrium, some $MalT_a$ is always present, and all the maltose genes are transcribed to some extent. This low-level expression of the maltose genes is necessary to allow the cells to take up and utilize any maltodextrins that might be present in the external medium. If maltotriose and ATP are present, the relative concentrations of $MalT_i$ and $MalT_a$ shift in the direction of $MalT_a$, and expression of the genes of the maltose regulon is increased 10- to 50-fold. Unlike many other DNA-binding proteins, $MalT_a$ recognizes an asymmetric site (GGA(T/G)GA) and binds as a monomer.

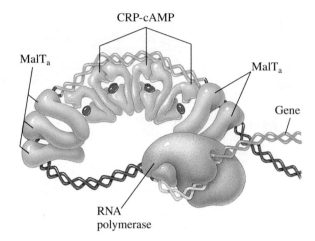

Figure 27·46
Organization of the MalT$_a$ transcription-initiation complex.

At some operons, there are multiple binding sites for MalT$_a$ and CRP-cAMP. Many of these sites are located a considerable distance from the promoter itself. Such sites are often referred to as **enhancers** because they enhance the rate of transcription initiation but are not obviously part of the promoter. MalT$_a$ and CRP-cAMP appear to interact with each other and contact RNA polymerase at the promoter by looping out the intervening DNA (Figure 27·46). One of the roles of CRP-cAMP in this complex of activators is to bend the DNA to form a loop. Enhancers used to be considered a unique feature of eukaryotic gene regulation, but there are now many known examples in bacteria.

27·12 Transcription of Some Genes Requires Specific σ Factors

The maltose regulon and the *lac* operon are transcribed by the common *E. coli* RNA polymerase holoenzyme containing σ^{70}. Other genes are transcribed by RNA polymerases that are bound to different σ factors. The genes controlling nitrogen metabolism in many bacteria, for example, are transcribed from promoters that are recognized by σ^{54}. Holoenzyme containing σ^{54} is incapable of forming an open complex at the promoter in the absence of additional activators. In this sense, these promoters resemble the eukaryotic promoters that require general transcription factors. Many prokaryotic genes require an activator protein, NR$_1$, also called NtrC, that interacts with σ^{54}-containing holoenzyme to promote transcription initiation. The activator does not affect the binding of RNA polymerase to the promoter, only formation of the open complex.

One of the characteristics of σ^{54} promoters is that there are usually multiple regulatory protein–binding sites located some distance upstream of the promoter. Interaction between the bound activator(s) and RNA polymerase often occurs through the formation of a DNA loop. NR$_1$-binding sites can be effective even when they are 1200 base pairs upstream of the promoter. Both orientations of the non-palindromic binding site are functional, even when positioned downstream of the promoter at the 3′ end of the gene. Binding sites or enhancers with these properties are common in eukaryotes as well.

Other σ factors in addition to σ^{70} and σ^{54} are found in bacteria. For example, a special class of genes is transcribed when *E. coli* cells are not growing and dividing. In such stationary-phase cells, σ^{38} is the predominant σ factor found associated with RNA polymerase core enzyme. The stationary-phase holoenzyme binds to promoters that have a recognizable TATA box but a different −35 region that has not been characterized. There is also a special σ factor (σ^{28}) that is specific for genes required for chemotaxis and for the synthesis of flagella.

The heat-shock genes in *E. coli* are under the control of a fifth σ factor named σ^{32}. σ^{32} is not active as a σ factor unless cells are stressed. Once it is activated in response to some trigger, σ^{32} directs RNA polymerase to the promoters of the heat-shock genes. Recall from Chapter 5 that many of the heat-shock genes encode chaperone proteins that aid in the recovery of misfolded proteins. The genes for these proteins are usually transcribed from a separate σ^{70} promoter at reduced rates in the absence of stress. There are many other examples of genes that have more than one promoter. Different promoters are used depending on the circumstances.

27·13 Bacteria Contain a Panoply of Repressors and Activators

There are dozens of different repressors and activators in the typical bacterial cell. Many of these are allosteric proteins like CRP-cAMP and *lac* repressor. Activators such as CRP-cAMP fall into category (a) in Figure 27·32. These are functional as activators when a ligand such as cAMP is bound. FadR is an example of an activator that falls into category (b). FadR regulates the genes that encode enzymes involved in fatty acid synthesis. These genes are normally transcribed by the combined action of FadR and RNA polymerase holoenzyme, but in the presence of exogenous fatty acids, FadR no longer serves as an activator because it binds acyl CoA derivatives that accumulate within the cell. In this way, expression of the *E. coli* genes is deactivated when an external source of fatty acids is available.

The *lac* repressor is an example of category (c) of Figure 27·32. It does not bind to DNA in the presence of the ligand allolactose. The *trp* repressor (TrpR) falls into category (d). The *trp* repressor controls transcription of the genes encoding enzymes required for the biosynthesis of tryptophan. These genes are repressed when free tryptophan is present inside the cell because tryptophan binds to the *trp* repressor and the ligand-repressor complex binds to the *trp* operator. Tryptophan is thus a corepressor of the *trp* operon. When tryptophan levels fall, the ligand is released and transcription of the operon is induced.

Not all activators and repressors are allosteric proteins. PhoB is an activator that regulates transcription of genes required for the transport and assimilation of phosphate. Only the phosphorylated form of PhoB functions as a specific DNA-binding protein and an activator; the nonphosphorylated form is not functional. Specific DNA-binding activity of the NR_1 activator is also regulated by phosphorylation. The methyltransferase encoded by the *ada* gene (Section 26·4A) is also a transcriptional activator whose activity is controlled by covalent modification. Recall that this protein repairs damage due to alkylating agents by transferring the alkyl group to an internal cysteine residue. The alkylated derivative becomes a transcriptional activator, stimulating transcription of its own gene.

LexA is an example of a repressor that is not an allosteric protein. LexA normally binds to promoter regions and represses transcription of many genes required for repair and recombination. In the presence of single-stranded DNA, which arises as a result of DNA damage, the repressor is inactivated and the repair genes are transcribed. This response is known as the SOS response. Inactivation of LexA is due to RecA, which when bound to single-stranded DNA functions as a specific protease, cleaving proteins such as LexA. In this case, inactivation is irreversible.

We have seen that CRP-cAMP can be both an activator and a repressor, depending on which gene is being controlled. It functions as an activator when its binding site is just upstream of the promoter, but it functions as a repressor when the binding site overlaps the promoter and CRP-cAMP competes with RNA polymerase in binding DNA. There are many similar examples of regulatory proteins that can be both repressors and activators; one well-studied protein is AraC, which regulates genes involved in the utilization of arabinose. The regulation of arabinose operons is complex; by binding to different sites on DNA, AraC functions as either a repressor (in the absence of arabinose) or an activator (when arabinose is available).

Finally, MerR is a simpler example of a regulatory protein that is both a repressor and an activator. The protein is required for the regulation of the *mer* operon, whose genes encode proteins that chelate mercury ions. MerR represses transcription of the *mer* operon by binding near the promoter. In the presence of mercury, a MerR-Hg^{2+} complex forms, and this complex acts directly as an activator at the same promoter.

27·14 Most Eukaryotic Genes Are Regulated by Transcriptional Activation

The formation of a transcription-initiation complex at eukaryotic promoters is also regulated by a variety of gene-specific activators and repressors that interact with RNA polymerase and with general transcription factors bound to DNA. There are a few examples of eukaryotic repressors, but most genes seem to be controlled by activators that stimulate transcription from poor promoters.

A. Some Hormone Receptors Are Transcriptional Activators

Estrogens, glucocorticoids, cortisol, testosterone, and progesterone are mammalian steroid hormones, and ecdysterone is an insect steroid hormone. Cells that respond to the presence of these hormones contain cytoplasmic receptors that bind the hormone once it traverses the plasma membrane. The hormone-receptor complex is then transported to the nucleus, where it acts as a transcriptional activator by binding to DNA sequences known as **hormone-response elements.**

The steroid-hormone receptors have several domains that act in DNA binding, ligand binding, and transcriptional activation. They are usually large proteins that associate with chaperones during transport from the cytoplasm, where they pick up hormone, to the nucleus, where they act as transcriptional activators. All of the steroid-hormone receptors are homologous.

The hormone-receptor complex usually binds to multiple hormone-response elements upstream of the TATA box. Each element exhibits dyad symmetry, and the receptors bind as dimers in a manner similar to that described for the interaction between a single operator and two *lac* repressor subunits. The bound receptors are thought to exert their effect as activators by interacting either directly with the initiation complex or through loops of DNA such as those described in the previous sections. Their effect can be profound. The ecdysterone receptor, for example, activates expression of many insect genes that are required for development of the adult. The testosterone, or androgen, receptor controls genes that must be activated for normal development of mammalian males. Human males who lack testosterone receptor develop into adults who appear to be female in every external aspect. The disorder is known as androgen-insensitivity syndrome or testicular feminization.

Steroid-hormone receptors have zinc-containing motifs of the C_2C_2 class (zinc atoms are chelated by four cysteine residues). The zinc-containing motifs occur in pairs, and each pair forms a single structure in which the side chains of only one α helix form contacts with the base pairs in DNA (Figure 27·47). The α helices from each subunit of the receptor dimer lie in adjacent major grooves on the same side of the DNA molecule (Figure 27·48). The zinc-containing DNA-binding domain of steroid-hormone receptors has a very different structure from that of the zinc fingers seen in TFIIIA, and the mechanism of binding is also different.

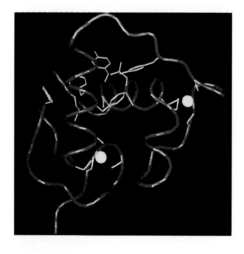

Figure 27·47
Zinc-containing DNA-binding domain of the human estrogen receptor. Zinc ions are shown as white spheres; zinc ligands are shown in yellow. Hydrophobic residues in the core of the protein are shown in blue. (Courtesy of John Schwabe.)

Thyroid hormones, retinoic acid, and vitamin D are believed to work in a manner similar to that of the steroid hormones. They bind to receptors in many cells and by doing so stimulate the transcription of responsive genes within those target cells.

B. Heat-Shock Induction in Eukaryotes Requires an Activator

Eukaryotic heat-shock genes, like those in prokaryotes, are transcribed in response to stresses such as high temperature. However, the mechanism of induction in eukaryotes is quite different. (Recall that transcription of the bacterial genes requires a rare σ factor.)

Heat-shock genes contain multiple heat-shock elements upstream of the TATA box. When cells are stressed, the protein heat-shock factor (HSF) becomes phosphorylated and binds to the heat-shock elements, activating transcription. In some species, the unphosphorylated HSF present in normal cells is already bound to the heat-shock element, and it is presumably phosphorylated in place. In other species, the heat-shock elements appear to be unoccupied in unstressed cells.

The heat-shock genes are found in all species. One of them, encoding the chaperone Hsp70 and its bacterial homologue DnaK, is the most highly conserved gene in all of biology. Although the genes themselves are conserved, the regulatory elements controlling their expression in bacteria and eukaryotes are not.

C. In Yeast, Galactose Metabolism Is Regulated by the Binding of the GAL4 Protein to an Upstream Activating Sequence

Genes involved in galactose uptake and metabolism are controlled by the GAL4 protein in the yeast *S. cerevisiae*. The dimeric GAL4 protein (M_r 99 000) regulates the transcription of at least five genes. Two of these, *GAL*1 and *GAL*10, encode galactose kinase and galactose epimerase, respectively, and are found side by side on chromosome II in the yeast genome. They are transcribed in opposite directions, and their promoters are contained within a 680 bp region that separates the genes. Also located in this region between the two promoters is an **upstream activating sequence** (UAS).

The galactose UAS is composed of four separate GAL4 binding sites, numbered I through IV (Figure 27·49). Each of the binding sites consists of a similar 17 bp sequence that displays dyad symmetry, and GAL4 probably binds as a dimer. Activation of transcription does not require that all four sites be bound by GAL4; however, experiments have shown that the rate of transcription of *GAL*1 and *GAL*10 is directly proportional to the number of GAL4 molecules bound to the UAS. Thus,

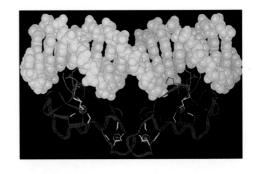

Figure 27·48
The DNA-binding domain of the human estrogen receptor in complex with DNA. DNA is shown as a white space-filling model. Amino acid side chains in contact with base pairs are shown in purple. Zinc ions and zinc ligands are shown in yellow. (Courtesy of John Schwabe.)

Figure 27·49
Galactose upstream activating sequence (UAS) in yeast. The *GAL*1 and *GAL*10 genes lie side by side and are transcribed from divergent promoters. In between the coding sequences of the two genes lies a 680 bp sequence that includes the galactose UAS. The galactose UAS contains four binding sites for the activator GAL4. Each binding site is composed of 17 base pairs and displays dyad symmetry. Activation of transcription does not require that all four sites be filled; however, the level of activation of transcription is proportional to the number of GAL4-binding sites occupied.

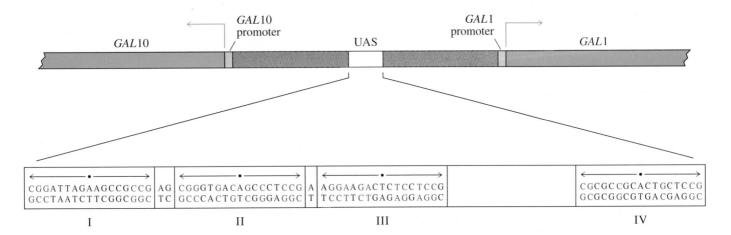

the effect of GAL4 binding is additive; each GAL4 dimer contributes independently to stimulation of transcription. Maximal activation increases the rate of transcription of both *GAL*1 and *GAL*10 over 1000-fold.

Like the prokaryotic operons encoding metabolic enzymes, the *GAL*1 and *GAL*10 genes are only transcribed in the presence of substrate, in this case galactose. In the absence of galactose, the galactose UAS is bound by GAL4 whose activation domain is in turn bound by a negative regulatory protein called GAL80 (M_r 48 300). Binding of GAL80 to GAL4 prevents GAL4 from activating transcription of *GAL*1 and *GAL*10. In the presence of galactose, however, galactose (or some other inducer derived from galactose) binds to GAL80 and causes it to dissociate from GAL4. GAL4 is then able to activate transcription of *GAL*1 and *GAL*10 by interacting with the transcription complex at the promoter. This interaction involves formation of a DNA loop (Figure 27·50).

Regulation of *GAL*1 and *GAL*10 is further complicated by the fact that both genes are also negatively regulated in the presence of glucose. This inhibition of transcription by glucose is analogous to catabolite repression in prokaryotes; when both sugars are present, *GAL*1 and *GAL*10 are transcribed at low basal rates. Glucose blocks transcription of *GAL*1 and *GAL*10 by preventing GAL4 from binding to the galactose UAS. It is not known whether glucose (or a metabolite derived from glucose) binds directly to GAL4.

GAL4 consists of two domains, one that binds to DNA and another, the activation domain, that stimulates transcription of downstream genes. The GAL4 DNA-binding domain contains chelated metal ions that can be either zinc or cadmium.

Figure 27·50
Regulation of transcription by GAL4. **(a)** In the absence of galactose, there is only basal transcription of the *GAL*1 gene. Although the GAL4 activator protein is bound to site IV of the UAS, it is unable to stimulate transcription due to the presence of the inhibitor GAL80. **(b)** In the presence of galactose (or some other inducer derived from galactose), GAL80 dissociates from GAL4, allowing GAL4 to interact with the transcription complex. As a result, transcription of the *GAL*1 gene increases.

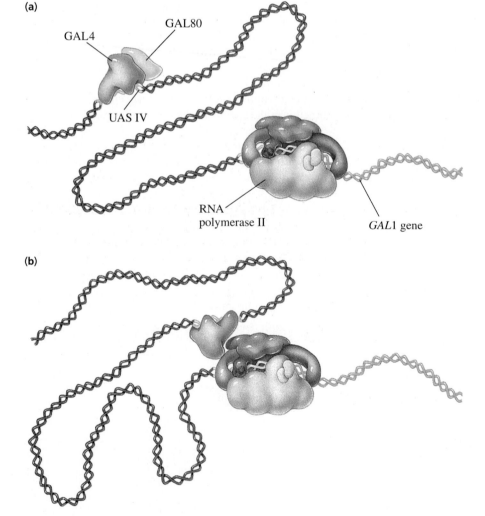

(a)

GAL4

GAL80

UAS IV

RNA polymerase II

*GAL*1 gene

(b)

The structure reveals a third type of zinc-containing motif that is unrelated to those in TFIIIA or steroid-hormone receptors. Six cysteine residues are involved in chelating two metal ions.

The GAL4 activation domain, like the activation domains of many other transcription factors, consists of a cluster of amino acid residues with negatively charged side chains (glutamate and aspartate). In the laboratory, the activation domain of GAL4 can be attached, or fused, to the DNA-binding domain of another protein, where it still functions as an activator of transcription initiation provided that the fusion protein binds near the promoter. Other proteins that contain activation domains with many negatively charged side chains include GCN4, a yeast transcription factor that activates genes involved in amino acid biosynthesis, and many of the hormone receptors, such as the glucocorticoid receptor.

The class II transcription factor SP1 is an example of a transcription factor with a different type of activation domain—one that is glutamine rich. Some activation domains, such as that of the class II factor CTF, are proline rich. Precisely how these activation domains stimulate transcription is not known, but it is likely that they make contact with RNA polymerase II or the general transcription factors at the promoter. In some cases, additional proteins are required.

D. Some Transcription Factors Contain Coiled-Coil Domains

In addition to the CCAAT box transcription factor, CTF (Section 27·8A), there are other mammalian transcription factors that bind to the core sequence CCAAT, although these other factors may differ in their specificity for flanking sequences. One of these proteins is called C/ERB (CCAAT enhancer binding). It is synthesized in liver, lung, and intestine shortly after birth. In these tissues, C/ERB activates transcription of genes that are expressed in terminally differentiated cells or cells that have stopped growing and dividing. Some of these genes encode enzymes involved in fatty acid metabolism and the insulin-responsive glucose transporter. Thus, C/ERB is a tissue-specific activator that is responsible for the presence of particular proteins in specialized tissues.

C/ERB is a dimer consisting of identical subunits that bind to adjacent CCAAT (or its inverse ATTGG) sequences in a manner analogous to the interaction of other dimeric binding proteins with symmetric binding sites. The two monomers bind to each other through contacts made by amphipathic α helices that form a coiled-coil structure such as that present in keratin and other fibrous proteins of the cytoskeleton and also in myosin. One of the characteristics of such amphipathic α helices is the presence of hydrophobic amino acid side chains at every seventh position. Frequently the residue is leucine; this has given rise to the name **leucine zipper** for this structural motif in transcription factors. Note that leucine zippers are also found in many other proteins; they are not directly related to protein-DNA interactions.

The yeast transcription factor GCN4 also contains leucine zippers that aid in the formation of the dimeric protein (Figure 27·51). There are several other examples of

Figure 27·51
Stereo view of a leucine-zipper motif of yeast GCN4 complexed with DNA. The protein is shown in blue; leucines that form the zipper are in purple. The DNA backbone is red; DNA bases are green. (Coordinates provided by T. E. Ellenberger and S. C. Harrison.)

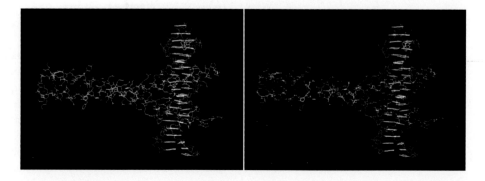

similar homodimers. However, the leucine-zipper motif can also be present on different polypeptides leading to the formation of a heterodimer. In mammals, there is a family of transcription factors, called the AP-1–Jun family, that are similar to GCN4 in amino acid sequence but that can only bind to DNA in the presence of an additional protein called Fos. Both Fos and AP-1–Jun polypeptides possess a leucine-zipper motif, and formation of the heterodimer is through a coiled-coil structure.

Jun, one of the members of the AP-1–Jun family, is encoded by *c-jun,* a proto-oncogene whose oncogenic form (*v-jun*) is found in certain retroviruses that cause cancer. Other cancer-causing retroviruses carry an oncogenic version of *c-fos.* Inappropriate expression of these transcription factors in infected cells leads to deregulated growth, or cancer, demonstrating the importance of transcription factors in regulating gene expression.

Leucine zippers represent just one of many types of protein-protein contacts within the initiation complex. The interaction of multiple transcription factors is a common theme in the regulation of transcription initiation in eukaryotes. In plants and animals, there are few genes that are regulated by a single transcription factor. Mammalian liver-specific genes that are regulated by C/ERB, for example, also require other factors that are not tissue specific. Heat-shock activation of the mammalian *hsp*70 genes requires CTF and SP1 as well as HSF, and the steroid-hormone receptors also interact with additional factors that bind near the promoters. Although there are examples of prokaryotic genes that require multiple activators (for example, the maltose regulon), in general, the regulation of transcription of eukaryotic genes involves more factors and a longer stretch of DNA to bind these factors than does the regulation of prokaryotic genes.

27·15 Transcription and Chromatin

It may seem that the presence of more specialized and complicated polymerases marks a significant difference between eukaryotic and prokaryotic transcription. However, although much remains to be learned about transcription in eukaryotes, it is clearly very similar to transcription in prokaryotes. In both prokaryotes and eukaryotes, transcription can be divided into initiation, elongation, and termination. In both types of organisms, each step is carried out by analogous reactions. Eukaryotic transcription differs significantly from prokaryotic transcription in the degree of complexity of the transcription apparatus and in the segregation of transcription, which occurs in the nucleus, from translation, which occurs in the cytoplasm.

Another distinction between the process of transcription in prokaryotes and eukaryotes is the state of the template. As discussed in Section 24·11, eukaryotic DNA is intimately associated with histones and other chromosomal proteins that package the DNA in a well-ordered structure. Neither replication nor transcription is compatible with the tightly packed chromatin structure found in metaphase chromosomes. During the late stages of meiosis and mitosis, when metaphase chromosomes are visible, transcription is inhibited. Transcription is typically associated with chromatin that has opened into the beads-on-a-string conformation. In all more densely packed conformations, including the 30-nm fiber, eukaryotic DNA appears to be relatively inaccessible to other DNA-binding proteins and hence to the transcription machinery.

We have visual evidence that the transition from a compact chromatin structure to the more open beads-on-a-string structure is related to transcriptional activity. In cells of the salivary glands from *D. melanogaster,* the chromosomes are extended and replicated in place, producing many copies of each chromosome, which are aligned to form a large structure called a **polytene chromosome.** When polytene chromosomes are examined, certain regions of the chromosome are seen as dark bands of condensed chromatin and other regions are expanded to form puffs. Puffs correspond to regions of transcriptional activity.

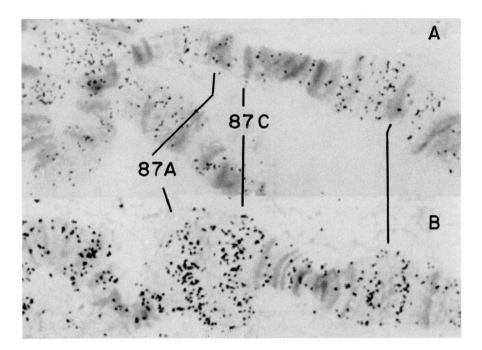

Figure 27·52
Formation of puffs on polytene chromosomes following heat shock. The autoradiograph shows grains at the sites of incorporation of ³H-uridine into RNA. Chromosome bands 87A and 87C contain several copies of *hsp*70 genes. (a) Before heat-shock induction, the 87A and 87C bands are condensed and no incorporation of ³H-uridine is observed. (b) After heat shock, the bands decondense to form large puffs, and the high density of grains demonstrates that the *hsp*70 genes are actively transcribed. (Courtesy of J. José Bonner.)

An example of polytene-chromosome puffs is shown in Figure 27·52. The bands labelled 87A and 87C correspond to the sites of the *hsp*70 heat-shock genes. In unstressed cells, these genes are in a compact, condensed region of chromatin, but following heat shock, the bands decondense to form large puffs as the genes are transcribed in response to stress. Chromatin decondensation is directly linked to activation of the genes by HSF.

A. One X Chromosome Is Inactivated in Mammalian Females by Condensation into Heterochromatin

The DNA within polytene-chromosome bands is condensed but nevertheless accessible to transcription factors. However, there are forms of chromatin, known as **heterochromatin,** that are much more highly condensed. *Constitutive heterochromatin* refers to chromosomes or parts of chromosomes that are heterochromatic in all cells of a given species. Examples of constitutive heterochromatin can be found in every multicellular eukaryote and can take the form of entire chromosomes or parts of chromosomes. For example, some maize cells contain multiple copies of a small, heterochromatic chromosome called chromosome B. In addition, between one-fourth and one-third of all DNA in *Drosophila* is found in heterochromatic regions near the centromeres.

Condensation of chromatin into heterochromatin is an effective mechanism of repressing eukaryotic gene expression and is best exemplified by the process of X-chromosome inactivation in mammalian females. The sex of a mammal is determined by the presence or absence of the male-specific Y chromosome. In humans, males normally have one X and one Y chromosome per somatic cell, whereas females normally have two X chromosomes per somatic cell. The X chromosome is quite large and contains a number of genes, most of which play no role in sex differences. Proper human development requires that only one X chromosome be fully active in each somatic cell of an adult. Thus, one of the X chromosomes in females is inactivated by condensation into heterochromatin (Figure 27·53, next page). Such condensed chromosomes are known as sex-chromosome bodies, or Barr bodies. X-chromosome inactivation is one example of the genetic phenomenon known as **dosage compensation** because it involves regulating the dosage of genes.

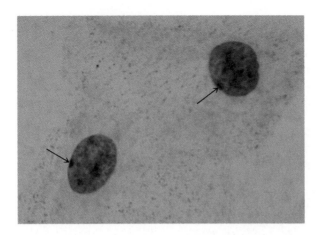

Figure 27·53
X-chromosome condensation in human epidermal cells. Female cells have a single condensed X chromosome (arrow) on the periphery of the nucleus. These condensed chromosomes are known as sex-chromosome bodies, or Barr bodies.

In human females, X-chromosome inactivation occurs very early in embryonic development, at about the 20-cell stage. Condensation of an X chromosome into heterochromatin appears to begin at a unique point, the *xist* gene, and proceed bi-directionally along the DNA. Inactivation is associated with extensive methylation of DNA. Once a specific X chromosome has been inactivated in a particular cell of the 20-cell embryo, the same X chromosome remains inactivated in all daughter cells descended from that precursor cell (Figure 27·54). In each human cell, either the maternal or the paternal X chromosome can be inactivated.

Figure 27·54
X-chromosome inactivation in human females. Each somatic cell of the embryo contains two X chromosomes, one inherited maternally (X_m) and one inherited paternally (X_p). Until about the 20-cell stage, genes on both X chromosomes are transcribed. In each cell of the 20-cell embryo, however, one X chromosome is inactivated at random by condensation of the DNA into heterochromatin. Once a certain X chromosome has been inactivated in a particular cell, the same X chromosome remains inactivated in all daughter cells descended from that precursor cell. Each of the cells in the 20-cell embryo goes on to produce many progeny cells. Since a mature female is composed of cells descended from more than one cell of the 20-cell embryo, human females are mosaics composed of some cells expressing the maternal X chromosome and some cells expressing the paternal X chromosome.

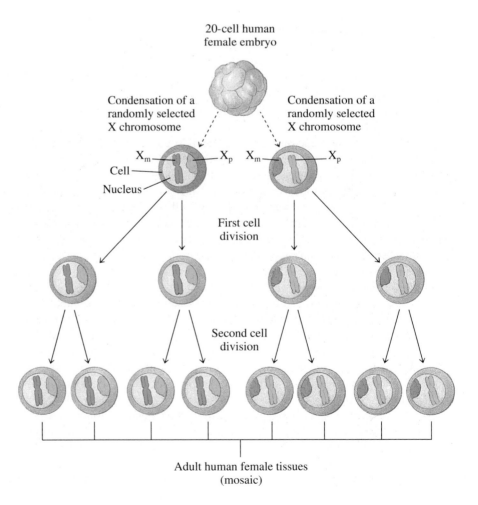

The frequencies of maternal and paternal X-chromosome inactivation vary among mammals. In female marsupials, for example, the paternal X chromosome appears to be preferentially inactivated. This observation indicates that the maternal and paternal X chromosomes are not identical and can be distinguished in the developing embryo. However, in most other mammals, including humans, the X chromosome that is condensed appears to be selected more or less at random. As a result, some of the cells in the mature organism contain an active maternal X chromosome, and some contain an active paternal X chromosome. Consequently, the organism is a mosaic composed of cells expressing different genetic information.

Sometimes cells containing active maternal X chromosomes can be physically distinguished from those containing active paternal X chromosomes. An example of such a visible mosaic is the calico cat, which has patches of orange and black fur (Figure 27·55). Calico cats are always female if they have normal X chromosomes. The patchiness results from random inactivation of X chromosomes in female cats in which the X chromosome inherited from one parent carries the gene for orange fur and the X chromosome inherited from the other parent carries the gene for black fur.

Genetic mosaicism due to X-chromosome inactivation also occurs in human females. For example, the gene for glucose 6-phosphate dehydrogenase is located on the X chromosome. If each chromosome carries a different allele, patches of cells will contain either one isoform or the other, depending on which X chromosome is inactivated. The theory of X-chromosome inactivation was developed in large part by Mary Lyon, and the process is sometimes referred to as Lyonization.

Figure 27·55
Calico cat with kittens. The mosaic of orange and black fur on the back is due to random X-chromosome inactivation. The white fur on the underside of the cat is due to expression of a different gene.

B. The Expression of Globin Genes Is Correlated with Changes in Chromatin Structure

Hemoglobin is a tetrameric protein, and each globin subunit is associated with a heme prosthetic group. In adults, the most abundant form of hemoglobin is composed of two α-globin and two β-globin subunits. α and β subunits are not the only types of subunits found in vertebrate hemoglobin, however. In embryos, a number of α-like and β-like subunits are produced. The common types of hemoglobin found in the different developmental stages of humans are shown in Table 27·7. In each case, the hemoglobins are composed of two α or α-like subunits and two β or β-like subunits. Each type of subunit is encoded by a separate globin gene whose expression is coordinated with that of other globin genes to ensure that no subunit is overproduced. The genes encoding the α and α-like subunits are clustered on one chromosome, and the genes encoding the β and β-like subunits are clustered on another chromosome. This clustered organization of hemoglobin-subunit genes is common among vertebrates; however, the gene clusters may vary in size and may contain pseudogenes (Chapter 32).

The gene cluster encoding the β and β-like subunits of chicken hemoglobin includes four genes (Figure 27·56). Two of the β-like genes, ε and ρ, are expressed predominantly during embryonic stages of development. Because avian erythrocytes retain their nuclei (unlike mammalian erythrocytes, which lose their nuclei before maturation), the chromatin structure of the globin genes can be analyzed at all developmental stages.

Table 27·7 Forms of hemoglobin found during human development

Stage of development	Hemoglobin subunit composition
Embryo	$\zeta_2\varepsilon_2$
Embryo	$\zeta_2\gamma_2,^1 \alpha_2\varepsilon_2$
Fetus	$\alpha_2\gamma_2^1$
Adult	$\alpha_2\beta_2, \alpha_2\delta_2^2$

[1] There are two types of human γ subunit, $^A\gamma$ and $^G\gamma$, which differ only in the amino acid at position 136.
[2] A minor form of hemoglobin, accounting for only 2% of the total hemoglobin present.

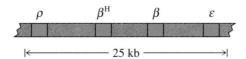

Figure 27·56
Organization of the gene cluster encoding the β and β-like subunits of chicken hemoglobin. This gene cluster contains four protein-coding genes. ρ and ε are expressed predominantly in embryos, whereas β and β^H are expressed predominantly in adults.

Figure 27·57
Procedure for probing chromatin structure with DNase I. Intact nuclei are isolated and treated with different concentrations of DNase I. DNA that is densely packaged in nucleosomes is relatively insensitive to digestion by DNase I, whereas DNA that is in the extended beads-on-a-string conformation is very sensitive to digestion by even low concentrations of DNase I. (Points of cleavage by DNase I are indicated by dashed arrows.) The nuclei are then lysed, the histones are removed from the DNA, and the DNA is digested with restriction endonucleases to yield fragments that can be resolved by gel electrophoresis. After separation, the DNA fragments are denatured and then transferred to a nitrocellulose filter paper. Single-stranded fragments derived from each gene are detected by the hybridization of a radioactive DNA fragment (probe) complementary to the gene of interest. The amount of probe bound to the filter indicates the relative abundance of each fragment. Genes that were inaccessible to DNase I bind similar amounts of radioactive probe regardless of the concentration of DNase I used to treat the nuclei, whereas genes that were accessible to DNase I bind decreasing amounts of radioactive probe as the DNase I concentration used to treat the nuclei increases.

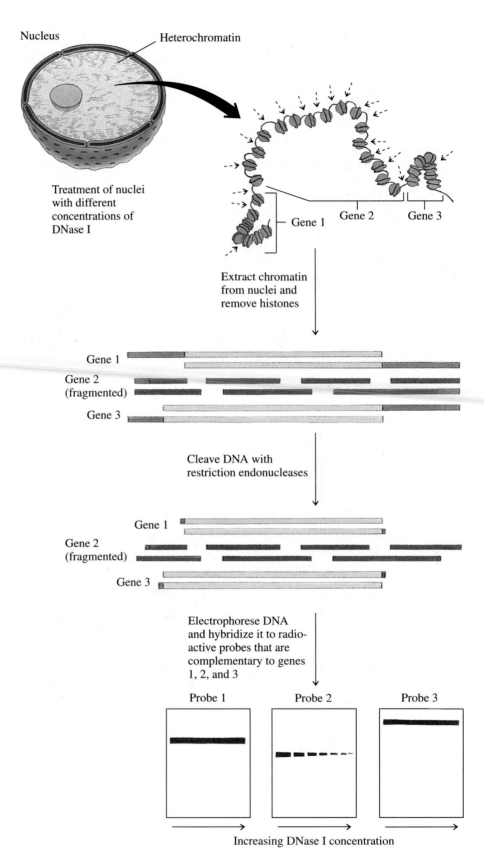

Nucleases such as DNase I can be used to probe the conformation of chromatin. DNase I poorly digests DNA that is either heterochromatic or in another condensed form, but even low concentrations of DNase I easily digest DNA that is in the more open beads-on-a-string conformation. Nuclei can be treated with DNase I, the chromatin isolated, the histones removed, and the DNA fragments separated by gel electrophoresis. The only genes isolated intact by this procedure are those that were packaged into transcriptionally inactive chromatin and hence not digested to small pieces by DNase I (Figure 27·57).

The genes from embryonic chicken erythrocytes can be grouped into different classes on the basis of their sensitivity to DNase I (Figure 27·58). The gene for chicken ovalbumin, which is never expressed in erythrocytes, is resistant to even high concentrations of DNase I in both 14-day-old and 5-day-old erythroid cells. The gene encoding the adult β subunit is slightly sensitive to DNase I in 5-day-old erythroid cells but is extremely sensitive to DNase I in 14-day-old cells. The reverse pattern is noted for the genes encoding embryonic β-like subunits; these genes are extremely sensitive in 5-day-old erythroid cells but only slightly sensitive in 14-day-old cells. This pattern of sensitivity to DNase I strongly suggests that developmental regulation of the expression of hemoglobin-subunit genes involves the decondensation of certain genes into an extended beads-on-a-string conformation and the condensation of other genes into transcriptionally inactive chromatin.

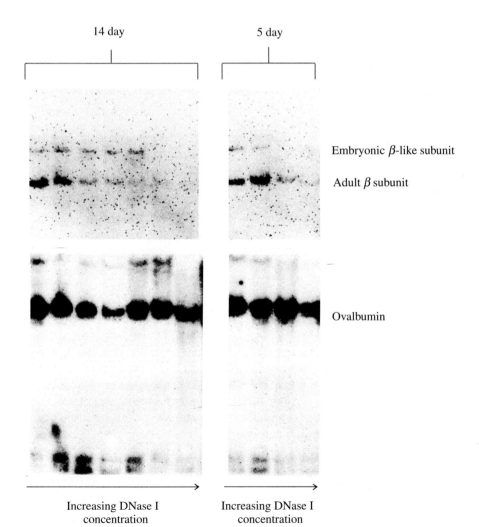

14 day   5 day

Embryonic β-like subunit

Adult β subunit

Ovalbumin

Increasing DNase I concentration   Increasing DNase I concentration

Figure 27·58
Relationship between chromatin structure and gene expression. Autoradiography reveals the sensitivity to DNase I of several genes in 14-day-old and 5-day-old chicken erythroid cells. The ovalbumin gene is never expressed in erythroid cells; in both the 14-day-old and the 5-day-old erythroid cells, the gene is relatively resistant to digestion by DNase I. In contrast, both the adult β and the embryonic β-like subunit genes are sensitive to digestion by DNase I. The genes for the embryonic β-like subunits are sensitive in 5-day-old cells but less sensitive in 14-day-old cells; the gene for the adult β subunit is less sensitive in 5-day-old cells but becomes more sensitive in 14-day-old cells. (Courtesy of Harold Weintraub.)

Figure 27·59
Transcription of chromatin. Note the presence of both nucleosomes (the DNA is in the beads-on-a-string conformation) and transcription complexes. (Courtesy of Victoria Foe.)

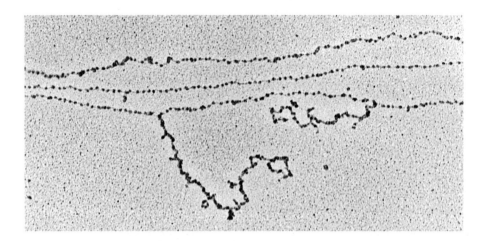

C. Nucleosomes Do Not Interfere with Transcription

Chromatin decondensation is necessary for transcription, but nucleosomes are still present when the DNA is organized as beads on a string. In spite of the presence of nucleosomes and the tight association of histones and DNA, the transcription complex is capable of transcribing DNA in the beads-on-a-string conformation (Figure 27·59).

It seems unlikely that RNA polymerase would track along the template strand, unwinding it and rotating 1.75 turns around each nucleosome. Instead, the histone-DNA complex probably dissociates to some extent, allowing RNA polymerase to pass through. One model proposes that each nucleosome splits in half and that each half remains bound to DNA. Splitting allows the DNA to unwind, which in turn permits passage of RNA polymerase.

As we discussed in Section 24·11B, the histones in decondensed chromatin are acetylated. This modification may alter the conformation of nucleosomes on genes that are actively transcribed. HMG proteins, which replace histone H1 in decondensed chromatin, may also play a role in decondensation. Transcription of regulated genes is preceded by a change in the chromatin organization to a form that is more accessible to transcription factors. The shift is mediated by proteins that bind to DNA sequences in the region of the gene. At the β-globin locus, for example, the entire region is decondensed prior to transcription; this decondensation requires sequences at both ends of the cluster.

Summary

Bacterial RNA polymerases consist of a multisubunit core and a σ factor that is required for initiation. The *E. coli* core enzyme contains two α subunits, a β subunit, and a β' subunit. The most common σ subunit is σ^{70}. Transcription begins at the 5′ end of a gene when the holoenzyme binds to the promoter to form a closed complex. The most common promoter sequence in *E. coli* is the one recognized by the σ^{70}-containing holoenzyme. The consensus sequence of this promoter includes a TATA box 10 base pairs upstream of the transcription start site and a −35 region. The next steps in initiation are unwinding of a small stretch of DNA to form an open complex, synthesis of a short RNA primer, and promoter clearance. The transition to the elongation phase of transcription is accompanied by release of the σ subunit and the association of additional factors such as NusA.

During transcription, RNA polymerase catalyzes a nucleotidyl-group–transfer reaction using nucleoside triphosphates as substrates. The nascent RNA is extended by addition of nucleoside monophosphates to the 3′ end, using one of the DNA strands as a template. A short stretch of RNA is base-paired to the template strand within the transcription bubble. Elongation is accompanied by continuous unwinding and rewinding of DNA.

Some nucleotide sequences cause the RNA polymerase to pause. Pausing is exaggerated when the nascent RNA forms a short hairpin structure. This structure can be recognized by NusA, preventing further transcription and leading to termination if the RNA-DNA hybrid is unstable. Termination can also occur when rho binds to free RNA, then comes in contact with the transcription complex. Rho-dependent termination is inhibited by antiterminators during ribosomal RNA synthesis.

There are three different eukaryotic RNA polymerases. RNA polymerase I transcribes large ribosomal RNA genes; RNA polymerase II transcribes most protein-encoding genes; and RNA polymerase III transcribes a variety of small genes. The eukaryotic RNA polymerases are related to each other, and they are more complex than the prokaryotic enzymes. Eukaryotic initiation is also more complex. Initiation on class II genes requires several transcription factors that assist in binding RNA polymerase II to the promoter. The TBP subunit of one of these factors, TFIID, binds to the TATA box. Initiation of class I and class III genes also requires specific transcription factors, but TBP is used by all three RNA polymerases. One of the class III transcription factors, TFIIIA, contains zinc-finger DNA-binding motifs.

Transcription initiation can be regulated negatively or positively. Negative regulation is mediated by repressors that bind DNA near the promoter and prevent initiation. Positive regulation is mediated by activators that bind near a weak promoter and assist initiation. Many repressors and activators are allosteric proteins whose activity is affected by ligand binding. For example, *lac* repressor binds allolactose, and CRP binds cAMP. Both of these proteins are involved in the regulation of the *lac* operon, and both contain a helix-turn-helix DNA-binding motif.

Transcription of some genes is regulated by multiple regulatory proteins that can associate with each other to form a complex at the promoter. In some cases, the regulatory proteins bind to DNA at sites that are far from the promoter, and the resulting complex forms a large loop of DNA. Transcription of many eukaryotic genes is inhibited because they are tightly associated with nucleosomes in heterochromatin. Transcription initiation is accompanied by decondensation of chromatin.

Selected Readings

General References

Alberts, B., Bray, D., Lewis, J., Raff, M., Roberts, K., and Watson, J. D. (1989). *Molecular Biology of the Cell,* 2nd ed. (New York: Garland Publishing).

Darnell, J., Lodish, H., and Baltimore, D. (1990). *Molecular Cell Biology,* 2nd ed. (New York: Scientific American Books).

Lewin, B. (1991). *Genes IV.* (New York: Oxford University Press).

Watson, J. D., Hopkins, N. H., Roberts, J. W., Steitz, J. A., and Weiner, A. M. (1987). *Molecular Biology of the Gene,* 4th ed. (Menlo Park, California: Benjamin/Cummings Publishing Company).

RNA Polymerases

Masters, B. S., Stohl, L. L., and Clayton, D. A. (1987). Yeast mitochondrial RNA polymerase is homologous to those encoded by bacteriophages T3 and T7. *Cell* 51:89–100.

Russo, F. D., and Silhavy, T. J. (1992). Alpha: the cinderella subunit of RNA polymerase. *J. Biol. Chem.* 267:14515–14518.

Willis, I. M. (1993). RNA polymerase III: genes, factors and transcriptional specificity. *Eur. J. Biochem.* 212:1–11.

Woychik, N. A., and Young, R. A. (1990). RNA polymerase II: subunit structure and function. *Trends Biochem. Sci.* 15:347–351.

Young, R. A. (1991). RNA polymerase II. *Annu. Rev. Biochem.* 60:689–715.

Transcription Initiation

Conaway, R. C., and Conaway, J. W. (1993). General initiation factors for RNA polymerase II. *Annu. Rev. Biochem.* 62:161–190.

Gill, G. (1992). Complexes with a common core. *Curr. Biol.* 2:565–567.

Helmann, J. D., and Chamberlin, M. J. (1988). Structure and function of bacterial sigma factors. *Annu. Rev. Biochem.* 57:839–872.

Lisser, S., and Margalit, H. (1993). Compilation of *E. coli* mRNA promoter sequences. *Nucleic Acids Res.* 21:1507–1516.

Transcription Elongation

Brewer, B. J. (1988). When polymerases collide: replication and the transcriptional organization of the *Escherichia coli* chromosome. *Cell* 53:679–686.

Daube, S. S., and von Hippel, P. H. (1992). Functional transcription elongation complexes from synthetic RNA-DNA bubble duplexes. *Science* 258:1320–1323.

French, S. (1992). Consequences of replication fork movement through transcription units in vivo. *Science* 258:1362–1364.

French, S. L., and Miller, O. L., Jr. (1989). Transcription mapping of the *Escherichia coli* chromosome by electron microscopy. *J. Bacteriol.* 171:4207–4216.

Kainz, M., and Roberts, J. (1992). Structure of transcription elongation complexes in vivo. *Science* 255:838–841.

Transcription Termination

Das, A. (1993). Control of transcription termination by RNA-binding proteins. *Annu. Rev. Biochem.* 62:893–930.

Richardson, J. P. (1993). Transcription termination. *Crit. Rev. Biochem.* 28:1–30.

Eukaryotic Transcription Factors

Braun, B. R., Bartholomew, B., Kassavetis, G. A., and Geiduschek, E. P. (1992). Topography of transcription factor complexes on the *Saccharomyces cerevisiae* 5S RNA gene. *J. Mol. Biol.* 228:1063–1077.

Greenblatt, J. (1991). RNA polymerase-associated transcription factors. *Trends Biochem. Sci.* 16:38–41.

Kim, J. L., Nikolov, D. B., and Burley, S. K. (1993). Co-crystal structure of TBP recognizing the minor groove of a TATA element. *Nature* 365:520–527.

Kim, Y., Geiger, J. H., Hahn, S., and Sigler, P. B. (1993). Crystal structure of a yeast TBP/TATA-box complex. *Nature* 365:512–520.

Pabo, C. O., and Sauer, R. T. (1992). Transcription factors: structural families and principles of DNA recognition. *Annu. Rev. Biochem.* 61:1053–1095.

Phillips, S. E. B. (1993). Saddling up the TATA box. *Curr. Biol.* 3:112–114.

Rhodes, D., and Klug, A. (1993). Zinc fingers. *Sci. Am.* 268(2):56–65.

Rigby, P. W. J. (1993). Three in one and one in three: it all depends on TBP. *Cell* 72:710.

Regulation of Transcription in Prokaryotes

Bushman, F. D. (1992). Activators, deactivators and deactivated activators. *Curr. Biol.* 2:673–675.

Kolb, A., Busby, S., Buc, H., Garges, S., and Adhya, S. (1993). Transcriptional regulation by cAMP and its receptor protein. *Annu. Rev. Biochem.* 62:749–95.

Lee, J., and Goldfarb, A. (1991). *lac* repressor acts by modifying the initial transcribing complex so that it cannot leave the promoter. *Cell* 66:793–798.

Richet, E., Vidal-Ingigliardi, V., and Raibaud, O. (1991). A new mechanism for coactivation of transcription initiation: repositioning of an activator triggered by the binding of a second activator. *Cell* 66:1185–1195.

Schwartz, M. (1987). The maltose regulon. In Escherichia coli *and* Salmonella typhimurium: *Cellular and Molecular Biology.* Vol. 2, F. C. Neidhardt, ed. (Washington, DC: American Society for Microbiology), pp. 1482–1502.

Regulation of Transcription in Eukaryotes

Alberts, B., and Sternglanz, R. (1990). Chromatin contract to silence. *Nature* 344:193–194.

Drapkin, R., Merino, A., and Reinberg, D. (1993). Regulation of RNA polymerase II transcription. *Curr. Opin. Cell Biol.* 5:469–476.

Dynan, W. S., and Gilman, M. Z. (1993). Transcriptional control and signal transduction: of soap-operas and reductionism. *Trends Genet.* 9:154–156.

Lyon, M. F. (1992). Some milestones in the history of X-chromosome inactivation. *Annu. Rev. Genet.* 26:1728.

Orkin, S. H. (1990). Globin gene regulation and switching: circa 1990. *Cell* 63:665–672.

Workman, J. L., and Buchman, A. R. (1993). Multiple functions of nucleosomes and regulatory factors in transcription. *Trends Biochem. Sci.* 18:90–95.

28

RNA Processing

Most RNA molecules are not released from the transcription complex in a fully functional form but must be extensively altered before they can adopt their mature structures and functions. These alterations are of three general types: 1) the removal of pieces of RNA from primary RNA transcripts; 2) the addition of RNA sequences not encoded by the corresponding genes; and 3) the covalent modification of certain nucleotides. The reactions that transform a primary RNA transcript into a mature RNA molecule are referred to collectively as **RNA processing.** RNA processing is an integral part of gene expression. RNA molecules can be processed while transcription is still under way (cotranscriptional processing) or after transcription termination (posttranscriptional processing).

Extensive RNA processing is generally associated with increased RNA stability, although stability also depends upon the degree of RNA secondary structure. At one extreme are the stable transfer RNA (tRNA) and ribosomal RNA (rRNA) molecules, whose half-lives are hours or days. These molecules have regions of secondary structure and a highly ordered tertiary structure that make them resistant to the ubiquitous nucleases found in cells. In both prokaryotes and eukaryotes, these RNA species are processed from primary transcripts. At the other extreme are the prokaryotic messenger RNA (mRNA) molecules, whose half-lives are generally on the order of minutes. These molecules seem to lack particular secondary structure and are rarely processed unless they are part of an operon containing tRNA or rRNA genes. In between these extremes are the eukaryotic mRNA molecules, whose half-lives are generally on the order of hours. Although eukaryotic mRNA molecules lack defined secondary structure, they undergo extensive processing in many organisms. Thus, to a first approximation, processing of RNA molecules is associated with increased stability.

 Related material appears in Exercise 8 of the computer-disk supplement *Exploring Molecular Structure.*

RNA processing has several functions in the cell. Some processing events increase the stability of an RNA molecule by altering its structure, thereby protecting it from degradation. Other processing events are required for subsequent recognition of the RNA molecule by cellular components. In eukaryotes, processing events, which occur in the nucleus, may be connected to transport of RNA to the cytosol, including passage through the nuclear pores. No matter what the function, RNA processing is an additional step following transcription and provides another opportunity for regulation of gene expression.

In this chapter, we will discuss specific processing reactions required for the maturation of tRNA, rRNA, and eukaryotic mRNA. An additional processing reaction, RNA editing, will be covered in the next chapter. The structure and organization of transcription units dictate whether processing events in vivo involve the removal of pieces of RNA from the primary transcript. Gene organization in different species, including the origin and evolution of introns, will be discussed in Chapter 32. In that chapter, we will also cover an unusual variant of RNA processing known as trans splicing.

28·1 Primary Transcripts of tRNA Genes Are Extensively Processed

Mature tRNA molecules are generated in both eukaryotes and prokaryotes by the processing of primary transcripts. In prokaryotes, such as *Escherichia coli,* most tRNA primary transcripts contain more than one tRNA precursor. About half of these primary transcripts also contain precursors of rRNA or mRNA. The monomeric tRNA precursors are cleaved from the large primary transcripts and trimmed to their mature lengths by a variety of ribonucleases (RNases). At least six ribonucleases are involved in the maturation of tRNA in *E. coli* (Table 28·1). The primary transcripts of eukaryotic tRNA molecules must also be processed extensively to yield mature tRNA. Eukaryotic cells contain an extraordinary variety of ribonucleases and other enzymes that participate in tRNA processing.

A. RNase P Cleavage Generates the 5′ Ends of Most Prokaryotic tRNA Precursors

The 5′ end of a tRNA precursor is generated by the activity of the endonuclease RNase P. Portions of the primary transcript that will become mature tRNA molecules exhibit tertiary structures that are recognized by RNase P. This endonuclease catalyzes the cleavage of the primary transcript on the 5′ side of each tRNA precursor,

Table 28·1 Ribonucleases involved in processing tRNA in *E. coli*

Ribonuclease	Activity	Function
RNase P	Endonuclease	Generation of 5′ end of tRNA
RNase D	$3′ \rightarrow 5′$ Exonuclease	Generation of 3′ end of tRNA
RNase BN	$3′ \rightarrow 5′$ Exonuclease	Generation of 3′ end of tRNA
RNase T	$3′ \rightarrow 5′$ Exonuclease	Generation of 3′ end of mature tRNA
RNase F	Endonuclease	Separation of tRNA precursors
RNase III	Endonuclease	Generation of monomeric precursors

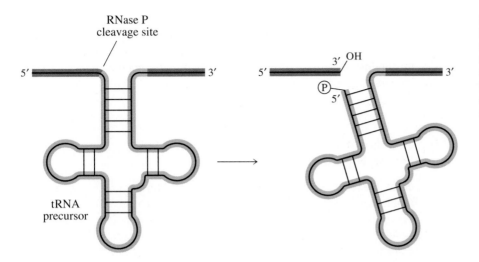

Figure 28·1
Release of tRNA precursors from a prokaryotic primary transcript. The mature 5′ terminus of each tRNA molecule is created by an endonucleolytic cleavage catalyzed by RNase P. The tRNA precursor is drawn schematically as a cloverleaf to indicate the secondary structure.

releasing tRNA precursors with mature 5′ ends (Figure 28·1). Digestion with RNase P in vivo is rapid and probably occurs while the transcript is still being synthesized, since tRNA molecules with immature 5′ ends generally cannot be isolated from living cells.

RNase P was one of the first ribonucleases studied in detail, and much is known about its structure. The enzyme is actually a ribonucleoprotein, composed in *E. coli* of a protein (M_r 18 000) and a 377-nucleotide RNA molecule (M_r 130 000) known as M1 RNA. The RNA component is catalytically active in the absence of protein; it was one of the first known examples of an RNA enzyme. RNase P has an absolute requirement for either Mg^{2+} or Mn^{2+} for catalytic activity. The cation is postulated to be hexagonally coordinated at the active site of M1 RNA through hydrogen bonds to the nucleotides. It may participate in cleavage by generating a nucleophile that attacks the phosphodiester linkage. The reaction catalyzed by RNase P has a substrate K_m of about 5×10^{-7} M; k_{cat} for the reaction is about $120\,s^{-1}$. These values are similar to those for protein enzymes.

RNase P is found in all organisms, and the sequence of the RNA component is conserved to some extent. The RNA contains extensive regions of secondary structure (Figure 28·2), and it presumably folds into a complex three-dimensional structure in vivo. The structure of the RNA seems to be more highly conserved than the sequence of individual residues, as is true of many protein enzymes. The protein component of RNase P helps to maintain the three-dimensional structure of the RNA and electrostatically shields the negatively charged catalytic RNA from its identically charged substrate.

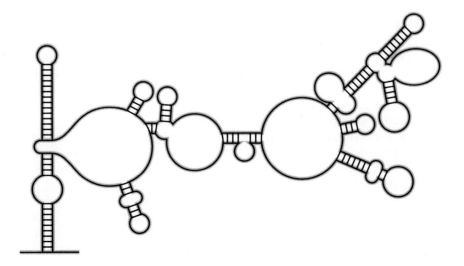

Figure 28·2
Proposed secondary structure of the M1 RNA from *E. coli* RNase P. This structure is one of several proposed by aligning regions of possible secondary structure that have been conserved among molecules from different species.

E. coli cells in which the genes for RNase P have been deleted still process tRNA accurately, which indicates that RNase P is not the only endonuclease that can catalyze the release of tRNA precursors. Several other endonucleases can compensate for the loss of activity of RNase P. Many of these endonucleases have not been well characterized.

B. The 3′ Ends of tRNA Precursors Are Generated by Exonucleolytic Digestion

In some cases, the primary tRNA transcript is short, and further processing involves only exonucleolytic digestion from the 3′ end. In other cases, a free 3′ end of a tRNA precursor can be generated by RNase P cleavage of an adjacent tRNA or by other endonucleases, such as RNase F, that release the monomeric tRNA precursor from a larger transcript. Processing the 3′ ends of tRNA precursors in *E. coli* involves several different exonucleases.

RNase D catalyzes the removal of individual nucleotide residues from the 3′ end of a monomeric tRNA precursor until it reaches the 3′ end of the tRNA sequence. The signal that halts digestion by RNase D has not been characterized. Like RNase P activity, RNase D activity can be lost without inhibiting the growth of a cell, which indicates that the activity can be replaced by the activity of other RNases. RNase BN and RNase T, for example, seem to carry out reactions that also generate mature 3′ ends.

All mature tRNA molecules, both prokaryotic and eukaryotic, possess the sequence CCA at their 3′ end. In some prokaryotic tRNA genes, this sequence is encoded in DNA, but the primary transcripts of other prokaryotic tRNA genes lack the 3′ CCA. This sequence is required for tRNA function. If it is not encoded by the gene, the CCA sequence must be added posttranscriptionally after all other types of processing at the 3′ end have been completed. The sequence CCA is added by the action of tRNA nucleotidyltransferase. This enzyme does not require a template. In *E. coli,* tRNA nucleotidyltransferase also catalyzes replacement of terminal adenylate residues that have been lost due to the activity of RNase T and other exonucleases. CCA is added posttranscriptionally to all eukaryotic tRNA genes in a similar reaction. A summary of the steps in the nucleolytic processing of the 3′ ends of prokaryotic tRNA precursors is shown in Figure 28·3.

C. Some Eukaryotic tRNA Precursors Contain Introns That Are Removed Through RNA Splicing

The tRNA precursors in some eukaryotes and in some archaebacteria require posttranscriptional processing to remove additional nucleotides not only from their 5′ and 3′ ends but also from within the transcript. Internal sequences that are present in the DNA and in the primary RNA transcript but do not appear in the mature RNA molecules are called **introns,** or intervening sequences. Since in Chapter 27 we defined a gene as a DNA sequence that is transcribed, introns are considered to be part of a gene even though they are removed during processing. Sequences that are present in both the DNA and the mature RNA are called **exons.** Introns are removed and exons are ligated to form a continuous RNA molecule in a process known as **RNA splicing.** RNA splicing is involved in the maturation of mRNA, tRNA, and rRNA, although the mechanisms of splicing differ in each case.

In a primary transcript, the intron separates two exons, one of which is located upstream, or 5′, of the intron, and one of which is located downstream, or 3′, of the intron. The region at the boundary between the 5′ exon and the intron is referred to as the **5′ splice site,** or the 5′ splice junction; the region at the boundary between the

(a) After the endonuclease RNase P generates mature 5′ ends by cleaving the tRNA primary transcripts, other endonucleases release the monomers by cleaving them at their 3′ ends.

Figure 28·3
Summary of the steps in the nucleolytic processing of the 3′ ends of prokaryotic tRNA precursors.

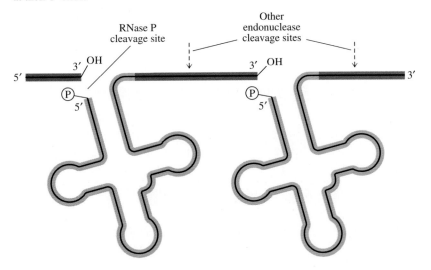

(b) The partially processed tRNA molecules are trimmed at their 3′ ends by the activity of the exonuclease RNase D.

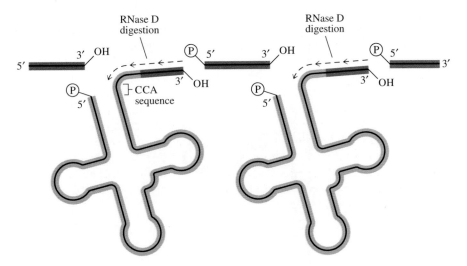

(c) If the 3′ terminal CCA sequence is present, digestion by the exonuclease RNase D produces a mature tRNA molecule. If the 3′ terminal CCA sequence is missing, it is added by the activity of tRNA nucleotidyltransferase.

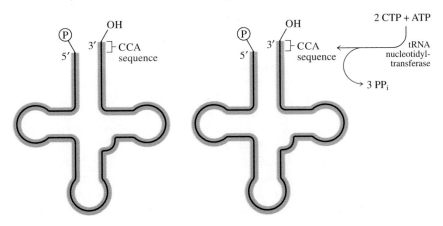

Figure 28·4
Nomenclature of splicing. Introns (dark brown) and exons (light brown) are shown. Splice sites are defined by their positions (5′ or 3′) relative to the intron. The blue arrow (labelled P for promoter) indicates the transcription start site. The site of termination (t) is also shown.

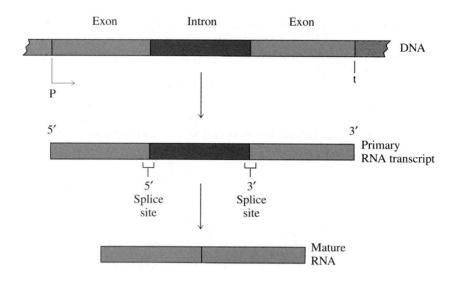

Figure 28·5
Secondary structures of two yeast tRNA precursor molecules. **(a)** Pre-tRNA^{Ile}. **(b)** Pre-tRNA^{Tyr}. Introns are shown in dark brown.

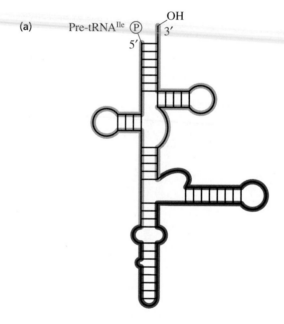

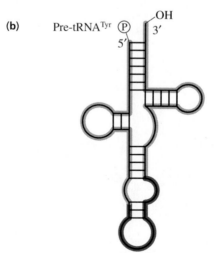

intron and the 3′ exon is referred to as the **3′ splice site,** or the 3′ splice junction (Figure 28·4). Each splice site contains at least one nucleotide on each side of the phosphodiester linkage broken during splicing, although some splice sites may contain many more nucleotides.

The yeast genome contains about 400 tRNA genes. About 10% of these genes contain a single intron each. The introns of tRNA precursors range from 14 to 46 base pairs in size and occur within at least partially self-complementary sequences that eventually form a secondary structure called the anticodon stem and loop (discussed further in Chapter 29). This stem-loop configuration (Figure 28·5) can be recognized by the enzyme or enzymes necessary for splicing.

Splicing of tRNA precursors in yeast involves two major steps. First, the precursor is cleaved at the junctions between the intron and the flanking exons, that is, at the 5′ and 3′ splice sites. Although the two exons are no longer covalently attached to one another after cleavage, they are held together by the stable tertiary structure of the tRNA precursor. In the second step, the two exons are covalently joined by the activity of an RNA ligase. The energy released by hydrolysis of the phosphodiester linkages in the first step is lost as heat, and the ligation reaction requires two molecules of ATP. A summary of the reactions involved in the splicing of a tRNA precursor in yeast is given in Figure 28·6.

Figure 28·6
Removal of an intron from the yeast tRNA precursor pre-tRNA^Tyr.

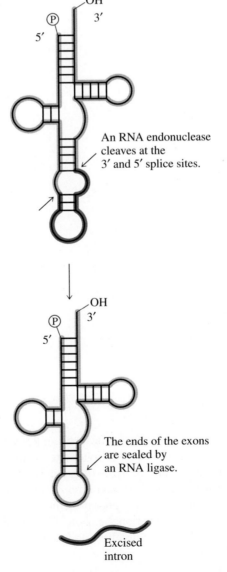

An RNA endonuclease cleaves at the 3′ and 5′ splice sites.

The ends of the exons are sealed by an RNA ligase.

Excised intron

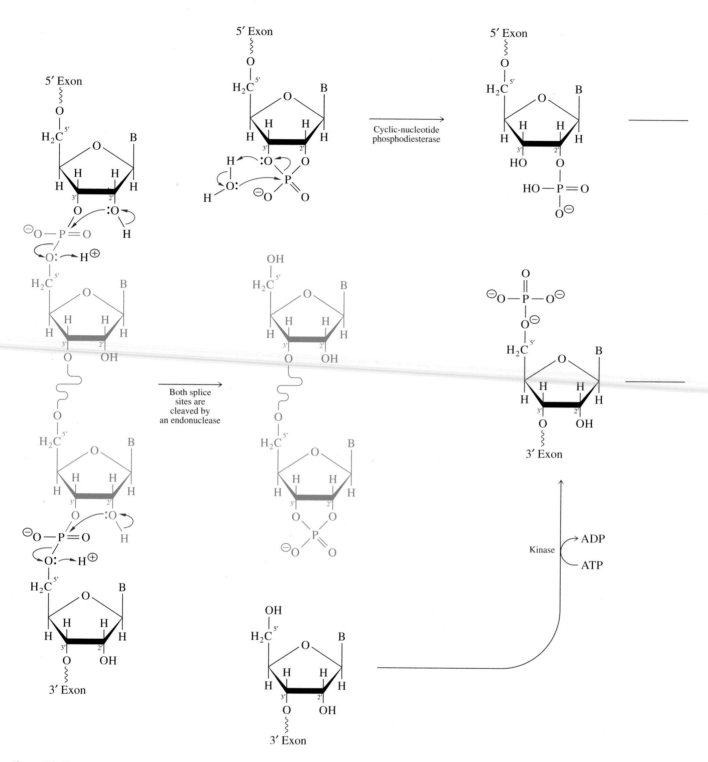

Figure 28·7
Mechanism of splicing of a tRNA precursor molecule. The intron is shown in blue, exons in black.

The cleavage step is catalyzed by an RNA endonuclease that recognizes tRNA precursors. Cleavage at the 5′ splice site leaves a 2′,3′-cyclic phosphate at the 3′ end of the first exon and a 5′-hydroxyl group at the 5′ end of the intron. A similar cleavage reaction leaves a hydroxyl group at the 5′ end of the second exon and a 2′,3′-cyclic phosphate at the 3′ end of the intron (Figure 28·7). The next steps are catalyzed by a multifunctional enzyme (M_r 90 000). It contains a cyclic-nucleotide phosphodiesterase activity that catalyzes the hydrolysis of the 2′,3′-cyclic phosphate

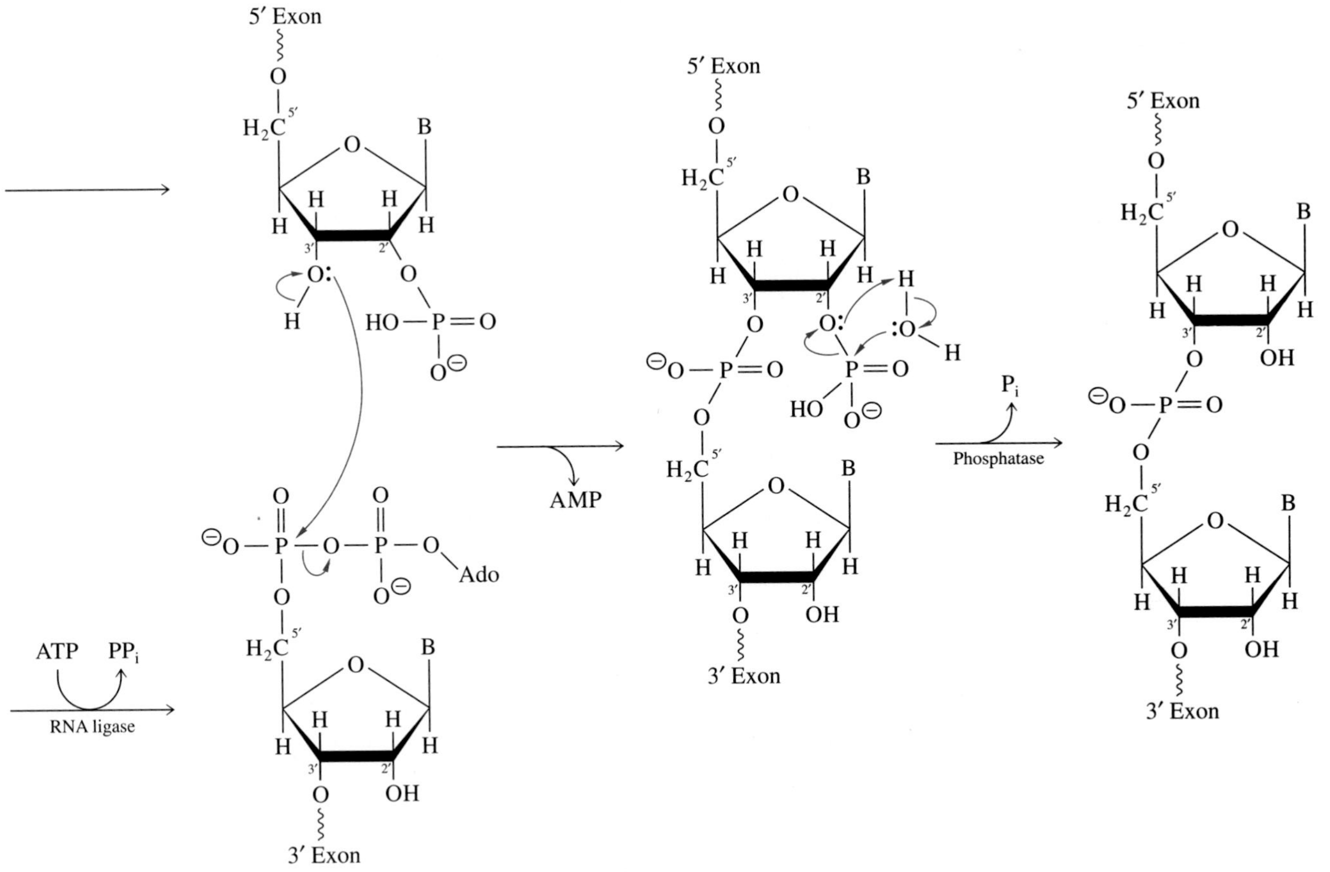

5′ Exon

B

H₂C

ATP   PPᵢ

RNA ligase

AMP

5′ Exon

B

B

3′ Exon

Pᵢ

Phosphatase

5′ Exon

B

B

3′ Exon

3′ Exon

of the first exon to produce a 2′-phosphate ester. The multifunctional enzyme also possesses kinase activity that catalyzes the transfer of a phosphoryl group from ATP to the 5′-hydroxyl group of the second exon.

The ligation step is catalyzed by an RNA ligase activity that is also part of the multifunctional splicing enzyme. This ATP-dependent enzyme catalyzes the transfer of AMP to the 5′-phosphate of the second exon. The transferred adenylyl group is next displaced by the free 3′-hydroxyl group of the first exon, thereby forming a 3′–5′ phosphodiester linkage between the two exons. (Note that the phosphate in this linkage is derived from ATP, not from the tRNA precursor.) A phosphatase then catalyzes the hydrolysis of the 2′-phosphate ester to complete the splicing reaction.

Not all eukaryotes exhibit precisely this mechanism for splicing exons of tRNA precursors. In human cells in culture, the initial cleavages at the intron boundaries leave 3′-phosphate groups and 5′-hydroxyl groups, and formation of the 2′,3′-cyclic phosphate requires an ATP-dependent cyclase activity. This reaction does not form a 2′-phosphate, and the subsequent ligation reaction leaves a 3′–5′ phosphodiester linkage in which the phosphate group is derived from the original tRNA precursor. In spite of these interspecies variations, normal intron removal always requires two separate steps: cleavage and ligation. The mechanism of splicing of tRNA precursors is very different from the other splicing mechanisms we will encounter.

The initial discovery of introns was quite a surprise. In 1967, the sequence of yeast tRNA^Tyr was determined directly from the RNA. In 1977, Benjamin Hall, Maynard Olson, and Howard Goodman isolated and sequenced four of the eight genes that encode this tRNA, using the then very new technology of DNA sequencing. When they aligned the newly derived DNA sequences and compared

Figure 28·8
Discovery of introns in tRNA genes. The first introns sequenced were from four yeast tRNA^{Tyr} genes. The sequences for three of these genes are shown here. Transcription of each of these genes yields an identical mature tRNA. A 14 bp sequence not present in the mature tRNA was found within each gene. (See Figure 28·9 for the structures of the modified nucleotides Ψ and i^{6}A.)

	30	40	50	60

tRNA^{Tyr} A A G A C U G Ψ A   A A Ψ C U U G A (with i^6 above)

pYT-A   A A G A C T G T A A T T T A T C A C T A C G A A A T C T T G A
       T T C T G A C A T T A A A T A G T G A T G C T T T A G A A C T

pYT-G   A A G A C T G T A A T T T A C C A C T A C G A A A T C T T G A
       T T C T G A C A T T A A A T G G T G A T G C T T T A G A A C T

pYT-C   A A G A C T G T A A T T T A C C A C T A C G A A A T C T T G A
       T T C T G A C A T T A A A T G G T G A T G C T T T A G A A C T

them to the sequence of the tRNA, they noted that the two were not colinear. The DNA of all four genes contained a sequence of 14 base pairs that was missing from the mature tRNA (Figure 28·8). This 14 bp sequence was the first intron whose existence was confirmed by sequencing, although by the mid-1970s there was good evidence for introns in rRNA genes and in adenovirus mRNA precursors.

D. Many Nucleotides in tRNA Molecules Are Covalently Modified

In both prokaryotes and eukaryotes, mature tRNA molecules exhibit a greater diversity of covalent modifications than any other class of RNA molecule. Creation of the 26 to 30 covalently modified nucleotides found in bacterial tRNA is catalyzed by about 45 enzymes. In *E. coli,* the genes for these enzymes may represent as much as 1% of the genome. Few of these enzymes from *E. coli* have been characterized. Like the other enzymes involved in the processing of tRNA precursors, enzymes involved in the covalent modification of tRNA precursors seem to recognize elements of both primary and tertiary structure near the target nucleotide residue.

Greater variation in covalent modification of tRNA occurs in prokaryotes than in eukaryotes, and only a few modifications are common to all organisms. Some covalent modifications of nucleotides are shown in Figure 28·9. Usually, each type of covalent modification occurs in only one location on each tRNA molecule, but a few can be found at more than one location. In both prokaryotes and eukaryotes, the targets of covalent modification are generally those regions of the tRNA molecule that do not form intramolecular base pairs.

28·2 Ribosomal RNA Molecules Are Produced by Processing Large Primary Transcripts

Ribosomal RNA molecules in all organisms are produced as large primary transcripts that require processing. The processing reactions include nucleotide modifications (mostly methylations) and cleavage by both endonucleases and exonucleases. In all living organisms, rRNA processing is coupled to the assembly of ribosomes, as we will see in Chapter 30.

The nucleotide sequences of rRNA molecules in different species are highly conserved. This conservation can be used to construct phylogenetic trees that connect living species with common ancestors that existed more than a billion years ago. The evolutionary conservation of rRNA molecules is also seen, to some extent, in the organization of rRNA genes in different species. We will expand on this point in Chapter 32. However, at the level of processing and modification of rRNA precursors, there are some important differences between prokaryotes and eukaryotes.

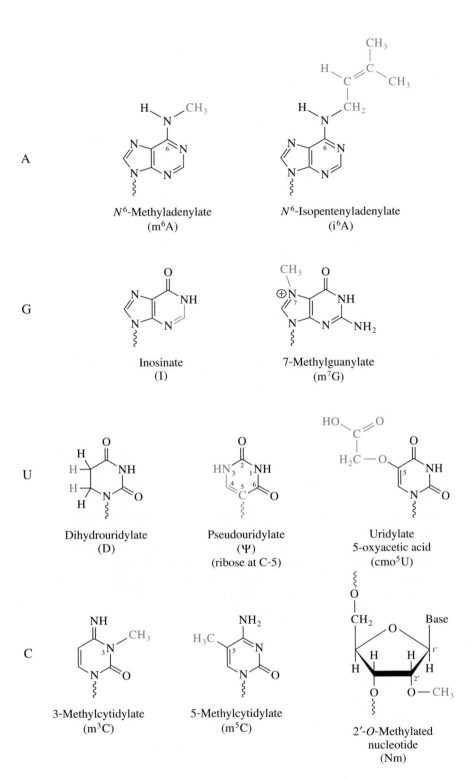

Figure 28·9
Examples of covalent modifications of nucleotides frequently found in tRNA molecules. Modifications are shown in blue.

A

N^6-Methyladenylate
(m^6A)

N^6-Isopentenyladenylate
(i^6A)

G

Inosinate
(I)

7-Methylguanylate
(m^7G)

U

Dihydrouridylate
(D)

Pseudouridylate
(Ψ)
(ribose at C-5)

Uridylate
5-oxyacetic acid
(cmo^5U)

C

3-Methylcytidylate
(m^3C)

5-Methylcytidylate
(m^5C)

2′-O-Methylated
nucleotide
(Nm)

A. Processing of Prokaryotic Ribosomal RNA Precursors

Bacterial ribosomes contain three species of rRNA: 16S, 23S, and 5S. (Note that S is the symbol for the Svedberg unit, a measure of the rate at which particles move in the gravitational field established in an ultracentrifuge. Large S values are associated with large masses. However, the relationship between S and mass is not linear; therefore S values are not additive.) The genes for the bacterial rRNA molecules are part of a large operon that is transcribed from a promoter at the 5′ end.

Figure 28·10
Ribosomal RNA operons in *E. coli* K12.
(a) Generalized map of the *E. coli* chromosome showing the locations of the rRNA operons. **(b)** The organization of genes within a typical operon. There may be one or two tRNA genes between the 16S and 23S genes. Following the 5S gene, some operons have no tRNA genes, whereas others have one or two. The transcribed regions at the ends of the operon and between genes are shown in gray.

(a)

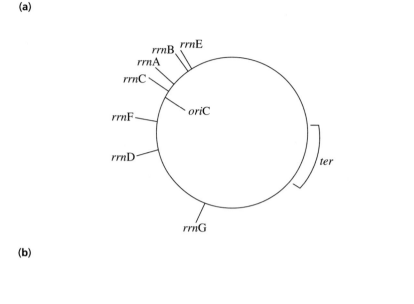

(b)

The order of the genes is 16S, 23S, 5S. There are usually multiple copies of these rRNA operons in the bacterial genome. Most *E. coli* strains, for example, have seven rRNA operons (*rrn*A–G) scattered throughout the chromosome (Figure 28·10). In all cases, the operons include tRNA genes, although the number, type, and position vary.

The primary transcript of the rRNA operon is a large 30S molecule of more than 6000 nucleotides. Within this primary transcript, the 5′ and 3′ ends of both 16S and 23S rRNA molecules form a base-paired region. In prokaryotes, an endonuclease binds to this region and cleaves both strands of the precursor. In *E. coli*, this endonuclease is RNase III (Figure 28·11).

Figure 28·11
Endonucleolytic cleavage of rRNA precursors in *E. coli*. The rRNA primary transcript contains a copy of each of the three rRNA species and may also contain several tRNA precursors. The 16S and 23S rRNA molecules are released from the large primary transcript by the action of RNase III and trimmed by the actions of the endonucleases M16 and M23, respectively. Cleavage by M5 releases 5S rRNA. (Slash marks indicate portions of the rRNA primary transcript that have been deleted for clarity.)

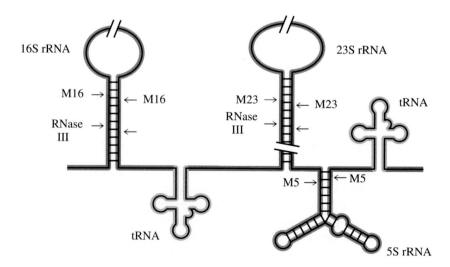

Figure 28·12
Arrangement of a typical rRNA gene repeat in vertebrates.

Further trimming of the 5′ and 3′ ends of the 16S and 23S precursors is carried out by RNases M16 and M23, respectively. These endonucleases remove about 10–20 nucleotides from the intermediate produced by RNase III cleavage. The smaller 5S rRNA is released from the precursor by endonuclease M5, which generates mature 3′ and 5′ ends.

B. Processing of Eukaryotic Ribosomal RNA Precursors

Eukaryotic ribosomes contain 18S, 28S, and 5S rRNA molecules that are homologous to the bacterial rRNA species. They also contain a separate small RNA, 5.8S, that is not found in prokaryotic ribosomes. The genes encoding 5S rRNA are usually found in clusters of numerous side-by-side, or tandem, repeats, and each gene is transcribed separately by RNA polymerase III, as we saw in Chapter 27. The separation of 5S from the rest of the rRNA genes is one of the significant differences between prokaryotic and eukaryotic rRNA genes. The remaining rRNA genes are arranged in the order 18S, 5.8S, 28S, and they are transcribed from a single promoter by RNA polymerase I. There are hundreds of copies of this arrangement in the eukaryotic genome. In some species, there are several clusters of rRNA genes at different loci, whereas in other species, all of the genes are found at a single locus. Each locus is a site where a nucleolus forms during transcription of the rRNA genes. Posttranscriptional processing of rRNA and assembly of ribosomes occurs in the nucleolus.

An example of a typical rRNA gene repeat in vertebrates is shown in Figure 28·12. The primary transcripts are usually more than 6500 nucleotides in length and contain a copy of each of three rRNA species: 18S, 5.8S, and 28S. The multimeric primary transcript is cleaved into smaller subunits by endonucleases in much the same manner as the bacterial transcripts. However, unlike bacterial processing, cleavage in eukaryotes does not begin until the primary transcript is completely synthesized and approximately 100 ribose groups within conserved sequences have been methylated. This is why such large transcripts are seen in electron micrographs of active eukaryotic rRNA genes (see Figure 27·27). Proper cleavage requires that each rRNA precursor bind to at least some of the proteins that will eventually be assembled with the mature rRNA in the ribosome.

28·3 Some RNA Precursors Contain Self-Splicing Introns

Introns are found in the nuclear rRNA precursors of some unicellular eukaryotes and in RNA precursors synthesized in mitochondria and chloroplasts. Several bacteriophage genes also contain introns, as do some genes in species of cyanobacteria and archaebacteria. These introns are excised by a mechanism distinct from the one described earlier for tRNA precursors (Section 28·1C). In fact, many of these introns are excised in a splicing reaction mediated by the RNA precursor itself. This self-splicing reaction occurs by two different mechanisms, as exemplified by the splicing of the nuclear rRNA precursor in the ciliated protozoan *Tetrahymena thermophila* and the cytochrome *c* oxidase mRNA precursor in yeast mitochondria.

A. Group I Introns Require a Guanosine Cofactor

The primary transcript for the large rRNA in *Tetrahymena* includes a 413-nucleotide intron. This intron is an example of **group I introns,** classified on the basis of conserved sequences and secondary structures. Other group I introns include the introns of chloroplast tRNA precursors in flowering plants, the introns of nuclear rRNA precursors in the fungus *Physarum polycephalum,* the introns encoded by several cytochrome *c* oxidase genes found in the mitochondria of fungi, bacteriophage introns, and the introns in the tRNA genes of cyanobacteria. Not all group I introns are self-splicing; perhaps those that are not share a common ancestor with self-splicing introns but have evolved a requirement for accessory proteins that help the RNA fold into the correct conformation in vivo.

Thomas Cech and his colleagues, while studying rRNA processing in *Tetrahymena,* discovered that intron excision in that organism occurs independently of protein and requires only Mg^{2+} and a guanosine cofactor. The splicing mechanism involves two separate transesterification reactions: one between the guanosine cofactor and the 5′ splice site and another between the 3′ end of the first exon and a guanylate residue at the 3′ splice site (Figure 28·13). The 5′ splice site is recognized and stabilized by formation of a double-stranded region between the end of the first exon and an internal guide sequence (IGS) in the intron (Figure 28·14).

Figure 28·13
Self-splicing of the *Tetrahymena* rRNA precursor. The intron is brown; the exons are purple. Dashed lines represent the remainder of the intron not directly involved in the splicing reaction.

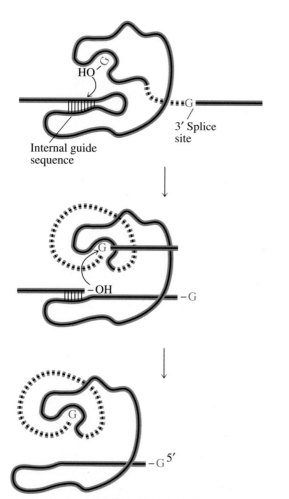

Internal guide sequence

3′ Splice site

A guanosine cofactor binds to the guanosine-binding site of the intron. The 3′-hydroxyl group of the guanosine cofactor attacks the phosphoryl group at the 5′ splice site, thereby cleaving the exon from the intron. The 5′ splice site remains base-paired to the internal guide sequence of the intron.

A guanylate residue at the 3′ splice site then binds to the guanosine-binding site. The 3′-hydroxyl group at the end of the first exon attacks the phosphoryl group at the 3′ splice site.

As a result, the exons are joined and the intron is released.

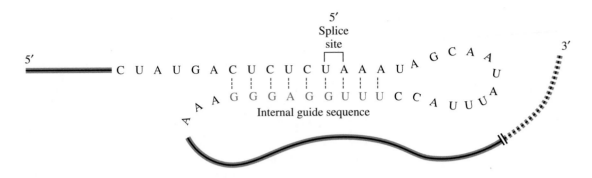

Figure 28·14
Model for alignment of the first exon of the *Tetrahymena* rRNA precursor and the internal guide sequence.

Most group I introns contain extensive regions of secondary structure (not shown in Figure 28·13) that are essential for efficient processing. The introns probably fold into a well-defined three-dimensional conformation as well. The guanosine cofactor binds to a guanosine-binding site formed by the folded intron. Nucleophilic attack by the guanosine cofactor results in cleavage at the 5′ splice site and covalent attachment of the guanosine cofactor to the 5′ end of the intron. Following this initial cleavage, the first exon remains bound to the IGS.

The second cleavage reaction requires that the ends of the two exons be brought together. This is accomplished when a guanylate residue at the 3′ splice site binds to the guanosine-binding site in the intron. A second transesterification reaction between the splice sites results in the joining of the exons. The intron is released but retains its secondary and tertiary structure.

Catalytic RNA molecules, such as the *Tetrahymena* rRNA intron, are often referred to as ribozymes to distinguish them from protein enzymes. This particular ribozyme efficiently carries out the self-splicing reaction.

B. Some Excised Introns Have Catalytic Activity

The self-splicing reaction of the intron in *Tetrahymena* is analogous to protein-mediated enzymatic reactions in a number of ways: the excision reaction proceeds orders of magnitude faster than basal transesterification, and the ribozyme is highly specific for its substrate. However, the self-splicing reaction is not truly enzymatic since the original catalyst is not renewed during the reaction. After its release from the rRNA precursor, the ribozyme still possesses autocatalytic activity, and it undergoes shortening at both ends to produce a form known as L-21. This form retains the guanosine-binding site and most of the IGS.

L-21 can act as a specific endonuclease, cleaving other RNA molecules at the sequence CCCUCU (Figure 28·15, next page). This reaction, analogous to the first step in the splicing reaction, requires a guanosine cofactor, and the substrate RNA binds to the IGS. The L-21 reaction is truly catalytic since the ribozyme remains intact. The rate constant for the cleavage step is $2.1 \times 10^3 \, \text{s}^{-1}$, which is 10^{11} times faster than uncatalyzed hydrolysis at the cleavage site. It should be noted, however, that the turnover number for this ribozyme is very low since the product remains tightly bound to the IGS. It is unlikely that this activity is significant in vivo. In vitro mutagenesis has been used to engineer derivatives of L-21 that are much more efficient or that can cleave other substrates.

Recall that the RNA component of RNase P is a true enzyme with biological significance. Telomerase (Section 25·11B) is also an enzyme with an associated RNA that provides a guide sequence, or template, but in this case, the RNA does not have catalytic activity.

Figure 28·15
Endonuclease activity of L-21. L-21 is dark brown; the substrate RNA is light brown.

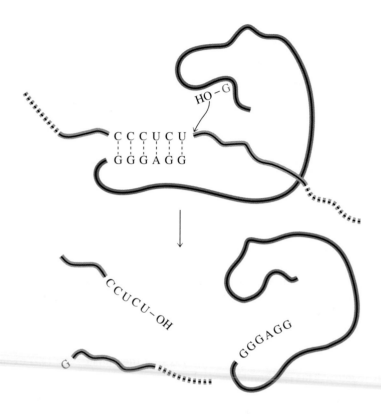

C. Group II Introns Include an Internal Branch Point

Group II introns occur in mitochondrial and chloroplast genes. The best-studied example is that found in the mRNA precursor for mitochondrial cytochrome c oxidase in yeast. (This gene also contains two group I self-splicing introns.) An intron is classified as group II on the basis of conserved sequences and secondary structure that juxtaposes the 3' and 5' splice sites and includes an internal branch point (Figure 28·16). Presumably, the intron has a well-defined tertiary structure as well. Splicing in vitro of group II introns requires magnesium but no guanosine cofactor. Like group I introns, not all group II introns are capable of self-splicing. Unlike group I introns, however, the self-splicing reaction of group II introns is sometimes reversible, leading to insertion of an intron into a mature mRNA.

Figure 28·16
Secondary structure of a group II intron.

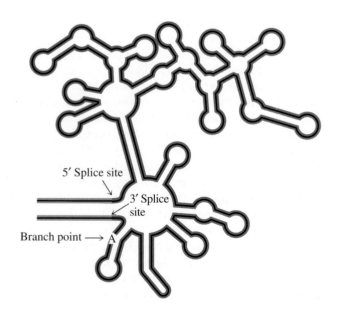

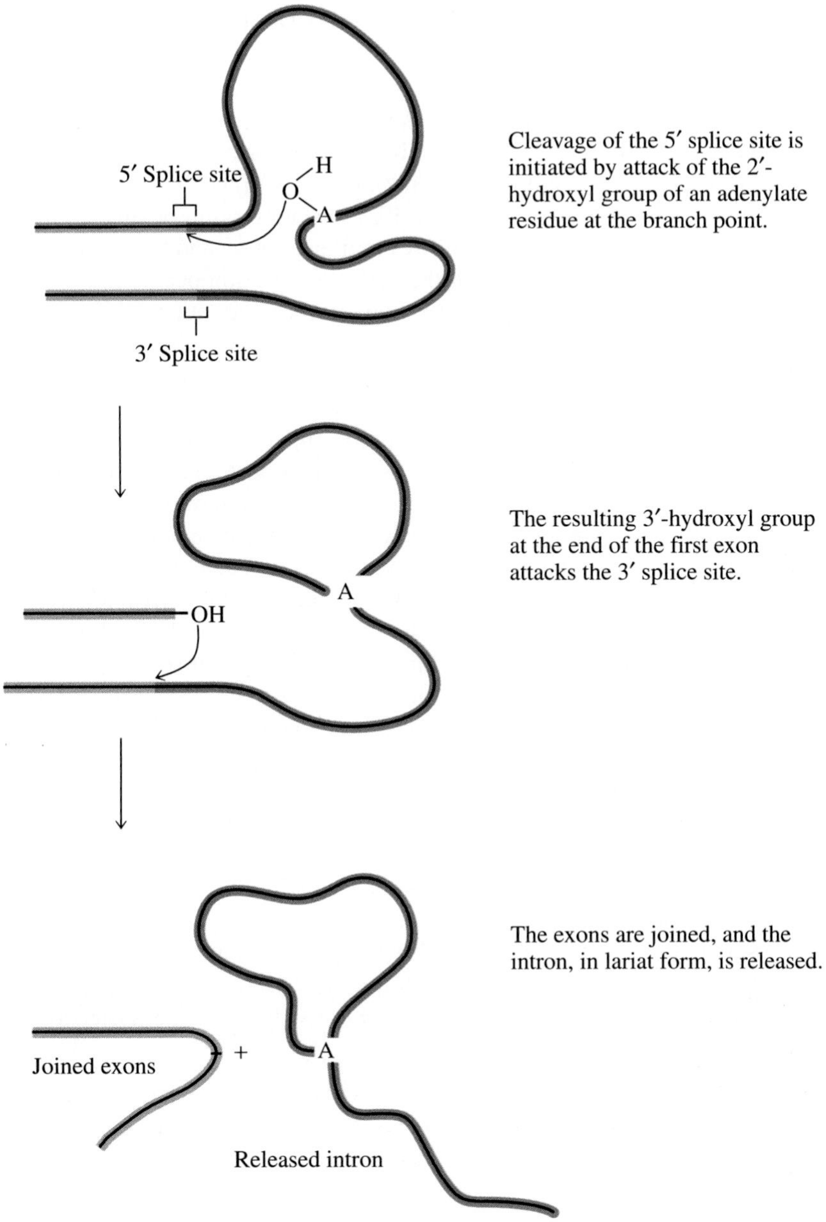

Figure 28·17
Self-splicing of group II introns.

5′ Splice site

3′ Splice site

Cleavage of the 5′ splice site is initiated by attack of the 2′-hydroxyl group of an adenylate residue at the branch point.

The resulting 3′-hydroxyl group at the end of the first exon attacks the 3′ splice site.

The exons are joined, and the intron, in lariat form, is released.

Joined exons

Released intron

Excision of group II introns, like excision of group I introns, begins with a transesterification reaction (Figure 28·17). However, since in this case no free nucleosides are used as cofactors, no 3′-hydroxyl groups are available to attack the 5′ splice site. Instead, the nucleophile is the 2′-hydroxyl group of an adenylate residue (the branch point) located inside a conserved hairpin structure within the intron. The 2′-hydroxyl group attacks the phosphorus at the 5′ splice site, releasing the first exon and a branched intermediate with a 2′–5′ phosphodiester linkage. The resulting free 3′-hydroxyl group on the first exon then attacks the phosphate at the 3′ splice site. The products of the reaction are the joined exons and a free intron. Group II introns are not excised as linear molecules but as branched, or lariat, structures. As in the self-splicing reactions of group I introns, the number of phosphodiester linkages is conserved, and no external energy source is required for these reactions.

The nucleotide sequences at the branch point and the 5′ and 3′ splice sites are conserved in all group II introns. As we will see, the presence of a branch point and the formation of a lariat structure are features that group II introns share with the more common introns in eukaryotic mRNA precursors.

(a)

(b)

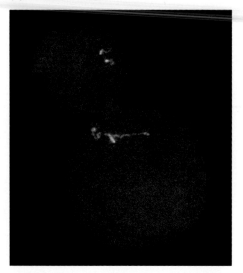

Figure 28·18
RNA tracks in the nucleus. **(a)** A fluorescent probe is used to detect the gene, which is present at a unique site in each homologous chromosome. **(b)** A probe is also used to detect RNA in the process of being transcribed and exported from the nucleus. Multiple copies of RNA form tracks that lead from the gene to the surface of the nucleus. In the upper nucleus, two tracks can be seen, one from each copy of the gene. (Courtesy of Jeanne B. Lawrence.)

28·4 Both Ends of Eukaryotic mRNA Precursors Are Modified

Extensive processing of eukaryotic mRNA precursors is one of the features that distinguish eukaryotes from prokaryotes. In bacteria, the primary mRNA transcript is usually translated directly, although in some cases tRNA precursors may first be excised. In fact, as we will see in Chapter 30, translation often begins before transcription is complete. In eukaryotes, however, transcription and translation are carried out in separate compartments of the cell: transcription occurs in the nucleus and translation in the cytosol. This separation allows eukaryotic mRNA precursors to be processed in the nucleus without interfering with translation.

All eukaryotic mRNA precursors undergo modifications that increase their stability and make them better substrates for translation. This processing involves the same kinds of steps that are responsible for producing mature tRNA and rRNA molecules, namely, cleavage of a precursor, covalent modification of nucleotides, addition and removal of terminal nucleotides, and splicing. One of the reasons eukaryotic mRNA precursors are modified is to protect them from exonucleolytic degradation during the time it takes for the mature mRNA to reach the cytosol.

The mechanism of transport of mRNA from the site of transcription to the nuclear pores has not been determined, but it is known to take about 30 minutes in mammalian cells. By specifically labelling one mRNA precursor, it is possible to detect tracks of RNA being transported from the gene to the surface of the nucleus (Figure 28·18). The appearance of these tracks suggests that the RNA is following a path defined by unknown elements of the nuclear matrix. (The nuclear matrix consists primarily of intermediate filaments that are composed of lamins.) Some of the processing events that occur in the nucleus may be required for efficient transport and export of mRNA.

In this section, we will concentrate on modifications of the ends of eukaryotic mRNA precursors. (Section 28·5 discusses the splicing of these precursors.) In addition, some internal nucleotides of eukaryotic mRNA molecules are covalently modified. The most common modification is methylation of adenylate to form N^6-methyladenylate. This modified nucleotide is also found in DNA (Section 24·5) and in all ribosomal RNA molecules.

A. The 5′ Ends of Eukaryotic mRNA Are Capped

One way to increase the stability of mRNA is to modify its ends so that it is no longer susceptible to cellular exonucleases that degrade RNA. The 5′ ends are modified before eukaryotic mRNA precursors are completely synthesized. The 5′ end of the primary transcript is a nucleoside triphosphate residue (usually a purine) that was the first nucleotide residue incorporated by RNA polymerase II. Modification of this end begins when the terminal phosphate group is removed by the activity of a phosphohydrolase (Figure 28·19). The resulting 5′-diphosphate group then reacts with the α-phosphorus of a GTP molecule to create a 5′–5′ triphosphate linkage. This reaction is catalyzed by the enzyme guanylyltransferase. The resulting structure is called a **cap.**

The cap structure is then covalently modified in several different ways. In some species of eukaryotes, most mRNA precursors are methylated at N-7 of the newly added guanylate by the addition of a methyl group from *S*-adenosylmethionine. This methylated structure is called cap 0. Further covalent modifications yield alternative cap structures, which are more prevalent in some eukaryotic species than in others. These additional modifications include methylation of the 2′-hydroxyl

Figure 28·19

Formation of a cap structure at the 5′ end of an mRNA precursor in eukaryotes. (1) Formation of the cap begins when a phosphohydrolase catalyzes removal of the terminal phosphate at the 5′ end of the precursor. (2) The 5′ end then receives a GMP group from GTP in a reaction catalyzed by guanylyltransferase. (3) The added GMP moiety is then methylated at N-7. (4) The 2′-hydroxyl group of the terminal ribose and of the penultimate ribose in the mRNA may also be methylated in consecutive steps. (The methyl-group donor, S-adenosylmethionine, is not shown in reactions 3 and 4.)

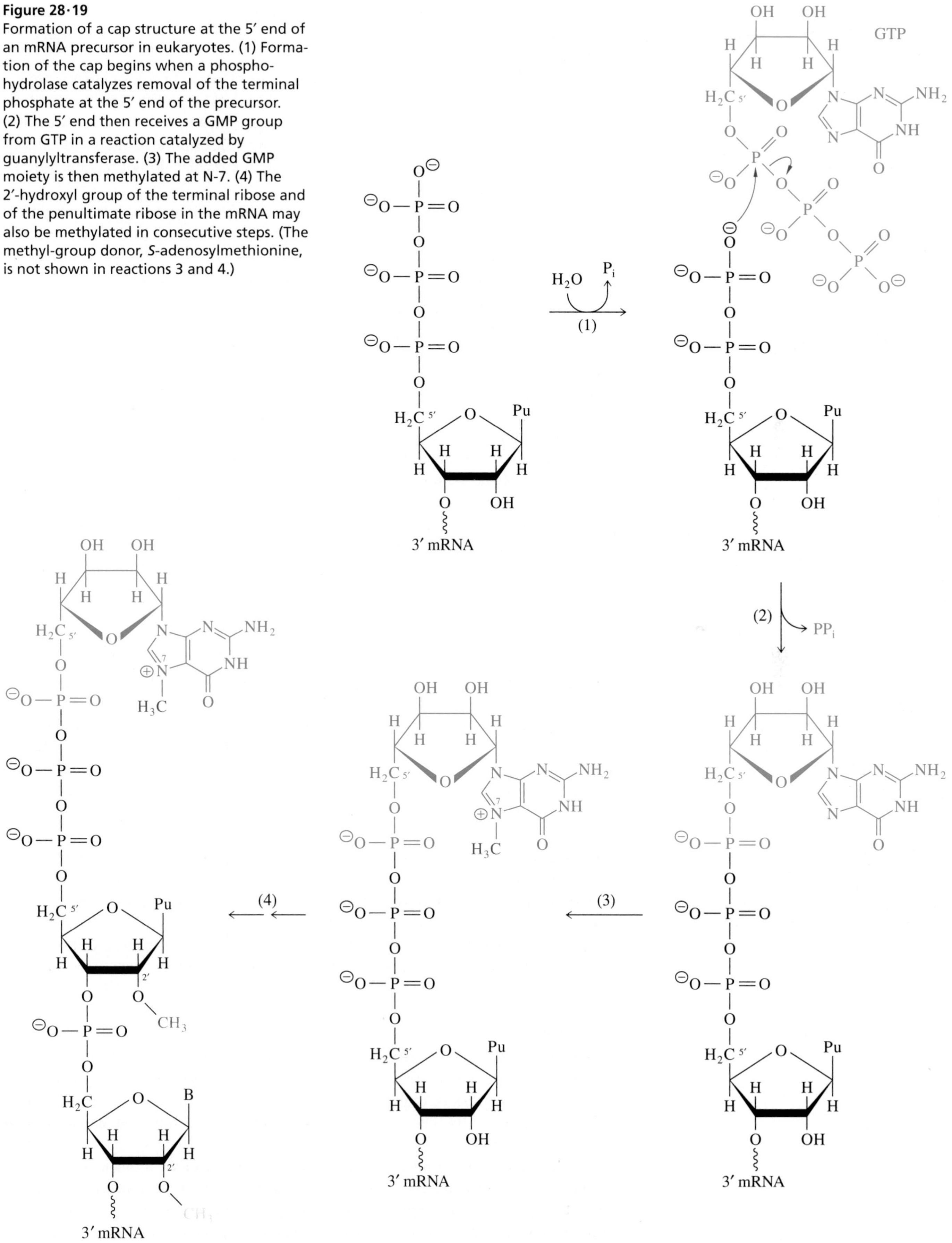

group of the last nucleotide of the original transcript (cap 1) or methylation of the 2′-hydroxyl groups of both of the last two nucleotides of the original transcript (cap 2).

Only mRNA precursors are modified in this manner; rRNA and tRNA are not capped, and the other small nuclear RNA molecules have a different cap structure. The specificity of the capping reaction is probably due to interaction between the capping enzymes and RNA polymerase II during transcription. Class I and class III transcripts are not modified because the capping enzymes do not recognize RNA polymerase I or RNA polymerase III.

Since the 5′–5′ triphosphate linkage leaves only a 3′-hydroxyl group exposed, the cap protects mRNA from 5′ exonucleases by blocking the 5′ end of the molecule. The cap also converts mRNA precursors into substrates for other processing steps in the nucleus, such as splicing. In addition, the cap is retained on mature mRNA, where it serves as the attachment site for ribosomes during protein synthesis (Chapter 30).

B. Eukaryotic mRNA Precursors Are Cleaved and Modified at Their 3′ Ends

Eukaryotic mRNA precursors are also modified at their 3′ ends. Once RNA polymerase II has transcribed past the 3′ end of the coding region of DNA, the newly synthesized RNA is cleaved by an endonuclease near a specific site whose consensus sequence is AAUAAA. This sequence is bound by a cleavage and polyadenylation specificity factor (CPSF), a protein that also interacts with the endonuclease and a polymerase (Figure 28·20). Cleavage generally occurs approximately 10 to 20 nucleotides downstream of this sequence and depends upon the presence of a downstream region that is rich in U or GU. The endonuclease then dissociates, and the newly generated 3′ end of the molecule serves as a primer for the addition of multiple adenylate residues. This addition is catalyzed by poly A polymerase (M_r 50 000–82 000, depending on the species), which adds adenylate residues distributively, using ATP as a substrate. Up to 250 nucleotides may be added to form a stretch of polyadenylate known as a **poly A tail.**

Polyadenylation is not coupled to the termination of transcription in most organisms; transcription may continue for hundreds of nucleotides beyond the terminal 3′ cleavage site. In the yeast *Saccharomyces cerevisiae,* however, the sites of termination of transcription are often close to the sites of polyadenylation, suggesting that in some organisms these two processes may be coupled. Yeast genes also have modified versions of the polyadenylation sequence. It makes sense that transcription termination and polyadenylation might be coupled in yeast since the yeast genome is relatively small and the genes are close together. If transcription were to continue much past the polyadenylation site, as it does in mammals, the adjacent gene would be transcribed.

With a few rare exceptions, all mature mRNA molecules in eukaryotes contain poly A tails. The poly A tail can vary in length from fewer than 50 to more than 250 residues, depending upon the species and perhaps upon other factors, such as the type of mRNA and the developmental stage of the cell. The length also depends upon the age of the mRNA, since the poly A tail is progressively shortened over time. In fact, the tail has already been shortened by 50 to 100 nucleotides by the time the mature mRNA reaches the nuclear pores. This shortening presumably arises from progressive digestion by exonucleases specific for 3′ ends. The presence of the poly A tail increases the time required for such exonucleases to reach the coding region.

The poly A tails of mRNA precursors and mature mRNA molecules are bound tightly to multiple copies of poly A binding protein (PAB I), which has a molecular weight of 70 000. The RNA-protein complex is presumed to stabilize the mRNA; the complex is not required for subsequent processing and is not involved directly in the translation of the mRNA.

(a) Polyadenylation begins when RNA polymerase II synthesizes a polyadenylation signal sequence whose consensus sequence is AAUAAA.

Figure 28·20
Polyadenylation of mRNA precursors.

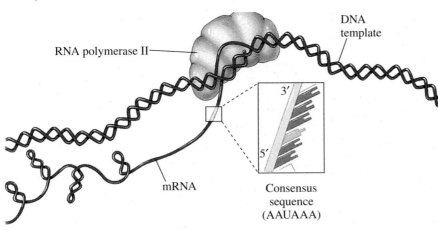

(b) The polyadenylation signal sequence is recognized by CPSF, which binds to the polyadenylation signal and forms a complex with the endonuclease and poly A polymerase. The endonuclease catalyzes cleavage of the transcript downstream of the AAUAAA sequence, producing a new 3' end.

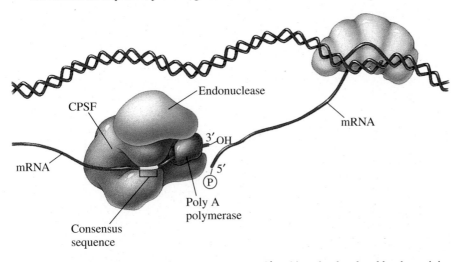

(c) Upon dissociation of the endonuclease, the new 3' end is polyadenylated by the activity of poly A polymerase.

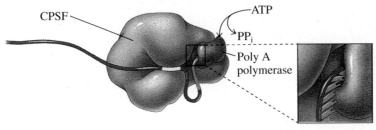

(d) Once the poly A tail has been added to the mRNA, the tail becomes associated with PAB I.

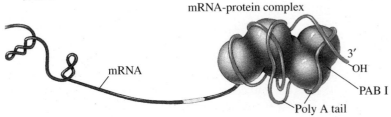

Figure 28·21

Purification of eukaryotic mRNA by affinity chromatography. Cells are briefly exposed to ³H-uridine, which radioactively labels newly synthesized RNA. The total cellular RNA is then isolated and passed through a column of immobilized oligo dT in a high-salt buffer. The column is then washed to remove any RNA that fails to bind. The amount of RNA exiting the column is determined by measuring the absorbance of the solutions in each tube, or fraction, of eluted buffer. (Recall from Section 24·3 that nucleic acids absorb ultraviolet light maximally at 260 nm.) Once no more mRNA is found in the eluted buffer (as indicated by the low absorbance at 260 nm), the bound mRNA containing poly A tails is removed from the column by washing with a buffer of low ionic strength. The peak of high radioactivity and low absorbance represents the fractions that contain the poly A mRNA. Note that although mRNA constitutes only 2–3% of the total RNA (as indicated by the relative areas underneath the absorbance peaks), it makes up 25–50% of the newly synthesized RNA (as indicated by the relative areas underneath the radio-activity peaks).

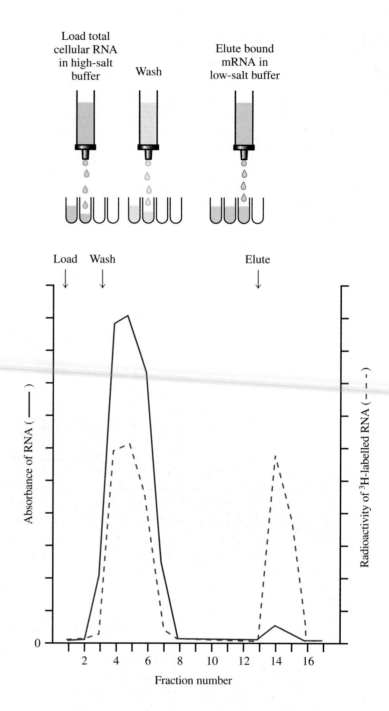

C. Purification of Poly A–Containing mRNA

The presence of poly A tails on most eukaryotic mRNA molecules has aided biochemists in purifying mRNA. Once the protein of the RNA-protein complex is removed, the poly A tail of the mRNA molecule can hybridize to a short stretch of thymidylate or uridylate residues (oligo dT or oligo U) that is covalently attached to a solid support (i.e., an affinity chromatography column). When an extract of cellular nucleic acids is passed over such a column under conditions that favor hybridization, only poly A–containing mRNA binds. All other RNA molecules, such as rRNA and tRNA, pass through the column. The bound mRNA can then be eluted by lowering the ionic strength of the buffer so that the mRNA-oligonucleotide hybrids are disrupted. Since only a small percentage of the total RNA in a cell is mRNA, this step can purify the mRNA 20- to 50-fold.

By separating mRNA from other types of RNA, this technique can be used to determine what fraction of the cell's RNA production is dedicated to the synthesis of mRNA (Figure 28·21). When radioactive uridylate is added to growing cells for a short period, the radioactivity is incorporated only into newly synthesized RNA. When this labelled RNA is applied to an oligo dT column, the column tightly binds 25–50% of the labelled RNA. This result indicates that mRNA synthesis occupies a large fraction of the capacity of the cell for RNA production. (Since mRNA is much less stable than other forms of RNA, however, it represents only a small fraction of steady-state RNA.)

(a)

28·5 Protein-Encoding Genes in Many Eukaryotes Contain Introns

Although splicing of mRNA precursors is rare in prokaryotes, it is the rule in vertebrates and flowering plants. In some unicellular eukaryotes, such as yeast and algae, only a few genes produce transcripts that are spliced. In insects, such as *Drosophila melanogaster,* genes containing introns are common, but many genes have no introns. There may be some eukaryotes, such as diplomonads, that have no intron-containing genes. These species are distantly related to most other eukaryotes.

Protein-encoding genes with introns were first identified in the mid-1970s. Splicing of mRNA precursors of adenovirus (a mammalian virus) was characterized first, followed by the discovery of splicing in globin genes and other similar genes that are expressed abundantly in certain tissues. Phillip Sharp and Richard Roberts (Figure 28·22) received a Nobel prize in 1993 for their discovery of introns in protein-encoding genes. Certain categories of genes rarely contain introns; these include the genes for heat-shock proteins, histones, and proteins specifically made in the male germ line of mammals. Other genes contain introns in almost every species examined. Most genes, however, have introns in some species and not in others.

(b)

A. Many Genes Contain Multiple Introns

Because of the loss of introns, mature mRNA is often a fraction of the size of its primary transcript. For example, the gene for triose phosphate isomerase from maize contains nine exons and eight introns and spans over 3400 base pairs of DNA, whereas the mature mRNA, which includes a poly A tail, is only one-third as large (Figure 28·23a, next page). The enzyme itself contains 253 amino acids.

Figure 28·22
(a) Phillip Sharp and (b) Richard Roberts, who received a Nobel prize in 1993 for their discovery of introns in protein-encoding genes.

(a)

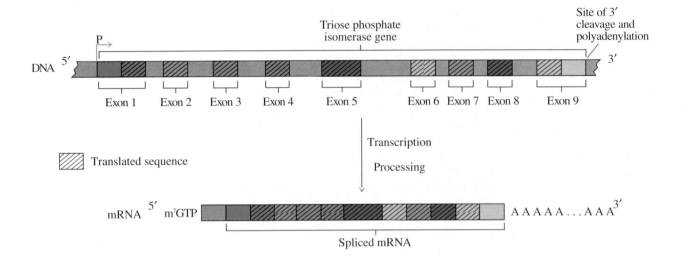

Transcription

Processing

Figure 28·23
Triose phosphate isomerase gene from maize and the encoded enzyme. **(a)** The gene contains nine exons and eight introns. Exons consist of both translated and untranslated sequences. **(b)** Three-dimensional structure of the protein showing the parts encoded by each exon. (Courtesy of Mark Marchionni.)

(b)

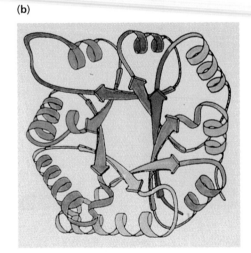

The introns in eukaryotic genes are not positioned randomly. They tend to be found in regions that encode loops on the surface of the protein, and they tend to be absent in parts of the gene that encode defined secondary structural elements such as α helices and β sheets (Figure 28·23b). Consequently, each exon often encodes a discrete domain of the protein with defined supersecondary structure. However, the frequency and importance of the correlation is often exaggerated. This organization contrasts sharply with the organization of most genes in prokaryotes. The gene for triose phosphate isomerase in *E. coli,* for example, has no introns, even though the enzymes from bacteria and from maize are homologous.

The gene for triose phosphate isomerase in maize is not unusual; many angiosperm and vertebrate genes contain multiple introns, some having more than 50. The organization of introns in a gene can be visualized using **R-looping,** a technique in which mature mRNA is hybridized to DNA corresponding to the gene. In an electron micrograph of the hybrid, long stretches of DNA that are not represented in the mRNA sequence appear as single-stranded loops, or R-loops. The 16 introns of the gene for chicken egg white conalbumin, for example, are seen as 16 loops in an electron micrograph of the mRNA-DNA hybrid (Figure 28·24).

B. Splice-Site Sequences Are Similar in All mRNA Precursors

The nucleotide sequences at splice sites are similar in all mRNA precursors. The vertebrate consensus sequences at the two splice sites are shown in Figure 28·25. All of the sequences at splice sites closely approximate the consensus sequences, although few match the consensus sequences exactly. An additional short consensus sequence is found within the intron near the 3′ end. This is the branch-point sequence, which plays the same role in splicing as the branch point in group II introns.

Introns in protein-encoding genes vary in length from as few as 65 base pairs to as many as 10 000 base pairs. Except for the short consensus sequences, the nucleotide sequence of an intron plays no known role in splicing or any other processing event. In fact, the sequences of introns in genes from closely related species are often quite different. This finding suggests that the sequences are not under selective pressure. Contrast this variability with the conservation of sequence and secondary structure seen in group I and group II introns.

Figure 28·24
Organization of the chicken egg white con-albumin gene visualized by R-looping. **(a)** An electron micrograph of a hybrid mRNA-DNA molecule reveals that the gene contains 17 exons and 16 introns. Exons form double-stranded regions with the single-stranded mature mRNA. The introns form large, single-stranded loops because there is no corresponding sequence in the mature mRNA to which they can bind.
(b) Tracing of the molecule in part (a), showing the positions of introns (A–P) and exons (1–17). (Courtesy of Pierre Chambon.)

(a)

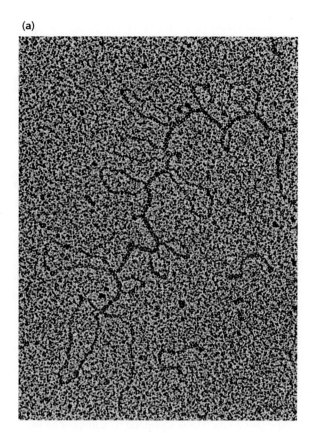

(b)

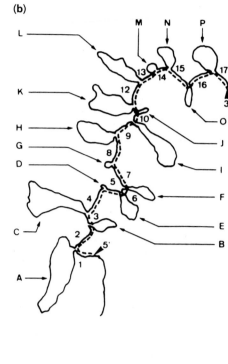

Figure 28·25
Consensus sequences at splice sites in vertebrates. The highly conserved nucleotides are under-lined. *Y* represents any pyrimidine (U or C), and *N* stands for any nucleotide. The splice sites, where the RNA precursors are cut and joined, are indicated by red arrows, and the intron is highlighted in blue.

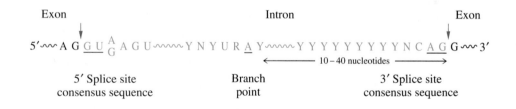

The consensus sequences of splice sites in plants and insects are very similar to the vertebrate consensus sequences, whereas the splice sites in yeast and some other eukaryotes differ somewhat. In spite of this variability, all conserved intron sequences share three important features that play a role in splicing.

1. In all eukaryotic genes without exception, the sequence GU is found at the 5′ splice site.

2. The dinucleotide sequence AG marks the 3′ splice site.

3. An adenylate residue is part of the branch point and is located 10–40 nucleotides upstream of the 3′ splice site.

Figure 28·26
Mutations that affect splicing in globin genes. Arched lines indicate excised introns.

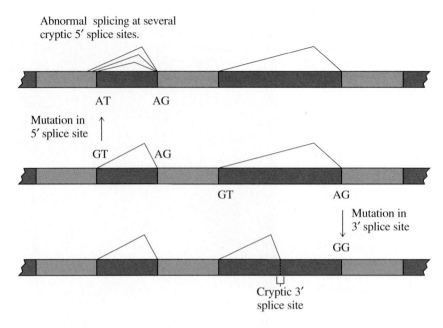

These common features give rise to the GT/AG rule that defines the ends of an intron in the DNA sequence of a gene. The importance of these nucleotides can be demonstrated experimentally by altering them and observing how the alteration affects splicing. Changing nucleotides within the consensus sequences interferes with or abolishes splicing. The highly conserved dinucleotide sequences at the ends of the intron are particularly sensitive, since any experimental alteration of these sequences invariably prevents correct splicing. Splicing is also altered by naturally occurring mutations in many species. These mutations either remove or modify one of the splice-site sequences and prevent normal processing of mRNA precursors. For example, several mutations in α-globin and β-globin genes of humans inhibit splicing and cause hereditary hemoglobin deficiency, or thalassemia. These mutations have been shown to occur within the vertebrate splice-site consensus sequences. Two examples of mutations in the β-globin gene are shown in Figure 28·26. In both cases, the mutation alters one of the conserved nucleotides at the splice sites. Normal splicing is abolished, and a cryptic splice site is used instead. This cryptic site partially matches the consensus sequence, but it is normally not involved in splicing.

C. Spliceosomes Catalyze Intron Removal

Eukaryotic mRNA precursors are spliced in a reaction that parallels the self-splicing reaction of group II introns. The intron is removed by two transesterification reactions: one between the 5′ splice site and a branch-point adenylate residue, and a second between the 5′ exon and the 3′ splice site. The products of these two reactions are the joined exons and the excised intron in the form of a lariat molecule. During splicing of mRNA precursors, however, these reactions are catalyzed not by the intron but by a large protein-RNA complex called the **spliceosome.** The spliceosome functions like the folded intron in group I and group II self-splicing reactions by holding the ends of the intron and the branch point in position during the

(a) The intron sequences are organized by the spliceosome so that the adenylate residue at the branch point is positioned near the 5′ splice site and the 2′-hydroxyl group of the adenylate can attack the 5′ splice site.

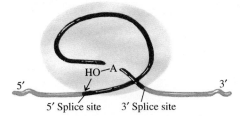

(b) Next, the 2′-hydroxyl group is attached to the 5′ end of the intron and the newly created 3′-hydroxyl group of the exon attacks the 3′ splice site.

(c) As a result, the ends of the exons are joined, and the intron, in the form of a lariat-shaped molecule, is released.

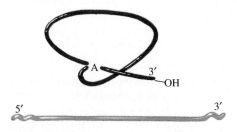

Figure 28·27
Intron removal in mRNA precursors. Splicing of mRNA precursors is catalyzed by the spliceosome, a RNA-protein complex. An adenylate residue within the intron is involved in the release of the 5′ exon, and the intron is excised as a lariat molecule, as in the group II mechanism. [Adapted from Sharp, P. A. (1987). Splicing of messenger RNA precursors. *Science* 235:766–771.]

reaction (Figure 28·27). The similarities in the reaction mechanisms for group II self-splicing and splicing catalyzed by the spliceosome strongly suggest that enzyme-catalyzed splicing of mRNA precursors is derived from a more primitive self-splicing reaction that was common in ancestral cells.

The spliceosome is a large, multisubunit complex of 3×10^6 Da. It is composed of 45 proteins and 5 molecules of RNA whose total length is about 5000 nucleotides. The RNA molecules are called **small nuclear RNA** (snRNA) molecules and are associated with protein to form **small nuclear ribonucleoproteins,** or snRNPs (pronounced "snurps"). snRNPs are important not only in the splicing of mRNA precursors but also in other cellular processes.

There are more than 100 000 copies of snRNA molecules per cell nucleus in vertebrates. These are divided into five types called U1, U2, U4, U5, and U6. (U stands for uracil, a common base in these small RNA molecules.) All five snRNA molecules are extensively base paired and contain modified nucleotides. Each snRNP is composed of one or two specific snRNA molecules plus a number of proteins that are bound to them. There are seven proteins that are found in all snRNPs, but each snRNP contains an additional two or three specific proteins.

Figure 28·28
Formation of a spliceosome. The snRNPs are shown in blue; additional protein factors are shown in purple.

(a) As soon as the 5′ splice site exits the transcription complex, a U1 snRNP binds to it.

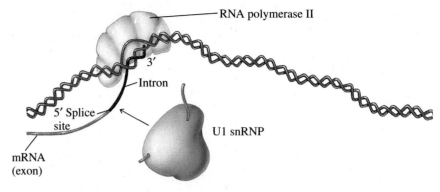

(b) Next, a U2 snRNP binds to the branch point within the intron.

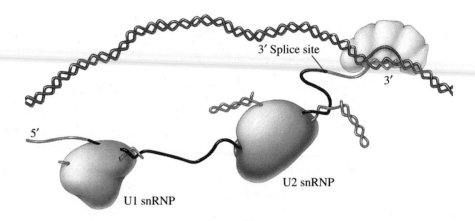

(c) When the 3′ splice site emerges from the transcription complex, a U5 snRNP binds and the complete spliceosome assembles around a U4/U6 snRNP.

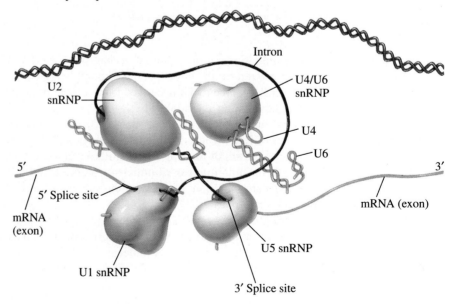

The various snRNPs assemble sequentially to form the spliceosome as shown in Figure 28·28. The formation of a spliceosome is initiated when a U1 snRNP binds to the newly synthesized 5′ splice site of the mRNA precursor. This interaction involves base pairing between the 5′ splice site and a complementary sequence near the 5′ end of U1 snRNA. A U2 snRNP then binds to the branch point of the intron, forming a stable complex that covers about 40 nucleotides. Next, a U5 snRNP associates with the 3′ splice site. Finally, U4/U6 snRNP joins the complex, and all snRNPs are drawn together to form the spliceosome. The assembly of the complete spliceosome also involves other protein factors in addition to the snRNPs.

The binding of U1, U2, and U5 snRNPs to consensus sequences at the 5′ splice site, branch point, and 3′ splice site, respectively, positions these three interactive sites properly for the splicing reaction. The spliceosome prevents the 5′ exon from diffusing away after cleavage and positions it to be joined to the 3′ exon.

Once a spliceosome has formed at an intron, it is quite stable and can be purified from cell extracts (Figure 28·29). Since the spliceosome, which is almost as large as a ribosome, is too large to fit through the nuclear pores, it prevents the mRNA precursor from leaving the nucleus before processing is complete.

The spliceosome does not form and splicing does not occur unless the 5′ and 3′ splice sites as well as the branch point are occupied by snRNPs. Since spliceosomes can be observed on nascent transcripts, it is thought that removal of the intron is the rate-limiting step in splicing. Although spliceosomes assemble during transcription, the intron is not usually removed until transcription is finished. Once the intron has been removed, the spliceosome disassembles.

Several human disorders, such as systemic lupus erythematosus, rheumatoid arthritis, and mixed connective tissue disease (MCTD), are known as **autoimmune diseases.** These diseases are so named because people afflicted with them produce antibodies against normal tissues and cellular components, such as DNA, immunoglobulins, kidney glomeruli, and neurons. The diseases are characterized by multiple symptoms, such as inflammation of bodily tissues and organ failure; they are often fatal.

Figure 28·29
Electron micrograph of a purified spliceosome. (Courtesy of Jack D. Griffith.)

Patients with autoimmune diseases frequently produce antibodies against nuclear components. In 1977, Joan Steitz and her coworkers analyzed the serum of a patient with MCTD to identify the particles bound by these antibodies. They found that the antibodies reacted with nuclear ribonucleoproteins. After five of these so-called snRNPs had been precipitated and characterized, they were recognized as assisting in intron removal. These newly discovered nuclear components were subsequently characterized further using sera from other patients with autoimmune diseases.

Why the sera of patients with autoimmune diseases contain antibodies against snRNPs is unknown, but there are at least two possible explanations. snRNPs are usually hidden from the immune system in the nucleus of a cell but could be released by cell lysis into the blood, where they may trigger the production of antibodies. Alternatively, the antibodies found by Steitz and coworkers may not be directed against snRNPs per se but, rather, against other proteins with similar structures, such as viral proteins.

D. mRNA Precursors Are Associated with Specific Proteins in the Nucleus

Splicing of free mRNA precursors does not occur. Rather, as soon as RNA polymerase II produces a sufficient length of RNA, the nascent transcript associates with specific proteins to form a **heterogeneous nuclear ribonucleoprotein complex** (hnRNP). This complex has a beads-on-a-string appearance reminiscent of chromatin (Section 24·11A); the particles are sometimes referred to as ribonucleosomes. The complex is composed of 8 to 10 proteins, whose molecular weights in mammals range from 32 000 to 120 000. These proteins bind to approximately 500 nucleotides of RNA.

The hnRNP may protect the primary transcript from endogenous nucleases and inhibit the formation of secondary structures. In fact, one of the proteins of the complex is a helix-destabilizing protein (HD protein) similar to proteins involved in DNA replication, such as SSB (Section 25·5). The HD protein denatures double-stranded RNA structures, such as hairpins. As we saw with chromatin, the packaging of nucleic acids with proteins does not seem to prevent complexes, such as snRNPs, from recognizing their binding sites. In fact, such interactions may be enhanced by the structure, since the subunits of the spliceosome interact with many of the proteins that are bound to RNA. These additional proteins, known as splicing factors, are required to ensure the positioning of snRNPs at the correct splice sites.

The mature mRNA in the cytoplasm also associates with proteins to form a messenger ribonucleoprotein (mRNP) particle. These proteins are not those that bind the mRNA precursor in the nucleus, however. The proteins probably exchange when mRNA is transported through the nuclear pore; this exchange may even be an essential part of the transport process.

E. Multiple Introns Are Not Always Removed Sequentially

The introns of some mRNA precursors, such as those for globin mRNA, are removed in the order they are transcribed; however, this is not the case for the introns in many other precursors. Since the nascent transcripts associate with protein to form hnRNPs, the order of intron removal is thought to be determined by the secondary and tertiary structure of the hnRNP. Excision of one intron may alter the secondary structure of the RNA, thereby making a second intron available for removal. The order of excision may also be kinetically determined; some introns may be removed within only a few seconds, whereas others may persist for 10 to 20 minutes. Complete splicing of the globin RNA precursor takes about five minutes.

Since the spliceosome is assembled during transcription, the 5′ splice-site sequence is recognized first, and the next 3′ splice site is the site of cleavage. In a gene with a large number of closely spaced introns, there is a risk that one 3′ splice-site sequence may be skipped and splicing may occur at the following one, leading to inappropriate excision of an exon. Such errors do occur, although they seem to be quite rare.

28·6 Gene Expression in Eukaryotes Can Be Regulated by Alternative Processing

The processing of mRNA precursors can be regulated at the level of splice-site selection. The use of alternative splice sites can give rise to variants of mRNA that may encode distinct proteins. In some cases, the variants are produced from a single primary transcript using alternative processing pathways. In other cases, the alternative pathways may utilize different primary transcripts of the same gene. In general, these alternative pathways are found in different tissues of multicellular organisms or at different stages of development in the same cell type. In most cases, the alternative pathways are controlled by protein factors within the cell. The use of alternative splice sites can also be controlled at the level of mRNA precursor production. Two or more promoters can be used for transcription of the same gene. The longer transcript may contain an additional exon not found in the shorter one. A similar situation may occur if two or more polyadenylation sites occur in the same gene.

There are many examples of alternative processing pathways. In a large number of cases, the alternative pathways involve regulation of multiple processing events involving several exons and introns.

A. Use of Two Promoters Leads to Alternative Processing Pathways

Some genes have two promoters. One promoter may be active in one cell type, whereas a second promoter may be active in other cells. If each promoter is associated with its own exon, alternative processing pathways are followed in each type of cell (Figure 28·30). When transcription begins at P_1, the primary transcript is processed by splicing between the 5′ splice site at the end of exon A and the next 3′ splice site, which is found at the beginning of exon C. Note that there is no 3′ splice site at the beginning of exon B, which instead has a second promoter. When transcription begins at promoter P_2, splicing occurs between the 5′ splice site at the end of exon B and the same 3′ splice site used in the first pathway.

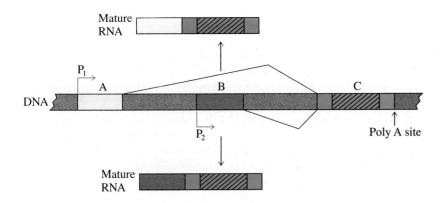

Figure 28·30
Alternative processing through the use of multiple promoters. Transcription from two different promoters results in two different primary transcripts that undergo alternative processing reactions. The protein-encoding region is cross-hatched. In this case, the same protein is produced from both mature transcripts.

In the example shown in Figure 28·30, exon C contains the protein-encoding region and part of the noncoding sequence at the 5′ end of the mature mRNA. Exons A and B do not contain protein-encoding sequences, so the alternative processing pathways result in production of the same protein. This is often the case with genes that have multiple promoters. For example, one of the mouse α-amylase genes is transcribed from one promoter in salivary glands and from another in liver, and the resulting primary transcripts are processed as shown here. Thus, transcription of the gene is regulated differently in different tissues, but the identical enzyme is produced as a result of alternative processing pathways.

B. Multiple Polyadenylation Sites Can Lead to Alternative Processing Pathways

Alternative processing pathways can also arise from the selective use of multiple polyadenylation sites. In the scheme shown in Figure 28·31, cleavage of the primary transcript can occur at polyadenylation site 1 or 2. The shorter transcript is spliced by removing the intron between the first and second exons, while the longer transcript is spliced by removing a larger sequence that includes the second exon. The second 3′ splice site is preferentially used in the longer transcript. The two different mRNA molecules produced by such alternative processing pathways often encode proteins whose primary sequences differ at the C-terminus.

Variants of the heavy chain of immunoglobulins are produced in this manner. When cleavage and polyadenylation occur at the first poly A site, the secreted, soluble form of the heavy chain found in immunoglobulins is produced. When the second poly A site is used, a longer protein containing a C-terminal, membrane-spanning region is made. This is part of membrane-bound immunoglobulin M.

C. Regulation of Processing Determines Sex in *Drosophila*

The determination of sex in *D. melanogaster* depends on the ratio of X chromosomes to the other chromosomes (autosomes). When two X chromosomes are present in normal diploid cells, the individual is female. Males have only one X chromosome and a much smaller Y chromosome that plays little role in sex determination. Different genes are expressed in males and females, and this differential expression is controlled by sex-determining genes that can be processed in several ways.

Figure 28·31
Alternative processing through the use of different polyadenylation sites. Protein-encoding regions are cross-hatched.

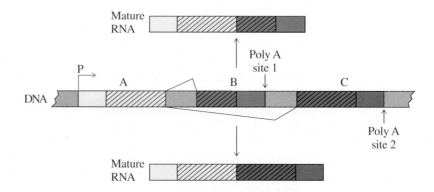

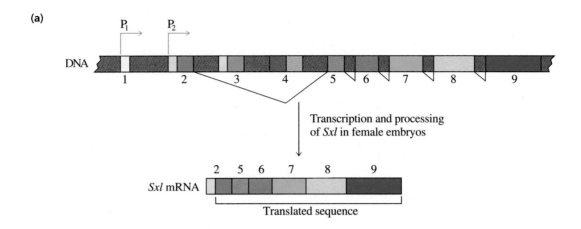

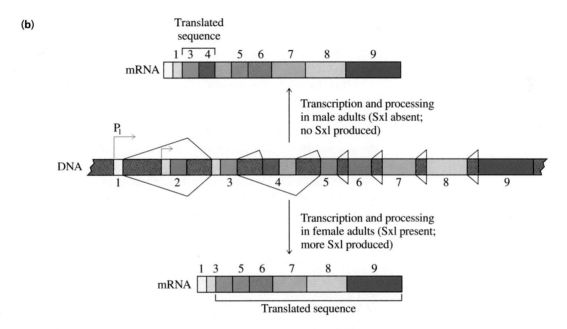

Figure 28·32
Transcription and processing of *Sxl*. (a) Transcription and processing of *Sxl* in female embryos. (b) Alternative processing pathways for *Sxl* in adults. The embryonic and adult forms of Sxl are slightly different at the N-terminus.

The first gene in the sex-determination pathway is *Sex-lethal (Sxl)*. It is transcribed from two different promoters (Figure 28·32). P_1 is a functional promoter in all cells of both males and females. P_2 is only active in embryos destined to be female due to the presence of specific transcription factors that are active only in the presence of two X chromosomes.

Transcription and processing in female embryos follows the pathway shown in Figure 28·32a. The mature mRNA is translated to produce Sxl. Sxl regulates processing of its own gene and other genes in adult tissues. The two alternative processing pathways for *Sxl* in adults are shown in Figure 28·32b. The male pathway is the default pathway that occurs in the absence of Sxl. In this pathway, exon 4 is included in the mature mRNA; exon 4 includes a stop codon, a three-nucleotide sequence that results in termination of translation. The resulting protein is nonfunctional. (Stop codons are discussed in more detail in Chapter 29.)

In females, exon 4 is skipped during processing. This results in a mature mRNA that can direct synthesis of more Sxl. Exon skipping is mediated directly by Sxl, which binds to the 3′ splice site adjacent to exon 4 and inhibits spliceosome assembly. When Sxl is present, the next available site is used, leading to excision of exon 4. Regulation of processing by Sxl ensures that Sxl is only produced in females.

Figure 28·33
Alternative processing pathways for *tra* transcripts.

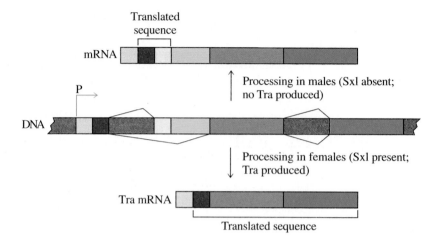

Transformer (tra) is the next gene in the sex-determination pathway. This gene is transcribed in males and females, but processing of the primary transcript is mediated by Sxl (Figure 28·33). The male pathway is followed in the absence of Sxl. Splicing of the first intron utilizes a 3′ splice site just upstream of a stop codon, and the mature mRNA does not encode a functional protein.

In females, Sxl binds to the polypyrimidine tract at the male-specific 3′ splice site (see Figure 28·25) and inhibits binding of a splicing factor that is required for spliceosome assembly, thereby forcing the use of an alternative site downstream of the stop codon within the exon. The same mechanism may be used to regulate processing of Sxl's own gene transcript. Mature mRNA encoding Tra is produced in females but not in males.

Tra regulates splicing of the third sex-determination gene, *doublesex (dsx)*. Tra binds to primary transcripts of the *dsx* gene and promotes female-specific splicing at a 3′ splice site. This leads to production of Dsx in females and the development of female organs and eggs. The action of Tra requires another protein, Tra-2, which is made constitutively in both males and females.

Tra-2 and Sxl both contain RNA-binding domains that are also found in other splicing factors. In addition, Tra, Tra-2, and several other proteins involved in regulation of splicing contain a serine- and arginine-rich domain that may play a role in association of these factors with the spliceosome.

Not all of the proteins that regulate splicing are splicing factors. One of the ribosomal proteins in yeast is encoded by a gene *(RPL32)* with a single intron. Normally, the primary transcript is spliced to produce mature mRNA, but in the presence of excess RPL32 protein, splicing is inhibited. RPL32 prevents spliceosome formation by binding directly to a region of secondary structure in the mRNA that resembles its rRNA binding site. This feedback mechanism ensures that RPL32 synthesis is blocked in the absence of ribosome assembly, when free RPL32 is available to bind its own mRNA.

D. Calcitonin and CGRP Are Produced by Differential Splicing

The gene in rats that encodes calcitonin, a thyroid hormone, also encodes a protein that is involved in taste perception and is localized in the brain. This other protein is called calcitonin gene–related protein (CGRP). Processing of the primary transcript involves the selective splicing of introns and alternative poly A sites (Figure 28·34).

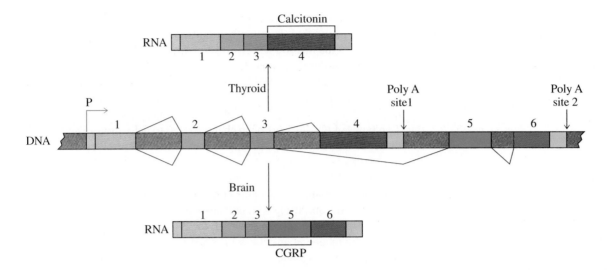

Figure 28·34
Differential expression of the calcitonin transcript. [Adapted from Rosenfeld, M. G., Mermod, J. J., Amara, S. G., Swanson, L. W., Sawchenko, P. E., Rivier, J., Vale, W. W., and Evans, R. M. (1983). Production of a novel neuropeptide encoded by the calcitonin gene via tissue-specific RNA processing. *Nature* 304:129–135.]

In the thyroid, the first polyadenylation site is used, and the primary transcript includes exons 1–4. This transcript is processed normally, and the polypeptide synthesized using the mature mRNA is cleaved to produce calcitonin. In the brain, the second poly A site is used, and the primary transcript contains exon 4, which encodes calcitonin. During processing, however, exon 4 is excised, and exons 3 and 5 are joined. CGRP is encoded by exon 5. This protein, like calcitonin, is generated by cleavage of the newly synthesized polypeptide chain.

The calcitonin pathway appears to be the default pathway. Neurons contain a factor that inhibits splicing at the splice site near the beginning of exon 4. There are many examples of such tissue-specific factors that influence processing in animals.

Summary

RNA processing includes chemical modification of nucleotides and the addition and deletion of nucleotides. Processing appears to be related, in part, to RNA stability. In addition, certain alterations refine RNA structure and allow the RNA to be recognized by other cellular components. Processing is required in all primary transcripts containing tRNA and rRNA and in eukaryotic mRNA molecules.

Precursor tRNA molecules in prokaryotes are part of longer primary transcripts that are cleaved by RNases into immature tRNA monomers. RNase P, which is made up of both protein and a catalytically active RNA molecule, is an endonuclease that cleaves the primary transcript once at each individual tRNA precursor to generate a mature 5′ end. Maturation of the 3′ end of a prokaryotic tRNA precursor involves several additional exonucleases, including RNases D and T. In organisms in which the 3′ terminal sequence CCA is not encoded by the tRNA gene itself, the sequence must be added posttranscriptionally by the enzyme tRNA nucleotidyltransferase.

Many primary transcripts contain sequences called introns that interrupt the functional part of the molecule. During maturation, introns are removed, and the structural portions of the precursor, called exons, are joined together. This process is called RNA splicing. In the case of yeast tRNA splicing, the reaction is mediated by a multifunctional enzyme and an RNA ligase and occurs in two steps. Excision of tRNA introns requires two molecules of ATP.

In both eukaryotes and prokaryotes, certain nucleotides of tRNA molecules are extensively modified. Covalently modified nucleotides tend to lie in regions of the tRNA molecule that are not involved in intrastrand base pairing.

The primary ribosomal RNA transcripts of both prokaryotes and eukaryotes contain more than one rRNA molecule. These individual rRNA species are separated from the long primary transcript by specific endonucleases that recognize double-stranded regions of the precursor. Some rRNA molecules contain introns. Two types of introns have been characterized to date: group I and group II. Some of these introns, called self-splicing introns, excise themselves without the assistance of proteins.

Group I introns are removed as linear molecules. Self-splicing by these introns requires Mg^{2+} and a guanosine cofactor. The guanosine cofactor is added to the 5′ end of the intron during a transesterification reaction, which releases the 5′ exon. Once the 5′ exon has been released, the 3′-hydroxyl group at its 3′ end attacks the 3′ exon boundary in another transesterification reaction. This reaction, like more conventional enzymatic reactions is reversible; these introns are examples of RNA enzymes, or ribozymes.

Self-splicing introns of group II are found in some mitochondrial rRNA and protein-encoding genes. They do not require guanosine cofactors. The mechanism of self-splicing in group II introns involves attack at the 5′ splice site by the 2′-hydroxyl group of an adenylate residue within the intron. Subsequent attack by the 3′-hydroxyl group of the 5′ exon at the 3′ splice site joins the exons and releases the intron as a lariat molecule.

Primary messenger RNA precursors in eukaryotes are processed by nucleotide modification (capping), polyadenylation, and, in some cases, splicing. The cap at the 5′ end protects the RNA from digestion during its long life span and also interacts with ribosomes during translation. mRNA precursors are modified at their 3′ ends by a complex consisting of a cleavage and polyadenylation specificity factor (CPSF), an endonuclease, and poly A polymerase. CPSF binds to a specific site that is 10 to 20 nucleotides upstream of the cleavage site. The endonuclease cleaves the precursor, and poly A polymerase adds a poly A tail consisting of up to 250 adenylate residues. This poly A tail binds to at least one protein that helps stabilize the RNA.

Introns in mRNA precursors range in size from 65 to 10 000 nucleotides. Except for short consensus sequences at the splice sites and the branch point, the sequences of introns share no similarities. Splicing of mRNA precursors is not self-catalyzed but involves a group of ribonucleoproteins (snRNPs) that assemble to form a spliceosome. The spliceosome correctly orients the reactants in the splicing reaction. Assembly of the spliceosome begins when a U1 snRNP binds to the 5′ splice site. This binding is mediated by base pairing between the U1 snRNA and the mRNA precursor. A U2 snRNP then binds at the branch point within the intron, while a U5 snRNP binds at the 3′ splice site. The three bound snRNPs move together, bind to the U4/U6 snRNP, and remove the intron. The actual reaction catalyzed by the spliceosome is similar to group II self-splicing reactions: in one transesterification reaction, nucleophilic attack by the 2′-hydroxyl group at the branch point opens the 5′ splice site; in the second transesterification reaction, the exons are joined and the intron is released as a lariat molecule.

Different mature mRNA molecules can be produced by alternative processing pathways. In some cases, alternative splice sites are used; in other cases, the processing pathways are coupled to the use of alternative promoters or polyadenylation sites.

Selected Readings

General References

Lewin, B. (1991). *Genes IV* (New York: Oxford University Press).

Watson, J. D., Hopkins, N. H., Roberts, J. W., Steitz, J. A., and Weiner, A. M. (1987). *Molecular Biology of the Gene,* Vol. 1, 4th ed. (Menlo Park, California: Benjamin/Cummings Publishing Company), pp. 360–381.

Processing of RNA

Abelson, J. (1992). Recognition of tRNA precursors: a role for the intron. *Science* 255:1390.

Aebi, M., and Weissman, C. (1987). Precision and orderliness in splicing. *Trends Genet.* 3:102–107.

Apirion, D., and Miczak, A. (1993). RNA processing in prokaryotic cells. *BioEssays* 15:113–120.

Baldi, M. I., Mattoccia, E., Bufardeci, E., Fabbri, S., Tocchini-Valentini, G. P. (1992). Participation of the intron in the reaction catalyzed by the *Xenopus* tRNA splicing endonuclease. *Science* 255:1404–1408.

Balvay, L., Libri, D., and Fiszman, M. Y. (1993). Pre-mRNA secondary structure and the regulation of splicing. *BioEssays* 15:165–169.

Belfort, M. (1991). Self-splicing introns in prokaryotes: migrant fossils? *Cell* 64:9–11.

Björk, G. R. (1987). Modification of stable RNA. In Escherichia coli *and* Salmonella typhimurium: *Cellular and Molecular Biology,* Vol. 1, F. C. Neidhardt, ed. (Washington, DC: American Society for Microbiology), pp. 719–731.

Cech, T. R. (1986). RNA as an enzyme. *Sci. Am.* 255(5):64–75.

Cech, T. R., Herschlag, D., Piccirilli, J. A., and Pyle, A. M. (1992). RNA catalysis by a group I ribozyme. *J. Biol. Chem.* 267:17 479–17 482.

Dreyfuss, G., Matunis, M. J., Piñol-Roma, S., and Burd, C. G. (1993). hnRNP proteins and the biogenesis of mRNA. *Annu. Rev. Biochem.* 62:289–321.

Ferat, J.-L., and Michel, F. (1993). Group II self-splicing introns in bacteria. *Nature* 364:358–361.

Fournier, M. J., and Maxwell, E. S. (1993). The nucleolar snRNAs: catching up with the spliceosomal snRNAs. *Trends Biochem. Sci.* 18:131–135.

King, T. C., and Schlessinger, D. (1987). Processing of RNA transcripts. In Escherichia coli *and* Salmonella typhimurium: *Cellular and Molecular Biology,* Vol. 1, F. C. Neidhardt, ed. (Washington, DC: American Society for Microbiology), pp. 703–718.

Kjems, J., and Garrett, R. A. (1988). Novel splicing mechanism for the ribosomal RNA intron in the archaebacterium *Desulfurococcus mobilis. Cell* 54:693–703.

Lamond, A. I. (1993). A glimpse into the spliceosome. *Curr. Biol.* 3:62–64.

Malter, J. S. (1989). Identification of an AUUUA-specific messenger RNA binding protein. *Science* 246:664–666.

Marchionni, M., and Gilbert, W. (1986). The triosephosphate isomerase gene from maize: introns antedate the plant-animal divergence. *Cell* 46:133–141.

McKeown, M. (1993). The role of small nuclear RNAs in RNA splicing. *Curr. Biol.* 5:448–454.

Phizicky, E. M., and Greer, C. L. (1993). Pre-tRNA splicing: variation on a theme or exception to the rule? *Trends Biochem. Sci.* 18:31–34.

Raghow, R. (1987). Regulation of messenger RNA turnover in eukaryotes. *Trends Biochem. Sci.* 12:358–360.

Reed, R., Griffith, J., and Maniatis, T. (1988). Purification and visualization of native spliceosomes. *Cell* 53:949–961.

Roger, A. J., and Doolittle, W. F. (1993). Why introns-in-pieces? *Nature* 364:289–290.

Steitz, J. A. (1988). "Snurps." *Sci. Am.* 258(6):56–63.

Takagaki, Y., Ryner, L. C., and Manley, J. L. (1989). Four factors are required for 3'-end cleavage of pre-mRNAs. *Genes Dev.* 3:1711–1724.

von Ahsen, U., and Schroeder, R. (1993). RNA as a catalyst: natural and designed ribozymes. *BioEssays* 15:299–307.

Witkowski, J. A. (1988). The discovery of "split" genes: a scientific revolution. *Trends Biochem. Sci.* 13:110–113.

Polyadenylation

Proudfoot, N. (1991). Poly(A) signals. *Cell* 64:671–674.

Wahle, E. (1992). The end of the message: 3′-end processing leading to polyadenylated messenger RNA. *BioEssays* 14:113–118.

Wahle, E., and Keller, W. (1992). The biochemistry of 3′-end cleavage and polyadenylation of messenger RNA precursors. *Annu. Rev. Biochem.* 61:419–440.

Wickens, M. (1990). How the messenger got its tail: addition of poly(A) in the nucleus. *Trends Biochem. Sci.* 15:277–281.

Alternative Splicing

Mattox, W., Ryner, L., and Baker, B. S. (1992). Autoregulation and multifunctionality among *trans*-acting factors that regulate alternative pre-mRNA processing. *J. Biol. Chem.* 267:19 023–19 026.

McKeown, M. (1992). Alternative mRNA splicing. *Annu. Rev. Cell Biol.* 8:133–155.

Valcárcel, J., Singh, R., Zamore, P. D., and Green, M. R. (1993). The protein Sex-lethal antagonizes the splicing factor U2AF to regulate alternative splicing of *transformer* pre-mRNA. *Nature* 362:171–175.

29

The Genetic Code and Transfer RNA

The structures of DNA and RNA make it easy to imagine a mechanism by which genetic information is transferred between the two: the hydrogen-bonding patterns of the bases allow a DNA strand to act as the template for synthesis of a complementary RNA strand. Much less apparent, however, is the mechanism by which DNA specifies the sequence of amino acids in a protein. Years ago, some mechanisms for protein synthesis were proposed in which DNA served as a template for direct synthesis of a protein, but we now know that these models are incorrect.

Early in the history of molecular genetics, two ideas were advanced to explain how the linear sequence of nucleotide residues in DNA indirectly specifies the linear sequence of amino acid residues in proteins. One hypothesis, developed by George Gamow, was that the information in DNA exists as a code that must be deciphered before proteins can be synthesized. The second hypothesis, developed by Francis Crick, suggested that the deciphering process depends upon the existence of adapter molecules that interact with both nucleic acids and amino acids. Crick hypothesized that the adapters are small nucleic acid molecules to which amino acids can attach and that at least twenty different adapters must exist, one for each kind of amino acid found in proteins.

Both of these ideas have proven to be correct. The sequence of residues in a nucleic acid specifies the amino acid sequence in a protein according to a **genetic code** that is almost universal among living organisms. Translation of the genetic code into an amino acid sequence is mediated by a class of relatively small RNA molecules called **transfer RNA** (tRNA). These molecules correspond to Crick's adapters; they form the interface between mRNA and proteins. One region of a tRNA molecule is covalently linked to a specific amino acid, and another region of

Related material appears in Exercise 10 of the computer-disk supplement *Exploring Molecular Structure.*

the same tRNA molecule interacts directly with an mRNA molecule by complementary base pairing. The progressive joining of amino acids that are sequentially specified by the mRNA template allows the precise synthesis of proteins.

An amino acid is attached to its specific tRNA molecule in a reaction catalyzed by an aminoacyl-tRNA synthetase. For instance, a tRNA^{Phe} molecule can be covalently linked to phenylalanine, forming a molecule referred to as phenylalanyl-tRNA^{Phe}. The reaction catalyzed by an aminoacyl-tRNA synthetase activates the amino acid in preparation for transfer to a growing peptide chain. Since aminoacyl-tRNA synthetases link amino acids to their corresponding tRNA molecules, they are essential to the transfer of genetic information from nucleic acid to protein. The accuracy and specificity of translation depend in part upon the fidelity of these enzymes.

29·1 The Genetic Code Consists of Nonoverlapping Triplet Codons

The specific code by which nucleic acids specify amino acids was not known until 1965. Ten years before that, George Gamow first proposed the basic structural units of the genetic code. He reasoned that, since the DNA "alphabet" consists of only four "letters" (A, T, C, and G) and since these four letters encode 20 amino acids, the genetic code might contain "words" made of a uniform length of three letters. Two-letter words constructed from any combination of the four letters produce a vocabulary of only 16 words (4^2), not enough for all 20 amino acids. In contrast, four-letter words produce a vocabulary of 256 words (4^4), far more than are needed. Three-letter words allow a vocabulary of 64 words (4^3), more than sufficient to specify each of the 20 amino acids, but not excessive.

In principle, a code made up of three-letter words can be either overlapping or nonoverlapping. In an overlapping code, a letter is part of more than one word; changing a single letter will change several words simultaneously. For example, in the sequence shown in Figure 29·1a, each letter is part of three different words in an overlapping code. In contrast, a particular letter is part of only one word in a nonoverlapping code (Figure 29·1b). In this case, changing a particular letter will change only one word. Studies have proven that the genetic codes used by all living organisms are nonoverlapping.

mRNA · · · A U G C A U G C A U G C · · ·

(a) Message read in   A U G
      overlapping   U G C
      triplet code   G C A
                 C A U
                  · · ·
                    · · ·

(b) Message read in   A U G
      nonoverlapping   C A U
      triplet code   G C A
                         U G C

Figure 29·1
Message read in (a) overlapping and (b) nonoverlapping three-letter codes. In an overlapping code, each letter is part of three different three-letter words; in a nonoverlapping code, each letter is part of only one three-letter word.

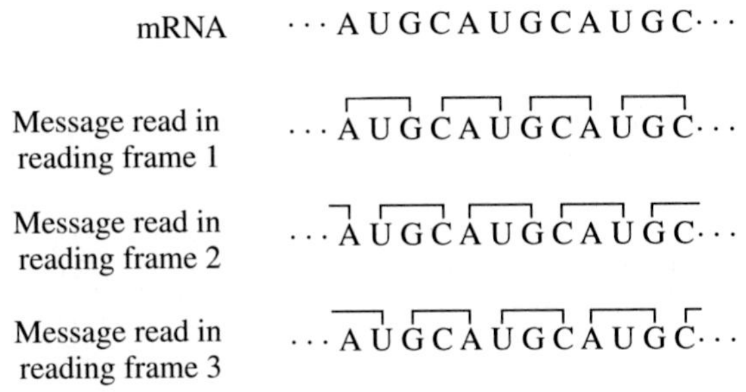

Figure 29·2
Reading frames in mRNA. The same string of letters read in three different reading frames can be translated as three different messages. Thus, translation of the correct message requires selection of the correct reading frame.

mRNA	$\cdots$ A U G C A U G C A U G C $\cdots$
Message read in reading frame 1	$\cdots$ AUG CAU GCA UGC $\cdots$
Message read in reading frame 2	$\cdots$ A UGC AUG CAU GC $\cdots$
Message read in reading frame 3	$\cdots$ AU GCA UGC AUG C $\cdots$

In a nonoverlapping code of three-letter words, the message translated from a particular sequence of letters depends upon the point at which translation begins; shifting the starting point for the reading of the code by one letter to the left or right changes the entire message. Each potential starting point defines a unique sequence of three-letter words, or **reading frame.** Thus, correct translation of the genetic code depends upon establishing the correct reading frame for translation. mRNA is translated in the same direction that it is synthesized, namely $5' \rightarrow 3'$, but since translation does not necessarily begin at the very beginning of the mRNA, there are three potential reading frames in each mRNA (Figure 29·2). Inside the cell, selection of the correct reading frame is determined by the translation machinery, as we will see in the next chapter.

In a series of elegant genetic experiments in the early 1960s, Francis Crick and Sydney Brenner demonstrated that the genetic code does in fact consist of three-letter words, or **codons.** They found that insertion or deletion of multiples of three nucleotides produced active proteins in bacteriophage T4. Insertion or deletion of one or two nucleotides, on the contrary, did not. The first major advance in deciphering the code came in 1961, when Marshall W. Nirenberg and J. Heinrich Matthaei demonstrated that a synthetic mRNA molecule could be translated in vitro. Using cell-free extracts that did not contain cellular mRNA, Nirenberg and Matthaei translated a synthetic mRNA, the homopolymer polyuridylate, or polyU. PolyU added to the cell-free extracts directed the synthesis of polyphenylalanine, suggesting that UUU is a codon for phenylalanine. Shortly thereafter, it was found that polycytidylate, polyC, is translated as polyproline, identifying CCC as a codon for proline. Similar experiments revealed that AAA is a codon for lysine and GGG is a codon for glycine.

In the mid-1960s, H. Gobind Khorana developed a method for synthesizing mRNA containing di- and trinucleotide repeats. Using a technique similar to that of Nirenberg and Matthaei, he was able to show that translation of polyAG yields a polypeptide with alternating arginine and glutamate residues, whereas translation of polyAGC yields a mixture of the polypeptides polyserine, polyalanine, and polyglutamine (different polypeptides are produced since the reading frames of the synthetic mRNA molecules are selected randomly). These studies not only helped crack the genetic code but also confirmed that codons are triplets and that the reading of codons is nonoverlapping and sequential. Eventually, these kinds of experiments allowed 61 of the 64 codons to be deciphered.

It took years of hard work to clarify the way that mRNA encodes proteins. The subsequent development of methods for sequencing genes and proteins has allowed the direct comparison of the amino acid sequences of proteins with the nucleotide sequences of their corresponding genes. Each time a new protein and its gene are characterized, the genetic code is confirmed, although there have been some surprises, as we will see in a moment.

Figure 29·3

The standard genetic code. The code is composed of 64 triplet codons whose mRNA sequences can be read from this chart. The left-hand column indicates the nucleotide found at the first (5′) position of the codon. The top row indicates the nucleotide found at the second (middle) position of the codon. The right column indicates the nucleotide found at the third (3′) position. Thus, the codon CAG encodes glutamine (Gln), and GGU encodes glycine (Gly). The codon AUG specifies methionine (Met) and is also used to initiate protein synthesis. STOP indicates a termination codon.

| First position (5′ end) | Second position | | | | Third position (3′ end) |
	U	C	A	G	
U	Phe	Ser	Tyr	Cys	U
	Phe	Ser	Tyr	Cys	C
	Leu	Ser	STOP	STOP	A
	Leu	Ser	STOP	Trp	G
C	Leu	Pro	His	Arg	U
	Leu	Pro	His	Arg	C
	Leu	Pro	Gln	Arg	A
	Leu	Pro	Gln	Arg	G
A	Ile	Thr	Asn	Ser	U
	Ile	Thr	Asn	Ser	C
	Ile	Thr	Lys	Arg	A
	Met	Thr	Lys	Arg	G
G	Val	Ala	Asp	Gly	U
	Val	Ala	Asp	Gly	C
	Val	Ala	Glu	Gly	A
	Val	Ala	Glu	Gly	G

29·2 The Genetic Code Is Degenerate and Almost Universal

By convention, the genetic code (like all nucleic acid sequences) is written in the 5′→3′ direction. Thus, UAC specifies tyrosine, and CAU specifies histidine. While the term *codon* usually refers to triplets of nucleotides in mRNA, it can also apply to triplets of nucleotides in the DNA sequence of a protein-encoding gene. For example, TAC is a DNA codon for tyrosine. As we discuss in the next chapter, codons are translated in the 5′→3′ direction.

The standard genetic code shown in Figure 29·3 is shared by nearly all living organisms. This code, like the alternate genetic codes we examine in Section 29·4, has several prominent features:

1. There are several codons for most amino acids. For example, there are six codons for serine, four for glycine, and two for lysine. Because of the existence of several codons for most amino acids, the genetic code is said to be **degenerate.** Different codons that specify the same amino acid are known as **synonymous codons.** The degeneracy of the genetic code minimizes the effects of mutations since the change of a single nucleotide often results in a codon that still specifies the same amino acid. In the standard genetic code, the only amino acids with single codons are methionine and tryptophan.

2. The first two nucleotides of a codon are often enough to specify a given amino acid. For example, the four codons for glycine all begin with GG: GGU, GGC, GGA, and GGG. Therefore, mutations that alter the 3′ position of these codons usually result in the same amino acid being incorporated into the protein.

3. Codons with similar sequences often specify chemically similar amino acids. For example, the codons for the amino acid threonine differ from four of the codons for serine by only a single nucleotide at the 5′ position, and the codons for the amino acids aspartate and glutamate all begin with GA. In addition, codons that have pyrimidines at their second position usually encode hydrophobic amino acids. Therefore, mutations that alter either the 5′ or 3′ position of these codons usually result in a chemically similar amino acid being incorporated into the protein.

4. Only 61 of the 64 codons specify amino acids. The three remaining codons (UAA, UGA, and UAG) are **termination codons,** or stop codons. In classical genetic terminology, these termination codons are referred to as ochre (UAA), opal (UGA), and amber (UAG) codons. Termination codons are not normally recognized by any tRNA molecule. Instead, they are recognized by specific proteins that cause newly synthesized peptides to be released from the translation machinery. The methionine codon, AUG, usually specifies the initiation site for protein synthesis and is often called the **initiation codon,** but other codons can also be initiation codons (we discuss the initiation of protein synthesis further in Chapter 30).

5. Each codon has only one meaning. (There are exceptions to this rule for initiation and termination codons; we examine these exceptions later in this chapter and in Chapter 30.)

29·3 Different Coding Regions Can Overlap

Although the genetic code is nonoverlapping (that is, individual codons in the same reading frame are composed of distinct, unshared nucleotides), an mRNA molecule may contain two or three functional coding regions that do overlap. Such RNA molecules are produced when regions of the chromosome that contain overlapping genes are transcribed.

As discussed in Chapter 27, many *E. coli* genes are organized into operons. In operons, the genes are close together, and the coding regions of adjacent genes may overlap. In the examples shown in Figure 29·4, the last nucleotide of the termination codon is also the first nucleotide of the next initiation codon. Short regions of gene overlap are common in bacteria.

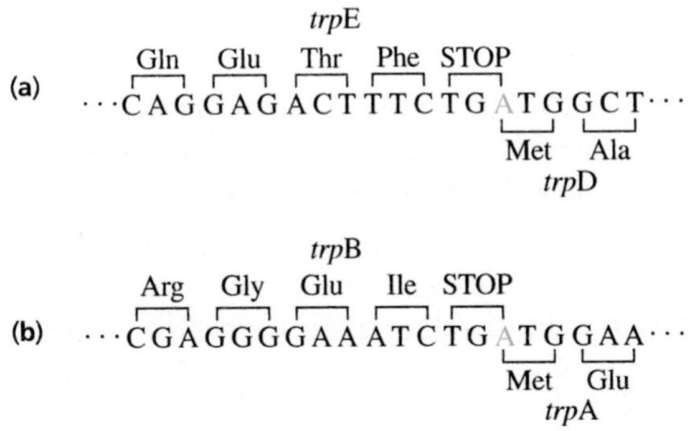

Figure 29·4
Overlapping genes in the *trp* operon of *E. coli*. (a) The last nucleotide in the reading frame of *trp*E overlaps the first nucleotide of the initiation codon of *trp*D. (b) The *trp*B and *trp*A genes also overlap by a single nucleotide.

Figure 29·5
Overlapping genes in φX174. **(a)** The organization of genes in φX174. **(b)** The nucleotide sequence at the junction of the K, A, and C genes. Only one strand is shown. The A residue shown in blue is used in all three reading frames.

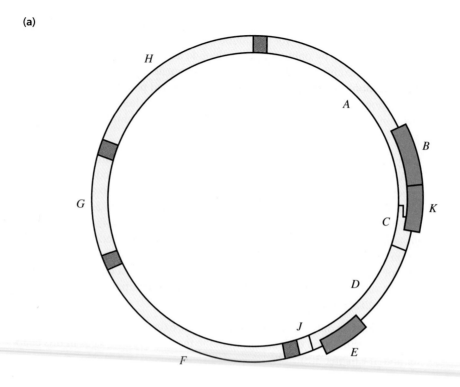

(a)

(b)

| Ile | Ser | Asp | Glu | Ser | | K |

T A T T T C T G A T G A G T C G A

Tyr Phe STOP   A

Met Ser Arg   C

More extensive regions of gene overlap are found in some viral genomes. The organization of genes in bacteriophage φX174 is shown in Figure 29·5. Gene *D* completely overlaps the shorter gene *E*, and gene *B* lies within gene *A*. (Here, we are using the word *gene* to mean coding region.) Genes *A* and *C* overlap by one nucleotide, and gene *K* overlaps the junction of genes *A* and *C*. All of the overlapping genes in φX174 are transcribed in the same direction; multiple proteins are synthesized by translating the single mRNA in more than one reading frame. In the case of genes *A, C,* and *K*, there is one nucleotide that is used in all three reading frames.

There are also examples of genes that encode two proteins with identical N-terminal amino acid sequences but different C-termini due to a shift in the reading frame during translation (these are discussed further in Section 30·10D). Finally, it is possible to have overlapping genes that are transcribed in opposite directions so that separate mRNA molecules are made from each gene. Any part of the chromosomal DNA is thus potentially capable of encoding six different proteins, although there are no known examples of such efficient use of genetic information.

29·4 There Are a Few Alternate Genetic Codes

The standard genetic code is found in nearly all organisms, but there are a few exceptions. In algae and some protozoa, for example, the codons UAA and UAG are

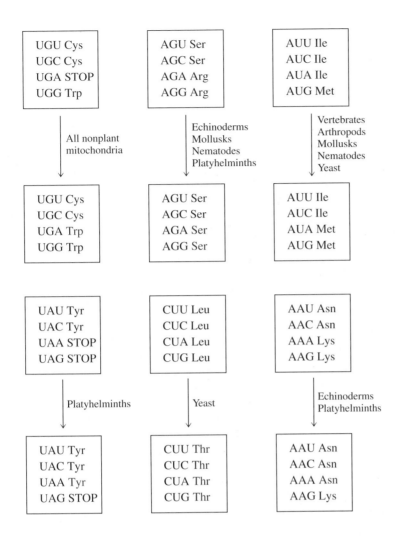

Figure 29·6
Alternate genetic codes found in nonplant mitochondria. [Adapted from Adams, R. L. P., Knowler, J. T., and Leader, D. P. (1992). *The Biochemistry of the Nucleic Acids,* 11th ed. (New York: Chapman and Hall), p. 525.]

not termination codons but instead specify glutamine. Similarly, in some prokaryotes, the termination codon UGA is used as a tryptophan codon. Such variants are rare in prokaryotes and in the nuclear genomes of eukaryotes.

Variation is more common in the mitochondria of some organisms. Recall that mitochondria and chloroplasts contain their own DNA. In plant mitochondria and in chloroplasts, the standard genetic code is used. But in the mitochondria of all other organisms, there are some deviations from the standard code (Figure 29·6). Mitochondrial genomes contain a limited number of tRNA genes, and it has been suggested that the altered genetic code allows translation to proceed efficiently with only this subset of tRNA molecules. Note, however, that there is no single mitochondrial genetic code; it appears that the variations have arisen independently in many lines of descent. Furthermore, mitochondria in plants and chloroplasts in both plants and algae use the standard code even though they also have a limited subset of tRNA genes. Thus, it is not obvious why there are alternate genetic codes in only some species.

Finally, the termination codon UGA sometimes specifies the rare amino acid selenocysteine in both prokaryotes and eukaryotes (Figure 29·7). Selenocysteine is found in only a few proteins, including formate dehydrogenases in *E. coli* and some peroxidases in mammals. Selenocysteine is not created by posttranslational modification but is incorporated into proteins in the same manner as the more common amino acids (selenocysteine insertion is discussed further in Section 30·12).

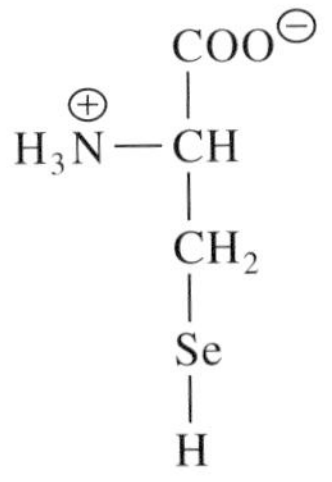

$$H_3\overset{\oplus}{N}-\underset{\underset{\underset{\underset{H}{|}}{Se}}{\underset{|}{CH_2}}}{\overset{\overset{COO^{\ominus}}{|}}{CH}}$$

Figure 29·7
Selenocysteine. Selenocysteine is a derivative of serine, with the oxygen of the hydroxyl group replaced by selenium (Se).

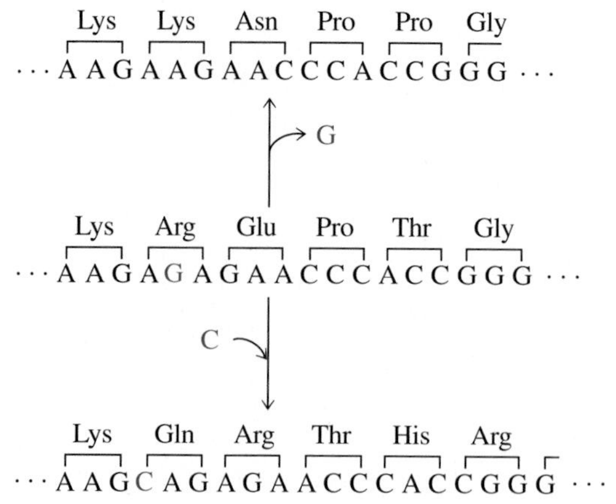

```
       Lys   Lys   Asn   Pro   Pro   Gly
      ┌─┐   ┌─┐   ┌─┐   ┌─┐   ┌─┐   ┌─┐
···A A G A A G A A C C C A C C G G G···
```

```
                    ↑→ G
```

```
       Lys   Arg   Glu   Pro   Thr   Gly
      ┌─┐   ┌─┐   ┌─┐   ┌─┐   ┌─┐   ┌─┐
···A A G A G A G A A C C C A C C G G G···
```

```
             C ↘
                    ↓
```

```
       Lys   Gln   Arg   Thr   His   Arg
      ┌─┐   ┌─┐   ┌─┐   ┌─┐   ┌─┐   ┌─┐
···A A G C A G A G A A C C C A C C G G G···
```

Figure 29·8
Changing the reading frame by frameshift mutation. Insertion or deletion of a residue changes the reading frame and thus the entire message downstream of the site of the mutation.

29·5 Mutations Can Alter the Sense of Codons

As we saw in Chapter 26, DNA is subject to a constant barrage of environmental influences that can cause mutations—alterations in the sequence of DNA. Errors in DNA replication are also important mutagenic events. Mutations can change the sense of codons by changing some nucleotides or by inserting or deleting others.

Some mutations involve the substitution of one nucleotide for another. When these substitution mutations alter the amino acid specified by the codon, they are called **missense mutations.** One example of missense mutation is the change from A to T in the middle of the GAA codon, which converts it from a codon for glutamate to a codon for valine. This is the mutation that changes normal hemoglobin to sickle-cell hemoglobin. Many mutations do not affect the sense of a codon because of degeneracy in the genetic code. Mutations that result in incorporation of the same amino acid are often referred to as neutral mutations because it is assumed that they have no effect on natural selection. This may not be a valid assumption, however, a point we will explore further when we talk about the use of synonymous codons in Section 29·7C.

When a mutation changes a codon from one that specifies an amino acid to a termination codon, the result is premature termination of protein synthesis. Such changes are called **nonsense mutations** and the resulting codon a **nonsense codon.** A nonsense mutation usually results in the synthesis of a truncated, nonfunctional polypeptide.

Some mutations involve the insertion or deletion of nucleotides. We noted earlier that the codons of a gene are all read in the same reading frame. Since a transition or transversion simply substitutes one nucleotide for another, it does not normally change the reading frame. However, the addition or deletion of a single nucleotide (or any number of nucleotides not divisible by three) changes the reading frame. These mutations are therefore called **frameshift mutations** (Figure 29·8). By adding or deleting nucleotides, frameshift mutations not only affect the codon in which they occur but also change the translation of the genetic code for all codons downstream of the mutation.

In most genes, only the reading frame that encodes the protein can be read for long stretches without encountering a termination codon. Such stretches are referred to as **open reading frames.** The other reading frames generally contain termination codons at frequent intervals. Thus, frameshift mutations usually result not only in the incorporation of incorrect amino acids but also in premature termination, both of which lead to production of inactive protein. One example of this occurs in the gene for the human glycosyltransferase enzymes, which are responsible for ABO blood types in humans (Section 12·5B). Individuals with type O blood carry a frameshift mutation at codon 86 of the gene for type A glycosyltransferase (Figure 29·9). As a result of this mutation, different amino acids are incorporated

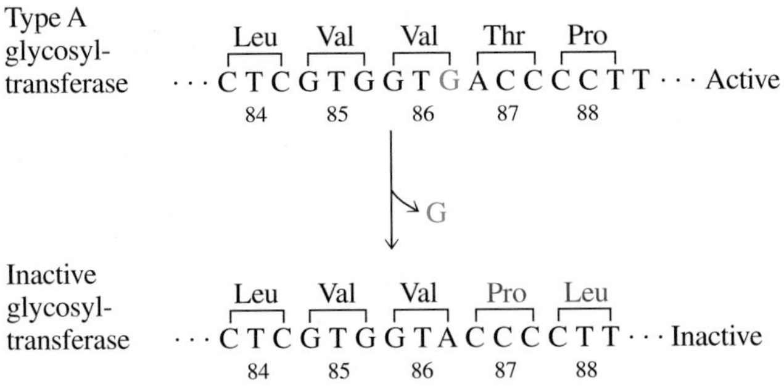

Type A
glycosyl-
transferase

Leu	Val	Val	Thr	Pro

···C T C G T G G T G A C C C C T T··· Active

84 85 86 87 88

↓ G

Inactive
glycosyl-
transferase

Leu	Val	Val	Pro	Leu

···C T C G T G G T A C C C C T T··· Inactive

84 85 86 87 88

Figure 29·9
Frameshift mutation in the gene for human type A glycosyltransferase. The deletion of a single G residue in codon 86 of the gene leads to an inactive protein. This change is responsible for the O blood group in humans.

into the protein after amino acid 86, and the protein truncates 30 amino acids later, when a termination codon is encountered in the new reading frame. The protein produced is only about one-third the size of type A glycosyltransferase and is inactive. Alternative splicing pathways can also result in altered reading frames in one of the products, which can again lead to premature termination of protein synthesis.

29·6 RNA Editing Can Change the Sense of a Message

In most cases, a protein-encoding gene is faithfully transcribed, and the mature mRNA product is translated without undergoing modifications that affect the reading frame. Exceptions occur when alternative processing pathways are used, but in such cases, the correct open reading frame is apparent from an examination of the nucleotide sequence of the gene and a knowledge of intron/exon boundaries and transcription termination sites. However, there are also several examples of direct RNA editing that involve posttranscriptional changes in the coding region of mRNA.

There are two general classes of RNA editing mechanisms: nucleotide modifications and insertions/deletions. Examples of nucleotide modifications include conversion of adenylate to guanylate or inosinate (I) in some vertebrate mRNA molecules. These modifications can alter the sense of the codons containing the original adenylate (I can base pair with U, A, or C). Such modifications are highly specific, with only particular adenylates modified.

Cytidylate-to-uridylate conversions have been found in mitochondrial and chloroplast mRNA molecules. Some of these conversions, particularly in chloroplasts, were originally thought to be examples of alternate genetic codes. For example, several chloroplast genes contain the arginine codon CGG at a position where a tryptophan codon was expected based on a comparison with homologous genes in nonplant species. It was therefore assumed that the chloroplast genetic code used CGG as a tryptophan codon. However, it turns out that a C-to-U conversion during RNA editing converts the CGG codon to the normal tryptophan codon, UGG.

The same type of conversion occurs in the mRNA for apolipoprotein B (apoB) in mammals. ApoB is a lipid-binding protein that exists in two forms: apoB-100 is one of the largest known proteins (4536 amino acid residues); apoB-48 contains only the N-terminal 2152 residues found in the larger form. ApoB-48 is produced from an mRNA molecule that contains a termination codon (UAA) at position 2153, a point at which the mRNA for apoB-100 contains a glutamine codon (CAA). The termination codon is created by posttranscriptional modification of cytidylate to uridylate, probably by deamination or transamination. The enzyme responsible for this modification of cytidylate binds to a specific RNA sequence on the 3′ side of the site of modification. Only one particular cytidylate is modified. In intestinal cells, about 90% of the apoB mRNA is edited, but in the liver, most of the mRNA is unedited (except in rodent liver, where equal amounts of edited and unedited forms are found).

(a)

(b)

Figure 29·10
RNA editing in mitochondrial RNA of *Try-panosoma brucei*. **(a)** The primary transcript. **(b)** The edited mRNA. Inserted U residues are highlighted in red.

Examples of insertions and deletions of nucleotides due to RNA editing are found in a few viruses and in some mitochondria. An extreme example is seen in some mRNA molecules in the mitochondria of trypanosomes, where hundreds of U residues are inserted into the mature mRNA. Such RNA editing requires the presence of a guide RNA (gRNA) that specifies the correct sequence of the edited mRNA. Figure 29·10 illustrates editing of the mRNA for cytochrome *c* oxidase subunit 2 found in the mitochondria of *Trypanosoma brucei*. The original transcript pairs with the guide RNA, which acts as a template for the insertion of U residues at specific sites. Pairing occurs at an "anchor" sequence adjacent to the site of insertion. The inserted uridylate residues are derived from the 3′ end of the gRNA, and the mechanism probably involves a transesterification reaction mediated by enzymes that form a complex with the gRNA and the unedited mRNA. The reaction is similar to splicing because the unedited transcript must be cut and ligated during insertion of nucleotides.

This form of RNA editing is rare and has no obvious function except to conserve space in the mitochondrial genome. The vast majority of organisms do not exhibit this form of RNA editing, suggesting it does not confer significant selective advantage. Insertions and deletions of nucleotides seem to be confined to viruses and the mitochondria of trypanosomes and slime molds. These organisms are only distantly related, and it is likely that this form of RNA editing has arisen independently in each lineage.

29·7 Transfer RNA Molecules Are Required for Protein Synthesis

Transfer RNA molecules serve as interpreters of the genetic code. They are the crucial link between the sequence of nucleotides in mRNA and the sequence of amino acids in a polypeptide. In order for tRNA to fulfill the role of adapter, every cell must contain at least twenty tRNA species (one for every amino acid), and each of these tRNA molecules must be able to recognize at least one mRNA codon.

A. All tRNA Molecules Have a Similar Three-Dimensional Structure

The nucleotide sequences of different tRNA molecules from many organisms have been determined. Despite diversity in their primary structures, the sequences of almost all analyzed tRNA molecules are compatible with the secondary structure shown in Figure 29·11. This cloverleaf structure is subdivided into a hydrogen-bonded stem and several arms, regions composed of a stem and a single-stranded

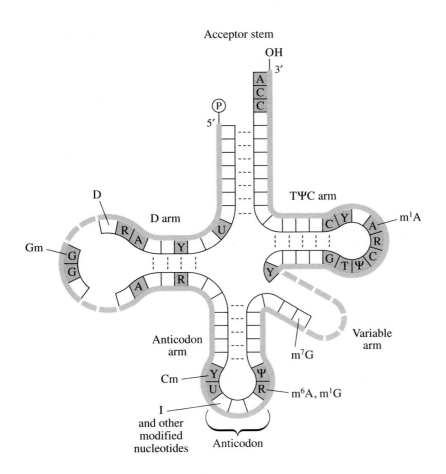

Acceptor stem

Figure 29·11
Cloverleaf secondary structure of tRNA. Watson-Crick base pairing is indicated by dashed lines between nucleotide residues. The molecule is divided into an acceptor stem and four arms. The acceptor stem is the site of amino acid attachment, and the anticodon arm is the region of the tRNA molecule that interacts with mRNA. The D and TΨC arms are named for modified nucleotides that are conserved within the arms. The number of nucleotide residues in each arm is more or less constant, except in the variable arm. Conserved bases (colored gray) and positions of commonly modified nucleotides are noted. Abbreviations other than standard nucleotides: R, a purine nucleotide; Y, a pyrimidine nucleotide; m^1A, 1-methyladenylate; m^6A, N^6-methyladenylate; Cm, 2'-*O*-methylcytidylate; D, dihydrouridylate; Gm, 2'-*O*-methylguanylate; m^1G, 1-methylguanylate; m^7G, 7-methylguanylate; I, inosinate; Ψ, pseudouridylate; and T, thymidylate.

loop. The double-stranded region of each arm forms a short, stacked, right-handed helix, similar to that of the A form of DNA.

The 5′ end and the region near the 3′ end of the tRNA molecule are base paired, forming a stem known as the **acceptor stem,** or amino acid stem. It is to this region that the amino acid is covalently attached. The carboxyl group of the amino acid is linked to either the 2′- or 3′-hydroxyl group of the ribose of the 3′ adenylate. (Recall from Section 28·1 that mature tRNA molecules are produced by processing a larger primary transcript and that the nucleotides at the 3′ end of a mature tRNA molecule are invariably CCA.) The nucleotide at the 5′ end of all tRNA molecules is phosphorylated.

The single-stranded loop opposite the acceptor stem is the **anticodon loop.** This loop contains the **anticodon,** the three-base sequence that binds to the complementary codon in mRNA. The arm of the tRNA molecule that contains the anticodon is referred to as the **anticodon arm.** Two of the arms of the tRNA molecule are named for the covalently modified nucleotides found within them (see Figure 28·9 for examples of these modified nucleotides). One of these arms contains thymidylate (T) and pseudouridylate (Ψ) followed by cytidylate (C) and is referred to as the **TΨC arm.** Dihydrouridylate (D) residues lend their name to the **D arm.** tRNA molecules also have a **variable arm** between the anticodon arm and the TΨC arm. The variable arm can range in length from about 3 to 21 nucleotides, and the length of the D arm is also somewhat variable. With a few rare exceptions, most tRNA molecules are between 73 and 95 nucleotides in length.

The presence of modified nucleotides in tRNA affects the secondary structure of the molecule. For example, dihydrouridine is more bulky than uridine due to the presence of extra hydrogen atoms (see Figure 28·9), and the ring atoms in this molecule are no longer coplanar. This in turn means that D residues are not stable in stacked, double-stranded regions. Thus, dihydrouridine is a helix breaker and is found exclusively in single-stranded loops.

Conservation of the nucleotide sequence in the D arm is more pronounced among eukaryotes than between prokaryotes and eukaryotes. This conservation is related to the regulation of transcription of eukaryotic tRNA genes. Recall from Chapter 27 that class III transcription factors, notably TFIIIC, bind to the internal control regions of eukaryotic tRNA genes. The TFIIIC binding site on the gene corresponds to the region that encodes the D loop in tRNA. The conservation of the internal control region gives rise to a conserved sequence of tRNA in these organisms.

The cloverleaf diagram of tRNA is a two-dimensional representation of a three-dimensional molecule. In three dimensions, the tRNA molecule is folded into an L shape as shown in Figures 29·12 and 29·13. The acceptor stem is at one end of the L-shaped molecule, and the anticodon loop is located at the opposite end. The

Figure 29·12
Relationship of the secondary and tertiary structures of tRNA. The ribose-phosphate backbone is indicated by a continuous ribbon. **(a)** Cloverleaf secondary structure. **(b)** The tertiary structure of tRNA involves base pairing and van der Waals interactions between the TΨC arm and the D arm and **(c)** between the variable arm and the D arm. Mature tRNA molecules are L shaped.

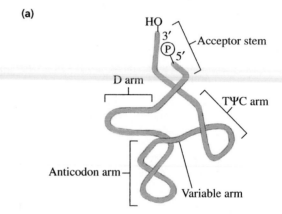

(a)

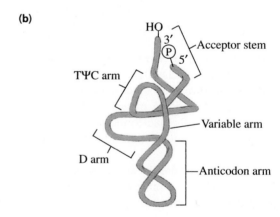

(b)

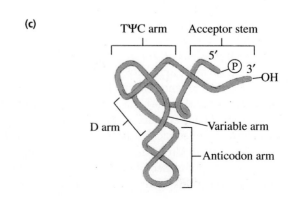

(c)

Figure 29·13
Stereo view of yeast tRNA^Phe. The residues are shown in green except for the anticodon, which is shown in blue. The colored ribbon highlights the sugar-phosphate backbone and the different regions of secondary structure. Color key for ribbon: acceptor stem, orange; TΨC arm, purple; D arm, red; variable arm, yellow; anticodon arm, blue. (Based on coordinates provided by E. Westhof and M. Sundaralingam.)

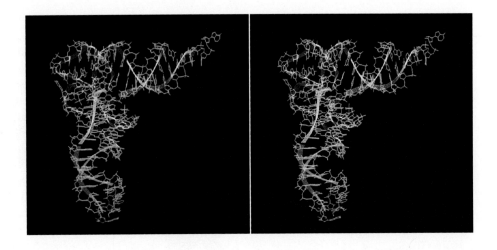

resulting structure is compact and very stable, in part because of hydrogen bonds between the nucleotides in the D, TΨC, and variable arms. This base pairing differs from the normal Watson-Crick base pairing that occurs between nucleotides in double-stranded regions (Figure 29·14). Most of the nucleotides in tRNA are part of

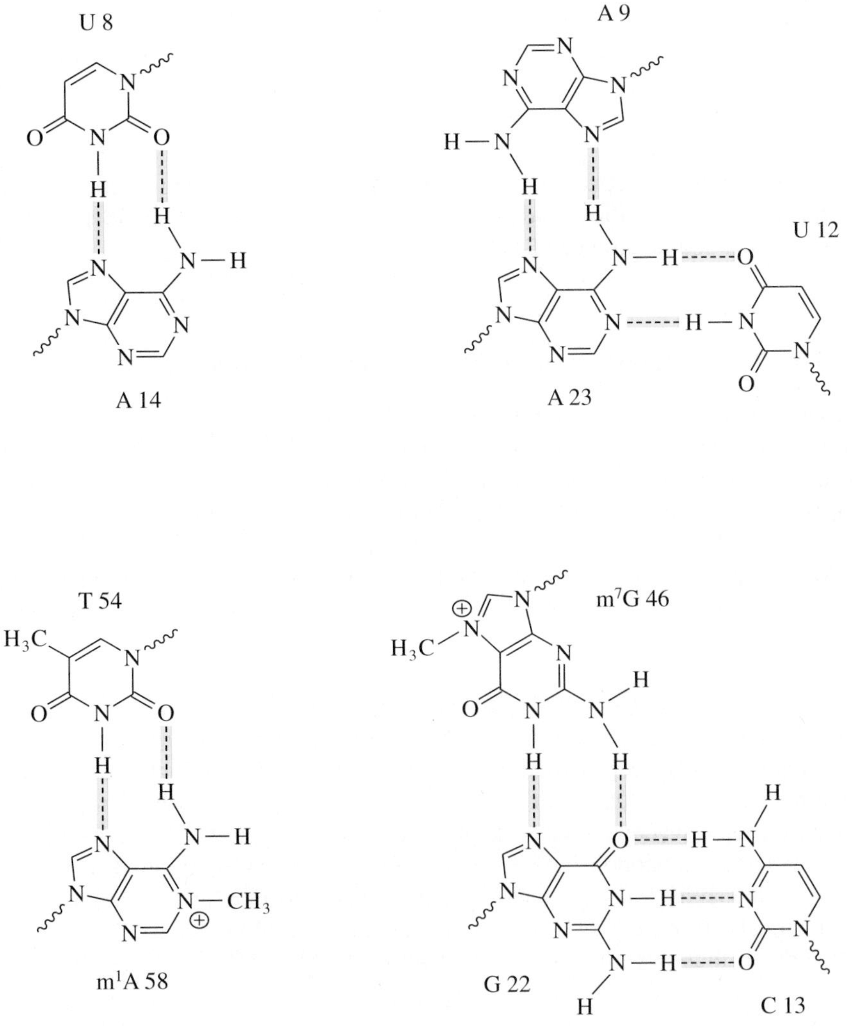

Figure 29·14
Examples of non–Watson-Crick base pairing in the tertiary structure of tRNA^Phe. Note that this type of base pairing differs from conventional Watson-Crick A/T and G/C base pairing. This non–Watson-Crick base pairing can involve bonding among three bases. The numbers next to the abbreviations for the nucleotides refer to the positions of the nucleotides in the primary structure of the yeast tRNA^Phe molecule. m^1A is 1-methyladenylate; m^7G is 7-methylguanylate.

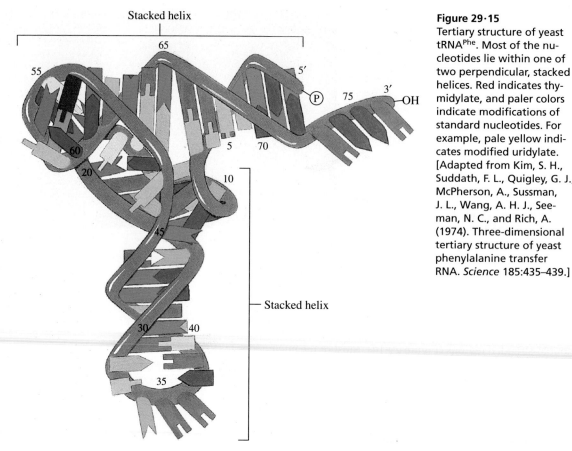

Stacked helix

Stacked helix

Figure 29·15
Tertiary structure of yeast tRNA[Phe]. Most of the nucleotides lie within one of two perpendicular, stacked helices. Red indicates thymidylate, and paler colors indicate modifications of standard nucleotides. For example, pale yellow indicates modified uridylate. [Adapted from Kim, S. H., Suddath, F. L., Quigley, G. J., McPherson, A., Sussman, J. L., Wang, A. H. J., Seeman, N. C., and Rich, A. (1974). Three-dimensional tertiary structure of yeast phenylalanine transfer RNA. *Science* 185:435–439.]

two perpendicular, stacked helices (Figure 29·15); the stacking interactions between the bases are additive and make a major contribution to the stability of tRNA, as they do in double-stranded DNA.

B. The tRNA Anticodons Base-Pair with mRNA Codons

tRNA and mRNA molecules interact through base pairing between anticodons and codons. The anticodon of a tRNA molecule therefore determines where the amino acid attached to its acceptor stem will be added to a growing polypeptide chain. A superscript is used to designate which amino acid the anticodon of the tRNA specifies. For instance, the tRNA molecule shown in Figure 29·15 has the anticodon GAA, which binds to the phenylalanine codon UUC. During aminoacylation, phenylalanine is covalently attached to the acceptor stem of this tRNA. The molecule is therefore designated tRNA[Phe].

Much of the base pairing between the codon and the anticodon is governed by the rules of Watson-Crick base pairing: A pairs with U, G pairs with C, and the strands in the base-paired region are antiparallel. There is some variation, however. In 1966, Francis Crick proposed that complementary, Watson-Crick base pairing is required for only two of the three base pairs formed when the codon interacts with the anticodon. The codon must form Watson-Crick base pairs with the 3′ and middle bases of the anticodon, but other types of base pairing are permitted at the 5′ position of the anticodon. This alternate pairing suggests that the 5′ position is conformationally flexible. Crick dubbed this flexibility **wobble.** The 5′ position of the anticodon is thus sometimes called the **wobble position.**

Table 29·1 Predicted base pairing between the 5′ (wobble) position of the anticodon and the 3′ position of the codon

Nucleotide at 5′ (wobble) position of anticodon	Nucleotide at 3′ position of codon
C	G
A	U
U	A or G
G	U or C
I*	U, A, or C

*I = Inosinate

Crick's prediction of the base pairs that could be formed between the wobble position of an anticodon and the 3′ position of an mRNA codon are shown in Table 29·1. When G is at the wobble position, for example, it can pair with either U or C, and similarly, when U is at the wobble position, it can pair with either G or A. The nucleotide at the wobble position of many anticodons is often covalently modified, permitting additional flexibility in codon recognition. For example, G at the 5′ anticodon position in several tRNA molecules is deaminated at C-2 to form I, which can hydrogen bond with A, C, or U (Figure 29·16). The presence of I at the 5′ position of the anticodon allows tRNA^Ala with the anticodon IGC to bind three codons specifying alanine: GCU, GCC, and GCA (Figure 29·17).

Figure 29·16
Inosinate base pairs. Inosinate (I) is often found at the 5′ (wobble) position of a tRNA anticodon. I can form hydrogen bonds with A, C, or U. This versatility in hydrogen bonding allows an anticodon to recognize more than one synonymous codon.

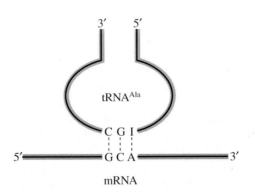

Figure 29·17
Base pairing at the wobble position. The tRNA^Ala molecule with the anticodon IGC can bind to any one of three codons specifying alanine (GCU, GCC, or GCA) because I can pair with U, C, or A. Note that the codon-anticodon interaction involves antiparallel RNA strands.

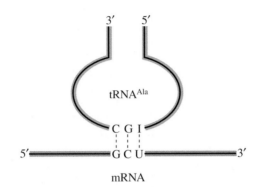

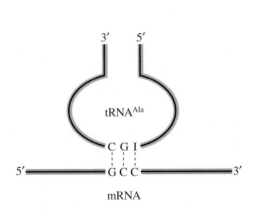

Although Table 29·1 indicates that A in the wobble position could potentially form a base pair with U, there are only a few examples of tRNA molecules with an A residue at the 5′ position of the anticodon. There is some evidence to suggest that when A occupies this position, it pairs with nucleotides other than U. In fact, A may be excluded from the wobble position because it discriminates poorly. Similarly, U is rarely found in the 5′ position of the anticodon, possibly because it may form base pairs with all nucleotides and not just A or G. However, modified U is common in the 5′ position; the modifications may enhance the specificity of base pairing so that usually A and sometimes G are recognized.

The base of the nucleotide on the 3′ side of the anticodon is almost always a purine, usually a modified purine (yeast tRNA^{Phe} is an exception). In some cases, the modified nucleotide gives rise to enhanced stacking interaction in the anticodon, which helps lock the nucleotides into a conformation that can form normal Watson-Crick base pairs with only the first two nucleotides of the mRNA codon.

C. Synonymous Codons Are Not Used Equally

Wobble allows some tRNA molecules to recognize more than one codon, but often several different tRNA molecules are required to recognize *all* synonymous codons. Different tRNA molecules that bind the same amino acid are called **isoacceptor tRNA molecules.** The term *isoacceptor* applies to tRNA molecules with different anticodons that bind the same amino acid, as well as to tRNA molecules with the same anticodon but different nucleotide sequences.

The genes that encode isoacceptor tRNA molecules are not expressed equally. For example, in *E. coli,* the alanine isoacceptor tRNA that recognizes the codon GCC is present at about one-tenth the concentration of the alanine isoacceptor tRNA that recognizes the codon GCU. Not only are the genes encoding isoacceptor tRNA molecules expressed unequally, but synonymous codons are used unequally. Consider the genes for the ribosomal proteins of *E. coli,* in which synonymous codons occur at remarkably different frequencies (Table 29·2). Among a total of

Table 29·2 Frequency of codon usage in *E. coli* ribosomal proteins

Arg	CGU	48	Val	GUU	54	Pro	CCU	3	His	CAU	3
	CGC	26		GUC	6		CCC	0		CAC	15
	CGA	0		GUA	40		CCA	4		Total	18
	CGG	0		GUG	16		CCG	36			
	AGA	1		Total	116		Total	43	Gln	CAA	9
	AGG	0								CAG	33
	Total	75	Ile	AUU	13	Thr	ACU	36		Total	42
				AUC	51		ACC	26			
Ser	UCU	18		AUA	0		ACA	3	Asn	AAU	3
	UCC	18		Total	64		ACG	0		AAC	42
	UCA	1					Total	65		Total	45
	UCG	1	Cys	UGU	1						
	AGU	1		UGC	6	Ala	GCU	93	Lys	AAA	90
	AGC	12		Total	7		GCC	10		AAG	24
	Total	51					GCA	45		Total	114
			Tyr	UAU	3		GCG	28			
				UAC	13		Total	176	Asp	GAU	17
Leu	UUA	1		Total	16					GAC	45
	UUG	2				Gly	GGU	49		Total	62
	CUU	4	Glu	GAA	61		GGC	34			
	CUC	3		GAG	16		GGA	0	Phe	UUU	10
	CUA	0		Total	77		GGG	0		UUC	23
	CUG	79					Total	83		Total	33
	Total	89	Trp	UGG	3						
Met	AUG	30									

176 codons specifying alanine, the codon GCU is used 93 times, the codon GCA 45 times, the codon GCG 28 times, and the codon GCC only 10 times. In general, the tRNA molecule corresponding to the codon used most often is present in considerable excess over its isoacceptors.

Although the reason some isoacceptor tRNA molecules are more abundant than others is not known, the effects of differential codon usage are apparent. If a gene contains a rare codon whose corresponding tRNA molecule is equally rare, the protein synthesis machinery may pause when translating that mRNA codon. In the case of prokaryotes, ribosomes normally follow closely behind the transcription complex. The slowing of protein synthesis might allow the transcription complex to pull ahead of the ribosome and thereby expose a rho binding site on the mRNA. Because the presence of a rare codon may cause premature termination of transcription in such cases, rare codons may be involved in the regulation of gene expression.

In general, the genes encoding abundant proteins contain codons recognized by the most abundant tRNA molecules. However, the most frequently used codons in one organism often differ from the most frequently used codons in another organism. For example, the codon usage tables for mammals and yeast are different from that shown for *E. coli*. Codon usage is also correlated with the overall base composition of the genome. Organisms containing DNA with a high A/T content, for example, preferentially use codons that are rich in A and T. These observations mean that not all synonymous codons are equal, and mutations that convert one codon into another synonymous codon are not necessarily neutral.

D. Some Suppressor tRNA Molecules Contain Altered Anticodons

Earlier we noted that there are three termination codons that normal tRNA molecules do not recognize. These termination codons, however, can be recognized by mutant tRNA molecules called **suppressor tRNA molecules** (because they can "suppress" the effect of nonsense mutations). A suppressor tRNA molecule may contain an altered anticodon that allows it to bind to a termination codon as if it were a codon for an amino acid. This in turn results in incorporation of the amino acid carried by the tRNA into the growing polypeptide chain. It is usually the genes for minor tRNA species that are mutated to suppressors.

The mutation that gives rise to a suppressor tRNA molecule not only allows the cell to overcome the effect of the nonsense mutation but also results in readthrough of some normal termination codons. Because of this readthrough, cells carrying suppressor tRNA genes are often less healthy than wild-type cells. In addition, any suppressor mutation also creates a tRNA species that can no longer recognize its usual codon. The functional loss of a tRNA species would be lethal unless the cell also contained an isoacceptor tRNA that could compensate for the loss.

Consider the recognition of the nonsense codon UAG by a suppressor tRNA molecule that allows insertion of tyrosine. The suppressor arises through a mutation in the anticodon of the normal tRNA^{Tyr} molecule that contains the anticodon GUA. The mutation results in a new tRNA^{Tyr} molecule with the anticodon CUA. A tRNA^{Tyr} so modified recognizes the nonsense codon UAG and inserts tyrosine into the growing protein. Thus, the suppressor tRNA^{Tyr} molecule prevents premature termination of protein synthesis. The existence of suppressor tRNA molecules illustrates the heart of the genetic code: the sense or nonsense of the information in DNA depends upon the machinery that assigns particular amino acids to particular codons—tRNA molecules. Equally important, however, are the enzymes that assign particular amino acids to particular tRNA molecules—aminoacyl-tRNA synthetases.

Table 29·3 Classification of aminoacyl-tRNA synthetases

Class I	Class II
Arg	Ala
Cys	Asn
Gln	Asp
Glu	Gly
Ile	His
Leu	Lys
Met	Phe
Trp	Pro
Tyr	Ser
Val	Thr

29·8 Aminoacyl-tRNA Synthetases Catalyze the Addition of Amino Acid Residues to tRNA Molecules

A particular amino acid is enzymatically attached to the 3′ end of each tRNA molecule. The product of this aminoacylation reaction is called an **aminoacyl-tRNA.** Because the aminoacyl-tRNA linkage is a high-energy linkage, the amino acid is said to be "activated" for subsequent transfer to a growing polypeptide chain. Similarly, aminoacyl-tRNA molecules are sometimes called "charged" or "activated" tRNA molecules. The specific type of aminoacylated tRNA molecule is indicated by a prefix; for instance, aminoacylated tRNA^{Ala} is called alanyl-tRNA^{Ala}. The enzymes that catalyze the aminoacylation reaction are called aminoacyl-tRNA synthetases.

Since 20 different amino acids are incorporated into proteins, at least 20 different aminoacyl-tRNA synthetases must be present in each cell. There are not many more than 20 aminoacyl-tRNA synthetases, however, because a single synthetase is able to recognize several isoacceptor tRNA molecules. For example, there are six codons for serine and several different isoacceptor tRNA^{Ser} molecules. However, each of these tRNA^{Ser} molecules is recognized by a single seryl-tRNA synthetase. The accuracy of protein synthesis depends on the ability of the aminoacyl-tRNA synthetases to catalyze the attachment of the correct amino acid to the appropriate tRNA species.

The aminoacyl-tRNA synthetases are divided into two classes (Table 29·3) based on nucleotide sequence and other structural similarities and on biochemical properties. The class I enzymes are very different from the class II enzymes, but within each class, the synthetases appear to have arisen from a common ancestor.

A. Aminoacyl-tRNA Molecules Are Synthesized in Two Steps

The activation of amino acids by aminoacyl-tRNA synthetases is similar in principle to the activation of fatty acids. The overall reaction is

$$\text{Amino acid} + \text{tRNA} + \text{ATP} \longrightarrow \text{Aminoacyl-tRNA} + \text{AMP} + \text{PP}_i \qquad \textbf{(29·1)}$$

The amino acid is attached to its tRNA molecule by formation of an ester linkage between the carboxylate group of the amino acid and the 2′- or 3′-hydroxyl group of the ribose at the 3′ end of the tRNA molecule. Recall that all tRNA molecules have the sequence –CAA at their 3′ termini, so the site of attachment is always the sugar moiety of an adenylate residue. In general, class I aminoacyl-tRNA synthetases catalyze a reaction that results in attachment of the amino acid to the 2′ carbon, whereas class II enzymes catalyze attachment to the 3′ position.

The aminoacylation reaction occurs in two discrete steps, as shown for a class II enzyme in Figure 29·18. In the first step, the aminoacyl-tRNA synthetase catalyzes formation of a reactive intermediate. The carboxylate group of the amino acid attacks the α-phosphorus of ATP, displacing pyrophosphate and producing an aminoacyl-adenylate.

$$\text{Amino acid} + \text{ATP} \longrightarrow \text{Aminoacyl-adenylate} + \text{PP}_i \qquad \textbf{(29·2)}$$

Formation of this high-energy intermediate activates the amino acid, and the aminoacyl-adenylate remains tightly but noncovalently bound to the aminoacyl-tRNA synthetase. Synthesis of the aminoacyl-adenylate is thermodynamically favored due to the subsequent hydrolysis of pyrophosphate inside the cell.

The second step of aminoacyl-tRNA synthesis is aminoacyl-group transfer from the intermediate to tRNA.

$$\text{Aminoacyl-adenylate} + \text{tRNA} \longrightarrow \text{Aminoacyl-tRNA} + \text{AMP} \qquad \textbf{(29·3)}$$

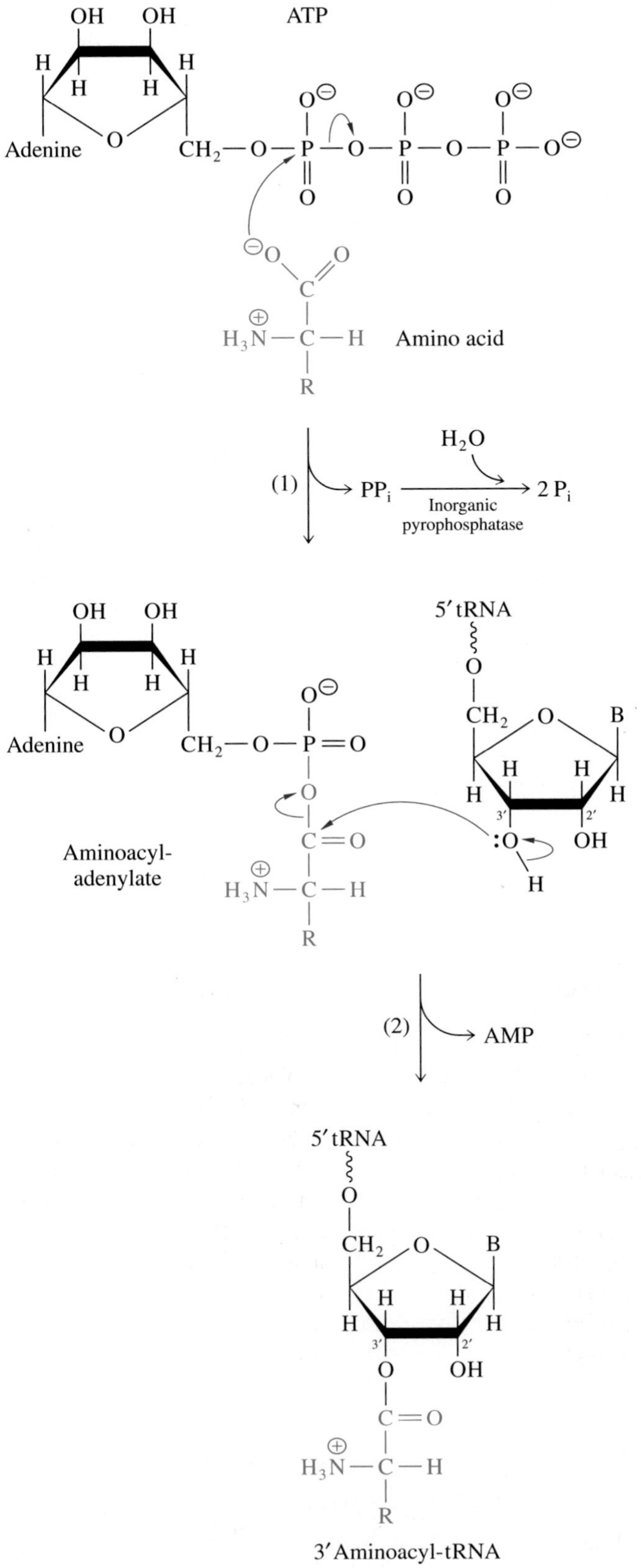

Figure 29·18
Synthesis of an aminoacyl-tRNA molecule catalyzed by a class II aminoacyl-tRNA synthetase. (1) The nucleophilic carboxylate group of the amino acid attacks the α-phosphorus atom of ATP, producing an aminoacyl-adenylate. (2) Nucleophilic attack of the 3′-hydroxyl group of the terminal residue of the tRNA leads to displacement of AMP and formation of an aminoacyl-tRNA molecule.

The mechanism of class I enzymes is similar to that shown for class II enzymes except that class I enzymes catalyze attachment of the amino acid to the 2′-hydroxyl group of the 3′-terminal adenylate residue of tRNA. In these cases, the amino acid is subsequently shifted to the 3′ position in a transesterification reaction. This is a crucial step; as we will see in the next chapter, the amino acid must be attached to the 3′ position for proper protein synthesis.

Under cellular conditions, the equilibrium for Reaction 29·1 lies overwhelmingly in the direction of aminoacyl-tRNA, and the intracellular concentration of free tRNA is very low. The energy stored in the aminoacyl-tRNA bond is ultimately used in the formation of a peptide bond during protein synthesis. Note that two high-energy bonds are consumed during each aminoacylation reaction.

B. Each Aminoacyl-tRNA Synthetase Is Highly Specific for One Amino Acid and Its Corresponding tRNA Molecules

Aminoacyl-tRNA synthetases must distinguish not only between amino acids but also between tRNA molecules. The attachment of a specific amino acid to its corresponding tRNA molecule is a critical step in the translation of information in nucleic acids. If it is not performed accurately, the wrong amino acid could be incorporated into a protein.

A given aminoacyl-tRNA synthetase first binds ATP and then selects the proper amino acid substrate based on the charge, size, and hydrophobicity of the amino acid. Amino acid selection initially involves exclusion of amino acids with inappropriate charges or hydrophobicity. For example, tyrosyl-tRNA synthetase readily distinguishes tyrosine from phenylalanine (see Section 7·16).

The aminoacyl-tRNA synthetase then selectively binds a specific tRNA molecule. The proper tRNA molecule is distinguished by features unique to its structure. In particular, the part of the acceptor stem that lies on the inside surface of the L of the tRNA molecule is implicated in the binding of tRNA to the corresponding

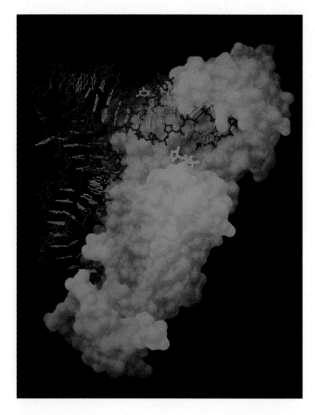

Figure 29·19
Structure of *E. coli* glutaminyl-tRNA synthetase bound to tRNA^{Gln}. The 3′ end of the tRNA is buried within a pocket on the surface of the enzyme. A molecule of ATP (green) is also bound at this site. The enzyme is interacting with the acceptor-stem region of the tRNA and also with the anticodon. (Courtesy of Thomas A. Steitz.)

aminoacyl-tRNA synthetase (Figure 29·19). In some cases, the synthetase recognizes the anticodon as well as the acceptor stem of the tRNA. In other cases, the binding of tRNA molecules to the appropriate aminoacyl-tRNA synthetase requires only interactions between the enzyme and the acceptor stem of the tRNA. In all cases, however, the net effect of the interaction is to position the 3′ end of the tRNA molecule in the active site of the enzyme.

Glutaminyl-tRNA synthetase, shown in Figure 29·19, is a class I enzyme. Since all class I enzymes have related structures, they probably bind tRNA molecules in a similar manner. The class II enzymes bind tRNA at the acceptor stem and at the anticodon as well, but they preferentially recognize the acceptor-stem region. Alanyl-tRNA synthetase, for example, binds and aminoacylates any tRNA with a G/U base pair at the third position in the acceptor stem, even when the anticodon loop has been deleted.

The way that aminoacyl-tRNA synthetases recognize tRNA molecules has been referred to as the second genetic code. In many cases, only a small number of specific nucleotides in tRNA are crucial for recognition, but in other cases, the aminoacyl-tRNA synthetase recognizes larger structural features of the cognate tRNA. The second genetic code does not appear to be a simple one, and it may not prove to be a useful concept.

C. Some Aminoacyl-tRNA Synthetases Have Proofreading Activity

Most aminoacyl-tRNA synthetases bind only a single amino acid; consequently, the error rate of aminoacylation is low for such enzymes. However, in a few cases, the enzyme is less specific. For example, isoleucine and valine are chemically similar amino acids, and both can be accommodated in the active site of isoleucyl-tRNA synthetase (Figure 29·20). Isoleucyl-tRNA synthetase mistakenly catalyzes the formation of the valyl-adenylate intermediate about one percent of the time. Based on this observation, we might expect valine to be attached to isoleucyl-tRNA and incorporated into protein in place of isoleucine about 1 time in 100. However, the substitution of valine for isoleucine in polypeptide chains occurs only about 1 time in 10 000. This lower rate of valine incorporation suggests that isoleucyl-tRNA synthetase also discriminates between the two amino acids *after* aminoacyl-adenylate formation. In fact, isoleucyl-tRNA synthetase catalyzes a proofreading step at the stage of the reaction in which the aminoacyl group is transferred from the aminoacyl-adenylate to the tRNA molecule. Although isoleucyl-tRNA synthetase may mistakenly catalyze the formation of valyl-adenylate, most of the time it catalyzes the hydrolysis of the incorrect valyl-adenylate to valine and AMP so that valyl-tRNA^{Ile} does not form.

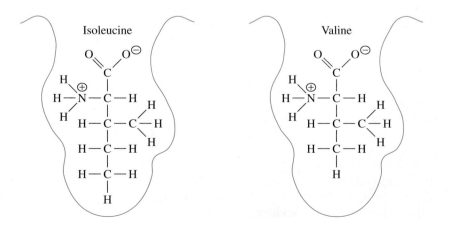

Figure 29·20
Model of the substrate binding site in isoleucyl-tRNA synthetase. Despite the similar size and charge of isoleucine and valine, isoleucyl-tRNA synthetase binds isoleucine about 100 times more readily than it binds valine. However, even this degree of accuracy is not sufficient for protein synthesis.

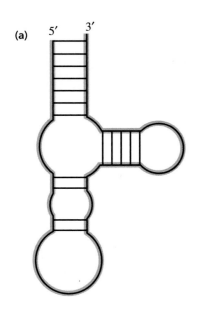

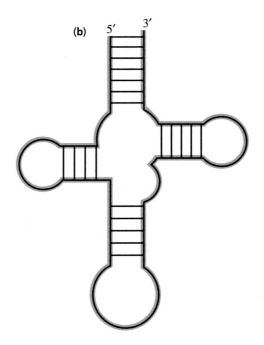

Figure 29·21
Proposed secondary structures of two unusual mitochondrial tRNA molecules based on the sequence of their genes. (a) tRNA^{Ser}. (b) tRNA^{Tyr}.

Of the 20 different synthetases, at least 7 have proofreading activity. Five synthetases lack proofreading activity, and the status of the remaining enzymes is unknown.

D. Mitochondria Have Their Own tRNA and Aminoacyl-tRNA Synthetases

As noted in Section 29·4, some mitochondria use a genetic code that is slightly different from that used by nuclear DNA of the same cell. In order to synthesize the few proteins encoded by mitochondrial DNA, mitochondria also possess unique tRNA molecules that recognize the codons of mitochondrial mRNA. In some cases, these mitochondrial tRNA molecules have unusual structures that do not resemble those of normal cytoplasmic tRNA molecules (Figure 29·21).

Some of the aminoacyl-tRNA synthetases that catalyze aminoacylation of mitochondrial tRNA molecules are found only in the mitochondrion. They are, however, encoded by the nuclear genome and are transported to the mitochondria after they are translated. In some cases, mitochondrial aminoacyl-tRNA synthetases are encoded by the same genes that encode cytoplasmic aminoacyl-tRNA synthetases, and the mitochondrial enzymes are generated as a result of differential splicing of the mRNA precursor. In other cases, the two forms of the enzyme are encoded by different genes.

E. Selenocysteine Is Synthesized from Seryl-tRNA

As mentioned earlier, selenocysteine is encoded by UGA, which is normally used as a termination codon. Cells contain a unique tRNA (tRNA^{Sec}) that is recognized by seryl-tRNA synthetase and charged with serine. This seryl-tRNA^{Sec} is subsequently modified by selenocysteine synthase, which converts the bound serine to selenocysteine. Thus, there is no specific selenocystyl-tRNA synthetase.

tRNA^{Sec} differs from the other seryl isoacceptor tRNA molecules in several important aspects. It has a longer acceptor stem, and it differs from all other tRNA molecules at a number of highly conserved positions. The variable arm is also very large compared to those of other tRNA molecules. None of these differences prevent it from being charged by seryl-tRNA synthetase, but they may ensure that it alone is modified by selenocysteine synthase.

Summary

The genetic code is nearly universal and consists of 64 nonoverlapping codons. The starting point for the reading of codons defines the reading frame of a gene; shifting the starting point for the reading of codons by a number of bases not divisible by three shifts the reading frame and changes the entire message.

The genetic code has several salient features. First, the code is degenerate; many synonymous codons specify the same amino acid, and the first two positions of a codon are often enough to specify a given amino acid. Mutations in the third position therefore often do not change the sense of the codons. Second, similar codons specify chemically similar amino acids. Third, special codons signal termination and initiation. Finally, each codon has only one meaning.

Mutations can alter the sense of codons. Missense mutations involve nucleotide substitutions that change the meaning of individual codons so that the wrong amino acid is incorporated into the protein during translation. Nonsense mutations produce termination codons called nonsense codons, which can result in premature termination of protein synthesis. Nonsense mutations can be suppressed by suppressor tRNA molecules, which insert amino acids at nonsense codons. Frameshift mutations involve insertions and deletions of nucleotides that change the reading frame of a gene.

tRNA molecules form the interface between proteins and nucleic acids; they carry activated amino acids to the ribosome for transfer to a growing peptide chain and interact with the mRNA by complementary base pairing. All tRNA molecules share certain structural features. tRNA molecules contain many conserved and many covalently modified nucleotides, some of which help stabilize the secondary and tertiary structures of the molecules. The secondary structure of a tRNA molecule resembles a cloverleaf with four Watson-Crick hydrogen-bonded stems and three loops. The acceptor stem is covalently attached to the amino acid residue, and the anticodon loop interacts with the codon in the mRNA. The tertiary structure of tRNA is L shaped, with the acceptor stem and the anticodon loop at opposite ends of the molecule. Non–Watson-Crick interactions help establish this tertiary structure.

The anticodon consists of three nucleotides that interact with the codon in mRNA by complementary base pairing. This interaction allows flexibility at the 5′ (wobble) position, where certain non–Watson-Crick base pairing is permitted. tRNA molecules that have different anticodons but bind to the same amino acid are called isoacceptor tRNA molecules. The genes that encode isoacceptor tRNA molecules are not equally expressed. In general, the codons used most often are those recognized by the most abundant isoacceptor tRNA molecules.

An amino acid is added to a tRNA molecule in a reaction catalyzed by an aminoacyl-tRNA synthetase. There are two classes of aminoacyl-tRNA synthetases. Class I enzymes catalyze attachment of the amino acid to the 2′-hydroxyl group of the terminal adenylate of the tRNA molecule, whereas class II enzymes catalyze attachment to the 3′-hydroxyl group. The two classes are distinguished on the basis of similarity in amino acid sequence and other aspects of structure.

Aminoacyl-tRNA molecules are synthesized in two steps. First, the amino acid is converted into an energy-rich aminoacyl-adenylate by adenylyl-group transfer from ATP. Second, the aminoacyl group of the energy-rich intermediate is transferred to a tRNA molecule. This aminoacylation of a given tRNA molecule determines which amino acid will be incorporated into the protein at a given codon.

Each aminoacyl-tRNA synthetase is highly specific for one amino acid and its corresponding tRNA molecules. Aminoacyl-tRNA synthetases discriminate among amino acid substrates on the basis of charge, size, and hydrophobicity. Some aminoacyl-tRNA synthetases exhibit a proofreading activity and can catalyze hydrolysis of an inappropriate substrate even after the aminoacyl-adenylate has formed.

Aminoacyl-tRNA synthetases also specifically bind to their corresponding tRNA molecules. The details of this recognition process are still unclear, but recognition is known to involve specific bases in the tRNA, including, in some cases, the anticodon.

Selected Readings

Genetic Code

Crick, F. H. C. (1966). Codon-anticodon pairing: the wobble hypothesis. *J. Mol. Biol.* 19:548–555.

Crick, F. H. C. (1968). The origin of the genetic code. *J. Mol. Biol.* 38:367–379.

Crick, F. H. C., Barnett, L., Brenner, S., and Watts-Tobin, R. J. (1961). General nature of the genetic code for proteins. *Nature* 192:1227–1232.

Lengyel, P., Speyer, J. F., and Ochoa, S. (1961). Synthetic polynucleotides and the amino acid code. *Proc. Natl. Acad. Sci. USA* 47:1936–1942.

Morgan, A. R. (1993). Base mismatches and mutagenesis: how important is tautomerism? *Trends Biochem. Sci.* 18:160–163.

Nirenberg, M. W., and Matthaei, J. H. (1961). The dependence of cell-free protein synthesis in *E. coli* upon naturally occurring or synthetic polyribonucleotides. *Proc. Natl. Acad. Sci. USA* 47:1588–1602.

Alternate Genetic Codes

Adams, R. L. P., Knowler, J. T., and Leader, D. P. (1992). *The Biochemistry of the Nucleic Acids,* 11th ed. (London: Chapman and Hall).

Farabaugh, P. J. (1993). Alternative readings of the genetic code. *Cell* 74:591–596.

RNA Editing

Benne, R. (1992). Guide RNA tails of the unexpected. *Curr. Biol.* 2:425–427.

Chan, L. (1993). RNA editing: exploring one mode with apolipoprotein B mRNA. *BioEssays* 15:33–41.

Structure of tRNA

Björk, G. R., Ericson, J. U., Gustafsson, C. E. D., Hagervall, T. G., Jönsson, Y. H., and Wikström, P. M. (1987). Transfer RNA modification. *Annu. Rev. Biochem.* 56:263–287.

Cigan, A. M., Feng, L., and Donahue, T. F. (1988). tRNA$_i^{Met}$ functions in directing the scanning ribosome to the start site of translation. *Science* 242:93–97.

Curran, J. F., and Yarus, M. (1987). Reading frame selection and transfer RNA anticodon loop stacking. *Science* 238:1545–1550.

Aminoacyl-tRNA Synthetases

Buechter, D. D., and Schimmel, P. (1993). Dissection of a class II tRNA synthetase: determinants for minihelix recognition are tightly associated with domain for amino acid activation. *Biochemistry* 32:5267–5272.

Carter, C. W., Jr. (1993). Cognition, mechanism, and evolutionary relationships in aminoacyl-tRNA synthetases. *Annu. Rev. Biochem.* 62:715–748.

Freist, W. (1989). Mechanisms of aminoacyl-tRNA synthetases: a critical consideration of recent results. *Biochemistry* 28:6787–6795.

Jakubowski, H., and Goldman, E. (1992). Editing of errors in selection of amino acids for protein synthesis. *Microbiol. Rev.* 56:412–429.

Kurland, C. G. (1992). Translational accuracy and the fitness of bacteria. *Annu. Rev. Genet.* 26:29–50.

Martinis, S. A., and Schimmel, P. (1993). Microhelix aminoacylation by a class I tRNA synthetase. *J. Biol. Chem.* 268:6069–6072.

Schimmel, P. (1989). Parameters for the molecular recognition of transfer RNAs. *Biochemistry* 28:2747–2759.

Selenocysteine

Böck, A., Forchhammer, K., Heider, J., and Baron, C. (1991). Selenoprotein synthesis: an expansion of the genetic code. *Trends Biochem. Sci.* 16:463–467.

Böck, A., Forchhammer, K., Heider, J., Leinfelder, W., Sawers, G., Veprek, B., and Zinoni, F. (1991). Selenocysteine: the 21st amino acid. *Mol. Microbiol.* 5:515–520.

Heider, J., Baron, C., and Böck, A. (1992). Coding from a distance: dissection of the mRNA determinants required for the incorporation of selenocysteine into protein. *EMBO J.* 11:3759–3766.

30

Protein Synthesis

We have now reached the final stage in the process of biological information flow: translation of mRNA and the polymerization of amino acids into proteins. Like DNA and RNA synthesis, protein synthesis is directed by a template (in this case, mRNA). Like DNA and RNA synthesis, protein synthesis can be divided into initiation, chain elongation, and termination. Finally, like DNA and RNA synthesis, protein synthesis is carried out by an elaborate complex, in this case composed of the ribosome and accessory protein factors as well as mRNA and charged tRNA molecules.

Since the association of an amino acid with a specific tRNA is an important step in decoding mRNA, we consider protein synthesis to begin with the activation of amino acids (Section 29·8). The initiation of polymerization involves the assembly of the translation complex at the first codon in the mRNA molecule. During polypeptide-chain elongation, the ribosomes and associated components move in the $5' \rightarrow 3'$ direction relative to the template mRNA, synthesizing the protein from the N-terminus to the C-terminus. Finally, when synthesis of the protein is complete, the translation complex is disassembled in a separate termination step.

The translation complex, like the replisome and the transcription complex discussed earlier, contains proteins that allow the complex to assemble and proteins that increase the efficiency of catalysis. However, the translation machinery differs profoundly from most other protein machines because two-thirds of the mass of its major component, the ribosome, is not protein but RNA. This ribosomal RNA (rRNA) is the heart of the ribosome; it is actively involved in many of the steps of protein synthesis. In fact, the ribosome may be considered a complex example of an RNA enzyme.

Many newly synthesized proteins are not fully functional until their polypeptide chains have been covalently modified and the proteins have been transported to particular locations inside (or even outside) the cell. Often, covalent modifications occur while proteins are still being synthesized on ribosomes or shortly afterward, and intracellular targeting is also coupled to translation in many cases. In this chapter, we discuss protein synthesis, transport, targeting, and some posttranslational modifications. We also discuss several mechanisms of translational regulation.

30·1 Formation of Peptide Bonds Occurs on Ribosomes

In the 1950s, the cytosol of a typical mammalian cell was found to contain large numbers of rRNA-protein particles—the ribosomes. Shortly thereafter, it was shown that these particles are the sites where radioactive amino acids are first incorporated into polypeptide chains. When cells are incubated with radioactive amino acids for about two seconds, most of the radioactivity appears associated with ribosomes. However, when the incubation is allowed to proceed for about a minute, much of the radioactive label moves into the cytosol. There, the radioactivity is found in proteins, which carry the radioactive label from one end of their polypeptide chains to the other. These results suggested that ribosomes are the sites of protein synthesis. We now know that in addition to the ribosomes themselves, other factors are required for accurate translation in vivo. We refer to the entire translation machine as a translation complex.

In a manner similar to other enzymes that catalyze polymerization reactions, ribosomes bring reactants (in this case, two activated amino acids aligned with mRNA codons) into close proximity and into the correct orientation for formation of a new bond (in this case, a peptide bond). However, ribosomes are also actively involved in other steps in protein synthesis. They align the mRNA during the initiation of protein synthesis and assist in selecting the proper codon for initiation. They also modulate the fidelity of translation by limiting the misreading of mRNA and the subsequent misincorporation of amino acids. In addition, they are involved in terminating translation and releasing the nascent polypeptide chain.

An *Escherichia coli* cell contains about 20 000 ribosomes, and many large eukaryotic cells contain several hundred thousand ribosomes. Large mRNA molecules can be translated simultaneously by many translation complexes, which form a polyribosome, or **polysome** (Figure 30·1). The number of ribosomes bound to an mRNA molecule depends on the length of the mRNA and the efficiency of initiation of protein synthesis. At maximal efficiency, the distance between each translation complex in the polysome is about 100 nucleotides. On average, each mRNA molecule in an *E. coli* cell is translated 30 times, thereby amplifying the information it contains 30-fold.

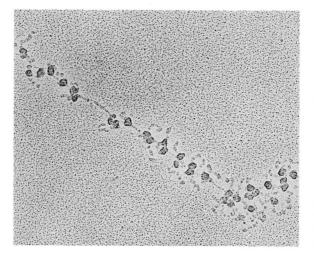

Figure 30·1
Translation of silk fibroin mRNA in the silkworm *Bombyx mori*. The mRNA is translated simultaneously by many translation complexes that form a large polysome. Translation begins at the 5′ end of the mRNA (upper left) and terminates at the 3′ end (lower right). Increasing lengths of fibroin are extruded from the ribosomes as protein synthesis proceeds. (Courtesy of S. L. McKnight and O. L. Miller, Jr.)

A. Ribosomes Are Composed of Both RNA and Protein

All ribosomes are composed of two subunits of unequal size. In *E. coli,* the small subunit is designated the 30S subunit, and the large subunit is called the 50S subunit (Figure 30·2). The 30S subunit is elongated and asymmetric, with overall dimensions of $5.5 \times 22 \times 22.5$ nm. A narrow neck separates the head from the base, and a protrusion extends from the base, forming a cleft where the mRNA molecule appears to rest. The 50S ribosomal subunit is fatter than the 30S subunit and has several protrusions; its dimensions are about $15 \times 20 \times 20$ nm. The 50S subunit contains a tunnel that is about 10 nm long and 2.5 nm in diameter. This tunnel, which

(a)
50S subunit

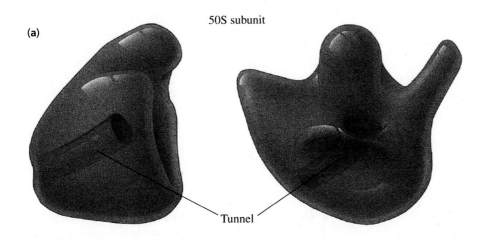

Tunnel

Figure 30·2
Structure of a 70S ribosome in *E. coli.* The 70S ribosome is composed of a 30S and a 50S subunit. **(a)** Two views of the large, 50S subunit. Note the tunnel that connects the site of active protein synthesis and the exterior surface. **(b)** Two views of the small, 30S subunit. **(c)** Arrangement of the two subunits in an intact ribosome.

(b)
30S subunit

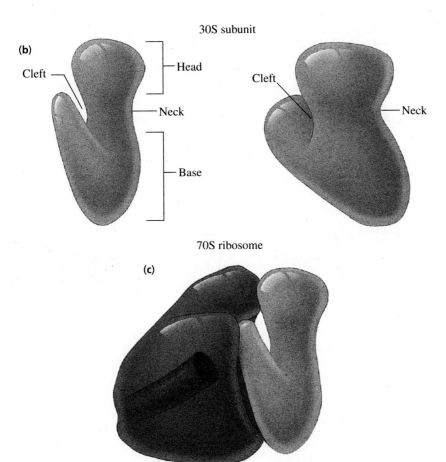

Cleft — Head

— Neck

Cleft — Neck

— Base

70S ribosome

(c)

connects the exterior of the ribosome with the site of peptide-bond formation, accommodates between 25 and 40 residues of a growing polypeptide chain during protein synthesis. The 30S and 50S subunits combine to form an active 70S ribosome.

In *E. coli,* the RNA component of the 30S subunit is a 16S rRNA molecule of 1542 nucleotides. Although 16S rRNA varies somewhat in length from species to species, it contains extensive regions of secondary structure whose sequences are highly conserved (Figure 30·3). In fact, comparisons of these conserved sequences from a variety of species led to construction of the phylogenetic trees that group all living organisms into three main branches (Section 2·3). The ribosomal RNA molecules of mitochondria and chloroplasts are also homologous to bacterial rRNA molecules, and comparisons of their sequences provide strong evidence that these organelles are derived from bacteria that invaded the primitive eukaryotic cell.

Ribosomal RNA makes up about two-thirds of the total mass of the ribosome. As we will see, many parts of the ribosomal RNA molecules serve as binding sites for factors during translation and as the catalytic center for peptide-bond formation.

Ribosomal RNA molecules are produced from a larger precursor by post-transcriptional processing, as discussed in Chapter 28. This processing is coupled to

Figure 30·3
Proposed secondary structure of 16S rRNA from *E. coli,* indicating predicted Watson-Crick base pairing. This structure is folded into a stable tertiary structure in the ribosome. Dots indicate G/U base pairs.

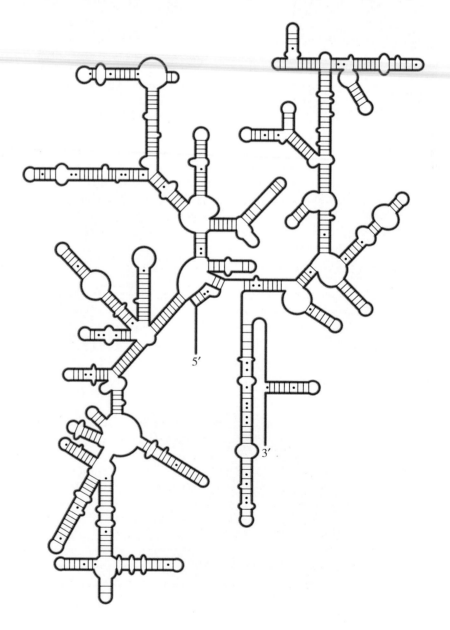

the assembly of ribosomes. We have seen that biological membranes and some protein complexes spontaneously self-assemble. Ribosomes also self-assemble; each subunit forms spontaneously in vitro from mixtures of pure proteins and rRNA, provided that the components are mixed in the proper order. Ribosomes assembled in vitro appear to function normally.

There are 21 ribosomal proteins in a complete 30S subunit. These proteins are named S1, S2, S3, and so on, where S stands for small subunit, and they range in mass from M_r 8500 to M_r 61 200. Many of these proteins form protein-RNA contacts and bind to specific regions of secondary structure on the 16S rRNA molecule, whereas others form protein-protein contacts and join the complex only after other ribosomal proteins are already bound. Assembly of the 30S subunit requires that the 16S rRNA precursor interact initially with at least six proteins—S4, S7, S8, S15, S17, and S20—and perhaps with a seventh, S13, to form a 21S particle (Figure 30·4). The 21S particle undergoes a conformational change, which allows the additional proteins to bind until all 21 are present. Assembly of the small ribosomal subunit occurs simultaneously with processing of the primary 16S rRNA transcript (Section 28·2A).

The 50S subunit of the *E. coli* ribosome contains two molecules of rRNA: one 5S rRNA molecule composed of 120 nucleotides and one 23S rRNA molecule composed of 2904 nucleotides. There are at least 31 different proteins (named L1, L2, and so on) associated with the 5S and 23S rRNA molecules in the mature 50S subunit. As with the 30S subunit, the proteins of the 50S subunit spontaneously assemble with the two rRNA molecules to form an active 50S subunit. About half the proteins bind the 23S and 5S rRNA molecules directly, and the others form protein-protein contacts.

Ribosomal proteins of a given species interact only with rRNA of the same or closely related species. For example, the 16S rRNA molecule of yeast does not form a functional ribosome with the ribosomal proteins of *E. coli*. This implies that although the gross morphology of ribosomes has been conserved during evolution, the specific protein-protein contacts and protein-rRNA interactions have not. In fact, the amino acid sequences of ribosomal proteins from *E. coli* and yeast are not highly conserved.

B. Prokaryotic and Eukaryotic Ribosomes Are Similar

The ribosomes of eubacteria and archaebacteria are remarkably similar in size, shape, and composition, although archaebacterial ribosomes have characteristic distinguishing features. Eukaryotic ribosomes are similar in shape to eubacterial ribosomes, but those of some eukaryotes, notably vertebrates and flowering plants, tend to be somewhat larger and more complex. The ribosomes of mitochondria and chloroplasts resemble bacterial ribosomes more closely than they resemble eukaryotic cytoplasmic ribosomes. All ribosomes contain a tunnel, although in some species, such as *E. coli*, it is difficult to detect.

Intact vertebrate ribosomes are designated 80S and are made up of 40S and 60S subunits (Figure 30·5, next page). The small, 40S subunit is analogous to the 30S subunit of the prokaryotic ribosome; it contains about 30 proteins and a single molecule of 18S rRNA. The large, 60S subunit contains about 40 proteins and three rRNA molecules: 5S rRNA, 28S rRNA, and 5.8S rRNA. The 5.8S rRNA molecule is about 160 nucleotides long, and its sequence is homologous to that of the 5′ end of prokaryotic 23S rRNA. This homology implies either that the gene encoding the largest rRNA in primitive organisms split into two genes in the line leading to eukaryotes or that two separate genes in primitive organisms became fused in bacteria. (Recall from Section 28·2B that the 5.8S and 28S rRNA molecules are transcribed from adjacent genes and that both are produced by processing a single large precursor.)

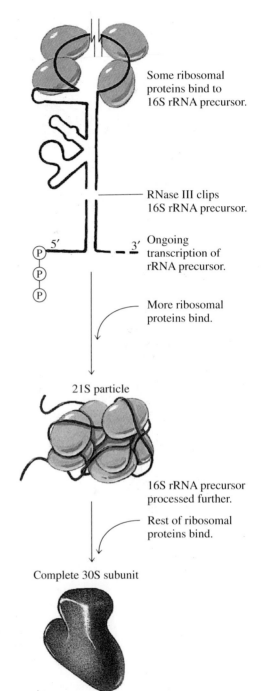

Some ribosomal proteins bind to 16S rRNA precursor.

RNase III clips 16S rRNA precursor.

Ongoing transcription of rRNA precursor.

More ribosomal proteins bind.

21S particle

16S rRNA precursor processed further.

Rest of ribosomal proteins bind.

Complete 30S subunit

Figure 30·4
Assembly of the 30S ribosomal subunit and maturation of 16S rRNA. Assembly of the 30S ribosomal subunit begins when six or seven ribosomal proteins bind to the 16S rRNA precursor as it is being transcribed, thereby forming a 21S particle. The 21S particle undergoes a conformational change, and the 16S rRNA molecule is processed to its final length. During this processing, the remaining ribosomal proteins of the 30S subunit bind.

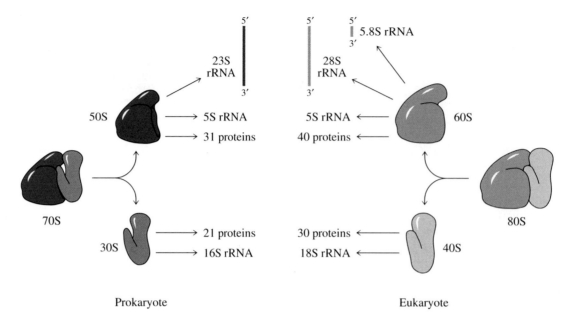

Prokaryote   Eukaryote

Figure 30·5
Comparison of eukaryotic ribosomes and prokaryotic ribosomes. Both consist of two subunits, each of which contains rRNA and proteins. The large subunit of the prokaryotic ribosome contains two molecules of rRNA: 5S and 23S. The large subunit of almost all eukaryotic ribosomes contains three molecules of rRNA: 5S, 5.8S, and 28S. The sequence of the eukaryotic 5.8S rRNA is similar to the sequence of the 5′ end of the prokaryotic 23S rRNA.

Prokaryotic and eukaryotic genomes both contain multiple copies of rRNA genes. The combination of a large number of copies and strong promoters for the rRNA genes allows cells to maintain a high level of ribosome synthesis. Eukaryotic rRNA genes, which are transcribed by RNA polymerase I, occur as tandem arrays of hundreds of copies. In most eukaryotes, these genes are clustered in the nucleolus, where the processing of rRNA and the assembly of ribosomes occur.

The ribosomal RNA molecules of both prokaryotes and eukaryotes share similar structures. For example, the secondary structures of 12S mitochondrial rRNA in humans, 18S rRNA in the frog *Xenopus laevis,* and bacterial 16S rRNA are quite similar. The 5S rRNA molecules of prokaryotes and eukaryotes also exhibit similarities (Figure 30·6). This conservation of ribosomal structure throughout evolution may reflect an ancient, conserved function in protein synthesis.

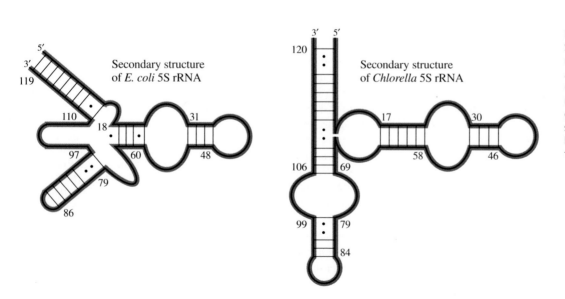

Figure 30·6
Secondary structures of 5S rRNA from the prokaryote *E. coli* and the eukaryote *Chlorella.* The prokaryotic molecule contains four regions of helical secondary structure, whereas the eukaryotic molecule contains five.

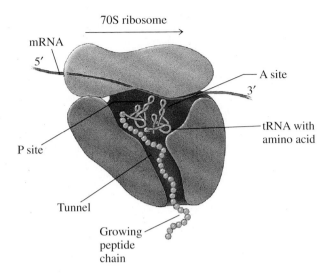

70S ribosome

mRNA

5′

A site

3′

tRNA with amino acid

P site

Tunnel

Growing peptide chain

Figure 30·7
Sites for tRNA binding on ribosomes. During protein synthesis, the P site is occupied by the tRNA molecule attached to the growing polypeptide chain, and the A site holds a second aminoacylated tRNA. The growing polypeptide chain extends through the tunnel of the large subunit.

C. Ribosomes Contain Two Aminoacyl-tRNA–Binding Sites

Ribosomes are enormous complexes. They hold the mRNA, the growing polypeptide chain, and two bulky aminoacyl-tRNA molecules, and they must also accommodate the binding of several protein factors. During protein synthesis, the ribosome aligns the two charged tRNA molecules so that their anticodons interact with the correct codons in mRNA and so that their aminoacylated ends are positioned at the site of peptide-bond formation.

The relative orientation of the two tRNA molecules during protein synthesis is shown in Figure 30·7. One tRNA molecule, to which the growing polypeptide chain is attached, is bound to the ribosome at the **peptidyl site** (P site). The second aminoacyl-tRNA is bound at the **aminoacyl site** (A site). As the polypeptide chain is synthesized, it extends through the tunnel of the large subunit.

30·2 Proteins Are Synthesized from the N-Terminus to the C-Terminus

The direction of growth of polypeptide chains was demonstrated early in the 1960s by H. M. Dintzis and his coworkers. They added radioactive amino acids to cells that were actively synthesizing proteins and determined where labelled amino acids first appeared in completed proteins. The cells they chose were rabbit reticulocytes—immature red blood cells that synthesize hemoglobin almost exclusively. The reticulocytes were incubated with labelled amino acids for intervals shorter than the time required to synthesize a complete molecule of hemoglobin. Under those conditions, only globin polypeptides that were already partially synthesized incorporated the labelled amino acids and were released into the cytosol. The location of the labelled amino acids in the polypeptide chain indicated which region of the molecule was synthesized last. When Dintzis and his coworkers digested newly synthesized hemoglobin with trypsin, they found the most radioactivity in peptides nearest the C-terminus of the protein and the least radioactivity in peptides nearest the N-terminus. This result indicates that the C-terminus of the protein is synthesized last, and consequently, that the direction of protein synthesis is from the N-terminus to the C-terminus.

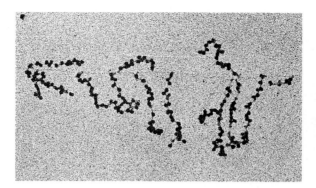

Figure 30·8
Coupled transcription and translation of an *E. coli* gene. The gene is being transcribed from left to right. Ribosomes bind to the 5′ end of the mRNA molecules as soon as they are synthesized. The large polysomes on the right will be released when transcription terminates. (Courtesy of Oscar L. Miller, Jr.)

30·3 mRNA Is Translated in the 5′→3′ Direction

The direction of translation of mRNA was determined experimentally by examining the polypeptides produced from synthetic oligonucleotides of known sequence. For example, the polymer 5′-AAAAAAAAA . . . AAC-3′ contains codons for lysine (AAA) and terminates with the codon for asparagine (AAC), if read in the 5′→3′ direction. When translated in vitro, this oligonucleotide directs the synthesis of a polypeptide from which the protease carboxypeptidase releases asparagine. This result demonstrates that mRNA is translated in the 5′→3′ direction. Since 5′→3′ is also the direction of mRNA synthesis, translation of mRNA in prokaryotes can begin before transcription is complete (Figure 30·8). Such coupling of transcription and translation is not possible in eukaryotes because these processes occur in different compartments of the cell (the nucleus and the cytoplasm, respectively).

30·4 Translation Begins with the Formation of an Initiation Complex

The initiation of protein synthesis involves the assembly of an initiation complex on mRNA at the beginning of the reading frame. The initiation complex consists of the ribosomal subunits, the mRNA template to be translated, a special initiator tRNA, and several accessory proteins called **initiation factors.** Part of the role of the initiation step is to ensure that the proper initiation codon and correct reading frame for the protein are selected before translation begins.

A. Initiation Complexes Assemble Only at Initiation Codons

Establishing the correct reading frame during initiation of translation is critical for the accurate transfer of information from mRNA to protein. Shifting the reading frame by even a single nucleotide alters the sequence of the entire peptide and results in a nonfunctional protein. The translation machinery must therefore be precise when locating the initiation codon that will serve as the start site for protein synthesis.

As mentioned in Section 29·2, the first codon translated in almost all mRNA messages is AUG. This initiation codon does not necessarily correspond to the first three nucleotides of the mRNA, however, and may in fact lie anywhere on the

(a)

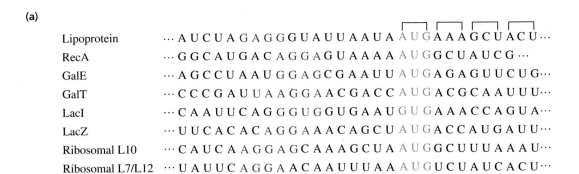

Lipoprotein	···A U C U A G A G G G U A U U A A U A A U G A A A G C U A C U···
RecA	···G G C A U G A C A G G A G U A A A A A U G G C U A U C G···
GalE	···A G C C U A A U G G A G C G A A U U A U G A G A G U U C U G···
GalT	···C C C G A U U A A G G A A C G A C C A U G A C G C A A U U U···
LacI	···C A A U U C A G G G U G G U G A A U G U G A A A C C A G U A···
LacZ	···U U C A C A C A G G A A A C A G C U A U G A C C A U G A U U···
Ribosomal L10	···C A U C A A G G A G C A A A G C U A A U G G C U U U A A A U···
Ribosomal L7/L12	···U A U U C A G G A A C A A U U U A A A U G U C U A U C A C U···

(b)

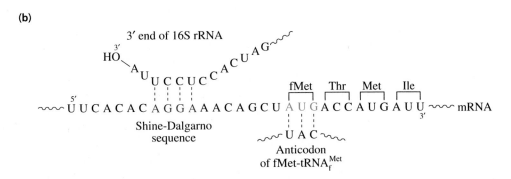

3' end of 16S rRNA

fMet   Thr   Met   Ile

~~~UUCACACAGGAAACAGCU AUGACCAUGAUU~~~ mRNA

Shine-Dalgarno
sequence

~UAC~
Anticodon
of fMet-tRNA$_f^{Met}$

Figure 30·9
Shine-Dalgarno sequences in *E. coli* mRNA.
(a) Ribosome-binding sites at the 5' end of mRNA for several *E. coli* proteins. The Shine-Dalgarno sequences (red) occur immediately upstream of initiation codons (blue).
(b) Complementary base pairing between the 3' end of 16S rRNA and the region near the 5' end of an mRNA. The Shine-Dalgarno sequence is highlighted in red. Binding of the 3' end of the 16S rRNA to the Shine-Dalgarno sequence helps establish the correct reading frame for translation by positioning the initiation codon at the P site.

mRNA template. Furthermore, the translation machinery does not initiate at every AUG codon; recall that AUG is also the codon for internal methionine residues. There must be some way of distinguishing between initiation and internal methionine codons.

In prokaryotes, the selection of a codon for initiation depends not only on the interaction between the tRNA anticodon and the mRNA codon but also on an interaction between the small subunit of the ribosome and the mRNA template. The 30S subunit binds to mRNA at a purine-rich region just upstream of the initiation codon. This region, called the **Shine-Dalgarno sequence,** is complementary to a pyrimidine-rich stretch at the 3' end of the 16S rRNA molecule. During formation of the initiation complex, the complementary nucleotides pair to form a double-stranded structure that binds the mRNA to the ribosome. Pairing of this untranslated segment of mRNA with the 16S rRNA molecule helps establish the correct reading frame for translation by positioning the initiation codon at the P site (Figure 30·9). Because Shine-Dalgarno sequences only occur immediately upstream of initiation codons, the initiation complex is assembled only at initiation codons and not at internal methionine codons.

B. Initiation Requires Special tRNA Molecules

In every cell, there are at least two different methionyl-tRNA^{Met} molecules that recognize the AUG codon. One of these is used exclusively at initiation codons and is called the **initiator tRNA.** The other recognizes only internal methionine codons. Although these two tRNA^{Met} molecules differ in nucleotide sequence and overall structure, both are aminoacylated by the action of the same methionyl-tRNA synthetase.

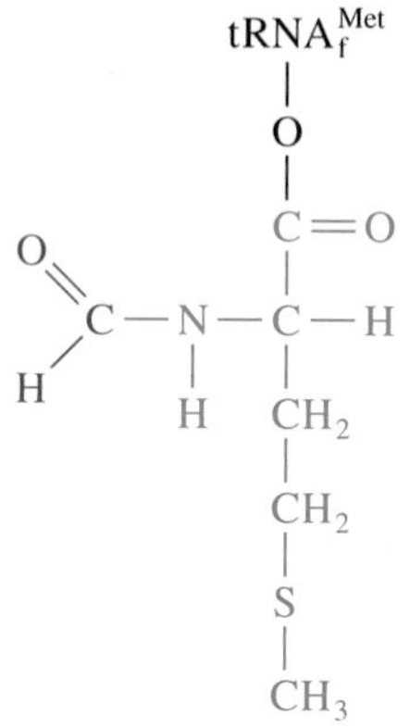

tRNA$_f^{Met}$

Figure 30·10
Structure of fMet-tRNA$_f^{Met}$. A formyl group (red) is added to the methionine (blue) of methionyl-tRNA$_f^{Met}$ in a reaction catalyzed by a formyltransferase.

In eubacteria, the initiator tRNA is called tRNA$_f^{Met}$. The charged initiator tRNA, called methionyl-tRNA$_f^{Met}$, is the substrate for a formyltransferase that catalyzes addition of a formyl group from 10-formyltetrahydrofolate to the methionine moiety, producing *N*-formylmethionyl-tRNA$_f^{Met}$ (fMet-tRNA$_f^{Met}$), shown in Figure 30·10. In eukaryotes and archaebacteria, the initiator tRNA is called tRNA$_i^{Met}$. The methionine that begins protein synthesis in eukaryotes is not formylated.

N-Formylmethionine in eubacteria or methionine in other organisms is the first amino acid incorporated into almost all proteins. After protein synthesis is underway, however, the N-terminal methionine of many proteins is either deformylated or removed from the polypeptide chain altogether.

Initiator tRNA molecules have the same anticodon (UAC) as the tRNA^{Met} used in elongation, but their structures have some unique features. For example, there are two different initiator tRNA molecules in *E. coli*, each encoded by a separate gene. Their structures are identical except for a single nucleotide substitution (7-methylguanylate for adenylate at position 47; Figure 30·11). Both these tRNA molecules lack the variable arm seen in most other tRNA molecules. Additional unique features of tRNA$_f^{Met}$ include unpaired bases at the end of the acceptor stem and three sequential G/C base pairs in the anticodon stem. The enhanced stacking interactions of these G/C base pairs alter the conformation of the anticodon nucleotides, making them somewhat less exposed to solvent than the nucleotides in most other tRNA molecules. This effect is enhanced by the absence of a modified purine on the 3′ side of the anticodon, which makes the anticodon less rigid. (Recall that the nucleotides of most anticodons are held in a stacked conformation due to the influence of an adjacent modified purine; Section 29·7B.) The presence of unique structural features enables formyltransferase to distinguish Met-tRNA$_f^{Met}$ from the charged tRNA used during elongation so that only the amino acid residue on the initiator tRNA is formylated. As we will soon see, the structure of the initiator tRNA also ensures that it is specifically used during assembly of the initiation complex.

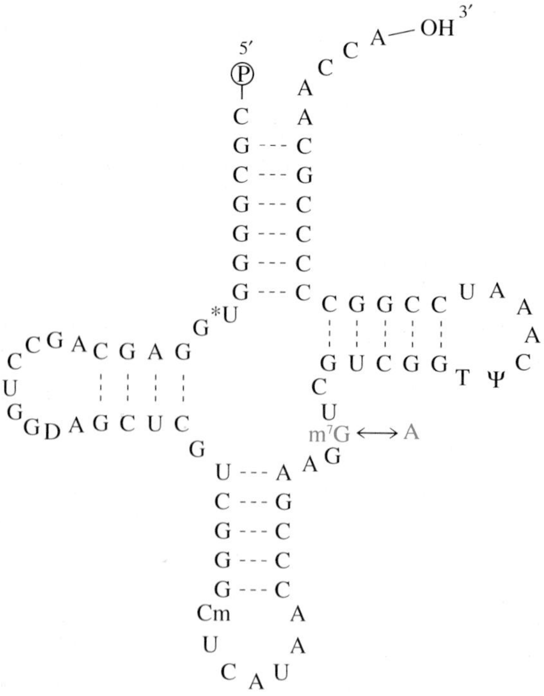

Figure 30·11
Initiator tRNA of *E. coli*. The two forms of initiator tRNA differ at a single position (shown in blue). The variable arm is not present in these molecules. Abbreviations other than standard nucleotides: m^{7}G, 7-methylguanylate; *U, 4-thiouridylate; Cm, 2′-*O*-methylcytidylate; Ψ, pseudouridylate; and D, dihydrouridylate.

C. Initiation Factors Assist in the Formation of the Initiation Complex

Formation of the initiation complex depends on the action of several initiation factors. In prokaryotes, there are three initiation factors: IF-1, IF-2, and IF-3. There are at least eight eukaryotic initiation factors, abbreviated eIF. In both prokaryotes and eukaryotes, the initiation factors promote the assembly of the initiation complex at the initiation codon.

IF-1 (M_r 8100) binds to the 30S subunit and facilitates the actions of both IF-2 and IF-3. One of the roles of IF-3 (M_r 20 600) is to maintain the ribosomal subunits in their dissociated state by binding to the small subunit. During initiation, the ribosomal subunits bind separately to the initiation complex, and the association of IF-3 to the 30S subunit prevents the 30S and 50S subunits from forming the 70S complex prematurely. IF-3 binds to a stem-loop structure in the 16S rRNA molecule, near a region that interacts with the 23S rRNA molecule in the 50S subunit. IF-3 also assists in positioning the 30S subunit at the initiation codon.

IF-2 (M_r 97 300), which has a binding site for GTP, selects the initiator tRNA from the pool of aminoacylated tRNA molecules in the cell. The IF-2–GTP complex binds to the 30S subunit, specifically recognizes the initiator tRNA, and rejects all other aminoacylated tRNA molecules. The presence of unpaired bases at the end of the acceptor stem in the initiator tRNA and the blocked α-amino group of *N*-formylmethionine allow IF-2 to distinguish an initiator tRNA from all others. It appears that IF-2–GTP binds first to the 30S subunit at the initiation codon and then to the initiator tRNA. However, IF-2–GTP may bind free fMet-tRNA$_f^{Met}$ to form a ternary complex that then associates with the 30S subunit.

One of the key roles of the initiation factors is to position fMet-tRNA$_f^{Met}$ and the initiation codon at the P site on the ribosome. This positioning is determined by the interaction of IF-2 with ribosomal proteins and by IF-3, which plays a role in recognizing the correct codon-anticodon base pairs at the P site. The initiation factors do not play a role in the initial binding of the 30S subunit to the Shine-Dalgarno sequence.

Once the 30S subunit has bound the initiation codon, forming the 30S pre-initiation complex, the 50S ribosomal subunit binds to the 30S subunit. At the same time, the GTP bound to IF-2 is hydrolyzed, P_i is released, and the initiation factors dissociate from the complex. IF-2–GTP is subsequently regenerated when the bound GDP is exchanged for GTP. The steps in the formation of the complete 70S initiation complex are summarized in Figure 30·12 (next page).

AUG is the most common initiation codon, but GUG is also used (about 8% of the time in *E. coli,* and even more frequently in certain other prokaryotic species, such as the archaebacterium *Methanobacterium thermoautotrophicum* and other methanogenic bacteria). Selection of the proper initiation codon depends on the presence of a Shine-Dalgarno sequence and the ability of the initiator tRNA and initiation factors, especially IF-3, to recognize and bind the initiation codon. Both AUG and GUG are recognized by the same initiator tRNA due to wobble in the last position of the anticodon (recall that the anticodon of tRNA$_f^{Met}$ has a less rigid conformation than other anticodons).

Codons other than AUG and GUG can also act as initiation codons in some cases. The initiation codon in IF-3 mRNA, for example, is AUU, an initiation codon not used in any other mRNA molecules that have been examined. IF-3 mRNA is translated very inefficiently under normal circumstances since IF-3 itself ensures that only interactions between AUG or GUG and the initiator tRNA are permitted in the P site of the pre-initiation complex. However, when the concentration of IF-3 falls relative to that of the 30S subunits and the other initiation factors, less stringent initiation complexes form. In the absence of IF-3, the initiation codon AUU can be used, and IF-3 mRNA is translated. Thus, synthesis of IF-3 is autoregulated at the level of translation. (We discuss regulation of gene expression at the level of translation further in Section 30·10.)

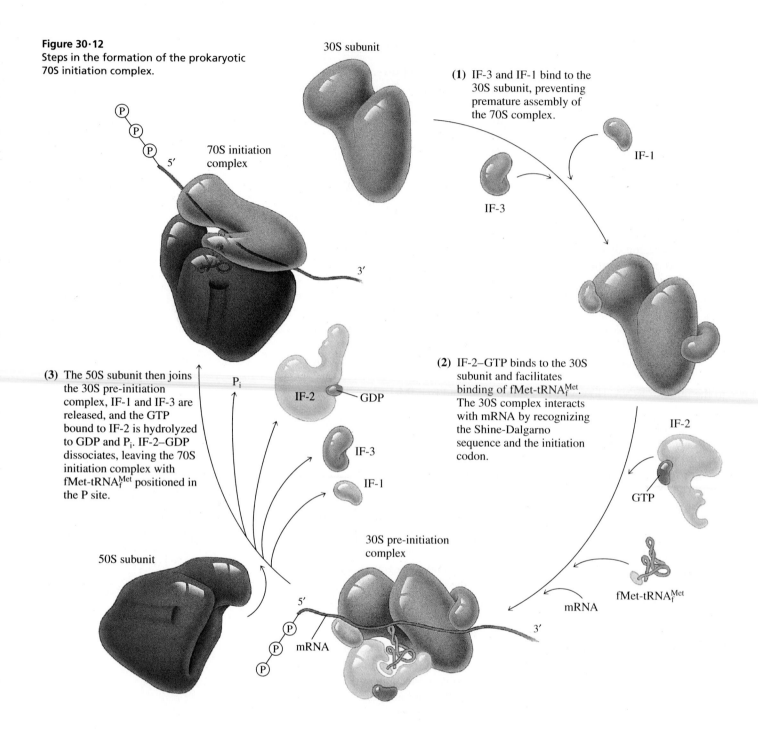

Figure 30·12
Steps in the formation of the prokaryotic 70S initiation complex.

30S subunit

(1) IF-3 and IF-1 bind to the 30S subunit, preventing premature assembly of the 70S complex.

IF-1

IF-3

70S initiation complex

5′

3′

(2) IF-2–GTP binds to the 30S subunit and facilitates binding of fMet-tRNA$_f^{Met}$. The 30S complex interacts with mRNA by recognizing the Shine-Dalgarno sequence and the initiation codon.

IF-2

GDP

IF-2

GTP

(3) The 50S subunit then joins the 30S pre-initiation complex, IF-1 and IF-3 are released, and the GTP bound to IF-2 is hydrolyzed to GDP and P$_i$. IF-2–GDP dissociates, leaving the 70S initiation complex with fMet-tRNA$_f^{Met}$ positioned in the P site.

P$_i$

IF-3

IF-1

fMet-tRNA$_f^{Met}$

mRNA

30S pre-initiation complex

5′

3′

mRNA

50S subunit

5′

mRNA

D. Initiation in Eukaryotes Differs Slightly from Initiation in Prokaryotes

As in prokaryotes, initiation of translation in eukaryotes requires several initiation factors (Figure 30·13). One of these is known as the **cap-binding protein** (CBP; also called eIF-4E in mammals). CBP, along with several other protein factors, interacts specifically with the 5′ 7-methylguanylate cap of eukaryotic mRNA (Figure 28·19). The mRNA-protein complex is then bound by the small ribosomal subunit, which is itself bound to several initiation factors. Next, this complex is joined by a ternary complex of eIF-2–GTP–Met-tRNA$_i^{Met}$, forming the 40S pre-initiation complex.

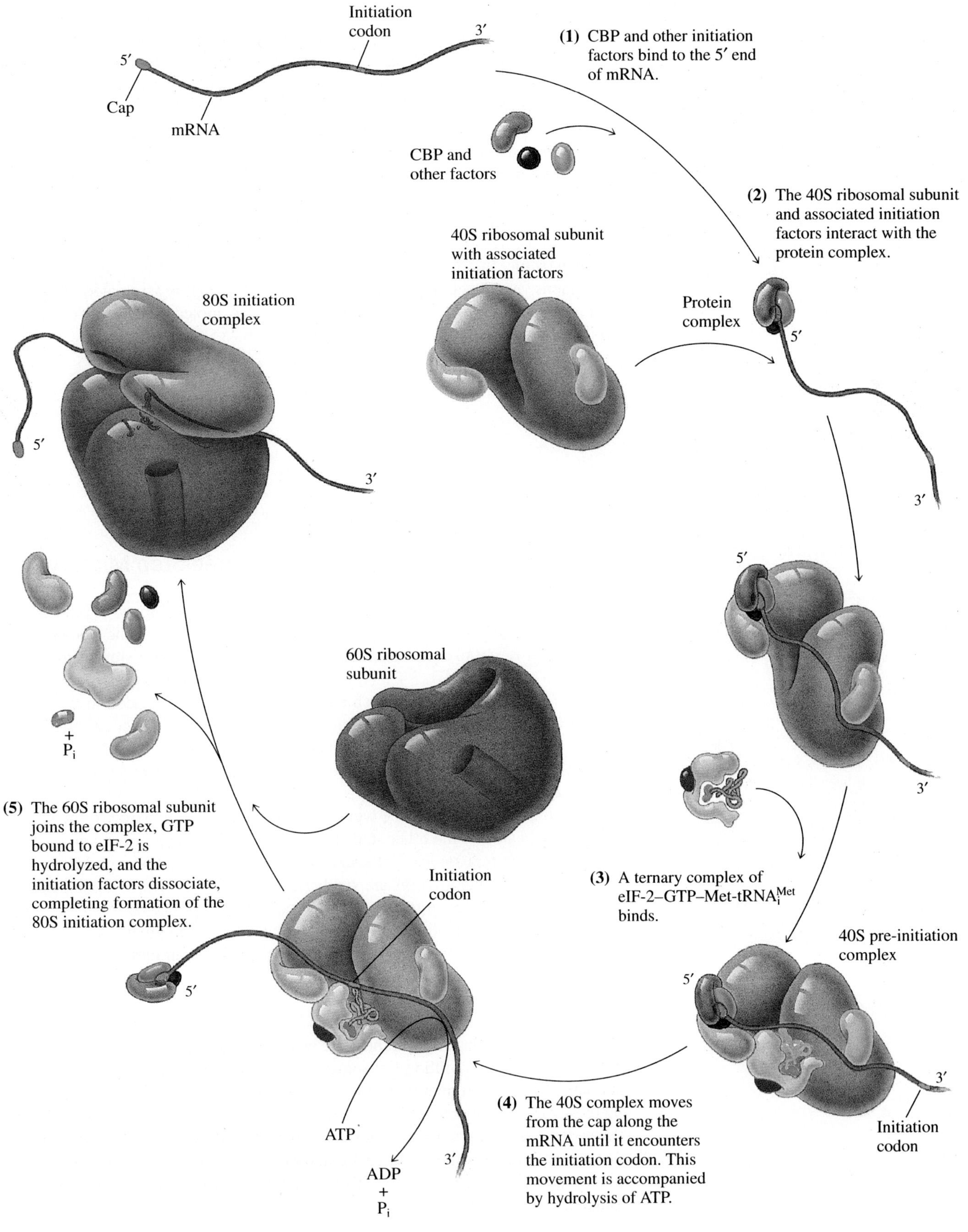

Figure 30·13
Assembly of the eukaryotic 80S initiation complex.

Initiation codon

3'

5'

Cap

mRNA

(1) CBP and other initiation factors bind to the 5' end of mRNA.

CBP and other factors

(2) The 40S ribosomal subunit and associated initiation factors interact with the protein complex.

40S ribosomal subunit with associated initiation factors

Protein complex

5'

80S initiation complex

5'

3'

+ P_i

60S ribosomal subunit

(5) The 60S ribosomal subunit joins the complex, GTP bound to eIF-2 is hydrolyzed, and the initiation factors dissociate, completing formation of the 80S initiation complex.

5'

3'

(3) A ternary complex of eIF-2–GTP–Met-tRNA$_i^{Met}$ binds.

40S pre-initiation complex

5'

3'

Initiation codon

Initiation codon

ATP

ADP + P_i

3'

(4) The 40S complex moves from the cap along the mRNA until it encounters the initiation codon. This movement is accompanied by hydrolysis of ATP.

Initiation codon

The 40S pre-initiation complex then moves along the mRNA molecule in the $5' \rightarrow 3'$ direction until it encounters an initiation codon. This unidirectional search is coupled to the hydrolysis of ATP, which may be required for movement of the ribosome or for the removal of secondary structure in the mRNA by the action of an RNA helicase. When the search is complete, the small ribosomal subunit is positioned so that Met-tRNA$_i^{Met}$ interacts with the initiation codon in the P site. In the final step, the 60S ribosomal subunit binds to complete the 80S initiation complex, and all the initiation factors dissociate. Dissociation of eIF-2 is accompanied by GTP hydrolysis, as is the case with IF-2.

Because the 40S ribosomal subunit binds to the 5′ end of the mRNA molecule and scans for the initiation codon, the first AUG codon in the message usually serves as the start site for protein synthesis in eukaryotes. Since this method of selecting the initiation codon permits only one initiation codon per mRNA, most mRNA molecules in eukaryotes encode only a single polypeptide and are said to be **monocistronic.** In contrast, prokaryotic mRNA molecules often have several coding regions, each beginning with an initiation codon that is associated with an upstream Shine-Dalgarno sequence. Messenger RNA molecules that encode several polypeptides are said to be **polycistronic.**

30·5 Chain Elongation Occurs in a Three-Step Microcycle

Following initiation, the second codon of the message is positioned to bind the second aminoacyl-tRNA. The initiator tRNA occupies the P site in the ribosome, and the A site is ready to receive an aminoacyl-tRNA in the first step of the chain-elongation reaction.

During chain elongation, each additional amino acid is added to the nascent polypeptide chain in a three-step, reiterative microcycle. The steps in this microcycle are 1) positioning the correct aminoacyl-tRNA in the A site of the ribosome, 2) forming the peptide bond, and 3) shifting the mRNA by one codon relative to the ribosome.

The translation machinery works relatively slowly compared to the enzyme systems that catalyze the other steps in the flow of biological information. For example, bacterial replisomes synthesize DNA at a rate of 1000 nucleotides per second (Chapter 25), but proteins are synthesized at a rate of only 18 amino acids per second. This difference in rates reflects, in part, the difference between polymerizing four types of nucleotides to make nucleic acids and polymerizing 20 types of amino acids to make proteins. Selecting the one correct aminoacyl-tRNA from a pool of 20 possibilities requires more time than selecting one out of four nucleotides, and protein synthesis is slower as a result. In spite of the fact that translation is slow, large amounts of new protein can be rapidly synthesized because the components of the translation machinery are abundant. For example, a large, eukaryotic cell may contain one million ribosomes, so protein synthesis in this cell can occur at a rate of 18 million peptide bonds every second.

The rate of transcription in prokaryotes averages 55 nucleotides per second (Section 27·2C), which corresponds to about 18 codons per second, or the same rate that the mRNA is translated. In bacteria, translation initiation occurs as soon as the 5′ end of an mRNA molecule is synthesized, but in eukaryotes, transcription and translation are not coupled.

A. Elongation Factors Dock an Aminoacyl-tRNA in the A Site

At the start of each chain-elongation microcycle, the A site is empty, and the P site is occupied by an aminoacylated tRNA. The tRNA molecule in the P site serves as the attachment point for the growing polypeptide chain. During each turn of the microcycle, the number of amino acid residues attached to the tRNA molecule in the P site increases by one. The tRNA molecule to which the growing peptide chain is attached is called the **peptidyl-tRNA.**

The first step in chain elongation is the insertion of the correct aminoacyl-tRNA into the A site of the ribosome. In bacteria, this step is catalyzed by an **elongation factor** called EF-Tu. EF-Tu is a monomeric protein (M_r 43 000) that contains a binding site for GTP. Once formed, EF-Tu–GTP associates with an aminoacyl-tRNA molecule to form a ternary complex that fits into the A site of a ribosome. Formation of such complexes is thermodynamically favored, and almost all aminoacylated tRNA molecules in vivo are found in ternary complexes. There are approximately 135 000 molecules of EF-Tu per cell, making it one of the most abundant proteins in *E. coli*. Two genes encode EF-Tu in the *E. coli* chromosome: *tuf*A and *tuf*B. The products of the two genes differ by a single amino acid residue and are indistinguishable in terms of function. Presumably two genes are needed to ensure an abundant supply of EF-Tu, sufficient to bind all the aminoacyl-tRNA molecules in a cell.

The EF-Tu–GTP complex recognizes common features of the tertiary structure of tRNA molecules and binds tightly to all aminoacyl-tRNA molecules except fMet-tRNA$_f^{Met}$. fMet-tRNA$_f^{Met}$ is distinguished by the distinctive secondary structure of its acceptor stem. The structure of EF-Tu is similar to that of IF-2 (which also binds GTP) and to that of the G proteins found in the membranes of eukaryotic cells (Section 12·9B), suggesting they all evolved from a common ancestral protein.

Ternary complexes of EF-Tu–GTP–aminoacyl-tRNA can diffuse freely into the A site of the ribosome. If the codon and anticodon do not match, the ternary complex is not stabilized in the A site and it leaves, soon to be replaced by another ternary complex. Given that the rate of chain elongation is 18 amino acids per second, one amino acid must be added to the polypeptide approximately every 50 ms. For every amino acid that is added, there are on average 10 unsuccessful attempts at fitting a ternary complex into the A site, suggesting that ternary complexes enter and leave the A site about every 5 ms.

If correct base pairs are formed between the anticodon of an aminoacyl-tRNA in a ternary complex and the codon in the A site, the complex is repositioned in a manner that allows EF-Tu–GTP to contact sites in the ribosome as well as the tRNA in the P site (Figure 30·14, next page). These contacts trigger hydrolysis of GTP to GDP and P$_i$, causing a conformational change in EF-Tu–GDP that releases the bound aminoacyl-tRNA. EF-Tu–GDP then dissociates from the chain-elongation complex. At the same time, the aminoacyl-tRNA is docked in the A site, where it is positioned for peptide-bond formation.

EF-Tu–GDP is unable to bind another aminoacyl-tRNA molecule until GDP is replaced by GTP. An additional elongation factor called EF-Ts (M_r 30 000) catalyzes exchange of bound GDP for GTP (Figure 30·15, Page 30·17). Once the EF-Tu–GTP complex has re-formed, it binds a new aminoacyl-tRNA molecule. Note that one GTP molecule is hydrolyzed for every aminoacyl-tRNA successfully docked at the A site by EF-Tu.

B. Ribosomal Peptidyl Transferase Catalyzes the Formation of Peptide Bonds

Once it is bound in the A site, the aminoacyl-tRNA is positioned such that the amino group of the amino acid (a nucleophile) can attack the carbonyl carbon of the ester formed by the nascent polypeptide chain and the 3′-hydroxyl group of the peptidyl-tRNA. Aminoacyl-group transfer results in the formation of a peptide bond by a nucleophilic displacement mechanism. During this reaction, the peptide chain, which has increased in length by one amino acid, is transferred from the tRNA in the P site to the tRNA in the A site (Figure 30·16, Page 30·18). The formation of the peptide bond requires the hydrolysis of the energy-rich aminoacyl-tRNA bond.

Figure 30·14
Insertion of tRNA by EF-Tu during chain elongation in *E. coli.*

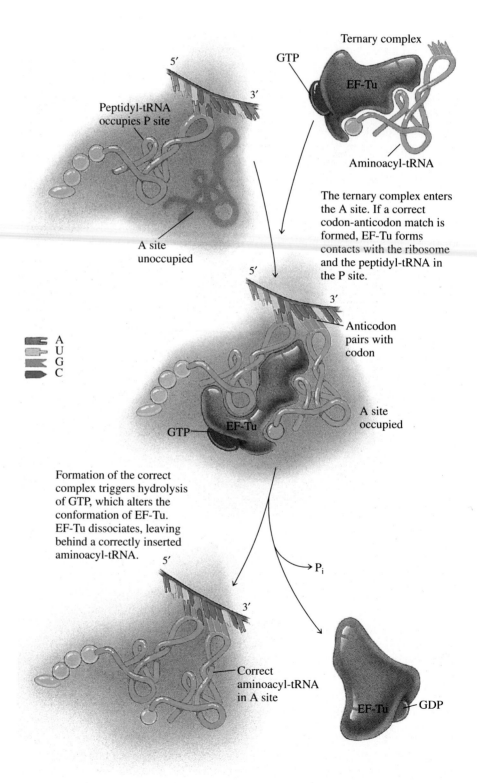

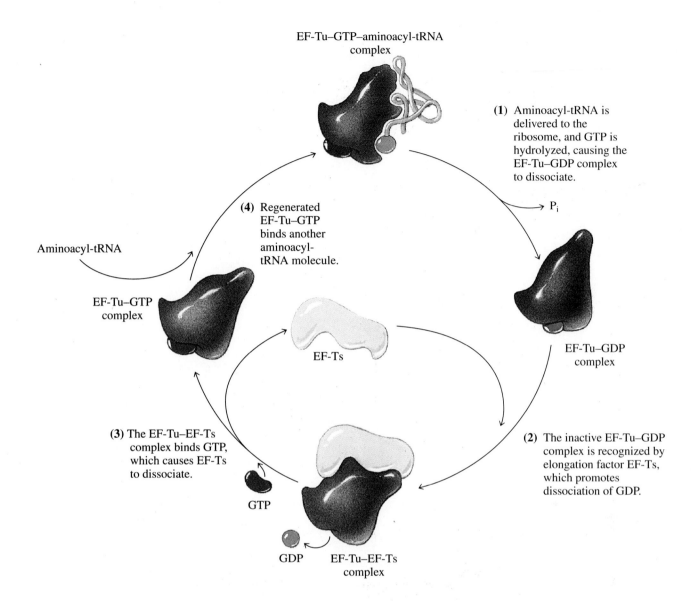

EF-Tu–GTP–aminoacyl-tRNA
complex

(1) Aminoacyl-tRNA is
delivered to the
ribosome, and GTP is
hydrolyzed, causing the
EF-Tu–GDP complex
to dissociate.

P_i

(4) Regenerated
EF-Tu–GTP
binds another
aminoacyl-
tRNA molecule.

Aminoacyl-tRNA

EF-Tu–GTP
complex

EF-Ts

EF-Tu–GDP
complex

(3) The EF-Tu–EF-Ts
complex binds GTP,
which causes EF-Ts
to dissociate.

(2) The inactive EF-Tu–GDP
complex is recognized by
elongation factor EF-Ts,
which promotes
dissociation of GDP.

GTP

GDP   EF-Tu–EF-Ts
complex

Figure 30·15
Cycling of EF-Tu–GTP.

The enzymatic activity responsible for the formation of the peptide bond is referred to as peptidyl transferase; this activity is contained within the large ribosomal subunit. Both the 23S rRNA molecule and the 50S ribosomal proteins contribute to the formation of the substrate-binding sites, but the catalytic activity is localized to the RNA component. Thus, peptidyl transferase is yet another example of an RNA molecule with enzymatic activity.

C. The Ribosome Is Repositioned During Translocation

After the peptide bond has formed, the newly created peptidyl-tRNA is partially in the A site and partially in the P site, as shown in Figure 30·17 (Page 30·19). The deaminoacylated tRNA has been displaced somewhat from the P site; it occupies a position on the ribosome that is referred to as the **exit site,** or E site. Before the next codon can be translated, the deaminoacylated tRNA must be released, and the peptidyl-tRNA must be completely transferred from the A site to the P site with concomitant movement of the mRNA relative to the ribosome by one codon. This third step in the chain-elongation microcycle is called **translocation.**

Figure 30·16

Formation of a peptide bond by an aminoacyl-group–transfer reaction. The carbonyl carbon of the peptidyl-tRNA undergoes nucleophilic attack by the amino group of the amino acid attached to the tRNA in the A site. This aminoacyl-group–transfer reaction results in the growth of the peptide chain by one residue and the transfer of the nascent peptide to the tRNA in the A site.

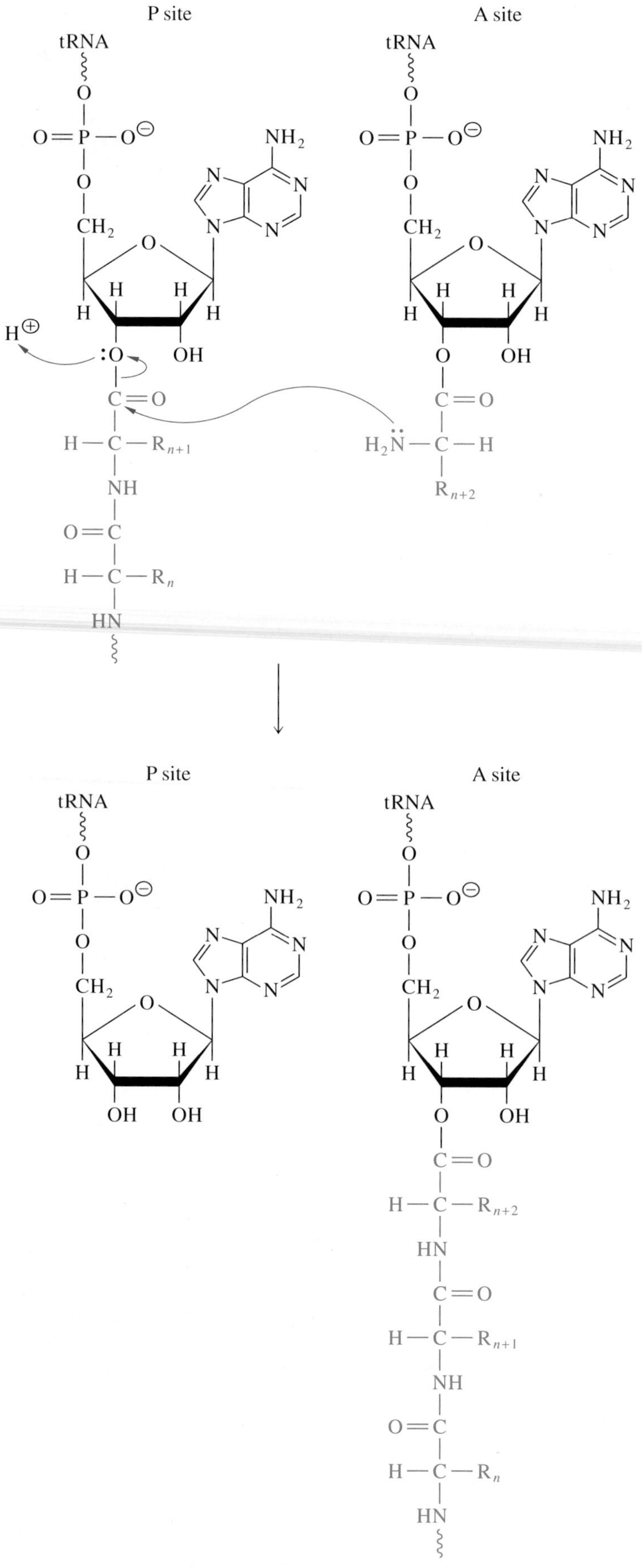

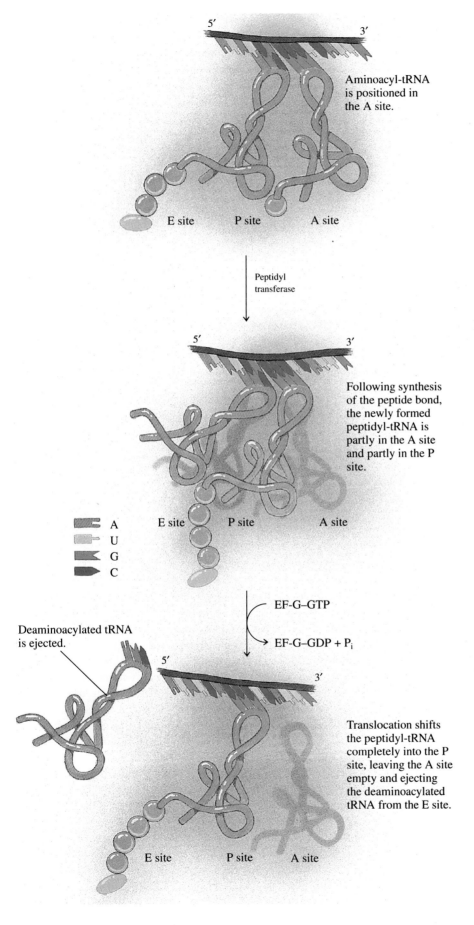

In prokaryotes, translocation requires the participation of a third elongation factor, EF-G. Like the other elongation factors, EF-G is an abundant protein: an *E. coli* cell contains approximately 20 000 molecules of EF-G, or roughly one for every ribosome. Like EF-Tu, EF-G has a binding site for GTP, and only the EF-G–GTP complex is active in translocation. Binding of EF-G–GTP to the ribosome releases the deaminoacylated tRNA from the E site and completes the translocation of the peptidyl-tRNA from the A site to the P site. EF-G itself is released from the ribosome only when its bound GTP is hydrolyzed to GDP, and P_i is released. The dissociation of EF-G–GDP leaves the ribosome free to begin another microcycle of chain elongation.

The polypeptide-chain–elongation reactions in eukaryotes are very similar to those described for *E. coli*. Three accessory protein factors participate in chain elongation in eukaryotes: EF-1α, EF-1β, and EF-2. EF-1α docks the aminoacyl-tRNA in the A site; its activity thus parallels that of *E. coli* EF-Tu. EF-1β acts analogously to EF-Ts, recycling EF-1α. EF-2 carries out translocation in eukaryotes.

In both prokaryotes and eukaryotes, the three-step microcycle is repeated as each new codon in mRNA is translated. This results in the synthesis of a polypeptide chain that may be hundreds or even thousands of residues long. Eventually, the translation complex reaches the end of the coding region, where it encounters a termination codon. At this point, elongation ceases and translation is terminated.

30·6 Release Factors Help Terminate Protein Synthesis

In *E. coli,* three **release factors,** designated RF-1, RF-2, and RF-3, participate in the termination of protein synthesis. After formation of the final peptide bond in a polypeptide chain, the peptidyl-tRNA, which holds the nascent protein, is translocated from the A site to the P site, as usual. The translocation also positions one of the three termination codons (UGA, UAG, or UAA) at the A site (Figure 30·18). After the termination codon in the A site is tested by ternary complexes of EF-Tu–GTP–aminoacyl-tRNA without success, one of the much less abundant release factors eventually diffuses into the A site. RF-1 binds UAA and UAG, and RF-2 binds UAA and UGA. The third release factor, RF-3, forms a heterodimer with either RF-1 or RF-2. RF-3 also binds GTP.

The binding of the heterodimer to the mRNA at the A site alters the activity of the peptidyl transferase, causing it to hydrolyze the ester of the peptidyl-tRNA. Release of the final polypeptide product is probably accompanied by GTP hydrolysis and dissociation of the release factors from the ribosome. At this point, the ribosomal subunits dissociate from the mRNA, and initiation factors bind to the 30S subunit in preparation for the next round of protein synthesis. Whereas three release factors are required during termination in prokaryotes, only one release factor (RF), which is GTP dependent, is required in eukaryotes.

Figure 30·18
Termination of protein synthesis in prokaryotes.

Termination
codon

5′   3′

tRNA
in P site

Nascent
polypeptide
chain

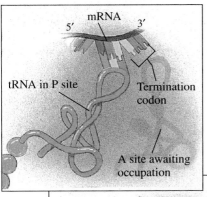

mRNA

5′   3′

tRNA in P site   Termination
codon

A site awaiting
occupation

After the final peptide bond in
a polypeptide chain is formed,
a termination codon is
positioned at the A site.

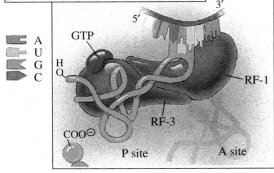

3′

5′

GTP

H
O

A
U
G
C

RF-1

RF-3

COO⊖

P site   A site

This codon is bound by a
heterodimer of release factors,
which alters the activity of the
peptidyl transferase, causing it
to catalyze hydrolysis of the
ester bond linking peptidyl-
tRNA and the nascent
polypeptide chain.

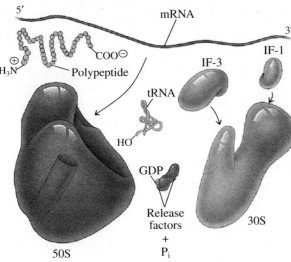

5′

mRNA

3′

IF-1

IF-3

COO⊖

H₃N⊕   Polypeptide

tRNA

HO

GDP

Release
factors
+
Pᵢ

50S   30S

As a result of this
hydrolysis, tRNA and the
polypeptide chain are
released from the ribosome.
Upon termination of chain
elongation, the 30S and 50S
subunits dissociate.
Initiation factors IF-3 and
IF-1 then bind to the 30S
subunit in preparation for a
new round of protein
synthesis.

30·7 Protein Synthesis Is Energetically Expensive

The relative cost of synthesis of the most important macromolecules in *E. coli* can be estimated. Although the rate of protein synthesis in *E. coli* is slow, many proteins are made simultaneously, and the overall reaction is costly, so protein synthesis requires almost 90% of the macromolecular biosynthetic capacity of a cell. When the cost of synthesizing small molecules is included, protein synthesis still seems to use 30 to 50% of all cellular ATP equivalents. Where does all this energy go?

For each amino acid added to a polypeptide chain, four phosphoanhydride bonds are cleaved: ATP is converted to AMP + 2 P_i during activation of the amino acid, and two GTP molecules are hydrolyzed to 2 GDP + 2 P_i during chain elongation. The hydrolysis of GTP is not associated with the formation of new covalent bonds but instead is coupled to conformational changes in the translation machinery. GTP hydrolysis appears to be necessary for the recycling of elongation factors and for intramolecular rearrangements. In this sense, GTP and GDP modulate the activity of elongation factors in much the same way GTP and GDP modulate the activity of the membrane-bound G proteins to which the elongation factors are related (Section 12·9B). The conformational changes that occur during protein synthesis are associated with the release of considerable energy.

The effects of GTP and GDP increase the fidelity of protein synthesis. When the EF-Tu–GTP–aminoacyl-tRNA complex enters the A site, EF-Tu is in one conformation. Once GTP has been hydrolyzed, EF-Tu adopts a second conformation. When EF-Tu is in either of these conformations, a correct aminoacyl-tRNA can interact strongly with the codon on the mRNA, but an incorrect aminoacyl-tRNA cannot. Therefore, the conformational change induced by GTP hydrolysis allows the ribosome to check the incoming aminoacyl-tRNA twice, under two sets of conditions. The eukaryotic elongation factors function in a similar manner.

The fidelity of protein synthesis is affected by the rate of GTP hydrolysis. If GTP were hydrolyzed quickly, the translation complex could check the incoming aminoacyl-tRNA only briefly, and the fidelity of protein synthesis would be lowered. If GTP hydrolysis were slow, there would be more time to discriminate among aminoacyl-tRNA molecules. The accuracy of protein synthesis would then increase, although its overall rate would decrease. The actual rate of GTP hydrolysis inside the cell is a compromise between the needs for accuracy and high rates of polypeptide-chain elongation.

The translation machinery couples the consumption of four high-energy phosphoanhydride bonds to the formation of each peptide bond. The consumption of four phosphodiester bonds represents a total standard free-energy change ($\Delta G^{\circ\prime}$) of $-120\,kJ\,mol^{-1}$ (-30×4), whereas the standard free-energy change for formation of a peptide bond is likely to be on the order of $+4\,kJ\,mol^{-1}$. This enormous difference in standard free-energy change ($-116\,kJ\,mol^{-1}$) makes it likely that synthesis of an entire protein is extremely exergonic and essentially irreversible in vivo. However, our estimate for the energetic cost of forming a peptide bond ($+4\,kJ\,mol^{-1}$) is probably not accurate since it does not consider that the sequence of amino acids generated is highly ordered. The creation of a chain of specific amino acid residues from an aqueous solution of 20 different kinds of free amino acids results in a large loss of entropy. In addition, entropy is lost when an amino acid is linked to a tRNA and when the aminoacyl-tRNA associates with a specific codon. These decreases in entropy are probably balanced by the release of energy stored in phosphoanhydride bonds.

30·8 Some Antibiotics Inhibit Protein Synthesis

Many microorganisms produce antibiotics, which they use as a chemical defense against competitors and predators. Some antibiotics prevent bacterial growth by inhibiting the formation of peptide bonds. For example, the antibiotic puromycin has a structure that closely resembles the 3′ end of an aminoacyl-tRNA molecule. Because of this structural similarity, puromycin can enter the A site of a ribosome. The peptidyl transferase then catalyzes the transfer of the nascent polypeptide to the free amino group of puromycin (Figure 30·19). The peptidyl-puromycin is bound weakly in the A site and soon dissociates from the ribosome, thereby terminating protein synthesis prematurely.

Although puromycin effectively blocks protein synthesis in prokaryotes, it is not medically useful in the treatment of disease because it also blocks protein synthesis in eukaryotes and is therefore poisonous to humans. Clinically important antibiotics must be specific for bacteria and have no effect on eukaryotic systems. Several examples of valuable antibiotics that can be used to inhibit prokaryotic protein synthesis are streptomycin, chloramphenicol, erythromycin, and tetracycline. These molecules can often be modified slightly to produce biochemically active derivatives.

The antibiotics shown in Figure 30·20 (next page) function at different steps in protein synthesis. Streptomycin inhibits the initiation of protein synthesis and also causes mRNA to be misread during elongation. For example, in the presence of streptomycin, the synthetic mRNA template polyU will occasionally be misread, allowing insertion of isoleucine into what would normally be polyphenylalanine. Streptomycin interacts with protein S12 in the 30S ribosomal subunit, but how this interaction alters protein synthesis is still not known. Chloramphenicol interacts

Figure 30·19
Formation of a peptide bond between puromycin at the A site of a ribosome and the nascent peptide bound to the tRNA in the P site. The product of this reaction is bound only weakly in the A site and soon dissociates from the ribosome, thus terminating protein synthesis and producing an incomplete, inactive peptide.

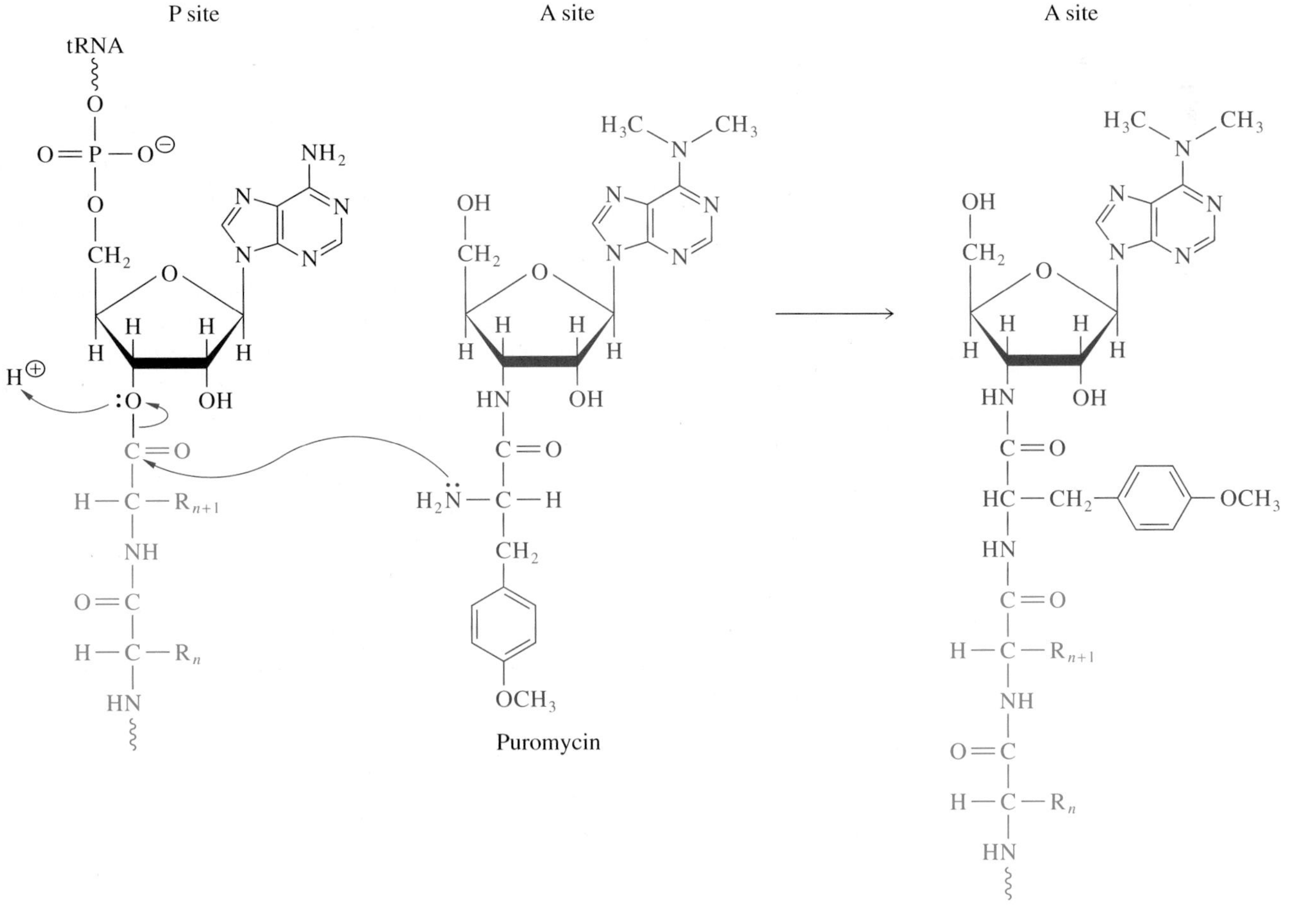

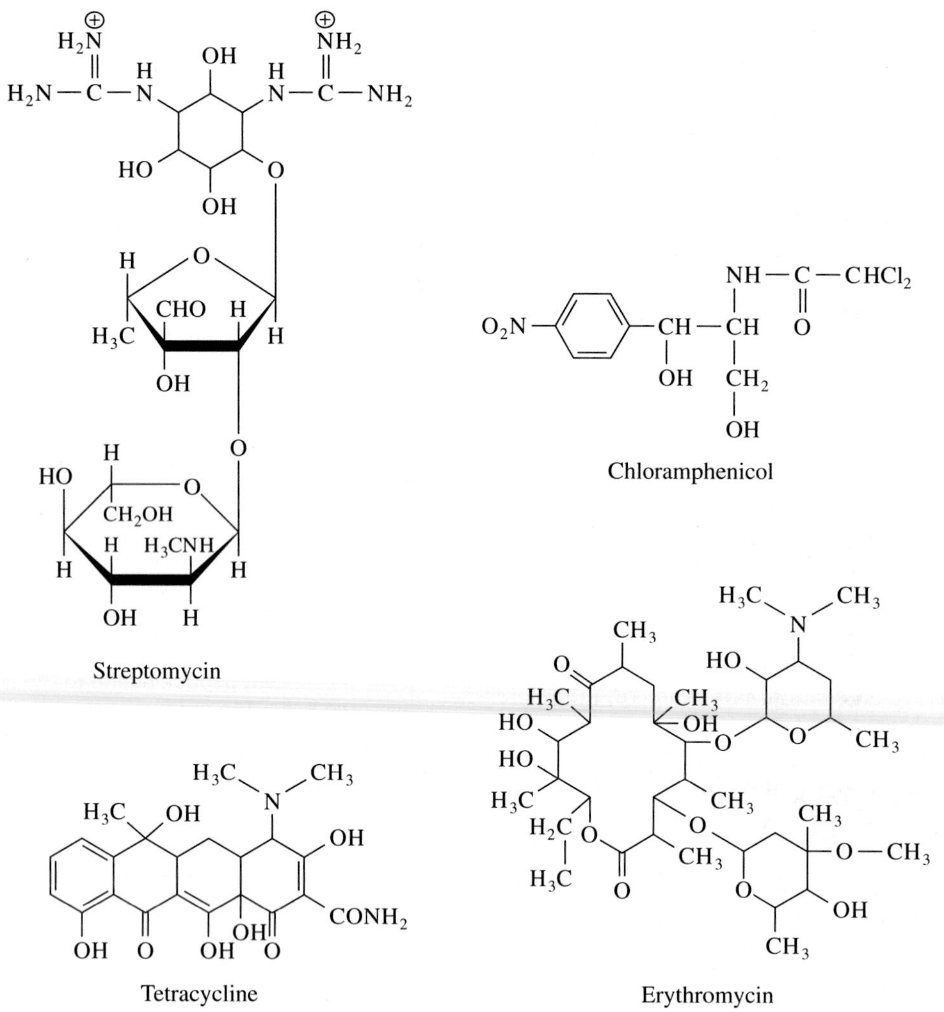

Figure 30·20
Four antibiotics that inhibit protein synthesis in prokaryotes.

Streptomycin

Chloramphenicol

Tetracycline

Erythromycin

with the 50S ribosomal subunit and inhibits peptidyl transferase and can be used to treat a broad range of bacterial infections. Tetracycline binds to the 30S subunit of prokaryotic ribosomes and prevents binding of aminoacyl-tRNA molecules. It also acts on eukaryotic ribosomes in vitro, but since it is not transported across the plasma membrane of eukaryotic cells, it does not inhibit eukaryotic protein synthesis in vivo. Erythromycin inhibits translocation by binding to the 50S subunit of prokaryotic ribosomes.

30·9 Diphtheria Toxin Inhibits Translocation of Ribosomes in Eukaryotes

Prior to mass immunization in this century, many children died of diphtheria. In 1884, the bacterial agent responsible for diphtheria, *Corynebacterium diphtheriae*, was identified, and four years later, it was found that this bacterium causes diphtheria by releasing a protein toxin. As it turns out, the toxin is not encoded by the bacterium but rather by a bacteriophage that infects the bacterium. Only those strains of *C. diphtheriae* carrying the phage cause the disease.

Diphtheria toxin consists of a single polypeptide chain (M_r 61 000) that specifically binds to ganglioside G_{M1} on the surface of the human plasma membrane (Section 11·6). The diphtheria toxin polypeptide is proteolytically cleaved into two pieces called the A fragment (M_r 21 000) and the B fragment (M_r 40 000). The B fragment appears to form a pore that traverses the plasma membrane, and the A fragment passes through the pore into the cell. This is the same type of entry mechanism exploited by cholera toxin.

The A fragment catalyzes ADP ribosylation of EF-2, leading to its inactivation. The target of ADP ribosylation is a covalently modified amino acid residue called diphthamide (Figure 30·21). ADP ribosylation of EF-2 prevents translocation of the growing peptide chain from the A site to the P site, thereby blocking protein synthesis. Since ADP ribosylation is an enzymatic reaction, a single A fragment can catalyze the ADP ribosylation of most of the 500 000 EF-2 molecules in a single eukaryotic cell. Thus, the presence of a single A fragment will cause cell death.

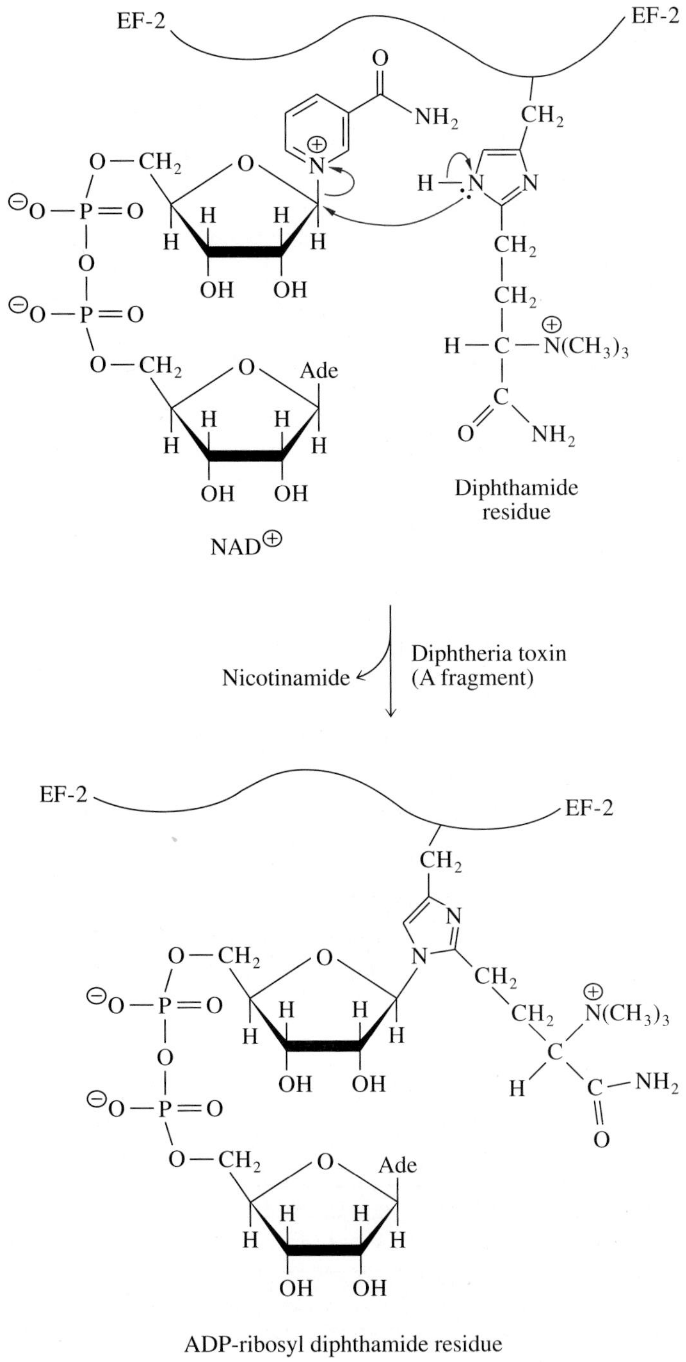

Figure 30·21
ADP ribosylation of EF-2 catalyzed by the A fragment of diphtheria toxin. ADP ribosylation of the diphthamide residue of EF-2 inactivates the elongation factor. *Ade* represents adenine.

The action of diphtheria toxin is analogous to the action of cholera toxin on membrane-bound G_s (see Figure 12·46). ADP-ribosylation of G_s by cholera toxin occurs at an arginine residue and leads to constitutive activity of the G protein, whereas ADP ribosylation of EF-2 blocks activity. As mentioned previously, the G proteins and the translation factors are related evolutionarily, and the different bacterial toxins that affect these related proteins may also have arisen from a common ancestor.

30·10 Gene Expression Can Be Regulated at the Level of Translation

As we saw with IF-3 synthesis, gene expression can be regulated by controlling the translation of mRNA into protein. Translational control can be exercised during initiation, elongation, or termination. In general, translational control of gene expression is used to regulate the production of those proteins that assemble into multisubunit complexes and those proteins whose expression in the cell must be strictly and quickly controlled to maintain viability.

As is the case with transcription, the rate of translation depends to some extent on the sequence of the template. An mRNA molecule containing an abundance of rare codons, for example, is translated less rapidly (and therefore less frequently) than one containing the most frequently used codons. In addition, the rate of translation initiation varies with the nucleotide sequence at the initiation site. A strong ribosome-binding site in bacterial mRNA leads to more efficient initiation. The use of AUG as an initiation codon instead of other initiation codons may also play a role

Figure 30·22
Some of the operons encoding *E. coli* ribosomal proteins. Each operon encodes one protein that inhibits translation of its own mRNA by binding near an initiation codon of one of the genes at the 5′ end of the mRNA transcript. The arrangement of *L22* and *S19* within the *S10* operon is not known. *P* respresents the promoter. α, β, and β' refer to genes that encode subunits of RNA polymerase. [Adapted from Nomura, M., Gourse, R., and Baughman, G. (1984). Regulation of the synthesis of ribosomes and ribosomal components. *Annu. Rev. Biochem.* 53:75–117.]

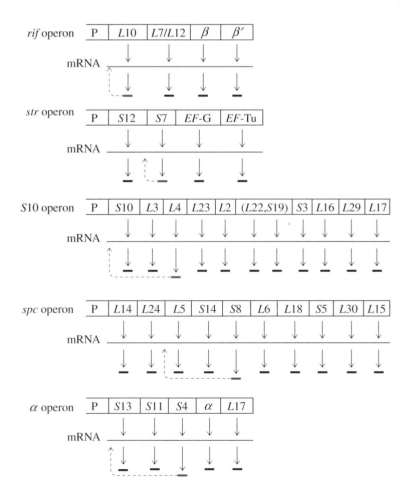

in regulating the rate of initiation and hence the amount of protein produced. There is also evidence that the nucleotide sequence surrounding the initiation codon in eukaryotic mRNA influences the rate of formation of the initiation complex.

One difference between initiation of translation and initiation of transcription is that the formation of a translation complex can be influenced by secondary structure in the message. The formation of intramolecular double-stranded regions in mRNA can, for example, mask ribosome-binding sites and the initiation codon. All factors that affect constitutive translation of mRNA molecules determine whether a given mRNA molecule is translated frequently or infrequently. However, *translational regulation* usually refers to examples where the frequency of translation of mRNA is affected by extrinsic factors.

A. The Synthesis of Ribosomal Proteins in *E. coli* Is Coupled to Ribosome Assembly

Every *E. coli* ribosome contains one molecule each of at least 52 different proteins and four molecules of the protein L7/L12. The genes encoding these ribosomal proteins are scattered throughout the genome of *E. coli*. They are found in 13 different operons and as seven isolated genes. When multiple copies of genes encoding some of these ribosomal proteins are inserted into *E. coli*, the concentration of mRNA encoding these proteins increases sharply, yet the net rate of ribosomal protein synthesis scarcely changes. Furthermore, the relative concentrations of ribosomal proteins remain unchanged, even though the various mRNA molecules for ribosomal proteins are present in unequal amounts. These findings suggest that the synthesis of ribosomal proteins is tightly regulated at the level of translation.

Translational regulation of ribosomal protein synthesis is crucial since ribosomes cannot assemble unless all proteins are present in the proper stoichiometry. The production of ribosomal proteins is controlled by regulating the efficiency with which the mRNA molecules encoding these proteins are translated. Each ribosomal protein operon encodes one ribosomal protein that inhibits translation of its own mRNA by binding near the initiation codon of one of the first genes of the operon (Figure 30·22).

The interactions between the inhibiting ribosomal proteins and their mRNA molecules may resemble the interactions between these proteins and the rRNA to which they bind when assembled into mature ribosomes. For example, ultraviolet light can be used to covalently cross-link 16S rRNA to ribosomal protein S7, demonstrating that S7 and the 16S rRNA contact each other in mature ribosomes. The mRNA transcript of the *str* operon, which includes the coding region for S7, contains some regions of RNA sequence that are identical to the S7-binding site in 16S rRNA. In addition, the proposed secondary structure of 16S rRNA in the region where it is covalently cross-linked to S7 resembles the proposed secondary structure of the *str* mRNA at the boundary between the coding regions for proteins S12 and S7 (Figure 30·23). S7 binds to this region of the *str* mRNA molecule and inhibits translation. It is likely that S7 recognizes analogous structural features in both RNA molecules. A similar mechanism is used in the translational inhibition of the other mRNA molecules shown in Figure 30·22.

The ribosomal proteins that inhibit translation bind more tightly to the specific site on their ribosomal RNA molecules than to the similar sites on their mRNA molecules. Thus, as long as newly synthesized ribosomal proteins are incorporated into ribosomes, the tighter binding of the inhibiting ribosomal proteins to ribosomal RNA ensures that the mRNA encoding ribosomal proteins continues to be translated. However, as soon as ribosome assembly slows and the concentration of free ribosomal proteins increases within the cell, the inhibiting ribosomal proteins bind to their own mRNA molecules and block additional protein synthesis. In this way, synthesis of ribosomal proteins is coordinated with ribosome assembly.

(a)

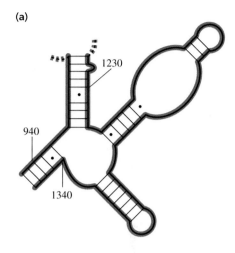

(b)

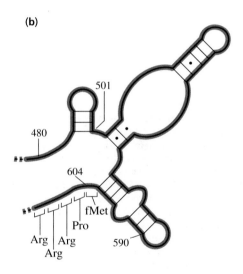

Figure 30·23
Comparison of proposed secondary structures of S7-binding sites. (a) S7-binding site on 16S rRNA. [Adapted from Watson, J. D., Hopkins, N. H., Roberts, J. W., Steitz, J. A., and Weiner, A. M. (1987). *Molecular Biology of the Gene,* 4th ed. (Menlo Park, California: Benjamin/Cummings Publishing Company), p. 396.] (b) S7-binding site on the *str* mRNA molecule at the boundary between the coding regions for proteins S12 and S7. The S12 coding region ends at nucleotide 501; the S7 coding region begins at nucleotide 604. [Adapted from Nomura, M., Yates, J. L., Dean, D., and Post, L. E. (1980). Feedback regulation of ribosomal protein gene expression in *Escherichia coli*: structural homology of ribosomal RNA and ribosomal protein mRNA. *Proc. Natl. Acad. Sci. USA* 77:7084–7088.]

Figure 30·24
Inhibition of translation by antisense RNA. *mic*F RNA (blue) is complementary to the 5′ region of OmpF mRNA (black). Formation of a double-stranded structure blocks ribosome assembly at the initiation codon, preventing translation.

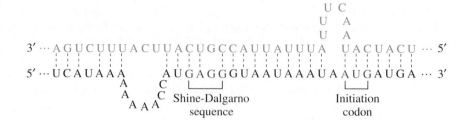

B. Translation Initiation Can Be Inhibited by Antisense RNA

RNA molecules in the cell can form double-stranded structures when they bind to complementary RNA sequences. Sometimes such binding occurs within an RNA molecule itself, but in other cases the binding occurs between RNA molecules. A variety of small RNA molecules called **antisense RNA** molecules have been characterized that bind to mRNA in vivo. When antisense RNA molecules bind to mRNA and form double-stranded regions near the initiation codon, translation can be inhibited. Some organisms use such antisense RNA molecules to control translation and thus regulate gene expression.

The synthesis of OmpF, a membrane protein in *E. coli,* is regulated by the production of antisense RNA. *mic*F is a gene for an antisense RNA in *E. coli.* The antisense RNA is complementary to the 5′ end of OmpF mRNA. Under certain conditions, transcription of *mic*F is stimulated, and the resulting product binds to the 5′ end of OmpF mRNA (Figure 30·24). Formation of the double-stranded RNA blocks access of the ribosomes to the initiation site, and production of OmpF ceases.

There are several other examples of naturally occurring regulation by antisense RNA. The production of antisense RNA can also be genetically engineered by placing an inverted gene next to a strong promoter. When such a construct is introduced into cells, the antisense RNA blocks translation of the endogenous mRNA derived from the normal gene. Constructs such as these have been used to repress synthesis of oncogenic proteins in cancer cells and show promise as possible therapies in several diseases where a protein is inappropriately expressed.

C. The Synthesis of Hemoglobin Subunits Is Coupled to the Availability of Heme

The synthesis of hemoglobin, which requires globin subunits and heme in stoichiometric amounts, is also controlled by regulation of translation initiation. Hemoglobin, the major protein in red blood cells, is initially synthesized in immature erythrocytes called rubriblasts. Mammalian rubriblasts lose their nuclei during subsequent maturation, eventually becoming reticulocytes, which are the immediate precursors of erythrocytes. Hemoglobin continues to be synthesized in reticulocytes, which are packed with processed, stable mRNA molecules that encode globin polypeptides.

The rate of globin synthesis in reticulocytes is determined by the concentration of heme: when the concentration of heme decreases, the translation of globin mRNA is inhibited. The effect of heme on globin mRNA translation is mediated by heme-controlled inhibitor (HCI), a protein kinase that, despite its name, is found in all mammalian cells. Active HCI catalyzes transfer of a phosphoryl group from ATP to the translation initiation factor eIF-2. Phosphorylated eIF-2 is unable to participate in translation initiation, and protein synthesis within the cell is inhibited.

During the initiation of translation, eIF-2 binds methionyl-tRNA$_i^{Met}$ and GTP. Recall from Section 30·4D that after formation of the 40S pre-initiation complex, methionyl-tRNA$_i^{Met}$ is transferred from eIF-2 to the initiation codon of the mRNA

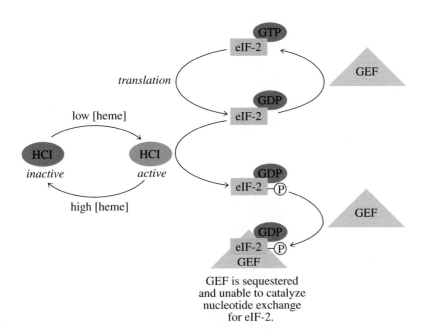

Figure 30·25
Inhibition of protein synthesis by phosphorylation of eIF-2 in reticulocytes. When the concentration of heme is high, HCI is inactive and translation proceeds normally. When the concentration of heme is low, HCI catalyzes the phosphorylation of eIF-2. Phosphorylated eIF-2 binds the limiting amounts of GEF in the cell very tightly, sequestering the GEF and preventing translation.

bound to the 40S ribosomal subunit (Figure 30·13). This transfer reaction is accompanied by hydrolysis of GTP and release of eIF-2 from the 40S ribosomal subunit as eIF-2–GDP. An enzyme called guanine nucleotide exchange factor (GEF) catalyzes the replacement of GDP with GTP on eIF-2 and the attachment of another methionyl-tRNA$_i^{Met}$ to eIF-2. GEF binds very tightly to phosphorylated eIF-2–GDP, inhibiting the exchange reaction. GEF is present in the cell in limiting amounts, and all GEF in the cell is complexed when only 30% of the cellular eIF-2–GDP is phosphorylated. When this happens, the initiation of protein synthesis is completely inhibited (Figure 30·25).

Heme regulates the synthesis of globin by interfering with the activation of HCI. When heme is abundant, HCI is inactive, and globin mRNA can be translated. When heme is scarce, however, HCI is activated, and translation of all mRNA within the cell is inhibited. Phosphorylation of eIF-2 appears to be a mechanism used to regulate the translation of mRNA in other mammalian cell types as well. For example, during infection of human cells by RNA viruses, the presence of double-stranded RNA leads to production of interferon, which, in turn, activates a protein kinase that phosphorylates eIF-2. This reaction inhibits protein synthesis in the virus-infected cell.

D. Protein Synthesis Can Be Regulated by a Shift in the Reading Frame During Translation

During the initiation of translation, an aminoacyl-tRNA molecule is aligned with an initiation codon, thereby establishing the correct reading frame for translation. A mistake in aligning a codon with its cognate aminoacyl-tRNA molecule prevents the synthesis of a functional protein since shifting the reading frame by even a single nucleotide alters the sequence of the entire protein. However, a few instances have been discovered in which synthesis of the protein *requires* a shift in the reading frame during translation. This phenomenon is called **translational frameshifting.**

Translational frameshifting regulates the synthesis of RF-2, one of the three *E. coli* release factors required for termination of translation. RF-2 binds to the ribosome at the A site when this site becomes occupied by a UGA or UAA termination codon (Section 30·6). RF-2 then assists in catalyzing the release of the newly synthesized polypeptide chain.

Figure 30·26
Translational frameshifting. Successful translation of the mRNA encoding RF-2 requires a shift in the reading frame at the 26th codon of the mRNA transcript. This codon, UGA, is a termination codon. Disassembly of the translation apparatus at a UGA codon requires the presence of RF-2. If the cellular concentration of RF-2 is low, the ribosome pauses and eventually bypasses the termination codon by shifting the reading frame by one nucleotide. The ribosome then translates the rest of the RF-2 mRNA using the new reading frame. In this way, synthesis of RF-2 is autoregulated.

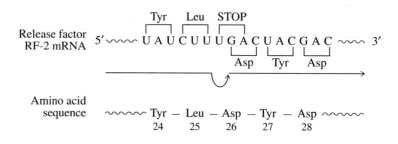

RF-2 (M_r 38 000) contains 315 amino acid residues. The first 25 amino acids are incorporated into a growing peptide chain when RF-2 mRNA is translated from the initiation codon AUG. The 26th codon of the mRNA, however, is a UGA termination codon. When the ribosome reaches the UGA termination codon, it pauses, as usual during termination. If the cellular concentration of RF-2 is relatively high, the release factor enters the A site and assists in catalyzing hydrolysis of the ester linkage of the peptidyl-tRNA, thereby prematurely terminating synthesis of more RF-2. If the cellular concentration of RF-2 is low, however, the ribosome eventually bypasses the termination codon by shifting the reading frame one nucleotide: the U is skipped and the next three nucleotides (GAC) are translated as aspartate. The ribosome then continues translating the RF-2 mRNA using the new reading frame (Figure 30·26). The synthesis of RF-2 is thus autoregulated by control of translation termination.

A similar mechanism is involved in the synthesis of the τ and γ subunits of DNA polymerase III (Section 25·3). Both proteins are encoded by the same gene (*dna*X), and the first 430 amino acids at their N-termini are identical. The common mRNA for these two proteins contains a short stretch of six adenylate residues followed by a region that can form a stable secondary structure (Figure 30·27). When translation through this region is unhindered, the τ protein (M_r 71 000) is produced, but about 40% of the time, the translation machinery shifts the reading frame by one nucleotide and promptly encounters a termination codon. When this happens, the shorter γ protein (M_r 52 000) is produced. Both the adenylate residues and the region of secondary structure are required for frameshifting. Presumably the translation machinery stalls at the stem-loop, and slippage is enhanced by the presence of the A-rich sequence.

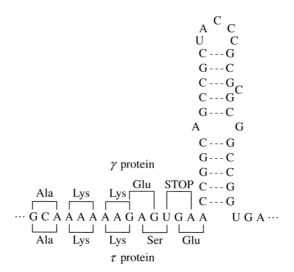

Figure 30·27
Translational frameshifting in the *dna*X gene. The *dna*X transcript contains a stem-loop structure that sometimes causes the translation machinery to pause. The mRNA is usually translated using the reading frame shown on the bottom to yield τ protein. About 40% of the time, however, the reading frame is shifted by one nucleotide, and translation terminates prematurely, yielding γ protein.

E. The *E. coli trp* Operon Is Regulated by Repression and Attenuation

The *trp* operon in *E. coli* encodes the proteins necessary for the biosynthesis of tryptophan. Most organisms synthesize their own amino acids but can also obtain them by degrading exogenous proteins. For this reason, most organisms have evolved mechanisms to repress the synthesis of enzymes required for de novo amino acid biosynthesis when the amino acid is available from exogenous sources. For example, in *E. coli,* tryptophan is a negative regulator of its own biosynthesis. In the presence of tryptophan, the *trp* operon is not expressed (Figure 30·28). Expression of the *trp* operon is inhibited in part by *trp* repressor, a tetramer composed of four identical subunits (subunit M_r 12 500). *trp* repressor is encoded by the *trp*R gene, which is located elsewhere on the bacterial chromosome and is transcribed separately. In the presence of excess tryptophan, *trp* repressor binds tryptophan, and the repressor-tryptophan complex binds to the operator *trp*O, which lies within the promoter. When bound to *trp*O, the repressor-tryptophan complex prevents RNA polymerase from binding to the promoter. Tryptophan is thus a corepressor of the *trp* operon.

Figure 30·28
Repression of the *E. coli trp* operon. The *trp* operon is composed of a leader region and five genes required for the biosynthesis of tryptophan from chorismate. The *trp*R gene, located upstream of the *trp* operon, encodes *trp* repressor, which is inactive in the absence of its corepressor, tryptophan. When tryptophan is present in excess, it binds to *trp* repressor, and the repressor-tryptophan complex binds to the *trp* operator (*trp*O). Once bound to the operator, the repressor-tryptophan complex prevents further transcription of the *trp* operon by excluding RNA polymerase from the promoter.

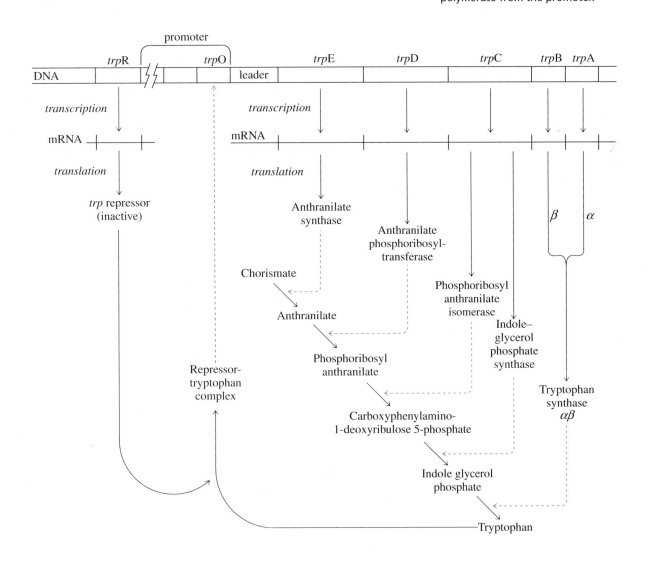

Note the difference between the effect of ligand on *lac* repressor (Section 27·10C) and on *trp* repressor. Allolactose signals the presence of the catabolic substrate and acts as an inducer of the *lac* operon: when *lac* repressor binds to allolactose, the repressor dissociates from the *lac* operators, and the *lac* operon is transcribed. In contrast, tryptophan signals an excess of end product and acts as a corepressor of the *trp* operon: when *trp* repressor binds to tryptophan, the repressor is able to bind to DNA, and transcription from the *trp* operon is repressed.

Regulation of the *E. coli trp* operon is supplemented and refined by a second, independent mechanism called **attenuation** that depends on translation and determines whether transcription of the *trp* operon proceeds or terminates prematurely. The movement of RNA polymerase from the promoter into the *trp*E gene is governed by a 162-nucleotide sequence that lies between the promoter and *trp*E. This sequence, called the **leader region,** includes a stretch of 45 nucleotides that encodes a 14–amino acid peptide called the **leader peptide.** The mRNA transcript of the leader region contains two consecutive codons specifying tryptophan near the end of the coding region for the leader peptide. In addition, the leader region contains four GC-rich sequences. The codons that specify tryptophan and the four GC-rich sequences regulate the synthesis of mRNA by affecting transcription termination.

When transcribed into mRNA, the four GC-rich sequences of the leader region can base-pair to form one of two alternative secondary structures (Figure 30·29). The first possible secondary structure involves the formation of two RNA hairpins. These hairpins form between the sequences labelled 1 and 2 in Figure 30·29a and between the sequences labelled 3 and 4. The 1-2 hairpin is a typical transcriptional pause site. The 3-4 hairpin is followed by a string of uridylate residues and is thus a typical rho-independent termination signal (Section 27·6). This particular termination signal is unusual, however, because it occurs upstream of the first gene in the *trp* operon. The other secondary structure involves formation of a single RNA hairpin between sequences 2 and 3. This hairpin, which is more stable than the 3-4 hairpin, forms only if sequence 1 is not available for hairpin formation with sequence 2.

During transcription of the leader region, RNA polymerase pauses when the 1-2 hairpin forms. While RNA polymerase pauses, a ribosome initiates translation of the mRNA encoding the leader peptide. This coding region begins just upstream of the 1-2 RNA hairpin. Sequence 1 encodes the C-terminal amino acids of the leader peptide and also contains a termination codon. As the ribosome translates sequence 1, it disrupts the 1-2 hairpin, thereby releasing the paused RNA polymerase, which then transcribes sequence 3. In the presence of tryptophanyl-tRNA^{Trp}, the ribosome and RNA polymerase move at about the same rate. When the ribosome encounters the termination codon of the *trp* leader mRNA, it dissociates, and the 1-2 hairpin re-forms. After the ribosome has disassembled, RNA polymerase transcribes sequence 4, which forms a termination hairpin with sequence 3. This termination signal causes the transcription complex to dissociate from the DNA template before the genes of the *trp* operon have been transcribed.

When tryptophan is scarce, however, the ribosome and RNA polymerase do not move in synchrony. If the concentration of cellular tryptophan is low, there is a deficiency of tryptophanyl-tRNA^{Trp} in the cell. Under these circumstances, the ribosome pauses when it reaches the two codons specifying tryptophan in sequence 1 of the mRNA molecule. RNA polymerase, which has already been released from the 1-2 pause site, transcribes sequences 3 and 4. While the ribosome is stationary and sequence 1 is covered, part of sequence 2 forms a hairpin loop with sequence 3. Since the 2-3 hairpin is more stable than the 3-4 hairpin, sequence 3 does not pair with sequence 4 to form the termination hairpin. Under these conditions, RNA

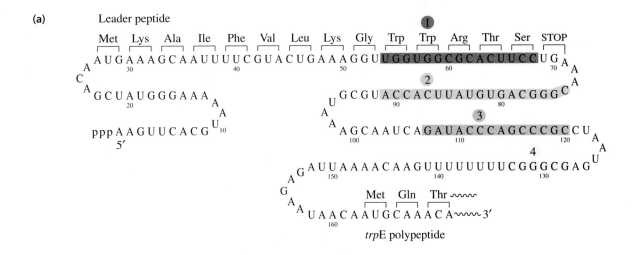

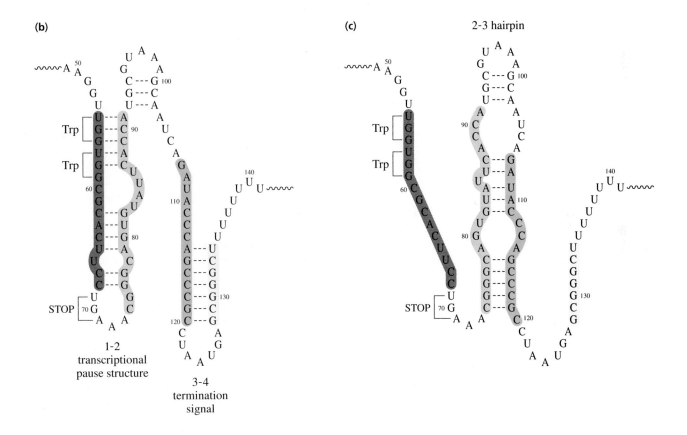

Figure 30·29

trp leader region. **(a)** mRNA transcript of the *trp* leader region. Included in this 162-nucleotide mRNA sequence are four GC-rich sequences and the coding region for a 14–amino acid leader peptide. This coding region includes two consecutive codons specifying tryptophan. The four GC-rich sequences can base-pair to form one of two alternative secondary structures. **(b)** Sequence 1 (red) and sequence 2 (blue) are complementary and, when base paired, form a typical transcriptional pause site. Sequence 3 (green) and sequence 4 (yellow) are complementary and, when base paired, form a rho-independent termination site. **(c)** Sequences 2 and 3 are also complementary and can form an RNA hairpin that is more stable than the 3-4 hairpin. This structure forms only if sequence 1 is not available for hairpin formation with sequence 2.

Figure 30·30
Attenuation of the *trp* operon in *E. coli.*

(a) During transcription of the *trp* leader region, RNA polymerase pauses at the 1-2 hairpin. While paused, a ribosome assembles at the initiation codon of the leader peptide.

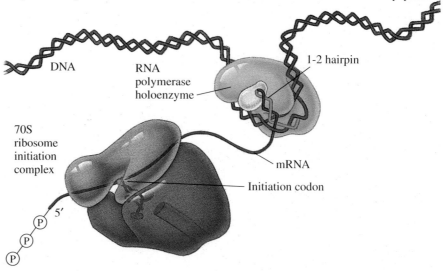

(b) The ribosome begins translation of the *trp* leader mRNA. The ribosome releases the paused RNA polymerase by disrupting the 1-2 hairpin as it translates sequence 1.

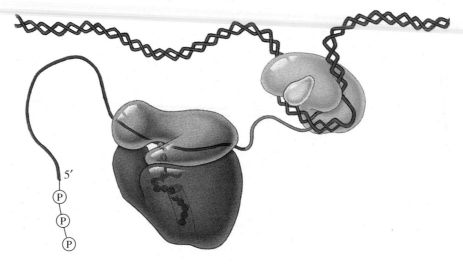

polymerase passes through the potential termination site, and the rest of the *trp* operon is transcribed (Figure 30·30).

Attenuation appears to be a regulatory mechanism that has evolved relatively recently and is found only in enteric bacteria, such as *E. coli.* (Attenuation cannot occur in eukaryotes because transcription and translation are separated by the nuclear membrane.) Several *E. coli* operons, including the *phe, thr, his, leu,* and *ile* operons, are regulated by attenuation mechanisms. Some operons, such as the *trp* operon, combine attenuation with repression, whereas others, such as the *his* operon, are regulated solely by attenuation. The leader peptides of operons whose genes are involved in amino acid biosynthesis may contain as many as seven codons specifying a particular amino acid.

(c) In the presence of tryptophanyl-tRNA^{Trp}, the ribosome and the RNA polymerase move at the same rate. The ribosome disassembles at the termination codon, allowing the 1-2 hairpin to re-form.

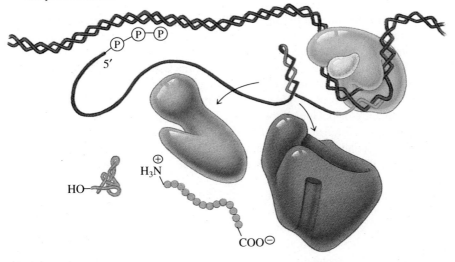

(d) RNA polymerase transcribes sequence 4. The formation of the 3-4 termination hairpin causes the transcription complex to dissociate, and transcription of the remainder of the *trp* operon is prevented.

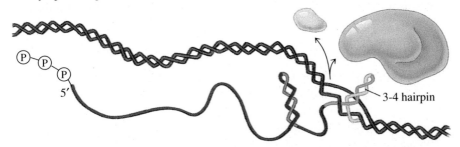

(e) In the absence of tryptophanyl-tRNA^{Trp}, the ribosome pauses over sequence 1 while the released RNA polymerase transcribes sequences 3 and 4. Since sequence 1 is covered by the ribosome, sequences 2 and 3 pair. The 2-3 hairpin is more stable than the 3-4 termination hairpin, and the rest of the *trp* operon is transcribed.

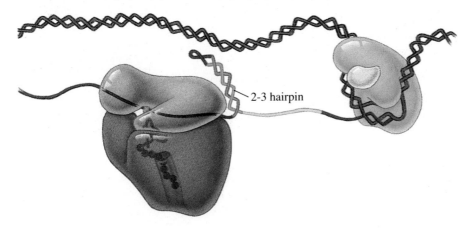

30·11 The Stringent Response Couples Transcription and Translation

When bacteria are starved for amino acids, uncharged tRNA molecules may enter the A site of the ribosome during protein synthesis. This triggers a pathway known as the **stringent response** that results in a decrease in transcription rates. The presence of an uncharged tRNA base paired with the codon at the A site activates a ribosome-associated enzyme known as stringent factor. This enzyme catalyzes the transfer of a pyrophosphate group from ATP to the 3′ position of GTP or GDP to produce pppGpp or ppGpp, respectively. These compounds are known as "magic spot" molecules since they were first detected as novel spots on chromatograms. The pentaphosphate compound can be converted to ppGpp by hydrolysis (Section 10·8D). It is the presence of this tetraphosphate derivative, known as magic spot 1, that signals amino acid starvation.

The magic-spot guanosine derivative inhibits transcription at the level of initiation and elongation. Synthesis of rRNA and tRNA is more severely affected than synthesis of mRNA. The net effect of the stringent response is to decrease the levels of RNA synthesis when amino acids are depleted and protein synthesis cannot proceed. Note that stringent factor specifically recognizes uncharged tRNA molecules bound to the ribosome. Since aminoacylation is a highly favorable reaction, the only time uncharged tRNA molecules are present within a cell is when amino acid concentrations are low. Thus, the stringent response provides a general sensing mechanism for amino acid starvation that couples translation and transcription; it is not a gene-specific mechanism.

30·12 Recognition of Selenocysteine Codons Involves a Unique mRNA Secondary Structure

Recall from Section 29·4 that the rare amino acid selenocysteine is sometimes incorporated at UGA codons, which normally serve as termination codons. This incorporation is specific and precise; only certain UGA codons encode selenocysteine, and selenocysteine is always inserted at these codons. How does the cell know when to read UGA as a termination codon and when to incorporate selenocysteine?

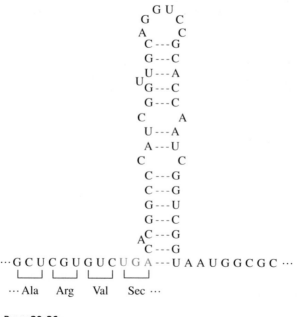

Figure 30·31
Secondary structure in the mRNA for formate dehydrogenase in *E. coli*. The UGA codon shown in blue encodes selenocysteine (Sec). Although not shown, the reading frame continues through the stem-loop region.

All proteins that contain selenocysteine are encoded by mRNA molecules that possess a defined secondary structure adjacent to the selenocysteine codon (Figure 30·31). In *E. coli,* insertion of selenocystyl-tRNA^{Sec} requires a special elongation factor, SelB, that is homologous to EF-Tu. SelB binds both selenocystyl-tRNA^{Sec} and the region of secondary structure conserved at all selenocysteine UGA codons. When a translation complex encounters a UGA codon at such a site, the seleno-cystyl-tRNA^{Sec} is inserted at the A site, and translation continues. Translation termination is inhibited by the presence of the mRNA secondary structure and because the SelB–selenocystyl-tRNA^{Sec} complex out-competes release factor for binding to UGA.

30·13 Proteins May Be Posttranslationally Modified and Transported

As translation proceeds, the nascent polypeptide chain grows in length. About 25 to 40 of the most recently polymerized amino acid residues remain in the tunnel of the ribosome, but amino acid residues closer to the N-terminus are extruded into the cytosol. These N-terminal residues consequently start folding into the native protein structure even before the C-terminus of the protein is synthesized. As these residues begin folding, they can be acted upon by enzymes that modify the nascent chain. Modifications that occur before chain elongation is complete are said to be **cotranslational,** and those that occur after elongation is complete are said to be **posttranslational.** Some examples from the hundreds of possible cotranslational and posttranslational modifications include deformylation of the N-terminal residue in prokaryotes, removal of the N-terminal methionine in prokaryotes and eukaryotes, disulfide-bond formation, acetylation of lysine residues (Section 24·11B), formation of diphthamide (Section 30·9), formation of desmosine (Section 5·8), and glycosylation of integral membrane proteins and secreted proteins (Section 30·15).

Among events that occur cotranslationally, one of the most important is the processing and transport of proteins through membranes. Protein synthesis occurs on cytosolic ribosomes, but the mature forms of many proteins are located on the noncytosolic side of membranes. For example, many receptor proteins in both bacteria and eukaryotes are embedded in the external membrane of the cell, with the bulk of the protein outside the cell. Other proteins are secreted from cells, and still others reside in lysosomes and other vesicles. In each of these cases, the cytosolically synthesized protein must be transported across a membrane barrier.

The best-characterized transport system is the one that carries proteins from the cytosol to the plasma membrane for secretion. In eukaryotes, proteins destined for secretion are transported across the membrane of the endoplasmic reticulum into the lumen. The lumen provides a network of channels through which newly synthesized proteins can travel on their way to the plasma membrane. Not all of the proteins passing through the endoplasmic reticulum are destined for secretion; some travel to other cellular organelles. In some cases, newly synthesized proteins contain markers that direct them to specific cellular locations, a phenomenon known as targeting.

30·14 Most Proteins Destined to Cross a Membrane Contain a Hydrophobic Signal Peptide

Eukaryotic proteins destined for secretion are carried in small vesicles from the endoplasmic reticulum to the Golgi apparatus. A Golgi apparatus is a stack of three to eight flattened, fluid-filled membrane vesicles, called Golgi sacs, or cisternae

(Section 2·4B). The various cisternae contain different enzymes and are the sites of distinct reactions. The side of the Golgi apparatus nearest the endoplasmic reticulum is called the *cis* face; the opposite side is called the *trans* face. From the *trans* face of the Golgi apparatus, secretory vesicles pinch off and move to the plasma membrane, where they release their contents from the cell by exocytosis (Figures 30·32 and 30·33). We can see from Figure 30·32 that the lumen of the endoplasmic reticulum is topologically equivalent to the outside of the plasma membrane, so that once a protein has crossed the endoplasmic reticulum membrane, it can be transported by secretory vesicles to the extracellular space without having to cross another membrane.

Figure 30·32
Secretory pathway. Proteins synthesized in the cytosol are transported into the lumen of the rough endoplasmic reticulum, processed further, and eventually secreted.

Proteins are transported across the membrane of the endoplasmic reticulum cotranslationally.

In the lumen of the endoplasmic reticulum, proteins may be covalently modified.

Proteins are transported to the Golgi apparatus in vesicles that bud off the endoplasmic reticulum. Additional posttranslational modifications occur within the Golgi apparatus.

Proteins to be secreted are carried from the Golgi apparatus to the plasma membrane in secretory vesicles. During transport, some of these proteins are activated by proteolytic cleavage.

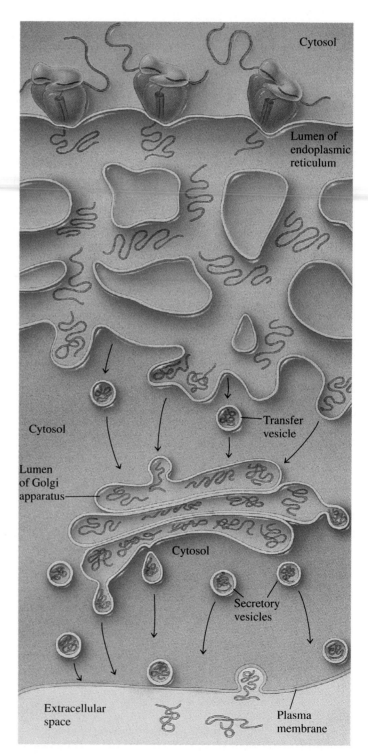

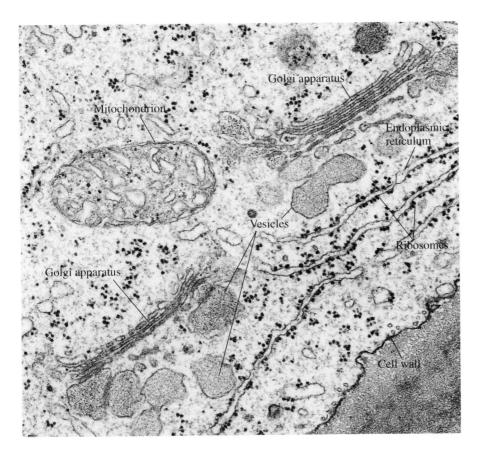

Figure 30·33
Secretory vesicles in a maize rootcap cell. Large secretory vesicles containing proteins are budding off the Golgi apparatus (center). Note the abundance of ribosomes bound to the endoplasmic reticulum. (Courtesy of Hilton H. Mollenhauer.)

The clue to the process by which many proteins cross the membrane of the endoplasmic reticulum appears in the first 20 or so residues of the nascent polypeptide chain. In most membrane-bound and secreted proteins, these residues are only present in the nascent polypeptide, not in the mature protein. The N-terminal sequence of residues that is proteolytically removed from the precursor is called the **signal peptide** since it is the portion of the precursor that signals the protein to cross a membrane. Signal peptides vary in length and composition, but they are typically from 16–30 residues in length and include 4–15 hydrophobic residues (Figure 30·34). The hydrophobicity of the signal peptide appears to be more important for

Figure 30·34
Signal peptides from secretory proteins. Hydrophobic residues are shown in blue, and arrows mark the sites where the signal peptide is cleaved from the precursor.

Prelysozyme

$H_3\overset{\oplus}{N}$-Met–Arg–Ser–Leu–Leu–Ile–Leu–Val–Leu–Cys–Phe–Leu–Pro–Leu–Ala–Ala–Leu–Gly↓–Gly〰

Preproalbumin

$H_3\overset{\oplus}{N}$-Met–Lys–Trp–Val–Thr–Phe–Leu–Leu–Leu–Leu–Phe–Ile–Ser–Gly–Ser–Ala–Phe–Ser↓–Arg〰

Alkaline phosphatase

$H_3\overset{\oplus}{N}$-Met–Lys–Gln–Ser–Thr–Ile–Ala–Leu–Ala–Leu–Leu–Pro–Leu–Leu–Phe–Thr–Pro–Val–Thr–Lys–Ala↓–Arg〰

Maltose-binding protein

$H_3\overset{\oplus}{N}$-Met–Lys–Ile–Lys–Thr–Gly–Ala–Arg–Ile–Leu–Ala–Leu–Ser–Ala–Leu–Thr–Thr–Met–Met–Phe–Ser–Ala–Ser–Ala–Leu–Ala↓–Lys〰

OmpA

$H_3\overset{\oplus}{N}$-Met–Lys–Lys–Thr–Ala–Ile–Ala–Ile–Ala–Val–Ala–Leu–Ala–Gly–Phe–Ala–Thr–Val–Ala–Gln–Ala↓–Ala〰

transport than its exact amino acid composition, since many residues can be substituted simultaneously in a signal peptide without loss of function. Secretion of proteins is not confined to eukaryotes. In fact, the role of signal peptides in protein export was first discovered in bacteria in experiments with maltose-binding protein and maltoporin.

In eukaryotes, many proteins destined for secretion appear to be translocated across the endoplasmic reticulum by the pathway shown in Figure 30·35. In the first step, an 80S initiation complex, composed of a ribosome, a Met-tRNA$_i^{Met}$ molecule, and an mRNA molecule, forms in the cytosol. In the next step, the ribosome begins translating the mRNA and synthesizing the signal peptide at the N-terminus of the precursor. Once the signal peptide is synthesized and extruded from the ribosome, it binds to a protein-RNA complex called a **signal-recognition particle** (SRP).

SRP is a small ribonucleoprotein complex composed of a 300-nucleotide RNA molecule called 7SL RNA and four proteins (two monomers of M_r 19 000 and 54 000 and two heterodimers of M_r 9000/14 000 and M_r 68 000/72 000). SRP recognizes and binds to the signal peptide as it emerges from the ribosome. When SRP binds, further translation is blocked. The complex then binds to an SRP receptor protein (sometimes known as docking protein) on the cytosolic face of the endoplasmic reticulum. The ribosome becomes anchored to the membrane of the endoplasmic reticulum by ribosome-binding proteins called **ribophorins,** and the signal peptide is inserted into the membrane at a pore that is part of the complex formed by the endoplasmic reticulum proteins at the docking site. Once the ribosome-SRP complex is bound to the membrane, the inhibition of translation is relieved, and SRP dissociates in a reaction coupled to GTP hydrolysis. Thus, the role of SRP is to recognize nascent polypeptides containing a signal peptide and to target the translation complex to the surface of the endoplasmic reticulum.

Once the translation complex is bound to the membrane, translation continues, and the polypeptide chain is passed through the membrane. The signal peptide is then cleaved by a **signal peptidase,** an integral membrane protein associated with the pore complex. The transport of proteins across the membrane is assisted by chaperones in the lumen of the endoplasmic reticulum. Chaperone proteins bind to the nascent polypeptide chain and help it fold properly. Chaperones are required for translocation, and their activity requires ATP hydrolysis. When protein synthesis terminates, the ribosome dissociates from the endoplasmic reticulum, and the translation complex is disassembled.

A modification of this process occurs in the synthesis of integral membrane proteins whose N-terminal sequences are on the luminal face of the endoplasmic reticulum and whose C-terminal portions reside either in the membrane or on the cytosolic face of the endoplasmic reticulum. These proteins contain a stop-transfer sequence, a series of amino acids that signals the transport machinery to stop passing the nascent chain through the membrane. In these proteins, the signal peptide directs the ribosome to the membrane and directs the N-terminus of the protein across the membrane, but the stop-transfer sequence assures that the C-terminus of the protein remains on the cytosolic side of the membrane.

Some integral membrane proteins are synthesized by a variation of the process just described, and others are inserted into the membrane by independent mechanisms. A few proteins are oriented in the membrane with their C-terminal residues on the luminal side and their N-terminal residues on the cytosolic side. In these proteins, the signal for transport across the membrane resides within the protein, not at the N-terminus. The protein is initially synthesized by cytosolic ribosomes, but once the internal signal peptide is synthesized, the nascent protein is moved to the membrane, where its C-terminus is synthesized and transported across the membrane. Other integral membrane proteins span the membrane more than once; perhaps these proteins have internal signal peptides that are not cleaved but that nevertheless direct the transport machinery to translocate the adjacent amino acid residues across the membrane.

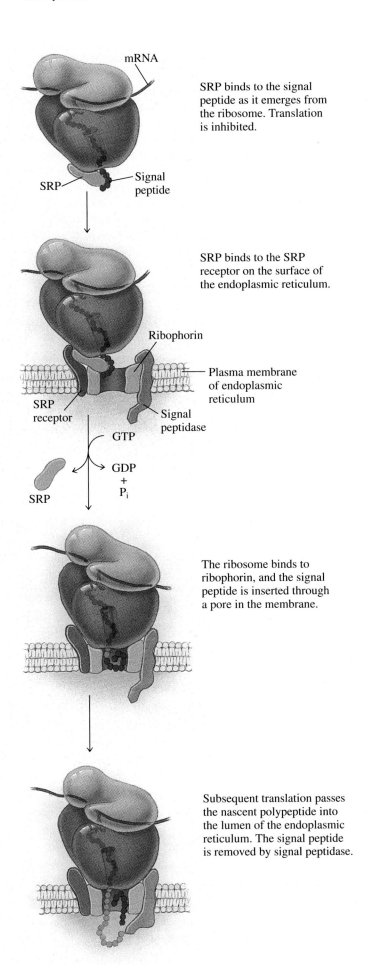

SRP binds to the signal peptide as it emerges from the ribosome. Translation is inhibited.

SRP binds to the SRP receptor on the surface of the endoplasmic reticulum.

The ribosome binds to ribophorin, and the signal peptide is inserted through a pore in the membrane.

Subsequent translation passes the nascent polypeptide into the lumen of the endoplasmic reticulum. The signal peptide is removed by signal peptidase.

Figure 30·35
Translocation of eukaryotic secretory proteins into the lumen of the endoplasmic reticulum.

Most membrane proteins of mitochondria and chloroplasts are encoded in the nucleus, synthesized in the cytosol, and transported across the membranes of the organelles. The transport of these proteins usually depends on a different N-terminal signal peptide and occurs posttranslationally with the assistance of chaperones both on the cytosolic side of the membrane and within the organelle.

There are two pathways for protein secretion in bacteria. The first pathway involves precursors that contain a signal peptide, and the mechanism of translocation is similar to that described for eukaryotes except that protein synthesis occurs on the plasma membrane. (Bacteria have a signal recognition particle that is also an RNA-protein complex and a signal peptidase that cleaves the signal peptide.) The second pathway is similar to the transfer of proteins across the mitochondrial and chloroplast membranes in that the action of several chaperones is required.

30·15 Some Proteins Are Glycosylated in the Lumen of the Endoplasmic Reticulum

Many integral membrane proteins and secretory proteins contain covalently bound oligosaccharide chains. The addition of these chains to proteins is called protein glycosylation. Protein glycosylation is one of the major metabolic activities of the lumen of the endoplasmic reticulum and of the Golgi apparatus. It can be regarded as an extension of the general process of protein biosynthesis and produces proteins known as glycoproteins. A glycoprotein can contain dozens, indeed hundreds, of monosaccharide units (Section 9·10). The mass of the carbohydrate portion of glycoproteins varies from less than 1% to more than 80% of the mass of the glycoprotein.

During their transit through the endoplasmic reticulum and the Golgi apparatus, proteins may be covalently modified in many ways, including formation of disulfide bonds and proteolytic cleavage. The oligosaccharide groups attached to the proteins are likewise modified during transit. By the time they have traversed the Golgi apparatus, the proteins and their oligosaccharides are usually fully modified.

A. The Biosynthesis of *O*-Linked Glycoproteins Requires Nucleotide Sugars and Specific Membrane-Bound Glycosyltransferases

The oligosaccharide chains of *O*-linked glycoproteins are generally attached to the side-chain hydroxyl groups of serine or threonine residues of the protein via C-1 of an *N*-acetylgalactosamine (GalNAc) residue (see Figure 9·41). After synthesis of the polypeptide backbone of an *O*-linked glycoprotein, GalNAc residues are transferred from UDP-GalNAc to one or more serine or threonine residues of the polypeptide in a reaction catalyzed by a specific GalNAc transferase. The attachment site is on the exterior surface of the protein, and there are always proline residues near the serine or threonine residues that serve as attachment sites. The remainder of the oligosaccharide chain is then assembled by the stepwise action of a battery of specific membrane-bound glycosyltransferases. These glycosyltransferases use nucleotide sugar donors (Section 8·4D) to build the oligosaccharide chain until the terminal sugar, usually a sialic acid residue, has been added.

Many glycosylation reactions involved in the synthesis of *O*-linked glycoproteins occur in the Golgi apparatus, whereas synthesis of most nucleotide sugars occurs in the cytosol. An adequate supply of nucleotide sugars is maintained within the Golgi apparatus by the action of transport proteins (Section 12·7). Many of

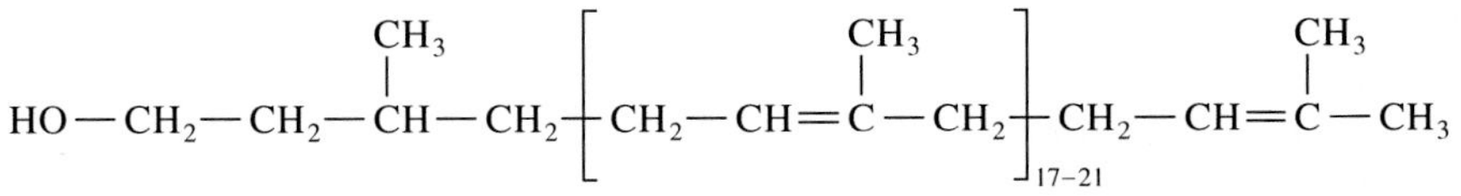

Figure 30·36
Structure of dolichol. The number of isoprene units in dolichol varies from approximately 17 to 21 in eukaryotes. The phosphate groups present in dolichol pyrophosphate–oligosaccharide are attached to the terminal hydroxyl group of dolichol.

these transport proteins employ an antiport mechanism in which the inward movement of the nucleotide sugar is accompanied by the outward movement of the corresponding nucleotide. In the Golgi apparatus, nucleotides are released from the corresponding nucleotide sugars during glycosylation reactions.

B. *N*-Glycosylation Occurs in the Endoplasmic Reticulum

The pathway of synthesis of *N*-linked glycoproteins employs a novel glycolipid, dolichol pyrophosphate–oligosaccharide, a compound not used in the pathway of synthesis of *O*-linked glycoproteins. Dolichol is a lipid composed of 17 to 21 repeating isoprenoid units (Figure 30·36). Dolichol pyrophosphate–oligosaccharide is shown in Figure 30·37. The first step in the synthesis of this compound is the phosphorylation of dolichol by a specific kinase. GlcNAc phosphate is then transferred from UDP-GlcNAc to dolichol phosphate, forming dolichol pyrophosphate–GlcNAc. Next, a second molecule of GlcNAc is donated by another molecule of UDP-GlcNAc. The other sugars of the oligosaccharide moiety are then added in a stepwise manner. Five mannose residues are added by sequential reactions employing GDP-mannose as the sugar donor. These initial reactions occur on the cytosolic side of the endoplasmic reticulum, with the dolichol moiety embedded in the membrane. Once synthesized, the dolichol pyrophosphate–heptasaccharide is translocated across the membrane, an event facilitated by the hydrophobic dolichol moiety. Further assembly of the dolichol pyrophosphate–oligosaccharide occurs on the luminal side of the endoplasmic reticulum. Four mannose residues and three glucose residues are added in reactions that employ dolichol phosphate–mannose and dolichol phosphate–glucose as donors, respectively, rather than nucleotide sugars. The assembled residues form two arms attached to the innermost mannose residue. One arm consists of five mannose residues, and the other arm consists of three mannose and three glucose residues. Because of the linkages involved, the first arm is referred to as the α-(1→6) arm, and the second as the α-(1→3) arm.

The entire oligosaccharide moiety ($Glc_3Man_9GlcNAc_2$—) of this complex glycolipid is transferred to the nitrogen of the side-chain amide group of specific asparagine residues in a reaction catalyzed by oligosaccharide transferase. The transfer occurs during synthesis of the polypeptide chain on membrane-bound polysomes (i.e., cotranslationally), and it is catalyzed by a complex of integral membrane proteins that includes ribophorin. The asparagine residues selected for

Figure 30·37
Structure of dolichol pyrophosphate–oligosaccharide. The two GlcNAc residues are donated by UDP-GlcNAc, and the five mannose residues indicated in red are donated by GDP-mannose. The other mannose residues and the three glucose residues are donated by dolichol phosphate–mannose and dolichol phosphate–glucose, respectively. See Figure 9·43a for details of the linkages in this structure.

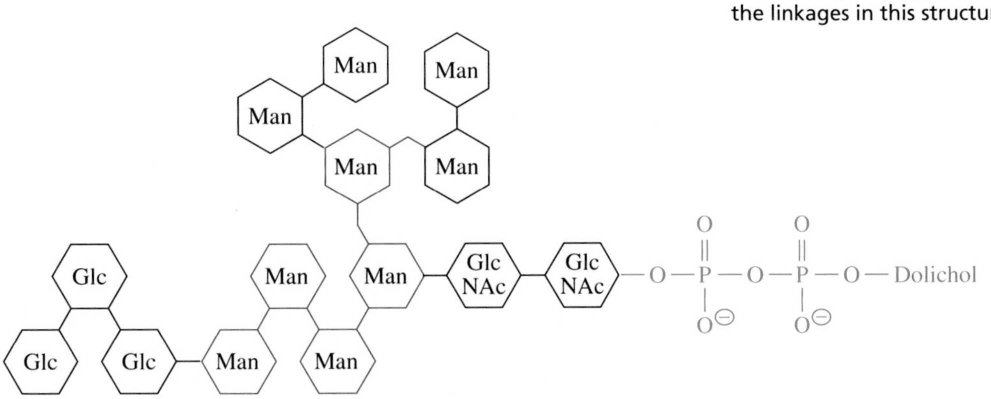

N-glycosylation are found in Asn-X-Ser or Asn-X-Thr tripeptide sequences, where *X* is any amino acid except proline and possibly aspartate. Not all asparagine residues located in such sequences are *N*-glycosylated.

Once the oligosaccharide is transferred to the polypeptide chain, a number of its sugar residues are removed by the action of specific hydrolases in the Golgi apparatus. Removal of the glucose residues generates a high-mannose oligosaccharide chain. Additional residues may be removed, leaving a core pentasaccharide to which new residues are subsequently added to produce a complex oligosaccharide chain. These modification reactions are collectively known as **oligosaccharide processing.**

Specific glycosidases and glycosyltransferases that use nucleotide sugars as glycosyl-group donors catalyze the steps of oligosaccharide processing. In certain cases, a glycoprotein may contain a hybrid oligosaccharide chain—a branched oligosaccharide chain in which one branch is of the high-mannose type and the other is of the complex type (see Figure 9·43). Certain oligosaccharide chains may contain three, four, or even more complex branches; these are called tri- or tetra-antennary structures. The same glycoprotein may also contain both *N*- and *O*-linked oligosaccharide chains.

C. Certain Enzymes Are Targeted to Lysosomes by a Mannose 6-Phosphate Signal

Certain lysosomal enzymes are directed toward that organelle by a unique signal that is attached covalently. The signal, mannose 6-phosphate, is generated in the *cis* Golgi by two sequential reactions. In the first step, GlcNAc 6-phosphate is transferred to one or more mannose residues attached to proteins destined to be lysosomal enzymes. The reaction is catalyzed by GlcNAc phosphotransferase.

$$\text{Lysosomal enzyme–Mannose } + \text{ UDP-GlcNAc}$$

$$\Big\downarrow \text{GlcNAc phosphotransferase} \qquad\qquad (30\cdot1)$$

$$\text{Lysosomal enzyme–Mannose 6-phosphate–GlcNAc } + \text{ UMP}$$

In the second step, the GlcNAc moiety is hydrolyzed by *N*-acetylglucosamine 1-phosphodiester α-*N*-acetylglucosaminidase, leaving mannose 6-phosphate attached to the lysosomal enzyme.

$$\text{Lysosomal enzyme–Mannose 6-phosphate–GlcNAc}$$

$$\Big\downarrow \begin{array}{l}\textit{N}\text{-Acetylglucosamine 1-phosphodiester} \\ \alpha\text{-}\textit{N}\text{-acetylglucosaminidase}\end{array} \qquad (30\cdot2)$$

$$\text{Lysosomal enzyme–Mannose 6-phosphate } + \text{ GlcNAc}$$

Next, the modified enzymes bind to mannose 6-phosphate receptor molecules in the lumen of the *trans* Golgi. The receptor molecules are transmembrane proteins of two distinct types with similar functions: one with a molecular weight of ~215 000, and the other with a molecular weight of ~45 000. Upon binding, specialized vesicles form, bud off, and fuse with a vesicle called the compartment of uncoupling of receptor and ligand (CURL). The low pH in CURL promotes dissociation of the prospective lysosomal enzymes from the mannose 6-phosphate receptors. Vesicles containing the enzymes but not the receptors then bud from this site and fuse with lysosomes, delivering the proteins to their destination. The mannose 6-phosphate receptors are then recycled, as is the case with a number of other receptors (e.g., asialoglycoprotein receptors and low density lipoprotein receptors).

In some individuals, mutations in the gene encoding GlcNAc phosphotransferase result in a defective enzyme. As a consequence, mannose 6-phosphate is not generated in these individuals, and lysosomal enzymes are not targeted to lysosomes. Instead, these enzymes are secreted from the cells into the circulation. If cells from such individuals are cultured, the enzymes may be detected in the medium, whereas little activity is detected in the medium from normal cells. Lysosomes from individuals deficient in GlcNAc phosphotransferase do not function properly, and various molecules accumulate that would normally be degraded. The accumulations are visible in the electron microscope as intracellular inclusions, and the resulting disorder is termed I cell disease (I for inclusion).

Summary

The general structure and function of ribosomes are conserved in all organisms. Ribosomes are composed of a large and a small subunit. In *E. coli,* the small subunit is the 30S subunit; it contains 21 proteins and one molecule of 16S rRNA. The large subunit is the 50S subunit; it contains at least 31 different proteins and two rRNA molecules, 5S and 23S. During protein synthesis, the two subunits combine to form an active 70S ribosome. The cytoplasmic ribosomes of eukaryotes are larger than prokaryotic ribosomes (80S, with subunits of 40S and 60S) and contain more proteins. The large subunit of eukaryotic ribosomes contains three molecules of rRNA: 28S, 5S, and 5.8S. The general structures of the rRNA molecules are conserved in all species, even though the exact nucleotide sequences of the molecules vary. Ribosome assembly is coupled to the processing of rRNA precursors in both prokaryotes and eukaryotes.

The formation of peptide bonds on the ribosome occurs in three distinct stages: initiation, elongation, and termination. During initiation in prokaryotes, an initiation complex is formed by the mRNA, the 30S subunit, an aminoacylated initiator tRNA (fMet-tRNA$_f^{Met}$), and the 50S subunit. Formation of the initiation complex is facilitated by three initiation factors that promote binding between the components of the complex and requires the consumption of one molecule of GTP. The proper codon for initiation is selected by an interaction between the initiation codon and the aminoacylated initiator tRNA; this interaction is augmented by base pairing between a region of the 16S rRNA and a complementary Shine-Dalgarno sequence in the mRNA molecule upstream of the initiation codon.

In the elongation stage of protein synthesis, an aminoacyl-tRNA binds initially at the A site, adjacent to a peptidyl-tRNA at the P site. Peptide-bond formation is followed by translocation of the new peptidyl-tRNA from the A site to the P site and ejection of the free tRNA from the E site. This process requires the participation of three elongation factors and the consumption of two molecules of GTP.

Termination of protein synthesis occurs at specific termination codons and requires the participation of release factors and the consumption of one molecule of GTP. Various steps of protein synthesis are inhibited by antibiotics.

Protein synthesis in eukaryotes is similar to protein synthesis in prokaryotes, particularly in chain elongation and termination. Initiation in eukaryotes requires the participation of at least eight initiation factors. One of these initiation factors interacts specifically with the 5′ cap on eukaryotic mRNA molecules. The initiation complex in eukaryotic translation is assembled at the end of the mRNA, and it moves to the first available initiation codon, where translation begins.

Translation can be regulated in several ways. Some proteins, such as ribosomal proteins, bind to the 5′ end of their mRNA and inhibit initiation. Translation can also be inhibited by antisense RNA. Modification of initiation factors can control translation initiation. Additional regulatory mechanisms include translational frameshifting and attenuation.

Many proteins are covalently modified during and after translation. Some of these modifications affect the transport of proteins to different cellular locations. A protein that is destined to be secreted or inserted into a plasma membrane often contains a hydrophobic signal peptide of 16–30 residues at its N-terminus. This peptide is recognized by signal recognition particle (SRP), which blocks further peptide-bond formation until the nascent polypeptide and ribosome are bound to the membrane of the endoplasmic reticulum. The newly synthesized protein is then translocated through a pore in the membrane into the lumen of the endoplasmic reticulum. A signal peptidase cleaves the signal peptide while protein synthesis is still in progress.

Once in the lumen of the endoplasmic reticulum, many secretory proteins are glycosylated. Oligosaccharides may be attached to serine or threonine residues (in *O*-linked glycoproteins) or to asparagine residues (in *N*-linked glycoproteins). In *N*-linked glycoproteins, the oligosaccharide chain is first synthesized as a 14-hexose chain attached to dolichol pyrophosphate. Protein-bound oligosaccharides may be modified in the Golgi apparatus. Some of these modifications direct the transport of glycoproteins to specific cellular locations.

Selected Readings

Ribosomes

Berkovitch-Yellin, Z., Bennett, W. S., and Yonath, A. (1992). Aspects in structural studies on ribosomes. *Crit. Rev. Biochem. Mol. Biol.* 27:403–444.

Hill, W. E., Dahlberg, A., Garret, R. A., Moore, P. B., Schlessinger, D., and Warner, J. R., eds. (1990). *The Ribosome: Structure, Function, and Evolution* (Washington, DC: American Society for Microbiology).

Noller, H. F. (1991). Ribosomal RNA and translation. *Annu. Rev. Biochem.* 60:191–227.

Noller, H. F. (1993). Peptidyl transferase: protein, ribonucleoprotein, or RNA? *J. Bacteriol.* 175:5297–5300.

Noller, H. F., Hoffarth, V., and Zimniak, L. (1992). Unusual resistance of peptidyl transferase to protein extraction procedures. *Science* 256:1416–1419.

Olsen, G., and Woese, C. R. (1993). Ribosomal RNA: a key to phylogeny. *FASEB J.* 7:113–123.

Yonath, A., Leonard, K. R., and Whittmann, H. G. (1987). A tunnel in the large ribosomal subunit revealed by three-dimensional image reconstruction. *Science* 236:813–816.

Yonath, A., and Whittmann, H. G. (1989). Challenging the three-dimensional structure of ribosomes. *Trends Biochem. Sci.* 14:329–335.

Protein Synthesis

Gualerzi, C. O., and Pon, C. L. (1990). Initiation of mRNA translation in prokaryotes. *Biochemistry* 29:5881–5889.

Hartz, D., McPheeters, D. S., and Gold, L. (1989). Selection of the initiator tRNA by *Escherichia coli* initiation factors. *Genes Dev.* 3:1899–1912.

Kozak, M. (1992). A consideration of alternative models for the initiation of translation in eukaryotes. *Crit. Rev. Biochem. Mol. Biol.* 27:385–402.

Kurland, C. G. (1992). Translational accuracy and the fitness of bacteria. *Annu. Rev. Genet.* 26:29–50.

Linder, P., and Pratt, A. (1990). Baker's yeast, the new work horse in protein synthesis studies: analyzing eukaryotic translation initiation. *BioEssays* 12:519–526.

Madden, T. (1993). The translational apparatus, from Berlin. *Trends Biochem. Sci.* 18:155–157.

McCarthy, J. E. G., and Gualerzi, C. (1990). Translational control of prokaryotic gene expression. *Trends Genet.* 6:78–85.

Salas, M., Smith, M. A., Stanley, W. M., Jr., Wahba, A. J., and Ochoa, S. (1965). Direction of reading of the genetic code. *J. Biol. Chem.* 240:3988–3995.

Schimmel, P. (1993). GTP hydrolysis in protein synthesis: two for Tu? *Science* 259:1264–1265.

Weijland, A., and Parmeggiani, A. (1993). Toward a model for the interaction between elongation factor Tu and the ribosome. *Science* 259:1311–1314.

Translational Regulation

Datta, B., Chakrabarti, D., Roy, A. L., and Gupta, N. K. (1988). Roles of a 67-kDa polypeptide in reversal of protein synthesis inhibition in heme-deficient reticulocyte lysate. *Proc. Natl. Acad. Sci. USA* 85:3324–3328.

Gold, L. (1988). Posttranscriptional regulatory mechanisms in *Escherichia coli. Annu. Rev. Biochem.* 57:199–233.

Kozak, M. (1992). Regulation of translation in eukaryotic systems. *Annu. Rev. Cell Biol.* 8:197–225.

Merrick, W. C. (1992). Mechanism and regulation of eukaryotic protein synthesis. *Microbiol. Rev.* 56:291–315.

Philippe, C., Eyermann, F., Bénard, L., Portier, C., Ehresmann, B., and Ehresmann, C. (1993). Ribosomal protein S15 from *Escherichia coli* modulates its own translation by trapping the ribosome on the mRNA initiation loading site. *Proc. Natl. Acad. Sci USA* 90:4394–4398.

Rhoads, R. E. (1993). Regulation of eukaryotic protein synthesis by initiation factors. *J. Biol. Chem.* 268:3017–3020.

Samuel, C. E. (1993). The eIF-2α protein kinases, regulators of translation in eukaryotes from yeasts to humans. *J. Biol. Chem.* 268:7603–7606.

Translational Frameshifting

Craigen, W. J., and Caskey, C. T. (1987). Translational frameshifting: where will it stop? *Cell* 50:1–2.

Craigen, W. J., Cook, J. R., Tate, W. P., and Caskey, C. T. (1985). Bacterial peptide chain release factors: conserved primary structure and possible frameshift regulation of release factor 2. *Proc. Natl. Acad. Sci. USA* 82:3616–3620.

Flower, A. M., and McHenry, C. S. (1990). The "gamma" subunit of DNA polymerase III holoenzyme of *Escherichia coli* is produced by ribosomal frameshifting. *Proc. Natl. Acad. Sci. USA* 87:3713–3717.

Attenuation

Landick, R., and Yanofsky, C. (1987). Transcription attenuation. In Escherichia coli *and* Salmonella typhimurium: *Cellular and Molecular Biology,* Vol. 2, F. C. Neidhardt, ed. (Washington, DC: American Society for Microbiology), pp. 1276–1301.

Platt, T. (1986). Transcription termination and the regulation of gene expression. *Annu. Rev. Biochem.* 55:339–372.

Yanofsky, C., and Crawford, I. P. (1987). The tryptophan operon. In Escherichia coli *and* Salmonella typhimurium: *Cellular and Molecular Biology,* Vol. 2, F. C. Neidhardt, ed. (Washington, DC: American Society for Microbiology), pp. 1453–1472.

Protein Transport and Glycosylation

Farquhar, M. G. (1985). Progress in unravelling pathways of Golgi traffic. *Annu. Rev. Cell Biol.* 1:447–488.

Hartl, F.-U., and Wiedmann, M. (1993). A signal recognition particle in *Escherichia coli? Curr. Biol.* 3:86–89.

High, S. (1992). Membrane protein insertion into the endoplasmic reticulum—another channel tunnel? *BioEssays* 14:535–540.

Hurtley, S. M. (1993). Hot line to the secretory pathway. *Trends Biochem. Sci.* 18:3–6.

Nothwehr, S. F., and Gordon, J. I. (1990). Targeting of proteins into the eukaryotic secretory pathway: signal peptide structure/function relationships. *BioEssays* 12:479–484.

Salmond, G. P. C., and Reeves, P. J. (1993). Membrane traffic wardens and protein secretion in Gram-negative bacteria. *Trends Biochem. Sci.* 18:7–12.

Sanders, S. L., and Schekman, R. (1992). Polypeptide translocation across the endoplasmic reticulum membrane. *J. Biol. Chem.* 267:13 791–13 794.

Siegel, V., and Walter, P. (1988). Each of the activities of signal recognition particle (SRP) is contained within a distinct domain. *Cell* 52:39–49.

31

Gene Expression and Development

The process of information flow discussed in the last few chapters results in the synthesis of the proteins and RNA needed by living organisms. Since no living organism needs all its gene products simultaneously and in the same amounts, the synthesis of most macromolecules is regulated. Regulation ensures that the resources of a cell are committed to the synthesis of particular gene products only when they are needed. Control of the synthesis of RNA and protein is called regulation of gene expression.

As we have already seen, gene expression can be regulated at any step in the pathway of biological information flow. For instance, the amount of mature RNA produced in a cell depends on the frequency of transcription initiation, the rate of RNA elongation, the efficiency of transcription termination, and the rates of the various RNA processing steps. Similarly, the amount of any protein produced by a cell depends on the stability of its mature mRNA, the frequency of translation initiation, the rate of polypeptide-chain elongation, the efficiency of translation termination, and the efficiency of posttranslational modification. Thus, gene expression is determined by many of the elements discussed in the preceding chapters, such as ribosome-binding sites, sequences of promoters, processing signals, and so on.

In this chapter, we focus on a subset of regulated genes called developmentally regulated genes. The expression of these genes is spatially or temporally regulated; that is, the genes are expressed only in certain cells or at certain times during the life of an organism according to a genetically programmed pattern. Developmentally regulated genes include those responsible for the proper embryonic development of multicellular organisms and those expressed in prokaryotes after infection by certain bacteriophages. Most of the examples of developmentally regulated genes that we encounter in this chapter involve controls at the level of transcription. In several cases, we examine networks of interacting genes and their products. In these cases, expression of one set of genes produces proteins that regulate expression of another set of genes whose products control expression of a third set of genes, and so on. Sequential gene expression of this sort is known as a cascade. We begin this chapter with examples of simple cascades in bacteriophages and viruses and then use these systems to introduce the more complex systems that regulate development in bacteria and multicellular eukaryotes.

31·1 Bacteriophage and Viral Genes Can Be Transcribed at Different Times During Infection

Recall from Section 27·12 that bacterial cells contain multiple σ factors that bind to RNA polymerase and control transcription of various sets of genes. For example, the σ^{70} subunit of *Escherichia coli* RNA polymerase recognizes the promoters of most housekeeping genes, whereas the σ^{32} and σ^{54} subunits recognize the promoters of heat-shock and nitrogen-metabolism genes, respectively. The rates at which prokaryotic genes are transcribed depend in part on the relative concentrations of the various σ factors in the cell. Thus, the transcription of prokaryotic genes can be regulated by controlling the synthesis of the various σ factors. The situation is slightly different in eukaryotes, which contain three different RNA polymerases and a panoply of transcription factors (Section 27·8). In this case, it is the relative concentrations of transcription factors that can be used to control gene expression.

When a bacteriophage or virus infects a cell, the host transcription machinery (be it prokaryotic or eukaryotic) must be used to express the bacteriophage or viral genes. Once expression has begun using the host machinery, the invader can use its initial gene products to subvert the host transcription apparatus and cause preferential expression of its own genes. There are two general strategies employed for such subversion: production of phage- or virus-specific RNA polymerases and synthesis of phage- or virus-specific transcription factors.

The control of gene expression during bacteriophage or viral infection provides simple examples of developmental regulation. In these cases, different genes are expressed at different times during infection. Early genes encode proteins that are required for viral DNA replication, subversion of the host transcription machinery, and expression of late genes. Late genes encode proteins required for the packaging of DNA into new infective particles and for release of these particles by lysis or extrusion. Ordered expression of these genes is essential for the proper development of the mature phage or viral particles.

A. A Phage RNA Polymerase Transcribes Late Genes in Bacteriophage T7

Bacteriophage T7 (Figure 31·1) is a phage that infects *E. coli*. The large T7 genome consists of a linear, double-stranded DNA molecule of 39 936 bp. The genome comprises 55 genes, which encode proteins required for transcription, DNA replication, and the production of new bacteriophage particles. The genes form three clusters (Figure 31·2). At one end are the class I genes, which are required for viral transcription and protection of the viral DNA. In the middle are the class II genes; most of these genes encode proteins required for DNA replication. At the other end are

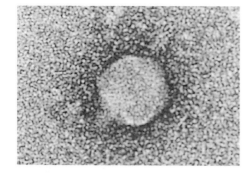

Figure 31·1
Electron micrograph of bacteriophage T7. (Courtesy of Jack Griffith.)

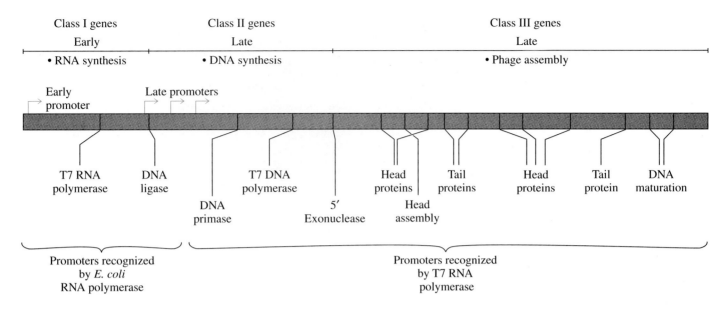

Figure 31·2
Abbreviated genetic map of bacteriophage T7. The T7 genome is 39 936 bp in size and consists of 55 genes (16 of which are shown here). Class I genes are required for viral transcription and protection of the viral DNA; they are transcribed immediately after infection. Class II genes encode mainly proteins required for DNA replication; they are transcribed later during infection. Class III genes encode proteins involved in the assembly of new phage particles; they are transcribed last. Promoters for all class I and some class II genes are recognized by *E. coli* RNA polymerase plus σ^{70}. Promoters for all other class II genes and all class III genes are recognized by T7 RNA polymerase, a product of a class I gene.

the class III genes, which encode proteins involved in the assembly of new phage particles. Class I genes are transcribed immediately after infection, class II genes are transcribed later during infection, and class III genes are transcribed last.

During infection, T7 binds to the surface of an *E. coli* cell and injects its linear DNA, beginning with the end on the left in Figure 31·2. This end contains genes that have promoters recognized by *E. coli* RNA polymerase containing σ^{70}. One of the products of these early genes is T7 RNA polymerase, an enzyme that in turn recognizes promoters of most of the T7 class II and all of the class III genes.

Bacteriophage T7 RNA polymerase is somewhat unusual in that it is a single polypeptide chain of M_r 98 000, whereas the other RNA polymerases we have examined so far are large, multisubunit enzymes. T7 RNA polymerase has some sequence similarity to eukaryotic mitochondrial RNA polymerases and DNA polymerases. As might be expected, these similarities predominantly involve amino acid residues in the vicinity of the active site for polymerization.

T7 RNA polymerase catalyzes all the reactions carried out by the more complex enzymes: searching for and binding the promoter, forming an open complex, winding and unwinding DNA, primer synthesis, elongation, and termination at RNA hairpins followed by runs of uridylate residues. Interestingly, the single-subunit enzyme does not require accessory factors for most of these reactions; for example, it does not need σ factors for promoter binding, nor does it need NusA for termination.

Genes that are expressed late during T7 infection have promoters with a highly conserved 23 bp consensus sequence. These promoters, although very different from those recognized by host σ factors, are high-affinity binding sites for T7 RNA polymerase. Transcription from late promoters occurs about six minutes after infection begins. This is enough time for the early genes to be transcribed by the host RNA polymerase and for the synthesis of T7 RNA polymerase. T7 RNA polymerase then catalyzes transcription of class II and III genes with T7 promoters. Phage DNA is injected slowly into the bacterial cell, so late genes nearer the end on the left in Figure 31·2 are transcribed before those near the right. This ensures that the class II genes are transcribed before the class III genes and that phage DNA is replicated before the assembly of new phage particles begins. Note that all T7 genes are transcribed from the same strand; in other words, all T7 genes have the same orientation (in fact, most of the late genes are part of a large operon). Transcription of the late genes can therefore begin before all the DNA has been injected into the cell.

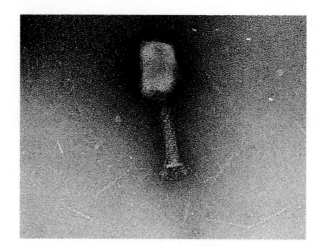

Figure 31·3
Electron micrograph of bacteriophage T4.
(Courtesy of Jack Griffith.)

B. A Bacteriophage T4 σ Factor Controls Late Gene Expression

Bacteriophage T4 is shown in Figure 31·3. The T4 genome is a linear, double-stranded DNA molecule of about 166 kb, four times the size of the T7 genome. Although the T4 chromosome is linear, the genetic map of T4 is usually shown as a circle (Figure 31·4). This is because each linear T4 chromosome contains duplicated ends (in other words, it contains one full chromosome, plus a bit more at one end). The duplicated sequences arise during DNA replication of a circular template and are not the same in every T4 chromosome (that is, different chromosomes have different ends). Thus, in order to properly explain all observed linkages among T4 genes, a circular map is required.

Figure 31·4
Abbreviated genetic map of bacteriophage T4. The genome is 166 kb in size and encodes about 200 proteins. Although the DNA in an infective phage is linear, the genome is drawn as a circular molecule to accurately represent genetic linkages. The numbers on the inside of the circle indicate kilobase pairs.

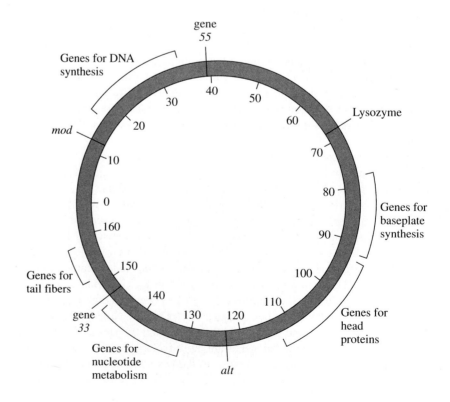

There are about 200 genes in the T4 genome. They encode a variety of proteins, including those responsible for DNA replication, transcription, recombination, and the synthesis of phage particles. As in T7 infection, the ordered expression of T4 genes is essential for the proper development of mature phage particles.

The first genes transcribed following injection of bacteriophage T4 DNA have promoters recognized by the bacterial holoenzyme containing σ^{70}. Among the first genes transcribed are the two genes *alt* and *mod*. After infection, the α subunits of all host RNA polymerase molecules are rapidly modified by ADP ribosylation mediated by the proteins encoded by *alt* and *mod*. The modified polymerase exhibits reduced affinity for σ^{70}; as we will see in a moment, this reduced affinity is important for late gene transcription.

Among the early genes transcribed by modified RNA polymerase are gene *55*, gene *33*, and a gene that encodes a small accessory protein of M_r 10 000. Gene *55* encodes a new σ factor, σ^{gp55} (*gp* stands for gene product). The protein encoded by gene *33* (gp33) and the small accessory protein bind to the ADP-ribosylated RNA polymerase as additional subunits of the holoenzyme. In doing so, these proteins further decrease the affinity of the enzyme for bacterial σ factors. When sufficient σ^{gp55} accumulates inside the cell, it competes successfully with σ^{70} for binding to the modified RNA polymerases and directs the enzyme to T4 late promoters having the consensus sequence TATAAATA (see Table 27·3). The combination of the small accessory protein, gp33, σ^{gp55}, and the covalently modified α subunits leads to reduced transcription of all bacterial and phage early genes by the subverted RNA polymerase and increased transcription of phage late genes (Figure 31·5).

Late genes are not transcribed until DNA replication is underway, and DNA replication cannot begin until many of the early gene products, including T4-specific replication factors, are synthesized. The coupling of late gene transcription and DNA replication is complex, involving several accessory factors that are integral parts of the replisome. These factors seem to stimulate the formation of an open complex by σ^{gp55}-containing RNA polymerase bound to a late promoter. Consequently, transcription initiation at the late promoters occurs only in the presence of a nearby, active replication complex. Furthermore, the replisome acts as an enhancer only at promoters whose genes are oriented in the same direction as the replication fork moves. This prevents head-on collisions between replication forks and transcription complexes.

C. A σ Cascade Regulates Expression of Bacteriophage SP01 Genes

The temporal expression of the genes of *Bacillus subtilis* bacteriophage SP01 (Figure 31·6) is also controlled by σ factors, but in this case there is sequential expression of two different factors, resulting in a cascade. The genes of SP01 can be

Figure 31·5
Late gene transcription in bacteriophage T4.

E. coli RNA polymerase is modified by the products of the *alt* and *mod* genes. The modified RNA polymerase binds to the promoter for gene *55*.

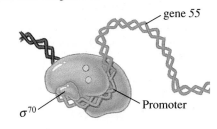

The transcript of gene *55* is translated to yield σ^{gp55}.

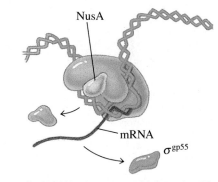

Modified RNA polymerase with bound gp33 and a small accessory protein binds σ^{gp55} and recognizes promoters for late genes.

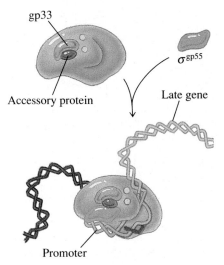

Figure 31·6
Electron micrograph of bacteriophage SP01. (Courtesy of Jack Griffith.)

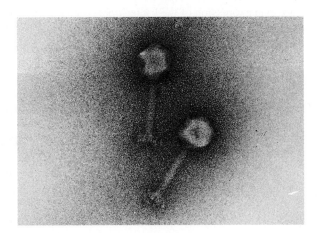

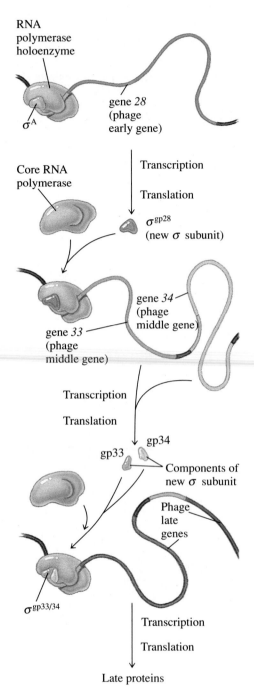

Figure 31·7

σ cascade in *B. subtilis* after infection by bacteriophage SP01. The phage early genes are transcribed by *B. subtilis* RNA polymerase holoenzyme (core + σ^A). One of these early genes encodes a phage-specific σ factor, σ^{gp28}. When σ^{gp28} associates with bacterial core RNA polymerase, the resulting holoenzyme recognizes the promoters of the phage middle genes and initiates transcription. Two of the phage middle genes encode the components of a second phage-specific σ factor, $\sigma^{gp33/34}$, which in turn interacts with bacterial core RNA polymerase. The resulting holoenzyme recognizes the promoters of the phage late genes.

Table 31·1 σ factors in bacteriophage SP01

Factor	Gene	Target genes	Consensus sequence*	
			−35	−10
σ^A	*sig*A	Housekeeping, phage early	TTGACA	TATAAT
σ^{gp28}	gene *28*	Phage middle	TNAGGAGA	TTTNTTT
$\sigma^{gp33/34}$	gene *33* gene *34*	Phage late	CGTTAGA	GATATT

*N represents any nucleotide.

divided into three groups—early, middle, and late—based on their order of transcription. Each of these groups possesses a promoter consensus sequence that is recognized by a different σ subunit (Table 31·1). All phage early genes possess promoters that are recognized by σ^A, the major σ factor of *B. subtilis*. One of the first phage early genes transcribed is gene *28,* which encodes the phage-specific σ factor σ^{gp28}. As soon as it is synthesized, σ^{gp28} associates with the bacterial core RNA polymerase, and the resulting holoenzyme recognizes the promoters of the phage middle genes and initiates transcription. Two of these middle genes, *33* and *34,* then encode the components of another phage-specific σ factor, $\sigma^{gp33/34}$. When this second σ factor associates with the bacterial core RNA polymerase, the resulting holoenzyme recognizes the promoters of the phage late genes and initiates transcription. Thus, the temporal regulation of gene expression in phage SP01 involves the sequential synthesis of two σ factors encoded by the phage genome. This mechanism of developmental regulation is termed a **σ cascade** (Figure 31·7). Studies of the temporal regulation of gene expression during SP01 infection provided some of the first evidence that different σ subunits could direct RNA polymerase to specific promoters.

D. A New Transcription Factor Is Synthesized During Adenovirus Infection

The adenoviruses are a group of related viruses that infect mammalian cells. Some of these viruses cause upper respiratory infection in humans, and others promote cancer. Adenovirus genomes consist of double-stranded DNA molecules of approximately 36 kb. Expression of adenovirus genes during infection is more complex than expression of the bacteriophage genes we just examined due to RNA processing, especially splicing. (Introns in protein-encoding genes were first discovered in adenovirus.) A simplified version of the adenovirus genome is shown in Figure 31·8. Promoters at one end of the viral DNA are recognized by host RNA polymerase II in combination with general class II transcription factors. The two early genes transcribed are *E1A* and *E1B*. The primary transcripts of these genes can be processed in several ways to yield multiple related proteins. The largest E1A protein (289 amino acid residues) is a transcriptional regulatory protein.

The largest E1A protein does not bind directly to DNA, but it does increase the rate of transcription at many class II promoters. E1A serves as a bridge between the transcription complex at the promoter and transcription factors bound to specific sites in the vicinity of the promoter. For example, E1A binds to a host transcription factor known as ATF-2. This factor is present in most cells, and there are ATF-2–binding sites near the promoters of many mammalian genes as well as promoters of

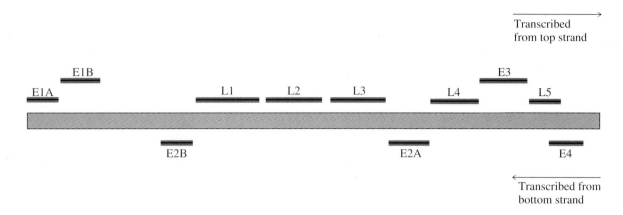

Figure 31·8
Abbreviated genetic map of adenovirus. The adenovirus genome is 36 kb in size. Unprocessed RNA transcripts encoded by the adenovirus genome are shown.

adenovirus genes. E1A also binds to the general transcription factor TFIID; this interaction probably leads to increased rates of initiation. The net effect of the presence of E1A is stimulation of transcription of all viral genes with ATF-2–binding sites.

In addition to enhancing viral-gene transcription, E1A also stimulates transcription of several host genes by interacting with additional transcription factors. One of the genes that is activated encodes proliferating cell nuclear antigen (PCNA). Recall that PCNA is required for eukaryotic DNA replication (Section 25·9). Increased levels of PCNA stimulate replication of adenovirus DNA.

Other eukaryotic viruses are also known to encode transcriptional activators. For example, *herpes simplex* DNA contains the gene for VP16. VP16 is contained within the viral particle and is injected into the host cell on infection. VP16 is a potent transcriptional activator that binds to specific sequences near *herpes* gene promoters. In the case of *herpes,* transcription requires an additional host factor called Oct-1.

Note the different strategies that exist for temporal expression of viral genes. In general, eukaryotic viruses produce activators that function in combination with host transcription factors (for example, E1A/ATF-2 and VP16/Oct-1). Such activators are common features of transcriptional regulation in eukaryotes, where they interact with RNA polymerase and the general class II transcription factors bound at the promoter (Section 27·8A). Bacteriophages T4 and SP01, on the contrary, direct the synthesis of new proteins (σ factors) that recognize different promoters, and bacteriophage T7 directs the synthesis of an entirely new RNA polymerase to transcribe phage late genes. In a functional sense, prokaryotic σ factors are analogous to the general transcription factors in eukaryotes, whereas E1A and VP16 are analogous to the activators MalT and CRP-cAMP (Sections 27·10 and 27·11).

31·2 Multiple σ Factors Control Sporulation in *B. subtilis*

A σ cascade is also used to control the temporal expression of the genes necessary for sporulation in *B. subtilis*. When exposed to unfavorable growth conditions, such as depletion of nitrogen or phosphate, *B. subtilis* differentiates to form a spore. The first step in sporulation involves the formation of a membrane, or polar septum, near the end of the bacterial cell. Four hours after initiation of sporulation, this polar septum divides the cell into a mother cell and a forespore. The mother cell then progressively engulfs the forespore, while the forespore synthesizes a cortex and an external coat. By 8 to 10 hours after initiation of sporulation, the endospore is fully

Figure 31·9
Life cycle of *B. subtilis.* In the presence of abundant nutrients, *B. subtilis* grows vegetatively, reproducing by binary fission. When nutrients become scarce, the vegetatively growing cell begins the process of sporulation.

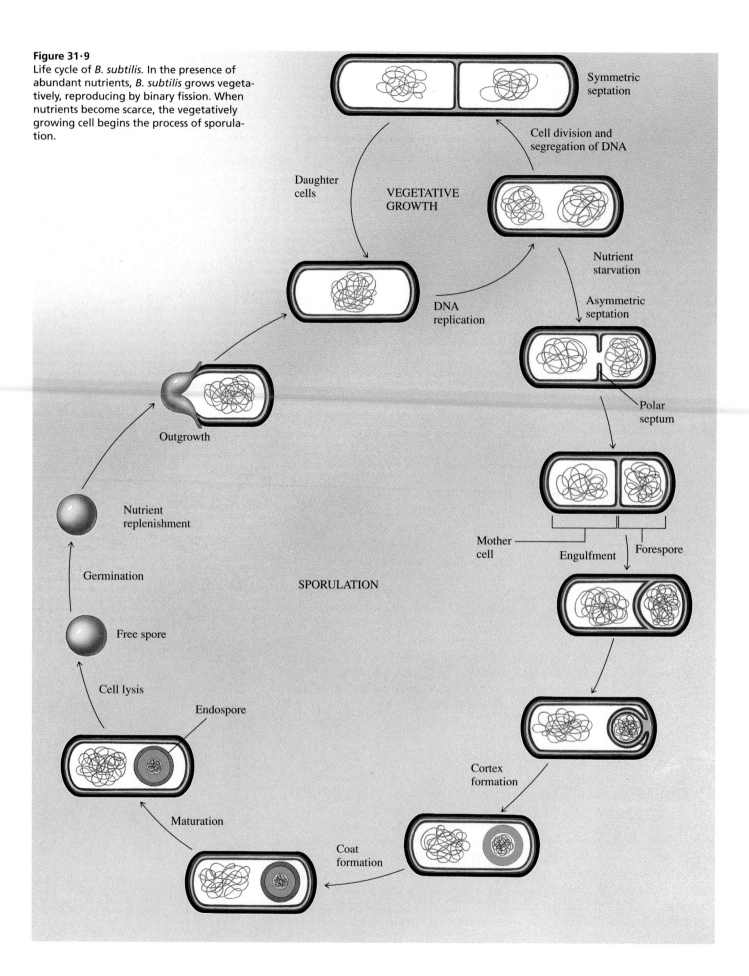

Symmetric septation

Cell division and segregation of DNA

Daughter cells

VEGETATIVE GROWTH

DNA replication

Nutrient starvation

Asymmetric septation

Outgrowth

Polar septum

Nutrient replenishment

Mother cell

Forespore

Germination

Engulfment

SPORULATION

Free spore

Cell lysis

Endospore

Cortex formation

Maturation

Coat formation

mature. Lysis of the mother cell releases a free spore. Under appropriate environmental conditions, the spore can germinate to produce another vegetatively growing bacterial cell (Figure 31·9). An electron micrograph of a mature endospore is shown in Figure 31·10.

The forespore and the mother cell are markedly different. Thus, sporulation in *B. subtilis* is an example of a developmental process that results in two differentiated cells (that is, cells expressing different genes) arising from a single progenitor. We don't usually think of bacteria as multicellular organisms, and we don't often realize that bacteria can form different cell types. However, sporulation is just one of many examples of prokaryotic differentiation.

Consideration of the mechanism that establishes differential gene expression in the forespore and the mother cell serves as a useful prelude to understanding differentiation in multicellular organisms. Expression of sporulation-specific genes must be properly timed to ensure successful spore formation. The temporal expression of these genes is mediated by several sporulation-specific σ factors that recognize the promoters of the genes. The consensus sequences recognized by these σ factors are shown in Table 31·2 and are compared to the consensus sequence for σ^A, the major σ factor in vegetatively growing *B. subtilis* cells.

In vegetatively growing cells, two sporulation-specific σ factors, σ^B and σ^C, are present but do not associate with core RNA polymerase. At the onset of sporulation, σ^B replaces σ^A on roughly 10% of the RNA polymerase holoenzymes, and transcription is initiated at the promoters of several early sporulation-specific genes. At the same time, σ^C, a minor species in the cell, replaces σ^A on other RNA polymerase holoenzymes, and transcription is initiated from the promoters of an additional, distinct class of early sporulation-specific genes.

Activation of σ^B and σ^C by nutrient starvation is a complex process. The replacement of σ^A by σ^B involves a number of other proteins, and the transcription of early sporulation-specific genes can be blocked by mutations in at least eight different genes. One of these genes encodes the protein δ (M_r 21 000). δ is one of the components of the *B. subtilis* transcription complex and must be present for the RNA polymerase holoenzyme to bind to promoters. However, the binding of δ to the holoenzyme has also been shown to destabilize the interaction between the core polymerase and the σ subunit. Thus, δ may assist in removing σ^A from the holoenzyme, thereby allowing competing σ factors, such as σ^B and σ^C, to bind core polymerase.

One of the genes expressed following activation of the early σ factors is *spo*IIAC, a gene that encodes an additional σ factor called σ^F. *spo*IIAC is part of an operon that includes the genes *spo*IIAA and *spo*IIAB. The product of the *spo*IIAB gene, SpoIIAB, is an inhibitor of σ^F. Synthesis of SpoIIAB ensures that the presence of

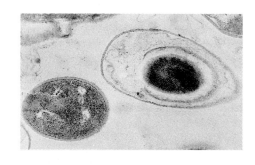

Figure 31·10
Electron micrograph showing a mature endospore in *B. subtilis*. The cortex and spore coat are clearly visible. A vegetatively growing cell is shown at the lower left. (Courtesy of JoAn S. Hudson.)

Table 31·2 Sporulation-specific σ factors in B. subtilis

Factor	M_r	Gene	Target genes	Consensus sequence[1] −35	−10
σ^A	43 000	*sig*A	Housekeeping	TTGACA	TATAAT
σ^B	30 000	*sig*B	Early sporulation	AGGNTT	GGNATTGNT
σ^C	32 000	*sig*C	Sporulation[2]	AAATC	TANTGNTTNTA
σ^E	24 000	*sig*E	Middle sporulation	TTNAAA	CCGATAT
σ^F	29 000	*spo*IIAC	Sporulation	[3]	[3]
σ^G	17 000	*spo*IIIG	Late sporulation	^T_CGNAT^A_G	CAN^T_ANTA
σ^K	27 000	*sig*K	Late sporulation	[3]	[3]

[1] *N* represents any nucleotide.
[2] This σ factor also directs the transcription of genes other than sporulation-specific genes.
[3] Consensus sequence is unknown.

σ^F does not alter the pattern of gene expression prematurely. When the polar septum forms, one of the integral membrane proteins within it is oriented so that its active site is in the forespore compartment. This membrane protein modifies SpoIIAA (the product of *spo*IIAA) in the forespore, which in turn blocks the action of SpoIIAB. This relieves inhibition of σ^F and leads to expression of a new set of genes that are transcribed only in the forespore. The net effect of this pathway is that σ^F is active in only one of the two cells formed during differentiation.

Regulation of sporulation-specific gene expression in *B. subtilis* also involves a σ cascade. Two early sporulation-specific genes transcribed by the core polymerase plus σ^B are *sig*E and *spo*IIGA. The product of the *sig*E gene is the protein pro-σ^E, an inactive precursor of the σ factor σ^E; *spo*IIGA encodes the inactive protease SpoIIGA. Activated σ^E is produced when the N-terminal 29 amino acids of

Figure 31·11
Regulation of sporulation-specific genes by a σ cascade. In vegetatively growing *B. subtilis*, most genes are transcribed by core polymerase + σ^A. At the onset of sporulation, σ^A is replaced by σ^B on some RNA polymerase holoenzymes and by σ^C on other RNA polymerase holoenzymes. These holoenzymes recognize the promoters of distinct classes of early sporulation-specific genes. Two of the early genes transcribed by core polymerase + σ^B are *sig*E and *spo*IIGA. *sig*E encodes the protein pro-σ^E, an inactive precursor of the σ factor σ^E. *spo*IIGA encodes the inactive protease SpoIIGA, which is localized to the polar septum and is activated by proteins in the forespore. Active SpoIIGA, along with three other proteins activated by septation, cleaves the N-terminal 29 amino acids of pro-σ^E to yield active σ^E. σ^E, which directs core RNA polymerase to the promoters of middle sporulation-specific genes, is therefore active in only the mother cell.

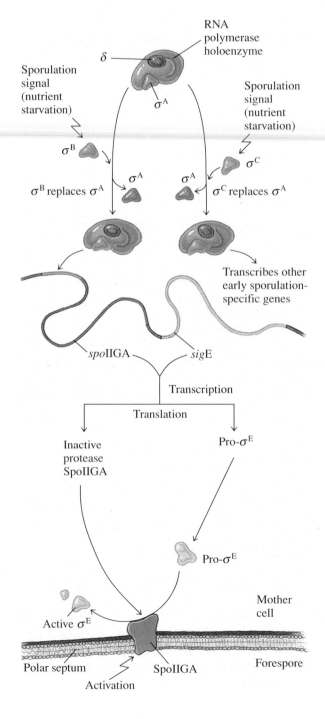

pro-σ^E are removed by the combined action of SpoIIGA and three other proteins. Cleavage of pro-σ^E to yield σ^E occurs approximately four hours after the onset of sporulation and, like σ^F activation, is coupled to formation of the polar septum. However, in this case, activation of σ^E occurs only in the mother cell. The protease SpoIIGA is an integral membrane protein that is localized to the polar septum. The active site of SpoIIGA lies in the mother cell, but the protease is only functional when stimulated by proteins that are produced in the forespore under the control of σ^F. In this way, cleavage of pro-σ^E is delayed until the septum forms and σ^F is active in the forespore (Figure 31·11). Such communication between cells mediated by integral membrane proteins is a common feature of differentiation in multicellular eukaryotes as well.

Once the polar septum has completely formed, the mother cell and the forespore become distinct compartments. From this point on, separate σ cascades control gene expression in the two cells. Some late sporulation-specific genes are expressed in only one of these two compartments. For example, the genes encoding cortex proteins are expressed only in the forespore, whereas the genes encoding coat proteins are expressed only in the mother cell. Transcription of these different classes of late sporulation-specific genes is mediated by different σ factors that are present in only one of the two cells: σ^G, encoded by the *spo*IIIG gene, is found exclusively in the forespore, whereas σ^K, encoded by the *sig*K gene, is found exclusively in the mother cell.

31·3 Bacteriophage λ Follows Either of Two Developmental Pathways

Up to this point, our discussion of the regulation of gene expression has focused on specific mechanisms by which genes are sequentially expressed. The examples we considered reveal the diversity of regulatory processes, but they run the risk of obscuring the interactive nature of regulation in an organism. In living cells, few regulatory mechanisms function in isolation. Rather, many different mechanisms act in concert to control both the magnitude and timing of gene expression.

The presence of multiple interacting mechanisms that control gene expression is beautifully illustrated by bacteriophage λ development. Regulation of gene expression during phage λ infection is better understood than the regulation of the other systems we have used as examples. The biochemistry and molecular biology of the λ life cycle have been subjects of intensive study for over 40 years. Consequently, many of the λ regulatory proteins serve as models of repressors, activators, and general DNA-binding proteins. Our analysis of gene expression during phage λ development will reveal the degree of fine-tuning and sophistication possible in a regulatory system.

Bacteriophage λ is shown in Figure 31·12. The phage consists of a head, or capsid, which holds the viral genome, and a tail through which the viral DNA is injected into a host cell. The genome of phage λ is a double-stranded DNA molecule of 48 502 bp. When packaged into the capsid, the molecule is linear, but it is converted into a covalently closed, circular molecule upon entering an *E. coli* cell.

After infecting an *E. coli* cell, bacteriophage λ can follow one of two alternate developmental pathways—the lytic pathway or the lysogenic pathway (Figure 31·13, next page). In either case, the initial infection of an *E. coli* cell by phage λ involves the insertion and circularization of the linear phage λ genome. After circularization, the phage immediate-early genes are transcribed, followed by the phage early genes. During expression of the early genes, a genetic switch is thrown that determines which of the two pathways of infection will be followed. Developmental regulation often involves the use of such genetic switches to mediate the selection of particular pathways.

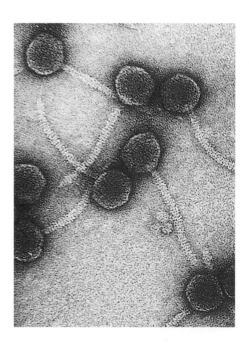

Figure 31·12
Electron micrograph of phage λ. (Courtesy of Roger W. Hendrix.)

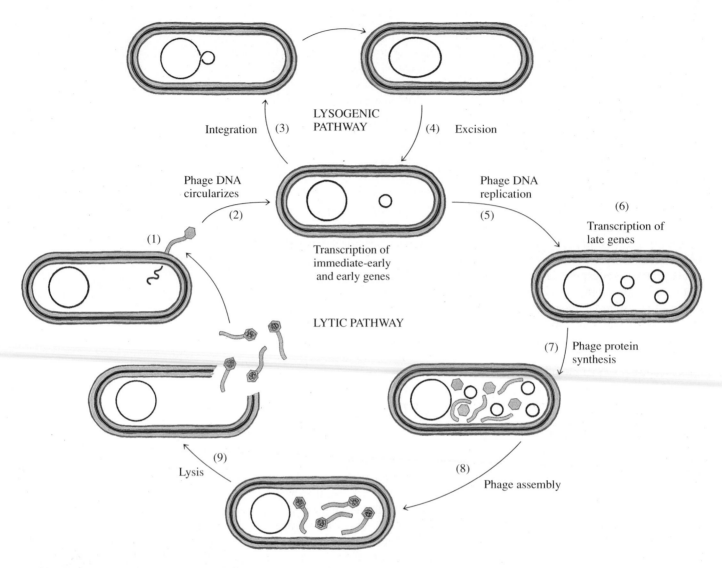

LYSOGENIC PATHWAY

Integration (3)

Excision (4)

Phage DNA circularizes (2)

(1)

Phage DNA replication (5)

(6)
Transcription of late genes

Transcription of immediate-early and early genes

LYTIC PATHWAY

(7) | Phage protein synthesis

(9)
Lysis

(8)
Phage assembly

Figure 31·13

Lytic and lysogenic pathways in bacterio-phage λ. (1) Infection begins when a phage attaches to the outer membrane of an *E. coli* cell and injects its DNA into the interior of the bacterium. (2) The phage DNA then circularizes, and bacterial RNA polymerase holoenzyme containing σ^{70} begins transcription of the immediate-early genes, followed by transcription of the early genes. During transcription of the early genes, phage λ development must take one of two alternative pathways. (3) If the lysogenic pathway is followed, the phage genome is integrated into the bacterial chromosome, where it can reside stably for many generations. (4) The lytic pathway can be entered directly after phage infection or after excision of the phage genome from the *E. coli* chromosome. Once in the lytic pathway, the circularized phage DNA is replicated (5), the late genes are transcribed (6), and the phage capsid and tail proteins are synthesized (7) and assembled with the phage DNA to form infective progeny (8). (9) Bacterial cell lysis releases multiple phage particles.

In the lysogenic pathway, the phage λ genome is integrated into the circular *E. coli* chromosome. Once integrated, the phage genome is known as a **prophage.** The prophage is replicated every time the *E. coli* chromosome is replicated, and a prophage segregates with each *E. coli* cell during cell division. During this phase of the life cycle, expression of most phage λ genes is repressed.

The integration of the phage genome into the *E. coli* chromosome reverses under certain conditions. When this happens, phage λ genes are derepressed, the prophage is excised, and phage development follows the lytic pathway. In the lytic pathway, the phage commandeers the synthetic machinery of the cell to transcribe phage genes and synthesize phage proteins. The lytic pathway can be entered either directly after phage λ infection or indirectly after prophage excision from an *E. coli* chromosome. During a lytic infection, the phage DNA is replicated, and the proteins necessary for forming new phage capsids and tails are synthesized. The newly replicated phage DNA is then packaged into phage capsids, and phage tails are added to produce infective progeny. These newly assembled phages are released by cell lysis and can begin a new round of infection.

A number of factors determine which of the two pathways of phage λ development is followed after infection of an *E. coli* cell. One factor is the multiplicity of infection (MOI), or how many phages on average infect a single bacterial cell. When the MOI is low, few bacterial cells in a population are infected by phages. Under these conditions, lytic growth is advantageous to the phage since there are many uninfected bacterial cells that can serve as hosts for the phage progeny released upon cell lysis. When the MOI is high, however, most bacterial cells in a population are infected by more than one phage, and few bacteria escape infection. Under these conditions, lysogeny is advantageous to the phage since a single cycle of lytic growth would produce progeny phages that could not find bacterial hosts for subsequent rounds of infection.

Another factor that determines which of the pathways is followed after phage λ infection is the metabolic state of the host cell. If the host cell is rapidly growing, lytic growth is advantageous because many phage progeny can be efficiently produced in a single round of infection. In contrast, if the host cell is starved for certain nutrients, lysogeny is advantageous since it allows the phage to persist until the bacterium encounters more favorable growth conditions.

31·4 Lytic Growth of Bacteriophage λ Is Controlled by an Antitermination Cascade

The sequence of gene expression in the phage λ life cycle is controlled by the temporal expression of genes encoding regulatory proteins. These genes must be expressed in a certain order and, sometimes equally importantly, repressed at appropriate times for proper phage development to occur. The immediate-early genes encode regulatory proteins that control the subsequent expression of the early genes. The early genes encode the proteins needed for DNA replication and for the integration of phage λ DNA into the *E. coli* chromosome, as well as regulatory proteins that control expression of the late genes. The late genes encode the proteins needed for new phage capsids and tails, for packaging the newly replicated DNA into new phage particles, and for lysis of the host cell. Figure 31·14 shows the organization of the phage λ genome.

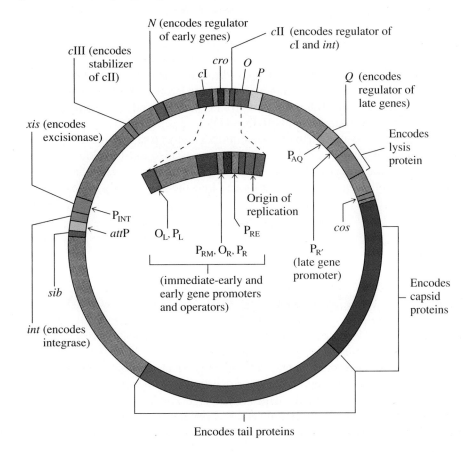

Figure 31·14
Organization of the phage λ genome. The phage λ control region is enlarged to show some of the immediate-early and early gene promoters and operators. The immediate-early, early, and late genes tend to be clustered into operons. In general, transcription in the clockwise direction leads to the production of proteins involved in lytic growth—proteins for phage λ DNA replication, cell lysis, and phage capsids and tails. Transcription in the counterclockwise direction leads to the production of proteins involved in lysogeny—proteins for integration of the phage λ genome.

Immediate-early, early, and late genes tend to be clustered into operons on the phage λ genome. The region containing the immediate-early genes is referred to as the control region. It includes the cI gene, which encodes the protein responsible for repressing transcription during the prophage stage. We will discuss this gene (and its product, cI) in more detail in Sections 31·5 and 31·6.

The lytic and lysogenic pathways are mutually exclusive, and entry into the lysogenic pathway can occur only at a specific point early in phage λ infection. Once the phage begins a productive lytic infection, the metabolism of the bacterial cell is so thoroughly disrupted that the cell is unlikely to survive even if the infection is aborted. In this section, we discuss the regulation of gene expression during a lytic infection. Later, we will return to discuss the lysogenic pathway.

A. One of the Phage λ Immediate-Early Genes Is an Antiterminator

Once phage λ DNA is injected into the *E. coli* cell, it is accessible to bacterial RNA polymerase. Regardless of which developmental pathway is eventually followed, the same phage λ genes are expressed immediately after infection. Bacterial RNA polymerase holoenzyme containing σ^{70} initially recognizes three promoters: P_L, P_R, and $P_{R'}$ (shown with blue arrows in Figure 31·15). P_L is a strong promoter; it initially directs transcription of the immediate-early gene *N*. In contrast, P_R is a weak promoter that initially directs transcription of the immediate-early gene *cro*. Because P_L is a stronger promoter than P_R, the rate of transcription of *N* initially exceeds that of *cro*. $P_{R'}$ is a strong promoter. At this stage in phage λ infection, transcription from $P_{R'}$ terminates at the strong terminator $t_{R'}$, producing a 194-nucleotide 6S RNA transcript that does not encode any proteins. (Transcription from $P_{R'}$ becomes more important during expression of the late genes in the lytic cycle, which we examine later.)

Transcription of immediate-early genes from both P_L and P_R stops at terminators located within a few thousand nucleotides of the transcription start sites. Transcription from P_L terminates at t_L, a strong terminator located just downstream of the *N* gene. Transcription from P_R terminates at either t_{R1} or t_{R2}. The first terminator encountered, t_{R1}, is a weak terminator located just downstream of the *cro* gene; about half of transcripts initiated at P_R pass through t_{R1} into the phage λ early genes *c*II, *O*, and *P*. The second terminator, t_{R2}, is a strong terminator located downstream of the *P* gene. Thus, initial transcription from P_L and P_R produces mainly N, some Cro, and even smaller amounts of cII, O, and P.

N is a regulatory protein that prevents termination of transcription at t_L, t_{R1}, and t_{R2}. Recall from Chapter 27 that such proteins are called **antiterminators**. Antitermination in the presence of N allows transcription complexes that initiate at P_L or P_R to proceed through the terminators and into the phage early genes (Figure

Figure 31·15

Transcription of phage λ immediate-early genes. After phage λ DNA is inserted into the *E. coli* cell, bacterial RNA polymerase holoenzyme containing σ^{70} initiates transcription from P_L, P_R, and $P_{R'}$. P_L is a strong promoter; transcription from P_L terminates at t_L and produces an mRNA transcript that encodes the protein N. P_R is a weak promoter; transcription from P_R terminates either at the weak terminator t_{R1} or at the strong terminator t_{R2}. Transcription from P_R that terminates at t_{R1} produces an mRNA transcript that encodes the repressor Cro. Transcription from P_R that terminates at t_{R2} produces an mRNA transcript that encodes Cro, cII, O, and P. $P_{R'}$ is a strong promoter; transcription from $P_{R'}$ terminates at $t_{R'}$ and yields a 6S RNA transcript that does not encode any proteins. The other promoters shown here (P_{INT}, P_{RM}, P_{RE}, and P_{AQ}) will be discussed later; they are shown here to indicate their relative positions.

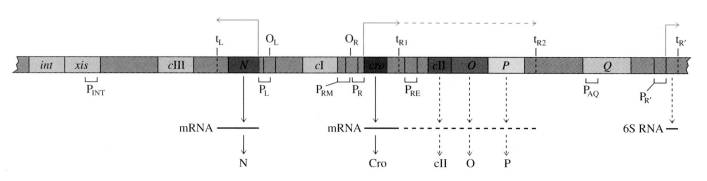

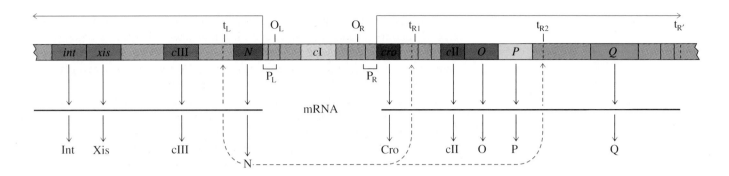

Figure 31·16
Transcription of phage λ early genes. In the presence of N, transcription from P_L and P_R continues past the terminators t_L, t_{R1}, and t_{R2} and into the phage early genes. Phage early genes include those for phage DNA integration and excision, those for replication of phage DNA, and those for several regulatory proteins.

31·16). Transcription of the immediate-early genes *N* and *cro* occurs very shortly after λ DNA is injected into the cell. The early genes (*cIII, xis, int, cII, O, P,* and *Q*) are only transcribed at significant levels after enough N protein accumulates to inhibit termination. This delay is usually a few minutes.

B. N Protein Is Part of the Transcription Complex

The antiterminator N does not interact directly with terminator sequences. Instead, it binds to the transcription complex and inhibits the ability of the complex to recognize terminators. N only associates with transcription complexes that initiate at P_L or P_R. This specificity depends on the presence of specific DNA sequences called *nut* sites (for N-utilization sites) in the vicinity of the promoter. Antitermination by N requires transcription of a *nut* site before the terminator is reached. There are two *nut* sites within the phage λ genome: *nut*$_R$ lies downstream of the *cro* gene, some 250 nucleotides away from P_R, and *nut*$_L$ lies just within the coding sequence of the *N* gene itself, some 50 nucleotides downstream of P_L (Figure 31·17). *nut* sites are composed of two separated sequences: a 17 bp, partially palindromic sequence called boxB, and an 8 bp sequence called boxA that is similar to the boxA sequences that specify antitermination at ribosomal RNA genes (Section 27·6D). Both sequences are required for antitermination in the presence of N.

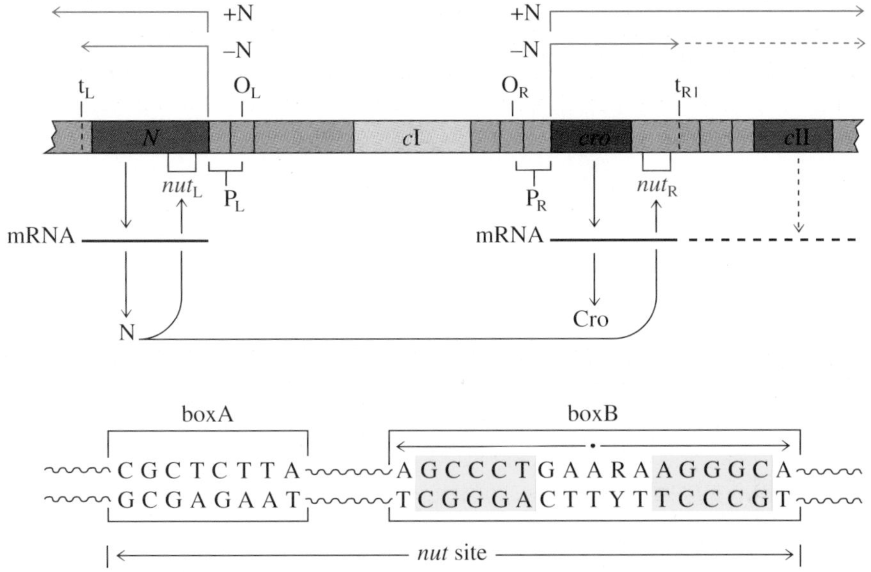

Figure 31·17
N-utilization (*nut*) sites. In the absence of the antiterminator N, transcription from P_L terminates at t_L, and transcription from P_R terminates at either t_{R1} or t_{R2} (not shown). In the presence of N, transcription from P_L and P_R proceeds through these terminators. Antitermination in the presence of N requires transcription of a *nut* site before the terminator is reached. The *nut* sites are composed of two separated sequences: an 8 bp sequence known as boxA and a 17 bp, partially palindromic sequence called boxB. The consensus sequence of these two elements is shown (abbreviations other than standard bases: R, a purine nucleotide; Y, a pyrimidine nucleotide).

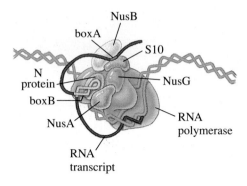

Figure 31·18
Antitermination transcription complex in cells infected by bacteriophage λ.

Transcription through a *nut* site produces an RNA transcript with a stem-loop structure due to the short region of dyad symmetry within boxB. N binds specifically to this stem-loop, and the N-RNA complex then associates with NusA in the elongation complex. This interaction is stabilized by NusG, NusB, and ribosomal protein S10 (Figure 31·18). (*Nus* stands for N-utilization substance; these factors were first discovered as proteins required for antitermination by N during bacteriophage λ infection.) NusG, NusB, and S10 are the same transcription factors required for antitermination at ribosomal RNA genes; they bind to boxA.

The antitermination complex formed by N and the nascent transcript remains associated with RNA polymerase during transcription. Since formation of the complex requires *nut* sites in the vicinity of the promoter, only transcription that begins at P_R or P_L is affected. Termination is inhibited at both rho-dependent and rho-independent termination sites. The mechanism of antitermination by N involves both the suppression of pausing (probably by interfering with the activity of NusA) and the blockage of interactions between the elongation complex and rho.

C. A Second Antiterminator Is Required for Late Gene Transcription

The genes required for lytic growth include some of the early genes and all of the late genes. The early genes include two genes whose products are needed to initiate DNA replication at the phage λ origin of replication. One of these, the *O* gene, encodes an origin-binding protein analogous to the *E. coli* protein DnaA (Section 25·6A). The O protein specifically recognizes the phage λ origin of replication and directs the components of the replisome to assemble on the phage λ genome.

The phage switches from transcribing the early genes to transcribing the late genes several minutes after infection. By this time, several rounds of DNA replication have occurred. The shift from early to late gene transcription is mediated by the products of the *Q* and *cro* genes, which are transcribed early in infection. The *Q* gene encodes Q, a second antiterminator whose presence is essential for transcription of the late genes. The *Q* gene is an early gene that is transcribed from P_R only in the presence of N. Like N, Q exerts its effect by interacting with RNA polymerase and altering the ability of the transcription complex to recognize terminators. But unlike N, this interaction does not involve NusG, NusB, or S10. Furthermore, the antitermination effect of Q is not mediated by bound RNA, as is the case with the N protein complexes. Instead, Q binds near the promoter $P_{R'}$, at a specific DNA sequence known as the *qut* site. The presence of Q at the site of transcription initiation results in modification of RNA polymerase so that it is not affected by pause sites or termination sequences.

Figure 31·19
Transcription of phage λ late genes. Transcription of *Q* from P_R is possible only in the presence of the antiterminator N. Q is itself an antiterminator; only in the presence of Q is complete transcription of the late operon from $P_{R'}$ possible. Transcription of the late genes thus involves an antitermination cascade.

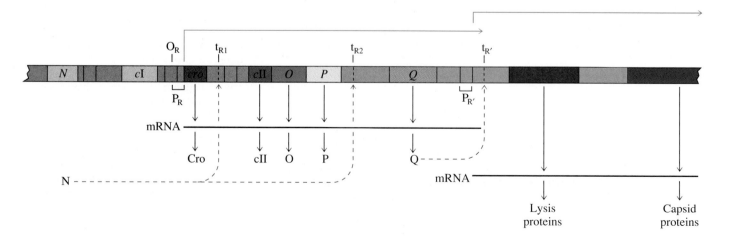

In the absence of Q, transcription from the σ^{70}-specific promoter $P_{R'}$ terminates at the strong terminator $t_{R'}$, which is located upstream of the late genes. Termination of transcription at $t_{R'}$ results in the production of a 6S RNA transcript that does not encode any proteins. In the presence of Q, however, the transcription complex is able to continue past $t_{R'}$, and the entire late operon of approximately 26 000 bp is transcribed. This operon includes genes for the lysis proteins and for the structural proteins of phage capsids and tails (Figure 31·19).

In the lysogenic pathway, the production of Q is inhibited, as we will see later. Because the antiterminator Q is only synthesized in the presence of the antiterminator N, Q and N form an antitermination cascade. This cascade, like the σ cascades we considered when reviewing bacteriophage SP01 infection and *B. subtilis* sporulation, ensures the correct temporal expression of genes required for phage development.

We have now encountered four examples of regulation by antitermination in prokaryotes: suppression of rho-dependent termination during transcription of ribosomal RNA genes (Section 27·6D); attenuation at the *trp* operon (Section 30·10E); N-mediated antitermination; and Q-mediated antitermination. All four mechanisms are different, although there are some similarities between N-mediated antitermination and the mechanism used for ribosomal RNA operons. There are also examples of antitermination that control gene expression in eukaryotic cells, although antitermination does not seem to be a common regulatory strategy in eukaryotes.

31·5 Some Genes of Bacteriophage λ Are Regulated by Repressors

The lytic pathway begins with the expression of the immediate-early genes *N* and *cro*. Antitermination by N leads to the expression of the early genes including *Q*. When enough Q accumulates, the late genes are expressed. The lysogenic pathway is quite different; there, the late genes and many of the early genes are repressed in order to prevent DNA replication and the formation of phage particles. Furthermore, the lysogenic pathway is characterized by expression of genes that promote the integration of λ DNA into the host chromosome. These genes are not expressed during the lytic pathway. In this section, we discuss the role of repressor proteins in regulating phage λ gene expression.

A. Two Repressors Bind the Bacteriophage λ Control Region

A map of the control region of phage λ is shown in Figure 31·20. The *cI* gene encodes the repressor cI (also known as λ repressor), which is responsible for maintaining lysogeny when the phage is integrated into the host chromosome. cI represses expression of *N* and *cro* by binding to the operators O_L and O_R. These operators partially overlap the promoters P_L and P_R, respectively. The presence of cI at these operators prevents RNA polymerase from initiating transcription at P_L and P_R.

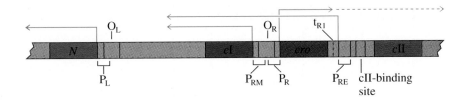

Figure 31·20
Phage λ control region. Transcription of the *cI* gene originates from either P_{RM} or P_{RE}; transcription of the *cro* gene originates from P_R; and transcription of the *N* gene originates from P_L.

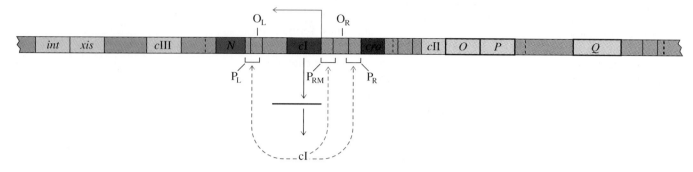

Figure 31·21

Action of cI. When the concentration of cI reaches a sufficient level, cI binds to O_L and O_R. When bound to O_L, cI inhibits transcription from P_L. When bound to O_R, cI simultaneously inhibits transcription from P_R and stimulates transcription from P_{RM}. Transcription of all phage λ genes other than cI itself is thus repressed. At high concentrations, multiple copies of cI bind to O_R and repress transcription from P_{RM}, thereby autoregulating cI synthesis.

Immediately following infection of an *E. coli* cell by phage λ, there is no cI present, and transcription occurs from P_L and P_R. As cI is synthesized, transcription from these promoters is repressed. If enough cI accumulates, expression of the early genes is prevented entirely. When this happens, the lytic pathway is aborted and lysogeny occurs.

Transcription of the early gene *c*I can originate from either one of two promoters—the promoter for repressor maintenance (P_{RM}) or the promoter for repressor establishment (P_{RE}). P_{RM} is used during lysogeny and when the phage DNA is stably integrated into the host chromosome (i.e., when the phage exists as a prophage), and P_{RE} is used for initial transcription of *c*I after infection. We will discuss transcription from P_{RM} here; transcription from P_{RE} is discussed in the next section.

Like *lac* repressor and CRP, which we considered in Chapter 27, cI can function as both a repressor and an activator. In the prophage, transcription of the *c*I gene is initiated at P_{RM}. P_{RM} is a relatively weak promoter, but when present at low concentrations, cI binds to O_R and increases transcription from P_{RM} 10-fold. At high concentrations, multiple copies of cI bind to O_R and repress transcription from P_{RM} as well as P_R. Thus, like production of CRP, production of cI is autoregulated (Figure 31·21).

Cro, the product of the *cro* gene, is also a repressor (*Cro* stands for *control of repressor and other things*). At low concentrations, Cro binds to O_R, where it represses transcription of the *c*I gene from P_{RM}. When this occurs, production of cI is inhibited, and development is directed along the lytic pathway. At high concentrations, multiple copies of Cro bind to O_R and O_L and repress transcription of the immediate-early and early genes from the promoters P_R and P_L (Figure 31·22). Thus, expression of *cro* is also autoregulated.

B. cI and Cro Bind to the Same Sites but with Different Affinities

Multiple copies of cI or Cro can bind at both O_R and O_L, and the two repressors have opposite and competing effects on transcription of the genes adjacent to these operators. The mechanism of regulation at these sites can be explained by examining the structures of the operators. Both O_R and O_L contain three protein-binding

Figure 31·22

Action of Cro. At low concentrations, Cro prevents transcription of cI from P_{RM}. At high concentrations, Cro inhibits transcription of all immediate-early and early genes.

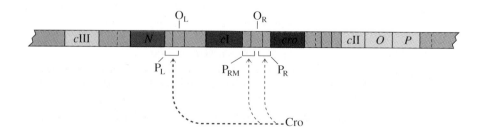

Figure 31·23
Nucleotide sequences of the protein-binding sites within O_R and O_L. The six 17–base pair sequences are similar and display dyad symmetry, as shown by the blue residues.

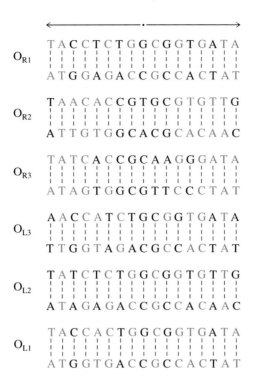

O_{R1}
```
TACCTCTGGCGGTGATA
ATGGAGACCGCCACTAT
```

O_{R2}
```
TAACACCGTGCGTGTTG
ATTGTGGCACGCACAAC
```

O_{R3}
```
TATCACCGCAAGGGATA
ATAGTGGCGTTCCCTAT
```

O_{L3}
```
AACCATCTGCGGTGATA
TTGGTAGACGCCACTAT
```

O_{L2}
```
TATCTCTGGCGGTGTTG
ATAGAGACCGCCACAAC
```

O_{L1}
```
TACCACTGGCGGTGATA
ATGGTGACCGCCACTAT
```

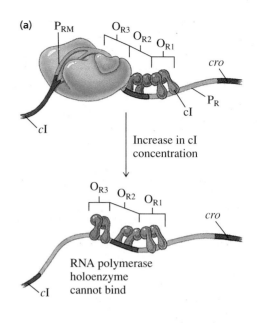

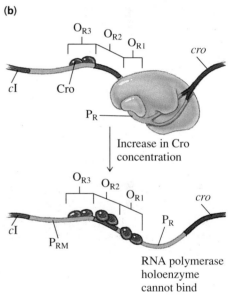

Figure 31·24
Effects of cI and Cro when bound at O_R. **(a)** cI binds cooperatively to O_{R1} and O_{R2}, the operator sites for which it has the highest affinity. When bound to these sites, cI simultaneously prevents transcription of the *cro* gene from P_R and stimulates transcription of the cI gene from P_{RM}. At high concentrations, cI binds noncooperatively to O_{R3}. When bound to O_{R3}, cI prevents transcription of the cI gene from P_{RM}. In this way, the production of cI is auto-regulated. **(b)** Cro does not display cooperative binding to DNA. When Cro is present in low concentrations, it binds to O_{R3}, the operator site for which it has the highest affinity. When bound to O_{R3}, Cro prevents transcription of the cI gene from P_{RM}. At high concentrations, Cro binds noncooperatively to O_{R1} and O_{R2}. When bound to these sites, Cro prevents transcription of the *cro* gene from P_R. Thus, Cro production is also autoregulated.

sites, designated O_{R1}, O_{R2}, and O_{R3}, and O_{L1}, O_{L2}, and O_{L3}, respectively. Each of these sites contains a similar sequence of 17 bp (Figure 31·23). The three sites of each operator can be bound by either Cro or cI, but the two repressors bind to each site with different affinities.

The affinity of cI for the different operator sites varies by a factor of 50. cI has the greatest affinity for O_{R1} ($K_{diss} \sim 8 \times 10^{-10}$ M), which is adjacent to *cro,* and only slightly less affinity for O_{R2}. Binding of cI at these two sites displays positive cooperativity; once cI is bound to one site, a second cI molecule binds more readily to the adjacent site. When the sites at O_{R1} and O_{R2} are occupied, cI simultaneously inhibits transcription from P_R and activates transcription from P_{RM}. Under these circumstances, the *cI* gene is transcribed while transcription of the *cro* gene is repressed. After the concentration of cI has increased sufficiently, cI binds noncooperatively to O_{R3}, which is adjacent to *cI,* and transcription of *cI* from P_{RM} is blocked.

Cro, on the other hand, binds preferentially to O_{R3}. This reversed affinity explains how cI and Cro can bind to the same operators but have opposite effects. The affinity of Cro for O_{R3} is an order of magnitude greater than that for O_{R1} or O_{R2}. In addition, binding of Cro is noncooperative: Cro binds to each operator binding site independently. Hence, at low concentrations of Cro, O_{R3} is bound but O_{R1} and O_{R2} remain unoccupied. When Cro is bound to O_{R3}, it prevents RNA polymerase from initiating transcription at P_{RM}, thereby blocking the synthesis of cI. Only when the concentration of Cro rises does it bind to its lower-affinity sites, O_{R1} and O_{R2}, thereby blocking transcription of *cro* from P_R. Both Cro and cI appear to function as repressors by preventing RNA polymerase from binding to the promoter and not by inhibiting transcription initiation, as does *lac* repressor. The competing and opposite actions of Cro and cI at O_R are summarized in Figure 31·24.

There is some evidence that efficient transcription of the phage λ late genes requires that the early genes be repressed. As we saw earlier, the decrease in early gene expression is mediated by Cro. At high concentrations, Cro binds to all three sites of both O_L and O_R, blocking transcription of all the immediate-early and early genes. Although at this point Cro is repressing transcription of both the *N* and *Q* genes, enough Q has already been synthesized to allow successful transcription of the late genes from $P_{R'}$.

Thus we see that the change from immediate-early to early to late gene transcription depends upon the concentrations of three gene products: N, Cro, and Q. None of these proteins is present at the beginning of infection. Transcription from P_R and P_L leads to the synthesis of Cro and N, respectively. When the concentration of N increases sufficiently, transcription switches from the immediate-early to the early genes. As transcription of the early genes continues, the concentrations of all three proteins continue to rise. Once the concentrations of Q and Cro increase to a sufficient level, transcription switches from the early genes to the late genes.

Note that this mechanism differs from the mechanism that regulates gene expression late in T4 or SP01 infection. In those cases, reduced expression of the early genes is not due to direct repression but to redirection of RNA polymerase to the late promoters by the synthesis of new σ factors.

C. The DNA-Binding Domains of cI, Cro, and CRP-cAMP Are Similar

The structures of cI and Cro have been solved, and both contain a helix-turn-helix motif. The cI repressor is a dimer of identical subunits (subunit M_r 26 000). Each subunit is composed of 236 amino acid residues and contains an N-terminal domain (residues 1–92) that binds to DNA and a C-terminal domain (residues 132–236) that is involved in the protein-protein interactions that lead to dimerization. The N-terminal domain of each subunit of the cI dimer contains five α helices. A model

(a)

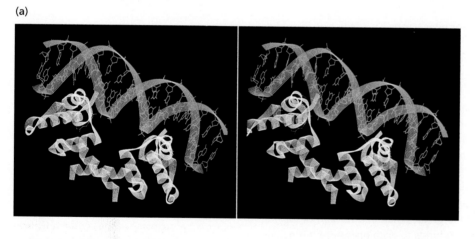

(b)

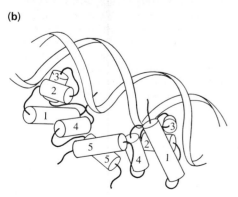

Figure 31·25
Interaction between the DNA-binding domains of cI and the operator site O_{R1}. (a) Stereo view of cI bound to O_{R1}. Each N-terminal domain of cI contains five α helices, shown here as helical ribbons. (Based on coordinates provided by Carl O. Pabo.) (b) Key showing the numbering of the α helices of cI.

of the symmetrical N-terminal domains of the dimer bound to the symmetrical halves of the operator site O_{R1} is shown in Figure 31·25. Helices 2 and 3 of each N-terminal domain contact the DNA most intimately, with helix 3 lying in the major groove of DNA and helix 2 oriented across the groove. These two α helices are joined by four amino acids that form a turn in the protein structure. Helix 3 of one subunit and helix 3 of the other subunit lie in adjacent major grooves of DNA so that the dimer lies on one side of the double helix and binds to successive turns of DNA.

The specificity of cI binding is determined by interactions between the amino acid side chains in the helices and specific base pairs at the binding sites. As shown in Figure 31·26, these interactions involve primarily hydrogen bonds, although van

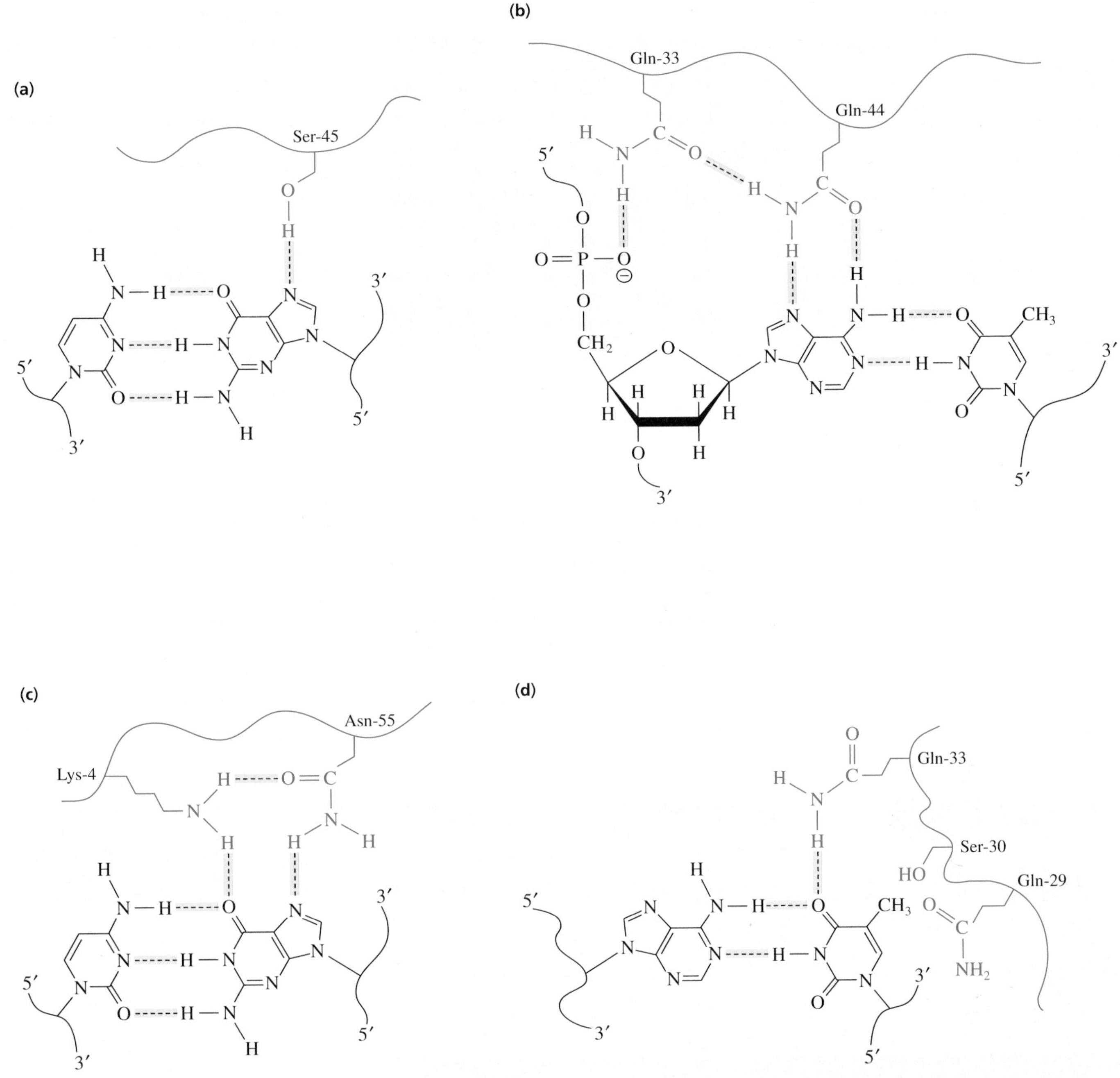

(a)

(b)

(c)

(d)

Figure 31·26
Interactions during the binding of two different cI repressors to DNA. **(a–c)** Interactions between the phage λ cI repressor and nucleotides at the operator binding site. **(d)** Interaction between the phage 434 repressor and nucleotides at its binding site. In all cases, the DNA-protein interactions involve primarily hydrogen bonds, but van der Waals contacts are also important.

der Waals contacts can be important as well. Hydrogen bonds can form between the amino acid residues and the bases (for example, between Ser-45 and a G residue or between Lys-4 and Asn-55 and a G/C base pair) or between the amino acid residues and the ribose-phosphate backbone of DNA (for example, between Gln-33 and a phosphate oxygen). The structure of the cI repressor from another lambdoid phage, 434, has also been determined. The 434 repressor is homologous to the phage λ cI repressor, but it recognizes operators with slightly different sequences. The interactions of Gln-33 and adjacent amino acid residues with an A/T base pair reveal that the two related repressors bind differently. In the 434 repressor, Gln-33 forms a hydrogen bond with thymine, and the side chains of Ser-30 and Gln-29 form van der Waals contacts with the methyl group of thymine.

(a)

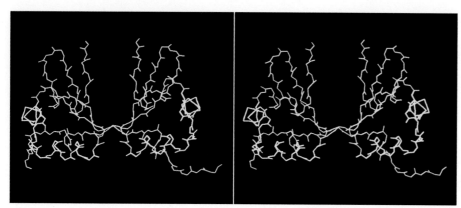

Figure 31·27
Structure of Cro. Cro is a dimer of identical subunits; each subunit contains three α helices. **(a)** Stereo view of Cro. Helices 2 and 3 of each subunit most intimately contact DNA and are highlighted in white. (Based on coordinates provided by B. W. Matthews.) **(b)** Key showing the numbering of the α helices of Cro.

(b)

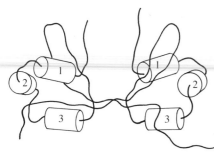

Cro is also a dimer of identical subunits (subunit M_r 7400); each subunit is composed of 66 amino acid residues and contains three α helices. The DNA-binding domains of Cro, like those of cI, contain a helix-turn-helix motif (Figure 31·27). Helix 3 of one subunit of Cro fits into the major groove of DNA and is separated by exactly one turn of the double helix from helix 3 of the second subunit, which also fits into the major groove.

Although the overall structures of cI and Cro are quite different, the two proteins share the helix-turn-helix structural motif within their DNA-binding domains. The structures of the helix-turn-helix motifs of cI, Cro, and CRP are very similar and can be superimposed with a deviation of only 0.11 nm per α carbon (Figure 31·28). The similarity is not surprising since all three proteins interact with the major groove of B-form DNA. However, the structural similarity of these three proteins does not extend beyond the helix-turn-helix motif of the DNA-binding domains. This is an important point since it suggests that DNA-binding motifs are similar as a result of convergent evolution and not because the genes for these three proteins are descended from a common ancestor.

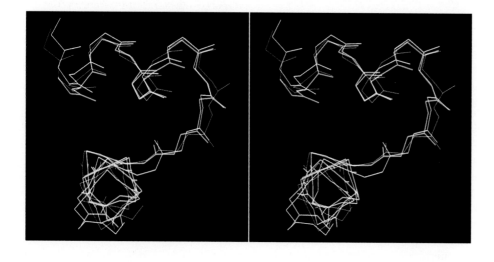

Figure 31·28
Stereo view of superimposed helix-turn-helix regions of cI (green), Cro (yellow), and CRP (red) DNA-binding domains. (Based on coordinates provided by Carl O. Pabo, B. W. Matthews, Thomas A. Steitz, and Irene T. Weber.)

cII-binding site

−35 region

−10 region

5′~~~~TTGCGTTTGTTTGCACGAACCATATGTAAGTATTTCCTTAG~~~3′

TTGACA σ^{70} consensus sequence   TATAAT

Figure 31·29
Sequence of P_{RE}. The promoter consensus sequence recognized by the σ^{70} subunit of RNA polymerase holoenzyme is also shown for comparison. The cII-binding site overlaps the −35 region of P_{RE} and includes a repeat of the nonpalindromic sequence TTGC. Transcription from P_{RE} can be initiated at either the A or the G residue, as shown.

31·6 The Genetic Switch In Bacteriophage λ Is Mediated by cII

Once N and Cro have been produced and efficient transcription of the early genes has begun, phage λ development follows one of the two developmental pathways. The lysogenic pathway requires transcription of the early gene *c*I because *c*I represses transcription of the immediate-early and early genes and prevents expression of the late genes. The lytic pathway, on the other hand, requires repression of *c*I transcription and expression of the late genes. The repression of *c*I transcription is mediated by Cro, as we saw in the previous section, and expression of the late genes requires the early gene product Q. Since we know that *cro* is one of the first genes transcribed, it is clear that a key step in the control of the genetic switch must be the regulation of transcription of *c*I.

A. Initial Expression of cI Requires the Activator cII

Because transcription of *c*I is not initiated at P_R or P_L, *c*I is not transcribed immediately after infection. Once lysogeny has been established, *c*I is transcribed from its own promoter, P_{RM}. But activation of transcription from P_{RM} requires that some *c*I already be present and bound to O_{R2}. The *c*I initially needed for the phage to enter the lysogenic pathway is produced by transcription from the promoter P_{RE}. The sequence of P_{RE} does not resemble the consensus sequence for promoters recognized by the σ^{70} subunit of bacterial RNA polymerase (Figure 31·29), and transcription from P_{RE} requires the presence of an activator, cII, which is encoded by the early gene *c*II. Efficient transcription of *c*II depends on the antiterminator N.

cII, a tetramer of identical subunits (subunit M_r 11 000), binds to sites at three different phage λ promoters—P_{RE}, P_{INT}, and P_{AQ} (Figure 31·30). The consensus sequence for the three binding sites is $TTGCN_6TTGC$, one of the few consensus sequences for DNA-binding proteins that contains a repeat of a nonpalindromic sequence. Transcription from P_{INT} and P_{AQ} is examined in Section 31·6E. When cII binds to P_{RE}, it activates transcription by increasing the rates of both the binding of RNA polymerase and the formation of the open complex. Transcription from P_{RE} increases the likelihood of lysogeny by allowing synthesis of *c*I.

Figure 31·30
Action of cII. cII activates transcription from three promoters—P_{INT}, P_{RE}, and P_{AQ}. Initiation at P_{RE} produces a transcript that encodes the repressor cI. Transcription from P_{RE} also interferes with transcription from the weak promoter P_R, since the same segment of DNA cannot be transcribed simultaneously by two converging RNA polymerases. Transcription from P_{INT} produces an mRNA molecule that encodes the enzyme integrase (see Section 31·6D). Transcription from P_{AQ} produces an antisense RNA molecule complementary to the 5′ end of the Q mRNA (Section 31·6E).

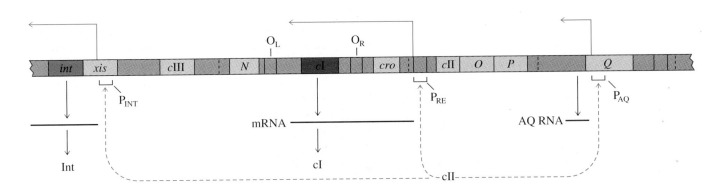

In addition to establishing synthesis of cI, transcription from P_{RE} also interferes with the production of Cro. RNA polymerase molecules that initiate at P_{RE} pass through the same region of DNA that is transcribed from P_R, but in the opposite direction. Since the same segment of DNA cannot be transcribed simultaneously by two converging RNA polymerases, transcription from P_{RE} interferes with transcription from the weak promoter P_R. Thus, cII simultaneously initiates the production of cI and decreases the production of Cro and the other proteins encoded by early genes transcribed from P_R.

Transcription of *cI* from P_{RE} is necessary for establishing lysogeny. If the concentration of cI reaches a sufficient level, cI binds to the operators O_R and O_L and represses transcription from P_R and P_L, respectively. Since cI binds cooperatively, complete repression of transcription from P_L and P_R is likely to occur once a critical threshold concentration of cI is reached. In addition to repressing transcription from P_R, the binding of cI to O_{R2} activates transcription from P_{RM}, thereby ensuring continued transcription of the *cI* gene even in the absence of cII. The continued synthesis of cI from P_{RM} may eventually lead to lysogeny and to the repression of transcription of all phage λ genes except *cI* itself.

B. The Choice of the Lytic Pathway or Lysogenic Pathway Depends upon the Metabolic State of the Host Cell

As previously mentioned (Section 31·3), the metabolic state of the host cell is one of the factors that determines which pathway is followed after phage λ infection. The metabolic state is monitored in part by cII. cII is an unstable protein, and the amount of cII in the cell is never very high. In addition, cII is extremely sensitive to cleavage by *E. coli* proteases whose activity is determined by the metabolic state of the cell. In cells that are rapidly growing, the proteases that degrade cII are active, and cII is rapidly destroyed. In cells that have encountered unfavorable growth conditions, however, these proteases are not very active, and cII concentrations are higher.

The activity of cII also depends upon the concentration of the product of the early gene *cIII*, whose transcription is initiated at P_L. The cIII protein helps protect cII from degradation, probably by interfering with the activity of the host proteases. In rapidly growing cells, the concentration of active host proteases is high, and not enough cIII is present to protect cII from degradation. When bacterial cells are growing more slowly, however, the same concentration of cIII is able to inhibit the few active host protease molecules, and cII escapes degradation. When several phage genomes simultaneously infect the same cell, they are all transcribed. The increased concentrations of cIII and cII that result from multiple infections overwhelm the activity of the host proteases even in rapidly growing cells. As a result, lysogeny is favored at high MOI.

Conditions that stabilize cII increase the frequency of lysogeny, whereas conditions that destabilize cII increase the frequency of lytic infections. The choice of lytic or lysogenic pathway depends upon the relative concentrations of Cro and cII. When cII is abundant, it activates transcription of *cI* from P_{RE}, and lysogeny is favored. When cII is degraded before a significant amount of cI is produced, Cro binds to O_L and O_R, and lytic growth is favored.

C. cII Also Controls Integration of the Phage λ Genome into the *E. coli* Chromosome

Repression of transcription from P_L and P_R by cI is necessary to establish lysogeny, but repression alone is not sufficient. Lysogeny also requires that the phage λ genome be inserted into the *E. coli* chromosome, where it can be passively replicated

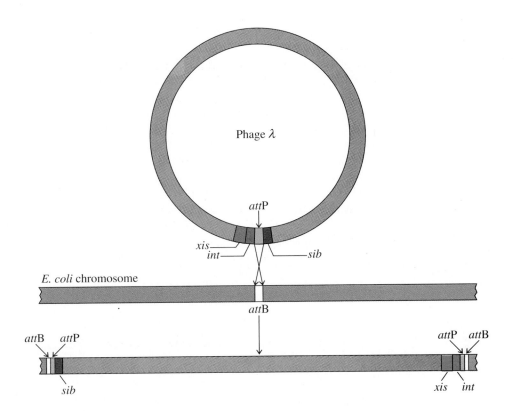

Figure 31·31
Site-specific integration of the phage λ
genome into the E. coli chromosome. The
circularized phage λ genome is integrated
into the E. coli chromosome by recombination between the attP and attB sites. The
DNA strands of both the host and the phage
are broken and rejoined to form hybrid attachment sites. Recombination results in the
formation of a DNA molecule that contains
both the bacterial and phage DNA.

during *E. coli* cell division. The establishment of lysogeny requires that integration of the phage λ genome and synthesis of cI from P_{RE} be coordinated.

Site-specific integration of the phage λ genome requires the enzyme integrase, which is the product of the phage *int* gene. Integration also requires at least one *E. coli* protein and two specific DNA sequences. One of these sequences, called *att*P (for attachment site phage), is located on the phage λ genome, and the other, called *att*B, is located on the bacterial chromosome. Integrase catalyzes the breaking of the DNA strands at both *att*P and *att*B and the subsequent ligation of phage λ DNA to bacterial DNA. The product of this reaction is a single, covalently closed, circular DNA molecule (Figure 31·31). This rearrangement of DNA molecules within the bacterium is an example of site-specific recombination. Note that in this case, recombination yields two hybrid attachment sites that are composed of DNA sequences taken from both the *att*P and *att*B sites.

The *int* gene is an early gene whose transcription is initiated at P_L and depends upon the presence of the antiterminator N. Since transcription of the early genes is eventually repressed when the phage enters the lysogenic pathway, expression of *int* during the establishment of lysogeny is supplemented by transcription from a second promoter, P_{INT}. P_{INT} is located upstream of the *int* gene, within the coding region of the *xis* gene. Transcription of *int* from P_{INT}, like transcription of *cI* from P_{RE}, is activated by cII. Thus, the expression of *int* is coordinated with the expression of *cI*; when the cellular concentration of cII rises sufficiently to activate transcription from P_{RE}, transcription from P_{INT} is also activated (see Figure 31·30). This ensures that integration and repression of transcription from P_L and P_R occur simultaneously.

D. Production of Integrase Is Inhibited During the Lytic Pathway

Integrase is detrimental during a lytic infection since it catalyzes integration of the phage genome into the host chromosome. Although the concentration of cII is too low during lytic infection to activate P_{INT}, the *int* gene is still transcribed from P_L.

Figure 31·32

Retroregulation of integrase mRNA. The stability of the mRNA transcript encoding integrase depends upon sequences at the 3′ end of the gene. The transcript initiated at P_{INT} during the establishment of lysogeny ends at a terminator located between attP and sib. Transcripts terminated at this site form a stable hairpin loop that resists degradation by nucleases. In contrast, the transcript initiated at P_L in the presence of N continues through the terminator and into the sib region. Transcripts that contain both int and sib form a stem-loop structure that is recognized and cleaved by RNase III. Cleavage by RNase III leaves the mRNA susceptible to further degradation by endogenous exonucleases, thereby allowing destruction of the integrase mRNA before a significant amount of integrase can be synthesized.

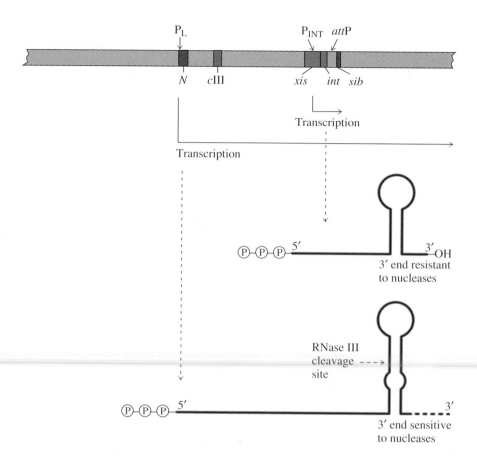

Under these conditions, the synthesis of integrase is blocked by the specific destruction of integrase mRNA. In the case of phage λ, the specific destruction of mRNA is known as **retroregulation.**

During lytic growth, the phage λ genome is not integrated into the *E. coli* chromosome, and the attP site is intact. Thus, transcripts initiated at P_L include xis, int, attP, and a sequence located downstream of attP called sib. When present in an mRNA transcript, sib forms a hairpin loop that is recognized and cleaved by the endonuclease RNase III (Figure 31·32). Once cleaved by RNase III, the newly formed 3′ end of the mRNA is recognized and degraded by endogenous exonucleases before significant amounts of integrase can be made. During the establishment of lysogeny, retroregulation does not inhibit synthesis of integrase because transcription complexes initiated at P_{INT} halt at a terminator sequence located between int and sib. Transcripts terminated at this site form a stable hairpin loop that resists degradation by nucleases. In the presence of N, transcription complexes initiated at P_L do not halt at this termination sequence since they have passed through nut_L and have become antitermination transcription complexes. Thus, sib is found only on transcripts that are synthesized from P_L in the presence of the antiterminator N. (The selective degradation of mRNA containing RNase III cleavage sites is common in *E. coli* when only a small amount of protein is needed by the cell.)

Retroregulation does not occur when phage λ DNA is integrated into the *E. coli* chromosome because recombination at the attP and attB sites separates the sib region of the DNA from the int gene (Figure 31·33). Thus, when transcription is initiated from P_L, the mRNA containing int does not contain sib and hence is not degraded. Since sib is reconnected to the operon transcribed from P_L only after the phage DNA has been excised from the host chromosome, retroregulation occurs only when the phage genome is in the covalently closed, circular configuration appropriate for lytic growth.

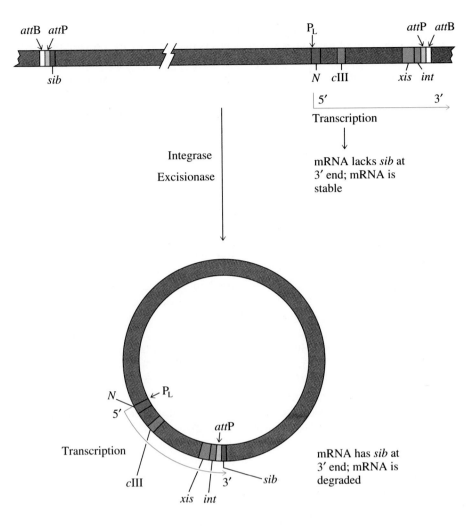

Figure 31·33
Inhibition of retroregulation by integration of the phage λ genome. When the phage λ genome is integrated into the *E. coli* chromosome, the *sib* sequence is separated from the *int* gene. Consequently, transcripts initiated at P_L do not contain *sib* and are relatively stable. Once the phage λ genome has been excised and has recircularized, *sib* is again present on the transcripts initiated at P_L, and retroregulation occurs.

E. cII Inhibits the Production of Q by Activating Transcription of an Antisense RNA

As mentioned earlier, activation of transcription from P_{RE} by cII interferes with transcription from P_R since two converging RNA polymerases cannot simultaneously transcribe the same segment of DNA (see Figure 31·30). This inhibition decreases the rate of transcription of all genes normally transcribed from P_R, including *cro, cII,* and *Q,* by about 50%. In this way, cII delays entry into the lytic pathway and the subsequent transcription of the late genes. This effect is consistent with the role of cII in encouraging lysogeny.

cII has a more direct effect on transcription of the late genes, however. A cII-dependent promoter called P_{AQ} is located within the *Q* gene. Transcription initiated from P_{AQ} produces an antisense RNA that inhibits *Q* expression. Translation of Q mRNA is inhibited because the antisense RNA binds the Shine-Dalgarno sequence of *Q,* preventing ribosome assembly (see Section 30·10B).

31·7 The Destruction of cI Repressor Leads to Induction of the Lytic Genes

As long as cI is present and active at sufficient concentrations in the cell, lysogeny is maintained. However, the lytic genes repressed by cI can be induced in the presence of a variety of agents that damage DNA. Unlike induction of the *lac* operon,

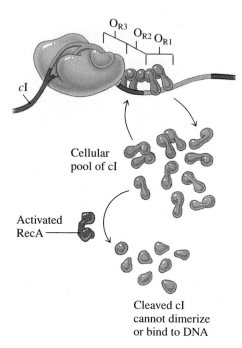

O_{R3} O_{R2} O_{R1}

cI

Cellular pool of cI

Activated RecA

Cleaved cI cannot dimerize or bind to DNA

Figure 31·34
Induction of the lytic genes by RecA. In the presence of fragments of single-stranded DNA, the *E. coli* protein RecA becomes an active protease. When RecA is activated, it cleaves cI between the N- and C-terminal domains. Cleaved cI cannot dimerize or bind cooperatively to DNA, and as the cellular pool of intact cI is depleted, fewer molecules are available to replace those that spontaneously dissociate from the operators. Eventually, transcription from P_L and P_R recommences and lytic growth becomes possible.

which is mediated by the binding of an inducer to bound *lac* repressor (Section 27·10C), induction of the phage λ lytic genes is mediated by the proteolytic cleavage of cI. The mechanism is the same as the one that mediates the SOS response in which the repressor LexA is cleaved by RecA protease activity (Section 27·13). In the presence of single-stranded DNA, RecA cleaves free cI between the N- and C-terminal domains (Figure 31·34). After the N- and C-terminal domains of cI have been separated, cI can no longer dimerize or bind cooperatively to DNA. As the cellular pool of intact cI is depleted, fewer molecules of cI are available to replace those that spontaneously dissociate from the operators. Eventually, transcription of *N* and *cro* begins from P_L and P_R, respectively. The induction by RecA of genes required for lytic growth allows phage λ to escape cells whose DNA has been severely damaged.

Just as the establishment of lysogeny requires coordination of cI production and integration, the switch from lysogeny to lytic growth requires coordination of induction and excision. The excision reaction is the reverse of the integration reaction: the DNA is cleaved at the two hybrid attachment sites and rejoined to re-form the *att*B and *att*P sites. This reaction is carried out by integrase; however, integrase alone cannot recognize the hybrid attachment sites. Instead, excision of the phage λ genome requires both integrase and another protein, excisionase, which is the product of the *xis* gene.

The *xis* gene is located immediately upstream of the *int* gene; both genes are coordinately expressed on transcripts initiated at P_L when phage λ DNA is integrated in the host genome. When cI is destroyed by RecA during induction, transcription of the immediate-early genes begins at P_L and P_R. Antitermination in the presence of N allows transcription of both the *int* and *xis* genes from P_L. The production of integrase and excisionase allows the phage λ genome to be excised from the *E. coli* chromosome. During excision, the phage λ genome is recircularized.

Note that although transcription of both *xis* and *int* is initiated from P_L, only integrase is present during the establishment of lysogeny. At that time, the phage λ genome is being integrated into the *E. coli* chromosome, and the presence of excisionase would be detrimental. During the establishment of lysogeny, transcription of *xis* is repressed because cI inhibits transcription from P_L. At the same time, cII activates transcription of *int* from P_{INT}. Excisionase is not encoded on the transcript that originates at P_{INT} because P_{INT} lies within the coding sequence of the *xis* gene itself.

31·8 Development and Differentiation in Multicellular Organisms Begins with a Single Cell

The timing of gene expression during the bacteriophage life cycle is similar in principle to the timing required during embryogenesis of multicellular organisms. But whereas sequential gene expression during phage infection takes place in a single cell, gene expression is tied to cell division and differentiation in multicellular organisms.

The regulation of gene expression during sporulation in *B. subtilis* is a simple example of differentiation linked to cell division. In this case, a single cell divides once to produce a forespore and a mother cell. Different genes are expressed in each cell, and the cells are morphologically and biochemically different. The pattern is the same during embryogenesis in multicellular eukaryotes, when each cell divides to produce two daughter cells.

In only a few cases have the fates of all of the cells in a multicellular organism been traced back to the original fertilized egg. One example is the nematode *Caenorhabditis elegans* (Figure 31·35). The adult is a small, wormlike creature about 1 mm in length. It contains about 1000 somatic cells plus germ cells (exactly

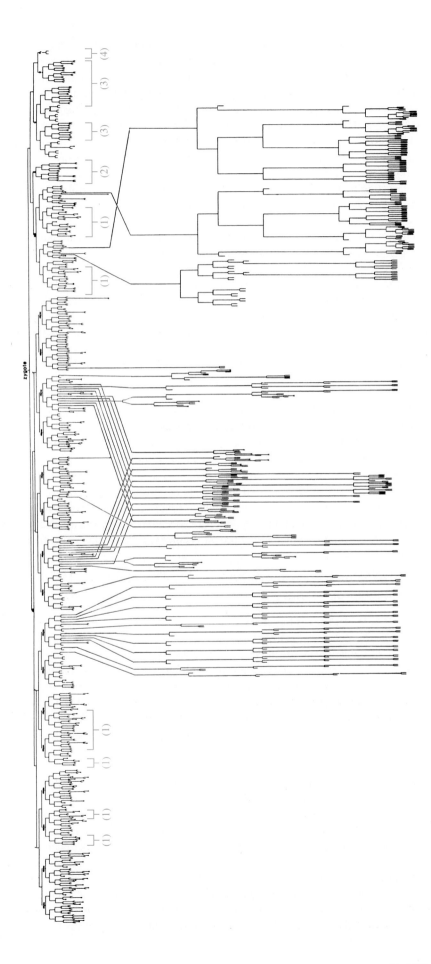

Figure 31·35
C. *elegans* lineage tree. The map shows the lineage of each cell, beginning with the fertilized egg, or zygote. At each step, two new cells are produced following mitosis. Each of these cells can in turn divide to produce two more cells. In some cases, cell division ceases after only a few generations, for example, division of the cells that form the pharynx, intestine, and body muscle. In other cases, cell division continues throughout embryogenesis, which lasts 50 hours. Cells marked with a small X die during embryogenesis. Only some of the tissues and differentiated cells are labelled. Differentiated tissues: (1) pharynx; (2) intestine; (3) body muscle; (4) germ line. (Courtesy of John E. Sulston.)

939 somatic cells in one sex and 1031 in the other). Each cell in the adult results from binary fission of a cell in the preceding generation. During embryogenesis, some cells die, others differentiate early, and still others continue dividing until very late in development. The cells that give rise to the germ line (eggs and sperm) appear very early; this is typical of development in multicellular organisms. The *C. elegans* fate map demonstrates that every cell in the adult is a direct descendant of the original zygote and is created by a program of sequential, binary cell division and differentiation. This is also true of larger organisms, such as flowering plants, arthropods, and vertebrates.

Interestingly, the process of differentiation is sometimes reversible. For example, it is common to grow a new plant from a cutting, and several decades ago, it was demonstrated that an entire plant could be reproduced from a single adult cell. Similarly, a new adult frog can be reproduced from the nucleus of an adult cell. These demonstrations indicate that differentiation is not usually accompanied by loss or rearrangement of genetic material; differentiated cells usually contain all the information needed to create a new organism. However, there are some instances of loss of genes during development; mammalian red blood cells, for example, do not have nuclei. We will see examples of genomic rearrangements that lead to deletion or destruction of genetic information in the next chapter. In such cases, differentiation is irreversible.

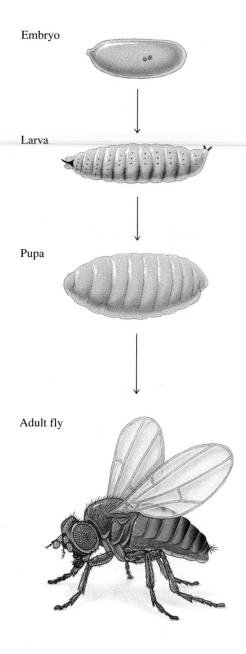

Embryo

Larva

Pupa

Adult fly

Figure 31·36
The four stages of *Drosophila* development. The structures are not to scale.

31·9 There Are Four Stages of Early Development in *Drosophila melanogaster*

Drosophila melanogaster is a typical dipteran insect with four stages of development (Figure 31·36). Embryogenesis takes place over a period of 20 hours (at 25°C) once the fertilized egg is laid. The larva that hatches from the egg feeds on decaying fruit and undergoes two molts, during which time its size increases severalfold. The third stage is pupation; in the pupa, most of the cells of the larva are destroyed, and the adult fly forms. Further development is completed after the adult emerges.

A. *Drosophila melanogaster* Development Depends on Specific Regulatory Genes

We know a great deal about the genes and mechanisms that control development in *D. melanogaster* because this insect has been studied intensively over the past century. Many of the genes involved encode repressors and activators of transcription. There are two categories of such genes: maternal and zygotic. Maternal genes are expressed in the mother, but their products act in the embryo. The proteins produced from maternal mRNA regulate gene expression during the first stages of embryogenesis. Zygotic genes are expressed only following fertilization. In general, zygotic genes are activated by the products of maternally expressed genes.

Early embryogenesis in *Drosophila* is characterized by rapid DNA replication and mitosis that is not accompanied by cell division (Figure 31·37). Within 1.5 hours after a fertilized egg has been laid, it contains approximately 1000 nuclei. This stage of embryogenesis is referred to as the syncytial stage. The absence of cell division means that maternal regulatory proteins present in the egg can interact with genes in the nuclei without having to cross plasma membranes. In the next stage, the nuclei migrate to the periphery of the egg where, after a few more divisions, cell membranes form. This gives rise to a single layer of about 6000 cells and a small number of special cells called pole cells that form at the posterior end. The pole cells are germ-line precursors. At this point, about 2.5 hours after the egg was laid, the developing embryo is called a blastoderm. In the blastoderm, zygotic genes are fully expressed and maternal mRNA has been degraded.

In the next stages of embryogenesis, the single layer of cells invaginates to form the gastrula and various larval structures. One of the most prominent features of the insect body plan is the presence of distinct segments that begin to form after about three hours of embryogenesis. As we will see, many genes are expressed only in one or a small number of segments, where their products regulate expression of other segment-specific genes. For example, the genes controlling the development of wings and legs are only expressed in the thoracic segments, and the genes controlling the development of antennae are only expressed in head segments.

Some of the best-known and most spectacular *Drosophila* mutations affect these segment-specific regulatory genes. For example, the *bithorax* complex (*BX-C*) contains a set of genes that control the development of wings and other features of the second segment of the thorax. Figure 31·38a shows a normal fly with a single set of wings attached to the second thoracic segment. Note the presence of small structures called halteres immediately behind the wings. The halteres are part of the third thoracic segment. Certain mutations cause the genes of the *bithorax* complex to be inappropriately expressed in the third thoracic segment, where they turn on expression of all of the genes that are normally transcribed only in the second thoracic segment. The result is a fly with two second thoracic segments and four wings (Figure 31·38b). Note that the halteres are not present on this mutant fly; they have been transformed into a second set of wings.

Mutations that result in the transformation of one structure into another are known as homeotic mutations, and the genes in which the mutations occur are called **homeotic genes.** The *BX-C* genes are segment-specific homeotic genes that act relatively late in development. Some genes of the *Antennapedia* complex (*ANT-C*) are also segment-specific homeotic genes. For example, the *Antennapedia*

Figure 31·37
Early embryogenesis in *D. melanogaster.* The original diploid nucleus, or zygotic nucleus, is formed from the fusion of sperm and oocyte nuclei. During the first few hours of embryogenesis, mitosis occurs, but cell membranes do not form. This gives rise to the syncytial stage, in which the cell contains about 1000 nuclei. Nuclei then migrate to the periphery of the embryo, forming a single layer. At this stage, the pole cells form at the posterior end of the cell; these give rise to the germ line. Cell membranes eventually form around all nuclei, producing a blastoderm.

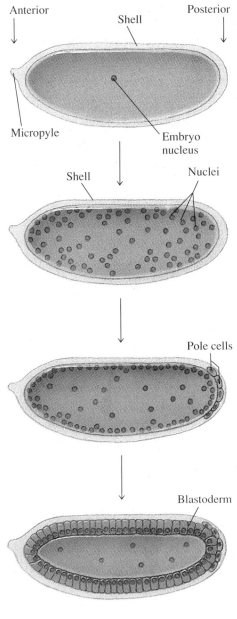

Figure 31·38
Effect of mutation in the *bithorax* complex. **(a)** Normal fly. Note the presence of halteres behind the wings. **(b)** Bithorax fly. This fly has two sets of wings and no halteres due to inappropriate expression of the *bithorax* complex genes in the third thoracic segment. (Courtesy of E. B. Lewis.)

(a)

(b)

gene controls gene expression in the first thoracic segment. Mutations in this gene can cause inappropriate expression in head segments, resulting in the development of legs where antennae usually grow.

Most of the regulatory genes we discuss in the rest of this section act much earlier in development than *BX-C* or *ANT-C*. We consider genes responsible for establishing the segmental compartments and for determining whether a cell will form part of the head or tail, or part of the dorsal or ventral surfaces.

B. Protein Gradients Control Anterior-Posterior Differentiation

One of the maternal genes that regulate embryonic gene expression is *bicoid*. This gene is transcribed in the nurse cells that surround the oocyte (the unfertilized egg), and Bicoid mRNA is transported out of the nurse cells and preferentially deposited

Figure 31·39
Production of the Bicoid gradient. The oocyte is surrounded by large nurse cells and smaller follicle cells. Bicoid mRNA is synthesized by the nurse cells and secreted into the anterior portion of the developing oocyte. Because Bicoid mRNA is anchored to the anterior cytoskeleton, Bicoid is synthesized at the anterior end of the embryo, and only traces of Bicoid are found at the posterior end.

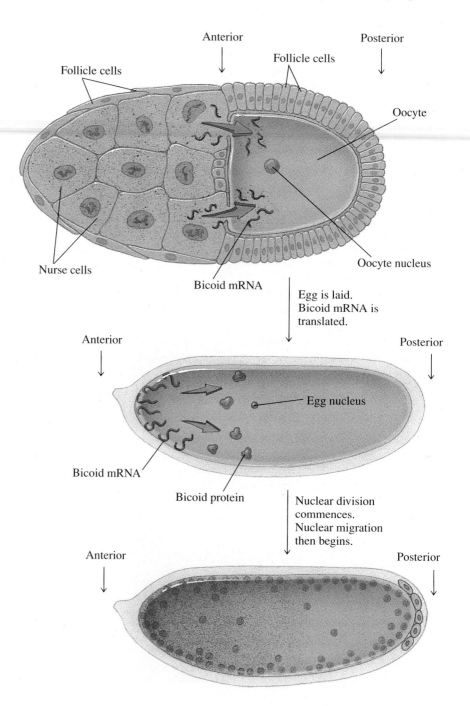

at the anterior end of the egg. Bicoid mRNA contains a long, untranslated region at its 3′ end that is bound by RNA-binding proteins. The mRNA-protein complex associates with the cytoskeleton at the anterior end of the egg, which prevents Bicoid mRNA from diffusing to the posterior end. When the mRNA is translated, Bicoid is exclusively synthesized at the anterior end of the embryo, and a gradient of Bicoid protein results (Figure 31·39). The steepness of the Bicoid gradient is further enhanced because Bicoid is unstable. The protein has a half-life of about 30 minutes, so most Bicoid is degraded before it can diffuse to the posterior end of the embryo. There is about a 1000-fold difference in Bicoid concentration between the anterior and posterior ends of the embryo.

Bicoid is a DNA-binding protein that acts as a transcriptional activator. The DNA-binding domain of Bicoid contains a motif known as a **homeodomain,** which consists of a stretch of 60 amino acid residues that form four α helices connected by turns. When Bicoid is bound to DNA, one of the helices lies in the major groove of the DNA, and various amino acid side chains within the helix contact specific nucleotide pairs in much the same manner as do the amino acids in the helix-turn-helix motifs of cI and Cro. The homeodomain is encoded by a DNA sequence known as the **homeobox.** Conserved homeobox sequences are found in genes for many *Drosophila* regulatory proteins and, in fact, in the genes for regulatory proteins of all other multicellular animals. Genes in the *Antennapedia* and *bithorax* complexes also contain homeoboxes.

Bicoid binds to sites on DNA with the consensus sequence TCTAATCC; several such sites are located near the promoters of some *Drosophila* genes. Because of the distribution of Bicoid in the embryo, the zygotic genes activated by Bicoid are expressed only in nuclei near the anterior end of the embryo. Several genes are positively regulated by Bicoid. One of these is *hunchback,* a gene whose product regulates gene expression at the anterior end of the embryo. The 5′ end of the *hunchback* gene contains multiple Bicoid-binding sites located from 60 to 300 bp upstream of the promoter (Figure 31·40). When Bicoid is bound at these sites, it interacts with RNA polymerase II and the general class II transcription factors to stimulate transcription initiation. Because of the distribution of Bicoid, Hunchback protein is synthesized only in the anterior end of the embryo.

Bicoid also activates some zygotic genes that are expressed only in the head, where they encode head-specific regulatory factors. One such gene is *Deformed;* the Deformed protein is found only at the extreme anterior end of the developing embryo. It appears that expression of *Deformed* requires much higher concentrations of Bicoid than those required for *hunchback* expression. This is probably related to the number and position of the Bicoid-binding sites. In the case of *hunchback,* it is likely that Bicoid binds cooperatively to the promoter region and that this allows activation in the presence of limited concentrations of Bicoid (such as those that occur near the middle of the embryo). This cooperative binding may be similar to the binding of phage λ repressor to O_{R1} and O_{R2}. In the case of *Deformed,* Bicoid binding may not be cooperative and a higher concentration of Bicoid may be needed to achieve gene activation. The net effect of the Bicoid gradient is to establish expression of a number of other regulatory proteins in nuclei at the anterior end.

There are two other components of the anterior-posterior differentiation system. The *nanos* gene is also expressed in follicle cells during oogenesis, but Nanos mRNA is deposited at the posterior end of the egg. During embryogenesis, Nanos forms a gradient opposite that of Bicoid. The primary role of Nanos is to prevent translation of Hunchback mRNA, thus ensuring that Hunchback is confined to the anterior end of the embryo.

Figure 31·40
Bicoid-binding sites near the *hunchback* gene. Strong binding sites are indicated by dark blue; weaker sites are indicated by light blue.

-300 -200 -100 *hunchback*

The third component of the anterior-posterior differentiation system consists of a tyrosine kinase receptor encoded by the *torso* gene and a ligand encoded by *torso-like*. The receptor is synthesized in the oocyte and translocated to the plasma membrane throughout the cell. Torsolike is made only by maternal cells at the extreme ends of the oocyte and is deposited at the ends of the egg in structures that lie outside the plasma membrane. When the egg is laid, Torsolike is released and binds to the Torso receptor, triggering a kinase cascade that activates genes in nuclei positioned near the poles. These activated genes are required for the formation of structures that are found only in the head and tail of the adult fly.

C. A Second Gradient Controls Dorsal-Ventral Differentiation

The differentiation of cells on the dorsal and ventral surfaces of the embryo is controlled by a transcription factor encoded by the maternal *dorsal* gene. Active Dorsal protein is localized to the ventral half of the developing embryo, forming a dorsal-ventral gradient similar to the anterior-posterior gradients of Bicoid and Nanos. Dorsal activates transcription of ventral-specific genes in those nuclei found on the ventral side of the embryo.

Formation of the Dorsal gradient involves a complex series of interactions between the nurse cells and the developing oocyte. Interest in this system stems from its similarity to processes that occur in mammals. For example, in the first part of the pathway, the developing oocyte secretes a factor that is detected by a receptor on the nurse cells near the ventral surface of the oocyte. The receptor, encoded by the *torpedo* gene, is homologous to mammalian epidermal growth-factor receptor, and the two receptors are thought to function in a similar manner. Another part of the pathway for establishing the Dorsal gradient requires the products of two genes, *snake* and *easter*, that encode serine proteases similar to those of the blood coagulation cascade (Section 6·11B). These serine proteases activate another protein that helps establish the dorsal-ventral gradient.

D. Segmentation Is Determined by a Network of Interacting Regulatory Factors

The *hunchback* gene is just one of a class of genes that is expressed in a localized region of the developing embryo due to the anterior-posterior gradients. The genes of this class are known as *gap* genes because mutations in these genes usually cause deletions of large parts of the differentiated larva, resulting in gaps. Other genes of this class include *Krüppel, knirps, giant, tailless,* and *huckebein.* With the exception of Giant, all of the products of the *gap* genes are DNA-binding, zinc-finger proteins (discussed in Section 27·8D) that act as repressors, activators, or both, by binding to specific sites near promoters.

Recall that expression of *hunchback* is regulated by Bicoid so that there are high levels of Hunchback protein at the anterior end of the egg. At high concentrations, Hunchback represses transcription of the *Krüppel* gene, which ensures that Krüppel protein is not synthesized at the anterior end. However, at lower concentrations, Hunchback activates expression of *Krüppel*. Thus, Krüppel accumulates at the position in the egg where Hunchback concentrations drop but do not fall off completely. The net effect of Hunchback on *Krüppel* expression is production of a band, or stripe, of Krüppel near the middle of the embryo. The border between Hunchback and Krüppel is sharp because Krüppel represses transcription of the *hunchback* gene.

(a)

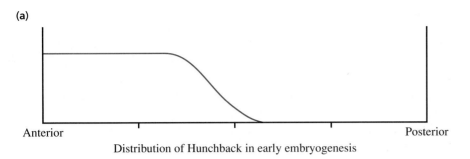

Anterior .... Posterior

Distribution of Hunchback in early embryogenesis

Figure 31·41
Distribution of *gap* gene products during embryogenesis along the length of the embryo. **(a)** The distribution of Hunchback in early embryogenesis is controlled by Bicoid. **(b)** By the blastoderm stage, the products of the *gap* genes are distributed in discrete regions of the embryo due to complex interactions that regulate *gap* gene expression. [Adapted from Lawrence, P. A. (1992). *The Making of a Fly: The Genetics of Animal Design* (Oxford, England: Blackwell Scientific Publications).]

(b)

Huckebein   Hunchback   Krüppel   Knirps   Huckebein
    Tailless   Giant   Giant   Tailless

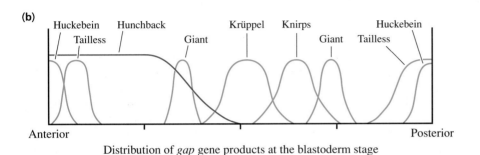

Anterior .... Posterior

Distribution of *gap* gene products at the blastoderm stage

A band of Knirps protein forms posterior to the Krüppel stripe. The *knirps* gene is activated by low concentrations of Krüppel and repressed by high concentrations in a manner analogous to the effects of Hunchback on the *Krüppel* gene. Furthermore, Knirps represses transcription of the *Krüppel* gene, thus sharpening the Krüppel-Knirps border.

Similar interactions among the remaining *gap* genes produce localized stripes or bands of protein along the length of the embryo (Figure 31·41). In the next stage of development, individual segments start to form under the influence of these *gap* genes and additional genes that are required specifically in each segment. Thus, anterior-posterior differentiation proceeds through successive stages in which smaller and smaller compartments are defined. Segmentation begins with broad regions defined by the presence of Hunchback and Deformed and progresses through formation of smaller regions defined by the *gap* gene products. The last stage involves expression of segment-specific genes.

There are also genes that are expressed in every second segment; these are known as **pair-rule** genes. One example is *fushi tarazu;* it is expressed in seven stripes in the embryo, corresponding to every other segment in the thorax and abdomen (Figure 31·42). Like the *gap* genes and *bicoid, fushi tarazu* encodes a DNA-binding protein that functions as a regulatory protein. Other examples of pair-rule genes are *hairy* and *eve*. The interactions between the products of the pair-rule genes and *gap* genes are complex. They involve combinations of regulatory proteins that bind to multiple sites, where they act as repressors and activators.

The final stage of embryogenesis in *Drosophila* is characterized by additional cascades of regulatory factors that lead to differentiation of nerve tissue, specialized organs, and large structures such as wings and legs. This is the stage at which the genes of the *bithorax* and *Antennapedia* complexes are transcribed.

While we have examined development in *Drosophila* in this chapter, similar mechanisms are known to control gene expression during embryogenesis in other multicellular organisms, including mammals and flowering plants. The elucidation of the early steps of development in these other organisms has lagged far behind the discoveries in *Drosophila* and *C. elegans,* but it is a major goal of scientists who work in the field of gene expression and development.

Figure 31·42
Distribution of the product of the pair-rule gene *fushi tarazu.* This gene product is visualized by the binding of antibodies labelled with a colored marker. The gene product is localized in seven stripes in the embryo. (Courtesy of Matthew Scott.)

Summary

The developmental regulation of gene expression involves many of the regulatory mechanisms described in earlier chapters. During infection by bacteriophages and viruses, the temporal expression of many genes is regulated at the level of transcription initiation. The T7 genome encodes a new RNA polymerase that is synthesized early in infection. This polymerase specifically transcribes T7 late genes. Late gene expression in T4 and SP01 is controlled by novel σ factors that are encoded by phage early genes. These σ factors direct host RNA polymerase to phage late-gene promoters. During adenovirus infection, a new transcriptional activator is produced.

Sporulation in *Bacillus subtilis* is an example of differentiation. Regulation of gene expression during sporulation is controlled by the sequential production of σ factors, which is called a σ cascade. In some cases, these σ factors are active either in the mother cell or in the forespore, leading to expression of different genes in the two cells. In these cases, production of active σ factors is coupled to formation of the polar septum.

Infection of *E. coli* by bacteriophage λ is a model of developmental regulation that involves cellular decision making. Bacteriophage λ can enter a lytic or lysogenic pathway after infecting a host cell. The expression of genes during a lytic infection is regulated in part by an antitermination cascade. The products of the *N* and *Q* genes modify the host transcription complex, enabling it to bypass terminators in the λ genome.

The lysogenic pathway of λ development is controlled mainly by cI and cII. The cI repressor activates transcription of its own gene and also represses transcription of phage early genes. cII acts at several promoters to increase transcription of genes required for lysogeny. One of these genes is *cI*. Activation of the *cI* gene by cII depends on the amount of cII present in an infected cell, and this ultimately determines whether lysogeny occurs.

The decision about which developmental pathway to enter is also regulated by competition between Cro and cI as they bind to operators near the early promoters. Both Cro and cI contain helix-turn-helix DNA-binding motifs.

Drosophila development is regulated by a cascade of activators and repressors that are synthesized during embryogenesis. Synthesis of these regulatory proteins is spatially and temporally regulated. The initial pattern of gene expression is defined by gradients of protein encoded by maternally expressed genes. The first round of embryonic gene expression is regulated by the presence or absence of the maternal gene products, and the embryonic gene products are themselves regulatory proteins that influence the expression of other embryonic genes.

Selected Readings

General Reference

McKnight, S. L., and Yamamoto, K. R., eds. (1992). *Transcriptional Regulation* (Cold Spring Harbor, New York: Cold Spring Harbor Laboratory Press).

Bacteriophage Development and σ Cascades

Helmann, J. D., and Chamberlin, M. J. (1988). Structure and function of bacterial sigma factors. *Annu. Rev. Biochem.* 57:839–872.

Kaiser, D., and Losick, R. (1993). How and why bacteria talk to each other. *Cell* 73:873–885.

Kunkel, B., Kroos, L., Poth, H., Youngman, P., and Losick, R. (1989). Temporal and spatial control of the mother-cell regulatory gene *spo*IIID of *Bacillus subtilis. Genes Dev.* 3:1735–1744.

Losick, R., Youngman, P., and Piggot, P. J. (1986). Genetics of endospore formation in *Bacillus subtilis. Annu. Rev. Genet.* 20:625–670.

Shapiro, L. (1993). Protein localization and asymmetry in the bacterial cell. *Cell* 73:841–855.

Sousa, R., Chung, Y. J., Rose, J. P., and Wang, B.-C. (1993). Crystal structure of bacteriophage T7 RNA polymerase at 3.3Å resolution. *Nature* 364:593–599.

Bacteriophage λ

Anderson, J. E., Ptashne, M., and Harrison, S. C. (1987). Structure of the repressor-operator complex of bacteriophage 434. *Nature* 326:294–295.

Greenblatt, J., Nodwell, J. R., and Mason, S. W. (1993). Transcriptional antitermination. *Nature* 364:401–406.

Steitz, T. A., Ohlendorf, D. H., McKay, D. B., Anderson, W. F., and Matthews, B. W. (1982). Structural similarity in the DNA-binding domains of catabolite gene activator and cro repressor proteins. *Proc. Natl. Acad. Sci. USA* 79:3097–3100.

Wolberger, C., Dong, Y., Ptashne, M., and Harrison, S. C. (1988). Structure of a phage 434 Cro/DNA complex. *Nature* 335:789–795.

Drosophila Development

Holland, P. (1992). Homeobox genes in vertebrate evolution. *BioEssays* 14:267–273.

Jäckle, H., and Sauer, F. (1993). Transcriptional cascades in *Drosophila. Curr. Biol.* 5:505–512.

Lawrence, P. A. (1992). *The Making of a Fly: The Genetics of Animal Design.* (Oxford, England: Blackwell Scientific Publications).

Sauer, F., and Jäckle, H. (1993). Dimerization and the control of transcription by Krüppel. *Nature* 364:454–457.

32

Genes and Genomes

As we pointed out in Chapter 27, there is no satisfactory biochemical definition of a *gene*. Our working definition has been "a DNA sequence that is transcribed," but different definitions are often encountered. One popular alternative, for example, is "a region of DNA or RNA that encodes a functional RNA molecule or polypeptide." Other definitions take into account all the regulatory sequences that influence transcription, including those that lie far away from the transcribed DNA.

All definitions have their strengths and weaknesses, but none work well in all cases. For example, is a bacterial operon a single gene, or is it composed of several genes? Are introns and transcribed noncoding regions parts of a gene? How do we define the limits of the genes for proteins with overlapping coding regions whose expression is regulated by frameshifting or alternative splicing? And what about large RNA or polypeptide precursors that are cleaved to create several smaller functional molecules; do these arise from one or several genes?

Overlooking semantic imprecision, we have presented many properties and functions of genes in the preceding chapters. Until now, we have assumed that a gene is unchanging, static over time. We have implied that a gene exists at a fixed locus in a chromosome and that its structure and function are not affected by rearrangement or recombination. In this chapter, we dash those notions. We will learn that genes and genomes are in fact dynamic. Among other topics, we examine genes that jump from one locus to another and genes that are rearranged before they are expressed. These examples will further challenge our concept of a gene.

For over a century, it has been known that genomes change over time encompassing many generations. This gradual change in genetic information is the basis of evolution. However, only in the past few decades have we recognized that genomes are dynamic structures that can change over much shorter periods of time. This conceptual revolution was driven by the discoveries of transposons and lysogenic bacteriophages by phage and bacterial geneticists and by the work of Barbara McClintock (Figure 32·1), who received the Nobel prize in 1983 for her demonstration of mobile genes in maize.

Genome is almost as difficult to define as *gene*. In practical terms, a genome is usually considered to be a single copy of the chromosomal DNA of an organism. The genome of *Escherichia coli*, for example, is a double-stranded, circular DNA

Figure 32·1
Barbara McClintock (1902–1992).

molecule, and the human genome consists of 23 double-stranded, linear chromosomes. The genome of an organism is usually exclusive of extrachromosomal DNA, such as plasmids. Nor is mitochondrial or chloroplast DNA in eukaryotic cells considered part of the genome, even though they may contain essential genes.

In this chapter, we examine the structure and organization of selected genomes and compare the organization of eukaryotic and prokaryotic chromosomes. We also address the fascinating problem of how to determine the number of different genes in a genome, and we examine why there is so much DNA in large genomes. Answers to these questions are likely to be known soon since the entire genomes of some organisms (including *E. coli, Haloferax volcanii, Caenorhabditis elegans, Drosophila melanogaster, Arabidopsis thaliana, Mus musculus,* and *Homo sapiens*) are currently being sequenced. These data will bring us closer to our goal of understanding the structure, organization, and evolution of genomes.

32·1 Some Bacteriophages and Viruses Can Integrate into Host DNA

We encountered an example of dynamic genomes in the previous chapter, when we explored bacteriophage λ integration and excision from the *E. coli* chromosome. *E. coli* is not unusual: the genomes of all organisms contain bacteriophages or viruses that remain stably integrated for many generations. The integration and excision of bacteriophage and viral DNA serve as models both for other recombinations between extrachromosomal DNA and genomes and for rearrangements within chromosomes.

Integration of bacteriophage λ is a classic example of site-specific recombination. Recall that bacteriophage λ recombines with *E. coli* DNA at a specific nucleotide sequence, the *att* site (Section 31·6C). Related phages also integrate at specific sites located elsewhere in the *E. coli* genome.

Normally, the phage λ genome is excised from the bacterial chromosome after a number of generations, but some phage λ mutants are incapable of excision and become permanent residents of the host genome. After many generations, these mutant phage genomes accumulate additional mutations in their unexpressed genes and become **cryptic phages.** Many strains of *E. coli* carry cryptic phages in their genomes, although in some cases, only a fragment of the original bacteriophage DNA remains. Cryptic phages are examples of extraneous, or "junk," DNA.

Like phage λ, bacteriophage Mu has a lysogenic stage during which it is integrated into the chromosome of its host. However, phage Mu differs from phage λ by recombining at many sites, showing little specificity. Another feature that distinguishes phage Mu from phage λ is that, once integrated, phage Mu can transpose to other sites in the genome. Transposition requires DNA replication, and the recombination mechanism involves the duplication of five base pairs of host DNA around the integration site. After transposition, there is a copy of the phage at the new integration site, and the original integrate is preserved.

Because phage Mu can insert almost randomly, it often integrates in the middle of genes. When phage Mu integrates into an essential gene, a lethal mutation results. The integration of lambdoid phages does not cause such mutations because the specific sites of integration are not within genes. Among those phage that integrate at specific sites, natural selection has presumably favored those that integrate outside of genes.

Retroviruses are RNA viruses that infect eukaryotic cells. During infection, the RNA genome is copied by the action of reverse transcriptase to make complementary, single-stranded DNA that can serve as a template for host DNA polymerase (Section 25·10C). The resulting double-stranded DNA can integrate into a host

Figure 32·2
Integrated retroviral DNA. The long terminal repeats (LTR) have the same orientation and contain short inverted repeats at their ends. The *gag, pol,* and *env* genes overlap. The retroviral insert is flanked by short repeats of the host DNA.

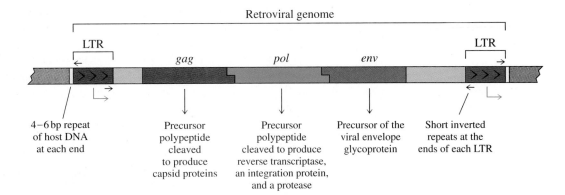

chromosome by a mechanism that causes a short duplication of the target site of four to six base pairs (depending on the type of retrovirus). Integration is catalyzed by a virally encoded integration protein and is not site specific.

The structure of an integrated retroviral genome is shown in Figure 32·2. The ends of the viral genome contain long terminal repeats (LTR) of several hundred base pairs. The two LTRs have the same orientation. This terminal redundancy is necessary for complete replication of the ends of unintegrated viral DNA and for the integration step. A short, inverted repeat segment of about 5 to 10 base pairs occurs at the ends of each LTR.

A retroviral genome contains at least three open reading frames, or "genes." The *gag* gene encodes a large polypeptide that is cleaved to produce the capsid proteins. The *pol* gene encodes a precursor polypeptide that is cleaved to produce reverse transcriptase, an integration protein, and a protease. In some retroviruses, the *pol* precursor is made by translating past a stop codon at the end of the *gag* open reading frame, whereas in others it is produced by translational frameshifting. The *env* gene encodes the viral envelope protein. The mRNA for the envelope protein is produced by transcribing the integrated viral DNA and splicing the transcript to remove the *gag* and *pol* sequences. The promoter for transcription of *gag, pol,* and *env* lies within the left-hand LTR, and an identical promoter is located within the right-hand LTR. The promoters have the same orientation, so initiation at the right-hand promoter results in transcription of adjacent host genes.

The integration of retroviral DNA into a eukaryotic genome can disrupt genes, but these disruptions are less likely to be lethal than phage Mu insertions in bacteria because most eukaryotic cells are diploid and integration affects only one copy of the gene. One well-known example of a mutation due to retroviral disruption of a gene occurs in mice. Insertion of a retrovirus at the *d* locus, which controls hair color, produces mice that exhibit dilute brown coat color rather than the normal dark brown.

Retroviral integration can also cause activation of downstream host genes due to transcription from the promoter in the right-hand LTR. In most cases, this does not affect the cell, but if the retrovirus integrates near a gene that regulates cell growth, overexpression of the gene can lead to cancer. For example, avian leukemia viruses infect chickens and other birds. When the virus integrates upstream of the proto-oncogene *c-myc, c-myc* is overexpressed and leukemia can result.

Eukaryotic genomes contain many copies of integrated retroviruses that have accumulated over generations. There are many different types of such endogenous retroviruses, and they are often nonfunctional, mutant versions, much like the cryptic phages found in bacteria. Mammalian genomes in particular appear to accumulate retroviruses, although the reason for this is unclear. The mouse genome, for example, contains thousands of copies of retroviruses that together account for up to 1% of the total DNA in mouse cells.

Figure 32·3
Prokaryotic class I transposons. **(a)** Insertion element. The inverted terminal repeats are indicated by arrows. The insertion element is flanked by a short repeat of host DNA. **(b)** Composite transposon *Tn*681. This class I transposon consists of two *IS*1 elements flanking a gene for a heat-stable enterotoxin.

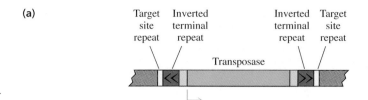

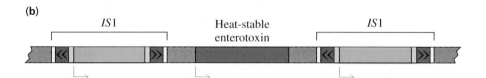

32·2 Transposons Can Recombine into Chromosomal DNA

In addition to bacteriophages and viruses, there are other examples of pieces of DNA that can integrate into and excise from the genome. **Transposons** are such DNA segments. These are mobile residents of the genome that can relocate within a chromosome or even between chromosomes within the same cell. Transposons typically encode the enzymes necessary to catalyze their relocation, or **transposition.** You will sometimes see transposons referred to as transposable elements, integrons, mobile genetic elements, or even jumping genes—all terms that emphasize the mobile nature of these DNA segments.

A. There Are Three Types of Bacterial Transposon

There are three classes of transposon in bacteria: class I, class II, and class III. Class I transposons consist of insertion elements (*IS* elements) that contain a transposase gene (Figure 32·3a). The ends of the transposon have short, inverted repeats. The transposase gene encodes the enzyme that catalyzes recombination between the *IS* element and the host DNA. Some of the most common *E. coli* insertion elements are listed in Table 32·1.

Table 32·1 *E. coli* insertion elements

Element	Length (bp)	Inverted terminal repeats (bp)	Target site repeats (bp)
*IS*1	768	23	9
*IS*2	1327	41	5
*IS*4	1428	18	11/12
*IS*5	1195	16	4
*IS*10	1329	22	9
*IS*50	1531	9	9
*IS*903	1057	18	9

[Adapted from Lewin, B. 1990. *Genes IV* (Oxford, England: Oxford University Press), p. 650.]

Table 32·2 Common composite transposons

Transposon	Flanking *IS* elements	Length (bp)	Captured genes
*Tn*5	*IS*50	5700	Kanamycin resistance
*Tn*9	*IS*1	2500	Chloramphenicol resistance
*Tn*10	*IS*10	9300	Tetracycline resistance
*Tn*681	*IS*1	2100	Heat-stable enterotoxin
*Tn*903	*IS*903	3100	Kanamycin resistance
*Tn*2571	*IS*1	23 000	Multiple resistance genes

Some class I transposons are known as composite transposons because they consist of two *IS* elements that flank a "captured" bacterial gene. The bacterial gene usually confers some selective advantage on the host, such as resistance to toxic agents or the ability to make toxins. For example, the transposon *Tn*681 consists of two *IS*1 elements flanking a gene for a heat-stable enterotoxin (Figure 32·3b). Common composite transposons are listed in Table 32·2.

During the excision of class I transposons, the transposase acts as an endonuclease, cleaving the DNA at the inverted terminal repeats on both sides of the transposon to create blunt ends. This excises the transposon, leaving a double-stranded break in the chromosome. Integration occurs at a **target site,** which is also recognized and cleaved by the transposase. In this case, cleavage by the transposase leaves staggered ends, like those generated by some restriction endonucleases (Figure 32·4). The transposon is then ligated to the single-stranded protruding ends of DNA at the target site, the resulting gap is repaired by DNA polymerase, and the nicks are sealed by DNA ligase. As a result of this mechanism, the target site at the ends of the integrated transposon is duplicated.

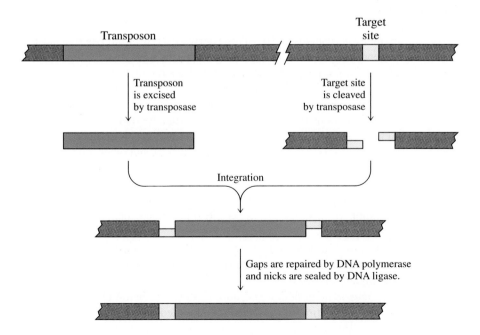

Figure 32·4
Mechanism of transposition of a class I transposon. Transposase catalyzes the excision of the transposon, leaving blunt ends. Transposase also cleaves the host DNA at a target site on the bacterial chromosome, producing a double-stranded break with single-stranded extensions. The transposon is inserted at the site of the staggered cut. The gaps are filled by the action of DNA polymerase, and the nicks are sealed by DNA ligase, leaving a short duplication at the ends of the insertion.

Figure 32·5
Class II transposon *Tn3*. This transposon contains a captured gene for β-lactamase, which confers resistance to the antibiotic ampicillin. Arrows indicate inverted terminal repeats. The duplicated target site is shown in yellow.

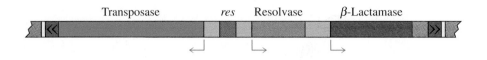

There are many target sites in bacterial genomes, and most transposons can integrate at a variety of sites. Some transposons show a preference for particular nucleotide sequences, whereas others preferentially integrate at sites that are within a few kilobase pairs of their existing integration sites.

Class II transposons are larger than class I transposons. They possess inverted terminal repeats of about 40 bp and carry genes encoding a minimum of two proteins: a transposase and a resolvase (Figure 32·5). The resolvase is required for integration; it catalyzes site-specific recombination between two transposons at the *res* site between the transposase and resolvase genes. The mechanism of integration of class II transposons is more complex than that of class I; it requires more extensive DNA synthesis and a recombination step catalyzed by the resolvase. This mechanism, called **replicative transposition,** is initiated by endonucleolytic cleavage at the target site and at sites at the ends of the transposon (Figure 32·6). One strand of the transposon is ligated to one strand at the target site, and the other strand of the transposon is ligated to the other strand at the target site. The resulting gaps are filled by the action of DNA polymerase, and the nicks are sealed by DNA ligase. When transposition involves two separate DNA molecules (as illustrated in Figure 32·6), it results in production of a double-sized chromosome referred to as a **cointegrate.** The cointegrate is resolved into two separate chromosomes by site-specific recombination between the *res* sites on the two transposons, catalyzed by resolvase. As a result of transposition, a new transposon is integrated at the target site, and the existing transposon still occupies its old location.

Most class II transposons are composite transposons that carry additional genes not required for integration and excision. The example shown in Figure 32·5 is *Tn3*, a transposon containing the gene for β-lactamase, which confers resistance to β-lactam antibiotics, such as ampicillin. This transposon is frequently inserted into plasmid DNA, and a small plasmid carrying *Tn3* and the ampicillin-resistance gene is the ancestor of many plasmids used in recombinant DNA technology (Chapter 33).

There are many derivatives of *Tn3*. These derivatives carry the *Tn3* transposase and resolvase genes and have the same terminal repeats as *Tn3*, but the β-lactamase gene is replaced by other genes. For example, *Tn21*, a member of the *Tn3* family, carries an entire bacterial operon that confers resistance to mercury ions (the genes encode proteins that convert toxic mercuric ions to metallic mercury). *Tn21* and its close relatives are widely distributed geographically and are found in many species of bacteria; in fact, *Tn21* was recently found in *Xanthomonas* growing deep in mercury mines in Central Asia. The wide distribution of *Tn21* and other transposons is evidence that transposons are able to move from one species of bacteria to another.

Class III transposons are bacteriophages such as phage Mu that behave as transposons following integration into the host chromosome. Phage Mu is the only well-studied example of this class. Like class II transposons, phage Mu integrates by replicative transposition.

Excision of most transposons is rare because the transposase gene is not transcribed at high rates. In some cases, transposition depends on whether DNA is methylated; methylated DNA is not cleaved by transposase, but hemimethylated DNA is. Since DNA is hemimethylated immediately following DNA replication (Section 24·10), transposition in these cases is coupled to DNA replication. One reason why this might be advantageous is that some transposition mechanisms, such as the one shown in Figure 32·4, involve a double-stranded break in the

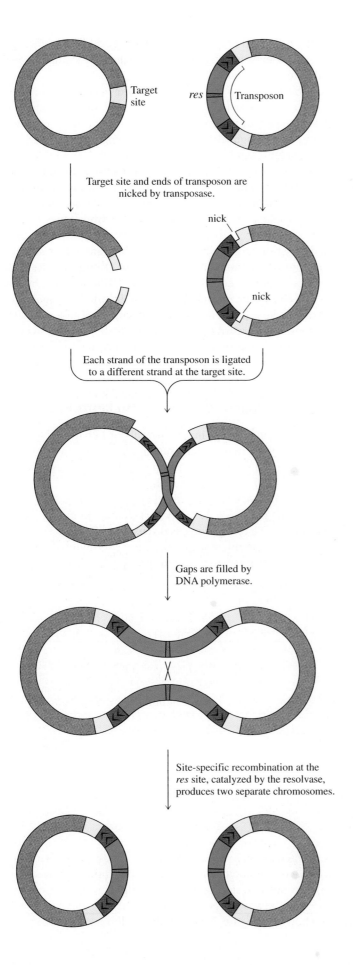

Target site

res Transposon

Target site and ends of transposon are nicked by transposase.

nick

nick

Each strand of the transposon is ligated to a different strand at the target site.

Gaps are filled by DNA polymerase.

Site-specific recombination at the *res* site, catalyzed by the resolvase, produces two separate chromosomes.

Figure 32·6
Replicative transposition of class II transposons. The transposase cleaves at the target site, generating overhanging ends, and at the ends of the integrated transposon, generating single-stranded nicks in the DNA. The two strands of the integrated transposon are ligated to the overhangs at the target site, and the resulting gaps are filled by the action of DNA polymerase. The resulting cointegrate is resolved by site-specific recombination at the *res* sites, catalyzed by transposon-encoded resolvase. The result of this sequence of reactions is duplication of the integrated transposon at a new site. In this figure, we show transposition between two separate DNA molecules, although it can also occur between sites on the same DNA molecule.

chromosome, which is lethal if not repaired. Without recent DNA replication, there would be no second chromosome to serve as a template for recombinational repair, and the cell would die.

All species of bacteria carry transposons in their genomes. Some have more than 100 copies of a single transposon, and others have multiple copies of many transposons. The frequency of transposition in bacteria is usually one event every 10^5 to 10^7 generations, although it can be higher in some cases. Bacteriophage Mu is a remarkable example; it can integrate into and excise from a host chromosome about 100 times during a single infection. However, because the rate of transposition of most transposons is low, the sites of integration of individual transposons can serve as relatively stable genetic markers over many generations. In *E. coli* strain W3310, for example, there are eight copies of *IS*1 whose sites have been precisely mapped.

There are a number of parallels between bacterial transposons and eukaryotic retroviruses. For example, the amino acid sequence of *IS*3 transposase is similar to that of the integration protein encoded by retroviruses. Furthermore, both prokaryotic transposons and eukaryotic retroviruses contain inverted terminal repeats, and in both cases the target site is duplicated during integration. These parallels suggest that bacterial transposons and retroviruses might be distantly related. As we will see in the next section, eukaryotic transposons and retroviruses are even more intimately related.

B. Many Eukaryotic Transposons Are Similar to Retroviruses

Eukaryotic transposons usually range in size from about 1 kb to over 8 kb. Like bacterial transposons, eukaryotic transposons contain inverted repeats at their termini, and integration is associated with the formation of a direct repeat of the target site. Some well-studied eukaryotic transposons are listed in Table 32·3. These can be divided into two distinct classes: retrovirus-like transposons and *Ac*-like transposons, the latter named after the prototype *Ac* (activator) element in maize. (A third class of eukaryotic transposon, the retroposons, is discussed in Section 32·4B).

The most common eukaryotic transposons are the retrovirus-like transposons, also known as integrating retroelements or retrotransposons. These transposons are

Table 32·3 Eukaryotic transposons

Transposon	Organism	Length (bp)	Inverted terminal repeats (bp)	Target site repeats (bp)
Retrovirus-like				
copia	Drosophila	5000	10	5
hobo	Drosophila	3100	12	?
gypsy	Drosophila	8300	10	4
Ty	Yeast	5900	2	5
IAP	Mouse	7100	4	6
Ac-like				
Ac	Maize	4563	10	8
Spm (En)	Maize	8287	13	3
Mu1	Maize	1367	213	9
P factor	Drosophila	2900	31	8

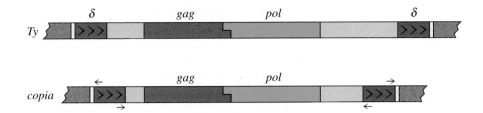

Figure 32·7
Eukaryotic retrovirus-like transposons *Ty* and *copia*. Retrovirus-like transposons contain LTRs, most with short, inverted repeats at their ends. These short repeats are absent in *Ty*, where the LTRs are called δ elements.

probably derived from retroviruses. They consist of pieces of DNA containing long terminal repeats similar to retroviral LTRs, and they contain genes that are homologous to the retroviral *gag* and *pol* genes. Some retrovirus-like elements carry a third gene similar to the retrovirus *env* gene, although these genes often contain multiple in-frame stop codons and usually cannot produce functional envelope proteins.

The *copia* transposon of *Drosophila* and the *Ty* element of yeast are typical examples of retrovirus-like transposons (Figure 32·7). The *copia* LTRs contain short, inverted repeats at the ends; these short repeats have apparently been lost in *Ty*, where the LTRs are referred to as δ elements. Both transposons contain *gag* and *pol* genes. The *IAP* transposon is an example of a mammalian retrovirus-like transposon. *IAP* is similar to *copia* in its structure, except that *IAP* also has a recognizable *env*-like gene.

Transcription of these transposons initiates from a promoter in the left-hand LTR and terminates at a site in the right-hand LTR. The transcript is a typical mRNA molecule that is polyadenylated and capped. The transposon mRNA can then be converted to double-stranded DNA in a series of reactions initiated by the transposon-encoded reverse transcriptase (one of the products of the *pol* gene). Occasionally, this double-stranded DNA is integrated into the chromosome in a reaction catalyzed by the transposon-encoded integration protein. Integration of the double-stranded DNA is accompanied by duplication of the target site. The net effect of this mechanism of transposition is duplication of the transposon at the new integration site.

Note that retrovirus-like transposons are not retroviruses because they cannot form infective viral particles. However, in some cases, noninfective virus-like particles containing transposon RNA have been observed in vivo. For example, *IAP* in mice produces abundant intracellular particles called intracisternal A particles.

Retrovirus-like transposons are common among eukaryotes. The genome of the house mouse *Mus musculus,* for example, contains as many as 1000 copies of *IAP*. *IAP* transposons are also found in other rodents, although it is not clear whether this is due to horizontal transfer (that is, transfer between species) of the transposon or whether the common ancestor of all living rodents contained *IAP*. There are abundant copies of *copia* in *D. melanogaster;* the genomes of individuals in most populations contain about 100 distinct inserts. In addition, *copia* is a member of a large family of transposons called *copialike* elements that includes *gypsy* and many others. This family accounts for over 5000 transposons in the genome of the typical fruit fly.

There are about 35 copies of *Ty* in the genome of most laboratory strains of *Saccharomyces cerevisiae*. In addition, yeast genomes contain many copies of single δ elements—remnants of homologous recombination between the ends of an integrated transposon that led to excision of the rest of the *Ty* element (Figure 32·8, next page). Homologous recombination can also occur between adjacent transposons or even between transposons on different chromosomes. Both of these events lead to genomic rearrangements, which are common in the genomes of organisms containing transposons.

Figure 32·8
Homologous recombination between δ elements of *Ty*. Such recombination leads to loss of the transposon but retention of one δ element in the genome.

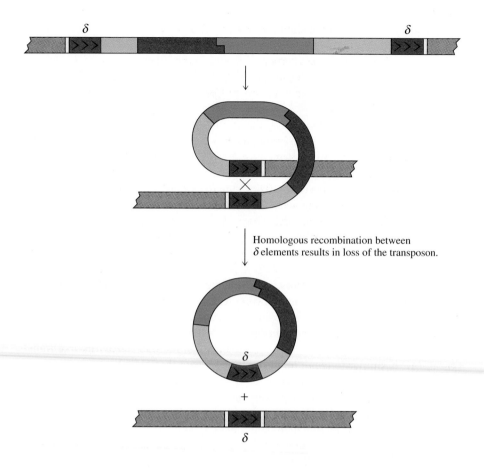

Homologous recombination between δ elements results in loss of the transposon.

The second class of eukaryotic transposons comprises the *Ac*-like elements. These transposons carry a transposase gene and are similar to bacterial class I transposons. Two typical transposons of this class, the *Drosophila P factor* and the maize *Ac* element, are shown in Figure 32·9. In both cases, the transposase genes contain introns, and the precursor mRNA transcribed from these genes is processed to a mature mRNA molecule that encodes the protein. The mechanism of transposition by *Ac*-like transposons is not well understood, but it is probably similar to that of bacterial class I transposons. Unlike retrovirus-like transposons, an RNA intermediate is not required.

The insertion of eukaryotic transposons can disrupt genes. There are many examples of such mutations in *S. cerevisiae* because transposition of *Ty* is frequent. Similarly, *P factor* insertions are associated with several mutations in *Drosophila*, including one in which the gene for red eye pigment is disrupted, leading to a white-eyed fly. This was one of the first mutations recognized in fruit flies. Transposon insertions cause several mutations in maize. In fact, the discovery that some mutations in maize are caused by mobile elements first pointed to the existence of

Figure 32·9
Eukaryotic *Ac*-like transposons *P factor* and *Ac*. In each case, the transposons consist of several exons (green) and introns (brown).

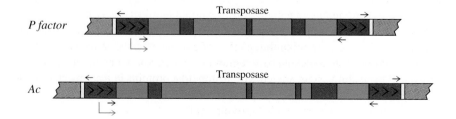

transposons. In the example shown in Figure 32·10, insertion of an *Spm* element into the gene whose product is responsible for synthesis of the purple pigment anthocyanin produces a colorless kernel. When the *Spm* element is excised during kernel formation, the activity of the gene is restored. The result is a spotted kernel containing some cells that still carry a disrupted gene and others that carry a restored gene.

In the pea plant, transposons similar to maize *Ac* elements can disrupt genes. For example, the gene for starch branching enzyme 1, *SBE1*, can be mutated by insertion of a transposon. This enzyme is required during pea seed formation for synthesis of amylopectin, a carbohydrate reserve used during germination. Normal pea seeds are smooth and spherical because they contain an abundant supply of amylopectin, but mutant seeds are wrinkled. These wrinkled pea seeds were one of the mutants studied by Gregor Mendel in the 1860s, when he did his pioneering work on the genetics of the pea plant.

The propagation of *P factor* among *Drosophila* species is an interesting example of the horizontal transfer of transposons. The genomes of modern populations of *D. melanogaster* now contain multiple copies of *P factor*. However, there is no evidence of *P factor* in wild-type populations of *D. melanogaster* isolated between 1910 and 1950, nor in the genomes of *Drosophila* species most closely related to *D. melanogaster*. However, the genomes of *Drosophila willistoni*, a more distantly related fly, not only contain *P factor*, but the nucleotide sequence of the *P factor* in these flies is almost identical to that now found in *D. melanogaster*. This suggests strongly that *P factor* was transferred recently from *D. willistoni* to *D. melanogaster*.

D. willistoni is found in South and Central America, and *D. melanogaster* originated in Western Africa. In the 1800s, *D. melanogaster* populations became established in the Americas, probably carried by ships between Africa and the West Indies during the slave trade. This was the first time the two species of *Drosophila* came into contact and the first opportunity for *P factor* to be transferred. The species do not interbreed, but exchange may have taken place via parasitic mites that feed on both species. Once *P factor* invaded *D. melanogaster*, it spread rapidly in the population, in part because *P factor* transposition is especially enhanced in germ-line cells.

Because transposons do not encode products vital to the survival of the host organism, deletions and point mutations in the integrated DNA accumulate relatively rapidly. Most eukaryotic genomes contain fragments of transposons and nonfunctional transposons, just as most bacteria carry cryptic bacteriophages.

(a)

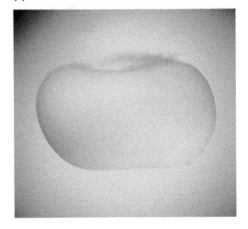

(b)

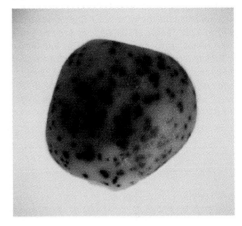

Figure 32·10
Effect of mobile transposons in maize. **(a)** A colorless kernel results when an *Spm* element integrates into the gene required for the synthesis of the pigment anthocyanin. **(b)** A spotted kernel results when the integrated *Spm* element is excised during development, restoring the pigment-producing gene in some cells. (Courtesy of Nina Fedoroff.)

32·3 Bacterial Genomes

The integration of transposons, bacteriophages, and viruses is an example of how genomes are routinely altered. We now turn to a description of the organization of bacterial and eukaryotic genomes as a prelude to discussing genomic rearrangements and gene expression. In this section, we examine the organization of the *E. coli* genome and establish a number of general features of all bacterial genomes. In the next section, we compare bacterial genomes to those of eukaryotes. One of the issues we address is how to determine the number of genes it takes to make an organism.

Bacterial DNA is localized to a region of the cell known as the nucleoid (see Figure 2·2). The DNA is associated with histonelike proteins in a compact structure similar to that of eukaryotic chromatin (Section 24·11C). The DNA is organized into large loops attached to an RNA-protein scaffold.

A. The *E. coli* Chromosome Is a Model for Prokaryotic Chromosomes

The *E. coli* genome consists of a double-stranded, circular DNA molecule 4600 kb in size. A simplified map of this chromosome is shown in Figure 32·11; a small region of the genome is expanded to show the detailed arrangement of transcription units. The expanded region contains several genes that encode proteins discussed in previous chapters. The genes for 16S, 23S, and 5S ribosomal RNA were introduced in Figure 28·10; recall that they are transcribed from a single promoter and that rRNA transcription units include at least one tRNA gene. In this example, the ribosomal RNA genes form part of the *rrn*B operon. Downstream of the *rrn*B operon is *tuf*B, which encodes the elongation factor EF-Tu (Section 30·5A); this gene is co-transcribed with four tRNA genes. *sec*E encodes a protein required for secretion of proteins by mechanisms discussed in Chapter 30; it is part of an operon that includes *nus*G, a transcription factor described in Sections 27·6D and 31·4B. The genes *L11*, *L1*, *L10*, and *L7/L12* encode proteins that form part of the large ribosomal subunit (Section 30·1A). Finally, *rpo*B and *rpo*C encode the largest subunits of RNA polymerase, β and β', respectively (Section 27·2A).

The organization of genes in the *rrn*B region is fairly typical of gene organization in bacteria, so we will use this as an example. The genes in this operon are close together, and there is little excess DNA between operons or between coding regions within an operon. In some cases, the coding regions actually overlap by a few nucleotides, as described earlier for reading frames in the *trp* operon (Section 29·3). The close packing of genes in bacterial genomes is due, in part, to the organization of prokaryotic genes into operons. With operons, there is no need for genes to have

Figure 32·11
Map of the *E. coli* chromosome, showing some of the genes mentioned in this book. An expanded view of the region near the *rrn*B operon is also shown. tRNA genes are identified by the one-letter abbreviation of the amino acid they carry. *tuf*B encodes the elongation factor EF-Tu; *sec*E encodes a protein required for protein secretion; *nus*G encodes the transcription factor NusG; *L11*, *L1*, *L10*, and *L7/L12* encode proteins that form part of the large ribosomal subunit; and *rpo*B and *rpo*C encode the β and β' subunits of RNA polymerase, respectively.

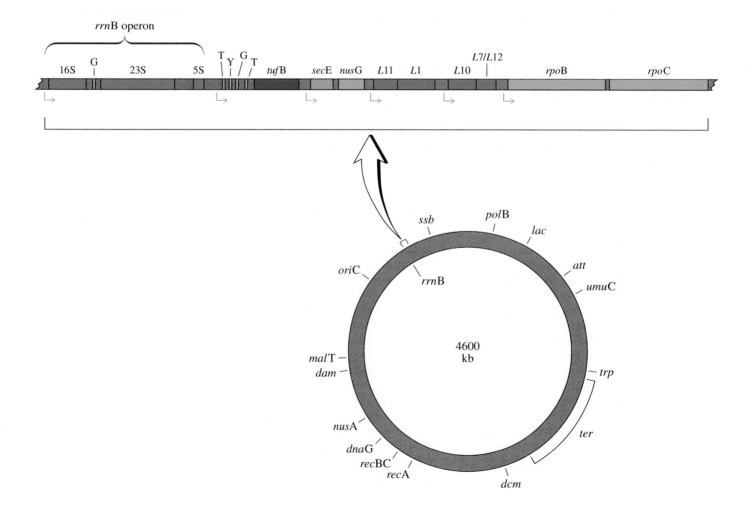

individual promoters and transcriptional regulatory regions. As we will see later, the dense packing of genes in bacterial genomes contrasts starkly with genome organization in multicellular eukaryotes.

In general, bacterial operons are composed of genes that are functionally related. For example, the genes for ribosomal proteins L11 and L1 are cotranscribed, as are several tRNA genes and *tuf*B. Similarly, the *trp* and *lac* operons contain genes for several enzymes in a single pathway. Having functionally related genes organized into operons allows their expression to be coordinately controlled. The selective advantage of this type of organization is strong enough that operons are preserved. However, not all genes encoding functionally related proteins must be in the same operon. In addition to the *L11/L1* operon, for example, there are several others that contain genes for ribosomal proteins (Figure 30·22), and the genes for the other subunits of RNA polymerase are located at different sites on the *E. coli* chromosome.

All the genes in the *rrn*B region are transcribed in the same direction. Since these genes are located clockwise of the origin of DNA replication, transcription of these genes and DNA replication occur in the same direction. While this gene organization is a general feature of the *E. coli* genome (as mentioned in Section 27·5), the presence of so many clustered genes oriented in the same direction is somewhat atypical. There are many genes that have the opposite orientation in other regions of the genome.

There is one other feature of the *E. coli* genome we should mention that is not apparent from Figure 32·11. With the exception of ribosomal RNA operons and some tRNA genes, it is unusual to find multiple copies of genes in bacterial genomes. Two known examples of duplicated protein-encoding genes in *E. coli* are the genes for elongation factor EF-Tu (*tuf*A and *tuf*B) and the genes for the DNA-packaging protein HU (*hup*A and *hup*B). Note that both pairs encode abundant proteins; presumably a single gene for these proteins is not sufficient.

The genetic map of *E. coli* is almost identical to that of *Salmonella typhimurium*, a closely related species that diverged from *E. coli* only a few million years ago. A detailed comparison of the organization of genes in the two species reveals that there have been very few changes since they diverged from a common ancestor. While there have been some localized insertions and deletions of DNA in one or the other species, such as deletion of the *lac* operon in *S. typhimurium*, large regions of the chromosomes of the two species show identical gene organization.

The overall organization of genomes in more distantly related species of bacteria is very different from the organization in *E. coli* and *S. typhimurium*. Nevertheless, in localized regions, the arrangements of genes have been conserved for over a billion years. For example, the genomes of all prokaryotes studied contain a region similar to the *rrn*B region of *E. coli*. This similarity includes not only the arrangement of the ribosomal RNA genes, which is highly conserved, but also the association of genes such as *tuf*B, *sec*E, *nus*G, the ribosomal protein genes, and the genes for the large subunits of RNA polymerase. The major differences seen in distantly related species are the inclusion of additional ribosomal protein genes in some operons and the splitting of the genes for the large RNA polymerase subunits into smaller genes (Section 27·2B). This localized conservation of gene organization is another general feature of bacterial genomes.

B. There Are About 3000 Genes in *E. coli*

There are 8 rRNA operons and about 80 tRNA genes in *E. coli*. The remaining genes—the vast majority of genes in the cell—are protein-encoding genes. A typical protein-encoding gene contains a coding region of about 1500 bp. This is sufficient to encode a protein of 500 amino acid residues, which corresponds to a relative molecular mass of about 55 000. Given the size of the *E. coli* genome (4600 kb),

Figure 32·12
Two-dimensional gel electrophoresis of
E. coli proteins. First, isoelectric focusing in
one dimension separates proteins by charge.
SDS-PAGE in the second dimension resolves
proteins by size. Each spot on the autoradiograph represents a protein from an extract
of *E. coli* cells grown in the presence of
^{35}S-methionine. There are over 1500 distinct
spots. (Courtesy of Patrick O'Farrell.)

Isoelectric focusing $\longrightarrow$

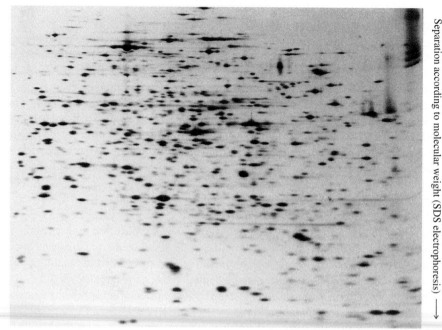

Separation according to molecular weight (SDS electrophoresis) $\longrightarrow$

about 3000 protein-encoding genes could be accommodated, and the cell could be
capable of making a comparable number of proteins. To date, about 1500 protein-
encoding genes in *E. coli* have been identified and mapped.

The proteins present in an *E. coli* cell can be visualized using two-dimensional
gel electrophoresis, which separates proteins by both size and charge (Figure
32·12). Over 1500 distinct proteins are detectable on such a gel, indicating that
there are at least that many genes. This is likely to be a low estimate, however, since
some proteins are only synthesized under special circumstances (for example, the
products of the *lac* operon) and others are present in amounts too small to be detected by such techniques (for example, *lac* repressor, which is present at about 10
molecules per cell).

An accurate count of the genes in *E. coli* will be available when the entire
genome has been sequenced. At the present time, more than half this task has been
completed. For example, the region around the origin of replication (a total of
325 kb of contiguous DNA) has been sequenced by Frederick Blattner and his colleagues. They have identified 293 genes by searching for open reading frames and
known tRNA and ribosomal RNA genes. Fewer than half of the genes identified
were previously known, and many of the newly discovered genes are unrelated to
known genes in other organisms. Extrapolating from this sample to the entire chromosome, we would expect a total of over 4000 genes in *E. coli*. However, there is
reason to believe that some regions of the *E. coli* chromosome (*ter,* for example)
contain a lower density of genes than the region around *ori*C, and there is some uncertainty about whether all identifiable open reading frames actually encode expressed, functional proteins. All things considered, it seems likely that *E. coli* has
about 3000 genes.

The sequence of the genome around *ori*C confirms some of the general points
we made earlier about prokaryotic genomes. The genes are closely packed and are
frequently part of operons, and two-thirds of the genes are oriented so that they are
transcribed in the same direction as replication-fork movement. In addition, analysis of the *ori*C region reveals multiple copies of short, repetitive sequences known
as repeated extragenic palindromic elements (REP elements). The consensus sequence of these repeats is

$$\text{GCC}^{\text{G}}_{\text{T}}\text{GATGNCG}^{\text{G}}_{\text{A}}\text{CG}^{\text{C}}_{\text{T}}\text{N}^{\text{G}}_{\text{A}}\text{CG}^{\text{C}}_{\text{T}}\text{CTTATC}^{\text{C}}_{\text{A}}\text{GGCCTAC} \qquad \textbf{(32·1)}$$

These sequences are the sites of binding of topoisomerase II (Section 24·6C), the enzyme that introduces negative supercoils into DNA loops and also relaxes super-coiled DNA during DNA replication and transcription. There are several hundred REP elements in the *E. coli* genome, distributed so there is at least one on each of the approximately 40 loops of DNA attached to the RNA-protein scaffold. As much as 1% of the *E. coli* genome is composed of REP elements. Other repetitive sequences are present, but they are not as prevalent.

The sequenced region around *oriC* also contains dozens of *chi* sites, which are required for RecBCD binding during recombination (Section 26·5B), and these are spaced at regular intervals of approximately 4 kb. When promoters and regulatory sites are included, less than 2% of this sequenced region contains unidentifiable features.

C. Bacterial Chromosomes Range Greatly in Size

The size of the *E. coli* genome (4600 kb) is average for bacteria, although the sizes of prokaryotic genomes range considerably (Table 32·4). The smallest prokaryotic genomes belong to *Chlamydia trachomatis* and to the mycoplasmas discussed in Section 2·2. *Chlamydia* is a parasitic bacterium first thought to be a virus. It lives within eukaryotic cells and lacks the capacity to encode enzymes for several metabolic processes, including oxidative phosphorylation. The chromosomes of *Mycoplasma* species are only about one-sixth the size of the *E. coli* genome. As pointed out in Chapter 2, the mycoplasmas are adapted to a parasitic lifestyle, which presumably means they do not need as many genes as free-living bacteria. Consider that the genome of bacteriophage T4 is 166 kb in size; the *Chlamydia* and *Mycoplasma* genomes are only about four times larger!

Some of the largest bacterial genomes are found in the myxobacteria such as *Stigmatella aurantiaca* and *Myxococcus xanthus*. These Gram-negative soil bacteria live on decaying organic matter. In response to adverse conditions, thousands of individual bacteria aggregate to form complex fruiting bodies on long stalks that project the fruiting bodies above the soil surface. The fruiting bodies contain cells that will differentiate into spores. The life cycle of these bacteria, with its multicellular stage, is complex, and the large genomes may reflect the presence of many genes expressed only during differentiation.

Nitrogen-fixing bacteria such as *Rhizobium meliloti*, *Bradyrhizobium japonicum*, and *Agrobacterium tumefaciens* also have larger genomes than *E. coli*. These species contain a number of genes required for nitrogen fixation and for the formation of symbiotic relationships with certain flowering plants (Section 21·1).

The largest known prokaryotic genome belongs to the cyanobacterium *Calothrix*. Like most cyanobacteria, this organism is photosynthetic and forms filamentous colonies that contain differentiated cells called heterocysts. Heterocysts carry out nitrogen fixation. Again, the complex lifestyle of this bacterium requires genes not found in simpler bacteria such as *E. coli*, although it is difficult to account for the size of the *Calothrix* genome when one considers that *Anabaena*, another cyanobacterium with a complex lifestyle, has a much smaller chromosome. The *Calothrix* genome is almost as large as the smallest eukaryotic genomes.

Many free-living bacterial species appear to have fewer genes than *E. coli*; 2000 seems to be the minimum number. About 600 different proteins are known to be needed for the metabolic reactions described in the first 23 chapters of this book, and at least 400 additional proteins and RNA molecules are required for the processes described in Chapters 24 through 30. Yet this totals only 1000 genes—less than half the number of genes in most bacteria. Obviously, many proteins and genes await discovery.

Table 32·4 Genome size in selected prokaryotes

Organism	Genome size* (kb)
Chlamydia trachomatis	600
Mycoplasma capricolum	724
Mycoplasma hominis	760
Mycoplasma pneumoniae	840
Borrelia burgdorferi	950
Methanobacterium thermoautotrophicum	1623
Hemophilus influenzae	1833
Brucella melitensis	3300 (2100, 1200)
Rhodobacter sphaeroides	3900 (3000, 900)
Haloferax volcanii	4100 (2900, 700, 400, 100)
Bacillus subtilis	4160
Escherichia coli	4600
Salmonella typhimurium	4800
Leptospira interrogens	4900 (4500, 400)
Agrobacterium tumefaciens	5100 (3100, 2000)
Bacillus cereus	5700
Rhizobium meliloti	6500 (3400, 1700, 1400)
Streptomyces coelicolor	~8000
Streptomyces lividans	8000
Bradyrhizobium japonicum	8700
Anabaena sp.	9100
Stigmatella aurantiaca	9350
Myxococcus xanthus	9454
Calothrix sp.	12 600

*For bacteria that contain more than one chromosome, individual chromosome sizes are indicated in parentheses.

Table 32·5 Number of chromosomes in selected eukaryotes

Organism	Haploid number
Amoeba (amoeba)	25
Dictyostelium discoideum (slime mold)	7
Caenorhabditis elegans (nematode)	6
Hydra vulgaris (hydra)	16
Planaria torva (flatworm)	8
Lumbricus terrestris (earthworm)	18
Drosophila melanogaster (fruit fly)	4
Musca domestica (house fly)	6
Bombyx mori (silkworm)	28
Culex pipiens (mosquito)	3
Cyprinus carpio (carp)	52
Rana pipiens (frog)	13
Alligator mississipiensis (alligator)	16
Gallus domesticus (chicken)	39
Oryctolagus ciniculus (rabbit)	22
Cavia cobaya (guinea pig)	32
Rattus norvegicus (rat)	21
Mus musculus (mouse)	20
Equus calibus (horse)	32
Felis domesticus (cat)	19
Canis familiaris (dog)	39
Bos taurus (cow)	30
Macaca mulatta (Rhesus monkey)	21
Homo sapiens (human)	23
Pongo pygmacus (orangutan)	24
Pan troglodytes (chimpanzee)	24
Gorilla gorilla (gorilla)	24
Arabidopsis thaliana (wall cress)	5
Allium cepa (onion)	8
Hordeum vulgare (barley)	7
Zea mays (maize)	10
Antirrhinum majus (snapdragon)	8
Lycopersicon esculentum (tomato)	12
Solanum tuberosum (potato)	24
Lilium (lily)	12
Pinus ponderosa (ponderosa pine)	12
Pisum sativum (pea)	7
Equisetum (horsetail fern)	108
Acetabularia mediterranea (green algae)	10
Chlamydomonas reinhardtii (green algae)	16
Ophioglossum reticulatum (fern)	>500
Aspergillis nidulans (mold)	8
Penicillium (penicillin mold)	4
Saccharomyces cerevisiae (yeast)	16
Schizosaccharomyces pombe (yeast)	3

D. Some Bacteria Contain More than One Chromosome

It was once thought that all bacteria contained a single, circular chromosome. However, we now know that *Borrelia burgdorferi* (the spirochete that causes Lyme disease) and several species of *Streptomyces* have linear chromosomes. Furthermore, as noted in Table 32·4, some bacteria contain more than one chromosome. *Rhizobium meliloti*, for example, has a large chromosome of 3400 kb and two smaller ones of 1700 kb and 1400 kb. The smaller chromosomes were once called megaplasmids before it became apparent that they were true chromosomes. Segregation of these chromosomes following DNA replication may involve spindlelike filaments, just as mitosis does in eukaryotic cells.

In practice, it is sometimes difficult to distinguish a plasmid from a chromosome. Theoretically, plasmids are nonessential; a bacterial cell can survive and grow under most conditions even when its plasmid DNA is eliminated. A circular, self-replicating piece of DNA that carries essential genes is called a chromosome and not a plasmid. A bacterial cell cannot survive when even one of its chromosomes is eliminated.

Size cannot be used to distinguish plasmids from chromosomes, since some plasmids are very large and some chromosomes are very small. For example, the F plasmids of *E. coli* (Section 25·10A) are over 100 kb in size and carry dozens of genes in addition to those that promote conjugation. In general, chromosomes are larger than 100 kb, although there are examples of chromosomes that are about this size (the smallest chromosome of *Haloferax volcanii,* for example, is only 100 kb).

32·4 Eukaryotic Genomes

Eukaryotic genomes are larger and more complex than prokaryotic genomes. Eukaryotic genomes are also much more variable than prokaryotic genomes in terms of their organization and the number of genes they contain. Eukaryotic genomic DNA is confined to the nucleus, where it is packaged in nucleosomes and arranged as large loops attached to an RNA-protein scaffold (Section 24·11).

Some of the differences between prokaryotic and eukaryotic genomes reflect the most noticeable distinction between them; namely, that the prokaryotic genome is densely packed and has little extraneous DNA, whereas the eukaryotic genome is usually loosely packed and does not seem to have faced strong selective pressure to conserve space. In fact, many eukaryotic genomes contain what appears to be large amounts of nonfunctional DNA.

A. Eukaryotic Genomes Contain Many Chromosomes

Eukaryotic chromosomes, unlike bacterial chromosomes, are linear pieces of DNA. Each chromosome contains a centromeric region, telomeres (Section 25·11B), and many origins of replication (Section 25·9C). Most eukaryotes are diploid; that is, they contain two complete sets of chromosomes. The haploid number of chromosomes ranges from 3 to more than 500 in different species (Table 32·5).

Closely related species tend to have similar numbers of chromosomes, although there can be considerable variation. One notable example of such variation can be seen in two closely related mammals, the Indian muntjac and the Reeves muntjac (Figure 32·13). The Indian muntjac has 23 pairs of chromosomes, which is about average for a mammal. In contrast, the Reeves muntjac has two large autosomes and a third chromosome that is fused to the X chromosome. The Y chromosome is separate.

Figure 32·13
Variation in chromosome number in Indian and Reeves muntjacs. (**a**) Indian muntjac. (**b**) Reeves muntjac. (**c**) Chromosomes of the Indian muntjac. This species has 23 pairs of chromosomes. (**d**) Chromosomes of the Reeves muntjac. This species, although closely related to the Indian muntjac, has only 4 pairs of chromosomes. (Parts (c) and (d) courtesy of Frank Johnston and R. B. Church.)

(a)

(b)

(c)
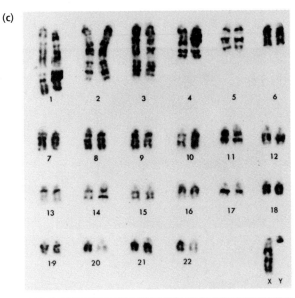

(d)
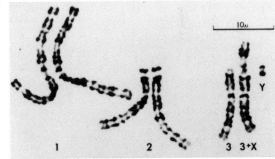

In spite of this difference in the number of chromosomes, the two types of muntjac can interbreed to produce viable hybrids, although the hybrids are infertile. This illustrates an important principle of eukaryotic genome organization: the number of chromosomes is not important for viability, although it does affect whether two species can produce viable offspring. The chromosomes of the Reeves muntjac represent an extreme example of chromosome fusions and rearrangements that have altered the total number of chromosomes. More frequently, such rearrangements occur within a chromosome, leading to speciation. For example, there are many species of *Drosophila* that have the same numbers of chromosomes but cannot produce fertile hybrids because of rearrangements within a chromosome. These rearrangements prevent proper pairing at meiosis in the hybrid.

Chromosomal rearrangements can be caused by homologous recombination between integrated transposons on different chromosomes. This results in the swapping of chromosome ends. Recombination between integrated transposons on the same chromosome can cause a chromosomal inversion. A bacterial example of such an inversion will be discussed in Section 32·6A.

As is the case in bacteria, the overall organization of genes in closely related eukaryotes is similar. Comparisons of the mouse and human genetic maps, for example, reveal about 50 large blocks of genes that, while similarly arranged, are located on different chromosomes or show evidence of chromosomal inversions. The similar arrangement of genes in the genomes of different organisms is known as **synteny.**

Most mammals appear to be descended from a common ancestor that lived approximately 100 million years ago, indicating that the organization of genes can be preserved for a considerable time. However, if one examines distantly related organisms, such as mammals and insects, one finds little evidence of synteny. Over

(a)

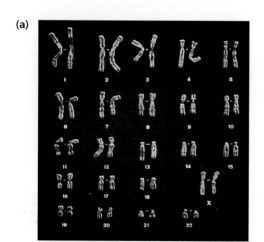

(b)

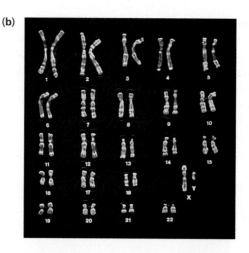

Figure 32·14
Karyotype of human chromosomes. (a) Female. (b) Male. Note the presence of the Y chromosome in the male.

longer periods of time, the genes in eukaryotic genomes appear to have been thoroughly shuffled. This is in marked contrast to the situation in prokaryotes, where species that have been separated for a billion years still preserve the original order of genes in some regions of the genome.

As an aside, it is interesting to note that, unlike genes on autosomes, genes on the X chromosome in mammals tend not to be transferred to other chromosomes. Ohno's Law states that because of the different dosage requirements of X-linked and autosomal genes (Section 27·15A), exchanges of DNA between the X chromosome and an autosome are likely to be lethal. Thus, all mammals should have the same X-linked genes. This is in fact true: although the order of genes on the X chromosome in different mammals is not the same, the same genes are located on the X chromosome in all cases. This is one of the few examples of selective pressure that favors chromosomal location in mammals.

The gross morphology of chromosomes can be detected by stains that bind specifically to certain regions of the chromosome. Such stains produce a characteristic banding pattern that is extremely reproducible. An example of a stained set of chromosomes, or **karyotype,** is shown in Figure 32·14. The banding pattern can be used to identify specific chromosomes and to detect chromosomal abnormalities. The presence and identity of extra chromosomes is readily determined using this technique.

The chromosomes of other primates have a banding pattern similar to that seen in humans, with one major exception. Within the ape family, the chimpanzee, gorilla, and orangutan all have 24 chromosomes, but humans have only 23. At some point within the past several million years, two chromosomes fused in the line leading to humans.

B. Eukaryotic Genomes Contain Repetitive Sequences

As we discussed in Section 32·3B, repetitive sequences are found in prokaryotes, but they generally constitute only a small percentage of the genome. For example, REP elements make up only 1% of the *E. coli* genome. In contrast, repetitive sequences are much more common in eukaryotes, in some cases accounting for 50% of all genetic material in the cell. There are several categories of repetitive sequences in these organisms.

One category of repetitive sequence consists of stretches of dinucleotide repeats (GA, CA, and AT are common). These stretches are usually found in the regions between genes and are not conserved in closely related species, suggesting that they have no function and that they arise and disappear rapidly on an evolutionary timescale.

Another category of repetitive sequence consists of larger repeats that appear to play some role in the cell. We have already encountered telomeric repeats of 4 to 8 bp (Section 25·11B), and the *chi* and REP sequences in *E. coli* are prokaryotic examples of repetitive DNA. The centromeric regions of eukaryotic chromosomes also contain large regions of highly repetitive sequence. The sizes of these repeats range from about 15 bp in *Drosophila* to 350 bp in other organisms. There are often thousands of copies of tandem repeats at each centromere. A certain number of these repeats must be present for proper chromosome segregation.

Some repetitive DNA consists of larger sequences that range from several hundred to several thousand base pairs in length. We now know that some of this repetition is due to the presence of multiple transposons in the genome. The genome of *S. cerevisiae,* for example, has hundreds of copies of *Ty* and δ elements, and the *Drosophila* genome contains many copies of *copialike* transposons.

The total amount of repetitive DNA at the centromere can be a significant percentage of the genome, as shown in Table 32·6. One might expect the amount to depend on the number of chromosomes, but no such correlation is evident. Closely related species, such as mouse and rat, which diverged only 30 million years ago, can differ significantly in the amount of repetitive DNA. This suggests that the repetitive region can expand and contract by internal recombination without affecting function. In other words, the amount of repetitive DNA in a species is due not to natural selection but to random genetic drift.

Even within a species, the amount of repetitive DNA can vary among individuals. Furthermore, the sequence of the repeats can differ due to the accumulation of mutations over several generations. This variation, or polymorphism, has been exploited in a technique that analyzes the unique patterns produced when DNA from different individuals is treated with restriction endonucleases. The technique, called **DNA fingerprinting,** has proven invaluable in the study of the evolution of many species. For example, it has been used to examine the dispersal of seeds of flowering plants and the relatedness of birds inhabiting a small forest. Patterns of restriction fragments have also been used to determine biological relationships among humans and to identify criminals through analysis of blood and tissue samples (Section 33·11D).

There are two other kinds of larger repetitive sequences in eukaryotes: short interspersed repeats (abbreviated SINES) and long interspersed repeats (abbreviated LINES). An example of each class is shown in Figure 32·15. SINES are about 100 to 500 bp in size and contain a region of DNA that is homologous to small RNA molecules (e.g., tRNA or 5S rRNA). The repetitive sequence is often a degenerative version of the normal RNA sequence; it frequently contains nucleotide substitutions, rearrangements, or deletions. SINES homologous to tRNA genes are found in many species, where they may be present in hundreds of copies. SINES homologous to the 7SL RNA of signal recognition particle (Section 30·14) are known in mammals. The human version of this SINE is referred to as the *Alu* sequence since it contains a sequence recognized by the *Alu*I restriction enzyme. *Alu* sequences are

Table 32·6 Percentage of centromeric repetitive DNA in the genome

Organism	Percentage
Land crab	21
Fruit fly	16
Mouse	8
Rat	~2
Cow	>23
African green monkey	~18
Human	~5

[Adapted from Singer, M., and Berg, P. (1991). *Genes and Genomes: A Changing Perspective* (Mill Valley, California: University Science Books).]

Figure 32·15
SINES and LINES. SINES are short, repetitive sequences that are homologous to small RNA molecules. LINES are larger repeated sequences that contain at least two open reading frames (ORF), including one that resembles the protein-encoding region of a retrovirus *pol* gene. Both SINES and LINES have a short, A/T-rich sequence at the 3′ end, and both are flanked by short, direct repeats of DNA.

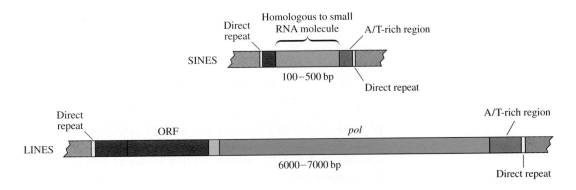

abundant in Old World primates, including humans. The human genome contains about one million dispersed copies of *Alu* sequences, which together account for about 9% of human DNA.

LINES are 6 to 7 kb in length and contain two open reading frames. One of these encodes a protein of unknown function, and the other encodes a protein similar to the reverse transcriptase of retrovirus-like transposons. In addition to being a type of repetitive DNA, LINES and SINES constitute a third class of eukaryotic transposon, referred to as retroposons or class II retrotransposons. Like other transposons, there is a short duplication of the target site at the ends of integrated LINES and SINES. Unlike other eukaryotic transposons, however, retroposons show no evidence of terminal repeats. Instead, one end contains an A/T-rich sequence that frequently consists of stretches of poly A or poly (TAA) on one strand. Most integrated LINES are degenerative versions of the complete version shown in Figure 32·15, which is why they were not at first recognized as transposons. The most frequent mutation is deletion of the 5′ end, producing an insert that is as short as 1 kb. There are about 10 000 copies of LINES in mammalian genomes. LINES are also found in *Drosophila,* where the various subclasses are referred to as *I elements, G elements, F elements,* and the *jockey* transposon.

Once integrated into chromosomes, LINES and SINES can spread to other sites. LINES appear to be transposons that jump by a mechanism that involves an RNA intermediate. Insertion into the coding regions of genes can cause mutations. For example, hemophilia *a* results when the gene for human coagulation factor VIII (Section 6·11B) is disrupted by a LINE.

C. Eukaryotic Genomes Contain Two Types of Pseudogene

Vertebrate chromosomes contain not only intact, expressed genes but also numerous stretches of DNA that are similar to genes but are not expressed. These **pseudogenes** share a common ancestor with a corresponding intact gene. Pseudogenes differ from expressed genes in that they often have mutations in their coding region and cannot produce functional proteins or RNA.

Pseudogenes of one type, called *processed pseudogenes,* contain stretches of adenine residues at their 3′ ends and lack introns. These features show that processed pseudogenes may represent DNA copies of mature mRNA molecules that were somehow synthesized and inserted into the genome (Figure 32·16). Because they have been produced from mRNA, processed pseudogenes are not associated with a promoter and are not transcribed. Free from selective pressures, these segments of DNA mutate without affecting the viability of the cell. As a result, many pseudogenes have over time accumulated mutations that would be detrimental in the gene from which they were derived.

The number and type of processed pseudogenes in eukaryotic genomes is highly variable. Some species, such as yeast and *Drosophila,* have very few processed pseudogenes, whereas most species of mammal (but not other vertebrates) have large numbers. In general, processed pseudogenes are derived from mRNA that is abundant in cells of the germ line. For example, there are 20 pseudogenes derived from actin and β-tubulin genes in humans. The human genome also contains about 20 glyceraldehyde 3-phosphate dehydrogenase pseudogenes, and there are more than 200 versions of this pseudogene in the mouse genome.

The other type of pseudogene is derived from gene duplication. Occasionally, genes are duplicated as a result of errors during homologous recombination. For example, when homologous chromosomes are aligned during meiosis, an incorrect crossover can occur if two similar regions lie near each other on the same chromosome (Figure 32·17). Local regions of homology can arise as the result of multiple transposon insertions or may be due to the presence of repetitive sequences such as *Alu* sequences or SINES.

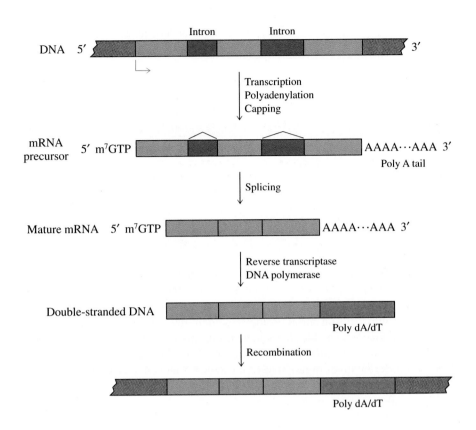

Figure 32·16
Possible origin of processed pseudogenes. An mRNA precursor is processed to produce a mature mRNA molecule. This molecule is converted to double-stranded DNA by the actions of reverse transcriptase and DNA polymerase. The double-stranded DNA is integrated into the chromosome to yield a processed pseudogene. Processed pseudogenes lack introns and contain stretches of poly dA/dT at their 3′ ends. These genes are not transcribed because they are not associated with promoter sequences.

Following gene duplication, the genome of one of the progeny contains duplicated genes (the other contains a deletion, which is likely to be lethal). Since there are two versions of the same gene, one of the genes may acquire mutations and become a pseudogene. This type of pseudogene can be distinguished from processed pseudogenes because it contains a promoter and other regulatory signals. It also contains introns if the original gene was interrupted. We will encounter an example of this type of pseudogene in the next section.

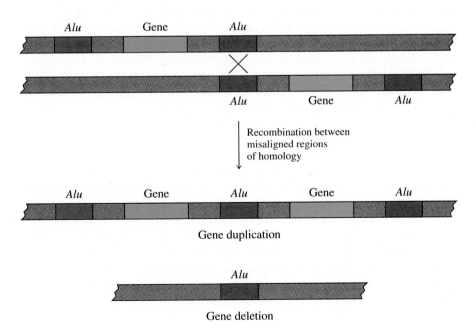

Figure 32·17
Gene duplication. During crossover at meiosis, the presence of repetitive sequences may cause homologous chromosomes to misalign. Subsequent crossover produces one daughter molecule with a gene deletion and the other with a gene duplication.

Figure 32·18
Globin gene loci in humans. The α-globin and β-globin genes form clusters on two separate chromosomes. Each of the genes encode similar but nonidentical globin proteins that are expressed at different times during development (see Table 27·7). ψ indicates a pseudogene. These pseudogenes have all arisen by gene duplication and subsequent mutation.

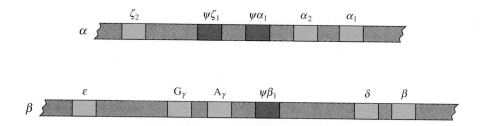

D. Families of Genes are Present in Eukaryotic Genomes

One of the differences between eukaryotic and prokaryotic genomes is that eukaryotes often contain gene families, sets of genes that encode closely related proteins that perform similar functions. The globin genes in humans represent one such family. The α-globin locus on chromosome 16 contains three functional α-globin genes and two pseudogenes, and the β-globin locus on chromosome 11 contains five functional genes and one pseudogene (Figure 32·18). The functional genes are expressed at different times during development (see Table 27·7), but the proteins differ at only a few positions.

The α-globin locus spans 25 kb, and the β-globin locus covers 35 kb. There are no other genes at these loci, and the density of genes is much less than in bacteria (the same amount of DNA in *E. coli* would accommodate 15 to 20 genes). The long intergenic regions contain repetitive sequences, such as *Alu* sequences, and unique DNA sequences whose function (if any) is unknown. The arrangement of globin genes in other mammals is similar to that seen in human chromosomes, but the sequence of the intergenic DNA is not conserved. This DNA is assumed to be nonfunctional, junk DNA. Much of the mammalian genome is composed of sequences such as this.

The vertebrate genes for the major histocompatibility proteins are also an example of a gene family. These proteins are found on the cell surface, where they bind antigen and present it to T cells that can stimulate an immune response. The genes for the major histocompatibility proteins are found in a large cluster called the *MHC* locus on the short arm of chromosome 6 in humans. This is one of the largest regions of the human genome to be analyzed in detail. The *MHC* locus covers 3800 kb, more than three-quarters the size of the entire *E. coli* chromosome. There are only about 100 genes in this region, and some of these are pseudogenes.

There are 30 functional members of the MHC gene family. Scattered among these genes is a variety of other genes not related to antigen presentation. They include genes for plasma proteins, heat-shock proteins, and a regulatory protein called tumor necrosis factor. In addition, there are genes that encode transport proteins that may play a role in the binding of peptides to MHC molecules.

As is the case in the globin loci, the density of genes at the *MHC* locus is much lower than that seen in prokaryotic genomes. The difference is again due to large stretches of intergenic DNA in the human genome. The arrangement of genes at the *MHC* locus is similar in most mammals and even in other vertebrates, but the sequence of the intergenic regions is highly variable, again suggesting that this DNA is nonfunctional.

The MHC- and globin-gene families illustrate a general rule of eukaryotic gene organization: members of gene families tend to form clusters in the genome. This is probably because the various genes that make up each family were generated by localized gene duplications that occurred relatively recently.

E. There Are Two Hypotheses for the Origin of Introns

In Section 28·5, we discussed the presence of introns in eukaryotic protein-encoding genes. We noted that introns are common in the genes of vertebrates and flowering plants but less common in the genes of other species. Any theory concerning the evolutionary origin of introns must satisfactorily explain why the vast majority of prokaryotic genes do not contain introns. It must also explain why intron-containing genes are more common in some eukaryotes than in others.

Suppose that the genes of ancestral organisms were made up of short pieces of DNA that encoded polypeptide domains with simple enzymatic activities. If true, complex proteins could be derived from the fusion of two or more of the primitive genes. The production of functional proteins from such fusions would be facilitated by a splicing mechanism, which would remove the need for precise joining of the short pieces of DNA in the genome. Joined DNA could then be transcribed as a single mRNA precursor, spliced, and the intervening RNA removed to produce an mRNA encoding a functional fused protein.

The hypothesized existence of splicing mechanisms in the primitive ancestor of eukaryotes and prokaryotes was originally supported by suggestions that eukaryotic exons encode distinct protein domains. This lent support to the idea that proteins evolved from fusions of domain-encoding DNA molecules. According to this hypothesis, the introns present in primitive cells have been lost from prokaryotic protein-encoding genes but are retained in many eukaryotes. This hypothesis, called the **introns-early hypothesis,** is consistent with the concept that the genomes of modern bacteria have undergone selection for compactness, whereas eukaryotic genomes can tolerate large amounts of excess DNA, including that found in introns.

However, examination of the increasing number of available protein structures fails to support the idea that exons consistently encode distinct domains of proteins. Furthermore, the introns-early hypothesis predicts that the positions of introns in eukaryotic genes should be highly conserved, although this does not seem to be the case. It now seems possible that introns in eukaryotic genes could have arisen recently as the result of insertions of DNA segments into protein-encoding genes. Such segments could be transposons or group II introns that have subsequently mutated so that they now require active removal by spliceosomes. This hypothesis, the **introns-late hypothesis,** is consistent with the concept that eukaryotic genomes are dynamic structures subject to considerable rearrangements, including the insertion and excision of transposons.

Note that whereas the introns-early hypothesis postulates that introns must have been lost from all prokaryotic genes and from all those eukaryotic genes that are missing them, the introns-late hypothesis postulates that introns must have been inserted into those eukaryotic genes that contain them. Is there any evidence of intron loss or intron insertion in living organisms? Unfortunately for those who seek a quick answer, there is evidence of both.

The evolution of rat insulin genes is an example of a rearrangement involving the loss of an intron. The rat insulin I gene contains two introns at the same locations as in the single insulin gene present in birds and all other mammals (Figure 32·19). The insulin II gene contains only one intron and has a stretch of poly A

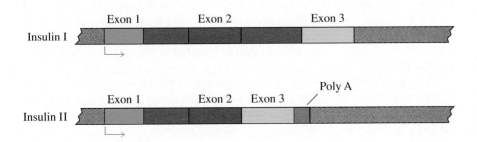

Figure 32·19
Structures of the genes for insulin in rats. Both genes encode the same protein and contain the same exon 1 and first intron. The insulin II gene lacks the second intron, which is found in the insulin I gene and in the insulin genes of birds and other mammals. The insulin II gene is derived from a partially processed RNA.

Figure 32·20
Trans-splicing. In trypanosomes, many mature mRNA molecules are created by the joining of a protein-encoding RNA molecule with a spliced leader (SL) sequence. The SL sequences are about 35 bp in size and are clustered in the trypanosome genome. *Trans-splicing* is similar to the splicing examined in Chapter 28, except that it involves the joining of separate RNA molecules, not the elimination of introns.

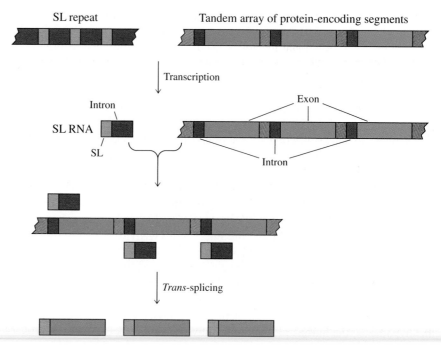

Mature *trans*-spliced mRNA

residues at its 3′ end. This appears to have arisen by the insertion of a partially processed RNA that had been converted into double-stranded DNA by the action of reverse transcriptase in a manner similar to that described for processed pseudogenes.

Unlike pseudogenes, the rat insulin II gene is expressed. It contains a promoter and regulatory regions that are homologous to those of the insulin I gene. Apparently, the rat insulin II gene arose from an aberrant transcript that initiated upstream of the rat insulin I gene (which is why the gene contains the rat insulin I promoter sequence).

There are also many examples of presumed intron insertion into protein-encoding genes. These are cases where the genes of only one particular group of eukaryotes contain an intron at a particular position. One example of presumed intron insertion is in a gene for one of the chaperones required for protein import into mitochondria. In eukaryotes, this gene is a nuclear gene, but sequence comparisons reveal that it is closely related to a bacterial gene, suggesting it was transferred from the mitochondrial genome. Consistent with this hypothesis is the observation that although this gene contains an intron in some protists, most eukaryotic versions of this gene do not contain introns.

There are some unusual examples of interrupted genes in trypanosomes and the nematode *Caenorhabditis elegans*. In these organisms, mature mRNA molecules consist of a transcript of the coding region covalently attached to a separately encoded RNA molecule called a 5′ spliced leader (SL) sequence. The SL RNA is attached to the mRNA precursor by *trans*-splicing (Figure 32·20). The mechanism is similar to normal splicing in that it involves snRNPs and there is an intermediate analogous to the lariat structure, while it differs in that the molecules spliced are not already covalently linked. All mature mRNA molecules in trypanosomes contain a spliced SL sequence, as do 10% of those in *C. elegans*. The leader sequence contains the capped structure required for assembly of the pre-initiation complex during translation (Section 30·4D).

SL RNA genes are present as large tandem arrays that are repeated 200 times in the trypanosome genome (there are fewer copies in *C. elegans*). The protein-encoding regions of those genes that are *trans*-spliced are also present as multiple copies arranged in tandem arrays. These regions are cotranscribed to produce a large precursor that is cleaved into individual mRNA molecules during *trans*-splicing. This unusual variation of the normal splicing reaction reveals that it is possible to create functional mRNA molecules from genes that are widely separated in the genome. The example also illustrates how difficult it is to define *gene* with any precision.

32·5 How Many Genes Are Present in Eukaryotic Genomes?

Eukaryotic cells are more complex than prokaryotic cells, so we expect even the simplest eukaryotes to contain more genes than do bacteria. Since no eukaryotic genome has been completely sequenced, we must again find ways to estimate the total number of genes in these organisms. However, this task is more difficult in eukaryotes than in well-characterized prokaryotes such as *E. coli*. Estimates based on the number of identified genetic loci are not accurate because in no eukaryote have enough genes been mapped to provide a reliable count of the total number of genes. Furthermore, we have already seen that genes in eukaryotes are widely separated, so it is not possible to guess the number of genes in eukaryotes by comparing the size of their genomes to that of *E. coli*. New strategies for estimating the numbers of genes are needed. We begin this section by examining the sizes of eukaryotic genomes and then presenting estimates of the numbers of genes in selected organisms.

A. The Size of Eukaryotic Genomes Is Highly Variable

The haploid genome sizes of selected eukaryotes are listed in Table 32·7. The genome sizes vary from 13 500 kb to 102 000 000 kb, a range that is proportionately much greater than that seen in bacteria. If you compare this table to Table 32·5, you will notice that there is no correlation between the number of chromosomes and the size of a genome in eukaryotes. For example, *Schizosaccharomyces pombe* is a yeast with a genome size similar to that of *S. cerevisiae*. However, *S. pombe* has only 3 chromosomes, whereas *S. cerevisiae* has 16. Note also that while there is some correlation between the size of the genome and the complexity of the organism (simpler organisms, such as the fungi, tend to have smaller genomes than complex organisms, such as mammals), the correlation is not absolute (the flowering plant *Arabidopsis thaliana* is a complex organism with a relatively small genome).

In some cases, species in the same phylum differ enormously in genome size. The difference between *Arabidopsis* and the lily *Fritillaria assyrinca* is an example. Differences in genome size between similar species is known as the **C-value paradox,** where C-value refers to the DNA content of a cell. Despite the term, this is not really a paradox since the difference is easily explained in one of two ways. First, the species with the larger genome is often a hybrid of two species, in which case it may still retain copies of both sets of chromosomes. Alternatively, the larger genome may contain increased numbers of repetitive DNA sequences.

B. The Yeast Genome Is Only Three Times Larger than Typical Prokaryotic Genomes

The yeast *S. cerevisiae* has a genome of 13 500 kb. This is only slightly larger than the largest prokaryotic genome (*Calothrix,* genome size 12 600 kb), and only three times larger than the *E. coli* genome. Such a small genome is unusual for a eukaryote and is only possible because the yeast genome is relatively compact and devoid of junk DNA. The compactness of the *S. cerevisiae* genome is consistent with yeast being a rapidly growing, single-celled organism. As is the case in bacteria, there has likely been selection for a small genome in this organism.

Table 32·7 Genome size in selected eukaryotes

Organism	Haploid genome size (kb)
Saccharomyces cerevisiae (yeast)	13 500
Schizosaccharomyces pombe (yeast)	14 000
Aspergillis nidulans (mold)	26 100
Dictyostelium discoides (slime mold)	70 000
Trypanosoma brucei (trypanosome)	80 000
Arabidopsis thaliana (wall cress)	64 000
Zea mays (maize)	7 000 000
Allium cepa (onion)	15 000 000
Fritillaria assyrinca (lily)	115 000 000
Caenorhabditis elegans (nematode)	80 000
Drosophila melanogaster (fruit fly)	165 000
Strongylocentratus purpuratus (sea urchin)	410 000
Bombyx mori (silkmoth)	500 000
Gallus domesticus (chicken)	630 000
Bos taurus (cow)	3 100 000
Homo sapiens (human)	3 200 000
Mus musculus (mouse)	3 300 000
Xenopus laevis (frog)	3 800 000
Triturus cristatus (newt)	31 000 000
Necturus maculosis (salamander)	50 000 000
Protopherus aethiopicus (lungfish)	102 000 000

There is little repetitive DNA in the yeast genome aside from that found at the centromeres and telomeres. About 2% of the genome consists of integrated *Ty* transposons and isolated δ elements. Many yeast genes do not contain introns, and those that do have much smaller introns than those typically found in mammals. Genes are close together in the yeast genome; in most cases, there is less than 2 kb of DNA between genes, and the promoter and regulatory regions of one gene often abut the termination region of an adjacent gene.

Chromosome III of *S. cerevisiae* has been completely sequenced. It is 315 kb in size and contains about 180 open reading frames, or potential genes. Extrapolating this gene density to the entire genome suggests that there are about 9000 genes in this organism, or three times the number in *E. coli.* Only half of the genes identified on chromosome III are homologous to known genes.

C. *Drosophila melanogaster* Has About 7000 Genes

The *Drosophila* genome is 165 000 kb in size. Repetitive DNA accounts for 30% of this amount, leaving 115 500 kb that could potentially encode proteins and RNA molecules. While this is sufficient to accommodate many more genes than are found in yeast, several experiments suggest that the number of genes in *D. melanogaster* is similar to the number in yeast.

The number of *Drosophila* genes has been estimated by measuring the amount of *Drosophila* DNA that can hybridize to expressed RNA. The percentage of DNA that hybridizes to RNA (assumed to be the percentage of DNA that is transcribed) is calculated and divided by the estimated size of the average *Drosophila* mRNA molecule. This calculation yields an estimate of 7000 genes in the fruit fly.

A technique called saturation mutagenesis has also been used to estimate the number of genes in particular regions of chromosomes. In such experiments, large numbers of mutations are introduced at a particular locus and then mapped to

Figure 32·21
Genetic analysis of the *D. melanogaster white* locus. A part of the polytene chromosome in the vicinity of the *white* (*w*) and *zeste* (*z*) genes is shown. Each of the chromosomal bands is labelled. This region has been saturated with mutations, and 16 genes have been identified. In most cases, each gene maps to a single band.

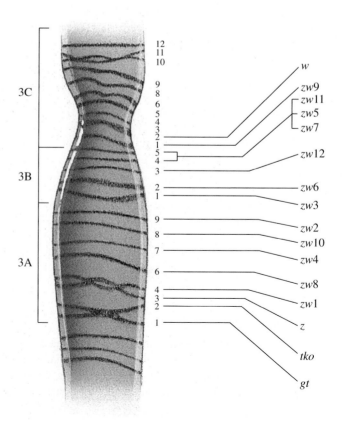

genes. The experiment is continued until many mutations in each gene have been isolated, indicating that further mutagenesis will not uncover additional genes. The data derived from such an experiment with *Drosophila* are shown in Figure 32·21. The region analyzed is the *white* locus near the tip of the *Drosophila* X chromosome. Within 15 bands visible in this region on polytene chromosomes are 16 genes. Most of these have been localized to a particular band. Mapping experiments at other loci give similar results. Since there are approximately 5000 bands in polytene chromosomes, this suggests there are approximately 5000 genes in the *Drosophila* genome.

Several large segments of *Drosophila* DNA have been sequenced. Sequence data indicate that the average gene is separated from its neighbors by 10 to 20 kb. Some genes contain very large introns, which account for a considerable amount of the DNA in the fruit-fly genome. Combining the results of all the available data suggests that there are 7000 to 10 000 genes in *Drosophila*. It is somewhat surprising that *Drosophila* does not contain many more genes that *S. cerevisiae,* since the fruit fly is a multicellular organism and therefore has a much more complex developmental pathway than unicellular *S. cerevisiae.*

D. Mammalian Genomes Contain at Least 50 000 Genes

Mammalian genomes are almost 700 times larger than that of *E. coli.* Were they packed as densely as the genome of *E. coli,* mammalian genomes might therefore accommodate over 2 million genes. One of the arguments against mammals having such a large number of genes was referred to in Section 26·1; if a large percentage of the mammalian genome were composed of genes, the genetic load would be high and the mutation rate due to DNA replication alone would be intolerable.

The analysis of cloned regions of mammalian genomes suggests that genes are on average 40 kb apart, although they may be more closely spaced in regions containing gene families. Such calculations suggest there are less than 80 000 genes in mammals, although even this estimate does not take into account large blocks of repetitive DNA at the centromeres and telomeres. Repetitive DNA accounts for about half the total DNA in the genome, and it appears that mammalian genomes may contain about 50 000 genes, more than five times the number in *Drosophila.*

32·6 Chromosomal Rearrangements and Gene Expression

We have seen that the organization of the genome can change as a result of evolution, but large-scale reorganizations are only evident after millions of years. We have also seen that the insertion and deletion of transposons take place on a much shorter timescale, although still one that spans many generations. In this section, we examine genomic rearrangements that occur much more frequently. These rearrangements are associated with the regulation of gene expression and involve recombination between specific sites in the genome.

A. Some Genes Can Be Flipped

In *Salmonella typhimurium,* there are two genes called *H*1 and *H*2 that encode the flagellar protein flagellin. The genes are found in different regions of the chromosome, and only one is expressed at a time. The two types of flagellin produced are recognized by different antibodies in an infected mammalian host. Invading bacteria can switch from expressing one type to expressing the other, a phenomenon known as **phase variation.** Phase variation helps the bacteria evade the host's immune response.

Figure 32·22
Phase variation in *S. typhimurium*. Inversion of a segment of DNA containing the *hin* gene occurs due to recombination between inverted repeats (IR). In one orientation (top), the *H2* and *rh*1 genes are transcribed. Following inversion, these genes lack a functional promoter and are not transcribed (bottom).

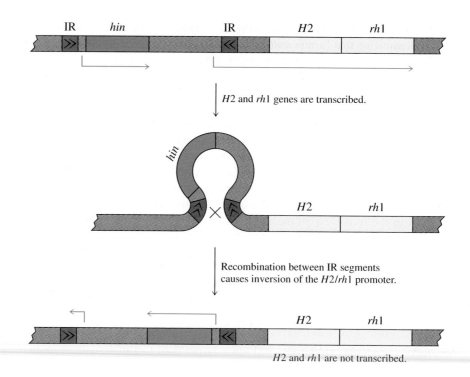

H2 and *rh*1 genes are transcribed.

Recombination between IR segments causes inversion of the *H2/rh*1 promoter.

H2 and *rh*1 are not transcribed.

The organization of genes at the *H2* locus is shown in Figure 32·22. The *H1* gene is located elsewhere in the genome. *H2* and *rh*1 are part of an operon transcribed from a single promoter. The *rh*1 gene encodes a repressor of *H1*. Transcription from the *H2/rh*1 promoter thus results in expression of *H2* and repression of *H1*.

The promoter is located on a fragment of DNA between two 26 bp inverted repeat (IR) segments. Also between these two IR segments lies the *hin* gene, whose product is a transposase that recognizes the IR segments. Phase variation occurs when Hin, the product of the *hin* gene, promotes recombination between the two IR segments, resulting in inversion of the DNA sequence containing *hin* and the promoter for *H2/rh*1. When the sequence is inverted, the *H2* and *rh*1 genes are not transcribed, and repression of *H1* is relieved. Thus, expression of the two types of flagellin is controlled by the rearrangement, or flipping, of a segment of the genome.

A system similar to the *Salmonella* Hin mechanism is found in bacteriophage Mu. In this case, the gene that is flipped encodes the phage tail-fiber proteins, and the two different orientations affect host specificity. In one orientation, the tail fibers produced bind to *E. coli* K12 strains, whereas in the other orientation, the tail fibers bind to other strains of *E. coli* and other species of bacteria. This switch is controlled by a protein called Gin.

Phase variation does not occur in most *E. coli* strains. However, some strains carry genes whose transcription is regulated by a mechanism that involves gene rearrangements. These rearrangements are under the control of a protein called Pin, and the genes regulated by Pin are found on integrated cryptic bacteriophages. It is likely that phase variation in these *E. coli* strains has been recently acquired from other bacteria via a phage vector. The Hin and Pin proteins are homologous, and both are related distantly to the bacteriophage Mu Gin protein. All three proteins are similar to the resolvase that catalyzes excision and integration of transposon *Tn3*.

B. Recombination Is Required for Mating-Type Switching in Yeast

The life cycle of *S. cerevisiae* consists of alternating haploid and diploid stages. In both stages, cells undergo mitosis followed by cell division, which takes the form of budding to form a small daughter cell. Haploid yeast cells are either mating type *a* or mating type *α*, although the cells can switch from one mating type to the other. When two haploid cells of opposite mating type come together, they form a diploid cell. This diploid cell can sporulate to produce four haploid spores.

Haploid *a* and *α* cells are very similar, but each expresses a distinct subset of genes. One type, for example, produces a pheromone that attracts the other, leading to fusion. Mating type–specific genes are regulated by the *a* and *α* genes, which are located on chromosome III (Figure 32·23). Both genes encode regulatory proteins that control expression of other mating type–specific genes. The *α* gene is located at the *HML* (homothalic, left) locus, and the *a* gene is located at *HMR* (homothalic, right). *HML* and *HMR* are silent loci; the genes they contain are not expressed, even though they have promoters and regulatory sequences. Repression is due to the formation of a heterochromatic region at these loci (Section 27·15). A single expressed gene, either *α* or *a*, is located at the *MAT* (mating-type) locus. Expression of the gene at this locus determines the mating type of the cell.

Cells can switch mating type as a result of recombination between one of the silent loci and *MAT*. Intrachromosomal pairing is mediated by homologous segments of DNA that flank the mating-type genes (*W*, *X*, and *Z*, as shown in Figure 32·23). The mechanism of recombination is similar to that of *Tn3* transposition in that it requires DNA replication, cleavage, and ligation, which results in excision of the original gene at the *MAT* locus. This is an example of site-specific recombination and gene conversion (Section 26·5E). Both *HML* and *HMR* can donate a copy of their genes to the *MAT* locus.

A similar example of substitution of alleles at a site of gene expression governs production of the variant surface glycoprotein (VSG) in trypanosomes. Trypanosomes cause sleeping sickness in humans, and during a parasitic infection, the organism frequently switches the type of surface glycoprotein expressed in order to evade the host's immune response. There are up to 1000 unexpressed copies of the VSG gene in the trypanosome genome, each one slightly different from the next. The expressed VSG gene is near the telomere of one of the trypanosome chromosomes. Recombination between one of the silent alleles and the expressed gene results in production of a different surface glycoprotein. Switching occurs up to 100 times during a single infection.

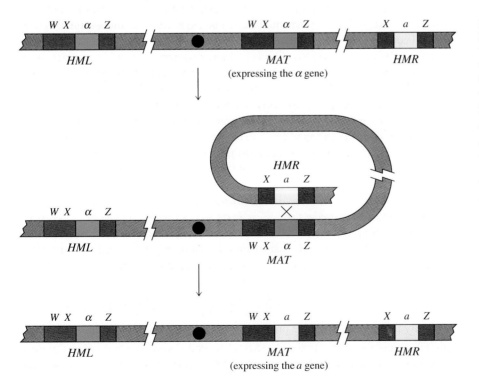

Figure 32·23
Mating-type switch in yeast. Yeast mating type is determined by expression of either an *a* or an *α* gene at the *MAT* locus of chromosome III. A single copy of *a* is found at the *HMR* locus, at one end of the chromosome, and a single copy of *α* is located at the *HML* locus, at the other end of the chromosome. The *α* and *a* genes are flanked by homologous DNA segments that allow recombination between *HMR* or *HML* and *MAT*. Recombination followed by gene conversion leads to replacement of one gene for the other at *MAT*.

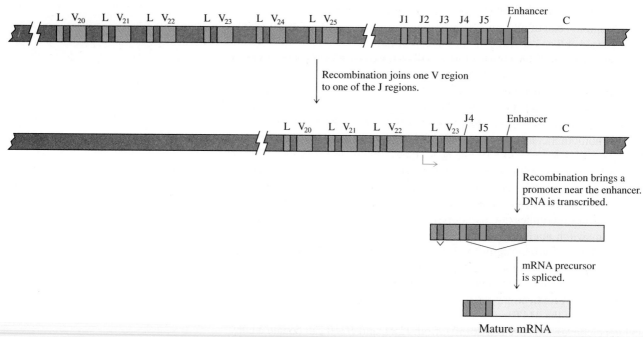

Figure 32·24
Genomic rearrangement of light-chain genes during lymphocyte differentiation. The immunoglobulin light-chain locus consists of several hundred leader (L) and variable (V) exons located upstream of the constant (C) exon. Near the C exon are several joining (J) exons. During differentiation, recombination between the 3′ end of a V exon and the 5′ end of a J exon occurs, creating a V-J fusion. The newly formed light-chain gene is transcribed from a promoter upstream of the L exon. Transcription occurs only after recombination because the promoter requires the close presence of the enhancer located between the J and C exons.

C. Rearrangements in Terminally Differentiated Cells

Phase variation in *Salmonella* and mating-type switching in yeast are examples of DNA rearrangements that do not irreversibly alter the genome; no DNA is deleted during these rearrangements, and the rearrangements are reversible. However, there are examples of genomic rearrangements during which pieces of DNA are lost. Because the genome is altered in these cases, the rearrangements are irreversible. Such rearrangements are confined to terminally differentiated cells, and the altered genome is not inherited.

Immunoglobulins, or antibodies, are Y-shaped molecules composed of light chains and heavy chains (Section 5·20). Each polypeptide has a constant C-terminus and a variable region at the N-terminus that binds to antigen. Immunoglobulins have identical constant regions, but there are hundreds of variations in the amino acid sequences of the variable regions. The J (joining) region between the two parts of the polypeptide exhibits hypervariation. Immunoglobulins are made only by circulating B cells in mammals.

The arrangement of exons encoding one of the light chains is shown in Figure 32·24. There are many copies of variable (V) region exons, each associated with a separate exon encoding a signal-peptide leader (L) sequence and a promoter. These are located upstream of the exon encoding the constant (C) region and the five exons encoding the short J region. During lymphocyte differentiation, recombination between one of the V regions and one of the J regions generates an intact gene that is transcribed from the V-region promoter. Any extra J regions are removed by splicing of the mRNA precursor.

Pairing during recombination is mediated by short, homologous repeats at the 3′ end of the V regions and the 5′ end of the J regions. These sequences are excised when the V and J segments are joined. Immunoglobulin gene rearrangement is an example of sloppy site-specific recombination. The actual site of V-J fusion can differ by up to three nucleotides, and this frequently results in an out-of-frame fusion and a nonfunctional gene. However, the sloppiness is also responsible for hypervariability in the J region of the protein, since different fusions sometimes generate different amino acid codons at this site.

The promoter of the light-chain gene is only active following rearrangement because the enhancer required for its transcription is located within the intron between the J regions and the C region. Before recombination, the promoters are located too far upstream of the enhancer to be affected by the regulatory protein that binds there. Heavy-chain genes are rearranged and regulated in a similar manner.

Recombination at the immunoglobulin gene loci sometimes results in chromosomal rearrangements. In such cases, recombination occurs between the immunoglobulin gene and a site on another chromosome. A frequently observed example occurs between a site on human chromosome 8 and either the heavy-chain locus on chromosome 14 or one of the light-chain loci on chromosomes 2 or 22. The chromosome 8 site is the location of the proto-oncogene *c-myc,* which can be activated by integration of the immunoglobulin gene and its enhancer. The resulting cancer is a type of lymphoma characterized by uncontrolled growth of B cells.

A similar rearrangement occurs during sporulation in *B. subtilis,* where the gene for σ^K, one of the sporulation-specific σ factors, exists as two separate segments in normal cells. During sporulation, the two segments are joined in the mother cell by a recombination event that deletes the DNA between the two halves of the gene. The rearrangement does not occur in the forespore.

Similarly, formation of heterocysts in cyanobacteria is associated with a genomic rearrangement that brings together several genes required for nitrogen fixation. A large segment of DNA is deleted. As a consequence, the genes for nitrogen fixation are only expressed in the heterocyst. Cyanobacteria form filamentous colonies in which only the occasional cell becomes a terminally differentiated heterocyst.

D. Some Genes Are Amplified During Development

In some organisms, gene expression is regulated by increasing the number of copies of a gene. This mechanism is known as **gene amplification.** One method of gene amplification involves multiple rounds of replication of specific DNA sequences. This particular method has been observed for certain genes whose products are required during development in some insects, protozoa, and amphibians. The amplification of certain genes in some cells facilitates the synthesis of large amounts of particular gene products. When gene amplification involves alteration of the genome, it is confined to terminally differentiated cells.

One example of gene amplification is found in *D. melanogaster.* Fruit flies lay a large number of eggs, which creates a heavy demand for the eggshell proteins, called chorion proteins. The chorion proteins are encoded by a family of genes that are clustered around an origin of replication. During eggshell formation, these genes are amplified by selective DNA replication that begins at the origin and extends outward for about 50 000 base pairs in either direction. The position of a given gene relative to the origin of replication determines the extent to which the gene is amplified. Each replication event not only generates additional genes encoding chorion proteins but also creates an additional origin of replication; thus, the number of chorion genes grows exponentially (Figure 32·25, next page). Typically, gene amplification produces 16 to 64 copies of the genes encoding chorion proteins. Amplification occurs in the nurse cells surrounding the oocyte, and the chorion proteins are deposited on the surface of the oocyte to form the eggshell.

Figure 32·25
Amplification of genes encoding chorion proteins in *D. melanogaster.* The chorion genes are part of a gene cluster centered around an origin of replication that is activated during development. DNA replication begins at the origin and simultaneously amplifies the genes on either side of the origin. Each replication event not only generates new genes encoding chorion proteins but also creates a new origin of replication; thus, the number of chorion genes grows exponentially.

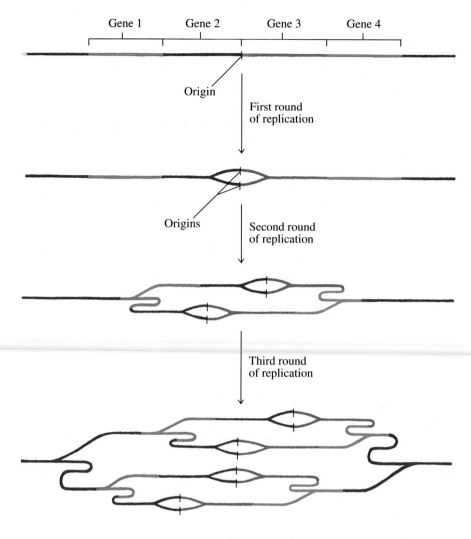

Another example of gene amplification involves the ribosomal RNA genes in the frog *Xenopus laevis.* During egg development in *X. laevis,* the number of copies of the genes encoding 5.8S, 18S, and 28S rRNA increases by a factor of several thousand, from about 500 copies to as many as 2 million copies. In this case, the additional copies of the rRNA genes exist on extrachromosomal pieces of DNA located in the nucleolus; the genome is not altered by amplification. The amplification of rRNA genes provides the egg with enough genes to synthesize the ribosomal RNA required to assemble the enormous number of ribosomes (about 10^{12}) needed for protein synthesis as the embryo develops.

Summary

Many bacteriophages and viruses can integrate into host chromosomes. Bacteriophage λ recombines with a specific site on the *E. coli* chromosome, but bacteriophage Mu can recombine at many sites. Bacteriophage Mu can also transpose to different sites by a mechanism that requires DNA replication and resolution of a cointegrate. Bacteriophage Mu is a class III transposon.

Retroviruses are RNA viruses that are copied into double-stranded DNA for integration into eukaryotic host chromosomes via nonspecific recombination. The ends of retroviral DNA consist of long terminal repeats (LTR) that contain the promoter for the retroviral genes *gag, pol,* and *env.*

Class I bacterial transposons are either simple *IS* elements or composite transposons with two *IS* elements flanking another gene. Class I transposons integrate at specific target sites recognized by the encoded transposase, leaving a short duplication of the host target site. Class II bacterial transposons contain a minimum of two genes: a transposase and a resolvase. Integration requires DNA replication and the resolution of a cointegrate by site-specific recombination.

There are three classes of eukaryotic transposon. The retrovirus-like transposons are probably derived from retroviruses because they contain genes homologous to *gag, pol,* and sometimes *env.* Transposition occurs when the transposon is transcribed, copied into double-stranded DNA, and then recombined with the host chromosome. The *Ac*-like transposons are similar to the bacterial class I transposons. They contain a transposase that promotes recombination. The third class of eukaryotic transposons are the retroposons.

Recombination between host DNA and a bacteriophage, virus, or transposon can result in disruption of a host gene. Prokaryotic genomes may include cryptic phages, the remnants of permanently integrated, mutated phage genomes. Both prokaryotic and eukaryotic chromosomes may contain multiple sequences derived from transposons.

Bacterial genomes are compact. There is little space between genes, and many genes are organized into operons. The sizes of bacterial genomes range from about 600 kb to over 12 000 kb. The *E. coli* genome, with about 3000 genes, is of average size for a prokaryote. There are very few examples of gene families or multiple copies of genes in bacterial chromosomes. Some bacteria contain more than one chromosome.

Eukaryotic genomes, which are larger and more complex than bacterial genomes, are organized into several linear chromosomes. The number of chromosomes varies considerably; even closely related species can have different numbers of chromosomes. There are many repetitive sequences in eukaryotic genomes. Some of these are large blocks of simple sequences found at centromeres and telomeres. Others are derived from transposons, such as the retroposons. Eukaryotic chromosomes often contain families of related genes and multiple copies of genes. In addition, the genomes of many multicellular eukaryotes contain pseudogenes and genes with introns. The available evidence suggests that introns in protein-encoding genes arose late in evolution.

The size of eukaryotic genomes is highly variable. Unicellular eukaryotes have compact genomes, much like those of bacteria. Yeast may have only three times as many genes as *E. coli.* The number of genes in more complex organisms such as *Drosophila* may not be much greater, but the distance between genes is greater. There may be about 50 000 genes in mammals.

Gene rearrangements are associated with expression and regulation of some genes. Recombination controls expression of flagellar genes in *Salmonella* and mating-type switching in the yeast *S. cerevisiae.* In some terminally differentiated cells, genomic rearrangements are required for expression; mammalian immunoglobulin genes are an example. Genes can also be selectively amplified during development.

Selected Readings

General References

Adams, R. L. P., Knowler, J. T., and Leader, D. P. (1992). *The Biochemistry of Nucleic Acids,* 11th ed. (London: Chapman and Hall).

Lewin, B. (1990). *Genes IV.* (New York: Oxford University Press and Cell Press).

Singer, M., and Berg, P. (1991). *Genes and Genomes: A Changing Perspective.* (Mill Valley, California: University Science Books).

Transposons

Engels, W. R. (1992). The origin of P elements in *Drosophila melanogaster. BioEssays* 14:681–686.

Kholodii, G. Ya., Yurieva, O. V., Lomovskaya, O. L., Gorlenko, Zh. M., Mindlin, S. Z., and Nikiforov, V. G. (1993). Tn5053, a mercury resistance transposon with integron's ends. *J. Mol. Biol.* 230:1103–1107.

McDonald, J. F. (1993). Evolution and consequences of transposable elements. *Curr. Opin. Genet. Dev.* 3:855–864.

Mizuuchi, K. (1992). Transpositional recombination: mechanistic insights from studies of Mu and other elements. *Annu. Rev. Biochem.* 61:1011–1051.

Stark, W. M., Boocock, M. R., and Sherratt, D. J. (1989). Site-specific recombination by Tn3 resolvase. *Trends Genet.* 5:304–309.

Wessler, S. R. (1988). Phenotypic diversity mediated by the maize transposable elements *Ac* and *Spm. Science* 242:399–405.

Whitcomb, J. M., and Hughes, S. H. (1992). Retroviral reverse transcription and integration: progress and problems. *Annu. Rev. Cell Biol.* 8:275–306.

Genomes

Bork, P., Ouzounis, C., Sander, C., Scharf, M., Schneider, R., and Sonnhammer, E. (1992). Comprehensive sequence analysis of the 182 predicted open reading frames of yeast chromosome III. *Prot. Sci.* 1:1677–1690.

Campbell, A. M. (1993). Genome organization in prokaryotes. *Curr. Opin. Genet. Dev.* 3:837–844.

Lambowitz, A. M., and Belfort, M. (1993). Introns as mobile genetic elements. *Annu. Rev. Biochem.* 62:587–622.

O'Brien, S. J., Seuánez, H. N., and Womack, J. E. (1988). Mammalian genome organization: an evolutionary view. *Annu. Rev. Genet.* 22:323–351.

Trowsdale, J. (1993). Genomic structure and function in the MHC. *Trends Genet.* 9:117–122.

Plunkett, G., Burland, V., Daniels, D. L., and Blattner, F. R. (1993). Analysis of the *Escherichia coli* genome. III. DNA sequence of the region from 87.2 to 89.2 minutes. *Nucl. Acid. Res.* 21:3391–3398.

Riley, M., and Krawiec, S. (1987). Genome organization. In Escherichia coli *and* Salmonella typhimurium: *Cellular and Molecular Biology,* F. C. Neidhardt, ed. (Washington, DC: American Society for Microbiology), pp. 967–981.

Chromosomal Rearrangements and Gene Expression

Haber, J. E. (1992). Mating-type gene switching in *Saccharomyces cerevisiae. Trends Genet.* 8:446–452.

Haselkorn, R. (1992). Developmentally regulated gene rearrangements in prokaryotes. *Annu. Rev. Genet.* 26:113–130.

Laurenson, P., and Rine, J. (1992). Silencers, silencing, and heritable transcriptional states. *Microbiol. Rev.* 56:543–560.

Schimke, R. (1984). Gene amplification in cultured animal cells. *Cell* 37:705–713.

Stark, G. R., and Wahl, G. M. (1984). Gene amplification. *Annu. Rev. Biochem.* 53:447–491.

33

Recombinant DNA Technology

Understanding the structure and function of the components of living organisms depends in part on the ability to isolate those components in pure form. Some cellular components are simple to isolate and purify due to their stability and abundance in certain cell types. For example, the enzyme lysozyme is found abundantly in egg white; hemoglobin composes over 95% of the protein inside red blood cells; and tRNA and rRNA are stable and abundant in all major cell types. Consequently, these components have been studied for many years. However, only since the early 1970s have *all* encoded components of a cell become potentially accessible to study due to the development of methods for isolating, manipulating, and amplifying identifiable sequences of DNA. These methodologies, collectively called **recombinant DNA technology,** have proven immensely powerful when applied to the study of biological systems, in particular the many previously uncharacterized components of eukaryotic species.

Recombinant DNA technology relies on an understanding of basic cellular biochemistry—a knowledge of the restriction enzymes of bacteria, the proteins involved in the synthesis of DNA, the proteins and enzymatic RNA involved in the processing of nucleic acids, and so on. The first engineered DNA molecules were

(a)

(b)

Figure 33·1
(a) Stanley N. Cohen. (b) Herbert Boyer. Cohen and Boyer conducted some of the pioneering experiments using engineered DNA.

composed of bacterial plasmids and fragments of bacterial DNA and were constructed in the early 1970s by Stanley N. Cohen and Herbert Boyer (Figure 33·1). Since then, recombinant DNA technology has led to increased understanding of such fundamental processes as gene expression and development and, coupled to the technology of DNA sequencing, is revealing the organization of the genomes of many species.

In addition to its most common role as a tool for basic research, recombinant DNA technology, popularly called genetic engineering, has numerous practical applications. Some people find these applications exciting and inspiring; others find them alarming. However, over the past 20 years, the benefits of this technology—increased knowledge of basic cellular processes, improved diagnostic and therapeutic products, and hardier crops—have generally been perceived to outweigh the risks, which include the possibility that engineered species or viruses may prove harmful or may upset the present ecosystem.

Recombinant DNA technology allows scientific questions that arise within the classic disciplines of biochemistry, genetics, biology, botany, and physiology to be pursued in the areas of molecular biology and chemistry. In this chapter, we describe the theory and a few details of the practice of recombinant DNA technology and give a few examples of current uses of the technology in agriculture, medicine, and industry.

33·1 DNA Is Manipulated by Six Basic Procedures

DNA molecules that are composed of DNA from different sources are called **recombinant DNA molecules,** or **recombinants.** The creation of recombinant DNA molecules occurs in nature more often than in the laboratory; for example, every time a bacteriophage or eukaryotic virus infects its host cell and integrates its DNA into the host genome, a recombinant is created. The ability of viruses to integrate has proven useful in genetic studies: in the 1950s, the tendency of some bacteriophages to pick up host DNA during excision from the host genome was used to study the operons that flank the sites of bacteriophage DNA integration. While this method is suitable for analyzing genes near viral integration sites, how do we locate and characterize a gene that is not near one of these sites? And how do we locate and characterize a gene expressed in only one cell or at only one time during development in a multicellular eukaryote? Clearly, in order to manipulate DNA to the point that we can identify and isolate a particular gene for study, we require more powerful technology.

The general steps in the generation of a recombinant DNA molecule in the laboratory are shown in Figure 33·2. A DNA fragment is liberated from a larger molecule and inserted into a particular site on a carrier molecule, or **vector.** Once this is done, the recombinant DNA can be introduced into a living host cell, where it is replicated and, in some cases, expressed. The replication of the recombinant DNA molecule, known as **cloning,** allows the production of millions of copies of the gene and, if the gene is expressed, millions of copies of its product.

Recombinant DNA technology is based on six basic procedures:

1. *Isolation of DNA.* DNA is separated by a variety of routine methods. We will not discuss these methods in detail here.

2. *Cleavage of DNA at particular sequences.* As we will see, the cleavage of DNA to generate fragments of defined length or with specific endpoints is crucial to recombinant DNA technology. In the laboratory, DNA is usually cleaved by treating it with commercially produced exonucleases and restriction endonucleases.

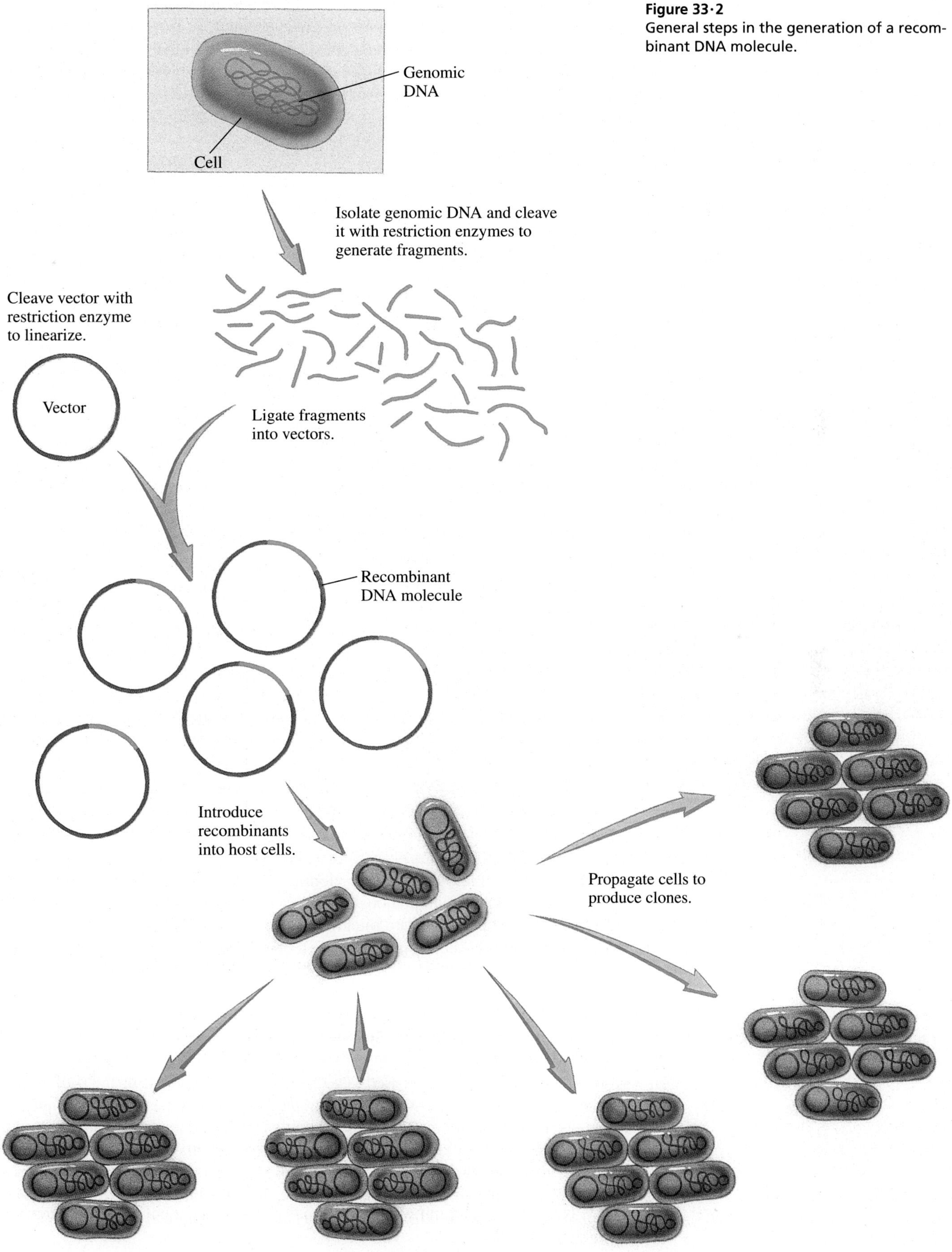

Figure 33·2
General steps in the generation of a recombinant DNA molecule.

3. *Ligation of DNA fragments.* The joining of DNA fragments also involves the use of commercially produced enzymes, notably DNA ligase. In the laboratory, a recombinant DNA molecule is usually formed by cleaving the DNA of interest to yield insert DNA and then joining the insert DNA to vector DNA. The vector DNA molecule is also referred to as the cloning vector, or cloning vehicle.

4. *Introduction of recombinant DNA into compatible host cells.* In order to be propagated, the recombinant DNA molecule (insert DNA and vector DNA joined together) must be introduced into a compatible host cell, where it can replicate. The introduction of foreign DNA into a cell is called **genetic transformation** (or just transformation) when the DNA is introduced directly and **transfection** when the DNA is introduced using a virus as a vehicle.

5. *Replication and in some cases expression of recombinant DNA within host cells.* Cloning vectors allow insert DNA to be replicated and, in some cases, expressed within a host cell. The ability to clone and express DNA efficiently depends on the choice of appropriate vectors and hosts.

6. *Identification of host cells that contain recombinant DNA of interest.* Vectors usually contain genetic markers that aid in the identification of cells that contain recombinants or, in some cases, specific insert DNA sequences. Identifying transformed or transfected cells involves the use of screens, selections, and probes, all of which we will discuss.

Before we begin our overview of recombinant DNA technology, we offer one brief note about nomenclature. The terms *recombinant* and *clone* are often used interchangeably to refer to a recombinant DNA molecule that has been transformed into a host cell. The term *clone*, however, should not be used for a recombinant molecule before it has been transformed, since the term properly refers to identical copies of an original, which are only produced within a host cell.

33·2 DNA Is Transported and Propagated in Cloning Vectors

Small fragments of free DNA do not survive long inside a cell. Thus, once a DNA fragment is liberated in vitro, it must be inserted into an appropriate vector that allows it to survive, to be transported among host cells, and to be easily identified.

Cloning vectors can be plasmids, bacteriophages, viruses, transposons, or even fragments of genomic DNA. Some cloning vectors contain sequences that allow them to be replicated autonomously within a compatible host cell, whereas others carry sequences that facilitate integration into the host genome. In addition, the most useful cloning vectors possess at least one unique cloning site, a sequence that can be cut by a restriction endonuclease to allow site-specific insertion of foreign DNA. Some cloning vectors contain several such sites grouped together in a multiple cloning site. Vectors that allow inserted DNA to be transcribed and translated into protein are called **expression vectors.** Table 33·1 lists the features of some common cloning vectors.

Plasmids are commonly used as vectors for DNA fragments up to 20 kb in size. Plasmids have the ability to replicate autonomously within a host and frequently carry genes conferring resistance to antibiotics such as tetracycline, ampicillin, or kanamycin. Such genes are known as **marker genes;** they can be used to distinguish between host cells that carry these vectors and those that do not (we will discuss selection of clones containing recombinant DNA in Section 33·5). Many of these marker genes are derived from transposons that have integrated into plasmid DNA.

Table 33·1 Common cloning vectors

Name	Type	Host	Selection/screening methods	Common uses
pBR322	Plasmid	*E. coli*	*amp*^{R}, *tet*^{R}	Library construction
pUC18	Plasmid	*E. coli*	*amp*^{R}, blue/white	Expression vector
M13mp18	Bacteriophage	*E. coli*	Blue/white	DNA sequencing, mutagenesis
pEMBL18	Plasmid	*E. coli*	*amp*^{R}, blue/white	Expression vector, DNA sequencing, mutagenesis
pcos2EMBL	Cosmid	*E. coli*	*amp*^{R}, *kan*^{R}	Library construction
λEMBL3	Bacteriophage	*E. coli*		Library construction
λgt11	Bacteriophage	*E. coli*	Blue/white	Expression vector
pMMB33	Cosmid	Gram-negative bacteria	*amp*^{R}, *kan*^{R}	Transfer of genomic DNA among gram-negative bacteria
pAH9	Plasmid	*E. coli*, yeast	*amp*^{R}, *LEU2*	Yeast expression vector, shuttle vector
pMH2	Plasmid	*E. coli*, Agrobacterium	*amp*^{R}, *neomycin*^{R}	Plant expression vector, shuttle vector
pSV2-DHFR	Plasmid	*E. coli*, monkey, Chinese hamster ovary cells	*amp*^{R}, *methotrexate*^{R}	Mammalian expression vector, shuttle vector

Plasmids are usually introduced as double-stranded DNA into bacterial cells, where they are maintained in either single or multiple copies. pBR322 (Figure 33·3) was one of the first plasmid vectors developed; it is the ancestor of many of the plasmid vectors used today. The most common differences between plasmid vectors lie in the number and type of restriction sites and selectable marker genes.

Some plasmid vectors have been developed that can replicate in either prokaryotic or eukaryotic cells. Such plasmids are known as **shuttle vectors** because they can be used to transfer recombinant DNA molecules between prokaryotes and eukaryotes. Shuttle vectors allow, for example, the initial cloning and manipulation of a mammalian gene to be carried out in a bacterial cell while the function and expression of the gene can be studied in a eukaryotic cell. This approach is useful for studying eukaryotic gene products; DNA is easily manipulated in *Escherichia coli*, but proper processing and glycosylation of an expressed protein require that it be produced in a eukaryote, such as the yeast *Saccharomyces cerevisiae*.

Another class of cloning vectors includes those derived from bacteriophages such as λ. The advantage of phage vectors over plasmid vectors is that transfection is much more efficient than transformation, especially for larger pieces of DNA. The disadvantages of phage vectors are that the recombinant DNA has to be packaged into phage particles in vitro, and it is not possible to isolate a line of cells that can propagate the recombinant DNA molecule (while propagating, the phages kill the host cells). Special vectors called cosmids have been developed that combine the advantages of plasmid and phage vectors (Section 33·6A). Cosmids accommodate large fragments of DNA and allow efficient transfection but permit propagation of the recombinant DNA molecule as a plasmid.

In eukaryotes, another way to clone large fragments of DNA is to construct artificial chromosomes that can be stably replicated. Such chromosomes must be linear and contain a eukaryotic origin of replication, a centromere, and telomeric sequences at the ends. An example of a yeast artificial chromosome (YAC) is shown

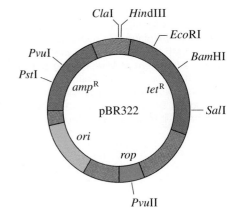

Figure 33·3
Plasmid vector pBR322. The plasmid contains an origin of replication (*ori*) and genes conferring resistance to the antibiotics ampicillin (*amp*^{R}) and tetracycline (*tet*^{R}). The *rop* gene (repressor of primer) regulates DNA replication so that there are about 20 copies of the plasmid per bacterial cell. There are also several restriction sites that are suitable for cloning (only a few common ones are shown). The entire plasmid contains only 4361 base pairs.

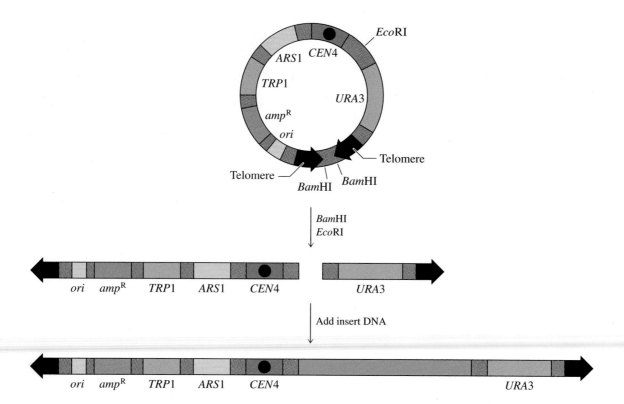

Figure 33·4

Cloning with a yeast artificial chromosome (YAC). This shuttle vector contains yeast centromeric DNA (*CEN*4) and a yeast origin of replication (*ARS*1). The plasmid also contains two marker genes for selection in yeast cells (*TRP*1 and *URA*3) and a marker gene for selection in *E. coli* (*amp*R). Large fragments of DNA (400–500 kb) can be inserted at the unique *Eco*RI site. Telomeric sequences are arranged in the plasmid in such a way that cleavage with the restriction enzyme *Bam*HI removes a fragment of DNA, leaving a linear vector with telomeric ends. Following cleavage of the recombinant DNA molecule by the action of *Bam*HI, the linear artificial chromosome is used to transform yeast cells.

in Figure 33·4. This shuttle vector contains yeast centromeric DNA (*CEN*4) and a yeast origin of replication (*ARS*1; Section 25·9C). The vector also contains two marker genes for selection in yeast cells (*TRP*1 and *URA*3, which encode proteins required in the biosynthetic pathways for tryptophan and uracil, respectively) and a marker gene for selection in *E. coli* (*amp*R).

YACs are introduced into yeast cells as linear molecules (unlike *E. coli*, yeast are transformed efficiently with linear DNA). Once introduced into a cell, YACs replicate at the same time as the normal yeast chromosomes, and they segregate at mitosis just as the other chromosomes do (some normal yeast chromosomes are about the size of YACs). YACs have been useful in the analysis of large genomes such as those of flowering plants and mammals because only about 5000 different YACs are needed to clone the entire genome of these organisms (fragments up to 500 kb can be cloned into YACs).

33·3 Recombinant DNA Molecules Are Constructed by Cutting and Splicing

DNA fragments can be generated by a variety of means. For example, DNA fragments can be produced by the mechanical shearing of long DNA molecules or by digesting DNA with type II restriction endonucleases (Section 24·10). Unlike shearing, which cleaves DNA randomly, restriction enzymes cleave DNA at specific sequences. For cloning purposes, this specificity offers extraordinary advantage.

The most useful restriction endonucleases are those that cleave to produce fragments with single-stranded extensions at their 3′ or 5′ ends. These ends, called sticky ends, can transiently base-pair to complementary termini on vector DNA and can be covalently joined to the vector in a reaction catalyzed by DNA ligase (Figure 33·5).

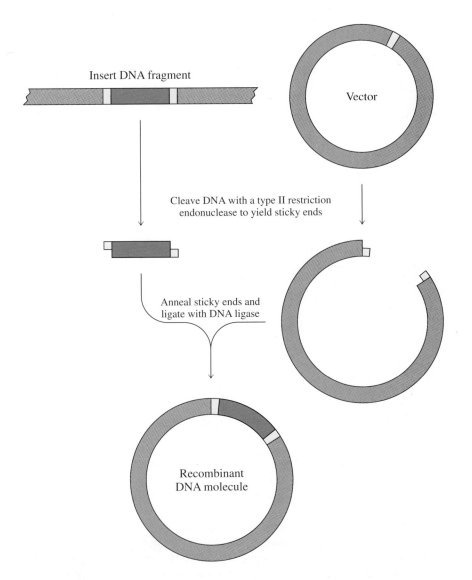

Insert DNA fragment

Vector

Cleave DNA with a type II restriction
endonuclease to yield sticky ends

Anneal sticky ends and
ligate with DNA ligase

Recombinant
DNA molecule

Figure 33·5
Use of restriction enzymes to generate recombinant DNA. The vector DNA and the insert DNA are cleaved in reactions catalyzed by restriction endonucleases to generate ends that can be joined together. In cases where sticky ends are produced, the two molecules join by annealing (base pairing) of the complementary ends. The molecules are then covalently attached to one another in a reaction catalyzed by DNA ligase.

Some DNA fragments are created without sticky ends. This happens, for example, when the DNA is cleaved with restriction enzymes that do not leave an overhang, or when DNA fragments are produced by mechanical shearing. It is possible to join blunt-ended insert DNA molecules to blunt-ended vectors by treatment with DNA ligase, but this reaction is much less efficient than ligation of ends that overlap by several bases. Sometimes it is easier to convert blunt ends into sticky ends by ligating **oligonucleotide linkers** to the ends of blunt-ended fragments. Oligonucleotide linkers are short, synthetic, double-stranded DNA molecules whose sequence contains a site recognized by a restriction endonuclease. When these linkers are ligated to the ends of blunt-ended fragments and then cut by the appropriate restriction endonuclease, blunt ends are converted to sticky ends (Figure 33·6, next page). The ends of the insert DNA can then be defined by the presence of the chosen restriction site (*Eco*RI in Figure 33·6), making subsequent removal of the insert from the vector much easier.

It is usually more efficient to ligate linkers to the blunt ends of DNA molecules than to ligate the ends of such molecules directly because the oligonucleotide linkers can be introduced in large excess relative to the ends of the DNA fragments. The ligation of multiple linkers onto a single end does not cause problems since any excess linkers are cleaved by subsequent treatment with the restriction enzyme before insertion of the DNA fragment into the vector.

Figure 33·6

Ligation of oligonucleotide linkers to DNA. The DNA sequence CAGCTG is recognized by the restriction endonuclease *Pvu*II and cleaved to yield two blunt-ended fragments. The blunt ends are then joined to oligonucleotide linkers in a blunt-end ligation catalyzed by DNA ligase. Each of these linkers contains an *Eco*RI restriction site and is thus called an *Eco*RI linker. Subsequent cleavage of the DNA catalyzed by *Eco*RI yields DNA with a sticky end. In this way, the *Pvu*II site is destroyed, and blunt-ended fragments are converted to sticky-ended fragments.

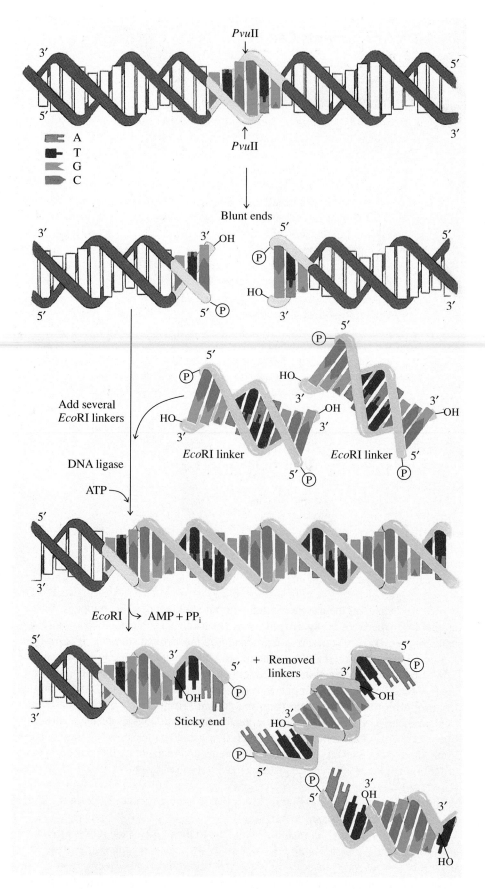

33·4 DNA Fragments Can Be Integrated into the Host Genome

In some cases, isolated DNA can be used directly to transform bacteria without being incorporated into a vector. For many decades, genetic studies in bacteria such as *Streptococcus pneumoniae* and *Bacillus subtilis* have relied on the ability of these organisms to take up DNA and integrate it by recombination into the host chromosome. More recently, direct integration has been studied in the yeast *S. cerevisiae* and in cultured mammalian cells.

Whether transformed directly or incorporated into a vector, foreign DNA integrated into the host genome is usually present as a single copy rather than as multiple copies (as is the case after transformation with plasmid DNA or transfection). Foreign DNA can sometimes be encouraged to integrate at a particular site on the host chromosome by site-specific recombination or by homologous recombination if the host gene and the foreign DNA share a similar DNA sequence. Site-specific integration offers a significant advantage when studying gene expression, especially in eukaryotic cells, since the foreign DNA is then in the same chromatin environment as the normal host gene.

Plasmid and phage vectors are frequently used in experiments involving integration into the host genome even though vector sequences are often not required for homologous recombination. Such vectors are convenient because they contain multiple cloning sites that make it easy to manipulate DNA fragments and because they carry genetic markers that make it easy to identify transformants.

33·5 Host Cells Containing Recombinant DNA Molecules Can Be Identified in a Variety of Ways

Once a cloning vector and insert DNA have been joined in vitro, the recombinant DNA molecule can be introduced into a host cell. In general, transformation is not very efficient; only a small percentage of host cells take up recombinant DNA. Therefore, transformed cells must be distinguished from the vast majority of untransformed cells. In addition, the DNA used in transformations often includes vectors that do not contain the insert DNA. If the vector in a cloning experiment has been prepared by cutting it with a single restriction endonuclease, the sticky ends of the vector DNA can reclose during ligation, thereby regenerating vector molecules without inserts. In some experiments, these reclosed vectors may be more abundant than vectors containing inserts. Cells containing recombinants must therefore also be distinguished from cells that contain only the vector. The identification of host cells containing recombinants requires either genetic selection or screening or both.

In a **selection,** cells are grown under conditions in which only transformed cells survive. In a **screen,** transformed cells are tested for the presence of the desired recombinant DNA. Normally, a number of colonies are first selected and then screened for colonies carrying the desired insert.

A. Selection Strategies Make Use of Marker Genes

Many of the selection strategies used in labs today involve selectable marker genes—genes whose presence can easily be detected or demonstrated. For example, many bacterial plasmid vectors carry the marker gene encoding β-lactamase, which catalyzes hydrolysis of β-lactam antibiotics, such as ampicillin. Only cells transformed with plasmids carrying and expressing this gene grow in media containing the antibiotic ampicillin; such cells are ampicillin resistant (amp^R). Bacteria that have not taken up the plasmid are killed when placed in medium containing ampicillin or spread on agar containing the antibiotic.

Figure 33·7
Yeast shuttle vector that can be propagated and selected in both *E. coli* and *S. cerevisiae*. Recombinants are selected in *E. coli* by the ability to grow in the presence of ampicillin and in *LEU2*-deficient yeast strains by the ability to grow in the absence of exogenous leucine.

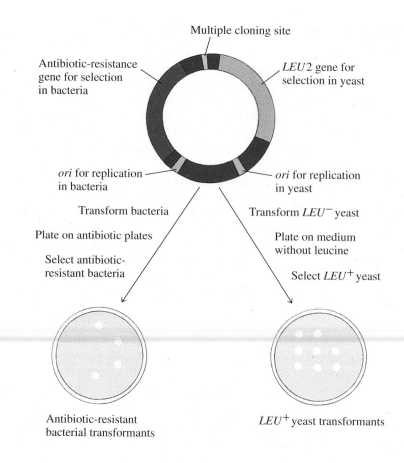

Similarly, many plasmids that grow in yeast carry the gene *LEU2*, which encodes the enzyme β-isopropylate dehydrogenase, an essential enzyme in the leucine biosynthesis pathway. Mutant yeast cells that lack a functional *LEU2* gene do not grow in the absence of leucine. If these cells are incubated with such a plasmid and grown in media lacking leucine, only transformants are able to grow. There are many other yeast marker genes commonly used in selections. Shuttle vectors contain prokaryotic genes for antibiotic resistance, allowing selection in bacterial hosts, and yeast genes for metabolite biosynthesis, allowing selection in yeast cells (Figure 33·7). Recombinants constructed from such vectors can be manipulated in *E. coli* and then transformed into yeast to express the cloned gene for analysis of its product.

Vectors for cloning in mammalian cells can also contain selectable markers. One common marker gene is the neo^R gene, which encodes the enzyme aminoglycoside phosphotransferase. This enzyme inactivates antibiotics such as neomycin and G418. A typical mammalian shuttle vector is shown in Figure 33·8. Mammalian cells transformed with recombinant DNA containing this vector can be treated with G418 so that only transformed cells survive. The vector shown in Figure 33·8 also contains a eukaryotic promoter (not shown) in front of the neo^R gene to ensure that the antibiotic-resistance gene is expressed in mammalian cells. In addition, the multiple cloning site is flanked by a promoter and a polyadenylation site so that the inserted DNA can be expressed in mammalian cells.

When the cloning vector is a bacteriophage, infected *E. coli* cells can be distinguished from those not infected because the phage either kills the cells or retards their growth. On culture plates, infected cells appear as either a clear plaque of dead cells or a turbid plaque of slowly growing cells on a densely packed layer of uninfected, rapidly growing bacterial cells (called a bacterial lawn).

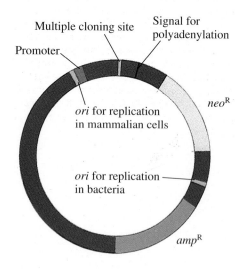

Figure 33·8
Example of a mammalian shuttle vector. This vector contains a bacterial origin of replication and the *amp*^{R} gene for growth and selection in *E. coli*. It also contains a mammalian origin of replication and the gene for resistance to the antibiotic G418 (*neo*^{R}).

B. Alkaline Phosphatase Can Be Used to Select for Recombinants

In addition to selecting for host cells carrying vector DNA molecules, many cloning strategies also select recombinant DNA molecules over vector DNA molecules carrying no insert. Circular DNA molecules transform many orders of magnitude more efficiently than do linear DNA molecules. This fact is often exploited in a technique that selects against vector molecules without inserts. When a restriction enzyme cleaves a DNA molecule, it leaves behind a 5'-monophosphate group and a 3'-hydroxyl group. The terminal 5'-monophosphate groups, required for ligation catalyzed by DNA ligase, can be removed by the action of the enzyme alkaline phosphatase. A circular, double-stranded plasmid cut with a restriction endonuclease and treated with alkaline phosphatase cannot be ligated to itself; however, it can be ligated to an insert DNA molecule that possesses a 5'-monophosphate group at each end. Ligation to such an insert results in a circular recombinant DNA molecule with a nick in one strand at one end of the insert and in the other strand at the other end of the insert (Figure 33·9). These nicks are repaired in vivo prior to replication. Any vector molecules not ligated to inserts remain linear and transform inefficiently.

C. Insertional Inactivation Can Be Used for Selection and Screening

Selection and screening can also be achieved using **insertional inactivation.** In this technique, a DNA fragment is inserted within the coding region of a functional gene on a vector, resulting in inactivation of that gene. If the gene encodes an easily detected product, insertional inactivation can be used as a method of screening or selection. For example, the plasmid pBR322 (see Figure 33·3) contains two genes conferring antibiotic resistance: one for ampicillin (amp^R) and one for tetracycline (tet^R). There are two commonly used cloning sites within the tet^R gene. Insertion of DNA at either of these sites inactivates the tet^R gene and gives rise to transformants sensitive to tetracycline (tet^S) but resistant to ampicillin (amp^R). Cells transformed by plasmids not bearing the insert are amp^R tet^R; those cells without any plasmid are amp^S tet^S.

Desired amp^R tet^S cells (those containing the inserted DNA) can be detected by selecting for amp^R cells and then screening for tet^S cells. In this method, the transformants are grown into colonies on a plate containing ampicillin. The colonies are then screened for sensitivity to tetracycline by growing them on medium containing tetracycline. Colonies that grow on the ampicillin-containing plate but not on the tetracycline-containing plate are amp^R tet^S and can be further examined.

The amp^R tet^S cells can also be selected directly by growing the cells for a short time after transformation in a medium containing tetracycline and cycloserine, a compound that kills only growing cells. In this medium, the unwanted tet^R cells begin to grow and are killed by the cycloserine, whereas growth of the desired tet^S cells is inhibited in the presence of tetracycline (the cells are not immediately killed by tetracycline). If surviving cells (amp^R tet^S and amp^S tet^S) are transferred to a medium containing only ampicillin, the desired amp^R tet^S cells grow (Figure 33·10, next page).

Visual markers can also be used to screen for hosts containing recombinant DNA molecules. For example, the *lac*Z gene of *E. coli* encodes the enzyme β-galactosidase, which catalyzes hydrolysis of the synthetic substrate 5-bromo-4-chloro-3-indolyl-β-D-galactoside (commonly known as X-gal). Cleavage of X-gal produces the blue dye 5-bromo-4-chloroindole, which identifies cells containing active β-galactosidase. In the absence of enzyme, X-gal is not cleaved, and no blue color is observed.

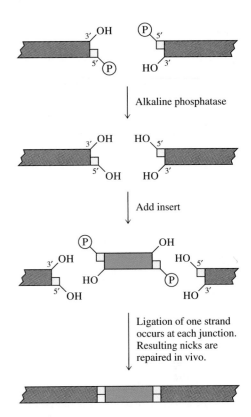

Figure 33·9
Ligation of insert DNA to vector DNA treated with alkaline phosphatase. Alkaline phosphatase catalyzes removal of the terminal 5'-monophosphate groups from the vector DNA molecule. Removal of these phosphate groups renders the vector an unsuitable substrate for DNA ligase, which requires a 5'-monophosphate group. As a result, vector molecules that re-anneal cannot be ligated. When vector molecules anneal with insert molecules, the insert molecules provide 5'-monophosphate groups for the ligation reaction, and DNA ligase catalyzes the covalent attachment of the insert DNA to the vector DNA through the formation of one phosphodiester bond at each junction.

Figure 33·10

Selection and screening by insertional inactivation. When DNA is cloned into pBR322 at the unique *Bam*HI site within the gene for tetracycline resistance (*tet*^R), a recombinant plasmid without a functional *tet*^R gene is produced. Cells containing such recombinants are resistant to ampicillin but sensitive to tetracycline (amp^R tet^S). Plasmids without inserts retain the *tet*^R gene and are thus amp^R tet^R; cells containing this parental plasmid grow on media containing either ampicillin or tetracycline. The two types of cells can be distinguished by either selection or screening. In selection, the transformants are grown in the presence of both cycloserine and tetracycline. The amp^R tet^R cells, which contain the undesired parental plasmid, start to grow but are killed by the cycloserine, whereas the desired amp^R tet^S cells lie dormant. When subsequently washed and plated onto media containing ampicillin, the amp^R tet^S recombinants are selected. In the screening method shown, the transformants are first selected by plating them on a medium containing ampicillin, where they grow into colonies. The colonies are then screened for tet^S cells by plating them on a medium containing tetracycline. Colonies that do not grow on the tetracycline plate are tet^S and can be isolated from the ampicillin plate for further use.

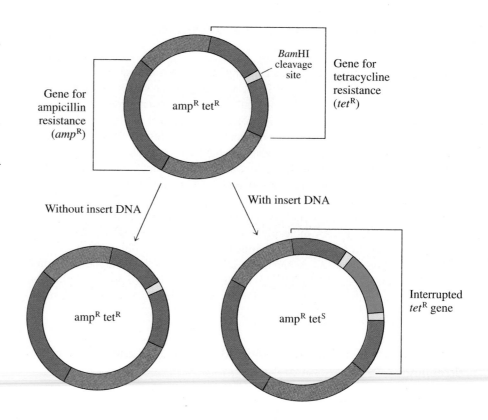

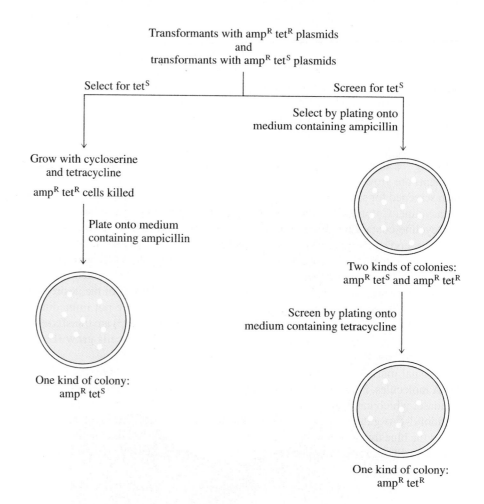

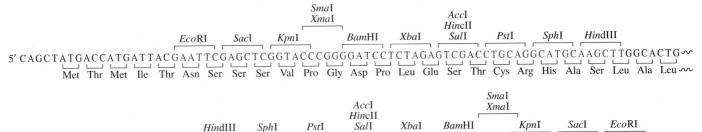

The *lac*Z gene is carried on many phage and plasmid cloning vectors. In these vectors, the 5' end of the gene has been reconstructed to include multiple restriction sites for cloning (sometimes called a restriction site cassette), while the reading frame of the gene has been preserved (Figure 33·11). Vectors without inserts produce functional β-galactosidase and therefore give rise to blue plaques (when the vector is a phage) or colonies (when the vector is a plasmid) in the presence of X-gal. If insert DNA is cloned into a site within the *lac*Z gene, however, the resulting recombinants do not produce functional β-galactosidase and therefore form colorless plaques or white colonies (Figure 33·12). This technique, also known as a blue/white screen, is quite useful since a single colorless plaque or white colony can often be distinguished from among 10⁴ blue ones.

Figure 33·11
Multiple restriction sites engineered into the 5' end of the β-galactosidase gene. A fragment of DNA containing multiple restriction sites has been inserted into the gene near the 5' end. Plasmid vectors containing both orientations of the multiple restriction site cassette are available. The reading frame of the gene is preserved, but if foreign DNA is inserted at one of the restriction sites, the gene is disrupted, and β-galactosidase is not produced.

D. Phage Packaging Selects for DNA Fragments of a Particular Size

It is possible to transform bacteria with plasmid vectors of any size up to about 25–30 kb. In contrast, there is often a lower limit to the size of a fragment that can be cloned using phage vectors. In vivo, infective bacteriophage λ is composed of a head and a tail (both made up of phage proteins) and DNA. The phage proteins and DNA are assembled into phage particles in a reaction that requires as a substrate long, linear concatemers of DNA (a linked series of complete λ genomes). These genomes are liberated by cleavage at specific sequences called *cos* (cohesive) sequences. The 12 bp *cos* sequence, or *cos* site, must be repeated every 45 to 50 kb for proper packaging to occur; if the distance between *cos* sites is outside this range, the DNA is not packaged into infective phages. Vectors are prepared by generating

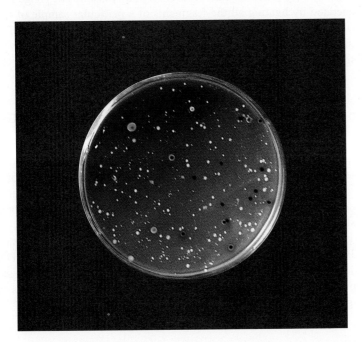

Figure 33·12
Blue/white screen. Many phage and plasmid vectors carry a restriction site cassette within the gene for β-galactosidase. X-gal is a chromogenic substrate cleaved by β-galactosidase to yield the blue dye 5-bromo-4-chloroindole. Blue colonies represent cells transformed with cloning vectors that do not contain inserts. In these cells, β-galactosidase is active. White colonies represent cells transformed with recombinants. In these cells, the β-galactosidase gene has been disrupted by the insert. (Courtesy of Philip L. Carl.)

Figure 33·13
Size selection in phage λ. During phage assembly, phage λ genomes are liberated by cleavage at appropriately spaced *cos* sites (shown in red). A phage λ vector is prepared by deleting a nonessential part of the phage genome. If the missing phage λ DNA is replaced by insert DNA of an appropriate size (green), infective phages are produced. If phage λ DNA is replaced by insert DNA that is too long or too short, infective phages are not produced. This technique can be used to select inserts of a particular size (usually 10–35 kb).

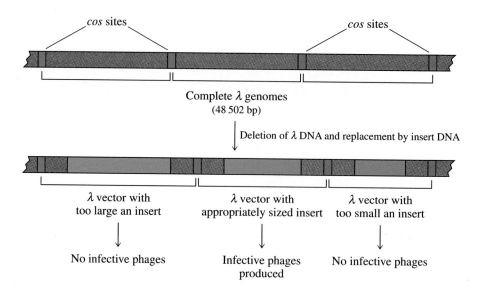

phage λ DNA that is missing a region of the genome not essential for phage replication. Such vectors are too short to be packaged without insert DNA, so only recombinant molecules containing DNA of the appropriate size are selected and packaged into phage particles (Figure 33·13). These phage particles then introduce the recombinant DNA molecules into host cells by transfection. This technique is particularly useful for cloning large DNA molecules (10 to 35 kb, depending on the phage vector used), such as those commonly generated when cloning eukaryotic genomic DNA.

33·6 DNA Libraries Contain Multiple Recombinant DNA Molecules

Cloning has proven exceptionally useful for isolating large quantities of specific DNA fragments from organisms with complex genomes. The process of cloning such fragments usually begins with the construction of a **DNA library,** or DNA bank. A DNA library consists of all the recombinant plasmids or phage DNA molecules generated by ligating all the fragments of a particular sample of DNA into vectors. The recombinant DNA molecules are then introduced into cells, where each recombinant replicates. Several different types of DNA libraries can be made, depending upon the nature of the insert, the type of vector used, and the type of the scientific problem to be investigated.

A. Genomic Libraries Include DNA Fragments from the Entire Genome

Genomic DNA libraries are constructed by fragmenting and cloning all the DNA from the genome of an organism. These libraries are frequently constructed using cloning vectors that can accommodate large inserts, such as phage λ vectors, so that fewer recombinants are required to hold the entire library and so that genes of interest—complete with important 5′ and 3′ flanking sequences—can be obtained intact in a single recombinant or a few different recombinants.

The general strategy for cloning genomic DNA in phage λ vectors is shown in Figure 33·14. The insert DNA is prepared by partially digesting genomic DNA with a restriction endonuclease, such as *Eco*RI. The resulting fragments are separated according to size by electrophoresis, and DNA of the appropriate size is removed from the gel for ligation into the vector.

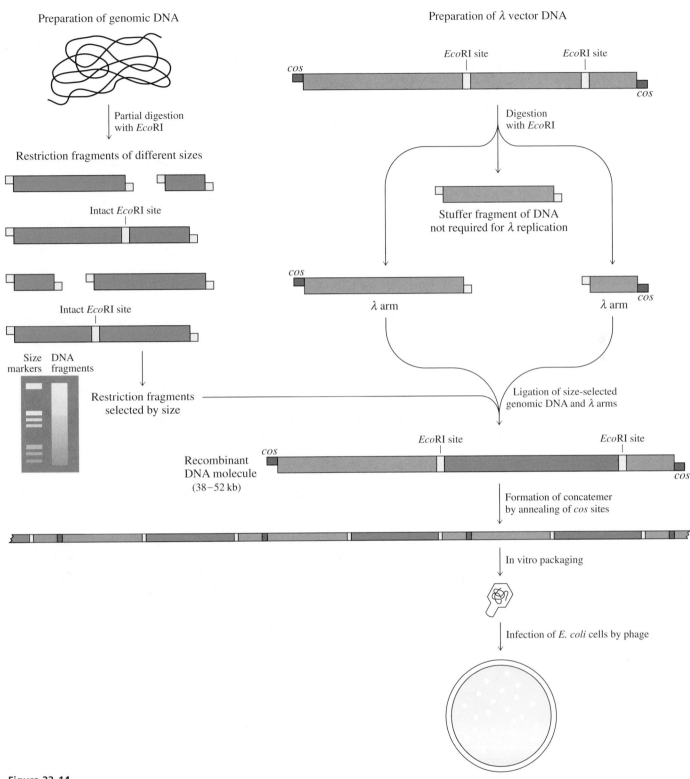

Preparation of genomic DNA

Preparation of λ vector DNA

Partial digestion with *Eco*RI

Restriction fragments of different sizes

Intact *Eco*RI site

Intact *Eco*RI site

Size markers

DNA fragments

Restriction fragments selected by size

*Eco*RI site

*Eco*RI site

cos

Digestion with *Eco*RI

Stuffer fragment of DNA not required for λ replication

cos

λ arm

λ arm

cos

Ligation of size-selected genomic DNA and λ arms

Recombinant DNA molecule (38–52 kb)

cos

*Eco*RI site

*Eco*RI site

cos

Formation of concatemer by annealing of *cos* sites

In vitro packaging

Infection of *E. coli* cells by phage

Figure 33·14
Use of a phage λ vector in constructing a genomic DNA library. Genomic DNA is partially digested with *Eco*RI (a partial digest leaves some *Eco*RI sites intact). The resulting fragments are electrophoresed through an agarose gel. (The fragments, which are in the right-hand lane, appear as a smear due to the range of their sizes.) DNA fragments of known size (size markers) are used to locate appropriately sized fragments, which are recovered from the gel and ligated to the arms of λ vector DNA. The λ vector DNA is prepared by digesting it with *Eco*RI, yielding the λ arms, each of which contains a half-*cos* site at one end. Since λ arms are assembled into phages only if ligated to insert DNA of an appropriate size, recombinant phages are selected.

The vector DNA is prepared by isolating DNA from phage λ particles and cutting it with *Eco*RI. The vector shown in Figure 33·14 is cleaved into three fragments: the two end fragments, called the λ arms, and a stuffer fragment. The two λ arms ligated together are too short to be packaged into a phage λ particle. During treatment with *Eco*RI, the stuffer fragment is digested into small fragments, leaving the two λ arms.

As we discussed in the previous section, the λ arms must be ligated to an appropriately sized insert fragment in order to be packaged. The λ arms each possess a 12-nucleotide, single-stranded extension at one end, generated by the cutting of λ DNA concatemers at the *cos* sites during in vivo packaging. When the arms are mixed with the prepared genomic DNA fragments, the sticky ends generated by the restriction endonuclease anneal, as do the 12-base sticky ends of the half-*cos* sites. The DNA is then sealed by the activity of DNA ligase into long concatemers, ready for packaging in vitro. The recombinant DNA molecules are packaged into phage particles that infect *E. coli* and produce plaques, each of which arises from a single recombinant phage.

The mechanism of phage DNA packaging was exploited in the creation of cosmid vectors. Cosmid vectors are small plasmids that contain a bacterial origin of replication, a selectable antibiotic-resistance gene, a *cos* sequence, and one or more cloning sites (Figure 33·15). Insertion of large fragments of DNA into a cosmid vector produces a recombinant cosmid that can be packaged in vitro into phage λ

Figure 33·15
Insertion of large DNA fragments into cosmid vectors. Large fragments of genomic DNA (35–45 kb) are generated by partial digestion with a restriction endonuclease. Cosmid vectors are cut with a compatible restriction endonuclease. The genomic DNA fragments are ligated to the linear cloning vectors to yield the concatemers of vectors and inserts as shown. Inserts that separate *cos* sites in the linear molecule by approximately 45 to 50 kb are selected by the packaging reaction; inserts of other sizes are not.

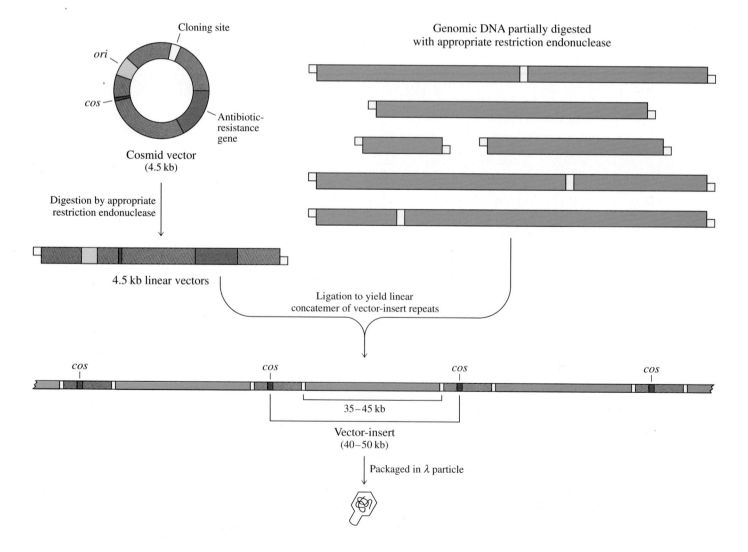

particles and introduced into host bacterial cells by transfection. Although the recombinants bear no resemblance to phage λ other than having *cos* sites, the in vitro packaging machinery only recognizes correctly spaced *cos* sites and does not discriminate between λ DNA and foreign DNA. Hence, correctly sized recombinant DNA is packaged in vitro into infective phage particles. Transfection of *E. coli* cells allows the recombinant DNA to be propagated as a plasmid. Cells containing the cosmid can be selected by growing the cells on media containing the appropriate antibiotic. Note that the phage particle serves only as an aid in transformation, and the transfected cells do not die since they do not contain viable phage. Because very large plasmids transform inefficiently, cosmid vectors are an excellent vehicle for introducing large plasmids into *E. coli*.

B. cDNA Libraries Are Made from Messenger RNA

A type of DNA library often used to study eukaryotic protein-encoding genes is known as a cDNA library. cDNA (complementary DNA) is double-stranded DNA synthesized from purified mRNA using reverse transcriptase.

The creation of a cDNA library starts with purification of mRNA. Most techniques for purifying mRNA molecules from eukaryotes take advantage of the poly A tails present on mature eukaryotic mRNA. Polyadenylation allows eukaryotic mRNA to be separated from the more abundant rRNA and tRNA molecules, which lack poly A tails (Section 28·5C). Poly A^+ mRNA (approximately 2% of total RNA) binds selectively to chromatography columns that contain oligodeoxythymidylate (oligo dT) covalently linked to an insoluble matrix.

The synthesis of double-stranded cDNA from poly A^+ mRNA is illustrated in Figure 33·16 (next page). In the first step, catalyzed by reverse transcriptase, the mRNA population is incubated with oligo dT, which hybridizes to the poly A sequence at the 3′ end of each mRNA molecule. The oligo dT fragment acts as a primer for synthesis of a complementary DNA strand. The resulting mRNA-cDNA hybrid is subsequently treated with RNase H, which nicks the RNA and leaves small RNA fragments base-paired to the DNA strand. These fragments are used as primers by DNA polymerase I, which degrades the RNA fragments with its 5′→3′ exonuclease activity and replaces the RNA with a continuous DNA strand as it proceeds along the template (a reaction analogous to the joining of Okazaki fragments during DNA replication; Section 25·4C). The single-stranded regions that remain at the termini of the duplex cDNA are removed by the 3′→5′ exonuclease activity of T4 DNA polymerase, which produces blunt-ended cDNA molecules. Blunt-ended cDNA molecules are often converted to sticky-ended cDNA molecules by the addition of oligonucleotide linkers. The generation of a cDNA library is completed by the ligation of the cDNA molecules to vectors and the introduction of the recombinants into a host cell (almost invariably *E. coli*).

A cDNA library takes advantage of the fact that different populations of mRNA molecules are found in different types of cells. Remember that although genomic DNA molecules are identical in almost all cells of an organism, the mRNA populations differ according to the specialized functions of each cell type. Thus, while liver and brain cells contain many identical mRNA molecules derived from housekeeping genes, each cell type also contains mRNA molecules that encode components necessary for specialized functions. Constructing a cDNA library from specific tissues in which the target protein is abundant increases the chances of successfully cloning the gene of interest. Note, however, that it is virtually impossible to clone the gene for a protein that is not expressed in the tissue from which the cDNA library was derived.

Figure 33·16

Steps in the synthesis of cDNA. (1) Total cellular poly A⁺ mRNA is isolated and purified. Oligo dT is annealed to the 3′ poly A tail on each mRNA molecule, where it serves as a primer for the copying of the RNA strand into DNA by reverse transcriptase. (2) The RNA in the resulting RNA-DNA duplex is nicked selectively by RNase H to generate a series of short RNA fragments that can act as primers for DNA synthesis by DNA polymerase I. (3) DNA polymerase I extends the primers and synthesizes the second strand of DNA. During synthesis of the second strand, the RNA fragments are degraded by the 5′→3′ exonuclease activity of the polymerase as it proceeds along the template. (4) The resulting duplex cDNA is made blunt ended by the 3′→5′ exonuclease activity of T4 DNA polymerase. (5) Blunt-ended, synthetic oligonucleotide linkers are added, and the fragments are digested with a restriction endonuclease to yield cDNA fragments with sticky ends. The fragments are then ligated into a vector.

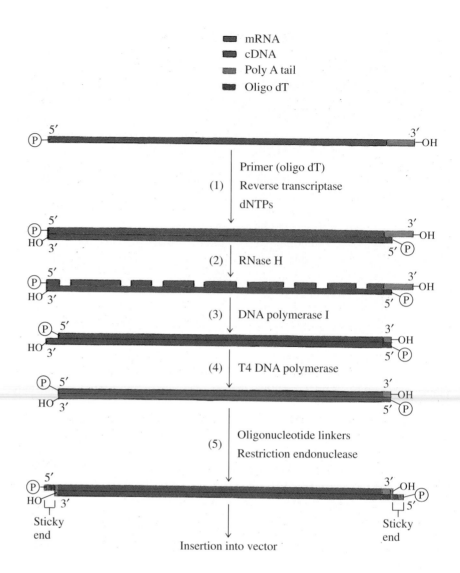

cDNA libraries are frequently constructed using both specialized phage λ vectors and plasmids. These libraries ideally represent all the genes expressed in a particular cell or tissue at a particular time during development. In addition, since cDNA libraries are derived from mature mRNA rather than from primary RNA transcripts, they do not include introns or flanking sequences and are therefore much less complex than genomic libraries. In high-quality cDNA libraries, the number of recombinant clones is great, the average size of the cDNA insert is large, and both abundant and rare mRNA molecules are represented. Clearly, the quality of the library depends upon the quality of the mRNA used at the outset, its size and purity, and the cumulative efficiency of the steps summarized in Figure 33·16. For these reasons, cDNA libraries that represent mRNA molecules of both high and low abundance are much more challenging to construct than genomic DNA libraries.

C. DNA Libraries Must Be Large to Be Useful

The most time-consuming and difficult step of cloning is frequently the isolation of the desired recombinant DNA from the million or so different recombinants present in a typical library. The probability that a library of a given size contains a particular clone is governed by the formula

$$P = 1 - (1-n)^N \quad \text{or} \quad N = \frac{\ln(1-P)}{\ln(1-n)} \qquad (33\cdot1)$$

where N is the number of recombinant clones in the library, P is the probability of finding a particular clone, and n is the frequency of occurrence of the desired clone. Usually, all DNA fragments are assumed to have been inserted into vectors and propagated with equal efficiency. The frequency of occurrence (n) depends upon the type of library. In a genomic DNA library, virtually all fragments generated from the genome have an equal probability of being ligated to vector and hence of being present in the library. Thus, in a genomic DNA library, n is the ratio of the size of the insert to the size of the genome. For example, to be 99% certain that a particular sequence from the human genome (genome size 3.2×10^9 bp) is present in a genomic library of insert size 20 kb, the library must contain approximately 8×10^5 recombinant clones (fewer recombinant clones would be needed in a YAC library, where the insert size can be as large as 500 kb). In a cDNA library, however, the probability of a desired clone being present depends upon the abundance of the desired mRNA and not on the genome size. Accordingly, in a cDNA library, n is the abundance of the mRNA of interest. Typically, moderately and highly abundant mRNA molecules (present at 20 to 1000 copies per cell) account for most of the mass of the approximately 2×10^5 mRNA molecules per cell, but rare mRNA molecules (<20 copies per cell and often as low as 0.001% of poly A^+ mRNA) account for most of the distinct species of mRNA. Thus, to be 99% certain that a cDNA library contains one clone of a particular mRNA representing 0.001% of the total mRNA, the library must contain approximately 5×10^5 recombinant clones.

Because up to 10^6 clones must be screened to isolate a single desired clone, it is not possible to screen each clone individually. In general, methods for screening DNA libraries for the desired clone must allow many thousands of recombinants to be analyzed simultaneously.

33·7 A Desired Clone Can Be Detected Using Probes

Screening a DNA library for a particular clone, also known as **probing,** requires some knowledge of the desired recombinant DNA. That information is used to generate a **probe,** a molecule that specifically recognizes the desired piece of DNA in a large library. The nature of the information available determines the kind of probe to be used and the method of screening. There are a variety of different types of probes, and screening methods have been developed that can detect cloned nucleic acid sequences either directly (by binding to the DNA of the clone) or indirectly (by detecting the presence of the gene product of the cloned DNA).

The same general procedure usually is followed whenever a library is screened using a probe (Figure 33·17, next page). Cells containing the cloned library are grown on petri plates into individually resolved colonies or plaques. The nucleic acids or proteins from the cells are liberated and transferred to a filter, where they are immobilized. The filter is then exposed to the probe, which is labelled so that the presence of bound probe on the filter can be detected. (The labels most commonly used are radioactive isotopes, but probes can also be labelled by coupling them to an enzyme whose activity produces a colored product when provided with

Figure 33·17
General technique for screening a DNA library with a probe. Cells containing recombinant molecules are plated on petri plates and grown into either colonies or plaques. A replica of the colonies or plaques is made by overlaying on the plate a filter disk. DNA and protein are released from the cells in situ and immobilized to the filter. The filter is then incubated with labelled probe under conditions in which the probe specifically recognizes the desired DNA or protein. Following washing to remove nonspecifically bound probe, specifically bound probe is detected by a method appropriate for the label used (in this case, autoradiography). Duplicate filters are used to distinguish false positives from true positives. False positives are present on only one filter; true positives are present on both filters. By aligning the filters with the original plates, cells or phages containing the recombinant of interest can be identified.

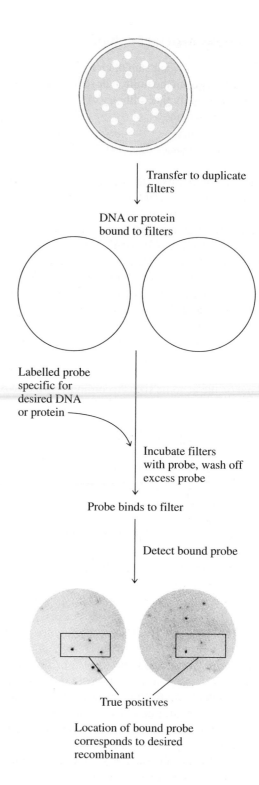

a suitable substrate.) After excess unbound probe has been removed by washing, the location of specifically bound probe is detected by the technique appropriate for the label (for example, autoradiography). The location on the filter of bound probe corresponds to the location on the petri plate of a colony or plaque containing the desired clone. Since no technique is 100% foolproof, screening is usually performed with duplicate filters to allow true positives (those present on both filters) to be distinguished from false positives (those present on only one of the two filters).

A. RNA and cDNA Probes Are Used to Detect Genes for Abundant Products

In certain cases, the RNA transcript of a gene may be so abundant and so easily purified that it can be labelled and used to screen a library directly. Some of the first eukaryotic genes isolated (those for globin, ovalbumin, and heat-shock proteins, for example) were detected in this manner. Unfortunately, this technique does not work for very many protein-encoding genes, although it is routinely used to isolate tRNA and rRNA genes.

It is more common to use cloned cDNA rather than mRNA as a probe for protein-encoding genes in genomic DNA libraries. The purification of a particular cDNA molecule using recombinant DNA technology is easier than the purification of an individual mRNA molecule from a cell, provided some information about the sequence of the protein is available (see Section 33·7C). cDNA clones from one type of organism can often be used to isolate the corresponding gene from another type of organism. In these cases, the DNA sequences need only be about 60 to 70% similar since the hybridization conditions (salt concentration, temperature) can be altered to allow the formation of DNA hybrids containing some mismatched base pairs (this is known as altering the stringency of the hybridization). For highly conserved genes, such as those for histones, the conserved sequences allow hybridization among even distantly related species.

B. Oligonucleotide Probes May Be Derived from Amino Acid Sequences

One development that revolutionized cloning was the advent of automated and efficient methods for synthesizing oligonucleotides and oligopeptides. Automated chemical synthesis allows the rapid and routine preparation of short, highly pure oligonucleotides up to 100 bp in length (Section 10·7).

The synthesis of an oligonucleotide probe suitable for screening a library requires information about the amino acid sequence of the protein of interest (or some portion of it). Using this information, all possible mRNA sequences that could encode a given portion of the protein can be derived. Figure 33·18 shows an example of an amino acid sequence from a protein and all the DNA sequences that could possibly encode it. Short stretches of DNA corresponding to each of these possible sequences can be synthesized and used collectively as a **mixed oligonucleotide probe.** The number of such possible sequences is known as the **degeneracy** of the probe. For example, the probe in Figure 33·18 contains two different nucleotides at each of three positions, resulting in a degeneracy of 2^3, or 8.

The most useful mixed oligonucleotide probes are those of maximum length and minimum degeneracy. Degeneracy is minimized by selecting regions of the protein sequence that contain amino acids encoded by few codons. In addition, unique oligonucleotides can be designed by considering the frequencies of codon usage in the particular organism and making educated guesses about which nucleotide to insert at each degenerate position. The need to synthesize multiple probes can be minimized further by using deoxyinosinate at degenerate sites (recall that deoxyinosinate forms stable base pairs with dA, dT, and dC).

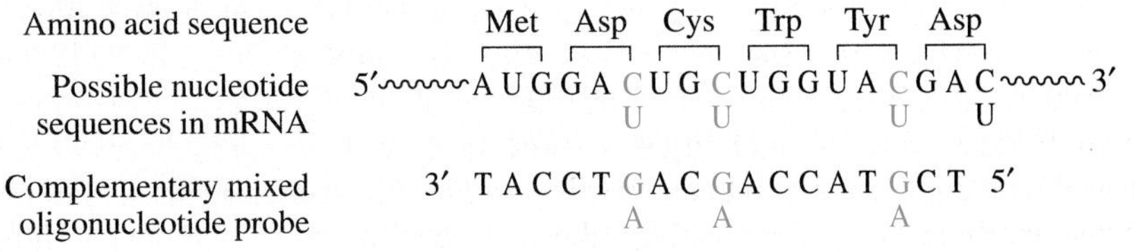

Amino acid sequence

Possible nucleotide sequences in mRNA

Complementary mixed oligonucleotide probe

Figure 33·18
Mixed oligonucleotide probe derived from the amino acid sequence of a protein. Because of the degeneracy of the genetic code, eight possible DNA sequences, each 17 bases long, could encode this peptide. The mixed oligonucleotide probe for the DNA encoding this amino acid sequence would contain all eight complementary oligonucleotides.

C. Complementation Can Be Used to Screen for Gene Products

Often all that is known about a desired gene is information about the function of its product. For example, a gene may encode an enzyme required for bacterial growth. In such a case, the gene from one organism can be selected by introducing it into another organism that lacks the gene and then growing the transformed cells under conditions where the gene product is required. This selection technique is known as **complementation.**

Complementation has been used to clone many bacterial genes and, somewhat surprisingly, certain eukaryotic genes that are serendipitously expressed in functional form in *E. coli.* The method requires that the host cell generate enough gene product for the organism to grow. Complementation with bacterial genomic clones is usually successful because the clones carry promoters and ribosome-binding sites that are recognized by RNA polymerase and the translation machinery of the bacterial host. One of the first successful complementation studies with a eukaryotic gene allowed the cloning of yeast *LEU2*, which encodes β-isopropylate dehydrogenase (discussed earlier as a selectable marker for yeast). This gene was isolated by complementing an *E. coli* strain defective in *leu*B (the corresponding gene in *E. coli*). The yeast gene was transcribed and translated in *E. coli* due to the fortuitous presence of an upstream yeast DNA sequence homologous to an *E. coli* promoter. Similar complementation studies subsequently allowed the isolation of yeast genes involved in the biosyntheses of uracil, histidine, and tryptophan.

D. Expression-Vector cDNA Libraries Can Be Screened with Antibodies

As mentioned earlier, one of the reasons that recombinant DNA technology has had such a large impact on biochemistry is that it has overcome many of the difficulties inherent in purifying low-abundance proteins and determining their amino acid sequences. Many proteins are discovered by their appearance in polyacrylamide gels or because they are recognized by antibodies, and this often occurs long before anything is known about either the structure or function of the new protein. Recombinant DNA technology allows us to purify such proteins without further characterization. Such purification begins with the overproduction of the protein in a cell using an expression vector.

Although some genes from yeast and other fungi can be expressed in prokaryotes directly from genomic clones, most eukaryotic gene products cannot be detected in prokaryotes unless they are produced from expression vectors. A classic cDNA cloning and expression vector is the λgt11 system. In this system, cDNA is cloned into a unique *Eco*RI site near the 3′ end of the *E. coli lac*Z gene (which encodes β-galactosidase). When *lac*Z expression is induced in bacterial cultures transfected with these phages, a fusion protein comprised of both β-galactosidase and the product of the cloned cDNA is produced. Note, however, that the fusion protein can be generated *only if* the protein encoded by the cloned cDNA has the same orientation and reading frame as *lac*Z. Thus, in general, this method detects only one in six clones containing cDNA from the desired gene. This does not usually present problems, however, since screening 600 000 recombinant λgt11 plaques requires little more work than screening 100 000 plaques.

Antibodies that recognize a protein (or fusion protein) can be used to screen clones from expression-vector cDNA libraries. This approach is especially useful for proteins still at an early stage of characterization since nothing more need be known about the protein beyond the position of its band in a polyacrylamide gel. The band can be excised and used to raise antibodies for screening. The resulting antibodies can be labelled and used as probes to identify recombinant cells that express the desired protein antigen.

Although antibody screening is a powerful technique, it may not be successful for several reasons. If the protein used to raise antibodies undergoes posttranslational modification, such as glycosylation or phosphorylation, the antibodies raised may not recognize the unmodified protein produced by bacterial cells. In addition, incomplete cDNA clones, which encode truncated proteins, may lack the antigenic regions of the protein and thus not be recognized by the antibody probe. For these reasons, expression-vector cDNA libraries are often more effectively screened with immune sera containing several different antibodies to the same protein (polyclonal antibodies) than with one type of antibody (a monoclonal antibody).

33·8 Chromosome Walking Is Used to Identify DNA Sequences Linked to Certain Sites

An elegant technique called **chromosome walking** is often used to order the DNA fragments in a genomic library. In this technique, overlapping recombinants are isolated in order to "walk" from one position, usually a previously cloned and analyzed gene, to another position nearby. Chromosome walking, outlined in Figure 33·19, is routinely used to isolate genes whose function is unknown but whose physical position on the chromosome has been mapped through detailed genetic studies. The *bithorax* and *Antennapedia* genes of *Drosophila* were isolated in this manner, for example, as was the human gene that, when mutated, leads to cystic fibrosis.

A chromosome walk begins with the isolation of a DNA fragment representing one terminus of the starting recombinant. This terminal fragment is labelled and used to probe a genomic library, where it identifies all recombinants that contain that fragment. These recombinants are then cleaved with restriction endonucleases and mapped to identify the one that extends furthest into the region adjacent to the starting recombinant. The terminal fragment of the newly isolated recombinant is then used as the probe for the second step in the walk. This process is continued until the walk reaches the desired destination.

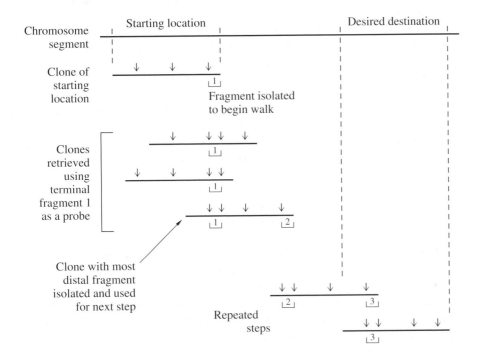

Figure 33·19
Chromosome walking. The restriction endonuclease sites (indicated by vertical arrows) are mapped on the starting recombinant. Based on this map, the terminal fragment (1) of the starting recombinant is isolated and used to probe a genomic library. The recombinants that hybridize to this fragment are restriction mapped and ordered to determine the recombinant that extends furthest into the region of the chromosome adjacent to the first recombinant. The process is then repeated using the restriction fragment furthest removed from the starting recombinant as the next probe.

Drosophila geneticists have also used chromosome walking to isolate a number of very large genes (>50 kb). These genes are often interrupted by such large introns that when a genomic library is screened with an mRNA or cDNA probe, two or more recombinants are identified, each containing complementary sequences but neither being contiguous on the chromosome or including the entire gene. Chromosome walking allows these noncontiguous recombinants to be linked, thereby establishing the complete structure of the gene.

33·9 The Polymerase Chain Reaction Amplifies Selected DNA Sequences

A simple and elegant technique known as the **polymerase chain reaction** (PCR) is reshaping molecular cloning experiments. The polymerase chain reaction provides a means of amplifying the amount of DNA present in a sample and also of enriching a population of mixed DNA molecules for the one of choice. The polymerase chain reaction is therefore extremely valuable for amplifying a particular DNA sequence when only small amounts of that DNA sequence are present. For developing PCR, Kary Mullis (Figure 33·20) shared the Nobel prize in chemistry in 1993.

The PCR technique is illustrated in Figure 33·21. Sequence information from both sides of the desired locus is used to construct oligonucleotide primers that flank the DNA sequence to be amplified. The oligonucleotide primers are complementary to opposite strands, and their 3′ ends are oriented toward each other. The DNA from the source (usually representing all the DNA in the cell) is denatured by heating in the presence of excess oligonucleotides. Upon cooling, the primers preferentially anneal to their complementary sites, which border the DNA sequence of interest. The primers are then extended using a heat-stable DNA polymerase. After one cycle of synthesis, the mixture is again heated to dissociate the strands and then cooled to reanneal the DNA with the oligonucleotides. The primers are then extended again. In this second cycle, two of the newly synthesized, single-stranded chains are precisely the length of the DNA between the 5′ ends of the primers. The cycle is repeated many times, with reaction time and temperature being carefully controlled. With each cycle, the number of DNA strands whose 5′ and 3′ ends are defined by the ends of the primers increases geometrically, whereas the number of DNA strands containing sequences outside the region bordered by the primers increases arithmetically. As a result, the desired DNA is preferentially replicated until, after 20 to 30 cycles, it constitutes virtually all the DNA. The target DNA sequence can then be cloned or analyzed directly by blotting or sequencing.

PCR is widely used in the study of molecular evolution, where the amplification and cloning of specific DNA sequences from a large number of individuals is necessary. Mitochondrial DNA sequences from humans around the world have been analyzed in order to trace the evolution of our species. The use of PCR technology avoids the need to take large samples of tissue in order to obtain enough DNA to sequence. In this case, DNA was isolated from the cells attached to a single hair follicle and amplified by PCR to allow analysis. The results show the evolutionary relationships of extant races and populations and have been used to infer that *Homo sapiens sapiens* originated in Africa approximately 100 000 years ago.

PCR has also allowed us to amplify (and hence sequence) the residual DNA in fossils that are millions of years old. The usual source of such DNA is tissue that is embedded in amber, such as the 40-million-year-old leaf shown in Figure 33·22 (Page 33·26), although DNA has also been recovered from younger air-preserved tissue, such as the skin of Egyptian mummies. The DNA from insects over 120 million years old has been amplified using PCR and sequenced. The resulting information allows us to construct more accurate phylogenetic trees.

Figure 33·20
Kary Mullis.

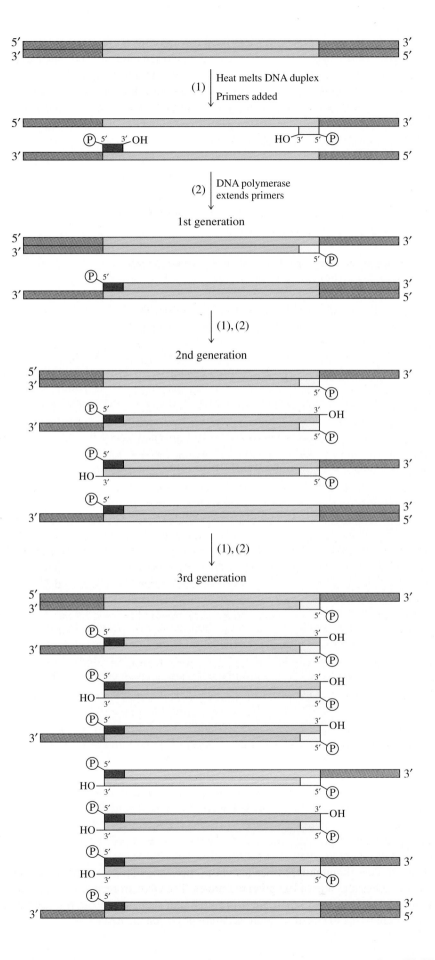

Figure 33·21
Three cycles of the polymerase chain reaction. The sequence to be amplified is shown in red. (1) The duplex DNA is melted by heating and cooled in the presence of a large excess of two primers that flank the region of interest. (2) A heat-stable DNA polymerase catalyzes the extension of these primers, copying each DNA strand. Successive cycles of heating and cooling in the presence of primer allow the desired sequence to be repeatedly copied until, after 20 to 30 cycles, it represents most of the DNA in the reaction mixture.

Figure 33·22
Forty-million-year-old leaf preserved in amber. PCR allowed the amplification and sequencing of DNA recovered from this fossil. (Courtesy of George O. Poinar, Jr.)

Figure 33·23 (next page)
Subcloning in M13 sequencing vectors. The two vectors differ only in the order of the restriction sites at the multiple cloning site. This feature allows a DNA fragment with termini generated by two different restriction endonucleases to be cloned in one orientation in one vector and in the opposite orientation in the sister vector (as indicated by the arrows inside the DNA fragment). Since the same strand of the vector is always packaged into the viral particles, the single-stranded regions of the insert produced by the two recombinant phages contain opposite DNA strands. In the manner, sequence information can be derived easily from both ends of the insert.

33·10 Cloned DNA Can Be Manipulated in a Variety of Ways

In order to pursue the answers to particular scientific questions, recombinant DNA can be manipulated in a variety of ways. In order to most easily sequence a gene, express a gene, or specifically alter a gene by mutagenesis, it is almost always necessary to move it from its original vector to new vectors designed for these purposes. Transferring cloned DNA between vectors is known as **subcloning.**

A. DNA Sequencing Vectors Allow Production of Single-Stranded DNA

The proof that a recombinant plaque or colony contains a desired DNA fragment requires evidence other than that used to identify the particular recombinant. Additional proof often comes from sequencing the DNA insert.

DNA is often subcloned into specialized vectors for DNA sequencing by the dideoxy method (Section 25·8). These vectors are derived from the single-stranded bacteriophage M13. The DNA of this phage exists in *E. coli* as a double-stranded, circular molecule that replicates autonomously (known as the replicative form, or RF). The phage particles that are released from *E. coli* cells, however, contain a single-stranded DNA molecule derived from one strand of the RF DNA. This single-stranded DNA can be isolated readily. Insert DNA can be subcloned into double-stranded RF DNA and transformed into *E. coli* as a plasmid. Following replication in the cell, the DNA can be recovered from isolated phage particles as single strands suitable for immediate sequencing.

M13 vectors have been engineered for convenience. For example, the vectors often come in pairs, with one vector containing a multiple cloning site oriented in one direction and the sister vector containing the multiple cloning site oriented in the opposite direction (Figure 33·23). The inclusion of oppositely oriented multiple cloning sites allows easy insertion of a DNA sequence in both orientations (recall Figure 33·11). For example, if one end of an insert were generated by the restriction endonuclease *Bam*HI and the other end by *Eco*RI, the insert could be cloned in both vectors at the multiple cloning site cut by both *Bam*HI and *Eco*RI. The resulting recombinant clones differ only in the orientation of the insert. The same strand of the vector is always packaged into the viral particle and exported from *E. coli,* so the single strands produced by one recombinant phage contain one strand of the insert, and the single strands produced by the sister recombinant phage contain the other strand of the insert. Subcloning the insert DNA into both vectors allows sequence information to be derived easily from both ends of the insert.

B. Prokaryotic Expression Vectors Optimize Protein Production in Bacteria

A cloned gene is often used to generate the corresponding gene product. Although some cloning vectors used to generate DNA libraries are also expression vectors (as described in Section 33·7C), their promoters are not usually strong enough to

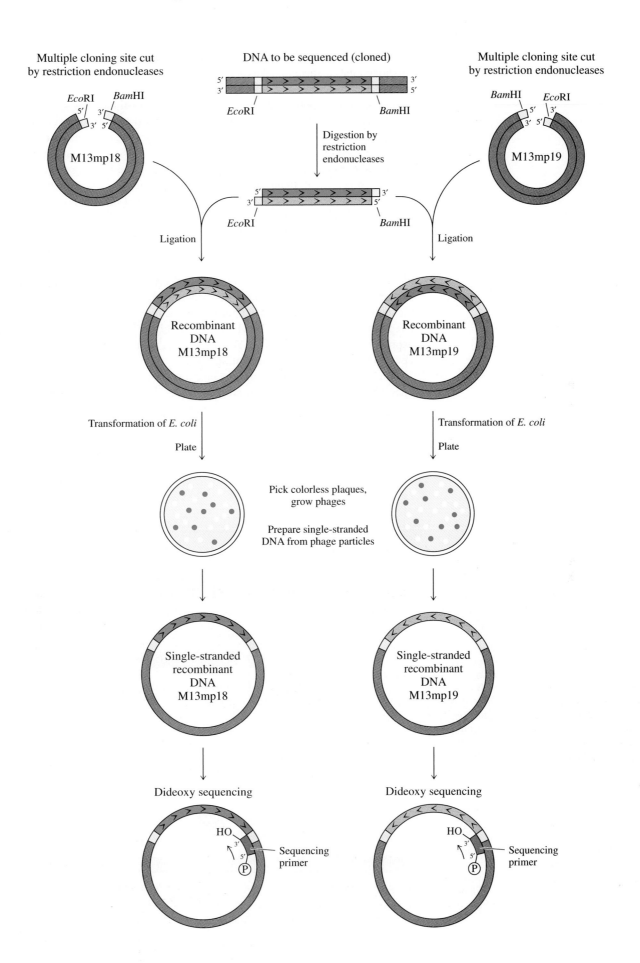

produce large quantities of protein for purification. Therefore, the gene is usually subcloned into a specialized expression vector. Expression vectors for prokaryotic hosts, such as *E. coli* or *B. subtilis,* are generally plasmids engineered to contain strong promoters, ribosome-binding sites, and useful cloning sites immediately downstream of these elements. For example, hybrid expression vectors have been created that contain the −35 promoter element from the *trp* operon, the TATA-box promoter element from the *lac*UV5 promoter, and a strong, synthetic ribosome-binding site. Because the spacing between the ribosome-binding site and the initiation codon is critical, many vectors contain cloning sites downstream of the initiation codon. When these vectors are used, the protein of interest is generated in high yield but with a few extra residues at the N-terminus. The extra amino acids do not usually present any problem for subsequent biochemical studies.

In some cases, it is necessary to regulate protein production by the expression vector. This is true, for example, when the protein being expressed is harmful to the host cell. To allow the production of harmful proteins, several prokaryotic expression vectors have been designed with inducible promoters. In these systems, operator sequences are engineered at appropriate locations on the expression vector, and a repressor gene is located elsewhere. Normally, the presence of repressor prevents expression of the harmful gene. However, at an appropriate time, the repressor can be inactivated to turn on transcription.

Inactivation of the repressor can take a variety of forms. Genes bearing the *lac* operators, for example, are derepressed by the addition of an inducer such as isopropyl β-D-thiogalactoside (IPTG), whereas genes controlled by a temperature-sensitive λ repressor are induced by a shift in temperature. Inducible promoters allow the production of proteins that are toxic to *E. coli* (the protease of human immunodeficiency virus I is an excellent example) since protein production can be

Figure 33·24
Inducible expression system using the promoter and RNA polymerase from bacteriophage T7. The insert DNA is cloned in the correct orientation downstream of the bacteriophage T7 promoter. Since this promoter is not recognized by *E. coli* RNA polymerase, the gene is not transcribed in transformants in the absence of T7 RNA polymerase. The transformants are then infected with a recombinant phage carrying the T7 RNA polymerase gene downstream of the inducible *lac* promoter. When no inducer is present, *lac* repressor binds the operators and prevents transcription of the T7 RNA polymerase gene. The insert DNA is not expressed. When an inducer is present, *lac* repressor dissociates, the T7 RNA polymerase gene is expressed, and the insert DNA is transcribed.

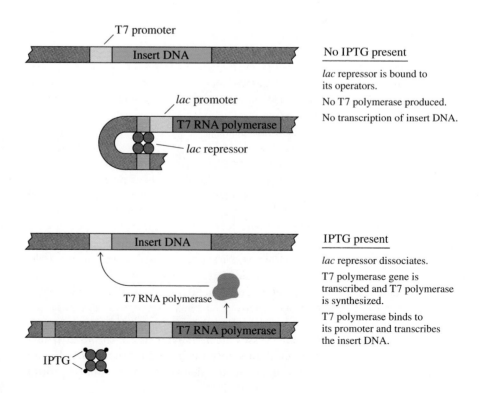

repressed until the cells have almost finished growing. Another advantage of inducible expression systems is that they protect proteins from degradation by endogenous proteases by allowing the protein to be produced just before the cells are harvested. Purification of the protein of interest may be made easier by fusing it to a carrier protein. In these cases, the fusion protein is purified on the basis of the properties of the carrier protein, and the protein of interest is eventually liberated from the rest of the fusion protein by site-specific chemical or proteolytic cleavage.

An example of an inducible system is diagrammed in Figure 33·24. The vector in this system uses a promoter recognized only by the RNA polymerase of bacteriophage T7. The gene whose product is of interest is cloned downstream of the T7 promoter and ribosome-binding site. This vector is placed in *E. coli* cells, which are then infected by phages carrying the T7 RNA polymerase gene under the control of the *lac* promoter. In the absence of IPTG, the *lac* operators are bound by repressor, no T7 RNA polymerase is produced, and the gene of interest is not expressed. Adding IPTG turns on the production of T7 RNA polymerase, which recognizes the T7 promoter upstream of the gene of interest. The addition of the antibiotic rifampicin to the medium inhibits the cellular RNA polymerase, thereby blocking all *E. coli* gene expression. In this way, *E. coli* cells are converted into factories that produce the desired protein at levels as high as 50% of total cellular protein.

C. Some Expression Vectors Function in Eukaryotes

Prokaryotic cells may be unable to produce functional proteins from eukaryotic genes even when all the signals necessary for gene expression are present. Furthermore, many eukaryotic proteins must be posttranslationally modified to be functional (for example, by glycosylation), and *E. coli* cells lack some of the enzymes needed to catalyze these modifications. Expressing some eukaryotic genes requires expression vectors that function in eukaryotes, and several such vectors have been developed. In addition to the features that allow them to be selected and replicated in *E. coli*, these vectors contain eukaryotic origins of replication, genes that can be selected in eukaryotes, transcription and translation control regions, and additional features required for efficient eukaryotic mRNA translation, such as polyadenylation signals and capping sites.

Yeast are eukaryotic organisms well suited to be hosts for the cloning and expression of eukaryotic genes. Since some yeast genes contain introns, yeast also have the enzymes necessary for splicing primary RNA transcripts to yield mature mRNA. For this reason, eukaryotic genes with introns can be cloned from genomic libraries and expressed directly in yeast without the need to isolate the clones from cDNA libraries. In addition, yeast are easily grown in the laboratory and are well understood at the molecular genetic level. These characteristics make them ideal recombinant DNA hosts.

Cloned recombinant DNA can be expressed in mammalian cells using similarly engineered vectors. It is also possible to create recombinant DNA molecules that can be integrated into the genome. Such recombinants usually contain a gene under the control of a promoter that is active in the host organism. If a zygote or an early embryo is transformed, an individual carrying the recombinant DNA in all cells can be created by subsequent crosses. Such individuals are known as **transgenic** organisms because they contain stably integrated foreign DNA. Figure 33·25 shows an example of a transgenic mouse that carries a gene for rat growth hormone.

Figure 33·25
Effect of an extra growth hormone gene in mice. The two mice shown are siblings. The mouse on the left is transgenic and carries a gene for rat growth hormone. The mouse on the right is normal. (Courtesy of Ralph L. Brinster.)

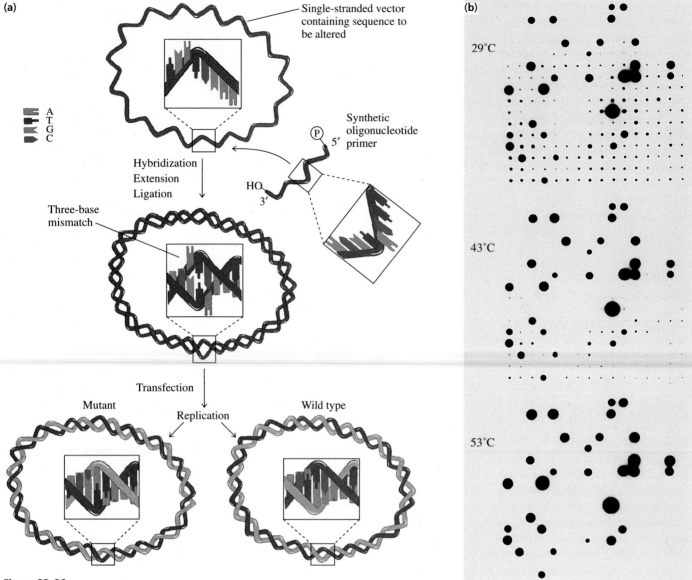

Figure 33·26
Oligonucleotide-directed, site-specific muta-genesis. **(a)** Schematic diagram showing a change of three bases. A synthetic oligo-nucleotide containing the desired changes is annealed to the single-stranded vector con-taining the sequence to be altered. The syn-thetic oligonucleotide is then used as a primer to synthesize a complementary strand. The double-stranded, circular hetero-duplex is transformed into *E. coli* cells, where replication produces mutant and wild-type DNA molecules. **(b)** The mutant DNA can be distinguished from wild-type DNA by hybridization with the radioactively labelled oligonucleotide used to generate the mutation. The oligonucleotide binds preferentially to the mutant rather than to the wild-type plasmids, where the three-base mismatch destabilizes the bind-ing of the probe. This differential binding becomes more evident (i.e., there is less non-specific binding) as the temperature of hy-bridization increases.

D. Site-Directed Mutagenesis of Cloned DNA Has Revolutionized Structure-Function Studies

One powerful technique that has had a great impact on structure-function studies of gene products is site-directed mutagenesis. In this technique, a desired mutation is engineered directly into a gene by synthesizing an oligonucleotide that contains the mutation flanked by sequences homologous to the target gene. If this oligo-nucleotide is used as a primer for in vitro DNA replication, the new copy of the gene will contain the desired mutation. Using this technique, alterations can be made at any position in a gene. Specific changes in proteins can thus be made, allowing di-rect testing of hypotheses about the functional role of key amino acid residues. (Of course, a mutant gene could also be synthesized completely in vitro, but site-directed mutagenesis is a much more rapid method of producing mutant proteins.) The technique, illustrated in Figure 33·26a, is commonly used to introduce single-nucleotide mutations into genes, although increasing the length of the homologous sequences flanking the mismatch allows larger mismatches and even drastic inser-tions or deletions to be introduced.

The desired mutant recombinants can be distinguished from wild-type recombinants by screening the transformants with the radioactively labelled oligonucleotide used to generate the mutation. Under highly stringent conditions of hybridization, this oligonucleotide, which now fully base-pairs with only the mutant sequence, hybridizes preferentially to the mutant sequence rather than to the wild-type sequence. An example of such a screen is shown in Figure 33·26b.

Much of the work in the development of site-directed mutagenesis was carried out by Michael Smith (Figure 33·27), who was awarded a Nobel prize in Chemistry in 1993 for his work in this area.

Figure 33·27
Michael Smith.

33·11 Recombinant DNA Technology Has Been Applied to Practical Problems

The vast increase in knowledge brought about by recombinant DNA technology is particularly evident when modern textbooks on genetics and biochemistry are compared with those of 10 years ago. Disciplines such as X-ray crystallography and biophysics are also benefiting from the availability of an ever-increasing number of purified proteins and DNA molecules.

The techniques described in the preceding sections, in addition to providing powerful tools to study fundamental biological questions, have numerous medical, ecological, industrial, and agricultural applications. The understanding to be gained by the in vitro and in vivo manipulation of DNA from all systems is limited now only by our understanding of the biology of those systems and by our ability to pose the right questions. Here we present a brief survey of the diverse areas that have benefited from recombinant DNA technology.

A. The Genes of Bacteria Have Been Engineered

Bacteria live under a bewildering variety of conditions of temperature, pressure, pH, nutrients, pollutants, and so on. Each bacterial species has evolved novel gene products to help it thrive under the particular conditions in which it lives. These novel genes are often found on plasmids, which can be isolated, analyzed, and introduced into other bacteria, thereby transferring the properties conferred by the gene product. As we saw in Chapter 32, this mechanism of gene transfer, particularly by plasmids with a broad host range, occurs naturally. However, recombinant DNA technology allows human intervention to accelerate this diversification of the prokaryotic world and to tailor it to meet specific needs not dictated by changes in the natural environment.

One remarkable example of the genetic engineering of bacteria is provided by the work of Ken Timmis and his colleagues. Their goal was to find biologically based ways to detoxify dangerous environmental pollutants, including chlorinated aromatic compounds. These researchers chose key enzymes from five different degradation pathways in three different soil bacteria (*Pseudomonas putida, Pseudomonas sp. B13, and Alcaligenes eutrophus*) and combined them by recombinant DNA techniques to generate a new bacterium. This organism can grow in the presence of and degrade hitherto lethal mixtures of chloroaromatics and methylaromatics such as chlorobenzenes, chlorophenols, toluene, and xylene. The enzymes of the hybrid metabolic pathway are not only stable in their new host, but their expression is specifically regulated in response to the presence of the various aromatic substrates.

B. The Genes of Plants Have Been Engineered

Plant-breeding techniques have been used for millennia to enhance certain desirable characteristics in important food crops. However, until recently, the study of the molecular biology of plants has been inhibited by the large genomes of plants and certain technical problems. Plant molecular biology has progressed in recent years largely through the exploitation of a bacterial plasmid called Ti. This plasmid, from the soil bacterium *Agrobacterium tumefaciens,* can transform many kinds of plants. The transforming DNA (T-DNA) of the Ti plasmid can be subcloned into an *E. coli*-compatible vector to yield plant–*E. coli* shuttle vectors. Next, foreign DNA can be inserted into the T-DNA at a nonessential region, and the recombinant plasmid can be transformed into *A. tumefaciens.* When the bacterium infects a plant, it passes the T-DNA into a plant cell, where the T-DNA becomes integrated into the genome of the host. The plant cell now contains the foreign gene, whose expression is controlled by the promoter upstream of the T-DNA cloning site (Figure 33·28). Although this technique results in the insertion of the foreign DNA into only one cell, it is still valuable for crop improvement: the foreign gene conferring the improvement can be transferred to an entire plant since plants can be regenerated from differentiated tissue. For example, a small piece of leaf taken from certain plants can de-differentiate, proliferate, and re-differentiate into an entire plant. When such

Figure 33·28
Introduction of foreign DNA into plants.
(a) The cloning vector is a plasmid carrying sequences essential for replication and selection in *E. coli* as well as Ti plasmid sequences required to mobilize the T-DNA. Foreign DNA can be inserted into this vector and ultimately integrated into the genome of the host plant. In the example shown, the luciferase gene from firefly is transformed into a tobacco plant using the Ti plasmid.
(b) Watering the plant with a solution of luciferin (the substrate for firefly luciferase) results in the generation of light by all plant tissues. (Courtesy of Donald R. Helinski.)

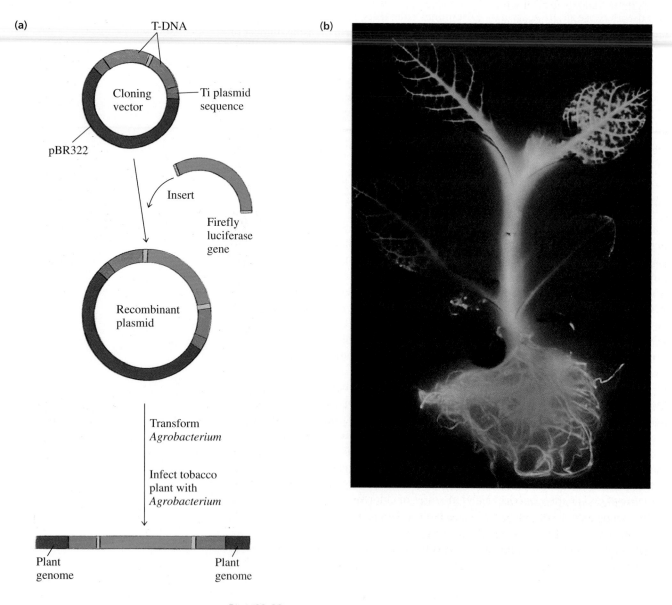

a leaf fragment is infected with a recombinant T-DNA plasmid or comes from an infected plant, every cell in the regenerated plant, including the cells that make pollen and seeds, is transformed.

The technology of T-DNA cloning is being used to engineer and stably transmit numerous genes for herbicide resistance into plant chloroplast and nuclear genomes. This approach may allow control of weeds using environmentally safe herbicides. Resistance to herbicides is engineered either by altering the target enzyme of the herbicide (frequently encoded by chloroplast DNA) or by introducing from resistant plant or bacterial sources foreign genes that catabolize the herbicide. For example, a subspecies of *Klebsiella pneumoniae* can use the herbicide bromoxynil (a benzonitrile derivative) as a source of nitrogen, thereby detoxifying it. This subspecies contains a plasmid that encodes a nitrilase highly specific for bromoxynil. Transfer of this gene to tomato plants yields transgenic plants resistant to the herbicide. Similarly, resistance to insect damage has been engineered by introducing into plants the gene for a toxin inimical to insects (Figure 33·29).

These manipulations are not currently possible with all types of plants. For example, such important world food crops as maize and wheat are resistant to infection by *A. tumefaciens*. Basic research in this area is progressing quickly, however, and the molecular biology governing resistance to infection will undoubtedly soon be understood, allowing resistance to be circumvented. The benefits of being able to engineer transgenic corn with, for example, increased levels of nutritionally important amino acids such as lysine would be immense. Similarly, the engineering of nonleguminous plants like wheat and corn to allow them to fix nitrogen would decrease their dependence on fertilizers and have overwhelming benefits for world food production.

Figure 33·29
Insect-resistant tomato plants. The plant on the left contains a gene that encodes a bacterial protein toxic to certain insects that feed on tomato plants. The plant on the right is a wild-type plant. Only the plant on the left is able to grow when exposed to the insects. (Courtesy of David A. Fischoff.)

C. Recombinant DNA Technology Helps Design Drugs

As we have seen, the overproduction of proteins is one of the major applications of recombinant DNA technology. Genes or cDNA molecules for a protein of interest can be isolated and subcloned into expression vectors, and the gene product manufactured in large quantity for further study or for therapeutic use. A number of therapeutic gene products—insulin, the interleukins, interferons, growth hormones, coagulation factor VIII—are now produced commercially from cloned genes. In some cases, these proteins are synthesized in transgenic animals, where they can be isolated from blood or meat. There are even transgenic mammals that synthesize foreign proteins in mammary glands. The proteins are conveniently secreted in their milk.

Recombinant DNA technology is especially powerful when used to isolate enzymes known to be the site of action of drugs and when applied to the identification, cloning, and characterization of potential new targets for drug intervention. Conventional approaches involving the random screening of large numbers of natural or synthetic compounds for therapeutic activity in order to locate new clinical agents are currently yielding diminishing returns. As a result, increasing attention is being given to the prospect of conducting the search in the opposite direction, that is, identifying a key gene product in a pathogen; isolating it in quantity via cloning, overexpression, and purification; determining its three-dimensional structure by X-ray crystallography; and then designing molecules to interfere with its function.

One objective of biotechnology is to design drugs that are more species-selective in their action. Many currently used drugs are toxic because they inhibit the target enzyme in the host organism as well as in the pathogen. Designed drugs can be made species-selective, and therefore less toxic, by selecting targets that are unique to the pathogenic organisms or by exploiting subtle structural differences between the pathogenic target enzyme and the host enzyme.

A similar approach is being used to develop selective drugs for the treatment of acquired immune deficiency syndrome (AIDS). Researchers have identified two key proteins, a protease and a reverse transcriptase, that are required for the reproduction of the human immunodeficiency virus. These proteins have been overproduced in *E. coli* by various sophisticated applications of recombinant DNA technology. They are currently the focus of intense efforts in drug design.

D. Recombinant DNA Technology Has Improved Diagnostic Medicine

The rapid ascendance of recombinant DNA technology has revolutionized many aspects of medicine. Based on information obtained from the study of cloned genes, new methods are being developed to diagnose genetic disorders.

Over 3000 human genetic diseases are currently known, many of which cause great suffering. Using traditional genetic techniques, the lesions causing these disorders have been mapped to specific chromosomes, and some have even been identified as mutations in specific genes and sequenced. These include mutations in the blood coagulation factors VIII and IX, which cause hemophilia; mutations in hypoxanthine-guanine phosphoribosyltransferase, which are associated with Lesch-Nyhan syndrome; and mutations in the hemoglobins, which cause diverse illnesses such as sickle-cell anemia and various thalassemias. Many of these genetic diseases may soon be treated by replacing the defective gene in key cells in the body, a process called **gene therapy.**

Figure 33·30
Detection of restriction fragment length polymorphisms (RFLP). **(a)** Schematic diagram of a region of DNA homologous to a probe. The homologous region in parent 1 is carried on a single, large restriction fragment; the same region in parent 2 is carried on a smaller fragment due to the presence of an extra restriction site in the DNA of parent 2. The offspring of these parents inherits one copy of the region from each parent. **(b)** The DNA from these individuals is digested with a restriction endonuclease and subjected to electrophoresis to separate the fragments according to size. The probe detects the homologous DNA region from each individual. DNA from parent 1 contains two copies of the large fragment. DNA from parent 2 contains two copies of the small fragment, and DNA from their child contains one DNA fragment of each size.

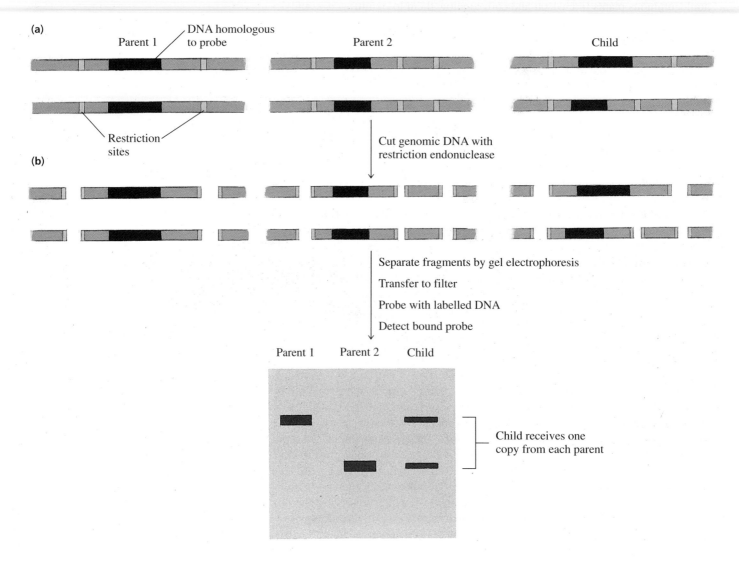

Even without cloned genes or an identified locus for a genetic disease, however, diagnosis is becoming easier through the use of genetic analyses of families
based on variations in the length of genomic restriction fragments. Such **restriction
fragment length polymorphisms,** or RFLPs, allow patterns of disease inheritance
to be traced through a family. An RFLP is detected by probing the restricted DNA
of many individuals with a cloned DNA probe and examining the pattern of the hybridizing fragments (Figure 33·30). Although the DNA sequences of different individuals are highly conserved, a difference of even one nucleotide can introduce or
abolish a restriction site and markedly change the pattern of hybridizing fragments.
The most valuable polymorphisms for genetic screenings are those caused by variations that occur in regions of the genome that are near an affected gene. Once an
associated polymorphism is detected using a particular set of probes, detection of
abnormalities is possible with a high degree of accuracy. Genetic screening can be
performed on an individual before symptoms appear, or it can be used when coun-
seling prospective parents about the probability of conceiving an afflicted child.
RFLP analysis followed by chromosome walking has led to the identification of the
mutations responsible for Duchenne muscular dystrophy and Huntington's chorea;
others are sure to follow as more DNA linkage studies are completed.

PCR and RFLP are also creating a revolution in forensic medicine, where they
are used as tools to distinguish one person from another using DNA from dried
blood, hair follicles, and semen found at the scenes of crimes. The polymerase
chain reaction, which requires extremely small quantities of DNA, is used to amplify a region of human DNA containing highly polymorphic sequence repeats. The
amplified DNA is then digested with restriction endonucleases, separated by electrophoresis, blotted to a filter, and probed with an oligonucleotide based on the
DNA repeat. The resulting pattern is unique to each individual human being (identical twins excepted) and may be used to identify perpetrators of crimes when the
pattern is compared to patterns arising from a molecular lineup of suspects.

Summary

Recombinant DNA technology allows scientific questions that arise within the classic disciplines of biochemistry, genetics, biology, botany, and physiology to be pursued in the areas of molecular biology and chemistry. A central technique of the
technology is that of molecular cloning, whereby DNA sequences of interest from
a particular organism are isolated and propagated to provide quantities of the DNA
suitable for subsequent studies.

Recombinant DNA technology builds upon a few basic techniques: isolation of
DNA, cleavage of DNA at particular sequences, ligation of DNA fragments, introduction of DNA into host cells, replication and expression of DNA, and identification of host cells that contain recombinants.

Cleavage of DNA usually involves the use of restriction endonucleases. Ligation of DNA fragments is usually accomplished using DNA ligase. During the generation of recombinant DNA molecules, DNA fragments of interest are inserted
into engineered cloning vectors. Vectors can be plasmids, bacteriophages, cosmids,
or genomic DNA molecules. Shuttle vectors are used to move DNA among different hosts, and expression vectors are used when a protein-encoding gene is to be
transcribed and translated in a host cell.

The introduction of DNA into host cells is called transformation when the
DNA is introduced directly and transfection when the DNA is introduced via a
phage vector. Once inside a host cell, recombinant DNA molecules are replicated.
Host cells containing recombinant DNA molecules are enriched using selections,

and populations of hosts that contain particular recombinant DNA molecules are identified using screens. Marker genes are often used in selections and screens.

DNA libraries are collections of populations of recombinant DNA molecules. DNA libraries can be made from genomic DNA or cDNA. Probes are used to identify particular cells that carry a recombinant DNA molecule of interest. Probes may be composed of nucleic acid, or they may be antibodies.

Chromosome walking is used to order fragments of genomic DNA. The polymerase chain reaction is used to amplify particular sequences of interest in DNA samples. Inducible expression vectors are used to overproduce harmful proteins in host cells. Additional techniques, such as DNA sequencing, chemical DNA synthesis, protein sequencing, and site-directed mutagenesis, allow molecular biologists to characterize virtually any gene or gene product in molecular terms. The ability to manipulate DNA in these ways has led to a rapidly increasing understanding of the basic biology of cells and to the production of materials useful for medicine, industry, and agriculture.

Selected Readings

General References

Micklos, D. A., and Freyer, G. A. (1990). *DNA Science: A First Course in Recombinant DNA Technology* (Cold Spring Harbor, New York: Cold Spring Harbor Laboratory Press; Burlington, North Carolina: Carolina Biological Supply Company).

Singer, M., and Berg, P. (1991). *Genes and Genomes: A Changing Perspective* (Mill Valley, California: University Science Books).

Watson, J. D., Gilman, M., Witkowski, J., and Zoller, M. (1992). *Recombinant DNA,* 2nd ed. (New York: W. H. Freeman and Company).

Watson, J. D., Hopkins, N. H., Roberts, J. W., Steitz, J. A., and Weiner, A. M. (1987). *Molecular Biology of the Gene,* 4th ed. (Menlo Park, California: Benjamin/Cummings Publishing Company).

Practical Aspects

Ausubel, F. M., Brent, R., Kingston, R. E., Moore, D. D., Seidman, J. G., Smith, J. A., and Struhl, K. (1987). *Current Protocols in Molecular Biology* (New York: John Wiley & Sons).

Colowick, S. P., Kaplan, N. O., Berger, S. L., and Kimmel, A. R., eds. (1987). *Methods in Enzymology: Guide to Molecular Cloning Techniques,* Vol. 152 (New York: Academic Press).

Subcloning and Expression

Reznikoff, W., and Gold, L. (1986). *Maximizing Gene Expression* (Boston: Butterworth).

Rodriguez, R. L., and Denhardt, D. T. (1987). *Vectors: A Survey of Molecular Cloning Vectors* (Boston: Butterworth).

Oligonucleotide and Oligopeptide Synthesis

Barany, G., and Merrifield, R. B. (1979). Solid phase peptide synthesis. In *The Peptides: Analysis, Synthesis, Biology,* Vol. 2, E. Gross and J. Meienhofer, eds. (New York: Academic Press).

Craik, C. S. (1985). *Use of Oligonucleotides for Site-Specific Mutagenesis Biotechniques* (San Francisco, California: Hormone Research Institute).

Itakura, K., Rossi, J. J., and Wallace, R. B. (1984). Synthesis and use of synthetic oligonucleotides. *Annu. Rev. Biochem.* 53:323–356.

PCR

Mullis, K. (1990). The unusual origin of the polymerase chain reaction. *Sci. Am.* 262(4):56–65.

Site-Directed Mutagenesis

Smith, M. (1986). Site-directed mutagenesis. *Phil. Trans. R. Soc. Lond.* A 317:295–304.

Drug Design

Bugg, C. E., Carson, W. M., and Montgomery, J. A. (1993). Drugs by design. *Sci. Am.* 269(6):92–98.

Index

(continued)

Illustration Credits

Cover, Front Matter, and Part Openers D. S. Goodsell.

Figure 1·1 The Bettmann Archive. **Figure 1·2** The Bettmann Archive.
Figure 1·3 The Bettmann Archive. **Figure 1·4** The Bettmann Archive.
Figure 1·8 T. A. Steitz. **Figure 1·10** Molecular Graphics and Modelling, Duke
University. **Figure 1·17** Coordinates by H. Drew and R. E. Dickerson; image by
Molecular Graphics and Modelling, Duke University. **Figure 1·19** L. A.
Shoemaker. **Figure 1·20** L. A. Shoemaker.

Figure 2·1 L. A. Shoemaker. **Figure 2·2** L. A. Shoemaker. **Figure 2·4** E. Angert
and N. R. Pace. **Figure 2·5** L. A. Shoemaker. **Figure 2·6** L. A. Shoemaker.
Figure 2·9 (a) L. A. Shoemaker; **(b)** K. R. Porter. **Figure 2·10** L. A. Shoemaker.
Figure 2·11 L. A. Shoemaker. **Figure 2·12** K. R. Porter. **Figure 2·13 (a)** L. A.
Shoemaker; **(b)** G. E. Palade. **Figure 2·14 (a)** L. A. Shoemaker; **(b)** K. R. Porter.
Figure 2·15 (a) L. A. Shoemaker; **(b)** K. R. Porter. **Figure 2·16 (a)** L. A.
Shoemaker; **(b)** A. D. Greenwood. **Figure 2·17** J. E. Heuser and M. Kirschner.
Figure 2·18 L. A. Shoemaker. **Figure 2·19** L. A. Shoemaker. **Figure 2·20** L. A.
Shoemaker. **Figure 2·21** L. A. Shoemaker. **Figure 2·22** L. A. Shoemaker.
Figure 2·23 D. S. Goodsell. From Goodsell, D. S. (1993). *The Machinery of Life*
(New York: Springer-Verlag). Reprinted with permission. **Figure 2·24** D. S.
Goodsell. **Figure 2·26** R. J. Feldmann. **Figure 2·27** L. A. Shoemaker.

Figure 3·1 (b) L. A. Shoemaker. **Figure 3·9** M. R. Dulude.

Figure 4·3 Molecular Graphics and Modelling, Duke University.

Figure 5·1 Carolina Biological Supply Company. **Figure 5·3** (**b**) E. Padlan. **Figure 5·4** (**a**) J. C. Kendrew; (**b**) The Bettmann Archive. **Figure 5·5** (**a**) Coordinates by H. C. Watson and J. C. Kendrew; image by Molecular Graphics and Modelling, Duke University; (**b**) Coordinates by T. Takano; image by Molecular Graphics and Modelling, Duke University; (**c**) Coordinates by J. C. Kendrew; image by R. J. Feldmann. **Figure 5·10** Molecular Graphics and Modelling, Duke University. **Figure 5·11** Molecular Graphics and Modelling, Duke University. **Figure 5·13** L. Pauling. **Figure 5·14** Coordinates by B. Shaanan; images by Molecular Graphics and Modelling, Duke University. **Figure 5·16** L. A. Shoemaker. **Figure 5·20** Coordinates by K. D. Hardman; image by R. J. Feldmann. **Figure 5·21** Carolina Biological Supply Company. **Figure 5·22** Molecular Graphics and Modelling, Duke University. **Figure 5·23** Coordinates by O. Epp and R. E. Huber; image by Molecular Graphics and Modelling, Duke University. **Figure 5·24** Coordinates by B. Brodsky and C. G. Long; image by Molecular Graphics and Modelling, Duke University. **Figure 5·25** R. J. Feldmann. **Figure 5·32** Coordinates by (**a**) T. Hamada, P. H. Bethge, and F. S. Mathews; (**b**) M. Legg, F. A. Cotton, and E. Hazen; (**c**) C. Kundrot and E. Richards; (**d**) J. Nachman and A. Wlodaver; (**e**) G. E. Schulz, C. W. Müller, and K. Diederichs; (**f**) M. Wilmanns, J. P. Priestle, and J. N. Jansonius; (**g**) W. J. Ray, J. B. Dai, Y. Liu, and M. Konno; (**h**) J. D. Griffith and M. Rossman. Images by Molecular Graphics and Modelling, Duke University. **Figure 5·34** T. C. Edgerton. **Figure 5·35** (**a**) Coordinates by H. C. Watson and J. C. Kendrew; image by Molecular Graphics and Modelling, Duke University; (**b**) Coordinates by J. C. Kendrew; image by R. J. Feldmann. **Figure 5·36** Coordinates by T. Takano; image by Molecular Graphics and Modelling, Duke University. **Figure 5·37** F. M. Richards. **Figure 5·40** Coordinates by J. Nachman and A. Wlodaver; image by Molecular Graphics and Modelling, Duke University. **Figure 5·42** Coordinates for Type I turn by B. W. Matthews and M. A. Holmes; coordinates for Type II turn by M. N. G. James and A. R. Sielecki; image by Molecular Graphics and Modelling, Duke University. **Figure 5·43** Coordinates by T. Takano; image by Molecular Graphics and Modelling, Duke University. **Figure 5·44** Molecular Graphics and Modelling, Duke University. **Figure 5·46** Coordinates by M. D. Yoder, N. T. Keen, and F. Jurnak; image by Molecular Graphics and Modelling, Duke University. **Figure 5·48** Coordinates by T. Alber, G. A. Petsko, and E. Lolis; images by Molecular Graphics and Modelling, Duke University. **Figure 5·49** Coordinates by G. E. Schulz, C. W. Müller, and K. Diederichs; images by Molecular Graphics and Modelling, Duke University. **Figure 5·50** Coordinates by I. Kamphuis and J. Drenth; image by Molecular Graphics and Modelling, Duke University. **Figure 5·51** K. R. Porter. **Figure 5·53** Coordinates by M. Perutz and G. Fermi; image by Molecular Graphics and Modelling, Duke University. **Figure 5·54** Coordinates by B. Shaanan; image by Molecular Graphics and Modelling, Duke University. **Figure 5·55** Coordinates by B. K. Vainshtein and E. H. Harutyunyan; image by Molecular Graphics and Modelling, Duke University. **Figure 5·57** Coordinates by T. Takano; images by Molecular Graphics and Modelling, Duke University. **Figure 5·62** L. A. Shoemaker. **Figure 5·65** S. Shyne. **Figure 5·66** L. A. Shoemaker. **Figure 5·67** (**b**) A. McPherson. **Figure 5·68** L. A. Shoemaker. **Figure 5·69** D. R. Davies.

Figure 6·4 (**a**) The Rockefeller Archive Center; (**b**) University of Toronto Library. **Figure 6·21** Coordinates for chymotrypsin by A. Tulinsky and R. Blevins; coordinates for chymotrypsinogen by D. Wang, W. Bode, and R. Huber; image by Molecular Graphics and Modelling, Duke University. **Figure 6·25** Coordinates by (**a**) A. Tulinsky and R. Blevins; (**b**) J. Walter, R. Huber, and W. Bode; (**c**) T. Prange and I. L. de la Sierra. Images by Molecular Graphics and Modelling, Duke University. **Figure 6·34** L. A. Shoemaker.

Figure 7·11 E. D. Getzoff, G. H. Liao, and C. L. Fisher. From Getzoff, E. D., Cabelli, D. E., Fisher, C. L., Parge, H. E., Viezzoli, M. S., Banci, L., and Hallewell, R. A. (1992). Faster superoxide dismutase mutants designed by enhancing electrostatic guidance. *Nature* 358:347–351. © 1992 Macmillan Magazines Limited. **Figure 7·18** Coordinates by **(a)** T. A. Steitz, C. Anderson, and R. Stenkamp; **(b)** W. Bennett, Jr. and T. A. Steitz. Images by Molecular Graphics and Modelling, Duke University. **Figure 7·19** Coordinates by H. R. Faber and B. W. Matthews; image by Molecular Graphics and Modelling, Duke University. **Figure 7·22** Coordinates by J. C. Cheetham, P. J. Artymink, and D. C. Phillips; image by Molecular Graphics and Modelling, Duke University. **Figure 7·26** Coordinates by A. Tulinsky and R. Blevins; image by Molecular Graphics and Modelling, Duke University.

Figure 8·2 Coordinates by Y. Babu, C. Bugg, and W. Cook; image by R. J. Feldmann. **Figure 8·18** Coordinates by U. M. Grant and M. G. Rossman; image by Molecular Graphics and Modelling, Duke University. **Figure 8·38** Coordinates by C. Bystroff, S. J. Oatley, and J. Kraut; image by Molecular Graphics and Modelling, Duke University. **Figure 8·51** K. J. Cournoyer. **Figure 8·59** Coordinates by S. K. Katti, D. M. LeMaster, and H. Eklund; image by Molecular Graphics and Modelling, Duke University.

Figure 9·1 C. May/Biological Photo Service. **Figure 9·3** Molecular Graphics and Modelling, Duke University. **Figure 9·6** Molecular Graphics and Modelling, Duke University. **Figure 9·14** Molecular Graphics and Modelling, Duke University. **Figure 9·15** Molecular Graphics and Modelling, Duke University. **Figure 9·16** Molecular Graphics and Modelling, Duke University. **Figure 9·28** Molecular Graphics and Modelling, Duke University. **Figure 9·31** Molecular Graphics and Modelling, Duke University. **Figure 9·32** Carolina Biological Supply Company. **Figure 9·52 (b)** J. A. Buckwalter. From Buckwalter, J. A., and Rosenberg, L. (1983). Structural changes during development in bovine fetal epiphyseal cartilage. *Collagen Rel. Res.* 3:489–504. Gustav Fischer Verlag, Stuttgart. Reprinted with permission.

Figure 10·14 Molecular Graphics and Modelling, Duke University. **Figure 10·17** Coordinates by G. E. Schulz, C. W. Müller, and K. Diederichs; images by Molecular Graphics and Modelling, Duke University. **Figure 10·21** Molecular Graphics and Modelling, Duke University.

Figure 11·4 Molecular Graphics and Modelling, Duke University. **Figure 11·7** M. C. Reedy. **Figure 11·10** R. J. Feldmann. **Figure 11·12** L. A. Shoemaker. **Figure 11·19** R. J. Feldmann. **Figure 11·22** R. J. Feldmann.

Figure 12·1 Biological Photo Supply. **Figure 12·2** L. A. Shoemaker. **Figure 12·3** L. A. Shoemaker. **Figure 12·4** L. A. Shoemaker. **Figure 12·5** V. T. Marchesi. **Figure 12·9** J. Schultz/Alaska Stock Images. **Figure 12·13** L. A. Shoemaker. **Figure 12·14** Coordinates by J. Deisenhofer; image by Molecular Graphics and Modelling, Duke University. **Figure 12·17** L. A. Shoemaker. **Figure 12·18** V. I. Kalnins. **Figure 12·21** L. A. Shoemaker. Adapted from Viitala, J., and Järnefelt, J. (1985). The red cell surface revisited. *Trends Biochem. Sci.* 14:392–395. **Figure 12·23 (b)** R. J. Feldmann. **Figure 12·24 (b)** L. A. Shoemaker. **Figure 12·25** L. A. Shoemaker. **Figure 12·26** L. A. Shoemaker. **Figure 12·27** L. A. Shoemaker. **Figure 12·28** Coordinates by M. Weiss and G. Schulz; image by Molecular Graphics and Modelling, Duke University. **Figure 12·29** L. A. Shoemaker. **Figure 12·32** L. A. Shoemaker. **Figure 12·33** L. A. Shoemaker.

Figure 12·34 L. A. Shoemaker. **Figure 12·35** L. A. Shoemaker. **Figure 12·37** L. A. Shoemaker. **Figure 12·38 (a)** M. M. Perry. **Figure 12·44** L. A. Shoemaker. **Figure 12·48** L. A. Shoemaker. **Figure 12·50** L. A. Shoemaker. **Figure 12·51** L. A. Shoemaker. **Figure 12·52** L. A. Shoemaker.

Figure 13·1 (a) Carolina Biological Supply Company; **(b)** K. J. Cournoyer; **(c)** Carolina Biological Supply Company. **Figure 13·2** L. A. Shoemaker. **Figure 13·3** L. A. Shoemaker. **Figure 13·6** L. A. Shoemaker. **Figure 13·7** L. A. Shoemaker. **Figure 13·15** Coordinates by R. Huber and J. Deisenhofer; images by Molecular Graphics and Modelling, Duke University. **Figure 13·18** L. A. Shoemaker. Adapted from Riddihough, G. (1993). Picture an enzyme at work. *Nature* 362:793. © 1993 Macmillan Magazines Limited. **Figure 13·23** L. A. Shoemaker.

Figure 14·2 (a) R. A. Tyrrell Photography; **(b)** Carolina Biological Supply Company. **Figure 14·10** Dembinsky Photo Associates.

Figure 15·2 The Nobel Foundation. **Figure 15·20** K. J. Cournoyer. **Figure 15·24** Coordinates by Y. Shirakihara and P. R. Evans; images by Molecular Graphics and Modelling, Duke University.

Figure 16·1 The Bettmann Archive. **Figure 16·7** A. Mattevi and W. G. J. Hol. **Figure 16·12** Coordinates by S. Remington, G. Weigand, and R. Huber; images by Molecular Graphics and Modelling, Duke University.

Figure 17·4 L. A. Shoemaker. Adapted from Sprang, S., Goldsmith, E., and Fletterick, R. (1987). Structure of the nucleotide activation switch in glycogen phosphorylase *a*. *Science* 237:1012–1019. **Figure 17·27** Coordinates by D. K. Wilson and F. A. Quiocho; image by Molecular Graphics and Modelling, Duke University.

Figure 18·2 L. A. Shoemaker. **Figure 18·3** Carolina Biological Supply Company. **Figure 18·4** The Bettmann Archive. **Figure 18·17** L. A. Shoemaker.

Figure 19·1 Carolina Biological Supply Company. **Figure 19·2 (a)** L. A. Shoemaker; **(b)** A. D. Greenwood. **Figure 19·6** L. A. Shoemaker. **Figure 19·7** L. A. Shoemaker. **Figure 19·11** L. A. Shoemaker. **Figure 19·12** L. A. Shoemaker. **Figure 19·14** Coordinates by D. Eisenberg; image by Molecular Graphics and Modelling, Duke University. **Figure 19·20** L. A. Shoemaker. **Figure 19·24** B. J. Miller/Biological Photo Service. **Figure 19·26** L. A. Shoemaker.

Figure 20·20 M. D. Lane.

Figure 21·1 Carolina Biological Supply Company.

Figure 24·1 Lawrence Livermore National Laboratory. **Figure 24·10** L. A. Shoemaker. **Figure 24·11** Coordinates by H. Drew and R. E. Dickerson; images by Molecular Graphics and Modelling, Duke University. **Figure 24·12** L. A. Shoemaker. **Figure 24·13** M. S. Webb. Adapted from a drawing by I. Geis in Dickerson, R. E. (1983). The DNA helix and how it is read. *Sci. Am.* 249(6):94–111. **Figure 24·16** L. A. Shoemaker. **Figure 24·17** Coordinates by U. Heinemann and H. Lauble; images by Molecular Graphics and Modelling, Duke University. **Figure 24·18** Molecular Graphics and Modelling, Duke University. **Figure 24·20** Coordinates by H. Drew and R. E. Dickerson; images by Molecular Graphics and Modelling, Duke University. **Figure 24·21** Coordinates by H. Drew, R. E.

Figure 27·2 L. A. Shoemaker. **Figure 27·5** O. L. Miller, Jr. From French, S. L., and Miller, O. L., Jr. (1989). Transcription mapping of the *Escherichia coli* chromosome by electron microscopy. *J. Bacteriol.* 171:4207–4216. Reprinted with permission. **Figure 27·8** D. J. Galas. **Figure 27·10** S. S. Judy/L. L. Murray. **Figure 27·12** L. A. Shoemaker. **Figure 27·13** S. S. Judy/L. L. Murray. **Figure 27·14** S. L. French. From French, S. L. (1992). Consequences of replication fork movement through transcription units in vivo. *Science* 258:1362–1365. © 1992 by the American Asssociation for the Advancement of Science. **Figure 27·17** L. A. Shoemaker/S. C. McQueen. **Figure 27·18** L. A. Shoemaker. Adapted from Platt, T. (1986). Transcription termination and the regulation of gene expression. *Annu. Rev. Biochem.* 55:339–372. **Figure 27·19** L. A. Shoemaker/S. C. McQueen. **Figure 27·23** Tomsich/Photo Researchers, Inc. **Figure 27·24** S. K. Burley. From Kim, J. L., Nikolov, D. B., Burley, S. K. (1993). Co-crystal structure of TBP recognizing the minor groove of a TATA element. *Nature* 365:520–527. © 1993 Macmillan Magazines Limited. **Figure 27·25** L. A. Shoemaker/S. C. McQueen. **Figure 27·26** L. A. Shoemaker/S. C. McQueen. **Figure 27·27** O. L. Miller, Jr. and B. R. Beatty. **Figure 27·28** L. A. Shoemaker/ S. C. McQueen. **Figure 27·30** J. Schwabe. **Figure 27·31** J. Schwabe. **Figure 27·32** L. A. Shoemaker/S. C. McQueen. **Figure 27·35** L. A. Shoemaker/ S. C. McQueen. **Figure 27·36** M. Amouyal. **Figure 27·40** L. A. Shoemaker/S. C. McQueen. **Figure 27·41** L. A. Shoemaker. **Figure 27·42** L. A. Shoemaker. **Figure 27·43** S. Schultz and T. A. Steitz. **Figure 27·44** L. A. Shoemaker/S. C. McQueen. **Figure 27·45** L. A. Shoemaker/S. C. McQueen. **Figure 27·46** L. A. Shoemaker/S. C. McQueen. **Figure 27·47** J. Schwabe. **Figure 27·48** J. Schwabe. **Figure 27·50** L. A. Shoemaker/S. C. McQueen. **Figure 27·51** Coordinates by T. E. Ellenberger and S. C. Harrison; image by Molecular Graphics and Modelling, Duke University. **Figure 27·52** J. J. Bonner. From Bonner, J. J. (1985) Mechanism of transcriptional control during heat shock. In *Changes in Eukaryotic Gene Expression in Response to Environmental Stress.* B. G. Atkinson and D. B. Walden, eds. (New York: Academic Press). **Figure 27·53** Carolina Biological Supply Company. **Figure 27·54** L. A. Shoemaker/L. L. Murray. **Figure 27·55** K. M. Dunlap. **Figure 27·57** L. A. Shoemaker. **Figure 27·58** H. Weintraub. **Figure 27·59** V. E. Foe. From Alberts, B., Bray, D., Lewis, J., Raff, M., Roberts, K., and Watson, J. D. (1983). *Molecular Biology of the Cell.* (New York: Garland Publishing). Reprinted with permission.

Figure 28·18 J. B. Lawrence. From Lawrence, J. B., Singer, R. B., and Marselle, L. M. (1989). Highly localized tracks of specific transcripts within nuclei visualized by in situ hybridization. *Cell* 57:493. © 1989 by Cell Press. **Figure 28·20** L. A. Shoemaker/S. C. McQueen. **Figure 28·22 (a)** P. A. Sharp; **(b)** R. Roberts. **Figure 28·23 (b)** M. Marchionni. From Marchionni, M., and Gilbert, W. (1986). The triosephosphate isomerase gene from maize: introns antedate the plant-animal divergence. *Cell* 46:138. © 1986 by Cell Press. **Figure 28·24** P. Chambon. **Figure 28·27** L. A. Shoemaker/S. C. McQueen. Adapted from Sharp, P. A. (1987). Splicing of messenger RNA precursors. *Science* 235:766–771. **Figure 28·28** L. A. Shoemaker/S. C. McQueen. **Figure 28·29** J. D. Griffith.

Figure 29·12 L. A. Shoemaker/L. L. Murray. **Figure 29·13** Coordinates by E. Westhof and M. Sundaralingam; image by Molecular Graphics and Modelling, Duke University. **Figure 29·15** L. A. Shoemaker/L. L. Murray. Adapted from Kim, S. H., Suddath, F. L., Quigley, G. J., McPherson, A., Sussman, J. L., Wang, A. H. J., Seeman, N. C., and Rich, A. (1974). Three-dimensional tertiary structure of yeast phenylalanine transfer RNA. *Science* 185:435–439. **Figure 29·19** T. A. Steitz.

Figure 30·1 S. L. McKnight and O. L. Miller, Jr. **Figure 30·2** L. A. Shoemaker. **Figure 30·4** L. A. Shoemaker/L. L. Murray. **Figure 30·7** L. A. Shoemaker. **Figure 30·8** O. L. Miller, Jr. From Miller, O. L., Jr., Hamkalo, B. A., and Thomas, C. A., Jr. (1970). Visualization of bacterial genes in action. *Science* 169:392. © 1970 American Association for the Advancement of Science. **Figure 30·12** L. A. Shoemaker. **Figure 30·13** L. A. Shoemaker. **Figure 30·14** L. A. Shoemaker. **Figure 30·15** L. A. Shoemaker/L. L. Murray. **Figure 30·17** L. A. Shoemaker. **Figure 30·18** L. A. Shoemaker. **Figure 30·30** L. A. Shoemaker. **Figure 30·32** L. A. Shoemaker. **Figure 30·33** H. H. Mollenhauer. **Figure 30·35** L. A. Shoemaker.

Figure 31·1 J. D. Griffith. **Figure 31·3** J. D. Griffith. **Figure 31·5** L. A. Shoemaker. **Figure 31·6** J. D. Griffith. **Figure 31·7** L. A. Shoemaker. **Figure 31·9** L. A. Shoemaker/S. C. McQueen. **Figure 31·10** J. S. Hudson. **Figure 31·11** L. A. Shoemaker. **Figure 31·12** R. W. Hendrix. From Hendrix, R. W., Roberts, J. W., Stahl, F. W., Weisberg, R. A. (1983). *Lambda II.* (Cold Spring Harbor, New York: Cold Spring Harbor Laboratories). **Figure 31·13** L. A. Shoemaker. **Figure 31·18** L. A. Shoemaker. **Figure 31·24** L. A. Shoemaker. **Figure 31·25 (a)** Coordinates by C. O. Pabo; image by Molecular Graphics and Modelling, Duke University. **(b)** M. S. Webb. **Figure 31·27 (a)** Coordinates by B. W. Matthews; image by Molecular Graphics and Modelling, Duke University. **(b)** M. S. Webb. **Figure 31·28** Coordinates by C. O. Pabo, B. W. Matthews, T. A. Steitz, and I. T. Weber; image by Molecular Graphics and Modelling, Duke University. **Figure 31·34** L. A. Shoemaker. **Figure 31·35** J. E. Sulston. **Figure 31·36** L. A. Shoemaker. **Figure 31·37** L. A. Shoemaker. **Figure 31·38** E. B. Lewis. **Figure 31·39** L. A. Shoemaker. **Figure 31·42** M. Scott.

Figure 32·1 The Nobel Foundation. **Figure 32·10** N. Fedoroff. From Banks, J. A., Masson, P., and Fedoroff, N. (1988). Molecular mechanisms in the developmental regulation of the maize *suppressor-mutator* transposable element. *Genes Dev.* 2:1364–1380. **Figure 32·12** P. O'Farrell. **Figure 32·13 (a, b)** Photo Researchers, Inc. **(c, d)** F. Johnston and R. B. Church. **Figure 32·14** Photo Researchers, Inc. **Figure 32·21** L. A. Shoemaker. **Figure 32·25** L. A. Shoemaker/B. D. Eller.

Figure 33·1 (a) S. N. Cohen. **(b)** H. Boyer. **Figure 33·2** L. A. Shoemaker. **Figure 33·6** L. A. Shoemaker/B. D. Eller. **Figure 33·7** L. A. Shoemaker/S. C. McQueen. **Figure 33·8** L. A. Shoemaker/S. C. McQueen. **Figure 33·12** P. L. Carl/K. J. Cournoyer. **Figure 33·16** L. A. Shoemaker/S. C. McQueen. **Figure 33·17** L. A. Shoemaker/S. C. McQueen. **Figure 33·20** K. Mullis. **Figure 33·22** G. O. Poinar, Jr. **Figure 33·25** R. L. Brinster. From Palmiter, R. D., Brinster, R. L., Hammer, R. E., Trumbauer, M. E., Rosenfeld, M. G., Birnberg, N. G., and Evans, R. M. (1982). Dramatic growth of mice that develop from eggs microinjected with metallothionein-growth hormone fusion genes. *Nature* 300:611–615. © 1982 Macmillan Magazines Limited. **Figure 33·26 (a)** L. A. Shoemaker/B. D. Eller. **Figure 33·27** M. Smith. **Figure 33·28 (a)** L. A. Shoemaker/ B. D. Eller; **(b)** D. R. Helinski. From Ow, D. W., Wood, K. V., Deluca, M., de Wet, J. R., Helinski, D. R., and Howell, S. H. (1986). Transient and stable expression of the firefly luciferase gene in plant cells and transgenic plants. *Science* 234:856. © 1986 by the American Association for the Advancement of Science. **Figure 33·29** D. A. Fischhoff. From Fischhoff, D. A., Bowdish, K. S., Perlak, F. J., Marrone, P. G., McCormick, S. M., Niedermeyer, J. G., Dean, D. A., Kusano-Kretzmer, K., Mayer, E. J., Rochester, D. E., Rogers, S. G., and Fraley, R. T. (1987). Insect tolerant transgenic tomato plants. *Bio/Tech.* 5:807–813. © 1987 Bio/Technology. **Figure 33·30** L. A. Shoemaker/B. D. Eller.